Modern Genetic Analysis

Modern Genetic Analysis

Integrating Genes and Genomes

SECOND EDITION

ANTHONY J. F. GRIFFITHS
University of British Columbia

WILLIAM M. GELBART
Harvard University

RICHARD C. LEWONTIN
Harvard University

JEFFREY H. MILLER
University of California, Los Angeles

W. H. FREEMAN AND COMPANY
NEW YORK

Acquisitions Editor: Jason Noe
Development Editor: Janet Tannenbaum
Marketing Director: John Britch
Project Editor: Jane O'Neill
Cover and Text Designers: Cambraia Fonseca Fernandes
and Maria Epes
Illustration Coordinator: Bill Page
Illustrations: Network Graphics
Photo Editor: Meg Kuhta
Photo Researcher: Elyse Rieder
Production Coordinator: Paul W. Rohloff
New Media and Supplements Editor: Joy Hilgendorf
Composition: Progressive Information Technologies
Manufacturing: RR Donnelley & Sons Company

Library of Congress Cataloging-in-Publication Data

Modern genetic analysis: integrating genes and genomes/Anthony J. F. Griffiths . . . [et al.].—2d ed.
p. cm.
Includes bibliographical references and index.
ISBN 0-7167-4382-5
1. Molecular genetics. I. Griffiths, Anthony J. F.
QH442 .M63 2002
572.8—dc21 2001059225

Printed in the United States of America
First printing 2002

W. H. Freeman and Company
41 Madison Avenue, New York, NY 10010
Houndmills, Basingstoke RG21 6XS, England

Contents in Brief

Contents

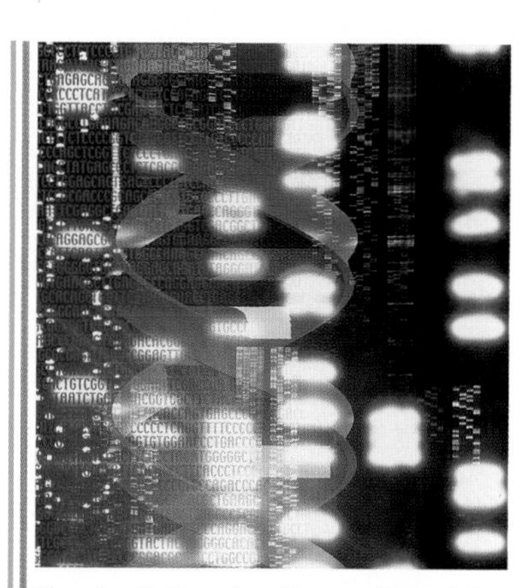

Chapter 2 Opening Figure Page 23

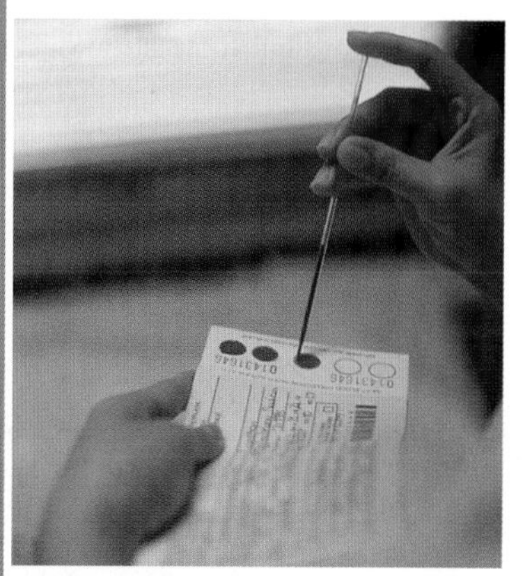

Figure 3-1b Page 56

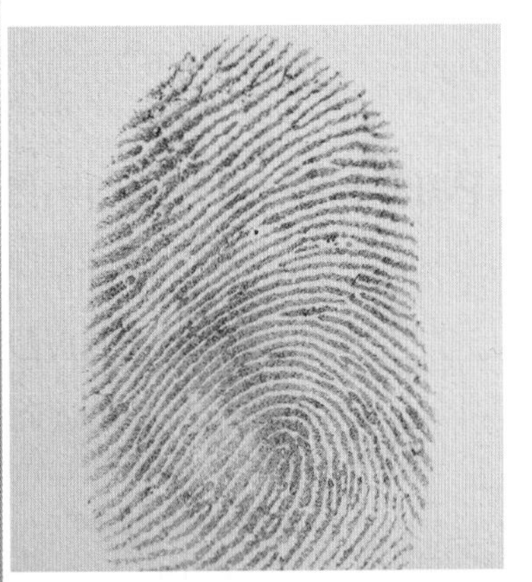

Chapter 8 Opening Figure 1
Page 213

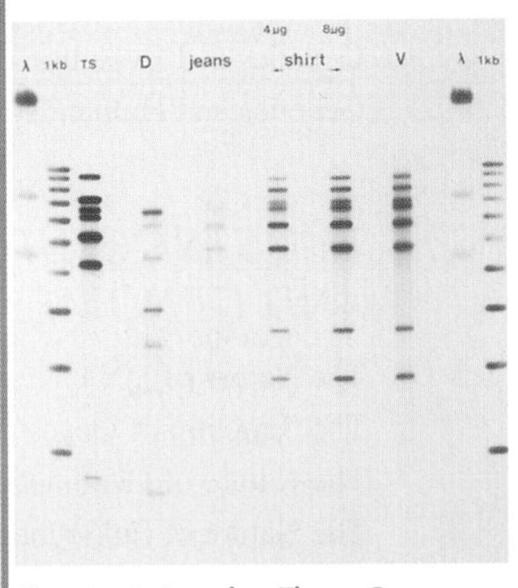

Chapter 8 Opening Figure 2
Page 213

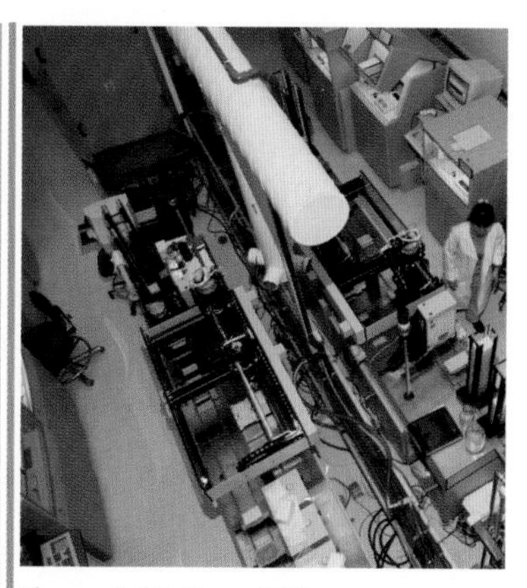

Figure 9-20 Page 286

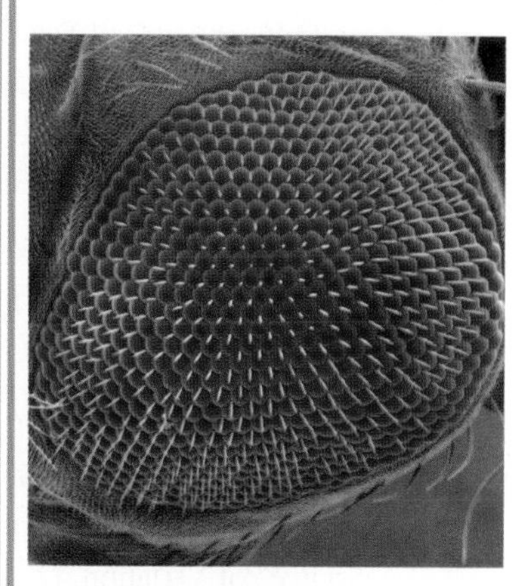

Figure 12-31 Page 407

Figure 14-14b Page 465

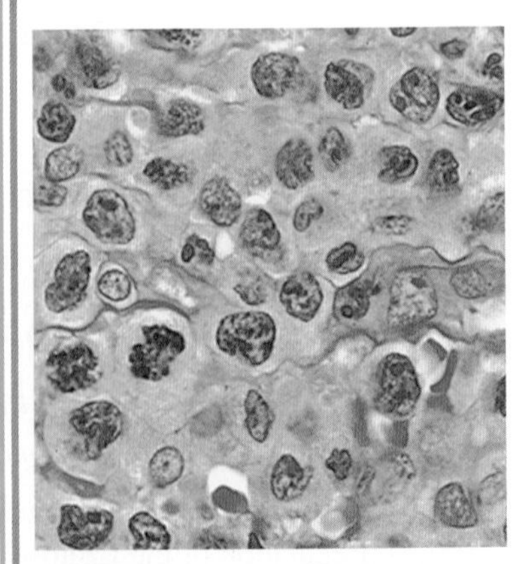

Figure 15-1b Page 484

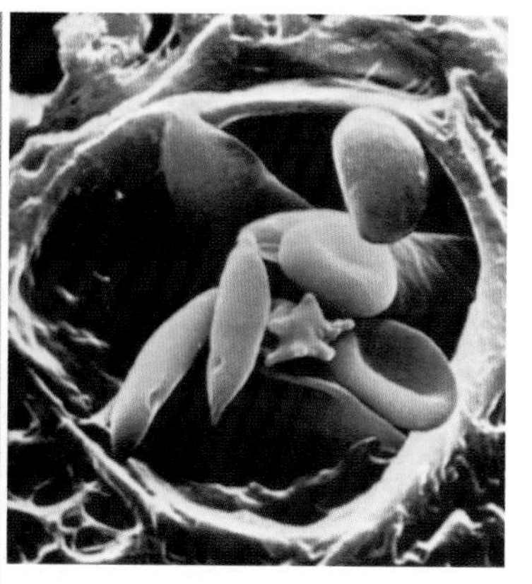

Figure 17-1a Page 554

Preface

As teachers of genetics and the authors of *An Introduction to Genetic Analysis*, we are aware of the quiet revolution occurring in the way genetics is being taught to beginning students. Many instructors are finding that a strictly chronological or historical approach no longer fits their method of teaching genetics, nor does it meet their students' needs. More and more, molecular genetics is being introduced earlier in the course and integrated with phenotypic and genotypic analysis.

Modern Genetic Analysis: Integrating Genes and Genomes, Second Edition, was written for instructors and students who need a textbook that supports an integrated "DNA early" approach. This departure from the traditional historical unfolding of genetics has had some significant side effects—chief among them, a more streamlined presentation in which genetic principles stand in bolder relief.

Regardless of whether the presentation is traditional or modern, it is essential that students learn to think like geneticists. Thus, as in *An Introduction to Genetic Analysis*, the focus in *Modern Genetic Analysis: Integrating Genes and Genomes,* Second Edition, is on teaching students to analyze data and draw conclusions.

A Modern Approach

We have divided this edition of *Modern Genetic Analysis* into four parts that reflect the broad divisions of modern genetics. Because of the way this edition is structured, not only is there an opportunity for the vertical integration of concepts at the organismal, cellular, chromosomal, protein, and DNA levels, but "classical genetics" is presented in a less abstract way that can be better grasped by students. We hope that this new approach will enable students to understand how genetics is done in the "real world," where classical and molecular approaches are not segregated but complement each other.

Part 1: Fundamentals of Gene Structure, Function, and Transmission deals with the fundamental themes of genetics. In this part, a student is immediately shown "the forest" before "the trees," a well-established learning method. After an opening chapter that gives an introductory overview of genetics, Chapter 2 introduces the student to gene and genome structure. Chapter 3 presents transcription and translation. Chapter 4 describes the molecular and cellular basis of gene

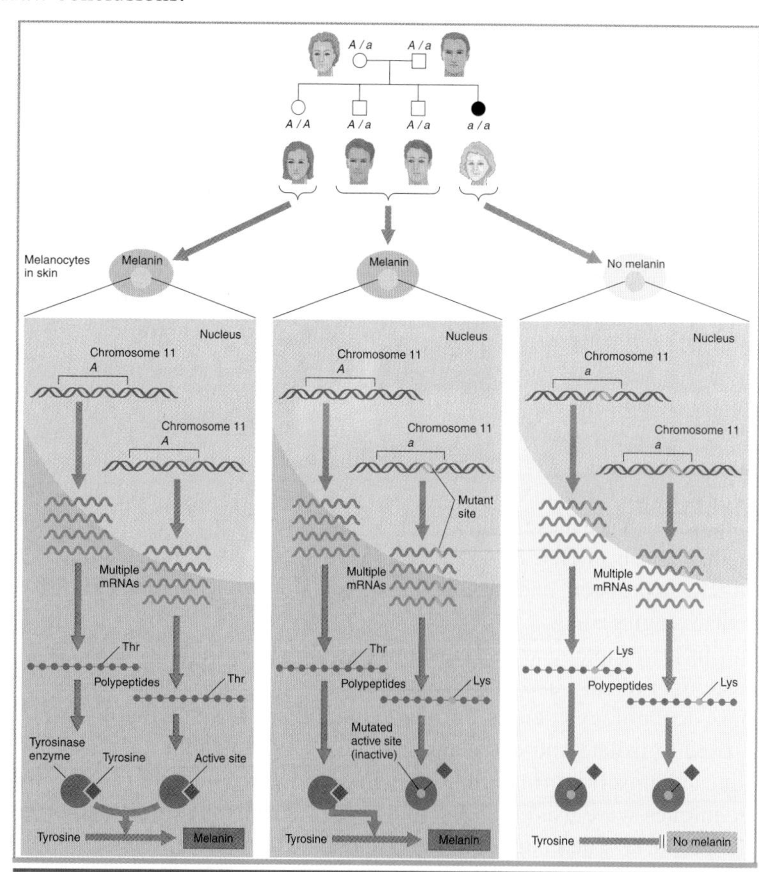

Figure 5-12 Page 132

transmission. Chapters 5 and 6 cover the principles of inheritance and recombination in eukaryotes, and Chapter 7 deals with the corresponding processes in prokaryotes. Thus, in this part, the core principles of inheritance are presented and integrated at the classical and molecular levels.

Part 2: Methods of Genetic Manipulation focuses on genetic manipulation. The presentation begins in Chapter 8 with a demonstration of the incisiveness of DNA technology as a means of manipulating individual genes; then Chapter 9 extends these approaches into the analysis of whole genomes—the subject of genomics. This part of the book then takes up the subject of mutation, examining gene mutation and repair in Chapter 10, continuing with chromosome mutation in Chapter 11, and culminating with a description of the power of mutation to manipulate and dissect genetic systems in Chapter 12.

Part 3: Systems Genetics: From Gene to Phenotype expands from the simple principles of genetics to the systems of complex interacting genetic components found in living cells. The part begins in Chapter 13 with an exquisite example of complex genetic interactions in the regulation of transcription. The theme of interaction is continued in Chapter 14 with an exploration of the complex relationships between genes and environment in controlling phenotype. Part 3 concludes with two topics that illustrate the wide variety of biological activities that result from gene interaction: cancer in Chapter 15 and development in Chapter 16.

Part 4: Genetic Analysis of Populations covers the special approaches used to analyze genes at the population level. Chapter 17 covers formal population genetics, and Chapter 18 lets quantitative genetics lead into evolutionary genetics in Chapter 19, the final chapter.

Focus on Principles

A primary goal in writing *Modern Genetic Analysis* was that the understanding and application of core genetic principles take priority over historical detail. We hope that students will more readily recognize and grasp fundamental principles and themes if their presentation is not encumbered by excessive detail. Thus, the text focuses on the overarching principles of genetics rather than on the historical experiments that generated them. The principles are used to explain to students how genetics is done today. For instance, in Chapter 3, we introduce the common themes of complementarity of nucleic acid sequences and specificity of protein–nucleic acid interactions. Students then see these themes again in subsequent chapters, applied in the analysis of DNA replication, protein synthesis, and regulation of gene expression, and exploited in the manipulations that underlie genetic engineering.

In all chapters, we stress the vertical relation among DNA, protein products, and phenotype. The chapter on recombinant DNA technology focuses on how recombinant DNA technology is used to isolate and characterize genes, rather than on the techniques themselves. Here students will see how recombinant DNA techniques were used to clone human genes such as the gene for alkaptonuria. The chapter on developmental genetics emphasizes the importance of signal transduction cascades in all aspects of a cell's or an organism's development and the important and varied switch mechanisms that underlie all developmental decisions.

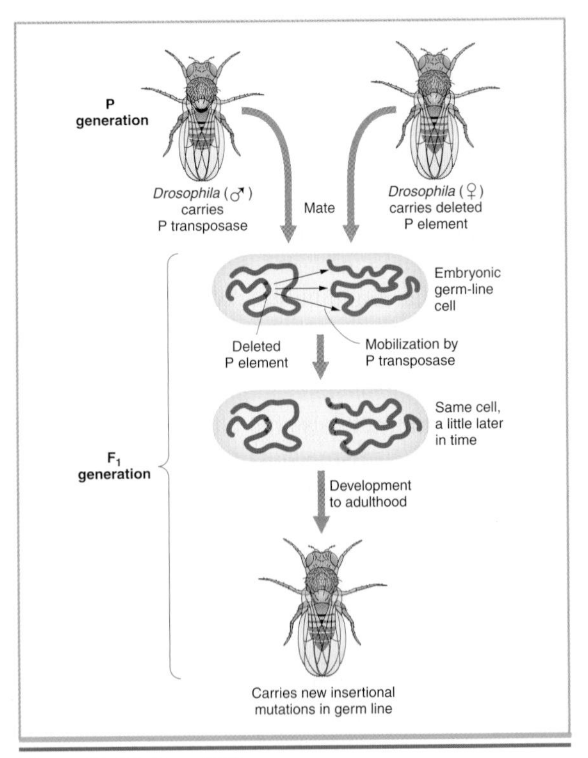

Figure 10-32 Page 336

Focus on Genetic Analysis

In *Modern Genetic Analysis* we focus on the questions that underlie much of modern genetics, such as, "How may genes affect this phenotypic difference? Are the genes linked? Are the mutations allelic? What is the cellular function of the gene?" In short, the focus is on the modes of inference geneticists use today. A well-rounded study of genetics should also expose students to some of the landmark experiments. These important investigations are set apart from the text in boxes called *FOUNDATIONS OF GENETICS.*

In these Foundations boxes, students will read about how Archibald Garrod inferred the nature of inborn errors of metabolism; the research that led Charles Yanofsky to deduce that gene and protein structure are colinear; Watson and Crick's model for the structure of DNA, including Watson's own description of the first assembly of the metal model; and Luria and Delbrück's method of deducing the random nature of mutation.

FOUNDATIONS OF GENETICS 3-2

Charles Yanofsky and the Colinearity of Gene and Protein Structure

In 1941, George Beadle and Edward Tatum originated the one-gene–one-enzyme hypothesis. When James Watson and Francis Crick deduced the structure of DNA in 1953, it seemed likely that there must be a linear correspondence between the nucleotide sequence in DNA and the amino acid sequence in protein (such as an enzyme). However, it was not until 1963 that an experimental demonstration of this colinearity was obtained, from two research groups, one of which was led by Charles Yanofsky at Stanford University.

Yanofsky had induced 16 mutant alleles of the *tryp A* gene of *Escherichia coli,* whose wild-type allele was known to code for the α protein subunit of the enzyme tryptophan syn-

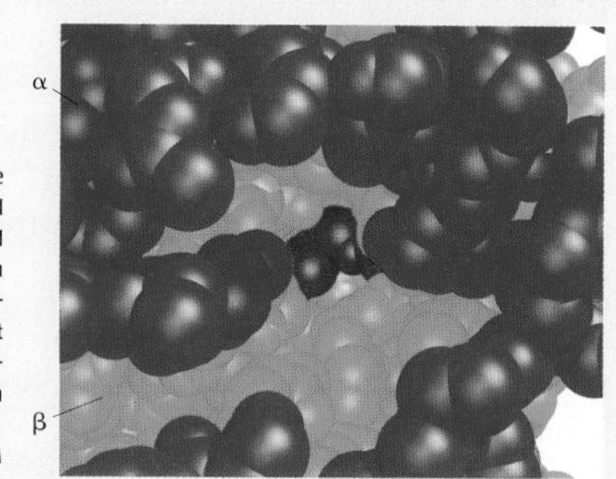

Model of tryptophan synthetase, with substrate in red.

Page 78

Added to this edition are experiments such as that of Hershey and Chase (presented in a solved problem that makes an interesting learning exercise for the student), the development of the Ames test, and Muller's experiment demonstrating that X rays are mutagenic. If your course favors a very heavy emphasis on landmark experiments, please visit Modern Genetics Online for Instructors (www.whfreeman.com/mga) to access our Landmark Experiments Resource Library.

Problems at the end of each chapter have been created to give students the chance to apply and exercise their analytical skills. Almost all problems have been class tested. Problems begin at the basic level and then proceed to a more challenging level. Included are *Pattern Recognition Problems,* which the authors designed to aid students in recognizing hereditary patterns in data, a key skill in genetic analysis. Most chapters also include a problem titled *Unpacking the Problem.* These exercises, written by Tony Griffiths, reveal the underlying levels of knowledge that must be applied for a problem to be solved constructively. In this edition, we now provide 20 Web-based *INTERACTIVE UNPACKING THE PROBLEM* exercises, which are indicated by the i icon next to each one. These exercises were written by Craig Berezowsky in consultation with Tony Griffiths, both of the University of British Columbia.

Many chapters also include several *Solved Problems* that walk students through the ways in which geneticists apply principles to experimental data. This type of problem will prepare students for solving problems on their own.

New to This Edition

The Latest Approaches

In keeping with the book's title, *we continue to emphasize the most current genetic research throughout the book.* More specifically, *CHAPTER 9, GENOMICS*, has been updated to show how genetics has migrated from recombination-based maps to sequence-level maps and to the sequencing of entire genomes. Beyond the acquisition of structural data, this chapter describes how bioinformatic and experimental approaches have been developed for analyzing genes and their products at the genomic level. It sets the stage for an integration of genomic approaches into most of the subsequent chapters, reflecting the fact that genetics and genomics have now become companion disciplines. In addition, through a class of Web-based tutorials new to this edition, entitled *EXPLORING GENOMES*, students can learn how to make use of the many genomic databases that have become an integral part of genetics today. (More on the specifics of these exercises can be found on the front endpapers of this text.)

A new *CHAPTER 19, EVOLUTIONARY GENETICS*, has been added in recognition of the inseparability of genetics and evolution in modern research. Some material from the first edition's Chapter 17, which briefly covered evolution along with population genetics, has been relocated to this chapter.

Reorganized, Expanded, and Updated

The material on gene inheritance (Chapter 4 in the first edition) has been reorganized into two chapters. In this edition, the first of these chapters, Chapter 4, focuses on the molecular and cellular basis of gene inheritance. Building on this, the second chapter, Chapter 5, then covers the genetic ratios that arise in the progeny of heterozygotes. Splitting this content into two chapters was done partly to shorten each chapter and partly to delineate the two complementary approaches.

The first edition had two chapters on genetic engineering and its applications (Chapters 10 and 11). These have now been streamlined into one chapter, Chapter 8, which has been completely reworked so that the two fundamental processes of base complementarity and sequence-specific interactions between DNA/RNA and proteins form its scaffold. The chapter takes a principles-based approach to genetic engineering.

The chapter on gene mutation in the first edition (Chapter 7) has now been expanded into two chapters: Chapter 10, on mechanisms of gene mutation and repair, as before, and Chapter 12, a new chapter on mutational dissection with a focus on the analytical process. Moving this topic to its new location after the chapters on genetic engineering and genomics allows the newer techniques of site-directed mutagenesis and targeted gene knockouts to be treated as part of the modern arsenal of techniques for mutation research.

Topics from the first edition Chapter 13, *Transposable Genetic Elements*, have here been separated and included in three chapters. Chapter 9, *Genomics*, examines the prominent representation of these mobile genetic elements as repetitive sequences in eukaryotic genomes. Chapter 10, *Gene Mutation*, discusses the mechanisms of transposition in both prokaryotic and eukaryotic systems, as well as the mutagenic potential of these elements. Chapter 12, *Mutational Dissection*, discusses the use of transposable elements as mutagens.

The first edition chapter on gene interaction (Chapter 6) has been redesigned and placed in Part 3 as Chapter 14, *From Gene to Phenotype*.

New Tools for Learning

Some instructors told us that in our attempt to focus on principles and avoid excessive detail in the first edition, we occasionally streamlined the text too much. In the second edition, we have addressed this feedback by fleshing out the explanations, providing additional background information where necessary, and adding new pedagogical diagrams to facilitate understanding of the more complex processes, such as DNA replication. While still focused on the essential principles, these enhanced explanations should make the material more accessible to students.

We also have taken the opportunity to add a number of new learning tools. Students want to and should learn about the exciting applications happening in genetics today and those that are poised to happen in the future. Each chapter now opens with a *REAL-WORLD STORY* that makes material even more relevant for students. Stories such as the existence of high frequencies of Huntington disease in certain human populations introduce the broader questions introductory students want answered concerning the inheritance and the cellular basis for such genetic diseases. Each chapter-opening story is revisited toward the end of the chapter in light of the principles that have been covered.

Pedagogically useful, conceptually rich illustrations are invaluable resources to help students develop genetic insight. In the second edition, the illustration program has been thoroughly expanded, updated, and revised with an emphasis on clarity and educational value in support of the text discussion. Over 60 new illustrations have been added with this theme in mind. Each of the remaining text illustrations has been carefully edited to ensure that its message is conveyed clearly. In addition, the colors of the entire art program have been enhanced so that images are more vibrant and will thus project better during lecture hall presentations.

In our commitment to focus on principles, we have developed for each new chapter a *CHAPTER OVERVIEW ILLUSTRATION* that serves as a clear and graphic road map for the chapter. These overviews were regarded by reviewers as an extremely useful learning tool and an innovative approach to presenting complex concepts.

As in the first edition, we continue to use study aids such as *Key Concepts, Message* boxes, end-of-chapter *Summaries, Concept Maps,* and *Unpacking the Problem* exercises.

As a result of user feedback, the end-of-chapter problem sets have been divided in this edition into a group of *BASIC PROBLEMS* followed by a group of *CHALLENGING PROBLEMS.* Each group of Basic Problems allows students to practice their mastery of the chapter and so gain confidence in their ability to apply principles learned. This arrangement should reduce any frustration some students might feel when dealing with a challenging problem later on.

Students should find two new appendixes very useful. Appendix A, *Genetic Nomenclature*, lists model organisms and their nomenclature. Appendix B, *Bioinformatic Resources for Genetics and Genomics*, builds on the theme of introducing students to the latest genetic research tools by providing valuable starting points for exploring the rapidly expanding universe of online resources for genetics and genomics.

Figures 5-1a and 5-1b Page 118

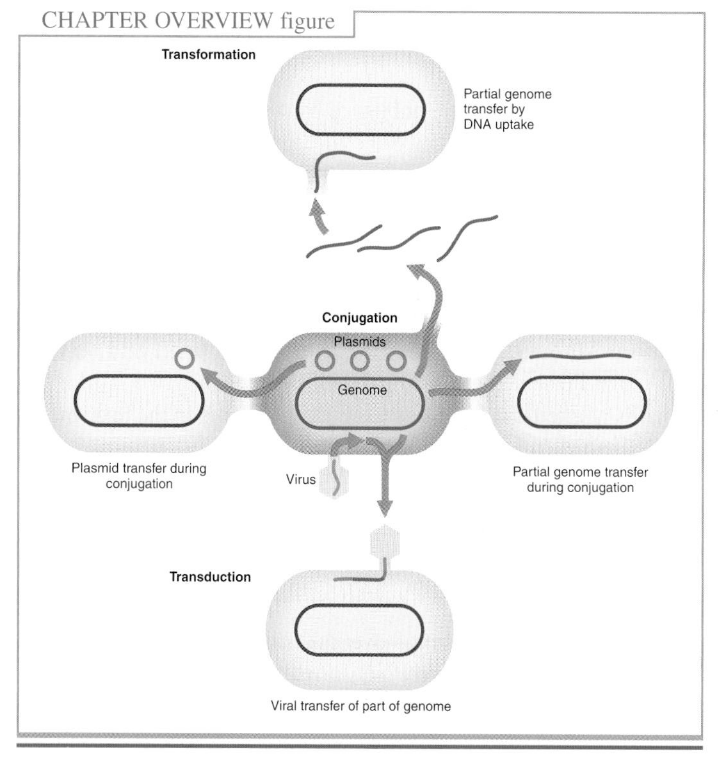

Figure 7-2 Page 185

Media and Supplements

For STUDENTS

NEW! Modern Genetics Online: www.whfreeman.com/mga

- *Exploring Genomes: Web-Based Bioinformatics Tutorials* by Paul G. Young of Queen's University. These 10 interactive tutorials teach students to use the vast genomic resources of the Web site of the National Center for Biotechnology Information. As a basis for the tutorials, Young draws from relevant material in the textbook, showing students how to use these bioinformatic resources as well as how to see the potential for their application in modern genetic analysis. For ease of use, the tutorials can be completed independently of the chapters in which they are located.
- Step-through and continuous-play FLASH™ Animations. These 45 animations were conceived by Tony Griffiths. In the text, each figure that has a related animation is tabbed for ready reference with this media icon: See the list inside the front cover of this text for descriptions of each animation and where each is connected to the text. The animations are also available on a student CD-ROM.
- Interactive Unpacking the Problem by Craig Berezowsky, University of British Columbia. As with the Unpacking the Problem exercises in the text, each series of these questions steps students through the thought processes needed to solve a problem. This Web-based and interactive version of Unpacking the Problem, indicated by an i next to one of the problems at the end of each chapter, features immediate feedback and grading functions.
- Online Sample Tests. In addition to practicing knowledge through the interactive Unpacking the Problem set of exercises, students can also test their fundamental understanding of key concepts and experiments through these online questions. All questions were chosen and often written by the textbook and test bank author teams to ensure that all questions cover what is core to each chapter. Questions are also keyed to the text so that students can easily review material needed for understanding.
- *The Human Genome*, published by *Nature* npg From the frontline of science reporting, and with a foreword by James Watson, here is the one book every student of genetics needs to understand—and to appreciate—one of the great revolutions in science. *The Human Genome* is the first comprehensive report on the Human Genome Project that combines essential background information, clear explanations, and expert analysis. Additionally, this invaluable resource evaluates the impact on science, medicine, and society—both opportunities and risks.

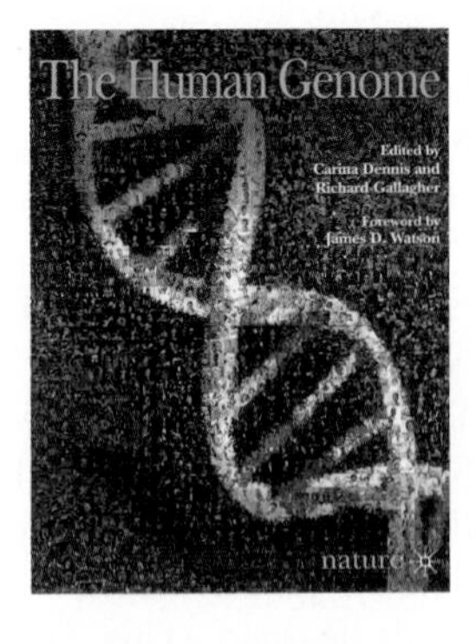

Solutions Manual, 0-7167-4703-0, **available in print and on the Web**
by William Fixsen, Harvard University.
Contains worked-out solutions to all the problems in the text, including the Unpacking the Problem exercises.

For INSTRUCTORS

NEW! Modern Genetics Online for Instructors
Available on the Web at www.whfreeman.com/mga or on CD-ROM, 0-7167-4498-8

- All images from the text—over 500 illustrations—in multiple downloadable formats including PowerPoint format. Also available in JPEG and PICT formats. All include enlarged labels that project more clearly for lecture hall presentation.

- PowerPoint lecture outlines
- 45 step-through and continuous-play **FLASH**™ Animations
- Landmark Experiments Resource Library (available on Web site only)
- Electronic Instructor's Resource Manual, Test Bank, and Student Solutions Manual

NEW! In partnership with NCBI Bookshelf, *Modern Genetic Analysis,* Second Edition, is now available as an online resource to assist instructors, researchers, and students while using PubMed. From a PubMed abstract, click on the "Books" link to view hyperlinked phrases that lead to sections from *Modern Genetic Analysis* for immediate online reference.

Overhead Transparency Set, 0-7167-4497-X
The full-color overhead transparency set contains 150 key illustrations from the text with enlarged labels that project more clearly for lecture hall presentation.

Instructor's Resource Manual and Test Bank, (printed) 0-7167-4496-1, (Dual Platform CD-ROM) 0-7167-4499-6, by Ewen J. Harrison, St. John's College, Santa Fe, and Sally Lyman Allen, University of Michigan.
The *Instructor's Resource Manual and Test Bank* contains over 700 test questions rated by difficulty in multiple-choice, true-false, and matching formats. It also contains extensive teaching notes, including ideas and tips for teaching the genetics course, learning objectives, vocabulary, teaching strategies, annotated problem solutions, expanded Unpacking the Problem exercises, and ethics/current-events topics. The easy-to-use electronic version of the test bank lets instructors edit and rearrange questions as well as add their own.

Acknowledgments

THANKS ARE DUE to the following people at W. H. Freeman and Company for their considerable support throughout the preparation of this edition: Janet Tannenbaum, Development Editor; Sara Tenney, Publisher; Jason Noe, Acquisitions Editor; Jane O'Neill, Project Editor; Joy Hilgendorf, Supplements Editor; Tanya Awabdy, Media Development Editor; Rawle Stoute II, Editorial Assistant; Cambraia Fonseca Fernandes, Designer; Paul W. Rohloff, Production Coordinator; Bill Page, Senior Illustration Coordinator; Meg Kuhta, Photo Editor; Elyse Rieder, Photo Researcher; and the Network Graphics art studio.

We also extend our thanks and gratitude to our colleagues who reviewed the manuscript and contributed to the project in numerous ways: their insights, advice, and efforts have been invaluable:

Sally Allen, *University of Michigan*
James O. Allen, *Florida International University*
Ruth Elizabeth Ballard, *California State University at Sacramento*
Craig Berezowsky, *University of British Columbia*
Daniel Bergey, *Montana State University*
Paul J. Bottino, *University of Maryland*
Kenneth C. Burtis, *University of California at Davis*
Pat Callie, *Eastern Kentucky University*
Rebecca L. Cann, *University of Hawaii*
Alan C. Christensen, *University of Nebraska*
Claire Cronmiller, *University of Virginia*
Diane Dodd, *University of North Carolina*
F. Paul Doerder, *Cleveland State University*
Dorothy B. Engle, *Xavier University*
William D. Fixsen, *Harvard University*
Rosemary H. Ford, *Washington College*
Denise K. Garcia, *California State University at San Marcos*
Dan Garza, *Novartis Pharmaceuticals Corporation*
Vaughn M. Gehle, *Southwest State University*
Elliott S. Goldstein, *Arizona State University*
Karen G. Hales, *Davidson College*
Susan R. Halsell, *James Madison University*
Ewen Harrison, *St. John's College, Santa Fe*
J. L. Henriksen, *Bellevue University*
Stanton F. Hoegerman, *College of William and Mary*
Margaret Hollingsworth, *State University of New York at Buffalo*
E. Jane Albert Hubbard, *New York University*
Kenton Ko, *Queen's University*
Jocelyn E. Krebs, *University of Alaska*
Nuran M. Kumbaraci, *Stevens Institute of Technology*
Carolyn R. Leach, *Adelaide University, Australia*
Heather E. Lorimer, *Youngstown State University*
David J. Matthes, *San Jose State University*
Bryant McAllister, *University of Texas, Arlington*
Kim S. McKim, *Rutgers University*
Brook Milligan, *New Mexico State University*
Annette Muckerheide, *College of Mount St. Joseph*
John C. Osterman, *University of Nebraska at Lincoln*
Carlos F. Quiros, *University of California at Davis*
Daniel J. Schoen, *McGill University*
Rodney J. Scott, *Wheaton College*
David T. Sullivan, *Syracuse University*
Beat Suter, *McGill University*
Tammy Tobin-Janzen, *Susquehanna University*
Frank Verley, *Northern Michigan University*
Linda Wells, *California State University at Bakersfield*
William Wellnitz, *Augusta State University*
Paul G. Young, *Queen's University*

Acknowledgments

GENETICS AND THE ORGANISM 1

Key Concepts

1 The hereditary material is DNA.

2 The functional units of DNA are genes.

3 DNA is a double helix composed of two intertwined nucleotide chains.

4 In the copying of DNA, the chains separate and serve as templates for making two identical daughter DNA molecules.

5 A gene consists of segments of DNA that specify the amino acid sequence of a polypeptide produced by the cell, as well as segments of DNA that are used to control the rate, time, and place of the polypeptide's production.

6 The polypeptide specified by a gene folds to make a protein. This folding depends on the amino acid sequence of the polypeptide and on the intracellular conditions during the folding process. The proteins so formed are the main determinants of the basic structural and physiological properties of an organism.

7 Hereditary variation is caused by variant forms of genes (alleles).

8 The tools of genetic analysis include controlled matings between organisms and statistical analyses of the resulting progeny, microscopic examination of cellular and tissue structures, the chemical and structural analysis of biological molecules, the identification of such molecules and their localization within cells, and analysis of DNA sequences.

9 The distinctive characteristics of a species are encoded by its genes, whereas variation within a species arises from the interaction of genetic variation, environmental variation, and random developmental noise.

10 Evolution is a consequence of genetic changes in the composition of a population over time and of the formation of new populations that are genetically isolated from their parental species.

Genetic variation in the color of corn kernels. Each kernel represents a separate individual with a distinct genetic makeup. This photograph symbolizes the history of humanity's interest in heredity. Humans were breeding corn thousands of years before the rise of the modern discipline of genetics. Extending this heritage, scientists still use corn today as an important research organism in classic and molecular genetics. *(William Sheridan, University of North Dakota; photo by Travis Amos.)*

Although we are ordinarily not aware of it, our everyday observations of living organisms present us with several genetic puzzles. Offspring resemble their parents, but not exactly (Figure 1-1). Humans give birth to humans, while dogs give birth to dogs. But all humans and all dogs are not alike. Children resemble their parents more than they resemble unrelated people, but that resemblance is far from perfect. Moreover, there are some very general resemblances between humans and dogs; for example, both have four limbs, two eyes, two ears, and teeth. These features of both humans and dogs appear during their development from a single fertilized egg cell into an adult. Their similarities are a consequence of the fact that both humans and dogs evolved from a common ancestral species in the remote past. What, then, are the mechanisms that explain the similarities and differences between ancestors and their descendants?

Figure 1-1 Offspring resemble their parents but are not identical to them. *(Paul Murphy/Unicorn Stock Photos.)*

THE SCIENCE OF GENETICS

The biggest scientific news of the year 2001 was the announcement, in scientific journals, newspaper headlines, and television features, that the sequence of the human genome had been determined. The London *Times* declared this accomplishment to be "a breathtaking moment for human health, even for philosophy. . . . The greatest scientific journey of this century starts here." The promise is that having the complete molecular description of human genes will not only lead to major advances in the diagnosis, treatment, and cure of disease, but will also revolutionize our understanding of what it is to be human.

Nor is the human genome the only one that has been sequenced. By the time the human genome was sequenced, complete gene sequences had also been obtained for more than thirty microorganisms; for a flowering plant, *Arabidopsis;* for a nematode worm that has become an important tool of genetic investigation, *Caenorhabditis elegans;* and for the fruit fly, *Drosophila melanogaster.* The immense efforts put into obtaining the DNA sequences of these organisms have been instigated not only by the desire to have more exact tools for genetic research, but also by the belief that such knowledge will lead to major advances in agricultural technology and in human health. Genetics has become the science at the center of public attention and excitement, just as nuclear physics was after the explosion of the first atomic bomb. Even for those who have no direct professional interest in genetics as an experimental science, it is obvious that a considerable knowledge of the facts and methods of genetics is essential, if only to have an informed view of the extraordinary claims and promises of the science.

The science of genetics is concerned with four specific puzzles about organisms that arise when we consider their similarities and differences. First, it is obvious that, in a very broad way, there is a resemblance between parents and their offspring. Humans give birth to other humans, dogs produce dogs, and there are no individuals that have some of the characteristics of each. How is this clear species identity passed from parents to their immediate offspring?

Second, the adults of humans, dogs, and even jellyfishes all develop from single fertilized egg cells that are seemingly quite similar to one another and show no hint of the very different sorts of organisms into which they will develop. How can a single cell, through its successive divisions, give rise during development to an enormously complex adult organism with anatomical and physiological features that are characteristic of its parental species? How does a single cell produce the immense diversity of cell types that make up a whole organism, arranged in tissues and organs that are like those of the parents that produced that cell?

Third, offspring do not resemble their parents exactly but differ from them in a variety of ways, depending on the characteristics being considered. In a sexually reproducing species like ours, in which there are clear anatomical differences between males and females, an individual offspring has the sexual anatomy of one of its two parents, while another child in the same family may have the other parent's sex. Yet this clear alternative between identity with one parent or with the other does not apply to other characteristics. A child may have her mother's eye color or a nose resembling her father's, and she may grow taller than either parent. What are the causes of these partial similarities to one parent or the other and differences from both?

Fourth, despite the fact that offspring resemble their parents, during the 3 billion years of the evolution of life on Earth, an immense variety of living forms have arisen from previously existing forms that were different from them. How can species as diverse as humans and jellyfishes, species that each reproduce their own kind, have descended from a remote common ancestor that differed from both of them?

All of these puzzles can be summarized by saying that genetics is a science concerned with explaining how cells and whole organisms can both resemble and differ from the cells and organisms that gave rise to them. As we shall see, the similarities and the differences are two aspects of the same underlying processes.

Genetics as a science had its origin in the desire of plant and animal breeders for a clear understanding of the inheritance of characteristics of economic importance in their orchards, fields, and flocks. In 1843 the Abbé Napp, who was the head of an Augustinian monastery in Moravia and an active amateur scientist, asked a teacher of physics to send him a bright student who might like to join the monastery and work on the problem of understanding the quantitative nature of variation in fruit trees. As a result, Gregor Mendel, pictured amid his fellow monks in Figure 1-2, was recruited into the monastery, where he carried out his now-famous experiments on garden peas (fruit trees, it turned out, were too large and slow-growing) that became the basis for the modern science of genetics. What Mendel showed was that inheritance is not, as previously thought, the result of the mixing of blood or some other continuous substance contributed by an individual's two parents. On the contrary, he found that for each characteristic that he studied, each individual carried two particles, which he called "factors," one "factor" having been inherited from its male parent and one from its female parent. Moreover, these "factors" kept their individuality in the offspring, and when that offspring produced gametes during its own reproduction, each gamete would contain one, and only one, of the two "factors," unmodified by its temporary passage through the individual.

Figure 1-2 Gregor Mendel (*standing, second from right*) with his fellow monks around 1862, seven years before he published his experiments on peas. He is holding a sprig of fuchsia, another plant on which he carried out hybridization experiments. *(Courtesy of the Moravian Museum, Brno.)*

Mendel's "factors" are now called genes, and modern genetics, since the beginning of the twentieth century, has been devoted to establishing the physical nature of genes, the cellular mechanisms by which genes are parceled out to separate gametes at reproduction, the pathways by which genes influence the development of the organisms that carry them, and the consequences of the entire apparatus of inheritance for evolution. Roughly the first third of the century was devoted to showing definitively that the genes, whatever their physical nature might turn out to be, are linearly arranged on threadlike bodies in cells, called chromosomes, and that the pattern of distribution of genes into gametes is a consequence of the physical behavior of the chromosomes during the cell divisions that give rise to gametes. During the second third of the century, the chemical nature of the genes was shown to be DNA (deoxyribonucleic acid), and it was shown how the chemistry and structure of this molecule allow three critical properties of genes to be realized.

- First, the cell manufactures copies of the DNA in such a way that the billions of cells produced during the development and reproduction of an individual carry identical copies of the original genes contained in the zygote from which the individual develops.
- Second, the cells read the information encoded in different bits and pieces of the extremely long DNA molecule—the different genes—and use it to direct the manufacture of the tens of thousands of different proteins of which the organism is made.
- Third, the diversity of cell forms and functions is possible because cells in different parts of the organism, at various times in development, read different parts of the DNA.

Research in the last part of the twentieth century, and continuing into our present time, has been devoted largely to elucidating the processes of spatial and temporal differentiation of cells and tissues by showing in detail how the molecules that result from reading and translating one part of the DNA molecule govern where and when in the organism other parts of the DNA will be read. Thus, the development of an organism is now understood as the successive reading of different genes at different times and in different places, as determined by proteins made from the information in other genes at previous times.

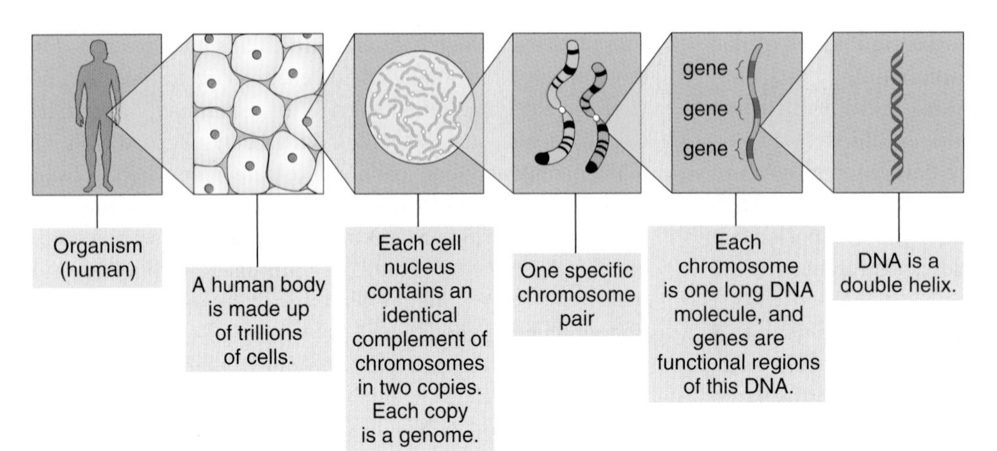

Figure 1-3 Successive enlargements bringing the genetic material of an organism into sharper focus.

THE PHYSICAL AND CHEMICAL BASIS OF HEREDITY

An organism's basic complement of DNA is called its **genome.** The somatic cells of most plants and animals contain two copies of their genomes (Figure 1-3); these organisms are **diploid.** The cells of most fungi, algae, and bacteria contain just one copy of the genome; these organisms are **haploid.** The genome itself is made up of one or more extremely long molecules of DNA that are organized into chromosomes. **Genes** are simply the regions of chromosomal DNA that are involved in the cell's production of proteins. Each chromosome in the genome carries a different array of genes. In diploid cells, each chromosome and its component genes are present twice. For example, human somatic cells contain two sets of 23 chromosomes, for a total of 46 chromosomes. Two chromosomes with the same gene array are said to be **homologous.** When a cell divides, all its chromosomes (its one or two copies of the genome) are replicated and then separated, so that each daughter cell contains the full complement of chromosomes.

To understand replication, we need to understand the basic nature of DNA. DNA is a linear, double-helical structure that looks rather like a molecular spiral staircase. The double helix is composed of two intertwined chains made up of building blocks called **nucleotides.** Each nucleotide consists of a phosphate group, a deoxyribose sugar molecule, and one of four different nitrogenous bases: adenine, guanine, cytosine, or thymine. Each of the four nucleotides is usually designated by the first letter of the base it contains: A, G, C, or T. Each nucleotide chain is held together by bonds between the sugar and phosphate portions of the consecutive nucleotides, which form the "backbone" of the chain. The two intertwined chains are held together by weak bonds between bases on opposite chains (Figure 1-4). There is a "lock-and-key" fit between the bases on the opposite strands, such that adenine pairs only with thymine and guanine pairs only with cytosine. The bases that form base pairs are said to be **complementary.** Hence a short segment of DNA drawn with arbitrary nucleotide sequence might be

···CAGT···

···GTCA···

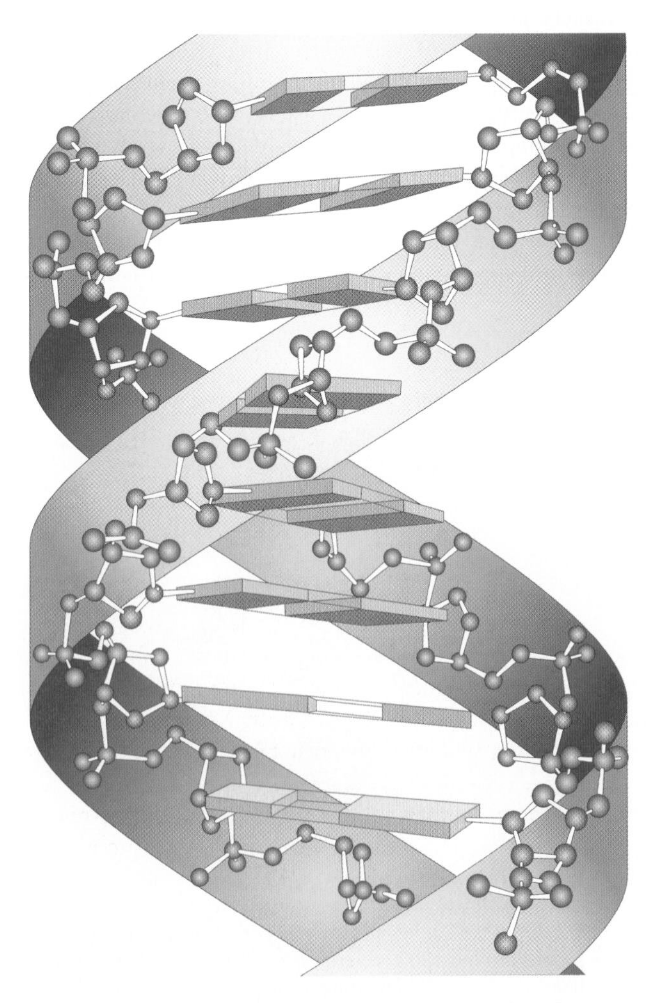

Figure 1-4 A representation of the DNA double helix. Blue = sugar-phosphate backbone; brown = base pairs.

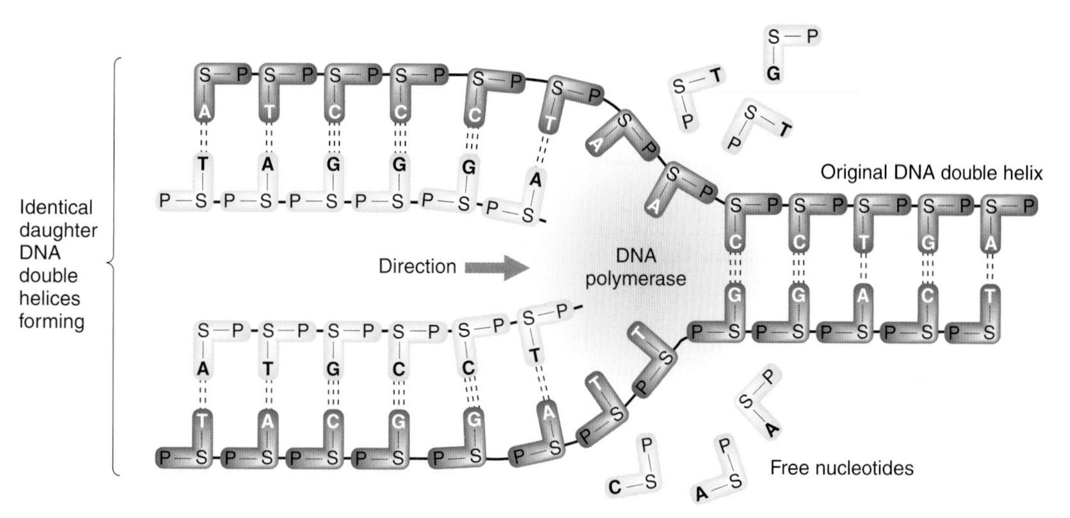

Figure 1-5 DNA replication in progress. Dark blue = nucleotides of the original double helix; light blue = new nucleotides being polymerized to form daughter chains. S = sugar; P = phosphate group.

MESSAGE

DNA is composed of two nucleotide chains held together by complementary pairing of A with T and G with C.

For replication of DNA to take place, the two strands of the double helix must come apart, rather like the opening of a zipper. The two exposed nucleotide chains then act as alignment guides, or templates, for the deposition of free nucleotides, which are then joined together by the enzyme DNA polymerase to form a new strand. The crucial point illustrated in Figure 1-5 is that, because of base complementarity, the two daughter DNA molecules are identical to each other and to the original molecule.

MESSAGE

DNA is replicated by the unwinding of the two strands of the double helix and the building up of a new complementary strand on each of the separated strands of the original double helix.

FROM GENE TO PROTEIN

The biological role of most genes is to carry information specifying the chemical composition of proteins and the regulatory signals that will govern their production by the cell. This information is encoded by the sequence of nucleotides. A typical gene contains the information for one specific protein. The collection of proteins an organism can synthesize, as well as the timing and amount of production of each protein, is an extremely important determinant of the structure and physiology of organisms. Proteins are important both as structural components—such as the proteins that constitute hair, skin, and muscle—and as active agents in cellular processes—such as enzymes and active transport proteins.

The primary structure of a protein is a linear chain of amino acids. This primary chain, called a **polypeptide,** is coiled and folded—and, in some cases, associated with other chains or small molecules—to form a functional protein. A given amino acid sequence may fold in a large number of stable ways. The final folded state of a protein depends both on the sequence of amino acids specified by its gene and on the physiology of the cell during folding.

MESSAGE

The sequence of nucleotides in a gene specifies the sequence of amino acids that is put together by the cell to produce a polypeptide. This polypeptide then folds under the influence of its amino acid sequence and other molecular conditions in the cell to form a protein.

The first step taken by the cell to make a protein is to copy, or **transcribe,** the nucleotide sequence in one strand of the gene into a complementary single-stranded molecule called **ribonucleic acid (RNA).** Like DNA, RNA is composed of nucleotides, but these nucleotides contain the sugar ribose instead of deoxyribose. Furthermore, in place of thymine, RNA contains uracil (U), which, like thymine, pairs with adenine. Hence the RNA bases are A, G, C, and U. The transcription process, which occurs in the cell nucleus, is very similar to the process for replication of DNA because the DNA strand serves as the template for making the RNA copy, which is called a **transcript.** The RNA transcript, which in many species undergoes some structural modifications, becomes a "working copy" of the information in the gene, a kind of "message" molecule called **messenger RNA (mRNA).** The mRNA then enters the cytoplasm, where it is used by the cellular machinery to direct the manufacture of a protein.

The determination of when and where the cell will make an RNA transcript of any particular gene, and the process of the final formation of the RNA "working copy"

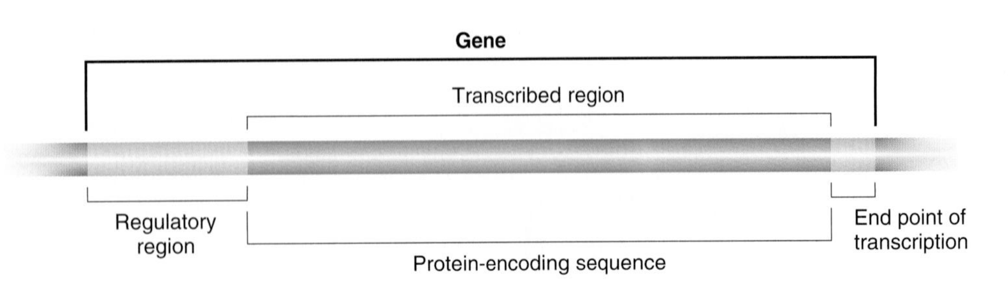

Figure 1-6 Generalized structure of a gene.

of the gene from such a transcript, depends on the structure of that gene. Figure 1-6 shows the general structure of a gene. At one end, there is a regulatory region to which various proteins involved in the regulation of the gene's transcription bind, causing the gene to be transcribed at the right time and in the right amount. A region at the other end of the gene signals the end point of the gene's transcription. Between these two end regions lies the DNA sequence that will be transcribed to specify the amino acid sequence of a polypeptide.

Some protein-encoding genes are transcribed more or less constantly; these are the "housekeeping" genes that are always needed for basic reactions. Other genes may be rendered unreadable or readable to suit the functions of the organism at particular times and under particular external conditions. The signal that masks or unmasks a gene may come from outside the cell; for example, from a steroid hormone or a nutrient. Or the signal may come from within the cell as the result of the reading of other genes. In either case, special regulatory sequences in the DNA are directly affected by the signal, and they in turn affect the transcription of the protein-encoding gene. The regulatory substances that serve as signals bind to the regulatory region of the target gene to control the synthesis of transcripts.

MESSAGE
During transcription, one of the DNA strands of a gene acts as a template for the synthesis of a complementary RNA molecule.

The process of producing a chain of amino acids based on the sequence of nucleotides in the mRNA is called **translation.** The nucleotide sequence of an mRNA molecule is "read" from one end of the mRNA to the other, in groups of three successive bases. These groups of three are called **codons.**

···	AUU	CCG	UAC	GUA	AAU	UUG	···
	codon	codon	codon	codon	codon	codon	

Because there are four different nucleotides, there are $4 \times 4 \times 4 = 64$ different possible codons, each one coding for an amino acid or a signal to terminate translation. Because only 20 kinds of amino acids are used in the polypeptides that make up proteins, more than one codon may correspond to the same amino acid. For instance, AUU, AUC, and AUA all encode isoleucine, while UUU and UUC code for phenylalanine, and UAG is a translation termination ("stop") codon.

Protein synthesis takes place on cytoplasmic organelles called **ribosomes.** A ribosome attaches to one end of an mRNA molecule and moves along the mRNA, catalyzing the assembly of the string of amino acids that will constitute the primary polypeptide chain of the protein. Each kind of amino acid is brought to the assembly process by a small RNA molecule called **transfer RNA (tRNA),** which is complementary to the mRNA codon that is being read by the ribosome at that point in the assembly.

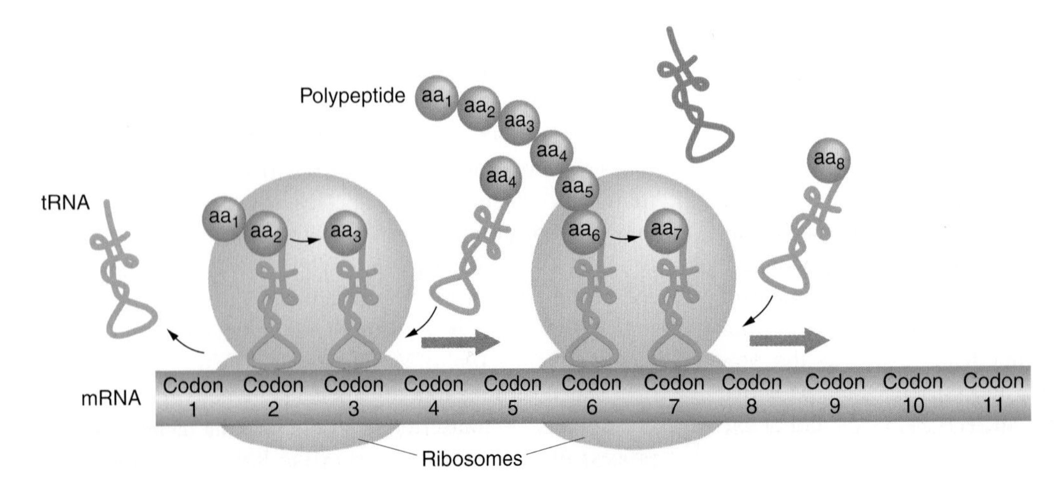

Figure 1-7 The addition of an amino acid (aa) to a growing polypeptide chain in the translation of mRNA. Multiple copies of the polypeptide are produced by a train of ribosomes following each other along the mRNA; two such ribosomes are shown.

Trains of ribosomes pass along an mRNA molecule, each member of a train making the same type of polypeptide. At the end of the mRNA, a termination codon causes the ribosome to detach and recycle to another mRNA. The process of translation is shown in Figure 1-7.

MESSAGE

> The information in genes is used by the cell in two steps of information transfer: DNA is transcribed into mRNA, which is then translated into the amino acid sequence of a polypeptide. The flow of information from DNA to RNA to protein is a central focus of modern biology.

GENETIC VARIATION

All members of the same species carry a set of genes that specifies the amino acid sequences, as well as the times and locations of production, of the set of different proteins that make up the individuals of that species. But any particular gene may exist in slightly different forms in different individuals. Different forms of the same gene, located in the same position along the chromosome in each individual, are called **alleles.** Some allelic DNA differences—for example, those that change one codon into another codon that still codes for the same amino acid—will not change the amino acid sequence of the protein specified by the gene. Other DNA differences, however, will change one or more of the amino acids in the protein, while yet others—those in regulatory regions—will alter the timing or location of production of the protein. This allelic variation is the basis for hereditary variation in the observed properties of organisms. In a population, for any given gene, there can be from one to many alleles; however, because most organisms carry only one or two chromosome sets per cell, an individual organism can carry only one or two alleles per gene.

Types of Variation

Until about 30 years ago, when methods were first invented to determine the actual nucleotide sequence of a gene isolated from an individual organism, the genetic variation studied by geneticists was inferred from differences in the appearance and physiology of the organism. These differences included variation among individuals in shape, size, fertility, longevity, behavior, and physiology, and in the chemical constitution and activity of their proteins. One of the difficulties of inferring genetic differences from such differences in morphology or physiology is that environmental variation, as well as genetic variation, contributes to the observed variation among individual organisms. The closest geneticists could get to observing variation in the genes themselves was to study differences in the shapes of chromosomes.

The characterization of organisms by their appearance or physiology sorts individuals into observable groups called **phenotypes.** "Blue eyes" is a phenotype, as is "5 feet 8 inches" or "blood type A" or "having sickle-cell anemia." The classification of individuals by allelic constitution, or **genotype,** was carried out by crossing individuals with different phenotypes and inferring their genetic composition from the phenotypes of the progeny that were produced. Now that sequencing the DNA of specific genes is a routine procedure, it is possible to sort individuals into genotypes by a direct description of their genes. By studying both the phenotype and the genotype of the same individuals, we can ask in a direct way how genetic variation influences the phenotypic differences among organisms—the problem that, in the end, is the motivation for the study of genetics.

The possibility of relating allelic variation in specific genes to specific phenotypic differences is greatly increased if there are only a small number of alternative phenotypic classes with no intermediates between them. In such cases it usually turns out that each phenotypic class corresponds to one or two genotypic classes formed by different alleles of a single gene. It is a great deal more difficult to determine the relationship between genotype and phenotype when there are many alternative phenotypic classes distinguished only by small, measurable differences. Sometimes these two kinds of phenotypic variation are described as **discontinuous,** or **qualitative,** variation and **continuous,** or **quantitative,** variation, respectively (Figure 1-8 on the next page), but these distinctions must not be taken too literally. For example, insects have large numbers of short bristles on their bodies. In a particular region of the abdomen, one fruit fly may have twelve bristles and another thirteen, but no fly has twelve and a half. Bristle number, although it may vary as widely as from, say, five to twenty, is a discontinuous character with no possibility of intermediates between adjacent classes, yet bristle number is a classic example of a "quantitative" character because there is a large range of phenotypes that differ from each other in small steps. Sometimes a character that is intrinsically continuous will nevertheless have one phenotypic class that is far removed in value from the rest of the distribution and can be treated as a qualitative difference, as, for example, a dwarf variety of a plant that also has normal variation in plant height. Such a dwarf may turn out to be the consequence of a single drastic allelic difference in one gene.

MESSAGE

> In many cases, an allelic difference in a single gene results in discrete phenotypic forms that make it easy to study the gene and its associated biological function.

Discontinuous variation. A good example of a genetically simple discontinuous character is albinism in humans. In most people, the cells of the skin can make a dark brown or

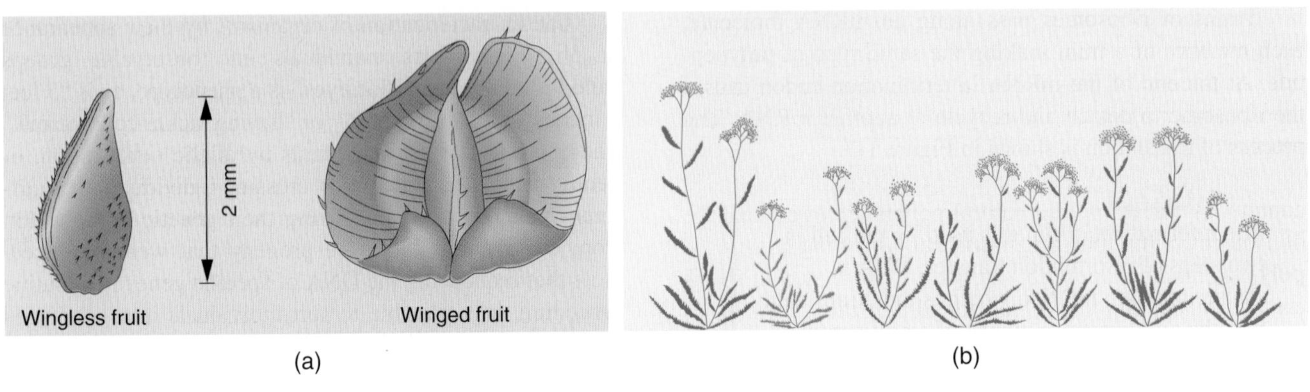

Figure 1-8 Examples of discontinuous and continuous variation in natural populations. (a) Fruits of the sea blush, *Plectritis congesta,* have one of two distinct forms. Any one plant has either all winged or all wingless fruits. (b) Variation in height, branch number, and flower number in the herb *Achillea.* *(Part b, Carnegie Institution of Washington.)*

black pigment called melanin, the substance that gives our skin its color. Human skin pigmentation ranges continuously from tan in people of European ancestry to brown or black in those of tropical and subtropical ancestry; this variation in skin pigmentation has proved very difficult to analyze genetically. However, rare albinos, who have totally pigmentless skin and hair, are found in all races (Figure 1-9). The difference between unpigmented skin and skin with any degree of pigmentation is caused by different alleles of a gene that encodes an enzyme involved in melanin synthesis.

The alleles of a gene are conventionally designated by letters. The allele that codes for the normal form of the enzyme involved in making melanin is called *A*, and the allele that codes for an inactive form of that enzyme (resulting in albinism) is designated *a*, to show that they are related. Because humans have two sets of chromosomes in each cell, a person's genotype can be A / A, A / a, or a / a (the slash shows that the two alleles are a pair). The phenotype of A / A is pigmented, that of a / a is albino, and that of A / a is pigmented. Because the presence of a single copy of the *A* allele is sufficient to allow the production of melanin, the allele *A* is said to be **dominant** over the allele *a*, which is said to be **recessive** (see Chapter 2).

Although allelic differences cause phenotypic differences such as pigmentation and albinism, this does not mean that only one gene affects skin color. It is known that there are several genes affecting skin color, although the identity and number of these genes are presently unknown. Nor do we know whether the pigmentation of Asians differs from that of Europeans by differences in the same genes that differ between Europeans and Africans. However, the difference between having some pigmentation, of whatever shade, and albinism is caused by an allelic difference in the one gene that determines the ability to make melanin; the allelic composition of other genes involved in skin pigmentation is irrelevant to this comparison.

In some cases of qualitative variation, there is a predictable one-to-one relation between allelic composition and phenotype under most conditions. In other words, the different phenotypes (and their underlying genotypes) can almost always be distinguished. In the albinism example, the *A* allele always allows some pigment formation, whereas the *a* allele always results in albinism when present in two copies. For this reason, discontinuous variation has been successfully used by geneticists to identify the alleles underlying quantitative characters and their roles in cellular functions. For most cases of "qualitative" variation, however, as we shall see below, the phenotypic classes may blend into each other under some environmental conditions but not others. If,

Figure 1-9 An albino. Such individuals lack the dark pigment melanin in the cells of the skin, hair, and retina. *(© Yves Gellie/Icône.)*

however, organisms are allowed to develop under controlled, uniform environmental conditions that allow an unambiguous classification of individuals into clear phenotypic classes, these variations too can be used successfully for simple genetic analysis.

Geneticists distinguish two categories of simple discontinuous variation. In a natural population, the existence of two or more common discontinuous variants is called **polymorphism** (Greek: "many forms"). The variant forms are called **morphs.** It is often found that different morphs are determined by different alleles of a single gene. Why do populations show genetic polymorphism? Special types of natural selection can explain a few cases, but in other cases, the morphs seem to be selectively neutral.

Rare, exceptional discontinuous variants are called **mutants,** whereas the more common, "normal" phenotype is called the **wild type.** Again, in many cases, the wild-type and mutant phenotypes are determined by different alleles of one gene. Both mutants and polymorphisms originally arise from rare changes in DNA **(mutations),** but somehow the mutant alleles of a polymorphism become common. Mutations (such as those that produce albinism) can occur spontaneously in nature, or they can be produced by treatment with mutagenic chemicals or radiation. Geneticists regularly induce mutations in experimental organisms and carry out genetic analyses on them because mutations that affect some specific biological function can identify the various genes that interact in that function.

Continuous variation. A continuously varying character shows an unbroken range of phenotypes in a population (see Figure 1-8b), and intermediate phenotypes are generally more common than extreme ones. In some cases, all the variation is environmental and has no genetic basis at all, as, for example, in the case of the slightly different regional dialects of a language spoken by different human groups. In other cases, such as that of the various shades of human eye color, the differences are caused by allelic variation in one or many genes, with no known effect of environmental variation. For most continuously variable traits, however, both genetic and environmental variation contribute to the observed phenotypic differences. In continuous variation, there is no one-to-one correspondence between genotype and phenotype. For this reason, little is known about the types of genes underlying continuous variation, and only recently have techniques become available for identifying and characterizing them.

Continuous variation is encountered more commonly than discontinuous variation in everyday life. We can all identify examples of continuous variation, such as variation in size or shape, in plant or animal populations that we have observed as well as in human populations. One area of genetics in which continuous variation is important is in plant and animal breeding. Many of the characters that are under selection in breeding programs, such as seed weight or milk production, have complex determination arising from many gene differences interacting with environmental variation and show continuous variation in populations. We shall return to the specialized techniques for analyzing continuous variation in Chapter 18, but, for the greater part of this book, we will be dealing with the genes underlying discontinuous variation.

The Molecular Basis of Simple Genetic Variation

Consider the difference between the pigmented and the albino phenotypes in humans. The dark pigment melanin has a complex structure that is the end product of a biochemical synthetic pathway (Figure 1-10 on the next page). Each step in the pathway involves a conversion of one molecule into another, leading to the formation of melanin in a step-by-step manner. Each step is catalyzed by a separate enzyme protein encoded by a specific gene. Most cases of albinism result from changes in one of these enzymes: tyrosinase. The enzyme tyrosinase catalyzes the last step of the pathway, the conversion of tyrosine into melanin.

To perform this task, tyrosinase binds to its substrate, a molecule of tyrosine, and facilitates the molecular changes necessary to produce melanin. There is a specific lock-and-key fit between tyrosine and a portion of the enzyme called the active site, which is a pocket formed by several crucial amino acids in the tyrosinase polypeptide. If a mutation changes the nucleotide sequence of the tyrosinase-encoding gene in such a way that one of these crucial amino acids is replaced by another or lost, then there are several possible consequences. First, the enzyme might still be able to perform its function, but in a less efficient manner. Such a change might have only a small effect at the phenotypic level, so small as to be difficult to observe, but it might lead to a reduction in the amount of melanin formed and, consequently, to a lighter skin color. Note that the tyrosinase protein is still more or less intact, but its ability to convert tyrosine into melanin has been compromised. Second, the enzyme might be incapable of any function, in which case the mutational event in the DNA would have produced a **null allele,** referred to earlier as the a allele. Hence a person of genotype a / a will be an albino. The genotype A / a results in normal pigmentation because transcription of one copy of the wild-type allele (A) can provide enough tyrosinase for synthesis of normal amounts of melanin.

In general, a mutation in a gene changes the amino acid composition of the encoded protein. The most important outcomes are changes in the shape and size of the folded protein. Such changes can result in no biological function (which would be the basis of a null allele) or reduced function. More rarely, mutation can lead to a new function for the protein produced (see Chapter 19).

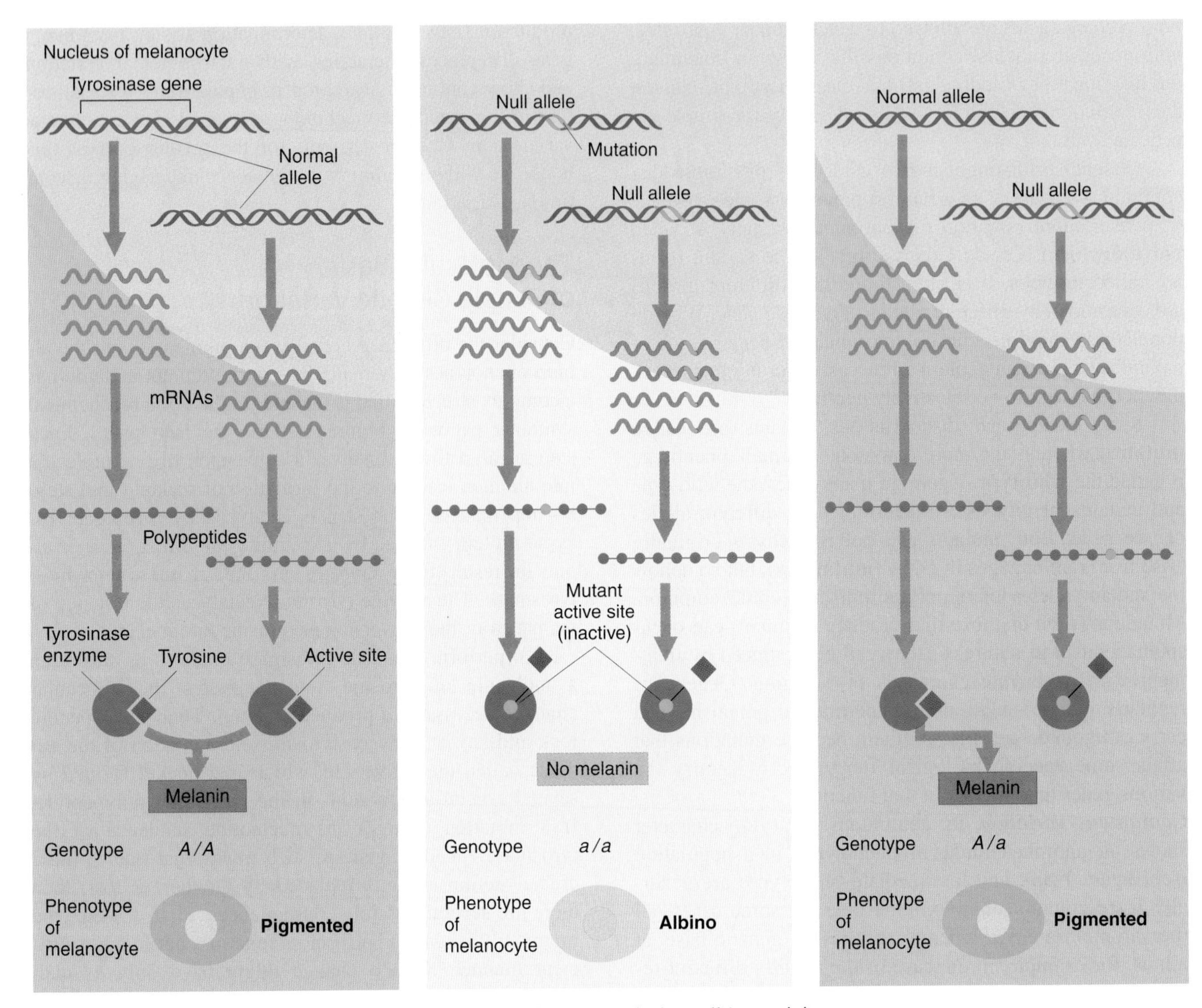

Figure 1-10 The molecular basis of albinism. Melanocytes (pigment-producing cells) containing two copies of the normal tyrosinase allele (*A*) produce the tyrosinase enzyme, which converts the amino acid tyrosine into the pigment melanin. Melanocytes containing two copies of the mutant null allele (*a*) are unable to produce any of the enzyme. Melanocytes containing one copy of the normal allele make enough tyrosinase to allow production of melanin and the pigmented phenotype.

METHODOLOGIES USED IN GENETICS

The study of genes has proved to be a powerful approach to understanding biological systems. Because genes affect virtually every aspect of the structure and function of an organism, being able to identify and determine the roles of genes and the proteins that they specify is an important step in charting the various processes that underlie a particular character under investigation. Geneticists study not only hereditary mechanisms, but all biological mechanisms. Many different methodologies are used to study genes and gene activities, and these methodologies can be summarized briefly as follows:

1. Isolation of mutations affecting the biological process under study. Each mutant gene reveals a genetic component of the process, and together they show the range of proteins that interact in that process.
2. Analysis of progeny of controlled matings ("crosses") between mutant and wild-type individuals or between other discontinuous variants. This type of analysis identifies genes and their alleles, their chromosomal locations, and their inheritance patterns.
3. Biochemical analysis of cellular processes that involve genes and the products of their transcription and

translation. Life is basically a complex set of chemical reactions, so studying the ways in which genes are relevant to these reactions is an important way of dissecting this complex chemistry. Mutant alleles underlying defective function (see method 1) are invaluable in this type of analysis. The basic approach is to find out how the cellular chemistry is disturbed in the mutant individual and, from this information, deduce the role of the product specified by the gene. The deductions from many genes are assembled to reveal the larger picture.

4. Microscopic analysis. The study of chromosome structure and movement has long been an integral part of genetics, but new technologies have provided ways of labeling genes and gene products so that their locations can be easily visualized under the microscope.
5. Direct analysis of DNA. Because the genetic material is composed of DNA, the ultimate characterization of a gene is the analysis of the DNA sequence itself. Many techniques, including gene cloning, are used to accomplish this. Cloning is a procedure by which an individual gene can be isolated and amplified (copied multiple times) to produce a pure sample for analysis. One way of doing this is by inserting the gene of interest into a small bacterial chromosome and allowing bacteria to do the job of copying the inserted DNA. After the clone of a gene has been obtained, its nucleotide sequence can be determined, and hence important information about its structure and function can be obtained. The sequencing of the entire genomes of organisms **(genomics)** and the extraction of information about the organization of genomes, their evolution, and the function of the proteins they encode from DNA sequences **(bioinformatics)** is now a major effort of molecular genetics.

GENES IN DEVELOPMENT

How can a single cell—a fertilized zygote—by successive cell divisions produce a whole organism with a myriad of different cell types, arranged into tissues and organs with different shapes and physiological functions, all in characteristic positions in the body? How does such heterogeneity arise from apparent homogeneity in a way that is reliably characteristic of a given species and reliably transmitted from parent to offspring? Our picture of the structure of genes and the way in which that structure is used by the cell machinery to produce proteins provides us with an outline of how the differentiation of form and function in a developing organism occurs.

The different shapes and physiologies of different cells and tissues are a consequence of the manufacture of different proteins at different times and in different places in the developing individual. This means that at different times and in different places, different genes must be read by the cells that contain them. Yet all the cells of the individual have the same genes. The key to understanding how different genes are read in different cells lies in the regulatory DNA sequences that are part of each gene. These sequences act as "switches" to determine whether a particular coding sequence will be read by the transcriptional apparatus of the cell. These switches are set to the "on" or "off" state by whether or not they are bound to regulatory molecules that are moving from place to place in the cell. The binding of such a molecule to the regulatory region of a gene's DNA is a reversible chemical reaction, so that a given gene in a given cell may be sometimes in a readable state and sometimes not, depending on the availability of the binding molecules and the stability of the binding. Thus some genes may be read at a very high rate for long periods, while others are read only for short periods of time or at a low rate.

The regulatory molecules that bind to a regulatory sequence are themselves the products of the reading of other genes. They may be the RNA molecules directly resulting from transcription of these other genes, or they may be proteins that have resulted from the folding of translated polypeptides. In the latter case, the folded form of the regulatory protein may be influenced by combination with small molecules such as hormones that have come into the cell from outside. This effect of small molecules on the folding of a regulatory protein is an important route through which conditions outside of an organism can affect the transcription of its genes and thus influence its development.

A very simple model of such an influence is the regulation of the production of the enzyme β-galactosidase, which allows the bacterium *E. coli* to use the sugar lactose as a source of carbon (Figure 1-11 on the next page). The enzyme is produced only when there is lactose in the bacterium's culture medium (an example of how the environment of a cell influences its phenotype). In the absence of lactose, a regulatory protein called a repressor binds to the regulatory sequence of the β-galactosidase gene and prevents the gene's transcription. When lactose molecules from the medium diffuse into the bacterial cells and bind to the repressor, they change its three-dimensional configuration in such a way that it, in turn, no longer binds to the regulatory sequence of the gene. It falls off the regulatory sequence, and the transcriptional machinery of the cell can now read the gene and produce the enzyme. But the repressor protein was itself produced by the reading of a different DNA sequence, so we see how the protein product of the reading of one gene can affect the reading of another gene.

The regulation of transcription of one gene by the products of other genes, sometimes modified by small molecules, provides us with a general model of how developmental differentiation occurs. Long chains of signals can be passed along as the product of one gene affects the reading of another, whose product affects the reading of yet another,

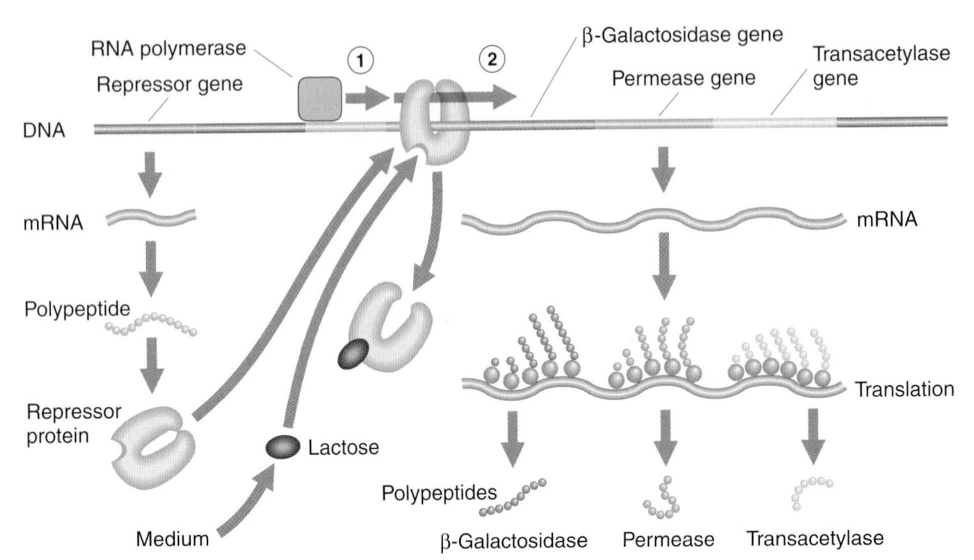

Figure 1-11 Regulation of the production of the enzymes β-galactosidase, permease, and transacetylase, which are part of a functional group in lactose metabolism in the bacterium *E. coli*. In the absence of lactose, a regulatory protein (the repressor) binds to the regulatory region of the β-galactosidase gene, preventing its transcription into RNA by the cell (blocked path 1). When lactose from the culture medium enters the cell, a lactose molecule associates with the repressor protein, changing its shape so that it can no longer bind to the regulatory region of the gene. The repressor protein falls off the DNA and allows transcription of the genes to proceed (unblocked path 2).

Animated Art • Regulation of the Lactose System in *E. coli*

and on and on, so that a sequence of proteins is produced. Some regulatory proteins are produced in the same cell in which they perform their regulatory function. Others enter the cell from neighboring cells, and still other regulatory molecules, such as hormones, may diffuse widely throughout the developing embryo. The ability of a signal to regulate the transcription of a particular gene in a given cell depends on the local cell conditions, including the permeability of the cell membrane to signaling molecules and the presence of other molecules that interact with them. Thus cells on the inside of the embryo will make different proteins from those on the outside, and cells very distant from the source of a protein signal will experience a different signaling level from those close to the source of the signal. Chapters 13 and 16 discuss in detail how the development of complex organisms from single cells is now understood in terms of these signaling pathways.

MESSAGE

The development of an organism is the consequence of the synthesis of different proteins in different cells at different times. The amino acid composition of these proteins is coded in the DNA sequences of different genes. The time, place, and rate of transcription and translation of these DNA sequences are dependent on the transcribed regulatory products of yet other DNA sequences.

GENES, THE ENVIRONMENT, AND THE ORGANISM

Organisms develop through the acquisition of materials from the world around them and the conversion of those materials into the various chemical forms—mostly proteins—that are characteristic of the organism. This process, as well as the maintenance of the individual throughout its life, requires energy, the sources of which are also external. While everyone recognizes these inputs into the organism, it would seem, at first sight, that the nature of the organism is internally determined by the particular form of the genes that it has inherited from its parents. Our discussion of genes in development hints that there are pathways by which the conditions of the external world may influence the outcome of genetic signaling pathways, but there is, in fact, very little concrete information at the molecular level about the interplay between gene and environment in development. In contrast, there is a large body of information at the level of phenotype—of shape, size, metabolism, and behavior—that shows a very strong gene-environment interdependence for many characters, even though we do not at present understand this interdependence at the cellular and molecular levels. The only generalization that we can now make is that for most characteristics of an individual, the information necessary to differentiate it from other individuals of its species is not completely contained in its genome. The degree to which particular characters in particular species are functions of internal and external factors varies from species to species and character to character. This variation gives rise to several different models of phenotypic determination.

MESSAGE

As an organism transforms developmentally from one stage to another, the DNA sequence of its genes and the environmental conditions in which it is developing jointly determine its future development at each moment of its life history.

Model I: Genetic Determination

It is clear that virtually all of the differences between species are determined by the differences in their genomes. There is no environment in which a lion will give birth to a lamb. An acorn develops into an oak, whereas the spore of a moss develops into a moss, even if both are growing side by side in the same forest. The two plants that result from these developmental processes resemble their parents and differ from each other, even though they have access to the same narrow range of materials from the environment.

Even within species, some variation is entirely a consequence of genetic differences that cannot be modified by any change in what we normally think of as environment. The children of African slaves brought to North America had dark skins, unchanged by the relocation of their parents from tropical Africa to temperate Maryland. The possibility of much of experimental genetics depends on the fact that many of the phenotypic differences between mutant and wild-type individuals resulting from allelic differences are insensitive to environmental conditions. The determinative power of genes is often demonstrated by differences in which one allele is normal and the other abnormal. The human inherited disease sickle-cell anemia is a good example. The underlying cause of the disease is a variant of hemoglobin, the oxygen-transporting protein molecule found in red blood cells. Normal people have a type of hemoglobin called hemoglobin A, the information for which is encoded in a gene. A single nucleotide change in the DNA of this gene, leading to a change in a single amino acid in the polypeptide, results in the production of a slightly changed hemoglobin, called hemoglobin S. In people possessing only hemoglobin S, the ultimate effect of this small change in DNA is severe ill health—sickle-cell anemia—and often death.

Such observations, if generalized, lead to a model, shown in Figure 1-12, of how genes and the environment interact. In this view, the genes act as a set of instructions for turning more or less undifferentiated environmental materials into a specific organism, much as blueprints specify what form of house is to be built from basic materials. The same bricks, mortar, wood, and nails can be made into an A-frame or a flat-roofed house, according to different plans.

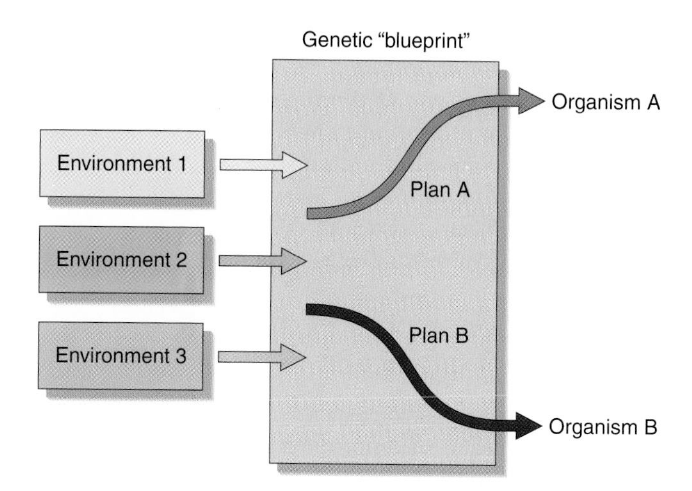

Figure 1-12 A model of phenotypic determination that emphasizes the role of genes.

Such a model implies that the genes are really the dominant elements in phenotypic determination; the environment simply supplies the undifferentiated raw materials.

Model II: Environmental Determination

Consider two monozygotic ("identical") twins, the products of a single fertilized egg that, at an early stage of development, divides to produce two complete babies with identical genes. Suppose that the twins are born in England but are separated at birth and taken to different countries. If one is reared in China by Chinese-speaking adoptive parents, she will speak Chinese, whereas her sister reared in Budapest will speak Hungarian. Each will absorb the cultural values and customs of her environment. Although the twins began life with identical genetic properties, the different cultural environments in which they live will produce differences between them (and differences from their biological parents). Obviously, the differences described in this case are due to the environment, and genetic effects are of little importance in determining those differences.

This example suggests the model shown in Figure 1-13, which is the converse of that shown in Figure 1-12. In

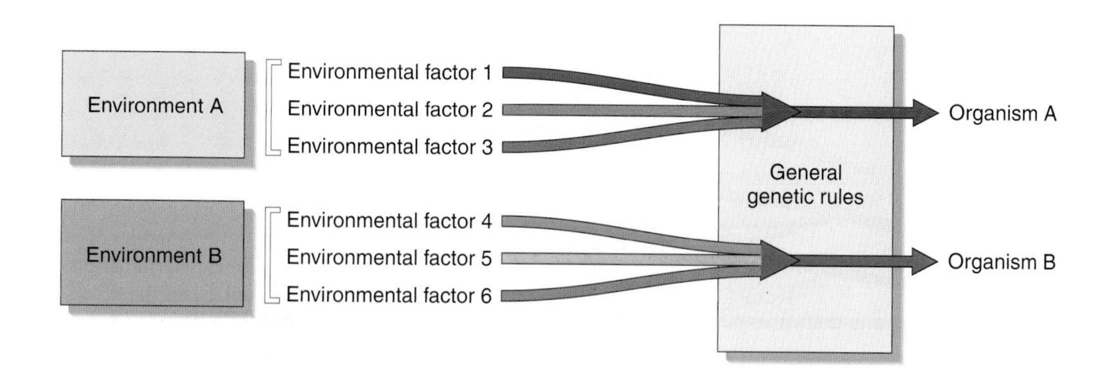

Figure 1-13 A model of phenotypic determination that emphasizes the role of the environment.

Figure 1-13, the genes impinge on the system, giving certain general signals for development, but the environment determines the actual course of development. Imagine a set of specifications for a house that simply call for "a floor that will support 300 pounds per square foot" or "walls with an insulation factor of 15 inches," allowing the actual appearance and other characteristics of the structure to be determined by the available building materials.

Model III: Genotype-Environment Interaction

Variations in most characteristics in most species conform to neither the genetic determination nor the environmental determination model. In general, to predict the phenotype of an organism, we need to know both the genetic constitution that it inherits from its parents and the historical sequence of environments to which it has been exposed. Every organism has a developmental history from conception to death. What an organism will become in the next moment depends critically on both its present state and the environment that it encounters during that moment. It makes a difference to an organism not only what environments it encounters, but in what sequence it encounters them. A fruit fly *(Drosophila melanogaster),* for example, develops normally at a temperature of 25°C. During that development, wings with a characteristic pattern of veins appear. There are a number of abnormal gene mutations that can disrupt this normal pattern. If the temperature is briefly raised to 37°C early in the pupal stage of development, the adult fly will be missing part of the normal vein pattern of its wings. However, if this "temperature shock" is administered just 24 hours later, the vein pattern will develop normally. This dependence of phenotype on both genotypic and environmental differences applies even to most single-gene mutations in an experimental organism such as *Drosophila.* Such environmental sensitivity makes mutants unsuitable for most experimental purposes, so only a small fraction of the mutations discovered in *Drosophila* have ever been used as experimental tools. A general model in which genes and the environment jointly determine (by some rules of development) the actual characteristics of an organism is depicted in Figure 1-14.

Norms of Reaction

For much of experimental genetics, for plant and animal breeding, and for trying to understand the relation between genetic variation and phenotypic differences among individuals within species in nature—for example, differential susceptibility to diseases—we need to do more than simply observe that phenotype is a consequence of the interaction between genes and environment. We need to make systematic descriptions of the consequences of that interaction that can be used to predict the phenotype in particular cases. How can we quantify the relation among the genotype, the environment, and the phenotype?

For a particular genotype, we could prepare a table showing the phenotype that would result from the development of that genotype in each possible environment. Such a set of environment-phenotype relations for a given genotype is called the **norm of reaction** of the genotype. In practice, we can make such a tabulation for only one or a few genes under study and one or a few particular aspects of the environment. For example, we might record the eye sizes that fruit flies have after developing at various constant temperatures; we could do this for several different eye-size genotypes to characterize their norms of reaction. Figure 1-15 represents just such norms of reaction for three eye-size genotypes in the fruit fly *Drosophila melanogaster.* The graph is a convenient summary of more extensive tabulated data. The size of the fly eye is measured by counting its individual units, called facets. The vertical axis of the graph shows the number of facets (on a logarithmic scale); the horizontal axis shows the constant temperature at which the flies developed.

Three norms of reaction are shown in Figure 1-15. When flies of the wild-type genotype that is characteristic of natural populations are raised at high temperatures, they develop eyes that are somewhat smaller than those of wild-type flies raised at cooler temperatures. The graph shows that wild-type phenotypes range from 700 to more than

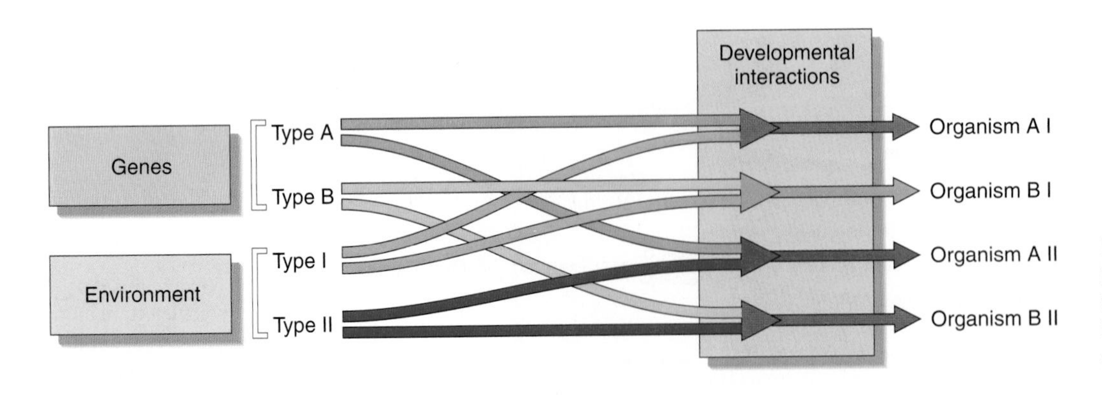

Figure 1-14 A model of phenotypic determination that emphasizes the interaction of genes and environment.

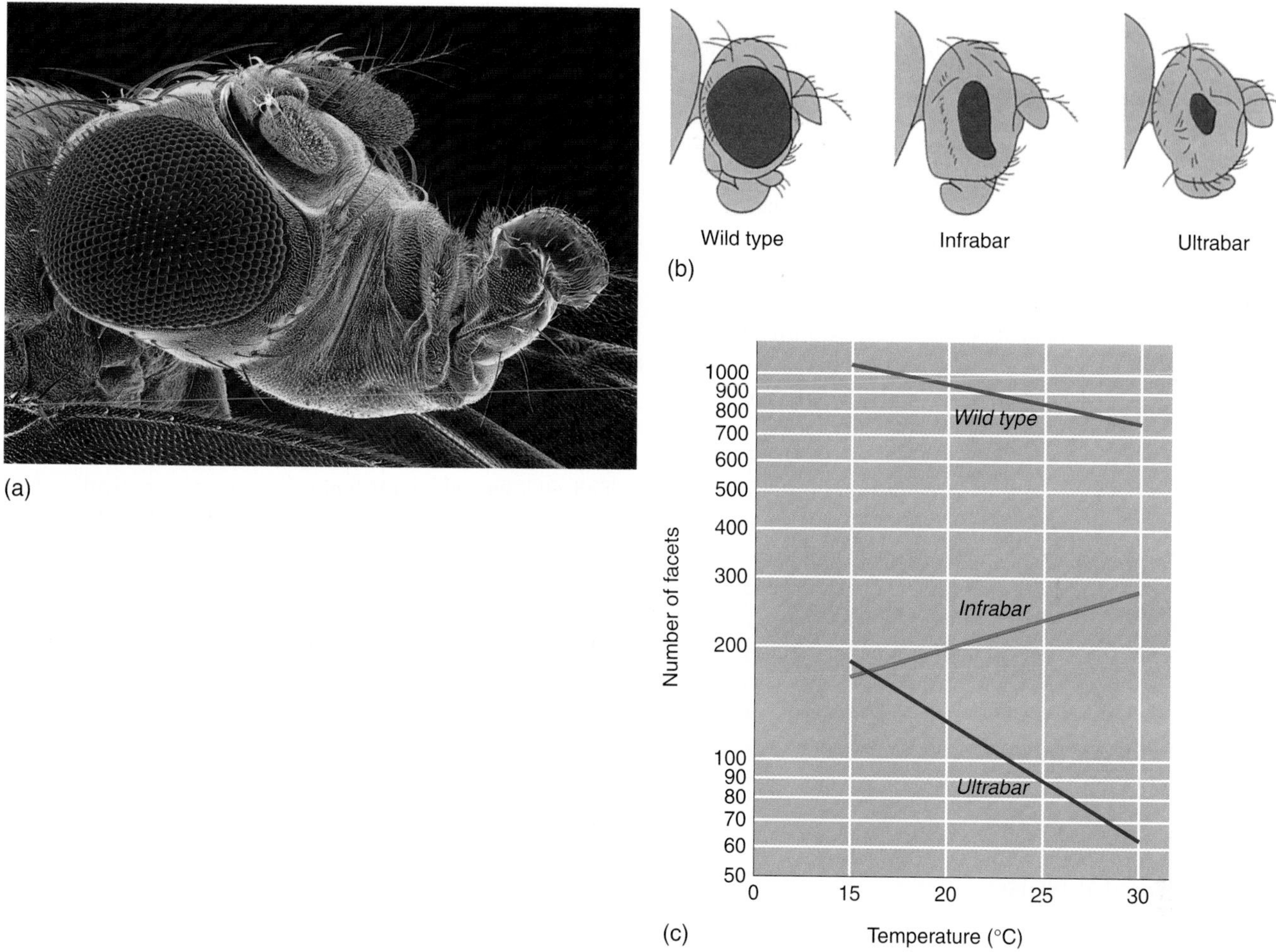

Figure 1-15 Norms of reaction to temperature during development for three different eye-size genotypes in *Drosophila.* (a) Close-up of a normal *Drosophila* eye, which comprises hundreds of units called facets. The number of facets determines eye size. (b) Relative eye sizes of wild-type, infrabar, and ultrabar flies raised at the higher end of the temperature range. (c) Norm of reaction curves for the three genotypes. *(Part a, Don Rio and Sima Misra, University of California, Berkeley.)*

1000 facets—the wild-type norm of reaction. A fly that has the *ultrabar* genotype has eyes that are smaller than those of wild-type flies, regardless of temperature during development. Temperatures during development have a stronger effect on *ultrabar* genotypes than on wild-type genotypes, as we can see by noticing that the *ultrabar* norm of reaction slopes more steeply than the wild-type norm of reaction. Any fly of the *infrabar* genotype also has smaller eyes than any wild-type fly, but temperatures have the opposite effect on flies of this genotype; infrabar flies raised at high temperatures tend to have larger eyes than those raised at lower temperatures. These norms of reaction indicate that the relation between genotype and phenotype is complex rather than simple.

If we know that a fruit fly has the wild-type genotype, this information alone does not tell us whether its eye has 800 or 1000 facets. Nor does the knowledge that a fruit fly's eye has 170 facets tell us whether its genotype is *ultrabar* or *infrabar.* We cannot even make a general statement about the effect of temperature on eye size in *Drosophila* because the effect is opposite in two different genotypes. We can see from Figure 1-15 that some genotypes do differ unambiguously in phenotype, no matter what the environment: any wild-type fly has larger eyes than any ultrabar or infrabar fly. But other genotypes overlap in their phenotypic expression: the eyes of an ultrabar fly may be larger or smaller than those of an infrabar fly, depending on the temperatures at which the individuals developed.

MESSAGE

A single genotype may produce different phenotypes, depending on the environment in which organisms develop. The same phenotype may be produced by different genotypes, depending on the environment.

It is important to understand that the norm of reaction of a genotype is simply a descriptive device to show the phenotypic outcome of experiencing different environments. It is not an explanation of the relationship among gene, environment, and phenotype, which can be provided only by a detailed understanding of the developmental and physiological pathways involved. Nevertheless, such observed norms of reaction are of immense predictive value in the practical applications of genetics, such as agriculture, pharmacology, and medicine, as well as in experimental genetics and in the application of genetic information to social policies.

At the present time, we do not know the norm of reaction of any human genotype for any character in any set of environments. To obtain a norm of reaction such as those shown in Figure 1-15, we must allow different individuals of identical genotype to develop in different environments. To carry out such an experiment, we must be able to obtain or produce many fertilized eggs with identical genotypes. To test a human genotype in 10 environments, for example, we would have to obtain genetically identical siblings and raise each of them in a different milieu. However, that is neither biologically nor ethically possible. Is it not clear how we could ever acquire norms of reaction for human characters without the unacceptable manipulation of human individuals.

Developmental Noise

Thus far, we have assumed that a phenotype is uniquely determined by the interaction of a specific genotype and a specific environment. But a closer look shows some further unexplained variation. According to Figure 1-15, a fruit fly of wild-type genotype raised at 16°C will have 1000 facets in each eye. In fact, this is only an average value; one fly raised at 16°C may have 980 facets and another may have 1020. Perhaps these variations are due to slight fluctuations in the local environment or slight differences in genotype. However, a typical count may show that a fly has, say, 1017 facets in the left eye and 982 in the right eye. In another wild-type fly raised at the same temperature, the left eye may have slightly fewer facets than the right eye. Yet the left and right eyes of the same fly are genetically identical. Furthermore, under typical experimental conditions, the fly develops as a larva (a few millimeters long), burrowing in homogeneous artificial food in a laboratory bottle, and then completes its development as a pupa (also a few millimeters long), glued vertically to the inside of the glass high above the food surface. Surely the environment does not differ significantly from one side of the fly to the other. But if the two eyes experience the same sequence of environments and are identical genetically, then why is there any phenotypic difference between the left and right eyes?

Differences in shape and size are partly dependent on the process of cell division that turns the zygote into a multicellular organism. Cell division, in turn, is sensitive to molecular events within the cell, and these events may have a relatively large random component. For example, the vitamin biotin is essential for *Drosophila* growth, but its average concentration is only one molecule per cell. Obviously, the rate of any process that depends on the presence of this molecule will fluctuate as the availability of biotin varies. But cells can divide to produce differentiated eye cells only within the relatively short developmental period during which the eye is being formed. Thus, we would expect random variation in such phenotypic characters as the number of eye facets, the number of hairs, the exact shape of small features, and the connections of neurons in a very complex central nervous system—even when the genotype and the environment are precisely fixed. Even such structures as the very simple nervous systems of nematodes, which contain

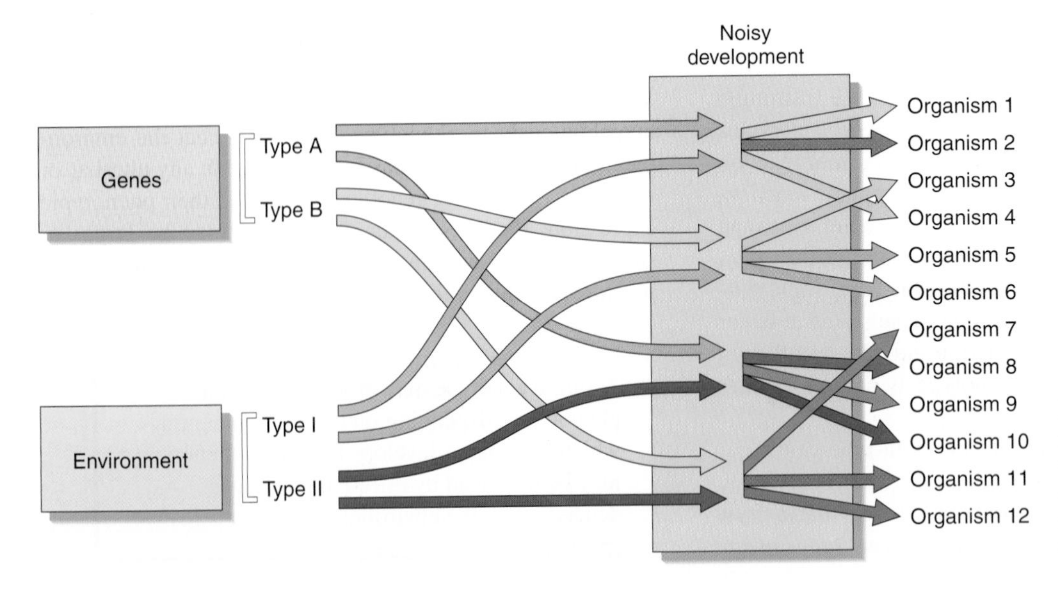

Figure 1-16 A model of phenotypic determination that shows how genes, environment, and developmental noise interact to produce a phenotype.

very few cells, vary at random for this reason. Random events in development lead to variations in phenotype called **developmental noise.**

Like noise in a verbal communication, developmental noise adds small random variations to the predictable developmental process governed by norms of reaction. Adding developmental noise to our model of phenotypic development, we obtain something like Figure 1-16. With a given genotype and environment, there is a range of possible outcomes for each developmental step. The developmental process does contain feedback systems that tend to hold the deviations within certain bounds, so that the range of deviation does not increase indefinitely through the many steps of development. However, this feedback is not perfect. For any given genotype developing in any given sequence of environments, there remains some uncertainty regarding the exact phenotype that will result.

MESSAGE

Developmental noise is a source of the observed phenotypic variation in some characters.

THREE LEVELS OF DEVELOPMENT

In the chapters of this book in which we discuss developmental genetics, we will be concerned with the way in which genes mediate development, but nowhere in those chapters will we consider the role of the environment or the influence of developmental noise. How can we, at the beginning of the book, emphasize the joint role of genes, environment, and developmental noise in influencing the phenotype, yet in our later considerations of the genetics of development ignore these factors? The answer is that modern developmental genetics is concerned with very basic processes of development that are common to all individual members of a species and, indeed, are common to animals as different as fruit flies and mammals. How does the front end of an animal become differentiated from the back end, the ventral from the dorsal side? How does the body become segmented, and why do limbs form on some segments and not on others? Why do eyes form on the head and not in the middle of the abdomen? Why do the antennae, wings, and legs of a fly look so different, even though they are derived by evolution from appendages that looked alike in the earliest ancestors of insects? At this first, most basic level of development, which is constant across individuals and species, normal environmental variation plays no role, and we can speak correctly of genes "determining" the phenotype. Precisely because the effects of genes can be isolated at this level of development, and because the processes seem to be general across a wide variety of organisms, they are easier to study than are characteristics for which environmental variation is important, and developmental genetics has concentrated on understanding them.

At a second level of development, there are variations on the basic developmental themes that differ between species but are constant within species. These variations too could be understood by concentrating on genes, although at the moment they are not part of the study of developmental genetics. For example, although both lions and lambs have four legs, one at each corner, lions always give birth to lions and lambs to lambs, and we have no difficulty in distinguishing between them in any environment. Again, we are entitled to say that genes "determine" the difference between the two species, although we must be more cautious here. Two species may differ in some characteristic because they live in different environments, and, until we can raise them in the same environment, we cannot always be sure whether environmental influences play a role. For example, two species of baboons in Africa, one living in the dry plains of Ethiopia and the other in more productive areas in Uganda, have very different food-gathering behaviors and social structures. Without actually transplanting colonies of the two species between the two environments, we cannot know how much of the difference is a direct response of these primates to the different food conditions in these different environments and what (if any) part of the difference is genetic.

It is at the third level of development—the one that results in differences in morphology, physiology, and behavior between individuals within species—that genetic, environmental, and developmental noise factors become intertwined, as discussed in this chapter. One of the most serious errors in the understanding of genetics by nongeneticists has been a confusion between variation at this level and variation at the higher levels. The experiments and discoveries to be discussed in Chapters 13 and 16 are not, and are not meant to be, models for the causation of individual variation. They apply directly only to those characteristics, deliberately chosen, that are general features of development and for which environmental variation appears to be irrelevant. There are, as yet, no experiments, or any ideas about how to perform them, that will bring together explanations for the three different levels of developmental variation.

GENETICS AND EVOLUTION

The first forms of life appeared on Earth approximately 3 billion years ago. Since that time, an immense number and variety of species have appeared as a result of modification and diversification of the progeny of those original forms. We owe our understanding of how such diversity could have arisen to the work of Charles Darwin, who, in 1859, provided an outline of the modern science of evolution in his revolutionary book *On the Origin of Species.*

The mechanism of evolution, as Darwin realized, rests on the following principles:

1. There is variation in morphology, physiology, and behavior among individuals within a species.
2. This variation may be heritable, so that offspring resemble their parents more than they resemble unrelated individuals.
3. New heritable variation is constantly arising within species, but that variation is independent of any adaptive properties that the new variants would confer on the organism. In other words, the variation is "random" with respect to the functional requirements of the organism.
4. Some variant types leave more offspring than others, with the result that there is a change over time in the frequency of different variants in the species.

The differential reproduction of some variants arises from two different sources. If the genetic variants result in differences in the morphology, physiology, and behavior of their carriers, then the different phenotypes may have different probabilities of survival and rates of reproduction. As a consequence, changes occur in the frequency of the genotypes in the species from generation to generation. This is the process of evolutionary change by **natural selection.** But nonselective change can also occur through sampling variation, which is the consequence of the reproduction of the next generation from a finite number of parents, each producing only a few offspring. Because of such random variation, the next generation will not have exactly the same proportion of different genetic variants as did the parental generation, and when this new generation reproduces there will again be a random fluctuation. This random change over time in the frequencies of genetic variants is the process of **random genetic drift.** It can result in the drifting apart of the genetic composition of isolated populations of the same species, leading eventually to the formation of new species.

The mechanism of evolution summarized in these principles is dependent on several important aspects of genetic processes, including the basic particulate, discontinuous nature of genes, the mechanism of transmission of genetic differences, the way in which genes enter the developmental pathways of organisms, and the sources of new genetic variation.

First, and most fundamental, is the discrete, particulate nature of genes as postulated by Mendel. If the mechanism of inheritance were a mixing of a continuous fluid, such as blood, then a mating between unlike individuals would result in offspring with intermediate characteristics, and the differences between the parents would never reappear in future generations, just as the mixing of paint of two different colors produces a fluid of intermediate color that can never again be separated into its original colors. The consequence of such a mixing mechanism of inheritance would be the dilution of any new variation that arose and the rapid homogenization of the population. Evolution based on the differential reproduction of different variants would not be possible. Although he knew nothing about Mendel's theory of inheritance, the existence of discrete heritable "factors" was essential to Darwin's theory of evolution.

Second, the relation between genotype and phenotype is of basic importance. If variation among individuals is not inherited, then differential reproduction of different variants cannot change the frequencies of those variants over time. If large individuals have more offspring than small ones, but the offspring of large individuals are no larger than the offspring of small ones, then no change in the composition of the population will occur. Only heritable variation can be the basis of evolutionary changes over time.

Third, the way in which new genetic variation arises in populations is critical to the mechanisms of evolutionary change. Already-existing allelic variation in different genes is reshuffled in the process of sexual reproduction to produce entirely new combinations. New allelic and genic variation is also introduced by mutations.

GENETICS AND HUMAN AFFAIRS

We began this chapter with the observation that offspring resemble their parents, but not exactly, and we saw how questions about this most basic observation form the core of genetics. We saw as well the agricultural origin of genetics in the desire of plant and animal breeders to understand the inheritance patterns of economically important characteristics. It was a concern with putting plant breeding on a firm quantitative footing that motivated Mendel's original experiments on garden peas—experiments that were the foundation of modern genetics. While the controlled breeding of domesticated animals and plants was at the center of the early interest in heredity, there was also a long-standing belief that various aspects of human temperament and abilities, as well as diseases, were "in the blood," as they were seen to run in families. The interest in the inheritance of useful agricultural traits and of various human abilities and disabilities has not diminished as modern scientific genetics has developed. On the contrary, it has come to play an increasingly dominant role in the research agendas of agricultural and health sciences and in the explanation of differences in social and individual life experiences. No day goes by without the announcement in newspapers of the discovery of a "gene for . . ." that is predicted to lead to an eventual ability to prevent or cure a disease, or of a new genetic technology that will feed the world's hungry. No more important reason can exist for a study of the science of genetics than to be able to evaluate such claims with a critical eye.

Genetics and Disease

More than 1000 distinct human disease syndromes are known to be caused by single-gene mutations. Nearly all of these are very rare conditions, and in the aggregate they account for a very small fraction of all human ill health and deaths, even in rich countries where most people are not suffering from severe malnutrition and infectious diseases. There are, however, some genetic diseases, such as cystic fibrosis, that are more common. Tay-Sachs disease, a lethal disorder involving a defect in lipid metabolism, can reach a frequency as high as 1 in 3500 births in some Jewish populations. The most extreme case is that of sickle-cell anemia, which reaches a frequency of 9 percent in some African populations. What is not clear is how identification and sequencing of the genes involved in these diseases will lead to their treatment. Indeed, the most successful and long-standing treatment of a metabolic disease syndrome is that for diabetes, a disease whose link to genes remains unclear, although some forms of diabetes are clearly familial. As yet, there is no case of a disease for which therapy has depended on knowledge of its genetic determination. It is hoped that it might even be possible to introduce a normal gene into the tissue of an individual with an inherited disorder and that the possession of this normal gene might cure the disorder. Several attempts at such gene therapy have been made, but so far none has provided a cure.

We now know that cancer is almost certainly the consequence of mutations in somatic cells of a number of genes, which have been identified. Some fraction of such mutations must also occur in the germ line and be passed on to offspring through the gametes. Again, however, it is not yet clear how such knowledge will be of use to human health, except to warn us against chemical pollutants or radiation that will increase the frequency of mutations. On the other hand, about 5 percent of breast cancer appears to be familial, and mutations in two genes, *BRCA-1* and *BRCA-2,* are associated with about two-thirds of these familial cancer cases. If a woman bears the *BRCA-1* mutation, her likelihood of developing breast cancer is very high, and a knowledge of her genetic constitution would be of importance in a variety of decisions she might make. In such cases, it is extremely important that a layperson be able to evaluate critically what is meant by statements such as "10 percent of cases of disease X are related to gene Y."

The most interesting and controversial area of disease etiology that concerns geneticists is mental health. There have been several reports of genes responsible for some fraction of cases of schizophrenia or bipolar (manic-depressive) disorder. All of these claims have been later questioned, and some have been retracted, because they were based on inadequate data. Again, the informed layperson needs to be able to make some judgment about the quality of such claims and about the meaning of statements linking some percentage of the cases of a disease to a reported gene. What are the observations? What is the methodology of the study? What accounts for the uncertainty and the contradictory reports? What is the meaning of quantitative statements about the heritability of a disorder? These are important issues for people planning to have children who may fear the presence of a heritable mental disorder.

Diabetes and mental disorders are examples of human characteristics with a complex causation that have become increasingly the subject of genetic investigation and genetic claims. We read of genes for obesity, genes for Alzheimer disease, genes for heart disease, for high cholesterol, for attention deficit disorder. No claim is being made that single-gene mutations account for these syndromes, but, more reasonably, it is claimed that genetic differences are somehow involved in the likelihood that people will display the characteristic under investigation. It must indeed be true that genetic differences have some relevance to these traits, but how are we to understand the various statements made, and how can we use that understanding to make decisions? These are issues that demand some detailed understanding of the concepts and methodologies of genetics.

Genetics and Social Policy

There is a long tradition, going back well before modern genetics, of claims about the inheritance and genetic fixity of various human personality and cognitive traits. The best-known episode in recent years has been debate on the inheritance of IQ. Repeated claims have been made that IQ is 70 to 80 percent "heritable." What is the exact meaning of such a statement? Does it mean that environmental measures (such as enrichment programs in schools) can erase only 20 to 30 percent of the differences in IQ among people? Does it mean that intelligence is mostly hard-wired? What are the observations on which such statements are based, and what methodologies have been used to generate such numbers?

A more recent example of a claim that is supposed to have social consequences is that heterosexuality and homosexuality are the consequence of different genes on the human X chromosome. Those who did the research say that since sexual preference is determined by genes, it is fixed rather than a matter of choice, and that therefore there can be no basis for social discrimination. Putting aside the ethical claims about the injustice of discrimination, does "genetic" mean "fixed" and unchangeable, and what sort of evidence has been presented that there are indeed genetic differences on the human X chromosome that account for differences in sexual behavior?

Biotechnology and Genetic Engineering

Increasingly, in agriculture and in the production of pharmaceuticals, genes from one species are being inserted into the genomes of other species. In the case of pharmaceuticals, genes coding for specific polypeptides are inserted into bacteria that can be cultured in mass quantities, so that the protein produced by the bacteria can be harvested in industrial quantities for sale as a drug. The gene coding for human insulin, for example, has been transferred to bacteria, which are then grown in large fermentation vats. The insulin can then be extracted so that, for the first time, human insulin is now available as a treatment for diabetes, in place of the insulin previously harvested from domesticated animals.

The applications of biotechnology in agriculture include the transfer into crop plants or animals of genes from other species that code for polypeptides not available in the genetic repertoire of the agricultural species. An example is the transfer into many crop plants of genes from *Bacillus thuringensis,* which provide resistance to insect pests by producing toxins normally made by these bacteria but not normally found in plants. The creation of such genetically modified organisms (GMOs) has created a storm of protest, especially in Europe, where governments have prohibited their importation. One biological claim on which these protests are based is that the transfer of genes from other species to make these transgenic hybrids may result in the accidental introduction into agricultural commodities of substances that are toxic to humans. While no such case has yet been discovered, it is certainly not an impossible consequence of transgenic transfer. It has also been claimed that such transfers may bring genes into crop species that will spread by hybridization into wild relatives of the crop plants and produce "superweeds." How likely are these scenarios? Are there biological (as opposed to economic) arguments against transgenic transfer?

In all these cases it is clear that the public cannot rely on reports published in the general media for the kind of critical evaluation needed to make informed personal and political decisions. Nor can it be left to the experts, who have their own biases and agendas. There is no substitute for acquiring the kind of basic knowledge of genetics that is essential to all informed decisions. The remaining chapters of this book will provide you with the necessary background information to distinguish which claims for the modern science of genetics are realistic and to think critically about the potential—both positive and negative—of modern genetic research.

SUMMARY

1. Genetics is the study of genes at all levels, from molecules to populations. As a modern discipline, it began in the 1860s with the work of Gregor Mendel, who first formulated the idea that discrete hereditary units exist. We now know that genes are composed of DNA.

2. DNA molecules constitute the hereditary material within a chromosome. A gene is a functional unit of a DNA molecule.

3. DNA is composed of four nucleotides, each containing a deoxyribose sugar molecule, a phosphate group, and one of four bases: adenine (A), thymine (T), guanine (G), or cytosine (C). DNA consists of two nucleotide chains, which form a double helix held together by bonding of A with T and G with C.

4. In replication of DNA, the two nucleotide chains separate, and their exposed bases are used as templates for the synthesis of two identical daughter DNA molecules.

5. Most genes consist of a region that encodes the amino acid sequence of a polypeptide and a control region that governs the amount and timing of the production of that polypeptide.

6. To make a protein, DNA is first transcribed into a single-stranded working copy called messenger RNA (mRNA). Only one of the DNA strands is used as a template for the mRNA. The nucleotide sequence in the mRNA is then translated into an amino acid chain that constitutes the primary chemical structure of a protein. These amino acid chains are synthesized on ribosomes. The polypeptide chain then folds to form a protein.

7. Genetic variation is of two types: discontinuous, showing two or more distinct phenotypes, and continuous, showing phenotypes with a wide range of quantitative values. Discontinuous variants are often determined by alternative forms (alleles) of a single gene.

8. The main tools of genetics are breeding analysis of variants, biochemistry, microscopy, and DNA cloning.

9. The relation of genotype to phenotype across a range of environmental conditions is called the norm of reaction. In the laboratory, geneticists can study discontinuous variants under conditions in which there is a one-to-one correspondence between genotype and phenotype. However, in natural populations, where environments vary, there is generally a more complex relation, and different genotypes can produce overlapping ranges of phenotypes. As a result, discontinuous variants have been the starting point for most experiments in genetic analysis.

10. The process of evolution is a change in the frequencies of different genotypes in populations and the genetic divergence of isolated populations to form new species. Only heritable variation can be the basis of biological evolution. New genes arise in evolution by the duplication of DNA within the genome of an organism or by the importation of DNA from other species. Genetic variation within a species is increased by mutation and by the shuffling of variation in sexual reproduction. Genetic changes in populations occur by natural selection or by random fluctuations in proportions as a result of reproduction by a finite number of parents.

CONCEPT MAP

Each chapter in this book contains an exercise on drawing concept maps. As you draw these maps, you will be organizing the knowledge that you have just acquired. As we learn, we must place our new knowledge into the overall conceptual framework of the discipline. An essential part of doing so is to discern the interrelations between new knowledge and previously acquired knowledge. It is not as easy as it sounds, and the mind can sometimes trick us into believing that we have the overall picture. Concept maps are a good way of proving to ourselves that we really do know how the various structures, processes, and ideas of genetics are interrelated. They can also help in identifying errors in our understanding.

In each concept map exercise, a selection of terms is given. The challenge is to draw lines or arrows between the terms you think are related, with a description on the arrow showing just what the relation is. Draw as many relations as possible. Force yourself to make connections, no matter how remote they might seem at first. The arrow description must be clear enough that another person reading the map can understand what you are thinking. Sometimes the maps reveal a simple series of consecutive steps, sometimes a complex network of interactions. There is no one correct answer; there are many correct variations. But you may make some incorrect connections, and finding these misunderstandings is part of the purpose of the exercise.

A sample concept map interrelating the following terms is shown in Figure 1-17:

base-pair substitution / recessive / albino / environment / melanin / metabolic pathway / active site / enzyme / gene

Now draw a concept map for this chapter, interrelating as many of the following terms as possible. Note that the terms are listed in no particular order.

genotype / phenotype / norm of reaction / environment / development / organism

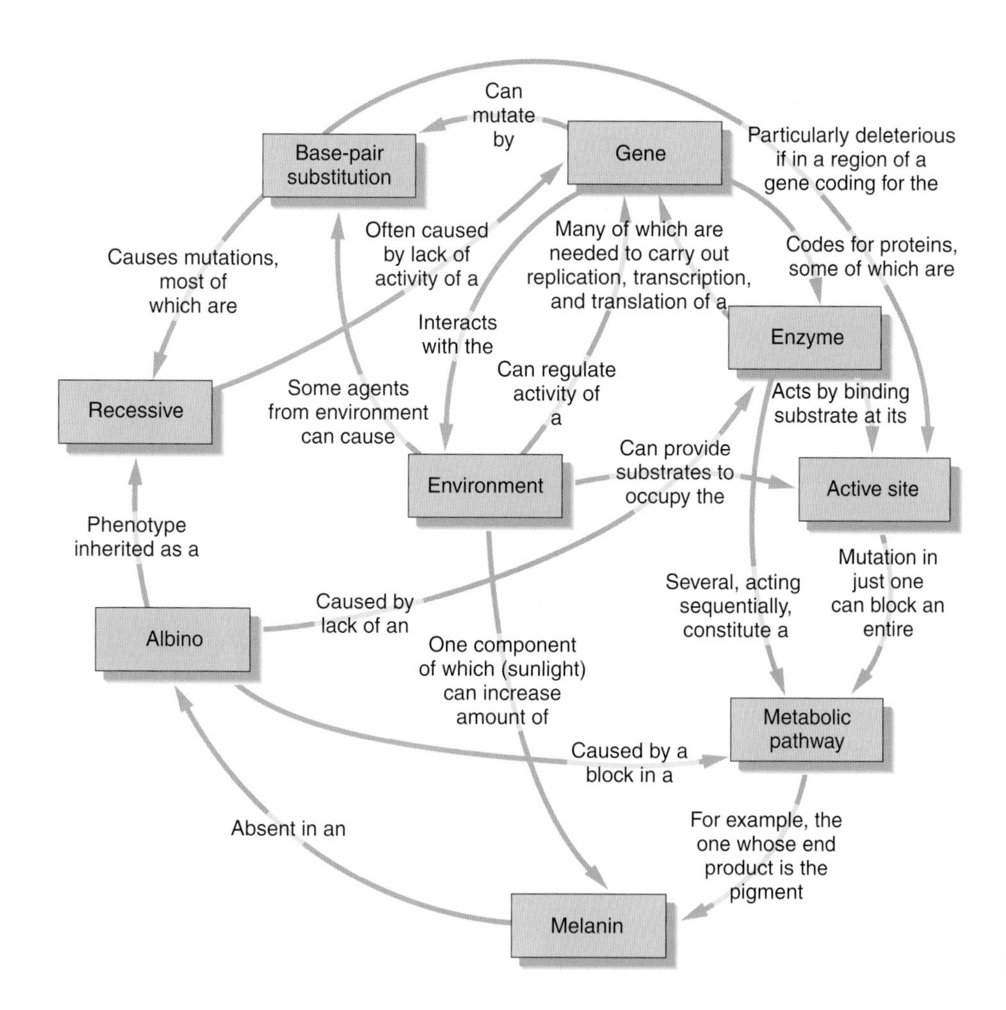

Figure 1-17 A sample concept map.

PROBLEMS

1. Define *genetics*. Do you think that ancient Egyptian racehorse breeders were geneticists? How might their approaches to breeding have differed from those of modern geneticists?

2. How does DNA dictate the general properties of a species?

3. What are the features of DNA that suit it for its role as a hereditary molecule? Can you think of alternative types of hereditary molecules that might be found in extraterrestrial life forms?

i 4. How many different DNA molecules 10 base pairs long are possible?

5. Draw a simple diagram of DNA showing the sugar-phosphate backbone and the complementary nucleotides.

6. Each cell of the human body contains 46 chromosomes.

   a. How many double-stranded DNA molecules does this statement represent?

   b. How many different types of DNA molecules does it represent?

7. A certain segment of DNA has the following nucleotide sequence in one strand:

   ATTGGTGCATTACTTCAGGCTCT

   What must be the sequence of the other strand?

8. In a single strand of DNA, is it ever possible for the number of adenines to be greater than the number of thymines?

9. In normal double-stranded DNA, is it true that

   a. A plus T will always equal G plus C?

   b. A plus C will always equal G plus T?

10. Suppose that the following DNA molecule replicates to produce two daughter molecules. Draw those daughter molecules by using black for previously polymerized nucleotides and red for newly polymerized nucleotides.

    TTGGCACGTCGTAAT

    AACCGTGCAGCATTA

11. In the DNA molecule in Problem 10, assume that the bottom strand is the template strand, and draw the mRNA that would be transcribed from it.

12. The most common elements in living organisms are carbon, hydrogen, oxygen, nitrogen, phosphorus, and sulfur. Which of them are not found in DNA?

13. What is a gene? What are some of the problems with your definition?

THE STRUCTURE OF GENES AND GENOMES 2

Key Concepts

1 The properties that characterize a species are based on its fundamental set of genetic information—its genome.

2 A genome is composed of one or more DNA molecules, each organized as a chromosome.

3 DNA is a double helix consisting of two intertwined antiparallel and complementary nucleotide chains.

4 Prokaryotic genomes are mostly single circular chromosomes.

5 Eukaryotic genomes consist of a large nuclear component and a much smaller organellar component.

6 The nucleus contains one or two sets of linear chromosomes (one or two nuclear genomes).

7 In a nuclear chromosome DNA is wound around histone proteins, resulting in a coiled, compact, linear structure.

8 A gene is a chromosomal region capable of making a functional transcript.

9 The genes of many eukaryotic species contain noncoding regions called *introns.*

10 Viruses, which are nonliving capsules parasitic on organisms, nevertheless also contain genetic material, which can be DNA or RNA.

11 Genes and genomes show many structural similarities across species.

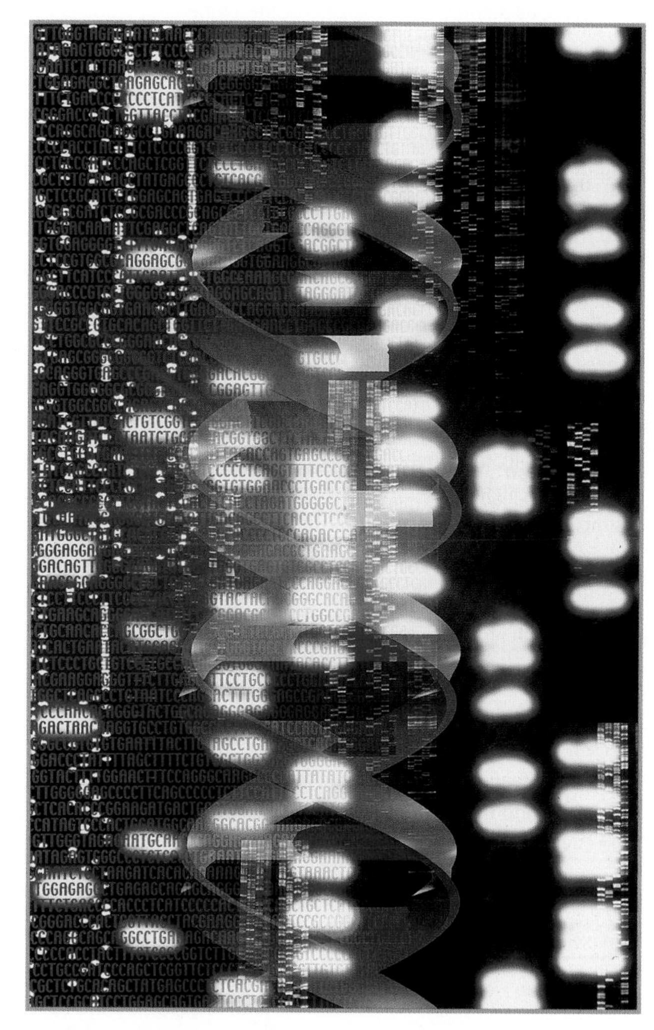

Views of the genome. At the center is a model of the double-helical structure of DNA. The right side shows various enlargements of bands on a sequencing gel used to determine nucleotide sequence. The left side shows some determined nucleotide sequences using letters representing the four bases found in nucleotides, A (adenine), T (thymine), G (guanine), and C (cytosine). *(Alfred Pasieka/Science Photo Library/Photo Researchers.)*

Duchenne muscular dystrophy is one of the many human diseases that are hereditary, in other words passed down through the generations in families. It is a particularly debilitating disease, leading to early death. Onset is usually before age six, followed by progressive wasting of certain sets of muscles, including those of the heart. Children affected with the disease, usually boys, are confined to a wheelchair by age 12 and generally die before age 17. The muscles of the calves are enlarged. Weakness of the hip muscles causes the spine to take on an unnatural curvature, thrusting the lower abdomen forward. These features are shown in Figure 2-1, taken from a paper by the nineteenth-century French physician Guillaume Duchenne, one of the pioneering workers on this disease. Another characteristic feature found in patients is blood serum concentrations of the enzyme creatine phosphokinase (CPK) hundreds of times higher than in normal individuals.

Interestingly, a hereditary disease with virtually identical symptoms, muscle wasting and elevated CPK concentration, is also found in golden retriever dogs, domestic cats, and mice.

This disease raises many questions about heredity. Since the disease is hereditary, presumably the genetic material (the DNA) of sufferers contains some type of alteration: But what specific part of the human DNA complement could be involved in such disease symptoms? This would be an important piece of information to have in order to understand the disease and to attempt therapy. To answer this question, we need to know the nature of the full DNA complements in living organisms generally, and in humans specifically, for this is the material that dictates the basic characteristic features of species that are handed down through generations. Then we must be able to zero in on the component or components of this genetic material that consistently differ between the normal and the disease states. Furthermore, how can we explain the parallel symptoms in other mammals? Are the similar symptoms just a coincidence, or are the genetic components altered in the disease really the same in the different species? To answer these questions, we need to be able to assess the general features of the genetic material in a range of different mammals, and then compare the disease-causing components of their DNA with those of humans.

Such issues are the theme of this chapter. We deal with the general nature of the genetic material, the variation between the disease and the nondisease states, and the differences between species.

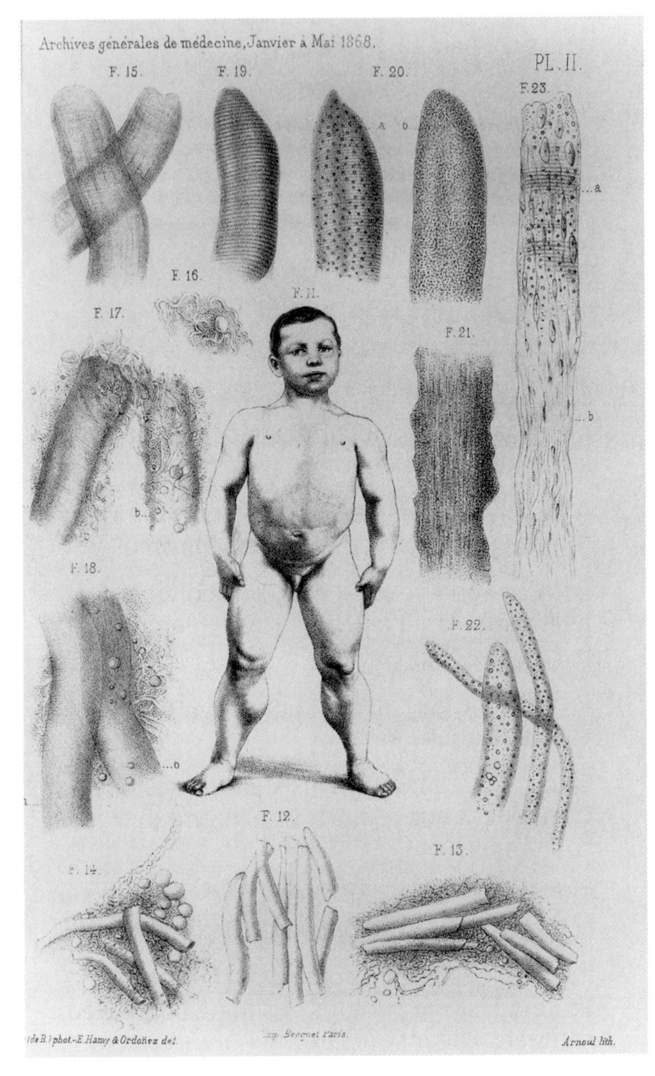

Figure 2-1 Nineteenth-century etching of a young boy with Duchenne muscular dystrophy, showing enlarged calf muscles and curvature of the spine. The border shows cellular structures connected to the disease symptoms. *(G. B. A. Duchenne, "Recherches sur la paralysie musculaire pseudo-hypertrophique ou paralysie myo-sclérosique,"* Archives Générales de Médicine *11, 1968, 5–25, 179–209, 307–321, 421–443, 552–588. Courtesy of the New York Academy of Medicine Library.)*

CHAPTER OVERVIEW

The core ideas of the chapter are illustrated in Figure 2-2. This figure is based on cells and virus particles. In living organisms, the keys to the questions of heredity are found at the level of the cell. The cell is the minimal unit of life. Organisms can be unicellular, such as bacteria and protists, or multicellular, such as plants, animals, and fungi. Multicellular organisms begin life as a single cell. The overall characteristics of an organism are determined by the structure, function, and spatial arrangement of its constituent cells. Therefore, to understand how the genetic material dictates the fundamental characteristics of different species of organisms and how they contribute to variation within a species, it is necessary to place the genetic material within the cellular context. In other words, we need to know the nature of the genetic material, its location within cells, and the way it performs its functions in the cellular setting. In contrast to organisms, viruses are generally considered not to be alive. They can propagate only by parasitizing living cells. Nevertheless, viruses do have their own individual characteristics and contain their own sets of genetic material.

This chapter provides an introduction to the nature and cellular organization of the genetic material in several categories of organisms and in viruses. At the end of the chapter, we will again look at Duchenne muscular dystrophy in the light of the material covered in the chapter. Chapter 3 extends the structural features of genes and genomes to the functional level.

CHAPTER OVERVIEW figure

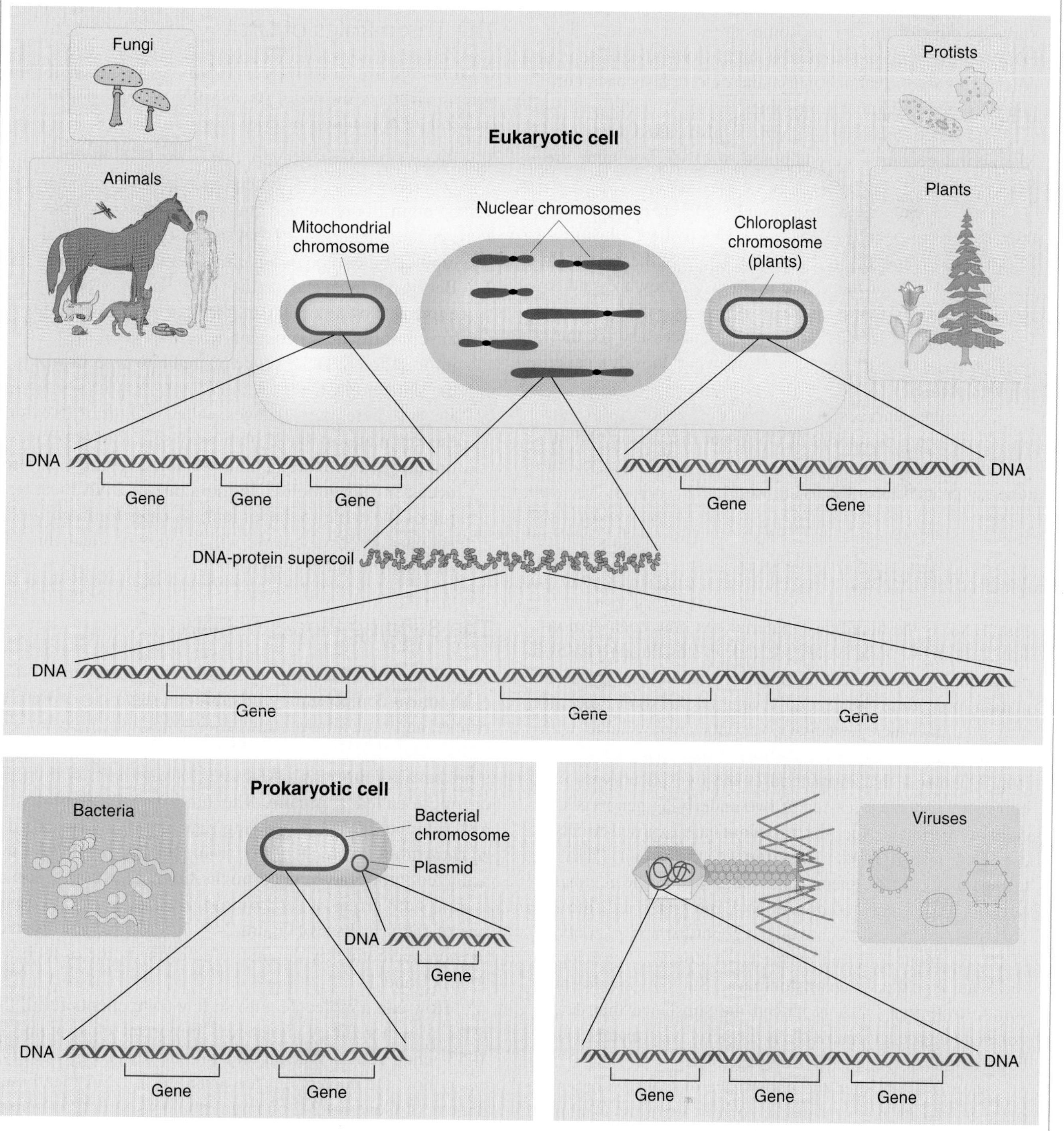

Figure 2-2 Overview of genomes and genes in living organisms and viruses.

How is DNA organized in a cell? One unique set of the DNA of an organism is called a **genome.** A genome is composed of long DNA molecules, which are, in turn, the main components of threadlike structures within the cell, called **chromosomes.** Each chromosome contains one DNA molecule carrying many genes. As we saw in Chapter 1, genes are regions of chromosomal DNA that can be transcribed into RNA. The genomes of most prokaryotic organisms consist of only one chromosome, whereas the genomes of eukaryotes consist of several to many chromosomes. In eukaryotes most of the chromosomes are in the nucleus. The DNA of these chromosomes is highly coiled and supercoiled. Eukaryotic mitochondria and chloroplasts each contain one small unique chromosome.

The genetic material of a virus is also called a *genome.* Many viral genomes are composed of DNA, but some are RNA.

Cells of many prokaryotes and some eukaryotes contain additional small DNA molecules called **plasmids.** Plasmids are generally not essential because the host cells can survive without them. For this reason they are said to be *extragenomic,* that is, not part of the essential genome. Plasmid DNA molecules contain genes necessary for their own propagation, so they are (somewhat loosely) called *plasmid genomes.*

Since the genomes of all organisms, all plasmids, and most viruses are composed of DNA, our discussion will begin with the structure of this molecule, which dictates the inherent properties of the living world.

THE NATURE OF DNA

That DNA is the hereditary material has now been demonstrated in many prokaryotes and eukaryotes through an experimental procedure called **transformation.** The transformation procedure is generally applied to some specific character in which hereditary variants are available with two different phenotypes, or forms, of that character. Recall from Chapter 1 that in such cases the two phenotypes can be viewed as the expression of two underlying genotypes. If cells of one phenotype (the **recipient**) are exposed to DNA extracted from another (the **donor**), the donor DNA is taken up by the recipient cells. Once inside the recipient, occasionally a piece of donor DNA integrates into the recipient's genome and changes the genotype and phenotype of the recipient into that of the DNA donor. The changed recipient is called a **transformant.** Such results clearly demonstrate that DNA is indeed the substance that determines genotype and therefore is the hereditary material (see Foundations of Genetics 2-1 on page 30).

Having established the importance of DNA, in order to place it into its proper cellular context we must establish where DNA resides in the cell. In eukaryotes, histochemical and physical techniques can be used to demonstrate that much of it resides in nuclear chromosomes. DNA-binding dyes such as Feulgen or DAPI primarily stain the nuclear chromosomes in cells and to a lesser extent also stain the mitochondria and chloroplasts. Furthermore, if a mass of cells is ground up and the different types of organelles separated experimentally, it becomes clear that the bulk of DNA can be isolated from the nuclei, and the remainder from mitochondria and chloroplasts.

The Three Roles of DNA

Even before the structure of DNA was elucidated, genetic studies clearly indicated three key properties that had to be fulfilled by hereditary material.

1. Because essentially every cell in the body has the same genetic makeup, it is crucial that the genetic material be faithfully replicated at every cell division. The structural features of DNA that allow such faithful replication will be considered later in this chapter.
2. Because it must encode the constellation of proteins expressed by an organism, the genetic material must have informational content. How the coded information in DNA is deciphered into protein will be the subject of Chapter 3.
3. Because hereditary changes, called *mutations,* provide the raw material that evolutionary selection operates on, the genetic material must be able to change on rare occasion. Nevertheless, the structure of DNA must be relatively stable so that organisms can rely on its encoded information. We will discuss the mechanisms of mutation in Chapter 10.

The Building Blocks of DNA

As a chemical, DNA is quite simple. It contains three types of chemical components: **phosphate,** a sugar called **deoxyribose,** and four nitrogenous **bases**—adenine, guanine, cytosine, and thymine. Two of the bases, adenine and guanine, have a double-ring structure characteristic of a type of chemical called a **purine.** The other two bases, cytosine and thymine, have a single-ring structure of a type called a **pyrimidine.** The chemical components of DNA are arranged into groups called **nucleotides,** each composed of a phosphate group, a deoxyribose sugar molecule, and any one of the four bases (Figure 2-3). It is convenient to refer to each nucleotide by the first letter of the name of its base: A, G, C, and T.

How can a molecule with so few components fulfill the roles of a hereditary molecule? Important clues came in 1953 when James Watson and Francis Crick showed precisely how the nucleotides are arranged in DNA (see Foundations of Genetics 2-2 on page 31). DNA structure is summarized in the next section.

Purine nucleotides

Phosphate

Nitrogen base (adenine, A)

Deoxyribose sugar

Deoxyadenosine 5′-phosphate (dAMP)

Guanine (G)

Deoxyguanosine 5′-phosphate (dGMP)

Pyrimidine nucleotides

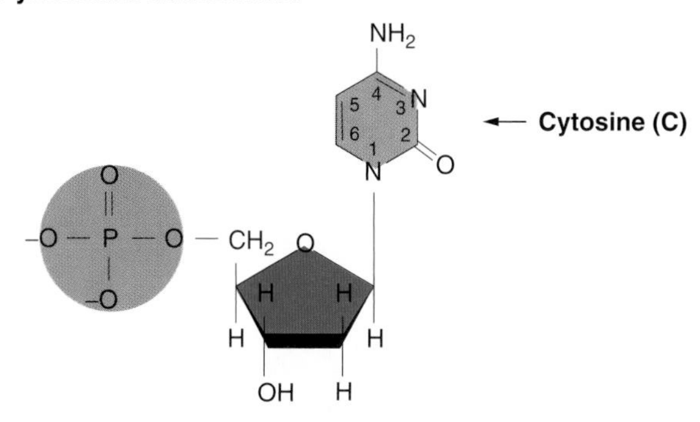

Deoxycytidine 5′-phosphate (dCMP)

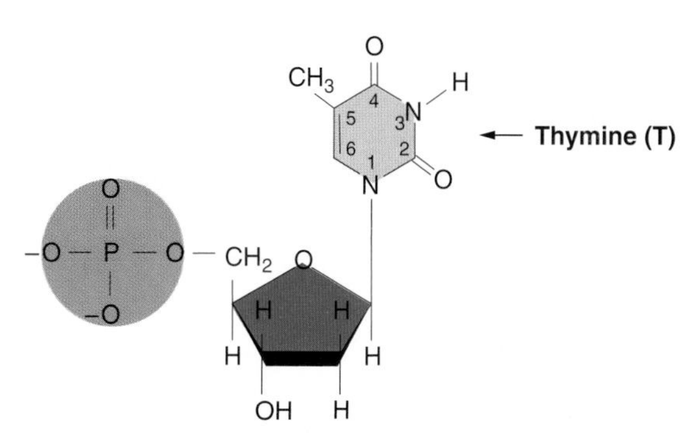

Deoxythymidine 5′-phosphate (dTMP)

Figure 2-3 Chemical structures of the four nucleotides (two with purine bases and two with pyrimidine bases) that are the fundamental building blocks of DNA. The sugar is called *deoxyribose* because it is a variation of a common sugar, ribose, which has one more oxygen atom.

The Double-Helical Nature of DNA

DNA is composed of two side-by-side chains ("strands") of nucleotides twisted into the shape of a double helix. The two nucleotide strands are held together by weak associations between the bases of each strand, forming a structure like a spiral staircase (Figure 2-4 on the next page). The backbone of each strand is a repeating phosphate–deoxyribose sugar polymer (Figure 2-5 on page 29). The sugar-phosphate bonds in this backbone are called **phosphodiester bonds.** The attachment of the phosphodiester bonds is used to describe how a nucleotide chain is organized. The carbons of the sugar groups are numbered 1′ through 5′. A phosphodiester bond links the 5′ carbon of a deoxyribose and the 3′ carbon of the adjacent deoxyribose. Thus, each sugar-phosphate backbone is said to have a 5′-to-3′ polarity, and understanding this polarity is essential in understanding how DNA fulfills its roles. In the double-stranded DNA molecule, the two backbones are in opposite, or **antiparallel,** orientation (see Figure 2-5). One strand is oriented 5′ → 3′; the other strand, though also 5′ → 3′, runs in the opposite direction, or, for convenience, can be viewed as oriented from 3′ → 5′.

The bases are attached to the 1′ carbon of each deoxyribose sugar in the backbone of each strand and face inward toward a base on the other strand. Interactions between pairs of bases, one from each strand, hold the two strands of the DNA molecule together. The bases of DNA interact according to a very straightforward rule, namely, that there are only two types of base pairs: A–T and G–C. The two bases in a base pair are said to be **complementary.** At any "step" of the stairlike double-stranded DNA molecule, the only base-to-base associations that can exist between the two strands without substantially distorting the double-stranded DNA molecule are A–T and G–C.

The association of A with T and G with C is by **hydrogen bonds.** The following is an example of a hydrogen bond:

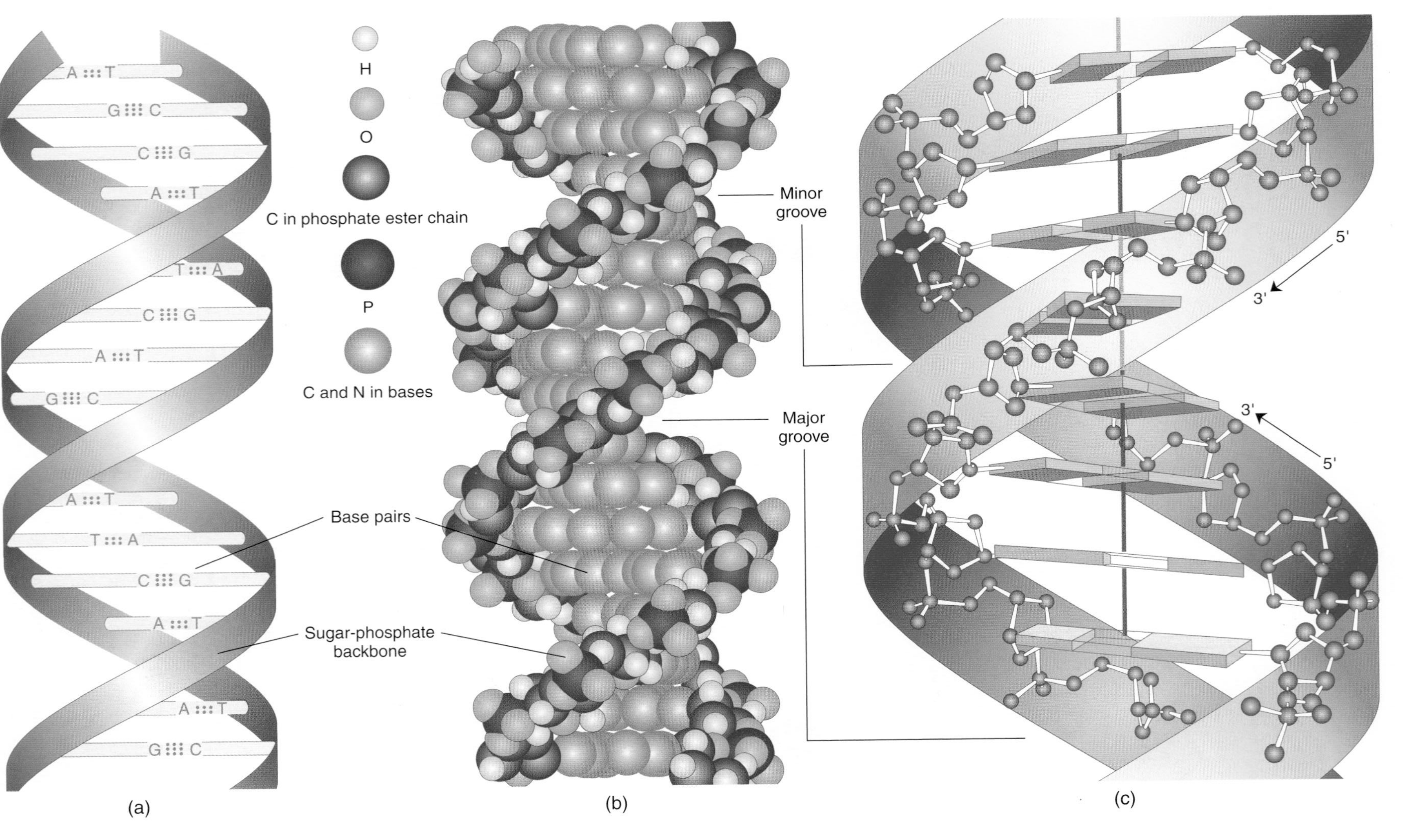

Figure 2-4 Three representations of the DNA double helix.

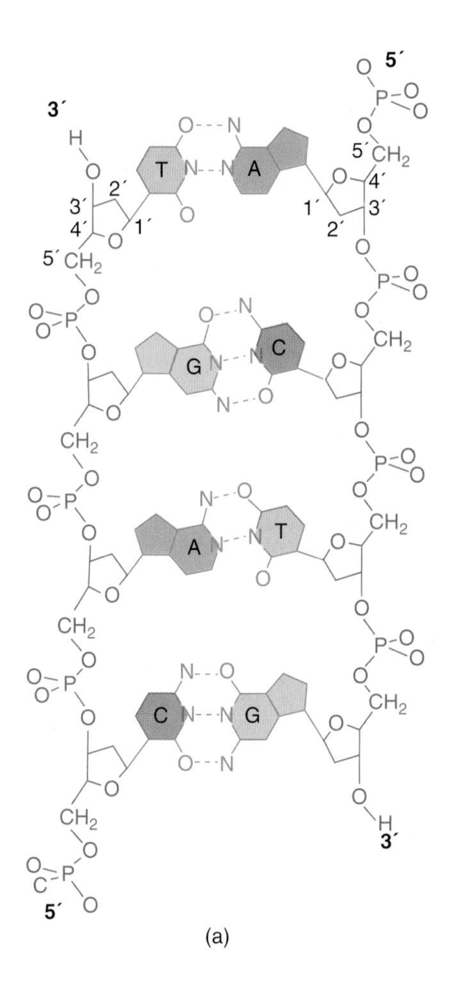

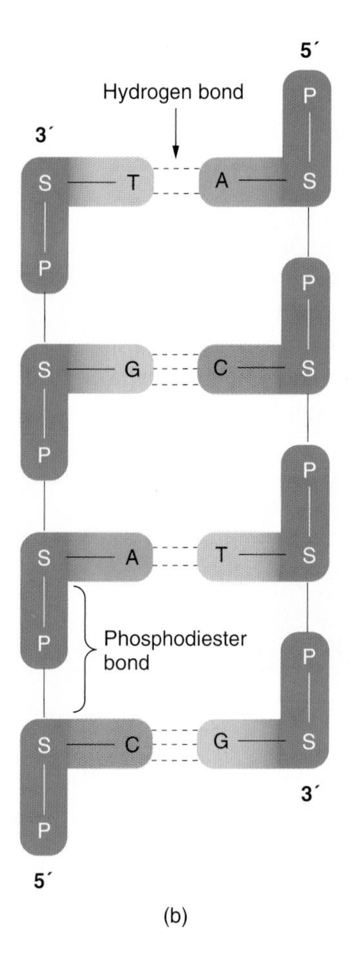

Figure 2-5 The arrangement of the components of DNA. A segment of the double helix has been unwound to show the structures more clearly. (a) An accurate chemical diagram showing the sugar-phosphate backbone in blue and the hydrogen bonding of bases in the center of the molecule. (b) A simplified version of the same segment emphasizing the antiparallel arrangement of the nucleotides, which are represented as L-shaped structures with 5′ phosphate "toes" and 3′ deoxyribose sugar "heels." S = deoxyribose sugar, P = phosphate, and G, C, A and T are bases.

Each hydrogen atom in the NH_2 group is slightly positive (δ^+) because the nitrogen atom tends to attract the electrons involved in the N—H bond, thereby leaving the hydrogen atom slightly short of electrons. The oxygen atom has six unbonded electrons in its outer shell, making it slightly negative (δ^-). A hydrogen bond forms between one slightly positive H and one slightly negative atom—in this example, O. Hydrogen bonds individually are quite weak (only about 3 percent of the strength of a covalent bond), but the large number of hydrogen bonds along the double helix provides the strength needed to hold its two halves together. An analogy here is Velcro, which binds tightly even though the individual links are weak and can be broken. This overall stability but potential for local separation is important in transcription and replication, when local separation of DNA strands is needed. One further important chemical fact: the hydrogen bond is much stronger if the participating atoms are "pointing at each other" (that is, if their linear structures are in alignment), as shown above.

Note that because the G–C pair has three hydrogen bonds, whereas the A–T pair has only two, one would predict that DNA containing many G–C pairs would be more stable than DNA containing many A–T pairs. This prediction is confirmed. Heat causes the two strands of the DNA double helix to separate (a process called DNA **melting** or DNA **denaturation**); it can be shown that DNAs with higher G + C content require higher temperatures to melt them because of the greater attraction of the G–C pairing.

Two nucleotide strands paired in an antiparallel manner automatically assume a double-helical configuration (see Figure 2-4), mainly through interaction of the base pairs. The base pairs, which are flat planar structures, stack on top of one another at the center of the double helix (see Figure 2-4c). Stacking adds to the stability of the DNA molecule by excluding water molecules from the spaces between the base pairs. The most stable form that results from base stacking is a double helix with two distinct sizes of grooves running around in a spiral. These are the **major groove** and the **minor groove,** which can be seen in the models. A single strand of nucleotides has no helical structure; the helical shape of DNA depends entirely on the pairing and stacking of the bases in antiparallel strands. DNA is a right-handed helix; in other words, it has the same structure as a screw that would be screwed into place using a clockwise turning motion.

The unit of measurement of DNA length is the **base pair.** A thousand base pairs is called one **kilobase** (1 kb) and a million is a **megabase** (1 Mb).

FOUNDATIONS OF GENETICS 2-1

Oswald Avery's Demonstration that the Hereditary Material Is DNA

In 1928 Frederick Griffith succeeded in permanently transforming a nonvirulent strain of the bacterium *Pneumococcus* into a virulent strain by adding an extract of dead cells of the latter type. The next step was to determine which chemical component of the dead donor cells had caused this transformation, because this substance had changed the genotype of the recipient strain and therefore might be a candidate for the hereditary material.

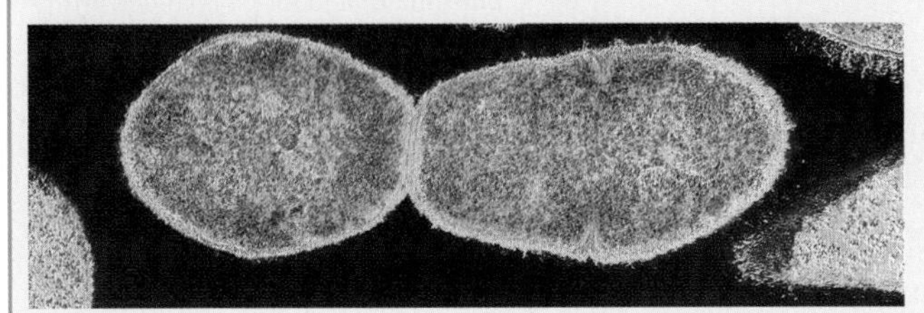

Pneumococcus. (Dr. Kari Lounatmaa/Science Photo Library/Photo Researchers, Inc.)

Oswald Avery's approach to this problem in 1944 was to chemically destroy all the major categories of chemicals in the extract of dead cells one at a time and find out if the transforming property was inactivated, too. The virulent cells had a smooth polysaccharide coat, whereas the avirulent cells did not; hence, polysaccharides were an obvious candidate for the transforming agent.

However, when polysaccharides were destroyed, the mixture could still transform. Proteins, fats, and ribonucleic acids (RNA) were all similarly shown not to be the transforming agent. However, when the donor mixture was treated with the enzyme deoxyribonuclease (DNase), which breaks up DNA, the transforming ability of the mixture was destroyed. (The DNase was extracted from dog intestinal mucosa or dog or rabbit serum.) When these mucosal or serum extracts were heated to denature the DNase, they were no longer able to inactivate the transforming property.

In a 1943 letter to his brother Roy, Avery wrote:

> For the past two years, first with MacLeod and now with Dr. McCarty, I have been trying to find out what is the chemical nature of the substance in the bacterial extract which induces this specific change. . . . Some job, full of headaches and heartbreaks. But at last perhaps we have it. . . . But today it takes a lot of well documented evidence to convince anyone that the sodium salt of deoxyribose nucleic acid, protein-free, could possibly be endowed with such biologically active and specific properties, and that is the evidence we are now trying to get. It's lots of fun to blow bubbles but it's wiser to prick them yourself before someone else tries to. *(From R. J. Dubos,* The Professor, the Institute, and DNA. *Rockefeller University Press, 1976, Appendix I.)*

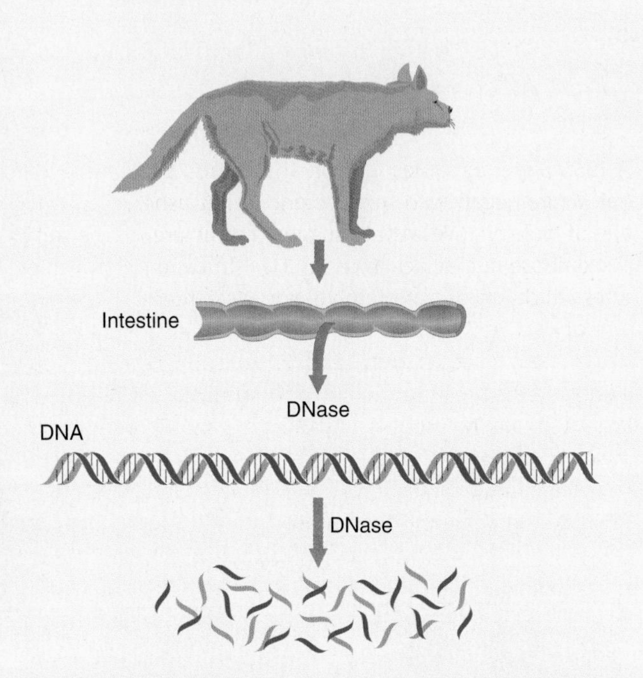

Therefore, by adding DNA alone, Avery could change the genotype and phenotype of bacterial cells, convincingly demonstrating that DNA is the genetic material. It is now known that fragments of the transforming DNA that confer virulence enter the bacterial chromosome and replace their counterparts that confer avirulence.

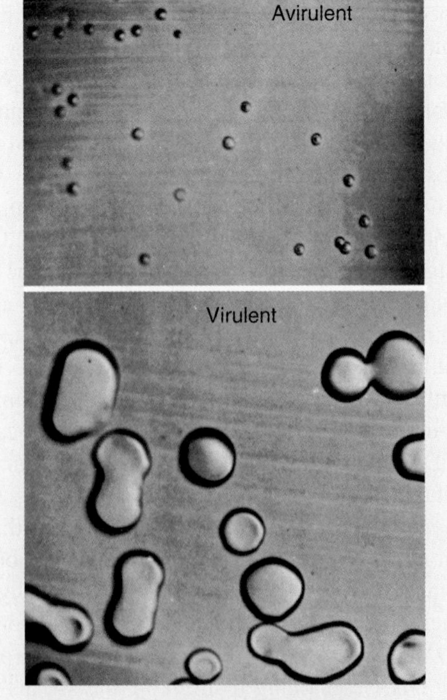

(From O. T. Avery, C. M. MacLeod, and M. McCarty, *J. Exp. Med.* 79, 1944, 158.)

FOUNDATIONS OF GENETICS 2-2

James Watson and Francis Crick Propose the Correct Structure for DNA

A 1953 paper by James Watson and Francis Crick in the journal *Nature* began with two sentences that ushered in a new age of biology: "We wish to suggest a structure for the salt of deoxyribose nucleic acid (D.N.A.). This structure has novel features which are of considerable biological interest." The structure of DNA had been a subject of intense debate since 1944, when Avery et al. showed that DNA is the hereditary substance. Although the general composition of DNA was known, it was not known how all the parts fit together. The structure had to fulfill the main requirements for a hereditary molecule: the ability to store information, the ability to be replicated, and the ability to mutate.

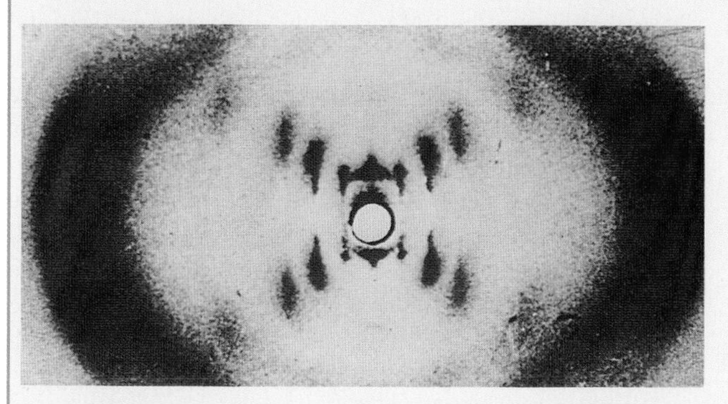

X-ray diffraction pattern of DNA. (Rosalind Franklin/Science Source/Photo Researchers.)

Watson (an American microbial geneticist) was a visitor in the Cavendish Laboratory in Cambridge, England, when he began his collaboration with Crick, an English physicist. Their DNA model was based on the evidence about structure available at the time. Maurice Wilkins and Rosalind Franklin had obtained X-ray diffraction patterns of DNA that suggested it had a helical structure. Erwin Chargaff, from chemical analyses of many different DNA samples, had devised several empirical rules concerning the four constituent bases, including that the amount of adenine (A) equals the amount of thymine (T), and guanine (G) equals cytosine (C).

From this information Watson and Crick pieced together a structure that seemed to fit the data. The structure they proposed was a double helix of two nucleotide chains running in opposite directions. The two chains were bound to each other by the pairing of specific bases, A with T, and G with C. This accounted satisfactorily for the equality of A with T and of G with C shown by Chargaff. Eventually they tested their ideas by building a scale model of DNA out of metal. Watson later described the excitement of assembling the model:

> I wandered down to see if the shop could be speeded up to produce the purines and pyrimidines later that afternoon. Only a little encouragement was needed to get the final soldering accomplished in the next couple of hours. The brightly shining metal plates were then immediately used to make a model in which for the first time all the DNA components were present. In about an hour I had arranged the atoms in positions which satisfied both the X-ray data and the laws of stereochemistry. The resulting helix was right-handed with the two chains running in opposite direction. Only one person can easily play with a model, and so Francis did not try to check my work until I backed away and said that I thought everything fitted. While one interatomic contact was slightly shorter than optimal, it was not out of line with several published values, and I was not disturbed. Another fifteen minutes' fiddling by Francis failed to find anything wrong, though for brief intervals my stomach felt uneasy when I saw him frowning. In each case he became satisfied and moved on to verify that another interatomic contact was reasonable. Everything thus looked very good when we went back to have supper. *From J. D. Watson,* The Double Helix, *Atheneum, New York, 1968, pp. 200–201.*

The double-helical structure proposed by Watson and Crick fulfilled the requirements for a hereditary substance: the nucleotide sequence in DNA could code for amino acid sequence in protein; replication was possible by strand separation and new synthesis directed by the specificity of the base pairing; and mutation was possible by base substitution.

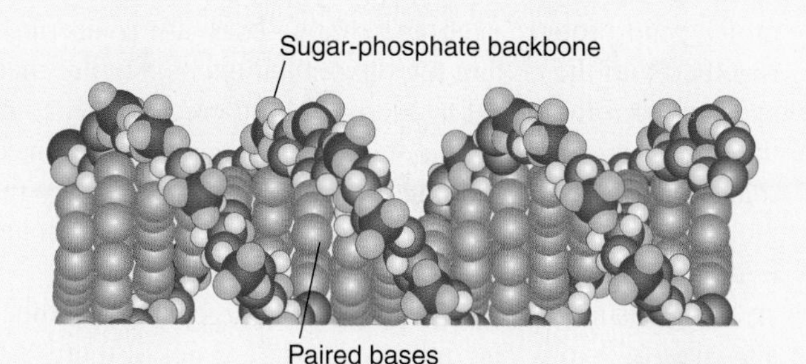

James Watson and Francis Crick with their DNA model. (Camera Press.)

DNA Structure Reflects Its Function

How does DNA structure fulfill the three requirements of a hereditary molecule to replicate, encode information, and mutate? First, replication. With the antiparallel orientation of the DNA strands, and the rules for proper base pairing, we can envision how DNA is faithfully replicated: each strand serves as an unambiguous **template** (alignment guide) for the synthesis of its complementary strand. If, for example, one strand has the base sequence AAGGCTGA (reading in the 5′-to-3′ direction), then we automatically know that its complementary strand can have only the sequence (written in the 3′-to-5′ direction) TTCCGACT. Replication is based on this simple rule. The two DNA strands (shown below in the darker blue) separate, and each serves as a template for the building of a new complementary strand (shown in light blue).

An enzyme called **DNA polymerase** is responsible for building new DNA strands, matching up each base of the new strand with the proper complement on the old, template strand. Thus, the complementarity of the DNA strands underlies the entire process of faithful duplication. This process will be examined and described more fully in Chapter 4.

The second requirement for DNA is that it have informational content. This informational requirement for DNA is fulfilled by its nucleotide sequence, which acts as a sequence of "letters" in a kind of coded language. The nucleotide sequence in DNA is copied into RNA, which then becomes the blueprint for the synthesis of the main component of biological form, protein. This will be the topic of Chapter 3.

The third requirement, the capacity for periodic change (mutation), is simply the occasional replacement, deletion, or addition of one or more nucleotide pairs, resulting in a change of the encoded information. The mechanisms of gene mutation will be the subject of Chapter 10.

MESSAGE

Double-stranded DNA is composed of two antiparallel, interlocked nucleotide chains, each consisting of a sugar-phosphate backbone with bases hydrogen-bonded to complementary bases of the other chain.

THE STRUCTURE OF GENES

We have seen that DNA fulfills the requirements of a hereditary molecule. Yet the functional units of the genome are the genes, which are simply regions along the DNA molecule. We need to understand the fundamental structure of genes to understand how a mere segment of a long DNA molecule can act as a functional unit. Genes are diverse in function and size, but despite this diversity, common topographical features can be delineated.

The Main Regions of a Gene

In Chapter 1 a gene was defined as a region of DNA capable of producing a functional RNA molecule (for most genes, an mRNA). However, the span of a gene is larger than the span of its RNA transcript. In addition to a DNA template for the RNA transcript, there must be regions of DNA that ensure the RNA is made at the correct time and in the correct place in the development of the organism. Only then is the gene properly functional. To regulate transcription, adjacent to the transcribed region there is a **regulatory region,** a segment of DNA that enables transcription of that gene to be turned on or off. The specific nucleotide sequence of the regulatory region promotes binding of special regulatory proteins produced in response to signals from outside the cell or from other parts of the cell. The protein that actually does the job of transcription also binds to this adjacent region. If the signals tell transcription to be turned on, the transcribing protein moves along the transcribed region, making RNA. At the other end of the gene there is a nucleotide sequence that signals that the transcript is complete.

MESSAGE

A gene is an operational region of the chromosomal DNA, part of which can be transcribed into a functional RNA at the correct time and place during development. Thus, the gene is composed of the transcribed region and adjacent regulatory regions.

Many eukaryotic genes contain mysterious segments of DNA, called **introns,** interspersed in the transcribed region of the gene. Introns do not contain information for functional gene product such as protein. They are transcribed together with the coding regions, called **exons,** but are then excised from the initial transcript. The correct sequence in the introns is necessary in order to generate a properly sized transcript. Hence, introns and the coding and regulatory regions must all be considered part of the overall functional unit—in other words, part of the gene (Figure 2-6).

Organisms differ a great deal in the average number and size of exons and introns in their genes. Figure 2-7

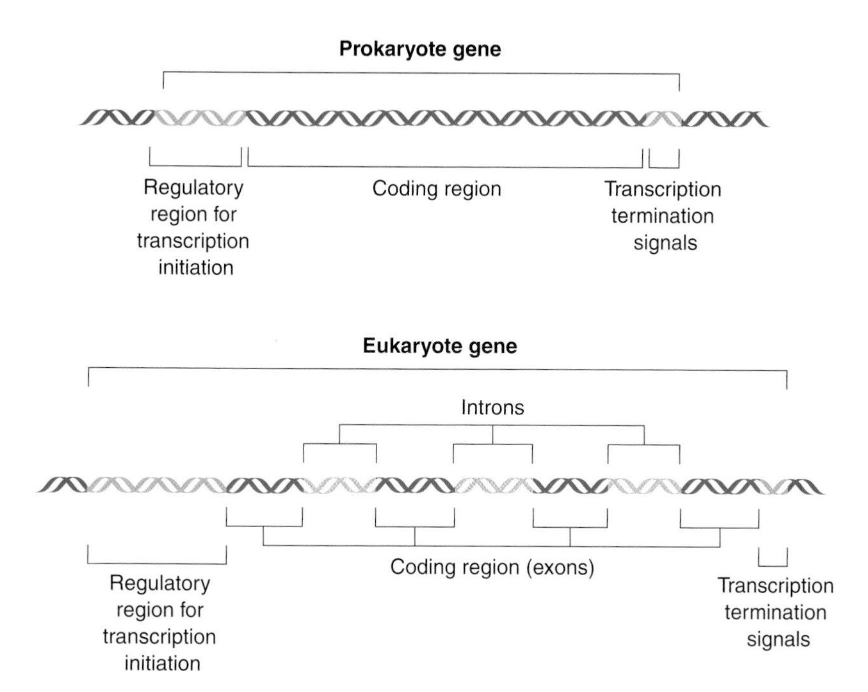

Figure 2-6 Generalized gene structure in prokaryotes and eukaryotes. The coding region is the region that contains the information for the structure of the gene product (usually a protein). The adjacent regulatory regions (lime green) contain sequences that are recognized and bound by proteins that make the gene's RNA and by proteins that influence the amount of RNA made. Note that in eukaryotic genes the coding region is often split into segments (exons) by one or more noncoding introns.

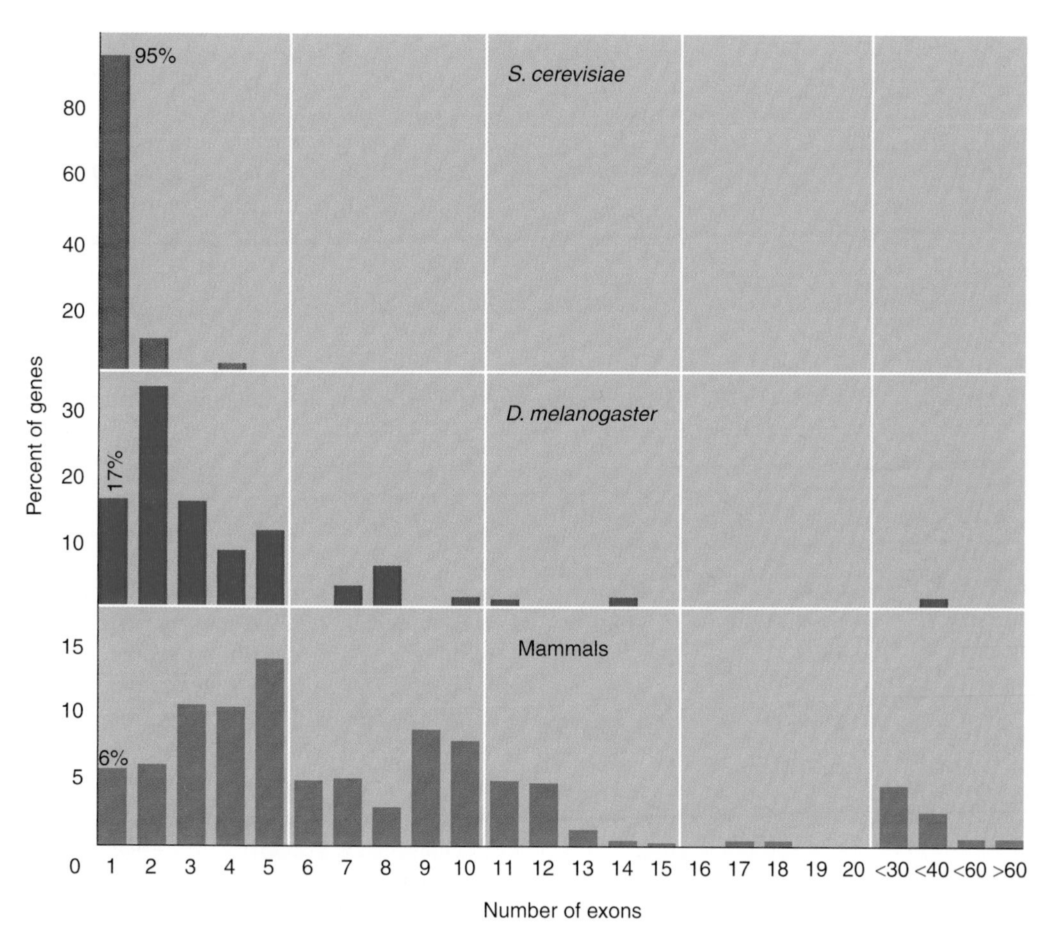

Figure 2-7 Distribution of exon numbers in the genes of three categories of organisms.

TABLE 2-1 The Relationship between Gene Size and mRNA Size

Species	Average Exon Number	Average Gene Length (kb)	Average mRNA Length (kb)
Hemophilus influenzae	1	1.0	1.0
Methanococcus jannaschii	1	1.0	1.0
S. cerevisiae	1	1.6	1.6
Filamentous fungi	3	1.5	1.5
Caenorhabditis elegans	4	4.0	3.0
D. melanogaster	4	11.3	2.7
Chicken	9	13.9	2.4
Mammals	7	16.6	2.2

Source: Based on B. Lewin, *Genes 5*, Table 2-2. Oxford University Press, 1994.

shows the number of exons in samples of genes from three eukaryotic groups, baker's yeast *(Saccharomyces cerevisiae)*, fruit flies *(Drosophila melanogaster)*, and mammals. As you can see, there are striking differences. In mammals, genes are generally very large and transcription units contain many exons (and, hence, introns). Table 2-1 shows average gene sizes of protein-coding genes in representative organisms and compares them with the size of the messenger RNA (which is the same size as all the exons combined). The coding regions of the genes in all these organisms are all in the 2–3-kb range, but the gene sizes are vastly different among these species, largely owing to differences in average intron size and intron number.

The Gene and Its Neighborhood

As is apparent from the previous discussion, many genes reside on the same DNA double helix. How are the genes that reside on the same molecule arranged relative to one another? Do they immediately abut their neighbors, or are they separated by DNA sequences that make no contribution to the genes in their vicinity? Earlier, we described the gene as a transcriptional unit composed of a transcribed region plus regulatory regions. However, in practice, it is difficult to identify and define all the regulatory elements associated with a given gene. Nevertheless, there appear to be many instances in which, once we account for all the regulatory information, there is clearly "spacer" DNA between adjacent genes. In many eukaryotes some of the DNA between genes is **repetitive DNA,** so called because it consists of a number of identical or nearly identical units repeated in the genome. Some of the repetitive DNA is dispersed; some is found in contiguous "tandem" arrays. Repetitive DNA is also found in some introns. The extent of this DNA is different in different species, and there is also variation of repeat number within species. A large

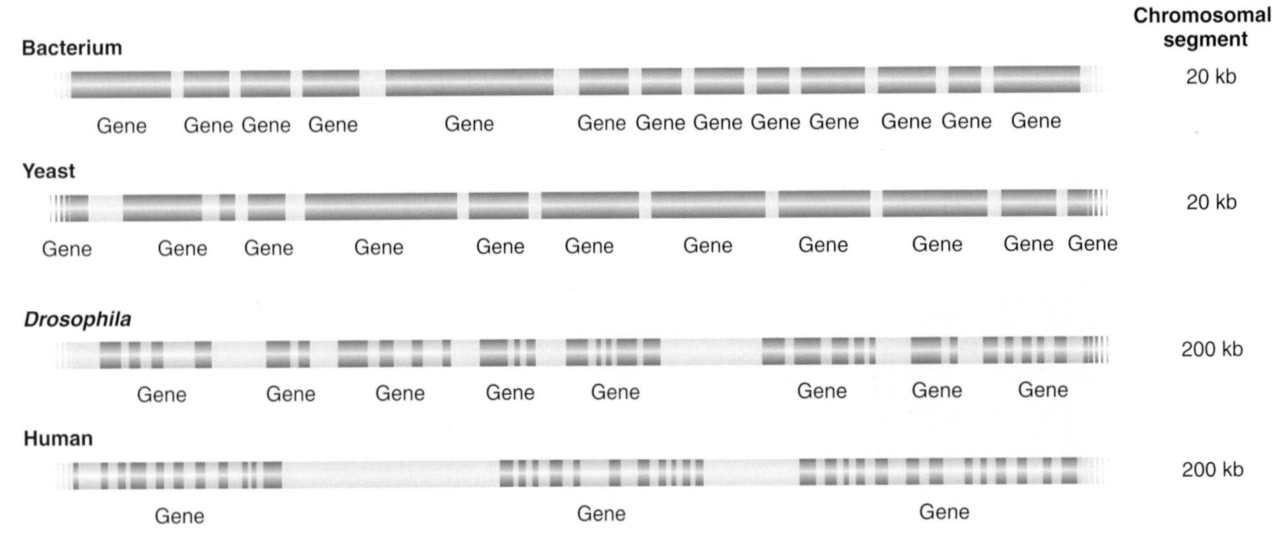

Figure 2-8 Differences in gene topography in four species. Light green = introns; dark green = exons; white = regions between the coding sequences (including regulatory region plus "spacer" DNA). Note the different scales of the top two and bottom two illustrations.

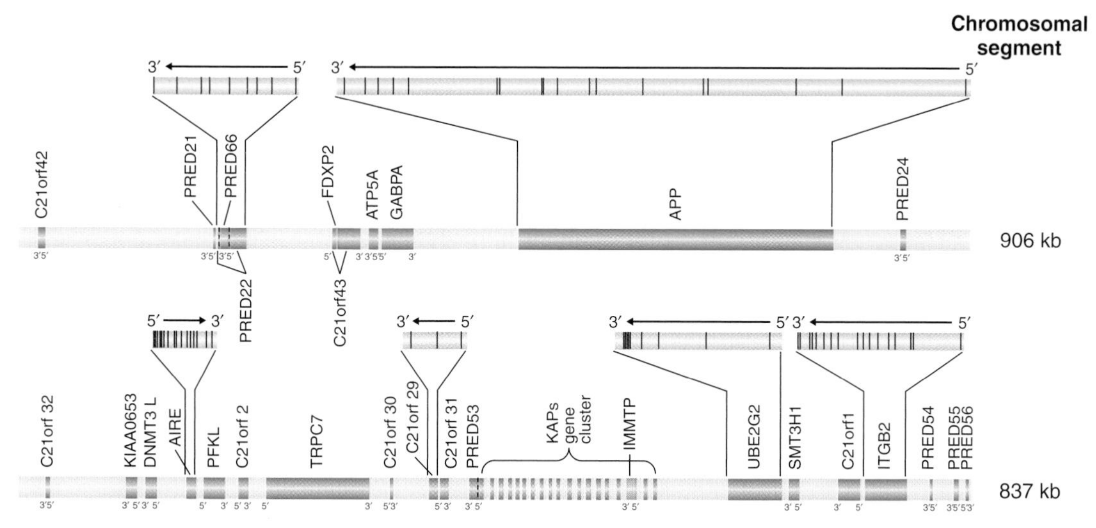

Figure 2-9 Transcribed regions of genes (green) in two segments of chromosome 21, based on the complete sequence for this chromosome. (Two genes, FDXP2 and IMMTP, have been colored orange so that they can be distinguished from neighboring genes.) Some genes are expanded to show exons (black bars and lines) and introns (light green). Vertical labels are gene names (some of known, some of unknown, function). The 5′ and 3′ labels show the direction of transcription of the genes. *(Modified from M. Hattori et al.,* Nature *405, 2000, 311–319.)*

proportion of repetitive DNA results from **mobile genetic elements,** DNA segments that can copy themselves and move within the genome. The function of repetitive DNA is still a mystery, but there are several hypotheses about its origin. (These topics will be addressed in more detail in Chapters 9 and 10.)

Some general chromosomal topologies in familiar organisms are illustrated in Figure 2-8. Note in this diagram that the segments of chromosomes from bacteria and yeast are only one-tenth the size of those from *Drosophila* and humans. Note also the large differences in gene size, intron number and size, and space between the transcribed regions of adjacent genes. Examples of some specific chromosomal regions from the human chromosome 21 are illustrated in Figure 2-9. Note the large differences in gene size and in intron number and size in these regions.

THE NATURE OF GENOMES

All organisms have a genome comprising DNA, and all organisms have genes. But layered beneath these generalities are many questions that naturally arise about genomes. How big are they? How many genes do they contain, and how does this vary between species? Which genes are on which chromosomes? How many chromosomes do genomes contain, and are there any special landmarks on the chromosomes? The study of the structure and function of entire genomes is called **genomics.** The ways in which genomes are analyzed will be covered in Chapter 9. However, for the present we need some simple genomic principles in order to understand basic genetics.

Understanding genome organization is particularly important for understanding the inheritance of genes. For the purposes of inheritance at cell division, genes need to be replicated, and then the copies need to be segregated to the two progeny cells. The mechanisms that govern this transmission of genes will be a major focus of this book. Many of the principles of inheritance from one generation to the next stem from the fact that genes are organized into chromosomes and from the structure of the chromosomes themselves. In theory, each gene could be on its own DNA molecule, but clearly this arrangement would present insurmountable problems in that each of this multitude of molecules would have to be replicated and parceled faithfully into the two daughter cells. In reality, linking genes together as chromosomes converts the genome into more manageable units that make for more reliable parceling of genetic material at cell division.

Genome Size

One basic genomic question that needs to be answered is "How big are genomes?" Figure 2-10 shows the ranges of total genome size in various groups of organisms. Notice that generally genome size increases with complexity of the group, but there is considerable variation (up to a thousandfold) in some of the groups. Table 2-2 extends these data by comparing genome sizes to the number of genes. The number of genes is roughly proportional to genome size. There are discrepancies in this proportionality, caused by repetitive DNA and the number and size of introns. Note that viruses, plasmids, mitochondria, and chloroplasts contain small numbers of genes.

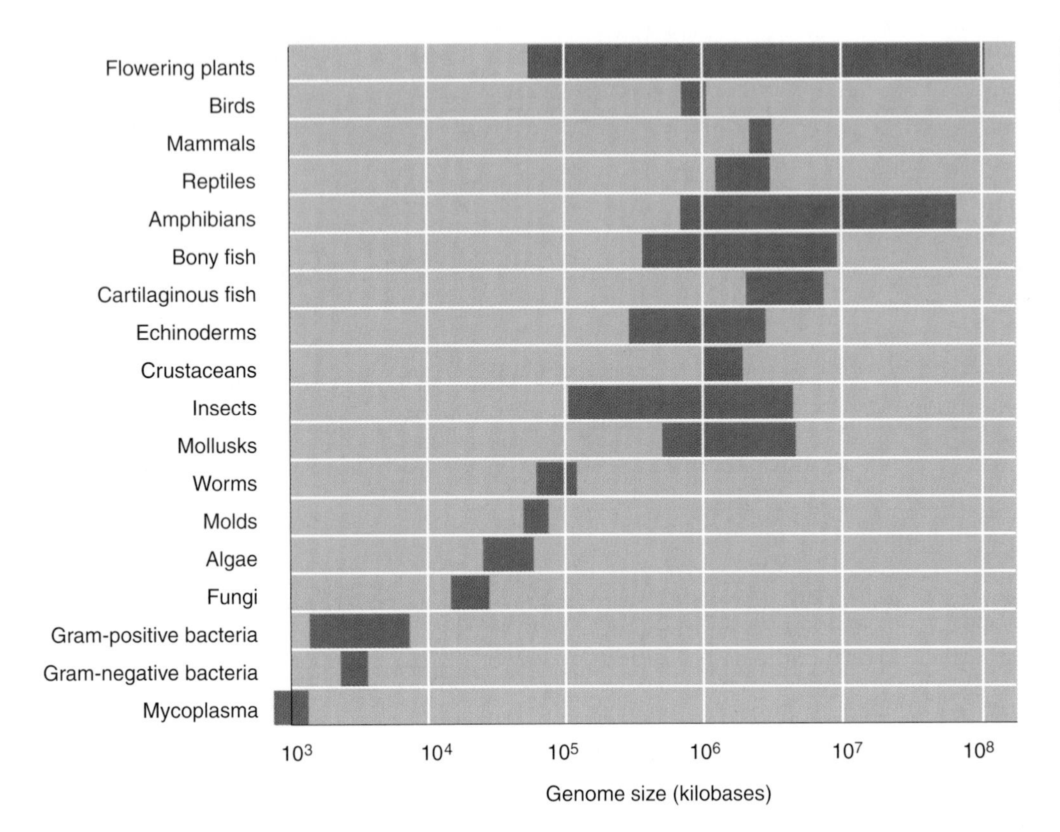

Figure 2-10 Amount of DNA in the genomes of various organisms. Note the log scale.

TABLE 2-2 Genomes: Sizes and Numbers of Genes

Genome	Group	Size (kb)*	Number of Genes
Eukaryotic nucleus			
Saccharomyces cerevisiae	Yeast	13,500 (L)	6,000
Caenorhabditis elegans	Nematode	100,000 (L)	13,500
Arabidopsis thaliana	Plant	120,000 (L)	25,000
Homo sapiens	Human	3,000,000 (L)	30,000 – 100,000
Prokaryote			
Escherichia coli	Bacterium	4,700 (C)	4,000
Hemophilus influenzae	Bacterium	1,830 (C)	1,703
Methanococcus jannaschii	Bacterium	1,660 (C)	1,738
Viruses			
T4	Bacterial virus	172 (L/C)	300
HCMV (herpes group)	Human virus	229 (L)	200
Eukaryotic organelles			
S. cerevisiae mitochondria	Yeast	78 (C)	34
H. sapiens mitochondria	Human	17 (C)	37
Marchantia polymorpha chloroplast	Liverwort	121 (C)	136
Plasmids			
F plasmid	In *E. coli*	100 (C)	29
Kalilo	In the fungus *Neurospora*	9 (L)	2

*C = circular DNA molecule; L = linear DNA molecule; L/C = linear in free virus, circular in cell.

We now turn to a detailed discussion of various genomes in order of increasing size.

Plasmids

Bacterial cells isolated from nature often contain small DNA elements that are not essential for the basic operation of the bacterial cell. These elements are called **plasmids.** Plasmids are symbiotic molecules that cannot survive at all outside of cells. Bacterial plasmids often contain genes that encode products extremely useful to the bacterial host, for example, those that promote sexual fusion, confer antibiotic resistance, or produce toxins. Since plasmids are not organisms, plasmid DNAs are strictly speaking not genomes; however, they are often referred to as such. Plasmid genomes generally are relatively small, and most plasmid genes do not contain introns. The amount of intergenic space is highly plasmid-specific.

Plasmids also are occasionally found in fungal and plant cells. Most are found inside mitochondria and chloroplasts, but some are found in nuclei or in the cytosol. Unlike the bacterial plasmids mentioned above, these eukaryotic plasmids seem to provide no benefits for their hosts—they seem to be molecular parasites, existing only by forcing their host cells to propagate them.

For their replication and maintenance, plasmids depend heavily on the general cellular machinery encoded by the host genome. Most bacterial plasmids consist of a closed circle or ring of DNA, with no 5′ or 3′ ends (Figure 2-11a). However, some bacteria have linear types, too. In fungi and plants, linear plasmids are most common, but circular plasmids are known in fungi.

Organellar DNA

The bulk of the eukaryotic genome is contained within the chromosomes in the nucleus. In addition, some cellular organelles—mitochondria and chloroplasts—contain an organelle-specific DNA molecule, called either an *organellar chromosome* or an *organellar genome.* Analysis of organellar genomes shows that they are inherently circular (Figure 2-11b), but linearized forms of these genomes can be detected in cells. Many organelle genes have introns. Generally there is little intergenic space.

Individual mitochondria and chloroplasts contain identical multiple copies of their chromosomes, and each cell contains several to many of one or both of these organelles. Therefore the copy number of these organelle chromosomes per cell can be quite high, often in the hundreds or thousands. Hence the DNA is relatively easy to extract from organelle fractions of disrupted cells. Organelle chromosomes contain genes specific to the functions of the organelle concerned, photosynthesis in chloroplasts and oxidative phosphorylation in mitochondria, but most of the biological functions that occur inside these organelles are

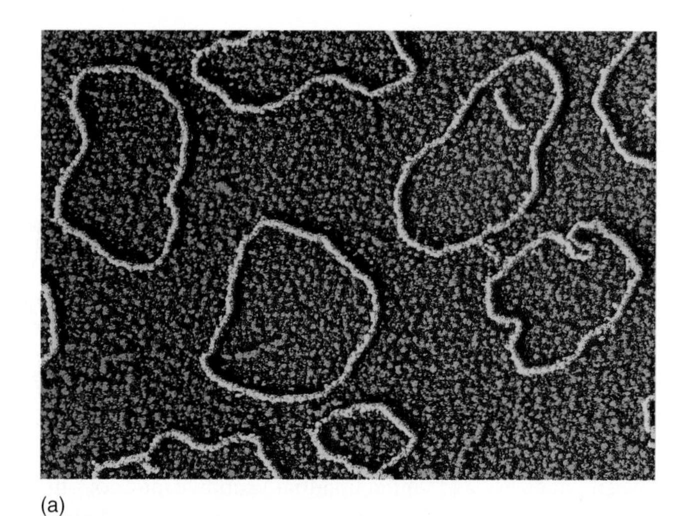

(a)

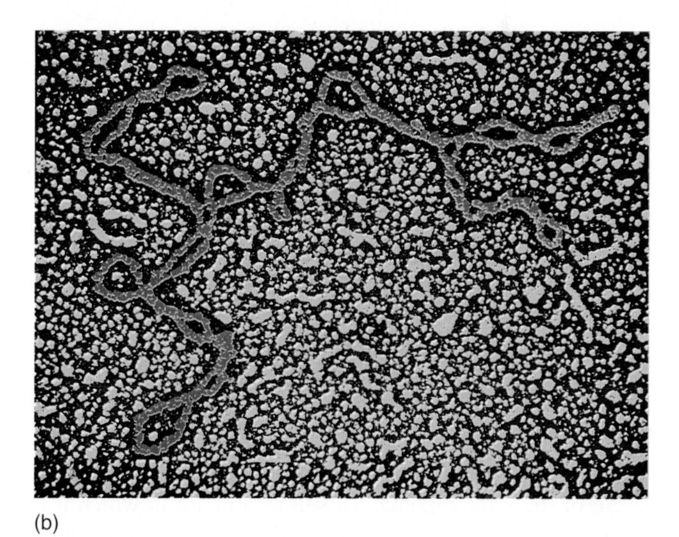

(b)

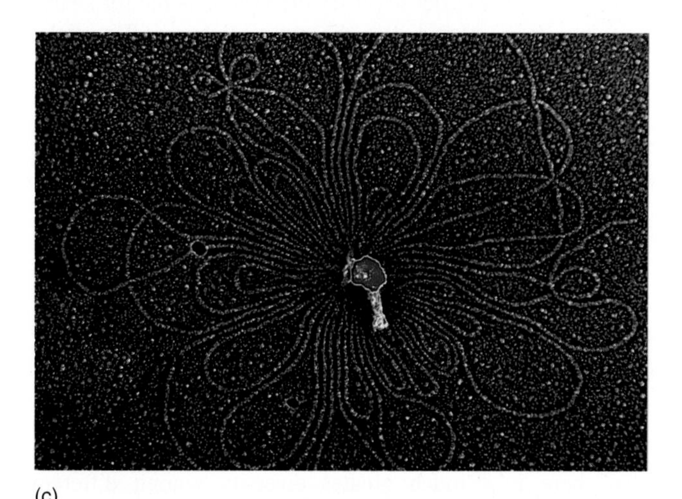

(c)

Figure 2-11 Electron micrographs of some small genomes. (a) Plasmids of *E. coli;* (b) mitochondrial DNA of humans; (c) T2 phage. *(Part a, Dr. Gopal Murti/Science Photo Library/Photo Researchers; part b, CNRI/Science Photo Library/Photo Researchers; part c, Biology Media/Science Photo Library/Photo Researchers.)*

specified by genes in the nuclear genome. There is almost no overlap between the genes in the organelle chromosome and those of the nuclear genome. According to the *endosymbiotic theory* of eukaryotic evolution, mitochondria and chloroplasts were originally prokaryotic cells that entered another cell and became indispensable to it. Throughout evolution most of the original prokaryotic genes were transferred to the nuclear genome or lost, but some were retained in the organelle. Mitochondria and their genomes can be experimentally eliminated in some organisms such as yeasts, which can produce energy from fermentation as an alternative to the oxidative phosphorylation mediated by mitochondria. However, most organisms cannot survive without them, so there is still mutual interdependence between nuclear and mitochondrial subdivisions of the genome. In some photosynthetic organisms chloroplasts can be experimentally eliminated, and the cells can survive by using chemicals instead of photosynthesis for their energy.

Viral Genomes

A virus is a nonliving particle that can reproduce itself only by infecting a living cell and subverting the cellular machinery of its host to generate progeny viral particles. Viruses of bacteria are also called bacteriophages (literally, "eaters of bacteria"). Viruses are composed of a protein coat and a core that contains the genome. During infection, the viral genome enters the cell, mediated by either a merger of the viral coat with the plasma membrane of the cell or by a syringelike injection process.

There is thus an intracellular phase to the replication cycle of the viral genome, as well as a phase in which it is packaged into its coat to form an infectious extracellular virus. Since it is easier to purify the viral genome for analysis when it is contained in the extracellular viral particle, viral genomes are usually described in terms of their structure as found in the viral particle. Viral genomes are quite varied. Many are composed of DNA. When packaged into the viral particle, in some viruses the DNA is double-stranded, in other cases single-stranded. Still other viruses, such as the retroviruses (HIV is an example), have RNA genomes in the viral particle. (The structure of RNA is described in Chapter 3.) Some RNA genomes are single-stranded; others are double-stranded. Some viral genomes contain linear DNA or RNA molecules, whereas others are circular (Figure 2-11c). Some viral genes have introns.

The genomes of all living cells contain one type of genetic material, double-stranded DNA. In contrast we see that there is a much greater diversity among different viruses in the kinds of molecules that form their genomes. This diversity is based partly on the various special strategies for packing the genome into infective viral particles and partly on the diverse evolutionary histories of viruses. Regardless of the nature of the packaged viral genome, all viruses have a phase of their life cycle in the host cell in which their genomes are composed of a double-stranded DNA molecule, or in some cases a double-stranded RNA molecule. Double-stranded DNA may be incorporated into a host chromosome or it may be a completely separate molecule, depending on the virus.

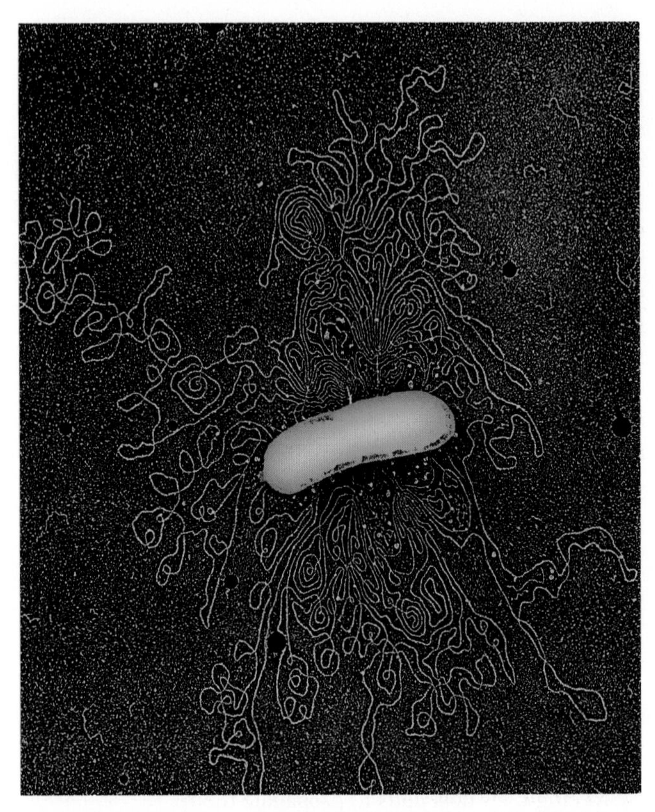

Figure 2-12 Electron micrograph of the genome of the bacterium *E. coli,* released from the cell by osmotic shock. *(Dr. Gopal Murti/Science Photo Library/Photo Researchers.)*

The genes of viruses are found close together with relatively little intergenic space. Linear viral genomes (and linear plasmids) have a special protein permanently attached to the 5′ phosphate group at each end; these proteins play a crucial role when the DNA replicates.

Prokaryotic Genomes

The genome of most prokaryotes is contained within a single chromosome that is a single, circular double helix of DNA. There are some exceptions, such as the bacterium *Borrelia burgdorfei,* in which the chromosome is a single linear DNA double helix. Within each bacterial cell there can be from one to several identical copies of the single chromosome type. A small minority of bacterial genomes consist of several different chromosomes.

Bacterial genes are arranged close together, and introns are extremely rare. In some regions of prokaryotic genomes, genes whose products are physiologically related are often located together as a group, and one molecule of mRNA is made from the entire unit; such a unit is termed an **operon.** For example, the genes concerned with lactose utilization in the colon bacterium *Escherichia coli* constitute an operon and are transcribed as one mRNA molecule. Operons are very rare in eukaryotes. The functional properties of operons will be discussed in Chapter 13.

In electron micrographs of bacterial cells, the DNA is seen arranged in a dense clump called a **nucleoid.** When cells are broken, the packing of the nucleoid is lost and DNA tumbles out in a disorganized skein (Figure 2-12). In each cell there is about 1 mm of DNA, whereas the cell is only 1 μm in diameter. Hence there is a clear need for effective packing of the DNA. Bacterial genomes have associated proteins that are thought to help package the genome into the nucleoid, but the precise functions of these DNA-associated proteins are not understood.

MESSAGE

In most species of prokaryotes, a cell has one or more copies of a single circular genome and, sometimes, one or more copies of plasmids.

Eukaryotic Nuclear Genomes

In eukaryotic organisms, the vast majority of genes are found in the chromosomes of the nucleus. Most eukaryotic species are classified as either **diploid,** carrying two sets of nuclear chromosomes (i.e., two nuclear genomes) in each body cell, or **haploid,** with only one chromosome set per cell. Most fungi and algae are haploids, whereas most other eukaryotes, including animals and flowering plants, are diploids. However, it is worth noting that diploid organisms produce haploid reproductive cells (such as ova and sperm in animals); conversely, haploid organisms produce specialized diploid cells during the sexual phase of their life cycles.

The letter ***n*** is used to designate the number of chromosomes in one nuclear genome, so the haploid condition is designated n (that is, $1 \times n$) and the diploid state is $2n$ (that is, $2 \times n$). The symbol n is called the **haploid chromosome number.** Conditions that are $3n$, $4n$, $5n$, $6n$, and so on are also known, especially in plants; these are called **polyploids,** to be discussed in Chapter 11. The number of chromosome sets (1, 2, 3, 4, 5, 6, etc.) is sometimes called the **ploidy** or **ploidy level.** Note that the ploidy level conventionally refers to the number of sets in a cell that has not entered cell division. It can be seen that the total number of chromosomes in a cell equals the haploid chromosome number times the ploidy level. In humans, for example, $n = 23$ and body (**somatic**) cells are diploid ($2n$) so there are $2 \times 23 = 46$ chromosomes in body cells. Chromosome numbers are determined simply by counting them in stained preparations of cells under a light microscope. Chromosomes are easier to see when they are in a condensed form, as during cell division. Figure 2-13 shows the nuclear genome of a species of small deer in which $2n = 6$. Notice from this figure that during division the chromosomes are compact and sausage-shaped, whereas in the nucleus caught between divisions the chromosomal material is spread throughout the nucleus in an extended state. Chromosome numbers of a variety of diploid organisms are given in Table 2-3 on the next page.

MESSAGE

The number of chromosomes in a eukaryotic nucleus equals the number of chromosomes in the chromosome set (the haploid chromosome number) multiplied by the number of sets (ploidy level).

In a diploid cell, since there are two chromosome sets, there are two chromosomes of each type—two of chromosome 1, two of chromosome 2, two of chromosome 3, and so on. The two members of a pair are called **homologous chromosomes,** or **homologs.** Members of a homologous pair are substantially alike in size and gene content, carrying the same genes in the same relative positions. Hence a

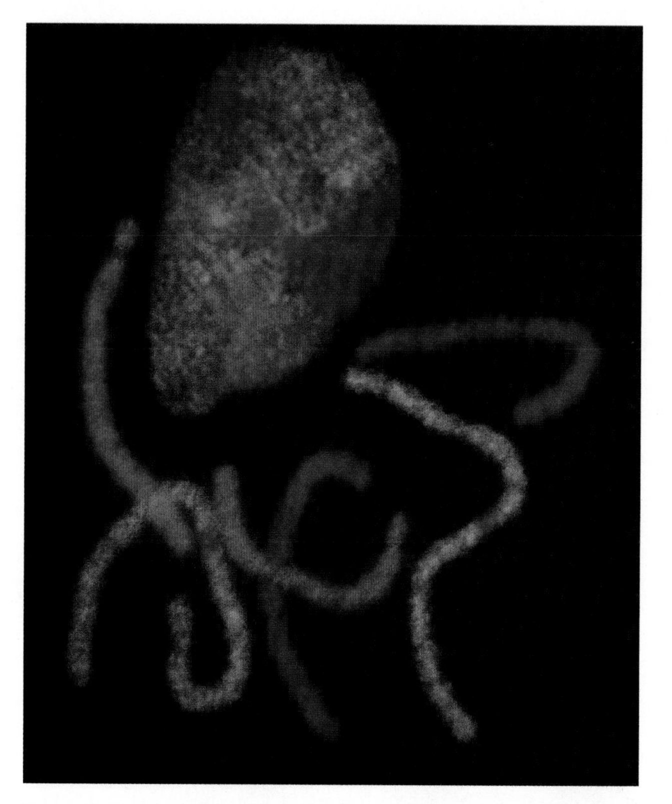

Figure 2-13 Nuclear genome in cells of a female Indian muntjac, a type of small deer ($2n = 6$). The six visible chromosomes are from a cell caught in the process of nuclear division. The three chromosomal types have been stained with chromosome-specific DNA probes, each tagged with a different fluorescent dye ("chromosome paint"). A nucleus derived from another cell is at the stage between divisions. Note that the chromosomes are compact when dividing; but in their extended state between divisions, they seem to "fill" the entire nucleus. *(Photo provided by Fengtang Yang and Malcolm Ferguson-Smith of Cambridge University. Appeared as the cover of* Chromosome Research, *vol. 6, no. 3, April 1998.)*

segment of a pair of homologous chromosomes in a diploid cell can be represented

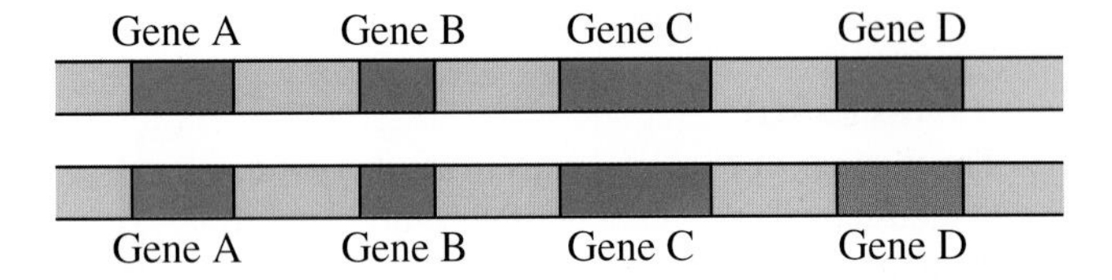

What about alleles, the variant forms of a gene? In a diploid individual carrying two different alleles, these alleles would both be located at the same relative position in the members of a pair of homologs. Even though some alleles can have dramatic impact at the level of phenotype, alleles represent minor differences between the homologs at the level of overall nucleotide sequence.

Experiments to study the precise locations of genes on chromosomes (Chapter 6) and to deduce the complete nucleotide sequences of chromosomes (Chapter 9) have shown that nonhomologous chromosomes in the genome contain largely nonoverlapping sets of genes. Furthermore, as we will see in the next section, chromosomes vary greatly in appearance under the microscope, and these unique diagnostic features are another manifestation of the different gene content of nonhomologous chromosomes. The differences between the chromosomes manifest them-

TABLE 2-3 Numbers of Pairs of Chromosomes in Different Species of Plants and Animals

Common Name	Scientific Name	Number of Chromosome Pairs	Common Name	Scientific Name	Number of Chromosome Pairs
Mosquito	*Culex pipiens*	3	Wheat	*Triticum aestivum*	21
Housefly	*Musca domestica*	6	Human	*Homo sapiens*	23
Garden onion	*Allium cepa*	8	Potato	*Solanum tuberosum*	24
Toad	*Bufo americanus*	11	Cattle	*Bos taurus*	30
Rice	*Oryza sativa*	12	Donkey	*Equus asinus*	31
Frog	*Rana pipiens*	13	Horse	*Equus caballus*	32
Alligator	*Alligator mississipiensis*	16	Dog	*Canis familiaris*	39
Cat	*Felis domesticus*	19	Chicken	*Gallus domesticus*	39
House mouse	*Mus musculus*	20	Carp	*Cyprinus carpio*	52
Rhesus monkey	*Macaca mulatta*	21			

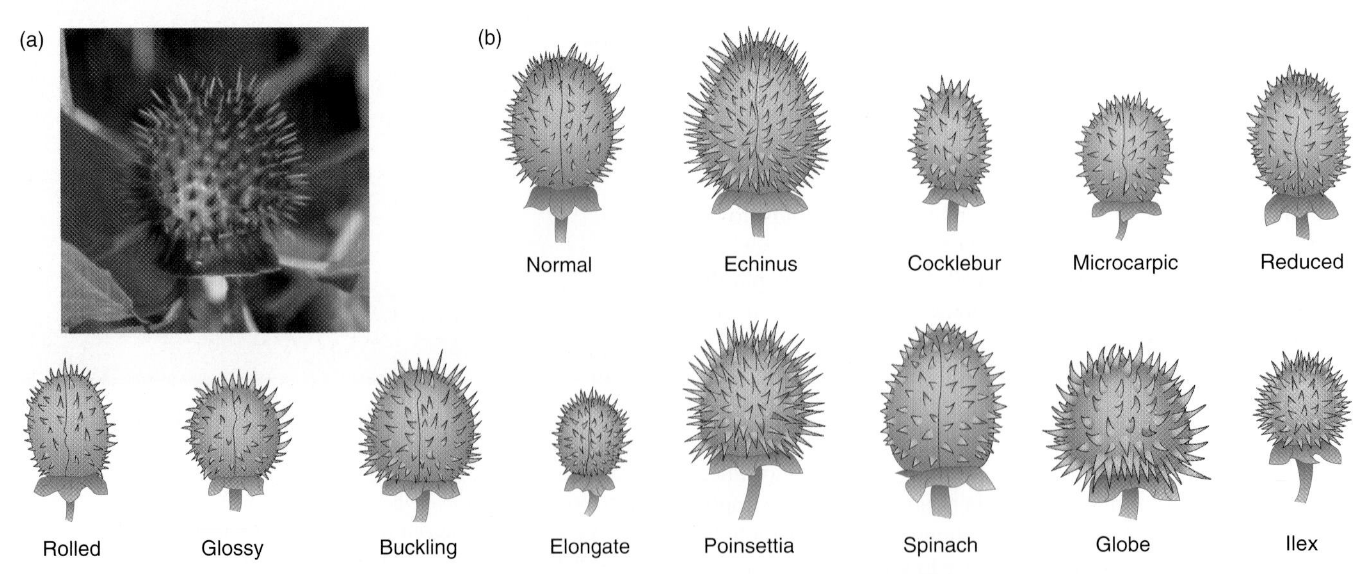

Figure 2-14 Fruits from *Datura* plants. (a) *Datura leichhardtii.* (b) Each *Datura* plant has one different extra chromosome. Their characteristic appearances suggest that each chromosome is different. *(Part a, Dr. G. W. M. Barendse/Nijmen University Botanical Garden; part b from E. W. Sinnott, L. C. Dunn, and T. Dobzhansky,* Principles of Genetics, *5th ed. McGraw-Hill Book Company, 1958, New York.)*

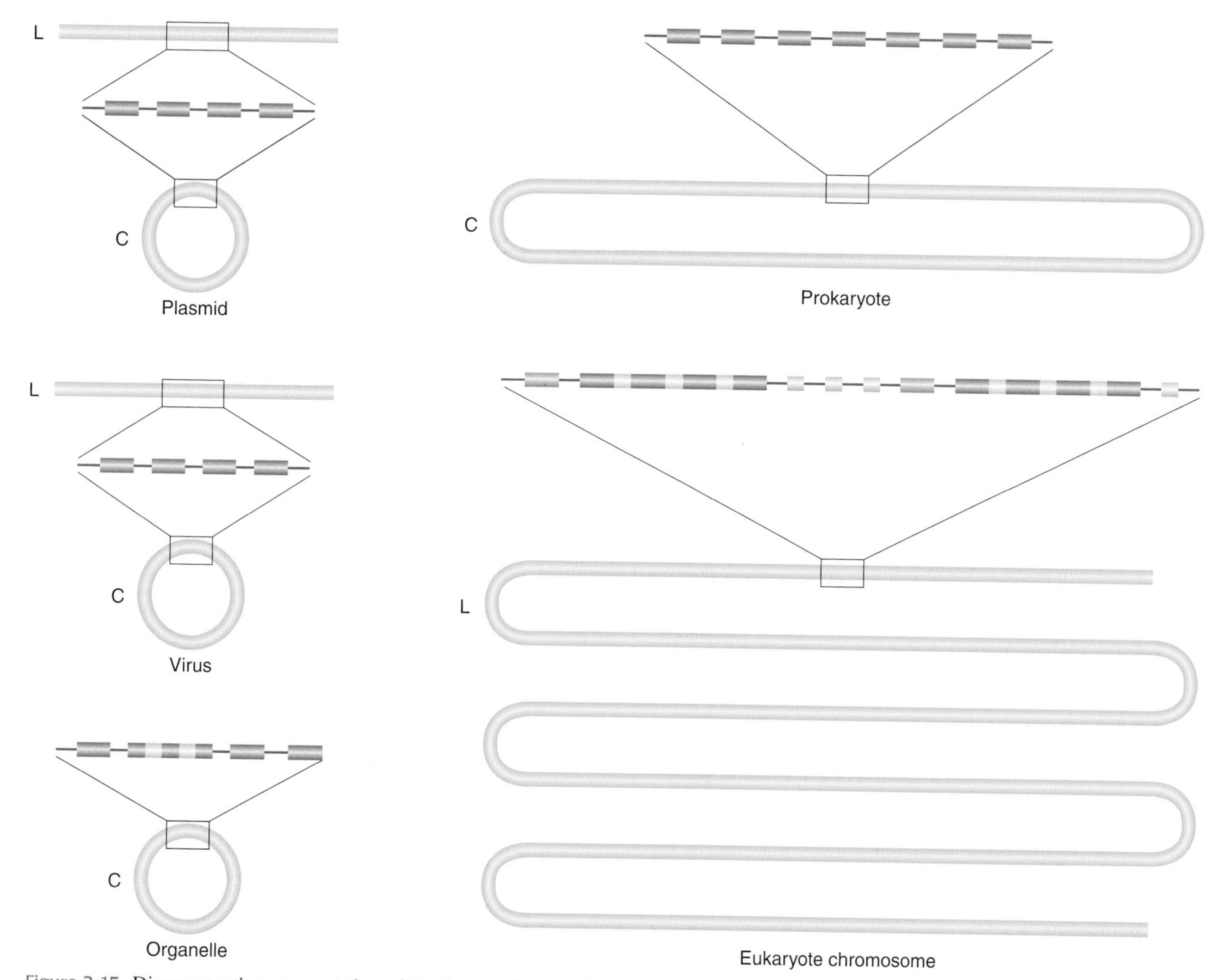

Figure 2-15 Diagrammatic representation of the form and approximate sizes of different types of genomes. The expanded segments show coding regions in green, introns in light green, and repetitive DNA in light blue and red.

selves in various ways at the anatomical level. For example, abnormal individuals that contain an extra copy of one specific chromosome show an abnormal appearance that is characteristic for that specific extra chromosome. An example from plants is shown in Figure 2-14.

MESSAGE

A eukaryotic cell has one or two sets of linear nuclear chromosomes, multiple copies of a circular mitochondrial chromosome, multiple copies of a circular chloroplast chromosome (plants only), and sometimes multiple copies of plasmids (some fungi and plants).

Figure 2-15 summarizes the main features of the broad categories of genomes discussed in the previous sections. Note the differences in size, shape, arrangement of genes, and intron content.

THE NATURE OF EUKARYOTIC NUCLEAR CHROMOSOMES

The set of chromosomes as viewed under the microscope is called a **karyotype** ("nuclear type"). The karyotype is defined by chromosome number and by other visible landmarks.

Visible Chromosomal Landmarks

The microscopic study of chromosomes and analysis of their genetic properties is called **cytogenetics,** a discipline that combines cytology with genetics. Since a great deal of space

TABLE 2-4 Human Chromosomes

Group	Number	Diagrammatic Representation*	Relative Length†	Centromeric Index‡
Large chromosomes				
A	1		8.4	48 (M)
	2		8.0	39
	3		6.8	47 (M)
B	4		6.3	29
	5		6.1	29
Medium chromosomes				
C	6		5.9	39
	7		5.4	39
	8		4.9	34
	9		4.8	35
	10		4.6	34
	11		4.6	40
	12		4.7	30
D	13		3.7	17 (A)
	14		3.6	19 (A)
	15		3.5	20 (A)
Small chromosomes				
E	16		3.4	41
	17		3.3	34
	18		2.9	31
F	19		2.7	47 (M)
	20		2.6	45 (M)
G	21		1.9	31
	22		2.0	30
Sex chromosomes				
	X		5.1 (group C)	40
	Y		2.2 (group G)	27 (A)

*Black dot indicates position of centromere on length of chromosome.
†Percentage of the total combined length of a haploid set of 22 autosomes.
‡Percentage of chromosome's length spanned by its short arm. The four most metacentric chromosomes are indicated by an (M), the four most acrocentric by an (A).

in this book will be devoted to the properties of chromosomes, we will begin with a look at the visible morphology of chromosomes, the features commonly used in cytogenetic analysis as landmarks for telling chromosomes apart.

Chromosomes can be viewed relatively easily under the microscope, but only just before, during, and immediately after cell division. When a cell divides, the nucleus and its chromosomes also divide. The topics of cell and nuclear division will be covered fully in Chapter 4. For the present discussion, we note that long, extended forms of chromosomes might tangle or break when moved around during division. Hence chromosomes need to become very compact in preparation for cell division. Thus, the components of a chromosome (including its DNA) become condensed from the extended form found in nondividing cells into a shorter, fatter form that can be easily handled by the division apparatus of the cell. It is the landmarks of these condensed chromosomes that are seen under the microscope, but it can be presumed that these landmarks are in the same relative positions on the extended chromosomes of nondividing cells.

Chromosome size. The chromosomes of a single genome may differ considerably in size. In the human genome, for example, there is about a fourfold range in size from chromosome 1 (the biggest) to chromosome 21 (the smallest), as shown in Table 2-4. In studying the chromosomes of a species, a cytogeneticist may have difficulty identifying individual chromosomes by size alone but may be able to group chromosomes of similar size. For example, in humans the chromosomes are placed into seven groups, labeled A (the largest chromosomes) through G (the smallest), as shown in the table.

Centromere position. The molecular strings (spindle fibers) that move chromosomes around during cell division attach to a specialized region of the chromosome called the **centromere.** The centromeric region usually appears as a constriction or neck at a specific position along the chromosome. This constriction divides the chromosome into two "arms." The shorter arm is called *p* (after the French word for "small," *petite*) and the longer arm *q*. The position of the constriction determines the **centromeric index,** defined as the length of the short arm over the total chromosome length. This index is a useful characteristic for identifying individual chromosomes (see Table 2-4). Chromosomes also can be categorized by the position of their centromeres; **telocentric** (at one end), **acrocentric** (close to one end), or **metacentric** (in the middle). Examine the chromosomes in Figure 2-13 and try to classify them according to these three terms.

The tips of linear chromosomes are called **telomeres.** Unlike centromeres, telomeres generally are not visibly distinct from the other parts of the chromosome. Both telomeres and centromeres have unique molecular structures that are crucial to normal chromosome behavior.

Position of nucleolar organizers. **Nucleoli** are spherical structures found at constrictions of the chromosomes called **nucleolar organizers** (Figure 2-16). Different organisms are differently endowed with nucleoli, which range in number from one to many per chromosome set. The diploid cells of many species have just a pair of nucleoli. Nucleolar organizers contain numerous tandem copies of the genes that code for ribosomal RNA, an untranslated RNA that is a component of the ribosomes. Ribosomal RNA is synthesized at the nucleolar organizers, deposited into the nucleoli, and later exported to the cytoplasm to become incorporated into ribosomes. The positions of nucleoli, like the positions of centromeres, are quite useful landmarks for cytogenetic analysis.

Variations in thickness. **Chromomeres** are beadlike, localized thickenings found along the chromosome during the early stages of nuclear division. Larger swellings found in some species are called **knobs.** The positions of these structures are constant and are the same in homologous chromosomes. Although they can be useful as markers, their molecular nature is not known.

Heterochromatin patterns. When chromosomes are treated with chemicals that react with DNA, such as Feulgen reagent, distinct regions with different staining characteristics are revealed. Densely staining regions are called **heterochromatin,** indicating a high degree of compactness; poorly staining regions are called **euchromatin** and indicate less tightly packed regions (see Figure 2-16). Most of the transcribed genes are in euchromatin. Heterochromatin is mostly untranscribed repetitive DNA.

Banding patterns. Special chromosome staining procedures have revealed intricate sets of euchromatin **bands** (transverse stripes) in many different organisms. The positions and sizes of the bands are highly chromosome-specific, making them useful landmarks. Commonly used banding patterns are **G bands** and **R bands.** The G bands are produced by treating the chromosomes with the protein-digesting enzyme trypsin followed by staining with the DNA-binding dye Giemsa. The dark-staining G bands are known to be relatively more compact than nonstaining regions at the time of staining (at nuclear division). G bands contain a high proportion of A–T base pairs. R ("reverse Giemsa") bands are produced by heat-treating the chromosomes in saline solution before staining with Giemsa. The R bands are rich in G–C pairs, and most active genes are located in these bands. The G banding patterns in two human chromosomes are shown in Figure 2-17.

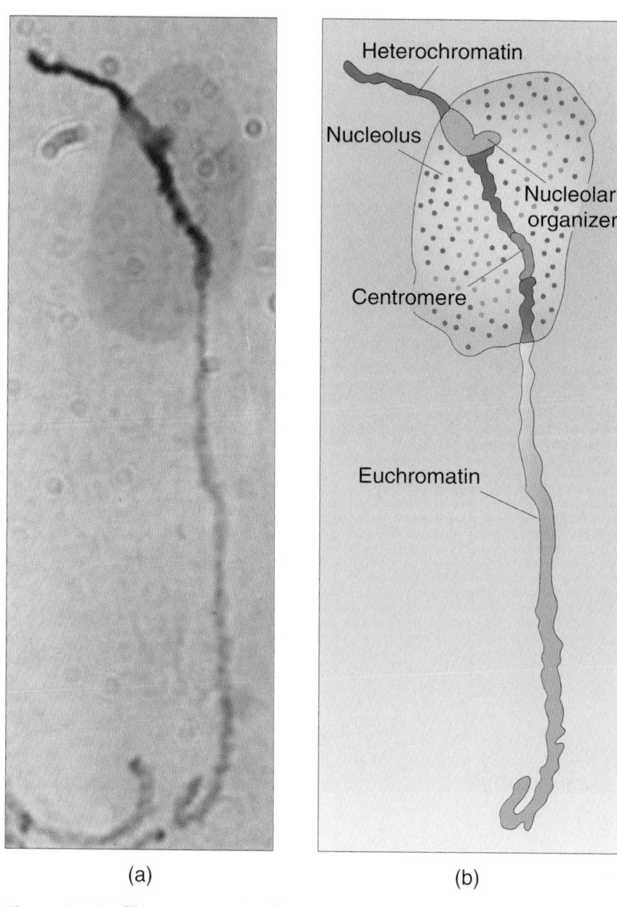

Figure 2-16 Chromosome 2 of tomato, showing the nucleolus and the nucleolar organizer: (a) photograph; (b) interpretation. *(Photo Peter Moens. From P. Moens and L. Butler, "The Genetic Location of the Centromere of Chromosome 2 in the Tomato,"* Can. J. Genet. Cytol. *5, 1963, 364–370.)*

A rather specialized kind of banding occurs in a few organisms whose chromosomes can replicate their DNA many times without separating. This produces giant chromosomes, which are in essence magnified versions of the unreplicated forms. Consequently, after DNA staining, the natural banding patterns of the chromosomes are highly conspicuous and are very useful as landmarks. These **polytene chromosomes** (*polytene* means "many-banded") are found in specialized

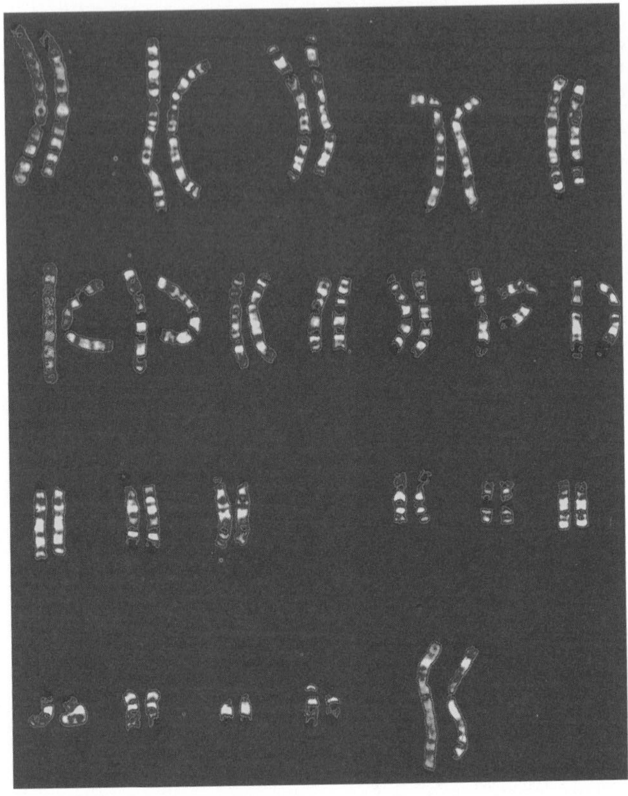

(a)

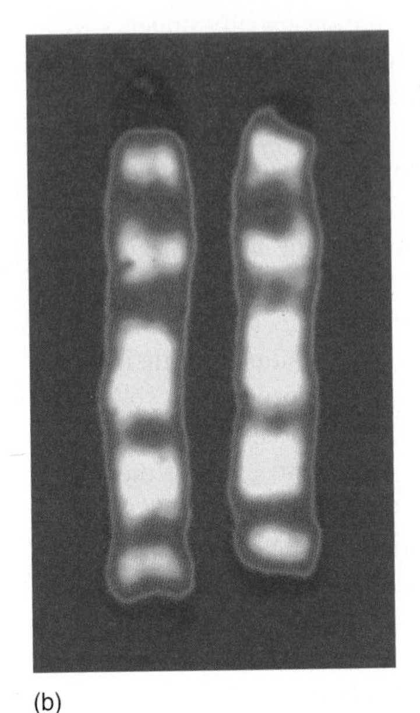

(b)

Figure 2-17 G-banding chromosomes of a human female. (a) Complete set (44A XX). The chromosomes are arranged in homologous pairs in order of decreasing size, starting with the largest autosome (chromosome 1) and ending with the relatively large X. (b) Enlargement of chromosome pair 13. *(L. Willatt, East Anglian Regional Genetics Service/Science Photo Library/Photo Researchers.)*

cells of the malpighian excretory tubules, rectum, gut, footpads, and salivary glands of the dipteran insects such as houseflies, mosquitoes, and fruit flies.

The fruit fly *Drosophila melanogaster* is a much-studied example. This insect (a diploid) has a $2n$ number of 8; these eight chromosomes (Figure 2-18a) are present in most cells. However, in the cells of the special organs that contain the polytene chromosomes, we see some interesting peculiarities (Figure 2-18b). First, there are only four polytene chromosomes per cell (not eight) because during the specialized replication process the members of each homologous pair unexpectedly unite with each other. Second, all four polytene chromosomes become joined at a structure called the **chromocenter,** a coalescence of the heterochromatic areas around the centromeres of all four chromosomes.

Along the length of a polytene chromosome, there are transverse bands (Figure 2-18c). Polytene bands are much more numerous than G or R bands, numbering in the hundreds on each chromosome. The bands differ in width and appearance, so that the banding pattern of each chromosome is unique and characteristic of that chromosome. Molecular studies have shown that in any chromosomal region of *Drosophila* there are more genes than there are polytene bands, so the bands do not represent genes.

Using all the available chromosomal landmarks together, cytogeneticists can distinguish each of the chromosomes of many species. Figure 2-19 shows a map of the chromosomal landmarks of the genome of corn.

MESSAGE

Such features as size, arm ratio, heterochromatin, number and position of swellings, number and location of nucleolar organizers, and banding pattern identify the individual chromosomes within the set that characterizes a species.

Three-Dimensional Structure of Nuclear Chromosomes

A human cell contains about 2 meters of DNA (1 m per chromosome set). The human body consists of approximately 10^{13} cells, and each cell is diploid; therefore, the body contains a total of about 2×10^{13} m of DNA. Some idea of the extreme length of this DNA can be obtained by comparing it with the distance from Earth to the sun, which is 1.5×10^{11} m. You can see that the DNA in your body could stretch to the sun and back almost 100 times. This peculiar fact makes the point that the DNA of eukaryotes is obviously efficiently packed. In fact, the packing occurs at the level of the nucleus, where the 2 m of DNA in a human cell is packed into 46 chromosomes, all inside a nucleus only 0.006 mm in diameter. How can such long molecules be packaged into the wormlike structures we call chromosomes? To answer this question we need to understand the three-dimensional structure of eukaryotic chromosomes.

If eukaryotic cells are broken, and the contents of their nuclei are examined under the electron microscope, chromo-

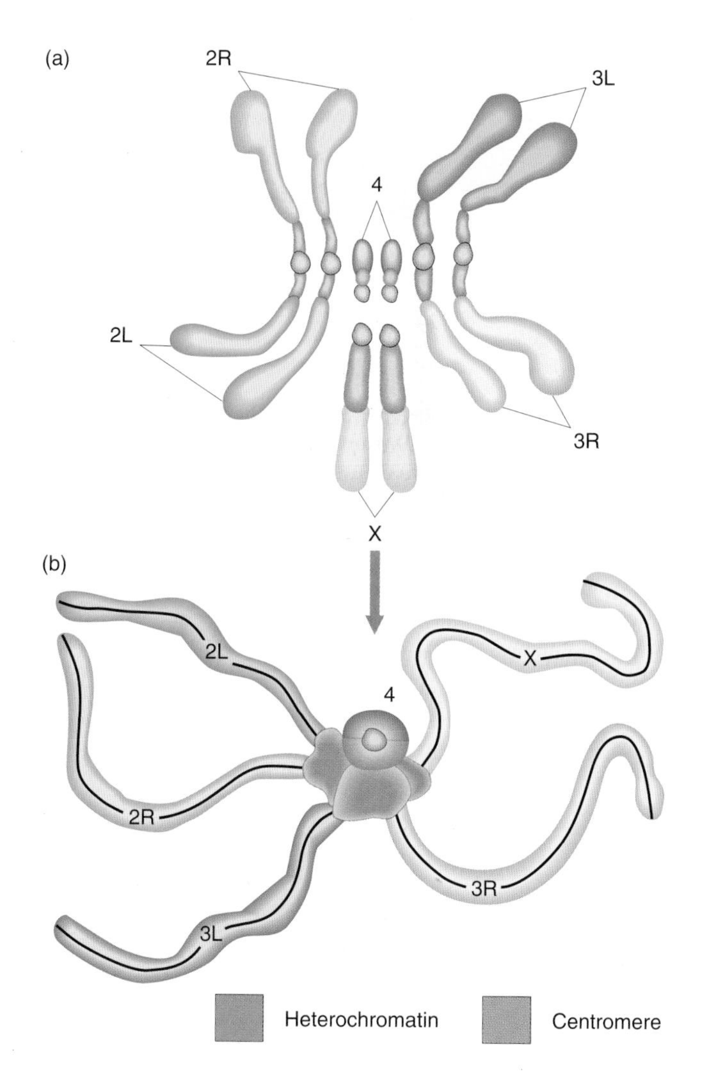

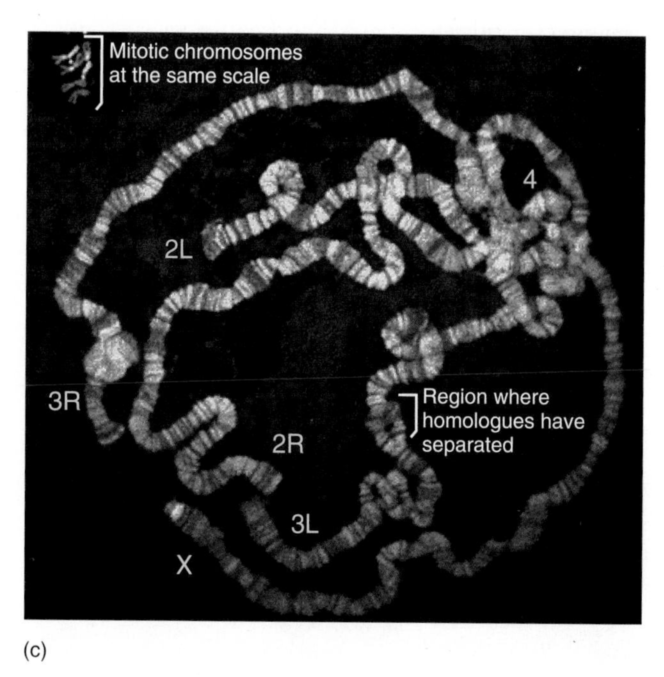

(c)

Figure 2-18 Polytene chromosomes form a chromocenter in a *Drosophila* salivary gland. (a) The basic chromosome set as seen in dividing cells, with arms represented by different shades. (b) In salivary glands, heterochromatin coalesces to form the chromocenter. (c) Photograph of polytene chromosomes. *(Courtesy of Brian Harmon and John Sedat, University of California, San Francisco.)*

somes appear as masses of spaghetti-like fibers with diameters of about 30 nm (Figure 2-20 on the next page). If such chromosomes are studied carefully, it becomes clear that for any internal segment of chromosome there are no ends protruding from the fibrillar mass. This suggests that each chromosome is one long, fine fiber folded up in some way. If the fiber has a DNA molecule as its core, then we arrive at the idea that each chromosome contains one densely folded DNA molecule. Single, long DNA molecules of chromosome size can be viewed using electron microscopy (Figure 2-21 on the next page).

Perhaps the best evidence that a chromosome consists of a single long DNA molecule comes from a technique called **pulsed field gel electrophoresis (PFGE).** If DNA from a large number of cells is extracted, purified with great care to avoid breaking the molecules, and placed on a gelatinous matrix under the influence of powerful, alternating, crossed electric fields, the DNA molecules of each chromosome move through the gel at speeds proportional to their size. All the DNA of one chromosomal type ends up in one position (called a *band*), so if the gel is stained with a dye that binds to DNA, the number of bands of DNA is always

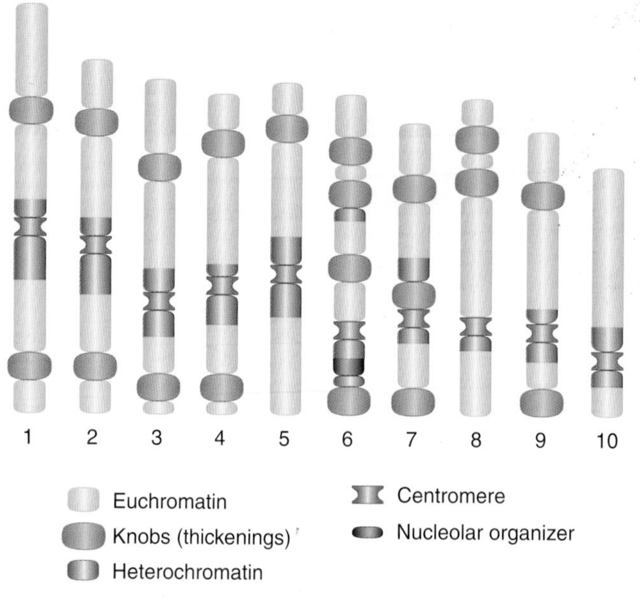

Figure 2-19 The landmarks that distinguish the chromosomes of corn.

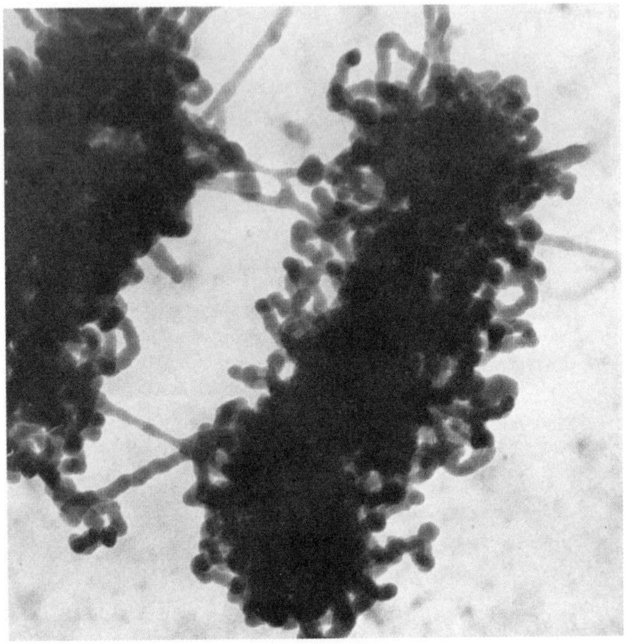

Figure 2-20 Electron micrograph of metaphase chromosomes from a honeybee. The chromosomes each appear to be composed of one continuous fiber 30 nm wide. *(From E. J. DuPraw,* Cell and Molecular Biology. *© 1968 by Academic Press.)*

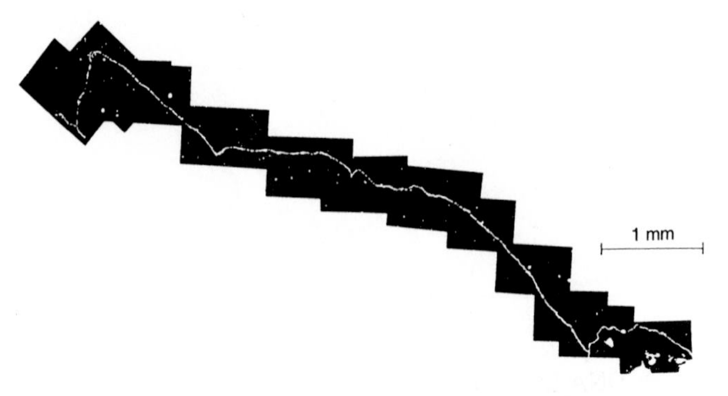

Figure 2-21 Composite electron micrograph of a single DNA molecule constituting one *Drosophila* chromosome. The overall length is 1.5 cm. *(From R. Kavenoff, L. C. Klotz, and B. H. Zimm,* Cold Spring Harbor Symp. Quant. Biol. *38, 1974, 4.)*

Animated Art • Three-Dimensional Structure of Nuclear Chromosomes

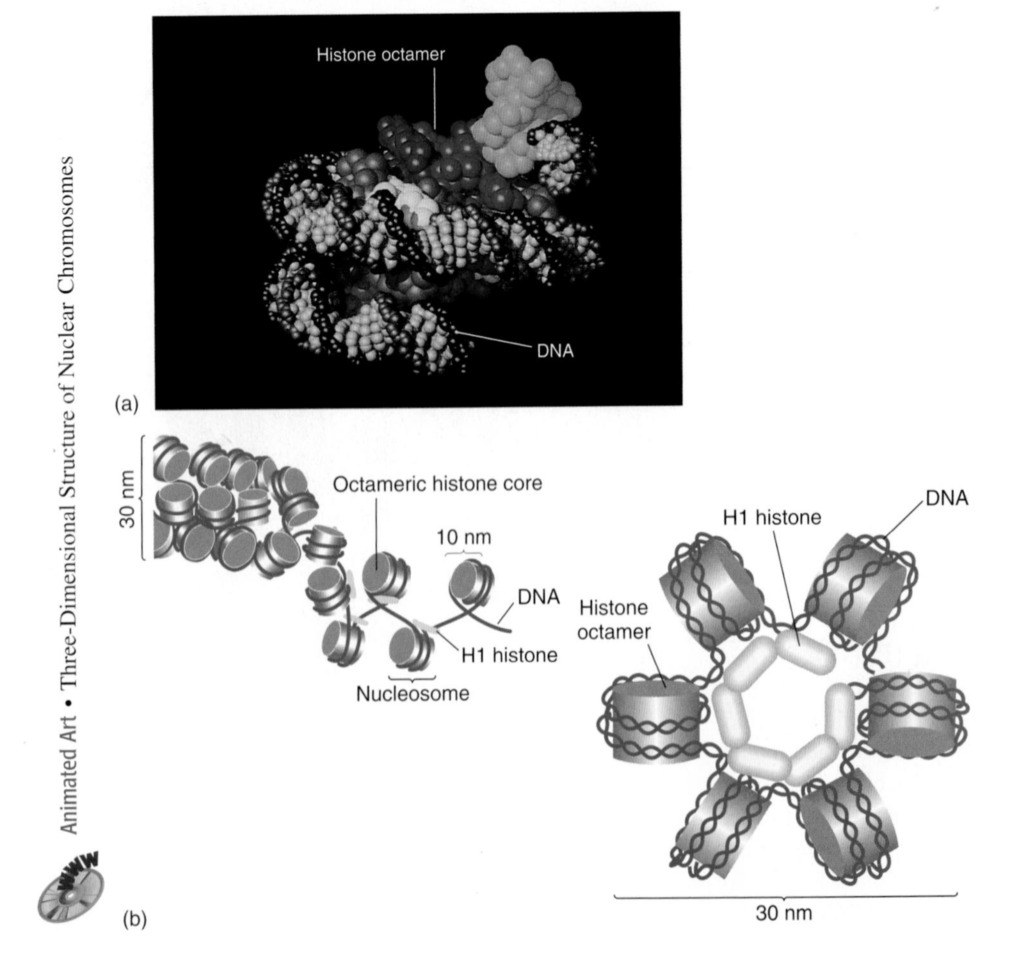

Figure 2-22 (a) Model of a nucleosome showing the DNA wrapped twice around a histone octamer. (b) Two views of a model of the 30-nm solenoid showing histone octamers as purple disks. (*Left*) Partially unwound lateral view. (*Right*) End view. The additional histone H1 is shown running down the center of the coil, probably acting as a stabilizer. With increasing salt concentrations, the nucleosomes close up to form a solenoid with six nucleosomes per turn. *(Part a, Alan Wolffe and Van Moudrianakis; part b adapted from H. Lodish, D. Baltimore, A. Berk, S. L. Zipursky, P. Matsudaira, and J. Darnell,* Molecular Cell Biology, *3d ed. © 1995 by Scientific American Books.)*

equal to the haploid chromosome number. If individual chromosomes consisted of more than one DNA molecule, PFGE would not be expected to yield the same number of bands as the haploid chromosome number.

MESSAGE

Each eukaryotic chromosome contains a single, long, folded DNA molecule.

Histone proteins. What are the precise mechanisms that pack DNA into chromosomes? How is the very long DNA thread converted into the chromosomes visible during cell division? The overall mixture of material that comprises chromosomes is given the general name **chromatin.** It is DNA and protein. If chromatin is extracted from nuclei and treated with differing concentrations of salt, it exhibits different degrees of compaction (condensation) when viewed under the electron microscope. With low salt concentrations, a structure about 10 nm in diameter is seen that resembles a beaded necklace. The string between the beads of the necklace can be digested away with the enzyme DNase, so the string can be inferred to be DNA. The beads on the necklace are called **nucleosomes;** these can be shown to be complexes of DNA and special chromosomal proteins called **histones.** Histone structure is remarkably conserved across the range of eukaryotic organisms, and nucleosomes are always found to contain an octamer of two units each of histones H2A, H2B, H3, and H4. The DNA is wrapped twice around the octamer, as shown in Figure 2-22a. When salt concentrations are higher, the nucleosome beaded necklace gradually assumes a coiled form called a **solenoid** (Figure 2-22b). This solenoid, produced in vitro, is 30 nm in diameter and probably corresponds to the in vivo spaghetti-like structures we saw in Figure 2-20. The solenoid is stabilized by another histone, H1, that runs down the center of the structure, as Figure 2-22b shows.

To achieve its first level of packaging, DNA winds onto histones, which act somewhat like spools. Further coiling results in the solenoid conformation. However, it takes one more level of packaging to convert the solenoids into the three-dimensional structure we call the chromosome.

Higher-order coiling. Many cytogenetic studies show that chromosomes are visibly coiled; Figure 2-23 shows a good example from the nucleus of a protozoan. Whereas the diameter of the solenoids is 30 nm, the diameter of these coiled coils, or **"supercoils,"** is the same as the diameter of the chromosome during cell division, often about 700 nm.

What produces the supercoils? One clue comes from observing chromosomes from which the histone proteins have been removed chemically. After such treatment, the chromosomes have a densely staining central core of nonhistone protein called the **scaffold,** as shown in Figure 2-24. Projecting laterally from this protein scaffold are **loops** of DNA. At high magnifications, it is clear from

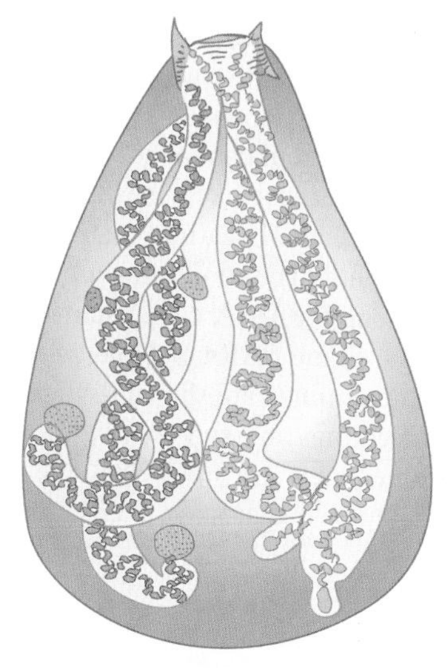

Figure 2-23 Coiling and supercoiling in chromosomes of a protozoan. Two large chromosomes are shown: one orange and the other yellow; the chromosomes are duplicated because cell division is about to occur. Two levels of coiling are visible; the coiled coils are referred to as supercoils. *(From L. R. Cleveland, "The Whole Life Cycle of Chromosomes and Their Coiling Systems,"* Trans. Am. Philosophical Soc. *39, 1949, 1.)*

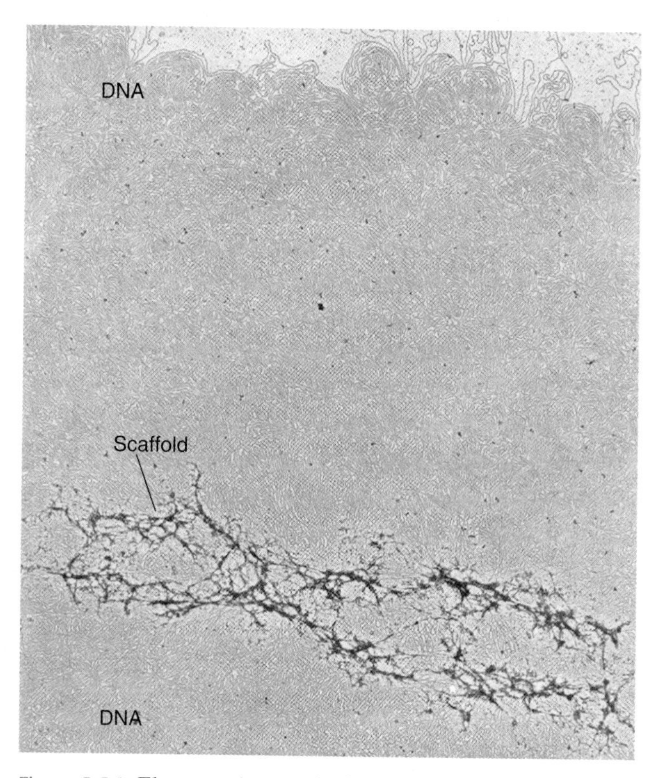

Figure 2-24 Electron micrograph of a chromosome from a dividing human cell. Note the central core, or scaffold, from which the DNA strands extend outward. No free ends are visible at the outer edge (*top*). At even higher magnification (not shown), it is clear that each loop begins and ends near the same region of the scaffold. *(From W. R. Baumbach and K. W. Adolph,* Cold Spring Harbor Symp. Quant. Biol., *Cold Spring Harbor Laboratory, Cold Spring Harbor, N.Y., 1977.)*

electron micrographs that each DNA loop begins and ends at the scaffold. It has been discovered that the central scaffold is largely composed of the enzyme topoisomerase II, an enzyme used during replication (see Chapter 4). The first level of condensation most likely is the looped solenoids emanating from the central scaffold matrix, which itself is in the form of a spiral. The loops attach to the scaffold by special regions along the DNA called **scaffold attachment regions,** or SARs. Some SARs of *Drosophila* whose positions have been determined are shown in Figure 2-25.

A model of coiling and supercoiling is shown in Figure 2-26. Chromosomes are much less condensed in the nondividing nucleus. At that stage they are essentially invisible at the light-microscopic level. This stage is shown on the right in Figure 2-26. In preparation for division, this whole structure becomes supercoiled, as shown on the left of Figure 2-26.

MESSAGE

In the progressive levels of chromosome packing

1. DNA winds onto nucleosome spools.
2. The nucleosome chain coils into a solenoid.
3. The solenoid loops, and the loops attach to a central scaffold.
4. The scaffold plus loops arrange into a giant supercoil.

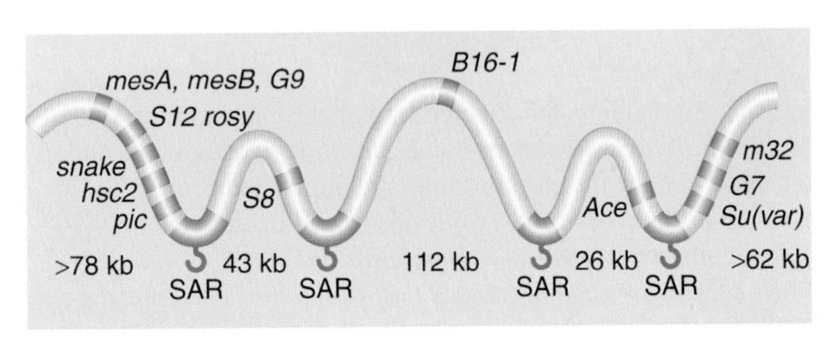

Figure 2-25 Some scaffold attachment regions (SARs) in *Drosophila*.

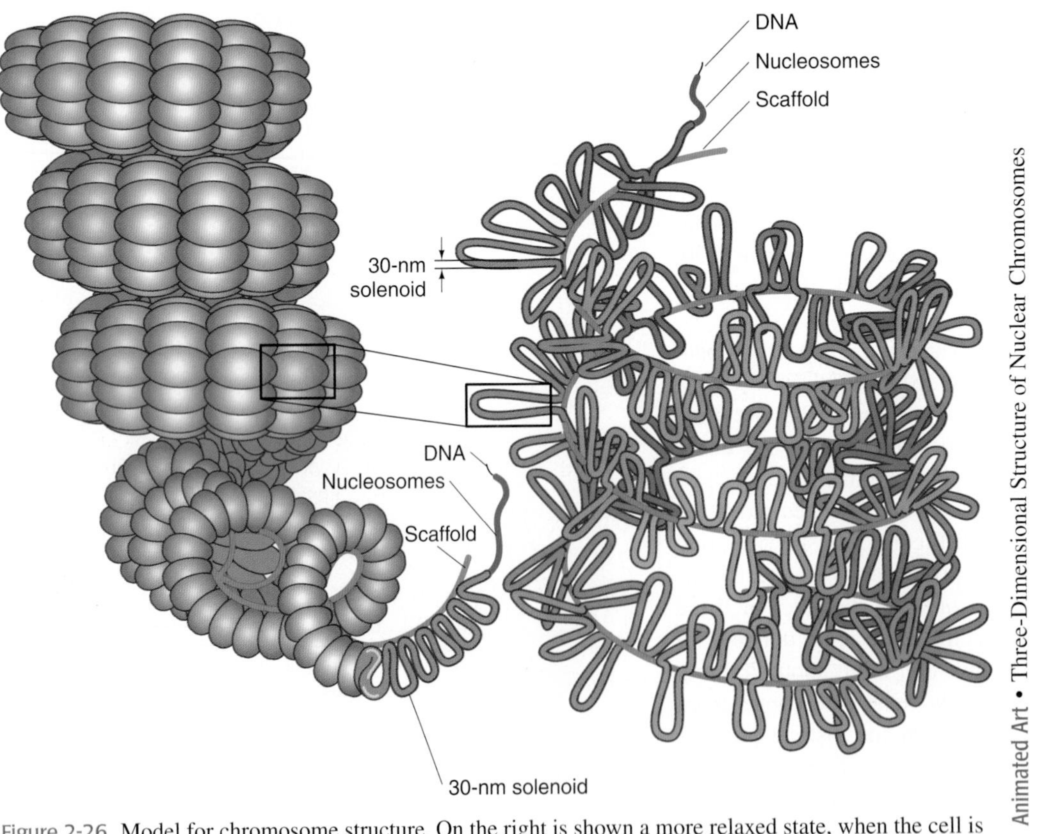

Animated Art • Three-Dimensional Structure of Nuclear Chromosomes

Figure 2-26 Model for chromosome structure. On the right is shown a more relaxed state, when the cell is not dividing. On the left, much tighter coiling is shown, representing a supercoiled chromosome during cell division: here the loops are so densely packed, only their tips are visible. At the free ends the solenoids are shown uncoiled to give an approximation of relative scale.

AN INTRODUCTION TO COMPARATIVE GENOMICS

The study of the similarities and differences between genomes is called **comparative genomics.** To do some basic comparative genomics, let us return to the questions we asked at the beginning of the chapter, which arose from a consideration of the human hereditary disease Duchenne muscular dystrophy (DMD). First, what is the genetic basis of the fundamental characteristics of a species such as *Homo sapiens?* We have seen that these characteristics are encoded in a specific set of genes called the *human genome.* Genetic analysis of DMD has revealed that the basis for the disease is a defect in the structure of a single unique gene located on the human X chromosome. The gene codes for a protein called **dystrophin,** which is located inside the membrane of certain cells, including muscle cells. The precise function of dystrophin is not known, but it appears to play an essential role in linking several different cellular components. Figure 2-27 shows dystrophin linking F-actin, a component of the cytoskeleton, to proteins in the plasma membrane.

In patients with DMD, the protein dystrophin is either absent or altered and nonfunctional, resulting in the various symptoms. The gene coding for dystrophin is one of the largest human genes, 2.5 Mb long (2500 kb), and has 78 introns (Figure 2-28). In fact the dystrophin gene occupies 1.5 percent of the length of the human X chromosome! (The average size of a human gene is 10–15 kb).

The DMD-like diseases in dogs and cats are also associated with a gene coding for dystrophin. At the nucleotide level, the genes are almost identical with that of humans. Indeed, the dog and cat dystrophin genes also reside on the X chromosomes of these species. Hence there is great genetic similarity between mammals in this gene, and indeed this principle extends to other genes. In general there is a striking similarity between the genomes of different mammals: they have the same genes, and

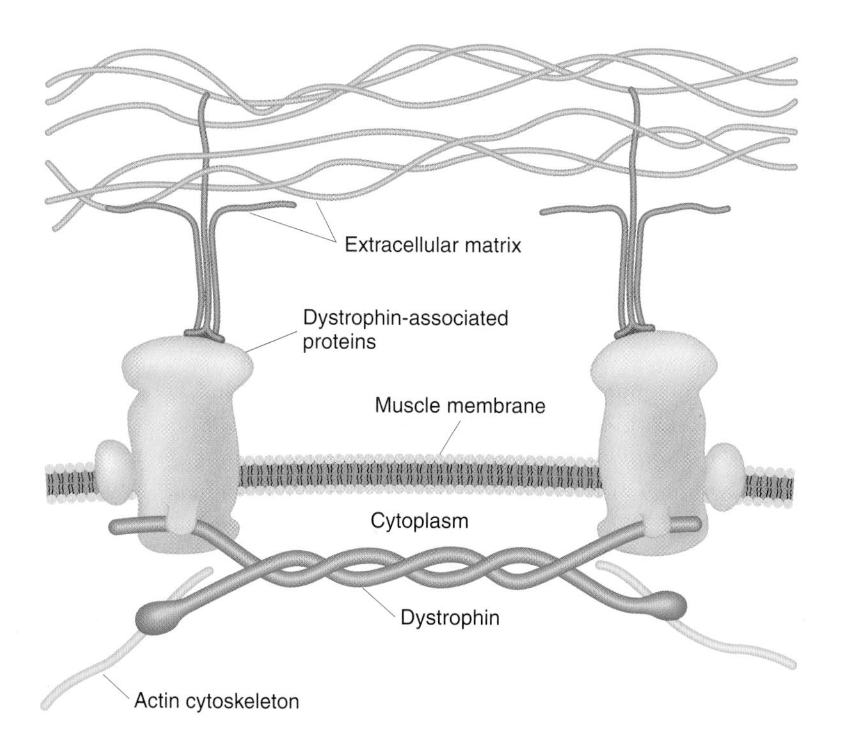

Figure 2-27 The proposed position and orientation of dystrophin, the protein altered in Duchenne muscular dystrophy, inside the membrane of a muscle cell.

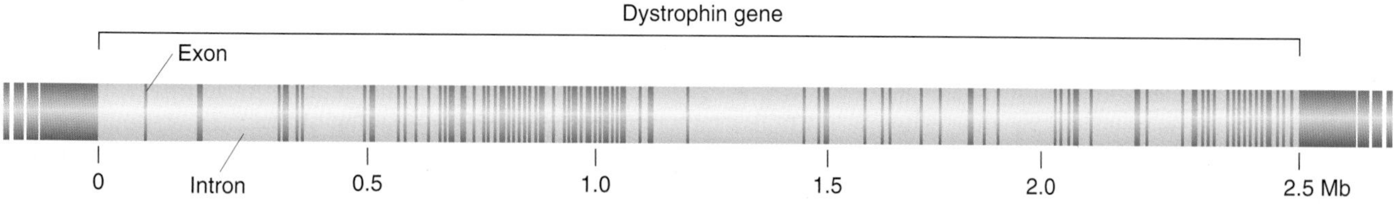

Figure 2-28 Structure of the gene for dystrophin, in which mutations give rise to Duchenne muscular dystrophy.

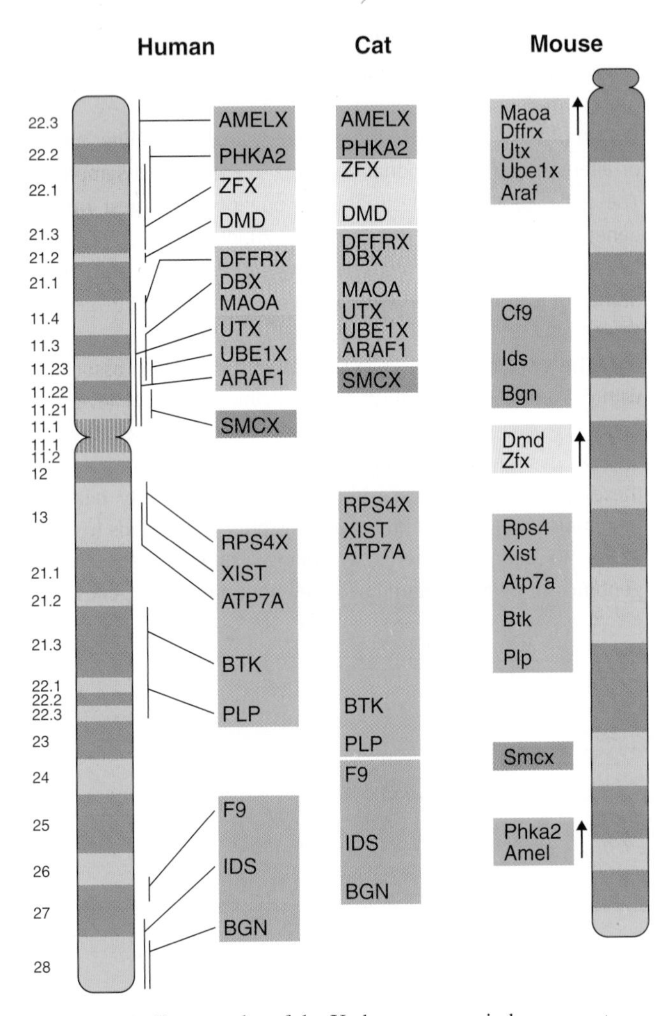

Figure 2-29 Topography of the X chromosomes in human, cat, and mouse, showing the similar gene content and position. The human chromosome is used as a standard, and its bands are numbered. Various blocks of genes are colored to aid comparisons. In the cat, the relative positions are the same as in humans, so no cat chromosome is shown. In the mouse, most of the blocks are the same, but some have been inverted (shown by arrows) or relocated during evolution. *(Modified from S. J. O'Brien et al.,* Science *286, 1999, 458–480.)*

often, in close evolutionary relatives, these genes are in the same positions in the genome (Figure 2-29).

More distantly related organisms show fewer but still striking gene similarities; for example, about half the genes of the fruit fly show similarity to specific mammalian genes. Fruit flies and nematodes even have genes similar to the dystrophin gene, as shown in Figure 2-30 (see "Duchenne MD^{+}-DMD" under "Neurological." Indeed the figure shows that many of the genes involved in human diseases are also found in these experimental organisms, usually performing similar functions. Thus, a process discovered and characterized in one experimental organism often finds application in other organisms, including humans.

Another interesting genomic comparison is between humans from different geographical regions. The genomes of sub-Saharan Africans, Europeans, Asians, and other major geographical groups (often referred to as "geographical races") are virtually identical, and no major discontinuities have been found. The genetic differences in skin pigmentation and hair, body, and facial morphology that characterize these geographical races represent only a minute fraction of the human genome.

SUMMARY

1. The phenotype of an organism is determined by the phenotype of its cells and their arrangement in tissues and organs. Therefore, to understand how genes dictate the characteristics of species, it is necessary to understand the nature of genomes at the cellular level.
2. The genetic material of all living organisms is DNA, a double-helical molecule composed of two intertwined, antiparallel nucleotide chains. There are always four nucleotides represented by the four bases A, T, G, and C. An A in one strand always pairs with T in the other because of complementary hydrogen-bond formation; similarly, G in one strand always pairs with C in the other. The nucleotide sequence in DNA represents a genetically encoded set of information ultimately necessary for specifying biological form.
3. A genome is the full set of DNA that codes for the characteristic features of that organism. A genome is composed of one or more DNA molecules in association with proteins such as histones; each DNA-protein unit is called a *chromosome.*
4. Prokaryotes (organisms lacking nuclei) have one circular chromosome present in one to several copies. Prokaryotic cells also commonly contain small extragenomic DNA molecules called *plasmids.*
5. Most of the DNA of eukaryotes is in the nuclear genome. The nuclear genome is generally much larger

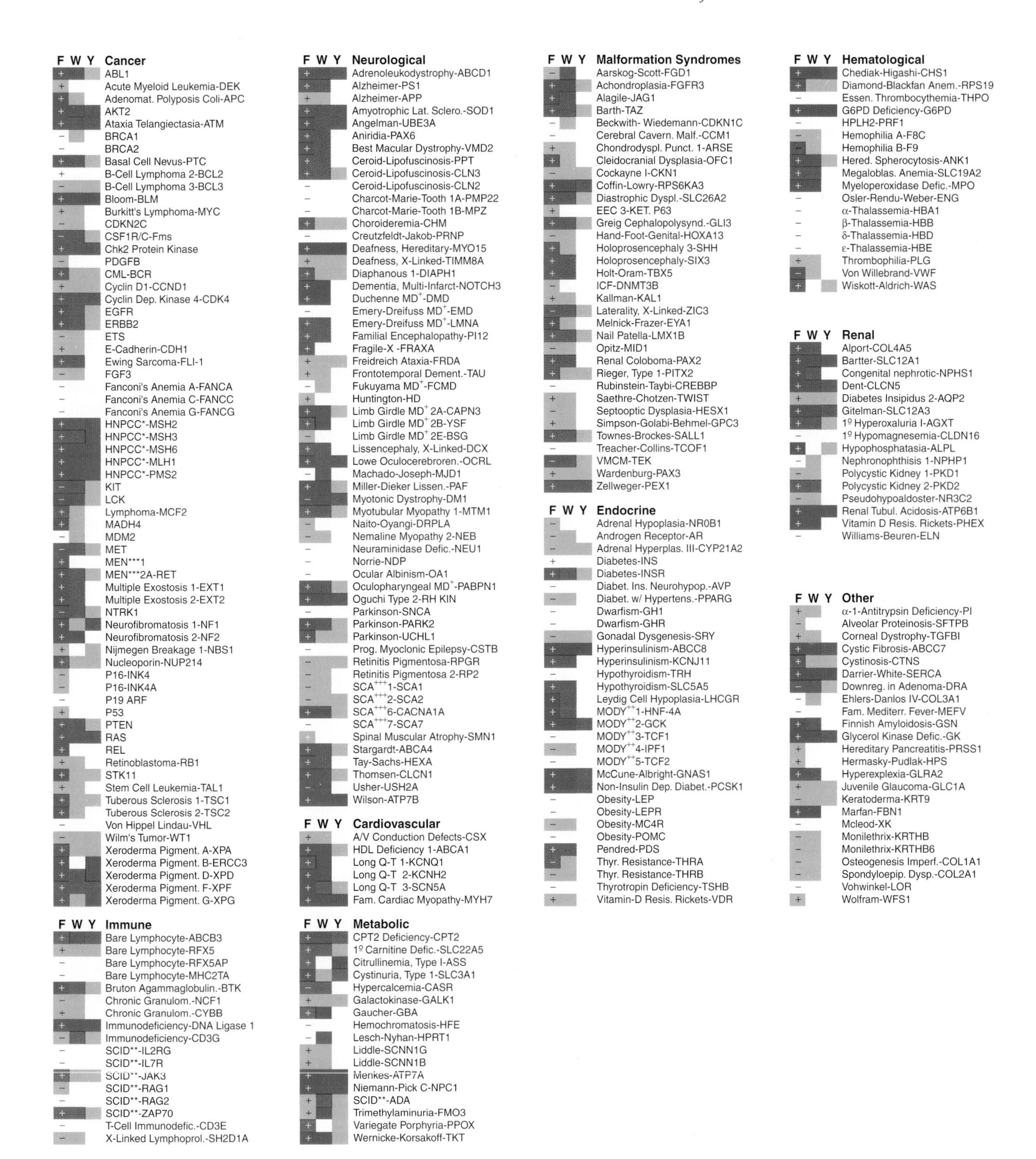

Figure 2-30 Genes of the fruit fly *Drosophila melanogaster* (F), the nematode worm *Caenorhabditis elegans* (W), and the yeast *Saccharomyces cerevisiae* (Y) showing similarity to human disease genes, which are grouped by function. Dark blue represents the strongest sequenced similarity, purple and light blue progressively less. White means no or weak similarity. A plus in the F column means that the gene functions in the same way in flies and humans; the minus means that no function comparable to a human function could be found for the gene in the fly. *(From G. M. Rubin et al.,* Science *287, 2000, 2204–2215.)*

than the prokaryotic genome. A smaller specialized component of eukaryotic genomes is found in chloroplasts (plants only) and in mitochondria. The few cases of eukaryotic plasmids have been found in mitochondria and chloroplasts.

6. The eukaryotic nuclear genome is composed of several to many linear chromosomes. There may be one or two nuclear genomes depending on whether the eukaryote is haploid or diploid. Mitochondria and chloroplasts contain one circular chromosome present in multiple copies.
7. A nuclear chromosome contains one long linear DNA molecule. The DNA is efficiently packed by coiling and supercoiling using spool-like structures composed of histone proteins and a scaffoldlike central core.
8. A gene is a region of the DNA that is capable of being transcribed to make a functional RNA product. The gene contains special sequences to signal the start and stop sites for transcription, and also sites for post-transcriptional processing of RNA. The gene also contains sequences that when transcribed into functional RNA can signal the beginning and end of the translation of the coded information into protein.
9. The genes of many eukaryotes contain intervening sequences called *introns,* which are not part of the final functional RNA product. The coding sequences that separate the introns are called *exons.* Although the size of the coding sequence of different genes shows some variation, there is great variation in the number and size of introns, which accounts for most of the differences in gene size within and between organisms.
10. Viruses, although not organisms, also have small genomes, which may be composed of DNA or RNA, single- or double-stranded, and linear or circular. When viruses parasitize cells, they always convert their viral genomes into a double-stranded DNA or RNA form.
11. Genomes show high levels of structural similarity. Many genes are found in virtually all species, and in closely related species the same genes are often found in similar positions. Mammalian genes are remarkably similar, and when damaged produce similar diseases in different species. Humans of all races are the same at the genomic level; racial differences are based on alleles of genes that represent a very small proportion of the genome.

CONCEPT MAP

Draw a concept map interrelating as many of the following terms as possible. Note that the terms are in no particular order.

DNA double helix / genome / nucleus / homologs / ploidy / *n* / haploid / diploid / chromosome

SOLVED PROBLEM

1. A molecule of double-helical DNA was found to have a purine:pyrimidine ratio of 1 : 4 in one nucleotide chain. What must be the purine:pyrimidine ratio in the other strand?

Solution

What facts about DNA are relevant here? We know from the general structure of DNA that a purine in one strand will always pair with a pyrimidine in the other strand (purine A pairs with pyrimidine T, and purine G pairs with pyrimidine C). In the strand in question for every 4 pyrimidines there is 1 purine, which means that the purines occupy $1/(1 + 4)$, or 1/5, of the sites in one strand. If 1/5 of the sites in one strand are purines, which we know pair only with pyrimidines, then 1/5 of the sites in the other strand must be pyrimidines. Likewise 4/5 of the sites in the other strand must be purines. Therefore, the purine:pyrimidine ratio in the other strand will be 4/5 : 1/5, which is 4 : 1. (Note that this is the inverse of the ratio in the opposite strand.)

SOLVED PROBLEM

2. The human genome contains 3,000,000 kb and according to some estimates contains 100,000 genes. If we make the simple assumption that the average size of the coding region of a gene (the part that codes for protein) is 2 kb, from these data:

a. What can be concluded about the average spacing between genes on chromosomal DNA (measured from the central point of one gene to the central point of the next)?

b. What can be concluded about the proportion of the genome that codes for protein?

c. What occupies the noncoding regions?

Solution

a. The average spacing between the central points of adjacent genes must be 3,000,000 divided by 100,000, which equals 30 kb.

b. Since the protein-coding segment is only 2 kb, it is clear that only 1/15, or about 6 percent, of the genome actually codes for protein.

c. The remaining sequences must be composed of regulatory regions adjacent to genes, as well as repetitive DNA and nonrepetitive spacer DNA between adjacent protein-coding regions and introns within the protein-coding regions.

PROBLEMS

Basic Problems

1. Draw a hierarchical classification diagram that shows the relationships of the terms *eukaryote, prokaryote, haploid, diploid, nuclear genome, chloroplast genome, mitochondrial genome, plasmids.*

2. In a single cell of a plant leaf, how many copies are there of the nuclear genome, the chloroplast genome, the mitochondrial genome (choose your answers from "many," "two," or "one")?

3. What does the word *homologous* mean?

4. Distinguish between the terms *chromosome, chromomere, chromocenter,* and *chromatin.*

5. a. Which of the following have a mitochondrial genome: a fish, a moss, a palm tree, baker's yeast, a bacterium?

b. Which of the following have a chloroplast genome: a diatom, a snake, a mushroom, the gut bacterium *E. coli,* mistletoe?

6. List the forces that hold the DNA double helix together as a stable unit.

7. If thymine makes up 15 percent of the bases in a certain DNA sample, what percentage of bases must be cytosine?

8. Somebody tells you that the G content of the DNA in a certain species is 55 percent. Why should you be suspicious of this statement?

9. If someone tells you that a certain DNA sequence is GTTAACGCT, what further information would you need in order to assess its orientation within the DNA double helix?

10. a. An organism has 5 chromosomes in its body cells. Can you tell from this if it is likely to be haploid or diploid?

b. An organism has 6 chromosomes in its body cells. Can you tell from this if it is likely to be haploid or diploid?

11. In a certain 1-kb DNA molecule the G+C content is 60 percent. How many hydrogen bonds hold the two strands of this molecule together?

12. If species A has more DNA per nucleus than species B, does A necessarily have more genes than B? Explain.

13. Write a sentence including the words *solenoid, histone, nucleosome, chromosome,* and *DNA.*

i **14.** If the G+C content of a DNA sample is 48 percent, what will be the proportions of the four different nucleotides?

15. Draw a simple diagram of DNA that makes it clear what 5′ and 3′ ends are.

16. Each cell of the human body (excluding gonads) contains 46 chromosomes. What are the ploidy level and haploid number?

17. How many DNA molecules are there in the nucleus of a human body cell?

18. A certain segment of DNA has the following nucleotide sequence in one strand:

5′ ATTGGCTCT 3′

What must be the sequence in the other strand (label its 5′ and 3′ ends)?

19. In normal double-helical DNA is it true that

a. A plus C will always equal G plus T?

b. A plus G will always equal C plus T?

c. Purines always equal pyrimidines?

d. Phosphate always equals deoxyribose sugar?

20. The most common elements in living organisms are carbon, hydrogen, oxygen, nitrogen, phosphorus, and sulfur.

a. Which of these is not found in DNA?

b. Which are not found in the sugar-phosphate backbone of DNA?

21. What is a gene? What are some of the problems with your definition?

22. The genome of the bacterium *Hemophilus influenzae* is 1830 kb in size, and sequencing has shown it has 1703 genes.

a. What is the average spacing between the central points of the genes?

b. If the average size of a coding region of a gene is assumed to be 1 kb, what is the average distance between genes?

c. What is in this region?

d. What proportion of the genome is expected to be introns?

e. What is the average space between coding regions?

23. The gene for the human protein albumin spans a chromosomal region 25,000 nucleotide pairs (25 kb) long from the beginning of the protein-coding sequence to the end of the protein-coding sequence, but the messenger RNA for this protein is only 2.1 kb long. What do you think accounts for this huge difference?

24. DNA is extracted from cells of *Neurospora,* a haploid fungus ($n = 7$); pea, a diploid plant ($2n = 14$); and housefly, a diploid animal ($2n = 12$). If the DNA is separated using pulsed field gel electrophoresis, how many bands will be produced by each of these three species?

25. Devise a formula that relates size of mRNA to gene size, number of introns, average size of introns, and size of the regulatory region.

26. Progress in genome sequencing has led to the estimate that there is a 1.7 percent difference between the genomes of humans and chimpanzees. Assuming the two genomes are approximately the same size, how many nucleotide differences are there between these two species?

27. In sentence form describe the difference between

a. The purine and pyrimidine bases in DNA

b. Adenine and guanine

c. Cytosine and thymine

28. What do the symbols 5′ and 3′ actually refer to in labeling the ends of DNA?

29. Draw the following pieces of DNA in the style of Figure 2-5b:

5′ ATGCT 3′

3′ TACGA 5′

3′ ATGCT 5′

5′ TACGA 3′

30. Using the figures of this chapter, estimate how many base pairs there are in one complete gyre of the DNA double helix. In other words, if you imagine you start walking up a vertical DNA double helix as you would walk up a spiral staircase, how many base pairs would you step on before you faced in the same relative direction in which you started?

31. In a certain insect, $2n = 18$. If polytene chromosomes form exactly as they do in *Drosophila,* how many polytenes would be found per cell in the salivary glands?

32. Arrange the following in order of increasing size of their genomes: *Saccharomyces cerevisiae, Escherichia coli,* herpes virus, *Homo sapiens, Arabidopsis thaliana.*

33. Mithramycin preferentially stains G–C base pairs. Would staining chromosomes with mithramycin produce G bands or R bands?

34. Chromosomes 21 and 22 of humans are almost the same size, 33.55 Mb for 21 and 33.46 Mb for 22. Each chromosome contains about 1 percent of the human genome. Chromosome 21 contains 225 genes, whereas 22 contains 545 (data from complete chromosome sequences). From these data estimate the number of genes in the human genome. What are some of the possible errors in your calculation?

35. On the human chromosome 22 there are 55,664 repetitive DNA elements totaling about 14 Mb. Look at the data in Problem 34 and deduce what proportion of chromosome 22 consists of repetitive DNA.

36. For chromosome 22 of humans, the average number of exons per gene is 5.4 (range 1–54). Average exon size is 266 bp, and mean gene size is 19.2 kb. Draw an average gene for chromosome 22, and using data from Problem 34, also show average spacing between genes.

Exploring Genomes: A Web-Based Bioinformatics Tutorial

Introduction to Genomic Databases

Where does a researcher turn to find information about a gene? Integrated genetic databases are maintained by a number of private and government organizations. In the first Genomics tutorial at www.whfreeman.com/mga, you will be introduced to the resources available through the National Center for Biotechnology Information (NCBI) in Washington, D.C.

GENE FUNCTION 3

Key Concepts

1. The functions of DNA and RNA are founded on two basic principles: base complementarity between single polynucleotide strands, and attachment of nucleic acid–binding proteins of various types.
2. Most genes code for proteins; a few code for untranslated RNAs that are functional components of the informational processing system.
3. In protein-coding genes, DNA is transcribed to make mRNA, which is translated to make a polypeptide chain.
4. Only one strand of DNA is used as a template for transcription of a given gene.
5. In RNA polymerization, new nucleotides are always added at the 3′ end of the growing molecule.
6. Protein function is based on the primary amino acid sequence, which is determined by nucleotide sequence.
7. The mRNA is translated three nucleotides (one codon) at a time; specificity arises through complementarity of the binding between the triplet codon and the anticodon of a tRNA bearing the appropriate amino acid.
8. Amino acids are added at the carboxyl end of the growing polypeptide chain.
9. Protein function is determined by shape, size, and binding properties, and all these are based on amino acid sequence.
10. Changes in DNA nucleotide sequence (mutations) may result in changes of amino acid sequence, which often result in reduced or no protein function.
11. In a diploid, if a gene is required in two copies for normal cellular function, a wild/mutant heterozygote will express a mutant phenotype.

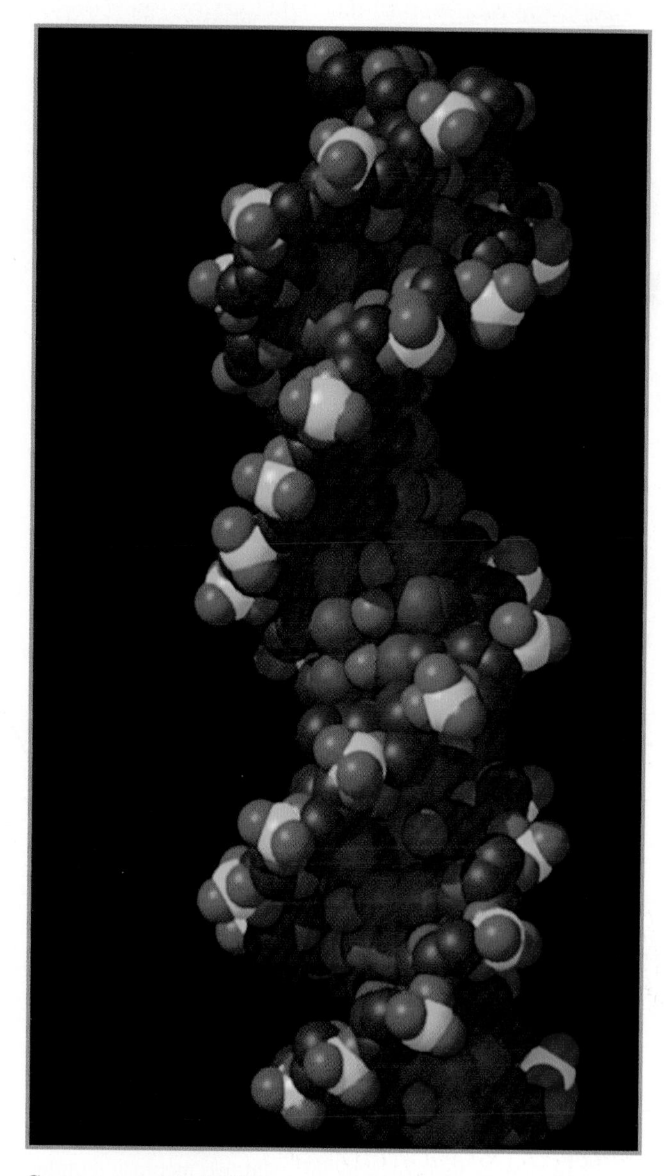

Computer model of DNA. *(J. Newdol, Computer Graphics Laboratory, University of California, San Francisco. © Regents, University of California.)*

(a)

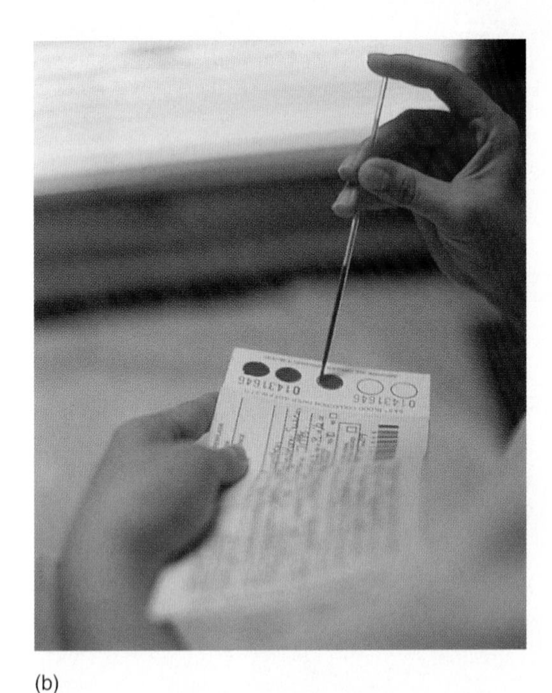

(b)

Figure 3-1 (a) Removing a blood sample from a baby's foot for the PKU test. (b) Applying the blood sample to the PKU test strip. *(Part a, Dan McCoy/Rainbow; part b, Custom Medical Stock Photo.)*

In Norway in 1934, a mother with two mentally retarded children consulted the physician Asbjørn Følling. In the course of the interview Følling learned that the urine of the children had a curious odor. This possible clue about the nature of the children's condition prompted him to focus on the chemistry of their urine. He tested their urine with ferric chloride and found that whereas normal urine gives a brownish color, the children's urine stained green. He deduced that the chemical responsible for the green color must be phenylpyruvic acid. Because of the chemical similarity of phenylpyruvic acid to the amino acid phenylalanine, it seemed likely that this substance had been formed from phenylalanine, and it was found that indeed the children had abnormally high levels of phenylalanine in their blood.

From a series of studies on similar families, Følling deduced that this disease, which came to be known as *phenylketonuria* (PKU), was hereditary; in fact, it seemed that it was caused by mutation of a single gene. The normal form of the gene provides a crucial function in human life. When this function is lacking, because an abnormal form of the gene cannot provide the necessary product, somehow the disease symptoms of abnormal chemical composition and mental retardation are the result. The PKU screening test (Figure 3-1) is now routinely applied to newborn babies and has become the most extensively applied genetic test in the world.

The role of genes in the function of the organism is the key to understanding this and the thousands of other hereditary diseases that can cause human illness and premature death. In this chapter we shall consider the mechanisms of gene function in organisms generally, and then return to phenylketonuria. We will examine the gene involved, its normal function, and the ways in which the defective form causes the multiple symptoms. We will see that understanding the disease at the functional level has led to a highly effective therapy. Indeed, a hope for such therapies underlies all attempts to understand hereditary disorders.

CHAPTER OVERVIEW

Figure 3-2 shows an overview of the main ideas of this chapter, using a eukaryotic system. The chapter concerns the transfer of information from genes to gene products, and how a mutation in a gene can cause cellular malfunction. Within the DNA sequences of any organism's genome is the encrypted information for each of the gene products that the organism can make. Not only do these DNA sequences encode the structure of these products, but they

CHAPTER OVERVIEW figure

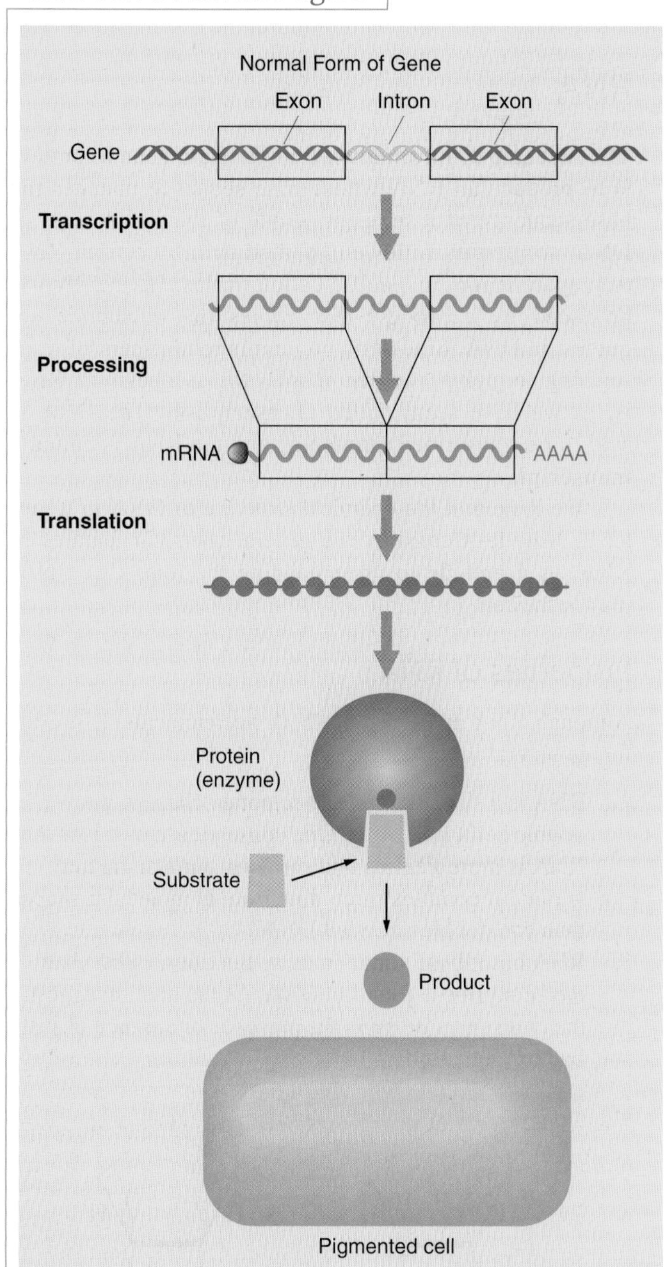

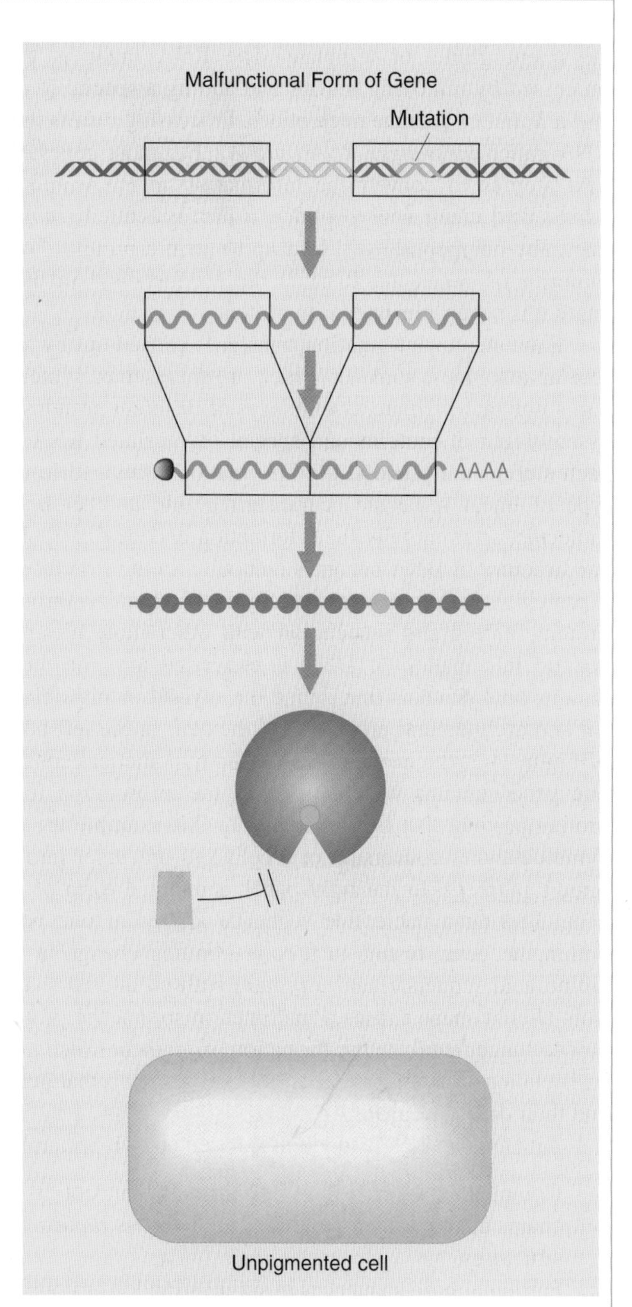

Figure 3-2 Overview of gene function and malfunction in a eukaryotic gene encoding a protein (the most common type of gene). The nucleotide sequence in DNA is transcribed to make single-stranded RNA. The initial transcript is processed to add a cap (the black dot) and a string of adenines (AAAA), and to excise introns. The result of processing is mRNA. The nucleotide sequence in mRNA is translated into an amino acid sequence (polypeptide chain), which specifies the basic properties of a protein. In this hypothetical, but typical, example, the protein is an enzyme that catalyzes the conversion of a colorless substrate into an orange pigment that is deposited in the cell. A defective form of the gene results in an amino acid change at a crucial position in the pocket necessary for the enzyme to engage with its substrate. Consequently no orange pigment is produced.

also contain the information for when, where, and how much of the product is made.

We can think of gene action as a process of copying and deciphering the encrypted information in the gene (Figure 3-2, left panel). The transfer of information from gene to gene product is in several steps. The first step is to copy *(transcribe)* the information into RNA using DNA as an alignment guide, or template. The RNA made is then processed by removing introns and adding a special 5′ cap and a 3′ tail of adenine nucleotides. Processing results in an RNA called *messenger RNA (mRNA)*. For most genes, the next step is to convert the information in RNA into an amino acid chain (polypeptide), a process called *translation.* The polypeptide will fold up to form a protein. For a minority of genes the RNA is the product, and in these cases it is never translated into protein.

Each step of information transfer is carried out by molecular machines, some of which are themselves proteins, and some of which are assemblages of different proteins or assemblages of proteins and special untranslated RNA. In each step of the process, the information contained in one type of linear molecule is transferred into another linear molecule, so eventually the information encoded in the linear structure of DNA becomes embodied in the structure of a protein, which is also essentially linear. Finally, through protein folding and association with other molecules, the genetic information of DNA is converted into the three-dimensional proteins that define the myriad of physiological and architectural properties of the cell. In the left panel of Figure 3-2, the normal form of the hypothetical eukaryotic gene contains within its exons the information for a protein product that is an enzyme. In this example the enzyme catalyzes conversion of a colorless precursor into an orange pigment. In the right panel, a mutated form of the gene, bearing a nucleotide sequence altered at one point within the gene, results in a corresponding change in the amino acid sequence, and a protein with an altered shape. This altered shape causes a malfunction so that the protein can no longer catalyze the formation of pigment. Such general relationships underlie the action of most normal genes and their defective forms.

All DNA and RNA function is based on two principles:

1. Complementary bases in single-stranded nucleotide chains can hydrogen-bond to form double-stranded structures.
2. Particular base sequences in single-stranded or double-stranded nucleic acids can be recognized by certain nucleic acid–binding proteins that specifically bind to these sequences and act upon them. (These proteins are themselves encoded by genes, but their function is to act on other genes.)

We shall see these two principles at work throughout the detailed discussions of transcription and translation that follow, and in chapters to come.

MESSAGE

The transactions of DNA and RNA are based on complementarity of base sequences and on attachment of various nucleic acid–binding proteins to specific sites.

RNA

The first step of information transfer from gene to protein is to produce an RNA whose base sequence matches the base sequence of a segment of one of the two strands of DNA, sometimes followed by modification of that RNA to prepare it for its specific cellular roles. Hence RNA is produced by a process that copies the nucleotide sequence in DNA. Since this process is reminiscent of transcribing (copying) written words, the synthesis of RNA is, as we have seen, called **transcription.** The DNA is said to be transcribed into RNA, and the RNA is called a **transcript.**

We looked at the chemical structure of DNA in Chapter 2. Now let's consider the general chemical features of RNA, as a prelude to understanding the roles that RNAs play in the cell.

Properties of RNA

Although RNA and DNA are both nucleic acids, RNA differs in several important ways:

1. RNA is a single-stranded nucleotide chain, not a double helix like DNA. One consequence of this is that RNA is more flexible and can form a much greater variety of complex three-dimensional molecular shapes than can double-stranded DNA.
2. RNA has **ribose sugar** in its nucleotides, rather than the deoxyribose found in DNA. As the names suggest, the two sugars differ in the presence or absence of just one oxygen atom.

Ribose   Deoxyribose

Analogous to the individual strands of DNA, there is a sugar-phosphate backbone to RNA, with a base covalently linked to the 1′ position on each ribose. Since the sugar-phosphate linkages are made at the 5′ and 3′ positions of the sugar, just as in DNA, an RNA chain will have a 5′ and a 3′ end.

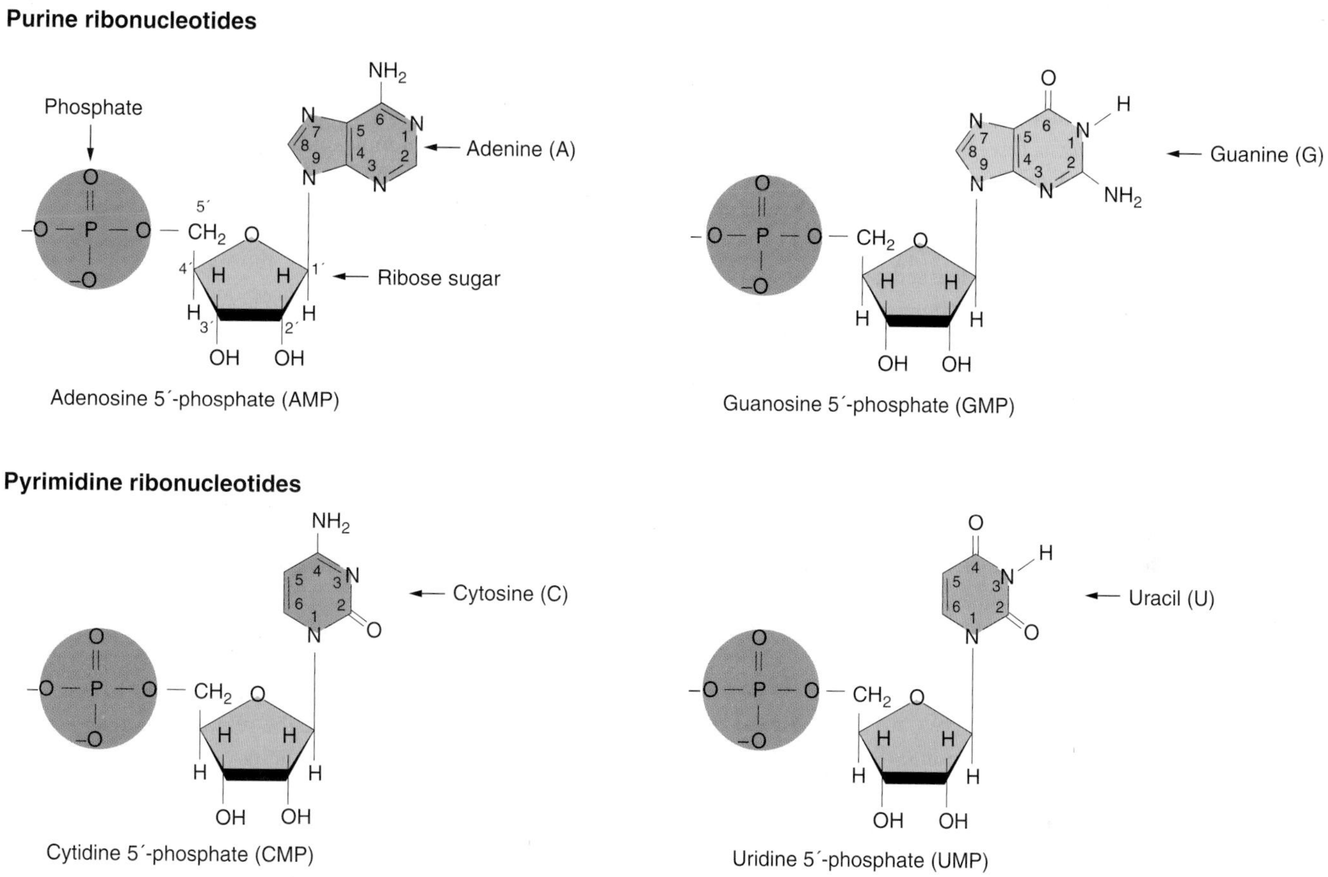

Purine ribonucleotides

Adenosine 5′-phosphate (AMP)

Guanosine 5′-phosphate (GMP)

Pyrimidine ribonucleotides

Cytidine 5′-phosphate (CMP)

Uridine 5′-phosphate (UMP)

Figure 3-3 The four ribonucleotides found in RNA.

3. RNA nucleotides carry the bases adenine, guanine, and cytosine, but the pyrimidine base **uracil** (abbreviated U) is found instead of thymine.

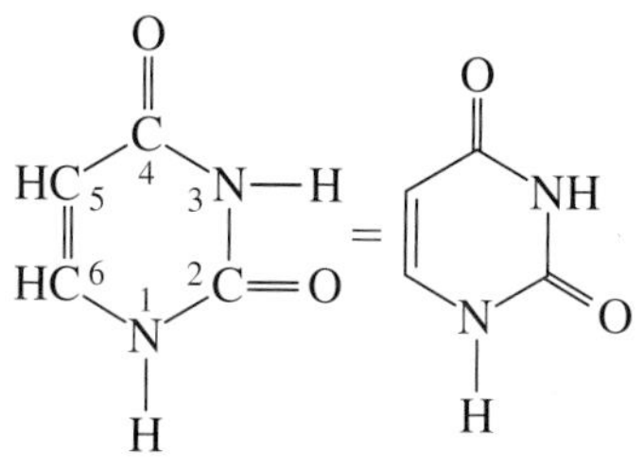

Uracil

However, uracil forms hydrogen bonds with adenine just as thymine does. Figure 3-3 shows the four ribonucleotides found in RNA.

Classes of RNA

RNAs can be grouped into two general classes. For protein-coding genes, RNAs are intermediaries in the process of decoding the genes into polypeptide chains. We will refer to these RNAs as "informational RNAs," since they bear a copy of the information in DNA. In the remaining minority of genes, the RNA itself is the final, functional product. We will refer to these RNAs as "functional RNAs."

Informational RNAs. The steps through which a gene influences phenotype are called *gene expression.* For the vast majority of genes, the RNA is only an intermediate necessary for the synthesis of the ultimate functional product that influences phenotype, which is a protein. The informational RNA of these genes is called **messenger RNA (mRNA).** Although gene expression is generally similar in most organisms, there are some differences. The most profound differences are between prokaryotes and eukaryotes, and presumably reflect the enormous amount of time since these two groups of organisms diverged during evolution. Let's compare prokaryotic and eukaryotic transcription and mRNA production. In prokaryotes, the transcript synthesized directly from the DNA (the **primary transcript**) is the mRNA. In eukaryotes, however, the primary transcript is modified before becoming the mRNA. This modification is called **processing.** The primary transcript, or **pre-mRNA,** is processed through modification of the 5′ and 3′ ends and by removal of the noncoding segments, the introns. When the processing of the primary transcript is complete, the product is the eukaryotic mRNA. We will discuss these steps in processing mRNAs later in this chapter.

The sequence of nucleotides in mRNA is "read out" as the sequence of amino acids in a polypeptide chain by a process called **translation.** In this connection the word *translation* is used in much the same way as we use it when we speak of translating a foreign language: the cell has a way of translating the language of RNA into the language of polypeptides. One or more polypeptide chains then fold up and associate together as a protein.

Functional RNAs. As more is learned about the intimate details of cell biology, it has become apparent that functional RNAs fall into a variety of classes that play diverse roles. Again, it's important to emphasize that functional RNAs are active as RNA; they are never translated into polypeptides. Each class of functional RNA is encoded by a relatively small number of genes (a few tens to a few hundred at most). However, while the genes that encode them are relatively few, these RNAs account for a large percentage of the RNA in the cell. This is because functional RNAs are more stable than informational RNAs, and also because abundance is necessary to carry out their functions.

The main classes of functional RNAs contribute to various steps in the informational processing of DNA to protein. Two such classes of functional RNAs are found in prokaryotes and eukaryotes: transfer RNAs and ribosomal RNAs.

Transfer RNA (tRNA) molecules act as transporters that bring amino acids to the mRNA during the process of translation. The tRNAs are general components of the translation machinery; they can bring amino acids to the mRNA of any protein-coding gene.

Ribosomal RNAs (rRNAs) are components of ribosomes, which are large macromolecular assemblies that act as guides to coordinate the assembly of the amino acid chain by the mRNA. Ribosomes are composed of several types of rRNA and about 100 different proteins. As in the case of tRNA, the rRNAs are general translational components that can be used to translate the mRNA of any protein-coding gene.

Two other classes of functional RNAs involved in information processing are specific to eukaryotes. **Small nuclear RNAs (snRNAs)** are part of the system that processes pre-mRNAs into messenger RNAs in the eukaryotic cell's nucleus. Several different snRNAs unite with several protein subunits to form the processing unit called a **small nuclear ribonucleoprotein particle (snRNP).** The second class, **small cytoplasmic RNAs (scRNAs),** direct protein traffic within the eukaryotic cell. For example, they ensure that polypeptides destined to be secreted from the cell are inserted into one of the membrane compartments of the cell (the rough endoplasmic reticulum). This begins the process of protein secretion.

MESSAGE

There are two types of genes, those coding for proteins (the majority) and those coding for functional RNAs.

MAKING AND PROCESSING TRANSCRIPTS

Exactly how is a gene-coding sequence, which is a relatively small segment of DNA embedded in a much longer DNA molecule (the chromosome), transcribed into a single-stranded RNA molecule of correct length and nucleotide sequence? In considering the several distinct steps involved, we will first examine the relationship between the two strands of the DNA double helix.

DNA as Transcription Template

Transcription relies on the complementary pairing of bases. In the chromosomal segment that constitutes a gene, the two strands of the DNA double helix separate, and one of the separated strands acts as a **template** for RNA synthesis. In the chromosome overall, both DNA strands are used as templates, *but in any one gene only one strand is used,* and in that gene it is always the same strand (Figure 3-4). Next, free ribonucleotides that have been chemically synthesized elsewhere in the cell form stable pairs with their complementary bases in the template. The free ribonucleotide A pairs with T in the DNA, G with C, C with G, and U with A. The enzyme **RNA polymerase** attaches to the DNA and moves along it linking the aligned ribonucleotides together to make an ever-growing RNA molecule, as shown in Figure 3-5a. Hence, already we see in action the two principles of base complementarity and nucleic acid–protein binding (in this case, the binding of RNA polymerase).

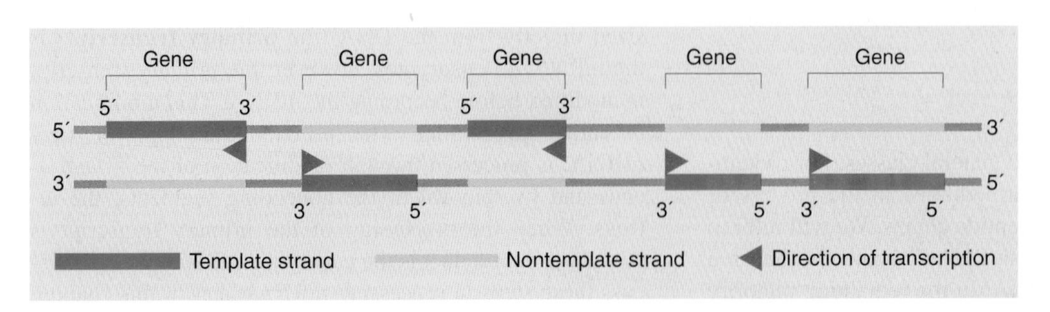

Figure 3-4 DNA strands used as transcriptional templates. The direction of transcription is always the same for any gene and starts from the 3′ end of the template. Hence genes transcribed in different directions use opposite strands of the DNA as templates.

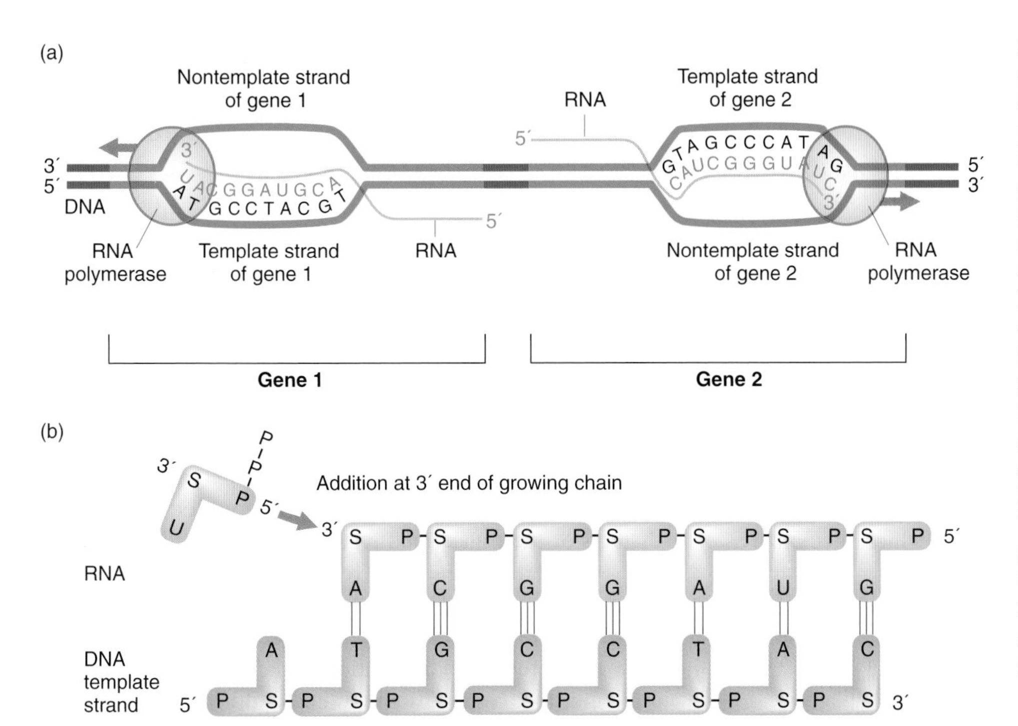

Figure 3-5 Transcription. (a) Transcription of two genes in opposite directions. RNA polymerase moves toward the left using the lower DNA strand of gene 1 *(left)* as template. The polymerase always adds ribonucleotides to the 3′ growing end of the RNA. The 5′ end of the RNA is released from the template as the transcription bubble closes behind the polymerase. RNA polymerase moves toward the right using the upper DNA strand of gene 2 *(right)* as template. (b) The growing 3′ end of the RNA on the DNA template strand for gene 1. The terminal two phosphate groups on the nucleotide being added to the 3′ end will be removed during the addition to the growing chain. Note that RNA growth is from 5′ to 3′.

Animated Art • Transcription

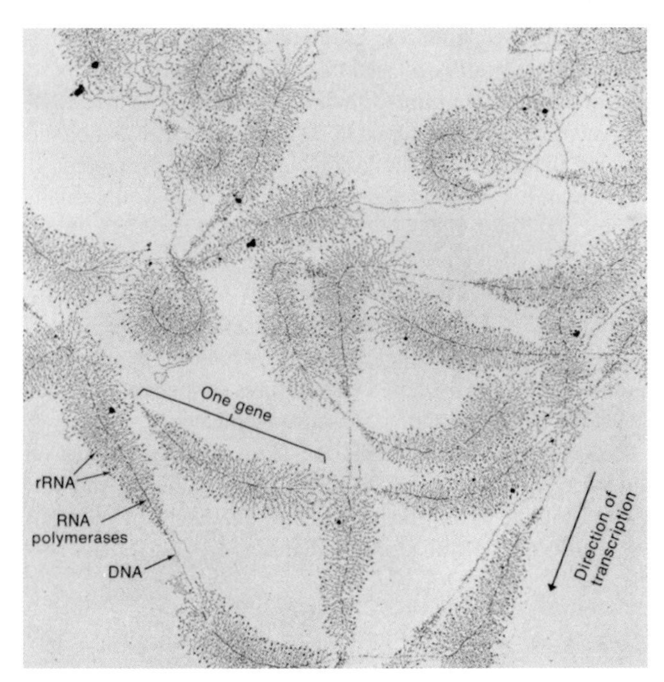

Figure 3-6 Tandemly repeated ribosomal RNA (rRNA) genes being transcribed in the nucleolus of *Triturus viridiscens* (an amphibian). (Ribosomal RNA is a component of the ribosome.) Along each gene, many RNA polymerase molecules are attached and transcribing in one direction. The growing RNA molecules appear as threads extending out from the DNA backbone. The shorter RNA molecules are nearer the beginning of transcription; the longer ones have almost been completed. The "Christmas tree" appearance is the result. *(Photograph from O. L. Miller, Jr., and Barbara A. Hamkalo.)*

We have seen that RNA has a 5′ and a 3′ end. During synthesis, RNA growth is always in the 5′ → 3′ direction; in other words, nucleotides are always added at a 3′ growing tip, as shown in Figure 3-5b. Because of the antiparallel nature of nucleotide pairing, the fact that RNA is synthesized 5′ → 3′ means that the template strand must be oriented 3′ → 5′. As RNA polymerase molecules move along the gene, the RNA molecule progressively lengthens. "Trains" of RNA polymerases, each synthesizing an RNA molecule, move along the gene, which allows the progressive enlargement of RNA to be visualized under the electron microscope (Figure 3-6).

Complementarity of bases in transcript and template dictates that the nucleotide sequence in the RNA be the same as that in the nontemplate strand of the DNA, except that the T's are replaced by U's, as shown in Figure 3-7 on the next page. When DNA base sequences are cited in scientific literature, by convention it is the sequence of the nontemplate strand that is given, because this is the same sequence as found in the RNA. This distinction is extremely important to keep in mind during discussions of transcription.

MESSAGE

Transcription is asymmetrical: only one strand of the DNA of a gene is used as a template for transcription. This strand is in 3′ → 5′ orientation, and RNA is synthesized in the 5′ → 3′ direction.

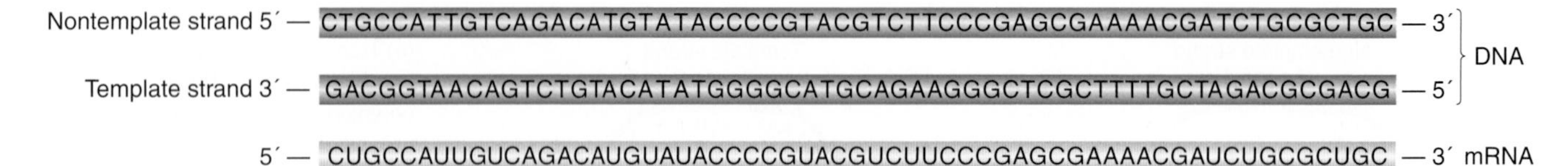

Figure 3-7 The mRNA sequence is complementary to the DNA template strand from which it is synthesized and therefore matches the sequence of the nontemplate strand (except that RNA has U where DNA has T). The sequence shown here is from the gene for the enzyme β-galactosidase, which is involved in lactose metabolism.

RNA Polymerases

In most prokaryotes, a single RNA polymerase does the job of transcribing all types of RNA. However, eukaryotes have three different RNA polymerases, which specialize as follows:

1. RNA polymerase I (Pol I) transcribes rRNA genes.
2. RNA polymerase II (Pol II) transcribes protein-coding genes, for which the ultimate transcript is mRNA.
3. RNA polymfFerase III (Pol III) transcribes other functional RNA genes (for example, tRNA genes).

In eukaryotes transcription of nuclear chromosomes takes place entirely within the nucleus, and the transcripts then move through nuclear pores out into the cytosol (the liquid phase of the cytoplasm), where translation occurs. Since prokaryotes have no nucleus, there is no comparable movement of transcripts, and translation can take place immediately, right on the growing transcript. In eukaryotic organelle genomes, transcription and translation take place within the organelle.

Since the DNA of a chromosome is a continuous unit, for RNA to be transcribed from a specific gene the transcriptional machinery must be directed to that gene to begin transcribing at the appropriate end, continue transcribing the length of the gene, and finally stop transcribing at the other end. These three distinct stages of transcription are called **initiation, elongation,** and **termination.** We will follow them using the system of the gut bacterium *E. coli* as an example.

Initiation

Initiation is based on binding of RNA polymerase to one specific DNA sequence called a **promoter,** located close to the end of the transcribed region. A promoter is an important part of the regulatory region of a gene. Remember, since the synthesis of an RNA transcript begins at its 5′ end and continues in the 5′ → 3′ direction, the convention is to draw and refer to the orientation of the gene in the 5′ → 3′ direction, too. Generally the 5′ end is drawn to the left and 3′ to the right. Using this view, since the promoter must be near the end of the gene where transcription begins, this is said to be at the 5′ end of the gene, and the regulatory region is called the 5′ regulatory region (Figure 3-8).

Figure 3-9 shows the promoter sequences of 13 different genes in the *E. coli* genome. The first-transcribed base, designated the *initiation site,* is numbered + 1. Since RNA polymerase binds to a specific DNA sequence, it is not surprising that there are similarities between the promoters. In particular, two regions of great similarity appear in virtually each case. These regions have been termed the − *35* (minus 35) and − *10 regions* because of their locations relative to the transcription initiation point. They are shown in yellow in Figure 3-9. In the analysis of similar DNA sequences of similar function, the nucleotides for which most sequences are in agreement are called a **consensus sequence.** In Figure 3-9c, the *E. coli* promoter consensus sequence is shown.

RNA polymerase scans the DNA for a promoter sequence, binds to the DNA at that point, then unwinds the DNA double helix and begins the synthesis of an RNA molecule at the transcriptional initiation site. Hence, we

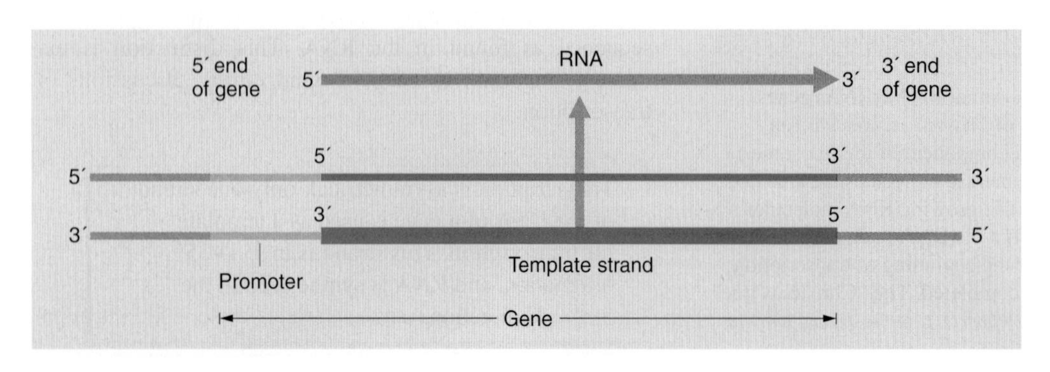

Figure 3-8 The convention for designating the 5′ and 3′ ends of a gene.

(a)

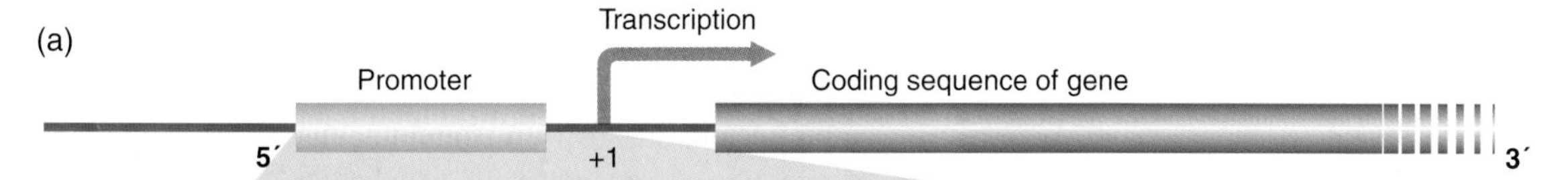

(b) Strong *E. coli* promoters

tyr tRNA	TCTCAACGTAACACTTTACAGCGGCG••CGTCATTTGATATGATGC•GCCCCGCTTCCCGATAAGGG
rrn D1	GATCAAAAAAATACTTGTGCAAAAAA••TTGGGATCCCTATAATGCGCCTCCGTTGAGACGACAACG
rrn X1	ATGCATTTTTCCGCTTGTCTTCCTGA••GCCGACTCCCTATAATGCGCCTCCATCGACACGGCGGAT
rrn $(DXE)_2$	CCTGAAATTCAGGGTTGACTCTGAAA••GAGGAAAGCGTAATATAC•GCCACCTCGCGACAGTGAGC
rrn E1	CTGCAATTTTTCTATTGCGGCCTGCG••GAGAACTCCCTATAATGCGCCTCCATCGACACGGCGGAT
rrn A1	TTTTAAATTTCCTCTTGTCAGGCCGG••AATAACTCCCTATAATGCGCCACCACTGACACGGAACAA
rrn A2	GCAAAAATAAATGCTTGACTCTGTAG••CGGGAAGGCGTATTATGC•ACACCCCGCGCCGCTGAGAA
λP_R	TAACACCGTGCGTGTTGACTATTTTA•CCTCTGGCGGTGATAATGG••TTGCATGTACTAAGGAGGT
λP_L	TATCTCTGGCGGTGTTGACATAAATA•CCACTGGCGGTGATACTGA••GCACATCAGCAGGACGCAC
T7 A3	GTGAAACAAAACGGTTGACAACATGA•AGTAAACACGGTACGATGT•ACCACATGAAACGACAGTGA
T7 A1	TATCAAAAAGAGTATTGACTTAAAGT•CTAACCTATAGGATACTTA•CAGCCATCGAGAGGGACACG
T7 A2	ACGAAAAACAGGTATTGACAACATGAAGTAACATGCAGTAAGATAC•AAATCGCTAGGTAACACTAG
fd VIII	GATACAAATCTCCGTTGTACTTTGTT••TCGCGCTTGGTATAATCG•CTGGGGGTCAAAGATGAGTG
	−35 −10 +1

(c) Consensus sequences for all *E. coli* promoters

−35 region: TTGACAT — 15–17 bp — −10 region: TATAAT

Figure 3-9 Promoter sequences. (a) The promoter lies "upstream" (toward the 5′ end) of the transcription initiation point and the coding sequences. (b) Promoter sites have regions of similar sequences, as indicated by the yellow region in the 13 different promoter sequences in *E. coli.* Spaces (dots) are inserted in the sequences to optimize the alignment of the common sequences. The name of the gene governed by each promoter sequence is indicated on the left. Numbering is given in terms of the number of bases before (−) or after (+) the RNA synthesis initiation point. Yellow shading indicates the consensus sequences shown in part c. (c) The consensus sequences for all *E. coli* promoters. *(From H. Lodish, D. Baltimore, A. Berk, S. L. Zipursky, P. Matsudaira, and J. Darnell,* Molecular Cell Biology, *3d ed. © 1995 by Scientific American Books, Inc. See W. R. McClure,* Ann. Rev. Biochem. *54, 1985, 171.)*

see again the principle of DNA binding in the interactions between the protein (here, the RNA polymerase) and a specific base sequence in the DNA. Note from Figure 3-9a that transcription starts *before* the protein-coding segment of the gene. Hence a transcript has what is called a **5′ untranslated region,** or **5′ UTR.**

Elongation

During elongation, the RNA polymerase moves along the DNA, maintaining a transcription "bubble" to expose the template strand, and catalyzes the 3′ elongation of the RNA strand. The polymerase monitors the binding of free ribonucleotide triphosphates with the next exposed base on the DNA template and, if there is a complementary match, adds it to the chain. The energy for the addition of a nucleotide is derived from splitting the high-energy triphosphate into the monophosphate and releasing the inorganic diphosphates, according to the following general formula:

$$\text{NTP} + (\text{NMP})_n \xrightarrow[\text{RNA polymerase}]{\text{DNA, } \text{Mg}^{2+}} (\text{NMP})_{n+1} + \text{PP}_\text{i}$$

Termination

The transcription of an individual gene is terminated beyond the protein-coding segment of the gene, creating a **3′ UTR** at the end of the transcript. Termination occurs when the RNA polymerase recognizes specific nucleotide sequences in the DNA that act as signals: at this point both the RNA strand and the polymerase are released from the DNA template. *E. coli* has two main mechanisms for termination, direct and indirect; we will look only at the direct termination mechanism here. The terminator sequence in the template strand consists of about 40 bp, ending in a GC-rich stretch that is followed by a run of six or more A's. Because G and C in the template will give C and G, respectively, in the transcript, the RNA in this region is also GC-rich and hence able to form complementary bonds with itself, resulting in a **hairpin loop** (Figure 3-10 on the next page). The loop is followed by the terminal run of U's that correspond to the A residues on the DNA template. The hairpin loop and section of U residues appear to serve as a signal for the release of RNA polymerase and termination of transcription.

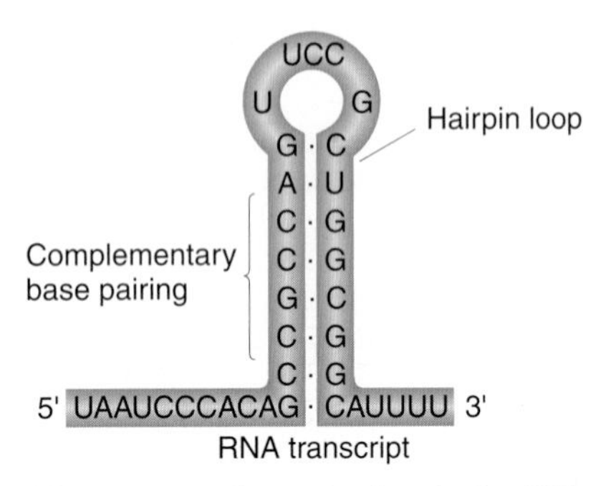

Figure 3-10 The structure of a termination site for RNA polymerase in bacteria. The hairpin structure forms by complementary base pairing *within* a GC-rich RNA strand. Most RNA base pairing is between G and C, but there is one single A–U pair.

RNA Processing in Eukaryotes

The mechanics of transcription works in much the same way in eukaryotes as in prokaryotes; that is, there are specific promoter sequences to which the RNA polymerase binds, and the polymerase moves along the gene synthesizing RNA in the 5′ → 3′ direction. However, in eukaryotes the primary RNA transcript is **processed** in several ways before its transport to the cytosol.

Processing 5′ and 3′ ends. Figure 3-11 depicts the processing of the 5′ and 3′ ends of the transcript of a protein-coding gene. First, during transcription, a **cap** consisting of a 7-methylguanosine residue is added to the 5′ end of the

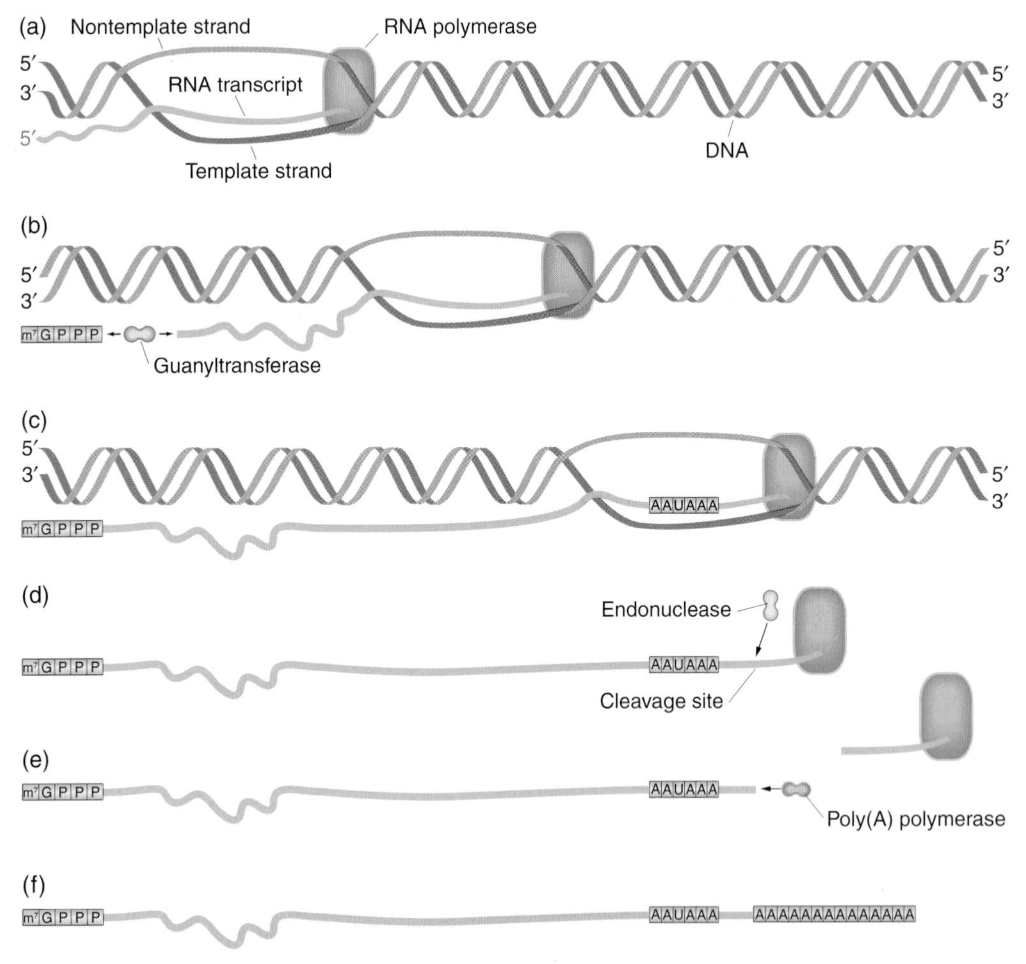

Figure 3-11 Processing the ends of a eukaryotic primary transcript. (a) Transcription is mediated by RNA polymerase. (b) Early in transcription an enzyme, guanyltransferase, adds 7-methylguanosine (m^7Gppp) to the 5′ end of the mRNA. (c) The sequence AAUAAA, near the 3′ end, helps signal a cleavage event (d) approximately 20 bp farther downstream by an endonuclease (an enzyme that can cut DNA strands). (e) Another enzyme, poly(A) polymerase, then adds a poly(A) tail, in reality composed of up to 150–200 adenosine residues, to the site of this cleavage at the 3′ end. (f) The resulting mRNA. *(From J. E. Darnell, Jr., "The Processing of RNA." © 1983 by Scientific American, Inc. All rights reserved.)*

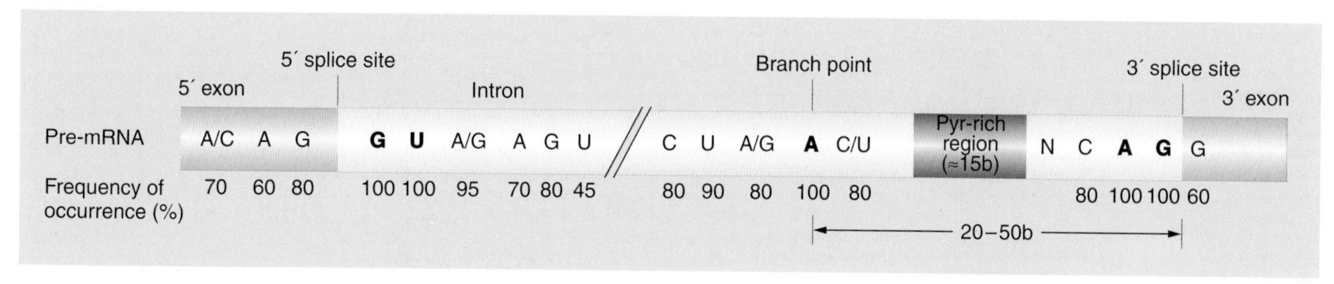

Figure 3-12 Conserved sequences related to intron splicing. The numbers below the nucleotides represent percent similarity among organisms. Of particular importance are the GU at the 5′ end, the AG at the 3′ end, and the central A labeled "branch point" (see Figure 3-15 for a view of the branch structure). N represents any base.

transcript, linked by three phosphate groups. Then an AAUAAA sequence near the 3′ end is recognized by an enzyme that cuts off the end of the RNA approximately 20 bases farther down. To this cut end, a stretch of 150 to 200 adenine nucleotides called a **poly(A) tail** is added at the cut 3′ end. Hence the AAUAAA sequence of protein-coding genes is called a *polyadenylation signal.*

Intron excision. Many eukaryotic genes contain introns. This is true not only for protein-coding genes but also for some cases of rRNA genes, and even tRNA genes. The removal of introns is called **splicing,** since it is reminiscent of the way in which videotape or movie film can be cut and rejoined to edit out some specific segment. In the same way, adjacent exons are spliced together. Introns are removed from the transcripts before transport of the transcript into the cytoplasm. In the case of protein-coding genes, after splicing of the capped and tailed transcript has taken place, pre-mRNA becomes mature mRNA. Splicing brings together the coding regions, the exons, so that the mRNA now contains a coding sequence that is completely colinear with the protein it codes.

Mechanism of exon splicing. Figure 3-12 shows the exon-intron junctions of pre-mRNAs, the splice sites where the splicing reactions occur. At these junctions, certain specific nucleotides are invariant, conserved across genes and across species, because they participate in the splicing reactions. Each intron will be cut at each end, and these ends almost always have a GU at the 5′ end and AG at the 3′ end (the **GU–AG rule**). Another invariant site is an A nucleotide in the central region of the intron. Other less well conserved nucleotides are found flanking these. These crucial nucleotides in the transcript are recognized by small nuclear ribonucleoprotein particles (snRNPs), complexes of protein and small nuclear RNA (snRNA). A team of snRNPs that constitutes a functional splicing unit is called a **spliceosome.** The spliceosome attaches to the intron, as shown in the electron micrograph in Figure 3-13 and the diagram in Figure 3-14 on the next page. The snRNPs help to align the splice sites by hydrogen bonding of their snRNAs to the conserved intron sequences. Then their proteins catalyze the removal of the intron through two consecutive splicing steps, labeled 1 and 2 in Figure 3-14. Splice 1 attaches one end of the intron to the conserved internal adenine, forming a structure the shape of a cowboy's lariat. Splice 2 releases the lariat and joins the two adjacent exons. Figure 3-15 on the next page portrays the chemistry behind intron excision. Chemically, steps 1 and 2 are transesterification reactions between the conserved nucleotides.

One exceptional case of RNA splicing occurs in the ciliate protozoan *Tetrahymena,* in which the splicing reaction is catalyzed not by a spliceosomal protein, but by the transcribed RNA molecule itself. This extraordinary

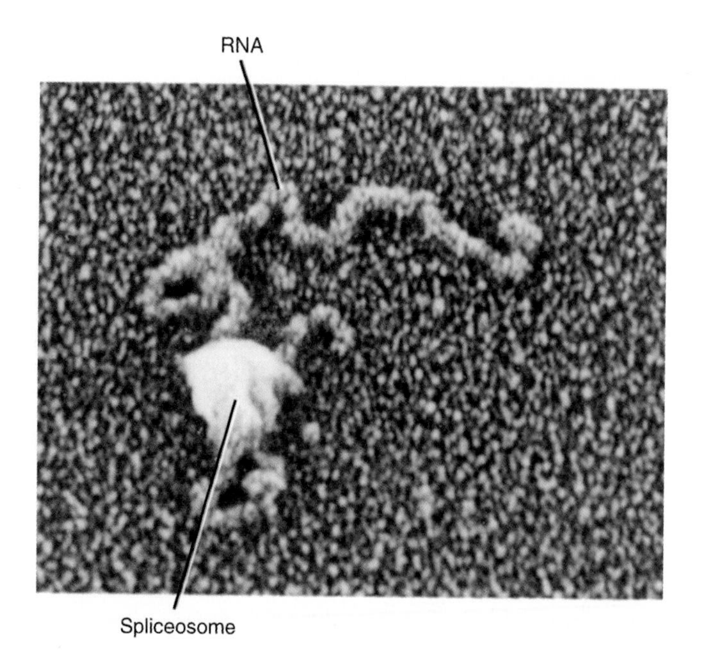

Figure 3-13 Electron micrograph of a spliceosome. *(From H. Lodish, D. Baltimore, A. Berk, S. L. Zipursky, P. Matsudaira, and J. Darnell,* Molecular Cell Biology, *3d ed. © 1995 by Scientific American Books, Inc.)*

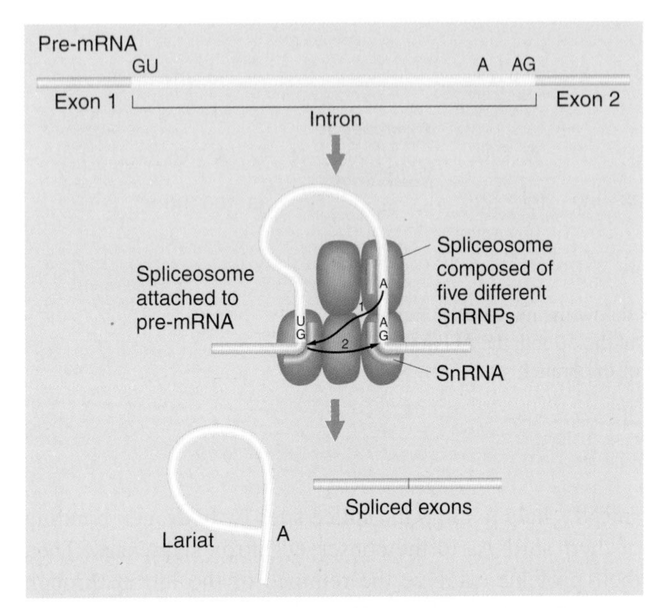

Figure 3-14 The structure and function of a spliceosome. The spliceosome is composed of several snRNPs that attach sequentially to the RNA, taking up positions roughly as shown. Alignment of the snRNPs results from hydrogen bonding of their snRNA molecules to the complementary sequences of the intron. In this way the reactants are properly aligned and the splicing reactions (1 and 2) can occur. The P-shaped loop, or lariat structure, formed by the excised intron is joined through the central adenine nucleotide.

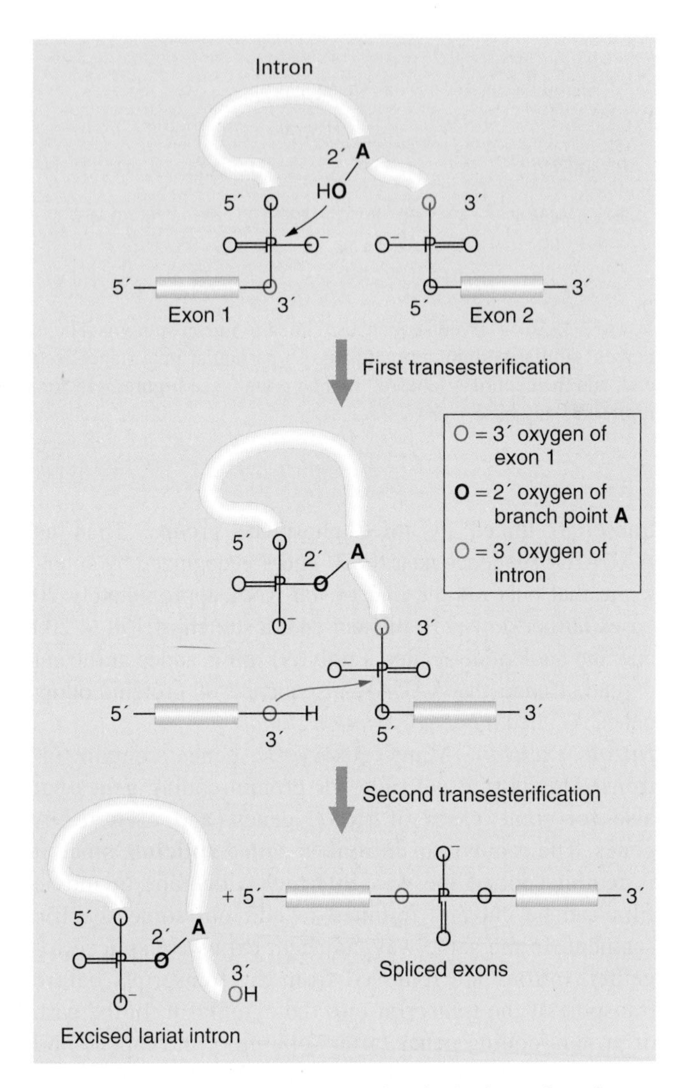

Figure 3-15 The reactions that take place in the formation of an intron lariat, as it is removed from RNA. Two transesterification reactions take place, first to join the GU end to the internal branch point (reaction 1 in Figure 3-14) and second to join the two exons together (reaction 2 in Figure 3-14). *(Modified from H. Lodish, A. Berk, S. L. Zipurski, P. Matsudaira, D. Baltimore, and J. Darnell,* Molecular Cell Biology, *4th ed. © 2000 by W. H. Freeman and Co.)*

finding was the first demonstration that an RNA molecule in some cases can act as an enzyme and catalyze a specific biological reaction. Such RNAs that have enzymatic activities have been termed **ribozymes.**

MESSAGE

Eukaryote mRNA arises by 5′ capping, 3′ polyadenylation, and splicing of the primary transcript.

In review, Figure 3-16 shows the general structure of a eukaryotic gene with some of the main landmarks relating to transcription, translation, and intron organization, and also summarizes the main steps of the processing of the primary transcript.

Once a primary transcript has been fully processed into a mature mRNA molecule, translation into protein can take place. Before discussing how proteins are made, it is necessary to understand protein structure.

PROTEIN

Proteins are the main determinants of biological form and function. The shape, color, size, behavior, and physiology of organisms are all heavily influenced by cellular proteins. Therefore understanding the nature of proteins is a key to understanding gene action.

Protein Structure

A protein is a polymer composed of monomers called **amino acids.** In other words, it is a chain of amino acids. Since amino acids were once called *peptides,* the chain is sometimes referred to as a **polypeptide.** Amino acids all have the general formula

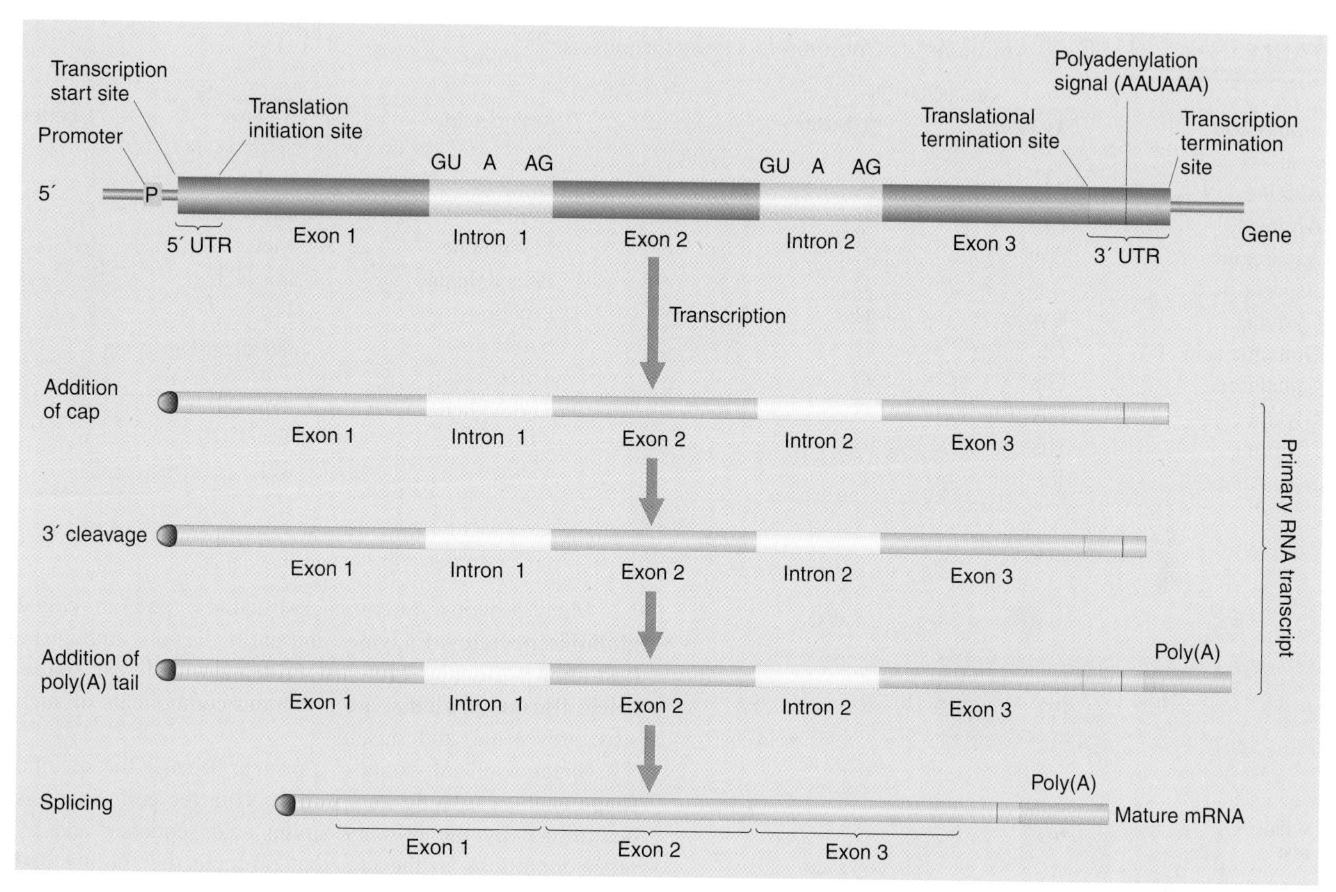

Figure 3-16 Transcriptional and translational landmarks in a eukaryotic gene with two introns (top line), and the processing of its transcript to make mRNA. Note that since the landmarks shown are relevant to RNA, U is given in the gene sequence instead of T.

$$H_2N-\overset{\displaystyle H}{\underset{\displaystyle R}{\overset{|}{\underset{|}{C}}}}-COOH$$

The side chain, or R (reactive) group, can be anything from a hydrogen atom (as in the amino acid glycine) to a complex ring (as in the amino acid tryptophan). There are 20 amino acids known to occur in proteins (Table 3-1 on the next page), each having a different R group that gives the amino acid its unique properties. In proteins the amino acids are linked together by covalent bonds called **peptide bonds.** A peptide bond is formed through a condensation reaction during which one water molecule is removed (Figure 3-17 on the next page). Because of the way in which the peptide bond forms, a polypeptide chain always has an **amino end** (NH_2) and a **carboxyl end** (COOH), as shown in Figure 3-17a.

Proteins have a complex structure that has four levels of organization, illustrated in Figure 3-18 on page 69. The linear sequence of the amino acids in a polypeptide chain constitutes the **primary structure** of the protein. The **secondary structure** of a protein is the specific shape the polypeptide chain takes on by folding. This shape arises from the bonding forces between amino acids that are close together in the linear sequence. Several types of weak bonds are involved, notably hydrogen bonds, electrostatic forces, and van der Waals forces. The most common secondary structures are an α helix and a pleated sheet. Different proteins show either one or the other, or sometimes both, within their structure. **Tertiary structure** is produced by folding the secondary structure. Some proteins have **quaternary structure,** being composed of two or more tertiary structures (separate folded polypeptides) joined together by weak bonds. The quaternary association can be between different types of polypeptides (resulting in a heterodimer) or between identical polypeptides (making a homodimer). Hemoglobin (Figure 3-18d) is an example of a heterodimer, composed of two copies each of two different polypeptides, shown in green and purple in the figure.

TABLE 3-1 The 20 Amino Acids Common in Living Organisms

Amino Acid	Abbreviation 3-Letter	Abbreviation 1-Letter	Amino Acid	Abbreviation 3-Letter	Abbreviation 1-Letter
Alanine	Ala	A	Leucine	Leu	L
Arginine	Arg	R	Lysine	Lys	K
Asparagine	Asn	N	Methionine	Met	M
Aspartic acid	Asp	D	Phenylalanine	Phe	F
Cysteine	Cys	C	Proline	Pro	P
Glutamic acid	Glu	E	Serine	Ser	S
Glutamine	Gln	Q	Threonine	Thr	T
Glycine	Gly	G	Tryptophan	Trp	W
Histidine	His	H	Tyrosine	Tyr	Y
Isoleucine	Ile	I	Valine	Val	V

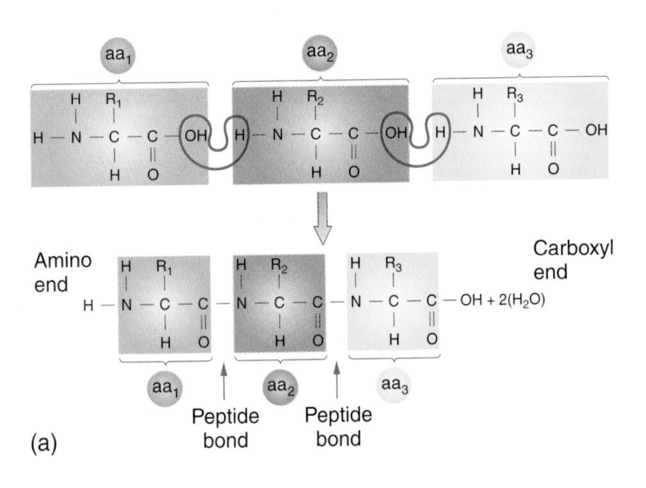

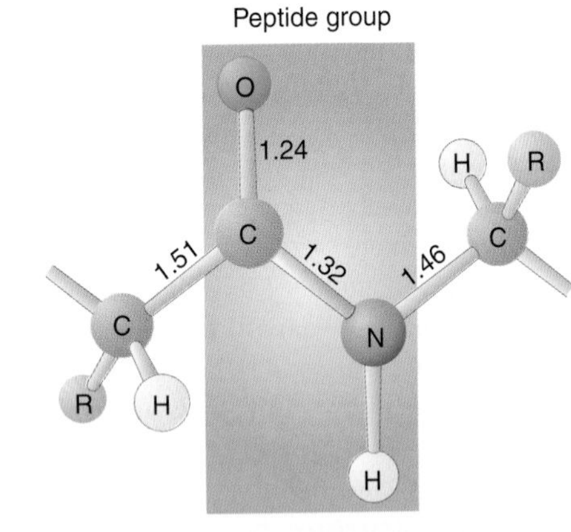

Animated Art • Translation: Peptide Bond Formation

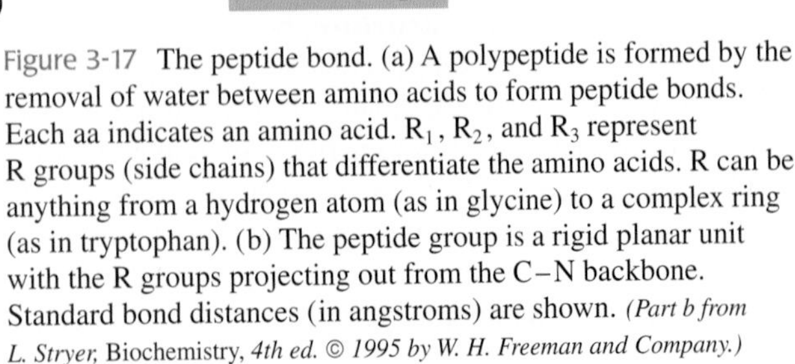

Figure 3-17 The peptide bond. (a) A polypeptide is formed by the removal of water between amino acids to form peptide bonds. Each aa indicates an amino acid. R_1, R_2, and R_3 represent R groups (side chains) that differentiate the amino acids. R can be anything from a hydrogen atom (as in glycine) to a complex ring (as in tryptophan). (b) The peptide group is a rigid planar unit with the R groups projecting out from the C–N backbone. Standard bond distances (in angstroms) are shown. *(Part b from L. Stryer,* Biochemistry, *4th ed. © 1995 by W. H. Freeman and Company.)*

Many proteins are compact structures; these are called **globular proteins.** Enzymes and antibodies are among the best-known globular proteins. Proteins with linear shape, called **fibrous proteins,** are important components of such structures as hair and muscle.

Shape is all-important to a protein because the specific shape enables it to do its specific job in the cell. Shape is determined by the primary amino acid sequence and by other conditions in the cell that promote the folding and bonding necessary for higher-level structures. The amino acid sequence also determines which R groups are present at specific positions and thus available to participate in binding with other specific cellular components. The **active sites** of enzymes are good illustrations. Each enzyme has a pocket called the *active site* into which its substrate or substrates can fit (Figure 3-19 on page 70). Within the active site, certain amino acid R groups are in key locations to aid in the chemical reactions undergone by the substrate.

At present, the rules by which primary structure is converted into higher-order structure are imperfectly understood. However, from knowledge of the primary amino sequence of a protein it is possible to predict the functions of specific regions. For example, some characteristic protein sequences are the contact points whereby the protein is positioned in phospholipid layers in membranes. Other characteristic domains act to bind the protein to DNA.

Armed with this brief introduction to protein structure, we can return now to protein synthesis.

Translation

How is the amino acid sequence of a protein determined? Quite simply, the amino acid sequence of a polypeptide is determined by the nucleotide sequence of the gene that encodes it. We have seen that the sequence of nucleotides in the DNA of a gene is transcribed into an equivalent

(a) **Primary structure**

Carboxyl end ... Amino end

$^{-}O-C(=O)-C(R_6)(H)-N(H)-C(=O)-C(H)(R_5)-N(H)-C(R_4)(H)-C(=O)-N(H)-C(H)(R_3)-N(H)-C(R_2)(H)-C(=O)-N(H)-C(H)(R_1)-{}^{+}NH_3$

(b) **Secondary structure**

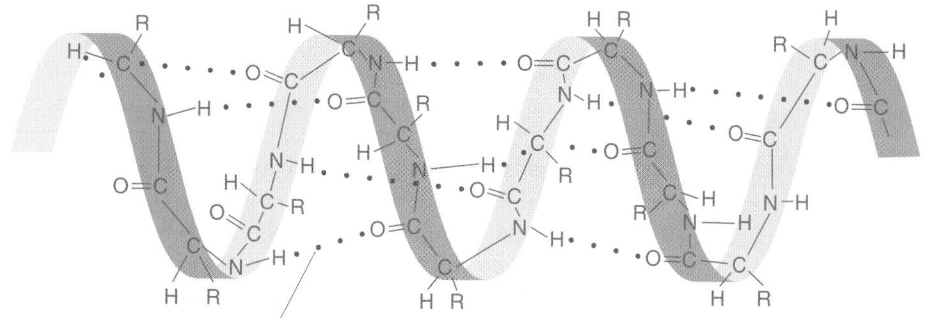

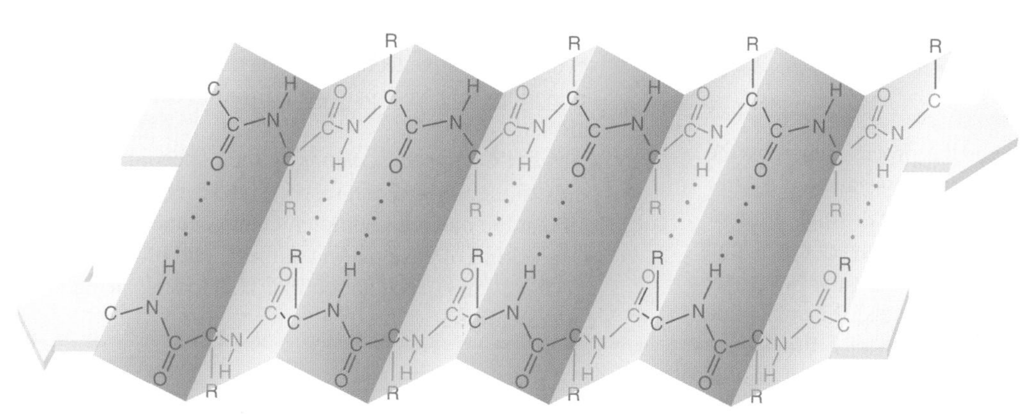

(c) **Tertiary structure**

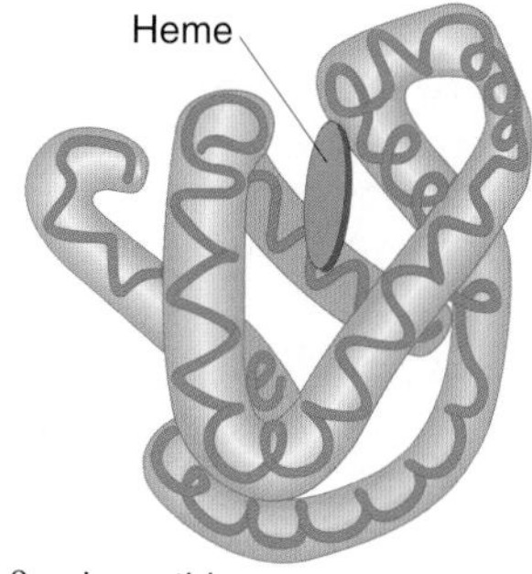

(d) **Quaternary structure**

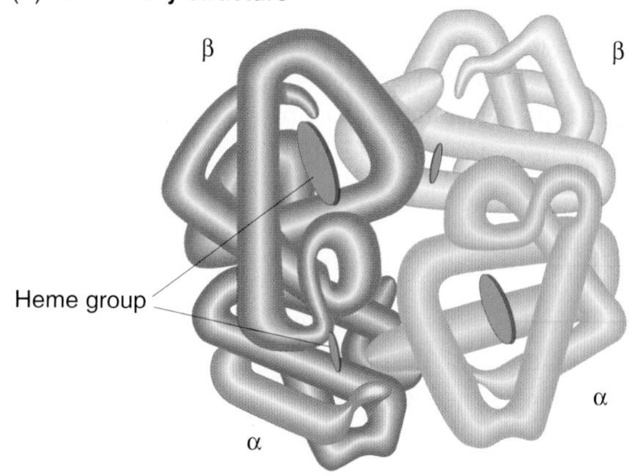

Figure 3-18 Different levels of protein structure. (a) Primary structure. (b) Secondary structure. The polypeptide can form a helical structure (an α helix) or a zigzag structure (a pleated sheet). The pleated sheet has two polypeptide segments arranged in opposite polarity, as indicated by the arrows. (c) Tertiary structure: the three-dimensional structure of myoglobin. The heme group is a nonprotein ring structure with an iron atom at its center. (d) Quaternary structure illustrated by hemoglobin, which is composed of four polypeptide subunits, two α subunits and two β subunits.

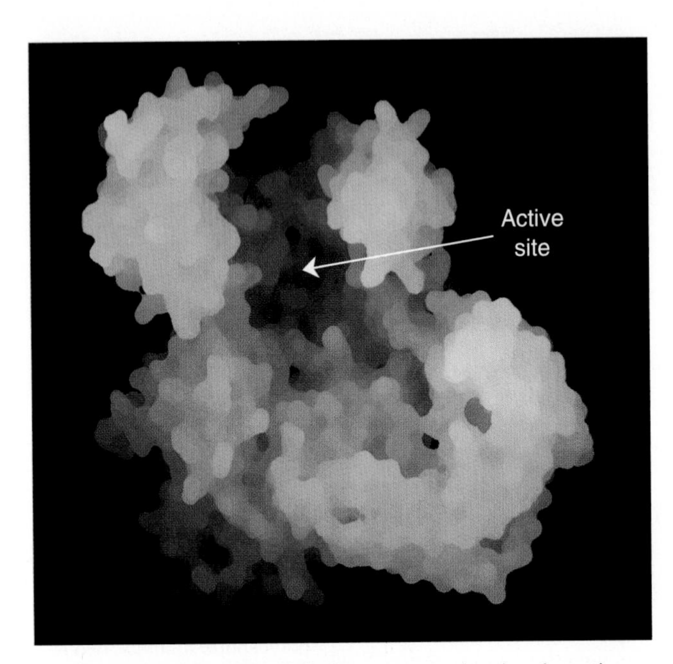

Figure 3-19 Structural model of an enzyme, showing the active site, in which the substrate is bound. *(From H. Lodish, D. Baltimore, A. Berk, S. L. Zipursky, P. Matsudaira, and J. Darnell,* Molecular Cell Biology, *3d ed. © 1995 by Scientific American Books, Inc.)*

sequence in mRNA. Translation is the process whereby the nucleotide sequence of mRNA directs the sequential assembly of amino acids into the polypeptide. A train of ribosomes moves along the mRNA, each starting at the 5′ end and proceeding along the entire length of the mRNA to the 3′ end. As a ribosome moves along, it "reads" the nucleotide sequence of the mRNA, three nucleotides at a time. Each group of three, called a **triplet codon,** stands for a specific amino acid. Since there are four different nucleotides in mRNA, there are $4 \times 4 \times 4 = 64$ different possible codons. These codons and the amino acids they stand for are shown in Figure 3-20. This same **genetic code** is used by virtually all organisms on the planet. (There are just a few exceptions in which a small number of the codons have different meanings, for example, in mitochondrial genomes.) Proteins contain only 20 common types of amino acids, so some amino acids are encoded by more than 1 of the 64 codons, as Figure 3-20 shows.

Codon Translation by tRNA

Another key component in translation is the set of **transfer RNA (tRNA)** molecules. Each amino acid becomes attached by an enzyme to a specific tRNA, which then brings that amino acid to the ribosome when the ribosome has reached that amino acid's own unique codon. The structure of tRNA holds the secret of the specificity between an mRNA codon and the amino acid it designates. A molecule of tRNA has a clover-leaf shape consisting of four double-helical stems and three single-stranded loops (Figure 3-21). The middle loop of each tRNA carries a nucleotide triplet called an **anticodon,** which has a sequence complementary to the codon for the amino acid carried by the tRNA. The anticodon in tRNA and the codon in the mRNA bind by specific RNA-to-RNA base pairing. (Again we see the principle of nucleic acid complementarity at work, this time in the binding of two different RNAs.) Since codons in mRNA are read in the 5′ → 3′ direction, anticodons are oriented and written in the 3′ → 5′ direction, as Figure 3-21 shows.

Amino acids are attached to tRNAs by enzymes called **aminoacyl-tRNA synthetases.** Each amino acid has a specific synthetase that links it only to those tRNAs that recognize the codons for that particular amino acid. An amino acid is attached at the free 3′ end of its tRNA, the amino acid alanine in the case shown in Figure 3-21.

We saw in Figure 3-20 that the number of codons for a single amino acid varies, ranging from one (tryptophan, UGG) to as many as six (serine, UCC, UCU, UCA, UCG, AGC or AGU). It is not exactly clear why the genetic code shows this variation, but two facts account for it:

1. Most amino acids can be brought to the ribosome by several alternative tRNA types having different anticodons that base-pair with different codons in the mRNA.
2. Certain tRNA species can bring their specific amino acids to the ribosome in response to several codons, not just the one with complementary sequence, through a loose kind of base pairing at the 3′ end of the codon and the 5′ end of the anticodon. This loose pairing is called **wobble.**

First letter	Second letter U		Second letter C		Second letter A		Second letter G		Third letter
U	UUU	Phe	UCU	Ser	UAU	Tyr	UGU	Cys	U
	UUC		UCC		UAC		UGC		C
	UUA	Leu	UCA		UAA	Stop	UGA	Stop	A
	UUG		UCG		UAG	Stop	UGG	Trp	G
C	CUU	Leu	CCU	Pro	CAU	His	CGU	Arg	U
	CUC		CCC		CAC		CGC		C
	CUA		CCA		CAA	Gln	CGA		A
	CUG		CCG		CAG		CGG		G
A	AUU	Ile	ACU	Thr	AAU	Asn	AGU	Ser	U
	AUC		ACC		AAC		AGC		C
	AUA		ACA		AAA	Lys	AGA	Arg	A
	AUG	Met	ACG		AAG		AGG		G
G	GUU	Val	GCU	Ala	GAU	Asp	GGU	Gly	U
	GUC		GCC		GAC		GGC		C
	GUA		GCA		GAA	Glu	GGA		A
	GUG		GCG		GAG		GGG		G

Figure 3-20 The genetic code. Notice that an amino acid can be coded by several different codons. A *stop* codon does not code for an amino acid, but instead signals to the ribosome that this is the end of the protein and that translation should cease.

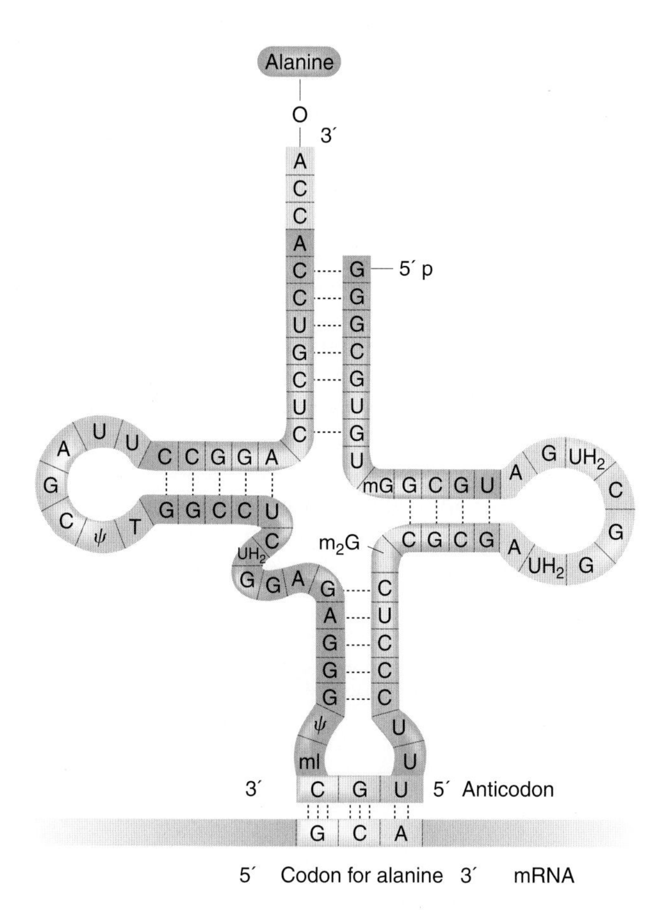

Figure 3-21 The structure of an alanine tRNA, showing the anticodon of the tRNA binding to its complementary codon in mRNA. The amino acid, in this case alanine, is attached to the 3′ end of the tRNA. The symbols ψ, mG, m_2G, mI, and UH_2 are abbreviations for the modified bases pseudouridine, methylguanosine, dimethylguanosine, methylinosine, and dihydrouridine, respectively. *(Adapted from L. Stryer,* Biochemistry, *4th ed. © 1995 by W. H. Freeman and Company.)*

Wobble is a situation in which the third nucleotide of an anticodon (at the 5′ end) can form two alignments (Figure 3-22). This third nucleotide can form hydrogen bonds not only with its normal complementary nucleotide in the third position of the codon but also with a different nucleotide in that position. There are "wobble rules" that dictate which nucleotides can and cannot form alternative hydrogen-bonded associations through wobble (Table 3-2). In the table, the letter *I* stands for inosine, one of the rare bases found in tRNA, often in the anticodon.

Table 3-3 on the next page lists all the codons for serine and shows how three different tRNAs ($tRNA^{Ser}{}_1$, $tRNA^{Ser}{}_2$, and $tRNA^{Ser}{}_3$) can pair with these codons. In some species there can be an additional tRNA species, which we could represent as $tRNA^{Ser}{}_4$, that has an anticodon identical with one of the three anticodons shown in Table 3-3 but differs in its nucleotide sequence elsewhere in the molecule. These four tRNAs are called **isoaccepting**

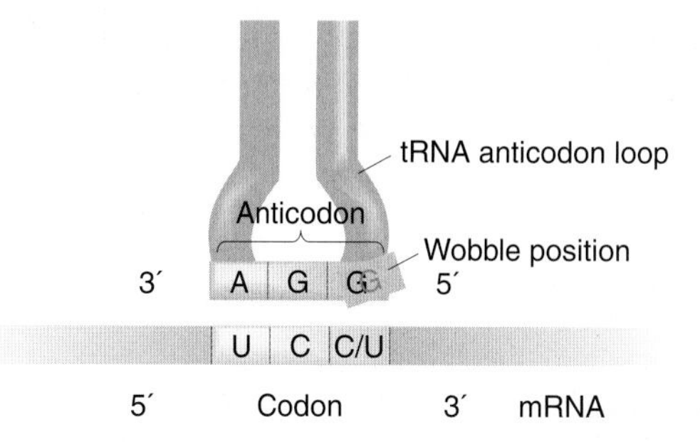

Figure 3-22 In the third site (5′ end) of the anticodon, G can take either of two positions through wobble, thus being able to pair with C (its usual partner) or U by wobble. This means that a single tRNA species carrying an amino acid (in this case, serine) can recognize two codons—UCC and UCU—in the mRNA.

tRNAs because they accept the same amino acid, but they are transcribed from different tRNA genes.

Ribosomes: Protein-Making Machines

The meeting place for amino acid–bound tRNAs and mRNA is the ribosome. Ribosomes are large macromolecular assemblies acting like complex subcellular machines. A ribosome consists of a large and a small subunit, each of which is composed of several rRNA types and as many as 50 proteins (Figure 3-23 on the next page). Ribosomes contain specific sites that enable them to bind to the mRNA, the tRNAs, and other specific protein factors, all required for protein synthesis. Let's look at a general picture of protein synthesis on the ribosome, and then examine each of the steps in the process in more detail.

The general process can be visualized as though viewing a frame from a movie. Figure 3-24 on page 73 shows a polypeptide caught in the process of being synthesized on the ribosome. We see that the mRNA is bound to the small ribosomal subunit. Two molecules of tRNA are bound to two different sites that span the large and small subunits. One of these sites, the **A** (*a*mino acid) **site,** is the entry site for the next aminoacyl-tRNA. A tRNA carrying the already synthesized amino acid chain (called peptidyl tRNA) is bound to the **P** (*p*olypeptide) **site.** Each new amino acid is added by reactions that transfer the growing chain to the

TABLE 3-2 Codon-Anticodon Pairings Allowed by the Wobble Rules

5′ End of Anticodon	3′ End of Codon
G	C or U
C	G only
A	U only
U	A or G
I	U, C, or A

TABLE 3-3 Different tRNAs That Can Service Codons for Serine

tRNA	Anticodon	Codon
$tRNA^{Ser}_1$	ACG + wobble	UCC UCU
$tRNA^{Ser}_2$	AGU + wobble	UCA UCG
$tRNA^{Ser}_3$	UCG + wobble	AGC AGU

aminoacyl-tRNA at the A site, forming a new peptide bond. The tRNA that has given up its amino acid is then released from the P site, and the ribosome moves one codon farther along the message, transferring the new polypeptide-holding tRNA to the P site, and leaving the A site vacant for the next incoming aminoacyl-tRNA. The growing end of the peptide chain (that is, the end closest to the ribosome) is the carboxyl end, and the free end (the first to be synthesized) is the amino end. Hence the amino end of the protein corresponds to the 5′ end of the mRNA, and the carboxyl end corresponds to the 3′ end of the mRNA.

The action of the ribosome during translation is divided into three distinct steps. **Initiation** is the recognition of the first triplet in the coding sequence and the placement of the first amino acid into position. **Elongation** is the growth of the chain by progressive amino acid addition. **Termination** is the process of stopping the chain elongation after the last amino acid has been added. (Note that these same terms were also applied to transcription,

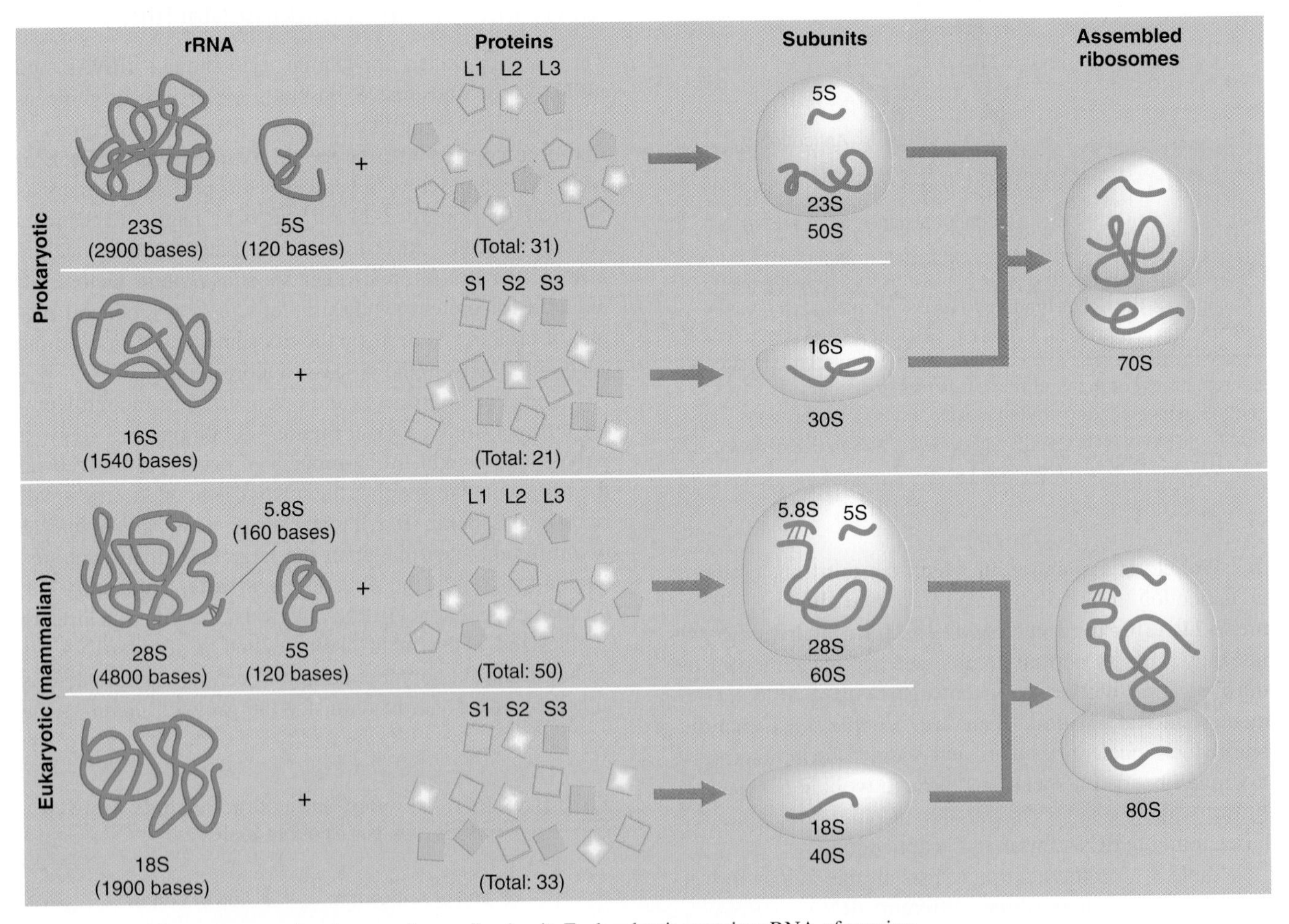

Figure 3-23 Ribosomes contain a large and a small subunit. Each subunit contains rRNA of varying lengths and a set of proteins. There are two principal rRNA molecules in all ribosomes (shown in the column on the left). Ribosomes from prokaryotes also contain one 120-base-long rRNA that sediments at 5S, whereas eukaryotic ribosomes have two small rRNAs: a 5S RNA molecule similar to the prokaryotic 5S, and a 5.8S molecule 160 bases long. The proteins of the large subunit are named L1, L2, etc., and those of the small subunit proteins S1, S2, etc. *(From H. Lodish, D. Baltimore, A. Berk, S. L. Zipursky, P. Matsudaira, and J. Darnell,* Molecular Cell Biology, *3d ed. © 1995 by Scientific American Books, Inc.)*

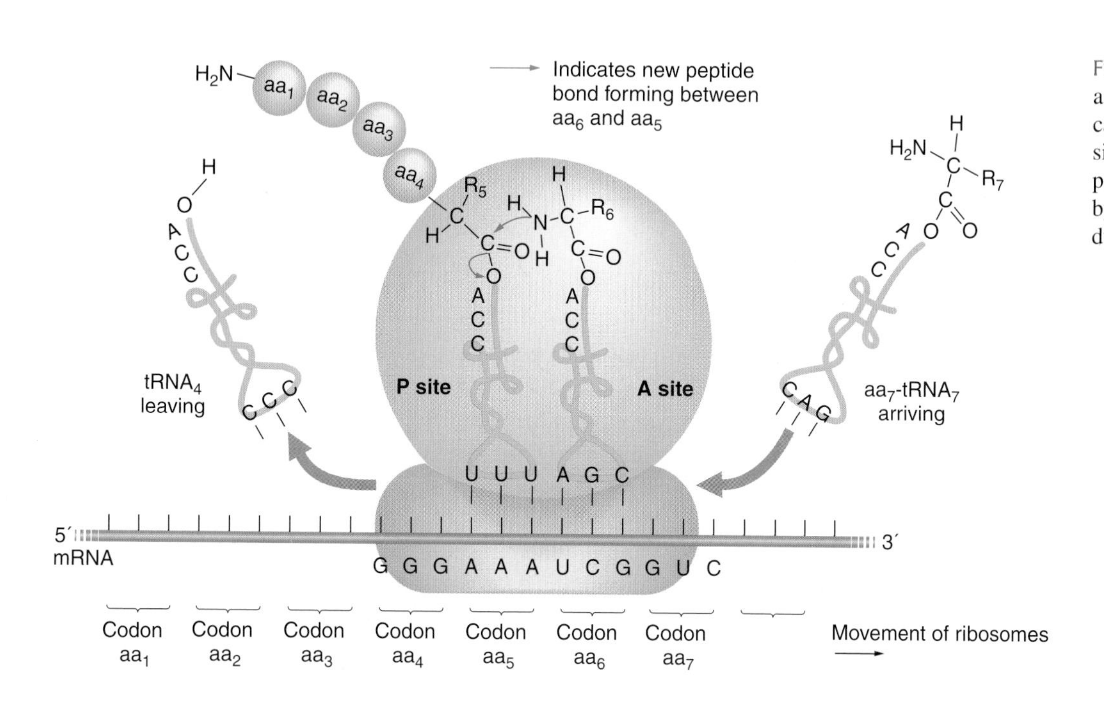

Figure 3-24 The addition of a single amino acid (aa_6), carried by the tRNA at the A site, to the growing polypeptide chain, tethered by the tRNA at the P site, during translation of mRNA.

so care must be taken to distinguish which topic is being discussed.)

Initiation. The main task of initiation is to place the first aminoacyl-tRNA in the P site of the ribosome. In most prokaryotes and eukaryotes, the first amino acid in any newly synthesized polypeptide is methionine, specified by the codon AUG. It is inserted not by $tRNA^{Met}$, but by a special tRNA called an *initiator,* symbolized $tRNA^{Met}{}_i$. In bacteria a formyl group is added to the methionine while attached to the initiator, making it *N*-formylmethionine. (The formyl group on *N*-formylmethionine is removed later.)

How does the translational machinery know where to begin? In other words, how is the initiation AUG codon selected from the many AUG codons in an mRNA molecule? Recall that in prokaryotes (Figure 3-9) and eukaryotes (Figure 3-16) an mRNA has a 5′ untranslated sequence consisting of the RNA between the transcriptional start site and the translational start site. In bacteria, initiation codons are preceded by special sequences called **Shine-Dalgarno sequences** that pair with the 3′ end of 16S rRNA in the ribosome, thereby positioning the ribosome properly next to the initiation codon, allowing the initiator tRNA and its *N*-formylmethionine to bind. There is a short but variable separation region between the Shine-Dalgarno sequence and the initiation codon.

Eukaryotic initiation is somewhat different. There is no Shine-Dalgarno sequence. Instead, a ribosome attaches to the mRNA under guidance of the 5′ cap and moves along the RNA; at the first AUG codon it encounters, an initiator tRNA carrying methionine is inserted at the P site. Conserved nucleotides nearby increase the likelihood of the initiator's binding to the first AUG. However, it is unclear if these sequences work through interactions with ribosomal RNAs.

Elongation. Elongation is assisted by several proteins, called *elongation factors,* that guide the binding and movement of tRNAs and the ribosome. The energy to drive the elongation process is provided by hydrolysis of GTP.

Termination. We noted in Figure 3-20 that some codons do not specify an amino acid at all. These codons, UAG, UGA, and UAA, are called **stop codons** or **termination codons.** They can be regarded as punctuation marks in the mRNA, ending the reading of the instructions for assembling a polypeptide. Stop codons often are called *nonsense codons* because they designate no amino acid, but this is misleading because punctuation is very much a part of the sense of any message. Interestingly, the three stop codons are not recognized by a tRNA, but instead by protein factors called **release factors.** When the peptidyl-tRNA is in the P site, the release factors bind to the A site in response to the chain-terminating codons. The polypeptide is then released from the P site, and the ribosomes dissociate into two subunits, ending translation.

MESSAGE

Translation is effected by ribosomes moving along mRNA in the 5′ → 3′ direction. A set of tRNA molecules brings amino acids to the moving ribosome, their anticodons binding to mRNA codons exposed on the ribosome. An incoming amino acid becomes bonded to the carboxyl end of the growing polypeptide chain on the ribosome.

PROTEIN FUNCTION AND MALFUNCTION IN CELLS

Transcription and translation ensure that a linear array of nucleotides in the DNA of a gene will be converted into a linear array of amino acids in the primary polypeptide chain. There is a precise correspondence of codons in DNA to amino acids in protein. This linear correspondence between gene and polypeptide product is called **colinearity.** Since the function of a protein depends on its shape, and shape is the result of precise folding of a specific chain of amino acids, it is clear that protein function depends absolutely on the DNA sequence of its coding gene.

Proteins can be classified into two broad types, *active proteins* and *structural proteins.* Good examples of active proteins are **enzymes,** the biological catalysts that make possible the thousands of chemical reactions that go on inside a living cell. Enzyme-encoding genes thus exert an enormous control over what goes on inside a cell. Other examples of active proteins are the microtubule proteins that produce movement of the internal components of the cell, and membrane-bound proteins that move ions and nutrients across the membrane. On the other hand, structural proteins, as their name suggests, contribute to the structural properties of the cell and the organism; human hair keratin and bone collagen are two structural proteins. Some proteins are both active and structural, for example, the contractile proteins actin and myosin, which comprise the great bulk of muscles and hence affect shape.

How Proteins Function

In this section we consider the molecular basis for individual protein function, and we look at an example of how several different proteins can act in concert to provide a specific cell function.

Active sites of enzymes. Enzymes do their job of catalysis by physically grappling with one or more substrate molecules, interacting with them to make or break chemical bonds. Enzymes are highly specific for the chemical reactions they catalyze, and the specificity lies in the precise fit between substrate and active site. The shape of the active site and of the substrates must have a complementary fit.

There are two basic models for enzyme-substrate fit: either the enzyme and substrate have rigid "lock-and-key" shapes, or the fit becomes precise only after the substrate binds, which is an "induced fit" (Figure 3-25). Figure 3-26a shows the active site of the enzyme carboxypeptidase, a digestive enzyme that cleaves polypeptides of food proteins between the amino acids glycine and tyrosine. As it binds its substrate, it undergoes a conformational shift that results in an induced fit necessary for catalysis (Figure 3-26b).

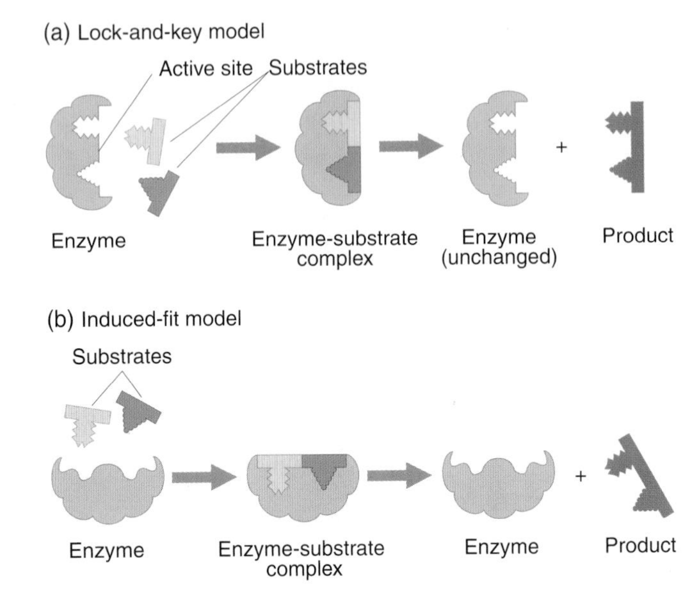

Figure 3-25 Schematic representation of the action of a hypothetical enzyme in putting two substrate molecules together. (a) In the lock-and-key mechanism the substrates have a fit complementary to the enzyme's active site. (b) In the induced-fit model, binding of substrates induces a conformational change in the enzyme.

Much of the globular structure of an enzyme is nonreactive material that simply supports the active site. However, the precise shape and binding properties of the active site are crucial for proper enzyme function, and a set of highly specific amino acids forms the lining of the active site. The codons for these amino acids tend to be clustered in the gene, but because of polypeptide folding some codons might be outside the cluster. In the following diagram the codons and amino acids for the active site of a hypothetical protein are shown in white; the remaining codons and amino acids of the protein, in black:

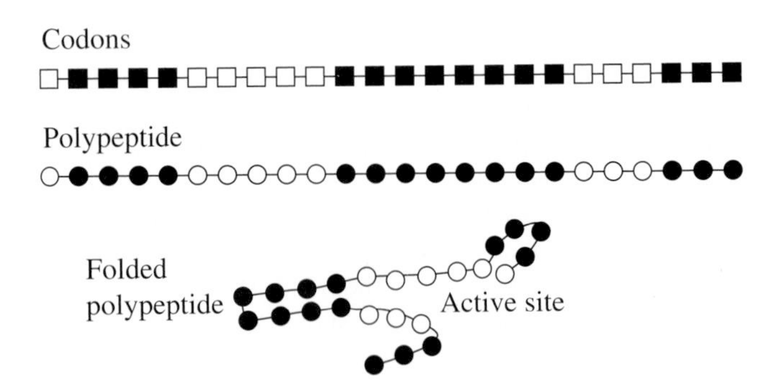

Protein domains. The active sites of different enzymes are good examples of localized functional regions of proteins. However, there are many other types of specialized functional regions in both active and structural proteins. Such regions are called **protein domains.** Some of these

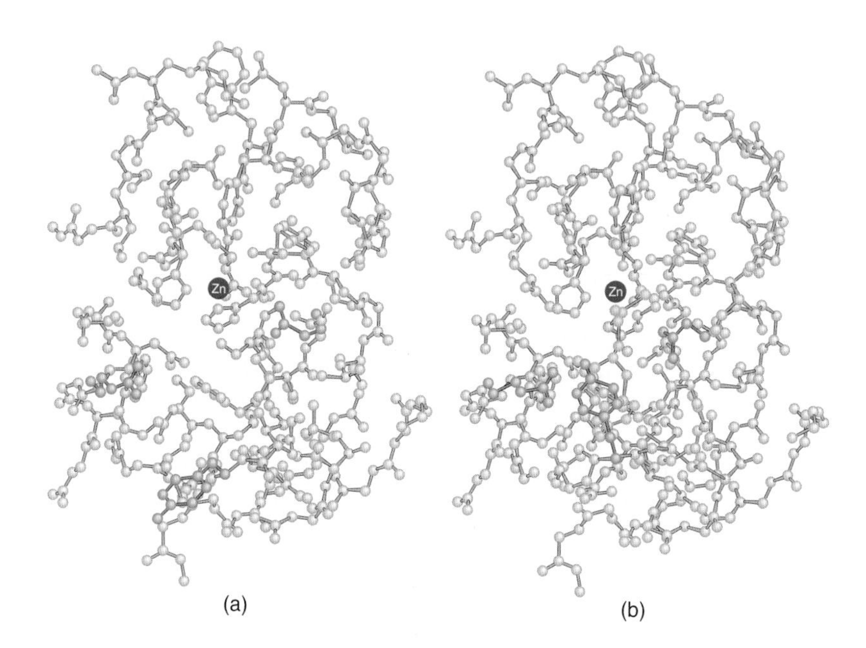

Figure 3-26 The active site of the digestive enzyme carboxypeptidase. (a) The enzyme without substrate. (b) The enzyme with its substrate (gold) in position. Three crucial amino acids (red) have changed positions to move closer to the substrate. Carboxypeptidase carves up proteins in the diet. *(From W. N. Lipscomb,* Proceedings of the Robert A. Welch Foundation Conferences on Chemical Research 15, *1971, 140–141.)*

show similarities of structure and function between proteins of similar function. For example, some domains are required for binding of proteins to nucleic acids. Some are required for anchoring the protein in a membrane. Others act as "address labels" targeting the protein to certain regions of the cell. The function of some domains can be regulated; for example, certain domains are active only when a phosphate group is attached to a specific amino acid within the domain. Furthermore, we can imagine how new proteins might evolve through acquisition or loss of specific domains by mutation or DNA rearrangement.

MESSAGE

Protein structure is the key to gene function. The specific amino acid sequence, together with cellular conditions, determines the shape, binding properties, and reactivity of the protein.

Enzyme pathways. Cells synthesize essential cellular substances step by step from compounds in the environment. This is done in a series of enzyme-mediated reactions called *biochemical pathways.* The haploid fungus *Neurospora,* like many microbes, can synthesize *all* its cellular molecules from inorganic components in the surroundings, using a myriad of biochemical pathways. For example, it synthesizes its own arginine (an amino acid) in a series of sequential biochemical steps. The immediate precursors in the pathway are ornithine and citrulline. Ornithine, which is made from earlier precursors, is converted into citrulline, which in turn is converted into arginine. These conversions are catalyzed by enzymes that we can designate enzymes 1 through 3, coded by three different genes that we can designate correspondingly 1 through 3:

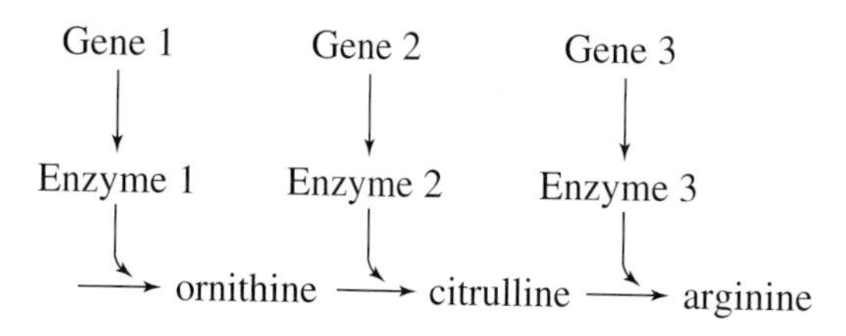

Hundreds of biochemical pathways involving thousands of reactions are active inside living cells. Collectively these reactions are termed *metabolism.* Each one of the metabolic reactions is catalyzed by an enzyme, whose structure is coded by a gene. Thus we see a glimmer of the pivotal position that genes play in controlling the chemistry of life.

Malfunctioning Alleles

Mutations and a short course in nomenclature. In genetics the standard organismal phenotype is called the **wild type,** because this is the type observed in the wild, in other words, in nature. All essential genes must be capable of producing their functional products in order to produce this wild-type phenotype. These normally functioning alleles are called **wild-type alleles. Mutations,** changes in the DNA sequence of a gene, occur spontaneously in nature. Since alleles are forms of genes, mutations by definition create new alleles. In genetic research, a gene is generally symbolized by a letter or several letters based on the word describing the phenotype produced by a mutant allele. Then the corresponding wild-type allele is designated by the addition of a superscript plus sign (+). For example, the wild-type eye color of the fruit fly *Drosophila* is red, but a mutant allele of a gene on the X chromosome produces white eyes, so the mutant allele is designated w and its wild-type allele is w^+.

As we saw earlier, the haploid fungus *Neurospora,* like most microbes, can synthesize all essential cellular compounds (such as amino acids) from simple inorganic substances in the growth medium. Therefore these organisms are said to be **prototrophs** ("self-feeders"). Some mutations render this fungus incapable of making specific essential substances. A mutant organism thus formed is called an **auxotroph** ("outside feeder") because it must obtain from the environment the essential compound it can no longer synthesize. Although such mutations would be lethal in nature, in the laboratory the mutant strains can be kept alive by supplying the missing substance in the growth medium. For example, some mutations result in the inability of a strain to make its own arginine, the synthesis of which was described in the previous section. A mutant allele is designated *arg* (standing for "arginine-requiring"), and its wild-type counterpart, which confers the ability to make arginine, is arg^+. Sometimes the + symbol is used by itself when the meaning is clear, so in the *Neurospora* example the two alleles would be designated *arg* and +, with the tacit understanding that the + corresponds to the wild-type allele of that particular gene.

An *arg* auxotroph will grow if arginine is added to the growth medium. Recall that the final portion of the arginine biosynthetic pathway is

Gene 1 Gene 2 Gene 3

$\longrightarrow$ ornithine $\longrightarrow$ citrulline $\longrightarrow$ arginine

An auxotrophic *arg* mutation could be in any one of the three genes controlling these conversions. The three possibilities can be distinguished. A mutation that codes for an inactive enzyme results in a blockage in the pathway, but this blockage can be bypassed by supplying any compound in the pathway past the blocked step. Hence a mutant blocked in gene 1 will be able to grow if supplied with arginine, citrulline, or ornithine. A mutant blocked in the middle step (gene 2) can be supplemented by arginine or citrulline. Finally, a mutant blocked in the last step (controlled by gene 3) would grow only if supplemented with arginine. This logic was used in the original elucidation of this pathway, as described in Foundations of Genetics 3-1. We could call the wild-type alleles of the three genes controlling the final portion of the arginine biosynthetic pathway $arg\text{-}1^+$, $arg\text{-}2^+$, and $arg\text{-}3^+$.

Types of mutations. The simplest mutant alleles show altered DNA sequence at only one spot in the gene, often a change of a single base pair. Such a mutant allele is composed of normal nucleotide sequence except in the area of change, known as the **mutant site** (Figure 3-27). Because during translation the genetic code is read in triplet codons, a mutation that changes one base pair to another often substitutes one amino acid for another. For example a change of TTT to TTA (UUU to UUA in mRNA) results in substitution of leucine for phenylalanine in the polypeptide. Furthermore, because of colinearity of gene and protein primary structure, a mutation in DNA results in an amino acid change in precisely the same relative position in the polypeptide. The research that first demonstrated this is described in Foundations of Genetics 3-2 on page 78.

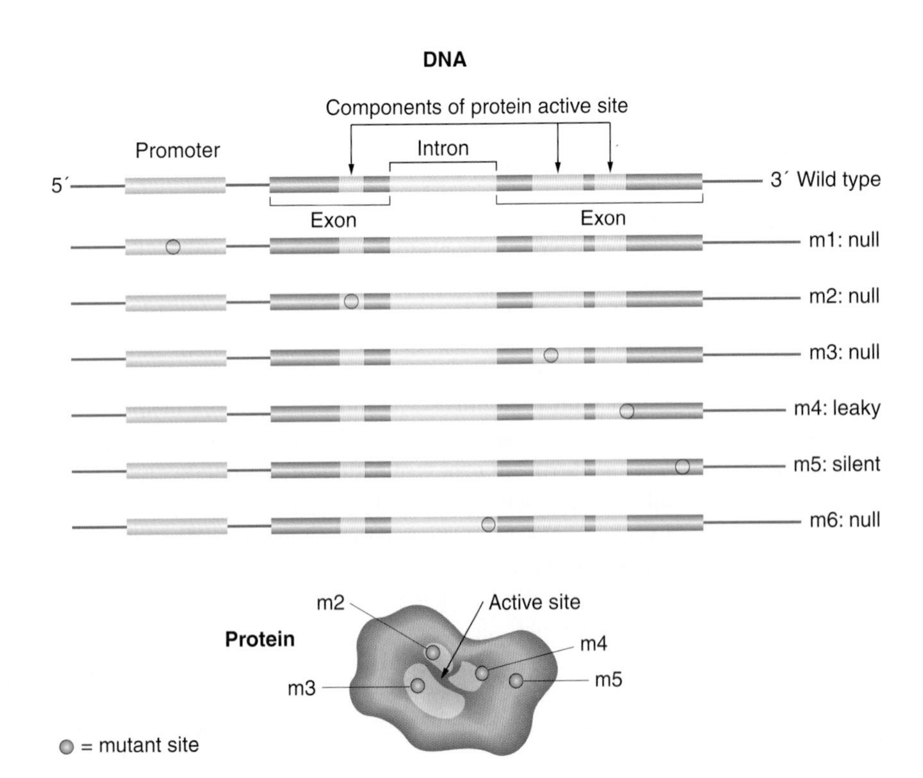

Figure 3-27 Representative positions of mutant sites and their functional consequences.

Beadle and Tatum and the Nature of Gene Action

In 1908 the English physician Archibald Garrod suggested that certain human diseases were caused by "inborn errors of metabolism." For example, the recessive disease alkaptonuria ("black urine disease") is characterized by the excretion in the urine of homogentisic acid, which turns black upon exposure to air. Garrod surmised that this unusual excretion was because those afflicted by the disease lacked an enzyme that normally broke this substance down, and it therefore built up in tissues and was excreted. This work was the dawn of the idea that the role of genes is to code for enzymes, which regulate the chemistry of the cell. The same idea arose in the literature several times over the next three decades, but it was the research of George Beadle and Edward Tatum on the fungus *Neurospora* that provided a clear statement of this mode of gene action.

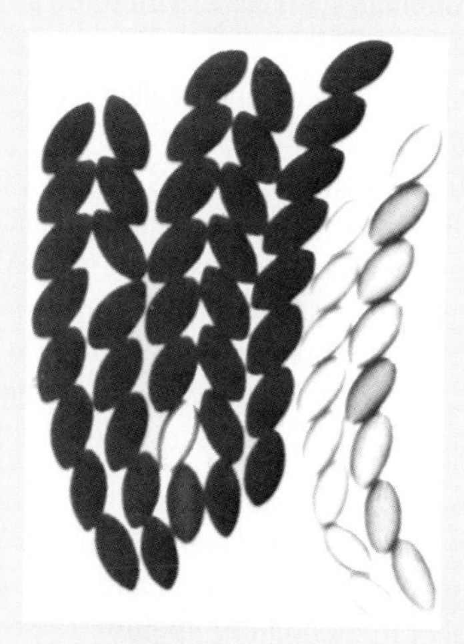

Sexual spores of *Neurospora.*

Beadle later wrote the following reminiscence:

> In 1940 we decided to switch from *Drosophila* to *Neurospora.* It came about in the following way. Tatum was giving a course in biochemical genetics and I attended the lectures. In listening to one of these—or perhaps not listening as I should have been—it suddenly occurred to me that it ought to be possible to reverse the procedure we had been following and instead of attempting to work out the chemistry of known genetic differences, we should be able to select mutants in which known chemical reactions were blocked. *Neurospora* was an obvious organism on which to try this approach, for its life cycle had been worked out by Dodge and Lindegren, and it probably could be grown in a culture of known composition. The idea was to select mutants unable to synthesize known metabolites such as vitamins and amino acids which could be supplied to the medium. In this way a mutant unable to make a given vitamin could be grown in the presence of that vitamin and classified on the basis of its differential growth response in media lacking or containing it.

Sure enough, after irradiating *Neurospora* cells to induce mutations in DNA, nutritionally defective mutants were easily found. Such mutants would grow only when the medium was supplemented with one of a number of essential cellular substances (for example, the amino acid arginine), implying that it was the synthesis of this substance that was defective in the mutant. Furthermore, known or suspected precursors of the essential substances could be tested. For example, possible precursors of arginine, such as ornithine and citrulline, could be tested in mutants that did not grow unless arginine was added to the medium. The arginine-requiring mutants could be classified into three groups on the basis of their different responses to ornithine and citrulline:

	Group 1	Group 2	Group 3
Ornithine	Growth	None	None
Citrulline	Growth	Growth	None
Arginine	Growth	Growth	Growth

Beadle and Tatum and their students deduced that groups 1 through 3 must each have a different mutation in one of three genes acting sequentially to make the end product:

$$\xrightarrow{\text{gene 1}} \text{ornithine} \xrightarrow{\text{gene 2}} \text{citrulline} \xrightarrow{\text{gene 3}} \text{arginine}$$

Only substances that normally were produced *after* the mutationally inactivated step could support growth of the mutants. From such experiments Beadle and Tatum devised the hypothesis that a gene generally acts by determining some specific enzyme activity—the "one-gene–one-enzyme" hypothesis. Subsequent work showed that this powerful catchphrase, which effectively united the two fields of biochemistry and genetics, could more accurately be stated "one-gene–one-polypeptide."

In the pathway that synthesizes orange carotenoid pigment in *Neurospora,* one of the intermediates is yellow, making the sequential process easier to appreciate.

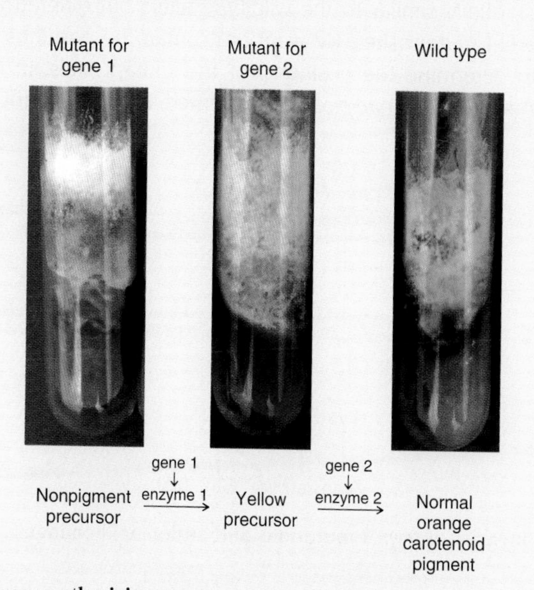

Pathway synthesizing orange carotenoid pigment in *Neurospora.*

FOUNDATIONS OF GENETICS 3-2

Charles Yanofsky and the Colinearity of Gene and Protein Structure

In 1941, George Beadle and Edward Tatum originated the one-gene–one-enzyme hypothesis. When James Watson and Francis Crick deduced the structure of DNA in 1953, it seemed likely that there must be a linear correspondence between the nucleotide sequence in DNA and the amino acid sequence in protein (such as an enzyme). However, it was not until 1963 that an experimental demonstration of this colinearity was obtained, from two research groups, one of which was led by Charles Yanofsky at Stanford University.

Yanofsky had induced 16 mutant alleles of the *tryp A* gene of *Escherichia coli,* whose wild-type allele was known to code for the α protein subunit of the enzyme tryptophan synthetase. Tryptophan synthetase is a tetramer composed of two α and two β subunits. It catalyzes the following reactions:

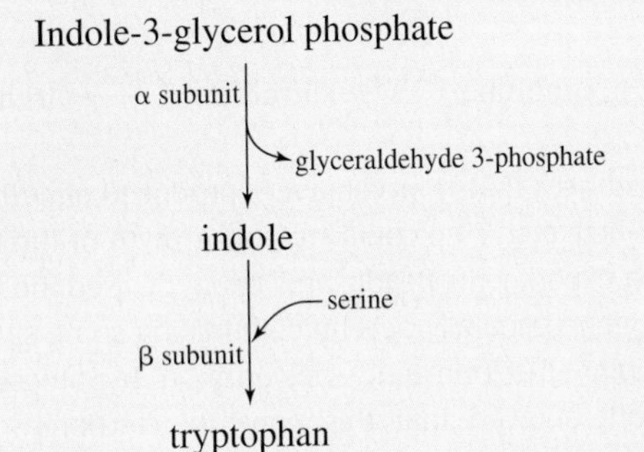

It is now known that the subunits of the enzyme act sequentially without letting go of the indole.

Yanofsky showed that the 16 mutant *tryp A* alleles, which produced inactive forms of the enzyme, were structurally all substantially similar to the wild-type allele but differed at 16 different mutant sites. He went on to map the positions, that is, to determine the positions, of the mutant sites. In 1963 DNA sequencing had not been invented, but the mutant sites could be mapped by a method called *recombination analysis,* which we will study in Chapter 5.

Model of tryptophan synthetase, with substrate in red.

Through biochemical analysis of the tryp A protein, it was shown that each of the 16 mutants resulted in an amino acid substitution at a different position in the protein. By comparing the relative positions of the mutant sites in the gene with the relative positions of the amino acid changes in the protein, it was shown that the two were colinear.

In a summary of his results in 1966, Yanofsky wrote:

> The concept of colinearity of gene structure and protein structure was examined with 16 mutants with alterations in one segment of the A gene and the A protein [now called α] of tryptophan synthetase. The results obtained demonstrate a linear correspondence between the two structures and further show that genetic recombination values [a measure of the distances between mutant sites] are representative of the distances between amino acid residues in the corresponding protein.

The results subsequently have been shown to be of general application to mutations in other proteins.

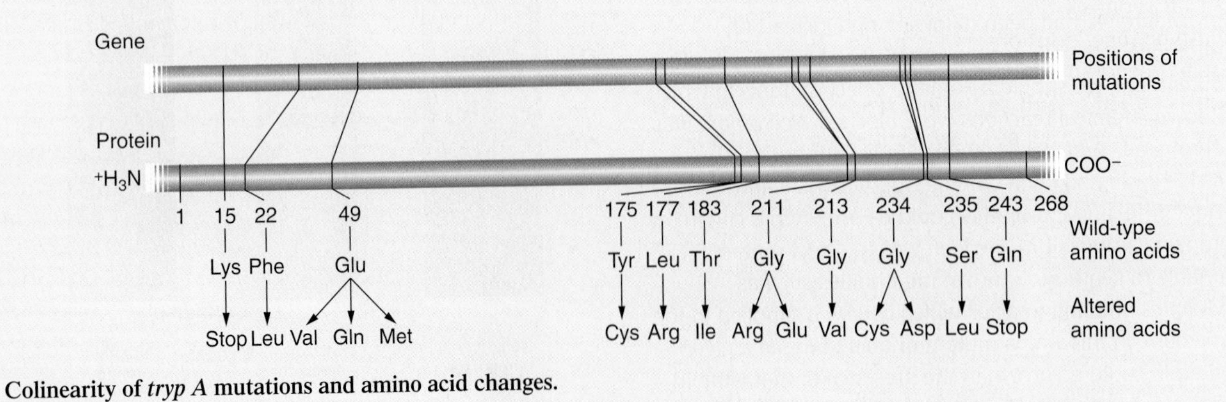

Colinearity of *tryp A* mutations and amino acid changes.

A change of a single nucleotide at a crucial functional site (such as the active site of an enzyme) can result in an amino acid replacement that leads to aberrant function. For example, a change from CCG to CTG replaces proline with leucine, an amino acid with quite different properties that could have adverse effects on protein function. The changes are summarized below:

	Wild type	Mutant
DNA	5′ CCG 3′	5′ CTG 3′
	3′ GGC 5′	3′ GAC 5′
mRNA	5′ CCG 3′	5′ CUG 3′
Amino acid	Proline	**Leucine**

Some DNA changes result in a premature stop codon. For example, one change in the tyrosine codon TAC results in the stop codon TAG (UAG in mRNA). This would cause premature termination of translation and a short, nonfunctional polypeptide.

Another common type of simple mutation is addition or deletion of single nucleotide pairs. These are called **frameshift mutations** because they cause the proper reading of the message to be off by one in every amino acid translationally "downstream" from the mutant site. For example, consider the effect of a deletion of a nucleotide on the following mRNA segment:

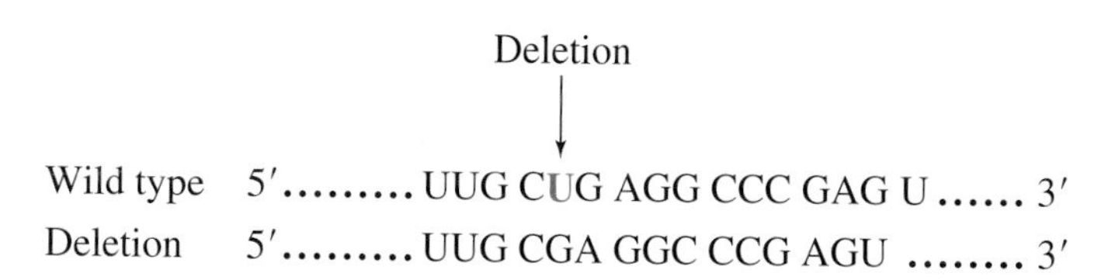

All the downstream codons will be changed, and most of the corresponding amino acids will be inappropriate for that protein. Hence frameshift mutations usually knock out protein function completely.

Effects of mutations on cellular function. Generally, mutations that show altered phenotypes produce their effect by reducing or eliminating protein function. A mutation with reduced function is called a **leaky mutation:** because partial wild-type function is observed, this seems to "leak" into the mutant phenotype. A mutation that results in no protein function is called a **null mutation** (the word *null* means "nothing"). Changes that do not affect the function of a protein are called **silent mutations.**

Some mutations alter the information-transfer process rather than directly altering the protein coding region of a gene. For example, some mutations produce malfunction not through any effect on amino acid sequence but by interfering with intron splicing. Recall that intron splicing depends on specific nucleotide sequences at the exon-intron boundaries and inside the intron (see Figure 3-12). If these sites are mutated, the intron cannot be excised and the proper mRNA will not be produced. Similarly, certain mutations can alter the regulation of the gene. For example, the promoter sequence to which the RNA polymerase binds is crucial, and if changes occur in this sequence, the gene might not be transcribed at all or might be transcribed at abnormally low (or high) rates. Finally, mutations in the untranslated 3′ tail of the transcript can have marked effects on mRNA stability and hence interfere with chemical balance in the cell. Some examples of these various types of simple mutations are shown in Figure 3-27.

MESSAGE

Mutations can lead to gene malfunction by changes in sequences that are protein-coding or important for information processing.

Genetic disease. We have seen that cells work through a myriad of metabolic reactions, arranged in biochemical pathways. Through these pathways, chemical structures are gradually built up or torn down. Each of the metabolic reactions is catalyzed by an enzyme, whose structure is dictated by a gene. A common cause of genetic disease in humans is enzyme deficiency resulting from mutation in one of these genes. Human cells carry two chromosome sets, so all genes are represented twice (gene pairs). Normally both copies of a gene pair are wild-type alleles. However, if an individual has a pair of defective alleles of a gene coding for an enzyme, there will be reduced or no enzyme activity. As a result, such an individual will show symptoms of the disease caused by deficiency of that enzyme. Table 3-4 on the next page gives some examples of genetic diseases caused by dysfunctional enzymes. Figure 3-28 on page 81 examines a corner of the human metabolic map and shows the genetic diseases that stem from the blockage of enzyme-catalyzed steps in a biosynthetic pathway.

The genetic basis of the disease phenylketonuria (PKU), with which we began the chapter, is revealed in part of Figure 3-28. A person with PKU has two defective alleles of the gene that codes for the enzyme phenylalanine hydroxylase. This results in a blockage in the pathway. The compound located just before the block, phenylalanine, builds up in concentration because it is constantly being supplied as a component of protein in food and cannot be converted to tyrosine if the pathway is blocked. At high concentrations phenylalanine is converted into phenylpyruvic acid. Normally this reaction does not take place because phenylalanine is processed immediately by phenylalanine hydroxylase. You will recall from earlier in the chapter that Følling originally detected this high level of phenylpyruvic acid in the urine of children with PKU. In an infant, phenylpyruvic acid interferes with the development of the nervous system, giving rise to mental retardation. Thus these symptoms are explained on the basis of this enzyme blockage. However, understanding the nature of this blockage has provided a treatment that

TABLE 3-4 Representative Examples of Enzymopathies: Inherited Disorders in Which Altered Activity (Usually Deficiency) of a Specific Enzyme Has Been Demonstrated in Humans

Condition	Enzyme with Deficient Activity*
Acatalasia	Catalase
Acid phosphatase deficiency	Acid phosphatase
Albinism	Tyrosinase
Aldosterone deficiency	18-Hydroxydehydrogenase
Alkaptonuria	Homogentisic acid oxidase
Angiokeratoma, diffuse (Fabry disease)	Ceramide trihexosidase
Apnea, drug-induced	Pseudocholinesterase
Argininemia	Arginase
Argininosuccinic aciduria	Argininosuccinase
Ataxia, intermittent	Pyruvate decarboxylase
Citrullinemia	Argininosuccinic acid synthetase
Crigler-Najjar syndrome	Glucuronyl transferase
Cystathioninuria	Cystathionase
Ehlers-Danlos syndrome, type V	Lysyl oxidase
Farber lipogranulomatosis	Ceramidase
Galactosemia	Galactose 1-phosphate uridyl transferase
Gangliosidosis, GM_1; generalized, type I, or infantile form	β-Galactosidase A, B, C
Gangliosidosis, GM_2; type II, or juvenile form	β-Galactosidase B, C
Gaucher disease	Glucocerebrosidase
Gout	Hypoxanthine-guanine phosphoribosyltransferase Phosphoribosyl pyrophosphate (PRPP) synthetase (increased activity)
Granulomatous disease	Reduced nicotinamide adenine dinucleotide phosphate (NADPH) oxidase
Hydroxyprolinemia	Hydroxyproline oxidase
Hyperlysinemia	Lysine-ketoglutarate reductase
Hypophosphatasia	Alkaline phosphatase
Immunodeficiency disease	Adenosine deaminase Uridine monophosphate kinase
Krabbe disease	Galactosylceramide β-galactosidase
Leigh necrotizing encephalomyelopathy	Pyruvate carboxylase
Maple-sugar urine disease	Keto acid decarboxylase
Niemann-Pick disease	Sphingomyelinase
Ornithinemia	Ornithine ketoacid aminotransferase
Pentosuria	Xylitol dehydrogenase (L-xylulose reductase)
Phenylketonuria	Phenylalanine hydroxylase
Refsum disease	Phytanic acid oxidase
Richner-Hanhart syndrome	Tyrosine aminotransferase
Sandhoff disease (GM_2 gangliosidosis, type II)	Hexosaminidase A, B
Tay-Sachs disease	Hexosaminidase A
Wolman disease	Acid lipase
Xeroderma pigmentosum	Ultraviolet-specific endonuclease

* The form of gout due to increased activity of PRPP is the only disorder listed that is due to *increased* enzymatic activity.
SOURCE: Victor A. McKusick, *Mendelian Inheritance in Man*, 4th ed. © 1975 by Johns Hopkins University Press.

alleviates the symptoms of PKU. If the high level of phenylpyruvic acid is detected soon after birth, in large-scale screening tests of newborns, the baby can be placed on a special low-phenylalanine diet. Hence phenylalanine does not accumulate in cells, no phenylpyruvic acid is made, and the child will develop without retardation. Notice from Figure 3-28 that other human hereditary disorders are caused by blockages in nearby steps in the same pathway.

The structure of the gene for phenylalanine hydroxylase is shown in Figure 3-29 on page 82. The gene spans approximately 90 kb of DNA, and contains 13 introns. This is one of the most thoroughly studied of all human genes. Many different mutations have been found, some of which are shown in the figure. Note that most are in the exons (represented approximately to scale as narrow vertical bars), thereby affecting the amino acid sequence. However some mutations are in the introns, and the action of these is to interfere with normal splicing, resulting in a defective mRNA.

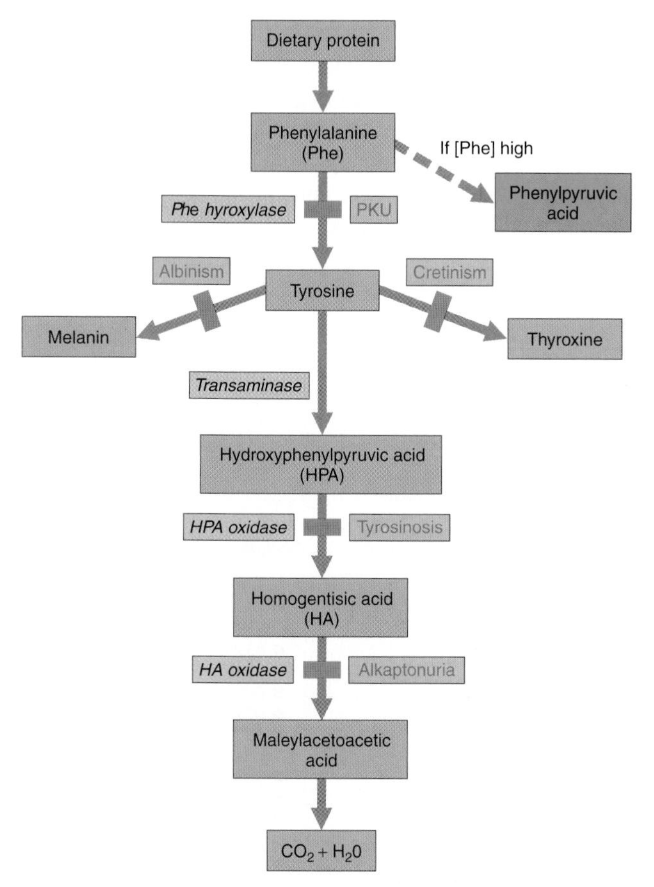

Figure 3-28 One small part of the human metabolic map, showing the consequences of various specific enzyme failures. The normal pathway is outlined in the purple boxes, with enzymes mediating the pathway in green boxes. Enzyme blockages are shown as red boxes, together with the names of the associated diseases. In the case of the disease PKU, the buildup of phenylalanine causes conversion into phenylpyruvic acid (orange), which interferes with development of the nervous system. *(After I. M. Lerner and W. J. Libby,* Heredity, Evolution, and Society, *2d ed. © 1976 by W. H. Freeman and Company.)*

We saw in Chapter 2 that dystrophin, the protein that is defective in Duchenne muscular dystrophy, is not an enzyme but an anchoring protein in the membrane. Hence genetic disease in humans can be caused by mutations in genes for any type of protein, enzymic or not.

DEFECTIVE PROTEINS AND DOMINANCE AND RECESSIVENESS

We saw above that phenylketonuria is caused by a defect in a single gene; such diseases are called *single-gene diseases* to distinguish them from genetic diseases of a more complex nature. PKU is said to be a **recessive** disease, as are the majority of single-gene diseases in humans. The term *recessive* means that both copies of the gene must be defective in order for the symptoms of the disease to develop. In a person carrying a normal allele on one chromosome and its defective allele on the homologous chromosome, the presence of that one normal allele produces enough protein to provide sufficient function for an apparently normal state. In this case the normal phenotype is said to be **dominant** because the normal phenotype is expressed even in the presence of one abnormal allele and thus seems to dominate it. Even though the terms *dominant* and *recessive* are strictly for describing phenotypes, they are also commonly applied to the respective causative alleles. In regard to PKU, the alleles we have discussed above are the dominant allele that codes for functional phenylalanine hydroxylase enzyme and the defective recessive allele that results in the disease PKU. Because one haploid "dose" of the normal allele provides sufficient protein for normal function, this general situation is termed **haplo-sufficiency.**

In conventional genetic nomenclature, recessive alleles are given lowercase letter symbols. If we consider a normal wild-type allele and a defective allele of any gene, there are three possible allelic combinations. A genotype with two copies of the dominant allele is called a **homozygous dominant,** with two copies of the recessive is a **homozygous recessive,** and with one of each is a **heterozygote.** The three possible genotypes and their associated phenotypes in a general case of haplo-sufficiency are represented in the following table. In the table, *a* is the mutant (defective) allele, and the wild-type counterpart is given a superscript "+" sign, a^+. [In the written genotypes, note the slash (/) between the allele pairs. This slash is used to designate homology in diploid organisms, and we will see extensions of its use in subsequent chapters.]

Dominance Relations in Genes Showing Haplo-Sufficiency

Genotype	Name of genotype	Phenotype
a^+ / a^+	Homozygous dominant	Normal
a^+ / a	Heterozygote	Normal
a / a	Homozygous recessive	Defective

In this case the phenotype of the heterozygote is the same as that of the homozygous dominant. Notice that the phenotype of the heterozygote is effectively a test of dominance. If the heterozygote is normal in phenotype, then the phenotype represented by the mutant allele must be recessive. Figure 3-30 on page 83 shows the molecular basis for haplo-sufficiency.

However, in heterozygotes for some genes, the normal allele *cannot* provide enough protein product to fulfill normal cell function; this situation is known as **haplo-insufficiency.** In these cases it is the defective (mutant) allele that is dominant. Uppercase letters are used to denote dominant mutations. As with recessive alleles, the wild-type allele is indicated by a superscript plus sign added to the letter. The general case for haplo-insufficiency can be written as follows, where *B* is a mutant allele:

Dominance Relations in Genes Showing Haplo-Insufficiency

Genotype	Name of genotype	Phenotype
B^+ / B^+	Homozygous recessive	Normal
B / B^+	Heterozygote	Defective
B / B	Homozygous dominant	Defective

Again we see that the test for dominance is the phenotype of the heterozygote. Here the heterozygote has the same phenotype as the homozygous mutant (defective), so the mutant allele is clearly dominant. A general rule of thumb is that there are many more haplo-sufficient genes than haplo-insufficient ones, but much depends on the organism and the genetic context.

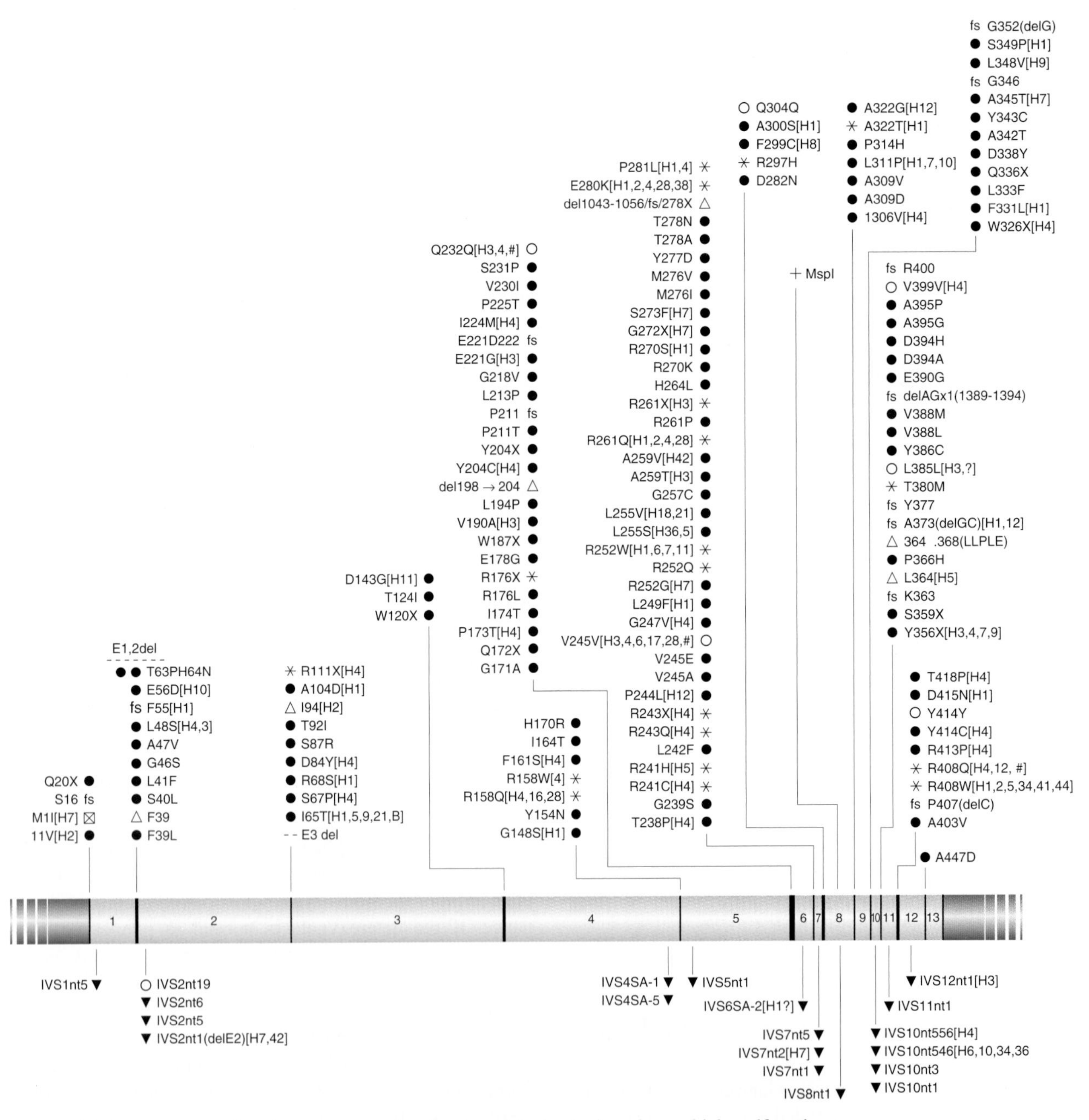

Figure 3-29 The structure of the human gene coding for phenylalanine hydroxylase, which malfunctions in the disease phenylketonuria (PKU). Exons are shown as black lines. Introns are in light green and numbered 1-13. Mutations resulting in malfunction for this gene can be in the protein-coding regions, the exons (mutations listed above the gene), or in the regions of the introns involved in splicing (listed below the gene). The various symbols represent different types of mutational changes. *(Modified from C. R. Scriver,* Ann. Rev. Genet. *28, 1994, 141–165.)*

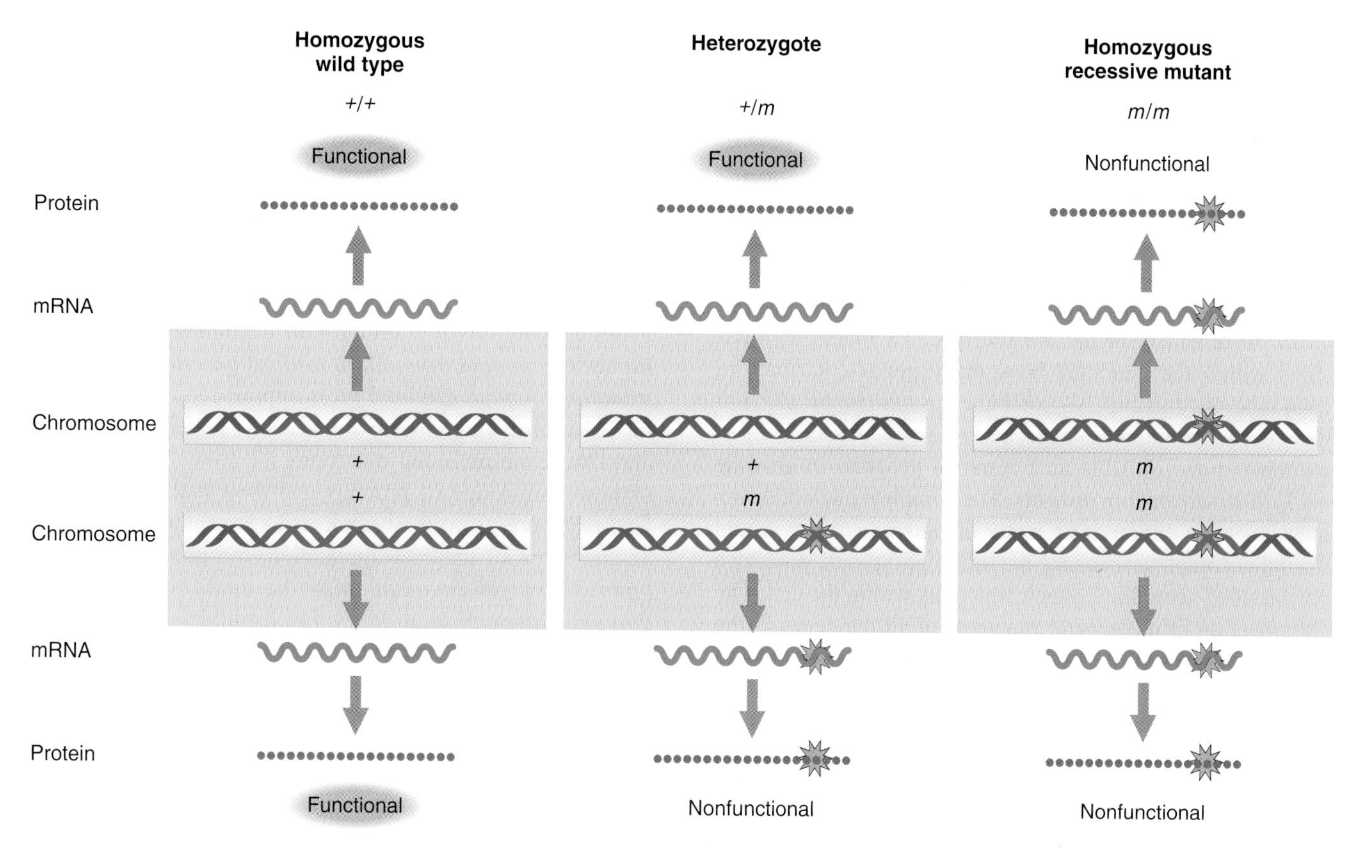

Figure 3-30 Recessiveness of a mutant allele of a haplo-sufficient gene. In the heterozygote, even though the mutated copy of the gene produces nonfunctional protein, the wild-type copy generates enough functional protein to produce the wild-type phenotype.

MESSAGE

In a normal/mutant heterozygote, if the single normal allele can provide adequate functional protein (haplo-sufficiency), the mutant allele is recessive; if the single normal allele cannot provide adequate functional protein (haplo-insufficiency), the mutant allele is dominant.

There are mechanisms other than haplo-insufficiency whereby mutant alleles can express dominance; we will discuss them only briefly here, then more extensively in later chapters. One way is by the mutant allele's taking on new cellular function. In other words, it can perform a function that the wild-type allele cannot do. These are called *gain-of-function dominants.* Another mechanism of dominance is shown by *dominant negative* mutations. Often dominant negative mutations are in proteins that form multimers. In a wild/mutant heterozygote the mutant protein subunit distorts the shape of the multimer and causes it to lose function even though some wild-type subunits are also present.

Haplo-sufficiency and -insufficiency imply the existence of some type of threshhold amount of gene activity in the cell. Above the threshhold we observe one phenotype, and below it another. However, for some genes, a wild/mutant heterozygote shows a phenotype intermediate between that of the homozygous wild type and the homozygous mutant. Neither the wild type nor the mutant allele can be considered dominant, and that is called a case of **incomplete dominance.** In such cases there seems to be no threshhold. What is important is the number of copies of a wild-type allele: 2, 1, or 0. There is proportionality between the number of alleles, the amount of gene product, and the phenotype.

In some situations geneticists instead use upper- and lowercase letters to designate alleles, such as *A* for dominant and *a* for recessive. In this symbolism there is no designation of wild type. The upper- and lowercase symbolism is appropriate in cases in which the wild-type allele is not known, for example, in extensively interbred lines of plants or animals. It is also useful in naming alleles that determine the morphs of a polymorphism because all the morphs are wild-type. There are also research situations in which knowledge of the wild-type state is unimportant.

A more complete discussion of variations on the theme of dominance will be found in Chapter 14.

FUNCTIONAL DIVISION OF LABOR IN THE GENE SET

So far, we have discussed function at the level of individual genes. However, function must also be considered at the level of the whole organism because the goal of genetics is to understand the way that genes influence the fundamental properties of entire organisms. Of particular interest is the nature of the complete set of genes that is needed to construct an organism. What are the types of functional divisions within the genome? How many genes contribute to each type of function? No complete answer can be given to this question at the present stage of research, but some information is now available as a result of progress in genome nucleotide-sequencing projects. The complete nucleotide sequence of the genomes of many organisms is known, and from this complete sequence the different types of genes can be classified according to their functions within the cell. The classification of the general functions of all the genes in the genome of the fruit fly (a favorite model genetic organism) is shown in Table 3-5. Notice that a large proportion of the genes code for enzymes; many of these are the enzymes of the basic cell metabolism. The same type of information, but for the model genetic plant *Arabidopsis thaliana,* is shown diagrammatically in Figure 3-31. Here, too, metabolic proteins are a large proportion of the total.

The set of protein-coding genes of an organism is called its **proteome.** Division of labor within the proteome is a topic of considerable interest in understanding the overall functional design features of organisms. Knowing which components of the proteome are active in specific tissues or during specific developmental phases provides great insight into the way that organisms work. For example, if we can show that in one cell type—say, a nerve cell—there are 50 genes expressed, and we can specify what proteins these genes encode, this is a very useful beginning for researching the function of that cell. In medical genetics the composition of the proteome is of great importance because of the insight it gives into disease and therapy. In the case of PKU and Duchenne muscular dystrophy we have seen examples of how knowledge of proteins and their malfunctions can lead to a cure, or at least point us in the right direction. In another type of medical application, the proteome can be "mined" for proteins that might be modified by disease-fighting pharmaceutical drugs. For example, if it is known which proteins are active in a cell type involved in some specific disease, each of these proteins becomes a target for development of a drug that can alter it.

From such accumulating information on genomes and proteomes, hazy outlines appear of the complete genetic blueprints for life on this planet, and the way in which it functions.

TABLE 3-5 Classification of All Genes of the Fruit Fly *Drosophila melanogaster*

Function	Number of Genes	Function	Number of Genes
Nucleic acid binding	1387	Enzyme inhibitor	68
DNA binding	919	Apoptosis inhibitor	15
DNA-repair protein	65	Signal transduction	622
DNA replication factor	38	Receptor	337
Transcription factor	694	Transmembrane receptor	261
RNA binding	259	G protein–linked receptor	163
Ribosomal protein	128	Olfactory receptor	48
Translation factor	69	Storage protein	12
Transcription factor binding	21	Cell adhesion	216
Cell cycle regulator	52	Structural protein	303
Chaperone	159	Cytoskeletal structural protein	106
Motor protein	98	Transporter	665
Actin binding	93	Ion channel	148
Defense/immunity protein	47	Neurotransmitter transporter	33
Enzyme	2422	Ligand binding or carrier	327
Peptidase	468	Electron transfer	124
Endopeptidase	378	Cytochrome *P450*	88
Protein kinase	236	Ubiquitin	11
Protein phosphatase	93	Tumor suppressor	10
Enzyme activator	9	Function unknown/unclassified	7576

SOURCE: G. M. Rubin et al., *Science* 287, 2000, 2204–2215.

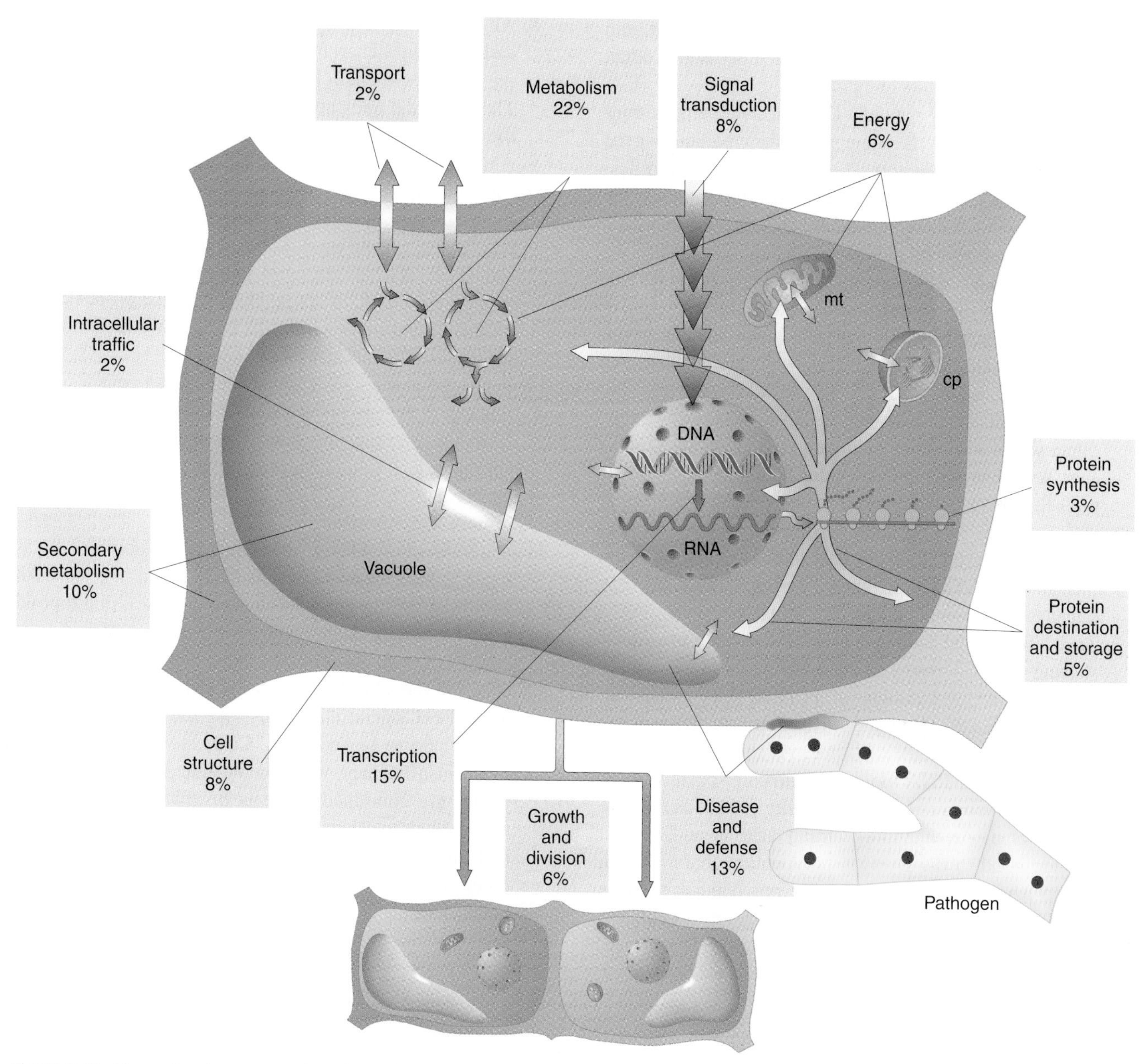

Figure 3-31 Sizes of various categories of protein-coding genes, estimated from currently known genes in the plant *Arabidopsis thaliana.*

SUMMARY

1. The three key macromolecules of genetics—DNA, RNA, and protein—are the main elements in the conversion of hereditary information into cellular form and function. The working of this system is based on two fundamental principles: complementary base pairing between single polynucleotide strands, and attachment of specific nucleic acid–binding proteins such as polymerases, transcription factors, and proteins associated with the ribosome. We see these two principles at work in transcription, pre-mRNA splicing, and translation.
2. Cellular form and function are largely determined by proteins, so for the great majority of genes their product is a protein. However, a minority of specialized genes code for RNAs that are never translated into protein; these RNAs act as recycling, information-processing components for all genes.
3. Protein-coding genes are transcribed to make RNA. In prokaryotes this RNA transcript is the messenger RNA (mRNA), which will be translated into protein. In eukaryotes, the initial transcript is processed in various

ways to add a 5′ cap, remove any introns present, and add a poly(A) tail at the 3′ end. The processed product is the mRNA.

4. Only one strand of a gene's DNA double helix is transcribed. It is always the same strand for a particular gene and is termed the *template strand.* Other genes (in different locations along the same chromosome) may use the other strand of the chromosomal DNA as their template.
5. Transcription of the template strand starts with the binding of RNA polymerase to a specific promoter sequence near the transcriptional start site. The polymerase then moves along the DNA from one end to the other, catalyzing a growing RNA molecule. RNA grows by the addition of new nucleotides at its 3′ end.
6. To function, a protein must have a precise three-dimensional structure, ultimately based on the amino acid sequence (the polypeptide chain), which constitutes the primary structure. The polypeptide chain folds and bends to give helices or pleated sheets, the secondary structure. The secondary structure may fold further to give tertiary structure, and several molecules may associate to give quaternary structure.
7. Translation of the mRNA sequence is also unidirectional starting at the 5′ end. The main translational machine is the ribosome, a complex of proteins and processing RNAs. It moves along the mRNA, translating it three nucleotides (a codon) at a time. As the ribosome progresses, transfer RNA molecules carrying specific amino acids bind their anticodons to specific codons as they are encountered. An incoming amino acid is added by a peptide bond to the growing polypeptide chain. During amino acid addition, the growing polypeptide chain is transferred to the tRNA of the incoming amino acid, and the previous tRNA, which is now "empty," is released.
8. All individual amino acids have an amino end and a carboxyl end. Therefore, when they are polymerized, the amino acid chain has an amino and a carboxyl end. The ribosome adds amino acids to the carboxyl end of the growing chain.
9. Assembly of the 20 common amino acids into polypeptide chains, as dictated by the nucleotide sequences of the many different genes, results in a vast array of protein types of different size, shape, and amino acid sequence. These different proteins have different properties, which, in conjunction with the proper physiological conditions within the cell, result in the different protein functions in a cell.
10. A change in the nucleotide sequence of a gene (a mutation), which can result from a chemical accident or from the presence of some reactive agent within the cell, may lead to an abnormal amino acid sequence and hence altered shape. Thus the mutation may cause the protein to malfunction.
11. At the functional level, genes may be classed broadly as haplo-sufficient or haplo-insufficient. In the case of haplo-sufficiency (the more common case), in a diploid cell, if a mutant allele coding for protein malfunction is combined with its normal allele, the single functional copy of the allele provides enough product to allow normal cell operation. In this case the dysfunctional allele is said to be recessive. Conversely, in the case of haplo-insufficiency, when the normal and nonfunctional alleles are combined, there is insufficient normal protein product to allow the cell to function, and the abnormal allele is said to be dominant. However, there are other mechanisms whereby an altered gene may become dominant, such as by the chance production of a new function.

CONCEPT MAP

Draw a concept map interrelating as many of the following terms as possible. Note that the terms are listed in no particular order.

gene / transcription / translation / RNA polymerase / dominant / ribosome / protein / mutation / haplo-insufficiency

SOLVED PROBLEM

1. Using the codon dictionary (Figure 3-20), show the consequences on subsequent translation of the addition of an extra adenine base to the beginning of the following coding sequence:

–CGA–UCG–GAA–CCA–CGU–GAU–AAG–CAU–

– Arg – Ser – Glu – Pro – Arg – Asp – Lys – His –

Solution

The addition of A at the beginning of the coding sequence constitutes a frameshift mutation, and a different set of amino acids is specified by the sequence, as shown here. (Note that a set of nonsense codons is encountered, which results in chain termination.)

–ACG–AUC–GGA–ACC–ACG–UGA–UAA–CGA–

– Thr – Ile – Gly – Thr – Thr – stp – stp –

SOLVED PROBLEM

2. Yeast (a haploid) can normally make its own proline. Six auxotrophic mutants were obtained that could not make proline and needed it to be supplied in the growth medium. Two other compounds structurally related to proline (GSA and glutamine) were added to the mutants to determine if they could cause growth instead of proline. The results were as follows, where "yes" means the compound permitted growth and "no" means it did not:

Mutant	Proline	Glutamine	GSA
1	Yes	No	Yes
2	Yes	No	No
3	Yes	Yes	Yes
4	Yes	No	No
5	Yes	No	Yes
6	Yes	No	Yes

a. How many different patterns of responses are there among the six mutants?

b. What do the different patterns of responses represent?

c. What is the metabolic relationship among the three compounds—proline, GSA, and glutamine—in yeast?

Solution

a. There are three different patterns, suggesting that the mutants fall into three groups:
Group A (mutants 2 and 4) can grow only if proline is supplied.
Group B (mutants 1, 5, and 6) can grow if proline *or* GSA is supplied.
Group C (mutant 3) can grow if proline, *or* GSA, *or* glutamine is supplied.

b. These three groups probably represent mutations in three different genes.

c. The different genes code for three enzymes (enzymes A, B and C, corresponding to genes *A*, *B*, and *C*) in a biochemical pathway for the synthesis of proline. The pathway must be as follows, because only this arrangement explains the pattern of restoration of growth by supplementation after a block in the pathway.

$$\xrightarrow{\text{C}} \text{glutamine} \xrightarrow{\text{B}} \text{GSA} \xrightarrow{\text{A}} \text{proline}$$

PROBLEMS

Basic Problems

1. a. In how many cases in the genetic code would you fail to know the amino acid specified by a codon if you know only the first two nucleotides of the codon?

b. In how many cases would you fail to know the first two nucleotides of the codon if you know which amino acid is specified by it?

2. a. Use the codon dictionary (Figure 3-20) to complete the following table. Assume that reading is from left to right and that the columns represent transcriptional and translational alignments.

<table>
<tr><td>C</td><td></td><td></td><td></td><td></td><td></td><td></td><td></td><td></td><td></td><td></td><td></td><td rowspan="2">DNA double helix</td></tr>
<tr><td></td><td></td><td></td><td></td><td></td><td></td><td>T</td><td>G</td><td>A</td><td></td><td></td><td></td></tr>
<tr><td></td><td>C</td><td>A</td><td></td><td></td><td></td><td>U</td><td></td><td></td><td></td><td></td><td></td><td>mRNA transcribed</td></tr>
<tr><td></td><td></td><td></td><td></td><td></td><td></td><td></td><td></td><td></td><td>G</td><td>C</td><td>A</td><td>Appropriate tRNA anticodon</td></tr>
<tr><td colspan="3"></td><td colspan="3">TRP</td><td colspan="3"></td><td colspan="3"></td><td>Amino acids incorporated into protein</td></tr>
</table>

b. Label 5′ and 3′ ends of DNA and RNA, and amino and carboxyl ends of protein.

3. Consider the following segment of DNA:

5′ GCTTCCCAA 3′
3′ CGAAGGGTT 5′

If the top strand is the template strand used by RNA polymerase,

a. draw the RNA transcribed.

b. label its 5′ and 3′ ends.

c. draw the corresponding amino acid chain.

d. label its amino and carboxyl ends.

Repeat, assuming the bottom strand is the template strand.

4. Invent symbols for the following gene mutations and their wild-type alleles (and explain your choices of symbols):

a. A *Neurospora* mutation auxotrophic for the amino acid histidine (wild type, prototrophic for histidine)

b. A recessive *Drosophila* mutation with yellow body (wild type, gray)

c. A dominant corn mutation that produces purple pigment (the wild type does not)

d. A null mutation in a haplo-insufficient gene that normally determines hairs on the stem of petunia plants

e. Write the genotype of a heterozygote in cases b through d.

5. List three examples of base complementarity in gene action.

6. List three examples of proteins that act on nucleic acids.

7. A linear plasmid contains only two genes, which are transcribed in opposite directions; each one from the end, toward the center of the plasmid. Draw diagrams that show

a. the plasmid DNA, showing 5′ and 3′ ends of the nucleotide strands.

b. the template strand for each gene.

c. the positions of the transcription initiation sites.

d. the transcripts, showing 5′ and 3′ ends.

e. the positions of the start and stop codons.

f. the encoded amino acid chains aligned with the transcripts and showing carboxyl and amino ends.

8. The following table shows the ranges of enzymatic activity (in units we need not worry about) observed for enzymes involved in two recessive metabolic diseases in humans. Similar information is available for many metabolic genetic diseases.

		RANGE OF ENZYME ACTIVITY		
Disease	Enzyme involved	Patients	Heterozygotes	Normal individuals
Acatalasia	Catalase	0	1.2–2.7	4.3–6.2
Galactosemia	Gal-1-P uridyl transferase	0–6	9–30	25–40

a. Of what use is such information to a doctor who is counseling prospective parents who know that these diseases have occurred in their families?

b. Indicate any possible sources of uncertainty in trying to advise prospective parents.

c. Discuss the concept of dominance in the light of such data.

9. In a certain plant, the flower petals are normally purple. Two recessive mutations arise in separate plants and are found to be in different genes. Mutation 1 (*m1*) gives blue petals when homozygous (*m1 / m1*). Mutation 2 (*m2*) gives red petals when homozygous (*m2 / m2*). Biochemists working on the synthesis of flower pigments in this species have already described the following pathway:

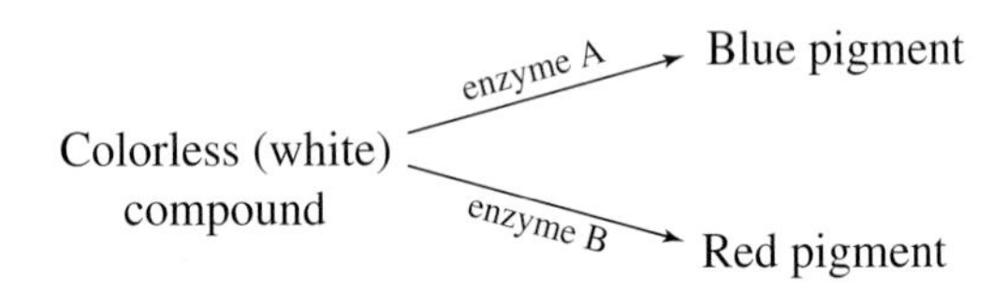

a. Which mutant would you expect to be deficient in enzyme A activity?

b. A plant has the genotype $m1^+$ / *m1* . $m2^+$ / *m2* (it is heterozygous for both genes). What would you expect its phenotype to be?

c. Why are these mutants recessive?

10. In sweet peas, the synthesis of purple anthocyanin pigment in the petals is controlled by two genes, *B* and *D*. The pathway is

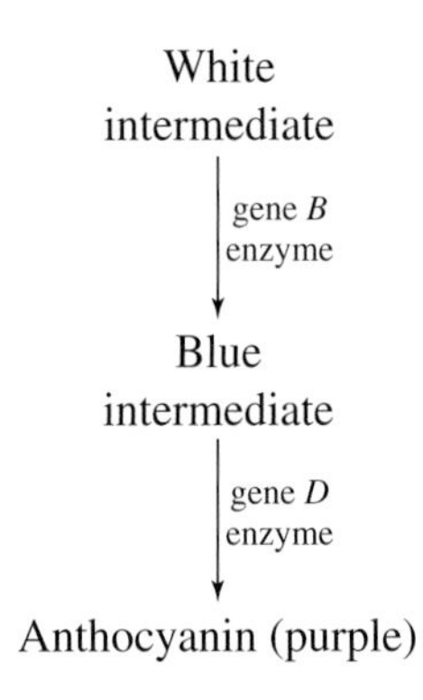

a. What petal color would you expect in a plant homozygous for a recessive mutation that renders it unable to catalyze the first reaction?

b. What petal color would you expect in a plant homozygous for a recessive mutation that renders it unable to catalyze the second reaction?

11. A certain *Drosophila* protein-coding gene has one intron. If a large sample of null alleles of this gene are examined, will any of the mutant sites be expected

a. in the exons?

b. in the intron?

c. in the promoter?

d. in the intron-exon boundary?

For each case in which your answer is yes, indicate whether any transcript, polypeptide, or both will be detectable.

12. Normally the thyroid growth hormone thyroxine is made in the body by an enzyme as follows:

$$\text{Tyrosine} \xrightarrow{\text{enzyme}} \text{thyroxine}$$

If the enzyme is deficient, the symptoms are called *genetic goiterous cretinism* (GGC), a rare syndrome consisting of slow growth, enlarged thyroid (called a *goiter*), and mental retardation.

a. If the normal allele is haplo-sufficient, would you expect GGC to be inherited as a dominant or a recessive phenotype? Explain.

b. Speculate on the nature of the GGC-causing allele, comparing its DNA sequence with the normal allele. Under your model show why it results in an inactive enzyme.

c. How might the symptoms of GGC be alleviated?

d. At birth infants with GGC are perfectly normal and develop symptoms only later. Why do you think this is so?

13. In Figure 3-5a, at which positions have former nucleotide triphosphates just liberated inorganic diphosphate?

14. Compare and contrast the process whereby *information* becomes *form* in an organism and in house construction.

15. Try to think of exceptions to the statement that "when you look at an organism, what you see is either a protein or something that has been made by a protein."

Challenging Problems

16. Which anticodon would you predict for a tRNA species carrying isoleucine? Is there more than one possible answer? If so, state any alternative answers.

17. The following data represent the base compositions of double-stranded DNA from two different bacterial species and their RNA products obtained in experiments conducted in vitro:

Species	(A + T)/(G + C)	(A + U)/(G + C)	(A + G)/(U + C)
Bacillus subtilis	1.36	1.30	1.02
E. coli	1.00	0.98	0.80

a. From these data, can you determine whether the RNA of these species is copied from a single strand or from both strands of the DNA? How? Drawing a diagram will make it easier to solve this problem.

b. Explain how you can tell whether the RNA itself is single-stranded or double-stranded.

18. Several yeast mutants are isolated, all of which require compound G for growth. The compounds (A to E) in the biosynthetic pathway to G are known, but not their order in the pathway. Each compound is tested for its ability to support the growth of each mutant (1 to 5). In the following table, "+" indicates growth and "−" indicates no growth:

	COMPOUND TESTED					
	A	B	C	D	E	G
Mutant 1	−	−	−	+	−	+
2	−	+	−	+	−	+
3	−	−	−	−	−	+
4	−	+	+	+	−	+
5	+	+	+	+	−	+

a. What is the order of compounds A to E in the pathway?

b. At which point in the pathway is each mutant blocked?

19. Twelve null alleles of an intronless *Neurospora* gene are examined, and it is found that all the mutant sites are clustered in a region occupying the central third of the gene. What might be the explanation for this finding?

20. An albino plant mutant is obtained that lacks red anthocyanin pigment, normally made by an enzyme P. Indeed the tissue of the mutant plant lacks all detectable activity for enzyme P. However, a study using antibodies that bind specifically to the enzyme shows that in the cells homozygous for the mutant the antibody still binds to a protein. How is this possible?

21. Explain fully the following observations on two genes in *Drosophila*. In all cases invent allele symbols.

a. Twenty units of enzyme E are normally found in homozygous wild-type cells. A homozygous mutant shows zero units. A heterozygote is mutant in appearance and has 10 units of E per cell.

b. Thirty units of enzyme F are normally found in homozygous wild-type cells. A homozygous mutant shows zero units. A heterozygote is wild-type in appearance and has 15 units of F per cell.

22. In a hypothetical diploid organism, squareness of cells is due to a threshold effect: more than 50 units of "square factor" per cell will produce a square phenotype, and fewer than 50 units will produce a round phenotype. Allele *sf* is a functional gene that causes the synthesis of the square factor. Each *sf* allele contributes 40 units of square factor; thus, *sf* / *sf* homozygotes have 80 units and are phenotypically square. A mutant allele *sn* arises; it is nonfunctional, contributing no square factor at all.

a. Which allele will show dominance, *sf* or *sn*?

b. Are functional alleles necessarily always dominant? Explain your answer.

c. In a system such as this one, how might a specific allele become changed in the course of evolution so that its phenotype shows recessive inheritance at generation 0 and dominant inheritance at a later generation?

23. In babies with PKU, a special diet low in phenylalanine allows development to continue without retardation. Indeed it has been found that after the child's nervous system has developed, the patient can be taken off the special diet. However, tragically, many women who had been born with PKU developed normally under the special diet but stopped the diet in adulthood and then gave birth to babies who were born mentally retarded. Giving the special diet to these babies had no effect on them.

a. Why do you think the babies of these mothers were born retarded?

b. Why did the special diet have no effect on these babies?

c. Explain the reason for the different response to the low-phenylalanine diet by the babies born with PKU and babies of mothers born with PKU.

d. Propose a treatment that might allow mothers with PKU to have unaffected children.

e. Write a short essay on PKU, integrating concepts at the genetic, diagnostic, enzymatic, and physiological levels.

24. A single nucleotide addition and a single nucleotide deletion approximately 15 sites apart in the DNA cause a protein change in sequence from

Lys—Ser—Pro—Ser—Leu—Asn—Ala—Ala—Lys

to

Lys—Val—His—His—Leu—Met—Ala—Ala—Lys

a. What are the old and the new mRNA nucleotide sequences? (Use the codon dictionary in Figure 3-20.)

b. Which nucleotide has been added and which has been deleted?

(Problem 24 is from W. D. Stansfield, *Theory and Problems of Genetics.* McGraw-Hill, 1969.)

25. Four yeast mutants were obtained that all lacked activity of a specific enzyme E, known to be encoded by one gene. All mutations were shown to be in that gene. Protein was extracted from the mutants and wild-type control and submitted to electrophoresis. (Electrophoresis is a technique in which macromolecules such as DNA, RNA, and protein are separated by size and charge in a gelatinous medium called a *gel,* using a strong electric field across the slab.) After electrophoretic separation, the gel was processed so that an antibody specific for enzyme E could be applied. The antibody was chemically attached to a colored dye. Thus the position of enzyme E could be visualized. The results are shown below:

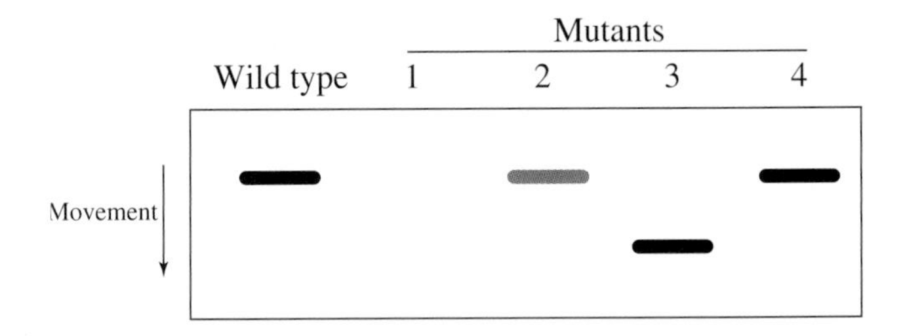

Provide a molecular explanation for the results in each lane (column of migration), and show how each is compatible with the observed lack of enzyme activity.

Exploring Genomes: A Web-Based Bioinformatics Tutorial

Learning to Use ENTREZ

The ENTREZ program at NCBI is an integrated search tool that links together a variety of databases that have different types of content. In the Genomics tutorial at Web site www.whfreeman.com/mga we will use it to look up the dystyrophin gene discussed in Chapter 2 and find research literature references, the gene sequence as well as conserved domains, the equivalent gene from a variety of organisms other than human, and the chromosome map of its location.

THE TRANSMISSION OF DNA AT CELL DIVISION 4

Key Concepts

1. DNA replication is a precursor to cell division in asexual and sexual cycles.
2. DNA is replicated by unwinding the double helix and using the exposed strands as templates for synthesis of new strands; hence, each daughter molecule is half old and half new.
3. The DNA-synthesizing enzyme adds nucleotides at 3′ growing points of the new DNA strands being synthesized.
4. On one template, DNA synthesis is continuous; on the other DNA synthesis is in short bursts in the opposite direction.
5. Replication of the ends of linear chromosomes requires special molecular devices to counteract the tendency of the DNA to shorten.
6. DNA replication underlies chromosome replication.
7. Chromosomes are partitioned into daughter cells during cell division by molecular strings, which attach to a specialized region of each chromosome.
8. When asexual eukaryotic cells divide, the accompanying nuclear division (mitosis) results in two genetically identical daughter cells.
9. Two consecutive divisions of a specialized diploid cell produce haploid sperm and eggs. Two nuclear divisions called meiosis accompany these cell divisions.
10. Meiosis produces haploid cells that show new combinations of genes.

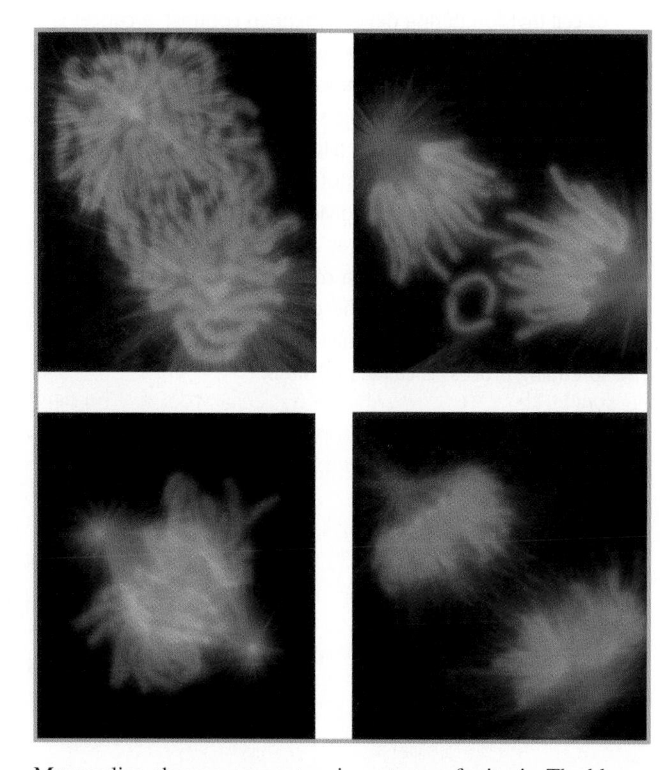

Mammalian chromosomes at various stages of mitosis. The blue color is from a DNA-binding dye, ethidium bromide, and the green color is from a fluorescent antibody that binds to the protein tubulin, a component of the spindle fibers. *(From J. C. Waters, R. W. Cole, and C. L. Rieder,* J. Cell Biol. *122, 1995, 361. Courtesy of C. L. Rieder.)*

Cancer is a disease that touches all our lives in some way. So important is this disease to medicine and genetics that we will later devote an entire chapter to it (Chapter 15). For now, though, we will consider how cancer arises in a population of normal cells. The key is cell division, the topic of this chapter. In normal cells, a genetic program tightly regulates division. Cell division has a major role in the development of an organism, in part because it is the main way that organisms grow. The growth and shaping of the parts of an organism, such as the organs of a human being, and the cessation of growth once the organs are complete present obvious examples of the need for tight regulation of cell division during development. In contrast, a cancer is a localized mass of cells produced by repeated division of cells that have lost this control system. Cancer begins with a series of mutations that accumulate in a single cell, enabling it to escape the normal checks on cell division (Figure 4-1). This mutant cell divides to produce two cells, both of which also carry the mutated alleles that confer the ability to divide in an uncontrolled manner. So both these cells also divide, as do their descendent cells, producing the mass of cells called a *cancer*, or *tumor*. A cell from the tumor may break away from the cell mass and travel to a new site in the body, where it can divide to produce another tumor. The spreading process is called *metastasis*.

The division and spreading process may be repeated many times to produce multiple tumors throughout the body. All the cellular descendents in all these tumors are genetically identical with the original cell in which the cancerous mutations took place.

In this process, the cancer cell capitalizes on one of the fundamental properties of a living cell, which is that it can accurately reproduce itself and its specific set of genetic material. In this case, a cancer cell, once formed, reproduces its cancer-causing property when it divides. In this chapter we will consider the ways in which cells accurately reproduce themselves and their genomes, and we will return to cancer at the end.

Down through the ages, one of the biggest mysteries of heredity has been reproduction of type. Reproduction takes place at many levels. Species reproduce: people always have babies, and cats always have kittens. Furthermore, specific variants within a species are reproduced; we all know examples, such as "she has inherited her father's nose" or "red hair runs in the family." A fertilized human egg reproduces itself by cell division so that all the cells of the fully formed body are genetically identical. In the same way, single-celled organisms such as yeasts and bacteria increase their numbers during population growth by reproducing genetically identical cells.

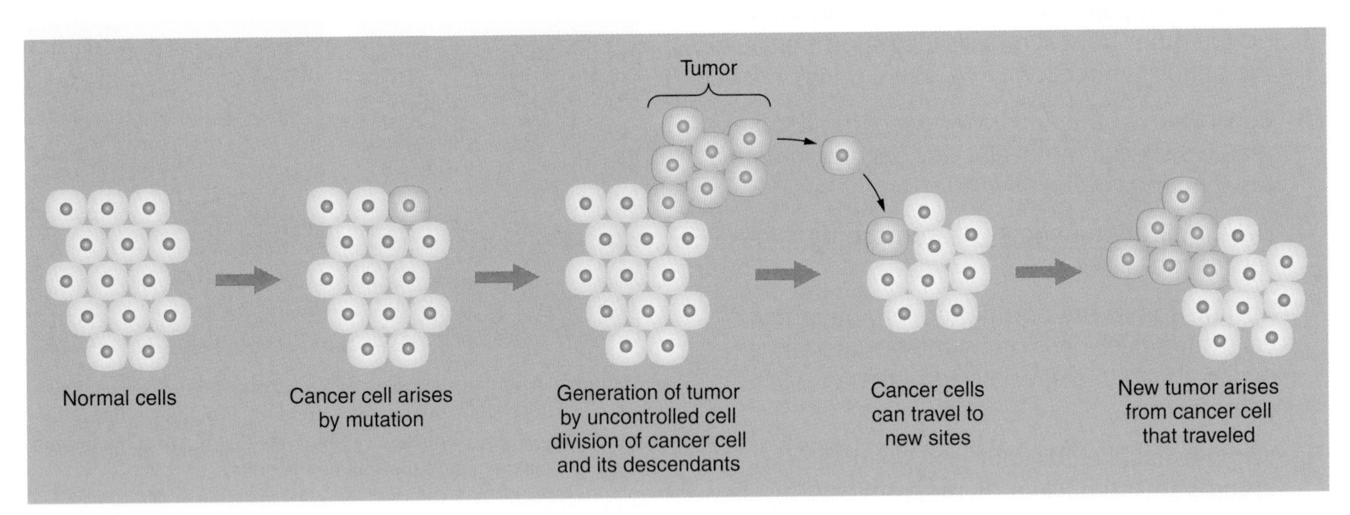

Figure 4-1 The faithful transmission of genotype at cell division, illustrated by the behavior of cancer cells. A cancer arises by mutation in the DNA of a single cell in the body. The cancerous genotype thus produced causes loss of normal control over cell division, resulting in continuous sequential division. The cancer-causing genotype is duplicated at every subsequent cell division, using the very mechanisms of DNA replication and cell division that maintain constancy of a normal cellular genotype. As a result, a tumor of cancer cells is produced. Eventually cancer cells from the original tumor spread throughout the body and form new tumors.

CHAPTER OVERVIEW

In this chapter we study the mechanisms that ensure high-fidelity reproduction at the level of individuals, cells, nuclei, chromosomes, and genes. The general basis for high-fidelity inheritance at all these levels is the division of the fundamental unit of life, the cell. Underlying all types of cell division, in both asexual and sexual cycles, we find nuclear division, chromosomal division, and ultimately the replication of DNA (Figure 4-2).

Replication is the process of making replicas or copies. A molecule of DNA can be replicated by the cell to become

CHAPTER OVERVIEW figure

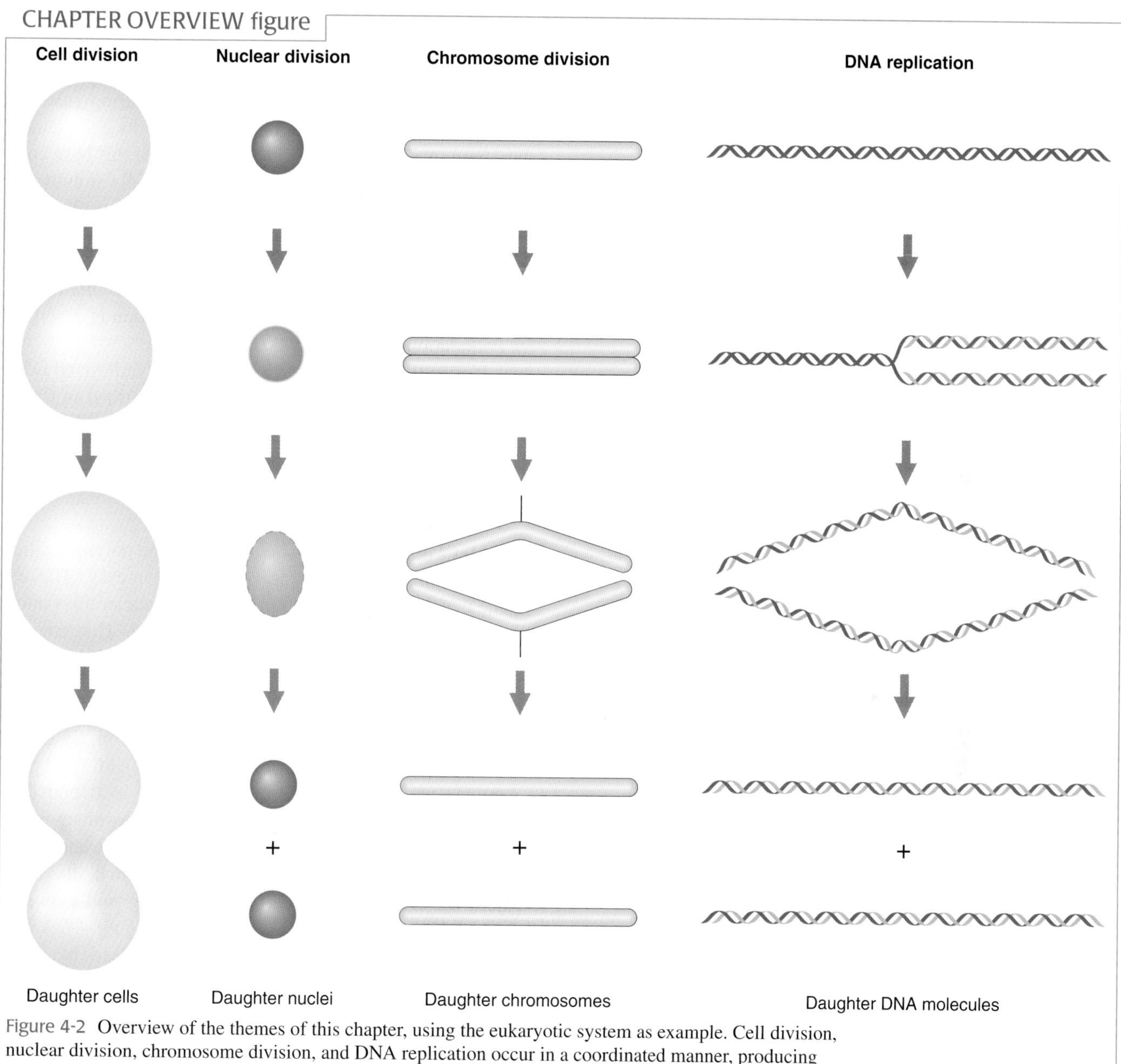

Figure 4-2 Overview of the themes of this chapter, using the eukaryotic system as example. Cell division, nuclear division, chromosome division, and DNA replication occur in a coordinated manner, producing daughter cells with identical genetic material.

two identical "daughter" molecules, and this is how chromosomes can be duplicated, how one cell can become two, how a new multicellular organism develops from a single cell, how populations grow, and how species survive through geological time. The DNA in every one of us is the product of a continuous series of replications going back billions of years.

DNA REPLICATION

In both prokaryotes and eukaryotes, DNA replication occurs as a prelude to cell division. This DNA replication phase is called the **S (synthesis) phase,** referring to the synthesis of copies of DNA. The two daughter DNA molecules formed from replication of a chromosome eventually become chromosomes in their own right in the daughter cells.

As with all the transactions of nucleic acids, the mechanism of DNA replication depends on base complementarity and on the ability of proteins to form specific interactions with DNA of specific sequences.

Semiconservative Replication

Replication of DNA in prokaryotes and eukaryotes is by a **semiconservative** mechanism; that is, the double helix of each daughter DNA molecule contains one strand from the original DNA molecule and one newly synthesized strand.

How is this done? Imagine that the double helix is like a zipper that becomes unzipped, starting at one end (the bottom end in Figure 4-3). We can see that the unwinding of the two strands exposes single bases on each strand, and each exposed base has the potential to pair with free nucleotides in solution. (These free nucleotides come from a pool of mononucleotide triphosphates that has been chemically synthesized in the cytoplasm.) Because base-pairing rules are strict, each exposed base can pair only with its complementary base, A with T and G with C. Because of this base complementarity, each of the two antiparallel single strands acts as a **template** (an alignment guide) to direct the assembly of complementary bases to re-form a double helix identical with the original. Thus, semiconservative replication is directly dependent on base complementarity. The first experimental support for semiconservative replication is described in Foundations of Genetics 4-1 on page 96.

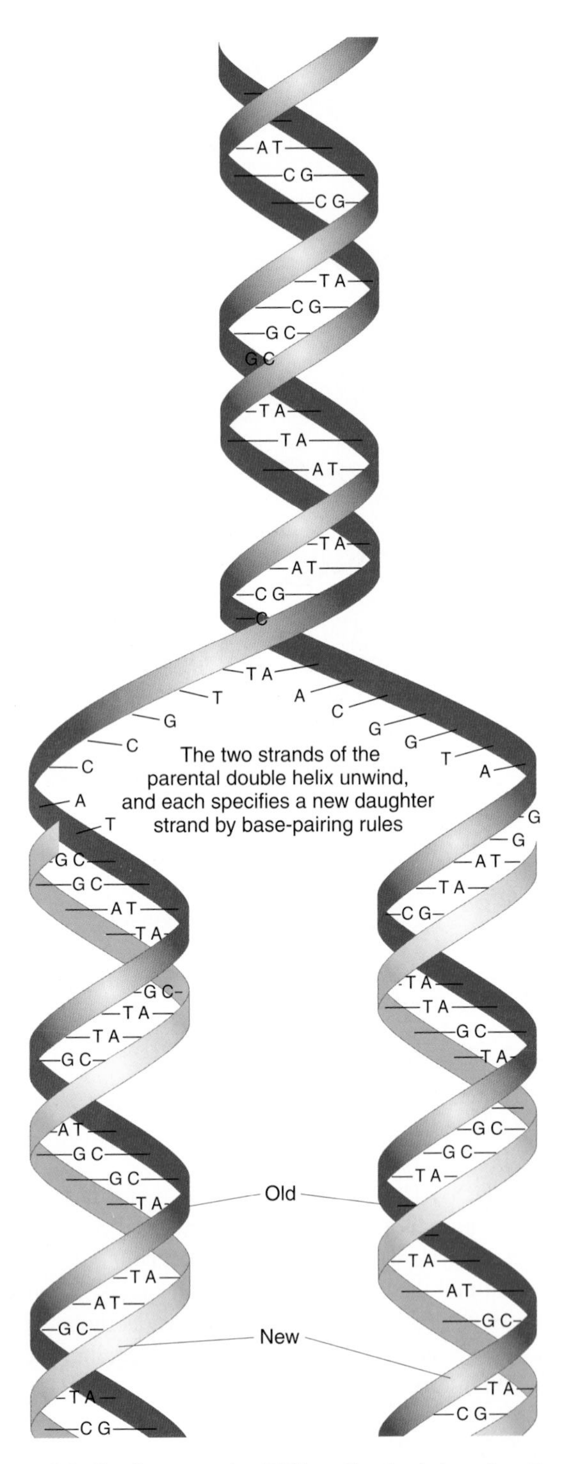

Figure 4-3 Semiconservative DNA replication is based on the hydrogen-bonding specificity of the base pairs. Parental strands, which serve as templates for polymerization, are shown in dark blue. Newly polymerized strands, with base sequence complementary to their templates, are shown in light blue.

Notice that the construction of newly synthesized DNA strands is analogous to RNA synthesis, described in Chapter 3. In each case a single strand of DNA serves as a template to direct the synthesis of a complementary strand.

The Polymerization Process

First, let's explore a general overview of the polymerization events occurring during the replication process. The enzyme that catalyzes the polymerization of nucleotides into a strand of DNA is called **DNA polymerase.** This enzyme works by adding deoxyribonucleotides to the 3′ end of a growing nucleotide chain, using for its template single-stranded DNA (Figure 4-4), which is exposed by localized unwinding of the DNA double helix. Recall from Chapter 3 that RNA polymerase acts in a similar way, adding ribonucleotides to an RNA molecule growing at it 3′ end. The substrates for DNA polymerase are the triphosphate forms of the deoxyribonucleotides, dATP, dGTP, dCTP, and dTTP. DNA polymerase acts at the **replication fork,** the zone where the DNA double helix is unwinding and exposing single DNA strands to act as templates (Figure 4-5a). Because the nucleotide polymerization catalyzed by DNA polymerase is always at the 3′ growing tip, only *one* of the two antiparallel strands can serve as a template for replication in the direction of the replication fork, which is moving toward the strand's 5′ end. For this strand, synthesis can occur in a smooth, continuous manner, in the direction of the fork; the new strand synthesized on this template is called the **leading strand** (Figure 4-5b). Synthesis on the other template also takes place at 3′ growing tips, but this synthesis is in the "wrong" direction, because for this strand, the 5′ → 3′ direction of synthesis is away from the replication fork (see Figure 4-5b). As we will see, the nature of the replication machinery requires that synthesis of *both* strands occur in the region of the replication fork. Therefore, synthesis moving away from the growing fork cannot go on for long. It must be in short segments: polymerase synthesizes a segment, then moves back to the segment's 5′ end, where the growing fork has exposed new template, and begins the process again. These short stretches of newly synthesized DNA are called **Okazaki**

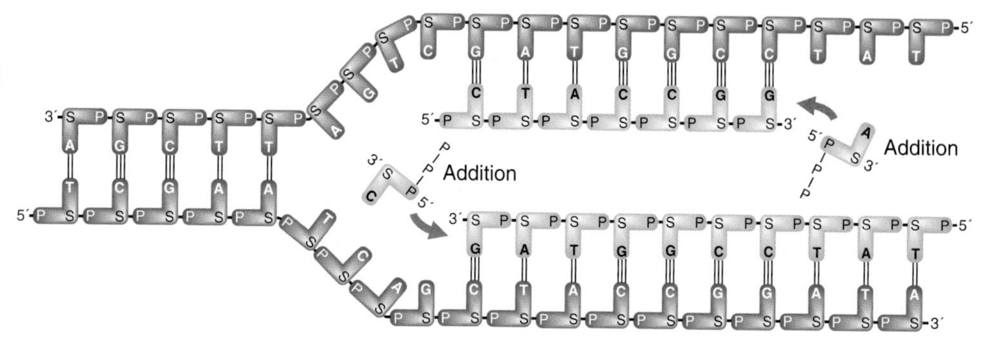

Figure 4-4 A DNA replication fork showing in simplified form how nucleotides are added to the 3′ growing tips on each template, but in opposite directions. See Figure 4-5 for more details.

Animated Art • DNA Replication: The Nucleotide Polymerization Process

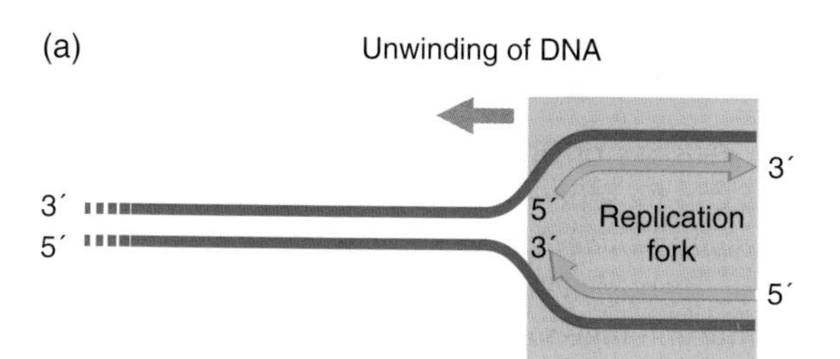

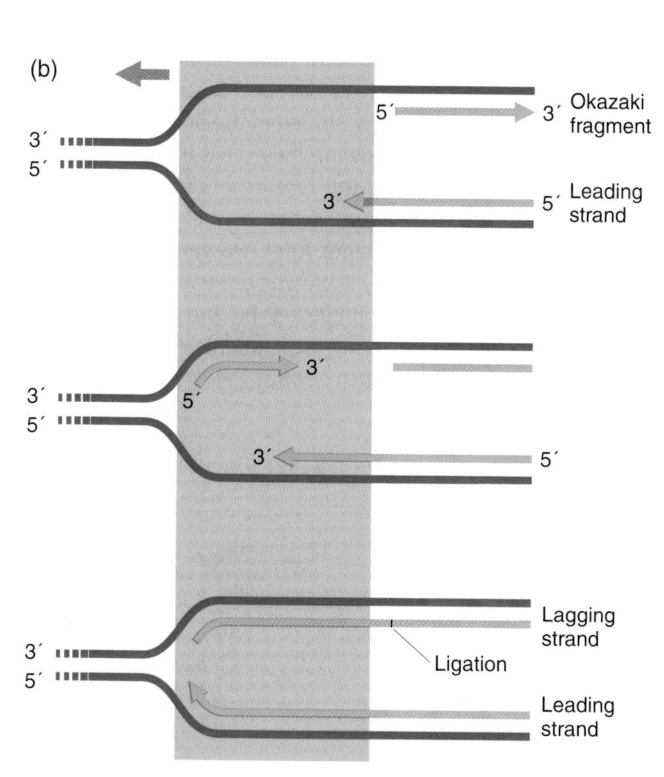

Figure 4-5 The movement of the replication fork (area in yellow box) during DNA synthesis. Part a shows the starting position. Part b shows the replication machinery has moved the length of one Okazaki fragment. 3′ synthesis is moving forward (toward the left) on the leading strand, but "backwards" (toward the right, away from the direction of DNA unwinding) on the lagging strand, forming an Okazaki fragment that will be ligated to a previous one. See the text for a detailed discussion.

fragments. An enzyme, DNA ligase, joins them into one long strand. The new strand thus formed is called the **lagging strand.**

Many active proteins are needed to carry out the general replication process just described. Some of the interacting components of replication in the bacterium *Escherichia coli* are shown in Figure 4-6 on page 97. The main polymerase is DNA polymerase III (pol III), which is the enzyme that catalyzes the 3′ nucleotide addition at the replication fork. At each replication fork there are two polymerase units attached to each other, oriented in opposite directions, to synthesize the two new strands. Polymerases of this type need to add nucleotides to an already existing nucleotide chain. Therefore at the beginning of replication of both the leading and the lagging strands, short stretches of *RNA* are synthesized to act as polymerization starters, or **primers.** On the leading strand, only one initial primer is needed because after the initial priming, continuous addition can use the growing DNA strand as the primer. However, on the lagging strand every Okazaki fragment needs its own primer (see Figure 4-6, a–d). The primers are synthesized by a set of proteins called a **primosome,** of which a central component is an enzyme **primase,** a type of RNA polymerase. Removal of the RNA primers and filling in of the resulting gaps with DNA are performed by a different DNA polymerase, pol I. After pol I has done its job, DNA ligase joins the 3′ end of the gap-filling DNA to the 5′ end of the downstream Okazaki fragment.

Figure 4-7 on page 97 shows some of the complexity of the events occurring at and near the replication fork. The movement of the replication fork is accomplished by the enzyme **helicase,** which breaks hydrogen bonds between the paired bases and unwinds the double helix ahead of the advancing DNA polymerase. The single strands of DNA so created are prevented from rejoining by single-strand binding proteins. As the DNA is unwound, it tends to become supercoiled, a process similar to the one we observe when trying to pull apart two strands of a piece of string or rope. The double helix is returned to its relaxed state by

FOUNDATIONS OF GENETICS 4-1

Meselson and Stahl Show that DNA Replicates Semiconservatively

In their 1953 paper first describing the structure of DNA, Watson and Crick ended with a brief, cryptic comment concerning the implications of complementary base pairing: "It has not escaped our notice that the specific pairing we have postulated immediately suggests a possible copying mechanism for the genetic material." To geneticists at the time, the meaning of this statement was clear: the two halves of the double helix must separate, and because of the specificity of base pairing, they both act as templates for the polymerization of new strands, thus forming two identical daughter double helices. This type of replication became known as *semiconservative.* A beautiful idea, but many beautiful ideas have been knocked down by ugly facts, so it was necessary to test the proposal. Matthew Meselson and Frank Stahl performed the first critical test.

Meselson and Stahl later wrote

> We met each other as graduate students at Woods Hole in 1954, the second summer after the Watson-Crick model had been announced. Our first conversations concerned the solutions to certain integrals describing cross-reactivation of UV-induced phages, which one of us was then studying. Later we got to talking about ways to test the prediction of semiconservative duplication of DNA. Perhaps the more intriguing ideas have been forgotten, but we remember talking about the possible use of density as a label for parental DNA molecules.

The idea was to allow parental DNA molecules containing components of one density to replicate in medium containing components of different density. If DNA replicated semiconservatively, the daughter molecules should be half old and half new and therefore of intermediate density. Initial attempts to use this approach on phage (bacterial viruses) using the heavy hydrogen isotope deuterium and the bromine-containing analog of thymine 5-bromouracil (5BU) gave uninterpretable results, but

> Finally we abandoned phage and 5BU and turned to bacteria *[E. coli]* and the heavy isotope of nitrogen (^{15}N). The second and third experiments along these lines worked beautifully; so we renumbered them 1 and 2 and began to write the paper.

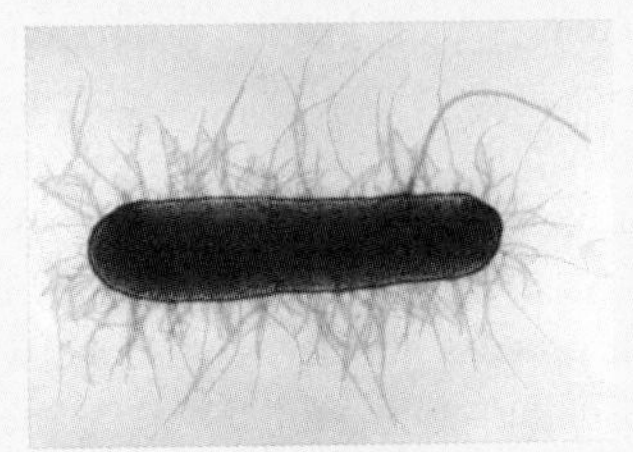

Escherichia coli. (Courtesy of S. Abraham and E. H. Beachey.)

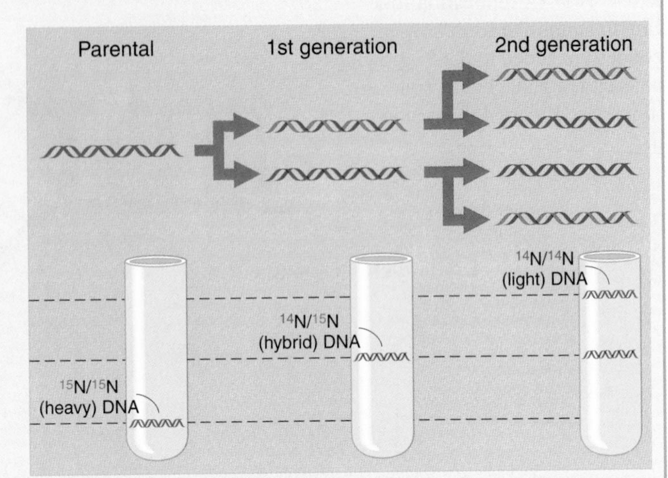

Predictions of semiconservative model.

Different densities of DNA can be distinguished because the molecules can be separated from each other by a procedure called *cesium chloride density gradient centrifugation.* In this procedure, the high gravitational forces generated in the high-speed ultracentrifuge make a density gradient of cesium chloride. DNA centrifuged with the cesium chloride forms a band at a position identical with its density in the gradient. DNA of different densities will form bands at different places. Cells initially grown in the heavy isotope ^{15}N showed DNA of high density. This DNA is shown in red in the first panel of the figure shown above. After growing these cells in the light isotope ^{14}N for one generation, the researchers found that the DNA was of intermediate density, shown half red (^{15}N) and half blue (^{14}N) in the central panel. After two generations both intermediate- and low-density DNA was observed (third panel of the figure), precisely as predicted by the Watson-Crick model. The paper reporting the experiments Meselson and Stahl had done was published in 1958.

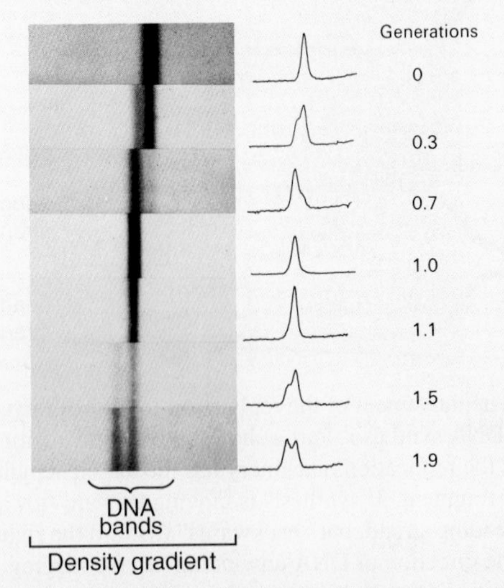

Bands of DNA in gradients; bottoms of tubes are toward the right. (From M. Meselson and F. W. Stahl, *Proc. Natl. Acad. Sci.* 44, 1958, 671.)

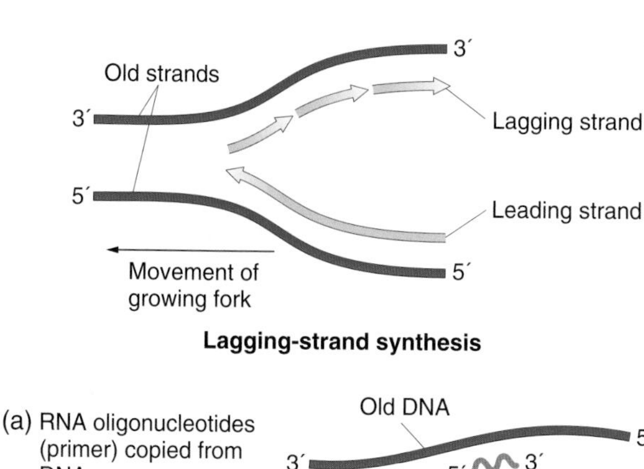

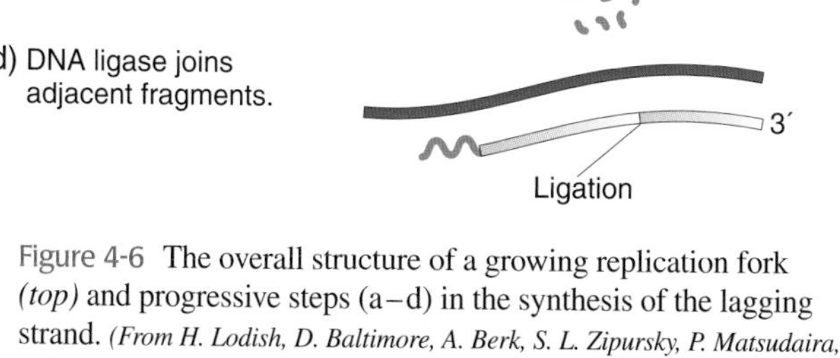

Animated Art • DNA Replication: Coordination of Leading- and Lagging-Strand Synthesis

Figure 4-6 The overall structure of a growing replication fork *(top)* and progressive steps (a–d) in the synthesis of the lagging strand. *(From H. Lodish, D. Baltimore, A. Berk, S. L. Zipursky, P. Matsudaira, and J. Darnell,* Molecular Cell Biology, *3d ed. © 1995 by Scientific American Books, Inc.)*

the action of another enzyme, **gyrase,** which is a type of **topoisomerase** (not shown in the diagram). This class of enzymes can cut and rejoin DNA strands, allowing them to "pass through" each other, like a magician interlocking and separating steel rings.

Note from Figure 4-7 that because the two DNA polymerase molecules are attached to each other, as the lagging-strand polymerase produces an Okazaki fragment by addition of nucleotides to the 3′ end of the lagging strand, it must push this fragment ahead of the polymerases, forming a loop, which is shown in the diagram. Once the Okazaki fragment has been completed, the loop is released and the polymerase starts synthesizing another Okazaki fragment, beginning at the next primer. In this way the attached polymerases synthesizing the leading and lagging strands can together follow the advancing replication fork.

Origins of Replication

In the genomes of prokaryotes and eukaryotes, replication begins from specific nucleotide sequences that are recognized by the replication apparatus; these are called **origins of replication.** Synthesis then proceeds *bidirectionally,* with two forks moving outward in opposite directions, as shown in Figure 4-8a on the next page. The replicated double helices that are being produced by each origin of replication elongate and eventually join each other. When replication of the two strands is complete, two identical **daughter molecules** of DNA result. In the replication of a eukaryotic chromosome, each daughter DNA molecule acquires its own set of histone proteins: at this stage the two copies, called **sister chromatids,** become visible under the microscope (Figure 4-8b). The term *chromatid* is applied only temporarily. Chromatids are in fact bona fide chromosomes, and each will be called a chromosome after cell division.

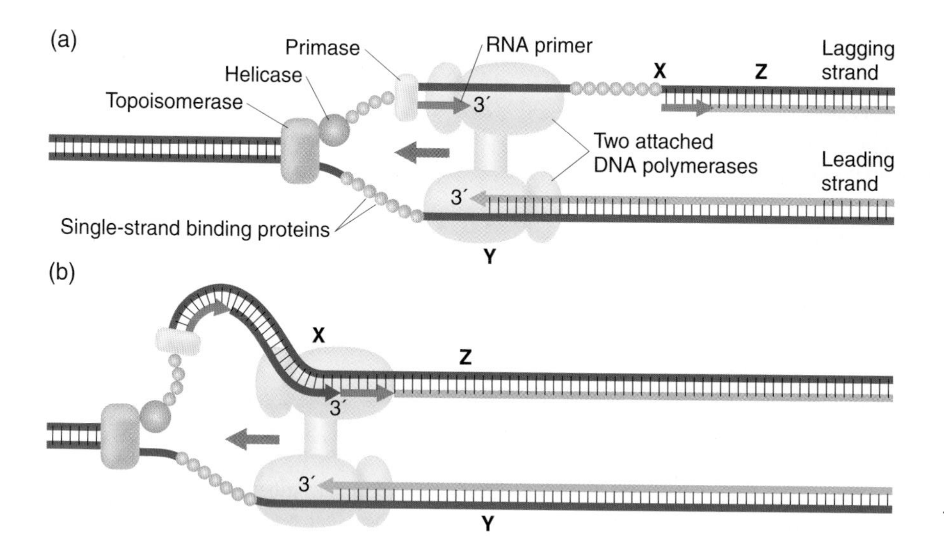

Figure 4-7 A DNA replication fork (a) at the beginning and (b) at the end of synthesis of a segment the size of an Okazaki fragment. Two DNA polymerase units attached to each other move with the replication fork, one producing the leading strand and one the lagging strand. On the lagging strand the growing Okazaki fragment is pushed out to form a loop that is later released, allowing the DNA polymerase to commence synthesis of the next Okazaki fragment. A helicase unwinds the double helix. A primase synthesizes an RNA segment to prime synthesis of an Okazaki fragment. Special binding proteins adhere temporarily to single-stranded regions of DNA. X, Y, and Z represent fixed reference points on the DNA. *(Adapted from H. Lodish et al.,* Molecular Cell Biology, *4th ed. © 2000 by W. H. Freeman and Co.)*

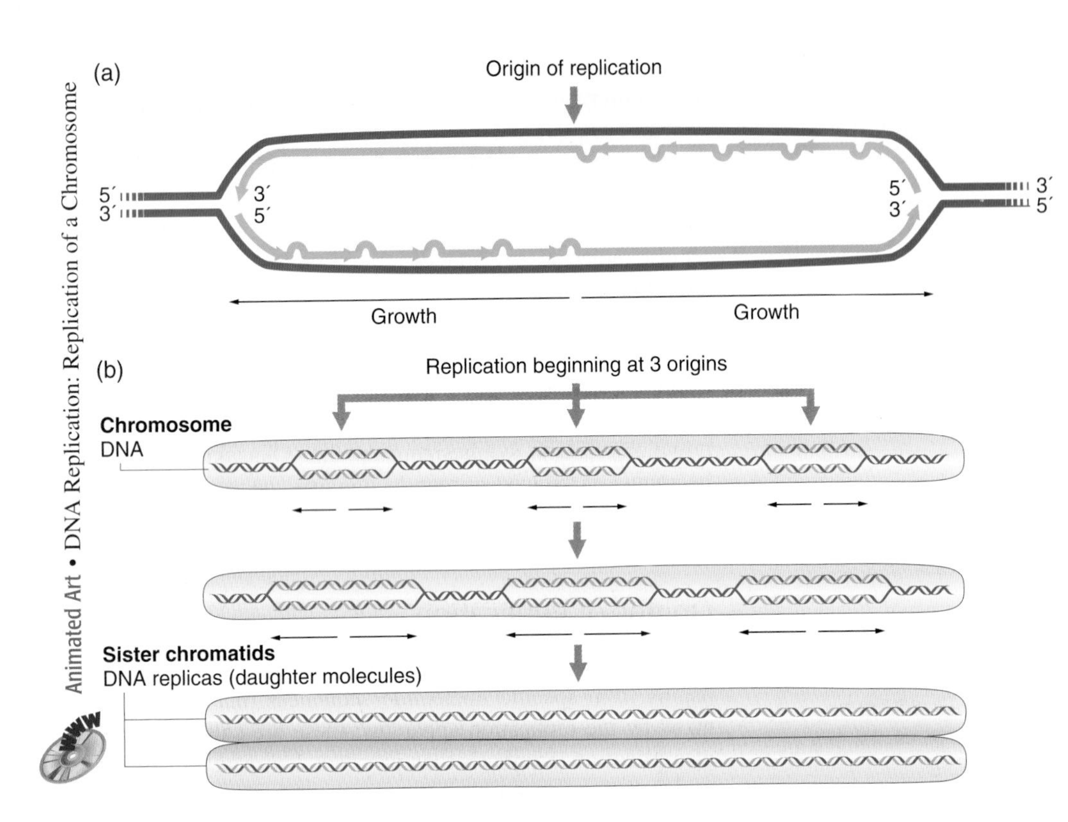

Figure 4-8 The bidirectional nature of DNA replication. Black arrows show the direction of growth of daughter DNA molecules. (a) Starting at the origin, DNA polymerases move outward in both directions. Long blue arrows represent leading strands and short joined blue arrows represent lagging strands. (b) How replication proceeds at the chromosome level. Three origins of replication are shown in this example.

MESSAGE

DNA replicates semiconservatively. With separated single strands of the double helix as templates, nucleotides are polymerized at the 3′ ends of new chains. Addition proceeds continuously in the leading strand but occurs in short bursts in the lagging strand.

Differences in the Replication of Linear and Circular DNA Molecules

In Chapter 2 we saw that in most bacteria the chromosome is a circular DNA molecule. Prior to bacterial cell division, bidirectional replication starting from one or more origins of replication proceeds until the complete circle has been replicated. Intermediates in the replication process can be visualized using specialized labeling techniques, and as expected the intermediates often have the appearance illustrated in Figure 4-9. These are called theta (θ) structures because of their similarity to the shape of that Greek letter.

Many plasmids (extragenomic elements) are also circular. In circular plasmids (and some other circular molecules), a mode of replication called **rolling circle replication** has been demonstrated (Figure 4-10). In this mode, one of the strands of the double helix remains circular and acts as the template strand for continuous replication. The 3′ continuous addition of new nucleotides displaces the 5′ end as a single-stranded tail, which acts as a template for discontinuous synthesis of Okazaki fragments. Synthesis often con-

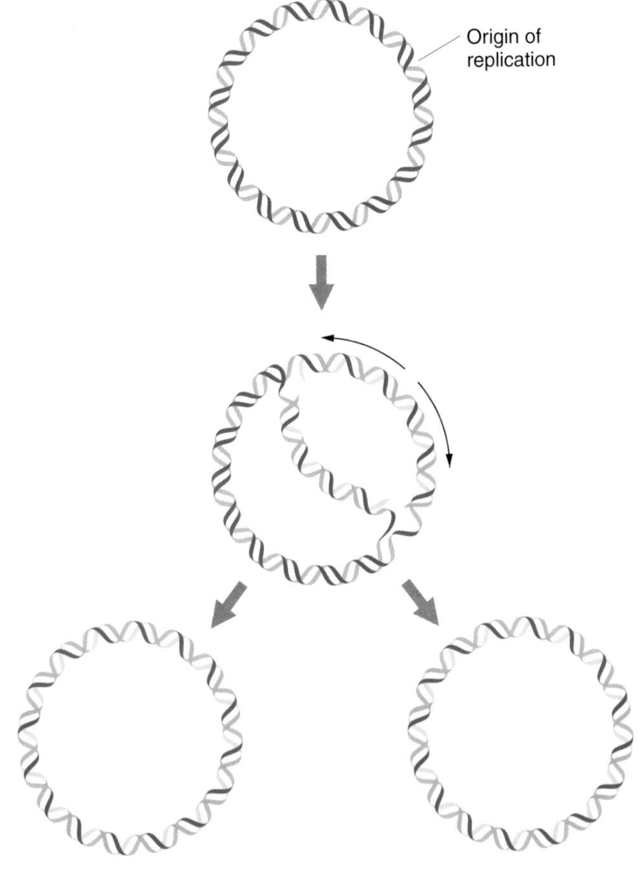

Figure 4-9 Bidirectional replication of a circular bacterial chromosome.

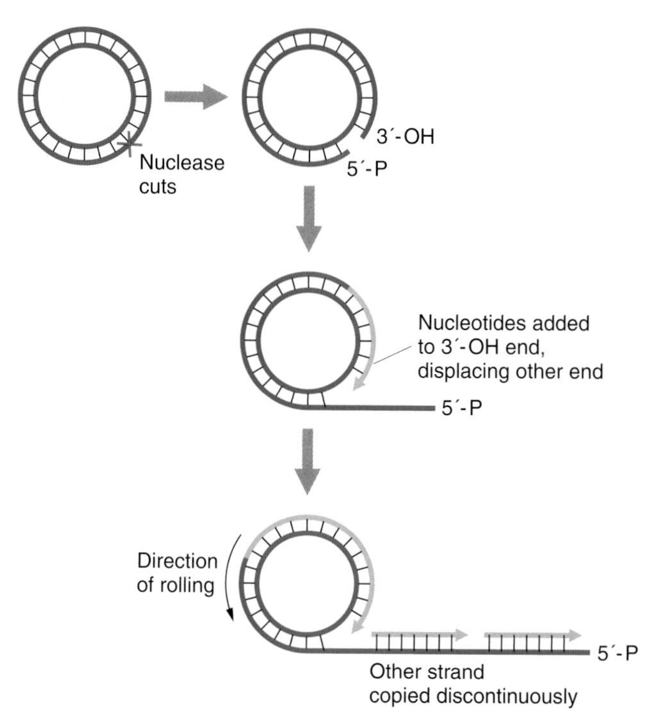

Figure 4-10 Rolling circle replication. Newly synthesized DNA is light blue. The displaced strand produced by 3′ addition around the circle is used as a template for discontinuous replication. *(After D. L. Hartl and E. W. Jones,* Genetics, Principles and Analysis, *4th ed. Jones and Bartlett.)*

tinues beyond a single chromosomal unit, resulting in a chain of head-to-tail copies. These are cut and rejoined to make new circular molecules.

Replication of the linear DNA molecule in a eukaryotic chromosome proceeds bidirectionally from numerous replication origins, as was shown in Figure 4-8. This process replicates most of the chromosomal DNA, but there is an inherent problem in replicating the two ends of linear DNA molecules, the regions called the **telomeres.** Continuous synthesis on the leading strand can proceed right to the very tip of the template. However, lagging-strand synthesis requires primers ahead of the process, so when the last primer is removed, a single-stranded tip remains in one of the daughter DNA molecules (Figure 4-11). If the daughter chromosome with this DNA molecule replicated again, this short single strand would become a shortened double-stranded molecule. At each replication cycle the telomere would continue to shorten, leading to loss of essential coding information.

Cells have devised a specialized system to prevent this loss. They add multiple copies of a simple noncoding sequence to the DNA at the chromosome tips in order to prevent shortening. For example, in the ciliate *Tetrahymena,* copies of the sequence TTGGGG are added to the 3′ end of each chromosome, and in humans it is TTAGGG.

The addition of this sequence is accomplished by the enzyme called **telomerase.** The telomerase protein carries a

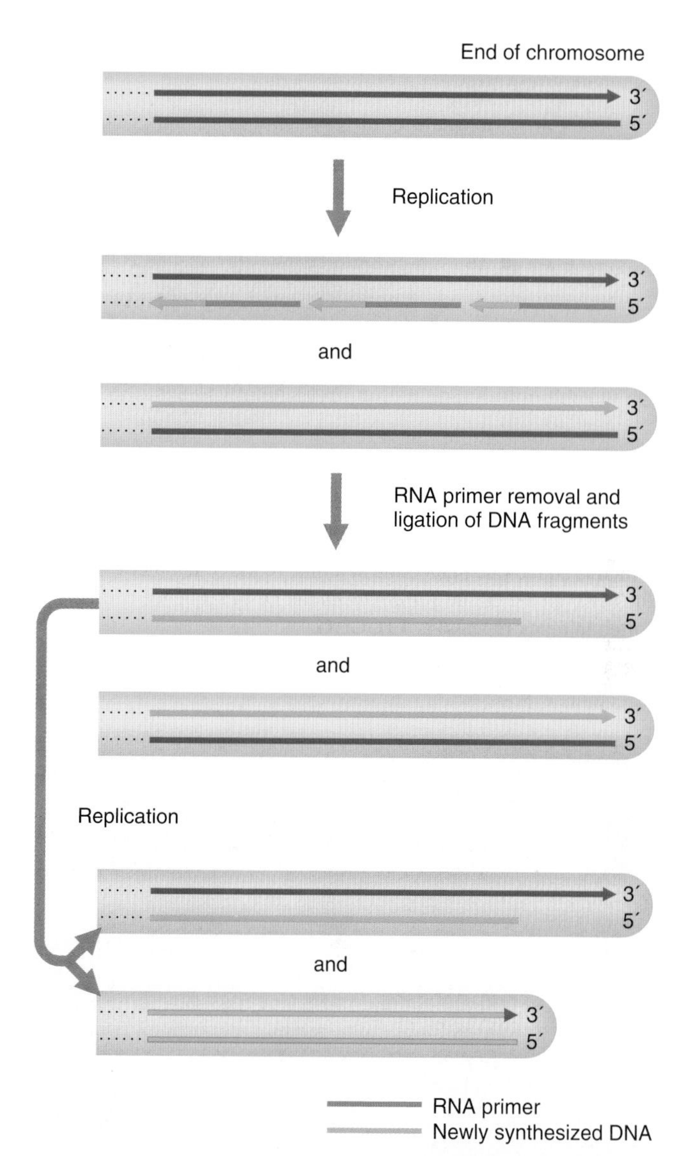

Figure 4-11 The replication problem inherent at the ends of linear chromosomes. Once the primer for the last section of the lagging strand is removed, there is no way to polymerize that segment, and a shortened chromosome would result when the chromosome containing the incomplete strand replicates.

small RNA molecule, part of which acts as a template for the polymerization of the telomeric repeat unit that is added to the 3′ end. In *Tetrahymena,* the RNA is 3′-AACCCC-5′, which acts as the template for the 5′-TTGGGG-3′ repeat unit (Figure 4-12 on the next page). Several methods have been proposed by which this additional DNA is able to act as a template for synthesis on the lagging strand. This lengthening of the DNA molecule by telomerase creates noncoding DNA that can be "sacrificed" to the replication process. Figure 4-13 on the next page demonstrates the positions of the telomeric DNA through a special chromosome labeling technique.

An age-dependent decline in telomere length has been found in several human somatic tissues. In addition, human fibroblast cells in culture show progressive shortening of

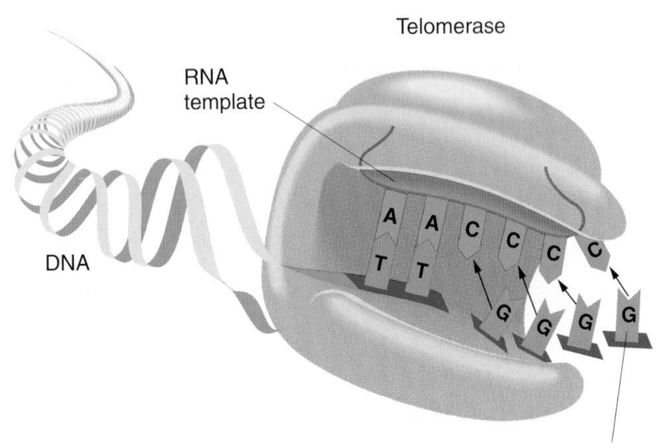

Figure 4-12 Telomerase carries a short RNA molecule that acts as a template for the addition of the complementary DNA sequence, which is added one nucleotide at a time to the 3′ telomeric end of the helix. In the ciliate *Tetrahymena* the DNA sequence added (often in many repeats) is TTGGGG.

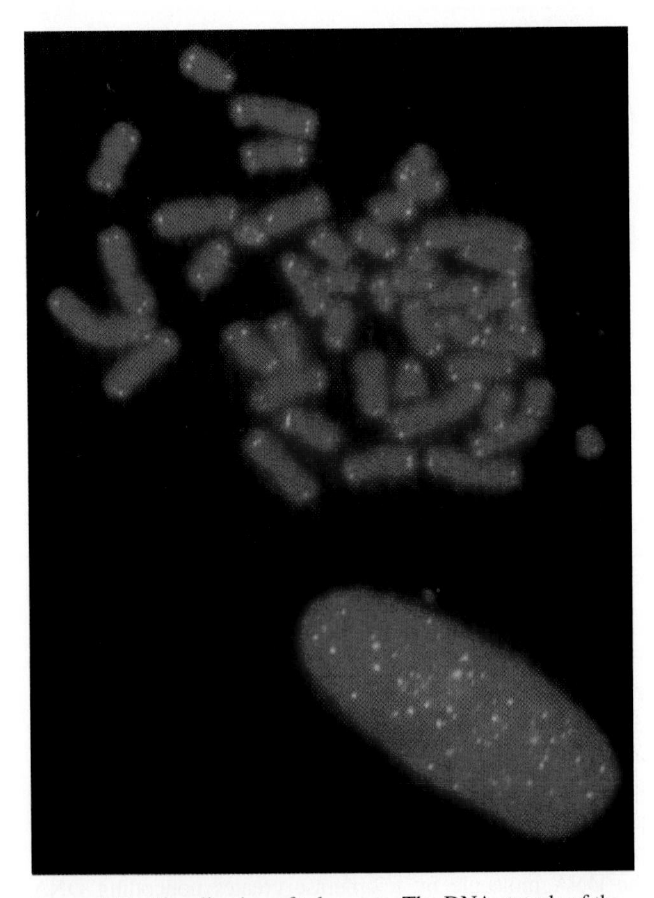

Figure 4-13 Visualization of telomeres. The DNA strands of the chromosomes are slightly separated and treated with a short segment of single-stranded DNA that specifically binds to the telomeres by complementary base pairing. The short segment has been coupled to a substance that can fluoresce yellow under the microscope. The chromosomes have formed sister chromatids. An unbroken nucleus is shown at the bottom. *(Robert Moyzis.)*

telomeres until the eventual death of the cells. Apparently the efficiency of telomerase declines with age. Such observations have led to a telomere theory for the mechanism of aging, in which shortening results in progressive loss of genes of essential function. The validity of this theory is now being tested.

CELL DIVISION: AN OVERVIEW

DNA replication must now be placed into a cellular context. Cell division is the basis for all organismal reproduction. In order for a cell to divide, the genome must also divide, so DNA replication is a necessary accompaniment to cell division. The main types of cell division are shown in Figure 4-14. There are two broad classes. The first is division of asexual cells, which occurs in both prokaryotes and eukaryotes. Such division produces two identical daughter cells containing the same genome as the original cell. Hence this type of division requires molecular mechanisms to conserve the genomic content of the cells. The other type of division is found only in the sexual cycles of eukaryotes. It is actually two consecutive cell divisions resulting in cells whose genomic content is halved. These two broad types of division are discussed in detail in the following sections.

CELL DIVISION THAT CONSERVES THE GENETIC MATERIAL

Prokaryotes

In prokaryotes, reproduction is by simple cell division. Formally this can be termed *asexual division,* since the cells are not part of any sexual cycle. Division produces two identical daughter cells from one progenitor cell (see Figure 4-14, top panel). Repeated cell duplication of this type is the basis of the exponential growth of bacterial cell populations, since the number of cells goes from 1 → 2 → 4 → 8, and so on. Prior to bacterial cell division, the bacterial DNA is replicated during S phase, and two full circular genomes result.

As the cell divides, the daughter cells each acquire one of the two daughter circular DNA molecules. The precise way in which the two daughter DNA molecules are partitioned into the two daughter cells is only now beginning to be understood. Recent studies in which the origin of replication is labeled with a fluorescent dye show that the pulling apart of the two daughter DNA molecules occurs well before the cell expands and begins division. Hence it has been proposed that there is an active motor system that drives the process along some type of track, as shown in Figure 4-15 on page 102.

Cell division that conserves the genetic content (asexual cell division)

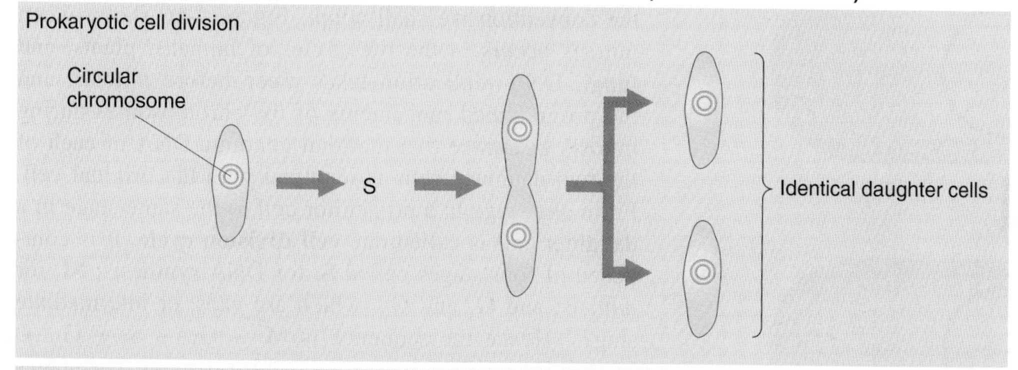

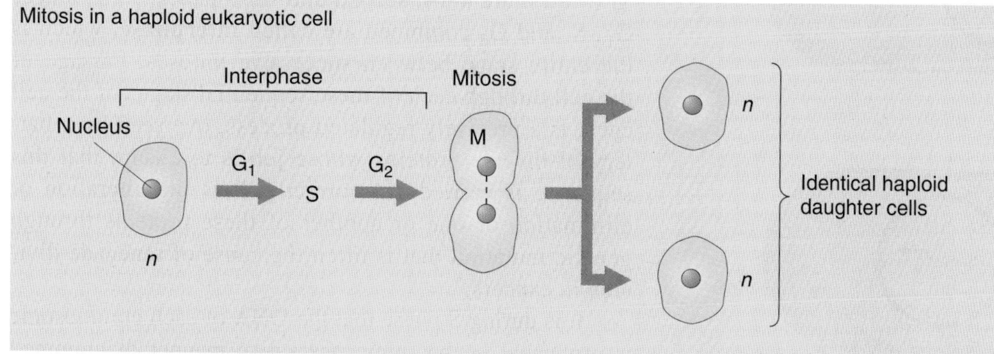

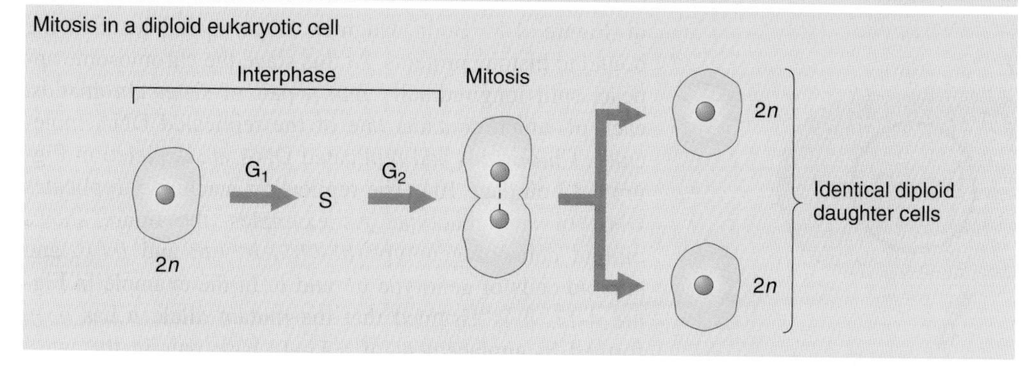

Consecutive cell divisions that halve the genetic content (sexual cell division)

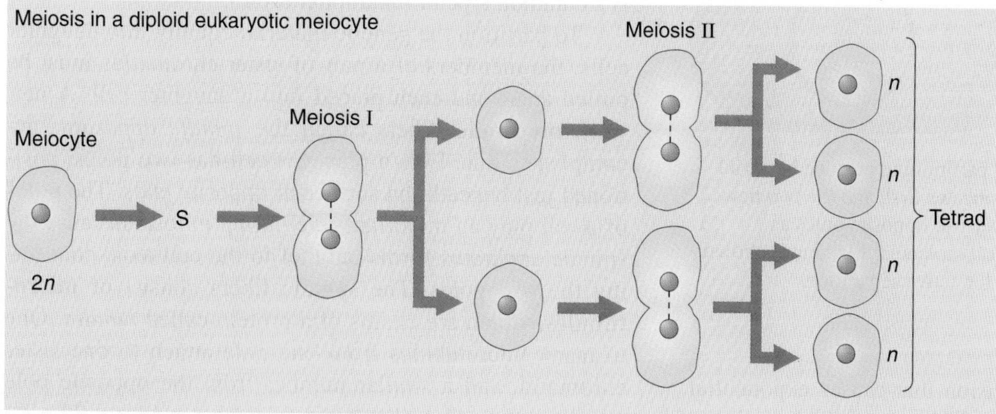

Figure 4-14 The different types of cell divisions and associated divisions of the genetic material. The first three panels show divisions that maintain the genomic content from one cell to its descendants, whereas the fourth panel shows two sequential divisions that halve the number of chromosomes. M = mitosis; S = DNA synthesis (replication); G_1 and G_2 are "gaps."

Eukaryotes

In eukaryotes the division of asexual cells also makes two identical daughter cells from one progenitor cell. Both haploid (n) and diploid ($2n$) cells can divide asexually, as shown in the second and third panels of Figure 4-14. This is the type of cell division that converts a single fertilized egg cell, a **zygote,** into two cells, then four, then eight, and so on until an organism composed of many cells is produced.

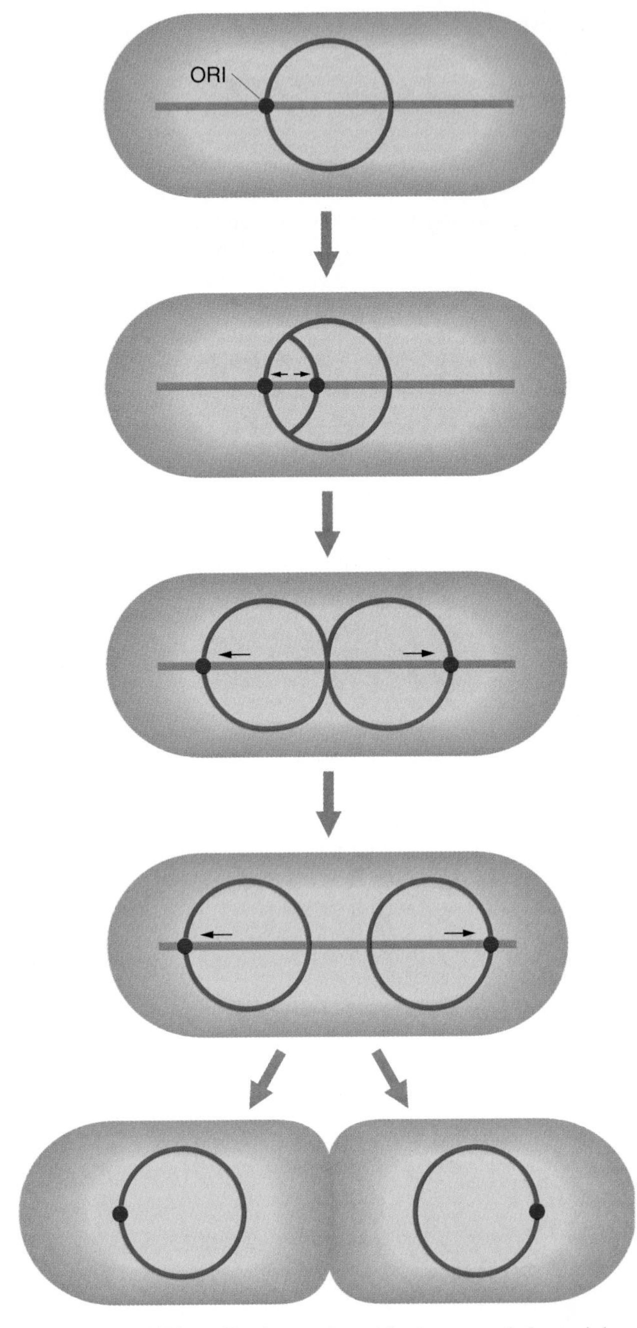

Figure 4-15 DNA replication and partitioning precede bacterial cell division. The origin (ORI) replicates first, and the two new origins can be shown to move rapidly to opposite poles as replication proceeds. A hypothetical molecular filament is shown to which the origins might attach for active separation.

It is also the type of cell division that causes exponential population growth of single-celled eukaryotic organisms, for example, yeasts and protozoans, as in prokaryotes.

When a eukaryotic cell divides asexually, the nucleus and its genetic contents divide in a process called **mitosis.** Geneticists use the term *mitosis* as shorthand to mean the cell division in which mitosis is occurring, and this is the convention we shall adopt. Figure 4-16 shows when mitosis occurs in the life cycles of humans, plants, and fungi. DNA replication takes place before mitosis, and the programmed movements of the chromosomes during mitosis guarantee that the total genomic DNA of each of the two daughter cells is identical with the original cell. From any stage in a progenitor cell to the same stage in a daughter cell is called one **cell division cycle.** It is composed of four stages called **S**, for DNA *s*ynthesis; **M**, for *m*itosis; and $\mathbf{G_1}$ and $\mathbf{G_2}$, which are *g*aps, or intermediate stages. Hence the sequence is $M \rightarrow G_1 \rightarrow S \rightarrow G_2 \rightarrow M$ (see Figure 4-14, second and third panels). The stages G_1, S, and G_2 combined are called **interphase,** which is the entire stage between successive mitoses. Passage of the cell through each of these sequential stages of the cell cycle is a precisely regulated process, overseen by a battery of diverse proteins whose job is to ensure that this sequence is carried out correctly. It is the alteration or elimination of one or another of these proteins through genetic mutation that is often the cause of renegade division in cancers.

It is during S phase that the DNA of each chromosome is replicated in the semiconservative manner diagrammed in Figure 4-8. Both daughter DNA molecules become bound to histone proteins. At this stage the chromosome appears split longitudinally into a pair of sister chromatids, each of which contains one of the replicated DNA molecules. Chromatids and replicated DNA are depicted in Figure 4-17 on page 104. The replication machinery replicates DNA of any genotype. As examples, the figure shows diploid cells of genotype b^+ / b^+, b^+ / b, and b / b, and haploid cells of genotype b^+ and b. In the example in Figure 4-17, it is assumed that the mutant allele b has been formed by replacement of a G–C base pair in the wild-type allele with an A–T base pair in the mutant allele (this is a common type of mutational event).

To partition the genetic material equally into daughter cells, the members of a pair of sister chromatids must be pulled apart and each placed into a daughter cell. A network of parallel fibers called the *spindle apparatus* accomplishes this. Like a planet, a cell has two poles, positioned just beneath the surface at opposite ends. The poles of a cell play an important role during mitosis because the spindle apparatus forms parallel to the cell axis, connecting the two poles. The spindle fibers consist of **microtubules,** which are chains of a protein called *tubulin.* One to many microtubules from one pole attach to one sister chromatid, and a similar number from the opposite pole attach to the other chromatid of a chromosome. The attachment point is a specific DNA sequence called the **centromere.** The centromere is replicated during the formation of sister chromatids, and each sister centromere acts as a binding site for a multiprotein complex called the

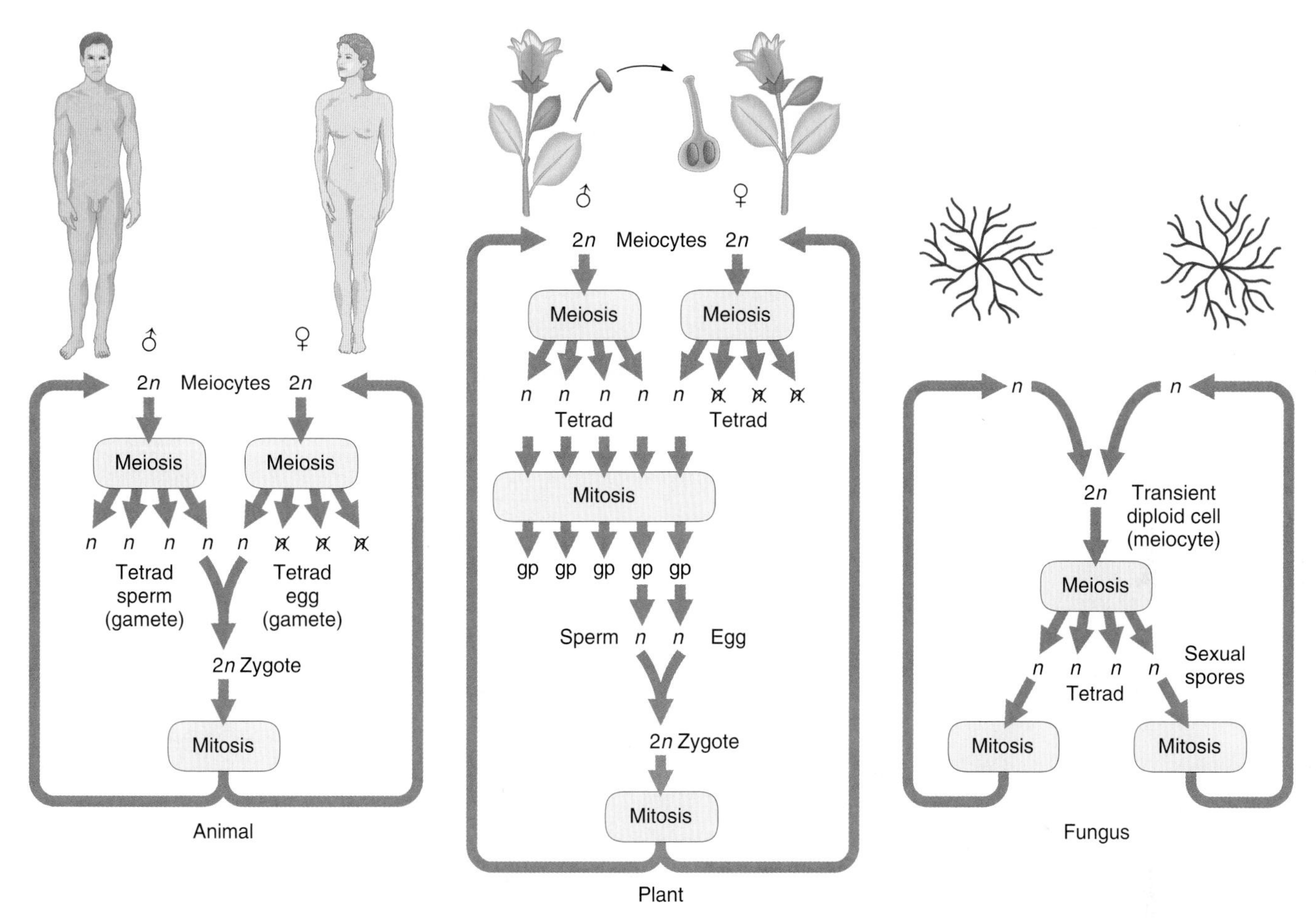

Figure 4-16 Life cycles of humans, plants, and fungi, showing the points in which mitosis and meiosis occur. Note that in females of humans and many plants, three cells of the meiotic tetrad abort. The abbreviation *n* indicates a haploid cell, *2n* a diploid cell; gp stands for "gametophyte," the name of the small structure composed of haploid cells that will produce gametes. In flowering plants the gametophyte stage is a small number of cells within the ovary or anther, but in others (such as mosses) the gametophyte is the green plant we see growing.

kinetochore (Figure 4-18 on page 105). The kinetochores in turn act as the sites for attachment of microtubules. The microtubules then pull sister chromatids to opposite poles. However, despite the ropelike nature of the microtubules, the pulling mechanism is not like pulling a rope. Instead, the tubulin units depolymerize at the kinetochores, resulting in shortening of the microtubule, and hence generating the motive force for pulling the sister chromatids apart (Figure 4-19 on page 105). Later, further separation of the sister chromatids is achieved by the action of "motor proteins" on another set of microtubules not connected to the kinetochores. These microtubules originate from the poles and overlap in the equatorial region. Here, molecular motors on microtubules from one pole attach to and "push" microtubules from the other pole, causing the two poles to separate, and thus further separating the chromatids.

In this way, each pole receives a copy of each chromosome from the parent cell, that is, a chromatid. The sets of chromatids at each pole become incorporated into the nuclei of the two daughter cells. These daughter nuclei are identical with each other and with the nucleus from which they were derived. In the daughter cells, chromatids are again called *chromosomes.* Note that it is the spindle apparatus and the kinetochore-centromere complex that determine the fidelity of nuclear division.

The S phase and the main events of mitosis are summarized in the left and central columns in Figure 4-20 on page 106. The figure uses a haploid cell of genotype *A* and a diploid of genotype *A* / *a* as examples. Although the figure shows just one chromosomal type, all other chromosomes in the nuclear genome would behave in the same manner. Note that mitosis generates two cells that have the same number of chromosomes, and also that the specific allelic constitution is also identical.

Having considered an overview of mitosis emphasizing the events important to genetics, we now turn to a more

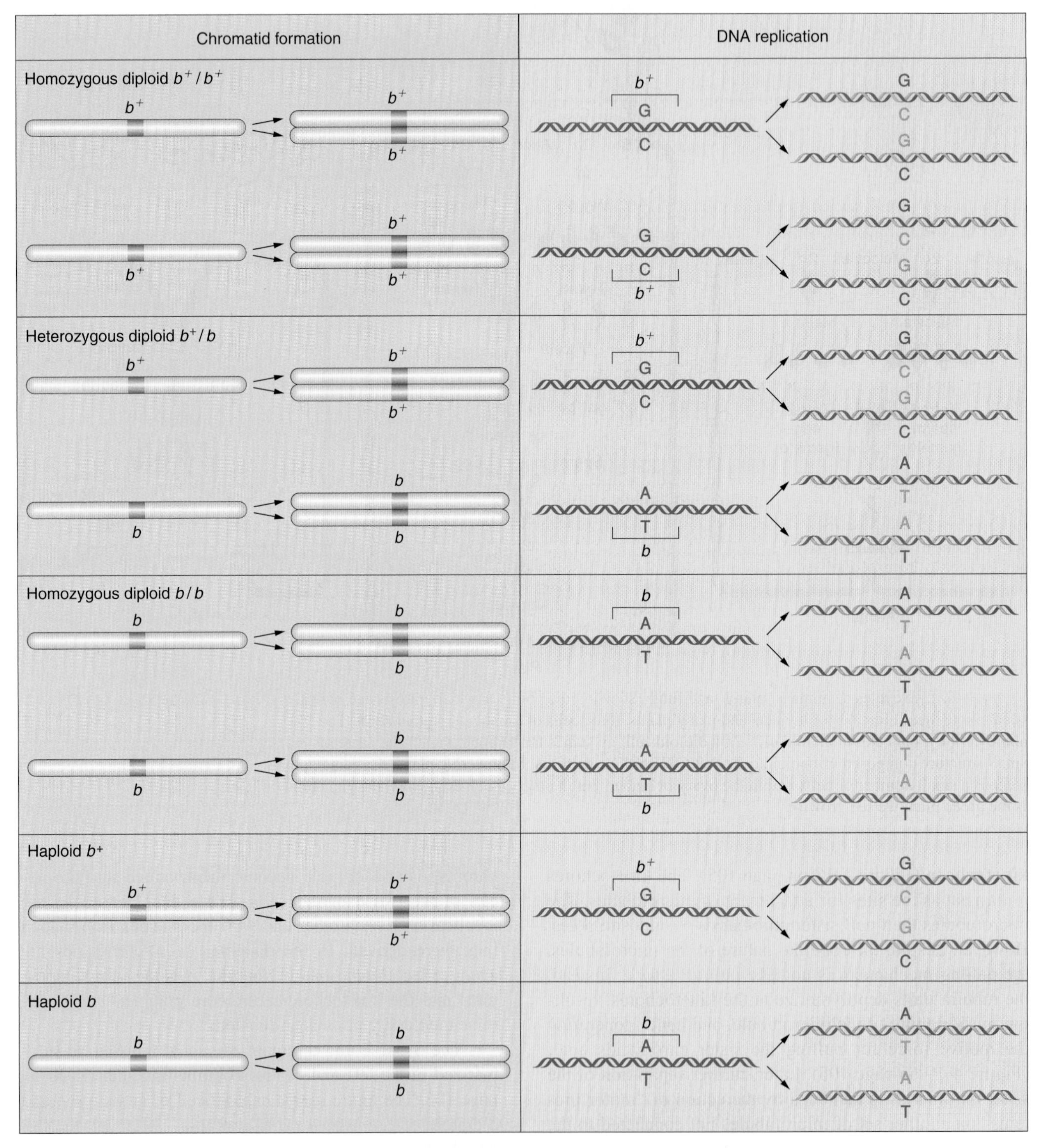

Figure 4-17 Chromatid formation and the underlying DNA replication. *(Left)* Each chromosome divides longitudinally into two chromatids; *(right)* at the molecular level, the single DNA molecule of each chromosome replicates, producing two DNA molecules, one for each chromatid. Also shown are various combinations of a gene with normal allele b^+ and mutant form b, caused by a change of a single base pair from GC to AT. Notice that at the DNA level the two chromatids produced when a chromosome replicates are always identical with each other and with the original chromosome.

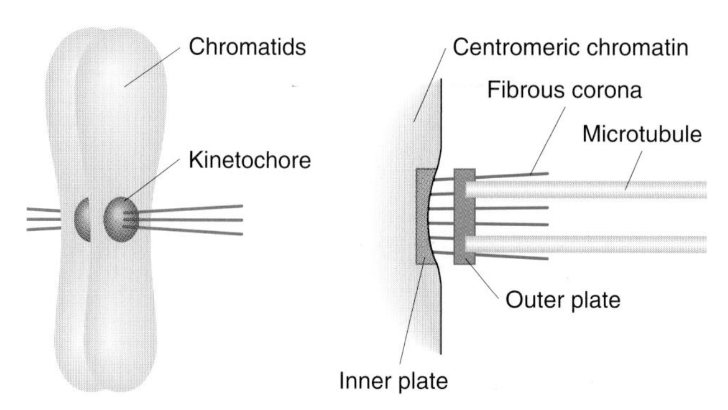

Figure 4-18 Microtubule attachment to the kinetochore at the centromere region of the chromatid in animal cells. The kinetochore is composed of an inner and outer plate and a fibrous corona. *(Adapted from A. G. Pluta et al.,* Science *270, 1995, 1592; taken from H. Lodish, A. Berk, S. L. Zipursky, P. Matsudaira, D. Baltimore, and J. Darnell,* Molecular Cell Biology, *4th ed. © 2000 by W. H. Freeman and Co.)*

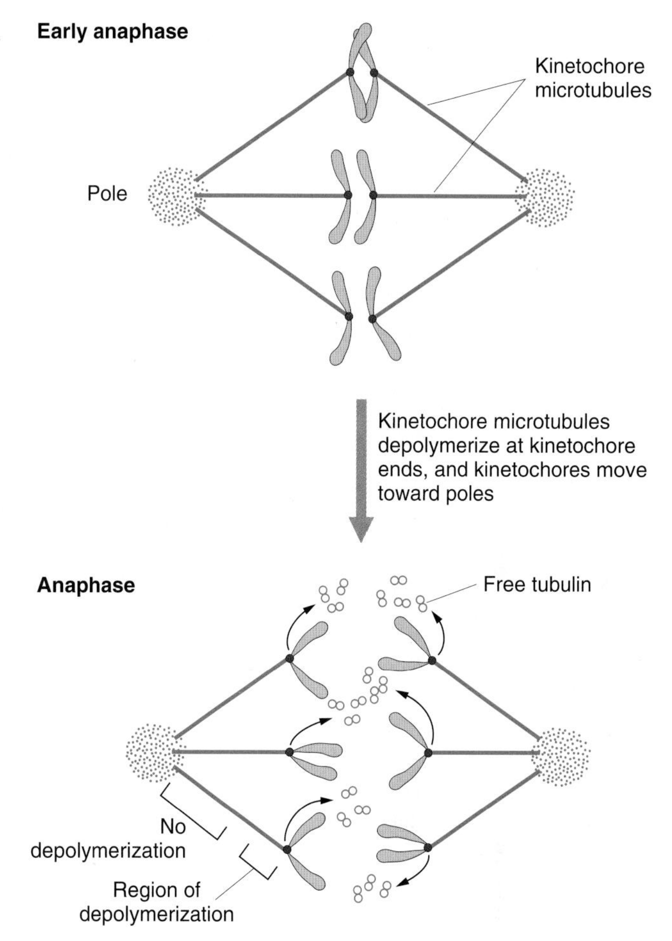

Figure 4-19 The microtubules exert pulling force on the chromatids by depolymerizing into tubulin subunits at the kinetochores. *(Adapted from G. J. Gorbsky, P. J. Sammak, and G. Borisy,* J. Cell Biol. *104, 1987, 9; and G. J. Gorbsky, P. J. Sammak, and G. Borisy,* J. Cell Biol. *106, 1988, 1185; modified from H. Lodish, A. Berk, S. L. Zipursky, P. Matsudaira, D. Baltimore, and J. Darnell,* Molecular Cell Biology, *4th ed. © 2000 by W. H. Freeman and Co.)*

detailed account of the microscopically visible stages of mitosis in the following section.

MESSAGE

During mitosis, daughter DNA molecules (visible as sister chromatids) are pulled apart to opposite poles of the cell, where they form two identical nuclei in the daughter cells resulting from cell division.

The Stages of Mitosis

Mitosis (M) usually takes up only a small proportion of the cell cycle, approximately 5 to 10 percent. The remaining time is the interphase, composed of G_1, S, and G_2 stages. Interphase used to be called the "resting period." However, this is a misnomer as cells are active in many ways during interphase, not the least of which, of course, is DNA replication. The chromosomes cannot be seen during interphase (Figure 4-21a on page 107), mainly because they are in an extended state and are intertwined with one another like a tangle of yarn.

For the sake of study, biologists divide mitosis into four stages called **prophase, metaphase, anaphase,** and **telophase** (see Figure 4-21). It must be stressed, however, that any nuclear division is a dynamic process and one stage blends into another.

Prophase. The onset of mitosis, called *prophase,* is heralded by the chromosomes' becoming visible microscopically for the first time (see Figure 4-21b). They get progressively shorter through a process of contraction, or condensation of the DNA and its associated histone proteins, into a series of spirals or coils, as we saw in Chapter 2. The coiling produces structures that are more easily moved around in the cell. As the chromosomes become visible, they are seen to be doubled, each chromosome being composed of two longitudinal halves, the sister chromatids (see Figure 4-21c). The chromatids become visible microscopically only during mitosis, but it is important to remember that the DNA replication that allowed division into chromatids took place during the premitotic S phase. In prophase the centromeric DNA has also doubled to form sister centromeres, but these are generally not apparent. The *nucleoli*—the large intranuclear spherical structures that contain ribosomal components—disappear at this stage. The nuclear membrane begins to break down, and the nucleoplasm and cytoplasm become one.

Metaphase. At metaphase, the spindle apparatus becomes prominent. The chromosomes move to the equatorial plane of the cell. Here one sister centromere of each chromatid pair becomes attached to a spindle fiber from one pole; the other sister centromere, to the other pole (see Figure 4-21d).

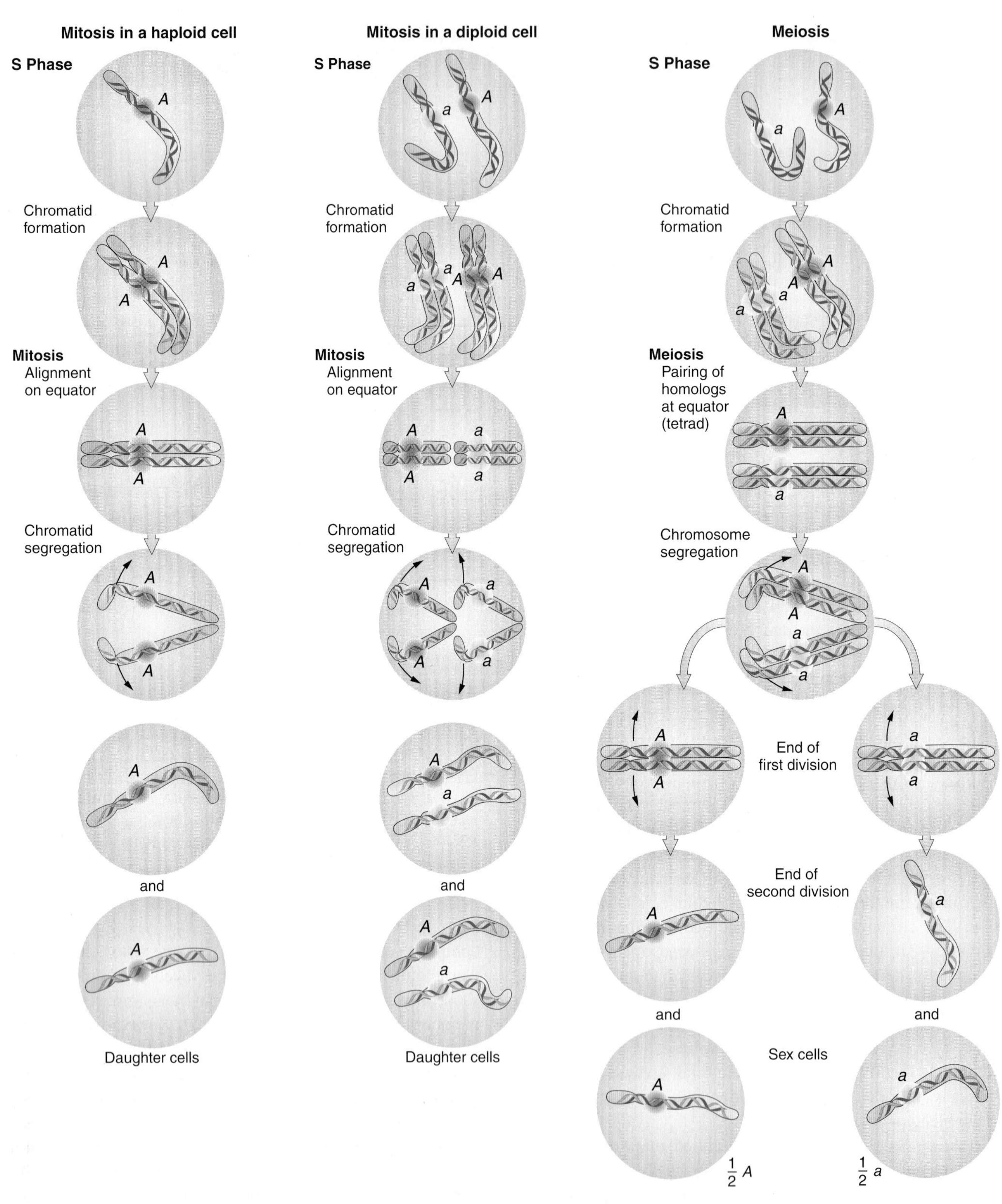

Figure 4-20 DNA and gene transmission during mitosis and meiosis in eukaryotes. S phase and the main stages of mitosis and meiosis are shown. Mitotic divisions (first two panels) conserve the genotype of the original cell. In the third panel, the two successive meiotic divisions that occur during the sexual stage of the life cycle have the net effect of halving the number of chromosomes. The diagrams emphasize the DNA content of each cell and chromosome. The first two rows in each column show DNA replication, which occurs during the S phase; the remaining panels show mitosis or meiosis. The alleles *A* and *a* of one gene are used to show how genotypes are transmitted during cell division.

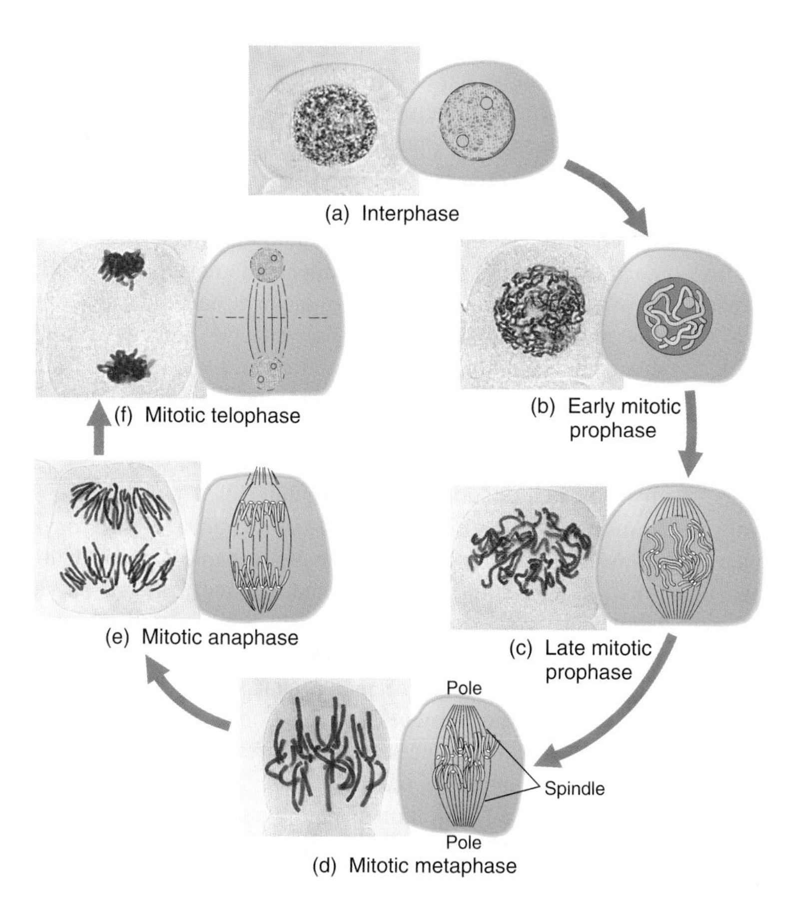

Figure 4-21 The stages of mitosis in the royal lily, *Lilium regale*. In each stage a photograph is shown at the left and an interpretive drawing at the right. For details, see the text. *(Modified from J. McLeish and B. Snoad,* Looking at Chromosomes. *© 1958, St. Martin's, Macmillan.)*

Animated Art • Mitosis

Anaphase. The anaphase begins when the pairs of sister chromatids separate, one from each pair moving to each pole (see Figure 4-21e). The sister centromeres, now clearly divided, separate first. As each chromatid moves, its two arms appear to trail its centromere; a set of V-shaped structures results, with the points of the V's directed at the poles. ***Telophase.*** At telophase a nuclear membrane re-forms around each daughter nucleus, the chromosomes uncoil, and the nucleoli reappear—all of this effectively re-forming the interphase nuclei (see Figure 4-21f). By the end of telophase, the spindle has dispersed and the cytoplasm has been divided into two by a new cell membrane, yielding two daughter cells. In each of the resultant daughter cells, the chromosome complement is identical with that of the original cell. What were referred to as *chromatids* now take on the role of full-fledged chromosomes in their own right.

CELL DIVISION THAT HALVES THE GENETIC CONTENT

In the preceding sections we have seen that both prokaryotic and eukaryotic cells need to reproduce themselves with high fidelity, conserving the precise genetic content of the original cell. However, specialized cells of eukaryotes alone show another type of cell division that halves the genetic content (chromosome number).

How Chromosome Number Is Halved

Most eukaryotic organisms can reproduce sexually (see Figure 4-16). Sex involves the fusion of cells from different individuals, at which time their nuclear contents combine. To compensate for the doubling of chromosome number that would result when fusion combines the entire genetic contents of two cells into one, at some stage of the cycle specialized cell divisions occur that halve the number of chromosome sets, returning cells to the original chromosome number. In plants and animals, the special cell divisions take place during the production of eggs and sperm (gametes). In fungi, these divisions occur during production of sexual spores such as ascospores.

The starting point for these specialized cell divisions in the sexual cycle is a diploid cell called a **meiocyte** (see Figure 4-14). In most complex organisms, such as animals and flowering plants, the cells of the organism are normally diploid and the meiocytes are simply a subpopulation of cells that are set aside for sexual division: for example, cells found in the testes and ovaries in animals. In haploid organisms a transient diploid meiocyte is constructed by fusion of two haploid cells as part of the normal reproductive cycle (see Figure 4-16).

The diploid meiocyte divides twice, resulting in four cells called a **tetrad.** The two nuclear divisions that accompany the two sexual cell divisions are called **meiosis.** As with mitosis, the term *meiosis* is used in a shorthand way to refer also to the two cell divisions in which meiosis takes place. Meiosis is preceded by the DNA synthesis (S) phase in the diploid meiocyte (see Figure 4-14). This achieves the same result as the S phase before a diploid mitosis; each chromosome (that is, each double-stranded DNA molecule) in the two sets is replicated into a pair of sister chromatids, precisely as we saw in Figure 4-17. Since the number of DNA molecules in the diploid meiocyte is doubled by DNA replication in the premeiotic S phase, the meiocyte becomes effectively $4n$ with regard to the number of DNA molecules present. Meiosis involves two divisions of the nucleus, reducing the number of DNA molecules to $2n$ after the first division and then to n after the second. Hence each of the cells of the tetrad that is produced at the second division is haploid.

Just before the first meiotic division, homologous chromosomes pair along their lengths, so that now for each chromosomal type there are two pairs of sister chromatids juxtaposed, making a bundle of four. This bundle of four is also called a *tetrad,* and this is significant because, as we shall see, each one of the members of the tetrad of *chromatids* ends up in one of the final tetrad of *cells.* At the tetrad stage a remarkable process occurs: paired *nonsister* chromatids exchange homologous sections of DNA through breakage and reunion of their arms at points called **crossovers.**

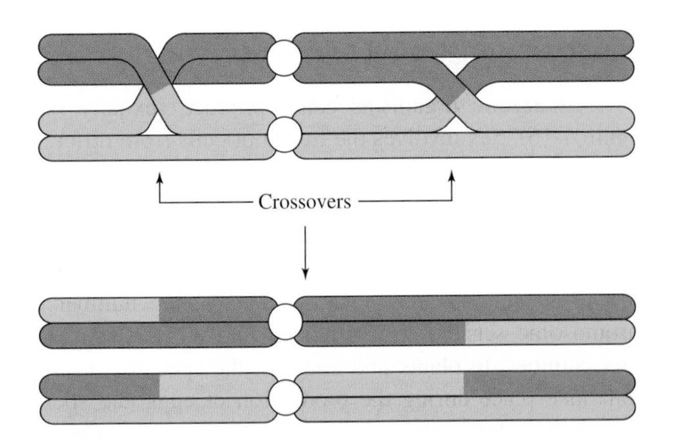

If a pair of homologous chromosomes is heterozygous for two or more genes, crossing-over can result in new allelic combinations, as we shall explore in Chapter 6. Hence crossing-over is an important generator of variation for natural selection to act on.

At the first division of meiosis, centromeres act as though they are still undivided, although it is known that replication of the centromeric DNA has occurred. However, unlike mitosis, the sister centromeres remain attached at this stage of meiosis. Spindle fibers (microtubules) from one pole attach to one centromere and spindle fibers from the other pole attach to the homologous centromere. These homologous centromeres and their attached chromatids are then segregated to opposite poles during the first division (see Figure 4-20). Therefore each of the two cells formed by the first division receives a pair of sister chromatids, still attached at the centromere. The number of centromeres in each of the two resulting cells appears to have been halved.

At the second division of meiosis it becomes apparent that at the DNA level the centromeres have divided. Spindle fibers attach and pull one sister chromatid to each pole. The resulting cells constitute the tetrad, the four final products of meiosis. Each product of meiosis contains one chromatid from each of the chromosome tetrads. In the products of meiosis, chromatids are once again called *chromosomes.* Figure 4-20 shows that if a diploid meiocyte is heterozygous (for example *A* / *a*), then half the haploid products of meiosis will carry the *A* allele and half will carry *a*. These alleles are said to **segregate** at meiosis because they separate into different haploid cells. The attachment of spindles to each chromosome pair is independent, resulting in **independent assortment** of chromosomes. Hence independent assortment produces many different possible chromosomal combinations in the meiotic products, contributing to the production of variation as possible raw material for natural selection to act on. We see that meiosis produces variation in two ways, by crossing-over and by independent assortment. Both these processes are the subject of Chapter 6.

MESSAGE

The process of meiosis starts with diploid meiocytes in the reproductive tissue and produces an array of haploid cells with diverse genotypes.

The detailed events of meiosis are described in the following section.

The Stages of Meiosis

Meiosis consists of two nuclear divisions distinguished as **meiosis I** and **meiosis II,** which take place in consecutive cell divisions. The events of meiosis I are quite different from those of meiosis II, and the events of both differ from those of mitosis (Figure 4-22). Each meiotic division is formally divided into prophase, metaphase, anaphase, and telophase. Of these stages, the most complex and lengthy is prophase I, which is divided into five stages. As with the stages of mitosis, keep in mind that these stages merge dynamically into one another with no clear borders.

Prophase I. For convenience, prophase I is divided into several stages: *leptotene, zygotene, pachytene, diplotene,* and *diakinesis* (see Figure 4-22, a–e). Much of what happens during these different stages of prophase I is a continuous

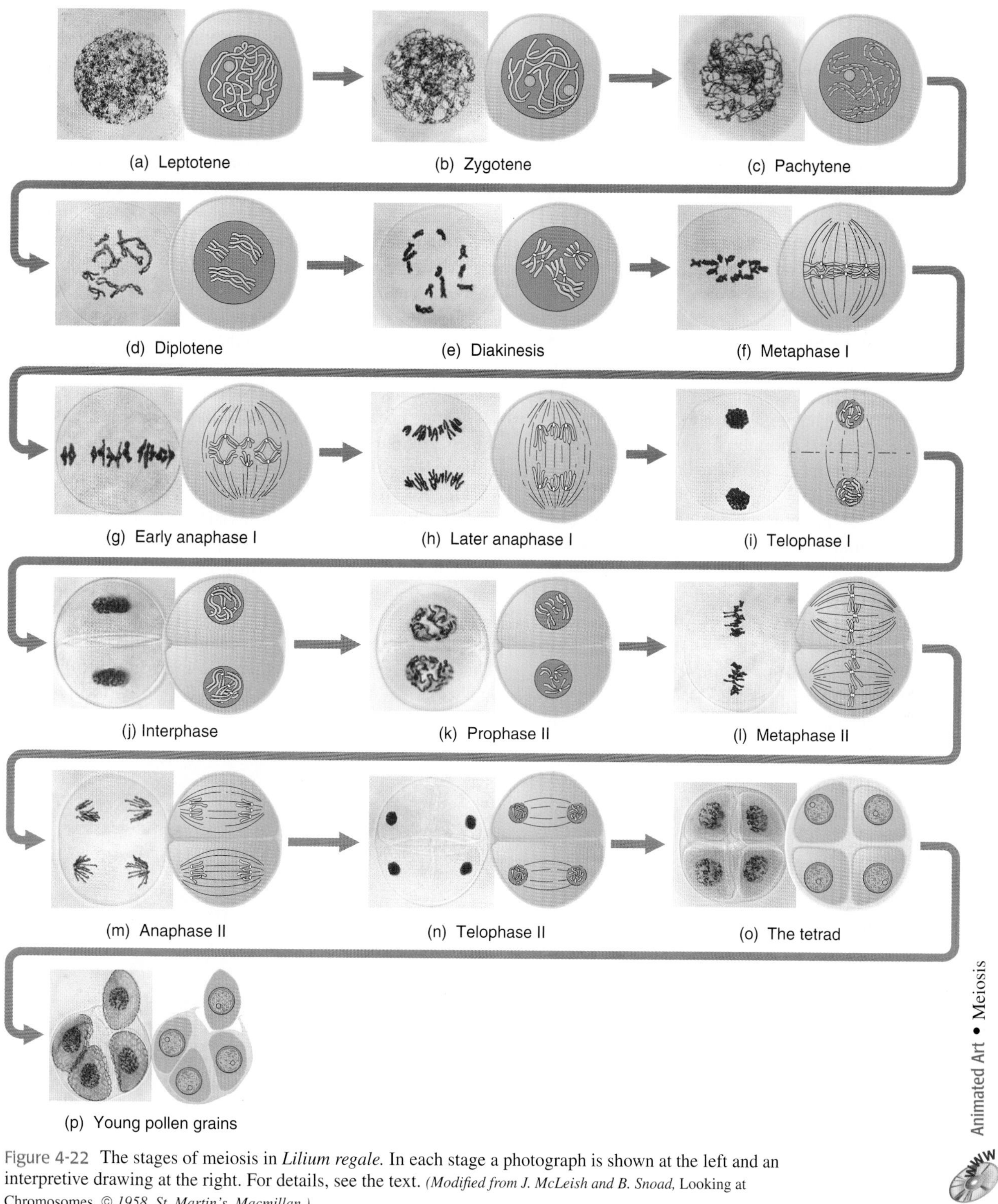

Figure 4-22 The stages of meiosis in *Lilium regale.* In each stage a photograph is shown at the left and an interpretive drawing at the right. For details, see the text. *(Modified from J. McLeish and B. Snoad,* Looking at Chromosomes. *© 1958, St. Martin's, Macmillan.)*

process involving the compaction of chromosomes and stabilization of homologous chromosome associations within the tetrad so that the tetrad can be divided in an orderly fashion at anaphase I.

The chromosomes become visible at **leptotene** as long, thin single threads (see Figure 4-22a). Although they appear single, each chromosome is composed of a pair of DNA molecules produced by replication during the premeiotic

S phase. The process of chromosome contraction continues during leptotene and throughout the entire prophase. During the leptotene stage, small areas of thickening called **chromomeres** develop along each chromosome, giving it the appearance of a necklace of beads.

Zygotene is a time of active pairing of homologous chromosomes (see Figure 4-22b). Thus, each chromosome has a homologous pairing partner, and the two homologs become progressively paired, or **synapsed,** along their lengths. Each pair is sometimes called a **bivalent.** It should be noted that pairing does not occur in mitosis: mitotic chromosomes lie randomly scattered around the equatorial plate, without regard to homologs.

How can homologs find each other during zygotene? The probable answer to this is that the ends of the chromosomes, the telomeres, are anchored in the nuclear membrane, and it is likely that homologous telomeres are anchored close to each other, and the pairing process begins from here. Although the mechanism for this pairing is not precisely understood, we know that it involves an elaborate structure composed of protein and DNA. This structure, called a **synaptonemal complex,** is always found sandwiched between homologs during synapsis. Figure 4-23 shows electron micrographs of several views of synaptonemal complexes. Part a shows meiosis in the silk moth ($2n = 62$). These chromosomes synapse into 31 pairs, each held together by a synaptonemal complex. The two homologs are seen as a diffuse material (DNA and histones) around the synaptonemal complex. In part b, on the left is an unpaired chromosome; note the dark scaffold through its center. This dark line is visible in duplicate in the homologous pair shown on the right. The two dark lines plus the material between them constitute the synaptonemal complex.

Pachytene is characterized by the presence of thick, fully synapsed chromosomes (see Figure 4-22c). Thus, the number of homologous pairs of chromosomes in the nucleus at this stage is equal to the number n. Nucleoli are often pronounced during pachytene. The beadlike chromomeres align precisely in the paired homologs, producing a distinctive pattern for each pair.

It is at **diplotene** that each chromosome is seen to have become a pair of sister chromatids. In other words, this is when the tetrads become visible (see Figure 4-22d). At diplotene, the pairing between homologs becomes less tight; in fact, they appear to repel each other, and as they separate slightly, cross-shaped structures called **chiasmata** (singular, *chiasma*) appear between nonsister chromatids. Each tetrad generally has one or more chiasmata. Chiasmata are the visible manifestations of crossovers that occurred earlier, probably during zygotene or pachytene. Studies performed on abnormal lines of organisms that undergo crossing-over very inefficiently, or not at all, show severe disruption of the orderly events that partition chromosomes into daughter cells at meiosis. Thus, crossovers have another role in addition to producing variation, which is to promote the proper segregation of chromosomes.

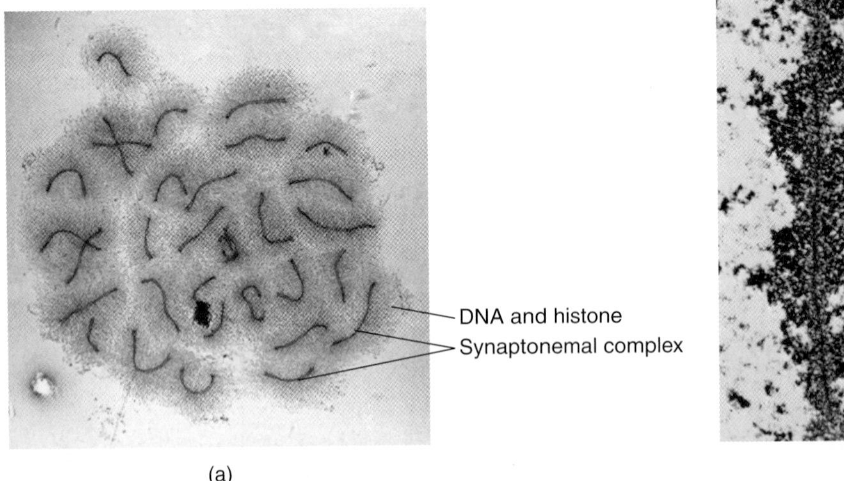

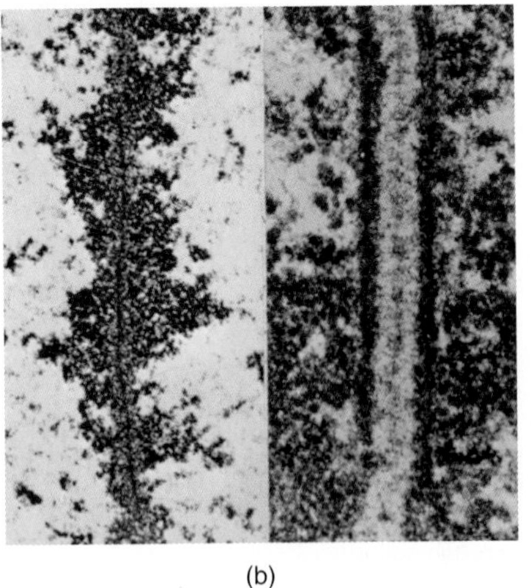

Figure 4-23 Synaptonemal complexes. (a) In *Hyalophora cecropia,* a silk moth, the normal male chromosome number is 62, giving 31 synaptonemal complexes. In the individual shown here, one chromosome *(center)* is represented three times; such a chromosome is termed *trivalent.* The DNA is arranged in regular loops around the synaptonemal complex. The black, dense structure is the nucleolus. (b) Regular synaptonemal complex in *Lilium tyrinum.* Note *(right)* the two lateral elements of the synaptonemal complex, corresponding to the scaffolding of each chromosome, and also *(left)* an unpaired chromosome, showing a central core corresponding to one of the lateral elements. *(Parts a and b courtesy of Peter Moens.)*

Diakinesis (see Figure 4-22e) differs only slightly from diplotene, in that chromosomes are further contracted. By the end of diakinesis, the long, filamentous chromosome threads of interphase have been replaced by compact units that are far more maneuverable.

Metaphase I. The nuclear membrane and nucleoli have disappeared by **metaphase I,** and each pair of homologs takes up a position in the equatorial plane (see Figure 4-22f). At this stage of meiosis, the two undivided homologous centromeres attach to spindle fibers from opposite poles. (This is a major difference from mitosis, in which it is *sister centromeres* on sister chromatids that attach to spindle fibers and separate.)

Anaphase I. As in mitosis, **anaphase** begins when chromosomes move directionally to the poles. The homologous chromosomes attached to homologous centromeres move to opposite poles (see Figure 4-22 g and h).

Telophase I. At **telophase I** (see Figure 4-22i) the chromosomes at each of the poles elongate and become diffuse, the nuclear membrane re-forms, and the cell divides. After telophase I, there is an "interphase," called **interkinesis** (Figure 4-22j). However, telophase I and interkinesis are not universal. In many organisms, these stages do not exist, no nuclear membrane re-forms, and after cell division the cells proceed directly to meiosis II. In other organisms, telophase I and the interkinesis are brief in duration. There is never DNA synthesis between telophase I and prophase II; thus each cell starts the second division of meiosis with each chromosome represented by a pair of sister chromatids, but no homolog.

Prophase II. The appearance of the haploid number of sister chromatid pairs in the contracted state characterizes **prophase II** (see Figure 4-22k).

Metaphase II. The pairs of sister chromatids arrange themselves on the equatorial plane during **metaphase II** (see Figure 4-22l). At this time the chromatids often partly dissociate from each other instead of being closely pressed together as they are in mitosis.

Anaphase II. During **anaphase II** the centromeres split and sister chromatids are pulled to opposite poles by the spindle fibers (see Figure 4-22m).

Telophase II. During **telophase II** the nuclei re-form around the chromosomes at the poles (see Figure 4-22n), and cell division occurs. Now in each cell each chromosome is represented by one of the four chromatids that formed the chromatid tetrad in prophase I, and hence each cell is haploid. Now these chromatids are called *chromosomes.* Cell division occurs in each of the cells that were the products of the first division, resulting in four cells.

The four products of meiosis (the tetrad) are shown in Figure 4-22o. Note again that in regard to any particular gene, the haploid product of meiosis will contain only one copy. Therefore for a gene A, if the genotype of the meiocyte is A / A, each meiotic product will be of genotype A. Furthermore, if the meiocyte is A / a, then half the products of meiosis will be A and half will be a (see the third column of Figure 4-20).

In the anthers of a flower, the four products of meiosis develop into pollen grains; these are shown in Figure 4-22p. In other organisms, differentiation produces other kinds of structures from the products of meiosis, such as sperm cells in animals. During oogenesis in animals and plants, one of the four meiotic products develops as the oocyte. The other three products degenerate as "polar bodies."

In review, some key features of meiosis are as follows:

1. There is one round of DNA replication but there are two rounds of nuclear division and two rounds of cell division. Thus, at the end of meiosis, the number of chromosomes per cell is halved.
2. At the first meiotic division, homologous chromosomes are pulled to opposite poles of the cell by spindle fibers. Because spindle attachment to one chromosome pair is independent of the attachment to other chromosome pairs, many different possible chromosomal combinations are found in the meiotic products. This is one source of the diversity generated by meiosis.
3. Early in the first meiotic division, there is exchange of chromosomal material between homologous chromatids as a result of crossing-over. Crossovers serve two roles. First, they are the other great source of genotypic diversity, since no two crossed-over chromatids are exactly the same. Second, the exchange events that occur between homologous chromatids in the tetrad serve to hold the tetrad together until the homologs pull apart at the end of the first division. This ensures proper chromosome segregation and prevents the formation of products of meiosis bearing abnormal chromosome numbers.
4. In contrast to mitosis, which achieves a *conservative* propagation of one genotype, meiosis not only halves the genomic content but also generates diversity in the haploid products. Meiosis shuffles allelic combinations by the two mechanisms of crossing-over and independent assortment of different chromosomes, which we shall discuss in detail in Chapter 6. Hence the products of meiosis (that is, the cells of the tetrads) will contain many different combinations of alleles. Meiosis is thought to be the main raison d'etre for the sexual cycle in that by generating variation it provides raw material for natural selection to act on.

In conclusion, the similarities and differences between meiosis and mitosis are summarized in Figure 4-24.

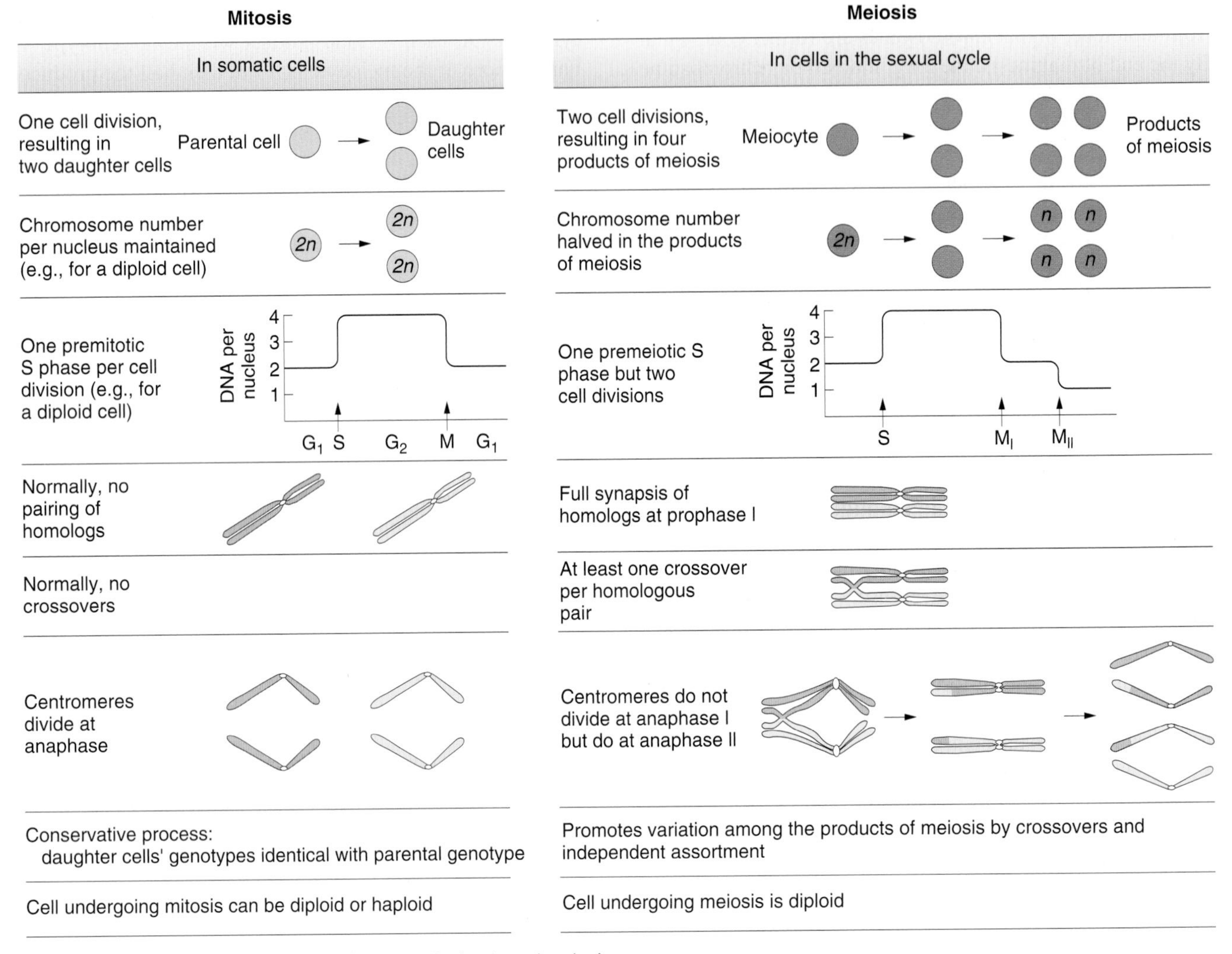

Figure 4-24 Comparison of the main features of mitosis and meiosis.

Let us revisit the topic with which we began this chapter, cancer. We will not discuss the mechanisms of cancer (for this, see Chapter 15), but the way in which cancer capitalizes on cell reproduction. The mutations that cause cancer are changes at the DNA level in cells of the body, not in cells that will become gametes. The mutations signal a cell to divide, usually a cell in a tissue in which cell division either is under strict regulation or has ceased. For cell division to occur, there must be a premitotic S phase at which DNA replication occurs. When DNA replicates, the cancer-causing mutational changes are replicated too. The cell divides and at the same time the nuclear contents divide by mitosis. The replicated mutational changes are segregated at mitotic anaphase on the chromatids on which they reside. Hence each daughter cell receives an identical copy of the altered nucleotide sequences that originally signaled the cell division. This process continues so that a population of identical cells is faithfully reproduced by the replication machinery and mitosis. This localized population of cells tends to stay together, as there is little tissue movement in an adult body. The resulting group of cells is the cancer, or tumor.

SUMMARY

1. Cell division is the basis for reproduction in prokaryotes and eukaryotes. When cells divide, the daughter cells must each receive a genome, so DNA replication is a requirement for cell division.

2. A DNA molecule replicates semiconservatively, producing two "half old–half new" daughter molecules. The DNA double helix is unwound at a replication fork, and the two single strands thus produced act as templates for the polymerization of free nucleotides.

3. Nucleotides are polymerized by the enzyme DNA polymerase, which adds new nucleotides only to the 3′ end of a growing nucleotide chain. DNA polymerases recognize and bind to a special sequence called a *replication origin,* and generate two replication forks moving outward bidirectionally until the whole molecule is replicated. Several other proteins are found acting at the replication fork, including helicase, single-strand binding proteins, and primase.
4. Because addition is only at 3′ ends, polymerization on one template is continuous, producing the leading strand, and on the other it is discontinuous in short stretches (Okazaki fragments), producing the lagging strand. Synthesis of the leading strand and of every Okazaki fragment is primed by a short RNA primer (synthesized by primase) that provides a 3′ end for deoxyribonucleotide addition. At each replication fork, two joined DNA polymerase units act on the lagging and leading strands. The action of the lagging-strand polymerase is to produce a loop containing the growing Okazaki fragment. The Okazaki fragments are joined into a continuous strand by the enzyme ligase.
5. The ends of linear chromosomes (telomeres) present a problem for the replication system because on one strand there is always a short stretch that cannot be primed. This chromosome would progressively shorten, thereby losing coding information. Adding a number of short repetitive elements to maintain length solves the problem. This is achieved by the enzyme telomerase, which carries a short RNA that acts as the template for synthesis of the telomeric repeats.
6. When a eukaryotic chromosome divides longitudinally to produce sister chromatids, the DNA content of each chromatid is one of the daughter DNA molecules produced by replication.
7. During cell division daughter DNA molecules are partitioned into daughter cells by molecular devices: spindle fibers made of tubulin in eukaryotes and an incompletely understood track system in prokaryotes.
8. Asexual cell division in prokaryotes and eukaryotes results in two cells with genetically identical genomes (diploid or haploid). This type of division is important in population growth of unicellular prokaryotes and eukaryotes, and in the development of multicellular eukaryotes.
9. In eukaryotes the sexual cycle requires the production of specialized haploid cells (for example, egg and sperm). This is achieved by DNA replication in a diploid meiocyte, followed by two successive cell divisions, resulting in a tetrad of four haploid products. The two divisions of the nucleus that produce the tetrad of haploid cells are called *meiosis.*
10. Meiosis halves the chromosome number from $2n$ to n. If the meiocyte contains heterozygous allele pairs, meiosis also creates new allelic combinations through crossing-over and independent assortment. This is an important source of variation as raw material for evolution.

CONCEPT MAP

Draw a concept map interrelating as many of the following terms as possible. Note that the terms are in no particular order.

meiosis / chromosome / chromatid / DNA replication / nucleotides / mitosis / haploid / diploid / spindle fibers

SOLVED PROBLEM

1. In 1952, Alfred Hershey and Martha Chase were looking for evidence on the nature of genetic material. They decided to use a simple system, bacteria-parasitizing virus called *T2*. The T2 virus is merely a protein shell carrying a DNA molecule. If a bacterial culture is infected with a few T2 viruses, the culture soon contains a large number of progeny viruses produced by replicating inside the bacteria and later bursting them. At the time of their experiment, the nature of genetic material had not been firmly established. Hershey and Chase asked the question "Is it the protein or the DNA of T2 that is its genetic material?" To answer this question, in separate experiments they made either the protein or the DNA of the infecting viruses radioactive and then observed in which case the emergent progeny viruses were radioactive. To make a specific viral molecule radioactive, one simply cultures the virus in a solution of the radioactive isotope of an atom found only in the molecule of interest. This is called radioactive *labeling,* since the radioactive atom acts as a tag, or label.

a. Which atom should be made radioactive to label the protein but not the DNA?

b. Which atom should be made radioactive to label the DNA but not the protein?

c. Which label would you expect to end up in the progeny T2 viruses?

Solution

a., b. DNA contains the atoms C, H, O, N, and P. Proteins contain the atoms C, H, O, N, and S. Hence it is clear that radioactive phosphorus would label DNA only and radioactive sulfur would label protein only.

Hershey and Chase used ^{32}P and ^{35}S in the separate experiments.

c. If protein is the genetic material, one would expect it to replicate; therefore, most of the radioactive sulfur label introduced as part of the original infecting viruses would end up in the progeny viruses. If DNA is the replicating genetic material, the radioactive phosphorus would end up in the progeny. The latter proved to be the result, showing that for T2 viruses DNA is the genetic material.

SOLVED PROBLEM

2. Consider the following segment of DNA that is part of a much larger molecule:

5′-ATTGCATTGCTTTTAGACGCTATACGTACG-3′

3′-TAACGTAACGAAAATCTGCGATATGCATGC-5′

Assume that the replication machinery passes from left to right as written. Also assume that this segment is the exact length of an Okazaki fragment plus a 6-base-pair RNA primer. (Note: In reality these are much larger.)

a. Which is the template for the leading strand?

b. Which is the template for the lagging strand?

c. What would be the sequence of the RNA primer?

d. In the center of the sequence there is a TAG stop codon. Assuming this is translationally in phase, would it present a problem to the DNA polymerase in replicating it?

Solution

a. Since the leading strand must grow from left to right and at a 3′ tip, the lower strand must be its template.

b. By similar logic, the upper strand must be the template for the lagging strand.

c. The primer must be placed at the right-hand end, and it must complement the top strand, so it is 3′-GCAUGC-5′.

d. Trick question. Replication will duplicate any template regardless of its protein-coding potential. So there will be no problem for the polymerase.

PROBLEMS

Basic Problems

1. Why are primers needed in DNA replication, and at what stage or stages are they used?

2. Write a sentence including the words *leading, lagging, discontinuous, continuous, replication, DNA.*

3. In a diploid cell in which $2n = 14$, how many telomeres are there during the following phases of the cell cycle:

a. G_1

b. G_2

c. Mitotic prophase

d. Mitotic telophase

4. In DNA replication, are both strands of the DNA double helix used as templates? Is your answer valid for the entire length of the molecule?

5. What are the similarities and differences between transcription and replication?

6. Assume a certain bacterial chromosome has one origin of replication. Under some conditions of rapid cell division, it is possible that replication can start from the origin before the previous replication cycle is complete. How many replication forks would be present under these conditions?

7. A molecule of composition

5′-AAAAAAAAAAA-3′

3′-TTTTTTTTTTT-5′

is replicated in a solution of adenine nucleoside triphosphate with all its phosphorus atoms in the form of the radioactive isotope ^{32}P. Will both the daughter molecules be radioactive? Explain. Then repeat the question for the molecule

5′-ATATATATATATAT-3′

3′-TATATATATATATA-5′

8. Draw a graph of DNA content against time in a diploid cell that first goes through mitosis and then through meiosis.

9. Why is DNA synthesis continuous on one template and discontinuous on the other?

10. In what way (if any) does the second division of meiosis differ from mitosis?

11. Draw mitosis in a diploid cell of genotype *A* / *a*.

12. A normal mitosis takes place in a cell that is heterozygous *A* / *a* for a gene on chromosome 1, and heterozygous for another gene B / b on chromosome 2. Draw the mitosis and show the genotypes of the two daughter cells.

13. What would have been the result of the Meselson and Stahl experiment if DNA replicated in a manner that generates two daughter double helices, one of which is identical with the parental molecule and one of which is all new (so-called conservative replication)?

14. The "politics" of nuclear division might be stated "mitosis is conservative; meiosis is liberal." Do you agree? State your reasons.

15. Consider the following segment of DNA, which is part of a much longer molecule constituting a chromosome:

5′.....ATTCGTACGATCGACTGACTGACAGTC.....3′

3′.....TAAGCATGCTAGCTGACTGACTGTCAG.....5′

If the DNA polymerase starts replicating this segment from the right,

a. which will be the template for the leading strand?

b. draw the molecule when the DNA polymerase is halfway along this segment.

c. draw the two complete daughter molecules.

d. is your diagram in b compatible with bidirectional replication from a single origin, the usual mode of replication?

16. The biologist Boveri once said, "The nucleus doesn't divide; it is divided." What was he getting at, do you think?

17. Specify the anatomical locations in which mitosis and meiosis take place in a fern, a frog, a moss, an onion, a pine tree, a mushroom, a snail.

18. Four of the following events are part of both mitosis and meiosis, but one is only meiotic. Which one?

Chromatid formation, spindle attachment, chromosome condensation, chromosome movement to the poles, chromosome pairing

19. State another event (in addition to your answer to Question 18) that takes place in meiosis and not in mitosis.

20. The DNA polymerases are positioned over the DNA segment below (which is part of a much larger molecule) and moving from right to left. If we assume that an Okazaki fragment is made from this segment, what will be its sequence? Label its 5′ and 3′ ends.

5′.....CCTTAAGACTAACTACTTACTGGGATC.....3′

3′.....GGAATTCTGATTGATGAATGACCCTAG.....5′

21. In Figure 4-4, why are only two nucleotides shown in triphosphate form?

Challenging Problems

22. If a mutation that inactivated the telomerase occurred in a cell (telomerase activity in the cell = zero), what do you expect to be the outcome?

23. Assume you have some diploid eukaryotic cells that will divide asexually in special nutrient solution in a test tube. Your starting point is cells that have not yet entered premitotic S phase (replication phase). You put the cells into nutrient solution that contains radioactive nucleotides that can be incorporated into newly synthesized DNA by DNA polymerase. Using red for a radioactive DNA strand and black for a nonradioactive one, draw the DNA double helix of a chromosome

a. at the start of the experiment (before premitotic S phase).

b. after chromatid formation.

c. before premitotic S phase in the daughter cells.

d. before premitotic S phase in *their* daughter cells (that is, after two divisions in the solution of radioactive nucleotides).

i **24.** Suppose meiosis occurs in a meiocyte in an anther of a plant in which $2n = 6$. What proportion of pollen grains will receive a complete set of centromeres from the plant's maternal parent? (Think of the maternal centromeres as being blue and the paternal ones as being red.)

25. When dividing asexual cells of haploid yeast are studied, it is found that 10 percent of the cells are in mitosis. Can this information be used to calculate the proportion of the cell cycle that mitosis (M) occupies? What assumptions do you need to make?

26. At mitosis, a spindle fiber by accident failed to attach to the kinetochore on one chromatid only. What might be the outcome of this accident?

27. In a certain multicellular organism, measurements are made of the DNA content of cells from various parts of the organism's body. In arbitrary units, these measurements were

a. 33 units.

b. 66 units.

c. 132 units.

What cells could these measurements have been made on (which types of cells, and which stages of division)? State whether there are several possible answers for each case.

Exploring Genomes: A Web-Based Bioinformatics Tutorial

Learning to Use BLAST

To compare one protein sequence to another, we most often use a computer program called BLAST. This program allows us to search using a protein sequence and to find sequences from other organisms that are similar to it. In the Genomics tutorial at www.whfreeman.com/mga, we will run BLAST on a small simple protein, insulin (Chapter 11), and on a large complex one, dystrophin (Chapter 2).

THE INHERITANCE OF SINGLE-GENE DIFFERENCES

5

Key Concepts

1. In matings of heterozygotes, precise phenotypic ratios (for example 3 : 1 and 1 : 1) are produced in descendents.
2. These phenotypic ratios in progeny are generated by the predictable movements of chromosome pairs at meiosis.
3. The members of a pair of heterozygous alleles segregate equally into the products of meiosis.
4. Understanding inheritance patterns allows progeny ratios to be predicted from parents of known genotype, or parental genotypes to be predicted from observed progeny ratios.
5. Dominance can be deduced from the phenotype of a known heterozygote.
6. In many organisms, sex is determined by sex chromosome constitution, for example, X and Y chromosome content.
7. Sex chromosomes contain genes for sex determination and many other genes unrelated to sex.
8. X-linked genes can show different phenotypic ratios in male and female progeny.
9. Inheritance patterns of single-gene differences in humans can be deduced from analysis of family pedigrees.
10. Organelle genes are inherited maternally.

The monastery garden of the monk Gregor Mendel, the "father of genetics," in the town of Brno (now part of the Czech Republic). Mendel's experiments in this small plot established most of the principles of inheritance covered in this chapter and some of those in the next chapter. A statue of Mendel is visible in the background. *(Anthony Griffiths.)*

Huntington disease (HD) is a rare inherited disease of the nervous system. Symptoms include convulsions, paralysis, and eventually death. The disease is unusual in that it shows late onset; in other words, people do not express the symptoms of the disease until the third, fourth, or fifth decade of life. Because of this late onset, parents often have had their children before they know that they themselves have the disease. There is a very high likelihood that the children of HD sufferers will have the disease. A large study on the inheritance pattern of this disease has been conducted in Venezuela (Figure 5-1a and b). The pattern in one particular family is shown in Figure 5-1c. This type of figure is called a *pedigree diagram.* We will discuss pedigrees in detail later in this chapter, but for now all we need to know is that round symbols indicate females, that squares indicate males, and that people with HD are designated by solid black symbols. We see that there are six marriages between people with HD and unaffected mates, highlighted in green in Figure 5-1c. In the total offspring of all these marriages, there are 13 with HD and 12 unaffected by the disease, a ratio that is approximately 1 : 1.

(a)

(b)

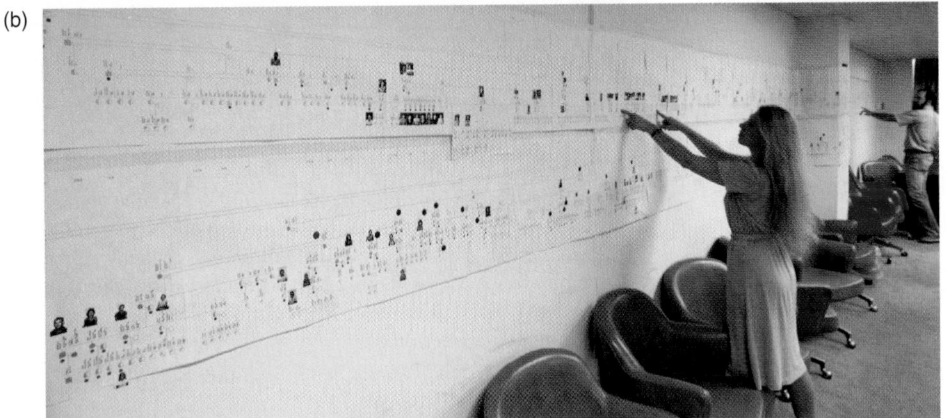

(c)

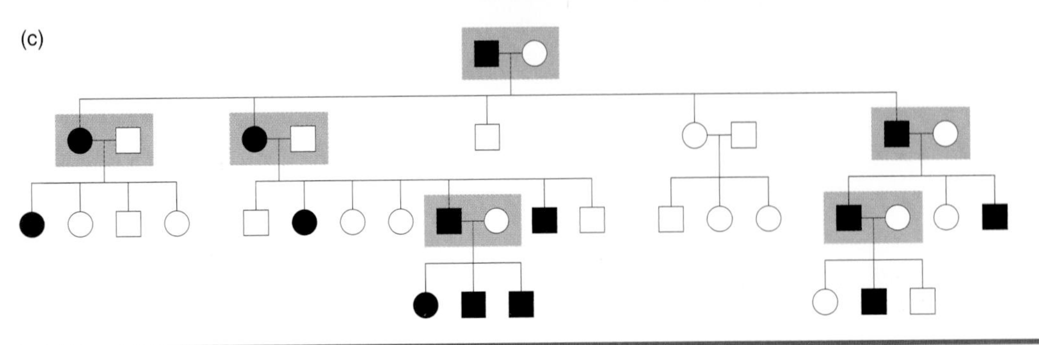

Figure 5-1 (a) Half the people in the Venezuelan town of Barranquitas have the neurodegenerative condition Huntington disease (HD). (b) A researcher examining a large HD pedigree of a family in Venezuela. The pedigree includes 10,000 people and is 100 feet long. (c) Family pedigree for HD. Black symbols represent people with the disease; white represents those who do not have the disease. Horizontal lines joining symbols represent marriages (those marriages involving HD are highlighted in color), and vertical lines connect to symbols for the children of the marriages. *(Part a, Ricardo Barbato/Black Star; part b, Steve Uzzell.)*

CHAPTER OVERVIEW

The above observations on HD raise many questions about this disease in particular and inherited diseases in general. What is the basis for the inheritance of the disease? What is the basis for the overall equal numbers of affected and normal children in marriages of HD and unaffected individuals? Why do the individual marriages show differences; for example, three children—all with HD—in one marriage and, in another marriage, two unaffected children and one with HD? Such questions are the topic of this chapter, which deals with the general nature of the inheritance patterns in organisms and the basis for the ratios, such as the 1 : 1 ratio of HD sufferers to unaffected individuals discussed above. We shall see that such simple inheritance ratios nearly always involve an allelic difference at one single gene, and that orderly chromosome behavior during meiosis, which we discussed in Chapter 4, is the process that generates the ratios. Figure 5-2 provides an overview of the main ideas in the chapter. It illustrates simple inheritance using an example from mice: two alleles of one gene, a dominant allele *B* which causes black coat, and *b* which causes brown. Here we see that meiosis in a heterozygote generates a 1 : 1 allelic ratio in gametes, and this 1 : 1 gametic ratio is converted into a 3 : 1 ratio of progeny phenotypes in the grid. (Note that any letter could be used for the allele symbols, but *B* and *b* are traditionally used in this case because of the match with the first letters of the names of the phenotypes.)

INHERITANCE PATTERNS

The Consequences of Chromosome Segregation at Meiosis

How are the inheritance patterns shown by individual genes dictated by the highly programmed movements of

CHAPTER OVERVIEW figure

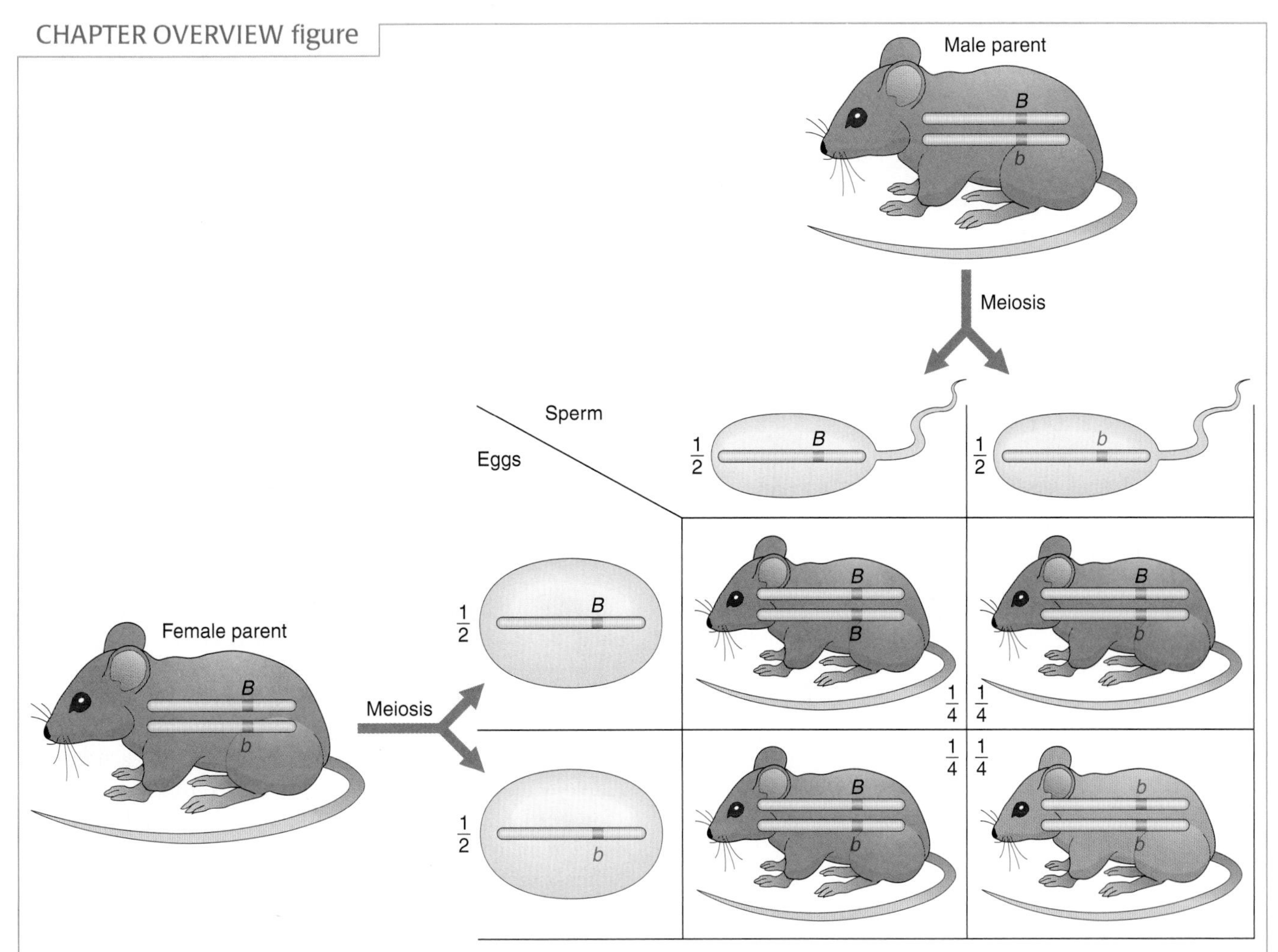

Figure 5-2 Overview of the chapter. The precise segregations of chromosomes at meiosis during gamete production lead to precise mathematical ratios of genotypes and phenotypes in the progeny. In the mice used as an example, the alleles B = black and b = brown. The grid represents the four different fusions of eggs with sperm.

chromosomes at meiosis? First let us review mitosis as a baseline for comparison. We will use the alleles *A* and *a* as "typical" dominant and recessive alleles of a gene. Haploid cells can be of genotype *A* or *a*; diploids can be homozygous, *A* / *A* and *a* / *a*, or heterozygous, *A* / *a*. As we saw in Chapter 4, because each of the chromosomes is replicated faithfully, and each daughter cell receives a full set of chromosomes, the genotypes of the daughter cells are identical with the original cell. We can represent these divisions as follows:

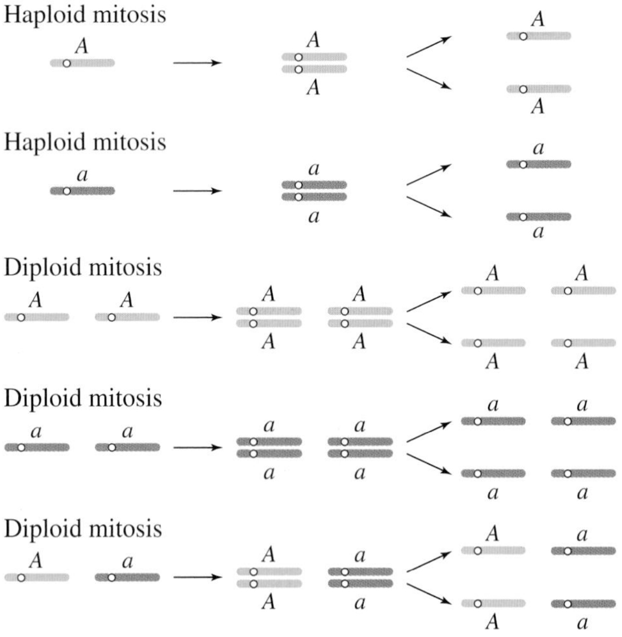

Haploid mitosis

Haploid mitosis

Diploid mitosis

Diploid mitosis

Diploid mitosis

Now let's review the outcomes of meiosis, using the same genotypes. Meiosis in *homozygous* meiocytes can produce only one genotype in the four haploid products of meiosis (the tetrad), as follows:

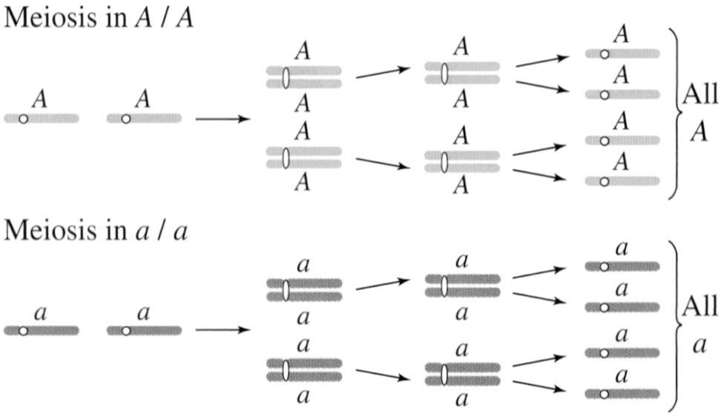

Meiosis in *A* / *A*

All *A*

Meiosis in *a* / *a*

All *a*

However, starting with a meiocyte of genotype *A* / *a*, meiosis produces four haploid cells, but there are two genotypes. Two of the haploid products are *A* and two are *a*; a ratio of 1 : 1.

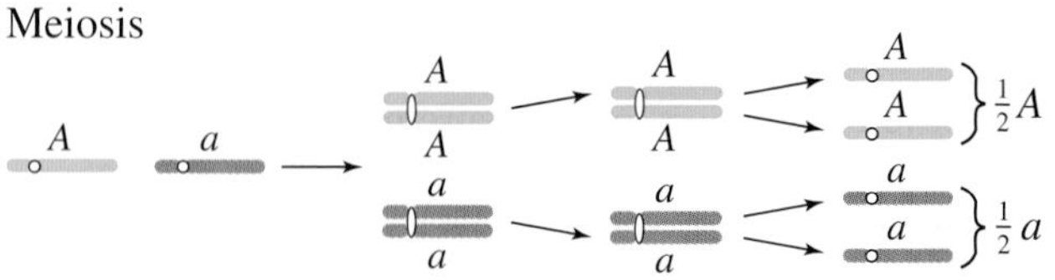

Meiosis

$\frac{1}{2}A$

$\frac{1}{2}a$

The basic reason for this ratio, as we saw in Chapter 4, is that in the *A* / *a* meiocyte, the *A* chromosome produces a pair of sister chromatids *A* / *A*, and the other homologous chromosome produces a pair *a* / *a*. These four chromatids bearing the four "copies" of the gene are separated by the two meiotic divisions and end up in separate meiotic products. The separation of the alleles of a heterozygote into equal numbers of meiotic products is called the law of **equal segregation.** It is stated formally in the following message:

MESSAGE

In the haploid products of a heterozygous meiocyte, half carry one allele and half the other allele. For example, a meiocyte *A* / *a* produces $\frac{1}{2}$ *A* and $\frac{1}{2}$ *a*.

Equal segregation was first observed by the monk Gregor Mendel in the middle of the nineteenth century, so the law of equal segregation is sometimes called **Mendel's first law.** However, Mendel did not know that such segregation is based on meiosis, since meiosis had yet to be discovered.

In animals and plants it is not possible to isolate a meiotic tetrad (the four cells that are the products of meiosis from a *single* meiocyte). The eggs and sperm that are released are the result of meioses in *many* meiocytes. But the law of equal segregation still holds, so that out of, say, 1000 sperm or eggs produced by many meiocytes in a heterozygote *A* / *a*, an average of 500 will be of genotype *A* and 500 will be *a*. (Of course, because of sampling error, in any real population of 1000 gametes there will not be precisely 500 of each type, just as there will not be precisely 500 heads and 500 tails in any particular set of 1000 coin tosses.) However, some fungi and unicellular algae do have sexual cycles that make it possible to isolate all four haploid products of a single meiocyte. Two parental haploid fungi (say, of genotypes *A* and *a*) unite sexually to form a transient diploid meiocyte that will be *A* / *a*. (Consult Figure 4-16 for the fungal sexual cycle.) Meiosis occurs and the products of meiosis in such cycles are called **sexual spores.** The four sexual spores that represent the products of meiosis from a single meiosis are held together as a tetrad. The ascomycete fungi derive their name from the fact that the meiotic tetrad is formed inside a sac called an **ascus.** In one well-known ascomycete fungus, baker's yeast, asci from such a parental pair will always contain 2 *A* and 2 *a* haploid

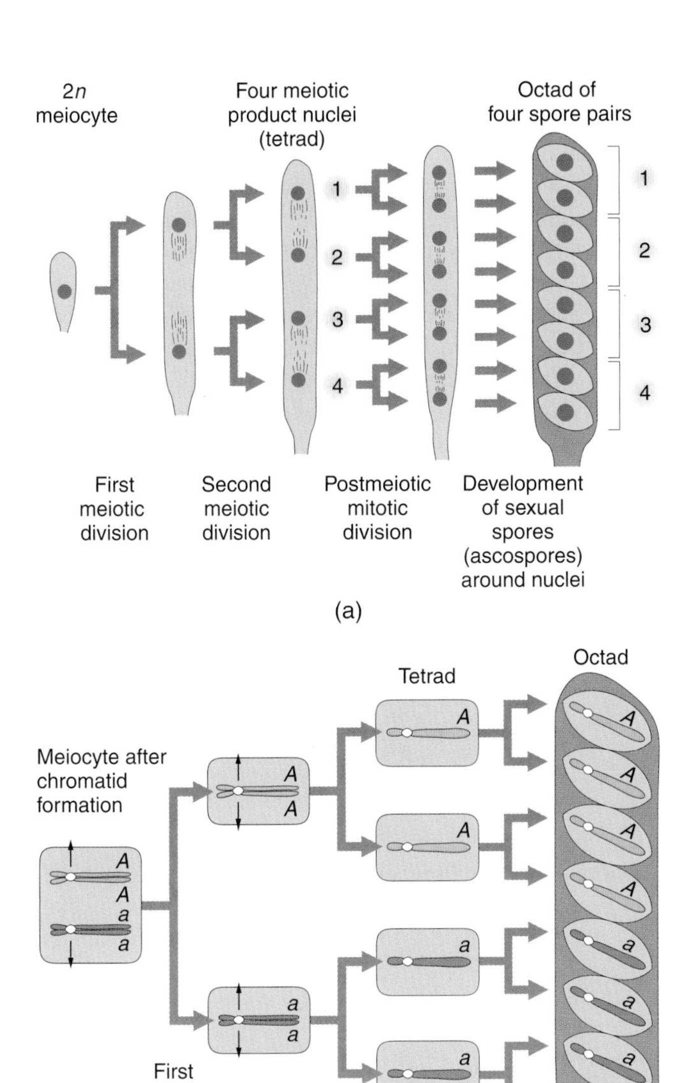

Figure 5-3 Certain fungi provide excellent systems for the study of allele segregation at meiosis. (a) In the fungus *Neurospora* (pink bread mold), meiosis produces four product nuclei (the tetrad), which then undergo mitosis to produce an octad. Since the meiotic and mitotic nuclear divisions all occur in a long, thin cell, nuclei never slip past each other, so the products form an ordered linear array. (b) An A / a meiocyte undergoes meiosis (and afterwards, mitosis), resulting in equal numbers of A and a products, a direct demonstration of the principle of equal segregation of alleles at meiosis (Mendel's first law). (Note: The meiosis illustrated happens to have no crossover in the region between the centromere and the gene under study. A crossover here would change the pattern of ascospores in the ascus but would not affect the 1 : 1 ratio of alleles in the meiotic products.)

sexual spores, showing the law of equal segregation directly in a single meiosis. In the ascomycete mold *Neurospora crassa* and related fungi, the four haploid nuclei resulting from meiosis undergo a mitotic division resulting in 8 sexual spores (an **octad**); in this case the law of equal segregation is expressed in the ascus as 4 A and 4 a (Figure 5-3). The sexual spores of ascomycetes are eventually released from the ascus for dispersal. If the released spores are sampled randomly, ratios are observed that are close to 1 : 1. Figure 5-4 on the next page shows the difference between analysis of the products of single meiocytes (as in fungi) and of randomly sampled products of many meiocytes (such as in plants and animals).

Crosses

A **cross** is a controlled mating between two specific organisms. The purpose of making crosses in genetics is either to obtain progeny of a specific genotype or, working the other way, to use the proportions of different phenotypes of progeny to deduce the genotypes of the parents. Some plants can be **selfed;** that is, they can be crossed to themselves: pollen is allowed to fall on the stigma in the same flower or another flower of the same plant. Most animals cannot be selfed, but matings between animals of identical genotypes are the equivalent of selfing. Crosses can be made in either diploid or haploid organisms.

Haploid crosses. The simplest crosses to analyze are those of haploid organisms, because each gene is present in only one copy in the parents. In genetics, a multiplication symbol, ×, is used to designate a cross. For a single gene with two known alleles, there are only three possible combinations of parents. These three types of crosses are shown below with their progeny:

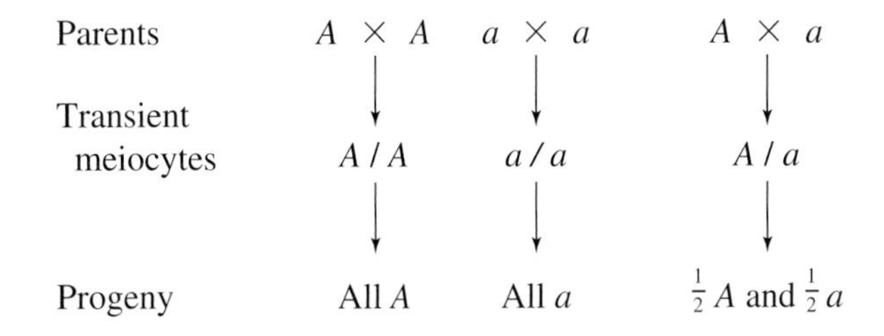

(Note that Mendel's first law of equal segregation is evident only in the progeny of the cross $A \times a$. However, equal chromosomal segregation is also occurring in the crosses $A \times A$ and $a \times a$. Because the crosses are homozygous, though, the effects of segregation are not visible at the phenotypic level.)

Diploid crosses. Considering again one gene with two known alleles, there are only three possible diploid genotypes, the homozygotes A / A and a / a, and the heterozygote A / a. Crosses between homozygotes have the simplest outcomes because only one type of meiotic product (shown below as gametes) is possible. Furthermore, each cross produces only one type of **zygote** (fertilized egg) and, hence, only one type of progeny. There are three possible types of crosses between homozygous parents:

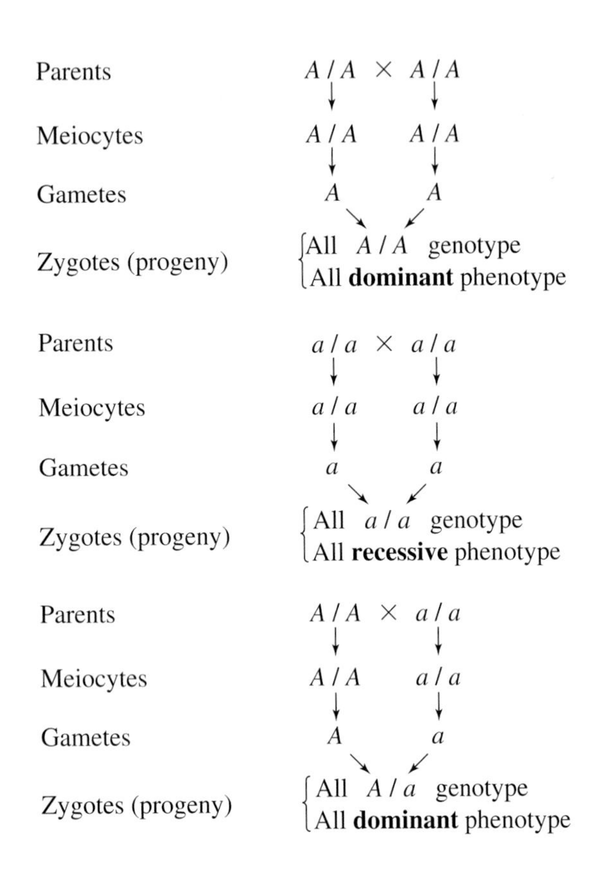

In the third cross, A / A × a / a, the progeny are genotypically A / a, but because of dominance they all show the dominant phenotype.

Crosses involving heterozygotes, A / a, are slightly more complex, but the law of equal segregation can be used to predict the proportions of gametes in these crosses as well. The following grids show graphically how the progeny are produced in the three different crosses involving heterozygotes. The gamete ratios produced by the two parents are drawn along the top and side of the grid, then the grid itself portrays the cells (the zygotes) that result from all the possible types of gamete fusions. (Note the useful symbolism A / –, in which the dash represents either an A or an a allele: this symbolism is used to represent the genotypes that give rise to the dominant phenotype in cases in which it is not necessary or possible to specify the exact genotype.)

A / a × a / a

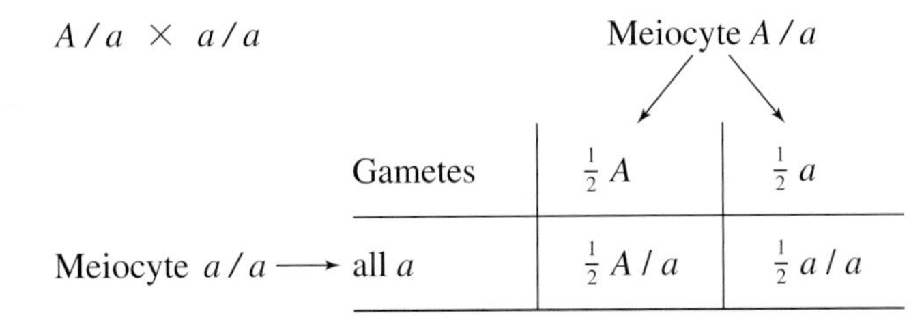

	Gametes	$\frac{1}{2}$ A	$\frac{1}{2}$ a
Meiocyte a / a ⟶	all a	$\frac{1}{2}$ A / a	$\frac{1}{2}$ a / a

Phenotypic ratio in progeny: $\frac{1}{2}$ **dominant,** $\frac{1}{2}$ **recessive**

1:1 segregation in products of single meiocyte

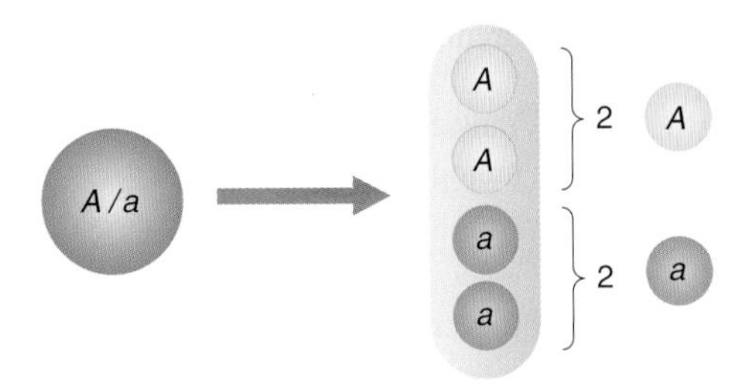

1:1 segregation in random meiotic products

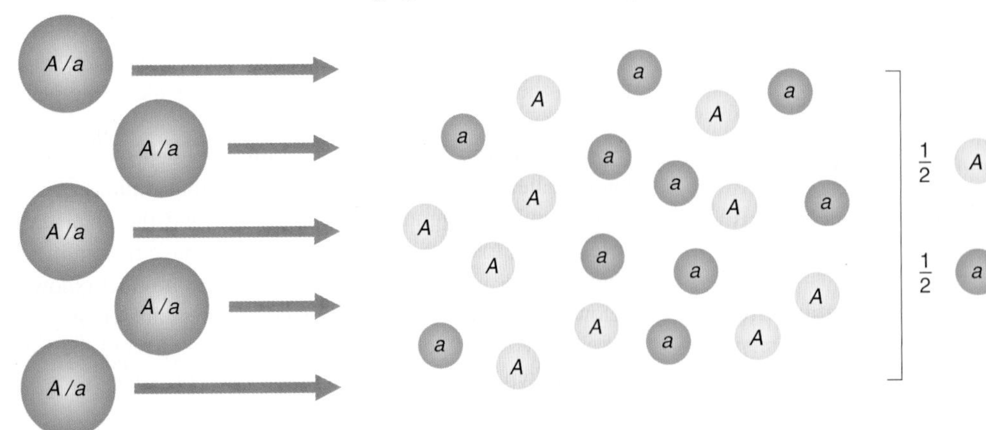

Figure 5-4 A representation of the different ways that the law of equal segregation is manifest using analysis of a single meiocyte, as in fungal tetrads, and of random meiotic products from many identical meiocytes, as in analysis of egg and sperm in animals.

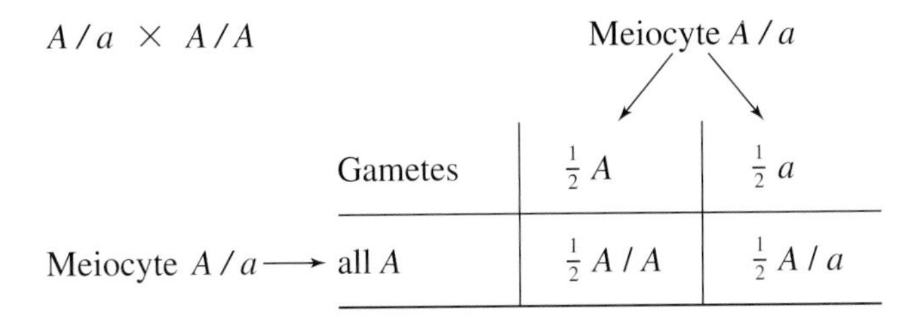

Phenotypic ratio in progeny: **All dominant A / –**

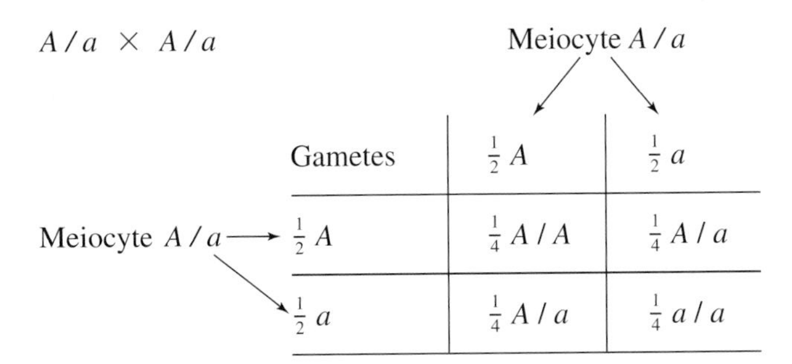

Phenotypic ratio in progeny: $\frac{3}{4}$ **dominant (A / –),**
$\frac{1}{4}$ **recessive (a / a)**

The progeny of these three crosses show different genotypic ratios. In the cross of A / a × a / a, the progeny show a 1 : 1 ratio of A / a : a / a, and this is directly evident in the 1 : 1 phenotypic ratio of dominant to recessive phenotypes. In the cross A / a × A / A, there is a 1 : 1 genotypic ratio of A / a : A / A, but because of dominance all the progeny show the dominant phenotype. In the cross A / a × A / a, the progeny show a genotypic ratio of 1 : 2 : 1 of A / A : A / a : a / a, but at the phenotypic level this is a 3 : 1 ratio of dominant to recessive.

Genotypic ratio	Phenotypic ratio
$\frac{1}{4}$ A / A, $\frac{1}{2}$ A / a	$\frac{3}{4}$ dominant A / –
$\frac{1}{4}$ a / a	$\frac{1}{4}$ recessive a / a

Some useful terms: a cross between two heterozygotes for a single gene is sometimes referred to as a **monohybrid cross;** a heterozygote is sometimes called a **carrier** of the unexpressed recessive allele.

In analyzing single genes, it is sometimes necessary to deduce whether an individual showing an established dominant phenotype is homozygous or heterozygous. A convenient way to do this is by a **testcross.** The individual of dominant phenotype is crossed to a homozygous recessive called a **tester** in this context. If there is roughly a 1 : 1 ratio of dominant to recessive phenotypes in the progeny, then we can infer that the strain under test must have been heterozygous. In fact, if we see *any* recessive phenotypes at all in the progeny, we know that the tested parent must have been a heterozygote because it must have contributed a recessive allele to the homozygous recessive progeny. For example, in mice the difference between black coat and brown coat is determined by a pair of alleles, B (black) and b (brown), and B is dominant to b. A black male of unknown ancestry is testcrossed to a brown female, who must be homozygous b / b, in order to determine the male's genotype. The two possible outcomes are shown below.

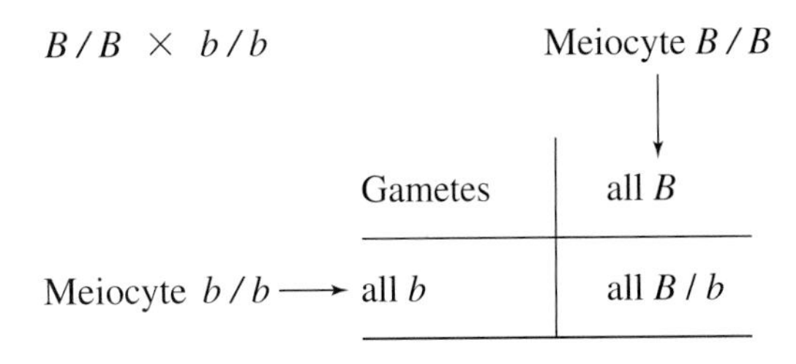

Phenotypic ratio in progeny: **All black**

B / b × b / b — Meiocyte B / b

Gametes	$\frac{1}{2}$ B	$\frac{1}{2}$ b
Meiocyte b / b → all b	$\frac{1}{2}$ B / b	$\frac{1}{2}$ b / b

Phenotypic ratio in progeny: $\frac{1}{2}$ **black (B / b),**
$\frac{1}{2}$ **brown (b / b)**

The cross results in eight offspring, of which three are black and five are brown. This result tells us the male must have been heterozygous B / b, because the brown offspring must be b / b and must have received a b allele from both their mother and their father. Therefore the cross was B / b × b / b, the second of the alternatives shown. Of course Mendel's law of equal segregation leads to an expectation of equal numbers of black and brown progeny (4 : 4), but a chance deviation such as 3 : 5 is likely in such a small sample size.

In plants an alternative way to decide if an individual is homozygous dominant or heterozygous is simply by selfing. If any progeny of recessive phenotype are observed, the individual must have been heterozygous.

Proving homozygosity for dominant alleles is important because animals and plants used for genetic analysis are often maintained as **pure lines** (also called *pure-breeding lines, stocks,* or *strains*). These are populations that when interbred or selfed will always produce progeny of the same phenotype. Such lines by definition must be homozygous. Pure lines are convenient because the animals or plants of the line can simply be allowed to breed randomly to propagate the genotype for research purposes, or for applications such as providing genetically identical seeds for crops. The genotypes *A* / *A* and *a* / *a* can both be bred as pure lines, whereas *A* / *a* cannot.

Using crosses to infer allelic differences and dominance and recessiveness. In deriving the above ratios, we used known parental genotypes involving known alleles with known dominance and recessiveness in order to make predictions about the phenotypic ratios expected in progeny. Working in the opposite direction, to infer allelic differences and their dominance when they are unknown, we have to make appropriate crosses and make deductions based on the progeny of these crosses.

Mendel was the first to establish this sort of experimental procedure, so we can do no better than to examine his work. As an example of the protocol, we shall use the character of seed color in peas, which was one of the cases studied by Mendel (Foundations in Genetics 5-1). The crossing procedure is shown in the upper panel of Figure 5-5. In describing this procedure, the symbol **P** is the conventional designation for the **parental generation,** the two different pure-breeding parents crossed to start the analysis. The progeny of such a cross is called the $\mathbf{F_1}$, or **first filial generation** (derived from the Latin words *filius,* "son," and *filia,* "daughter"). If the F_1 individuals are allowed to interbreed, their progeny are the $\mathbf{F_2}$, or **second filial generation.** In the P generation, we cross two pure-breeding lines of peas, one yellow and one green, to produce the F_1. An F_2 is then derived by selfing or intercrossing the F_1 plants. We find that the F_1 plants all have yellow seeds, and that of the F_2 plants $\frac{3}{4}$ have yellow seeds and $\frac{1}{4}$ have green seeds, a 3 : 1 phenotypic ratio. From these results two deductions can be made (see lower panel of Figure 5-5):

1. The green phenotype disappears in the F_1 but reappears in the F_2, so we can deduce that yellow must have dominated green in the F_1. Therefore yellow is the dominant phenotype and green the recessive.
2. We know that a 3 : 1 ratio (as found in the F_2) is diagnostic of a monohybrid cross, a cross between two heterozygotes for a single gene. Therefore individuals in the F_1 must have been heterozygotes of a genotype we can represent by *Y* / *y*, where *Y* stands for a dominant allele determining the yellow phenotype and *y* stands for a recessive allele determining the green phenotype.

MESSAGE

Genetic analysis works in two directions:

1. Crossing parents of known genotypes and dominance relations to obtain specific progeny types
2. Observing progeny phenotypes and ratios to infer parental genotypes and dominance

The analytical technique invented by Mendel has revealed the genotype, inheritance patterns, and dominance of phenotypes in many different organisms. Some more examples follow.

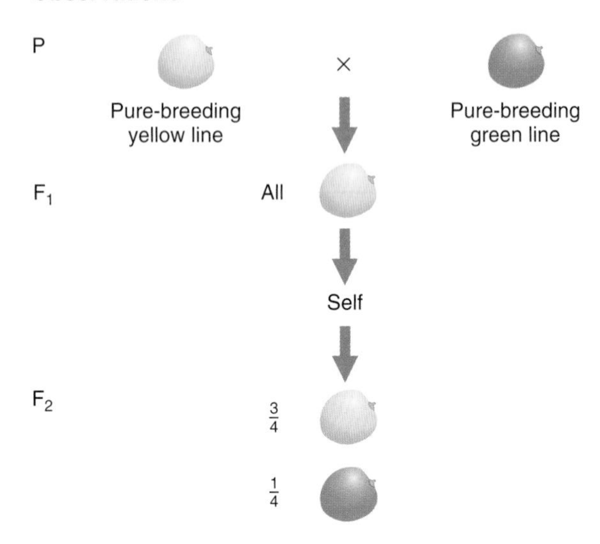

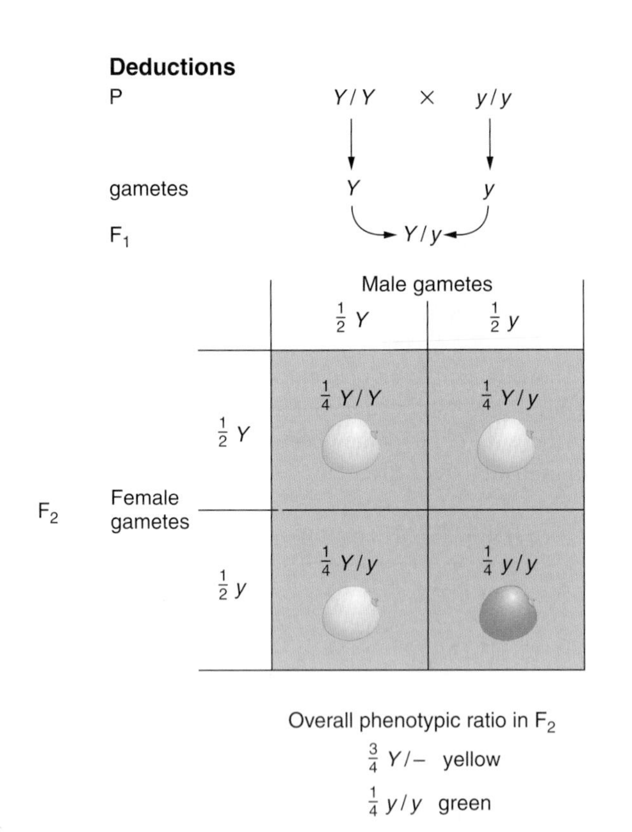

Figure 5-5 Using pure-breeding lines to deduce genotypes and dominance and recessiveness.

In yeast (haploid) the colonies are normally white in color. However, a rare strain arose that had pink colonies. A cross was made between a white and a pink strain and a tetrad analysis was performed. All tetrads showed 2 white spores and 2 pink spores. This result shows the 1 : 1

FOUNDATIONS OF GENETICS 5-1

Mendel, Originator of the Gene Concept

In his 1866 paper "Experiments on Plant Hybridization," a Moravian monk, Gregor Mendel, laid down the basis for the entire subject of genetics by postulating for the first time the existence of the discrete hereditary determinants we now call *genes.* The introductory statement of his paper summarized the purpose of Mendel's study:

> Artificial fertilization undertaken on ornamental plants to obtain new color variants initiated the experiments to be described here. The striking regularity with which the same hybrid form always reappeared whenever fertilization between like species took place suggested further experiments whose task it was to follow the development of hybrids and their progeny.

Gregor Mendel (Moravian Museum, Brno.)

Mendel's choice of an experimental organism was the garden pea. This was a choice based largely on practical convenience. Peas have large flowers that are easily self-pollinated or cross-pollinated. Also, they are annual plants that take up relatively little space in an experimental plot. Even more important, peas were available to Mendel in many different pure-breeding lines. A pure line of plants is a stock of seeds all of the same genetic constitution, so that a cross of any two will always produce the same phenotype. (We now know—but Mendel at first did not know—that they breed in this way because they are genetically homozygous.) These pure lines showed contrasting phenotypes for certain characters. For example, one line had yellow seeds and another had green seeds, contrasting phenotypes of the character *seed color.*

Green and yellow peas. (Leonard Lessin, FPBA.)

Upon crossing these lines he obtained progeny that were called "hybrids." A hybrid is an individual produced from genetically different parents. The hybrid seeds were all yellow. However, when the hybrids were grown up and self-pollinated, $\frac{3}{4}$ of the seeds in the next generation were yellow and $\frac{1}{4}$ were green, the now famous 3:1 ratio that we have seen is produced in a monohybrid cross.

Another character Mendel used was flower color. Two pure lines, one with purple flowers and the other with white, gave hybrids that were all purple. When these hybrid purple-flowered plants were selfed, there was a 3:1 ratio of purple- to white-flowered plants in the next generation. Mendel repeated the analysis with a total of seven different characters and obtained 3:1 ratios in each case. Hence by "following the development of hybrids" (see quote above) Mendel had arrived at a repeatable and consistent hereditary pattern, which demanded an explanation.

To explain the 3:1 ratios, Mendel devised the following hypothesis:

1. The difference between yellow and green is caused by differences in discrete hereditary determinants he called *factors* (what we call "genes" today).
2. The factors exist in pairs, one pair for each character.
3. One of the pairs of phenotypes was dominant over the other in each character studied. For example, in seed color yellowness is dominant over greenness, which is recessive, and in flower color purple is dominant to white.
4. At gamete formation the members of a pair separate, each into $\frac{1}{2}$ of the gametes.
5. Male and female gametes fuse randomly.

Pea flowers. (John Kaprielian/Photo Researchers.)

(continued on next page)

(continued from preceding page, FOUNDATIONS OF GENETICS 5-1 ***Mendel, Originator of the Gene Concept)***

In his analysis he represented genes with letters, the beginning of the practice still used today. He let *Y* stand for the yellow-determining allele, and *y* for the green-determining allele. Using this symbolism, he represented the components of his hypothesis in the following way: the original pure lines were *Y / Y* and *y / y* (points 1 and 2), and the hybrid was *Y / y* and yellow (point 3); the hybrids produce gametes that were $\frac{1}{2}$ *Y* and $\frac{1}{2}$ *y* [point 4 (now called "Mendel's first law")]. Fusion of male and female gametes of these two types (point 5) gave the 3:1 ratio of dominant to recessive phenotypes in the next generation. Thus, Mendel had succeeded not only in identifying the existence of genes, but also in establishing a mechanism of their inheritance.

He also made crosses between pure lines differing in *two* characters, for example a green-seeded, purple-flowered line and a yellow-seeded, white-flowered line. We shall follow this cross more carefully in Chapter 6. Briefly, he used the results of such crosses to show that the genes for different characters behaved independently at gamete formation, a principle now called "Mendel's second law."

Without knowing it, Mendel had devised rules of inheritance that were applicable not only to peas, but to virtually all eukaryotic organisms. His work was little appreciated at the time, and it was only when his laws were independently rediscovered in the early twentieth century that he received the credit for them. He is credited not only with the discovery of hereditary laws, but also with the establishment of a rigorous scientific protocol for identifying specific genes and their alleles, which is still used today.

segregation expected of heterozygous meiocytes. Therefore we can deduce that the phenotypic difference was determined by two alleles of one gene. Since in this case we know the wild-type phenotype (white), we can switch to the symbolism that uses a "+" for the wild-type allele. Recall that in such cases the gene is named on the basis of its mutant phenotype. Recessive mutant alleles are given lowercase letter symbols, and dominant alleles are shown in uppercase. However, in this example we do not know if the mutant allele is dominant or recessive, because yeast is haploid and the dominance test cannot be easily done. Nevertheless we can make the assumption that the mutant allele would be recessive and let p = pink and p^+ = white (wild type). Then the cross must have been as follows:

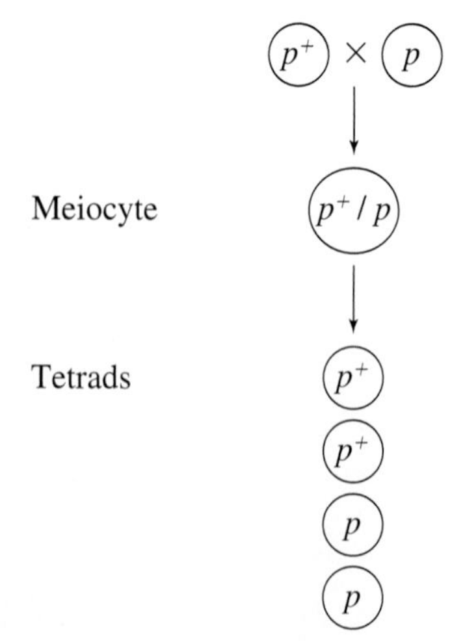

Fruit flies *(Drosophila)* normally have relatively large wings. In the laboratory population a fly was found that had miniature wings. When this fly was crossed to the wild type, the F_1 all had normal large wings. When F_1 flies were allowed to interbreed, thus producing an F_2, 400 flies were counted and, of these, 305 had normal wings and 95 had miniature wings. This result tells us that:

1. The two phenotypes must be determined by two alleles of one gene, because the observed 3:1 F_2 ratio is based on 1:1 meiotic segregation in a heterozygote.
2. Large wings must be dominant to miniature wings, because the F_1 were all large-winged. Since we know the mutant allele is recessive, we can use a lowercase symbol, letting m = miniature wings and m^+ = large.

Now we can deduce the cross must have been

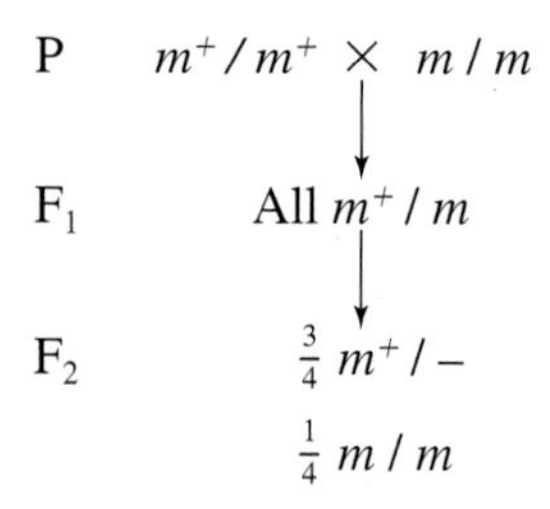

Autosomal and Sex-Linked Genes

In many organisms, sex is determined by special **sex chromosomes.** This is true for most animals, such as humans and fruit flies, and dioecious species of plants, those with male and female sex organs on separate plants. In such cases it is necessary to distinguish between the sex chromosomes and the remaining "regular" chromosomes, which are called **autosomes.** Generally there is only one pair of sex chromosomes. In human females the sex chromosomes are a pair of chromosomes called **X chromosomes,** and in males there is nonidentical pair, an X and a **Y chromosome.** Hence, using a letter A to represent an autosome, a female is designated 44A XX and a male is 44A XY. The set of

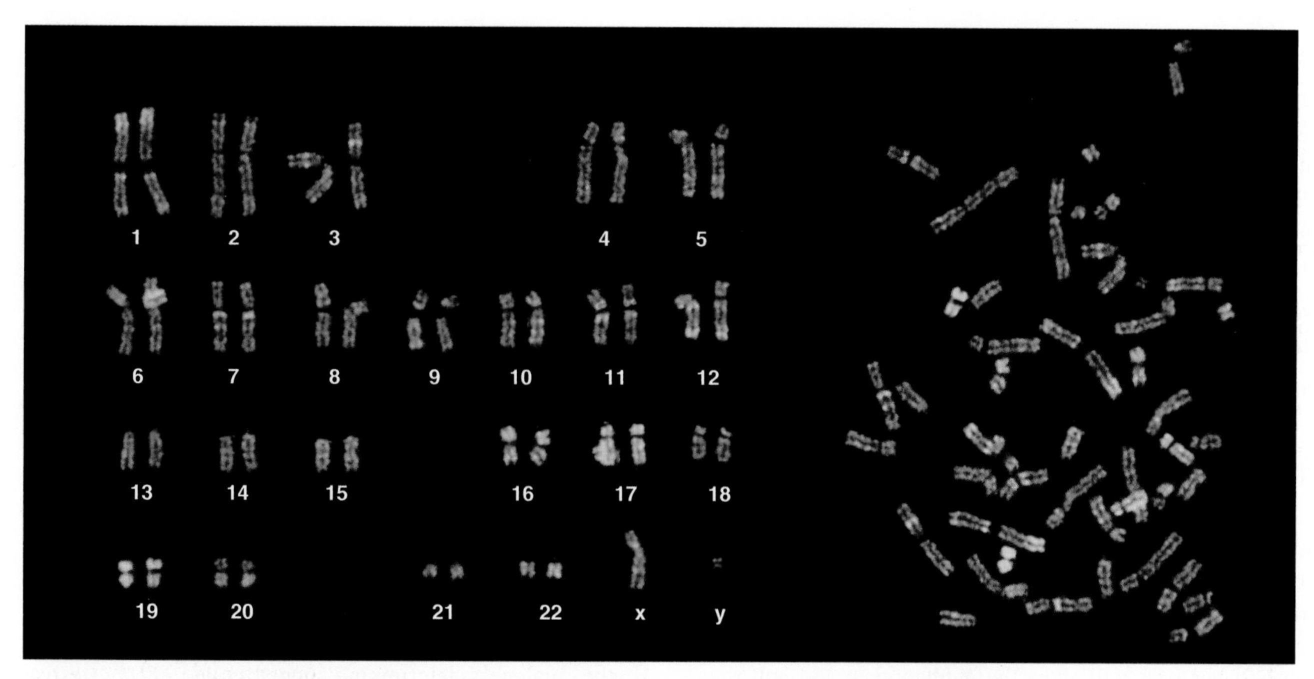

Figure 5-6 Stained human chromosomes. Under the microscope the chromosomes appear as a jumbled cluster, as shown to the right. This array is photographed; the individual chromosomes are cut out of the photograph and then grouped by size and banding pattern, as shown at the left. The chromosome set of a male is shown.

human chromosomes is shown in Figure 5-6. In females, the pair of homologous X chromosomes segregates at meiosis, so all eggs have a single X. In males, the X and the Y chromosomes are homologous only in a small region at one end, but at meiosis they pair in this region and segregate equally so that $\frac{1}{2}$ the sperm bear an X and $\frac{1}{2}$ bear a Y. When these sperm fuse randomly with the X-bearing eggs, 50 percent male (XY) and 50 percent female (XX) progeny result.

	Gametes	$\frac{1}{2}$ X (Sperm)	$\frac{1}{2}$ Y (Sperm)
Eggs ⟶	all X	$\frac{1}{2}$ XX (female)	$\frac{1}{2}$ XY (male)

Hence we see that the common observation that there are equal numbers of males and females is based on equal segregation of the X and Y chromosomes at meiosis.

In humans, sex is determined by the presence or absence of the Y chromosome. However, in fruit flies sex is determined by the number of X chromosomes in relation to the number of autosomes. In Table 5-1 the individuals with irregular numbers of sex chromosomes demonstrate this subtle difference in chromosomal sex determination.

The key subdivisions of the sex chromosomes in humans and in a dioecious plant are shown in Figure 5-7. Notice that for most of its length the X chromosome has

TABLE 5-1 Chromosomal Determination of Sex in *Drosophila* and Humans

Species	SEX CHROMOSOMES XX	XY	XXY	X0*
Drosophila	♀	♂	♀	♂
Humans	♀	♂	♂	♀

* "0" indicates that the chromosome is absent.

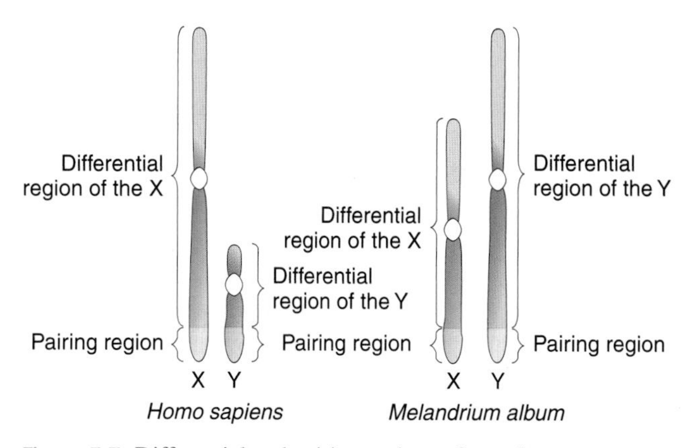

Figure 5-7 Differential and pairing regions of sex chromosomes of humans and of the plant *Melandrium album.* The regions were located by observing where the chromosomes paired during meiosis and where they did not. Genes on the differential regions show X-linked or Y-linked inheritance patterns.

no counterpart on the Y. This region of the X, called the **differential region,** contains many genes with alleles that produce distinct phenotypes. These genes for the most part are not involved in sex determination. Genes on this differential region of the X are called **X-linked genes.** An X-linked gene will have two allelic representatives in females, since females have two X chromosomes, but only one representative in males, since males have one X, and the Y chromosome lacks these genes. Thus, the genotype of a female can be *A* / *A* or *A* / *a* or *a* / *a*, but males can be only *a* or *A*. We will adopt a special symbolism to designate X-linked genes, showing the X chromosome with an allele written as a superscript; hence, the genotypes in the previous sentence become X^AX^A, X^AX^a, X^aX^a (females) and X^AY and X^aY (males).

X-linked genes show a pattern of inheritance called **X-linked inheritance.** This pattern is based on two key facts. First, male offspring inherit their Y chromosome from their fathers and thus must inherit their X chromosome from their mothers. Hence, if a mother is a heterozygote X^AX^a, then half her sons will show the A phenotype and half the a phenotype, irrespective of the genotype of her mate. Second, daughters inherit X-linked genes equally from mother and father. Hence the X genotype of the father can influence the genotypes of his daughters but not of his sons. These differences are illustrated in the example below. Note that the two effects just described result in different phenotypic ratios among sons (a 1 : 1 ratio) and daughters (a 1 : 0 ratio).

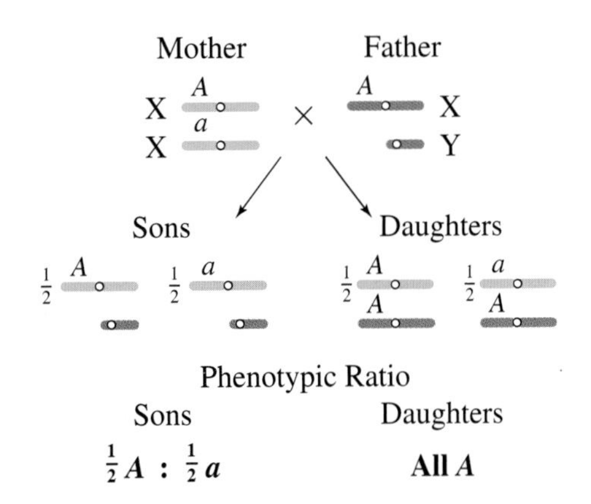

In contrast, remember that autosomal genes show exactly the same phenotypic ratios in male and female progeny. For example, in an autosomal monohybrid cross *A* / *a* × *A* / *a* both male and female progeny show a 3 : 1 ratio of *A* / – : *a* / *a*. Indeed the observation of *identical* phenotypic ratios in male and female progeny is an experimental diagnostic for telling that a gene is autosomal in location. Conversely, observing *different* ratios between the two sexes of progeny is one diagnostic for inferring that the genes concerned have a sex chromosome location.

A typical case of X-linked inheritance is illustrated by the red and white alleles of an eye-color gene in *Drosophila* (Figure 5-8). Note the crosses in which the male and female progeny show different phenotypic ratios (the diagnostic for inferring X-linked inheritance). For example, in the F_2 of the first cross, there is a 3 : 1 ratio of red to white, but unlike cases of autosomal gene inheritance, the white flies are all males. Furthermore, in the second cross the female progeny in the F_1 are all red but the males are all white.

There are few genes in the short homologous region of the X and Y chromosomes (see Figure 5-7). We will not consider the inheritance patterns of these genes other than to say that they follow essentially an autosomal pattern, and therefore this region is sometimes called the *pseudoautosomal region.* However, for the segregation of the X and Y chromosomes to opposite poles to take place, the homologous regions are crucial: they must pair, and furthermore at least one crossover must occur.

The differential region of the Y contains few genes and even fewer with known alleles that cause distinct phenotypic variants. Genes on the differential region of the Y are called **Y-linked genes,** and these are expressed only in males. One important Y-linked gene in humans is the maleness-determining gene, called *TDF* (testis-determining factor). The role of this gene in sex determination will be discussed in Chapter 16.

The bearer of the pair of sex chromosomes that differ from each other is called the **heterogametic sex.** Hence, in humans and fruit flies, as in most cases of chromosomal sex determination, the male is the heterogametic sex. However, in chickens and moths, the female is the heterogametic sex.

To summarize, we see that genes on the sex chromosomes present an inheritance pattern different from that of genes on the autosomes **(autosomal genes).** One key difference is that in crosses of some sex chromosome genotypes there are large differences in the phenotypic ratios among the male and female progeny. Indeed the observation of *identical* phenotypic ratios in male and female progeny is an experimental diagnostic for telling that a gene is autosomal in location. Conversely, observing *different* ratios between the two sexes of progeny is one diagnostic for inferring that the genes concerned have a sex chromosome location. The precise pattern of inheritance is determined by exactly where a gene is located on the sex chromosomes.

HUMAN PEDIGREE ANALYSIS

In humans, controlled experimental crosses cannot be made. Furthermore, the number of offspring from any mating is quite small. Therefore geneticists must resort to scrutinizing family records over several generations, including a variety of relatives, in the hope that informative matings have been made that can be used to deduce dominance and distinguish autosomal from X-linked inheritance. The investigator traces the history of some variant phenotype (such as phenylketonuria) back through the history of the

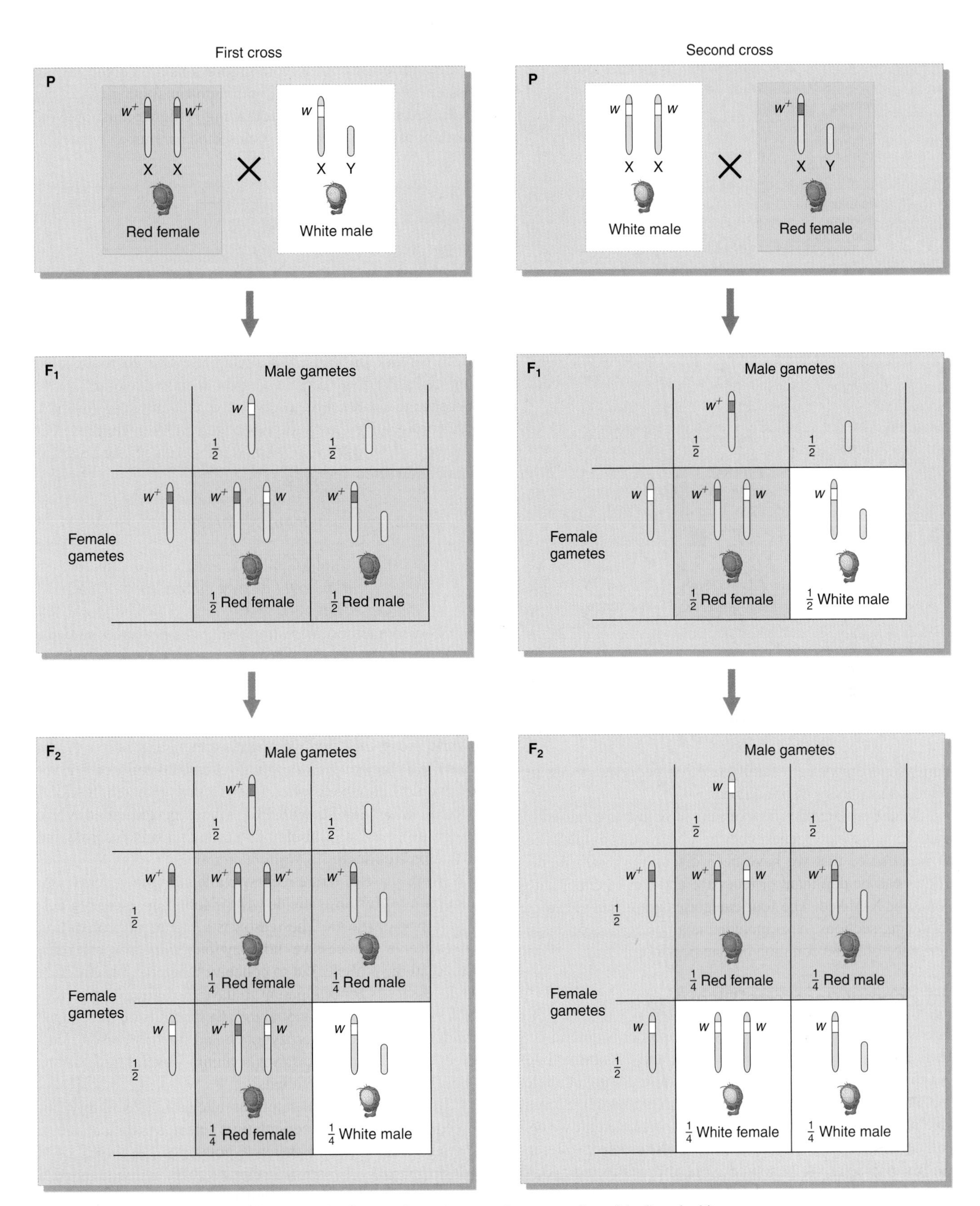

Figure 5-8 Explanation of the different results from reciprocal crosses between red-eyed (red) and white-eyed (white) *Drosophila*. The alleles are X-linked, and the inheritance of the X chromosome explains the phenotypic ratios observed, which are different from those of autosomal genes.

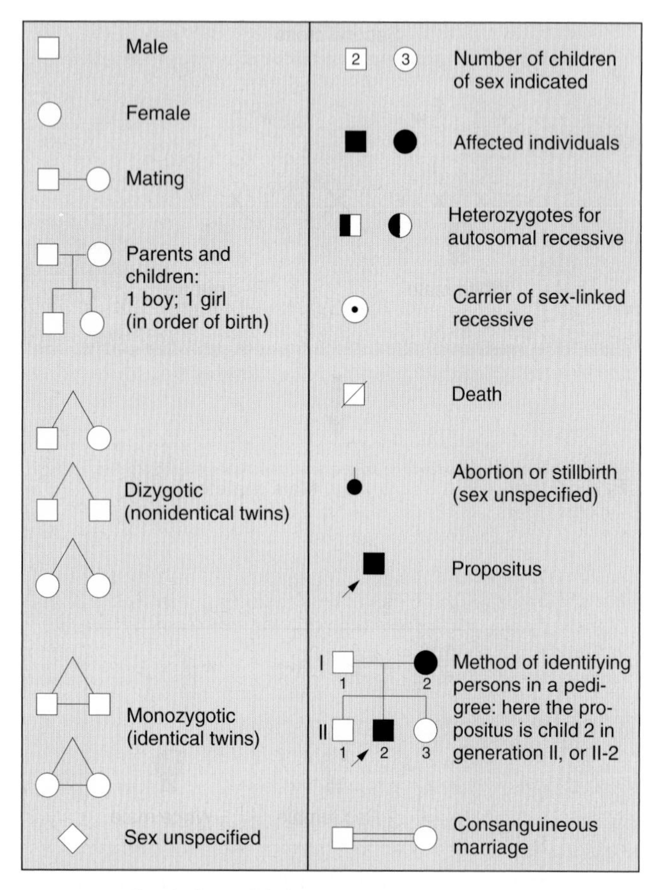

Figure 5-9 Symbols used in human pedigree analysis. A propositus is the first case brought to the attention of the geneticist who drew the pedigree. *(After W. F. Bodmer and L. L. Cavalli-Sforza,* Genetics, Evolution, and Man. *© 1976 by W. H. Freeman and Company.)*

family and draws up a family tree, or **pedigree,** using the standard symbols given in Figure 5-9.

Many genetic disorders of human beings are caused by mutations of single genes, and these show the simple types of inheritance that we have been discussing. Such mutant alleles can be dominant or recessive and can be either autosomal or X-linked. The four categories are discussed in the following sections. Also included is a description of the inheritance of some common human polymorphisms.

Autosomal Recessive Disorders

The mutant phenotype of a recessive disorder is determined by homozygosity for a recessive allele, and the normal phenotype is determined by the corresponding dominant allele. In Chapter 3 we saw that phenylketonuria (PKU) is a recessive phenotype. PKU is determined by an allele that we can call p, and the normal condition by P. Therefore, sufferers of this disease are of genotype p / p, and unaffected people are either P / P or P / p. What patterns in a pedigree would reveal such an inheritance? Two key features of the pedigree are, first, the affected progeny include both males and females equally and, second, generally the disease appears in the progeny of unaffected parents. When we observe that male and female phenotypic proportions are equal, we can assume that we are dealing with autosomal inheritance, not X-linked inheritance. The following pedigree illustrates the birth of affected children to unaffected parents:

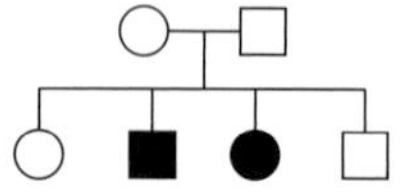

From this pattern we can deduce autosomal inheritance, with the recessive allele responsible for the exceptional phenotype (indicated in black). Furthermore, we can deduce that the parents must both be heterozygotes, P / p. Both parents must have a p allele because each contributed one to each affected child, and both must have a P allele because the parents themselves are phenotypically normal. We can identify the genotypes of the children (in the order shown) as $P / -$, p / p, p / p, and $P / -$. Hence, the pedigree can be rewritten

$$P / p \times P / p$$

$$P/- \quad p/p \quad p/p \quad P/-$$

Notice another interesting feature of pedigree analysis: even though Mendelian rules are at work, Mendelian ratios are rarely observed in single families, because the sample sizes are too small. In the above example, we see a 1:1 phenotypic ratio in the progeny of what is clearly a monohybrid cross, in which we might expect a 3:1 ratio of $P / -$ to p / p. If the couple were to have, say, 20 children, the ratio would undoubtedly be closer to the predicted 15 unaffected children and 5 with PKU (the expected monohybrid 3:1 ratio), but in a sample of four any ratio is possible and all ratios are found.

In the case of a rare recessive allele, in the general population most of these alleles will be found in heterozygotes, not in homozygotes. The reason is a matter of probability: to conceive a recessive homozygote, *both* parents must have had the p allele, but to conceive a heterozygote all that is necessary is for *one* parent to have the allele. The formation of an affected individual usually depends on the chance union of unrelated heterozygotes, and for this reason pedigrees of autosomal recessives spanning several generations show few affected individuals.

If there are rare deleterious alleles in the family, inbreeding (mating between relatives) increases the chance of matings between two heterozygotes. An example of a cousin marriage is shown in Figure 5-10. Individuals III-5 and III-6 are first cousins and produce two affected children. Hence these two parents must have been heterozygotes. You can see from the figure that an ancestor who is a heterozygote may produce many descendants who are also

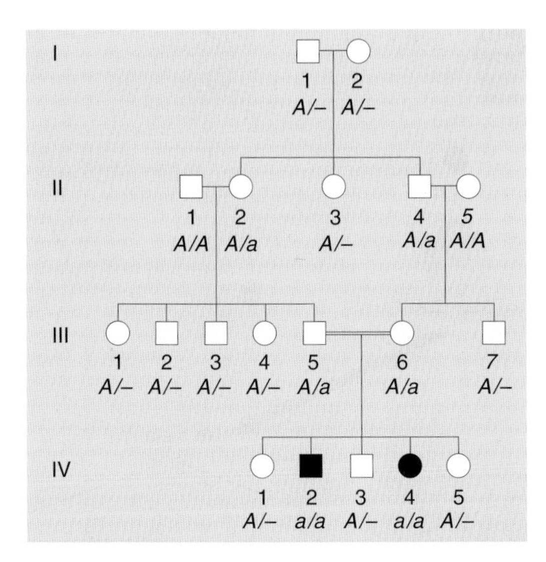

Figure 5-10 Pedigree of a rare recessive phenotype determined by a recessive allele *a*. Gene symbols normally are not included in pedigree charts, but genotypes are inserted here for reference. Note that individuals II-1 and II-5 have married into the family; they are assumed not to carry the recessive allele *a* because the heritable condition under scrutiny and its causative allele are rare. Note also that it is not possible to be certain of the genotype in some individuals with normal phenotype; such individuals are indicated by A / –.

heterozygotes. Even though many are shown as *A* / –, some of these will be *A* / *a*. Matings between relatives thus run a higher risk of producing abnormal homozygous recessives than do matings between nonrelatives. It is for this reason that first cousin marriages are responsible for a large portion of individuals with autosomal recessive disorders in human populations.

Albinism (Figure 5-11) is another rare condition that is inherited in a Mendelian manner as an autosomal recessive phenotype in many animals, including humans. The striking "white" phenotype is caused by a defect in an enzyme that synthesizes melanin, the pigment responsible for most black and brown coloration of animals. In humans, such coloration is most evident in hair, skin, and retina, and its absence in albinos (who have the homozygous recessive genotype *a* / *a*) leads to white hair, white skin, and eye pupils that are pink because of the unmasking of the red hemoglobin pigment in blood vessels in the retina. Since its genetic and molecular bases are well known, albinism affords a useful opportunity to examine the phenotype at several different organizational levels, and these are presented in Figure 5-12 on the next page. Albinism has the same biological basis in virtually all animals.

MESSAGE

In pedigrees, an autosomal recessive disorder is revealed by the appearance of the phenotype in both the male and female progeny of unaffected individuals, who may be inferred to be heterozygous carriers.

Autosomal Dominant Disorders

In autosomal dominant disorders, the normal allele is recessive and the abnormal allele is dominant. It might seem paradoxical that a rare disorder can be dominant, but remember that dominance and recessiveness are simply reflections of how alleles interact in the heterozygous condition and are not defined in terms of predominance in the population. An example of a rare autosomal dominant phenotype is achondroplasia, a type of dwarfism (Figure 5-13 on page 133). In this case, people with normal stature are genotypically *d* / *d*, and the dwarf phenotype in principle could be *D* / *d* or *D* / *D*. However, it is believed that in *D* / *D* individuals the two "doses" of the *D* allele produce such a severe effect that this genotype is lethal. If true, all achondroplastics are heterozygotes. Thus, paradoxically, a mutation that shows a dominant phenotypic effect on stature is a recessive with respect to lethality.

In pedigree analysis, the main clues for identifying an autosomal dominant disorder are, first, that affected fathers and affected mothers transmit the phenotype to both sons and daughters, and second that the phenotype tends to appear in every generation of the pedigree. Again, the representation of both sexes equally among the affected offspring

Figure 5-11 An albino. The phenotype is caused by homozygosity for a recessive allele, say, *a* / *a*. The dominant allele *A* determines one step in the chemical synthesis of the dark pigment melanin in the cells of skin, hair, and eye retinas. In *a* / *a* individuals this step is nonfunctional, and the synthesis of melanin is blocked. *(© Yves Gellie/Icône.)*

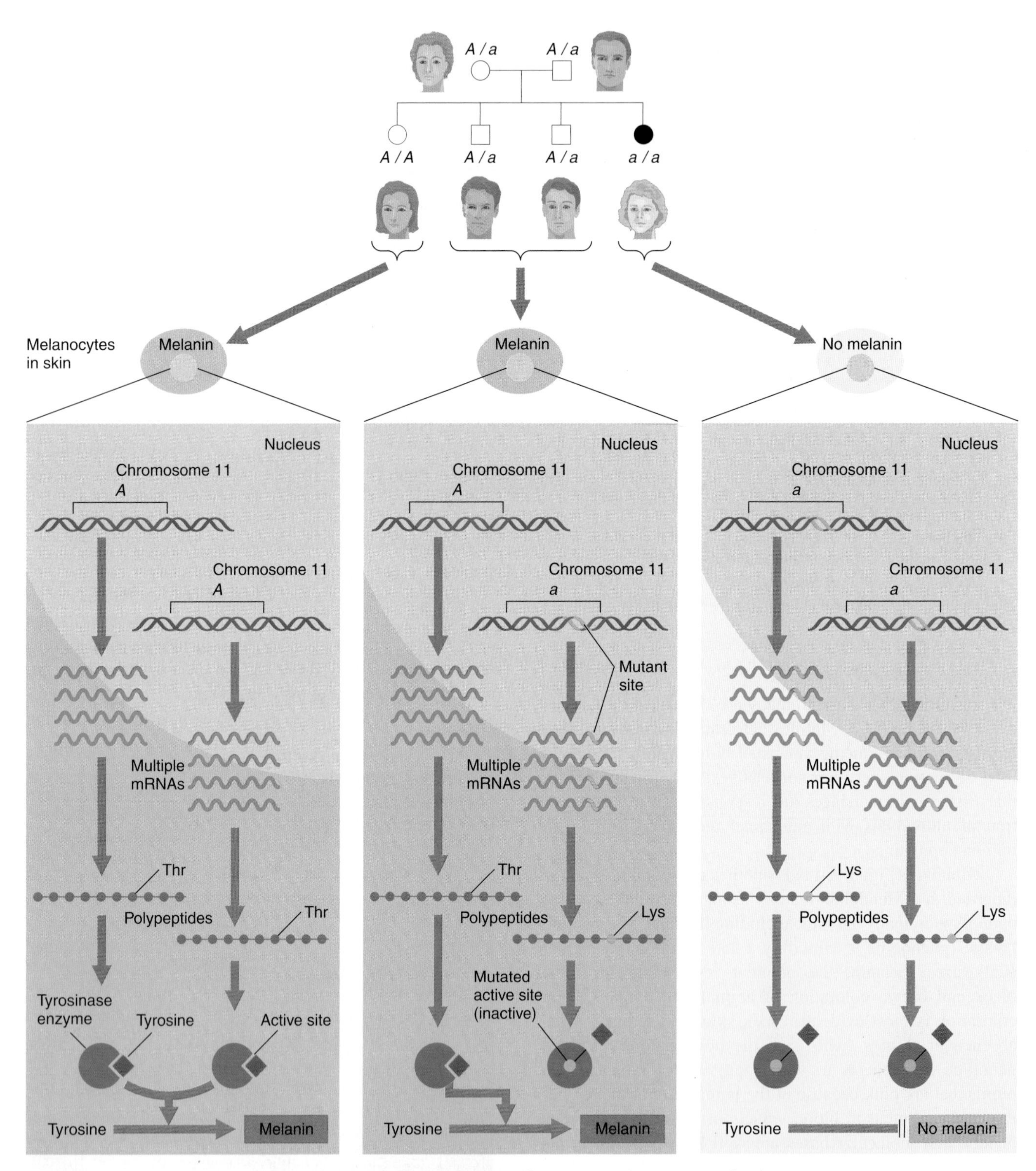

Figure 5-12 Genetics and the molecular biology of albinism. In the pedigree, parents heterozygous for the recessive albinism allele produce three *A* / – progeny, who have melanin in their cells, and one *a* / *a* female, who is albino. The three panels at the bottom of the figure show why this is so. The gene for the *A* allele codes for a polypeptide chain that constitutes tyrosinase, an enzyme necessary to convert tyrosine to melanin. A mutation in the gene for the *a* allele causes the replacement of a threonine with a lysine. The resulting polypeptide is unable to catalyze melanin production. The single *A* allele in heterozygotes encodes sufficient functional polypeptide to create melanin. The *a* / *a* individual produces no functional polypeptide and, hence, no melanin.

Figure 5-13 The human achondroplasia phenotype, illustrated by a family of five sisters and two brothers. The phenotype is determined by a dominant allele, which we can call D, that interferes with bone growth during development. Most members of the human population can be represented as d / d in regard to this gene. This photograph was taken upon the arrival of the family in Israel after the end of the Second World War. *(UPI/Bettmann News Photos.)*

argues against X-linked inheritance. The phenotype appears in every generation because generally an abnormal dominant allele carried by any given individual must have come from a parent in the previous generation who probably also expressed it. An exception is mutation. Abnormal dominant alleles can arise anew by mutation. Mutation is relatively rare, but for dominant disorders in which the disorder affects reproduction, mutation is the source of most cases in the population.

A typical pedigree for a dominant disorder is shown in Figure 5-14. Once again, notice that precise Mendelian ratios are not necessarily observed in families of limited size.

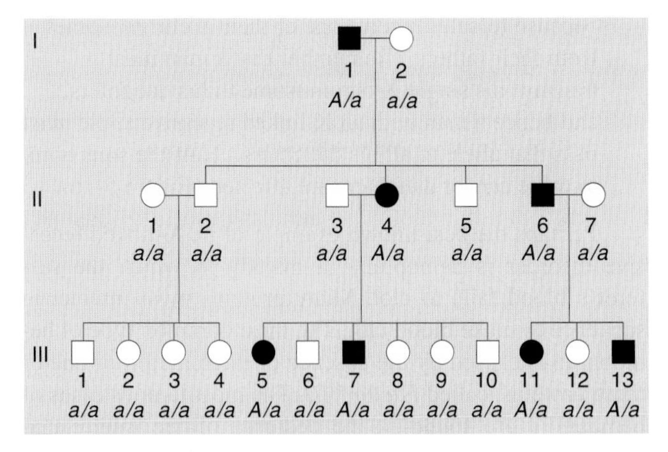

Figure 5-14 Pedigree of a dominant phenotype determined by a dominant allele A. In this pedigree, all the genotypes have been deduced.

As with recessive disorders, individuals bearing one copy of the rare allele (A / a) are much more common than those bearing two copies (A / A), so most affected people are heterozygotes, and virtually all matings involving dominant disorders are $A / a \times a / a$. Therefore, when the progeny of such matings are totaled, a 1 : 1 ratio is expected of unaffected (a / a) to affected individuals (A / a).

Huntington disease, with which we began this chapter, is another example of an autosomal dominant disorder. Hence we can let H be the disease-causing mutant allele and h the normal recessive counterpart. Since the dominant disease allele is rare, most people who have HD are heterozygotes. Hence if such a heterozygote marries an unaffected person, the mating is always genetically H / h (HD) $\times$ h / h (normal) and the law of equal segregation dictates that $\frac{1}{2}$ the progeny will have the HD genotype. Because disease symptoms generally develop after parents have their children, a person with Huntington disease does not have the opportunity to exercise reproductive choice, and an affected parent knows that the disease has a 50% chance of appearing in his or her child. This tragic pattern has led to a drive to find ways of identifying people who carry the abnormal allele before they experience the symptoms of the disease, so that if they do carry the mutant allele, they can decide whether or not to have children. The discovery of DNA variants adjacent to the H allele that act as "markers" were of great help in this sort of diagnosis. Most recently, finding the DNA sequence of H alleles has led to the ability to diagnose the alleles directly.

MESSAGE
Pedigrees of autosomal dominant disorders show affected males and females in each generation and also show affected men and women transmitting the condition to equal proportions of their sons and daughters.

Human Polymorphisms

In human populations there are many examples of polymorphisms (generally dimorphisms) in which the common alternative phenotypes of the character are determined by alleles of a single gene. Some examples are the dimorphisms for chin dimple versus none, for pendulous earlobes versus attached, and for widow's peak hairline versus straight. The interpretation of pedigrees for polymorphisms is somewhat different from those for rare disorders, because by definition the morphs in a dimorphism are common.

Let's look at a pedigree for an interesting human dimorphism. Most human populations are dimorphic for the ability to taste the chemical phenylthiocarbamide (PTC): people can either detect it as a foul, bitter taste or—to the great surprise and disbelief of tasters—cannot taste it at all. From the pedigree in Figure 5-15, we can see that two tasters (for example, I-3 and I-4) sometimes produce nontaster children. This makes it clear that the allele for ability to taste is dominant, and that the allele for nontasting is recessive because it must have been masked in the parents I-3 and I-4. Knowing this, we can designate the mating between I-3 and I-4 as $T / t \times T / t$, and their children II-4, II-5, and II-6 as t / t, $T / -$, and T / t.

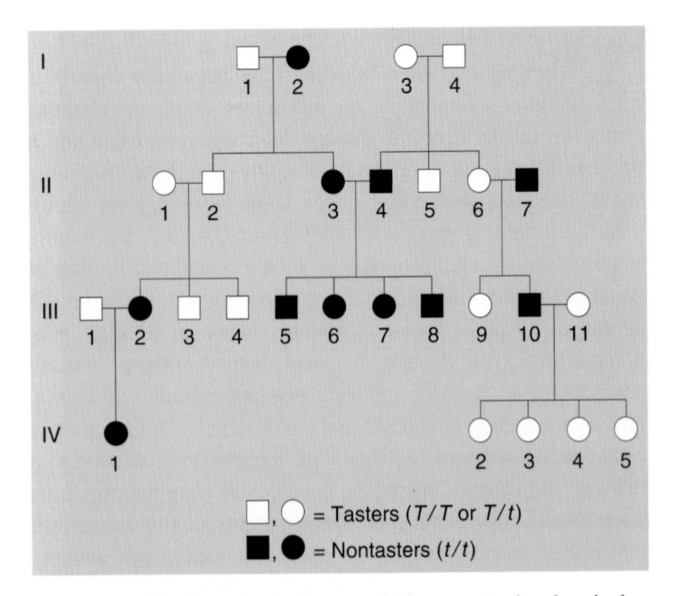

Figure 5-15 Pedigree for the human ability to taste the chemical PTC. These phenotypes constitute a genetic dimorphism, defined as the presence of two *common* forms of a character. Hence the assumptions about rarity of the alleles are quite different from those in a pedigree for a rare disorder such that in Figure 5-14.

Notice, however, that almost all people who marry into this pedigree carry the recessive allele for inability to taste PTC, either in homozygous or in heterozygous condition. (The heterozygotes can be pinpointed as tasters producing nontaster children, such as II-1.) Hence such a pedigree is quite different from those of rare recessive disorders, for which it is conventional to assume that all who marry into a family are homozygous normal. As both PTC alleles are common, it is not surprising that all but one of the family members in this pedigree married individuals with at least one copy of the recessive allele.

MESSAGE
In a polymorphism the contrasting morphs are often determined by alleles of a single autosomal gene.

X-Linked Recessive Disorders

Phenotypes with X-linked recessive inheritance typically show the following patterns in pedigrees:

1. Many more males than females show the phenotype. This is because a female showing the phenotype can result only from a mating in which both the mother *and* the father bear the allele (for example, $X^AX^a \times X^aY$). If the recessive allele is rare, it is extremely unlikely that both parents carry it. In contrast, a male with the disorder can be produced when only the mother carries the mutant allele. Even though such carrier mothers are rare, they are much more common than matings of two individuals with the mutant allele. Hence almost all individuals showing recessive X-linked disorders are males.
2. None of the offspring of an affected male are affected, but all his daughters must be heterozygous carriers because females receive one of their X chromosomes from their fathers. (Remember that a man must transmit his single X chromosome to his daughters, and hence a man with an X-linked recessive allele *must* pass that allele to all his daughters.) Half the sons born to these carrier daughters are affected (Figure 5-16).

Perhaps the best-known example of an X-linked recessive disorder is hemophilia, a malady in which the sufferer's blood fails to clot. Many proteins must interact in sequence to make blood clot. The most common type of hemophilia is caused by the absence or malfunction of one of these proteins, called *Factor VIII.* The most famous cases of hemophilia are found in the pedigree of the interrelated royal families of Europe (Figure 5-17). The original hemophilia allele in the pedigree arose spontaneously (as a mutation), probably in the reproductive cells of Queen Victoria's

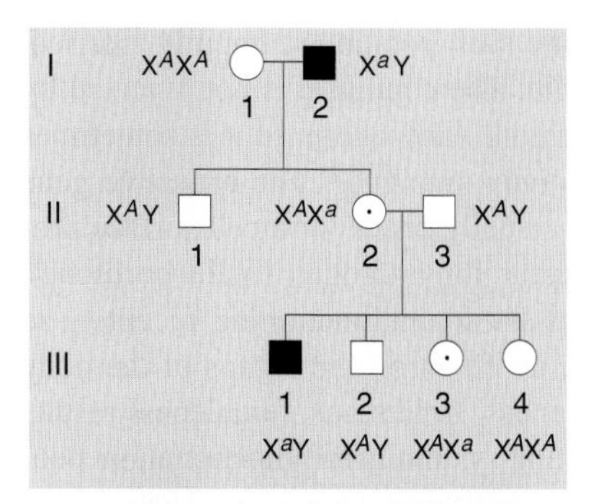

Figure 5-16 Pedigree showing that X-linked recessive alleles expressed in males are then carried unexpressed by their daughters in the next generation, to be expressed again in their sons. Note that III-3 and III-4 could not be distinguished phenotypically, but III-3, like her mother (II-2), has been drawn as a carrier and has the recessive allele on one of her X chromosomes. Therefore III-3 could produce affected children but III-4 could not.

(a)

(b)

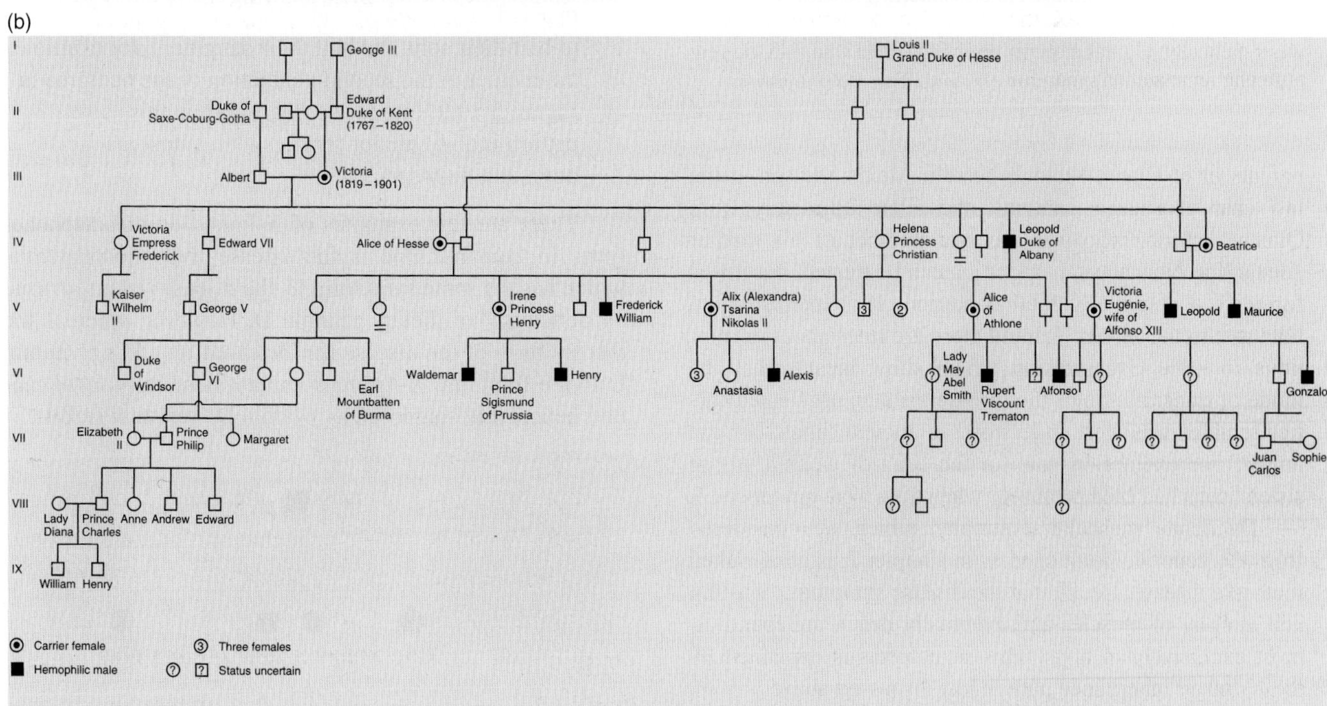

Figure 5-17 The inheritance of the X-linked recessive condition hemophilia in the royal families of Europe. A recessive allele causing hemophilia (failure of blood clotting) arose in the reproductive cells of Queen Victoria, or one of her parents, through mutation. This hemophilia allele spread into other royal families by intermarriage. (a) A painting showing Queen Victoria surrounded by her numerous descendants. (b) This partial pedigree shows affected males and carrier females (heterozygotes). Most spouses marrying into the families have been omitted from the pedigree for simplicity. From examination of this pedigree, do you think it is possible that the present British royal family harbors the recessive allele? *(Part a, Royal Collection, St. James's Palace. © Her Majesty Queen Elizabeth II.; part b modified from C. Stern,* Principles of Human Genetics, *3d ed. © 1973 by W. H. Freeman and Company)*

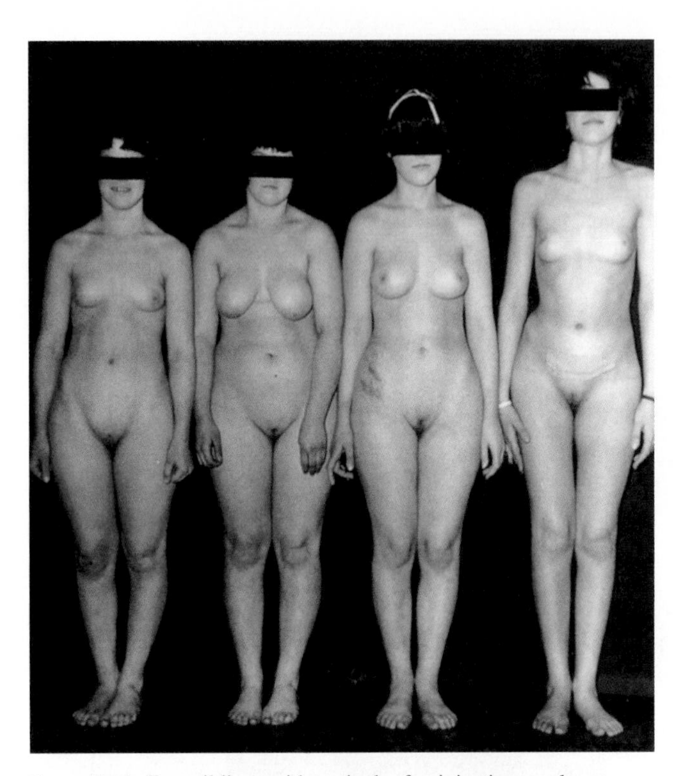

Figure 5-18 Four siblings with testicular feminization syndrome (congenital insensitivity to androgens). All four subjects in this photograph have 44 autosomes plus an X and a Y, but they have inherited the recessive X-linked allele conferring insensitivity to androgens (male hormones). One of their sisters (not shown) who was genetically XX was a carrier and bore a child who also showed testicular feminization syndrome. *(Leonard Pinsky, McGill University.)*

parents or of Queen Victoria herself. Alexis, the son of the last czar of Russia, inherited the allele ultimately from Queen Victoria, who was the grandmother of his mother Alexandra. Nowadays, hemophilia can be treated, but it was formerly a potentially fatal condition. It is interesting to note that in the Jewish Talmud there are rules about exemptions to male circumcision which show clearly that the mode of transmission of the disease through unaffected carrier females was well understood in ancient times. For example, one exemption was for the sons of women whose sisters' sons had bled profusely when they were circumcised.

Duchenne muscular dystrophy, which we considered from the genomic point of view in Chapter 2, is an X-linked recessive disease. Recall that the disease symptoms, wasting and atrophy of muscles and eventually death, are found almost exclusively in boys. This sexual bias is explained by the X-linked inheritance pattern just discussed above.

Testicular feminization syndrome, a rare X-linked recessive phenotype (about 1 in 65,000 male births), provides an interesting insight into sexual differentiation. People afflicted with this syndrome are chromosomally males, 44A XY, but they develop superficially as females (Figure 5-18). They have female external genitalia and a vagina, but no uterus. Testes may be present either in the labia or in the abdomen. Although many such people are happily married, they are, of course, sterile. The condition is not reversed by treatment with male hormone (androgen), so it is sometimes called *androgen insensitivity syndrome.* The causative gene on the X chromosome normally codes for a cell surface protein that acts as a receptor for androgen (male hormone). The mutant allele codes for a malfunctioning receptor, so male hormone can have no effect on the organs of the body that are involved in maleness. In humans, femaleness results when the male-determining system is not functional, so people who are androgen-insensitive develop as females.

X-Linked Dominant Disorders

Pedigrees of rare X-linked dominant phenotypes show the following characteristics:

1. Affected males pass the condition on to all their daughters but to none of their sons (see the grandfather in the pedigree of Figure 5-19). This point directly reflects the transmission of the male X chromosome only to daughters. (Note for comparison that in the case of autosomal dominants, fathers transmit the condition to half their sons and half their daughters. If a disorder is passed from father to son, that disorder cannot be caused by an X-linked allele.)
2. Affected females are mostly heterozygotes, so when married to unaffected males they pass the condition on to half their sons and half their daughters, for example, the mother in the second generation of the pedigree in Figure 5-19. (Note for comparison that this same pattern is also true for females with autosomal dominant disorders.)

There are few examples of X-linked dominant phenotypes in humans. One is the disease hypophosphatemia, which has the same symptoms as the disease rickets, caused by deficiency of dietary vitamin D. However, whereas the dietary form of the disease can be cured by adding vitamin D to the food, the X-linked dominant hereditary form cannot; hence, it is diagnosed as vitamin D–resistant rickets.

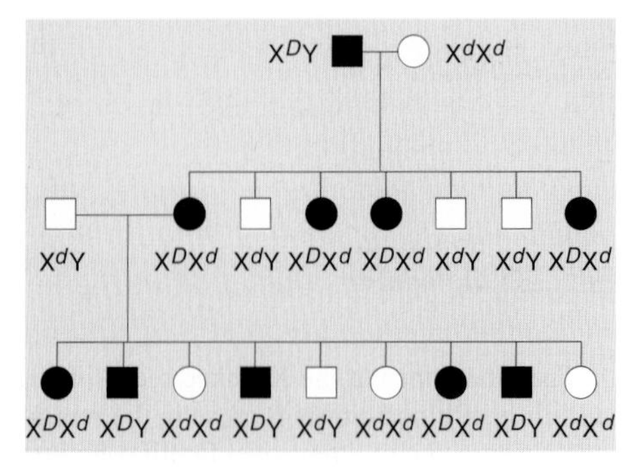

Figure 5-19 Pedigree of an X-linked dominant disorder. Note the typical features: the disorder appears in each generation, males transmit only to females, and females transmit to males and females.

In mammals (including humans) the mechanisms of X-linked dominance and recessiveness are somewhat complicated by the phenomenon of X chromosome inactivation. This topic will be covered in detail in Chapter 16. Briefly, in females one of the X chromosomes is randomly inactivated in each cell. Hence the body of a female heterozygous for an X-linked gene is composed of a mixture of cells, some expressing one allele and some the other. Since the phenotype of the heterozygote is the key for diagnosing dominance and recessiveness, X inactivation means that the phenotype of the heterozygous female must be assessed with great care. For gene products that circulate in the body, the same types of dominance mechanisms that pertain to the cell also can apply to the overall body. For example, female heterozygotes with one normal allele and one hemophilia allele have normal clotting throughout their bodies because cells in which the X containing the normal allele is active will release normal clotting factor that circulates throughout the body.

Calculating Risks in Pedigree Analysis

When a disease allele is known to be present in a family, knowledge of simple gene transmission patterns can be used to calculate the probability of prospective parents' having a child with the disorder. For example, a newly married husband and wife find out that each had an uncle with Tay-Sachs disease. This is a severe autosomal recessive disease caused by malfunction of the enzyme hexosaminidase A. The defect leads to build up of fatty deposits in nerve cells, causing paralysis followed by an early death. The pedigree is as follows:

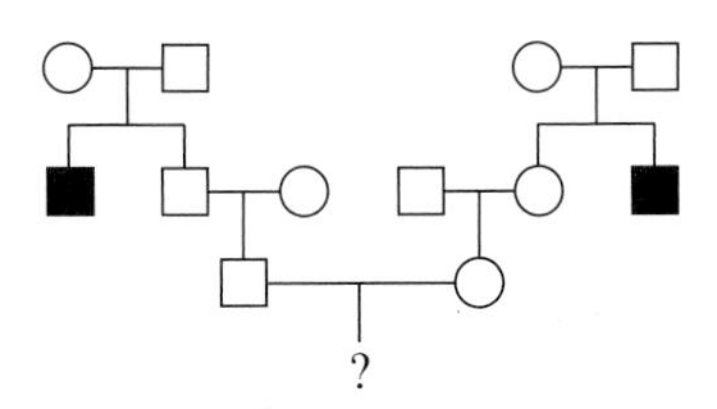

The probability of their having a child with Tay-Sachs can be calculated in the following way. Because neither of the couple has the disease, they can be only homozygous, normal or heterozygous. If they are both heterozygotes, then they stand a chance of both passing the recessive allele on to a child, who would then have Tay-Sachs disease. Hence we must calculate the probability of their being heterozygotes, and then, if so, the probability of passing the deleterious allele on to a child.

1. The husband's grandparents must have both been heterozygotes T / t because they produced a t / t child (the uncle). Therefore, they effectively constituted a monohybrid cross. The husband's father could be T / T or T / t, but we know that the relative probabilities of these genotypes must be $\frac{1}{4}$ and $\frac{1}{2}$, respectively (because the expected progeny ratio in a monohybrid cross is $\frac{1}{4}$ T / T, $\frac{1}{2}$ T / t, and $\frac{1}{4}$ t / t). Therefore, there is a $\frac{2}{3}$ probability that the father is a heterozygote ($\frac{2}{3}$ is the proportion of unaffected progeny who are heterozygotes: $\frac{1}{2}$ divided by $\frac{3}{4}$).
2. The husband's mother is assumed to be T / T, since she married into the family and disease alleles generally are rare. Thus *if* the father is T / t, then the mating to the mother was a cross $T / t \times T / T$ and the expected proportions in the progeny (which includes the husband) are $\frac{1}{2}$ T / T and $\frac{1}{2}$ T / t.
3. The overall probability of the husband's being a heterozygote must be calculated using a statistical rule called the **product rule,** which states

   The probability that two independent events will both occur is the product of their individual probabilities.

   Since gene transmissions in different generations are independent events, we can calculate that the probability of the husband's being a heterozygote is the probability of his father's being a heterozygote *times* the probability of this heterozygous father having a heterozygous son, which is $\frac{2}{3} \times \frac{1}{2} = \frac{1}{3}$.
4. Likewise the probability of the wife's being heterozygous is also $\frac{1}{3}$.
5. If the husband and wife are both heterozygous (T / t), this is a standard monohybrid cross and so the probability of their having a t / t child is $\frac{1}{4}$.
6. Overall, the probability of the couple's having an affected child is the probability of their both being heterozygous and then having an affected child. Again these are independent events, so we can calculate the overall probability as $\frac{1}{3} \times \frac{1}{3} \times \frac{1}{4} = \frac{1}{36}$. In other words, there is a 1 in 36 chance of their having a child with Tay-Sachs disease.

Note that in some Jewish communities the Tay-Sachs allele is not as rare as in the general population. In such cases it cannot be assumed that unaffected people marrying into the family are T / T. If the frequency of T / t heterozygotes in the community is known, this can be factored into the product rule calculation. Nowadays, molecular diagnostic tests for Tay-Sachs alleles are available, and the judicious use of these tests has drastically reduced the frequency of the disease in some communities.

INHERITANCE OF ORGANELLE GENES

Mitochondria and chloroplasts are specialized organelles located in the cytoplasm. They contain a defined subset of the total cell genome (mitochondrial genes are shown in

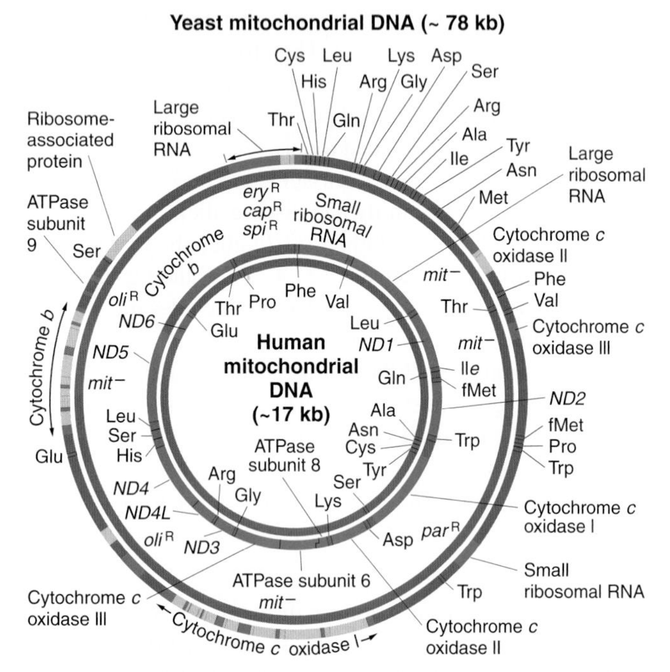

Figure 5-20 Maps of yeast (outer circle) and human (inner circle) mitochondrial DNAs (mtDNAs). Each map is shown as two concentric circles corresponding to the two strands of the DNA helix. Note that the mutants used in yeast mtDNA analysis are shown opposite their corresponding structural genes. Green = exons and uninterrupted genes, red = tRNA genes, yellow = URFs ("unassigned reading frames," probable genes but of unknown function), and blue = absence of an open reading frame. tRNA genes are shown by their amino acid abbreviations; ND genes code for subunits of NADH dehydrogenase. (Note that the human mtDNA map is not drawn to the same scale as the yeast map.)

Figure 5-20). These genes show their own special mode of inheritance. In a cross, both parents contribute equally to the nuclear genome of the zygote. However, the cytoplasmic contribution of the male and the female parent is generally unequal; the egg contributes the bulk of the cytoplasm and the sperm essentially none. Because organelles reside in the cytoplasm, the organelle genes generally show strictly **maternal inheritance.** In other words, the female parent contributes the organelles along with the cytoplasm, and essentially none of the organelle DNA in the zygote is from the male parent. A simple example is seen in the inheritance of the *Neurospora* slow-growing mutant phenotype poky, which is caused by a defect in one of the mitochondrial genes. *Neurospora* can be crossed in such a way that one parent acts as the maternal and the other the paternal parent, and hence the progeny receive only maternal cytoplasm. In the cross of a poky female with a normal male, the progeny are all poky (Figure 5-21a), the precise inheritance pattern expected from a mitochondrial gene. In the reciprocal cross of normal female by poky male (Figure 5-21b), the progeny receive only normal cytoplasm and therefore grow normally.

Figure 5-22 demonstrates maternal inheritance of a mutation in a chloroplast gene in the four-o'clock plant. This mutation blocks synthesis of the green pigment chlorophyll in a chloroplast. The color of the chloroplasts determines the color of cells and hence the color of the branches and leaves composed of those cells. Variegated branches are mosaics of all-green and all-white cells. Flowers can develop on green, white, or variegated branches, and the cytoplasm of the flower's cells will be that of the branch on which it grows. Hence, when crossed, it is the maternal gamete within the flower—the egg cell—that

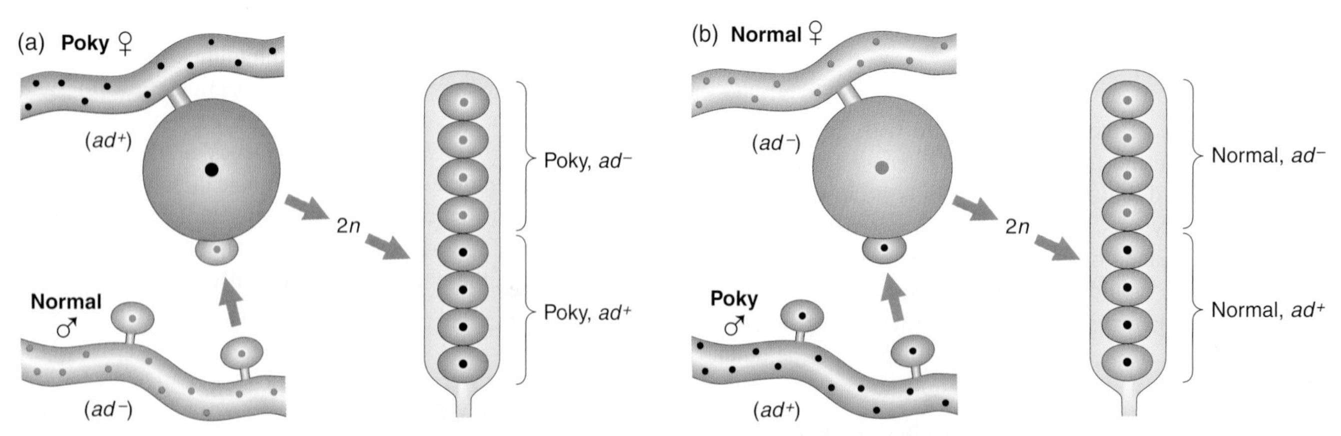

Figure 5-21 Explanation of the different results from reciprocal crosses of poky and normal *Neurospora*. The parent contributing most of the cytoplasm of the progeny cells is called *female*. Brown shading represents cytoplasm with mitochondria containing the poky mutation, and green shading cytoplasm with normal mitochondria. Note that in (a) the progeny will all be poky, whereas in (b) the progeny are all normal. Hence both crosses show maternal inheritance. The nuclear gene with the alleles ad^+ (black) and ad^- (red) is used to illustrate the segregation of the nuclear genes in the 1 : 1 Mendelian ratio expected for this haploid organism.

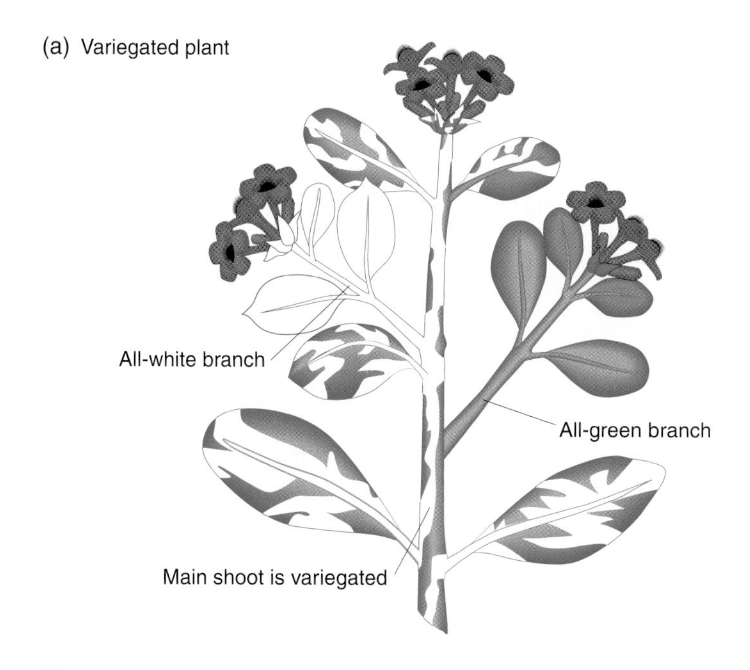

(b) Results of crosses between branches

Egg cell of female (*n*)	Pollen cell of male (*n*)	Zygote constitution (2*n*)
White ♀	Any ♂	White
Green ♀	Any ♂	Green
Variegated ♀	Any ♂	
Egg type 1		White
Egg type 2		Green
Egg type 3		Cell division → Variegated

Figure 5-22 Leaf variegation in *Mirabilis jalapa,* the four-o'clock plant.
(a) Flowers may form on any branch (variegated, green, or white), and these flowers may be used in crosses. (b) In crosses between flowers on different-colored branches, the color of the zygote and resulting plant is determined by maternal inheritance. The first two crosses shown here exhibit strict maternal inheritance: white females produce white zygotes and plants, and green females produce green zygotes and plants. If the maternal branch is variegated, the pattern of inheritance is more complex. Three types of zygotes can result, depending on whether the egg cell produced by a flower on a variegated branch contains only white, only green, or both green and white chloroplasts. Maternal inheritance still applies in all three types, but in the variegated zygotes, a process of cytoplasmic segregation during subsequent cell divisions produces a mosaic of all-green and all-white cells and, hence, a variegated plant.

determines color in the resulting plant. For example, if an egg cell is from a flower on the white branch, it will have white chloroplasts and, regardless of the origin of the pollen, the resulting plant will be white, thus showing maternal inheritance. The variegated zygotes (bottom of Figure 5-22) come from eggs that are cytoplasmic mixtures of two chloroplast types. Interestingly, when such a zygote divides, the white and green chloroplasts often sort themselves into separate cells, yielding the distinct green and white sectors that cause the variegation in the branches.

MESSAGE

Variant phenotypes caused by mutations in organelle DNA are generally inherited maternally.

SUMMARY

1. Crosses involving individuals that bear different alleles of a single gene can produce descendents that show predictable mathematical ratios of the phenotypes determined by the alleles. For example, the cross $A / a \times A / a$ produces a 3:1 ratio of the A phenotype to the a phenotype, and $A / a \times a / a$ produces a 1:1 ratio. Precise mathematical *genotypic* ratios such as the 1:1 ratio of A / a and A / A resulting from the cross $A / A \times A / a$ are not expressed as a phenotypic ratio because of dominance.
2. These mathematical ratios are based on the predictable separation (segregation) of homologous chromosomes at the first division of meiosis.
3. The equal segregation of the alleles of a heterozygote into the gametes is the fundamental mechanism underlying the precise ratios in progeny. For example, the heterozygote A / a produces meiotic products, $\frac{1}{2}$ of which are A and $\frac{1}{2}$ a. This principle of equal segregation has been called Mendel's first law. If both parents are heterozygous, the 1:1 ratios in male and female gametes combine randomly, resulting in a 3:1 ratio of $A / -$ to a / a, another direct manifestation of equal segregation.
4. An understanding of progeny ratios can be used in two directions. First, knowing the genotypes of two *parents* makes it possible to predict the progeny genotypes and phenotypes and their expected proportions. Second, observing phenotypes and their proportions among *offspring* allows the genotypes of the parents to be deduced.
5. *Dominance* is a term that describes the expression of one allele over another. It is defined as the phenotype that is expressed in a known heterozygote. The term can be applied to phenotypes and alleles.
6. The genome can be divided into autosomes—the regular chromosomes—and sex chromosomes, whose inheritance is associated with the sex of the individual. For example, in humans there are 44 autosomes in each sex. However, males carry X and Y sex chromosomes and females carry two X's. The X and Y chromosomes are radically different in size and gene content.
7. Sex determination is largely by one or a small number of genes. For example, in humans a single gene on the Y determines maleness, and its absence femaleness. The human Y contains few other genes. The genes on the X chromosome are mostly unrelated to sex.
8. Genes on the X chromosome show an inheritance pattern correlated with sex. In other words, it is often observed that the two sexes of progeny show different genotypic and phenotypic ratios. This principle can be used to deduce that a gene of previously unknown location is on the sex chromosome.
9. In human genetics, controlled crosses cannot be made. Therefore, analysis of human inheritance patterns makes use of information about the transmission of phenotypes in several family generations. This is called *pedigree analysis.* It can be used to infer single-gene inheritance, dominance and recessiveness, and autosomal or sex linkage.
10. In most organisms, the cytoplasm of the progeny comes from the maternal parent only. Hence the genes in the cytoplasmic organelle's mitochondria and chloroplasts are not transmitted through the male sperm, but they are transmitted in the female egg. Therefore if there is a single-gene difference between parents for an organelle gene, the offspring will all inherit only the allele from the mother (maternal inheritance).

CONCEPT MAP

Draw a concept map interrelating as many of the following terms as possible. Note that the terms are listed in no particular order.

equal segregation / heterozygote / 1:1 ratio / 3:1 ratio / 1:2:1 ratio / genotype / phenotype / products of meiosis / fertilization

SOLVED PROBLEM

1. You have four mice. Numbers 1 (male), 2 (female), and 3 (female) are agouti in color ("mousy" gray), and number 4 (male) is black. The following crosses were made and progeny obtained:

$$1 \times 2 \longrightarrow \text{all agouti}$$

$$1 \times 3 \longrightarrow \tfrac{3}{4}\text{ agouti and }\tfrac{1}{4}\text{ black}$$

$$2 \times 4 \longrightarrow \text{all agouti}$$

a. Which phenotype is dominant?

b. List the genotypes of all four mice.

c. Predict the progeny types and proportions of a cross of 3×4.

Solution

a. It is a good idea to look over all the data before attempting a solution. Are there any big clues? The most informative is the cross of 1×3 because here we learn that some black mice are born to agouti parents; therefore, agouti must be the dominant phenotype, and black recessive. This cross also establishes that the two phenotypes are determined by alleles of one gene because the 3:1 ratio is a single-gene ratio. The cross 2×4 is also informative and confirms the dominance of agouti.

b. Now filling in the genotypes is relatively straightforward. Let A be the allele for agouti and a the allele for black. We can deduce that 1 and 3 must both be $A\,/\,a$, because the 3:1 progeny ratio is produced only by a cross between heterozygotes. Number 2 must be homozygous $A\,/\,A$ because if it were $A\,/\,a$, it would have produced some black progeny in the cross with 1. Since number 4 is black, it must be homozygous recessive ($a\,/\,a$) because black is recessive. The cross of 2×4 confirms that the genotypes are $A\,/\,A$ and $a\,/\,a$, because all the progeny are expected to be $A\,/\,a$ (agouti), which was observed. In summary,

$$1 = A\,/\,a$$

$$2 = A\,/\,A$$

$$3 = A\,/\,a$$

$$4 = a\,/\,a$$

c. The cross of 3×4 is $A\,/\,a \times a\,/\,a$, and since the heterozygote will produce equal numbers of A and a gametes, the progeny are predicted to be $\frac{1}{2}\,A\,/\,a$ and $\frac{1}{2}\,a\,/\,a$.

SOLVED PROBLEM

2. Phenylketonuria (PKU) is a rare autosomal recessive disease. A couple wants to have children but consults a genetic counselor because the man has a sister with PKU and the woman has a brother with PKU. There are no other known cases in their families. They ask the genetic counselor to determine the probability that their first child will have PKU. What is this probability?

Solution

What can we deduce? If we let the allele causing the PKU phenotype be p and the respective normal allele be P, then the man's sister and woman's brother must have been $p\,/\,p$. In order to produce these affected individuals, all four grandparents of the future child must have been heterozygous normal ($P\,/\,p$). The pedigree can be summarized as follows:

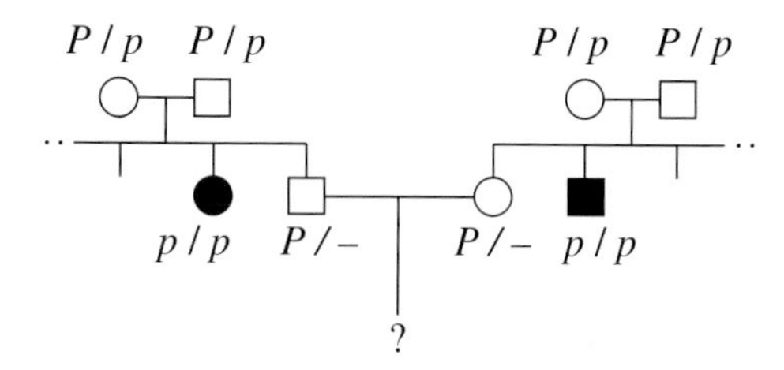

The only way the couple can have a child with PKU is if both of them are heterozygotes (it is obvious that they themselves do not have the disease). Both the grandparental matings are simple monohybrid crosses ($P\,/\,p \times P\,/\,p$) expected to produce progeny in the following proportions:

$$\left.\begin{array}{l}\tfrac{1}{4}P\,/\,P\\ \tfrac{1}{2}P\,/\,p\end{array}\right\}\ \tfrac{3}{4}\text{ normal}$$

$$\tfrac{1}{4}p\,/\,p\ \}\ \tfrac{1}{4}\text{ PKU}$$

We know that the man and the woman are normal, so for each of them the probability of being a heterozygote is $\frac{2}{3}$, because within the normal $P\,/\,-$ class, $\frac{2}{3}$ are $P\,/\,p$ and $\frac{1}{3}$ are $P\,/\,P$. Therefore, using the product rule, the probability of both the man and the woman being heterozygotes is $\frac{2}{3} \times \frac{2}{3} = \frac{4}{9}$. If they are both heterozygous, then one-quarter of their children would have PKU, so the probability that their first child will have PKU is $\frac{1}{4}$, and the probability of their being heterozygous *and* of their first child having PKU is $\frac{4}{9} \times \frac{1}{4} = \frac{4}{36} = \frac{1}{9}$, which is the answer to the question.

PROBLEMS

Basic Problems

1. A cross is made: $A / A \times a / a$.

a. What will be the genotype and phenotype of the F_1? Give proportions.

b. If an F_1 individual is crossed to the recessive parent, what will be the genotypes and phenotypes of the progeny? Give proportions.

c. If an F_1 individual is crossed to the dominant parent, what will be the genotypes and phenotypes of the progeny? Give proportions.

d. If an F_1 individual is selfed, what will be the genotypes and phenotypes of the progeny? Give proportions.

2. In nasturtiums, orange and red petals are determined by alleles of one gene. A pure-breeding red line is crossed to a pure-breeding orange line, and the progeny are all orange.

a. Explain this result, using allele symbols of your own choosing.

b. If one of these progeny is selfed, what progeny are expected and in what proportions?

3. In a wild population of foxgloves, a rare variant is discovered that has double flowers (normal wild-type plants have single flowers). Pollen from the double plant is used to fertilize several different wild-type plants, and the result is the same in each case, a 1 : 1 ratio of double to single. Explain this result using allele symbols of your own choosing.

4. Two first cousins marry, and their first child has the rare autosomal recessive disease galactosemia (inability to process galactose, leading to muscle, nerve, and kidney malfunction). With the aid of a pedigree diagram, show how their child might have inherited both copies of the recessive allele from a single ancestor who did not have galactosemia.

5. If you had a fruit fly *(Drosophila melanogaster)* that was of phenotype $A / -$, what genetic test would you make to determine if it was A / A or A / a? (Specify what other genotype or genotypes you would need.)

6. Two black guinea pigs were mated and over several years produced 29 black and 9 white offspring. Explain these results, giving the genotypes of both parents and progeny.

7. Holstein cattle normally are black and white. A superb black and white bull, Charlie, was purchased by a farmer for \$100,000. The progeny sired by Charlie were all normal in appearance. However, certain pairs of his progeny, when interbred, produced red and white progeny at a frequency of about 25 percent. Charlie was soon removed from the stud lists of the Holstein breeders. Explain precisely why, using symbols.

8. The plant blue-eyed Mary grows on Vancouver Island and on the lower mainland of British Columbia. The populations are dimorphic for purple blotches on the leaves—some plants have blotches, and others don't. Near Nanaimo, one plant in nature had blotched leaves. This plant, which had not yet flowered, was dug up and taken to a laboratory, where it was allowed to self. Seeds were collected and grown into progeny. One randomly selected (but typical) leaf from each of the progeny is shown in the figure below.

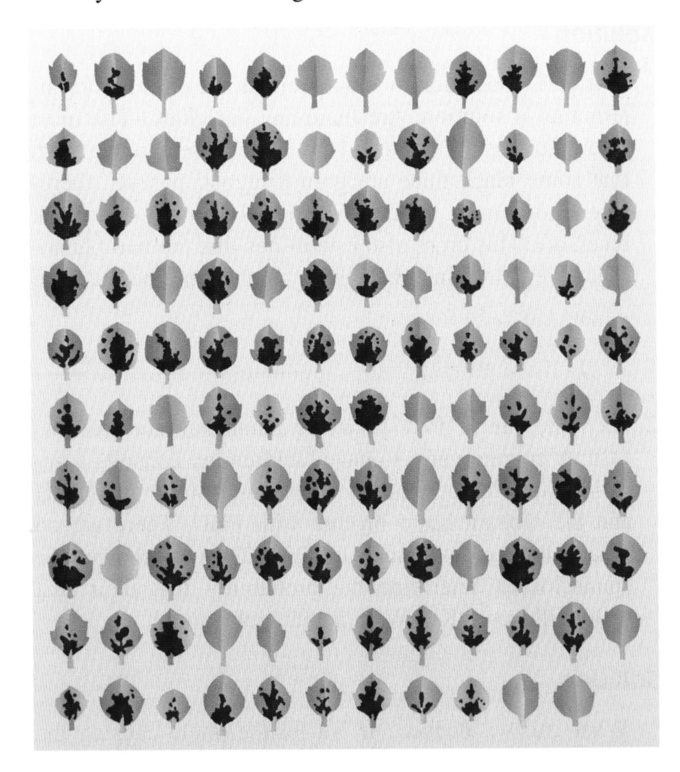

a. Formulate a concise genetic hypothesis to explain these results. Explain all symbols and show all genotypic classes (and the genotype of the original plant).

b. How would you test your hypothesis? Be specific.

Unpacking the Problem

a. Briefly sketch the life cycle of a flowering plant.

b. Define the word *dimorphic.*

c. How does the term *dimorphic* relate to the term *polymorphism*?

d. What is a population?

e. Name a population of plants or animals that you know about near where you live.

f. In the present analysis, were any plants found that had both blotched and unblotched leaves on the same plant? Explain the reason for your answer.

g. What character do you think is under study in this analysis?

h. How many phenotypes were apparent for this particular character?

i. Name the phenotypes from your answer to Question h.

j. In the diagram, do you think the variation in the blotch patterns is relevant to this analysis? Explain.

k. Do you think the blotch patterns on all the leaves of one plant would be exactly the same or approximately the same? Sketch two possible leaves from one plant.

l. Why was it important that the plant dug up had not flowered?

m. What role does counting have in genetic analysis?

n. What counting analyses can be done on the data in this question?

o. Can the numbers in your counts be expressed as ratios?

p. What is the importance of ratios in genetic analysis?

q. List three ratios discussed in this chapter.

r. Do the numbers you have counted match any of these ratios?

s. What is the difference between a genotypic ratio and a phenotypic ratio? Give an example from this chapter.

t. Is there any relationship between the number of identifiable phenotypes and the number of alleles?

u. What is the definition of *dominance?*

v. What is a self? How would you make a self of the plant in question?

w. In a self, what does the observation of progeny with phenotype different from the parent tell you about dominance?

x. Can you tell which is the wild-type phenotype in the present analysis?

y. How does the issue of designating wild type relate to the allele symbols you choose in analyzing this data set?

9. In nature, the plant *Plectritis congesta* is dimorphic for fruit shape; that is, individual plants bear either wingless or winged fruits:

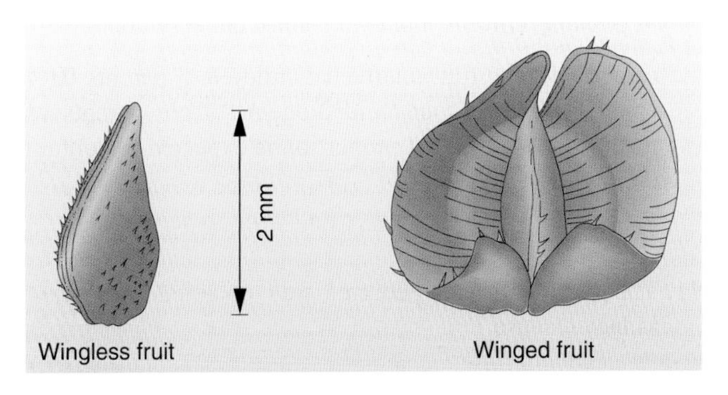

Plants were collected from nature before flowering and were crossed or selfed with the following results:

	NUMBER OF PROGENY	
Pollination	Winged	Wingless
Winged (selfed)	91	1*
Winged (selfed)	90	30
Wingless (selfed)	4*	80
Winged × wingless	161	0
Winged × wingless	29	31
Winged × wingless	46	0
Winged × winged	44	0
Winged × winged	24	0

* The phenotypes of progeny marked by asterisks probably have a nongenetic explanation.

Interpret these results and derive the mode of inheritance of these fruit-shape phenotypes, using symbols. What do you think the explanation is for the results indicated by the asterisk?

10. The accompanying pedigree is for a rare, but relatively mild, hereditary disorder of the skin.

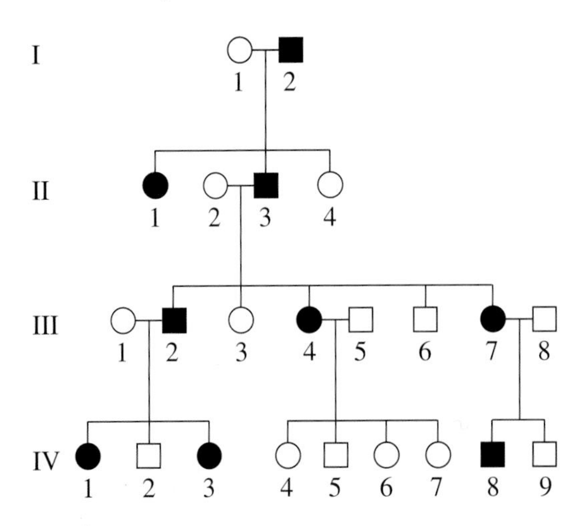

a. Is the disorder inherited as a recessive or a dominant phenotype? State reasons for your answer.

b. Give genotypes for as many individuals in the pedigree as possible. (Invent your own defined allele symbols.)

c. Consider the four unaffected children of parents III-4 and III-5. In all four-child progenies from parents of these genotypes, what proportion is expected to contain all unaffected children?

11. The accompanying pedigree was obtained for a rare kidney disease.

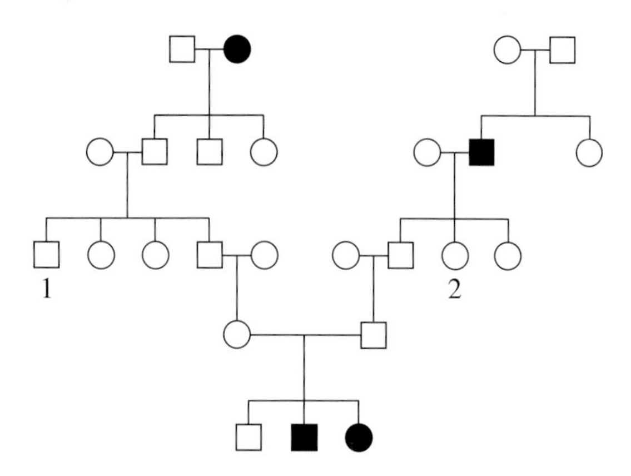

a. Deduce the inheritance of this condition, stating your reasons.

b. If individuals 1 and 2 marry, what is the probability that their first child will have the kidney disease?

12. A cross is made between a leucine-requiring haploid yeast strain of genotype *leu* and a wild-type strain that can make its own leucine (leu^+).

a. What will be the genotype of the transient meiocyte?

b. What progeny genotypes will be produced and what will be their proportions?

c. Draw diagrams that show a suggestion for the nature of the mutant allele at the DNA level, and the corresponding protein product.

13. The recessive allele *s* causes *Drosophila* to have small wings, and the s^+ allele causes normal wings. This gene is known to be X-linked. (Note: It is not the same gene discussed earlier in the chapter.) If a small-winged male is crossed with a homozygous wild-type female, what ratio of normal to small-winged flies can be expected in each sex in the F_1? If F_1 flies are intercrossed, what F_2 progeny ratios are expected? What progeny ratios are predicted if F_1 females are crossed to their father?

Pattern Recognition Problems

In Problems 14 through 18, diagrams show phenotypes and the results of breeding analyses. Deduce the genotypes of the individuals shown in each diagram as far as possible, using allelic symbols that take into account any evidence of dominance in the data.

14. Pure-breeding diploid × Pure-breeding diploid

All

Self

$\frac{3}{4}$

$\frac{1}{4}$

15.

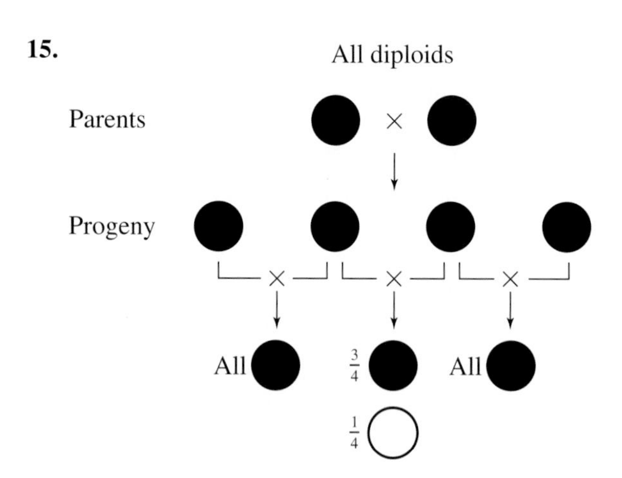

16.

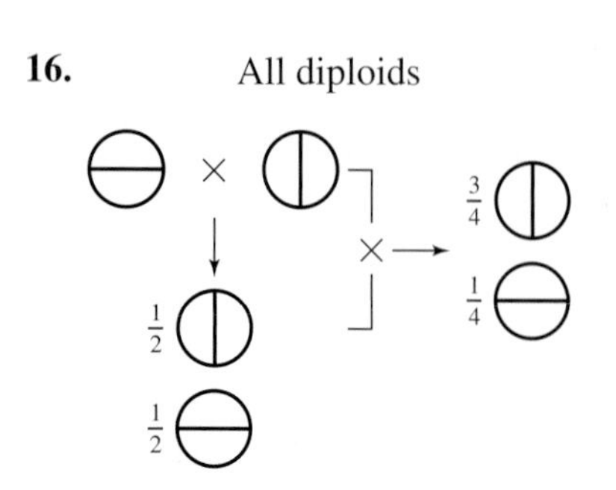

17.

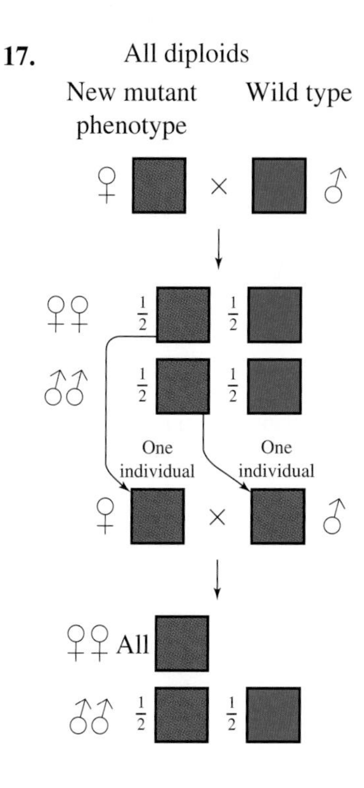

18.

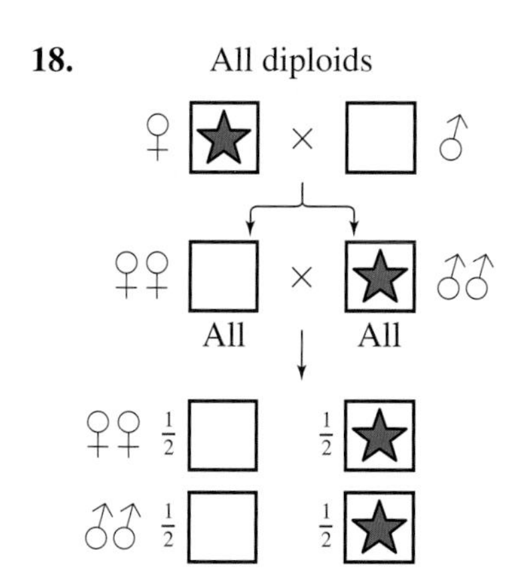

Challenging Problems

19. Duchenne muscular dystrophy is X-linked and usually affects only males. Victims of the disease become progressively weaker, starting early in life.

a. What is the probability that a woman whose brother has the disease will have an affected child?

b. If your mother's brother (that is, your uncle) had Duchenne muscular dystrophy, what is the probability that you have received the allele?

c. If your father's brother had the disease, what is the probability that you have received the allele?

20. The accompanying pedigree concerns a certain rare disease that is incapacitating but not fatal.

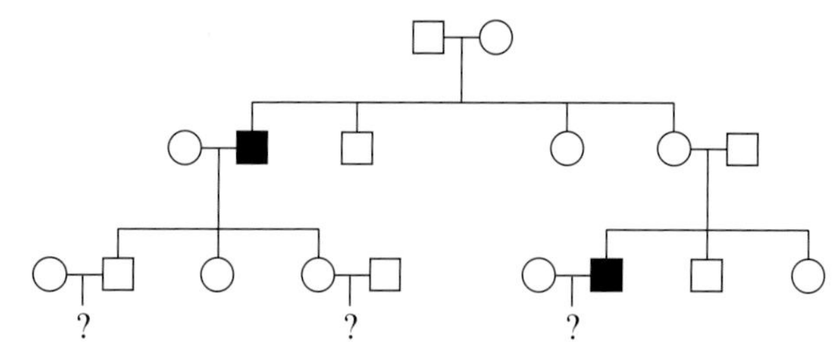

a. Determine the mode of inheritance of this disease.

b. Write the genotype of each individual according to your proposed mode of inheritance.

c. If you were this family's doctor, how would you advise the three couples in the third generation about the likelihood of having an affected child?

21. A mutant allele in mice causes a bent tail. Six pairs of mice were crossed. Their phenotypes and those of their progeny are given below. N is normal phenotype; B is bent phenotype. Deduce the mode of inheritance of this phenotype.

Cross	Parents		Progeny	
	♀	♂	♀	♂
1	N	B	All B	All N
2	B	N	$\frac{1}{2}$ B, $\frac{1}{2}$ N	$\frac{1}{2}$ B, $\frac{1}{2}$ N
3	B	N	All B	All B
4	N	N	All N	All N
5	B	B	All B	All B
6	B	B	All B	$\frac{1}{2}$ B, $\frac{1}{2}$ N

a. Is it recessive or dominant?

b. Is it autosomal or sex-linked?

c. What are the genotypes of all parents and progeny?

22. A new mutant strain of the haploid fungus *Neurospora* was found. It showed erratic stop-start growth and, so, was called "stopper" (*Neurospora* usually grows continuously). The stopper strain was crossed to another mutant strain, which had a yellow color instead of the usual orange. The cross was done reciprocally, using each strain as either maternal or paternal parent. The results were as follows:

Cross 1

Parents: Stopper, orange female × continuous, yellow male

Progeny: $\frac{1}{2}$ stopper, orange
$\frac{1}{2}$ stopper, yellow

Cross 2

Parents: Continuous, yellow female × stopper, orange male

Progeny: $\frac{1}{2}$ continuous, yellow
$\frac{1}{2}$ continuous, orange

a. What can you conclude about the location of these genes in the cell?

b. Explain clearly the difference between the results of the reciprocal crosses.

23. A man's grandfather had galactosemia. This is a rare autosomal recessive disease caused by inability to process galactose, leading to muscle, nerve, and kidney malfunction. The man married a woman whose sister had galactosemia. The woman is now pregnant with their first child.

a. Draw the pedigree as described.

b. What is the probability that this child will have galactosemia?

c. If the first child does have galactosemia, what is the probability a second child will have it?

24. A man whose grandfather on his mother's side had favism (a rare X-linked recessive sensitivity to broad beans) marries a woman whose uncle on her mother's side also had favism.

a. Draw the pedigree as described.

b. What is the probability that their first child will have favism?

c. If the first child does have favism, what is the probability a second child will have favism?

25. The following pedigree is of a rare hereditary disease of the nervous system:

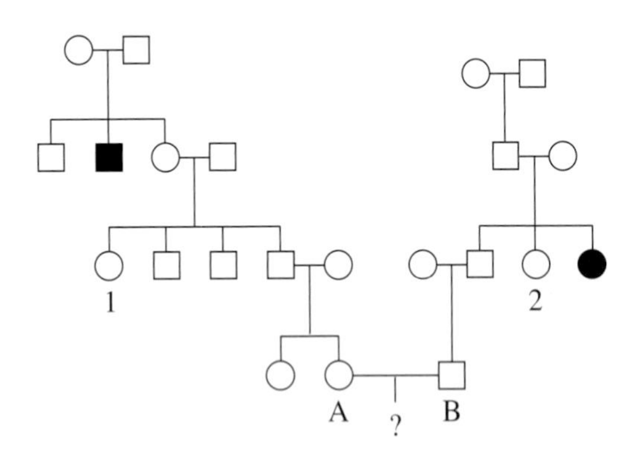

a. Do you think the disease is recessive or dominant? (Explain.)

b. Do you think it is autosomal or X-linked? (Explain.)

c. Under your scheme, what is the probability that the first child of the couple A and B will have the disease?

26. In *Mimulus* (the monkey flower) plant stems can be either smooth or hairy, and this difference is determined by a simple allelic difference. Examine the following crosses and give the genotypes of the parental and progeny plants in each cross. (Invent your own gene symbols and define their meaning.)

Cross	Parents	Progeny
1	Plant 1 (hairy) × plant 2 (hairy)	All hairy
2	Plant 1 (hairy) × plant 3 (hairy)	$\frac{3}{4}$ hairy $\frac{1}{4}$ smooth
3	Plant 1 (hairy) × plant 4 (smooth)	$\frac{1}{2}$ hairy $\frac{1}{2}$ smooth
4	Plant 2 (hairy) × plant 4 (smooth)	All hairy

27. Represent the progeny of the cross $A/a \times A/a$ both as a grid and as a pedigree diagram. What are the strengths and weaknesses of each representational method?

28. Consider the following pedigrees for two different rare diseases:

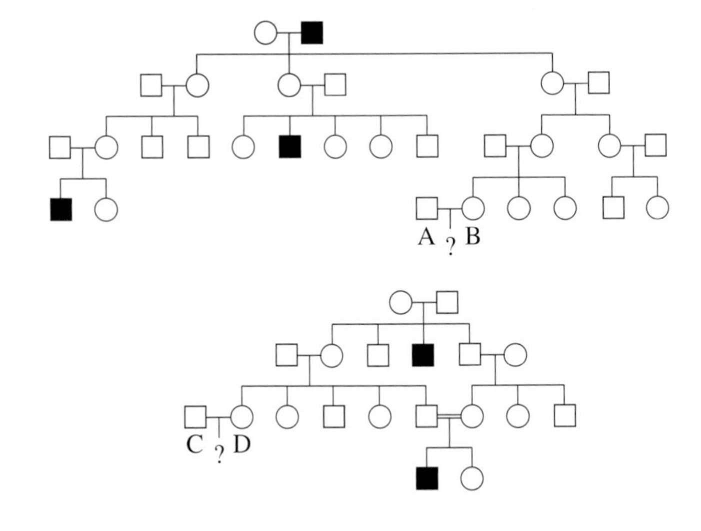

a. Deduce the most likely mode of inheritance of these two diseases and show your reasoning.

b. Calculate the respective probabilities of couple A and B and of couple C and D having an affected child.

Exploring Genomes: A Web-Based Bioinformatics Tutorial

Using BLAST to Compare Nucleic Acid Sequences

The BLAST algorithm is also able to search for and compare nucleic acid sequences. In the Genomics tutorial at www.whfreeman.com/mga, we will find that comparing transfer RNA sequences (Chapter 3) among species is a good way to explore this capability.

GENETIC RECOMBINATION IN EUKARYOTES 6

Key Concepts

1. During meiosis, two different processes generate recombinant products (those with new combinations of parental input alleles).
2. One source of recombination is through the independent segregation of allelic differences located on nonhomologous chromosomes.
3. The other source of recombination is through crossing-over, the exchange of DNA segments between homologous chromatids.
4. In a dihybrid meiosis, if 50 percent of products are recombinant, this indicates either that two genes are on different chromosomes or that they are far apart on the same chromosome.
5. If two genes are near each other on the same chromosome, less than 50 percent of meiotic products will be recombinant.
6. Crossing-over occurs after premeiotic S phase, and a crossover can involve any two nonsister chromatids of the prophase I tetrad.
7. Gene loci on a chromosome can be mapped using crossovers; that is, their locations relative to one another can be determined through an analysis of crossover frequencies.
8. Crossover analysis results in linear maps, consistent with the linear nature of double-stranded DNA, the genetic material of chromosomes.
9. Crossing-over occurs by the formation of a DNA segment that is a heteroduplex—a strand from one parental chromatid and a strand from the other.

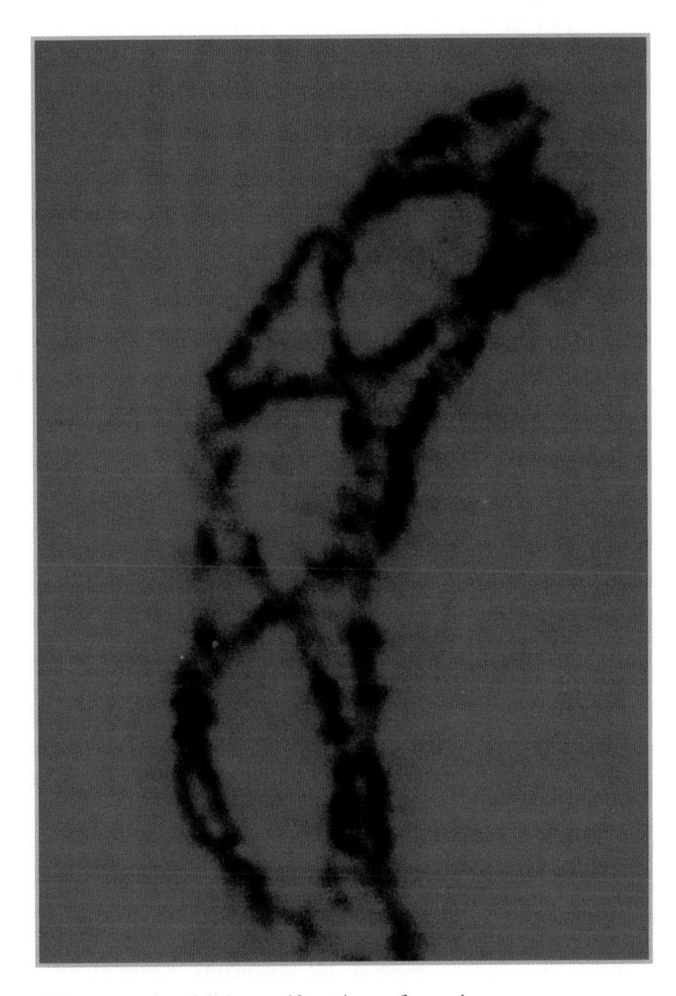

Chiasmata, the visible manifestations of crossing-over, photographed during meiosis in a grasshopper testis. *(John Cabisco/Visuals Unlimited.)*

Most eukaryotic organisms undergo sexual reproduction, the type of reproduction in which different individuals mate for the purpose of producing progeny. It is common knowledge that plants and animals reproduce sexually by making and uniting male and female gametes. Other eukaryotes, such as many fungal species, do not have a sexual cycle, but nevertheless engage in a *parasexual* cycle, which mimics some of the processes and outcomes that take place in true sex. As we shall see in Chapter 7, prokaryotes, too, engage in a type of sexual reproduction. The widespread occurrence of sex suggests that it has great adaptive significance. One of the most obvious outcomes of sexual reproduction is the production of progeny, but this same effect could be achieved with asexual reproduction. Most biologists believe that the reason for the sexual cycle is that it produces progeny that show different combinations of the phenotypes of the parents, and this is the aspect of sex that has adaptive value. Figure 6-1 is an example from humans, the family of Gerald Ford, former president of the United States. The two parents have many contrasting phenotypic differences of height, shape, and color. It is clear that the children show variation. Furthermore, not only are they different from their parents, but they have combinations of many of the attributes from each parent.

Figure 6-1 Children show combinations of the physical features of their parents, as shown in the family of President and Mrs. Gerald Ford. (Gerald R. Ford Library.)

Examine various facial features, and for each compare children with parents. (One useful approach is to view the photograph upside down.) Of course, age and environment can influence some features, but the uniqueness of an individual's appearance is in large part the result of a combination of parental contributions.

CHAPTER OVERVIEW

Variation among individuals provides the raw material for evolution. The variation that results from sexual reproduction results in a greater rate of evolution because there are many more variants among which to select. Natural selection is based on relative reproductive rates of different genotypes in a population. In the struggle for existence, some genotypes are more successful and produce more offspring; hence, their alleles become represented at a higher frequency in the next generation. In this way evolution gradually shapes the genetic structure of a population.

Two processes are responsible for genetic variation, *recombination* (the subject of this chapter) and *mutation.* We have seen that mutation is a change in the DNA sequence of a gene. Mutation is the ultimate source of evolutionary change; new alleles arise in all organisms, some spontaneously, others as a result of exposure to radiation and chemicals in the environment (a topic covered in detail in Chapter 10). The new alleles produced by mutation become the raw material for a second level of variation, effected by recombination. As its name suggests, recombination is the outcome of cellular processes that cause alleles of different genes to become grouped in new combinations. To use an analogy, mutation produces new playing cards, and then recombination shuffles them and deals them out as different hands.

In most familiar eukaryotes, essentially all recombination takes place at meiosis. The processes underlying meiotic recombination shuffle heterozygous allele pairs and deal them out in different combinations into the products of meiosis (such as the gametes of plants and animals). If, for example, we assume a diploid organism is heterozygous at only 10 genes (a gross underestimate in most organisms), then each allele pair will produce two types of gametes, and if we extend this idea to all 10 allele pairs, the total possible number of gamete genotypes will be $2 \times 2 \times 2 \times 2 \times 2 \times 2 \times 2 \times 2 \times 2 \times 2 = 2^{10} = 1024$. In contrast, without recombination there would be only two different types. The cellular processes responsible for recombination also occur when alleles are homozygous, that is, when the DNA sequences of the two homologous genes are identical, but then no recombinants are produced.

Although in nature recombination produces hereditary variation for natural selection to act on, geneticists make use of recombination for experimental genetic analysis. One important use of recombination is to create specific genotypes needed for experimental purposes, by combining particular sets of alleles present in different established genetic stocks. (Sometimes these interesting alleles have been isolated from nature, but more often they have been induced experimentally using mutation-causing agents such as high-energy radiation.) Another experimental use of recombination is to find out which specific chromosome is the location of a gene of interest, and then to locate ("map") the specific position of the gene in relation to other genes. These applications of recombination, which we introduce in this chapter, are central aspects of genetics.

How can we define *recombination?* We have seen that it is the bringing together of alleles in new combinations. However, geneticists define it in a precise way that can be applied experimentally. Recombination in eukaryotes is largely a meiotic process, although recombination does occur much more rarely at mitosis. For the present we shall define recombination in terms of meiosis. *Meiotic recombination* is defined as the production of haploid products of meiosis with genotypes differing from both the haploid genotypes that originally combined to form the diploid meiocyte. The product of meiosis so generated is called a **recombinant.**

This is very much an operational definition of recombination. Figure 6-2 shows the way it works. This diagram illustrates the important point that we detect recombination by comparing the *output* genotypes of meiosis with the *input* genotypes. The inputs are the two haploid genotypes that combined to form the meiocyte, the diploid cell that undergoes meiosis. In a plant or animal the inputs are the two haploid gametes that united to form the diploid zygote. Within the developed zygote, a subset of cells become meiocytes. The gametes produced by the meiocytes are the output haploids.

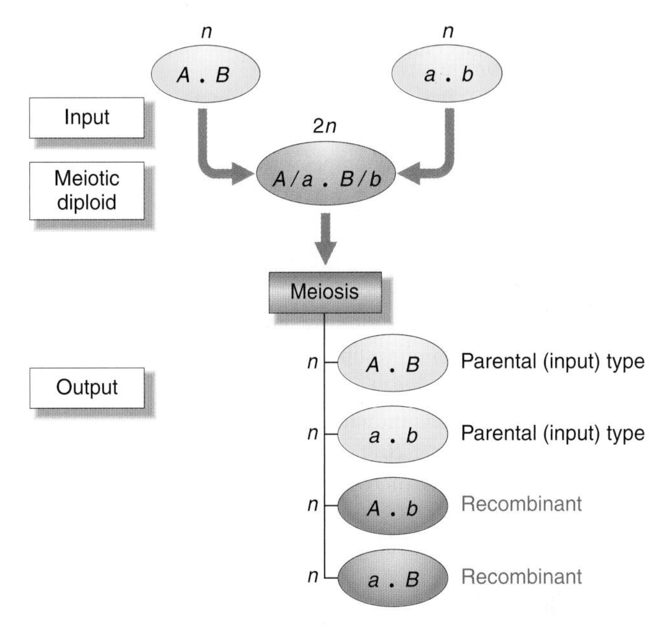

Figure 6-2 Recombinants are those products of meiosis with allelic combinations different from those of the haploid cells that formed the meiotic diploid.

There are two different mechanisms of meiotic recombination: *independent assortment* of heterozygous genes on different chromosomes and *crossing-over* between heterozygous genes on the same chromosome. These two processes, forming the basis for the analyses in this chapter, are shown in Figure 6-3. In this figure one input genotype carried all the dominant alleles, and the other input contained all recessive alleles. A typical meiotic product is

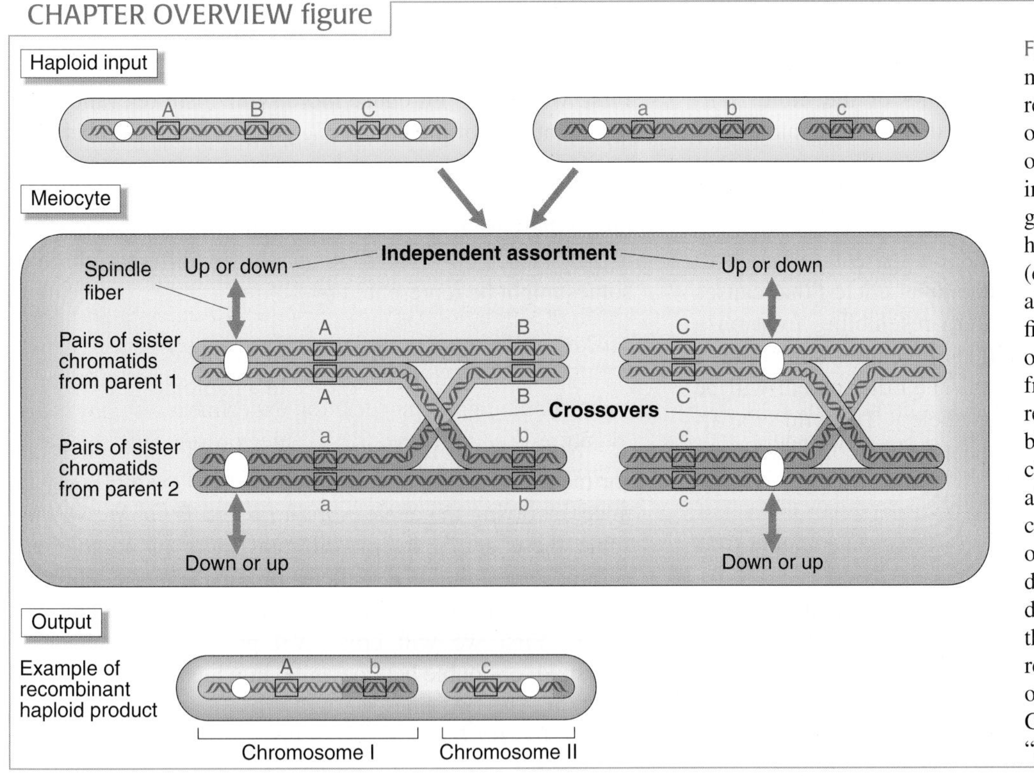

Figure 6-3 The two mechanisms of meiotic recombination are crossing-over (breakage and reunion of nonsister chromatids) and independent assortment of genes on different homologous chromosomes (caused by random attachment of the spindle fibers to one homolog or the other). Chromosomal DNA from one parent is shown in red and from the other in blue. The parental chromosomes carry the alleles *A*, *B*, and *C* in one case and *a*, *b*, and *c* in the other. "Up" and "Down" are directions on the page, designating opposite poles of the cell. An example of a recombinant meiotic product of genotype *A b c* is shown. Gametes are one example of "haploid input."

shown, which, using the above definition of recombination, is clearly a recombinant.

In Figure 6-2, the process of recombination is illustrated by two allele pairs, A / a and B / b. For the first time in the book we need to write genotypes involving two genes. The definition of recombination makes no assumptions about whether the genes are on the same or different chromosomes. Hence we need a way of representing this uncertainty symbolically. There is no universally agreed-upon convention for this situation, but in this book we will use a *period,* or *dot,* between the genes to represent this uncertainty. Therefore in the example in Figure 6-2 the input genotypes are written $A\ .\ B$ and $a\ .\ b$. The output genotypes are written $A\ .\ B$, $a\ .\ b$, $A\ .\ b$, and $a\ .\ B$.

Now we see how the operational definition of recombination is applied. In the example in Figure 6-2, the two input genotypes $A\ .\ B$ and $a\ .\ b$ might be gametes from two pure-bred lines $A/A\ .\ B/B$ and $a/a\ .\ b/b$. All we have to do is compare the output (which might be gametes produced by the diploid) to these inputs. We see that two output genotypes, $A\ .\ b$ and $a\ .\ B$, are different from the input genotypes, so these are by definition recombinants.

RECOMBINATION ANALYSIS

The goal of recombination analysis is to detect and quantify recombinants at meiosis. Since generally it is not possible to study recombination directly, it must be done indirectly, by making a cross between the parents of suitable genotype. This process is simply illustrated using the life cycle of haploid fungi. As an example, let's make a cross between two different mutants of the mold *Neurospora.* Recall from Figure 4-16 that in the haploid sexual cycle two haploid parents unite to form a transient diploid, which then undergoes meiosis. Therefore, to make a cross between haploids we simply unite two parental strains. In our example, one parental strain will be frost, a mutant phenotype due to an allele f that causes a highly branched growth pattern resembling frost on a windowpane. The wild-type allele f^+ causes spreading, a less branched pattern of growth. The other parent will be the mutant yellow, caused by an allele y. The wild-type allele y^+ promotes orange pigment production. The cross can be designated

$$f\ .\ y^+ \text{ (frost, orange)} \times f^+ .\ y \text{ (spreading, yellow)}$$

Cells of these cultures, which represent meiotic input in this type of cycle, unite to form a transient diploid meiocyte. Meiosis takes place and results in the production of haploid products of meiosis called *ascospores,* according to the standard haploid life cycle of this fungus. These ascospores represent meiotic output. The phenotypes of the cultures growing from the ascospores will be

Frost, orange	$f\ .\ y^+$
Spreading, yellow	$f^+ .\ y$
Spreading, orange	$f^+ .\ y^+$
Frost, yellow	$f\ .\ y$

The definition of recombination tells us that the spreading, orange ($f^+ .\ y^+$) and frost, yellow ($f\ .\ y$) progeny are clearly recombinants—they represent new combinations of alleles when compared with the meiotic input genotypes contributed by the cross parents. Now that we know which genotypes are recombinants, we can ask whether the two genes are on the same or different chromosomes. As we shall see, the key lies in the frequency of recombinants. In the *Neurospora* example no numbers of progeny were given, so we cannot tell. However, in order to understand the significance of recombinant frequencies we must consider the two meiotic processes that cause recombination, independent assortment and crossing-over, because they generally result in different frequencies of recombinants. In a case in which the chromosomal locations of genes are unknown, the frequency of recombinants can be used to deduce whether they are on different chromosomes or on the same chromosome. The two mechanisms of recombination and their associated recombinant frequencies are discussed in the following two sections.

INDEPENDENT ASSORTMENT

First, we need to introduce more symbolism relevant to the representation of more than one gene. A semicolon between two genes is widely used by geneticists to represent two genes on different chromosomes. We have already introduced the slash (/) between two alleles to show that they are a pair. Hence two genes, A and B, on different chromosomes might be represented by

$A/a\ ;\ B/b$	In a diploid
$A\ ;\ b$	In a haploid

Now we can illustrate the process of independent assortment. Let us start with a hypothetical example, say, in mice, involving the allele pairs A / a and B / b. We will assume these genes are on different pairs of homologous chromosomes, the A or a alleles on a pair of long chromosomes and the B or b alleles on a pair of short chromosomes. First we will cross two pure lines to create a **dihybrid,** or double heterozygote, in which to study recombination. The two lines we will cross to create a dihybrid are $A/A\ ;\ B/B$ and $a/a\ ;\ b/b$:

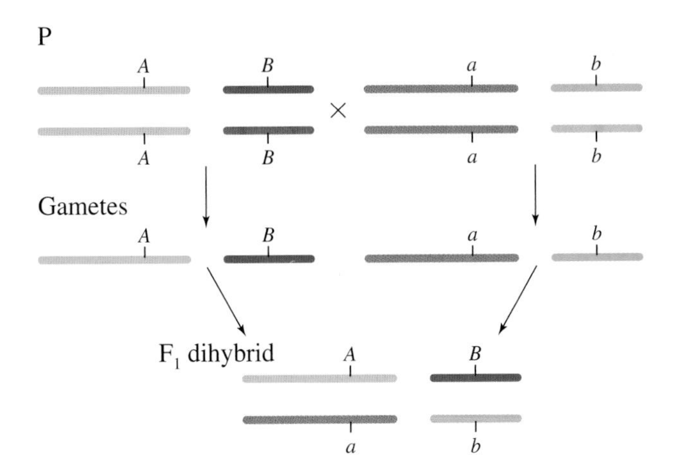

Now we can use the law of equal segregation to predict the gametic genotypes produced by the dihybrid F_1, and also their proportions. We know that $\frac{1}{2}$ the gametes will be *A* and $\frac{1}{2}$ will be *a*. We also know that $\frac{1}{2}$ the gametes will be *B* and $\frac{1}{2}$ will be *b*. These two ratios must be multiplied to determine the overall gametic genotypes. Why? Recall that the product rule (Chapter 5) states that the probability of independent events is the product of their individual probabilities. Since the two genes in question are on different chromosomes, the attachment of spindle fibers to the pair of homologous centromeres of one chromosome pair is independent of the attachment of the fibers to the centromeres of the other pair. Hence the two chromosome pairs will segregate independently, and it is appropriate to apply the product rule and multiply probabilities. The expected gametic *proportions* of $\frac{1}{2}$ can be thought of as *probabilities* of $\frac{1}{2}$. Therefore, the proportions of gametic types produced by the dihybrid can be calculated by multiplying the proportions produced by each allele pair. The procedure can be represented graphically as in the following, called a *branch diagram:*

Gametes

$\frac{1}{2} A$ → $\frac{1}{2} B$ → $\frac{1}{4} A \,;\, B$

$\frac{1}{2} A$ → $\frac{1}{2} b$ → $\frac{1}{4} A \,;\, b$

$\frac{1}{2} a$ → $\frac{1}{2} B$ → $\frac{1}{4} a \,;\, B$

$\frac{1}{2} a$ → $\frac{1}{2} b$ → $\frac{1}{4} a \,;\, b$

One way of interpreting the top line branch diagram (as an example) is that the *A* gametes make up half the total gametes (the other half are *a*), and half of these will be *B*. One half of a half is a quarter, a simple application of the product rule.

Figure 6-4 shows how chromosome movements at meiosis produce this ratio. Basically it is because there are two different but equally frequent combinations in which the spindle fibers attach to the centromeres:

- In half of the meiocytes, *A* and *B* are pulled to one pole, and *a* and *b* are pulled to the other pole (panel 4 in Figure 6-4).
- In the other half of the meiocytes, *A* and *b* are pulled to one pole, and *a* and *B* to the other (panel 4′).

The general principle is known as **independent assortment** of allele pairs. This principle is based on the fact that the equal segregation of one allele pair into the meiotic products is *independent* of the equal segregation of the other allele pair, simply because they are on different chromosomes. Independent assortment was first observed by Gregor Mendel. Although he did not know that the independent

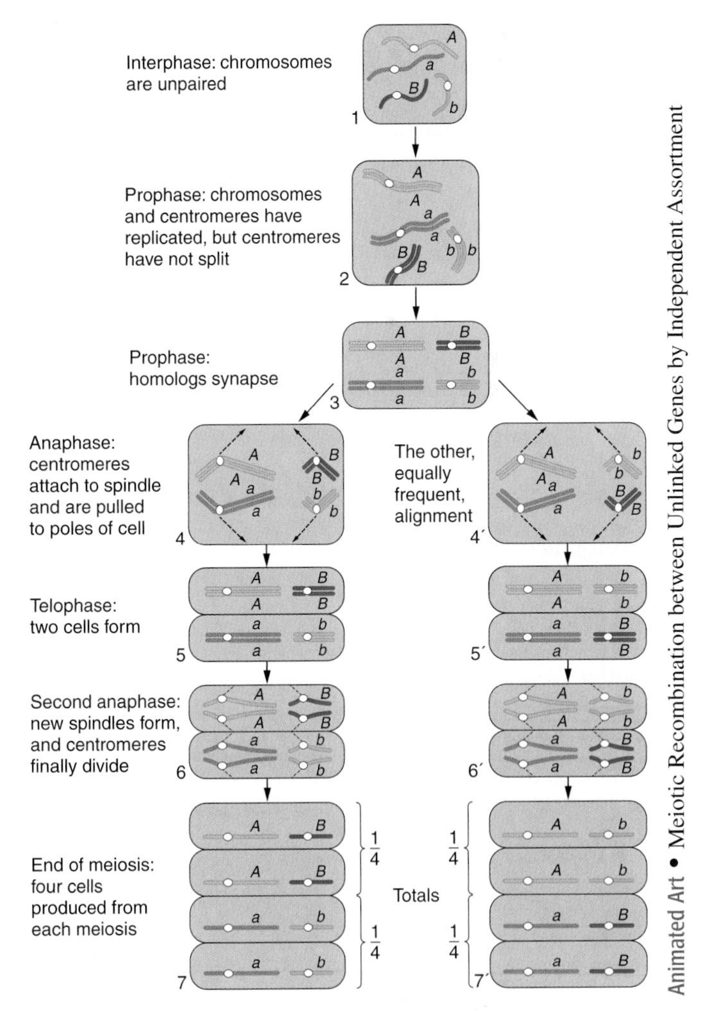

Figure 6-4 Meiosis in a diploid dihybrid cell of genotype *A* / *a* ; *B* / *b*, showing how the segregation and independent assortment of different chromosome pairs give rise to the 1 : 1 : 1 : 1 gametic ratio. The diagram splits into two channels at stage 4 (one designated by prime); these channels represent the two equally frequent possible combinations of spindle fiber attachments.

Animated Art • Meiotic Recombination between Unlinked Genes by Independent Assortment

assortment is based on the independent behavior of different chromosome pairs, the principle is sometimes called **Mendel's second law** in his honor.

> **MESSAGE**
> Allele pairs on different chromosome pairs assort independently.

Which of the gametes shown in panels 7 and 7′ of Figure 6-4 are recombinants? It is simply a matter of applying the definition of recombination: compare the meiotic input genotypes with the meiotic output genotypes. The input genotypes were the genotypes of the parental gametes that fused to form the dihybrid F_1, and we know they were *A* ; *B* and *a* ; *b*. The output is the set of gametes produced by meiosis in the F_1, as shown in Figure 6-4. The genotypes *A* ; *B* and *a* ; *b* are the same as the parental input, so these types are not recombinant. However, the gametic genotypes *A* ; *b* and *a* ; *B* are not the same as the input, so by definition they are recombinants. In summary, the gametes are of the following proportions:

Gametes			
$\frac{1}{4}$	*A*	*B*	Parental
$\frac{1}{4}$	*a*	*b*	Parental
$\frac{1}{4}$	*a*	*B*	*Recombinant*
$\frac{1}{4}$	*A*	*b*	*Recombinant*

Hence all four genotypes are equally likely, and the total **recombinant frequency (RF)** must be $\frac{1}{4} + \frac{1}{4} = \frac{1}{2} = 50$ percent. This RF value of 50 percent is always observed for genes on different chromosome pairs.

The analysis will be exactly the same if instead of starting with the pure-breeding strains *A* / *A* ; *B* / *B* and *a* / *a* ; *b* / *b*, we start with the pure-breeding strains *A* / *A* ; *b* / *b* and *a* / *a* ; *B* / *B*. Here the parental gametes are *A* ; *b* and *a* ; *B*, and the F_1 is *A* / *a* ; *B* / *b* as before. Again, because of the law of segregation, there will be the same types of gametes in the same proportions:

$\frac{1}{4}$ *A* ; *B*
$\frac{1}{4}$ *A* ; *b*
$\frac{1}{4}$ *a* ; *B*
$\frac{1}{4}$ *a* ; *b*

Now the recombinants will be *A* ; *B* and *a* ; *b*, because these are not input haploid genotypes; nevertheless, they will still be at a frequency of 50 percent.

> **MESSAGE**
> In a dihybrid *A* / *a* ; *B* / *b* involving two genes on two different pairs of chromosomes, the recombinant frequency will always be 50 percent.

So far we have not addressed the problem of how to identify the genotypes of the products of meiosis issuing from a dihybrid meiocyte. In haploid organisms this is simple (as we saw in the *Neurospora* example above) because the products are the sexual spores (such as ascospores in fungi), and these can easily be grown up into haploid organisms in which the phenotype will correspond to the genotype. However, in a diploid organism such as an animal the gametes generally cannot be observed directly, so the individual under investigation must be crossed, to test the gametes *indirectly* by looking at the phenotypes of the progeny that result. There are two types of crosses that can be used to measure recombination in a diploid, a testcross and a self.

Testcross of a Dihybrid

The best way to determine the genotypes of the gametes of a dihybrid diploid is to make a cross to a **tester,** an individual that carries only recessive alleles for the genes under investigation. Such a cross is called a **testcross.** A tester must be fully homozygous recessive, for example, *a* / *a* ; *b* / *b*. Since the gametes of the tester carry only recessive alleles, the genotypes of the gametes of the dihybrid will be directly reflected in the phenotypes of the testcross progeny. For example an *A* ; *b* gamete from the dihybrid will combine with an *a* ; *b* gamete from the tester to produce an *A* / *a* ; *b* / *b* zygote that will have phenotype A ; b, revealing the genotype of the dihybrid's gamete. The general nature of the testcross is illustrated in Figure 6-5. If the genes are on different chromosomes, independent assortment will produce a 1 : 1 : 1 : 1 ratio of gametes, which in turn will produce a 1 : 1 : 1 : 1 phenotypic ratio in the progeny:

A* / *a ; ***B* / *b*** (dihybrid) × *a* / *a* ; *b* / *b* (tester)

Gametes	Gametes *a* ; *b*	Progeny proportions
$\frac{1}{4}$ ***A*** ; ***B***	***A*** / *a* ; ***B*** / *b*	$\frac{1}{4}$
$\frac{1}{4}$ ***A*** ; ***b***	***A*** / *a* ; ***b*** / *b*	$\frac{1}{4}$
$\frac{1}{4}$ ***a*** ; ***B***	***a*** / *a* ; ***B*** / *b*	$\frac{1}{4}$
$\frac{1}{4}$ ***a*** ; ***b***	***a*** / *a* ; ***b*** / *b*	$\frac{1}{4}$

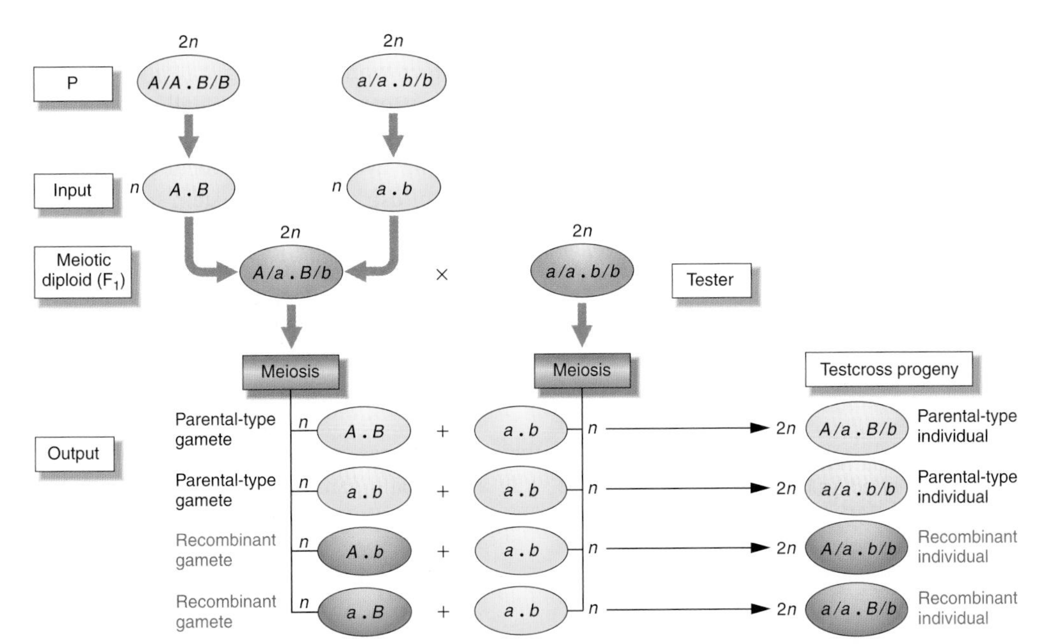

Figure 6-5 The detection of recombination in diploid organisms. Recombinant products of a diploid meiosis are most readily detected in a cross of a heterozygote to a recessive tester. A dihybrid with parents of known genotype is testcrossed to a doubly recessive homozygote. The phenotypes of the progeny are the same as the genotypes of gametes from the dihybrid, because the tester contributes only recessive alleles. Hence progeny can be categorized as parental (referring to the P generation that produces the dihybrid) or recombinant.

The chromosomal assortment that leads to this result is summarized in Figure 6-6. The analysis represented in this figure hinges on the F_1 dihybrid, and its input and output genotypes. Note that the word *parental* is sometimes used to signify the nonrecombinant genotypes. The term refers to the parents that provided the haploid inputs to the dihybrid, not to the parents of the testcross.

In Figure 6-6 we have illustrated independent assortment with genes *known* to be on different chromosomes. However, if we did not know whether the two genes were on different chromosomes, obtaining the standard 1 : 1 : 1 : 1 ratio in the testcross progeny ratio can be used to *infer* that genes are assorting independently and are likely to be on different chromosomes. We can illustrate this with a cross used by Mendel. Recall from Chapter 5 that Mendel deduced that the pea color phenotypic difference yellow versus green was determined by the alternative alleles of one gene: *Y* (yellow) and *y* (green). From other similar analyses he knew that a phenotypic difference in pea shape, round versus wrinkled, was determined by alternative alleles of another gene: *R* (round) and *r* (wrinkled). From his cross of a pure-breeding round, yellow plant ($R / R \ . \ Y / Y$) with a pure-breeding wrinkled, green one ($r / r \ . \ y / y$) Mendel obtained a dihybrid F_1 that was (as expected) round, yellow ($R / r \ . \ Y / y$). When these F_1 individuals were testcrossed to an $r / r \ . \ y / y$ tester, the following progeny were produced:

31 round, yellow	$R / r \ . \ Y / y$
26 wrinkled, green	$r / r \ . \ y / y$
26 round, green	$R / r \ . \ y / y$
27 wrinkled, yellow	$r / r \ . \ Y / y$

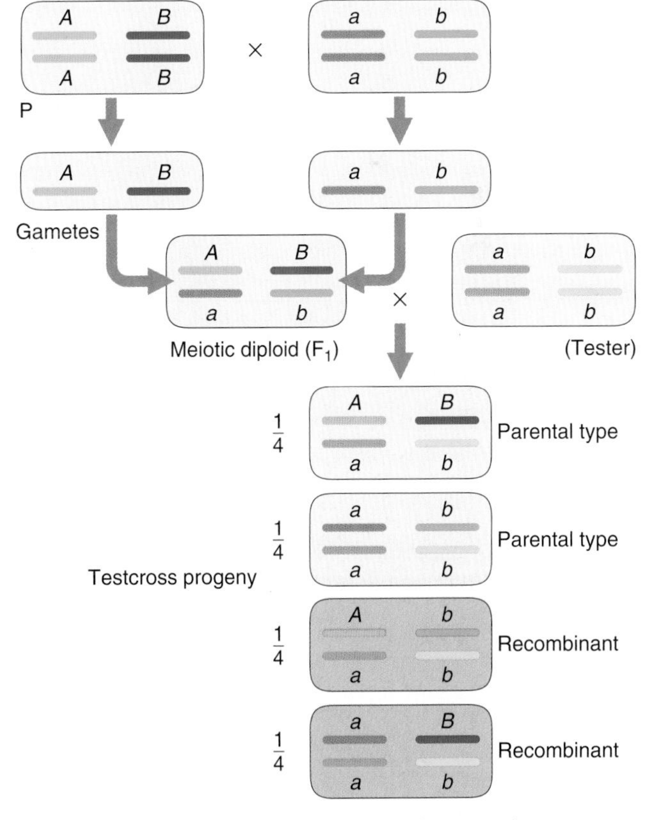

Figure 6-6 Independent assortment, which always produces a recombinant frequency of 50 percent. This diagram shows two chromosome pairs of a diploid organism, with *A* and *a* on one pair and *B* and *b* on the other. A dihybrid with parents of known genotype is testcrossed to a doubly recessive homozygote. The chromosomes of the tester are shown as the lower chromosomes in each progeny type. The phenotypes of the testcross progeny are the same as those of the genotypes of the gametes produced by the dihybrid.

Mendel saw that these numbers were very close to a $\frac{1}{4}:\frac{1}{4}:\frac{1}{4}:\frac{1}{4}$ ratio, so he deduced that the two pairs of alleles were assorting independently. We now know that this is because they are in fact located on different chromosome pairs. We would now rewrite the genotype of the dihybrid as R/r ; Y/y.

Self of a Dihybrid

A testcross is the best way to study recombination in a dihybrid, but sometimes a tester is not available. Nevertheless, independent assortment can be demonstrated by *selfing* a dihybrid, in other words, crossing it to itself. In the plant kingdom selfing is easily achieved by allowing pollen from a plant to fertilize its own ovules. In animals, selfing is achieved by crossing two genetically identical animals. As an example we can use a dihybrid system from Gregor Mendel's work (Figure 6-7, top panel). From a cross of two pure lines, R/R ; y/y (round, green) and r/r ; Y/Y (wrinkled, yellow), a dihybrid F_1 of genotype R/r ; Y/y was produced. The dihybrid was selfed:

R/r ; Y/y (round, yellow) × R/r ; Y/y (round, yellow)

We know that if the genes are on different chromosome pairs, the male and female gametes produced by the dihybrid will be in the ratio

$\frac{1}{4}$ R ; Y

$\frac{1}{4}$ R ; y

$\frac{1}{4}$ r ; y

$\frac{1}{4}$ r ; Y

The fusion of the male and female gametes to form the F_2 can be represented by a 4 × 4 grid, called a **Punnett square,** as shown in the lower panel of Figure 6-7. Each of the 16 squares in the grid represents $\frac{1}{16}$ of the total progeny outcomes, so the overall phenotypic ratio in the progeny is

$\frac{9}{16}$ $R/-$; $Y/-$	Round, yellow
$\frac{3}{16}$ $R/-$; y/y	Round, green
$\frac{3}{16}$ r/r ; $Y/-$	Wrinkled, yellow
$\frac{1}{16}$ r/r ; y/y	Wrinkled, green

This 9:3:3:1 ratio is typical of selfed dihybrids showing independent assortment. We derived this ratio in a way that mimics the biology of a dihybrid self, that is, by considering all the gametes and their possible pairings. However, the ratio can be derived merely by combining two 3:1 ratios at random. We know that both allele pairs considered separately should give monohybrid ratios; in the present example $\frac{3}{4}$ $R/-$:$\frac{1}{4}$ r/r and $\frac{3}{4}$ $Y/-$:$\frac{1}{4}$ y/y. We could combine those using a branch diagram, and multiply their proportions in an application of the product rule:

$\frac{3}{4}$ $R/-$ → $\frac{3}{4}$ $Y/-$ → $\frac{9}{16}$ $R/-$; $Y/-$
$\frac{3}{4}$ $R/-$ → $\frac{1}{4}$ y/y → $\frac{3}{16}$ $R/-$; y/y

$\frac{1}{4}$ r/r → $\frac{3}{4}$ $Y/-$ → $\frac{3}{16}$ r/r ; $Y/-$
$\frac{1}{4}$ r/r → $\frac{1}{4}$ y/y → $\frac{1}{16}$ r/r ; y/y

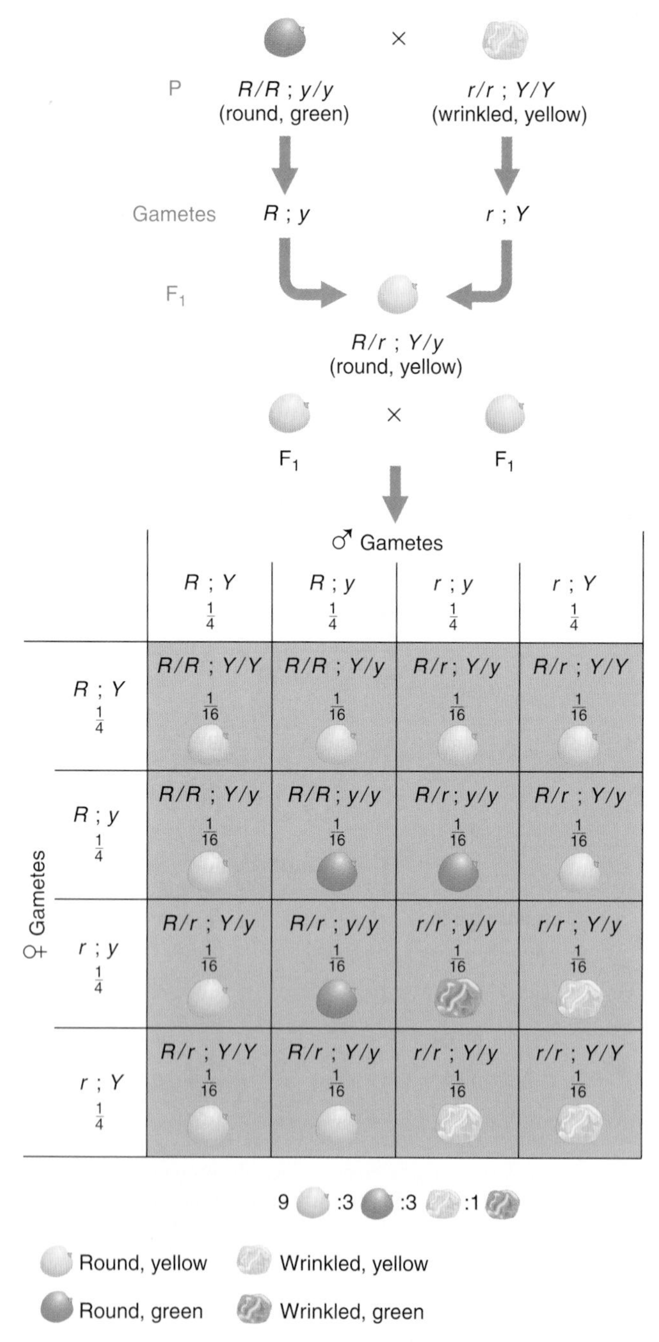

Figure 6-7 Punnett square showing predicted genotypic and phenotypic constitution of the F_2 generation from a dihybrid cross.

Figure 6-8 A 9 : 3 : 3 : 1 ratio in the phenotypes of kernels of corn. Each kernel represents a progeny individual. The progeny result from a self on an individual of genotype A/a ; B/b, where A = purple, a = yellow, B = smooth, and b = wrinkled. *(Anthony Griffiths.)*

In general the phenotypic ratio in the progeny of a self of a dihybrid for genes on separate chromosomes can be written:

	PHENOTYPE	
	First gene	Second gene
$\frac{9}{16}$	Dominant	Dominant
$\frac{3}{16}$	Dominant	Recessive
$\frac{3}{16}$	Recessive	Dominant
$\frac{1}{16}$	Recessive	Recessive

A photograph of a 9 : 3 : 3 : 1 phenotypic ratio in corn *(Zea mays)* is shown in Figure 6-8.

MESSAGE

A 1 : 1 : 1 : 1 ratio in a testcross of a dihybrid and a 9 : 3 : 3 : 1 ratio in a self of a dihybrid both reflect a gametic ratio of 1 : 1 : 1 : 1, which shows the allele pairs are assorting independently and that the RF is 50 percent.

Calculating Phenotypic and Genotypic Ratios for Independently Assorting Genes

In Chapter 5 we calculated the phenotypic ratios expected from progeny of crosses involving alleles of single genes. For example,

$$A/a \times A/a \longrightarrow \tfrac{3}{4}\,A/- \text{ and } \tfrac{1}{4}\,a/a$$
$$A/a \times a/a \longrightarrow \tfrac{1}{2}\,A/a \text{ and } \tfrac{1}{2}\,a/a$$

This knowledge makes it possible to predict genotypes and phenotypes involving several genes *if* they are assorting independently. Consider this cross:

$$A/a \;;\; b/b \times A/a \;;\; B/b$$

We might want to calculate the expected phenotypic ratio in the progeny. From our knowledge of simple single-gene ratios derived in Chapter 5, we know that for the first gene the ratio is

$$\tfrac{3}{4}\,A/- : \tfrac{1}{4}\,a/a$$

and for the second gene it will be

$$\tfrac{1}{2}\,B/b : \tfrac{1}{2}\,b/b.$$

Therefore, since these genes assort independently, we can combine the two phenotypic ratios randomly by drawing a branch diagram. The progeny ratios are derived simply by multiplying the proportions according to the product rule, as shown below:

$A/a \;;\; b/b \times A/a \;;\; B/b$

- $\tfrac{3}{4}\,A/-$
  - $\tfrac{1}{2}\,B/b \longrightarrow \tfrac{3}{8}\,A/- \;;\; B/b$
  - $\tfrac{1}{2}\,b/b \longrightarrow \tfrac{3}{8}\,A/- \;;\; b/b$
- $\tfrac{1}{4}\,a/a$
  - $\tfrac{1}{2}\,B/b \longrightarrow \tfrac{1}{8}\,a/a \;;\; B/b$
  - $\tfrac{1}{2}\,b/b \longrightarrow \tfrac{1}{8}\,a/a \;;\; b/b$

The same progeny ratio could be derived using a grid, which also portrays random associations. Note that the axes of this grid show *phenotypic* proportions in the progeny, as opposed to the *gametic* proportions that we have been showing up to now. A grid is simply a device for displaying random combinations of any kind.

A/a gene / B/b gene	$\tfrac{3}{4}\,A/-$	$\tfrac{1}{4}\,a/a$
	Progeny	
$\tfrac{1}{2}\,B/b$	$\tfrac{3}{8}\,A/- \;;\; B/b$	$\tfrac{1}{8}\,B/b \;;\; a/a$
$\tfrac{1}{2}\,b/b$	$\tfrac{3}{8}\,A/- \;;\; b/b$	$\tfrac{1}{8}\,a/a \;;\; b/b$

In a more complex cross, say,

$$A/a \;;\; B/b \;;\; C/c \times a/a \;;\; B/b \;;\; C/c,$$

we might want to predict the proportion of some specific progeny genotype that we want to use for an experiment, perhaps the genotype $a / a \ ; \ b / b \ ; \ c / c$ for use as a tester. Since we know the genotypic proportions for individual genes will be

$$A / a \times a / a \longrightarrow \tfrac{1}{2} A / a + \tfrac{1}{2} \mathbf{a / a}$$
$$B / b \times B / b \longrightarrow \tfrac{1}{4} B / B + \tfrac{1}{2} B / b + \tfrac{1}{4} \mathbf{b / b}$$
$$C / c \times C / c \longrightarrow \tfrac{1}{4} C / C + \tfrac{1}{2} C / c + \tfrac{1}{4} \mathbf{c / c}$$

if the genes assort independently, we can use the product rule simply to multiply all the proportions of the desired genotypes (shown in bold) to obtain the expected proportion (probability) of $a / a \ ; \ b / b \ ; \ c / c$. It will be

$$\tfrac{1}{2} \times \tfrac{1}{4} \times \tfrac{1}{4} = \tfrac{1}{32}.$$

Therefore, if we need this genotype, we would have to obtain more than 32 progeny to stand a reasonable chance of obtaining one that is $a / a \ ; \ b / b \ ; \ c / c$. These simple calculations can be used to predict phenotypic, genotypic, or gametic proportions in crosses if it is known or assumed that the genes are assorting independently. However, as we saw above and in Chapter 5, genetic analysis works in two directions, so we can also use specific progeny proportions to make deductions about the genotypes and phenotypes of the parents if these are not known. For example, what could we deduce if selfing some plant with the normal wild-type phenotype of blue, large petals produced the following numbers of phenotypes in the progeny?

Blue, large	182
Blue, small	60
White, large	57
White, small	21

We would note that the ratio of blue:white is approximately 3:1 (242:78), that the ratio of large:small is also approximately 3:1 (239:81), and that the four combined phenotypes represent a very close fit to a 9:3:3:1 ratio. Therefore we could deduce the following:

1. The parent must have been heterozygous for two genes, one affecting petal color (blue or white) and the other petal size (large or small).
2. Because of the 3:1 ratios, blue petal is dominant to white and large petal is dominant to small. We could invent allele symbols w^+ = blue and w = white, and s^+ = large and s = small.
3. Because we know that the 9:3:3:1 ratio represents a random combination of two 3:1 ratios, the two genes are assorting independently. One possible reason is that they are on different chromosome pairs. If true, the selfed plant must have been of the genotype $w^+ / w \ ; \ s^+ / s$. (However, we show below that genes far apart on the same chromosome can show RF values up to 50 percent, which gives another possibility.)

CROSSING-OVER

The second way of producing recombinants is crossing-over between genes on the same chromosome. This is a breakage-and-rejoining process between homologous DNA double helices. To track the process of crossing-over, we must go back to a stage just before meiosis. As we saw in Chapter 4, in each diploid meiocyte, before meiosis begins, there is a replication of all the chromosomal DNA molecules. This is the premeiotic S, or synthesis, phase. Now each chromosomal type is represented in four copies, in other words, two pairs of sister chromatids. The two pairs of sister chromatids align, constituting a tetrad, or group of four. It is at this stage that crossing-over takes place. For any particular tetrad in any particular meiocyte, there can be from one to several crossovers. The crossovers can occur at any position along the chromatids, and the positions are different in different meiocytes. Furthermore, crossovers occur only between nonsister chromatids. If we designate the sister chromatids from one parent 1 and 2, and from the other parent 3 and 4, crossovers can occur between 1 and 3, 1 and 4, 2 and 3, and 2 and 4. Some examples are shown in Figure 6-9. What is the basis for the precision of breakage and reunion between the nonsisters? There are still many gaps in our understanding of this precision, but some ideas will be discussed later in the chapter.

These events that take place between DNA molecules are not visible under the microscope; however, the crossover becomes visible later in prophase I of meiosis. When homologous pairs of sister chromatids are visible in their "stacked" configuration during the diplotene stage,

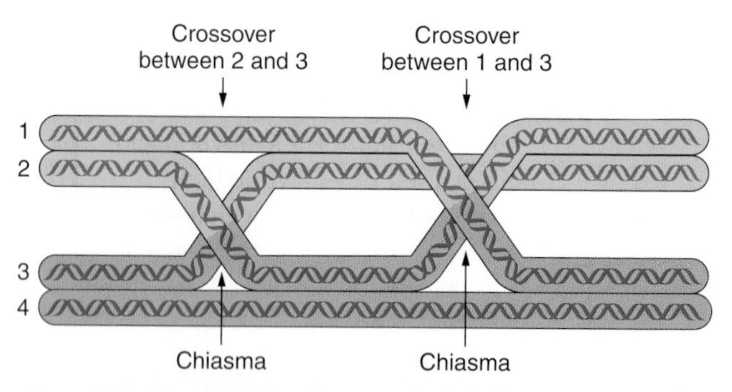

Figure 6-9 Diagrammatic representation of chiasmata at meiosis. Microscopically, a crossover shows up as a cross-shaped structure (a chiasma) between the paired homologs. Crossovers can be between any two nonsister chromatids. In this example one crossover happens to be between chromatids 2 and 3, and the other is between 1 and 3, but other combinations are possible.

cross-like structures called **chiasmata** (singular, *chiasma*) are regularly seen between nonsister chromatids. Each homologous pair shows one to several chiasmata. Chiasmata can be anywhere along the length of a pair but are seen in different positions in different meiocytes. Indeed the chiasmata seem to occur more or less randomly. These chiasmata are the positions of the crossovers, the points at which a breakage-and-reunion event has taken place between homologous nonsister chromatids. The chapter-opening photograph shows a total of four chiasmata on one chromosome pair. In a photograph of the same chromosome pair from another meiocyte, the crossovers would be at different positions, and there might be a number other than four.

Genes on one chromosome are said to be **linked** for the obvious reason that they are physically linked, or joined together, by the segment of chromosome between them. The position of alleles of one gene on a chromosome is always the same, and it is called the gene's **locus,** a term that comes from the Latin word meaning "a place" (plural, *loci*). Therefore, we can say that two genes whose loci are on the same chromosome are linked. In a dihybrid in which the genes are linked, a crossover can lead to the production of recombinants. In order to illustrate this process, we need some more terms. First, there are two kinds of dihybrids involving linked genes. The difference lies in whether or not the recessive alleles of the two genes are on the same parental homolog. If the recessive alleles are on the same homolog, it is a **cis dihybrid.** If they are on different homologs, it is a **trans dihybrid.** The two arrangements are as follows:

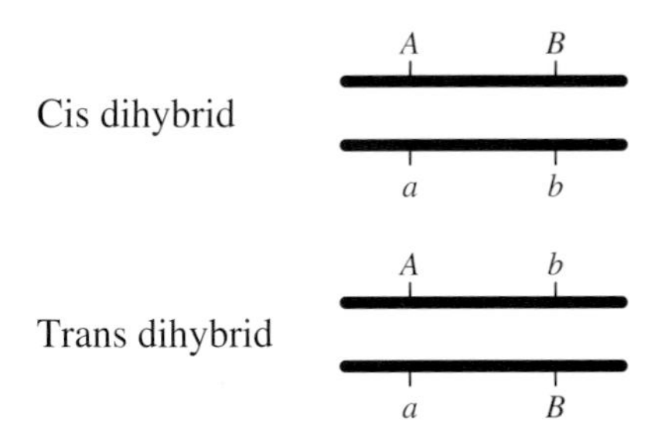

Linked genes are conventionally symbolized by listing to the left of a slash the alleles on one homolog, and to the right of the slash the alleles on the other homolog. The genes are listed in the same order on both sides of the slash. Hence the cis and trans dihybrids above would be represented as *A B* / *a b* and *A b* / *a B*, respectively. (Note there is no punctuation symbol between alleles on the same homolog.)

How can we tell which of these two arrangements (cis or trans) we are working with—they contain the same alleles and look exactly the same at the phenotypic level. There are two ways to tell them apart. First, when the two parents of the dihybrid are pure-breeding strains, this automatically dictates the allelic arrangement in their offspring. For example, if the two parents of a dihybrid are *A b* / *A b* and *a B* / *a B*, their gametes must be *A b* and *a B*, respectively, and therefore the dihybrid must be *A b* / *a B*, a trans configuration. Conversely, if the parents are *A B* / *A B* and *a b* / *a b*, the dihybrid offspring must be in cis, *A B* / *a b*. Another way to determine whether a dihybrid for linked genes is in cis or trans is to work backwards from the phenotypic ratios in the progeny when dihybrids are crossed. We shall see examples below.

Using a cis dihybrid, let's see how a crossover leads to recombinants. Because crossovers occur anywhere along the paired homologs in the population of meiocytes undergoing meiosis, some will have a crossover between the two genes we are studying and some will not. The set of four meiotic products will be different in each case, as shown in Figure 6-10. In meioses with a crossover, two of the haploid meiotic products will be recombinant, of genotypes *A b* and *a B*. (If we had used a trans dihybrid, the recombinants would be *A B* and *a b*.) Since crossover meioses produce equal numbers of recombinant and parental chromosomes, and since both crossover and noncrossover meioses are occurring, the frequency of recombinants must be less than 50 percent ($<\frac{1}{2}$). Each recombinant genotype will be at a frequency of $<\frac{1}{4}$.

Figure 6-10 In dihybrids for linked genes, recombinants arise from meioses in which nonsister chromatids cross over between the genes under study.

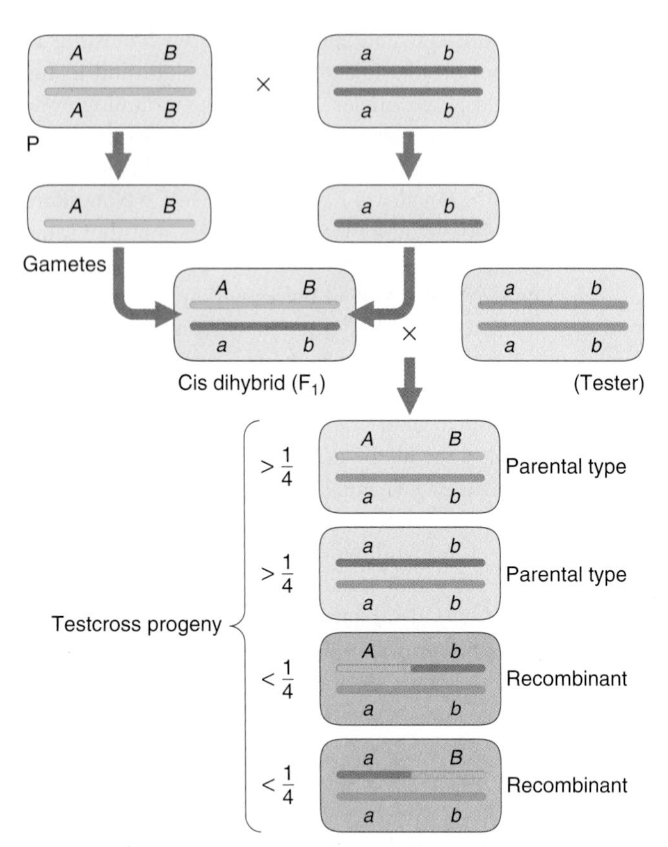

Figure 6-11 Recombinants produced by crossing-over. Notice that the frequencies of the recombinants add up to less than 50 percent.

A typical analysis of a cis dihybrid is shown in Figure 6-11, which should be compared with Figure 6-6, the comparable situation for independent assortment. The genotypes of the two pure parental lines dictate the nature of the meiotic "input" and also dictate that, in this example, the dihybrid F_1 will be in cis configuration. The dihybrid is testcrossed to a fully recessive tester to reveal the proportions of its gametes, the "meiotic output." The testcross progeny proportions show that the recombinant frequency is <50 percent. Note that if we had not known the genotypes of the parental lines, we could have deduced that the dihybrid was in cis configuration because the two most common output genotypes are cis; therefore, the homologs of the dihybrid must have been *A B* and *a b*.

Foundations of Genetics 6-1 examines one of the experiments that showed that recombination between linked loci is a result of crossing-over between chromosomes.

LINKAGE MAPS

Analysis of recombinant frequency can reveal a case of two genes linked on the same chromosome. However, when many such analyses are compared and integrated, the assembled data can be used to construct a **linkage map,** showing the loci of many genes in relation to one another. The groups of linked genes are called **linkage groups,** each corresponding to a part of a chromosome. This exercise to establish linkage groups is important because geneticists are fundamentally interested in knowing where a gene is located within the genome. There are many reasons for this interest. A gene is a segment of a much larger DNA molecule that represents the chromosome. A gene's locus tells us its neighbors and its adjacent regulatory regions, all fundamental aspects of its biological function. An accurate genetic location also makes it feasible to isolate and proceed with a molecular biological analysis of the gene, and eventually to learn what protein the gene encodes and how the gene is activated. In addition, knowledge of gene location is useful in constructing special strains for specific experimental purposes. Maps of the gene loci on chromosomes can be constructed by crossover analysis, and it is to the techniques of crossover-based mapping that we now turn.

The recombinant frequency (RF) is different in crosses involving different linked loci. However, for any two loci the RF value is characteristic and experimentally reproducible. The RF for two loci is roughly proportional to the physical length of DNA between them on the chromosome. This makes sense because crossovers occur more or less randomly along the chromosome, so if two loci are far apart, we would expect that the proportion of meiocytes in which a crossover occurs between the loci would be greater than in cases in which two loci are very close together. (If we think of crossovers as a random scatter of pellets fired from a shotgun, it is intuitive that there will be a greater chance of hitting a large target—a large chromosomal distance—than a small one.) Thus, by determining the frequency of recombinants, we should be able to obtain a measure of the distance between the genes. In fact, one genetic **map unit (m.u.)** is defined as that distance between genes for which one product of meiosis out of 100 is recombinant. Put another way, a recombinant frequency of 1 percent is defined as 1 m.u., and so on. A map unit is sometimes referred to as a **centimorgan (cM)** in honor of Thomas Hunt Morgan, a pioneer in the field of chromosome mapping.

Let's look at a simple calculation of map distance using the haploid fungus *Neurospora.* We will make a cross involving the following pairs of alleles:

ad	Needs adenine to grow
ad^+	Needs no adenine
al	Albino spores
al^+	Black spores

We want to know if these two loci are linked and, if so, how many map units apart. Assume that two haploid strains of

FOUNDATIONS OF GENETICS 6-1

Linking Gene Recombination and Chromosomal Rearrangements

In 1931 most geneticists suspected that recombination between linked genes resulted from chromosomal exchanges dubbed *crossovers,* an idea originally suggested by Thomas Hunt Morgan. Indeed the microscopically visible structures called *chiasmata* seemed to be candidates for the sites at which crossing-over occurs. However, there was no proof of the connection between allelic recombination and chromosome exchange.

In 1931 two papers appeared that convincingly demonstrated this connection. One of these papers, by two maize geneticists, Harriet Creighton and Barbara McClintock, began with the following statement that laid down the basis for their elegant experimental approach:

> The analysis of the behavior of homologous or partially homologous chromosomes, which are morphologically distinguishable at two points, should show evidence of cytological crossing-over. It is the aim of the present paper to show that cytological crossing-over occurs and that it is accompanied by genetical crossing-over.

Controlled pollination of a corn plant. (Chuck O'Rear/Woodfin Camp & Associates.)

Creighton and McClintock used a special strain of maize in which one chromosome 9 was morphologically normal, but the other chromosome 9 had a knob at one end and an extra piece they called an *interchange* at the other. The knob and interchange acted as structural markers by which crossing over could be followed microscopically, and then correlated with recombination of genes. This pair of chromosomes also carried two heterozygous genes *C* / *c* (*C* = colored, or pigmented, *c* = unpigmented) and *Wx* / *wx* (*Wx* = nonwaxy, *wx* = waxy). The arrangement was as follows:

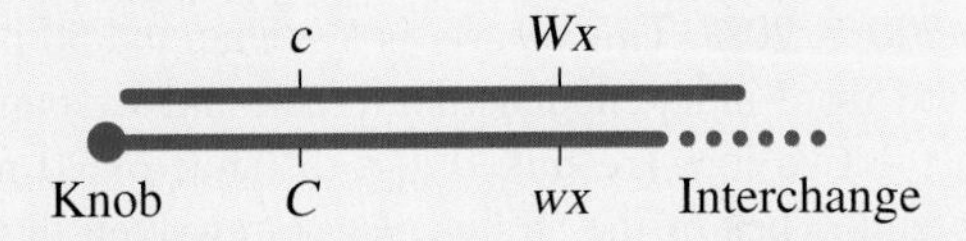

This strain was crossed to a tester with normal chromosomes. They examined recombinant progeny derived from gametes *c wx* and *C Wx*. If recombination is a result of crossing-over, the first type of recombinant should show absence of a knob and presence of the interchange, and the second type should show presence of a knob but absence of an interchange. Both these predictions were observed, as shown in the following diagram (the tester chromosome has been omitted for simplicity):

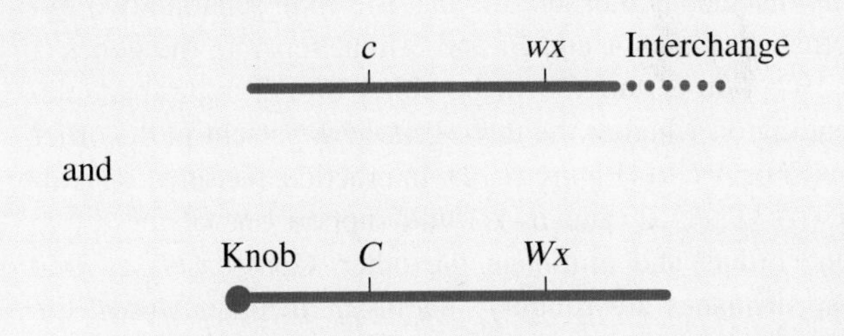

The paper ended with the statement "The foregoing evidence points to the fact that cytological crossing-over occurs and is accompanied by the expected types of genetic cross-over."

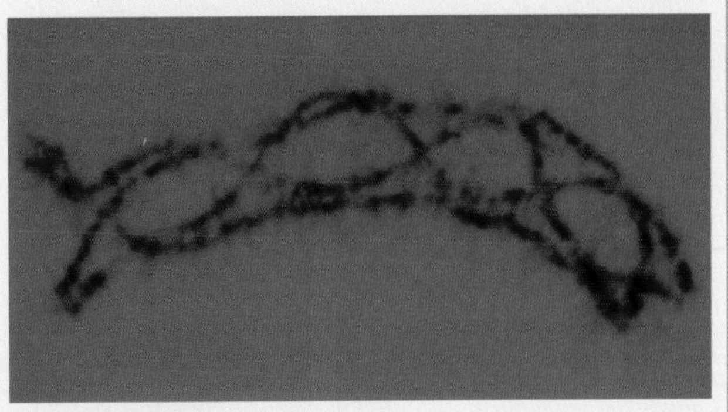

Chiasmata. (John Cabisco/Visuals Unlimited.)

genotype $ad^+ \,.\, al$ and $ad \,.\, al^+$ are crossed and 200 progeny isolated, with the following results:

Parents	$ad^+ \,.\, al \times ad \,.\, al^+$	
Meiocyte	$ad^+/ad \,.\, al^+/al$	
Progeny	$ad^+ \,.\, al$	78
	$ad \,.\, al^+$	82
	$ad^+ \,.\, al^+$	19
	$ad \,.\, al$	21

By considering the haploid input that the two parents provide, we can tell that the last two progeny genotypes are recombinant. They are at a combined frequency of $(19 + 21)/200 = 20\%$. This clearly is less than the 50 percent expected from independent assortment, and we can conclude that the two loci are linked at a distance of 20 m.u. apart. The genotypes of the strains can be rewritten in the proper symbolism, omitting the dots that indicate unknown linkage. We can draw a simple linkage map showing the relative positions of these two genes, using the mutant alleles to designate the locus of each gene:

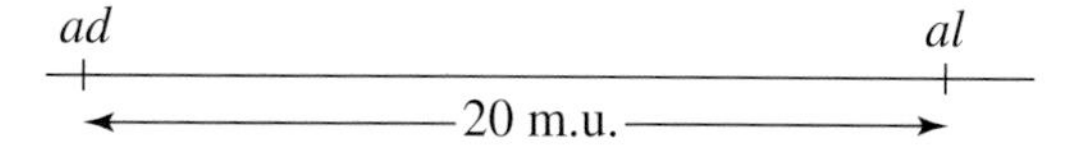

In mapping generally, a direct consequence of the way map distance is measured is that if 5 m.u. separate two hypothetical genes *A* and *B*, and 3 m.u. separate the genes *A* and *C*, then *B* and *C* will be either 8 or 2 m.u. apart, depending on whether the three genes are present in the order *C A B* or *A C B* (Figure 6-12). In practice, the three RF values (*A*–*B*, *A*–*C*, and *B*–*C*) will support one of these two gene orders and eliminate the other. Geneticists say that map distances are roughly "additive." In other words, the RF analysis is compatible with the fact that genes are arranged in linear order on the chromosome, which in turn reflects that they are part of a large linear DNA molecule (see Foundations of Genetics 6-2).

Note that by itself a simple linkage map of the type developed in Figure 6-12 does not place the gene loci at specific places on the chromosome; all the RF analysis does is determine the positions of genes *relative to one another.* The small cluster of three genes could in principle be anywhere on the actual chromosome. However, as more and more recombination analyses are done with many more genes, a linkage map comes to span an entire chromosome.

We shall see in Chapters 8, 9, and 11 that molecular and cytological analysis can also be used to determine gene position on specific chromosomes. These types of analyses are used to complement linkage maps based on recombinant frequency.

As we have seen, genetic analysis works in two directions, and this is true for linkage analysis as well. In addition to using RF measurements to draw a genetic map, we can work in the other direction and use a map we already have to predict RF values and hence frequencies of progeny in different genotypic classes. For example, assume that we know that the *Drosophila* autosomal genes for eye color (p^+ = red, p = purple) and wing size (v^+ = long, v = vestigial) are linked on the same chromosome and are 13 m.u. apart. If we testcross a trans dihybrid

$$p^+\, v\, /\, p\, v^+ \times p\, v\, /\, p\, v$$

we can predict that there will be 13 percent recombinant progeny. Furthermore, there should be two equally frequent types of recombinant progeny, because they are *reciprocal* products of the same type of crossover event. Hence each should be at a frequency of $\frac{13}{2} = 6.5$ percent. The remaining

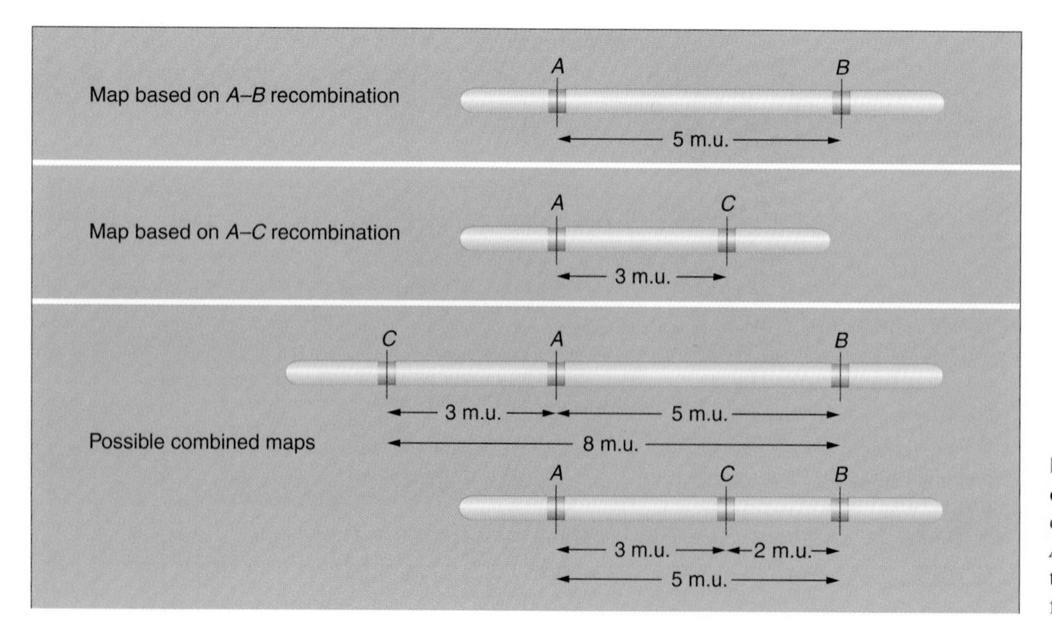

Figure 6-12 Because map distances are additive, calculation of the *A*–*B* and *A*–*C* distances leaves us with the two possibilities shown for the *B*–*C* distance.

FOUNDATIONS OF GENETICS 6-2

The First Chromosome Map

Alfred Sturtevant began his career in genetics in 1911 as an undergraduate in the lab of the great geneticist Thomas Hunt Morgan. Morgan assigned Sturtevant the task of making sense of some data showing various allelic combinations arising from *Drosophila* crosses of genes showing X chromosome–linked inheritance.

Drosophila melanogaster. (Robert Calentine/Visuals Unlimited.)

In one evening Sturtevant made sense of all the data and in doing so developed the analytical method for mapping genes on chromosomes that is still used today. In a later recollection he said:

> In the latter part of 1911, in conversation with Morgan, I suddenly realized that the variations in the strength of linkage [recombinant frequency], already attributed by Morgan to the spatial separation of genes, offered the possibility of determining sequences in the linear dimension of a chromosome. I went home and spent most of the night (to the neglect of my undergraduate homework) in producing the first chromosome map.

The analysis was published in 1913. The introduction to the paper states:

> The parallel between the behavior of the chromosomes in reduction, and that of Mendelian factors in segregation was first pointed out by Sutton (1902) although earlier in the same year Boveri (1902) had referred to a possible connection.

The paper went on to map five X-linked genes using the following logic:

> The proportion of "crossovers" could be used as an index of the distance between any two factors. Then by determining the distances (in the above sense) between A and B and between B and C, one should be able to predict AC. For, if proportion of crossovers really represents distance, AC must be approximately either AB plus BC or AB minus BC, and not any intermediate value.

In Sturtevant's map of the X chromosome (the first genetic map) the gene locus at one end was given the position zero, and the map positions of the other four loci, based on the recombinant frequencies observed, were shown in relation to this one, as in this diagram.

Black body 0 · White eye 1.0 · Vermilion eye 30.7 · Rudimentary wing 33.7 · Miniature wing 57.6

Note that (for example) the map distance between the rudimentary wing and miniature wing genes is 57.6 minus 33.7, or 23.9 map units.

Sturtevant's conclusion follows:

> They [the results] form a new argument in favour of the chromosome view of inheritance, since they strongly indicate that the factors investigated are arranged in a linear series, at least mathematically.

Since Sturtevant's first map, extensive linkage maps of *Drosophila* have been made based on recombinant frequency, and these maps have been shown to represent specific chromosomes. One example is shown in the following RF map.

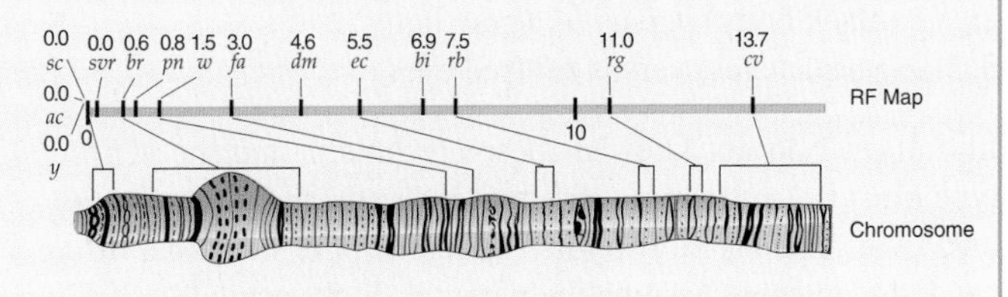

87 percent will be nonrecombinant, and again of equal frequency, at 43.5 percent each. In summary, we expect

Recombinant	$p^+ v^+ / p\ v$	6.5%
Recombinant	$p\ v / p\ v$	6.5%
Parental	$p^+ v / p\ v$	43.5%
Parental	$p\ v^+ / p\ v$	43.5%

MESSAGE

Recombination between linked genes can be used to map their distance apart on the chromosome. The map unit (1 m.u.) is defined as a recombinant frequency of 1 percent.

In a dihybrid for linked loci, the recombinant frequency does not exceed 50 percent, no matter how far apart the loci are. This is counterintuitive because one imagines that as chromosomal distance increases, more crossovers can occur and more recombinants will be produced. However, the multiple crossovers that occur over large chromosomal distances occur among all four chromatids and essentially uncouple the loci under study, causing them to assort independently and therefore to show an RF of 50 percent. (This is rather like taking scissors and glue and repeatedly cutting and cross-joining nonsister chromatids throughout the chromosome.)

MESSAGE

In a dihybrid of linked genes, the RF will be between 0 percent and 50 percent.

MAPPING USING A TRIHYBRID TESTCROSS

So far we have considered the analysis of recombination in dihybrids. By compiling map distances obtained in many dihybrid testcrosses, it is possible to build up entire linkage maps that correspond to maps of loci on chromosomes. The same methodology can be applied faster to more than two loci at once. Sometimes multiple heterozygotes (trihybrids, tetrahybrids, etc.) are available to the researcher, and in these cases it is possible to perform several map-distance calculations by making one testcross to a multiple recessive tester. Trihybrid crosses also enable an unambiguous determination of gene order. We will follow an example using a testcross of a trihybrid, sometimes called a **three-point testcross.**

The example analyzes alleles at three loci in *Drosophila:*

1. v (vestigial wings) versus v^+ (long wings)
2. b (black body) versus b^+ (gray body)
3. p (purple eyes) versus p^+ (red eyes)

Initially we do not know if these are linked, and the genotype must be written in a way that indicates that the order is unknown. To make a trihybrid we cross pure lines that differ at three genes. Assume the parental stocks available to us are the following, which are crossed to form the trihybrid:

Pure lines crossed

$$v^+/v^+ \;.\; b/b \;.\; p/p \times v/v \;.\; b^+/b^+ \;.\; p^+/p^+$$

Gametes $\quad v^+ \;.\; b \;.\; p \qquad v \;.\; b^+ \;.\; p^+$

Trihybrid progeny $\quad v/v^+ \;.\; b/b^+ \;.\; p/p^+$

We will take females of this genotype and cross them to tester males of genotype $v/v \;.\; b/b \;.\; p/p$. For each allele pair of the trihybrid, two genotypes of gametes will be produced (for example, v^+ and v), so since there are three allele pairs, there are $2 \times 2 \times 2 = 8$ gametic genotypes possible. As with simpler testcrosses, these gametic genotypes determine the eight progeny phenotypes from this testcross because the tester contributes only recessive alleles. For simplicity the progeny will be written in terms of the gametic genotypes, and the $v \;.\; b \;.\; p$ contribution from the tester is omitted. The data are shown below as numbers of progeny out of a total sample of 1448 flies from an actual cross.

Trihybrid $\qquad$ Tester

$$v^+/v \;.\; b^+/b \;.\; p^+/p \times v/v \;.\; b/b \;.\; p/p$$

Progeny

v	b^+	p^+	580
v^+	b	p	592
v	b	p^+	45
v^+	b^+	p	40
v	b	p	89
v^+	b^+	p^+	94
v	b^+	p	3
v^+	b	p^+	5
			1448

Notice that, as expected, there are eight genotypes. For each heterozygous locus, the two alleles are each found in $\frac{1}{2}$ the progeny. If all three loci show independent assortment, the eight progeny genotypes will all be of frequency $(\frac{1}{2})^3 = \frac{1}{8}$. This is clearly not the case, so we can infer that there is some linkage. To determine the precise situation, the most straightforward approach is to measure all possible recombinant frequencies. Therefore we must compare input and output genotypes for all pairs of loci. Note that the parental input genotypes for the trihybrid are $v \;.\; b^+ \;.\; p^+$ and $v^+ \;.\; b \;.\; p$; we must take this into consideration when we decide what constitutes a recombinant.

Starting with the v and b loci, we know that the recombinants must be of genotype $v \;.\; b$ and $v^+ \;.\; b^+$ and that there are $45 + 40 + 89 + 94 = 268$ of these. Out of a total of 1448 flies, this gives an RF of 18.5 percent, showing that the v and b loci are linked 18.5 m.u. apart.

For the v and p loci, the recombinants are $v \;.\; p$ and $v^+ \;.\; p^+$. There are $89 + 94 + 3 + 5 = 191$ of these among 1448 flies, so the RF $=$ 13.2 percent and v and p are linked 13.2 m.u. apart. Now we know all the loci are linked. The remaining RF determines their order.

For p and b, the recombinants are $b \;.\; p^+$ and $b^+ \;.\; p$. There are $45 + 40 + 3 + 5 = 93$ of these among the 1448, so that RF $=$ 6.4 percent and p and b are 6.4 m.u. apart.

Because the v and b loci show the largest RF value, they must be farthest apart. The map distances from p to v and from p to b approximately total the v to b distance, so the p locus must be in the middle. A map can be drawn as follows:

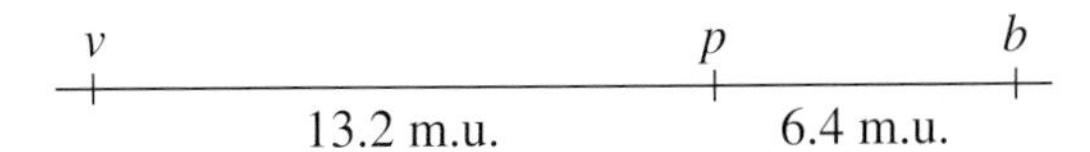

Hence, using a single testcross of a trihybrid, we have established that the three genes are linked and that p is between v and b, and we have determined the distances between the loci in map units. Note that we have arbitrarily placed v to the left and b to the right; the map could equally well be inverted, b–p–v instead of v–p–b.

Note that the two smaller map distances, 13.2 m.u. and 6.4 m.u., add up to 19.6 m.u., which is a distance greater than 18.5 m.u., the distance calculated for v and b. Why is this so? The answer to this question lies in the way in which we have analyzed the two rarest classes in our classification of recombination for the v and b loci, the classes $v\ p\ b^+$ and $v^+\ p^+\ b$. Now that we have the map, we can see that these two rare classes are in fact *double recombinants,* recombinant for the v and p loci and for the p and b loci. They must have arisen from two crossovers (Figure 6-13). However, we did not count the $v\ p\ b^+$ and $v^+\ p^+\ b$ genotypes when we calculated the RF value for v and b; after all, with regard to v and b they are parental combinations ($v\ b^+$ and $v^+\ b$). In the light of our map, however, failure to include these two rare classes led to an underestimate of the distance between the v and b loci. The reason is that the method of mapping by RF depends on proportionality of chromosomal distance and crossover frequency. Therefore not only should we have counted these two rarest classes, we should have counted each of them *twice* because each represents a *double*-recombinant class. Hence, we can correct the value by adding the numbers $45 + 40 + 89 + 94 + 3 + 3 + 5 + 5 = 284$. Out of the total of 1448, this is exactly 19.6 percent, which is identical with the sum of the two component values.

It is worth noting that double and other multiple crossovers lead to RF values that underestimate physical distance (that is, length of DNA). In the above analysis, if the p locus had not been heterozygous, we would have underestimated the v–b map distance as 18.5 m.u. Therefore, the most accurate recombination-based map distances are those based on small RF values because in such short distances multiple crossovers are less likely. Larger distances can be measured more accurately by summing many smaller intervals, as shown in the above three-point cross.

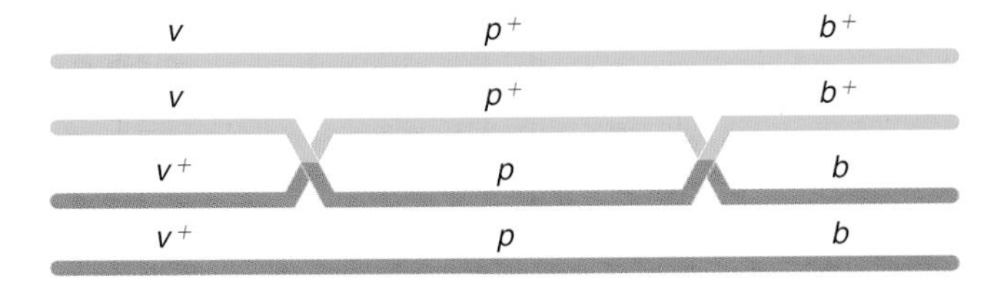

Figure 6-13 An example of one possible type of double crossover in the cross analyzed. Notice that a double crossover can produce double-recombinant chromatids that have the parental allelic combinations at the outer loci.

Now that we have had some experience with the data of this cross, we can look back at the progeny listing and see that it is usually possible to deduce gene order by inspection, without a recombinant-frequency analysis. Only three gene orders are possible, each with a different gene in the middle position (Figure 6-14, left column). The double-recombinant classes are the smallest ones, so only one gene order should be compatible with the smallest classes having been formed by double crossovers (Figure 6-14, right column). Only one gene order gives double recombinants of genotype $v\ p\ b^+$ and $v^+\ p^+\ b$.

MESSAGE

Testcrosses of multiple heterozygotes enable linkage between several loci to be evaluated in one cross.

Interference

The detection of double-recombinant classes shows that double crossovers must occur. The existence of double crossovers permits testing the assumption we have been making that crossovers occur randomly along the chromosome. For example, we might ask if a crossover in one region affects the likelihood of there being a crossover in an adjacent region. We might surmise that a crossover is a large-scale distortion of the chromatids that could decrease the likelihood of another crossover in an adjacent region. It turns out that often it is possible to detect such interaction

Possible gene orders	Double recombinant chromatids
$v\ p^+\ b^+$ / $v^+\ p\ b$	$v\ p\ b^+$ / $v^+\ p^+\ b$
$p^+\ v\ b^+$ / $p\ v^+\ b$	$p^+\ v^+\ b^+$ / $p\ v\ b$
$p^+\ b^+\ v$ / $p\ b\ v^+$	$p^+\ b\ v$ / $p\ b^+\ v^+$

Figure 6-14 With three genes, only three gene orders are possible. Double crossovers create unique double-recombinant genotypes for each gene order. Only the first possibility is compatible with the data in the text, in which $v\ p\ b^+$ and $v^+\ p^+\ b$ are the smallest (and hence double-recombinant) classes.

between adjacent regions, and the effect is appropriately called **interference** because a crossover in one region seems to interfere with the possibility of a crossover nearby.

The analysis can be approached in the following way. If the crossovers in the two regions are independent, then according to the product rule, the frequency of double recombinants would equal the product of the recombinant frequencies in the adjacent regions. In the $v\text{–}p\text{–}b$ recombination data, the $v\text{–}p$ RF value is 13.2 percent and the $p\text{–}b$ value is 6.4 percent, so if there is independence, double recombinants might be expected at the frequency $0.132 \times 0.064 = 0.0084$ (0.84 percent). In the sample of 1448 flies, a total of $0.0084 \times 1448 = 12$ double recombinants are expected. But the data show that only 8 were actually observed. If this deficiency of double recombinants were consistently observed, and not just an effect of sampling, it would show us that the two regions are not independent and suggest that the distribution of crossovers favors single recombinants at the expense of double recombinants. In other words, there *is* interference: a crossover reduces the probability of a crossover in an adjacent region.

Interference (I) is quantified by first calculating a term called the **coefficient of coincidence (c.o.c.),** which is the ratio of observed to expected double recombinants. This value is then subtracted from 1. Hence,

$$\text{Interference} = 1 - \text{c.o.c.}$$
$$= 1 - \frac{\text{observed double recombinants}}{\text{expected double recombinants}}$$

In our example $\text{I} = 1 - (\frac{8}{12}) = \frac{4}{12} = \frac{1}{3}$, or 33 percent.

In some regions, there are never any observed double recombinants. In these cases, c.o.c. = 0, so I = 1 and interference is complete. Most of the time, the interference values that are encountered in mapping chromosome loci are between 0 and 1, but for some chromosomal regions in some organisms, observed double crossovers exceed expected numbers, giving negative interference values.

Examples of Linkage Maps

Linkage maps are an important part of the experimental genetic study of any organism because, like any type of map, they represent an essential aid to finding your way around—in this case, finding your way in the genome. In other words, they are a component of the general goals of the study of genomes: to find out what genes there are, how they are arranged, and how they function. Many eukaryotic organisms have had their chromosomes intensively mapped in this way. Some examples are baker's yeast; the mold *Neurospora crassa;* the plants *Arabidopsis thaliana,* corn, rice, wheat, and tomato; the nematode *Caenorhabditis elegans;* the fruit fly *Drosophila melanogaster;* and the mouse. The resultant maps represent a vast amount of genetic analysis generally achieved by integrating the efforts of research groups throughout the world.

Let us illustrate the early stages of recombination-based genome mapping using data from the screwworm *(Cochliomyia hominivorax).* The larval stage of this insect—the "worm"—is parasitic on mammalian wounds and is a costly pest of livestock in some parts of the world. In order to control this insect, it is advantageous to be able to understand its biological cycle, and a key part of this is to characterize the genes that are relevant to its parasitic lifestyle. A crucial aspect of finding and manipulating such genes is to map them.

The animal has six chromosome pairs ($2n = 12$). Genetic analysis starts by finding and analyzing as many variant phenotypes as possible in order to discover as many genes as possible, including those that might be directly relevant to agriculture control of the insect. Although the genes underlying most of these phenotypes considered would not be directly involved in a fly-control project, they are needed as markers, or signposts, to build up the chromosome map.

The adult stage of this insect is a fly, and we will examine phenotypic variants of adult screwworm flies, which of course have the same genome as the larvae. We will consider six different eye-color genes, all different from the brown-eyed wild-type flies (Figure 6-15a), and five variants

(a)

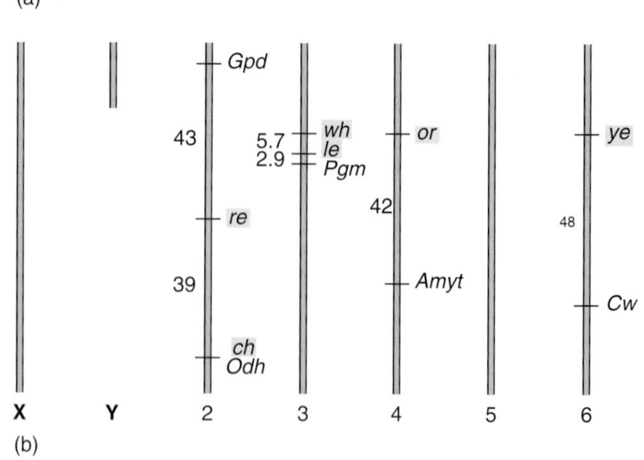

(b)

Figure 6-15 (a) Wild-type adult of screwworm and six flies whose eye colors are determined by alleles at six different autosomal loci. (b) Linkage maps of the six eye-color loci (highlighted in blue) and five other loci of screwworms. The numbers between the loci give the recombinant frequencies. *(D. B. Taylor, USDA.)*

for other characters. In total, 11 mutant alleles are involved. In mapping these genes, pure lines of each phenotype were intercrossed to generate dihybrid F_1s, and then these were testcrossed or selfed according to the methods discussed earlier. The analysis of these mutants reveals a set of four linkage groups (the chromosomes carrying the genes highlighted in blue in Figure 6-15b). Notice that the *ye* and *Cw* loci are is not significantly different from 50 percent.

Because *Drosophila* was one of the first model genetic organisms, a dense recombination-based map has been developed for it. The map in Figure 6-16 shows only a fraction of the known *Drosophila* loci. Tomatoes, also, have been interesting from the perspectives of both basic and applied genetic research, and the tomato genome is one of the best characterized of plants (Figure 6-17 on the next page). The different panels of Figure 6-17 illustrate some of the stages of understanding through which research arrives at a comprehensive map. First, although chromosomes are visible under the microscope, there is initially no way to locate genes on them. However, the chromosomes can be individually identified and numbered, using their inherent landmarks such as staining patterns and centromere positions (see Figure 6-17a and b). Next, analysis of recombinant frequencies generates a set of linkage groups that must correspond to chromosomes, but specific correlations cannot necessarily be made with the numbered chromosomes. At some stage molecular and cytogenetic analysis must be performed to enable the linkage groups to be assigned to specific chromosomes, by techniques we will cover in Chapters 8 and 11. Part c of Figure 6-17 shows some of the loci on a tomato map. Each locus is represented by the two alleles used in the recombination experiments. Today the map contains hundreds of loci.

Molecular Markers

We have seen that in linkage mapping it is important to have a supply of alleles to act as markers that span all regions of the genome. These genes may not themselves be interesting in any particular experiment, but they act as milestones useful for mapping the interesting genes. Likewise, with geographical maps, if your destination is city Z, the map must tell you where cities such as D, M, and R are located, even though they are not destinations, because they will orient you toward Z. Morphological variants provide useful chromosome markers, but molecular technology has provided many more markers for development of recombination-based maps. This technology has provided molecular "probes" that can detect sites of neutral variation in the DNA base sequence that results in no detectable phenotypic variation. There is a wide spectrum of such neutral DNA variation, ranging from single base-pair

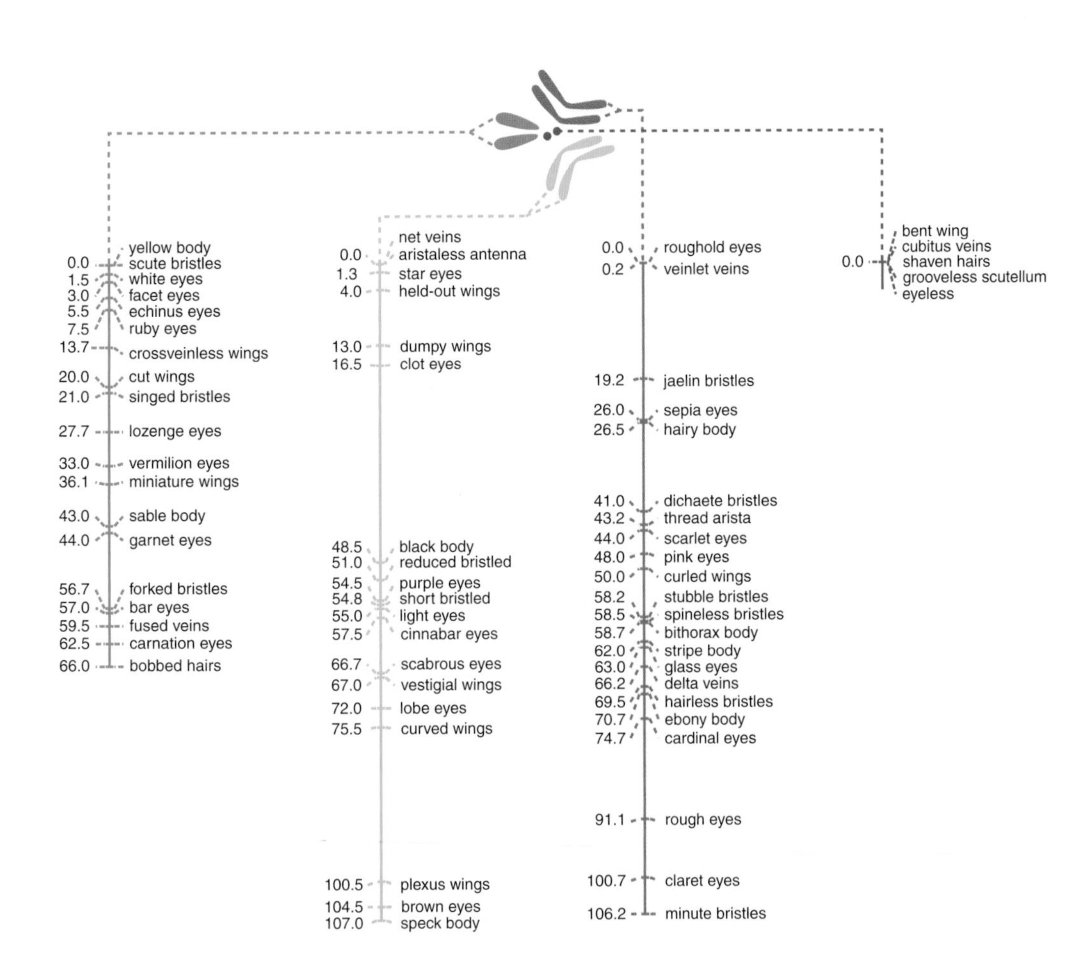

Figure 6-16 A genetic map of the *Drosophila* genome, showing how each linkage group corresponds to one chromosome pair. Values are given in map units measured from the gene closest to one end. The values > 50 m.u. were calculated as sums of individual intervals between loci. *(From E. W. Sinnott, L. C. Dunn, and T. Dobzhansky,* Principles of Genetics, *5th ed. © 1962 by McGraw-Hill.)*

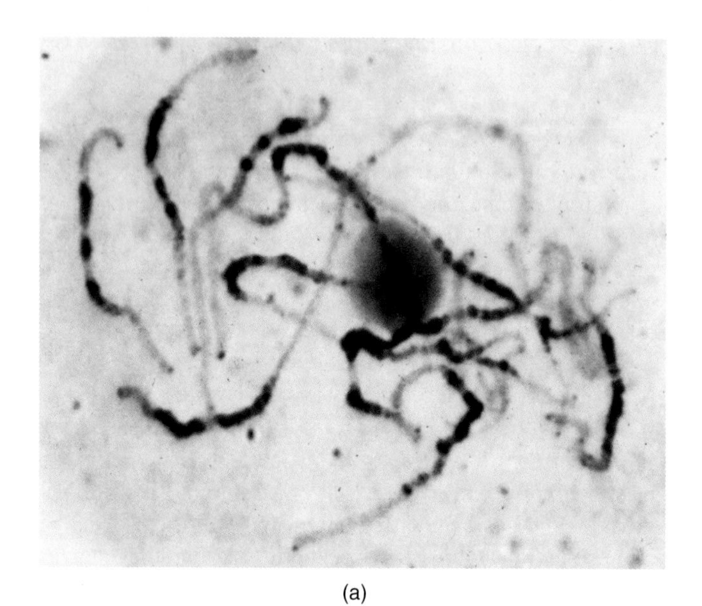

(a)

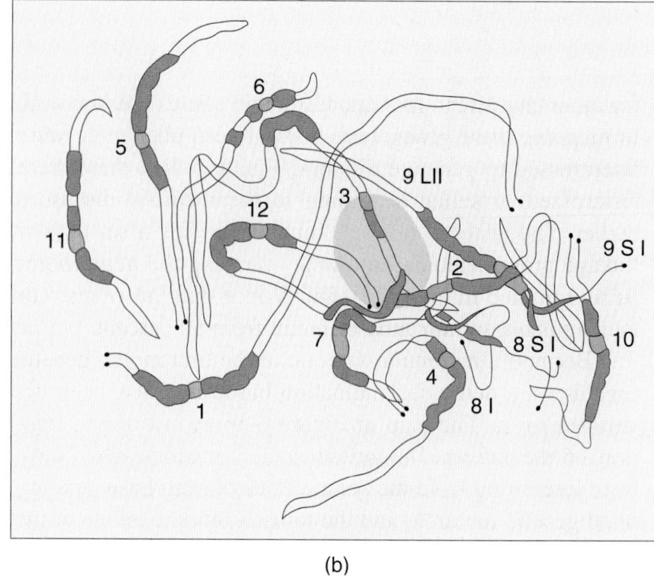

(b)

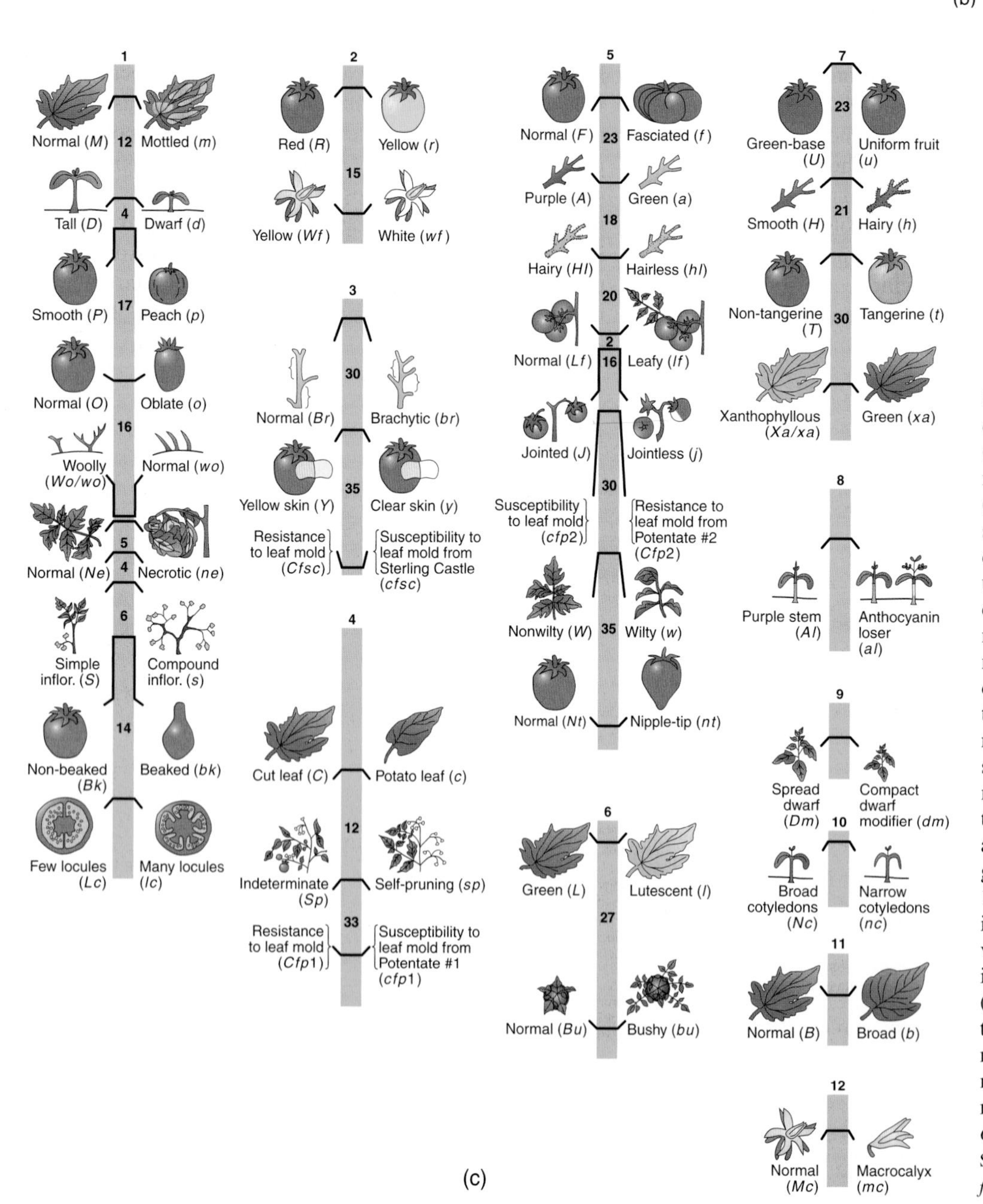

(c)

Figure 6-17 Mapping the chromosomes of tomatoes. (a) Photomicrograph of a meiotic prophase I (pachytene) from anthers, showing the 12 pairs of chromosomes as they appear under the microscope. (b) The currently used chromosome numbering system, using both numerals and letters. The centromeres are orange, and the flanking, densely staining regions (heterochromatin) are shown in green. (c) A linkage map made in 1952 showing the genes known at that time arranged into 12 linkage groups, corresponding to the 12 chromosomes. Each locus is flanked by drawings of the variant phenotype that first identified that genetic locus (generally recessive, and to the right) and the appropriate normal phenotype. Interlocus map distances are shown in map units. *(Parts a and b from C. M. Rick, "The Tomato,"* Scientific American, *1978; part c from L. A. Butler.)*

differences to different numbers of repeats of specific DNA segments Such variation is sometimes called *molecular polymorphism.* If a pair of homologous chromosomes differ at a site of neutral variation—that is, one chromosome shows variant a^1 at the neutral site and its partner shows variant a^2—then this heterozygous locus becomes a **molecular marker** that can be used in mapping in exactly the same way as a *morphological* marker such as the purple-eye locus of *Drosophila* (p^+ / p) can be mapped. Because they are segments of the DNA of chromosomes, molecular markers segregate and recombine like genes. They form valuable chromosomal landmarks for fleshing out a chromosome map, as will be discussed further in Chapter 9.

THE χ^2 TEST

In linkage analysis, the question often arises "Are these two genes linked?" Sometimes the answer is obvious, and sometimes not. But in either situation it is helpful to apply an objective statistical test that can support or not support one's intuitive feeling. The χ^2 test is a statistical test that is useful in science generally. It finds widespread application in genetic analysis, including helping to decide if two genes are linked.

In which situations is the χ^2 test applicable? In research generally, experimental results often fall into several distinct classes or categories, such as red, blue, male, female, lobed, unlobed, and so on. Furthermore, it is often necessary to compare the observed numbers of items in the different categories with numbers that are predicted on the basis of some hypothesis. This is the general situation in which the χ^2 test is useful. In a simple genetic example, suppose you have bred an animal that you hypothesize on the basis of previous analysis to be a heterozygote, A / a. To test this hypothesis you would make a cross to a tester of genotype a / a and count the numbers of $A / -$ and a / a phenotypes in the progeny. Then you would need to assess whether the numbers you obtain constitute the expected 1:1 ratio. If there is a close match, then the hypothesis is deemed consistent with the result, while if there is a poor match, the hypothesis is rejected. As part of this process a judgment has to be made about whether the observed numbers are close *enough* to those expected. Very close matches and blatant mismatches generally present no problem, but inevitably there are gray areas in which the match is not obvious.

The χ^2 test is simply a way of quantifying the various deviations expected by chance if a hypothesis is true. Take the simple hypothesis predicting a 1 : 1 ratio, for example. Even if the hypothesis is true, we would not always expect an exact 1 : 1 ratio. We can model this with a barrel full of equal numbers of red and blue marbles. If we blindly remove samples of 100 marbles, on the basis of chance we would expect samples to show small deviations, such as 52 red : 48 blue, quite commonly, and larger deviations, such as 60 red : 40 blue, less commonly. Even 100 red marbles is a possible outcome, at a probability of $(\frac{1}{2})^{100}$. The χ^2 test gives a simple way to calculate the probability of chance deviations from expectations if the hypothesis is true. But if all levels of deviation are expected with different probabilities even if the hypothesis is true, how can we ever reject a hypothesis? It has become a general scientific convention that if there is a probability of less than 5 percent of observing a deviation from expectations this large, or larger, then this is taken as a criterion for rejecting the hypothesis. The hypothesis might still be true, but we have to make a decision somewhere and the 5 percent is the conventional decision line. The implication is that although results this far from expectations are anticipated 5 percent of the time even when the hypothesis is true, we will mistakenly reject the hypothesis in only 5 percent of cases and we are willing to take this chance of error.

How is the χ^2 test applied to linkage? We have learned in this chapter that we can infer that two genes are linked on the same chromosome if the RF is less than 50 percent. But how much less? It is not possible to test for linkage directly because *a priori* we do not have a precise linkage distance to use in the hypothesis. Are the genes 1 m.u. apart? 10 m.u.? 45 m.u.? The only genetic criterion for linkage with which we can make a precise prediction is the presence or absence of independent assortment. Consequently, it is necessary to test the hypothesis of the *absence* of linkage. If the observed results cause rejection of the hypothesis of *no linkage,* then we can infer linkage. This type of hypothesis, called a **null hypothesis,** is generally useful in χ^2 analysis because it provides a precise experimental prediction that can be tested.

Let's test a specific set of data for linkage using χ^2 analysis. Assume that we have crossed pure-breeding parents of genotypes $A / A \;.\; B / B$ and $a / a \;.\; b / b$ and obtained a dihybrid $A / a \;.\; B / b$, which we testcross to $a / a \;.\; b / b$. A total of 500 progeny are classified as follows (written as gametes from the dihybrid):

142	$A \;.\; B$	Parental
133	$a \;.\; b$	Parental
113	$A \;.\; b$	Recombinant
112	$a \;.\; B$	Recombinant
Total 500		

From these data the recombinant frequency is $\frac{225}{500} = 45$ percent. On the face of it, this seems like a case of linkage because the RF is less than the 50 percent expected from independent assortment. However, it is possible that the recombinant classes are less than 50 percent merely on the basis of chance. Therefore, we need to perform a χ^2 test and calculate the likelihood of this result under a hypothesis of independent assortment. The statistic χ^2 is always calculated from actual numbers, not from percentages, proportions, or fractions. Sample size is therefore very important in the χ^2 test, as it is in most considerations of chance phenomena. The sample is composed of four phenotypic classes. The letter O is used to represent the observed number in a class, and E the expected number for the same class based on the predictions of the hypothesis. The general formula for calculating χ^2 is as follows:

$$\chi^2 = \sum \frac{(O - E)^2}{E} \text{ for all classes}$$

The problem, then, is to calculate the expectations E for each class. As we saw above, the hypothesis that must be tested in this case is one of independent assortment of the two loci, because the hypothesis of linkage cannot provide a precise set of expected proportions. We might assert that if the allele pairs of the dihybrid are assorting randomly (that is, there is no linkage), there should be a 1 : 1 : 1 : 1 ratio of gametic types. Therefore it seems reasonable to use $\frac{1}{4}$ as the expected proportion of each gametic class. Note that if there is equal viability of all genotypes, we expect allelic proportions of $\frac{1}{2}$ for each allele of each gene, which are multiplied to give us a $\frac{1}{4}:\frac{1}{4}:\frac{1}{4}:\frac{1}{4}$ ratio, as discussed earlier in the chapter. However, this expectation of equal viability of all genotypes is often not the case because some mutations or combinations of mutations are less vigorous than their wild-type counterparts and do not survive to adulthood. A better test is whether or not the two allele ratios are combined randomly in the meiotic products, because this does not assume that the two alternative alleles at each locus have equal effects on viability. The observed allele ratios in the present example can be multiplied using a grid. Let's start with the observed genotypic classes and rearrange them in a grid to see the allele ratios more clearly.

Observed values

		Segregation of A and a		
		A	a	Totals
Segregation of B and b	B	142	112	254
	b	113	133	246
	Totals	255	245	500

We see that the allele proportions are $\frac{255}{500}$ for A, $\frac{245}{500}$ for a, $\frac{254}{500}$ for B, and $\frac{246}{500}$ for b. Now we calculate the values expected under independent assortment by multiplying these allelic proportions. For example, to find the expected value of $A\ B$ if the two ratios are combined randomly, we simply multiply as follows:

$$\text{Expected value } (E) \text{ for } A\ .\ B = \left(\tfrac{255}{500}\right) \times \left(\tfrac{254}{500}\right) \times 500$$
$$= 129.54.$$

Using this approach, we can complete the entire grid of E values:

Expected values

		Segregation of A and a		
		A	a	Totals
Segregation of B and b	B	129.54	124.46	254
	b	125.46	120.56	246
	Totals	255	245	500

The value of χ^2 is calculated as follows:

Genotype	O	E	$(O - E)^2/E$
$A\ .\ B$	142	129.54	1.19
$a\ .\ b$	133	120.56	1.29
$A\ .\ b$	113	125.46	1.24
$a\ .\ B$	112	124.46	1.25
Total (which equals the χ^2 value)			4.97

The obtained value of χ^2 (4.97) is used to find a corresponding probability value p, using a χ^2 table (Table 6-1).

To do this we need to compute the number of **degrees of freedom (df)** used in the χ^2 calculation. Generally in a statistical test the number of degrees of freedom is the number of nondependent values. Working through the following "thought experiment" will show what this means. In the grids we used to show the combined segregations of A / a and B / b (called *contingency tables*), the column and row totals are given from the experimental results, so specifying any one value within the grid automatically dictates the other three values. Thus there are three dependent values but only one nondependent value and, therefore, only 1 degree of freedom. A rule of thumb useful for larger contingency tables is that the number of degrees of freedom is equal to the number of classes represented in the rows minus 1, times the number of classes represented in the columns minus 1. According to the rule for calculating degrees of freedom, in the present example

$$\text{df} = (2 - 1) \times (2 - 1) = 1$$

as stated earlier.

Therefore, using Table 6-1, we look along the row corresponding to 1 degree of freedom until we find our χ^2 value of 4.97. Not all values of χ^2 are shown in the table, but 4.97 is close to the value 5.024. Hence the corresponding probability value is very close to 0.025, or 2.5 percent. This p value is the required probability, that of obtaining a deviation from expectations this large or

TABLE 6-1 Critical Values of the χ^2 Distribution

df \ p	0.995	0.975	0.9	0.5	0.1	0.05	0.025	0.01	0.005	df
1	.000	.000	0.016	0.455	2.706	3.841	5.024	6.635	7.879	1
2	0.010	0.051	0.211	1.386	4.605	5.991	7.378	9.210	10.597	2
3	0.072	0.216	0.584	2.366	6.251	7.815	9.348	11.345	12.838	3
4	0.207	0.484	1.064	3.357	7.779	9.488	11.143	13.277	14.860	4
5	0.412	0.831	1.610	4.351	9.236	11.070	12.832	15.086	16.750	5
6	0.676	1.237	2.204	5.348	10.645	12.592	14.449	16.812	18.548	6
7	0.989	1.690	2.833	6.346	12.017	14.067	16.013	18.475	20.278	7
8	1.344	2.180	3.490	7.344	13.362	15.507	17.535	20.090	21.955	8
9	1.735	2.700	4.168	8.343	14.684	16.919	19.023	21.666	23.589	9
10	2.156	3.247	4.865	9.342	15.987	18.307	20.483	23.209	25.188	10
11	2.603	3.816	5.578	10.341	17.275	19.675	21.920	24.725	26.757	11
12	3.074	4.404	6.304	11.340	18.549	21.026	23.337	26.217	28.300	12
13	3.565	5.009	7.042	12.340	19.812	22.362	24.736	27.688	29.819	13
14	4.075	5.629	7.790	13.339	21.064	23.685	26.119	29.141	31.319	14
15	4.601	6.262	8.547	14.339	22.307	24.996	27.488	30.578	32.801	15

larger. Since this probability is less than 5 percent, the hypothesis of independent assortment must be rejected. Thus, having rejected the hypothesis of no linkage, we are left with the inference that indeed the loci are probably linked.

MESSAGE

In genetic analysis, the χ^2 test is used to test observed numbers in various observed classes against the expectations derived from a predictive hypothesis. The test generates the probability of obtaining by chance a specific deviation at least as great as the one observed, assuming that the hypothesis is correct.

THE MECHANISM OF MEIOTIC CROSSING-OVER

Crossing-over is a remarkably precise process. Some kind of cellular machinery takes two huge molecular assemblages (homologous nonsister chromatids), breaks them in the same relative position, and then rejoins them in a new arrangement so that no genetic material is lost or gained in either. The exact molecular mechanism of crossing-over is not known, but several models have been proposed. In all these models there is general agreement that crossing-over must involve the formation of **heteroduplex DNA,** a hybrid type of DNA molecule that is composed of a single strand from a chromatid derived from one parent and a single strand from a chromatid derived from the other parent. The formation of heteroduplex DNA is the way in which a homologous alignment of the two chromatids is assured.

MESSAGE

Heteroduplex formation, the association of complementary strands from the two parental DNA molecules, explains the molecular precision of cross-over events.

Most of the evidence in favor of models based on heteroduplexes comes from the study of fungal tetrads and octads. Octads are particularly informative in pointing to the existence of heteroduplexes in crossing-over. We saw in Chapter 5 that in fungi a cross *A* × *a* will create a monohybrid meiocyte *A* / *a* that is expected to segregate in a 1 : 1 ratio in the meiotic products according to the law of equal segregation. Indeed this is found in most meiocytes. In fungi with ordered octads, the four nuclei that represent the four products of meiosis undergo an extra mitotic division to produce four ascospore pairs, which stay together in the octad sac (see Figure 5-3). Therefore the expected octad ratio from a monohybrid meiocyte is 4 : 4, and this is seen in most cases. However, in rare meiocytes (generally on the order of 0.1 percent to 1 percent) any one of five types of aberrant ratios can be found, and these give the clues needed to build a heteroduplex crossover model. The aberrant types are as follows:

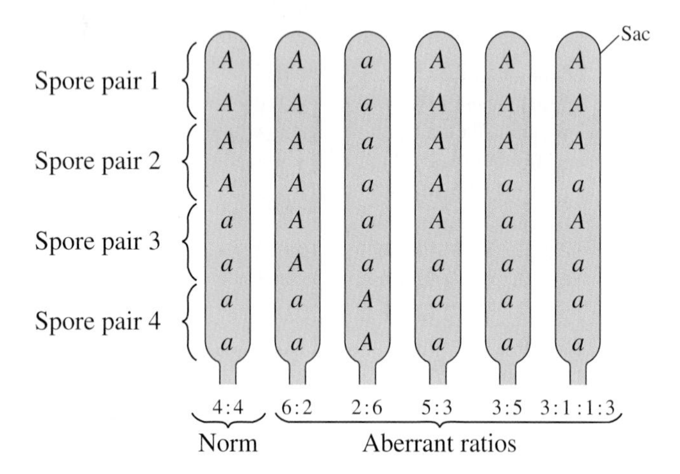

All these aberrant ratios need to be explained, but the key types that led to the heteroduplex model are the odd number–containing ratios 5:3, 3:5, and 3:1:1:3, because these contain spore pairs with *nonidentical sister spores.* (Note that the postmeiotic round of mitosis is expected to produce identical sister genotypes by replication of each of the products of meiosis.) Nonidentical sister spore genotypes must reflect heteroduplex DNA in the meiotic product that gave rise to the nonidentical pair, that is, DNA with a segment in which one strand is the nucleotide sequence of the *A* allele and one strand is the nucleotide sequence of the *a* allele. After mitosis the two sister cells resulting from division of such a heteroduplex-containing nucleus will be different, one *A* and one *a*. This process is shown in the following sketch; *A* and *a* represent the site in the gene at which the alleles differ:

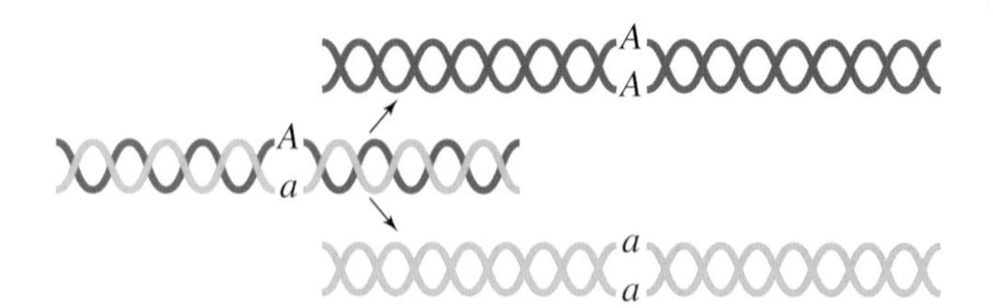

Accordingly, in the octads showing aberrant ratios, heteroduplexes must have been responsible for spore pair 3 in the 5:3 octad, spore pair 2 in the 3:5 octad, and spore pairs 2 and 3 in the 3:1:1:3 octad. Aberrant ratios are often found in octads or tetrads in which there is a crossover in the vicinity, leading to the idea that the underlying heteroduplexes are part of the crossover mechanism.

Figure 6-18 shows one model of how heteroduplexes might form, the so-called Meselson-Radding model. In the following discussion it is important to refer closely to this fig-

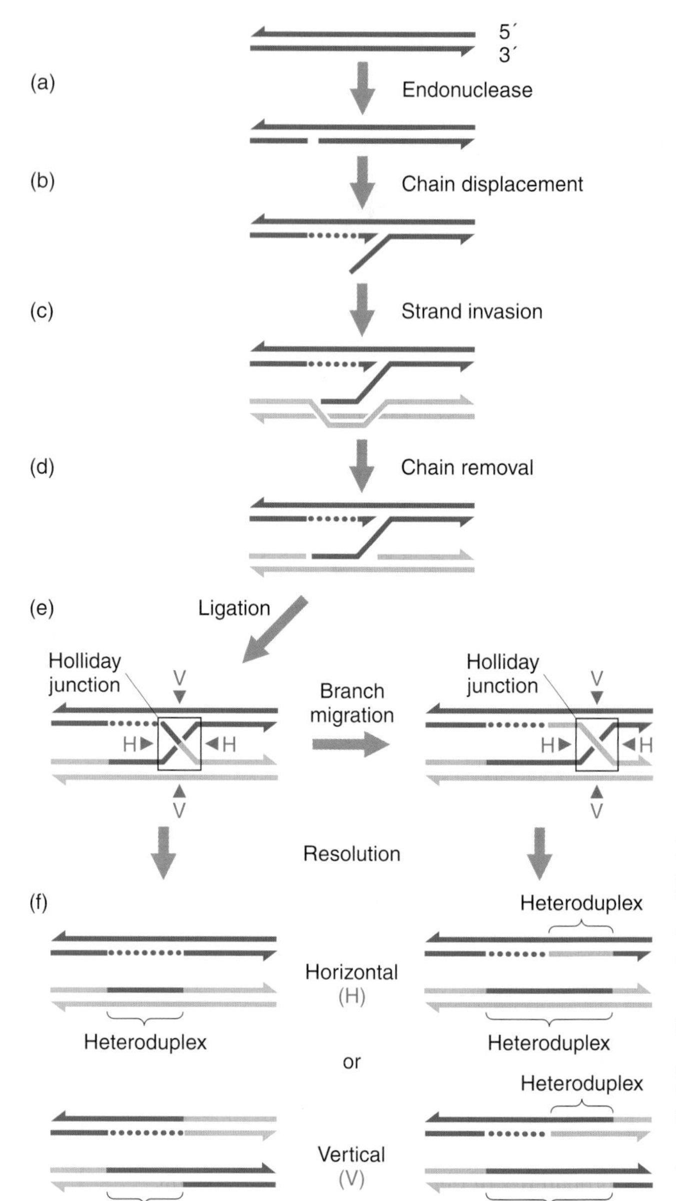

Figure 6-18 The Meselson-Radding heteroduplex model. Blue represents one parental DNA and yellow the other. In total, at meiosis there are two blue and two yellow molecules corresponding to the two pairs of sister chromatids, but nonparticipating chromatids have been omitted for simplicity. (a) In one chromatid (assumed to be a blue one), one chain of the double-stranded helix is cut. (b) DNA polymerase synthesizes a new strand (dotted) and displaces one original strand. (c) The displaced strand invades a double helix of the other parental type (yellow) and displaces its counterpart. (d) This displaced yellow chain is enzymatically digested. (e) The two free ends join (ligation), and this creates a Holliday junction (left column). Only one of the two duplexes has a region of heteroduplex DNA (blue/yellow). If the junction migrates, heteroduplex DNA can arise on both duplexes (right column). (f) Cutting and rejoining of the Holliday junction in two alternative ways (depicted in part e by H and V) creates either a crossover chromatid (V, vertical cut) or a noncrossover chromatid (H, horizontal cut). *(Modified from F. W. Stahl, "The Holliday Junction on Its Thirtieth Anniversary,"* Genetics *138, 1994, 241–246.)*

Animated Art • A Mechanism of Crossing-Over: Meselson-Radding Heteroduplex Model

ure. Note that for simplicity the diagram shows only the two nonsister chromatids participating in the crossover. A strand from the DNA molecule of one chromatid (blue) breaks, unwinds, and "invades" a nonsister chromatid (yellow), displacing a segment of yellow strand, which is removed. One of the immediate products of strand invasion is a blue/yellow heteroduplex region (on the lower strand in Figure 6-18). New strand synthesis replaces the invading blue strand. Joining of free ends (ligation) results in a peculiar type of half-helix crossover called a **Holliday junction.** If the Holliday junction moves along the paired chromatids ("migrates") by unwinding and rewinding of the helix in one direction (assumed to be to the right in the figure), then *two* aligned blue/yellow heteroduplexes are formed, one on each chromatid. These are in addition to the initial heteroduplex formed on the lower chromatid during strand invasion.

What will happen to the Holliday junction? It has been shown that a Holliday junction can resolve itself in two different ways, "horizontally" or "vertically." To picture these resolutions, imagine holding a pair of scissors horizontally and cutting both strands of the Holliday junction (see Figure 6-18e and f, upper panel). Then paste the cut ends together, giving two noncrossover chromatids (although the single heteroduplex formed on the lower strand during strand invasion remains). Alternatively, hold the scissors vertically and at the Holliday junction cut the two strands not in the single-stranded crossover (see Figure 6-18e and f, lower panel). Rejoin the top cut ends to the bottom cut ends, thereby making a double-stranded crossover. Hence you can see that the vertical resolution produces the crossover chromatids that the model started out to explain. Note that in the model the heteroduplex is an essential precursor of a crossover. However, the heteroduplex persists either in the presence or the absence of the crossover. Now we must see what happens to the heteroduplexes.

First consider the case where there is no migration of the Holliday junction (left-hand part of Figure 6-18f). The tetrad would contain the following four DNA molecules:

- Two DNA molecules that did not engage in heteroduplex formation, one of each parental type (the chromatids not shown in Figure 6-18); one blue DNA double helix and one yellow
- One blue and yellow heteroduplex molecule (the lower, "invaded" chromatid in Figure 6-18f)
- Another parental DNA molecule—one blue double helix—which was the source of the invading strand

If the heteroduplex spans a heterozygous gene locus, then after the postmeiotic mitosis, the resulting octad would show a 5:3 phenotypic ratio because the wholly blue and yellow DNA molecules will replicate themselves, giving 4 blue and 2 yellow, whereas the heteroduplex will form an unequal sister spore pair, 1 blue and 1 yellow. Figure 6-19 on the next page summarizes creation of an asymmetric octad for a heterozygous locus, showing the role of heteroduplex.

However, a heteroduplex is an unstable structure containing a mismatch of DNA. If the mutant site is an AT pair whereas in the wild type it is GC, then depending on which strand invades, the possible mismatches would be a GT pair or a CA pair. Enzymes are known that recognize such mismatches and occasionally cut out one of the mispaired strands and replace it with one containing a properly hydrogen-bonded nucleotide (Figure 6-20 on the next page). This type of mismatch repair of the simple unmigrated structure could change the heteroduplex in two ways. For example, let's assume that the mismatch in the heteroduplex is GT, as shown in Figure 6-20: this could be repaired to GC (wild type) or to AT (mutant). Hence, an octad that without repair would have been a 5:3 ratio, such as the one explained above, will show either a 6:2 ratio or a normal 4:4 ratio, depending on whether the repair went from wild to mutant or mutant to wild, respectively:

Chromatid 1			G C		
Chromatid 2			G C		
Chromatid 3	A T	← ←	G T	→ →	G C
Chromatid 4			A T		

In cases where there are two heteroduplexes resulting from migration of the Holliday junction (see Figure 6-18f, right-hand side), if there is no repair, the result is a 3:1:1:3 ratio. Repair can produce any of the other aberrant ratios.

The Meselson-Radding model is only one of a number of models based on heteroduplex formation. The recombination mechanism in this model is initiated by a single-strand break in DNA. In another model, the recombination mechanism is initiated by a double-strand break, and there is support for this model from studies on yeast recombination. However, there is no model that is generally agreed to be applicable to all eukaryotes.

MESSAGE

The observation of rare cases of nonidentical ascospore pairs in fungal octads suggests that DNA heteroduplexes can form that are the basis of crossovers. The extent of the heteroduplex region and the repair of DNA mismatches in the heteroduplexes can explain all the aberrant ratios observed.

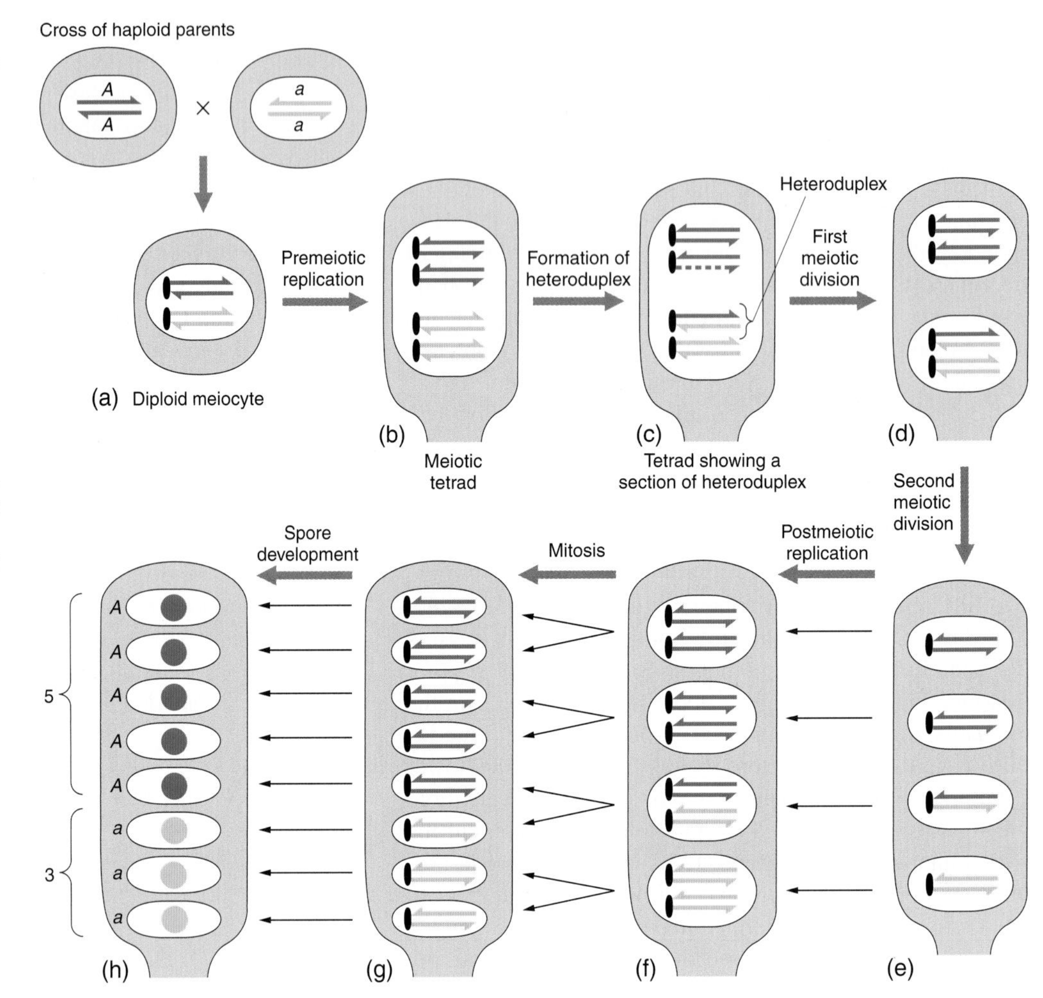

Figure 6-19 The role of a heteroduplex in creating an abberant 5:3 octad for a heterozygous locus in *Neurospora*. (a) Two haploid parents unite sexually to form a diploid meiocyte. Each colored line represents a portion of one strand of a DNA double helix containing a heterozygous locus for the alleles *A* and *a*. (b) Replication and the beginning of meiosis produce pairing of the homologous chromosomes, yielding a tetrad of the four chromatids. Centromeres are shown, to keep track of sister and nonsister chromatids. (c) The heteroduplex is formed (as in Figure 6-18). (d, e) Two meiotic divisions occur, producing four haploid nuclei, one of which contains heteroduplex DNA. When this heteroduplex DNA is replicated, two sister chromatids lacking the heteroduplex are formed (f). Mitosis yields eight nuclei (g), and spore development produces the octad (h).

Animated Art • A Mechanism of Crossing-Over: Genetic Consequences of the Meselson-Radding Model

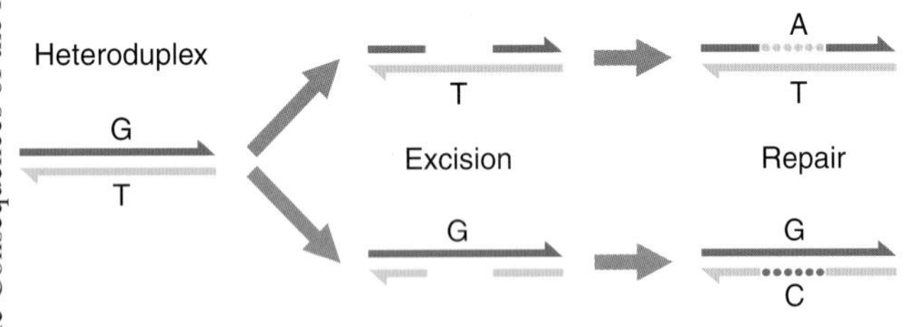

Figure 6-20 Repair of mispaired nucleotides in heteroduplex DNA to either AT or GC. Heteroduplex DNA containing mismatched base pairs is unstable. The mispaired nucleotides can produce a distortion in the DNA at these points that will be recognized by a repair system.

It is worth emphasizing that the aberrant meiotic ratios are rare cases in which a researcher stumbles accidentally upon a situation in which a heterozygous locus happens to be in a region that forms a DNA heteroduplex. The importance of the aberrant ratios is that in explaining them, an explanation was deduced for the molecular mechanism of crossovers in general, most of which would *not* involve a heteroduplex spanning a heterozygous allele pair that is under study.

RECOMBINATION BETWEEN ALLELES OF A GENE

The molecular events that generate a crossover can occur between genes or *within* a gene. If the crossover is within a gene, it can produce recombination between the mutant sites within that gene. Meiosis in a diploid that is heterozygous for two different mutant alleles of the same gene (say, a^1 / a^2) would generally produce meiotic products half of which are a^1 and half a^2, according to the law of segregation. However, it is likely that these two mutant alleles will have their mutant sites in different places within the gene, and if this is the case, there is a finite probability that a crossover will occur between the two sites. The probability will be low because the two sites are within a gene and therefore very close together on the chromosome. The products will be a wild-type chromatid and a mutant with two mutant sites:

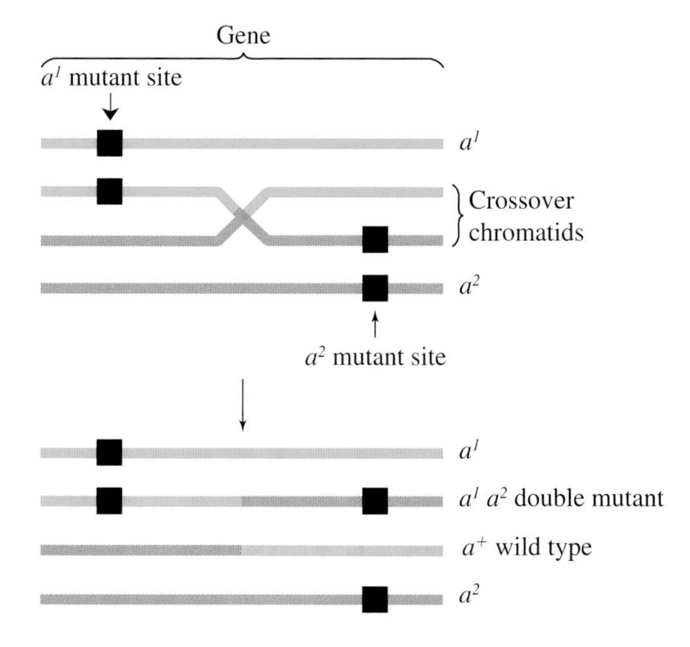

These rare recombinants that occur at frequencies of 10^{-3} to 10^{-6} can be detected in some systems. For example, in fungi if the mutant alleles determine auxotrophy (for example, an arginine requirement), the wild types can be detected by simply spreading the meiotic products on medium containing no arginine—only the wild-type recombinants (which are prototrophic) will be able to grow to form colonies. Indeed mutant sites within a gene can be mapped by this method in an approach essentially identical with that used in chromosome mapping.

Let us return to the human family illustrated in Figure 6-1. Can this family be analyzed for recombination using the techniques covered in this chapter? The answer is in general yes, but we must be cautious in doing so. Normal human phenotypic variation has a complex genetic basis involving many genes affecting a wide range of characters, and some of these genes interact to influence one another's expression (as we shall see in Chapter 14). Furthermore, human matings are generally not between the genotypes that we need to make clear inferences about linkage; for example, matings that would be the equivalent of a testcross are rare. Hence, whereas we might be very interested in trying to analyze phenotypic variation in such a family (perhaps our own family) on the basis of allelic recombination, this is fraught with difficulty. However, molecular analysis is more reliable because the molecular "phenotypes", such as DNA or protein variants, are much more cut-and-dried. Hence linkage analysis of human phenotypes, especially disease phenotypes, is heavily dependent on the use of linked molecular markers, as we shall see in Chapter 9.

SUMMARY

1. Most organisms have a sexual cycle, whose main outcome is variant offspring resulting from meiotic recombination. Recombination produces genotypes with new combinations of parental alleles. Essentially, recombination is a process that takes two parental haploid *input* genotypes (such as the two gametes that form a diploid organism) and creates haploid *output* genotypes (such as gametes) that have new combinations of the input alleles. Such new combinations are thought to be important as a source of evolutionary flexibility.

2. In a dihybrid for genes on two different chromosomes, the genes assort independently because each chromosome pair attaches to the spindle independently and undergoes anaphase I movement. The resulting independent assortment produces recombinant meiotic products.

3. In a dihybrid of genes on the same chromosome (linked genes), the genes cannot assort independently at meiosis because they are attached by the intervening segment of chromosome. However, since a pair of homologous chromosomes can undergo crossing-over by a breakage and exchange of nonsister chromatids, recombinant allele combinations can be produced by crossovers occurring between two loci.

4. Dihybrids of genes on different chromosomes produce recombinants at frequencies of 50 percent.

5. Recombinant frequencies between 0 and 50 percent indicate linked genes. The precise value depends on the proximity of the two loci. Note that RF values of 50 percent can be explained by postulating loci on different chromosomes or loci far apart on the same chromosome.

6. Crossing-over takes place early in prophase I of meiosis, after a premeiotic round of DNA replication (S phase). Hence at the time of crossing-over each chromosome is represented by two pairs of sister chromatids (two pairs of sister DNA molecules), one pair from each parent (the four-chromatid stage). Any particular crossover can take place between any two nonsister chromatids.

7. Since recombinant frequency between linked loci is roughly proportional to their physical distance apart, the relative positions of gene loci on a chromosome can be mapped relative to one another by measuring the frequencies of recombinants produced by crossing-over between them. One map unit (m.u.) is defined as a recombinant frequency of 1 percent.

8. Over short chromosomal segments, map distances are roughly additive. Therefore if the distance between *A* and

C is 10 m.u, and B is between these genes, the distance from A to B plus the distance from B to C will equal 10 m.u. Deviations from additivity are caused by multiple crossovers (mostly doubles), which lead to an underestimate of map distance because they produce some parental chromatids. The maps produced are linear, reflecting the fact that they are maps of linear DNA molecules.

9. Most models of the molecular mechanism of crossing-over involve an intermediate recombinant DNA molecule that is a heteroduplex, composed of a DNA strand from one parental chromatid and a strand from the other. If a heterozygous locus happens to be spanned by this heteroduplex, then, in a fungal octad analysis, aberrant allele ratios such as 3:1:1:3, 5:3, and 6:2 can result.

CONCEPT MAP

Draw a concept map interrelating as many of the following terms as possible. Note that the terms are listed in no particular order.

recombinant / chromatid / 50 percent RF / meiosis / crossing-over / independent assortment / mitosis / heteroduplex DNA / linkage

SOLVED PROBLEM

1. Consider three yellow, round peas, labeled A, B, and C. Each was grown into a plant and crossed to a plant grown from a green, wrinkled pea. Exactly 100 peas issuing from each cross were sorted into phenotypic classes as follows:

A	51	Yellow, round
	49	Green, round
B	100	Yellow, round
C	24	Yellow, round
	26	yellow, wrinkled
	25	green, round
	25	green, wrinkled

What were the genotypes of A, B, and C? (Use gene symbols of your own choosing; be sure to define each one.)

Solution

Notice that each of the crosses is

Yellow, round × green, wrinkled

Because A, B, and C were all crossed to the same plant, all the differences between the three progeny populations must be attributable to differences in the underlying genotypes of A, B, and C. How much we can deduce from the data? What about dominance? The key cross for deducing dominance is B. Here, the inheritance pattern is

Yellow, round × green, wrinkled
↓
All yellow, round

In the progeny, the green and wrinkled phenotypes disappear completely, so yellow and round must be dominant phenotypes. Now we know that the green, wrinkled parent used in each cross must be fully recessive; we have a very convenient situation because it means that each cross is a testcross, which is generally the most informative type of cross.

Turning to the progeny of A, we see a 1:1 ratio for yellow to green. This is a demonstration of Mendel's first law (equal segregation) and shows that for the character of color, the cross must have been heterozygote × homozygous recessive. Letting Y = yellow and y = green, we have

$$Y/y \times y/y$$
↓
$\frac{1}{2}\,Y/y$ (yellow)
$\frac{1}{2}\,y/y$ (green)

For the character of shape, because all the progeny are round, the cross must have been homozygous dominant × homozygous recessive. Letting R = round and r = wrinkled, we have

$$R/R \times r/r$$
↓
R/r (round)

Combining the two characters, we have for A

$$Y/y\ ;\ R/R \times y/y\ ;\ r/r$$
↓
$\frac{1}{2}\,Y/y\ ;\ R/r$
$\frac{1}{2}\,y/y\ ;\ R/r$

Now, cross B becomes crystal clear and must have been

$$Y/Y \; ; \; R/R \times y/y \; ; \; r/r$$
$$\downarrow$$
$$Y/y \; ; \; R/r$$

because any heterozygosity in pea B would have given rise to several progeny phenotypes, not just one.

What about C? Here, we see a ratio of 50 yellow : 50 green (1 : 1) and a ratio of 49 round : 51 wrinkled (also 1 : 1). So both genes in pea C must have been heterozygous, and cross C was

$$Y/y \; ; \; R/r \times y/y \; ; \; r/r$$
$$\downarrow$$

$$\tfrac{1}{2}\, Y/y \begin{cases} \tfrac{1}{2}\, R/r \longrightarrow \tfrac{1}{4}\, Y/y \; ; \; R/r \\ \tfrac{1}{2}\, r/r \longrightarrow \tfrac{1}{4}\, Y/y \; ; \; r/r \end{cases}$$

$$\tfrac{1}{2}\, y/y \begin{cases} \tfrac{1}{2}\, R/r \longrightarrow \tfrac{1}{4}\, y/y \; ; \; R/r \\ \tfrac{1}{2}\, r/r \longrightarrow \tfrac{1}{4}\, y/y \; ; \; r/r \end{cases}$$

which is a good demonstration of Mendel's second law (independent assortment of different genes).

How would a geneticist have analyzed these crosses? Basically, the same way we just did but with fewer intervening steps. Possibly something like this: "yellow and round dominant; single-gene segregation in A; B homozygous dominant; independent two-gene segregation in C."

SOLVED PROBLEM

2. A human pedigree shows people affected with the rare nail-patella syndrome (misshapen nails and kneecaps) and also gives the ABO blood group genotype of each individual. (Note: ABO blood types are determined by three alleles I^A, I^B, and i. I^A determines group A, I^B group B, and i group O. The I^A/I^B heterozygote gives group AB.) Both loci concerned are autosomal. To answer the following questions, study the accompanying pedigree.

a. Is the nail-patella syndrome a dominant or recessive phenotype? Give reasons to support your answer.

b. Is there evidence of linkage between the nail-patella gene and the gene for ABO blood type, as judged from this pedigree? Why or why not?

c. If there is evidence of linkage, draw the alleles on the relevant homologs of the grandparents. If there is no evidence of linkage, draw the alleles on two homologous pairs.

d. According to your model, which descendants represent recombinants?

e. What is the best estimate of RF?

f. If man III-1 marries a normal woman of blood type O, what is the probability that their first child will be blood type B with nail-patella syndrome?

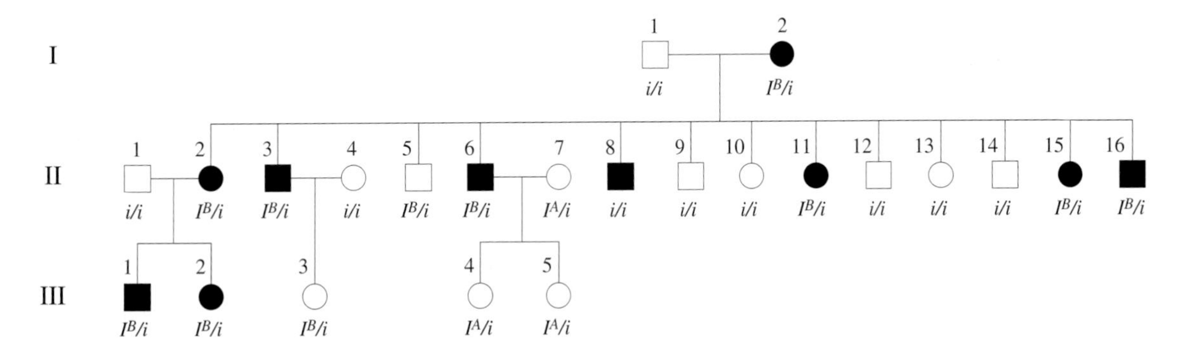

Solution

a. Nail-patella syndrome is most likely dominant. We are told that it is a rare abnormality, so it is unlikely that the unaffected people marrying into the family carry a presumptive recessive allele for nail-patella syndrome. Let N be the causative allele. Then all people with the syndrome are heterozygotes N/n because all (probably including the grandmother, too) result from a mating to an n/n normal individual. Notice that the syndrome appears in all three successive generations—another indication of dominant inheritance.

b. There is evidence of linkage. Notice that most of the affected people—those that carry the N allele—also carry the I^B allele; most likely, these alleles are linked on the same chromosome.

c.

$$\frac{n \quad i}{n \quad i} \times \frac{N \quad I^B}{n \quad i}$$

(The grandmother must carry one recessive allele for each gene in order to produce offspring of genotype i / i and n / n.)

d. Notice that the grandparental mating is equivalent to a testcross, so the recombinants in generation II are

II-5: $n\ I^B$ / $n\ i$ and II-8: $N\ i$ / $n\ i$

whereas all others are nonrecombinants, either $N\ I^B$ / $n\ i$ or $n\ i$ / $n\ i$.

e. Notice that the grandparental cross and the first two crosses in generation II are (disregarding gender) identical and, of course, are all testcrosses. Three of the total sixteen progeny are recombinant (II-5, II-8, and III-3). This gives a recombinant frequency of RF = 18.8 percent. (We cannot include the cross of II-6 × II-7, because the progeny cannot be designated as recombinant or not.)

f.

(III-1 ♂) $\frac{N \quad I^B}{n \quad i} \times \frac{n \quad i}{n \quad i}$ (normal type O ♀)

↓

Gametes

81.2%	$N\ I^B$	40.6%	← Nail-patella, blood type B
	$n\ i$	40.6%	
18.8%	$N\ i$	9.4%	
	$n\ I^B$	9.4%	

The two parental classes are always equal, and so are the two recombinant classes. Since RF = 18.8, the parental classes must comprise 81.2 percent of the offspring. Hence, the probability that the first child will have nail-patella syndrome and blood type B is 40.6 percent.

PROBLEMS

Basic Problems

1. If the genes assort independently, what will be the progeny genotypes and their proportions in a cross A / a ; B / b × A / a ; b / b?

2. In mice, *black* coat is dominant to *brown,* and *intense* (high) pigment concentration is dominant to *dilute* (low) pigment concentration. One mouse with dilute, black pigment is crossed to another with intense, brown pigment. The progeny are

$\frac{1}{4}$ dilute, black

$\frac{1}{4}$ dilute, brown

$\frac{1}{4}$ intense, black

$\frac{1}{4}$ intense, brown

a. What are the genotypes of parents and progeny?

b. Draw the alleles involved on their respective chromosomes.

3. Look at the grid in Figure 6-7.

a. How many genotypes are there in the 16 squares of the grid?

b. What is the genotypic ratio underlying the 9:3:3:1 phenotypic ratio?

c. Can you devise a simple formula for the calculation of the number of progeny genotypes in dihybrid, trihybrid, etc., crosses? Repeat for phenotypes.

d. Mendel predicted that within all but one of the F_2 phenotypic classes there should be several different genotypes. In particular, he performed many crosses to identify the underlying genotypes of the round, yellow phenotype. Show two different ways that could be used to identify the various genotypes underlying the round, yellow phenotype. (Remember, all the round, yellow peas look identical.)

4. The following testcross of a plant is made:

$$\frac{A \quad B}{a \quad b} \times \frac{a \quad b}{a \quad b}$$

If the two loci are 10 m.u. apart, what proportion of the progeny will be $A\ B$ / $a\ b$?

5. The A locus and the D locus are so tightly linked that no recombination is ever observed between them. If $A\ d$ / $A\ d$ is crossed to $a\ D$ / $a\ D$, and the F_1 is intercrossed, what phenotypes will be seen in the F_2 and in what proportions?

6. A strain of *Neurospora* with the genotype *H . I* is crossed with a strain with the genotype *h . i*. Half the progeny are *H . I* and half are *h . i*. Explain how this is possible.

7. A female animal with genotype *A / a . B / b* is crossed with a double-recessive male *a / a . b / b*. Their progeny include

   442 *A / a . B / b*

   458 *a / a . b / b*

   46 *A / a . b / b*

   54 *a / a . B / b*

Explain these proportions and draw the chromosomes of the dihybrid parent, showing the positions of the genes and alleles.

8. If *A / A . b / b* is crossed to *a / a . B / B*, and the F_1 is testcrossed, what percent of the testcross progeny will be *a / a . b / b* if the two genes are (a) unlinked, (b) completely linked (no crossing-over at all), (c) 12 map units apart, (d) 24 map units apart?

9. A fruit fly of the genotype *B R / b r* is testcrossed to *b r / b r*. In 84 percent of the meioses, there are no chiasmata between the linked genes; in 16 percent of the meioses, there is one chiasma between the genes. Is the proportion of the progeny that will be *B r / b r* (a) 50 percent, (b) 4 percent, (c) 84 percent, (d) 25 percent, (e) 16 percent?

10. In dogs, dark coat color is dominant over albino and short hair is dominant over long hair. Assuming these effects are caused by two independently assorting genes, write the genotypes of the parents in each of the crosses shown below, where D and A stand for the dark and albino phenotypes, respectively, and S and L stand for the short-hair and long-hair phenotypes.

Parental phenotypes	NUMBER OF PROGENY			
	D, S	D, L	A,S	A, L
a. D, S × D, S	89	31	29	11
b. D, S × D, L	18	19	0	0
c. D, S × A, S	20	0	21	0
d. A, S × A, S	0	0	28	9
e. D, L × D, L	0	32	0	10
f. D, S × D, S	46	16	0	0
g. D, S × D, L	30	31	9	11

Use the symbols *C* and *c* for the dark and albino coat-color alleles and the symbols *S* and *s* for the short-hair and long-hair alleles, respectively. Assume homozygosity unless there is evidence otherwise.

(Problem 10 reprinted by permission of Macmillan Publishing Co., Inc., from *Genetics* by M. Strickberger. © 1968 by Monroe W. Strickberger.)

11. In tomatoes, two alleles of one gene determine the phenotypes purple (P) versus green (G) stems and two alleles of a separate, independent gene determine the phenotypes "cut" (C) versus "potato" (Po) leaves. The results for five matings of tomato plant phenotypes are shown below:

Mating	Parental phenotypes	NUMBER OF PROGENY			
		P, C	P, Po	G, C	G, Po
1	P, C × G, C	321	101	310	107
2	P, C × P, Po	219	207	64	71
3	P, C × G, C	722	231	0	0
4	P, C × G, Po	404	0	387	0
5	P, Po × G, C	70	91	86	77

a. Determine which phenotypes are dominant.

b. What are the most probable genotypes for the parents in each cross?

(Problem 11 modified from A. M. Srb, R. D. Owen, and R. S. Edgar, *General Genetics*, 2d ed. © 1965 by W. H. Freeman and Company.)

12. From four crosses of the general type *A / A . B / B* × *a / a . b / b* the F_1 individuals of type *A / a . B / b* were testcrossed to *a / a . b / b*. The results are shown below:

Testcross of F_1 from cross	TESTCROSS PROGENY			
	A/a . B/b	*a/a . b/b*	*A/a . b/b*	*a/a . B/b*
1	310	315	287	288
2	36	38	23	23
3	360	380	230	230
4	74	72	50	44

For each set of progeny, use the χ^2 test to decide if there is evidence of linkage.

13. In the following pedigree, the vertical lines stand for protan color blindness and the horizontal lines stand for deutan color blindness. These are separate conditions causing different misperceptions of colors; each is determined by a separate gene.

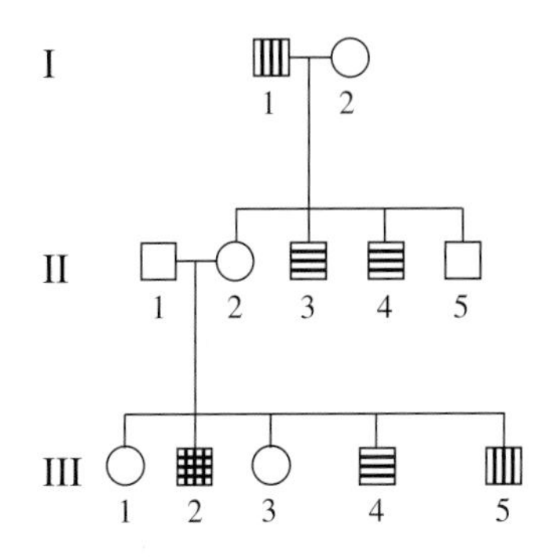

a. Does the pedigree show any evidence that the genes are linked?

b. If there is linkage, does the pedigree show any evidence of crossing-over?

Explain both your answers with the aid of the diagram.

c. Can you calculate a value for the recombination between these genes? Does this recombination arise from independent assortment or crossing-over?

14. A three-point testcross was made in corn. The results and a partial recombination analysis are shown in the following display, which is typical of three-point testcrosses (p = purple leaves, $+$ = green; v = virus-resistant seedlings, $+$ = sensitive; b = brown midriff to seed, $+$ = plain). Study the display and answer the questions below.

P	$+/+ \,.\, +/+ \,.\, +/+ \times p/p \,.\, v/v \,.\, b/b$	
Gametes	$+ \,.\, + \,.\, +$	$p \,.\, v \,.\, b$
F_1	$+/p \,.\, +/v \,.\, +/b$	
Testcross	$+/p \,.\, +/v \,.\, +/b$ (F_1)	$\times\; p/p \,.\, v/v \,.\, b/b$ (Tester)

Class	Progeny phenotypes	F_1 gametes	Numbers	RECOMBINANT FOR p-b	p-v	v-b
1	gre sen pla	$+ . + . +$	3,210			
2	pur res bro	$p . v . b$	3,222			
3	gre res pla	$+ . v . +$	1,024		R	R
4	pur sen bro	$p . + . b$	1,044		R	R
5	pur res pla	$p . v . +$	690	R		R
6	gre sen bro	$+ . + . b$	678	R		R
7	gre res bro	$+ . v . b$	72	R	R	
8	pur sen pla	$p . + . +$	60	R	R	
			Total 10,000	1,500	2,200	3,436

a. Determine which genes are linked.

b. Draw a map that shows distances in map units.

c. Calculate interference if appropriate.

Unpacking the Problem

a. Sketch cartoon drawings of the parent (P), F_1, and tester corn plants, and use arrows to show exactly how you would perform this experiment. Show where all meioses occur and where seeds are collected.

b. Why do all the +'s look the same, even for different genes? Why does this not cause confusion?

c. How can a phenotype be purple and brown (for example) at the same time?

d. Is it significant that the genes are written in the order p–v–b in the problem?

e. What is a tester and why is it used in this analysis?

f. What does the column marked "progeny phenotypes" represent? In class 1, for example, state exactly what "gre sen pla" means.

g. What does the line marked "gametes" represent, and how is this different from the column marked "F_1 gametes"? In what way is comparison of these two types of gametes relevant to recombination?

h. Which meiosis is the main focus of study? Label it on your drawing.

i. Why are the gametes from the tester not shown?

j. Why are there only eight phenotypic classes? Are there any classes missing?

k. What classes (and in what proportions) would be expected if all the genes are on separate chromosomes?

l. To what do the four class sizes (two very big, two intermediate of equal size, two intermediate of another equal size, two very small) correspond?

m. What can you tell about gene order from inspecting the phenotypic classes and their frequencies?

n. What would the expected phenotypic class distribution be if only two genes are linked?

o. What does the word *point* refer to in a "three-point testcross"? Does this word usage imply linkage? What would a four-point testcross be like?

p. What is the definition of *recombinant,* and how is it applied here?

q. What do the "recombinant for" columns mean?

r. Why are there only three "recombinant for" columns?

s. What do the R's mean? How are they determined?

t. What do the column totals signify? How are they used?

u. What is the diagnostic test for linkage?

v. What is a map unit? Is it the same as a centimorgan?

w. In a three-point testcross such as this one, why aren't the F_1 and the tester considered to be parental in calculating recombination? (They *are* parents in one sense.)

x. What is the formula for interference? How are the "expected" frequencies calculated in the coefficient of coincidence formula?

y. Why does part c of the problem say "if appropriate"?

z. How much work is it to obtain such a large progeny size in corn? Approximately how many progeny are represented by one corncob?

15. R. A. Emerson crossed two different pure-breeding lines of corn and obtained a phenotypically wild-type F_1 that was heterozygous for alleles of three different genes that determine recessive phenotypes: *an* determines the mutant phenotype anther; *br*, brachytic; and *f*, fine. He testcrossed the F_1 to a tester that was homozygous recessive for the three genes and obtained these progeny phenotypes:

Anther	355
Brachytic, fine	339
Completely wild type	88
Anther, brachytic, fine	55
Fine	21
Anther, brachytic	17
Brachytic	2
Anther, fine	2

a. What were the genotypes of the parental lines?

b. Draw a linkage map for the three genes (include map distances).

c. Calculate the interference value.

16. The year is 1868. You are a skilled young lens maker working in Vienna. With your superior new lenses, you have just built a microscope that has better resolution than any others available. During your testing of this microscope, you have been observing the cells in the testes of grasshoppers and have been fascinated by the behavior of strange elongated structures you have seen within the dividing cells. One day in the library you read a recent journal paper by G. Mendel on hypothetical "factors" that he claims explain the results of certain crosses in peas. In a flash of revelation, you are struck by the parallels between your grasshopper studies and Mendel's ratios (1 : 1 : 1 : 1, 9 : 3 : 3 : 1, etc.), and you resolve to write him a letter. What do you write?

(Problem 16 based on an idea by Ernest Kroeker.)

Pattern Recognition Problems

In Problems 17 through 21 diagrams show phenotypes and the results of breeding analyses. Deduce the genotypes of the individuals shown in each diagram as far as possible.

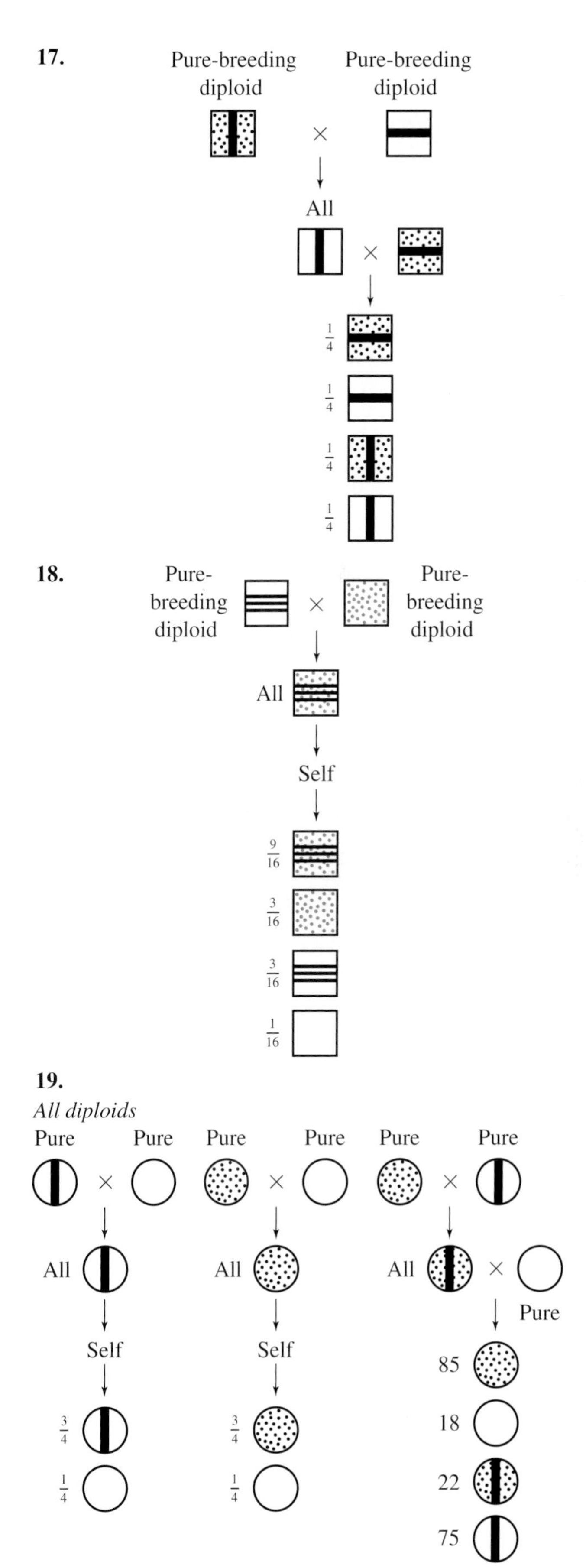

20.

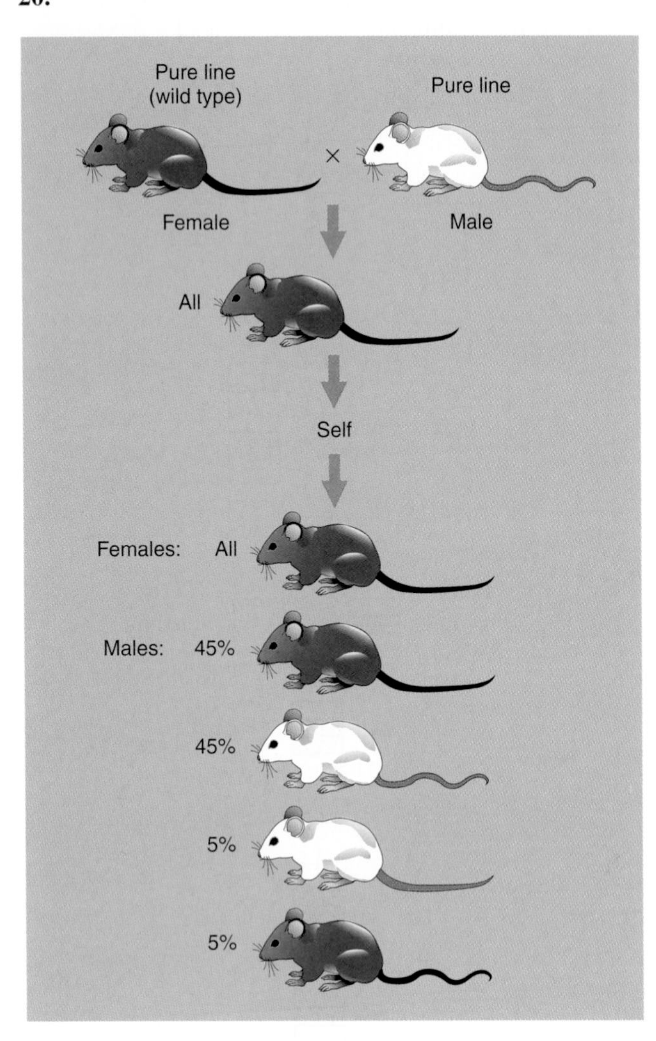

21.

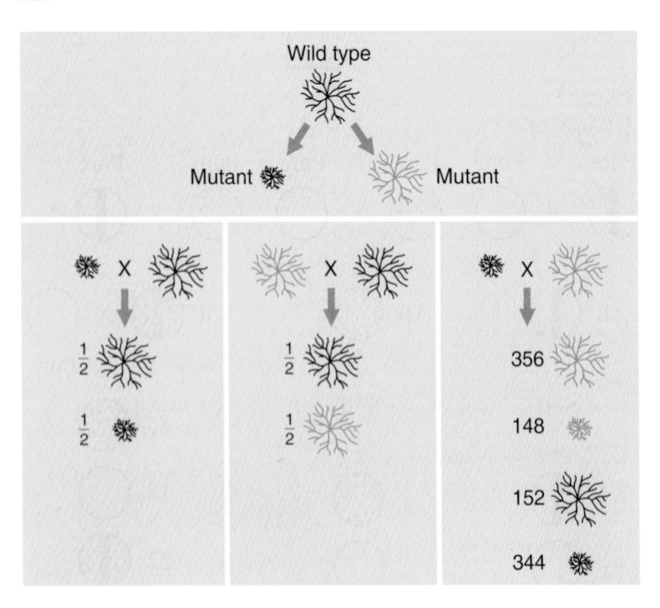

Challenging Problems

22. We have dealt mainly with independent assortment of only two genes, but the same principles hold for more than two genes. Consider the cross

$$A/a \ ; B/b \ ; C/c \ ; D/d \ ; E/e \times a/a \ ; B/b \ ; c/c \ ; D/d \ ; e/e$$

a. What proportion of progeny will *phenotypically* resemble (1) the first parent, (2) the second parent, (3) either parent, and (4) neither parent?

b. What proportion of progeny will be *genotypically* the same as (1) the first parent, (2) the second parent, (3) either parent, and (4) neither parent?

23. From the five sets of data given in the following table, determine the order of genes by inspection—that is, without calculating recombination values. Recessive phenotypes are symbolized by lowercase letters, and dominant phenotypes by pluses.

Phenotypes observed in 3-point testcross	DATA SET				
	1	2	3	4	5
+ + +	317	1	30	40	305
+ + c	58	4	6	232	0
+ b +	10	31	339	84	28
+ b c	2	77	137	201	107
a + +	0	77	142	194	124
a + c	21	31	291	77	30
a b +	72	4	3	235	1
a b c	203	1	34	46	265

24. In the plant *Arabidopsis thaliana* the loci for pod length (L = long, l = short) and fruit hairs (H = hairy, h = smooth) are linked 16 map units apart on the same chromosome. The following crosses were made:

$$L\,H/L\,H \times l\,h/l\,h \longrightarrow F_1$$

$$L\,h/L\,h \times l\,H/l\,H \longrightarrow F_1$$

If the F_1s from above are crossed,

a. what proportion of the progeny are expected to be $l\,h/l\,h$?

b. what proportion of the progeny are expected to be $L\,h/l\,h$?

25. Assume in an ascomycete fungus that heteroduplex DNA occurs at a heterozygous locus A/a (consult Figure 6-18). Assume that an A strand invades an a helix at the same frequency as the invasion of A by a. Assume no branch migration occurs. Also assume heteroduplexes are repaired to A 80 percent of the time and to a 20 percent of the time. Out of all the cases of heteroduplex DNA at that site, what proportion of heteroduplexes will become converted into aberrant ratios of the types 6:2, 2:6, 5:3, and 3:5?

26. The plant *Haplopappus gracilis* is diploid, and $2n = 4$. There are one long pair and one short pair of chromosomes. The diagrams at the bottom of the page represent anaphases ("pulling-apart" stages) of individual cells during meiosis or mitosis in a plant that is genetically a dihybrid (*A / a* ; *B / b*) for genes on different chromosomes. The lines represent chromosomes or chromatids, and the points of the "V's" represent centromeres. In each case, determine whether the diagram represents a cell in meiosis I, meiosis II, or mitosis. If a diagram shows an impossible situation, indicate this.

27. The *Neurospora hist-1* gene, which is on the right arm of chromosome 1, is involved in the biochemical pathway for the synthesis of histidine. Two mutations of the *hist-1* gene were obtained in separate experiments. Both left the strains auxotrophic—that is, they would not grow unless histidine was added to the growth medium. These mutant strains were known never to back-mutate to wild type. The two mutants were crossed together (*hist-1*^{i} × *hist-1*^{ii}), and thousands of ascospores were spread onto medium lacking histidine (minimal medium). Most ascospores did not grow, but three colonies were observed. Draw a diagram to account for the origin of these rare prototrophic colonies.

28. In *Drosophila* the genes for wing size (phenotypes large or vestigial) and bristle length (phenotypes long or short) are inherited independently. The following crosses are made:

	PROGENY			
Parental crosses	Large, long	Large, short	Vestigial, long	Vestigial, short
large, long × vestigial, short	246	0	250	0
large, long × vestigial, long	302	98	297	101
large, long × large, short	150	148	48	51
large, short × vestigial, long	89	93	90	94

a. Deduce the dominance relations of the phenotypes involved.

b. Deduce the genotypes of the parents in the four crosses.

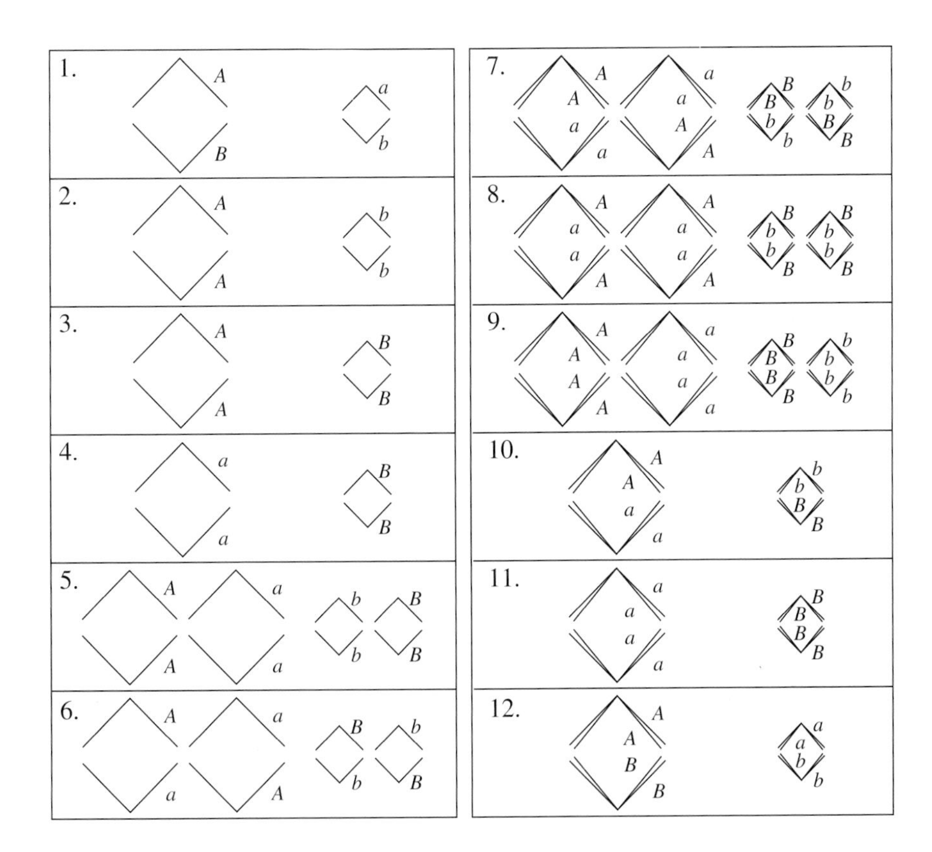

29. A plant of genotype $Q/q \cdot R/r \cdot T/t$ produces the following gamete genotypes in the proportions shown:

$\frac{1}{8}\ Q \cdot R \cdot T$

$\frac{1}{8}\ Q \cdot R \cdot t$

$\frac{1}{8}\ q \cdot R \cdot T$

$\frac{1}{8}\ Q \cdot r \cdot T$

$\frac{1}{8}\ q \cdot r \cdot t$

$\frac{1}{8}\ Q \cdot r \cdot t$

$\frac{1}{8}\ q \cdot R \cdot t$

$\frac{1}{8}\ q \cdot r \cdot T$

Draw labeled meiosis diagrams to explain clearly how these gametic proportions were produced. Show the chromosomes

a. before premeiotic S phase (replication phase).

b. after chromatid formation.

c. during pairing.

d. at anaphase I.

e. at anaphase II.

30. In beans, tall *(T)* is dominant to short *(t)*, red flowers *(R)* is dominant to white *(r)*, and wide leaves *(W)* is dominant to narrow *(w)*. The following cross is made and progeny obtained as shown:

Cross	tall, red, wide × short, white, narrow	
Progeny	tall, white, wide	478
	tall, red, wide	21
	short, white, wide	19
	short, red, wide	482

a. Explain why these progeny phenotypes were obtained and in the proportions observed (list all genotypes and show chromosomal positions).

b. Under your hypothesis, if the tall, red, wide parent is selfed, what will be the proportion of short, white, wide progeny?

Exploring Genomes: A Web-Based Bioinformatics Tutorial

Learning to Use PubMed

PubMed provides a searchable database of the world's scientific literature. In the Genomics tutorial at www.whfreeman.com/mga, you will learn to do a literature search to find the first report of a gene sequence and subsequent papers demonstrating the function of the gene.

RECOMBINATION IN BACTERIA AND THEIR VIRUSES 7

Key Concepts

1. Some *E. coli* cells carry a circular plasmid called the *sex factor, F.*
2. The F factor can exist in a free state in the cytoplasm, or it can be integrated into the circular bacterial chromosome.
3. In the nonintegrated state, F can pass into F^- free cells during cell conjugation.
4. When F is integrated, the bacterial chromosome can be transferred linearly to an F-free cell during conjugation.
5. Bacteriophages can carry (transduce) bacterial DNA from one cell to another.
6. During generalized transduction, random fragments of the bacterial genome are incorporated into the heads of phages and transferred to other bacteria by infection.
7. During specialized transduction, specific genes near the sites on the bacterial chromosome where certain phages can integrate are sometimes incorporated into the phage genome when it exits the chromosome. The phages carry these specific genes to other cells.
8. DNA from the medium can enter a bacterial cell and integrate into the chromosome, thereby transforming the genotype.
9. The different methods of gene transfer in bacteria generate partial diploids that permit the study of recombination, dominance, and gene interaction.

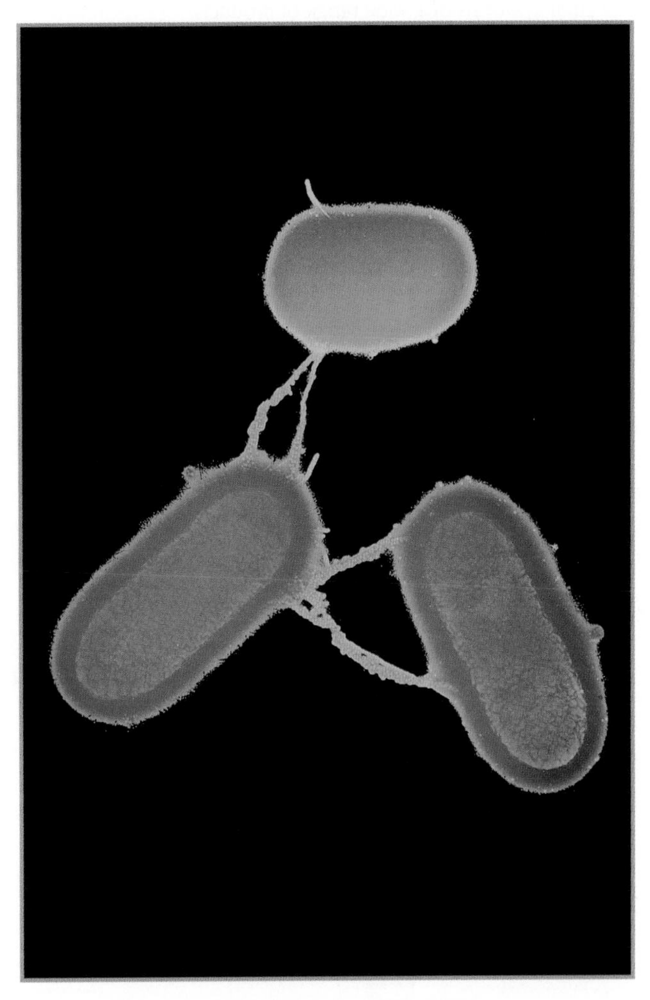

Cells of *Escherichia coli* that have become attached by "pili" prior to DNA transfer between donor and recipient cell types.
(Dr. L. Caro/Science Photo Library/Photo Researchers.)

Although bacteria are generally invisible to the human eye, they play key roles in biology. Estimates of their total biomass on the planet vary, but there is general agreement that bacteria represent one of the largest biomasses of any category of organism. Given the small size of their cells, they are the most numerous organisms on our planet. They contribute to the recycling of nutrients such as nitrogen, sulfur, and carbon in ecosystems. Furthermore, they interact with humans in many ways; for example, some bacteria live symbiotically inside our mouths and intestines. Others are agents of human, animal, and plant disease. In addition, many types of bacteria are used by humans for industrial synthesis of a wide range of products.

A familiar example of the industrial use of bacteria is in the manufacture of cheese and other dairy products. Let us begin our discussion of bacterial genetics with a genetical case history concerning *Lactococcus lactis*, a bacterium used in making dairy products. Antibiotics usually kill the wild-type form of this bacterium, but in 1993 a strain of *L. lactis* was isolated from yogurt that was resistant to three antibiotics: streptomycin, tetracycline, and chloramphenicol. Investigation of these resistances showed that they were determined not by genes in the *L. lactis* genome, but by genes on a plasmid called *pK214*. (Recall from Chapter 2 that plasmids are extragenomic DNA molecules, mostly circular.) Base sequencing in pK214 revealed that the plasmid is composed of DNA sequences from several other species of bacteria, as shown in Figure 7-1. The streptomycin and chloramphenicol resistances were caused by genes 19 and 22, from *Staphylococcus aureus*, a bacterium that can infect and cause various disease symptoms in humans, including pneumonia. The tetracycline resistance (gene 28) was from *Listeria monocytogenes*, a bacterium that causes meningitis-like symptoms in humans.

The existence of antibiotic-resistant bacteria is the result of a two-step process. First, spontaneous mutations occur to confer resistance to an antibiotic. Second, if the environment contains significant concentrations of that antibiotic, the cells carrying the resistance mutations are better able to survive than are cells lacking these mutations, and they eventually dominate the population (a process called *selection*). Human activities provide many opportunities for selection of antibiotic resistance. We expose ourselves to antibiotics medically. Also, many animals that we eat are fed antibiotics, for example, farmed fish, chickens, and (most relevant to the present example involving dairy products) cattle. Bacteria are omnipresent in our human environment and in our food chain. Since our own bodies and our food are exposed to antibiotics, it is easy to imagine how resistant bacteria arise and move up and down the food chain.

What is the significance of the finding that resistance genes in the strain of *L. lactis* are all carried on a plasmid? The basis for this situation can be deduced from the analysis of bacterial recombination, the topic of this chapter.

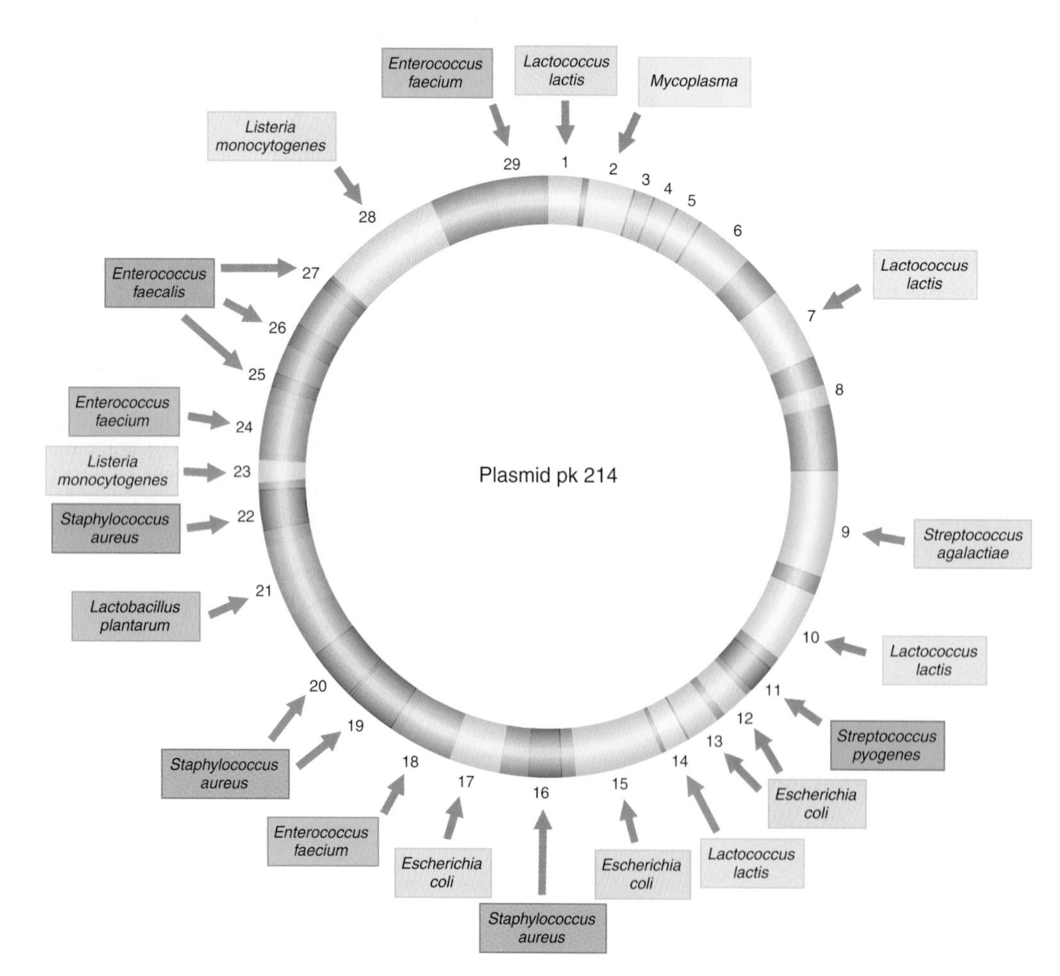

Figure 7-1 Origins of genes of the *Lactococcus lactis* plasmid pK214. *(Data from Table 1 in V. Perreten, F. Schwarz, L. Cresta, M. Boeglin, G. Dasen, and M. Teuber,* Nature *389, 1997, 801–802.)*

CHAPTER OVERVIEW

The evolutionary divergence of eukaryotes and prokaryotes was quite ancient. Because of their common ancestry, the two groups share many common features in their cell biology; for example, both are based on a DNA-RNA-protein information system. However, there are some profound differences between them, a good example being the way they accomplish genetic recombination, the topic of this chapter. Let's consider plasmid pK214, described above. This plasmid has obtained a large proportion of its genes from other bacteria, so it can be thought of as a live record of past recombinations between the DNA of these bacteria. What mechanisms can account for this exchange of DNA? We shall see that bacteria regularly exchange genes with other bacteria, by a variety of routes summarized in Figure 7-2. Bacterial cells sometimes fuse, a process known as **conjugation,** and then separate. Conjugation provides a way for plasmids to be transmitted from one cell to another (cell at left in Figure 7-2). Occasionally these transferred plasmids have picked up part of the bacterial genome and carry this along with them. Also during conjugation, segments of the bacterial genome can be transmitted directly, not as part of a plasmid (cell at right in Figure 7-2). Bacterial viruses **(bacteriophages)** can pick up a fragment of a bacterial genome and deliver it to another cell, a process called **transduction** (lower cell in Figure 7-2). Finally, DNA released from a cell into the extracellular medium can be taken up by another cell, in a process called **transformation** (upper cell in Figure 7-2). (Recall from Chapter 2 that the observation of transformation was one of the key experiments that showed DNA is the genetic material; see Foundations of Genetics 2-1.) All four of these processes provide a way to get a fragment of the genome of one bacterial cell into another. Once transferred, the "new" fragment may recombine with the genome of its new host and thus become part of the main chromosome or the plasmid.

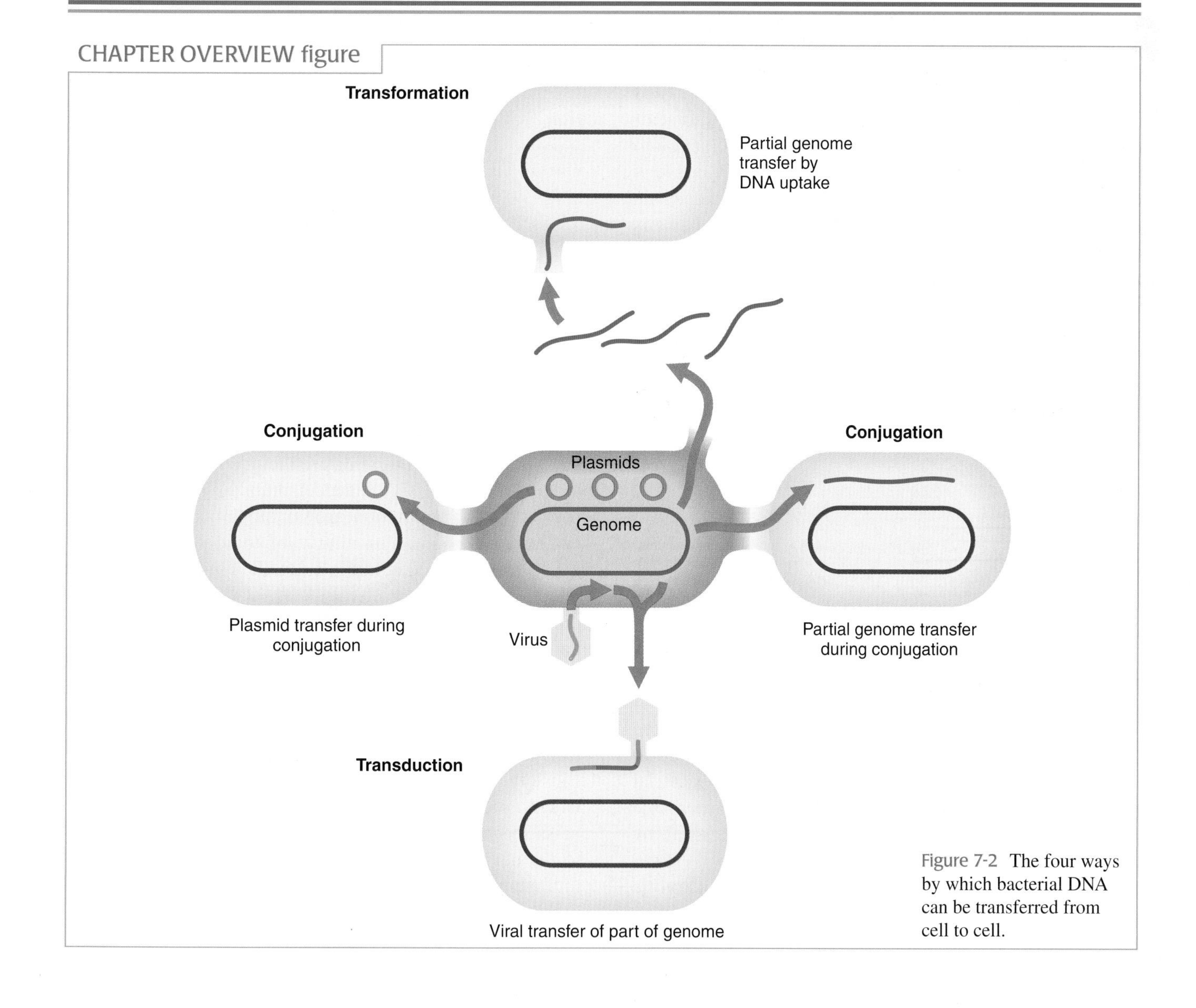

Figure 7-2 The four ways by which bacterial DNA can be transferred from cell to cell.

DETECTING RECOMBINATION IN BACTERIA

A large part of the science of genetics is concerned with bacteria. Genetic analysis of bacteria has been a source of key insights into the nature and structure of the genetic material, the genetic code, mutation, and DNA technology. Bacteria and phages have relatively small and simple genomes that make them excellent model organisms for genetics. Because bacteria are prokaryotes, they do not undergo meiosis. However, there are several ways in which bacterial genomes can unite and recombine. Therefore the approach to the genetic analysis of recombination in these organisms is remarkably similar to that for eukaryotes—namely, place two different genomes in the same cell to afford an opportunity for recombination, and then look for recombinants in the descendant cells.

The union of bacterial DNA for recombination is quite different from the comparable process in eukaryotes. In eukaryotes two parental cells contribute *equally* to a cell in which recombination will occur. However, in the bacterial processes that lead to recombination (conjugation, transduction, and transformation) the gene transfer is *partial* and *unidirectional.* In other words, only a *part* of the genome of one organism is transferred and becomes incorporated into the complete genome of another. The bacterial cell that contributes the partial genomic fragment is called the **donor,** and the one that contributes the complete genome is the **recipient.** The donor fragment is called the **exogenote,** and the recipient genome is the **endogenote.** A cell containing both an endogenote and an exogenote is a partial diploid, or **merozygote.** The merozygote provides an opportunity for genetic recombination to occur between any genes that are heterozygous in the partially diploid segment, such as the genes in the following example.

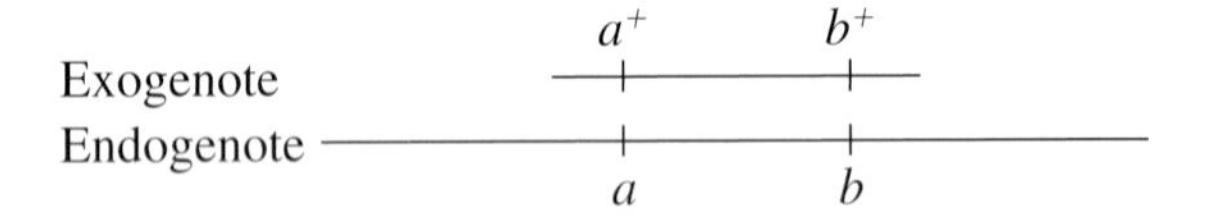

The merozygote also provides opportunities for other types of genetic tests not possible in haploids, such as tests for dominance.

Working with Microorganisms

How can organisms as small as bacteria be manipulated for genetic analysis? In particular, how is it possible to study bacterial phenotypes? Although individual bacterial cells are invisible to the unaided eye, bacteria are visible as colonies, opaque masses of cells, each derived by asexual cell division from a single progenitor cell (lower panel in Figure 7-3). This division is the prokaryotic equivalent of mitotic cell division in eukaryotes, so it results in daughter cells of the same genotype (Chapter 4). Since the cells in a colony are genetically identical, it is legitimate to use the phenotype of the colony as a manifestation of the genotypes of the individual bacterial cells.

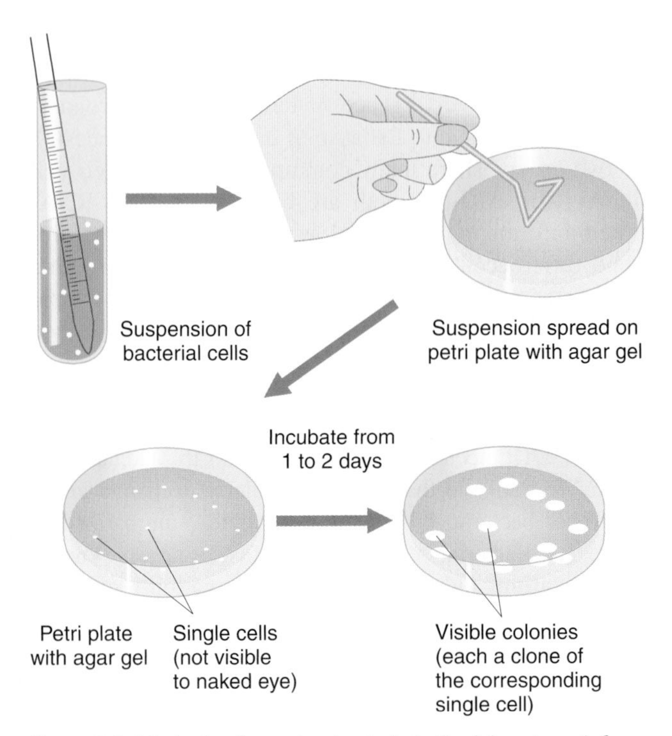

Figure 7-3 Methods of growing bacteria in the laboratory. A few bacterial cells that have been grown in a liquid medium containing nutrients can be spread on an agar medium also containing the appropriate nutrients. Each of these original cells will divide many times by binary fission and eventually give rise to a colony. All cells in a colony, being derived from a single cell, will have the same genotype and phenotype.

Bacteria can be cultured either in a liquid medium or on the surface of a medium made semisolid with gelled agar. In both cases nutrients are included. The bacteria divide by binary fission, and cell number increases exponentially until population growth stops because of nutrient exhaustion or accumulation of toxic waste products. In solid-medium culture, a small amount of liquid medium containing actively dividing bacterial cells is pipetted into a petri plate containing a solid agar medium, and the cells are spread evenly on the agar surface with a sterile spreader (the solution is sufficiently dilute to prevent clumping of cells). This process is called **plating** (see Figure 7-3). Each cell then divides exponentially until a large number of descendant cells are produced. Because the cells are immobilized on the gel surface, all the daughter cells remain together in a clump. When this mass reaches approximately 10^7 cells, it becomes visible to the naked eye as a colony. A colony will grow in size until nutrients are exhausted, but colony phenotype can be observed and categorized at any convenient time.

Phenotypes

We have seen above that examining a bacterial colony provides a convenient way of studying the properties of bacteria. What are some specific colony phenotypes that can be studied? One category concerns the ability to synthesize essential biochemical compounds. Wild-type bacteria are **prototrophic** ("self-feeding"): they can produce colonies on **minimal medium**—a solution containing only inorganic salts, a source of energy such as a sugar, and water. From these simple substances, the bacteria synthesize all the molecules of which they are composed. Some mutant clones can be identified because they are **auxotrophic** ("outside-feeding"): they will not grow unless one or more specific nutrients—say, adenine, threonine, or biotin—are added to the minimal medium. Such a phenotype is indicative of the loss of the cell's ability to make that specific compound using the components of the minimal medium.

The two alternative phenotypes of prototrophy and auxotrophy for specific nutrients have proved to be very useful markers for genetic analysis. As with discrete phenotypic variants in eukaryotes, these alternative bacterial phenotypes are often determined by a pair of alleles. For example, ad^+ stands for an allele coding for ability to synthesize adenine, making the strain prototrophic, and ad^- stands for inability to synthesize adenine, making the strain auxotrophic.

A second character used by bacterial geneticists also concerns nutrition but in a fundamentally different way. Bacteria need to be opportunists in using energy sources in the natural environment. For example, wild-type bacteria can utilize many different sugars as an energy source, including the sugar galactose. However, mutations arise that inactivate the ability to use galactose, so other energy sources must be found. Here the determining alleles are gal^+(galactose-utilizing) and gal^- (galactose-nonutilizing).

In comparing these two different types of phenotypes, note the different meaning of the allele symbols with negative superscripts; ad^- means cannot *make* adenine, whereas gal^- means cannot *utilize* galactose. (There is no way to distinguish these meanings just by looking at the symbols themselves.)

Another useful phenotype concerns resistance. Whereas wild types are susceptible to various inhibitors, such as the antibiotic streptomycin, resistant mutants differ in that they can form colonies in the presence of the inhibitor. Hence, in this case, the two alternative phenotypes that can be studied genetically are resistance and sensitivity to the inhibitor, determined in this example by the alleles str^r and str^s, respectively.

Note that in all the foregoing phenotypes, the alternative phenotypes analyzed by bacterial geneticists are growth and absence of growth under particular conditions. This is quite different from most of the alternative phenotypes analyzed, say, in a fruit fly or a pea plant, which tend to be expressed in most laboratory conditions.

Individual bacterial strains are generally maintained as populations of cells growing in nutrient-rich medium in test tubes. The phenotype of a culture can be tested by inoculating a few cells from the culture onto the surface of an appropriate medium using a sterile needle, and recording whether or not a colony appears. Table 7-1 lists some examples of bacterial phenotypes and their genetic symbols.

Selective Systems

We have seen that genetic analysis in eukaryotes and prokaryotes is based on genetic variants, called *mutations*. The rarity of mutations is a problem if an investigator is trying to find one or to amass a collection of a specific type of mutation for genetic study. One solution is to use a selective system, an experimental protocol designed to allow the desired mutant types to survive and propagate, but not the wild types. In this way rare mutants can be found in a population consisting mainly of wild types. Bacterial phenotypes are well suited to finding mutants by selection.

Selection of mutations to antibiotic resistance provides a good example (Figure 7-4a, on the next page). Wild-type cells sensitive to some antibiotic such as streptomycin are plated in large numbers on medium containing streptomycin. The antibiotic will kill most wild-type cells, but if there are rare mutant cells resistant to streptomycin, they will each divide and form a colony.

Sometimes it is necessary to obtain mutations that cause a reverse change from mutant to wild type; such mutations are called *revertants*. The change from auxotrophy back to prototrophy provides a simple selective system for obtaining revertants. As an example, let's look at an adenine-requiring auxotroph (Figure 7-4b). To obtain cells

TABLE 7-1 Some Genotypic Symbols Used in Bacterial Genetics

Symbol	Character or Phenotype Associated with Symbol
bio^-	Requires biotin added as a supplement to minimal medium
arg^-	Requires arginine added as a supplement to minimal medium
met^-	Requires methionine added as a supplement to minimal medium
lac^-	Cannot utilize lactose as a carbon source
gal^-	Cannot utilize galactose as a carbon source
str^r	Resistant to the antibiotic streptomycin
str^s	Sensitive to the antibiotic streptomycin

NOTE: Minimal medium is the basic synthetic medium for bacterial growth without nutrient supplements.

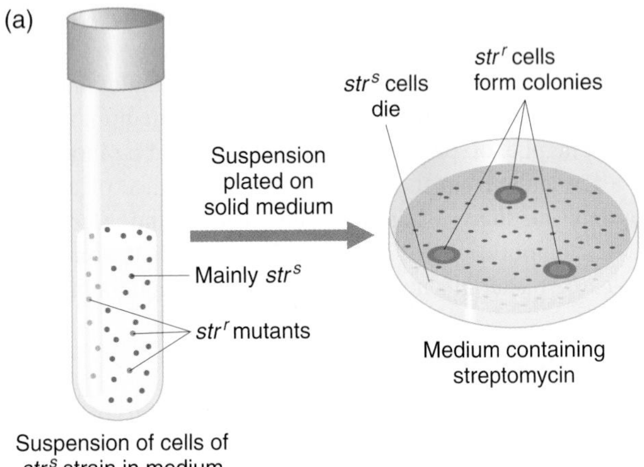

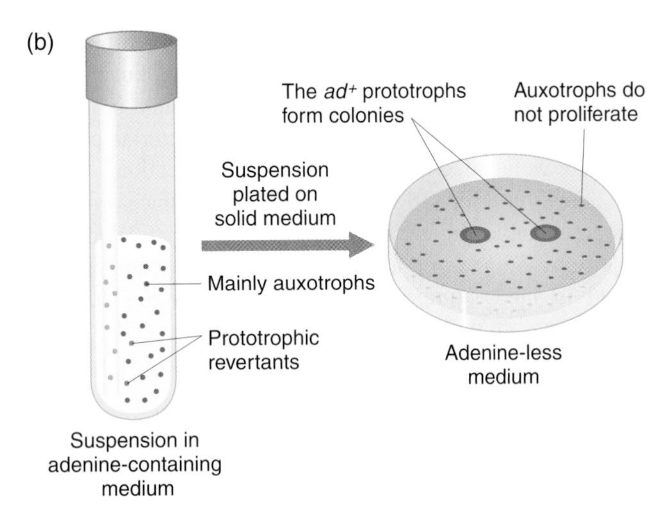

Figure 7-4 Selective systems for finding bacterial mutants. (a) Selection for mutant cells resistant to streptomycin. (b) Selection for prototrophic revertants of auxotrophic adenine-requiring mutants.

to work with, the strain is grown in an adenine-containing medium. Then a large number of these cells are plated on a solid nutritive medium containing no adenine (minimal medium). The only cells that can grow, divide, and hence form colonies on this medium are adenine prototrophs, which must have arisen by reverse mutation in the original culture, symbolically $ad^- \rightarrow ad^+$.

For mutations that change the wild-type state of prototrophy to auxotrophy, a selection technique is also available. Many species of bacteria are highly sensitive to the antibiotic penicillin (isolated from the fungus *Penicillium*). This sensitivity is expressed only in actively dividing cells. If penicillin is added to a suspension of cells in liquid minimal medium, all the prototrophic cells are killed because they start to divide, but the auxotrophic mutants survive because they cannot divide without supplementation. After treatment, the penicillin can be removed by washing the cells on a filter. There will be many different types of auxotrophic mutants in the culture but, if the washed cells are plated on a range of different media, each supplemented with one specific chemical, the specific requirement of any given auxotroph can be deduced. For example, we might get the following results when we plated the washed cells; a "+" means growth and a "−" no growth in response to various additives:

	COMPOUND ADDED TO MEDIUM			
Auxotrophic mutant	None	Adenine	Histidine	Inositol
1	−	−	−	+
2	−	−	−	−
3	−	−	+	−

We can deduce that mutant 1 is an inositol-requiring auxotroph and 3 is histidine-requiring. However, the data do not tell us what type of auxotroph mutant 2 is, and more tests adding a larger range of compounds would be required to determine that.

Armed with these techniques for manipulating bacteria, we now return to the analysis of bacterial recombination, starting with the process of conjugation.

BACTERIAL CONJUGATION

Conjugation starts with the contact of two compatible bacterial cells, followed by the formation of a bridge that unites their cytoplasms. Bringing two cell types together and allowing them to conjugate is the way of making a cross. Our discussion of conjugation will center on the gut bacterium *Escherichia coli (E. coli),* which is the most-studied bacterial species. Conjugation and gene transfer in *E. coli* are driven by a circular DNA plasmid called the **fertility factor** or **sex factor (F).** This genetic element is found in some but not all *E. coli* cells. Hence to understand how to make a cross in *E. coli,* we have to understand the properties of F.

The Remarkable Properties of the F Plasmid

Cells carrying the F plasmid are designated $\mathbf{F^+}$; those lacking it are $\mathbf{F^-}$. The F plasmid contains approximately 100 genes, which give the plasmid many important properties:

1. The F plasmid DNA can be replicated inside the bacterial cell, using the replication machinery of the bacterium. This allows the plasmid to be maintained in a dividing cell population (Figure 7-5a).
2. Cells carrying the F plasmid promote the synthesis of **pili** (singular, *pilus*) on the bacterial cell surface. Pili are minute proteinaceous tubules that allow the F^+ cells to attach to F^- cells to conjugate (Figure 7-5b).
3. F^+ and F^- cells can conjugate. When conjugation occurs, the F^+ cells can act as F donors. The circular

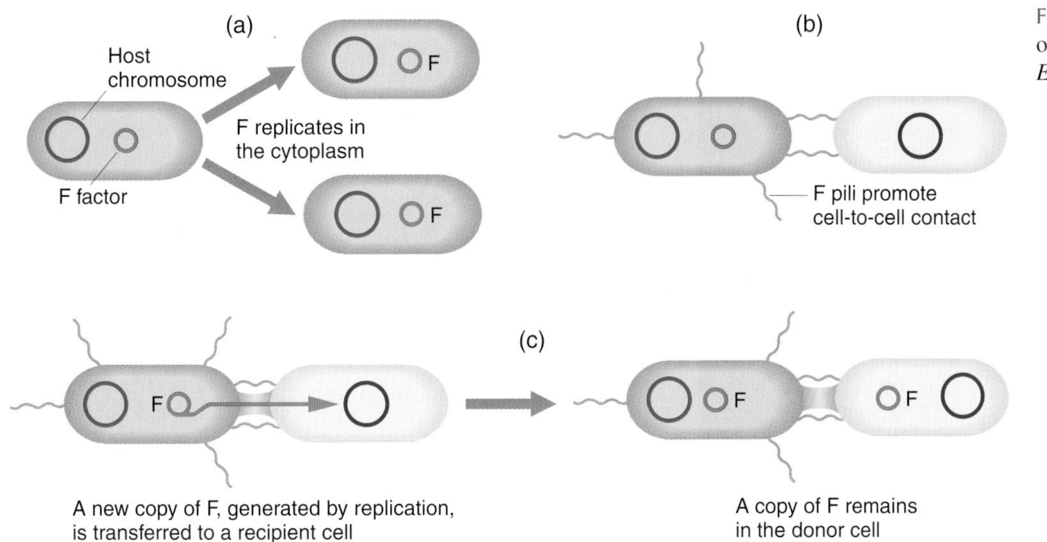

Figure 7-5 Some properties of the fertility (F) factor of *E. coli.*

F plasmid DNA replicates by producing a linear single-stranded copy of itself, and this copy is transferred to the F^- recipient (Figure 7-5c). However, an F plasmid always remains behind in the donor cell. In the recipient cell the single-stranded F is converted into a double-stranded circular form, and this converts the cell into an F^+. Because the transfer of the F plasmid from F^+ to F^- is rapid, the plasmid can spread like wildfire throughout an initially F^- population.

4. F^+ cells are usually inhibited from making contact with other F^+ cells; therefore, the F plasmid is not transferred from F^+ to F^+.
5. Typically F carries within its genome one or more **IS (insertion-sequence)** elements, which are repeated segments of DNA that can move from place to place within the host chromosome or between chromosome and plasmid. The presence of an IS element both in the plasmid *and* at various sites in the bacterial chromosome provides sites at which crossing-over between homologous sequences can occasionally occur. A crossover between these two circular DNAs leads to the integration of the plasmid into the bacterial chromosome (Figures 7-6 and, on the next page, 7-7a). The integrated form of F can promote the transfer of a single-stranded copy of the entire host chromosome into a recipient cell, along with its own integrated F DNA (Figure 7-7b).

This transfer of F and the associated host genome has some interesting features. First, in any population of cells containing the F factor, F will integrate into the chromosome only in a small fraction of cells (Figure 7-7c). These few cells can now transfer all or part of their chromosomes to a second strain. This transfer produces a merozygote, and the donor and recipient alleles can recombine to produce genetic recombinants (see Figure 7-7c). Note in the figure that recombination in a merozygote is by double crossovers, by which a piece of the donor chromosome is inserted into the chromosome of the recipient.

The rare cells in which the F factor is integrated into the host chromosome can be isolated, and pure strains grown from each. In such strains with an integrated F plasmid, *every* cell has the F factor in its chromosome and consequently donates chromosomal DNA during conjugation. Therefore in crosses using these strains the frequency of recombinants is much higher than when using the original population. Strains with an integrated F factor are termed **high frequency of recombination (Hfr)** strains to distinguish them from normal F^+ strains, which contain only a few rare Hfr cells and thus display only a low frequency of recombination for the strain as a whole. Because they transfer chromosomal markers efficiently, Hfr strains are the ones used for genetic mapping, as we shall see later on.

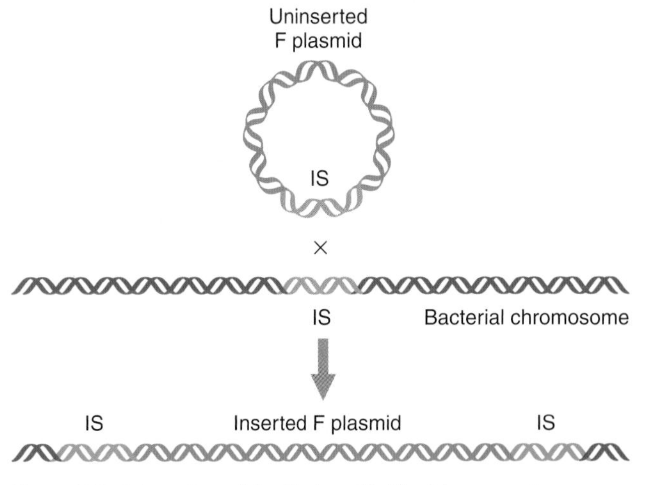

Figure 7-6 Integration of the F plasmid. The × represents a crossover.

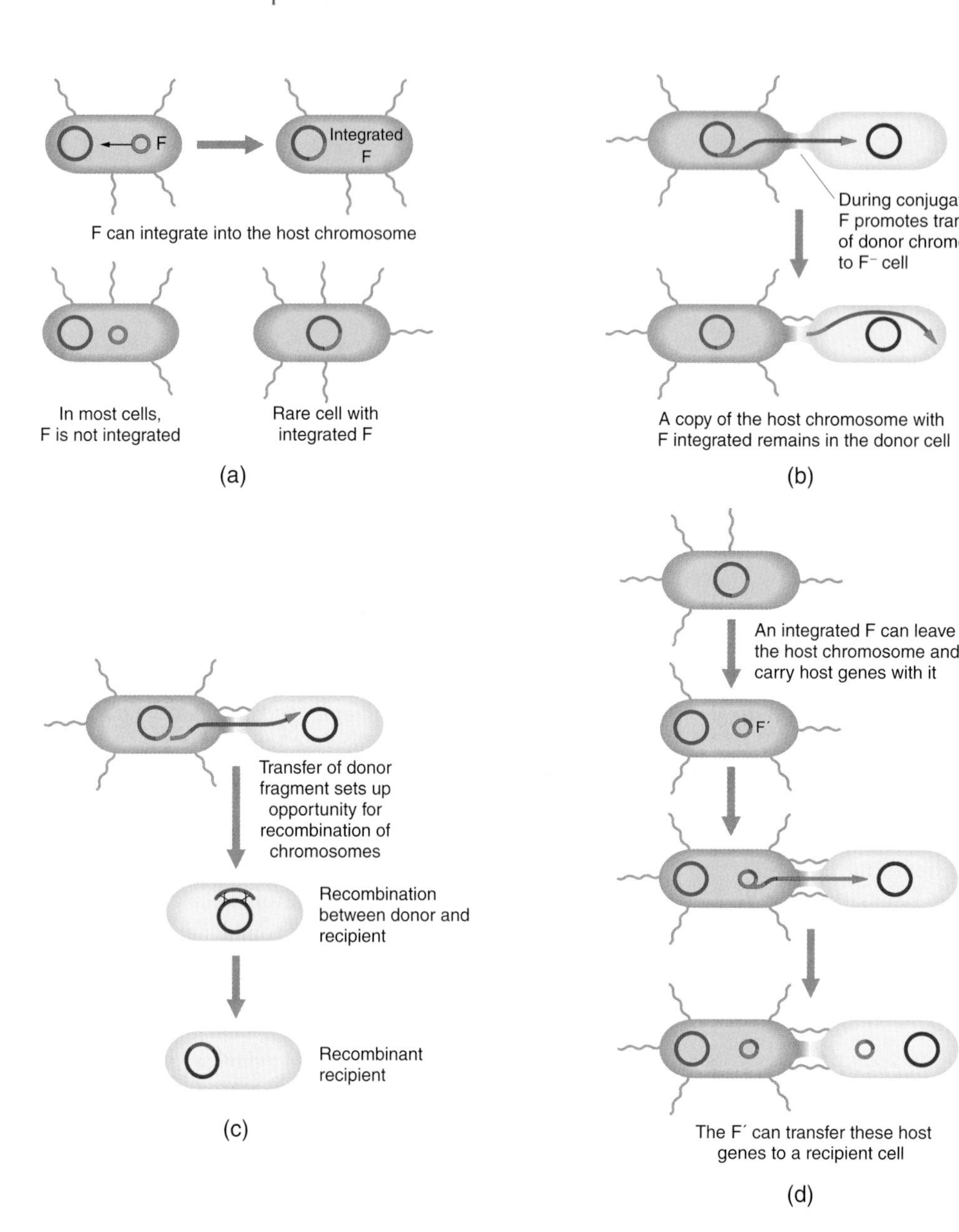

Figure 7-7 The transfer of *E. coli* chromosomal markers mediated by F. (a) Occasionally, the independent F factor combines with the *E. coli* chromosome. (b) When the integrated F transfers to another *E. coli* cell during conjugation, it carries along any *E. coli* DNA that is attached, thus transferring host chromosomal markers to a new cell. (c) In a population of F^+ cells, the few cells that have F integrated into the chromosome can transfer chromosomal markers. Therefore, when a population of F^+ cells is mixed with a population of F^- cells, a few F^- cells will acquire markers from the donor F^+ cells and undergo recombination. (d) Occasionally, the integrated F can leave the chromosome and return to the cytoplasm. In rare cases, F can carry host genes with it, incorporating them into the circular F, which is now termed an F′. When a copy of the F′ is transferred to a recipient cell during conjugation (see Figure 7-5), these incorporated host genes are also transferred at high efficiency because they are part of the F′ genome.

6. The integrated F factor occasionally leaves the chromosome of an Hfr cell and moves back to the cytoplasm. In some rare cases it carries a few host chromosomal genes along with it (Figure 7-6d). This modified F, called **F′** (read as "F prime"), can now transfer these specific donor genes to a recipient (F^-) cell in an infectious manner, in the same way that F is spread. Thus, the recipient cell now contains two copies of the same gene—one resident copy on its bacterial chromosome and one copy on the newly transferred cytoplasmic F′ factor. Since the F′ can replicate, its presence establishes a stable partial diploid.

Recombination between Donor and Recipient DNA

In bacterial genetics, all conjugations ("crosses") are, by definition, of the type Hfr (donor) × F^- (recipient). After cell fusion, the Hfr chromosome replicates by rolling circle replication (see Figure 4-10), which produces a single-stranded copy of the Hfr chromosome. This copy is then transferred linearly into the F^- cell. The replication and transfer begin at one side of the integrated F, called the **origin (O).** Genes close to the origin are transferred first, and the remaining genes are transferred in the order in which they are situated on the chromosome. The integrated F factor would be transferred last; however, in most conjugations, the chromosomal transfer process stops before F enters (Figure 7-8). After separation of the conjugating pair, the F^- cell is known as an **exconjugant.**

Once inside the F^- cell, the linear single-stranded DNA molecule acts as a polymerization template and is converted into a DNA double helix. The resulting linear donor fragment is the exogenote, and the resident F^- chromosome is the endogenote. As a free molecule, the exogenote cannot replicate and would become lost. However,

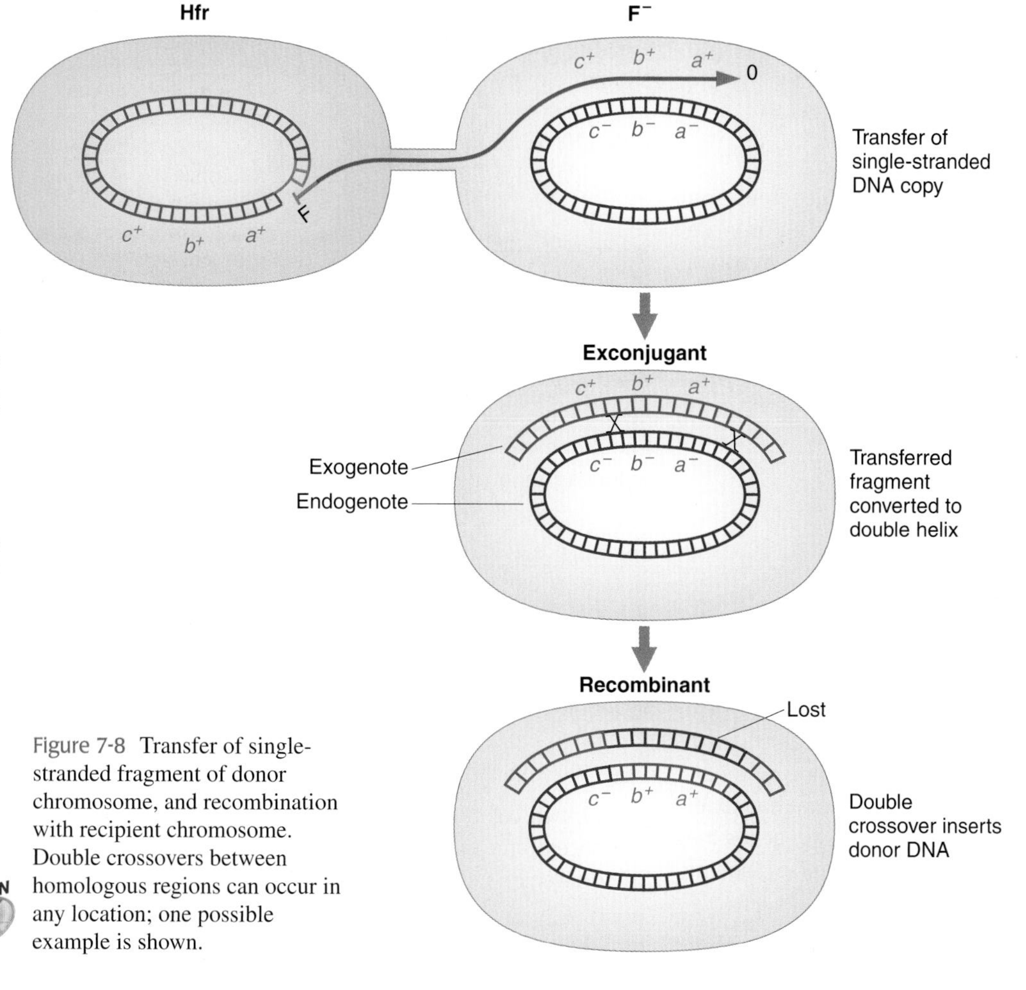

Animated Art • Bacterial Conjugation and Mapping by Recombination

Figure 7-8 Transfer of single-stranded fragment of donor chromosome, and recombination with recipient chromosome. Double crossovers between homologous regions can occur in any location; one possible example is shown.

because exogenotes and endogenotes are homologous, crossing-over can take place between them. A single crossover between a linear molecule (the exogenote) and a circular one (the endogenote) would produce a single long molecule that would be inviable because circularity is needed for replication. However, *two* crossovers integrate a part of the donor genome into the recipient. It is in this way that recombination takes place (see Figure 7-8), and donor DNA becomes part of a stable recombinant chromosome in the exconjugant. Note that, although such integrative exchanges *look like* double crossovers in the formal genetic sense, at the DNA level what takes place is a *single* integration event in which a long donor segment replaces the equivalent segment in the recipient.

Historically, the detection of recombinants was the evidence from which it was deduced that bacteria do in fact engage in a type of sexual union (Foundations of Genetics 7-1).

Since bacterial crosses can be made, and recombination detected, chromosome mapping is possible. There are two main mapping methods based on conjugation: mapping by interrupted conjugation, which produces a low-resolution map of large parts of the genome, and mapping by recombinant frequency, which produces higher-resolution maps of smaller regions.

Mapping by Interrupted Conjugation

Hfr strains transmit their chromosome to F^- in a linear manner at a rate of about 1 percent of the bacterial chromosome per minute. This feature of chromosome transfer is exploited to map the chromosome in the technique called **interrupted conjugation.** To map by interrupted conjugation, Hfr donor and F^- recipient cells are mixed, and conjugation allowed to proceed. Then, at fixed times, the F^- cells are sampled and growth tests performed on them in order to determine which donor alleles have entered. Agitation of conjugating cells in a kitchen blender disrupts the connections between the joined pairs, thus interrupting the conjugation. After interruption, the Hfr cells are selectively killed so that the analysis can focus on the exconjugants. Then the exconjugants are tested to see which of the donor alleles have entered and stably recombined with the endogenote. The times are calculated at which various donor alleles can first be detected in the exconjugants. If a donor allele a^+ enters the recipient at 5 minutes after conjugation began and allele b^+ enters at 8 minutes, the two genes are inferred to be 3 minutes apart on the chromosome. Therefore the map units in this case are minutes.

FOUNDATIONS OF GENETICS 7-1

Lederberg and Tatum Discover Genetic Recombination in Bacteria

In eukaryotes, the sexual cycle brings together genomes from two parents into one zygotic cell; then, in this cell or in descendant cells, meiosis takes place, which allows the recombination of parental alleles.

Eukaryotic sex.

This process has been known since the time of Mendel's paper in the 1860s. However, it was not until 1946 that a short paper by Joshua Lederberg and Edward Tatum appearing in *Nature* reported the discovery of "sex" in bacteria. The experiments that they used did not detect sexual union directly under the microscope, but indirectly, using a genetic method. They started with two strains of *E. coli* that had different nutritional deficiencies caused by mutations in genes that normally synthesize biotin, cysteine, leucine, phenylalanine, thiamine, and threonine. They wrote

> . . . single nutritional requirements were established as single mutational steps under the influence of X-ray or ultra-violet. By successive treatments, strains with several requirements have been obtained. *(J. Lederberg and E. L. Tatum,* Nature *158, 1946, 558.)*

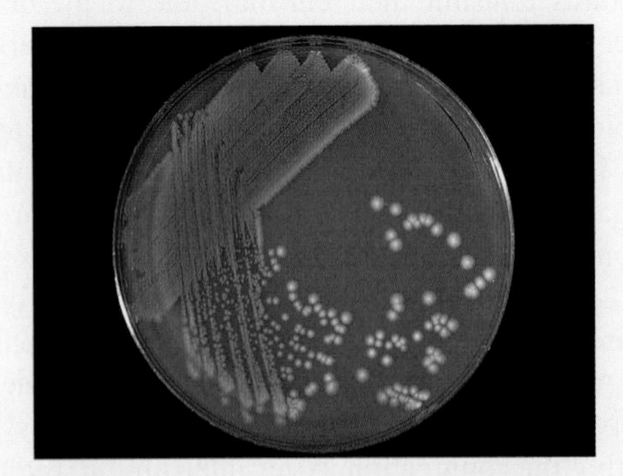

Bacterial colonies. (Biophoto Associates/Science Source/Photo Researchers.)

The strains that they used were both triple mutants, as shown below, where the genes are listed alphabetically:

Strain Y10:	+	+	*leu*	+	*thi*	*thr*
Strain Y24:	*bio*	*cys*	+	*phe*	+	+

Strains Y10 and Y24 were grown together in culture medium containing all six supplements. After a period of coincubation, cells were plated and recombinant genotypes were detected. Of the recombinants, the easiest to obtain and work with were wild types, which could be selected by plating cells on minimal medium. Wild types arose at a frequency of about one in a million cells. Wild types must have been of the following genotype:

+ + + + + +

As a control experiment, the parental triple mutant strains were shown *never* to revert to wild type when grown individually; hence, the wild types were the product of coculturing in some way.

Lederberg and Tatum concluded that there were only two likely explanations for these results:

1. The two cell types fused in some type of bacterial sexual process and engaged in recombination of alleles by "assortment in new combinations."

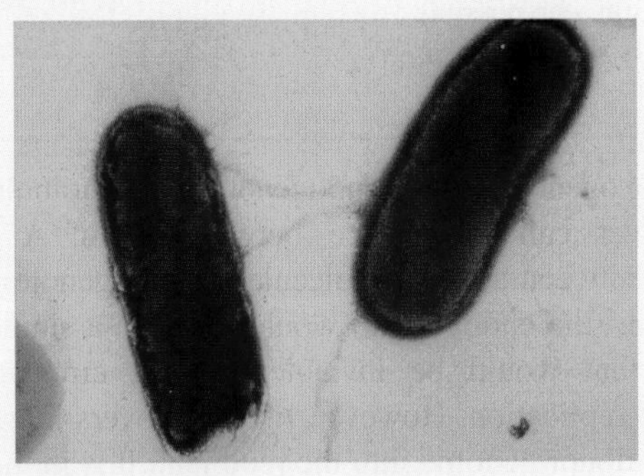

Bacterial cells attached by pili. (Fred Marsik/Visuals Unlimited.)

2. Something was passing through the medium from one cell type to another, transferring hereditary material. However, they found that cell-free filtrates of one culture could not "transform" the other into the wild-type genotype, so they concluded that "these experiments imply the occurrence of a sexual process in the bacterium *Escherichia coli*."

Let's analyze a typical cross in which the order and map positions of the genes under study are not known initially. In this particular cross, the genes by which the parents differ will be *azi* (resistance or sensitivity to sodium azide), *gal* (ability or inability to utilize galactose as an energy source), *lac* (ability or inability to utilize lactose as an energy source), and *ton* (resistance or sensitivity to bacteriophage T1). A streptomycin-sensitivity allele (str^s) in the Hfr along with a streptomycin-resistance allele (str^r) in the recipient are used to selectively kill the Hfr cells after conjugation. Selective killing is accomplished by adding streptomycin to the mixture of cells after interrupting the conjugation. It is advantageous if the allele used to kill the Hfr enters close to last, because then it will only rarely enter the F$^-$ and make the exconjugant sensitive. In other words, the *str*^{s} allele should be close to the integrated F factor. (This must be established in previous experiments.) The parents of the cross under consideration here are as follows, where the unmapped genes are written in alphabetical order:

Hfr	*str*^{s}	*azi*^{r}	*gal*$^+$	*lac*$^+$	*ton*^{r}
F$^-$	*str*^{r}	*azi*^{s}	*gal*$^-$	*lac*$^-$	*ton*^{s}

The results of the interrupted-mating experiment are shown in Figure 7-9. The *azi*^{r} gene is the first to be detected, entering at 8 minutes, followed by *ton*^{r}, *lac*$^+$, and *gal*$^+$ in that order. Therefore, not only is gene order on the chromosome map established, but map distances in minutes also are obtained, as shown in Figure 7-10 on the next page.

Note from Figure 7-9 that alleles transferred early, such as *azi*^{r}, are found in a high percentage of F$^-$ exconjugants, but the alleles that enter later, such as *gal*$^+$, are found in only a small proportion. The reason for this difference is either that transfer spontaneously stops or that the chromosome breaks, leaving the later genes inside the Hfr. However, this result does not affect the time-of-entry calculations.

The relative positions of the *azi, ton, lac,* and *gal* genes are established by this experiment. However, the chromosomal region containing these loci might be only a small proportion of the entire chromosome. A complete map is obtained from many such interrupted-conjugation experiments, involving different Hfr donors. In Hfr strains isolated at different times, the integrated F factor is typically in different locations and may be inserted in either of two orientations, pointing "clockwise" or "counterclockwise." Examples of the positions and orientations of F in different Hfrs, and the different orders of gene entry resulting, are shown in Figure 7-11 on the next page.

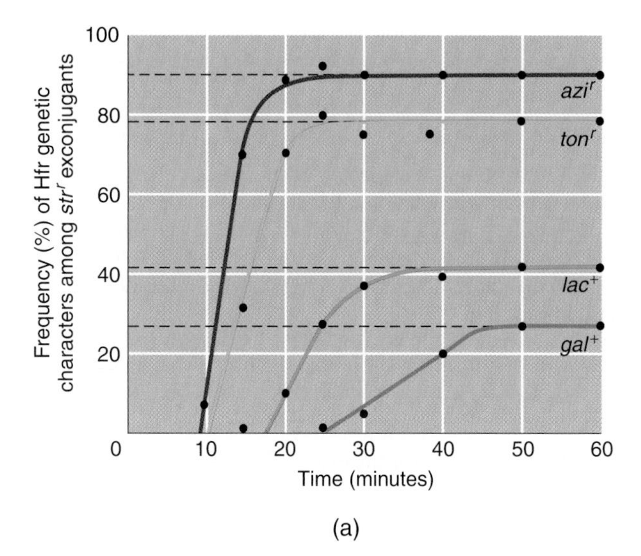

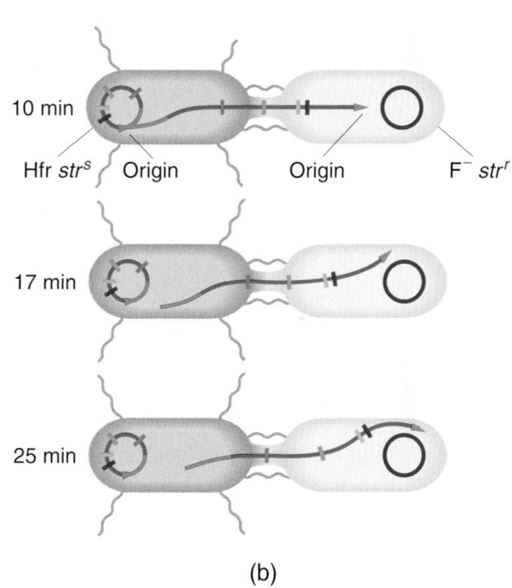

Figure 7-9 Interrupted-mating conjugation experiments with *E. coli.* F$^-$ cells that are *str*^{r} are crossed with Hfr cells that are *str*^{s}. The F$^-$ cells carry a number of mutations, *azi, ton, lac,* and *gal.* However, the Hfr cells carry the wild-type alleles of all these genes. At different times after the cells have been mixed, samples are withdrawn, disrupted in a blender to break conjugation between cells, and plated on media containing streptomycin. The antibiotic kills the Hfr cells but allows the F$^-$ cells to grow and to be tested for their ability to carry out the four metabolic steps. (a) A plot of the frequency of exconjugants containing donor alleles for each gene, as a function of time after mating. (b) A schematic view of the transfer of donor alleles (shown in different colors) over time. *(Part a modified from E. L. Wollman, F. Jacob, and W. Hayes,* Cold Spring Harbor Symposia on Quantitative Biology *21, 1956, 141.)*

High-Resolution Mapping by Recombinant Frequency

Interrupted-conjugation experiments provide a rough set of gene locations over the entire map. As we learned, the genes are mapped by time of entry. In such experiments, to be able to detect an exconjugant carrying a donor gene, the exogenote from the donor must integrate by a double-crossover event (see Figure 7-7c). Nevertheless the mapping method is not based on any measurement of recombinant frequencies: all we measure is *when* any individual gene enters. However, over smaller distances the time-of-entry method is not accurate, especially for genes that are close together. To provide a higher-resolution method for measuring the sizes of smaller map distances, recombinant frequencies between two or more genes are used. The following example will show how the method provides higher resolution over small distances.

Suppose that we decide to map three genes—*met, arg,* and *leu*—by recombinant frequency. To measure recombination between these genes, we must set up a merozygote

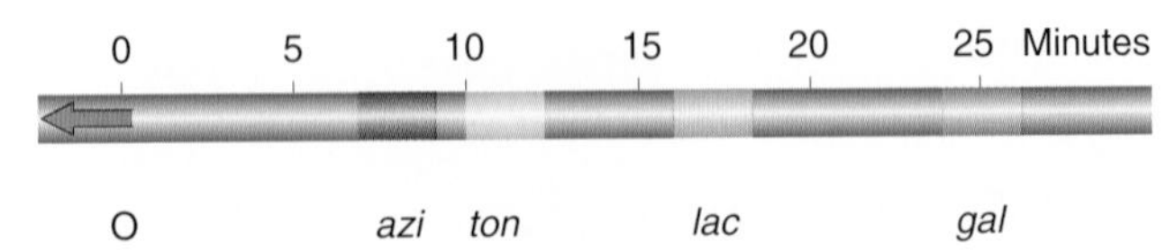

Figure 7-10 Chromosome map based on Figure 7-9. A linkage map can be constructed for the *E. coli* chromosome from interrupted-mating studies, by using the time at which the donor alleles first appear after mating. The units of distance on the map are given in minutes. The arrow at left indicates the direction of transfer of the donor alleles.

that is heterozygous for all three. This can be accomplished if we can establish which of these three genes enters last during conjugation, which we can do by an interrupted-conjugation analysis. Exconjugants that have received the donor allele of the last entering gene are selected by plating on medium on which only cells carrying that donor allele will grow. Knowing that we have selected the last of the three donor genes we want to map assures us that the other two donor genes must have been received by the merozygote. In the cross

$$\text{Hfr } met^+ \; arg^+ \; leu^+ \times \text{F}^- \; met^- \; arg^- \; leu^-$$

assume that we have established that the gene order is *met* first, followed by *arg*, and then *leu*.

The last gene to enter is leu^+; therefore, we select initially for leu^+ exconjugants by plating the cells on medium containing no leucine. This medium must contain methionine and arginine because some of the leu^+ recombinants we have selected might carry the recipient alleles met^- and arg^-. (Note: it is important to distinguish between alleles that must have entered the recipient as an exogenote—leu^+, arg^+, and met^+ in this case—and those that recombine into the recipient's chromosome. Only the latter can become part of the stable exconjugant genotype because an exogenote cannot replicate.) Then the merozygote must have been as follows:

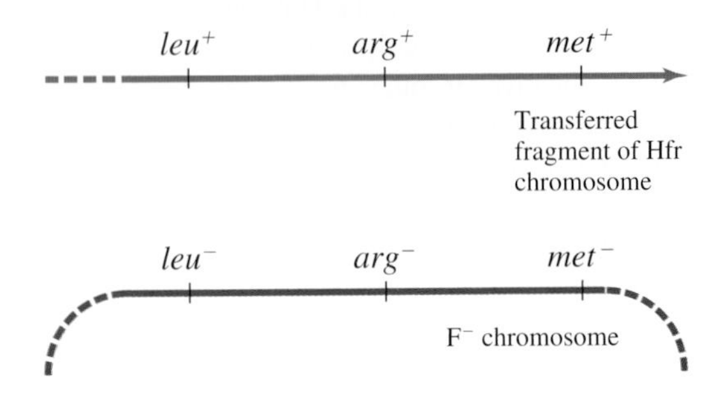

Now we can proceed to calculate map distance in the standard way by using a map unit equal to a recombinant frequency of 1 percent. In practice, this calculation is done by measuring the proportion of the total leu^+ exconjugants that also carry arg^+, or met^+, or both, or neither. The recombination events needed to produce these recombinant genotypes are shown in Figure 7-12. We know that a double crossover must have occurred to integrate leu^+: one crossover is at the left of the *leu* gene, but the other can be in various positions to its right. Hence the genotype that arises from recombination between *leu* and *arg* will be leu^+ arg^- met^- (see Figure 7-12a): so the percentage of bacteria with this genotype in the leu^+ exconjugants will give us our recombinant frequency value for the *leu*-to-*arg* interval. The leu^+ exconjugants arising from recombination between *met* and *arg* will be leu^+ arg^+ met^- (see Figure 7-12b). The percentage of bacteria with this leu^+ subgenotype will provide the map distance between the *met* and *arg* genes.

If the second crossover occurs to the right of *met*, giving the genotype leu^+ arg^+ met^+ (see Figure 7-12c), this will tell us nothing about recombination in the *leu–met* region, so the size of this class is not informative. The leu^+ arg^- met^+ recombinants would require four crossovers instead of two (see Figure 7-12d). These recombinants would be relatively rare, so we will not consider these further.

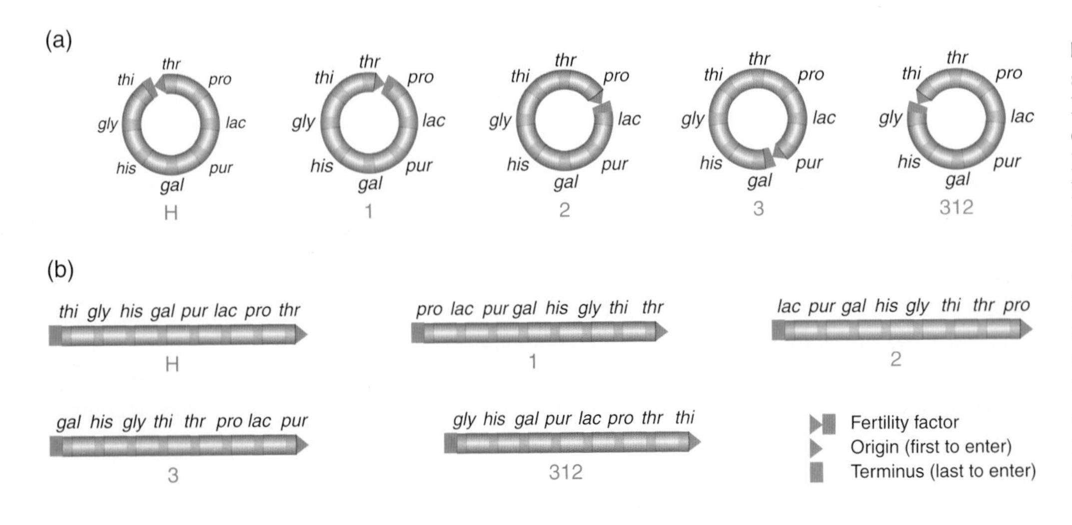

Figure 7-11 Origin of Hfr strains of *E. coli* by F insertion into the chromosome. (a) Different Hfr strains (H, 1, 2, 3, 312) have the fertility factor inserted into the chromosome at different points and in different directions. The insertion point is shown for each strain. (b) The linear order of transfer of markers for each Hfr strain. Arrowheads indicate the origin and direction of transfer.

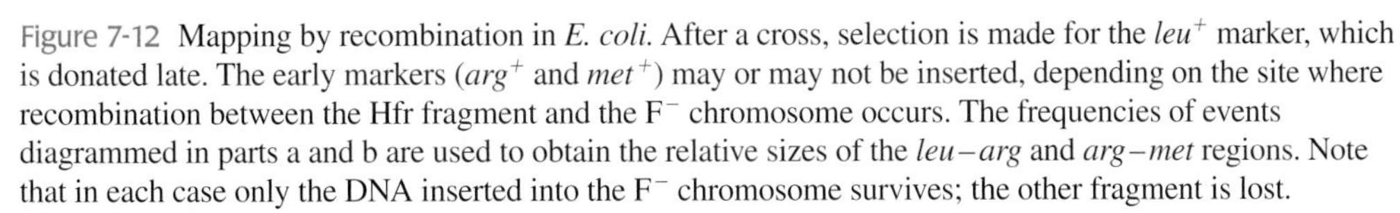

Figure 7-12 Mapping by recombination in *E. coli.* After a cross, selection is made for the leu^+ marker, which is donated late. The early markers (arg^+ and met^+) may or may not be inserted, depending on the site where recombination between the Hfr fragment and the F^- chromosome occurs. The frequencies of events diagrammed in parts a and b are used to obtain the relative sizes of the *leu–arg* and *arg–met* regions. Note that in each case only the DNA inserted into the F^- chromosome survives; the other fragment is lost.

Let us examine some data from this cross. The percentages of the three main genotypes obtained after testing leu^+ exconjugants are

$leu^+\ arg^-\ met^-$	4%
$leu^+\ arg^+\ met^-$	9%
$leu^+\ arg^+\ met^+$	87%

From these results, we can conclude that the *leu–arg* distance is 4 map units and that the *arg–met* distance is 9 map units.

MESSAGE

Time-of-entry measurements in interrupted conjugation can generate a broad-scale map of the bacterial chromosome. Recombinant frequencies among exconjugants can be used in fine-scale mapping.

F Factors Carrying Bacterial Genes

Occasionally, the integrated F factor of an Hfr strain exits from the bacterial chromosome. Usually this event is a clean excision regenerating an intact F plasmid. However, in some cases, the excision event uses an IS element in a position different from the one used for entry. Hence exit is not a precise reversal of the original insertion, and a part of the bacterial chromosome is incorporated into the liberated plasmid. Figure 7-13, on the next page, shows incorporation of a nearby *lac* gene into the plasmid. The precise gene incorporated depends on where the F factor had originally integrated in the particular Hfr (see Figure 7-13a). Such plasmids carrying bacterial genes are called **F′**. They are named for the gene that they carry: F′-*lac,* as in the case illustrated in Figure 7-13, or F′-*gal,* F′-*trp,* and so forth. Any specific F′ can be detected by looking for rapid infectious transfer of a gene that would normally be transferred late in the particular Hfr strain used.

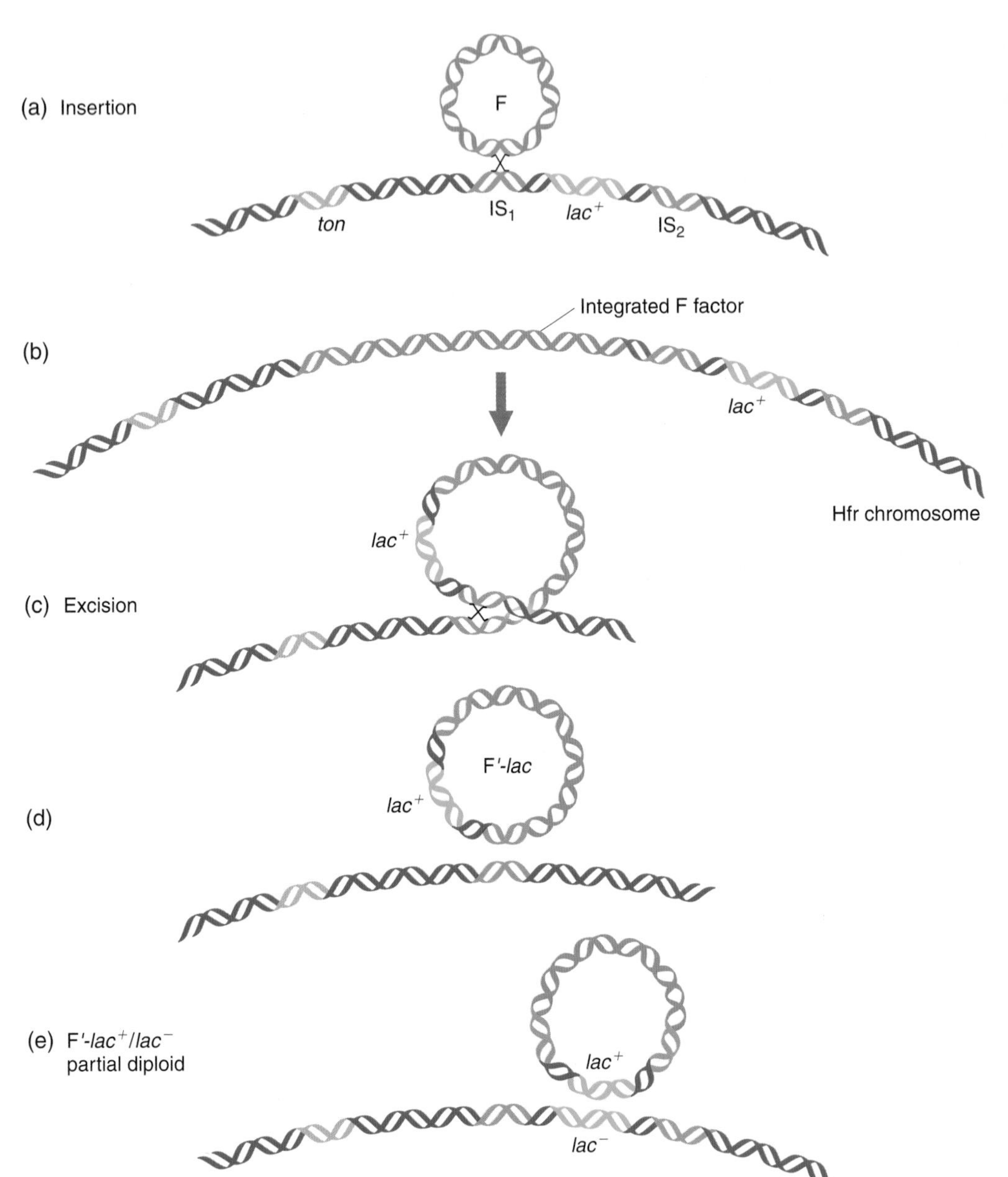

Figure 7-13 Origin of an F′ factor. (a) F is inserted in an Hfr strain at a repetitive element labeled IS_1 between the *ton* and lac^+ alleles. (b) The inserted F factor. (c) Abnormal "outlooping" by crossing-over with a different element IS_2, to include the *lac* locus. (d) The resulting F′-lac^+ particle. (e) An F′-lac^+ / lac^- partial diploid is produced by the transfer of the F′-lac^+ particle to an F^- lac^- recipient.

If an F′ plasmid is transferred upon conjugation with an F^- strain, the recipients generated are stable merozygotes, because the donor fragment is carried on a plasmid, which can replicate (see Figure 7-13e). The process of creating a merozygote by an F′ element is called **sexduction** or **F′-duction.** Such stable partial diploids are extremely valuable because they can be used for genetic studies usually possible only in a diploid cell, such as determination of dominance, or to see if two different mutations are alleles of the same gene. For example, if a lac^+ donor is used to create an F′-lac^+ plasmid and this plasmid is transferred to an F^- recipient that carries the allele lac^-, the partial diploid is heterozygous lac^+ / lac^-. Therefore these cells can be used to determine which allele is dominant (lac^+ turns out to be dominant in this case).

BACTERIAL TRANSFORMATION

In Chapter 2 we learned that the genotype of cells of the bacterium *Streptococcus* can be altered by the addition of exogenous DNA extracted from a donor culture of a different genotype, a process called **transformation.** Recipient bacterial cells can be transformed only if they have been put into an appropriate physiological state called *competence.* Then the exogenous DNA can pass through the cell membrane and integrate into the bacterial chromosome. Transformation can be demonstrated in many different kinds of bacteria, and even in many eukaryotic cells, too. Transformation has some key uses in bacterial genetics (and in eukaryotic genetics).

Bacterial transformation can be demonstrated by using the genes for drug resistance. Transforming DNA from a str^r culture is added to cells of a recipient culture of genotype str^s, which are then plated on streptomycin. The frequency of resistant colonies (transformants) is found to be proportional to the amount of transforming DNA used (Figure 7-14). Transformants are stable because, once inside the recipient cell, the exogenous DNA inserts into the bacterial chromosome by a double-crossover process analogous to that which produces recombinants in Hfr × F^- crosses, depicted in Figure 7-8. (However, note that, in *conjugation,* DNA is transferred from one living cell to another through cell fusion, whereas in *transformation* a cell takes up isolated pieces of DNA from the external environment.)

Transformation can be used to assess the tightness of linkage. When the DNA from the donor is extracted for use in a transformation experiment, the isolation procedure causes random breakage of the donor DNA molecules into smaller pieces. If two donor genes are located close together on the chromosome, it is unlikely that a break will occur between them during DNA extraction. Then there is a good chance that they will be carried on the same piece of transforming DNA, and hence there will be frequent **cotransformation** (transformation by both genes). Conversely, if genes are very far apart on the chromosome, when DNA is extracted, most of the time a break will occur between them so the genes will be carried on separate DNA segments. There will still be cotransformants, but these will most likely be cases in which the two genes enter on separate DNA fragments. Hence, since the individual transformations will be independent, in this case the frequency of cotransformants will equal the product of the single-gene transformation frequencies. Thus, it should be possible to test for close linkage by testing for a departure from the product rule. Suppose DNA from an $a^+ b^+$ donor is used to transform $a^- b^-$ recipient cells; if a and b are closely linked, the proportion of $a^+ b^+$ cotransformants should exceed the product of the proportions of single a^+ and b^+ transformants. Therefore, if there is a total of T transformants involving genes a and b, the diagnostic for linkage is

$$a^+ b^+ / \mathrm{T} > a^+ b^- / \mathrm{T} \times a^- b^+ / \mathrm{T}$$

Furthermore, the relative distances between linked genes can be obtained from the proportions of cotransformants, in an approach similar to that used in mapping by transduction, considered later in this chapter.

Transformation can be used to map genes on defined segments of the bacterial genome, rather than random fragments as used above. As we will see in Chapter 8, DNA technology allows DNA to be cut at defined positions in the chromosome and the resulting cut fragments to be isolated. This makes it possible to test various fragments of a bacterial chromosome for the presence of specific donor alleles in transformation tests. If the relative positions of the individual fragments are known, it is possible to build up a map of the loci of genes on the DNA.

MESSAGE

A bacterial genome can be transformed into a new stable genotype by the entry and integration of DNA from another strain. The frequency of cotransformation of two donor genes can be used to map the distance between the two genes.

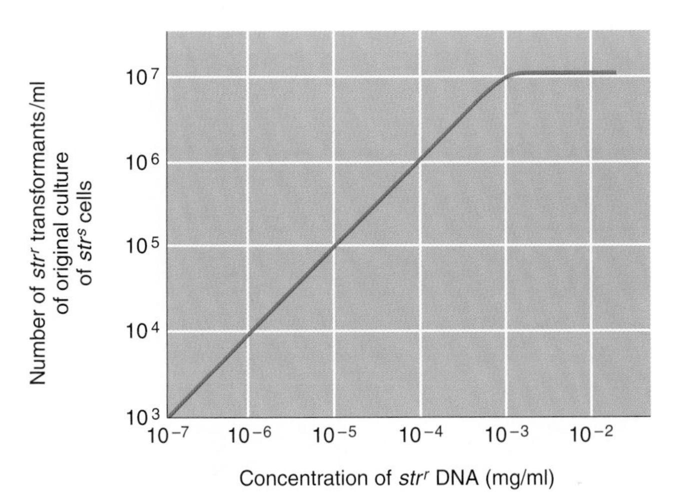

Figure 7-14 Transformation of streptomycin sensitivity (str^s) to streptomycin resistance (str^r) in *E. coli*. The frequency of str^r transformants among str^s cells depends on the concentration of str^r DNA used. *(From G. S. Stent and R. Calendar,* Molecular Genetics, *2d ed. © 1978 by W. H. Freeman and Company, New York.)*

BACTERIOPHAGE GENETICS

The word *bacteriophage,* which is a name for bacterial viruses, means "eater of bacteria." These viruses, which parasitize and kill bacteria, can be used in two different types of genetic analysis, which we will examine in this chapter. First, two distinct phage genotypes can be crossed to measure recombination and hence map the *viral* genome. We will examine this process in this section. Second, they can be used as a way of bringing *bacterial* genes together for linkage and other genetic studies. We will study this in the next section. In addition, as we shall see in Chapter 8, phages are used in DNA technology as carriers, or vectors, of foreign DNA inserts from any organism. Before we can understand phage genetics, we must examine the infection cycle of phages.

Infection of Bacteria by Phages

Most species of bacteria are susceptible to attack by one or more types of bacteriophage. A phage consists of a nucleic acid chromosome (generally DNA) surrounded by a coat of protein molecules. The T phages—T1, T2, T3, T4, and so forth—are an extensively studied class of phage strains; phage T4 is a good example (Figure 7-15).

Figure 7-16 shows the complicated structure of phage T4. The head contains the phage DNA, and the remaining parts are concerned with injecting this DNA into the bacterial cell. During infection, a phage attaches to a bacterium and injects its genetic material through the cell wall into the bacterial cytoplasm (Figures 7-16 and 7-17a). The phage genetic information then takes over the machinery of the bacterial cell by turning off the synthesis of bacterial components and redirecting the bacterial synthetic machinery to make phage components (Figure 7-17b). The use of the word *information* is most appropriate in this context; it literally means "to give form." That is precisely the role of the genetic material: to provide blueprints for the construction of form. In the present context, the form is the elegantly symmetrical structure of the new phages. Ultimately, many phage descendants are released when the bacterial cell wall breaks open. This bursting process is called **lysis.**

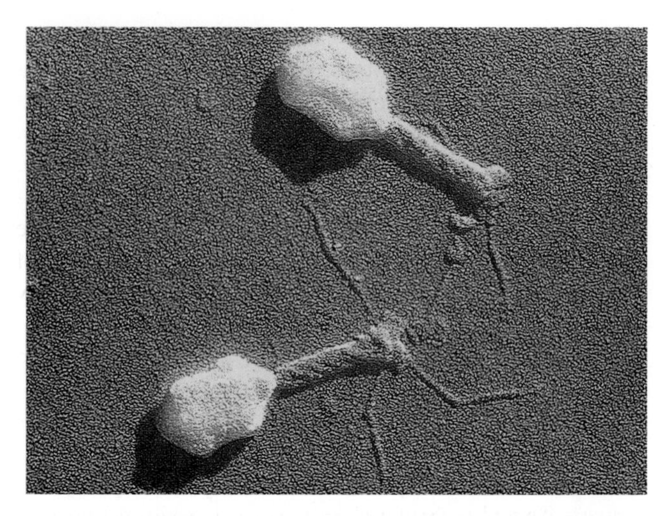

Figure 7-15 Mature particles of the *E. coli* phage T4 (× 97,500). *(M. Wurtz/Biozentrum, University of Basel/SPL/Photo Researchers.)*

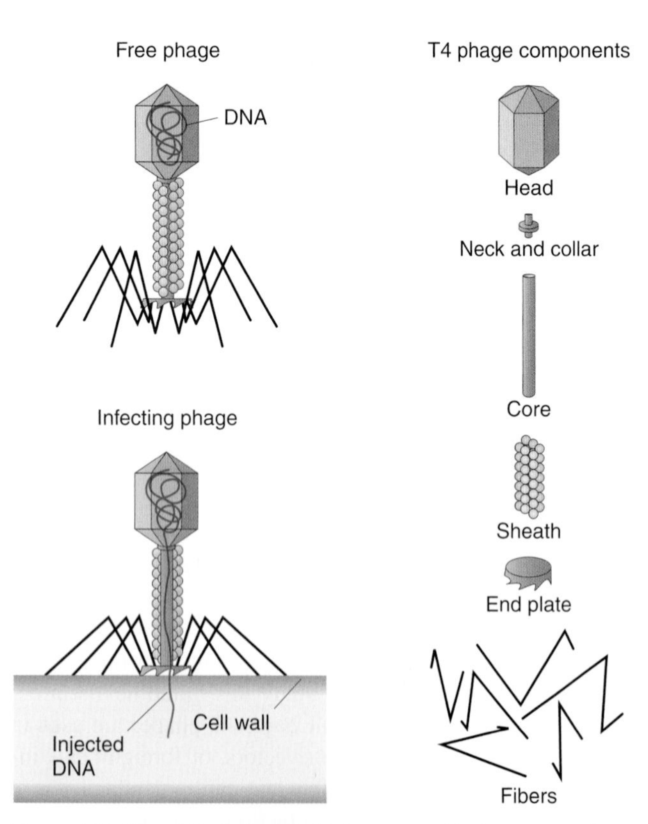

Figure 7-16 Phage T4, depicted in its free state and in the process of infecting an *E. coli* cell. The infecting phage injects DNA through its core structure into the cell. On the right, a phage has been diagrammatically dissected to show its highly ordered three-dimensional structure. *(Modified from R. S. Edgar and R. H. Epstein, "The Genetics of a Bacterial Virus." © 1965 by Scientific American, Inc. All rights reserved.)*

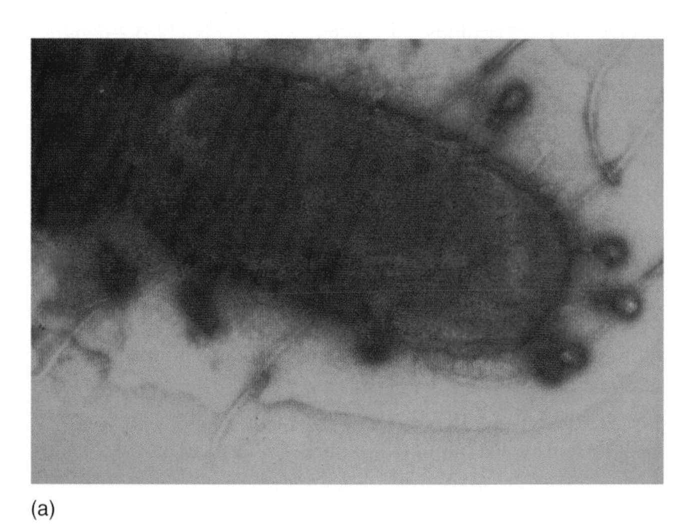

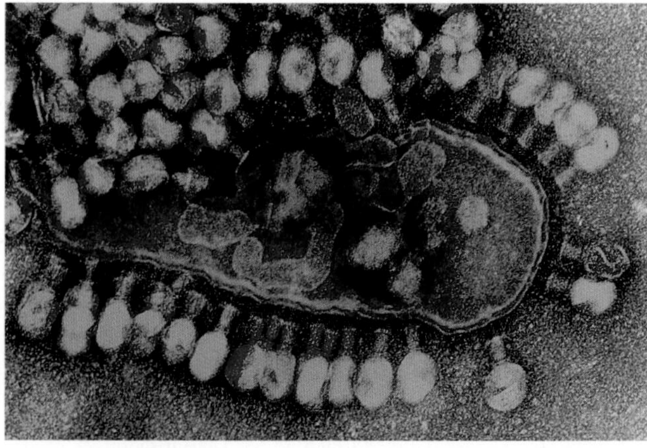

Figure 7-17 (a) Bacteriophage λ attached to an *E. coli* cell and injecting its genetic material. (b) Progeny particles of phage λ maturing inside an *E. coli* cell. *(Part a, Fred Hossler/Visuals Unlimited; part b, Lee D. Simon/Science Source/Photo Researchers.)*

But what phage phenotypes can be studied, given that they are even smaller than bacteria, and visible only under the electron microscope? In the study of phages, we cannot produce a visible colony by plating, but we can produce a visible phenotype by taking advantage of several phage properties. Let's look at the consequences of a phage's infecting a bacterial culture. Figure 7-18 shows the sequence of events in the infectious cycle that leads to the release of progeny phages from the lysed cell. After lysis, the progeny phages infect neighboring bacteria, and these bacteria subsequently lyse and the new generation of progeny phage then reinfects. The reinfection cycle results in an exponential increase in the number of lysed cells. Within 15 hours after the start of an experiment of this type, the effects are visible to the naked eye as a clear area, or **plaque,** in an opaque "lawn" of bacterial culture growing on the surface of a plate of solid medium (Figure 7-19 on the next page). Such plaques can be large or small, fuzzy- or sharp-edged, and so forth, depending on the phage type. Thus, *plaque morphology* is a phage character that can be analyzed. Another phage character that we can analyze genetically is *host range,* because phages may differ in the spectrum of bacterial strains that they can infect and lyse. For example, certain strains of bacteria are resistant to attachment or injection by specific phages.

The Phage Cross

Genetic analysis of any organism is accomplished by bringing together two genetically different genomes into the same cell and giving them an opportunity to engage in recombination. In an analogous manner, two distinct phage genomes can be united, but, because of the parasitic nature of phages, this union must be inside a bacterial cell. In other words, a phage cross must be made by a double infection of the same bacterial cell by at least one phage of each of the two types.

The procedure can be illustrated by a cross of T2 phages. The genotypes of the two parental strains are $h^- \, r^+$ and $h^+ \, r^-$. The gene symbol h stands for host range. Phages with the allele h^- can infect *two* different *E. coli* strains (which we can call strains 1 and 2). Those with h^+ can infect only one strain (strain 1). Phages with the allele r^- *rapidly* lyse cells, and so many cells are lysed per unit time that they produce large plaques. Those with r^+ *slowly* lyse cells, thus producing smaller plaques. To make the cross, *E. coli* strain 1 is infected with both parental T2 phage genotypes at a phage : bacteria ratio that is high enough to ensure that a large percentage of cells are simultaneously infected by both phage types (Figure 7-20 on the next page). The phage lysate (the progeny phages) is then analyzed by spreading it onto a bacterial lawn composed of a mixture of *E. coli* strains 1 and 2. Phage particles with the allele h^- will infect both hosts, forming a clear plaque, whereas h^+ results in a cloudy plaque because one host is not infected. Four plaque types are then distinguishable (Figure 7-21 and Table 7-2, both on the next page). These four genotypes can be scored easily as parental ($h^- \, r^+$ and $h^+ \, r^-$) and recombinant ($h^+ \, r^+$ and $h^- \, r^-$), and a recombinant frequency can be calculated as follows:

$$\text{RF} = \frac{(h^+ \, r^+) + (h^- \, r^-)}{\text{total plaques}}$$

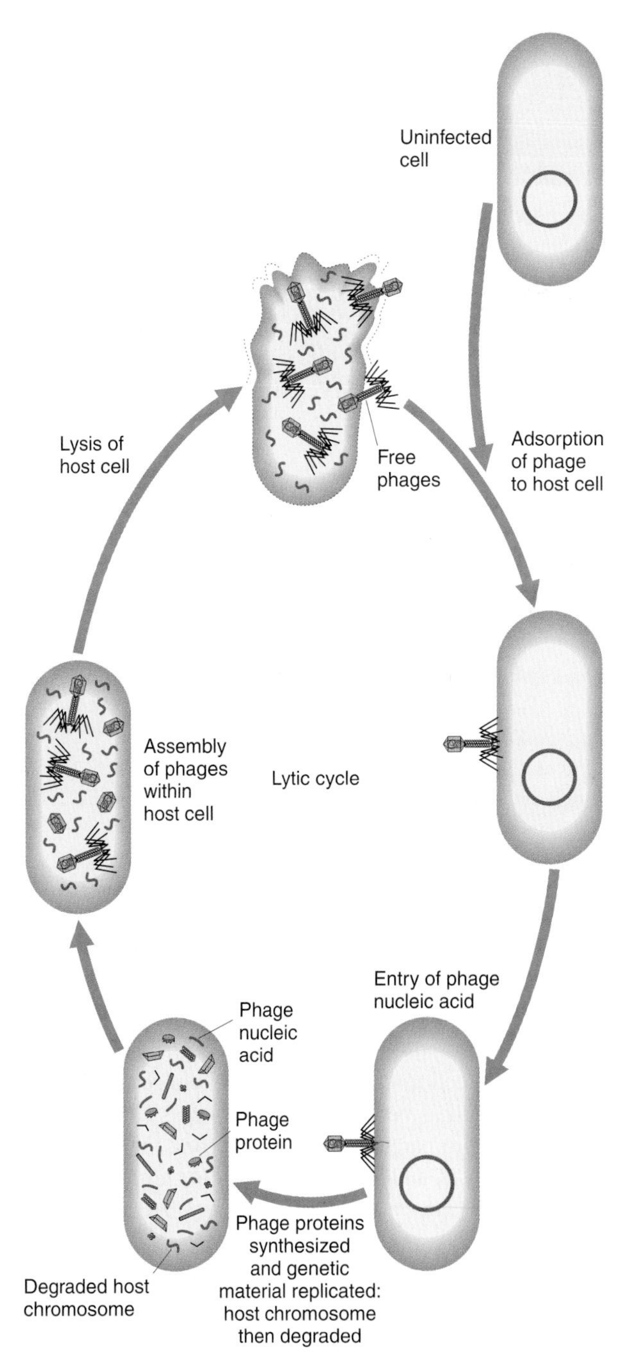

Figure 7-18 A generalized bacteriophage lytic cycle. *(Adapted from J. Darnell, H. Lodish, and D. Baltimore,* Molecular Cell Biology. *© 1986 by W. H. Freeman and Company, New York.)*

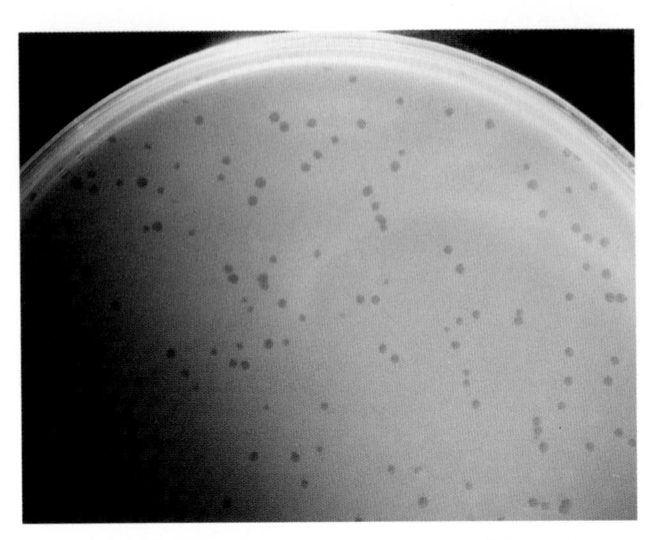

Figure 7-19 The appearance of phage plaques. Individual phages are spread on an agar medium that contains a fully grown "lawn" of *E. coli*. Each phage infects one bacterial cell, producing 100 or more progeny phages that burst the *E. coli* cell and infect neighboring cells. They in turn are exploded with progeny, and the continuing process produces a clear area, or plaque, where cells in the opaque lawn of bacterial have lysed. *(Barbara Morris, Novagen.)*

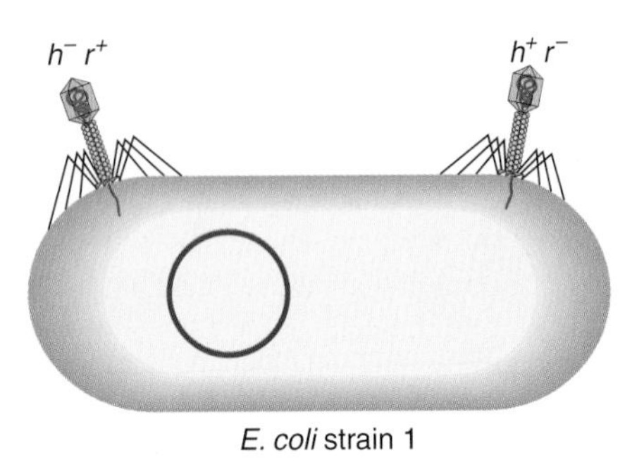

Figure 7-20 A double infection of *E. coli* by two phages.

Analysis of phage crosses is subject to complications. First, several rounds of recombination can potentially occur within one host cell: a recombinant produced by exchange of DNA between two phage chromosomes shortly after infection may undergo additional recombination later with other phage chromosomes present in the infected cell. Second, double infection can occur between phages of the same genotype as well as between different types. Thus, if we let P_1 and P_2 refer to parental phage genotypes, then crosses of the type $P_1 \times P_1$ and $P_2 \times P_2$ occur in addition to the informative cross $P_1 \times P_2$. The crosses between similar genotypes produce no detectable recombinants. For both of these reasons, recombinants from phage crosses are a consequence of a *population* of events rather than defined, single-step exchange events as in bacterial conjugation or eukaryotic meiosis. Nevertheless, so long as the input phage ratios and concentrations are appropriate, the RF calculation is a valid index of map distance in phages.

MESSAGE
Phage genomes can be mapped by analysis of recombinant frequency in lysates produced from double infections.

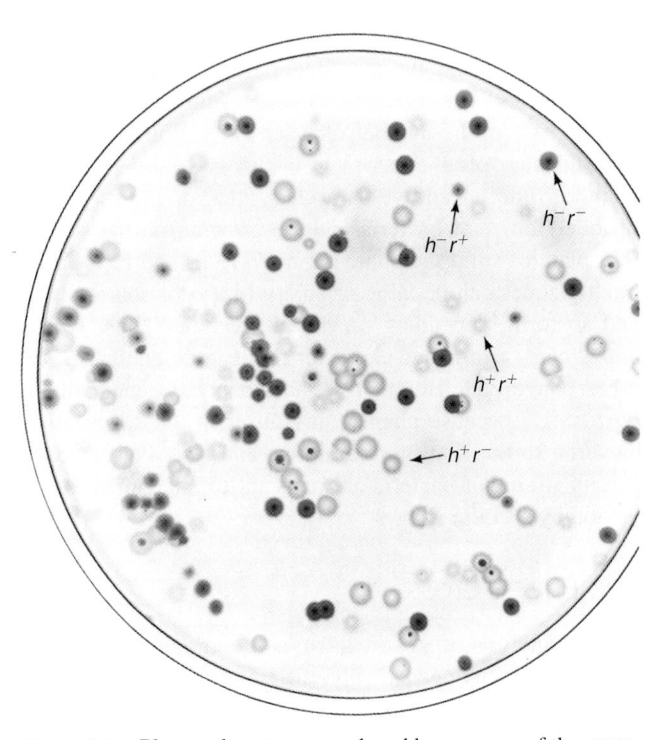

Figure 7-21 Plaque phenotypes produced by progeny of the cross $h^- r^+ \times h^+ r^-$. Enough phages of each genotype are added to ensure that most bacterial cells are infected with at least one phage of each genotype. After lysis, the progeny phages are collected and added to an appropriate *E. coli* lawn. Four plaque phenotypes can be differentiated, representing two parental types and two recombinants. *(From G. S. Stent,* Molecular Biology of Bacterial Viruses. *© 1963 by W. H. Freeman and Company, New York.)*

TABLE 7-2 Progeny Phage Plaque Types from Cross $h^- r^+ \times h^+ r^-$

Phenotype	Inferred Genotype
Clear, small	$h^- r^+$
Cloudy, large	$h^+ r^-$
Cloudy, small	$h^+ r^+$
Clear, large	$h^- r^-$

NOTE: Clearness is produced by the h^- allele, which allows infection of *both* bacterial strains in the lawn; cloudiness is produced by the h^+ allele, which limits its infection to the cells of strain 1.

PHYSICAL INTERACTION OF PHAGE AND BACTERIAL GENOMES

We now turn to the types of genetic analysis in which phages are used to help analyze bacterial genomes. Phage DNA physically interacts with bacterial DNA during infection, and this interaction can be used in bacterial genetics, as shown in the following sections.

Lysogeny

Phages are categorized into two types. **Virulent phages** are always **lytic;** that is, they infect and lyse the host cell, resulting in progeny phages. In contrast, **temperate phages** may undergo a lytic cycle under some circumstances, but more often the phage chromosome integrates into the bacterial chromosome, in some cases in a manner resembling F plasmid insertion. Henceforth the inserted phage is replicated along with the bacterial chromosome. In this condition, the phage is referred to as a **prophage** and the bacterial host is said to be in a **lysogenic** state. A lysogenic bacterium, or **lysogen,** is resistant to subsequent infection, because an "immunity" is conferred by the presence of the prophage. The lysogenic state can be transmitted genetically through many bacterial generations. However, a prophage can occasionally spontaneously excise from the bacterial chromosome and enter a lytic cycle, leading to lysis of its host cell. The lytic cycle can be induced experimentally using ultraviolet radiation. The progeny phages in the lysate can infect and cause lysis in any nonlysogenic cells present in the same culture. (The word *lysogenic* means "lysis-causing"; the ability to lyse other genotypes was the way this condition was first identified.)

MESSAGE

Virulent phages cannot become prophages; they are always lytic. The chromosome of a temperate phage can integrate into the bacterial chromosome as a prophage, allowing the host cell to survive in the lysogenic state. Prophages occasionally exit the bacterial chromosome and enter the lytic cycle.

Further properties of lysogenic phage can be illustrated with the use of a temperate phage of *E. coli* called **lambda (λ).** Phage λ has become the most intensively studied and best-characterized phage. If a lysogenic Hfr conjugates with a nonlysogenic F^- recipient, the donor chromosome is introduced linearly into the recipient, as expected. However when the section of chromosome carrying the λ prophage enters the nonimmune cell, it immediately triggers the prophage into a lytic cycle; this event is called **zygotic induction** (Figure 7-22a). But, in the cross Hfr(λ) $\times$ $F^-(\lambda)$, conjugation proceeds normally, no lysis takes place,

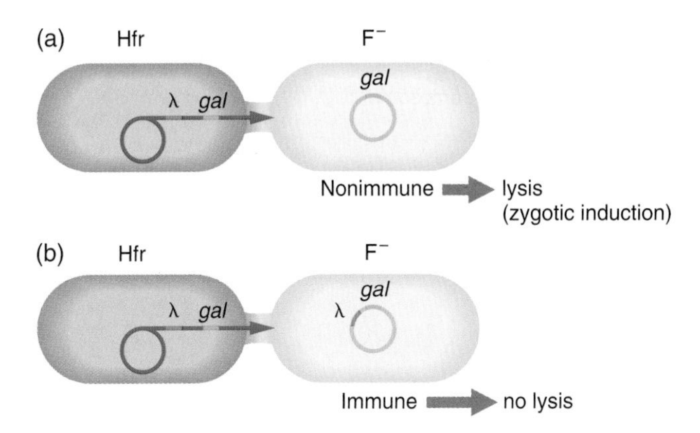

Figure 7-22 Zygotic induction.

and bacterial recombinants are readily recovered. In other words, there is no zygotic induction followed by lysis (Figure 7-22b); the cells are immune to λ. The cytoplasm of the F^- cell exists in two different states, depending on whether the cell contains a prophage. Contact between an entering prophage and the cytoplasm of a nonlysogenic cell immediately induces the lytic cycle. It is known that a gene product encoded by the prophage represses the multiplication of the virus. Entry of the prophage into the environment of a nonlysogenic cell immediately dilutes this repressing factor, and therefore the virus reproduces. (One might wonder why, if it specifies the repressing factor, the virus doesn't shut itself off again? Clearly, it can: a large proportion of infected cells do become lysogenic. There is a race between the λ gene signals for reproduction and those specifying a shutdown.) The model of a phage-directed cytoplasmic repressor nicely explains the immunity of the lysogenic bacteria, because another phage infecting later would immediately encounter a repressor and be inactivated.

The phenomenon of lysogeny is a way for a temperate phage to avoid eating itself out of house and home. Lysogenic cells can perpetuate and carry the phages relatively innocuously.

Phage Integration

How does the prophage integrate into the bacterial genome? The free form of the phage chromosome that has entered the bacterial cell is circular at this stage. There is alignment of a specific site in λ, the λ **attachment site,** and a homologous site in the bacterial chromosome located between the genes *gal* and *bio* (Figure 7-23 on the next page). Then a crossover takes place between the circular phage and the circular bacterial chromosome, resulting in the integration of the phage in the *E. coli* chromosome. The integration of the phage to become a prophage increases the genetic distance between the bacterial genes that flank the attachment site, as can be seen in Figure 7-23 for *gal* and *bio.* This integration is promoted by proteins encoded by

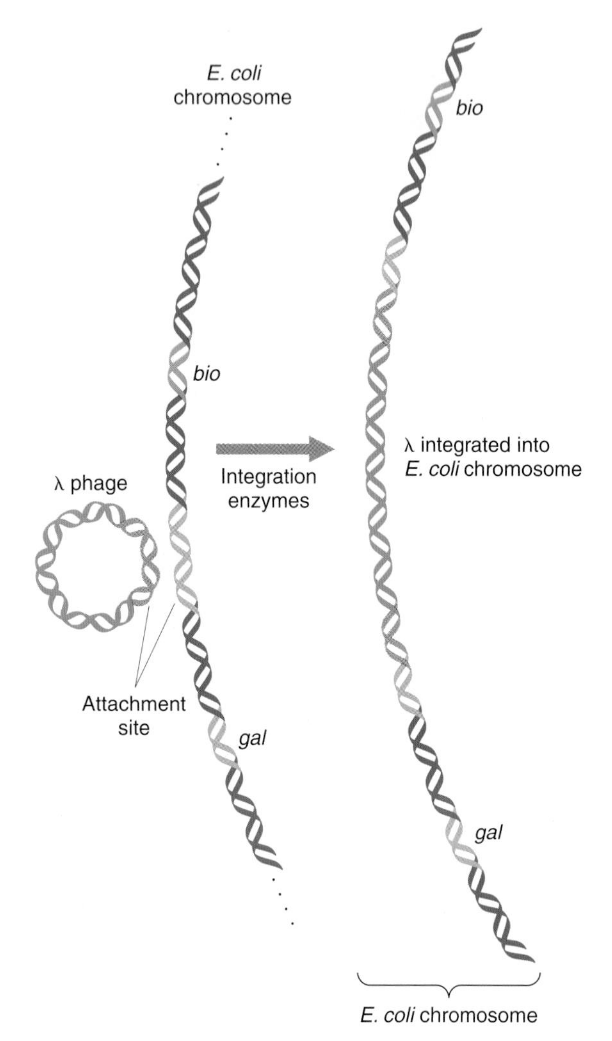

Figure 7-23 Model for the integration of phage λ into the *E. coli* chromosome. Reciprocal recombination takes place between a specific attachment site on the circular λ DNA and a specific region on the bacterial chromosome between the *gal* and *bio* genes.

phage genes that promote recombination between homologous segments of bacterial DNA.

Transduction

Phages infect bacteria so that they can reproduce. However, during their infection cycle some phages accidentally pick up bacterial DNA and carry it from one bacterial cell to another. This process is called **transduction** ("leading across"). Geneticists use transduction to study bacterial gene linkage and to deliberately manipulate bacterial genotypes for special experimental purposes. Phages are allowed to infect one bacterial genotype (the donor), and then, after lysis, progeny phage particles are collected. Most of these phage progeny carry chromosomes that are exact duplicates of the original infecting phage. However, a small fraction of them carry, instead, fragments of DNA from the donor bacterial strain. The progeny phage population is used to infect another bacterial genotype (the recipient). Many of the phages that carry donor DNA fragments will infect a recipient. Hence, in these cases, a temporary donor-recipient merozygote is created. Subsequently, the transduced bacterial genes may be incorporated into the bacterial chromosome by recombination (Figure 7-24). The presence of transduced donor alleles can be detected by plating on selective medium, in a procedure similar to that used for detecting recombinants after bacterial conjugation.

There are two kinds of transduction: generalized and specialized. Some phages can carry any part of the bacterial chromosome; these are **generalized transducing phages.** In contrast, **specialized transducing phages** carry only *specific* parts of the bacterial chromosome.

Generalized transduction. Phages P1 and P22 both show generalized transduction. These temperate phages have different fates in the cell: P22 inserts into the host chromosome as a prophage, whereas P1 remains free like a large plasmid. However, both act as generalized transducers because occasionally, in the course of lysis, bacterial DNA is accidentally "stuffed" into the phage head instead of phage DNA. Inside the recipient cell, the transducing fragments integrate by double crossover, utilizing the host recombination system, as we saw in Figure 7-24.

Specialized transduction. Phage λ is a good example of a specialized transducer. We learned earlier that λ always inserts at an attachment site between the *gal* and *bio* genes in the *E. coli* host chromosome (see Figure 7-23). The prophage normally excises by an outlooping mechanism that is a precise reversal of its mechanism of integration. It is able to transduce only when an occasional faulty outlooping produces a λ chromosome that includes the *gal* or *bio* genes in the phage DNA circle. (This process is like the one that produces F′ plasmids; see Figure 7-13c.) Let's trace the details of λ transduction.

In Figure 7-25a, we see a schematic representation of the integration of the λ chromosome into the bacterial chromosome to produce a lysogen, or lysogenic bacterium. The lytic cycle can be induced experimentally by ultraviolet light to produce a lysate. The normal outlooping of the prophage restores the original phage chromosome (Figure 7-25b, i). These phage chromosomes can integrate normally (as in Figure 7-25a) upon subsequent infection of a bacterial strain that is not lysogenic for λ.

Faulty outlooping can occur if the phage recombines at a site other than the attachment site. This results in a phage chromosome that contains a segment of bacterial DNA carrying the *gal* gene and that lacks a segment of λ chromosome (Figure 7-25b, ii). The phage particles carrying these chromosomes are *defective* in that some phage genes have been left behind in the host; consequently, they are called **λdgal** (that is, λ-defective gal). The λdgal particle has a λ protein coat and tail fibers and can infect bacteria, but because it lacks some phage functions, it cannot recombine with the bacterial genome. Therefore, efficient integration cannot take place if λdgal alone infects in *E. coli.* However, coinfection with a wild-type λ phage results in efficient integration of λdgal (Figure 7-25c, i); the wild-type phage supplies the missing functions and acts as a **helper phage.**

The experimental sequence for specialized transduction is that a lysate of λ is produced (see Figure 7-25b) and used to infect a gal^- recipient culture that is nonlysogenic for λ. These infected cells are plated on a minimal medium

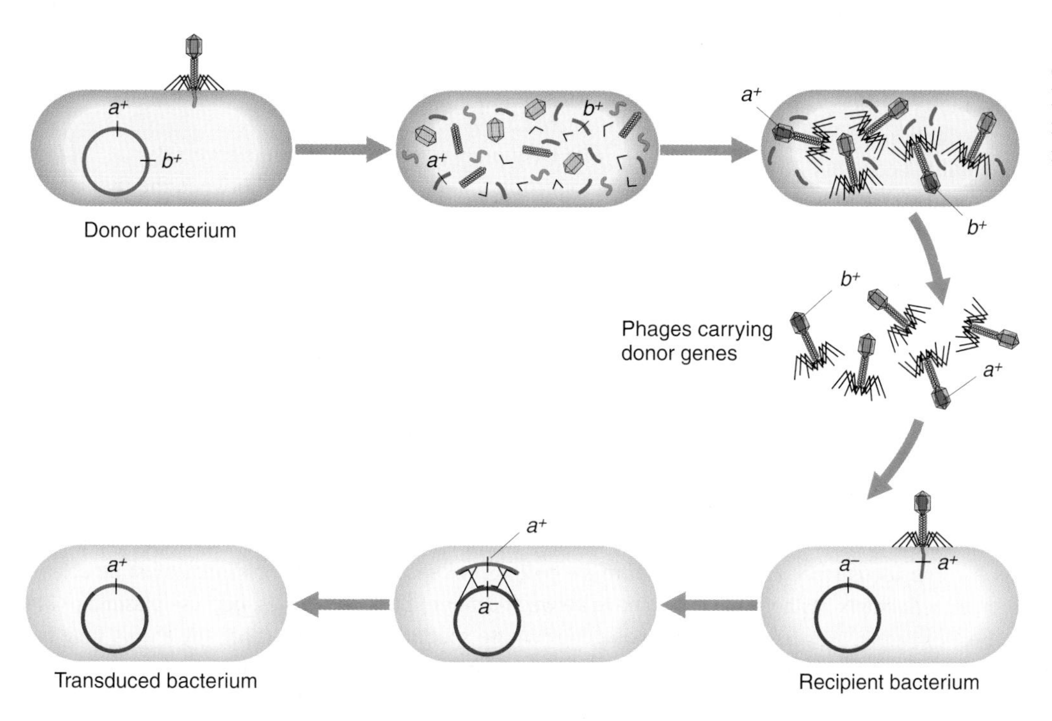

Figure 7-24 The mechanism of generalized transduction. In reality, only a very small minority of phage progeny (1 in 10,000) carry donor genes.

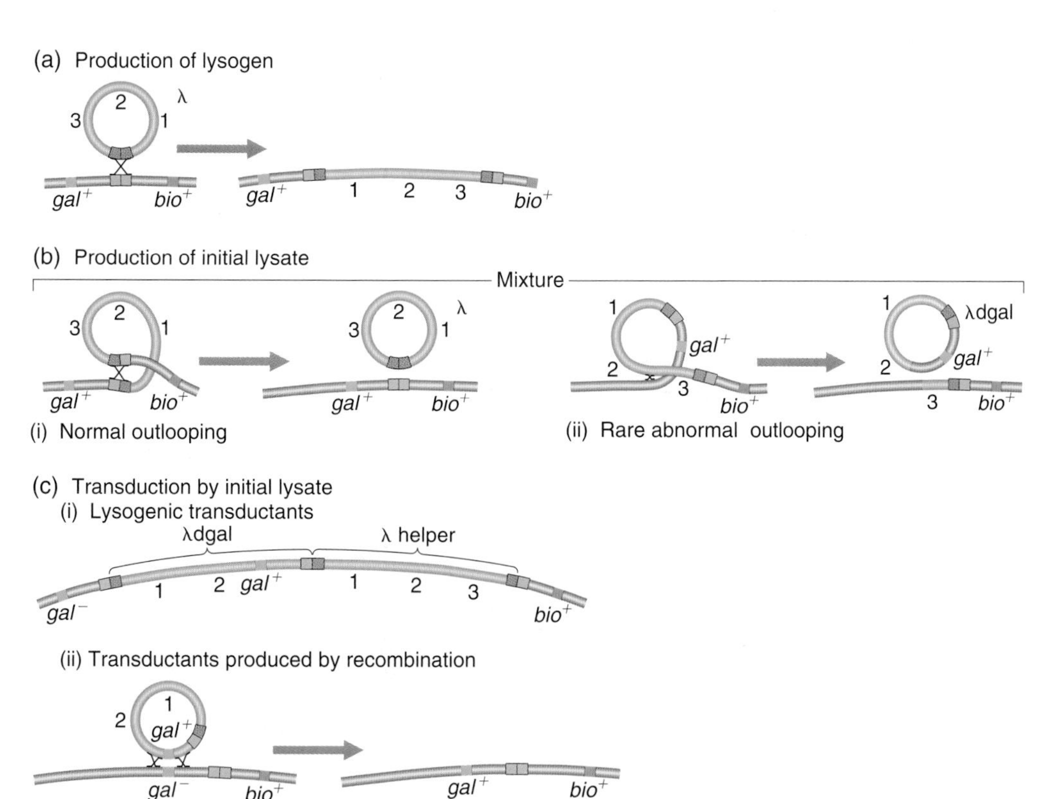

Figure 7-25 Specialized transduction mechanism in phage λ. (a) The production of a lysogenic bacterium takes place by crossing-over at a specialized attachment site. (b) The lysogenic bacterial culture can produce (i) normal λ or, rarely, (ii) an abnormal particle, λdgal, which is the transducing particle. (c) Transduction by the mixed lysate of λ and λdgal can produce gal^+ transductants by the coincorporation of λdgal and λ (acting as a helper) or, more rarely, by crossovers flanking the *gal* gene. The blue double boxes are the bacterial attachment site, the purple double boxes are the λ attachment site, and the pairs of blue and purple boxes are hybrid integration sites, derived partly from *E. coli* and partly from λ.

to select for rare gal^+ transductants. Most of the transductants (see Figure 7-25c, i) are double lysogens produced by integration of λdgal and λ. (A small percentage of transductants result from a different mechanism; recombination between the *gal* regions of the λdgal phage particle and the recipient chromosome, as shown in Figure 7-25c, ii.)

Specialized transduction involves only small regions of the bacterial genome that flank the prophage. Phage λ can transduce the *gal* gene by faulty outlooping to one side of the normal integration site, as shown above, but it can also transduce the *bio* gene, by faulty outlooping to the other side. Hence specialized transduction by λ is particularly useful when it is experimentally necessary to move these specific genes from one bacterial strain to another.

Determining Linkage from Transduction

Generalized transduction can be used to derive linkage information about bacterial genes in cases in which genes are close enough that the phage can pick them up and transduce them as a single piece of DNA, an outcome known as **cotransduction.** The closer two genes are, the more frequently are they cotransduced. (Note that this is the same way that transformation is used to analyze linkage.) The approach is to select transductants for one locus (the *selected locus*) and then analyze those for transductants at the other locus (the *unselected locus*). For example, suppose we wanted to measure the linkage between *met* and *arg* in *E. coli.* A generalized transducing phage such as phage P1 could be used. We would allow phage P1 to infect a donor met^+ arg^+ strain, then collect P1 progeny and use them to infect a recipient strain of genotype met^- arg^-. Then we would select for one donor gene—say, met^+—by plating cells on medium that lacks methionine and isolating met^+ colonies. These colonies would then be tested for the *unselected* marker (arg^+) to see what proportion of them also received arg^+; in this way, we could determine the percentage of met^+ arg^+ cotransductants. Distances between the two genes are expressed as percentages of cotransduction (Figure 7-26). For example, if 20 percent of met^+ colonies are also found to be arg^+ (that is, 20 percent are met^+ arg^+ and 80 percent are met^+ arg^-), then the cotransduction frequency is 20 percent.

Note that, in this type of mapping, there is an inverse relation quite different from that obtained by using map units based on recombinant frequencies: the *greater* the cotransduction frequency, the closer two genetic markers are and the *shorter* is the map distance (see Figure 7-26).

MESSAGE

The frequency of cotransduction is inversely proportional to the map distance between the two genes concerned.

Let us return to *Lactococcus lactis* plasmid pK214, introduced at the beginning of the chapter. Now in retrospect we can begin to understand the ways in which this plasmid obtained its diverse set of genes. Plasmids can pick up genes from bacterial genomes, as we have seen in the case of the F′ plasmids. Furthermore, clearly plasmids can be transferred easily from one strain to another during conjugation. What is more surprising from the pK214 DNA sequences is that plasmids must be able to move between different bacterial species, because plasmid pK214 has genes whose specific DNA sequence clearly shows they are from species other than *L. lactis.* Indeed multiple drug resistances obtained in this way are a large problem in medicine. When plasmids that have picked up a variety of antibiotic-resistance genes from a series of different hosts enter their next hosts, they bestow upon these bacteria resistance to a large proportion of the antibiotic agents that can be directed at them. As bacteria develop multiple resistances, there is serious concern that we will soon run out of antibiotics to combat bacterial infections.

The recombination processes of bacteria are summarized diagrammatically in Figure 7-27.

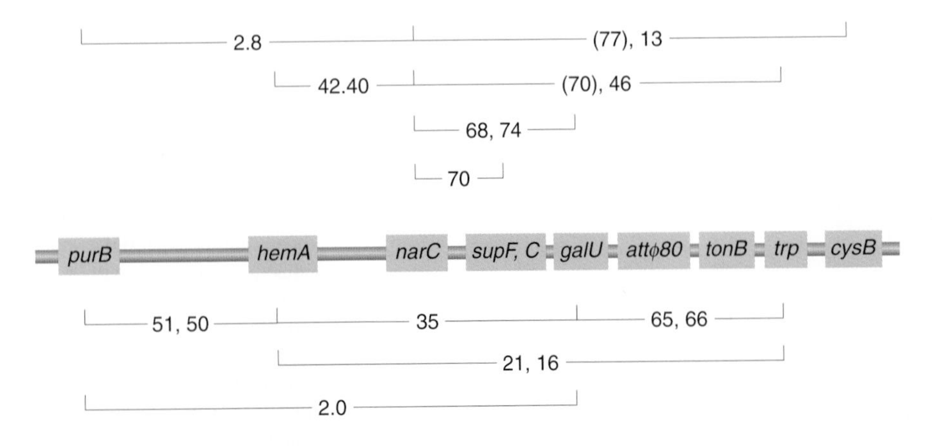

Figure 7-26 Genetic map of the *purB–cysB* region of *E. coli* determined by P1 cotransduction. The numbers given are the averages in percent for cotransduction frequencies obtained in several experiments. The values in parentheses are considered unreliable. *(Redrawn from J. R. Guest,* Molecular and General Genetics *105, 1969, 285.)*

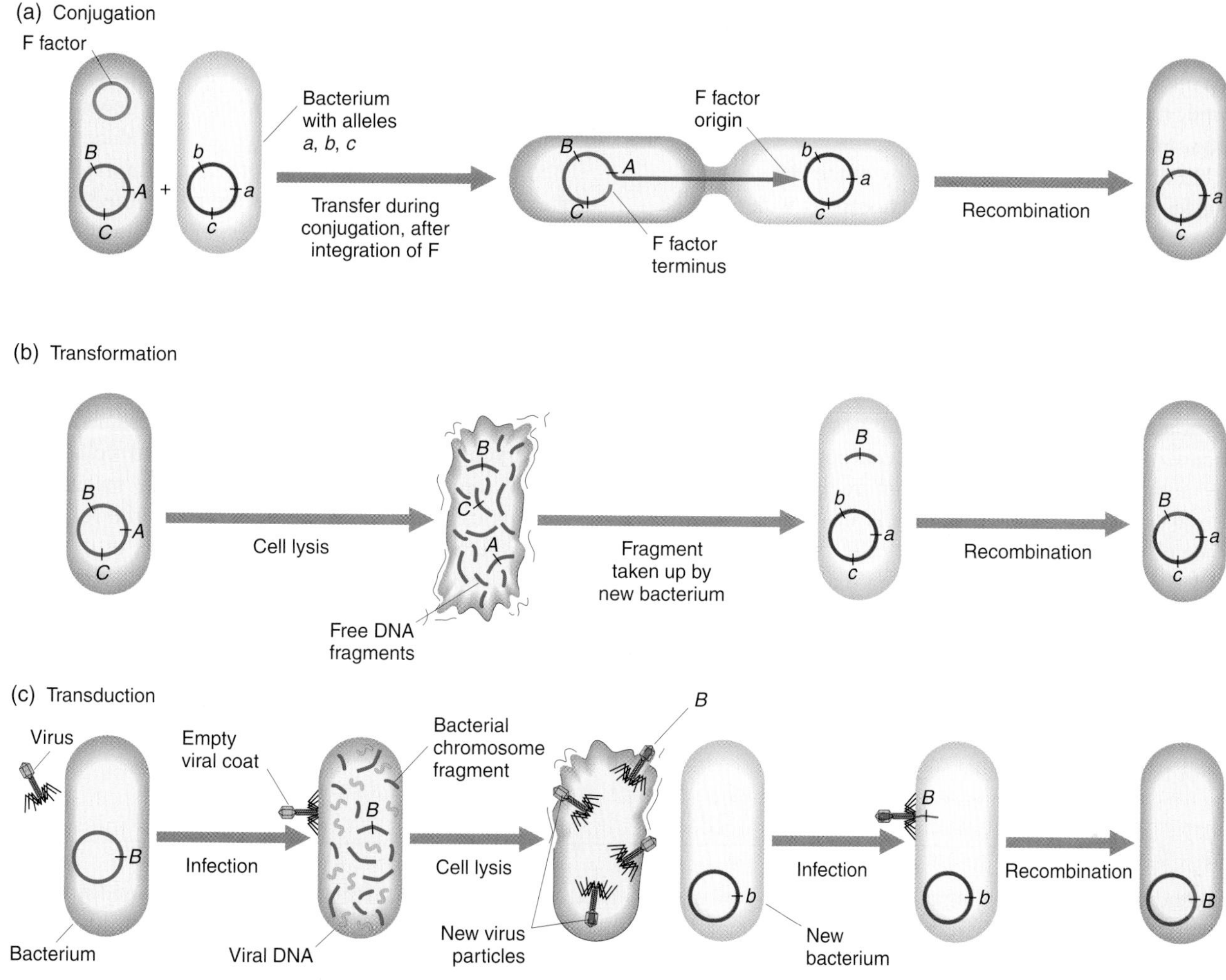

Figure 7-27 Recombination processes in bacteria. Bacterial recombination requires that a bacterial cell receive an allele obtained from another cell. (a) In conjugation, a fertility factor (F) integrates into the chromosome of a bacterial cell. During cell-to-cell contact, the integrated factor can transfer part of or all that chromosome to another cell. The transferred segment recombines with the homologous segment in the recipient cell's chromosome. In the example shown here, allele *B* thereby replaces allele *b*. (b) In transformation, a recipient bacterial cell takes up a DNA segment released into the environment by another cell that acts as donor. The fragment can be inserted by homologous recombination. In some cases this causes a replacement of a recipient allele (*b*) by an allele from the donor (*B*). (c) In transduction, after a phage has infected a bacterial cell, some of the newly forming phage particles pick up a bacterial DNA segment instead of viral DNA. When one of these phage particles infects another cell, it injects its bacterial DNA, which recombines with a homologous segment in the second cell, sometimes replacing the recipient allele (*b*) with that from the donor (*B*).

SUMMARY

1. The gut bacterium *Escherichia coli* has been used as a model organism for studying bacterial recombination. A key to understanding recombination in *E. coli* is a large circular plasmid called F, the sex factor. F is present in some cells, which then act as donors of genes in a unilateral exchange leading to genetic recombination. If F is absent (F^- type), the cell acts as a recipient.

2. The F plasmid can exist in two states. It can replicate as a free extragenomic element in the cytoplasm of the bacterial cell; such cells are called F^+. F can also integrate into the bacterial genomic circle; such a cell is called Hfr (high-frequency recombination). Integration of F is achieved by crossing-over between the plasmid and genomic circles at one of several small regions of homology.

3. If an F^+ cell fuses with an F^- cell, a copy of the F plasmid can move rapidly into the F^- during conjugation, converting it to an F^+. If there are genomic genes on the plasmid, these are also transferred rapidly, in a process called *sexduction.*
4. If an Hfr cell fuses with an F^-, a copy of the Hfr genome is transferred into the F^- cell linearly, starting with genes next to the inserted F plasmid and ending with F itself. The progress of this conjugative transfer can be interrupted at different times; and by measuring which specific Hfr alleles are in the F^-, time of entry can be used to map the entire bacterial genome. After interrupted conjugation, the transferred fragment integrates into the F^- genome by double crossover. The positions and frequencies of the double crossovers can be used to map genes on the fragment by recombinant frequency.
5. Bacterial viruses (phages) can pick up parts of the genome of an infected cell and carry the fragment to an unrelated cell, where the fragment can integrate into the genome. This process is called *transduction.*
6. In generalized transduction, some phages incorporate random fragments of the bacterial genome and introduce them into the cells they infect. The frequency of cotransduction of two alleles in this manner can be used as an index of their physical distance apart.
7. Some phages integrate into the bacterial genome by crossing-over. This process is reversible through loop formation and crossing-over. If there is imprecise outlooping, then parts of the adjacent bacterial genome can become incorporated into the phage and transduce an unrelated cell. This is called *specialized transduction,* because only specific adjacent genes are transmitted.
8. Bacteria can exchange parts of their genomes without the aid of plasmids or phages, simply by incorporating genomic DNA fragments present in the extracellular medium. This process, called *transformation,* can also be used to obtain mapping information by measuring frequency of cotransformation.
9. Bacterial recombination is conceptually the same as that in eukaryotes (which recombine in diploid cells), but there are some significant differences in the mechanism and analytical approach. Most cases of bacterial recombination, whether by conjugation, transduction, or transformation, arise from crossovers between the complete genome of a recipient cell and a part of a genome from a donor cell. These partial diploids can be stably maintained in some cases; they can be used to study dominance and to determine whether two mutations are alleles.

CONCEPT MAP

Draw a concept map interrelating as many of the following terms as possible. Note that the terms are listed in no particular order.

bacteria / conjugation / recombination / F plasmid / Hfr / F′ / donor / recipient / interrupted mating / chromosome map / merozygote / gene

SOLVED PROBLEM

1. We saw in Chapter 6 how recombination takes place by breakage and reunion of chromosomes. Suppose a bacterial cell were unable to carry out homologous recombination (it has the mutant genotype rec^-). How would this cell behave as a recipient in generalized and specialized transduction? First compare each type of transduction and then determine the effect of the rec^- mutation on the inheritance of genes by each process.

Solution

Generalized transduction involves the incorporation of chromosomal fragments into phage heads, which then infect recipient strains. Fragments of the chromosome are randomly incorporated into phage heads, so any marker on the bacterial host chromosome can be transduced to another strain by generalized transduction. In contrast, specialized transduction involves the integration of the phage at a specific point on the chromosome and the rare incorporation of chromosomal markers near the integration site into the phage genome. Therefore, only those markers that are near the specific integration site of the phage on the host chromosome can be transduced.

Inheritance of markers occurs by different routes in generalized and specialized transduction. A generalized transducing phage injects a fragment of the donor chromosome into the recipient. This fragment must be incorporated into the recipient's chromosome by recombination, using the recipient recombination system. Therefore, a rec^- recipient will not be able to incorporate fragments of DNA and cannot inherit markers by generalized transduction. On the other hand, the major route for the inheritance of markers by specialized transduction includes integration of the specialized transducing particle into the host chromosome at the specific phage integration site. This integration, which sometimes requires an additional wild-type (helper) phage, is mediated by a phage enzyme system that is independent of the bacterial recombination enzymes. Therefore, a rec^- recipient can still inherit genetic markers by specialized transduction.

SOLVED PROBLEM

2. In *E. coli,* four Hfr strains donate the following genetic markers shown in the order donated:

Strain 1:	*Q*	*W*	*D*	*M*	*T*
Strain 2:	*A*	*X*	*P*	*T*	*M*
Strain 3:	*B*	*N*	*C*	*A*	*X*
Strain 4:	*B*	*Q*	*W*	*D*	*M*

All these Hfr strains are derived from the same F^+ strain. What is the order of these markers on the circular chromosome of the original F^+?

Solution

Here the relevant principle is that each Hfr strain donates genetic markers from a fixed point on the circular chromosome, which causes the markers donated earliest to be donated with the highest frequency. Because not all markers are donated by each Hfr, only the early markers must have been donated for each Hfr. Each strain allows us to draw the following circles:

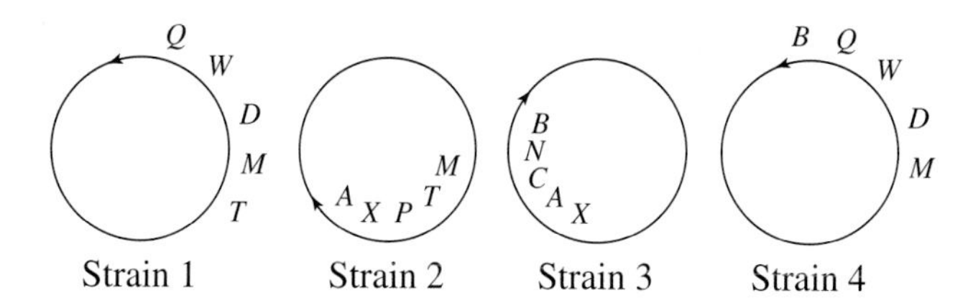

From this information, we can consolidate each circle into one circular linkage map of the order *Q, W, D, M, T, P, X, A, C, N, B, Q.*

SOLVED PROBLEM

3. Compare the mechanism of transfer and inheritance of donor lac^+ genes in crosses using Hfr, F^+, and F'-lac^+ strains. How would an F^- cell that cannot undergo normal homologous recombination (rec^-) behave in crosses with each of these three strains? Would the cell be able to inherit the lac^+ genes?

Solution

Each of these three strains donates genes by conjugation. In regard to the Hfr and F^+ strains, the lac^+ genes on the host chromosome are donated. In regard to the Hfr strain, the F factor is integrated into the chromosome in every cell, so a chromosomal marker can be donated efficiently, particularly if the marker is near the integration site of F and is donated early. Therefore, for *lac,* we expect high frequencies of transfer. The F^+ cell population contains only a small percentage of Hfr cells in which F is integrated into the chromosome. These rare cells are the ones responsible for the gene transfer displayed by cultures of F^+ cells. Hence, for the *lac* gene, we expect a low frequency of transfer. In regard to Hfr- and F^+-mediated gene transfer, inheritance requires the incorporation of a transferred fragment into the F^- chromosome by recombination (recall that two crossovers are needed). Therefore, any F^- strain that is rec^- (recombination-defective) cannot integrate the donor fragment by recombination and hence cannot inherit donor chromosomal markers even though they are transferred from the donor cells. Because these donor fragments do not possess the ability to replicate within the F^- cell, they are rapidly diluted out in cell division.

F' cells transfer chromosomal genes on the F' plasmid, a process that does not require *chromosome* transfer. We have seen that, in F'-*lac* cells, lac^+ genes are transferred on the plasmid at a high efficiency. In the F^- cell, no recombination is required for inheritance of the lac^+ allele in the recipient population, because the F'-lac^+ plasmid can replicate and be maintained in the dividing F^- cell population. Therefore, the lac^+ genes are inherited even in a rec^- strain.

PROBLEMS

Basic Problems

1. Describe the state of the F factor in strains of type Hfr, F^+ and F^-.

2. How does a culture of F^+ cells transfer markers from the host chromosome to a recipient?

3. Compare and contrast the mode of intercellular gene transfer and the maintenance of the transferred gene in the recipient cell in the following pairs of cases:

a. Hfr conjugation and generalized transduction

b. F' derivatives such as F'-*lac* and specialized transduction

4. Why can generalized transduction transfer any gene, but specialized transduction only a small, restricted set?

5. A microbial geneticist isolates a new mutation in *E. coli* and wishes to map its chromosomal location. She uses interrupted-mating experiments with Hfr strains and generalized transduction experiments with phage P1. Explain why each technique, by itself, is insufficient for accurate mapping.

6. In *E. coli,* four Hfr strains donate the following markers, shown in the order donated:

Strain 1:	*M*	*Z*	*X*	*W*	*C*
Strain 2:	*L*	*A*	*N*	*C*	*W*
Strain 3:	*A*	*L*	*B*	*R*	*U*
Strain 4:	*Z*	*M*	*U*	*R*	*B*

All these Hfr strains are derived from the same F^+ strain. What is the order of these markers on the circular chromosome of the original F^+?

7. Four *E. coli* strains of genotype a^+ b^- are labeled 1, 2, 3, and 4. Four strains of genotype a^- b^+ are labeled 5, 6, 7, and 8. The two genotypes are mixed in all possible combinations and (after incubation) are plated to determine the frequency of a^+ b^+ recombinants. The following results are obtained, where M = many recombinants, L = low numbers of recombinants, and 0 = no recombinants:

	1	2	3	4
5	0	M	M	0
6	0	M	M	0
7	L	0	0	M
8	0	L	L	0

On the basis of these results, assign a sex type (either Hfr, F^+, or F^-) to each strain.

Pattern Recognition Problem

8. Deduce the genotypes of the following four *E. coli* strains:

Minimal
1 2
3 4

Minimal plus arginine
1 2
3 4

Minimal plus methionine
1 2
3 4

Minimal plus arginine and methionine
1 2
3 4

9. An Hfr strain of genotype a^+ b^+ c^+ d^- str^s is mated with an F^- strain of genotype a^- b^- c^- d^+ str^r. At various times, the culture is shaken vigorously to separate mating pairs. The cells are then plated on agar of the following three types, where nutrient A allows the growth of a^- cells; nutrient B, of b^- cells; nutrient C, of c^- cells; and nutrient D, of d^- cells (a plus indicates the presence of streptomycin or the indicated nutrient, and a minus indicates its absence):

Medium	Str	A	B	C	D
1	+	+	+	−	+
2	+	−	+	+	+
3	+	+	−	+	+

a. What donor genes are being selected on each type of agar?

b. The following table shows the number of colonies on each type of agar for samples taken at various times after the strains are mixed. Use this information to determine the order of the genes *a*, *b*, and *c*.

Sampling time (min)	NUMBER OF COLONIES		
	Medium 1	Medium 2	Medium 3
0	0	0	0
5	0	0	0
7.5	100	0	0
10	200	0	0
12.5	300	0	75
15	400	0	150
17.5	400	50	225
20	400	100	250
25	400	100	250

c. From each of the 25-minute plates, 100 colonies are picked and transferred to a dish containing agar with all the nutrients except D. The numbers of colonies that grow on this medium are 89 for the sample from medium 1, 51 for the sample from medium 2, and 8 for the sample from medium 3. Using these data, fit gene *d* into the sequence of *a*, *b*, and *c*.

d. At what sampling time would you expect colonies to first appear on agar containing C and streptomycin but no A or B?

(Problem 9 is from D. Freifelder, *Molecular Biology and Biochemistry.* © 1978 by W. H. Freeman and Company, New York.)

10. In an interrupted-conjugation experiment in *E. coli,* it is established that the *pro* gene enters after the *thi* gene. A pro^+ thi^+ Hfr is crossed with a pro^- *thi* F^- strain, and exconjugants are plated on medium containing thiamine but no proline. A total of 360 colonies are observed, and they are isolated and cultured on fully supplemented medium. These cultures are then tested for their ability to grow on medium containing no proline or thiamine (minimal medium), and it is found that 320 of the cultures can grow but the remainder cannot.

a. Deduce the genotypes of the two types of cultures.

b. Draw the crossover events required to produce these genotypes.

c. Calculate the distance between the *pro* and *thi* genes in recombination units.

Unpacking the Problem

a. What type of organism is *E. coli?*

b. What does a culture of *E. coli* look like? (Sketch one.)

c. On what sort of substrates does *E. coli* generally grow in its natural habitat?

d. What are the minimal requirements for *E. coli* cells to divide?

e. Define the terms *prototroph* and *auxotroph.*

f. Which cultures in this experiment are prototrophic and which are auxotrophic?

g. Given some strains of unknown genotype regarding thiamine and proline, how would you test their genotypes? Give precise experimental details, including equipment.

h. What kinds of chemicals are proline and thiamine? Does this matter in this experiment?

i. Draw a schematic diagram showing the full set of manipulations performed in the experiment.

j. Why do you think the experiment was done?

k. How was it established that *pro* enters after *thi?* Give precise experimental steps.

l. In what way does the interrupted-mating experiment differ from the experiment described in this problem?

m. What is an exconjugant? How do you think that exconjugants were obtained? (This might involve genes not described in this problem.)

n. When the *pro* gene is said to enter after *thi,* does it mean the *pro* allele, the pro^+ allele, either, or both?

o. What is "fully supplemented medium" in the context of this question?

p. Some exconjugants did not grow on minimal medium. On what medium would they grow?

q. State the types of crossovers that are involved in Hfr × F^- recombination. How do these crossovers differ from crossovers in eukaryotes?

r. What is a recombination unit in the context of the present analysis? How does it differ from the map units used in eukaryote genetics?

s. Now try to solve the problem.

i **11.** A cross is made between two *E. coli* strains:

$$\text{Hfr } arg^+\ bio^+\ leu^+ \times F^-\ arg^-\ bio^-\ leu^-$$

Interrupted-mating studies show that arg^+ enters the recipient last, so arg^+ recombinants are selected on a medium containing *bio* and *leu* only. These recombinants are tested for the presence of bio^+ and leu^+. The following numbers of individuals are found for each genotype:

Genotype	Number
$arg^+\ bio^+\ leu^+$	320
$arg^+\ bio^+\ leu^-$	8
$arg^+\ bio^-\ leu^+$	0
$arg^+\ bio^-\ leu^-$	48

a. What is the gene order?

b. What are the map distances in recombination units?

12. A particular Hfr strain normally transmits the pro^+ marker as the last one during conjugation. In a cross of this strain with an F^- strain, some pro^+ recombinants are recovered early in the mating process. When these pro^+ cells are mixed with F^- cells, the majority of the F^- cells are converted into pro^+ cells that also carry the F factor. Explain these results.

13. F′ strains in *E. coli* are derived from Hfr strains. In some cases, these F′ strains show a high rate of integration back into the bacterial chromosome of a second strain. Furthermore, the site of integration is often the same site that the sex factor occupied in the original Hfr strain (before production of the F′ strains). Explain these results.

14. You have two *E. coli* strains, $F^-\ str^r\ ala^-$ and Hfr str^s ala^+, in which the F factor is inserted close to ala^+. Devise a screening test to detect strains carrying F'-ala^+.

15. Five Hfr strains A through E are derived from a single F^+ strain of *E. coli.* The following chart shows the entry times of the first five markers into an F^- strain when each is used in an interrupted-conjugation experiment:

A	B	C	D	E
mal^+ (1)	ade^+ (13)	pro^+ (3)	pro^+ (10)	his^+ (7)
str^s (11)	his^+ (28)	met^+ (29)	gal^+ (16)	gal^+ (17)
ser^+ (16)	gal^+ (38)	xyl^+ (32)	his^+ (26)	pro^+ (23)
ade^+ (36)	pro^+ (44)	mal^+ (37)	ade^+ (41)	met^+ (49)
his^+ (51)	met^+ (70)	str^s (47)	ser^+ (61)	xyl^+ (52)

a. Draw a map of the F^+ strain, indicating the positions of all genes and their distances apart in minutes.

b. Show both the insertion point and orientation of the F plasmid in each Hfr strain.

c. In using each of these Hfr strains, state which gene you would select to obtain the highest proportion of Hfr exconjugants.

16. *Streptococcus pneumoniae* cells of genotype $str^s\ mtl^-$ are transformed by donor DNA of genotype $str^r\ mtl^+$ and (in a separate experiment) by a mixture of two DNAs with genotypes $str^r\ mtl^-$ and $str^s\ mtl^+$. The following table shows the results:

	PERCENTAGE OF CELLS TRANSFORMED INTO		
Transforming DNA	$str^r\ mt^-$	$str^s\ mtl^+$	$str^r\ mtl^+$
$str^r\ mtl^+$	4.3	0.40	0.17
$str^r\ mtl^- + str^s\ mtl^+$	2.8	0.85	0.0066

a. What does the first line of the table tell you? Why?

b. What does the second line of the table tell you? Why?

17. A generalized transduction experiment uses a $metE^+\ pyrD^+$ strain as donor and $metE^-\ pyrD^-$ as recipient. $MetE^+$ transductants are selected and then tested for the $pyrD^+$ allele. The following numbers were obtained:

$metE^+\ pyrD^-$	857
$metE^+\ pyrD^+$	1

Do these results suggest that these loci are closely linked? Explain. What other explanations are there for the lone "double"?

18. In a generalized transduction system using P1 phage, the donor is $pur^+\ nad^+\ pdx^-$ and the recipient is $pur^-\ nad^-\ pdx^+$. The donor allele pur^+ is initially selected after transduction, and 50 pur^+ transductants are then scored for the other alleles present. The results follow:

Genotype	Number of colonies
$nad^+\ pdx^+$	3
$nad^+\ pdx^-$	10
$nad^-\ pdx^+$	24
$nad^-\ pdx^-$	13

a. What is the cotransduction frequency for *pur* and *nad?*

b. What is the cotransduction frequency for *pur* and *pdx?*

c. Which of the unselected loci is closest to *pur?*

d. Are *nad* and *pdx* on the same side or on opposite sides of *pur?* Explain. (Draw the exchanges needed to produce the various transformant classes under either order to see which requires the minimum number to produce the results obtained.)

19. You have infected *E. coli* cells with two strains of T4 virus. One strain is minute (*m*), rapid-lysis (*r*), and turbid (*tu*); the other is wild-type for all three markers. The lytic products of this infection are plated and classified. Of 10,342 plaques, the following numbers are classified as each genotype:

m r tu	3467	*m* + +	520
+ + +	3729	+ *r tu*	474
m r +	853	+ *r* +	172
m + *tu*	162	+ + *tu*	965

a. Determine the linkage distances between *m* and *r*, between *r* and *tu*, and between *m* and *tu*.

b. What linkage order would you suggest for the three genes?

c. What is the coefficient of coincidence in this cross, and what does it signify?

(Problem 19 is reprinted with the permission of Macmillan Publishing Co., Inc., from Monroe W. Strickberger, *Genetics*. © 1968 by Monroe W. Strickberger.)

Challenging Problems

20. Linkage maps in an Hfr bacterial strain are calculated in units of minutes (the number of minutes between genes indicates the length of time that it takes for the second gene to follow the first during conjugation). In making such maps, microbial geneticists assume that the bacterial chromosome is transferred from Hfr to F^- at a constant rate. Thus, two genes separated by 10 minutes near the origin end are assumed to be the same physical distance apart as two genes that are separated by 10 minutes near the F attachment end. Suggest a critical experiment to test the validity of this assumption.

21. A transformation experiment is performed with a donor strain that is resistant to four drugs: A, B, C, and D. The recipient is sensitive to all four drugs. The recipient cell population to which DNA has been added is divided up and plated on media containing various combinations of the drugs. The following table shows the results:

Drug(s) added	Number of colonies	Drug(s) added	Number of colonies
None	10,000	BC	51
A	1,156	BD	49
B	1,148	CD	786
C	1,161	ABC	30

(continued on next page)

Drug(s) added	Number of colonies	Drug(s) added	Number of colonies
D	1,139	ABD	42
AB	46	ACD	630
AC	640	BCD	36
AD	942	ABCD	30

a. One of the genes obviously is quite distant from the other three, which appear to be tightly (closely) linked. Which is the distant gene?

b. What is the probable order of the three tightly linked genes?

(Problem 18 is from Franklin Stahl, *The Mechanics of Inheritance*, 2d ed. © 1969, Prentice-Hall, Englewood Cliffs, New Jersey. Reprinted by permission.)

22. Although most λ-mediated gal^+ transductants are inducible lysogens, a small percentage of these transductants in fact are not lysogens (that is, they contain no integrated λ). Control experiments show that these transductants are not produced by mutation. What is the likely origin of these types?

23. An ade^+ arg^+ cys^+ his^+ leu^+ pro^+ bacterial strain is known to be lysogenic for a newly discovered phage, but the site of the prophage is not known. The bacterial map is

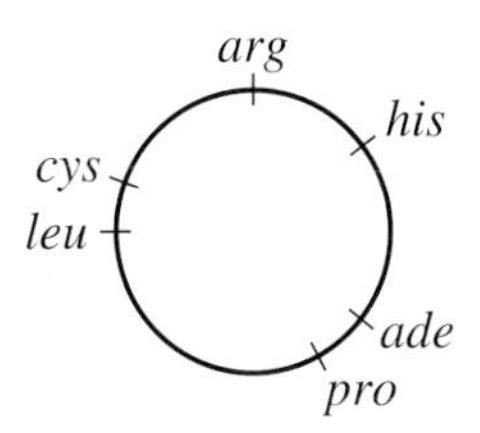

The lysogenic strain is used as a source of the phage, and the phages are added to a bacterial strain of genotype ade^- arg^- cys^- his^- leu^- pro^-. After a short incubation, samples of these bacteria are plated on six different media, with the supplementations indicated in the following table. The table also shows whether colonies were observed on the various media. (Note: A plus sign indicates the presence of a nutrient supplement; a minus sign indicates supplements are not present; N indicates that no colonies are present; and C indicates that colonies are present.)

Medium	NUTRIENT SUPPLEMENTATION *ade*	*arg*	*cys*	*his*	*leu*	*pro*	Presence of colonies
1	−	+	+	+	+	+	N
2	+	−	+	+	+	+	N
3	+	+	−	+	+	+	C
4	+	+	+	−	+	+	N
5	+	+	+	+	−	+	C
6	+	+	+	+	+	−	N

a. What genetic process is at work here?

b. What is the approximate locus of the prophage?

24. You have two strains of λ that can lysogenize *E. coli;* the following figure shows their linkage maps:

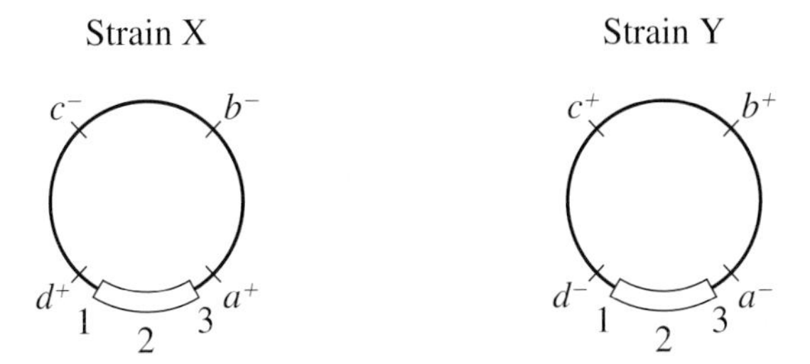

The segment shown at the bottom of the chromosome, designated 1–2–3, is the region responsible for pairing and crossing-over with the *E. coli* chromosome. (In answering the following questions, keep the markers on all your drawings.)

a. Diagram the way in which λ strain X is inserted into the *E. coli* chromosome (so that the *E. coli* bacterium is lysogenized).

b. It is possible to superinfect the bacteria that are lysogenic for strain X by using strain Y. A certain percentage of these superinfected bacteria become "doubly" lysogenic (that is, lysogenic for both strains). Diagram how this will occur. (Don't worry about how double lysogens are detected.)

c. Diagram how the two λ prophages pair.

d. It is possible to recover crossover products between the two prophages. Diagram a crossover event and the consequences.

25. A generalized transducing phage is used to transduce an a^- b^- c^- d^- e^- recipient strain of *E. coli* with an a^+ b^+ c^+ d^+ e^+ donor. The recipient culture is plated on various media with the results shown in the following table. (Note that a^- determines a requirement for A as a nutrient, and so forth. Plus and minus signs indicate presence or absence of colonies, respectively.) What can you conclude about the linkage and order of the genes?

Compounds added to minimal medium	Colonies
C D E	−
B D E	−
B C E	+
B C D	+
A D E	−
A C E	−
A C D	−
A B E	−
A B D	+
A B C	−

26. In a generalized transduction experiment, phages are collected from an *E. coli* donor strain of genotype cys^+ leu^+ thr^+ and used to transduce a recipient of genotype cys^- leu^- thr^-. Initially, the treated recipient population is plated on a minimal medium supplemented with leucine and threonine. Many colonies are obtained.

a. What are the possible genotypes of these colonies?

b. Strains from these colonies are then inoculated onto three different media: (1) minimal plus threonine only, (2) minimal plus leucine only, and (3) minimal. What genotypes could, in theory, grow on these three media?

c. It is observed that 56 percent of the original colonies grow on medium 1, that 5 percent grow on medium 2, and that no colonies grow on medium 3. What are the actual genotypes of the colonies on these three media?

d. Draw a map showing the order of the three genes and which of the two outer genes is closer to the middle gene.

Exploring Genomes: A Web-Based Bioinformatics Tutorial

OMIM and Huntington Disease

The Online Mendelian Inheritance in Man (OMIM) program collects data on the genetics of human genes. In the Genomics tutorial at www.whfreeman.com/mga, you will learn how to search OMIM for data on gene locus and inheritance pattern for conditions such as Huntington disease.

RECOMBINANT DNA AND GENETIC ENGINEERING

8

Key Concepts

1. Recombinant DNA technology exploits the fundamental principles of molecular biology: complementarity of antiparallel nucleic acid sequences and sequence-specific DNA–protein interactions.
2. Recombinant DNA technology also requires an understanding of the enzymology of DNA replication to create novel DNA molecules and an understanding of the biology of a recipient organism so that these engineered DNA molecules can be introduced and replicated within this recipient's cells.
3. Recombinant DNA is made by splicing a foreign DNA fragment into a small replicating DNA molecule (a vector), which can then replicate and amplify the foreign DNA fragment along with itself, producing a molecular clone of the inserted DNA.
4. Restriction enzymes that cut DNA at specific target sites to produce defined fragments with ends suitable for insertion into a vector that has been cut open by the same enzyme are crucial tools in genetic engineering.
5. PCR, the polymerase chain reaction, can be used for amplification of specific DNA segments in a test tube.
6. Cloning is the replication and amplification of a specific recombinant DNA molecule in a population of identical microbial cells.
7. Labeled single-stranded DNA or RNA can be used as a probe to fish out individual sequences from among complex-sequence populations by nucleic acid complementarity.
8. Restriction enzyme target sites can be mapped, providing useful markers for DNA manipulation.
9. The structure of any portion of the genome can be reconstructed by producing maps of overlapping clones; this reconstruction process is called a chromosome walk.

(continued on next page)

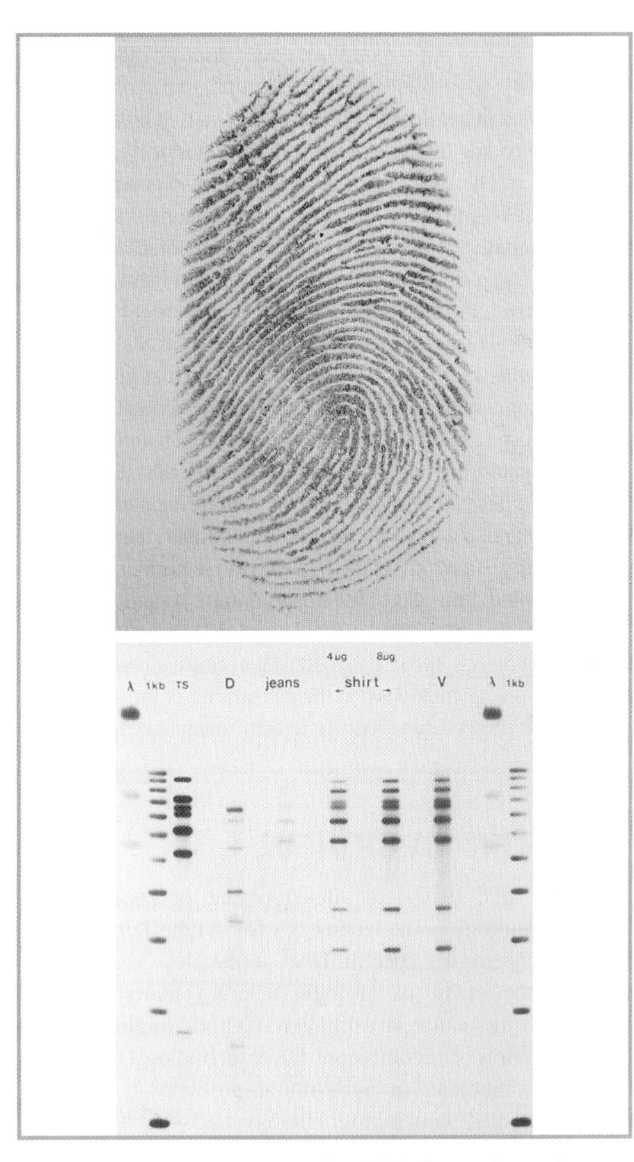

Classical true fingerprint and modern DNA fingerprint (gel pattern). Both of these images show the results of applying techniques used by forensic scientists for unique individual identification. The DNA fingerprint is an example from an actual murder case. D = defendant, V = victim; patterns from blood stains on the defendant's shirt and jeans are shown. The remaining lanes are various controls. *(Classical fingerprint from Dennie Cody/FPG; modern DNA fingerprint courtesy of Cellmark Diagnostics, Inc., Germantown, Maryland.)*

10 The sequence of any amplified DNA segment can be determined, and the sequence can be used to study gene function.

11 Recombinant DNA molecules can be used as probes to assess genotype and to predict the risk of having acquired a heritable genetic disorder.

12 Transgenes are recombinant DNA molecules introduced into higher eukaryotes.

13 Gene therapy is the application of transgenic technology to the treatment of heritable disorders.

A boy is born with a disease that completely compromises his immune system. Diagnostic testing determines that he has a recessive genetic disorder called SCID (severe combined immunodeficiency disease), more commonly known as *bubble-boy disease.* This disease is caused by a mutation in the gene coding for the blood enzyme adenosine deaminase (ADA). As a result of the loss of this enzyme, the precursor cells that give rise to one of the cell types of the immune system are missing. Because this boy has no ability to fight infection, he has to live in a completely isolated and sterile environment—that is, a bubble in which the air is filtered for sterility (Figure 8-1). No pharmaceutical or other conventional therapy is available to treat this disease. Tissue-transplantation therapies such as giving him the precursor cells obtained from another person would not work, because such cells would end up creating an immune response against the boy's own tissues (graft vs. host disease). In the past two decades, techniques have been developed that offer the possibility of a different kind of transplantation therapy—**gene therapy**—in which, in the present case, a normal ADA gene is "transplanted" into cells of the boy's immune system, thereby permitting their survival and normal function. This type of transplantation has become a possibility because a revolution in molecular biology took place in the last quarter of the twentieth century that allowed geneticists to directly manipulate individual genes in test tubes and return them to intact organisms by employing the techniques of genetic engineering. We will consider the basic methods and applications of these techniques and then return at the end of the chapter to consider in more detail how gene therapy is being attempted to cure SCID.

Figure 8-1 A boy with SCID living in a protective bubble. *(UPI/Bettmann/Corbis.)*

CHAPTER OVERVIEW

Since the mid-1970s, the development of **recombinant DNA technology**—the techniques for synthesizing, amplifying, and purifying specific DNA molecules—has revolutionized the study of biology, opening many areas of research to molecular investigation. **Genetic engineering**—the application of recombinant DNA technology to specific biological, medical, or agricultural problems—is now a well-established branch of technology, and—as we will see in Chapter 9—**genomics** is the ultimate extension of recombinant DNA technology to the global analysis of the nucleic acids present in a nucleus, a cell, an organism, or a group of related species. The techniques of recombinant DNA are pervasive in modern biological research, and various applications of genetic engineering will underlie many of the topics that we consider throughout the remainder of this book.

Recombinant DNA technology exploits (1) the basic principles of molecular biology and (2) the behavior of nonessential "accessory" chromosomes (such as plasmid chromosomes of the sort discussed in Chapter 7) either in a test tube or inside simple unicellular organisms, usually prokaryotes. The principles of molecular biology as described in detail in Chapters 2 and 3 (Figure 8-2a and b) permit the chopping apart and gluing together of DNA molecules according to the needs and ingenuity of the investigator (hence the term *recombinant* DNA). We will see over and over again that recombinant DNA technology depends on two basic principles:

- The ability of antiparallel complementary single-stranded DNA or RNA sequences to form double-stranded molecules.

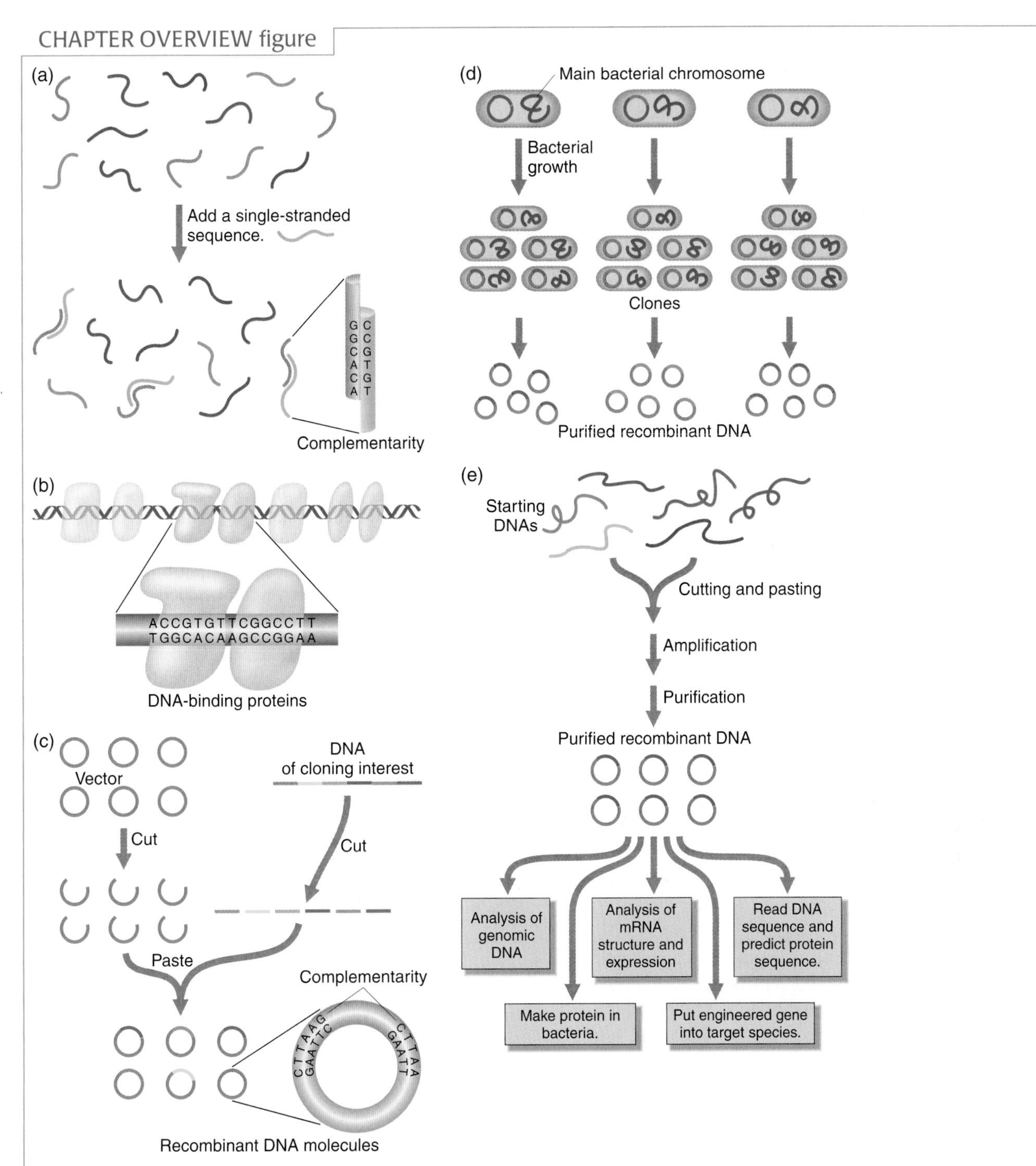

Figure 8-2 The basic principles of molecular biology employed in recombinant DNA research (a and b) and their application to the production of recombinant DNA molecules (c–e). (a) The ability of complementary single-stranded nucleic acid segments to hybridize together. In a complex mixture of single-stranded molecules (different colors represent different sequences), a double-stranded molecule forms by hybridization of a single-stranded nucleotide sequence (green) to another, complementary single-stranded sequence (red) within the complex mixture. (b) Specific DNA sequences are docking sites for sequence-specific DNA binding proteins. In this hypothetical DNA segment, the DNA is decorated with many different DNA-binding proteins according to nucleotide sequence present. An example of two proteins that bind to different DNA sequences is shown in the enlargement. (c and d) An overview of the technique for producing and amplifying recombinant molecules. (c) In this example, the starting materials are a circular accessory chromosome (vector) from a bacterium and a linear DNA molecule of interest. The accessory chromosome is cut open to turn it into a linear molecule. The DNA molecule of interest is chopped up to convert it into a series of smaller fragments, with each fragment represented by a different colored line. The linearized accessory chromosomes are mixed with the DNA fragments under conditions in which the fragments become attached to the DNA of the linearized accessory chromosome. In this way, new *recombinant* circular accessory chromosomes are produced, each of which has a different inserted fragment from the DNA molecule of interest. (d) The recombinant accessory chromosomes are then reintroduced into different bacteria, each of which is cultured as a separate clone to amplify a different fragment of the DNA molecule of interest. (e) A summary of the overall pathway of cloning and some of the major ways in which these clones are used for research and genetic engineering.

- The ability of specific proteins to recognize and bind to specific sequences (i.e., specific runs of base pairs within the DNA double helix).

Recombinant DNA technology uses these principles for both cutting DNA molecules apart and then *pasting* them together in novel combinations (Figure 8-2c). Restriction endonucleases—the class of sequence-specific DNA-binding proteins that enzymatically cleave the sugar-phosphate backbone of each of the two strands of the double helix at the site of binding—are very important molecular "scissors" used to cut a DNA molecule into many fragments of manageable and appropriate size for manipulations in a test tube. Pasting typically depends on the ability of short single-stranded regions at the ends of DNA fragments to hybridize to other short complementary sequences (see enlargement in Figure 8-2d). Short single-stranded complementary sequences are thus the "glue" that is used to join different DNA molecules together to produce recombinant DNAs. Environmental conditions (usually salt and temperature conditions) can be adjusted in a test tube so that even very short (from three to four complementary antiparallel base pairs) are sufficient to join two separate DNA molecules.

Accessory chromosome behavior permits the amplification of many identical copies of each individual cut-and-pasted DNA molecule during the growth and division of the bacterial cell in which it resides (Figure 8-2d). Because this amplification results in a *clone* of identical cells, each containing the recombinant DNA molecule, the technique of obtaining and isolating a specific recombinant molecule is called **cloning.** In general, recombinant DNA technologists "trick" the replication machinery (in a test tube or in a cell) into replicating the recombinant molecule. In the examples that we will encounter, tricking the replication machinery in such a way relies on an understanding of the biology of the recipient cell and of the nature and role of the accessory chromosome. In thinking about cloning systems, we will consider several facets of the process:

- The necessary components of the accessory chromosome.
- The origins of the donor DNA that becomes fused with the accessory chromosome to form the recombinant molecule.
- Techniques for getting recombinant DNA molecules into recipient cells.
- Techniques for selectively growing only those recipient cells containing a recombinant DNA molecule.

With the application of the principles of molecular biology and cloning, individual recombinant DNA molecules can be amplified in sufficient quantities to permit purification and detailed analysis of their structures and functions. The availability of these purified molecules enables a researcher to understand the structure of a gene, analyze its DNA sequence, or reintroduce it into another genome (Figure 8-2e). As we will see in several subsequent chapters, recombinant DNA technologies can be merged with other genetic techniques to uncover the genetic components contributing to a disease state, to deduce the nature and function of the protein product of the recombinant DNA segment, and to serve as a source of insight into the biological process(es) of which this protein is a part (Figure 8-3). How these techniques are merged to produce such insight will be the subject matter of much of the remainder of this book. In this chapter, we will focus on the basic elements of recombinant DNA technology. We will examine the production of recombinant molecules by cutting and pasting first, and then we will look at the techniques necessary to clone many identical DNA molecules. Finally, we will examine some of the ways in which these cloned recombinant molecules are analyzed and manipulated.

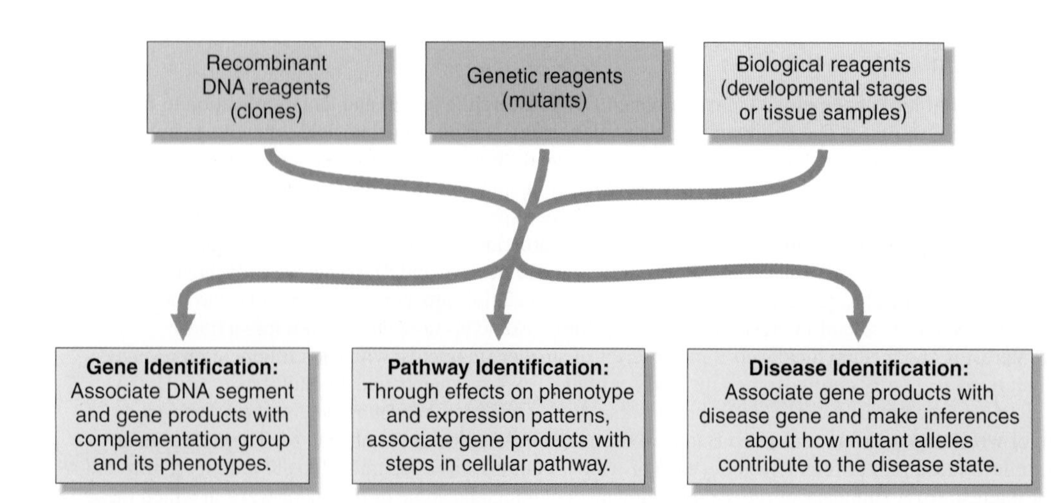

Figure 8-3 The merger of recombinant DNA technology and other techniques is a source of insight into biological processes.

GENERATING RECOMBINANT DNA MOLECULES

How can we make recombinant DNA molecules? We can divide the process into several steps:

- Where does the *donor DNA* come from?
- How does the donor DNA get chopped up?
- How do we glue the donor DNA segments to other segments, including the accessory chromosome DNA?

We'll consider each of these steps of the process in turn.

The Source of the Donor DNA

The organism under study, which will be used to donate DNA for the analysis, is called the **donor organism.** Essentially any organism's DNA can be purified in a test tube by straightforward isolation procedures in which the cells of the donor are broken apart and DNA is separated from other molecular constituents of the cell, such as proteins, RNA, lipids, and carbohydrates. Regardless of their source, the donor DNAs all behave similarly because they are double helices of complementary antiparallel strands of deoxyribonucleic acid. Three sources of DNA segments can be considered:

1. Genomic DNA. These DNA sequences exist in the chromosomes of the organism under study (i.e., they are already present in its genome) and thus are the most straightforward source of DNA.
2. cDNA. **Complementary DNA,** or **cDNA,** is essentially a double-stranded DNA version of an mRNA molecule. cDNA is made from mRNA with the use of a special enzyme called *reverse transcriptase* originally isolated from retroviruses. With the use of an mRNA molecule as a template, reverse transcriptase synthesizes a single-stranded DNA molecule that can then be used as a template for double-stranded DNA synthesis (Figure 8-4). cDNAs are especially useful because RNAs are inherently less stable than DNA, and techniques for routinely amplifying and purifying individual RNA molecules do not exist. For this reason, we use the enzymatically synthesized double-stranded cDNA replicas of the original mRNA. Researchers frequently need to understand the structure of a transcript and, in particular, to characterize an mRNA to predict the polypeptide sequence that it encodes. This characterization is especially important in eukaryotes, because direct prediction of a polypeptide sequence from a genomic sequence is very difficult—especially owing to the presence of introns in the genomic sequence, which are removed from the pre-mRNA by splicing to produce the mature mRNA (see Figure 3-16).
3. Chemically synthesized DNA. Sometimes, a researcher needs to include in a recombinant DNA molecule a specific sequence that cannot be isolated from available natural genomic or cDNAs. Highly automated techniques for the chemical synthesis of oligonucleotides (DNA segments typically ranging in length from 15 to 100 nucleotides) have been developed to produce DNA sequences for this purpose.

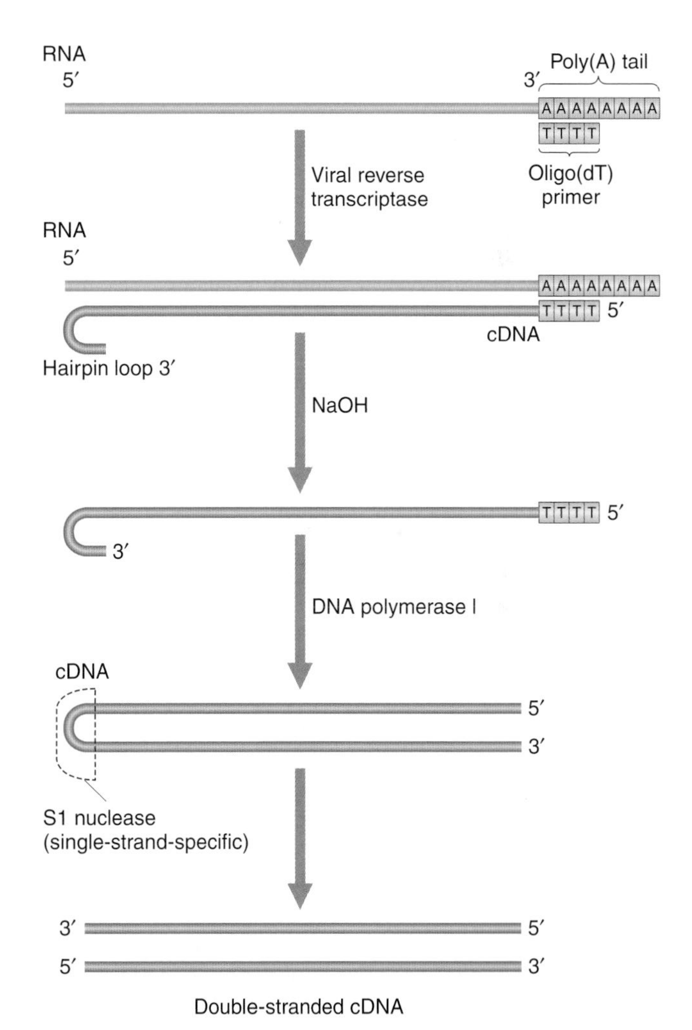

Figure 8-4 The synthesis of double-stranded cDNA from mRNA. A short oligo(dT) chain is hybridized to the poly(A) tail of an mRNA strand. The oligo(dT) segment serves as a primer for the action of viral reverse transcriptase, an enzyme that uses the mRNA as a template for the synthesis of a complementary DNA strand. The resulting cDNA ends in a hairpin loop. When the mRNA strand has been degraded by treatment with NaOH, the hairpin loop becomes a primer for DNA polymerase I, which completes the paired DNA strand. The loop is then cleaved by S1 nuclease (which acts only on the single-stranded loop) to produce a double-stranded cDNA molecule. *(From J. D. Watson, J. Tooze, and D. T. Kurtz,* Recombinant DNA: A Short Course. *© 1983 by W. H. Freeman and Company.)*

Chopping Up the Donor DNA

Many kinds of donor DNA are too large for easy manipulation in a test tube. Even when donor DNA size is not important, intact donor DNA molecules may not have the single-stranded regions at their ends that will permit them to be glued to accessory chromosomes by base-pair complementarity to construct recombinant DNA molecules. DNA fragments from a donor genome can be produced in several ways, according to the needs of the particular experiment. DNA can be broken into small fragments by physically shearing it (e.g., by agitating it in a blender) or by enzymatically digesting it with endonucleases that cleave at specific sites.

The most important technique for endonuclease digestion uses **restriction enzyme** digestion. Restriction enzymes cut at specific DNA target sequences, and this property is one of the key features that make them suitable for DNA manipulation. Purely by chance, any DNA molecule, be it derived from virus, fly, or human, contains restriction enzyme target sites. Thus, in the presence of the appropriate restriction enzyme, the DNA will be cut into a set of **restriction fragments** according to the locations of the restriction sites.

Look at an example: the restriction enzyme *Eco*RI (from *E. coli*) recognizes the following six-nucleotide-pair sequence in the DNA of any organism:

5′-GAATTC-3′
3′-CTTAAG-5′

This type of segment is called a DNA **palindrome,** which means that both strands have the same nucleotide sequence but in antiparallel orientation. (Rotate the sequence 180 degrees within the plane of the paper and you will see that the sequence is identical. For this reason, we say that palindromes have *rotational symmetry*.) Many different restriction enzymes recognize and cut specific palindromes; however, different restriction enzymes cut at different palindromic sequences. Sometimes the cuts are in the middle of the palindrome on each of the two antiparallel strands. However, in the most useful restriction enzymes, the cuts are offset or staggered. The enzyme *Eco*RI produces such staggered cuts. *Eco*RI recognizes and cuts only within the GAATTC palindrome sequence. The cuts are between the G and the A nucleotides on each strand of the palindrome:

↓
5′-GAATTC-3′
3′-CTTAAG-5′
↑

These staggered cuts leave a pair of identical five-base-long single-stranded "sticky ends." The ends are called *sticky* because, being single stranded, they can hybridize through base-pair hydrogen bonding (i.e., stick) to a complementary sequence. The production of these sticky ends is another feature of many restriction enzymes that makes them suitable tools for recombinant DNA technology. If two different DNA molecules are cut with the same sticky-end-producing restriction enzyme, the fragments of each will have the same complementary sticky ends, enabling them to hybridize with each other under the appropriate salt and temperature conditions in a test tube. Figure 8-5 (top left) illustrates the restriction enzyme *Eco*RI making a single cut in a circular DNA molecule such as a plasmid; the cut opens up the circle, and the resulting linear molecule has two sticky ends. If such a molecule is mixed with a different DNA molecule (e.g., a donor DNA fragment cut with *Eco*RI, as shown in Figure 8-5, top right), the two can then hybridize to each other through their complementary sticky ends to form a recombinant molecule (Figure 8-5, bottom).

Dozens of restriction enzymes with different sequence specificities are now known; some of them are listed in Figure 8-6. All the target sequences are palindromes, meaning

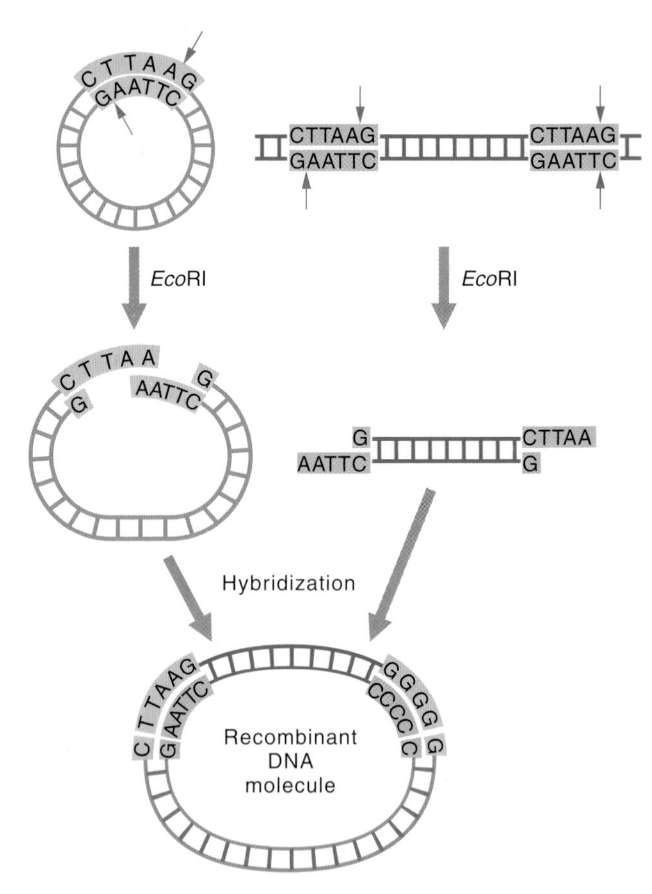

Figure 8-5 The restriction enzyme *Eco*RI cuts a circular DNA molecule bearing one target sequence, resulting in a linear molecule with single-stranded sticky ends. Because of complementarity, other linear molecules with *Eco*RI-cut sticky ends can hybridize with the linearized circular DNA, forming a recombinant DNA molecule.

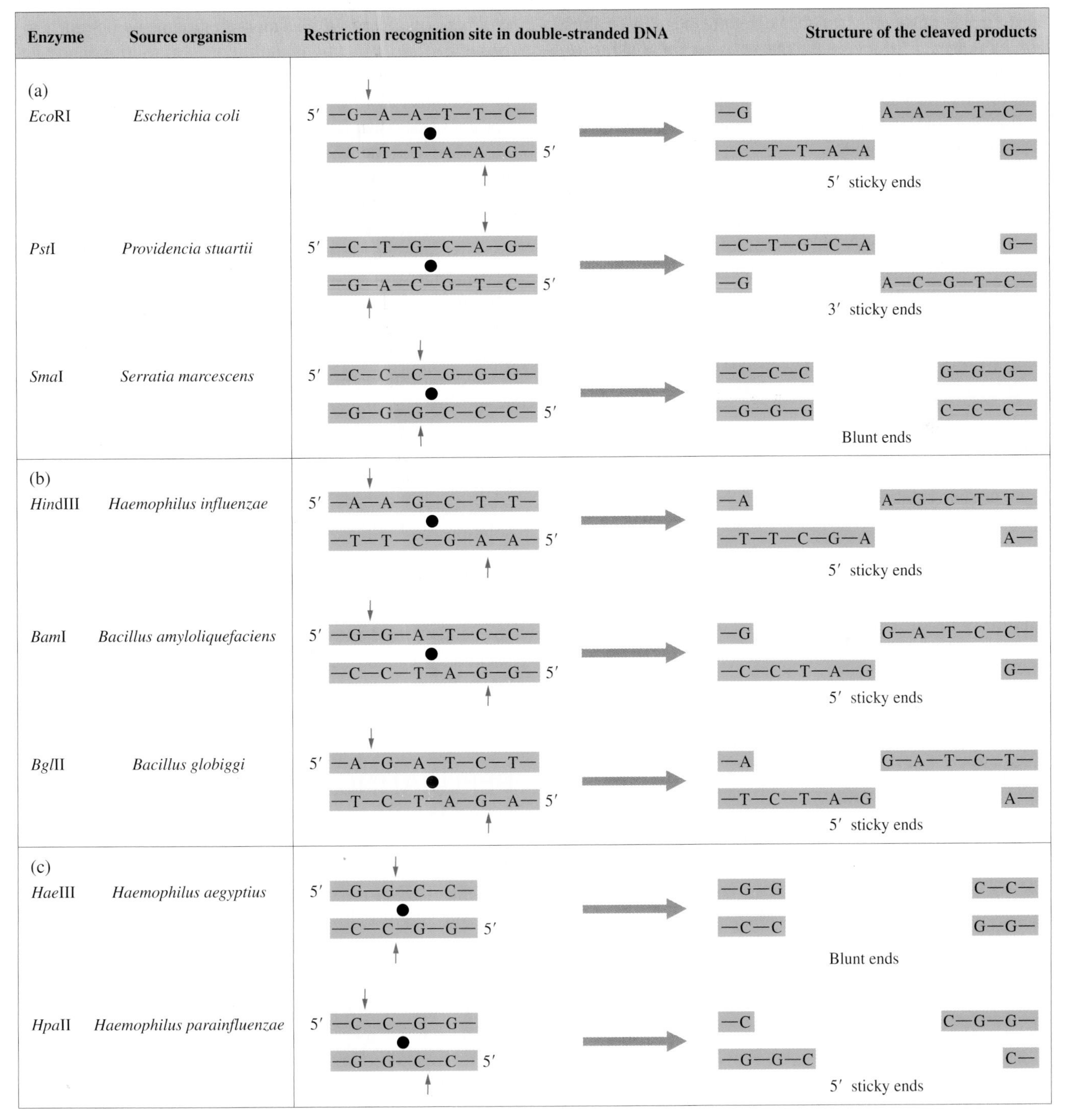

Figure 8-6 The specificity and results of restriction enzyme cleavage. The 5′ end of each DNA strand and the site of cleavage (small red arrows) are indicated. The large dot indicates the site of rotational symmetry of each recognition site. Note that the recognition sites differ for different enzymes. In addition, the positions of the cut sites may differ for different enzymes, producing single-stranded overhangs (sticky ends) at the 5′ or 3′ end of each double-stranded DNA molecule or producing blunt ends if the cut sites are not offset. (a) Three hexanucleotide (six-cutter) recognition sites and the restriction enzymes that cleave them. Note that one site produces a 5′ overhang, another a 3′ overhand, and the third a blunt end. (b) Some other examples of six-cutter enzymes showing different recognition sequences from those in part a. (c) Some examples of enzymes that have tetranucleotide recognition sites (four-cutters).

that they have rotational symmetry. Some enzymes, such as *Eco*RI or *Pst*I, make staggered cuts, whereas others, such as *Sma*I, make flush cuts and leave blunt ends. Even flush cuts, which lack sticky ends, can be used for making recombinant DNA with the use of enzymes that join blunt ends together or with the use of other special enzymes that synthesize short single-stranded sticky ends on the exposed 3′ strand of the blunt end.

MESSAGE

Restriction enzymes have two properties useful in recombinant DNA technology. First, they cut DNA into fragments of a size suitable for manipulation in a test tube. Second, many restriction enzymes make staggered cuts, generating single-stranded sticky ends conducive to the formation of recombinant DNA.

Gluing the Donor DNA to Other DNA

There are two reasons why the donor DNA must be attached to other DNA segments to form useful recombinant molecules. First, on its own, a fragmented donor DNA segment typically does not have the necessary DNA sequences, such as an origin of replication, to enable it to be replicated in a test tube or inside a host organism. The donor DNA must be physically attached to other DNA segments that can support replication in a test tube or inside a host cell. Second, an experiment may demand that multiple fragments be glued together to form a functional unit (e.g., a transcriptionally active gene). Even such synthetic multicomponent DNAs must be eventually inserted into other DNA segments (accessory chromosomes) that can support replication.

Most commonly, both donor and accessory DNA are digested with the use of a restriction enzyme that produces sticky ends and then mixed in a test tube to allow the sticky ends of vector and donor DNA to bind to each other and form recombinant molecules. Figure 8-7a shows a bacterial plasmid DNA that carries a single *Eco*RI restriction site; so digestion with the restriction enzyme *Eco*RI converts the circular DNA into a single linear molecule with sticky ends. Donor DNA from any other source (say, *Drosophila*) also is treated with the *Eco*RI enzyme to produce a population of fragments carrying the same sticky ends. When the two populations are mixed at the proper salt and temperature conditions, DNA fragments from the two sources can hybridize, because double helices form between their sticky ends. There are many opened-up plasmid molecules in the solution, as well as many different *Eco*RI fragments of donor DNA. Therefore a diverse array of plasmids recombined with different donor fragments will be produced. At this stage, the hybridized molecules do not have complete covalently joined sugar-phosphate backbones. However, the backbones can be sealed by the addition of the enzyme **DNA ligase,** which creates phosphodiester bonds at the junctions (Figure 8-7b).

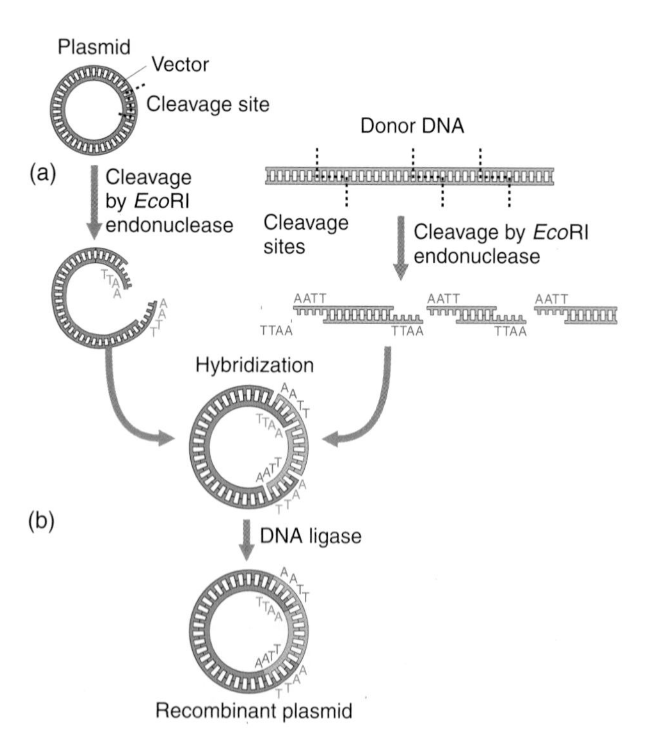

Figure 8-7 Method for generating a recombinant DNA plasmid containing genes derived from donor DNA. *(From S. N. Cohen, "The Manipulation of Genes." © 1975 by Scientific American, Inc. All rights reserved.)*

REPLICATING RECOMBINANT MOLECULES

To do further manipulations and analyses of individual recombinant DNA molecules, we must have many copies of the molecule, usually in purified form. There are two general ways to produce large quantities of a single recombinant DNA molecule:

- Chemically. Selectively replicate a recombinant DNA molecule in a test tube.
- Biologically. Co-opt selective replication machinery of bacterial or simple eukaryotic cells to do the job for us, "tricking" those cells into replicating a recombinant DNA molecule of interest.

We'll consider each of these alternatives.

Selective Replication in a Test Tube: The Polymerase Chain Reaction

If we know enough about a recombinant molecule, such as the sequence of at least some parts of the molecule, we can amplify it in a test tube. To do so, we employ the **polymerase chain reaction (PCR).** The basic strategy of PCR is outlined in Figure 8-8. Suppose, for example, that we start with exactly one linear double-stranded DNA molecule that we wish to amplify. With the use of two primers of defined sequence chosen to be complementary to the opposite ends of the linear DNA molecule (recall that DNA polymerase requires a primer—a DNA or RNA with an exposed 3′-OH end—as a starting point for replication), each of the two strands of the DNA are replicated.

The great advantage of the technique is that no cloning procedures are necessary because base-pair complementarity to the sequence of the primers determines the specificity of the DNA segment that is amplified. If the sequences corresponding to the primers are each present only once in the genome and are sufficiently close together (maximum distance, about 2 kb), the *only* DNA segment that can be amplified is the one between the two primers. This will be true even if this DNA segment is present at very low levels (e.g., one part in a million) in a complex mixture of DNA fragments such as might be generated from a preparation of human genomic DNA.

To get enough material for analysis, multiple rounds of PCR amplification need to be carried out. We start with a solution containing the recombinant DNA molecule, the

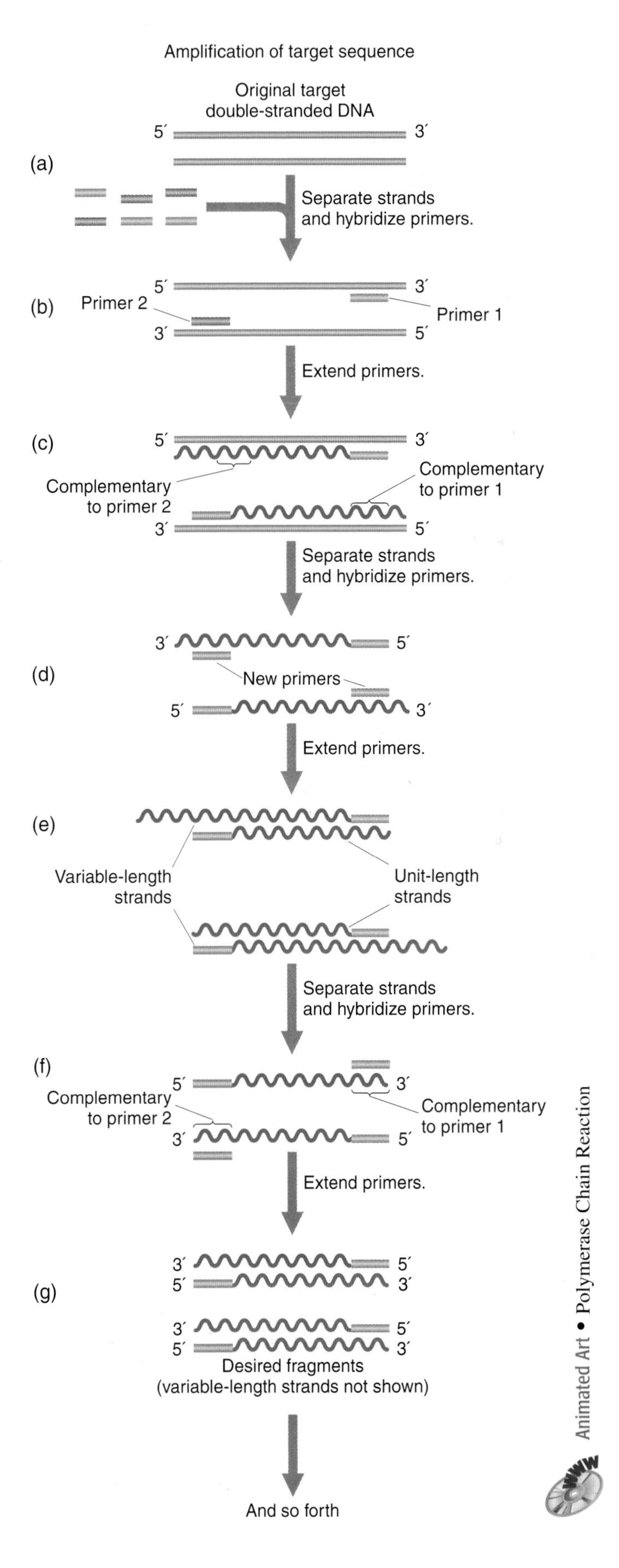

Figure 8-8 The polymerase chain reaction. (a) Double-stranded DNA containing the target sequence. (b) Two chosen or created primers have sequences complementing primer-binding sites at the 3′ ends of the target gene on the two strands. The strands are separated by heating, allowing the two primers to anneal to the primer-binding sites. Together, the primers thus flank the targeted sequence. (c) *Taq* polymerase then synthesizes the first set of complementary strands in the reaction. These first two strands are of varying length, because they do not have a common stop signal. They extend beyond the ends of the target sequence as delineated by the primer-binding sites. (d) The two duplexes are heated again, exposing four binding sites. (For simplicity, only the two new strands are shown.) The two primers again bind to their respective strands at the 3′ ends of the target region. (e) *Taq* polymerase again synthesizes two complementary strands. Although the template strands at this stage are variable in length, the two strands just synthesized from them are precisely the length of the target sequence desired. This precise length is achieved because each new strand begins at the primer-binding site, *at one end of the target sequence,* and proceeds until it runs out of template, *at the other end of the sequence.* (f) Each new strand now begins with one primer sequence and ends with the primer-binding sequence for the other primer. Subsequent to strand separation, the primers again anneal and the strands are extended to the length of the target sequence. (The variable-length strands from part c are also producing target-length strands, which, for simplicity, is not shown.) (g) The process can be repeated indefinitely, each time creating two double-stranded DNA molecules identical with the target sequence. *(From J. D. Watson, M. Gilman, J. Witkowski, and M. Zoller,* Recombinant DNA, *2d ed. © 1992 by Scientific American Books.)*

Animated Art • Polymerase Chain Reaction

synthesized primers, the four deoxyribonucleotide triphosphates, and a special DNA polymerase. By modulating the temperature of this solution, we accomplish the following steps for each round of PCR amplification:

- Single-stranded DNA is produced by heat denaturing the double-stranded DNA molecules.
- Primers anneal (hybridize) to their complementary sequences in cooled solutions containing the denatured (single-stranded) DNA molecules.
- The special DNA polymerase replicates the single-stranded DNA segments to which primers have annealed.

If temperature conditions are properly controlled, there can be many rounds of amplification. After 20 rounds of amplification, for example, we would have about 1,000,000 identical molecules of the desired sequence.

Figure 8-8 illustrates the principle of the technique. A temperature-resistant DNA polymerase, *Taq* polymerase, from the bacterium *Thermus aquaticus* is used to catalyze growth from DNA primers. (This bacterium normally grows in thermal vents and so has evolved proteins that are extremely heat resistant. This characteristic turns out to be very important because heat denaturation of the DNA duplex requires temperatures that would denature and inactivate DNA polymerase from most species.) Through replication starting from the locations of the pairs of primers on opposite strands, the single-stranded DNA molecules are extended toward each other, forming a double-stranded DNA molecule identical with each starting one. After completion of the replication of the segment between the two primers (one cycle), the two new duplexes are heat denatured to generate single-stranded templates, and a second cycle of replication is carried out by lowering the temperature in the presence of all the components necessary for the polymerization. Repeated cycles of denaturation, annealing, and synthesis result in an exponential increase in the number of segments replicated. Amplifications by as much as a millionfold can be readily achieved within 1 to 2 hours.

Because PCR is a very sensitive technique, it has many other applications in biology. Target sequences that are in extremely low copy number in a sample can be amplified, as long as primers specific to this rare sequence are present. For example, in criminal investigations, where physical samples are often minute, segments of human DNA can be amplified from the few follicle cells surrounding a single pulled-out hair.

Although PCR's sensitivity and specificity are clear advantages, the technique does have some significant limitations. To design PCR primers for selective amplification of a particular DNA segment, at least some sequence information must be available for the piece of DNA that is to be amplified; in the absence of such information, PCR amplification cannot be applied. Practical limitations of the enzymatic activity of the polymerase ensure amplification of DNA segments of <2 kb by PCR. Thus, PCR is best used for relatively small fragments of recombinant DNA.

MESSAGE

The polymerase chain reaction uses specially designed primers for direct amplification of specific short regions of DNA in a test tube.

Selective Replication inside a Host Cell: Cloning of Recombinant Molecules

Unlike PCR, the cloning of recombinant DNA molecules inside host cells can permit the amplification of molecules of many different sizes from a few hundred base pairs to a few hundred kilobase pairs of DNA. Many of the basic recombinant DNA systems depend on amplification in bacterial cells. Prokaryotic genetic mechanisms, including those of bacterial transformation, plasmid replication, and bacteriophage growth discussed in Chapter 7, underlie the techniques for cloning recombinant molecules. Typically, as a result of amplification, the resulting colony of bacteria will contain billions of copies of the single donor DNA insert fused to an appropriate accessory chromosome; the latter is usually referred to as a **cloning vector.** This set of amplified copies of the single donor DNA fragment within the cloning vector is the recombinant DNA clone. Figure 8-9 illustrates the cloning of a donor DNA segment within a bacterial plasmid cloning vector.

Four basic questions must be considered in thinking about cloning in bacteria:

1. What are the appropriate cloning vectors for amplifying particular recombinant molecules?
2. How do the recombinant molecules enter the bacterial cell?
3. How do the introduced recombinant molecules replicate?
4. How is the amplified recombinant DNA recovered from the clone?

The answers to these questions depend on the nature of the vector DNA molecules into which the donor DNA of interest is inserted.

Cloning vectors—properties and strategies. Vectors must be relatively small molecules for the convenience of manipulation. They must be capable of prolific replication in a living cell, thereby enabling the amplification of the inserted donor fragment. Another important requirement is that there be convenient restriction sites that can be used for insertion of the DNA to be cloned. Unique sites (sites present exactly once in the vector) are most useful because then donor DNA treated with the restriction enzyme that cuts at this site will insert only at the precise desired location in the vector. (If there are multiple restriction sites in the vector, then, in the process of the fragments being glued together by sticky-end hybridization, some restriction frag-

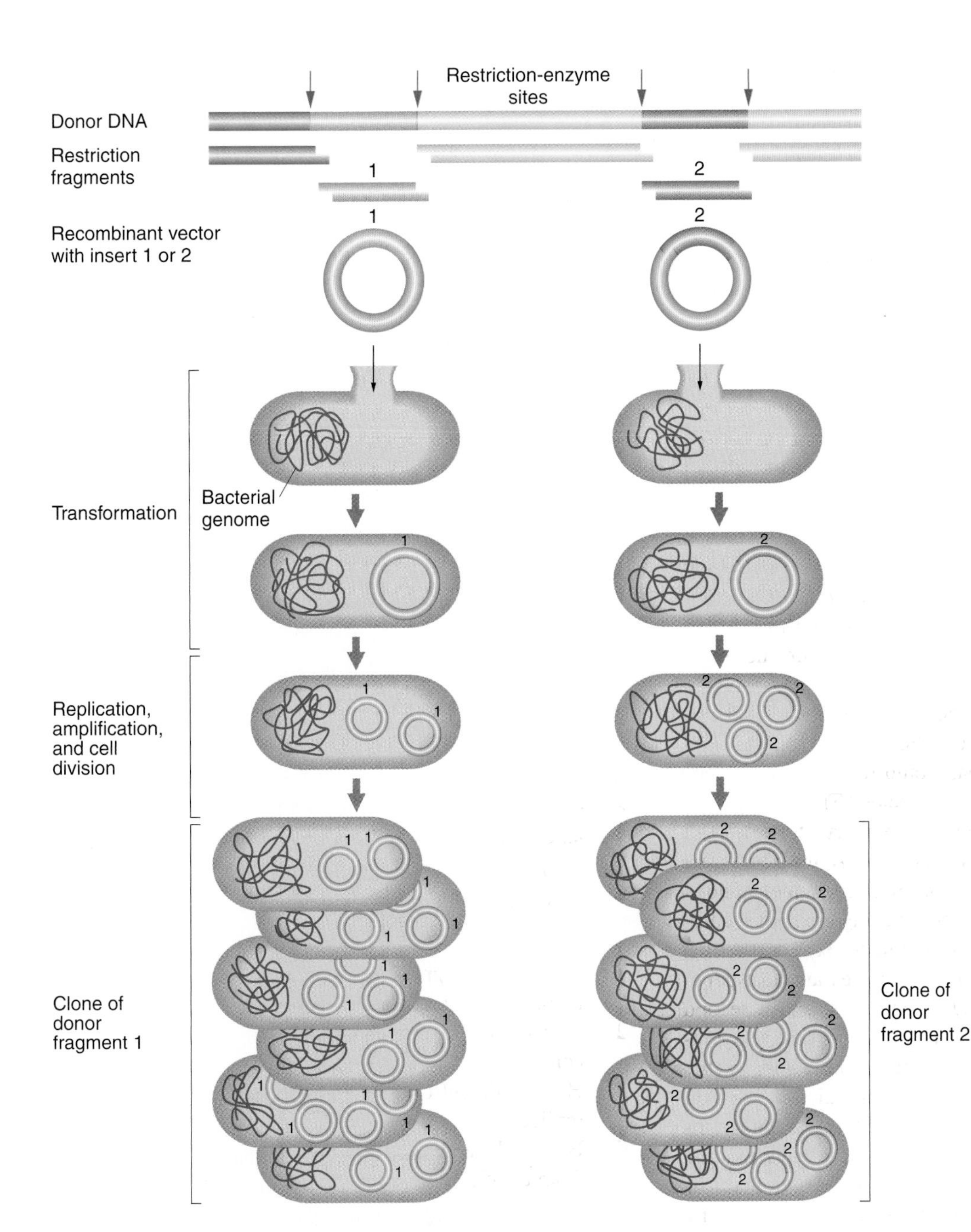

Figure 8-9 How amplification works. Restriction enzyme treatment of donor DNA and vector allows the insertion of single fragments into vectors. A single vector enters a bacterial host, where replication and cell division result in a large number of copies of the donor fragment.

Animated Art • Finding Specific Cloned Genes by Functional Complementation: Making a Library of Wild-Type Yeast DNA

ments of the vector may be in the wrong order or may be entirely lost.) It is also important that there be a mechanism for easy identification and recovery of the recombinant molecule. There are numerous cloning vectors in current use, and the choice among them generally depends on the size of the DNA segment that needs to be cloned or the use to which the clone will be put. Cloning systems for smaller inserts are generally easier to manipulate, but modern requirements, particularly for genome-level analysis, demand cloning vectors that can tolerate much larger sized inserts. Some general classes of cloning vectors follow.

Plasmid vectors. As described earlier, bacterial plasmids are small circular DNA molecules that are distinct from, as well as additional to, the main bacterial chromosome. They replicate their DNA independently of the bacterial chromosome. In Chapter 7, we encountered the F plasmid, which confers certain types of conjugative behavior to cells of *E. coli.* The F plasmid can be used as a vector for carrying large donor DNA inserts. However, the plasmids that are routinely used as vectors are those that carry genes for drug resistance. The drug-resistance genes are useful because the drug-resistant phenotype can be used to select for cells transformed by plasmids, as shown at left in Figure 8-10 on the next page. Plasmids are also an efficient means of amplifying cloned DNA because there are many copies per cell, as many as several hundred for some plasmids. Examples of specific plasmid vectors are shown in Figure 8-10.

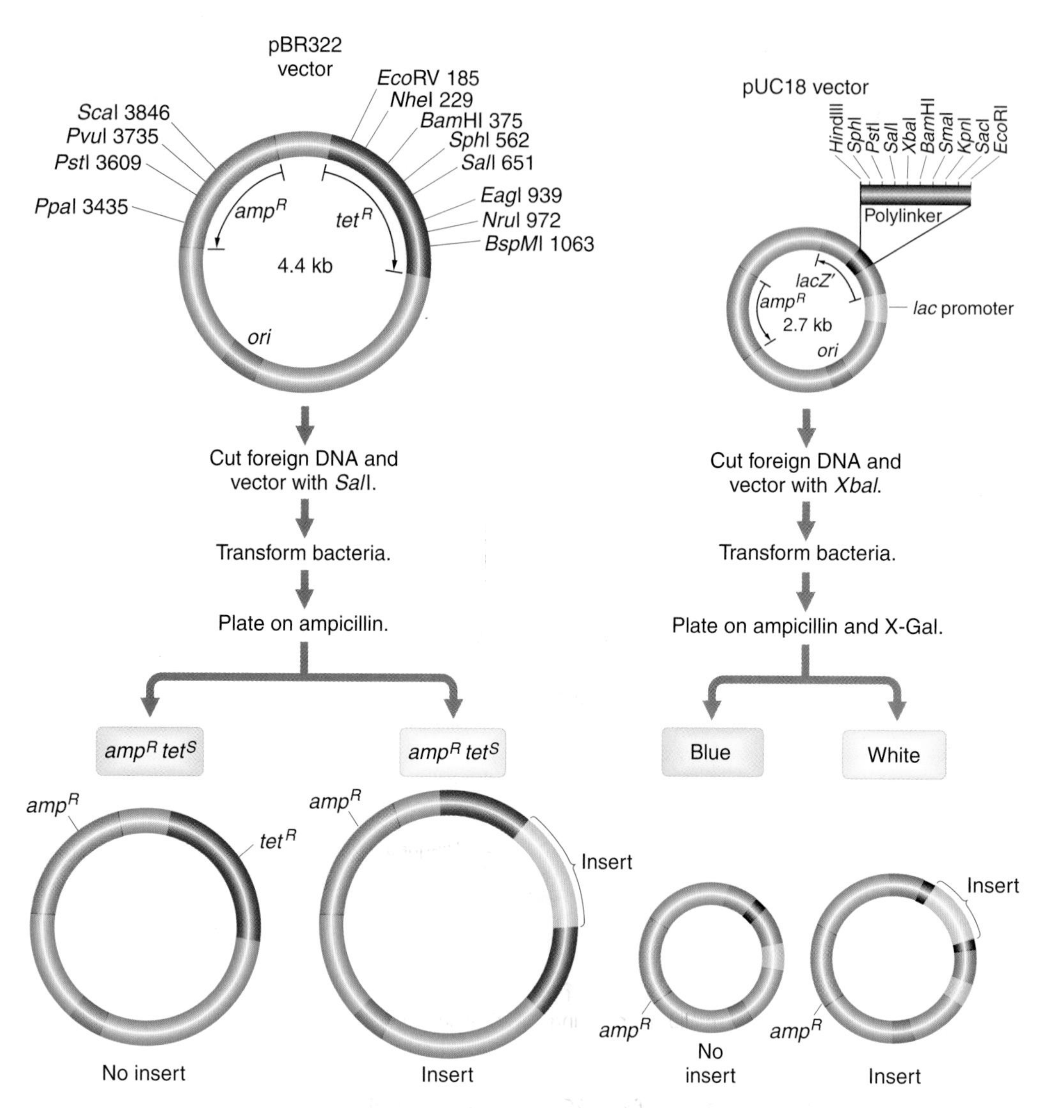

Figure 8-10 Two plasmids designed as vectors for DNA cloning, showing general structure and restriction sites. Insertion into pBR322 is detected by inactivation of one drug-resistance gene (*tet*^{R}), indicated by the *tet*^{S} (sensitive) phenotype. Insertion into pUC18 is detected by inactivation of the β-galactosidase function of *Z*′, resulting in an inability to convert the artificial substrate X-Gal into a blue dye.

Bacteriophage vectors. There are different classes of bacteriophage vectors, depending on (1) whether the chromosomal DNA inside the bacteriophage is single stranded or double stranded, and (2) the size of the donor DNA insert that can be harbored within the vector chromosome. A given bacteriophage has a standard amount of DNA that can be inserted, or "packaged," inside the phage particle. If too little or too much DNA is packaged, the resulting structurally abnormal bacteriophage particle is noninfective and hence does not replicate; such an abnormal phage is therefore of no use in making a recombinant clone. Some of the genes of the bacteriophage are absolutely necessary for replication of that bacteriophage chromosome, whereas others are dispensable. A typical bacteriophage cloning vector consists of the essential DNA sequences, and the size of the donor DNA segment that can be fused to the vector to produce a "packagable" recombinant bacteriophage chromosome depends on the size of the dispensable sequences that were removed in building the vector.

Some bacteriophage vectors, such as bacteriophage M13, are useful because the DNAs that they package are single stranded (i.e., contain only one of the two complementary DNA strands). Some DNA-sequencing techniques require single-stranded DNA as substrate, and so these single-stranded bacteriophages can provide the required DNA. Typically, they package DNA in lengths less than about 1 to 2 kb.

Bacteriophage λ (lambda) is used as a cloning vector for double-stranded DNA inserts as long as about 15 kb. Lambda phage heads can package DNA molecules no larger than about 50 kb in length (the size of a normal λ chromosome). The central part of the phage genome is not required for replication or packaging of λ DNA molecules in *E. coli,* and so the central part can be cut out by using restriction enzymes and discarded. The two remaining "arms" at either end of the linear λ chromosome are ligated to restriction-digested donor DNA (Figure 8-11). The recombinant molecules can be introduced into *E. coli* by transfor-

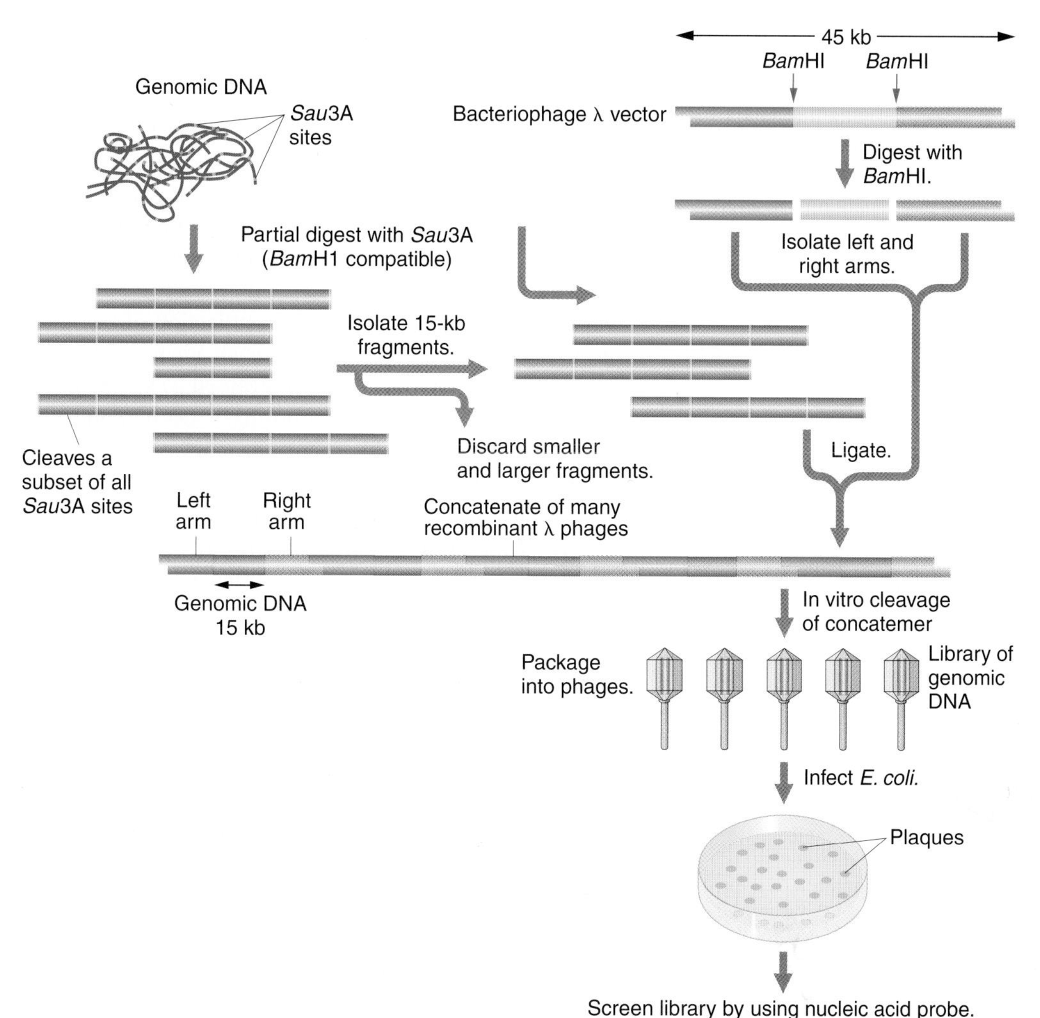

Figure 8-11 Cloning in phage λ. A nonessential central region of the phage chromosome is discarded, and the ends are ligated to random 15-kb fragments of donor DNA. A linear multimer (concatenate) forms, which is then stuffed into phage heads one monomer at a time by using an in vitro packaging system. *(From J. D. Watson, M. Gilman, J. Witkowski, and M. Zoller,* Recombinant DNA, *2d ed. © 1992 by Scientific American Books.)*

mation and recovered in λ phage particles after infecting the transformed bacteria with phage λ. The infecting phage and the recombinant phage chromosomes are both replicated and packaged in progeny phage heads, and they are released from the infected bacterial cell when it lyses. Alternatively, the recombinant molecules can be directly packaged into phage heads in vitro. In the in vitro system, DNA and phage-head components are mixed together, and infective λ phages form spontaneously. In either method, recombinant molecules with 10- to 15-kb inserts are the ones that will be most effectively packaged into phage heads, because an insert of this size substitutes for the deleted central part of the phage genome and brings the total chromosome size to 50 kb. (If smaller inserts are desired, a smaller dispensable central fragment of the λ phage chromosome can be discarded, thus leaving less "room" for insert DNA in a "packagable" chromosome.) Therefore the presence of a phage plaque on the bacterial lawn automatically signals the presence of recombinant phage bearing an insert.

Vectors for larger DNA inserts. The standard plasmid and phage λ vectors just described have upper limits ranging from 25 to 30 kb for the size of the donor DNA that can be inserted. However, many purposes require inserts well in excess of this upper limit. To meet these needs, other prokaryotic vectors have been engineered. In all cases, bacteriophages serve as the delivery systems for introducing the large DNA molecules intact into the bacterial cells. **Cosmids** are vectors that can carry 35- to 45-kb inserts. The cosmid vector sequences are hybrids of λ phages and plasmids that can be inserted into λ particles. The λ phage is used as the "syringe" to introduce these big pieces of recombinant DNA into the recipient *E. coli* cell. However, because the vector lacks the essential phage sequences necessary to form progeny λ phages, the recombinant DNA molecule depends on the plasmid sequences in the cosmid. Once in the cell, these hybrids form circular molecules that replicate extrachromosomally in the same manner as plasmids do. PAC (P1 artificial chromosome) vectors are based on the bacteriophage P1 and can tolerate inserts ranging from 80 to 100 kb; they, too, replicate as very large plasmids. The largest prokaryotic inserts use the BAC (bacterial artificial chromosome) vector system, (Figure 8-12 on the next page), which is based on the 7-kb F plasmid and can be used for inserts ranging from 150

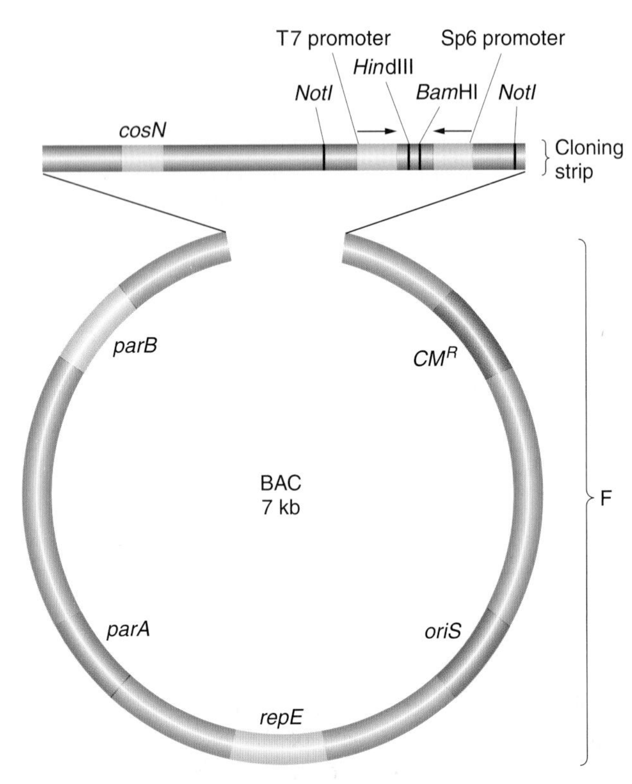

Figure 8-12 Structure of a bacterial artificial chromosome (BAC), used for cloning large fragments of donor DNA. CM^R is a selectable marker for chloramphenicol resistance. *oriS, repE, parA,* and *parB* are F genes for replication and regulation of copy number. *cosN* is the *cos* site from λ phage. *Hin*dIII and *Bam*HI are cloning sites at which donor DNA is inserted. The two promoters are for transcribing the inserted fragment. The *NotI* sites are used for cutting out the inserted fragment.

to 300 kb. BACs have proved to be particularly important as the "workhorse" cloning vectors of large-scale genome-sequencing projects (discussed in Chapter 9).

Finally, for inserts larger than 300 kb, YACs (yeast artificial chromosomes)—a eukaryotic vector system based on independently segregating yeast chromosomes introduced into yeast cells by transformation—can be used to clone recombinant molecules of several hundred kilobases in length. Although the YAC system works for some applications, it turns out to be technically more challenging to purify the YAC DNA molecule away from the normal yeast chromosomes than it is to purify cosmids, PACs, or BACs, and so the prokaryotic systems are typically used unless extremely large DNA segments need to be cloned.

Entry into bacterial cells. Exogenous DNA molecules can enter a bacterial cell by two basic paths (described in Chapter 7): (1) transformation and (2) delivery by a transducing bacteriophage (Figure 8-13). If recombinant DNA molecules are introduced by bacterial transformation, bacteria are in essence bathed in a solution containing the desired recombinant DNA molecule, which then forms a plasmid chromosome (Figure 8-13a). If bacteriophage form the delivery system for injecting the recombinant DNA into the host bacterial cells, the recombinant molecule is packaged into the appropriate bacteriophage particles. These engineered bacteriophages are then mixed with the bacteria and the bacteriophages inject their DNA cargo into the bacterial cells. Whether the results of injection will be the formation of a recombinant plasmid chromosome (Figure 8-13b) or the production of progeny phages carrying the recombinant DNA molecule (Figure 8-13c) depends on the vector system.

Replication in bacterial cells. The replication of recombinant molecules exploits the normal mechanisms in the bacterial cell for replication of chromosomal DNA. One requirement is the presence of an origin of DNA replication (as described in Figure 4-8a). For plasmid-based and most phage-based replication systems, the recombinant DNA must exist in the cell as a closed circular double-stranded molecule. With these requirements fulfilled, the bacterial replication system will copy the inserted DNA along with the vector sequences, thereby enabling massive amplification of the original DNA molecule.

Recovery of amplified recombinant molecules. Recombinant molecules that are packaged into phage particles can be recovered from progeny phage particles released from lysed bacteria originally infected with single recombinant phage. The recombinant DNA is chemically extracted from the protein coat of the bacteriophage particle. For plasmids and other recombinant molecules that are replicating extrachromosomally, the bacteria are chemically or mechanically broken apart. The recombinant DNA molecule is separated from the much larger main bacterial chromosome by centrifugation or chromatography techniques on the basis of size differential or other physical characteristics.

The basic approaches for the biological cloning of recombinant DNA molecules have been described in the preceding sections. Above all, we should realize that these techniques are based not only on an understanding of genetic mechanisms that operate in the recipient cells, but also on tricks that allow researchers to subvert these mechanisms for the purpose of amplifying specific DNA molecules.

MESSAGE

Recombinant DNA cloning is carried out through the introduction of single recombinant DNA molecules into recipient cells, followed by the amplification of these molecules, thanks to the replication properties of the vector DNAs into which donor DNA is inserted.

EXPLOITING RECOMBINANT MOLECULES

Now that we have considered the basics of how to generate and amplify individual recombinant DNA molecules, we

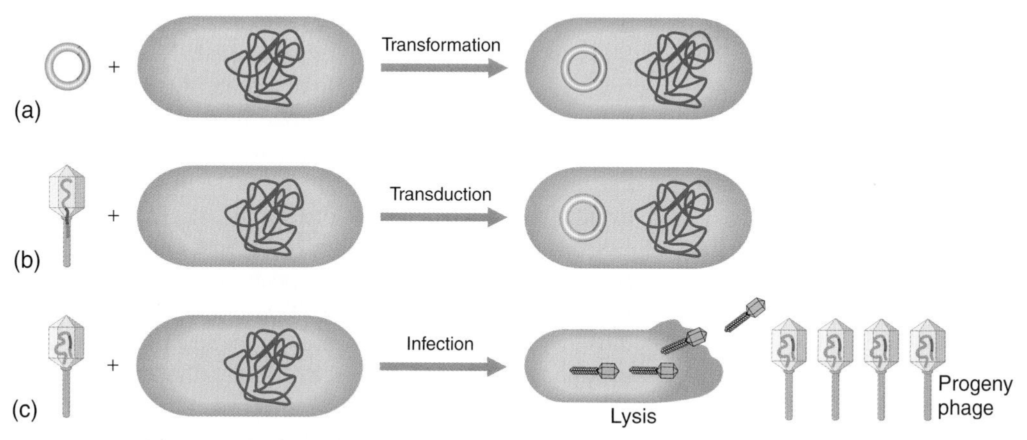

Figure 8-13 The modes of delivery of recombinant DNA into bacterial cells. (a) A plasmid vector is delivered by DNA-mediated transformation. (b) Certain vectors such as cosmids are delivered within bacteriophage heads (transduction); however, after having been injected into the bacterium, they form circles and replicate as large plasmids. (c) Bacteriophage vectors such as phage λ infect and lyse the bacterium, releasing a clone of progeny phage, all carrying the identical recombinant DNA molecule within the phage genome.

turn our attention to the techniques that can be used to analyze these molecules.

Creating Representative Sets of Recombinant DNA Molecules: Genomic and cDNA Libraries

We have seen how to make and amplify individual recombinant DNA molecules. Any one clone represents a small part of the genome or only one of thousands of mRNA molecules that an organism can synthesize. Often, the identity of a DNA segment in which we are interested is not known to us ahead of time. To ensure that we have cloned this DNA segment, we make collections of DNA segments that are all-inclusive. For example, we take all the DNA from a genome, break it up into segments of the right size for the cloning vector that we are using, and insert each segment into a different copy of the vector, thereby creating a collection of recombinant DNA molecules that, taken together, represent the entire genome. We can then transform or transduce these molecules into separate bacterial recipient cells, where they are amplified. The resulting collection of recombinant DNA-bearing bacteria or bacteriophages is called a **genomic library.** If we are using a cloning vector that accepts an average insert size of 10 kb and if the entire genome is 100,000 kb in size (the approximate size of the genome of the nematode *Caenorhabditis elegans*), then 10,000 independent recombinant clones would represent one genome's worth of DNA. Partly by chance, and because some sequences from other genomes replicate less well in bacteria than do others, one genome's worth of clones would in reality contain multiple copies of some parts of the genome and no copies of other parts. To ensure that all sequences of the genome that can be cloned are contained within a collection, genomic libraries typically represent an average segment of the genome at least five times (and so, in our example, there would be 50,000 independent clones in the genomic library). This multifold representation makes it highly unlikely that, by chance, a sequence would not be represented at least once in the library.

Similarly, representative collections of cDNA inserts require tens or hundreds of thousands of independent cDNA clones; these collections are **cDNA libraries.** Because we do not know how many mRNAs are produced from an organism's genome, it is impossible to give a solid estimate of the number of clones constituting a representative library. Further, not all mRNAs are expressed at any given time in the life of an organism; so sampling mRNA from different tissues, different developmental stages, and organisms grown in different environmental conditions is a necessary prerequisite for collecting cDNA representations of the totality of mRNAs encoded by a genome.

The choice between genomic DNA and cDNA library construction depends on the situation. If a specific gene that is active in a specific type of tissue in a plant or animal is being sought, then it makes sense to use that tissue to prepare mRNA to be converted into cDNA and then make a cDNA library from that sample. For example, for identifying cDNAs corresponding to hemoglobin mRNAs, mRNAs from erythrocytes (the red blood cells where hemoglobin protein is in abundance) are the appropriate source of material. Such a cDNA library should be enriched for the gene in question. A cDNA library represents a subset of the transcribed regions of the genome; so it will inevitably be smaller than a complete genomic library, which by definition contains clones representing the entire genome. Although genomic libraries are bigger, they do have the benefit of containing genes in their native form, including introns and untranscribed regulatory sequences. If the purpose of constructing the library is a prelude to cloning an entire gene or an entire genome, then a genomic library is necessary at some stage. There are methods for enriching genomic libraries for one particular fraction of the genome. For example, there are techniques for purifying individual chromosomes. Chromosome-specific "subgenomic" libraries can then be constructed from DNA segments isolated from these purified chromosomes.

MESSAGE

The task of isolating a clone of a specific gene begins with making a library of genomic DNA or cDNA—if possible, enriched for sequences containing the gene in question.

Identifying DNA Molecules of Interest

The production of a library is sometimes referred to as "shotgun" cloning because the experimenter clones a large sample of fragments and hopes that one of the clones will contain a "hit"—the desired gene. The task then is to find that particular clone. How to do so is what we will consider next.

Finding specific clones by using probes. The library, which might contain as many as hundreds of thousands of cloned fragments, must be screened to find the recombinant DNA molecule of interest—typically, a molecule represents a gene of interest to a researcher. Such screening is accomplished by using a specific **probe** that will find and mark the clone for the researcher to identify. Broadly speaking, there are two types of probes: those that recognize a specific DNA sequence and those that recognize part of a specific protein.

Probes for finding DNA. Probes for finding specific DNA sequences depend on the ability of a single strand of nucleic acid to find and hybridize to another single strand with a complementary base sequence. A probe that is itself DNA, when denatured (made single stranded by unwinding the two halves of the double helix), will therefore find and bind to other similar denatured DNAs in the library. As exemplified in the diagram below, probes as small as 15 to 20 base pairs will specifically hybridize to complementary sequences within much larger cloned DNAs. Thus, probes can be thought of as "bait" to identify much larger "prey."

The identification of a specific clone in a library is a two-step procedure (Figure 8-14). First, colonies or plaques of the library on a petri dish are transferred to an absorbent membrane (often nitrocellulose) by simply laying the membrane on the surface of the medium. The membrane is peeled off, colonies or plaques clinging to the surface are lysed in situ, and the DNA is denatured. Next, the membrane is bathed with a solution of a probe that is specific for the DNA being sought. The probe must be labeled either with a radioactive isotope or a fluorescent dye. Generally, the probe is itself a cloned piece of DNA that has a sequence homologous to that of the desired gene. The probe DNA must be denatured; it will then bind only to the DNA of the clone being sought. The position of a positive clone will become clear from the position of the concentrated radioactive or fluorescent label. For radioactive labels, the membrane is placed on a piece of X-ray film, and the decay of the radioisotope produces subatomic particles that "expose" the X-ray film, producing a dark spot on the film adjacent to the location of the radioisotope concentration. Such an exposed film is called an **autoradiogram** because the membrane itself, instead of an outside source of X-rays, exposes the film. Alternatively, the label can be a fluorescent dye. The membrane containing hybridized dye-labeled probe is exposed to the correct wavelength of light to activate the dye's fluorescence, and a photograph is taken of the resulting fluorescence on the membrane. Regardless of the type of label, the position of the spot of label indicates the location of DNA segments containing sequences complementary to the probe.

Where does the DNA to make a probe come from? The DNA can be from one of several sources. One source is cDNA from tissue that expresses a gene of interest. Another source of DNA for a probe might be a homologous gene from a related organism. This method depends on the evolutionary conservation of DNA sequences through time. Even though the probe DNA and the DNA of the desired clone might not be identical, they are often similar enough to promote hybridization. The method is jokingly called "clone by phone" because, if you can phone a colleague who has a clone of your gene of interest from a related organism, then your job of cloning is made relatively easy. If the protein product of the gene of interest is known, it can provide another way to generate a probe. After an amino acid sequence of the protein product has been obtained, synthetic DNA probes can be designed by using the table of the genetic code in reverse (from amino acid to codon); so an amino acid sequence merely has to be back-translated to obtain the DNA sequence that encoded it. However, because the genetic code is degenerate—that is, most amino acids are encoded by multiple codons—several possible DNA sequences could in theory encode the protein in question (but only one of these DNA sequences is present in the gene that actually encodes the protein). To get around this problem, a short stretch of amino acids with minimal de-

```
Probe                          3′- AAGCCTATTTATGGGCAAT -5′
Hydrogen bonds                     |||||||||||||||||||
Clone           5′- . . . AGCTAGGGATCTTCGGATAAATACCCGTTACGTACTATTGGAAGGA . . . -3′
```

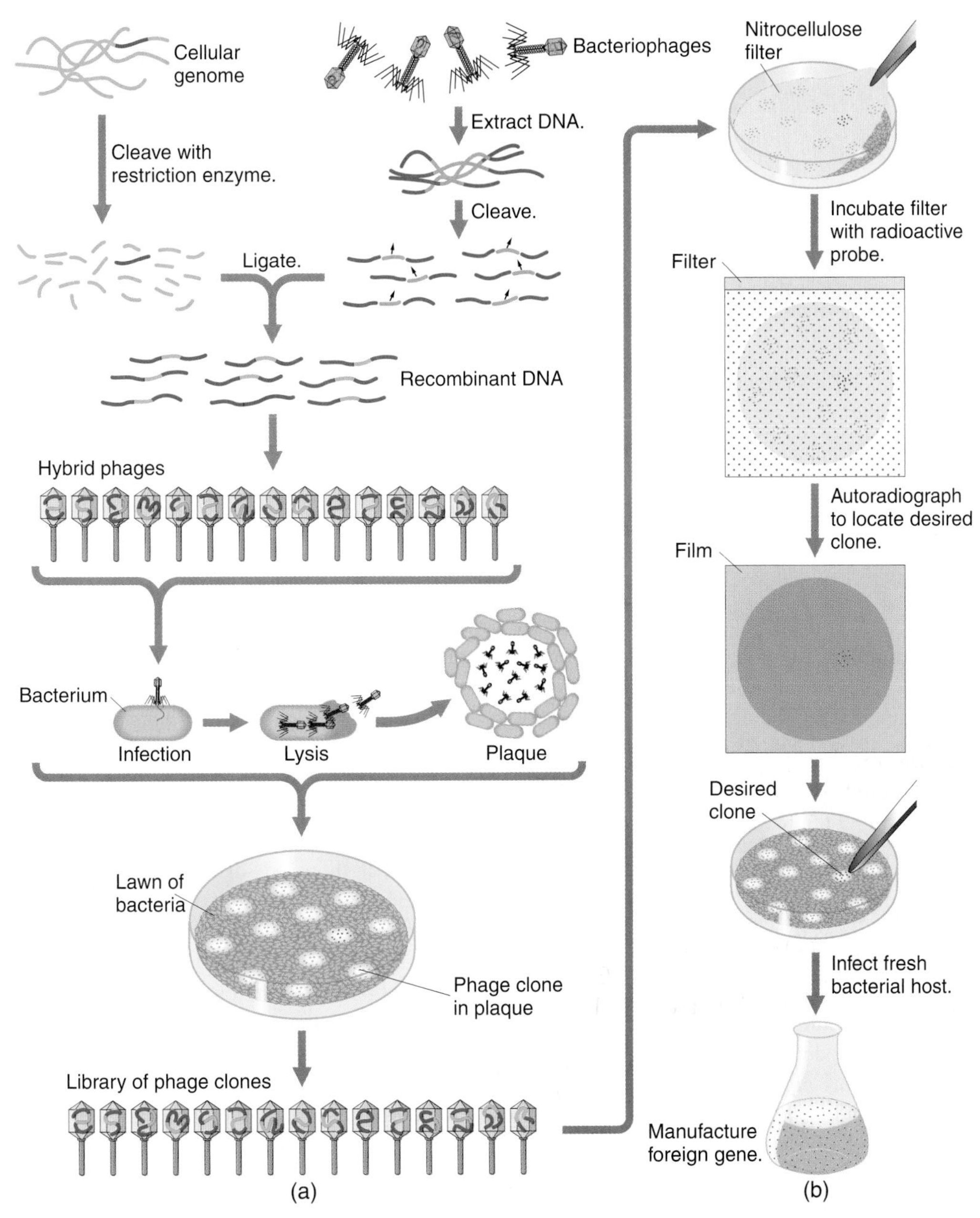

Figure 8-14 (a) A genomic library can be made by cloning genes in λ bacteriophages. When a lawn of bacteria on a petri dish is infected by a large number of different hybrid phages, each plaque in the lawn is inhabited by a single clone of phages descended from the original infecting phage. Each clone carries a different fragment of cellular DNA. The problem now is to identify the clone carrying a particular gene of interest (dark blue) by probing the clones with DNA or RNA known to be related to the desired gene. (b) The plaque pattern is transferred to a nitrocellulose filter, and the phage protein is dissolved, leaving the recombinant DNA, which is then denatured so that it will stick to the filter. The filter is incubated with a radioactively labeled probe—that is, a DNA copy of the messenger RNA representing the desired gene. The probe hybridizes with any recombinant DNA incorporating a matching DNA sequence, and the position of the clone having the DNA is revealed by autoradiography. Now the desired clone can be selected from the corresponding spot on the petri dish and transferred to a fresh bacterial host so that a pure gene can be manufactured. *(Adapted from R. A. Weinberg, "A Molecular Basis of Cancer," and P. Leder, "The Genetics of Antibody Diversity." © 1983, 1982 by Scientific American, Inc. All rights reserved.)*

generacy is selected. A mixed set of probes containing all possible DNA sequences that can encode this amino acid sequence is then designed by using the codon dictionary. The chemical DNA synthesis reaction is a step-by-step process, so wherever in the sequence there are alternative nucleotides, a mixture of those alternative nucleotides is fed into the reaction and all possible DNA strands are synthesized. An example in which there are five positions of

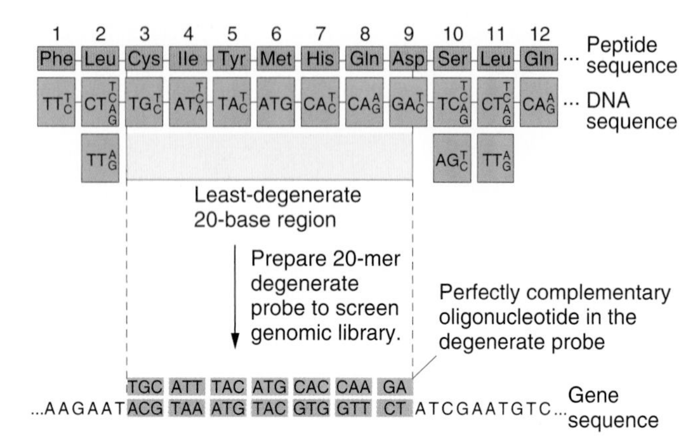

Figure 8-15 A short sequence of a protein is used to design a set of redundant oligonucleotides for use as a probe to recover the gene that encoded the protein. One of the set of probes will be a perfect match for the gene. *(From H. Lodish, D. Baltimore, A. Berk, S. L. Zipursky, P. Matsudaira, and J. Darnell,* Molecular Cell Biology, *3d ed. © 1995 by Scientific American Books.)*

degeneracy for amino acids 2 through 8, showing two, three, two, two, and two alternatives at the degenerate positions, respectively, appears in Figure 8-15. The reaction would make $2 \times 3 \times 2 \times 2 \times 2 \times 2 = 48$ **oligonucleotide** strands at the same time. This "cocktail" of oligonucleotides would be used as a probe. The correct strand within this cocktail would find the gene of interest. As noted previously, 20 nucleotides embody enough specificity to hybridize to one unique complementary DNA sequence in the library. One final possibility is to use labeled free RNA as a probe. This type of probe is possible only when a nearly pure population of identical molecules of RNA can be isolated, such as rRNA or fractionated tRNAs.

Probes for finding proteins. If the protein product of a gene is known and isolated in pure form, then this protein can be used to detect the clone of the corresponding gene in a library. The process is described in Figure 8-16. An **antibody** (a protein made by the immune system that binds with high affinity to a given molecule) to the protein is prepared, and this antibody is used to screen an expression library. These libraries are special cDNA libraries made by using **expression vectors,** vectors designed to express high levels of a recombinant protein. To make the library, cDNA is inserted into the vector in the correct triplet reading frame with a bacterial protein (in this case, β-galactosidase), and the cells containing the recombinant expression insert produce a "fusion" protein that is partly a translation of the cDNA insert and partly a portion of the normal β-galactosidase. A membrane is laid over the surface of the medium and removed with some of the cells of each colony now attached to the membrane at locations reflective of their positions on the original petri dish (see Figure 8-16). The imprinted membrane is then dried and bathed in a solution of the antibody, which will bind to the imprint of any colony that contains the fusion protein of interest. Positive clones are revealed by making an antibody to the first antibody; the second antibody is labeled by a radioactive isotope or a chemical that will fluoresce or become a colored dye. By detecting the correct protein, the antibody effectively identifies the clone containing the gene that must have synthesized that protein and therefore contains the cDNA whose recovery was wanted.

MESSAGE

A cloned gene can be selected from a library by using probes for the gene's DNA sequence or for the gene's protein product.

Probing to find a specific nucleic acid in a mixture. In the course of gene and genome manipulation, it is often necessary to detect and isolate a specific DNA molecule from among a complex mixture of restriction fragments. For example, recall that, in cloning with the use of λ phage, it is necessary to separate the two chromosome arms from the unwanted central region of the λ genome. There are several ways of detecting and isolating particular restriction fragments, but the most extensively used method is **gel electrophoresis**. If a mixture of linear DNA molecules is placed in a well cut into an agarose gel and the well is placed near the cathode of an electric field, the molecules will migrate through the gel to the anode, at speeds inversely dependent on their size; in other words, the smallest pieces migrate most rapidly (Figure 8-17 on page 232). Therefore, if there are distinct size classes in the mixture, these classes will form distinct bands on the gel. The bands can be visualized by staining the DNA with ethidium bromide, which causes the DNA to fluoresce in ultraviolet light. A comparison of the migration of the visualized restriction fragments with a set of control size standards (a set of fragments of accurately measured sizes) allows the absolute size of each restriction fragment in the mixture to be determined. If the bands are well separated, an individual band can be cut from the gel, and the DNA sample can be purified from the gel matrix. Therefore DNA electrophoresis can be either diagnostic (showing sizes and relative amounts of the DNA fragments present) or preparative (useful in isolating specific DNA fragments).

Restriction enzyme digestion of genomic DNA results in so many fragments that a stained gel of a digest that has been separated by electrophoresis shows a continuous smear of DNA rather than discrete bands. A probe can identify one fragment in this mixture, with the use of a technique developed by E. M. Southern called **Southern blotting** (Figure 8-18 on page 233). Just as in clone identification (see Figure 8-14b), this technique entails getting an imprint of DNA molecules on a membrane (in this case, the imprint reflects the geographic location of the DNA on the gel after electrophoresis is complete) and hybridizing with labeled probe to identify the locations of DNA sequences on the membrane that are complementary to the probe. After DNA

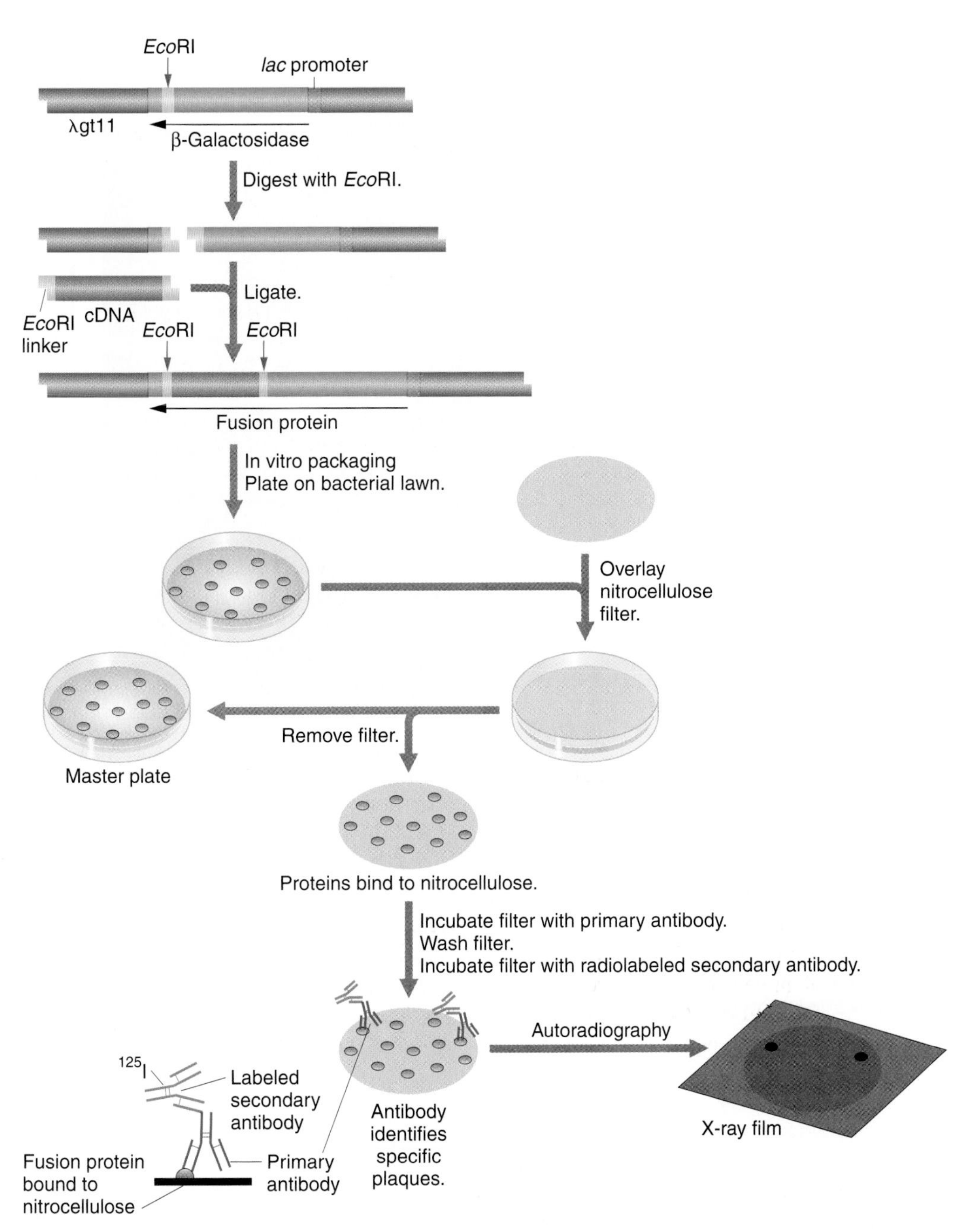

Figure 8-16 Finding the clone of interest by using antibody. An expression library made with special phage λ vector called λgt11 is screened with a protein-specific antibody. After the unbound antibodies have been washed off the filter, the bound antibodies are visualized through the binding of a radioactive secondary antibody. *(From J. D. Watson, M. Gilman, J. Witkowski, and M. Zoller,* Recombinant DNA, *2d ed. © 1992 by Scientific American Books.)*

fragments are separated by gel electrophoresis, an absorbent membrane is laid over the gel, and the DNA bands are transferred ("blotted") onto the membrane by capillary action. The DNA is then treated to denature it and anchor it to the locations of its bands on the membrane. Thus, when transferred to the membrane, the DNA bands stay in the same relative positions as on the gel. The membrane is bathed in a labeled single-stranded DNA or RNA probe (both work) and an autoradiogram or a photograph of fluorescent bands is used to reveal the presence of any bands on the gel that are complementary to the probe. If appropriate, those bands can be cut out of the gel and further processed. With the use of size-calibration standards, the sizes of any interesting fragments in the experimental sample can be determined.

The Southern-blotting technique can be extended to detect a specific *RNA* molecule from a mixture of RNAs fractionated on a gel. This technique is called **Northern blotting** (thanks to some scientist's sense of humor) to contrast it with the Southern technique for *DNA* analysis. The fractionated RNA is blotted onto a membrane and probed in the same way as DNA is blotted and probed for Southern blotting. One application of Northern analysis is in determining whether a specific gene is transcribed in a certain tissue or under certain environmental conditions. RNA is extracted from the appropriate cell sample and then electrophoresed, blotted, and probed with the cloned gene in question. A positive signal shows the presence of the transcript. Hence we see that cloned DNA finds widespread application as a probe, for detecting a specific clone, DNA fragment, or RNA. In all these cases, the ability of nucleic acids with *complementary* nucleotide sequences to find and bind to each other is being exploited.

MESSAGE

Recombinant DNA techniques that depend on complementarity to a cloned DNA probe include blotting and hybridization systems for the identification of specific clones, restriction fragments, or mRNAs or for measurement of the size of specific DNAs or RNAs

Finding specific clones by functional complementation. Transformation is now possible in many species, as discussed later in this chapter. For organisms in which transformation is possible, specific clones in a bacterial or phage library can be detected through their ability to confer a missing function on a mutant line of the donor organism. This procedure is called **functional complementation** or **mutant rescue.** The general outline of the procedure is as follows:

Make a bacterial or phage library containing wild-type a^+ recombinant donor DNA inserts.

↓

Transform cells of recessive mutant cell-line a^- by using the DNA from individual clones in the library.

↓

Identify clones from the library that produce transformed cells with the dominant a^+ phenotype.

↓

Recover the a^+ gene from the successful bacterial or phage clone.

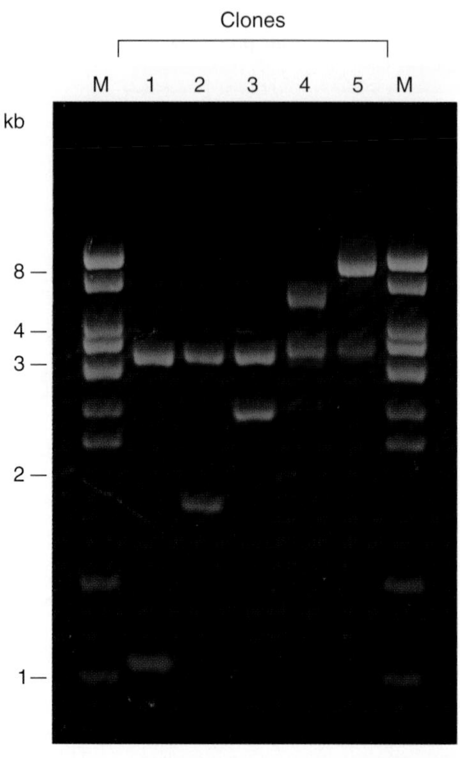

Figure 8-17 Mixtures of different-sized DNA fragments separated electrophoretically on an agarose gel. In this case, the samples are five recombinant vectors treated with *Eco*RI. The mixtures are applied to wells at the top of the gel, and fragments move under the influence of an electric field to different positions dependent on size (and, therefore, number of charges). The DNA bands have been visualized by staining with ethidium bromide and photographing under UV light. (*M* represents lanes containing standard fragments acting as markers for estimating DNA length.) *(From H. Lodish, D. Baltimore, A. Berk, S. L. Zipursky, P. Matsudaira, and J. Darnell,* Molecular Cell Biology, *3d ed. © 1995 by Scientific American Books.)*

Finding specific clones on the basis of genetic-map location—positional cloning. Information about a gene's position in the genome can be used to circumvent the hard work of assaying an entire library to find the clone of interest. **Positional cloning** is a term that can be applied to any method that makes use of such information. A common starting point is the availability of a landmark cloned sequence known to be closely linked on a genetic map to the gene being sought. The linked sequence is a marker for the target gene and serves as the departure point in a process called **chromosome walking** that will terminate at the target gene.

Figure 8-19 on page 234 summarizes the procedure of chromosome walking. The basic idea is to use the first cloned landmark sequence as a probe to identify a second set of clones that contain inserts from the same location in the donor genome as that of the landmark clone but extend out from the landmark clone in one of two directions (toward the target or away from the target). If the overlaps between the landmark clones and the new clones can be measured and if the ends of the new clones farthest from the landmark clone can be identified, then these new end fragments can be used as probes for identifying a third set of overlapping clones from the genomic library. In this *step-by-step* fashion, a set of clones representing the region of the genome extending out from the landmark clone can be arrayed until one obtains clones that can be shown to include the target gene (e.g., by identifying mutant DNA sequences in individuals mutant for the target gene). Because the process consists of steps, the term *chromosome walking* has been applied to it.

The key to chromosome walking is therefore to have the ability to order the array of clones that hybridize to a given probe. Such ordering is done by creating restriction maps of the clones. A **restriction map** is a linear map of the order and distances of restriction endonuclease cut sites in a segment of DNA. An example of one method to create a restriction map

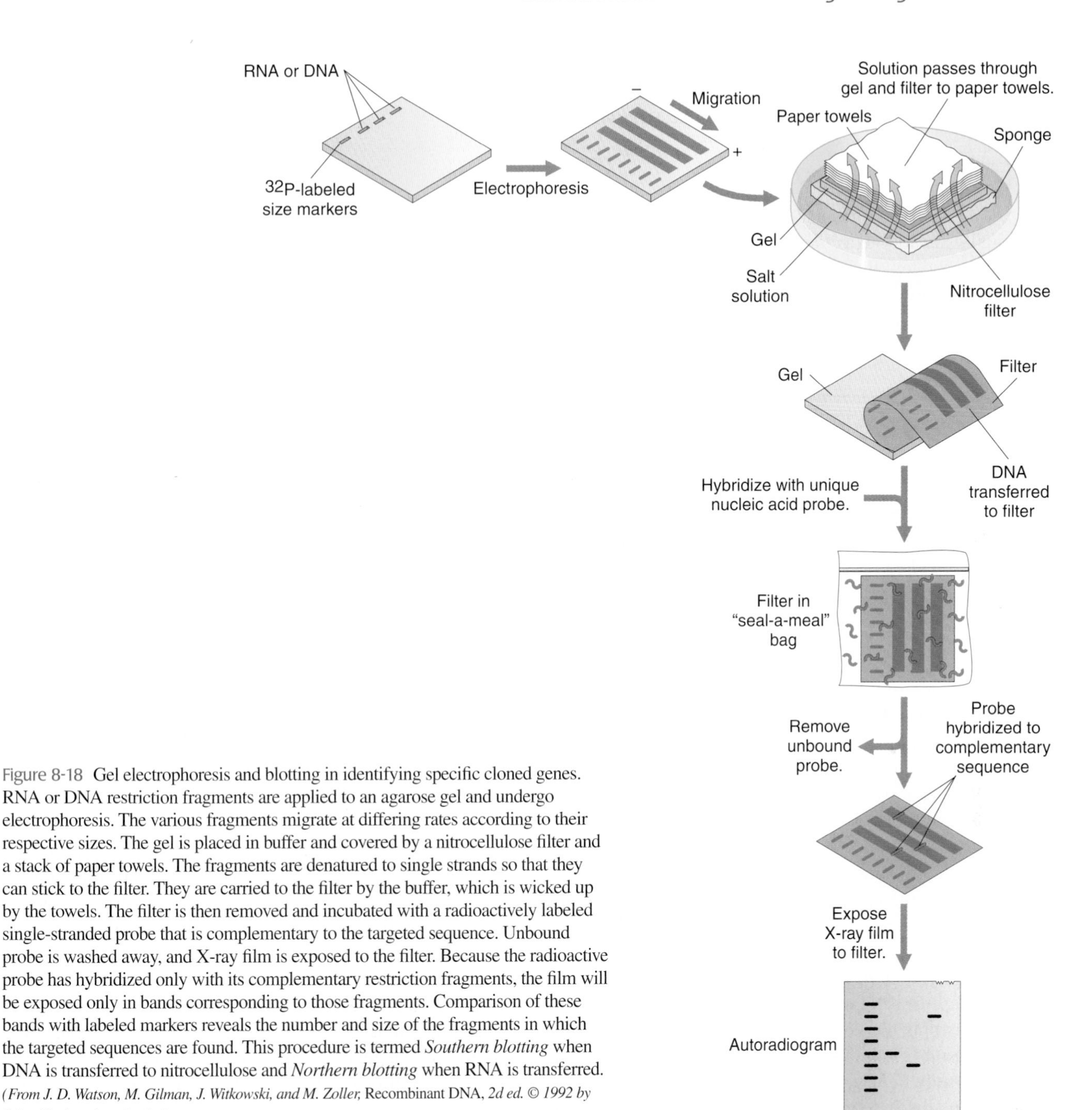

Animated Art • Finding Specific Cloned Genes by Functional Complementation: Using the Cloned GAL Gene as a Probe for GAL mRNA

Figure 8-18 Gel electrophoresis and blotting in identifying specific cloned genes. RNA or DNA restriction fragments are applied to an agarose gel and undergo electrophoresis. The various fragments migrate at differing rates according to their respective sizes. The gel is placed in buffer and covered by a nitrocellulose filter and a stack of paper towels. The fragments are denatured to single strands so that they can stick to the filter. They are carried to the filter by the buffer, which is wicked up by the towels. The filter is then removed and incubated with a radioactively labeled single-stranded probe that is complementary to the targeted sequence. Unbound probe is washed away, and X-ray film is exposed to the filter. Because the radioactive probe has hybridized only with its complementary restriction fragments, the film will be exposed only in bands corresponding to those fragments. Comparison of these bands with labeled markers reveals the number and size of the fragments in which the targeted sequences are found. This procedure is termed *Southern blotting* when DNA is transferred to nitrocellulose and *Northern blotting* when RNA is transferred. *(From J. D. Watson, M. Gilman, J. Witkowski, and M. Zoller,* Recombinant DNA, *2d ed. © 1992 by Scientific American Books.)*

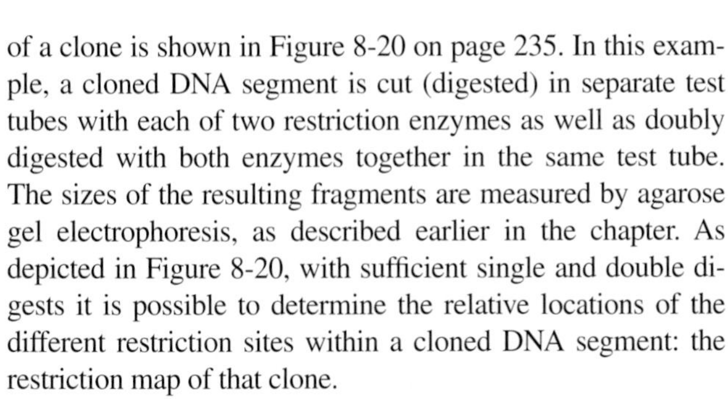

of a clone is shown in Figure 8-20 on page 235. In this example, a cloned DNA segment is cut (digested) in separate test tubes with each of two restriction enzymes as well as doubly digested with both enzymes together in the same test tube. The sizes of the resulting fragments are measured by agarose gel electrophoresis, as described earlier in the chapter. As depicted in Figure 8-20, with sufficient single and double digests it is possible to determine the relative locations of the different restriction sites within a cloned DNA segment: the restriction map of that clone.

The resulting restriction maps of the DNA of the clones can then be used to determine the most efficient way to take the next step in the walk. Once the restriction maps of the several overlapping clones from one step in the walk are determined, the order and degree of overlap of the clones can be established, and the end fragment of the clone that extends this step in the walk the farthest can be determined and used as the probe for the next step. In this way, every stride in this step-by-step procedure can be as long as possible.

It is important to recognize that chromosome walking is only one of many applications of restriction mapping. In a sense, the restriction map is a partial sequence map of a DNA segment, because every restriction site is one at

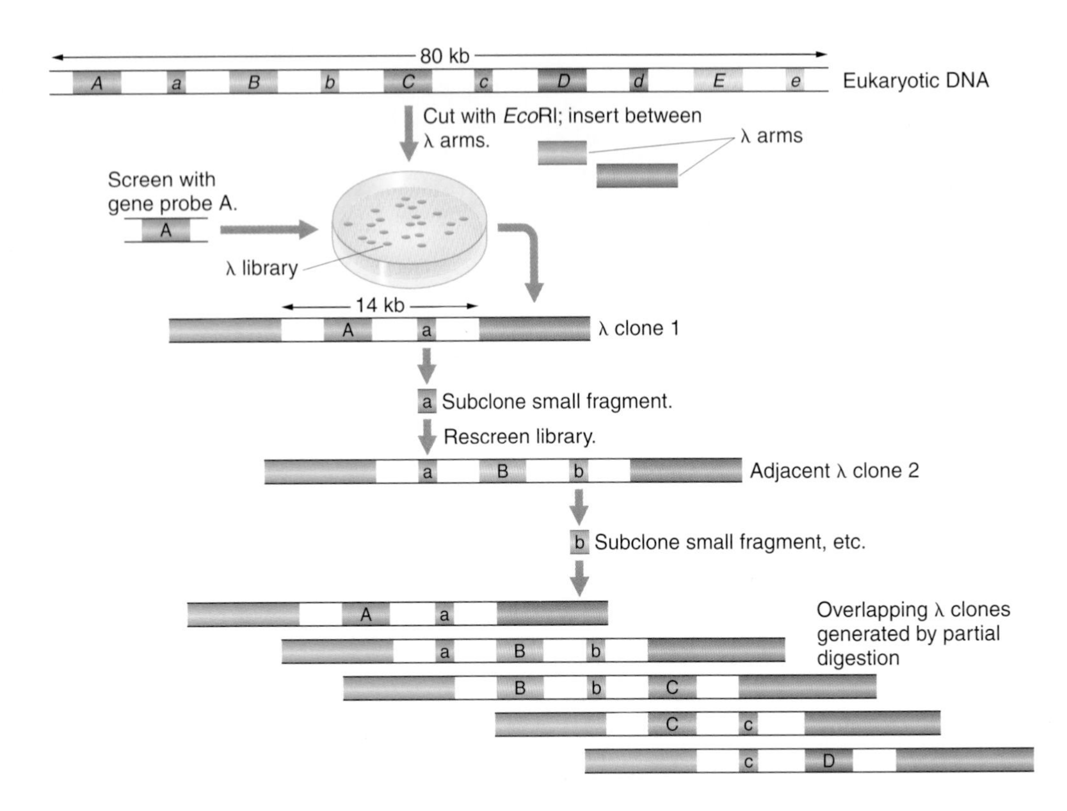

Figure 8-19 Chromosome walking. One recombinant phage obtained from a phage library made by the partial *Eco*RI digestion of a eukaryotic genome can be used to isolate another recombinant phage containing a neighboring segment of eukaryotic DNA, as described in the text. This walk illustrates how to start at molecular landmark A and get to target gene D. *(From J. D. Watson, J. Tooze, and D. T. Kurtz,* Recombinant DNA: A Short Course. © *1983 by W. H. Freeman and Company.)*

which a particular short DNA sequence resides (depending on which restriction enzyme cuts at that site). We will see in Chapter 9 an application of restriction mapping in determining the locations of genes whose mutant alleles contribute to genetic disease. In addition, the task of creating recombinant DNA molecules for a particular segment of the genome is simplified by knowing the restriction map of the segment so that the appropriate restriction enzymes can be chosen to produce a restriction fragment of desired size for cloning into a specific vector.

Finding specific clones by using mutant landmarks—gene tagging. In **gene tagging,** a piece of DNA of defined sequence disrupts the normal sequence of the gene by becoming inserted within the gene, thereby disrupting its normal function. The specific sequence then becomes a landmark (really, a sequence complementary to a probe) that can be used to identify DNA sequences corresponding to the gene. The approach is summarized in Figure 8-21. When the insertional mutation (a mutation caused by fusion of an exogenous DNA segment into the DNA sugar-phosphate backbone corresponding to the gene) has been identified, a genomic library from the strain containing the insertional mutation is prepared. In the simplest case—the only copy of the insertion sequence in the genome resides within the gene of interest—all that need be done is to probe the genomic library with labeled single-stranded insertion-sequence DNA and identify clones containing DNA corresponding to this sequence. Because of the presence of the genetically defined mutation, these clones would then be inferred to also contain sequences from the disrupted gene. When the sequence surrounding the inserted sequence has been cloned, it can be characterized for its genetic content, such as being searched for hybridization to known mRNAs and sequenced to predict protein products. The same basic approach can be taken even when there are multiple copies of the insertion sequence in the genome; in this case, there are more clones to slog through before one can determine which clones include DNA from the gene of interest (rather than DNA adjacent to one of the other insertions in this genome).

One type of tag is *transforming DNA.* When exogenous DNA is introduced into a cell by transformation, viral or phage infection, or other methods such as injection, it can insert into the genome and become part of the chromosome. Ectopic insertion occurs throughout the genome, and apparently no segment of chromosomal DNA is completely immune to integration. When integration takes place within or near a gene, the inserting fragment acts as a mutagen, disrupting the function of the interrupted gene. This property can be used to good advantage. Suppose that we are interested in cloning a gene called *A* in the fungus *Aspergillus.* Suppose also that we have a different cloned gene x^+ that we will use as our "tag DNA" (see Figure 18-21). By DNA transformation, we now have transformed x^- *Aspergillus* cells into x^+. Many of the x^+ transformants will be mutant for the genes into which the transforming DNA has inserted ectopically. If we're lucky or if we've generated a very large number of transformants, a few of such x^+ transformants will be mutant for gene *A*, the gene that we're interested in cloning, and will be of phenotype A^-. After we have discovered the A^- mutations among the x^+ transformants, the next step is to cross these transformed *Aspergillus* cells with $x^- A^+$ cells to determine whether A^- and x^+ cosegregate. If they do, then the A^- mutation is likely to have been caused by the insertion of the x^+ transforming DNA into gene A. A genomic library from this mutant transformed *Aspergillus* strain is generated, and DNA corresponding to gene x^+ is used as a probe to recover the clone of the disrupted A gene. To recover the intact wild-type A gene, a fragment of the cloned disrupted A gene sequence is used in another round of probing, this time with a library made from an A^+ strain.

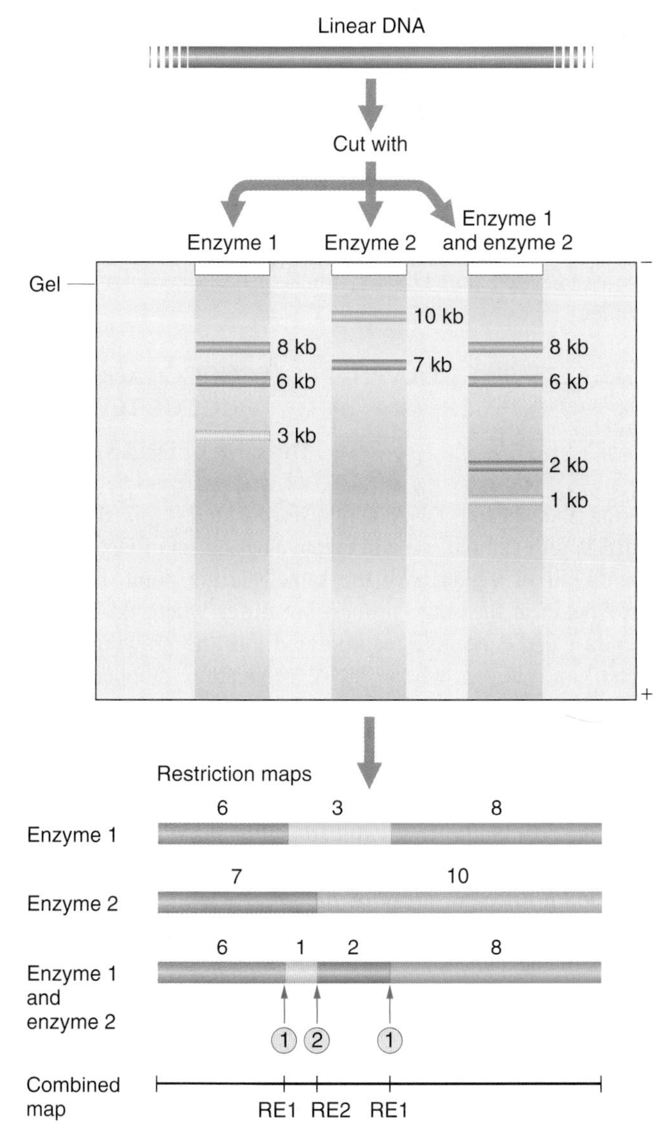

Figure 8-20 Restriction mapping by comparing electrophoretic separations of single and multiple digests. In this simplified example, digestion with enzyme 1 shows that there are two restriction sites for this enzyme but does not reveal whether the 3-kb segment generated by this enzyme is in the middle or on one of the ends of the digested sequence, which is 17 kb long. Combined digestion by both enzyme 1 (RE1) and enzyme 2 (RE2) leaves the 6- and 8-kb segments generated by enzyme 1 intact but cleaves the 3-kb fragment, showing that enzyme 2 cuts at a site within the 3-kb fragment. If the 3-kb segment were at one of the ends of the 17-kb sequence, digestion of the 17-kb sequence by enzyme 2 alone would yield a 1- or 2-kb fragment by cutting at the same site at which this enzyme cut to cleave the 3-kb fragment in the combined digestion by enzymes 1 and 2. Because this is not the case, of the three restriction fragments produced by enzyme 1, the 3-kb fragment must lie in the middle. That the RE2 site lies closer to the 6-kb section than to the 8-kb section can be inferred from the 7- and 10-kb lengths of the enzyme 2 digestion.

A similar approach uses transposons as tags. *Transposons* are naturally occurring mobile DNA elements—that is, DNA segments that can be excised from one position in the genome and inserted at other genomic locations. We will consider their molecular properties and their uses as mutagens in detail in Chapters 9, 10, and 12. When mobilized, a given transposon can insert at one of many different locations in the genome. If it inserts within a gene, it can disrupt that gene's normal function—often causing null mutations. In a line containing an active transposon, mutants for a gene that we are interested in cloning are selected. Typically, these mutants will be caused by the insertion of the transposon into the gene of interest. We can then clone the DNA of the gene in a manner completely analogous to that described in the preceding paragraph for gene A, except that the transposon DNA is the first probe, the probe used to recover DNA sequences corresponding to the mutated gene (see Figure 8-21).

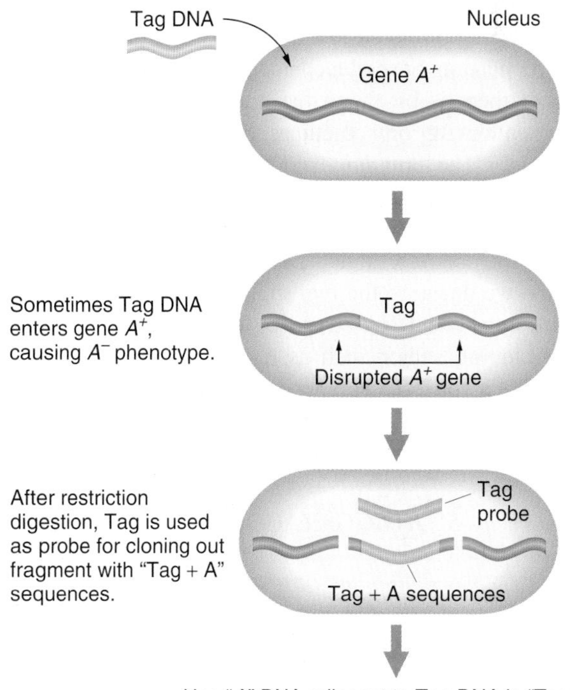

Figure 8-21 Using DNA insertion as a tag for marking and recovering a gene from the genome. The tag DNA can be transforming DNA or an endogenous transposon (movable element).

MESSAGE

Mutating a gene by the insertion of transforming DNA or a transposon allows the gene to be tagged as a prelude to its isolation.

Determining the Base Sequence of a DNA Segment

The ultimate language of the genome is composed of strings of A's, T's, C's, and G's. Although we know a fair bit about how to comprehend the meaning of some of the word strings, such as protein-coding sequences, our ability to decipher other strings of this four-letter alphabet is still rather limited. Nonetheless, the scientific literature is full of examples in which the sequence of genes and genomes has been extremely informative in understanding protein function, pathways, and disease. Therefore, obtaining the complete nucleotide sequence of a segment of DNA is often a very important part of understanding the organization and regulation of a gene, its relation to other genes, or the function of its encoded RNA or protein. Indeed, for the most part, it is simpler to translate the nucleic acid sequence of a cDNA to discover the amino sequence of its encoded polypeptide chain than to directly sequence the polypeptide itself. In this section, we will consider the techniques used to read the nucleotide sequence of DNA.

As with other recombinant DNA technologies, DNA sequencing exploits base-pair complementarity together with an understanding of the basic enzymology and biochemistry of DNA replication. Several techniques have been developed, but one of them is by far in the greatest use. It is called **dideoxy sequencing** or, sometimes, **Sanger sequencing** after its inventor. The term *dideoxy* comes from a special modified nucleotide that is key to this technique.

The logic of dideoxy sequencing is straightforward. To read the sequence of a cloned DNA segment of, say, 5000 base pairs, we denature the two strands of this segment (or start with only one DNA strand by cloning into a single-stranded cloning vector such as bacteriophage M13). We create a primer for DNA synthesis that will hybridize to exactly one location on the cloned DNA segment and then add a special "cocktail" of DNA polymerase, normal nucleotide triphosphates (dATP, dCTP, dGTP, and dTTP), and a small amount of a special dideoxynucleotide for one of the four bases (e.g., dideoxy adenosine triphosphate, abbreviated ddATP). As synthesis of the complementary DNA strand proceeds, if the dideoxynucleotide triphosphate is incorporated into the growing DNA chain in place of the normal nucleotide triphosphate, the chain stops growing at that point. Suppose the DNA sequence of the DNA segment that we're trying to sequence is:

5′ ACGGGATAGCTAATTGTTTACCGCCGGAGCCA 3′

We would then start DNA synthesis from a complementary primer:

5′ ACGGGATAGCTAATTGTTTACCGCCGGAGCCA 3′
3′ CGGCCTCGGT 5′

← Direction of DNA synthesis

Using the special DNA synthesis cocktail "spiked" with ddATP, for example, we will create a nested set of DNA fragments, all of which have the same starting point, because they begin at the same primer, but differ where ddATP instead of dATP became incorporated into the growing DNA chain and hence where DNA replication stopped. The array of different ddATP-arrested DNA chains looks like the diagram below (*A indicates the dideoxynucleotide).

We can generate an array of such fragments for each of the four possible dideoxynucleotide triphosphates in four separate cocktails (one spiked with ddATP, one with ddCTP, one with ddGTP, and one with ddTTP). Each will produce a different array of dideoxynucleotide-terminated fragments, with no two spiked cocktails producing fragments of the same size. Further, if we add up the results of all four cocktails, we will see that the fragments can be ordered in length, with the lengths increasing by one base at a time. The final steps of the process are:

- Display the fragments in size order. This is done by gel electrophoresis, using the logic described earlier in this chapter.
- Label the newly synthesized strands so that they can be visualized after they have been separated according

5′	ATGGGATAGCTAATTGTTTACCGCCGGAGCCA	3′	Template DNA clone
3′	CGGCCTCGGT	5′	Primer for synthesis
	←		Direction of DNA synthesis
3′	*ATGGCGGCCTCGGT	5′	Dideoxy fragment 1
3′	*AATGGCGGCCTCGGT	5′	Dideoxy fragment 2
3′	*AAATGGCGGCCTCGGT	5′	Dideoxy fragment 3
3′	*ACAAATGGCGGCCTCGGT	5′	Dideoxy fragment 4
3′	*AACAAATGGCGGCCTCGGT	5′	Dideoxy fragment 5
3′	*ATTAACAAATGGCGGCCTCGGT	5′	Dideoxy fragment 6
3′	*ATCGATTAACAAATGGCGGCCTCGGT	5′	Dideoxy fragment 7
3′	*ACCCTATCGATTAACAAATGGCGGCCTCGGT	5′	Dideoxy fragment 8

to size by gel electrophoresis. Do so by either radioactively or fluorescently labeling the primer (initiation labeling) or the individual dideoxynucleotide triphosphate (ddNTP) (termination labeling).

The result of such dideoxy sequencing reactions is shown in Figure 8-22. Because the end result of the sequencing reactions is a ladder of labeled DNA chains increasing in length by one, all we need do is read up the gel to read the DNA sequence of the synthesized strand in the 5′-to-3′ direction.

Until now, the answers to two questions have remained a mystery: What is a dideoxynucleotide triphosphate (generically, a ddNTP)? And how does it block continued DNA synthesis? A dideoxynucleotide lacks the 3′-hydroxyl group as well as the 2′-hydroxyl group, which is also absent in a deoxynucleotide (Figure 8-23 on the next page). DNA synthesis requires a DNA polymerase-catalyzed

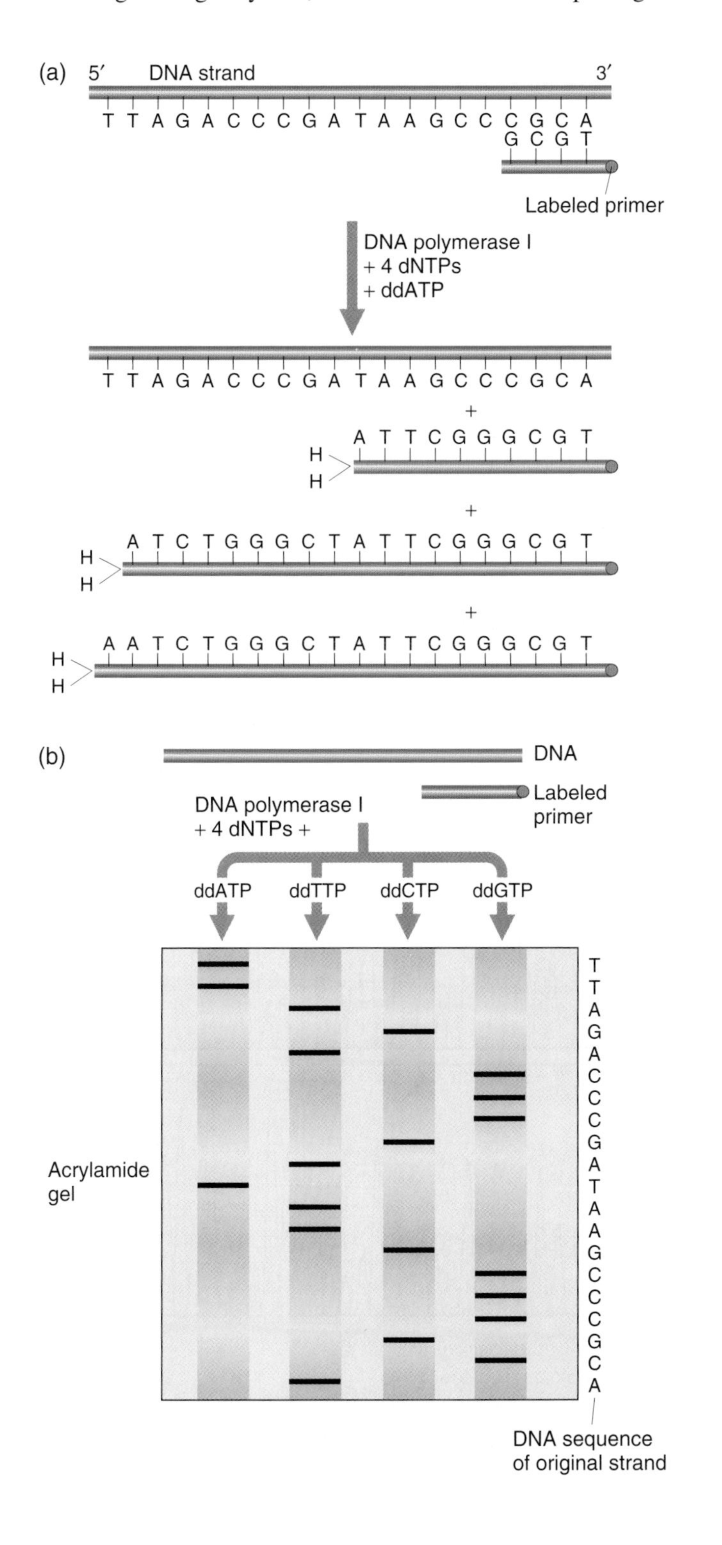

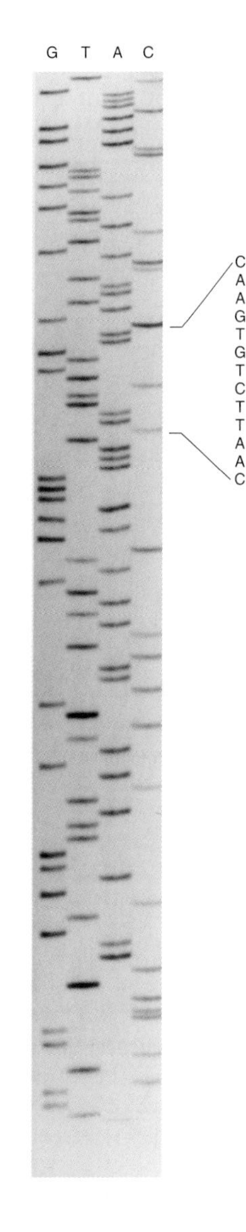

Figure 8-22 The dideoxy sequencing method. (a) A labeled primer (designed from the flanking vector sequence) is used to initiate DNA synthesis. The addition of four different dideoxy nucleotides (ddATP is shown here) randomly arrests synthesis. (b) The resulting fragments are separated electrophoretically and subjected to autoradiography. (c) Sanger sequencing gel. The inferred sequence is shown at the right. *(Parts a and b from J. D. Watson, M. Gilman, J. Witkowski, and M. Zoller,* Recombinant DNA, *2d ed. © 1992 by Scientific American Books; part c is from Loida Escote-Carlson.)*

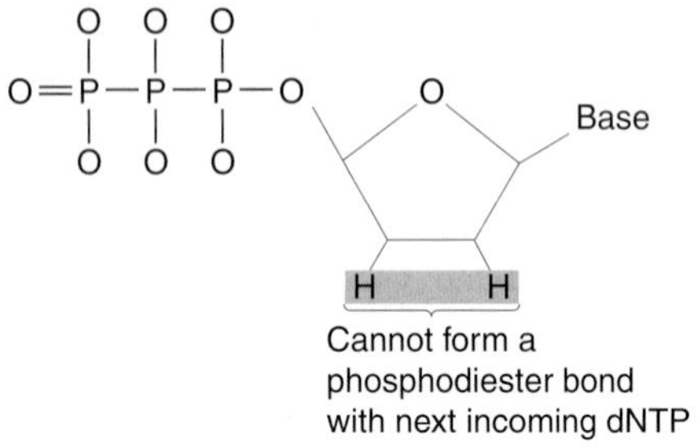

O=P—P—P—O
Base
H H
Cannot form a phosphodiester bond with next incoming dNTP

Figure 8-23 The structure of 2′,3′-dideoxynucleotides, which are employed in the Sanger DNA-sequencing method.

condensation reaction between the 3′-hydroxyl group of the last nucleotide added to the growing chain and the 5′-phosphate group of the next nucleotide to be added, releasing water and forming a phosphodiester covalent bond with the 3′-carbon atom of the adjacent sugar. Because a dideoxynucleotide lacks the 3′-hydroxyl group, this reaction cannot take place, and therefore DNA synthesis is blocked at the point of addition.

Instead of a radioactive label, the tag can be a fluorescent dye. If four different fluorescent color emitters are used for each of the four ddNTP reactions, then the four reactions can take place in the same test tube and the four sets of nested DNA chains can be electrophoresed together. Thus, four times as many sequences can be produced in the same space as can be produced by running the reactions separately. This logic is used in the fluorescence detection of automated DNA-sequencing machines. These machines perform electrophoresis on flat gel sheets (slab-gel electrophoresis), where adjacent lanes now contain reaction ladders from different clones, or in gel-containing capillary tubes, where each reaction ladder is physically isolated from its neighbors. For technical reasons of the ease and reliability of data capture, the capillary machines have proved far superior to the slab-gel machines. Thanks to these machines and the automation that they permit, DNA sequencing can proceed at a massive level, and sequences of whole genomes can be obtained by scaling up the procedures discussed in this section. Figure 8-24 illustrates a readout of automated sequencing. Each colored peak represents a different size fragment of DNA whose fluorescent base was detected by the fluorescent scanner of the automated DNA sequencer; the four different colors represent the four bases of DNA. Applications of automated sequencing technology on a genomewide scale will be a major focus of Chapter 9.

MESSAGE

A cloned DNA fragment can be sequenced by generating a set of labeled single-stranded DNAs that are complementary to one strand of the fragment, differ in length by one nucleotide, and terminate with a nucleotide of known identiry. When these DNAs are electrophoresed, the nucleotide sequence can be read directly from an autoradiogram of the gel.

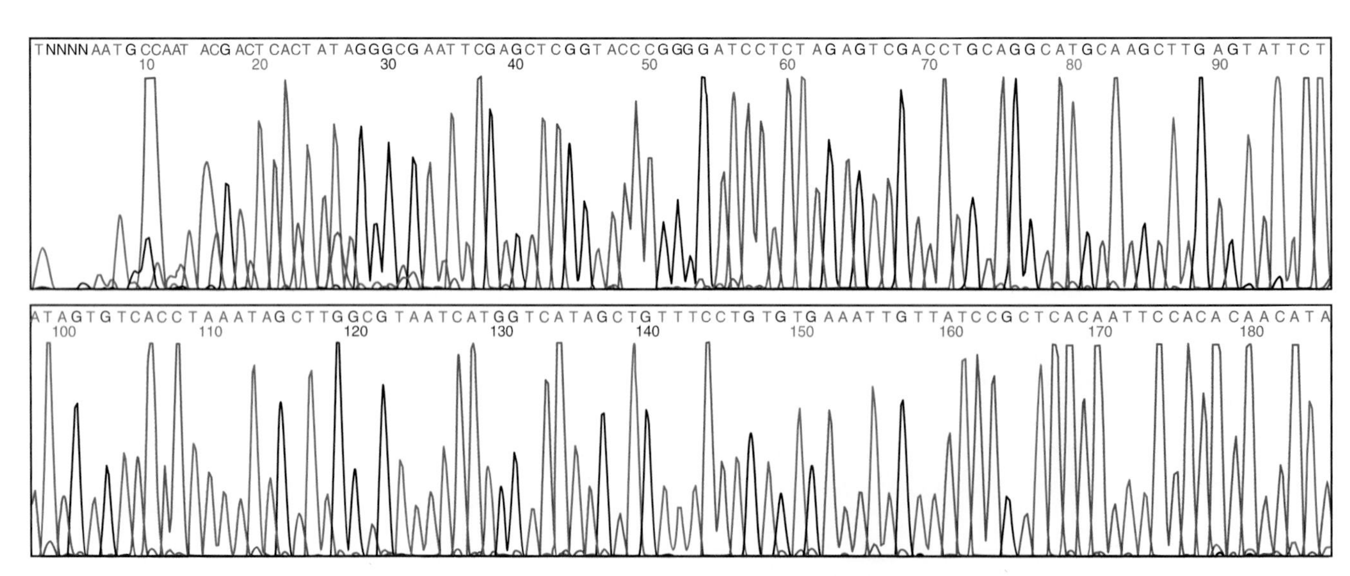

Figure 8-24 Printout from an automatic sequencer that uses fluorescent dyes. Each of the four colors represents a different base. N represents a base that cannot be assigned, because peaks are too low. Note that if this were a gel as in Figure 8-22c, each of these peaks would correspond to one of the dark bands on the gel; in other words, these colored peaks represent a different readout of the same sort of data as are produced on a sequencing gel.

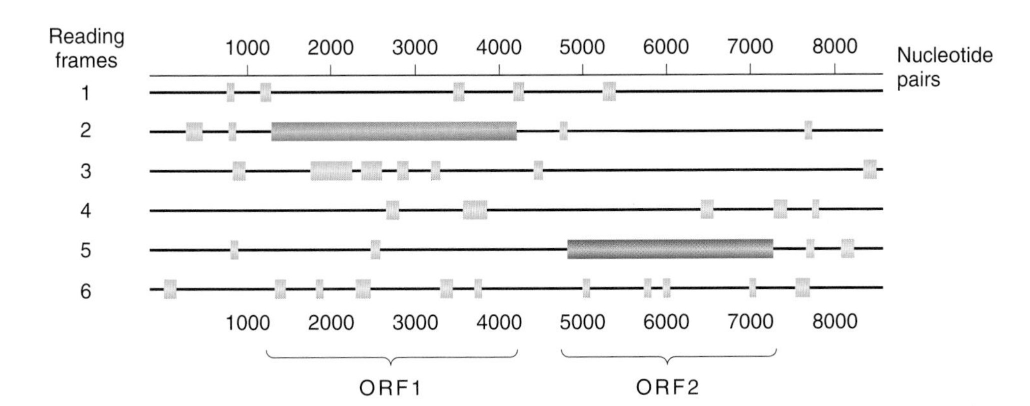

Figure 8-25 Any piece of DNA has six possible reading frames, three in each direction. Here the computer has scanned a 9-kb fungal plasmid sequence in looking for ORFs (potential genes). Two large ORFs, 1 and 2, are the most likely candidates as potential genes.

The nucleotide sequence of a cloned DNA fragment can be examined to find the gene or genes that it contains. The nucleotide sequence is fed into a computer, which then scans all six reading frames (three in each direction) in the search for possible protein-coding regions that begin with an ATG initiation codon, end with a stop codon, and are long enough that it is unlikely that a polypeptide of this length could not arise by chance. These stretches are called **open reading frames (ORFs).** They represent sequences that are candidate genes. Figure 8-25 shows such an analysis in which two candidate genes have been identified as ORFs. It is interesting to note that ORF detection and segregation of allelic differences, although poles apart in their approaches, are both ways of identifying genes.

The preceding sections have introduced the fundamental techniques that have revolutionized genetics. The remainder of the chapter will focus on the application of these techniques to diagnosis and genetic engineering. Foundations of Genetics 8-1 on the next page describes the molecular genetic analysis of the disease alkaptonuria, which incorporates and integrates many of the techniques introduced in this chapter.

Early Detection of Disease-Associated Alleles by Using Cloned DNA: Molecular Genetic Diagnostics

Recessive mutant phenotypes that follow single-gene inheritance are an important component of more than 500 human genetic diseases. For families at risk for such diseases (e.g., those who had previous cases among close relatives, with an inheritance pattern that could lead to an affected offspring), early fetal detection of homozygous recessive genotypes is important, for reasons of traditional genetic counseling, for early intervention of drug or dietary therapeutics, or, in the future, possibly for gene therapy.

To detect homozygosity for these defective alleles, fetal cells can be taken from the amniotic fluid, separated from other components, and cultured to allow the analysis of chromosomes, proteins, enzymatic reactions, and other biochemical properties. This process, **amniocentesis** (Figure 8-26), can identify a number of known disorders; Table 8-1 on page 242 lists some examples. **Chorionic villus sampling (CVS)** is a related technique in which a small sample of cells from the placenta is aspirated out with a long syringe. CVS can be performed earlier in the pregnancy than can amniocentesis, which must await the development of a large enough volume of amniotic fluid.

Testing physiological properties or enzymatic activity in cultured fetal cells limits the screening procedure to those disorders that affect characters or proteins expressed in the cultured cells. However, recombinant DNA technology improves the screening for genetic diseases in utero, because the DNA can be analyzed directly. In principle, the fetal gene being tested for the presence of a mutation that would produce the disease in question could be cloned and its sequence compared with that of a cloned normal gene to see if the fetal gene is normal. However, this procedure would be lengthy and impractical, and so shortcuts have to be devised to allow more rapid screening. Several of the useful techniques that have been developed for this purpose are explained in this section.

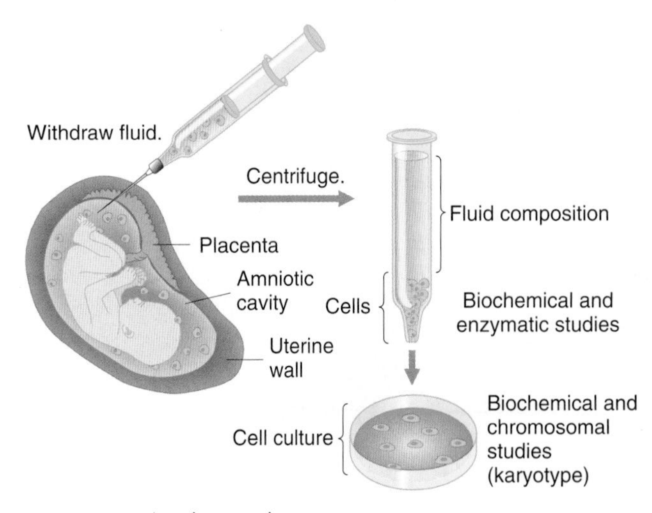

Figure 8-26 Amniocentesis.

Diagnosis of mutations on the basis of restriction-site differences. Sometimes mutations responsible for a specific disease happen to remove a restriction site normally present at a particular location in the DNA. Conversely, occasionally a mutation associated with a disease alters the normal sequence such that a restriction site is created. If either such feature is produced by a mutation that is commonly associated with a disease, the presence or absence of the restriction site becomes a convenient assay for determining the relevant genotype of an individual.

For example, sickle-cell anemia is a genetic disease that is commonly caused by a well-characterized mutational alteration. Affecting approximately 0.25 percent of African Americans, the disease results from an altered hemoglobin in which the amino acid valine replaces glutamic acid at position 6 in the β-globin chain. The GAG-to-GTG change that is responsible for the Glu-to-Val amino acid replacement eliminates a cut site for the restriction enzyme *Mst*II, which cuts the sequence CCTNAGG (in which N represents any of the four bases). The change from CCTGAGG to CCTGTGG can thus be recognized by Southern analysis using labeled β-globin cDNA as a probe, because the DNA derived from persons with sickle-cell disease lacks one fragment contained in the DNA of normal

FOUNDATIONS OF GENETICS 8-1

A Century of Genetic Research on Alkaptonuria

Alkaptonuria is a human disease with several symptoms, of which the most conspicuous is that the urine turns black when exposed to air. In 1898, an English doctor named Archibald Garrod showed that the substance responsible for the black color is homogentisic acid, which is excreted in abnormally large amounts into the urine of alkaptonuria patients. In 1902, early in the post-Mendelian era, Garrod suggested, on the basis of pedigree patterns, that alkaptonuria is inherited as a Mendelian recessive. Soon after, in 1908, he proposed that the disorder was caused by the lack of an enzyme that normally splits the aromatic ring of homogentisic acid to convert it into maleylacetoacetic acid. Because of this enzyme deficiency, he reasoned, homogentisic acid accumulates. Thus alkaptonuria is among the earliest proposed cases of an "inborn error of metabolism," an enzyme deficiency caused by a defective gene. There was a 50-year delay before it was shown by others that, in the liver of patients with alkaptonuria, activity for the enzyme that normally splits homogentisic acid, an enzyme called homogentisate 1,2-dioxygenase (HGO), is indeed totally absent. It seemed likely that the enzyme HGO was normally encoded by the alkaptonuria gene. In 1992, the alkaptonuria gene was mapped genetically to band 2 of the long arm of chromosome 3 (band 3q2).

Archibald Garrod. (Courtesy of Dr. Alexander G. Beran, American Philosophical Society.)

In 1995, Jose Fernandez-Canon and colleagues characterized a gene coding for HGO activity in the fungus *Aspergillus nidulans,* and, in 1996, they used the deduced amino acid sequence of this gene to do a computer search through a large number of sequenced fragments of a human cDNA library. They identified a positive clone that contained a human gene coding for 445 amino acids, which showed 52 percent similarity to the *Aspergillus* gene. When the human gene was expressed in an *E. coli* expression vector, its product had HGO activity. Furthermore, when the gene was used as a probe in a Northern analysis of liver RNA, a single RNA of the expected size was hybridized.

When the cloned gene was used as a probe for hybridization to chromosomes in which the DNA had been partly denatured (in situ hybridization—see Chapter 9), the probe bound to band 3q2, showing that it was indeed the gene for alkaptonuria.

The cDNA clone was used to recover the full-length gene from a genomic library. The gene was found to have 14 exons and spanned a total of 60 kb. A family of seven in which three children suffered from alkaptonuria was tested for mutations in this gene. PCR analysis was used to amplify all the exons individually. The amplified products were sequenced. One parent was heterozygous for a proline → serine substitution at position 230 in exon 10 (mutation P230S). The other parent was heterozygous for a valine → glycine substitution at position 300 in exon 12 (mutation V300G). All three children with alkaptonuria were of the constitution P230S/V300G, as expected if these positions were the mutant sites inactivating the HGO enzyme.

(Foundations of Genetics based on an idea by Charles Scriver.)

persons and contains a large (uncleaved) fragment not seen in normal DNA (Figure 8-27 on the next page).

Diagnosis of mutations on the basis of sequence differences. Most mutations of diagnostic interest are not associated with restriction-site changes. Rather, techniques for distinguishing mutant and normal alleles on the basis of any single base substitution that makes them differ are very important for diagnostics. Synthetic oligonucleotide probes can be designed to identify such differences. A good example is α_1-antitrypsin deficiency, which leads to a greatly increased probability of developing pulmonary emphysema. The condition results from a single base change at a known nucleotide position. A synthetic oligonucleotide probe that contains the wild-type sequence in the relevant region of the gene can be used in Southern-blot analysis to determine whether the DNA contains the wild-type or the mutant sequence. At higher temperatures, a complementary sequence will hybridize, whereas a sequence containing even a single mismatched base will not.

Diagnosis with the use of PCR tests. Because PCR allows an investigator to zero in on any desired sequence, it can be used to look specifically at a potentially defective DNA sequence in the diagnosis of diseases in which the presence of a specific mutational site is in question. For example,

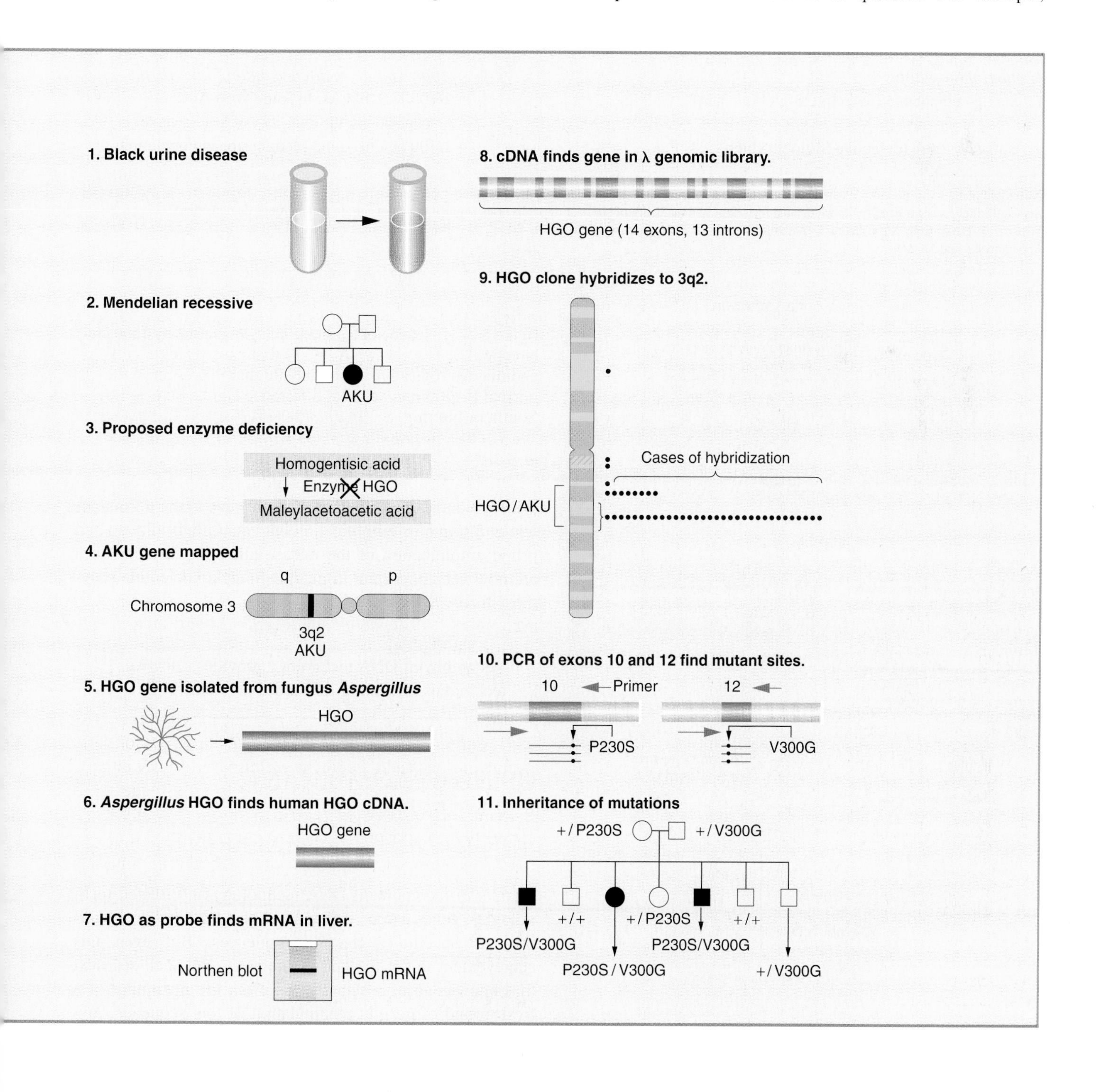

TABLE 8-1 Some Common Genetic Diseases

Inborn Errors of Metabolism	*Approximate Incidence among Live Births*
1. Cystic fibrosis	1/1600 Caucasians
2. Duchenne muscular dystrophy	1/3000 boys (X linked)
3. Gaucher disease (defective glucocerebrosidase)	1/2500 Ashkenazi Jews; 1/75,000 others
4. Tay-Sachs disease (defective hexosaminidase A)	1/3500 Ashkenazi Jews; 1/35,000 others
5. Essential pentosuria (a benign condition)	1/2000 Ashkenazi Jews; 1/50,000 others
6. Classic hemophilia (defective clotting factor VIII)	1/10,000 boys (X linked)
7. Phenylketonuria (defective phenylalanine hydroxylase)	1/5000 Celtic Irish; 1/15,000 others
8. Cystinuria (mutated gene unknown)	1/15,000
9. Metachromatic leukodystrophy (defective arylsulfatase A)	1/40,000
10. Galactosemia (defective galactose 1-phosphate uridyl transferase)	1/40,000

Hemoglobinopathies	*Approximate Incidence among Live Births*
1. Sickle-cell anemia (defective β-globin chain)	1/400 U.S. blacks. In some West African populations, the frequency of heterozygotes is 40%.
2. β-Thalassemia (defective β-globin chain)	1/400 among some Mediterranean populations

Note: Although a vast majority of more than 500 recognized recessive genetic diseases are extremely rare, in combination they constitute an enormous burden of human suffering. As is consistent with Mendelian mutations, the incidence of some of these diseases is much higher in certain racial groups than in others.
SOURCE: J. D. Watson, M. Gilman, J. Witkowski, and M. Zoller, *Recombinant DNA*, 2d ed. Scientific American Books. © 1992 by J. D. Watson, M. Gilman, J. Witkowski, and M. Zoller.

Type of Hb	Amino acid sequence Nucleotide sequence
A	—Pro—Glu—Glu— —CCT—GAG—GAG— *Mst*II
S	—Pro—Val—Glu— —CCT—GTG—GAG

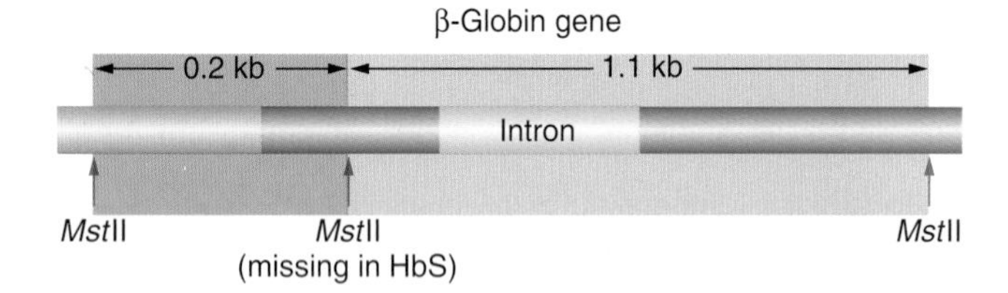

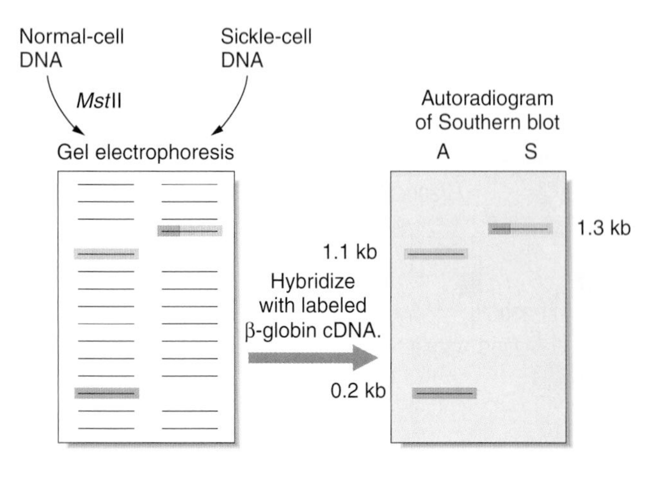

Figure 8-27 Detection of the sickle-cell globin gene by Southern blotting. The base change (A → T) that causes sickle-cell anemia destroys an *Mst*II target site that is present in the normal β-globin gene. This difference can be detected by Southern blotting. *(Modified from* Recombinant DNA, *2d ed. Scientific American Books. © 1992 by J. D. Watson, M. Gilman, J. Witkowski, and M. Zoller.)*

primers can be designed that can hybridize to the normal allele and prime its amplification but cannot hybridize to and prime amplification of the mutant allele. Such techniques are now very important in mapping single nucleotide polymorphisms (SNPs).

MESSAGE
Recombinant DNA technology provides sensitive techniques for testing for defective alleles.

USING RECOMBINANT DNA TECHNOLOGY FOR GENETIC ENGINEERING

Thanks to recombinant DNA technology and related technologies, genes can be isolated in a test tube and characterized as specific nucleotide sequences. But even this achievement is not the end of the story. We shall see next that knowledge of a sequence is often the beginning of a fresh round of genetic manipulation of that sequence, giving rise to sophisticated new approaches for altering an organism's phenotype. The use of sophisticated recombinant DNA techniques to alter the genotype and phenotype of an organism is termed **genetic engineering.**

Techniques for gene manipulation and cloning were first developed in bacteria but are now applied routinely in a variety of eukaryotes. The genomes of eukaryotes are larger and more complex than those of bacteria, so modifications of the techniques are needed to handle the larger amounts of DNA and the array of different cells and life cycles of eukaryotes. Although eukaryotic genes are typically cloned and sequenced in bacterial hosts, it is often desirable to reintroduce such **transgenes** (engineered genes) back into the original eukaryotic host or into another eukaryote —in other words, to make a **transgenic** eukaryote.

Transgenesis is applied to a great many experimental problems. For example, to identify the regulatory sequences in a segment of *Drosophila* DNA that induce the expression of the genes that they regulate in eye tissue, that DNA segment can be placed adjacent to a transcription unit (the DNA sequences encoding an mRNA) that has a readout in eye tissue that is easily detected. Such a transcription unit is called a "reporter gene." In this example, the reporter gene might be the *E. coli* β-*galactosidase* gene. The regulatory sequence–β-*galactosidase* fusion is transferred back into a recipient *Drosophila.* The eyes of the fly are then stained with a dye called X-gal. Where β-*galactosidase* is present, the enzyme converts colorless X-gal into a blue precipitate, thus revealing transgenic cells that expressed the reporter. In this manner, the regulatory elements within the segment of eukaryotic DNA can be observed, and the locations of these elements can be mapped; we'll look at this topic in detail in Chapter 13.

A transgenic "rescue" construct can be used to determine whether a piece of cloned genomic DNA can rescue a mutant phenotype to wild type. That is, it tests whether the function lacking in the mutant can be supplied by the cloned DNA, which would indicate that the cloned DNA segment corresponds to all or part of the gene that bears the mutation. The "rescue transgene" can be manipulated by site-directed mutagenesis of the transgene before it is introduced into a eukaryotic host. The mutated transgene can then be assayed for its biological activity, such as the ability to rescue an endogenous mutant allele.

As already discussed, transgenes may insert into ectopic positions in the genome. Alternatively, as we will see, some transgenes can be targeted to specific sites on the basis of sequence identity. In each case, the transgene can act as a special kind of mutagen.

As a final example, a transgene can be expressed at high levels in particular tissues; examples are transgenes encoding proteins that are important pharmaceutical drugs. Regulatory sequences that can drive high levels of tissue-specific expression are found by using the techniques described in the first example in this section. Some proteins are not active when expressed in bacteria, because the conditions and the available enzymes cannot modify and fold the proteins properly. Expressing these proteins in eukaryotes, especially in secretory tissues such as cow mammary glands, allows the production of large amounts of active protein. We will have numerous opportunities throughout the text to see how such transgenes are employed for genetic research.

Let us now investigate how such transgenes are produced. As with cloning in bacteria, the techniques for creating transgenic eukaryotes exploit naturally occurring molecular mechanisms specific to a given species. DNA is introduced into a eukaryotic cell by a variety of techniques, including transformation, injection, bacterial or viral infection, or bombardment with DNA-coated tungsten particles (Figure 8-28). When exogenously added DNA that is originally from that organism inserts into the genome, it can either replace the resident gene or insert ectopically (i.e., at other locations in the genome). If the DNA is a transgene from another species, it typically inserts ectopically.

MESSAGE

Transgenesis is in a sense molecular hitchhiking in that it exploits naturally occurring DNA infection and transfer phenomena to incorporate engineered DNA molecules into eukaryotic cells.

The transgenic modification of eukaryotes such as plants and animals (including humans) opens up many new approaches to research because genotypes can be genetically

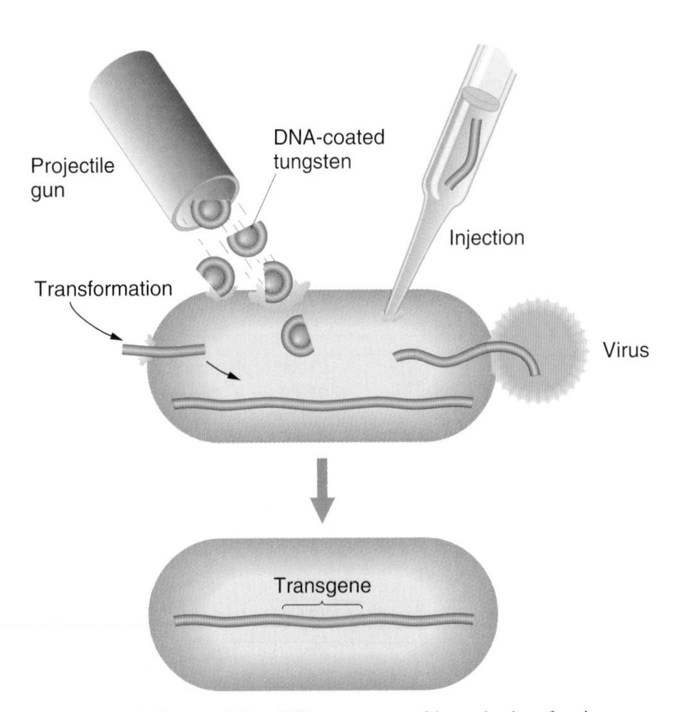

Figure 8-28 Some of the different ways of introducing foreign DNA into a cell.

engineered to make them suitable for some specific experiment. Furthermore, because plants, animals, fungi, and the products extracted from them are of major economic and biomedical importance, transgenic "designer" genotypes are finding extensive use in applied research.

> **MESSAGE**
>
> Transgenesis, the design of a specific genotype by the addition of exogenous DNA to the genome, has increased the range of gene-transmission experiments in basic genetic research and in commercial applications.

We now turn to some of the specialized approaches to genetic engineering used in fungi, plants, and animals and to its applications in attempts at human gene therapy.

Genetic Engineering in Fungi

Cloning systems have been established for many types of fungi used in genetic analysis. The first and perhaps the most extensive cloning systems have been established for baker's and brewer's yeast, *Saccharomyces cerevisiae.* It is fair to say that *S. cerevisiae* is the most sophisticated eukaryotic model for recombinant DNA technology. One of the main reasons is that the transmissional genetics of yeast is extremely well understood, and the stockpile of thousands of mutants affecting hundreds of different phenotypes is a valuable resource when yeast is used for genetic engineering. In yeast, another important advantage is the availability of a circular 6.3-kb natural yeast plasmid that segregates into daughter cells at meiosis and mitosis. This plasmid, which has a circumference of 2 μm, has become known as the "2 micron" plasmid. It forms the basis for several sophisticated cloning vectors.

Integrative vectors. The simplest yeast vectors are yeast integrative plasmids (YIps), derivatives of bacterial plasmids into which the yeast DNA of interest has been inserted (Figure 8-29a). When transformed into yeast cells, these plasmids insert into yeast chromosomes, generally by homologous recombination with the resident gene and either a single or a double crossover (Figure 8-30). As a re-

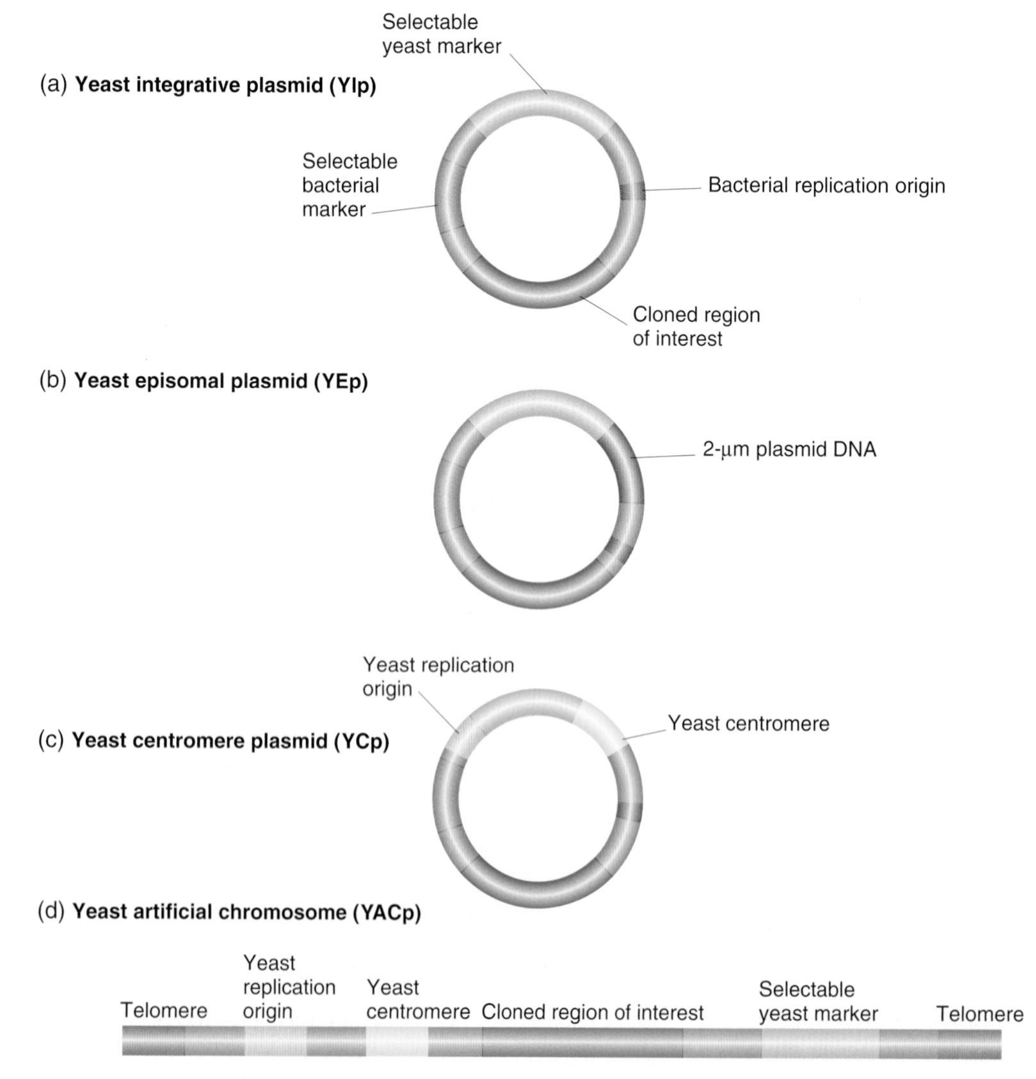

Figure 8-29 Simplified representations of four different kinds of plasmids used in yeast. Each is shown acting as a vector for some genetic region of interest, which has been inserted into the vector. The function of such segments can be studied by transforming a yeast strain of suitable genotype. Selectable markers are needed for the routine detection of the plasmid in bacteria or yeast. Origins of replication are sites needed for the bacterial or yeast replication enzymes to initiate the replication process. (DNA derived from the 2-μm natural yeast plasmid has its own origins of replication.)

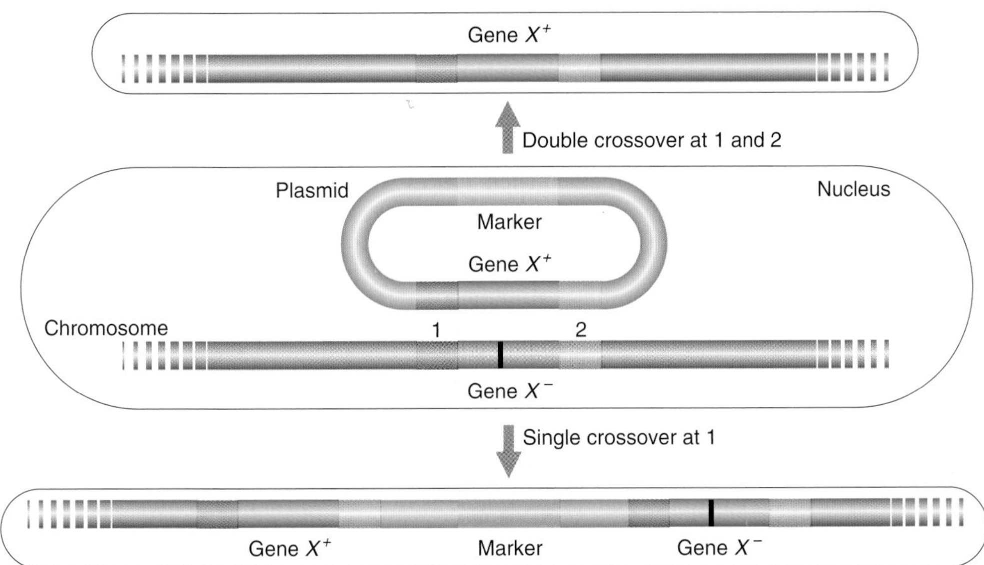

Figure 8-30 Two ways in which a recipient yeast strain bearing a defective gene (X^-) can be transformed by a plasmid bearing an active allele (gene X^+). The mutant site of gene X^- is represented as a vertical black bar. Single crossovers at position 2 also are possible but are not shown.

sult, either the entire plasmid is inserted or the targeted allele is replaced by the allele on the plasmid. Such integrations can be detected by plating cells on a medium that selects for the allele on the plasmid. The bacterial origin of replication is different from eukaryotic origins, and so bacterial plasmids do not replicate in yeast. Because of the failure to replicate in yeast, integration is the only way to generate a stable modified genotype with the use of bacterial plasmid vectors.

Autonomously replicating vectors. If the 2-μm plasmid is used as the basic vector and other bacterial and yeast segments are spliced into it, then a yeast episomal plasmid (YEp) construct with several useful properties is obtained (Figure 8-29b). First, the 2-μm segment confers the ability to replicate autonomously in the yeast cell, and integration into a yeast chromosome is not necessary for a stable transformation. Second, after genes have been introduced into yeast and their effects studied in that organism, the plasmid can be recovered and put back into *E. coli,* provided that a bacterial replication origin and a selectable bacterial marker are on the plasmid. Such **shuttle vectors** are very useful in the routine cloning and manipulation of yeast genes, because growing and purifying the DNA in bacteria is much easier than in yeast, though the genetic assays must be performed in the yeast recipient cells.

Yeast artificial chromosomes. With any autonomously replicating plasmid, there is the possibility that a daughter cell will not inherit a copy, because the partitioning of plasmid copies to daughter cells depends on where the plasmids are in the cell when the new cell wall is formed. However, if the section of yeast DNA containing a centromere and replication origins is added to the plasmid (Figure 8-29c), then the nuclear spindle that ensures the proper segregation of chromosomes will treat the resulting yeast centromere plasmid (YCp) in somewhat the same way as it would treat a chromosome and partition it into daughter cells at cell division. The addition of a centromere is one step toward the creation of an artificial chromosome. A further step has been made by linearizing a plasmid containing a centromere and adding the DNA from yeast telomeres to the ends (Figure 8-29d). If this construct contains yeast replication origins (also called autonomous replication sequences), then it constitutes a **yeast artificial chromosome (YAC),** which behaves in many ways like a small yeast chromosome at mitosis and meiosis. For example, when two haploid cells—one bearing a ura^+ YAC and another bearing a ura^- YAC—are brought together to form a diploid, many tetrads will show the clean 2:2 segregations expected if these two elements are behaving as regular chromosomes.

Yeast artificial chromosomes have been extensively used as cloning vectors for large sections of eukaryotic DNA. Consider, for example, that the size of the region encoding blood-clotting factor VIII in humans is known to span about 190 kb and that the gene for Duchenne muscular dystrophy spans more than 1000 kb. Furthermore, the large size of mammalian genomes in general means that libraries made in bacterial vectors would be huge. Yeast artificial chromosomes carry much longer inserts, as many as 1000 kb, and the library size is correspondingly smaller. We return to this topic in Chapter 9.

MESSAGE

Yeast vectors can be integrative, can autonomously replicate, or can resemble artificial chromosomes, allowing genes to be isolated, manipulated, and reinserted in molecular genetic analysis.

Some foreign sequences are unable to replicate in bacterial cloning vectors: some of these sequences are thought to form structures that cannot be replicated or properly packaged by the bacterial chromosome-replication machinery. In some cases, these sequences can be cloned and studied in yeast.

Genetic Engineering in Plants

Because of their economic significance in agriculture, many plants have been the subject of genetic analysis aimed at developing improved varieties. Recombinant DNA technology has introduced a new dimension to this effort because the genome modifications made possible by this technology are almost limitless. No longer is genetic diversity confined to selecting variants within a given species. DNA can now be introduced from other species of plants, animals, or even bacteria. The new possibilities created by the ability to clone genes into plants has led to a great deal of concern by a sector of the public in regard to possible unanticipated health problems produced by the introduction of **GMOs** (genetically modified organisms) into the food supply. The concern about GMOs is a current example of public debate that has been going on for some time between different sectors of the public: namely, if new genetic technologies are to be accepted by regulatory agencies and by consumers, the complex public health, safety, ethical, and educational issues need to be addressed as well as the much more straightforward scientific and engineering issues.

The Ti-plasmid system. A major vector routinely used to produce transgenic plants is the **Ti plasmid,** derived from a soil bacterium called *Agrobacterium tumefaciens.* This bacterium causes what is known as *crown gall disease,* in which the infected plant produces uncontrolled growths (tumors, or galls), normally at the base (crown) of the stem of the plant. The key to tumor production is a large (200-kb) circular DNA plasmid—*the Ti (tumor-inducing) plasmid.* When the bacterium infects a plant cell, a part of the Ti plasmid—a region called *T-DNA* for transfer-DNA—is transferred and inserted, apparently more or less at random, into the genome of the host plant (Figure 8-31). The structure of a Ti plasmid is shown in Figure 8-32. The genes whose products catalyze this T-DNA transfer reside in a region of the Ti plasmid separate from the T-DNA region itself. The T-DNA region encodes several interesting functions that contribute to the bacterium's ability to grow and divide inside the plant cell, including enzymes contributing to the production of the tumor (and hence more infected cells) and other proteins that direct the synthesis of compounds called *opines* (important substrates for the bacterium's growth). Opines are actually synthesized by the infected plant cells by expression of the opine-synthesizing genes located in the transferred T-DNA region and are imported into the bacterium and metabolized by enzymes encoded by the bacterium's opine-utilizing genes on the Ti plasmid.

The natural behavior of the Ti plasmid makes it well suited to the role of a plant vector. If the DNA of interest could be spliced into the T-DNA, then the whole package would be inserted in a stable state into a plant chromosome. This system has indeed been made to work essentially in this way but with some necessary modifications. Let us examine one protocol.

Ti plasmids are too large to be easily manipulated and cannot be readily made smaller, because they contain few

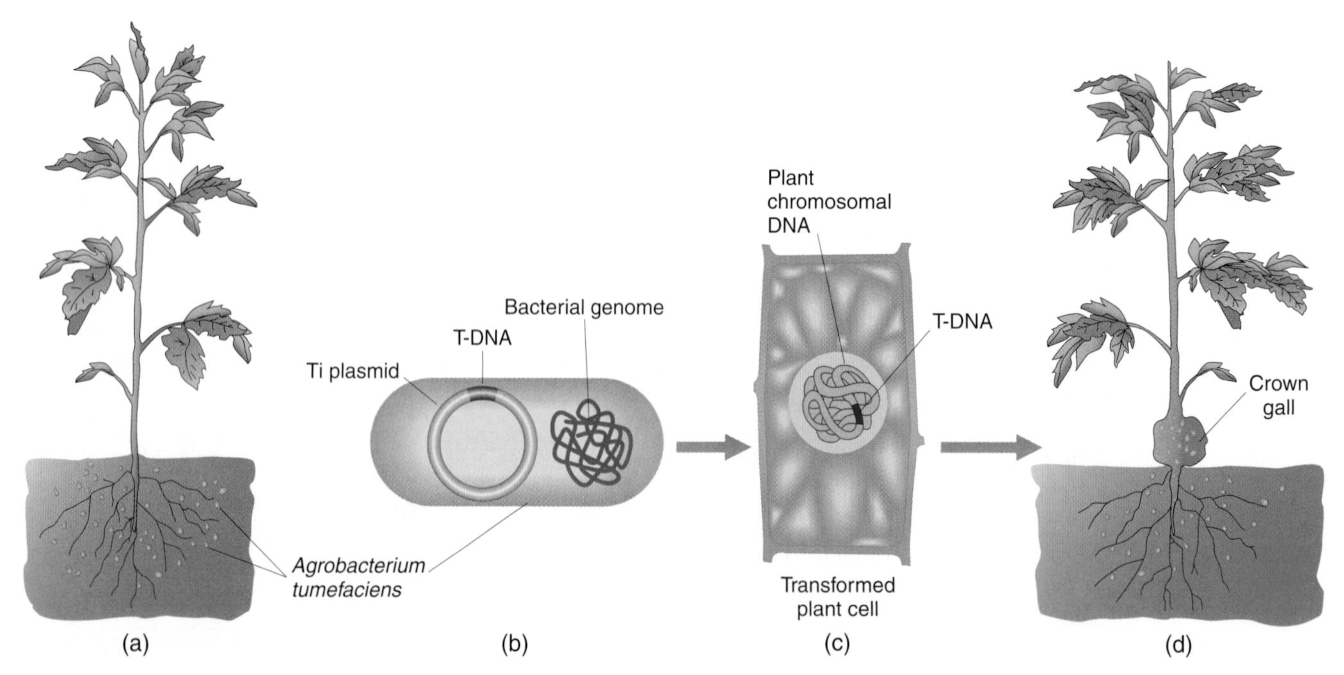

Figure 8-31 In the process of causing crown gall disease, the bacterium *Agrobacterium tumefaciens* inserts a part of its Ti plasmid—a region called T-DNA—into a chromosome of the host plant.

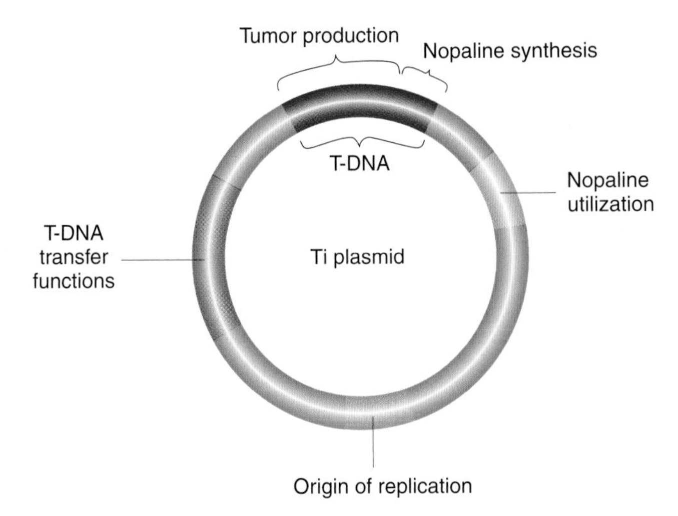

Figure 8-32 Simplified representation of the major regions of the Ti plasmid of *A. tumefaciens.* The T-DNA, when inserted into the chromosomal DNA of the host plant, directs the synthesis of nopaline, which is then utilized by the bacterium for its own purposes. T-DNA also directs the plant cell to divide in an uncontrolled manner, producing a tumor.

unique restriction sites and much of the plasmid is necessary for either its replication or for the infection and transfer process. Therefore, cloning and modification of the Ti plasmid to appropriately engineer inserts of interest into the T-DNA region are performed in steps. The first cloning steps take place in *E. coli,* producing an intermediate vector considerably smaller than Ti. The intermediate vector initially receives the insert of interest and the various other genes and segments necessary for plasmid recombination, replication, and antibiotic resistance. When engineered with the desired gene elements, this intermediate vector can then be inserted into the Ti plasmid, forming a *cointegrate plasmid* that can be introduced into a plant cell by *Agrobacterium* infection and transformation.

Figure 8-33a on the next page shows one method of creating the cointegrate. A modified version of the Ti plasmid is used as a cloning vector in *Agrobacterium.* This modified Ti plasmid is attenuated (incapable of producing a tumor), because all we want the *Agrobacterium* to do for us is to be the delivery system to bring the transgenic T-DNA into the plant cell and enable its integration into the DNA of a plant chromosome. (We don't want tumor-forming infections to occur, because they may interfere with things that we may ultimately want to do with these plants, such as growing morphologically normal plants.) The attenuated Ti plasmid lacks the entire right-hand part of the T-DNA region and is thus missing its tumor genes and nopaline-synthesis genes, thereby rendering it incapable of tumor formation. It retains the left-hand portion (L) of the T-DNA region, which will be used as the crossover site for incorporation (also called *cointegration*) of the intermediate vector.

The intermediate vector that will be engineered in *E. coli* has several important features. It has a segment called a **polylinker,** so-called because it contains several unique restriction sites that can be used to produce sticky ends for inserting the DNA segments to be cloned. In our example, the gene of interest has been inserted into the intermediate vector at one of the polylinker's restriction sites (see Figure 8-33a). The vector contains an origin of replication that functions inside the *E. coli* cell but no origin that is functional in *Agrobacterium.* Also contained in the vector are a selectable gene (spc^R) for spectinomycin resistance that is used for selection of bacteria containing this plasmid; a bacterial kanamycin-resistance gene (kan^R), engineered for expression in plants so that plant cells containing the integrated transgene can be selected for in plant cell culture; and two segments of T-DNA. One T-DNA segment carries the nopaline-synthesis gene *(nos)* plus the right-hand T-DNA border sequence (R). The second T-DNA segment comes from near the left-hand border (L) and provides a section for recombination with an identical sequence of region L that is present in the attenuated Ti plasmid. After the intermediate vector has been constructed in *E. coli,* it is then introduced into *Agrobacterium* cells containing the attenuated Ti plasmids by conjugation with *E. coli.* Plasmid recombinants *(cointegrates)* form by a single crossover at the point of sequence identity of the L segments of the T-DNA region contained on both plasmids. The cointegrates are selected by plating on spectinomycin. The selected bacterial colonies will contain only the cointegrate Ti plasmid because the intermediate vector itself cannot replicate in *Agrobacterium.*

As Figure 8-33b shows, after spectinomycin selection for the cointegrates, bacteria containing the cointegrant plasmid are then used to infect cut segments of plant tissue, such as punched-out leaf disks. If bacterial infection of plant cells takes place, any genetic material between the left and right T-DNA border sequences can be inserted into the plant chromosomes. If the leaf disks are placed on a medium containing kanamycin, the only plant cells that will undergo cell division are those that have acquired the kan^R gene from T-DNA transfer. The growth of such cells results in a clump, or callus, which is an indication that transformation has taken place. These calli can be induced to form shoots and roots, at which time they are transferred to soil where they develop into transgenic plants (see Figure 8-33b). Typically, only a single copy of the T-DNA region inserts into a given plant genome, where it segregates at meiosis like a regular Mendelian allele (Figure 8-34 on page 249). The insert can be detected on the basis of transgenic genetic markers, or molecularly by screening purified DNA with a T-DNA probe in a Southern hybridization, or by the presence of nopaline in the transgenic tissue.

The gene gun. Another approach to producing transgenic plants is to engineer the gene of interest in standard bacterial vectors and to essentially shoot it into the plant cell by

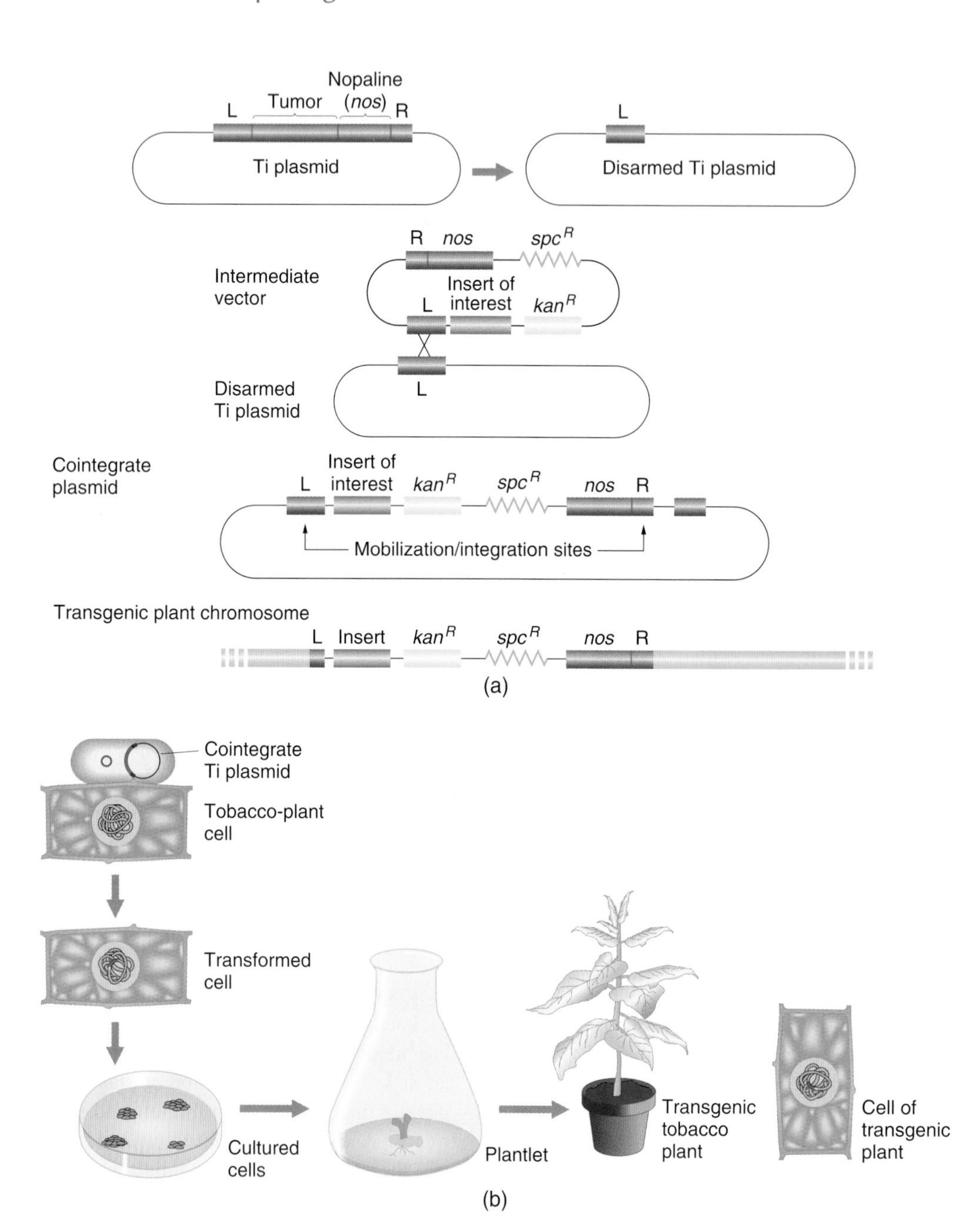

Figure 8-33 (a) To produce transgenic plants, an intermediate vector of manageable size is used to clone the segment of interest. In the method shown here, the intermediate vector is then recombined with an attenuated ("disarmed") Ti plasmid to generate a cointegrate structure bearing the insert of interest and a selectable plant kanamycin-resistance marker between the T-DNA borders, which is all the T-DNA that is necessary to promote insertion. (b) The generation of a transgenic plant through the growth of a cell transformed by T-DNA.

means of high-velocity micropellets (see Figure 8-28). Here, as in other cases of nonbiologically mediated introduction of DNA (such as direct injection into cells), the DNA tends to be incorporated as multicopy tandem arrays, in which copy number can be quite variable from one transgenic integration event to the next. The variable copy number and structure of these arrays sometimes presents problems in the subsequent analysis of the transgenes. In contrast, the Ti-plasmid system incorporates one copy of the T-DNA–borne transgene, albeit at ectopic positions in the genome.

Transgenic plants carrying any one of a variety of foreign genes are in current use, and many more are in development. Not only are the qualities of plants themselves being manipulated, but, like microorganisms, plants are also being used as convenient "factories" to produce proteins encoded by foreign genes.

Genetic Engineering in Animals

There are many animal model systems for which transgenic technologies are now being employed. We will focus on the three animal models most heavily used for basic genetic research: the nematode *Caenorhabditis elegans,* the fruit fly *Drosophila melanogaster,* and the mouse *Mus musculus.* Versions of many of the techniques considered so far can also be applied in these animal systems.

***Transgenesis in* Caenorhabditis elegans.** Transgenics in *C. elegans* are produced by direct injection of DNAs, typically DNAs cloned in bacteria (plasmids, cosmids, etc.).

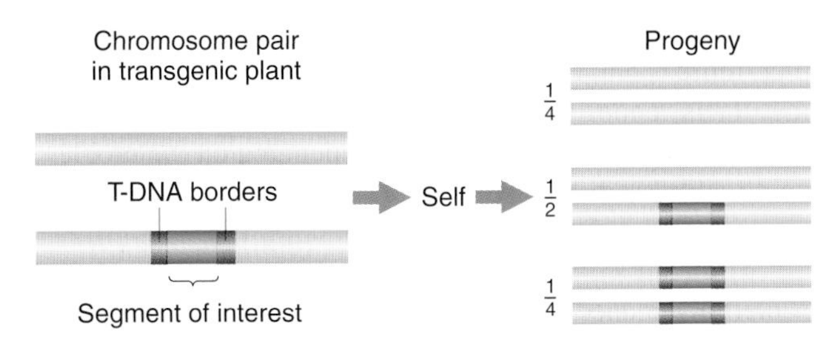

Figure 8-34 T-DNA region and any DNA inserted into a plant chromosome in a transgenic plant are transmitted in a Mendelian pattern of inheritance.

The strategy for where to inject depends on an understanding of the reproductive biology of this worm. The gonads of the worm are syncitial, meaning that there are many nuclei within the same gonadal cell. One syncitial cell is a large proportion of one arm of the gonad, and the other syncitial cell is the bulk of the other arm (Figure 8-35a). The nuclei do not form individual cells until meiosis, when formation of the individual egg or sperm begins. A solution of DNA is injected into the syncitial region of one of the arms, thereby exposing >100 germ-cell precursor nuclei to the transforming DNA. By chance, a few of these nuclei will incorporate the DNA in their mitotic divisions (remember, the nuclear membrane breaks down in the course of mitosis and meiosis, and so the cytoplasm into which the DNA is injected becomes continuous with the nucleoplasm for part of each division cycle). Typically, the transgenic DNA forms extrachromosomal multicopy arrays (Figure 8-35b) that replicate and show sufficiently regular mitotic disjunction (one replicated copy to each of the two mitotic poles), and so the arrays are retained in the genome for many cell divisions. Occasionally, the transgenes will become integrated into an ectopic position in a chromosome, but such integration is a rarer event.

***Transgenesis in* Drosophila melanogaster.** Transgenesis in *D. melanogaster* proceeds by a mechanism different from those discussed so far. It is based on the properties of a **transposable element** called the P element, a DNA segment that, catalyzed by a protein that it encodes, called P transposase, is capable of excision from one location in the genome and integration at many, many other genomic positions. We will consider the molecular biology and the mutational properties of transposable elements in detail later in the book (Chapters 9, 10, and 12). As with *C. elegans,* the DNA is injected into a syncitium—in this case, the early *Drosophila* embryo (Figure 8-36a). The DNA is specifically injected into the posterior pole of the syncitial egg, because it is known that the posterior pole is the site of germ-cell formation. Successfully produced transgenic adults from these injections will contain some transgenic germ cells, but their soma will typically not contain the transgene.

The production of transgenic *Drosophila* requires the coinjection of two separate bacterial recombinant plasmids.

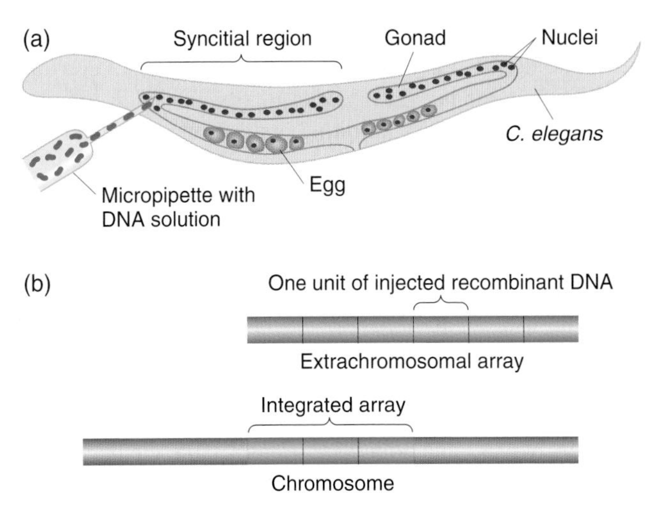

Figure 8-35 Creation of *C. elegans* transgenes. (a) Method of injection. (b) The two main types of transgenic results: extrachromosomal arrays and arrays integrated in ectopic chromosomal locations.

One, the P element vector, contains the ends of the P element (only 200 bp from each end of the 2912-bp P element are needed for a modified P element to be transposable); the piece of cloned DNA that will be incorporated as a transgene into the fly genome is inserted into the two ends. The other, the P-element helper, contains the gene for P transposase but lacks one end of the P-element DNA; it supplies the P transposase enzyme required to catalyze the integration of the P-element vector into the genome. A DNA solution containing both of these plasmids is injected into the posterior pole of the syncitial embryo. Catalyzed by the P transposase expressed from the injected P helper plasmid, insertion of the P-element vector transgene occurs.

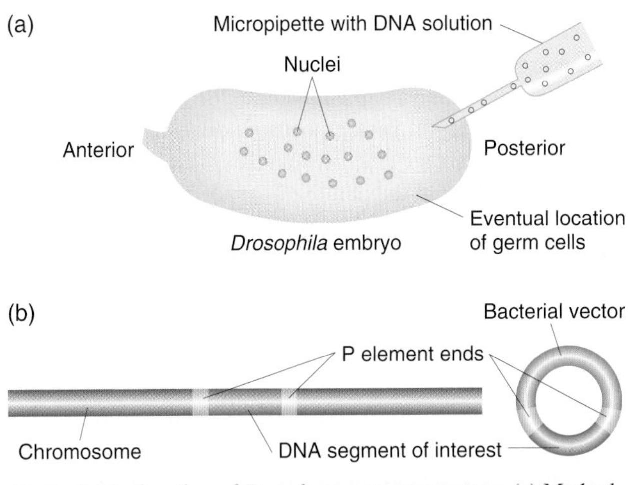

Figure 8-36 Creation of *D. melanogaster* transgenes. (a) Method of injection. (b) The circular P-element vector *(right)* and a typical integration event at an ectopic chromosomal location *(left)*. Note that the bacterial vector sequences do not become integrated into the genome; rather, in integration exactly one copy of the DNA segment is contained between the P-element ends.

Typically, the progeny that develop from gametes that receive the cloned DNA as a result of these events are detected among the population of F_1 progeny of injected individuals as transgenics exhibiting expression of a dominant wild-type transgenic allele of a gene for which the recipient strain carries a recessive mutant endogenous allele. Because of the enzymology of the transposase reaction, only a single copy of the element inserts at a given location (Figure 8-36b).

The use of transposable elements as vectors and helpers for transgenesis is very widespread. In *Drosophila,* other transposable-element systems have been developed into transgenic cloning systems: they include the Hobo, Mariner, Minos, and PiggyBac transposable-element systems. Further, some of these systems work in other organisms as well, such as in *Anopheles gambiens,* the mosquito that carries the protozoan that causes malaria. Other, related transposable elements have been shown to be transgenic cloning vectors in other systems. For example, the Activator transposable-element system first described in *Zea mays* (corn), which is a molecular relative of the *D. melanogaster* Hobo element, has been developed into a transgenic cloning system in many plants.

***Transgenesis in* Mus musculus.** Mice are the most important models for mammalian genetics. Furthermore, much of the technology developed in mice is potentially applicable to humans. There are two strategies for transgenesis in mice: ectopic insertions (similar to what happens in transgenesis in *C. elegans, D. melanogaster,* and plant systems) and gene targeting (similar to one of the modes of transgenesis in *S. cerevesiae,* which inserts the cloned gene in place of the gene present in the recipient genome). There are pros and cons to each approach. Important advantages of gene targeting are that it enables us to use transgenic constructs to eliminate or modify the function encoded by a gene in its normal chromosomal environment. In one application, a mutant allele can be repaired through **gene replacement**—substituting a wild-type allele for a mutant one in its normal chromosomal location. Considerations of chromosome environment are very important, because genes inserted at ectopic positions are at risk of being expressed in different patterns according to the nature of the regulatory sequences near the location of insertion of these ectopic inserts; such location-specific changes in gene expression are called **position effects.** (We will consider regulatory sequences in detail in Chapter 13.) Further, targeted insertions are typically single copy, and so complexities of multicopy arrays are averted—such as scrambling of the sequences within the arrays or unexpected patterns of gene expression.

However, sometimes we desire to express genes in tissues where they are not normally expressed, and the best way to accomplish this expression is to obtain transgene insertions in foreign environments where, by chance, the regulatory sites near the site of insertion will drive interesting patterns of ectopic expression that can be valuable. Further, the technology associated with ectopic integration is simpler, more rapid, and much less expensive than gene targeting. Thus, choosing between these options is a matter of scientific and practical judgment, depending on the problem being addressed. We will consider the technology of these two approaches in turn.

For ectopic insertions, the procedure is simply to inject bacterially cloned DNA solutions into the nuclei of early-stage embryos, before germ cells have developed (Figure 8-37a). On occasion, the descendants of the injected cells form part of the germ line, and so the adults developing from the injected embryos will produce germ cells containing the transgene array inserted at some ectopic position in one of the mouse chromosomes (Figure 8-37b). The location, size, and structure of the arrays will be different for each integration event.

Gene targeting is a rare event and, to be able to select for cells that have integrated the transgene at the homologous target site, a multistep process is needed. The targeting procedure itself is performed in cultured mouse germ-line precursor cells under conditions in which the cells can be selected for the presence of a transgene construct that includes a drug-resistance gene. Then, those identified targeted transgenic cells are transplanted back into genetically distinct host embryos in a such a manner that the transplanted cells will contribute to the germ line. The adults from such targeted embryonic stem-cell (ES-cell) transplantations are crossed with normal mates, and the progeny are assayed with molecular probes by PCR or Southern hybridization for sequences unique to the targeted copy of the gene. Progeny mice containing the targeted transgene in each of their cells are identified molecularly and maintained for further genetic analysis.

Embryonic stem cells have the ability to form any and all parts of a mouse—that is, they are **totipotent.** The process of gene targeting can be exemplified by one of its outcomes—namely, the substitution of an inactive gene for the normal gene. Such a targeted inactivation is called a

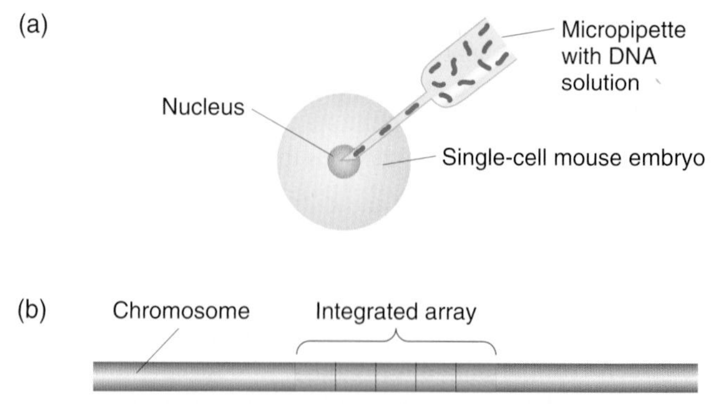

Figure 8-37 Creation of *M. musculus* transgenes inserted in ectopic chromosomal locations. (a) Method of injection. (b) A typical ectopic integrant, with multiple copies of the recombinant transgene inserted in an array.

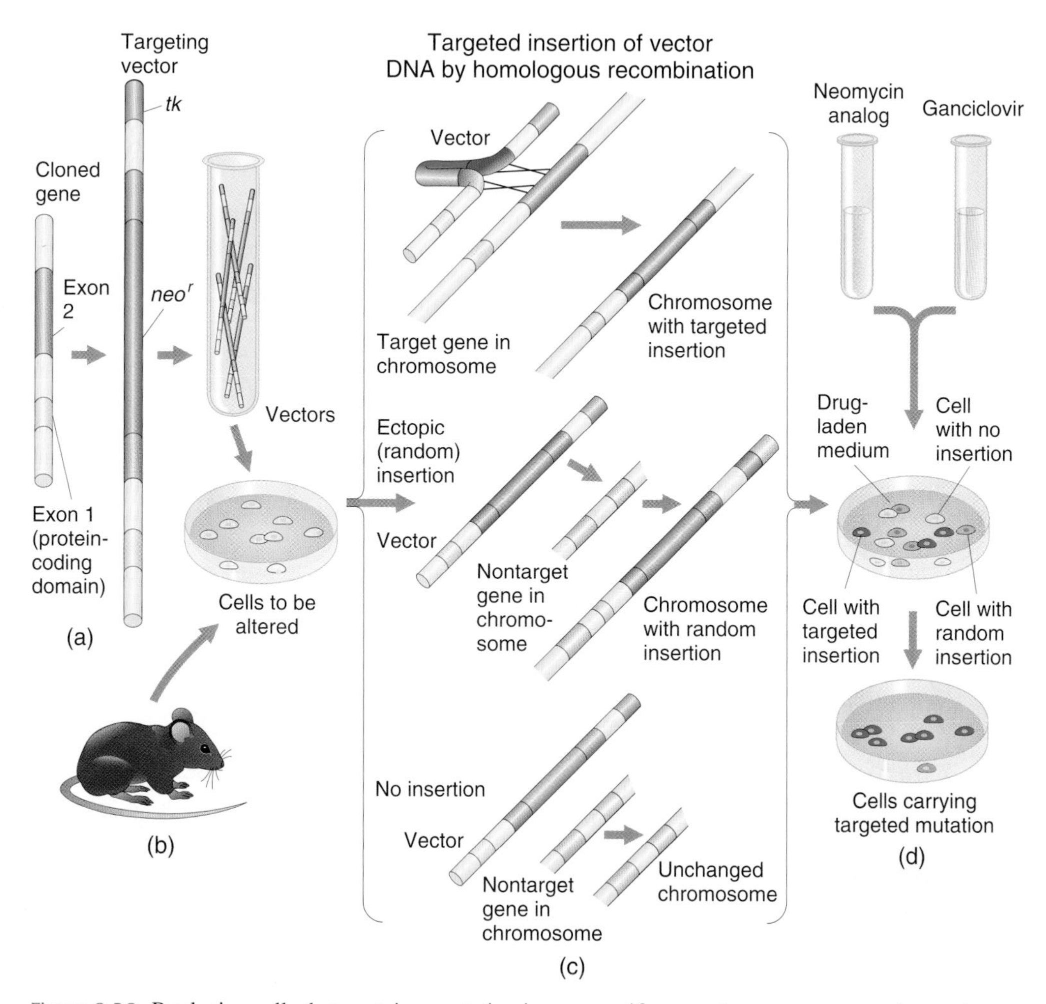

Figure 8-38 Producing cells that contain a mutation in one specific gene, known as a targeted mutation or a gene knockout. (a) Copies of a cloned gene are altered in vitro to produce the targeting vector. The gene shown here has been inactivated by the insertion of the neomycin-resistance gene *(neo^r)* into a protein-coding region (exon 2) of the gene and had been inserted into a vector. The *neo^r* gene will serve later as a marker to indicate that the vector DNA took up residence in a chromosome. The vector has also been engineered to carry a second marker at one end: the herpes *tk* gene. These markers are standard, but others could be used instead. (b) When a vector, with its dual markers, is complete, it is introduced into cells isolated from a mouse embryo. (c) When homologous recombination occurs *(top)*, the homologous regions on the vector, together with any DNA in between but excluding the marker at the tip, take the place of the original gene. This is important because the vector sequences serve as a useful tag for detecting the presence of this mutant gene. In many cells, though, the full vector (complete with the extra marker at the tip) inserts ectopically *(middle)* or does not become integrated at all *(bottom)*. (d) To isolate cells carrying a targeted mutation, all the cells are put into a medium containing selected drugs—here a neomycin analog (G418) and ganciclovir. G418 is lethal to cells unless they carry a functional *neo^r* gene, and so it eliminates cells in which no integration of vector DNA has taken place (yellow). Meanwhile, ganciclovir kills any cells that harbor the *tk* gene, thereby eliminating cells bearing a randomly integrated vector (red). Consequently, virtually the only cells that survive and proliferate are those harboring the targeted insertion (green). *(Redrawn from M. R. Capecchi, "Targeted Gene Replacement." © 1994 by Scientific American, Inc. All rights reserved.)*

gene knockout. First, a cloned, disrupted gene is used to produce ES cells containing a gene knockout (Figure 8-38a and b). Although recombination of the defective part of the gene into nonhomologous (ectopic) sites is much more frequent than its recombination into homozygous sites (Figure 8-38c), selections for site-specific recombinants and against ectopic recombinants can be used, as shown in Figure 8-38d. Second, the ES cells that contain one copy of the disrupted gene of interest are injected into an early embryo (Figure 8-39a on the next page). The resulting progeny are chimeric, having tissue derived from either recipient or transplanted ES lines. Chimeric mice are then mated with their siblings to produce homozygous mice with the knockout in each copy of the gene (Figure 8-39b).

MESSAGE

Germ-line transgenic techniques have been developed for all well-studied eukaryotic species and depend on an understanding of the reproductive biology of the recipient species.

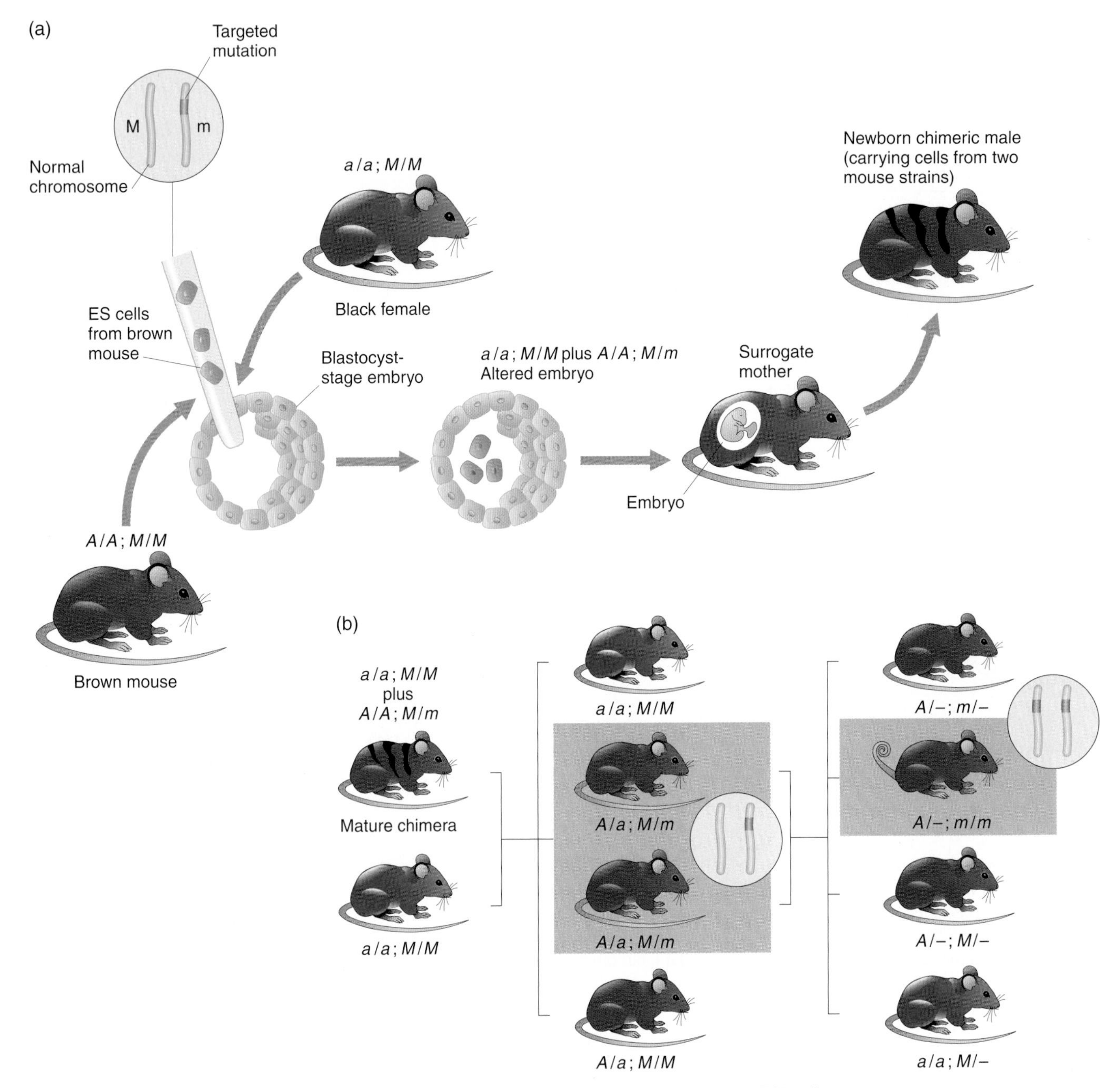

Figure 8-39 Producing a knockout mouse carrying the targeted mutation. (a) Embryonic stem (ES) cells are isolated from an agouti (brown) mouse strain *(A / A)* and altered to carry a targeted mutation *(m)* in one chromosome. The ES cells are then inserted into young embryos, one of which is shown. Coat color of the future newborns is a guide to whether the ES cells have survived in the embryo. Hence, ES cells are typically put into embryos that, in the absence of the ES cells, would acquire a totally black coat. Such embryos are obtained from a black strain that lacks the dominant agouti allele *(a / a)*. The embryos containing the ES cells grow to term in surrogate mothers. Agouti shading intermixed with black indicates those newborns in which the ES cells have survived and proliferated. (Such mice are called *chimeras* because they contain cells derived from two different strains of mice.) Solid black coloring, in contrast, would indicate that the ES cells had perished, and these mice are excluded. *A* represents agouti, *a* black; *m* is the targeted mutation, and *M* is its wild-type allele. (b) Chimeric males are mated with black (nonagouti) females. Progeny are screened for evidence of the targeted mutation (green in *inset*) in the gene of interest. Direct examination of the genes in the agouti mice reveals which of those animals *(boxed)* inherited the targeted mutation. Males and females carrying the mutation are mated with one another to produce mice whose cells carry the chosen mutation in both copies of the target gene *(inset)* and thus lack a functional gene. Such animals *(boxed)* are identified definitively by direct analyses of their DNA. The knockout in this case results in a curly-tail phenotype. *(Redrawn from M. R. Capecchi, "Targeted Gene Replacement." © 1994 by Scientific American, Inc. All rights reserved.)*

Considerations for Human Gene Therapy

The general goal of **gene therapy** is to attack the genetic basis of disease at its source: "curing" or correcting the abnormal condition by means of transgenic DNA from a wild-type allele introduced into the cells of an individual with a mutant allele that causes a heritable disease. This technique of transgene-mediated phenotypic rescue has been successfully applied in many experimental organisms. The technique has the potential in humans to correct some hereditary diseases, particularly those associated with single genetic differences (monogenic traits). Although attempts at gene therapy have been attempted for several genetic diseases, thus far there are no clear instances of success. However, the implications of gene therapy are so far-reaching that the approach merits consideration here.

To understand the approach, consider an example from mice in which a growth-hormone deficiency was corrected (Figure 8-40). The recessive mutation *little (lit)* results in dwarf mice. Even though a mouse's growth-hormone gene is present and apparently normal, no mRNA for this gene is produced. The initial step in correcting this deficiency was to inject homozygous *lit / lit* eggs with about 5000 copies of a 5-kb linear DNA fragment that contained the rat growth-hormone structural gene *(RGH)* fused to mouse metallothionein gene regulatory sequences, which lead to the expression of any immediately adjacent gene in the presence of heavy metals. The eggs were then implanted

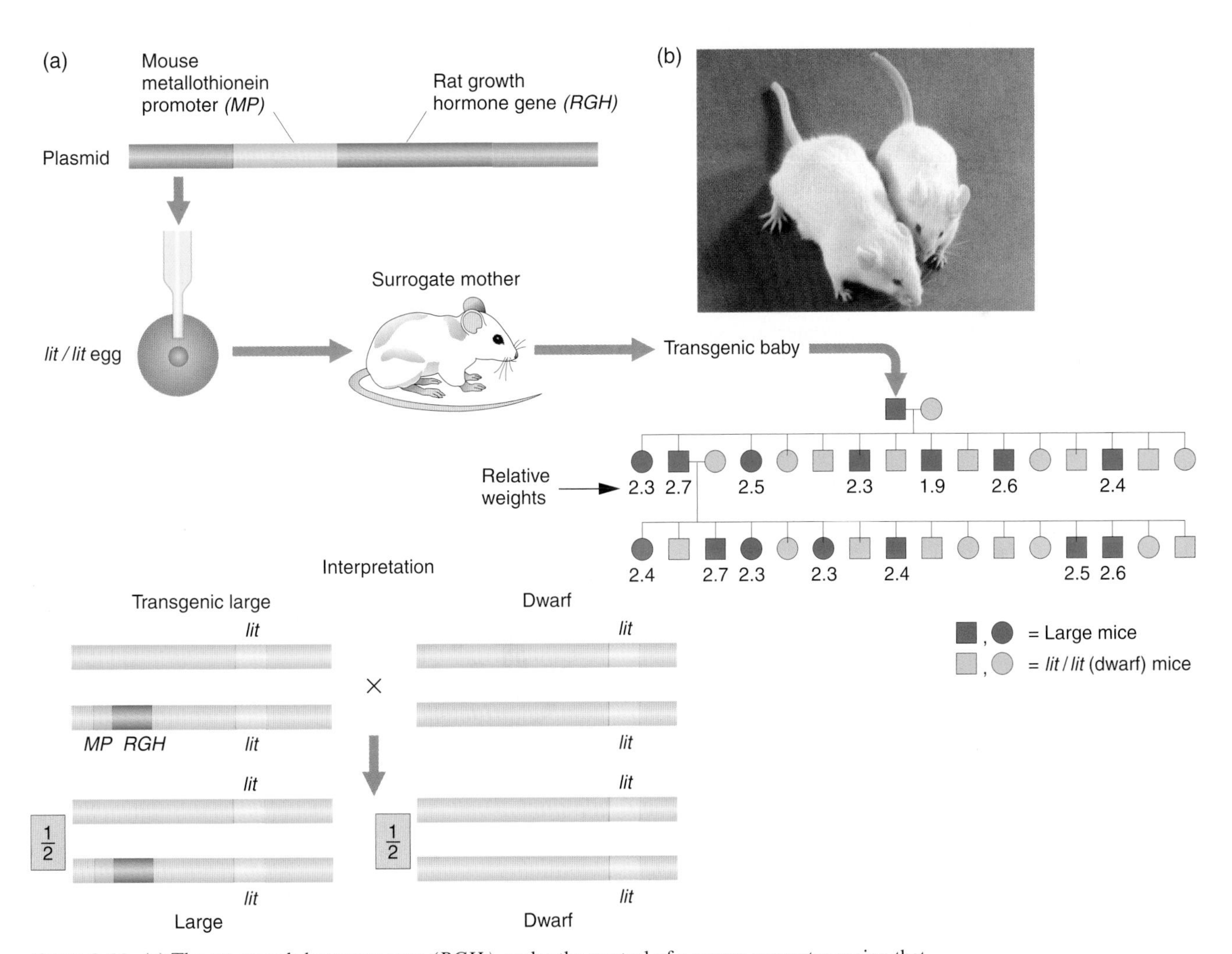

Figure 8-40 (a) The rat growth-hormone gene *(RGH)*, under the control of a mouse promoter region that is responsive to heavy metals, is inserted into a plasmid and used to produce a transgenic mouse. *RGH* compensates for the inherent dwarfism *(lit / lit)* in the mouse. *RGH* is inherited in a Mendelian dominant pattern in the ensuing mouse pedigree. (b) Transgenic mouse. The mice are siblings, but the mouse on the left was derived from an egg transformed by injection with a new gene composed of the mouse metallothionein promoter fused to the rat growth-hormone structural gene. (This mouse weighs 44 g, and its untreated sibling weighs 29 g.) The new gene is passed on to progeny in a Mendelian manner and so is proved to be chromosomally integrated. *(From R. L. Brinster.)*

into the uteri of surrogate mother mice, and the baby mice were born and raised. About 1 percent of these babies turned out to be transgenic, showing increased size when heavy metals were administered in the course of development. A representative transgenic mouse was then crossed with a homozygous *lit* / *lit* female. The ensuing pedigree is shown in Figure 8-40a. Here we see that mice two to three times the weight of their *lit* / *lit* relatives are produced in subsequent generations, with the rat growth-hormone transgene acting as a dominant allele, always heterozygous in this pedigree. Thus, the phenotype derived from the defective *lit* gene mutation has been "cured" by the introduction of the *RGH* transgene.

In an extension of the logic of this approach to conditions of human disease, there are both technical and ethical considerations. A thorough treatment of the ethical issues is beyond the scope of this book. However, we can at the very least enumerate some of them:

- What are the implications for the human gene pool of attempting germ-line (and hence transmissible) gene therapy? What are the long-term effects on genetic population structure and human health of introducing transgenes in the germ lines of reproductively active humans?
- How do we define *genetic disease?* When is something a *characteristic* rather than a *disease?* Ought we to provide the technology for transgenic intervention in things that are characteristics rather than diseases, such characteristics as sex; height; intellectual, athletic, or artistic ability; skin or hair color? Is there a gray area where it is hard to distinguish between a characteristic and a disease? If we develop the technology for clear examples of *genetic disease,* should there be rules to regulate its application for the transgenic alteration of personal characteristics? If so, what should be the rules and who should determine them? How much of the decision making should be left to parents or the person in question?
- How do we assess benefit versus risk of inadvertant mutation induction as a by-product of transgene incorporation into the genome? How does the potential risk gibe with the medical dictum "Do no harm"?

Let us now turn to the status of various technical approaches. Two basic types of gene therapy can be applied to humans: germ line and somatic. The goal of **germ-line gene therapy** (Figure 8-41a) is the more ambitious: to introduce transgenic cells into the germ line as well as into the somatic-cell population. Not only would this therapy achieve a cure of the person treated, but gametes and thus children of the person who received the therapy also will contain the therapeutic transgene. The example of recombinant *RGH* cure of the mouse *lit* recessive defect is germ-line gene therapy. At present, these technologies depend on chance ectopic integration or gene replacement events, and these events are sufficiently infrequent to make germ-line gene therapy currently impractical.

Somatic gene therapy (Figure 8-41b) focuses only on the body (soma). The approach is to attempt to correct a disease phenotype by treating *some* somatic cells in the affected person. At present, it is not possible to render an entire body transgenic, and so the method addresses diseases whose phenotype is caused by genes that are expressed predominantly in one tissue. In such cases, it is likely that not all the cells of that tissue need to become transgenic; a portion of cells being transgenic can ameliorate the overall disease symptoms. The method proceeds by removing some

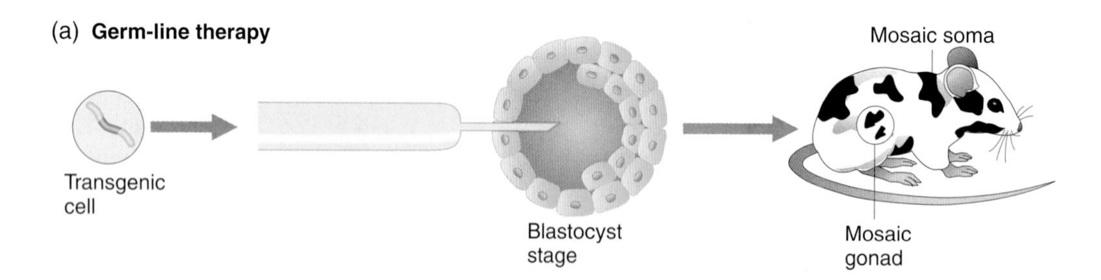

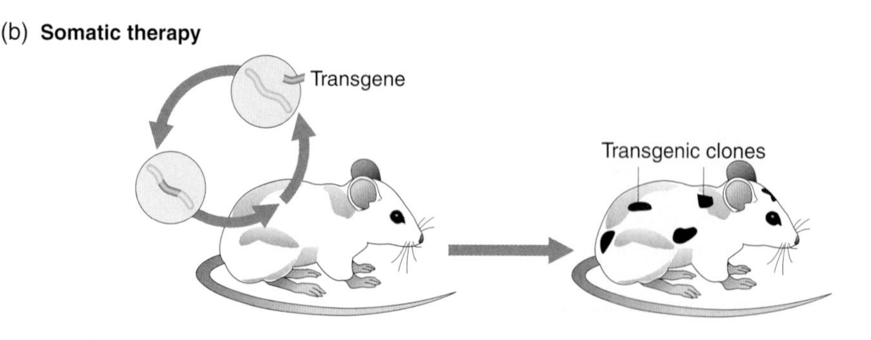

Figure 8-41 Types of gene therapy in mammals.

cells from a patient with the defective genotype and making these cells transgenic through the introduction of copies of the cloned wild-type gene. The transgenic cells are then reintroduced into the patient's body, where they provide normal gene function. Because somatic gene therapy by definition does not alter the germ line, one set of ethical issues—introducing transgenes into the general human population through standard human reproduction by cured persons—are bypassed. The others still apply.

Let us return to the boy with severe combined immunodeficiency disease described at the beginning of this chapter. In his case, the defect is in stem cells of the immune system. These cells can be isolated from bone marrow. If the defect in adenosine deaminase in these stem cells can be repaired by the introduction of a normal ADA gene, then the progeny of these repaired cells will populate his immune system and cure the SCID condition. Because only a small set of stem cells needs to be repaired to cure the disease, SCID is ideally suited for gene therapy. How has gene therapy been attempted for SCID?

The method uses a "disarmed" retrovirus with the normal ADA transgene spliced into its genome, replacing most of the viral genes; this retrovirus is unable to form progeny viruses and is thus avirulent, or disarmed. The natural cycle of retroviruses includes the integration of the viral genome at some location in one of the host cell's chromosomes. The recombinant retrovirus will carry the ADA transgene along with it into the chromosome. Blood stem cells are removed from the bone marrow of the person who has SCID, the retroviral vector containing the ADA transgene is added, and the transgenic cells are reintroduced into the blood system. Thus far, no case of long-term cure has been achieved, but there have been some encouraging results (Figure 8-42).

The retroviral vector poses a potential problem, because the integrating virus can act as an insertional mutagen and inactivate some unknown resident gene; thus it has the potential to cause a mutation. Another problem with this type of vector is that a retrovirus infects only proliferating cells, such as blood cells, and thus cannot be used in the context of the many heritable disorders that affect other tissues in which cells rarely or never divide.

Another vector used in human gene therapy is the adenovirus. This virus normally infects respiratory epithelia, injecting its genome into the epithelial cells. The viral genome does not integrate into a chromosome but persists extrachromosomally in the cells, which eliminates the problem of insertional mutagenesis by the vector. Another advantage of the adenovirus as a vector is that it attacks nondividing cells, making most tissues susceptible in principle. Inasmuch as cystic fibrosis is a disease of the respiratory epithelium, adenovirus is an appropriate choice of vector for treating this disease, and gene therapy for cystic fibrosis is currently being attempted with the use of this vector. Viruses bearing the wild-type cystic fibrosis allele are introduced through the nose as a spray.

Although there is some reason to believe that the technical hurdles of somatic gene therapy will be overcome, these hurdles are considerable. One hurdle concerns targeting the transgenic delivery system to the appropriate tissue for a given disease. Another is how to build the transgene to ensure sufficiently and consistently high levels of expression to produce a phenotypic cure. Still another is how to protect against potentially harmful side effects of the therapy, such as might be caused by misexpression of the transgenic gene. These are major areas of gene-therapy research.

MESSAGE

The technologies of transgenesis are currently being applied to humans with the specific goal of applying gene therapy to the correction of certain heritable disorders. The technical, societal, and ethical challenges of these technologies are considerable and are active areas of research and debate.

Figure 8-42 Ashanti de Silva, the first person to receive gene therapy. She was treated for SCID, and her symptoms have been ameliorated. *(Courtesy of Van de Silva.)*

SUMMARY

1. Genetics focuses on the nature of genes, and a major goal is to characterize their structure and function. Recombinant DNA technology has allowed individual purified and amplified genes to be isolated in a test tube and then characterized at the molecular level. The methodologies of recombinant DNA are completely reliant on the two fundamental principles of molecular biology: (1) hydrogen bonding of complementary antiparallel nucleotide sequences and (2) interactions between specific proteins and specific nucleotide sequences. Examples of the applications of the principles are numerous. We exploit complementarity to glue together DNA fragments with complementary sticky ends; to probe for specific sequences in colony, Southern, and Northern hybridizations; and to prime cDNA synthesis, PCR, and DNA-sequencing reactions. The specificity of interactions between proteins and nucleotide sequences enables the cutting action of restriction endonucleases to be at specific target recognition sites, and the transposition reactions catalyzed by transposases to act on specific transposons.
2. Recombinant DNA technology also exploits basic DNA biochemistry and the basic life history and reproductive biology of organisms. Various DNA synthetic or degradatory enzymes are important tools in the recombinant DNA toolbox. DNA ligase is used to form covalent phosphodiester bonds to fuse recombinant molecules that initially are held together by a few hydrogen bonds at their sticky ends. Restriction endonucleases are able to chop up DNA at specific sites. Special heat-resistant DNA polymerases, which evolved in bacteria that live in high-temperature thermal vents, are able to remain enzymatically active during the thermal cycling of PCR. Plasmid DNAs have a natural role in conferring antibiotic resistance on host bacteria. An understanding of how the germ line develops in fly, worm, and mouse is essential in determining how to deliver transgenic DNAs to the right locations in the right cell types.
3. Recombinant DNA is the result of cutting and pasting of DNA segments such that the synthesized molecules are able to be replicated and amplified, either in a test tube or in an appropriate host cell. It is a mix-and-match process whereby we choose DNA fragments that together have the right combination of properties to fulfill the needs of a particular experiment or genetic-engineering task. One fragment may contribute a bacterial antibiotic-resistance gene, another might provide a polylinker, another an origin of replication, another the ends of a transposon, and still another a wild-type gene that is expressed in the eyes of a fruit fly. Glued together by sticky-end complementarity and covalently fused with DNA ligase, these fragments form a plasmid that can be cloned in bacteria and that can be used as a cloning vector for making transgenic *Drosophila.*
4. Restriction enzymes are a fundamental and invaluable tool of recombinant DNA technology. Each restriction enzyme recognizes a specific DNA sequence and cuts the sugar-phosphate backbone. Those restriction enzymes that make staggered cuts produce single-stranded ends that can be hybridized to other fragments that have complementary single-stranded sequences. The large number of restriction-enzyme and target-recognition sites make restriction enzyme-mediated cutting and pasting of recombinant DNA fragments a versatile and broadly applicable technique.
5. PCR is a powerful method for the direct amplification of a relatively small sequence of DNA without the need of a host cell and without the need of very much starting material. The selection of the DNA to be amplified from a given sample is determined by the sequence specificity of the primer. Multiple rounds of priming and synthesis lead to an exponential amplification of the sequence of interest. When the exact sequence of the primer is uncertain, degenerate primers are used to amplify the desired PCR product.
6. The cloning of DNA segments in host cells is quite versatile, with vectors that can enable the cloning of segments that are small (a few kilobases or less) or as large as several hundred kilobases. Sometimes, cloning focuses on constructing one specific DNA molecule; alternatively, it is used to construct an entire library of sequences that represents, in aggregate, the entirety of the genome or expressed transcripts in a cell, a tissue, or an individual organism.
7. Labeled single-stranded DNA or RNA probes are important "bait" for fishing out similar or identical sequences from complex mixtures of molecules in genomic or cDNA libraries and in Southern and Northern blotting.
8. Because restriction enzyme target sites are short, specific DNA sequences, they are landmarks that can be used to describe the structure of a cloned DNA molecule. A restriction map of the location and distance between restriction sites is produced by ordering such markers. Knowledge of restriction maps is essential in the design of engineered molecules.
9. It is important to be able to use the structure and overlap of clones to develop an overall map of a part of the genome, which is accomplished by comparing the restriction maps or the full sequence of overlapping clones. With the use of chromosome-walking techniques, a megabase-sized region can be mapped even though the individual clones that contain DNA segments from that region are much smaller. Chromosome-walking techniques can also be used to identify genes that map near

known molecular markers; this application of chromosome walking is called *positional cloning*.

10. After a DNA molecule has been amplified, its sequence can be determined by standard methods. The sequence can be used to predict the locations of DNA sequences that encode transcripts and proteins, as well as the structures of these encoded products. The order of bases in a nucleotide segment can be considered the ultimate genetic map of that segment.

11. Recombinant DNA molecules can be used as diagnostics for genotyping experimental organisms or, in a clinical context, for assessing the risk of a genetic disease. Some heritable diseases are associated with a single mutant allele that, for some reason, has spread in the population. In such cases, diagnostics can be developed to identify the presence of that mutant allele or its wild-type counterpart or both.

12. Transgenes—engineered molecules that are introduced and expressed in eukaryotic cells—are important in several ways. They can be used to demonstrate an association between a transmissible mutation and a specific DNA sequence (and a specific transcript's protein) by rescuing that recessive mutation with a wild-type transgene. Transgenes can also be employed to engineer a novel mutation or to study the regulatory sequences that constitute part of a gene. They can be introduced as extrachromosomal molecules or as integrants in ectopic chromosomal locations or can be targeted to the location of the normal (endogenous) gene, depending on the system.

13. Gene therapy is the clinical extension of transgenic technology to the treatment of certain human diseases. As with many applied aspects of genetic engineering, gene therapy raises a complex set of social and ethical issues.

CONCEPT MAP

Draw a concept map interrelating as many of the following terms as possible. Note that the terms are listed in no particular order.

restriction enzyme / recombinant DNA / DNA clone / probe / transgene / vector / sequencing / chromosome walk / library / PCR / gene therapy

SOLVED PROBLEM

1. In Chapter 3, we studied the structure of tRNA molecules. Suppose that you want to clone a fungal gene that encodes a certain tRNA. You have a sample of the purified tRNA and an *E. coli* plasmid that contains a single *Eco*RI cutting site in a tet^R (tetracycline-resistance) gene, as well as a gene for resistance to ampicillin (amp^R). How can you clone the gene of interest?

Solution

You could use the tRNA itself or a cloned cDNA copy of it to probe for the DNA containing the gene. One method is to digest the genomic DNA with *Eco*RI and then mix it with the plasmid, which you also have cut with *Eco*RI. After transformation of an amp^S tet^S recipient, Amp^R colonies are selected, indicating successful transformation. Of these Amp^R colonies, select the colonies that are Tet^S. These Tet^S colonies will contain vectors with inserts in the tet^R gene, and a great number of them are needed to make the library. Test the library by using the tRNA as the probe. Those clones that hybridize to the probe will contain the gene of interest.

Alternatively, you can subject *Eco*RI-digested genomic DNA to gel electrophoresis and then identify the correct band by probing with the tRNA. This region of the gel can be cut out and used as a source of enriched DNA to clone into the plasmid cut with *Eco*RI. You would then probe these clones with the tRNA to confirm that these clones contain the gene of interest.

SOLVED PROBLEM

2. The restriction enzyme *Hin*dIII cuts DNA at the sequence AAGCTT, and the restriction enzyme *Hpa*II cuts DNA at the sequence CCGG. On average, how frequently will each enzyme cut double-stranded DNA? (In other words, what is the average spacing between restriction sites?)

Solution

We need consider only one strand of DNA, because both sequences will be present on the opposite strand at the same site owing to the symmetry of the sequences. The frequency of the six-base-long *Hin*dIII sequence is $(1/4)^6 = 1/4096$, because there are four possibilities at each of the six positions. Therefore, the average spacing between *Hin*dIII sites is approximately 4 kb. For *Hpa*II, the frequency of the four-base-long sequence is $(1/4)^4$, or 1/256. The average spacing between *Hpa*II sites is approximately 0.25 kb.

SOLVED PROBLEM

3. A yeast plasmid carrying the yeast *leu2*$^+$ gene is used to transform haploid *leu2*$^-$ yeast cells. Several *leu*$^+$-transformed colonies appear on a medium lacking leucine. Thus, *leu2*$^+$ DNA presumably has entered the recipient cells, but now we have to decide what has happened to it inside these cells. Crosses of transformants to *leu2*$^-$ testers reveal that there are three types of transformants, A, B, and C, representing three different fates of the *leu2*$^+$ gene in the transformation. The results are:

Type A × *leu2*$^-$ ⟶ $\frac{1}{2}$ *leu*$^-$
$\frac{1}{2}$ *leu*$^+$ × standard *leu2*$^+$
⟶ $\frac{3}{4}$ *leu*$^+$
$\frac{1}{4}$ *leu*$^-$

Type B × *leu2*$^-$ ⟶ $\frac{1}{2}$ *leu*$^-$
$\frac{1}{2}$ *leu*$^+$ × standard *leu2*$^+$
⟶ 100% *leu*$^+$
0% *leu*$^-$

Type C × *leu2*$^-$ ⟶ 100% *leu*$^+$

What three different fates of the *leu2*$^+$ DNA do these results suggest? Be sure to explain *all* the results according to your hypotheses. Use diagrams if possible.

Solution

If the yeast plasmid does not integrate, then it replicates independently of the chromosomes. In meiosis, the daughter plasmids would be distributed to the daughter cells, resulting in 100 percent transmission. This percentage was observed in transformant type C.

If one copy of the plasmid is inserted, the resulting offspring from a cross with a *leu2*$^-$ line would have a ratio of 1 *leu*$^+$: 1 *leu*$^-$. This ratio is seen in type A and type B.

When the resulting *leu*$^+$ cells are crossed with standard *leu2*$^-$ lines, the data from type A cells suggest that the inserted gene is segregating independently of the standard *leu2*$^+$ gene, and so the *leu2*$^+$ transgene has inserted ectopically into another chromosome.

Type A: *leu2*$^-$ | *leu2*$^+$

When this is crossed with a standard wild-type strain,

leu2$^+$

then the following segregation results:

$\frac{1}{2}$ *leu2*$^-$ ⟶ $\frac{1}{2}$ *leu2*$^+$ ⟶ $\frac{1}{4}$ +
$\frac{1}{2}$ *leu2*$^-$ ⟶ $\frac{1}{2}$ no allele ⟶ $\frac{1}{4}$ −
$\frac{1}{2}$ *leu2*$^+$ ⟶ $\frac{1}{2}$ *leu2*$^+$ ⟶ $\frac{1}{4}$ +
$\frac{1}{2}$ *leu2*$^+$ ⟶ $\frac{1}{2}$ no allele ⟶ $\frac{1}{4}$ +

The data from type B cells suggest that the inserted gene has replaced the standard *leu2*$^+$ allele at its normal locus.

Type B: *leu2*$^+$

When crossed with a standard wild type, all the progeny will be *leu*$^+$.

PROBLEMS

Basic Problems

1. Calculate the average distances (in nucleotide pairs) between the restriction sites in organism X for the following restriction enzymes, assuming an AT:GC ratio of 50:50:

*Alu*I	5′ AGCT 3′
	3′ TCGA 5′
*Eco*RI	5′ GAATTC 3′
	3′ CTTAAG 5′
*Acy*I	5′ G Pu CG Py C 3′
	3′ C Py GC Pu G 5′

(NOTE: Py = any pyrimidine; Pu = any purine.)

2. A circular bacterial plasmid (pBP1) has a single *Hin*dIII restriction enzyme site in the middle of a tetracycline-resistance gene (*tet*^{R}). Fruit fly genomic DNA is digested with *Hin*dIII, and a library is made in the plasmid vector pBP1. Probing reveals that clone 15 contains a specific *Drosophila* gene of interest. Clone 15 is studied by restriction analysis with *Hin*dIII and another restriction enzyme, *Eco*RV. The ethidium bromide-stained electrophoretic gel shows bands as in the accompanying diagram (the control was the plasmid pBP1 vector without an insert). The sizes of the bands (in kilobases) are shown alongside. (NOTE: Circular molecules do not give intense bands on this type of gel; so you can assume that all bands represent linear molecules.)

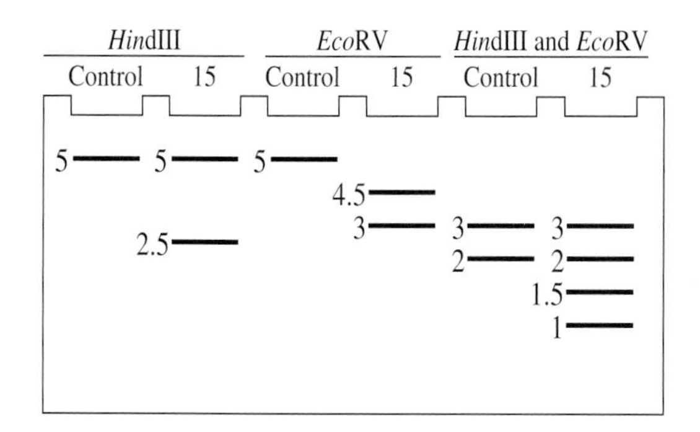

a. Draw restriction maps for plasmid pBP1 with and without the insert, showing the sites of the target sequences and the approximate position of the tet^R gene.

b. If the same tet^R gene cloned in a completely nonhomologous vector is made radioactive and used as a probe in a Southern blot of this gel, which bands do you expect to appear radioactive on an autoradiogram?

c. If the same gene of interest from a fly closely related to *Drosophila* has been cloned in a nonhomologous vector and this clone is used as a probe for the same gel, what bands do you expect to see on the autoradiogram?

Unpacking the Problem

a. Which plasmid described in this chapter seems closest in type to pBP1?

b. What is the importance of the single *Hin*dIII restriction site?

c. Why is it important that the single site is in a resistance gene? Would it be useful if not?

d. What is the effect of the insertion of donor DNA into the resistance gene? Is this effect important for this problem?

e. What is a library? What type was used in this experiment and does it matter for this problem?

f. What kind of probing would have shown that clone 15 contains the gene of interest? and is it relevant to the present problem?

g. What is an electrophoretic gel?

h. What function does ethidium bromide serve in this experiment?

i. Does the gel shown represent a Southern blot or a Northern blot or neither?

j. Generically, what types of molecules are visible on the gel?

k. How many fragments are produced if a circular molecule is cut once?

l. How many fragments are produced if a circular molecule is cut twice?

m. Can you write a simple formula relating the number of restriction enzyme sites in a circular molecule to the number of fragments produced?

n. If one enzyme produces n fragments and another produces m fragments, how many fragments are produced if both enzymes are used?

o. In the diagram, at what positions were the DNA samples loaded into the gel?

p. Why are the smaller-molecular-weight fragments at the bottom of the gel?

q. What is the total molecular weight of the fragments in all the lanes? What patterns do you see?

r. Is it a coincidence that the 3- and 2-kb fragments together equal the 5-kb fragment in size?

s. Is it a coincidence that the 1.5- and 1-kb fragments together equal the 2.5-kb fragment in size?

t. If a fragment produced by one enzyme disappears when the DNA is treated with that same enzyme plus another enzyme, what does the disappearance signify?

u. What determines whether a probe will hybridize to a DNA blot (denatured)?

v. In part c, why is it stressed that a nonhomologous vector is used?

Now attempt to solve the problem.

3. You have a purified DNA molecule, and you wish to map restriction enzyme sites along its length. After digestion with *Eco*RI, you obtain four fragments: 1, 2, 3, and 4. After digestion of each of these fragments with *Hin*dII, you find that fragment 3 yields two subfragments (3_1 and 3_2) and that fragment 2 yields three (2_1, 2_2, and 2_3). After digestion of the entire DNA molecule with *Hin*dII, you recover four pieces: A, B, C, and D. When these pieces are treated with *Eco*RI, piece D yields fragments 1 and 3_1, A yields 3_2 and 2_1, and B yields 2_3 and 4. The C piece is identical with 2_2. Draw a restriction map of this DNA.

4. After *Drosophila* DNA has been treated with a restriction enzyme, the fragments are attached to plasmids and selected as clones in *E. coli*. With the use of this "shotgun" technique, every DNA sequence of *Drosophila* in a library can be recovered.

a. How would you identify a clone that contains DNA coding for the protein actin, whose amino acid sequence is known?

b. How would you identify a clone coding for a specific tRNA?

5. You have isolated and cloned a segment of DNA that is known to be a unique sequence in the human genome. It maps near the tip of the X chromosome and is about 10 kb in length. You label the 5′ ends with ^{32}P and cleave the molecule with *Eco*RI. You obtain two fragments: one is 8.5 kb long; the other is 1.5 kb. You split a solution containing the 8.5-kb fragment into two samples, partly digesting one with *Hae*III and the other with *Hin*dII. You then separate each sample on an agarose gel. By autoradiography, you obtain the following results:

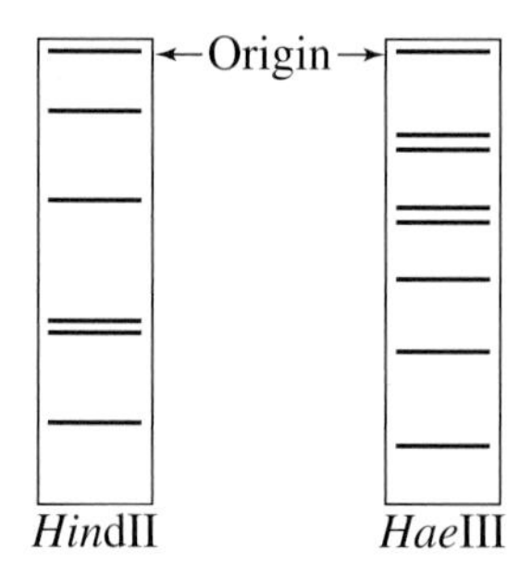

Draw a restriction enzyme map of the complete 10-kb molecule.

6. A linear fragment of DNA is cleaved with the individual restriction enzymes *Hin*dIII and *Sma*I and then with a combination of the two enzymes. The fragments obtained are:

*Hin*dIII	2.5 kb, 5.0 kb
*Sma*I	2.0 kb, 5.5 kb
*Hin*dIII and *Sma*I	2.5 kb, 3.0 kb, 2.0 kb

a. Draw the restriction map.

b. The mixture of fragments produced by the combined enzymes is cleaved with the enzyme *Eco*RI, resulting in the loss of the 3-kb fragment (band stained with ethidium bromide on an agarose gel) and the appearance of a band stained with ethidium bromide representing a 1.5-kb fragment. Mark the *Eco*RI cleavage site on the restriction map.

(Problem 6 courtesy of Joan McPherson. From A. J. F. Griffiths and J. McPherson, *100+ Principles of Genetics*. W. H. Freeman and Company, 1989.)

7. The gene for β-tubulin has been cloned from *Neurospora* and is available. List a step-by-step procedure for cloning the same gene from the related fungus *Podospora,* using as the cloning vector the pBR *E. coli* plasmid shown here, where *kan* = kanamycin and *tet* = tetracycline:

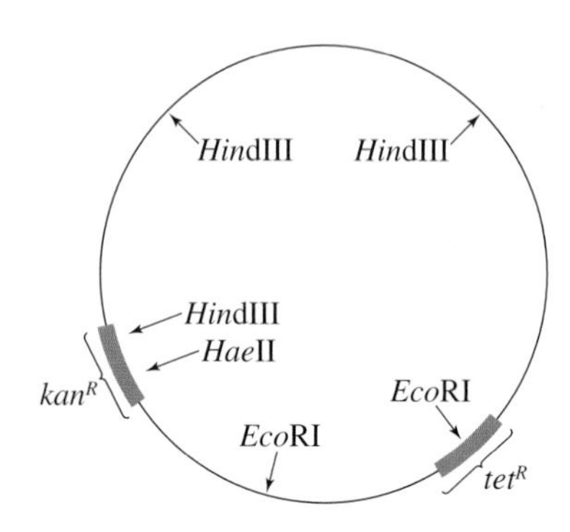

8. In any particular transformed eukaryotic cell (say, of *Neurospora*), how could you tell if the transforming DNA (carried on a circular bacterial vector)

a. replaced the resident gene of the recipient by double crossing-over or single crossing-over?

b. was inserted ectopically?

9. In an electrophoretic gel across which is applied a powerful electric alternating pulsed field, the DNA of the haploid fungus *Neurospora crassa* (n = 7) moves slowly but eventually forms seven bands, which represent DNA fractions that are of different sizes and hence have moved at different speeds. These bands are presumed to be the seven chromosomes. How would you show which band corresponds to which chromosome?

10. The protein encoded by the alkaptonuria gene is 445 amino acids long, yet the gene spans 60 kb (see Foundations of Genetics 8-1). How is this possible?

11. In yeast, you have sequenced a piece of wild-type DNA and it clearly contains a gene, but you do not know what gene it is. Therefore, to investigate further, you would like to find out its mutant phenotype. How would you use the cloned wild-type gene to do so? Show your experimental steps clearly.

Challenging Problems

12. A circular bacterial plasmid containing a gene for tetracycline resistance was cut with restriction enzyme *Bgl*II. Electrophoresis showed one band of 14 kb.

a. What can be deduced from this result?

The plasmid was cut with *Eco*RV and electrophoresis produced two bands, one of 2.5 kb and the other 11.5 kb.

b. What can be deduced from this result?

Digestion with both enzymes together resulted in three bands of 2.5, 5.5, and 6 kb.

c. What can be deduced from this result?

Plasmid DNA cut with *Bgl*II was mixed and ligated with donor DNA fragments, also cut with *Bgl*II, to make recombinant DNA molecules. All recombinant clones proved to be tetracycline sensitive.

d. What can be deduced from this result?

One recombinant clone was cut with *Bgl*II, and fragments of 4 and 14 kb were observed.

e. Explain this result.

The same clone was treated with *Eco*RV and fragments of 2.5, 7, and 8.5 were observed.

f. Explain these results by showing a restriction map of the recombinant DNA.

13. a. A fragment of mouse DNA with *Eco*RI sticky ends carries the gene *M*. This DNA fragment, which is 8 kb long, is inserted into the bacterial plasmid pBR322 at the *Eco*RI site. The recombinant plasmid is cut with three different restriction enzyme treatments. The patterns of ethidium bromide fragments, after electrophoresis on agarose gels, are shown in this diagram:

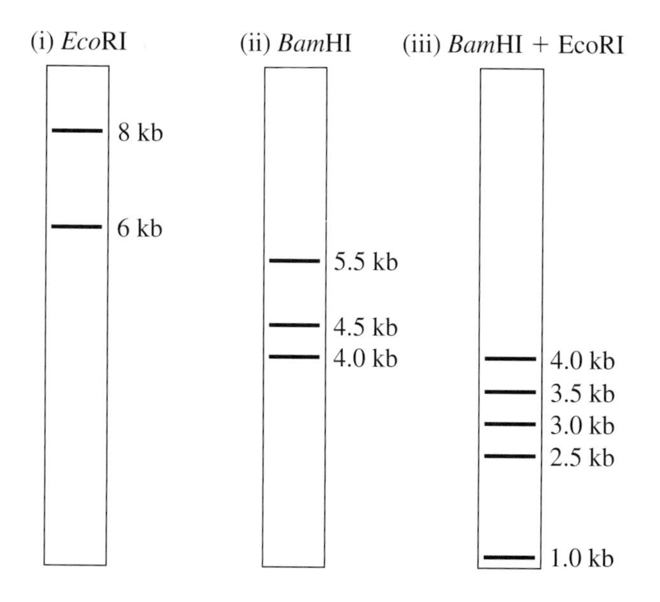

A Southern blot is prepared from gel iii. Which fragments will hybridize to a probe (^{32}P) of pBR plasmid DNA?

b. Gene *X* is carried on a plasmid consisting of 5300 nucleotide pairs (5300 bp). Cleavage of the plasmid with the restriction enzyme *Bam*HI gives fragments 1, 2, and 3, as indicated in the following diagram. (*B* = *Bam*HI restriction site). Tandem copies of gene *X* are contained within a single *Bam*HI fragment. If gene *X* encodes a protein X of 400 amino acids, indicate the approximate positions and orientations of the gene *X* copies.

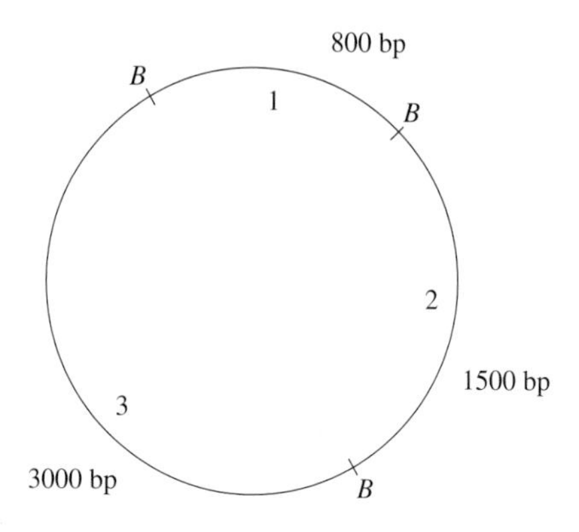

(Problem 13 courtesy of Joan McPherson. From A. J. F. Griffiths and J. McPherson, *100+ Principles of Genetics*. W. H. Freeman and Company, 1989.)

14. Prototrophy is often the phenotype selected to detect transformants. Prototrophic cells are used for donor DNA extraction; then this DNA is cloned and the clones are added to an auxotrophic recipient culture. Successful transformants are identified by plating the recipient culture on minimal medium and looking for colonies. What experimental design would you use to make sure that a colony that you hope is a transformant is not, in fact,

a. a prototrophic cell that has entered the recipient culture as a contaminant?

b. a revertant (mutation back to prototrophy by a second mutation in the originally mutated gene) of the auxotrophic mutation?

15. Two children are investigated for the expression of a gene (*D*) that encodes an important enzyme for muscle development. The results of the studies of the gene and its product follow.

For child 2, the enzyme activity of each stage was very low and could be estimated only at approximately 0.1 unit at ages one, two, three, and four.

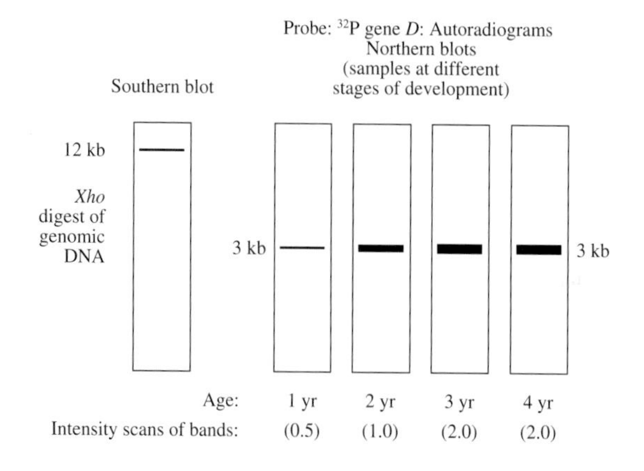

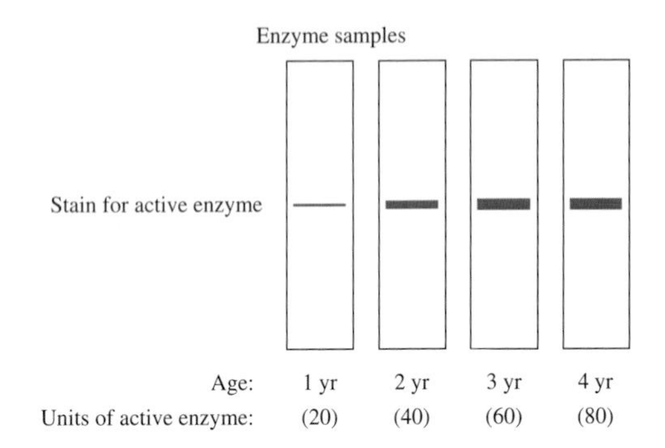

Child 2

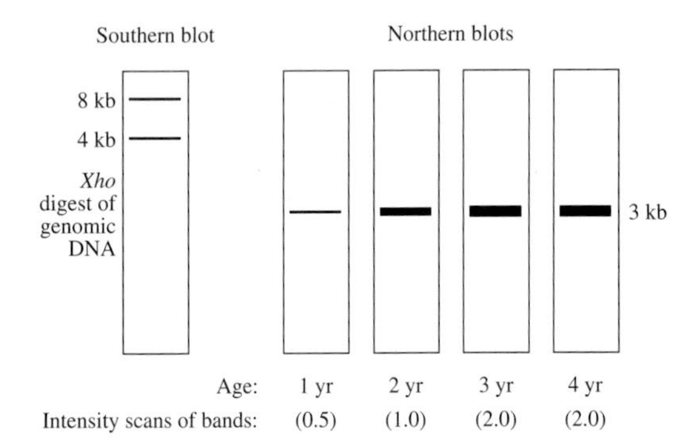

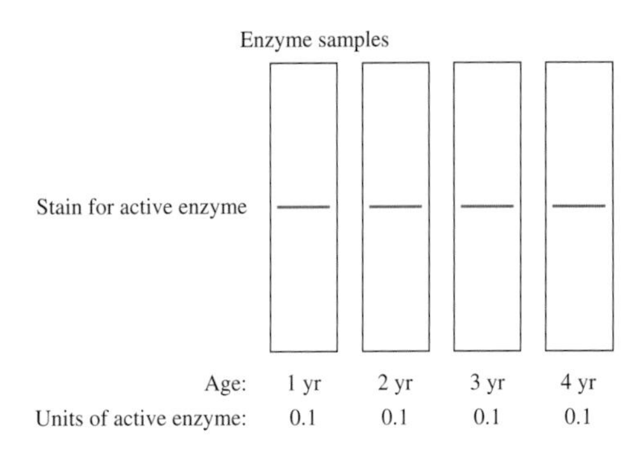

a. For both children, draw graphs representing the developmental expression of the gene. (Fully label both axes.)

b. How can you explain the very low levels of active enzyme for child 2? (Protein degradation is only one possibility.)

c. How might you explain the difference in the Southern blot for child 2 compared with that for child 1?

d. If only one mutant gene has been detected in family studies of the two children, define the individual children as either homozygous or heterozygous for gene *D*. (Problem 15 courtesy of Joan McPherson. From A. J. F. Griffiths and J. McPherson, *100+ Principles of Genetics*. W. H. Freeman and Company, 1989.)

16. A cloned fragment of DNA was sequenced by using the dideoxy method. A part of the autoradiogram of the sequencing gel is represented here.

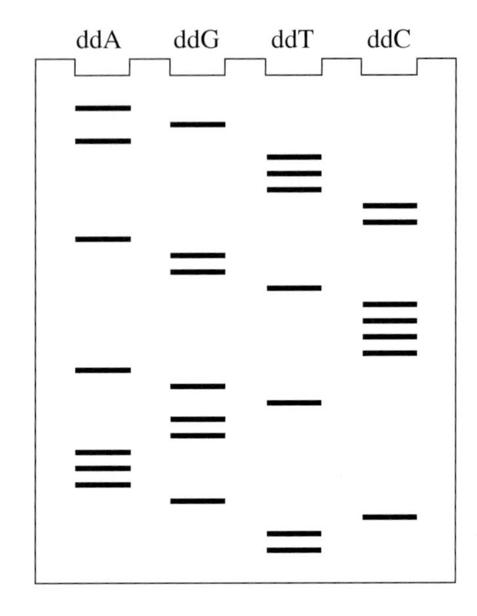

a. Deduce the nucleotide sequence of the DNA nucleotide chain synthesized from the primer. Label the 5′ and 3′ ends.

b. Deduce the nucleotide sequence of the DNA nucleotide chain used as the template strand. Label the 5′ and 3′ ends.

c. Write out the nucleotide sequence of the DNA double helix (label the 5′ and 3′ ends).

d. How many of the six reading frames are "open" as far as you can tell?

17. The cDNA clone for the human gene encoding tyrosinase was radioactively labeled and used in a Southern analysis of *Eco*R1-digested genomic DNA of wild-type mice. Three mouse fragments were found to be radioactive (were bound by the probe). When albino mice were used in this Southern analysis, no genomic fragments bound to the probe. Explain these results in relation to the nature of the wild-type and mutant mouse alleles.

18. Transgenic tobacco plants were obtained in which the vector Ti plasmid was designed to insert the gene of interest plus an adjacent kanamycin-resistance gene. The inheritance of chromosomal insertion was followed by testing progeny for kanamycin resistance. Two plants typified the results obtained generally. When plant 1 was backcrossed with wild-type tobacco, 50 percent of the progeny were kanamycin resistant and 50 percent were sensitive. When plant 2 was backcrossed with the wild type, 75 percent of the progeny were kanamycin resistant, and 25 percent were sensitive. What must have been the difference between the two transgenic plants? What would you predict about the situation regarding the gene of interest?

19. A cystic fibrosis mutation in a certain pedigree is due to a single nucleotide-pair change. This change destroys an *Eco*RI restriction site normally found in this position. How would you use this information in counseling members of this family about their likelihood of being carriers? State the precise experiments needed. Assume that you find that a woman in this family is a carrier, and it transpires that she is married to an unrelated man who is also a heterozygote for cystic fibrosis, but, in his case, it is a different mutation in the same gene. How would you counsel this couple about the risks of a child's having cystic fibrosis?

20. Bacterial glucuronidase converts a colorless substance called X-Gluc into a bright-blue indigo pigment. The gene for glucuronidase also works in plants if given a plant promoter region. How would you use this gene as a reporter gene to find out in which tissues a plant gene that you have just cloned is normally active? (Assume that X-Gluc is easily taken up by the plant tissues.)

21. A *Neurospora* geneticist is interested in the genes that control hyphal extension. He decides to clone a sample of these genes. It is known from previous mutational analysis of *Neurospora* that a common type of mutant has a small-colony ("colonial") phenotype on plates, caused by abnormal hyphal extension. Therefore, he decides to do a tagging experiment by using transforming DNA to produce colonial mutants by insertional mutagenesis. He transforms *Neurospora* cells by using a bacterial plasmid carrying a gene for benomyl resistance *(ben-R)* and recovers resistant colonies on benomyl-containing medium. Some colonies show the colonial phenotype being sought, and so he isolates and tests a sample of these colonies. The colonial isolates prove to be of two types as follows:

Type 1 *col . ben-R* × wild type (+ . *ben-S*)

Progeny	1/2 *col . ben-R*
	1/2 + . *ben-S*

Type 2 *col . ben-R* × wild type (+ . *ben-S*)

Progeny	1/4 *col . ben-R*
	1/4 *col . ben-S*
	1/4 + . *ben-R*
	1/4 + . *ben-S*

a. Explain the difference between these two types of results.

b. Which type should he use to try to clone the genes affecting hyphal extension?

c. How should he proceed with the tagging protocol?

d. If a probe specific for the bacterial plasmid is available, which progeny should be hybridized by this probe?

Unpacking the Problem

a. What is hyphal extension? and why do you think anyone would find it interesting?

b. How does the general approach of this experiment fit in with the general genetic approach of mutational dissection?

c. Is *Neurospora* haploid or diploid? and is this property relevant to the problem?

d. Is it appropriate to transform a fungus (a eukaryote) with a bacterial plasmid? Does it matter?

e. What is transformation? and in what way is it useful in molecular genetics?

f. How are cells prepared for transformation?

g. What is the fate of transforming DNA if successful transformation takes place?

h. Draw a successful plasmid entry into the host cell, and draw a representation of a successful stable transformation.

i. Does it make any difference to the protocol to know what benomyl is? What role is the benomyl-resistance gene playing in this experiment? Would the experiment work with another resistance marker?

j. What does the word *colonial* mean in the present context? Why did the experimenter think that finding and characterizing colonial mutations would help in understanding hyphal extension?

k. What kind of "previous mutational studies" do you think are being referred to?

l. Draw the appearance of a typical petri dish after transformation and selection. Pay attention to the appearance of the colonies.

m. What is tagging? How does it relate to insertional mutagenesis? How does insertion cause mutation?

n. How are crosses made and progeny isolated in *Neurospora?*

o. Is recombination relevant to this question? Can an RF value be calculated? What does it mean?

p. Why are there only two colonial types? Is it possible to predict which type will be more common?

q. What is a probe? How are probes useful in molecular genetics? How would the probing experiment be done?

r. How would you obtain a probe specific to the bacterial plasmid?

22. The plant *Arabidopsis thaliana* was transformed by using the Ti plasmid into which a kanamycin-resistance gene had been inserted in the T-DNA region. Two kanamycin-resistant colonies (A and B) were selected, and plants were regenerated from them. The plants were allowed to self-pollinate, and the results were as follows:

Plant A selfed ⟶ $\frac{3}{4}$ progeny resistant to kanamycin
$\frac{1}{4}$ progeny sensitive to kanamycin

Plant B selfed ⟶ $\frac{15}{16}$ progeny resistant to kanamycin
$\frac{1}{16}$ progeny sensitive to kanamycin

a. Draw the relevant plant chromosomes in both plants.

b. Explain the two different ratios.

23. Two different circular yeast plasmid vectors (YP1 and YP2) were used to transform leu^- cells into leu^+. The resulting leu^+ cultures from both experiments were crossed with the same leu^- cell of opposite mating type. Typical results were as follows:

YP1 $leu^+ \times leu^-$ ⟶ all progeny leu^+ and the DNA of all these progeny showed positive hybridization to a probe specific to the vector YP1

YP2 $leu^+ \times leu^-$ ⟶ $\frac{1}{2}$ progeny leu^+ and hybridize to vector probe to YP2
$\frac{1}{2}$ progeny leu^- and do not hybridize to YP2 probe

a. Explain the different actions of these two plasmids during transformation.

b. If total DNA is extracted from YP1 and YP2 transformants and digested with an enzyme that cuts once within the vector (and not within the insert), predict the results of electrophoresis and Southern analyses of the DNA; use the specific plasmid as a probe in each case.

24. A linear 9-kb *Neurospora* plasmid, mar1, has the following restriction map:

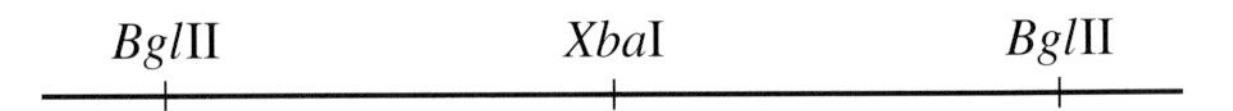

This plasmid was suspected to sometimes integrates into genomic DNA. To test this idea, the central large *Bgl*II fragment was cloned into a pUC plasmid vector, and this was used as a probe in a Southern analysis of *Xba*I-digested genomic DNA from a mar1-containing strain. Predict what the autoradiogram will look like if

a. the plasmid never integrates.

b. the plasmid occasionally integrates into genomic DNA.

Exploring Genomes: A Web-Based Bioinformatics Tutorial

Finding Conserved Domains

Conserved protein sequences are a reflection of conservation of amino acid residues necessary for structure, regulation, or catalytic function. Oftentimes, groups of residues can be identified as being a pattern or signature of a particular type of enzyme or regulatory domain. In the Genomics tutorial at www.whfreeman.com/mga, you will learn how to find the conserved domains in a complex protein.

GENOMICS 9

Key Concepts

1. Genomics is the molecular mapping and characterization of whole genomes and whole sets of gene products.
2. The first step in whole genome mapping is the development of high-resolution genetic maps in which classical genetic markers and phenotypically silent polymorphic DNA differences are interdigitated.
3. High-resolution genetic maps can be based on ordering of markers by meiotic recombination or by co-localization of genes to the same fragments of broken chromosomes.
4. High-resolution genetic mapping techniques are indispensable in the localization of genes contributing to heritable diseases.
5. The second step in whole genome mapping is the production of physical maps, overlapping clones that together represent the entire genome.
6. The final and highest-resolution level of genome mapping is a complete DNA sequence.
7. The choice of genome sequencing strategy in part depends on the size of the genome and the array and distribution of repetitive sequences within it.
8. There are two approaches to assembling millions of sequencing reads into a reconstruction of an intact genome: clone-by-clone and whole genome assembly strategies.
9. With the availability of a complete genomic sequence, one can use computational analysis to predict the full array of RNA and protein products encoded by a genome, and to group them into structural and functional families of related molecules.
10. Functional genomics explores the expression and interaction of the products encoded by the genome and the phenotypes that these products elicit.

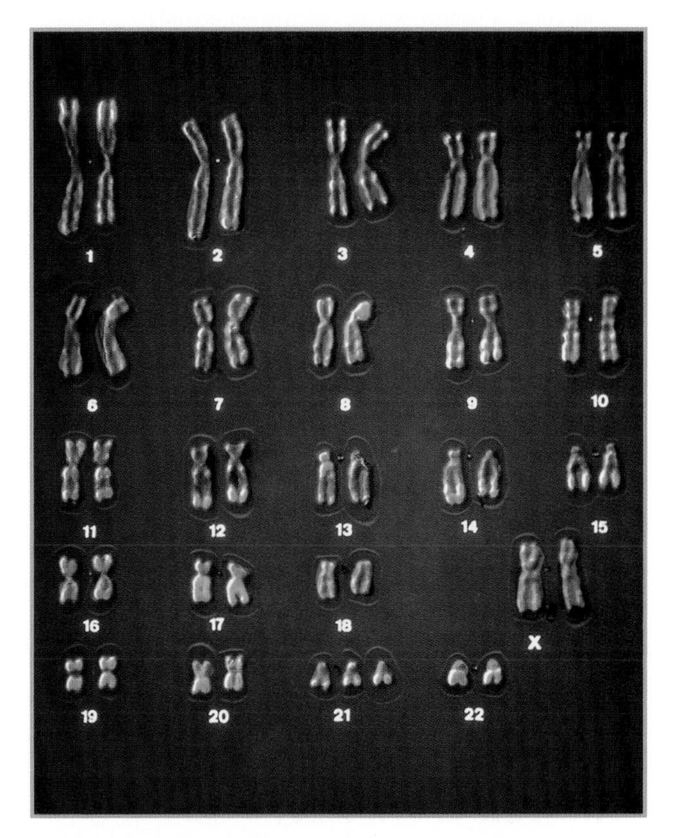

The karyotype of a female with three copies of human chromosome 21 (trisomy-21), which causes a genetic disease called *Down syndrome*. With the sequencing of the human genome nearing completion, the array of gene products on chromosome 21 that might be responsible for Down syndrome can be determined. *(CNRI/Science Photo Library/Science Source/Photo Researchers.)*

In May 2000, with the imminent release of the first "draft" sequence of a human genome, Ewan Birney of the Sanger Centre and European Bioinformatics Institute in Hinxton, England, and others set up a pool in which scientists wagered on the total number of protein-coding genes in the human genome. The pool is ongoing, and the winner will be determined in 2002, roughly at the target date of the release of the "finished" human genome sequence. Each bettor is permitted one bet per calendar year between 2000 and 2002, for a dollar a bet in 2000, increasing to $20 a bet in 2002 as more data emerge and less speculation is needed. (Scientists are notorious as high rollers!) At the appointed hour in 2002, a winner will be determined according to criteria that will be set up at a specified human genome meeting earlier that year. The person with the bet closest to the final number wins the entire pool, winner-take-all, with the award to be given at the 2003 international conference on the human genome at Cold Spring Harbor Laboratories on Long Island in New York. The current distribution of 165 bets (Figure 9-1) ranges from 27,462 to 153,478, with a mean wager of 61,710 genes. What's your bet? Friendly suggestion—don't place your bet until we get to the end of the chapter and have seen how gene prediction analysis is performed in eukaryotes and where the current results of that analysis stand.

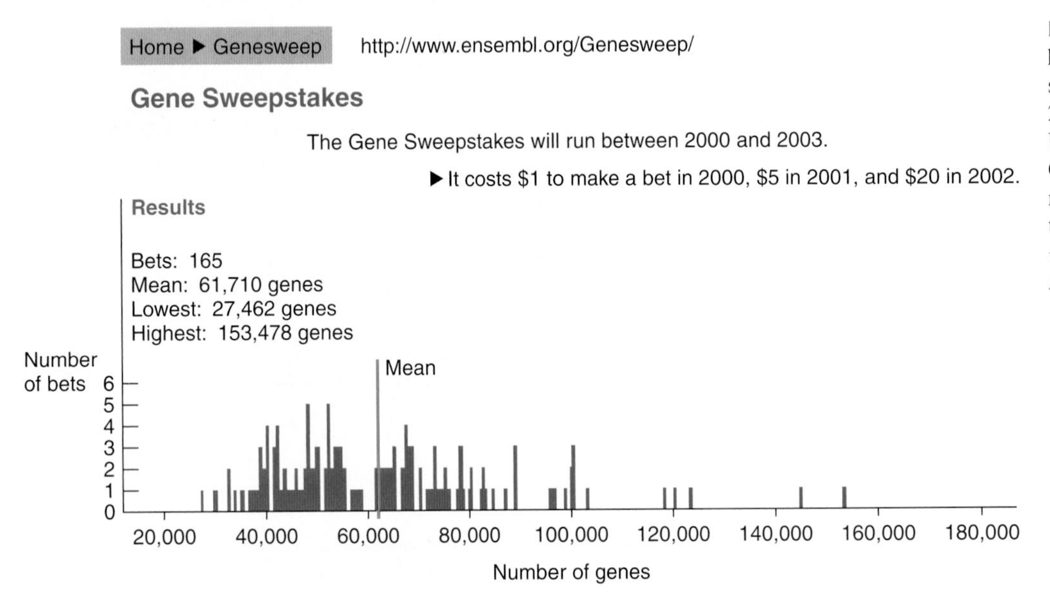

Figure 9-1 The status of the human gene count sweepstakes as of April, 2001. Go to the Web site at URL http://www.ensembl.org/Genesweep/ to see the contest rules and the latest view of the sweepstakes. *(Reproduced with permission of EBI and Ewan Birney.)*

CHAPTER OVERVIEW

Geneticists typically study genes singly or in small groups: a region of the genome within which resides one or more genes of phenotypic or functional interest, genes encoding a group of structurally related proteins (a protein family), the genes encoding products that form sequential steps in a biochemical pathway, and so on. **Genomics** researchers are interested in the same things—understanding gene organization, studying protein families, characterizing biochemical pathways—but do so at the level of the entire genome or for all expressed transcripts or polypeptides. In other words, *genetics is local; genomics is global.*

What are the relative advantages of the local genetic and global genomic approaches? The local approach is most helpful in the analysis of systems where there are a limited number of components and they are all known. In such cases, all effort can go into the detailed analysis of these components. However, often we do not know all the components, or even if we do, their numbers are quite large (such as all genes expressed in a hepatocyte—a liver cell). Global approaches permit the identification of all genes within the genome contributing to a system of interest according to some shared molecular property—structural similarity, common time or anatomical site of expression, common change in expression levels when cells are exposed to an environmental agent, same chromosomal location, etc. These approaches are not mutually exclusive, but rather are complementary. Once a global genomic approach is applied to a particular problem, the in-depth local analysis of the identified group of genes is pursued. Some of the general techniques of mutational analysis are described in Chapter 12, and some specific applications of such analyses are described in Chapters 13, 15, and 16.

Since genomics is just an extension of techniques we've already discussed under the heading of genetic analysis, why does genomic analysis merit special attention? The answer is twofold:

1. Special experimental techniques have been devised to carry out the difficult task of manipulating and characterizing large numbers of genes and large amounts of DNA. For example, the sequencing of an entire complex genome involves handling and sequencing millions of clones, with the same experimental manipulations carried out on each clone. The automation of such repetitive procedures by computer-driven **robotics** has several advantages: (1) the process flow rate can be much more rapid, allowing these projects to be **high-throughput,** (2) personnel can focus on the more individualized and interpretive aspects of the genome projects, and (3) samples can be tracked more accurately (that is, there will be more assurance that a sequence read truly comes from a particular clone stored in a freezer). Robotic engineering would not be possible without sufficient computational ability. Thus, genome projects also rely heavily on bioinformatics for robotics and sample tracking, and also for subsequent data analysis. Programs have to be written to regulate the movements of the robots, and large databases must maintain the description of each clone and what has been done with it. As with any experimental data set, any individual sequence "read" is subject to experimental error. Interactive computer programs are required to identify those sequences that overlap one another and statistically analyze their overlaps to determine the most likely correct sequence (the **consensus sequence**).
2. The analysis of whole genomes gives us new insights into global organization, expression, regulation, and evolution of the hereditary material. Once the consensus sequences have been produced, the fun has just begun. The interpretation of the informational content of the consensus sequence of a genome is a major ongoing research effort. It requires programs that compare a new consensus sequence with databases of many different kinds of sequences. Such comparisons look for statistically significant levels of sequence similarity at the level of encoded proteins or DNA landmarks called **motifs.** Motifs are short strings of base pairs characteristic of sites regulating particular events in gene expression or chromosome replication, such as 5′ splice sites or origins of replication. In **comparative genomics,** other programs examine the overall structure of a genome and compare it to that of taxonomically related species to understand the evolutionary events that have occurred subsequent to the divergence of these species (discussed in Chapter 19).

Genomics as a discipline follows logically from recombinant DNA analysis, applying the techniques of genetic engineering on a large scale to address the problems geneticists are interested in. More broadly, *genomics* is becoming used as an umbrella term that refers to techniques that globally apply to a given type of analysis. Figure 9-2 on the next page describes the major data sets that are being subject to genomic, high-throughput analysis. The steps in the flow of information from the genome to the production of a phenotype are described in Figure 9-2a. The first series of steps, as we already know, lead to the production of the active RNAs and proteins in the cell. These individual molecules then interact with one another, or with segments of DNA, or with both, to produce multimeric complexes (from dimeric proteins and single protein subunits binding to an RNA molecule or a DNA segment to macromolecular complexes such as ribosomes, spliceosomes, and DNA polymerase). These complexes then interact with other molecules in the cell in sequential reactions to produce pathways and processes, and these in aggregate form the phenotype of the individual.

In this chapter, we will consider the main branches of genomics, and the data sets that are being compiled within each branch (Figure 9-2b). Most of our discussion will focus on **whole genome mapping,** the creation of molecular views of the genome at different levels of resolution. High-resolution genetic maps are used to place molecularly defined differences (such as the presence or absence of a restriction enzyme cut site) on the sorts of linkage maps already discussed (Chapter 6) or on cytogenetic maps (as will be discussed in Chapter 11). These maps provide molecular landmarks for building the higher-resolution physical and sequence maps, and also provide molecular entry points for researchers interested in cloning genes with interesting phenotypes (positional cloning—see Chapter 8). **Physical maps** provide a view of how the clones from genomic clone libraries are distributed throughout the genome. These maps support positional cloning efforts, since in essence they are chromosome walks throughout the genome (see Chapter 8 for a description of chromosome walks), and they also provide logical criteria for selection of clones for production of the highest-resolution maps. The highest-resolution maps are of course the **nucleotide sequence maps** of the genome. There are two major routes of production of sequence maps: **clone-based genome sequencing,** in which a subset of clones from the physical map is fully sequenced and the overlaps between sequences of the individual clones are used to assemble the entire sequence map of the genome. The other is **whole genome shotgun (WGS) sequencing,** in which the portions of the inserts adjacent to the junction points with vector sequences are sequenced from a great many random clones throughout the genome, and the overlapping sequence information is used to assemble the sequence of the entire genome and to reconstruct the physical map of the clones. We will discuss these mapping techniques and the basic procedures used to identify the parts list—the RNAs and proteins—encoded within a genome.

Finally, we will turn to **functional genomics,** the global study of the structure, expression patterns, interactions, and regulation of the RNAs and proteins encoded by

CHAPTER OVERVIEW figure

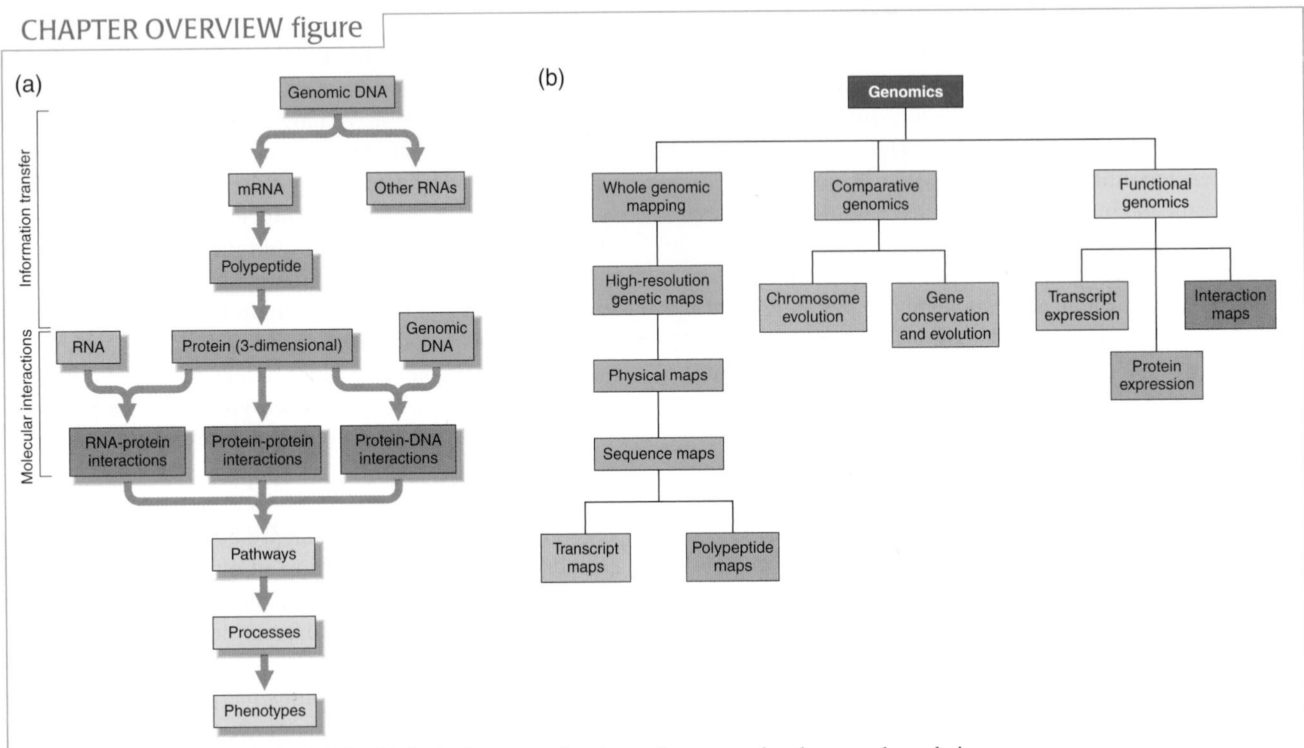

Figure 9-2 Genomic analysis. (a) The biological macromolecules and macromolecular complexes being subjected to high-throughput global "genomic" analysis. One goal of these studies is for these global data sets to serve as a foundation for understanding molecular pathways, biological processes, and phenotypes. (b) A hierarchical view of genomic analysis. (Comparative genomics will be discussed in Chapter 19.)

the genome. The strings of A's, T's, C's, and G's that make up the sequences of genomes are of little value if the information that is enciphered within these strings cannot be intelligibly read. By analogy, knowing the 26 letters of the English alphabet is not of very much use unless we know how to string these letters together to form words, and string the words together to form sentences, etc. Thus, the ways that we can experimentally and computationally decode such information is also an important part of our discussion. Much of this analysis is focused on deciphering the structure and expression of gene products, which, we have already noted, is called *functional genomics.* The field of functional genomics can be broken down into categories on the basis of how many steps separate the products being assessed from the gene itself. The direct products of genes are the RNAs, and in turn, the mRNAs encode the proteins. Thus, characterizations of the structures of the transcripts and the polypeptides are of primary importance. The genome encodes information determining when, where, and how much of each protein is expressed. Characterizing the expression patterns of the RNAs and proteins is then the second level of functional analysis. Finally, individual RNA and polypeptide molecules typically do not function by themselves, but rather they interact within complexes with other molecules or more copies of themselves. Techniques exist to identify such interactions, and these form the third level of functional genomic analysis.

We have reached the point when most of the DNA sequence of the genome of any organism can be determined. For prokaryotes with genomes of at most a few million base pairs (megabase pairs, or Mbp) of DNA, such sequencing efforts are within the capabilities of a single laboratory, but for complex eukaryotes with genomes in the range of several tens to some thousands of megabase pairs, such sequencing requires a complex strategy and coordination of the efforts of many scientists and engineers; these are referred to as **genome projects.** These coordinated efforts for many microbes, for several of the important model genetic organisms, and for humans have culminated in first drafts or complete views of their genomic sequence. Some of the intermediary products that have helped move complex genomes to the point of complete sequence-level analysis have proved to be extremely valuable. Recombinational maps of differences within short DNA sequences dispersed throughout the genome have provided high-resolution maps that have helped in cloning genes of interest, including genes that have alleles contributing to human disease. Physical maps provide clone resources for positional cloning. New genome projects are being initiated at an increasing rate, and we can look forward over the next few years to having an array of genome sequences available as important data sets in genetic and genomic analysis.

MAPPING WHOLE GENOMES

Before the advent of genomic analysis, the genetic basis of our knowledge of an organism usually included relatively low-resolution chromosomal maps of genes producing known mutant phenotypes. We call this *low-resolution* because we now know that phenotypic analysis (even when it is quite thorough) identifies only a small subset of the genes in the genome (about one-third to one-fifth). Starting with these genetic linkage maps, whole genome molecular mapping generally proceeds through several steps of increasing resolution (Figure 9-3a):

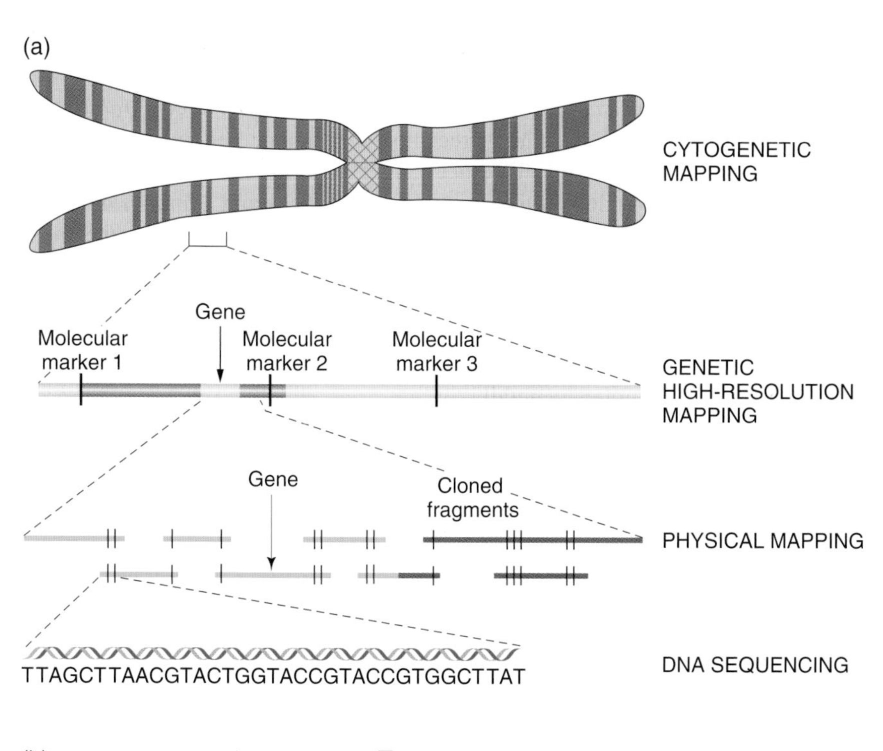

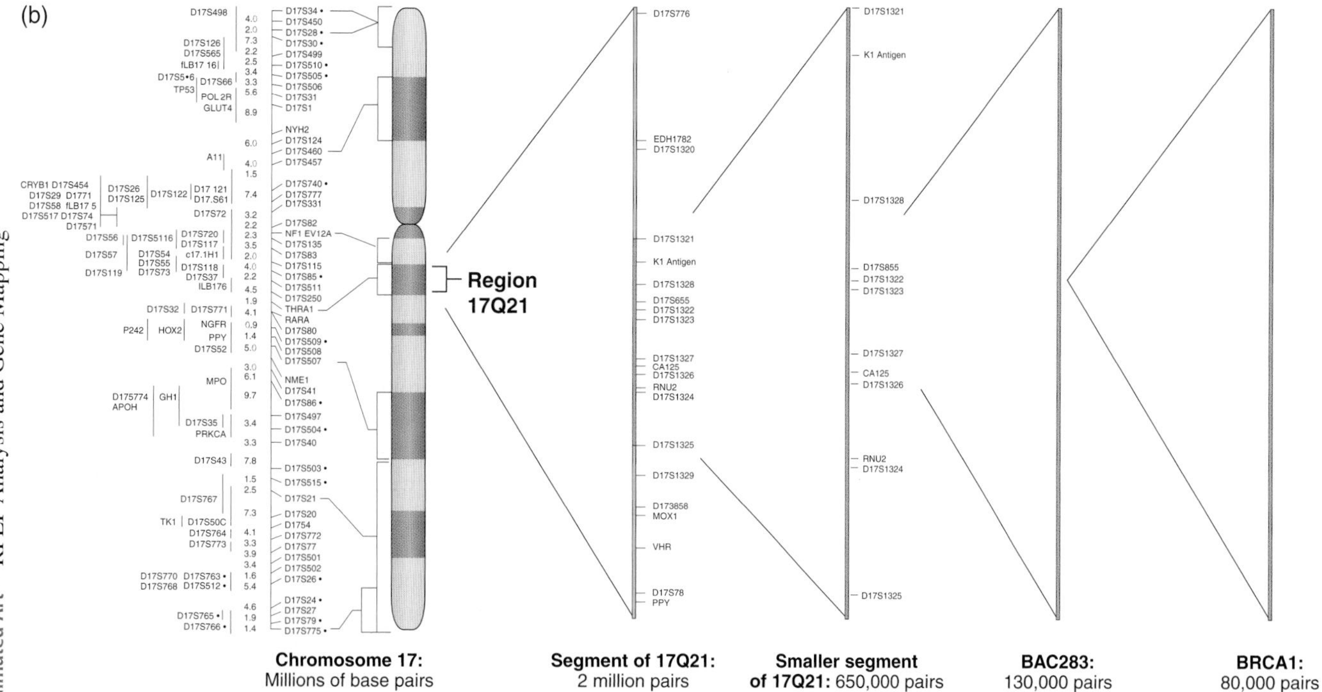

Animated Art • RFLP Analysis and Gene Mapping

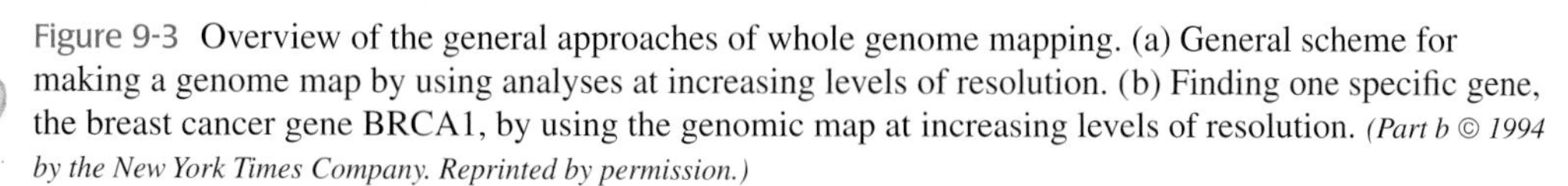
Figure 9-3 Overview of the general approaches of whole genome mapping. (a) General scheme for making a genome map by using analyses at increasing levels of resolution. (b) Finding one specific gene, the breast cancer gene BRCA1, by using the genomic map at increasing levels of resolution. *(Part b © 1994 by the New York Times Company. Reprinted by permission.)*

1. Position genes and molecular markers on high-resolution genetic (recombinational and cytogenetic) maps of each chromosome.
2. Physically characterize and position individual cloned DNA fragments relative to one another to create a synthetic clone-based view of each chromosome. During this process, the high-resolution genetic map or maps of the genome will be anchored to the physical map.
3. Conduct large-scale genomic DNA sequence analysis to produce a complete sequence map of each chromosome. The genetic and physical maps can then be anchored to the sequence map.

In the process of creating these several maps, important reagents for cross-comparison and evaluation of consistency are obtained. These reagents are valuable because any one of the maps has some possibility of being incorrect in its description of a particular region of the genome, and only by cross-comparisons can these errors be detected and then sorted out.

The progressively increasing resolution of analysis in mapping the genome is paralleled by the increasing resolution of analysis needed to find any specific gene (Figure 9-3b). Increasing the level of resolution helps researchers focus in on a region of the genome sufficiently small that it can be fully characterized. Without the ability to progressively focus in on such a small region, gene identification is essentially looking for a needle in a haystack. We will consider how analysis proceeds at these different levels and then come back to consider what we now know about the information content and sequence organization of the genome.

MESSAGE

Whole genome mapping proceeds in several steps, each increasing in level of resolution, culminating in a full sequence map of each chromosome of the species.

GENOME PROJECTS

Genome projects have been undertaken in a range of different organisms, including humans and representatives of every kingdom of life. The initial genomes to have been completely sequenced were chosen for the practical reason of their small size: first the individual chromosomes of viral genomes and then the circular chromosomes of mitochondria and chloroplasts. Next, full bacterial genome sequences began to emerge. Here, some of the genomes were chosen for their genetic interest, others for analyzing evolutionary diversity within prokaryotes, and still others because their organisms were important human pathogens. In addition to the human genome itself, the model eukaryotes that were first selected for genome scale analysis are the same ones that are intensively studied using standard genetic analysis (Table 9-1); in these organisms, an extensively described genome could be best exploited because of the availability of a rich set of mutant strains and other genetic resources. The first eukaryotic model organism genome project was that of a unicellular fungus (the yeast *Saccharomyces cerevisiae*), then a worm (the roundworm *Caenorhabditis elegans*), an insect (the fruit fly *Drosophila melanogaster*), a mammal (the house mouse *Mus musculus*), and a plant (the mustard weed *Arabidopsis thaliana*). Very recently, new major genome projects are expanding to other model systems, including the genetic models *Danio rerio* (the zebrafish) and *Rattus norvegicus* (the laboratory rat), and the carrier of malaria, the mosquito *Anopheles gambiae.* While a great many resources are invested in genome projects, there is a great deal of scientific value that emerges. Let's explore some of the payoffs from genome projects before considering the techniques in detail.

First, some of the intermediate steps along the way are of tremendous importance. Obtaining a set of ordered clones of an entire chromosome or an entire genome is an invaluable baseline for future molecular studies of any type. For example, these clones can be used for finding and manipulating individual genes of interest. (These clones ultimately serve as the raw material for generating the genomic DNA sequence.)

Second, genomic DNA is the blueprint of a species, the information needed to build a living cell and a living organism. Hence knowledge of its specific nucleotide sequence lays the groundwork for eventually providing answers to one of the basic questions of biology: why is an organism the way it is and different from other organisms? Just knowing the sequence of an entire genome will not in itself answer such a question, but it is the prelude to the functional round of analysis, which attempts to understand the functions of individual genes and how they interact as a set.

Third, although a great deal is now known about the structure and function of individual genes, little is known about the *principles* by which DNA segments are organized into whole genomes. This is another level of inquiry that can be addressed through analysis of whole genome sequences. Part of the inquiry will be evolutionary and can be attacked by comparing the entire DNA sequences of different organisms: How were genomes assembled and scrambled through the processes of evolution? Are the precise positions of genes on chromosomes unimportant, or must genes be in specific locations to ensure proper function? It is known that there is conservation of gene position between related taxa, but the significance of this conservation has not been established. Another question concerns the numbers and types of genes needed to specify an individual species and how the numbers and types differ between taxa. There are undoubtedly many other regularities to genome

TABLE 9-1 Some of the Major Eukaryotes Studied in Advanced Genome Projects

Organism	Megabases Sequenced	Percentage of Euchromatin Sequenced	Haploid No. of Chromosomes	Estimated Gene Number
S. cerevisiae	12	100	16	6,800
C. elegans	97	100	6	19,099
D. melanogaster	116	97	4	13,601
A. thaliana	115	100	5	25,498
H. sapiens	2693	90	23	?

structure and function that cannot even be imagined without full sequences from several organisms.

Fourth, human genomics has particular significance to us as humans, not only because of its value as a source of insight into human biology, but also because of the unique applications in medicine. The human genome project has already speeded up the identification and cloning of mutations and genetic variants associated with many human diseases. An understanding of the genetic alterations underlying heritable diseases is an essential step in the path leading to better disease diagnosis and the development of therapeutic treatments.

MESSAGE

Characterizing whole genomes is important to a fundamental understanding of the design principles of living organisms and for the discovery of new genes such as those that are involved in human genetic disease.

HIGH-RESOLUTION GENETIC MAPS

Let's consider how one analyzes a whole genome. Generally, as starting material, we have low-resolution genetic maps derived from the recombinational mapping of heritable differences (Chapters 6 for eukaryotes and 7 for prokaryotes) or cytogenetic mapping techniques (as we will discuss in Chapter 11). Now, what we need to do is to layer onto the low-resolution genetic maps the locations of **DNA polymorphisms**—that is, molecularly defined differences between individuals. Such DNA polymorphisms are far more frequent than gene mutations affecting observable phenotypes on the morphology or viability of an organism, and so the potential density of the markers on these maps is much higher than that of classical genetic maps.

How do we determine the position of a DNA marker on the chromosome? Several different methods are used in localizing genes or markers, as we shall now discuss.

High-Resolution Meiotic Recombination Maps

Meiotic linkage mapping is based on the principles covered in Chapter 6, in other words, on analyzing recombinant frequency in dihybrid and multihybrid crosses. In experimental organisms such as *Saccharomyces, Neurospora, C. elegans, Drosophila,* and *Arabidopsis,* the genes that determine qualitative phenotypic differences can be mapped in a straightforward way because of the ease with which controlled experimental crosses (such as testcrosses) can be made. Therefore, in these organisms, the low-resolution maps built over the years appear full of genes with known phenotypic effect, all mapped to their respective loci.

This is not the case for humans, because, first, there is a lack of informative crosses. Second, progeny sample sizes are too small for accurate statistical determination of linkage. Third, the human genome is enormous, with 24 linkage groups (X, Y, and 22 autosomes), some of which are larger than the entire *Drosophila* or *Arabidopsis* genome. In fact, it was a difficult task even to assign a human disease gene to an individual autosome by linkage analysis.

Even in those organisms in which the maps appeared to be "full" of loci of known phenotypic effect, measurements showed that the chromosomal intervals between the mapped genes had to contain vast amounts of DNA. These intervals could not be mapped by linkage analysis, because there were no markers in those regions. What was needed was large numbers of additional differential markers (that is, genetic differences) that fall in the gaps to provide a higher-resolution map. This need has been met by the exploitation of various kinds of polymorphic DNA markers, typically a site of heterozygosity for some type of **neutral DNA sequence variation.** Neutral variation is that which is not associated with any measurable phenotypic variation. Such a "DNA locus," when heterozygous, can be used in mapping analysis just as with a conventional heterozygous allele pair. DNA markers are easily detected on Southern blot hybridization or by PCR amplification (see Chapter 8) and, it turns out, are so numerous in a genome that, when enough of them are genetically mapped by recombination or cytogenetic analysis, many of them fill the voids between genes of known phenotype. Note that, in mapping, the biological significance of the DNA marker (if any) is not of relevance; the heterozygous site is merely a convenient reference point that will be useful in navigating the genome. In this way, markers are being used just as milestones were used by travelers in previous centuries. Travelers were not interested in the milestones (markers) per se, except as pointers to their final destinations. We will now discuss some of the molecular markers that are employed to produce high-resolution meiotic maps.

Restriction fragment length polymorphisms. Restriction fragment length polymorphisms (RFLPs) are restriction enzyme recognition sites (discussed in Chapter 8) that are present in some strains and absent in others. Consider the progeny of an individual heterozygous for an RFLP backcrossed to an individual homozygous for one of the RFLP variants (presence or absence of the restriction site). Southern blots of restriction-digested genomic DNA from these progeny are hybridized with a probe that will distinguish the various genotypes for an RFLP. The probe in this case is a cloned DNA fragment that uniquely comes from only one DNA segment of the genome (called a **single-copy DNA** probe) and that overlaps the restriction site. By measuring the recombination frequency between this RFLP with other RFLP markers, a detailed RFLP map of the genome can be produced.

RFLP mapping is often performed on a defined set of strains or individuals that become "standards" for mapping that species. For example, in the fungus *Neurospora,* two wild-type strains, Oak Ridge and Mauriceville, are known to show many RFLP differences, so these strains have become standards used in RFLP mapping. The RFLPs can be mapped relative to one another or to genes of known phenotypic expression. Figure 9-4 exemplifies such a cross, in this case with a phenotypic mutant *m*, and RFLP loci 1 through 5 bearing either the Oak Ridge (O) or Mauriceville (M) "alleles." A cross can be made of the type

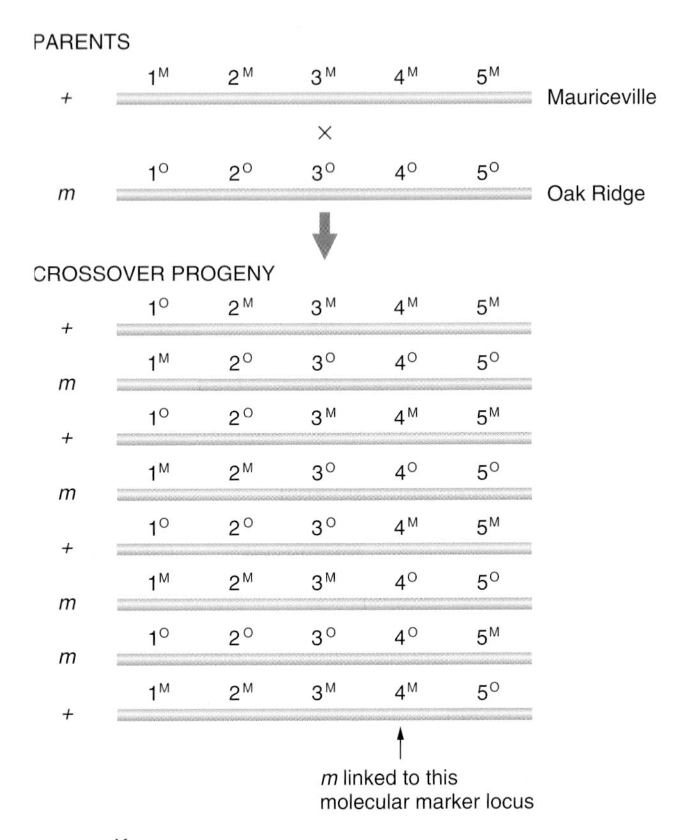

Figure 9-4 Mapping a gene *(m)* by RFLP analysis in *Neurospora.* The two parental strains show many RFLPs ranged along the chromosomes; their loci are labeled 1 through 5. The two parental strains are from Oak Ridge (Tennessee) and Mauriceville (Texas), and their RFLP alleles are labeled O and M. Many different progeny types are recovered, and some of the more common types are shown. The results show that the + allele always segregates with 4^M and the *m* allele always segregates with 4°, suggesting linkage of *m* to RFLP locus 4.

$$m\ .\ 1^O\ .\ 2^O\ .\ 3^O\ .\ 4^O\ .\ 5^O\ \times\ m^+\ .\ 1^M\ .\ 2^M\ .\ 3^M\ .\ 4^M\ .\ 5^M$$

In *Neurospora,* the resulting heterozygote forms haploid progeny spores that can be isolated and grown to determine their genotype at all six loci. The phenotypic mutant or wild-type genotype is observed directly. The RFLP allelic states are determined by Southern blot analysis with five single-copy DNA probes, one overlapping each of the five individual RFLPs. Once the progeny genotypes are determined and grouped according to parental and recombinant types, recombinant frequencies are then calculated in the usual way (Chapter 6). In this case, the complete cosegregation of *m* with RFLP 4 demonstrates the close linkage of these two markers. In this way, the RFLP marker map and the low-resolution phenotypic marker map can be integrated to form a high-resolution genetic map.

An analogous approach has been used in human genome mapping by collecting DNA from a defined set of individuals in 61 families with an average of eight children per family and making this DNA available throughout the world to provide a standard for RFLP mapping. Figure 9-5 shows an example of linkage of a human disease allele to an RFLP locus and the potential for using this information in diagnostics. Because of the close linkage, future generations of persons showing the RFLP morph 1 (another way of saying allele 1 of the RFLP locus) can be predicted to have a high chance of inheriting the disease allele *D*. With more progeny in a pedigree, the position of the RFLP and

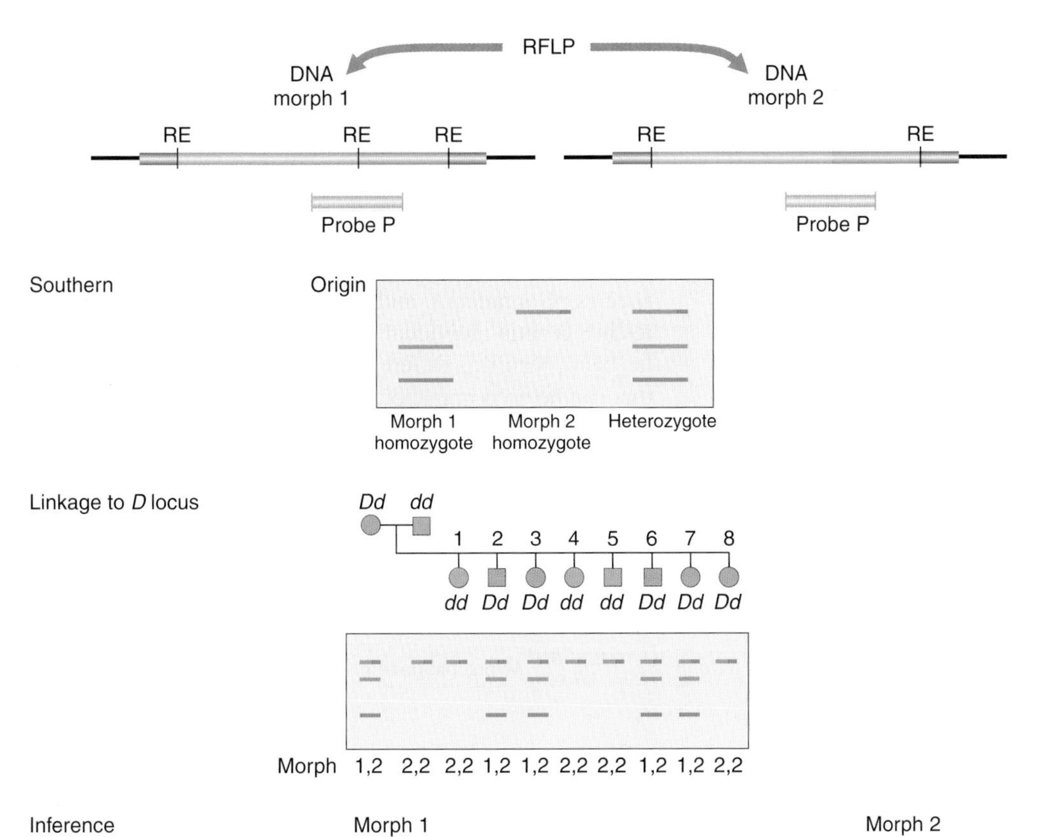

Figure 9-5 The detection and inheritance of a restriction fragment length polymorphism (RFLP). A probe P detects two DNA morphs when the DNA is cut by a certain restriction enzyme (RE). The pedigree of the dominant disease phenotype D shows linkage of the *D* locus to the RFLP locus; only child 8 is recombinant.

Animated Art • RFLP Analysis and Gene Mapping

the disease gene can be measured with considerable accuracy (the eight progeny in Figure 9-5 are far too few for a very accurate determination). Thus, RFLP and other DNA markers can be used to help map human disease genes. Further, the probes used to detect these nearby DNA markers can themselves be used as starting points for chromosome walks to clone the disease genes. This approach has succeeded in cloning hundreds of disease-associated genes in humans.

Now that we've examined one example of DNA marker mapping in detail, let's look at some of the other kinds of DNA markers that are in common use.

DNA markers based on variable numbers of short-sequence repeats. Although RFLPs were the first DNA markers to have been generally used in genomic characterization, in the analysis of animal and plant genomes, they have now been largely replaced by markers based on variation in the number of short tandem repeats. (A tandem repeat is a sequence that is repeated two or more times in the same orientation: for example, ATCTCATCTCATCTCATCTC is a fourfold tandem repeat of the sequence ATCTC.) These markers are collectively called **simple-sequence length polymorphisms (SSLPs).**

SSLPs have two basic advantages over RFLPs. First, RFLPs usually have only one or two "alleles," or morphs, in a pedigree or population under study. This small number limits their usefulness; it would be better to have a larger number of alleles that could act as specific tags so that the input alleles from both parents (four total alleles, two from each parent) can all be tracked in a pedigree. The SSLPs fill this need because multiple allelism is much more common, and as many as 15 alleles have been found for an SSLP locus. Second, the heterozygosity for RFLPs can be low; in other words, if one allele is relatively uncommon in relation to the other, the proportion of heterozygotes (the crucial individuals useful in mapping) will be low. However, SSLPs show much higher levels of heterozygosity. This makes them more useful in mapping because the likelihood of having an SSLP difference located in a region of interest in a particular pedigree is quite high. Two types of SSLPs are now routinely used in genomics, minisatellite and microsatellite markers.

Minisatellite markers. **Minisatellite markers** are based on variation of the number of **VNTRs (variable number tandem repeats)**. The VNTR loci in humans are 1- to 5-kb sequences consisting of variable numbers of a repeating unit from 15 to 100 nucleotides long. If a VNTR probe is available and the total genomic DNA is cut with a restriction enzyme that has no target sites within the VNTR arrays, then a Southern blot will reveal a large number of different-sized fragments that are bound by the probe. Because of the variability in the number of tandem repeats from person to person, the set of fragments that are revealed by Southern blot analysis is highly individualistic. In fact, these patterns are sometimes called **DNA fingerprints.** Figure 9-6 shows a human VNTR obtained from a quadruple repeat of a 33-bp sequence found within the first intron of the myoglobin gene. When some of the VNTRs in the DNA fingerprint were cloned and sequenced, the reason for hybridization of the probe to the VNTRs was a 13-bp core sequence common to all the repeats and to the probe. The longer sequences into which the core was embedded were not necessarily similar. Because VNTRs are common, DNA fingerprints can be made in many different species of organisms, and the technique has great value in testing genetic individuality. If parents differ for a particular band, then this difference becomes a heterozygous site that can be used in mapping. A simple example is shown in Figure 9-7.

Microsatellite markers. **Microsatellite markers** are dispersed regions of the genome composed of variable numbers of dinucleotides repeated in tandem. The most

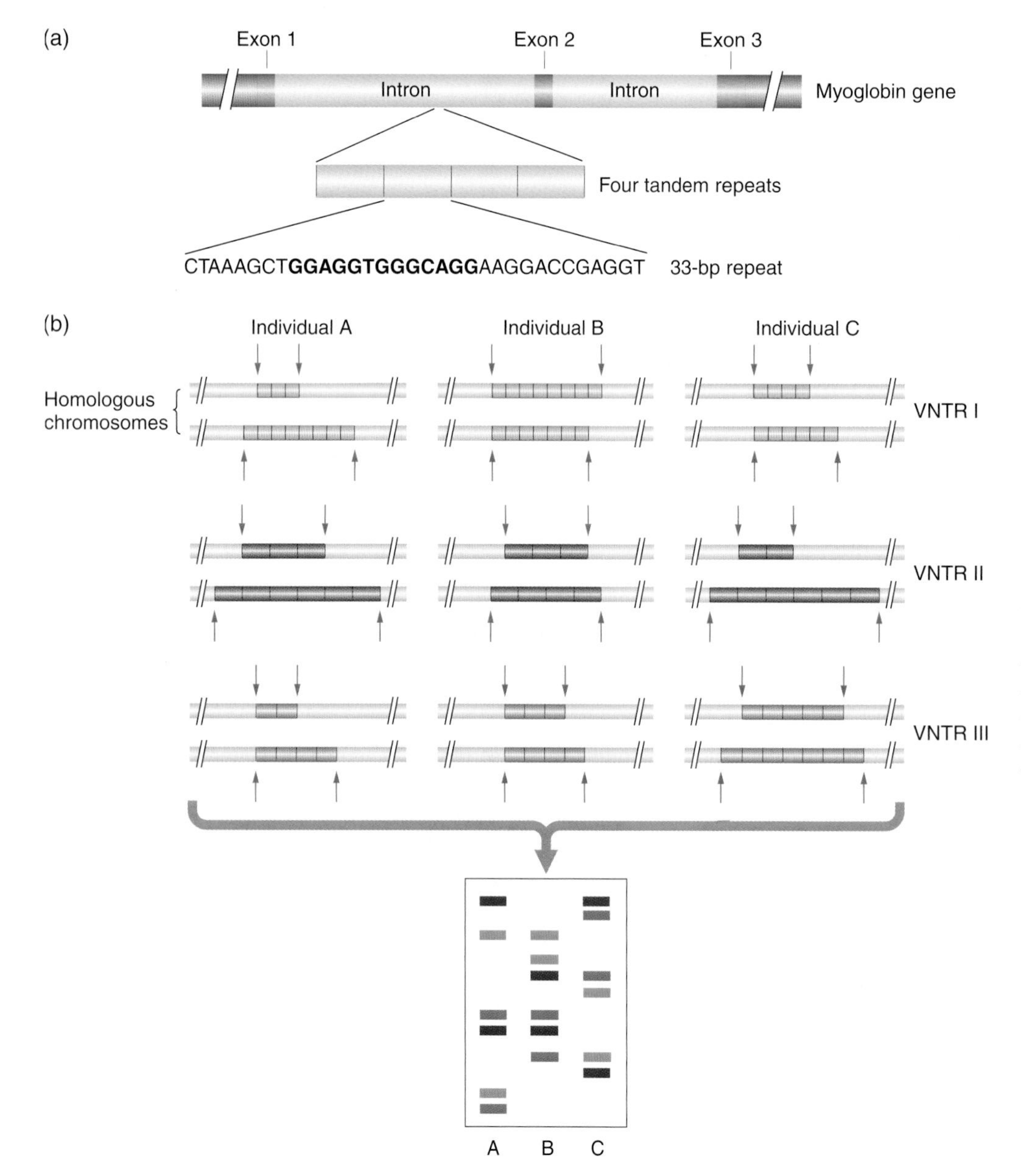

Figure 9-6 Obtaining a DNA fingerprint by using a VNTR probe. (a) Preparation of the probe. The first intron of the myoglobin gene has four repeats of the 33-bp sequence shown, which contains a 13-bp core sequence (shown in boldface). This core sequence is found at other VNTR loci, labeled VNTR I, II, and III in this simple diagrammatic representation. (b) The number of repeats at the three VNTR loci with the core sequence. The Southern blot has been probed with the 33-bp repeat detailed in (a) and shows the VNTR fingerprints of three people. Small red arrows indicate restriction enzyme target sites. *(From J. D. Watson, M. Gilman, J. Witkowski, and M. Zoller,* Recombinant DNA, *2d ed. © 1992 by James D. Watson, Michael Gilman, Jan Witkowski, and Mark Zoller.)*

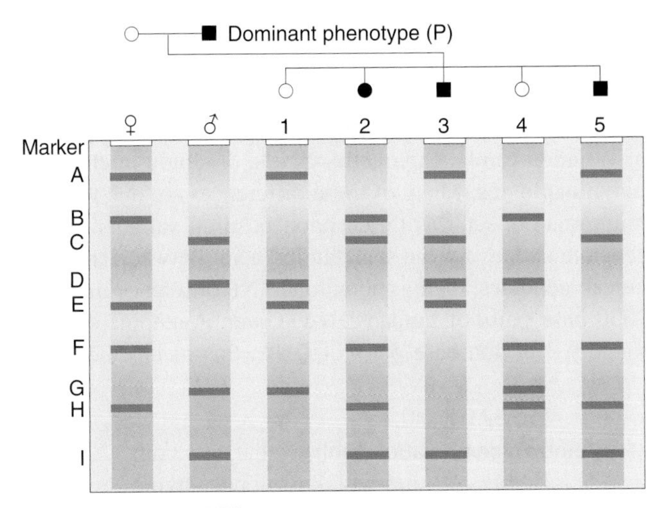

ANALYSIS EXAMPLES
F and H Always inherited together — linked?
A and B In progeny, always either A *or* B — "allelic"?
A and D Four combinations; A and D, A, D, or neither — unlinked?
F, H, and E Always *either* F and H *or* E — closely linked in trans?
Allele *P* Possibly linked to I and C.

Figure 9-7 Using DNA fingerprint bands as DNA markers in mapping. Simplified fingerprints are shown for parents and five progeny. Examples illustrate methods of linkage analysis.

common type is a repeat of CA and its complement GT, as in the following example:

5′ C-A-C-A-C-A-C-A-C-A-C-A-C-A-C-A 3′

3′ G-T-G-T-G-T-G-T-G-T-G-T-G-T-G-T 5′

Probes for detecting DNA regions surrounding individual microsatellite repeats are made with the help of the polymerase chain reaction (PCR; see Chapter 8). For example, digestion of human DNA with the restriction enzyme *Alu*I results in fragments with an average length of 400 bp. These fragments can then be cloned into a sequencing vector. Vectors with genomic inserts that include $(CA)_n/(GT)_n$ are identified by Southern blot hybridization analysis using a $(CA)_n/(GT)_n$ probe. Positive clones are sequenced, and PCR primer pairs are designed that will recognize single-copy DNA sequences flanking the repetitive microsatellite tract:

Primer 1

Alu ——— $(CA)_n$ ——— *Alu*

Alu ——— $(GT)_n$ ——— *Alu*

Primer 2

The primers are used in a PCR amplification with genomic DNA as the substrate. An individual primer pair will recognize the single-copy DNA sequences flanking a specific repetitive tract, and any microsatellite size variants of that tract can be determined by gel electrophoresis of the DNAs from different individuals. Thus, all size variations of the microsatellite repeat will be detectable. A high proportion of microsatellite PCR primer pairs reveal at least three marker "alleles" of different-sized amplification products. An example of the microsatellite mapping technique is shown in Figure 9-8. Thousands of microsatellite primer pairs can be made that likewise detect thousands of marker loci.

Note some differences in the convenience of RFLP and SSLP analyses. RFLP analysis requires a specific cloned single-copy DNA probe to be on hand in the laboratory for the detection of each individual marker locus. Microsatellite analysis requires a single-copy DNA PCR primer pair for each marker locus. Minisatellite analysis on the other hand requires just one probe that detects the core sequence of the repetitive element simultaneously at all sites of that repetitive element anywhere in the genome.

RAPDs: DNA markers based on random PCR amplification. A single PCR primer designed at random will often by chance amplify several different regions of the genome.

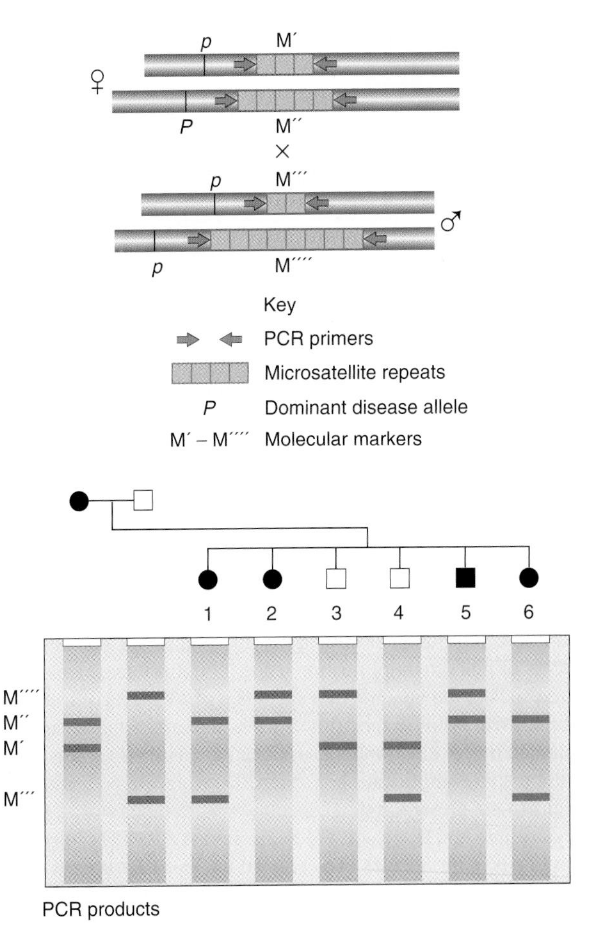

Figure 9-8 Using microsatellite repeats as molecular markers for mapping. A hybridization pattern is shown for a family with six children, and this pattern is interpreted at the top of the illustration with the use of four different-sized microsatellite "alleles," M′ through M⁗, one of which (M″) is probably linked in cis configuration to the disease allele *P*.

The single sequence amplifies only those DNA segments bracketed by nearby inverted copies of the primer sequence (typically a few hundred base pairs apart). The result is a set of different-sized amplified bands of DNA (Figure 9-9). The set of amplified DNA fragments is called **randomly amplified polymorphic DNA (RAPD,** pronounced "rapid"). In a cross, some of the amplified bands may be unique to one parent, in which case they can be treated as heterozygous loci and used as DNA markers in mapping analysis.

Together, the discovery of RFLP, SSLP, and RAPD markers has enabled the construction of a human genetic map with about 1 centimorgan (cM) marker density. Although this density permits high-resolution mapping and is a remarkable achievement, a centimorgan of human DNA is still a huge segment, estimated as 1 megabase (1 Mb = 1 million base pairs, or 1000 kb). A DNA marker map of human chromosome 1 is shown in Figure 9-10.

It has become clear that even with all these DNA markers, there still are not sufficient markers at sufficient density for some purposes. With the availability of a whole genome sequence, as we will describe below, an even richer source of variation can be exploited—**SNPs,** or **single-nucleotide polymorphisms.** Many of these differences are due to neutral variation, such as third codon position variants in degenerate codons for the same amino acid. Between any two human genomes, there is about one SNP difference in every 1000 base pairs of human DNA. Given a genome size of roughly 3 billion base pairs of DNA, this means there are

MESSAGE

Meiotic recombination analysis that uses both the locations of genes with known phenotypic effect and the locations of DNA markers has produced high-resolution genetic maps.

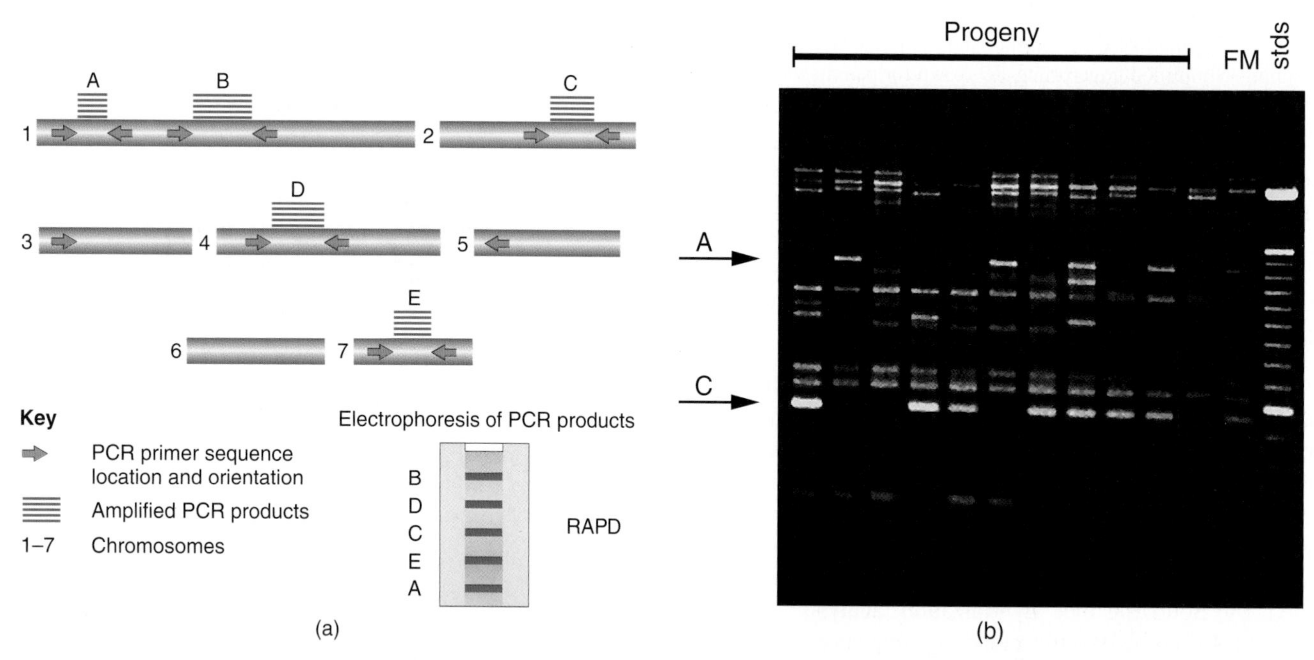

Figure 9-9 Randomly amplified polymorphic DNA analysis (RAPD analysis). (a) A representation of the process of random amplification using a PCR primer that is able to hybridize to five appropriately oriented paired sites in the genome of a specimen of a species of tree. The five amplified DNAs (A–E) are of different size because the primers differ in their distances apart. Thus, the amplified DNAs can be separated by gel electrophoresis. Locations with only one site of primer binding (chromosomes 3 and 5) do not display amplification. For RAPD primers to be useful, not all of the paired sites present in one specimen are present in other specimens—in other words, presence or absence of an amplified band is polymorphic in the species. (b) Segregation of RAPD molecular markers in a cross between polymorphic individuals in the species of tree. A 10-nucleotide primer was used to amplify regions of genomic DNA in the parental trees [the lanes marked F (female) and M (male)]. Similarly, the amplified bands in the progeny are displayed. The differential bands among the progeny reflect the segregation of two heterozygous markers in one of the parents (bands marked by the arrows). Since only the male parent showed these bands but not all the progeny did, the male must have been heterozygous for the presence (+ allele) and absence (− allele) of bands at both loci. Thus, the male could be designated C^+ / C^- . A^+ / A^-, and the female C^- / C^- . A^- / A^-, where C and A refer to the molecular locations of each of the two RAPD sites in the genome, analogous to the designations in part a. DNA standards for size calibration are shown in the right lane, marked "stds." *(Part b from John E. Carlson.)*

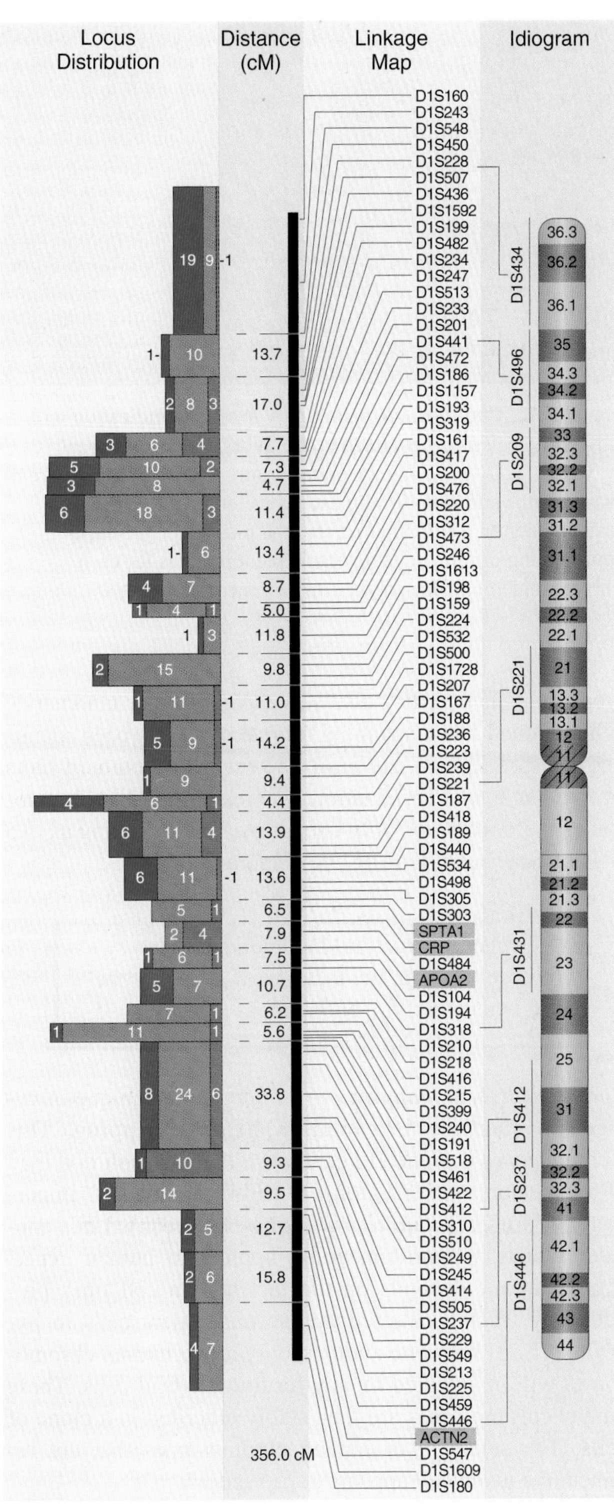

Key

Short sequence length polymorphisms
Other DNA polymorphisms } DNA markers
Genes
Genes included on the linkage map

Figure 9-10 Linkage map of human chromosome 1, correlated with chromosome banding pattern. The diagram shows the distribution of all genetic differences that had been mapped to chromosome 1 at the time this diagram was drawn. Some markers are genes of known phenotype (their numbers are represented in green), but most are polymorphic DNA markers (the numbers in mauve and blue represent two different classes of molecular markers). A linkage map displaying a well spaced out set of these markers, based on recombinant frequency analyses of the type described in this chapter, is in the center of the figure. Map distances are shown in centimorgans (cM). At a total length of 356 cM, chromosome 1 is the longest human chromosome. Some markers have also been localized on the chromosome 1 cytogenetic map (right-hand map, called an *idiogram*), using techniques discussed later in this chapter. Having common landmark markers on the different genetic maps permits the locations of other genes and molecular markers to be estimated on each map. Most of the markers shown on the linkage map are molecular, but several genes (highlighted in light green) also are included. *(B. R. Jasney et al., Science, Sept. 30, 1994.)*

about 3 million differences between any two of our genomes. Extensive projects to develop high-density SNP maps for humans and some model organisms have been carried out.

High-Resolution Cytogenetic Maps

High-resolution cytogenetic maps can be produced in a variety of ways, by relating the locations of DNA markers to cytogenetic landmarks such as chromosome bands and puffs, or to disruptions to the chromosomal integrity (rearrangement breakpoints). Let's look at several techniques in turn.

In situ hybridization mapping. If a cloned DNA sequence is available, then it can be used to make a labeled probe for hybridization to chromosomes in situ. The logic of this approach is identical with any filter hybridization technique such as Southern blotting, except that here, largely intact chromosomes are the target for probe hybridization (rather than DNAs extracted onto a membrane). In this technique, the denatured and labeled probe is hybridized to preparations on microscope slides in which cells have been broken open and their chromosomes spread out. The DNA of the chromosomes has been denatured such that the labeled probe will hybridize to the sites of homologous sequences within the chromosomal DNA. If the individual chromosomes of the genomic set are recognizable through their morphology (banding patterns, size, centromere location, or other cytological features), the probe sequence can be mapped to the approximate position on the chromosome to which it hybridizes. The term *approximate* is used, because this technique does not have the resolving power of recombinational mapping. For example, two genes 5 cM apart in the human genome will have indistinguishable positions by in situ hybridization mapping.

Commonly used labels for probes are radioactivity and fluorescence. In the process of **fluorescent in situ hybridization (FISH),** the cloned DNA is labeled with a fluorescent dye, and a denatured chromosome preparation is bathed in this probe. When the probe binds to the chromosome in situ, the location of the cloned fragment is revealed by a bright fluorescent spot (Figure 9-11). An extension of FISH is **chromosome painting.** Rather than chromosome morphology landmarks, this technique uses a standard control set of probes homologous to known locations to establish the cytogenetic map. Sets of cloned DNA known to be from specific chromosomes or specific chromosome regions are labeled with different fluorescent dyes. These dyes then "paint" specific regions and identify them under the microscope when the set of labeled clones are used as probes in FISH (Figure 9-12). If a probe consisting of a cloned sequence of unknown location is labeled with yet another dye, then its position can be established in the painted array.

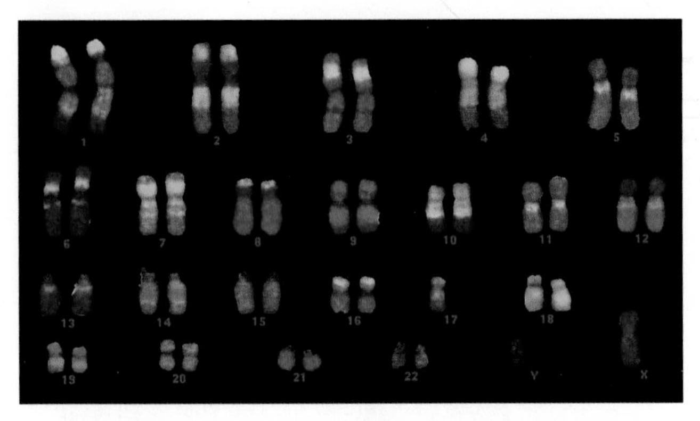

Figure 9-12 Chromosome painting by in situ hybridization with different labeled probes. Each probe fluoresces at a different wavelength. By exposing the slide serially to each of the different wavelengths and then turning each wavelength into a different virtual color on a computer screen, the multicolored images representing different characteristic paint patterns for each chromosome in the karyotype can be generated. *(Applied Imaging, Hylton Park, Wessington, Sunderland, U.K.)*

Rearrangement breakpoint mapping. In Chapter 11, we shall discuss chromosomal rearrangements, a class of mutations that result from the severing of the DNA backbone of a chromosome at one location and its rejoining with another similarly derived **DNA breakpoint** (broken end). That is, a segment of chromosome now has new neighboring sequences. One helpful feature about rearrangement breakpoints is that they also serve as molecular landmarks. When cloned DNA spanning a breakpoint has been identified, the breakpoints are easily detected on Southern blots as two bands of hybridization, whereas in normal chromosomes there would be only one, or by FISH mapping as two sites of labeling instead of one (Figure 9-13).

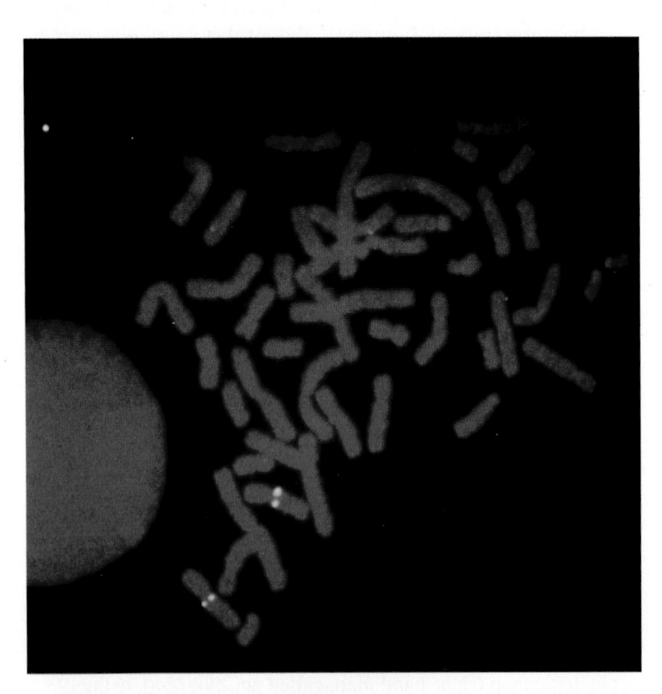

Figure 9-11 FISH analysis. Chromosomes probed in situ with a fluorescent probe specific for a gene present in a single copy in each chromosome set—in this case, a muscle protein. Only one locus shows yellow fluorescence, which corresponds to the probe bound to the muscle protein gene. We see four spots corresponding to this one locus because there are two homologous mitotic prophase chromosomes, each of which contains two chromatids. *(From Peter Lichter et al.,* Science *247, 1990, 64.)*

MESSAGE

By correlating structural landmarks on chromosomes with the location of cloned probe DNA, extensive cytogenetic maps are produced.

Radiation hybrid mapping. An important technique of cytogenetic mapping is **radiation hybrid mapping.** This technique was designed to generate a higher-resolution map of molecular markers along a chromosome, and importantly, it does not require marker heterozygosity. Let's consider radiation hybrid mapping applied to human genes. The technique depends on the fact that in cell lines produced by forcing cultured human and rodent cells to hybridize (fuse) with one another, only a few human chromosomes will be retained by any resulting hybrid cells. These human chromosomes are then stably inherited in a clone of cells. The selection of human chromosomes that are retained in a given cell appears to be random.

In radiation hybrid mapping, instead of whole chromosomes, each hybrid cell line contains a random set of human chromosome fragments. The procedure is to irradiate human cells with 3000 rads of X rays to fragment the chromosomes, and then fuse the irradiated cells with rodent cells to form a radiation hybrid mapping panel—a series of clones, each containing a different random assortment of fragments of human chromosomes, as diagrammed in Figure 9-14. Typically, the fragments are integrated into the

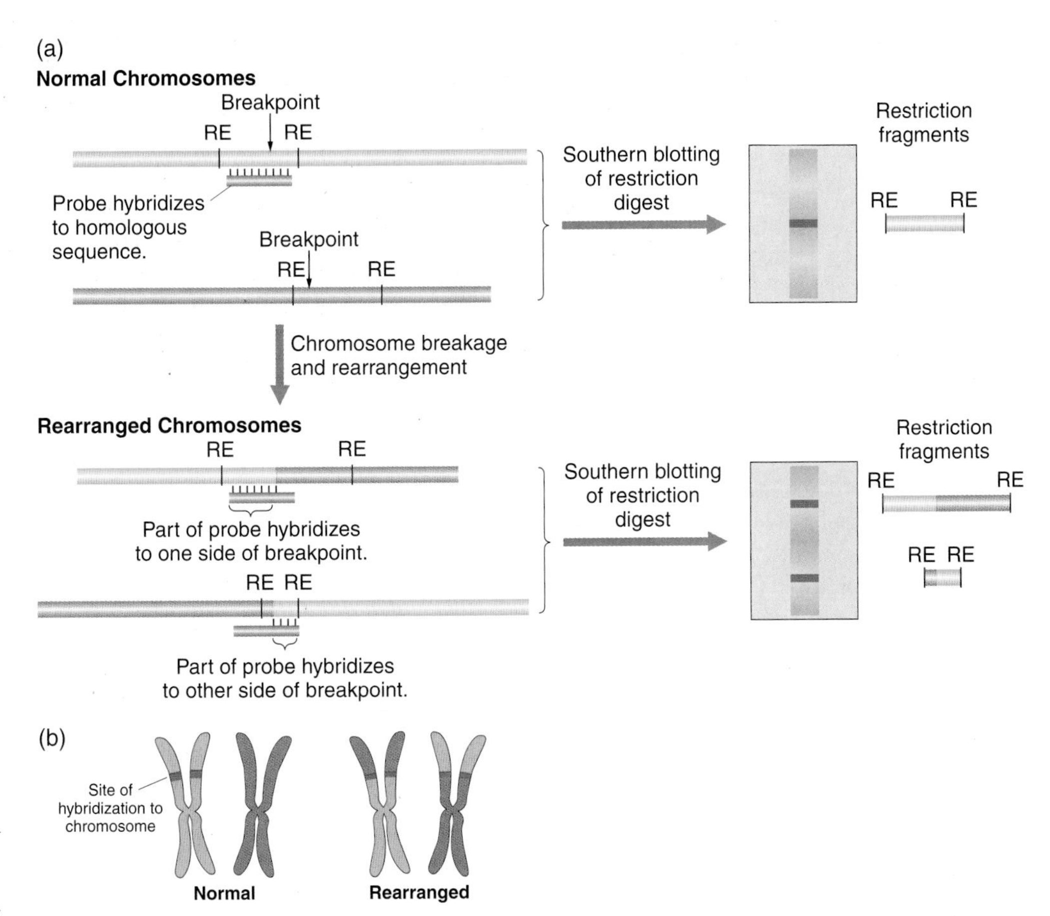

Figure 9-13 Chromosomal rearrangement breakpoints are detected by additional sites of hybridization to a probe that spans the DNA breakpoint. (a) On Southern blots of restriction digests, a chromosomal rearrangement leads to an extra band of hybridization by a probe spanning the breakpoint. (b) By FISH, a probe that spans the breakpoint hybridizes to two spots in the rearranged karyotype as opposed to the single spot in a normal karyotype. RE = sites cut by restriction enzymes.

rodent chromosomes by X-ray-induced chromosome breakage and rejoining.

DNA from each cell line in the radiation hybrid mapping panel is isolated, placed in separate spots on membranes, and denatured. A labeled single-copy human DNA probe is hybridized to such membranes, and sites of labeling identify cell lines containing a human chromosome fragment corresponding to the DNA homologous to the probe.

After accumulating probe-hybridization information for many probes, the data are analyzed for co-retention of DNA markers in a manner analogous to that for mapping bacterial genes by cotransduction (see Chapter 7). Distant DNA markers and markers on different chromosomes should be present in the same hybrid cell line at frequencies close to the product of their individual frequencies in the entire set of cell lines in the mapping panel. However, if two DNA markers map relatively near each other on the same chromosome, they will be co-retained unless an X-ray-induced breakpoint occurred in the interval between them, thereby leaving the two markers in different DNA fragments. The probability that a radiation-induced break will occur between the loci is roughly proportional to the molecular distance separating them. As a quantitative measure

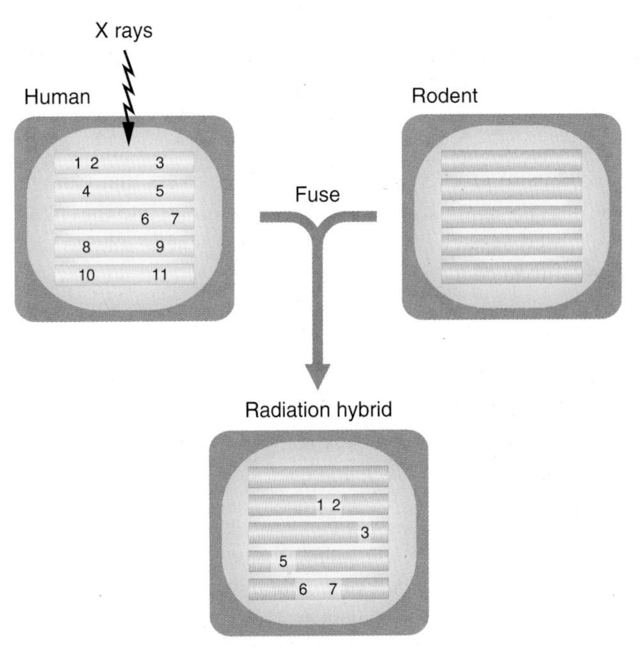

Figure 9-14 Making radiation hybrids using X rays. Fragments of human chromosomes integrate into rodent chromosomes. A panel of different radiation hybrids is analyzed for cotransfer of human markers, which can indicate linkage.

of this distance, we calculate the frequency of coincidental retention (co-retention) of pairs of human molecular markers divided by the sum of their independent retention in the mapping panel. Thus, the higher the frequency of co-retention of two human DNA markers, the closer the two markers map on the same human chromosome. A mapping unit of the probability of a rearrangement between the two DNA markers, cR_{3000}, is calculated as

$$100(1 - \text{frequency of co-retention})$$

The cR_{3000} unit reflects the fact that the frequency will depend on the dose of X-irradiation to which the human cells are exposed; the subscript 3000 indicates a radiation dosage of 3000 rads. 1.0 cR_{3000} has been estimated to be approximately the equivalent of 0.1 cM on the human linkage map.

An advantage of this method is in sample size; a large number of radiation hybrids can be amassed relatively easily. It has been estimated that a standard panel of only 100 to 200 hybrids is sufficient to obtain a high-resolution cR_{3000} map of the human genome—in other words, a map with tenfold increased resolution over the centimorgan map.

MESSAGE

The co-retention of different human markers in radiation hybrid mapping panels allows high-resolution mapping of the chromosomal loci of the DNA markers.

In this section on DNA marker mapping we have encountered techniques based on widely differing biological phenomena—for example, meiotic crossover frequency and radiation-induced breakage. The outcomes of these diverse mapping techniques are high-resolution maps that can be integrated with one another and with low-resolution genetic maps to provide a dense array of markers that can be used directly as starting points in gene cloning and that also can be used as anchor points to relate the genetic map to the physical map of a series of overlapping clones representing the genome. We will now consider in detail the process of physical mapping of the genome.

PHYSICAL MAPS OF THE GENOME

Physical maps are maps of *physically* isolated pieces of the genome—in other words, maps of cloned genomic DNA. A complete physical map of a genome consists of a series of maps for each chromosome in the haploid chromosome set. For each chromosome, a complete map consists of a series of continuous overlapping cloned genomic DNA segments extending from one telomere of the chromosome to the other. We can think of these continuous overlapping clone sets that represent each chromosome as giant chromosome walks (discussed in Chapter 8).

Physical maps (that is, an array of clones covering the genome) have two main values. First, previously cloned genes and DNA markers on a genetic map can be localized to a specific clone on a physical map by hybridization techniques such as Southern blotting or PCR. This gives us a way to measure DNA locations and distances between these markers—something that genetic maps cannot provide by themselves. If for example, you find that an uncloned human disease gene maps between two DNA markers and the physical map between these markers is complete, then you can be sure that one or more of the clones in this portion of the physical map must contain your gene—this simplifies the process of positional cloning. Second, as we will see below, one of the main strategies for sequencing an entire genome requires a complete physical map. Thus, the availability of the physical map provides an intermediary in moving from genetic maps to complete sequence maps of the genome.

In preparing physical maps of genomes, vectors that can carry very large inserts are the most useful because their use in cloning requires breaking the genome up into fewer pieces than would be necessary for other vectors (meaning that there are fewer clones to keep track of in trying to reconstruct the genome). Cosmids, YACs (yeast artificial chromosomes), BACs (bacterial artificial chromosomes), and PACs (phage P1-based artificial chromosomes) have been the main types of such vectors used (see Chapter 8 for a description of these vectors). Although the maximum insert sizes of BACs (300 kb) and PACs (100 kb) are not as large as those of YACs (1000 kb), the former types have important advantages over YACs. They can be amplified in bacteria and isolated and manipulated simply with basic bacterial plasmid technology. Also, BACs and PACs form fewer hybrid inserts than YACs do. By *hybrid inserts,* we mean cloning artifacts composed of several fragments of the genome instead of just one. *Cloning artifacts* refers to clone inserts that do not accurately represent the material their DNA insert is derived from—in this case, the genome. The presence of cloning artifacts can thwart attempts to create a physical map that is a true reflection of the genome because the hybrid clones can create a clone overlap even though the sequences actually come from different regions of the genome.

The goal of physical mapping is to develop a clone-overlap map that truly reflects the organization of the intact genome. These vectors that carry large inserts are useful, but the task of creating a physical map is nonetheless a daunting one. Even so-called small genomes contain huge amounts of DNA. Consider, for example, the 100-Mb genome of the tiny nematode *Caenorhabditis elegans.* Because an average cosmid insert is about 40 kb, at least 2500 cosmids would be required to embrace this genome, and many more to be sure that all segments of the genome were represented. If you had a *C. elegans* BAC library with an average insert site of 200 kb, your task would be fivefold simplified.

Creating a complete physical map begins by amassing a large number of randomly cloned inserts. In some manner, sequence identities between end regions of the clones must be identified; these sequence identities are used to infer that the ends of the two clones overlap and thus the clones come from adjacent regions of the genome. A set of overlapping clones is called a **contig.** In the early phases of a genome project, contigs are numerous and represent separate cloned segments of the genome. But, as more and more clones are characterized, clones are found that overlap two previously separate contigs, and these "joining clones" then permit the merger of the two contigs into one larger one. This process of contig merging continues until eventually a set of contigs is built that is equal to the number of chromosomes. At this point, if the contigs in this set extend out to the telomeres of the chromosomes, the physical map is complete.

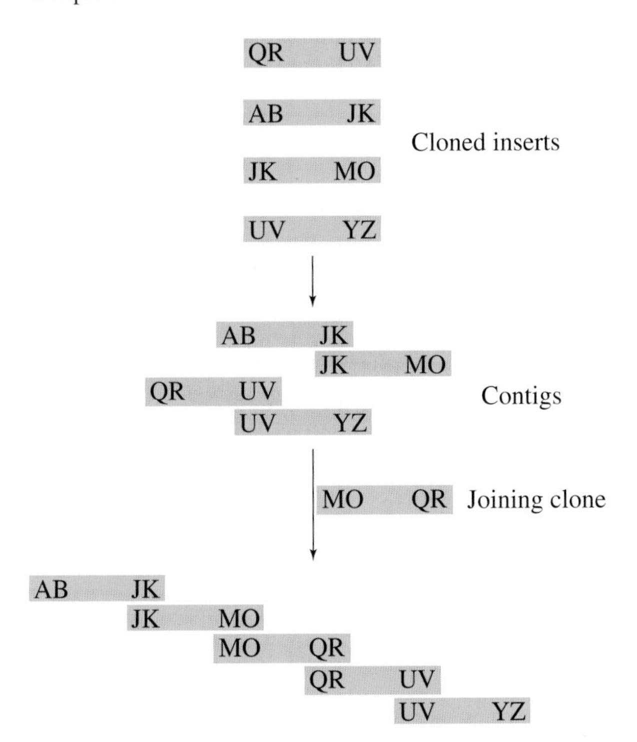

MESSAGE

Genomic cloning proceeds by assembling clones into overlapping groups called *contigs*. As more data accumulate, the contigs extend the length of whole chromosomes.

Let's now look at techniques for identifying clone overlaps—in other words, techniques for turning a set of randomly isolated clones into a contig.

Ordering by Clone Fingerprints

The genomic insert carried by a vector has its own unique sequence that can be used to generate a DNA fingerprint. A multiple-restriction-enzyme digestion will generate a set of bands whose number and positions are a unique fingerprint of that clone. The pattern of bands generated by each separate clone can be digitized, and the bands from different clones can be aligned by computer to determine if there is any overlap between the inserted DNAs. What is determined is the proportion of bands that are shared between two clones. The order of the bands is not determined in this technique, just the proportion of shared bands. A set of overlapping cloned inserts is thus determined (Figure 9-15 on the next page). Depending on issues like genome size, the experimenters determine the proportion of shared bands that indicate a true overlap; this is usually in the range of 25–50 percent of the bands. If sufficient clones to cover any region of the genome ten times are fingerprinted, the fingerprint of any given clone will be found to overlap several other clones. If many of the clones that overlap clone A, say, themselves overlap clone B, it means that it is very unlikely that these clones overlap for some incorrect reason (such as a cloning artifact coming from different regions of the genome). These multiply overlapping matches then provide the basis for building a highly reliable complete physical map. Fingerprinting has proved to be an extremely important technique in developing physical maps of the *C. elegans,* human, and mouse genomes.

Ordering by Sequence-Tagged Sites

Short unique sequences that can be amplified using defined PCR primers are called **sequence-tagged sites (STSs).** STSs derive from sequenced regions of the genome and can be used as landmarks for clone classification in establishing a physical map of a genome. The logic of the technique is that clones that share STSs must overlap each other. The more STSs that they share, the more they must overlap. The resulting physical maps are called **STS content maps.**

Development of an STS content map proceeds according to the following steps:

1. Collect a large number of usable STS sequences, for example by sequencing the ends of a random set of clones from a genomic library.
2. Develop two PCR primers per STS such that a given pair of primers will uniquely amplify a single STS.
3. Use PCR to screen individually all the clones from a large insert genome library. All clones containing the particular STS that is amplified by the primer pair used will yield a PCR amplification product. The PCR product is detectable because it incorporates labeled nucleotide triphosphates during its synthesis. Samples of the DNA in the PCR reaction mixtures for each of the clones after STS amplification with a specific

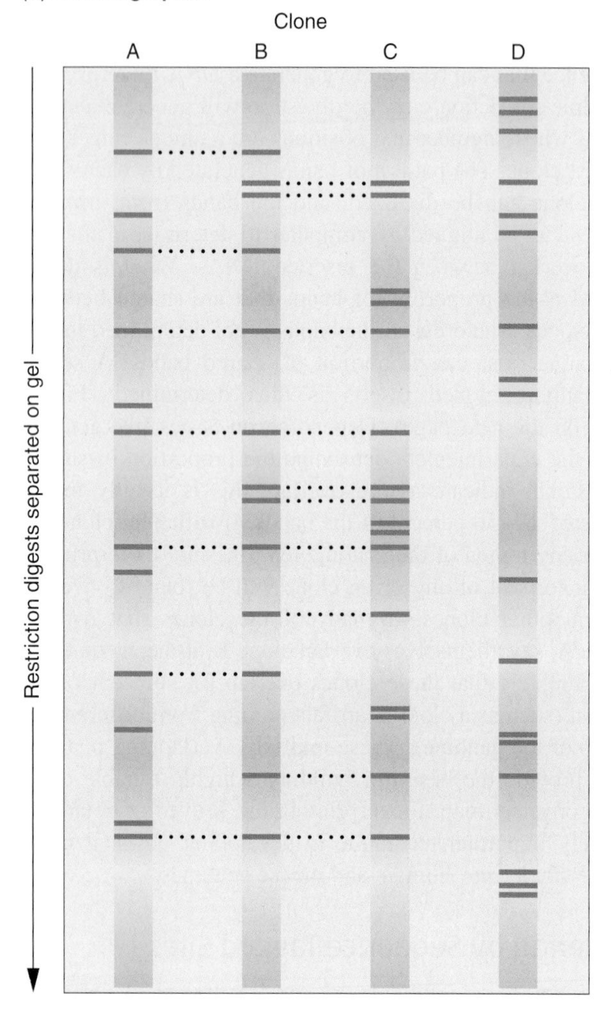

Figure 9-15 Creating a physical map by clone fingerprint mapping. (a) Four clones are digested with multiple restriction enzymes, and the resulting complex mixture of restriction fragments is separated on the basis of size by gel electrophoresis. The bands containing the fragments are stained to show their location. The number of identically sized bands for each pair of digests is determined. A and B digests share > 50 percent of the bands, as do the B and C digests, indicating that they come from overlapping regions of the genome. No other pair of digests shares a significant number of bands. (b) The physical map derived from the data in part (a). Clones A and C both overlap clone B but do not overlap each other. Clone D is from somewhere else in the genome since it doesn't overlap any of the other three clones.

primer pair are examined for labeled DNA products—either radioactive or fluorescent—by complementarity-detection techniques similar to those used for Southern blotting or any other hybridization technique.

Consider several STS markers that are being used to create a BAC STS content map (Figure 9-16). Even though the locations of these STSs in the genome are not known initially, a panel of many STSs can be used to characterize the BAC clones. From this characterization, the clones that are shown to have specific STSs in common must have overlapping inserts and therefore can be aligned into physical map contigs.

The combination of fingerprinting and STS content mapping has resulted in the complete or near-complete physical maps of several organisms. For example, the *C. elegans* genome is available as sets of cosmid or YAC contigs. Furthermore, the DNA of the contigs has been arranged on nitrocellulose filters in ordered arrays. So, to find out where a specific piece of DNA of interest lies in the genome, that DNA is used as a probe on the contig filters, and a positive hybridization signal will announce the precise location of the DNA (Figure 9-17).

DATA (STS content of BACs)

STSs (Sequence-tagged sites)

BACs	1	2	3	4	5	6	7	8	9	10	11	12
A	−	+	−	−	−	−	−	−	−	−	+	+
B	−	−	−	+	+	+	+	−	−	−	−	−
C	+	+	−	−	−	−	−	+	−	+	−	−
D	−	−	+	+	−	−	+	−	−	−	−	+
E	+	−	−	−	−	−	−	+	+	−	−	−

CONTIG
STS map: 9 1 8 10 2 11 12 3 4 7 5 6

BAC coverage
E 9 1 8
C 1 8 10 2
A 2 11 12
D 12 3 4 7
B 4 7 5 6

Order uncertain

Figure 9-16 Using sequence-tagged sites (STSs) to order overlapping clones (BACs, in this example) into a contig. Five different BACs are tested to determine which STSs they contain *(top panel)*, and these data are used to assemble a physical map (STS content map) *(bottom overlapping red bars)*. STS markers that completely co-map (map to the same clones, such as markers 1 and 8) cannot be ordered on the physical map (the order could be 1, 8 or 8, 1 for these two markers). STS pairs whose respective order cannot be determined here are marked by brackets.

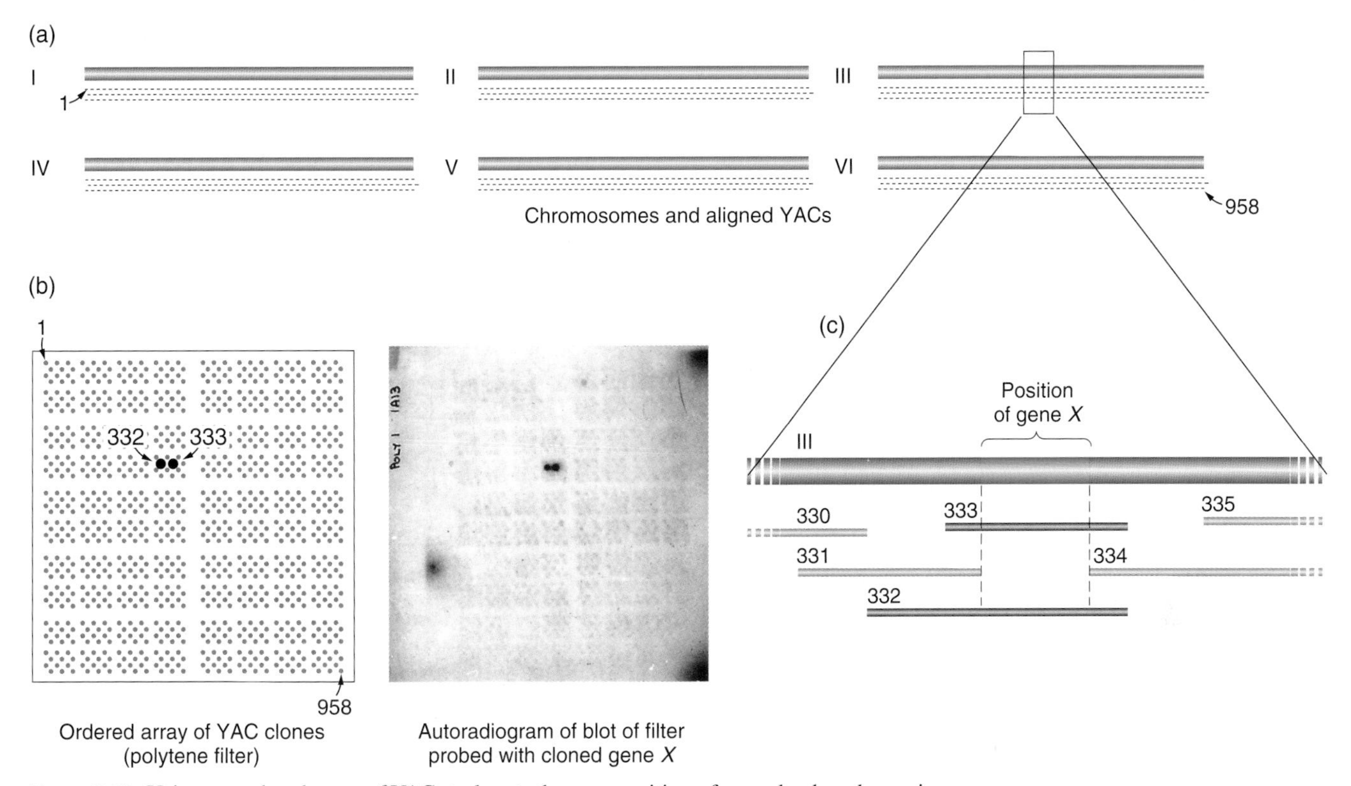

Figure 9-17 Using an ordered array of YACs to locate the map position of a newly cloned gene in *Caenorhabditis elegans.* (a) The 958 YACs, indicated as a series of dashes, have been ordered into six contigs that correspond to chromosomes I through VI. (b) The YAC clones are placed in order on a membrane filter. DNA on this filter is probed with the cloned gene *X,* giving an autoradiogram with two positive spots corresponding to two adjacent YACs numbered 332 and 333, which have hybridized with the cloned gene *X.* (c) Because of the hybridization pattern obtained, the location of the cloned gene can be narrowed down to a small region on chromosome III that corresponds to the overlapping section of YACs 332 and 333. *(Autoradiogram from Alan Coulson.)*

MESSAGE

Physical maps can be developed by matching DNA fingerprints or by matching unique sequences within cloned segments.

Subdividing the Genome: Simplifying the Physical Mapping Task

Compiling a physical map of an entire genome is a considerable task, and even with fingerprinting or STS content mapping techniques, there are biological or technical challenges to creating maps that are valid reflections of the genome. For example, in humans, there are megabase-size regions of the genome that exist as essentially perfectly duplicate copies on two different chromosomes (a biological challenge). In addition, some regions of the genome do not clone efficiently, leading to gaps in the physical map corresponding to the regions that do not clone in standard vectors (a technical challenge). Some of these challenges can be circumvented if there are ways to subdivide the entire genome into smaller entities. If we need assemble sets of clones corresponding only to subregions of the genome, then the number of clones required to construct a complete physical map of a particular subregion is far fewer than that required for the genome as a whole. We will next consider two techniques for subdividing the genome for physical mapping: chromosome isolation and mapping by FISH.

Chromosome-specific libraries: subdividing the genome. One way to subdivide the genome is by separating the actual DNA molecules of the genome—those contained within each of the specific chromosomes. **Pulse field gel electrophoresis** (**PFGE**, discussed in Chapter 2) is an electrophoretic technique that can be used to isolate individual chromosomes if they are small (such as yeast chromosomes) or—for large chromosomes—to isolate chromosome fragments generated by cutting with "rare cutter" enzymes such as *Not*I (which has an 8-base-pair restriction recognition sequence and thus on average cuts only once in every 64,000 base pairs). The technique is a modification of standard gel electrophoresis that adjusts conditions to permit separation of large DNA molecules.

Flow sorting is another option for preparing DNA of a specific chromosome. Chromosomes (such as human chromosomes) can be flow-sorted by **fluorescence-activated chromosome sorting** (**FACS;** Figure 9-18). In this procedure, cells are disrupted to liberate whole metaphase chromosomes into liquid suspension. The metaphase chromosomes released in this suspension are stained with two dyes, one of which binds to AT-rich regions (regions that are composed of long strings of A's and T's), and the other to GC-rich regions. Every chromosome in a genome has its own characteristic ratio of AT-rich to GC-rich regions, and these characteristic ratios are used to distinguish between the various chromosomes. The suspension is first diluted to an appropriate chromosome concentration so that it can be converted into a spray of droplets that each contain one chromosome. The droplets flow in a tube past laser beams tuned to excite the fluorescence. Each chromosome produces its own characteristic fluorescence signal, which is recognized electronically, and two deflector plates direct the droplets containing the specific chromosomes needed into each of several collection tubes, according to the ratio of the AT-rich and GC-rich dyes. Each collection tube then contains a different chromosome from the genome.

Once individual chromosomes are isolated, their DNA can be purified and genomic libraries with inserts from this DNA can be prepared. These libraries then serve as a source of clones for fingerprint or STS content physical mapping. One question is how to tell which chromosome is which. The most straightforward way is by hybridization with probes known to map to each of the chromosomes. This allows identification of each of the isolated chromosomes from a genome, and permits development of their physical maps.

Ordering by FISH: confirming physical map order. If good chromosomal landmarks are known, FISH analysis can be used to locate the approximate positions of the cloned DNAs. Figure 9-19 shows results of a FISH analysis of human BACs and PACs by individual in situ hybridization of each clone to human chromosomes. Many clones in situ hybridize to the same landmark regions (using human chromosome banding techniques), and thus this technique puts clones into one of a number of cytogenetic regions within a given chromosome. These clones can then be evaluated by fingerprint analysis or STS content mapping to produce the physical map of this region of the genome.

In addition to reducing the number of clones that need to be organized into a complete physical map of a given region of the genome, FISH mapping also provides an independent way of corroborating the results of physical map production by clone fingerprint or STS content mapping. Suppose that we have organized several clones into one small region of the physical map of the genome. One prediction of this is that each of these clones will FISH map to the same cytogenetic location. If the FISH map agrees with this prediction, it is supporting evidence that the physical map has been correctly compiled. If it does not, then it contradicts the validity of the physical map assembly.

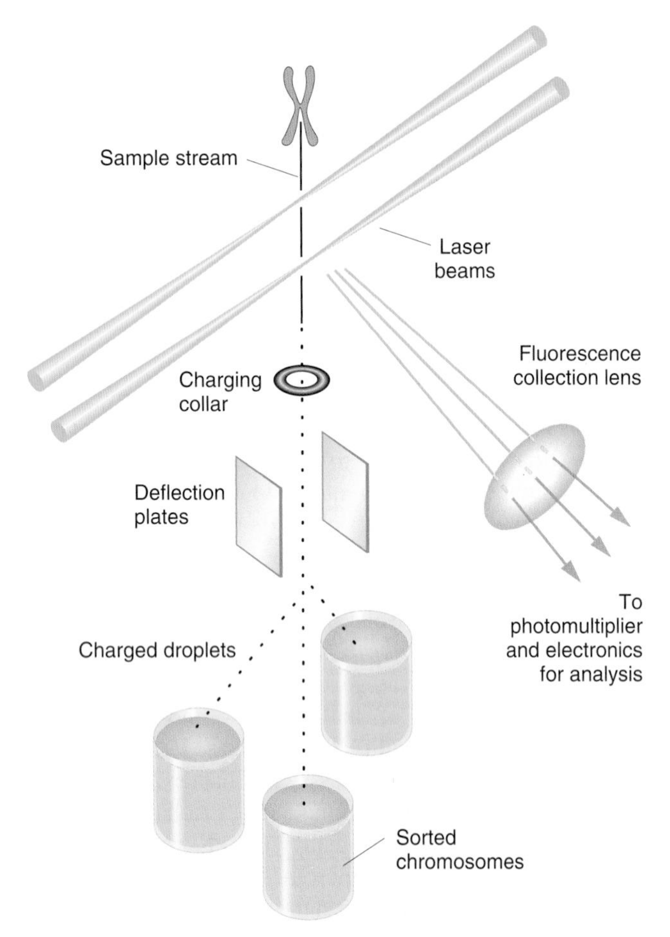

Figure 9-18 Chromosome purification by flow sorting. Chromosomes stained with two fluorescent dyes are passed through laser beams. Each time, the amount of fluorescence at each of the two dye wavelengths is measured, and the chromosome is deflected accordingly. The chromosomes are then collected as droplets.

THE SEQUENCE OF THE GENOME

The highest-resolution genomic map is its complete DNA sequence—that is, the complete base pair sequence of A's, T's, C's, and G's. Establishing a complete sequence map of a genome is a massive undertaking, of a sort not seen before in biology. Several different strategies have been used to determine the sequence of an entire genome; these depend on high levels of robotics and automation, as exemplified by the automated sample preparation and sequencing production line of a genome center.

Why is a high level of robotics and automation necessary to create a sequence map of a genome? To understand this, let's consider the human genome, which contains about

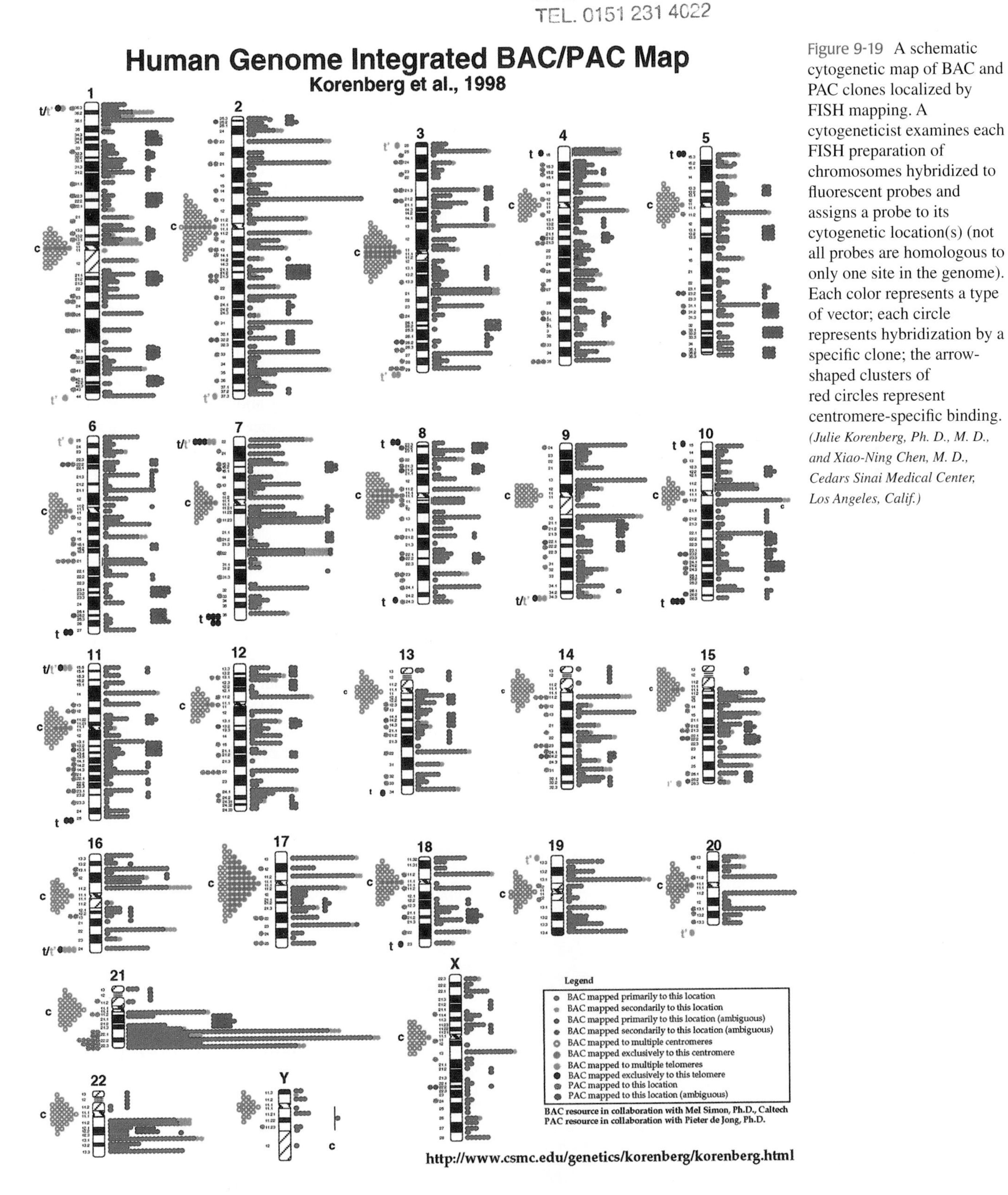

Figure 9-19 A schematic cytogenetic map of BAC and PAC clones localized by FISH mapping. A cytogeneticist examines each FISH preparation of chromosomes hybridized to fluorescent probes and assigns a probe to its cytogenetic location(s) (not all probes are homologous to only one site in the genome). Each color represents a type of vector; each circle represents hybridization by a specific clone; the arrow-shaped clusters of red circles represent centromere-specific binding. *(Julie Korenberg, Ph. D., M. D., and Xiao-Ning Chen, M. D., Cedars Sinai Medical Center, Los Angeles, Calif.)*

3×10^9 bp (base pairs) of DNA. If we could purify the DNA intact from each of the 24 different human chromosomes (X, Y, and the 22 autosomes), separately put each of these 24 DNA samples into a sequencing machine and read their sequences directly from one telomere to the other, creating a complete sequence map would be utterly straightforward, like reading a books with 24 chapters—albeit a very, very long book with 3 billion characters! Unfortunately, such a sequencing machine does not exist. Rather, automated fluorescence sequencing of the sort we discussed in Chapter 8 is the current state of the art in DNA sequencing technology. Individual sequencing reactions (called *sequencing reads*) generally fall in the 600-bp range. Obviously, such lengths are a small portion of a single chromosome. For example, an individual read is only 0.0002 percent of the longest human chromosome (about 3×10^8 bp of DNA) and only about 0.00002 percent of the entire human genome. Even within a 300,000-bp BAC clone, an individual read is only about 0.2 percent of the length of that clone. Thus, a major challenge facing a genome project is **sequence assembly**—that is, how to build the individual reads into a **consensus sequence** that is an authentic representation of the sequence of each of the DNA molecules in that genome.

Let's look at these numbers in a somewhat different way to understand the scale of the problem. As with any experimental observation, automated sequencing machines do not always give perfectly accurate sequence reads. The error rate is not constant; it depends on such factors as the dyes that are attached to the sequenced molecules, the purity and homogeneity of the starting DNA sample, the actual run of base pairs in the DNA sample. Thus, the convention genome projects use is to require 10 independent sequence reads for each base pair in a genome. This provides 10 independent attempts to read the true sequence of any region and ensures that chance errors in the reads do not give us a false reconstruction of the consensus sequence. Given an average sequence read of about 600 bases of DNA and a human genome of 3 billion base pairs, 50 million successful independent reads are required to give us our 10-fold average coverage of each base pair. However, not all reads are successful; there is a failure rate of about 20 percent; thus the real numbers are about 60 million attempted reads to cover the human genome. Thus, there is an enormous amount of information and material to be tracked. To try to minimize both human error and the need for people to carry out highly repetitive tasks, genome project laboratories have implemented automation wherever practical.

For these reasons, preparation of clones, DNA isolation, electrophoresis, and sequencing protocols have all been adapted to automation. For example, one of the recent breakthroughs has been the development of production-line sequencing machines that run around the clock without human intervention, producing about 96 reads every 3 hours (Figure 9-20). A genome center with 200 of these capillary sequencing machines can produce about 150,000 reads in a single day. Thus, there is the capacity for a single sequencing center to generate enough reads to assemble the sequence of a mammal [about 3 billion base pairs or gigabase pairs (Gbp) of DNA] in about 2 years.

What are the goals of sequencing the genome? Ultimately, we strive to produce a consensus sequence that is a true and accurate representation of the genome as it exists in the individual from whom the DNA being sequenced was first cloned. This sequence will then serve as a reference sequence for the species. We now know that there are many DNA sequence differences between the genomes of different individuals within a species, and even between the maternally and paternally contributed genomes within a

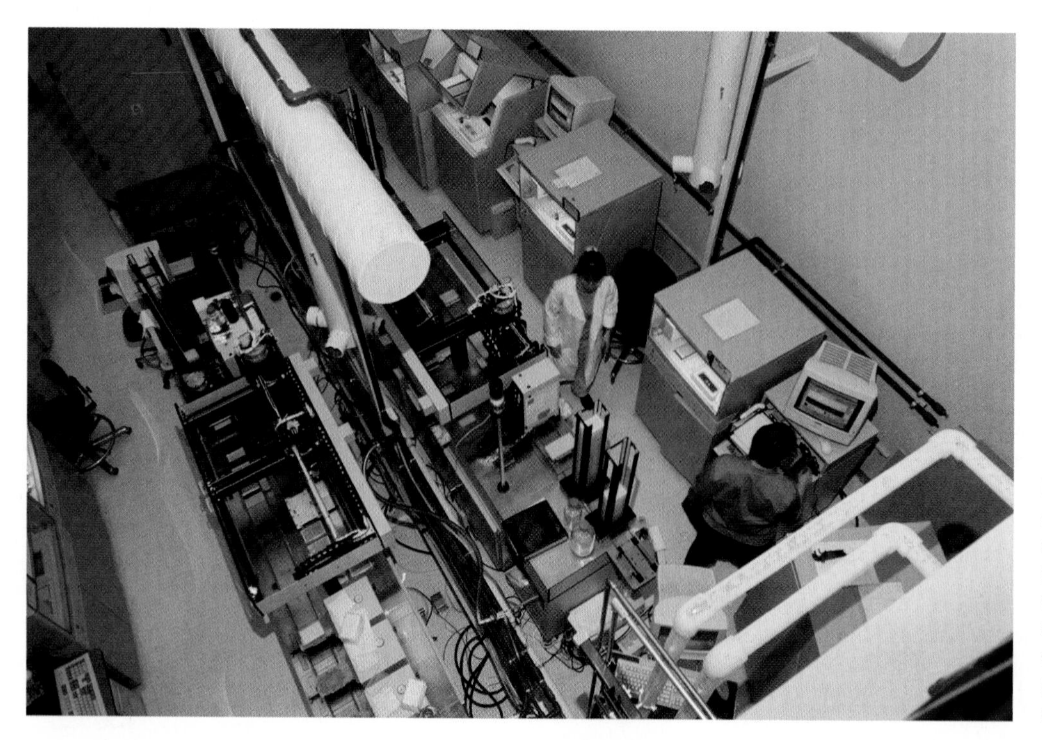

Figure 9-20 Part of the automated production line of a major human genome sequencing center. All this equipment is used for high-throughput processing of clones for DNA sequencing. (*© Bethany Versoy; all rights reserved.*)

single diploid individual. Thus, no one genome sequence truly represents the genome of the entire species, but nonetheless, the genome sequence serves as a standard or reference sequence with which other sequences can be compared, and which can be analyzed to determine the information encoded within the DNA—such as the array of encoded RNAs and polypeptides.

All current genome sequencing strategies are clone-based. That is, first you make a clone or subclone library, and then you sequence all or part of the inserts of the individual clones or subclones in this library. There are two basic ways that a consensus sequence of a genome is assembled. One is called *ordered clone sequencing* and the other is called *whole genome shotgun sequencing.*.

In **ordered clone sequencing,** a complete physical map of the genome is first produced, and then from the set of clones that make up the physical map, an ordered subset of minimally overlapping clones is selected for sequencing. Once a consensus sequence for each clone is assembled, these clone-consensus sequences are assembled into an overall consensus sequence for the genome according to the known order of these clones on the physical map.

In **whole genome shotgun sequencing,** sequence reads are obtained from randomly selected clones from a whole genome library (such a library is called a *shotgun library;* hence, a sequencing strategy using random clones from a shotgun library is called *whole genome shotgun sequencing*), without any information on where these clones map in the genome. The homologous sequences shared by reads from overlapping clones allow the sequencers to assemble these sequences into consensus sequences covering the whole genome.

Let's now consider these two strategies in more detail.

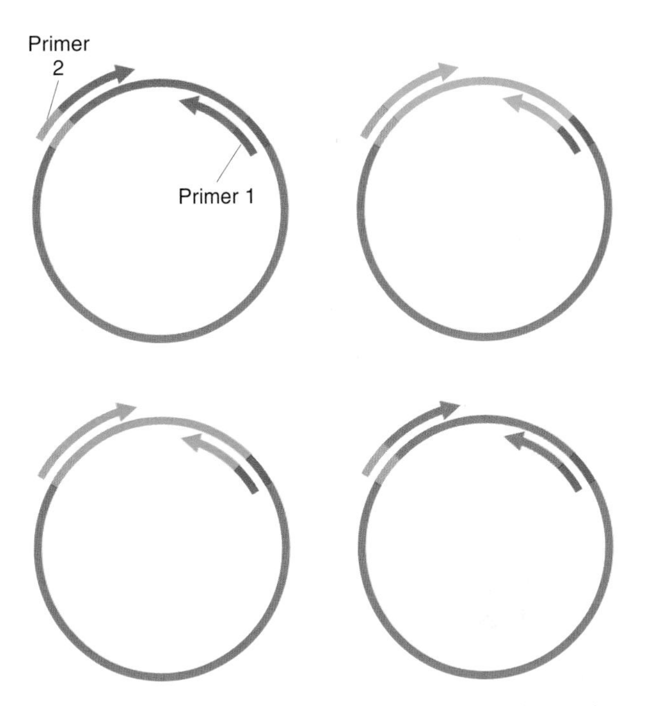

Figure 9-21 The production of terminal insert sequencing reads. The last 600 or so base pairs at each end of the genomic insert are sequenced. Two different sequence priming sites are used, one at each end of the vector. Because vector sites are used for initiating dideoxy sequencing (Chapter 8), the primers hybridize to vector sequences and thus can be used to sequence several hundred base pairs from the ends of any insert within the library. If both ends of the same clone are sequenced, the two resulting sequence reads are called *paired-end reads.*

Sequencing Strategies in Bacteria

Bacterial DNA is essentially single-copy. This means that any given DNA sequence read will come from one unique place in the bacterial genome. In addition, a typical bacterial genome is only a few megabase pairs of DNA in size. Owing to this simpler system, whole genome shotgun sequencing can be effectively applied to bacterial genomes. Clones from a shotgun library are randomly sequenced, using primers based on adjacent vector DNA to sequence short regions at the ends of the genomic inserts (Figure 9-21). Because so many random short sequences are obtained, together their homologous sequence overlaps can be used to build a consensus sequence covering the great majority of the bacterial genome. Occasional gaps are encountered because a region of the genome was not found in the shotgun library—some DNA fragments do not grow well in particular cloning vectors. Such gaps are filled in by techniques such as **primer walking**—that is, by using the end of a cloned sequence as a primer to sequence into adjacent uncloned fragments. Using this basic approach, by January 2001, about 40 bacterial species had been sequenced. Because a large genome center can sequence a bacterial genome in only 1 to 2 days, the number of such fully sequenced prokaryotic genomes is increasing rapidly.

If whole genome shotgun sequencing works so well in bacteria, why is it not the obvious choice for sequencing eukaryotic genomes? The answer is that the eukaryotic genome is not composed solely of single-copy DNA. While a considerable proportion of all genomes are single-copy, there are also substantial segments of the genome that are composed of repetitive sequences—that is, identical sequence strings present many times in a given genome. Because these repetitive sequences are often larger than the size of an individual sequence read, their presence provides a major technical hurdle to the assembly of a correct consensus sequence. If a sequence is repeated at different locations in a genome, how do we determine where a particular sequence read that includes it actually belongs? Before considering how eukaryotic whole genome shotgun and ordered clone sequencing strategies address this hurdle, we need to understand the nature of these repetitive sequences.

The Many Classes of Repeated Genomic Sequences

As we just noted, a major stumbling block in reassembling a consensus sequence of a eukaryotic genome is the existence of numerous classes of repeated sequences. What are these sequences and why are they problems for genome sequencing? There are two main classes of such sequences, tandem repeat arrays and mobile genetic elements, which we will examine here.

Tandem repeat arrays. **Tandem repeats,** sequences that are present in multiple copies adjacent to one another, come in several flavors. Some are very large and repeated many times; others, just a few times. We have already discussed tandem repeats that are interspersed throughout the genome: minisatellites or VNTRs, and microsatellites or dinucleotide repeats. In addition to the problem of reassembling repeated sequences, tandem repeats present an additional problem for genome sequencing: DNA fragments containing high-copy-number tandem repeats are difficult to maintain stably in clones. When tandemly repeated sequences are grown in bacterial clones, the repeat units, which are homologous to one another, tend to recombine asymmetrically with one another, leading to changes in the number of copies of the tandem repeat in the clone. Because smaller clones often replicate faster, the clone DNA inserts that have lost many of their tandem repeat copies replicate fastest and become the predominant DNA molecules in the clone. Such clones are obviously not representative of the portion of the genome from which they originate. A correct consensus sequence for these regions of the genome is extremely difficult, or often impossible, to reconstruct. In the next sections we'll examine some of the classes of tandem repeats.

Tandemly repeated genes. Some genes are present in multiple tandem copies, with essentially identical sequence from one copy to another. Why do we encounter such tandem arrays of identical genes? Cells need large amounts of the products of some genes, such as those that encode components of the translation or the replication machinery, and the solution that has evolved for making a great deal of these products is to have many copies of these genes in the genome. Also, we know in some cases that the clustering of such tandemly arrayed genes is essential for ensuring that they are expressed at the same time. The tandem array organization seems to provide an opportunity to ensure that the copies of the genes remain identical by using the sequence from one copy as a template for repairing mutations that might arise in another. As one example, the genes for the histone chromosomal proteins that make up nucleosomes are typically arranged in tandem arrays of hundreds of copies in higher eukaryotes (Figure 9-22). Typically, these identical histone genes are expressed early in the cell cycle, when DNA replication demands doubling of the number of nucleosomes in the cell.

Noncoding tandem repeats—telomeres. There are specialized structural features of chromosomes that are crucial to their biological role. The best understood of these are telomeres, the specialized structures at the tips of chromosomes. As we saw in Chapter 4, telomeres have tandem arrays of simple repeating units that serve a structural role essential to chromosome replication. In some cases, these sequences can extend for many kilobases of DNA.

Noncoding tandem repeats—heterochromatin. When genomic DNA is centrifuged in a cesium chloride density gradient, the "satellite" band is often visible at a position different from that of the main DNA band. On isolation, such **satellite DNA** is found to consist of multiple tandem repeats of short nucleotide sequences, stretching to hundreds of kilobases in length. Because the satellite repeats are based on small repeat units of DNA, they constitute a nonrepresentative sample of the genomic DNA. They consequently can have nucleotide proportions that are significantly different from the bulk of the DNA. This is why this repeat DNA forms a separate satellite band in a cesium chloride gradient.

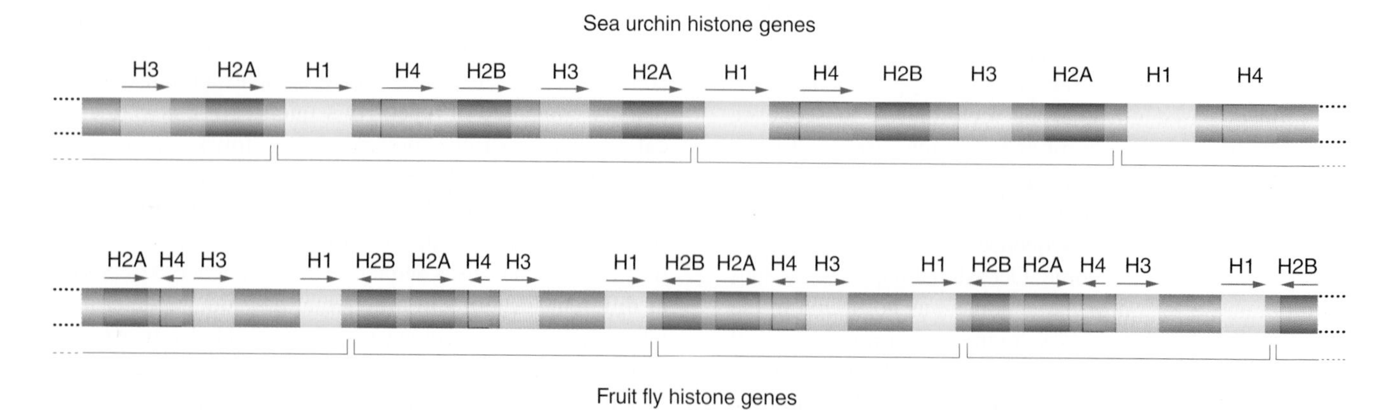

Figure 9-22 Tandem repeats of histone genes in sea urchin and fruit fly. Only a small fraction of the repeats is shown. Arrows indicate direction of transcription.

When probes are prepared from such simple-sequence DNA and used in chromosomal in situ labeling experiments, the great bulk of the satellite DNA is found to reside in the regions of **heterochromatin,** densely staining, highly compact regions flanking the centromeres. In situ labeling by a mouse satellite DNA that hybridizes in the centric heterochromatin of all mouse chromosomes is shown in Figure 9-23. There can be either one or several basic repeating units, but usually they are less than 10 bases long. For example, in *Drosophila melanogaster,* the sequence AATAACATAG is found in tandem arrays around all centromeres; other short repetitive sequences are found near only some of the centromeres. Similarly, in guinea pigs, tandem arrays of the shorter sequence CCCTAA flank the centromeres. Typical regions and extents of centric heterochromatin are depicted in Figure 9-24; note that the heterochromatin can be a substantial proportion of the length of a chromosome. There is no demonstrable function for centric heterochromatin. Some organisms have staggering amounts of this DNA; for example, as much as 50 percent of kangaroo DNA can be centromeric satellite DNA!

Mobile genetic elements: dispersed repeats. One component of genomes consists of dispersed repetitive elements that have propagated within the genome by making copies of themselves, which can move into new locations, a process termed **transposition.** Particularly in higher eukaryotes, these repetitive elements make up a large portion of their genomes. Collectively, we term these repetitive elements **mobile genetic elements.** We will discuss the mutagenic consequences of transposition in Chapter 10. Here we will consider mobile genetic elements from a structural point of view. The problem that these repetitive elements pose for genome sequence analysis is the one that we raised previously—that they are longer than a sequence read and thus there is a problem of identifying the true location of a mobile element sequence read in the genome.

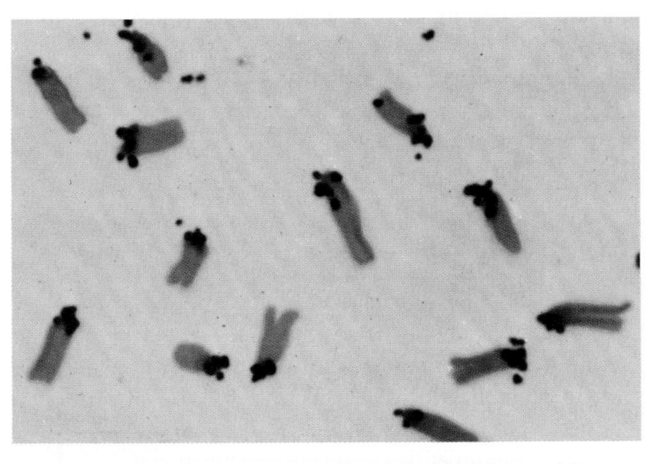

Figure 9-23 Autoradiographic localization of highly repetitive simple sequence mouse DNA to the heterochromatin surrounding the centromeres. A radioactive probe for simple-sequence DNA was added to chromosomes whose DNA had been denatured. *(From M. L. Pardue and J. G. Gall,* Science *168, 1970, 1356.)*

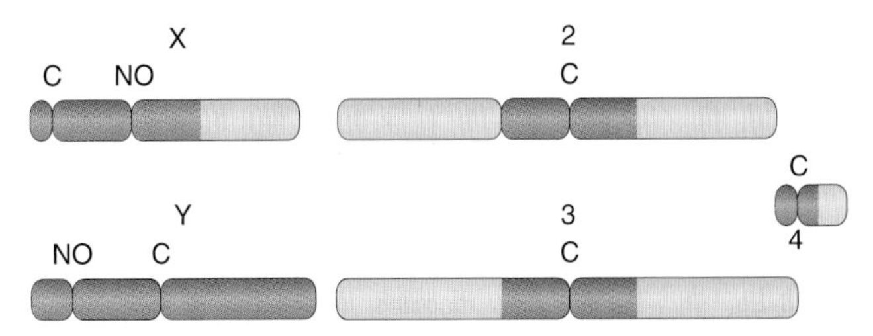

Figure 9-24 Position of heterochromatin in *Drosophila melanogaster.* C represents the centromeres, and NO the nucleolar organizers. The darkly colored regions are heterochromatin. *(From A. Hilliker and C. B. Sharp, in J. P. Gustafson and R. Appels, eds.,* Chromosome Structure and Function, *Plenum, 1988, pp. 91–95.)*

Elements that move in the form of DNA are called *transposons.* Many genomes have multiple copies of such elements or truncated versions of them dispersed throughout the genome. The full-length copies of these elements encode enzymes called **transposases** that are necessary for their transposition (Figure 9-25 on the next page). When we consider the problem these elements pose for genome sequencing, we should be aware that these elements can be present tens to hundreds of times in a genome, with sequences that are very similar or identical to the members of a given family of elements.

Another general class of transposed DNA sequences, called **retrotransposons,** move in the form of RNA (Figure 9-26 on the next page). One type within this category comprises repetitive sequences whose structures are similar to those of retroviruses (which replicate by reverse transcriptase) and may be derived from such viruses. Retrotransposons propagate through the action of the reverse transcriptase they encode. Reverse transcriptase is an enzyme that synthesizes a DNA strand from an RNA template, and the resulting DNA molecules can move through RNA intermediates to new locations in the genome. Examples of retroviral-like retrotransposons are the *copia* elements of *Drosophila* (5-kb sequences present at about 50 copies per genome) and the Ty elements of yeast (6-kb elements with about 30 full-length copies per genome). By contrast, the **long interspersed elements,** or **LINEs,** of mammals (1- to 5-kb elements present in 20,000 to 40,000 copies per human genome) move by retrotransposition but lack some structural features of retrovirus-like elements.

The *Alu* repetitive sequence in the human genome, so named because it contains a target site for the *Alu* restriction enzyme, is an example of a class of retrotransposon that is not similar to retroviruses. The human genome contains hundreds of thousands of whole and partial *Alu* sequences, scattered between genes and within introns and making up about 5 percent of human DNA. The full *Alu* sequence is about 200 nucleotides long and bears remarkable resemblance to 7SL RNA, an RNA that is part of a complex

Classes of interspersed repeat in the human genome

Element	Transposition	Structure	Length	Copy number	Fraction of genome
LINEs	Autonomous	ORF1 ORF2 (*pol*) AAA	1–5 kb	20,000–40,000	21%
SINEs	Nonautonomous	A B AAA	100–300 bp	1,500,000	13%
Retrovirus-like elements	Autonomous	*gag* *pol* (*env*)	6–11 kb	450,000	8%
	Nonautonomous	*gag*	1.5–3 kb		
DNA transposons	Autonomous	*transposase*	2–3 kb	300,000	3%
	Nonautonomous		80–3000 bp		

Figure 9-25 The general classes of transposable elements found in the human genome. Note the enormous copy numbers of these elements, which total nearly half the entire genome sequence. Also note that many of the elements are termed "nonautonomous." This refers to the fact that many of the transposable elements are missing some of the genes required for transposition; however, these elements can still move because other copies of the element in the genome encode the necessary gene products. Genes that may or may not be present in an element are indicated by parentheses. *(Reprinted by permission from* Nature *vol. 409: 880 (15 Feb 2001), "Initial Sequencing and Analysis of the Human Genome," The International Human Genome Sequencing Consortium. © 2001 Macmillan Magazines Ltd.)*

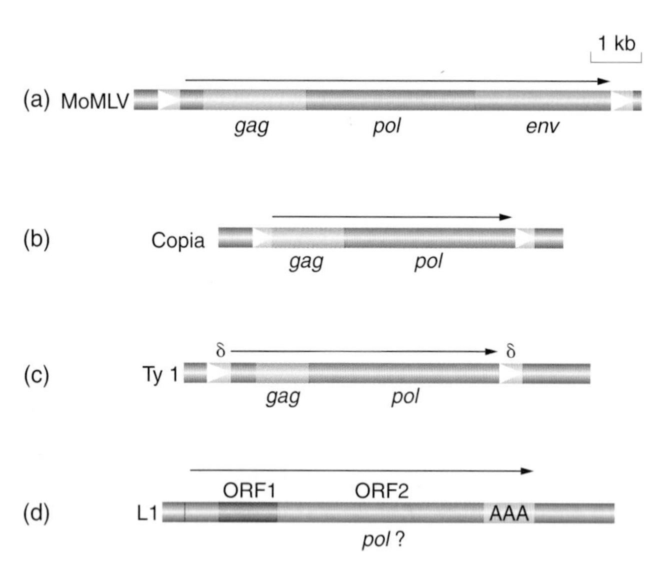

Figure 9-26 The general structure of some retrotransposons found in eukaryotic genomes. *Gag, pol,* and *env* = retroviral genes; ORF = open reading frame; AAA = a poly(A) tail. Unshaded triangles represent direct repeats (repeats oriented in the same 5′-to-3′ direction) at the termini of the retrotransposon. Arrows show direction of transcription. (a) A retrovirus, Moloney murine leukemia virus of mice. (b) Copia (*Drosophila* retrotransposon). (c) Ty 1 (yeast retrotransposon). (d) L1, a human LINE. *(Modified from J. R. S. Fincham,* Genetic Analysis: Principles, Scope and Objectives. *Blackwell, London, 1994.)*

by which newly synthesized polypeptides are secreted through the endoplasmic reticulum. Presumably, the *Alu* sequences originate as reverse transcripts of these RNA molecules. Repeats such as *Alu* sequences are collectively called **SINEs,** for **short interspersed elements.** Other examples of this class of moderately repetitive elements are the many scattered **pseudogenes** that have clearly been created by reverse transcription of spliced mRNAs, because they do not contain the introns that are found in the original functional gene.

Figure 9-27 is an example of the diversity of mobile elements present in the human genome, based on the identification of elements in the vicinity of a single human gene. There is considerable variation between species in the

MESSAGE

The landscape of eukaryotic chromosomes includes a panoply of repetitive DNA segments. Some, like telomeres, clearly have a structural role in replication. Others, such as centric heterochromatic repeats, have an uncertain role. Still other repetitive segments are functional genes or pseudogenes. Finally, many repeats are mobile genetic elements.

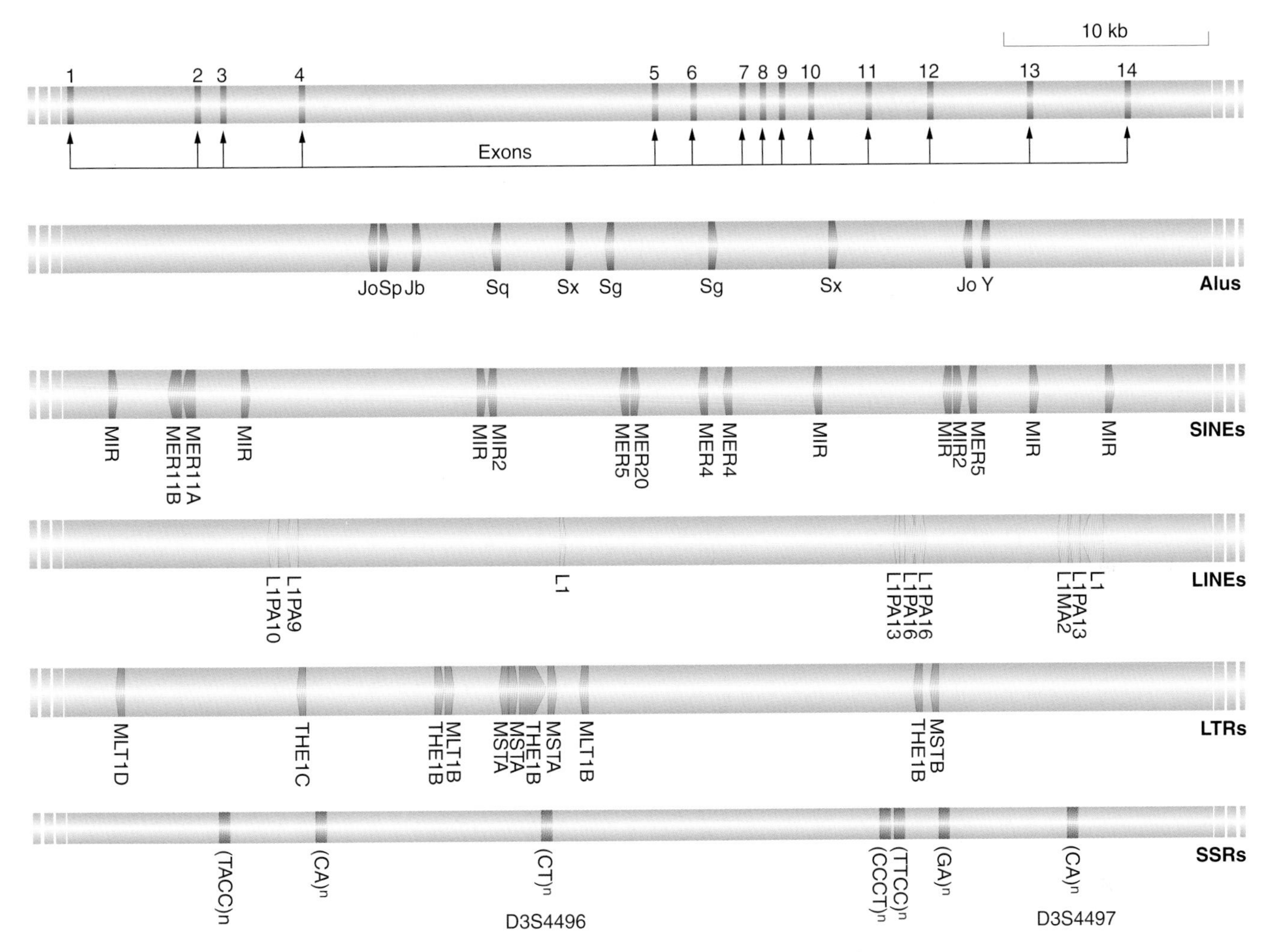

Figure 9-27 Repetitive elements found in the human gene *(HGO)* coding for homogentisate 1,2-dioxygenase, the enzyme whose deficiency causes alkaptonuria. The first line diagrams the position of the *HGO* exons. The location of *Alus* (blue), SINEs (purple), LINEs (yellow), retrotransposon-derived sequences (LTRs, red), and short-sequence repeats (SSRs, maroon) in the *HGO* sequence are indicated in the five rows below; note that these repetitive elements all fall into the introns of the *HGO* gene. *(Modified from B. Granadino, D. Beltrán-Valero de Bernabé, J. M. Fernández-Cañón, M. A. Peñalva, and S. Rodríguez de Córdoba, "The Human Homogentisate 1,2-Dioxygenase* (HGO) *Gene,"* Genomics *43, 1997, 115.)*

amounts of nonfunctional repetitive DNA; for example, there is little in yeasts and other fungi.

We thus see that there are considerable complexities to the organization of higher eukaryotic genomes. A summary view of the organization of a higher eukaryotic chromosome is depicted in Figure 9-28.

Tackling Genomes with Repetitive Sequences

Now that we have seen the diversity of repetitive elements, we can appreciate the problems they raise for producing a consensus sequence assembly for a eukaryotic genome. As

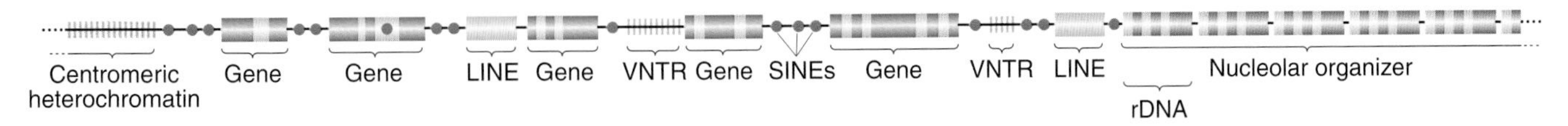

Figure 9-28 General depiction of a eukaryotic chromosomal landscape. The small region of a chromosome shown happens to have five protein-coding genes, one end of a nucleolar organizer, and one end of centromeric heterochromatin. Various kinds of repetitive DNAs are shown. (Each chromosome would normally have several thousand genes.)

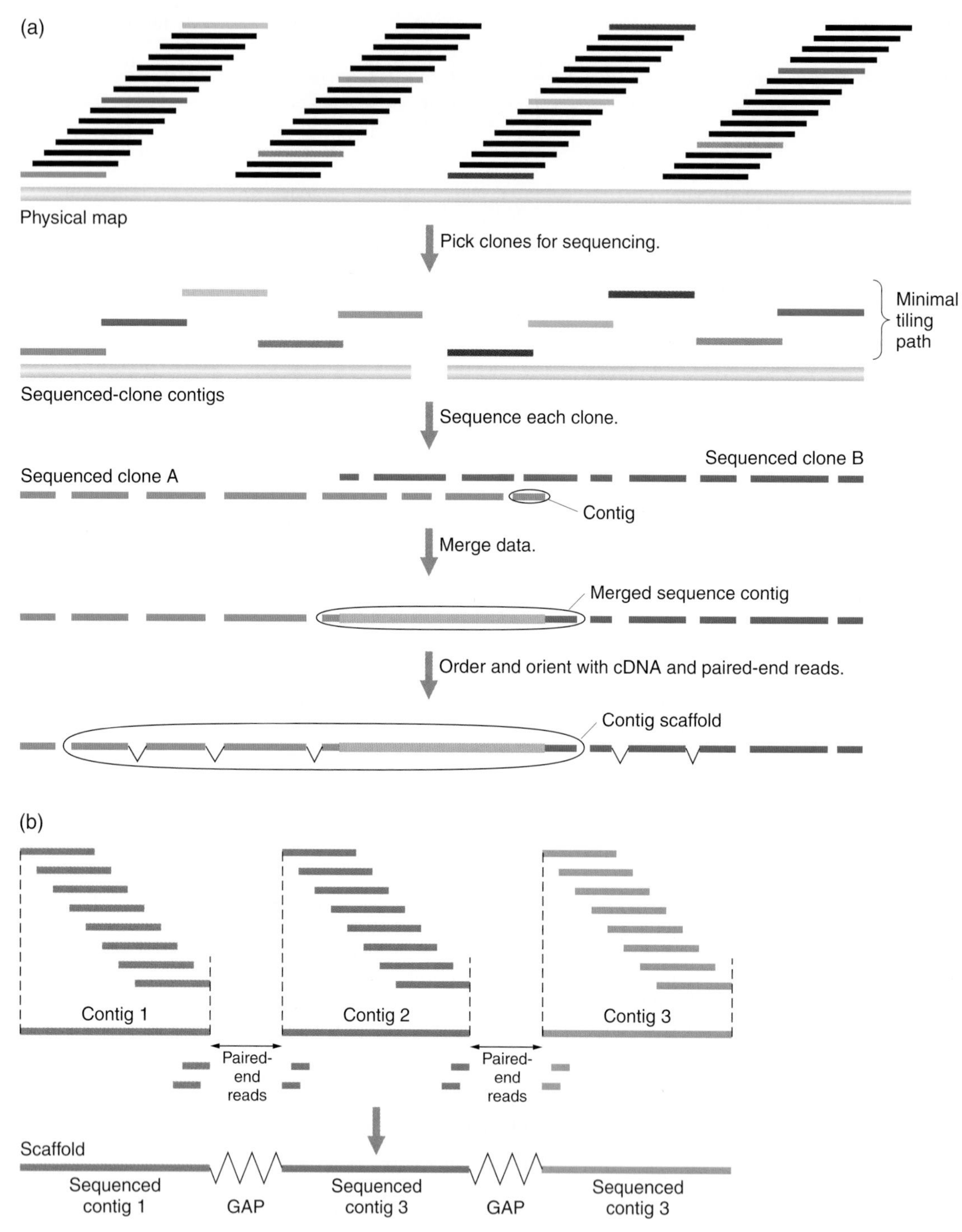

Figure 9-29 A comparison of the steps in ordered clone sequencing and whole genome shotgun sequencing assembly of a minimum tiling path of clones from a complete physical map of a genome. (a) By physical mapping, a series of clones that minimally overlap are identified; this series is called a *minimum tiling path,* and the alternating clones chosen as the "steps" of this path are represented in the various colors. The physical map data confirmed by the sequence overlaps between the ends of adjacent clones allow reconstruction of the genome into contigs. These contigs are ordered and oriented by means of aligning them relative to paired-end sequence reads and to cDNA sequences. Gaps in the sequenced-clone contigs indicate missing bits of sequence. Introns in the cDNA are indicated by V-shaped black lines connecting the colored exons. (b) Whole genome shotgun sequencing assembly. First, the unique sequence overlaps between sequence reads are used to build contigs. Paired-end reads are then used to span gaps and to order and orient the contigs into larger units called *scaffolds. (Part a reprinted by permission from* Nature *vol. 409: 870 (15 Feb 2001), "Initial Sequencing and Analysis of the Human Genome," The International Human Genome Sequencing Consortium. © 2001 Macmillan Magazines Ltd.)*

we shall see, ordered clone sequencing and whole genome shotgun sequencing each attempt to deal with repetitive DNAs in different ways. Whole genome shotgun sequence assembly connects the single-copy sequences on either side of a repetitive element but ignores the sequence of the repetitive element itself, while ordered clone sequencing allows straightforward assembly of many of the dispersed repeats, since they are present only once in an individual clone. Let's examine these approaches and see how they circumvent the repetitive DNA issue.

Assembling a sequence from ordered clones. As an example of the approach of assembling a sequence from ordered clones, we will consider the sequencing of the nematode *Caenorhabditis elegans* genome. As a first step, genomic cosmid clones were fingerprinted and thus placed into a complete physical map of the *C. elegans* genome. For reasons of economy and efficiency, it was desirable to sequence clones with as little overlap as possible. Thus, as outlined in Figure 9-29a, a subset of the cosmid clones in the physical map with clear but minimal overlap was selected for sequencing; such a subset of clones is called a **minimum tiling path,** since it is the minimum number of clones that represents the entirety of the genome. The minimum tiling path clones were subcloned into sequencing cloning vectors that accept inserts of up to 2 kb, and the inserts were then sequenced using a dideoxynucleotide sequence priming site within the vector DNA immediately adjacent to the genomic insert, the approach we previously discussed in Figure 9-21. This primer produced sequence reads of the adjacent region of the genomic insert DNA. Recall that the same priming site could be used universally for all inserts, a great simplification when many millions of individual sequence reads are required to reconstruct the consensus sequence of the entire genome. The next step was to assemble the sequence reads of individual clones into a clone-consensus sequence, with multiple reads of each and every base in the clone. While there are many mobile element families in the *C. elegans* genome, they are dispersed. Thus, it was rare that two or more members of a given mobile element family were encountered in the same cosmid. In other words, by creating the consensus sequence clone by clone, these mobile element families are single-copy with regard to their frequency within a cosmid, and hence their assembly into a consensus sequence for that particular cosmid is straightforward. This is one of the major advantages of ordered clone assembly. Since the order of the clones had been determined thanks to the physical mapping project, once the individual clone-consensus sequences were assembled it was straightforward to merge the individual consensus sequences with those of their neighbors, eventually producing an overall consensus sequence for the entire genome. The ability to rely on the physical map to order and orient the clone sequences is the other major advantage of the ordered clone approach.

Whole genome shotgun assembly. As an example of whole genome shotgun sequencing, we will consider the sequencing of the fruit fly *Drosophila melanogaster* genome, for which the initial sequence was produced by this method. Libraries of genomic clones of different sizes (2 kb, 10 kb, 150 kb) were sequenced. Sequence reads from the two ends of genomic clones were obtained and aligned by a logic identical with that used for bacterial sequencing. Through this logic, sequence reads of single-copy sequences were aligned by their homologous sequence overlaps, and contigs—consensus sequences for these single-copy stretches of the genome—were produced. However, unlike the situation in bacteria, where there is only single-copy DNA, the contigs would eventually run into a repetitive DNA segment such as a mobile genetic element that would prevent unambiguous assembly of the sequence reads (unambiguous in the sense of knowing that there is a unique location of that sequence in the genome). Because of the average spacing of mobile elements that are in the *D. melanogaster* genome, contigs had an average size of about 150 kb. The challenge then was how to glue the thousands of contigs together in their correct genomic order and orientation.

The solution to this problem was to obtain sequences from both ends of the inserts in the sequenced clones. The two ends of the inserted clones were sequenced by use of two universal primers, one from either end of the vector sequence. In this way, **paired-end sequences** were obtained, that is, sequences from the portion of the insert adjacent to each end of the plasmid cloning vector (see Figure 9-21). If one paired end was a single-copy sequence read that was part of one contig, and the other paired end was a sequence read that was from a second contig, then the two contigs were clearly near each other. Indeed, since the size of the clone was known (that is, it came from a library containing genomic inserts of uniform size, either the 2-kb, 100kb, or 150-kb library), the distance between the paired-end reads was known. Further, because the sequence reads are oriented relative to each other, by aligning the sequences of the two contigs to the paired-end sequences from the genomic insert, the relative orientation of the two contigs could be determined. In this manner, single-copy contigs could be joined together, albeit with gaps where the repetitive elements reside. These gapped collections of joined-together contigs are called **scaffolds.** Because most *Drosophila* repeats are large (3 to 8 kb) and widely spaced (one repeat approximately every 150 kb), this technique was extremely effective at producing a correctly assembled sequence of the single-copy DNA. A summary of the logic of this technique is shown in Figure 9-29b.

What are the advantages and disadvantages of whole genome shotgun assembly? The advantage is that overall assembly occurs without the need for any physical mapping of clones. Shotgun sequencing reads are very efficient, so an overview of the structure of most genes can be obtained very rapidly compared with ordered clone sequencing. There are two chief disadvantages. First, in some genomes, the repetitive elements are so numerous and frequent that even with paired-end reads, average scaffold size is less than 100 kb. Second, in order to fill in the gaps in the sequence, individual clones need to be identified and sequenced. Thus, the finished product is actually a mixture of whole genome shotgun and ordered clone sequencing. Even with these cautionary notes, however, whole genome shotgun assembly has proved to be a powerful way of getting a high-quality first look at the single-copy consensus sequence of a complex genome.

MESSAGE

The two basic approaches to genome sequencing are minimum tiling path ordered clone sequencing from physical maps and whole genome shotgun sequencing.

Filling sequence gaps. For both whole genome shotgun and ordered clone sequencing, in practice there are gaps that remain in the genome. Sometimes, these are due to sequences from the genome that are incapable of growth inside of the bacterial cloning host. Special techniques must be used to fill the holes in the sequence assemblies left by these gaps. If the gaps are short, PCR fragments can be generated from the ends of the assemblies adjacent to the gaps, and the PCR fragments can then be directly sequenced without a cloning step. If the gaps are longer, attempts can be made to grow clones containing these sequences in a different cloning host, such as yeast. If this fails, then the gaps in the sequence may remain. The other source of these gaps is the presence of nucleotide sequences that interfere with sequencing reactions in the test tube, or that lead to unusual electrophoresis properties of the DNA such that the sequence traces (an example of a trace is shown in Figure 8-24) from the sequencing machines are unreadable.

SUCCESS STORIES: THE WEED, THE WORM, THE FLY, AND THE HUMAN GENOMES

Even with all of the problems that can get in the way of accurate sequence assemblies, many of them can be circumvented. Thus, the sequencing projects for the major genetically accessible eukaryotes are well under way, including the genomic sequencing of the yeast *Saccharomyces cerevisiae,* the nematode *Caenorhabditis elegans,* the fruit fly *Drosophila melanogaster,* the mustard weed *Arabidopsis thaliana,* the laboratory mouse *Mus musculus,* the zebrafish *Danio rerio,* and humans.

For the genomic sequencing projects that are largely complete, different strategies were used for sequencing. For *Caenorhabditis elegans* and *Arabidopsis thaliana,* clone-by-clone methods were employed, in which a minimum tiling path of overlapping cosmid *(C. elegans)* and BAC *(A. thaliana)* clones formed the major backbones of the sequences. As discussed, for *D. melanogaster,* the bulk of the consen-sus sequence was obtained from whole genome shotgun assembly.

For the human genome, repeats are generally shorter but much more frequent than in other sequenced genomes. As we saw from the diversity and frequency of repetitive elements found in even a single human gene (see Figure 9-27), the frequency, size, and similarity of these repeats in the human genome make whole genome shotgun assembly as it was done in *Drosophila* quite challenging for the human sequence. Two approaches to producing a genome assembly were taken. One relied purely on an ordered clone sequencing approach using BAC clone libraries but employed some paired-end data to jump over repeats within a single BAC clone and connect consensus sequences on either side. The other used a combination of paired-end whole genome shotgun data and a distillation of the ordered clone data from the other project to create its consensus sequence. These sorts of hybrid strategies seem to offer the best approach at present for sequencing genomes as large and complex as those of mammals.

Both approaches to sequencing the human genome allowed production of consensus sequences amounting to about 2.6 billion base pairs of DNA, out of the estimated 3 billion base pairs in the human genome. (Much of the remaining 0.4 billion base pairs probably comprises repeats that could not be assembled.) While the overall size of the assemblies produced by the two approaches is the same, there were nonetheless many differences between the two consensus sequences and in the gene predictions that were made from them. At this point, it is unclear which of the assemblies and gene prediction sets is more accurate, and considerable evaluation of these two first "drafts" of the human genome sequence will be necessary to understand and resolve data conflicts.

BIOINFORMATICS: GLEANING MEANING FROM GENOMIC SEQUENCE

The availability of "draft" or completed genomic sequences has proved to be a treasure trove, providing the first opportunity for a global evaluation of the information encoded in the DNA of the various organisms whose genomes have been sequenced, and for overlaying of this information on the detailed genetic maps of these species. However, just because we have the raw four-letter sequence of DNA does not mean that we are in a position to read this information from beginning to end in the way we would read a novel. Rather, much of the information content of these sequenced genomes remains indecipherable by our current understanding of the various genetic codes.

The Information Content of DNA

In one sense, DNA can be viewed as a scaffold containing a series of docking sites for different proteins and RNAs. Some of these docking sites function in the DNA itself, while others encoded in the DNA function only in the RNA transcribed from that DNA sequence (Figure 9-30). It is the relative location of those docking sites to one another that permits genes to be transcribed, spliced, and translated properly and in specific spatial and temporal patterns. For example, for transcription to occur, multiple docking sites on the DNA must be appropriately spaced so that RNA polymerase can simultaneously contact them all. At the RNA levels, the locations of docking sites for RNAs and proteins of spliceosomes at eukaryotic 5′ and 3′ splice sites will determine where introns are removed. As another example at the RNA level, translation in prokaryotes requires that ribosome-binding sites be a certain fixed distance from initiation AUG codons at which *N*-formylmethionyl-tRNA will bind. Indeed, every codon in the translated region of an mRNA is a docking site for a specific aminoacyl-tRNA. Even stop codons are locations at which proteins—those that terminate translation—bind to the mRNA. Regardless of whether a docking site actually functions as such in DNA or RNA, the site must be encoded in the DNA. In order to be able to recognize the presence of such sites in the DNA sequence, we need to understand the rules by which they operate.

Hurdles to Deciphering the Information Content of DNA

It may seem surprising that the words and languages of DNA are not well understood after 50 years of research by many tens of thousands of scientists. Just as we have learned how to cope with imprecision in a spoken or written language in which we are fluent, cells have evolved mechanisms to cope with imprecision in the languages of DNA. At this point, we don't understand how cells deal with such imprecision; we can only infer from the reliable properties of cells that indeed they do cope. We will examine three of the major sources of imprecision that we have to face in thinking about how to decipher the languages of DNA.

Not all words describing functions encoded in DNA have been discovered. We now know many (but perhaps not all) of the classes of docking sites encoded within genomic DNA. These include those that participate in DNA replication, transcription initiation and termination, pre-mRNA splicing, translation initiation and termination, and—as we will discuss in Chapter 13—those that act as regulatory protein binding sites necessary to determine when, where, and at what level a gene will be expressed. While we know many of the kinds of regulatory docking sites that DNA can contain, we do not know all the specific DNA sequences that encode the thousands of docking sites for DNA or RNA-binding regulatory proteins in the genome of a higher eukaryote (Figure 9-31a on the next page). This is analogous to saying that while we can generally define the role of a verb in a sentence, we do not have a complete (or even near complete) dictionary of all the verbs in a given language—there may be many combinations of letters that spell verbs, but we are unable to recognize them as verbs, or perhaps as words at all.

One word means more than one thing. A major problem in deciphering the DNA sequence is that we must use context-dependent interpretation. Just as one word in English (or any other language) may have multiple meanings, the same is true for DNA "words"; a DNA nucleotide sequence may have different meaning, depending on where it located within the DNA (see Figure 9-31b). TTAATTGCA on one strand of DNA located in an exon that will form part of the translated region of an mRNA will be translated as leucine-isoleucine-alanine, if it is on the correct DNA strand and in the correct reading frame. However, if

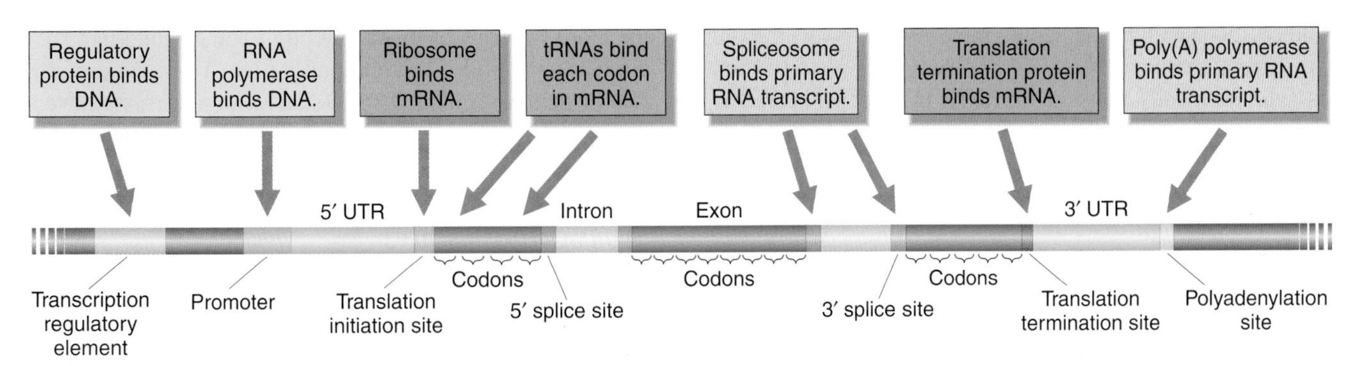

Figure 9-30 A depiction of a gene within DNA as a series of docking sites for proteins and RNAs.

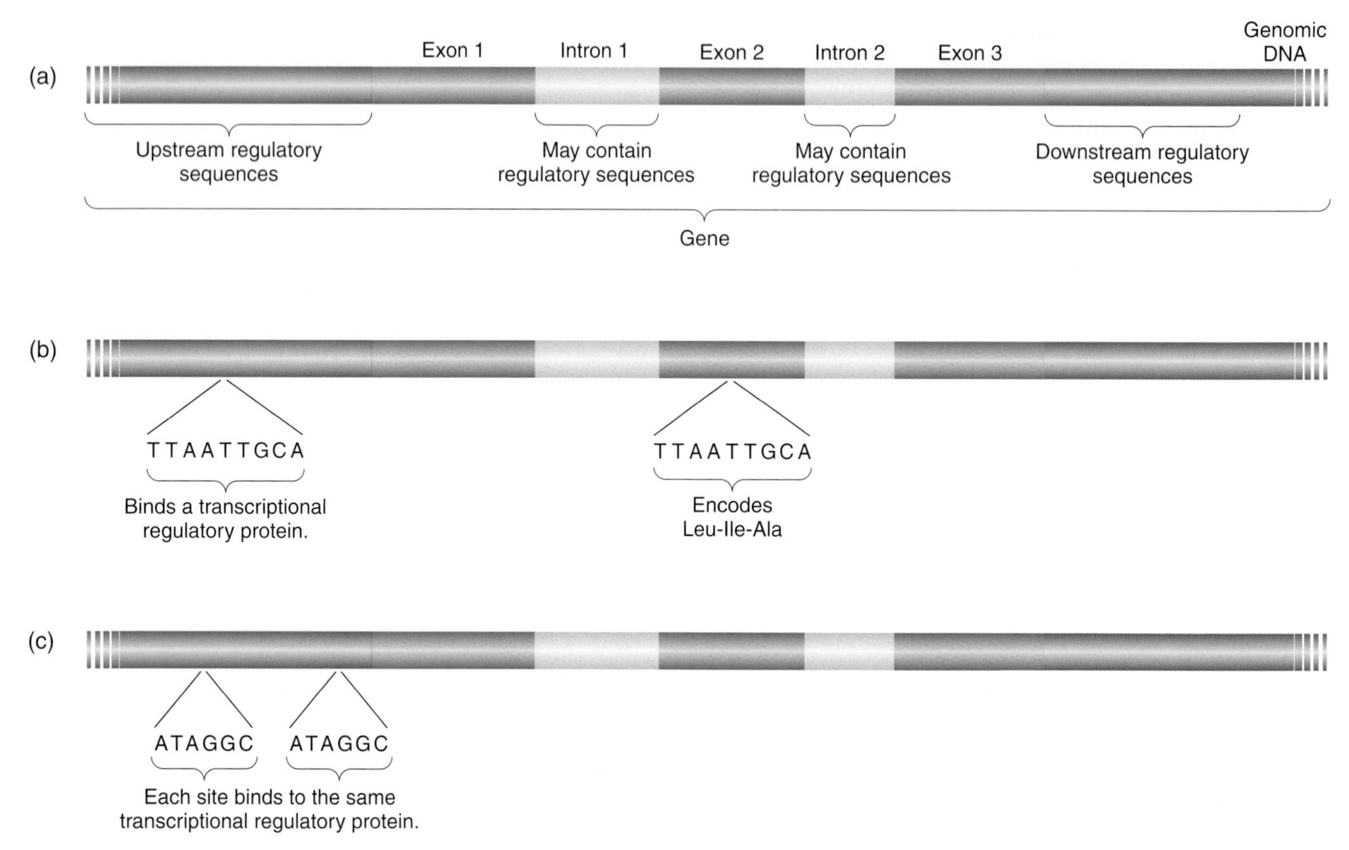

Figure 9-31 Some of the issues in interpreting DNA sequence. (a) Many of the docking sites within genes have not yet been identified. Thus, while regulatory elements generally are found upstream and sometimes downstream of transcription and within introns, many of the hundreds or thousands of regulatory protein binding sites have not yet been discovered. (b) The same nucleotide sequence string can carry different information, depending on where it is located within the gene. (c) Multiple sequences can be bound by the same regulatory protein.

TTAATTGCA is located in a nontranscribed region of the genome, it can act as a binding site for a protein, which in turn forms a protein complex with other components of the transcriptional machinery to control certain aspects of transcription of a nearby gene. Our knowledge of DNA words is limited, and so we can't rule out the possibility of still other interpretations of TTAATTGCA. Even with the triplet code for amino acids, we see context-dependence. Most codons are universally interpreted by the translation machinery as encoding the same amino acid. However, there are exceptions. For example, UGA, which is a stop codon in cytoplasmically translated mRNAs in humans, is read as tryptophan in translation of mitochondrial genes in the same human cells. These are but a couple of many examples of context dependence.

More than one word means the same thing. Another issue in deciphering the DNA sequence is that DNA has the equivalent of synonyms in English or other human languages. We have already pointed out that we can view DNA as a string of docking sites for sequence-specific DNA or RNA-binding proteins. Sometimes, a group of sites act in concert, cooperatively serving as binding sites for several subunits of a multiprotein complex. Often, there do not seem to be absolute rules for unique protein-DNA binding. Rather, we find multiple sequences that can be bound by the same protein (see Figure 9-31c). Thus, instead of being able to describe one and only one specific DNA sequence that serves as a particular docking site, we often have to refer to a consensus sequence for a protein-binding motif in DNA, and we assign a certain probability to each base in the motif in terms of its likelihood of being found in a specific instance of this protein-binding site. Thus, the description of many words in DNA is statistical; that is, we have a frequency distribution of bases for some or all sites within the consensus sequence.

Bioinformatics and the Analysis of Gene Structure

Because the proteins present in a cell largely determine cell shape, role, and physiological properties, one of the first orders of business in genome analysis is to try to determine a list of all of the polypeptides encoded by an organism's genome. This list of the polypeptides encoded by a genome is sometimes termed the organism's **proteome** and can be considered a parts list for the cell. To determine the list of polypeptides, the structure of each mRNA encoded by the genome must be deduced. Because of intron splicing, this is particularly challenging in higher eukaryotes, where the presence of introns is the rule, and, for example, in humans, where an average gene has about 10 exons. Furthermore, many genes encode alternative exons—that is, some exons are included in some versions of a processed mRNA encoded by the gene but are not included in others. Often, the alternatively processed mRNAs encode polypeptides sharing much, but not all, of their amino acid sequence. Even though we have a great many examples of completely sequenced genes and mRNAs, it has not yet been possible to identify 5′ and 3′ splice sites from sequence-gazing with a high degree of accuracy, which means we cannot be certain which sequences are introns. Predictions of alternatively used exons are even more error-prone. Many other bits of information contained in the genome are completely obscured at our current level of understanding. A major problem then is to use our knowledge of information transfer from DNA to RNA to protein to deduce the total parts list of polypeptides in higher eukaryotes. The computational analysis of sequence information to predict mRNA and polypeptide sequences is an important part of the discipline called **bioinformatics.**

How does bioinformatics approach the problem of obtaining information from the sequences of complex genomes? It largely is a problem in code-breaking: decrypting enciphered information using approaches similar to those used by cryptographers, linguists, and archeologists to break coded messages, develop language recognition systems, and reveal the meaning of ancient languages. All these disciplines use both experimental data sets (dictionaries of known words) and statistical prediction techniques to turn encoded message into useful information, and as we shall see, this is precisely the approach that bioinformaticians use to decipher the languages of DNA. We purposefully use the plural "languages" here because we should recognize that indeed many different kinds of languages are encoded in DNA—the triplet code for amino acid sequence, and the codes that underlie the specificity of nucleotide sites at which proteins bind to begin transcription, to replicate DNA, to regulate gene expression, to pull chromosomes to the spindle poles during mitosis and meiosis, are examples.

Bioinformatics uses several independent sets of information to predict the most likely sequence for mRNA and polypeptide coding regions:

- cDNA sequences
- Docking site sequences marking the start and end points for the events in information transfer (transcription, pre-mRNA splicing, translation)
- Sequences of related polypeptides
- Species-specific usage preferences for some codons over others encoding the same amino acid

We'll now consider these four sources of evidence in turn.

Direct cDNA evidence. cDNA sequences are extremely valuable in understanding transcript and polypeptide structure (Figure 9-32). Because cDNAs are DNA copies of mRNAs, if cDNAs are aligned with their corresponding genomic sequence, the regions that align are identified as exons, and the regions in between areas of alignment are introns. cDNA sequences can also be examined to reveal **motifs** (sequence strings that are characteristic for DNA sites that function in specific roles) that are predictive of initiation codons and for likely translated regions termed *ORFs* (*o*pen *r*eading *f*rames). Full-length cDNA evidence is taken as the gold-standard proof for identification of the sequence of a transcription unit, determination of how it is

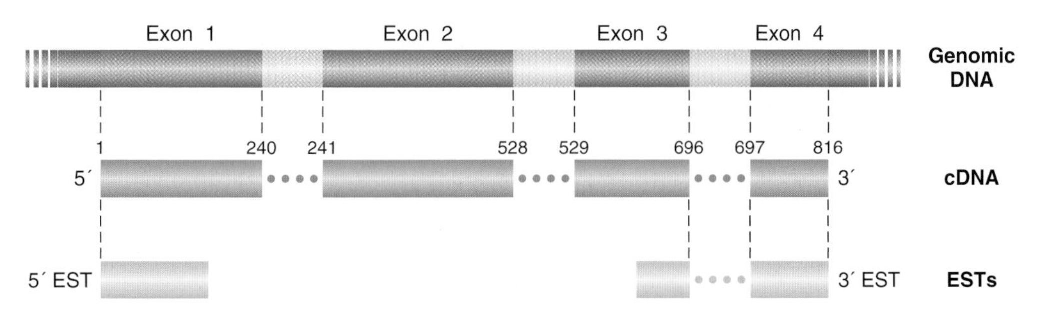

Figure 9-32 The alignment of fully sequenced cDNAs and ESTs with genomic DNA. The solid lines indicate regions of alignment; for the cDNA, these are the exons of the gene. The dots between segments of cDNA or ESTs indicate regions in the genomic DNA that do not align with cDNA or EST sequences; these are the locations of the introns. The numbers above the cDNA line indicate the base coordinates of the cDNA sequence, where base 1 is the 5′-most base and base 816 is the 3′-most base of the cDNA. For the ESTs, only a short sequence read from either the 5′ or 3′ end of the corresponding cDNA is obtained. This establishes the boundaries of the transcription unit, but it is not informative about the internal structure of the transcript unless the EST sequences cross an intron (as is true for the 3′ EST depicted here).

processed, and localization of the ORF it encodes. In addition to full-length cDNA sequences, there are large data sets of cDNAs for which only the 5′ or the 3′ ends, or both, have been sequenced. These short cDNA sequence reads are called *ESTs* for *e*xpressed *s*equence *t*ags. ESTs can be aligned with genomic DNA and thereby used to determine the 5′ and 3′ ends of transcripts—in other words, to determine the boundaries of the transcript—and thus aid in predicting mRNA structure.

Predictions of mRNA and open-reading-frame structures. As we have discussed, a gene consists of a segment of DNA that encodes a transcript, as well as the regulatory signals that determine when, where, and how much of that transcript is made. In turn that transcript has the signals necessary to determine its processing into mRNA and the translation of that mRNA into a polypeptide (Figure 9-33). There are now statistical "gene-finding" computer programs that include predictions of the various docking sites for transcription start sites, 3′ and 5′ splice sites, and translation initiation codons within genomic DNA. These predictions are based on consensus motifs for such known sequences, but they are by no means perfect. For mammalian genomes such as human, where exons are very small (about 150 bases) and introns are numerous (on average 9), it is relatively infrequent that gene-finding programs make completely accurate mRNA and ORF predictions in the absence of cDNA evidence.

Polypeptide similarity evidence. Proteins with related function or of related ancestry typically have much more amino acid sequence similarity than do unrelated proteins. Sometimes this similarity extends over the entire length of the polypeptide sequences, and sometimes it is localized to specific domains within the protein sequence, such as the amino acid sequences corresponding to the active site of an

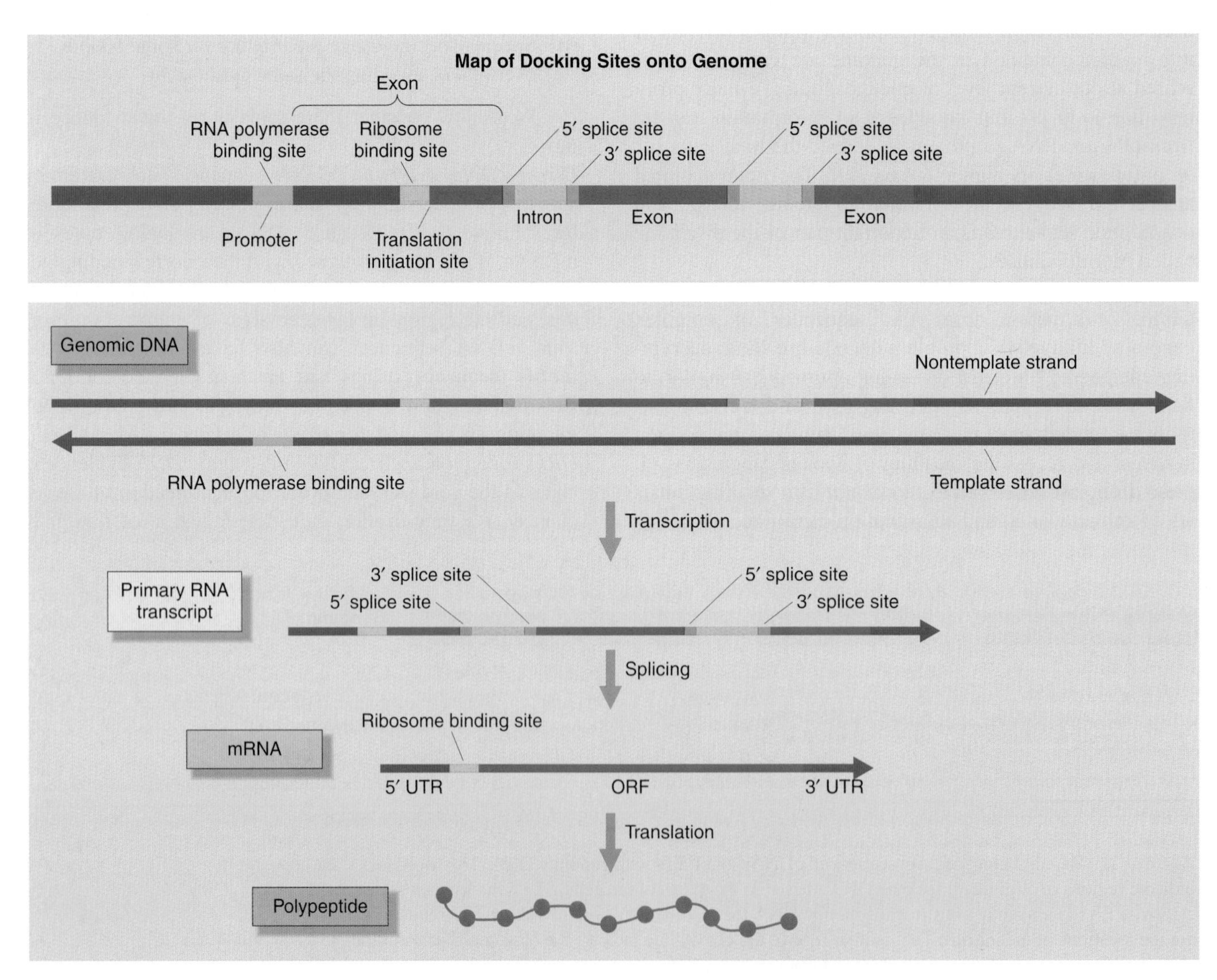

Figure 9-33 Eukaryotic information transfer from gene to polypeptide chain, with emphasis on the DNA and RNA "docking sites" that are bound by protein complexes to initiate the events of transcription, splicing, and translation.

enzyme. Statistical tests have been devised that can identify many such related proteins. A level of 35 percent or more amino acid identity at comparable positions in two polypeptides is indicative of a common three-dimensional structure and often suggests that the two polypeptides have at least some aspects of their function in common.

Instead of comparing two known polypeptides, one can use the detection methods of polypeptide similarity to detect likely protein-coding genes within a sequenced genome. Suppose that we wanted to identify all the genes involved in oxygen transport in the bloodstream—that is, genes encoding hemoglobins. We can use the known amino acid sequence of a hemoglobin (a query sequence) as a computational probe and compare that sequence with the amino acid sequences predicted for all possible polypeptides encoded by the nucleotide sequence of the genome in question. One way to do this is to compare the query sequence to all polypeptide sequences predicted by the sort of gene prediction programs described in the previous section. The common statistical tool for carrying out such searches is called BLAST (*b*asic *l*ocal *a*lignment *s*earch *t*ool), which is accessed through the Worldwide Web (see Appendix B). Figure 9-34 on the next page shows the results of a BLASTp search (one kind of BLAST search) for predicted proteins in the human genome related to the β chain of mouse hemoglobin (our query sequence). In this case, several human proteins related to mouse β-hemoglobin are detected.

A variation on this theme is to compare the query sequence with the six-frame translation (the three possible reading frames in each direction along the double-stranded DNA molecule) of the entire genome. This variation has the advantage that it does not depend on the accuracy of the gene prediction programs, but it can miss similarities if they are spread out over too many small exons. This logic can be extended through statistical methods to ask if a hypothetical amino acid sequence significantly matches any polypeptide in a data bank of all known proteins. The hypothetical amino acid sequence can come from translation of the open reading frame of a predicted mRNA or from six-frame translation of a portion of the genome. The mRNA ORF translation makes a best guess at the amino acid sequence of joined exons. The six-frame translation makes no assumption about the correct reading frame but requires matches on an exon-by-exon basis.

If significant levels of similarity are observed between the query amino acid sequence and one or more known polypeptides, this is considered strong evidence that these identified sequences are contained within the ORF of an mRNA. Often, the similarity is strong in some regions of the protein and absent in others, so similarity evidence cannot be used to infer the entire structure of an ORF.

Predictions based on codon bias. Recall from Chapter 3 that the triplet code for amino acids is degenerate; that is, most amino acids are encoded by two or more codons (see Figure 3-20). The multiple codons for a single amino acid are termed *synonymous codons*. In a given species, not all synonymous codons for an amino acid are used with equal frequency. Rather, certain codons are present much more frequently in ORFs. For example, in *Drosophila melanogaster*, of the two codons for cysteine, UGC is used 73 percent of the time, while UGU is used 27 percent. Of the four codons for valine, GUG is used 48 percent of the time, GUC 26 percent, GUU 18 percent, and GUA only 8 percent. In other organisms, this "codon bias" pattern is quite different. Codon biases are thought to reflect the relative abundance of the tRNAs complementary to these various codons in a given species. Presumably the rate of translation of an mRNA is higher if its codon bias is skewed toward codons complementary to the most abundant tRNAs. Statistical comparisons of the codon usage of a predicted ORF in a species with that species' known codon usage pattern can reveal significant agreement or disagreement with codon bias. Agreement in codon bias can then be taken as supporting evidence for the accuracy of a gene prediction.

Putting it all together. A summary of how these different sources of information are combined to create the best possible mRNA and ORF predictions is depicted in Figure 9-35 on page 301. These different kinds of evidence are complementary and can cross-validate one another. For example, the structure of a gene may be inferred from protein similarity evidence within a region of genomic DNA bounded by 5′ and 3′ ESTs. Even in the absence of cDNA sequence or evidence from protein similarities, proper codon bias in a hypothetical ORF based on docking site prediction programs would be supporting evidence for the presence of a gene that contained the predicted ORF.

Using the approaches outlined in the previous sections, predictions of the expressed genes from the whole genome can be assessed. An example of the outcome of these predictions is shown in Figure 9-36 on page 302. These predictions are being revised continually as new data and new computer programs become available. The current state of the predictions can be viewed and queried at many Web sites, most notably at the public DNA data banks in the United States and Europe (see Appendix B). These are current best guesses of the protein-coding genes present in the sequenced species and, as such, are works-in-progress. Over the next few years, these best guesses will no doubt become closer to the correct answer for at least three reasons:

1. The genomic sequences will gradually go from "draft" to "finished" stages as scientists refine the sequence analysis on a clone-by-clone basis, with all sequence reads of high quality and all gaps filled.
2. cDNAs for many more genes will be identified and sequenced.
3. Computational algorithms for identifying sequence motifs and similarities at both the nucleotide and polypeptide levels will improve as our understanding

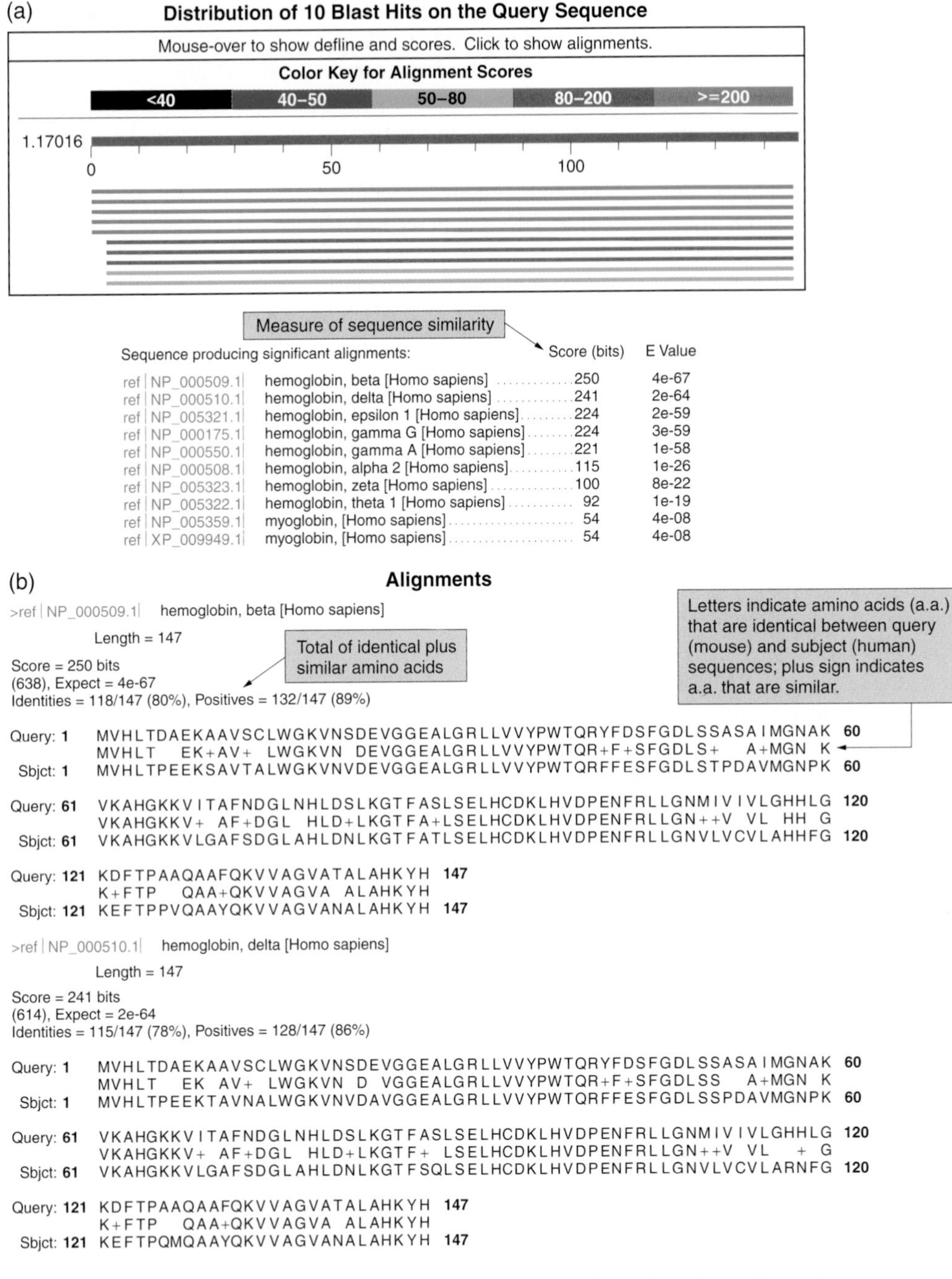

Figure 9-34 An example of a BLASTp output. A mouse hemoglobin 147-amino acid protein sequence, having the identifier number AAA37791.1 in the public sequence data banks (essentially the social security number of the protein), was used as a query sequence looking for matches in the proteins predicted from the human genome sequencing project. (See Appendix B for more information about BLASTp and other BLAST tools.) (a) A list of human proteins that have statistically significant levels of similarity to the query sequence (the mouse hemoglobin), along with their unique identifier numbers (blue), names, and their level of similarity to the 17,016-bp-long query sequence. The degree of sequence similarity is indicated by the "score," and the probability of obtaining this match by chance is indicated by the "E value," the probability expressed to the base *e*. The diagram above the list depicts the degree of alignment and where the similarities of each of these proteins align to the query sequence. (b) Alignment of the query sequence with two of the protein "hits" listed in part a is shown using the one-letter abbreviations for the 20 amino acids. Bold numbers indicate amino acid number in the sequence. (Report generated using the public NCBI BLAST server at the URL http://www.ncbi.nlm.nih.gov/BLAST/.)
(Reprinted with permission of the National Center for Biotechnology Information.)

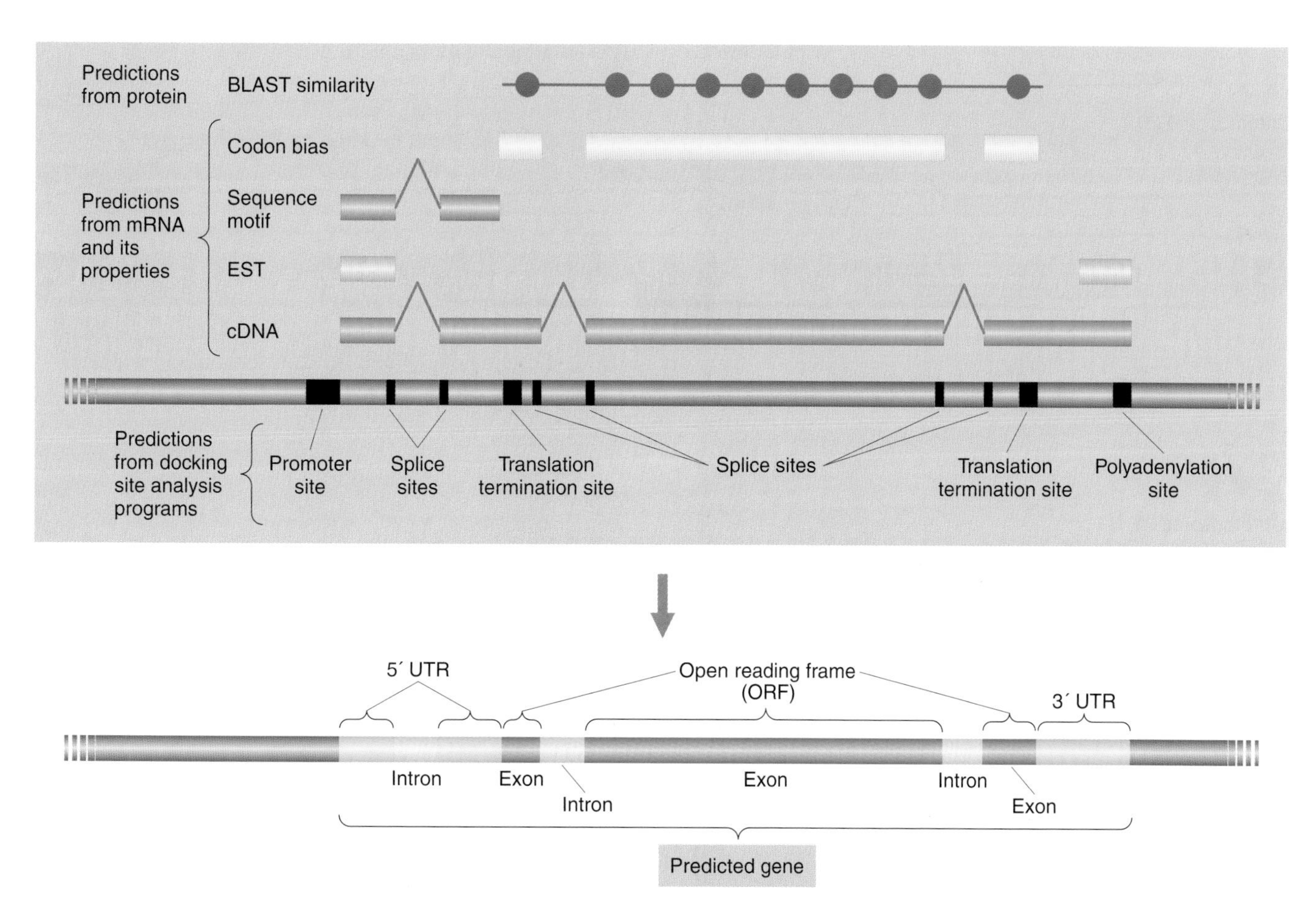

Figure 9-35 The different forms of gene product evidence—cDNAs, ESTs, BLAST similarity hits, codon bias, and motif hits—are integrated to make gene predictions. Where multiple classes of evidence are found to be associated with a particular genomic DNA sequence, there is greater confidence in the likelihood that a gene prediction is accurate.

of how to decipher the codes embedded in DNA becomes clearer.

MESSAGE

Predictions of mRNA and polypeptide structure from genomic DNA sequence depend on an integration of information from cDNA sequence, docking site predictions, polypeptide similarities, and codon bias.

Take-Home Lessons from the Genomes

There is now such a flood of genomic information that any attempt to summarize it in a few paragraphs will fall far short of the mark. Instead, let's consider a few of the insights from our first view of the overall structure of chromosomes and genomes and the global parts list of the species whose genomes have been sequenced. We will view the genome sequences by introspection—that is, what can we learn by looking at a single genome by itself? We will use the human genome as our example.

Many of the analytical methods that are used in sieving through genome, mRNA, and protein sequence data are based on comparisons between a given sequence and a data bank of other known sequences, such as those using the BLAST alignment tool (see Figure 9-34 and Appendix B). Using this statistical analytical tool, we can identify polypeptides with related structure and, very likely, with related function. If one of these polypeptides has a known function, we can use guilt-by-association logic to infer a similar function for the related polypeptides. Considerable experience in such sequence comparisons has borne out the validity of this approach, and it has proved to be a powerful way to classify polypeptides. (As the adage goes, "If it walks like a duck and talks like a duck, it probably is a duck!")

The Structure of the Human Genome

In describing the overall structure of the human genome, one of the first things we wish to delineate is its repeat structure. A considerable fraction of the human genome,

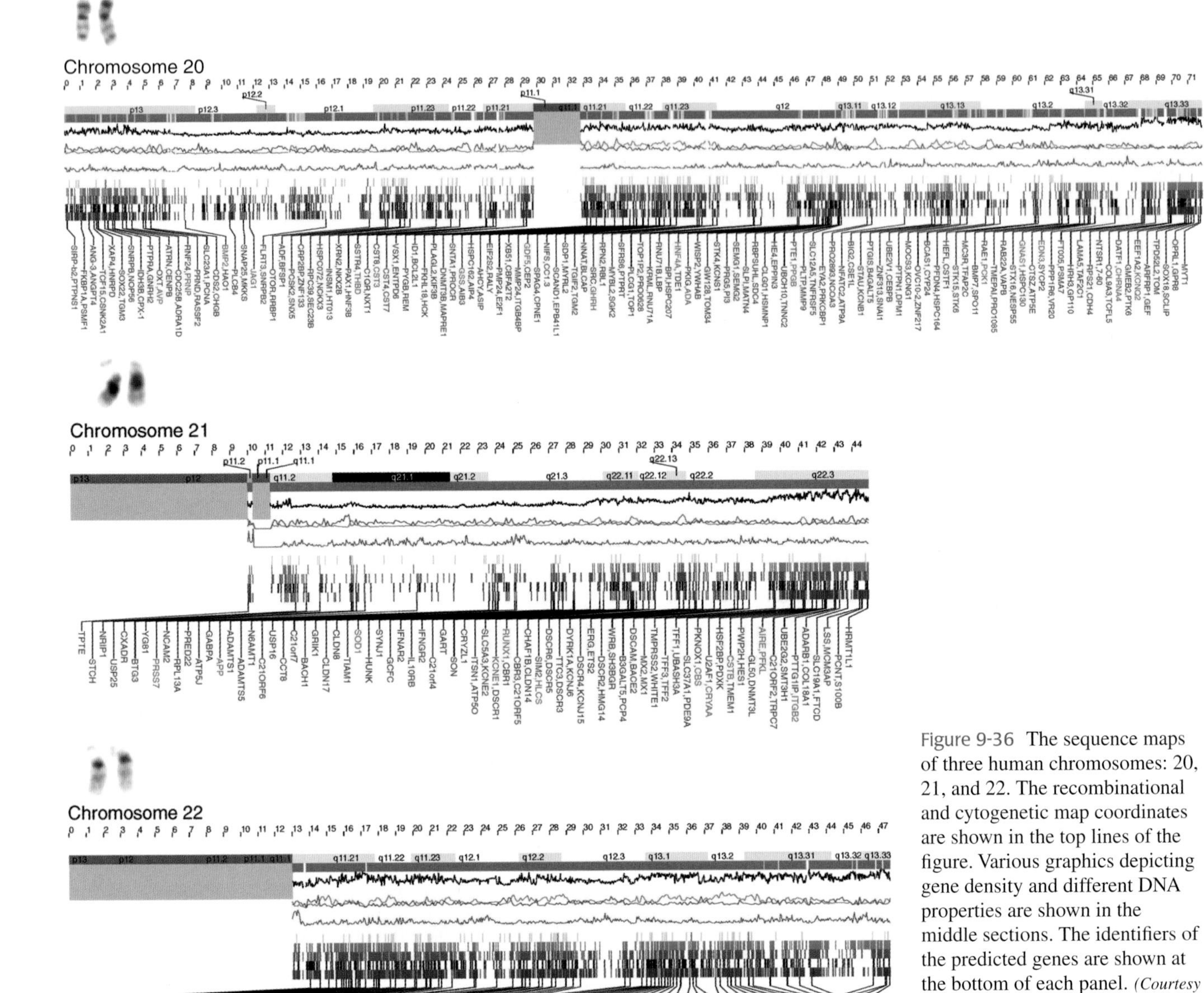

Figure 9-36 The sequence maps of three human chromosomes: 20, 21, and 22. The recombinational and cytogenetic map coordinates are shown in the top lines of the figure. Various graphics depicting gene density and different DNA properties are shown in the middle sections. The identifiers of the predicted genes are shown at the bottom of each panel. *(Courtesy of Jim Kent, Ewan Birney, Darryl Leja, & Francis Collins. Adapted from Initial Sequencing and Analysis of the Human Genome. The International Human Genome Sequencing Consortium.* Nature *409:860–921, 2001.)*

about 45 percent, is repetitive and composed of recognizable transposable elements. Indeed, even within the remaining single-copy DNA, a fraction has sequences suggesting that they might be descended from ancient transposable elements that are now immobile and have accumulated random mutations, causing them to diverge in sequence from the canonical transposable elements. Thus, it appears that the majority of our genome is composed of genetic hitchhikers.

There are larger-scale repeat structures involving the standard protein-coding genes of the human genome. Some regions of the genome look very similar to one another. That is, some regions that are over tens of kilobases long are >99 percent identical. It is presumed that these nearly identical copies arose by some sort of duplication mechanism. Thus, the genome contains footprints of previous expansion events.

Only a small portion of the human genome is polypeptide-coding; that is, somewhat less than 5 percent of the human genome encodes exons of mRNAs. Exons are typically small (about 150 bases), while introns are large, many extending more than 1000 bases and some extending more than 100,000 bases. Transcripts are composed of an average of 10 exons, although of course many have substantially more exons. Finally, considerable added diversity in mRNA and polypeptide sequence is generated through splicing variation. Based on current cDNA and EST data, it is likely that 60 percent of human genes have two or more splice variants, and most of these variants are in protein-coding exons. On average, there are about three splice variants per gene. Hence the number of distinct proteins encoded by the human genome is about threefold greater than the number of recognized genes.

Proteins can be grouped into structurally and functionally related families on the basis of amino acid sequence similarities. For a given protein family that is known in many taxa, family size is larger in humans than in invertebrates whose genomes have been sequenced. Proteins are thought of as composed of modular domains, with modules mixed and matched to carry out different roles. Many domains are indicative of specific biological functions. The number of modular domains per protein also seems to be higher in humans than in the other sequenced organisms. In succeeding chapters in the book, we will encounter numerous of these families, and the roles they play in various biological processes. The degree of similarity parallels the confidence we have in relating the function of a known family member to the other unknown members. We will see examples where members of a protein family clearly are "brothers" or "sisters" in the sense that they share all protein domains and act similarly in parallel processes, or act in some coordinated or cooperative mode in the same biological process. We will also encounter examples where the family members are "half-brothers" or "half-sisters," meaning that they share some protein domains (and some functional attributes) but not others. Finally, there are many cases where the level of protein similarity is such that we are in a statistical gray zone, where we don't know if we're really looking at significant relationships (perhaps "second or third cousins") or not. At present, it is very difficult to make sense of such cases from the primary amino acid sequence comparisons, but it is hoped that comprehensive studies of the three-dimensional structures of all protein families (a project sometimes called **structural genomics**) will shed light on these weak relationships by determining the degree to which weakly related primary sequences reflect conserved three-dimensional shape and, by inference, conserved function.

FUNCTIONAL GENOMICS

While the culmination of the genome project in the finished sequence of a genome is indeed cause for scientific celebration, it is truly the point at which things get *really* interesting. Staring (virtually) at the sequence of a genome, we have before our eyes the entire parts list of RNAs and proteins that the organism can produce, the instruction manual for when, where, and at what level to express those RNAs and proteins, and for how each chromosome operates as a subcellular object that goes through rounds of condensation, decondensation, replication, and segregation.

As we have already discussed, we don't yet know how to decipher this molecular instruction manual, but experimental and computational methods for doing so are getting better all the time, and we can expect one day, perhaps in the next few years, to be rather fluent in the languages of the genome. Once we have reached this point of fluency, we will be in the position of having a full list of the genes (the genomic DNA sequence) and gene products (the complete repertoire of RNAs and polypeptides). Then we will want to go to the next level, understanding how these molecules cooperate and interact (under a given set of environmental conditions) to effect all the processes and phenotypes that make up a biological system. As with other aspects of genetic and genomic analysis, genetic studies of the expression and interactions of gene products have been going on *locally* for the past 25 years—that is, a scientist might study one or a few genes at a time. However, now we have the opportunity to explore these subjects *globally,* using genomic approaches to study some aspect of all gene products simultaneously. This global approach to the study of the expression and interaction of gene products is termed **functional genomics.**

-Ome, Sweet -Ome

Just as we talk about the genome as DNA sequences that we want to understand at a global level, there are other complete experimental data sets that we want to create. Following from the example of the genome, where "gene" plus "ome" becomes a word for "all genes," genomics researchers have coined a number of terms to describe other global data sets that they are working on acquiring. This -ome wish list includes:

A description of the transcriptome: The sequence and expression patterns of all transcripts (where, when, how much)

A description of the proteome: The sequence and expression patterns of all proteins (where, when, how much)

A description of the interactome: (a) The complete set of physical interactions between all proteins and all DNA segments, (b) the complete set of physical interactions between all proteins and all RNA segments, and (c) the complete set of physical interactions among all proteins

A description of the phenome: The description of the complete set of phenotypes produced by inactivation of gene function for each gene in the genome

We will not discuss all of these -omes in this section but focus on some of the global techniques that are beginning to be exploited to harvest these data sets.

(a)

(b)

Figure 9-37 Fluorescence detection of binding to DNA microarrays: (a) Array of 1046 cDNAs probed with fluorescently labeled cDNA made from bone-marrow mRNA. Level of hybridization signal follows the colors of the spectrum, with red highest and blue lowest. (b) Affymetrix GeneChip 65,000 oligonucleotide array representing 1641 genes, probed with tissue-specific cDNAs. *(Part a courtesy of Mark Scheria, Stanford University. Image appeared in* Nature Genetics, *vol. 16, June 1997, p. 127, Fig. 1a. Part b courtesy of Affymetrix Inc. Santa Clara, CA. Image generated by David Lockhart. Affymetrix and GeneChip are U.S. registered trademarks used by Affymetrix. This image appeared in* Nature Genetics, *vol. 16, June 1997, p. 127, Fig. 1b.)*

The Use of DNA Chips to Study the Transcriptome and the Interactome

Recombinant DNA analysis relies heavily on membrane hybridization techniques to detect nucleotide sequences homologous to probe DNAs or RNAs, or subject to binding by specific proteins (see Chapter 8). The logic of these probing techniques has been adapted to the needs of genomics for automation and miniaturization of assay methods. DNA chips are the results of this adaptation. The chips contain samples of DNA laid out as a series of microscopic spots bound to a glass "chip" the size of a microscope cover slip. One chip can contain spots of DNA corresponding to all the genes of a complex genome. DNA chips, functional genomics' moral equivalent of a hybridization filter, may revolutionize genetics by permitting the straightforward assay of all gene products in a single experiment. Before considering how DNA chips are used, let's briefly examine how they are constructed.

One protocol is as follows. Robotic machines with multiple printing tips resembling miniature fountain pen nibs deliver microscopic droplets of DNA solution to specific positions (addresses) on the chip. The DNA is dried and treated so that it will bind to the glass. Thousands of samples can be applied to one chip. In one protocol, the

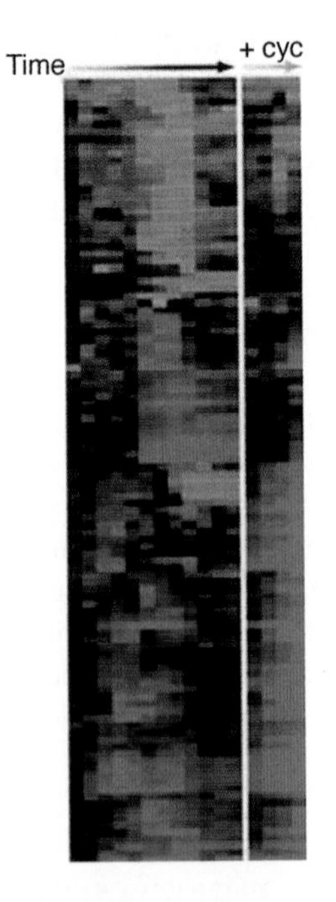

Figure 9-38 Display of gene expression patterns detected by DNA microarrays. Each row is a different gene and each column a different time point. Red means that transcript levels for the gene are higher than at the initial time point; green means transcript levels are lower. The four columns labeled +cyc are from cells grown on cycloheximide, meaning that no protein synthesis was occurring in these cells. *(Mike Eisen and Vishy Iyer, Stanford University. Image appeared in* Nature Genetics, *vol. 18, March 1998, p. 196, Fig. 1.)*

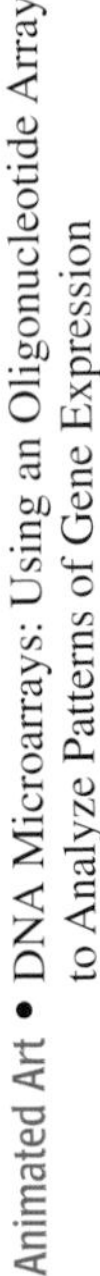

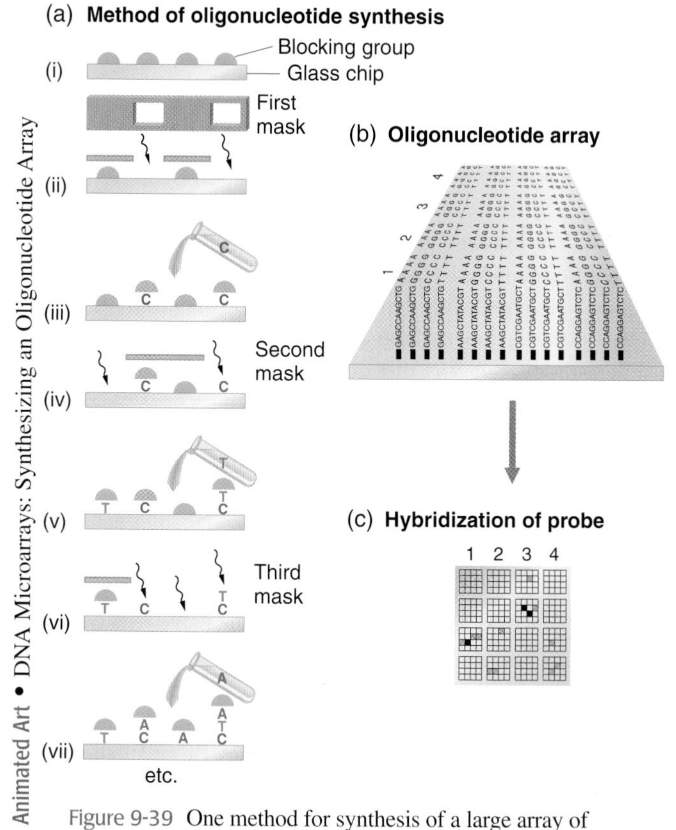

Figure 9-39 One method for synthesis of a large array of oligonucleotides on a glass chip in situ. Nucleotides are deposited one at a time at addresses activated by light shining through a pattern of holes in a mask. Each nucleotide carries a blocking group that prevents further polymerization unless activated by light.

array of DNAs consists of known cDNAs from different genes. Such chips are exposed to a probe, for example, one consisting of the RNA extracted from a particular cell type at a specific stage in development. Fluorescent labels are attached to the probe, and the binding of the probe molecules to the homologous DNA spots on the glass chip is monitored automatically with the use of a laser beam-illuminated microscope. A typical result is shown in Figure 9-37a. In this way, the genes that are active at any stage of development or under any environmental condition can be assayed. Once again, the idea is to identify protein networks that are active in the cell at any particular stage of interest. Figure 9-38 shows an example of a developmental expression pattern generated by assaying this kind of chip.

Another protocol loads the chip with an array of oligonucleotides chemically synthesized one nucleotide at a time on the chip itself (Figure 9-39). The glass slide is first covered with protecting groups that prevent DNA deposition. A mask is placed on the glass with holes corresponding to the sites of deposition. Then laser beams are shone onto the holes where synthesis is to begin. The light knocks off the protecting groups. Then the glass is bathed in the first nucleotide to be deposited. Each nucleotide carries its own protection group, which can be knocked off for the second round of deposition. Hence, by sequential application of the appropriate masks and bathing sequences, arrays of different nucleotides can be built up. For studying genomic function, these oligonucleotides could be unique sequences that are essentially bar codes identifying each gene in the genome. As before, the completed chip is bathed in fluorescent probe isolated at some developmental stage. Binding to such an oligonucleotide chip is shown in Figure 9-37b. These chips are hybridized and analyzed with automated laser beam illuminated microscopy in an analogous fashion to the analysis described in Figure 9-38.

Note that these DNA array methods basically take an approach to genetic dissection that is an alternative to mutational analysis. Under either method the goal is to define the set of genes or proteins that are involved in any specific process under study. Traditional mutational analysis accomplishes this objective by amassing mutations that disrupt a specific process under study; chip technology does it by detecting the specific mRNAs that are expressed in that process.

DNA chips can also be used to detect protein-DNA interactions. For example, a DNA-binding protein can be fluorescently tagged and bound to DNA sequences on a chip to identify specific binding sites within the genome.

The Study of the Interactome Using the Yeast Two-Hybrid System

The yeast two-hybrid system investigates interaction between proteins. The basis for the test is the yeast GAL4 transcriptional activator. This protein has two domains, a DNA-binding domain that binds to the site of transcriptional activation (the GAL4 start of transcription) and an activation domain that will activate transcription but cannot itself bind to DNA to do so. Thus, both must be in close proximity in order for transcription from the GAL4 gene to occur. In the two-hybrid system, the gene for the GAL4 transcriptional activator is divided between two plasmids so that one contains the portion encoding the GAL4 DNA-binding domain and the other the portion encoding the activation domain. On one plasmid, a gene for one protein under investigation is spliced next to the GAL4 DNA-binding domain, and the fusion protein that is produced acts as "bait." On another plasmid, a gene for another protein under test is spliced to the activation domain; the resulting fusion protein is said to be the "target" (Figure 9-40 on the next page). The two hybrid plasmids are then introduced into the same yeast cell. One way of doing this is to mate haploid cells containing bait and target plasmids, and to look for activation of GAL4 transcription by use of a **reporter gene** (the gene for an easily detected protein) that has the GAL4 transcription initiation region fused to it. The only way that the GAL4 binding and activation domains can come together and activate transcription of the reporter gene is if the bait and target proteins

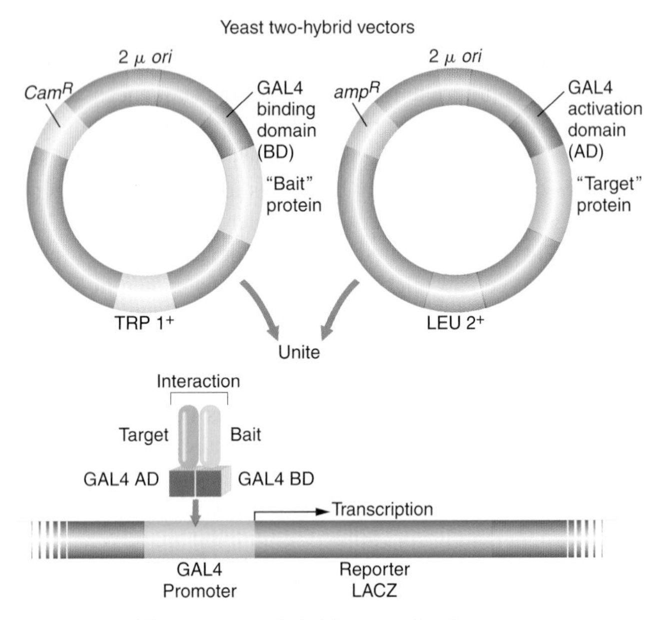

Figure 9-40 The yeast two-hybrid system for detecting gene interaction. The system uses the binding of two proteins under test to restore the function of the GAL4 protein, which activates a reporter gene.

bind to each other, demonstrating a physical interaction between the two proteins being tested. The two-hybrid system can be automated to facilitate large-scale hunting for protein interactions throughout the proteome.

THE GENE CONTENT OF THE HUMAN GENOME

Let's return now to our wager on the human genome project. How many genes does a human have? We still have a couple of years remaining to place our bets, and in the meantime, the genome is moving from its draft status to being completely sequenced at high quality. Further, the quality of gene prediction programs will improve. Because of changes to the sequence as it is completed as well as the improved ability to predict genes, we might find that the current gene estimates might be considerably off from reality. Nonetheless, right now, the numbers hover in the range of 27,000 to 35,000, and have been arrived at by several independent approaches. Many researchers were surprised (and perhaps humbled!) that the predicted human gene count is only greater than the worms, flies, and mustard weed by a factor of 2 or so (compare the data in Table 9-1 with the human gene estimate). Does this mean that a human is only about twice as complex as a little roundworm that's only a millimeter long? The truth is that we don't know how to evaluate this question. For example, with the much greater amount of DNA in the human genome, it is clear that human genes are larger and more complex than *C. elegans* genes. However, the genomic dust will have to settle before we have a clear answer to this numerological quandary.

SUMMARY

1. Genomic analysis takes the approaches of genetic analysis and applies them to the collection of global data sets to fulfill goals such as the mapping and sequencing of whole genomes and the characterization of all transcripts and proteins. Genomic techniques involve high-throughput processing of large sets of experimental material and depends on extensive automation.

2. Whole genome mapping begins with high-resolution genetic maps. In these maps, classical genetic mapping data are merged with data from the localization of molecular markers, generally polymorphisms such as RFLPs, SSLPs, RAPDs, and SNPs.

3. One approach to producing high-resolution genetic maps is by meiotic recombinational mapping. As with purely genetic mapping, the closer molecular markers are to each other or to genetic markers, the smaller the physical interval that needs to be searched for the gene of interest. The alternative to meiotic recombinational mapping is cytogenetic mapping relative to landmarks on the chromosomes, or the use of radiation hybrids to test proximity.

4. High-resolution genetic mapping techniques are indispensable in the localization of genes contributing to heritable diseases. The availability of human high-resolution genetic maps has enabled the identification of candidate genes for many heritable diseases.

5. Physical maps are an important intermediate between high-resolution genetic maps and a complete sequence map of the genome. These maps create overlapping sets of clones that represent the entire genome.

6. The final and highest-resolution level of the genome is a complete map of its DNA sequence. From physical maps, a minimum tiling path of slightly overlapping clones can be generated as substrate for the clone-by-clone sequencing strategy. These clones are also important for taking whole genome assemblies, which essentially are draft assemblies, to the standards of complete high-quality sequence assemblies.

7. The repeat structure of a genome is a major determinant of sequencing strategy. Repeats longer than a few hundred base pairs that are present in nearly identical copies dispersed throughout the genome are a major roadblock to accurate sequence assembly. Clone-by-clone sequencing strategies are able to deal more effectively with repeated sequences, but even here, frequent interspersed repeats can create difficulties in sequence assembly.

8. By current techniques, an individual sequencing reaction results in a "read" of several hundred base pairs at most, a size orders of magnitude smaller than the sizes of genomes (millions to billions of base pairs). This necessitates complex strategies to assemble the individual reads into a representation of the sequence of an entire genome. Whole genome shotgun assembly and clone-by-clone sequence assembly are the two fundamental strategies. They each have made major contributions to genome sequencing.

9. With the availability of a compete genomic sequence, bioinformatic approaches can be used to predict the full array of RNA and protein products encoded by a genome, and to group these products into structural and functional families of related molecules. The techniques of gene prediction and decoding of the information content within the genome are imperfect, and it is likely to be a few years before we have fully determined the transcriptome and proteome of complex genomes.

10. Functional genomics exploits high-throughput global technologies to describe the expression profiles and molecular interactions of the transcriptome and the proteome, and to characterize the phenotypes due to perturbation of all the genes in the genome. Such global studies are in their infancy. We will see applications of functional genomics throughout the remainder of the text.

CONCEPT MAP

Draw a concept map interrelating as many of the following terms as possible. Note that the terms are listed in no particular order.

contig / physical map / BAC / RFLP / SSLP / STS / EST / FISH / recombinant frequency / molecular marker / whole genome shotgun assembly / BLAST

SOLVED PROBLEM

1. A *Neurospora* geneticist has just isolated a new mutation that causes aluminum insensitivity *(al)* in a strain of Oak Ridge background (see Figure 9-4). She wishes to clone the gene by positional cloning and therefore needs to map it. For reasons that we do not need to go into, she suspects that it is located near the tip of the right arm of chromosome 4. Luckily, there are three RFLP markers (1, 2, and 3) available in that vicinity, so the following cross is made:

al (Oak Ridge background)
$\times$ *al*$^+$ (Mauriceville background)

One hundred progeny are isolated and tested for *al* and the six RFLP alleles 1^O, 2^O, 3°, 1^M, 2^M, and 3^M. The results were as follows, where O and M represent the RFLP alleles, and *al* and + represent *al* and *al*$^+$:

RFLP 1	O	M	O	M	O	M
RFLP 2	O	M	M	O	O	M
RFLP 3	O	M	M	O	M	O
al locus	*al*	+	*al*	+	*al*	+
Total of genotype	34	36	6	4	12	8

a. Is the *al* locus in fact in this vicinity?

b. If so, to which RFLP is it closest?

c. How many map units separate the three RFLP loci?

Solution

This is a mapping problem, but with the twist that some of the markers are traditional types (which we have encountered in preceding chapters) but others are molecular markers (in this case, RFLPs). Nevertheless, the principle of mapping is the same as we used before; in other words, it is based on recombinant frequency. In any recombination analysis, we must be clear about the genotype of the parents before we can classify progeny into recombinant classes. In this case, we know that the Oak Ridge parent must contain all O alleles, and Mauriceville all M alleles; therefore, the parents were

$$al\ 1^O\ 2^O\ 3^O \times al^+\ 1^M\ 2^M\ 3^M$$

and knowing this makes determining recombinant classes easy. We see from the data that the parental classes are the two most common (34 and 36). We first of all notice that the *al* alleles are tightly linked to RFLP 1 (all progeny are *al* 1^O or + 1^M). Therefore, the *al* locus is definitely on this part of chromosome 4. There are 6 + 4 = 10 recombinants between RFLP 1 and 2, so these loci must be 10 map units apart. There are 12 + 8 = 20 recombinants between RFLP 2 and 3; that is, they are 20 map units apart. There are 6 + 4 + 12 + 8 = 30 recombinants between RFLP 1 and 3, showing that these loci must flank RFLP 2. Therefore, the map is

al RFLP 1 10 m.u. RFLP 2 20 m.u. RFLP 3

There are evidently no double recombinants, which would have been of the type M O M and O M O.

Notice that there is no new principle at work in the solution of this problem; the real challenge is to understand the nature of RFLPs and to translate this understanding into genotypes from which to study recombination. If you still don't understand RFLPs, you might ask yourself how we assess which RFLP alleles are present.

SOLVED PROBLEM

2 Duchenne muscular dystrophy (DMD) is an X-linked recessive human disease affecting muscles. Six small boys had DMD, together with various other disorders, and they were found to have small deletions of the X chromosome, as shown here:

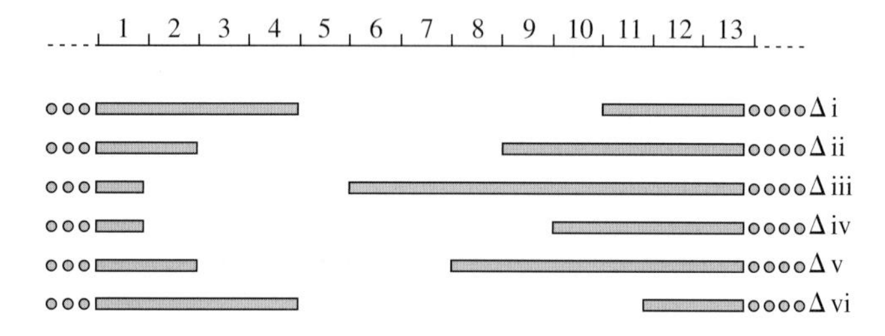

a. On the basis of this information, which chromosomal region most likely contains the gene for DMD?

b. Why did the boys show other symptoms in addition to DMD?

c. How would you use DNA samples from these six boys and DNA from unaffected boys to obtain an enriched sample of DNA containing the gene for DMD, as a prelude to cloning the gene?

Solution

a. The only region that all the deletions are lacking is the chromosomal region labeled 5, so this presumably contains the gene for DMD.

b. The other symptoms probably result from the deletion of the other regions surrounding the DMD region.

c. If the DNA from all the DMD deletions is denatured (i.e., its strands separated) and bound to some kind of filter, the normal DNA can be cut by shearing or by restriction-enzyme treatment, denatured, and passed through the filter containing the deleted DNA. Most DNA will bind to the filter, but the region-5 DNA will pass through. This process can be repeated several times. The filtrate DNA can be cloned and then used in a FISH analysis to see if it binds to the DMD X chromosomes. If not, it becomes a candidate for the DMD-containing sequence.

PROBLEMS

Basic Problems

1. A cloned gene from *Arabidopsis* is used as a radioactive probe against DNA samples from cabbage (which is in the same plant family) digested by three different restriction enzymes. For enzyme 1, there were three radioactive bands on the autoradiogram; for enzyme 2, there was one band; and, for enzyme 3, there were two bands. How can these results be explained?

2. Five YAC clones of human DNA (YAC A through YAC E) were tested for sequence-tagged sites STS1 through STS7. The results are shown in the following table, in which a plus sign shows that the YAC contains that STS:

	STS						
YAC	1	2	3	4	5	6	7
A	+	−	+	+	−	−	−
B	+	−	−	−	+	−	−
C	−	−	+	+	−	−	+
D	−	+	−	−	+	+	−
E	−	−	+	−	−	−	+

a. Draw a physical map showing the STS order.

b. Align the YACs into a contig.

3. You have two strains of the ascomycete fungus *Aspergillus nidulans* that are from different continents and are likely to have accumulated many different genetic variations in the time since they became geographically isolated from each other. You are studying polymorphisms in this strain by RAPD analysis. A RAPD primer amplified two bands in haploid strain 1 and no bands in strain 2. These strains were crossed, and seven progeny were analyzed. The results were as follows:

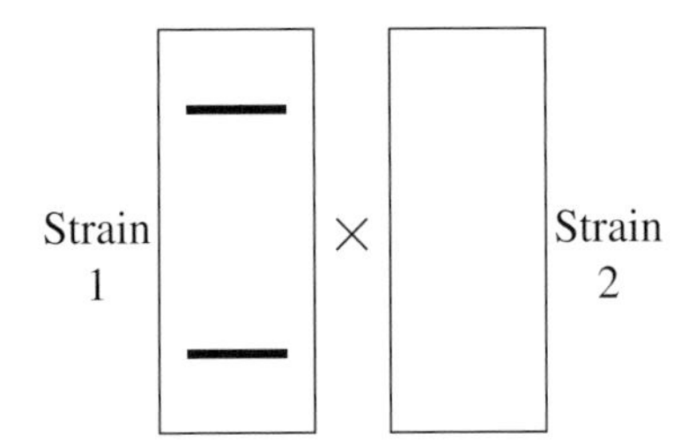

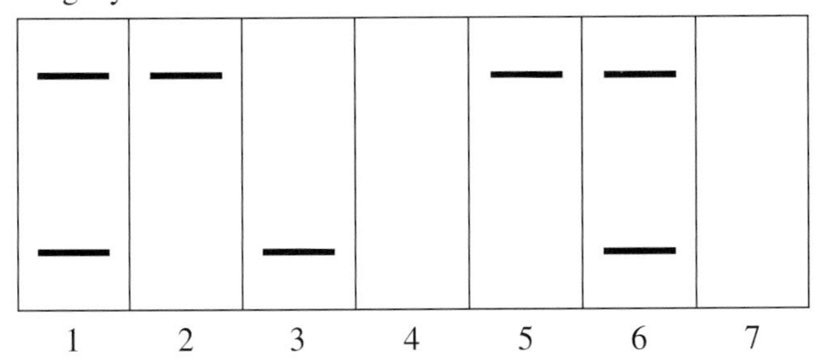

a. Draw diagrams that explain the difference between the parents.

b. Explain the origin of the progeny and their relative frequencies.

c. Draw an example of a single tetrad from this cross, showing RAPD bands.

4. A certain disease is inherited as an autosomal dominant *N*. It was noted that some patients carry chromosomal rearrangements that share the property of always having a chromosomal breakpoint in band 3q3.1 on chromosome 3. Four molecular probes (a through d) are known to hybridize in situ to this band, but their order is not known. In the rearrangements, only probe c hybridizes to one side of the chromosome 3 breakpoint, and probes a, b, and d always hybridize to the other side of the breakpoint.

a. Draw diagrams that illustrate the meaning of these findings.

b. How would you use this information for positional cloning of the normal allele *n*?

c. Once *n* is cloned, how would you use this clone to investigate the nature of the mutations in patients with the same disease but who do not have a chromosomal rearrangement with a breakpoint in 3q3.1?

5. Place the following terms in the order in which these techniques would be employed in taking a genome sequencing project from low to highest resolution. (Not all terms necessarily need to be used.)

a. STS content mapping

b. Clone contig assembly

c. Microsatellite mapping

d. DNA fingerprint mapping

e. DNA sequencing of BAC clones

f. Mapping of phenotypic markers

g. Clone scaffold assembly

h. Paired-end reads

6. Place the following terms in the order in which these techniques would be employed in taking a genome sequencing project from low to highest resolution. (Not all terms necessarily need to be used.)

a. STS content mapping

b. Whole genome shotgun contig assembly

c. Microsatellite mapping

d. DNA fingerprint mapping

e. Whole genome shotgun sequencing

f. Mapping of phenotypic markers

g. Whole genome shotgun scaffold assembly

h. Paired-end reads

7. You have the following sequence reads from a genomic clone of the *Drosophila melanogaster* genome:

Read 1: TGGCCGTGATGGGCAGTTCCGGTG
Read 2: TTCCGGTGCCGGAAAGA
Read 3: CTATCCGGGCGAACTTTTGGCCG
Read 4: CGTGATGGGCAGTTCCGGTG
Read 5: TTGGCCGTGATGGGCAGTT
Read 6: CGAACTTTTGGCCGTGATGGGCAGTTCC

Use these six sequence reads to create a sequence contig of this portion of the *D. melanogaster* genome.

8. In whole genome shotgun sequencing, paired-end reads are used to join contigs together into scaffolds. You have two contigs, called *contig A* and *contig B*. Contig A is 4833 nucleotides long, and contig B is 3320 nucleotides long. Paired-end reads have been made from two ends of a clone containing a 2000-bp genomic insert. The sequencing read from one end of this clone is

210 bp long, and it aligns with nucleotides 4572–4781 of contig A. The sequencing read from the other end of the clone is 342 nucleotides long and aligns with nucleotides 245–586 of contig B. From this information, draw a map of the scaffold containing contig A and contig B, indicating the overall size of the scaffold and the size of the gap between contig A and contig B.

9. Sometimes, cDNAs turn out to be "monsters," that is, fusions of DNA copies of two different mRNAs accidentally inserted adjacent to each other in the same clone. You are suspicious that a cDNA clone from the nematode *Caenorhabditis elegans* is such a monstrosity because the sequence of the cDNA insert predicts a protein with two structural domains not normally observed in the same protein. How would you use the availability of the entire genomic sequence to assess if this cDNA clone is a monster or not?

10. You have sequenced the genome of the bacterium *Salmonella typhimurium,* and you are using BLAST analysis to identify similarities within the *S. typhimurium* genome to known proteins. You find a protein that is 100 percent identical in the bacterium *Escherichia coli.* When you compare nucleotide sequences of the *S. typhimurium* and *E. coli* genes, you find their nucleotide sequences are only 87 percent identical.

a. Explain this observation.

b. What do these observations tell you about the merits of nucleotide versus protein similarity searches in identifying related genes?

Challenging Problems

11. Seven human-rodent radiation hybrids were obtained and tested for six different human genome molecular markers A through F. The results are shown here, where a plus sign indicates the presence of a marker.

	RADIATION HYBRIDS						
Markers	1	2	3	4	5	6	7
A	−	+	−	−	+	+	−
B	+	−	+	−	−	−	−
C	+	−	+	+	−	+	−
D	−	+	−	+	+	+	−
E	+	−	−	+	+	−	+
F	+	−	−	+	+	−	+

a. What marker linkages are suggested by these results?

b. Is there any evidence of markers being on separate chromosomes? Justify your answer.

12. From in situ hybridizations, five different YACs containing genomic fragments were known to hybridize to one specific chromosome band of the human genome. Rare-cutting restriction enzymes, which have 8-bp recognition sequences and which cut on average once in every 64,000 bp, can be used to digest genomic DNAs. Genomic DNA was digested with a rare-cutter restriction enzyme, and radioactively labeled YACs were each hybridized to blots of the digest. The autoradiogram was as follows:

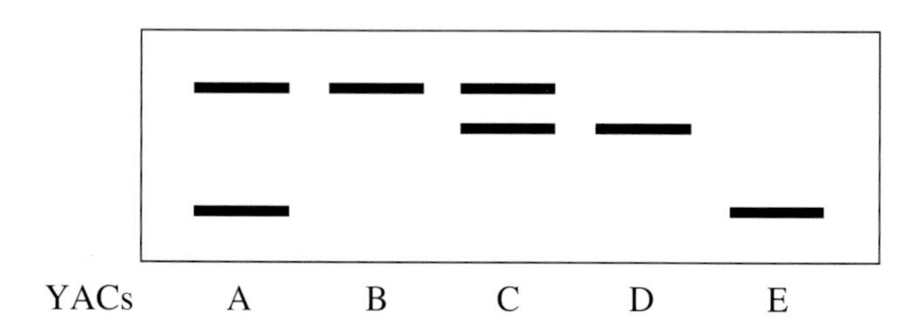

a. Use these results to order the three hybridized restriction fragments.

b. Show the locations of the YACs in relation to the three genomic restriction fragments in part a.

Unpacking the Problem

a. State two types of hybridization used in genetics. What types of hybridizations are used in this question, and what is the molecular basis for such hybridizations? (Draw a rough sketch of what happens at the molecular level during hybridization.)

b. How are in situ hybridizations done in general? How would the specific in situ hybridizations in this question be done (as in the first sentence of the problem)?

c. What is a YAC?

d. What are chromosome bands, and what procedure is used to produce them? Sketch a chromosome with some bands and show how the in situ hybridizations would look.

e. How would five different YACs have been shown to hybridize to one band?

f. What is a genomic fragment? Would you expect the five YACs to contain the same genomic fragment or different ones? How do you think these genomic fragments were produced (what are some general ways of fragmenting DNA)? Does it matter how the DNA was fragmented?

g. What is a restriction enzyme?

h. What is a rare cutter?

i. Why were the YACs radioactively labeled? (What does it mean to radioactively label something?)

j. What is an autoradiogram?

k. Write a sentence that uses the terms *DNA, digestion, restriction enzyme, blot,* and *autoradiogram.*

l. Explain exactly how the pattern of dark bands shown in the question was obtained.

m. Approximately how many kilobases of DNA are in a human genome?

n. If human genomic DNA were digested with a restriction enzyme, roughly how many fragments would be produced? Tens? Hundreds? Thousands? Tens of thousands? Hundreds of thousands?

o. Would all these DNA fragments be different? Would most of them be different?

p. If these fragments were separated on an electrophoretic gel, what would you see if you added a DNA stain to the gel?

q. How does your answer to the preceding question compare with the number of autoradiogram bands in the diagram?

r. Part a of the problem mentions "three hybridized restriction fragments." Point to them in the diagram.

s. Would there actually be any restriction fragments on an autoradiogram?

t. Which YACs hybridize to one restriction fragment and which YACs hybridize to two?

u. How is it possible for a YAC to hybridize to two DNA fragments? Suggest two explanations, and decide which makes more sense in this problem. Does the fact that all the YACs in this problem bind to one chromosome band (and apparently to nothing else) help you in deciding? Could a YAC hybridize to more than two fragments?

v. Distinguish the use of the word *band* by cytogeneticists (chromosome microscopists) from the use of the word *band* by molecular geneticists. In what way do these uses come together in this problem?

13. You have the following sequence reads from a genomic clone of the *Homo sapiens* genome:

Read 1: ATGCGATCTGTGAGCCGAGTCTTTA
Read 2: AACAAAAATGTTGTTATTTTTATTTCAGATG
Read 3: TTCAGATGCGATCTGTGAGCCGAG
Read 4: TGTCTGCCATTCTTAAAAACAAAAATGT
Read 5: TGTTATTTTTATTTCAGATGCGA
Read 6: AACAAAAATGTTGTTATT

a. Use these six sequence reads to create a sequence contig of this portion of the *H. sapiens* genome.

b. Translate the sequence contig in all possible reading frames.

c. Go to the BLAST page of NCBI (http://www.ncbi.nlm.nih.gov/BLAST/—see Appendix B) and see if you can identify the gene that this sequence is part of, by using each of the reading frames as a query for protein-protein comparison (BLASTp).

14. A *Neurospora* geneticist wanted to clone the gene *cys-1,* which was believed to be near the centromere on chromosome 5. Two RFLP markers (RFLP1 and RFLP2) were available in that vicinity, so he made the cross

Oak Ridge *cys-1* × Mauriceville *cys-1*$^+$

Then 100 ascospores were tested for RFLP and *cys-1* genotypes, and the following results were obtained:

RFLP 1	O	M	O	M	O	M
RFLP 2	O	M	M	O	M	O
cys locus	*cys*	+	+	*cys*	*cys*	+
Total of genotype	40	43	2	3	7	5

a. Is *cys-1* in this region of the chromosome?

b. If so, draw a map of the loci in this region, labeled with map units.

c. What would be a suitable next step in cloning the *cys-1* gene?

15. It turns out that some sizable regions of different chromosomes of the human genome are >99 percent nucleotide-identical with one another. These were overlooked during the production of the draft genome sequence of the human genome because of their high level of similarity. Of the mapping techniques discussed in this chapter, which would allow genome researchers to identify the existence of such duplicate regions?

16. A *Caenorhabditis* contig for one region of chromosome 2 is as follows, where A through H are cosmids:

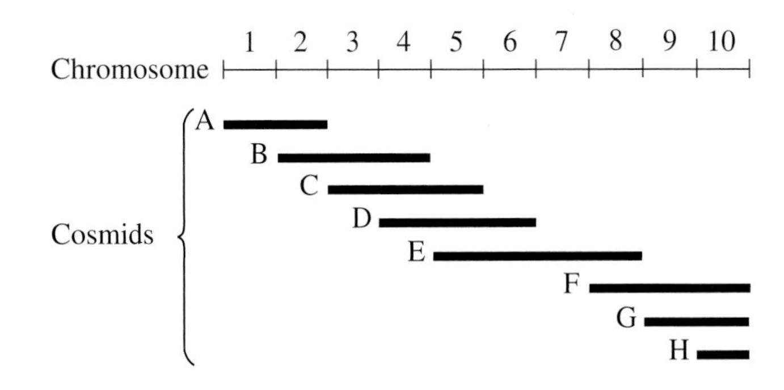

a. A cloned gene *pBR322-x* hybridized to cosmids C, D, and E. What is the approximate location of this gene *x* on the chromosome?

b. A cloned gene *pUC18-y* hybridized only to cosmids E and F. What is its location?

c. Explain exactly how it is possible for both probes to hybridize to cosmid E.

17. The gene for the autosomal dominant disease shown in this pedigree is thought to be on chromosome 4, so five RFLPs (1–5) mapped on chromosome 4 were tested in all family members. The results of the testing are shown below each individual listed in the pedigree. Vertical lines represent the two homologous chromosomes, and the superscripts represent different alleles of the RFLP loci.

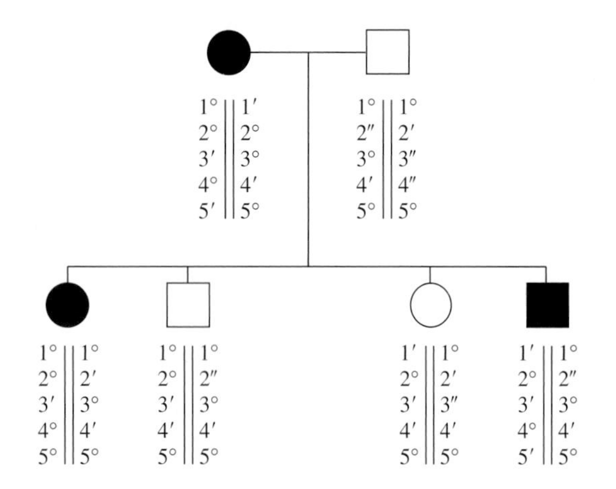

a. Explain how this experiment was carried out.

b. Decide which RFLP locus is closest to the disease gene (explain your logic).

c. How would you use this information to clone the disease gene?

18. Three techniques for physical mapping are clone fingerprinting, STS content mapping, and radiation hybrid mapping.

a. Briefly describe each of these techniques.

b. How might the presence of repetitive transposable element sequences in a genome interfere with the use of each of these techniques?

c. Suggest "work-arounds" for each of these techniques—that is, ways in which the techniques can be used to establish valid physical maps even though a portion of the genome is composed of dispersed repetitive sequences such as transposable elements.

19. Cystic fibrosis is a disease that shows autosomal recessive inheritance. A couple has three children with cystic fibrosis (CF), as shown in the pedigree below. Their oldest son has recently married his second cousin. He has molecular testing done to determine if there is a chance that he may have children with CF. Three probes detecting RFLPs known to be very closely linked to the *CF* gene were used to assess the genotypes in this family. Answer the following questions, describing your logic.

a. Is this man homozygous normal or a carrier?

b. Are his three normal siblings homozygous normal or carriers?

c. From which parent did each carrier inherit the disease allele?

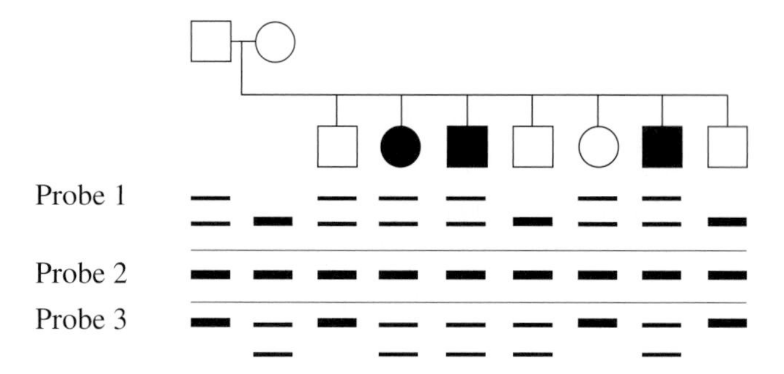

(Problem 19 is from Tamera Western.)

20. Some exons in the human genome are quite small (less than 75 bp long). Identification of such "micro-exons" is difficult because these distances are too short to reliably use open reading frame identification or codon bias to determine if small genomic sequences are truly part of an mRNA and a polypeptide. What techniques of "gene finding" can be used to try to assess if a given region of 75 bp constitutes an exon?

Exploring Genomes: A Web-Based Bioinformatics Tutorial

Determining Protein Structure

Protein function depends on 3D structure, which is in turn dependent on the primary sequence of the protein. Protein structure is determined experimentally by X-ray crystallography or NMR. In the absence of direct experimental information, powerful programs are used to try to fit primary amino acid sequence data to a 3D model. We'll try one in the Genomics tutorial at www.whfreeman.com/mga.

GENE MUTATION: ORIGINS AND REPAIR PROCESSES

10

Key Concepts

1. Mutation is the process whereby genes change from one allelic form to another.
2. Mutational variation consists of the differences that underlie the study of genetics.
3. Mutations can be induced by mutagens or arise spontaneously.
4. Point mutations include single base-pair substitutions, additions, or deletions.
5. The expansion of the number of copies of trinucleotide repeats in some genes can lead to mutant phenotypes.
6. Transposable-element insertion can disrupt gene integrity and cause gene mutation.
7. Cells have sophisticated surveillance and repair pathways that can reverse DNA damage directly or that can exploit complementarity to indirectly replace damaged DNA segments with restored normal counterparts.

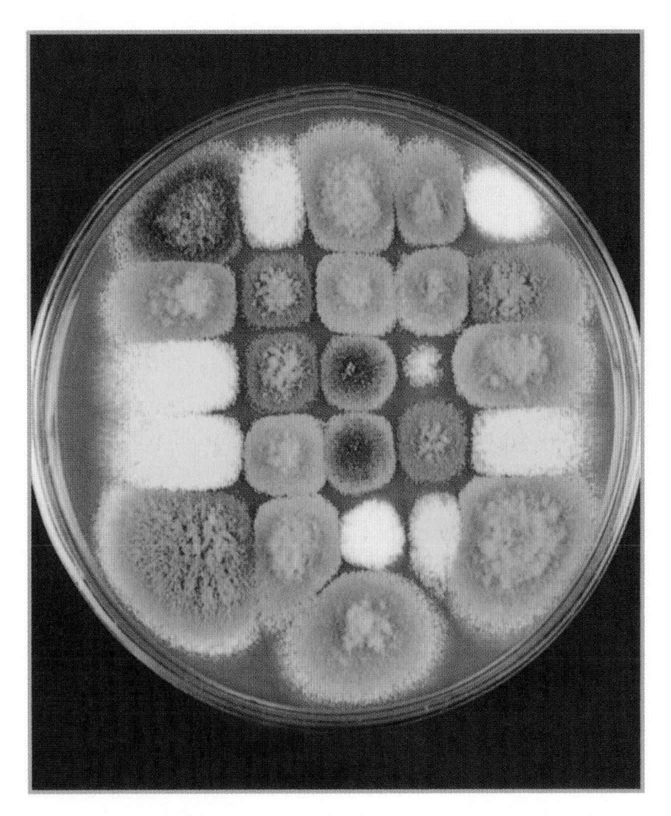

Mutant colonies of the mold *Aspergillus.* The mutations are in genes that control the type and the amount of several different pigments synthesized by this fungus, whose normal color is dark green. *(Courtesy of J. Peberdy, Department of Life Sciences, University of Nottingham, England.)*

A young patient develops a great many small, freckle-like, precancerous skin growths, and is extremely sensitive to sunlight (Figure 10-1). She also has sensory and cognitive neurological deficiencies. A family history is taken, and she is diagnosed with an autosomal recessive disease called xeroderma pigmentosum (XP). Throughout her life, she will be prone to developing pigmented skin cancers. How can we understand the pleiotropic effects exerted by the XP mutation? Several different genes can be mutated to generate the xeroderma pigmentosum phenotype. In a person without the disease, each of these genes contributes to the biochemical processes in the cell that respond to chemical damage to DNA, and *repair* this damage before it leads to the formation of new mutations. We shall first consider the nature of mutation and repair and then turn to genetic diseases such as xeroderma pigmentosum.

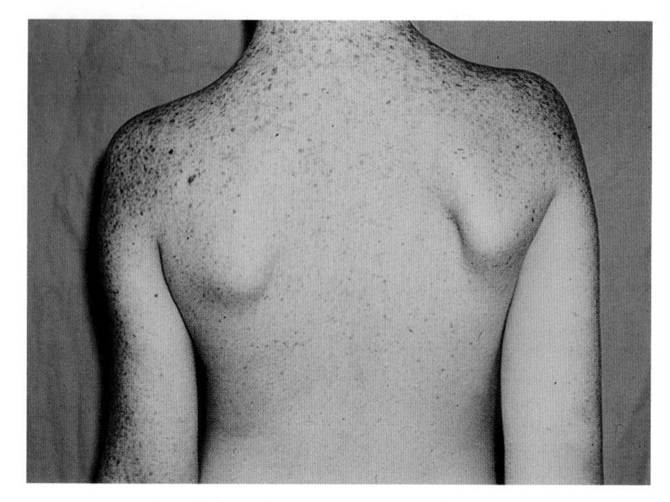

Figure 10-1 Skin cancer in xeroderma pigmentosum. This recessive hereditary disease is caused by a deficiency in an enzyme that helps correct damaged DNA. This enzyme deficiency leads to the formation of skin cancers on exposure of the skin to the UV rays in sunlight. *(Ken Greer/Visuals Unlimited.)*

CHAPTER OVERVIEW

In large part, genetics is the study of inherited differences. Thus, genetic analysis would not be possible without *variants*—organisms that show phenotypic differences in one or more particular characters. In previous chapters we performed many analyses of the inheritance of such variants; now we consider their origin. How do genetic variants arise?

The simple answer to this question is that, in the cellular environment, DNA molecules are not utterly and absolutely stable; each base pair in a DNA double helix has a certain probability of undergoing change. Such hereditary change is called **mutation.** As we shall see, the term *mutation* covers a broad array of different kinds of changes. In the next chapter, we shall consider mutational changes that affect entire chromosomes or large pieces of chromosomes. In the present chapter, we focus on mutational events that take place *within* individual genes. We call such events **gene mutations.**

DNA can be viewed as being subjected to a dynamic tug of war between the chemical processes that damage DNA and that can lead to new mutations and the cellular repair processes that constantly monitor DNA for such damage and that correct these premutational events. We will consider this tug of war by examining three main topics, summarized in Figure 10-2:

- We will examine the kinds of gene alterations that can occur within DNA molecules and the consequences of these various changes on gene structure and function. As indicated at the bottom of Figure 10-2, these events can be as simple as the swapping of one base pair for another. Alternatively, some mutations entail a change in the number of copies of a repeated sequence (such as AGCAGCAGC becoming AGCAGCAGCAGCAGCAGCAGC). Still other mutations are due to the integration of segments of DNA from elsewhere in the genome into the DNA backbone of a gene. Often, these exogenous segments of DNA have specially evolved the capability to integrate elsewhere in the genome and hence have been called *mobile elements* or *transposable* elements.
- We will look at the molecular mechanisms whereby such mutations arise. We will consider the action of certain agents, **mutagens,** that increase the rate at which mutations occur. We will also consider possible molecular mechanisms for the much less frequent (and hence harder to study) but evolutionarily more important mutations that occur "spontaneously." We will see that there are a host of different molecular mechanisms that underlie mutation ranging from the production of highly reactive cellular components through cell metabolism or through the introduction of environmental agents (reactive chemicals or high-energy radiation such as X rays) to mistakes in the DNA replication process or to the enzymatic attack of the DNA backbone by enzymes encoded by transposable elements. As we will see, the consequences of these events for mutational outcome can be quite different.
- We will explore the sophisticated systems that cells have evolved to identify and repair damaged DNA,

CHAPTER OVERVIEW figure

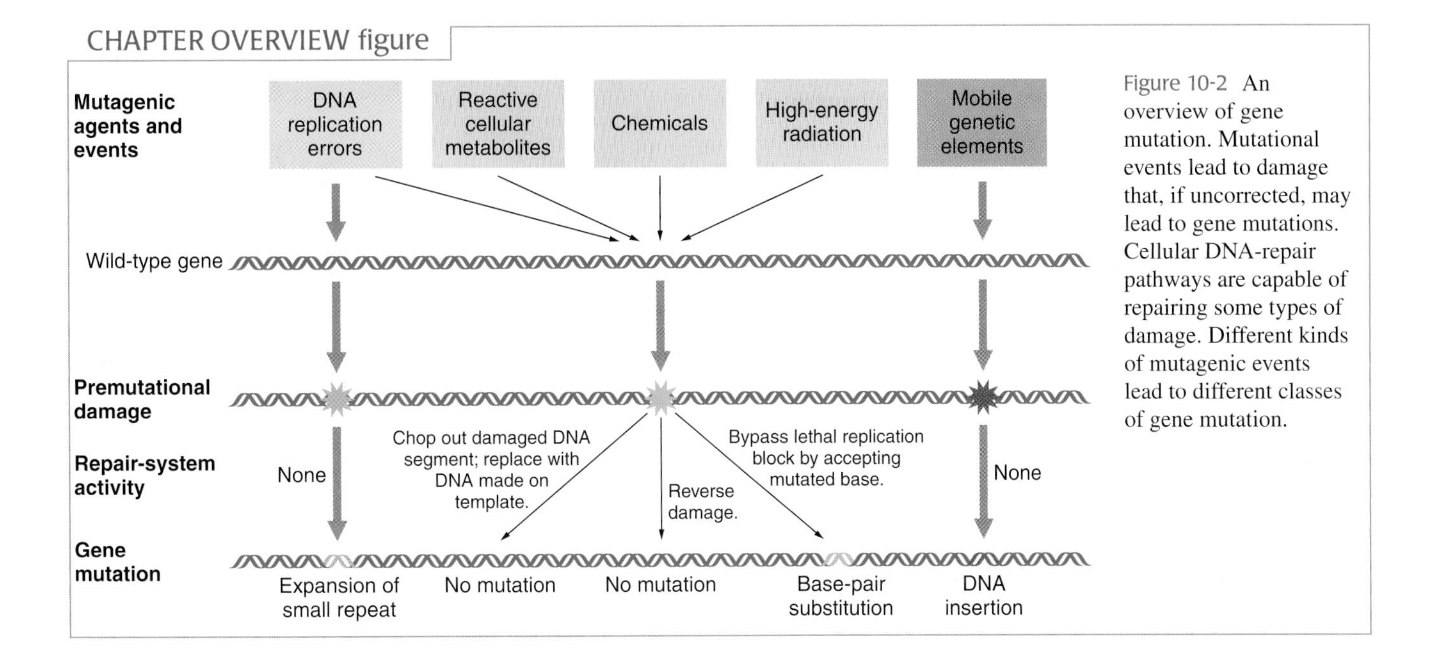

Figure 10-2 An overview of gene mutation. Mutational events lead to damage that, if uncorrected, may lead to gene mutations. Cellular DNA-repair pathways are capable of repairing some types of damage. Different kinds of mutagenic events lead to different classes of gene mutation.

thereby preventing the occurrence of mutations. Most notably, we will see that there are a variety of repair systems and that most of them rely on DNA complementarity to carry out effective and reliable repair by using one DNA strand as a template for the correction of DNA damage that has occurred on the other strand. As we will see, much of the premutational damage that occurs is corrected before it can lead to mutations—changes in the DNA sequence. However, some damage cannot be repaired, and such damage leads to the results that we detect phenotypically as mutational events.

MOLECULAR CATEGORIES OF GENE MUTATIONS

We are going to consider several general classes of gene mutation:

- Mutations affecting single base pairs of DNA. Essentially, these "point" mutations are the minimum changes that can be produced—changing only one "letter" in the "book of DNA." The two general classes of such point mutations are:
    - mutations in which one base pair is swapped for another, and
    - mutations in which an extra base pair is inserted or in which one of the normal base pairs is removed.
- Mutations altering the number of copies of a small repeated sequence within a gene.
- Mutations in which a large block of foreign DNA is incorporated into the normal DNA sequence of a gene.

POINT MUTATIONS

Point mutations typically refer to alterations of single base pairs of DNA or of a small number of adjacent base pairs—that is, mutations that map to a single location, or "point," within a gene. Here we will focus on the point mutations that alter one base pair at a time.

The constellation of possible ways in which point mutations could change a wild-type gene is very large, and it varies according to the particular structure and sequence of the gene. However, it is always true that such mutations are more likely to reduce or eliminate gene function (thus they are loss-of-function mutations) than to enhance it. The reason is simple: it is much easier to break a machine than to alter the way that it works by randomly changing or removing one of its components. Conversely, mutations that increase or alter the type of activity of the gene or change where within the body of a multicellular organism it is expressed (gain-of-function mutations) are much rarer.

The Origin of Point Mutations

Newly arising mutations are categorized as *induced* or *spontaneous*. **Induced mutations** are defined as those that arise after purposeful treatment with **mutagens,** environmental agents that are known to increase the rate of mutations. **Spontaneous mutations** are those that arise in the absence of *known* mutagen treatment. They account for the "background rate" of mutation and are presumably the ultimate source of natural genetic variation that is seen in populations.

The frequency at which spontaneous mutations occur is low, generally in the range of one cell in 10^5 to 10^8.

Therefore, if a large number of mutants is required for genetic analysis, mutations must be induced. The induction of mutations is accomplished by treating cells with mutagens. The most commonly used mutagens are high-energy radiation or specific chemicals; examples of these mutagens and their efficacy are given in Table 10-1. The greater the dose of mutagen, the greater the number of mutations induced, as shown in Figure 10-3. Note that Figure 10-3 shows a *linear* dose response, which is often observed in the induction of point mutations. The molecular mechanisms whereby mutagens act will be covered in subsequent sections.

Recognize that the distinction between induced and spontaneous is purely operational. If we are aware that an organism was **mutagenized,** then we surmise that the bulk of the mutations that arise after this **mutagenesis** treatment (i.e., exposure to a mutagen) were induced by the mutagen applied to the organism. However, this is not true in an absolute sense. The mechanisms that give rise to spontaneous mutations also are in action in this mutagenized organism. In reality, there will always be a subset of mutations recovered after mutagenesis that arose independently of the action of the mutagen. The proportion of mutations that fall into this subset depends on how potent a mutagen is. The higher the rate of induced mutations, the lower the proportion of recovered mutations that are actually "spontaneous" in origin.

Induced and spontaneous mutations arise by generally different mechanisms, and so they will be covered separately. After considering these mechanisms, we shall explore the subject of biological mutation repair. Without these repair mechanisms, the rate of mutation would be so high that cells would accumulate too many mutations to remain viable and capable of reproduction. Thus, the mutational events that do occur are those rare events that have somehow been overlooked or bypassed by the repair processes.

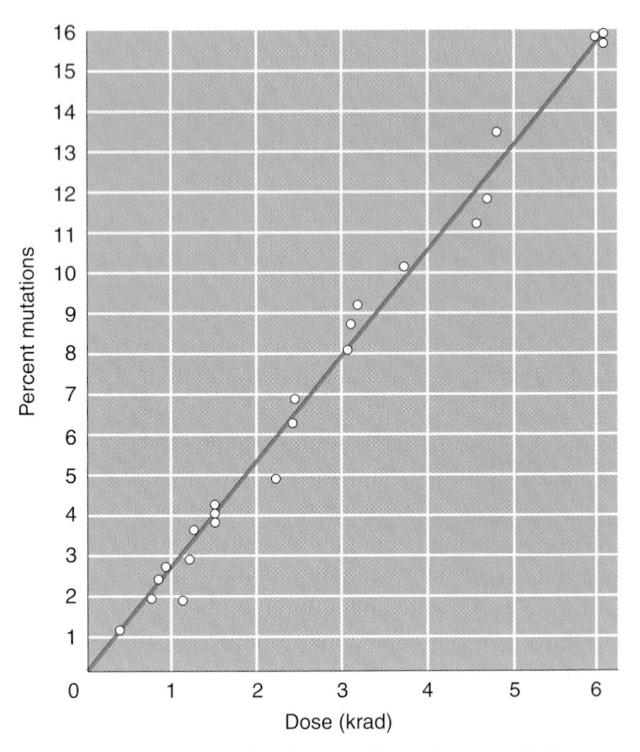

Figure 10-3 Linear relation between X-ray dose to which *Drosophila melanogaster* were exposed and the percentage of mutations (mainly sex-linked recessive lethals).

The Molecular Anatomy of Point Mutations

Point mutations are classified in molecular terms in Table 10-2, which shows the main types of DNA changes and their functional effects at the protein level when such changes occur within the protein-coding region of a gene.

There are two main types of point mutational changes in DNA: *base substitutions* and *base additions* or *deletions.*

TABLE 10-1 Mutation Frequencies Obtained with Various Mutagens in *Neurospora*

Mutagenic Treatment	Exposure Time (minutes)	Survival (%)	Number of *ad-3* Mutants per 10^6 Survivors
No treatment (spontaneous rate)	–	100	~0.4
Amino purine (1–5 mg/ml)	During growth	100	3
Ethyl methanesulfonate (1%)	90	56	25
Nitrous acid (0.05 M)	160	23	128
X rays (2000 r/min)	18	16	259
Methyl methanesulfonate (20 mM)	300	26	350
UV rays (600 erg/mm²/min)	6	18	375
Nitrosoguanidine (25 mM)	240	65	1500
ICR-170 acridine mustard (5 mg/ml)	480	28	2287

Note: The assay measures the frequency of *ad-3* mutants. It so happens that such mutants are red, so they can be detected against a background of white *ad-3*$^+$ colonies.

TABLE 10-2 Point Mutations at the Molecular Level

Type of Mutation	Result and Example(s)
At DNA level	
Transition	Purine replaced by a different purine, or pyrimidine replaced by a different pyrimidine: A·T → C·G C·G → A·T C·G → T·A T·A → C·G
Transversion	Purine replaced by a pyrimidine, or pyrimidine replaced by a purine: A·T → C·G A·T → T·A G·C → T·A G·C → C·G T·A → G·C T·A → A·T C·G → A·T C·G → G·C
Indel	Addition or deletion of one or more base pairs of DNA (inserted or deleted bases are underlined): AAGACTCCT → AAGAGCTCCT AAGACTCCT → AAACTCCT
At protein level	
Synonymous mutation	Codons specify the same amino acid: AGG → CGG Both encode Arg.
Missense mutation	Codon specifies a different amino acid.
Conservative missense mutation	Codon specifies chemically similar amino acid: AAA → AGA Changes basic Lys to basic Arg; does not alter protein function in many cases.
Nonconservative missense mutation	Codon specifies chemically dissimilar amino acid: UUU → UCU Hydrophobic Polar Phenylalanine Serine
Nonsense mutation	Codon signals chain termination: CAG → UAG Change from a codon for Gln to an amber termination codon
Frameshift mutation	AAG ACT CCT → AAG AGC TCC T... one-base-pair addition (underlined) or AAG ACT CCT → AAA CTC CT... one-base-pair deletion (underlined)

Base substitutions are mutations in which one base pair is replaced by another. Base substitutions can be divided into two subtypes: transitions and transversions. To describe these subtypes, we consider how a mutation alters the sequence on one DNA strand (the complementary change will take place on the other strand). A **transition** is the replacement of a base by the other base of the same chemical category (purine replaced by purine: either A to G or G to A; pyrimidine replaced by pyrimidine: either C to T or T to C). A **transversion** is the opposite—the replacement of a base of one chemical category by a base of the other (pyrimidine replaced by purine: C to A, C to G, T to A, T to G; purine replaced by pyrimidine: A to C, A to T, G to C, G to T). In describing the same changes at the double-stranded level of DNA, we must represent both members of a base pair in the same relative location. Thus, an example of a transition would be G·C → A·T; that of a transversion would be G·C → T·A.

Addition or **deletion mutations** are actually of *nucleotide* pairs; nevertheless, the convention is to call them *base*-pair additions or deletions. Collectively, they are termed **indel mutations.** The simplest of these mutations are single-base-pair additions or single-base-pair deletions. (There are examples in which mutations arise through the simultaneous addition or deletion of multiple base pairs at once. As we shall see later in this chapter, mechanisms that selectively produce certain kinds of multiple-base-pair additions or deletions are the cause of certain human genetic diseases.)

The Molecular Consequences of Point Mutations on Gene Structure and Expression

What are the functional consequences of these different types of point mutations? First, consider what happens when a mutation arises in a polypeptide-coding part of a gene. For single-base substitutions, there are several possible outcomes, which are direct consequences of two aspects of the genetic code: degeneracy of the code and the existence of translation termination codons (Figure 10-4a).

- **Synonymous mutations.** The mutation changes one codon for an amino acid into another codon for that same amino acid. Synonymous mutations are also referred to as silent mutations.
- **Missense mutations.** The codon for one amino acid is changed into a codon for another amino acid. Missense mutations are sometimes referred to as nonsynonymous mutations.
- **Nonsense mutations.** The codon for one amino acid is changed into a translation termination (stop) codon.

Synonymous substitutions never alter the amino acid sequence of the polypeptide chain. The severity of the effect of missense and nonsense mutations on the polypeptide differs on a case-by-case basis. For example, if a missense mutation causes the substitution of a chemically similar amino acid, referred to as a **conservative substitution,** then it is likely that the alteration will have a less-severe effect on the protein's structure and function. Alternatively, chemically different amino acid substitutions, called **nonconservative substitutions,** are more likely to produce severe changes in protein structure and function. Nonsense mutations will lead to the premature termination of translation. Thus, they have a considerable effect on protein function. The closer the nonsense mutations are to the 3′ end of the open reading frame, the more plausible it is that the resulting proteins might possess some biological activity. However, nonsense mutations often produce completely inactive protein products.

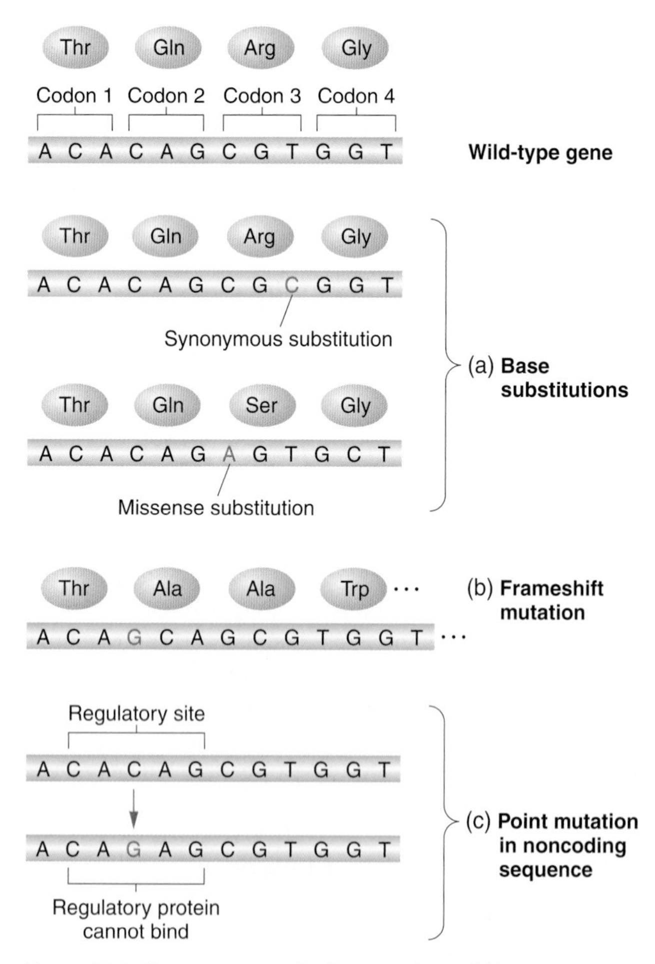

Figure 10-4 Consequences of point mutations within genes. (a) Effects of base substitutions on coding. (b) Effects of frameshift mutations on coding. (c) Effects of point mutations on noncoding sequences.

Like nonsense mutations, indel mutations (base-pair additions or deletions) have consequences on polypeptide sequence that extend far beyond the site of the mutation itself (Figure 10-4b). Because the sequence of mRNA is "read" by the translational apparatus in register, three bases (one codon) at a time, the addition or deletion of a single base pair of DNA changes the reading frame from the site of the base-pair addition or deletion for the remainder of the translation process, until a stop codon in the alternative reading frame is reached. Hence, these lesions are called **frameshift** mutations. These mutations cause the entire amino acid sequence translationally downstream of the mutant site to bear no relation to the original amino acid sequence. Thus, frameshift mutations typically exhibit complete loss of normal protein structure and function.

Now let's turn to those mutations that occur in regulatory and other noncoding sequences (Figure 10-4c). Those parts of a gene that do not directly encode a protein contain many crucial docking sites, interspersed among sequences that are nonessential to gene expression or gene activity. At the DNA level, the docking sites include the sites to which RNA polymerase and its associated factors bind, as well as sites to which specific transcription-regulating proteins must bind. At the RNA level, additional important docking sites include the ribosome-binding sites of bacterial mRNAs, the 5′ and 3′ splice sites for exon joining in eukaryotic mRNAs, and sites that regulate translation and localize the mRNA to particular areas and compartments within the cell.

The ramifications of mutations in parts of a gene other than the polypeptide-coding segments are much harder to predict. In general, the functional consequences of any point mutation (substitution or addition or deletion) in such a region depend on its location and on whether it disrupts (or creates) a docking site. Mutations that disrupt these sites have the potential to change the expression pattern of a gene in regard to the amount of product expressed at a certain time, or in response to certain environmental cues, or in certain tissues. We shall see numerous additional examples of such target sites as we explore mechanisms of gene regulation later on (Chapters 13 through 16). It is important to realize that such regulatory mutations will affect the amount of the protein product of a gene but will not alter the structure of the protein. Alternatively, some docking-site mutations might completely obliterate a required step in normal gene expression (such as polymerase binding or mRNA splicing) and hence totally inactivate the gene product or block its formation.

It is important to keep in mind the distinction between the occurrence of a gene mutation—that is, a change in the DNA sequence of a given gene—and the detection of such an event at the phenotypic level. Many point mutations within noncoding sequences elicit little or no phenotypic change; these phenotypically silent sites may be functionally irrelevant or may overlap in function with other sites within the gene. Nonetheless, with the ever more powerful techniques of genome sequencing, we can imagine coming to the point at which we can survey all gene mutations purely on the basis of changes in genomic sequence, rather than relying on phenotypic manifestations to identify mutational events.

Mechanisms of Point-Mutation Induction

When we examine the array of mutations induced by different mutagens, we see a distinct specificity that is characteristic of each mutagen. Such **mutational specificity** was first noted at the *rII* locus of the bacteriophage T4. Specificity arises from a given mutagen's "preference" both for a certain *type* of mutation (e.g., G·C → A·T transitions) and for certain mutational *sites* called **hot spots.**

Mutagens act through at least three different mechanisms. They can *replace* a base in the DNA, *alter* a base so that it specifically mispairs with another base, or *damage* a base so that it can no longer pair with any base under normal conditions.

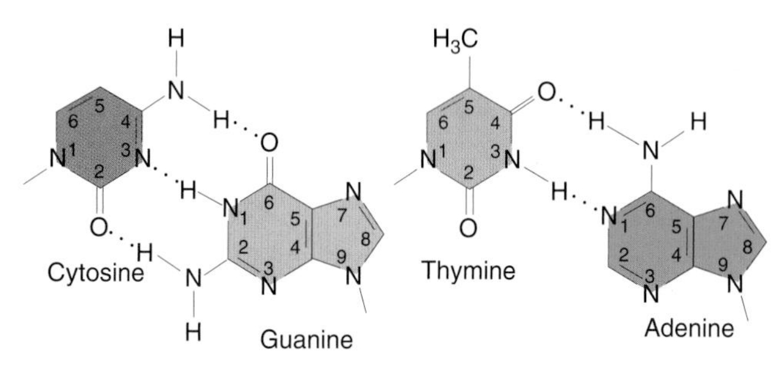

Figure 10-5 Pairing between the normal (keto) forms of the bases.

Base replacement. Some chemical compounds are sufficiently similar to the normal nitrogen bases of DNA that they are occasionally incorporated into DNA in place of normal bases; such compounds are called **base analogs.** Many of these analogs have pairing properties unlike those of the normal bases; thus they can produce mutations by causing incorrect nucleotides to be inserted in the course of replication. To understand the action of base analogs, we must first consider the natural tendency of bases to assume different forms.

All of the bases in DNA can exist in one of several chemical forms, called **tautomers,** which are isomers that differ in the positions of their atoms and in the bonds between the atoms. The forms are in equilibrium. The distribution of tautomers of bases is skewed so that the **keto** form of each base is normally found in DNA (Figure 10-5), whereas the **enol** forms of the bases are rare. A base that is in the enol configuration will spontaneously shift back to the keto tautomer, and, reciprocally, a keto base will very occasionally shift to an enol tautomer. The complementary base pairing of the enols is different from that of the keto forms. Figure 10-6 demonstrates the possible mispairs resulting from

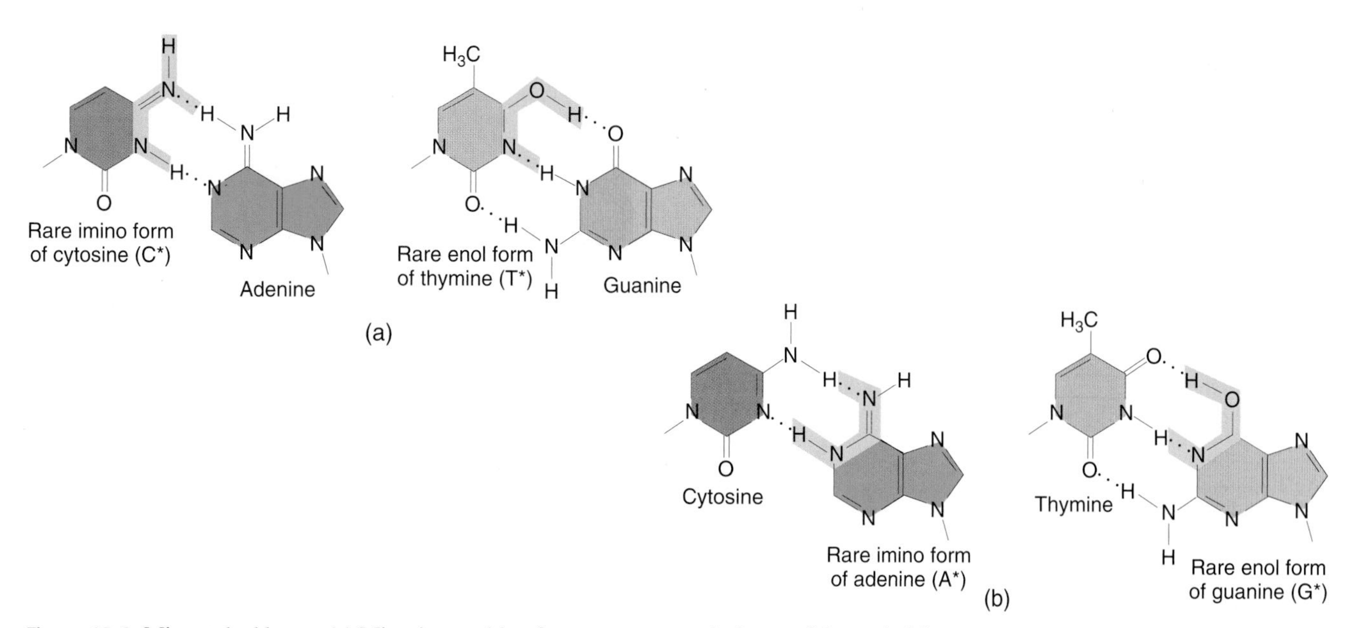

Figure 10-6 Mismatched bases. (a) Mispairs resulting from rare tautomeric forms of the pyrimidines; (b) mispairs resulting from rare tautomeric forms of the purines.

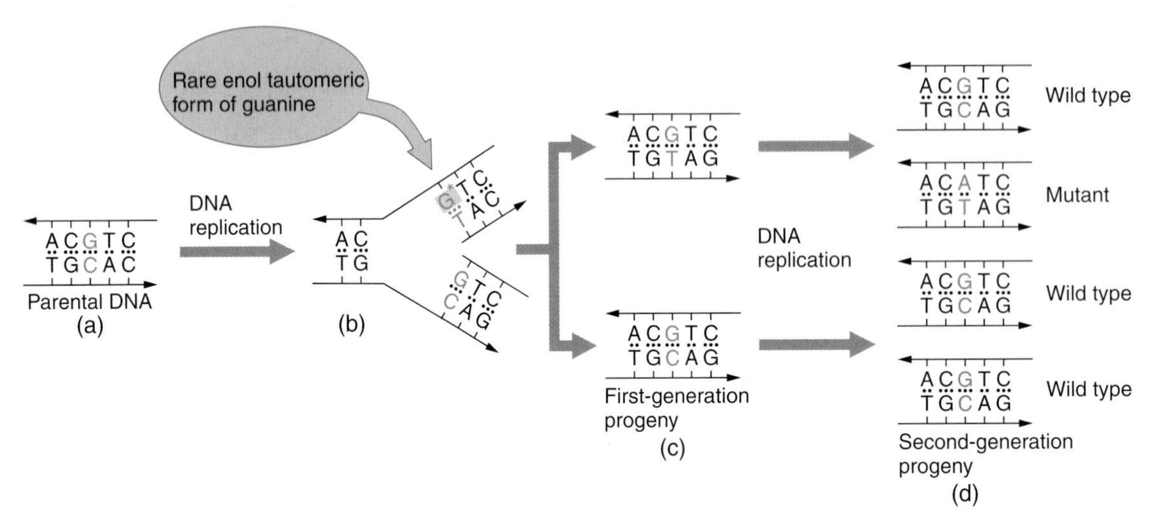

Figure 10-7 Mutation by tautomeric shifts in the bases of DNA. (a) In the example diagrammed, a guanine residue undergoes a tautomeric shift to its rare enol form (G*) at the time of replication. (b) In its enol form, it pairs with thymine. (c and d) In the next replication, the guanine residue shifts back to its more stable keto form. The thymine residue incorporated opposite the enol form of guanine, seen in part b, directs the incorporation of adenine in the subsequent replication, shown in parts c and d. The net result is a G · C → A · T mutation. *(From E. J. Gardner and D. P. Snustad,* Principles of Genetics, *5th ed. © 1984 by John Wiley & Sons, New York.)*

tautomeric shifts to the enol form. Because of such mispairing, the enols are one source of *rare* spontaneous mutations. (Note that, in Figure 10-6 and throughout the text, an asterisk is used to identify a chemically modified base, such as G* for a chemically altered form of guanine.)

For example, assume that a guanine base in DNA changes into its enol form at the moment at which it is copied in the course of replication (it changes back into its keto form soon after). The enol form will bind to an incoming thymine. Hence we can represent the mutagenic process as follows, in which G* is the enol form of guanine:

G · C ⟶ G* · C: first round of replication ⟶ G · T and G · C

G · T ⟶ second round of replication ⟶ G · C and **A · T**

The appearance of the A—T pair represents a G · C → A · T base-pair transition. The same process in the double helix is diagrammed in Figure 10-7.

If, in the course of replication, an *incoming* free base (say, guanine) temporarily enolizes, it will pair with thymine in the DNA, and the consequence is an A · T → G · C transition as follows:

A · T ⟶ first round of replication ⟶ A · T and G* · T

G · T ⟶ second round of replication ⟶ **G · C** and A · T

Mispairs can also arise when bases become spontaneously **ionized.** The mutagen **5-bromouracil (5-BU)** is an analog of thymine that has bromine at the carbon-5 position in place of the CH_3 group found in thymine (Figure 10-8a). Its mutagenic action is based on enolization and ionization. In 5-BU, the bromine atom is not in a position in which it can hydrogen-bond during base pairing. Thus the keto form of 5-BU pairs with adenine, as would thymine; this pairing is shown in Figure 10-8a. However, the presence of the bromine atom significantly alters the distribution of electrons in the base ring; so 5-BU can *frequently* change to either the enol form or an ionized form. The ionized form of 5-BU pairs with guanine (Figure 10-8b). 5-BU causes G · C → A · T or A · T → G · C transitions in the course of

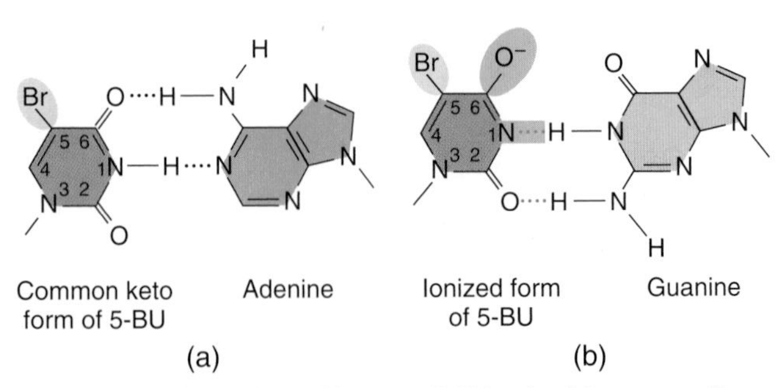

Figure 10-8 Alternative pairing possibilities for 5-bromouracil (5-BU). An analog of thymine, 5-BU can be mistakenly incorporated into DNA as a base. It has a bromine atom in place of the methyl group. (a) In its normal keto state, 5-BU mimics the pairing behavior of the thymine residue that it replaces, pairing with adenine. (b) The presence of the bromine atom, however, causes a relatively frequent redistribution of electrons; so 5-BU can spend part of its existence in the rare ionized form. In this state, it pairs with guanine, mimicking the behavior of cytosine and thus inducing mutations in the course of replication.

replication, depending on whether 5-BU has been enolized or ionized within the DNA molecule or as an incoming base. Hence the action of 5-BU as a mutagen is due to the fact that the molecule spends more of its time in the enol or ion form.

Another base analog widely employed as a mutagen is **2-aminopurine (2-AP),** which is an analog of adenine that can pair with thymine (Figure 10-9a). When protonated, 2-AP can mispair with cytosine (Figure 10-9b). Therefore, when 2-AP is incorporated into DNA by pairing with thymine, it can generate A · T → G · C transitions by mispairing with cytosine in subsequent replications. Or, if 2-AP is incorporated by mispairing with cytosine, then G · C → A · T transitions will result when 2-AP pairs with thymine in subsequent replications. Genetic studies have shown that 2-AP, like 5-BU, is highly specific for transitions.

Base alteration. Some mutagens are not incorporated into the DNA but instead alter a base, causing specific mispairing. Certain commonly used **alkylating agent mutagens,** such as ethyl methanesulfonate (EMS) and nitrosoguanidine (NG), operate by this pathway.

$H_5C_2-O-S(=O)_2-CH_3$

EMS

$O=N-N(CH_3)-C(=N-H)-N(H)-NO_2$

NG

Such agents add alkyl groups (an ethyl group in EMS and a methyl group in NG) to many positions on all four bases. However, the mutagenicity of alkylating agents (their ability to cause mutations) is best correlated with an addition to the oxygen at position 6 of guanine to create an *O*-6-alkylguanine. This alkylation leads to direct mispairing with thymine, as shown in Figure 10-10, and results in G · C → A · T transitions in the next round of replication. Alkylating agents can also modify the bases of incoming nucleotides in the course of DNA synthesis.

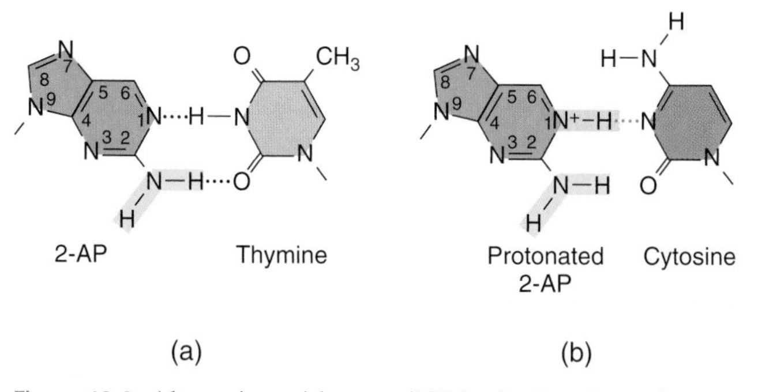

Figure 10-9 Alternative pairing possibilities for 2-aminopurine (2-AP), an analog of adenine. Normally, 2-AP pairs with thymine (a), but, in its protonated state, it can pair with cytosine (b).

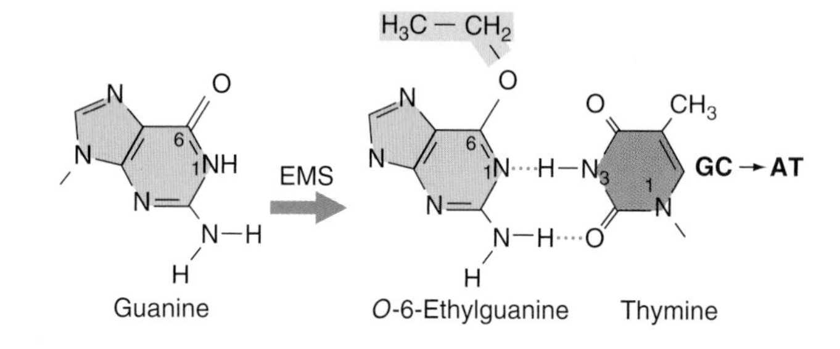

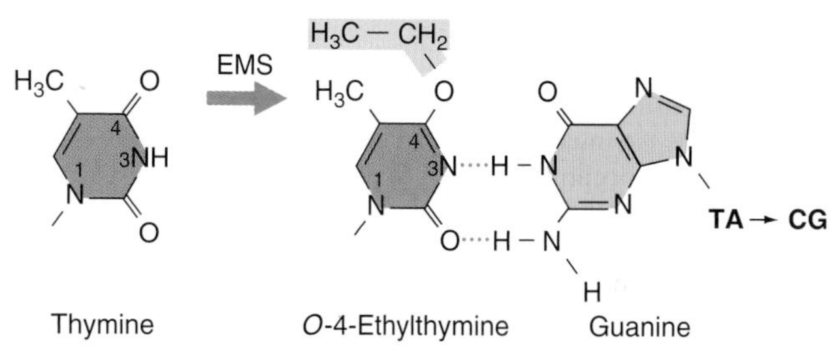

Figure 10-10 Alkylation-induced specific mispairing. The alkylation (in this case, EMS-generated ethylation) of the *O*-6 position of guanine, as well as the *O*-4 position of thymine, can lead to direct mispairing with thymine and guanine, respectively, as shown here. In bacteria, where mutations have been analyzed in great detail, the principal mutations detected are G · C → A · T transitions, indicating that the *O*-6 alkylation of guanine is most relevant to mutagenesis.

The **intercalating agents** are another important class of DNA modifiers. This group of compounds includes **proflavin, acridine orange,** and a class of chemicals termed **ICR compounds** (Figure 10-11a on the next page). These agents are flat planar molecules that mimic base pairs and are able to slip themselves in *(intercalate)* between the stacked nitrogen bases at the core of the DNA double helix (Figure 10-11b). In this intercalated position, an agent can cause single-nucleotide-pair insertions or deletions. Intercalating agents may also stack between bases in single-stranded DNA; in so doing, they may stabilize bases that are looped out during frameshift formation, as depicted in Figure 10-12 on the next page.

Base damage. A large number of mutagens *damage* one or more bases; so no specific base pairing is possible. In other words, a base is chemically modified in such a way that it has no complementary base with which it can hydrogen-bond to form a planar base pair with the approximate molecular dimensions of G · C or A · T base pairs. The result is a replication block, because DNA polymerase cannot continue DNA synthesis past such a damaged template base. In bacterial cells (and probably in eukaryotic cells as well), such replication blocks can be *bypassed* by inserting nonspecific bases. The process requires the activation of a special system, the **SOS system.** The name *SOS* comes from the idea that this system is induced as an emergency response to prevent cell death in the presence of significant DNA damage.

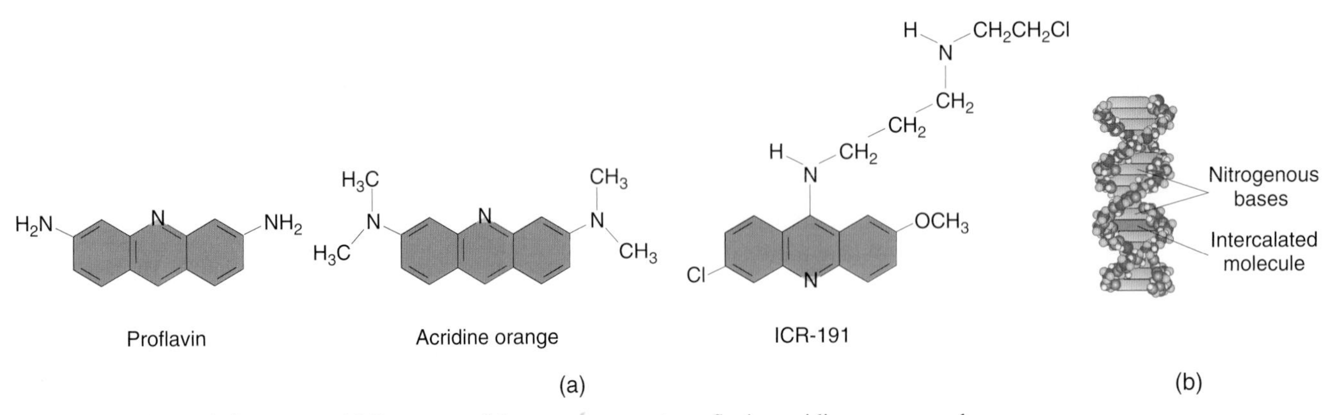

Figure 10-11 Intercalating agents. (a) Structures of the common agents proflavin, acridine orange, and ICR-191. (b) An intercalating agent slips between the nitrogenous bases stacked at the center of the DNA molecule. This occurrence can lead to single-nucleotide-pair insertions and deletions. *(From L. S. Lerman,* Proceedings of the National Academy of Sciences U.S.A. *49, 1963, 94.)*

SOS induction is a mechanism of last resort, allowing the cell to trade death for a certain level of mutagenesis.

Exactly how the SOS bypass system functions is not clear, although in *E. coli* it is known to be dependent on at least three genes, *recA* (which also has a role in general recombination), *umuC,* and *umuD.* Current models for SOS bypass suggest that the UmuC and UmuD proteins combine with the polymerase III DNA replication complex to suppress its "proofreading" function, which evaluates the molecular "quality" of each base pair as synthesis proceeds. In this way, the normally strict specificity is relaxed, a random base is inserted, and replication proceeds past such damaged template bases.

Figure 10-13 shows a model for the SOS bypass system that operates after DNA polymerase III stalls at a type of damage called a T–C photodimer, in which the adjacent bases on the same DNA strand become cross-linked. Because replication can restart downstream from the dimer, a single-stranded region of DNA is generated (Figure 10-13a).

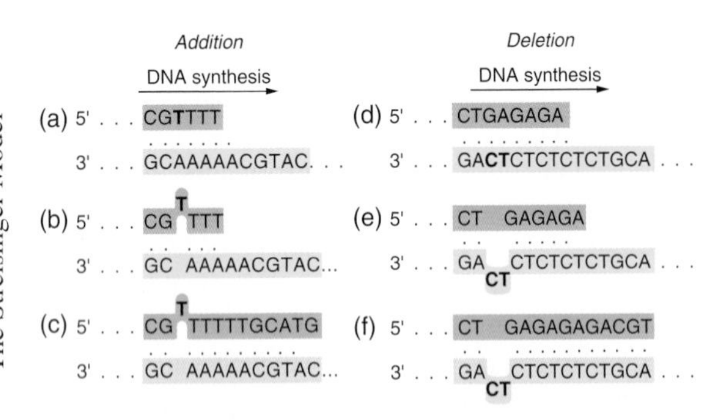

Figure 10-12 A model for indel mutations, resulting in frameshifts. (a–c) In DNA synthesis, the newly synthesized strand slips, looping out one or several bases. This loop is stabilized by the pairing afforded by the repetitive-sequence unit (the adenine bases, in this case). An addition of one base pair, A · T, will result at the next round of replication in this example. (d–f) If, instead of the newly synthesized strand, the template strand slips, then a deletion results. Here the repeating unit is a CT dinucleotide. After slippage, a deletion of two base pairs (C · G and T · A) would result at the next round of replication.

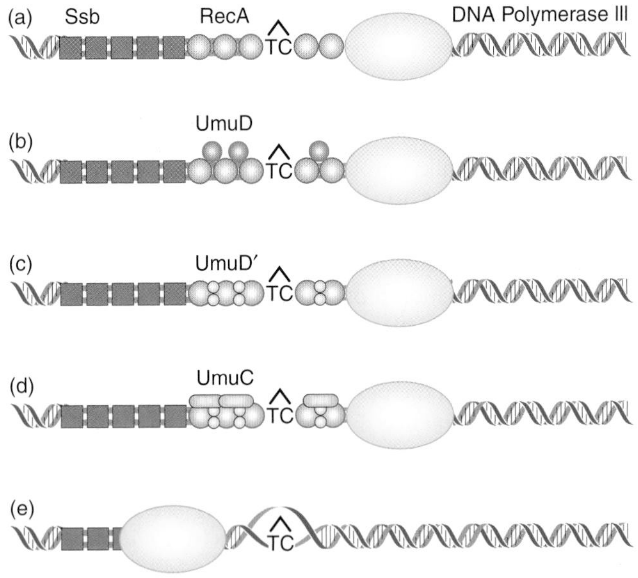

Figure 10-13 (a) DNA polymerase III, shown in blue, stops at a noncoding lesion, such as the T · C photodimer shown here, generating single-stranded regions that attract the Ssb protein (dark purple) and RecA protein (light purple), which forms filaments. (b) The presence of RecA filaments helps to signal the cell to synthesize the UmuD protein (red circles), which is cleaved by RecA to yield UmuD′ (pink circles, part c). (d and e) The UmuC protein (yellow ovals) is recruited to form a complex with UmuD′ that permits DNA polymerase III to proceed past the blocking lesion, resulting in the incorporation of random nucleotides opposite the positions where the template strand is nonfunctional.

This region attracts a stabilizing protein, called single-strand-binding (Ssb) protein, as well as the RecA protein, which forms filaments and signals the cell to synthesize the UmuC and UmuD proteins. The UmuD protein binds to the filaments (Figure 10-13b) and is cleaved by the RecA protein to yield a shortened version termed UmuD′ (Figure 10-13c). In turn, UmuD′ then recruits the UmuC protein to form a complex (Figure 10-13d). This complex allows DNA polymerase III to continue past the dimer, adding bases across from the dimer with a high error frequency (Figure 10-13e).

Mutagens that create bases unable to form stable base pairs are thus dependent on the SOS system for their mutagenic action, because the incorporation of incorrect nucleotides requires the activation of the SOS system. The category of SOS-dependent mutagens is important, because it includes most cancer-causing agents (carcinogens), such as ultraviolet (UV) light and aflatoxin B_1. Indeed, a great deal of work has been done on the relation of mutagens to carcinogens (see Foundations of Genetics 10-1 on the next page). The relation of mutation to cancer will be discussed in detail in Chapter 15.

Because the SOS system targets only single-stranded DNA, it lowers the fidelity of DNA replication only in regions of the genome with otherwise irreparable DNA damage. Consistent with this conclusion, a series of different SOS-dependent mutagens have markedly different mutational specificities; that is, each mutagen induces a unique distribution of mutations. This finding is consistent with the idea that the SOS-associated mutations are generated in response to specific kinds of damaged base pairs. The type of lesion differs in many cases. Some of the most widely studied SOS-inducing lesions include UV photoproducts and apurinic sites.

Ultraviolet light generates a number of photoproducts in DNA. Two different lesions that unite adjacent pyrimidines in the same strand have been most strongly correlated with mutagenesis. These lesions are the cyclobutane pyrimidine photodimer and the 6-4 photoproduct (Figure 10-14). These lesions interfere with normal base pairing; hence, induction of the SOS system is required for mutagenesis. The insertion of incorrect bases across from UV photoproducts is at the 3′ position of the dimer and, more frequently, for 5′-CC-3′ and 5′-TC-3′ dimers. The C → T transition is the most frequent mutation, but other base substitutions (transversions) and frameshifts also are induced by UV light, as are larger duplications and deletions.

Aflatoxin B_1 (AFB_1) is a powerful carcinogen originally isolated from fungal-infected peanuts. Aflatoxin

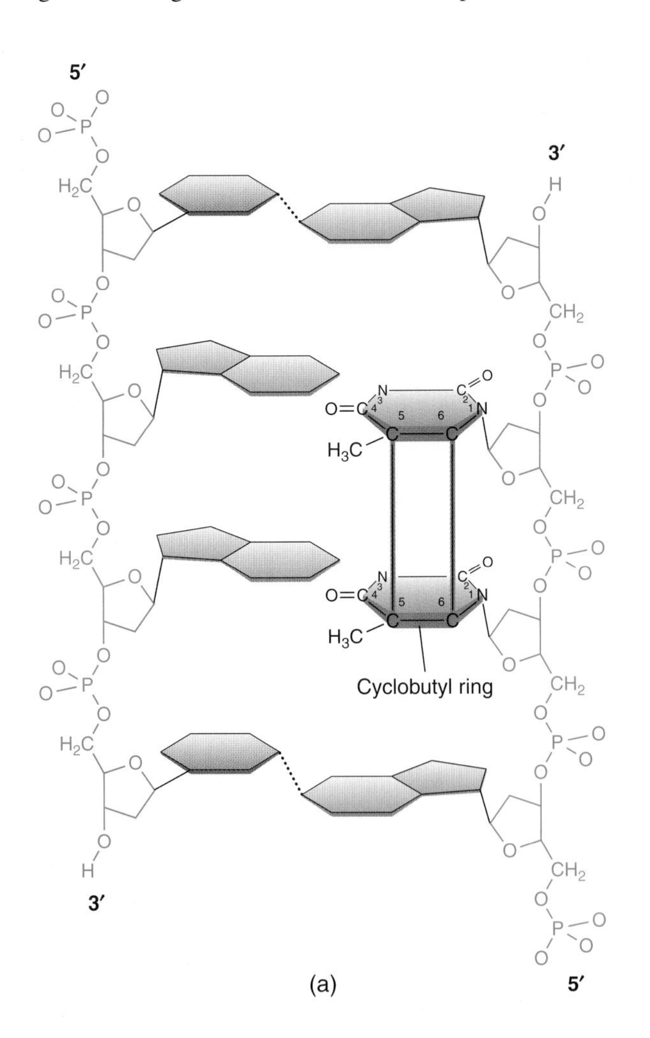

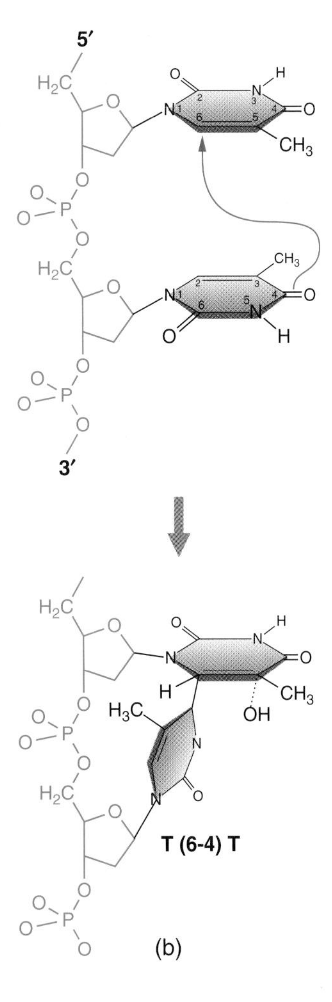

Figure 10-14 (a) Structure of a cyclobutane pyrimidine dimer. Ultraviolet light stimulates the formation of a four-membered cyclobutyl ring (green) between two adjacent pyrimidines on the same DNA strand by acting on the 5,6 double bonds. (b) Structure of the 6-4 photoproduct. The structure forms most prevalently with 5′-CC-3′ and 5′-TC-3′, between the C-6 and C-4 positions of two adjacent pyrimidines, causing a significant perturbation in local structure of the double helix. *(Part a adapted from E. C. Friedberg,* DNA Repair. *© 1985 by W. H. Freeman and Company. Part b from J. S. Taylor et al.)*

FOUNDATIONS OF GENETICS 10-1

Bruce Ames Develops the Ames Test for Evaluating Carcinogens and Mutagens

A huge number of chemical compounds have been synthesized, and many have possible commercial applications. We have learned the hard way that the potential benefits of these applications have to be weighed against health and environmental risks. Thus, having efficient screening techniques to assess some of the risks of a large number of compounds is essential.

An important risk factor of many compounds is as cancer-causing agents (carcinogens). Thus, having valid model systems in which the carcinogenicity of compounds can be efficiently and effectively evaluated is very important. However, using a model mammalian system such as mouse is very slow, time-consuming, and expensive.

In the 1970s, Bruce Ames recognized that there was a strong correlation between the carcinogenicity and the mutagenicity of compounds. He surmised that measurement of mutation rates in bacterial systems would be an effective model for evaluating the mutagenicity of compounds as a first level of detection of potential carcinogens. However, it became clear that the carcinogens themselves were not mutagenic; rather, the carcinogens' metabolites produced in the body were actually the mutagenic agents that acted as carcinogens as well. Typically, these metabolites are produced in the liver, and the enzymatic reactions that converted the carcinogens into the bioactive metabolites did not take place in bacteria.

Ames realized that he could overcome this problem by injecting the test compound into rats and, after an appropriate period of time, removing the livers of the injected rats and treating special strains of the bacterium *Salmonella typhimurium* with extracts of these livers. The special strains of *S. typhimurium* had one of several mutant alleles of a gene responsible for histidine synthesis that were known to "revert" (i.e., return to wild-type phenotype) only by certain kinds of additional mutational events. For example, an allele called TA100 could be reverted to wild type only by a base-substitution mutation, whereas TA1538 could be reverted only by indel mutations resulting in a protein frameshift.

By growing the treated bacteria of each of these strains on petri plates containing medium lacking histidine, only revertant individuals containing the appropriate base substitution or frameshift mutation would grow. The number of colonies on each plate and the total number of bacteria tested would be determined, allowing Ames to measure the frequency of reversion. Compounds that yielded metabolites inducing elevated levels of reversion relative to untreated control liver extracts were then clearly mutagenic and were possible carcinogens. The Ames test thus provided an important way of screening thousands of compounds and evaluating one aspect of their risk to health and the environment. It is still in use today as an important tool for the evaluation of the safety of chemical compounds.

Summary of the procedure used for the Ames test.

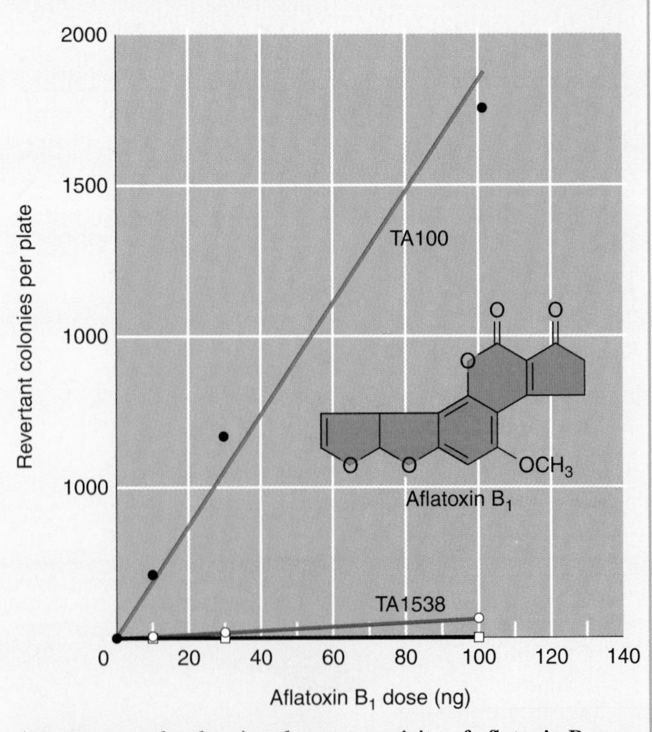

Ames test results showing the mutagenicity of aflatoxin B_1 or its metabolites by base substitution. (From J. McCann and B. N. Ames, in *Advances in Modern Toxicology,* vol. 5. Edited by W. G. Flamm and M. A. Mehlman. © by Hemisphere Publishing Corp., Washington, DC.)

forms an addition product at the N-7 position of guanine (Figure 10-15). This product leads to the breakage of the bond between the base and the sugar, thereby liberating the base and resulting in an **apurinic site** (Figure 10-16). The results of studies with apurinic sites generated in vitro have demonstrated that the SOS bypass of these sites leads to the preferential insertion of an adenine residue across from an apurinic site. This finding allows the prediction that agents that cause depurination at guanine residues should preferentially induce G · C → T · A transversions. For example, with 0 (zero) representing an apurinic site,

$$G \cdot C \longrightarrow 0 \cdot C \longrightarrow 0 \cdot A \text{ plus } G \cdot C$$

$$0 \cdot A \longrightarrow 0 \cdot A \text{ plus } T \cdot \mathbf{A}$$

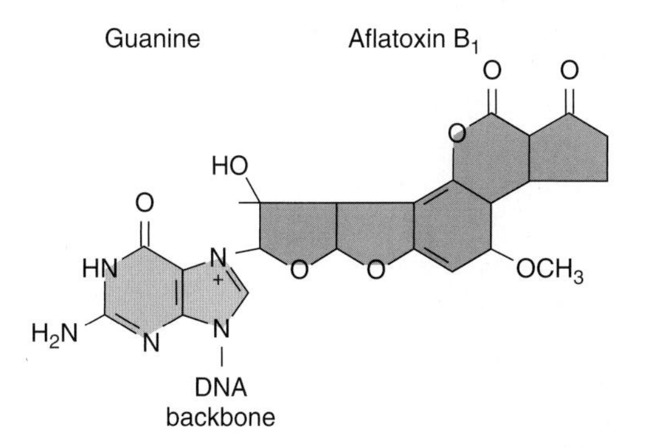

Figure 10-15 The binding of metabolically activated aflatoxin B_1 to DNA.

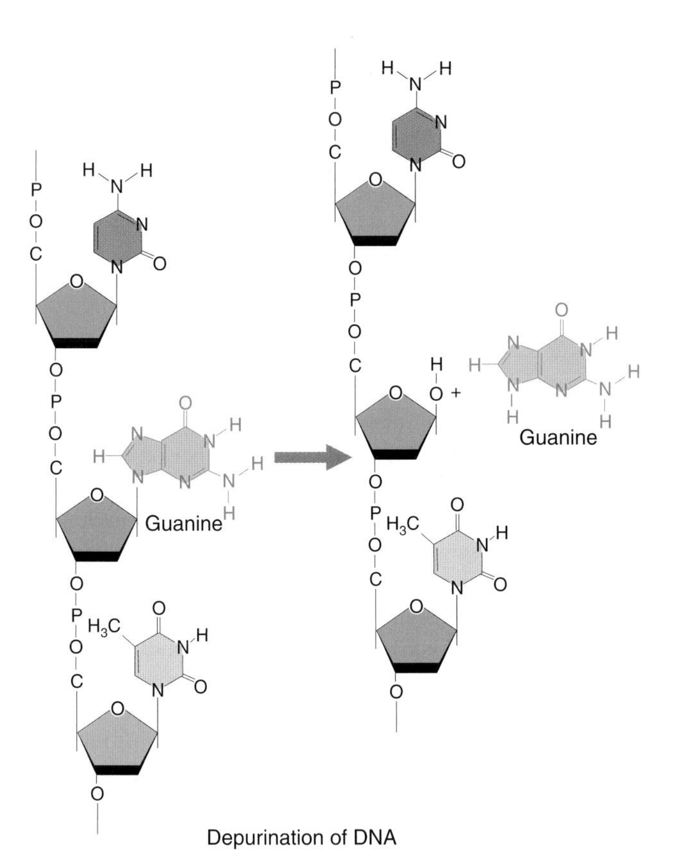

Figure 10-16 The loss of a purine residue (guanine) from a single strand of DNA. The sugar-phosphate backbone is left intact.

MESSAGE

Mutagens induce mutations by a variety of mechanisms. Some mutagens mimic normal bases and are incorporated into DNA, where they can mispair. Others damage bases and either cause specific mispairing or destroy pairing by causing nonrecognition of bases.

Mechanisms of Spontaneous Mutation

The origin of spontaneous hereditary change has always been a topic of considerable interest (see Foundations of Genetics 10-2 on the next page). We now know that spontaneous mutations arise from a variety of sources, including errors in DNA replication, spontaneous lesions, and transposable elements.

Spontaneous mutations are very rare, making it difficult to determine the underlying mechanisms. What sources of insight do we then have into the processes governing spontaneous mutation? Even though they are rare, some selective systems allow spontaneous mutations to be obtained and then characterized at the molecular level—for example, their DNA sequences can be determined. From the nature of the sequence changes, inferences can be made about the processes that have led to the spontaneous mutations.

Spontaneous lesions. Naturally occurring damage, called **spontaneous lesions,** to DNA can generate mutations. Two of the most frequent spontaneous lesions are depurination and deamination, the former being more common.

We learned earlier that aflatoxin induces **depurination;** however, depurination also occurs spontaneously. A mammalian cell spontaneously loses about 10,000 purines from its DNA in a 20-hour cell-generation period at 37°C. If these lesions were to persist, they would result in significant genetic damage because, during replication, the apurinic sites cannot specify any kind of base, let alone the correct one. However, as mentioned earlier in the chapter, under certain conditions, a base can be inserted across from an apurinic site, frequently resulting in a mutation.

The **deamination** of cytosine yields uracil (Figure 10-17a on the next page). Unless corrected, uracil residues will pair with adenine in the course of replication, resulting in the conversion of a G · C pair into an A · T pair (a G · C → A · T transition). Deaminations at certain cytosine positions have been found to be one type of mutational hot spot. DNA sequence analysis of hot spots for G · C → A · T transitions in the *lacI* gene has shown that 5-methylcytosine residues are present at the position of each hot spot.

FOUNDATIONS OF GENETICS 10-2

Salvador Luria and Max Delbrück Show That Bacterial Mutations Are Random

Bacteria and phages have played key roles in genetic research, including the study of mutation. One experiment by Luria and Delbrück in 1943, with the use of bacteria and bacteriophages, was particularly influential in shaping our understanding of the nature of mutation not only in bacteria, but in organisms generally. In the adjoining electron micrograph of some bacteria, the bacterial chromosomes are shown in red. The structure of bacteriophage T1 is shown in the diagram. Note that the DNA is packaged in the head and that the tail fibers anchor the phage to the bacterial membrane. The tail is a syringelike structure that is used to inject the bacteriophage DNA into the bacterium.

It was known at the time that, if *Escherichia coli* bacteria are spread on a plate of nutrient medium in the presence of

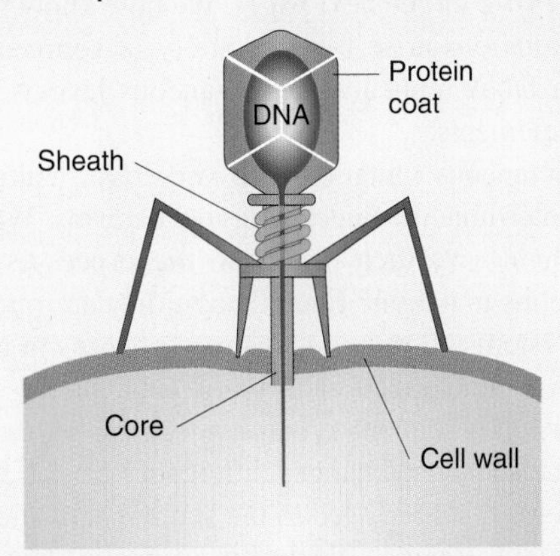

The structure of bacteriophage T1.

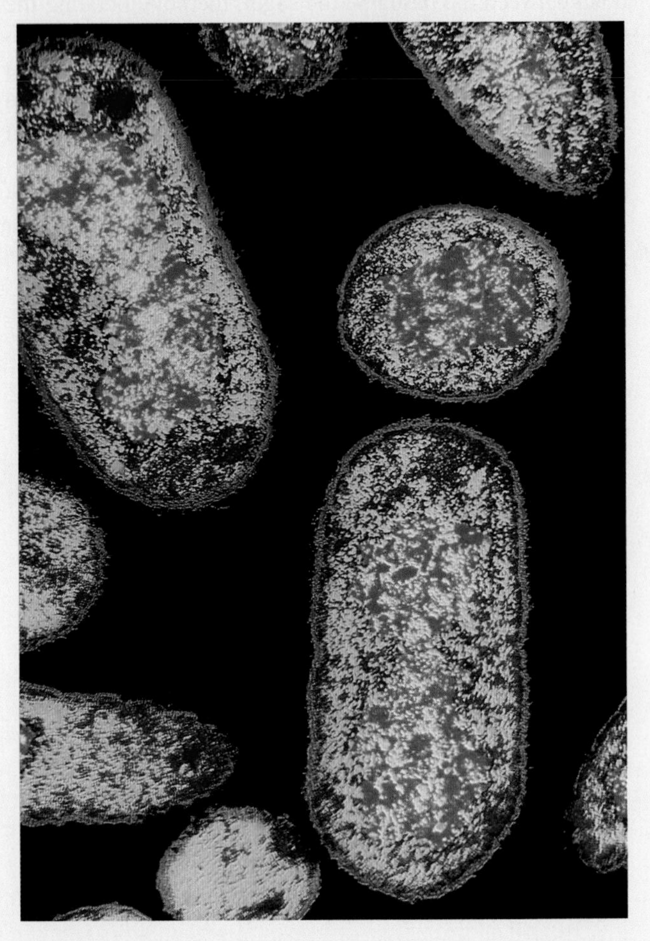

An electron micrograph of *E. coli*. (A. B. Dowsett/Science Photo Library/Photo Researchers.)

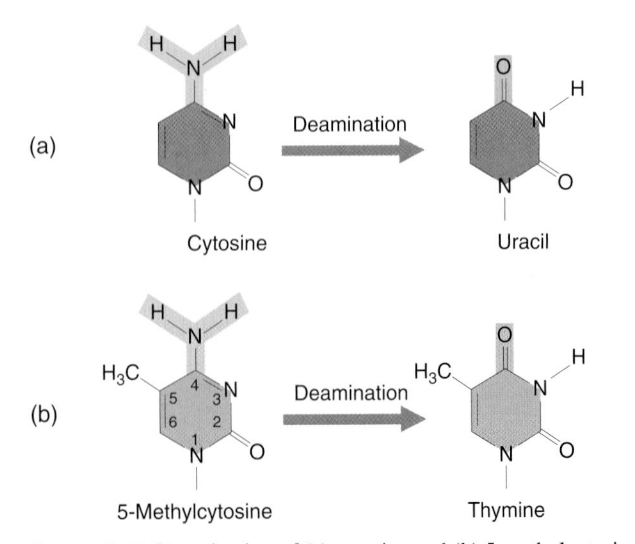

Figure 10-17 Deamination of (a) cytosine and (b) 5-methylcytosine.

(Certain bases in prokaryotes and eukaryotes are normally methylated.) Some of the data from this *lacI* study are shown in Figure 10-18. The height of each bar on the graph represents the frequency of mutations at each of a number of sites. It can be seen that the positions of 5-methylcytosine residues correlate nicely with the most mutable sites. How can 5-methylcytosine residues lead to mutations? The deamination of 5-methylcytosine (Figure 10-17b) generates thymine (5-methyluracil). Thus, C → T transitions generated by deamination are seen frequently at 5-methylcytosine sites.

Oxidatively damaged bases constitute a third type of spontaneous lesion implicated in mutagenesis. Active oxygen species, such as superoxide radicals (O_2), hydrogen peroxide (H_2O_2), and hydroxyl radicals ($OH\cdot$), are produced as by-products of normal aerobic metabolism. These oxygen species can cause oxidative damage to DNA, as well as to precursors of DNA (such as GTP), resulting in

phage T1, the bacteria soon become parasitized and killed by the phages. However, rarely but regularly, colonies were seen that were resistant to phage attack; these colonies were stable and so appeared to be genuine mutants. However, it was not known whether these mutants were produced spontaneously but randomly in time or were induced by the presence of the phage.

Luria reasoned that, if mutations occurred spontaneously, then the mutations might be expected to occur at different times in different cultures; so the resulting numbers of resistant colonies per culture should show high variation (or "fluctuation" in his words). He later claimed that he obtained the idea while watching the fluctuating returns obtained by colleagues gambling on a slot machine at a faculty-dance in a local country club.

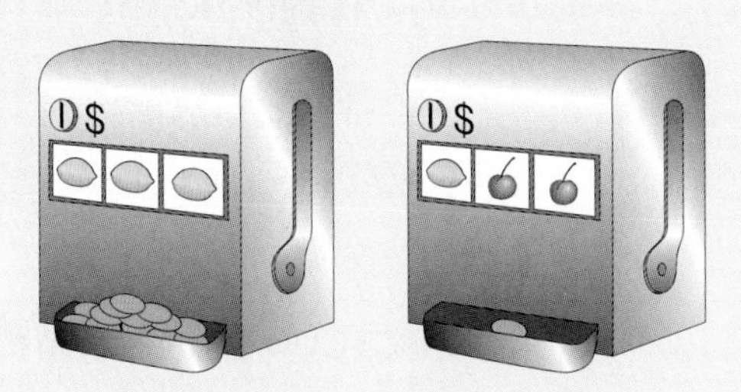

Luria and Delbrück designed their "fluctuation test" as follows. They inoculated 20 small cultures, each with a few cells, and incubated them until there were 10^8 cells per milliliter. At the same time, a much larger culture also was inoculated and incubated until there were 10^8 cells per milliliter. The 20 individual cultures and 20 aliquots of the same size from the large culture were plated in the presence of phage. The 20 individual cultures showed high variation in the number of resistant colonies: 11 plates had 0 resistant colonies, and the remainder had 1, 1, 3, 5, 5, 6, 35, 64, and 107 per plate. The 20 aliquots from the large culture showed much less variation from plate to plate, all in the range of 14 to 26. If the phage were inducing mutations, there was no reason why fluctuation should be higher on the individual cultures, because all were exposed to phage similarly. The best explanation was that mutation was occurring randomly in time: the early mutations gave the higher numbers of resistant cells because they had time to produce many resistant descendants. The later mutations produced fewer resistant cells. This result led to the reigning "paradigm" of mutation; that is, whether in viruses, bacteria, or eukaryotes, mutation can occur in any cell at any time.

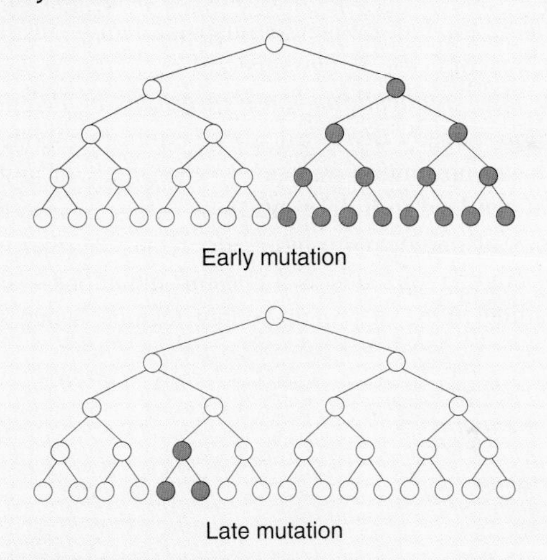

The fluctuation test data also allowed calculation of the rate of mutations per cell division, and the approach is still used today in microbial genetics.

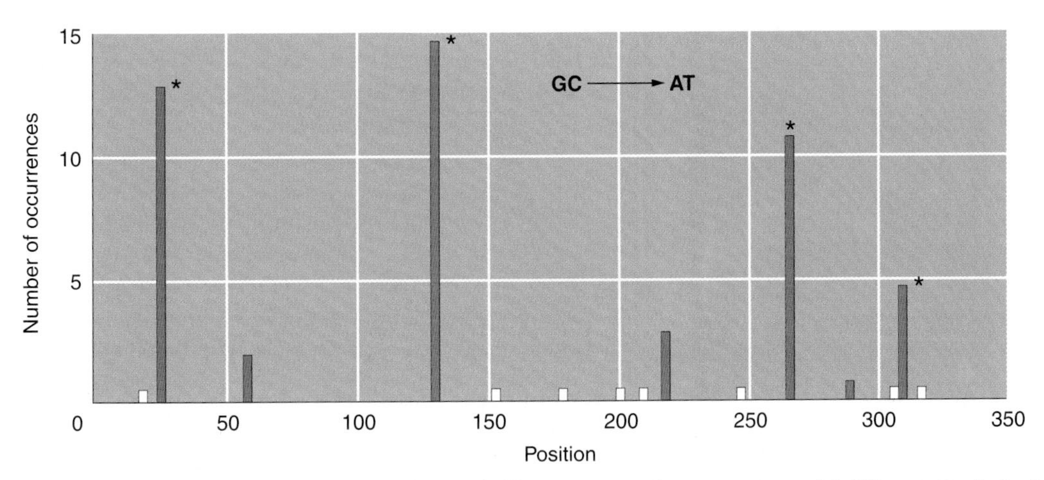

Figure 10-18 5-Methylcytosine hot spots in *E. coli*. Nonsense mutations occurring at 15 different sites in *lacI* were scored. All result from the G · C → A · T transition. The asterisks (*) mark the positions of 5-methylcytosines. The white bars depict sites at which the G · C → A · T change could be detected but at which no mutations occurred in this particular collection. It can be seen that 5-methylcytosine residues are hot spots for the G · C → A · T transition. Of 50 independently occurring mutations, 44 were at the four 5-methylcytosine sites and only 6 were at the 11 unmethylated cytosine residues. *(From C. Coulondre, J. H. Miller, P. J. Farabaugh, and W. Gilbert,* Nature *274, 1978, 775.)*

Thymidine glycol

8-Oxo-7-hydrodeoxyguanosine (8-oxo dG)

Figure 10-19 DNA damage products formed after attack by oxygen radicals. dR = deoxyribose.

mutation. Such mutations have been implicated in a number of human diseases. Figure 10-19 shows two products of oxidative damage. The 8-oxo-7-hydrodeoxyguanosine (8-oxo dG, or "GO") product frequently mispairs with A, resulting in a high level of G → T transversions.

Errors in DNA replication—base insertions or deletions. Although, as already discussed, some errors in replication produce base-substitution mutations, other kinds of replication errors can lead to **indel mutations**—that is, insertions or deletions of one or more base pairs. One way that indel mutations are recognized is as frameshift mutations—shifts in the reading frame of protein-coding regions (when such mutations add or subtract a number of bases not divisible by three). The nucleotide sequence surrounding frameshift-mutation hot spots was determined in the lysozyme gene of phage T4. These mutations often occur at repeated bases. The prevailing model (see Figure 10-12) proposes that indels arise when loops in single-stranded regions are stabilized by the "slipped mispairing" of repeated sequences in the course of replication. In the *E. coli lacI* gene, certain hot spots result from repeated sequences, just as predicted by this model. Figure 10-20 depicts the distribution of spontaneous mutations in the *lacI* gene. Note how one or two mutational sites dominate the distribution. In *lacI,* a four-base-pair sequence (CTGG) repeated three times in tandem in the wild type is the cause of the indel hot spots (for simplicity, only one strand of the DNA is shown):

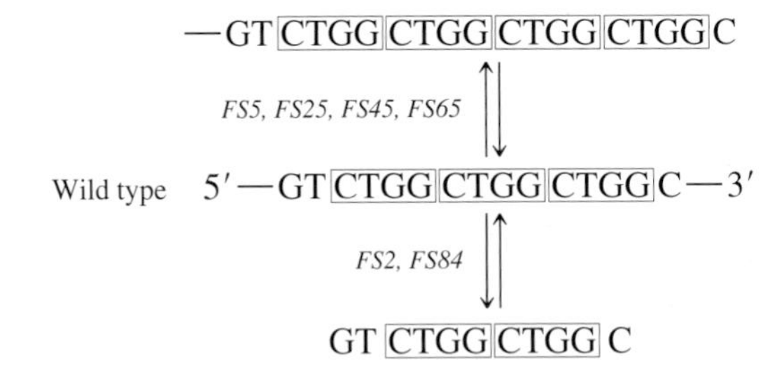

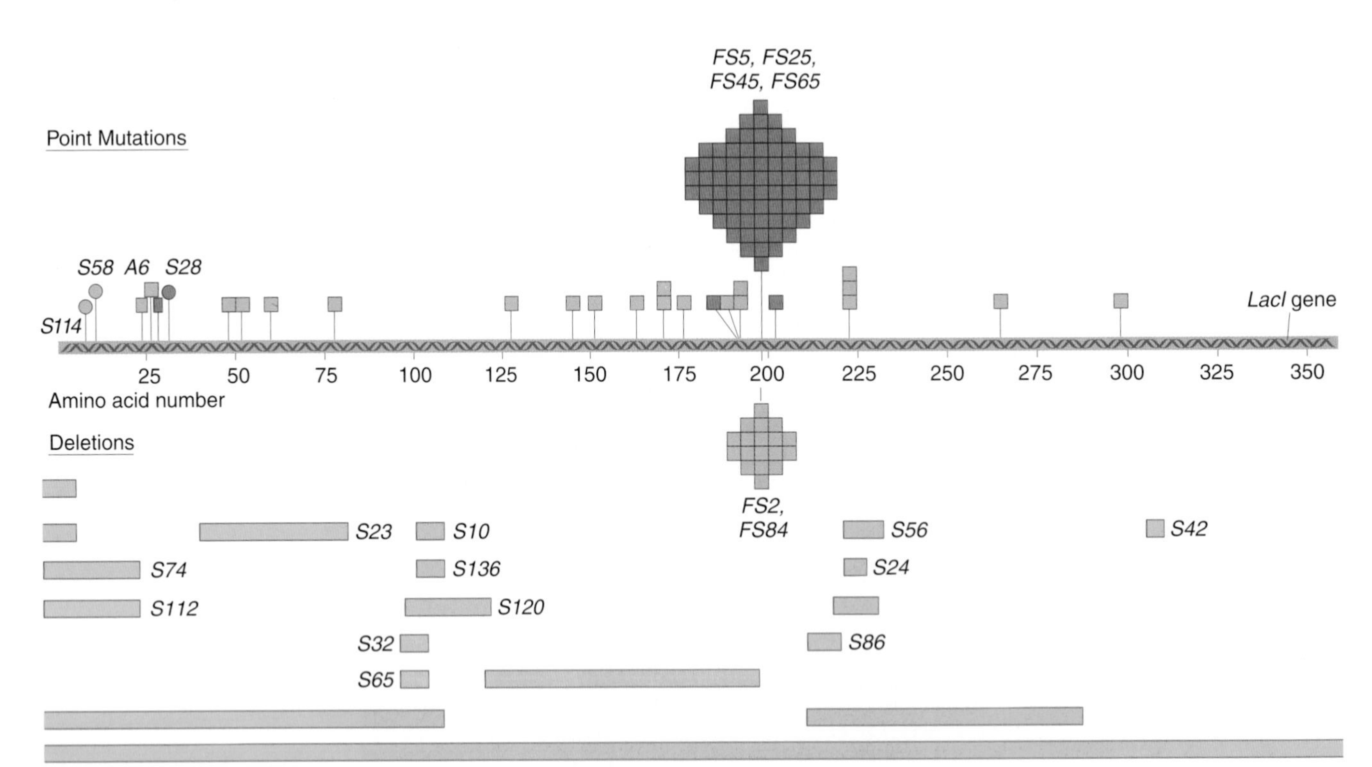

Figure 10-20 The distribution of 140 spontaneous mutations in *lacI.* Each occurrence of a point mutation is indicated by a box. Red boxes designate fast-reverting mutations. Deletions (gold) are represented below. The *lacI* map is given in terms of the amino acid number in the corresponding Lac repressor that *lacI* encodes. Allele numbers refer to mutations that have been analyzed at the DNA sequence level. The mutations *S114* and *S58* (blue circles) result from the insertion of transposable elements. *S28* (red circle) is a duplication of 88 base pairs. *(From P. J. Farabaugh, U. Schmeissner, M. Hofer, and J. H. Miller,* Journal of Molecular Biology *126, 1978, 847.)*

S74, S112 75 bases

CAATTCAGGGTGGTGAATGTGAAACC------CGCGTGGTGAACCAGG

Site (no. of bp)	Sequence repeat	No. of bases deleted	Occurrences
20 to 95	GTGGTGAA	75	2 *S74, S112*
146 to 269	GCGGCGAT	123	1 *S23*
331 to 351	AAGCGGCG	20	2 *S10, S136*
316 to 338	GTCGA	22	2 *S32, S65*
694 to 707	CA	13	1 *S24*
694 to 719	CA	25	1 *S56*
943 to 956	G	13	1 *S42*
322 to 393	None	71	1 *S120*
658 to 685	None	27	1 *S86*

Figure 10-21 Deletions in *lacI.* Deletions occurring in *S74* and *S112* are shown at the top of the diagram. As indicated by the gold bars, one of the sequence repeats (green) and all the intervening DNA have been deleted, leaving one copy of the repeated sequence. All mutations were analyzed by direct DNA sequence determination. *(From P. J. Farabaugh, U. Schmeissner, M. Hofer, and J. H. Miller,* Journal of Molecular Biology *126, 1978, 847.)*

The major hot spot, represented here by the mutations *FS5, FS25, FS45,* and *FS65,* results from the addition of one extra set of the four bases CTGG. The minor hot spot, represented here by the mutations *FS2* and *FS84,* results from the loss of one set of the four bases CTGG.

How can we explain these observations? The model predicts that the frequency of a particular indel depends on the number of base pairs that can form during the slipped mispairing of repeated sequences. The wild-type sequence shown for the *lacI* gene can slip out one CTGG sequence and stabilize this structure by forming nine base pairs (apply the model in Figure 10-12 to the sequence shown for *lacI*). Whether a deletion or an insertion is generated depends on whether the slippage is on the template or on the newly synthesized strand, respectively.

Larger deletions (more than a few base pairs) constitute a sizable fraction of observed spontaneous mutations, as shown in Figure 10-20. Most, although not all, of the deletions are of repeated sequences. Figure 10-21 shows 12 deletions analyzed at the DNA sequence level in the *lacI* gene of *E. coli.* The results of further studies have shown that the longer repeats constitute hot spots for deletions. Duplications of segments of DNA have been observed in many organisms. Like deletions, they often occur at sequence repeats.

How do these deletions and duplications form? Several mechanisms could account for their formation. Deletions may be generated as replication errors. For example, an extension of the frameshift model of slipped mispairing could explain why deletions predominate at short repeated sequences. Alternatively, deletions and duplications could be generated by offset homologous recombination between copies of the repeats.

Spontaneous mutations in humans—trinucleotide repeat diseases. DNA sequence analysis has revealed the gene

MESSAGE

Spontaneous mutations can be generated by several different processes. Spontaneous lesions and replication errors generate most of the base-substitution and indel mutations, respectively.

mutations contributing to numerous human hereditary diseases. Many are of the expected base-substitution or single-base-pair indel type. However, some mutations are more complex but reminiscent of the previously discussed bacterial mutations, allowing us to infer mechanisms that cause these human disorders. A number of these disorders are due to duplications of short repeated sequences.

A common mechanism responsible for a number of genetic diseases is the expansion of a three-base-pair repeat, as in the fragile X syndrome (Figure 10-22 on the next page). For this reason, they are termed **trinucleotide repeat** diseases. Fragile X syndrome is the most common form of inherited mental retardation, occurring in close to 1 of 1500 males and 1 of 2500 females. It is manifested cytologically by a fragile site in the X chromosome that results in breaks in vitro. Fragile X syndrome results from changes in the number of a $(CGG)_n$ repeat in the transcribed but not translated region of the *FMR-1* gene. How does repeat number correlate with the disease phenotype? Humans normally show a considerable variation in the number of CGG repeats in the *FMR-1* gene, ranging from 6 to 54, with the most frequent allele containing 29 repeats. Sometimes, unaffected parents and grandparents give rise to several offspring with fragile X syndrome. The offspring with the symptoms of the disease have enormous repeat numbers, ranging from 200 to 1300. Their unaffected ancestors have also been found to contain increased copy numbers of the repeat, but ranging from only 50 to 200. For this reason, these ancestors have been said to carry *premutations.* The repeats in these premutation alleles are not sufficient to cause the disease phenotype, but they are much more unstable (i.e., readily expanded) than normal alleles, and so they lead to even greater expansion in their offspring. (In general, it appears that the more expanded the repeat number, the greater the instability.)

The proposed mechanism for these repeats is a slipped mispairing in the course of DNA synthesis, just as discussed previously for the one-step expansion of the four-base-pair sequence CTGG at the *lacI* hot spot. However, the extraordinarily high frequency of mutation at the trinucleotide repeats in fragile X syndrome suggests that in human cells, after a threshold level of about 50 repeats, the replication machinery cannot faithfully replicate the correct sequence, and large variations in repeat numbers result.

Other diseases, such as Huntington disease (see Chapter 5) also have been associated with the expansion of trinucleotide repeats in the HD gene. Several general themes apply to these diseases. The wild-type HD gene includes a

Figure 10-22 Expansion of the CGG triplet in the *FMR-1* gene seen in the fragile X syndrome. Normal persons have from 6 to 54 copies of the CGG repeat, whereas those from susceptible families display an increase (premutation) in the number of repeats: normally transmitting males (NTMs) and their daughters are phenotypically normal but display 50 to 200 copies of the CGG triplet; the number of repeats expands to some 200 to 1300 in persons showing full symptoms of the disease.

repeated sequence, often within the protein-coding region, and mutation correlates with a considerable expansion of this repeat region. The severity of the disease correlates with the number of repeat copies.

Huntington disease and Kennedy disease (also called X-linked spinal and bulbar muscular atrophy) result from the amplification of a three-base-pair repeat, CAG. Normal persons have an average of 19 to 21 CAG repeats, whereas affected patients have repeats averaging about 46. In Kennedy disease, which is characterized by progressive muscle weakness and atrophy, the expansion of the trinucleotide repeat occurs in the gene that encodes the androgen receptor.

Myotonic dystrophy, the most common form of adult-onset muscular dystrophy, is yet another example of sequence expansion causing a human disease. Susceptible families display an increase in the severity of the disease in successive generations; this increased severity is caused by the progressive amplification of a CTG triplet at the 3′ end of a transcript. Normal people possess, on average, five copies of the CTG repeat; mildly affected people have approximately 50 copies; and severely affected people have more than 1000 repeats of the CTG triplet.

Properties common to some trinucleotide-repeat diseases suggest a common mechanism by which the abnormal phenotypes are produced. First, many of these diseases seem to include neurodegeneration—that is, cell death within the nervous system. Second, in such diseases the trinucleotide repeats fall within the open reading frames of the transcripts of these genes, leading to expansions or contractions of polypeptide repeats of a single amino acid (e.g., CAG repeats encode a polyglutamine repeat). Thus, it is no accident that these diseases entail expansions of codon-size three-base-pair units. These repeats are thought to encode repeated stretches of a single amino acid (such as polyglutamine for CAGCAGCAG . . .) that serve as interaction domains; that is, they are local regions of the protein that interact with specific domains of certain other proteins. The expansion of the numbers of these single amino acid repeats increases the strength of these interactions or creates opportunities for interactions with proteins with which the normal repeat domains do not interact. These abnormal interactions then, through an unknown cascade of cellular events, lead to the cellular damage that eventually causes neurodegeneration.

If this is so, why don't all of these diseases display identical phenotypes? The protein interactions of these gene products may be different from one another, and certainly the genes are not all expressed in exactly the same tissues, at the same time in the life of the individual person, and in the same amount. Thus, the individual differences in the diseases is due to the specific structures of the encoded proteins and to their patterns of expression within the nervous system.

This explanation cannot hold for all trinucleotide-repeat diseases. For example, in fragile X syndrome, the trinucleotide expansion occurs in the 5′ mRNA-coding region that is untranslated (the 5′ untranslated region). Thus, we cannot ascribe the phenotypic abnormalities of these mutations to their effect on protein repeat motifs. The trinucleotide-repeat basis of the fragile X phenotype remains to be understood.

MESSAGE

Trinucleotide-repeat diseases arise through the expansion of the number of copies of a three-base-pair sequence normally present in several copies, often within the coding region of a gene.

MOBILE ELEMENTS AND GENE MUTATION

The last example of mutation that we will consider is mutation induction through the integration of foreign DNA segments into a gene, thereby disrupting its function. In Chapter 9, we saw that a significant proportion of any genome consists of nomadic DNA sequences that are present at different genomic locations in different isolates of a given species. A mobile element (also called a transposable element) encodes products that catalyze the element's transposition from one location in the genome to another. Mobile elements, which vary in size from a few hundred to a few thousand base pairs of DNA, can produce major alterations in gene expression when they are inserted within a gene (Figure 10-23). Within the coding region of a gene, they typically completely disrupt the gene's function. Within adjacent regulatory sequences, they may partly or fully inactivate the gene, depending on how they interfere with the necessary interactions between regulatory sites and the promoter of the gene. If, as is true for some mobile elements, they carry their own tissue-specific regulatory elements, an insertion of such an element can cause a gene to be expressed in a tissue where it is not ordinarily expressed. Such expression can lead to a gain-of-function dominant mutant phenotype (more on this topic in Chapter 13). For now, we will consider the general properties of mobile elements and the mechanisms by which they move and cause gene mutations.

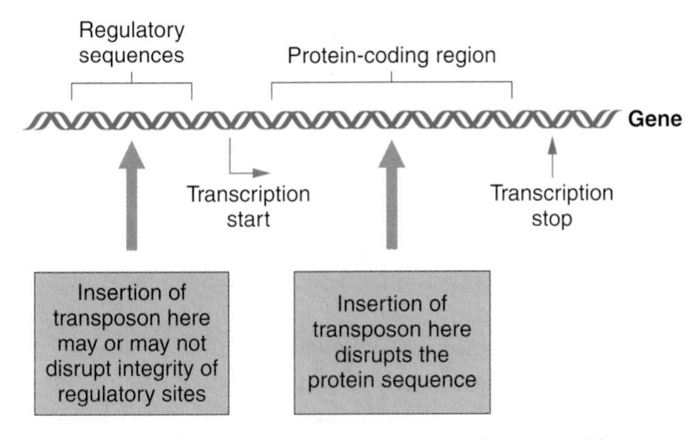

Figure 10-23 The phenotypic consequences of transposable-element insertion depend on where the DNA element inserts within a gene.

Examples of Mobile Genetic Elements

The molecular nature of transposable genetic elements was first understood in bacteria and phages. Therefore, we shall begin with the information derived from the original studies in these prokaryotes.

Insertion sequences. One of the structurally simplest classes of mobile element consists of **insertion sequence (IS) elements.** These elements are found only in prokaryotes. IS elements can jump from one location in an *E. coli* chromosome to another or from one chromosome to another (e.g., from an F plasmid to the standard *E. coli* chromosome). When an IS element appears in the middle of a gene, it interrupts the coding sequence and inactivates the expression of that gene. If one denatures DNA of an allele of a gene in which an IS element is inserted and mixes it with denatured DNA of a wild-type allele of that gene, **heteroduplex,** or hybrid DNA molecules, can renature in which one DNA strand comes from the IS-element-bearing allele and the complementary strand comes from the wild-type allele. When examined under an electron microscope, such heteroduplex DNA shows a single-stranded buckle, or loop (Figure 10-24) corresponding to the IS element, which in this case is about 800 bp long. This sort of experiment constitutes proof that IS mutations insert foreign DNA into the mutated gene.

There are different families of IS elements based on size and DNA sequence. Some of them are listed in Table 10-3 on the next page. The genome of one of the standard wild-type *E. coli* strains is rich in IS elements: it contains eight copies of IS1, five copies of IS2, and copies of other

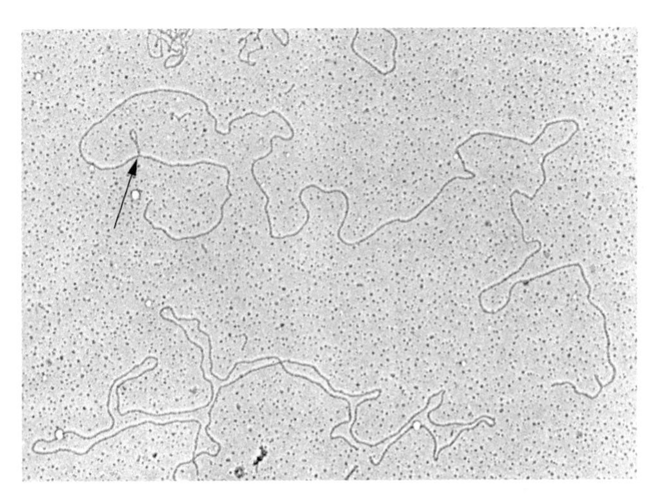

Figure 10-24 Electron micrograph of a $\lambda dgal^+$ / $\lambda dgal^m$ DNA heteroduplex. The single-stranded loop (arrow) is caused by the presence of an insertion sequence in $\lambda dgal^m$. *(From A. Ahmed and D. Scraba, "The Nature of the* gal3 *Mutation of* Escherichia coli," Molecular and General Genetics *136, 1975, 233.)*

TABLE 10-3 Prokaryotic Insertion Elements

Insertion Sequence	Normal Occurrence in *E. Coli*	Length (bp)	Inverted Repeat (bp)*
IS1	5–8 copies on chromosome	768	18–23
IS2	5 on chromosome; 1 on F	1327	32–41
IS3	5 on chromosome; 2 on F	1400	32–38
IS4	1 or 2 copies on chromosome	1400	16–18
IS5	Unknown	1250	Short

*The numbers represent the lengths of the 5′ and 3′ copies of the imperfect inverted repeats.
SOURCE: M. P. Calos and J. H. Miller, *Cell* 20, 1980, 579–595.

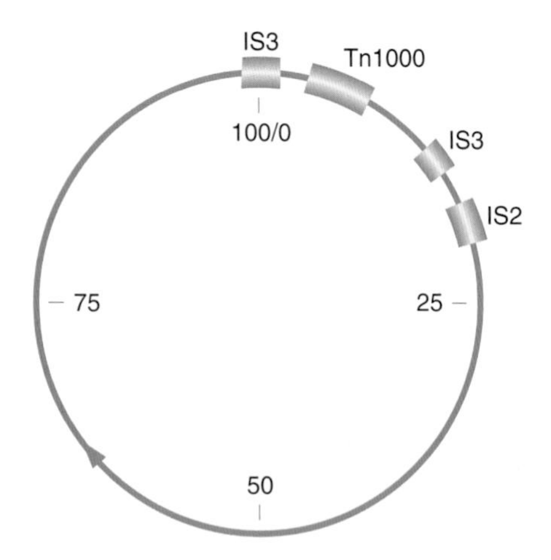

Figure 10-25 Genetic and physical map of the F factor. The positions of the resident insertion sequences, IS2, IS3, and Tn1000 *(boxes),* are shown relative to the map coordinates from 0 to100. The arrow indicates the origin and direction of F DNA transfer during conjugation.

less well studied IS types. IS elements are also portable regions of sequence identity, which can mediate site-specific recombination. For example, recombination between the F factor plasmid and the *E. coli* chromosome to form Hfr strains is mediated by single crossovers between IS elements located on each. Figure 10-25 shows an example of the IS element distribution on one F factor plasmid.

MESSAGE

Prokaryotic genomes contain segments of DNA, termed *IS elements,* that can move from one position on the chromosome to a different position on the same chromosome or on a different chromosome.

Bacterial transposons. A frightening ability of pathogenic bacteria was discovered in Japanese hospitals in the 1950s. Bacterial dysentery is caused by bacteria of the genus *Shigella.* This bacterium initially proved to be sensitive to a wide array of antibiotics that were used to control the disease. In the Japanese hospitals, however, *Shigella* isolated from patients with dysentery proved to be simultaneously resistant to many of these drugs, including penicillin, tetracycline, sulfanilamide, streptomycin, and chloramphenicol. This multiple-drug-resistance phenotype was inherited as a single genetic package, and it could be transmitted in an infectious manner—not only to other antibiotic-sensitive *Shigella* strains, but also to other related species of bacteria. This talent is an extraordinarily useful one for the pathogenic bacterium, and its implications for medical science were terrifying. From the point of view of the geneticist, however, the situation is very interesting. The vector carrying these resistances from one cell to another proved to be a self-replicating element similar to the F factor considered in Chapter 7. These **R factors** (for "resistance") are transferred rapidly on cell conjugation, much like the F particle in *E. coli.*

These R factors proved to be just the first of many similar F-like factors to be discovered. These elements, which exist in the plasmid state in the cytoplasm, have been found to carry many different kinds of genes in bacteria. What is the mode of action of these plasmids? How do they acquire their new genetic abilities? How do they carry them from cell to cell? It turns out that the drug-resistance genes reside between two identical IS-like elements oriented in opposite directions within the plasmid chromosome; because these elements are in opposite orientation, they are called **inverted repeat (IR) sequences** (Figure 10-26). As one example, the element called IS10 is present at the ends of the region carrying the genes for tetracycline resistance (Figure 10-27). The IR sequences together with their contained genes have been collectively called a **transposon (Tn).** Transposons are therefore longer than IS elements (usually a few kilobases in length), inasmuch as they contain extra protein-coding genes. Although IS elements and transposons as defined here are prokaryotic mobile elements, their properties typify many kinds of mobile elements that are found in eukaryotes as well.

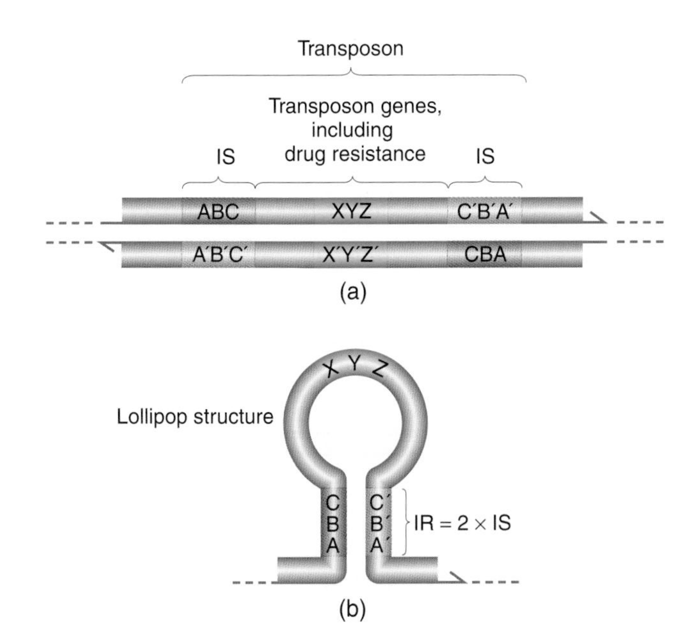

Figure 10-26 The structure of a transposon. (a) The transposon in its double-stranded form before denaturing. Note the presence of oppositely oriented copies of an insertion sequence (IS), which constitute an inverted repeat (IR). (b) The lollipop structure formed by sequence alignment of the two IS regions that form the IR of the transposon; in this view, the transposon genes are carried in the lollipop loop.

MESSAGE

Transposons were originally detected as mobile genetic elements that confer drug resistance. Many of these elements consist of recognizable IS elements flanking a gene that encodes drug resistance. IS elements and transposons are now grouped together under the single term *transposable elements.*

A transposon can jump from a plasmid to a bacterial chromosome or from one plasmid to another plasmid. In this manner, multiple-drug-resistant plasmids are generated. This then raises the question of how such **transposition** or mobilization events occur. We will consider this question next.

Mechanisms of Transposition

Several different mechanisms of transposition are employed by prokaryotic mobile elements. And, as we shall see later, some eukaryotic elements employ the same mechanisms, but other eukaryotic elements utilize different mechanisms of transposition. In *E. coli,* we can identify **replicative** and **conservative** (nonreplicative) modes of transposition (Figure 10-28 on the next page). In the replicative pathway, a new copy of the transposable element is generated in the transposition event (Figure 10-28a). The results of the transposition are that one copy appears at the new site and one copy remains at the old site. In the conservative pathway, there is no replication. Instead, the element is excised from the chromosome or plasmid and is integrated into the new site (Figure 10-28b).

Replicative transposition. When transposition is from one locus to a second locus for certain transposons, a copy of the transposable element is left behind at the first locus. As Figure 10-28a illustrates, such transposons are said to mobilize through **replicative transposition** because transposition is associated with the production of two copies of the transposon from an initial single copy.

Replicative transposition is mediated by an enzyme called a **transposase** that is encoded by the transposon itself. Genetic analysis demonstrates that transposase mutations are recessive. One replicative transposon is Tn3 (Figure 10-29 on the next page). (Replicative transposition defective mutants are described in Figure 10-29.) The transposase acts on an inserted copy of the transposon and mediates its conversion into an intermediate form (diagrammed in Figure 10-30 on the next page) consisting of a double plasmid, with both donor and recipient plasmid being fused together. The combined circle resulting from the fusion of two circular elements is termed a **cointegrate.** The cointegrate then resolves by a recombination-like event that turns a cointegrate into two smaller single-element circles (see Figure 10-30), where one copy remains at the original location of the element, whereas the other is integrated at a new genomic position. Cis-acting mutations (i.e., mutations acting only on the copy of the transposon in which they are located) that interfere with transposition delete a region of the

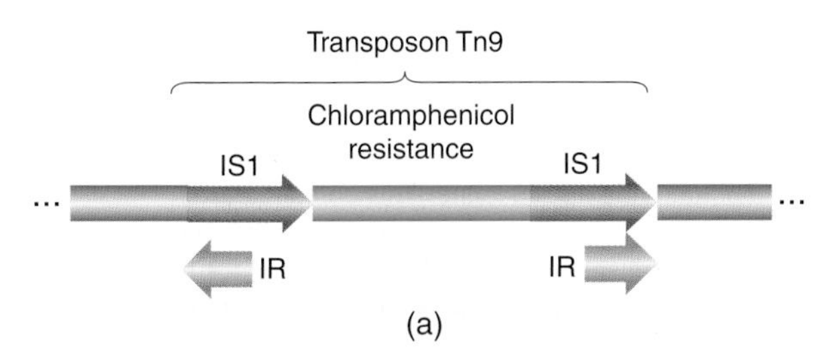

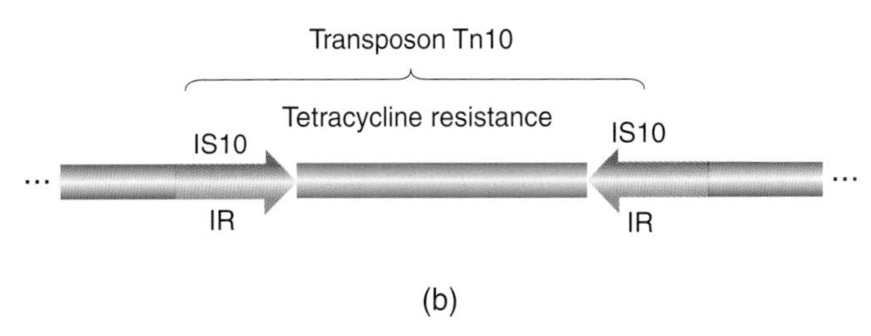

Figure 10-27 Two different transposons having different IR regions and carrying different drug-resistance genes. (a) Tn9 has a short IR region, because the two IS1 elements are in the same orientation and each element has a short inverted repeat. (b) Tn10 has a large IR region because the two IS10 components have opposite orientations, and the entire IS10 sequence constitutes the inverted repeat.

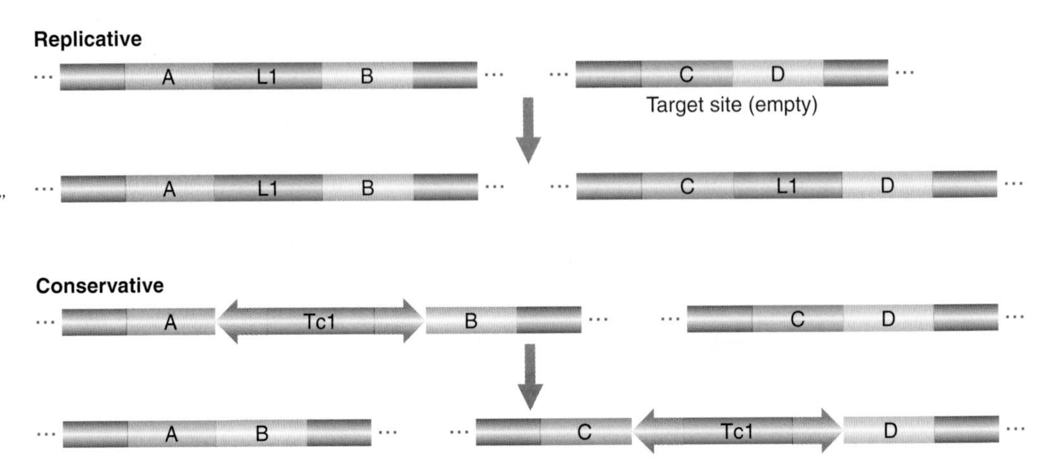

Figure 10-28 Two general modes of mobile-element transposition. *(Adapted with permission from* Nature Reviews: Genetics, *vol. 1, no. 2, page 138, Figure 3, Nov. 2000, "Mobile Elements and the Human Genome," E. T. Luning Prak and H. H. Kazazian, Jr. © 2000 Macmillan Magazines Ltd.)*

transposon responsible for cointegrate recombination. This region, called the **internal resolution site (IRS),** is noted in Figure 10-29.

The finding of a cointegrate structure as an intermediate in transposition helped elucidate the mechanism for the replicative mode of transposition for certain elements. In Figure 10-30, note how the transposable element is duplicated during the fusion event and how the recombination event that resolves the cointegrate into two single transposon-containing circles leaves one copy of the transposable element in each circular chromosome (in this case, circular plasmid chromosomes).

Conservative transposition. Some transposons, such as Tn10, excise from the chromosome and integrate into the target DNA. In these cases, DNA replication of the element does not take place, and the element is lost from the site of the original chromosome.

Molecular consequences of transposition. On integration into a new target site, transposable elements generate a repeated sequence of the target DNA in both replicative and conservative transposition. Figure 10-31 depicts the integration of IS1 into a gene. In the example shown, the integration event results in the repetition of a nine-base-pair target sequence. The target repeat sequence for each transposable element in prokaryotes (and in eukaryotes as well) has a characteristic length, as small as two base pairs for some mobile elements. The repeated sequence is generated in the process of integration itself, presumably catalyzed by an element-specific transposase that generates staggered double-strand breaks (not unlike the staggered breaks catalyzed by restriction endonucleases in the sugar-phosphate

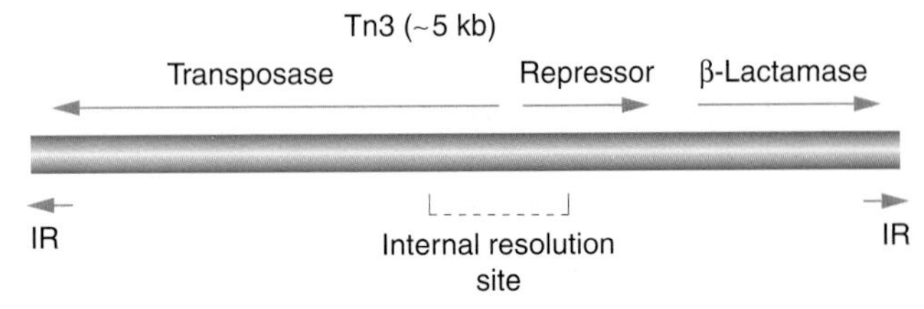

Figure 10-29 The structure of Tn3. Transposon Tn3 contains 4957 base pairs and codes for three polypeptides: the transposase is required for transposition, the repressor is a protein that regulates the transposase gene (see Chapter 13), and β-lactamase confers ampicillin resistance. Tn3 is flanked by inverted repeats (IRs) of 38 base pairs and contains a site designated the *internal resolution site* necessary for the resolution of Tn3 cointegrates.

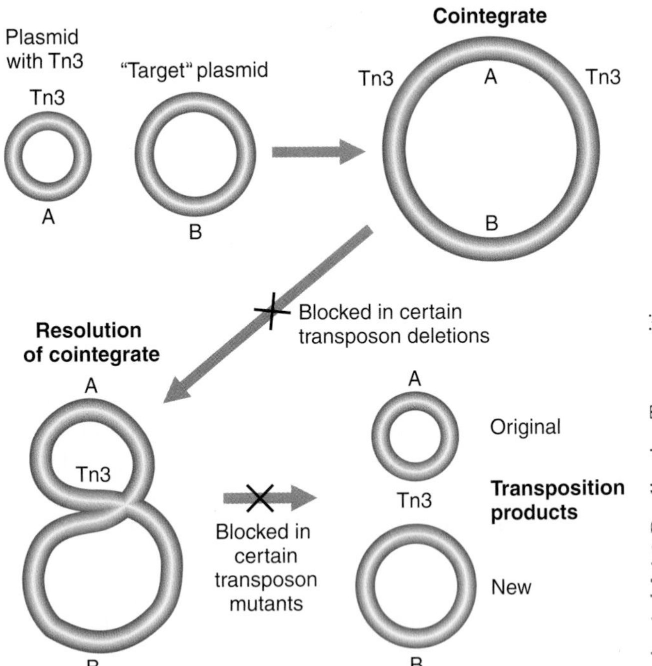

Figure 10-30 Replicative transposition of Tn3 takes place through a cointegrate intermediate. This cointegrate intermediate persists in some mutants that have internal deletions in the transposon gene. The explanation for this observation is that the cointegrate is the intermediate in Tn3 transposition and its resolution to form the two products of the transposition event is blocked if the internal deletion has removed the internal resolution site (IRS), where resolution takes place by recombination. *(From F. Heffron, in J. A. Shapiro, ed.,* Mobile Genetic Elements, *pp. 223–260. © 1983 by Academic Press.)*

Animated Art • Replicative Transposition

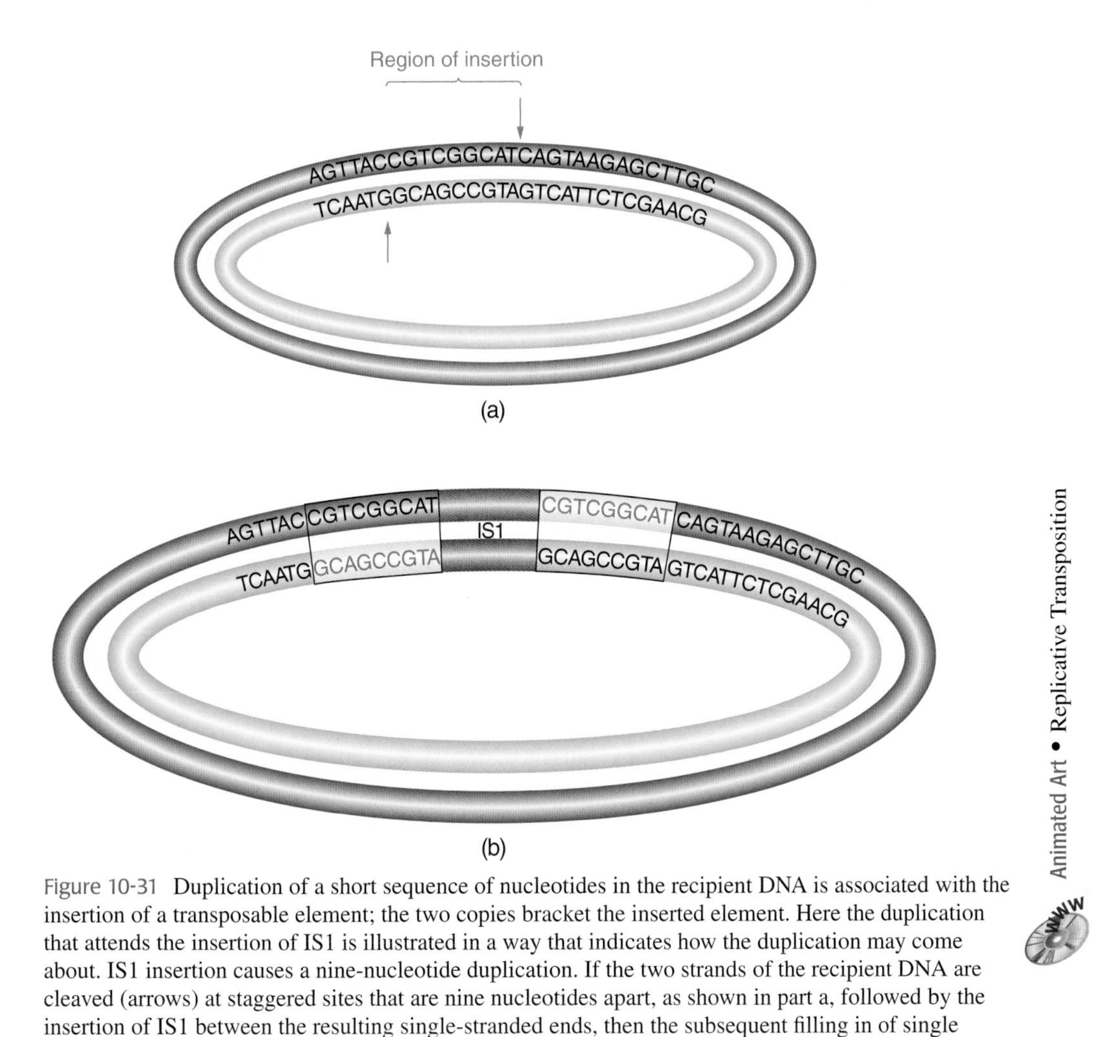

Figure 10-31 Duplication of a short sequence of nucleotides in the recipient DNA is associated with the insertion of a transposable element; the two copies bracket the inserted element. Here the duplication that attends the insertion of IS1 is illustrated in a way that indicates how the duplication may come about. IS1 insertion causes a nine-nucleotide duplication. If the two strands of the recipient DNA are cleaved (arrows) at staggered sites that are nine nucleotides apart, as shown in part a, followed by the insertion of IS1 between the resulting single-stranded ends, then the subsequent filling in of single strands on each side of the newly inserted element, indicated by red letters in part b, with the right complementary nucleotides could account for the duplicated sequences *(boxes)*. *(From S. N. Cohen and J. A. Shapiro,* "Transposable Genetic Elements." © 1980 by Scientific American, Inc. All rights reserved.)

Animated Art • Replicative Transposition

WWW

backbone of DNA), in which the single-strand overhangs are used as templates for creating a second complementary strand. The transposon itself is inserted between these repeated oligomers.

> **MESSAGE**
>
> In prokaryotes, transposition occurs by at least two different pathways. Some transposable elements can replicate a copy of the element into a target site, leaving one copy behind at the original site. In other cases, transposition consists of the direct excision of the element and its reinsertion into a new site.

Eukaryotic Mobile Elements

Historically, mobile elements were first detected in eukaryotes as unstable mutations that seemed to map at different places in the genome in different strains. Some of the most important early work emerged from studies in corn (see Foundations of Genetics 10-3 on pages 338–339).

Several classes of mobile elements are found in eukaryotes, some of which appear to transpose by mechanisms similar to those in bacteria. Briefly, among the families of eukaryotic mobile elements, several structural classes can be distinguished, as discussed in Chapter 9. One class called *SIR* (short inverted repeat) elements, typified by the **Ac (Activator) element** in corn (described in Foundations of Genetics 10-3) and the **P element** in *Drosophila* (described in Chapter 9), contains short (for Ac, 11 bp; for the P element, 31 bp) inverted repeats bracketing a DNA sequence that encodes the element's transposase. Many SIR elements in a typical genome have deletions removing some or all of the coding information for the transposase. Such deleted elements are considered nonautonomous; that is, they are not capable of encoding a functional transposase, but they can be mobilized by the transposase contributed by another, autonomous copy of the element in the same genome. By doing crosses in which a source of SIR transposase is introduced into a genome containing only deleted elements, one can induce nonautonomous element mobilization on command. In this way, a researcher can exploit the biological properties of this element as a mutagen (Figure 10-32). Such transposable-element-induced mutations are especially useful because, inasmuch as the sequence of these elements is known, they provide molecular tags for cloning any gene into which they are inserted. They are widely used in genetic research for exactly this reason.

Retroviral-like Mobile Elements in Eukaryotes

Unique to eukaryotes are elements that transpose through RNA intermediates. One major class comprises the ***copia*-like elements,** which include many different families of elements (each family identified by a different canonical sequence) ranging in size from about 4 kb to 9 kb (see also Chapter 9). In *Drosophila,* for example, members of about 50 different families each appear at positions ranging in number from 10 to 100 in a single genome. Each *copia*-like element carries a **long terminal repeat (LTR)** in the same (uninverted) orientation (Figure 10-33). The LTR typically makes up about 5 percent of the sequence length of the element. The other class of elements that transposes through RNA intermediates lacks LTRs. This class includes the most frequently encountered transposable elements in mammals: the LINEs (long interspersed elements) and SINEs (short interspersed elements) discussed in Chapter 9. More is understood about the transposition of the *copia*-like elements, and so we will focus on them here.

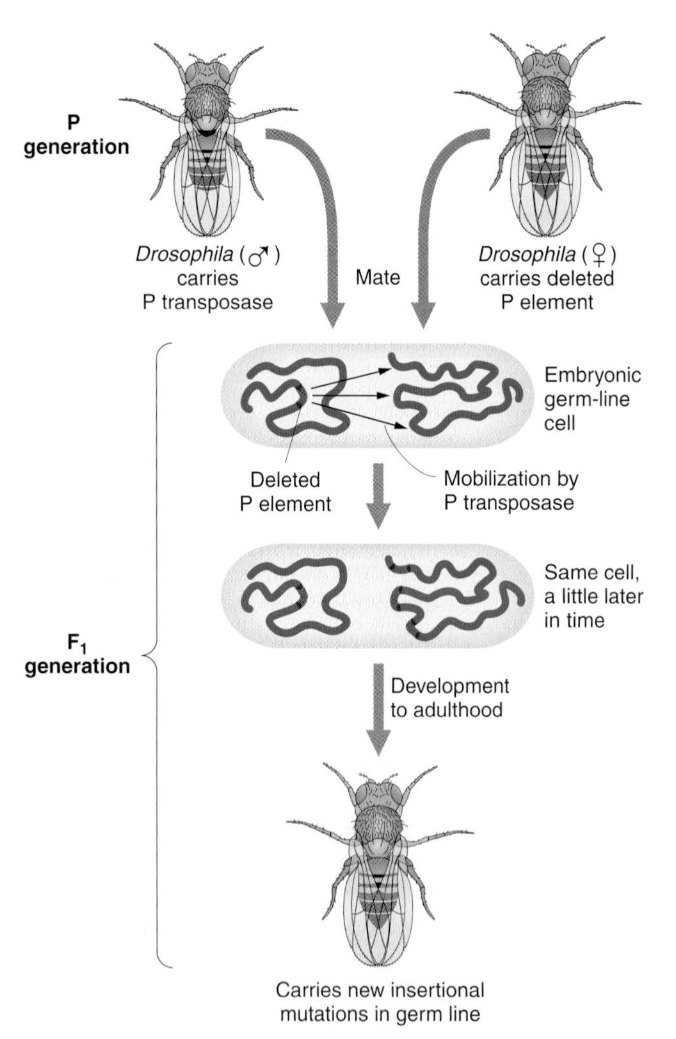

Figure 10-32 Crosses of P-transposase-bearing with defective P-element-bearing *Drosophila* produce P-transposable-element mutations in the germ line of F_1 progeny.

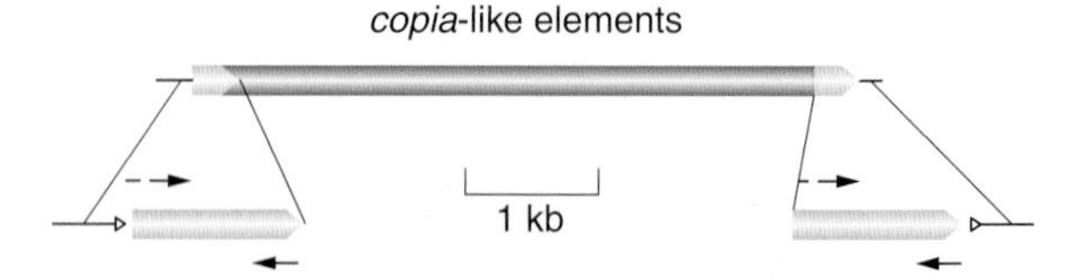

Figure 10-33 Summary of the structures of the *copia*-like elements of *Drosophila.* The *copia*-like elements carry long direct terminal repeats (LTRs, in yellow). These repeats are shown on an expanded scale below the element to illustrate the presence of a few base pairs of duplicate target sequence (triangle pointing right) flanking the element after insertion. Such duplication of the target sequence is a universal feature of transposable-element insertion.

Copia-like elements are similar in some ways to retroviruses, a class of viruses that infect animal cells. **Retroviruses** are single-stranded RNA viruses that employ a double-stranded DNA intermediate for replication. The RNA is copied into DNA by the enzyme **reverse transcriptase** (recall this enzyme from the discussion of cDNA synthesis in Chapter 8). The life cycle of a typical retrovirus is shown in Figure 10-34. Some retroviruses, such as mouse mammary tumor virus (MMTV) and Rous sarcoma virus (RSV), are responsible for the induction of cancerous tumors. When integrated into host chromosomes as double-stranded DNA, these retroviruses are termed **proviruses.** Certain proviruses can be considered transposable elements, because they can, in effect, transpose from one location to another.

Figure 10-35 summarizes the structural similarities between a provirus genome and two retroviral-like mobile elements: Ty 912, from *Saccharomyces cerevisiae,* and *copia,* from *Drosophila melanogaster.* The ends of the proviruses have long terminal repeats reminiscent of the *copia*-like LTRs. In addition, the integration of a provirus results in the duplication of a short target sequence in the host chromosome. The mechanisms of transposition of the proviruses and the *copia*-like elements also are identical. As do retroviruses, *copia*-like elements transpose through an RNA intermediate, utilizing a reverse transcriptase encoded by the *pol* gene.

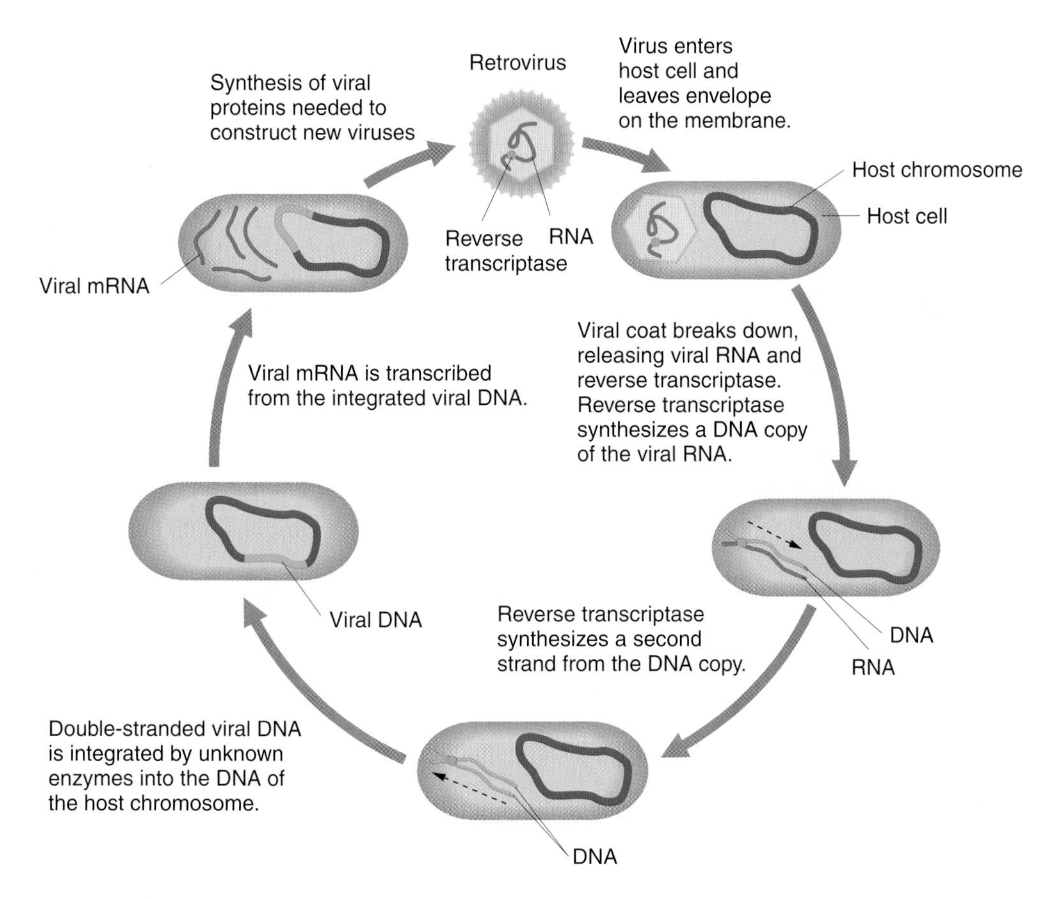

Figure 10-34 The life cycle of a retrovirus. Viral RNA is shown in red; DNA, in blue.

MESSAGE

Retrovirus-like transposable elements exploit the enzyme reverse transcriptase to produce an RNA intermediate that is capable of integrating a DNA copy of itself at a new position in the genome.

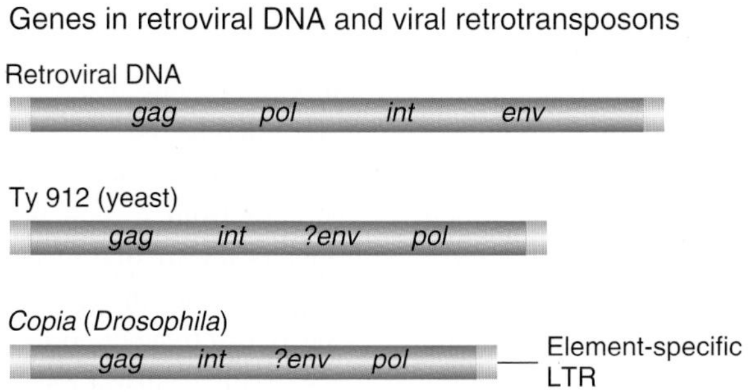

Genes in retroviral DNA and viral retrotransposons

Retroviral DNA

gag *pol* *int* *env*

Ty 912 (yeast)

gag *int* *?env* *pol*

Copia (*Drosophila*)

gag *int* *?env* *pol* — Element-specific LTR

Figure 10-35 A comparison of the genes of integrated retrovirus DNA, the yeast Ty elements, and *Drosophila copia* elements. The four functions encoded by the retroviral DNA have counterparts in the yeast and *Drosophila* elements shown here. The LTRs are represented by the yellow ends of the elements. *(From H. Lodish, D. Baltimore, A. Berk, S. L. Zipursky, P. Matsudaira, and J. Darnell,* Molecular Cell Biology, *3d ed., p. 329. © 1995 by Scientific American Books.)*

BIOLOGICAL REPAIR MECHANISMS

Living cells have evolved a series of enzymatic systems that repair DNA damage in a variety of ways. The low rate of spontaneous mutation is indicative of the efficiency of these repair systems. We can think of the spontaneous-mutation rate as being at a balance point between the rate at which premutational damage arises through mechanisms such as those discussed in the preceding section and the capability of the repair systems to recognize this damage and restore the normal base sequence.

Failure of these systems can lead to a higher mutation rate, as we shall see later. Let's first examine some of the characterized repair pathways and then consider how the cell integrates these systems into an overall strategy for repair. For the error-free repair pathways, one of two things happens:

- The repair pathway chemically reverses the damage to the DNA base.
- The repair pathway takes advantage of antiparallel complementarity and uses an existing complementary sequence as a template. In this way the restoration of the damaged DNA sequence to its normal counterpart is ensured.

FOUNDATIONS OF GENETICS 10-3

Rhoades and McClintock Describe the First Examples of Transposable Elements in Maize

In 1938, Marcus Rhoades analyzed an ear of Mexican black corn. The ear came from a selfing of a pure-breeding pigmented genotype, but it showed a surprising modified Mendelian dihybrid segregation ratio of 12:3:1 among pigmented, dotted, and colorless kernels. Analysis showed that two events had occurred at unlinked loci. At one locus, a pigment gene A_1 had mutated to a_1, an allele for the colorless phenotype; at another locus, a dominant allele *Dt (Dotted)* had appeared. The effect of *Dt* was to produce pigmented dots in the otherwise colorless phenotype of a_1 / a_1. Thus, the original line was very probably A_1 / A_1 ; dt / dt, and the mutations generated an A_1 / a_1 ; Dt / dt plant, which on selfing gave the observed ratio of progeny.

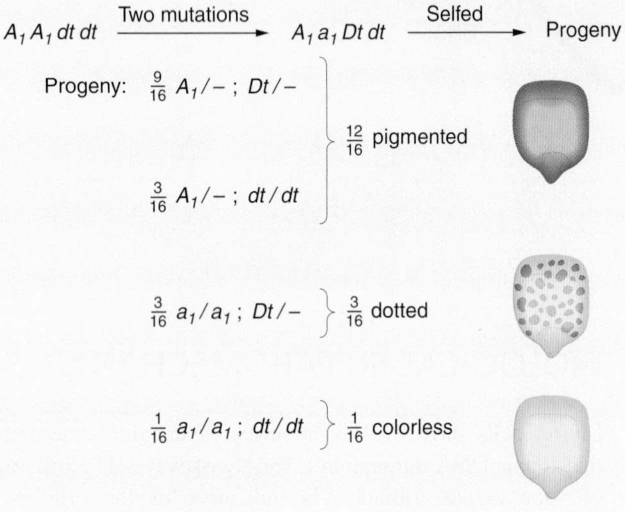

But what was causing the dotted phenotype? A reverse mutation of $a_1 \rightarrow A_1$ in somatic cells would be an obvious possibility, but the large numbers of dots in the Dotted kernels would require extremely high reversion rates. Using special stocks, Rhoades was able to find anthers in the flowers of a_1 / a_1 ; Dt / – plants that showed patches of pigment. He reasoned that these anthers might contain pollen grains bearing the reverted pigment genotype, and he used the pollen from these anthers to fertilize a_1 / a_1 tester females. Sure enough, some of the progeny were completely pigmented, showing that each dot in the parental plants was in fact the phenotypic manifestation of a genetic reversion event. Thus, a_1 is one of the first known examples of an **unstable mutant allele**—an allele for which reverse mutation occurs at a very high rate. However, the allelic instability is dependent on the presence of the unlinked *Dt* gene. Once the reverse mutations occur, they are stable; the *Dt* gene can be crossed out of the line with no loss of the A_1 character. This finding would not be surprising if the a_1 phenotype arose from the insertion of a defective transposable element that by itself is unable to move. In the presence of a transfactor produced by the Dt locus, however, the element could move, yielding reversion to A_1. The A_1 allele would remain stable in the absence of *Dt*.

In the 1950s, Barbara McClintock demonstrated an analogous situation in another study of corn. She found a genetic factor *Ds (Dissociation)* that causes a high tendency toward chromosome breakage at the location at which it appears. These breaks can be located cytologically or they can be located genetically by the uncovering of recessive genes. This action of *Ds* is another kind of instability. Once again, the instability proved to be dependent on the presence of an un-

$A_1 A_1$ *dt dt* —Two mutations→ $A_1 a_1$ *Dt dt* —Selfed→ Progeny

Progeny: $\frac{9}{16}$ A_1/– ; *Dt*/–
$\frac{3}{16}$ A_1/– ; *dt*/*dt*
} $\frac{12}{16}$ pigmented

$\frac{3}{16}$ a_1/a_1 ; *Dt*/– } $\frac{3}{16}$ dotted

$\frac{1}{16}$ a_1/a_1 ; *dt*/*dt* } $\frac{1}{16}$ colorless

We can divide repair pathways into several categories:

- Reversal of damage
- Excision repair
- Postreplication repair

Direct Reversal of Damaged DNA

The most straightforward way to repair a lesion is to reverse it directly, thereby regenerating the normal base. Reversal is not always possible, because some types of damage are essentially irreversible. In a few cases, however, lesions can be repaired in this way. One case is a mutagenic photodimer caused by UV light (see Figure 10-14). The cyclobutane pyrimidine photodimer can be repaired by a **photolyase.** The enzyme binds to the photodimer and splits it, in the presence of certain wavelengths of visible light, to regenerate the original bases (Figure 10-36 on page 340). This enzyme cannot operate in the dark, and so other repair pathways are required to remove UV damage in the absence of visible light.

Alkyltransferases also are enzymes that directly reverse lesions. They remove certain alkyl groups that have been added to the *O*-6 positions of guanine (see Figure 10-10) by such mutagens as nitrosoguanidine and ethyl methanesulfonate. The methyltransferase from *E. coli* has been well studied. This enzyme transfers the methyl group from *O*-6-methylguanine to a cysteine residue on the protein. When this happens, the enzyme is inactivated; so this repair system can be saturated if the level of alkylation is high enough.

Homology-Dependent Repair Systems

One of the overarching themes of genetics is the power of nucleotide sequence complementarity. The properties of antiparallel complementarity are exploited by important repair

linked gene, *Ac (Activator),* in the same way that the instability of a_1 is dependent on *Dt.*

McClintock found it impossible to map *Ac.* In some plants, it mapped to one position; in other plants of the same line, it mapped to different positions. As if this were not enough of a curiosity, the *Ds* locus itself was constantly changing position on the chromosome arm, as indicated by the differing phenotypes of the variegated sections of the seeds (as different recessive gene combinations were uncovered in a system such as the one illustrated below). We now know that *Ac* and *Ds* are respectively the transposase-encoding and internally deleted versions of a mobile element with overall structural similarity to the P element of *Drosophila*—namely, the presence of short inverted terminal repeats (11 base pairs). As has been the case for the P element in *Drosophila, Ac* has proved to be a very powerful tool for producing "transposon tagged" insertional mutations in a variety of plants.

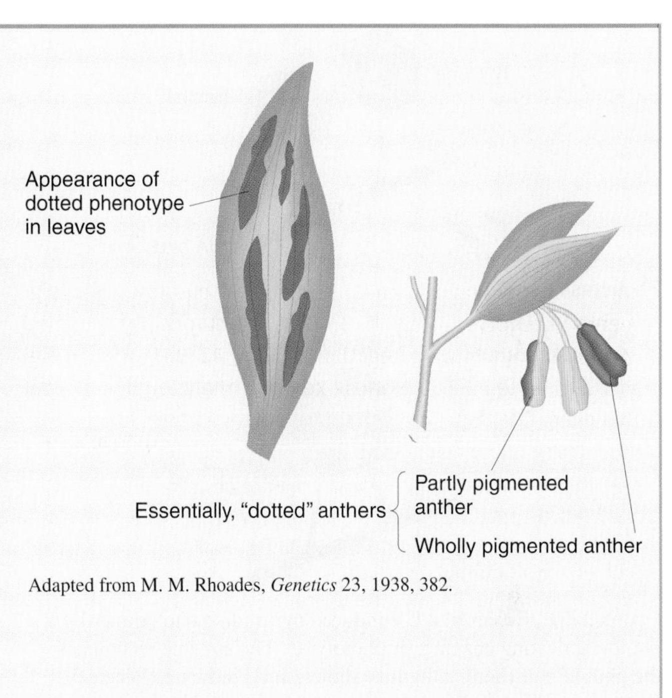

Adapted from M. M. Rhoades, *Genetics* 23, 1938, 382.

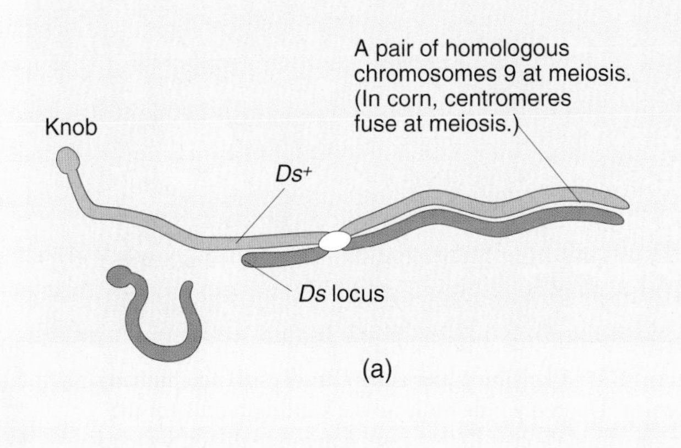

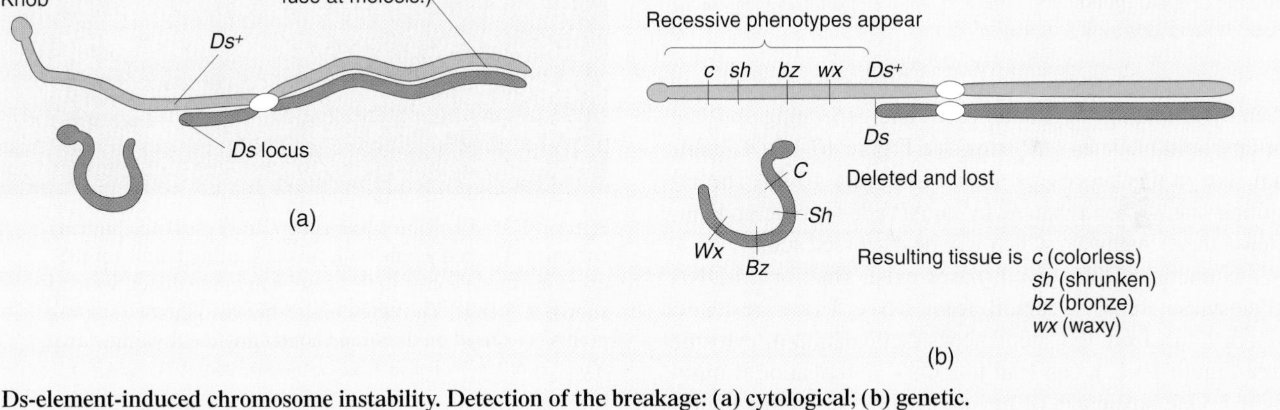

Ds-element-induced chromosome instability. Detection of the breakage: (a) cytological; (b) genetic.

systems to restore damaged DNA segments back to their initial, nondamaged state. In these systems, a segment of a DNA chain is removed and replaced with a newly synthesized nucleotide segment complementary to the opposite template strand. Because these systems depend on the complementarity or homology of the template strand to the strand being repaired, they are called homology-dependent repair systems. Because repair takes place through a template, the rules of DNA replication ensure that repair is accomplished with high fidelity—that is, it is error free. There are two major homology-dependent error-free repair systems. One system (excision repair) deals with damage that has been detected before replication. The other (postreplication repair) deals with damage that is detected in the course of or after the replication process.

Excision-repair pathways. The **nucleotide excision-repair system** recognizes an abnormal base and enzymatically breaks a phosphodiester bond a few base pairs away on either side of the lesion, on the same strand. The excision enzyme removes this segment of a few bases, including the lesion itself. The removal leaves a gap that is then filled by repair DNA synthesis, in which the template strand is used to produce an accurate copy of the original DNA sequence. A DNA ligase seals the breaks by connecting the newly replicated base sequence to the adjacent sequences by formation of new phosphodiester bonds. In prokaryotes, the excision enzyme removes 12 or 13 nucleotides, whereas, in eukaryotes, from 27 to 29 nucleotides are eliminated. Figure 10-37 on the next page depicts the starting points for excision by endonuclease typical of both prokaryotes and eukaryotes.

Certain lesions are too subtle to cause a distortion large enough to be recognized by the general excision-repair system. Thus, additional **specific excision pathways** are necessary. **Base-excision repair** is carried out by **DNA glycosylases** that cleave *N*-glycosidic (base–sugar) bonds,

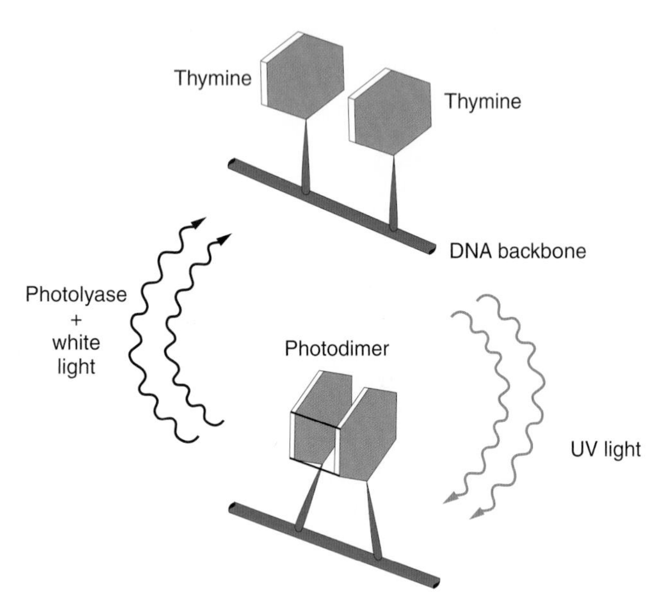

Figure 10-36 Repair of a UV-induced pyrimidine photodimer by a photoreactivating enzyme, or photolyase. The enzyme recognizes the photodimer (here, a thymine dimer) and binds to it. When light is present, the photolyase uses its energy to split the dimer into the original monomers. *(After J. D. Watson,* Molecular Biology of the Gene, *3d ed. © 1976 by W. A. Benjamin.)*

thereby liberating the altered bases and generating apurinic or apyrimidinic sites (AP sites; see Figure 10-16). The initial step in this process is shown in Figure 10-38. The resulting site is then repaired by an AP site-specific endonuclease repair pathway, which will be discussed soon.

Numerous DNA glycosylases exist. One, uracil-DNA glycosylase, removes uracil from DNA. Uracil residues, which result from the spontaneous deamination of cytosine (see Figure 10-17), can lead to a C → T transition if unrepaired. One advantage of having thymine (5-methyluracil), rather than uracil, as the natural pairing partner of adenine in DNA is that spontaneous cytosine-deamination events can be recognized as abnormal and then excised and repaired. If uracil were a normal constituent of DNA, such repair would not be possible.

All cells have endonucleases that attack the sites remaining after the spontaneous loss of single purine or pyrimidine residues. The **AP endonucleases** are vital to the cell, because, as noted earlier, spontaneous depurination is a relatively frequent event. These enzymes introduce chain breaks by cleaving the phosphodiester bonds at AP sites. This bond cleavage initiates an excision-repair process mediated by three further enzymes—an exonuclease, DNA polymerase I, and DNA ligase (Figure 10-39).

Owing to the efficiency of the AP endonuclease repair pathway, it can be the final step of other repair pathways. Thus, if damaged base pairs can be excised, leaving an AP site, the AP endonucleases can complete the restoration to the wild type, which is what happens in the DNA glycosylase repair pathway.

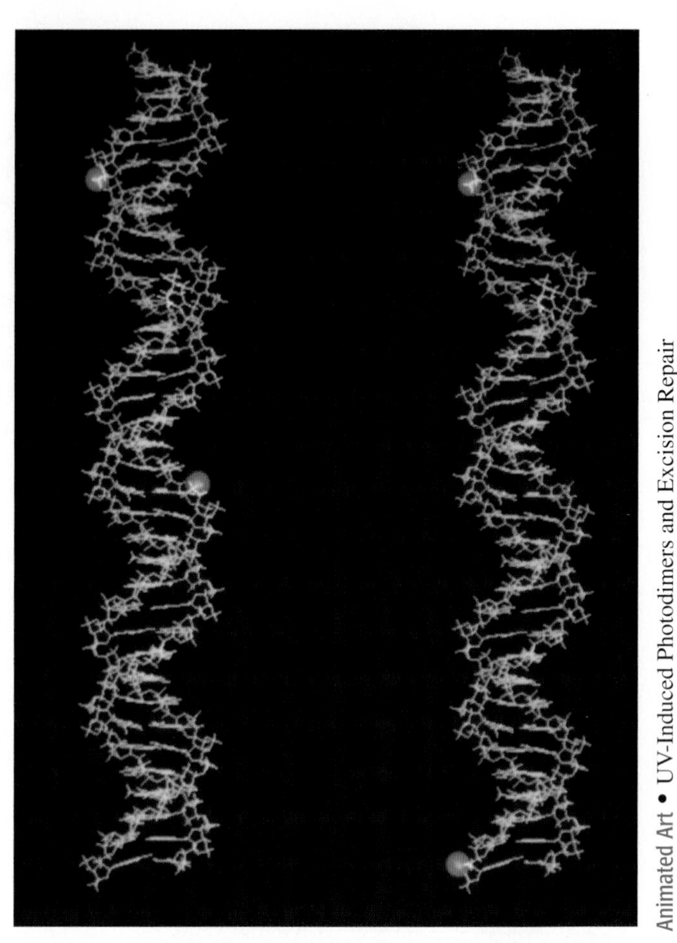

Figure 10-37 Excision patterns by *E. coli (left)* and human enzymes. The red points indicate the starting points for the excision of a lesion—in this case, a thymine dimer, which is shown in orange. The entire region between the two starting points is excised on the strand containing the thymine dimer. *(Courtesy of J. E. Hearst, in A. Sancar,* Science *266, 1974, 1954.)*

Animated Art • UV-Induced Photodimers and Excision Repair

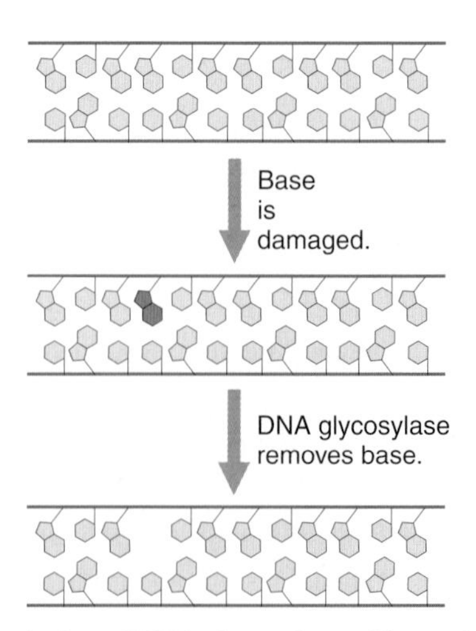

Figure 10-38 Action of DNA glycosylases. Glycosylase removes an altered base and leaves an AP site. The AP site is subsequently excised by the AP endonucleases diagrammed in Figure 10-39. *(After B. Lewin,* Genes. *© 1983 by John Wiley.)*

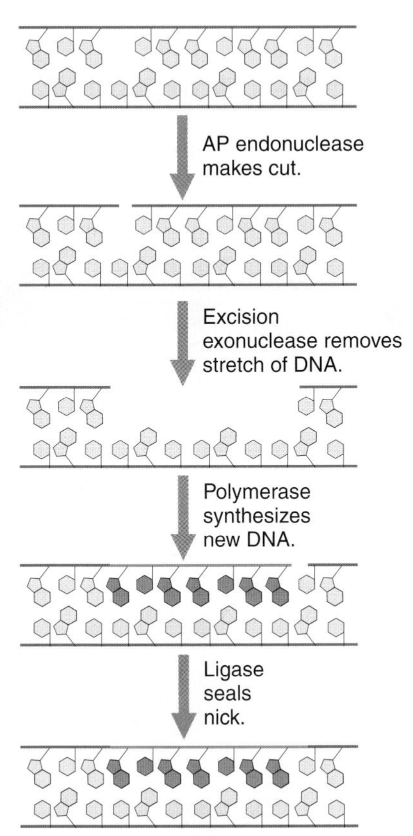

Figure 10-39 Repair of AP (apurinic or apyrimidinic) sites. AP endonucleases recognize AP sites and cut the phosphodiester bond. A stretch of DNA is removed by an exonuclease, and the resulting gap is filled in by DNA polymerase I and DNA ligase, by using the complementary strand as template. *(After B. Lewin,* Genes. *© 1983 by John Wiley.)*

Postreplication repair. Some repair pathways are capable of recognizing errors even after DNA has already undergone replication. One such pathway, termed the **mismatch-repair** system, can detect such mismatches. Mismatch-repair systems have to do at least three things:

- Recognize mismatched base pairs
- Determine which base in the mismatch is the incorrect one
- Excise the incorrect base and carry out repair synthesis

The second property is the crucial one of such a system. Unless it is capable of discriminating between the correct and the incorrect bases, the mismatch-repair system cannot determine which base to excise to prevent a mutation from arising. If, for example, a G · T mismatch occurs as a replication error, how can the system determine whether G or T is incorrect? Both are normal bases in DNA. But replication errors produce mismatches on the newly synthesized strand, and so it is the base on this strand that must be recognized and excised.

To distinguish the old, template strand from the newly synthesized strand, the mismatch-repair system, best characterized in bacteria, takes advantage of a delay in the methylation of the following sequence (methylation that normally takes place after replication):

5′—G—A—T—C—3′
3′—C—T—A—G—5′

The methylating enzyme is **adenine methylase,** which creates 6-methyladenine on each strand. However, it takes adenine methylase several minutes to recognize and modify the newly synthesized GATC stretches. During that interval, the mismatch-repair system can operate because it can now distinguish the old strand from the new one by the methylation pattern. Methylating position 6 of adenine does not affect base pairing, and it provides a convenient tag that can be detected by other enzyme systems. Figure 10-40 on the next page shows the replication fork during mismatch correction. Note that only the old strand is methylated at GATC sequences right after replication. After the mismatched site has been identified, the mismatch-repair system corrects the error.

The *E. coli recA* gene, one of the genes of the SOS bypass system (see Figure 10-13), also takes part in postreplication repair. Here the DNA replication system stalls at a UV photodimer or other blocking lesions and then restarts past the block, leaving a single-stranded gap. The RecA product takes part in **recombinational repair,** a process in which the gap is patched by DNA cut from the parental strand contained in the sister molecule (Figure 10-41 on the next page). This process is not error prone; that is, it does not lead to the production of new mutations.

Revisiting the SOS System

Let us consider the SOS system discussed earlier in the chapter (see Figure 10-13) in the current context of the repair of damaged DNA. The repair systems that directly reverse DNA damage or that use a template strand to direct restoration of the damaged sequence are quite error free—that is, again, they do not lead to the production of new mutations. In contrast, we can think of the SOS bypass system as an error-prone repair pathway. Given the options of "change or die," change (i.e., mutate) seems like not such a bad choice. Change is the choice that the SOS pathway allows. DNA damage that blocks the ability of a cell to divide and produce viable progeny is restored to an undamaged state, even though the original sequence is likely to be altered. Because SOS repair leads to new mutations, this system is induced only under conditions of blocked DNA replication. In contrast, error-free repair systems are always "on" as part of the normal surveillance machinery that searches out and corrects damaged DNA before it becomes new mutations.

Figure 10-40 Steps in *E. coli* mismatch repair. (1) MutS binds to mispair. (2) MutH and MutL are recruited to form a complex. MutH cuts the newly synthesized (unmethylated) strand, and exonuclease degradation proceeds past the point of the mismatch, leaving a gap in the newly synthesized strand. (3) Single-strand-binding protein (Ssb) protects the single-stranded region across from the gap. (4) Repair synthesis and ligation fill in the gap. *(From J. Jiricny,* Trends in Genetics, *vol. 10. Elsevier Trends Journals, Cambridge, UK, 1995.)*

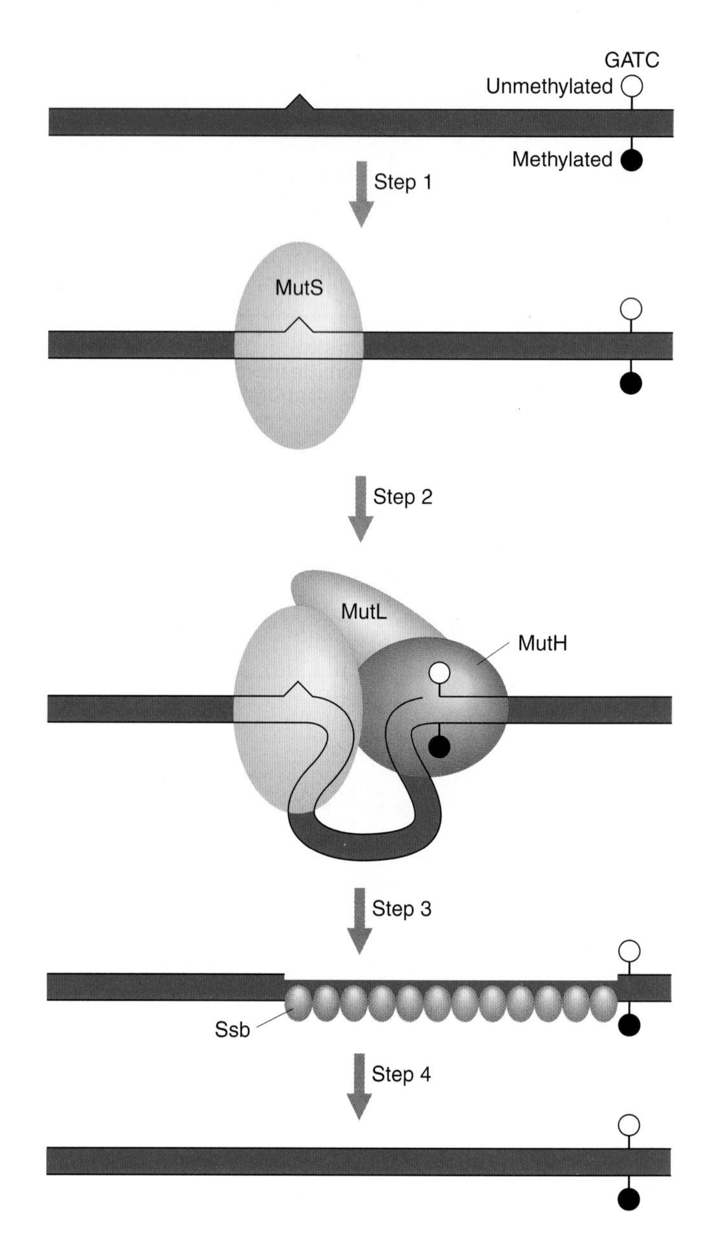

Strategies for Repair

We can now assess the overall repair-system strategy used by the cell. The many different repair systems available to the cell are summarized in Table 10-4. It would be convenient if enzymes could be used to directly reverse each specific lesion. However, sometimes that is not chemically possible, and not every possible type of DNA damage can be anticipated. Therefore, a general excision-repair system is used to remove any type of damaged base that causes a recognizable distortion in the double helix. When lesions are too subtle to cause such a distortion, specific excision systems, such as glycosylases, or removal systems are designed. To eliminate replication errors, a postreplication mismatch-repair system operates; finally, postreplication recombinational systems eliminate gaps across from blocking lesions that have escaped the other repair systems.

MESSAGE

Repair enzymes play a crucial role in reducing genetic damage in living cells. The cell has many different repair pathways at its command to eliminate potentially mutagenic errors.

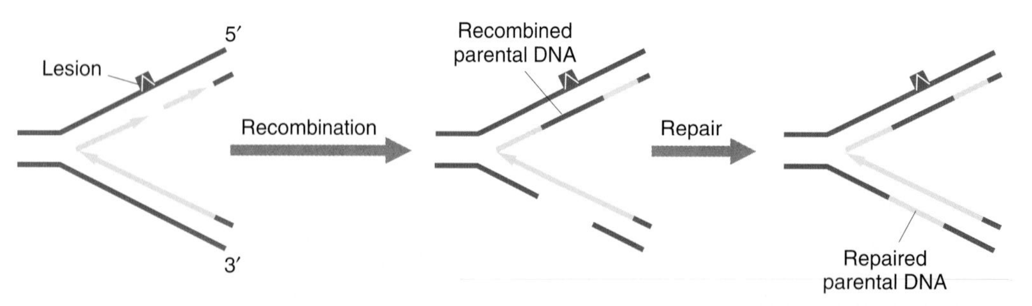

Figure 10-41 Scheme for postreplication recombinational repair. In this repair mechanism, replication jumps across a blocking lesion, leaving a gap in the new strand. A *recA*-directed protein (not shown) then fills in the gap, by using a piece from the opposite parental strand. (Because of DNA complementarity, this filler will supply the correct bases for the gap.) Finally, the RecA protein repairs the gap that was created in the parental strand. Note that the original blocking lesion is still there and must be repaired by some other repair pathway. *(Adapted from A. Kornberg and T. Baker,* DNA Replication, *2d ed. © 1992 by W. H. Freeman and Company.)*

TABLE 10-4 Repair Systems in *E. coli*

General Mode of Operation	Example	Type of Lesion Repaired	Mechanism
Direct removal of lesions	Alkyltransferases	*O*-6-Alkylguanine	Transfers alkyl group from *O*-6-alkylguanine to cysteine residue on transferase
	Photolyase	6-4 photoproduct	Breaks 6–4 bond and restores bases to normal
	Photolyase	UV photodimers	Splits dimers in the presence of white light
General excision	*uvrABC*-encoded exonuclease system	Lesions causing distortions in double helix, such as UV photoproducts and bulky chemical additions	Makes endonucleolytic cut on either side of lesion; resulting gap is repaired by DNA polymerase I and DNA ligase
Specific excision	AP endonucleases	AP sites	Makes endonucleolytic cut; exonuclease creates gap, which is repaired by DNA polymerase I and DNA ligase
	DNA glycosylases	Deaminated bases (uracil, hypoxanthine), certain methylated bases, ring-opened purines, oxidatively damaged bases, and certain other modified bases	Removes base, creating AP site, which is repaired by AP endonucleases
	GO system	8-oxo dG	A glycosylase removes 8-oxo dG from DNA; another glycosylase converts any remaining 8-oxo dG · A mispairs into 8-oxo dG · C pairs, and the first glycoslyase then removes the 8-oxo dG
Postreplication	Mismatch-repair system	Replication errors resulting in base-pair mismatches	Recognizes newly synthesized strand by detecting nonmethylated adenine residues in 5′-GATC-3′ sequences; then excises bases from the new strand when a mismatch is detected
	Recombinational repair	Lesions that block replication and result in single-stranded gaps	Recombinational exchange
	SOS system	Lesions that block replication	Allows replication bypass of blocking lesion, resulting in frequent mutations across from lesion

MUTAGENESIS, REPAIR, AND HUMAN DISEASE

This chapter began with a brief description of a disease called xeroderma pigmentosum. Recall that the disease syndrome includes cognitive and sensory deficiencies, sensitivity to the sun, high rates of skin cancer, and darkly pigmented spots on the skin. On the basis of the results of pedigree analyses and somatic-cell hybridization studies, eight different complementation groups (for our purposes, eight different genes) have been identified as mutated in different xeroderma pigmentosum patients. Different genes that can be mutated to similar phenotypes usually constitute steps in the same biochemical pathway. This is certainly true for xeroderma pigmentosum, for which these eight genes have been shown to encode different proteins participating in the excision repair of damaged DNA. With this knowledge, let us consider the symptoms of the disease. The sensitivity to the sun is most likely due to a defective repair response to ultraviolet irradiation (Figure 10-42). We have already seen that there is a correlation between mutagenicity and the carcinogenic properties of many mutagens (see Foundations of Genetics 10-1). As we will explore in depth in Chapter 15, cancer is essentially a genetic disease of somatic cells. Thus, elevations in the mutation rate due to defective repair processes elevate the frequency of mutation overall, including that of mutations in those genes that can contribute to tumor formation. The connections between defects in DNA repair and the neurological deficiencies in xeroderma pigmentosum are somewhat less clear but are probably related to an elevated sensitivity of neuronal cells to abnormalities in the replication process.

The link among DNA repair, elevated mutation rate, and cancer is very strong. Several other diseases affecting different DNA repair pathways are now known, including ataxia telangiectasia, Bloom syndrome, Cockayne syndrome, and Fanconi anemia. For each disease, an elevated risk of cancer is found. The study of such genetic diseases of DNA-repair systems not only has revealed much important information for understanding and treating these diseases, but has been very informative in regard to understanding the basic biology of DNA repair as well.

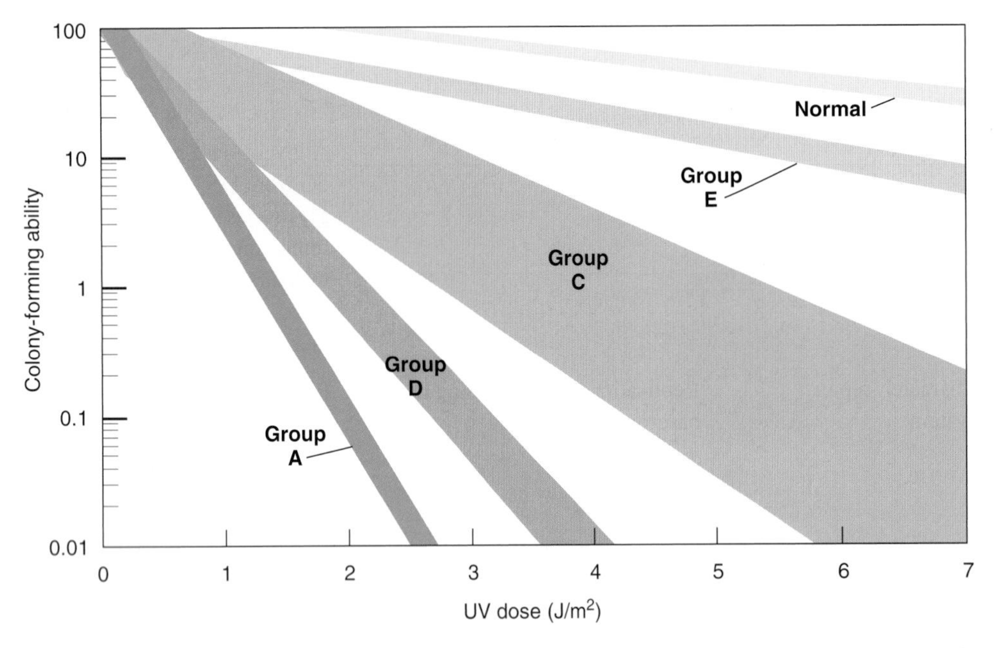

Figure 10-42 Hypersensitivity to UV radiation of XP cells in culture. Here the cells from a number of complementation groups are shown. The complementation groups vary but all are more sensitive to UV radiation than are normal cells. *(Adapted from J. E. Cleaver and K. H. Kraemer, "Xeroderma Pigmentosum," in C. R. Scriver et al., eds.* The Metabolic Basis of Inherited Disease. *© 1989 by McGraw-Hill Book Co., New York.)*

SUMMARY

1. Gene mutation is a change in the DNA sequence from one allelic form of a gene to another. The consequences of a particular change in the DNA sequence are very much context dependent. That is, the consequences of a mutation on gene expression depend not only on the kind of change to the DNA, but also on where within the gene the change has occurred.
2. Genetics is the study of differences. Thus, it is important to understand how such heritable differences arise. In the absence of new mutation, genetic difference would not occur. Fortunately, through a dynamic equilibrium between chemical modifications that create new potential mutations and repair processes that fix the problems before they turn into mutations, a mutation rate appropriate to a given species is achieved.
3. Spontaneous and mutagen-induced mutations arise by a variety of molecular mechanisms. Because these mechanisms are so varied, the arrays of mutational events produced by treatment with different mutagens can be very different from one another.
4. Many gene mutations are point mutations; that is, they alter individual base pairs of DNA. The phenotypic consequences of point mutations differ, depending on whether they occur in coding or noncoding sequences and on exactly where in the open reading frame they occur.
5. Certain genes that contain trinucleotide repeats—especially those that are expressed in neural tissue—become mutated through the expansion of these repeats and can thus cause disease. For some of these diseases, it is clear that the formation of mono-amino acid repeats within the polypeptides encoded by these genes is responsible for the mutant phenotypes.
6. Transposable elements are a rich source of spontaneous gene mutations. A variety of molecular mechanisms lead to the mobilization and movement of these elements around the genome. As mutagens, transposable elements have the additional attribute of molecularly "tagging" the gene into which the insertion has occurred.
7. Biological repair processes have evolved to cope with many different types of DNA damage. The basal error-free repair systems, which are always present in the cell, utilize complementarity to ensure that the DNA sequence present before the damaging event is restored. In instances in which the basal repair systems are unable to repair a particular type of damage, the SOS system is activated. The kinds of damage that activate the SOS system would be lethal if not repaired, because DNA replication could not be completed, and hence intact progeny chromosomes could not be produced. On activation, SOS proteins relax the high-fidelity constraints on ordinary repair replication. By relaxing these constraints, mutations arise as the cost of repair events that ensure the successful completion of DNA replication.

CONCEPT MAP

Draw a concept map interrelating as many of the following terms as possible. Note that the terms are listed in no particular order.

mutation / wild type / transition / frameshift / transversion / mutagen / repair / genetic disease / IS element / transposons / inverted repeats / phenotype / tagging

SOLVED PROBLEM

1. In Chapter 3, we learned that UAG and UAA codons are two of the chain-terminating nonsense triplets. On the basis of the specificity of aflatoxin B_1 and ethyl methanesulfonate (EMS), describe whether each mutagen would be able to revert these codons to wild type.

Solution

EMS induces primarily G·C → A·T transitions. UAG codons could not be reverted to wild type, because only the UAG → UAA change would be stimulated by EMS and that generates a nonsense (ochre) codon. UAA codons would not be acted on by EMS. Aflatoxin B_1 induces primarily G·C → T·A transversions. Only the third position of UAG codons would be acted on, resulting in a UAG → UAU change (on the mRNA level), which produces tyrosine. Therefore, if tyrosine were an acceptable amino acid at the corresponding site in the protein, aflatoxin B_1 could revert UAG codons. Aflatoxin B_1 would not revert UAA codons, because no G·C base pairs appear at the corresponding position in the DNA.

SOLVED PROBLEM

2. Explain why mutations induced by acridines in phage T4 or by ICR-191 in bacteria cannot be reverted by 5-bromouracil.

Solution

Acridines and ICR-191 induce mutations by deleting or adding one or more base pairs, which results in a frameshift. However, 5-bromouracil induces mutations by causing the substitution of one base for another. This substitution cannot compensate for the frameshift resulting from ICR-191 and acridines.

SOLVED PROBLEM

3. A mutant of *E. coli* is highly resistant to mutagenesis by a variety of agents, including ultraviolet light, aflatoxin B_1, and benzo(*a*)pyrene. Explain one possible cause of this mutant phenotype.

Solution

The mutant might lack the SOS system and perhaps carries a defect in the *umuC* gene. Such strains would not be able to bypass replication-blocking lesions of the type caused by the three mutagens listed. Without the processing of premutational lesions, mutations would not be recovered in viable cells.

SOLVED PROBLEM

4. Transposable elements have been referred to as "jumping genes" because they appear to jump from one position to another, leaving the old locus and appearing at a new locus. In light of what we now know concerning the mechanism of transposition, how appropriate is the term "jumping genes" for bacterial transposable elements?

Solution

In bacteria, transposition occurs by two different modes. The conservative mode results in true jumping genes, because, in this case, the transposable element excises from its original position and inserts at a new position. A second mode is termed the *replicative mode.* In this pathway, transposable elements move to a new location by replicating into the target DNA, leaving behind a copy of the transposable element at the original site. When operating by the replicative mode, transposable elements are not really jumping genes, because a copy does remain at the original site.

PROBLEMS

Basic Problems

1. Differentiate between the elements of the following pairs:

a. Transitions and transversions

b. Synonymous and neutral mutations

c. Missense and nonsense mutations

d. Frameshift and nonsense mutations

2. Why are frameshift mutations more likely than missense mutations to result in proteins that lack normal function?

3. Diagram two different mechanisms for deletion formation. How do DNA sequencing experiments suggest these possibilities?

4. Describe two spontaneous lesions that can lead to mutations.

5. Compare the mechanism of action of 5-bromouracil (5-BU) with ethyl methanesulfonate (EMS) in causing mutations. Explain the specificity of mutagenesis for each agent in light of the proposed mechanism.

6. Compare the two different systems required for the repair of AP sites and in the removal of photodimers.

7. Describe the repair systems that operate after replication.

8. A certain compound that is an analog of the base cytosine can become incorporated into DNA. It normally hydrogen-bonds just as cytosine does, but it quite often isomerizes to a form that hydrogen-bonds as thymine does. Do you expect this compound to be mutagenic and, if so, what types of changes might it induce at the DNA level?

9. Describe the repair systems that operate after depurination and deamination.

10. Explain the difference between the replicative and the conservative modes of transposition.

Challenging Problems

11. Describe the model for indel formation. Show how this model can explain mutational hot spots in the *lacI* gene of *E. coli.*

12. **a.** Why is it impossible to induce nonsense mutations (represented at the mRNA level by the triplets UAG, UAA, and UGA) by treating wild-type strains with mutagens that cause only A—T → G—C transitions in DNA?

b. Hydroxylamine (HA) causes only G—C → A—T transitions in DNA. Will HA produce nonsense mutations in wild-type strains?

c. Will HA treatment revert nonsense mutations?

13. Suppose that you want to determine whether a new mutation in the *gal* region of *E. coli* is the result of an insertion of DNA. Describe a physical experiment that would allow you to demonstrate the presence of an insertion.

14. Several auxotrophic point mutants in *Neurospora* are treated with various agents to see if reversion will take place. The following results were obtained (a plus sign indicates reversion; HA causes only G—C → A—T transitions).

Mutant	5-BU	HA	Proflavin	Spontaneous reversion
1	–	–	–	–
2	–	–	+	+
3	+	–	–	+
4	–	–	–	+
5	+	+	–	+

a. For each of the five mutants, describe the nature of the original mutation event (not the reversion) at the molecular level. Be as specific as possible.

b. For each of the five mutants, name a possible mutagen that could have caused the original mutation event. (Spontaneous mutation is not an acceptable answer.)

c. In the reversion experiment for mutant 5, a particularly interesting prototrophic derivative is obtained. When this type is crossed with a standard wild-type strain, the progeny consist of 90 percent prototrophs and 10 percent auxotrophs. Give a full explanation for these results, including a precise reason for the frequencies observed.

15. You are using nitrosoguanidine to "revert" mutant *nic-2* (nicotinamide-requiring) alleles in *Neurospora.* You treat cells, plate them on a medium without nicotinamide, and look for prototrophic colonies. You obtain the following results for two mutant alleles. Explain these results at the molecular level, and indicate how you would test your hypotheses.

a. With *nic-2* allele 1, you obtain no prototrophs at all.

b. With *nic-2* allele 2, you obtain three prototrophic colonies A, B, and C, and you cross each separately with a wild-type strain. From the cross prototroph A × wild type, you obtain 100 progeny, all of which are prototrophic. From the cross prototroph B × wild type, you obtain 100 progeny, of which 78 are prototrophic and 22 are nicotinamide requiring. From the cross prototroph C × wild type, you obtain 1000 progeny, of which 996 are prototrophic and 4 are nicotinamide requiring.

16. Fill in the following table, using a plus sign (+) to indicate that the mutagenic lesion (base damage) induces the indicated base change and a minus sign (–) if it does not.

Base change	*O*-6-Methyl G	8-Oxo dG	C—C Photodimer
A—T to G—C			
G—C to T—A			
G—C to A—T			

17. You are working with a newly discovered mutagen, and you wish to determine the base change that it introduces into DNA. Thus far you have determined that the mutagen chemically alters a single base in such a way that its base-pairing properties are altered permanently. In order to determine the specificity of the alteration,

you examine the amino acid changes that take place after mutagenesis. A sample of what you find is shown below:

Original:	Gln-His-Ile-Glu-Lys
Mutant:	Gln-His-Met-Glu-Lys
Original:	Ala-Val-Asn-Arg
Mutant:	Ala-Val-Ser-Arg
Original:	Arg-Ser-Leu
Mutant:	Arg-Ser-Leu-Trp-Lys-Thr-Phe

What is the base-change specificity of the mutagen?

18. You now find an additional mutant from the experiment in Problem 17:

Original:	Ile-Leu-His-Gln
Mutant:	Ile-Pro-His-Gln

Could the base-change specificity in your answer to Problem 17 account for this mutation? Why or why not?

19. Describe the generation of multiple-drug-resistance plasmids.

20. Explain how the properties of P elements in *Drosophila* make gene-tagging experiments in this organism possible.

CHROMOSOME MUTATIONS 11

Key Concepts

1 In organisms with multiple chromosome sets (polyploids), an even number of sets is more likely to result in fertility than is an odd number of sets.

2 A cross between two different species of plants followed by the doubling of the chromosome number in the hybrid produces a special kind of fertile interspecific polyploid.

3 Variants differing from the wild type by parts of chromosome sets show phenotypic abnormalities caused by an upset gene balance.

4 Variants in which a single chromosome has been gained or lost generally arise by abnormal chromosome segregation at meiosis.

5 Chromosome segments are sometimes lost, duplicated, or relocated to a new position within the set. Owing to the strong meiotic pairing affinity of homologous chromosomes, diploids with one standard chromosome set and one altered set produce meiotic pairing structures that have shapes and properties unique to that type of alteration.

6 If a single segment takes up an opposite orientation within a chromosome, it leads to reduced recombination between genes spanning the segment, and also to reduced fertility.

7 If nonhomologous chromosomes exchange segments, the result is 50 percent sterility and linkage of genes on the exchanged segments.

8 Deletion of a chromosome segment is generally harmful because it results in gene imbalance and in the unmasking of deleterious alleles in the other chromosome set.

9 Duplication of chromosome segments can lead to gene imbalance, but it can also provide extra material for evolutionary divergence.

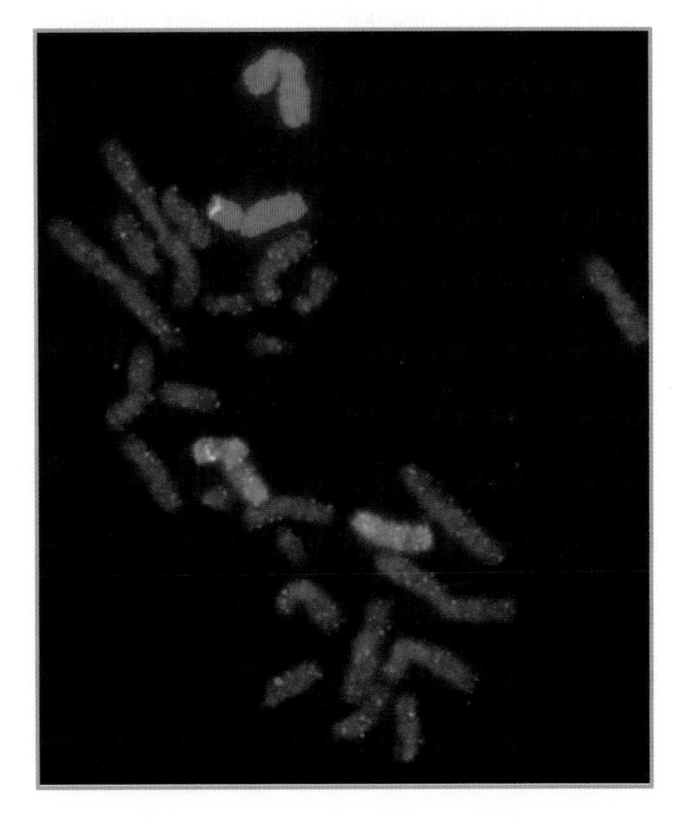

A reciprocal translocation demonstrated by a technique called *chromosome painting*. A suspension of chromosomes from many cells is passed through an electronic device that sorts them by size. DNA is extracted from individual chromosomes, denatured, joined to one of several fluorescent dyes, and then added to partially denatured chromosomes on a slide. The fluorescent DNA "finds" its own chromosome and binds along its length by base complementarity, thus "painting" it. In this preparation, a bright blue and a pink dye have been used to paint different chromosomes. The preparation shows one normal pink chromosome, one normal light blue, and two that have exchanged their tips. *(Lawrence Berkeley Laboratory)*

A young couple is planning to have children. The husband knows that his grandmother had a child with Down syndrome by a second marriage. Down syndrome is a set of physical and mental disorders caused by the presence of an extra chromosome 21 (Figure 11-1). No records of the birth, which occurred early in the twentieth century, are available, but the couple knows of no other cases of Down syndrome in their families.

The couple has heard that Down syndrome results from a rare chance mistake in egg production and therefore decide that they stand only a low chance of having such a child. They decide to have children. Their first child is unaffected, but the next conception aborts spontaneously (a miscarriage), and their second child is born with Down syndrome. Was this a coincidence, or is it possible that there is a connection between the genetic makeup of the man and that of his grandmother that led to their both having Down syndrome children? Was the spontaneous abortion significant? What tests might be necessary to investigate this situation? The analysis of such questions is the topic of this chapter.

Figure 11-1 Children with Down syndrome. *(Bob Daemmrich/The Image Works.)*

CHAPTER OVERVIEW

We saw in Chapter 10 that gene mutations are one source of genomic change. However, the genome can also be remodeled on a larger scale by alterations to chromosome structure or by changes in the number of copies of chromosomes in a cell. These large-scale variations are termed **chromosome mutations** to distinguish them from gene mutations. Broadly speaking, gene mutations are defined as changes that take place within a gene, whereas chromosome mutations are changes at the multigene level. Chromosome mutations can be detected by microscopic examination, genetic analysis, or both. In contrast, gene mutations are never detectable microscopically; a chromosome bearing a gene mutation looks the same under the microscope as one carrying the wild-type allele. Chromosome

CHAPTER OVERVIEW figure

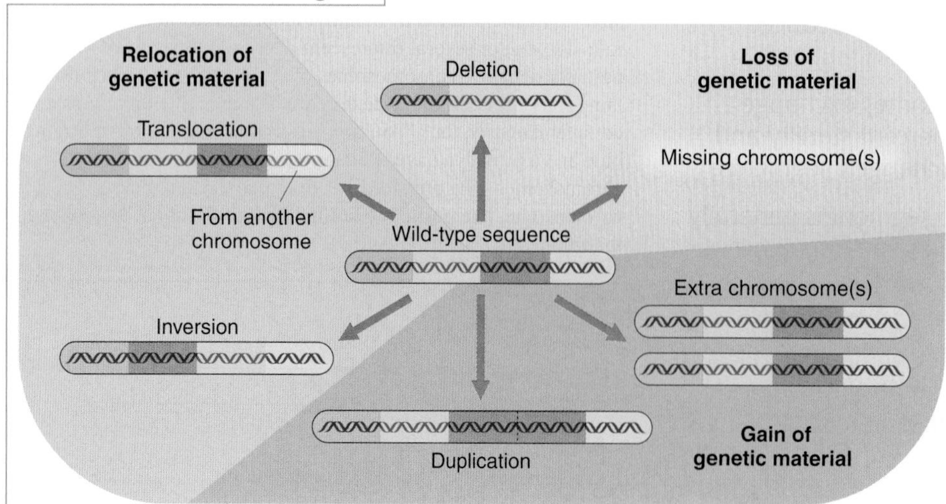

Figure 11-2 Overview of chromosome mutations. The wild-type chromosome (*center*) has been divided into four colored regions to depict the main types of chromosome mutations that can occur. These involve the loss, gain, or relocation of entire chromosomes or chromosome segments. Note that in translocation the chromosome shown has exchanged material with another one (not shown), losing its blue segment in exchange for a red one.

mutations have been best characterized in eukaryotes, and all the examples in this chapter are from that group.

Chromosome mutations are important from several biological perspectives. First, they can be sources of insight into gene function on a genomic scale. Second, they reveal several important features of meiosis and chromosome architecture. Third, they constitute useful tools for experimental genomic manipulation. Fourth, they are sources of insight into evolutionary processes.

Many chromosome mutations cause abnormalities in cell and organismal function. Most of these abnormalities stem from changes in *gene number* or *gene position.* In some cases, a chromosome mutation results from chromosome breakage, and the break may occur within a gene, thereby leading to functional *disruption* of that gene.

For our purposes, we shall divide chromosome mutations into two groups: changes in chromosome *number* and changes in chromosome *structure.* These two groups represent two fundamentally different kinds of events. Changes in chromosome number are not associated with structural alterations of any of the DNA molecules of the cell. Rather, it is the *number* of these DNA molecules that is changed, and this change in number is the basis of their genetic effects. Changes in chromosome structure, on the other hand, result in novel sequence arrangements of one or more DNA double helices. These two types of chromosome mutations are illustrated in Figure 11-2, which is a summary of the topics of this chapter. We begin by exploring the nature and consequences of changes in chromosome number.

CHANGES IN CHROMOSOME NUMBER

In genetics as a whole there are few topics that impinge on human affairs quite so directly as that of changes in the number of chromosomes present. Foremost is the fact that a large proportion of genetically determined health problems in humans are caused by a small group of common genetic disorders resulting from the presence of an abnormal number of chromosomes. Also of relevance to humans is the manipulation of chromosome number by plant breeders to improve commercially important agricultural crops.

Changes in chromosome number are of two basic types: changes in *whole* chromosome sets, resulting in a condition called aberrant euploidy, and changes in *parts* of chromosome sets, resulting in a condition called aneuploidy.

Aberrant Euploidy

Organisms with multiples of the basic chromosome set (genome) are referred to as **euploid.** We learned in earlier chapters that familiar eukaryotes such as plants, animals, and fungi carry in their cells either one chromosome set (haploid) or two chromosome sets (diploid). In these species, the haploid and diploid states are both cases of normal euploidy. Organisms that have more or less than the normal number of sets are aberrant euploids. **Polyploids** are individual organisms in which there are more than two chromosome sets. They can be represented by $3n$ **(triploid),** $4n$ **(tetraploid),** $5n$ **(pentaploid),** $6n$ **(hexaploid),** and so forth. (Recall that the number of chromosome sets is called the ploidy or ploidy level.) An individual of a normally diploid species that has only one chromosome set (n) is called a **monoploid** to distinguish it from an individual of a normally haploid species (also n). Examples of these conditions are shown in the first four rows of Table 11-1.

Monoploids. Male bees, wasps, and ants are monoploid. In the normal life cycles of these insects, males develop by **parthenogenesis** (the development of a specialized type of unfertilized egg into an embryo without the need for fertilization). In most other species, however, monoploid zygotes fail to develop. The reason is that virtually all individuals in

TABLE 11-1 Chromosome Constitutions in a Normally Diploid Organism with Three Chromosomes (Labeled A, B, and C) in the Basic Set

Name	Designation	Constitution	Number of Chromosomes
Euploids			
Monoploid	n	A B C	3
Diploid	$2n$	AA BB CC	6
Triploid	$3n$	AAA BBB CCC	9
Tetraploid	$4n$	AAAA BBBB CCCC	12
Aneuploids			
Monosomic	$2n - 1$	A BB CC	5
		AA B CC	5
		AA BB C	5
Trisomic	$2n + 1$	AAA BB CC	7
		AA BBB CC	7
		AA BB CCC	7

a normally outbreeding diploid species carry a number of deleterious recessive mutations, together called a "genetic load." The deleterious recessive alleles are sheltered by wild-type alleles in the diploid condition. These alleles are automatically expressed in a monoploid derived from a diploid. Monoploids that do develop to advanced stages are abnormal. If they survive to adulthood, their germ cells cannot proceed through meiosis normally because the chromosomes have no pairing partners. Thus, monoploids are characteristically sterile. (Male bees, wasps, and ants bypass meiosis; in these groups, gametes are produced by *mitosis.*)

Polyploids. Polyploidy is very common in plants but rare in animals (for reasons that we will consider later). Indeed, polyploidization has been an important factor in the origin of new plant species. The evidence for this is shown in Figure 11-3, which displays the frequency distribution of haploid chromosome numbers in dicotyledonous plant species. Above a haploid number of about 12, even numbers are much more common than odd numbers. This pattern is a consequence of the polyploid origin of many plant species, because doubling and redoubling of a number can only give rise to even numbers. Animal species do not show such a distribution, owing to the rareness of polyploidy in animals.

In aberrant euploids, there is often a correlation between the number of copies of the chromosome set and the size of the organism. A tetraploid organism, for example, typically looks very similar to its diploid counterpart in its proportions, except that the tetraploid is bigger, both as a whole and in its component parts. The higher the ploidy level, the larger the size of the organism (Figure 11-4).

MESSAGE

Polyploid plants are often larger and have larger component parts than their diploid relatives.

In the realm of polyploids, we must distinguish between **autopolyploids,** which have multiple chromosome sets originating from within one species, and **allopolyploids,** which have sets from two or more different species. Allopolyploids form only between closely related species; however, the different chromosome sets are only **homeologous** (partially homologous), not fully homologous as they are in autopolyploids.

Autopolyploids. Triploids are usually autopolyploids. They arise spontaneously in nature, and they can be constructed by geneticists from the cross of a $4n$ (tetraploid) and a

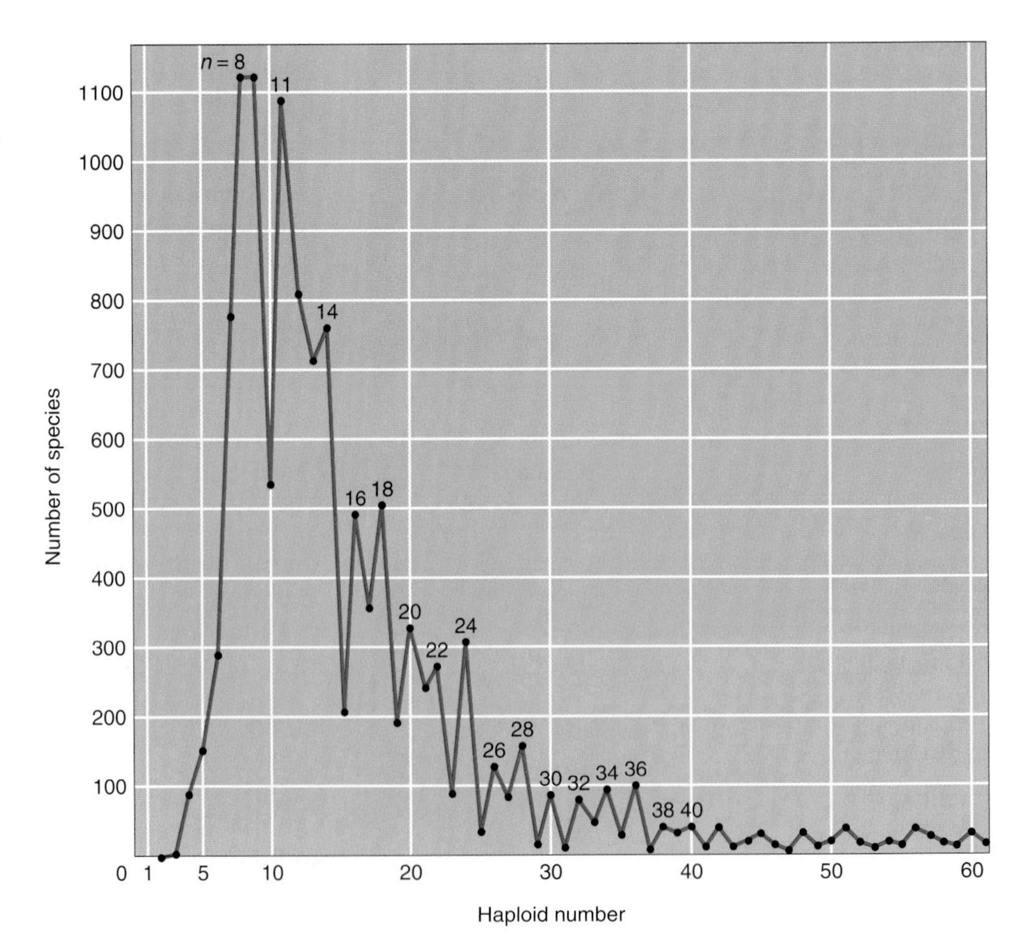

Figure 11-3 Frequency distribution of haploid chromosome number in dicot plants. Notice the excess of even-numbered values in the higher ranges, suggesting ancestral polyploidization. *(Adapted from Verne Grant,* The Origin of Adaptations, *Columbia University Press, 1963.)*

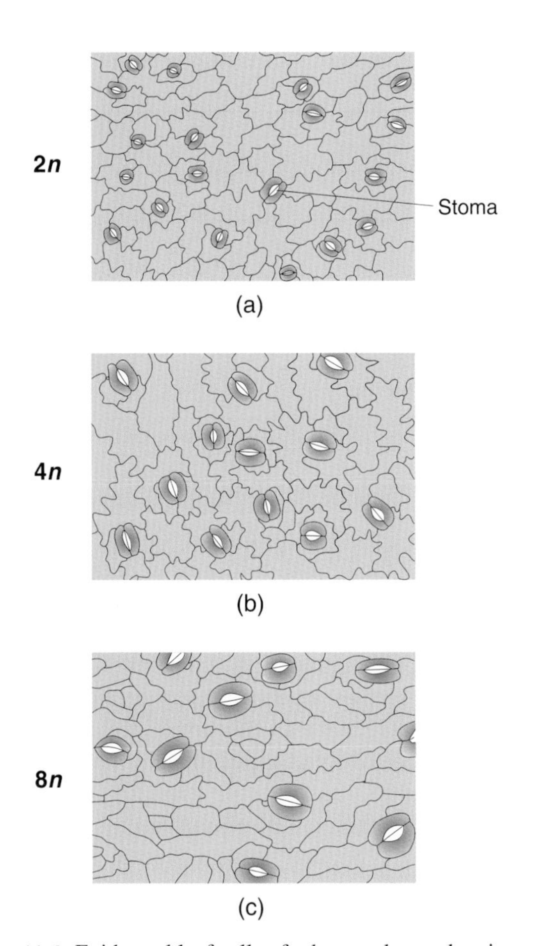

Figure 11-4 Epidermal leaf cells of tobacco plants, showing an increase in cell size, particularly evident in stoma size, with an increase in ploidy. (a) Diploid; (b) tetraploid; (c) octoploid. *(From W. Williams,* Genetic Principles and Plant Breeding. *Blackwell Scientific Publications, Ltd.)*

$2n$ (diploid). The $2n$ and the n gametes produced by the tetraploid and the diploid, respectively, unite to form a $3n$ triploid. Triploids are characteristically sterile. The problem, as in monoploids, lies in pairing at meiosis. The molecular mechanisms for synapsis, or true pairing, dictate that pairing can take place between only two of the three chromosomes of each type (Figure 11-5). Paired homologs **(bivalents)** segregate to opposite poles, but the unpaired homologs **(univalents)** pass to either pole randomly. In the case of a **trivalent,** the paired centromeres segregate as a bivalent and the unpaired one as a univalent. These segregations take place for every chromosome threesome, so for any chromosomal type, the gamete could receive either one or two chromosomes. It is unlikely that a gamete will receive two for *every* chromosomal type, or that it will receive one for *every* chromosomal type. Hence the likelihood is that meiotic products will have chromosome numbers intermediate between the haploid and diploid number; such genomes are of a type called **aneuploid** ("not euploid").

Aneuploid gametes generally do not give rise to viable offspring. There are a couple of reasons for this. First, in

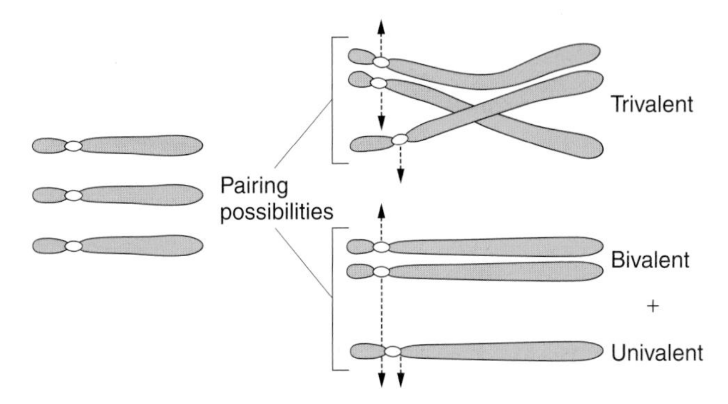

Figure 11-5 Two possibilities for the pairing of three homologous chromosomes before the first meiotic division in a triploid. Notice that the outcome will be the same in both cases: one resulting cell will receive two chromosomes and the other will receive just one. The probability that the latter cell will become a functional haploid gamete is very small, however, because to do so it would also have to receive only one of the three homologous chromosomes of every other set in the organism. (Remember that each chromosome is actually a pair of chromatids.)

plants, pollen cells are very sensitive to aneuploidy, so aneuploid pollen grains are generally inviable and hence unable to fertilize the female gamete. Second, zygotes that do result from the fusion of a haploid and an aneuploid gamete will themselves be aneuploid, and typically these zygotes also are inviable. We will examine the underlying reason for the inviability of aneuploids when we consider gene balance later in the chapter.

MESSAGE

Polyploids with odd numbers of chromosome sets are sterile or highly infertile because their gametes and offspring are aneuploid.

Autotetraploids arise by the doubling of a $2n$ complement to $4n$. This doubling can occur spontaneously, but it can also be induced artificially through the application of chemical agents that disrupt microtubule polymerization. We saw in Chapter 4 that chromosome segregation is powered by spindle fibers, which are polymers of the protein tubulin. Hence disruption of microtubule polymerization blocks chromosome segregation. The chemical treatment is normally applied to vegetative tissue during the formation of spindle fibers in cells undergoing division. The resulting polyploid tissue (such as a polyploid branch) can be detected cytologically. Such a branch can be removed and used as a cutting to generate a polyploid plant or allowed to produce flowers, which would be selfed to produce polyploid offspring. A commonly used anti-tubulin agent is colchicine, an alkaloid drug extracted from the autumn crocus. In colchicine-treated cells, an S phase of the cell cycle occurs, but not chromosome segregation or cell division. As the treated cell enters telophase, a nuclear membrane forms around the entire doubled set of chromosomes. Thus, treating diploid ($2n$) cells for one cell cycle leads to tetraploids

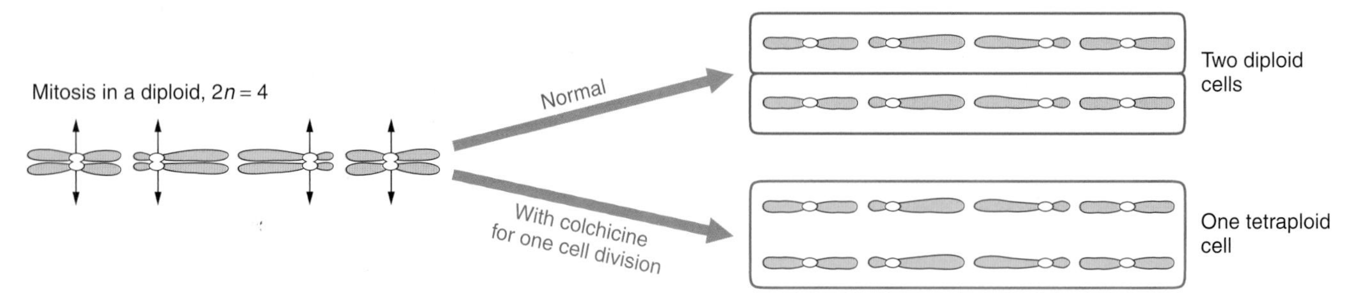

Figure 11-6 The use of colchicine to generate a tetraploid from a diploid. Colchicine added to mitotic cells during metaphase and anaphase disrupts spindle fiber formation, preventing the migration of chromatids after the centromere is split. A single cell is created that contains pairs of identical chromosomes that are homozygous at all loci.

($4n$) with exactly four copies of each type of chromosome (Figure 11-6). Treatment for an additional cell cycle produces octaploids ($8n$), and so forth. This method works in both plant and animal cells, but generally plants seem to be much more tolerant of polyploidy. Note that all alleles in the genotype are doubled. Therefore, if a diploid cell of genotype A / a ; B / b is doubled, the resulting autotetraploid will be of genotype $A / A / a / a$; $B / B / b / b$.

Because 4 is an even number, autotetraploids can have a regular meiosis, although this is by no means always the case. The crucial factor is how the four chromosomes of each set pair and segregate. There are several possibilities, as shown in Figure 11-7. In cases in which the chromosomes pair as bivalents or quadrivalents, the normal meiotic segregation processes result in diploid gametes, which upon fusion regenerate the tetraploid state. If trivalents form, segregation leads to nonfunctional aneuploid gametes, and hence sterility.

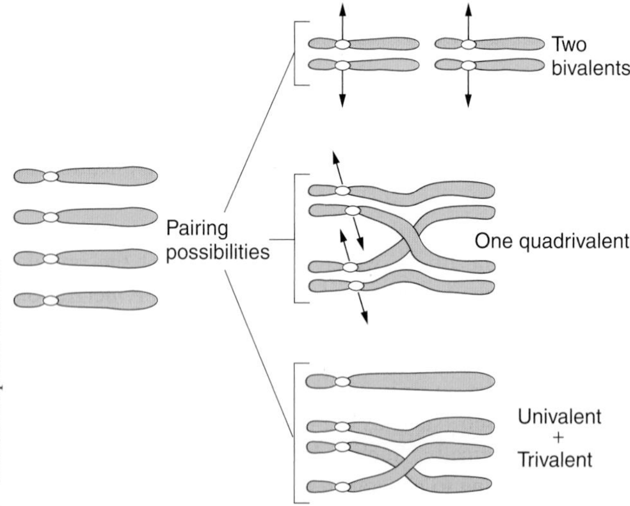

Animated Art • Autotetraploid Meiosis

Figure 11-7 Meiotic pairing possibilities in tetraploids. (Remember that each chromosome is actually a pair of chromatids.) The four homologous chromosomes may pair as two bivalents or as a quadrivalent. Both possibilities can yield functional gametes. However, the four chromosomes may also pair in a univalent-trivalent combination, yielding nonfunctional gametes. A specific tetraploid species can show one of these pairings predominantly.

Allopolyploids. An allopolyploid is a plant that is a hybrid of two or more species, containing two or more copies of each of the input genomes. The prototypic allopolyploid was an allotetraploid synthesized by G. Karpechenko in 1928. He wanted to make a fertile hybrid that would have the leaves of the cabbage *(Brassica)* and the roots of the radish *(Raphanus),* because these were the agriculturally important parts of each plant. Each of these two species has 18 chromosomes, so $2n_1 = 2n_2 = 18$, and $n_1 = n_2 = 9$. The species are related closely enough to allow intercrossing. Fusion of an n_1 and an n_2 gamete produced a viable hybrid progeny individual of constitution $n_1 + n_2 = 18$. However, this hybrid was functionally sterile because the 9 chromosomes from the cabbage parent were different enough from the radish chromosomes that pairs did not synapse and segregate normally at meiosis, and thus the hybrid could not produce functional gametes.

Eventually, one part of the hybrid plant produced some seeds. On planting, these seeds produced fertile individuals with 36 chromosomes. All of these individuals were allopolyploids. They had apparently been derived from spontaneous, accidental chromosome doubling to $2n_1 + 2n_2$ in one region of the sterile hybrid, presumably in tissue that eventually became a flower and underwent meiosis to produce gametes. In $2n_1 + 2n_2$ tissue, there is a pairing partner for each chromosome, and functional gametes of the type $n_1 + n_2$ are produced. These gametes fuse to give $2n_1 + 2n_2$ allopolyploid progeny, which also are fertile. This kind of allopolyploid is sometimes called an **amphidiploid,** or doubled diploid (Figure 11-8). Treating a sterile hybrid with colchicine greatly increases the chances of doubling of the chromosome sets. Amphidiploids are now synthesized routinely in this manner. (Unfortunately for Karpechenko, his amphidiploid had the roots of a cabbage and the leaves of a radish.)

When Karpechenko's allopolyploid was crossed with either parental species—the cabbage or the radish—sterile

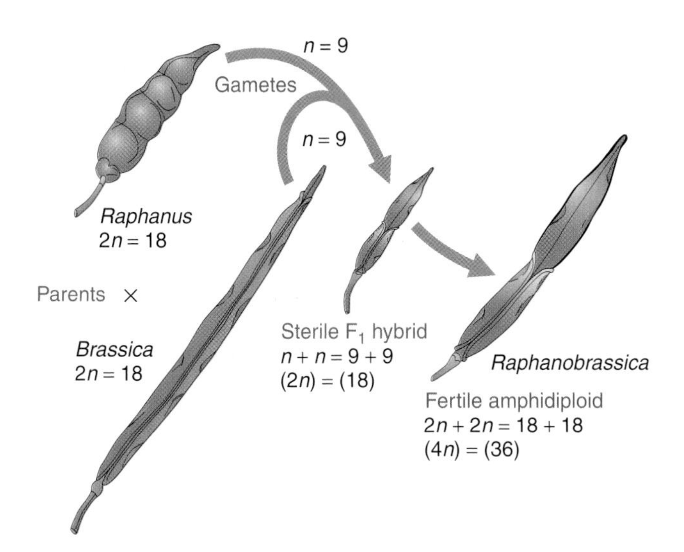

Figure 11-8 The origin of the amphidiploid *(Raphanobrassica)* formed from cabbage *(Brassica)* and radish *(Raphanus)*. The fertile amphidiploid arose in this case from spontaneous doubling in the $2n = 18$ sterile hybrid. Colchicine can also be used to promote such doubling *(From A. M. Srb, R. D. Owen, and R. S. Edgar,* General Genetics, *2d ed. © 1965 by W. H. Freeman. Adapted from G. Karpechenko,* Z. Indukt. Abst. Vererb. *48, 1928, 27.)*

offspring resulted. The offspring of the cross with cabbage were $2n_1 + n_2$, constituted from an $n_1 + n_2$ gamete from the allopolyploid and an n_1 gamete from the cabbage. The n_2 chromosomes had no pairing partners; hence, a normal meiosis could not take place, and sterility resulted. Consequently, Karpechenko had effectively created a new species, with no possibility of gene exchange with either cabbage or radish. He called his new plant *Raphanobrassica.*

In nature, allopolyploidy seems to have been a major force in the speciation of plants. One extensive example is shown by the genus *Brassica,* as illustrated in Figure 11-9. Here three different parent species have hybridized in all possible pair combinations to form new amphidiploid species.

A particularly interesting natural allopolyploid is bread wheat, *Triticum aestivum* ($6n = 42$). By studying its wild relatives, geneticists have reconstructed a probable evolutionary history of this plant. Figure 11-10 on the next page, shows that bread wheat is composed of two sets of each of three ancestral genomes. At meiosis, pairing is always between homologs from the same ancestral genome. Hence, in bread wheat meiosis, there are always 21 bivalents.

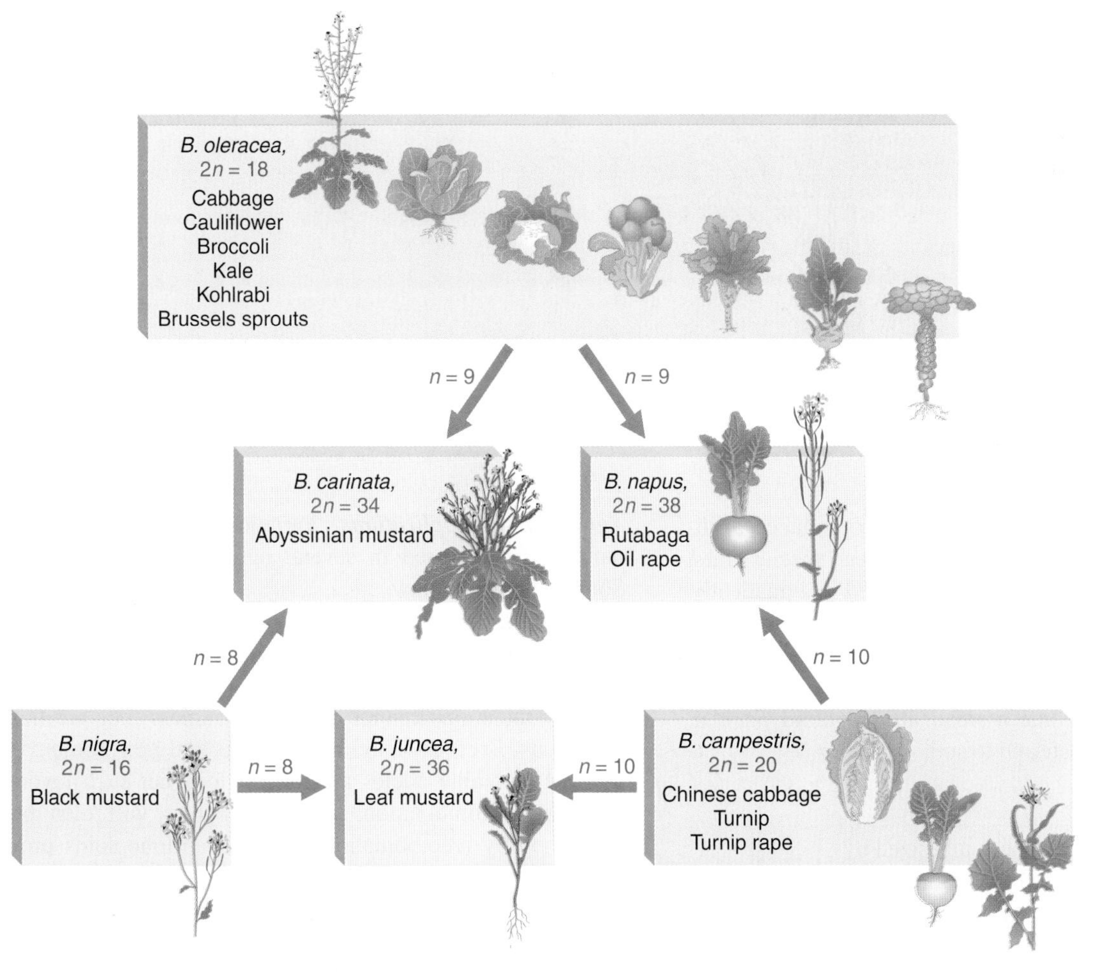

Figure 11-9 Three species of *Brassica* (blue boxes) and their allopolyploids (pink boxes), showing the importance of allopolyploidy in the production of new species.

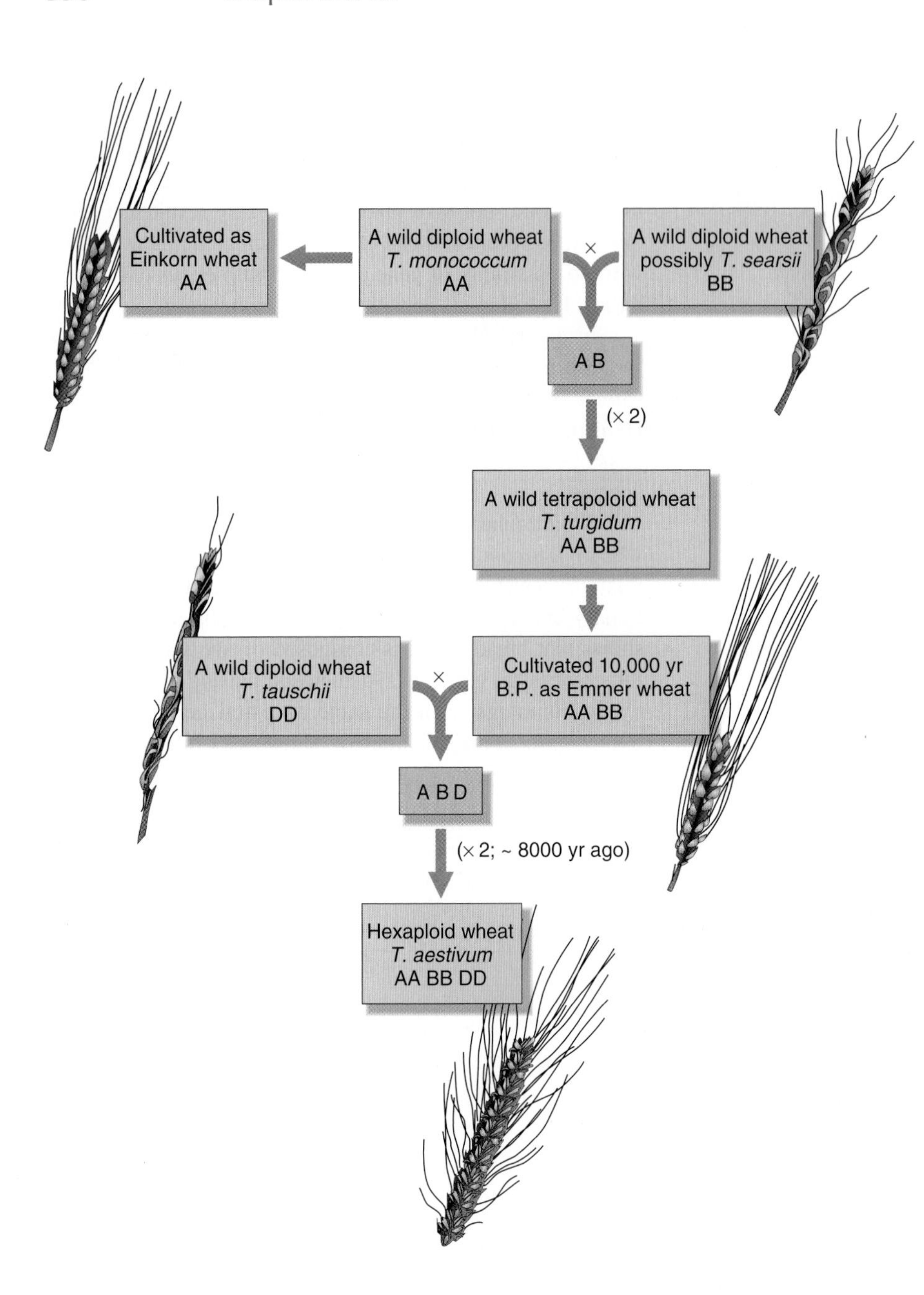

Figure 11-10 Diagram of the proposed evolutionary history of modern hexaploid wheat, in which amphidiploids are produced at two points. A, B, and D are different chromosome sets.

Allopolyploid plant cells can also be produced artificially by fusing diploid cells from different species. First the walls of two diploid cells are removed enzymatically, and the membranes of the two cells fuse and become one. The nuclei often fuse, too, resulting in the polyploid. If the cell is nurtured with the appropriate hormones and nutrients, it divides to become a small allopolyploid plantlet, which can then be transferred to soil.

MESSAGE

Allopolyploid plants can be synthesized by crossing related species and doubling the chromosomes of the hybrid or by fusing diploid cells.

Agricultural applications. Variations in chromosome number can be used in several agricultural applications. Some examples follow.

Monoploids. Diploidy is an inherent nuisance for plant breeders. When they want to induce and select new recessive mutations that are favorable for agricultural purposes, the new mutations cannot be detected unless they are homozygous. Breeders may also want to find new combinations of favorable alleles at different loci, but such favorable allelic combinations in heterozygotes will then be broken up by recombination at meiosis. Monoploids provide a way around some of these problems.

Monoploids can be artificially derived from the products of meiosis in the plant's anthers. A cell destined to become a pollen grain can instead be induced by cold treat-

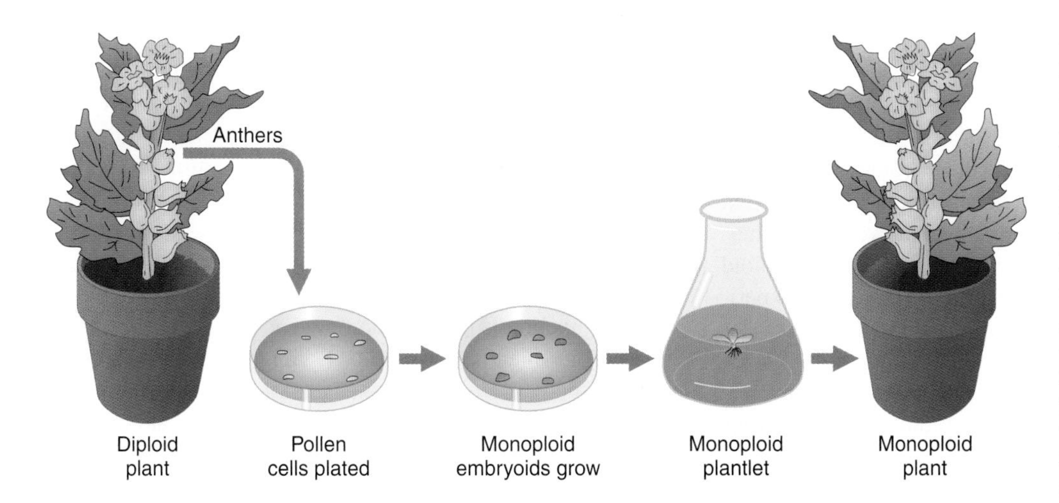

Figure 11-11 Generating monoploid plants by tissue culture. Immature pollen grains (which are haploid) are treated so that they form monoploid embryoids, which are placed on agar plates containing certain plant hormones. Under these conditions, the embryoids will grow into monoploid plantlets. After being moved to a medium containing different plant hormones, these plantlets will grow into mature monoploid plants with roots, stems, leaves, and flowers.

ment to grow into an **embryoid,** a small dividing mass of monoploid cells. The embryoid can be grown on agar to form a monoploid plantlet, which can then be potted in soil and allowed to mature (Figure 11-11).

Plant monoploids can be exploited in several ways. In one approach, they are first examined for favorable allelic combinations that have arisen from heterozygosity already present in the diploid parent. Hence from a parent that is A / a ; B / b might come a favorable monoploid combination a ; b. The monoploid can then be subjected to chromosome doubling, through application of microtubule inhibitors such as colchicine, to produce homozygous diploid cells, a / a ; b / b, that are capable of normal reproduction.

Another approach is to treat monoploid cells basically as a population of haploid organisms in a mutagenesis-and-selection procedure. A population of monoploid cells is isolated, their walls are removed by enzymatic treatment, and they are exposed to a mutagen. They are then plated on a medium that selects for some desirable phenotype. This approach has been used to select for resistance to toxic compounds produced by a plant parasite as well as to select for resistance to herbicides being used by farmers to kill weeds. Resistant plantlets eventually grow into monoploid plants, whose chromosome number can then be doubled using colchicine. This treatment produces diploid tissue, and eventually, by taking a cutting or by selfing a flower, a fully resistant diploid plant can be obtained.

These powerful techniques can circumvent the normally slow process of meiosis-based plant breeding. They have been successfully applied to important crop plants such as soybeans and tobacco.

MESSAGE

To create new plant lines, geneticists produce monoploids with favorable genotypes and then double their chromosomes to form fertile, homozygous diploids.

Autotriploids. The bananas that are widely available commercially are sterile triploids with 11 chromosomes in each set ($3n = 33$). The most obvious expression of the sterility of bananas is the absence of seeds in the fruit that we eat. (The black specks in bananas are not seeds; banana seeds are rock hard—real tooth-breakers.) Another example of the commercial exploitation of triploidy in plants is the production of triploid watermelons, which also are seedless, a phenotype favored by some for its convenience.

Autotetraploids. Many autotetraploid plants have been developed as commercial crops because of their increased size (Figure 11-12). Large fruits and flowers are particularly favored.

Allopolyploids. Allopolyploidy has been important in the production of modern crop plants. New World cotton is a natural allopolyploid that occurred spontaneously, as is wheat. Allopolyploids also are synthesized artificially to combine the useful features of parental species into one type. Only one synthetic amphidiploid has ever been widely used commercially: *Triticale,* an amphidiploid between

Figure 11-12 Diploid *(left)* and tetraploid *(right)* grapes. *(© Leonard Lessin/Peter Arnold Inc.)*

wheat (*Triticum*, $6n = 42$) and rye (*Secale*, $2n = 14$). Hence, for *Triticale*, $2n = 2 \times (21 + 7) = 56$. This novel plant combines the high yields of wheat with the ruggedness of rye.

Polyploid animals. Polyploidy is more common in plants than in animals, but there are cases of naturally occurring polyploid animals. Polyploid species of flatworms, leeches, and brine shrimps reproduce by parthenogenesis. Triploid and tetraploid *Drosophila* have been synthesized experimentally. However, examples are not limited to these so-called lower forms. Naturally occurring polyploid amphibians and reptiles are surprisingly common. They have several modes of reproduction: polyploid species of frogs and toads participate in sexual reproduction, whereas polyploid salamanders and lizards are parthenogenetic. The Salmonidae (the family of fishes that includes salmon and trout) provide a familiar example of the numerous animal species that appear to have originated through ancestral polyploidy. Genomics has been successful in revealing cases of such doubled genomes.

The sterility of triploids has been commercially exploited in animals as well as in plants. Triploid oysters have been developed, and such oysters have a commercial advantage over their diploid relatives. The diploids go through a spawning season, when they are unpalatable, but triploids, because of their sterility, do not spawn and are palatable year-round.

Aneuploidy

Aneuploidy is the second major category of chromosomal aberrations in which the chromosome number is abnormal. An aneuploid is a individual organism whose chromosome number differs from the wild type by part of a chromosome set. Generally, the aneuploid chromosome set differs from the wild type by only one chromosome or by a small number of chromosomes. An aneuploid can have a chromosome number either greater or smaller than that of the wild type. Aneuploid nomenclature (see Table 11-1) is based on the number of copies of the specific chromosome in the aneuploid state. For example, the aneuploid condition $2n - 1$ is called **monosomy** (meaning "one chromosome") because there is only one copy of some specific chromosome present instead of the usual two found in the diploid progenitor. For autosomes in diploid organisms, the aneuploid $2n + 1$ is **trisomic,** $2n - 1$ is **monosomic,** and $2n - 2$ (the -2 represents the loss of two homologs) is **nullisomic.** In haploids, $n + 1$ is **disomic.** Special notation has to be used to describe sex chromosome aneuploids because we are dealing with two different chromosomes (X and Y), and the homogametic and heterogametic sexes have different sex chromosome compositions even in euploid individuals. This notation merely lists the copies of each sex chromosome, such as XXY, XYY, XXX, or XO (the "O" stands for absence of a chromosome and is included to show that the single X symbol is not a typographical error).

Nondisjunction. The cause of most aneuploidy is **nondisjunction** in the course of meiosis or mitosis. *Disjunction* is another word for the normal segregation of homologous chromosomes or chromatids to opposite poles at meiotic or mitotic divisions. Nondisjunction is a failure of this process, in which two chromosomes or chromatids go to one pole and none to the other.

Mitotic nondisjunction during development results in aneuploid sections of the body (aneuploid *sectors*). *Meiotic* nondisjunction is more commonly encountered. It results in aneuploid meiotic products, leading to descendants in which the entire organism is aneuploid. In meiotic nondisjunction, the chromosomes may fail to disjoin at either the first or the second meiotic division (Figure 11-13). Either way, $n + 1$ and $n - 1$ gametes are produced. If an $n - 1$ gamete is fertilized by an n gamete, a monosomic ($2n - 1$) zygote is produced. The fusion of an $n + 1$ and an n gamete yields a trisomic $2n + 1$.

MESSAGE

Aneuploid organisms result mainly from nondisjunction during a parental meiosis.

Nondisjunction occurs spontaneously; like most gene mutations, it is an example of a chance failure of a basic

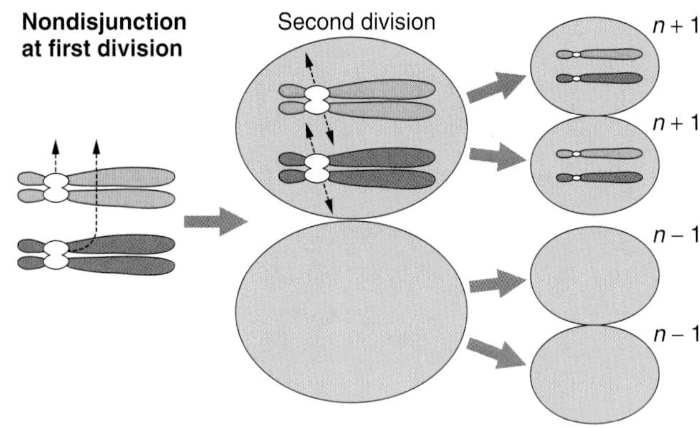

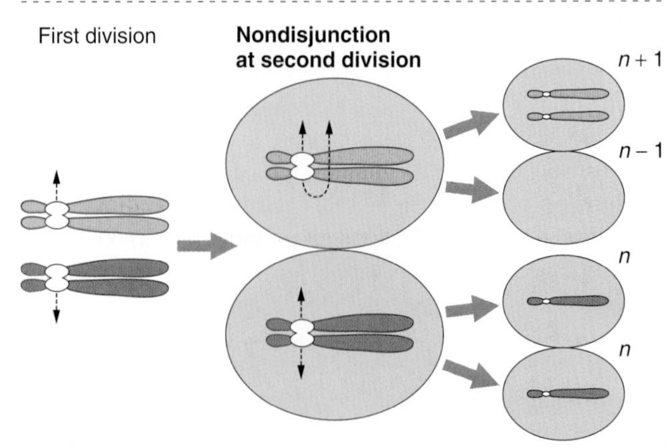

Figure 11-13 The origin of aneuploid gametes by nondisjunction at the first or second meiotic division.

Animated Art • Meiotic Nondisjunction

cellular process. The precise molecular processes that fail are not known, but in experimental systems, the frequency of nondisjunction can be increased by interference with microtubule polymerization action, thereby inhibiting normal chromosome movement. It appears that disjunction is more likely to go awry in meiosis I. This may not be surprising, because normal anaphase I disjunction requires that proper associations between the homologous chromosomes of the tetrad be maintained during prophase I and metaphase I, and also requires crossovers. In contrast, proper disjunction at anaphase II or at mitosis requires that the centromere split properly but does not require chromosome pairing or crossing-over. Meiosis I nondisjunction can thus be viewed as failure to form or maintain the tetrad until anaphase I.

Crossovers are required in the normal disjunction process. Somehow the formation of a chiasma in a chromosome pair helps to hold the tetrad together and assures that the members of a pair will go to opposite poles. In most organisms, the amount of crossing-over is sufficient to ensure that all tetrads will have at least one exchange per meiosis. In *Drosophila,* many of the nondisjunctional chromosomes seen in disomic gametes are nonrecombinant, showing that they arise from meioses with no crossing-over on that chromosome. Similar observations have been made in human trisomies. In addition, in several different experimental organisms, mutations that interfere with recombination have the effect of massively increasing the frequency of meiosis I nondisjunction. All of these observations provide evidence for the role of crossing-over in maintaining chromosome associations in the tetrad; in the absence of these associations, chromosomes are vulnerable to anaphase I nondisjunction.

MESSAGE

Crossovers are needed to maintain the intact tetrad until anaphase I. If crossing-over fails for some reason, first-division nondisjunction occurs.

Monosomics (2n − 1). In most diploid organisms, monosomic chromosome complements are deleterious. In humans, monosomics for any of the autosomes die in utero. Many X chromosome monosomics also die in utero, but some are viable. A human chromosome complement of 44 autosomes + 1 X (XO) produces a condition known as ***Turner syndrome.*** Affected persons have a characteristic phenotype: they are sterile females, short in stature, and often have a web of skin extending between the neck and shoulders (Figure 11-14). Although their intelligence is near normal, some of their specific cognitive functions are defective. About 1 in 5000 female births show Turner syndrome.

Geneticists have used viable plant monosomics to identify the chromosomes that carry the loci of newly discovered recessive mutant alleles. For example, one can make a set of monosomic lines, each of which is known to lack a different chromosome. Homozygotes for the new mutant allele are crossed with each monosomic line, and the progeny of each cross are inspected for the recessive phenotype. The appearance of the recessive phenotype identifies the chromosome that is monosomic as the one the gene is normally located on. The test works because half the gametes of a fertile monosomic will be $n - 1$, and when an $n - 1$ gamete is fertilized by a gamete bearing a new mutation on the homologous chromosome, the mutant allele will be hemizygous and hence will be expressed.

As an example, let's assume that a gene A / a is on chromosome 2. Crosses of a / a to monosomics for chromosome 1 and chromosome 2 illustrate the method (the chromosomes are abbreviated chr1 and chr2):

$$\text{chr1 / chr1 ; } a/a \times \text{chr1 / 0 ; } A/A \longrightarrow \text{progeny all } A/a$$

$$\text{chr1 / chr1 ; } a/a \times \text{chr1 / chr1 ; } A/0 \longrightarrow \text{progeny } \tfrac{1}{2}\,A/a,\ \tfrac{1}{2}\,a/0$$

Trisomics (2n + 1). In diploid organisms generally, the trisomic condition is also one of chromosomal imbalance and can result in abnormality or death. However, there are many examples of viable trisomics. Furthermore, trisomics can be fertile. When cells from some trisomic organisms are observed under the microscope at the time of meiotic chromosome pairing, the trisomic chromosomes are seen to

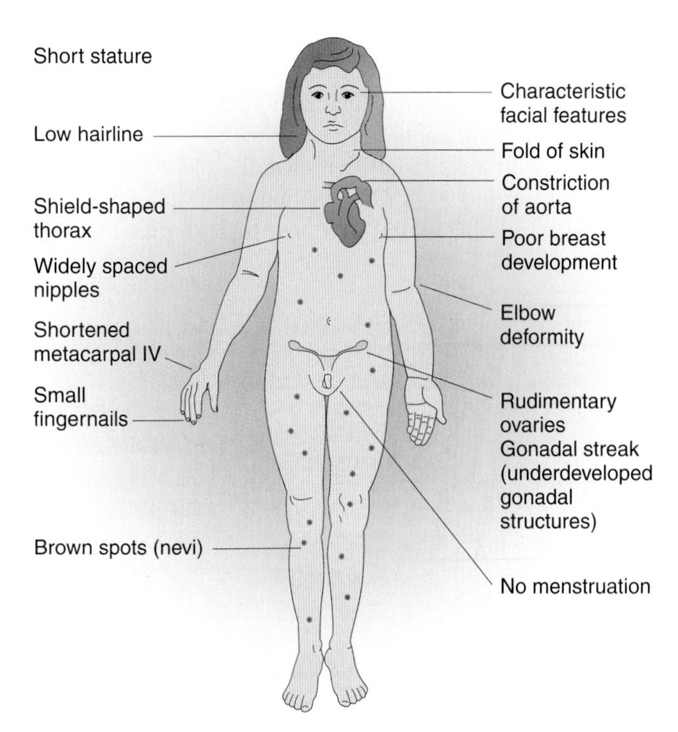

Figure 11-14 Characteristics of Turner syndrome, which results from the presence of a single X chromosome (XO). *(Adapted from F. Vogel and A. G. Motulsky,* Human Genetics. *Springer-Verlag, 1982.)*

form an associated group of three (a trivalent), whereas the other chromosomes form regular pairs.

What genetic ratios might we expect for genes on the trisomic chromosome? Let us consider a gene A that is close to the centromere on that chromosome, and let us assume that the genotype is $A\,/\,a\,/\,a$. Furthermore, if we postulate that at anaphase I the two paired centromeres in the trivalent pass to opposite poles and that the other centromere passes randomly to either pole, then we can predict the three equally frequent segregations shown in Figure 11-15. These segregations result in an overall gametic ratio as shown in the six compartments of Figure 11-15; that is,

$$\tfrac{1}{6}\,A$$
$$\tfrac{2}{6}\,a$$
$$\tfrac{2}{6}\,A\,/\,a$$
$$\tfrac{1}{6}\,a\,/\,a$$

If a set of lines is available, each carrying a different trisomic chromosome, then a gene mutation can be located to a chromosome by determining which of the lines gives a trisomic ratio of the above type.

There are several examples of viable human trisomies. Several types of sex chromosome trisomics can live to adulthood. Each of these types is found at a frequency of about 1 in 1000 births of the relevant sex. (In considering human sex chromosome trisomies, recall that mammalian sex is determined by the presence or absence of the Y chromosome.) The combination XXY results in **Klinefelter syndrome.** Persons with this syndrome are males with lanky builds, and are mentally retarded and sterile (Figure 11-16). Another abnormal combination, XYY, has a controversial history. Attempts have been made to link the XYY condition with a predisposition toward violence. However, it is now clear that an XYY condition in no way guarantees such behavior. Males with XYY are usually fertile. Meioses

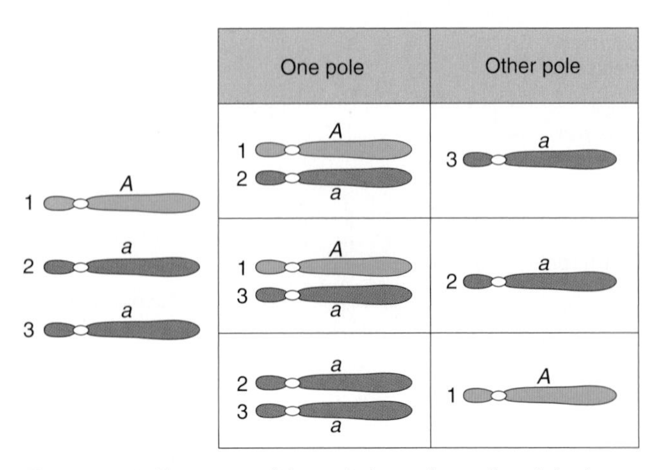

Figure 11-15 Genotypes of the meiotic products of an $A\,/\,a\,/\,a$ trisomic. The three segregations shown are equally likely.

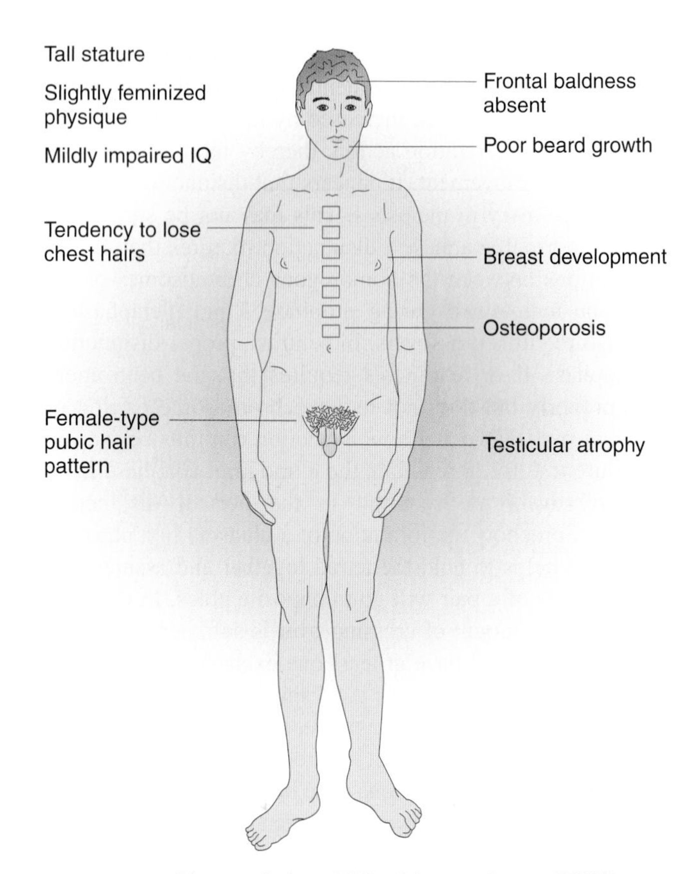

Figure 11-16 Characteristics of Klinefelter syndrome (XXY) *(Adapted from F. Vogel and A. G. Motulsky,* Human Genetics. *Springer-Verlag, 1982.)*

show normal pairing of the X with one of the Y's; the other Y does not pair and is not transmitted to gametes. Therefore the gametes contain either X or Y, never YY or XY. Triplo-X trisomics (XXX) are phenotypically normal and fertile females. Meiosis shows pairing of only two X chromosomes; the third does not pair. Hence eggs bear only one X and, as in the case of XYY individuals, the condition is not passed on to progeny.

Of human trisomies, the most familiar type is **Down syndrome** (Figure 11-17), which we discussed briefly at the beginning of the chapter. Down syndrome occurs at a frequency of about 0.15 percent of all live births. Most affected individuals show trisomy 21 caused by nondisjunction of chromosome 21 in a parent who is chromosomally normal. In this *sporadic* type of Down syndrome, there is no family history of aneuploidy. Some rare types of Down syndrome arise from translocations (a type of chromosomal rearrangement discussed later in the chapter); in these cases, as we shall see, Down syndrome recurs in the pedigree because of the transmission of the translocation.

The combined phenotypes that make up Down syndrome include mental retardation (with an IQ in the 20-to-50 range); a broad, flat face; eyes with an epicanthic fold; short stature; short hands with a crease across the middle; and a large, wrinkled tongue. Females may be fertile and

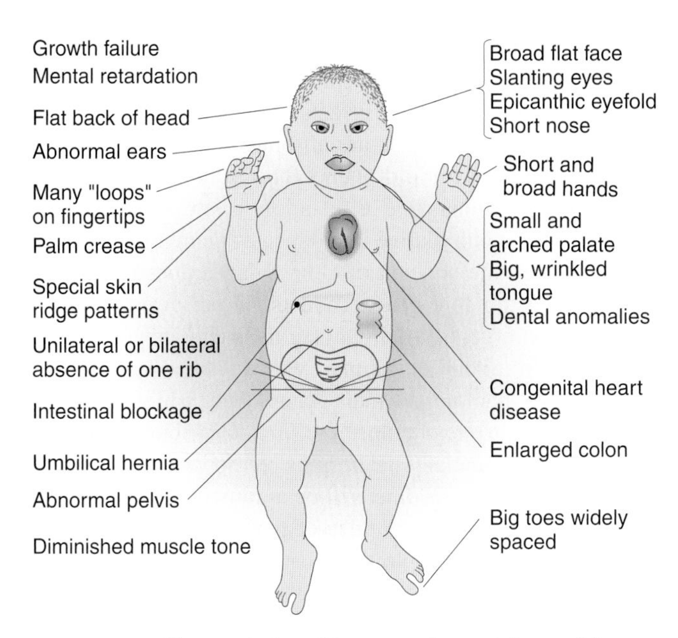

Figure 11-17 Characteristics of Down syndrome (trisomy 21). *(Adapted from F. Vogel and A. G. Motulsky,* Human Genetics. *Springer-Verlag, 1982.)*

may produce normal or trisomic progeny, but males do not reproduce. Mean life expectancy is about 17 years, and only 8 percent of persons with Down syndrome survive past age 40.

Down syndrome is related to maternal age; older mothers run a greatly elevated risk of having a child with Down syndrome (Figure 11-18). For this reason, fetal chromosome analysis (by amniocentesis or by chorionic villus sampling) is now recommended for older mothers. A less pronounced paternal-age effect also has been demonstrated.

Even though the maternal-age effect has been known for many years, its cause is still not known. Nonetheless, there are some interesting biological correlations. It is possible that one aspect of the strong maternal-age effect on nondisjunction is an age-dependent decrease in the probability of keeping the chromosome tetrad together during prophase I of meiosis. Meiotic arrest of oocytes (female meiocytes) in late prophase I is a common phenomenon in many animals. In female humans, all oocytes are arrested at diplotene before birth. Meiosis resumes at each menstrual period, which means that proper chromosome associations in the tetrad must be maintained for as long as several decades. If we speculate that these associations have an increasing probability of breaking down by accident over time, we can envision a mechanism contributing to increased maternal nondisjunction with age. Consistent with this speculation, most nondisjunction related to the effect of maternal age is due to nondisjunction at anaphase I, not anaphase II.

The only other human autosomal trisomics to survive to birth are those with trisomy 13 (Patau syndrome) and trisomy 18 (Edwards syndrome). Both show severe physical and mental abnormalities. The phenotypic syndrome of trisomy 13 includes a harelip; a small, malformed head; "rockerbottom" feet; and a mean life expectancy of 130 days. That of trisomy 18 includes "faunlike" ears, a small jaw, a narrow pelvis, and rockerbottom feet; almost all babies with trisomy 18 die within the first few weeks after birth. All other trisomics die in utero.

The concept of gene balance. In considering aberrant euploidy, we noted that an increase in the number of full chromosome sets correlates with increased organism size, but that the general shape and proportions of the organism remain very much the same. In contrast, autosomal aneuploidy typically alters the organism's shape and proportions in characteristic ways.

Plants tend to be somewhat more tolerant of aneuploidy than are animals. Studies in jimsonweed *(Datura stramonium)* provide a classic example of the effects of aneuploidy and polyploidy. In jimsonweed, the haploid chromosome number is 12. As expected, the polyploid jimsonweed is proportioned like the normal diploid, only larger. In contrast, each of the 12 possible trisomics is disproportionate, but in ways different from one another, as exemplified by changes in the shape of the seed capsule (see Figure 2-14). The 12 different trisomies lead to 12 different and characteristic shape changes in the capsule. Indeed, these and other characteristics of the individual trisomics are so reliable that the phenotypic syndrome can be used to identify plants carrying a particular trisomy. Similarly, the 12 monosomics are themselves different from one another and from each of the trisomics. In general, a monosomic for a particular chromosome is more severely abnormal than is the corresponding trisomic.

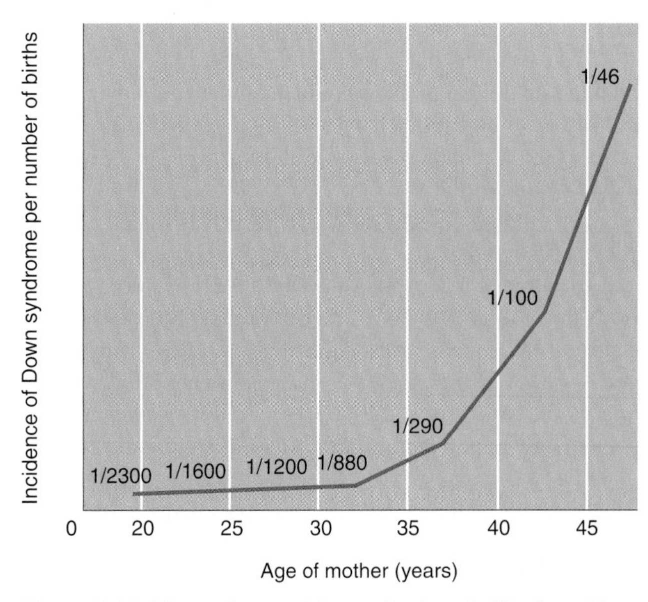

Figure 11-18 Maternal age and the production of offspring with Down syndrome. *(From L. S. Penrose and G. F. Smith,* Down's Anomaly. *Little, Brown and Company, 1966.)*

We see similar trends in aneuploid animals. In the fruit fly *(Drosophila),* the only autosomal aneuploids that survive to adulthood are trisomics and monosomics for chromosome 4, which is the smallest *Drosophila* chromosome, representing only about 1 to 2 percent of the genome. Trisomics for chromosome 4 are only very mildly affected and are much less abnormal than are monosomics for chromosome 4. In humans, no autosomal monosomic survives to birth, but as we have seen, three types of autosomal trisomics can do so. As is true of aneuploid jimsonweed, these three trisomics each show unique phenotypic syndromes because of the special effects of altered dosages of each of these chromosomes.

Why are aneuploids so much more abnormal than polyploids? Why does aneuploidy for each chromosome have its own characteristic phenotypic effects? And why are monosomics typically more severely affected than are the corresponding trisomics? The answers seem certain to be a matter of **gene balance.** In a euploid, the ratio of genes on any one chromosome to the different genes on other chromosomes is 1:1, regardless of whether we are considering a monoploid, diploid, triploid, or tetraploid. In contrast, in an aneuploid, the ratio of genes on the aneuploid chromosome to genes on the other chromosomes differs from the wild type by 50 percent (50 percent for monosomics; 150 percent for trisomics). Thus, we can see that the aneuploid genes are out of balance. How does this help us to answer the questions raised?

In general, the amount of transcript produced by a gene is directly proportional to the number of copies of that gene in a cell. That is, for a given gene, the rate of transcription is directly related to the number of DNA templates available. Thus, the more copies of the gene, the more transcripts are produced and the more of the corresponding protein product is made. This relationship between the number of copies of a gene and the amount of the gene's product made is called a **gene-dosage effect.**

We can infer that normal physiology in a cell depends on the proper ratio of gene products in the euploid cell. This ratio is the normal gene balance. If the relative dosage of certain genes changes—for example, because of the removal of one of the two copies of a chromosome (or even a segment thereof)—physiological imbalances in cellular pathways can arise.

In some cases, the imbalances of aneuploidy result from the effects of a few "major" genes whose dosage has changed, rather than from changes in the dosage of all the genes on a chromosome. Such genes can be viewed as *haplo-abnormal* (resulting in an abnormal phenotype if present only once) or *triplo-abnormal* (resulting in an abnormal phenotype if present in three copies) or both. They contribute significantly to the aneuploid phenotypic syndromes. For example, the study of persons trisomic for only part of chromosome 21 has made it possible to localize genetic determinants specific to Down syndrome to various regions of chromosome 21; the results hint that some aspects of the phenotype might be due to trisomy for single major genes in these chromosome regions. In addition to these major gene effects, other aspects of aneuploid syndromes are likely to result from the cumulative effects of aneuploidy for numerous genes whose products are all out of balance. Undoubtedly, the entire aneuploid phenotype results from a combination of the imbalance effects of a few major genes, together with a cumulative imbalance of many minor genes.

However, the concept of gene balance does not tell us why having too few gene products (monosomy) is much worse for an organism than having too many gene products (trisomy). In a parallel manner, we can ask why there are many more haplo-abnormal genes than triplo-abnormal ones. A key to explaining the extreme abnormality of monosomics is that any deleterious recessive alleles present on a monosomic autosome will be automatically expressed.

How do we apply the idea of gene balance to cases of sex chromosome aneuploidy? Gene balance holds for sex chromosomes as well, but we also have to take into account the special properties of the sex chromosomes. In organisms with X-Y sex determination, the Y chromosome seems to be a degenerate X chromosome in which there are very few functional genes other than some involved in sex determination itself, in sperm production, or in both. The X chromosome, on the other hand, contains many genes involved in basic cellular processes ("housekeeping genes") that just happen to reside on the chromosome that eventually evolved into the X chromosome. X-Y sex determination mechanisms have probably evolved independently from 10 to 20 times in different taxonomic groups. Thus, there appears to be one sex determination mechanism for all mammals, but it is completely different from the mechanism governing X-Y sex determination in fruit flies.

In a sense, X chromosomes are naturally aneuploid. In species with an X-Y sex determination system, females have two X chromosomes, whereas males have only one. Nonetheless, it has been found that the X chromosome's housekeeping genes are expressed to approximately equal extents per cell in females and in males. In other words, there is **dosage compensation.** How is this accomplished? The answer depends on the organism. In fruit flies, the male's X chromosome appears to be hyperactivated, allowing it to be transcribed at twice the rate of either X chromosome in the female, giving the XY male *Drosophila* an X gene dosage equivalent to that of an XX female. In mammals, in contrast, the rule is that no matter how many X chromosomes are present, there is only one transcriptionally active X chromosome in each somatic cell, which gives the XX female mammal an X gene dosage equivalent to that of an XY male. Dosage compensation in mammals is achieved by random **X chromosome inactivation.** (A person with two X chromosomes, for example, is a mosaic of two cell types in which one or the other X is active.) Thus, XY and XX individuals produce the same amounts of X chromosome housekeeping-gene products. X chromosome inactivation also explains why triplo-X humans are phenotypically normal, inasmuch as only one of the three X chromosomes is transcriptionally active in a given cell. Similarly, an XXY male is only moderately affected because only one of his two X chromosomes is active in each cell.

Why are XXY individuals abnormal at all, given that triplo-X individuals are phenotypically normal? It turns out that a few genes scattered throughout an "inactive X" are still transcriptionally active. In XXY males, these genes are transcribed at twice the level they are in XY males. In XXX females, on the other hand, the few transcribed genes are active at only 1.5 times the level that they are in XX females. This lower level of "functional aneuploidy" in XXX than in XXY, plus the fact that the active X genes appear to lead to feminization, may explain the feminized phenotype of XXY individuals. The severity of Turner syndrome (XO) may be due to the deleterious effects of monosomy and to the lower activity of the transcribed genes of the X (compared with XX females). As is usually observed for aneuploids, monosomy for the X chromosome produces a more abnormal phenotype than does having an extra copy of the same chromosome (triplo-X females or XXY males).

Gene dosage is also important in the phenotypes of polyploids. Human polyploid zygotes do arise through various kinds of mistakes in cell division. Most die in utero. Occasionally, triploid babies are born, but none survive. This fact seems to violate the principle that we have been discussing—namely, that polyploids are more normal than aneuploids. The explanation for this violation seems to lie with X chromosome dosage compensation. Part of the rule for gene balance in organisms that have a single active X seems to be that there must be one active X for every two copies of the autosomal chromosome complement. Thus, some cells in triploid mammals are found to have one active X, whereas others, surprisingly, have two. Neither situation is in balance with autosomal genes. Presumably this functional underrepresentation (in $3n$ cells with one active X) or functional overrepresentation (in $3n$ cells with two active X's) of housekeeping genes leads to substantial functional aneuploidy and the inviability of triploid individuals.

MESSAGE

Aneuploidy is nearly always deleterious because of gene imbalance—the ratio of genes is different from that in euploids, and this difference interferes with the normal function of the genome.

CHANGES IN CHROMOSOME STRUCTURE

Changes in chromosome structure, called **rearrangements,** encompass several major classes of events. A chromosome segment can be lost, constituting a **deletion,** or doubled to form a **duplication.** The direction of a segment within the chromosome can be reversed, constituting an **inversion.** Or a segment can be moved to a different chromosome, constituting a **translocation.** Each of these events can be caused by the breakage of DNA double helices at two different locations, followed by a rejoining of the broken ends to produce a new chromosomal arrangement (Figure 11-19a on the next page). Chromosomal rearrangements by breakage can be induced artificially by using ionizing radiation. This kind of radiation, of which X rays and gamma rays are the domains most commonly used, is highly energetic and causes numerous double-stranded breaks in DNA.

To understand how chromosomal rearrangements are produced by breakage, several points should be kept in mind:

1. Each chromosome is a single double-stranded DNA molecule.
2. The first event in the production of a chromosomal rearrangement is the generation of two or more double-stranded breaks in the chromosomes of a cell (see Figure 11-19a, top row at left).
3. Double-stranded breaks are potentially lethal, unless they are repaired.
4. Repair systems in the cell correct the double-stranded breaks by joining broken ends back together (see Chapter 10 for a detailed discussion of DNA repair).
5. If the two ends of the same break are rejoined, the original DNA order is restored. If the ends of two different breaks are joined together, however, one or another type of chromosomal rearrangement is produced.
6. The only recoverable chromosomal rearrangements are those that produce DNA molecules that have one centromere and two telomeres. If a rearrangement produces a chromosome that lacks a centromere, such an **acentric** chromosome will not be dragged to either pole at anaphase of mitosis or meiosis and will not be incorporated into either progeny nucleus. Therefore acentric chromosomes are not inherited. If a rearrangement produces a chromosome with two centromeres (a **dicentric**), it will often be simultaneously dragged to opposite poles at anaphase, forming an **anaphase bridge.** Anaphase bridge chromosomes typically will not be incorporated into either progeny cell, depending on the organism under consideration. If a chromosome break produces a chromosome lacking a telomere, that chromosome cannot replicate properly. Recall from Chapter 4 that telomeres are needed to prime proper DNA replication at the ends (see Figure 4-11).
7. If a rearrangement duplicates or deletes a segment of a chromosome, gene balance may be affected. The larger the segment that is lost or duplicated, the more likely it is that the change will cause phenotypic abnormalities because of gene imbalance.

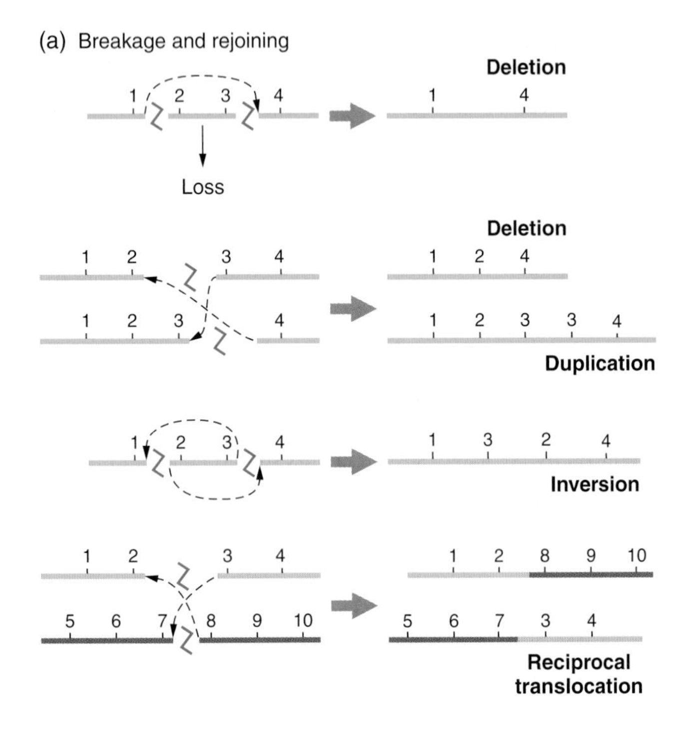

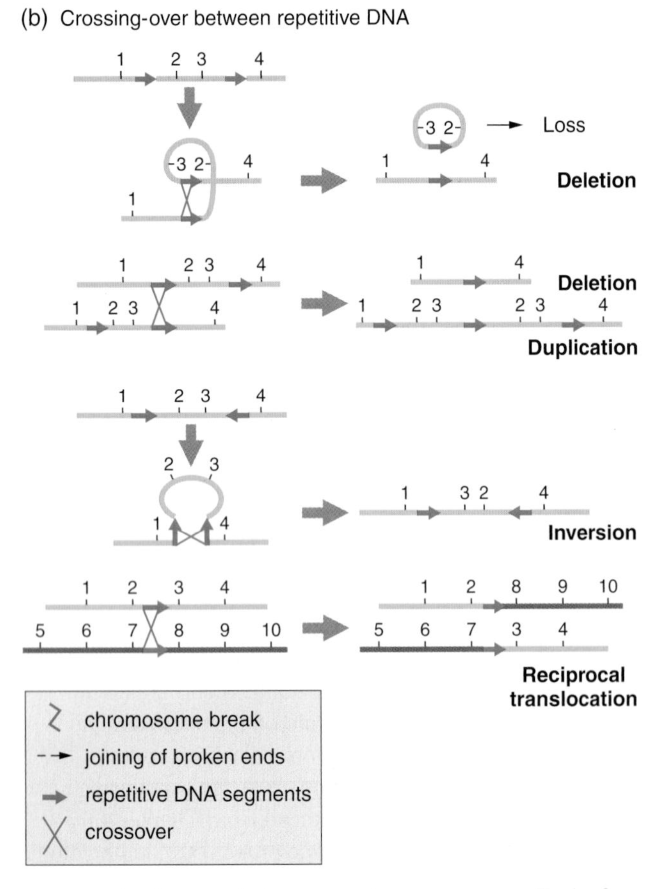

Figure 11-19 Origins of chromosomal rearrangements. Each of the four types of chromosomal rearrangements can be produced by either of two basic mechanisms: (a) chromosome breakage and rejoining or (b) crossing-over between repetitive DNA. Chromosome regions are numbered 1 through 10. Homologous chromosomes are the same color.

In organisms with repeated short DNA sequences—homologous repetitive segments within one chromosome or on different chromosomes—there is some ambiguity in which of the repeats will pair with each other at meiosis. If pairing occurs between sequences that are not in the same relative positions on the homologs, crossing-over can lead to aberrant chromosomes. Deletions, duplications, inversions, and translocations can all be produced by such crossing-over (see Figure 11-19b). Thus, in addition to chromosome breakage, crossing-over between repetitive segments probably constitutes a significant source of chromosomal rearrangements.

There are two general types of rearrangements, balanced and unbalanced. **Balanced rearrangements** change the chromosomal gene order but do not remove or duplicate any of the DNA of the chromosomes. The two simple classes of balanced rearrangements are inversions and reciprocal translocations.

An **inversion** is a rearrangement in which an internal segment of a chromosome has been broken twice, flipped 180 degrees, and rejoined:

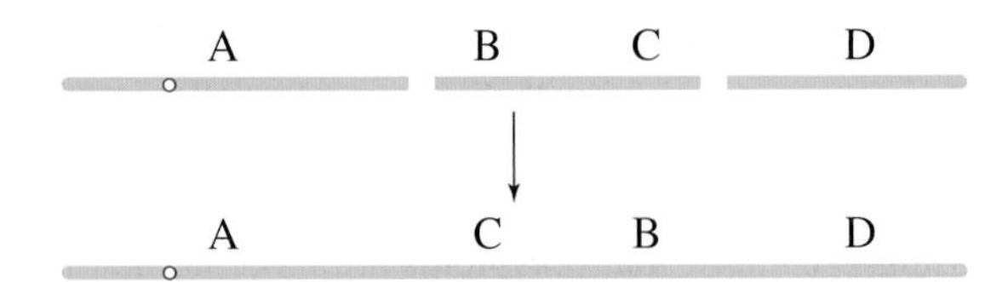

A **reciprocal translocation** is a rearrangement in which two nonhomologous chromosomes are each broken once, creating acentric fragments, which then trade places:

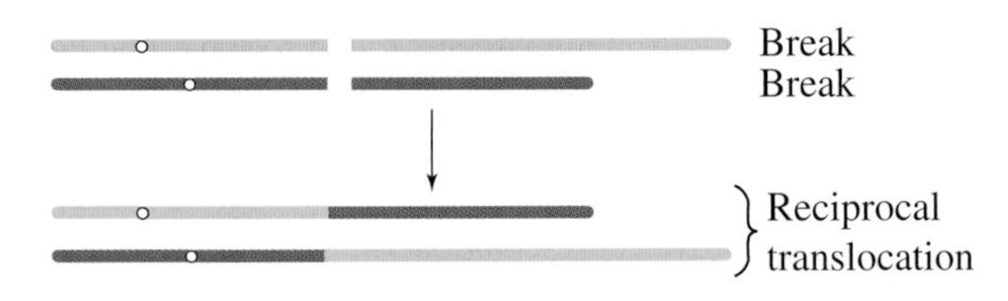

Note that, for both inversions and translocations, no chromosomal material is gained or lost—there is simply a change in the relative locations of genes on the rearranged chromosomes.

It is important to realize that, in addition to the effects of the rearrangement itself, the DNA molecules are disrupted at the location of each break. Sometimes these breaks occur *within* genes. When they do, they generally disrupt gene function because part of the gene moves to a new location and no complete transcript can be made. In addition, the DNA sequences on either side of the rejoined ends of a rearranged chromosome are ones that are not normally juxtaposed. Sometimes the junction occurs in such a way that a novel gene fusion is produced.

Unbalanced rearrangements change the gene dosage of a part of the affected chromosome. As with aneuploidy for whole chromosomes, the loss of one copy or the addition of an extra copy of a chromosome segment can disrupt normal gene balance. The two simple classes of unbalanced rearrangements are deletions and duplications. A **deletion** is the loss of a segment within one chromosome arm and the juxtaposition of the two segments on either side of the deleted segment, as in this example, which shows loss of segment C-D:

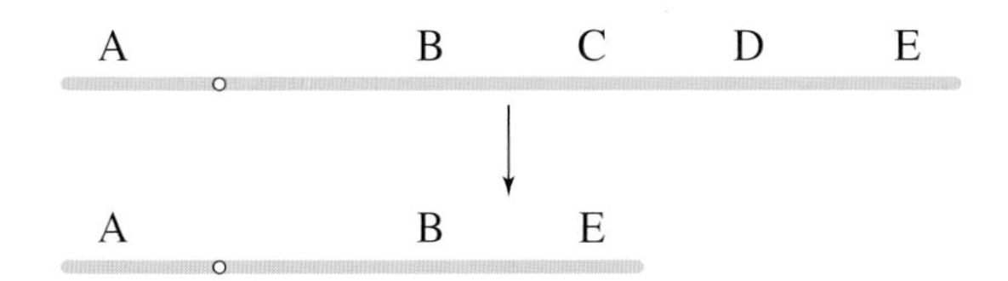

A **duplication** is a repetition of a segment of a chromosome arm. In the simplest type of duplication, the two segments are adjacent to each other (a tandem duplication), as in this duplication of segment C:

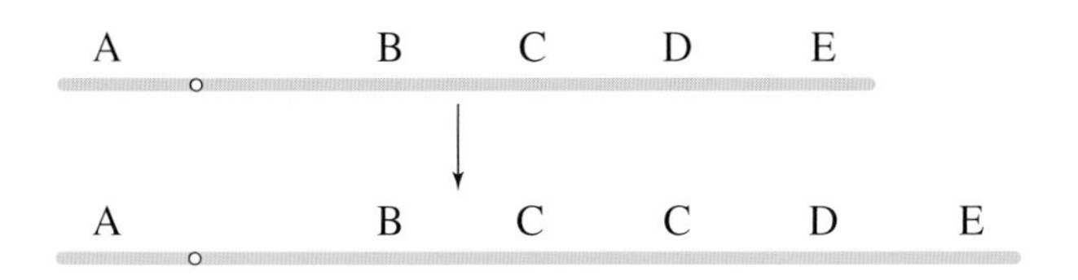

However, the duplicate segment can end up at a different position on the same chromosome, or even on a different chromosome.

The following sections consider the properties of these balanced and unbalanced rearrangements.

Inversions

Inversions are of two basic types. If the centromere is outside the inversion, the inversion is said to be **paracentric;** inversions spanning the centromere are **pericentric.**

Normal sequence A B C D E F

Paracentric A B C E D F

Pericentric A D C B E F

Because inversions are balanced rearrangements, they do not change the overall amount of genetic material, so they do not result in gene imbalance and generally do not lead to inviability nor show abnormalities at the phenotypic level. In some cases, however, one of the chromosome breaks is within a gene, which disrupts that gene and thus produces a mutation that may be phenotypically detectable. If the gene has an essential function, then the breakpoint acts as a lethal mutation linked to the inversion. In such a case, the inversion cannot be bred to homozygosity. However, many inversions can be made homozygous, and furthermore, inversions can be detected in haploid organisms; in these cases, the breakpoints of the inversion are clearly not in essential regions. Some of the possible consequences of inversion at the DNA level are shown in Figure 11-20 on the next page.

Most analyses of inversions are carried out on diploid cells that contain one normal chromosome set plus one set carrying the inversion. This type of cell is called an **inversion heterozygote,** but note that this designation does not imply that any gene locus is heterozygous, but rather that one normal and one abnormal chromosome set are present. Microscopic observation of meiosis in inversion heterozygotes reveals the location of the inverted segment because one chromosome twists once at the ends of the inversion to pair with the other, untwisted chromosome; in this way the paired homologs form an **inversion loop** (Figure 11-21 on page 367).

At meiosis, crossing-over within the inversion loop of a heterozygous paracentric inversion connects homologous centromeres in a **dicentric bridge** while also producing an **acentric fragment** (Figure 11-22 on page 367). Then, as the chromosomes separate during anaphase I, the centromeres remain linked by the bridge. The acentric fragment cannot align itself or move, and consequently it is lost. Tension eventually breaks the dicentric bridge, forming two chromosomes with terminal deletions. The gametes containing such chromosomes may be inviable, but even if they are viable, the zygotes that they eventually form will probably be inviable. Hence, a crossover event, which normally generates the recombinant class of meiotic products, instead produces lethal products. The overall result is a drastically lower recombinant frequency. In fact, for genes within the inversion, the RF is close to zero. (It is not exactly zero because double crossovers involving only two chromatids—which are rare—are viable.) For genes flanking the inversion, the RF is reduced in proportion to the size of the inversion, because for a longer inversion, there is a greater probability of a crossover occurring within it, and hence of an inviable meiotic product.

Inversions affect recombination in another way, too. Inversion heterozygotes often have mechanical pairing problems in the region of the inversion. The inversion loop causes a large distortion that can extend beyond the loop itself. This distortion reduces the opportunity for crossing-over in the neighboring regions.

The net genetic effect of a heterozygous pericentric inversion is the same as that of a paracentric inversion—crossover products are not recovered—but the reasons are different. In a pericentric inversion, because the centromeres are contained within the inverted region, the chromosomes that have engaged in crossing-over separate in the normal fashion, without the creation of a bridge. However, the crossover produces chromatids that contain a duplication and a deletion for different parts of the chromosome (Figure 11-23 on page 368). In this case, if a gamete carrying a crossover chromosome is fertilized, the zygote dies

Breakpoints *between* genes

Normal sequence

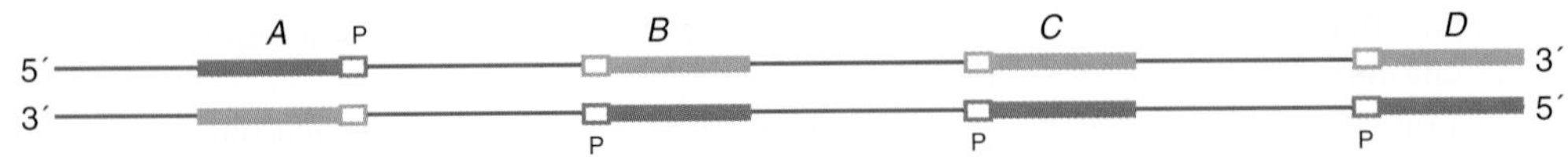

Breaks in DNA

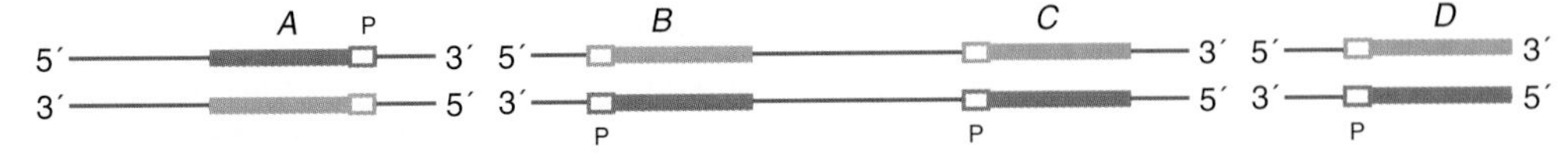

Inverted alignment

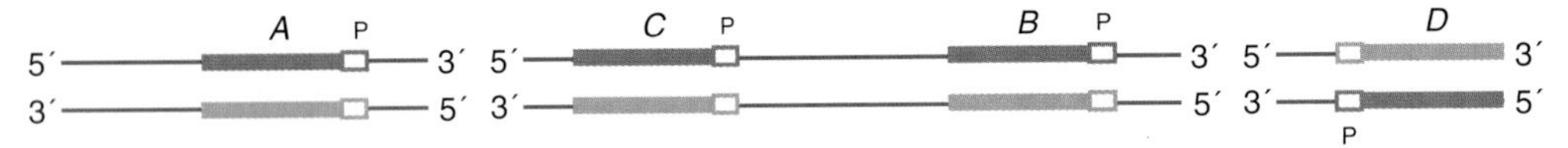

Joining of breaks to complete inversion

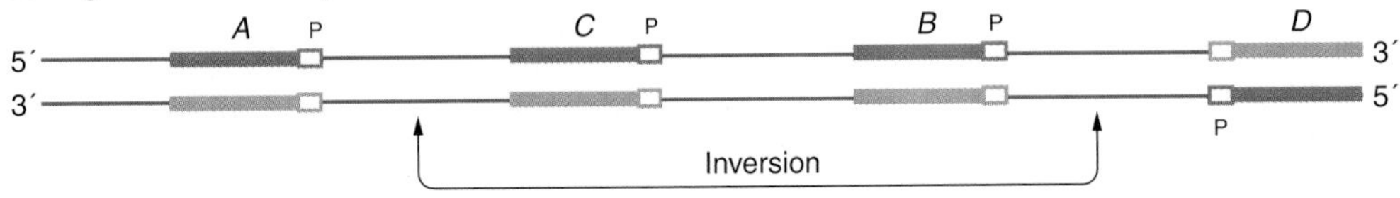

One breakpoint *between* genes
One *within* gene *C* (*C* disrupted)

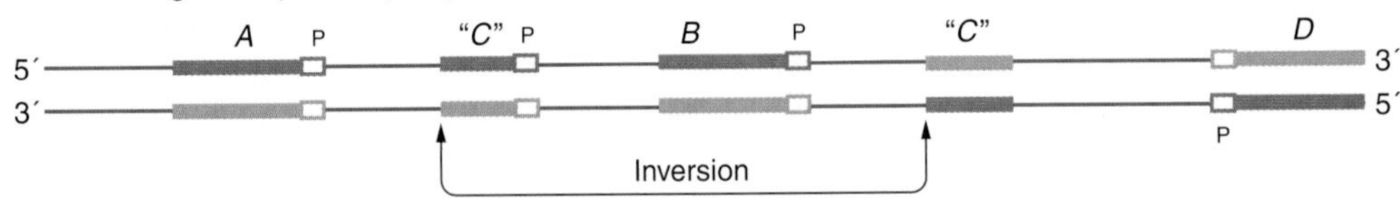

Breakpoints *in* genes *A* and *D*
Creating gene fusions

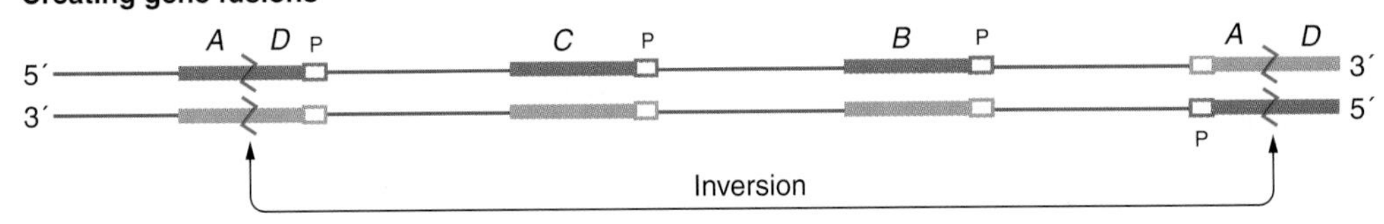

Figure 11-20 Effects of inversions at the DNA level. Genes are represented by *A*, *B*, *C*, and *D*. Template strand is dark green; nontemplate strand is light green; jagged lines indicate where breaks in the DNA produced gene fusions (*A* with *D*) after inversion and rejoining. The letter P stands for promoter; arrows indicate the positions of the breakpoints.

because of its gene imbalance. Again, the result is the selective recovery of noncrossover chromatids in viable progeny, which greatly lowers the RF value.

Let us consider an example of the effects of an inversion on recombinant frequency. A wild-type *Drosophila* specimen from a natural population is crossed with a homozygous recessive laboratory stock *dp cn* / *dp cn*. (The *dp* allele codes for dumpy wings and *cn* codes for cinnabar eyes. The two genes are known to be 45 map units apart on chromosome 2.) The F_1 generation is wild-type. When an F_1 female is crossed with the recessive parent, the progeny are

250	wild type	+ + / *dp cn*
246	dumpy cinnabar	*dp cn* / *dp cn*
5	dumpy	*dp* + / *dp cn*
7	cinnabar	+ *cn* / *dp cn*

In this cross, which is effectively a dihybrid testcross, 45 percent of the progeny are expected to be dumpy or cinnabar (they constitute the crossover classes), but only 12 out of 508, about 2 percent, are obtained. Something is reducing crossing-over in this region, and a likely explanation is an inversion spanning most of the *dp-cn* region. Because the expected RF was based on measurements made on laboratory strains, the wild-type fly from nature was the most likely source of the inverted chromosome. Hence chromosome 2 in the F_1 can be represented as follows:

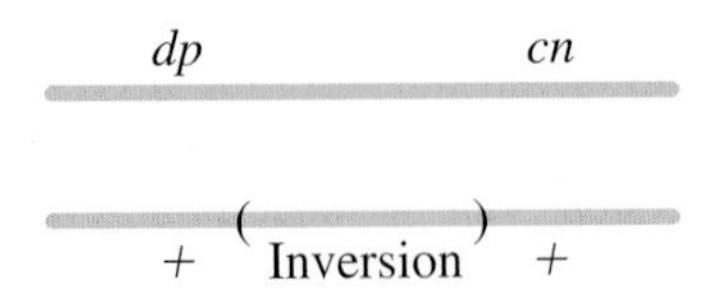

Animated Art • Chromosome Rearrangements: Formation of Paracentric Inversions

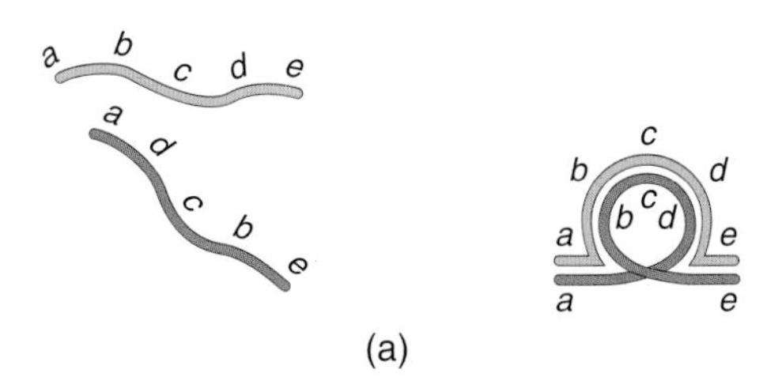

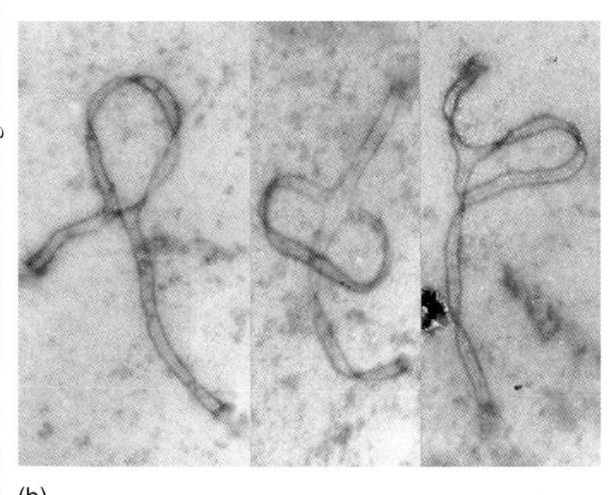

Figure 11-21 The chromosomes of inversion heterozygotes pair in a loop at meiosis. (a) Diagrammatic representation (remember that each chromosome is actually a pair of sister chromatids). (b) Electron micrographs of synaptonemal complexes at prophase I of meiosis in a mouse heterozygous for a paracentric inversion. Three different meiocytes are shown. *(Part b from M. J. Moses, Department of Anatomy, Duke Medical Center.)*

Pericentric inversions also can be detected microscopically through new arm ratios. Consider the following pericentric inversion:

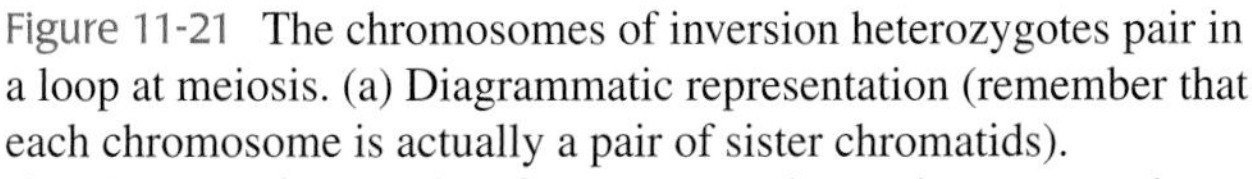

Note that the length ratio of the long arm to the short arm has been changed from about 4:1 to about 1:1 by the inversion. Paracentric inversions do not alter the arm ratio, but they may be detected microscopically if banding or other chromosomal landmarks are available.

MESSAGE

The main diagnostic features of heterozygous inversions are inversion loops, reduced recombinant frequency, and reduced fertility because of unbalanced or deleted meiotic products.

In some experimental systems, notably the fruit fly *(Drosophila)* and the nematode *(Caenorhabditis elegans),* inversions are used as balancers. A **balancer** chromosome contains *multiple* inversions, so that when it is combined with the corresponding wild-type chromosome, there can be no viable crossover products. In some analyses, it is important to keep all the alleles on one chromosome together; combining such chromosomes with a balancer eliminates crossovers, and only parental combinations survive. Balancer chromosomes are marked with a dominant morphological mutation. The marker allows the geneticist to track the segregation of the entire balancer or its normal homolog by following the presence or absence of the marker.

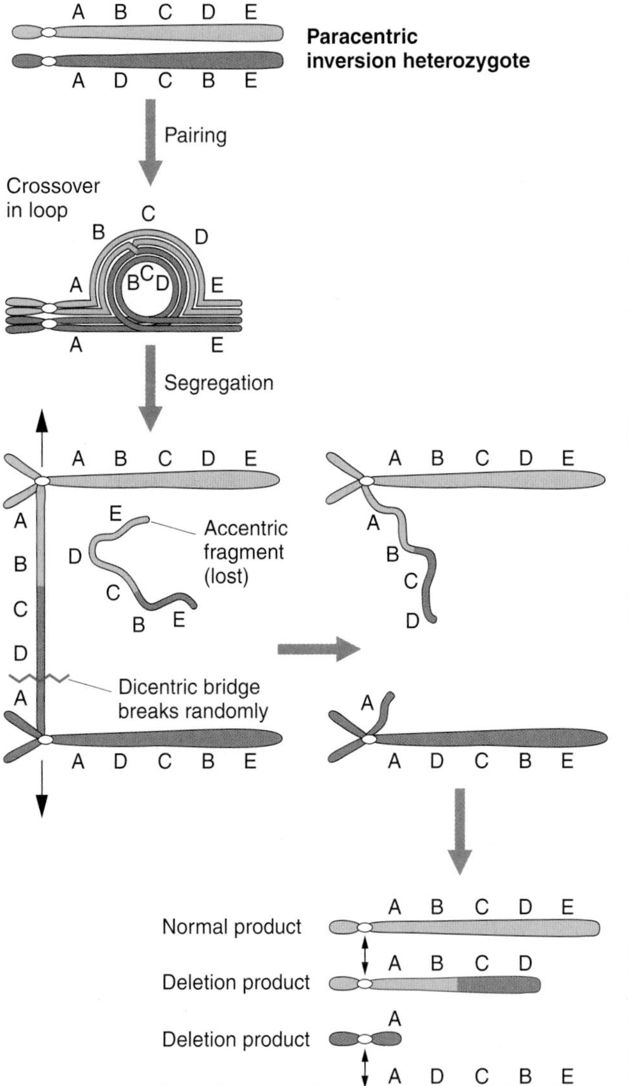

Figure 11-22 Meiotic products resulting from a single crossover within a paracentric inversion loop. Two nonsister chromatids cross over within the loop.

Figure 11-23 Meiotic products resulting from a single crossover within a pericentric inversion loop.

Reciprocal Translocations

There are several types of translocations, but only reciprocal translocations (the simplest type) will be illustrated here. Meiosis in heterozygotes having two translocated chromosomes and their normal counterparts causes some important genetic and cytological effects. Again, as with other rearrangements, the pairing affinities of homologous regions dictate a characteristic configuration for chromosomes synapsed in meiosis. Figure 11-24, which illustrates meiosis in a heterozygote for a reciprocal translocation, shows that the pairing configuration is cross-shaped. Because the law of independent meiotic assortment is still in force, there are two common patterns of segregation. Let us use N_1 and N_2 to represent the normal chromosomes, and T_1 and T_2 the translocated chromosomes. The segregation of each of the structurally normal chromosomes with one of the translocated ones ($T_1 + N_2$ and $T_2 + N_1$) is called **adjacent-1 segregation.** Both meiotic products are duplicated and deficient for different arms of the cross. These products are inviable. On the other hand, the two normal chromosomes may segregate together, as will the reciprocal parts of the translocated ones, to produce $N_1 + N_2$ and $T_1 + T_2$ products. This segregation pattern is called **alternate segregation.** These products are both balanced and viable.

As a result of the equal numbers of adjacent-1 and alternate segregations, half the overall population of gametes will be nonfunctional, a condition known as **semisterility** or "half sterility." Semisterility is an important diagnostic tool for identifying translocation heterozygotes. However, semisterility is defined differently for plants and animals. In plants, the 50 percent unbalanced meiotic products from the adjacent-1 segregation generally abort at the gametic stage (Figure 11-25). In animals, the duplication-deletion products are viable as gametes but lethal to the zygotes they produce upon fertilization.

Remember that heterozygotes for inversions may also show some reduction in fertility, but by an amount dependent on the size of the affected region; the precise 50 percent reduction in viable gametes or zygotes is usually a reliable diagnostic clue for a translocation.

Genetically, genes on translocated chromosomes act as though they are linked if their loci are close to the translocation breakpoint. Figure 11-26 shows a translocation heterozygote that has been established by crossing an $a\,/\,a$; $b\,/\,b$ individual with a translocation homozygote bearing the wild-type alleles. On testcrossing the heterozygote, the only viable progeny are those bearing the parental genotypes, so linkage is seen between loci that were originally on different chromosomes. Apparent linkage of genes

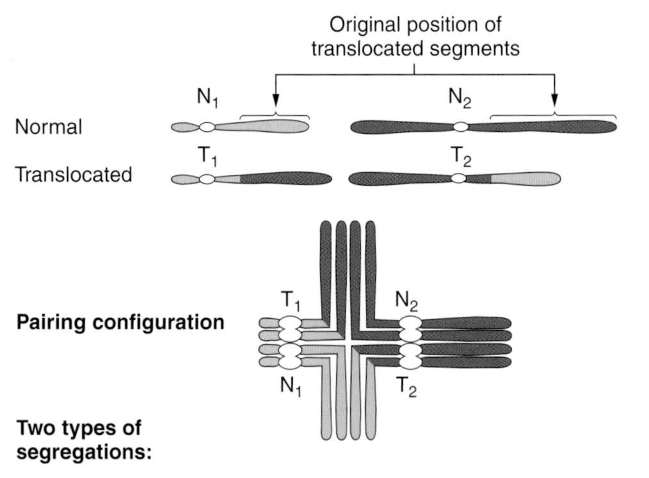

Adjacent-1		Products	
Up	$T_1 + N_2$	Duplication of purple, deletion of orange translocated segment	Often inviable
Down	$T_2 + N_1$	Duplication of orange, deletion of purple translocated segment	
Alternate			
Up	$T_1 + T_2$	Translocation genotype	Both complete and viable
Down	$N_1 + N_2$	Normal	

Figure 11-24 The meiotic products resulting from the two most commonly encountered chromosome segregation patterns in a reciprocal translocation heterozygote. N_1 and N_2 = normal nonhomologous chromosomes; T_1 and T_2 = translocated chromosomes.

Figure 11-25 Photomicrograph of normal and aborted pollen of a semisterile corn plant. The clear pollen grains contain chromosomally unbalanced meiotic products of a reciprocal translocation heterozygote. The opaque pollen grains, which contain either the complete translocation genotype or normal chromosomes, are functional in fertilization and development. *(William Sheridan.)*

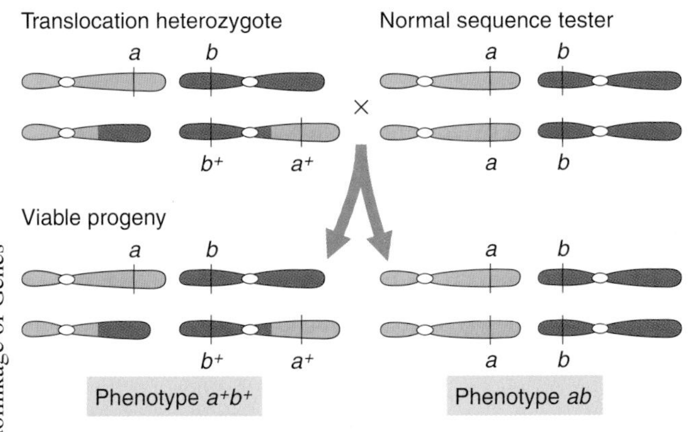

Figure 11-26 When a translocated fragment carries a marker gene, this marker can show linkage to genes on the other chromosome because the recombinant genotypes (in this case, $a^+ . \, b$ and $a \, . \, b^+$ tend to be in duplication-deletion gametes and do not survive.

Animated Art • Chromosome Rearrangements: Pseudolinkage of Genes

known normally to be on separate nonhomologous chromosomes—sometimes called **pseudolinkage**—is a genetic diagnostic clue to the presence of a translocation.

MESSAGE

Heterozygous reciprocal translocations are diagnosed genetically by semisterility and by the apparent linkage of genes whose normal loci are on separate chromosomes.

Applications of Inversions and Translocations

Translocations and inversions have proved to be useful genetic tools; some examples of their uses follow.

Gene mapping. Translocations and inversions are useful for the mapping and subsequent isolation of specific genes. The gene for human neurofibromatosis was isolated in this way. The critical information came from people who not only had the disease, but also carried chromosomal translocations. All the translocations had one breakpoint in common, in a band close to the centromere of chromosome 17. Hence it appeared that this must be the locus of the neurofibromatosis gene, which had been disrupted by the translocation breakpoint. Subsequent analysis showed that the chromosome 17 breakpoints were not at identical positions; however, since they must have been within the gene, the range of their positions revealed the segment of the chromosome that constituted the neurofibromatosis gene. Isolation of DNA fragments from this region eventually led to the recovery of the gene itself.

Synthesizing specific duplications or deletions. Translocations and inversions are routinely used to delete and duplicate specific chromosome segments. Recall, for example, that both translocations and pericentric inversions generate products of meiosis that contain a duplication *and* a deletion (see Figures 11-23 and 11-24). If the dimensions of the parental rearrangement are such that the duplicated or the deleted segment is very small, then the duplication-deletion meiotic products are tantamount to deletions or duplications, respectively. Duplications and deletions are useful for a variety of experimental applications, including the mapping of genes and the varying of gene dosage for the study of regulation, as we shall see in the following sections.

Another approach uses unidirectional *insertional* translocations, in which a segment of one chromosome is removed and inserted into another. In an insertional translocation heterozygote, if the chromosome with the insertion segregates along with the normal copy, a duplication results.

Position-effect variegation. Gene action can be affected by proximity to the densely staining chromosome regions called *heterochromatin,* and translocations or inversions can be used to study this effect. The locus for white eye color in *Drosophila* is near the tip of the X chromosome. Consider a translocation in which the tip of an X chromosome carrying w^+ is relocated next to the heterochromatic region of, say, chromosome 4 (Figure 11-27a, top section, on the next page). **Position-effect variegation** is observed in flies that are heterozygotes for such a translocation, in which the normal X chromosome carries the recessive allele w. The eye phenotype is expected to be red because the wild-type allele is dominant to w. However, in such cases, the observed phenotype is a variegated mixture of red and white eye facets (Figure 11-27b). How can we explain the white areas? The w^+ allele is not always expressed because the heterochromatin boundary is somewhat variable: in some cells it engulfs and inactivates the w^+ gene, thereby allowing the expression of w. If the position of the w^+ and w alleles is exchanged by a crossover, then position-effect variegation is not detected (Figure 11-27a, lower section).

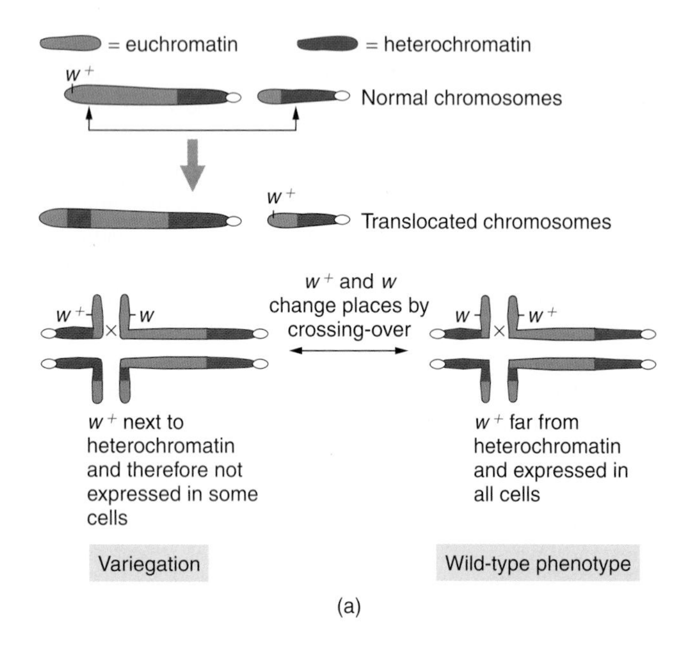

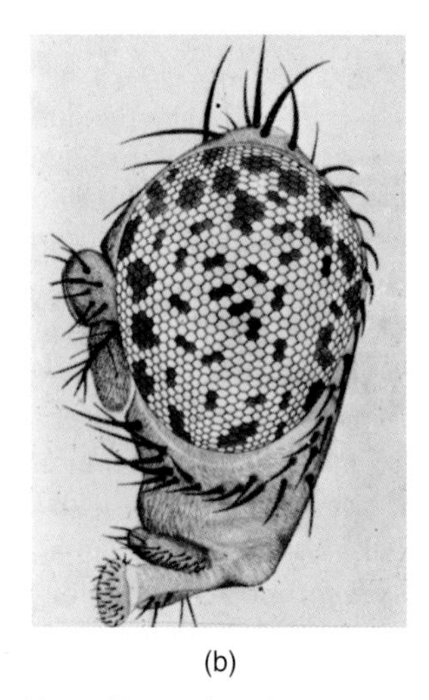

Figure 11-27 Position-effect variegation. (a) The translocation of w^+ to a position next to heterochromatin causes the w^+ function to fail in some cells, producing position-effect variegation. (b) A *Drosophila* eye showing position-effect variegation. *(Part b from Randy Mottus.)*

Position-effect variegation can be used to study the regulatory effects of heterochromatin and thereby the effects of chromosome condensation (coiling), a key feature of chromosome structure.

Deletions

A deletion is simply the loss of a part of one chromosome arm. The process of deletion requires two chromosome breaks to cut out the intervening segment. The deleted fragment has no centromere; consequently, it cannot be pulled to a spindle pole in cell division and is lost. The effects of deletions depend on their size. A small deletion *within* a gene, called an **intragenic deletion,** inactivates the gene and has the same effect as other null mutations of that gene. If the homozygous null phenotype is viable (as, for example, in human albinism), then the homozygous deletion also will be viable. Intragenic deletions can be distinguished from mutations caused by single nucleotide changes because deletions are nonrevertible.

For most of this section, we shall be dealing with **multigenic deletions,** which have more severe consequences than do intragenic deletions. If such a deletion is made homozygous by inbreeding (that is, if both homologs have the same deletion), then the combination is always lethal. This fact suggests that all regions of the chromosomes are essential for normal viability and that complete elimination of any segment from the genome is deleterious. Even an individual organism heterozygous for a multigenic deletion—that is, having one normal homolog and one that carries the deletion—may not survive. Principally, this lethal outcome is due to disruption of normal gene balance. Another cause is the expression of deleterious recessive alleles uncovered by the deletion.

MESSAGE

The lethality of large heterozygous deletions can be explained by gene imbalance and the expression of deleterious recessives.

Small deletions are sometimes viable in combination with a normal homolog. Such deletions may be identified by cytogenetic analysis. If meiotic chromosomes are examined, the region of the deletion can be identified by the failure of the corresponding segment on the normal homolog to pair, resulting in a **deletion loop** (Figure 11-28a). In dipteran insects, deletion loops are also detected in the polytene chromosomes, in which the homologs are tightly paired and aligned (Figure 11-28b). A deletion can be assigned to a specific chromosome location by examining polytene chromosomes microscopically and determining the position of the deletion loop.

Another clue to the presence of a deletion is that deletion of a segment on one homolog sometimes unmasks recessive alleles present on the other homolog, leading to their unexpected expression. Consider, for example, the deletion shown in the following diagram:

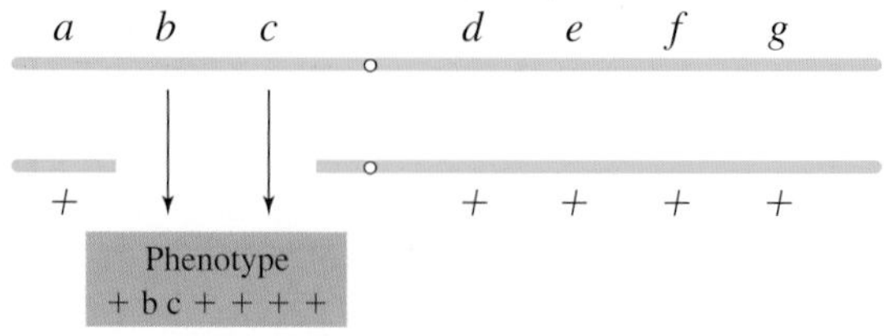

In this case, none of the six recessive alleles is expected to be expressed; however, if b and c are expressed, then a deletion spanning the b^+ and c^+ genes has probably occurred on the other homolog. Because in such cases it seems as if recessive alleles are showing dominance, the effect is called **pseudodominance.**

(a) Meiotic chromosomes

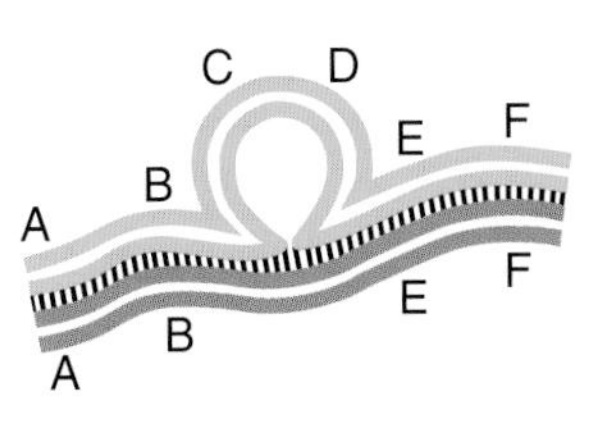

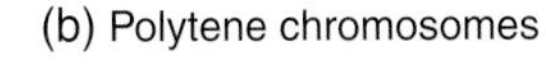

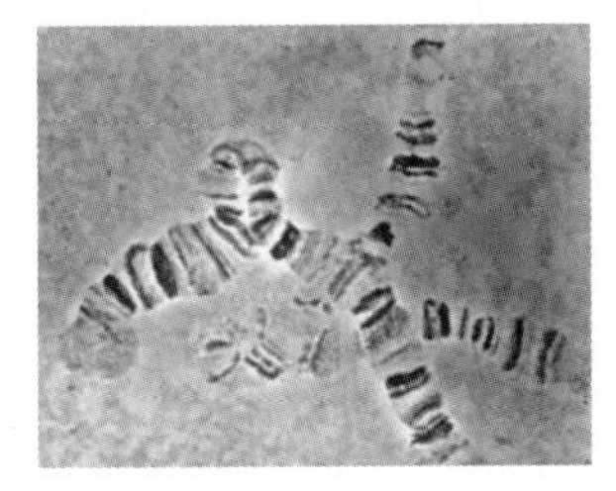

Figure 11-28 Looped configurations in a *Drosophila* deletion heterozygote. (a) In meiotic pairing, the normal homolog forms a loop. The genes in this loop have no alleles with which to synapse. (b) Because polytene chromosomes in *Drosophila* have specific banding patterns, we can infer which bands are missing from the homolog with the deletion by observing which bands appear in the loop of the normal homolog. *(Part b from William M. Gelbart.)*

The pseudodominance effect can also be applied in the opposite direction by using a set of defined overlapping deletions to locate the map positions of mutant alleles. This procedure is called **deletion mapping.** An example from the fruit fly *(Drosophila)* is shown in Figure 11-29. In this diagram, the recombination map is shown at the top, marked with distances in map units from the left end. The horizontal red bars below the chromosome show the extent of the deletions listed at the left. Each deletion is paired with each mutation under test, and the phenotype is observed to see if the mutation is pseudodominant. The mutation prune *(pn)*, for example, shows pseudodominance only with deletion 264-38, and this determines its location in the 2D-4 to 3A-2 region. However, *fa* shows pseudodominance with all but two deletions (258-11 and 258-14), so its position can be pinpointed to band 3C-7, which is the region that all but two deletions have in common.

MESSAGE

Deletions can be recognized by deletion loops and pseudodominance.

Clinicians regularly find deletions in human chromosomes. The deletions are usually relatively small, but they do have an adverse phenotypic effect, even though heterozygous. Deletions of specific human chromosome regions cause unique syndromes of phenotypic abnormalities. An example is cri du chat syndrome, caused by a heterozygous deletion of the tip of the short arm of chromosome 5 (Figure 11-30 on the next page). The specific bands deleted in cri du chat syndrome are 5p15.2 and 5p15.3, the two most distal bands identifiable on 5p. The most characteristic phenotype in the syndrome is the one that gives it its name, the distinctive catlike mewing cries made by affected infants. Other phenotypic manifestations of the syndrome are microencephaly (abnormally small head) and a moonlike face. Like syndromes caused by other deletions, cri du chat syndrome also includes mental retardation. Fatality rates are low, and many persons with this deletion reach adulthood.

Most human deletions, such as those that we have just considered, arise spontaneously in the gonads of a normal parent of an affected person; thus no signs of the deletions are generally found in the chromosomes of the parents. Less commonly, deletion-bearing individuals can arise in the offspring of parents with an undetected balanced rearrangement. For example, adjacent segregation of a reciprocal translocation heterozygote or recombination within a

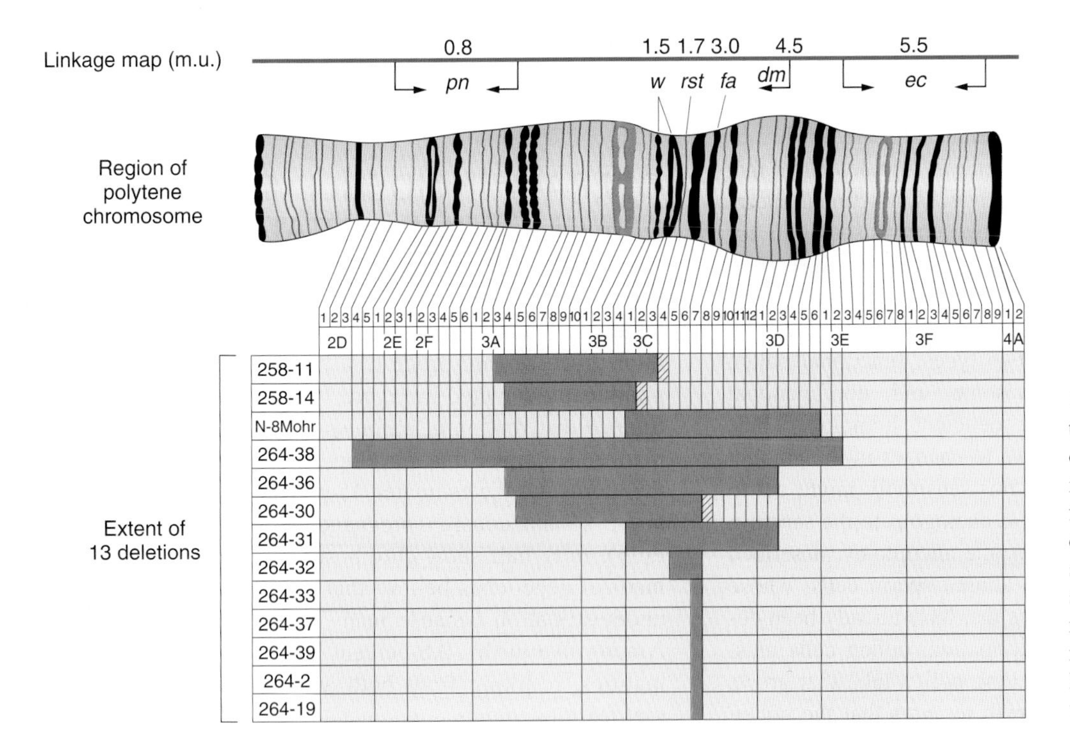

Figure 11-29 Locating genes to chromosome regions by observing pseudodominance in *Drosophila* heterozygous for deletion and normal chromosomes. The red bars show the extent of the deleted segments in 13 deletions. All recessive alleles in the same region that is deleted in a homologous chromosome will be expressed.

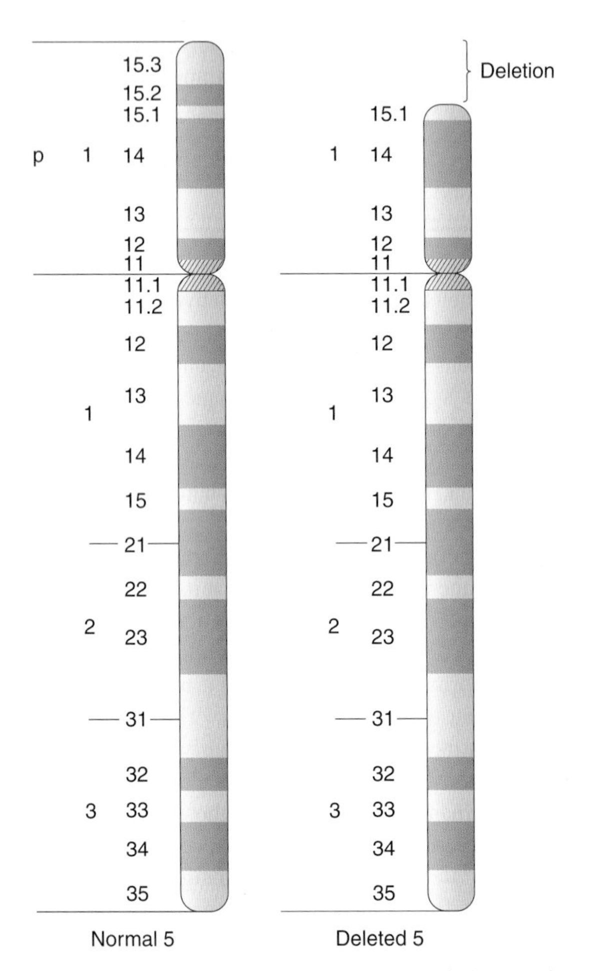

Figure 11-30 The cause of cri du chat syndrome in humans is loss of the tip of the short arm of one of the homologs of chromosome 5.

pericentric inversion heterozygote produces deletions. Cri du chat syndrome can result from a parent heterozygous for such a translocation.

Animals and plants show differences in the survival of gametes or offspring that bear deletions. A male animal that is heterozygous for a deletion produces functional sperm carrying one or the other of the two chromosomes in approximately equal numbers. In other words, sperm seem to function to some extent regardless of their genetic content. In diploid plants, on the other hand, the pollen produced by a deletion heterozygote is of two types: functional pollen carrying the normal chromosome and nonfunctional (aborted) pollen carrying the deficient homolog. Thus, pollen cells seem to be sensitive to changes in the amount of chromosomal material, and this sensitivity might act to weed out deletions. This effect is analogous to the sensitivity of pollen to whole-chromosome aneuploidy, described earlier in this chapter. Unlike animal sperm cells, whose metabolic activity uses enzymes that have already been deposited in them during their formation, pollen cells must germinate and then produce a long pollen tube that grows to fertilize the ovule. This growth requires that the pollen cell manufacture large amounts of protein, thus making it sensitive to genetic abnormalities in its own nucleus. Plant ovules, in contrast, are quite tolerant of deletions, presumably because of the nurturing effect of the surrounding maternal tissues.

Duplications

The processes of chromosome mutation sometimes produce an extra copy of some chromosome region. In considering a haploid organism, we can easily see why such a product is called a *duplication*—because the region is now present in duplicate. The duplicate regions can be located adjacent to each other—called a **tandem duplication**—or one copy can be in its normal location and the other in a novel location on a different part of the same chromosome or even on another chromosome—called an **insertional duplication.** In a diploid organism, the chromosome set containing the duplication is generally present together with a standard chromosome set. The cells of such an organism will have three copies of the chromosome region in question, but nevertheless such cells are generally referred to as duplication heterozygotes because they carry the product of one duplication event. In meiotic prophase, tandem duplication heterozygotes show a loop representing the unpaired extra region.

Synthetic duplications can be used for mapping genes by duplication coverage. In haploids, for example, by crossing a chromosomally normal strain carrying a new recessive mutation *m* to strains bearing a number of duplication-generating rearrangements (for example, translocations and pericentric inversions), *m*-bearing progeny can be obtained with a range of duplicated wild-type segments. If the progeny are "m" in phenotype, then the particular wild-type segment added to create the duplication does not span gene *m*, but, if the strain is wild-type, then *m* must be in that equivalent segment.

OVERALL INCIDENCE OF HUMAN CHROMOSOME MUTATIONS

Chromosome mutations arise surprisingly frequently in human sexual reproduction, showing that the relevant cellular processes are prone to a high level of error. This is graphically depicted in Figure 11-31, which shows the estimated distribution of chromosome mutations among human conceptions that develop sufficiently to implant in the uterus. Of the estimated 15 percent of conceptions that abort spontaneously (pregnancies that terminate naturally), fully half show chromosomal abnormalities. Some medical geneticists believe that even this high level is an underestimation because many cases are never detected. Among live births, 0.6 percent have chromosomal abnormalities, resulting from both aneuploidy and chromosomal rearrangements.

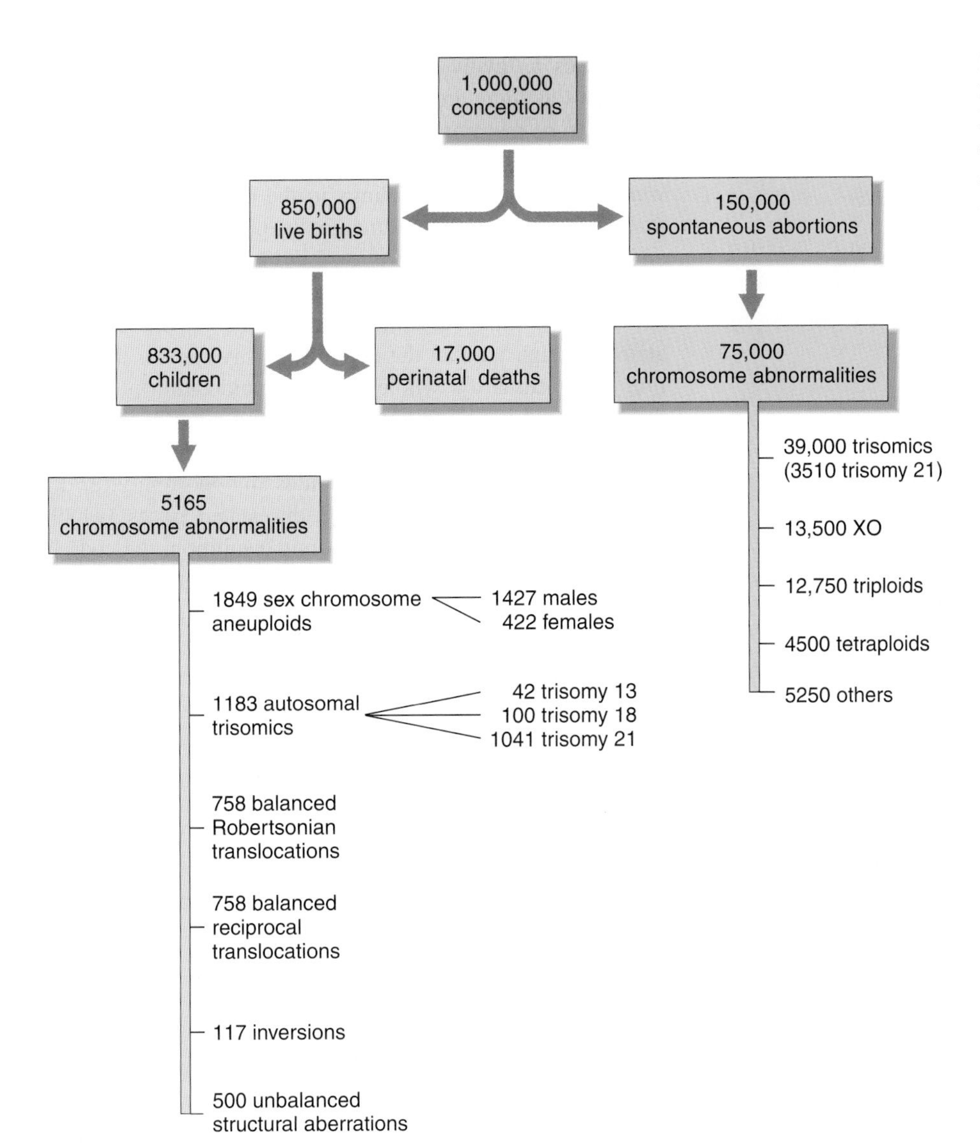

Figure 11-31 The fates of a million implanted human zygotes. (In Robertsonian translocations, centromeres fuse or dissociate.) *(From K. Sankaranarayanan,* Mutation Research *61, 1979.)*

EVOLUTION OF THE GENOME

Different phylogenetic groups differ in chromosome number and structure. In this section, we consider some of the ways that chromosome mutations contribute to the evolutionary divergence of genomes, and we explore the comparative molecular anatomy of genomes.

Chromosomal Polymorphism

Some natural populations show two or more chromosomal forms, a phenomenon called **chromosomal polymorphism.** *Drosophila,* because of its polytene chromosomes, has been a favorite organism for the study of natural chromosomal variation. In particular, *Drosophila* shows abundant polymorphism for chromosomal inversions—specifically, paracentric inversions (Figure 11-32 on the next page).

Inversion polymorphism is so common that determining the "wild-type" gene order in a species of *Drosophila* is quite arbitrary. There are many different gene orders, depending on the inversions present in a given individual fly. Because of the strong reduction in RF caused by inversions, the genes in the inverted region segregate as a unit—a so-called supergene—and they evolve to interact favorably with one another. The commonness of inversion polymorphisms in *Drosophila* is a consequence of two peculiarities of *Drosophila* meiosis that prevent the formation of inviable recombinant gametes. First, there is no crossing-over in *Drosophila* males, so the deleterious effects of crossing-over in an inversion are not encountered. Second, in oogenesis, if there is crossing-over inside the inversion, the bridge between the recombinant strands holds the two nuclei that are formed in the middle of a line of four nuclei produced during meiosis. Only the two end nuclei can become included in an egg, and since they were not involved in the crossover within the inversion, they do not contain a duplication-deletion. Other kinds of chromosomal rearrangement polymorphism seem to be rarer in *Drosophila* because they tend to be more deleterious than inversions and because they allow the formation of inviable gametes from crossing-over.

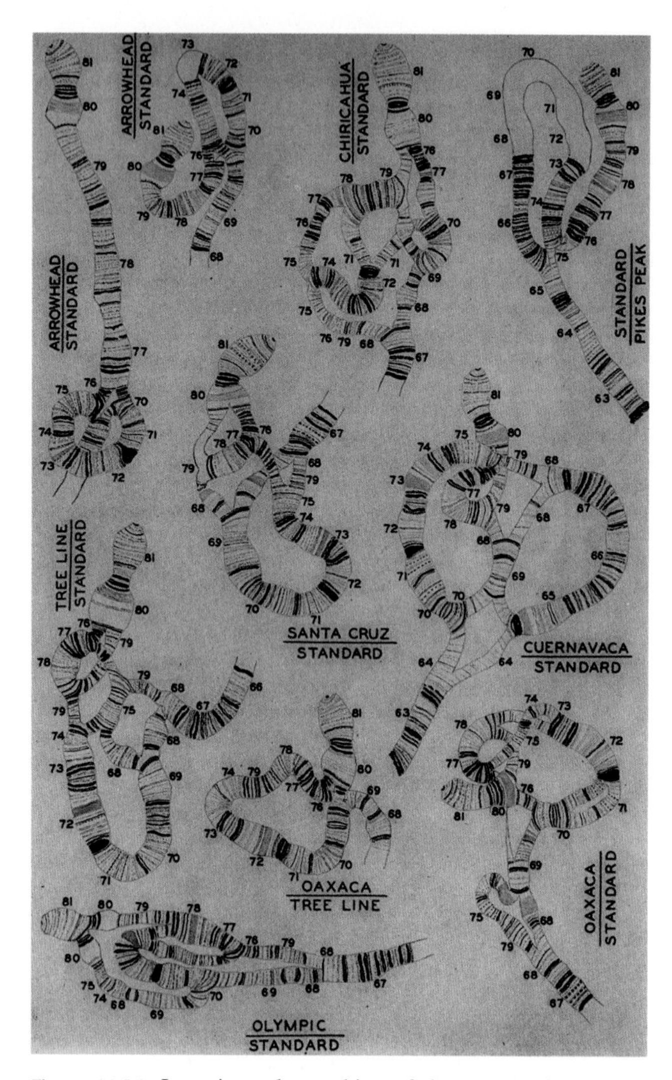

Figure 11-32 Inversion polymorphism of chromosome 3 among natural populations of *Drosophila*. Each chromosome 3 is given a name based on its geographical origin. Pairings of different chromosomes show loops in the polytene chromosomes, revealing the locations of inversions. *(Th. Dobzhansky, Chromosomal Races in Drosophila pseudoobscura and Drosophila persimilis, pp. 47–144, 1944. Carnegie Institution of Washington.)*

Chromosomal Changes and Speciation

How do chromosome mutations contribute to speciation? No general answer can be given to this question at present. However, we can describe the kinds of differences that we see between species, and these differences provide clues about the types of events that might underlie speciation.

First, consider gene families. As we saw in Chapter 9, some types of genes are found in several nearly identical copies per haploid genome; such groups of clearly related genes are called gene families. In some phylogenetic comparisons, the number of genes in certain gene families varies in interesting ways. For example, there are frequently four times as many members of a given gene family in mammals as there are in primitive chordates (a group considered to be mammalian ancestors). This finding has led to the suggestion that the primitive chordate genome underwent two successive doublings in the evolutionary lineage that eventually formed the mammals, providing another example of how polyploidization might have contributed to speciation.

Second, many chromosome regions appear to have arisen through tandem duplication. In any case of duplication (and certainly wholesale polyploidization), one of the duplicates is free to undergo gene mutation because the necessary basic gene functions will be provided by the other copy. Duplication thus provides an opportunity for divergence in the function of the duplicated genes, which could be advantageous in genomic evolution. Indeed, it seems certain that some groups of adjacent genes with closely related functions, such as the human globin genes, arose as tandem duplicates (see the discussion of the evolution of globin genes in Chapter 19).

MESSAGE

Duplication and polyploidization supply additional genetic material capable of evolving new functions.

Chromosomal Synteny

Inversions and translocations also are notable in comparisons of different species. Even in closely related species, a group of genes that forms a block in one species will be in a different position in another species, sometimes elsewhere on the same chromosome but often on a different chromosome. Thus the genome becomes scrambled in its order during evolution. Compare the genomes of humans and mice. The human genome includes 23 pairs of chromosomes; the mouse has only 20 pairs. Mapping of the same genes in mice and in humans has shown that many blocks of genes are the same in both. Such blocks of genes are said to be **syntenic.** A synteny map shows how the chromosome locations of the blocks in one species relate to those of the comparable blocks in another (Figure 11-33). However, a great deal of gene scrambling is also apparent. Between 50 and 100 different inversion and translocation events must have mixed and swapped these blocks to enable them to reside in their present arrangements in humans and mice.

MESSAGE

The study of synteny shows that chromosomal rearrangements have been involved in evolution at the chromosomal level.

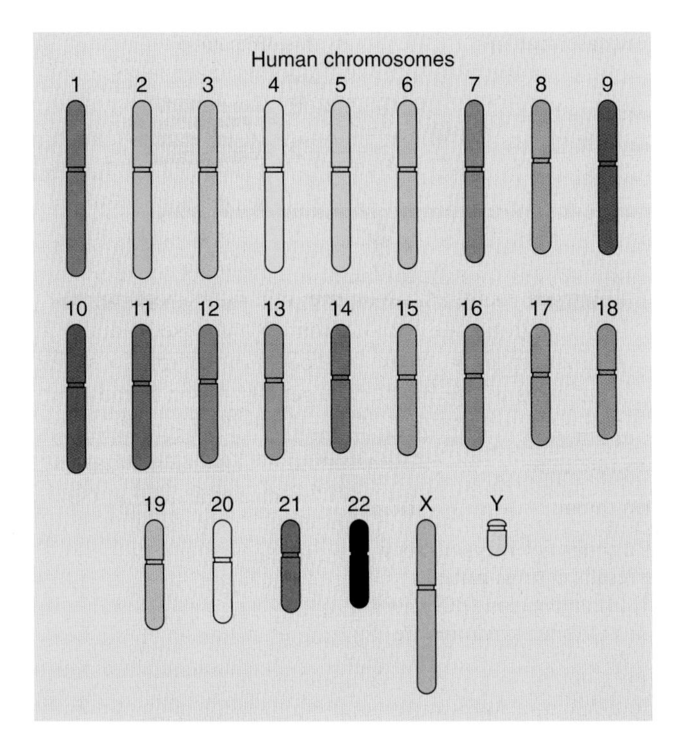

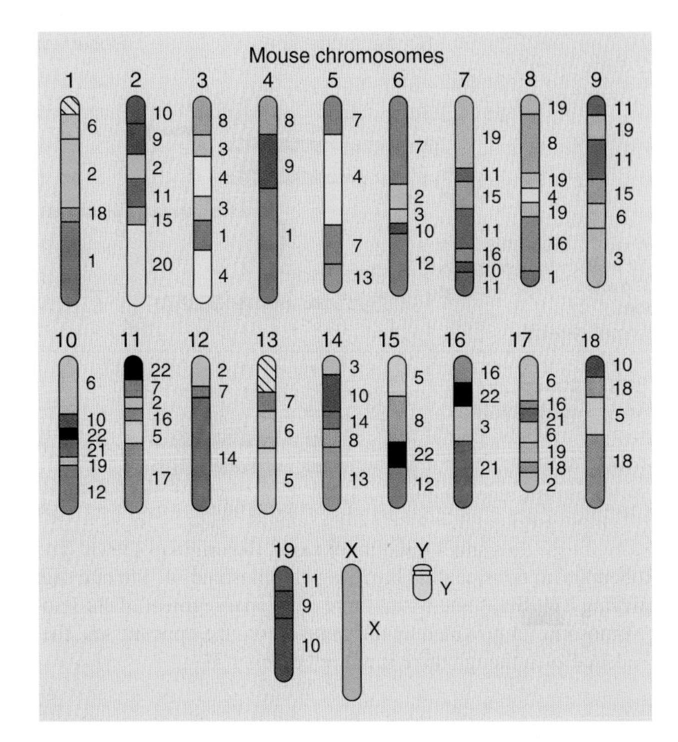

Figure 11-33 Synteny of human and mouse chromosomes. Each human chromosome has been given a different color. Gene mapping has revealed the locations of some of the same blocks of genes in the mouse. Since evolutionary divergence, multiple rearrangements have placed homologous blocks of genes in different combinations. *(From Lisa Stubbs, LLNL.)*

IDENTIFYING CHROMOSOME MUTATIONS BY GENOMICS

DNA microarrays (see Figure 9-45) can be made so that the DNA samples are laid out in chromosomal position. This has made it possible to detect increases or decreases in the representation of a given DNA segment in the genome of a mutant under test. This procedure is relevant to changes in whole chromosomes or to changes in parts of chromosomes. The technique is called *comparative genomic hybridization.* The total DNA of the wild type and of the mutant are labeled with two different fluorescent dyes that emit distinct wavelengths of light. These labeled DNAs are added to a cDNA microarray together, and they both hybridize to the array. The array is then scanned with a detector tuned to one fluorescent wavelength, and then again for the other wavelength. The ratio of values for each cDNA is calculated. Ratios for mutant/wild type substantially greater than 1 represent regions that have been amplified. A ratio of 2 points to a duplication, and a ratio of less than 1 points to a deletion. Some examples are shown in Figure 11-34.

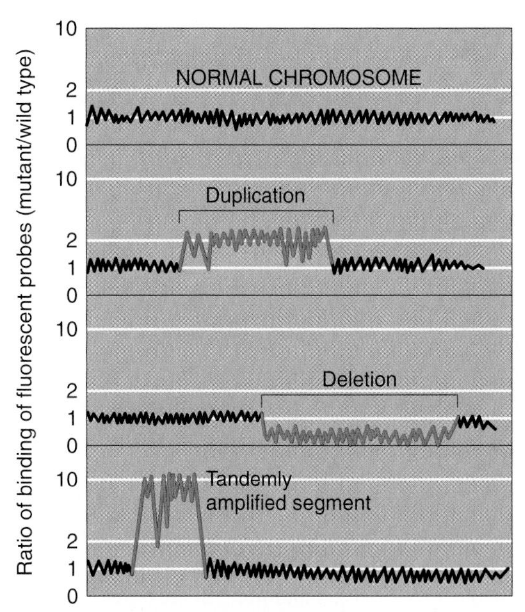

Figure 11-34 Detecting changes in the amount of chromosomal DNA by comparative genomic hybridization. Mutant and wild-type genomic DNA is tagged with dyes that fluoresce at different wavelengths. These are added to cDNA clones arranged in chromosomally ordered microarrays, and the ratio of bound fluorescence at each wavelength is calculated for each clone. The expected results for a normal genome and three types of mutants are illustrated.

Now that we have covered a number of analyses bearing on chromosome mutations, let's return to the family with the Down syndrome child, introduced at the beginning of the

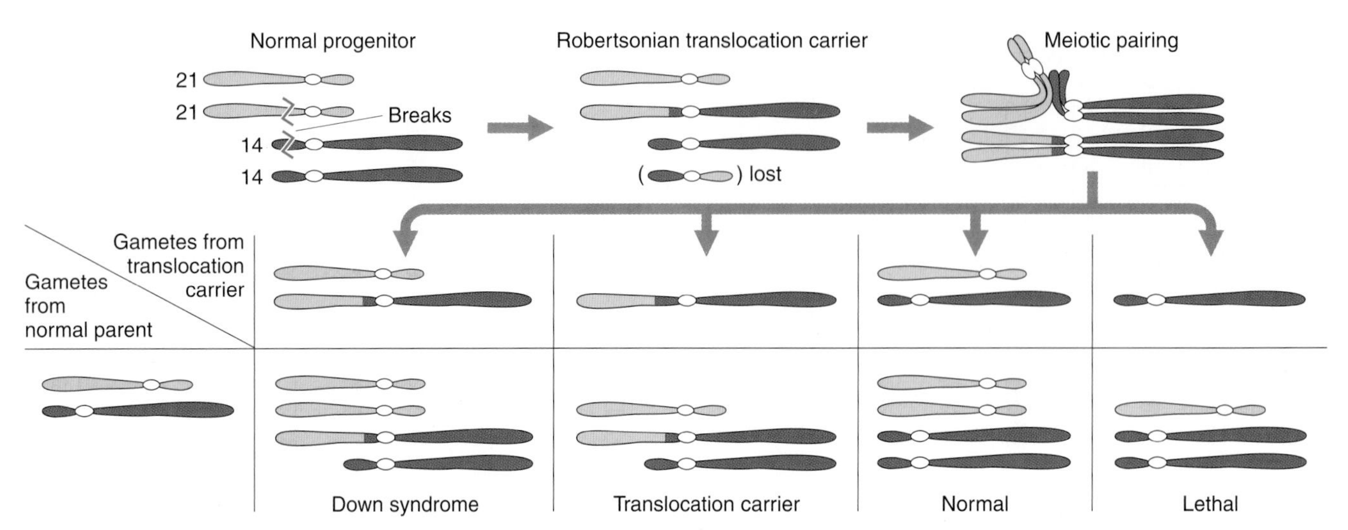

Figure 11-35 In a small minority of cases, the origin of Down syndrome is a parent heterozygous for a Robertsonian translocation involving chromosome 21. Meiotic segregation results in some gametes carrying a chromosome with a large additional segment of chromosome 21. In combination with a normal chromosome 21 provided by the gamete from the opposite sex, the symptoms of Down syndrome are produced even though there is not full trisomy 21.

chapter. It is possible that the birth is indeed a coincidence—after all, coincidences do happen. However, the miscarriage gives a clue that something else might be going on. Recall that a large proportion of spontaneous abortions carry chromosomal rearrangements, so perhaps that is the case in this example. If so, the couple may have had two conceptions with chromosome mutations, which would be very unlikely unless there was a common cause. It is known that a small proportion of Down syndrome cases result from a translocation in one of the parents. We have seen that translocations can produce progeny that have extra material from part of the genome, so a translocation involving chromosome 21 can produce progeny that have extra material from that chromosome. In Down syndrome, the translocation responsible is of a type called a *Robertsonian translocation.* It produces progeny carrying an almost complete extra copy of chromosome 21. The translocation and its segregation are illustrated in Figure 11-35. Note that not only Down syndrome–causing complements are produced, but other aberrant chromosome complements, too, most of which abort. In our example, the man may have this translocation, which he may have inherited from his grandmother. To confirm this, his chromosomes would be checked. His unaffected child might have normal chromosomes or might have inherited his translocation.

SUMMARY

1. Polyploidy is an abnormal condition in which there is a larger than normal number of chromosome sets. Polyploids such as triploids ($3x$) and tetraploids ($4x$) are common among plants and are represented even among animals. An odd number of chromosome sets makes an organism sterile because there is not a partner for each chromosome at meiosis. Unpaired chromosomes pass randomly to the poles of the cell during meiosis, leading to unbalanced sets of chromosomes in the resulting gametes. Such unbalanced gametes do not result in viable progeny. In polyploids with an even number of sets, each chromosome has a potential pairing partner and hence can produce balanced gametes and progeny. Polyploidy can result in an organism of larger dimensions; this discovery has permitted important advances in horticulture and in crop breeding.

2. In plants, allopolyploids (polyploids formed by combining chromosome sets from different species) can be made by crossing two related species and then doubling the progeny chromosomes through the use of colchicine or through somatic cell fusion. These techniques have potential applications in crop breeding because allopolyploids combine the features of the two parental species.

3. When mutations change parts of chromosome sets, aneuploids result. Aneuploids are important in the engineering of specific crop genotypes, although aneuploidy per se usually results in an unbalanced genotype with an abnormal phenotype. Examples of aneuploids include monosomics ($2n - 1$) and trisomics ($2n + 1$). Aneuploid conditions are well studied in humans; Down syndrome (trisomy 21), Klinefelter syndrome (XXY), and Turner syndrome (XO) are well-documented examples. The spontaneous level of aneuploidy in humans is quite high and accounts for a large proportion of genetically based ill health in human populations. The phenotype of an aneuploid organism depends very much on the particular chromosome involved. In some cases, such as human trisomy 21, there is a highly characteristic constellation of associated phenotypes.

4. Most instances of aneuploidy result from accidental chromosome mis-segregation at meiosis (nondisjunction). The error is spontaneous and can occur in any particular meiocyte at the first or second division. In humans there is a maternal-age effect associated with nondisjunction of chromosome 21, resulting in a higher incidence of Down syndrome in the children of older mothers.
5. The other general category of chromosome mutations is structural rearrangements, which include deletions, duplications, inversions, and translocations. Chromosomal rearrangements are an important cause of ill health in human populations and are useful in engineering special strains of organisms for experimental and applied genetics. In organisms with one normal chromosome set plus a rearranged set (heterozygous rearrangements), there are unusual pairing structures at meiosis resulting from the strong pairing affinity of homologous chromosome regions. For example, heterozygous inversions show loops, and reciprocal translocations show cross-shaped structures. Segregation of these structures results in abnormal meiotic products unique to the rearrangement.
6. An inversion is a 180-degree turn of a part of a chromosome. In the homozygous state, inversions may cause little problem for an organism unless heterochromatin brings about a position effect or one of the breaks disrupts a gene. On the other hand, inversion heterozygotes show inversion loops at meiosis, and crossing-over within the loop results in inviable products. The crossover products of pericentric inversions, which span the centromere, differ from those of paracentric inversions, which do not, but both show reduced recombinant frequency in the affected region and often result in reduced fertility.
7. A translocation moves a chromosome segment to another position in the genome. A simple example is a reciprocal translocation, in which parts of nonhomologous chromosomes exchange positions. In the heterozygous state, translocations produce duplication-deletion meiotic products, which can lead to unbalanced zygotes. New gene linkages can be produced by translocations. Random segregation of centromeres in a translocation heterozygote results in 50 percent unbalanced meiotic products, and hence 50 percent sterility (called semisterility).
8. Deletions are losses of a section of chromosome, either because of chromosome breaks followed by loss of the intervening segment or because of segregation in other heterozygous translocations or inversions. If the region removed in a deletion is essential to life, a homozygous deletion is lethal. Heterozygous deletions may be nonlethal, or they may be lethal because of chromosomal imbalance or because they uncover recessive deleterious alleles. When a deletion in one homolog allows the phenotypic expression of recessive alleles in the other, the unmasking of the recessive alleles is called pseudodominance.
9. Duplications are generally produced from other rearrangements or by aberrant crossing-over. They also unbalance the genetic material, producing a deleterious phenotypic effect or death of the organism. However, duplications can be a source of new material for evolution because function can be maintained in one copy, leaving the other copy free to evolve new functions.

CONCEPT MAP

Draw a concept map interrelating as many of the following terms as possible. Note that the terms are listed in no particular order.

pairing / inversion / translocation / loop / duplication / crossing-over / deletion / sterility / recombinant frequency

SOLVED PROBLEM

1. A corn plant is obtained that is heterozygous for a reciprocal translocation and therefore is semisterile. This plant is crossed to a chromosomally normal strain that is homozygous for the recessive allele brachytic (*b*), located on chromosome 2. A semisterile F_1 plant is then backcrossed to the homozygous brachytic strain. The progeny obtained show the following phenotypes:

NONBRACHYTIC		BRACHYTIC	
Semisterile	Fertile	Semisterile	Fertile
334	27	42	279

a. What ratio would you expect to result if the chromosome carrying the brachytic allele is not involved in the translocation?

b. Do you think that chromosome 2 is involved in the translocation? Explain your answer, showing the conformation of the relevant chromosomes of the semisterile F_1 and the reason for the specific numbers obtained.

Solution

a. We should start with the methodical approach and simply restate the data in the form of a diagram, where

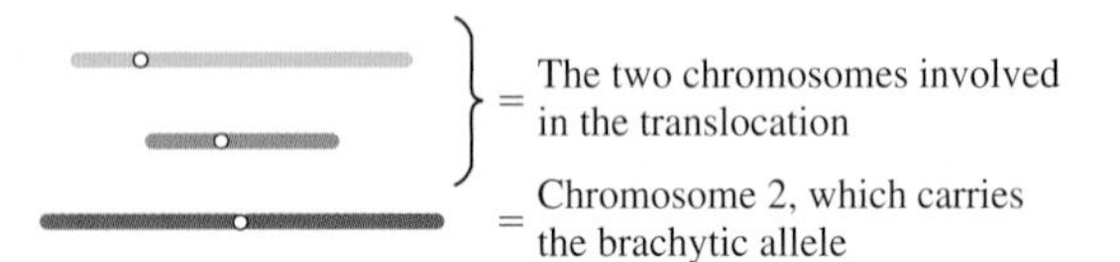

To simplify the diagram, we do not show the chromosomes divided into chromatids (although they would be at this stage of meiosis):

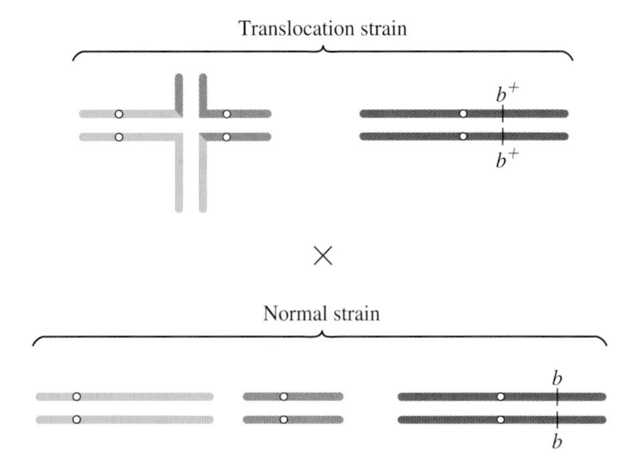

All the progeny from this cross will be heterozygous for the chromosome carrying the brachytic allele, but what about the chromosomes involved in the translocation? In this chapter, we have seen that only alternate-segregation products survive, and that half these survivors will be chromosomally normal and half will carry the two rearranged chromosomes. The rearranged combination will regenerate a translocation heterozygote when it combines with the chromosomally normal complement from the normal parent. These latter types—the semisterile F_1s—are diagrammed as part of the backcross to the parental brachytic strain:

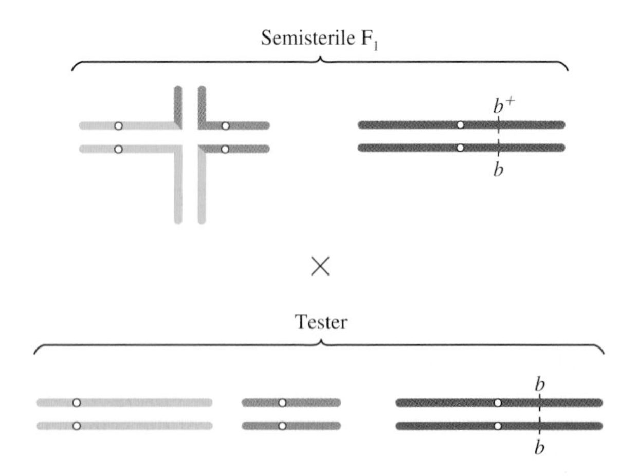

In calculating the expected ratio of phenotypes from this cross, we can treat the behavior of the translocated chromosomes independently of the behavior of chromosome 2. Hence, we can predict that the progeny will be

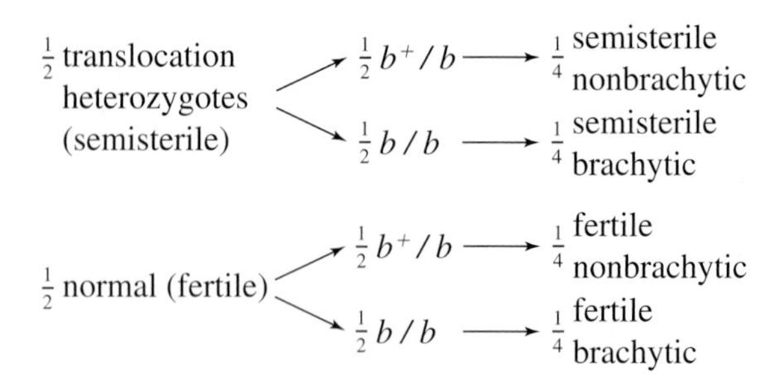

This predicted 1 : 1 : 1 : 1 ratio is quite different from that obtained in the actual cross.

b. Because we observe a departure from the expected ratio based on the independence of the brachytic phenotype and semisterility, it seems likely that chromosome 2 *is* involved in the translocation. Let's assume that the brachytic locus (b) is on the orange chromosome. But where? For the purpose of the diagram, it doesn't matter where we put it, but it does matter genetically because the position of the b locus affects the ratios in the progeny. If we assume that the b locus is near the tip of the piece that is translocated, we can redraw the pedigree:

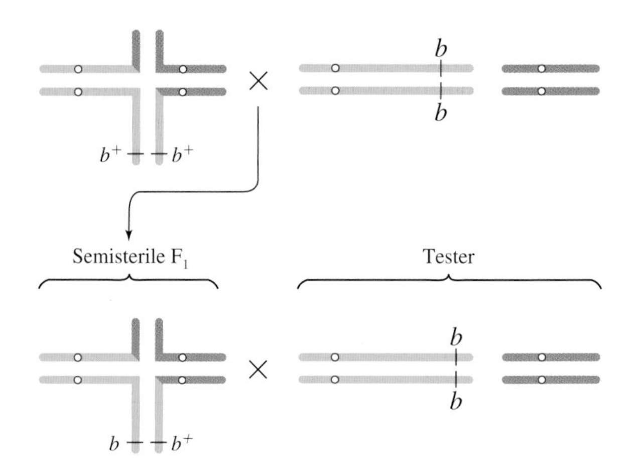

If the chromosomes of the semisterile F_1 segregate as diagrammed here, we could then predict

$\frac{1}{2}$ fertile brachytic

$\frac{1}{2}$ semisterile nonbrachytic

Most progeny are certainly of this type, so we must be on the right track. How are the two less frequent types produced? Somehow, we have to get the b^+ allele onto the normal orange chromosome and the b allele onto the translocated chromosome. This must be achieved by crossing-over between the translocation breakpoint (the center of the cross-shaped structure) and the brachytic locus. To represent this crossover, we must show chromatids, because crossing-over occurs at the chromatid stage:

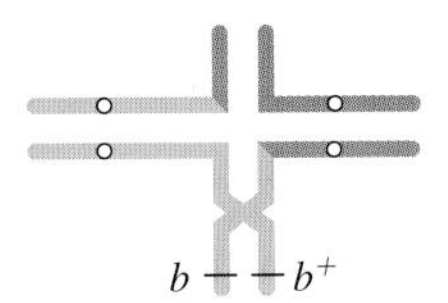

The recombinant chromosomes produce some progeny that are fertile and nonbrachytic and some that are semisterile and brachytic (these two classes together constitute 69 progeny of a total of 682, or a frequency of about 10 percent). We can see that this frequency is really a measure of the map distance (10 m.u.) of the brachytic locus from the breakpoint. (The same basic result would have been obtained if we had drawn the brachytic locus in the part of the chromosome on the other side of the breakpoint.)

SOLVED PROBLEM

2. We have lines of mice that breed true for two alternative behavioral phenotypes that we know are determined by two alleles at a single locus: *v* causes a mouse to move with a "waltzing" gait, whereas *V* determines a normal gait. After crossing the true-breeding waltzers and normals, we observe that most of the F_1 is normal, but, unexpectedly, there is one waltzer female. We mate the F_1 waltzer with two different waltzer males and note that she produces only waltzer progeny. When we mate her with normal males, she produces normal progeny and no waltzers. We mate three of her normal female progeny with two of their brothers, and these mice produce sixty progeny, all normal. When, however, we mate one of these same three females with a third brother, we get six normals and two waltzers in a litter of eight. By thinking about the parents of the F_1 waltzer, we can consider some possible explanations of these results:

a. A dominant allele may have mutated to a recessive allele in her normal parent.

b. In one parent there may have been a dominant mutation in a second gene to create an allele that acts to prevent *V*'s expression, leading to waltzing. (This phenomenon, called epistasis, is discussed in detail in Chapter 14.)

c. Meiotic nondisjunction of the chromosome carrying *V* in her normal parent may have given a viable aneuploid.

d. There may have been a viable deletion spanning *V* in the meiocyte from her normal parent.

Which of these explanations are possible, and which are eliminated by the genetic analysis? Explain in detail.

Solution

The best way to answer the question is to take the explanations one at a time and see if each fits the results given.

a. Mutation *V* to *v*

This hypothesis requires that the exceptional waltzer female be homozygous *v* / *v*. This assumption is compatible with the results of mating her both to waltzer males, which would, if she is *v* / *v*, produce all waltzer offspring (*v* / *v*), and to normal males, which would produce all normal offspring (*V* / *v*). However, brother-sister matings within this normal progeny should then produce a 3:1 normal-to-waltzer ratio. Because some of the brother-sister matings actually produced no waltzers, this hypothesis does not explain the data.

b. Epistatic mutation *s* to *S*

Here the parents would be *V* / *V* . *s* / *s* and *v* / *v* . *s* / *s*, and a germinal mutation in one of them would give the F_1 waltzer the genotype *V* / *v* . *S* / *s*. When we crossed her with a waltzer male, who would be of the genotype *v* / *v* . *s* / *s*, we would expect some *V* / *v* . *s* / *s* progeny, which would be phenotypically normal. However, we saw no normal progeny from this cross, so the hypothesis is already overthrown. Linkage could save the hypothesis temporarily if we assumed that the mutation was in the normal parent, giving a gamete *V S*. Then the F_1 waltzer would be *V S* / *v s*, and, if linkage were tight enough, few or no *V s* gametes would be produced, the type that are necessary to combine with the *v s* gamete from the male to give *V s* / *v s* normals. However, if this were true, the cross with the normal males would be *V S* / *v s* × *V s* / *V s*, and this would give a high percentage of *V S* / *V s* progeny, which would be waltzers, none of which were seen.

c. Nondisjunction in the normal parent

This explanation would give a nullisomic gamete that would combine with *v* to give the F_1 waltzer the hemizygous genotype *v*. The subsequent matings would be:

- *v* × *v* / *v* gives *v* / *v* and *v* progeny, all waltzers. This fits.
- *v* × *V* / *V* gives *V* / *v* and *V* progeny, all normals. This also fits.
- First intercrosses of normal progeny: *V* × *V*. This gives *V* and *V* / *V*, which are all normal. This fits.
- Second intercrosses of normal progeny: *V* × *V* / *v*. This gives 25 percent each of *V* / *V*, *V* / *v*, *V* (all normals), and *v* (waltzers). This also fits.

This hypothesis is therefore consistent with the data.

d. Deletion of *V* in normal parent

Let's call the deletion D. The F_1 waltzer would be D / *v*, and the subsequent matings would be:

- D / *v* × *v* / *v*. This gives *v* / *v* and D / *v*, all of which are waltzers. This fits.
- D / *v* × *V* / *V*. This gives *V* / *v* and D / *V*, all of which are normal. This fits.
- First intercrosses of normal progeny: D / *V* × D / *V*. This gives D / *V* and *V* / *V*, all normal. This fits.

Second intercrosses of normal progeny: D / *V* × *V* / *v*. This gives 25 percent of each of *V* / *V*, *V* / *v*, D / *V* (all normals), and D / *v* (waltzers). This also fits.

Once again, the hypothesis fits the data provided, so we are left with two hypotheses that are compatible with the results, and further experiments are necessary to distinguish them. One obvious way of doing this would be to examine the chromosomes of the exceptional female under the microscope; aneuploidy should be relatively easy to distinguish from deletion.

PROBLEMS

Basic Problems

1. List the diagnostic features (genetic or cytological) that are used to identify these chromosomal alterations:

a. Deletions

b. Duplications

c. Inversions

d. Reciprocal translocations

2. The normal sequence of nine genes on a certain *Drosophila* chromosome is 123 · 456789, where the dot represents the centromere. Some fruit flies were found to have aberrant chromosomes with the following structures:

a. 123 · 476589

b. 123 · 46789

c. 1654 · 32789

d. 123 · 4566789

Name each type of chromosomal rearrangement and draw diagrams to show how each would synapse with the normal chromosome.

3. The two loci *P* and *Bz* are normally 36 m.u. apart on the same arm of a certain plant chromosome. A paracentric inversion spans about one-fourth of this region but does not include either of the loci. What approximate recombinant frequency between *P* and *Bz* would you predict in plants that are

a. Heterozygous for the paracentric inversion?

b. Homozygous for the paracentric inversion?

4. As we saw in Solved Problem 2, certain mice called waltzers have a recessive mutation that causes them to execute bizarre steps. W. H. Gates crossed waltzers with homozygous normals and found, among several hundred normal progeny, a single waltzing female mouse. When mated with a waltzing male, she produced all waltzing offspring. When mated with a homozygous normal male, she produced all normal progeny. Some males and females of this normal progeny were intercrossed, and there were no waltzing offspring among their progeny. T. S. Painter examined the chromosomes of waltzing mice that were derived from some of Gates's crosses and that showed a breeding behavior similar to that of the original, unusual waltzing female. He found that these individuals had 40 chromosomes, just as in normal mice or the usual waltzing mice. In the unusual waltzers, however, one member of a chromosome pair was abnormally short. Interpret these observations as completely as possible, both genetically and cytologically.

(Problem 4 is from A. M. Srb, R. D. Owen, and R. S. Edgar, *General Genetics*, 2d ed. W. H. Freeman and Company, 1965.)

5. Six bands in a salivary gland chromosome of *Drosophila* are shown in the following figure, along with the extent of five deletions (Del 1 to Del 5):

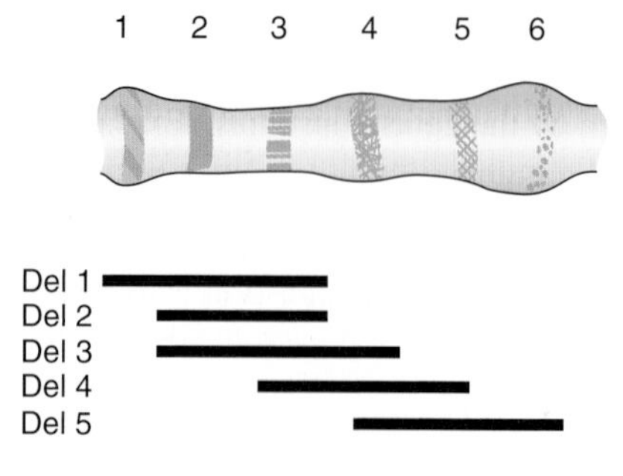

Recessive alleles *a*, *b*, *c*, *d*, *e*, and *f* are known to be in the region, but their order is unknown. When the deletions are combined with each allele, the following results are obtained:

	a	*b*	*c*	*d*	*e*	*f*
Del 1	–	–	–	+	+	+
Del 2	–	+	–	+	+	+
Del 3	–	+	–	+	–	+
Del 4	+	+	–	–	–	+
Del 5	+	+	+	–	–	–

In this table, a minus sign means that the deletion is missing the corresponding wild-type allele (the deletion uncovers the recessive) and a plus sign means that the corresponding wild-type allele is still present. Use these data to infer which salivary band contains each gene.
(Problem 5 is from D. L. Hartl, D. Friefelder, and L. A. Snyder, *Basic Genetics*. Jones and Bartlett, 1988.)

6. A fruit fly was found to be heterozygous for a paracentric inversion. However, it was impossible to obtain flies that were homozygous for the inversion even after many attempts. What is the most likely explanation for this inability to produce a homozygous inversion?

7. Orangutans are an endangered species in their natural environment (the islands of Borneo and Sumatra), so a captive breeding program has been established using orangutans currently held in zoos throughout the world. One component of this program is research into orangutan cytogenetics. This research has shown that all orangutans from Borneo carry one form of chromosome 2, as shown in the following diagram, and all orangutans from Sumatra carry the other form. Before this cytogenetic difference became known, some matings were carried out between animals from different islands, and 14 hybrid progeny are now being raised in captivity.

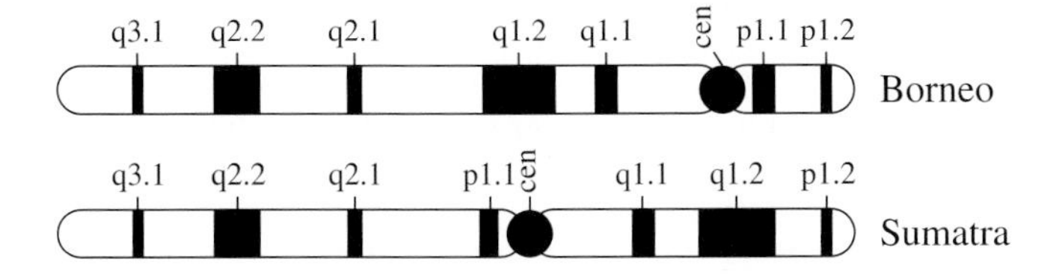

a. What term or terms describe the differences between these chromosomes?

b. Draw the chromosomes 2, paired during the first meiotic prophase, of such a hybrid orangutan. Be sure to show all the landmarks indicated in the preceding diagram, and label all parts of your drawing.

c. In 30 percent of meioses, there will be a crossover somewhere in the region between bands p1.1 and q1.2. Draw the gamete chromosomes 2 that would result from a meiosis in which a single crossover occurred within band q1.1.

d. What fraction of the gametes produced by a hybrid orangutan will give rise to viable progeny, if these are the only chromosomes that differ between the parents?
(Problem 7 is from Rosemary Redfield.)

8. In corn, the genes for tassel length (alleles T and t) and rust resistance (alleles R and r) are known to be on separate chromosomes. In the course of making routine crosses, a breeder noticed that one $T/t\ ;\ R/r$ plant gave unusual results in a testcross with the double-recessive pollen parent $t/t\ ;\ r/r$. The results were

Progeny:	$T/t\ ;\ R/r$	98
	$t/t\ ;\ r/r$	104
	$T/t\ ;\ r/r$	3
	$t/t\ ;\ R/r$	5

Corn cobs: Only about half as many seeds as usual

a. What key features of the data are different from the expected results?

b. State a concise hypothesis that explains the results.

c. Show genotypes of parents and progeny.

d. Draw a diagram showing the arrangement of alleles on the chromosomes.

e. Explain the origin of the two classes of progeny having three and five members.

Unpacking the Problem

a. What do "a gene for tassel length" and "a gene for rust resistance" mean?

b. Does it matter that the precise meaning of the allelic symbols T, t, R, and r is not given? Why or why not?

c. How do the terms *gene* and *allele,* as used here, relate to the concepts of locus and gene pair? (A concept map would be one way of answering this question.)

d. What prior experimental evidence would give the corn geneticist the idea that the two genes are on separate chromosomes?

e. What do you imagine "routine crosses" are to a corn breeder?

f. What term is used to describe genotypes of the type $T/t\ ;\ R/r$?

g. What is a "pollen parent"?

h. What are testcrosses, and why do geneticists find them so useful?

i. What progeny types and frequencies might the breeder have been expecting from the testcross?

j. Describe how the observed progeny differ from expectations.

k. What does the approximate equality of the first two progeny classes tell you?

l. What does the approximate equality of the second two progeny classes tell you?

m. What were the gametes from the unusual plant, and what were their proportions?

n. Which gametes were in the majority?

o. Which gametes were in the minority?

p. Which of the progeny types seem to be recombinant?

q. Which allelic combinations appear to be linked in some way?

r. How can there be linkage of genes supposedly on separate chromosomes?

s. What do these majority and minority classes tell us about the genotypes of the parents of the unusual plant?

t. What is a corn cob?

u. What does a normal corn cob look like? (Sketch one and label it.)

v. What do the corn cobs from this cross look like? (Sketch one.)

w. What exactly is a kernel?

x. What effect could lead to the absence of half the kernels?

y. Did half the kernels die? If so, was the female or the male parent the reason for the deaths?

Now try to solve the problem.

9. A yellow body in *Drosophila* is caused by a mutant allele *y* of a gene located at the tip of the X chromosome (the wild-type allele causes a gray body). In a radiation experiment, a wild-type male was irradiated with X rays and then crossed with a yellow-bodied female. Most of the male progeny were yellow, as expected, but the scanning of thousands of flies revealed two gray-bodied (phenotypically wild-type) males. These gray-bodied males were crossed with yellow-bodied females, with the following results:

	Progeny
Gray male 1 × yellow female	Females all yellow Males all gray
Gray male 2 × yellow female	$\frac{1}{2}$ females yellow $\frac{1}{2}$ females gray $\frac{1}{2}$ males yellow $\frac{1}{2}$ males gray

a. Explain the origin and crossing behavior of gray male 1.

b. Explain the origin and crossing behavior of gray male 2.

10. In corn, the allele *Pr* stands for green stems, *pr* for purple stems. A corn plant of genotype *pr / pr* that has standard chromosomes is crossed with a *Pr / Pr* plant that is homozygous for a reciprocal translocation between chromosomes 2 and 5. The F_1 is semisterile and phenotypically Pr. A backcross with the parent with standard chromosomes gives 764 semisterile Pr; 145 semisterile pr; 186 normal Pr; and 727 normal pr. What is the map distance between the *Pr* locus and the translocation point?

11. Distinguish among Klinefelter, Down, and Turner syndromes.

12. Show how one could make an allotetraploid between two related diploid plant species in both of which $2n = 28$.

13. In *Drosophila,* trisomics and monosomics for the tiny chromosome 4 are viable, but nullisomics and tetrasomics are not. The *b* locus is on this chromosome. Deduce the phenotypic proportions in the progeny of the following crosses of trisomics.

a. $b^+ / b / b \times b / b$

b. $b^+ / b^+ / b \times b / b$

c. $b^+ / b^+ / b \times b^+ / b$

14. A woman with Turner syndrome is found to be colorblind (X-linked recessive phenotype). Both her mother and her father have normal vision.

a. Explain the simultaneous origin of Turner syndrome and colorblindness by the abnormal behavior of chromosomes at meiosis.

b. Can your explanation distinguish whether the abnormal chromosome behavior occurred in the father or the mother?

c. Can your explanation distinguish whether the abnormal chromosome behavior occurred at the first or second division of meiosis?

d. Now assume that a colorblind Klinefelter man has parents with normal vision, and answer parts a, b, and c.

15. a. How would you synthesize a pentaploid ($5x$)?

b. How would you synthesize a triploid ($3x$) of genotype *A / a / a*?

c. You have just obtained a rare recessive mutation a^* in a diploid plant, which Mendelian analysis tells you is A / a^*. From this plant, how would you synthesize a tetraploid ($4x$) of genotype $A / A / a^* / a^*$?

d. How would you synthesize a tetraploid of genotype *A / a / a / a*?

16. Suppose you have a line of mice that has cytologically distinct forms of chromosome 4. The tip of the chromosome can have a knob (called 4^K) or a satellite (4^S) or neither (4). Here are sketches of the three types:

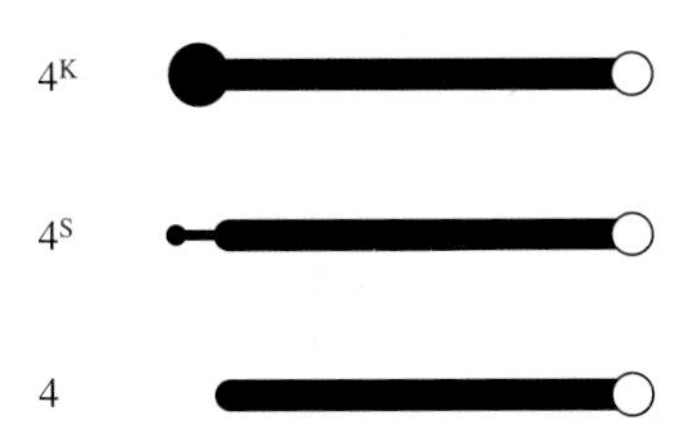

You cross a 4^K / 4^S female with a 4 / 4 male and find that most of the progeny are 4^K / 4 or 4^S / 4, as expected. However, you occasionally find some rare types as follows (all other chromosomes are normal):

a. 4^K / 4^K / 4

b. 4^K / 4^S / 4

c. 4^K

Explain the rare types that you have found. Give, as precisely as possible, the stages at which they originate, and state whether they originate in the male parent, the female parent, or the zygote. (Give reasons briefly.)

17. A cross is made in tomatoes between a female plant that is trisomic for chromosome 6 and a normal diploid male plant that is homozygous for the recessive allele for potato leaf (*p* / *p*).

a. A trisomic F_1 plant is backcrossed to the potato-leaved male. What is the ratio of normal-leaved plants to potato-leaved plants when you assume that *p* is located on chromosome 6?

b. What is the ratio of normal-leaved to potato-leaved plants when you assume that *p* is not located on chromosome 6?

18. A tomato geneticist attempts to assign five recessive genes to specific chromosomes by using trisomics. She crosses each homozygous mutant (2*n*) to each of three trisomics, involving chromosomes 1, 7, and 10. From these crosses, the geneticist selects trisomic progeny (which are less vigorous) and backcrosses them to the appropriate homozygous recessive. The *diploid* progeny from these crosses are examined. Her results, in which the ratios are wild type:mutant, are as follows:

Trisomic chromosome	GENE *d*	*y*	*c*	*h*	*cot*
1	48:55	72:29	56:50	53:54	32:28
7	52:56	52:48	52:51	58:56	81:40
10	45:42	36:33	28:32	96:50	20:17

Which of the genes can the geneticist assign to which chromosomes? (Explain your answer fully.)

Challenging Problems

19. The *Neurospora un-3* locus is near the centromere on chromosome 1, and crossovers between *un-3* and the centromere are very rare. The *ad-3* locus is on the other side of the centromere of the same chromosome, and crossovers occur between *ad-3* and the centromere in about 20 percent of meioses (no multiple crossovers occur).

a. What types of linear asci (see Chapter 5) do you predict, and in what frequencies, in a normal cross of *un-3 ad-3* × wild type? (Specify genotypes of spores in the asci.)

b. Most of the time such crosses behave predictably, but in one case, a standard *un-3 ad-3* strain was crossed with a wild type isolated from a field of sugarcane in Hawaii. The results follow:

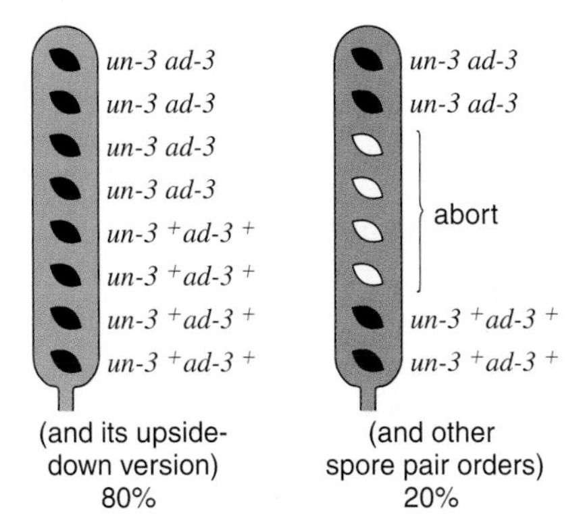

Explain these results, and state how you could test your idea. (Note: In *Neurospora,* ascospores with extra chromosomal material survive and are the normal black color, whereas ascospores lacking any chromosome region are white and inviable.)

20. Two mutations in *Neurospora, ad-3* and *pan-2,* are located on chromosomes 1 and 6, respectively. An unusual *ad-3* line arises in the laboratory, giving the following results:

	Ascospore appearance	RF between *ad-3* and *pan-2*
1. Normal *ad-3* × normal *pan-2*	All black	50%
2. Abnormal *ad-3* × normal *pan-2*	About $\frac{1}{2}$ black and $\frac{1}{2}$ white (inviable)	1%

3. Of the black spores from cross 2, about half were completely normal and half repeated the same behavior as the original abnormal *ad-3* strain.

Explain all three results with the aid of clearly labeled diagrams. (Note: In *Neurospora,* ascospores with extra chromosomal material survive and are the normal black color, whereas ascospores lacking any chromosome region are white and inviable.)

21. Deduce the phenotypic proportions in the progeny of the following crosses of autotetraploids in which the a^+ / a locus is very close to the centromere. (Assume that the four homologous chromosomes of any one type pair randomly two-by-two and that only one copy of the a^+ allele is necessary for the wild-type phenotype.)

a. $a^+ / a^+ / a / a \times a / a / a / a$

b. $a^+ / a / a / a \times a / a / a / a$

c. $a^+ / a / a / a \times a^+ / a / a / a$

d. $a^+ / a^+ / a / a \times a^+ / a / a / a$

22. The New World cotton species *Gossypium hirsutum* has a $2n$ chromosome number of 52. The Old World species *G. thurberi* and *G. herbaceum* each have a $2n$ number of 26. Hybrids between these species show the following chromosome pairing arrangements at meiosis:

Hybrid	Pairing arrangement
G. hirsutum × *G. thurberi*	13 small bivalents + 13 large univalents
G. hirsutum × *G. herbaceum*	13 large bivalents + 13 small univalents
G. thurberi × *G. herbaceum*	13 large univalents + 13 small univalents

Draw diagrams to interpret these observations phylogenetically, clearly indicating the relationships between the species. How would you go about proving that your interpretation is correct?

(Problem 23 is adapted from A. M. Srb, R. D. Owen, and R. S. Edgar, *General Genetics,* 2d ed. W. H. Freeman and Company, 1965.)

23. Several kinds of sexual mosaicism are well documented in humans. Suggest how each of the following examples may have arisen by nondisjunction at *mitosis:*

a. XX / XO (that is, there are two cell types in the body, XX and XO)

b. XX / XXYY

c. XO / XXX

d. XX / XY

e. XO / XX / XXX

24. In *Drosophila,* a cross (cross 1) was made between two mutant flies, one homozygous for the recessive mutation bent wing (b) and the other homozygous for the recessive mutation eyeless (e). The mutations e and b are alleles of two different genes that are known to be very closely linked on the tiny autosomal chromosome 4. All the progeny had a wild-type phenotype. One of the female progeny was crossed with a male of genotype $b\ e\ /\ b\ e$; we shall call this *cross 2.* Most of the progeny of cross 2 were of the expected types, but there was also one rare female of wild-type phenotype.

a. Explain what the common progeny are expected to be from cross 2.

b. Could the rare wild-type female have arisen by (1) crossing-over? (2) nondisjunction? Explain.

c. The rare wild-type female was testcrossed to a male of genotype $b\ e\ /\ b\ e$ (cross 3). The progeny were

$\frac{1}{6}$ wild type

$\frac{1}{6}$ bent, eyeless

$\frac{1}{3}$ bent

$\frac{1}{3}$ eyeless

Which of the explanations in part b are compatible with this result? Explain the genotypes and phenotypes of progeny of cross 3 and their proportions.

25. In the fungus *Ascobolus* (similar to *Neurospora*), ascospores are normally black. The mutation *f,* producing fawn-colored ascospores, is in a gene just to the right of the centromere on chromosome 6, whereas mutation *b,* producing beige ascospores, is in a gene just to the left of the same centromere. In a cross of fawn and beige parents ($+ f \times b +$), most octads showed four fawn and four beige ascospores, but three rare exceptional octads were found, as shown below. In the sketch, black is the wild-type phenotype, a vertical line is fawn, a horizontal line is beige, and an empty circle represents an aborted (dead) ascospore.

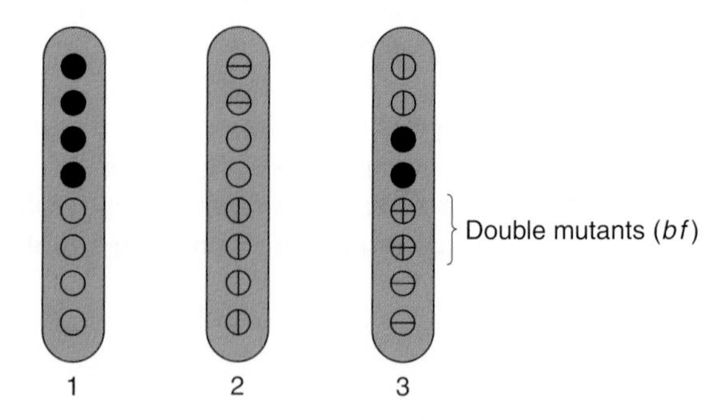

a. Provide reasonable explanations for these three exceptional octads.

b. Diagram the meiosis that gave rise to octad 2.

MUTATIONAL DISSECTION 12

Key Concepts

1 Mutational analysis is a way of discovering genes that encode products involved in specific biological processes.

2 Genes involved in a process can be identified initially by relevant mutant phenotypes or by properties of the RNAs or proteins that they encode.

3 Phenotypes of interest can be identified by treating organisms with general mutagens; a variety of general mutagens can produce different sorts of alterations, ranging from point mutations to deletions and other events of chromosomal breakage.

4 In some organisms, phenotypic analysis of a gene can be initiated by targeting that gene for mutational inactivation or modification by genetic engineering.

5 The phenotypic contribution of a cloned gene can also be studied by phenocopy techniques that inactivate its RNA or protein product (thereby mimicking the effect of mutating the gene) but do not alter the gene itself.

6 One way to assay for mutations is by genetic selection, in which only individuals with a desired phenotype survive the assay; such selections are the most powerful way to assay large numbers of mutagenized individuals for rare mutational events.

7 Genetic screens can survey for any phenotype of interest; they can identify mutations by directly observable phenotypes or their interactions with mutations in other genes.

8 Mutations can eliminate or reduce normal gene function; such mutations are called loss-of-function.

9 Mutations can increase gene activity or even create a novel activity for a gene; such mutations are called gain-of-function.

10 To use mutant phenotypes to understand the role of a wild-type gene in a process, it is necessary to know whether the mutations are loss-of-function or gain-of-function.

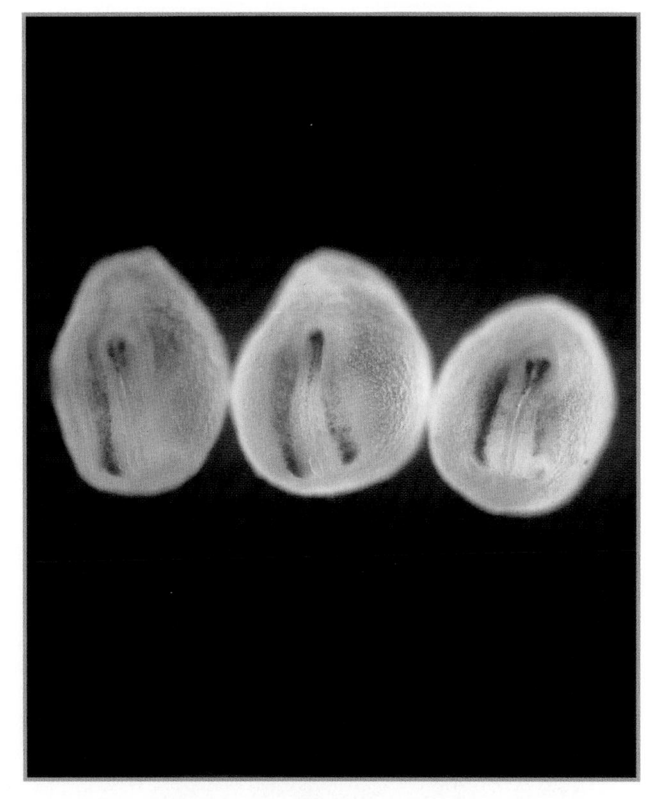

The expression pattern of a gene that controls left-right symmetry of the body plan. The mouse embryos on both sides show the asymmetric pattern of this gene in a wild-type genotype embryo. The embryo in the middle shows symmetric expression in a mutant genotype whose internal organs will be arranged in mirror-image to wild type. *(Courtesy of D. Norris and E. Robertson, Harvard University.)*

The jury has certainly not come to a final verdict, but current estimates are that there are about 30,000–35,000 genes in the human genome (as explained in Chapter 9). Focusing on the numerology is interesting, but regardless of the exact count of human genes, the much more difficult problem is to ascertain the function and biological role of each of the RNA and protein products encoded by the human genome. This is especially challenging in humans, where, for important ethical considerations, many of the experimental approaches useful in genetic model systems cannot be applied. For example, there are about 35 genes of one family of proteins involved in sending signals from one cell to its neighbors (Figure 12-1a), a topic that will be covered in more detail in Chapters 15 and 16. How do we begin to understand the roles that these proteins play in humans, given that, for many of the genes that encode these proteins, naturally occurring genetic variants do not exist? In one exceptional case, genetic variation does exist and the phenotypic effects of the lack of a functional copy of this gene are profound. When both copies

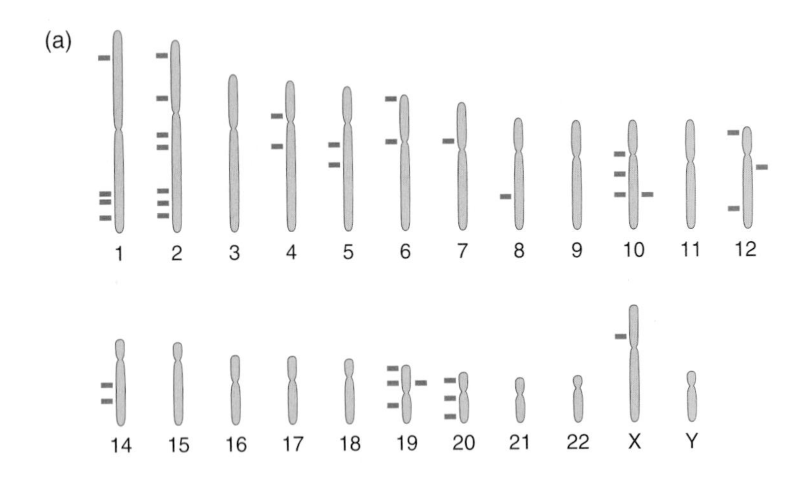

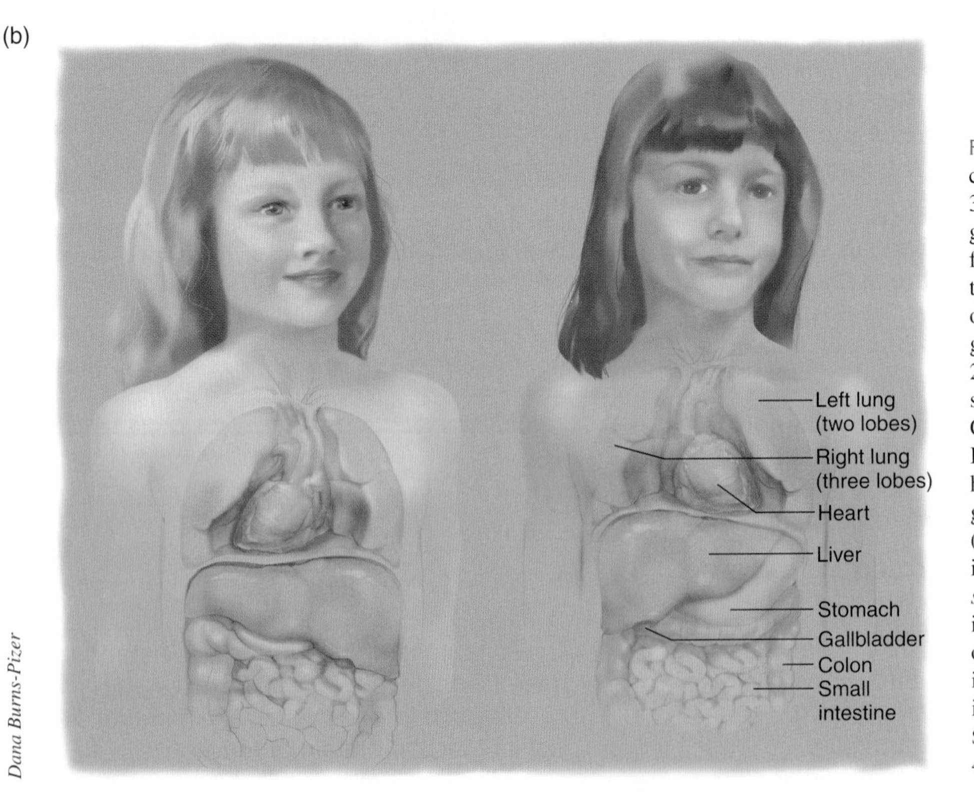

Figure 12-1 (a) The chromosomal distribution of 35 genes of the transforming growth-factor-beta gene family as determined by translational BLAST analysis of the compiled human genome sequence (as of May 2001), from the BLAST Web site provided by the National Center for Biotechnology Information (the URL is http://www.ncbi.nlm.nih.gov/genome/seq/HsBlast.html). (b) Internal organ placement in normal *(left panel)* and *situs inversus (right panel)* individuals. One of the causes of this condition is mutations in one of the genes depicted in part a. *(From J.C.I. Belmonte,* Scientific American *280(6), 1999, 46–47.)*

of the gene are mutant, the phenotype is a syndrome called *situs inversus,* in which all of the internal organs of the body are in mirror-image left-right orientation (Figure 12-1b). This remarkable phenotype clearly warrants study of the other members of this protein family. Genome analysis shows tremendous conservation of gene and protein structure across major branches of the tree of life. By analyzing mutations in experimental model organisms, we can get hints of the likely functions of the "homologous" proteins in humans (that is, proteins derived from the same gene in a common evolutionary ancestor). Let's consider how such mutational analysis is carried out in experimental systems and then return at the end of the chapter to some examples of the insights that can be gained.

CHAPTER OVERVIEW

The purpose of mutational analysis is to understand normal biological function by genetically disrupting normal gene activity, and then analyzing the phenotypes of the resulting mutant organisms. This chapter will explore the basic approaches that geneticists use to produce mutations in genes of interest, summarized in Figure 12-2. In some cases, the geneticist wants to study a particular biological process, for example, brain development, and wants to survey the genome for all the genes that contribute to this process. In such cases, the challenge is to sieve through a collection of individuals with mutagenized genomes and identify the few with phenotypes suggestive of a role in brain development. In other cases, the geneticist already knows of a phenotype produced by mutations in a gene but wants a broader range of mutations in the gene to understand its effects fully. In such cases, the collection of mutagenized individuals is sieved in different ways to focus attention on the one gene of interest. Once the genes of interest have been identified on the basis of phenotype and map location, the geneticist studies the genes and their mutant alleles by various techniques, including the sequence-level analysis of the cloned genes and the identification of their encoded products. These cases can be thought of as **forward genetics,** because the identification of heritable differences and description of phenotype (the "genetics") precedes the molecular analysis of the products encoded by the wild-type and mutant genes.

In a third type of case, one that appears more and more frequently with the rapid sequencing of entire genomes, analysis goes in the reverse direction. A researcher has identified a protein or an RNA and wants to know what the phenotype is when the gene encoding this product is mutated. This approach, starting by studying a molecule and then mutating the gene that encodes it, is called **reverse genetics.**

This chapter will explore both classical and recently developed techniques for forward and reverse genetics. The classical approaches to mutational analysis, which have been practiced for about 100 years, relied on mutagens that would mutate any gene at a low rate; the challenge to the geneticist is to identify the proverbial needle-in-the-haystack, that is, the rare mutant that arises in the gene(s) that he or she is interested in. This approach is still extremely important in forward genetics, since it is often true that a geneticist does not know which genes contribute to the biological process that he or she wishes to understand. In the past two decades, the classical approach has been made *neoclassical* by the addition of naturally occurring or transgenic insertional DNA elements to the arsenal of mutagens at its disposal. Unlike classical chemical and radiation mutagens, these *insertional mutagens* not only disrupt gene function, but also molecularly tag the DNA of the gene so

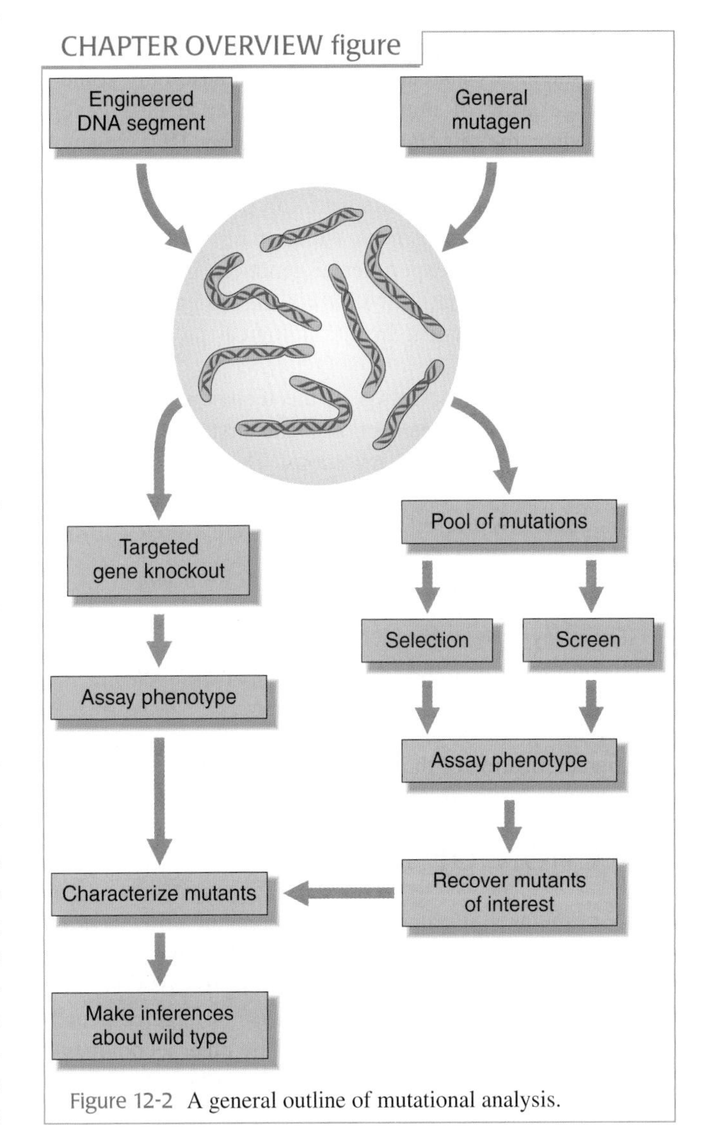

Figure 12-2 A general outline of mutational analysis.

that it can be cloned fairly easily. This chapter will also examine directed techniques, in which a gene of interest can be mutated in a test tube and reintroduced as a mutant transgene or in which the gene of interest can be targeted for disruption.

There are various approaches to recovering mutant forms of a gene. This chapter describes the different general approaches to mutational dissection. Later chapters will show how these concepts and approaches are used to address major biological questions.

THE COMPONENTS OF MUTATIONAL DISSECTION

There is a large chasm between the kinds of mutations that *could* arise in a gene and the ones that actually can be recovered after mutagenesis. Several factors are crucial in the recovery of mutations. The choice of mutagen is important, because different mutagens produce different arrays of mutations. The phenotypes that are used to identify mutant-bearing individuals also are important, because only some of the huge array of mutations that can arise in a gene might produce the relevant mutant phenotype. Thus, in any given mutagenesis procedure, every gene will have its own *target size.* It is possible that a gene can present a large target in one kind of mutagenesis but a small one in another. The challenge for the experimental geneticist is then to design a mutagenesis that will *saturate* the system—that is, identify every component in the biological process being studied.

The following description of the techniques of mutational analysis is in three sections. The first two focus on how to harvest the desired mutations. The third addresses what to do with the mutant strains once they are isolated. The three sections are:

- The selection of mutagen according to the kinds of lesions that are desired
- The assay system that identifies the relevant mutations according to the kinds of processes that one wishes to study
- The first phase of genetic and phenotypic characterization of the recovered mutations

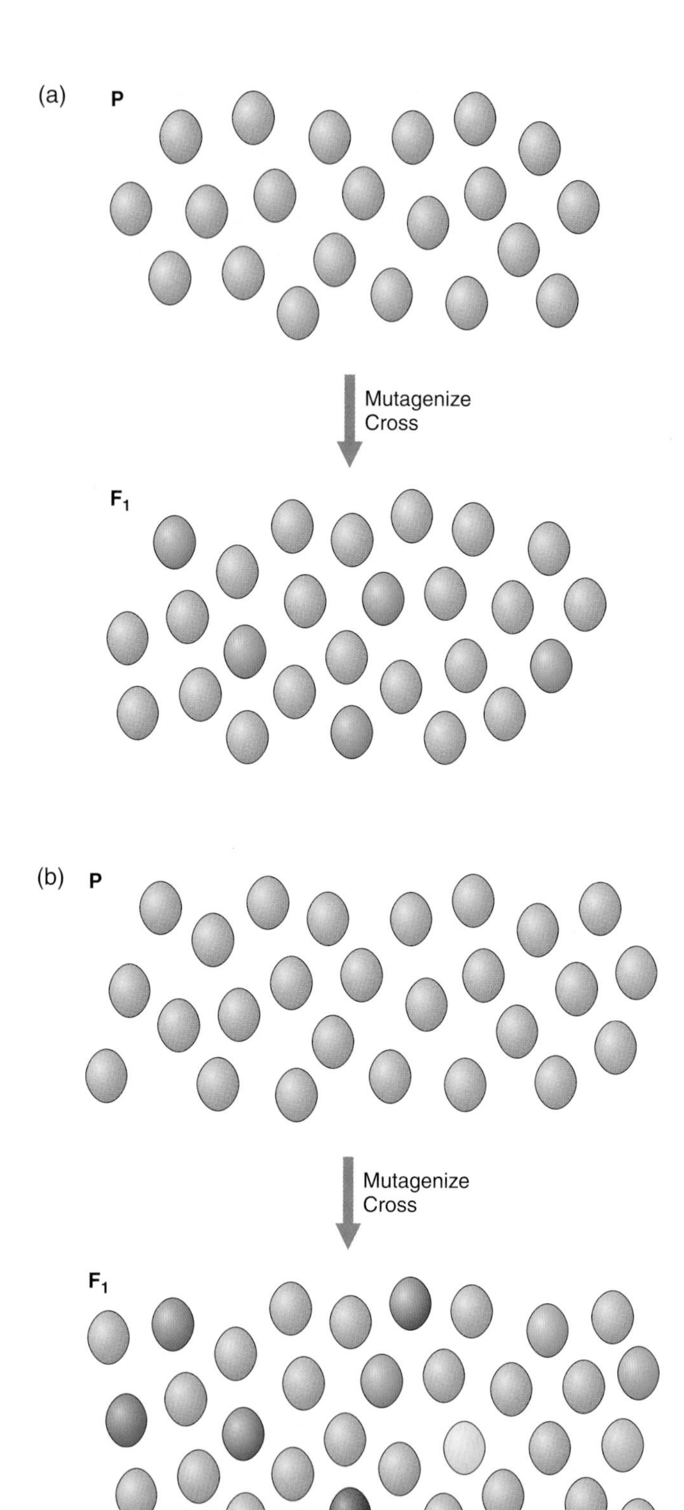

Figure 12-3 Targeted versus general mutagenesis. (a) Targeted mutagenesis produces mutations (represented by the red circles) in only the targeted gene. (b) General mutagenesis produces a variety of mutations (represented by the different-colored circles). Any given mutation is a rare event, even after mutagenesis, and so a mutagenized population will contain mostly normal individuals with occasional independent mutations.

SELECTION OF MUTAGEN

Choice of mutagen is crucial to the outcome of a mutagenesis. Mutagens can be classified in several ways. As described in Chapter 10, mutagens can be agents of random damage in DNA. Alternatively, in the case of targeted gene knockouts (described in Chapter 8) and by other techniques described in this chapter, nucleotide sequences can be introduced into a cell in such a way that they are very likely to disrupt the structure or activity of a sequence-homologous gene in the genome. In other words, the effects of these nucleotide agents is directed, rather than random.

What are the advantages of the random and directed approaches? Directed approaches, if they have been developed for a given organism, have considerable advantages for reverse genetics; in these cases, the researcher knows the sequence of the gene and wishes to see what happens when he or she mutates its function. On the other hand, many mutageneses are fishing expeditions. In these, the geneticist hopes to create a population of mutagenized organisms that includes different individuals with mutations in each of the genes in the genome (Figure 12-3). For such random mutageneses, general mutagens are required.

Choice of General Mutagenic Agents

Ideally, general mutagens should mutate all genes at a constant frequency and produce a broad array of different mutational events. This would make it straightforward to *saturate* the genome for mutations of a certain class, since we would just have to screen enough individuals to be confident of getting *at least* one mutation in each gene. However, several other factors must be considered as well.

Genes come in different physical sizes—for example, the smallest human genes are a few hundred base pairs in size, whereas the largest genes are about two million base pairs! The physical size of genes can thus differ by several orders of magnitude. The mutational target size is harder to calculate (Figure 12-4). The basic problem is that a typical mutagenesis does not identify all of the mutational changes to the DNA that can arise from treatment with a specific mutagen. Rather, the mutagenesis identifies the subset of changes that lead to the abnormal phenotype that is being assayed (see the section on mutagenesis assay systems, below). Many mutations that arise in a gene will not alter it to produce the sought-after mutant phenotype. Just considering base-substitution point mutations, the proportion of mutations in two genes that give rise to the mutant phenotype might be very different. It depends on (1) the frequency of amino-acid substitutions or protein truncations due to nonsense mutations that make each protein sufficiently abnormal to induce the desired mutant phenotype and (2) the frequency of non-protein-coding mutations that alter the expression of each gene sufficiently to produce the desired mutant phenotype. As every mutagen has its own characteristic pattern of nucleotide alterations, in reality (1) and (2)

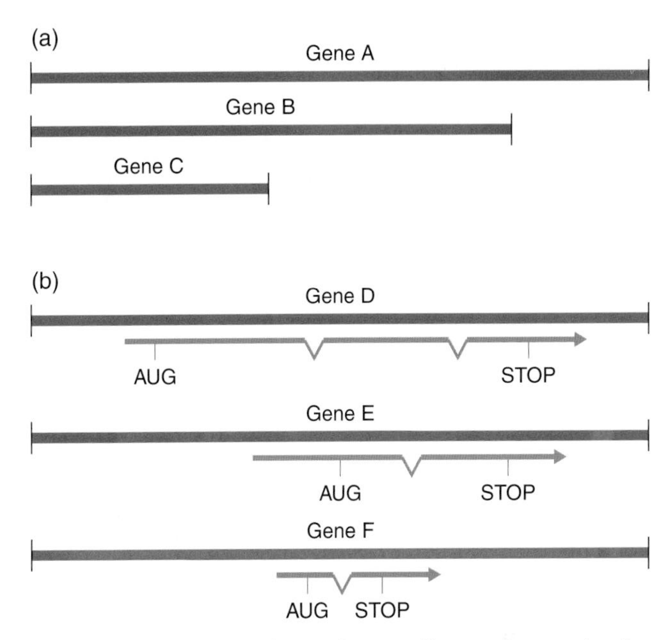

Figure 12-4 Some of the factors that contribute to the mutational target size of a gene. (a) The size of the entire gene, including the transcribed region and the regulatory regions, is one contributing factor. (b) The size of the protein-coding region of the gene is another. Particularly for point mutagens, the size of the protein-coding region correlates with the proportion of mutations that inactivate the protein.

need to be calculated in accordance with the spectrum of mutational events generated by a given mutagen.

Let's turn now to some of the commonly used general mutagens. The nature of some of them was described in Chapter 10, so they are just briefly reviewed here (Figure 12-5 on the next page). The mutagenicity of a compound depends on several factors. It has to be taken up by the organism (and the germ line) in sufficient quantity to cause mutation. It cannot be metabolized rapidly in the body before it reaches the germ line. Most mutagens are cytotoxic (that is, they kill cells). Effective mutagens are those that can be delivered at sufficient dosages to cause a high frequency of mutations without being severely cytotoxic. Presumably because of physiological differences between different species (and even between the germ lines of the two sexes), some agents are effective mutagens in one species but not in another. For example, the alkylating agent ethyl methanesulfonate is an effective mutagen in *Drosophila* males but not in females.

MESSAGE
The detailed molecular function of a gene product as well as the chosen mutagenesis can have a big impact on its observed target size.

MESSAGE
The mutagenicity of an agent depends on many factors, including how it is taken up and metabolized and how it interacts with the physiology of the mutagenized organism.

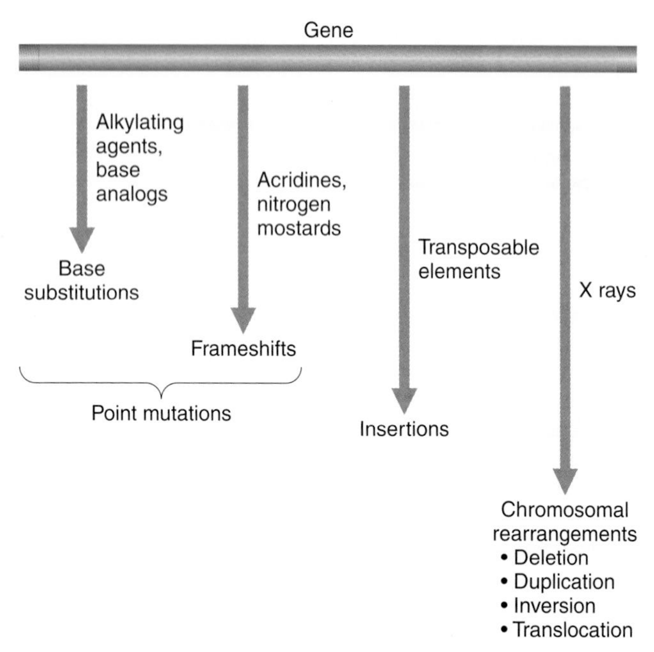

Figure 12-5 Some of the outcomes of general mutagenesis with different classes of mutagen.

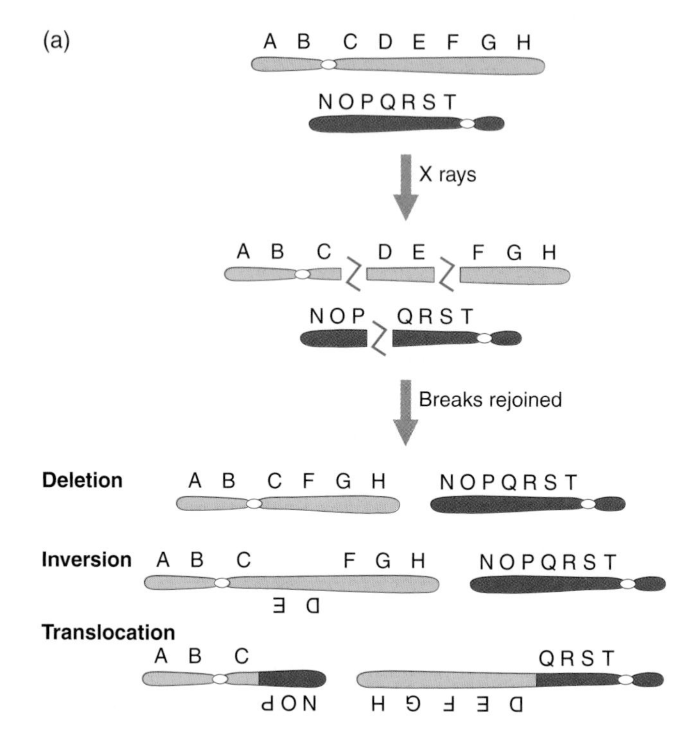

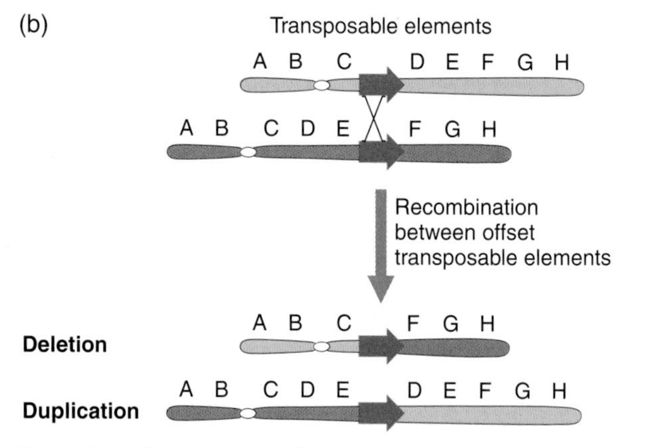

Figure 12-6 Some classes of chromosomal rearrangements that are produced by (a) X-ray mutagenesis or (b) recombination between offset copies of members of the same family of transposable elements.

Base-substitution mutagens. Many mutagens are known that produce base substitutions (see Chapter 10). Typically, they cause transitions at much higher rates than transversions. The alkylating agents ethyl methanesulfonate, ethyl nitrosourea, and nitrosoguanidine are widely used as mutagens. The base analogs 5-bromo-uracil and 2-amino-purine are also widely used in microbial systems.

Indel mutagens. The indel mutagens largely produce insertions or deletions of single base pairs of DNA. In microbial systems, proflavin is a very potent one. Some quinacrine mustards, such as ICR-170, are effective indel mutagens in some eukaryotes.

Insertional mutagens: natural and engineered transposable elements. Transposable elements, which also have been described in Chapter 10, can be very potent mutagens. In a number of genetic model organisms, the biology of transposition is sufficiently understood so that transposition events can be manipulated experimentally to occur only in a single generation of a series of controlled crosses. In this way, new transposition events can occur, producing insertions at new sites in the genome.

Mutagens that induce chromosomal rearrangements. Chromosomal rearrangements include the deletions, duplications, inversions, and translocations described in Chapter 11 (Figure 12-6). Chemical agents and certain kinds of high-energy radiation can cause such rearrangements. Generally, the action of these agents leads to double-stranded chromosome breaks, and this potentially lethal damage must be repaired. When multiple broken chromosomes are present in a cell, the repair mechanisms sometimes rejoin the double-stranded breaks improperly, leading to rearrangements. For example, chemical cross-linking compounds, such as formaldehyde, that can create covalent bonds between bases on complementary strands of DNA indirectly lead to double-strand breaks. X rays, gamma rays, and other forms of high-energy radiation also are potent chromosome-breaking agents (see Foundations of Genetics 12-1 on pages 392–393). The resulting mutations can be very small deletions or duplications (e.g., as small as a few base pairs deleted or duplicated), or they can be macro-events that may be detected merely by examining chromosomes under a microscope.

Another source of new rearrangements are the excision and recombinational events associated with the mobilization of transposable elements. Mistakes during the excision of a transposable element from a location in the genome can delete sequences next to a transposable element. Recombinational events can also occur between copies of the same transposable element located at different positions on a chromosome; such events can produce deletions, duplications, and inversions. Additional events can be caused by recombination between copies of elements located on different chromosomes; these produce translocations. Thus, transposable elements can be used as a mutagen for chromosomal rearrangements as well as for new insertions.

Directed Mutations and Phenocopies

For most of the twentieth century, researchers saw the ability to target or direct mutations to a specific gene as the holy grail of genetics. As it turns out, an understanding of the basic molecular biology of cellular processes such as DNA replication, DNA repair, and information transfer during gene expression has enabled the development of such techniques. These targeting techniques enable us to evaluate the phenotype exhibited by individuals lacking the function of the targeted gene. There are two broad ways of achieving this goal:

- Inactivate the gene by targeted changes to the DNA.
- Leave the gene intact but block the activity of one of the gene products—the mRNA or the protein.

Targeted gene knockouts. Targeted gene knockouts replace some of the sequences of the *endogenous* (that is, preexisting) chromosomal copy of a gene with corresponding sequences from an *ectopic* (introduced) DNA segment engineered in such a way that the introduced sequences inactivate the gene. The techniques for performing targeted gene knockouts in yeast and in the mouse were described in Chapter 8. Basically, this technique performs a mutational inactivation of natural genes at their normal location in the chromosome. A mutated copy of a gene is recombined into the chromosome by a homologous-recombination-like mechanism, replacing the corresponding normal sequences with the mutant gene (Figure 12-7).

Targeted gene knockouts can only be carried out in a small subset of organisms: in bacteria by DNA transformation or transduction and in a very few eukaryotes, most notably yeast and mouse. Even among very well-studied eukaryotes, not all species have proved susceptible to such manipulations. For this reason, alternative methods of accomplishing that end have been pursued.

Site-directed mutagenesis and related techniques. In many circumstances, it is not possible to replace the normal gene. However, the next best thing is to produce a null mutant of the endogenous gene by a general mutagenesis technique, and then to introduce a normal or mutated transgene as the only candidate for interaction with that mutant. Thus,

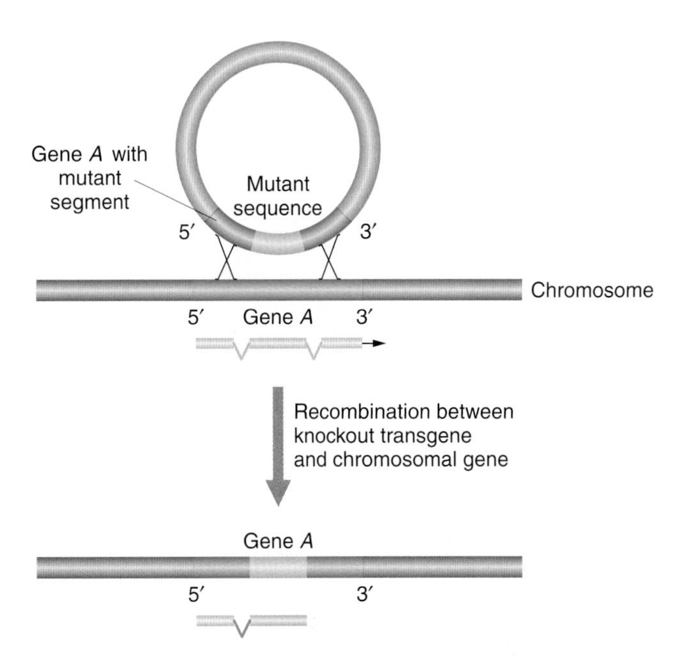

Figure 12-7 The basic molecular event in targeted gene knockouts. A transgene containing sequences from two ends of a gene but with a different segment of DNA in between is introduced into a cell. Double recombination between the transgene and a normal chromosomal gene produces a recombinant chromosomal gene that has incorporated the abnormal segment. This segment has been selected to ensure that the resulting gene would be nonfunctional.

the ability to alter the sequences of a transgene in vitro is a very powerful way to perform targeted mutagenesis.

One of the best-established techniques for altering such transgenes is **site-directed mutagenesis.** This method can create mutations at any specific site in a gene that has been cloned and sequenced. Information about the gene sequence is used to chemically synthesize short DNA segments (**oligonucleotides**). Through single-strand hybridization, an oligonucleotide can be directed to any site in the gene.

In one approach, the gene of interest is inserted into a single-stranded bacteriophage vector, such as the phage M13. A synthetic oligonucleotide containing the desired mutation is designed. This oligonucleotide is allowed to hybridize to the mutant site by complementary base pairing. The oligonucleotide serves as a primer for the in vitro synthesis of the complementary strand of the M13 vector (Figure 12-8a on page 394). Any desired specific base change can be programmed into the sequence of the synthetic primer. Although there will be a mispaired base when the synthetic oligonucleotide hybridizes with the complementary sequence on the M13 vector, a few mismatched bases can be tolerated when hybridization takes place at a low temperature and a high salt concentration. After DNA synthesis by DNA polymerase in vitro, the M13 DNA is allowed to replicate in *E. coli,* and many of the resulting phages will be the desired mutant. The synthetic oligonucleotide can be used as a labeled probe to distinguish wild-type from mutant phages. At low temperature, the mismatched base does not prevent the primer from

FOUNDATIONS OF GENETICS 12-1

Muller Demonstrates That X Rays Are Mutagenic

So far, the text has taken for granted that there are environmental agents that act as mutagens. In reality, one of the first applications of mutational analysis was to assess whether outside agents could increase the mutation rate. In the 1920s, geneticists began to address the question of whether the environment could modify the mutation rate. The first attempts to answer this question, however, got negative answers. In part, it was because geneticists selected the wrong environmental agents, and in part because they didn't have assay systems that would allow them to compile enough for statistically significant results. H. J. Muller recognized these underlying flaws and worked out a system in *Drosophila* that allowed him to report in 1927 an unequivocal proof that X rays were mutagenic.

Muller had recognized that he needed to identify a phenotype that he could classify objectively and that was shared by a large number of genes so that he could amass a statistically significant amount of data in a reasonable period of time. He knew that one of the most common phenotypes in *Drosophila* was recessive lethality and that several hundred sex-linked genes could very likely be mutated to this state. He then designed a mating scheme that would allow him to look for recessive (actually, hemizygous) X-linked lethality by the absence of grandsons in the F_2 of mutagenized grandfathers. To do this, he designed the first *balancer* chromosome (see Chapter 11), an X chromosome balancer that he called *ClB*. The "*C*" stood for "crossover-suppressor," because it contained two inversions that prevented recombinants from forming between *ClB* and a normal X homologue; the "*l*" stood for the recessive lethality due to a mutation on the balancer chromosome; and the "*B*" stood for the dominant morphological mutation *B (Bar eye)*, which was carried on *ClB*.

Muller reasoned that he could use the *ClB* balancer chromosome to track simultaneously new mutations arising in any of hundreds of different genes, specifically that he could observe forward mutations changing wild-type genes to recessive lethal alleles. (We now estimate that about one in four genes in the *Drosophila* genome can mutate to recessive lethality, but Muller didn't know this in 1927.) To make use of the *ClB* system, Muller set up a series of crosses as follows (in these crosses, the *w* mutation is the X-linked recessive white-eye color mutation, and Y refers to the Y chromosome).

He knew that he must be starting with an X chromosome that did not carry any recessive lethal mutation, as the X chromosome is viable as the hemizygous X in the P-generation males. So, if this X chromosome produces an inviable phenotype such as the hemizygous X in the F_2 males, then it must have picked up a recessive lethal mutation during mutagenesis. To measure the rate at which such mutations were induced, for each dose of X rays as well as the unirradiated control, Muller set up several hundred individual F_1 crosses in parallel, each with a single female (that is, with a single mutagenized X chromosome). Because the *ClB* chromosome is

hybridizing with both types of phage, but at high temperature, the primer hybridizes only with the mutant phage. Oligonucleotides with deletions or insertions will cause similar mutations in the resident gene. The site-directed method can also be used on genes cloned in double-stranded vectors if their DNA is first denatured.

A knowledge of restriction sites is also useful in modifying a cloned gene. For example, a small deletion can be made by removing the fragment liberated by cutting at two restriction sites (Figure 12-8b). With a similar double cut, a fragment, or *cassette*, can be inserted at a single restriction cut to create a duplication or other modification (Figure 12-8c). Another approach is to erode enzymatically a cut end created by a restriction enzyme to create deletions of various lengths (Figure 12-8d).

MESSAGE

Site-directed mutagenesis and other such techniques allow various types of directed changes to be made at specific sites in a cloned DNA molecule.

The polymerase chain reaction (PCR) also can be used (Figure 12-8e). A primer containing a mutation is used in a first round of PCR. The product of the first round is used as a primer for a second round of PCR, whose product is the mutant gene.

Targeted gene knockouts and site-directed mutagenesis are the two mutational methods of inactivating specific genes in a directed fashion. Often, however, directed gene inactivation by one of these techniques is not possible. Because of this, researchers have looked for more general ways of doing the next best thing: inactivating the gene product rather than the gene itself. The advantage of inactivating the gene itself is that mutations will be passed on from one generation to the next, so that manipulation of the cell, as in injection of engineered DNA segments, needs to be done only once. On the other hand, inactivation of the gene product—the general term for the process is **phenocopying** (mimicking a mutant phenotype by manipulating something in the interior environment of the cell or its surroundings)—can be applied to a great many organisms regardless of how well developed the genetic technology is for a given species. The next three sections exam-

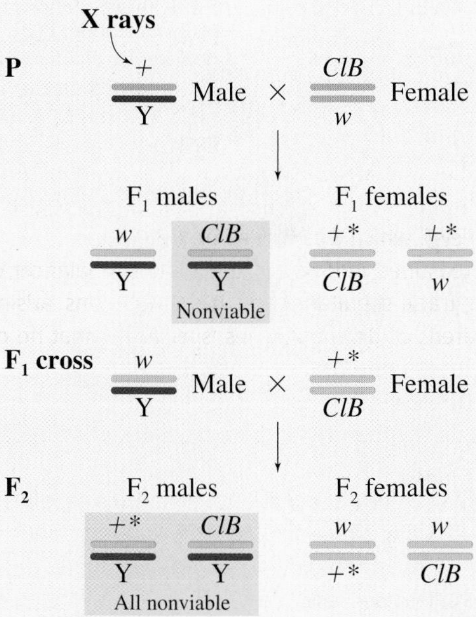

X rays

P: $+$ / Y Male × ClB / w Female

F_1 males: w / Y; ClB / Y (Nonviable)

F_1 females: $+*$ / ClB; $+*$ / w

F_1 cross: w / Y Male × $+*$ / ClB Female

F_2 males: $+*$ / Y; ClB / Y (All nonviable)

F_2 females: w / $+*$; w / ClB

lethal in hemizygous males (remember the "*l*" in *ClB* stands for recessive lethality of this balancer), if there is also a newly induced recessive lethal on the mutagenized "+" X chromosome, then neither X chromosome is hemizygous viable and there will be no grandsons of the original mutagenized P-generation male. For this reason, the technique was sometimes called the *grandsonless technique.*

In his several hundred unirradiated control cultures, Muller recovered no recessive lethal (grandsonless) crosses. Thus, the spontaneous mutation rate to recessive lethality was quite low. In contrast, Muller found that about 1000r of X rays (a unit of total exposure) gave about 2.5 percent recessive lethal crosses, and that about 2000r gave about 5 percent. From this, it was very obvious that X rays were a potent mutagen and that the mutation rate per gene appeared to be linear with dose.

Muller's observations were exceedingly important. For the first time, geneticists were able to accelerate the rate of mutant production, greatly enhancing their ability to isolate mutations of interest. Fourteen years later (1941), Charlotte Auerbach and colleagues used the same technique to demonstrate the first example of chemical mutagenesis. This work, on the chemical-warfare agent mustard gas, was classified by the British government during World War II and not published until 1947. In short order, many other compounds were then identified as mutagens.

At the time of Muller's original work, X rays were already being used extensively in medicine for diagnosis and therapy, and his findings struck a cautionary note to limit exposure to medical X rays. Indeed, Muller's experience with X rays (one form of high-energy radiation) later led him to be an early and outspoken critic of nuclear weapons production and testing. The work served as a model for how to measure mutation rates in any experimental organism, and its logic is still followed today in the assessment of environmental agents as mutagens.

ine three such phenocopying techniques. These techniques have been very effective in some systems and less so in others, for reasons that are not always understood or predictable. Nonetheless, they are part of the growing arsenal of techniques for studying the phenotype that arises from shutting down the activity of a gene product.

Antisense RNA. Translation of an mRNA can be inhibited by introducing RNA molecules with sequences complementary to a portion of that mRNA. Because the mRNA is considered the sense strand (it encodes the sequences that are translated into the polypeptide product), the RNA that is complementary to the mRNA and intended to antagonize its action is called **antisense RNA.**

Antisense RNA is typically introduced into target cells by mechanical or chemical means similar to the ways that transgenic DNA constructs are introduced, except that the goal is to get the antisense RNA into the cytoplasm, not the nucleus. Once inside the cell, antisense RNA can hybridize with the mRNA and either prevent its translation or cause degradation of the mRNA-antisense RNA complex. Regardless of the mechanism, the end result is that the protein product of the mRNA is reduced or eliminated (Figure 12-9 on page 395).

What must be considered when using antisense RNA as a phenocopying tool? It works best when the protein product of the gene is unstable, so that the pool of preexisting protein will be degraded rapidly. When proteins are stable, mutant phenotypes may not be apparent even if the antisense RNA reduces functional mRNA levels very effectively. The antisense RNAs themselves may be degraded by ribonucleases (RNAses) in the cell. This problem is typically addressed by synthesizing the antisense RNA with chemically modified ribonucleotides that resist RNAse digestion. Specificity is important, but many genes have multiple members with very similar protein-coding regions. To address this, antisense RNAs are often synthesized against the sequences of the 3′ or 5′ untranslated regions (3′ UTR or 5′ UTR) of the mRNA. Because the 3′ UTR and 5′ UTR are typically more divergent than the coding region of the mRNA, these antisense RNAs can be directed much more specifically than those complementary to translated regions. In addition, it is now known that the 5′ UTR and 3′ UTR domains of the mRNA must interact with each other (probably through proteins that bind to each) as part of the initiation of translation. Thus, antisense

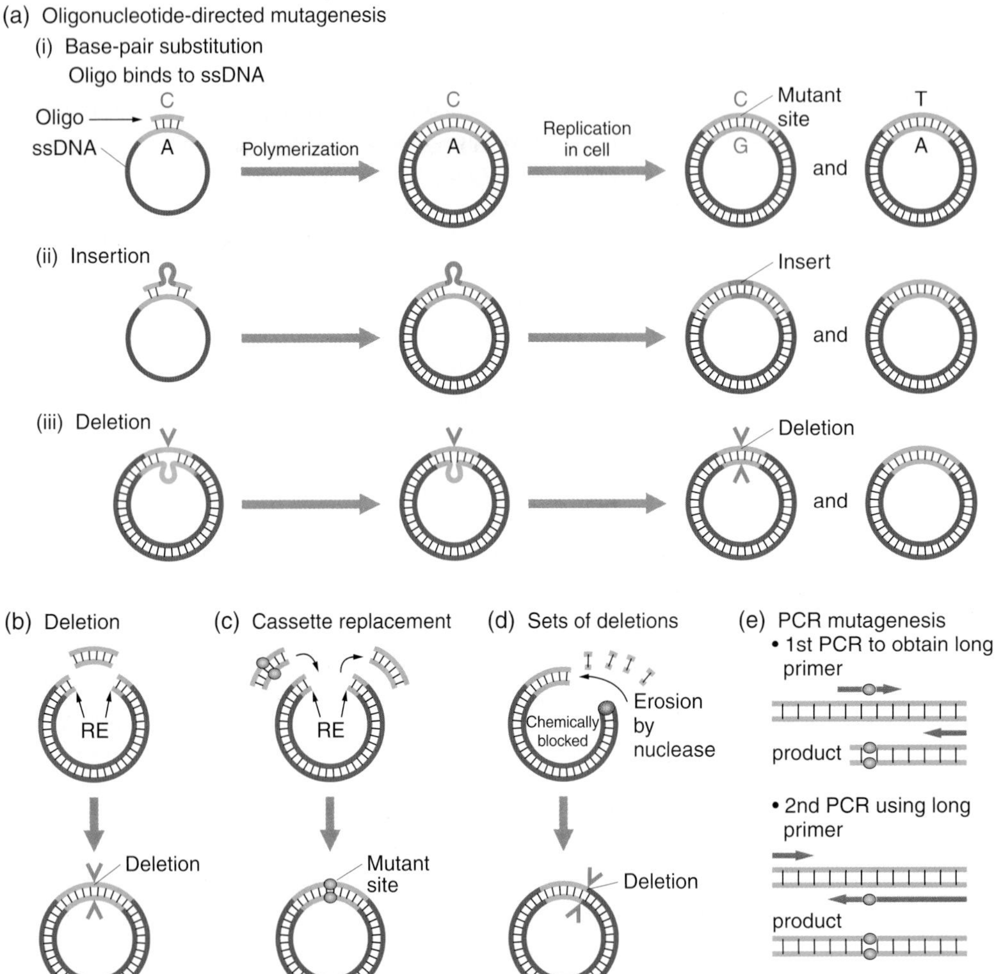

Figure 12-8 Site-directed mutagenesis. (Oligo = oligonucleotide; PCR = polymerase chain reaction; RE = restriction enzyme; ssDNA = single-stranded DNA.)

RNAs directed at one of the UTRs can be very effective in blocking message translation.

Double-stranded RNA interference. The introduction into a cell (for RNA) or into the genome (for DNA) of additional nucleotide sequences homologous with an endogenous gene can shut down the expression of that endogenous gene, even though the gene itself has not been mutated. It is thought that such phenomena reflect a "search-and-destroy" activity of cells that has evolved as a mechanism to protect them from infection by viral genomes. Among the best characterized and readily manipulable of these phenomena is **double-stranded-RNA interference (dsRNAi).**

In dsRNAi, a double-stranded RNA with sequences homologous to part of an mRNA is synthesized (Figure 12-10) and introduced into a cell. The result is a major reduction of mRNA levels that lasts for hours or days. It has been applied successfully in several systems, including *Caenorhabditis elegans, Drosophila,* and several plant species. It has been especially successful in applications to *C. elegans.* In a few cases, the mechanism of action of dsRNAi in *C. elegans* has been worked out (Figure 12-11 on page 396). The introduced dsRNAi is chopped into 21 to 23 sequences (it is unclear whether they are single- or double-stranded) and bound to a complex called the *RISC (RNA-induced silencing complex).* The RNA component of the complex, called the *guide RNA,* helps the complex find homologous RNAs by standard base-pair complementarity. Once the complex binds to the target mRNA, the nuclease activities of the complex degrade it. Phenocopies of mutant phenotypes can then ensue, depending upon the level and length of the RNA inhibition.

Interestingly, the chance discovery of dsRNAi led to the identification of the RISC complex. Because a general biological phenomenon underlies dsRNAi, it is expected to be applicable to many genes and many species. This seems to be true, but there is still considerable gene-to-gene variability in the level of reduction in activity caused by dsRNAi. Some of the considerations regarding antisense RNA also hold for dsRNAi.

Chemical-library screening. The last point in the information transfer process that can be targeted for phenocopying is the protein itself. A genomic scale, high-throughput technique has been developed called **chemical genetics,** in

MESSAGE
A variety of techniques that disrupt phenotype by interfering with steps in information transfer from the gene to the protein, but that do not impair the heritable gene itself, offer another approach for reverse genetics.

which libraries of tens or hundreds of thousands of related synthesized small molecules are tested for their ability to bind to a specific protein with high affinity (that is, bind very tightly). A small molecule that shows specific high-affinity binding to a single protein in a proteome can then be introduced into a cell by physical or chemical means and tested for its ability to inhibit the activity of the protein. If it does inhibit the protein activity sufficiently, then a phenocopy of the mutant phenotype for that protein may be achieved by treating a cell or an organism with that chemical compound.

Even though the name "chemical genetics" implies that this is a genetic technique (that is, one involving inheritance), it is not. Rather, it is a systematic extension of the long-standing use of inhibitory drugs to inactivate a protein involved in a specific biochemical process in the cell. The problem with most inhibitory drugs is that specificity is incomplete and so, inadvertently, multiple proteins and multiple biochemical processes in the organism are inhibited by the drug. Chemical genetics offers two advantages over classical drug inhibition.

- First, chemists have developed chemical libraries—tens or hundreds of thousands of related synthetic compounds—that can be assayed by high-throughput robotic techniques for their ability to bind a specific protein (Figure 12-12a on the next page) or to phenocopy a specific cellular phenotype (Figure 12-12b).
- Second, once a compound from one of these libraries is identified, it can be tested by high-throughput robotic techniques for specificity—that is, the ability to bind uniquely to a single protein among many of the proteins in the proteome of a cell or an organism. We are not yet at the point at which the entire proteome of complex genomes can be assayed, but functional genomics is rapidly developing such techniques for applications to higher eukaryotes.

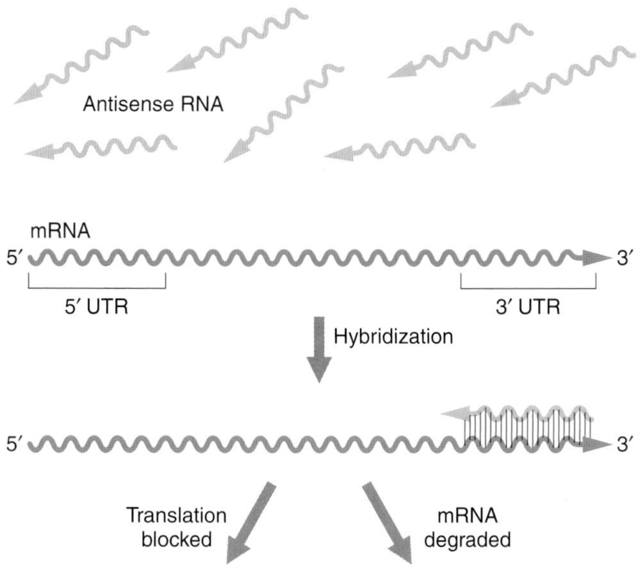

Figure 12-9 The general action of antisense RNA. The antisense RNA is complementary to some portion of the mRNA. Often, the sequence chosen for complementarity comes from the untranslated region of the mRNA. This is because, even if the genome includes related genes (members of the same gene family), their untranslated regions are typically very different from one another. In this way, the antisense RNA can be designed to bind specifically to the mRNA produced by a single gene. There are two mechanisms by which antisense mRNA may act. It may interfere with translation (both the 5′ and 3′ untranslated regions are known to be very important for initiating translation), or it could produce a double-stranded RNA segment that is vulnerable to digestion by special double-stranded-RNA endonucleases in the cell, resulting in degradation of the mRNA.

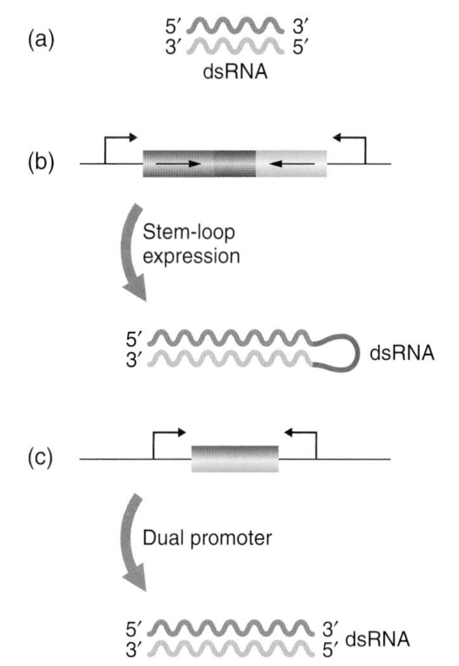

Figure 12-10 Three ways to introduce dsRNAi into a cell. (a) The double-stranded RNA can be synthesized in vitro and injected into a cell. (b) A transgene containing a reverse repeat copy of a DNA segment from a gene is introduced into the genome. When the transgene is expressed, it forms a self-complementary stem and loop with a double-stranded region. (c) A transgene containing two promoters is introduced into the genome. When both promoters are expressed in the same cell, the complementary RNA molecules can hybridize to form dsRNAi. *(Reprinted with permission from S. Hammond, A. Caudy, and G. Hannon,* Nature Reviews: Genetics *2, 2001, 116.)*

As with the other phenocopying techniques, there are both advantages and disadvantages to the use of chemical libraries. As with the other methods, these techniques can be applied in the absence of genetic analytical tools. In a sense, this is also the disadvantage, because the lack of heritability means that the inhibitor identified by screening a chemical library has to be inserted into a cell (or every cell of the organism) every time it is used. Sometimes, this is not a practical limitation, particularly in the study of cultured cell lines or unicellular microorganisms, but its application to multi-tissued higher organisms in vivo remains uncertain.

MESSAGE

Several techniques are available for the directed phenocopy analysis of the mutant phenotype of the products of a specific gene.

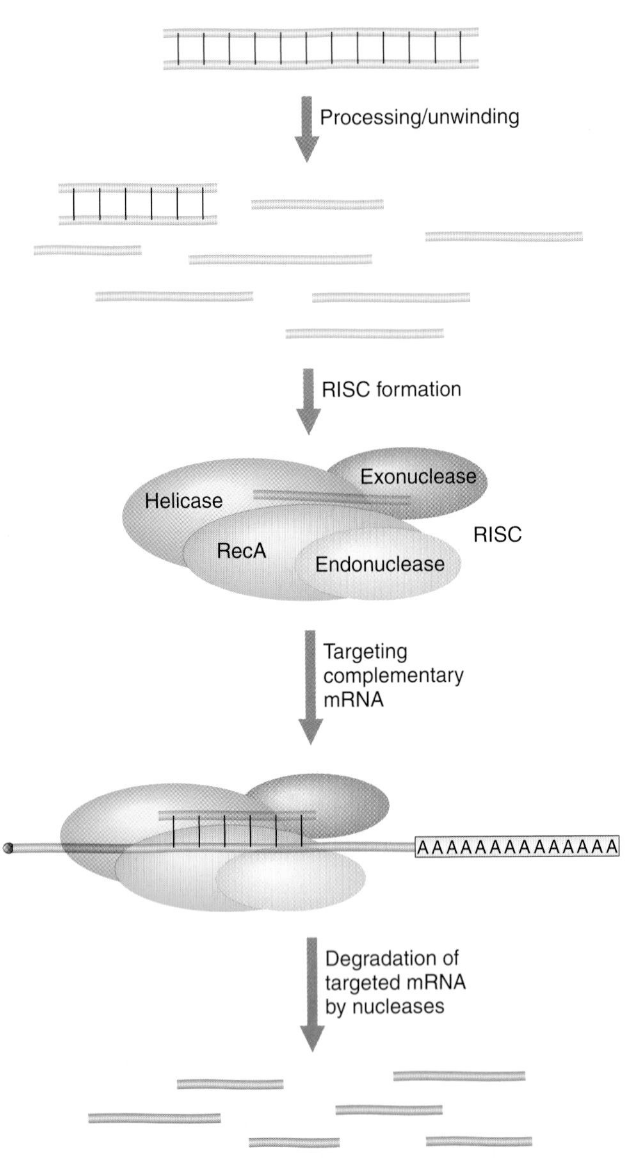

Figure 12-11 A proposed mechanism of action for dsRNAi. In this mechanism, the dsRNAi specifically interacts with a complex of DNA/RNA binding proteins (RecA), nucleases, and unwinding proteins (helicase). This RNA/protein complex is then able to bind to and degrade the complementary mRNA. It is thought that dsRNAi exploits a mechanism that has evolved to protect cells from attack by double-stranded RNA viruses. *(Modified from S. Hammond, A. Caudy, and G. Hannon,* Nature Reviews: Genetics *2, 2001, 115.)*

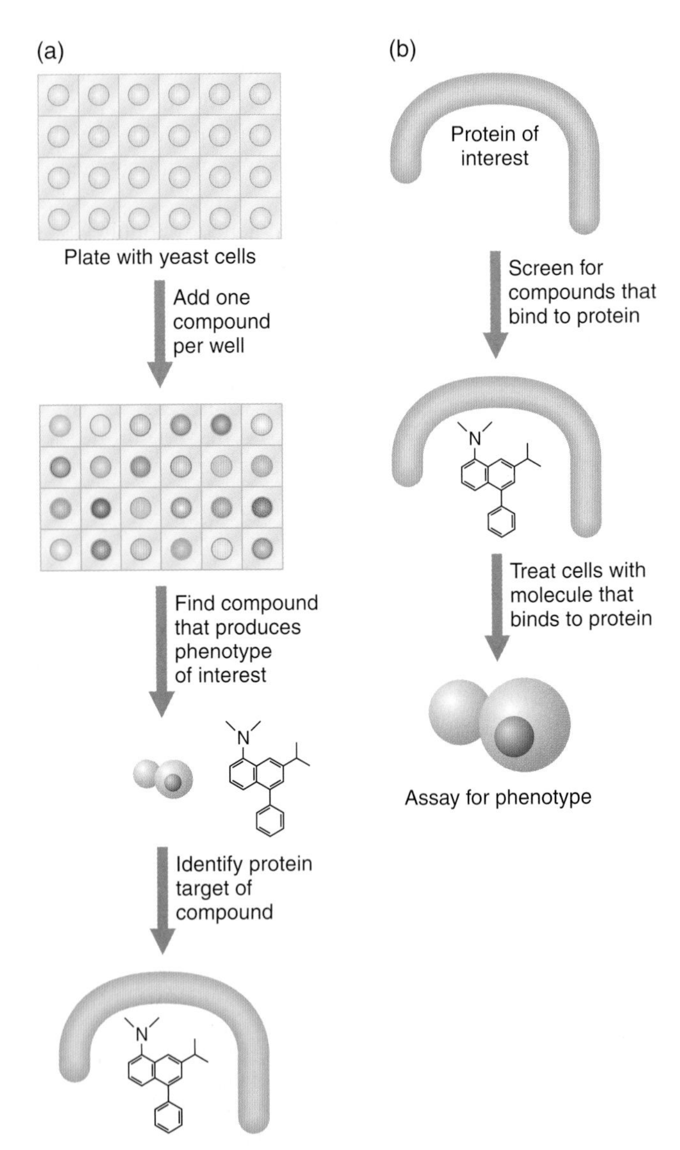

Figure 12-12 Chemical genetics. (a) An example of *forward* chemical genetics, in which small molecules are directly tested on yeast cells to identify one that produces a phenotype of interest. (b) An example of *reverse* chemical genetics, in which a small molecule is first shown to bind to a protein of interest and is subsequently tested for its phenotypic effect when applied to cells. *(From B. Stockwell,* Nature Reviews: Genetics *1, 2000, 117.)*

THE MUTATIONAL ASSAY SYSTEM

Let's now turn to the other part of harvesting mutations—the mutational assay system.

The key to mutational analysis is to wed the appropriate mutagen to the appropriate system for detecting mutations—that is, for generating candidate individuals with a phenotype of interest. What are the elements of an appropriate mutation-detection system? Many factors come into play, as described in the following sections.

Somatic versus Germ-Line Mutation

One question in the design of mutational analysis is the type of tissue or cells to examine. In multicellular organisms, genes can mutate in either somatic or germinal tissue, and these changes are called **somatic mutations** and **germ-line** (also called **germinal**) **mutations,** respectively. These two different types are shown diagrammatically in Figure 12-13. If a *somatic mutation* occurs in a single cell in developing somatic tissue, that cell is the progenitor of a population of identical mutant cells, all of which are descended from the cell that mutated. A population of identical cells derived asexually from one progenitor cell is called a **clone.** Because the members of a clone tend to stay close to one another during development, an observable outcome of a somatic mutation is often a patch of phenotypically mutant cells called a **mutant sector.** The earlier in development the mutation event, the larger the mutant sector will be (Figure 12-14). Mutant sectors can be identified visually only if their phenotype contrasts with the phenotype of the surrounding wild-type cells (Figure 12-15).

In diploids, any *dominant* somatic mutation is expected to show up in the phenotype of the cell or clone of cells containing it. On the other hand, a *recessive* mutation will not be expressed, because it is masked by a wild-type allele that is by definition dominant to the recessive mutation. A second mutation could create a homozygous recessive mutation, but this event would be extremely rare.

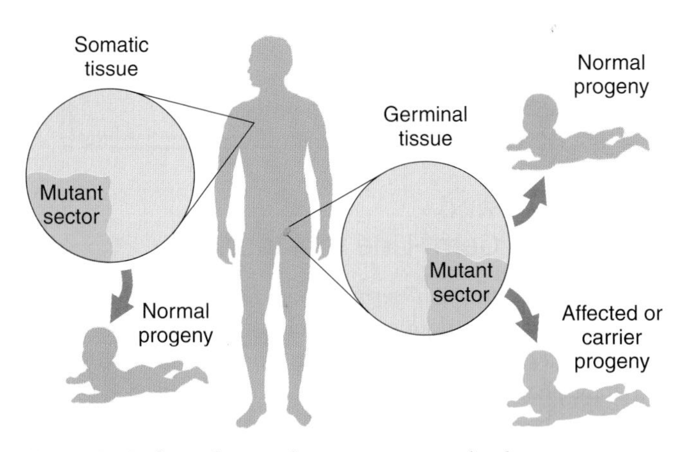

Figure 12-13 Somatic mutations are not transmitted to progeny, but germinal mutations may be transmitted to some or all progeny.

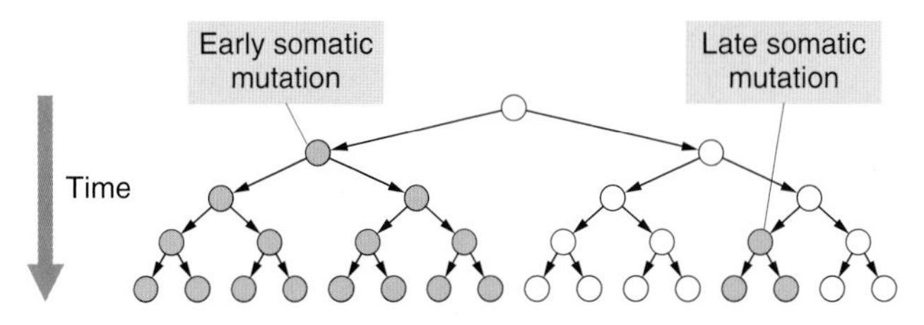

Figure 12-14 Early mutation produces a larger proportion of mutant cells in the growing population than does later mutation.

What would be the consequences of a somatic mutation in a cell of a fully developed organism? If the mutation is in tissue in which the cells are still dividing, then it is possible that a mutant clone will arise. If the mutation is in a postmitotic cell—that is, one that is no longer dividing—then the effect on phenotype is likely to be negligible. Even when a dominant mutation renders a cell either dead or defective, this loss of function will be compensated for by other normal cells in that tissue. However, mutations that give rise to cancer are a special case. Mechanisms of cancer will be considered in Chapter 15.

Are somatic mutations ever passed on to progeny? This is impossible, because somatic cells by definition are those that are never transmitted to progeny. However, note that, if we take a plant cutting from a stem or leaf that includes a mutant somatic sector, the plant that grows from the mutant cutting may develop germinal tissue that contains the mutation. Put another way, a branch bearing flowers (that is, germinal tissue) can grow out of the mutant somatic sector.

Figure 12-15 Somatic mutation in the red delicious apple. The mutant allele determining the golden color arose in a flower's ovary wall, which eventually developed into the fleshy part of the apple. The seeds would not be mutant and would give rise to red-appled trees. (Note that, in fact, the golden delicious apple originally arose as a mutant branch on a red delicious apple tree.) *(Anthony Griffiths.)*

Figure 12-16 A mutation producing an allele for white petals that arose originally in somatic tissue but eventually became part of germinal tissue and could be transmitted through seeds. The mutation arose in the primordium of a side branch of the rose. The branch grew long and eventually produced flowers. *(From Harper Horticultural Slide Library.)*

Figure 12-18 A mutation to an allele determining curled ears arose in the germ line of a normal straight-eared cat and was expressed in progeny such as the cat shown here. This mutation arose in a population in Lakewood, California, in 1981. It is an autosomal dominant. *(From R. Robinson,* Journal of Heredity *80, 1989, 474.)*

Figure 12-17 Germinal mutation determining white petals in viper's bugloss *(Echium vulgare)*. A recessive germinal mutation, a, arose in an A / A blue plant of the preceding generation, making its germinal tissue A / a. Upon selfing, the mutation was transmitted to progeny, some of which were a / a and expressed the mutant phenotype. *(Anthony Griffiths.)*

Hence, what arose as a somatic mutation can be transmitted sexually. An example is shown in Figure 12-16.

A *germinal mutation* arises in the **germ line,** special tissue that is set aside during development to form gametes. If a mutant gamete participates in fertilization, then the mutation will be passed on to the next generation. However, an individual of normal phenotype and of normal ancestry can harbor undetected mutant gametes. These mutations can be detected only if they are included in a zygote (Figures 12-17 and 12-18). Recall from Chapter 5 that the X-linked hemophilia mutation in the European royal families is thought to have arisen in the germ cells of Queen Victoria or one of her parents (see Figure 5-17). The mutation was expressed only in her male descendants.

MESSAGE

Mutations can occur in somatic cells or in the germ line. In somatic cells, mutations can be detected in the same individual that they occcur in. In germ-line cells, mutations can be detected through their transmission to progeny.

Dominant and Recessive Germ-Line Mutations

Let us focus now on germ-line mutations. We know that mutations can be dominant or recessive. Mutageneses can be carried out so that only F_1 individuals are assayed for the mutant phenotype. In haploid organisms (Figure 12-19) or for sex chromosomes in diploid organisms (Figure 12-20), both dominant and recessive mutations can be identified in the F_1 of mutagenized individuals. For autosomal genes in

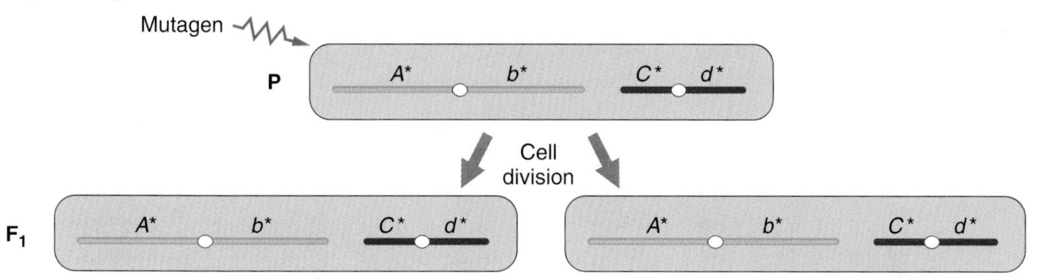

Figure 12-19 A haploid mutagenesis scheme. Note that crosses are not necessary to produce progeny with appropriate mutant genotypes. Some of the genes on two chromosomes are shown here. Note that dominant or recessive mutations in any of these genes can be recovered after the mutagenesis.

diploid organisms, such F_1 screens would reveal only dominant mutations (see Figure 12-20). Ordinarily, autosomal recessive mutations can be recovered only in the F_2 or F_3 of diploid organisms, depending upon whether the organism is monoecious (self-fertilizes) or dioecious (has two cross-fertilizing sexes) and how the mutagenesis is carried out.

> **MESSAGE**
> Haploidy permits very effective detection of both dominant and recessive mutations. In diploid species, detection of dominant mutations is also very efficient.

Detecting autosomal recessive mutations. In a self-fertilizing species (such as many plant species), a new autosomal recessive mutation in the F_1 is expected to be homozygous in 25 percent of the F_2 individuals (Figure 12-21a on the next page).

In cross-fertilizing species, in order to recover homozygous autosomal recessive mutations in any gene in the genome, F_3 individuals have to be analyzed (Figure 12-21b). In the F_1, the mutation is heterozygous in one of the two sexes. From backcrosses of this mutation-bearing individual with wild-type mates, F_2 heterozygotes for the mutation will constitute one class of F_2 progeny. Finally, through intercrosses of the F_2, the recessive mutation can

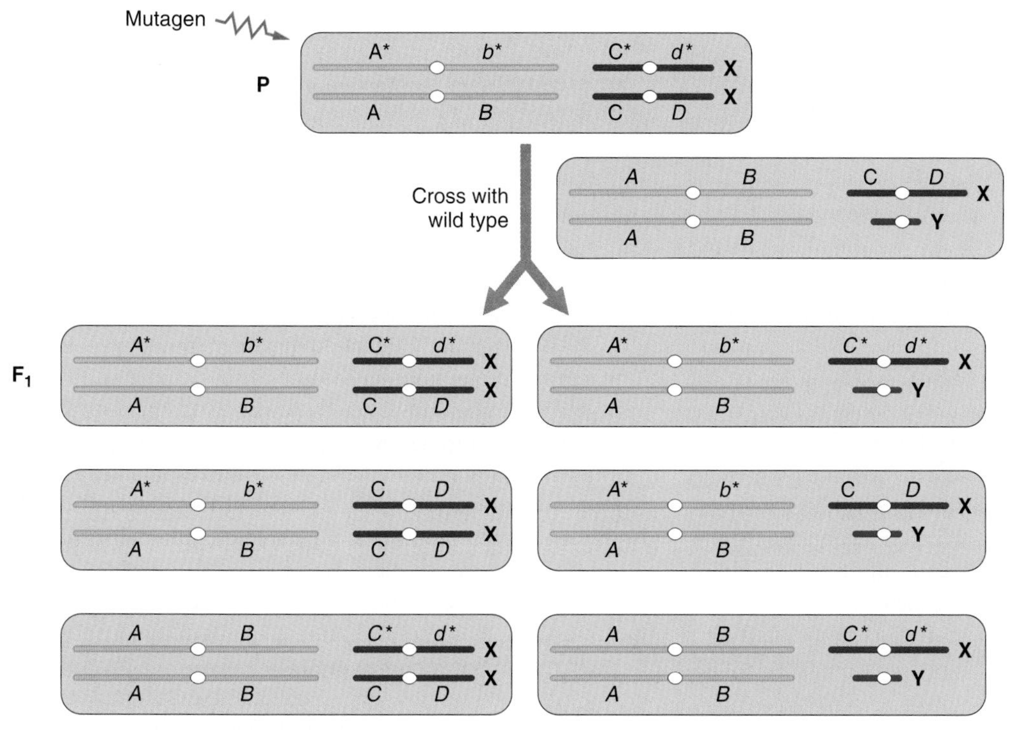

Figure 12-20 A diploid dominant F_1 mutagenesis scheme. The genome of one of the two sexes is mutagenized and crossed with unmutagenized mates. The heterozygous F_1 can be surveyed for newly arising dominant mutations (autosomal and sex-linked) and possibly for recessive sex-linked mutations (if the mutagenized individual carries two copies of the sex chromosome and will therefore transmit a mutagenized sex chromosome to the hemizygous offspring).

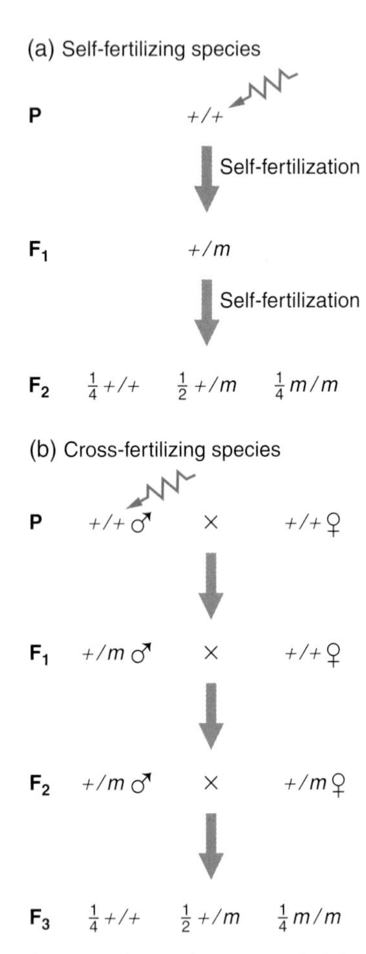

Figure 12-21 Autosomal recessive screens in (a) a self-fertilizing species and (b) a cross-fertilizing species. Note that in both cases, 25 percent of the final generation's offspring should be homozygous for the new mutation *m*. However, in self-fertilizing species these homozygous mutants can be detected in the F_2, whereas in cross-fertilizing species, they cannot be detected until the F_3.

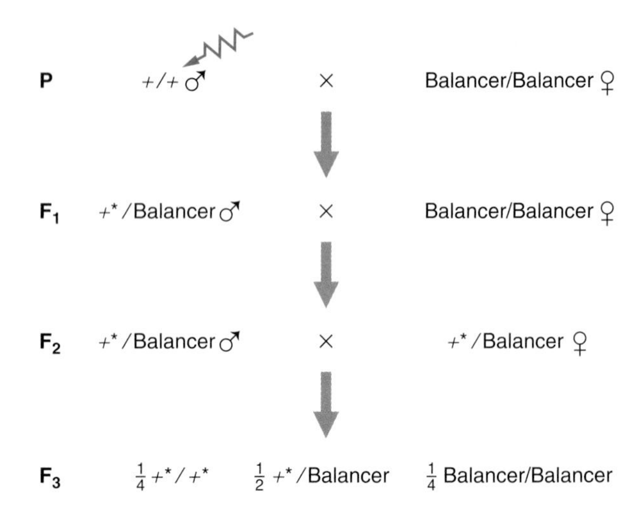

Figure 12-22 The use of balancer chromosomes in *Drosophila* to make an F_3 recessive scheme more efficient. Typically, a balancer chromosome contains a dominant morphological marker and a series of inversions that prevent crossing-over between it and a normal homologue. Thus, any fly displaying the balancer's dominant morphological marker will carry the balancer chromosome, and the flies lacking this dominant morphological marker are homozygous for a mutagenized autosome.

then be made homozygous in the F_3, permitting its recessive phenotype to be observed. Note that, at best, this is a tedious procedure. However, such F_3 screens can be made much more efficient if the mutagenized chromosome can be tracked, such as by use of balancer chromosomes in *Drosophila,* in which an entire chromosome behaves as a single indivisible unit of inheritance (Figure 12-22).

In cross-fertilizing species, recessive mutations of a specific gene may be recovered in the F_2 if the F_1 mutagenized progeny are test-crossed with individuals carrying a recessive mutant allele of the gene of interest (Figure 12-23). However, in this case, rather than any recessive mutation in any gene in the genome being detectable in the F_2 (the case for self-fertilizing species), only recessive mutations in the genes (loci) that are mutant in the tester strain can be recovered in the second generation after mutagenesis. For this reason, the protocol exemplified in Figure 12-23 is called a **specific locus test.**

MESSAGE

Ordinarily, detection of autosomal recessive mutations requires transmission through at least two generations of progeny after the mutation has been induced.

Accelerating the identification of autosomal recessive mutants. Sometimes, genetic tricks can be used to accelerate the identification of autosomal recessive mutations. For example, in the zebrafish *(Danio rerio),* applying pressure to unfertilized eggs undergoing meiosis can activate the gamete nucleus, bypass the need for fertilization, and create an instant haploid (Figure 12-24 on page 402). By applying this technique to the oocytes that would, if fertilized, ordinarily become the F_1 diploid progeny, homozygosity for mutations in any gene in the zebrafish genome can be observed in the F_1. However, these haploid zebrafish survive only partway through development and so, while the mutation can be observed in the F_1, it still takes another generation (the F_2) to isolate it by crosses of the standard diploid siblings of the adult-inviable haploid mutant individuals. Nonetheless, every generation of crosses that can be eliminated in mutant detection means that many labor-intensive crosses need not be done and, therefore, that more mutagenized genomes can be surveyed.

A second method of accelerated autosomal mutant identification depends upon the identification of mutations in clones of somatic cells. For example, in *Drosophila* it is possible to make clones homozygous for a portion of a

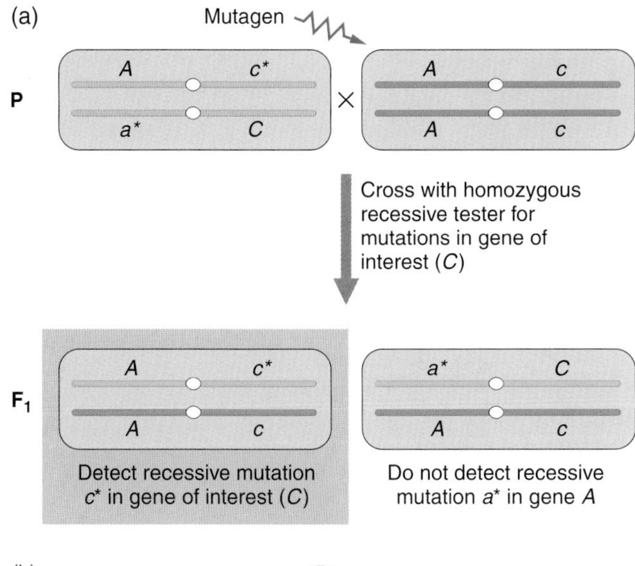

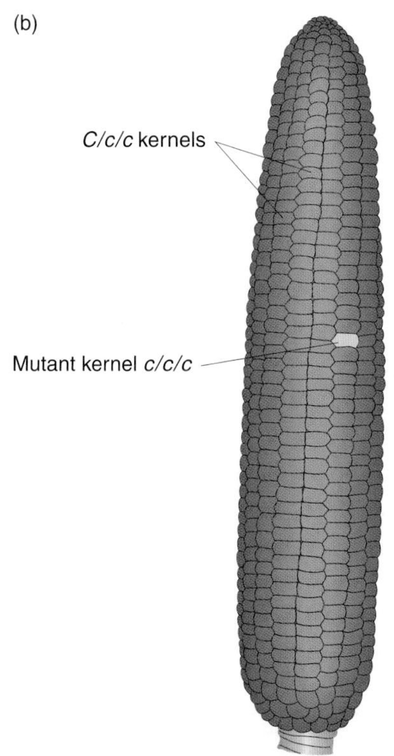

Figure 12-23 The detection system for mutations at a specific locus of corn. (a) The C allele determines the presence of a purple pigment in kernels, whereas c results in none. Mutations in other genes such as from $A \rightarrow a^*$ go undetected in this procedure. The geneticist makes the cross $c \,/\, c\cdot \;\times\; C \,/\, C\cdot$, and $C \rightarrow c$. (b) Mutations in the male germ line show up as unpigmented kernels on the cobs. (Note: The colored layer of the corn seed is triploid, formed by two identical female nuclei plus a male nucleus.)

chromosome arm (Figure 12-25 on page 403). In this technique, a mitotic crossover event occurs on a chromosome arm and, as a result, a progeny cell can be produced that is homozygous for all genes *distal* (that is, on the telomere side) of the location of the mitotic crossover event. If such a crossover occurs near the centromere of the chromosome arm, then that entire arm can be made homozygous in this progeny cell. When that cell divides several times, a local clone of mutant tissue is produced, which can be identified by being homozygous for a genetic marker on the portion of the chromosome that is homozygous because of the somatic crossover. This clone can then be examined for an autosomal recessive mutant phenotype. If mitotic clones are generated in the developing F_1 progeny of mutagenized individuals, and a morphological mutant phenotype can be observed, then the corresponding recessive autosomal mutations can be identified in the F_1. However, mutant recovery will still require two more generations of crosses. Even so, as with the zebrafish example, identification is accelerated, and this translates into many more mutagenized genomes that can be examined for the same amount of effort.

MESSAGE

In some experimental systems, procedures are available to detect some classes of autosomal recessive mutations directly in the F_1 of mutagenized individuals.

Forward and Reverse Mutations

Mutageneses can be set up so that they start with a wild-type allele and look for changes to mutant or, alternatively, start with a mutant allele and look for changes back to wild type. A process leading to any change *away from* the wild-type allele is called **forward mutation;** a process leading to any change back *toward* the wild-type allele is called **reverse mutation** (also called **reversion** or **back mutation**). For example,

$$\left.\begin{array}{l} a^+ \longrightarrow a \\ D^+ \longrightarrow D \end{array}\right\} \text{Forward mutation}$$

$$\left.\begin{array}{l} a \longrightarrow a^+ \\ D \longrightarrow D^+ \end{array}\right\} \text{Reverse mutation}$$

Note that these terms—"forward mutation" and "reverse mutation"—have to do with the functional state of the starting allele being mutated and are unrelated to the terms "forward genetics" and "reverse genetics" introduced earlier in the chapter.

Genetic Selections versus Genetic Screens

Some mutagenesis schemes are designed to kill off all individuals who do not have the mutation that the geneticist is seeking—such mutageneses are called **genetic selections** (Figure 12-26 on page 404). The alternative is mutageneses

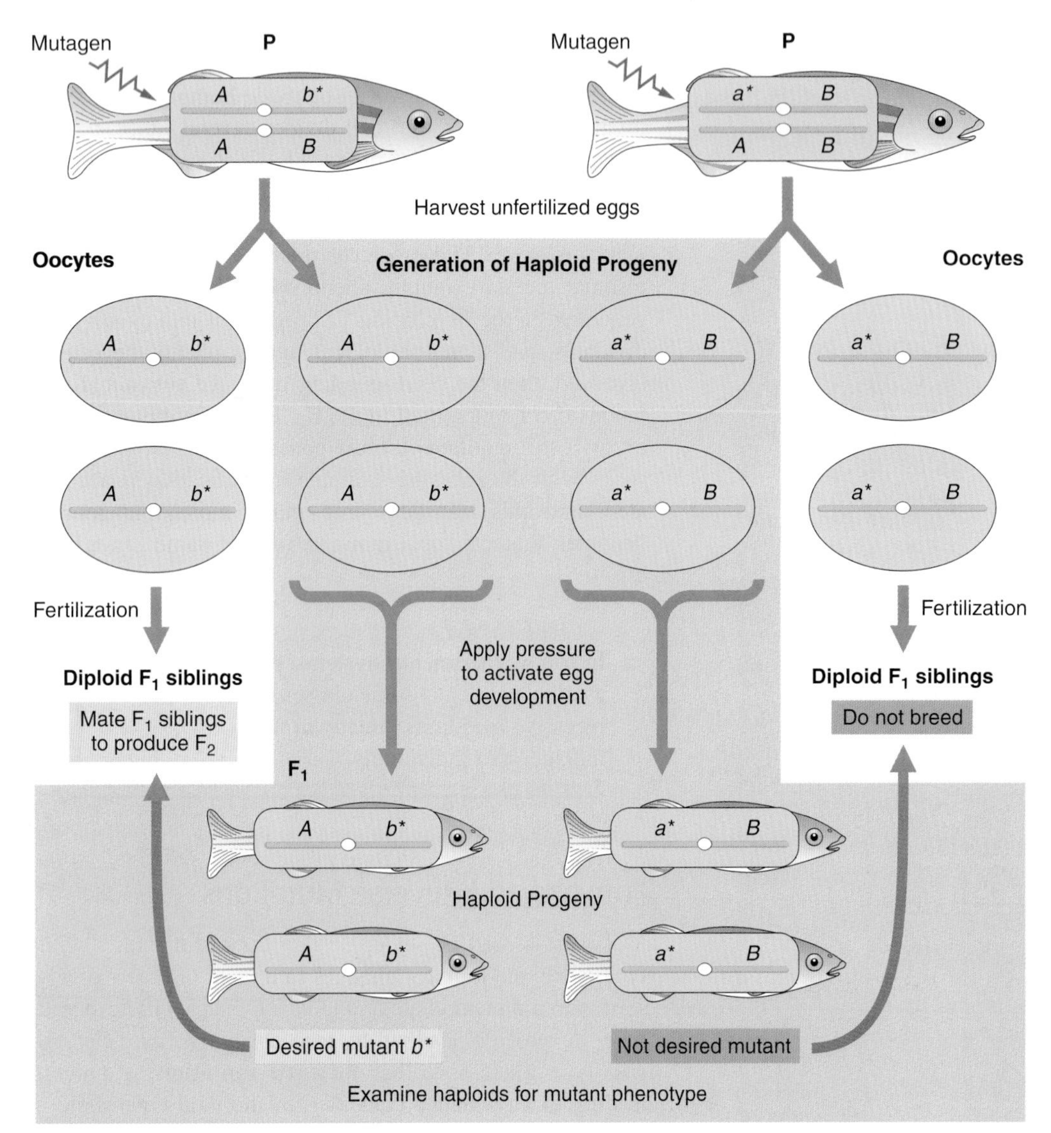

Figure 12-24 Generating haploids in the F_1 of a zebrafish *(Danio rerio)* mutagenesis.

in which both mutated and nonmutated individuals are recovered, but the individuals carrying the mutation of interest are identified because they (or some of their progeny) display a phenotype associated with this mutation—these mutageneses are called **genetic screens** (see Figure 12-26).

What are the advantages of one approach over the other? Genetic selections have one major advantage. Because mutational events are typically quite rare, the ability to kill off 99.9999 percent of the progeny makes it possible to identify a mutational event that might occur only 0.0001 percent of the time (once in a million) or even more rarely! In other words, genetic selections allow geneticists to put their effort into scaling up the size of the mutagenesis, because every offspring that survives is a "hit." Alternatively, in a genetic screen, the advantage is that any imaginable phenotype can be sought. Here, geneticists' effort goes into examining every individual for the phenotype; thus, because it is labor-intensive, far fewer individuals can be examined in a genetic screen than in a genetic selection experiment.

MESSAGE

Genetic selections are incredibly efficient but cannot be applied to the detection of many kinds of mutant phenotypes. Genetic screens are more laborious but far more adaptable to the detection of many classes of mutant phenotypes.

Genetic selections. Genetic selections can best be performed in microbial systems (bacteria, fungi, and cultured somatic cells) that can be grown on defined media. These systems provide many more targets for genetic selection than multi-tissued eukaryotes, because so many of a microbe's genes involve metabolic functions. How can genetic selections work? Figure 12-27 on page 404, shows a few examples.

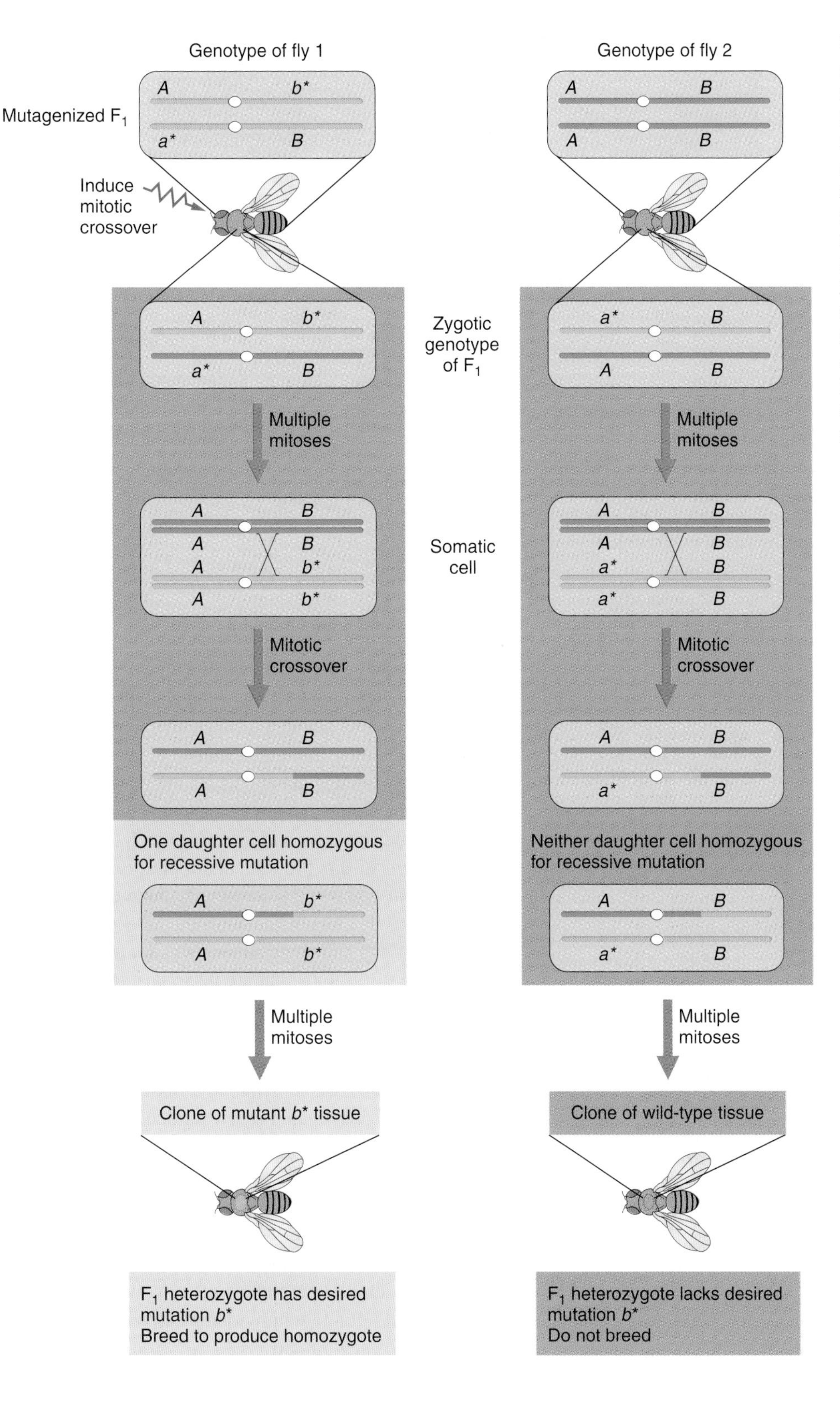

Figure 12-25 Somatic recombination in *Drosophila* as a tool for accelerating the identification of autosomal recessive mutations. Mitotic crossovers are induced in the F_1 of such mutageneses. A crossover followed by mitotic DNA replication and appropriate segregation leads to a clone of cells homozygous for the mutant-bearing segment of a chromosome. Note that it is the chromosome material *distal* to the crossover that becomes homozygous. ("Distal" in this case means on the telomere side of the crossover.)

- **Reversion to prototrophy.** Simple reversion (reverse mutation) of auxotrophic alleles in a haploid fungus or bacterium is often possible. The auxotrophic mutation can be grown on a minimal medium supplemented with a compound that is downstream of the enzymatic block in the affected biochemical pathway. Very large numbers of cells can be plated then on minimal medium lacking this supplement, and the rare prototrophic revertants will announce themselves as colonies able to grow in the absence of the supplement

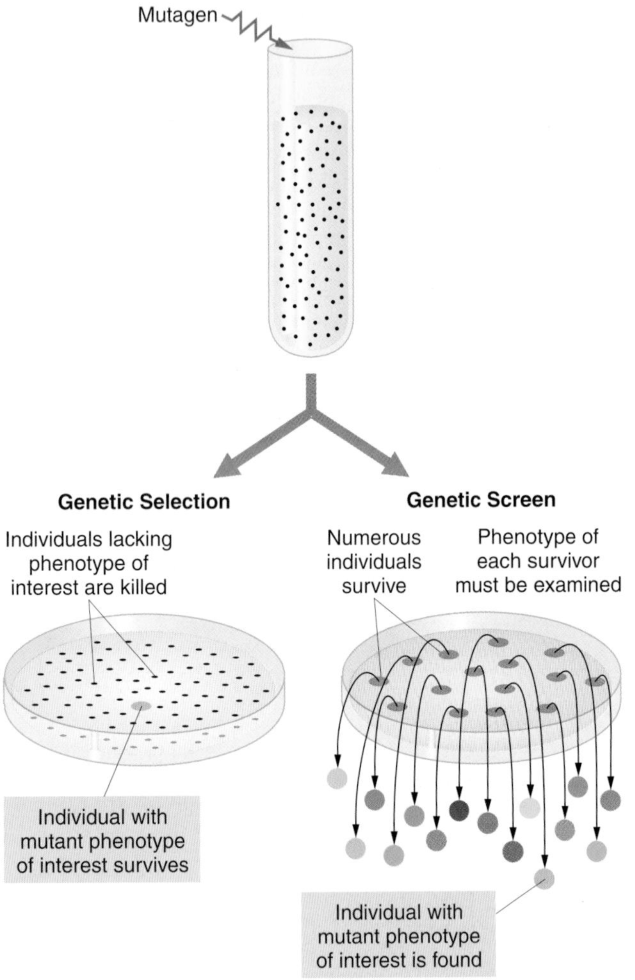

Mutagen

Genetic Selection

Individuals lacking phenotype of interest are killed

Individual with mutant phenotype of interest survives

Genetic Screen

Numerous individuals survive

Phenotype of each survivor must be examined

Individual with mutant phenotype of interest is found

Animated Art • Screening and Selecting for Mutations

Figure 12-26 Genetic selections permit highly efficient recovery of mutations because individuals die if they lack a newly induced (and rare) mutation that affects the phenotype in question. Genetic screens are less efficient but are quite flexible in permitting the investigator to recover mutations with just about any kind of phenotype.

(in other words, there is a built-in selection system). For example, to detect revertants of an auxotrophic mutation at the *leu-3* locus of yeast, large numbers of cells are plated on leucine-free medium on which only the prototrophic *leu-3*$^+$ revertants can grow. This is an example of positive selection.

- **Suppression of a mutant defect.** Extragenic suppressor mutations can also be identified by genetic selections. For example, a nonsense mutation in an auxotrophic gene can be suppressed by a mutation in a gene for an appropriate tRNA that changes the anticodon into one that binds to the nonsense codon in the mutation. Such nonsense-suppressor mutations can be detected by the ability of cells that contain them to grow on medium lacking the supplement needed to sustain the auxotroph.

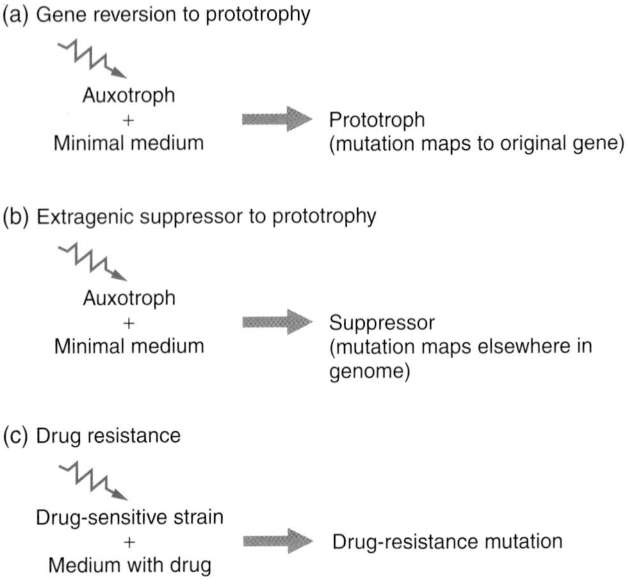

(a) Gene reversion to prototrophy

Auxotroph + Minimal medium → Prototroph (mutation maps to original gene)

(b) Extragenic suppressor to prototrophy

Auxotroph + Minimal medium → Suppressor (mutation maps elsewhere in genome)

(c) Drug resistance

Drug-sensitive strain + Medium with drug → Drug-resistance mutation

Figure 12-27 Three types of genetic selection

- **Drug resistance.** Mutations that can grow in the presence of a drug that kills normal sensitive individuals can be recovered by growth in normally toxic levels of the drug. For example, *E.coli* mutations can be recovered that are resistant to the antibiotic streptomycin. Streptomycin binds to a protein subunit of the *E. coli* ribosome and interferes with translation. Mutant forms of this protein cannot bind to streptomycin and so cells containing them can survive otherwise toxic doses of the antibiotic. Drug resistance can also arise by other means, such as the mutational inactivation of proteins that are necessary for the cell to take in the drug, or mutations that lead to rapid enzymatic degradation of the drug. In addition, for drugs in which an enzyme is required to change the drug from a nontoxic substrate to a toxic product, mutations that deactivate the relevant enzyme would be selected for drug resistance.

MESSAGE

Genetic selections depend either on reversion to prototrophy or on the exposure of mutagenized individuals to environmental agents or conditions that kill off all but a few individuals carrying a special class of new mutations.

Genetic screens: general phenotypes. Genetic screens can be applied to any problem, depending only upon the ingenuity of the researcher in coming up with a phenotype that will reveal a class of mutations in which he or she is interested. Here are some examples of common phenotypes:

- **Biochemical (auxotrophic) mutations.** Microbial cultures are convenient material for the study of biochemical mutations. Wild-type microorganisms are prototrophic, existing on a substrate of simple inorganic salts and an energy source. From these simple raw materials, the microorganisms synthesize all necessary compounds by using their many biochemical pathways. In contrast, biochemical mutants are often auxotrophic—they must be supplied certain additional nutrients if they are to grow. The chemicals that will restore growth are those presumed to be missing in the mutant cells. Therefore analyzing the precise set of chemicals that restores growth is instructive in piecing together the relevant biochemical pathway. The practical method of testing for the auxotrophic or prototrophic phenotype is shown in Figure 12-28. Although microbial cultures are used for the experimental induction of biochemical mutations, we should note that many human hereditary diseases are biochemical mutations defective in some step of cellular chemistry. The expression "inborn errors of metabolism" is sometimes used to describe such biochemical disorders. Phenylketonuria and alkaptonuria are two examples.
- **Morphological mutations.** *Morph* means "form." Morphological mutations affect the outwardly visible properties of an organism, such as shape, color, or size. Albino ascospores in *Neurospora,* curly wings in *Drosophila,* and dwarf stature in peas are morphological mutations. Additional examples are shown in Figure 12-29 on the next page.
- **Lethal mutations.** Lethal mutations cause the premature death of mutant individuals. Dominant lethal mutations occur, but by their very nature, they cannot be maintained as a strain, since *mutant* / + heterozygotes are inviable. Recessive lethal mutations, on the other hand, are a frequent mutational class that is readily recovered in genetic screens. Sometimes a primary cause of death from a recessive lethal mutation is easy to identify by some clear anatomical or physiological defect. But often the cause of death is unclear, and the mutant allele is recognizable *only* by its effects on viability. The analysis of recessive lethal mutations is of considerable importance in the genetic dissection of development (Chapter 16).
- **Conditional mutations.** Not all mutations reliably produce a mutant phenotype regardless of environmental conditions. Indeed, conditional mutations have been a gold mine for genetic analysis. A **conditional mutant** allele causes a mutant phenotype only in a certain environment, called the **restrictive condition,** but produces a wild-type phenotype in some different environment, called the **permissive condition.** Geneticists have studied many temperature-conditional mutations. Conditional mutants are useful because they can be grown under permissive conditions and then shifted to restrictive conditions for study. For example, one problem in haploid genetics is that mutations in genes whose only phenotype is recessive lethality (an abundant phenotypic class) cannot be recovered as heterozygotes. Conditional mutations offer a solution to this dilemma. Special alleles of recessive lethal mutations can be screened for, such as mutations that permit survival at a lower temperature but are lethal (don't support growth) at a higher temperature. Such **temperature-sensitive mutations** are of great value in microbial genetics, where the predominant part of the life cycle is haploid. For example, Chapter 15 will show that the mutations that have been so helpful for understanding the cell cycle are a special class of

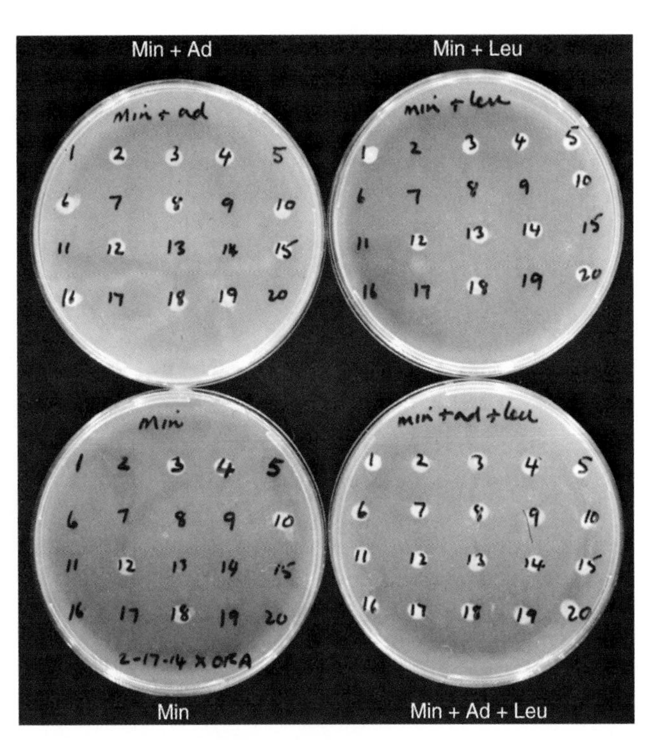

Figure 12-28 Testing strains of *Neurospora crassa* for auxotrophy and prototrophy. In this example, the test uses 20 progeny from a cross of an adenine-requiring auxotroph *(ad)* with a leucine-requiring auxotroph *(leu)*. Genotypically, the cross is $ad\ .\ leu^+ \times ad^+\ .\ leu$, and the progeny can carry any of the four possible combinations of these alleles. To test the progeny, the geneticist attempts to grow cells on various kinds of gelled media in petri dishes. The media are minimal medium (Min) with either adenine (Ad, *top left*), leucine (Leu, *top right*), neither *(bottom left),* or both *(bottom right).* Growth appears as a small circular colony (white in the photograph). Any culture growing on minimal medium must be $ad^+\ .\ leu^+$, one growing on adenine and no leucine must be leu^+, and one growing on leucine and no adenine must be ad^+. All should grow on adenine plus leucine; it is a kind of control to check viability. For example, culture 8 must be $ad\ .\ leu^+$, 9 must be $ad\ .\ leu$, 10 must be $ad^+\ .\ leu^+$, and 13 must be $ad^+\ .\ leu$. *(Anthony Griffiths.)*

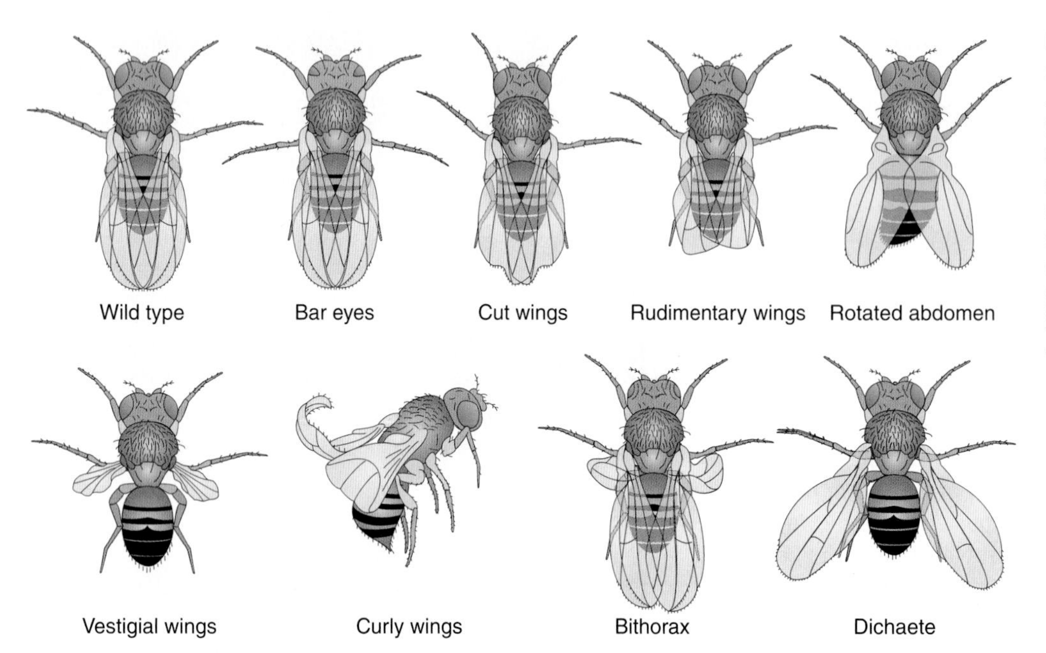

Figure 12-29 Eight morphological mutants of *Drosophila,* with the wild type for comparison. Most of the mutant phenotypes are self-explanatory; bithorax is an abnormality of the thorax featuring small wings instead of balancers; the most prominent feature of dichaete is that the wings are held at 45 degrees to the body.

temperature-sensitive yeast mutations. When shifted to a restrictive temperature, cells containing these mutations stop dividing. A given mutant has a characteristic cell-division-arrest phenotype, with all of the cells blocked at the same point in the cell cycle. There are many other examples in microbial genetics of the value of temperature-sensitive mutations in the study of cellular processes.

- **Behavioral mutations.** Geneticists can screen for mutations that affect the behavior of the organism just as they can screen for morphological mutations. In *Drosophila,* for example, wild-type flies are positively phototropic; that is, they migrate toward light. Mutants that are unable to migrate toward light are recovered by putting the flies into a T-maze in which one arm of the T-maze is illuminated and the other is in the dark (Figure 12-30). Other behavioral traits for which mutants have been screened include circadian (daily) rhythms, learning, courtship behavior, smell, taste, visual acuity, and substance abuse.

MESSAGE

Many mutant phenotypes can be detected in genetic screens.

Genetic screens: secondary screens. Recently, screens have focused on obtaining mutations in many genes affecting a given biological process. Some of them can be recovered directly by looking for mutations with a specific morphological phenotype. For example, as Chapter 16 will show, a great deal has been learned about developmental biology by looking for mutations that affect some physical aspect or pattern, such as the regular hexagonal array of facets in the *Drosophila* eye (Figure 12-31). However, not all of the genes that affect a process like eye development can be identified through direct phenotypic screens for mutations that affect the shape, number, or cell types of the facets of the eye. Why not? For one reason, some of these mutations are recessive and will affect the eye shape only in homozygotes; however, the mutants themselves are also recessive lethal because of other demands on this gene during development. Another reason to miss some genes is that their function overlaps that of other genes in the genome, and mutations in one gene are masked by the "redundant" activity of the product of the other. Here are some examples of secondary screens that provide alternative routes for identifying these "missing" genes.

- **Modifier mutations.** A modifier is a mutation in one gene that ameliorates (suppresses) or worsens (enhances) the phenotype of mutations in a second gene. In modifier-mutational screens, the mutagenized genotype is *sensitized* by introducing a known mutation into a gene that affects the biological process of interest (Figure 12-32). Modifier mutations by themselves might not detectably affect the process, but they can be detected in combination with the sensitizing mutation. The interaction between the mutations in these two genes is often the first hint that the modifier gene contributes to the same process as the gene whose phenotype it modifies. In addition, modifier mutations can often be recovered on the basis of their dominant interaction with a sensitized genotype, but in a wild-type genetic background, their effects on the process of interest would be recessive. Thus, the modifier genes can be identified as dominant mutations in a much more efficient F_1 screen instead of in a recessive F_3 screen. (We should realize that the

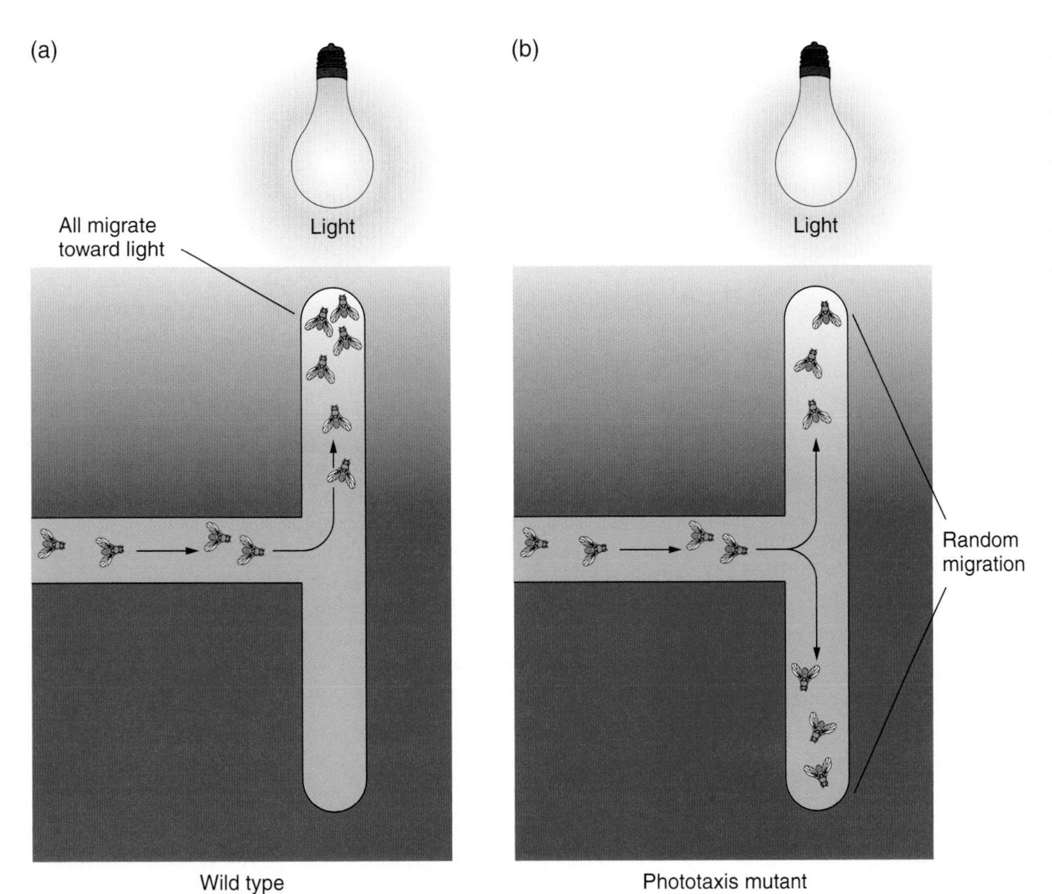

Figure 12-30 An example of a behavioral mutant screen in *Drosophila:* a T-maze for identifying mutants that are unable to orient and travel toward light. (a) Wild-type flies display positive phototaxis and all accumulate on the illuminated end of the T-maze. (b) Phototaxis-defective mutants go to either the light or dark end of the T-maze with equal probability.

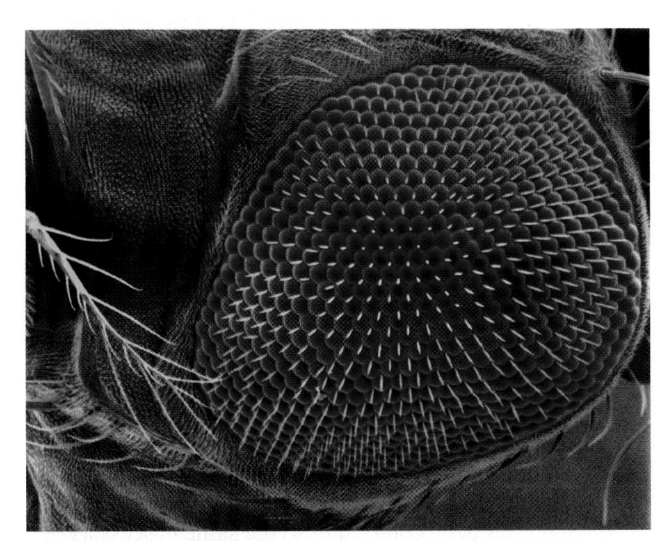

Figure 12-31 The *Drosophila* compound eye, as seen under a scanning electron microscope. Each of the approximately 800 facets is built of the same set of eight primary photoreceptor cells and several accessory cells. Many mutations that affect the number of facets or the number of cells in each facet have been identified. Because mutations that affect the eye do not affect the survival of the fly under laboratory conditions, it has been a favorite object of study for *Drosophila* geneticists. *(Dennis Kunkel Microscopy.)*

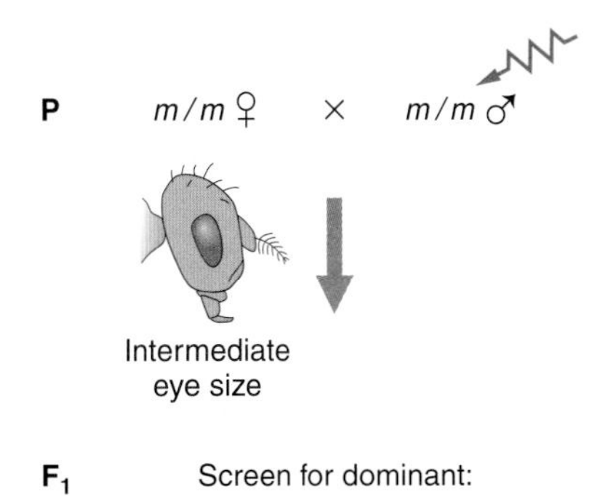

(a) Mutants with more severe phenotype (enhancer mutations)

(b) Mutants with more normal phenotype (suppressor mutations)

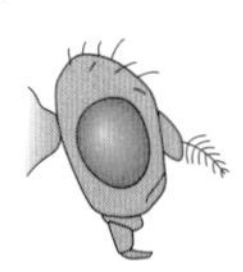

Figure 12-32 An example of a modifier screen for mutations affecting the *Drosophila* eye. Starting with a genotype (m / m) in which the eye has an intermediate number of facets, mutations in other genes that dominantly (a) *enhance* or (b) *suppress* the eye phenotype can be recovered.

same effort to screen hundreds of thousands of mutagenized F_1 individuals in a *Drosophila* cross, for example, would permit the analysis of only a thousand or so F_3 mutagenesis crosses.)

- **Mutations detected in somatic mosaics.** Consider a *Drosophila* gene that contributes to eye development, a process that occurs fairly late in development. Suppose that, because this gene also contributes to some earlier developmental process, mutants in the gene die before the stage at which eye development can be studied. Can we screen for mutants in this gene and learn that it is involved in eye development? In some experimental systems, including *Drosophila,* the answer is yes. The trick relies on mitotic crossing-over, which was described earlier in this chapter (see Figure 12-25). We perform a cross between mutagenized flies and normal spouses and, in the F_1 heterozygotes, generate clones of eye tissue homozygous for a section of one of the mutagenized chromosomes by mitotic crossing over (Figure 12-33). If such a clone shows a mutant eye phenotype, then one of the chromosomes in that fly's genome contains a mutation in a gene that contributes in some manner to eye development.
- **Screens based on gene expression.** In higher eukaryotes, many DNA regulatory sequences control transcription (more on this in Chapter 13). These regulatory elements activate transcription of any gene whose transcription start site is nearby. We can build a transgene that has a transcription start site and a "reporter" gene whose protein product can be visualized directly (such as green fluorescent protein—GFP) or indirectly through an enzymatic reaction that forms a colored precipitate (such as beta-galactosidase—β-gal). If we conduct genetic crosses

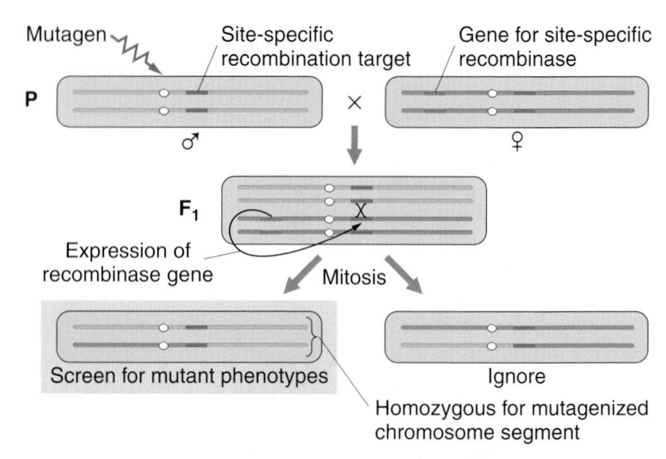

Figure 12-33 A screen that detects recessive morphological mutations in mosaic clones. Efficient systems for generating mosaic clones are possible. They exploit site-specific recombination by using a recombinase (a recombination-promoting enzyme) and the DNA target that the recombinase acts upon. Because the somatic recombination event in the outlined scheme occurs only at the target site, all resulting somatic clones are homozygous for the same segment of the genome (the segment of the chromosome arm *distal* to the target site).

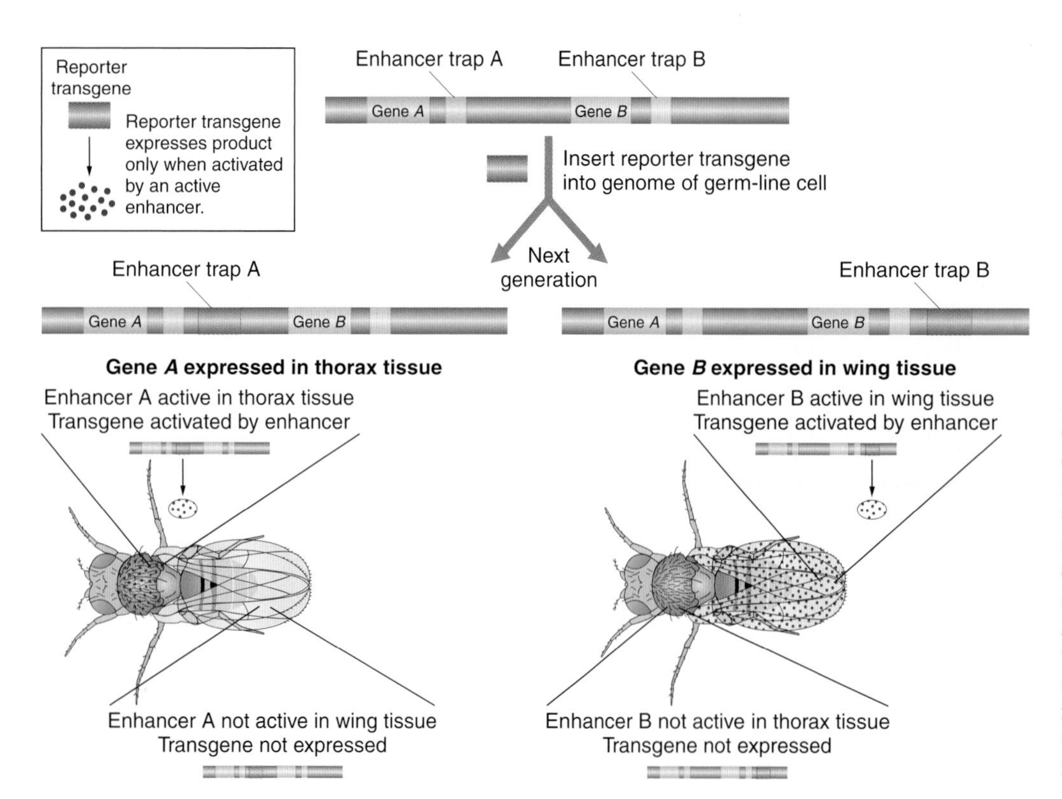

Figure 12-34 Different enhancer traps lead to expression of reporter genes in different tissues. Insertion of the enhancer-trap construct near enhancer elements that drive expression in the thorax *(left)* and wing *(right)* of *Drosophila* are shown. Note that the enhancer traps reveal some aspects of the normal regulation of the genes in which the constructs are inserted.

that mobilize this reporter transgene so that it inserts itself at various places in the genome and then we visualize the distribution of the reporter protein product, we can identify the locations of regulatory elements that drive a particular tissue and temporal pattern of gene expression (Figure 12-34). Suppose that some of these reporter-transgene insertions (also called enhancer traps because regulatory sequences that activate transcription are called enhancer elements) are expressed in developing *Drosophila* eye tissue. It is likely then that a gene that resides in the vicinity of this eye regulatory element is expressed in the eye. Thus, the immediately neighboring genes are candidates to be involved in some aspect of eye development or cellular function

MESSAGE

Some genetic screens are indirect—they detect new mutations because of their interactions with preexisting mutations in the same genome.

ANALYSIS OF THE RECOVERED MUTATIONS

Once mutations have been detected and isolated, it is important to be able to evaluate them and draw conclusions about their properties. Much of Chapters 13 to 16 explains in detail examples of how to analyze a series of mutations that affect a biological process. At a certain point, the analysis requires an understanding of the process under consideration, and those aspects of the analysis are deferred to those chapters. However, some aspects of the analysis are generic and will be explained here.

Mutations: A Mini-Review

One way that scientists gain insight into a process is by disrupting the process in various ways, including mutagenesis, observing the consequences in each case, and using this information to try to understand the step-by-step mechanism of the normal process. The analysis is facilitated by an understanding of what has gone awry in these mutations. Let's review some of the facts about mutant alleles:

- Single genetic differences (that is, mutant versus wild type) can be identified and mapped by comparing the inheritance patterns of these differences with those of standard genetic markers.
- Some mutant phenotypes are recessive to wild type; others are dominant.
- Multiple mutant alleles of the same gene can arise; these may fall into different phenotypic classes.
- Crosses between two recessive mutants that share a common phenotype reveal whether they are mutations in the same gene or in different genes. If they are mutations in the same gene, then the heterozygote for the two mutations will show a mutant phenotype; if mutations in two different genes, then the heterozygote will be wild type in phenotype. (This subject, called complementation analysis, is covered in depth in Chapter 14.)
- Genes can be cloned by recombinant DNA techniques, and the molecular differences between mutant and wild-type genes can be identified by comparing the sequence of the mutant and wild-type alleles.
- By overlaying the sequence-level mutational map and the transcript and protein maps, inferences can be made about the effect of a given mutation on the expression or structure of gene products.
- Some mutations are in protein-coding regions of genes, and their phenotypes can be understood in terms of their effects on the structures of the encoded proteins.
- Other mutations fall outside of coding regions, and their phenotypes are due to alterations in the levels, places, or times of expression of encoded gene products.

Classification Systems for Mutations

The preceding list of facts is not enough to draw inferences about a mutational event from the resulting phenotype. One of the elements that is missing is being able to classify the *kind* of alteration to gene function that occurs in a given mutation. To understand why this is a key element, think about trying to understand how a series of gears mesh to cause an axle to rotate (Figure 12-35):

- One thing that we could do is just remove any one of the gears. If we did that, then we would see that all the

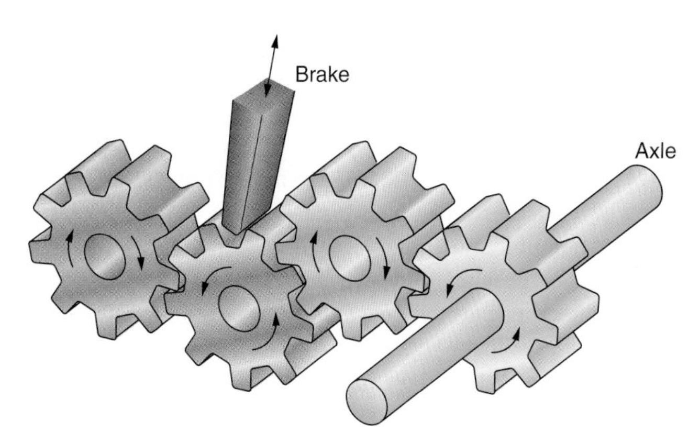

Figure 12-35 A gear mechanism that moves an axle. We can think of such a mechanism as the equivalent of a cellular pathway, with gears that act positively to turn the axle and a braking mechanism that acts negatively to prevent the axle from turning. The effects of alterations in these basic components are the equivalent of mutations in the genes that encode the components of a cellular pathway.

"downstream" gears stopped rotating. We could accomplish the same thing by leaving the gears in place, but stripping all of the gear teeth from one of the gears. If this was a biological process being studied by mutational analysis, we would say that the gear removal or gear-tooth stripping were *loss-of-function mutations* because we have simply eliminated a step in the process.

- Suppose that there is a brake that can be applied to one of the gears to slow it down. If we remove the brake pad so it doesn't work any more, then it can't slow down the gears. This too is analogous to a *loss-of-function mutation,* but in a step that *negatively* regulates the process.
- Other things we might do would be (1) to speed up the rotation of one of the gears, or (2) increase the tooth size of one of the gears, and in each case then observe the effects on gear rotation and meshing. In cases (1) and (2), rather than eliminating a step, we have given one of the gears different properties. If we were talking about mutational analysis, we would refer to such alterations as *gain-of-function mutations* because the step in the process is still active, but with different characteristics from before, so that the outcome is an increase in the speed of rotation of the axle.

Suppose that we couldn't look directly at the gears because they were hidden inside a gear box, but we could look at the resulting rotation of an axle. It is important to know how many steps there are in the process of rotating the axle. It is also important to know how the mutations in a given step alter the process. If we see that the loss-of-function "phenotype" for a given step is slowing down or stopping axle rotation (and the gain-of-function "phenotype" is increased rotational speed), then we can infer that the function of the wild-type component is to promote rotation of the downstream gears. Alternatively, if the loss-of-function "phenotype" is increased axle rotation speed, then we will infer that the contribution of the wild-type component is negative (to slow it down or stop it). Thus, using the individual mutations to learn about the process requires knowing whether any given case is a loss-of-function or gain-of-function mutation. How do we get this information? In our gear analogy, we have no way of doing this without peering into the gear box; but in mutational analysis, we shall see that we can infer loss- versus gain-of-function by phenotypic analysis of different dosages of mutant and wild-type genes.

MESSAGE

An important way of classifying mutations is as loss-of-function or gain-of-function.

Counting the genes in a biological process. A typical mutational screen or selection recovers a large number of mutations that represent multiple "hits" in a smaller number of genes. How many genes are represented by this mutant collection? The answer is to use genetic transmissional and complementation analysis (Figure 12-36). In transmissional analysis, if two mutations map to different locations on the same chromosome or on different chromosomes, then they are clearly mutations in different genes. In complementation analysis (see Chapter 14 for a more detailed description), if two *recessive* mutations *m1* and *m2* complement one another (that is, if *m1* / *m2* is wild type in phenotype), then we infer that the two recessive mutations represent different genes. On the other hand, if two recessive mutations fail to complement (that is, if *m1* / *m2* is mutant in phenotype), then we infer that the two mutants are allelic mutations in the same gene. (This analysis doesn't work with dominant mutations—by the definition of dominance, their

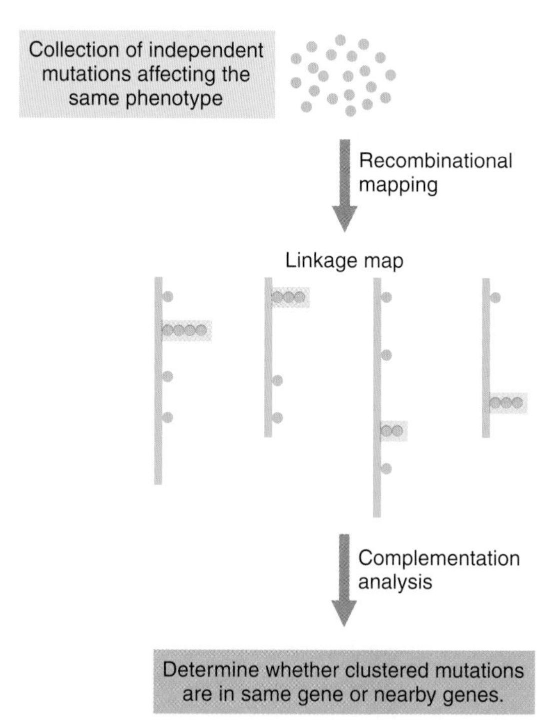

Figure 12-36 Recombinational mapping and complementation analysis can resolve the number of genes mutated in a given collection of mutations that share a common phenotype. Recombinational analysis helps us know where mutations map, and this is very important for subsequent analysis, both genetic and molecular. (For example, knowing where a gene maps allows us to use *positional cloning* techniques to isolate the DNA that corresponds to the gene.) Mutations that map to the same region of the genome may represent multiple hits in the same gene or may represent lesions in a nearby group of genes. Complementation analysis can resolve these two possibilities. The clusters of mutations (light blue boxes) must be examined by complementation analysis to determine whether mutations in the same cluster are in the same gene.

MESSAGE

New mutations with a common phenotype can be grouped according to the mutated genes by genetic mapping and, if the mutations are recessive, by testing them for allelism.

phenotype will be mutant regardless of the mutant state of the other allele.)

Diagnostics for loss of function versus gain of function. Both gain-of-function and loss-of-function mutations can be dominant or recessive. We already know how to tell dominant from recessive, but how do we tell gain-of-function from loss-of-function? Let's consider dominant mutations.

Knowing where any of the mutations map in the genome, we can determine whether chromosomal deletions or duplications of the gene exist (recall that deletions and duplications are two classes of chromosomal aberrations—see Chapter 11 and Figure 12-37). If they do, they can indicate definitively whether a dominant mutation is loss-of-function or gain-of-function (Figure 12-38 on the next page).

In the case of dominant loss-of-function, a single copy of the wild-type allele is insufficient to make enough gene product to generate a wild-type phenotype (Table 12-1 on page 413). For this reason, a loss-of-function dominant mutation is also called *haplo-insufficient.* This property leads to two predictions (see Figure 12-38a). First, a deletion of the gene *(Df)* heterozygous with the wild-type allele of the gene should also show a mutant phenotype (*Df* / + is mutant). Second, the dominant mutant phenotype should be "cured" by adding a duplicate copy of the wild-type gene (Dp^+) in a duplication (*Mut* / + / Dp^+ is wild type).

Gene dosage also can be used to distinguish between levels of loss of function. Some loss-of-function mutations completely remove the activity of the gene product; these are called **null mutations.** Others only decrease the activity of the gene product; these are called hypomorphic mutations. As Figure 12-38a shows, for a haplo-insufficient null mutation, the mutant heterozygote with wild type would be expected to have a phenotype identical with that of the deletion heterozygote with wild type. We can represent this symbolically as *Mut* / + = *Mut* / *Df*. For a haplo-insufficient hypomorphic mutation, the mutant with wild-type heterozygote should have more gene activity than the deletion with wild-type heterozygote (see Figure 12-38b). Symbolically, *Mut* / + > *Df* / + (where ">" means "more normal than").

MESSAGE

Loss-of-function mutant phenotypes can be due to partial or complete elimination of the activity of a gene's encoded product.

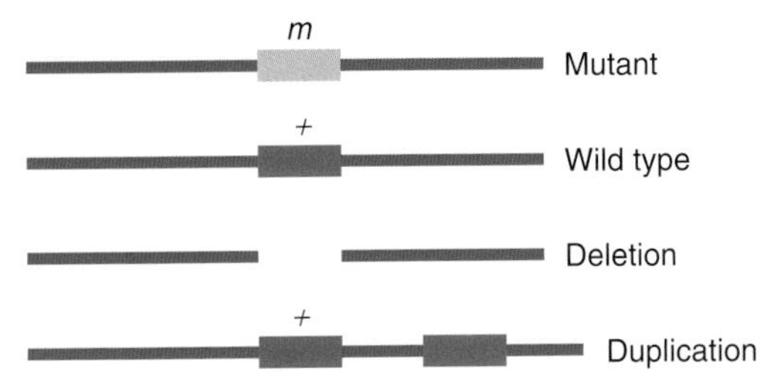

Figure 12-37 Chromosomes used in determining the loss-of-function or gain-of-function status of mutations. In addition to a dominant mutation and its recessive wild-type allele, a chromosome lacking the gene entirely (a deletion) or carrying an extra copy of the gene (a duplication) are needed for this analysis.

Let's now compare these predictions with those for dominant gain-of-function mutations. The term "gain-of-function" is an umbrella word for several different classes of mutation (recall the different types of gain-of-function changes in the gear analogy). Here are two examples of dominant gain-of-function mutations: **hypermorph** and **neomorph.**

Hypermorph: A hypermorph is a mutation that produces more gene activity per gene dose than wild type, but in all other respects the gene product is normal (see Figure 12-38c). For dominant hypermorphs, the *Mut* / + mutant phenotype is due to the extra gene activity of the *Mut* allele. If we remove the + allele with a deletion *(M / Df),* we reduce the combined activity of the two alleles and thus the phenotype should become more normal—symbolically, *M* / *Df* > *M* / +. On the other hand, if we increase the dosage of wild type with a duplication, then the phenotype should become more mutant—that is, *Mut* / + > *Mut* / + / Dp^+.

Neomorph: A neomorph is a mutation that produces some novel gene activity that is not a characteristic of the wild type. For example, if the coding sequences of two genes are fused in the right open reading frame in the mutant, a novel protein can be produced that may have cellular activities different from those of either parental protein. Alternatively, the same gene product as in the wild type is misregulated, so that it is expressed in a tissue in which the wild-type gene product is never expressed. This *ectopically expressed* gene product may then affect the biochemical pathways of these ectopic cells and thereby produce a completely unpredictable and novel phenotype. How is a neomorphic mutation identified? Because the gene product or site of action is novel, a neomorphic mutation is insensitive to dosage of the wild-type allele (see Figure 12-38d). Having zero, one, or two copies of the wild-type allele in a genotype with the corresponding neomorphic mutation produces the same dominant mutant phenotype. Symbolically, for a neomorphic mutation, *Mut* / + = Mut / *Df* = *Mut* / + / Dp^+.

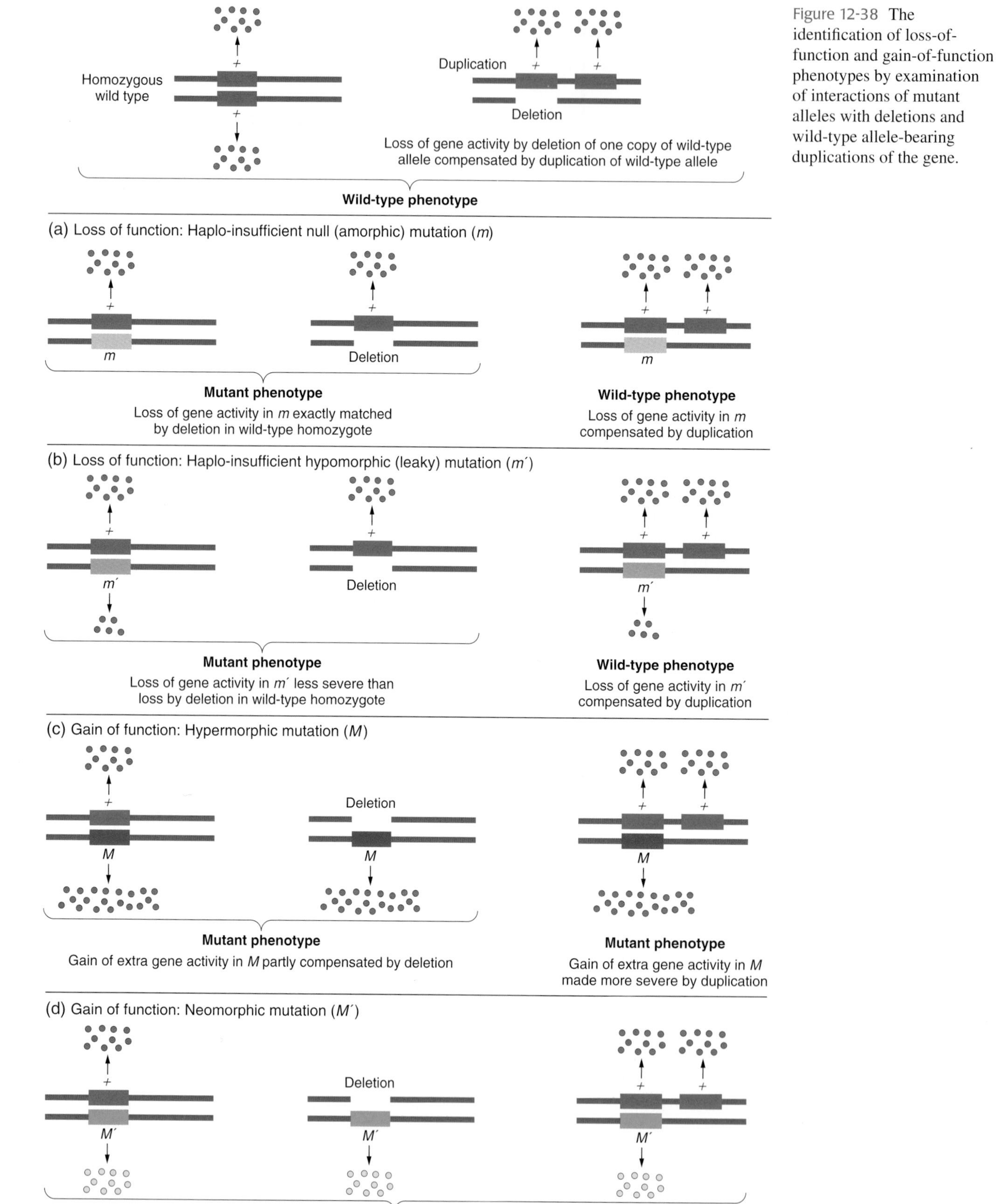

Figure 12-38 The identification of loss-of-function and gain-of-function phenotypes by examination of interactions of mutant alleles with deletions and wild-type allele-bearing duplications of the gene.

TABLE 12-1 Gene Dosage Analysis of Different Classes of Dominant Mutations

Mutant Class	Deficiency Comparison[1, 2, 3]	$Mut / + / Dp^+$ Comparison
Loss-of-function:	$+ / + > Df / +$	
Amorph (Null)	$Mut / + = Df / +$	$Mut / + / Dp^+ = + / +$
Hypomorph (Leaky)	$Mut / + > Df / +$	$Mut / + / Dp^+ = + / +$
Gain-of-function	$+ / + = Df / +$	
Hypermorph	$Mut / Df > Mut / +$	$Mut / + > Mut / + / Dp^+$
Neomorph	$Mut / Df = Mut / +$	$Mut / + = Mut / + / Dp^+$

[1]The symbol "=" means "has the same phenotype as".
[2]The symbol ">" means "is more normal in phenotype than".
[3]Df = Deficiency removing the gene; Dp^+ = Duplication with a wild-type copy of the gene

For recessive mutations, gene dosage can be used to make similar distinctions between loss-of-function and gain-of-function alterations.

MESSAGE

Gain-of-function mutant phenotypes can be due to an elevation in the activity or amount of a gene's encoded product, or to a change in the product or its site of expression so that it has a completely novel function.

Understanding the loss-of-function/gain-of-function status of each of a set of new mutations is the beginning of understanding pathways and processes. How to take this further by examining interactions between mutations in different genes will be the subject of Chapter 14. From there, specific examples of pathways that underlie the regulation of cell number and the regulation of development will be described in Chapters 15 and 16.

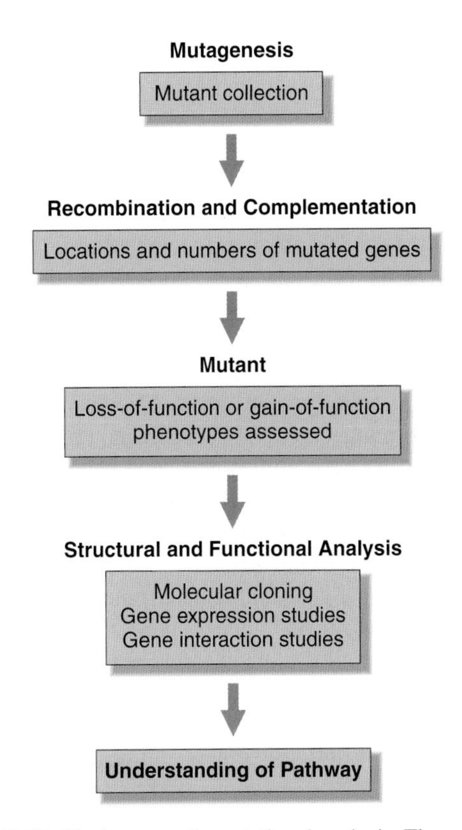

Figure 12-39 Further steps in mutational analysis. These further steps (genetic and molecular structural and functional analysis) will be the subject of the next several chapters.

FURTHER ASPECTS OF MUTATIONAL ANALYSIS

What are the next steps once the basic genetic and phenotypic analysis of a set of mutations is accomplished? Certainly, some of the steps involve other experimental approaches that have already been described (Figure 12-39). These approaches, including recombinational mapping for positional cloning and insertional mutagenesis for molecular "tagging," enable the molecular identification of genes and their mRNA and protein products. With these products in hand, the next questions are where and when they are expressed, what other products they interact with, and how their expression is regulated. These topics will be the subject of the next four chapters.

Let's return now to how analysis of mutations in experimental model organisms can help inform our understanding of human genetics and human biology. Simply put, the underpinnings of understanding human biochemical and cellular pathways depends upon their elucidation in model systems. The experimental approaches available in model organisms allow the manipulation of their genes in ways that we cannot use with their counterparts in humans (Figure 12-40). For example, targeted gene knockout in the mouse reveals that one of the members of the family of signaling molecules described in Figure 12-1 is required for kidney development. Secondary genetic screens in the fly identify the gene products involved in converting these signals into changes in the transcription patterns in cells that receive the signals. How to exploit these mutations will be considered in the next chapters, and especially in an evolutionary context in Chapter 19, the final chapter of this book.

Figure 12-40 The phenotype of a mutation in one of the transforming-growth-factor-beta genes in the mouse. The middle section is of the excretory organs of a wild-type mouse (prenatal). Note the large kidney capsules at the top of the dissection. The outer dissections are of the corresponding excretory organs of a homozygous targeted gene knockout of this transforming-growth-factor-beta family member. Note that the kidneys are missing or rudimentary. *(Courtesy of E. Robertson, Harvard University).*

SUMMARY

1. Mutational analysis is an important discovery tool for identifying gene products involved in biological processes. We can view the discovery of mutations of interest as a way of sieving through the many thousands of genes in a genome and "purifying" for those genes that contribute to a process of interest.

2. One way to identify genes involved in a process of interest is to mutagenize the genome and then search for mutations that have a phenotype that suggests involvement in that process. This is sometimes called forward genetics. Recombinational and complementation analysis are then used to identify those mutations that are lesions in the same gene and to determine the number and location of the affected genes. These genes can then be subjected to detailed molecular and genetic analysis.

3. For such phenotypic surveys, various mutagens are available. Some of the point mutagens primarily generate base substitutions, others frameshift mutations. Transposable elements can be used to disrupt the continuity of the gene and at the same time insert a molecular tag into the gene that facilitates its cloning. Still other mutagens, such as X rays, can be used to produce various mutational events, including chromosomal rearrangements; these rearrangements are of great value in genetic analysis. The selection of mutagen can dramatically affect the kinds of mutations that can be recovered.

4. With many fully or partly sequenced genomes available, we are often in a position of having identified genes of interest based on some molecular property, but not knowing any mutations in these genes. Among the molecular properties of interest are the similarity of the protein product of the gene to other proteins of known function, the interaction of the protein product with other proteins in the cell, and the tissue pattern of expression of the gene products. The genetic characterization of such genes is called reverse genetics, because we proceed from the molecule to the gene (rather than the reverse). Mutations in these cloned genes can sometimes be identified by general mutageneses, but only if their mutant phenotypes match the selection or screening criteria. An alternative route is to mutate the cloned gene directly using any of several genetic engineering techniques, most notably targeted gene inactivation.

5. Another route for assessing the phenotypic consequences of mutation in a cloned gene is to bypass entirely the need for creating a mutation, but instead to inactivate the RNA or protein product of that gene. Because these effects are transient, only lasting as long as the inhibitory molecule is present in the cell, these techniques are said to *phenocopy* (mimic the phenotype of) loss-of-function mutations in the cloned gene. Several phenocopy techniques have been developed, including antisense RNA, double-stranded RNA inhibition, and chemical genetics.

6. Genetic selection is the most efficient way to identify mutations that possess a specific phenotype. In such selections, all individuals lacking that phenotype fail the selection and are eliminated from the population, with survival restricted only to the rare individuals that have the desired phenotype. Although selection for a specific phenotype is a very efficient method, most phenotypes of interest do not lend themselves to selection techniques.

7. Genetic screens can be designed to identify any mutation with a discernible phenotype. Dominant mutations can be screened in the first generation after mutagenesis. Recessive mutations require two to three generations of crosses after mutagenesis before screening. Genetic screens are more versatile but also less efficient than genetic selections.
8. Gene-dosage experiments can identify those mutations that are loss-of-function. Deletions and duplications of the gene can be introduced into appropriate genotypes with the mutation and/or its wild-type allele. Loss-of-function mutations mimic the phenotype of gene deletions and are suppressed by the presence of extra copies of the wild-type gene. Some loss-of-function mutations are amorphs, meaning that they remove all phenotypically detectable gene activity. Others are hypomorphs, meaning that they have partly retained gene activity.
9. Similarly, gene-dosage experiments can identify those mutations that are gain-of-function. Gain-of-function phenotypes are not mimicked by gene deletions and are not suppressed by extra copies of the wild-type gene. Some gain-of-function mutations are hypermorphs, meaning that they show an increase in gene activity over the wild-type allele. Other gain-of-function alleles are neomorphs, meaning that they produce a novel sort of gene activity, that is, one that is completely distinct from the wild-type.
10. An understanding of whether a particular mutation is loss-of-function or gain-of-function is crucial to any inference about the role of the wild-type gene and its encoded products. These inferences are themselves among the first steps in organizing phenotypic information into a view of biochemical or cellular pathways.

CONCEPT MAP

Draw a concept map interrelating as many of the following terms as possible. Note that the terms are listed in no particular order.

gain-of-function / morphological mutation / dsRNAi / genetic screen / neomorph / base substitution / enhancer trap / auxotroph / targeted gene inactivation / X rays / complementation analysis

SOLVED PROBLEM

1. You want to study the development of the olfactory (smell-reception) system in the mouse. You know that the cells that sense specific chemical odors (odorants) are located in the lining of the nasal passages of the mouse. Describe some approaches for doing both forward and reverse genetics to study olfaction.

Solution

There are many approaches that can be imagined.

a. For forward genetics, the first trick is developing an assay system. One such assay system would be behavioral. For example, we could identify odorants that either attracted or repelled wild-type mice. We could then perform a genetic screen after mutagenesis, looking for mice that do not recognize the presence of this odorant, by using a maze in which a stream of air containing the odorant is directed at the mice from one end of the maze. This assay has the advantage of allowing us to process lots of mutagenized mice without any anatomical analysis.

As a first step, you might want to identify mutations that act dominantly to affect odorant reception, simply because these mutations could be identified in an F_1 screen. If this didn't work, you might need to resort to an F_2 screen for X-linked recessives or an F_3 screen for autosomal recessive mutations. Given that you don't know anything about the number or target size of genes that might produce this phenotype, you would want to use a highly mutagenic agent, such as a base-substitution mutagen, that would be able to generate mutations very efficiently in most protein-coding genes.

b. For reverse genetics, we would want to identify candidate genes that are expressed in the lining of the nasal passages. Given the techniques of functional genomics (Chapter 9), this could be accomplished by purifying RNA from isolated nasal-passage-lining cells and using this RNA as a probe of DNA chips containing sequences that correspond to all known mRNAs in the mouse. For example, you may choose first to examine mRNAs that are expressed in the nasal passage lining but nowhere else in the mouse as important candidates for a specific role in olfaction. (Many of the important molecules may also have other jobs elsewhere in the body, but you have to start somewhere!) Alternatively, you may choose to start with those genes whose protein products are located in the cell membrane as candidate proteins for binding the odorants themselves. Regardless of your choice, the next step would be to engineer a targeted knockout of the gene that encodes each mRNA or protein of interest, or to use antisense or dsRNAi injection to attempt to phenocopy the loss-of-function phenotype of each of the candidate genes.

PROBLEMS

Basic Problems

1. A certain species of plant produces flowers with petals that are normally blue. Plants with the mutation *w* produce white petals. In a plant of genotype *w* / *w*, one *w* allele reverts in the development of a petal. What detectable outcome would this reversion produce in the resulting petal? Would this mutation be inherited?

2. How would you select revertants of the yeast allele *pro-1?* This allele confers an inability to synthesize the amino acid proline, which can be synthesized by wild-type yeast and which is necessary for growth.

3. Suppose that you want to determine whether caffeine induces mutations in higher organisms. Describe how you might do so (include control tests).

4. One of the jobs of the Hiroshima-Nagasaki Atomic Bomb Casualty Commission was to assess the genetic consequences of the blast. One of the first things that they studied was the sex ratio in the offspring of the survivors. Why do you suppose they did this?

5. Cells of a haploid wild-type *Neurospora* strain were mutagenized with EMS. Large numbers of these cells were plated and grown into colonies at 25°C on complete medium (containing all possible nutrients). These strains were tested on minimal medium and complete medium at both 25°C and 37°C. There were several mutant phenotypes, as shown in the accompanying sketch. The large circles represent luxuriant growth, the spidery symbols represent weak growth, and a blank means no growth at all. How would you categorize the types of mutants represented by isolates 1 through 5?

	Minimal		Complete	
Strain	25°C	37°C	25°C	37°C
Mutant 1			●	●
Mutant 2	●		●	
Mutant 3	✱	✱	●	●
Mutant 4	●	✱	●	✱
Mutant 5	●		●	●
Wild type (control)	●	●	●	●

6. Devise imaginative screening procedures for detecting

a. nerve mutants in *Drosophila.*

b. mutants lacking flagella in a haploid unicellular alga.

c. supercolossal-size mutants in bacteria.

d. mutants that overproduce the black compound melanin in normally white haploid fungus cultures.

e. individual humans (in large populations) whose eyes polarize incoming light.

f. negatively phototrophic *Drosophila* or unicellular algae.

g. UV-sensitive mutants in haploid yeast.

7. A man and a woman with no record of genetic disease in their families have one child with neurofibromatosis (autosomal dominant) and another child who is unaffected. The penetrance of neurofibromatosis is close to 100 percent.

a. Explain the birth of the affected child.

b. How would you counsel the parents if they contemplated having another child?

8. Describe three different methods used to generate phenocopies. What is the purpose of generating a phenocopy?

9. What is the benefit of using a balancer chromosome?

10. What is the difference between forward and reverse mutation?

11. What are the advantages and disadvantages of genetic screens versus selections?

Challenging Problems

12. A haploid strain of *Aspergillus nidulans* carried an auxotrophic *met-8* mutation conferring a requirement for methionine. Several million asexual spores were plated on minimal medium, and two prototrophic colonies grew and were isolated. These prototrophs were crossed sexually with two different strains, with the progeny shown in the body of the following table, where met^+ means that methionine is not required for growth and met^- means that methionine is required for growth.

	Crossed with a wild-type strain	Crossed with a strain carrying the original *met-8* allele
Prototroph 1	All met^+	1/2 met^+ 1/2 met^-
Prototroph 2	3/4 met^+ 1/4 met^-	1/2 met^+ 1/2 met^-

a. Explain the origin of both of the original prototrophic colonies.

b. Explain the results of all four crosses, using clearly defined gene symbols.

Unpacking the Problem

Before you try to solve this problem, follow the instructions and answer the questions that pertain to the experimental system.

a. Draw a labeled diagram that shows how this experiment was done. Show test tubes, plates, and so forth.

b. Define all the genetic terms in this problem.

c. Many problems show a number next to the auxotrophic mutation's symbol—here the number *8* next to *met*. What does the number mean? Is it necessary to know in order to solve the problem?

d. How many crosses were actually made? What were they?

e. Represent the crosses by using genetic symbols.

f. Is the question about somatic mutation or germinal mutation?

g. Is the question about forward mutation or reversion?

h. Why was such a small number of prototrophic colonies (two) found on the plate?

i. Why didn't the several million asexual spores grow?

j. Do you think any of the millions of spores that did not grow were mutant? Dead?

k. Do you think the wild type used in the crosses was prototrophic or auxotrophic? Explain.

l. If you had the two prototrophs from the plates and the wild-type strain in three different culture tubes, could you tell them apart just by looking at them?

m. How do you think the *met-8* mutation was obtained in the first place? (Show with a simple diagram. NOTE: *Aspergillus* is a filamentous fungus.)

n. Is the concept of recombination relevant to any part of this problem?

o. What progeny do you predict from crossing *met-8* with a wild type? What about *met-8* × *met-8*? What about wild type × wild type?

p. Do you think this is a random meiotic progeny analysis or a tetrad analysis?

q. Draw a simple life-cycle diagram of a haploid organism showing where meiosis takes place in the cycle.

r. Consider the 3/4 : 1/4 ratio. In haploids, crosses heterozygous for one gene generally give progeny ratios based on halves. How can this idea be extended to give ratios based on quarters?

13. Every mutagen produces a certain characteristic type of mutational event. Explain your answers to the following questions.

a. Would you expect hypomorphic mutations to be more frequent among all mutations produced by base-substitution mutagens or by frameshift mutagens?

b. Would you expect amorphic mutations to be more frequent among all mutations produced by base-substitution mutagens or by frameshift mutagens?

14. Neomorphic mutations can be reverted by treatment with standard mutagens. When the revertants are examined, they typically turn out to be recessive loss-of-function mutations. Explain this observation.

15. You are trying to identify all mutations that affect development of the dorsal fin in the zebrafish. You do an F_3 mutagenesis analysis for recessive mutations that cause loss of the dorsal fin. By recombination and complementation analysis, you find that 40 mutations that you've isolated represent mutations in 5 genes, with 12 mutations in one gene, 10 mutations in each of two others, 7 mutations in a fourth gene, and only 1 mutation in a fifth.

a. Is it surprising that you recovered so many mutations in each of four of the genes and only one in the fifth? Justify your answer.

b. Would you expect that this screen has identified all genes that contribute to dorsal fin development? Why or why not? If you think there should be other classes of mutations, propose some experiments to identify them.

16. You are studying proteins involved in translation in the mouse. By BLAST analysis of the predicted proteins of the mouse genome, you identify a set of mouse genes that encode proteins with sequences similar to those of known eukaryotic translation-initiation factors. You are interested in determining the phenotypes associated with loss-of-function mutations of these genes.

a. Would you use forward or reverse genetics approaches to identify these mutations?

b. Briefly outline two different approaches you might use to look for loss-of-function phenotypes in one of these genes.

17. Normal ("tight") auxotrophic mutants will not grow at all in the absence of the appropriate supplementation to the medium. However, in mutant hunts for auxotrophic mutants, it is common to find some mutants (called "leaky") that grow very slowly in the absence of the appropriate supplement but normally in the presence of the supplement. Propose an explanation for the molecular action of the leaky mutants in a biochemical pathway.

18. A botanist interested in the chemical reactions whereby plants capture light energy from the Sun decided to dissect this process genetically. She decided that leaf fluorescence would be a useful mutant phenotype to select because it would show something wrong with the process whereby electrons normally are transferred from chlorophyll. Therefore four fluorescent (*fl*) mutants were obtained in the plant *Arabidopsis* after mutagenesis. All were found to be recessive in the simple Mendelian manner. Homozygous stocks of the mutants were intercrossed, and the F_1s were each testcrossed to a strain that was homozygous recessive for all the genes involved in that cross. The results are shown in the table below.

Cross	Percentage of wild types in F_1	Percentage of wild types in progeny of testcross of F_1
1 × 2	100	25
1 × 3	100	25
1 × 4	0	0
2 × 3	100	10
2 × 4	100	25
3 × 4	100	25

a. How many genes are represented by these mutants?

b. What can you deduce about the chromosomal location of the genes?

c. Use your own gene symbols to explain the F_1 and testcross results.

19. As was described in Chapter 9, the entire genome of the yeast *Saccharomyces cerevisiae* has been sequenced. This sequencing has led to the identification of all the open reading frames (ORFs) in the genome (gene-sized sequences with appropriate translational initiation and termination signals). Some of these ORFs are previously known genes with established functions; however, the remainder are unassigned reading frames (URFs). To deduce the possible functions of the URFs, they are being systematically, one at a time, converted into null alleles by in vitro knockout techniques. To date, the results are as follows:

- 15 percent are lethal when knocked out.
- 25 percent show some mutant phenotype (altered morphology, altered nutrition, and so forth).
- 60 percent show no detectable mutant phenotype at all and resemble wild type.

Explain the possible molecular-genetic basis of these three mutant categories, inventing examples where possible.

20. In *Drosophila*, the genes for ebony body (*e*) and stubby bristles (*s*) are linked on the same arm of chromosome 2. Flies of genotype $+\ s\ /\ e\ +$ develop predominantly as wild type but occasionally show two different kinds of unexpected abnormalities on their bodies. The first abnormality is the presence of pairs of adjacent patches, one with stubby bristles and the other with ebony color. The second abnormality is the presence of solitary patches of ebony color.

a. Draw diagrams to show the likely origin of these two types of unexpected abnormalities.

b. Explain why there are no single patches that are stubby.

21. A strain of *Aspergillus* was subjected to mutagenesis by X rays, and two tryptophan-requiring mutants (A and B) were isolated. These tryptophan-requiring strains were plated in large numbers to obtain revertants to wild type. You failed to recover any revertants from mutant A and recovered one revertant from mutant B. This revertant was crossed with a normal wild-type strain. What proportion of the progeny from this cross would be wild type if

a. the reversion precisely reversed the original change that produced the trp^- mutant allele?

b. the revertant phenotype was produced by a mutation in a second gene located on a different chromosome (the new mutation suppresses trp^-)?

Propose an explanation of why no revertants from mutant A were recovered.

REGULATION OF GENE TRANSCRIPTION 13

Key Concepts

1. An important mode of gene regulation occurs via proteins that sense environmental signals and appropriately raise or lower the transcription rates of sets of genes.
2. In prokaryotes, coordinate gene control is achieved by clustering coordinated structural genes on the chromosome so that they are transcribed into multigenic mRNAs.
3. Negative regulatory control is exemplified by the *lac* system, in which a repressor protein blocks transcription by binding to DNA at a site termed the *operator.*
4. Positive regulatory control requires protein factors to activate transcription.
5. Many regulatory proteins have common structural features.
6. In eukaryotes, regulatory sites on DNA can act at considerable distance from the transcription start site to modulate gene expression by interacting with specific regulatory proteins.
7. Some regulation occurs at the epigenetic level, in which gene expression is modulated according to the parentage and cell division history of the gene.
8. Genome-wide approaches are being applied to the analysis of transcriptional regulation.

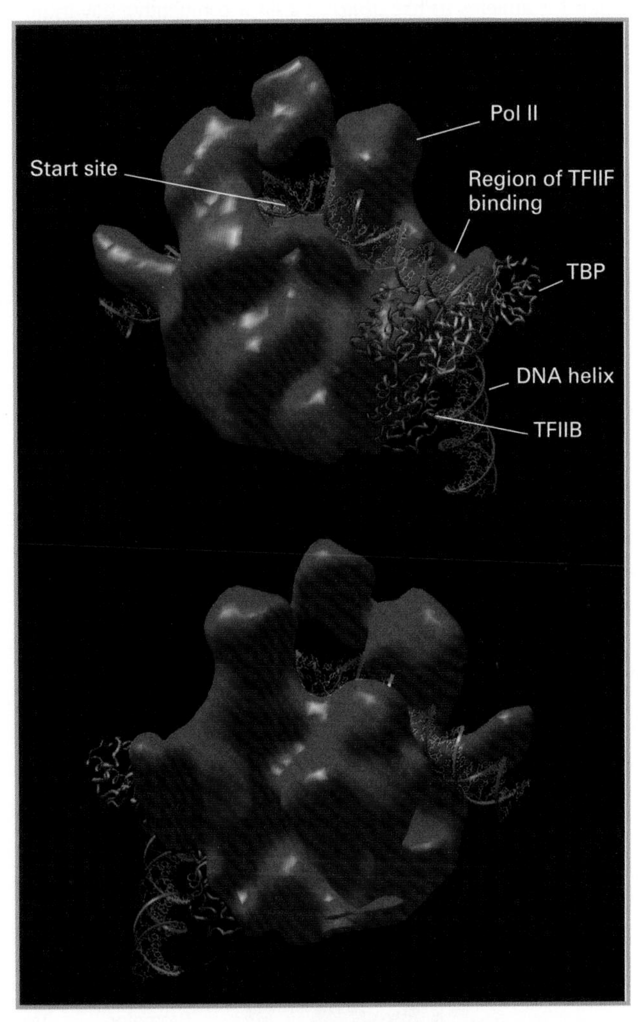

A three-dimensional model of promoter DNA complexed with eukaryotic RNA polymerase (Pol II) and other transcription initiation proteins (TBP, TFIIB). *(From H. Lodish, D. Baltimore, A. Berk, S. L. Zipursky, P. Matsudaira, and J. Darnell,* Molecular Cell Biology, *4th ed. © 2000 by Scientific American Books, p. 383. Adapted from T.-K. Kim et al.,* Proc. Natl. Acad. Sci. U.S.A. *94, 1997, 12268.)*

Performance-enhancing drugs have been a major issue in world-class athletic events (Figure 13-1). At every Olympics, questions are raised about the possible use of such drugs by athletes who suddenly emerge at the top of their events, as well as by long-term stars. One of the classes of drugs that have been at the center of controversy is the anabolic steroids, natural hormones which increase the activity of certain enzymatic pathways which lead to the expansion of muscle tissue and increased muscle mass. (Some of these same drugs are given to beef cattle to increase muscle mass and hence their sale price when brought to market.) Not only are there issues of whether it is fair for athletes to use drugs to gain a competitive advantage, but also there are valid reasons to be concerned about long-term health risks to athletes who make chronic use of these drugs.

Anabolic steroids, like all steroids, act as environmental signals to modulate transcription patterns in cells that receive these signals. Ordinarily, they are produced under the proper circumstances by some of the body's glands. The long-term health risks come from taking large doses of these hormones—much larger than the body normally produces—over long periods of time. To understand these risks, we need to understand the processes that govern transcription, one of the major regulatory steps in gene expression. Such regulation will be the subject of this chapter.

Figure 13-1 A men's track event. *(Duomo/CORBIS.)*

CHAPTER OVERVIEW

The biological properties of each cell are largely determined by the active proteins expressed in it. Proteins determine much of the cell's architecture, its enzymatic activities, its interactions with its environment, and many other physiological properties. At any given time in the life history of a cell, only a small proportion of the proteins that its genome encodes are expressed. Moreover, some of the expressed proteins are present at relatively low levels, whereas others are quite abundant. How is the regulation of protein expression accomplished? For one protein or another, we find examples of regulation at each of the steps leading from the initiation of gene transcription to the formation of the final protein. Transcription patterns directly reflect the primary readout of the genome's instruction manual, since the first step in information transfer is transcription itself. This has made transcriptional regulation more accessible to experimental analysis than is true for some of the other steps, and regulation at this level will therefore be the focus of this chapter.

We can divide the genome's contribution to transcription into a transcriptional parts list and an instruction manual for transcription (Figure 13-2). The parts list consists of template domains—the regions of the genome that encode RNAs, most of which in turn encode proteins. The instruc-

CHAPTER OVERVIEW figure

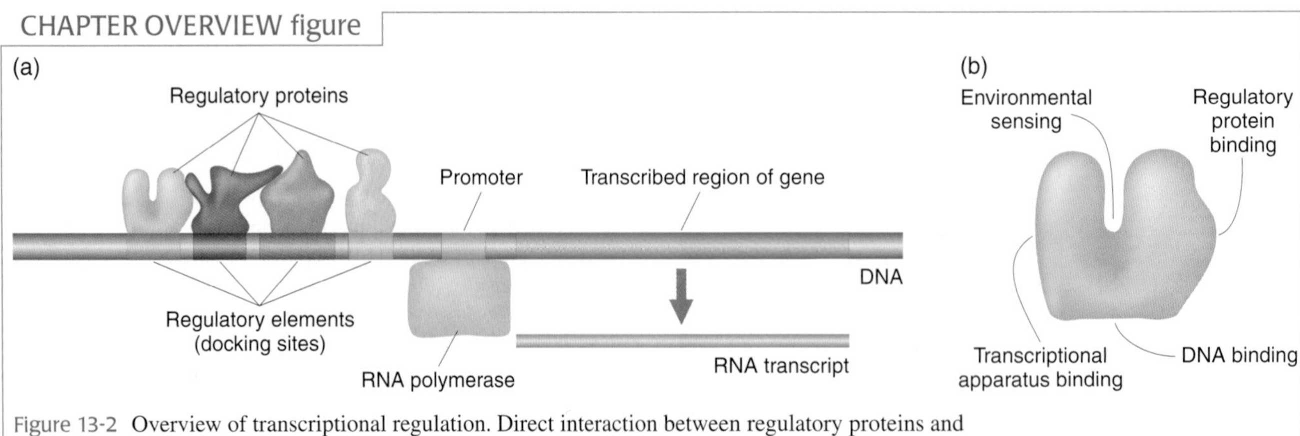

Figure 13-2 Overview of transcriptional regulation. Direct interaction between regulatory proteins and specific sequences on the DNA determines the location of transcription initiation and when, where, and how much transcription occurs for a given gene. (a) Interaction of regulatory proteins and their docking sites. (b) The domains of a regulatory protein.

tion manual of transcription consists of the genes' regulatory elements—the docking sites for transcriptional regulatory proteins that determine when, where, and how much of each RNA (and indirectly, each protein) is expressed. In order to modulate transcription appropriately, these regulatory proteins usually comprise several regulatory domains:

1. A sequence-specific DNA-binding surface that can recognize the correct docking site
2. A surface that can interact with one or more proteins of the basal transcriptional apparatus
3. A surface that interacts with proteins that bind to nearby docking sites, such that they can act cooperatively to regulate transcription
4. A surface that acts as a sensor of environmental conditions within the cell

Having all these domains ensures that the transcription regulatory proteins of the cell are correctly wired into the metabolic and other biochemical circuitry of the cell. With the proper circuits in place, the regulatory proteins will bind only to their docking sites and activate transcription under the appropriate set of environmental conditions. When these conditions are not present, the circuit is not closed and the regulatory proteins are toggled to the "off" position. Indeed, this connection between the cell circuitry and transcriptional activity is so crucial to normal cell function that when this connection is disrupted, major diseases such as cancer can result—as discussed in Chapter 15.

Studying transcription can be a source of insight into the general theme of how regulatory information is encoded in the genome. We shall see that some of the aspects of transcription regulation are common to prokaryotes and eukaryotes. In higher eukaryotes, regulatory events are necessarily more complex to coordinate the production of a wide variety of cells with different arrays of proteins in space (tissue-specific gene expression) and time (development). As we shall see toward the end of this chapter, some of the challenges we face are finding ways to "read" the transcriptional instruction manual of the genome and to predict a gene's expression pattern from the sequence of its surrounding DNA. But to begin, we shall examine some well-understood examples of transcriptional regulation in bacteria.

THE LOGIC OF PROKARYOTIC GENE REGULATION

Bacteria have a profound need to regulate the expression of their genes. Enzymes taking part in sugar metabolism provide an example. Metabolic enzymes are required to break down different carbon sources to yield energy. However, there are many different types of compounds that bacteria could use as carbon sources, including sugars such as lactose, glucose, galactose, and xylose. A different enzyme is required to allow each of these sugars to enter the cell. Further, a different set of enzymes is required to break down each of these sugars. If a cell were to simultaneously synthesize all the enzymes that it might possibly need, it would cost the cell much more energy to produce the enzymes than it could ever derive from breaking down any of the prospective carbon sources. The cell has devised mechanisms to shut down (repress) transcription of all the genes encoding enzymes that are not needed at a given time and to turn on (activate) those genes encoding enzymes that are required. For example, if lactose is the primary carbon source in the environment, the shutdown of transcription of the relevant genes ensures that no enzymes are present for the import and metabolism of glucose, galactose, xylose, and other sugars. Conversely, the initiation of transcription of the appropriate genes ensures that the enzymes for lactose import and metabolism are available when needed. Such selective repression and activation mechanisms must fulfill two criteria:

1. They must have the ability to reversibly turn on or switch off transcription of each specific gene or group of genes.
2. They must be responsive to environmental conditions in which transcription of the relevant gene or genes should be activated or repressed.

Let's preview the current model for prokaryotic transcriptional regulation and then use a well-understood example—the regulation of the genes involved in the metabolism of the sugar lactose—to examine it in detail.

THE BASICS OF PROKARYOTIC TRANSCRIPTIONAL REGULATION

To understand how prokaryotic transcriptional regulation is accomplished, we shall consider a series of controlled, sequence-specific DNA-protein interactions at DNA target sites located near the beginning of transcription of a gene. Two types of DNA-protein interactions are required for regulated transcription.

One required DNA-protein interaction determines where transcription begins. It involves the DNA segment called the **promoter** and the protein **RNA polymerase.** When RNA polymerase binds to the promoter DNA, transcription can initiate a few bases away from the promoter site. Every gene must have a promoter or it cannot be transcribed.

The other type of required DNA-protein interaction regulates whether or not promoter-driven transcription occurs. DNA segments near the promoter serve as protein-binding sites—most of these sites are termed **operators**—

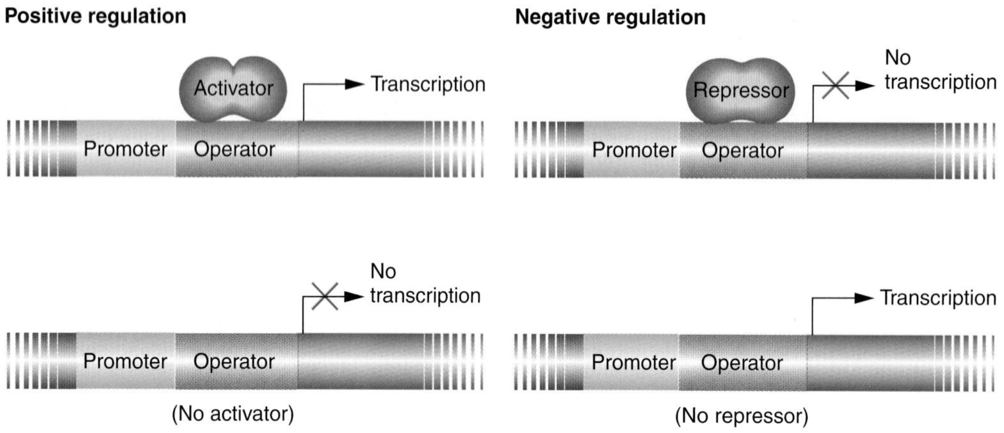

Positive regulation

Activator
Transcription
Promoter Operator

No transcription
Promoter Operator
(No activator)

Negative regulation

Repressor
No transcription
Promoter Operator

Transcription
Promoter Operator
(No repressor)

Figure 13-3 The different consequences of activator and repressor binding to operator DNA. Activator binding is required for transcription. Repressor binding blocks transcription.

for regulatory proteins called **activators** and **repressors.** For some genes, the binding of an activator protein to its target DNA site is a necessary prerequisite for transcription to begin. Such instances are sometimes referred to as *positive regulation* because it is the *presence of the bound protein* that is required for transcription (Figure 13-3). For other genes, preventing the binding of a repressor protein to its target site is a necessary prerequisite for transcription to begin. Such cases are sometimes termed *negative regulation* because it is the *absence of the bound protein* that allows transcription to occur. How do activators and repressors regulate transcription? Often, a DNA-bound activator protein acts at the level of transcription initiation, by physically helping to tether RNA polymerase to its nearby promoter. A DNA-bound repressor protein typically acts either by physically interfering with the binding of RNA polymerase to its promoter (blocking transcription initiation) or by impeding the movement of RNA polymerase along the DNA chain (blocking transcription elongation).

MESSAGE

Genes must contain two kinds of binding sites to permit regulated transcription. First, binding sites for RNA polymerase must be present. Second, binding sites for activator or repressor proteins must be present in the vicinity of the promoter.

For activator or repressor proteins to do their job, each must be able to exist in two states: one that can bind its DNA targets and one that cannot. The binding state must be in accord with the cellular environment, that is, be appropriate for a given set of physiological conditions. For many activator or repressor proteins, DNA binding is regulated through the interaction of two different sites in the three-dimensional structure of the protein. One site is the **DNA-binding domain.** The other site, the **allosteric site,** acts as the toggle switch that sets the DNA-binding domain in one of two modes: functional or nonfunctional. The allosteric site interacts with small molecules called *allosteric effectors.* In our earlier example of lactose metabolism, the metabolite lactose would be an allosteric effector. An **allosteric effector** binds to the allosteric site of the regulatory protein in such a way that it changes the structure of the DNA-binding domain. Some activator or repressor proteins must bind to their allosteric effectors before they can bind DNA. Others can bind DNA only in the absence of their allosteric effectors. Two of these situations are shown in (Figure 13-4).

MESSAGE

Allosteric effectors work like toggle switches to control the ability of activator or repressor proteins to bind to their DNA target sites.

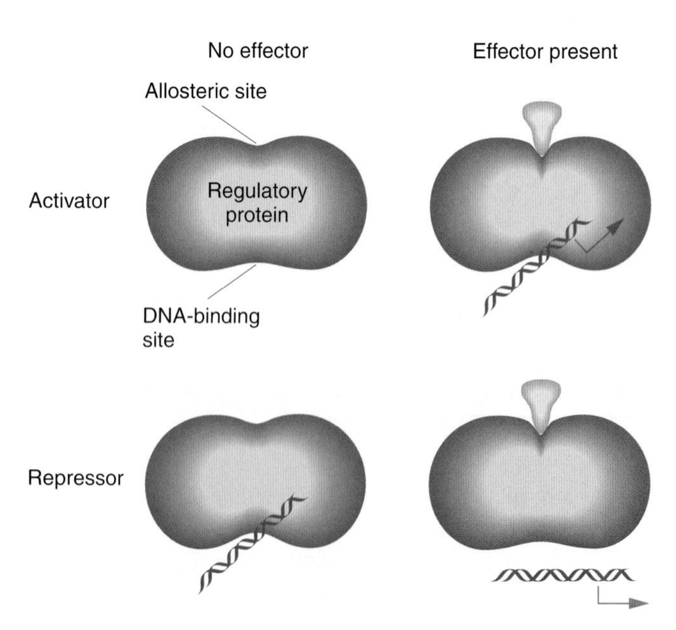

Figure 13-4 The influence of allosteric effectors on the DNA-binding activities of activators and repressors.

REGULATION OF THE LACTOSE SYSTEM

The salient features of transcriptional regulation in prokaryotes can be seen in the regulation of expression of the enzymes necessary for lactose metabolism in the bacterium *Escherichia coli.* We should recognize that the existing system is one that, through a long evolutionary process, has been selected to operate in an optimal fashion for the energy efficiency of the bacterial cell. Presumably because of energy-efficiency considerations, two environmental conditions have to be satisfied for the lactose metabolic enzymes to be expressed.

One condition that must be met is that lactose must be present in the environment. This condition makes sense, because it would be inefficient for the cell to produce the lactose metabolic enzymes in circumstances where there is no substrate to metabolize. We shall see that the cell's recognition that lactose is present is accomplished by a repressor protein.

The other condition is that glucose cannot be present in the cell's environment. Because glucose metabolism yields more usable energy to the cell than does lactose metabolism, mechanisms have evolved that prevent the synthesis of the enzymes for lactose metabolism in the presence of glucose. The repression of transcription of the lactose-metabolizing genes in the presence of glucose is an example of **catabolite repression.** The transcription of proteins necessary for the metabolism of many different sugars is repressed by catabolites. We shall see that catabolite repression works through an activator protein.

A First Look at the *lac* Regulatory Circuit

Thanks to the pioneering work of François Jacob and Jacques Monod, we have learned a lot about how lactose metabolism is regulated. First, we will deal with the system used to regulate lactose metabolism according to the presence or absence of lactose. Figure 13-5 is a simplified view of the components of this system. The cast of characters for *lac* regulation includes protein-coding genes and sites on the DNA that are targets for DNA-binding proteins.

The *lac* Structural Genes

The metabolism of lactose requires two enzymes: a permease to transport lactose into the cell and β-galactosidase to cleave the lactose molecule to yield glucose and galactose (Figure 13-6 on the next page). Permease and β-galactosidase are encoded by two contiguous genes, *Z* and *Y*, respectively. A third gene, the *A* gene, encodes an additional enzyme, termed *transacetylase,* but this enzyme is not required for lactose metabolism, and so we will focus on *Z* and *Y*. All three genes are transcribed into a single, multigenic messenger RNA (mRNA) molecule. Regulation of the production of this mRNA coordinates the regulation of the synthesis of all three enzymes.

MESSAGE

If the genes encoding proteins of a given pathway are joined into a single transcription unit, the expression of all these genes will be coordinately regulated.

Regulatory Components of the *lac* System

There are three components in addition to RNA polymerase to consider in understanding the basics of regulation of the lactose metabolic system. These include a transcription regulatory protein and two docking sites—one for the regulatory protein and one for RNA polymerase.

1. *The gene for the Lac repressor:* A fourth gene, the *I* gene, encodes the Lac repressor protein, so named

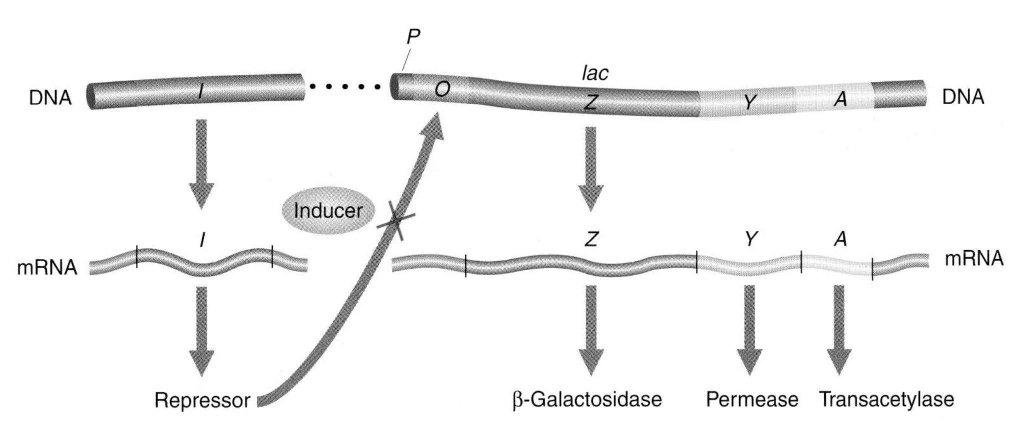

Figure 13-5 A simplified *lac* operon model. The three genes *Z*, *Y*, and *A* are coordinately expressed. The product of the *I* gene, the repressor, blocks the expression of the *Z*, *Y*, and *A* genes from the promoter (*P*) by interacting with the operator (*O*). The inducer can inactivate the repressor, thereby preventing interaction with the operator. When this happens, the operon is fully expressed.

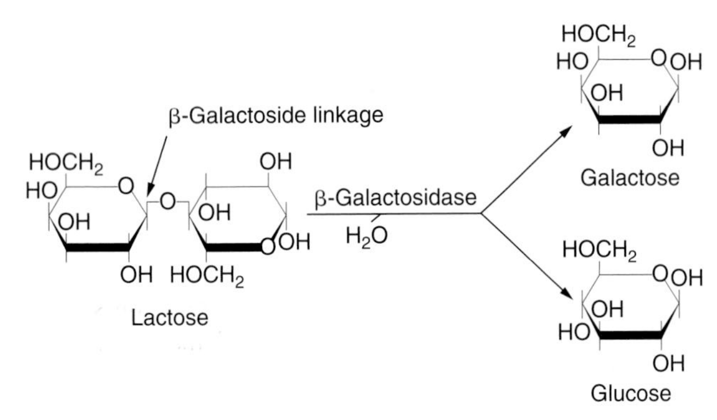

Figure 13-6 The metabolism of lactose. The enzyme β-galactosidase catalyzes a reaction in which water is added to the β-galactoside linkage to break lactose into separate molecules of galactose and glucose. The enzyme lactose permease (not shown) is required to transport lactose into the cell.

because it can block the expression of the *Z*, *Y*, and *A* genes. The *I* gene happens to map fairly near the *Z*, *Y*, and *A* genes, but this proximity does not seem to be important to its function.

2. *The lac promoter site:* The promoter (*P*) is the site on the DNA to which RNA polymerase binds to initiate transcription of the *lac* structural genes (*Z*, *Y*, and *A*).
3. *The lac operator site:* The operator (*O*) is the site on the DNA to which the Lac repressor binds. It is located between the promoter and the *Z* gene near the point at which transcription of the multigenic mRNA begins.

The *lac* Operon: Assaying the Presence or Absence of Lactose through the Lac Repressor

The *P*, *O*, *Z*, *Y*, and *A* segments shown in Figure 13-7 constitute an **operon,** which is a genetic unit of coordinate expression. The interaction between the *lac* operator site on the DNA and the Lac repressor is crucial to proper regulation of the *lac* operon. The Lac repressor is a molecule with two recognition sites—a *DNA-binding* site that can recognize the specific operator DNA sequence for the *lac* operon and an *allosteric site* that binds the allosteric effector lactose and similar molecules (analogs of lactose).

The DNA-binding site of the Lac repressor is able to bind with high affinity to only one DNA sequence in the entire *E. coli* genome—the *lac* operator. The specificity of high-affinity DNA binding ensures that the repressor will bind only to the site on the DNA near the genes that it is controlling and not to random sites distributed throughout the chromosome. By binding to the operator, the repressor prevents transcription by RNA polymerase that has bound to its *lac* promoter site.

As already mentioned, the allosteric site of the Lac repressor binds to lactose and structurally similar molecules (lactose analogs). We shall see later in the chapter that some lactose analogs are very useful tools for experimentally inducing *lac* operon expression. When the repressor protein binds to lactose or its analogs, the protein undergoes a conformational (allosteric) change; this slight alteration in shape changes the DNA-binding site so that the repressor no longer has high affinity for the operator. Thus, in response to binding lactose, the repressor falls off the DNA, which satisfies one requirement for such a control system—the ability to recognize conditions under which it is worthwhile to activate expression of the *lac* genes.

The relief of repression for systems such as *lac* is termed **induction;** lactose and its analogs that allosterically inactivate the repressor and lead to expression of the *lac* genes are termed **inducers.** We now know that the Lac repressor is a protein consisting of four identical subunits, each with a molecular weight of approximately 38,000. Each tetrameric Lac repressor molecule contains four

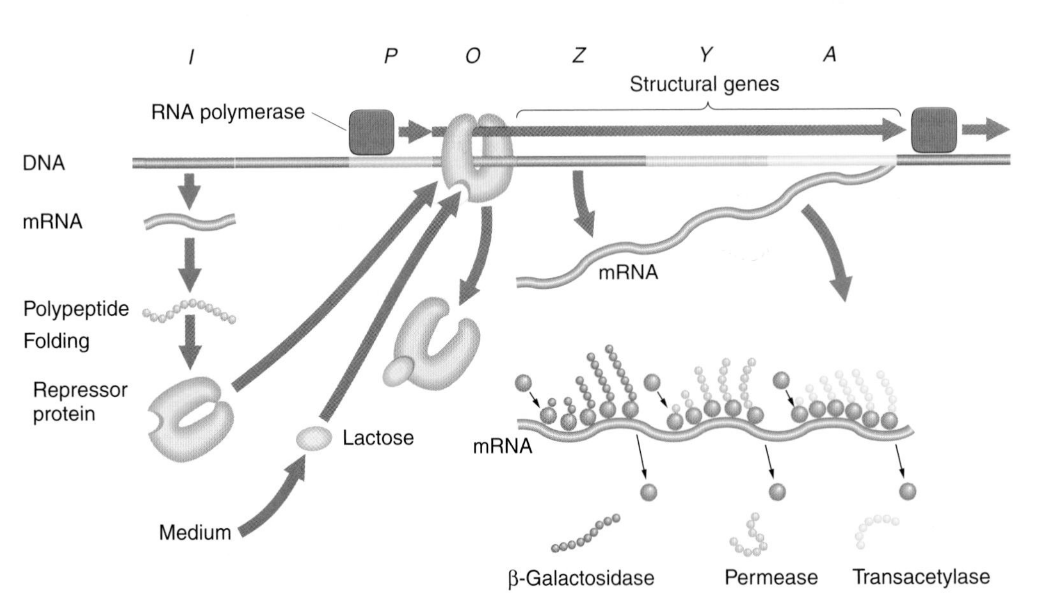

Figure 13-7 Regulation of the *lac* operon. The *I* gene continually makes repressor. The repressor binds to the *O* (operator) region, preventing the RNA polymerase bound to *P* (the promoter region) from transcribing the adjacent structural genes. When lactose is present, it binds to the repressor and changes its shape so that the repressor no longer binds to *O*. The RNA polymerase is then able to transcribe the *Z*, *Y*, and *A* structural genes, so the three enzymes are produced.

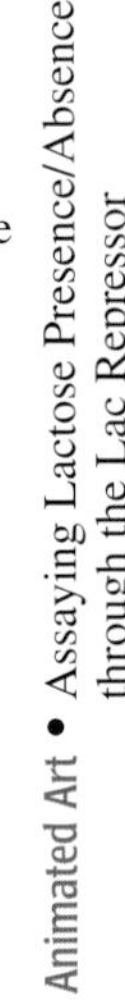

inducer-binding allosteric sites. (A more detailed description of the repressor is given later in the chapter.)

Let's examine the implications of the last few paragraphs. In the absence of an inducer (lactose or an analog), the Lac repressor binds to the *lac* operator site and prevents transcription of the *lac* operon. Most of the effect of the Lac repressor's binding to the operator is to block the progression of RNA polymerase transcription. In this sense, the Lac repressor acts as a roadblock on the DNA. Consequently, all the structural genes of the *lac* operon (the *Z*, *Y*, and *A* genes) are repressed, and there is no β-galactosidase, β-galactoside permease, or transacetylase in the cell. In contrast, when an inducer is present, it binds to the allosteric site of each Lac repressor subunit, thereby inactivating the operator DNA-binding site of the Lac repressor protein. This inactivation permits the induction of transcription of the structural genes of the *lac* operon and, through the translation of the multigenic mRNA, the enzymes β-galactosidase, β-galactoside permease, and transacetylase now appear in the cell in a coordinated fashion.

Before going any further, however, we should note that there is more to the regulation of *lac* operon transcription. Recall that the induction of transcription of the *lac* operon also requires a second environmental condition—namely, that glucose is *not* present in the environment of the cell. We shall consider the reasons for this condition and the mechanisms governing glucose repression next.

Catabolite Repression of the *lac* Operon: Choosing the Best Sugar to Metabolize

An additional control system is superimposed on the repressor-operator system. This control system is thought to have evolved because the cell can capture more energy from the breakdown of glucose than it can from the breakdown of other sugars. If both lactose and glucose are present, the synthesis of β-galactosidase is not induced until all the glucose has been utilized. Thus, the cell conserves its energy pool used, for example, to synthesize the Lac enzymes by metabolizing any existing glucose before going through the energy-expensive process of creating new machinery to metabolize lactose.

Studies indicate that a breakdown product of glucose (the identity of this catabolite is as yet unknown) prevents activation of the *lac* operon by lactose—this is the catabolite repression referred to earlier. The glucose catabolite modulates the level of an important cellular constituent—**cyclic adenosine monophosphate (cAMP).** When glucose is present in high concentrations, the cell's cAMP concentration is low; as the glucose concentration decreases, the cellular concentration of cAMP increases correspondingly. A high concentration of cAMP is necessary for activation of the *lac* operon. Mutants that cannot convert ATP into cAMP cannot be induced to produce β-galactosidase, because the concentration of cAMP is not great enough to activate the *lac* operon. In addition, there are other mutants that do make cAMP but cannot activate the Lac enzymes, because they lack yet another protein, called **CAP (catabolite activator protein),** made by the *crp* gene. By itself, CAP cannot bind to the CAP site of the *lac* operon. However, by binding to its allosteric effector, cAMP, CAP is able to bind to the CAP site. The DNA-bound CAP is then able to interact physically with RNA polymerase and essentially increase the affinity of RNA polymerase for the *lac* promoter. In this way, the catabolite repression system contributes to the selective activation of the *lac* operon (Figure 13-8).

MESSAGE

The *lac* operon has an added level of control so that the operon remains inactive in the presence of glucose even if lactose also is present. An allosteric effector, cyclic adenosine monophosphate (cAMP), binds to the activator CAP, to permit the induction of the *lac* operon. However, high concentrations of glucose catabolites produce low concentrations of cAMP, thus failing to produce cAMP–CAP and thereby failing to activate the *lac* operon.

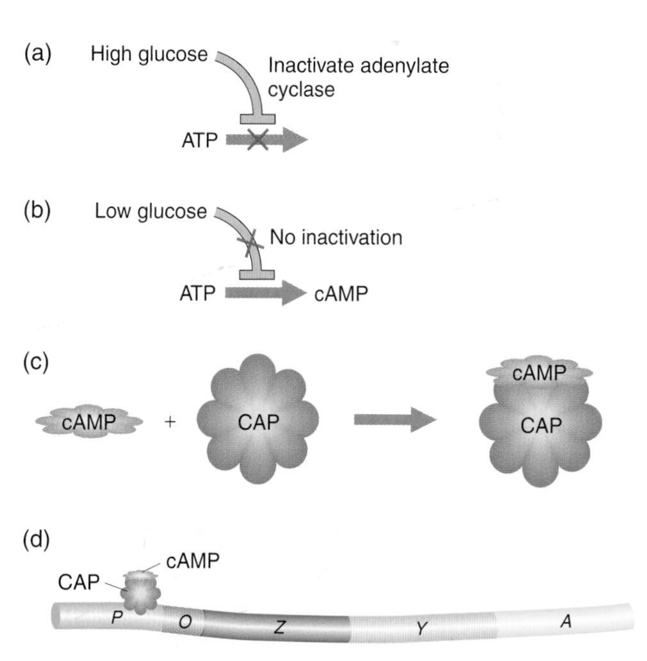

Figure 13-8 Catabolite control of the *lac* operon. The operon is inducible by lactose to the maximal levels when cAMP and CAP form a complex. (a) Under conditions of high glucose, a glucose-breakdown product inhibits the enzyme adenylate cyclase, preventing the conversion of ATP into cAMP. (b) Under conditions of low glucose, there is no breakdown product, and therefore adenylate cyclase is active and cAMP is formed. (c) When cAMP is present, it acts as an allosteric effector, complexing with CAP. (d) The cAMP–CAP complex acts as an activator of *lac* operon transcription by binding to a region within the *lac* promoter. (CAP = catabolite activator protein; cAMP = cyclic adenosine monophosphate.)

Genetic Aspects of the *lac* Operon Model

One way to be certain that we understand how the *lac* operon works is by considering the genetic consequences of mutations in the various components of the *lac* operon. The properties of mutations in the structural genes and the regulatory elements of the *lac* operon are quite different. Indeed, the phenotypes of these mutations in homozygotes and hemizygotes, as well as their complementation behavior, were important clues for Jacob and Monod in unraveling the mechanisms of gene regulation in bacteria. Here, we shall examine the kinds of mutations that occur and how the *lac* operon model can explain them.

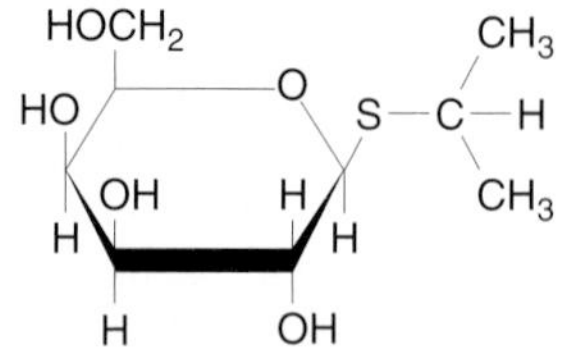

HOCH$_2$ HO O S C H CH$_3$ CH$_3$ OH H H H H OH

Isopropyl-β-D-thiogalactoside (IPTG)

Figure 13-9 Structure of IPTG, an inducer of the *lac* operon. The β-D-thiogalactoside linkage is not cleaved by β-galactosidase, allowing manipulation of the intracellular concentration of this inducer.

Techniques for Mutational Analysis

For studying the Lac repressor and the *lac* operator, we shall consider enzymatic assays for β-galactosidase and the permease under two sets of environmental conditions. In both conditions, no glucose will be present, so we do not have to consider the effects of catabolite repression. Genotypes will be assayed in the *uninduced* state, in which no inducer molecule is present. Tests of uninduced bacteria permit us to assess whether the genetic circuitry necessary for repression is present. Genotypes will also be assayed in the *induced* state, that is, in the presence of inducers in the culture medium. Natural inducers, such as lactose, are not optimal for these experiments, because they are hydrolyzed by β-galactosidase; the inducer concentration decreases during the experiment, and so the measurements of enzyme induction become quite complicated. Rather, for such experiments, we use synthetic inducers, such as isopropyl-β-D-thiogalactoside (IPTG; Figure 13-9), that bind to the allosteric site of the Lac repressor but are not hydrolyzed by β-galactosidase. By assaying wild-type and mutant genotypes in the induced state, we determine whether the genetic circuitry necessary to overcome repression is functioning normally.

We shall see that several different classes of mutations can alter the expression of the structural genes of the *lac* operon. Genetic complementation tests, in which we construct bacterial genotypes heterozygous for various *lac* operon mutations, are essential for distinguishing between these different mutant classes. Ordinarily, bacteria are haploid, making such complementation analysis difficult. However, by using F′ factors (see Chapter 7) carrying the *lac* region of the genome, we can produce bacteria that are diploid and heterozygous for the desired *lac* mutations. We shall see that complementation tests allow us to distinguish mutations in the *lac* operator from mutations in the *I* gene (encoding the Lac repressor).

Results of Genetic Analysis

Let's begin to look at the effects of various combinations of mutations, in the induced and noninduced state, on the production of β-galactosidase and permease. The first thing that we learn from examining mutations that inactivate the structural genes for β-galactosidase and permease (designated Z^- and Y^-, respectively) is that Z^- and Y^- are recessive to their respective wild-type alleles (Z^+ and Y^+). For example, strain 2 in Table 13-1 is inducible for β-galactosidase (like the wild-type haploid strain 1 in this table), even though it is heterozygous for mutant and wild-type *Z* alleles. This demonstrates that the Z^+ allele is dominant to its Z^- counterpart.

Mutations in the repressor and operator cause global misregulation of the *lac* operon structural genes. Let's first consider operator mutations, O^C mutations that make the operator nonfunctional—that is, that make the operator incapable of binding to repressor. Such O^C mutations cause the *lac* operon structural genes to be constitutive in expres-

TABLE 13-1 Synthesis of β-Galactosidase and Permease in Haploid and Heterozygous Diploid Operator Mutants

		β-GALACTOSIDASE (Z)		PERMEASE (Y)		
Strain	**Genotype**	**Noninduced**	**Induced**	**Noninduced**	**Induced**	**Conclusions**
1	$O^+Z^+Y^+$	−	+	−	+	Wild type is inducible
2	$O^+Z^+Y^+/\text{F}'\ O^+Z^-\ Y^+$	−	+	−	+	Z^+ is dominant to Z^-
3	$O^CZ^+Y^+$	+	+	+	+	O^C is constitutive
4	$O^+Z^-\ Y^+/\text{F}'\ O^CZ^+Y^-$	+	+	−	+	Operator is cis-acting

Note: Bacteria were grown in glycerol (no glucose present) with and without the inducer IPTG. The presence or absence of enzyme is indicated by + or −, respectively. All strains are I^+.

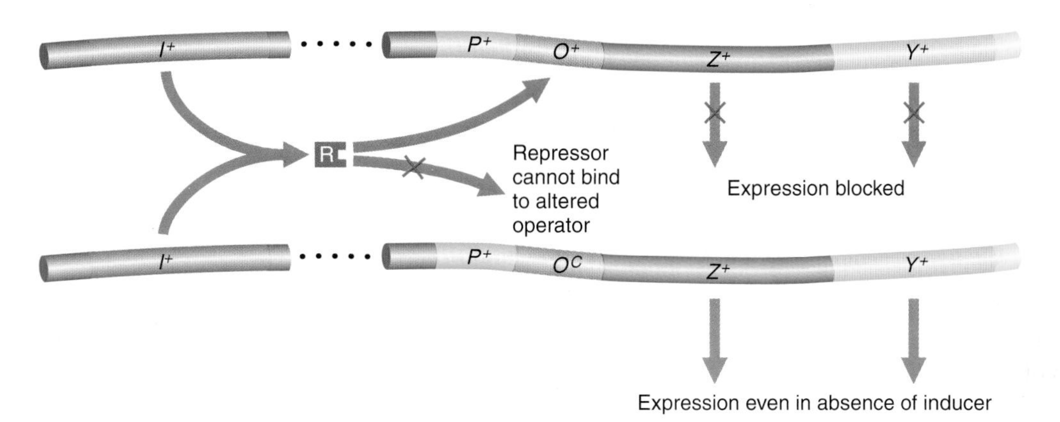

Figure 13-10 O^+ / O^C heterozygotes demonstrate that operators are cis-acting. Because a repressor cannot bind to O^C operators, the *lac* structural genes linked to an O^C operator are expressed even in the absence of an inducer. However, the *lac* genes adjacent to an O^+ operator are still subject to repression.

sion (see Table 13-1, strain 3). That is, regardless of whether inducer is present, these mutations lead to the transcription and translation of the *lac* operon genes. Furthermore, the constitutive effects of O^C mutations are restricted solely to those *lac* structural genes on the same chromosome; for this reason, we say that operators are **cis-acting.** (More generally, we use the term *cis-acting* to refer to genetic interactions that are restricted to elements on the same DNA molecule.) This is demonstrated by the phenotype of strain 4 in Table 13-1. Here, because the wild-type permease (Y^+) gene is cis to the wild-type operator, permease activity is inducible. In contrast, the wild-type β-galactosidase (Z^+) gene is cis to the O^C mutant operator; hence, β-galactosidase expression is constitutive. How can we understand the cis-acting nature of the operator? The explanation is that the operator acts solely as a protein-binding site—it makes *no* gene product. The operator-binding site thus regulates only the expression of the structural genes linked to it (Figure 13-10).

Now let's consider the effects of constitutive I^- mutations (Table 13-2). In I^- mutations, the DNA-binding site of the repressor has been mutated, so no functional operator-binding repressor protein is made. Thus, unlike the inducible phenotype of the wild-type I^+ (strain 1), I^- mutations are constitutive (strain 2). In addition, we see that the inducible phenotype of I^+ is dominant to the constitutive phenotype of I^- (strain 3). This tells us that the amount of wild-type repressor encoded by one copy of the gene is sufficient to regulate both copies of the operator in a diploid cell. Finally, we see that the I^+ gene product is **trans-acting** (strain 4), which means that the gene product regulates *all* structural *lac* operon genes, both those in cis and those in trans (residing on different DNA molecules). How do we explain the trans action of the I^+ gene product? The *I* gene is a standard protein-coding gene. The protein product of the *I* gene is able to diffuse and act on all operators in the cell (Figure 13-11 on the next page).

MESSAGE

Mutations in a target site for DNA binding reveal that such a site is cis-acting; that is, the target site regulates the expression of an adjacent transcription unit on the same DNA molecule. In contrast, mutations in the gene encoding an activator or repressor protein reveal that this protein is trans-acting; that is, it can act on any copy of the target DNA site in the cell.

Genetic Evidence for Allostery

Another class of repressor mutations reveals the importance of allostery. Recall that the Lac repressor has to inhibit transcription of the *lac* operon in the absence of an inducer

TABLE 13-2 Synthesis of β-Galactosidase and Permease in Haploid and Heterozygous Diploid Strains Carrying I^+ and I^-

		β-GALACTOSIDASE (Z)		PERMEASE (Y)		
Strain	**Genotype**	**Noninduced**	**Induced**	**Noninduced**	**Induced**	**Conclusions**
1	$I^+Z^+Y^+$	−	+	−	+	I^+ is inducible
2	$I^-Z^+Y^+$	+	+	+	+	I^- is constitutive
3	$I^+Z^-Y^+$/F′ $I^-Z^+Y^+$	−	+	−	+	I^+ is dominant to I^-
4	$I^-Z^-Y^+$/F′ $I^+Z^+Y^-$	−	+	−	+	I^+ is trans-acting

Note: Bacteria were grown in glycerol (no glucose present) and induced with IPTG. The presence of the maximal level of the enzyme is indicated by a plus sign; the absence or very low level of an enzyme is indicated by a minus sign. (All strains are O^+.)

Animated Art • I^- Lac Repressor Mutations

Figure 13-11 The recessive nature of I^- mutations demonstrates that the repressor is trans-acting. Although no active repressor is synthesized from the I^- gene, the wild-type (I^+) gene provides a functional repressor that binds to both operators in a diploid cell and blocks *lac* operon expression (in the absence of an inducer).

but must permit transcription when the inducer is present. This is accomplished through a second site on the repressor protein, the allosteric site, which binds to the inducer. When bound to the inducer, the repressor is changed in overall structure such that the DNA-binding site can no longer function.

I^S mutations (super-repressors) cause repression even in the presence of an inducer (compare strain 2 in Table 13-3 with the inducible wild-type strain 1). Unlike the case for I^-, I^S mutations are dominant to I^+ (see Table 13-3, strain 3). I^S mutations alter the stereospecific allosteric site such that it can no longer bind to an inducer. In the absence of an ability to bind an inducer, I^S-encoded repressor protein continually binds to the operator—preventing transcription of the *lac* operon even when the inducer is present in the cell. On this basis, we can see why I^S is dominant to I^+. Mutant I^S protein will bind to all operators in the cell, even in the presence of an inducer and regardless

TABLE 13-3 Synthesis of β-Galactosidase and Permease by the Wild Type and by Strains Carrying Different Alleles of the *I* Gene

		β-GALACTOSIDASE (Z)		PERMEASE (Y)		
Strain	**Genotype**	**Noninduced**	**Induced**	**Noninduced**	**Induced**	**Conclusions**
1	$I^+Z^+Y^+$	−	+	−	+	I^+ is inducible
2	$I^S Z^+ Y^+$	−	−	−	−	I^S is always repressed
3	$I^S Z^+ Y^+ / F' I^+ Z^+ Y^+$	−	−	−	−	I^S is dominant to I^+

Note: Bacteria were grown in glycerol (no glucose present) with and without the inducer IPTG. Presence of the indicated enzyme is represented by +; absence or low levels, by −.

Animated Art • I^S Lac Super-Repressor Mutations

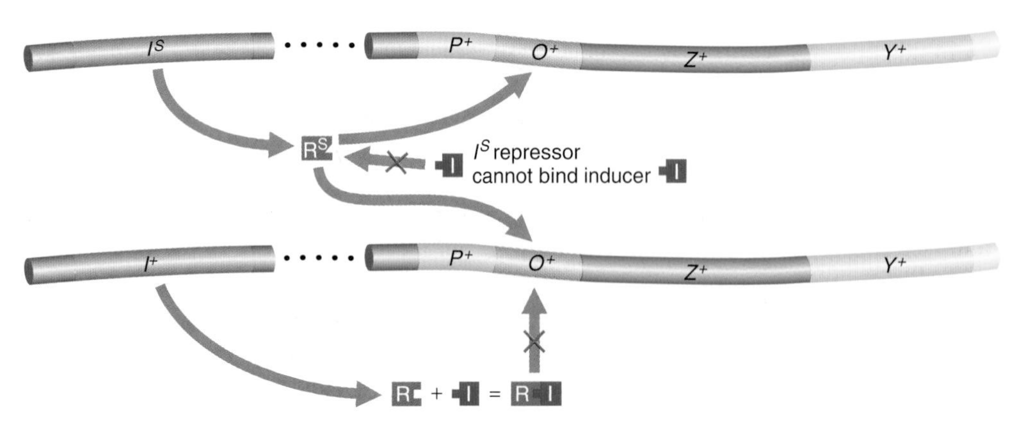

Figure 13-12 The dominance of I^S mutations is due to the inactivation of the allosteric site on the Lac repressor. In an I^S / I^+ diploid cell, none of the *lac* structural genes are transcribed, even in the presence of an inducer. In contrast with the wild-type repressor, the I^S repressor lacks a functional lactose-binding site (the allosteric site) and thus is not inactivated by an inducer. Thus, even in the presence of an inducer, the I^S repressor binds irreversibly to all operators in a cell, thereby blocking transcription of the *lac* operon.

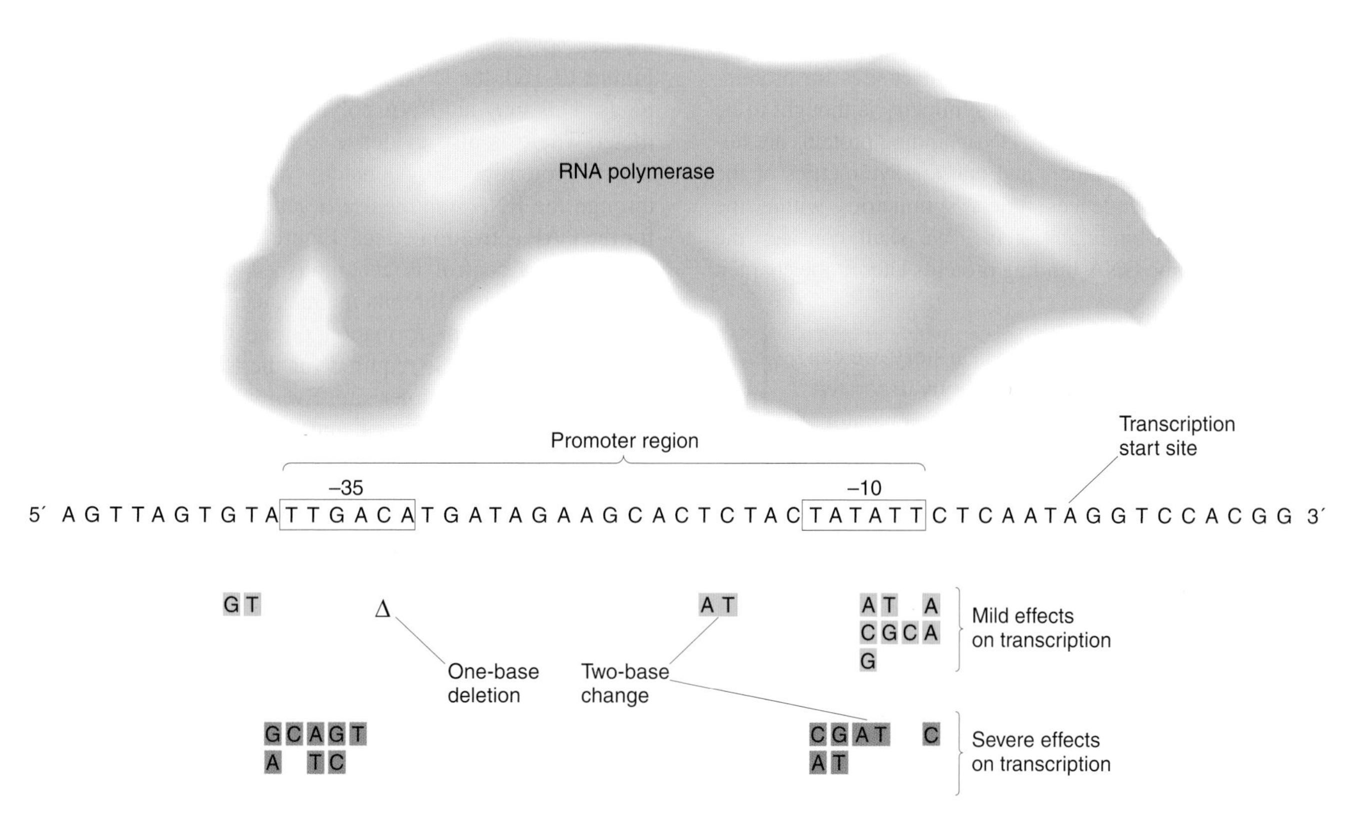

Figure 13-13 Specific DNA sequences are important for efficient transcription of *E. coli* genes by RNA polymerase. The boxed sequences at approximately 35 and 10 nucleotides before the transcription start site are highly conserved in all *E. coli* promoters, an indication of their role as contact sites on the DNA for RNA polymerase binding. Mutations in these regions have mild (gold) and severe (brown) effects on transcription. The mutations may be changes of single nucleotides or pairs of nucleotides, or a deletion (Δ) may occur. *(From J. D. Watson, M. Gilman, J. Witkowski, and M. Zoller,* Recombinant DNA, *2d ed. © 1992 by James D. Watson, Michael Gilman, Jan Witkowski, and Mark Zoller.)*

of the fact that I^+ protein may be present in the same cell (Figure 13-12).

Genetic Analysis of the *lac* Promoter

Genetic experiments demonstrated that an element essential for *lac* transcription is located between *I* and *O*. This element, termed the **promoter** *(P),* serves as the initiation site for transcription. There are two binding regions for RNA polymerase in a typical prokaryotic promoter, shown in Figure 13-13 as the two highly conserved regions at −35 and −10. Promoter mutations affect the transcription of all structural genes in the operon in a similar manner. The dominance of promoter mutations is cis-acting because promoters, like operators, are sites on the DNA molecule that are bound by proteins.

The Structures of Target DNA Sites

The DNA sequences to which the Lac repressor and the CAP–cAMP complex bind are now known. These sequences (Figure 13-14) are very different from each other, and these differences underlie the specificity of binding by these very different DNA-binding proteins. One property that they do share, and which is common to many other DNA-binding sites, is rotational twofold symmetry. In other words, if we rotate the DNA sequence 180° within the plane of the page, the sequence of the highlighted bases of the

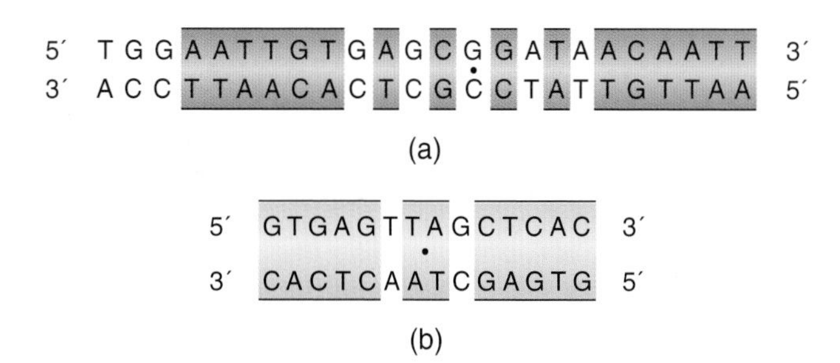

Figure 13-14 The DNA base sequences of (a) the *lac* operator, to which the Lac repressor binds, and (b) the CAP-binding site, to which the CAP–cAMP complex binds. Sequences exhibiting twofold rotational symmetry are indicated by the colored boxes and by a dot at the center point of symmetry. *(Part a from W. Gilbert, A. Maxam, and A. Mirzabekov, in N. O. Kjeldgaard and O. Malløe, eds.,* Control of Ribosome Synthesis. *Academic Press, 1976. Used by permission of Munksgaard International Publishers, Ltd., Copenhagen.)*

binding sites will be identical. The highlighted bases are thought to constitute the important contact sites for protein-DNA interactions. This rotational symmetry is thought to be reflective of the fact that many DNA-binding proteins are homodimers or homotetramers and that the symmetries of the target-site DNA sequences reflect symmetries within the multimeric DNA-binding proteins. We shall consider the structures of some DNA-binding proteins later in the chapter.

MESSAGE

Generalizing from the *lac* operon story, we can envision the chromosome as heavily decorated by regulatory proteins binding to the operator sites that they control. The exact pattern of decorations will depend on which genes are turned on or off and whether activators or repressors regulate particular operons.

A Summary of the *lac* Operon

We can now fit the known repressor, CAP–cAMP, and RNA polymerase binding sites into the detailed model of the *lac* operon, as shown in Figures 13-15 and 13-16. In Figure 13-16d, the DNA is shown as being bent when CAP binds. This may aid RNA polymerase's binding to the promoter. There is also evidence to suggest that CAP makes direct protein-protein contacts with RNA polymerase, through the RNA polymerase α subunit, that are important for the CAP activation effect (Figure 13-16e).

Glucose control is accomplished because a glucose-breakdown product inhibits maintenance of the high cAMP levels necessary for formation of the CAP–cAMP complex, which in turn is required for the RNA polymerase to attach at the *lac* promoter site. Even when there is a shortage of glucose catabolites and CAP–cAMP forms, the mechanism for lactose metabolism will be implemented only if lactose is present. This level of control is accomplished because lactose must bind to the repressor protein to remove it from the operator site and permit transcription of the *lac* operon. Thus, the cell conserves its energy and resources by producing the lactose-metabolizing enzymes only when they are both needed and useful.

Inducer-repressor control of the *lac* operon is an example of repression, or negative control, in which expression

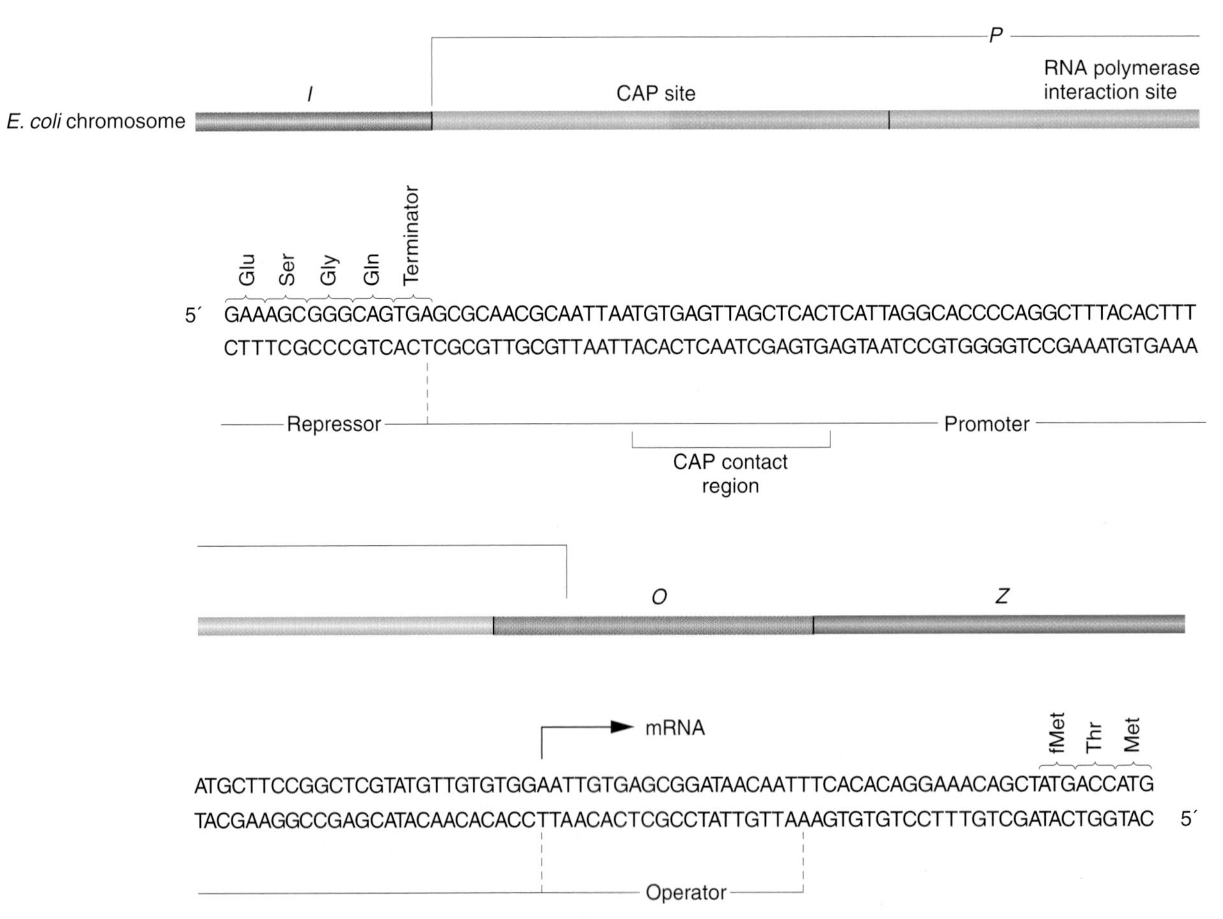

Figure 13-15 The base sequence and the genetic boundaries of the control region of the *lac* operon, with partial sequences for the structural genes. *(After R. C. Dickson, J. Abelson, W. M. Barnes, and W. S. Reznikoff, "Genetic Regulation: The Lac Control Region,"* Science *187, 1975, 27. © 1975 by the American Association for the Advancement of Science.)*

(a) Glucose present (cAMP low); no lactose; no *lac* mRNA

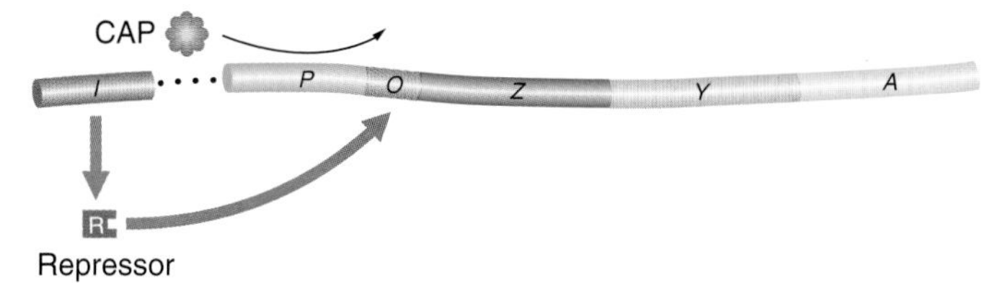

(b) Glucose present (cAMP low); lactose present

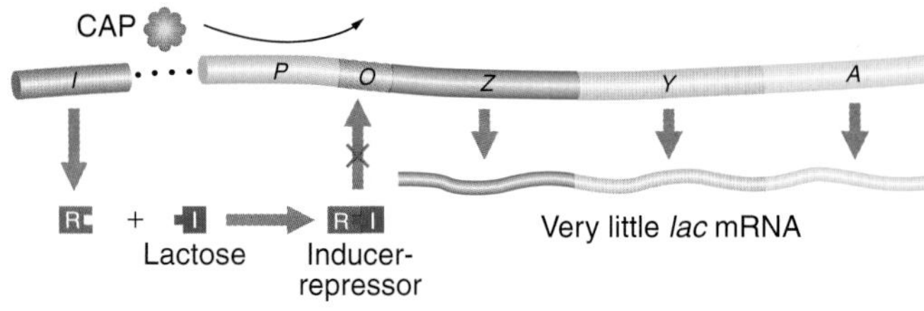

(c) No glucose present (cAMP high); lactose present

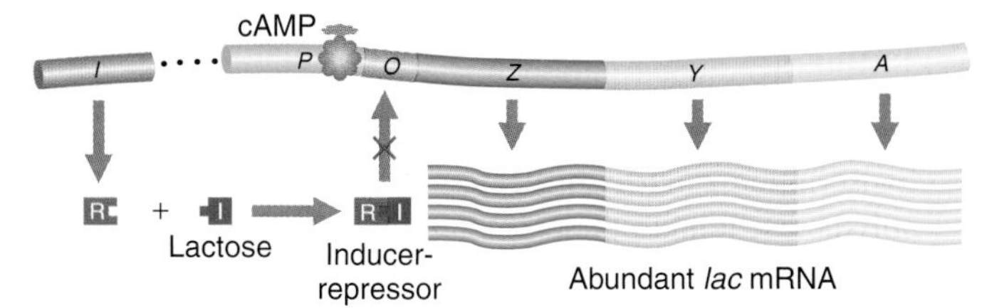

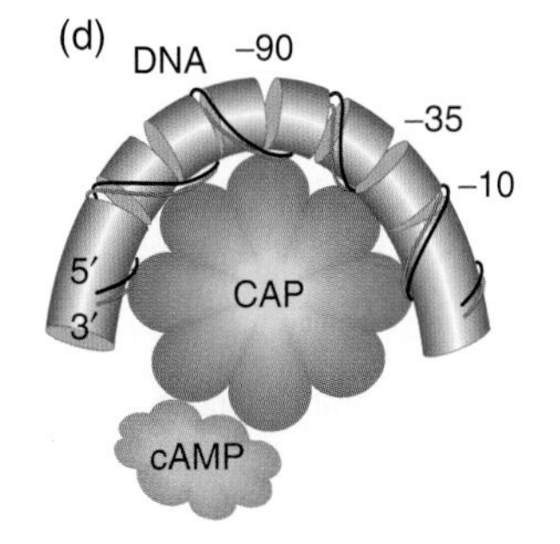

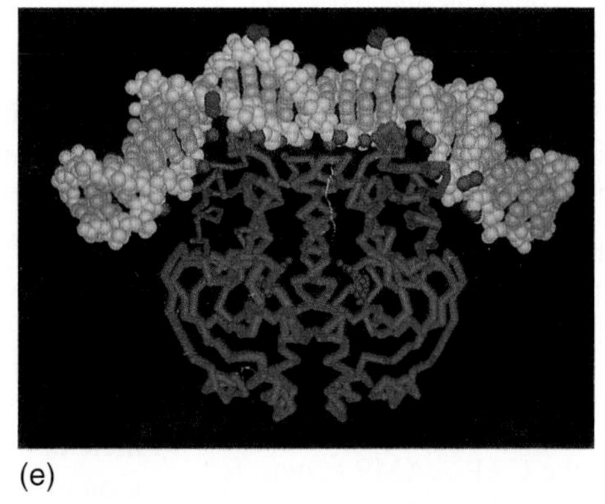

(e)

Figure 13-16 Negative and positive control of the *lac* operon by the Lac repressor and catabolite activator protein (CAP), respectively. (a) In the absence of lactose to serve as an inducer, the Lac repressor is able to bind the operator; regardless of the levels of cAMP and the presence of CAP, mRNA production is repressed. (b) With lactose present to bind the repressor, the repressor is unable to bind the operator; however, only small amounts of mRNA are produced because the presence of glucose keeps the levels of cAMP low, and thus the CAP–cAMP complex does not form and bind the promoter. (c) With the repressor inactivated by lactose and with high levels of cAMP present (owing to the absence of glucose), cAMP binds CAP. The CAP–cAMP complex is then able to bind the promoter; the *lac* operon is thus activated, and large amounts of mRNA are produced. (d) When CAP binds the promoter, it creates a bend greater than 90° in the DNA. Apparently, RNA polymerase binds more effectively when the promoter is in this bent configuration. (e) CAP bound to its DNA recognition site. This part is derived from the structural analysis of the CAP–DNA complex. *[Parts a–d redrawn from B. Gartenberg and D. M. Crothers,* Nature *333, 1988, 824. (See H. N. Lie-Johnson et al.,* Cell *47, 1986, 995.) Adapted from H. Lodish, D. Baltimore, A. Berk, S. L. Zipursky, P. Matsudaira, and J. Darnell,* Molecular Cell Biology, *3d ed. © 1995 by Scientific American Books. Part e from L. Schultz and T. A. Steitz.]*

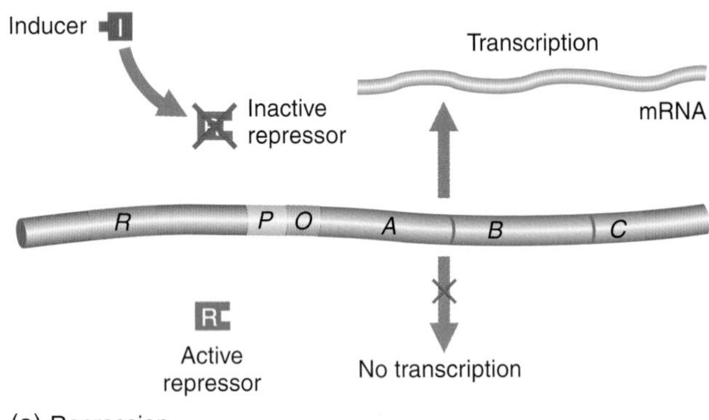

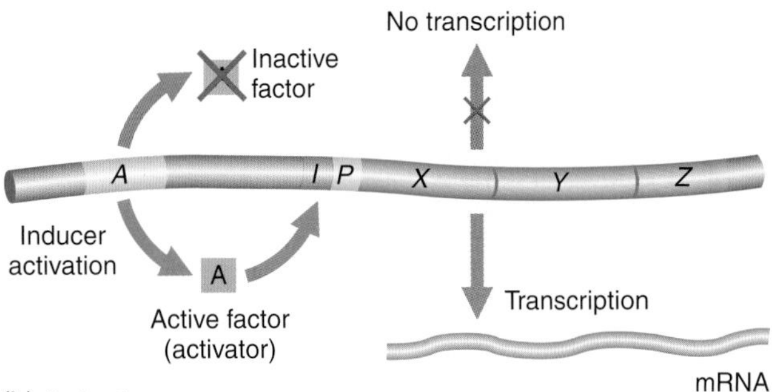

Figure 13-17 Comparison of repression and activation.
(a) In repression, an active repressor (encoded by the *R* gene in the example shown here) blocks gene expression of the *A*, *B*, *C* operon by binding to an operator site *(O)*. An inactive repressor allows gene expression. The repressor can be inactivated either by an inducer or by mutation. (b) In activation, a functional activator is required for gene expression, as shown for the *X*, *Y*, *Z* operon here. Small molecules can convert a nonfunctional activator into a functional one, as in the case of cyclic AMP and the CAP protein in the *lac* operon system. A nonfunctional activator results in no gene expression. The activator binds to the control region of the operon, termed *I* in this case. (The positions of both *O* and *I* with respect to the promoter *P* in the two examples are arbitrarily drawn.)

is normally blocked. In contrast, the CAP–cAMP system is an example of activation, or positive control, because its expression requires the presence of an activating signal—in this case, the interaction of the CAP–cAMP complex with the CAP site. Figure 13-17 outlines these two basic types of control systems.

MESSAGE

The *lac* operon is a cluster of structural genes that specify enzymes taking part in lactose metabolism. These genes are controlled by the coordinated actions of cis-acting promoter and operator regions. The activity of these regions is, in turn, determined by repressor and activator molecules specified by separate regulator genes.

DUAL POSITIVE AND NEGATIVE CONTROL: THE ARABINOSE OPERON

Prokaryotic control of transcription seems to mix and match different aspects of positive and negative regulation in different ways. Thus the general themes to derive from our detailed examination of the lactose operon should focus on how positive and negative regulation can work through the activity of sequence-specific, allosterically regulated DNA-binding proteins. Even in the metabolism of other sugars, we see transcriptional regulatory mechanisms that are certainly different in detail. The arabinose operon is such an example (Figure 13-18).

The structural genes (*araB, araA,* and *araD*) that encode the metabolic enzymes that break down arabinose are transcribed as a multigenic mRNA. Transcription is activated at *araI,* the **initiator** region, which contains both an operator site and a promoter. The *araC* gene encodes an activator protein that, when bound to arabinose, activates transcription of the *ara* operon, perhaps by helping RNA polymerase bind to the promoter, located within in the *araI* region. An additional activation event is mediated by the same CAP–cAMP catabolite repression system that regulates *lac* operon expression.

In the presence of arabinose, both the CAP–cAMP complex and the AraC–arabinose complex must bind to the initiator region in order for RNA polymerase to bind to the promoter and transcribe the *ara* operon (Figure 13-19a). In the absence of arabinose, the AraC protein assumes a different conformation and represses the *ara* operon by binding both to *araI* and to a second operator region, *araO,* thereby forming a loop (Figure 13-19b) that prevents transcription. Thus, the AraC protein has two conformations, one that acts as an activator and the other that acts as a repressor. The two conformations, dependent on whether the allosteric effector has bound to the protein, also differ in their abilities to bind a specific target site in the *araO* region of the operon.

MESSAGE

Operon transcription can be regulated by both activation and repression. Operons regulating the metabolism of similar compounds, such as sugars, can be regulated in quite different ways.

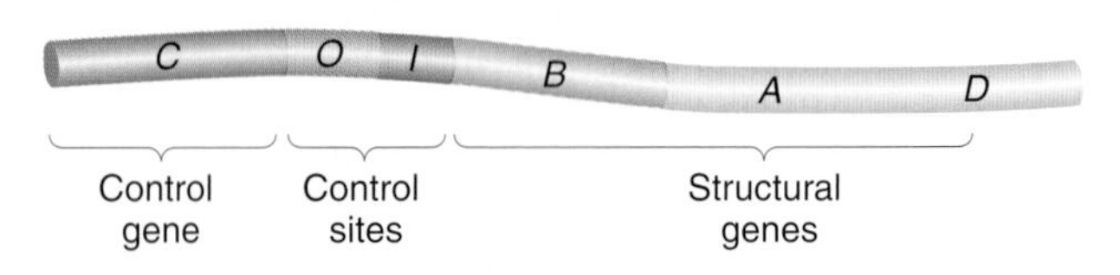

Figure 13-18 Map of the *ara* region. The *B*, *A*, and *D* genes together with the *I* and *O* sites constitute the *ara* operon.

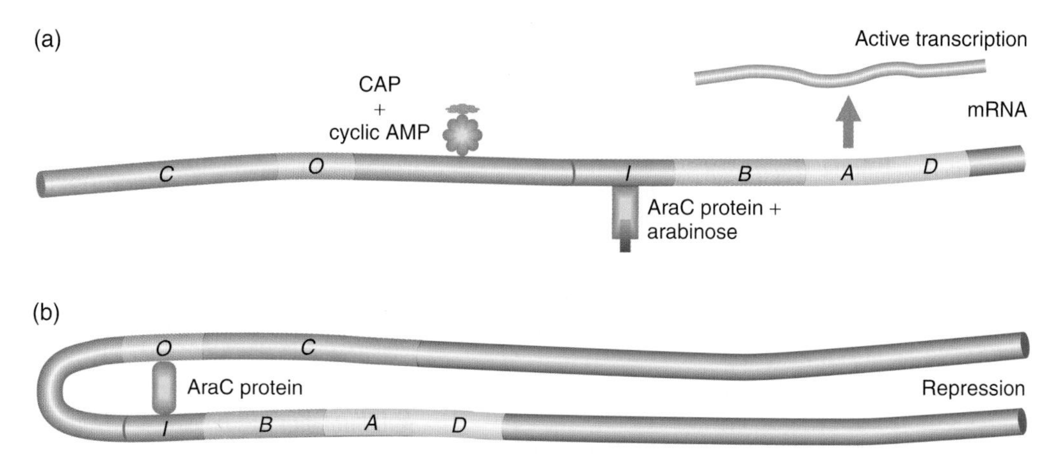

Figure 13-19 Dual control of the *ara* operon. (a) In the presence of arabinose, the AraC protein binds to the *araI* region and, when bound to cAMP, the CAP protein binds to a site adjacent to *araI*. This stimulates the transcription of the *araB, araA,* and *araD* genes. (b) In the absence of arabinose, the AraC protein binds to both the *araI* and *araO* regions, forming a DNA loop. This prevents transcription of the *ara* operon.

METABOLIC PATHWAYS

Coordinate control of genes through operon-level regulation in bacteria and bacteriophage is widespread. In the preceding section, we looked at examples of regulation of pathways for the metabolism of specific sugars. In fact, most coordinated gene function in prokaryotes acts through operon mechanisms. For example, clustered genes organized into operons, complete with multigenic mRNAs, encode enzymes in many biosynthetic pathways. Furthermore, in cases where the sequence of catalytic activity is known, there is a remarkable congruence between the sequence of genes on the chromosome and the sequence in which their products act in the metabolic pathway. This congruence is strikingly illustrated by the organization of the tryptophan operon in *E. coli* (Figure 13-20).

MESSAGE

In prokaryotes, genes that encode enzymes that are in the same metabolic pathways are generally organized into operons.

TRANSCRIPTIONAL REGULATION IN EUKARYOTES

Eukaryotes face the same basic tasks of coordinating gene expression as prokaryotes do but in a much more intricate

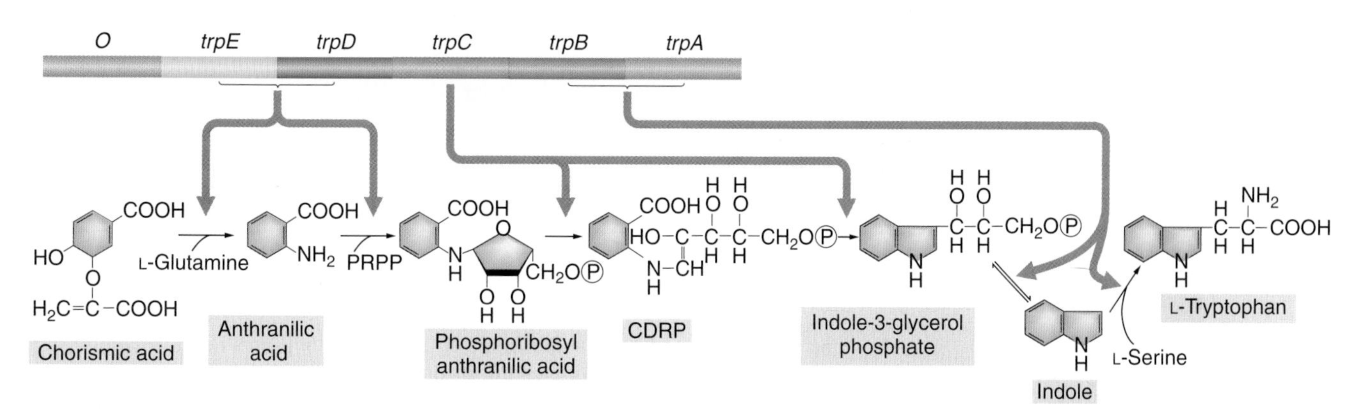

Figure 13-20 The chromosomal order of genes in the *trp* operon of *E. coli* and the sequence of reactions catalyzed by the enzyme products of the *trp* structural genes. The products of genes *trpD* and *trpE* form a complex that catalyzes specific steps, as do the products of genes *trpB* and *trpA*. Tryptophan synthetase is a tetrameric enzyme formed by the products of *trpB* and *trpA*. It catalyzes a two-step process leading to the formation of tryptophan. [PRPP = phosphoribosylpyrophosphate; CDRP = 1-(*o*-carboxyphenylamino)-1-deoxyribulose 5-phosphate.] *(After S. Tanemura and R. H. Bauerle,* Genetics *95, 1980, 545.)*

way. Some genes have to respond to changes in physiological conditions. Many others are parts of developmentally triggered genetic circuits that organize cells into tissues and tissues into an entire organism (except for unicellular eukaryotes). In these cases, the signals controlling gene expression are the products of developmental regulatory genes, rather than signals from the external environment.

Most eukaryotic genes are controlled at the level of transcription, and the mechanisms are similar in concept to those found for bacteria. Trans-acting regulatory proteins work through sequence-specific DNA binding to their cis-acting regulatory target sequences. Because of the much more complex regulation that is required to coordinate proper gene activity throughout the lifetime of a multicellular organism, there are some considerable novelties as well. We shall see examples of these novelties in this and subsequent chapters.

Typically, eukaryotes have many more genes than prokaryotes do, sometimes by several orders of magnitude. The genes of higher organisms also tend to be larger, owing to the facts that cis-acting sequences on the DNA can be located tens of thousands of base pairs away from the transcription start site and that a battery of regulatory factors is sometimes needed to bring about proper regulation of certain genes.

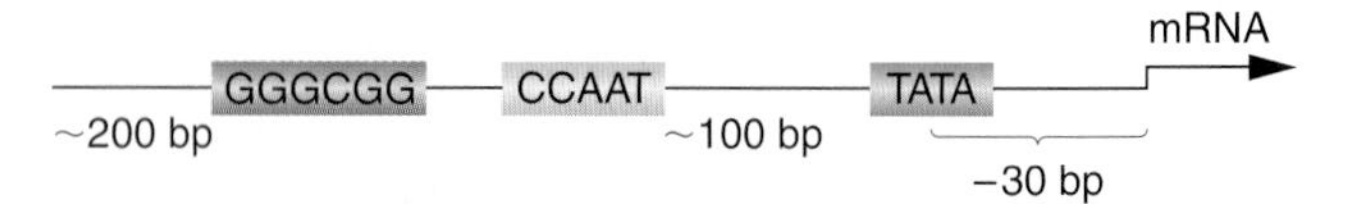

Figure 13-21 The region upstream of the transcription start site in higher eukaryotes. The TATA box is located approximately 30 base pairs from the mRNA start site. Usually, two or more promoter-proximal elements are found 100 and 200 bp upstream of the mRNA start site. The CCAAT box and the GC-rich box are shown here. Other upstream elements include the sequences GCCACACCC and ATGCAAAT.

Cis-Acting Sequences in Transcriptional Regulation

As mentioned in Chapter 3, eukaryotes have three different classes of RNA polymerase (distinguished by roman numerals I, II, and III). All mRNA molecules are synthesized by RNA polymerase II, and the rest of this chapter will focus on the transcription of mRNAs. To achieve maximal rates of transcription by RNA polymerase II, the cooperation of multiple cis-acting regulatory elements is required. We can distinguish three classes of elements on the basis of their relative locations. Near the transcription initiation site are the core promoter (the RNA polymerase II–binding region) and promoter-proximal cis-acting sequences that bind to proteins that in turn assist in the binding of RNA polymerase II to its promoter. Additional cis-acting sequence elements can act at considerable distance—these elements, which are independent of distance, are termed *enhancers* and *silencers*. Often, an enhancer or silencer element will act only in one or a few cell types in a multicellular eukaryote. The promoters, promoter-proximal elements, and distance-independent elements are all targets for binding by different trans-acting DNA-binding proteins.

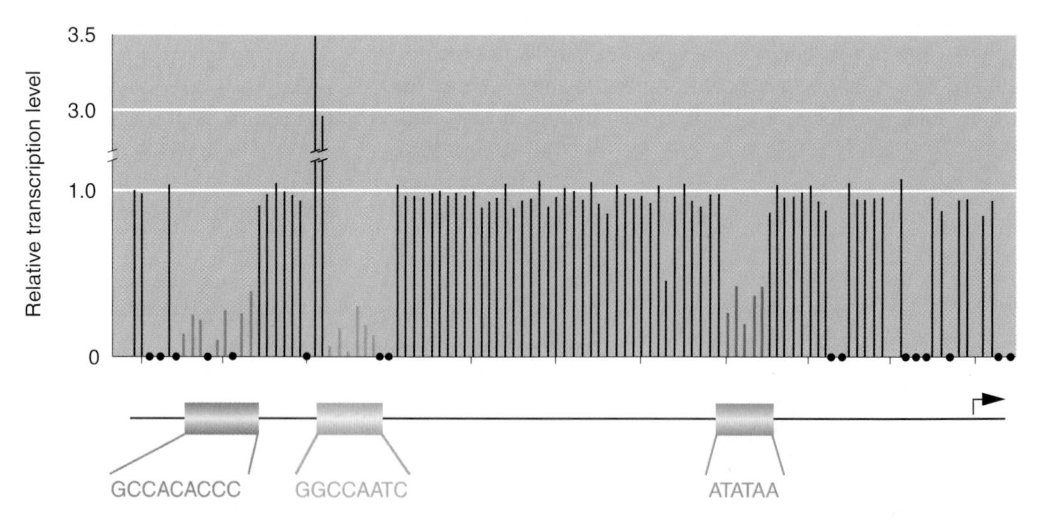

Figure 13-22 Consequences of point mutations in the promoter for the β-globin gene. Point mutations throughout the promoter region were analyzed for their effects on transcription rates. The height of each line represents the transcription level, relative to a wild-type promoter, that results from promoters with base changes at that point. A level of 1.0 means that the rates are equal to the wild-type rate; reductions in transcription rates yield levels less than 1.0. Almost every nucleotide throughout the promoter was tested, except for the positions shown with black dots. The diagram below the bar graph shows the position of the TATA box and two upstream elements of the promoter. Only the base substitutions that lie within the three promoter elements change the level of transcription. *(From T. Maniatis, S. Goodbourn, and J. A. Fischer,* Science *236, 1987, 1237.)*

```
             1                              2
3′  TTGGTCGACACCTTACACACAGTCAATCCCACACCTTTCAGGGGTCCGAGGGGTCGT
5′  AACCAGCTGTGGAATGTGTGTCAGTTAGGGTGTGGAAAGTCCCCAGGCTCCCCAGCA

            3         4          5
     CCGTCTTCATACGTTTCGTACGTAGAGTTAATCAGTCGTTGGTC 5′
     GGCAGAAGTATGCAAAGCATGCATCTCAATTAGTCAGCAACCAG 3′
```

Figure 13-23 Organization of the SV40 enhancer. SV40 is a DNA virus that infects primates, and its regulatory sequences interact with the eukaryotic cell's transcriptional regulatory machinery. Boxed sequences 1 to 5 indicate the sequences that are required for maximum levels of enhancer activity. *(From T. Maniatis, S. Goodbourn, and J. A. Fischer,* Science *236, 1987, 1237.)*

The core promoter and promoter-proximal elements. Figure 13-21 is a schematic view of the core promoter and promoter-proximal sequence elements. The **core promoter** usually refers to the region from the transcription start site upstream about 50–150 bp, including the TATA box, which resides approximately 30 bp upstream of the transcription initiation site. This core promoter is unable to mediate efficient transcription by itself. Some important elements near the promoter, the **promoter-proximal elements,** are found within 100–200 bp of the transcription initiation site. The CCAAT box functions as one of these promoter-proximal cis-acting sequences, and a GC-rich segment often functions as another. An example of the consequences of mutating these sequence elements on transcription rates is shown in Figure 13-22.

Distance-independent cis-acting elements. In eukaryotes, we distinguish between two classes of cis-acting elements that can exert their effects at considerable distance from the promoter. **Enhancers** are cis-acting sequences that can greatly increase transcription rates from promoters on the same DNA molecule; thus, they act to activate, or positively regulate, transcription. Silencers have the opposite effect. **Silencers** are cis-acting sequences that are bound by repressors, thereby inhibiting activators and reducing transcription. Enhancers and silencers are similar to promoter-proximal regions in that they are organized as a series of cis-acting sequences that are bound by trans-acting regulatory proteins. However, they are distinguished from promoter-proximal elements by being able to act at a distance, sometimes 50 kb or more, and by being able to operate either upstream or downstream from the promoter they control. Enhancer and silencer elements are intricately structured. Figure 13-23 shows the DNA sequence of the SV40 (simian virus 40) enhancer, which is required for high-level expression of SV40 transcripts. Within the enhancer, there are five sequence elements required for maximal enhancement of transcription. Enhancers that are themselves composed of multiple copies of a DNA-binding element are common. Different DNA sequences serve as target-recognition sites for specific trans-acting regulatory proteins.

How do enhancer and silencer elements many thousands of base pairs away regulate transcription? Most models for such action at a distance include some type of DNA looping. Figure 13-24 details a DNA-looping model for activation of the initiation complex (see the next section). In this model, a DNA loop brings activator proteins bound to distant enhancer

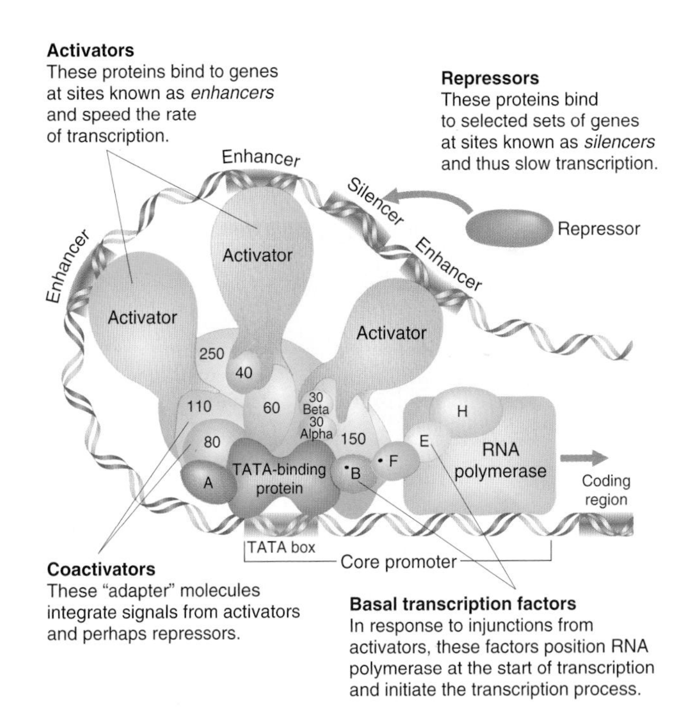

Figure 13-24 The molecular apparatus controlling transcription in human cells consists of four kinds of components. (The numbered proteins are the names of the subunits of RNA polymerase II. Each subunit is named according to its molecular mass in kilodaltons.) Basal transcription factors (labeled A, B, F, E, H) are essential for transcription but cannot by themselves increase or decrease its rate. That task falls to regulatory molecules known as activators and repressors. Activators, and possibly repressors, communicate with the basal factors through coactivators–proteins that are linked in a tight complex to the TATA-binding protein, the first of the basal transcription factors to land on the core promoter. *(From R. Tjian, "Molecular Machines That Control Genes."*

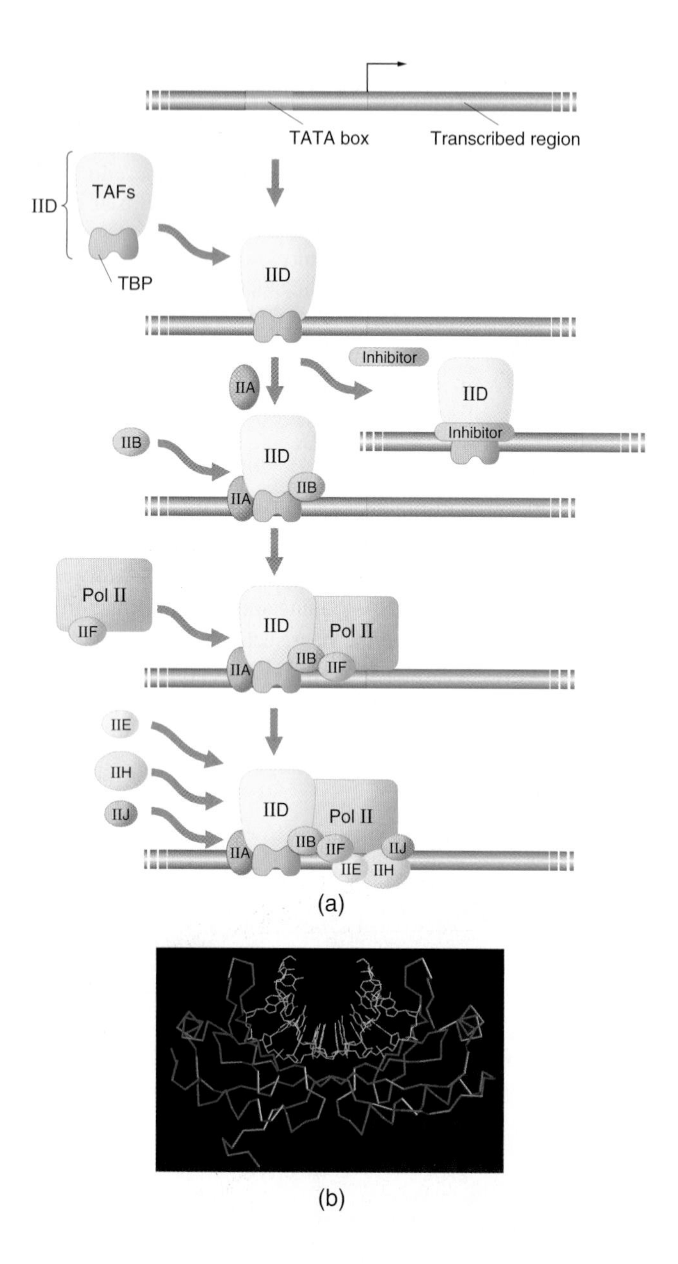

Figure 13-25 (a) Assembly of the RNA polymerase II initiation complex begins with the binding of transcription factor IID (TFIID) to the TATA box. TFIID is composed of one TATA box-binding subunit called *TBP* and more than eight other subunits (TAFs), represented by one large symbol (TAFs). Inhibitors can bind to the TFIID-promoter complex, blocking the binding of other general transcription factors. Binding of TFIIA to the TFIID-promoter complex (to form the D–A complex) prevents inhibitor binding. TFIIB then binds to the D–A complex, followed by binding of a preformed complex between TFIIF and RNA polymerase II. Finally, TFIIE, TFIIH, and TFIIJ must add to the complex, in that order, for transcription to be initiated. (b) TATA-binding protein (blue) is a remarkably symmetrical, saddle-shaped molecule. Its underside rides on DNA and seems to bend it. This bending may somehow facilitate assembly of the complex that initiates transcription. *(Part a from H. Lodish, D. Baltimore, A. Berk, S. L. Zipursky, P. Matsudaira, and J. Darnell,* Molecular Cell Biology, *3d ed. © 1995 by Scientific American Books; part b courtesy of J. L. Kim and S. K. Berley, from* Nature *365, 1993, 520.)*

elements into proximity to protein complexes associated with promoter-proximal cis-acting sequences.

MESSAGE

Eukaryotic enhancers and silencers can act at great distance.

Trans Control of Transcription

A large number of trans-acting regulatory proteins have now been identified in eukaryotic cells. Like their counterparts in prokaryotes, these regulatory proteins act by binding to specific target DNA sequences.

Regulatory proteins that bind the core promoter and promoter-proximal elements help RNA polymerase II to initiate transcription, and together with the polymerase they form an initiation complex, as pictured in Figure 13-25a. Several different transcription factor complexes (TFII complexes) interact with RNA polymerase II. For example, the TFIID complex consists of a TATA box–binding protein (TBP) and more than eight additional subunits (TAFs). The TFII complexes are often referred to as *basal* or *general* transcription factors, because they are the minimal requirement for RNA polymerase II to initiate transcription (usually very weakly) at a promoter. Figure 13-25b shows the structure of the TATA box–binding protein binding to DNA. CCAAT and GC boxes are recognized by additional DNA-binding proteins.

Some of the proteins that bind distance-independent elements also have been identified. The protein encoded by the yeast *GCN4* gene is an example of a trans-acting enhancer-binding protein. It operates on enhancers called *upstream activating sequences (UASs).* GCN4 activates the transcription of many yeast genes that encode enzymes of amino acid biosynthetic pathways. In response to amino acid starvation, the level of GCN4 protein rises and, in turn, increases the expression levels of the amino acid biosynthetic genes. The UASs recognized by GCN4 contain the principal recognition sequence element ATGACTCAT.

MESSAGE

The proper constellation of proteins binding to the core promoter, promoter-proximal elements, and distance-independent elements is required for effective transcription inititation.

Tissue-Specific Regulation of Transcription

Many enhancer elements in higher eukaryotes activate transcription in a tissue-specific manner—that is, they induce expression of a gene in one or a few cell types. For example, antibody genes are flanked by powerful enhancers that operate only in the B lymphocytes of the immune system. Many enhancers are integral components of complex tissue-specific genetic circuits that underlie complex events in development in higher eukaryotes. Tissue specificity is conferred in one of two ways. An enhancer can act in a tissue-specific manner if the activator that binds to it is present in only some types of cells. Alternatively, a tissue-specific repressor can bind to a silencer element located very near the enhancer element, making the enhancer inaccessible to its transcription factor.

In some genes, regulation can be controlled by simple sets of enhancers. For example, in *Drosophila,* vitellogenins are large egg yolk proteins made in the female adult's ovary and fat body (an organ that is essentially the fly's liver) and transported into the developing oocyte. Two distinct enhancers located within a few hundred base pairs of the promoter regulate the vitellogenin gene, one driving expression in the ovaries and the other in the fat body.

The array of enhancers for a gene can be quite complex, controlling similarly complex patterns of gene expression. The *dpp* (decapentaplegic) gene in *Drosophila,* for example, encodes a protein that mediates signals between cells (as we shall see in Chapter 15). It contains numerous enhancers, perhaps numbering in the tens or hundreds, dispersed along a 50-kb interval of DNA. Some of these enhancers are located 5′ (upstream) of the transcription initiation site of *dpp,* others are downstream of the promoter, some are in introns, and still others are 3′ of the polyadenylation site of the gene. Each of these enhancers regulates the expression of *dpp* in a different site in the developing animal. Some of the better-characterized *dpp* enhancers are shown in Figure 13-26.

The requirement for multiple enhancer elements to regulate tissue-specific expression helps to explain the large size of genes in higher eukaryotes. The tissue-specific regulation of a gene may be quite complex, requiring the action of numerous, distantly located enhancer elements.

Dissecting Eukaryotic Regulatory Elements

An important part of modern genetics is the identification and characterization of distantly located regulatory elements by means of transgenic constructs, in which recombinant DNA molecules are inserted into the genome of an organism (see Chapter 8). In these constructs, isolated pieces of a gene are incorporated to determine what tissue and temporal regulatory patterns it can control. By means of such slicing and dicing experiments, it is possible to home in on the locations of specific regulatory elements. We can also exploit these regulatory elements to develop new ways to identify genes of interest. In the following sections, we shall see how such experiments are carried out, by using studies of transcriptional enhancers as examples. Remember, however, that the same techniques and logic can be applied to any other classes of regulatory elements, some of which are considered in Chapter 16.

Enhancers of a cloned gene are typically identified by means of transgenic **reporter genes** (Figure 13-27). In reporter-gene constructs, pieces of cis-regulatory DNA are fused (usually by restriction-enzyme-based "cutting and pasting" of recombinant DNA molecules) near a transcription unit that can express a reporter protein—that is, a protein whose presence can be monitored. Our old friend, the *E. coli* β-galactosidase enzyme encoded by the *lacZ* gene, is a very popular reporter protein. It is very easy to detect the presence of β-galactosidase histochemically by adding a synthetic substrate, X-gal, to the medium and observing which tissues turn blue, an effect that happens only where β-galactosidase is present to convert X-gal to a blue product. The reporter-gene-construct transcription unit contains

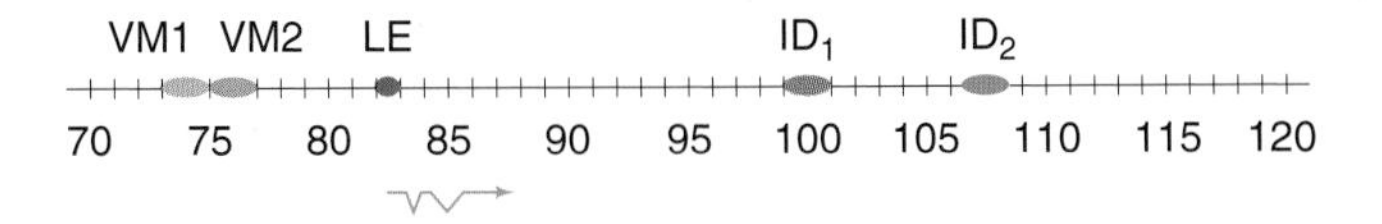

Figure 13-26 A molecular map of a complex gene—the *dpp* gene of *Drosophila.* Units on the map are in kilobases. The basic transcription unit of the gene is shown below the map coordinate line. The abbreviations above the line mark the sites of a few of the many tissue-specific enhancers that regulate this transcript in different stages of development. Tissue-specific expression patterns conferred by these enhancers are shown in Figure 13-28, and the abbreviations are explained there.

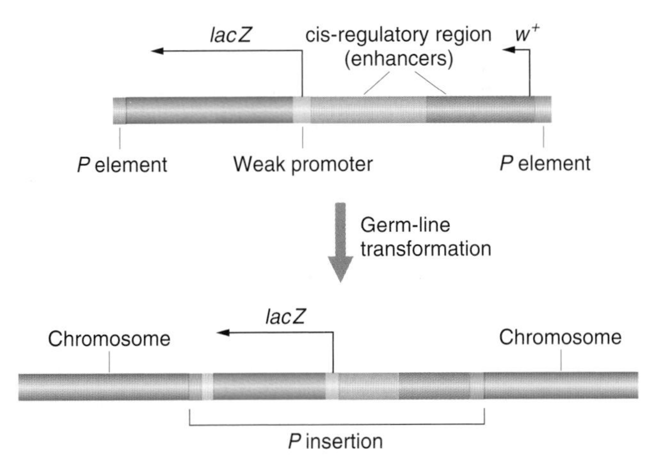

Figure 13-27 Use of a reporter-gene construct in *Drosophila* to identify enhancers. The top line represents a part of a plasmid, bracketed by *P*-element ends so that the material in between can be inserted into the genome by *P*-element transformation (see Chapter 8). Between these *P* elements a region of DNA thought to contain one or more enhancers is inserted immediately adjacent to a "weak" promoter—that is, a promoter that by itself cannot initiate transcription. The promoter is joined to the *E. coli lacZ* structural gene (the reporter gene), which encodes β-galactosidase. When the construct is introduced to fly genome, if there are any enhancers in the construct, they will induce tissue-specific expression of *lacZ*. W^+ is the wild-type allele of the *Drosophila* white gene, used to detect the presence of the reporter-gene construct in the fly. The embryos and imaginal disks in Figures 13-28 and 13-29 are examples of expression from reporter-gene constructs in *Drosophila* and in mice.

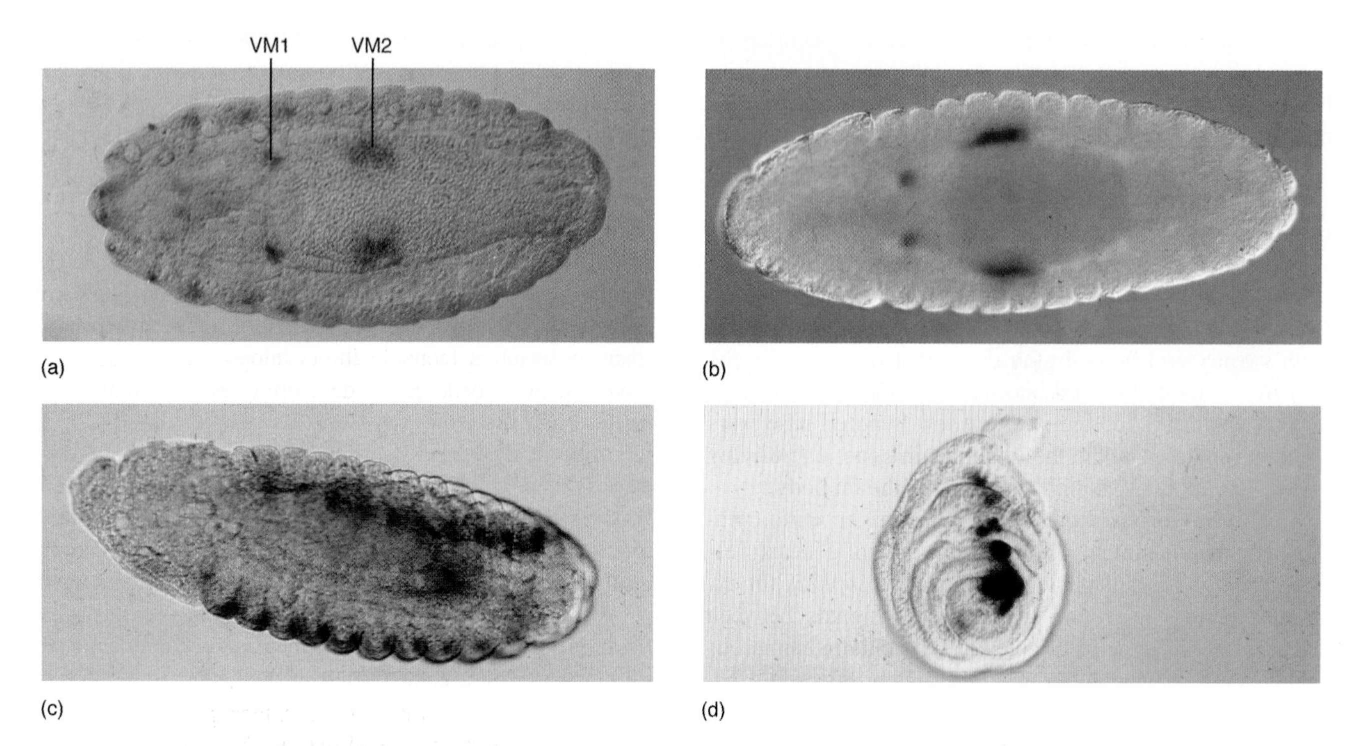

Figure 13-28 Examples of the complex tissue-specific regulation of the *dpp* gene. In parts a, c, and d, the blue staining is due to a histochemical assay for *E. coli* β-galactosidase activity (the protein encoded by the *lacZ* reporter gene). The map positions of the *dpp* enhancers responsible for the staining patterns seen here are shown in Figure 13-26. (a) Reporter-gene assay for expression of the *Drosophila dpp* gene in two parts of the embryonic visceral mesoderm, the precursor of the gut musculature. The left-hand block of blue staining resulted from reporter-gene expression activated by the enhancer VM1, which drives anterior visceral mesoderm expression; the right-hand block of staining indicates activation by the enhancer VM2, which drives posterior visceral mesoderm expression. (b) RNA in situ hybridization assay of *dpp* expression in the embryonic visceral mesoderm. Note that the blue reporter-gene expression pattern in part a is the same as the brown *dpp* RNA expression pattern shown here, confirming the reliability of the reporter-gene assay. (c) Reporter-gene expression driven by a different enhancer (LE) of *dpp* in the lateral ectoderm of an embryo. (d) Reporter-gene expression driven by ID, one of many enhancer elements driving imaginal disk expression of *dpp*. (An imaginal disk is a flat circle of cells in the larva that gives rise to one of the adult appendages.) A blue sector of *dpp* reporter-gene expression in a leg imaginal disk is shown. *(Parts a and b courtesy of D. Hursh, part c courtesy of R. W. Padgett, and part d courtesy of R. Blackman and M. Sanicola.)*

a "weak" promoter—one that cannot initiate transcription without the assistance of an enhancer. The construct is then introduced by DNA transformation into the germ line of a host organism, and appropriate cells are histochemically assayed for the presence of the reporter protein. Figure 13-28 shows assays for enhancers of the *dpp* gene in *Drosophila;* Figure 13-29, for a mouse enhancer expressed in muscle precursor cells.

Reporter-protein expression in a tissue indicates the presence of one or more enhancers within the tested piece of DNA. When a piece of DNA has been found to act as an enhancer, the enhancer can be further localized by testing smaller and smaller subfragments of the original DNA segment, using the same reporter-gene assay.

Ultimately, the DNA sequence of the enhancer can be identified by whittling down the piece of cis-regulatory DNA. With this sequence known, the next question of importance is the identity of the transcription-factor proteins that bind to the enhancer. Methods now exist to identify enhancer-binding proteins and to clone the genes that encode these proteins. When these genes have been cloned, they can be characterized by genetic and molecular techniques. With the use of these approaches, it is possible to build detailed circuit diagrams of the genetic pathways that regulate gene expression in eukaryotes.

MESSAGE

Reporter-gene techniques can be used to isolate individual regulatory elements of genes.

Regulatory Elements and Dominant Mutations

The properties of regulatory elements help us to understand the nature of gain-of-function dominant mutations (dis-

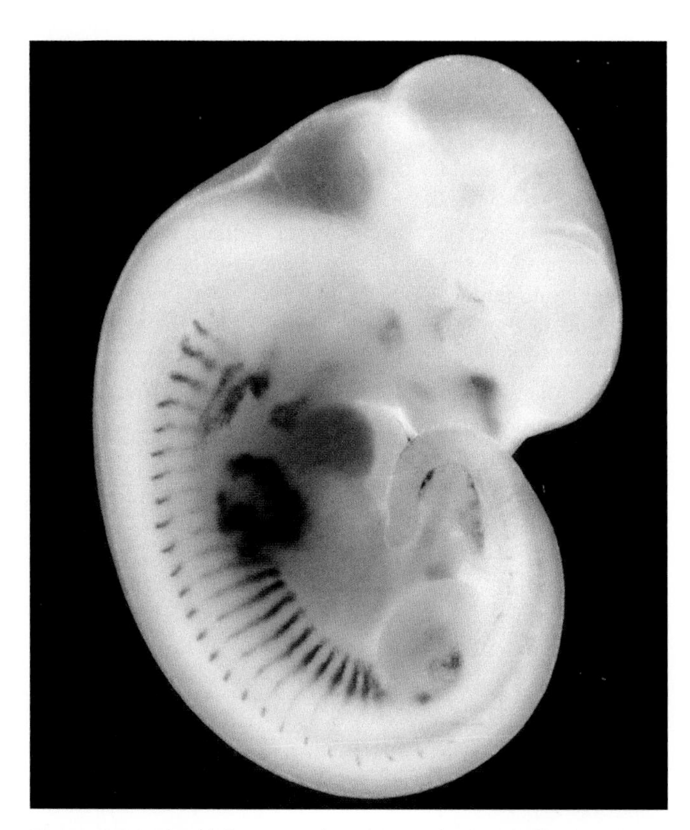

Figure 13-29 An 11.5 transgenic mouse embryo contains recombinant DNA composed of a 258-bp mouse DNA sequence fused to the *E. coli lacZ* gene, which encodes the enzyme β-galactosidase. The 258-bp mouse DNA contains all the cis-regulatory sequences necessary to direct expression of *lacZ* in muscle precursor cells, as revealed by the blue histochemical stain for β-galactosidase enzyme activity. The stained muscle precursors include the somites, limb buds, and branchial arches of the embryo. *(Reproduced from D. J. Goldhamer, B. P. Brunk, A. Faerman, A. King, M. Shani, and C. P. Emerson, Jr.,* Development *121, 1995, 644.)*

cussed in Chapter 12). Recall that *gain-of-function dominant mutations* refer to those mutations in which the dominant phenotype is due to some new property of the mutant gene, not to a reduction in its normal activity.

One way that gain-of-function dominant mutations arise is through the fusion of the regulatory elements of one gene to the structural RNA or protein-coding sequences of another. (Be aware that there are many other mechanisms for producing a gain-of-function dominant mutation.) Such fusions can occur at the breakpoints of chromosomal rearrangements such as inversions, translocations, duplications, or deletions (see Chapter 11). Because enhancers can act at long distance and can activate many different promoters, misregulation of a gene can occur if a chromosomal rearrangement juxtaposes enhancers of one gene and a transcription unit of another gene. In such cases, the enhancers of the gene at one breakpoint can now regulate the transcription of a gene near the other breakpoint. Often, this misregulation leads to the misexpression of the mRNA encoded by the transcription unit in question. Such fusions can lead to gain-of-function dominant mutant phenotypes, depending on the nature of the protein product of the misexpressed mRNA and the havoc it wreaks in the tissues in which it is misexpressed.

The classic *Bar* dominant eye-shape mutation in *Drosophila* is an example of such misregulation through gene fusion. In the *Bar* mutation, cis-regulatory elements that promote expression in the developing eye are fused to a gene that is ordinarily not expressed in the eye. This latter gene encodes a transcription factor, and misexpression of that transcription factor in the developing eye leads to the death of many cells of the developing eye and thus to the small-eye Bar phenotype.

In a few cases, the basis for the misexpression in such gene fusions is quite well understood. One such case is the *Tab (Transabdominal)* mutation in *Drosophila. Tab* causes part of the thorax of the adult fly to develop instead as tissue normally characteristic of the sixth abdominal segment (A6) (Figure 13-30). *Tab* is associated with a chromosomal inversion. One breakpoint of the inversion is within an enhancer region of a different gene, the *sr* (striped) gene. These enhancers of the *sr* gene induce gene expression in certain parts of the thorax of the fly. The other breakpoint is near the transcription unit of the *Abd-B (Abdominal-B)* gene. The *Abd-B* gene encodes a transcription factor that normally is expressed only in posterior regions of the animal, and this *Abd-B* transcription factor is responsible for conferring an abdominal phenotype on any tissues that

Figure 13-30 The *Tab* mutation. The fly on the left is a wild-type male. The fly on the right is a *Tab* / + heterozygous mutant male. In the mutant fly, part of the thorax (the black tissue) is changed into tissue normally found in the dorsal part of one of the posterior abdominal segments. The rest of the thorax is normal. *(From S. Celniker and E. B. Lewis,* Genes and Development *1, 1987, 111.)*

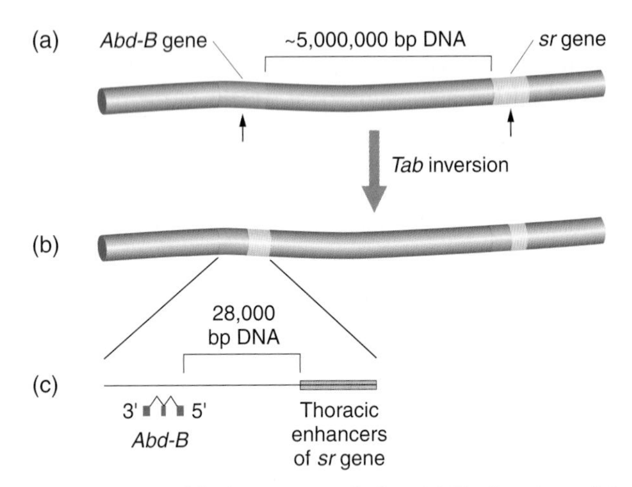

Figure 13-31 *Tab* is due to a gene fusion. (a) The locations of the *Abd-B* and *sr* genes on a map of a normal chromosome 3 are depicted. The two genes are (very) approximately 5 million base pairs apart, and there are hundreds of genes in between them. (b) The *Tab* inversion has one chromosomal breakpoint within the *Abd-B* gene and the other within *sr* (small arrows in part a). As a result of misfusion of these breakpoints, an inversion arises in which pieces of the *Abd-B* and *sr* genes are fused at either end of the inversion. (c) A magnified view of the centromere-proximal breakpoint of the inversion. The inversion breakpoints have produced a DNA molecule in which the promoter region of the *Abd-B* gene is now only 28,000 base pairs away from enhancer elements of the *sr* gene. This causes the *Abd-B* transcript to be ectopically expressed in some parts of the fly's thorax, where the wild-type *Abd-B* gene is not transcribed.

express it. (We shall have more to say about genes such as *Abd-B* in the treatment of homeotic genes in Chapter 16.) In the *Tab* inversion, the *sr* enhancer elements controlling thoracic expression are juxtaposed to the *Abd-B* transcription unit, causing the *Abd-B* gene to be activated in exactly those parts of the thorax where *sr* would ordinarily be expressed (Figure 13-31). Because of the function of the *Abd-B* transcription factor, its activation in these thoracic cells changes their fate to that of posterior abdomen. In this way, we can understand the molecular basis of a dominant mutation.

Gene fusions are an extremely important source of genetic variation. Through chromosomal rearrangements, novel patterns of gene expression can be generated. In fact, we can imagine that such fusions might play an important role in the shifts of gene expression pattern in the divergence and evolution of species (discussed in Chapter 19). In addition to affecting development (discussed in Chapter 16), such mutations can play a pivotal role in the formation and progression of many cancers (discussed in Chapter 15).

MESSAGE

The fusion of tissue-specific enhancers to genes not normally under their control can produce dominant gain-of-function mutant phenotypes.

STRUCTURE AND FUNCTION OF TRANSCRIPTION FACTORS

The activities of transcription factors in eukaryotes need to be responsive to the cellular environment, just as in prokaryotes. Many of the eukaryotic signals, particularly in multitissue higher eukaryotes, are not substrates or products of metabolic pathways, but rather are intrinsic signals synthesized by the organism itself to send global or regional signals between cells and tissues. Here, we will consider the case of global signaling—sending signals from a master command tissue (a gland) through the bloodstream to be received and interpreted in appropriate ways by the relevant target tissues. In Chapters 15 and 16, we will consider regional or nearby signals—those that take place between neighboring cells of the same tissue—in the contexts of understanding how cell numbers are locally regulated and how cell populations make complex division-of-labor decisions during development.

DNA-Binding Specificity of Regulatory Proteins

Throughout this chapter, we have seen that proteins such as Lac repressor, CAP, or TATA-binding protein are crucial to gene regulation. Such sequence-specific DNA-binding proteins are vital to transcriptional regulation in all organisms. We need to consider how DNA sequence-specific binding takes place.

Protein sequence analyses and structural comparisons indicate that DNA-binding regulatory proteins have important features in common. Many consist of a DNA-binding domain, located at one end of the protein, that protrudes from the main "core" of the protein. In certain cases, the core protein contains the allosteric effector site. This arrangement holds for the Lac repressor (Figure 13-32) and CAP, as well as for many other regulatory proteins such as

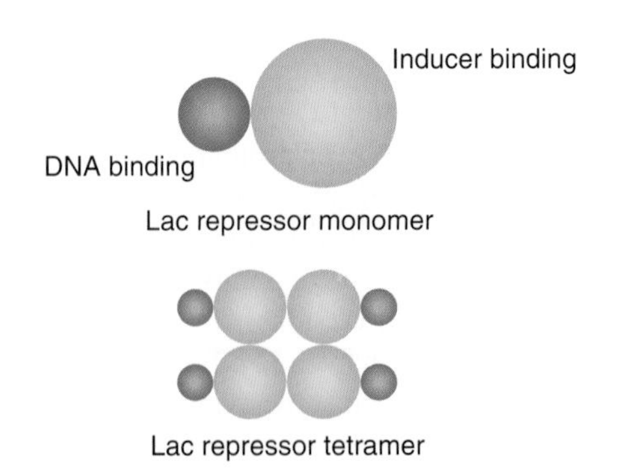

Figure 13-32 Schematic representation of the arrangement of domains in the Lac repressor. All mutations affecting DNA and operator binding result in alterations in the amino-terminal end of the protein, whereas mutants defective in inducer binding or aggregation of monomers into the tetramer result in alterations in the remainder of the protein.

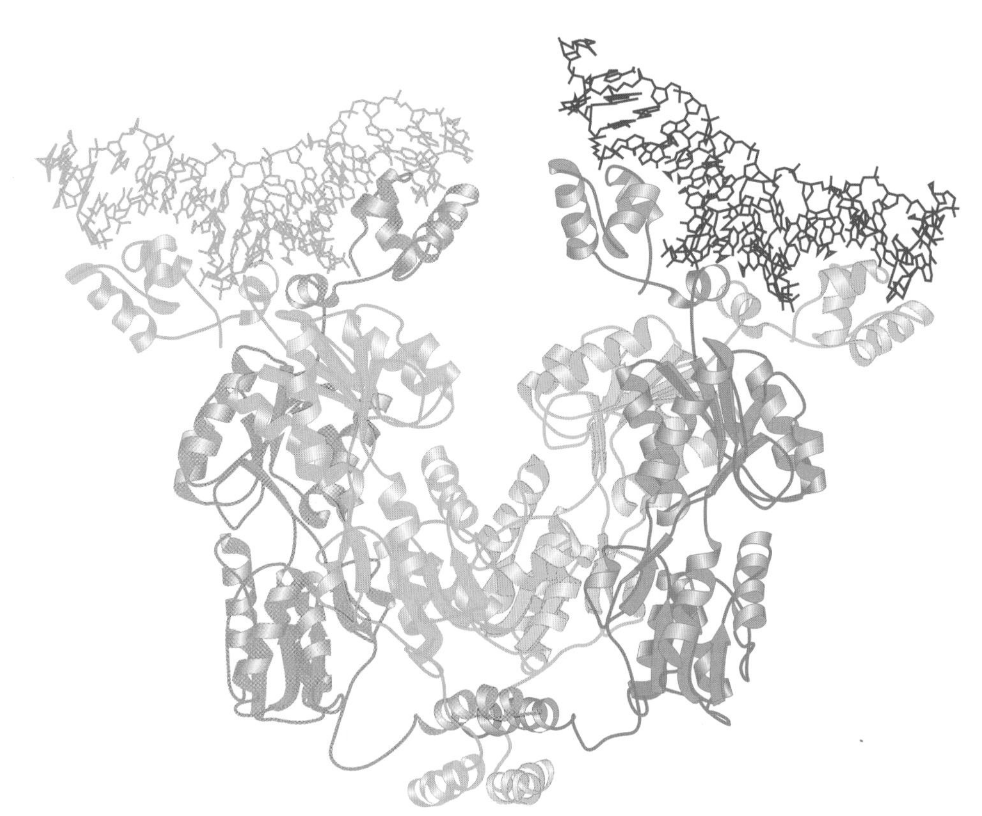

Figure 13-33 View of the Lac repressor–DNA complex, as determined by X-ray crystallography. Here the repressor tetramer is shown binding to two operators on two separate DNA molecules. Each dimer binds to one operator. The operators are 21 base pairs long here and are shown in dark and light blue. The monomers are colored green, pink, red, and yellow. Note how the amino-terminal headpiece on one monomer (pink) crosses the other monomer (green) to bind the first part of the operator, whereas the second monomer headpiece (green) crosses over to bind the second part of the operator. Thus, the recognition helix from each of two subunits binds to the consecutive parts of the same operator. *(Illustration supplied by M. Lewis, Geoffrey Chang, N. C. Horton, M. A. Kercher, H. C. Pace, M. A. Schumacher, R. G. Brennan, and P. Lu, University of Pennsylvania.)*

the steroid receptors. For such proteins, certain protruding α helices fit into the major groove of the DNA. This fit has been visualized by determining the three-dimensional structures of various protein-DNA complexes, such as the Lac repressor-operator complex (Figure 13-33). Here two α helices, each one from a different monomer of the repressor protein, interact with two consecutive turns of the major groove of the DNA at the operator site. Each helix in the major groove is connected by a turn in the protein secondary structure to another helix. This helix-turn-helix motif (Figure 13-34) is common to many regulatory proteins. However, many other DNA-binding motifs abound as well. For example, Figure 13-35 shows the structure of part of a *zinc-finger* protein, in which a zinc atom is conjugated to four amino acids [two cysteines (C) and two histidines (H)] of a small part of a polypeptide chain. Zinc-finger proteins generally have several such zinc fingers, each of which seems able to interact with a specific DNA sequence.

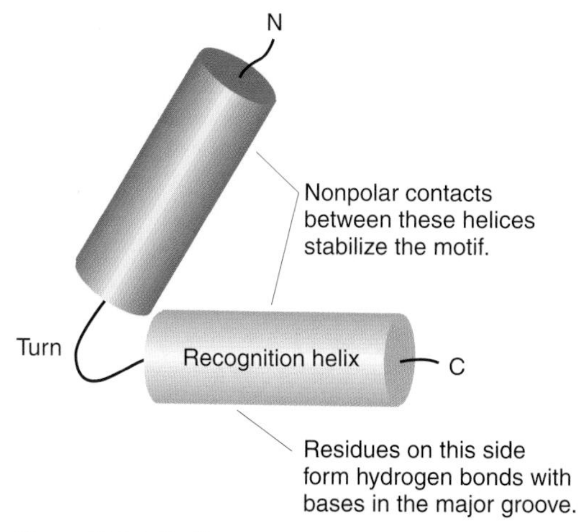

Figure 13-34 Helix-turn-helix motif of DNA-binding proteins. Each monomer of these dimeric proteins contains a helix-turn-helix; the two units are separated by 34 Å—the pitch of the DNA helix. *(From L. Stryer,* Biochemistry, *4th ed. © 1995 by Lubert Stryer.)*

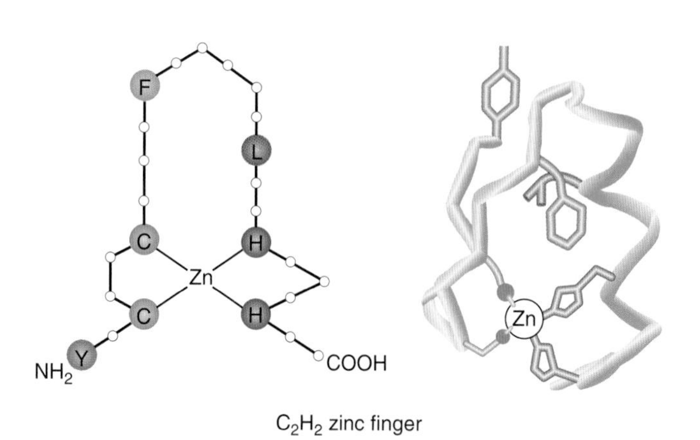

Figure 13-35 Structural model for the zinc-finger DNA-binding domain. *(From J. D. Watson, M. Gilman, J. Witkowski, and M. Zoller,* Recombinant DNA, *2d ed. © 1992 by James D. Watson, Michael Gilman, Jan Witkowski, and Mark Zoller.)*

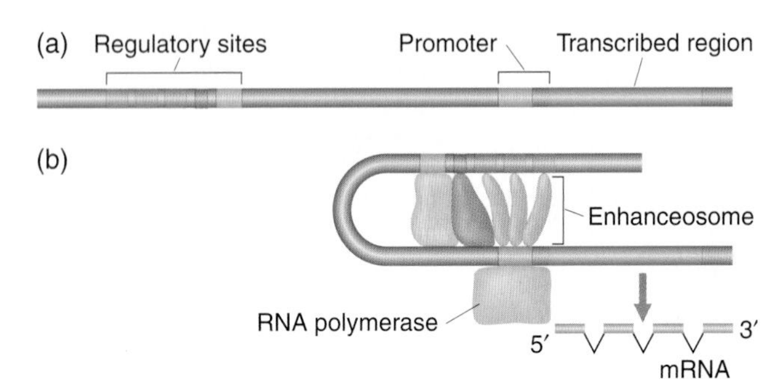

Figure 13-36 Enhanceosome binding to a cluster of transcription factor-bound enhancer elements. (a) Gene structure represented in extended linear orientation. (b) Enhanceosome interaction with enhancer elements leads to initiation of transcription.

Cooperative Interactions and DNA-Binding Activity

Frequently, there are multiple copies of the same type of docking site near one another. The reason for this appears to be that having several copies of a particular transcription factor binding to adjacent sites (sites that are the correct distance apart) can lead to an amplified, or superadditive, effect of these transcription factors on activating transcription. It is now thought that the presence of multiple bound sites can catalyze the formation of an **enhanceosome,** a large protein complex that can mediate the interactions between distant enhancers and the basal transcriptional machinery (Figure 13-36).

MESSAGE

The structures of DNA-binding proteins help us understand how they contact specific DNA sequences through polypeptide domains that fit into the major groove of the DNA double helix.

EPIGENETIC INHERITANCE

We now have a general view of transcriptional regulation that can account for most observations that geneticists have made in the past century. However, there are still some phenomena that beg for explanation. Important sets of phenomena, termed *epigenetic inheritance,* seem to be due to heritable alterations in which the DNA sequence itself is unchanged. Indeed, it is likely that these phenomena constitute another, poorly understood level of gene control. Paramutation and parental imprinting are two examples of epigenetic inheritance in which the activity state of a gene depends on its genealogical history.

Paramutation

The phenomenon of **paramutation** has been described in several plant species, most notably in corn (Figure 13-37). Paramutation was observed at only a few genes in corn. In this phenomenon, certain special but seemingly normal alleles, called *paramutable alleles,* suffer irreversible changes after having been present in the same genome as another class of special alleles, called *paramutagenic alleles.* The *B-I* gene in corn encodes an enzyme in the pathway of anthocyanin pigments in various tissues in corn. Ordinary null *b* alleles lack these pigments, and these *b* alleles are completely recessive to *B-I.* There is a special

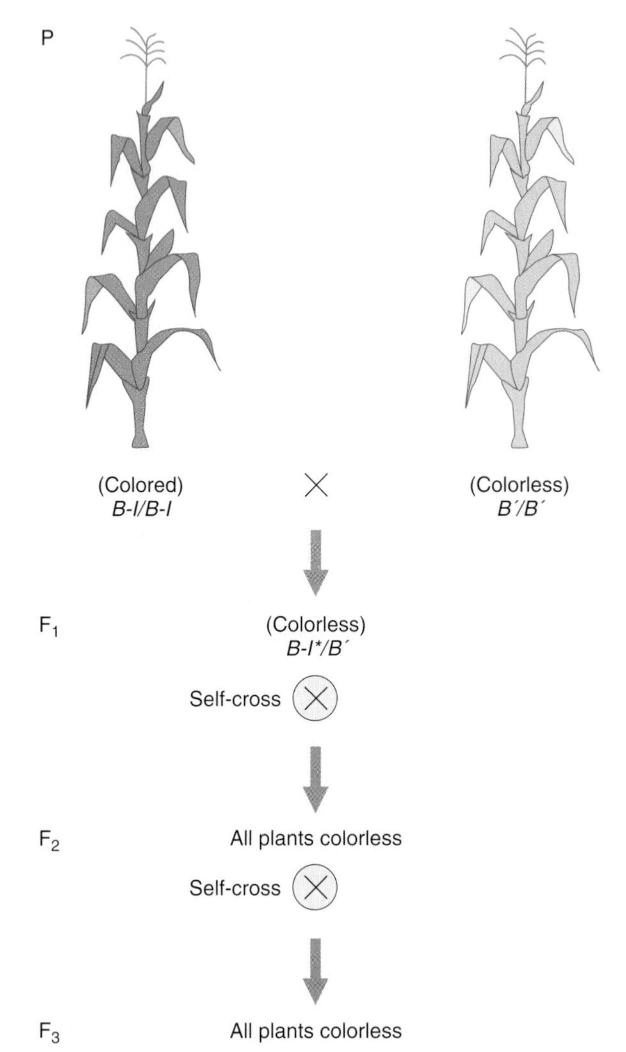

Figure 13-37 A series of crosses depicting paramutation. The *B-I* mutation produces pigmented plants, whereas the *B′* mutation produces nearly unpigmented plants. Normally, when *B-I* is crossed with recessive colorless alleles of the *b* gene, the resulting plants are pigmented. However, when *B-I* and *B′* plants are intercrossed, the F_1 plants are essentially unpigmented, like the *B′* homozygotes. Thus, *B-I* is altered by being in the same genome as *B′*, indicated by the *B-I** designation. If this outcome were due simply to the dominance of *B′* to *B-I,* then a self-cross of the F_1 plants should generate *B-I*-colored homozygotes as approximately $\frac{1}{4}$ of the F_2 progeny. Instead, no F_2 are pigmented. Intercrosses of the F_2 and of further generations do not restore the pigmented phenotype. Thus, *B-I* is said to have been paramutated by virtue of being in the same nucleus with the *B′* allele.

paramutagenic allele, called B', that confers the ability to make only a small amount of anthocyanin pigment. In crosses of *B-I* with B' homozygotes, the resulting heterozygotes are weakly pigmented, thus appearing indistinguishable from the B' homozygous plants. This result would seemingly suggest that *B-I* is recessive to B'. If this simple explanation were true, self-crosses of these heterozygous plants would generate homozygous *B-I* plants. However, instead, only B' alleles appear in the next (and subsequent) generations, indicating that the *B-I* allele has been paramutated. Somehow, by virtue of having been exposed to the paramutagenic B' allele by being in the same genotype for but a single generation, the *B-I* allele has been permanently crippled in its activity.

Parental Imprinting

Another example of epigenetic inheritance, discovered about 15 years ago in mammals, is **parental imprinting.** In parental imprinting, certain autosomal genes have seemingly unusual inheritance patterns. For example, the mouse *Igf2* gene is expressed in a mouse only if it was inherited from the mouse's father. It is said to be maternally imprinted, inasmuch as a copy of the gene derived from the mother is inactive. Conversely, the mouse *H19* gene is expressed only if it was inherited from the mother; *H19* is paternally imprinted. The consequence of parental imprinting is that imprinted genes are expressed as if they were hemizygous, even though there are two copies of each of these autosomal genes in each cell. Furthermore, when these genes are examined at the molecular level, no changes in their DNA sequences are observed. Rather, the only changes that are seen are extra methyl ($—CH_3$) groups present on certain bases of the DNA of the imprinted genes. Occasional bases of the DNA of most higher organisms are methylated (an exception being *Drosophila*). These methyl groups are enzymatically added and removed, through the action of special methylases and demethylases. The level of methylation generally correlates with the transcriptional state of a gene: active genes are less methylated than inactive genes. However, whether altered levels of DNA methylation cause epigenetic changes in gene activity or whether altered methylation levels arise as a consequence of such changes is unknown.

Note that parental imprinting can have profound effects on pedigree analysis. Because the allele inherited from one parent is inactive, a mutation in the allele inherited from the other parent will appear to be dominant, whereas it is really displaying a hemizygous phenotype—not due to the presence of only one chromosome (as would be true for the mammalian X chromosome in males), but because only one of the two homologs is active for this gene. Genome-wide screens for imprinted genes suggest that there may be a hundred or so such parentally imprinted autosomal genes in the genome. Some human genetic diseases, such as Prader-Willi syndrome (PWS), are due to parentally imprinted genes. In PWS, patients have short stature, mild mental retardation, and poor muscle tone, and they are compulsive eaters. The gene, located on human chromosome 15, is maternally imprinted such that the maternal copy of the gene is inactive (Figure 13-38). A minority of PWS individuals turn out to have both chromosome 15s derived maternally,

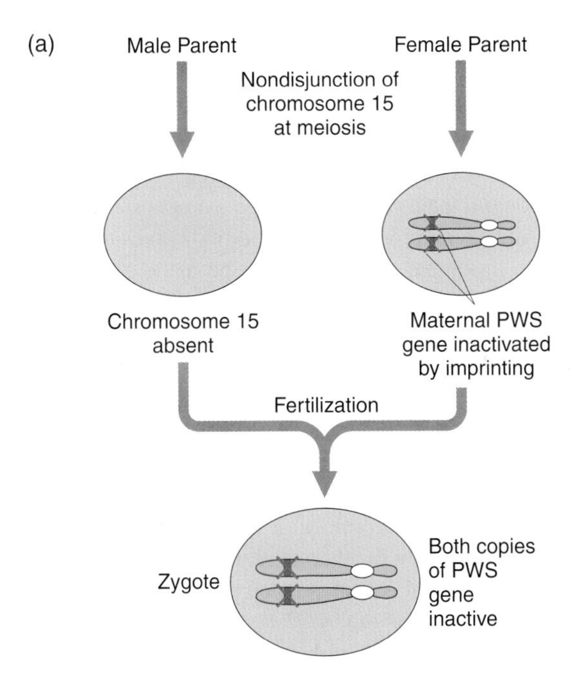

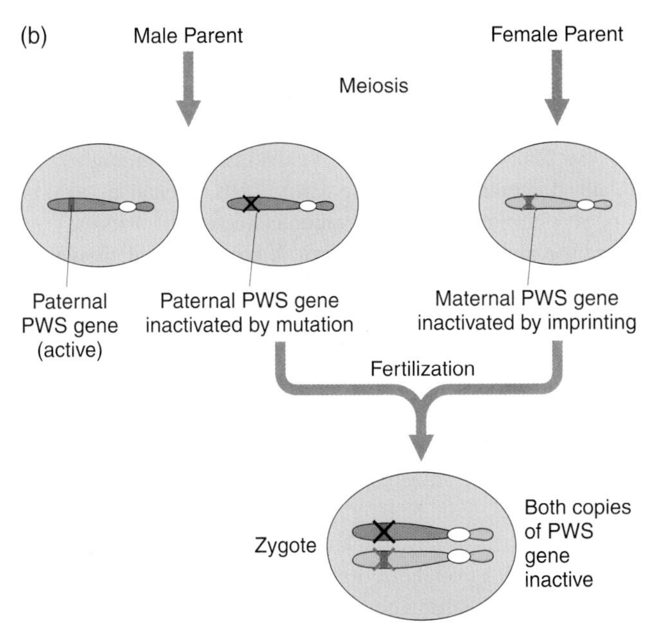

Figure 13-38 The genetic origin of Prader-Willi syndrome (PWS). (a) PWS caused by nondisjunction and maternal inheritance of two imprinted chromosome 15s. (b) PWS caused by mutation of the paternally derived PWS gene and imprinting of the maternally derived PWS allele.

owing to chromosome 15 nondisjunction in each sex; homozygous inactivation of the maternal PWS genes through imprinting then leads to PWS. In most cases, however, PWS is associated with mutational inactivation of the paternal gene copy. Most PWS mutations turn out to be deletions, and while the entire region shared by these deletions has been sequenced, it is not known which gene (if it is just one) is responsible for PWS itself. It is clear, however, that the syndrome is a loss-of-function phenotype, with inactivation of the PWS gene or genes because of a combination of mutation and imprinting.

X-Inactivation in Mammals

In Chapter 11, we discussed the phenomenon of dosage compensation in mammals. In dosage compensation, the amount of most gene products from the two copies of the X chromosome in female mammals must be made equivalent to the single dose of the X chromosome in males. This is accomplished by inactivating one of the two X chromosomes in each cell at an early stage in development, and then propagating this inactive state to all progeny cells. (In the germ line, the second X becomes reactivated during oogenesis.)

Two aspects of this phenomenon reflect imprinting (Figure 13-39a). First, the selection of the X chromosome is not always random. In protherian mammals (marsupials), it is always the paternal X that is inactivated—that is, there is paternal imprinting of the X. Paternal X inactivation is true in some cell types in eutherian mammals (all other mammals) as well—the cells that form the amnion, chorion, and placenta, the extraembryonic membranes of the embryo. Only in the eutherian mammalian embryo proper is the decision made randomly, at about the 64-cell stage of development.

Second, as we have said, the decision is "remembered" in all descendants of these early cells. Because of this clonal inheritance, large continuous sectors of a tissue can share the inactivation of the same X chromosome, as is true for calico cats (Figure 13-40). Part of the memory system is methylation, with the inactive X being methylated to a much greater extent than the active X. Another correlation with X-inactivation is the lower level of acetylation of a special lysine residue in one of the histone subunits on the inactive X. A special RNA, called **Xist,** is synthesized by the inactive X chromosome and is thought to act as an inactivation center. This RNA appears to spread from this center, where it is synthesized, and is found along the length of the inactive X throughout development (see Figure 13-39b). While there is a strong correlation between inactivation and the presence of this special RNA, which does not appear to encode any protein, how it might contribute to the inactivation process remains unclear. We should note that not all of the X is inactivated, however. The pseudoautosomal region contains several genes that remain active on both chromosomes. In addition, some genes interspersed among the inactive blocks also somehow escape inactivation. How any of the molecular events—methylation, histone acetylation,

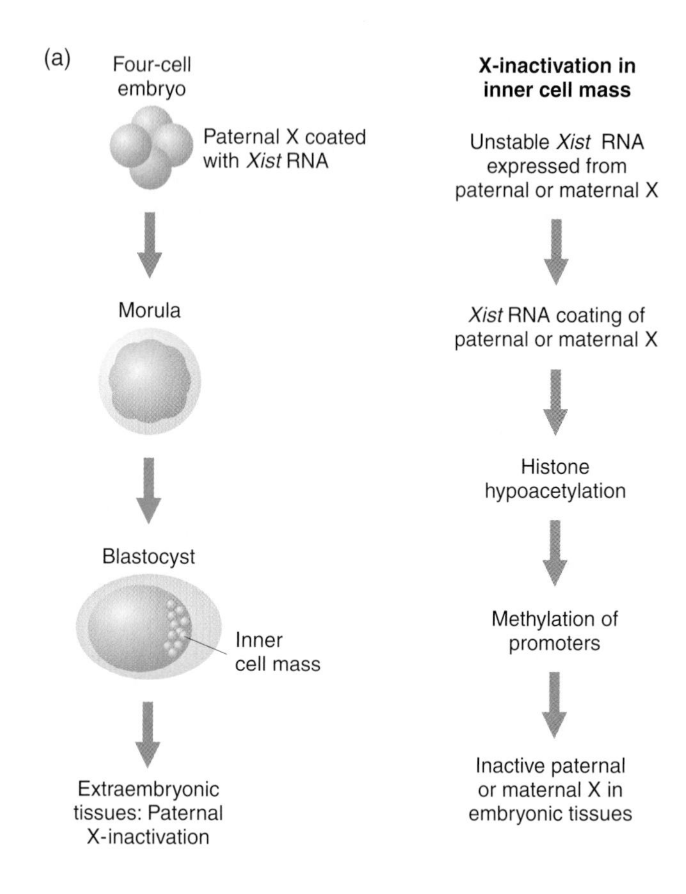

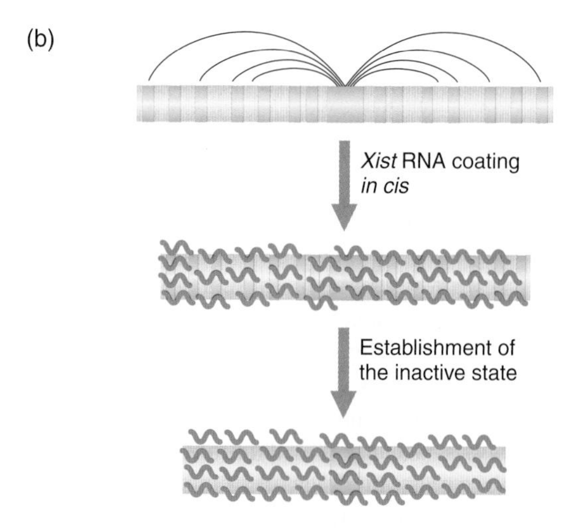

Figure 13-39 (a) Imprinting during X-inactivation. In the cells that form the extraembryonic membranes, the paternally derived X is always inactivated. In cells of the inner cell mass, that form the embryo itself, the maternal or paternal X will be randomly inactivated in each XX cell. (b) *Xist* spreads from its site of transcription in the middle of the X chromosome, leading to establishment of X-inactivation. *(From P. Avner and E. Heard,* Nature Reviews: Genetics *2, 2001, 64–65.)*

Figure 13-40 A calico cat. *(Anthony Griffiths.)*

Xist RNA localization—contribute to X inactivation remains to be determined.

What do these examples of epigenetic inheritance have in common? The main thread is that, somehow, a piece of a chromosome can be labeled as different on the basis of its ancestry or depending on which other genes were in the same genome. For many of these examples, differences in DNA methylation have been associated with differences in gene activity. Nonetheless, the underlying mechanisms and rationales for why such systems evolved still seem rather mysterious.

TRANSCRIPTIONAL REGULATION IN THE ERA OF GENOMICS

We have seen in Chapter 9 that eukaryotic genomes contain some tens of thousands of genes. However, much of the genome does not code for protein. (Only about 5 percent of the human genome does!) Some unknown fraction of the rest consists of the instruction manual of the genome—enhancers and other regulatory elements. How can we identify these elements on a genome-wide basis and learn to "read" the information they encode?

One approach combines comparative genomics with expression studies. The general idea is that the sequencing of related genomes (such as human and mouse) will identify conserved sequence elements within noncoding regions, and these conserved sequences are candidates to be regulatory elements. However, we know that some conservation may be by chance. How can we recognize which ones are meaningful? Perhaps correlation of these conservations with transcriptional expression arrays (Chapter 9), in which RNA expression of each of the genes of the genome can be assayed, will help. The idea is to computationally cluster those genes that display similar expression profiles in a given cell type, and then to identify conservations among their noncoding sequences (Figure 13-41).

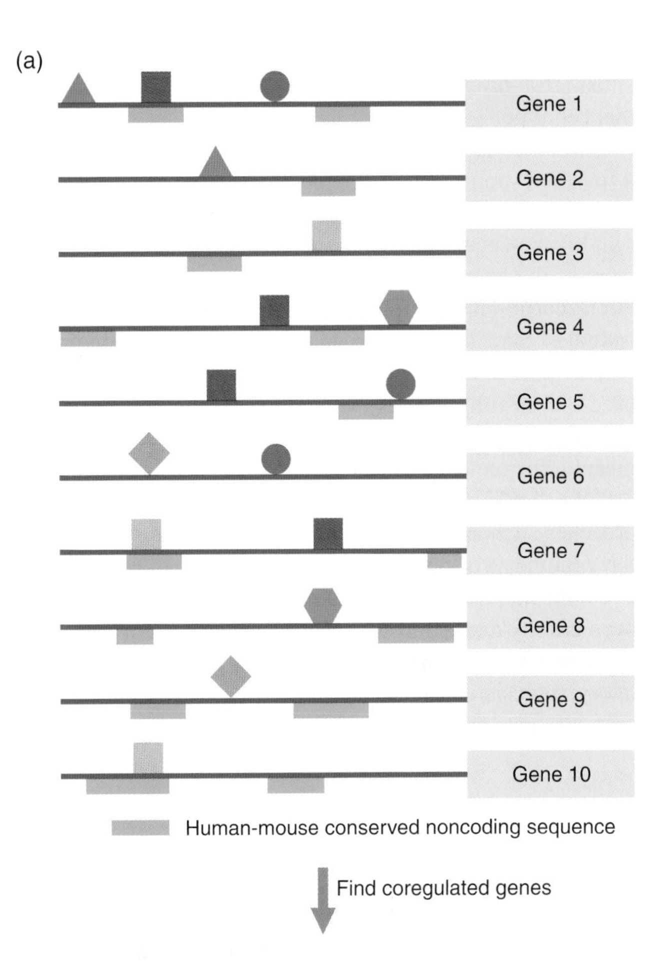

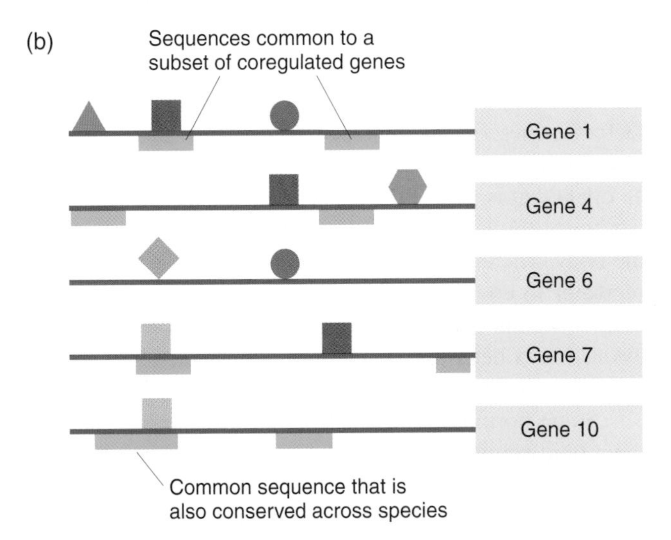

Figure 13-41 Correlating expression array and comparative genomic data. Conserved noncoding regions between the equivalent genes in human and mouse (termed **orthologs**) are identified (each marked by a different geometric symbol). Of these, a subset of the genes (gene 1, 4, 6, 7, and 10) are found to be coregulated in a particular tissue. If they have certain conserved noncoding regions in common, these regions are candidates to be regulatory elements controlling the shared expression pattern. *(From L. Pennacchio and E. Rubin,* Nature Reviews: Genetics *2, 2001, 106.)*

These conservations would then be candidate regulatory elements for driving the observed patterns of expression in that cell type.

Another approach is to measure directly the binding of a purified transcription factor to all sequences of the genome. In principle, microarrays or DNA chips can represent all the sequences of the genome, or at least of a segment of the genome (depending on the genome's size). The fluorescence-labeled transcription factor can then be allowed to bind to the microarray or chip under conditions where specific binding is required. This approach has been used successfully in yeast, where the sites of binding can be correlated with clustered expression array data to identify meaningful sites of binding.

We don't yet know if these approaches, or other approaches such as computational analysis of the sequence of the genome, will be the ones that enable us to read the instruction manual of the genome. A strong foundation has been cast by dissections of specific gene regulatory regions. Moving forward from this foundation, genome-wide regulatory analysis is a major area of current research, and we can anticipate a great deal of progress in the next few years.

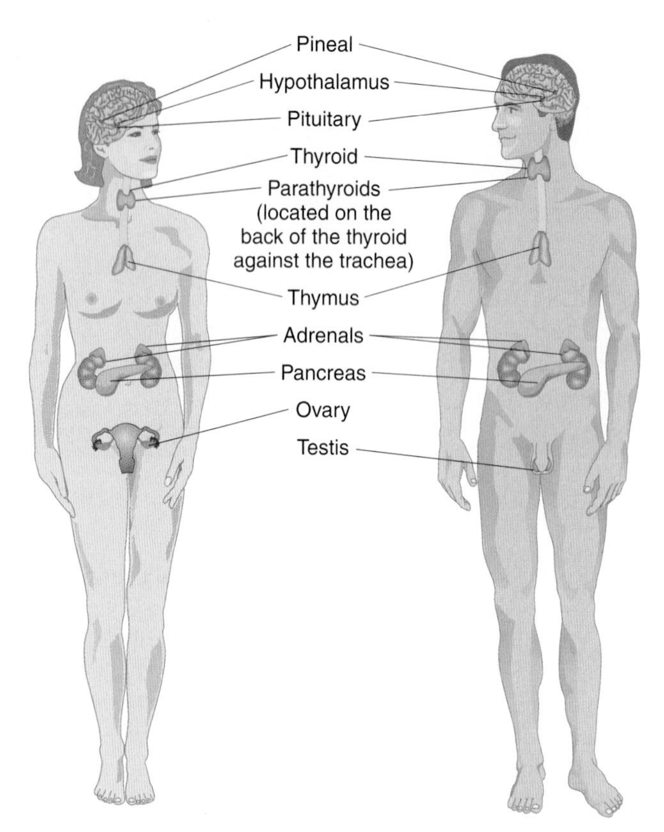

Figure 13-42 The human endocrine organs. *(Reprinted from W. K. Purves, G. H. Orians, and H. C. Heller,* Life, The Science of Biology, *4th ed. Sinauer Associates, Inc., and W. H. Freeman and Company, 1995.)*

ENDOCRINE REGULATION OF TRANSCRIPTION FACTOR ACTIVITY

Just as it was crucial in prokaryotes to wed the activity of transcriptional regulatory proteins to the physiological state of the bacterium, the tie between transcriptional regulation and physiology is crucial in eukaryotes as well. Sometimes, the regulatory signal that activates eukaryotic transcription factors comes from a very distant source in the body. For instance, hormones released into the circulatory system by an organ that is part of the endocrine system (Figure 13-42) can travel through the circulation to essentially all parts of the body. The endocrine system can thus serve as a master regulator to coordinate changes in transcription in cells of many different tissues. This mechanism underlies the basic events in sex determination in mammals, which will be discussed in Chapter 16.

Some hormones are small molecules that, because of their lipid-solubility properties, can pass directly through the plasma membrane of the cell—examples are various steroid hormones, such as glucocorticoids, testosterone, and estrogen. Once in the cell, steroid hormones bind to and regulate specific transcription factors in the nucleus. In this respect, we can think of steroids as being analogous to allosteric effectors regulating some bacterial operons.

One example is the female sex hormone estrogen. In chicken oviducts, the egg white protein ovalbumin is specifically synthesized in response to estrogen, which causes increased transcription of the ovalbumin gene. The estrogen molecule activates transcription by binding to a

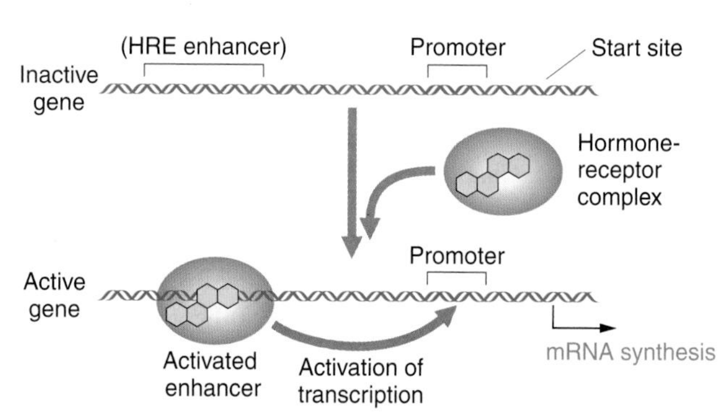

Figure 13-43 The action of steroid hormones at enhancer sequences. A steroid hormone (yellow) binds to a soluble receptor protein (blue). This complex, in turn, binds to enhancer sequences and enables them to stimulate the transcription of hormone-responsive genes. *(From L. Stryer,* Biochemistry, *4th ed. © 1995 by Lubert Stryer.)*

protein receptor molecule that first recognizes the estrogen molecule in the cytoplasm and transports it to the nucleus. The receptor molecule is a transcription factor that then binds to the DNA at an enhancer site called an *HRE* (hormone-response element). The general process of steroid-receptor activation is depicted in Figure 13-43. The anabolic steroids discussed at the beginning of this chapter are another example.

MESSAGE

Just as in prokaryotes, eukaryotic transcriptional regulatory proteins must contain domains that interact with molecular signals of the physiological state of the cell.

Let's return now to our initial discussion of anabolic steroids as performance-enhancing drugs. By analogy with other steroids, the anabolic steroids act through specific nuclear hormone receptors to alter the transcriptional activity of target genes. For the anabolic steroids, among the genes that are activated are genes that contribute to muscular development—hence, the increase in muscle mass associated with taking these drugs. We can also see why chronic use of these drugs is dangerous. If taken over long periods of time, a drug that has broad effects on gene expression will alter gene expression in ways that run counter to the natural regulatory mechanisms of the body. While the muscle mass effects are desirable for an athlete, undoubtedly there are many other genes being overexpressed in many different tissues as a result of steroid treatment. A variety of health problems, including higher cancer rates, are a consequence of abuse of this drug.

A general take-home lesson is that any drug needs to be understood in terms of its effects on the normal biology of the individual, in addition to its effects on the target of interest. This is why part of drug testing is to assess general toxicity and side effects. One of the challenges of pharmacology is to identify truly specific drugs that interact with only one molecular species within an individual.

SUMMARY

1. Gene regulation is often mediated by proteins that react to environmental signals by raising or lowering the transcription rates of specific genes. The logic of this is straightforward. In order for regulation to operate appropriately, the regulatory proteins have to have built-in sensors that continually monitor cellular conditions. The activities of these proteins would then be dependent on the right set of environmental conditions.
2. In prokaryotes, coordinate gene control is achieved by clustering coordinated structural genes together into operons on the chromosome so that they are transcribed into multigenic mRNAs. This simplifies the task for bacteria. Rather than evolving mechanisms by which parallel regulation is achieved (for example, having the same regulatory element next to every coding sequence), one cluster of regulatory sites per operon is sufficient to regulate expression of all the operon's genes.
3. Negative regulatory control is exemplified by the *lac* system, in which a repressor protein blocks transcription by binding to DNA at a site termed the *operator.* Since the *lac* system requires that the genes be shut down in the absence of appropriate sugars in the environment, negative regulation is one very straightforward way to achieve that.
4. Positive regulatory control requires protein factors to activate transcription. Some prokaryotic gene control, such as for catabolite repression, and many eukaryotic gene regulatory events operate through positive gene control.
5. Many regulatory proteins have common structural features, such that they fall into families of proteins that share very similar DNA-binding motifs. Other parts of the proteins, such as their protein-protein interaction domains, tend to be less similar.
6. In eukaryotes, regulatory sites on DNA, such as enhancers, can act at considerable distance from the transcription start site to modulate gene expression by interacting with specific regulatory proteins. The distance- and orientation-independent action of enhancer elements has permitted evolution of large, highly fragmented genes with complex arrays of regulatory proteins to drive complex patterns of tissue- and stage-specific gene expression.
7. Some regulation occurs at the epigenetic level, in which gene expression is modulated according to the lineage ancestry of the gene. In one class of epigenetic regulation, called *paramutation,* cellular mechanisms permanently modify one allele's ability to be expressed if it was present in a nucleus with another specific allele of that gene. In the classes of epigenetic regulatory phenomena that are known as *parental imprinting* and *X-inactivation,* cellular mechanisms can distinguish whether a gene has been introduced from the father or the mother. At present, the mechanisms underlying these phenomena are very incompletely understood.
8. At present, we can identify regulatory elements only by experimentation. One goal of genomics research is to enable researchers to globally identify and interpret the regulatory information encoded in an organism's DNA. Approaches currently underway involve expression microarrays and microarray assays of transcription-factor binding, combined with comparative genomics.

CONCEPT MAP

Draw a concept map interrelating as many of the following terms as possible. Note that the terms are listed in no particular order.

environment / promoter / operator / gene / operon / RNA polymerase / mRNA / enhancer / trans-acting factors / regulation / steroids / reporter genes

SOLVED PROBLEMS

This set of four solved problems, which are similar to Problem 4 in the Basic Problems at the end of this chapter, is designed to test understanding of the operon model. Here we are given several diploids and are asked to determine whether Z and Y gene products are made in the presence or absence of an inducer. Use a table similar to the one in Problem 4 as a basis for your answers, except that the column headings will be as follows:

	Z GENE		*Y* GENE	
Genotype	No inducer	Inducer	No inducer	Inducer

1. $\dfrac{I^- \; P^- \; O^C \; Z^+ \; Y^+}{I^+ \; P^+ \; O^+ \; Z^- \; Y^-}$

Solution

One way to approach these problems is first to consider each chromosome separately and then to construct a diagram. The following illustration diagrams this diploid:

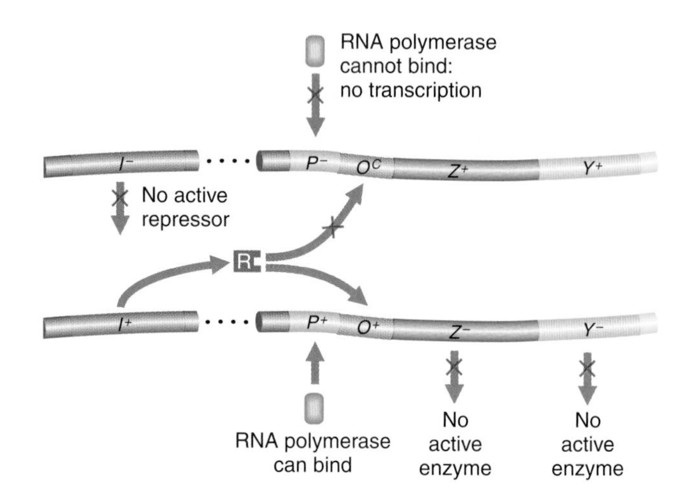

The first chromosome is P^-, so transcription is blocked and no Lac enzyme can be synthesized from it. The second chromosome (P^+) can be transcribed, and thus transcription is repressible (O^+). However, the structural genes linked to the good promoter are defective; thus, no active Z product or Y product can be produced. The symbols to add to your table are "−, −, −, −."

SOLVED PROBLEM

2. $\dfrac{I^+ \; P^- \; O^+ \; Z^+ \; Y^+}{I^- \; P^+ \; O^+ \; Z^+ \; Y^-}$

Solution

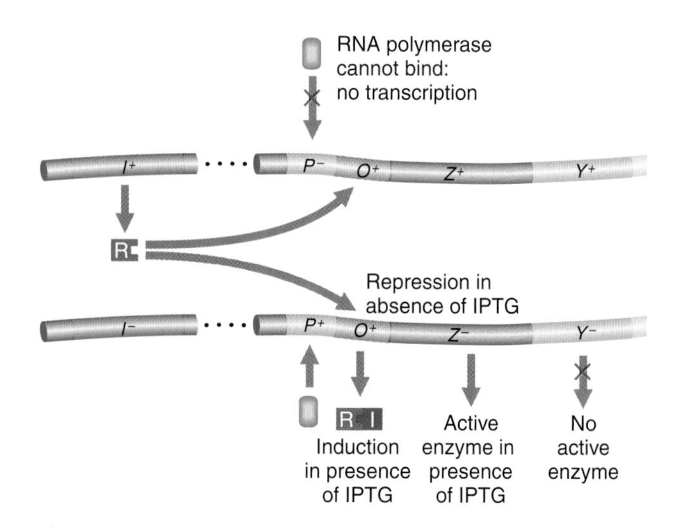

The first chromosome is P^-, so no enzyme can be synthesized from it. The second chromosome is O^+, so transcription will be repressed by the repressor supplied from the first chromosome, which can act in trans through the cytoplasm. However, only the Z gene from this chromosome is intact. Therefore, in the absence of an inducer, no enzyme will be made; in the presence of an inducer, only the Z gene product, β-galactosidase, will be produced. The symbols to add to the table are "−, +, −, −."

SOLVED PROBLEM

3. $\dfrac{I^+ \; P^+ \; O^C \; Z^- \; Y^+}{I^+ \; P^- \; O^+ \; Z^+ \; Y^-}$

Solution

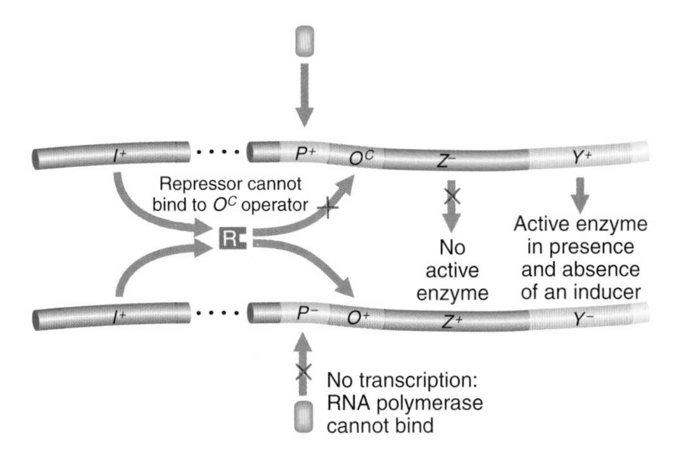

Because the second chromosome is P^-, we need consider only the first chromosome. This chromosome is O^C, so enzyme is made in the absence of an inducer, although only the *Y* gene is active. The entries in the table should be "−, −, +, +."

SOLVED PROBLEM

4. $\dfrac{I^S \; P^+ \; O^+ \; Z^+ \; Y^-}{I^- \; P^+ \; O^C \; Z^- \; Y^+}$

Solution

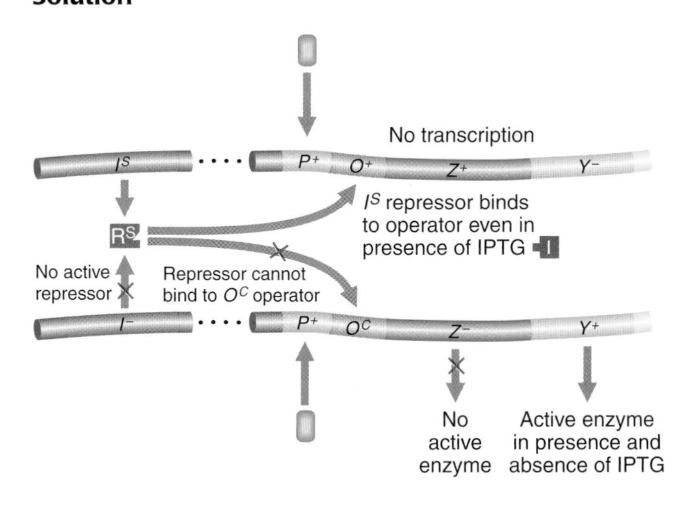

In the presence of an I^S repressor, all wild-type operators will be shut off, both with and without an inducer. Therefore, the first chromosome will be unable to produce any enzyme. However, the second chromosome has an altered (O^C) operator and can produce enzyme in both the absence and the presence of an inducer. Only the *Y* gene is active on this chromosome, so the entries in the table should be "−, −, +, +."

PROBLEMS

Basic Problems

1. Explain why I^- mutations in the *lac* system are normally recessive to I^+ mutations and why I^+ mutations are recessive to I^S mutations.

2. What do we mean when we say that O^C mutations in the *lac* system are cis-acting?

3. The genes shown in the following table are from the *lac* operon system of *E. coli.* The symbols *a*, *b*, and *c* represent the repressor *(I)* gene, the operator *(O)* region, and the structural gene *(Z)* for β-galactosidase, although not necessarily in that order. Furthermore, the order in which the symbols are written in the genotypes is not necessarily the actual sequence in the *lac* operon.

	ACTIVITY (+) OR INACTIVITY (−) OF Z GENE	
Genotype	Inducer absent	Inducer present
$a^- b^+ c^+$	+	+
$a^+ b^+ c^-$	+	+
$a^+ b^- c^-$	−	−
$a^+ b^- c^+ / a^- b^+ c^-$	+	+
$a^+ b^+ c^+ / a^- b^- c^-$	−	+
$a^+ b^+ c^- / a^- b^- c^+$	−	+
$a^- b^+ c^+ / a^+ b^- c^-$	+	+

a. State which symbol (*a*, *b*, or *c*) represents each of the *lac* genes *I*, *O*, and *Z*.

b. In the table, a superscript minus sign on a gene symbol merely indicates a mutant, but you know that some mutant behaviors in this system are given special mutant designations. Use the conventional gene symbols for the *lac* operon to designate each genotype in the table.

(Problem 3 is from J. Kuspira and G. W. Walker, *Genetics: Questions and Problems.* © 1973 by McGraw-Hill.)

4. The map of the *lac* operon is

I P O Z Y

The promoter *(P)* region is the start site of transcription through the binding of the RNA polymerase molecule before actual mRNA production. Mutationally altered promoters (P^-) apparently cannot bind the RNA polymerase molecule. Certain predictions can be made about the effect of P^- mutations. Use your predictions and your knowledge of the lactose system to complete the following table. Insert a "+" where an enzyme is produced and a "−" where no enzyme is produced. The first one has been done as an example.

	β-GALACTOSIDASE		PERMEASE	
Genotype	No lactose	Lactose	No lactose	Lactose
$I^+ P^+ O^+ Z^+ Y^+ / I^+ P^+ O^+ Z^+ Y^+$	−	+	−	+
a. $I^- P^+ O^C Z^+ Y^- / I^+ P^+ O^+ Z^- Y^+$				
b. $I^+ P^- O^C Z^- Y^+ / I^- P^+ O^C Z^+ Y^-$				
c. $I^S P^+ O^+ Z^+ Y^- / I^+ P^+ O^+ Z^- Y^+$				
d. $I^S P^+ O^+ Z^+ Y^+ / I^- P^+ O^+ Z^+ Y^+$				
e. $I^- P^+ O^C Z^+ Y^- / I^- P^+ O^+ Z^- Y^+$				
f. $I^- P^- O^+ Z^+ Y^+ / I^- P^+ O^C Z^+ Y^-$				
g. $I^+ P^+ O^+ Z^- Y^+ / I^- P^+ O^+ Z^+ Y^-$				

5. Explain the fundamental differences between negative control and positive control.

6. Mutants that are *lacY*$^-$ retain the capacity to synthesize β-galactosidase. However, even though the *lacI* gene is still intact, β-galactosidase can no longer be induced by adding lactose to the medium. How can you explain this?

7. What analogies can you draw between transcriptional trans-acting factors that activate gene expression in eukaryotes and the corresponding factors in prokaryotes? Give an example.

8. Compare the arrangement of cis-acting sites in the control regions of eukaryotes and prokaryotes.

a. $R^+ O^+ A^+$

b. $R^- O^+ A^+ / R^+ O^+ A^-$

c. $R^+ O^- A^+ / R^+ O^+ A^-$

9. What is meant by the term *epigenetic inheritance?* What are two examples of such inheritance?

10. Explain how models for bacterial operons such as *ara* relate to eukaryotic trans-acting proteins and their mechanism of action.

Challenging Problems

11. One interesting mutation in *lacI* results in repressors with 100-fold increased binding to both operator and nonoperator DNA. These repressors display a "reverse" induction curve, allowing β-galactosidase synthesis in the absence of an inducer (IPTG) but partly repressing β-galactosidase expression in the presence of IPTG. How can you explain this? (Note that when IPTG binds a repressor, it does not completely destroy operator affinity, but rather it reduces affinity 1000-fold. Additionally, as cells divide and new operators are generated by the synthesis of daughter strands, the repressor must find the new operators by searching along the DNA, rapidly binding to and dissociating from nonoperator sequences.)

12. You are studying a mouse gene that is expressed in the kidneys of male mice. You have already cloned this gene. Now you wish to identify the segments of DNA that control the tissue-specific and sex-specific expression of this gene. Describe an experimental approach that would allow you to do so.

13. In *Neurospora,* all mutants affecting the enzymes carbamyl phosphate synthetase and aspartate transcarbamylase map at the *pyr*-3 locus. If you induce *pyr*-3 mutations by ICR-170 (a chemical mutagen), you find that either both enzyme functions are lacking or only the transcarbamylase function is lacking; in no case is the synthetase activity lacking when the transcarbamylase activity is present. (ICR-170 is assumed to induce frameshifts.) Interpret these results in regard to a possible operon.

14. Certain *lacI* mutations eliminate operator binding by the Lac repressor but do not affect the aggregation of subunits to make a tetramer, the active form of the repressor. These mutations are partially dominant to wild type. Can you explain the partially dominant I^- phenotype of the I^- / I^+ heterodiploids?

15. You are examining the regulation of the lactose operon in the bacterium *Escherichia coli.* You isolate seven new independent mutant strains that lack the products of all three structural genes. You suspect that some of these mutations are $lacI^S$ mutations and that other mutations are alterations that prevent binding of RNA polymerase to the promoter region. Using whatever haploid and partial diploid genotypes you think are necessary, describe a set of genotypes that will permit you to distinguish between the *lacI* and *lacP* classes of uninducible mutations.

16. You are studying the properties of a new kind of regulatory mutation of the lactose operon. This mutation, called *S*, leads to the complete repression of the *lacZ, lacY,* and *lacA* genes, regardless of whether or not inducer (lactose) is present. Studies of this mutation in partial diploids demonstrate that this mutation is completely dominant to wild type. When you treat bacteria of the *S* mutant strain with a mutagen and select for mutant bacteria that can express the enzymes encoded by *lacZ, lacy,* and *lacA* genes in the presence of lactose, some of the mutations map to the *lac* operator region and others to the *lac* repressor gene. Based on your knowledge of the lactose operon, provide a molecular genetic explanation for all these properties of the *S* mutation. Include an explanation of the constitutive nature of the "reverse mutations."

17. The *trp* operon in *E. coli* encodes enzymes essential for the biosynthesis of tryptophan. The general mechanism for controlling the *trp* operon is similar to that observed with the *lac* operon: when the repressor binds to the operator, transcription is prevented; when the repressor does not bind the operator, transcription proceeds. The regulation of the *trp* operon differs from the regulation of the *lac* operon in the following way: the enzymes encoded by the *trp* operon are not synthesized when tryptophan is present but, rather, when it is absent. In the *trp* operon the repressor has two binding sites: one for DNA and one for the effector molecule, tryptophan. The *trp* repressor must first bind to a molecule of tryptophan before it can bind effectively to the *trp* operator.

a. Draw a map of the tryptophan operon, indicating the promoter (*p*), operator (*o*), and the first structural gene of the tryptophan operon (*trpA*). In your drawing, indicate where on the DNA the repressor protein binds when it is bound to tryptophan.

b. The *trpR* gene encodes the repressor; *trpO* is the operator; *trpA* encodes the enzyme tryptophan synthetase. A $trpR^-$ repressor cannot bind tryptophan; a $trpO^-$ operator cannot be bound by the repressor; and the enzyme encoded by a $trpA^-$ mutant gene is completely inactive. Do you expect to find active tryptophan synthetase in each of the following mutant strains when the cells are grown in the presence of tryptophan? In its absence?

i. $R^+ \ O^+ \ A^+$ (wild type)

ii. $R^- \ O^+ \ A^+ \ / \ R^+ \ O^+ \ A^-$

iii. $R^+ \ O^- \ A^+ \ / \ R^+ \ O^+ \ A^-$

18. The activity of the enzyme β-galactosidase produced by wild-type cells grown in media supplemented with different carbon sources is measured. In relative units, the following is found:

Glucose	Lactose	Lactose + glucose
0	100	1

Predict the relative levels of β-galactosidase activity in cells grown under similar conditions when the cells are $lacI^-$; $lacI^S$; $lacO^-$; and crp^-.

19. The diagram below represents the structure of a gene in *Drosophila melanogaster;* blue segments are exons, and yellow segments are introns.

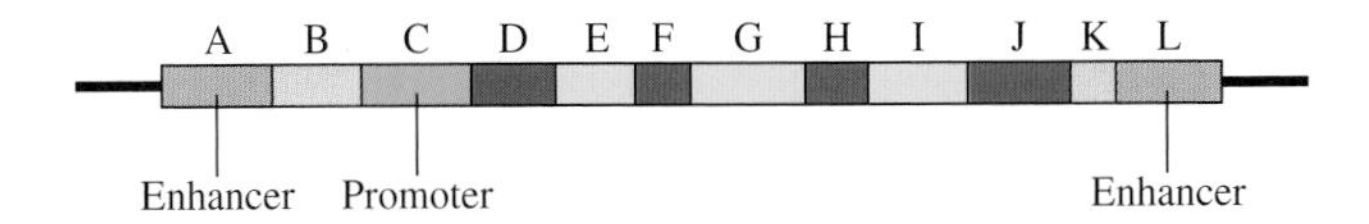

a. Which segments of the gene will be represented in the initial RNA transcript?

b. Which segments of the gene will be removed by RNA splicing?

c. Which segments would most likely bind proteins that interact with RNA polymerase?

20. You wish to find the cis-regulatory DNA elements responsible for the transcriptional responses of two genes, *c-fos* and *globin.* Transcription of the *c-fos* gene is activated in response to fibroblast growth factor (FGF), but it is inhibited by cortisol (Cort). On the other hand, transcription of the *globin* gene is not affected by either FGF or cortisol, but it is stimulated by the hormone erythropoietin (EP). To find the cis-regulatory DNA elements responsible for these transcriptional responses, you use the following clones of the *c-fos* and *globin* genes, as well as two "hybrid" combinations (fusion genes), as diagrammed below. A is the intact *c-fos* gene, D is the intact *globin* gene, and B and C are the *c-fos/globin* gene fusions. The c-fos and globin exons (E) and introns (I) are numbered. For example, E3(f) is the third exon of the *c-fos* gene and I2(g) is the second intron of the globin gene. (These labels are provided to help you make your answer clear.) The transcription start sites (↓) and polyadenylation sites (colored arrows) are indicated.

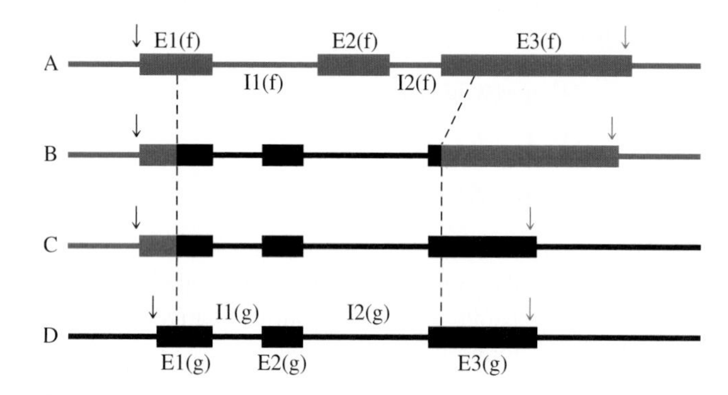

You introduce all four of these clones simultaneously into tissue culture cells and then stimulate individual aliquots of these cells with one of the three factors. Gel analysis of the RNA isolated from the cells gives the following results. The level of transcripts produced from the introduced genes in response to various treatments are shown; the intensity of these bands is proportional to the amount of transcript made from a particular clone. (Where no band appears, it indicates that the level of transcript is undetectable.)

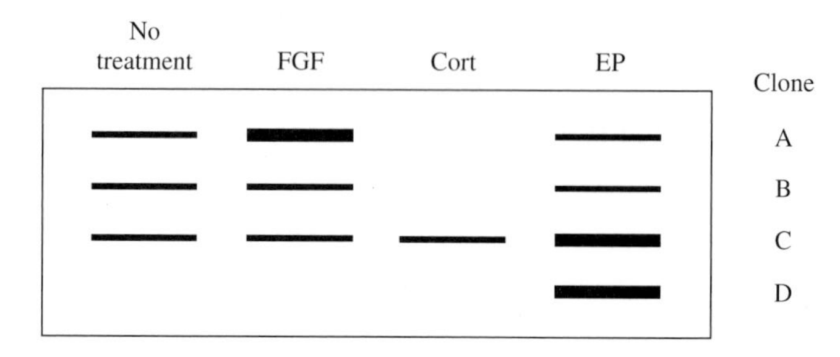

a. Where is the DNA element that permits activation by FGF?

b. Where is the DNA element that permits repression by Cort?

c. Where is the DNA element that permits induction by EP? Explain your answer.

FROM GENE TO PHENOTYPE 14

Key Concepts

1. Different alleles of a gene may cause phenotypic differences, but an allele cannot determine a phenotype by itself: many other genetic and environmental contributions are needed to produce a phenotype.
2. When two different recessive mutant alleles are united in the same cell and the phenotype is mutant, the mutations must be in the same gene.
3. Conversely, when recessive mutant alleles are combined in one cell and the phenotype is wild-type, the mutations must be in different genes.
4. The phenotypes of some heterozygotes reveal types of dominance other than full dominance.
5. Some mutant alleles can kill the organism.
6. Most characters are determined by sets of genes that interact with one another and with the environment.
7. Modified monohybrid ratios reveal allelic interactions such as incomplete dominance.
8. Modified dihybrid ratios reveal gene interactions such as gene collaboration and modulation or suppression of expression.
9. Lack of expression of an allele, or expression to varying degrees, can be caused by the influence of other genes or of the environment.

The colors of peppers are determined by the interaction of several genes. An allele *Y* promotes early elimination of chlorophyll, whereas *y* does not. *R* determines red and *r* yellow carotenoid pigments. Alleles *c1* and *c2* of two different genes down-regulate the amounts of carotenoids causing the lighter shades. Orange is down-regulated red. Brown is green plus red. Pale yellow is down-regulated yellow. *(Anthony Griffiths.)*

By analyzing the results of crosses, geneticists can often attribute discrete phenotypic differences in a specific character to different alleles of a single gene. Indeed, much of the success of genetics can be attributed to its ability to identify genes in this way. However, there is a natural tendency to view alleles as somehow *determining* phenotypes. Although this assumption is a useful mental shorthand, we must now examine the relationship between genes and phenotypes more carefully. Traditional genetic analysis deals only with differences; therefore, a more accurate statement might be that a discrete phenotypic *difference* in a character can be determined by an allelic *difference* in a single gene. Hence, alleles do not determine phenotypes; they determine phenotypic differences. This might seem to be splitting hairs, but it is an important point. It reflects the fact that there is no way a gene can act in a vacuum. For a gene to influence a phenotype, it must act in concert with many other genes and with the environment. In this chapter we examine the ways in which these interactions take place. Even though these interactions represent a higher level of complexity in genetic analysis, there are standard analytical approaches that can help elucidate the type of interaction occurring in a particular case.

Let us introduce the topic using phenylketonuria (PKU), the human disease that we first encountered in Chapter 3. Recall that PKU was described as a simple autosomal recessive disease caused by a defective allele of the gene coding for the liver enzyme phenylalanine hydroxylase (PAH). In the absence of normal PAH, the phenylalanine entering the body in food is not broken down, and hence accumulates. Under such conditions, phenylalanine is converted into phenylpyruvic acid, which is transported via the bloodstream to the brain, where it impedes normal development, leading to mental retardation. Because PKU is associated with elevated concentrations of phenylalanine, it is sometimes referred to as a case of *hyperphenylalaninemia.*

This simple model of PKU has been extremely useful medically, mainly because (as we saw) it has provided a successful treatment and therapy for the disease. However, there are some interesting exceptions to, or extensions of, this model that draw attention to the underlying complexity of the genetic system involved. For example, some cases of hyperphenylalaninemia and its symptoms are associated not with the PAH locus, but with

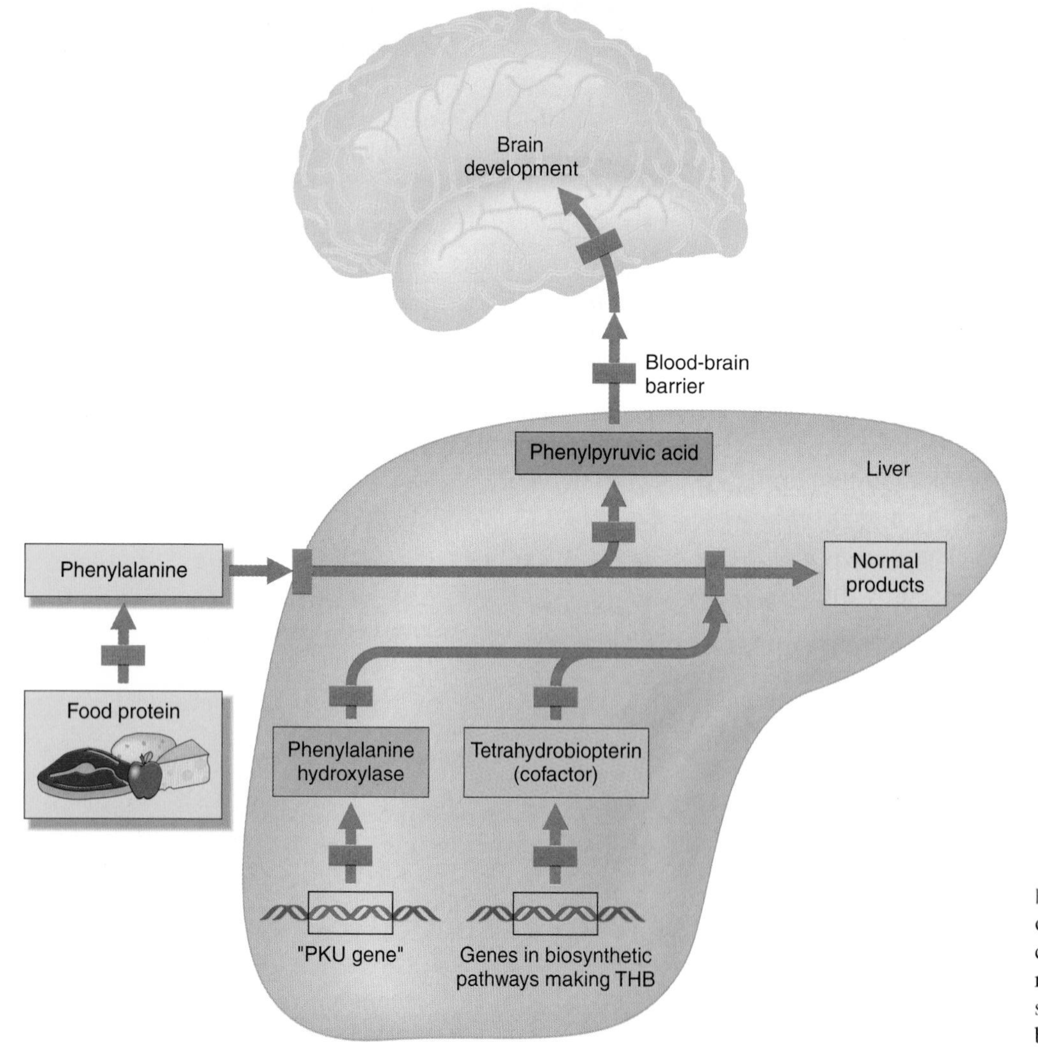

Figure 14-1 The determination of the disease PKU involves a complex series of steps. The red rectangles indicate those steps where variation or blockage is possible.

other genes. Also, some people who have PKU and its associated hyperphenylalaninemia do not show abnormal cognitive development. The reason behind these apparent exceptions to the model is that the expression of the symptoms of PKU depends not only on the PAH locus, but also on many other genes and on the environment. This complex situation is summarized in Figure 14-1. There are many steps in the pathway from dietary phenylalanine to impaired cognitive development, and any one of them can show variation. First, the amount of phenylalanine in the diet is obviously of key importance. Then the phenylalanine must be transported to the appropriate sites in the liver, the "chemical factory" of the body. In the liver, PAH must act in concert with its cofactor, tetrahydrobiopterin. If excess phenylpyruvic acid is produced, it must be transported to the brain in the bloodstream and then pass through the blood-brain barrier. Inside the brain, developmental processes must be susceptible to the detrimental action of phenylpyruvic acid. Each of these multiple steps is a possible site at which genetic or environmental variation may act. Hence what seems to be a simple "monogenic" disease is actually based on quite a complex set of processes. We also see how exceptions to the simple model can be explained; for example, we see how mutations in genes other than the gene for PAH—such as the gene for tetrahydrobiopterin—can cause hyperphenylalaninemia.

Gene action is a term that embraces the complete set of events that begins with the gene's DNA sequence and leads all the way to the observed phenotype. As the PKU example shows, these steps generally constitute a long and complex network involving interactions between genes and the environment.

CHAPTER OVERVIEW

To date, there is no case in which we understand the steps of gene action all the way from the level of expression of a single gene to the level of an organism's phenotype. However, one general goal of genetic analysis is to identify all the genes that affect a particular character and to understand their genetic, cellular, developmental, and molecular roles. To do this, geneticists start by finding mutations affecting the character under study. Next they must deduce which mutations are in which genes, and then the genes must be sorted according to their apparent functions. Finally the genes are assembled into a model showing the components of the system and how they interact. In the first section of this chapter, we shall see how we can use genetic analysis to determine whether two mutant phenotypes are caused by mutations in the same gene (that is, whether the mutations are alleles) or in different genes. In the rest of the chapter we shall consider how genetic analysis can be used to make inferences about gene interactions in developmental and biochemical pathways.

Some of the general ways in which genes interact with each other and with the environment are summarized by the hypothetical model in Figure 14-2 on the next page. Often there is a "gene of interest" that is the center of the study. The structural aspects of this gene itself are obviously important in determining the function of its protein product. Transcription of the gene of interest may be turned on or off by other genes encoding regulatory proteins that bind to its 5′ region. In due course, the protein encoded by the gene of interest may have to bind to other proteins to form an active complex necessary for function. Furthermore, modification of the protein of interest by other proteins or by the physiological conditions of the cell may activate or deactivate its function. Addition of phosphate groups is one way in which a protein can be activated or deactivated, and the enzyme that mediates its phosphorylation is encoded by yet another gene. If the protein of interest is itself an enzyme, its substrate may be supplied by the environment. The environment can also supply signals that set a chain of gene action in motion. Here we see the importance of gene-environment interaction.

How can such complex systems be studied? There are several general approaches:

1. The activity of individual proteins and the elements they interact with can be studied at a purely biochemical level using appropriate purification techniques and chemical assays for function.
2. Functional genomics (Chapter 9) provides powerful ways of defining the set of genes that participate in particular systems. For example, using microarrays, the genes specific to some developmental process can be deduced by comparing the transcripts present at one developmental stage with those at another. The yeast two-hybrid test can also be used to study pairwise interaction of proteins.
3. In the genetic approach, which will be the main focus of this chapter, interacting components are studied by analyzing mutants. Single-gene mutations identify the important components in the system. Interactions between different mutant alleles of a single gene can reveal aspects of the gene's normal function. Normal interactions *between* genes (usually two) are deduced by bringing their mutant alleles together in one genotype. The phenotype of the double mutant, and the phenotypic ratios produced when it is crossed, are the keys to such inference.

Certain commonly encountered types of interactions between mutations of different genes have been given formal names; Figure 14-2 can be used to illustrate some of those covered in this chapter. If mutation of one gene prevents the expression of another, the former gene is said to be *epistatic*. An example would be a mutation of the gene encoding a regulatory protein because, if its protein product is defective, no allele of the gene it regulates could be

CHAPTER OVERVIEW figure

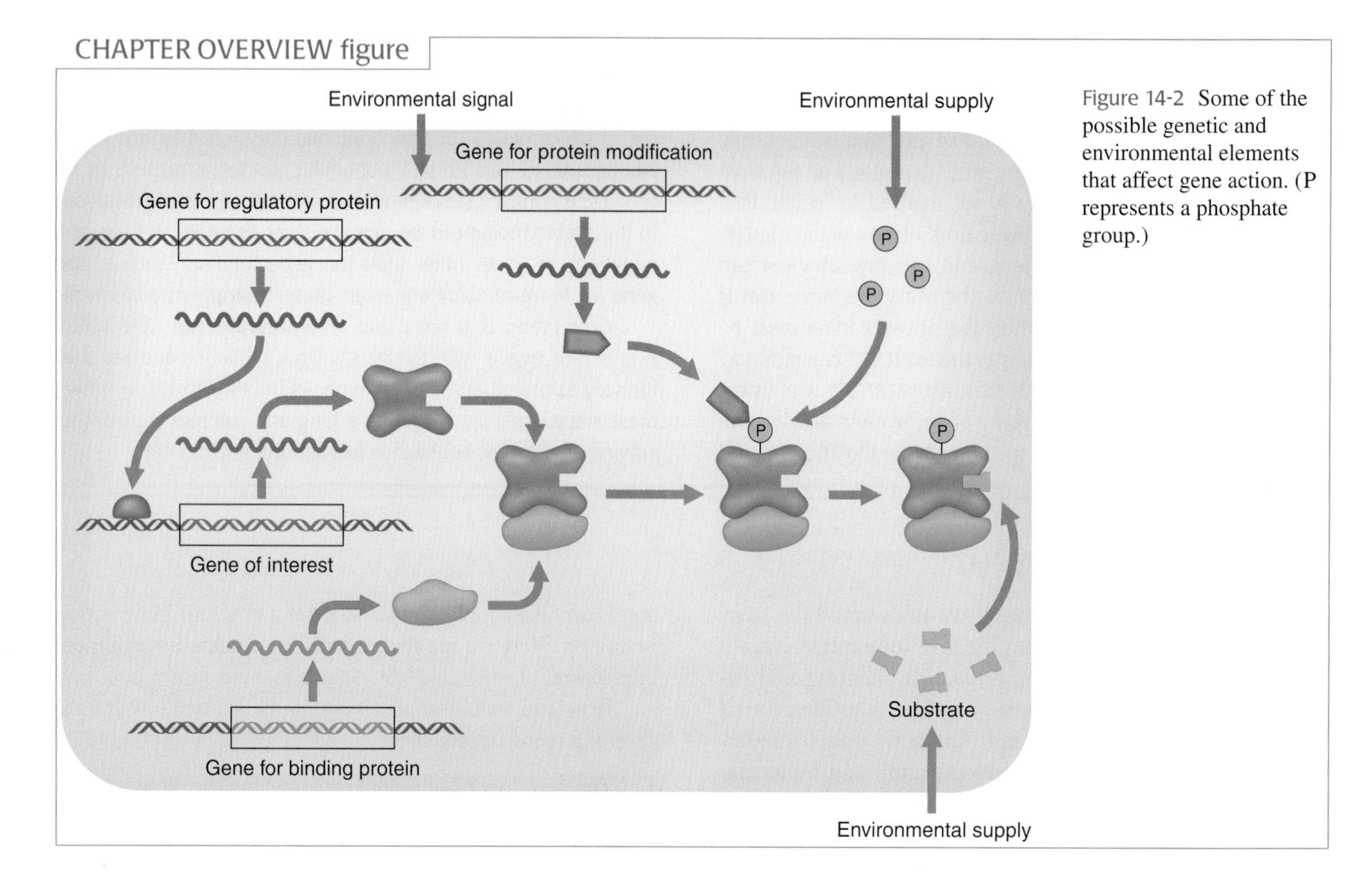

Figure 14-2 Some of the possible genetic and environmental elements that affect gene action. (P represents a phosphate group.)

transcribed properly. Sometimes mutation in one gene can restore wild-type expression to a mutant allele of another gene; in this case the restoring mutation is said to be a *suppressor.* An example might be seen in the interaction of a gene product with a binding protein. A mutation causing a shape change in the product might lead to malfunction because it could no longer bind to the binding protein. However, a mutation in the gene for the binding protein might cause a shape change that would now allow the mutant product to bind again, hence restoring an active complex. As we shall see, both epistasis and suppression can be detected by unique dihybrid F_2 ratios.

A DIAGNOSTIC TEST FOR ALLELES

In Chapter 12 we saw that the determination of which genes contribute to any particular biological process begins with the analysis of a collection of related mutant phenotypes affecting that process. For example, a geneticist who was interested in the genes determining locomotion in a nematode worm would begin by isolating a set of different mutants with defective locomotion. An important task is to determine how many different genes are represented by the alleles that determine these related phenotypes, because this information defines the set of genes that affect the process under study. Hence it is necessary to have a test to find out if the mutant alleles are alleles of one gene or of different genes.

The allelic test that finds widest application is the **complementation test,** which is illustrated in the following example. Consider a species of harebell in which the wild-type flower color is blue. Let's assume that by application of mutagenic radiation we have induced three white-petaled mutants, and that they are all homozygous pure-breeding strains. We can call the mutant strains \$, £, and ¥, using currency symbols to avoid prejudicing our thinking concerning dominance. When crossed to wild type, each mutant gives the same results in the F_1 and F_2, as follows:

White \$ × blue ⟶ F_1, all blue ⟶ F_2, $\frac{3}{4}$ blue, $\frac{1}{4}$ white
White £ × blue ⟶ F_1, all blue ⟶ F_2, $\frac{3}{4}$ blue, $\frac{1}{4}$ white
White ¥ × blue ⟶ F_1, all blue ⟶ F_2, $\frac{3}{4}$ blue, $\frac{1}{4}$ white

In each case, the results show that the mutant condition is determined by the recessive allele of a single gene because the 3:1 ratio is diagnostic for a monohybrid cross. However, are these three conditions caused by three alleles of one gene, or of two or three genes? This question can be answered by asking whether the mutants complement each other. **Complementation** is defined as the production of a wild-type phenotype when two recessive mutant alleles are brought together in the same cell. In a diploid organism, the complementation test is performed by intercrossing homozygous recessive mutants two at a time and observing whether or not the progeny have the wild-type phenotype.

If recessive mutations represent alleles of the same gene, then obviously they will not complement each other, because they both represent loss of the same gene function. Such alleles can be labeled a' and a'', using primes to distinguish between two different mutant alleles of a gene whose wild-type allele is a^+. These alleles could have different mutation sites within the same gene but be functionally identical. The heterozygote a' / a'' would be

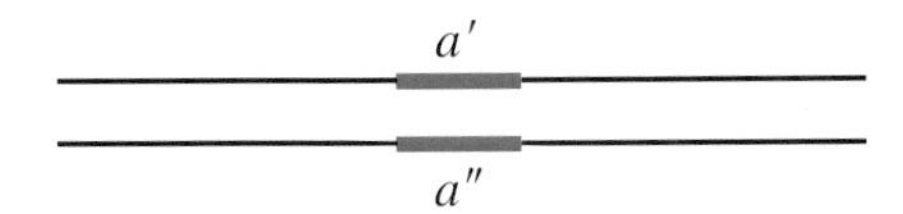

However, two recessive mutations in different genes would have wild-type function, provided by the respective wild-type alleles. Here we can name the different genes *a1* and *a2*, after their mutant alleles. Heterozygotes would be $a1$ / + ; + / $a2$ (unlinked genes) or $a1$ + / + $a2$ (linked genes), and we can diagram them as follows:

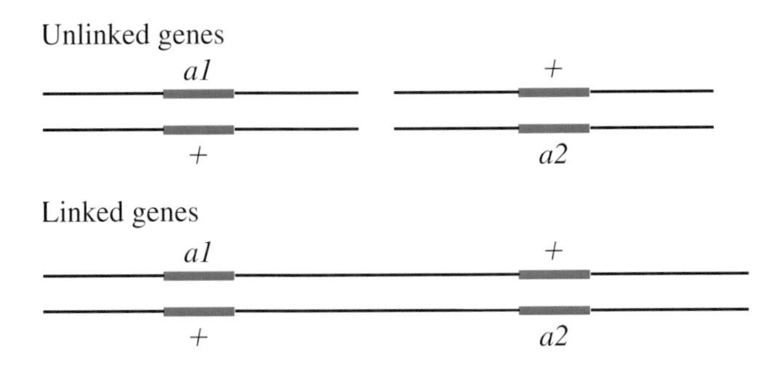

Let us return to the harebell example and intercross the mutants to test for complementation. Assume that the results of intercrossing mutants \$, £, and ¥ are as follows:

White \$ × white £ ⟶ F_1, all white
White \$ × white ¥ ⟶ F_1, all blue
White £ × white ¥ ⟶ F_1, all blue

From this set of results we would conclude that mutants \$ and £ must be caused by alleles of the same gene (say, *w1*) because they do not complement; but ¥ must be caused by a mutant allele of another gene (*w2*).

How does complementation work at the molecular level? The first key point is that although it is conventional to say that *mutants* complement each other, in fact the active agents in complementation are the proteins produced by the *wild-type* alleles. That is, in an $a1$ / + . $a2$ / + dihybrid, it is the two plus symbols that permit the wild-type phenotype. Complementation is generally merely a collaboration of different proteins in a biochemical or developmental pathway. The harebell example provides a simple case. The normal blue color of the flower is caused by a blue pigment called *anthocyanin.* Pigments are chemicals that absorb certain parts of the visible spectrum; in the case of the harebell, the anthocyanin absorbs all wavelengths except blue, which is reflected into the eye of the observer. However, the anthocyanin is made from chemical precursors that are not pigments; that is, they do not absorb light of any specific wavelength and simply reflect back the white light of the sun to the observer, giving a white appearance. The blue pigment is the end product of a series of biochemical conversions of nonpigments. Each step is catalyzed by a specific enzyme encoded by a specific gene. We can accommodate the results with a pathway as follows:

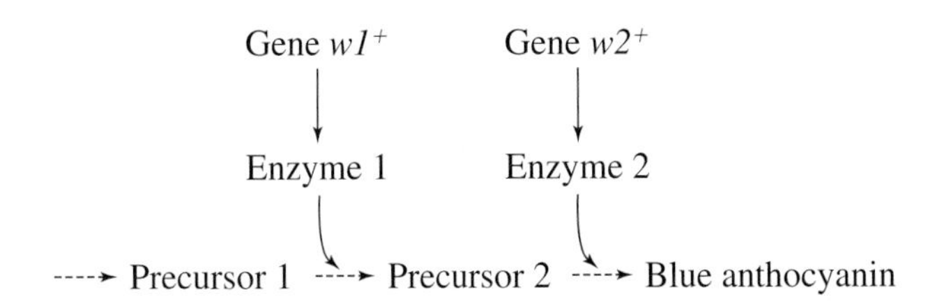

A homozygous mutation in either of the genes will lead to the absence of the enzyme it encodes and hence the accumulation of a precursor, which will simply make the flower white. Now, the mutant designations can be written as follows:

\$ $w1_{\$} / w1_{\$} \ . \ w2^+ / w2^+$
£ $w1_{£} / w1_{£} \ . \ w2^+ / w2^+$
¥ $w1^+ / w1^+ \ . \ w2_{¥} / w2_{¥}$

In practice, however, the subscript symbols would be dropped and the genotypes written

\$ $w1 / w1 \ . \ w2^+ / w2^+$
£ $w1 / w1 \ . \ w2^+ / w2^+$
¥ $w1^+ / w1^+ \ . \ w2 / w2$

Hence an F_1 from \$ × £ will be

$$w1 / w1 \ . \ w2^+ / w2^+$$

which will have two defective alleles for *w1* and in which anthocyanin production will therefore be blocked at step 1. Even though enzyme 2 is fully functional, it has no substrate to act on, so no anthocyanin will be produced, and the phenotype will be white.

However, the F_1s from the other crosses will have the wild-type alleles for both the enzymes needed to take the intermediates to the final anthocyanin product. Their genotypes will be

$$w1^+ / w1 \ . \ w2^+ / w2$$

Hence we see why complementation is actually a result of the cooperative interaction of the *wild-type* alleles of the two genes. Figure 14-3 on the next page shows a summary diagram of the interaction of the complementing and noncomplementing white mutants.

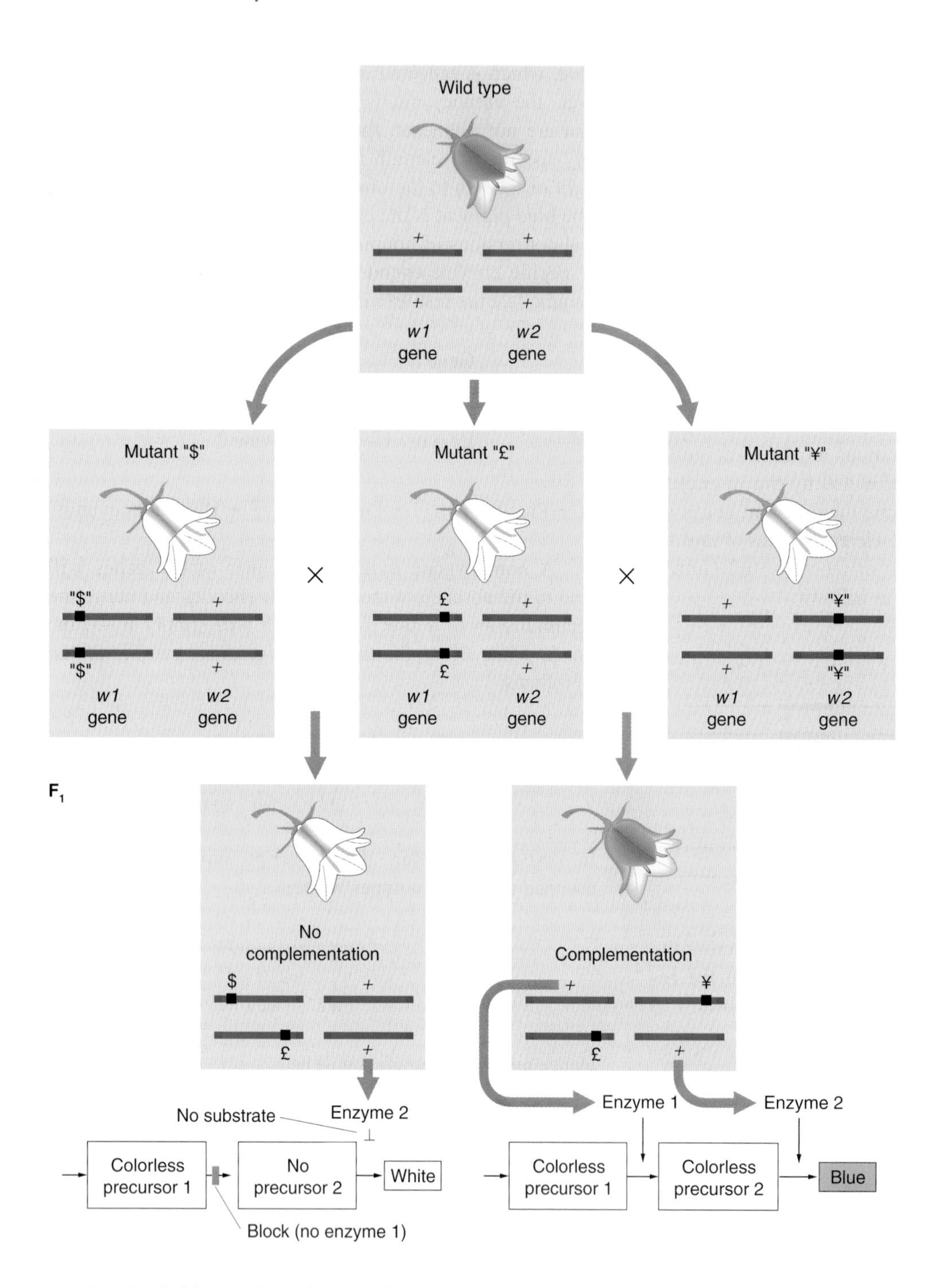

Figure 14-3 The molecular basis of genetic complementation. Three phenotypically identical white mutants, $, £, and ¥, are intercrossed to form heterozygotes whose phenotypes reveal whether or not the mutations complement one another. (Only two of the three possible crosses are shown here.) If two mutations are in different genes (such as £ and ¥), then complementation between the wild-type alleles of each gene results in the completion of the biochemical pathway, because both enzymes are produced and the end product of the pathway (a blue pigment in this example) is made. If the mutations are in the same gene (such as $ and £), no complementation occurs because the biochemical pathway is blocked at the step controlled by the enzyme product of that gene, and the colorless intermediates in the pathway are not converted to blue pigment, leaving the flowers white. (What would you predict to be the result of crossing $ and ¥?)

In a haploid organism, the complementation test cannot be performed by intercrossing. In fungi, an alternative way to test for complementation is to make a **heterokaryon** (Figure 14-4). Fungal cells fuse readily, and when two different strains fuse, the nuclei from the different strains come to occupy one cell, which is called a heterokaryon (Greek: "different kernels"). The nuclei in a heterokaryon generally do not fuse. In one sense, it merely mimics a diploid condition.

Assume that in different haploid strains we have mutations in two different genes conferring the same mutant phenotype—for example, an arginine requirement. We can call these genes *arg-1* and *arg-2*. The two strains, whose genotypes can be represented as *arg-1* . $arg\text{-}2^+$ and $arg\text{-}1^+$. *arg-2*, can be fused to form a heterokaryon with the two nuclei in a common cytoplasm:

Nucleus 1: *arg-1* . $arg\text{-}2^+$
Nucleus 2: $arg\text{-}1^+$. *arg-2*

Because gene expression takes place in a common cytoplasm, the two wild-type alleles can exert their dominant effect and cooperate to produce a heterokaryon of wild-type phenotype. In other words, the two mutations complement each other, just as they would in a diploid. If the mutations had been alleles of the same gene, no complementation would have occurred.

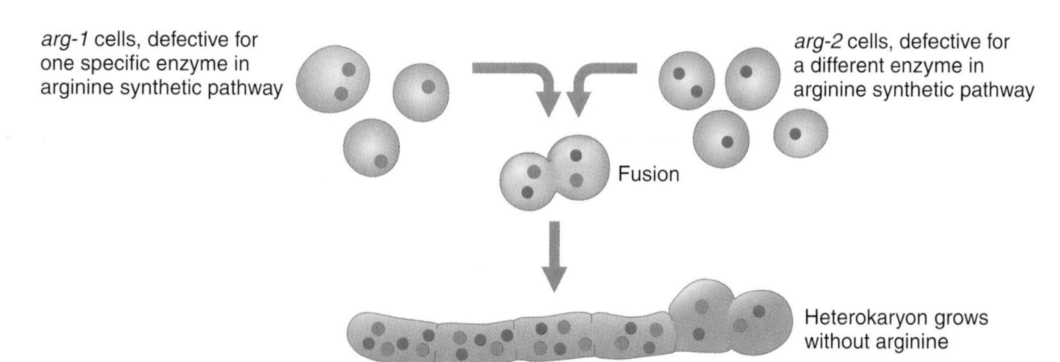

Figure 14-4 Formation of a heterokaryon of *Neurospora,* demonstrating both complementation and recessiveness. Vegetative cells of this normally haploid fungus can fuse, allowing the nuclei from the two strains to intermingle within the same cytoplasm. If each strain is blocked at a different point in a metabolic pathway, as are *arg-1* and *arg-2* mutants, all functions are present in the heterokaryon and the *Neurospora* will grow without arginine; in other words, complementation takes place. The growth must be due to the presence of the wild-type gene; hence, the mutant genes *arg-1* and *arg-2* must be recessive.

MESSAGE

When two independently derived recessive mutant alleles producing similar recessive phenotypes fail to complement each other, the alleles must be of the same gene. If they do complement each other, they must be of different genes.

INTERACTIONS BETWEEN THE ALLELES OF ONE GENE

Some of the complexity encountered in analyzing variant phenotypes affecting one biological process is due to interactions of alleles of a single gene. The alleles of one gene can interact in several different ways at the functional level, resulting in variations in the type of dominance and markedly different phenotypic effects in different allelic combinations.

Incomplete Dominance

Four-o'clocks are plants native to tropical America. Their name comes from the fact that their flowers open in the late afternoon. When a wild-type four-o'clock plant with red petals is crossed to a pure line with white petals, the F_1 has pink petals. If an F_2 is produced by selfing the F_1,

$\frac{1}{4}$ of the progeny have red petals
$\frac{1}{2}$ of the progeny have pink petals
$\frac{1}{4}$ of the progeny have white petals

From this 1:2:1 *phenotypic* ratio in the F_2, we can deduce an inheritance pattern based on two alleles of a single gene. We know that generally this is the *genotypic* ratio from selfing a monohybrid, and that the "2" represents the heterozygotes. If we postulate that the heterozygotes (all of the F_1 and half of the F_2 progeny) are intermediate in phenotype—that is, pink rather than red or white—we can explain the results. An intermediate phenotype in the heterozygotes would suggest an incomplete type of dominance. Inventing allele symbols, we can list the genotypes of the four-o'clocks in this experiment as

c^+ / c^+ (red)
c / c (white)
c^+ / c (pink)

The crosses can then be represented as

Parents c^+ / c^+ (red) $\times$ c / c (white)
↓
F_1 c^+ / c (pink)
↓
F_2 $\frac{1}{4}$ red c^+ / c^+
$\frac{1}{2}$ pink c^+ / c
$\frac{1}{4}$ white c / c

Incomplete dominance describes the general situation in which the phenotype of a heterozygote is intermediate between those of the two homozygotes on some quantitative scale of measurement. Figure 14-5 on the next page summarizes the theoretical positions on such a scale. At the molecular level, incomplete dominance is generally caused by a quantitative effect of the number of "doses" of a wild-type allele: two doses produce the largest amount of transcript, and therefore the largest amount of functional protein product. One dose produces less transcript and product, whereas zero doses produce no functional transcript or product. In some cases of full dominance, the wild-type/mutant heterozygote has half the normal amount of transcript and product, but this amount is adequate for normal function. As we saw earlier in Chapter 3, this situation is termed haplosufficiency (see Figure 3-30). In other cases of full dominance, transcription of the single wild-type allele is increased to bring the concentrations of transcript and product up to the same levels as in the homozygous wild type.

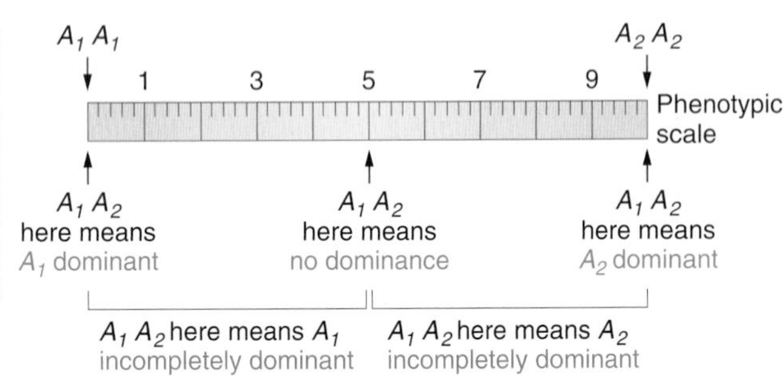

Figure 14-5 Summary of dominance relationships. The ruler represents any quantitative phenotypic measurement, such as amount of pigment.

Codominance

The four human ABO blood types are determined by three alleles of one gene that show several types of interaction. The allelic series includes three major alleles, i, I^A, and I^B, but of course any person can have only two of those three alleles (or two copies of one of them). There are six different possible genotypes, the three homozygotes and three different types of heterozygotes:

Genotype	Blood type
I^A / I^A or I^A / i	A
I^B / I^B or I^B / i	B
I^A / I^B	AB
i / i	O

In this allelic series, the alleles are concerned with an antigen, a cell-surface molecule that can be recognized by the immune system. The alleles I^A and I^B determine two different forms of this antigen, which is deposited on the surface of red blood cells. However, the allele i results in no antigen of this type. In the genotypes I^A / i and I^B / i, the alleles I^A and I^B are fully dominant to i. However, in the genotype I^A / I^B, each of the alleles produces its own form of the antigen, so the two alleles are said to be **codominant.** Formally, codominance is defined as the expression in a heterozygote of *both* the phenotypes normally shown by the two alleles.

The human disease sickle-cell anemia gives some interesting insights into dominance. The gene concerned affects the hemoglobin molecule, which transports oxygen and is the major constituent of red blood cells. The three genotypes have different phenotypes:

Hb^A / Hb^A	Normal; red blood cells never sickle
Hb^S / Hb^S	Severe, often fatal anemia; abnormal hemoglobin causes red blood cells to have a sickle shape (Figure 14-6)
Hb^A / Hb^S	No anemia; red blood cells rarely sickle, generally doing so only at low oxygen concentrations

In regard to the presence or absence of anemia, the Hb^A allele is obviously dominant. In regard to blood cell shape, however, there is incomplete dominance. Finally, as we shall now see, in regard to hemoglobin itself, there is codominance. The alleles Hb^A and Hb^S actually code for two different forms of hemoglobin that differ by a single amino acid, and both these forms are synthesized in the heterozygote.

The different hemoglobin forms can be visualized using **electrophoresis,** which separates proteins or nucleic acids on the basis of their different charges or sizes (Figure 14-7). Because the A and S forms of hemoglobin have different charges, they can be separated using this technique (Figure 14-8). We see that homozygous normal people have one type of hemoglobin (type A) and people with sickle-

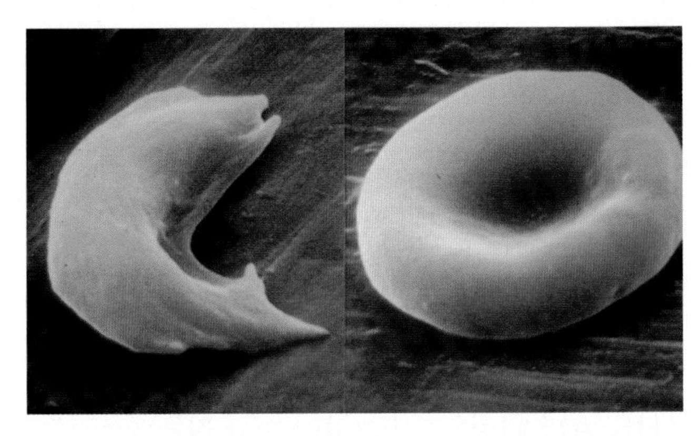

Figure 14-6 Electron micrograph of a red blood cell from an individual with sickle-cell anemia *(left)* and one from a normal individual *(right)*. *(© Stan Flegler/Visuals Unlimited.)*

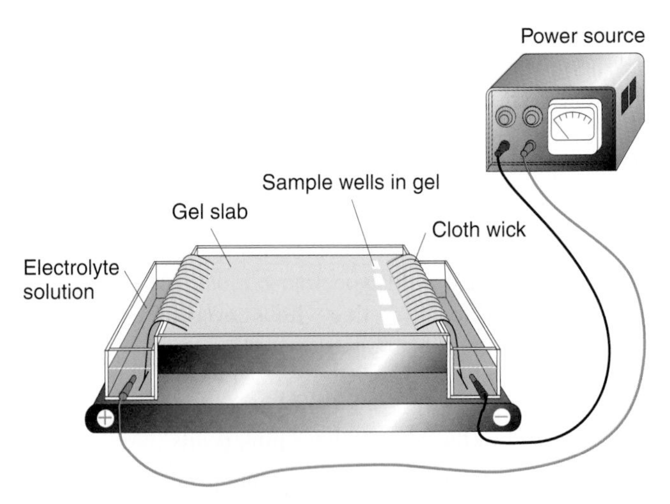

Figure 14-7 Apparatus for electrophoresis. Each sample is placed in a well in a gelatinous slab (a gel). When an electric current is applied, the molecules in the samples migrate different distances on the gel owing to their different electric charges. Several samples can be tested at the same time (one in each well). The positions to which the molecules have migrated are later revealed by staining.

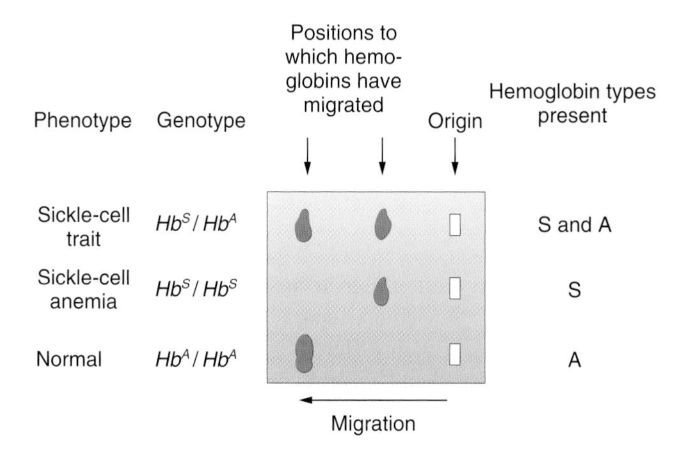

Figure 14-8 Results of electrophoresis of hemoglobin from an individual with sickle-cell anemia, a heterozygote (a condition called *sickle-cell trait*), and a normal individual. The smudges show the positions to which the hemoglobins have migrated on the gel.

cell anemia have another (type S), which moves more slowly in the electric field. Heterozygotes have both types, A and S. In other words, there is codominance at the molecular level.

Sickle-cell anemia illustrates that the terms *dominance, incomplete dominance,* and *codominance* are somewhat arbitrary. The type of dominance inferred depends on the phenotypic level at which the observations are being made—organismal, cellular, or molecular. Indeed, the same caution can be applied to many of the categories that scientists use to classify structures and processes; these categories are devised by humans for convenience of analysis.

In nature, white clover *(Trifolium repens)* shows considerable variation among individuals in the curious V or chevron pattern on the leaves. The different chevron forms (and the absence of chevrons) are determined by a series of alleles of one gene (Figure 14-9). The figure shows the many different types of interactions that are possible, even for one allele. For example, V^b is dominant to v, codominant with V^l, and recessive to V^f.

MESSAGE

A gene can have several different states or forms, called alleles. The alleles of one gene are said to constitute an allelic series. The members of an allelic series can show various degrees of dominance over one another. The type of dominance is determined by the molecular functions of the alleles of a gene and by the investigative level of analysis.

Lethal Alleles

Many mutant alleles are capable of causing the death of an organism; such alleles are called **lethal alleles.** A good example is an allele of a coat color gene in mice. Normal wild-type mice have coats with a dark overall pigmentation. A mutation called *yellow* (which results in a lighter coat color) shows a curious inheritance pattern. If a yellow mouse is mated to a homozygous wild-type mouse, a 1:1 ratio of yellow to wild-type mice is always observed in the progeny. This observation suggests (1) that a single gene with two alleles determines these phenotypic alternatives, (2) that the yellow mouse was heterozygous for these alleles, and (3) that the allele for yellow is dominant to an allele for normal color. However, if two yellow mice, which from the above logic would seem to be heterozygotes, are crossed, the result is always as follows:

$$\text{yellow} \times \text{yellow} \longrightarrow \tfrac{2}{3}\ \text{yellow},\ \tfrac{1}{3}\ \text{wild type}$$

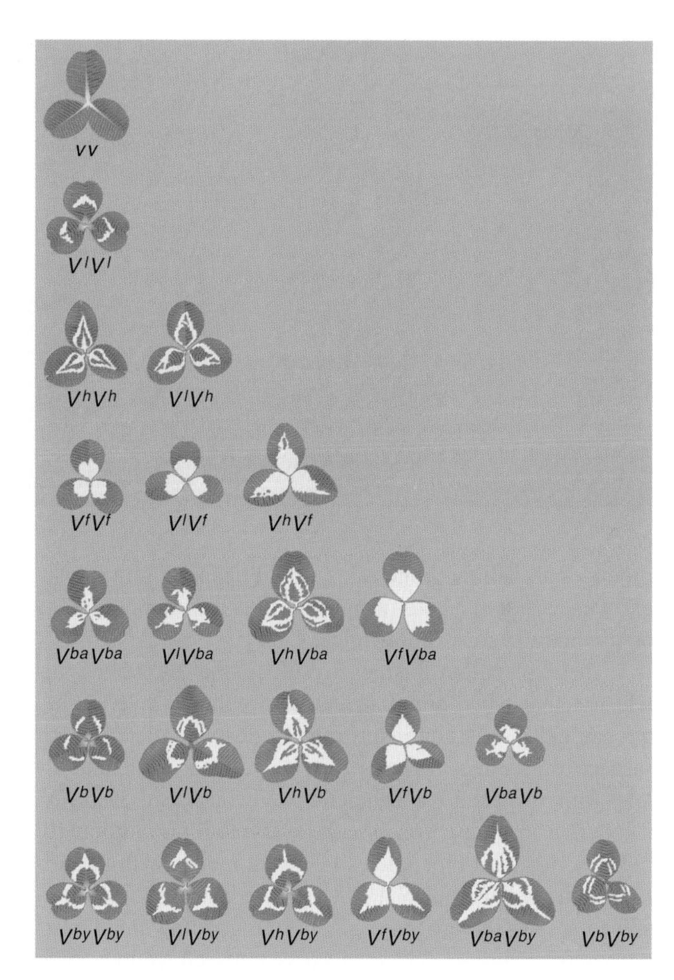

Figure 14-9 Multiple alleles determine the chevron pattern on the leaves of white clover. The genotype of each plant is shown below it. *(Adapted from photo by W. Ellis Davies.)*

Figure 14-10 on the next page shows a typical litter from a cross between yellow mice. Note two interesting features in these results. First, the 2:1 phenotypic ratio is a departure from the expectations for a monohybrid cross. Second, because no cross of yellow × yellow ever

Figure 14-10 A mouse litter from two parents heterozygous for the yellow coat color allele, which is lethal in a double dose. The larger mice are the parents. Not all progeny are visible. *(Anthony Griffiths.)*

produces all yellow progeny, as would happen if either parent were a homozygote, it appears that it is impossible to obtain homozygous yellow mice.

The explanation for these results is that the yellow allele is lethal when homozygous. The yellow allele, which we can call A^Y, is dominant to the wild-type allele A with respect to its effect on color, but it is a recessive lethal allele with respect to a character we would call *viability.* (Note that in mouse genetics the + symbol is traditionally not used for wild-type alleles.) Thus, a mouse with the homozygous genotype A^Y / A^Y dies before birth and is not observed among the progeny. All surviving yellow mice must be heterozygous A^Y / A, so a cross between yellow mice will always yield the following results:

$$A^Y / A \times A^Y / A$$

Progeny	$\frac{1}{4}\ A^Y / A^Y$	Lethal
	$\frac{1}{2}\ A^Y / A$	Yellow
	$\frac{1}{4}\ A / A$	Wild type

The expected monohybrid ratio of 1:2:1 would be found among the zygotes, but it is altered to a 2:1 ratio in the progeny actually seen at birth because zygotes with a lethal A^Y / A^Y genotype do not survive to be counted. This hypothesis is supported by observations of the progeny before birth. If the uteri from pregnant females of the yellow × yellow cross are removed for observation, one-fourth of the embryos are found to be dead.

The A^Y allele produces effects on two characters: coat color and survival. It is thus said to be **pleiotropic.** It is entirely possible, however, that both effects of the A^Y pleiotropic allele result from the same basic cause, which results in yellowness of coat in a single dose and death in a double dose. Most mutant alleles have pleiotropic effects to greater or lesser degrees, a reflection of the complexity of biological systems and the interaction of their components.

The tailless Manx phenotype in cats (Figure 14-11) is also produced by an allele that is lethal in the homozygous state. A single dose of the Manx allele M^L severely interferes with normal spinal development, resulting in the absence of a tail in the M^L / M heterozygote. But in M^L / M^L homozygotes, the double dose of the gene produces such an extreme abnormality in spinal development that the embryo does not survive.

There are many different types of lethal alleles. Some lethal alleles produce a recognizable phenotype in the heterozygote, as in the yellow mouse and Manx cat. Some lethal alleles are fully dominant and kill in one dose in the heterozygote. Most, however, confer no detectable effect on the heterozygote at all, and the lethality is fully recessive. Furthermore, lethal alleles differ in the developmental stage at which they express their effects. Human lethals illustrate this very well. It has been estimated that we are all heterozygous for a small number of recessive lethals. The

Figure 14-11 Manx cat. All such cats are heterozygous for a dominant allele that causes no tail to form. The allele is lethal in homozygous condition. The dissimilarity of this cat's eyes is unrelated to its taillessness. *(Gerard Lacz/NHPA.)*

lethal effect is expressed in the homozygous progeny of a mating between two people who by chance carry the same recessive lethal in the heterozygous condition. Some lethals are expressed as deaths in utero, where they either go unnoticed or are noticed as spontaneous abortions. Other lethals, such as those responsible for Duchenne muscular dystrophy, cystic fibrosis, and Tay-Sachs disease, exert their effects in childhood. The time of death can even be in adulthood, as in Huntington disease. The total of all the deleterious and lethal genes that are present in the individuals of a population is called **genetic load,** a kind of genetic burden that the population has to carry.

Exactly what goes wrong in lethal mutations? In many cases it is possible to trace the complex cascade of events that leads to death. Many lethal alleles cause a deficiency in some enzyme that catalyzes an essential chemical reaction. The human disease PKU is a good examples of this kind of deficiency, as we saw in Chapter 3. In other cases the allele results in a structural defect. For example, a lethal allele of rats determines abnormal cartilage protein, and the effect of this abnormality is expressed phenotypically in several different organs, resulting in lethal symptoms, as shown in Figure 14-12. Sickle-cell anemia, discussed above, is another example.

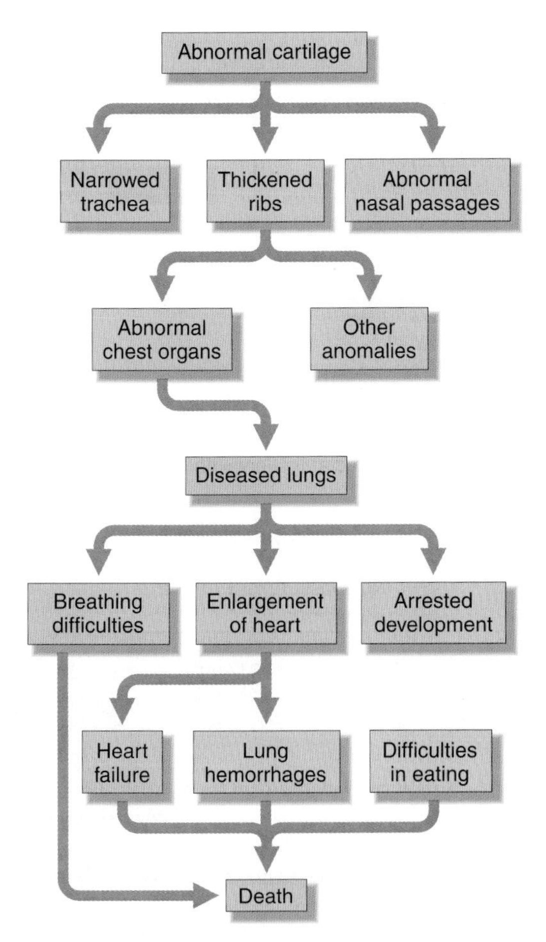

Figure 14-12 Diagram showing how one specific lethal allele causes death in rats. *(From I. M. Lerner and W. J. Libby,* Heredity, Evolution, and Society, *2d ed. W. H. Freeman, 1976; after H. Gruneberg.)*

Whether an allele is lethal or not often depends on the environment in which the organism develops. Whereas certain alleles would be lethal in virtually any environment, others are viable in one environment but lethal in another. Human hereditary diseases provide some examples. Cystic fibrosis and sickle-cell anemia are diseases that would be lethal without treatment, and individuals with PKU would undoubtedly not survive in a natural setting in which the special diet they need would be impossible. Likewise, many of the alleles favored and selected by animal and plant breeders would almost certainly be eliminated in nature as a result of competition with the members of the natural population. Modern high-yielding dwarf varieties of grain provide good examples; only careful nurturing by the farmer has maintained such alleles for our benefit.

Geneticists commonly encounter situations in which expected phenotypic ratios are consistently skewed in one direction by reduced viability caused by one allele. For example, in the cross $A / a \times a / a$, we might predict a progeny phenotypic ratio of 50 percent A / a and 50 percent a / a, but we might consistently observe ratios such as 55%:45% or 60%:40%. Because the deviation from expectations is consistently in one direction, it is unlikely to be due to chance. In such cases, the recessive phenotype is said to be *subvital,* or *semilethal,* because the lethality is expressed in only some individuals. Thus, lethality may range from 0 to 100 percent, depending on the gene itself, the rest of the genome, and the environment.

MESSAGE

Many alleles that result in lost, reduced, or altered cellular function are potentially lethal. In diploids, a recessive lethal allele (l) can survive in a $+ / l$ heterozygote. Lethal alleles can be detected via modified monohybrid ratios.

INFERRING GENE INTERACTION FROM DIHYBRID RATIOS

We have examined above the analysis of the interaction between alleles of one gene. Genetic analysis can also identify multiple genes that interact in a biological system. Mutations in two interacting genes produce diagnostic *dihybrid ratios*—for example, F_2 ratios. There are various types of gene interactions, which lead to a range of different ratios. An important initial distinction is between interacting genes in the same biological pathway and those in different pathways.

Interacting Genes in the Same Pathway

To look at the effects of interaction between genes in the same biological pathway, we need only return to the example of petal color in harebells, used to introduce the idea of gene complementation (see right side of Figure 14-3). Recall that the anthocyanin pathway terminated in a blue pigment, and that the intermediates were all colorless. Two different white-petaled homozygous lines of harebells were crossed, and the F_1 was blue-flowered, showing complementation and thus demonstrating that the two mutations must have been in two different genes. What will the F_2 resulting from crossing these F_1 plants be like?

If the two genes assort independently, the F_2 will show both blue and white plants in a ratio of 9:7. How can these results be explained? The 9:7 ratio is clearly a modification of the dihybrid 9:3:3:1 ratio, with the 3:3:1 combined to make 7. The cross of the two white lines and subsequent generations can be represented as follows:

P
$w1/w1\ ;\ w2^+/w2^+$ (white) $\times$ $w1^+/w1^+\ ;\ w2/w2$ (white)

F_1 $\quad w1^+/w1\ ;\ w2^+/w2$ (blue)

$w1^+/w1\ ;\ w2^+/w2 \times w1^+/w1\ ;\ w2^+/w2$

F_2
9 $w1^+/-\ ;\ w2^+/-$ (blue) 9
3 $w1^+/-\ ;\ w2/w2$ (white) }
3 $w1/w1\ ;\ w2^+/-$ (white) } 7
1 $w1/w1\ ;\ w2/w2$ (white) }

The results show that homozygosity for the recessive mutant allele of *either* or *both* genes causes a plant to have white petals. To have the blue phenotype, a plant must have at least one dominant wild-type allele of both genes, because both are needed to complete the sequential steps in the pathway (see Figure 14-3).

In the anthocyanin example, all the intermediate compounds are colorless. However, if one or more intermediates produced color, different F_2 ratios would be produced because in the single mutants, different-colored pigments would accumulate. In the following example, taken from the plant blue-eyed Mary *(Collinsia parviflora),* the pathway is

$$\text{Colorless} \xrightarrow{\text{gene } w^+} \text{magenta} \xrightarrow{\text{gene } m^+} \text{blue}$$

The w and m genes are not linked. If homozygous white and magenta plants are crossed, the F_1 and F_2 are as follows:

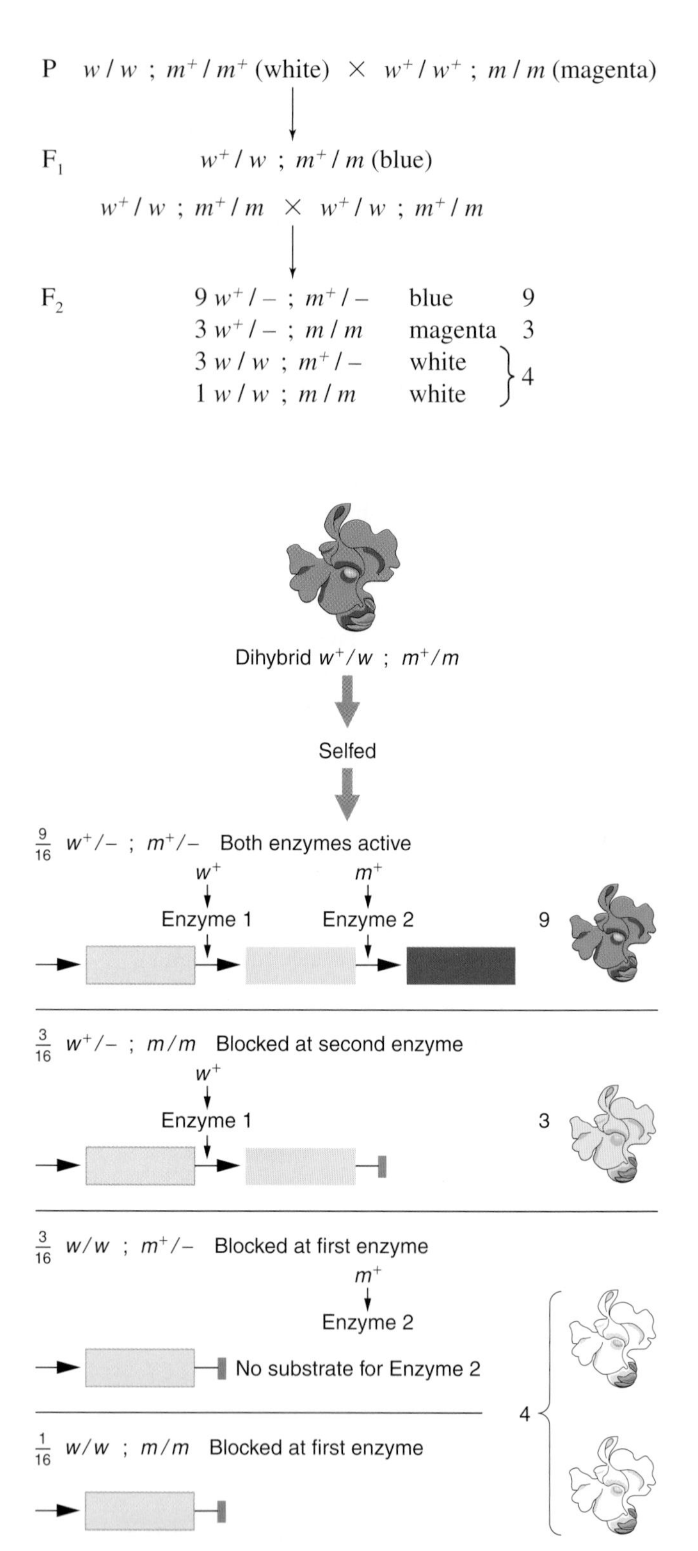

Figure 14-13 A molecular mechanism for recessive epistasis in *Collinsia.* Two genes code for enzymes catalyzing successive stages in the synthesis of a blue petal pigment. The substrates for these enzymes are colorless and pink, respectively, so null alleles of these genes result in colorless (white) or pink petals. The epistasis is revealed in the double mutant because it shows the phenotype of the earlier of the two blocks in the pathway (that is, white). Hence the mutation in the earlier-acting gene precludes expression of any allele of the later-acting gene.

Overall, a 9:3:4 phenotypic ratio is produced. This kind of gene interaction is called **epistasis.** The word *epistasis* literally means "standing on"; in other words, an allele of one gene masks expression of alleles of another gene and expresses its own phenotype instead. In this example, the w allele is epistatic on m^+ and m because no matter which allele of the later-acting gene is present, it will not be expressed. Conversely, m^+ and m can be expressed only in the presence of w^+. Because in this example a recessive allele (w) is epistatic, this is a case of **recessive epistasis.**

Note that the distinction between the 9:7 ratio in harebells ("complementation") and the 9:3:4 ratio in *Collinsia* ("epistasis") is arbitrary. Both reveal a type of epistasis because in both cases an early-acting gene cancels expression of a later-acting gene. In one case the epistasis is made clear by the different-colored intermediates. Furthermore, in both cases, the F_1 dihybrid clearly shows complementation, defined as the production of a wild-type phenotype when different mutant alleles are united in one cell. Even though the F_2 ratios are different, the underlying genetic interaction is exactly the same in both cases.

MESSAGE

Epistasis is inferred when an allele of one gene eliminates the expression of alleles of another gene and expresses its own phenotype instead.

In general, every time one gene acts earlier than others in some biochemical pathway, we would expect there to be an epistatic effect of a defective allele on alleles of genes later in the sequence. Therefore, finding a case of epistasis (for example, a 9:3:4 modified dihybrid ratio) can provide insight about the sequence in which genes act. This principle can be useful in piecing together biochemical pathways. We can piece together the genetic and biochemical systems acting in *Collinsia* as shown in Figure 14-13.

Another case of recessive epistasis with a somewhat different underlying cellular basis is the yellow coat color of some Labrador retriever dogs. Two alleles, B and b, produce black and brown melanin, respectively, but the allele e of another gene is epistatic on these, giving a yellow coat (Figure 14-14). Therefore, the genotypes $B/-\ ;\ e/e$ and $b/b\ ;\ e/e$ both produce a yellow phenotype, whereas $B/-\ ;\ E/-$ and $b/b\ ;\ E/-$ are black and brown, respectively. This case of epistasis is *not* caused by an upstream block in a pathway leading to dark pigment. Yellow dogs can make black or brown melanin, as can be seen in their noses and lips. The action of the allele e is to prevent deposition of the pigment in the hairs. In this case, the epistatic gene is *developmentally downstream;* it represents a kind of developmental target that has to be in the E state before pigment can be deposited.

MESSAGE

Epistasis points to an interaction of genes in the same biochemical or developmental pathway.

Interacting Genes in Different Pathways

A simple, yet striking, example of interacting genes in different pathways is the inheritance of skin coloration in corn snakes. The snake's natural color is a repeating black-and-orange camouflage pattern, as shown in Figure 14-15a on the next page. The phenotype is produced by two separate pigments, both of which are under genetic control. One gene determines the orange pigment, and the alleles we

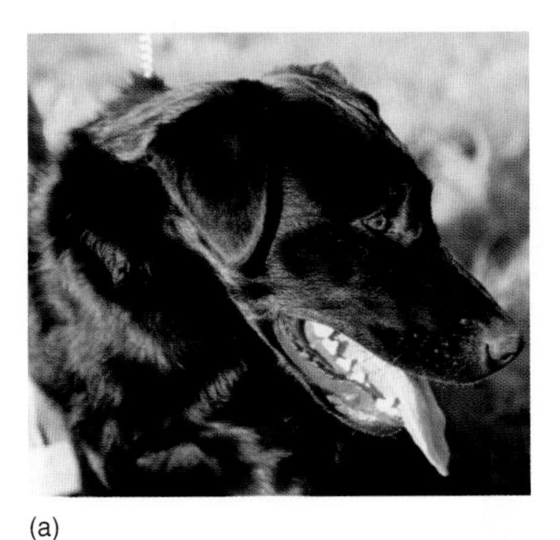
(a)

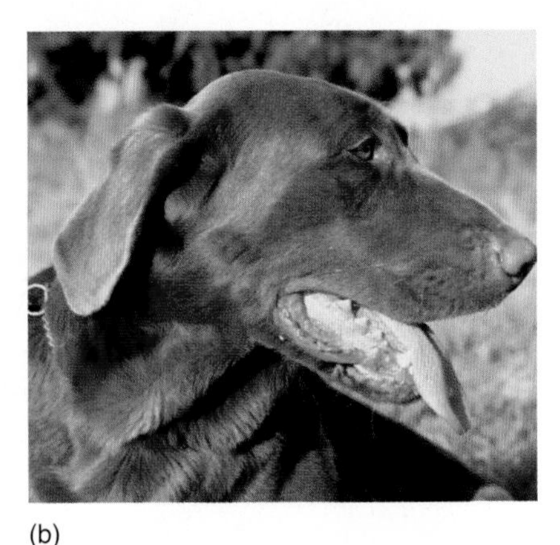
(b)

(c)

Figure 14-14 Coat color inheritance in Labrador retrievers. Two alleles of a pigment gene, B and b, determine (a) black and (b) brown, respectively. At a separate gene, $E/-$ allows pigment deposition in the coat and e/e prevents deposition, resulting in (c) the yellow phenotype. This is a case of recessive epistasis. Thus, the three homozygous genotypes are (a) $B/B\ ;\ E/E$, (b) $b/b\ ;\ E/E$, and (c) $B/B\ ;\ e/e$ or $b/b\ ;\ e/e$. The dog in (c) is most likely $B/B\ ;\ e/e$; the animal still has the ability to make black pigment, as witnessed by its black nose and lips, but not to deposit this pigment in the hairs. (A dog of genotype $b/b\ ;\ e/e$ would be yellow with brown nose and lips.) The progeny of a dihybrid cross would produce a 9:3:4 ratio of black : brown : yellow. *(Anthony Griffiths.)*

shall consider are o^+ (presence of orange pigment) and o (absence of orange pigment). Another gene determines the black pigment, with alleles b^+ (presence of black pigment) and b (absence of black pigment). These two genes are unlinked. The wild-type pattern is produced by the genotype $o^+ / - \; ; \; b^+ / -$. A snake that is $o / o \; ; \; b^+ / -$ is black because it lacks orange pigment (Figure 14-15b), and a snake that is $o^+ / - \; ; \; b / b$ is orange because it lacks black pigment (Figure 14-15c). The double homozygous recessive $o / o \; ; \; b / b$ is albino (Figure 14-15d). Notice, however, the faint pink color of the albino snake, which comes from yet another pigment, the hemoglobin of the blood that is visible through its skin when the other pigments are absent. The albino snake also shows clearly that there is another element to the camouflage pattern in addition to pigment: the repeating motif in and around which pigment is deposited. (If we had mutations affecting that property, it, too, could be analyzed genetically.)

Since there are two genes in this system, and since their products are synthesized in different biochemical pathways, we obtain a typical dihybrid inheritance pattern, and the four unique phenotypes form a 9 : 3 : 3 : 1 ratio in the F_2. A typical analysis might be as follows:

female $o^+ / o^+ \; ; \; b / b$ × male $o / o \; ; \; b^+ / b^+$
(orange) (black)

F_1 $o^+ / o \; ; \; b^+ / b$
(camouflaged)

female $o^+ / o \; ; \; b^+ / b$ × male $o^+ / o \; ; \; b^+ / b$
(camouflaged) (camouflaged)

F_2
9 $o^+ / - \; ; \; b^+ / -$ (camouflaged)
3 $o^+ / - \; ; \; b / b$ (orange)
3 $o / o \; ; \; b^+ / -$ (black)
1 $o / o \; ; \; b / b$ (albino)

(a)

(b)

(c)

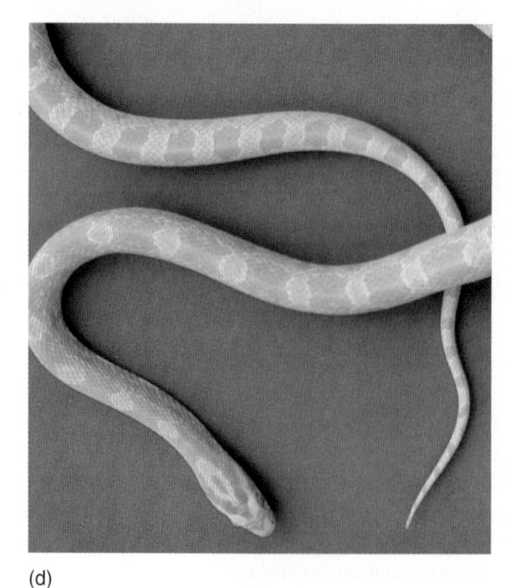
(d)

Figure 14-15 Analysis of the genes for skin pigment in the corn snake. The wild type (a) has a skin pigmentation pattern made up of a black and an orange pigment. The gene o^+ determines an enzyme in the synthetic pathway for orange pigment; when this enzyme is deficient (o / o), no orange pigment is made, and the snake is black (b). Another gene, b^+, determines an enzyme for black pigment; when this enzyme is deficient (b / b), the snake is orange (c). When both enzymes are deficient, the snake is albino (d). Hence the four homozygous genotypes are (a) $o^+ / o^+ \; ; \; b^+ / b^+$, (b) $o / o \; ; \; b^+ / b^+$, (c) $o^+ / o^+ \; ; \; b / b$, and (d) $o / o \; ; \; b / b$. A cross of (a) × (d) or (b) × (c) would give a dihybrid wild-type F_1 and a 9 : 3 : 3 : 1 ratio of the four phenotypes in the F_2. *(Anthony Griffiths.)*

In summary, the 9:3:3:1 dihybrid F_2 ratio is produced because

1. the two loci assort independently.
2. the two gene products are synthesized independently in two distinct biochemical pathways:

$$\begin{array}{l} \text{Precursors} \xrightarrow{b^+} \text{black pigment} \searrow \\ \qquad\qquad\qquad\qquad\qquad\qquad \longrightarrow \text{camouflaged} \\ \text{Precursors} \xrightarrow[o^+]{} \text{orange pigment} \nearrow \end{array}$$

3. the black and orange pathway products do not interact physically. The interaction of these compounds that determines the wild-type camouflaged pattern is really just their co-presence in the skin.

Another important type of interaction between genes in different pathways is **suppression.** Here the two components of the system *do* interact physically. A suppressor is a mutant allele of one gene that reverses the effect of a mutation of another gene, resulting in the normal (wild-type) phenotype. For example, assume that an allele a^+ produces the normal phenotype, whereas a recessive mutant allele a results in abnormality. A recessive mutant allele s at another gene suppresses the effect of a, so the genotype $a\,/\,a\ .\ s\,/\,s$ will have the wild-type (a^+-like) phenotype. Suppressor alleles sometimes have no effect by themselves; in such a case, the phenotype of $a^+\,/\,a^+\ .\ s\,/\,s$ would be wild-type. In other cases the suppressor might have its own phenotype. Like epistasis, suppressors produce modified dihybrid ratios.

Let's look at a real-life example from *Drosophila.* A recessive allele *su* has no detectable phenotype itself, but suppresses the unlinked recessive purple-eye-color allele *pd*. Hence $pd\,/\,pd\ ;\ su\,/\,su$ is wild-type in appearance and has red eyes. The following analysis illustrates the inheritance pattern. A homozygous purple-eyed fly is crossed to a homozygous red-eyed stock carrying the suppressor:

P $pd\,/\,pd\ ;\ su^+\,/\,su^+$ (purple) $\times$ $pd^+\,/\,pd^+\ ;\ su\,/\,su$ (red)

$\downarrow$

F_1 All $pd^+\,/\,pd\ ;\ su^+\,/\,su$ (red)

$pd^+\,/\,pd\ ;\ su^+\,/\,su$ (red) $\times$ $pd^+\,/\,pd\ ;\ su^+\,/\,su$ (red)

$\downarrow$

$$F_2 \quad \left.\begin{array}{ll} 9\ pd^+/-\ ;\ su^+/- & \text{red} \\ 3\ pd^+/-\ ;\ su/su & \text{red} \\ 1\ pd/pd\ ;\ su/su & \text{red} \end{array}\right\} 13 \qquad 3\ pd/pd\ ;\ su^+/-\ \ \text{purple}\}\ 3$$

The overall ratio in the F_2 is 13 red : 3 purple. This ratio is characteristic of a recessive suppressor acting on a recessive mutation. Both recessive and dominant suppressors are found. If a recessive suppressor had its own phenotype, the F_2 ratio would be $\frac{9}{16}$ wild type, $\frac{3}{16}$ suppressor phenotype, $\frac{3}{16}$ mutant phenotype under study, and another $\frac{1}{16}$ wild type due to suppression.

How do suppressors work at the molecular level? There are many possible mechanisms. One well-researched type is **nonsense suppressors,** which act on mutations that produce stop ("nonsense") codons in the middle of the protein-coding sequence (see Chapter 10). Such a nonsense codon results in premature amino acid chain termination because there is no tRNA with an anticodon to bind to it. However, a mutation in a tRNA anticodon that allows the tRNA to bind and insert an amino acid at the nonsense codon will suppress the effect of the mutation by allowing protein synthesis to proceed past the site of the mutation in the mRNA. The amino acid inserted will not necessarily be the "correct" one, but it may be capable of wild-type function in that site. Since there are often repetitive copies of tRNA genes, a suppressor mutation in one copy will be viable because the remaining unmutated copies preserve the original specificity.

Another type of suppression is possible in cases of protein-protein interactions. If two proteins normally fit together to provide some type of cellular function, then when a mutation causes a shape change in one of the proteins, it no longer fits together with the other, and hence the function is lost. However, a suppressor mutation that causes a compensatory shape change in the second protein can restore fit and hence normal function (Figure 14-16).

Finally, in situations in which a mutation causes a block in a metabolic pathway, the suppressor finds some way of bypassing the block; for example, by rerouting into

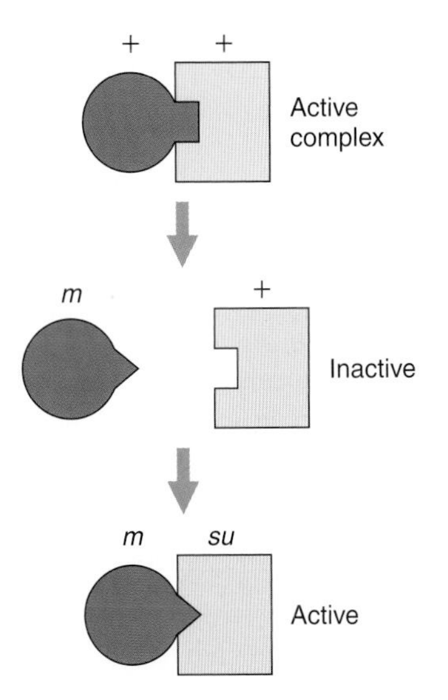

Figure 14-16 A molecular mechanism for suppression. The protein products of two genes normally bind to form a complex needed for some cellular function. Mutation *m* in one gene results in a protein that does not fit with the other product. Mutation *su* in the second gene acts as a suppressor because it produces a complementary shape change so that the proteins can once again bind to each other to form an active complex.

the blocked pathway intermediates similar to those beyond the block. In the following example, the suppressor provides an intermediate B to circumvent the block.

No suppressor

A ⟶|⟶ ~~B~~ ⟶ ~~product~~

With suppressor

A ⟶|⟶ B ⟶ product
B ↗

Because of the demonstrable interaction of a suppressor with its target gene, geneticists deliberately seek suppressors as another way of piecing together the set of interacting genes that affect a biological process or structure. The approach is relatively easy: perform a large-scale mutagenesis experiment starting with a mutant line (say, genotype *m*), and then simply look for rare individuals that are wild-type. Most of these will be m^+ reverse mutations, but some will be suppressed (*m . su*), and these can be distinguished by the dihybrid ratios produced on crossing. This procedure can be very easily applied in haploid organisms. For example, if large numbers of cells of an arginine-requiring mutant (*arg*) are spread on a plate of growth medium lacking arginine, most cells will not grow; however, reverse mutations to the true wild-type allele (arg^+) and suppressor mutations (*arg . su*) will grow and announce their presence by forming visible colonies. The suppressed colonies can be distinguished by crossing to wild type because arginine-requiring progeny will be produced:

$$arg \;.\; su \times arg^+ \;.\; su^+$$

↓

$arg^+ \;.\; su^+$	non-arginine-requiring
$arg^+ \;.\; su$	non-arginine-requiring
$arg \;.\; su$	non-arginine-requiring
$arg \;.\; su^+$	arginine-requiring

MESSAGE

A suppressor mutation cancels the effect of a mutant allele of another gene, resulting in the normal wild-type phenotype.

Coat Color in a Mammalian Model, the Mouse

The analysis of coat color in mammals has provided a beautiful example of how different genes cooperate in the determination of the overall phenotype. Because of the evolutionary relatedness of all mammals, there is a large degree of overlap in the sets of coat color genes found in different mammals; hence what is learned in one mammal is often widely applicable. The mouse has become the model organism for mammalian genetics because it is easy to maintain in the laboratory and its reproductive cycle is short. For similar reasons, it is the best-studied mammal with regard to the genetic determination of coat color. At least five major genes interact to determine the coat color of mice; these genes are labeled *A*, *B*, *C*, *D*, and *S*.

The A gene. The *A* gene determines the distribution of pigment in the hair. The wild-type allele *A* produces a phenotype called *agouti*, an overall grayish coat color with a brindled, or "salt-and-pepper," appearance. It is a common color of mammals in nature. The effect is produced by a band of yellow on the otherwise dark hair shaft. In the nonagouti phenotype (determined by the allele *a*), the yellow band is absent, so there is solid dark pigment throughout the hair (Figure 14-17). The lethal allele A^Y, discussed in an earlier section, is another allele of this gene; it makes the entire shaft yellow. Still another allele a^t results in a "black-and-tan" effect, a yellow belly with dark pigmentation elsewhere.

The B gene. The *B* gene determines the color of melanin. There are two major alleles: *B*, coding for black melanin, and *b*, for brown. The allele *B* gives the normal agouti color in combination with *A* but gives solid black with *a / a*. The genotype *A / –* ; *b / b* gives a streaked brown color called *cinnamon*, and *a / a* ; *b / b* gives solid brown. In horses, the breeding of domesticated lines seems to have eliminated the *A* allele that determines the agouti phenotype, although certain wild relatives of the horse do have this allele. The color we have called *brown* in mice is called *chestnut* in horses, and, as in mice, chestnut is recessive to black. Note that we have already encountered these same alleles in the determination of coat color in Labrador retrievers.

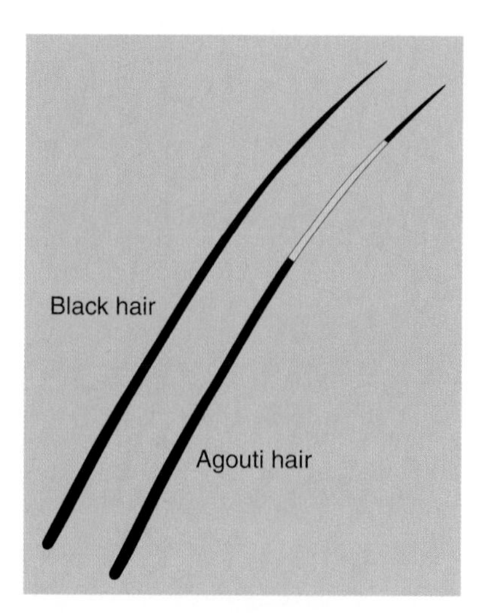

Figure 14-17 Individual hairs from an agouti and a black mouse. The yellow band on each hair gives the agouti pattern its brindled appearance.

The C gene. The wild-type allele C permits color expression, and the allele c prevents color expression. The c / c constitution is epistatic to the other color genes. A c / c animal is, of course, an albino, as we have already discussed in Chapter 5. Albinos are common in many mammalian species and have also been reported among birds, snakes, and fish (Figure 14-18). In most cases, the C gene codes for the enzyme tyrosinase, which is involved in the production of melanin.

In rabbits, an allele of the C gene, the c^h (Himalayan) allele, determines that melanin will be deposited only at the body extremities. In mice, the same allele produces a similar phenotype, also called *Himalayan,* and in cats the same allele produces the phenotype called *Siamese* (Figure 14-19 on the next page). The allele c^h can be considered a version of the c allele with heat-sensitive expression. It is only in the colder body extremities that the protein product of c^h is functional and can make pigment, because at lower temperatures the protein undergoes a conformational change that enables its active site to assume the correct shape. In warmer parts of the body it is expressed just like the albino allele c (that is, the protein is not functional). The c^h allele shows clearly how the expression of an allele can depend on the environment.

(a)

(b)

Figure 14-18 Albinism in reptiles and birds. In each case, the phenotype is produced by a recessive allele that determines an inability to produce the dark pigment melanin in skin cells. (The normal allele determines the ability to synthesize melanin.) (a) In the rattlesnake, the normal dark coloration is due entirely to melanin, so the albino allele results in a completely unpigmented appearance. (b) In the penguin, melanin normally makes the dorsal feathers black, but the reddish-orange colors in the head feathers and beak are due to another pigment chemically unrelated to melanin. The recessive albino allele results in no melanin, but the reddish parts are unaffected and retain their normal coloration. *(Part a, K. H. Switak/NHPA; part b, A.N.T./NHPA.)*

The D gene. The D gene controls the intensity of the coloration specified by the other coat color genes. The genotypes D / D and D / d permit full expression of color in mice, but d / d "dilutes" the color, making it look milky. The effect is due to an uneven distribution of pigment in the hair shaft. Dilute agouti, dilute cinnamon, dilute brown, and dilute black coats are all possible. A gene that affects the expression of another gene in such a way is called a **modifier gene.** The dilution allele has also been identified in horses, but in contrast to the one in mice, it shows incomplete dominance, so that two doses of the allele produce a milkier effect than one. Figure 14-20 on the next page shows how dilution affects the chestnut color and another coat color called bay. Cases of dilution in the coats of domestic cats also are commonly seen.

Modifier gene action can be based on many different molecular mechanisms. One example involves regulatory genes whose products bind to the upstream region of another gene near the promoter and affect the level of its transcription. Positive regulators increase ("up-regulate") transcription rates, and negative regulators decrease ("down-regulate") them. Consider the regulation of a gene G. G is the normal allele coding for an active protein, whereas g is a null allele (caused by a base-pair substitution) that codes for an inactive protein. At an unlinked locus, R codes for a regulatory protein that causes high levels of transcription at the G locus, whereas r yields a protein that allows only a basal level of transcription. If a dihybrid G / g ; R / r is selfed, a 9:3:4 ratio of protein activity is produced, as follows:

		Transcription	Protein activity
9	$G/-$; $R/-$	High	High
3	$G/-$; r/r	Low	Low
3	g/g ; $R/-$	High	None
1	g/g ; r/r	Low	None

(Note that a null mutation of R would probably act in an epistatic manner, shutting down expression of G completely; it would cause *no* transcription rather than *low* transcription. Hence different mutant alleles of one gene can act as modifiers or in an epistatic manner, depending on the mutation.)

The S gene. The S gene controls the presence or absence of spots by controlling the migration of clumps of melanocytes

Figure 14-19 The temperature-sensitive Himalayan allele (c^h) of the C gene results in similar phenotypes in several different mammals. In homozygous animals, there is reduced or no synthesis of the dark pigment melanin in the skin covering warmer parts of the body. At lower temperatures, such as those found at the body extremities, melanin is synthesized, producing darker snout, ears, tail, and feet. The Siamese cat (a), Himalayan mouse (b), and Himalayan rabbit (not shown) are all of genotype c^h / c^h. *(Part a, Walter Chandoha; part b, Anthony Griffiths.)*

(pigment-producing cells) across the surface of the developing embryo. The genotype $S / -$ results in no spots, and s / s produces a spotting pattern called *piebald* in both mice and horses. This pattern can be superimposed on any of the coat colors discussed earlier, with the exception of albino, of course. Piebald mutations are also known in humans.

Figure 14-21 illustrates some of the pigment patterns found in mice. We see that the normal coat color of wild-type mice is produced by a complex set of interacting genes determining the presence or absence of pigment, pigment type, pigment distribution in the individual hairs, and pigment distribution over the animal's body. Such interactions can be deduced from dihybrid ratios. Interacting genes such as these determine most characters in any organism.

MESSAGE

Different kinds of modified dihybrid ratios point to different ways in which genes can interact with one another to determine phenotype.

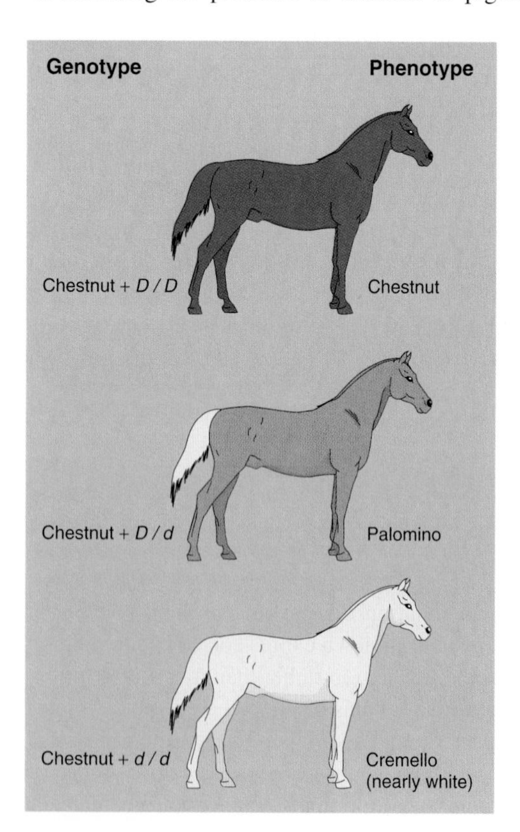

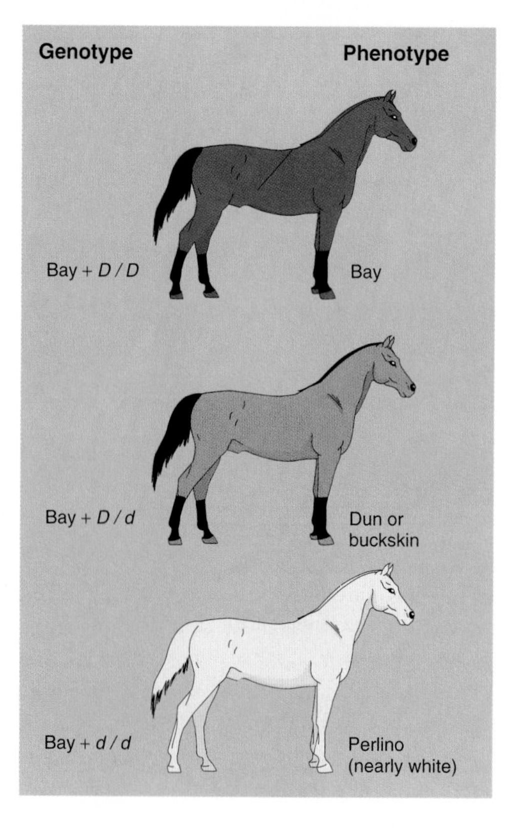

Figure 14-20 The modifying effect of the dilution allele on basic chestnut and bay genotypes in horses. Note the incomplete dominance shown by D. *(From J. W. Evans et al.,* The Horse. *W. H. Freeman and Company, 1977.)*

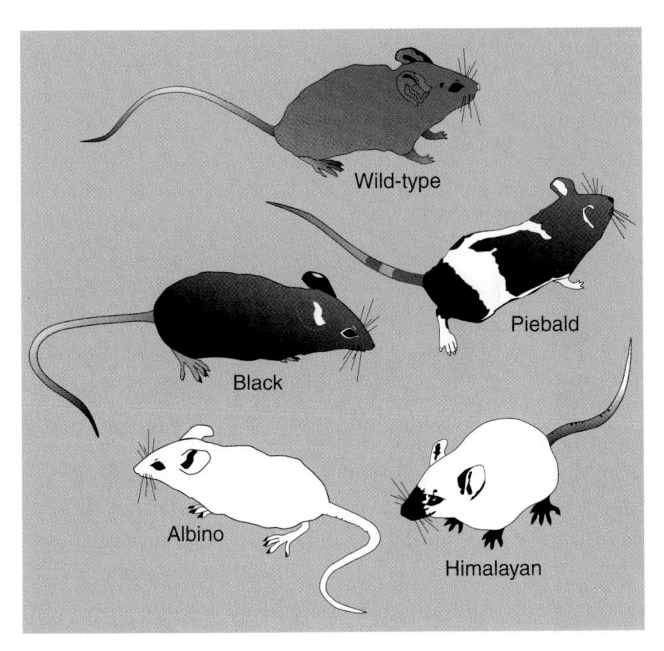

Figure 14-21 Some coat color phenotypes in mice. Genotypes written as homozygotes are

Wild type	$A/A \;.\; B/B \;.\; C/C \;.\; D/D \;.\; S/S$
Black	$a/a \;.\; B/B \;.\; C/C \;.\; D/D \;.\; S/S$
Albino	$-/- \;.\; -/- \;.\; c/c \;.\; -/- \;.\; -/-$
Himalayan	$a/a \;.\; B/B \;.\; c^h/c^h \;.\; D/D \;.\; S/S$
Piebald	$a/a \;.\; B/B \;.\; C/C \;.\; D/D \;.\; s/s$

PENETRANCE AND EXPRESSIVITY

In the previous examples, the genetic basis of the dependence of one gene on another was deduced from clear genetic ratios. However, only a relatively small proportion of genes in the genome have properties that make them favorable for such analysis. One important requirement for doing such an analysis is that the mutation not exhibit decreased viability or fertility relative to the wild type, so that the ratio of mutant and wild-type classes is not skewed.

Another requirement is that the difference between mutant and wild type in the norm of reaction (see Chapter 1) be dramatic enough that there is no overlap of their reaction curves, making it possible to distinguish mutant and wild-type genotypes with 100 percent certainty. In such cases, we say that the mutation is 100 percent **penetrant.** However, many mutations show incomplete penetrance. For example, an organism may have a particular allele but may not express the corresponding phenotype because of modifiers, epistatic genes, or suppressors in the rest of the genome or because of a modifying effect of the environment. Penetrance is thus defined as the *percentage of individuals* with a given allele that exhibit the phenotype associated with that allele.

Another useful term for describing the range of phenotypic expression is **expressivity.** Expressivity measures the *extent to which* a given allele is expressed at the phenotypic level. Different degrees of expression in different individuals may be due to variation in the allelic constitution of the rest of the genome or to environmental factors. Figure 14-22 illustrates the distinction between penetrance and expressivity. Like penetrance, expressivity is integral to the concept of the norm of reaction. An example of variable expressivity in dogs is found in Figure 14-23 on the next page.

Any kind of genetic analysis, such as human pedigree analysis and genetic counseling, can be made substantially more difficult by the phenomena of incomplete penetrance and variable expressivity. For example, if a disease-causing allele is not fully penetrant (as often is the case), it is difficult to give a clean genetic bill of health to any individual in a disease pedigree (for example, individual R in Figure 14-24 on the next page). On the other hand, pedigree analysis can sometimes identify individuals who do not express, but almost certainly do have, a disease genotype (for example, individual Q in Figure 14-24 on the next page). Similarly, variable expressivity can confound diagnosis.

MESSAGE

Only a subset of mutations are fully penetrant and show consistent expressivity. These experimentally reliable mutations are the most suitable for testing gene interaction patterns.

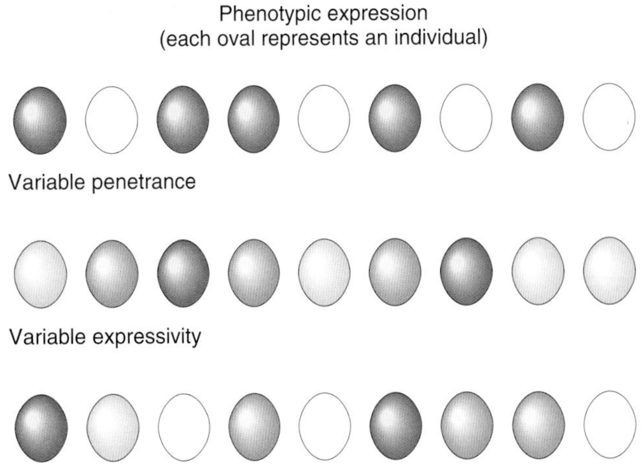

Figure 14-22 The difference between penetrance and expressivity for a hypothetical character called "pigment intensity." In each row, all individuals have the same allele—say, P—giving them the same "potential to produce pigment." However, effects deriving from the rest of the genome and from the environment may suppress or modify pigment production in an individual.

Figure 14-23 Variable expressivity as shown by 10 grades of piebald spotting in beagles. Each of these dogs has S^P, the allele responsible for the piebald pattern in dogs. *(Adapted from Clarence C. Little,* The Inheritance of Coat Color in Dogs, *Cornell University Press, 1957, and from Giorgio Schreiber,* Journal of Heredity *9, 1930, 403.)*

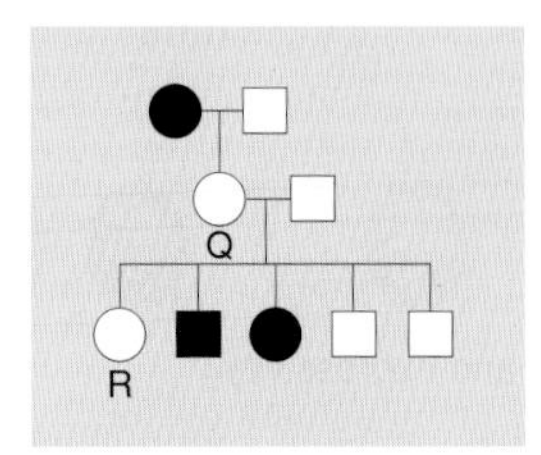

Figure 14-24 Lack of penetrance illustrated by a pedigree for a rare dominant allele. Individual Q must have the allele because it was passed on to her progeny, even though it was not expressed in her phenotype (her husband is an unlikely source of the allele because it is rare). An individual such as R cannot be sure that her genotype lacks the allele.

Let us return to the phenylketonuria system used to introduce this chapter. In the light of what we have learned, what can we say about the system? A person who is homozygous null for the phenylalanine hydroxylase gene should normally have PKU, but we saw that in some cases such persons do not. Although little is known definitely, we can make sense of this anomaly in several different ways. For example, it is possible that there is a defect in the synthesis of phenylpyruvic acid or in its transport to the brain. Such a defect could involve one of many different processes. Alternatively, in some people, there might be variants affecting the ability to transport dietary phenylalanine into a liver cell. All these defects would act as suppressors of PKU, enabling a person with the PKU mutation to avoid the disease. Then what could explain those persons who suffer the symptoms of PKU without carrying a mutation of the phenylalanine hydroxylase locus? Three percent of people with hyperphenylalaninemia show reduced levels of tetrahydrobiopterin, the cofactor of the phenylalanine hydroxylase enzyme. This deficiency results from defects in genes in the pathways that produce tetrahydrobiopterin: the most common such defect is in the gene for dihydrobiopterin synthetase, which makes one of the component structures of the tetrahydrobiopterin molecule. Alternatively, there could be other enzymes "downstream" of PAH that, if defective, could cause a backup of phenylalanine concentration. Such a mutation would act essentially as a functional duplicate of the PAH gene.

Hence, overall, we see that genetic systems are inherently complex because they reflect the complexity of the cellular machinery. We can identify individual components of the system using genetic and other analyses, but we must not be seduced into believing that the component we have identified genetically is the main component. This caution comes back to the idea of norms of reaction, introduced in Chapter 1. Each allele of a gene has a norm of reaction, which is determined (to varying degrees) by the state of all the other genes in the genome and the range of environments that the organism or cell is exposed to during its development.

SUMMARY

1. Segregation analysis in crosses can show that two alternative phenotypes are caused by different alleles of one gene. However, a gene or its alleles cannot act alone: the effect of a gene on the phenotype depends on its interaction with other genes and with the external and internal environments.
2. Genetic dissection of a biological system starts with the collection of mutations affecting the process or structure in question. The next step is to sort these mutations into genes and their alleles. Hence a reliable test for allelism is required. The most commonly used test is the complementation test, in which two mutant haploid genomes are introduced into one cell. If the phenotype of the cell is mutant, the paired mutations must be allelic.

3. Conversely, if the two mutations genomes together produce a wild-type phenotype (i.e., they complement), the mutations must be in different genes, either unlinked or linked. Thus key genes in the system may be identified.
4. Alleles show a wide range of interactions with other alleles at the same locus. In some cases heterozygotes show full dominance, in others intermediate or incomplete dominance, and in others joint expression or codominance. These categories of dominance are mere conveniences: for example, a heterozygote may show full dominance at the observable phenotypic level but codominance at the protein level.
5. Many mutant alleles that drastically alter function are lethal because they remove an essential function. Others have a less drastic or sublethal effect. In some cases a mutant allele can have an observable phenotypic effect in just one dose (heterozygous with wild type) but may be lethal in two doses (homozygous).
6. A biological property of an organism is the result of the actions of many genes in combination with the environment. Genetic dissection can identify individual components of such a system, generally by the severe effects of mutations on that system. Then, by intercrossing mutants in a pairwise manner, researchers can infer various types of interactions, gradually piecing together a picture of the intact system. Examples of the types of interactions revealed are collaboration of genes, epistasis (a mutation of one gene taking phenotypic precedence over mutations at another), modification of one gene's effect by another, and suppression of one mutation by another, resulting in the wild-type phenotype.
7. In a specific heterozygous combination at one locus, interaction between the alleles is revealed by monohybrid ratios such as 3:1 (full dominance), 1:2:1 (incomplete dominance or codominance) or 2:1 (recessive lethal).
8. Interactions between two specific genes are generally revealed as modified 9:3:3:1 dihybrid ratios. In cases of interaction, the 9, 3, 3, and 1 components of the ratio are combined in various ways to yield the modified ratios. An example is a 9:3:4 ratio, which points to recessive epistasis. In many such cases, the two genes act sequentially in some type of cellular pathway, so that when the earlier-acting gene is mutant, it does not matter what allele is present at a later-acting gene. Hence the 4 in the 9:3:4 ratio all have the same phenotype because they all bear the homozygous recessive epistatic genotype.
9. In some pedigrees, an individual that is known to bear a mutant phenotype does not express it—a case of lack of penetrance. In other cases, the degree of expression of a particular allele can show wide variation—a case of variable expressivity. Both effects are thought to be the result of genetic or environmental modification.

CONCEPT MAP

Draw a concept map interrelating as many of the following terms as possible. Note that the terms are in no particular order.

dihybrid / 9:3:4 ratio / independent assortment / biochemical pathway / recessive epistasis / gene interaction / meiosis / wild type

SOLVED PROBLEM

1. Beetles of a certain species may have green, blue, or turquoise wing covers. Virgin beetles were selected from a polymorphic laboratory population and mated to determine the inheritance of wing-cover color. The crosses and results were as follows:

Cross	Parents	Progeny
1	Blue × green	All blue
2	Blue × blue	$\frac{3}{4}$ blue : $\frac{1}{4}$ turquoise
3	Green × green	$\frac{3}{4}$ green : $\frac{1}{4}$ turquoise
4	Blue × turquoise	$\frac{1}{2}$ blue : $\frac{1}{2}$ turquoise
5	Blue × blue	$\frac{3}{4}$ blue : $\frac{1}{4}$ green
6	Blue × green	$\frac{1}{2}$ blue : $\frac{1}{2}$ green
7	Blue × green	$\frac{1}{2}$ blue : $\frac{1}{4}$ green : $\frac{1}{4}$ turquoise
8	Turquoise × turquoise	All turquoise

a. Deduce the genetic basis of wing-cover color in this species.

b. Write the genotypes of all parents and progeny as completely as possible.

Solution

a. These data seem complex at first, but the inheritance pattern becomes clear if we consider the crosses one at a time. A general principle of solving such problems, as we have seen, is to begin by looking over all the crosses and by grouping the data to bring out the patterns. One clue that emerges from an overview of the data is that all the ratios are one-gene ratios (3:1, 1:2:1, 1:1); there is no evidence of two separate genes being involved. How can such variation be explained with a single gene? The obvious answer is multiple allelism. Perhaps there are three alleles of one gene; let's call the gene *w* (for wing-cover color) and represent the alleles as w^g, w^b, and w^t. Now we have an additional problem, which is to determine the dominance of these alleles.

Cross 1 tells us something about dominance because the progeny of a blue × green cross are all blue; hence, blue appears to be dominant to green. This conclusion is supported by cross 5 because the green determinant must have been present in the blue parental stock to appear in the progeny. Cross 3 informs us about the turquoise determinants, which must have been present in the green parental stock because there are turquoise wing covers in the progeny. So green must be dominant to turquoise. Hence, we have formed a model in which the order of dominance is $w^b > w^g > w^t$. Indeed, the inferred position of the w^t allele at the bottom of the dominance series is supported by the results of cross 7, in which turquoise shows up in the progeny of a blue × green cross.

b. Notice that the question states that the parents were taken from a polymorphic population; this means that they could be either homozygous or heterozygous. A parent with blue wing covers, for example, might be homozygous (w^b / w^b) or heterozygous (w^b / w^g or w^b / w^t).

Here, a little trial and error and common sense are called for, but by this stage, the question has essentially been answered and all that remains is to "cross the t's and dot the i's." The following genotypes explain the results. (A dash indicates that the genotype may be homozygous or heterozygous with a second allele farther down the dominance series.)

Cross	Parents	Progeny
1	$w^b / w^b \times w^g / -$	w^b / w^g or $w^b / -$
2	$w^b / w^t \times w^b / w^t$	$\frac{3}{4}\ w^b / - : \frac{1}{4}\ w^t / w^t$
3	$w^g / w^t \times w^g / w^t$	$\frac{3}{4}\ w^g / - : \frac{1}{4}\ w^t / w^t$
4	$w^b / w^t \times w^t / w^t$	$\frac{1}{2}\ w^b / w^t : \frac{1}{2}\ w^t / w^t$
5	$w^b / w^g \times w^b / w^g$	$\frac{3}{4}\ w^b / - : \frac{1}{4}\ w^g / w^g$
6	$w^b / w^g \times w^g / w^g$	$\frac{1}{2}\ w^b / w^g : \frac{1}{2}\ w^g / w^g$
7	$w^b / w^t \times w^g / w^t$	$\frac{1}{2}\ w^b / - : \frac{1}{4}\ w^g / w^t : \frac{1}{4}\ w^t / w^t$
8	$w^t / w^t \times w^t / w^t$	all w^t / w^t

SOLVED PROBLEM

2. The leaves of pineapples can be classified into three types: spiny, spiny tip, and piping (nonspiny). In crosses between pure strains followed by intercrosses of the F_1, the following results were obtained:

	PHENOTYPES		
Cross	Parental	F_1	F_2
1	spiny tip × spiny	spiny tip	99 spiny tip : 34 spiny
2	piping × spiny tip	piping	120 piping : 39 spiny tip
3	piping × spiny	piping	95 piping : 25 spiny tip : 8 spiny

a. Assign gene symbols. Explain these results in terms of the genotypes produced and their ratios.

b. Using the model you developed in part a, give the phenotypic ratios you would expect if you crossed (1) the F_1 progeny from piping × spiny with the spiny parental stock, and (2) the F_1 progeny of piping × spiny with the F_1 progeny of spiny × spiny tip.

Solution

a. First, let's look at the F_2 ratios. We have clear 3:1 ratios in crosses 1 and 2, indicating single-gene segregations. Cross 3, however, shows a ratio that approximates a 12:3:1 ratio. This could be a modification of a 9:3:3:1 dihybrid ratio. Therefore, it looks as if we are dealing with a two-gene interaction. This seems the most promising place to start; we can go back to crosses 1 and 2 and try to fit them in later.

Any dihybrid self-cross involving two independently assorting genes (with complete dominance at both) gives the proportions 9:3:3:1. Since we are not told which phenotype is the wild type, we cannot use the wild-type symbols, so we just use the generic *A* / *a* and *B* / *b* symbols to represent the two interacting loci. The genotypes must be grouped as follows to obtain a 12:3:1 ratio:

$$\left.\begin{array}{l} 9\ A/- \ ;\ B/- \\ 3\ A/- \ ;\ b/b \end{array}\right\}\ 12 \text{ piping}$$

$$3\ a/a \ ;\ B/- \quad 3 \text{ spiny tip}$$

$$1\ a/a \ ;\ b/b \quad 1 \text{ spiny}$$

So, without worrying about the name of the type of gene interaction we are seeing (we are not asked to supply this anyway), we can already define our three pineapple leaf phenotypes in terms of the proposed allelic pairs *A*, *a* and *B*, *b*:

Piping	$A / -$ (*B*, *b* irrelevant)
Spiny tip	$a / a \ ;\ B / -$
Spiny	$a / a \ ;\ b / b$

NOTE: We could have designated these alleles, reversing the letters A and B, as follows: $B / -$ (*A*, *a* irrelevant), $b / b \ ;\ A / -$, and $b / b \ ;\ a / a$—it makes no difference. What about the parents of cross 3? The spiny parent must be $a / a \ ;\ b / b$, and because the *B* gene is needed to produce F_2 spiny tip individuals, the piping parent must be $A / A \ ;\ B / B$. (Note that we are told that all parents are pure, or homozygous.) The F_1 must therefore be

$A/a\ ;\ B/b$. Without further thought, we can write out cross 1 as follows:

$$a/a\ ;\ B/B \times a/a\ ;\ b/b \longrightarrow$$

$$a/a\ ;\ B/b \begin{cases} \frac{3}{4}\ a/a\ ;\ B/b \\ \frac{1}{4}\ a/a\ ;\ b/b \end{cases}$$

Cross 2 can also be written out partially without further thought, using our arbitrary gene symbols:

$$A/A\ ;\ -/- \times a/a\ ;\ B/B \longrightarrow$$

$$A/a\ ;\ B/- \begin{cases} \frac{3}{4}\ A/-\ ;\ -/- \\ \frac{1}{4}\ a/a\ ;\ B/- \end{cases}$$

We know that the F_2 of cross 2 shows single-gene segregation, and it seems certain now that the A/a allelic pair is involved. But the B allele is needed to produce the spiny tip phenotype, so all individuals must be homozygous B/B:

$$A/A\ ;\ B/B \times a/a\ ;\ B/B \longrightarrow$$

$$A/a\ ;\ B/B \begin{cases} \frac{3}{4}\ A/-\ ;\ B/B \\ \frac{1}{4}\ a/a\ ;\ B/B \end{cases}$$

Notice that the two single-gene segregations in crosses 1 and 2 do not show that the genes are *not* interacting. What is shown is that the two-gene interaction is not *revealed* by these crosses—only by 3, in which the F_1 is heterozygous for both genes, which segregate and produce all four phenotypes in the F_2.

b. Now it is simply a matter of using the laws of segregation and independent assortment to predict the cross outcomes:

(1) $A/a\ ;\ B/b \times a/a\ ;\ b/b \longrightarrow$

(independent assortment in a standard testcross)

$\frac{1}{4}\ A/a\ ;\ B/b$ } piping
$\frac{1}{4}\ A/a\ ;\ b/b$ } piping
$\frac{1}{4}\ a/a\ ;\ B/b$ spiny tip
$\frac{1}{4}\ a/a\ ;\ b/b$ spiny

(2) $A/a\ ;\ B/b \times a/a\ ;\ B/b \longrightarrow$

$\frac{1}{2}\ A/a$ → $\frac{3}{4}\ B/-$ → $\frac{3}{8}$ } $\frac{1}{2}$ piping
$\frac{1}{2}\ A/a$ → $\frac{1}{4}\ b/b$ → $\frac{1}{8}$ } $\frac{1}{2}$ piping
$\frac{1}{2}\ a/a$ → $\frac{3}{4}\ B/-$ → $\frac{3}{8}$ spiny tip
$\frac{1}{2}\ a/a$ → $\frac{1}{4}\ b/b$ → $\frac{1}{8}$ spiny

PROBLEMS

Basic Problems

1. Both recombination and complementation can take two different recessive mutants and from them produce a wild type. How would you explain the difference between these two processes?

2. Erminette fowls have mostly light-colored feathers with an occasional black one, which gives them a flecked appearance. A cross of two erminettes produced a total of 48 progeny, consisting of 22 erminettes, 14 blacks, and 12 pure whites. What genetic basis of the erminette pattern is suggested? How would you test your hypothesis?

3. In the multiple allelic series that determines coat color in rabbits, c^+ codes for agouti, c^{ch} for chinchilla (a beige coat color), and c^h for Himalayan. Dominance is in the order $c^+ > c^{ch} > c^h$. In a cross of $c^+/c^{ch} \times c^{ch}/c^h$, what proportion of progeny will be chinchilla?

4. Black (B), sepia (S), cream (C), and albino (A) are all coat colors found among laboratory guinea pigs. Individual animals (not necessarily from pure lines) showing these colors were intercrossed; the results are tabulated as follows.

		PHENOTYPES OF PROGENY			
Cross	Parental phenotypes	B	S	C	A
1	B × B	22	0	0	7
2	B × A	10	9	0	0
3	C × C	0	0	34	11
4	S × C	0	24	11	12
5	B × A	13	0	12	0
6	B × C	19	20	0	0
7	B × S	18	20	0	0
8	B × S	14	8	6	0
9	S × S	0	26	9	0
10	C × A	0	0	15	17

a. Deduce the inheritance of these coat colors, using gene symbols of your own choosing. Show all parent and progeny genotypes.

b. If the black progeny of crosses 7 and 8 were crossed, what progeny proportions would you predict using your model?

5. In a maternity ward, four babies have been accidentally mixed up. The ABO types of the four babies are known to be O, A, B, and AB. The ABO types of the four sets of parents have also been determined. Indicate which baby belongs to each set of parents: (a) AB × O, (b) A × O, (c) A × AB, (d) O × O.

6. Two fruit flies of wild-type phenotype were crossed, and the progeny consisted of 202 females and 98 males.

a. What is unusual about this result?

b. Provide a genetic explanation.

c. Provide a test of your hypothesis.

7. A pure-breeding strain of squash that produced disk-shaped fruits (see the figure below) was crossed to a pure-breeding strain that produced long fruits. The F_1 had disk fruits, but the F_2 showed a new phenotype, sphere, and consisted of the following proportions:

Disk	270
Sphere	178
Long	32

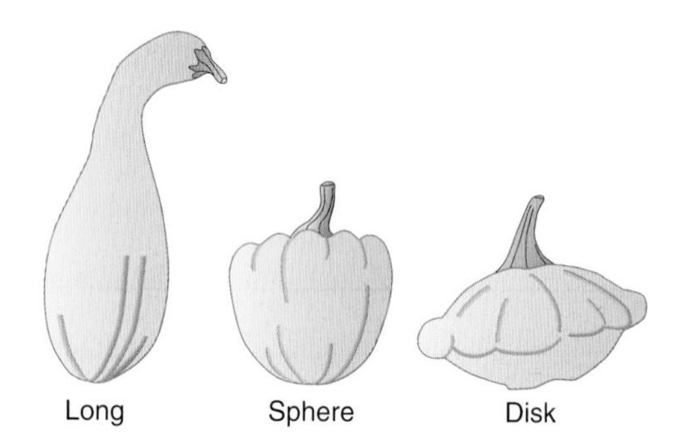

Propose an explanation for these results, and show the genotypes of the P, F_1, and F_2 generations.

(Illustration from P. J. Russell, *Genetics*, 3d ed. Harper-Collins, 1992.)

8. Wild-type snapdragons *(Antirrhinum)* make a red anthocyanin pigment and have red petals. Two pure anthocyaninless mutant lines of *Antirrhinum* were developed, one in California and one in Holland. Both mutant lines had no red pigment at all, producing identical-looking white (albino) flowers. However, when petals from the two lines were ground up together in buffer in the same test tube, the solution, which appeared colorless at first, gradually turned red.

a. What control experiments should an investigator conduct before proceeding with further analysis?

b. What could account for the production of the red color in the test tube?

c. Based on your answer to part b, what would be the genotypes of the two lines?

d. If the two white lines were crossed, what would you predict the phenotypes of the F_1 and F_2 to be? What genetic principle would be illustrated by the F_1 results? The F_2 results?

9. Four homozygous recessive mutant lines of *Drosophila melanogaster* (labeled 1 through 4) showed abnormal leg coordination, which made their walking highly erratic. These lines were intercrossed; the phenotypes of the F_1 flies are shown in the following grid, in which "+" represents wild-type walking and "−" represents abnormal walking:

	1	2	3	4
1	−	+	+	+
2	+	−	−	+
3	+	−	−	+
4	+	+	+	−

a. What type of test does this analysis represent?

b. How many different genes were mutated in creating these four lines?

c. Invent wild-type and mutant symbols and write out full genotypes for all four lines and for the F_1s.

d. Do these data tell us which genes are linked? If not, how could linkage be tested?

e. Do these data tell us the total number of genes involved in leg coordination in this animal?

10. Three independently isolated tryptophan-requiring mutants of haploid yeast are called *trpB, trpD,* and *trpE.* Cell suspensions of each are streaked on a plate of nutritional medium supplemented with just enough tryptophan to permit weak growth for a *trp* strain. The streaks are arranged in a triangular pattern so that they do not touch one another. Luxuriant growth is noted at both ends of the *trpE* streak and at one end of the *trpD* streak (see the figure below).

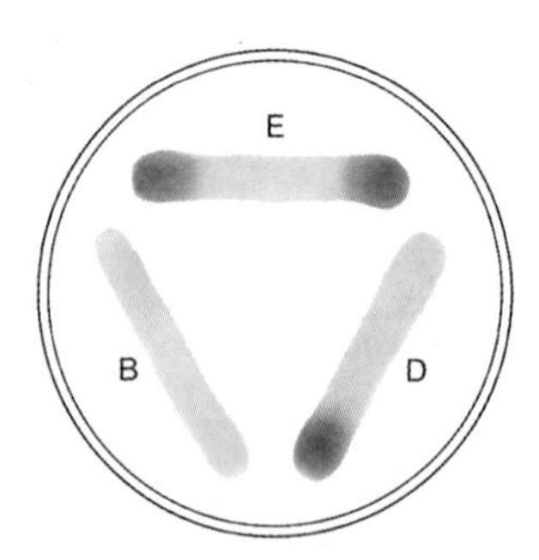

a. Do you think complementation is involved?

b. Briefly explain the pattern of luxuriant growth.

c. In what order in the tryptophan-synthesizing pathway are the enzymatic steps that are defective in *trpB, trpD,* and *trpE?*

d. Why was it necessary to add a small amount of tryptophan to the medium in order to demonstrate such a growth pattern?

11. The frizzle fowl is much admired by poultry fanciers. It gets its name from the unusual way that its feathers curl up, giving the impression that it has been (in the memorable words of animal geneticist F. B. Hutt) "pulled backwards through a knothole." Unfortunately, frizzle fowls do not breed true; when two frizzles are intercrossed, they always produce progeny that are 50 percent frizzles, 25 percent normal, and 25 percent with peculiar woolly feathers that soon fall out, leaving the birds naked.

a. Give a genetic explanation for these results, showing genotypes for all phenotypes, and provide a statement of how your explanation works.

b. If you wanted to mass-produce frizzle fowls for sale, which genotypes would be best to use as breeding pairs?

12. Marfan's syndrome is a disorder of the fibrous connective tissue characterized by many symptoms, including long, thin digits; eye defects; heart disease; and long limbs. (Flo Hyman, the American volleyball star, suffered from Marfan's syndrome. She died soon after a game from a ruptured aorta.)

a. Use the pedigree shown below to propose a mode of inheritance for Marfan's syndrome.

b. What genetic phenomenon is shown by this pedigree?

c. Speculate on a reason for such a phenomenon.

(Illustration from J. V. Neel and W. J. Schull, *Human Heredity,* University of Chicago Press, 1954.)

13. A woman owned a fine purebred albino female poodle (an autosomal recessive phenotype) and wanted to mate her to produce a litter of white puppies. She took the dog to a breeder, who said he would mate the female with an albino stud male, also from a pure stock. When six puppies were born, they were all black, so the woman sued the breeder, claiming that he had replaced the stud male with a black dog, giving her six unwanted puppies. You are called in as an expert witness, and the defense asks you whether it is possible to produce black offspring from two pure-breeding recessive albino parents. What testimony do you give?

14. The petals of the plant *Collinsia parviflora* are normally blue, giving the species its common name, "blue-eyed Mary." Two pure-breeding lines were obtained from color variants found in nature; the first line had pink petals and the second line had white petals. The following crosses were made between pure lines, with the results shown below.

Parents	F_1	F_2
Blue × white	Blue	101 blue, 33 white
Blue × pink	Blue	192 blue, 63 pink
Pink × white	Blue	272 blue, 121 white, 89 pink

a. Explain these results genetically. Define the allele symbols you use and show the genetic constitution of parents, F_1, and F_2.

b. A cross between a certain blue F_2 plant and a certain white F_2 plant gave progeny of which $\frac{3}{8}$ were blue, $\frac{1}{8}$ were pink, and $\frac{1}{2}$ were white. What must the genotypes of these two F_2 plants have been?

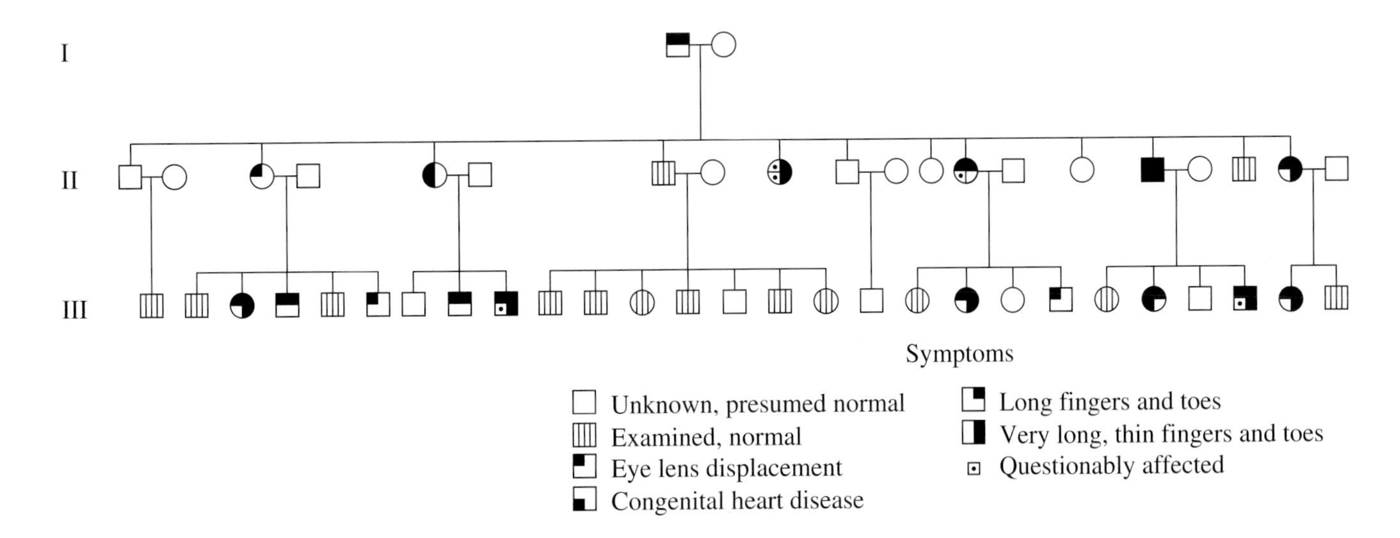

Unpacking the Problem

a. What is the character being studied?

b. What is the wild-type phenotype?

c. What is a variant?

d. What are the variants in this problem?

e. What does "in nature" mean?

f. In what way would the variants have been found in nature? (Describe the scene.)

g. At which stages in the experiments would seeds be used?

h. Would writing "blue × white" (for example) mean the same as writing "white × blue"? Would you expect similar results from the two crosses? Why or why not?

i. In what way do the first two rows in the table differ from the third row?

j. Which phenotypes are dominant?

k. What is complementation?

l. Where does the blueness come from in the progeny of the pink × white cross?

m. What genetic phenomenon does the production of a blue F_1 from a pink and a white parent represent?

n. List any ratios you can see.

o. Are there any monohybrid ratios?

p. Are there any dihybrid ratios?

q. What does observing monohybrid and dihybrid ratios tell you?

r. List four modified Mendelian ratios that you can think of.

s. Are there any modified Mendelian ratios in the problem?

t. What do modified Mendelian ratios indicate in general?

u. What do the specific modified ratio or ratios in this problem indicate?

v. Draw chromosomes representing the meioses in the parents in the cross blue × white, and meiosis in the F_1.

w. Repeat for the cross blue × pink.

15. Most flour beetles are black, but several color variants are known. Crosses of pure-breeding parents produced the following results in the F_1 generation, and intercrossing the F_1 from each cross gave the ratios shown for the F_2 generation. The phenotypes are black (Bl), brown (Br), yellow (Y), and white (W).

Cross	Parents	F_1	F_2
1	Br × Y	Br	3 Br : 1 Y
2	Bl × Br	Bl	3 Bl : 1 Br
3	Bl × Y	Bl	3 Bl : 1 Y
4	W × Y	Bl	9 Bl : 3 Y : 4 W
5	W × Br	Bl	9 Bl : 3 Br : 4 W
6	Bl × W	Bl	9 Bl : 3 Y : 4 W

a. From these results, deduce and explain the inheritance of these colors.

b. Write out the genotypes of each of the parents, the F_1, and the F_2 in all crosses.

c. Arrange the genes and pigments into a biochemical pathway or pathways.

16. An allele *A* that is not lethal when homozygous causes rats to have yellow coats. The allele *R* of a separate gene that assorts independently produces a black coat. Together, *A* and *R* produce a grayish coat, whereas *a* and *r* produce a white coat. A gray male is crossed to a yellow female, and the F_1 is $\frac{3}{8}$ yellow, $\frac{3}{8}$ gray, $\frac{1}{8}$ black, and $\frac{1}{8}$ white. Determine the genotypes of the parents.

17. The genotype r/r ; p/p gives fowls a single comb, $R/-$; $P/-$ gives a walnut comb, r/r ; $P/-$ gives a pea comb, and $R/-$; p/p gives a rose comb (see the figure below). Note that the genes are unlinked.

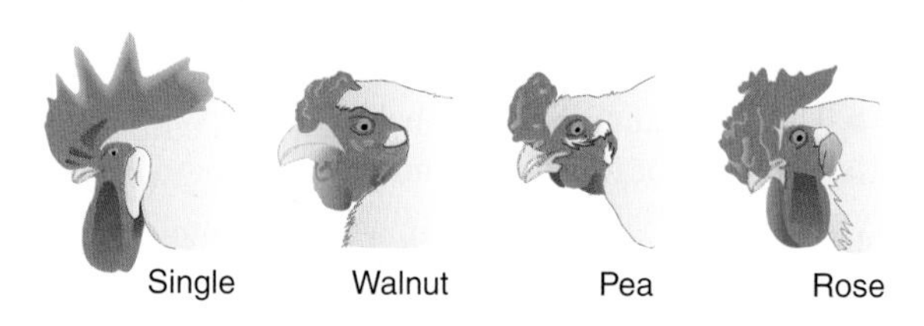

a. If single-combed birds are crossed to birds of a true-breeding walnut strain, what comb types will appear in the F_1 and in the F_2 in what proportions?

b. What are the genotypes of the parents in a walnut × rose mating from which the progeny are $\frac{3}{8}$ rose, $\frac{3}{8}$ walnut, $\frac{1}{8}$ pea, and $\frac{1}{8}$ single?

c. What are the genotypes of the parents in a walnut × rose mating from which all the progeny are walnut?

d. How many genotypes produce a walnut phenotype? Write them out.

Challenging Problems

18. The allele *B* gives mice a black coat, and *b* gives a brown one. The genotype e/e for another, independently assorting gene prevents expression of *B* and *b*, making the coat color beige, whereas $E/-$ permits expression of *B* and *b*. Both genes are autosomal. In the following pedigree, black symbols indicate a black coat, pink symbols indicate brown, and white symbols indicate beige:

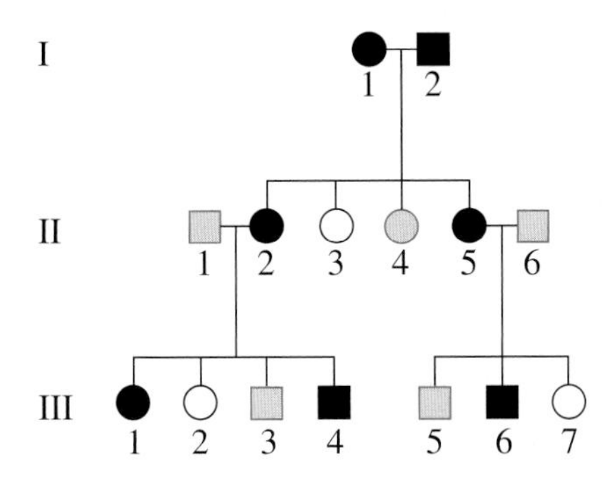

a. What is the name given to the type of gene interaction in this example?

b. What are the genotypes of the individuals in the pedigree? (If there are alternative possibilities, state them.)

19. Assume that two pigments, red and blue, mix to give the normal wild-type purple color of petunia petals. Separate biochemical pathways synthesize the two pigments, as shown in the top two rows of the accompanying diagram. "White" refers to compounds that are not pigments. (Total lack of pigment results in a white petal.) Red pigment forms from a yellow intermediate that normally is at a concentration too low to color petals. A third pathway whose compounds do not contribute pigment to petals normally does not affect the blue and red pathways, but if one of its intermediates ($white_3$) builds up in concentration, it can be converted to the yellow intermediate of the red pathway. In the diagram, A–E represent enzymes; their corresponding genes, all of which are unlinked, may be symbolized by the same letters.

Pathway I $\cdots \longrightarrow \text{white}_1 \xrightarrow{E} \text{blue}$

Pathway II $\cdots \longrightarrow \text{white}_2 \xrightarrow{A} \text{yellow} \xrightarrow{B} \text{red}$

(C: $\text{white}_3 \dashrightarrow \text{yellow}$)

Pathway III $\cdots \longrightarrow \text{white}_3 \xrightarrow{D} \text{white}_4$

Assume that wild-type alleles are dominant and code for enzyme function, and that recessive alleles represent lack of enzyme function. Deduce which combinations of true-breeding parental genotypes could be crossed to produce F_2 progenies in the following ratios:

a. 9 purple : 3 green : 4 blue

b. 9 purple : 3 red : 3 blue : 1 white

c. 13 purple : 3 blue

d. 9 purple : 3 red : 3 green : 1 yellow

e. Which of the mutations is acting as a suppressor?

f. Which of the mutations is showing recessive epistasis?

(NOTE: Blue mixed with yellow makes green; assume that no mutations are lethal.)

20. The following pedigree shows the inheritance of deaf-mutism:

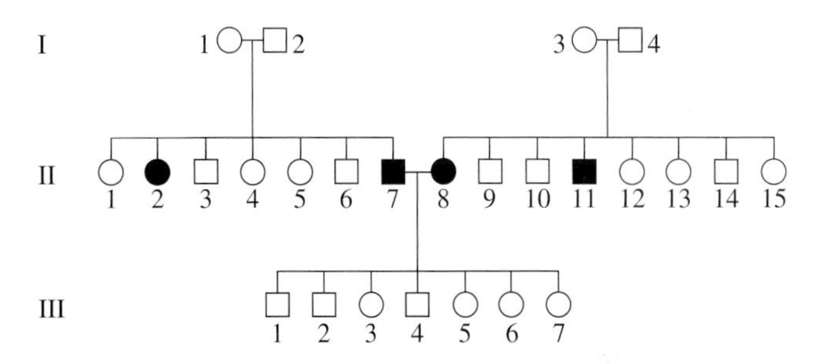

a. Provide an explanation for the inheritance of this rare condition in the two families in generations I and II, showing the genotypes of as many individuals as possible using symbols of your own choosing.

b. Provide an explanation for the presence of only normal individuals in generation III, making sure your explanation is compatible with your answer to part a.

21. The following pedigree shows the inheritance of blue sclera (the thin outer wall of the eye) and brittle bones:

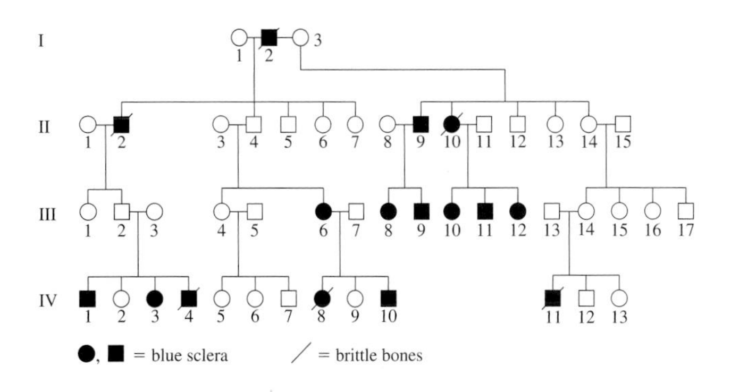

a. Are these two disorders caused by the same gene or by two different genes? State the reasons for your answer clearly.

b. Is the allele or alleles involved autosomal or sex-linked?

c. Does the pedigree show any evidence of incomplete penetrance or expressivity? If so, make the best calculation that you can of penetrance.

22. In minks, the wild type has an almost black coat. Breeders have developed many pure lines of color variants for the mink coat industry. Two such pure lines are platinum (blue-gray) and aleutian (steel-gray). These lines were used in crosses, with the following results:

Cross	Parents	F_1	F_2
1	Wild × platinum	Wild	18 wild, 5 platinum
2	Wild × aleutian	Wild	27 wild, 10 aleutian
3	Platinum × aleutian	Wild	133 wild 41 platinum 46 aleutian 17 sapphire (a new silvery color)

a. Devise a genetic explanation for the results of these three crosses. Show genotypes for parents, F_1, and F_2 in the three crosses, and make sure you show the alleles of each gene you hypothesize in every individual.

b. Predict the F_1 and F_2 phenotypic ratios that would result from crossing sapphire with platinum and aleutian pure lines.

23. A geneticist is interested in the genes whose products interact to synthesize leucine (an amino acid) in the haploid filamentous fungus *Aspergillus*. He treats haploid cells with ultraviolet light to increase the mutation rate and obtains five haploid leucine-requiring mutants (*a* to *e*), all of which need to have leucine added to their medium in order to grow. (Without leucine, none of them will grow.)

Experiment I. The geneticist first makes heterokaryons between the mutant strains to check on their functional relationships. He obtains the following results (where "+" indicates that the heterokaryon grew, and "−" that it did not grow, on a medium lacking leucine):

	a	*b*	*c*	*d*	*e*
a	−	+	+	+	−
b		−	+	+	+
c			−	+	+
d				−	+
e					−

Experiment II. The geneticist then intercrosses the mutants in all possible combinations. From each cross, he tests 500 ascospore progeny by placing them on a medium lacking leucine. The results are shown in the following table. The numbers represent the number of progeny (out of 500) that grew on the medium (showing that they were able to synthesize their own leucine):

	a	*b*	*c*	*d*	*e*
a	0	125	128	126	0
b		0	124	2	125
c			0	124	127
d				0	123
e					0

a. In both experiments, leucine-independent strains are created. Explain the different origins of these strains in the two experiments.

b. Explain the pattern of + and − growth in the heterokaryons in the different pairings in experiment I.

c. Explain the different frequencies of strains found in experiment II that are independent of leucine. (Note that the two leucine-independent progeny of the *b* × *d* cross were found *not* to be due to mutation back to wild type.)

d. Draw a diagram that summarizes the results of both experiments.

24. Wild-type *Neurospora* (a haploid fungus) can synthesize its own adenine from inorganic components in the growth medium. In a mutational analysis of the synthetic pathway for making adenine, 10 adenine-requiring mutations of *Neurospora* were obtained, and heterokaryons were made in all pairwise combinations. The results are shown in the following table, in which a plus sign means that the heterokaryon grew and a minus sign, that it did not grow on a medium lacking adenine:

	1	2	3	4	5	6	7	8	9	10
1	−	−	+	+	+	+	−	+	+	+
2		−	+	+	+	+	−	+	+	+
3			−	+	−	+	+	−	+	+
4				−	+	−	+	+	−	−
5					−	+	+	−	+	+
6						−	+	+	−	−
7							−	+	+	+
8								−	+	+
9									−	−
10										−

a. How many genes were involved in these mutations?

Mutants 1, 3, and 4 were tested for growth on the compounds CAIR, AIR, and SAICAR, all chemically related to adenine. The results were as follows:

	Adenine	CAIR	AIR	SAICAR
1	+	−	−	+
3	+	−	−	−
4	+	+	−	+

b. Explain these results.

Mutants 1, 3, and 4 were intercrossed, and 1000 ascospores from each cross were placed on a medium lacking adenine to select for wild types. The results were as follows:

	1	3	4
1	0	4	245
3	7	0	260
4	255	252	0

c. Explain these results and integrate them with the previous results to provide a summary statement about this genetic system.

25. In a plant, the gene e^+ codes for a certain enzyme. The gene r^+ codes for a regulatory protein that must bind to the regulatory region of gene e^+ for its transcription to occur. A recessive null mutation is obtained for each of the two genes, *r* and *e*. A dihybrid e^+ / e ; r^+ / r is allowed to self. In the progeny, what will be the proportion of individuals with the enzyme? (Assume that lack of the enzyme is not lethal and that the genes are on different chromosomes.)

26. It is known that a certain *Drosophila* protein (which we shall call P) is necessary for normal flight. This protein is known to be composed of only one polypeptide chain. Two pure lines of mutants (lines 1 and 2) were obtained that could not fly. When crossed to wild type, they both gave the same result:

Nonflyer × wild type ⟶ F_1 wild type ⟶ F_2 $\frac{3}{4}$ wild type : $\frac{1}{4}$ nonflyer

When the two pure lines were intercrossed, the results were

Nonflyer 1 × nonflyer 2 ⟶ F_1 all wild type ⟶ F_2 $\frac{9}{16}$ wild type : $\frac{7}{16}$ nonflyer

Protein from *all* members of the F_2 was run on electrophoretic gels and stained using an antibody that is specific for protein P. It was found that the F_2 individuals could be grouped into six different types as follows:

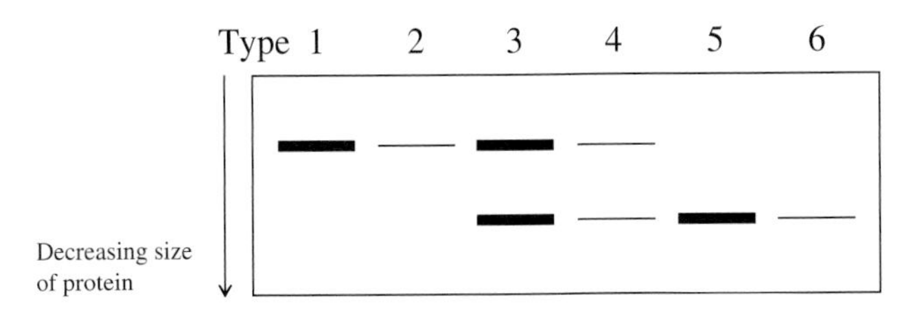

(The thin bands represent very weak antibody staining.)

a. Provide a general explanation for the results of the electrophoresis. In particular, state the difference in the molecular nature of the two original mutations.

b. According to your explanation, what are the genotypes of the six F_2 types found?

c. What is the expected frequency of type 4?

d. Which of the six F_2 types represent the $\frac{9}{16}$ of the F_2 that are phenotypically wild-type?

e. Draw the expected gel pattern of the two parental mutants and the F_1.

Pattern Recognition Problems

In Problems 27 through 32, the diagrams show phenotypes and the results of breeding analyses. Deduce the genotypes of the individuals shown in each diagram as far as possible. All organisms are diploid.

27. Lines 1, 2, and 3 are three independently obtained pure-breeding mutant lines.

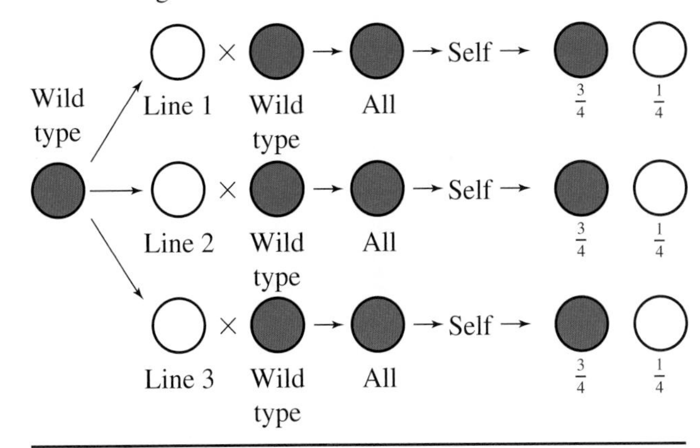

28.

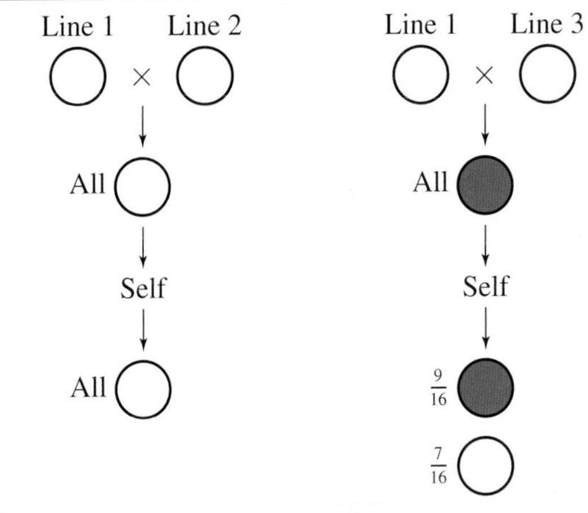

29.

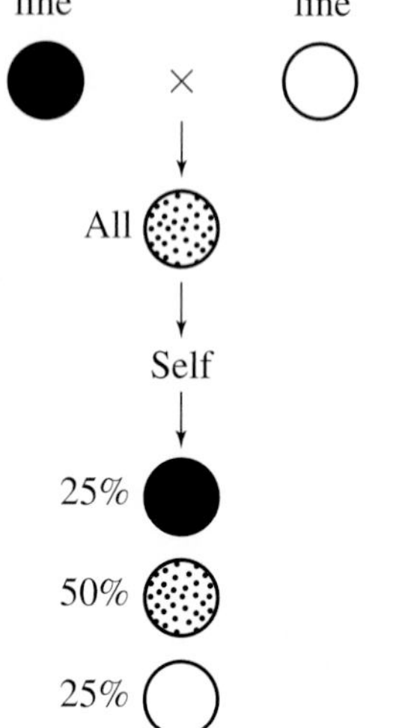

30.

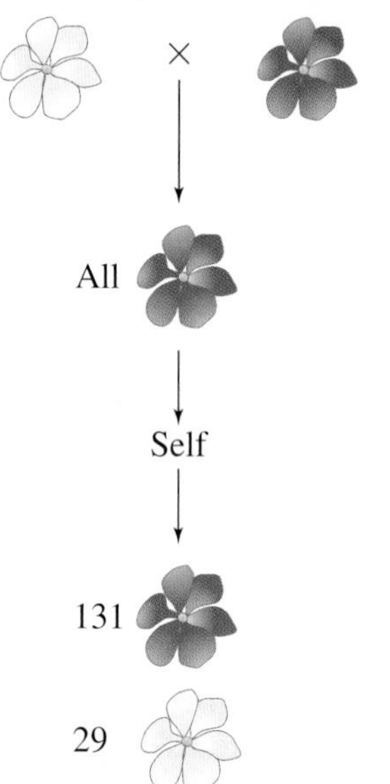

31.

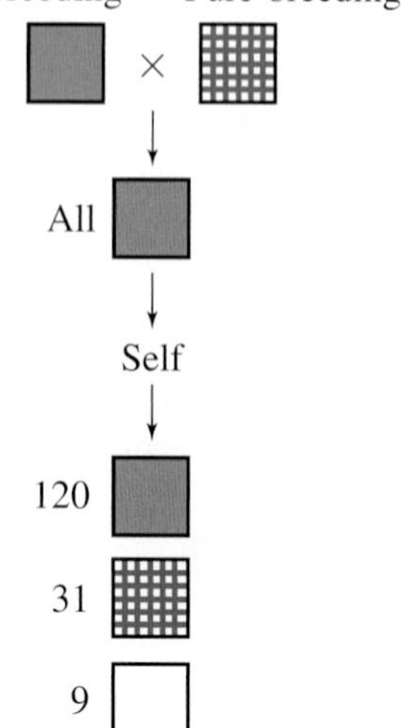

32. Using the following diagram, deduce genotype and dominance and predict the genotypes in the F_2s.

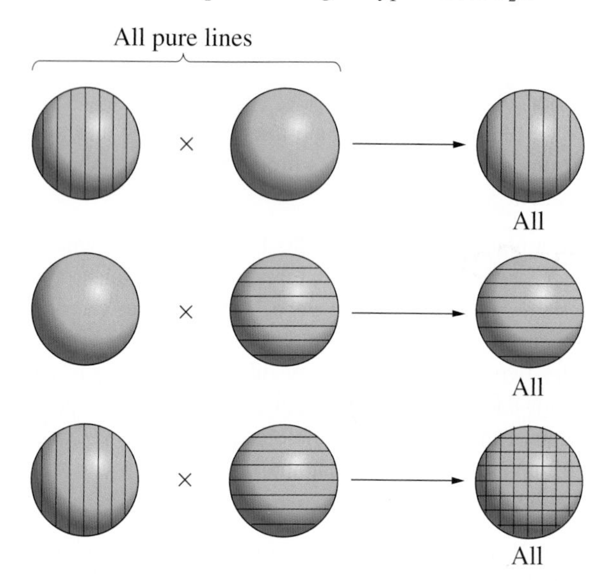

33. The wing covers of grain beetles exhibit one of three phenotypes, showing a diamond, a spot, or a stripe. The following crosses were made between various strains that were not necessarily pure-breeding:

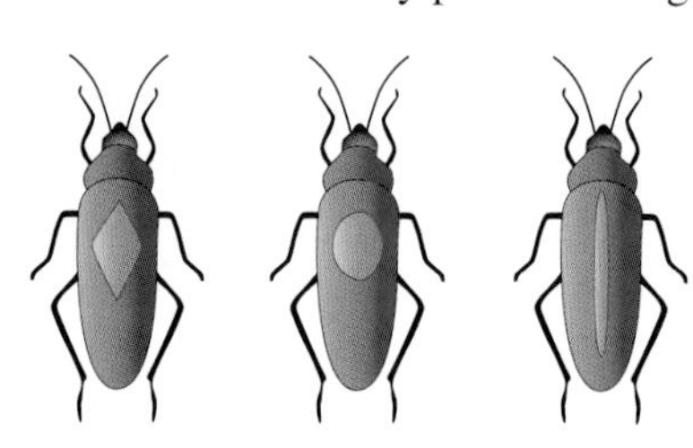

Cross	Progeny
a. Diamond × diamond	$\frac{3}{4}$ diamond
	$\frac{1}{4}$ stripe
b. Spot × spot	$\frac{6}{8}$ spot
	$\frac{1}{8}$ diamond
	$\frac{1}{8}$ stripe
c. Spot × stripe	$\frac{1}{2}$ spot
	$\frac{1}{4}$ diamond
	$\frac{1}{4}$ stripe

a. Invent allele symbols, explain clearly the action of all the alleles involved, then write the full genotypes of the parents and progeny in all three crosses. (If your explanation involves more than one gene, make sure *all* the genes are represented in the genotypes in every cross.)

b. What phenotypic ratio would result from crosses between the spotted progeny in cross 3?

REGULATION OF CELL NUMBER: NORMAL AND CANCER CELLS

15

Key Concepts

1. Higher eukaryotic cells have evolved mechanisms that control their survival and their ability to proliferate.
2. These controls are highly integrated and depend on the continual evaluation of the state of the cell and the continual communication of information among neighboring cells and between different tissues.
3. Normal cell proliferation is modulated by regulation of the cell cycle.
4. Apoptosis is a normal self-destruction mechanism that eliminates damaged and potentially harmful cells as well as cells needed only transiently during the program of development.
5. Intercellular signaling systems permit proliferation and apoptosis to be coordinated within a population of cells.
6. In cancer, cells proliferate out of control and avoid self-destruct mechanisms through the accumulation of a series of tumor-promoting mutations in the same somatic cell.
7. Many of the classes of genes in which mutations can cause cancers normally contribute directly or indirectly to growth control and differentiation.
8. Functional genomics is being used in diagnostic tests and in the identification of drug targets in studies to improve the detection and treatment of cancer.

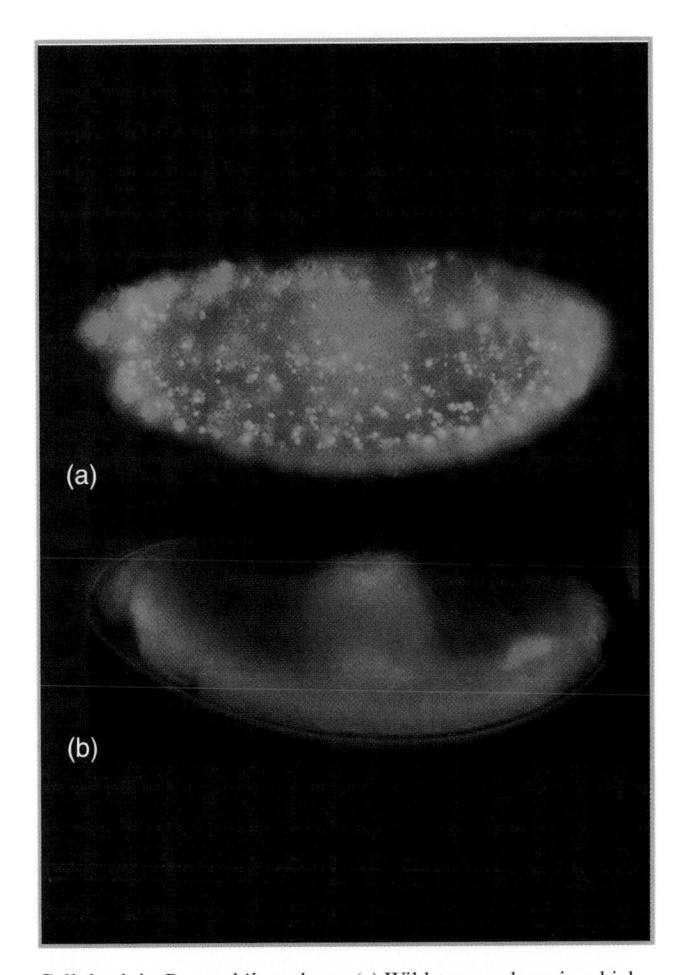

Cell death in *Drosophila* embryo. (a) Wild-type embryo in which the bright spots are cells carrying out a genetic program to die (apoptosis). (b) Mutant embryo in which this genetic program does not proceed. *(Kristin White, Massachusetts General Hospital and Harvard Medical School.)*

Lymphomas and leukemias are cancers of the white blood cells—the cells that make up the immune system. In these diseases, certain white blood cells massively overproliferate, leading to a complete imbalance in immune cells and, ultimately, a failure of the immune system. One class of lymphoma is called diffuse large B-cell lymphoma (DLBCL), the most common form of non-Hodgkin's lymphoma. In the United States alone, about 25,000 new cases of DLBCL are diagnosed every year. The diagnosis is based on a characteristic set of symptoms and on the histology (microscopic examination of cell and tissue morphology) of affected lymph nodes (Figure 15-1). Standard chemotherapy cures about 40 percent of DLBCL patients, but other patients do not respond to this regimen. Why this dichotomy among the patient population? It could be differences in genetic or environmental factors, differences in stages of the disease, or any of many other factors. Without understanding the molecular basis of the disease, it is difficult to home in on the right explanation and adjust treatments accordingly. Before discussing DLBCL further, let's explore how the body normally regulates cell numbers and then see how these regulatory mechanisms are disrupted in cancer—a group of diseases of massive cell overproliferation of cells.

(a)

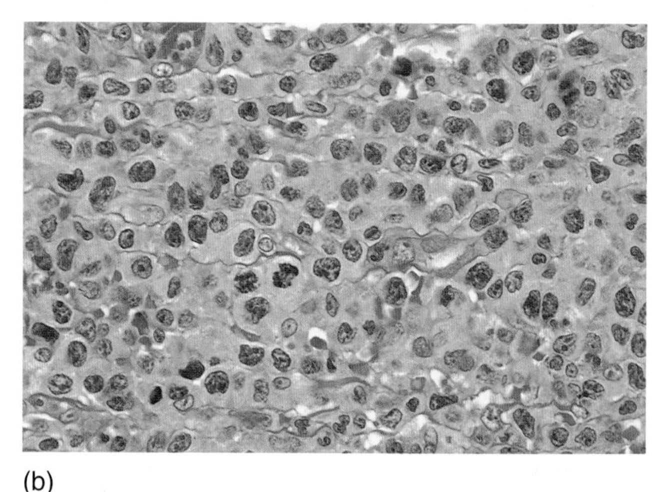

(b)

Figure 15-1 Histology of (a) a normal lymph node and (b) diffuse large B-cell lymphoma (DLBCL). The darkly stained cells are lymphoid cells. Note that the lymphocytes in the normal lymph node are small and uniformly shaped, but the lymphoma cells are large and more varied in shape. *(G. W. Willis, M.D. / Visuals Unlimited.)*

CHAPTER OVERVIEW

Chapter 13 described some ways that a cell compares its own condition with its environment and responds accordingly. For example, by using certain metabolites as allosteric effectors of transcriptional regulatory proteins, an *E. coli* cell can make decisions about which sugar metabolic pathways to implement at any given time. In analogous ways, metazoa (multi-tissued animals) use steroids and other low-molecular-weight hormones as allosteric effectors of transcriptional regulatory molecules to coordinate appropriate responses of different organs to a particular physiological event.

It is important to remember that cells have evolved mechanisms that modulate the activity of key target proteins by relatively minor modifications—in the two preceding examples, by forming complexes of the proteins with allosteric effectors. Much of genetics, indeed much of the biology of a cell, depends on such modulations, in which key proteins are toggled between active and inactive states.

This chapter will show how such modulations achieve proper control of cell numbers and how the systems can be undermined by certain classes of mutations, producing uncontrolled proliferation—the diseases that we call cancer (Figure 15-2).

We'll begin with normal regulation of cell numbers. Cell proliferation is necessary for both development and maintenance of adult life. For example, in most of our tissues, cells die all the time by chance, and mechanisms to monitor a shortage of cells and recruit replacements are necessary for tissues to remain intact. Many tissues contain **stem cells,** undifferentiated cells that can divide and give rise to several different cell types depending upon local environmental cues. Other cues regulate the ability of stem cells to undergo mitosis. Cell death results from various kinds of cell damage. Internal and external cues indicate when a cell is damaged and instruct it to self-destruct. This is important because the presence of damaged cells can be more harmful to the organism as a whole than the loss of energy and material due to replacing the damaged cell. Thus, the cell-death program is essential to the health of an organism and is the counterpoint to cell-proliferation mech-

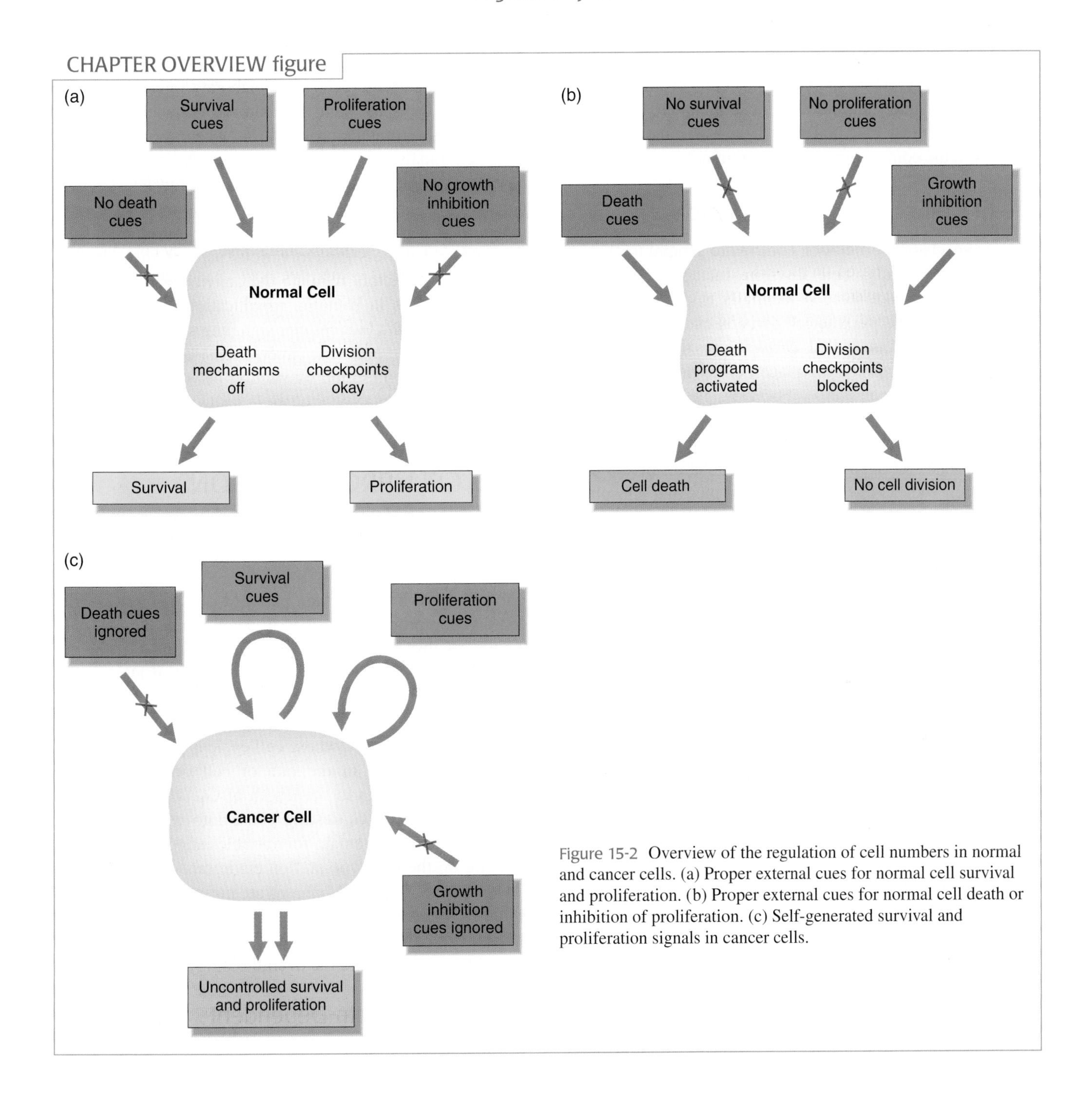

Figure 15-2 Overview of the regulation of cell numbers in normal and cancer cells. (a) Proper external cues for normal cell survival and proliferation. (b) Proper external cues for normal cell death or inhibition of proliferation. (c) Self-generated survival and proliferation signals in cancer cells.

anisms. For normal proliferation and normal programmed cell death, fail-safe mechanisms monitor the status of important components of the cell and permit proliferation or cell survival only if the fail-safe criteria are passed.

Cancers develop through the accumulation of mutations in a clone of somatic cells that ignore these normal fail-safe control mechanisms. Hundreds of mutational events that can lead to **neoplasia** (cancer formation) have now been identified, and certain themes emerge. Instead of depending upon the proper local cues to determine proliferation and cell survival, malignant cells continuously produce their own proliferation and survival signals. The second half of this chapter will explore this topic in detail.

THE BALANCE BETWEEN CELL PROLIFERATION AND CELL ELIMINATION

Cancer is now clearly understood as a genetic disease of somatic cells. In cancer, the fail-safe mechanisms that ensure that cell number remains balanced to the needs of the whole organism are subverted, and cancerous cells proliferate out of control. To understand how cells can mutate to a cancerous state, we must first understand the basic mechanisms that govern the control of normal cell numbers.

The Machinery of Cell Proliferation

Certain aspects of proliferation control are general to all eukaryotes. Universally, cell division includes numerous events that must take place sequentially to produce viable progeny cells. Moreover, the cell division cycle has evolved fail-safe mechanisms, called *checkpoints,* that prevent a subsequent event from taking place before the prerequisite events have been completed. For example, it would be lethal for mitosis to occur before DNA replication was completed. Mechanisms have evolved that prevent such cellular disasters. We shall explore the regulation of the eukaryotic cell cycle. Protein kinases, enzymes that phosphorylate certain specific amino acid residues on target proteins, and protein phosphatases, enzymes that remove phosphate groups from such amino acid residues, modulate the activities of key proteins in the cell cycle. These phosphorylation-dephosphorylation pathways ultimately converge to determine which key proteins are active for some fraction of the entire cell cycle. Put another way, it is the cyclical variations in these key proteins that determine which parts of the cell cycle are currently active.

The Machinery of Cell Death

Some aspects of cell control appear to have evolved only in multicellular organisms. To develop and maintain themselves normally, multicellular organisms must properly balance the numbers of the cell types in their various tissues. Almost all of these cell types are somatic—that is, they do not contribute to the germ line. Loss of such somatic cells is not a problem for propagation of the species, as long as proliferation of the remaining cells of that type in a particular tissue of the organism compensates for the cells that are eliminated. Furthermore, abnormal cells can do considerable harm. Thus, mechanisms have evolved for eliminating certain cells—through a process called **programmed cell death** or **apoptosis.** A cascade of enzymes called caspases kill a cell by disrupting numerous structural and functional systems within it. Once the cell is dead, its carcass is removed by scavenger cells.

Linking Cell Proliferation and Death to the Environment

The cell-proliferation and cell-death machinery must be interconnected so that each is activated only under the appropriate environmental circumstances. In adult organs, for example, maintenance of proper cell numbers requires a balance between the birth of new cells and the loss of existing ones. Eukaryotic cells have elaborate intercellular signaling pathways as status indicators of the environment. Some signals stimulate proliferation, whereas others inhibit it. Other signals can activate apoptosis, whereas still others block activation. Intercellular signaling pathways typically consist of several components: the signals themselves, the receptors that receive the signals, and the transduction systems that relay the signal to various parts of the cell. Just as allosteric effectors regulate the activity of many DNA-binding proteins in bacteria, modifications of the various components of the intercellular signaling systems—protein phosphorylation, allosteric interactions between proteins and small molecules, and interaction between protein subunits—control the activity of these pathways.

THE CELL-PROLIFERATION MACHINERY: CELL CYCLE REGULATION

The cell cycle has four main parts: M phase—mitosis, the cell division process described in detail in Chapter 4—and three parts that are components of interphase—G_1, the period between the end of mitosis and the start of DNA replication; S, the period of DNA synthesis; and G_2, the period that follows DNA replication and precedes the mitotic prophase. In mammals, whose cell cycle is particularly well studied, differences in the rate of cell division are due largely to differences in the length of G_1. This variation is due to an optional G_0 resting phase into which G_1-phase cells can shunt and remain for variable lengths of time, depending on the cell type and on environmental conditions. Conversely, S, G_2, and M phases are normally quite fixed in duration. This section will consider the molecules that drive the cell cycle. A later section will consider how these molecules are integrated into the overall biology of the cell.

Cyclins and Cyclin-Dependent Protein Kinases

The engines that drive the cell cycle from one step to the next are a series of protein complexes composed of two subunits: a **cyclin** and a **cyclin-dependent protein kinase** (abbreviated **CDK**). Every eukaryote has a family of structurally and functionally related cyclin proteins. Cyclins are so named because each is found only during a particular segment or segments of the cell cycle. The appearance of a specific cyclin is due to cell cycle–controlled transcription, in which the activity of the preceding cyclin-CDK complex leads to the activation of a transcription factor for the new cyclin. The disappearance of a cyclin depends on three factors: rapid inactivation of the transcription activator for the cyclin's gene (so that no new mRNA is produced), a high

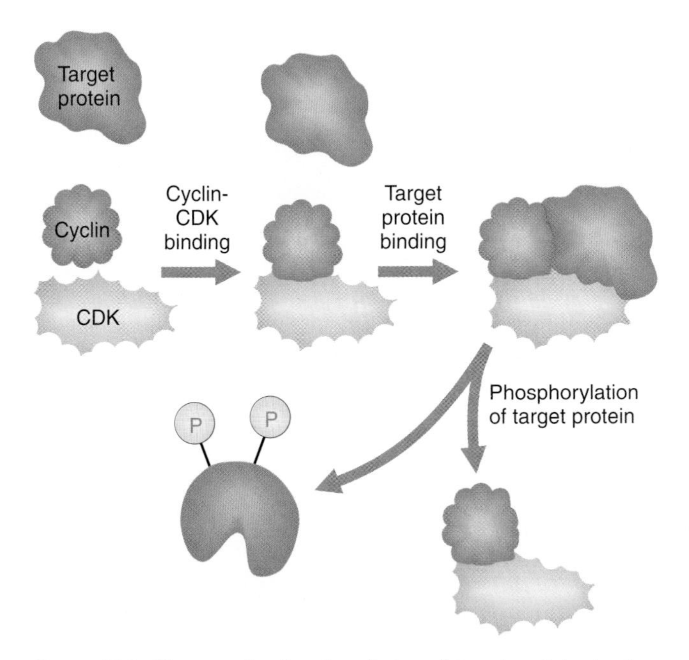

Figure 15-3 The steps in phosphorylation of target proteins by the cyclin-CDK complex. First, a cyclin and a CDK subunit bind to form an active cyclin-CDK complex. Then, the target protein binds to the cyclin part of the complex, putting the target phosphorylation sites close to the active site of CDK. The target protein is then phosphorylated; it is then no longer able to bind to cyclin and is released from the complex.

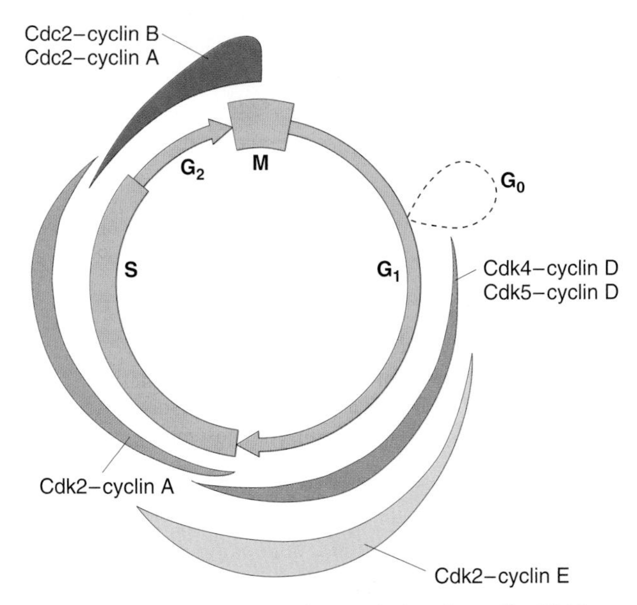

Figure 15-4 A current view of the variations in cyclin-CDK activities throughout the cell cycle of a mammalian cell. The widths of the bands indicate the relative kinase activities of the various cyclin-CDK complexes. Note that several different cyclins (A, B, D, and E) and several different CDKs (Cdc2, Cdk2, Cdk4, and Cdk5) can bind to one another to form different complexes, increasing the array of combinations of cyclin-CDK complexes that can form in the course of the cell cycle. *(Modified from H. Lodish, A. Berk, S. L. Zipursky, P. Matsudaira, D. Baltimore, and J. Darnell,* Molecular Cell Biology, *4th ed. © 2000 by W. H. Freeman and Company.)*

degree of instability of the cyclin mRNA (so that the existing pool of mRNA is quickly eliminated), and a high degree of instability of the cyclin itself (so that the pool of cyclin protein is quickly destroyed).

Cyclin-dependent protein kinases are another family of structurally and functionally related proteins. **Kinases** are enzymes that add phosphate groups to target substrates; for protein kinases, such as CDKs, the substrates are proteins. CDKs are so named because their activities are regulated by cyclins and because they catalyze the phosphorylation of specific serine and threonine residues of target proteins.

The target proteins for CDK phosphorylation are determined by the associated cyclin, because the cyclin tethers the target protein so that the CDK can phosphorylate it (Figure 15-3), thereby changing the activity of the target protein. Because different cyclins are present at different phases of the cell cycle (Figure 15-4), each phase is characterized by the phosphorylation of different target proteins. The phosphorylation events are transient and reversible. When the cyclin-CDK complex disappears, the phosphorylated substrate proteins are rapidly dephosphorylated by protein phosphatases.

CDK Targets

How does the phosphorylation of some target proteins control the cell cycle? Phosphorylation initiates a chain of events that culminates in the activation of certain transcription factors. These transcription factors promote the transcription of certain genes whose products are required for the next stage of the cell cycle. Much of our knowledge of the cell cycle comes both from genetic studies in yeast (see Foundations of Genetics 15-1 on page 489) and from biochemical studies of cultured mammalian cells. Indeed, this work was recognized with the Nobel Prize for medicine and physiology in 2001. A well-understood example is the Rb-E2F pathway in mammalian cells. Rb is the target protein of a CDK-cyclin complex called Cdk2-cyclin A, and E2F is the transcription factor that Rb regulates (Figure 15-5 on the next page). From late M phase through the middle of G_1, the Rb and E2F proteins are combined in a protein complex that does not promote transcription. In late G_1, however, the Cdk2–cyclin A complex is produced and it phosphorylates the Rb protein. This phosphorylation produces a change in the shape of Rb so that it can no longer bind to the E2F protein. The free E2F protein is then able to promote the transcription of certain genes that encode enzymes vital for DNA synthesis. This allows the next phase of the cell cycle—S phase—to proceed.

Rb and E2F are in fact representatives of two families of related proteins. In mammals, different cyclin-CDK complexes are thought to selectively phosphorylate different proteins of the Rb family, each of which in turn releases the specific E2F family member to which it is bound. The different E2F transcription factors then promote the

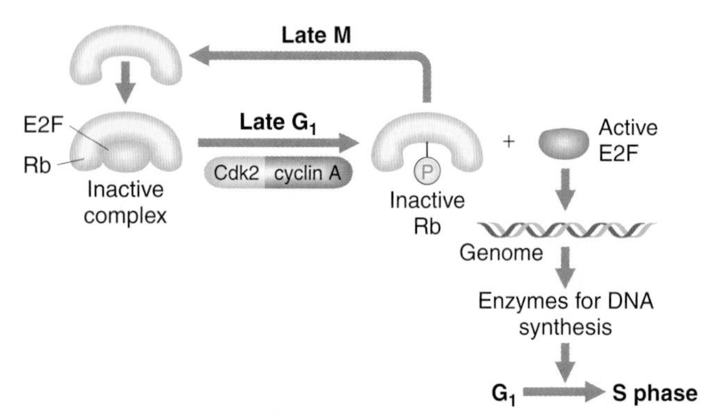

Figure 15-5 The contributions of the Rb and E2F proteins in the regulation of the G_1-to-S-phase transition in a mammalian cell. *(From H. Lodish, D. Baltimore, A. Berk, S. L. Zipursky, P. Matsudaira, and J. Darnell,* Molecular Cell Biology, *3d ed. © 1995 by Scientific American Books.)*

transcription of different genes that execute different aspects of the cell cycle.

MESSAGE

Sequential activation of different CDK-cyclin complexes ultimately controls progression of the cell cycle.

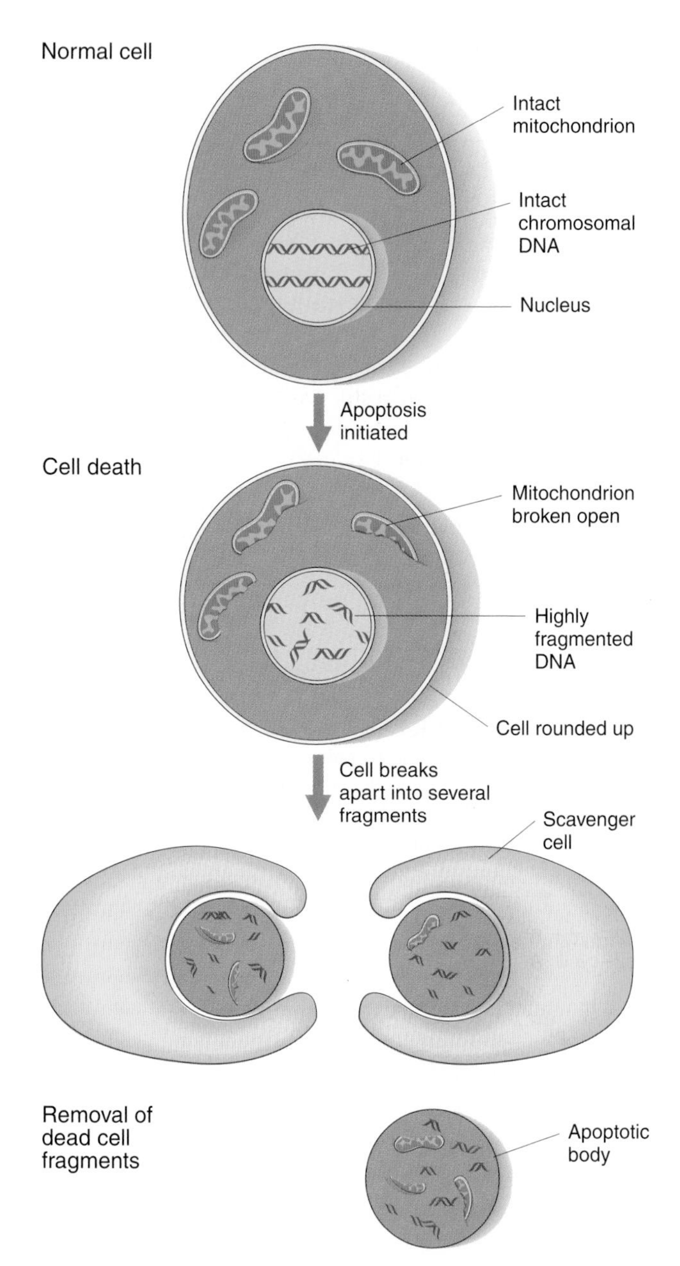

Figure 15-6 The sequence of events in apoptosis. First, the membranes of organelles such as mitochondria are disrupted and their contents leak into the cytoplasm, chromosomal DNA breaks into small pieces, and the cell loses its normal shape. Then the cell breaks apart into small fragments, which are disposed of by phagocytotic scavenger cells.

THE MACHINERY OF PROGRAMMED CELL DEATH: THE APOPTOSIS PATHWAY

In multicellular organisms, systems have evolved to eliminate damaged (and hence potentially harmful) cells by a self-destruct and disposal mechanism called **programmed cell death,** or **apoptosis.** This self-destruct mechanism can be activated under many different circumstances, includinjg elimination of cells no longer needed for development. In all cases, however, the events in apoptosis seem to be the same (Figure 15-6). First, the DNA of the chromosomes is fragmented, organelle structure is disrupted, and the cell loses its normal shape (apoptotic cells become spherical). Then, the cell breaks up into small fragments called *apoptotic bodies,* which are phagocytosed (literally, eaten up) by motile scavenger cells.

This section briefly describes the molecules responsible for carrying out apoptosis. The next section will describe how these responses are regulated within the cell.

The engines of self-destruction are a series of enzymes called **caspases** (cysteine-containing aspartate-specific proteases—see Foundations of Genetics 15-2 on page 490). Proteases are enzymes that cleave other proteins. Each caspase is a protein rich in cysteines: when activated, it cleaves certain target proteins at specific aspartate residues in the polypeptide chains. Every organism has a family of caspase proteins, related to each other by polypeptide sequence; in humans, for example, 10 caspases have been identified so far. In normal cells, each caspase is in an inactive state, called the **zymogen** form. In general, a zymogen is an inactive precursor form of an enzyme that contains a longer polypeptide chain than the final active enzyme does. To turn the zymogen form into the active caspase, a part of the polypeptide is removed by enzyme cleavage (also known as proteolysis).

FOUNDATIONS OF GENETICS 15-1

Yeast as a Model for the Cell Cycle

Work initiated by Leland Hartwell and his associates on the cell cycle genetics of the budding yeast *Saccharomyces cerevisiae* revealed a large array of genetic functions that maintain the proper cell cycle. These functions were identified as a specific subset of temperature-sensitive (ts) mutations called *cdc* (cell division cycle) mutations. When grown at low temperature, yeasts with these *cdc* mutations grew normally. When shifted to higher, restrictive temperatures, the yeasts stopped growing. What made these *cdc* mutations novel among the more general class of ts mutations was that a particular *cdc* mutant would stop growing at a specific time in the cell cycle, and all the yeast cells would look alike. Consider some examples of *S. cerevisiae,* a yeast that divides through budding, a process in which a mother cell develops a small outpocketing, a "bud." The bud grows and mitosis occurs in such a way that one spindle pole is in the mother cell and the other in the bud. The bud continues to grow until it is as big as the mother cell, and the mother cell and bud then separate into two daughter cells. Any run-of-the-mill ts mutations in *S. cerevisiae,* when shifted to restrictive temperature, stop growing at various times in the cycle of bud formation and growth. In contrast, after a shift to restrictive temperature, one *S. cerevisiae cdc* mutation produces yeast cells with only tiny buds, whereas another produces yeast cells exclusively with larger buds, half the size of the mother cell. Such different cdc phenotypes indicate different defects in the machinery required to execute specific events in the cell cycle.

With the sequencing of the *S. cerevisiae* genome (see Chapter 9) now complete, we are able to identify the entire array of proteins of the cyclin and CDK families (22 and 5 proteins, respectively). Their genes are now being systematically mutagenized and genetically characterized to reveal how each contributes to the cell cycle.

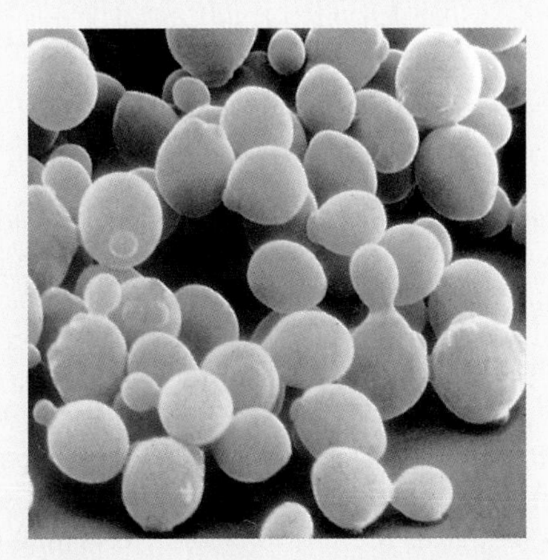

Scanning electron micrograph of *S. cerevisiae* cells at different points in the cell cycle, as indicated by different bud sizes. (Courtesy of E. Schachtbach and I. Herskowitz.)

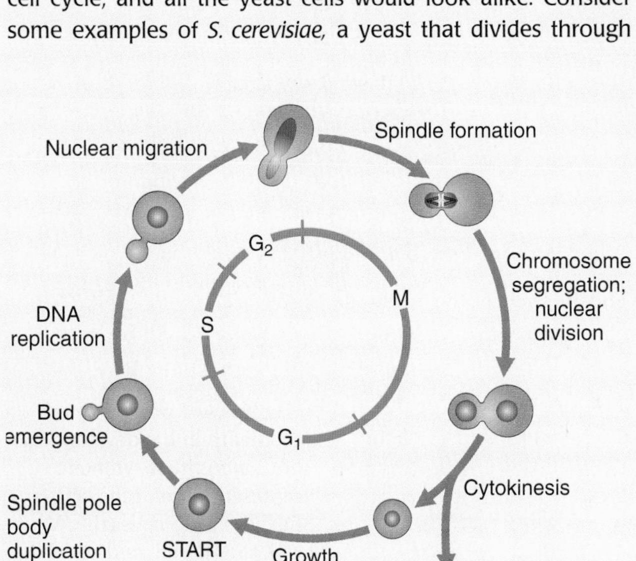

Cell cycle of *S. cerevisiae* (From H. Lodish, D. Baltimore, A. Berk, S. L. Zipursky, P. Matsudaira, and J. Darnell, *Molecular Cell Biology,* 3d ed. © 1995 by Scientific American Books.)

The current view is that there are two classes of caspases: *initiators* and *executioners.* Exactly how they are organized into a cascade of cleavage events is still unclear. One suggested scenario is that the initiator caspases are cleaved in response to activation signals coming from other classes of proteins. They in turn cleave one of the executioner caspases, which in turn cleaves another, and so forth.

MESSAGE

Programmed cell death is mediated by a sequential cascade of proteolysis events that activate enzymes that destroy several key targeted cellular components.

How do the executioner caspases carry out the cellular sentence of death? In addition to activating other caspases, executioner caspases enzymatically cut the target proteins (Figure 15-7 on page 491). One target is a "sequestering" protein that forms a complex with a DNA endonuclease, thereby holding (sequestering) the endonuclease in the cytoplasm. On cleavage of the sequestering protein, the endonuclease is free to enter the nucleus and chop up the cell's DNA. Another target is a protein that, when cut by the caspases, cleaves actin, a major component of the cytoskeleton, causing disruption of actin filaments and thus leading to a loss of normal cell shape. In similar fashion, all other aspects of apoptosis are thought to be mediated by caspase-activated proteases.

FOUNDATIONS OF GENETICS 15-2

The Nematode *Caenorhabditis elegans* as a Model for Programmed Cell Death

Programmed cell death has been described in various systems. However, genetic studies by Robert Horvitz and his associates in the past 15 years in the nematode (roundworm) *Caenorhabditis elegans* have propelled the field forward. Researchers have mapped the entire series of somatic cell divisions that produce the 1000 or so cells of the adult worm. For some of the embryonic and larval cell divisions, particularly those that will contribute to the worm's nervous system, a progenitor cell gives rise to two progeny cells, one of which then undergoes programmed cell death (see the adjoining figure). These divisions, in which a progenitor cell gives rise to only one viable progeny cell, are necessary for the progeny cell to fulfill its proper developmental role.

A set of mutations has been identified in the worm that block this cell-death phenotype. Some of these mutations knock out genes that encode caspases—an example is *ced-3* (cell-death gene number 3)—clearly implicating these caspases in the apoptosis process. The analysis of other genes with mutant cell-death phenotypes is being carried out in worms and other experimental systems and is uncovering other key players in this process.

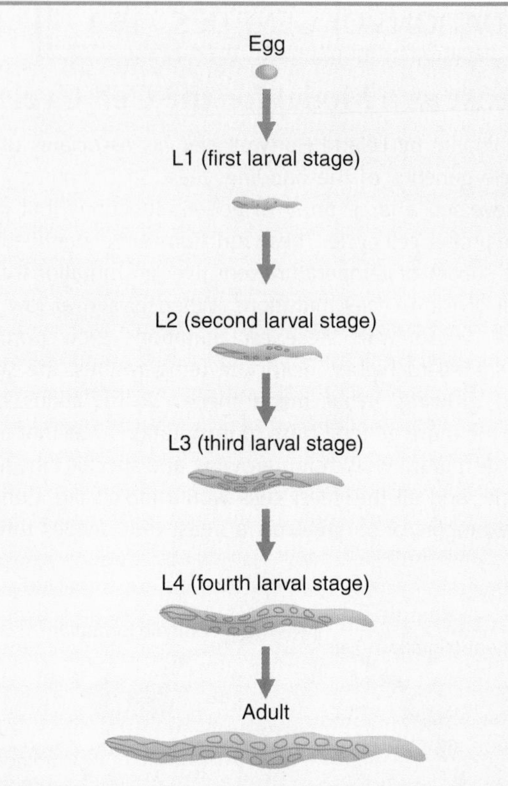

The life cycle of the nematode *C. elegans*.

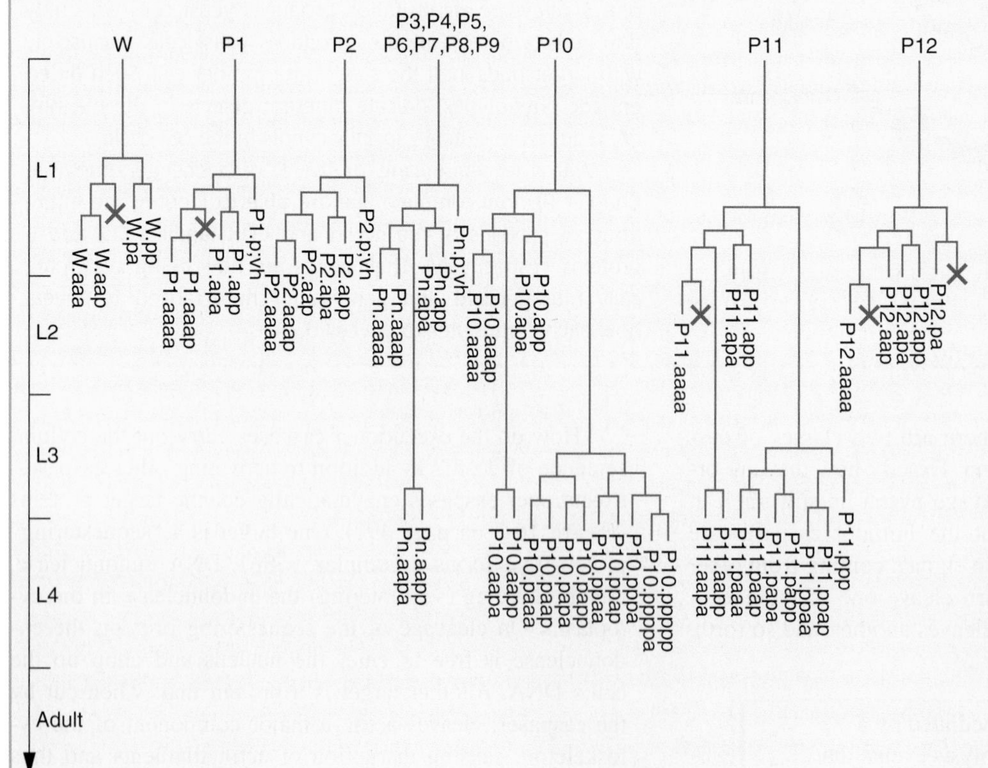

Examples of programmed cell death in the development of *C. elegans*. A symbolic representation of the cell lineages of 13 cells (the W cell, the P1 cell, and so forth) produced during embryogenesis. The vertical axis is developmental time, beginning with the hatching of the egg into the first larval stage (L1). In each lineage, a vertical blue bar connects the various division events (horizontal blue bars). The names of the final cells are shown, such as W.aaa or P1.apa. In several cases, a cell division produces one viable cell and one cell that undergoes apoptosis. A cell that undergoes programmed cell death is indicated with a blue × at the end of a branch of a lineage. In homozygotes for mutations such as *ced-3*, these cell deaths do not occur.

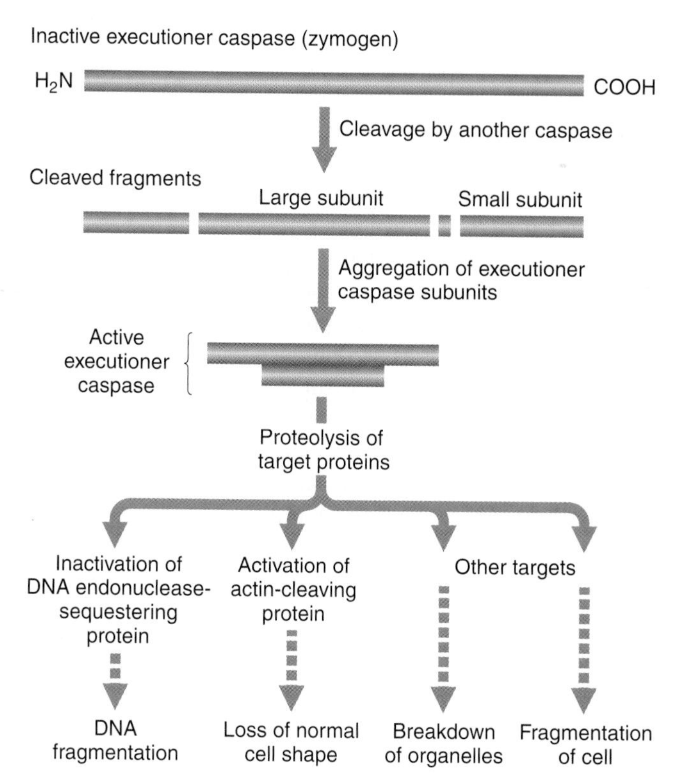

Figure 15-7 The role of executioner caspases in apoptosis. A cascade of caspase activation leads to the activation of the executioner caspases. Cleavage of the zymogen (inactive precursor) form of the executioner caspase by another caspase takes place at several aspartate residues, producing several protein fragments. Two of these fragments, the large and small subunits, bind and form the enzymatically active caspase. Through cleavage of a series of target proteins (also by cleavage at aspartate residues), the various cellular breakdown events take place, leading to cell death and removal.

CONTROLLING CELL PROLIFERATION AND DEATH MACHINERY

We have used the term *engine* to describe the role of the cyclin-CDK complex and the caspase cascade in cell proliferation and programmed cell death. To continue the analogy, ignition switches and accelerators (positive controls) start up the engines and get these processes moving, and brakes (negative controls) slow down or halt the processes when necessary. Like the cell cycle and apoptosis, the positive and negative controls result from a series of modulations of protein activity through protein-protein interactions and protein modifications.

Intracellular Signals

Some of the elements of the positive and negative control loops consist of signals that originate within the cell.

The cell cycle: negative intracellular controls. Through activation of proteins that can inhibit the protein kinase activity of CDK-cyclin complexes, the cell cycle can be held in check until various monitoring mechanisms give a "green light," indicating that the cell is properly prepared to proceed to the next phase of the cycle.

One example of how this "checkpoint" system operates begins with damaged DNA (Figure 15-8). When DNA is damaged during G_1 (for example, by X-irradiation), CDK activity of CDK-cyclin complexes is inhibited. The inhibition seems to be mediated by a protein called p53. Part of the p53 protein recognizes certain kinds of DNA mismatches. In the presence of such mismatches, p53 is able to activate another protein, p21. When its levels are high, p21 binds to the CDK-cyclin complex and inhibits its protein kinase activity. In the absence of its protein kinase activity, the target proteins of CDK are not phosphorylated, and the cell cycle is unable to progress. When the DNA mismatches have been repaired, the inhibiting processes are reversed. This reversal is accomplished by a post-DNA-repair drop in p53 levels and a cessation of inhibition of CDK-cyclin protein kinase activity, which leads to removal of the G_1-to-S checkpoint block.

In this manner, checkpoints monitor the status of DNA replication, the spindle apparatus, and other key components of the cell cycle and can operate as braking systems when necessary. The key is the existence of regulatory proteins that can modulate the protein kinase activity of the cyclin-CDK complex.

> **MESSAGE**
> Fail-safe systems (checkpoints) ensure that the cell cycle does not progress until the cell has completed all prior events necessary to assure its survival through the next steps.

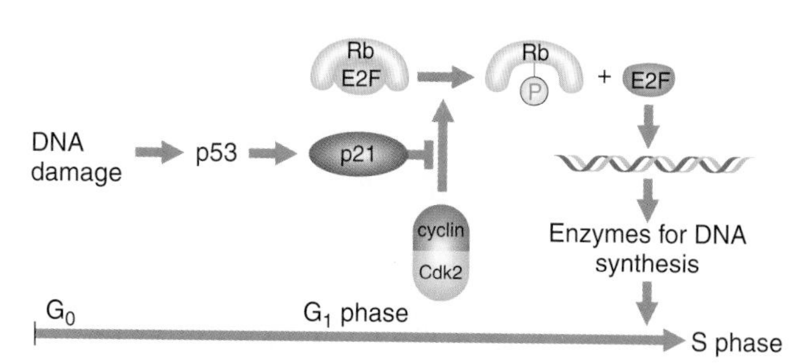

Figure 15-8 An example of inhibitory control of the progression of the cell cycle. In mammals, the transition from the G_1 to the S phase requires the phosphorylation of Rb protein by the Cdk2-cyclin complex. In the presence of damaged DNA, p53 protein is induced, which in turn induces p21 protein. The elevated levels of p21 inhibit the protein kinase activity of the Cdk2-cyclin complex. When the damaged DNA has been repaired, p53 levels drop. In turn, p21 levels decrease, and the inhibition of the Cdk2-cyclin protein kinase activity is relieved, which allows Rb to be phosphorylated and E2F to become an active transcription factor, permitting the cell to enter S phase. *(Adapted from C. J. Sherr and J. M. Roberts,* Genes and Development *9, 1995, 1150.)*

The cell cycle: positive intracellular controls. It is necessary not only to release the cell cycle "brake" but also to engage the "transmission" and the "engine" to advance the cell cycle. Once the brake has been released, independent signals from within or outside the cell induce a cascade of protein kinases that phosphorylate the appropriate cyclin-CDK complex, thereby activating the complex. This activation in turn allows the complex to phosphorylate its target proteins.

Apoptosis: positive intracellular controls. It has been known for several years that, in some manner, many forms of cellular damage trigger leakage of material from mitochondria and that this leakage somehow induces the apoptotic response. Indeed, it now appears that one of the ignition switches for apoptosis is cytochrome *c*, one of the mitochondrial proteins that normally takes part in cell respiration. Leakage of cytochrome *c* into the cytoplasm is detected and triggers the activation of initiator caspases. This detection is thought to happen through the binding of cytochrome *c* to another protein called Apaf (apoptotic-protease-activating factor). The cytochrome *c*-Apaf complex then binds to and activates the initiator caspase.

Apoptosis: negative intracellular controls. The irreversibility of cell death has probably been the compelling factor in the evolution of backup systems to make sure that the apoptosis pathway remains "off" under normal conditions. Proteins such as Bcl-2 and Bcl-x in mammals accomplish this. Among the possible actions of these Bcl proteins is that they block the release of cytochrome *c* from mitochondria (possibly by making it more difficult for mitochondria to burst) and bind to Apaf, preventing its interaction with the initiator caspase.

Extracellular Signals

A cell in a multicellular organism continually assesses its own internal status regarding proliferation and survival. Nonetheless, the proliferative and survival abilities of a cell must be subservient to the needs of the population of cells of which it is a member (populations such as the entire early embryo, a tissue, or a body part such as a limb or an organ). For example, in many adult organs, stem cells divide to produce replacement cells only when there is a depletion of cell numbers. Without such homeostatic mechanisms, organs would not be proportioned appropriately for the size of a given individual organism.

Mechanisms for cell-cell communication. Many kinds of signals need to be transmitted between cells to coordinate virtually all aspects of the development and physiology of complex multicellular organisms. The major routes of cell-cell communication are briefly outlined here.

All systems for intercellular communication have several components. A molecule called a **ligand** is produced by secretion from signaling cells (Figure 15-9). Some ligands, called hormones, are long-range **endocrine signals** that are released from endocrine organs into the circulatory system, which transmits them throughout the body. (Recall the description of steroid hormones and their receptors in Chapter 13.) Hormones can act as master control switches for many different tissues, which can then respond in a coordinated fashion. Other secreted ligands act as **paracrine signals;** that is, they do not enter the circulatory system but act only locally, in some cases only on adjacent cells. Chapter 16 will describe paracrine and endocrine signals in further detail. Some ligands are proteins, whereas others are small molecules such as steroids or vitamin D. Most (but not all) endocrine signals are small molecules, such as the mammalian steroid hormones that are responsible for male (androgen) or female (estrogen) sex-specific phenotypes. In contrast, most paracrine signals are proteins. Here the focus is on paracrine signaling through protein ligands.

Protein ligands act as signals by binding to and thereby activating transmembrane receptors, proteins that are embedded in the plasma membrane at the surface of the cell. These ligand-receptor complexes release chemical signals into the cytoplasm just inside the plasma membrane. Such signals are passed through a series of intermediary molecules until they finally alter the structure of transcription factors in the nucleus, activating the transcription of some genes and the repression of others. Transmembrane receptors have one part (the extracellular domain) outside of the cell, a middle part that passes once or several times through the plasma membrane, and another part (the cytoplasmic domain) inside the cell (Figure 15-10).

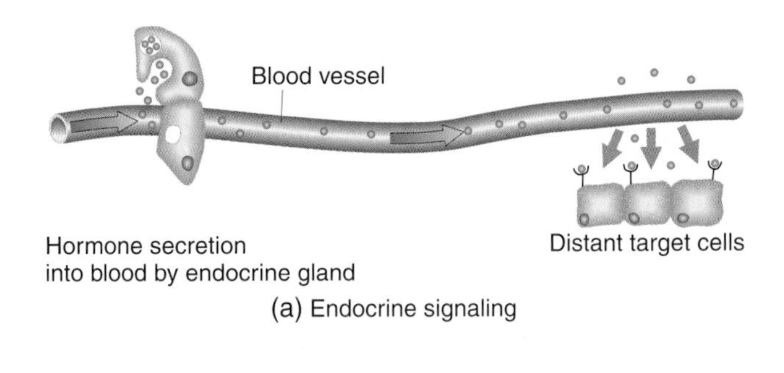

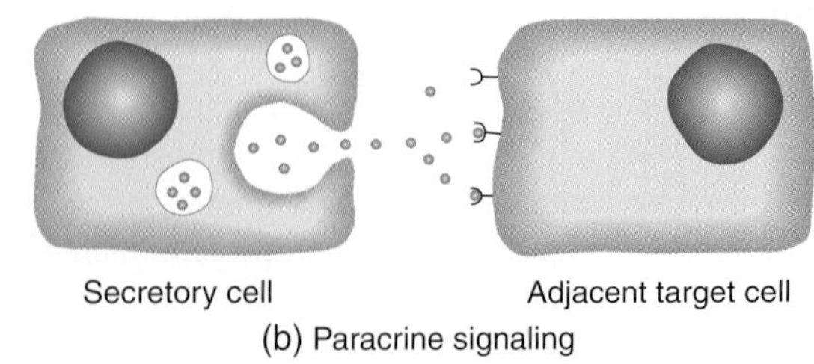

Figure 15-9 Modes of intercellular signaling. (a) Endocrine signals enter the circulatory system and can be received by distant target cells. (b) Paracrine signals act locally and are received by nearby target cells. *(From H. Lodish, D. Baltimore, A. Berk, S. L. Zipursky, P. Matsudaira, and J. Darnell,* Molecular Cell Biology, *3d ed. © 1995 by Scientific American Books.)*

The extracellular domain of the receptor is the site to which the ligand binds. Many polypeptide ligands are dimers and can simultaneously bind two receptor monomers. This simultaneous binding brings the cytoplasmic domains of the two receptor subunits into close proximity and activates their signaling activity. Some receptors for polypeptide ligands are receptor tyrosine kinases (RTKs, see Figure 15-10b). Their cytoplasmic domains, when activated, have the ability to phosphorylate certain tyrosine residues on target proteins. Others are receptor serine/threonine kinases. Still other receptors have no enzymatic activity, but conformational changes in these receptors (when a ligand binds) cause conformational changes in (and activation of) receptor-bound cytoplasmic proteins.

Perhaps the best understood of the receptors for polypeptide ligands are the receptor tyrosine kinases (Figure 15-11 on the next page). RTK is a monomer "floating" in the plasma membrane. When ligand and RTK bind to form a ligand-RTK complex, two RTK monomers bind to form a dimer. RTK dimerization activates the protein kinase

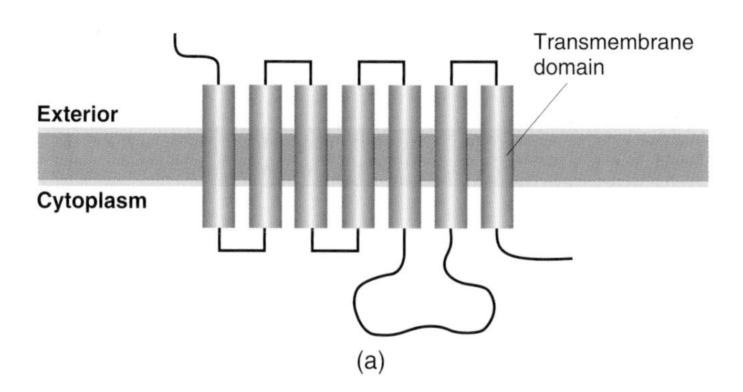

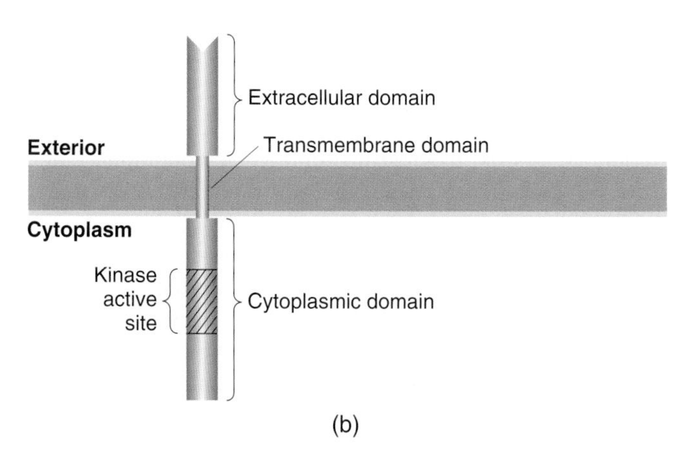

Figure 15-10 Examples of transmembrane receptors. (a) A receptor that passes through the cell membrane seven times. (b) Receptor tyrosine kinase (RTK), which has a single transmembrane domain. The extracellular domain binds to ligand. The active site of the tyrosine kinase is in the cytoplasmic domain. *(Adapted from H. Lodish, D. Baltimore, A. Berk, S. L. Zipursky, P. Matsudaira, and J. Darnell,* Molecular Cell Biology, *3d ed. © 1995 by Scientific American Books.)*

enzymatic activity of the cytoplasmic domain of the RTK. The first phosphorylation targets of the kinase are several tyrosines in the cytoplasmic domain of the RTK itself; this process is called *autophosphorylation,* because the kinase acts upon itself. Autophosphorylation initiates a signal transduction cascade in which, sequentially, modification of the conformation of one protein leads to modification of the conformation of others. Eventually, the signal transduction cascade leads to the modification of transcriptional activators and repressors and hence to changes in the activities of many genes in the target cell.

RTK autophosphorylation activates signal transduction cascades in two ways. In one process, phosphorylated sites on the RTK are targets for binding by various *adaptor* proteins (see Figure 15-11a). Multiple adaptor proteins "dock" on phosphorylated sites on the RTK in the vicinity of one another. These adaptor proteins in turn have affinity for other proteins—elements of signal transduction cascades. By bringing these other transduction elements close together, protein-protein interactions lead to activation of the cascades. In the other process, the phosphorylated RTK is conformationally changed so that its tyrosine kinase active site phosphorylates other target proteins (see Figure 15-11b). The conformation of these phosphorylated proteins is then changed, allowing them to participate in a signal transduction cascade. By these two processes, activation of one RTK can lead to the simultaneous activation of multiple signal transduction pathways.

Quite often, the next step in propagating the signal is to activate a G-protein. G-proteins cycle between being bound by GDP (the inactive state) and being bound by GTP (the activated state). Propagation of the signal from the RTK leads to the activation of a protein that binds to the inactive GDP-bound G-protein, changing its conformation so that it then binds to a molecule of GTP (Figure 15-12 on page 495). The specific G-protein called Ras is especially important in carcinogenesis, as will be explained later.

The activated GTP-bound G-protein then binds to a cytoplasmic protein kinase, in turn changing its conformation and activating its kinase activity. This protein kinase then phosphorylates other proteins, including other protein kinases. (In the example in Figure 15-13 on page 495, the protein kinases farther down the cascade are called Raf, MEK, and MAP kinase.) The targets of some of these protein kinases are transcriptional activators and repressors. The phosphorylation of the transcription factors changes their conformation, leading to the transcription of some genes and the repression of others (see Figure 15-13).

Cell-cell signaling depends on conformational changes. The steps in ligand-receptor binding and in signaling within the cell depend on conformational changes. For example, the conformational changes caused by the binding of ligands to receptors activate the signaling pathways. Likewise, conformational changes in protein kinases enable them to phosphorylate specific amino acids on specific proteins, and

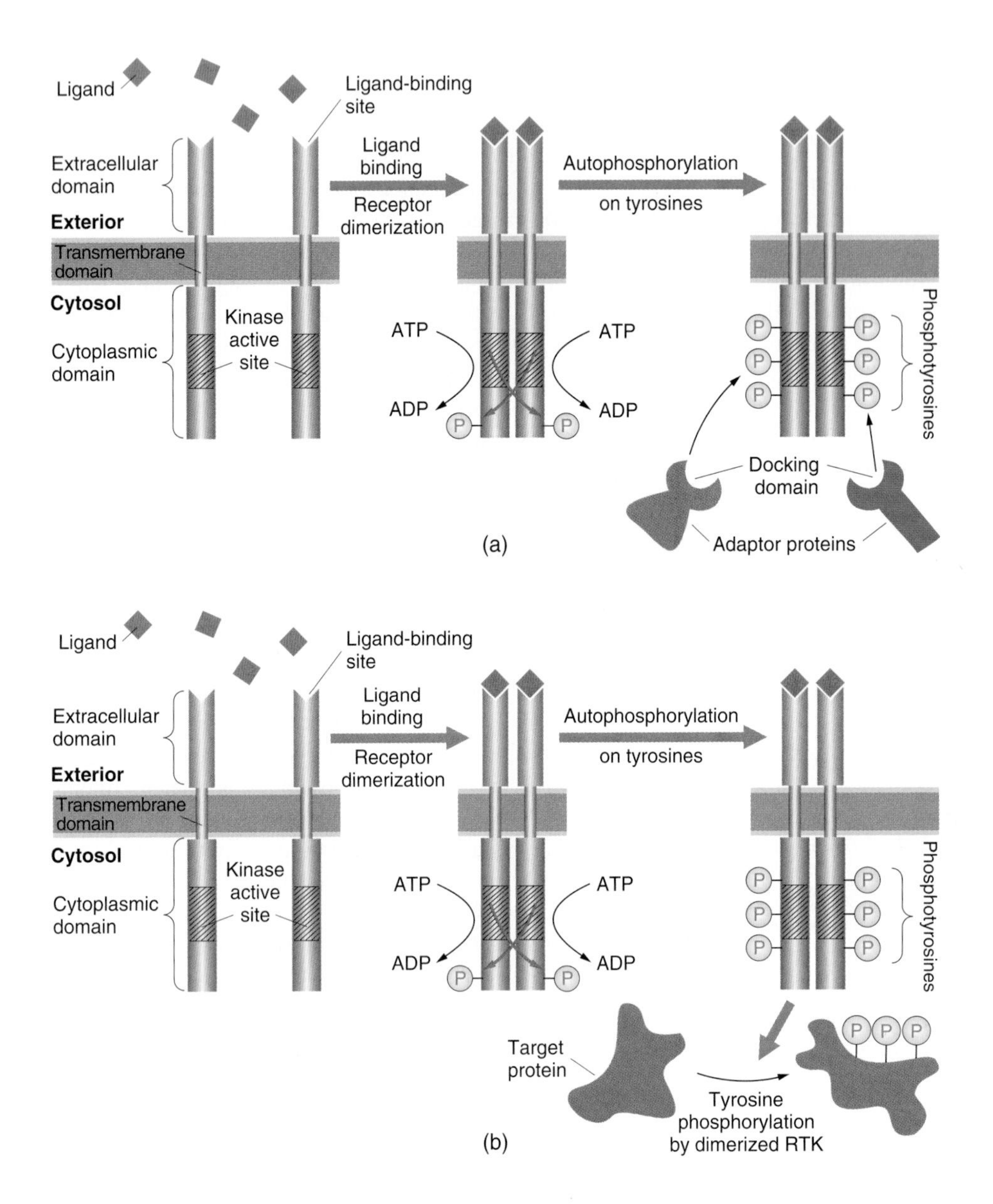

Figure 15-11 The consequences of ligand binding on RTK activity and the initiation of the signal-transduction cascade. On dimerization of the RTK, autophosphorylation occurs. Activated RTK initiates signal transduction by two different routes: (a) the activated RTK serves as a docking site for some signal-transduction proteins and (b) the activated RTK phosphorylates other proteins. (a) Autophosphorylation creates binding sites for *adaptor,* or "docking," proteins that bind to sites that contain specific phosphorylated tyrosine residues. Through the docking of different adaptor proteins close to one another, protein-protein conformational changes lead to induction of the signal-transduction activities of the proteins. (b) The other consequence of dimerization and autophosphorylation is that the RTK directly phosphorylates *target* proteins and, through this phosphorylation, induces their signal-transduction activities. *(Adapted from H. Lodish, D. Baltimore, A. Berk, S. L. Zipursky, P. Matsudaira, and J. Darnell,* Molecular Cell Biology, *3d ed. © 1995 by Scientific American Books.)*

other proteins undergo conformational changes when they bind to GTP. Not only do these conformational changes permit rapid response to an initial signal, but they also are readily reversible, enabling signals to be shut down rapidly and permitting the components of the signaling system to be recycled so that they are ready to receive further signals.

The cell cycle: positive extracellular controls. Cell division is promoted by the action of **mitogens,** polypeptide ligands released usually from a paracrine (nearby) source. Many mitogens, also called **growth factors,** such as EGF (epidermal growth factor), activate RTKs and initiate exactly the sort of signal transduction pathway described above.

The cell cycle: negative extracellular controls. Certain secreted proteins are known to inhibit cell division. One example is TGF-β, a ligand that is thought to be secreted in a variety of tissues under growth-inhibitory conditions. The TGF-β ligand binds to the TGF-β receptor and activates the receptor's serine/threonine kinase activity. This activation in turn leads to phosphorylation of proteins called SMADs, which cause changes in transcriptional activities, and perhaps to phosphorylation of other substrates as well. As a result of this signal transduction cascade, the phosphorylation and inactivation of the Rb protein are eventually blocked. Recall, from earlier in the chapter, the role that Rb plays in regulating the cell cycle by preventing activation of the E2F transcription factor. Blocking Rb inactivation thus keeps E2F off and blocks progression of the cell cycle.

Apoptosis: positive extracellular controls. Often, the command for self-destruction comes from a neighboring cell. For example, within the immune system, only a small percentage of B cells and T cells mature to make functional

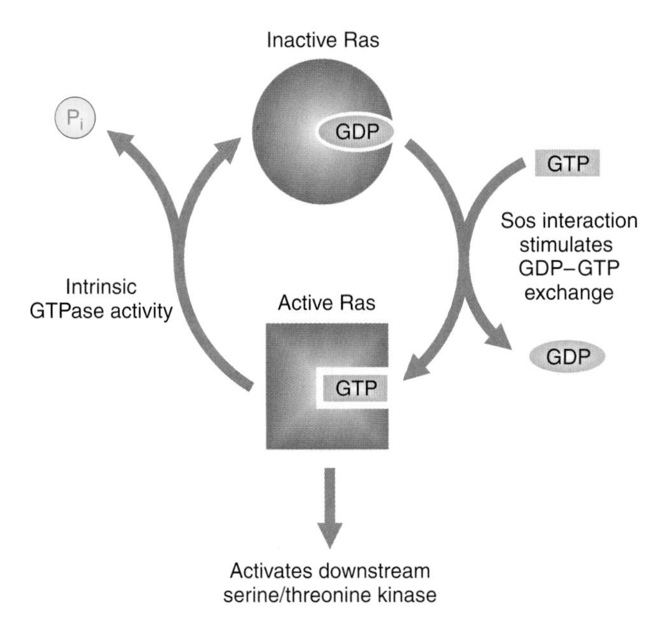

Figure 15-12 An example of the G-protein activity cycle. Ras is a member of the G-protein family. When GDP is bound to Ras, it does not signal. Through direct interactions with Ras, another protein called Sos causes conformational changes in Ras so that it preferentially binds GTP. The Ras-GTP complex is able to interact in turn with a cytoplasmic serine/threonine kinase, activating its activity, and thus transmits the signal to the next step in the signal transduction pathway. When Ras-GTP is released from Sos, it hydrolyzes GTP to GDP and reassumes the inactive Ras-GDP state. *(Adapted from J. D. Watson, M. Gilman, J. Witkowski, and M. Zoller,* Recombinant DNA, *2d ed. © 1992 by James D. Watson, Michael Gilman, Jan Witkowski, and Mark Zoller.)*

antibody or receptor protein, respectively. If the nonfunctional immature B cells and T cells were not eliminated by induced self-destruction, the numbers of these unnecessary cells would clog up the immune system. The self-destruction signal is activated through the Fas system (Figure 15-14 on the next page). A cell-surface membrane-bound protein called FasL (Fas ligand) binds to Fas cell-surface receptors on an adjacent cell. This induces trimerization of the ligand-receptor complex and trimerization of a cytoplasmic domain of the Fas transmembrane receptor. This in turn, directly or indirectly, activates a molecule such as Apaf (described earlier in this chapter), which activates an initiator caspase and thus the caspase cascade.

Apoptosis: negative extracellular controls. Negative secreted factors that are necessary to block activation of the apoptosis pathway also exist, and they are sometimes called **survival factors.** How they influence the apoptosis pathway is not now clear.

MESSAGE

Intercellular signaling systems communicate instructions to proliferate or to arrest the cell cycle and to initiate or postpone self-destruction.

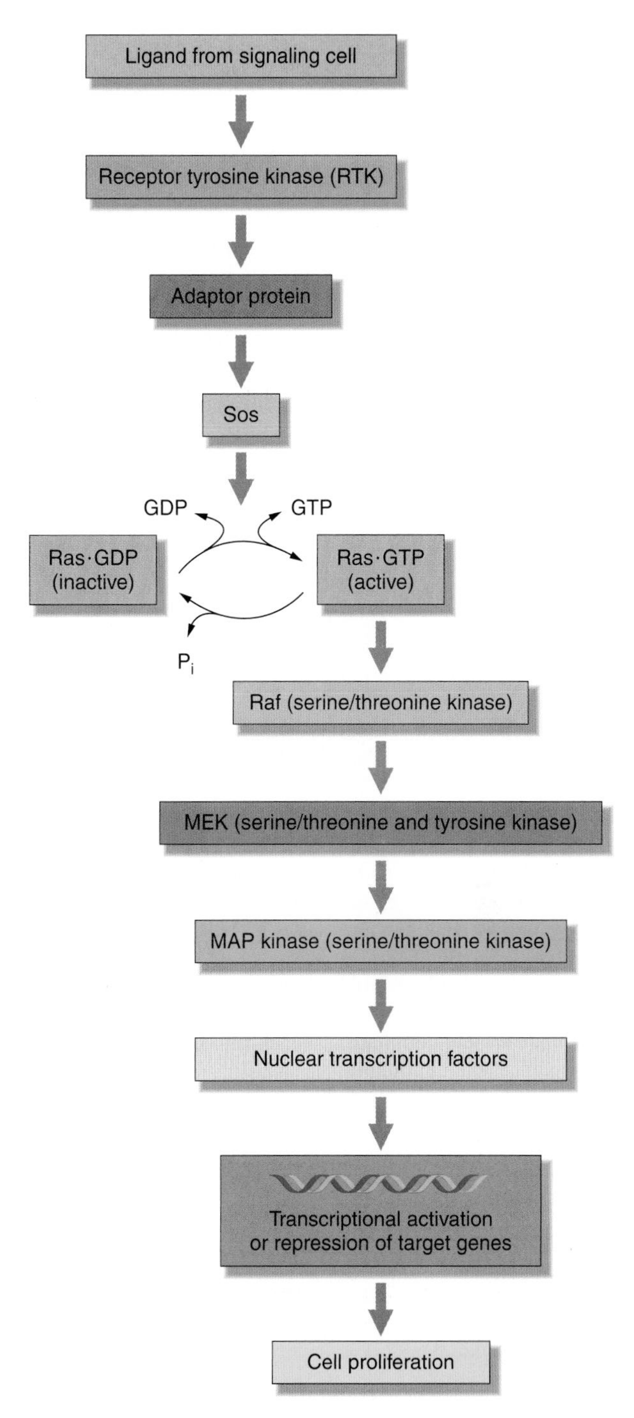

Figure 15-13 One pathway for RTK signaling. Raf, MEK, and MAP kinase are three cytoplasmic protein kinases that are sequentially activated in the signal transduction cascade. *(Adapted from H. Lodish, D. Baltimore, A. Berk, S. L. Zipursky, P. Matsudaira, and J. Darnell,* Molecular Cell Biology, *3d ed. © 1995 by Scientific American Books.)*

An Integrated View of the Control of Cell Numbers

The preceding sections have described numerous ways to modulate cell number. The general theme is that there

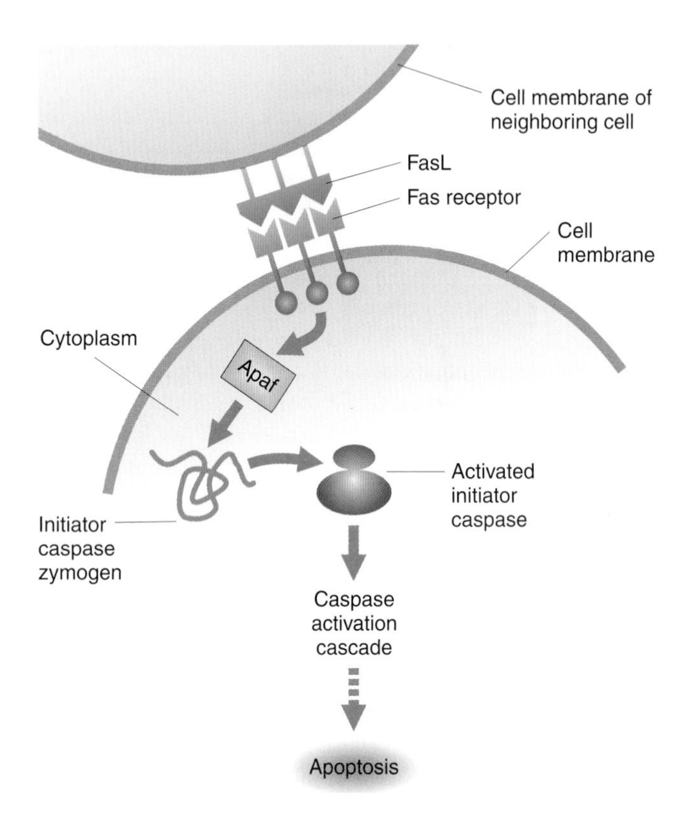

Figure 15-14 Positive extracellular control of apoptosis. A molecule such as Apaf is activated by the binding of FasL (Fas ligand), anchored on the outside of an adjacent cell. FasL binds to the Fas transmembrane receptor on the cell that will undergo apoptosis. Apaf activation in turn causes proteolysis and activation of the initiator caspase. A series of caspases are then proteolysed and activated in turn, ultimately leading to apoptosis of the cell. *(Adapted from S. Nagata,* Cell *88, 1997, 357.)*

are pathways for controlling cell proliferation and self-destruction and that activation of these pathways requires the correct array of positive inputs and the absence of negative, or inhibitory, inputs. Not only are cells able to assess their own status regarding proliferation or viability, but neighboring cells can play instructive roles through cell-cell signaling (Figure 15-15).

CANCER: THE GENETICS OF ABERRANT CELL CONTROL

A basic article of faith in genetic analysis is that we learn a great deal about normal biology and about the disease state by studying the properties of mutations that disrupt normal processes. This has certainly been true in regard to cancer. It has become clear that virtually all cancers of somatic cells are due to a series of special mutations that accumulate in a cell. These mutations fall into a few major categories: those that increase the ability of a cell to proliferate, decrease the susceptibility of a cell to apoptosis, or increase the general mutation rate of the cell, so that all mutations, including those that affect regulation of proliferation or apoptosis, are more likely to occur.We are getting the first glimmers of hope that these insights into the basic events in cancer biology will translate into improved diagnosis, treatment, and control of this widespread and devastating group of diseases.

How Cancer Cells Differ from Normal Cells

Malignant tumors, or cancers, are clonal. **Cancers** are aggregates of cells, all derived from an initial aberrant founder cell that, although surrounded by normal tissue, no

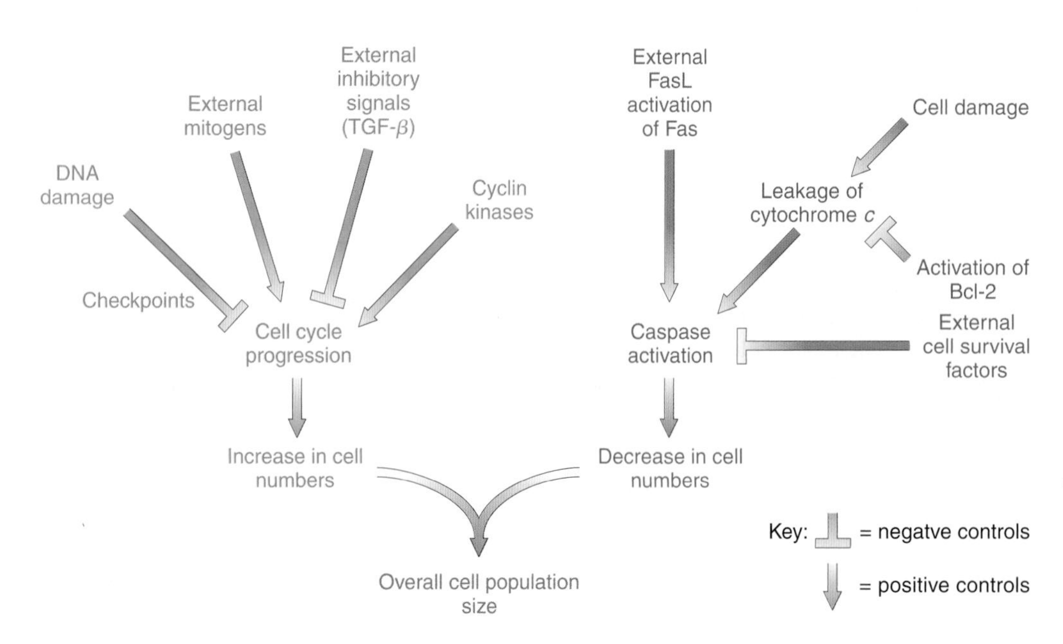

Figure 15-15 The inputs into control of cell number.

longer participates in that environment. Cancer cells typically differ from their normal neighbors by a host of phenotypic changes, such as rapid division rate, invasion of new cellular territories, high metabolic rate, and abnormal shape. For example, when cells from normal epithelial cell sheets are placed in cell culture, they can grow only when anchored to the culture dish itself. In addition, normal epithelial cells in culture divide until they form a continuous monolayer. Then, they somehow recognize that they have formed a single epithelial sheet and stop dividing. In contrast, malignant cells derived from epithelial tissue continue to proliferate, piling up on one another (Figure 15-16b).

Clearly, the factors regulating normal cell differentiation have been altered. What, then, is the underlying cause of cancer? Many different cell types can be converted into a malignant state. Is there a common theme, or does each arise in a quite different way? We can think about cancer in a general way: as due to the accumulation of multiple mutations in a single cell that cause it to proliferate out of control. Some of those mutations may be transmitted from the parents through the germ line. Others arise de novo in the somatic-cell lineage of a particular cell.

Evidence for the Genetic Origin of Cancers

Several lines of evidence point to a genetic origin of the transformation of cells from the benign to the cancerous state. Most carcinogenic agents (chemicals and radiation) are also mutagenic. Occasionally, certain cancers are inherited as highly penetrant single Mendelian factors; an example is familial retinoblastoma. Perhaps more general are less-penetrant susceptibility alleles that increase the probability of developing a particular type of cancer. In the past few years, several susceptibility genes have been recombinationally mapped and molecularly cloned and localized with the use of RFLP mapping or related techniques. **Oncogenes,** dominant mutant genes that contribute to cancer in animals, have been isolated from tumor viruses—viruses that can transform normal cells in certain animals into tumor-forming cells. Such dominant oncogenes can also be isolated from tumor cells by cell-culture assays that can distinguish between some types of benign and malignant cells. A tumor does not arise as a result of a single genetic event but rather as the result of multiple hits, in which several mutations must arise within a single cell for it to become cancerous. In some of the best-studied cases, the progression of colon cancer and astrocytoma (a brain cancer) has been shown to entail the accumulation of several different mutations in the malignant cells (Figure 15-17 on the next page). The next sections will consider further the genetic origin of cancers and the nature of the proteins that are altered by cancer-producing mutations. Many of these proteins take part in intercellular communication and the regulation of the cell cycle and apoptosis.

> **MESSAGE**
> Tumors arise from a sequence of mutational events that lead to uncontrolled proliferation.

Mutations in Cancer Cells

Two general kinds of mutations are associated with tumors: oncogene mutations and mutations in tumor-suppressor genes. Oncogenes are mutated in such a way that the proteins that they encode are activated in tumor cells that carry the dominant mutant allele. A typical tumor cell will

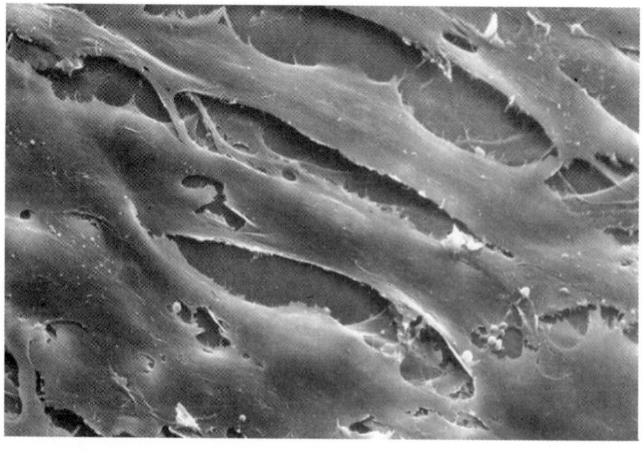

(a)

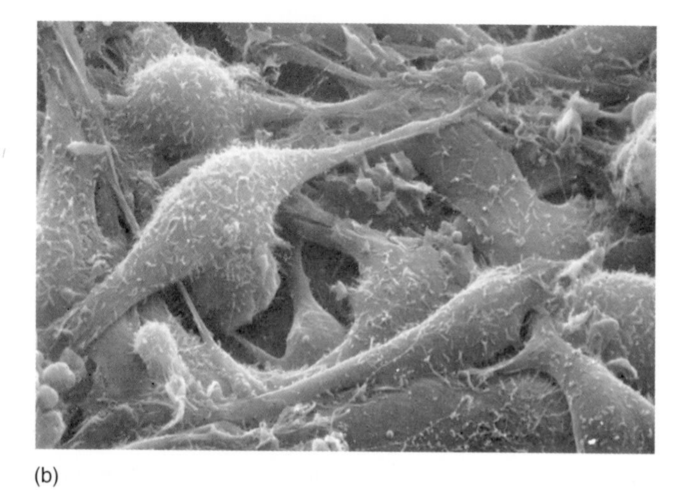

(b)

Figure 15-16 Scanning electron micrographs of (a) normal cells and (b) cells transformed by Rous sarcoma virus, which infects cells with the *src* oncogene. (a) A normal cell line called 3T3. Note the organized monolayer structure of the cells. (b) A transformed derivative of 3T3. Note how the cells are rounder and piled up on one another. *(From H. Lodish, A. Berk, S. L. Zipursky, P. Matsudaira, D. Baltimore, and J. Darnell,* Molecular Cell Biology, *4th ed. © 2000 by W. H. Freeman and Company. Courtesy of L.-B. Chen.)*

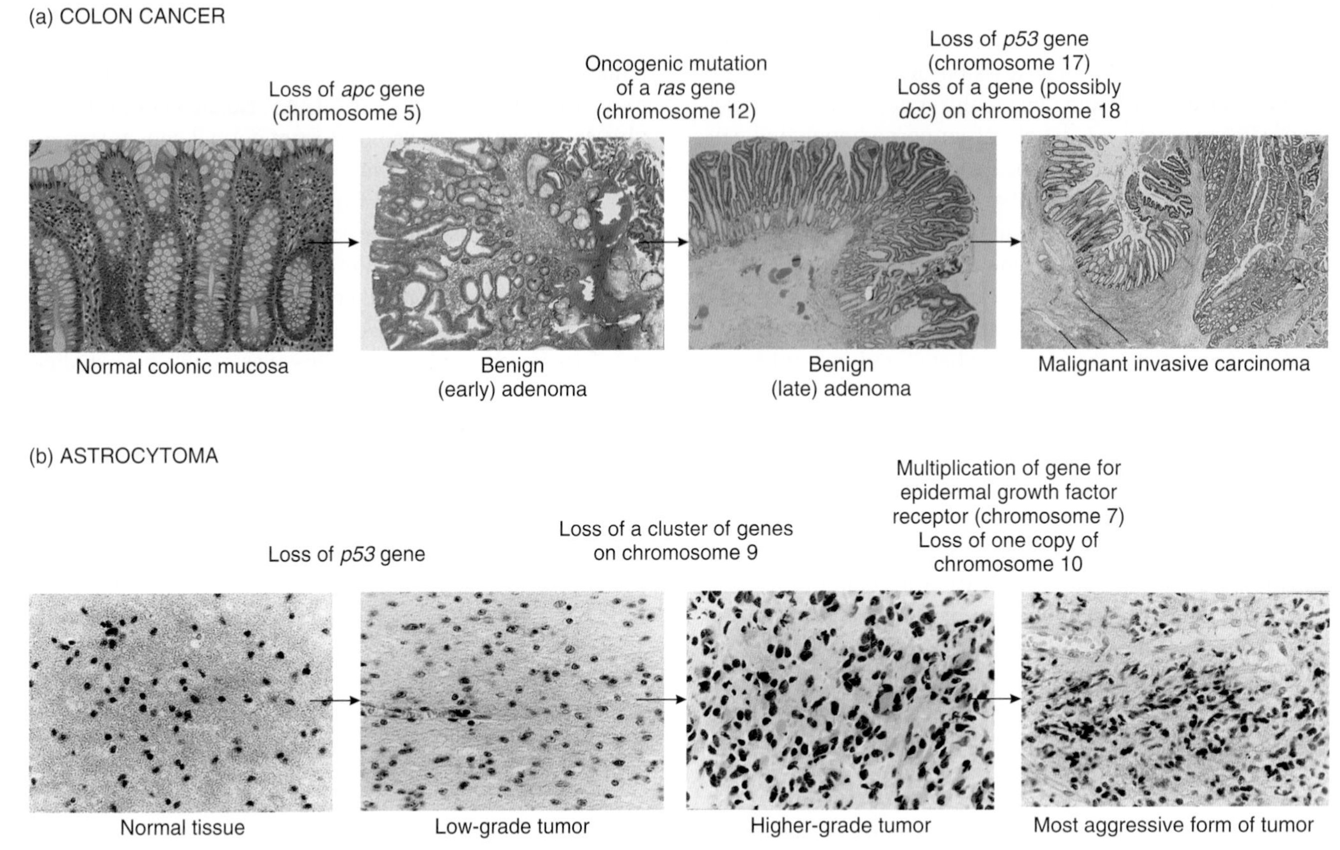

Figure 15-17 The multistep progression to malignancy in cancers of the colon and brain. Several histologically distinct stages can be distinguished in the progression of these tissues from the normal state to benign tumors to a malignant cancer. (a) A common sequence of mutational events in the progression to colon cancer. Note that the tissue becomes more disorganized as the tumor progresses toward malignancy. (b) A different characteristic series of mutations marks the progression toward a malignant astrocytoma, a form of brain cancer. *(Micrographs by E. R. Fearon and K. Cho. From W. K. Cavanee and R. L. White,* Scientific American, *March 1995, 78–79.)*

be heterozygous or an oncogene mutation and its normal allelic counterpart. Tumor-promoting mutant alleles of **tumor-suppressor genes** inactivate the proteins that they encode. In such mutations, the tumor cell lacks any copy of the corresponding wild-type allele; in essence, tumor-suppressor mutations that are found in a tumor cell are recessive.

How have tumor-promoting mutations been identified? Several approaches have been used. It is well known that certain types of cancer can "run in families." With modern pedigree analysis techniques, familial tendencies toward certain kinds of cancer can be mapped and matched with molecular markers such as microsatellites; in several cases, this has led to the successful identification of the mutated genes. Cytogenetic analysis of tumor cells themselves also has proved invaluable. Many types of tumors have characteristic chromosomal translocations or deletions of particular chromosomal regions. In some cases, these chromosomal rearrangements are so reliably a part of a particular cancer that they can be used for diagnostic purposes. For example, 95 percent of patients with chronic myelogenous leukemia (CML) have a characteristic translocation between chromosomes 9 and 22. This translocation, called the *Philadelphia chromosome* after the city where this translocation was first described, is a critical part of the CML diagnosis. The Philadelphia chromosome will be considered in more detail later in this chapter. Other translocations characterize other sorts of tumors; diagnostic translocations are most often associated with cancers of the white blood cells—leukemias and lymphomas. However, not all tumor-promoting mutations are specific to a given type of cancer. Rather, the same mutations seem to be promote tumors in a variety of cell types and thus are seen in many different cancers.

MESSAGE

Tumor-promoting mutations can be identified in various ways. When located, they can be cloned and studied to learn how they contribute to the malignant state.

It is obvious why mutations that increase the rate of cell proliferation cause tumors. It is not so obvious why mutations that decrease the chances that a cell will undergo apoptosis cause tumors. The reason seems to be twofold: (1) a cell that cannot undergo apoptosis has a much longer lifetime in which to accumulate proliferation-promoting mutations and (2) the sorts of damage and unusual physiological changes that occur inside a tumor cell would otherwise induce the self-destruction pathway.

Whether a change in an element of the cell cycle or the apoptosis pathway is due to a dominant oncogene mutation or to a recessive tumor-suppressor gene mutation depends on how that normal protein contributes to the regulation of cell proliferation or programmed cell death (Table 15-1). Genes encoding proteins that positively control (turn on) the cell cycle or negatively control (block) apoptosis can typically be mutated to become oncogenes; these tumor-promoting alleles are *gain-of-function* mutations because the mutant proteins have an altered function and now are active even in the absence of the appropriate activation signals. On the other hand, genes encoding proteins that negatively regulate the cell cycle or positively regulate apoptosis are found in the tumor-suppressor class; in these cases, the tumor-promoting alleles are *loss-of-function* mutations.

TABLE 15-1 Functions of Wild-Type Proteins and Properties of Tumor-Promoting Mutations That Can Arise in the Genes Encoding Them

Wild-Type Protein Function	Properties of Tumor-Promoting Mutations
Promotes cell cycle progression	Oncogene (gain-of-function)
Inhibits cell cycle progression	Tumor-suppressor mutation (loss-of-function)
Promotes apoptosis	Tumor-suppressor mutation (loss-of-function)
Inhibits apoptosis	Oncogene (gain-of-function)
Promotes DNA repair	Tumor-suppressor mutation (loss of function)

Classes of Oncogenes

Roughly a hundred different oncogenes have been identified (some examples are listed in Table 15-2). How do their normal counterparts, **proto-oncogenes,** function? Proto-oncogenes generally encode a class of proteins that are active only when the proper regulatory signals allow them to be activated. As mentioned, many proto-oncogene products are elements positive-control pathways of the cell cycle, including growth-factor receptors, signal transduction proteins, and transcriptional regulators. Other proto-oncogene products regulate the apoptotic pathway negatively. However, in an oncogene mutation, the activity of the mutant oncoprotein has been uncoupled from its normal regulatory pathway, leading to its continuous unregulated expression. Several categories of oncogenes have been identified

TABLE 15-2 Some Well-Characterized Oncogenes and the Functions of the Proteins That They Encode

	PROPERTIES OF PROTEIN	
Oncogene	**Location**	**Function**
Nuclear Transcription Regulators		
jun	Nucleus	Transcription factor
fos	Nucleus	Transcription factor
erbA	Nucleus	Member of steroid-receptor family
Intracellular Signal Transducers		
abl	Cytoplasm	Protein tyrosine kinase
raf	Cytoplasm	Protein serine kinase
gsp	Cytoplasm	G-protein α subunit
ras	Cytoplasm	GTP/GDP–binding protein
Mitogen		
sis	Extracellular	Secreted growth factor
Mitogen Receptors		
erbB	Transmembrane	Receptor tyrosine kinase
fms	Transmembrane	Receptor tyrosine kinase
Apoptosis Inhibitor		
bcl2	Cytoplasm	Upstream inhibitor of caspase cascade

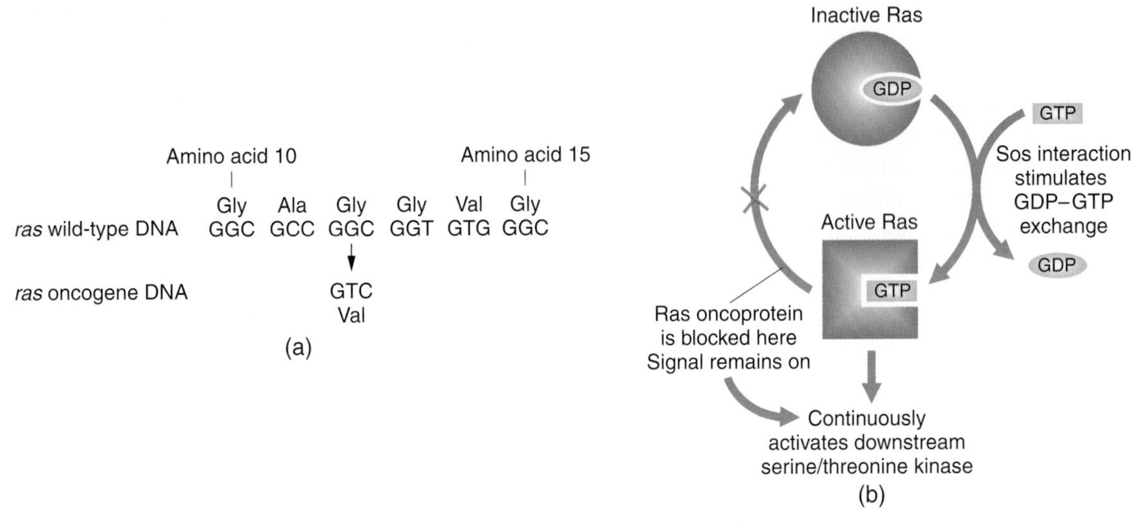

Figure 15-18 The Ras oncoprotein. (a) The *ras* oncogene differs from the wild type by a single base pair, producing a Ras oncoprotein that differs from the wild type in one amino acid, at position 12 in the *ras* open reading frame. (b) The effect of this missense mutation is to create a Ras oncoprotein that cannot hydrolyze GTP to GDP (compare with the normal Ras cycle depicted in Figure 15-12). Because of this defect, the Ras oncoprotein remains in the active Ras-GTP complex and continuously activates the downstream serine/threonine kinase (see Figure 15-13).

according to the different ways in which the regulatory functions have been uncoupled.

MESSAGE

Oncogenes encode deregulated forms of oncoproteins whose wild type participates in the positive control of the cell cycle or in the negative control of apoptosis.

Types of Oncogene Mutations

This section will describe the range of mutations that can produce dominant oncogenes. A variety of lesions, from point mutations to gene fusions, can produce germ-line gain-of-function dominant phenotypes. The same is true for dominant oncogene mutations. Indeed, we can think of these as the somatic version of gain-of-function dominant mutations, where the dominant phenotype is enhanced cell survival or cell proliferation.

Point mutations. The change from normal protein to oncoprotein often includes structural modifications of the protein itself, such as those caused by simple point mutation. A single base-pair substitution that converts glycine into valine at amino acid number 12 of the Ras protein, for example, creates the oncoprotein found in human bladder cancer (Figure 15-18a). Recall that the normal Ras protein is a G-protein subunit that takes part in signal transduction and, as described earlier in this chapter, normally functions by cycling between the active GTP-bound state and the inactive GDP-bound state (see Figure 15-12). The change caused by the *Ras* oncogene missense mutation produces an oncoprotein that always binds GTP (Figure 15-18b), even in the absence of the normal signals, such as phosphorylation of Ras, required for such binding by a wild-type Ras protein. In this way, the Ras oncoprotein continuously propagates a signal that promotes cell proliferation.

Loss of protein domains. Structural alterations that produce an oncoprotein can also be due to the deletion of parts of a normal protein. The v-*erbB* oncogene (a mutated gene in a tumor-producing virus that infects birds) encodes a mutated form of an RTK known as the EGFR, a receptor for the epidermal growth factor (EGF) ligand (Figure 15-19). The mutant form of the EGFR lacks the extracellular, ligand-binding domain as well as some regulatory components of the cytoplasmic domain. The result of these deletions is that the truncated v-*erbB*-encoded EGFR oncoprotein is able to dimerize even in the absence of the EGF ligand. The constitutive EGFR oncoprotein dimer is always autophosphorylated through its tyrosine kinase activity and so continuously initiates a signal transduction cascade.

Gene fusions. Perhaps the most remarkable type of structurally altered oncoprotein is caused by a gene fusion. The classic example emerged from studies of the Philadelphia chromosome, which, as already mentioned, is a translocation between chromosomes 9 and 22 that is a diagnostic feature of chronic myelogenous leukemia (CML). Recombinant DNA methods have shown that the breakpoints of the Philadelphia chromosome translocation in different

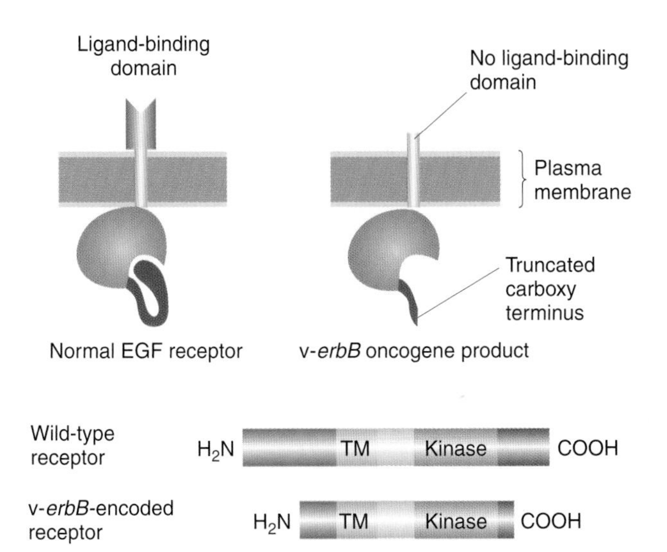

Figure 15-19 An oncogenic mutation that affects signaling between cells. EGFR, the normal receptor for epidermal growth factor (EGF), has a ligand-binding domain outside the cell, a transmembrane (TM) domain that allows the protein to span the plasma membrane, and an intracellular domain that has tyrosine-specific protein kinase activity. Normally, the kinase activity is activated only when EGF binds to the ligand-binding domain, and this activity is only transitory. The erythroblastosis tumor virus carries the v-*erbB* oncogene, which encodes a mutant form of EGFR that lacks pieces at both ends. These deletions allow the mutant protein to dimerize constitutively, leading to continuous autophosphorylation, which results in continuous transduction of a signal from the receptor. *(Adapted from J. D. Watson, M. Gilman, J. Witkowski, and M. Zoller,* Recombinant DNA, *2d ed. © 1992 by James D. Watson, Michael Gilman, Jan Witkowski, and Mark Zoller.)*

CML patients are quite similar and cause the fusion of two genes, *bcr1* and *abl* (Figure 15-20). The *abl* (Abelson) proto-oncogene encodes a cytoplasmic tyrosine-specific protein kinase, the Bcr1-Abl fusion oncoprotein, which has a permanent protein kinase activity that is responsible for its oncogenic effect.

Some oncogenes produce an oncoprotein that is identical in structure to the normal protein. In these cases, the mutation induces misexpression of the protein—that is, it is expressed in cell types from which it is ordinarily absent. Several oncogenes that cause misexpression are also associated with chromosomal translocations diagnostic of various B-lymphocyte tumors. B lymphocytes and their descendants, plasma cells, are the cells that synthesize antibodies, or immunoglobulins. In these B-cell oncogene translocations, no protein fusion results; rather, the chromosomal rearrangement causes a gene near one breakpoint to be turned on in the wrong tissue. In follicular lymphoma, 85 percent of patients have a translocation between chromosomes 14 and 18 (Figure 15-21). Near the chromosome 14 breakpoint

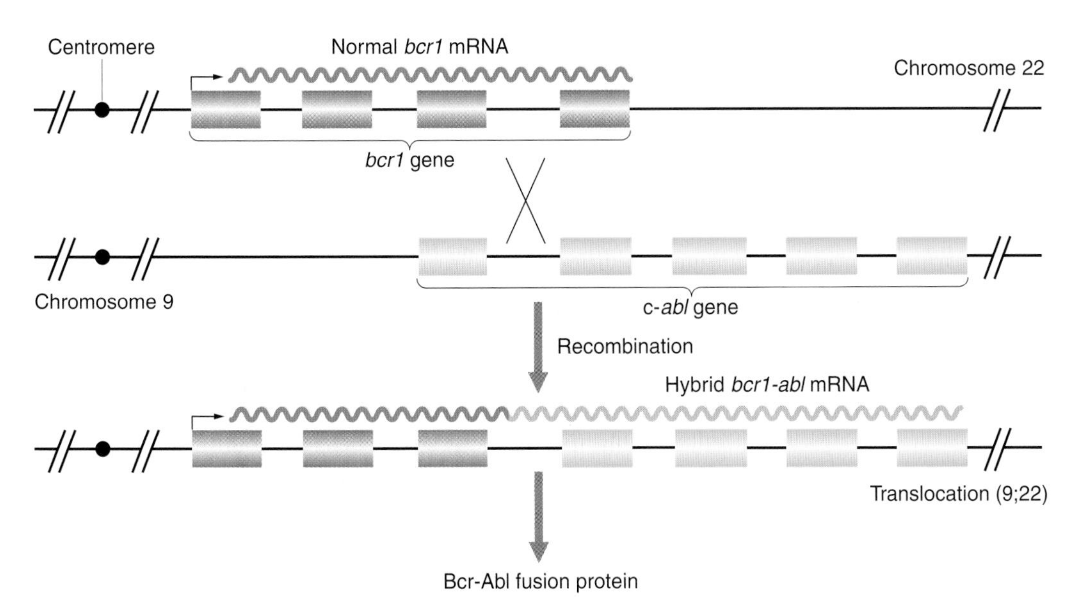

Figure 15-20 The chromosome rearrangement in CML. The Philadelphia chromosome, which is diagnostic of CML, is a translocation between chromosomes 9 and 22. The translocation breakpoints are in the middle of the c-*abl* gene, which encodes a cytoplasmic protein tyrosine kinase, and the *bcr1* gene, which is also thought to be a protein kinase. The translocation produces a hybrid Bcr1-Abl protein that lacks the normal controls for repressing c-*abl*-encoded protein tyrosine kinase activity. Only one of the two rearranged chromosomes of the reciprocal translocation is shown. *(Adapted from J. D. Watson, M. Gilman, J. Witkowski, and M. Zoller,* Recombinant DNA, *2d ed. © 1992 by James D. Watson, Michael Gilman, Jan Witkowski, and Mark Zoller.)*

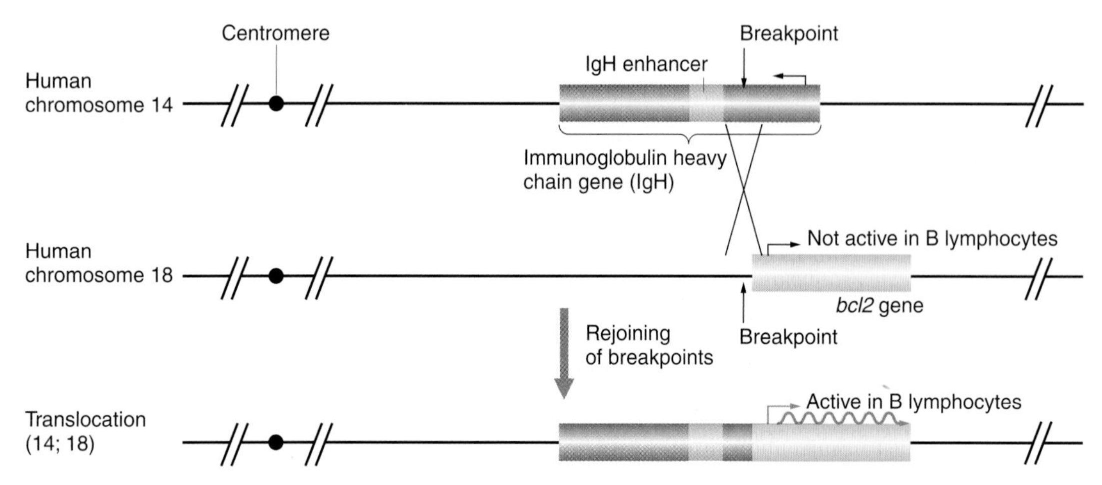

Figure 15-21 The chromosomal rearrangement in follicular lymphoma. The translocation fuses the transcriptional enhancer element of a gene, on chromosome 14, that makes one protein subunit of an antibody (the IgH subunit, also called the immunoglobulin heavy chain) with the transcription unit of a gene, on chromosome 18, that encodes Bcl2, a negative regulator of apoptosis. In this way, the Bcl2 protein is produced in antibody-producing cells, preventing any self-destruction signals from inducing apoptosis in those cells.

is a transcriptional enhancer from one of the immunoglobulin genes. This translocated enhancer element is fused with the *bcl2* gene, which is a negative regulator of apoptosis. This enhancer-*bcl2* fusion causes large amounts of Bcl2 to be expressed in B lymphocytes. This effectively blocks apoptosis in these lymphocytes and gives them an unusually long lifetime in which to accumulate mutations that promote cell proliferation. There are strong parallels between this sort of dominant oncogene mutation and the dominant gain-of-function phenotypes caused by the fusion of the enhancer of one gene to the transcription unit of another in producing the *Tab* allele of the *Abd-B* gene (see Chapter 13). In each case, the introduction of an enhancer produces a dominant gain-of-function phenotype by misregulation of the transcription unit. Mutations such as *Tab* arise in the germ line and are transmitted from one generation to the next, whereas most oncogene mutations arise in somatic cells and are thus not inherited by offspring.

MESSAGE

Dominant oncogenes contribute to the oncogenic state by causing a protein to be expressed in an activated form or in the wrong cells.

Classes of Tumor-Suppressor Genes

The normal functions of tumor-suppressor genes fall into categories complementary to those of proto-oncogenes (see Table 15-1). Some tumor-suppressor genes encode negative regulators of the cell cycle, such as the Rb protein or elements of the TGF-β signaling pathway. Others encode positive regulators of apoptosis (at least part of the function of p53 falls into this category). Still others are indirect players in cancer, with a normal role in the repair of damaged DNA. We shall consider two examples here.

Inheritance of the tumor phenotype. In retinoblastoma, the gene that encodes the Rb protein has mutated. In this cancer, which typically affects young children, retinal cells that lack a functional *RB* gene proliferate out of control. These *rb* null cells are either homozygous for a single mutant *rb* allele or are heterozygous for two different *rb* mutations. Most patients have one or a few tumors localized to one site in one eye, and the condition is sporadic—in other words, there is no history of retinoblastoma in the family and the affected person does not transmit it to his or her offspring. Retinoblastoma is not transmitted, in this case, because the *rb* mutation or mutations that inactivate both alleles of this autosomal gene arise in a somatic cell whose descendants populate the retina (Figure 15-22). Presumably, the mutations arise by chance at different times in development in the same cell lineage.

A few patients, however, have an inherited form of the disease, called hereditary binocular retinoblastoma (HBR). Such patients have many tumors, and the retinas of both eyes are affected. Paradoxically, even though *rb* is a recessive trait at the cellular level, the transmission of HBR is as an autosomal dominant (see Figure 15-22). How do we resolve this paradox? In the presence of a germ-line mutation that knocks out one of the two copies of the *RB* gene, the mutation rate for *RB* makes it virtually certain that at least some of the retinal cells of patients with HBR will have acquired an *rb* mutation in place of the single remaining nor-

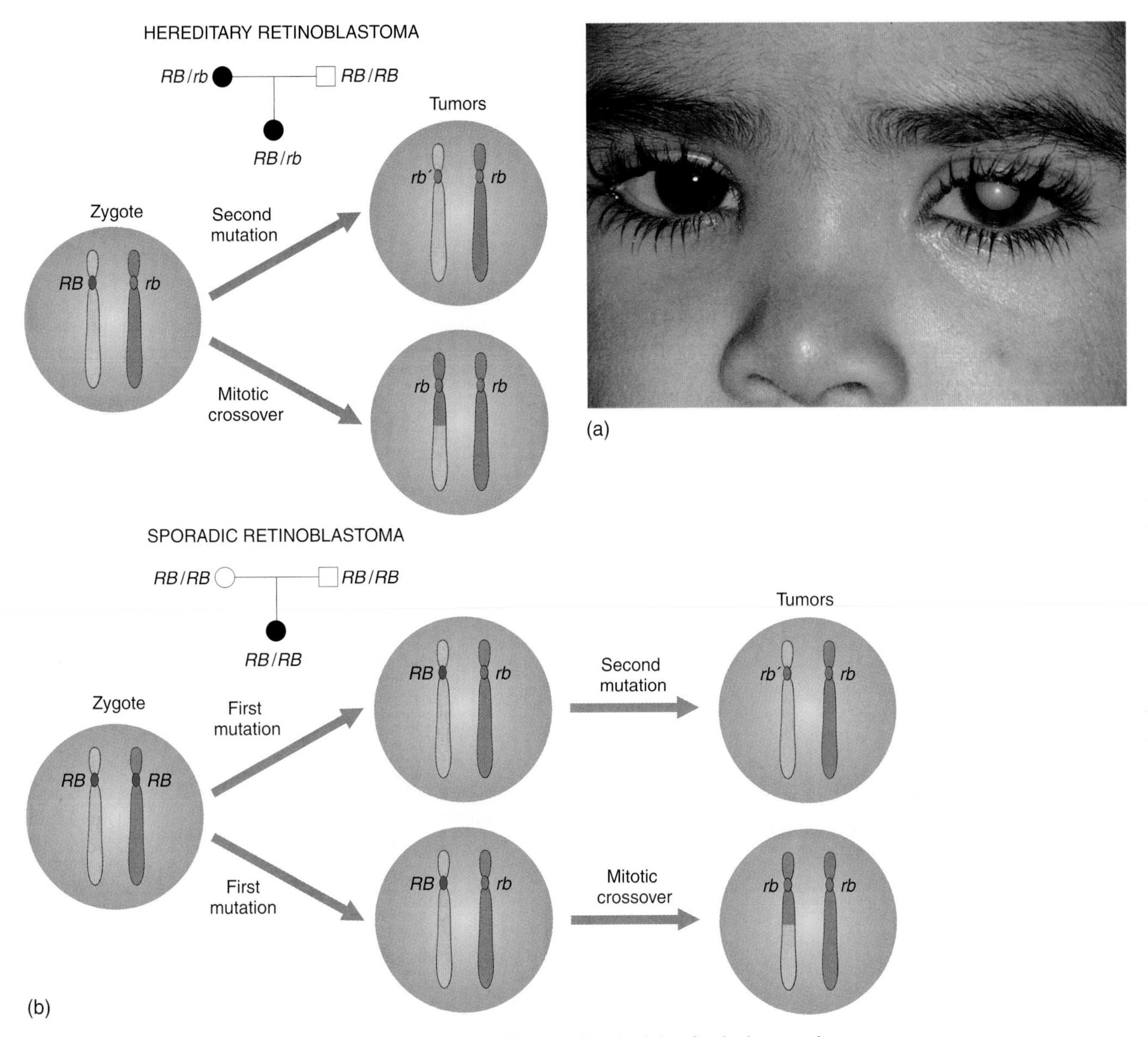

Figure 15-22 (a) Retinoblastoma, a cancer of the retina. (b) The mutational origin of retinal tumors in hereditary and sporadic retinoblastoma. Recessive *rb* alleles of the *RB* gene lead to tumor development. *(Part a from Custom Medical Stock.)*

mal *RB* gene, thereby producing cells with no functional Rb protein.

Why does the absence of *RB* promote tumor growth? Recall from our consideration of the cell cycle that Rb protein functions by binding the E2F transcription factor. Bound Rb prevents E2F from promoting the transcription of genes whose products are needed for S-phase functions such as DNA replication. An inactive Rb is unable to bind E2F, and so E2F can promote the transcription of S-phase genes. In homozygous null *rb* cells, Rb protein is permanently inactive. Thus, E2F is always able to promote S phase, and the arrest of normal cells in late G_1 (shortening their time in G_0) does not occur in retinoblastoma cells.

***The p53* tumor-tuppressor gene: a link between the cell cycle and apoptosis.** Another very important recessive tumor-promoting mutation has identified the *p53* gene as a tumor-suppressor gene. Mutations in *p53* are associated with many types of tumors, and estimates are that 50 percent of human tumors lack a functional *p53* gene. The active p53 protein is a transcriptional regulator that is activated in response to DNA damage. Activated wild-type p53 serves double duty, preventing progression of the cell cycle until the DNA damage is repaired and, under some circumstances, inducing apoptosis. In the absence of a functional *p53* gene, the p53 apoptosis pathway does not become activated, and the cell cycle progresses even in the absence of

DNA repair. This progression elevates the overall frequency of mutations, chromosomal rearrangements, and aneuploidy and thus increases the chances that other mutations that promote cell proliferation or block apoptosis will arise. Other recessive tumor-promoting genes that have been identified recently also are implicated in the repair of DNA damage. Recent research suggests that null mutations able to produce the phenotype of an elevated mutation rate are very important contributors to the progression of tumors in humans. Such recessive tumor-suppressor mutations that interfere with DNA repair promote tumor growth indirectly, because the elevated mutation rate makes it much more likely that a series of oncogene and tumor-suppressor mutations will arise, corrupting the normal regulation of the cell cycle and programmed cell death.

MESSAGE

Mutations in tumor-suppressor genes, like mutations in oncogenes, act directly or indirectly to promote the cell cycle or block apoptosis.

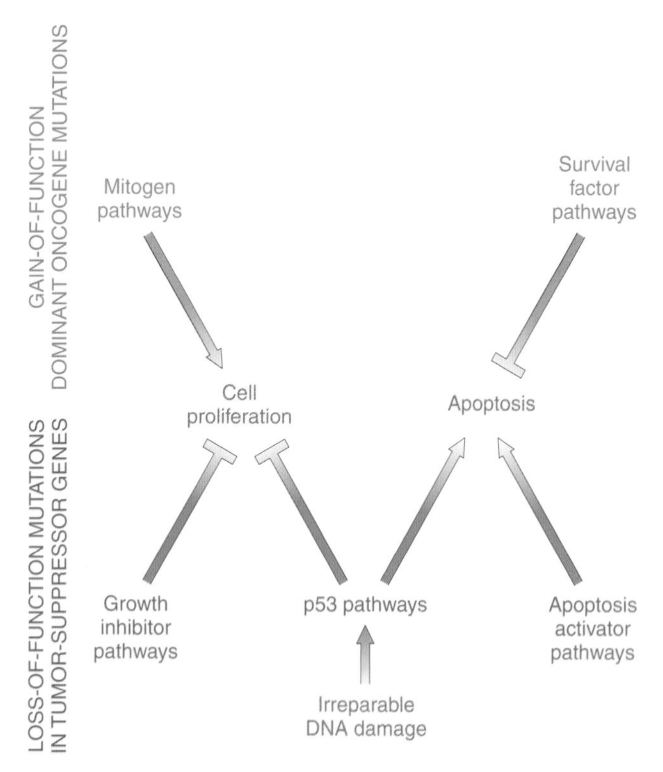

Figure 15-23 The major pathways in which mutations contribute to cancer formation and progression. The main events that contribute to tumor formation are increased cell proliferation and cell survival (decreased apoptosis). The pathways in red are susceptible to gain-of-function oncogene mutation. The pathways in blue are susceptible to loss-of-function tumor-suppressor gene mutation.

The Complexities of Cancer

Numerous mutations that promote tumor growth can arise. These mutations are thematically related and can be understood in relation to the ways in which they alter the normal processes that govern proliferation and apoptosis (Figure 15-23). In some instances, such as colon cancer (see Figure 15-17), we are even able to identify a series of independent mutations that contribute to the progression of a cell from a normal state through various stages of a benign tumor to a truly malignant state. The story does not stop there, however. Even among malignant tumors, their rates of proliferation and their abilities to invade other tissues, or metastasize, are quite different. Undoubtedly, even after a malignant state is achieved, more mutations accumulate in the tumor cell that further promote its proliferation and invasiveness. Thus, there is a considerable way to go before we have a truly comprehensive view of how tumors arise and progress.

Even so, there is light at the end of the tunnel. Overexpression of the Abelson tyrosine kinase through the Philadelphia chromosome gene fusion was completely correlated with chronic myelogenous leukemia (CML) in the early 1990s. Chemists then developed a compound, called ST1571, that bound to the ATP-binding site of the Abelson tyrosine kinase and thereby inhibited its ability to phosphorylate tyrosines on target proteins. ST1571 was shown to inhibit cell proliferation induced by *bcr-abl* in cell lines, and then was tested as the drug *Gleevec* in human clinical trials, where it has had dramatic effects on CML patients; more than 90 percent of treated patients have recovered normal blood counts. Even more striking, more than half show no evidence of cells containing the Philadelphia chromosome in their bloodstream. (No other treatment for CML has ever been this effective.) Because this drug has been in testing for only about two years, it remains to be determined whether relapses will occur or if there are serious side effects of long-term use of Gleevec. Regardless, this is the first clear case in which an understanding of the molecular biology of a malignancy has been translated into a targeted treatment.

With a general understanding of the types of molecular events that contribute to cancer, let's go back to our original question about DLBCL (diffuse large B-cell lymphoma) patients and why only 40 percent of them respond to treatment. A breakthrough has come from the transcriptional profiling of malignant cells from 40 patients, using microarray technology (Figure 15-24). In this study, two different patterns of gene expression can be observed. One pattern, called the germinal-center B-cell-like pattern (GCB), correlates with a much higher probability of patient survival after standard chemotherapy than the other pattern, called activated B-cell-like DLBCL. (These patterns are named for their similarities in gene expression to those of certain

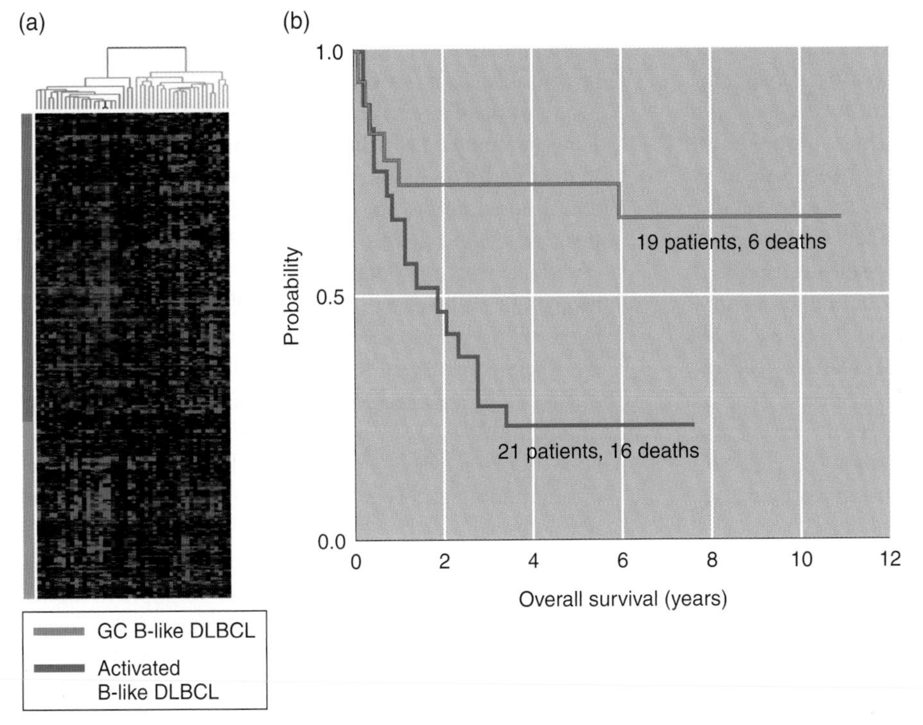

Figure 15-24 Functional genomic analysis of DLBCL. (a) Microarray analysis of malignant cells indicates that patients possess one of two basic RNA-expression profiles. The data come from microarray experiments as described in the Functional Genomics section of Chapter 9. Each row represents a cDNA sample for a different gene. Each column represents an mRNA sample from malignant cells of different DLBCL patients; these mRNA samples were used to probe one copy of the microarray. Green represents genes that were expressed at least twofold-lower levels in the DLBCL malignant cells than the mRNA of control nonmalignant cells. Red represents genes that were expressed at least twofold-higher levels in the DLBCL malignant cells than the control mRNA sample. The black areas represent genes whose expression levels were not significantly different in DLBCL and control-cell mRNA preparations. Note that there are two markedly different DLBCL types in terms of the overall pattern of gene expression—one termed GC B-like and the other Activated B-like. (b) Higher patient survival correlates with the GC B-like RNA profile. *(Reprinted by permission from L. Liotta and E. Petricoin,* Nature Reviews: Genetics *1, 2000, 53. Macmillan Magazines Ltd.)*

normal B-lymphocyte cell types.) This suggests that even though DLBCL is diagnosed histologically as one disease, at a molecular level it is more like two different diseases. This raises the hope of more accurate diagnosis and of treatment that can be focused on each specific disease. Further, microarray experiments such as these are beginning to reveal unexpected RNAs that differ in level either between normal and tumor cells, or between different stages of malignancy; the protein products of these RNAs are then additional candidate targets for specific drug therapy in cancer treatment.

SUMMARY

1. Higher eukaryotic cells have evolved mechanisms that control their survival and their ability to proliferate. These mechanisms are essential for carrying out normal integrative functions of tissues and organ systems.
2. These controls are all highly integrated and depend on the continual evaluation of the state of the cell and the continual communication of information between neighboring cells and among different tissues. Higher eukaryotes are heavily invested in a "division of labor" life strategy, in which somatic cells are integrated into a variety of tissues, each with a specialized role in the physiology of the organism. Regulation of cell numbers is essential to maintain proper physiological balance between the various tissues and cell types.
3. Normal cell proliferation is modulated by regulation of the cell cycle. This achieves multiple ends. First, regulation at checkpoints of the cell cycle ensures that progression can occur only when a set of preconditions that ensure proper chromosome replication and segregation are fulfilled. Second, because the cell cycle is wired into the

intercellular signaling system, proliferation can be halted until the proper signals for it are received.

4. Apoptosis is a normal self-destruction mechanism that eliminates damaged and potentially harmful cells as well as cells that have only temporary functions during development. We now know that cell survival requires direct survival signals that inhibit activation of the execution program and the caspase cascade that lead to self-destruction. Having a built-in cellular self-destruct system is very valuable in the context of the higher eukaryotic life strategy, in which the cost of replacing discarded cells is far less than compromising the health of the entire organism.
5. Intercellular signaling systems permit proliferation and apoptosis to be coordinated in a population of cells. Signaling involves secretion of a signal from sending cells, receipt of that signal by surface receptors on target cells, and transduction of that signal from the transmembrane receptor to the interior (usually the nucleus) of the target cells.
6. In cancer, cells proliferate out of control and avoid self-destruct mechanisms through the accumulation of a series of tumor-promoting mutations in the same somatic cell. These mutations have the net effect of converting the cell from a state that relies on input from its neighbors to survive and to proliferate, to an alternative state where it is produces the necessary survival and proliferation signals itself. In other words, instead of depending upon paracrine signaling for survival and proliferation, it now uses autocrine signals.
7. Many of the classes of genes that promote cancers when they are mutated are important components of the cell that directly or indirectly contribute to growth control and differentiation. Basically, mutations in pathways that positively or negatively control survival and proliferation contribute to tumor development. Genes that normally accelerate survival or proliferation are found to be mutated to dominant oncogenes. Genes that normally accelerate apoptosis or inhibition of mitosis are found as recessive tumor-suppressor mutations. In addition, mutants in repair pathways, as well as those that increase the general mutation rate, increase the likelihood of tumor-promoting mutations and hence indirectly promote tumor development.
8. Functional genomics is being used in diagnostic tests and in drug-target identification in studies to improve the detection and treatment of cancer. Instead of the former method of screening for specific genetic contributions to tumor development, we are now able to survey virtually the entire genome for changes in gene activity associated with tumor development. Even if we don't understand the nature of all the changes, functional genomics reveals gene expression "fingerprints" that allow different disease states to be identified.

CONCEPT MAP

Draw a concept map interrelating as many of the following terms as possible. Note that the terms are listed in no particular order:

DNA repair / enhancer / oncogene / cell proliferation / gain-of-function mutations / signal transduction pathway / cyclin-dependent protein kinase / apoptosis / checkpoints / RTK / tumor-suppressor gene

SOLVED PROBLEM

In the inherited form of retinoblastoma, an affected child is heterozygous for an *rb* mutation, which is either passed on from a parent or has newly arisen in the sperm or the oocyte nucleus that gave rise to the child. The heterozygous *RB* / *rb* cells are nonmalignant, however. The *RB* allele of the heterozygote must also be knocked out in the developing retinal tissue to create a tumorous cell. Such a knockout can occur through an independent mutation of the *RB* allele or by mitotic crossing-over such that the original *rb* mutation would now be homozygous.

a. If retinoblastoma is passed on to other siblings as well, could we determine whether the original mutation was derived from the mother or the father? How?

b. Could we determine whether the *rb* mutation was maternally or paternally derived if it arose de novo in a germ cell of one parent?

Solution

a. If the trait is inherited, we can determine from which parent it came. The most straightforward approach is to identify DNA polymorphisms, such as restriction fragment length polymorphisms (RFLPs), that map within or near the *RB* gene. *RB* has the curious property of being inherited as an autosomal dominant, even though it is recessive on a cellular basis. Given the dominant pattern of inheritance, by finding DNA differences in the parental genomes that map near each of the four parental alleles, we should be able to determine which allele has been passed on to all affected offspring. That allele is the mutant one.

b. If the trait arose de novo in the sperm or the oocyte, we can still possibly determine from which parent it came, but with considerable difficulty. One way to do so would be to use recombinant DNA cloning techniques to isolate each of the two copies of the *RB* gene from normal cells. Only one of these alleles should be mutant. When the gene has been cloned, by DNA sequencing of the two alleles we may be able to identify the mutation that inactivates *RB*. If it arose de novo in the sperm or the oocyte, this mutation would not be present in the somatic cells of the parents. If sequencing also reveals some polymorphisms (for example, in restriction-enzyme recognition sites) that distinguish the alleles, we should then be able to go back to the parents' DNA to find out whether the mother or the father carries the polymorphisms that were found in the cloned mutant allele. (Whether this approach will work depends on the exact nature of the parental alleles and the mutation.)

If the mutation arose by mitotic crossing-over, additional tools are available. In this case, the entire region around the *rb* gene will be homozygous for the mutant chromosome. By examining DNA polymorphisms known to map in this region, we should be able to determine whether this chromosome derived from the mother or from the father. This becomes a standard exercise in DNA fingerprinting similar to that described in part a.

PROBLEMS

Basic Problems

1. Describe three lines of evidence that cancer is a genetic disease.

2. Describe three mechanisms used to control the activities of the proteins in the cell cycle and cell-death pathways.

3. What are the two roles of cyclins? How are the levels of cyclins regulated?

4. What are three major categories of mutations that lead to the development of cancer?

5. What are the roles of Apaf and Bcl proteins in apoptosis?

6. What are two mechanisms by which translocations can lead to oncogene formation?

7. How is the activity of caspases controlled?

8. Give an example of an oncogene. Why is the mutation dominant?

9. Give an example of a tumor-suppressor gene. Why is the mutation recessive?

10. How do the following mutations lead to cancer?

a. v-*erbB*

b. *ras* oncogene

c. Philadelphia chromosome

11. For each of the following, describe how a mutation could lead to the formation of an oncogene.

a. growth-factor receptor

b. transcriptional regulator

c. G-protein

Challenging Problems

i 12. Cancer is thought to be caused by the accumulation of two or more "hits"—that is, two or more mutations that affect cell proliferation and survival in the same cell. Many of these oncogenic mutations are dominant: one mutant copy of the gene is sufficient to change the proliferative properties of a cell. Which of the following general types of mutations have the potential to create dominant oncogenes? Justify each answer.

a. a mutation that increases the number of copies of a transcriptional activator of cyclin A

b. a nonsense mutation located shortly after the beginning of translation in a gene that encodes a growth-factor receptor

c. a mutation that increases the level of FasL

d. a mutation that disrupts the active site of a cytoplasmic tyrosine-specific protein kinase

e. a translocation that joins a gene encoding an inhibitor of apoptosis to an enhancer element for gene expression in the liver

13. Many of the proteins that participate in the progression pathway of the cell cycle are reversibly modified, whereas, in the apoptosis pathway, the modification events are irreversible. Rationalize these observations in relation to the nature and end result of the two pathways.

14. Tumor-promoting mutations are described as being either gain-of-function or loss-of-function. Describe the effects of both classes of tumor-promoting mutations on the cell cycle and apoptosis.

15. Normally, FasL is present on cells only when a message needs to be sent to neighboring cells instructing them to undergo apoptosis. Suppose that you have a mutation that produces FasL on the surface of all liver cells.

    a. If the mutation were present in the germ line, would you predict it to be dominant or recessive?

    b. If such a mutant arose in somatic tissues, would you expect it to be tumor-promoting? Why or why not?

16. Some genes can be mutated to oncogenes by increasing the copy number of the gene. This, for example, is true of the Myc transcription factor. On the other hand, oncogenic mutations of *ras* are always point mutations that alter the protein structure of Ras. Rationalize these observations in relation to the roles of normal and oncogenic versions of Ras and Myc.

17. We now understand that mutations that inhibit apoptosis are found in tumors. Because proliferation per se is not induced by the inhibition of apoptosis, explain how this inhibition might contribute to tumor formation.

18. Suppose that you had the ability to introduce normal copies of a gene into a tumor cell in which mutations in the gene caused it to promote tumor growth.

    a. If the mutations were in a tumor-suppressor gene, would you expect that these normal transgenes would block the tumor-promoting activity of the mutations? Why or why not?

    b. If the mutations were of the oncogene type, would you expect that the normal transgenes would block their tumor-promoting activity? Why or why not?

19. Insulin is a protein that is secreted by the pancreas (an endocrine organ) when the level of blood sugar is high. Insulin acts on many distant tissues by binding and activating a receptor tyrosine kinase (RTK), leading to a reduction in blood sugar by appropriate storage of the products of sugar metabolism. Diabetes is a disease in which the level of blood sugar level remains high because some part of the insulin pathway is defective. One kind of diabetes (call it type A) can be treated by giving the patient insulin. Another kind of diabetes (call it type B) does not respond to insulin treatment.

    a. Which type of diabetes is likely to be due to a defect in the pancreas, and which type is likely to be due to a defect in the target cells? Justify your answer.

    b. Type B diabetes can be due to mutations in any of several different genes. Explain this observation.

20. Irreparable DNA damage can have consequences for both the cell cycle and apoptosis. Explain what the consequences are, as well as the pathways by which the cell implements them.

21. Retinoblastomas arise through mutations in the *RB* gene. In HBR (hereditary binocular retinoblastoma), the gene is inherited from parent to offspring as a simple autosomal dominant. Nonetheless, *RB* is thought to be a recessive tumor-suppressor gene. In HBR, there are typically multiple tumors in both eyes. In contrast, in the form of retinoblastoma that is not inherited (called sporadic retinoblastoma), there are fewer tumors and they are usually restricted to one eye. Account for all of these observations in terms of your understanding of how tumor-suppressor genes mutate to cause neoplasia.

22. The wild-type Rb protein functions to sequester E2F in the cytoplasm. At the appropriate time in the cell cycle, E2F is released so that it can act as a functional transcription factor.

    a. In cells homozygous for a loss-of-function *rb* mutation, where in the cell do you expect to find E2F?

    b. If a cell were doubly homozygous for loss-of-function mutations in both the *RB* and *E2F* genes, would you expect tumor growth? Explain.

    c. How would a mutation in the RB protein that irreversibly binds E2F affect the cell cycle?

Exploring Genomes: A Web-Based Bioinformatics Tutorial

Exploring the Cancer Anatomy Project

The Cancer Anatomy project at NCBI is a specialized database focused on a particular subset of data. It brings together, in a graphical and searchable format, all known information about individual genes, map positions, and chromosomal location for genes associated with cancer. In the Genomics tutorial at www.whfreeman.com/mga, we will explore gene expression data, gene mutations, and chromosomal aberrations associated with the various tumor types.

THE GENETIC BASIS OF DEVELOPMENT 16

Key Concepts

1. A programmed set of instructions in the genome of a higher organism establishes the developmental fates of cells with respect to the major features of the basic body architecture.
2. Developmental pathways consist of a sequence of various regulatory steps.
3. The zygote is totipotent, giving rise to every adult cell type; as development proceeds, successive decisions restrict each cell and its descendants (a lineage) to its particular fate.
4. In the egg, gradients of maternally derived regulatory proteins establish polarity along the major body axes (head to tail; front to back); these proteins control the local transcriptional activation of genes that encode master regulatory proteins in the zygote.
5. Many proteins that act as master regulators of early development are transcription factors; others are components of pathways that mediate signaling between cells. Consequently, many fateful decisions require communication and collaboration between cells.
6. The same basic set of genes identified in *Drosophila* and the regulatory proteins that they encode are conserved in mammals and appear to govern major developmental events in many—perhaps all—higher animals.
7. The molecules that underlie regulation of development in plants are different from those in animals, but many of the same themes are seen to direct the fate of cells in plants.

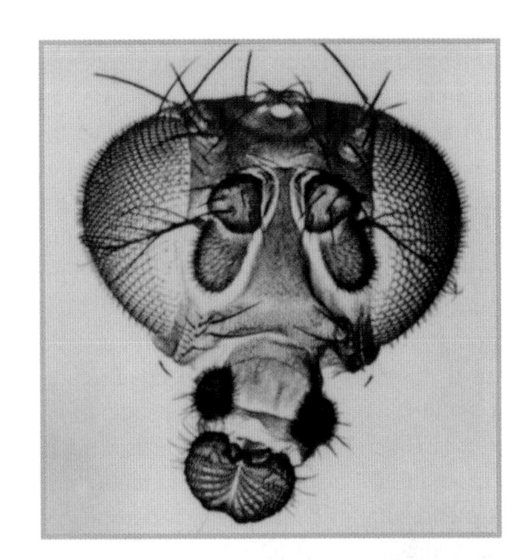

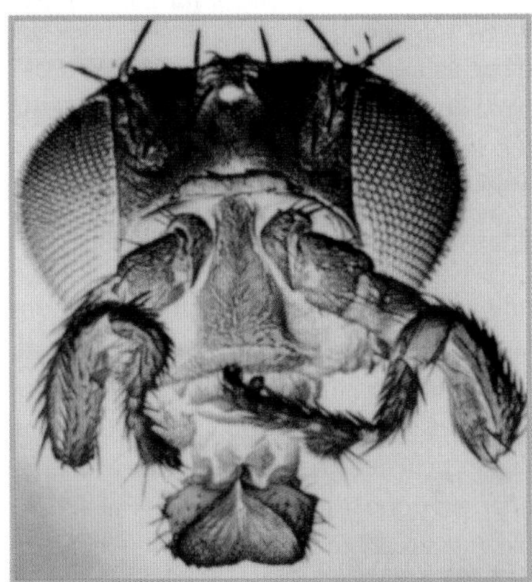

A homeotic mutation that alters the basic *Drosophila* body plan. Homeosis is the replacement of one body part by another. In place of the normal antennae (shown at the top), an *Antennapedia* mutation causes the antennal precursor cells to develop into a leg. *(F. R. Turner/BPS.)*

A child is born, seemingly normal except for a shorter great toe and thumb. By the time the child is two, her ability to turn her head is somewhat impaired. When she is four, she develops a big painful red swelling on her back. The swelling goes away in a few weeks but, six months later, bone appears at the site of the swelling. Over time, this swelling-bone deposition cycle continues, with more ectopic bone appearing until, as a young adult, she is effectively encased in an ectopic skeleton. Ultimately, such individuals are literally frozen in one position for the rest of their lives and require around-the-clock care.

This process of bone deposition describes the onset of a dominant autosomal genetic disorder called *FOP, Fibrodysplasia Ossificans Progressiva*. The basic event is the conversion, over time, of connective tissue to bone (Figure 16-1). This conversion is accelerated at sites of even mild injury such as an intramuscular injection, and the exact locations where the ectopic bones will develop are not predictable. The great toe and thumb phenotypes are due to a missing segment (called a *phalanges*) in each digit. Why this digit defect correlates with the ectopic bone phenotype of FOP is completely unclear. Because few individuals with FOP are affected mildly enough to be able to reproduce, the genetic analysis of the disease is based on very limited pedigree data. This is a rare disorder, and its frequency is about the same in all ethnic groups—about 1 case per 1.5 million individuals. This is consistent with most cases of FOP being due to new mutations.

Several lines of experimental evidence suggest that the genetic program for bone development in FOP individuals has gone awry; somehow, this program becomes activated in connective tissue as well as in normal bone tissue. (FOP patients have a complete normal skeleton, except for the missing phalange of each of the first digits.) Knowing this has not in itself produced a treatment. Surgical intervention to free joints frozen by ectopic bone is not possible, because surgery is injurious and actually accelerates ectopic bone deposition at the sites of surgery. It seems clear that one key to developing a treatment is to understand the molecular basis of the disease, by identifying the mutated gene or genes responsible for the condition. However, in the absence of sufficient pedigree data, how can this be accomplished? The alternative to pedigree analysis is the **candidate gene approach.** In this approach, one uses known biological information to make plausible guesses about the nature of the mutated gene and tests those guesses by cloning and sequencing candidate alleles from affected individuals, looking for meaningful mutations in these alleles. (We use the term *meaningful* because many DNA differences between allelic genes occur naturally but do not affect gene activity; see Chapter 17.) For this approach to be effective, however, there must be a reasonable basis for identifying a list of candidates. What might that candidate list be for FOP and bone development? Let's consider how development in general is thought to work and then return to the candidate gene approach for FOP.

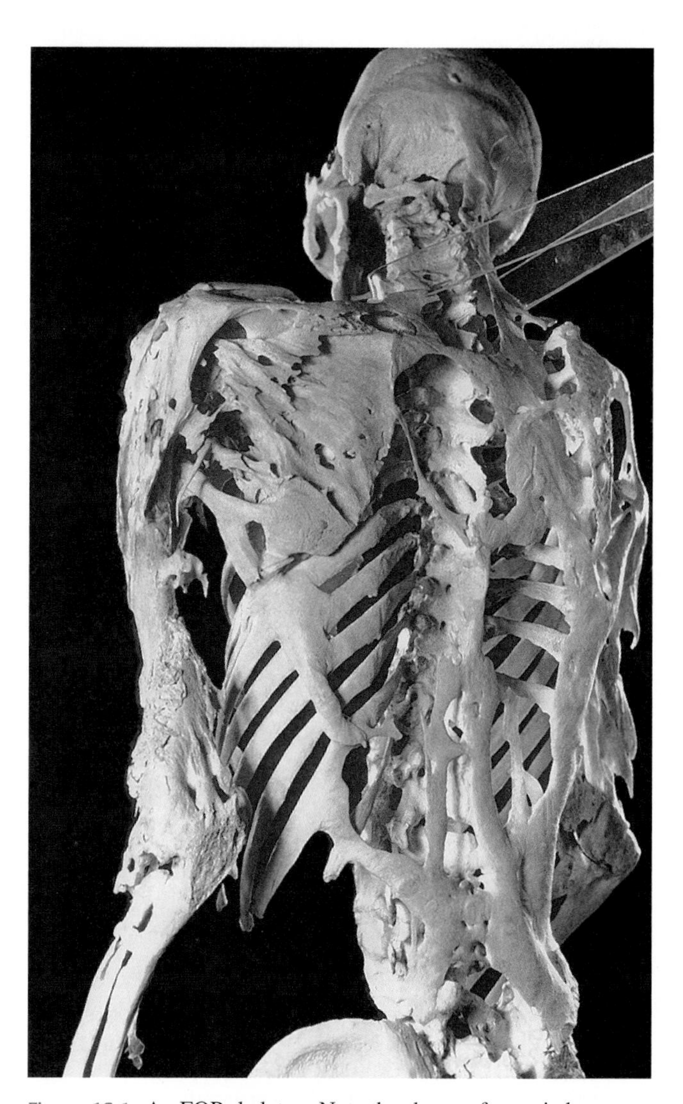

Figure 16-1 An FOP skeleton. Note the sheets of ectopic bone that are connected to the normal skeleton. (*"Overexpression of an Osteogenic Morphogen in Fibrodysplasia Ossificans Progressiva" by A. B. Shafritz, E. M. Shore, F. H. Gannon, M. A. Zasloff, R. Taub, M. Muenke, and F. S. Kaplan, 1996, Massachusetts Medical Society.*)

CHAPTER OVERVIEW

The general body plan (the overall architecture) is common to all members of a species and, indeed, is common to many very different species. All mammalian species have four limbs, whereas all insects have six. But all mammals and insects must, in the course of their development, differentiate the anterior from the posterior end and the dorsal from the ventral side. Eyes and legs always appear in the appropriate places. The basic body plan of a species appears to be quite robust—that is, the internal genetic program produces the same body plan within a broad range of environmental conditions. We should not forget, however, that the study of the genetic determination of these basic developmental processes does not provide an explanation of the phenotypic differences between individuals. This chapter focuses on the processes that underlie pattern formation, the construction of complex form, and how these processes operate reliably to execute the developmental program for the basic body plan.

This chapter will show that the production of the body plan is a stepwise process of cell commitment, in which the outline of the major body subdivisions is first painted with broad strokes and then refined until fine-grained pattern is finally laid down (Figure 16-2). This is a quite sensible approach to building pattern. Create a few different populations of cells, and then let them assort in stepwise fashion into a larger and larger set of coherently distributed cell types. At each step in the process, there is the opportunity to adjust cell number and cell commitment if the previous step didn't work quite right. Only after these mid-course corrections have been accomplished, go on to the next step. In a sense, this is analogous to the checkpoint regulation of the cell cycle, but on a multicellular level.

The development of a pattern involves toggle switches that turn certain developmental pathways on or off; other "radio" switches are permitted to select only one option among many. The coherence of the body plan is produced by wiring the transcriptional regulatory apparatus within each cell to intercellular signaling systems. In this way, cells can communicate and coordinate their commitments as a community effort, ensuring representation of all of the necessary cell types in the proper spatial deployment for building the tissues, organs, and appendages of the mature body plan. Thus, this chapter will not present new types of molecular functions. Rather, we will encounter the same cast of characters that take part in gene regulation and intercellular signaling but in more highly integrated and coordinated contexts. This should not be surprising because one of the basic themes of biology is that natural selection exploits a limited set of existing tools to solve new problems, such as the construction of complex body plans.

CHAPTER OVERVIEW figure

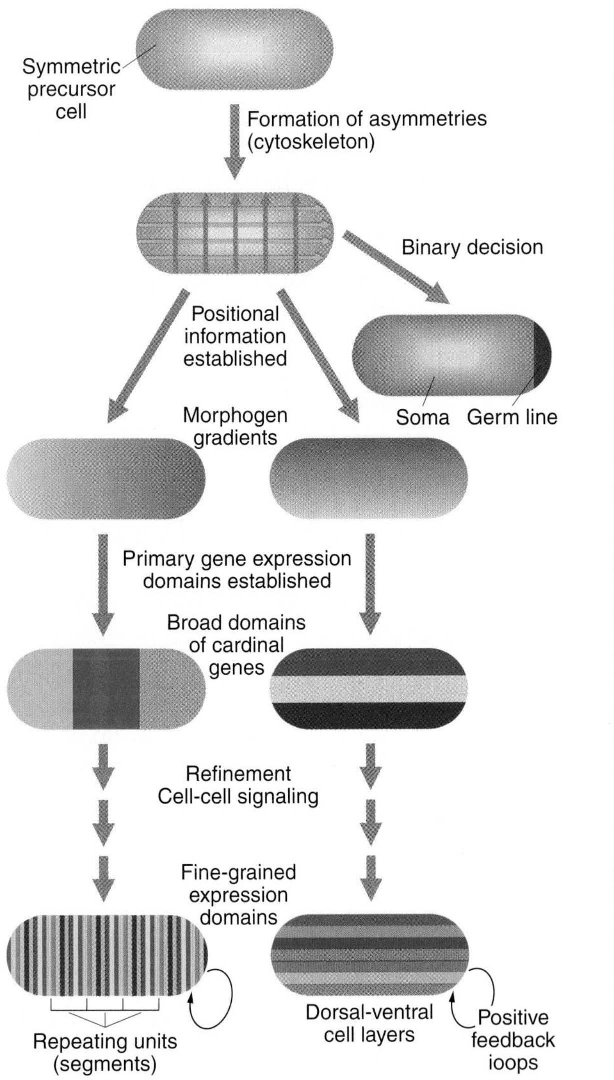

Figure 16-2 An overview of developmental strategies.

THE LOGIC OF BUILDING THE BODY PLAN

In all higher organisms, life begins as a single cell, the newly fertilized egg. It reaches maturity as a population of thousands, millions, or even trillions of cells combined into a complex organism with many integrated organ systems. The goal of developmental biology is to unravel the fascinating and mysterious processes that achieve the transfiguration of egg into adult. Because we know more about development in animal systems than in plants or multicellular fungi, we shall focus our attention on these organisms.

The different cell types of the body are distinguished by the variety and amounts of the proteins that they express—the protein profile of each cell (that is, the quantitative and qualitative array of proteins that it contains). The protein profile of a cell in a multicellular organism is the end result of a series of genetic regulatory decisions that determine the "when, where, and how much" of gene expression. Thus, for a particular gene, a geneticist is interested in which tissues and at what developmental times the gene is transcribed and how much of the gene product is synthesized. From that point of view, all developmental programming that controls an organism's protein profiles is determined by the regulatory information encoded in the

DNA. We can look at the genome as a parts list of all the gene products (RNAs and polypeptides) that can possibly be produced and as an instruction manual of when, where, and how much of these products are to be expressed. Thus, one aspect of developmental genetics is to understand how this instruction manual operates to send cells down different developmental pathways, ultimately producing a large constellation of characteristic cell types.

This is not all that there is to developmental genetics and production of cellular diversity, however. How are these different cell types deployed coherently and constructively—in other words, how do they become organized into organs and tissues, and how are those organ systems and tissues organized into an integrated, coherently functioning individual organism? We shall explore these questions by examining the formation of the basic animal body plan.

During the elaboration of the body plan, cells commit to specific **cell fates;** that is, the capacity to differentiate into particular kinds of cells. The commitments have to make sense in regard to the location of the cell, because all organs and tissues are made up of many cells, and the structure of an organ or tissue requires a cooperative division of labor among the participating cells. Thus, somehow, cell position must be identified, and fate assignments must be parceled out among a cooperating group of cells—called a **developmental field.**

Positional information is generally established through protein signals that emanate from a localized source within a cell (the initial one-cell zygote) or within a developmental field. This is the molecular equivalent of establishing the rules for geographical longitudes and latitudes. Just as we need longitudes and latitudes to navigate on Earth, cells need positional information to determine their location within a developmental field and to respond by executing the appropriate developmental program. When that positional information has been received, generally a few intermediate cell types are created within a field. Through further processes of cell division and decision making, a population of cells with the necessary final diversity of fates will be established.

These further processes—**fate refinement**—can be of two types. Sometimes, through asymmetric divisions of one of the intermediate types of cells, descendants are created that have received different regulatory instructions and therefore become committed to different fates. This can be thought of as a cell lineage–dependent mechanism for partitioning fates. Often, such fate decisions are made by committee—that is, the fate of a cell depends on input from, and feedback to, neighboring cells through paracrine signaling mechanisms (see Chapter 15). Such neighborhood-dependent decisions are extremely important, because the chemical dialogue between cells ensures that all fates have been allocated and that the pattern of allocation is coherent. The cell neighborhood–dependent mechanisms also provide for a certain developmental flexibility. Developmental mechanisms need to be flexible so that an organism can compensate for accidents such as the unplanned death of some cell. If some cells are lost by accident, the normal paracrine intercellular communication is then aborted and the surviving neighbors can become reprogrammed to divide and instruct a subset of their descendants to adopt the fates of the deceased cells. Indeed, the regeneration of severed limbs, as occurs in some animals, manifests the power of stepwise specification of cell fate through local intercellular interactions versus hard-wired determination in building pattern.

MESSAGE

Cells within a developmental field must be able to identify their geographic locations and make developmental decisions in the context of the decisions being made by their neighbors.

The consequence of the preceding scenario for development is that the process of commitment to a particular fate is a gradual one. A cell does not go in one step from being totally uncommitted, or totipotent, to becoming earmarked for a single fate. Each major patterning decision is, in actuality, a series of events in which multiple cells that are at the same level of commitment to their fate are, step by step, assigned different fates. These events unfold, generally, along with cell proliferation. Thus, if we examine a cell lineage—that is, the family tree of a somatic cell and its descendants—we see that parental cells in the tree are less committed than their descendants.

MESSAGE

As cells proliferate in the developing organism, decisions are made that specify more and more precisely the options for the fate of cells of a given lineage.

The Major Decisions in Building the Embryo

A variety of developmental decisions are undertaken in the early embryo to give cells their proper identities and to build the body plan. Some of them are simple binary decisions:

- Separation of the germ line (the gamete-forming cells) from the soma (everything else).
- Establishment of the sex of the organism. (Ordinarily, all cells of the body make the same choice.)

These binary decisions tend to be made at one developmental stage and, as we shall see later, are examples of irreversible fate determination.

The other major decisions involve multiple fate options and far more intricate decision-making pathways; they lead to the complex elements of the body plan, composed almost entirely of the somatic cells. Most of these pathways are specified by decisions taken by local populations of cells:

- Establishment of the positional information necessary to orient and organize the two major body axes of the embryo: anterior-posterior (head to tail) and dorsal-ventral (back to front)
- Subdivision of the embryonic anterior-posterior axis into a series of distinct units called **segments,** or metameres, and assignment of distinct roles to each segment according to its location in the developing animal
- Subdivision of the embryonic dorsal-ventral axis into the outer, middle, and inner sheets of cells, called the **germ layers,** and assignment of distinct roles to each of these layers
- Production of the various organs, tissues, systems, and appendages of the body through the coordinated and cooperative action of localized groups of cells characteristic of each segment and germ layer

MESSAGE

Among developmental decisions, the simpler ones tend to involve irreversible commitment to one of two options, whereas the more complex ones involve selections among multiple options.

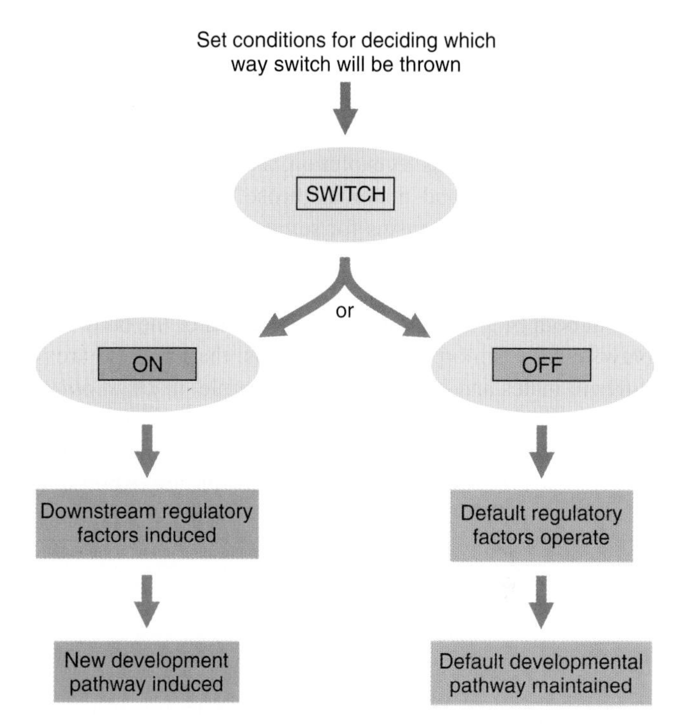

Figure 16-3 Decision making in developmental pathways. In these pathways, cells have to achieve different roles through a series of binary (on-off) decisions. Conditions within the cell allow a master switch to be regulated. When the master switch is activated, it sets in motion a cascade of "downstream" regulatory events that eventually lead the cell down a new developmental pathway. Without the activation of the master switch, a set of default signals remain in place and maintain the cell in the default pathway.

Applying Regulatory Mechanisms to Developmental Decisions

Chapter 13 described how transcriptional regulation of biochemical pathways controls the production of specific proteins at the correct time, in the correct place, and in the correct amounts. In a biochemical pathway, the regulatory "switch" that activates or blocks the synthesis of the enzymes of that pathway is usually some nutrient being supplied to the organism externally. In the developmental pathway, by contrast, the regulatory switch depends on some key molecule that is produced internally by the organism itself: either a molecule synthesized by the very cell that makes the decision or a molecule produced by other cells. In a simple developmental pathway, the concentration of such a key molecule determines whether the "on" or the "off" binary choice is made (Figure 16-3). Above some threshold level, one decision will be taken; below this threshold, the opposite decision will be made. The "off" decision implies that development will proceed along the path that had been programmed by previous decisions in the history of that cell; this is usually called the *default* pathway. The "on" decision shunts the cell into an alternative pathway.

Many pathways also have a maintenance mode that locks the "on" or "off" decision permanently in place for a given cell and its descendants. Making a pathway decision and subsequently remembering that decision are keys to the commitment of cells to their fate.

In dissecting a developmental pathway, its regulatory decisions might be found at any level in the process of transferring information from the gene to the active protein (Figure 16-4 on the next page). This chapter will describe this process at levels ranging from transcriptional regulation to protein modification and subunit interactions that control the active protein profiles within cells.

BINARY FATE DECISIONS: PATHWAYS OF SEX DETERMINATION

In many species, sex determination is associated with the inheritance of a heteromorphic chromosome pair in one sex. However, this does not imply that all such species have evolved from a common ancestor that possessed such a heteromorphic sex chromosome set. Rather, XX-XY mechanisms for determining sex appear to have arisen independently many times in evolution. The XX-XY sex chromosomes of flies and mammals arose independently, and the underlying mechanisms for sex determination are quite different. The main features of mechanisms that determine sex in flies and mammals are contrasted in Table 16-1.

In both flies and mammals, many areas of the body display sexually dimorphic characteristics; that is, these areas differ in males and in females. For example, in *Drosophila,* the two sexes differ in the structure of the sex organs themselves and in pigmentation of the abdomen. The mechanisms of sex determination and how they ensure expression of such sexually dimorphic characteristics will be the main focus of the following pages.

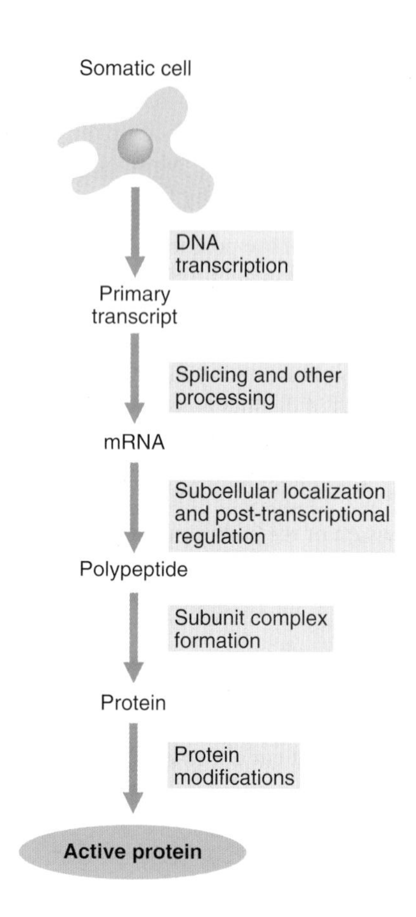

Figure 16-4 Regulation of protein activity. In a cell, many steps transfer information between the gene and the active protein that it encodes. In development, regulatory mechanisms at any of these levels can control the production of active protein products.

TABLE 16-1 A Comparison of the Features of Sex Determination in *Drosophila* and Mammals

Aspect of Sex Determination	*Drosophila*	Mammals
Chromosomal basis of XX vs. XY	X : A ratio	Presence or absence of Y
Molecular basis of XX vs. XY	Numerator : denominator transcription factor subunit interactions	*SRY* (human) ~ *sry* (mouse) gene for testis determination located on Y chromosome encodes transcription factor
Decision-making process	Cell by cell	Testis as male-determining organ that instructs other tissues by release of testosterone
Switch	Transcription of *Sxl (Sex lethal)* in early embryo	Presence or absence of *SRY* (human) ~ *Sry* (mouse) gene (on Y chromosome)
Level of downstream regulation	Alternative mRNA splicing	Transcription
Positive feedback mechanism	Positive feedback loop of SXL protein on *Sxl* mRNA splicing	Continued presence of testis throughout development
Elaboration of sex-specific gene	Presence of male (M)- or female (F)-specific form of *dsx (doublesex)*-encoded protein. Generally, DSX-F is a repressor of expression of male-specific genes, and DSX-M is a repressor of female-specific genes. For a few genes, DSX-F or DSX-M acts as an activator of female-specific or male-specific gene expression, respectively.	Testosterone-mediated activation of androgen receptor in target tissues. Testosterone-androgen receptor complex regulates transcription by binding to testosterone-responsive elements (enhancers and silencers).

DROSOPHILA SEX DETERMINATION: EVERY CELL FOR ITSELF

Every cell lineage in *Drosophila* makes its own sexual decision. One of the best ways to demonstrate this is by analyzing XX-XY mosaic flies; that is, individual flies containing a mixture of XX and XY cells. Such mosaics show a mixture of male and female phenotypes, according to the genotype of each individual cell. The interpretation of this difference is that every cell in *Drosophila* determines its sex independently.

Phenotypic Consequences of Different Ratios of X Chromosomes to Autosomes

In *Drosophila,* sex determination is due to the ratio of X chromosomes to sets of autosomes. Recall that, in *Drosophila,* $n = 4$: one sex chromosome and three separate autosomes. Hence, one autosomal set, which we shall represent as A, comprises the three separate autosomes and, in a diploid fly, A = 2. The effect of this X:A ratio can best be seen by examining sex-chromosome aneuploids (Table 16-2). A normal XX *Drosophila* diploid (XX AA) has an X:A ratio of 1.0 and is phenotypically female. An XY diploid (XY AA) has an X:A ratio of 0.5 and is male; an XO diploid also is male (although sterile). Triploids with three X chromosomes (XXX AAA) are females, those with one X (XYY AAA) are male, and those with two Xs (XXY AAA) are "in between" (intersexes).

The Basics of the Regulatory Pathway

An overview of the regulatory pathway for sex determination in *Drosophila* is shown in Figure 16-5. The X:A ratio in the early embryo establishes whether a fly will become male or female. This directive for establishing the sexual phenotype is carried out by a master regulatory switch and several downstream sex-specific genes. The "off" position of the *Sxl* switch produces the male pathway of determination, whereas the "on" position shunts cells into the female

TABLE 16-2 Effect of Sex-Chromosome Dosage on Somatic Sexual Phenotype in Diploid *Drosophila* and Humans

	SOMATIC SEXUAL PHENOTYPE	
Sex-Chromosome Constitution	***Drosophila***	**Humans**
Euploidy		
XX	♀	♀
XY	♂	♂
Aneuploidy		
XXY	♀	♂
XO	♂	♀

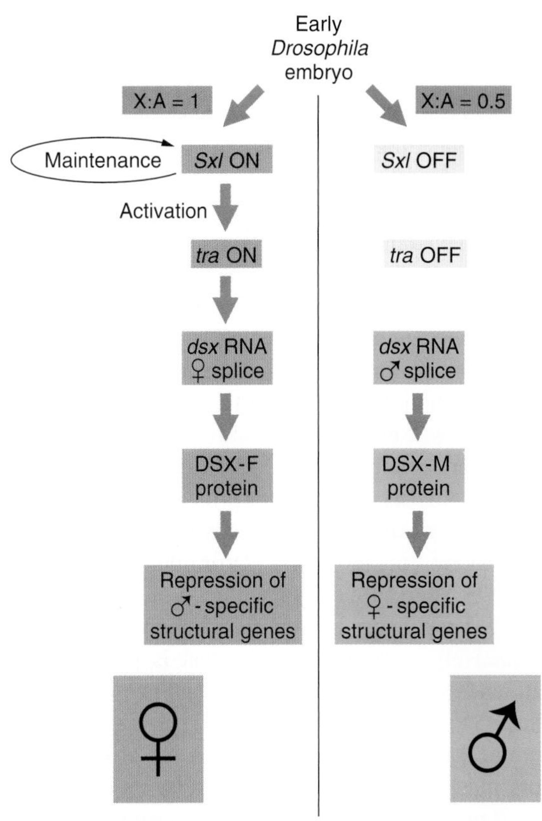

Figure 16-5 The pathway of sex determination and differentiation in *Drosophila.* The X:A ratio is evaluated through the interaction of numerator and denominator monomeric protein subunits to produce an active complex referred to as NUM-NUM transcription factor (see Figure 16-6). The level of active numerator transcription factor determines whether *Sxl (Sex lethal)* is to be permanently turned on or is to remain off. If *Sxl* is on, then the female sex-differentiation pathway is turned on, ultimately causing splicing of a form of the *dsx* mRNA that produces the DSX-F transcription factor, which represses male-specific genes. If *Sxl* is off, then the sex-differentiation pathway is not activated and the default *dsx (doublesex)* splicing pattern creates an mRNA encoding the DSX-M transcription factor, which represses female-specific genes.

mode of sexual determination. The choice of pathway is initiated by differential transcription of the *Sxl* and *tra (transformer)* genes, the direction of the switch is maintained by an *Sxl* autofeedback loop, and the decision is propagated along the developmental pathway by differential RNA splicing of the *dsx (doublesex)* gene. The default mode of the pathway culminates in the production of male-specific transcription factors, whereas the alternative shunt culminates in the production of female-specific transcription factors.

The Regulatory Switch

Genetic analysis has shown that the master regulatory switch is the activity of a gene called *Sxl (Sex lethal).* In a fly with an X:A ratio of 1, SXL protein is synthesized and the fly develops as a female. In a fly with an X:A ratio of 0.5, no functional SXL protein is produced, and consequently the fly develops as a male.

Setting the switch in the "on" or "off" position. The X:A ratio sets in motion the sex-determination pathway by an interaction between the protein products of a series of X-chromosomal, zygotically expressed *numerator* genes and autosomal, maternally and zygotically expressed *denominator* genes. At least some of the numerator and denominator genes encode transcription factors of a type called **basic helix-loop-helix (bHLH) proteins.** bHLH proteins are known to function as transcription factors only when two bHLH monomers complex to form a dimeric protein. Here, NUM will indicate the X chromosome–encoded bHLH numerator proteins and DEM will indicate the autosomally encoded bHLH denominator proteins.

These transcription factors have only one role: in a narrow time window in the early *Drosophila* embryo—roughly 2 to 3 hours after fertilization—they determine whether the *Sxl* regulatory switch gets flipped on. The *Sxl* gene is effectively a "toggle switch" that is permanently locked into an "on" position in females and an "off" position in males (Figure 16-6a). To set the *Sxl* switch in the "on" position, the level of the active X:A NUM transcription factors must be high (owing to an X:A ratio of 1.0). With high (female) levels, the X:A transcription factors present in the early embryo bind to enhancers of the *Sxl* gene, activating its transcription from the *Sxl* early promoter. The transcript made from the early promoter then produces active SXL protein.

In contrast, if the levels of the NUM factors are too low, as is the case when the X:A ratio is 0.5, then there is insufficient transcription factor to activate *Sxl* transcription and no SXL protein is made.

The NUM proteins very likely measure the X:A ratio by competing for dimer formation with the DEM proteins to form active NUM transcription-factor dimeric protein complexes (Figure 16-6b). Although how this works is not known for sure, here is a plausible mechanism.

- The NUM monomers have a sequence-specific DNA-binding site, whereas the DEM proteins lack any DNA-binding site.
- The DNA-binding site of NUM recognizes an enhancer sequence that regulates transcription from the promoter of the *Sxl* regulatory switch gene. As the next section will show, transcription from this promoter is required for establishment of *Sxl* gene expression in the early embryo.
- Both NUM and DEM polypeptides are synthesized at levels proportionate to the number of copies of each bHLH-encoding numerator or denominator gene in the cell. In this way, embryos with an X:A ratio of 1.0 have twice as much NUM polypeptide per cell as do embryos with an X:A ratio of 0.5. In contrast, regardless of the X:A ratio, these cells have the same level of DEM.
- All possible combinations of dimers can form, in proportion to the concentrations of NUM and DEM monomers in a cell: NUM-NUM homodimers, NUM-DEM heterodimers, and DEM-DEM homodimers.
- To be an active transcription factor, both subunits of a bHLH dimer must possess sequence-specific DNA-binding sites. This is true only for NUM homodimers. In a sense, then, when present in the same dimer, the DEM monomers inhibit the transcriptional activity of the NUM subunits.

The conclusion is that the higher the NUM:DEM ratio, the more active the NUM-NUM transcription factor in a cell. Thus, in early embryos with an X:A ratio of 1.0, we can expect that much more active numerator transcription factor will accumulate than in embryos with an X:A ratio of 0.5.

MESSAGE

Protein-protein interactions, such as competition between normal and inhibitory subunits for dimer formation, can be triggers for controlling developmental switches.

Maintaining the switch in a stable position. The *Sxl* gene has two promoters. The early promoter is the only one that is activated by the NUM-NUM transcription factors. The early promoter (P_E) is active only early in embryogenesis. Later in embryogenesis and for the remainder of the life cycle, the *Sxl* gene is transcribed from the late promoter (P_L) regardless of the X:A ratio or any other condition. This late promoter is active in every cell in the animal, beginning with mid-embryogenesis and persisting for the lifetime of the organism. The primary transcript produced by *Sxl* transcription from the late promoter is much larger than the primary transcript from the early promoter, and it is subject to alternative mRNA splicing, depending on the presence or absence of preexisting active SXL protein in the cell. The SXL protein is an RNA-binding protein that alters the splicing of the nascent *Sxl* transcript coming from this late promoter. When mRNA splicing occurs in the presence of bound SXL protein, splicing of the *Sxl* primary transcript produces an mRNA that encodes additional active SXL RNA-binding protein. This SXL protein in turn binds to more *Sxl* primary transcript from the late promoter, creating the spliced form of the mRNA that encodes functional SXL protein, and so forth. Thus a feedback, or autoregulatory, loop, controlled at the level of RNA splicing, maintains SXL activity throughout development in flies with an X:A ratio of 1.0.

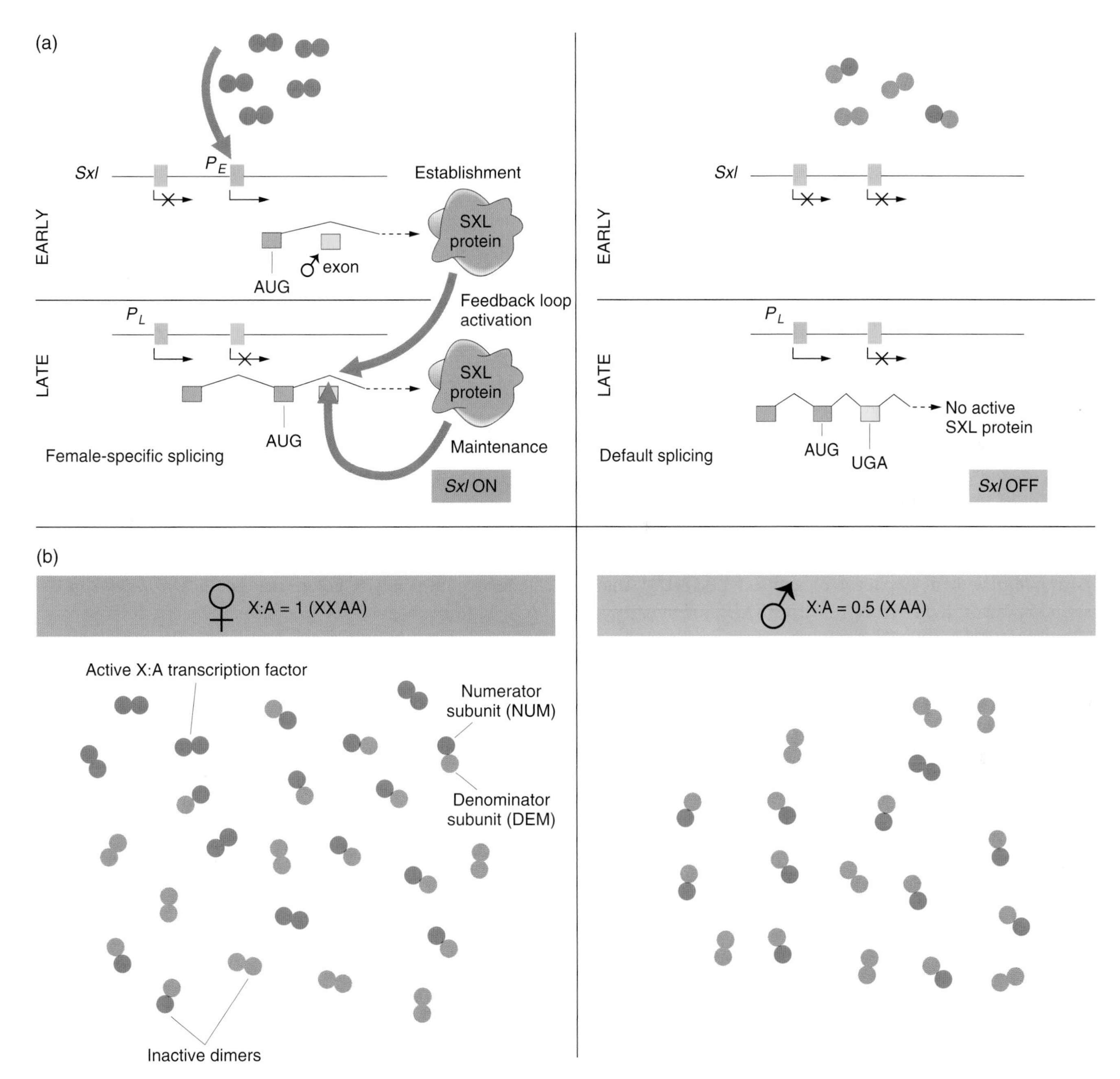

Figure 16-6 The initiation and maintenance of the *Sxl* switch. (a) The *Sxl* switch. (*left panel*) High levels of the NUM-NUM transcription-factor dimers (red circles) activate transcription from the early promoter (P_E) of *Sxl*. By mid-embryogenesis, P_E is turned off, and *Sxl* is transcribed in all cells of the animal from the constitutively active late promoter (P_L). Through binding to the primary P_L transcript and regulating its splicing pattern (preventing the male exon from being included in the final mRNA), preexisting active SXL protein ensures the further production of active SXL protein, which continues the splicing pattern that leads to its own formation, thus creating a positive autoregulatory loop. (*right panel*) On the other hand, when the X:A ratio is 0.5, P_E is not activated. Thus, no SXL protein is present in the early embryo, and the RNA-splicing pattern of the P_L transcript generates an mRNA that includes the male exon. This exon contains a stop codon (UGA), which terminates the SXL polypeptide prematurely. The short SXL protein made in males is completely nonfunctional. AUG denotes the location of the translation-initiation codon for the SXL polypeptide. (b) A plausible mechanism for the molecular basis of the X:A ratio. During early embryogenesis, the numerator and denominator genes are expressed. NUM subunits (red circles) encoded by X chromosome genes and DEM polypeptide subunits (green circles) encoded by autosomal genes form dimers at random. Only NUM-NUM dimers form active transcription factor. If the X:A ratio is 1.0, high levels of these NUM-NUM dimers form, bind to the P_E enhancer, and activate transcription from P_E. If the X:A ratio is 0.5, most NUM subunits are part of NUM-DEM heterodimers, which do not function as active transcription factors.

MESSAGE

The autoregulatory loop exemplifies how an early developmental decision can be "remembered" for the rest of development, even after the initial signals that established the decision have long disappeared.

In contrast, when the X:A ratio is 0.5, the *Sxl* switch is set in the "off" position. The early promoter is not activated early in embryogenesis and hence the early X:A = 0.5 embryo has no SXL protein. As a consequence, in the absence of any active SXL protein, the primary *Sxl* transcript of the late, constitutive *Sxl* promoter is processed in the default mRNA-splicing pattern. This default *Sxl* mRNA is nonfunctional, in the sense that it encodes a stop codon shortly after the translation-initiation codon of its protein-coding region. The small protein produced from this male-specific spliced mRNA has no biological activity. Thus, in *Drosophila* with a low level of active NUM-NUM transcription factor, the absence of active SXL protein early in development predestines that there will be no SXL activity throughout the remainder of development.

Propagating the decision. Not only must SXL have an autoregulatory maintenance function, but it must be capable of activating the shunt pathway that will lead to female-specific gene expression. It accomplishes this through the same RNA-binding activity. Only in the presence of SXL protein is the primary *tra (transformer)* transcript spliced to produce an mRNA that encodes active TRA protein (Figure 16-7a). In turn, TRA protein is an RNA-binding protein that produces female-specific splicing of the *dsx (doublesex)* nascent RNA. The mRNA produced by this splicing pattern encodes a DSX-F protein, a transcription factor that globally represses male-specific gene expression (Figure 16-7b).

In the absence of active SXL protein, the splicing pattern of *tra* primary transcript produces an mRNA that does not encode functional TRA protein. In the absence of active TRA protein, splicing of the *dsx* primary transcript leads to the production of a DSX-M transcription factor that represses female-specific gene expression (Figure 16-7b).

The way that genetic analysis has contributed to this understanding of sex determination in *Drosophila* is described in Foundations of Genetics 16-1.

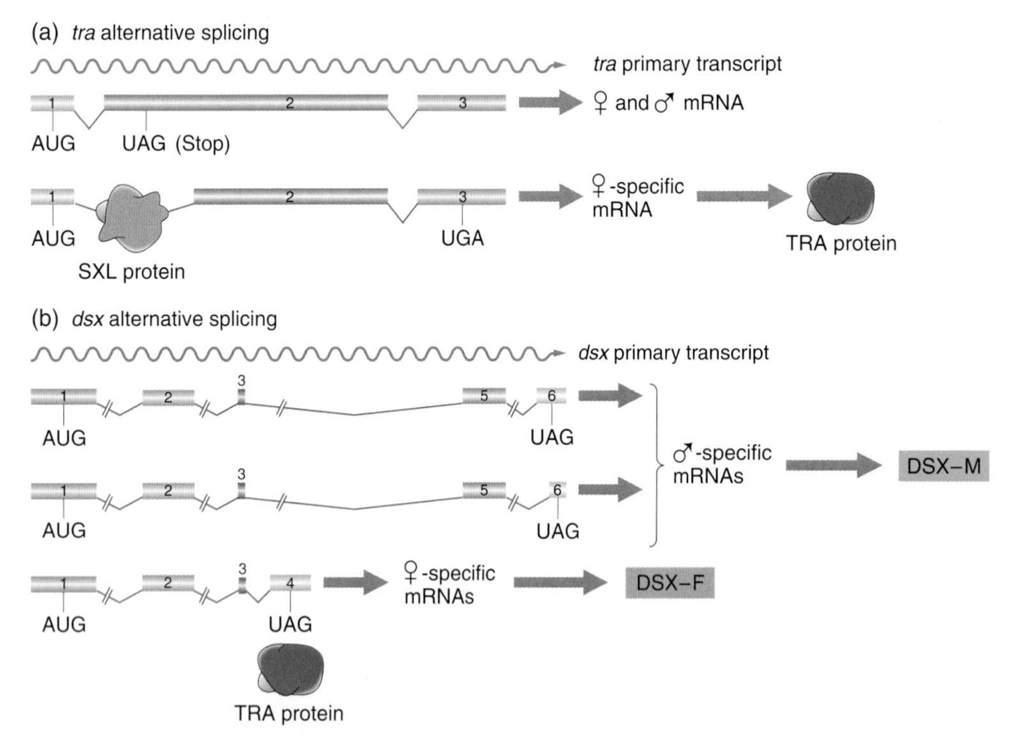

Figure 16-7 Alternative splicing of *tra* and *dsx* transcripts. (a) Two forms of *tra* mRNA are produced. One form is present in both sexes and, because of a stop codon (UAG) in exon 2, does not encode a functional protein. The other form is female specific and encodes the active TRA polypeptide. (b) Different *dsx* mRNAs are produced in both sexes. In males, exon 4 is not included in two related male-specific mRNAs. Both male-specific mRNAs make related polypeptides, DSX-M, that act to repress transcription of female-specific genes. In females, exons 5 and 6 are not included in the *dsx* mRNA. The DSX-F polypeptide that is produced acts to repress transcription of male-specific genes. *(Modified from M. McKeown,* Current Opinion in Genetics and Development *2, 1992, 301.)*

FOUNDATIONS OF GENETICS 16-1

The Mutational Analysis of *Drosophila* Sex Determination

Thomas Hunt Morgan, the founding father of *Drosophila* genetics, was quoted as saying: "Treasure your exceptions." This has been the guiding principle of all modern genetic analysis of any biological process. This approach, studying the properties of rare mutant individual organisms and using these observations to make inferences about what the wild-type process is doing, has greatly enhanced our understanding of sex determination in several species from very different taxonomic groups.

Insights into *Drosophila* sex determination have emerged through molecular and genetic analysis of mutations that alter the phenotypic sex of the fly, especially studies by Thomas Cline, Bruce Baker, and their colleagues. What kinds of mutations have been encountered? In regard to sexually dimorphic phenotypes, the effects of null mutations in several of the genes in the pathway are to transform females into phenotypic males. Males homozygous for these mutations are completely normal. These mutated genes include *sis-b (sisterless-b), Sxl (Sex-lethal),* and *tra (transformer).* These genes are dispensable in males because the male developmental pathway seems to be the default state of the developmental switch. In other words, the sex-determination pathway in *Drosophila* is constructed so that the activities of several gene products are needed to shunt the animal from the default state into the female developmental pathway. The *sis-b* gene, a numerator gene that encodes a bHLH protein, must be active to achieve an X:A ratio of 1.0. The mRNA-splicing regulators—the RNA-binding proteins encoded by the *Sxl* and *tra* genes—must be active for female development to ensue. The *sis-b, Sxl,* and *tra* genes are ordinarily off in males anyway, so it is of no consequence to male development to have mutations knocking out the functions of these genes.

The exceptional gene is *dsx (doublesex).* The knockout of the *dsx* gene leads to the production of flies that simultaneously have male and female attributes. The reason for this phenotype is that each of the two alternative *dsx* proteins, DSX-F and DSX-M, represses the gene products that produce the phenotypic structures characteristic of the other sex. In the absence of repression, the gene products that build the structures characteristic of each of the two sexes operate simultaneously, and a fly that is simultaneously male and female develops.

[An aside: You may wonder why *Sxl* is called *Sex-lethal* even though phenotypic sex is a dispensable trait. The answer is that the phenomenon of dosage compensation—equalizing the expression of X-linked genes in 2X females and 1X males—also operates through the numerator/denominator balance and through *Sxl* (but not through *tra* or *dsx*). When proper dosage compensation is impaired, lethality ensues. Special genetic tricks that circumvent this lethality problem are used to be able to study the sex determination–specific aspects of *Sxl*.]

SEX DETERMINATION IN MAMMALS: COORDINATED CONTROL BY THE ENDOCRINE SYSTEM

An analysis of sex-chromosome aneuploids demonstrates that mammalian sex determination and differentiation are quite different from those of *Drosophila* (see Table 16-2). An XXY human is phenotypically male, with a syndrome of moderate abnormalities (Klinefelter syndrome; see Figure 11-16). XO humans have a number of abnormalities (Turner syndrome; see Figure 11-14), including short stature, mental retardation, and mere traces of gonads, but they are clearly female in morphology. These data are consistent with a mammalian sex-determination mechanism based on the presence or absence of a Y chromosome. Without a Y, the person develops as a female; with it, as a male.

Mammalian Reproductive Development and Endocrine Organ Control

In contrast with flies, where all the signals that regulate sex determination are generated within each cell, each individual human cell does not make an independent determination of its sex. Rather, the human mosaic for XX and XY tissues typically has a general appearance characteristic of one or the other sex. The observation of nonautonomy in mammalian sex determination can be understood in view of the biology of the mammalian reproductive system: sex-specific phenotypes are driven by the presence or absence of the testes.

The gonad forms within the first 2 months of human gestation. Primordial germ cells migrate into the genital ridge, which sits atop the rudimentary kidney. The chromosomal sex of the germ cells determines whether they will migrate superficially or deeply into the genital ridge and whether they will organize it into a testis or an ovary

(a)

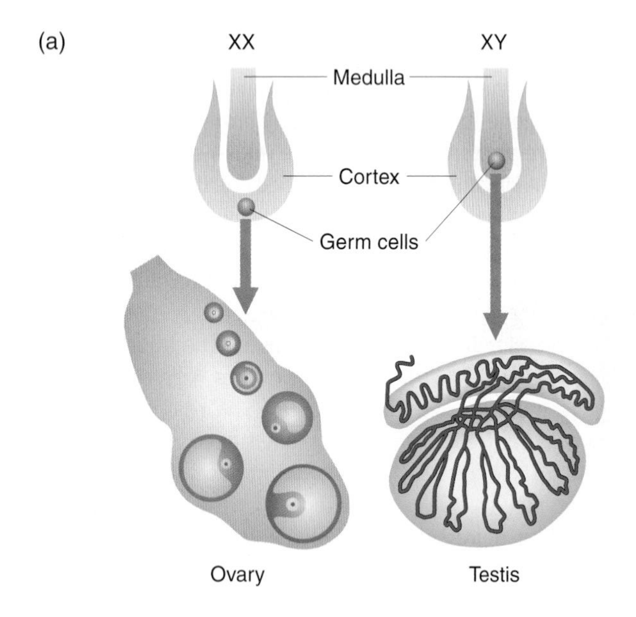

(b)

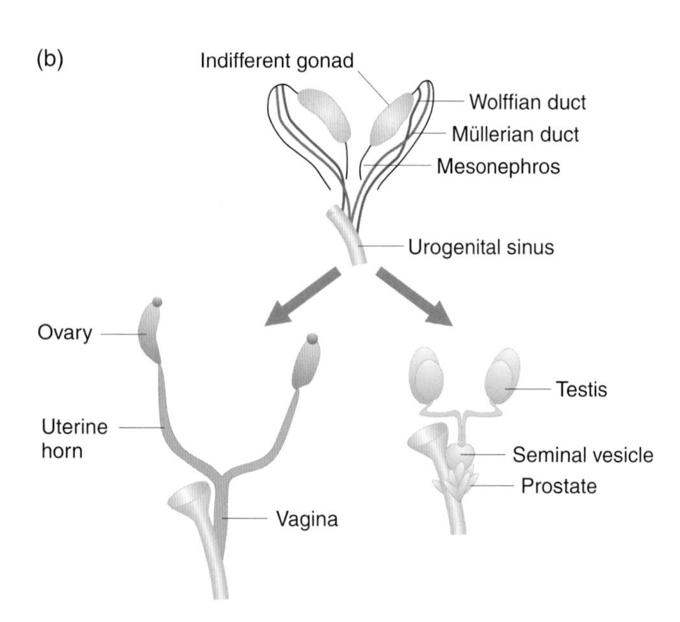

Figure 16-8 Development of the mammalian urogenital system. (a) The embryonic genital ridge consists of a medulla surrounded by a cortex. Female germ cells migrate into the cortex, which becomes organized into an ovary. Male germ cells migrate into the medulla, which becomes organized into a testis. (b) In the initial urogenital organization at the indifferent gonad stage, precursors of both male (Wolffian) and female (Müllerian) ducts are present. If a testis is present, it secretes two hormones, testosterone and a polypeptide hormone called Müllerian-inhibiting substance (MIS, or anti-Müllerian hormone). This causes the Müllerian ducts to regress and the Wolffian ducts to develop into the male reproductive ducts. If an ovary is present, testosterone and MIS are absent and the opposite happens: the Wolffian ducts regress and the Müllerian ducts develop into the female reproductive ducts. *(From U. Mittwoch,* Genetics of Sex Differentiation. *© 1973 by Academic Press.)*

(Figure 16-8). If they form a testis, the Leydig cells of the testis secrete testosterone, which is an androgenic (male-determining) steroid hormone. (Recall the discussion of steroid hormone receptor transcription factors in Chapter 13.) This hormone binds to androgen receptors. These receptors function as transcription factors; their transcription-factor activity, however, depends on binding to their cognate hormone. Thus, the androgen-receptor complex binds to androgen-responsive enhancer elements, leading to the activation of male-specific gene expression. In chromosomally female embryos, no Leydig cells form in the gonad, no testosterone is produced, androgen receptor is not activated, and the embryos continue along the default female pathway of development. Hence, it is the presence or absence of a testis that determines the sexual phenotype, through the endocrine release of testosterone. Indeed, in XY embryos lacking the androgen receptor, development proceeds along a completely female pathway even though the embryos have testes.

Setting the Switch in the On or Off Position

What initiates the sex-determination pathway? Molecular genetic analysis has focused on identifying the locus on the Y chromosome that drives testis formation. This hypothetical gene has been called the testis-determining factor on the Y chromosome (*TDF* in humans, *Tdy* in mice) and is now known to be identical to the *SRY* (human) ~ *Sry* (mouse) gene (different nomenclature for human and mouse genes), first identified through its gain-of-function dominant sex-reversal effect, in which heterozygous mutant XX individuals develop as phenotypic males (see Foundations of Genetics 16-2 on page 522). Furthermore, because the wild-type *SRY* ~ *Sry* gene is on the Y chromosome, we can easily understand how the on-off switch is set. The wild-type XY individual has an *SRY* ~ *Sry* gene, which causes the male shunt pathway to be activated, whereas the normal XX individual lacking *SRY* ~ *Sry* remains in the female default pathway.

How does *SRY* ~ *Sry* contribute to sex determination? The SRY ~ Sry protein is a transcription factor and is expressed in the primitive male gonad. Exactly how the SRY ~ Sry protein initiates testis formation is not understood. However, with the SRY ~ Sry protein sequence in hand, many avenues for answering this and other age-old questions about the biological basis of sexual phenotype can be pursued.

MESSAGE

In mammals, a Y chromosome gene encodes a transcription factor that causes the gonad to become a testis. The testis serves as a command organ that, through the release of testosterone, directs male phenotypic development throughout the body. In the absence of testosterone signaling, a female phenotype develops.

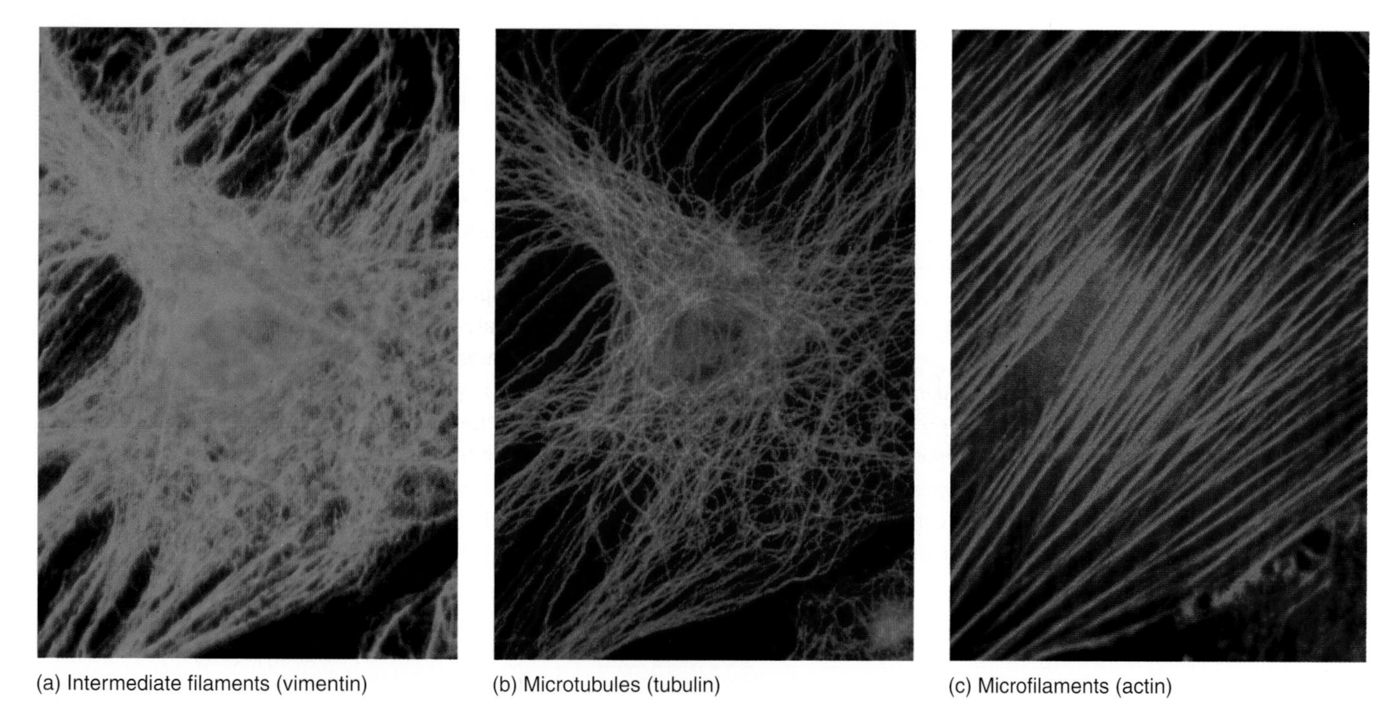

Figure 16-9 Different cytoskeletal systems in the same cell. The distribution of (a) the intermediate filament protein vimentin, (b) the microtubulin protein tubulin, and (c) the microfilament protein actin are shown. *(Courtesy of V. Small. Reprinted from H. Lodish, D. Baltimore, A. Berk, S. L. Zipursky, P. Matsudaira, and J. Darnell,* Molecular Cell Biology, *3d ed. © Scientific American Books, 1995.)*

BINARY FATE DECISIONS: THE GERM LINE VERSUS THE SOMA

In animal development, the earliest developmental decision is the separation of the germ line from the soma. Once this separation occurs, it is irreversible. Germ cells do not contribute to somatic structures. Somatic cells cannot form gametes, and thus their descendants never contribute genetic material to the next generation. This early separation means that genetic or regulatory modifications of somatic cells can occur in the course of development without affecting the genetic information in the germ line.

In making this decision of germ line versus soma, the embryo exploits its machinery for creating intracellular asymmetries—the **cytoskeleton,** the girders or "roadways" that support and shape the cell—to localize a germ-line determinant to a subset of early embryonic cells and unidirectional motors that move up or down the linear cytoskeletal roadways. Before directly addressing the question of how the germ-line-versus-soma decision is made, we need to consider the nature of cytoskeletal and cellular asymmetries.

The Cytoskeleton of the Cell

The cytoskeleton consists of several networks of very highly organized structural rods that run through each cell: microfilaments, intermediate filaments, and microtubules (Figure 16-9). Each has its own architecture formed of unique sets of protein subunits and proteins that promote production or disassembly of the rods. Each type of rod forms higher-order networks through different sets of proteins that reversibly cross-link the individual rods to one another.

Several roles of the cytoskeleton are relevant to the formation of assymetries: control of the location of the mitotic cleavage plane within the cell, control of cell shape, and directed transport of molecules and organelles within the cell. All of these roles depend on the fact that the cytoskeletal rods are polar structures (Figure 16-10). The contributions

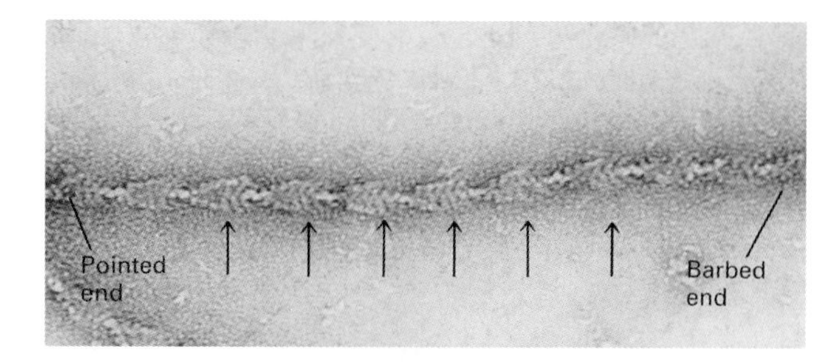

Figure 16-10 The polarity of subunits in an actin microfilament. An actin microfilament does not ordinarily have this appearance, but the microfilament has been coated with a protein that binds in a fashion that reveals the underlying polarity of the actin microfilament itself. *(Courtesy of R. Craig. Reprinted from H. Lodish, D. Baltimore, A. Berk, S. L. Zipursky, P. Matsudaira, and J. Darnell,* Molecular Cell Biology, *3d ed. © Scientific American Books, 1995.)*

FOUNDATIONS OF GENETICS 16-2

The Mutational Analysis of Mammalian Sex Determination

The Y chromosome testis-determining gene was identified through mapping and characterization by Robin Lovell-Badge and Peter Goodfellow of a genetic syndrome common to mice and humans that almost certainly affects this factor (see molecular map). This syndrome is called sex reversal. Sex-reversed XX individuals are phenotypic males and have been shown to carry a fragment of the Y chromosome in their genomes. In general, these Y chromosome duplications arise by an illegitimate recombination between the X and Y chromosomes that fuses a piece of the Y chromosome to a tip of one of the X chromosomes. The part of the Y chromosome that includes these duplications was cloned; by subsequent molecular analysis, Lovell-Badge and Goodfellow identified from this region a transcript that is expressed in the appropriate location of the developing kidney capsule.

The gene encoding this transcript was named the *sex reversal on Y* gene (*SRY* in humans, *Sry* in mice), because it was identified on the basis of the sex-reversal syndrome, but it is certainly the same gene as *TD* ~ *Tdy.* Lovell-Badge and Goodfellow used a transgene to provide spectacular evidence in support of this identity (see photographs). A cloned 14-kb genomic fragment of the mouse Y chromosome, including the *Sry* gene, was inserted into the mouse genome by germ-line transformation. An XX offspring containing this inserted *Sry* ~ *Sry* DNA (the transgene) was completely male in external and internal phenotype and, as predicted, possessed the somatic tissues of the testis—including the Leydig cells that

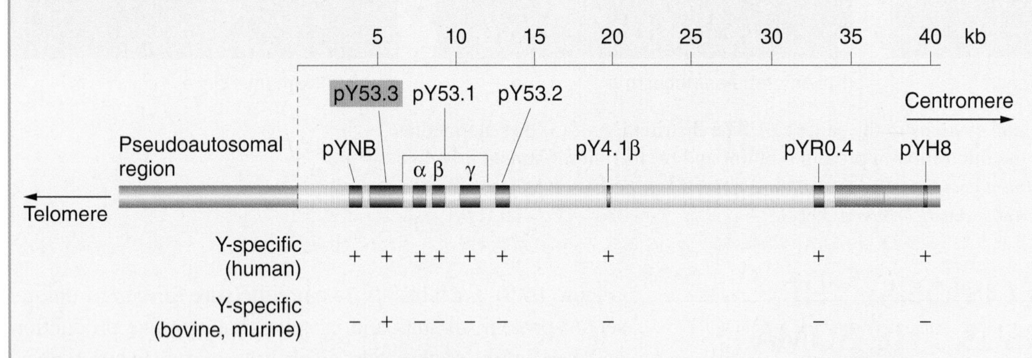

A molecular map of the distal part of the short arm of the human Y chromosome. The gene for testis-determining factor *(TDF)* was localized to a 34-kb region on the basis of the following logic: to the left of this region, the DNA sequences are the same on the X and Y chromosomes. (These sequences to the left are called the *pseudoautosomal region.*) Pieces of the Y chromosome starting at the telomere and ending at position 34 (kb) of the map are sufficient to cause the sex-reversal syndrome. Within this region, much of the DNA is repeated many times throughout the genome, but eight blocks of DNA unique to the Y chromosome were identified (the black regions). Cross-hybridization tests showed that only one of these (pY53.3) is present in bovine and mouse Y chromosomes. This fragment encodes SRY. In some humans with sex reversal, point mutations in *SRY* have been found. (Modified from A. H. Sinclair et al., *Nature* 346, 1990, 240.)

to pattern formation of microfilaments and microtubules—polymers of actin and tubulin subunits, respectively—are better documented than those of intermediate filaments, and so we shall focus on these two classes of cytoskeletal elements.

The Intrinsic Asymmetry of Cytoskeletal Filaments

The polarity of microfilaments and microtubules is crucial to their roles as intracellular highways. The ability of other molecules to move up and down these highways is an important aspect of all of their cellular roles. Microfilaments and microtubules are linear polymers with polarity—conceptually like the 5′-to-3′ polarity of DNA and RNA strands, even though the molecular basis of polarity is quite different. Furthermore, the polarity of the cytoskeletal elements can be organized within a cell. Consider microtubules. Near the center of most cell types are found all the "−" (minus) ends of the microtubules (Figure 16-11 on page 524). This location is called the microtubule organizing center (MTOC). The "+" (plus) ends of microtubules are located at the periphery of the cell. Very much as an automobile uses the combustion of gasoline to create energy that is then transduced into motion, special

make testosterone. (It should be noted, however, that this mouse was sterile. The sterility is probably a consequence of having two X chromosomes in a male germ cell, because XXY male mice are similarly sterile.) Thus, a single genetic unit was directly shown to greatly alter the mammalian sexual phenotype, completely consistent with the role of *SRY* ~ *Sry* as the gene that determines testis development.

The role of the androgen receptor in receiving the testosterone signal and establishing the male secondary sexual characteristics was elucidated through the study of rare *Tfm* mice lacking this receptor. Chromosomally XY mice hemizygous for the X-linked *Tfm (Testicular feminization)* mutation develop as phenotypic females (see Figure 5-18), except that they are infertile and are typically diagnosed at puberty because of their failure to menstruate. *Tfm* XY mice have testes, but the target cells that must decide between alternative pathways regarding sexually dimorphic characteristics lack androgen receptors and so are completely insensitive to the presence of testosterone. Thus, these mice develop along the default developmental pathway, which leads to phenotypic feminization.

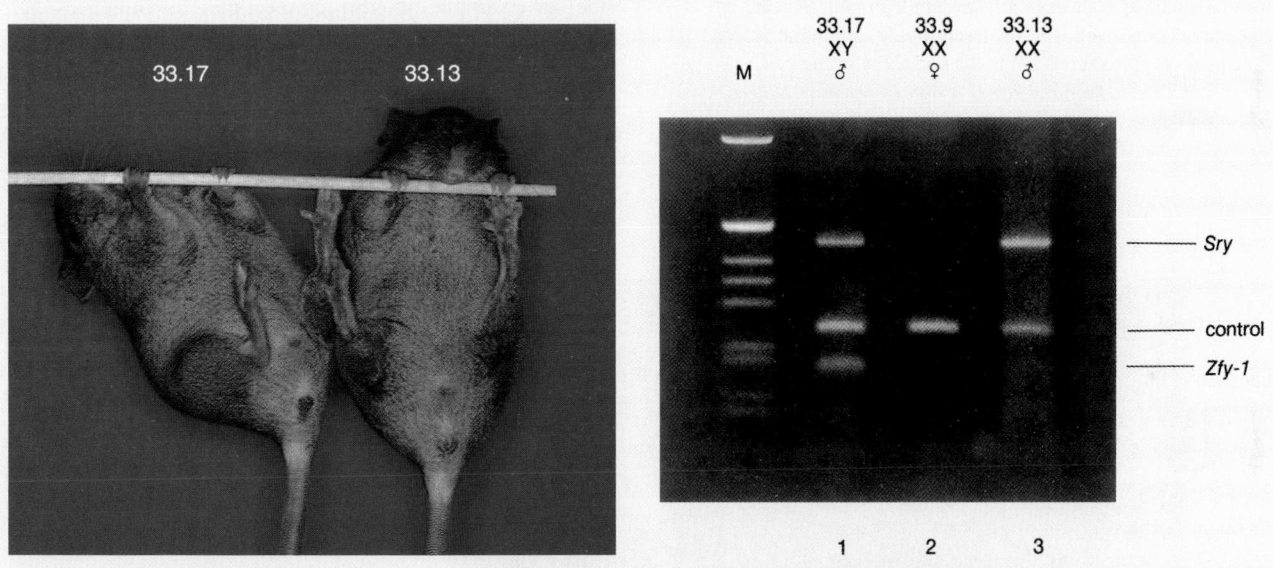

A transgenic mouse that proves that *Sry* can cause the sex-reversal syndrome. The external genitalia of sex-reversed XX transgenic mouse 33.13 are indistinguishable from those of a normal XY male sib (33.17), demonstrating that the *Sry* gene is sufficient to cause the sex-reversal phenotype. Gel pattern of genomic DNA amplified by the polymerase chain reaction (PCR) shows that mouse 33.13 lacks a DNA marker for the presence of a Y chromosome *(Zfy-1)* but that it does contain the *Sry* transgene. M = marker bands. (From P. Koopman, J. Gubbay, N. Vivian, P. Goodfellow, and R. Lovell-Badge, *Nature* 351, 1991, 117.)

"motor" proteins hydrolyze ATP for energy that is utilized to propel movement along a microtubule. For example, a protein called kinesin is able to move in a − to + direction along microtubules, carrying cargoes such as vesicles from the center of the cell to its periphery (Figure 16-12a and b on the next page). The motor—the part of the kinesin protein that directly interacts with the microtubule rod—is contained in the globular heads of the protein (Figure 16-12c). The tail of kinesin is thought to be where the cargo is attached. These cargoes might be individual molecules, organelles, or other subcellular particles to be towed from one part of the cell to another. (Comparable motors exist for actin microfilaments.)

What is the value of having multiple independent cytoskeletal systems? A part of the answer is probably division of labor. Just as cities have complex grids of intersecting streets to permit travel from a starting point to any other location, cells use multiple cytoskeletal systems to move cargo from one part of the cell to any other.

MESSAGE

The cytoskeleton serves as a highway system for the directed movement of subcellular particles and organelles.

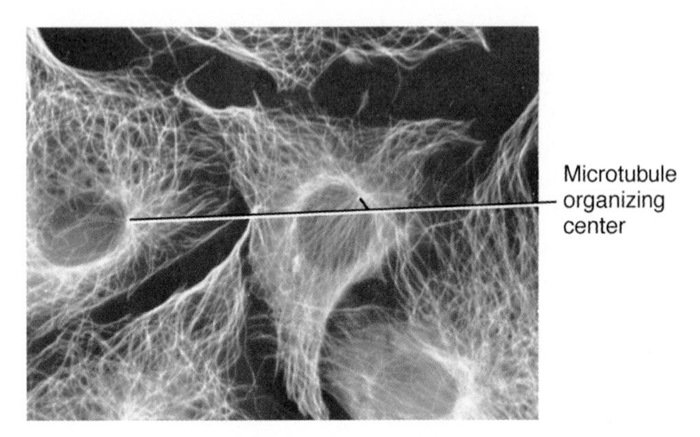

Figure 16-11 Fluorescence micrograph showing the distribution of tubulin in an interphase animal fibroblast. Notice that the microtubules radiate out from a microtubule organizing center. The negative (minus) ends of the microtubules are in the center, and the positive (plus) ends are at the periphery of the cell. *(Courtesy of M. Osborn. Reprinted from H. Lodish, D. Baltimore, A. Berk, S. L. Zipursky, P. Matsudaira, and J. Darnell,* Molecular Cell Biology, *3d ed. © Scientific American Books, 1995.)*

Localizing Determinants through Cytoskeletal Asymmetries: The Germ Line

In many organisms, a visible particle is asymmetrically distributed to the cells that will form the germ line. These particles—called P granules in *Caenorhabditis elegans,* polar granules in *Drosophila,* and nuage in frogs—are thought to be transport vehicles that ride on specific cytoskeletal highways to deliver the attached germ-cell determinants (regulatory molecules) to the appropriate cell. In *C. elegans* and *Drosophila,* the evidence relating germ-line determination to mechanisms that depend on the cytoskeleton is particularly strong. Let's consider both of these cases.

The early cell divisions of the *C. elegans* zygote provide an example of how cytoskeletal asymmetries help form the germ line. One of the favorable properties of *C. elegans* as an experimental system is that the same pattern of cell divisions occurs from one animal to another—a pattern that can be readily followed under the microscope. A

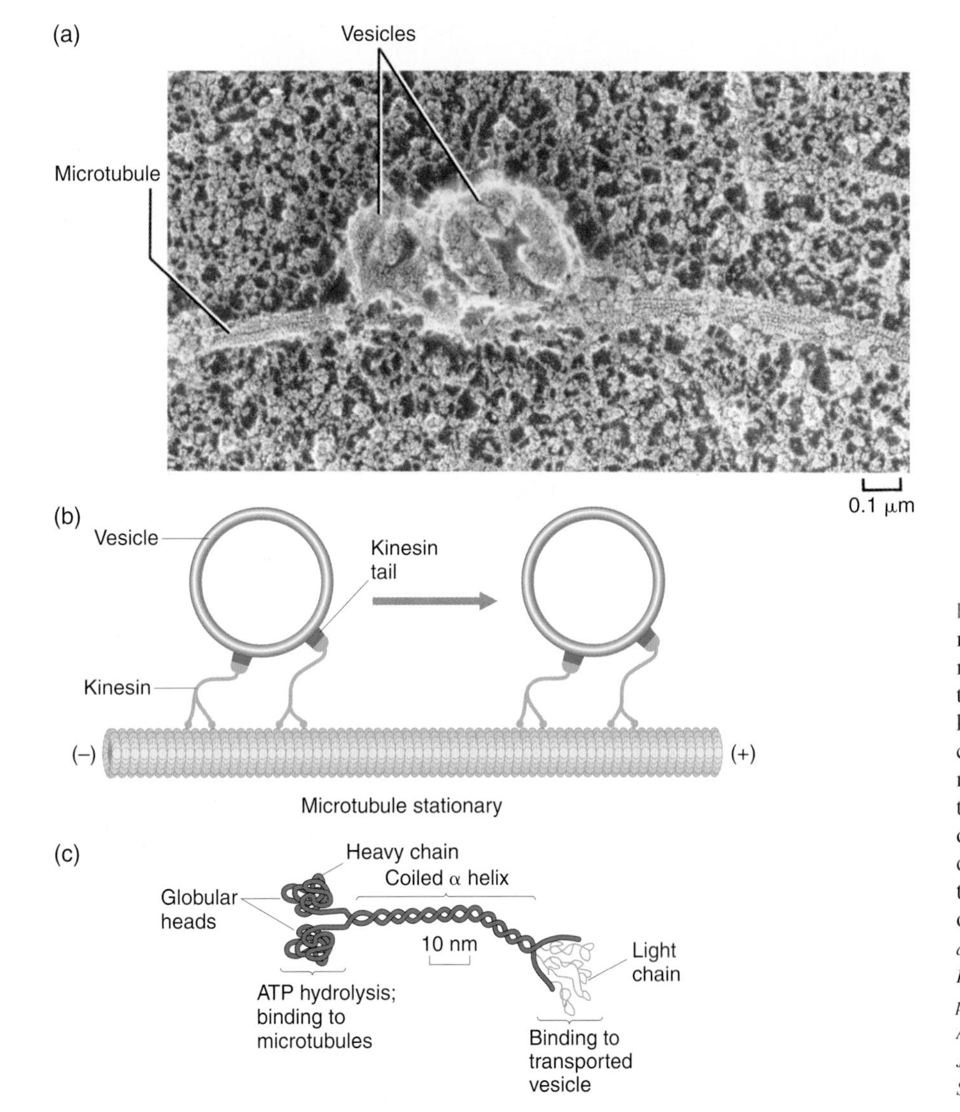

Figure 16-12 Movement of vesicles along microtubules. (a) A scanning electron micrograph of two small vesicles attached to a microtubule. (b) A diagram of how kinesin is thought to attach to cellular cargoes such as vesicles at its tail and to move the cargoes along the microtubule in the • to + direction by using the motor domain in the kinesin heads. (c) A diagram of the kinesin protein showing the functions associated with various parts of the molecule. *(Part a from B. J. Schnapp et al.,* Cell *40, 1985, 455. Courtesy of B. J. Schnapp, R. D. Valle, M. P. Sheetz, and T. S. Reese. All parts: Reprinted from H. Lodish, D. Baltimore, A. Berk, S. L. Zipursky, P. Matsudaira, and J. Darnell,* Molecular Cell Biology, *3d ed. © Scientific American Books, 1995.)*

lineage tree can then be constructed that traces the descent of each of the thousand or so somatic cells of the worm (see the lineage diagrams in Foundations of Genetics 15-2).

The one-cell zygote of *C. elegans* that is produced on fertilization is called the P_0 cell. It is an ellipsoidal cell that divides asymmetrically across its long axis to produce a larger, anterior AB cell and a smaller, posterior P_1 cell (Figure 16-13). This is a very important division in that it already sets up specialized roles for the descendants of these first two cells. Descendants of the AB cell will produce most of the skin cells of the worm (the hypoderm) and most of the neurons of the nervous system, whereas most of the muscles, all the digestive system, and the germ-line cells will come from the P_1 cell.

The germ-cell fate in the earliest divisions of the P_0 cell and its posterior descendants (P_1, P_2, and so forth) correlates with the distribution of certain fluorescent cytoplasmic particles called P granules. These granules are incorporated exclusively into the P_1 cell at the first division. When the P_1 cell divides, also asymmetrically, the P granules are incorporated into the progeny P_2 cell and, similarly, at the next division into the P_3 cell and so forth. Only the P_x cell that has these P granules becomes the germ line of the worm—all other cells are somatic. The asymmetric distribution of P granules is microfilament dependent. When applied at the right time to the P_0 cell, drugs (such as cytochalasin) that disrupt actin subunit polymerization into microfilaments produce a symmetrical distribution of the P granules between the two progeny cells. (Presumably because other fate determinants are abnormally distributed owing to the actin disruption, the resulting embryos are quite "confused" and die as masses of cells that look nothing like a normal worm.)

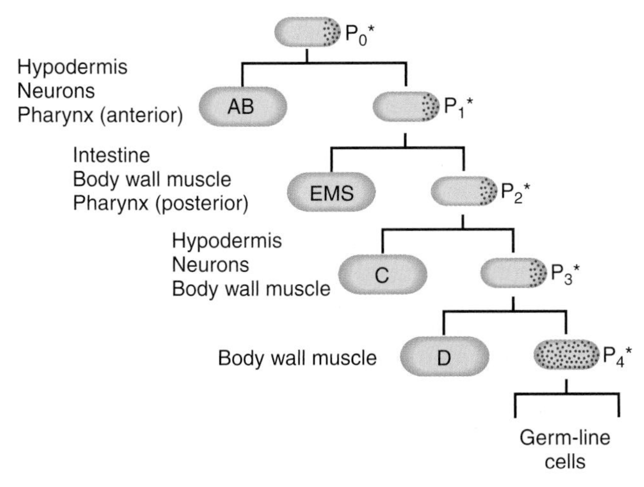

Figure 16-13 The early development of *C. elegans,* showing the early lineage divisions of the zygote. The mature cell types that arise from the various daughter cells of the early divisions are indicated. Note that the entire germ line comes from the P_4 cell. Each of the P cell divisions indicated by an asterisk is asymmetric, and each of the posterior daughter cells inherits all the P granules, which are thought to be germ-line determinants in the worm. The letters (for example, AB, EMS) are symbols for names of daughter cells.

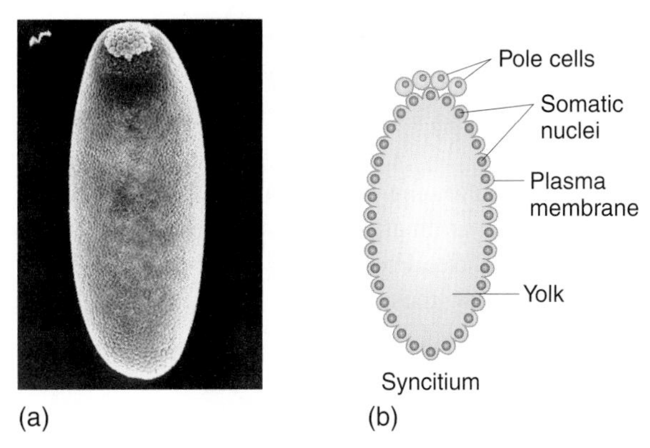

Figure 16-14 Pole-cell formation at the syncitial stage of the early *Drosophila* embryo. (a) A scanning electron micrograph of a *Drosophila* embryo with the egg shell removed. Note that the pole cells (the cap of cells on top of the embryo) lie outside the somatic syncitium. (b) A diagram of a longitudinal section through the embryo in part a, showing that the germ-line cells—the pole cells—have formed, whereas the soma still consists of syncitial nuclei. *(Modified from F. R. Turner and A. P. Mahowald,* Developmental Biology *50, 1976, 95.)*

In early *Drosophila* development, the cytoskeleton also is exploited to localize the structure containing the germ-line determinants: the polar granules. During oogenesis in the ovary of the mother, the polar granules are constructed and become tethered to the posterior pole of the oocyte by virtue of their attachment to one end of the microtubules. They remain in this location throughout early embryogenesis until nuclear division 9, when a few nuclei migrate to the posterior pole. (An unusual feature of early *Drosophila* development is that the first 13 mitoses are nuclear divisions without concomitant cytoplasmic division, making the early embryo a syncitium—a multinucleate cell.) After nuclear division 9, the plasma membrane of the oocyte evaginates at the posterior pole to surround each nucleus and pinches off some of the polar-granule-containing cytoplasm. This creates the pole cells, the first mononucleate cells of the embryo and the cells that will uniquely form the fly's germ line (Figure 16-14).

How do the polar granules get tethered to the posterior pole of the oocyte? Again, the subcellular localization is accomplished by one of the cytoskeletal networks. In contrast with *C. elegans,* in which the actin-based microfilaments seem to form the polar structure to which the P granules attach, here the tubulin-based microtubules provide the essential asymmetry. Probably this is just an accident of the evolutionary history of these organisms. In each case, the germ-line determinant evolved to co-opt any cytoskeletal system that had the appropriately oriented asymmetry.

FORMING COMPLEX PATTERN: ESTABLISHING POSITIONAL INFORMATION

This section and following ones about the early development of *Drosophila* summarize the results of a good deal of mutational analysis. See Foundations of Genetics 16-3 for a description of how these mutants were recovered and analyzed.

Cytoskeletal Asymmetries and the *Drosophila* Anterior-Posterior Axis

As we shall see here, not only is the *Drosophila* germ line established through a localized determinant anchored to microtubules, but the same is also true for formation of the anterior-posterior (A-P) axis of the soma. Positional information along the A-P axis of the syncitial *Drosophila* embryo is initially established through the creation of concentration gradients of two transcription factors: the BCD and HB-M proteins. The BCD protein, encoded by the *bicoid (bcd)* gene, is distributed in a steeper gradient in the early embryo, whereas the HB-M protein, encoded by one of the mRNAs of the *hunchback (hb)* gene, is distributed in a shallower but longer gradient (Figure 16-15). Both gradients have their high points at the anterior pole. In somewhat different ways, the gradients of both these proteins depend on the diffusion of protein from a localized origin: localized translation of two mRNA species, one tethered to microtubules at the anterior pole, and the other at the posterior pole of the syncitial embryo (see discussion on page 528).

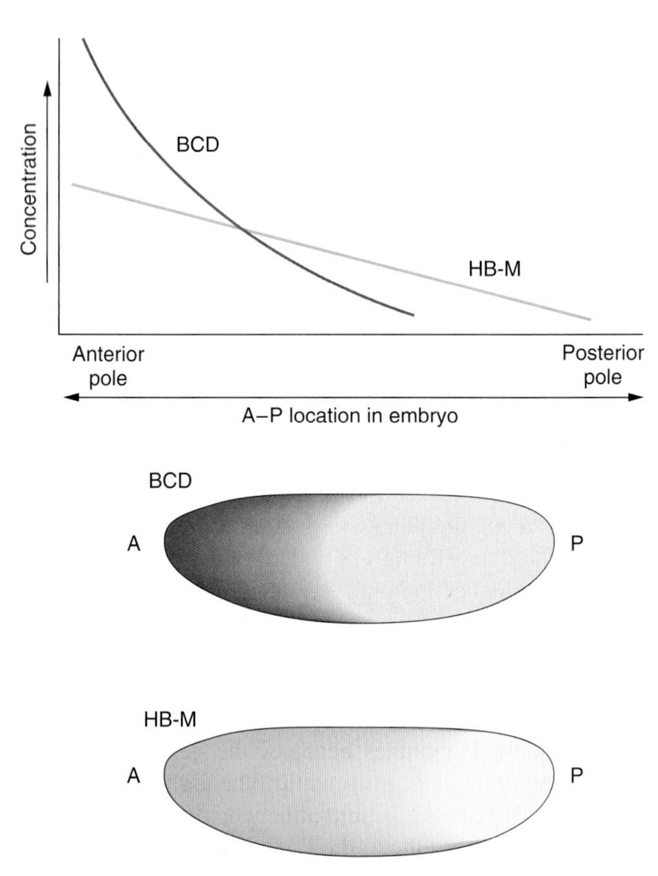

Figure 16-15 The concentration gradients of BCD and HB-M proteins. The BCD gradient is steeper, and the BCD protein is not detectable in the posterior half of the early *Drosophila* embryo. The HB-M gradient is shallower, and HB-M protein can be detected well into the posterior half of the embryo.

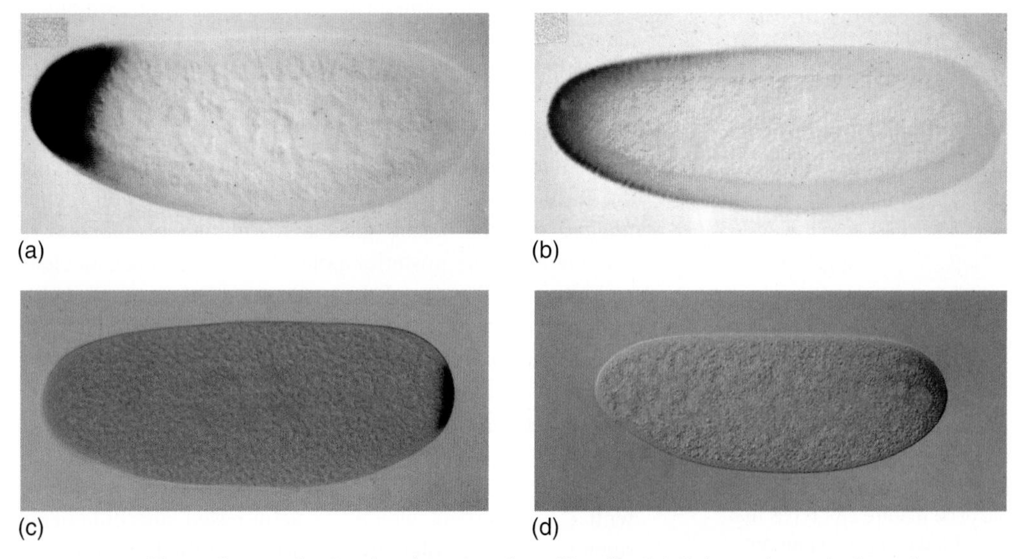

Figure 16-16 Photomicrographs showing the expression of localized A-P determinants in the embryo. (a) By in situ hybridization to RNA, the localization of *bcd* mRNA to the anterior (left) tip of the embryo can be seen. (b) By antibody staining, a gradient of BCD protein (brown stain) can be visualized, with its highest concentration at the anterior tip. (c) Similarly, *nos* mRNA localizes to the posterior (right) tip of the embryo, and (d) NOS protein is in a gradient with a high point at the posterior tip. *(Parts a and b from Christiane Nüsslein-Volhard,* Development, *Suppl. 1, 1991, 1. Parts c and d provided by E. R. Gavis, L. K. Dickinson, and R. Lehmann, then of Massachusetts Institute of Technology.)*

The origin of the BCD gradient is quite straightforward. The maternal *bcd* mRNA, packaged during oogenesis into the developing oocyte, is tethered to the − (minus) ends of microtubules, which are located at the anterior pole (Figure 16-16a). Translation of BCD protein begins midway through the early nuclear divisions of the syncitial embryo. The protein diffuses in the common cytoplasm of the syncitium. Because the protein is a transcription factor, it

FOUNDATIONS OF GENETICS 16-3

Mutational Analysis of Early *Drosophila* Development

The initial insights into the genetic control of pattern formation have emerged from studies of the fruit fly *Drosophila melanogaster.* The reason that *Drosophila* development has proved to be a gold mine to researchers is that developmental problems can be simultaneously approached by the use of genetic and molecular techniques. Let's consider the basic genetic and molecular techniques that are employed.

The *Drosophila* embryo has been especially important in understanding the formation of the basic animal body plan. One important reason is that the formation of the larval exoskeleton in the *Drosophila* embryo lends itself to easy identification of body plan mutant phenotype. The exoskeleton of the *Drosophila* larva is laid down as a mosaic in the embryo. Each structure of the exoskeleton is built by the epidermal cell or cells underlying that structure. With its intricate pattern of hairs, indentations, and other structures, the exoskeleton provides numerous landmarks to serve as indicators of the fates assigned to the many epidermal cells. In particular, there are many anatomical structures that are distinct along the anterior-posterior (A-P) and dorsal-ventral (D-V) axes. Furthermore, because all the nutrients necessary to develop to the larval stage are prepackaged in the egg, mutant embryos in which the A-P or D-V cell fates are drastically altered can nonetheless develop to the end of embryogenesis and produce a mutant larva. The exoskeleton of such a mutant larva mirrors the mutant fates assigned to subsets of the epidermal cells and can thus identify genes worthy of detailed analysis.

Researchers, most notably Christiane Nüsslein-Volhard, Eric Wieschaus, and their colleagues, have performed extensive mutational screens, essentially saturating the genome for mutations that alter the A-P or D-V patterns of the larval exoskeleton. These mutational screens identified two broad classes of genes that affect the basic body plan: zygotically acting genes and maternal-effect genes (see diagrams). The zygotically acting genes are those zygotic genes whose expression in the zygote is the sole source of their gene products that contribute to early development. They are part of the DNA of the zygote itself and are the "standard" sorts of genes that we are used to thinking about. Recessive mutations in zygotically acting genes elicit mutant phenotypes only in homozygous mutant animals. The alternative category—the maternal-effect genes—affects early development through contributions of gene products from the ovary of the mother to the developing oocyte. In maternal-effect mutations, the phenotype of the offspring depends on the genotype of the mother, not of the offspring, because the source of the gene products is the mother's genes. A recessive maternal-effect mutation will produce mutant animals only when the mother is a mutant homozygote.

Equally important to the mutational identification of genes that affect the body plan is the ease with which these genes can be cloned and characterized at the molecular level. Any *Drosophila* gene can be cloned, if its chromosomal map location has been well established, by using recombinant DNA techniques such as those described in Chapter 8. The analysis of the cloned genes often provides valuable information on the function of the protein product—usually by identifying close relatives in amino acid sequence of the encoded polypeptide through comparisons with all the protein sequences stored in public databases. In addition, one can investigate the spatial and temporal pattern of expression of (1) an mRNA, by using histochemically tagged single-stranded DNA sequences complementary to the mRNA to perform RNA in situ hybridization, or (2) a protein, by using histochemically tagged antibodies that bind specifically to that protein.

Extensive use is also made of in vitro mutagenesis techniques. P elements are used for germ-line transformation in *Drosophila* (see Chapter 8). A cloned pattern-formation gene is mutated in a test tube and put back into the fly. The mutated gene is then analyzed to see how the mutation alters the gene's function.

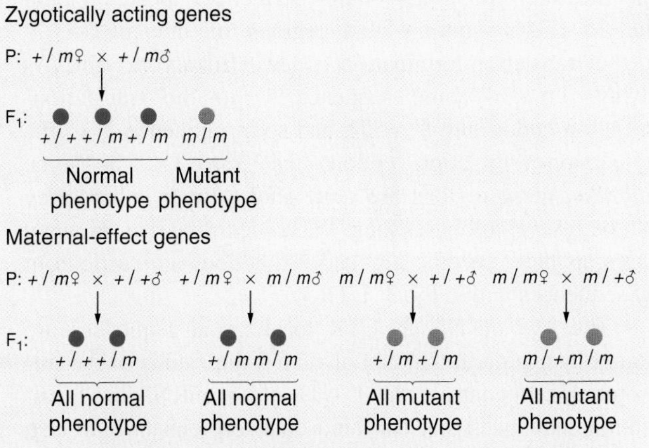

Zygotically acting genes

P: + / *m* ♀ × + / *m* ♂

F_1: + / + + / *m* + / *m* *m* / *m*

Normal phenotype | Mutant phenotype

Maternal-effect genes

P: + / *m* ♀ × + / + ♂ | + / *m* ♀ × *m* / *m* ♂ | *m* / *m* ♀ × + / + ♂ | *m* / *m* ♀ × *m* / + ♂

F_1: + / + + / *m* | + / *m* *m* / *m* | + / *m* + / *m* | *m* / + *m* / *m*

All normal phenotype | All normal phenotype | All mutant phenotype | All mutant phenotype

The genetic distinction between recessive zygotically acting and maternal-effect mutations.

contains signals to become localized in nuclei. By diffusion, those nuclei nearer to the source of translation (the anterior pole) incorporate a higher concentration of BCD protein than do those farther away; this difference results in the steep BCD protein gradient (Figure 16-16b).

The origin of the HB-M protein gradient is more complex. The HB-M protein gradient is produced by means of post-transcriptional regulation of *hb-m* mRNA. The *hb-m* mRNA is maternal in origin, being packaged during oogenesis into the developing oocyte. Unlike the *bcd* mRNA, though, the *hb-m* RNA is uniformly distributed throughout the oocyte and the syncitial embryo. However, translation of *hb-m* mRNA is blocked by a translational repressor protein—the NOS protein product, encoded by the *nanos (nos)* gene. Like *bcd* mRNA, *nos* mRNA is maternal in origin. However, in contrast with *bcd* mRNA, *nos* mRNA is localized at the posterior pole, through its association with the + (plus) ends of microtubules (Figure 16-16c). When translation of *nos* mRNA begins midway through the syncitial stage of early embryogenesis, NOS protein becomes distributed by diffusion in a gradient opposite that of BCD. The NOS gradient has a high point at the posterior pole and drops down to background levels in the middle of the A-P axis of the embryo (Figure 16-16d). NOS protein has the ability to specifically inhibit translation of *hb-m* mRNA, and the level of inhibition is proportional to the concentration of NOS protein. Through this ability, the posterior-to-anterior concentration gradient of the NOS translation repress-or produces the shallow anterior-to-posterior gradient of HB-M protein.

MESSAGE

Localization of mRNAs within a cell is accomplished by anchoring the mRNAs to one end of polarized cytoskeleton chains.

How do the maternal *bcd* and *nos* mRNAs get tethered to opposite ends of the polarized microtubules of the oocyte and early syncitial embryo? The answer is that there are specific microtubule-association sequences located within the 3′ UTRs—untranslated regions of the mRNA 3′ to the translation-termination codon. (Eukaryotic mRNAs always contain some sequence 5′ to the translation-initiation codon, the 5′ UTR, and some sequence 3′ to the translation-termination codon, the 3′ UTR. In some mRNAs, these regions are quite short, but in others they can be several kilobases long. We are learning that, in many cases, as here, specific regulatory functions are carried out by sequences within the 3′ UTR.)

The *bcd* mRNA 3′ UTR localization sequences are bound by a protein that can also bind the • ends of the microtubules. In contrast, the 3′ UTR of *nos* mRNA has localization sequences that can bind other proteins, which also bind to the + ends of microtubules. (In actuality, there are more intermediary steps in anchoring *nos* mRNA at the posterior end, with the 3′ UTR localization sequences of *nos* mRNA being anchored to the molecules in the polar granules, which are in turn anchored to the + end of the microtubules.)

How can we demonstrate that the 3′ UTRs of the mRNAs are where the localization sequences reside? This has been determined in part by "swapping" experiments. For example, when a synthetic transgene that produces an mRNA with the 5′ UTR and protein-coding regions of the normal *nos* mRNA glued to the 3′ UTR of the normal *bcd* mRNA is inserted into the fly genome, this fused *nos-bcd* mRNA will be localized at the anterior pole of the oocyte. This causes a double gradient of NOS: one from anterior to posterior (due to the transgene's mRNA) and one from posterior to anterior (due to the normal *nos* gene's mRNA). This procedure produces a very weird embryo, with two mirror-image posteriors and no anterior (Figure 16-17). This double-abdomen embryo arises because NOS protein is now present throughout the embryo and translationally represses *hb-m* mRNA (it also represses translation of *bcd* mRNA, although it is not clear that this is its normal function in wild-type animals). For more detailed information about how A-P positional information is generated, see Foundations of Genetics 16-4 on pages 530–531.

MESSAGE

The positional information of the *Drosophila* A-P axis is generated by protein gradients. The gradients ultimately depend on diffusion of newly translated protein from localized sources of specific mRNAs anchored by their 3′ UTRs to ends of cytoskeletal filaments.

Cell-Cell Signaling and the *Drosophila* Dorsal-Ventral Axis

In the examples considered thus far, the determinants were intracellular products: mRNAs or larger macromolecular assemblies packaged into the oocyte. In many circumstances, though, positional information depends on extracellular proteins secreted from a localized subset of cells within a developing field. These secreted proteins diffuse in the extracellular space to form a concentration gradient of ligand that binds to receptors on target cells. The ligand then activates the target cells in a concentration-dependent fashion through a receptor-signal transduction system.

An example of such a mechanism for position information is the establishment of the dorsal-ventral (D-V) axis in the early *Drosophila* embryo. The proximate effect of the D-V positional information is to create a gradient of DL protein activity in cells along the D-V axis. The DL protein is a transcription factor encoded by the *dorsal (dl)* gene. It exists in two forms: (1) active transcription factor located in the nucleus and (2) inactive protein located in the cyto-

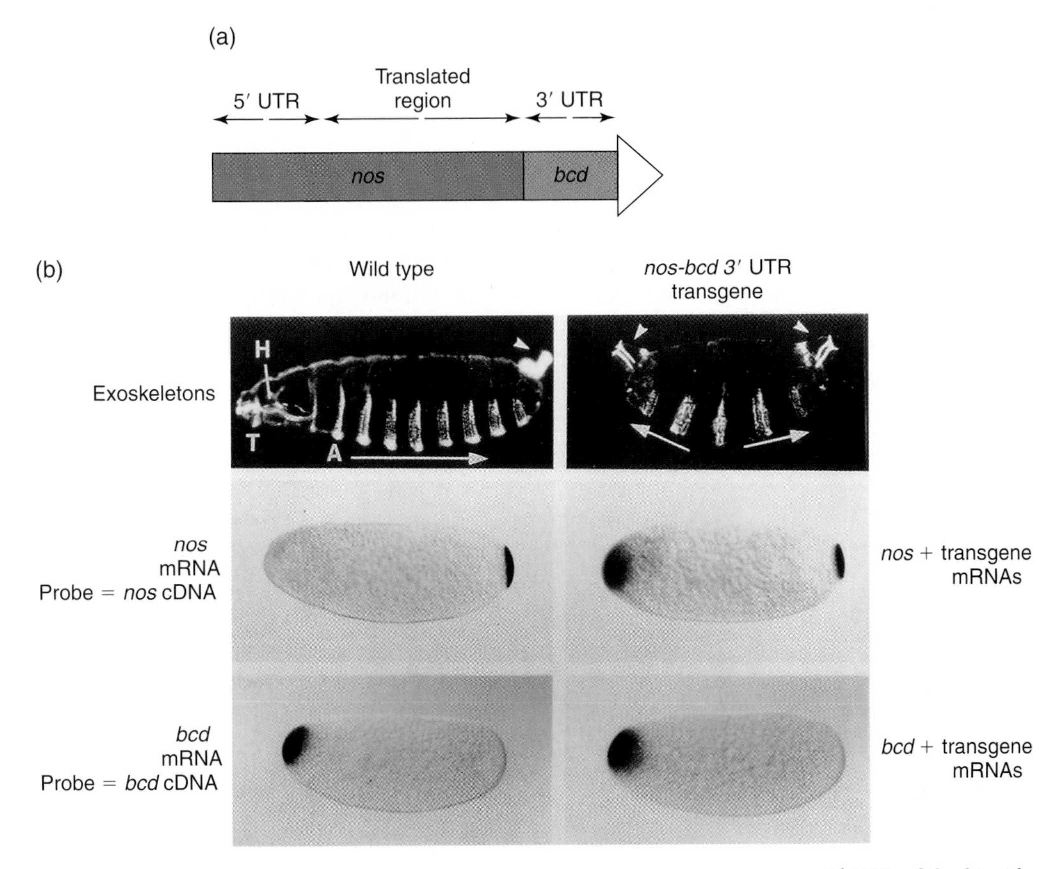

Figure 16-17 The effect of replacement of the 3′ UTR of the *nanos* mRNA with the 3′ UTR of the *bicoid* mRNA on mRNA localization and embryonic phenotypes. (a) The structure of the *nos-bcd* 3′ UTR transgene. (b) The effects on embryonic development of the transgene. The embryos and larva in the left column are derived from wild-type mothers. Those in the right column are derived from transgenic mothers. All embryos are shown with anterior to the left and posterior to the right. The exoskeletons are shown for comparison. The transgene causes a perfect mirror-image double abdomen. In the embryo from a transgenic mother, mRNAs coding for NOS protein are now present at both poles of the embryo. NOS protein will inhibit translation of *hb-m* mRNA (and, actually, of *bcd* mRNA as well). *(From E. R. Gavis and R. Lehmann,* Cell *71, 1992, 303.)*

plasm, where it is sequestered in a complex bound to the CACT protein encoded by the *cactus (cact)* gene. A concentration gradient of active DL protein determines cell fate along the D-V axis. Both *dl* mRNA and DL protein are distributed uniformly in the oocyte and the very early embryo. However, late in the syncitial embryo stage, there develops a gradient of active DL protein, with its high point at the ventral midline of the embryo (Figure 16-18).

How does D-V positional information generate the gradient of active DL protein? The key events take place in oogenesis, through an interaction between the oocyte itself and the layer of surrounding somatic cells—the follicle cells, which also make the eggshell (Figure 16-19 on the next page). The follicle cells on the ventral side of the oocyte-follicle cell complex synthesize and secrete some proteins that lead to a gradient of activation of a secreted precursor of the SPZ ligand, encoded by the *spaetzle (spz)* gene. On the inner boundary of the eggshell is the vitelline membrane (Figure 16-20a on page 532). The SPZ ligand is

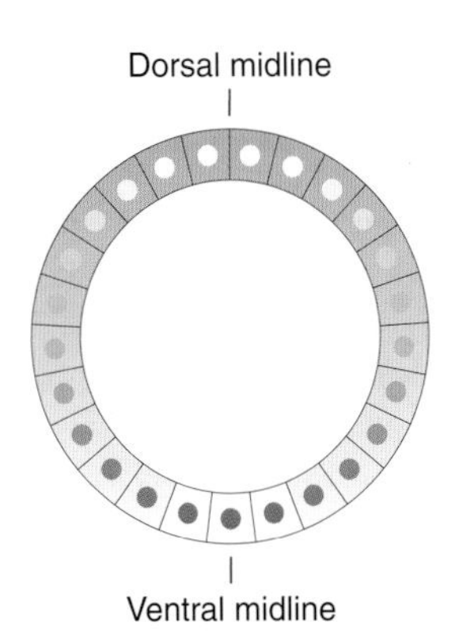

Figure 16-18 The distribution of DL shown in a cross section of the cellular blastoderm-stage *Drosophila* embryo. The dorsal midline is at the top and the ventral midline at the bottom. Note that the DL protein is in the nucleus (where it is active) ventrally, throughout the cell laterally, and in the cytoplasm (where it is inactive) dorsally. *(Adapted from S. Roth, D. Stein, and C. Nüsslein-Volhard,* Cell *59, 1989, 1196.)*

FOUNDATIONS OF GENETICS 16-4

Studying the BCD Gradient

How do we know that molecules such as BCD and HB-M contribute A-P positional information? Let's consider the example of BCD in detail. First, genetic changes in the *bcd* gene alter anterior fates. Embryos derived from *bcd* homozygous null mutant mothers lack anterior segments (see micrographs). If, on the other hand, we overexpress *bcd* in the mother, by increasing the number of copies of the bcd^+ gene from the normal two to three, four, or more, we "push" fates that ordinarily appear in anterior positions into increasingly posterior zones of the resulting embryos (as the first diagram shows). These observations suggest that BCD protein exerts global control of anterior positional information.

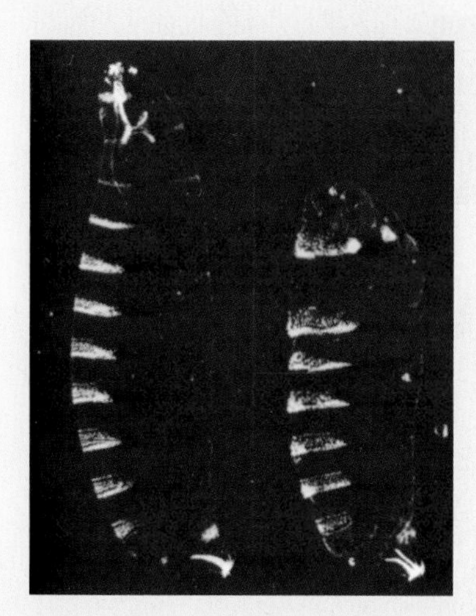

Photomicrographs of the exoskeletons of larvae derived from wild-type and *bcd* maternal-effect-lethal mutant mothers. These photomicrographs are in darkfield, so dense structures appear white, as in a photographic negative. Note the bright, segmentally repeated denticle bands present on the ventral side of the embryo. Maternal genotypes and larval phenotypes (and class of mutation) are as follows: (*left*) wild-type, normal phenotype; (*right*) *bcd (bicoid)*, anterior head and thoracic structures missing (anterior). (From C. H. Nüsslein-Volhard, G. Frohnhöfer, and R. Lehmann, *Science* 238, 1987, 1678.)

Second, *bcd* mRNA can completely substitute for the anterior determinant activity of anterior cytoplasm (see the second diagram). If one removes the anterior cytoplasm from a punctured syncitial embryo, anterior segments (head and thorax) are lost (not shown). Injection of anterior cytoplasm from another embryo into the anterior region of the anterior cytoplasm-depleted embryo restores normal anterior segment formation, and a normal larva is produced. Similarly, synthetic *bcd* mRNA can be made in a test tube and injected into the anterior region of an anterior cytoplasm-depleted embryo. Again, a normal larva is produced. Unlike that of anterior cytoplasm, transplantation of cytoplasm from middle or posterior regions of a syncitial embryo does not restore normal anterior formation. Thus, the anterior determinant should be located *only* at the anterior end of the egg. As already discussed, this is exactly where *bcd* mRNA is found.

Third, also as has already been explained, the BCD protein shows the predicted asymmetric and graded distribution required to fulfill its role of establishing positional information.

temporarily bound to structures in the vitelline membrane, which sequester it until near the end of the syncitial stage of early embryogenesis, when it is released. Active SPZ ligand (with its highest concentration at the ventral midline) then binds to the TOLL transmembrane receptor, encoded by the *Toll* gene, present uniformly in the oocyte plasma membrane (see Figure 16-20a). In a concentration-dependent manner, the SPZ-TOLL complex triggers a signal transduction pathway that ends up phosphorylating the inactive DL and CACT cytoplasmic proteins of the DL-CACT complex (Figure 16-20b). Phosphorylation of DL and CACT causes conformational changes that break apart the cytoplasmic complex. The free phosphorylated DL protein is then able to migrate into the nucleus, where it serves as a transcription factor activating genes necessary for establishing the ventral fates.

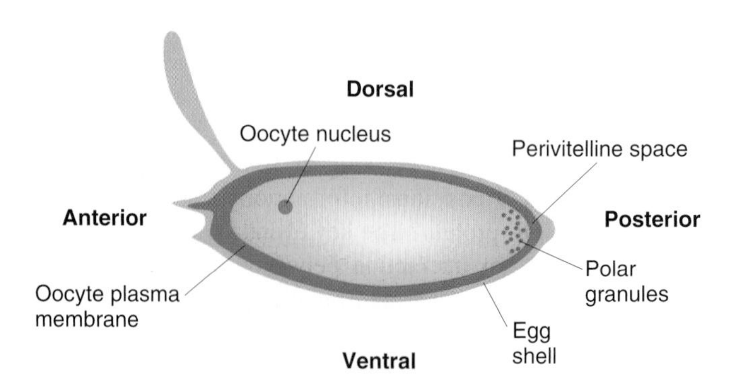

Figure 16-19 A mature oocyte. The follicle cells that built and surrounded the eggshell have been discarded, and the plasma membrane of the oocyte is interior to the eggshell. Note the polar granules at the posterior end of the oocyte. *(Modified from Christiane Nüsslein-Volhard,* Development, *Suppl. 1, 1991, 1.)*

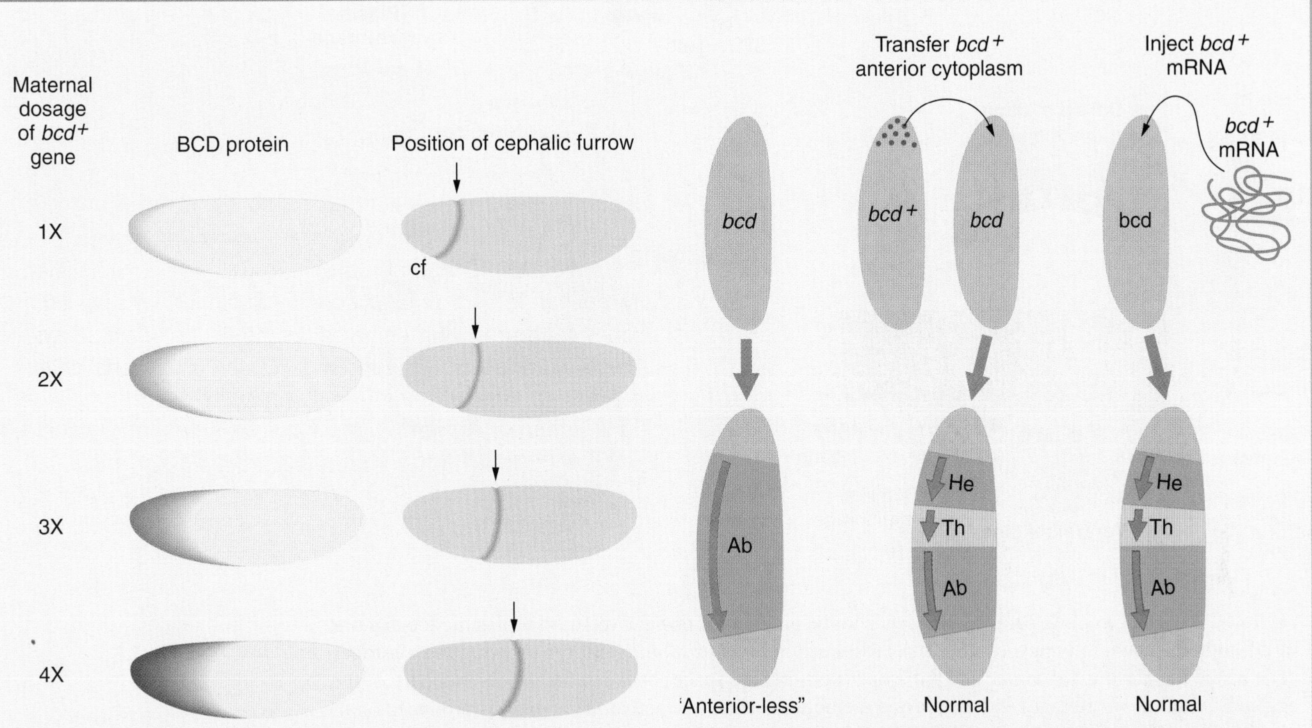

The concentration of BCD protein affects A-P cell fates. The amount of BCD protein can be changed by varying the number of copies of the *bcd*$^+$ gene in the mother. Embryos derived from mothers carrying from one to four copies of *bcd*$^+$ have increasing amounts of BCD protein. (Two copies is the normal diploid gene dose.) Cells that invaginate to form the cephalic furrow (cf) are determined by a specific concentration of BCD protein. In the progression from one maternal copy to four copies of *bcd*$^+$, this specific concentration is present more and more posteriorly in the embryo. Thus, the position of the cephalic furrow (marked dorsally by the arrow) arises farther toward the posterior according to *bcd*$^+$ gene dosage. (Adapted from W. Driever and C. Nüsslein-Volhard, *Cell* 54, 1988, 100.)

The *bcd* "anterior-less" mutant phenotype can be rescued by wild-type cytoplasm or purified *bcd*$^+$ mRNA. In embryos derived from *bcd* mothers, the anterior (head and thoracic) segments do not form, producing the anterior-less phenotype. If anterior cytoplasm from an early wild-type donor embryo is injected into the anterior of a recipient embryo derived from a *bcd* mutant mother, a fully normal embryo and larva are produced. Cytoplasm from any other part of the donor embryo does not rescue. Injection of *bcd*$^+$ mRNA into the anterior of an embryo derived from a *bcd* mutant mother also rescues the wild-type segmentation pattern. (Ab = abdomen; He = head; Th = thorax.) (Adapted from C. H. Nüsslein-Volhard, G. Frohnhöfer, and R. Lehmann, *Science* 238, 1987, 1678.)

MESSAGE

Positional information can be established by cell-cell signaling through a concentration gradient of a secreted signaling molecule.

The Two Classes of Positional Information

To summarize this section, the most important message is that the specific examples that we considered fall into two general categories of positional information (Figure 16-21 on the next page).

- *Localization of mRNAs within a cell.* This type of positional information can be utilized only in cases where the developmental field begins as a single cell. It is used to form gradients of positional information in unusual cases such as *Drosophila* early embryogenesis, because, at this time, the embryo is a unicellular syncitium. More generally, it is used as a way of asymmetrically distributing local determinants to progeny cells.
- *Formation of a concentration gradient of an extracellular diffusible molecule.* This type of positional information can be employed in multicellular developmental fields, because the gradient is extracellular and therefore is not limited by the boundaries of the individual cells. Indeed, we now know of several cases where concentration gradients of secreted protein ligands that activate receptors and signal transduction systems fulfill the properties expected of classical developmental **morphogens**—literally, concentration-dependent determinants of form.

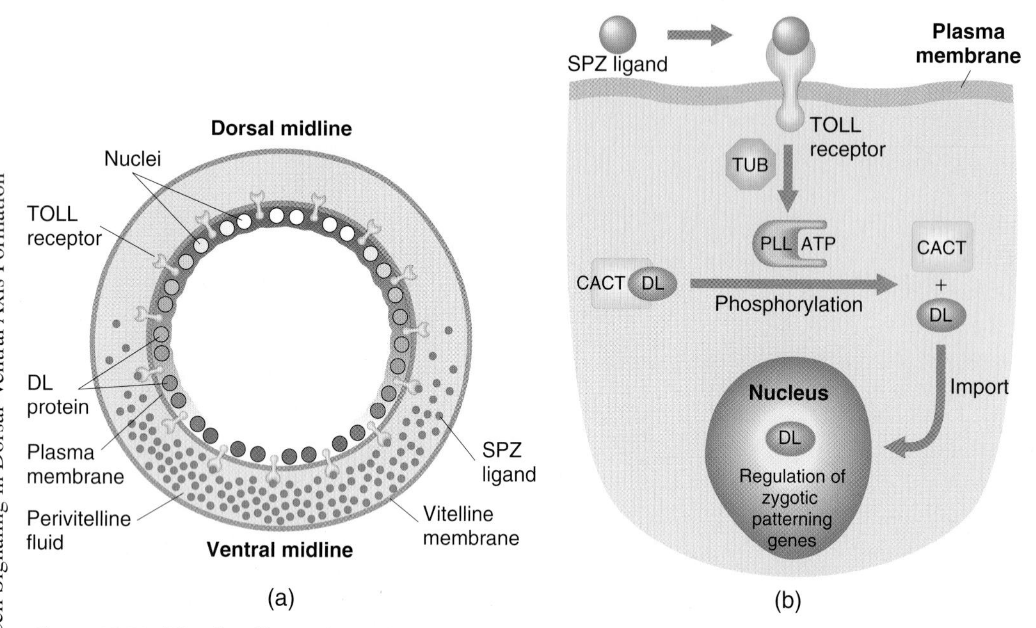

Animated Art • Gene Interactions in *Drosophila* Embryogenesis: Cell Signaling in Dorsal-Ventral Axis Formation

Figure 16-20 The signaling pathway that leads to the gradient of nuclear versus cytoplasmic localization of DL protein shown in Figure 16-18. (a) A cross section of a *Drosophila* embryo showing the blastoderm cells inside the plasma membrane and the space (perivitelline space) between the inside boundary of the eggshell (vitelline membrane) and the plasma membrane, where the active SPZ ligand is produced on the ventral side of the embryo. (b) The SPZ ligand binds to the TOLL receptor, activating a signal transduction cascade through two proteins called TUB and PLL, leading to the phosphorylation of DL and its release from CACT. DL then is able to migrate into the nucleus, where it serves as a transcription factor for D-V cardinal genes. *(Adapted from H. Lodish, D. Baltimore, A. Berk, S. L. Zipursky, P. Matsudaira, and J. Darnell,* Molecular Cell Biology, *3d ed. Scientific American Books, 1995.)*

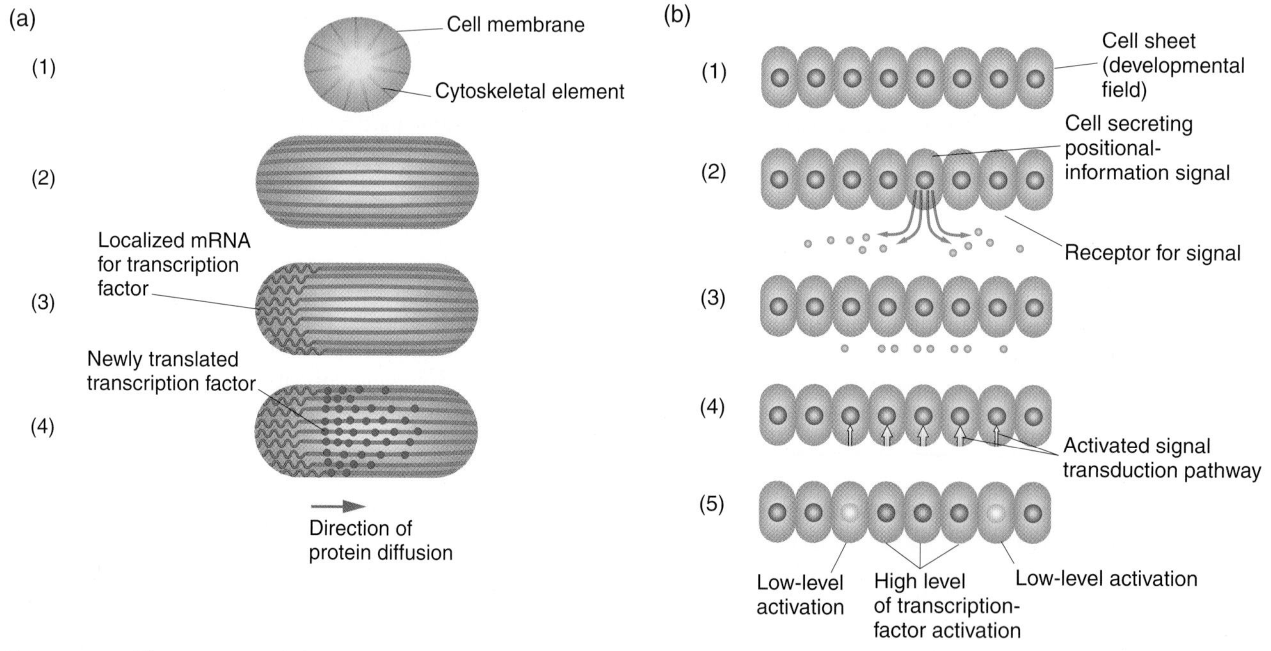

Figure 16-21 The two general classes of positional information. (a) Asymmetric organization of the cytoskeletal system permits localization of mRNA encoding a transcription factor that will provide positional information. Translation of the anchored mRNA will lead to diffusion of the newly translated transcription factor and the formation of a transcription-factor gradient with a high point near the site of the mRNA. (b) Secretion of a positional-information signaling molecule from a localized source activates the signal-transduction apparatus and the target transcription factors according to the level of signaling molecule that binds to its transmembrane receptor.

FORMING COMPLEX PATTERN: UTILIZING POSITIONAL INFORMATION TO ESTABLISH CELL FATES

When positional information has been provided, it can be useful only if a system that can interpret the positional data is present. To use a geographical analogy, it is not sufficient to have a system of longitudes and latitudes; we also need equipment that can receive longitude and latitude information—special instruments to read the positions of stars or receivers that can triangulate signals transmitted from radio beacons. In the same way, the developmental positional-information system requires that the signals transmitted be interpretable by elements within the cell.

The Initial Interpretation of Positional Information

As described earlier, two very different kinds of positional signal can be produced. However, both lead to the same outcome: a gradient in the concentration of one or more specific transcription factors within the cells of the developmental field. In the examples of localized mRNAs in *Drosophila* A-P axis development, this outcome is a direct consequence of the fact that the positional-information gradients are of transcription factors themselves (BCD and HB-M). For diffusible extracellular sources of positional information, this outcome requires several intermediary steps. Such cases typically involve a gradient of a secreted protein ligand that binds to a transmembrane receptor in a concentration-dependent fashion. In turn, this binding activates a signal transduction pathway in proportion to the level of receptor activation. Eventually, this signal transduction pathway proportionately activates the key transcription factor(s) in the target cells. This is exactly what we saw in regard to the *Drosophila* D-V axis, where the SPZ extracellular gradient leads to a graded activation of the DL transcription factor.

Given that positional information leads to a gradient of transcription-factor activities, we would naturally expect that the receivers are regulatory elements (enhancer and silencer elements) of genes whose protein products can begin the gradual process of specifying cell fate. This is exactly what we see. The genes targeted by the A-P and D-V transcription factors are zygotically expressed genes collectively known as the **cardinal genes** because they are the first genes to respond to the maternally supplied positional information. For an example, let's consider the A-P cardinal genes (Table 16-3). (The logic by which the D-V axis is divided initially into three domains through the action of the DL transcription-factor activity gradient and then into numerous finer subdivisions is identical with that described below for the A-P axis.) Before considering how cell fates are assigned along the A-P axis, we need to review a bit of *Drosophila* embryology.

TABLE 16-3 Examples of *Drosophila* A-P Axis Genes That Contribute to Pattern Formation

Gene Symbol	Gene Name	Protein Function	Role(s) in Early Development
hb-z	*hunchback-zygotic*	Transcription factor—Zn-finger protein	Gap gene
Kr	*Krüppel*	Transcription factor—Zn-finger protein	Gap gene
kni	*knirps*	Transcription factor—steroid receptor–type protein	Gap gene
eve	*even-skipped*	Transcription factor—homeodomain protein	Pair-rule gene
ftz	*fushi tarazu*	Transcription factor—homeodomain protein	Pair-rule gene
opa	*odd-paired*	Transcription factor—Zn-finger protein	Pair-rule gene
prd	*paired*	Transcription factor—PHOX protein	Pair-rule gene
en	*engrailed*	Transcription factor—homeodomain protein	Segment-polarity gene
ci	*cubitus-interruptus*	Transcription factor—Zn-finger protein	Segment-polarity gene
wg	*wingless*	Signaling WG protein	Segment-polarity gene
hh	*hedgehog*	Signaling HH protein	Segment-polarity gene
fu	*fused*	Cytoplasmic serine/threonine kinase	Segment-polarity gene
ptc	*patched*	Transmembrane protein	Segment-polarity gene
arm	*armadillo*	Cell-cell junction protein	Segment-polarity gene
lab	*labial*	Transcription factor—homeodomain protein	Segment-identity gene
Dfd	*Deformed*	Transcription factor—homeodomain protein	Segment-identity gene
Antp	*Antennapedia*	Transcription factor—homeodomain protein	Segment-identity gene
Ubx	*Ultrabithorax*	Transcription factor—homeodomain protein	Segment-identity gene

After the early embryonic syncitial stage, all somatic nuclei migrate to the surface of the egg and cellularize (Figures 16-22 and 16-23a). A few hours later, the first morphological manifestations of segmentation are apparent. At the end of 10 hours of development, the embryo is already externally divided into 14 segments from anterior to posterior—3 head, 3 thoracic, and 8 abdominal segments (Figure 16-23b). By this time, each segment has developed a unique set of anatomical structures, reflective of its special identity and role in the biology of the animal. At the end of 12 hours, organogenesis occurs. At 15 hours, the exoskeleton of the larva begins to form, with its specialized hairs and other external structures. Only 24 hours after development began at fertilization, a fully formed larva hatches out of the eggshell (Figure 16-23c). Of special note in considering the A-P pattern, the segmental arrangement of spikes, hairs, and other sensory structures on the larval exoskeleton makes each segment distinct and recognizable. Now let's return to the A-P cardinal genes.

The A-P cardinal genes are also known as **gap genes,** because flies that have mutations in these genes lack a sequential series of larval segments, producing a gap in the normal segmentation pattern (look at the phenotypes of mutations in two gap genes, *Krüppel* and *knirps* in Figure 16-24). BCD or HB-M protein or both bind to enhancer elements of the promoters of the gap genes, thereby regulating their transcription. For example, transcription of one gene, *Krüppel (Kr),* is repressed by high levels of the BCD transcription factor but is activated by low levels of BCD and HB-M. In contrast, the *knirps (kni)* gene is repressed by the presence of any BCD protein but does require low levels of the HB-M transcription factor for its expression. These en-

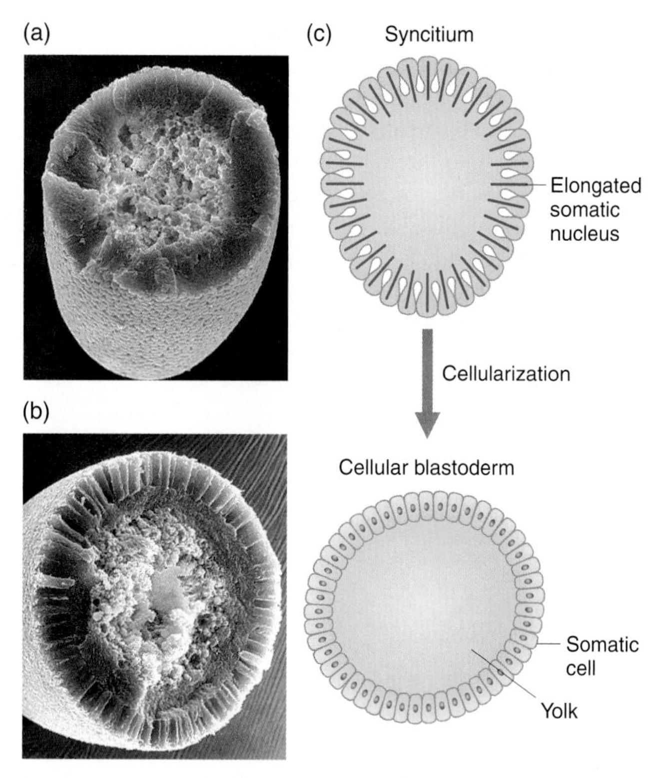

Figure 16-22 Cellularization of the *Drosophila* embryo. The embryo begins as a syncitium; cellularization is not completed until there are about 6000 nuclei. (a) and (b) Scanning electron micrographs of embryos removed from the eggshell. (a) A syncitium-stage embryo, fractured to reveal the common cytoplasm toward the periphery and the central yolk-filled region. The bumps on the outside of the embryo are the beginning of cellularization, in which the plasma membrane of the egg folds inward from the outside. (b) A cellular blastoderm embryo, fractured to reveal the columnar cells that have formed by cell membranes being drawn down between the elongated nuclei to create some 6000 mononucleate somatic cells. (c) Schematic diagram of the changes taking place during cellularization. *(Modified from F. R. Turner and A. P. Mahowald,* Developmental Biology *50, 1976, 95.)*

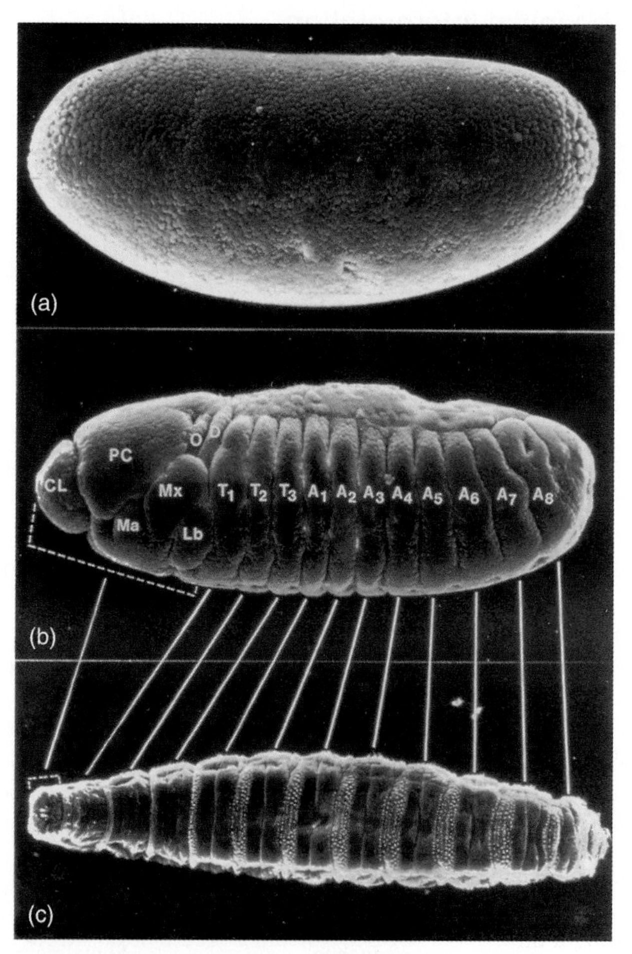

Figure 16-23 Scanning electron micrographs of (a) a 3-hour *Drosophila* embryo, (b) a 10-hour embryo, and (c) a newly hatched larva. Note that outlines of individual cells are visible by 3 hours, and by 10 hours the segmentation of the embryo is obvious. Lines are drawn to indicate that the segmental identity of cells along the A-P axis is already fated early in development. The abbreviations refer to different segments of the head (CL, PC, O, D, Mx, Ma, Lb), thorax (T_1 to T_3), and abdomen (A_1 to A_8). *(T. C. Kaufman and F. R. Turner, Indiana University.)*

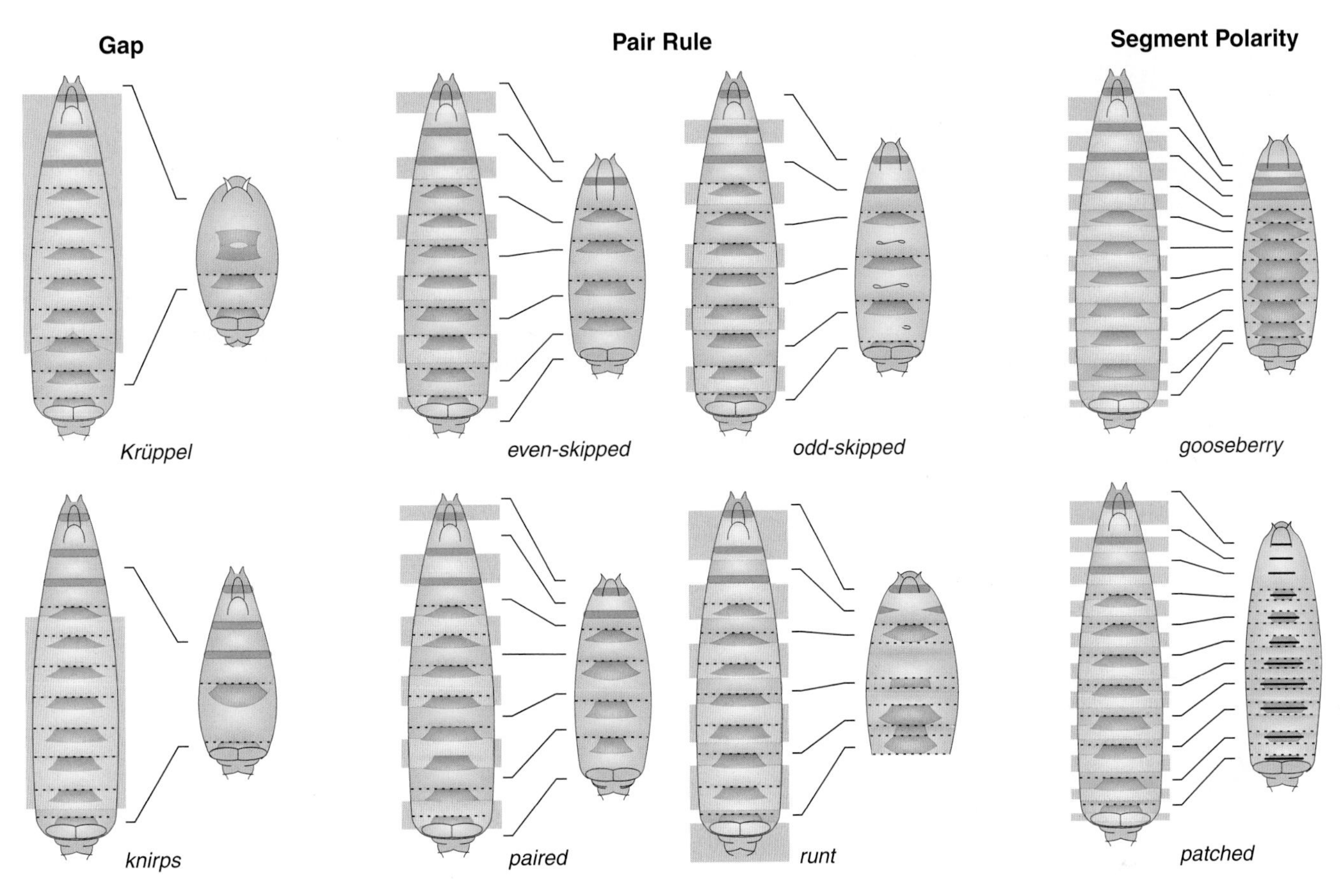

Figure 16-24 Diagrams depicting representative mutants in each class of mutant larval phenotypes due to mutations in the three classes of zygotically acting genes controlling segment number in *Drosophila.* The denticle belts, shown as red trapezoids, are segmentally repeating swatches of dense projections on the ventral surface of the larval exoskeleton. The boundary of each segment is indicated by a dotted line. The left diagram of each pair depicts a wild-type larva, and the right diagram the indicated A-P mutant. The pink regions on the wild-type diagram indicate the A-P domains of the larva that are missing in the mutant.

hancer and promoter properties thus ensure that the *kni* gene is expressed more posteriorly than is *Kr* (Figure 16-25a on the next page). By having promoters that are differentially sensitive to the concentrations of the transcription factors of the A-P positional information system, the gap genes can be expressed in a series of distinct domains; these domains then become different developmental fields. That is, the cells in the different domains become committed to different A-P fates.

These commitments of the domains to different fates occur because each of the gap genes encodes a different transcription factor and thereby has the capability of regulating a different set of downstream target genes necessary to refine (create finer-grained) A-P segmental fate.

Refining Fate Assignments through Transcription-Factor Interactions

The gap-gene expression pattern slices up the A-P axis into several domains. However, gap genes are expressed too coarsely to allocate all the A-P cell fates that are needed. Further, the end of the gap-gene expression stage and the beginning of refinement closely coincide with the time when the syncitial embryo becomes fully cellularized. When cellularization has established separate cells, the cytoplasm of each cell then contains a particular concentration of one or perhaps two adjacent gap-gene-encoded proteins, which enter its nucleus. Essentially all further decisions are driven by the particular A-P gap proteins present in the nucleus of a given cell.

The A-P developmental pathway downstream of the gap genes bifurcates. Each of the two branches illustrates strategies for refining the pattern. One branch establishes the correct number of segments. The other assigns the proper identity to each segment. (These different identities are manifest in the unique patterns of spikes and hairs on each segment of the larva, as described earlier.) The existence of two branches means that there are two different sets of target genes for regulation by the gap-gene-encoded transcription factors.

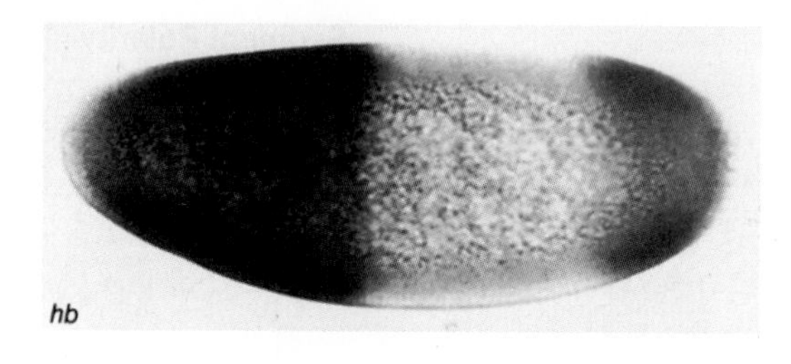

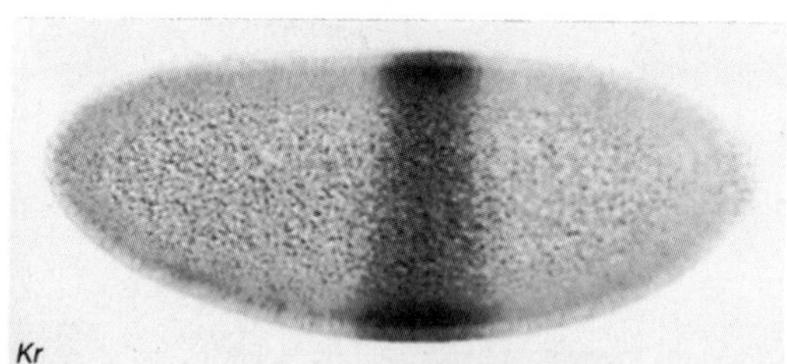

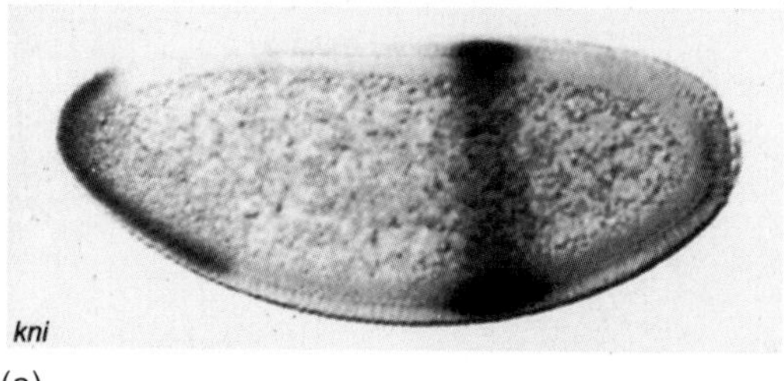

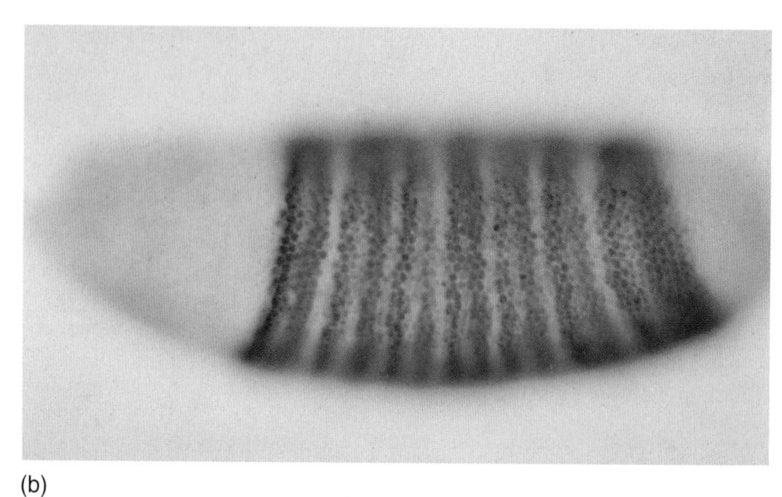

Figure 16-25 Photomicrographs showing the early embryonic expression patterns of gap and pair-rule genes. All embryos are shown with anterior to the left and posterior to the right. (a) Early blastoderm expression patterns of proteins from three gap genes: *hb-z*, *Kr*, and *kni*. (b) Late blastoderm expression patterns of proteins from two pair-rule genes: *ftz* (stained gray) and *eve* (stained brown). Note the localized gap-gene expression patterns compared with the reiterated pair-rule gene expression pattern. *(Part a from M. Hulskamp and D. Tautz,* BioEssays *13, 1991, 261. Part b from Peter Lawrence,* The Making of a Fly. *© 1992 by Blackwell Scientific Publications.)*

First, let's briefly consider the formation of segment number. (Refer to Figure 16-24 for a description of the mutant phenotypes that the different classes of segment-number genes produce.) The gap genes activate a set of secondary A-P patterning genes called the **pair-rule genes,** which encode transcription factors in a repeating pattern of seven stripes (see Figure 16-25b). There are several different pair-rule genes, and each of them produces a slightly offset pattern of stripes. There is a hierarchy within the pair-rule gene class. Some of these genes, called primary pair-rule genes, are regulated directly by the gap genes, whereas others are activated by proteins encoded by the primary pair-rule genes (these proteins are also transcription factors).

The products of the pair-rule genes then act combinatorially (several of their proteins are expressed within a given cell) to regulate the transcription of the segment-polarity genes, which are expressed in offset patterns of 14 stripes. Thus, the hierarchy of transcription-factor regulation extends all the way from the positional information system to the repeating pattern of segment-polarity gene expression. The products of the segment-polarity genes then permit the 14 segments to form and define the individual A-P rows of cells within each segment.

MESSAGE

Through a hierarchy of transcription-factor regulation patterns, positional information leads to the formation of the correct number of segments. In the readout of positional information, transcription factors act combinatorially to create the proper segment-number fates.

How do the primary pair-rule genes become activated in a repeating pattern by the asymmetrically expressed gap proteins? The key is that the regulatory elements for the primary pair-rule genes are quite complex. For primary genes such as *eve (even-skipped),* separate enhancer elements regulate the activation of each of the seven *eve* stripes. The enhancer for the first *eve* stripe is activated by high levels of the HB-Z gap transcription factor, the enhancer for the second *eve* stripe by low levels of HB-Z but high levels of the KR gap transcription factor, and so forth.

MESSAGE

The complexity of regulatory elements of the primary pair-rule genes turns an asymmetric (gap-gene) expression pattern into a repeating one.

Next, let's briefly consider the establishment of segmental identity. The gap genes target a clustered group of genes known as the **homeotic gene complexes.** They are called gene complexes because several of the genes are clustered together on the DNA. *Drosophila* has two homeotic gene clusters. The ANT-C (*Antennapedia* complex) is largely responsible for segmental identity in the head and anterior thorax, whereas the BX-C (*Bithorax* complex) is responsible for segmental identity in the posterior thorax and abdomen.

Homeosis, or homeotic transformation, is the development of one body part with the phenotype of another. Three examples of body-part-conversion phenotypes due to homeotic gene mutations are (1) the loss-of-function *bithorax* class of mutations that cause the entire third thoracic segment (T_3) to be transformed into a second thoracic segment (T_2), giving rise to flies with four wings instead of the normal two (Figure 16-26); (2) the gain-of-function dominant *Tab* mutation described in Chapter 13 (see Figures 13-30 and 13-31) that transforms part of the adult T2 segment into the sixth abdominal segment (A6); and (3) the gain-of-function dominant *Antennapedia (Antp)* mutation that transforms antenna into leg (see the photograph on the first page of this chapter). Note that, in each of these cases, the number of segments in the animal remains the same; the *only* change is to the identity of the segments. The study of these homeotic mutations has revealed much about how segment identity is established.

The domains of expression of the various gap proteins activate the target homeotic genes initially in a series of overlapping domains (Figure 16-27 on the next page). These homeotic genes encode transcription factors of a class called homeodomain proteins. **Homeodomain proteins** interact with the regulatory elements of the homeotic genes in a specific pattern such that expression patterns of the homeotic genes become mutually exclusive. (We shall consider the relations of structure and function within the homeotic gene complexes later in this chapter, in the context of the evolution of developmental mechanisms.) These homeodomain proteins also regulate downstream target

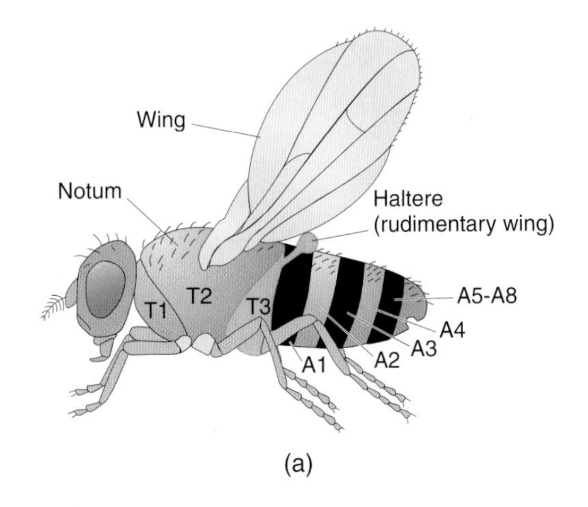

Figure 16-26 The homeotic transformation of the third thoracic segment (T3) of *Drosophila* into an extra second thoracic segment (T2). (a) Diagram showing the normal thoracic and abdominal segments; note the rudimentary wing structure (haltere) normally derived from T3. Most of the thorax of the fly, including the wings and the dorsal part of the thorax, comes from T2. (b) A wild-type fly with one copy of T2 and one of T3. (c) A *bithorax* triple mutant homozygote completely transforms T3 into a second copy of T2. Note the second dorsal thorax and second pair of wings (T2 structures) and the absence of the halteres (T3 structures). *(From E. B. Lewis,* Nature *276, 1978, 565. Photographs courtesy of E. B. Lewis. Reprinted by permission of* Nature. *© 1978 by Macmillan Journals Ltd.)*

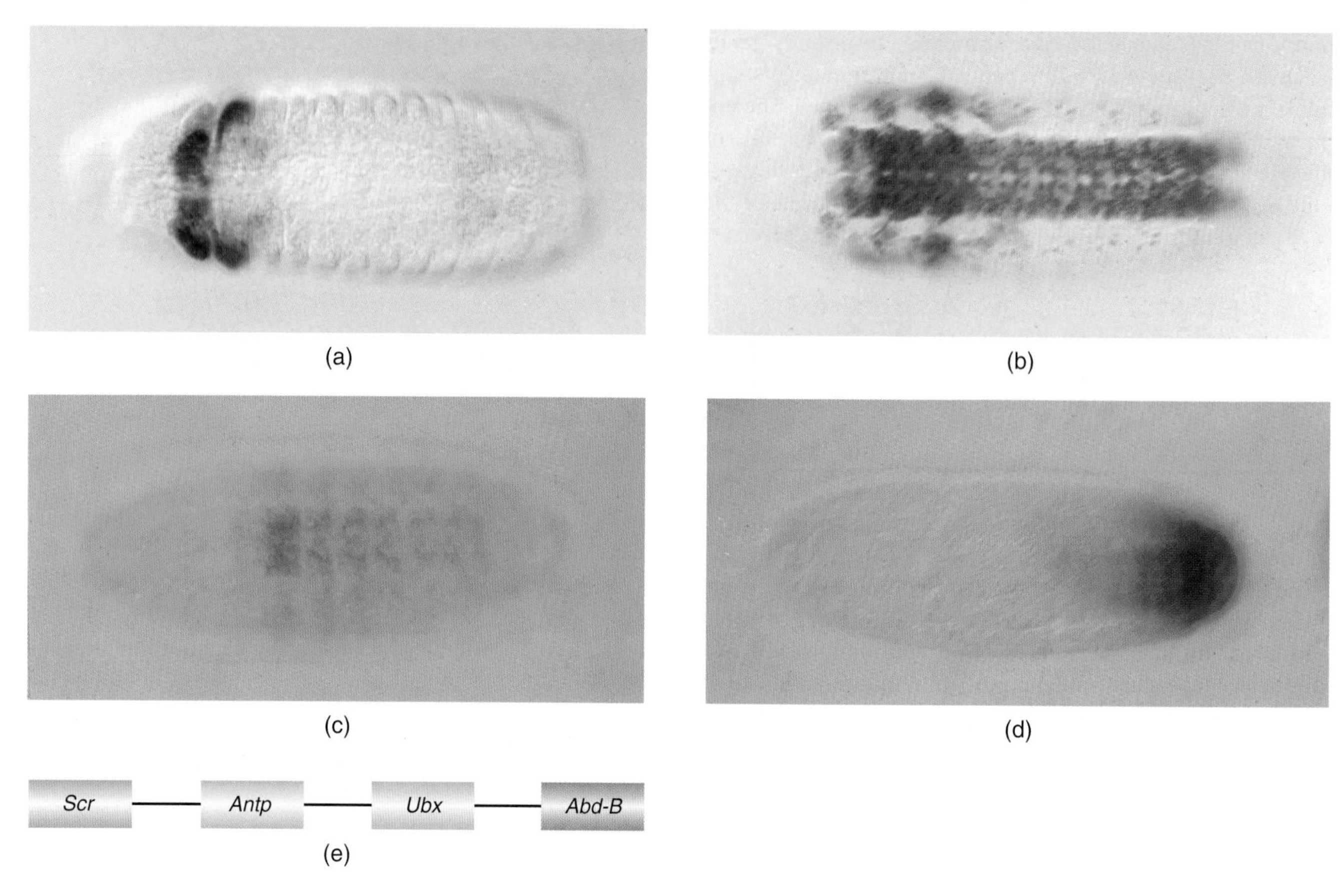

Figure 16-27 Photomicrographs of embryos that exhibit protein expression patterns encoded by homeotic genes in *Drosophila.* (a) SCR; (b) ANTP; (c) UBX; (d) ABD-B. Note that the anterior boundary of homeotic gene expression is ordered from SCR (most anterior) to ANTP, UBX, and ABD-B (most posterior). (e) This order is matched by the linear arrangement of the corresponding genes along chromosome 3. *(Parts a and b from T. C. Kaufman, Indiana University. Parts c and d from S. Celniker and E. B. Lewis, California Institute of Technology.)*

genes that are responsible for conferring specific functions and identities on different regions of each segment.

MESSAGE

Segment identity is established through asymmetric gap-gene expression patterns that deploy an asymmetric pattern of homeotic gene expression.

A Cascade of Regulatory Events

As we have seen, A-P patterning of the *Drosophila* embryo occurs through a sequential triggering of regulatory events. Positional information establishes different concentrations of transcription factors along the A-P axis, and target regulatory genes are then deployed accordingly to execute the increasingly finer subdivisions of the embryo, establishing both segment number and segment identity (Figure 16-28).

ADDITIONAL ASPECTS OF PATTERN FORMATION

The principles delineated in the preceding sections lay out initial fates, but additional mechanisms must be in place to ensure that all aspects of patterning are elaborated. Some of these mechanisms are considered in this section.

Memory Systems for Remembering Cell Fate

Patterning decisions frequently need to be maintained in a cell lineage for the lifetime of the organism. This requirement is certainly true of the segment-polarity and homeotic gene expression patterns that are set up by the A-P patterning system. Such maintenance is accomplished through intracellular or intercellular positive-feedback loops.

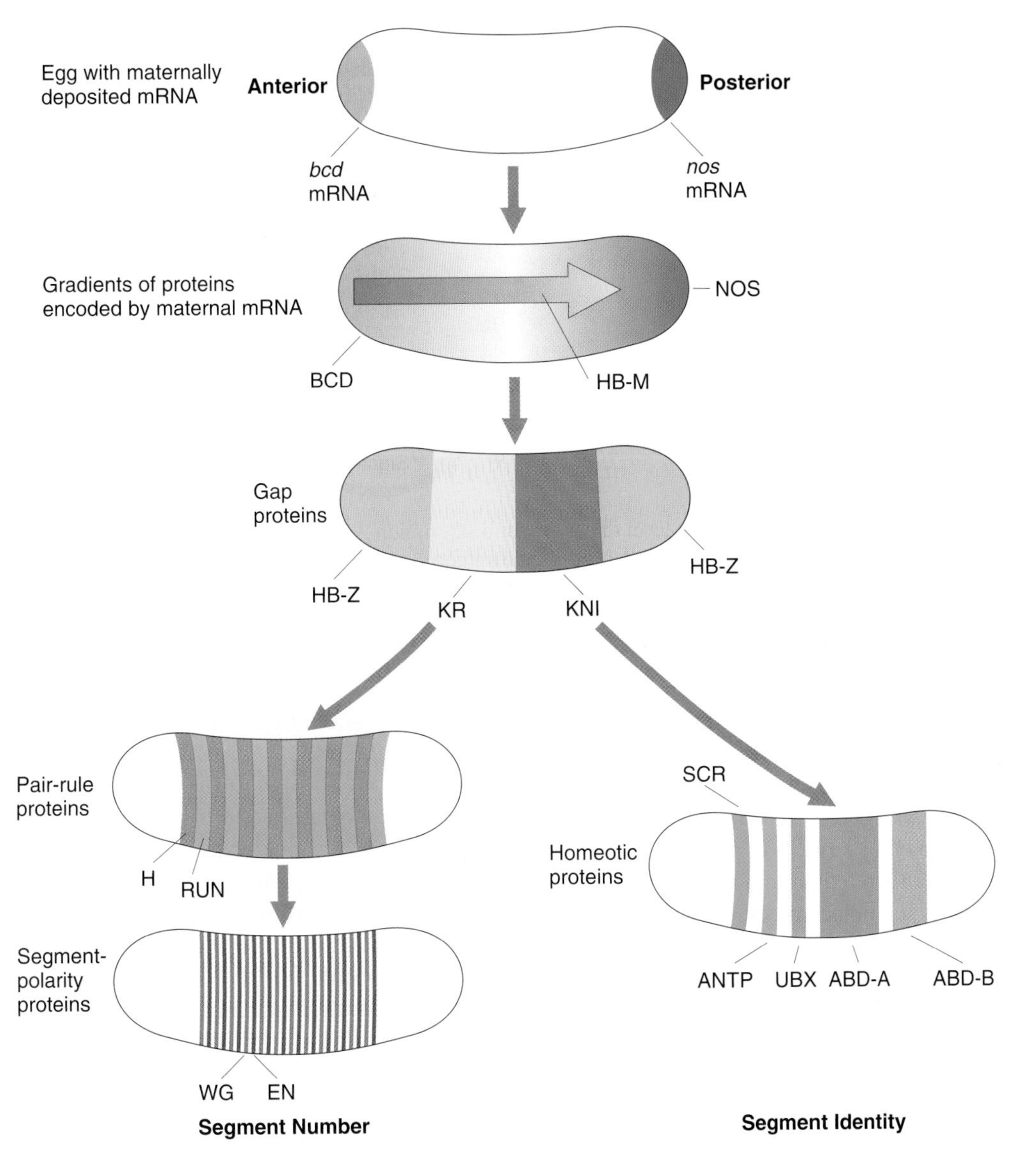

Figure 16-28 The hierarchical cascade that activates the elements forming the A-P segmentation pattern in *Drosophila*. The maternally derived *bcd* and *nos* mRNAs are located at the anterior and posterior poles, respectively. Early in embryogenesis, these mRNAs are translated to produce a steep anterior-posterior gradient of BCD transcription factor. The posterior-anterior gradient of NOS inhibits translation of *hb-m* mRNA, thereby creating a shallow anterior-posterior gradient of HB-M transcription factor (shown as an arrow). The gap genes, which are the A-P cardinal genes, are activated in different parts of the embryo in response to the anterior-posterior gradients of the two factors BCD and HB-M. (The posterior band of HB-Z expression gives rise to certain internal organs, not to segments.) The correct number of segments is determined by activation of the pair-rule genes in a zebra-stripe pattern in response to the gap-gene-encoded transcription factors. The segment-polarity genes are then activated in response to the activities of the several pair-rule proteins, leading to further refinement of the organization within each segment. The correct identities of the segments are determined by expression of the homeotic genes due to direct regulation by the transcription factors encoded by the gap genes. *(Modified from J. D. Watson, M. Gilman, J. Witkowski, and M. Zoller,* Recombinant DNA, *2d ed. © 1992 by James D. Watson, Michael Gilman, Jan Witkowski, and Mark Zoller.)*

In several tissues, positive-feedback loops are established in which the homeodomain protein that is expressed binds to enhancer elements in its own gene, ensuring that more of that homeodomain protein will continue to be produced (Figure 16-29a on the next page). (This is reminiscent of the positive-feedback loop for *Sxl* splicing in the female developmental pathway, described earlier in this chapter.)

The other memory system requires cell-cell interactions (see Figure 16-29b). For example, among the segment-polarity genes, adjacent cells express the WG and

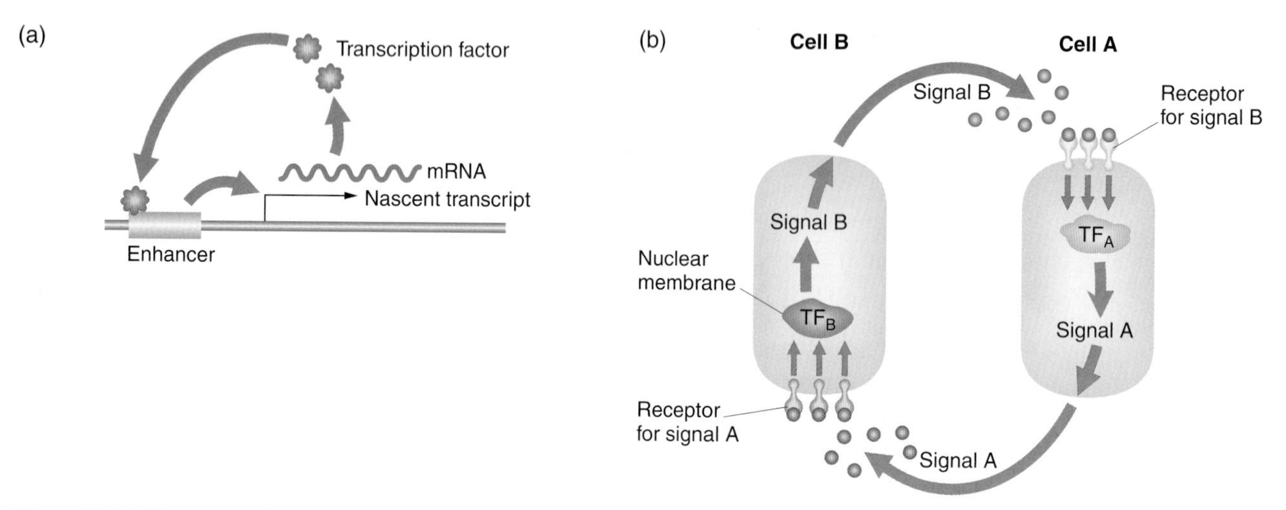

Figure 16-29 Two types of positive-feedback loops that maintain the level of activity of transcription factors that determine cell fate. (a) The transcription factor binds to an enhancer of its own gene, maintaining its transcription. (b) Each adjacent cell sends out a signal (different signals from each cell) that activates receptors, signal transduction pathways, and transcription-factor (TF) expression in the other cell. This activation leads to a mutual positive-feedback loop between the cells.

EN proteins. The EN protein is a transcription factor that activates HH in the same cells. HH is a secreted protein-signaling molecule that induces a receptor-mediated signal transduction cascade in the WG-expressing cell, activating *wg (wingless)* gene expression and causing more WG protein to be expressed. WG similarly is a secreted protein that activates *en (engrailed)* expression in the adjacent cell, inducing more EN protein in that cell.

MESSAGE

Once the fate of a cell lineage has been established, it must be remembered. This is accomplished by intracellular or intercellular positive-feedback loops.

Ensuring That All Fates Are Allocated: Decisions by Committee

Ultimately, for a developmental field to mature into a functional organ or tissue, cells must be committed in appropriate numbers and locations to the full range of fates that are needed. Cell-cell interactions ensure that these proper allocations are made. We should be aware of two types of interactions, both of which operate in the development of the vulva, the opening to the outside of the reproductive tract of the nematode *C. elegans* (Figure 16-30). One type is the ability of one cell to induce a developmental commitment in one neighbor of many, and the other is the ability of one cell to inhibit its neighbors from adopting its fate.

Vulva development has been studied in detail through the analysis of *C. elegans* mutants that have either no vulva

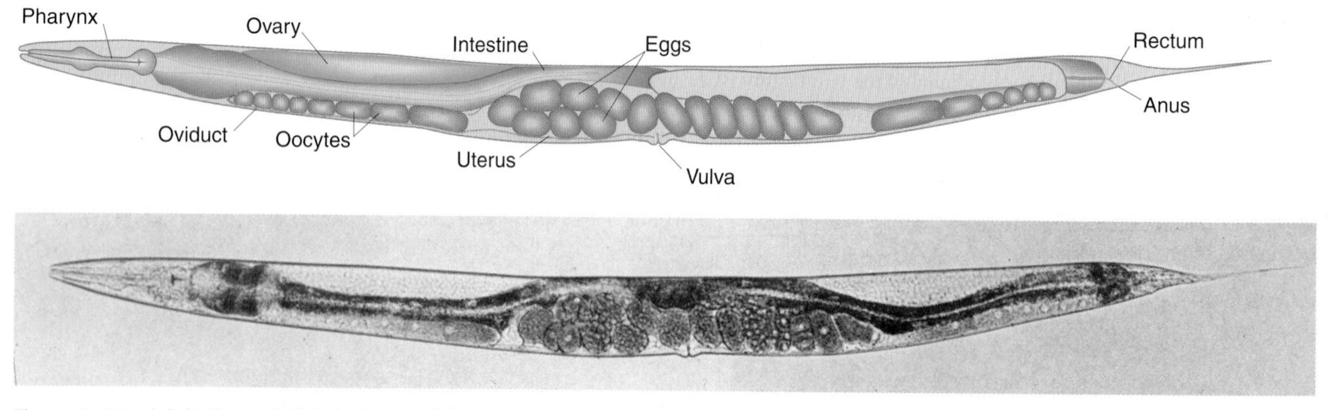

Figure 16-30 Adult *Caenorhabditis elegans.* Photomicrograph and drawing of an adult hermaphrodite, showing various organs readily identified by their location. Note the position of the vulva midway along the anterior-posterior axis of the worm. *(From J. E. Sulston and H. R. Horvitz,* Developmental Biology *56, 1977, 111.)*

or too many (i.e., more than one). Within the hypoderm (the body wall of the worm), several cells have the potential to build certain parts of the vulva. Initially, all these cells can adopt any of the required roles and so are called an **equivalence group**—in essence, a developmental field. To make an intact vulva, one of the cells must become the primary vulva cell, and two others must become secondary vulva cells; yet others become tertiary cells that contribute to the surrounding hypoderm (Figure 16-31a and b).

The key to allocating the different roles to these cells is another single cell, called the **anchor cell,** which lies underneath the cells of the equivalence group (Figure 16-31c). The anchor cell secretes a polypeptide ligand that binds to a receptor tyrosine kinase (RTK) present on all the cells of the equivalence group. Only the cell that receives the highest level of this signal (the equivalence group cell nearest the anchor cell) activates the signal transduction pathway at a sufficient level to activate the transcription factors necessary for that cell to become a primary vulva cell (Figure 16-31d). Thus we can say that the anchor cell operates through an **inductive interaction** to commit a cell to the primary vulva fate.

Having acquired its fate, the primary vulva cell sends out a different paracrine signal to its immediate neighbors in the equivalence group, inhibiting those cells from similarly interpreting the anchor-cell signal, thus preventing them from also adopting the primary role. This process of **lateral inhibition** leads these neighboring cells to adopt the secondary fate. The remaining cells of the equivalence group develop as tertiary cells and contribute to the hypoderm surrounding the vulva. For each of the three cell types into which the equivalence group develops, there is a specific constellation of transcription factors that are activated and that typify the state of the cell: primary, secondary, or tertiary. Thus, through a series of paracrine intercellular signals, a group of equivalent cells can develop into the three necessary cell types.

MESSAGE

Fate allocation can be made through a combination of inductive and lateral inhibitory interactions between cells.

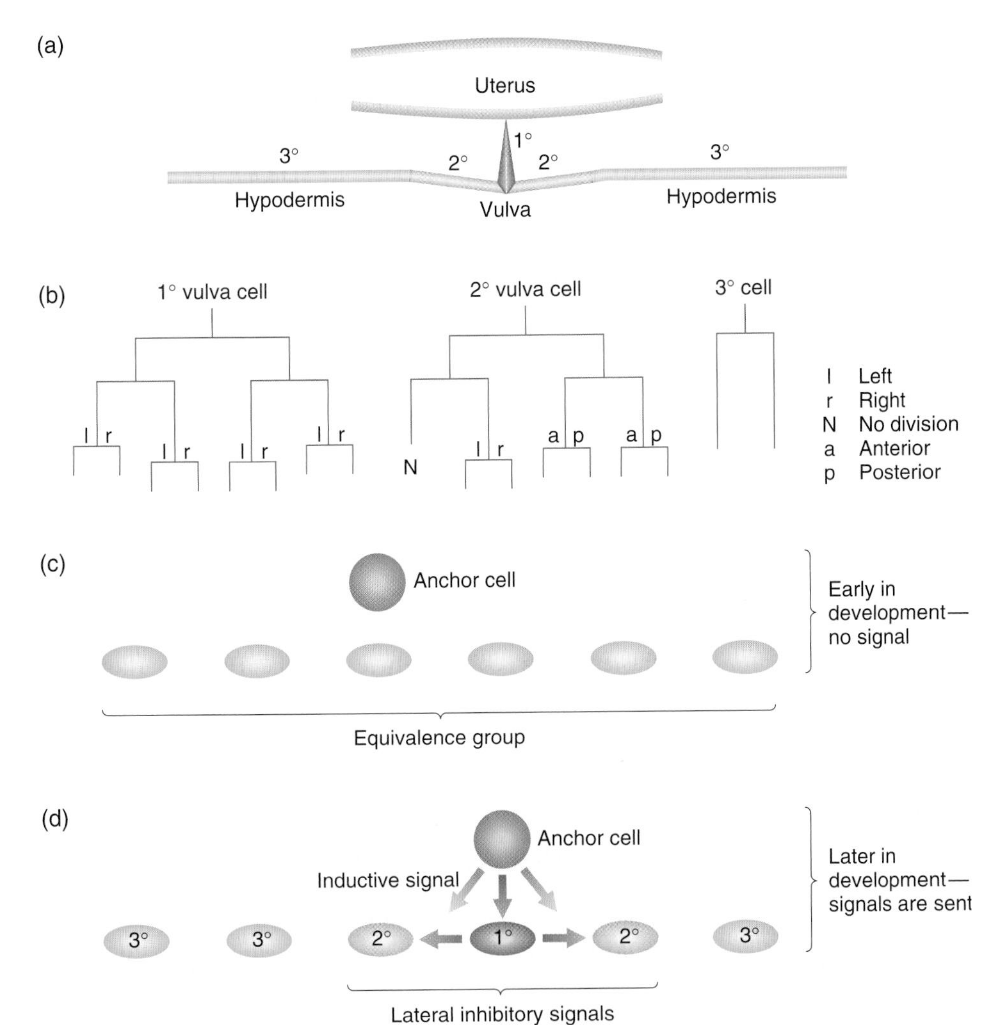

Figure 16-31 The production of the *C. elegans* vulva from the equivalence group by cell-cell interactions. (a) The parts of the vulva anatomy occupied by the descendants of cells in the equivalence group. (b) The primary, secondary, and tertiary cell types are distinguished by the cell division patterns that they undergo. (c) Early in development, there is no signal from the anchor cell, and all the equivalence group cells are in the default tertiary cell state. (d) Later in development, the anchor cell sends a signal that activates a receptor tyrosine kinase signal transduction cascade. The equivalence group cell nearest the anchor cell receives the strongest signal and becomes the primary vulva cell. It then sends out lateral inhibition signals to its neighbors, preventing them from also becoming primary vulva cells and shunting them into the secondary vulva cell pathway. *(Part b from R. Horvitz and P. Sternberg,* Nature *351, 1991, 357. Parts a, c, and d adapted from I. Greenwald,* Trends in Genetics *7, 1991, 366.)*

Developmental Pathways Are Composed of Plug-and-Play Modules

Many developmental pathways are under active investigation in model organisms. From these studies, it is clear that the components of developmental pathways contribute over and over again to the development of a given species. Rather few gene products take part in pattern formation that contributes to only one developmental decision. Instead, bits and pieces of pathways are combined in different ways to determine different outcomes. It is as if, once an effective solution to a particular developmental problem evolved, it was then applied to solve many other problems. This can be thought of as a molecular correlate of the general evolutionary point of view that new structures arise by modifications and adaptations of existing structures rather than by the invention of something totally new.

MESSAGE
Components of one developmental pathway are also found in many others but are often mixed and matched as if they are reusable cartridges.

Typically, a part of a pathway has different inputs, in regard to the signals that regulate it, and sometimes the outputs are different as well (Figure 16-32). As an example of different inputs, the *dpp (decapentaplegic)* gene, whose complex structure was described in Chapter 13, is regulated during embryonic D-V pattern formation by the DL transcription factor, whereas, during visceral mesoderm development, it is regulated by the UBX transcription factor.

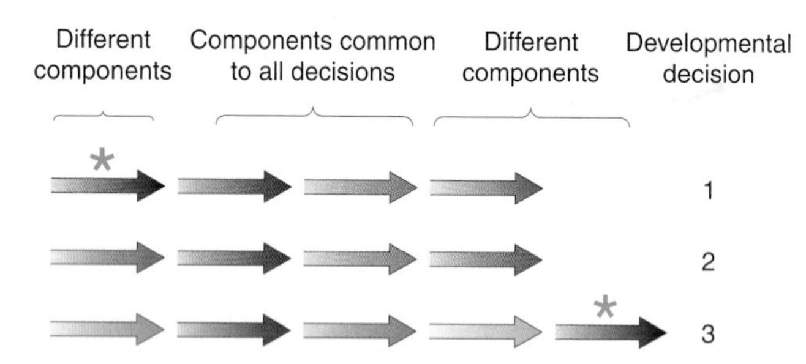

Figure 16-32 Different developmental decisions are made by using mixed and matched combinations of pathway components. Each colored arrow represents a different component. These pathways are joined together in different combinations to achieve different developmental decisions in different cell lineages or in the same cell lineage at different developmental stages. Some components of these decision processes are common; others are unique. It is also possible that some components (denoted by an asterisk) are used in multiple decision processes but occupy different positions in each one.

In regard to different outputs, in maintenance of proper segment-polarity gene expression and cell fates, for example, the EN-HH component of the positive-feedback loop activates the gene that encodes WG in adjacent cells (described earlier). During the larval development of the *Drosophila* adult wing, the EN-HH component activates transcription of a gene that encodes a different secreted signaling protein—DPP—in cells adjacent to those expressing EN-HH. As another example of output differences, the DPP signal transduction pathway leads to activation of completely different transcription factors in D-V patterning, wing, and visceral mesoderm development. Indeed, the complicated regulatory element structure common to many higher eukaryotic genes is probably a necessary consequence of the need to respond to many different tissue-specific inputs.

MESSAGE
Differences in the developmental context of different cell lineages—that is, the transcription factors active in these cells—permit different inputs to, and outputs from, a given developmental circuit.

THE MANY PARALLELS IN VERTEBRATE AND INSECT PATTERN FORMATION

How universal are the developmental principles uncovered in *Drosophila?* Until recently, the type of genetic analysis possible in *Drosophila* was not feasible in most other organisms, at least not without a huge investment to develop comparable genetic tools. However, in the past few years, recombinant DNA technology has provided the tools for addressing the generality of the *Drosophila* findings (see Foundations of Genetics 16-5). Some of the most spectacular advances have come from studying early mouse development.

Perhaps the most striking case is the similarity between the clusters of mammalian homeotic genes called the Hox complexes and the insect ANT-C and BX-C homeotic gene clusters, now collectively called the HOM-C (homeotic gene complex) (Figure 16-33). The ANT-C and BX-C clusters, which are far apart on chromosome 3 of *Drosophila,* are in one cluster in more primitive insects such as the flour beetle *Tribolium castaneum.* This indicates that the general case in insects is that there is only one homeotic gene cluster—HOM-C—and that, in the evolution of the *Drosophila* lineage, it was separated into two clusters. Moreover, as was noted in Figure 16-27e, the genes of the HOM-C cluster are arranged on the chromosome in an order that is colinear with their spatial pattern of expression:

FOUNDATIONS OF GENETICS 16-5

Finding Cognates of *Drosophila* Developmental Genes

With the discovery that there were numerous homeobox genes within the *Drosophila* genome, similarities among the DNA sequences of these genes could be exploited in treasure hunts for other members of the homeotic gene family. These hunts depend on DNA base-pair complementarity. For this purpose, DNA hybridizations were carried out under *moderate stringency conditions,* in which some mismatch of bases between the hybridizing strands could occur without disrupting the proper hydrogen bonding of nearby base pairs. Some of these treasure hunts were carried out in the *Drosophila* genome itself, looking for more family members. Others searched for homeobox genes in other animals, by means of *zoo blots* (Southern blots of restriction-enzyme-digested DNA from different animals), by using radioactive *Drosophila* homeobox DNA as the probe. This approach has led to the discovery of homologous homeobox sequences in many different animals, including humans and mice. (Indeed, it is a very powerful approach to go "fishing" for relatives of almost any gene in your favorite organism.) Some of these mammalian homeobox genes are very similar in sequence to the *Drosophila* genes.

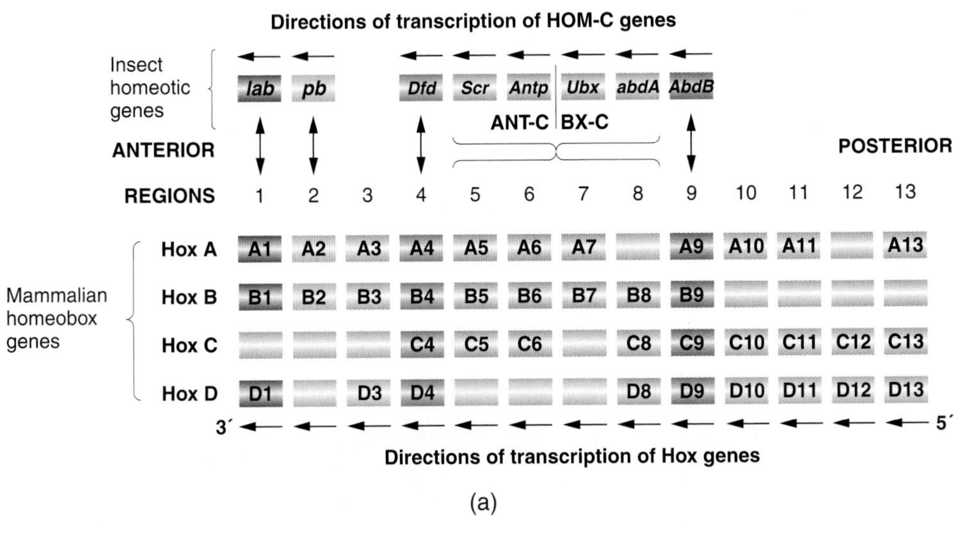

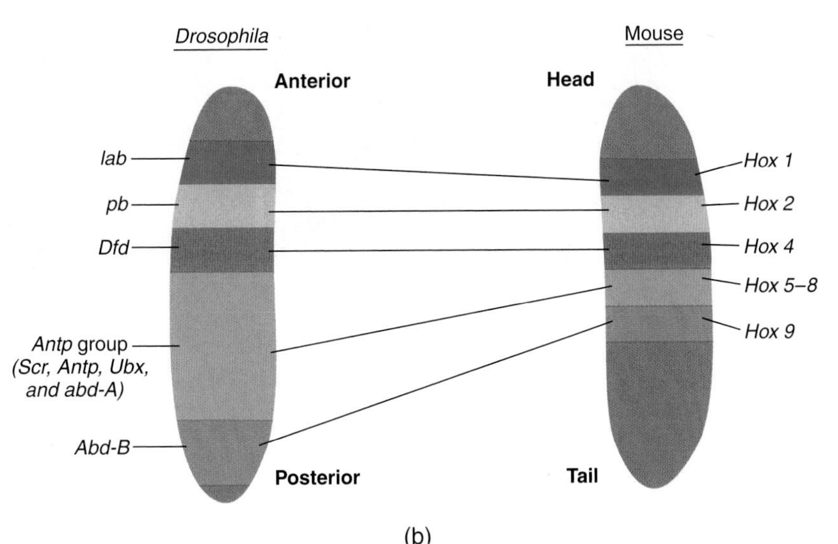

Figure 16-33 Comparison of the structures and functions of the insect and mammalian homeotic genes. (a) The comparative anatomy of the HOM-C (insect) and Hox (mammalian) gene clusters. The genes of the HOM-C are shown at the top. Each of the four paralogous (see text) Hox clusters maps on a different chromosome. Genes shown in the same color are most closely related to one another in structure and function. (b) The expression domains and regions of the *Drosophila* and mouse embryos that require the various HOM-C and Hox genes. The color scheme parallels that in part a. Note that the order of domains in the two embryos is the same. *(Adapted from H. Lodish, D. Baltimore, A. Berk, S. L. Zipursky, P. Matsudaira, and J. Darnell,* Molecular Cell Biology, *3d ed. © Scientific American Books, 1995.)*

the genes at the left-hand end of the complex are transcribed near the anterior end of the embryo; rightward along the chromosome, the genes are transcribed progressively more posteriorly (compare Figure 16-33a and b).

We still do not know why the insect genes are clustered or organized in this colinear fashion, but, regardless of the roles of these features, the same structural organization—clustering and colinearity—is seen for the equivalent genes in mammals, which are organized into the Hox clusters (see Figure 16-33a). The major difference between flies and mammals is that there is only one HOM-C cluster in the insect genome, whereas there are four Hox clusters, each located on a different chromosome, in mammals. These four Hox clusters are **paralogous,** meaning that the structure (order of genes) in each cluster is very similar, as if the entire cluster had been quadruplicated in the course of vertebrate evolution. The genes near the left end of each Hox cluster are quite similar not only to each other, but also to one of the insect HOM-C genes at the left end of the cluster. Similar relations hold throughout the clusters. Finally, and most notably, the Hox genes are expressed in a segmental fashion in the developing somites and central nervous system of the mouse (and presumably the human) embryo. Each Hox gene is expressed in a continuous block beginning at a specific anterior limit and running posteriorly to the end of the developing vertebral column (Figure 16-34). The anterior limit differs for different Hox genes. Within each Hox cluster, the leftmost genes have the most anterior limits. These limits proceed more and more posteriorly in the rightward direction in each Hox cluster. Thus, the Hox gene clusters appear to be arranged and expressed in an order that is strikingly similar to that of the insect HOM-C genes (see Figure 16-33b).

The correlations between structure and expression pattern are further strengthened by consideration of mutant phenotypes. In vitro mutagenesis techniques permit efficient gene knockouts in the mouse. Many of the Hox genes have now been knocked out, and the striking result is that the phenotypes of the homozygous knockout mice are thematically parallel to the phenotypes of homozygous null HOM-C flies. For example, the *Hoxc-C8* knockout causes ribs to be produced on the first lumbar vertebra, L1, which ordinarily is the first nonribbed vertebra behind those vertebrae-bearing ribs (Figure 16-35). Thus, when *Hox C8* is knocked out, the L1 vertebra is homeotically transformed to the segmental identity of a more anterior vertebra. To use geneticists' jargon, *Hox C8*$^-$ has caused a fate shift toward anterior. Clearly, this Hox gene seems to control segmental fate in a manner quite similar to the HOM-C genes, because, for example, the absence of the *Drosophila Ubx* gene also causes a fate shift toward anterior in which T3 and A1 are transformed to T2.

How can such disparate organisms—fly, mouse, human—have such similar gene sequences? (The same is true for the worm *C. elegans.*) The simplest interpretation is that the Hox and HOM-C genes are the vertebrate and insect descendants of a homeobox gene cluster present in a common ancestor some 600 million years ago. The evolu-

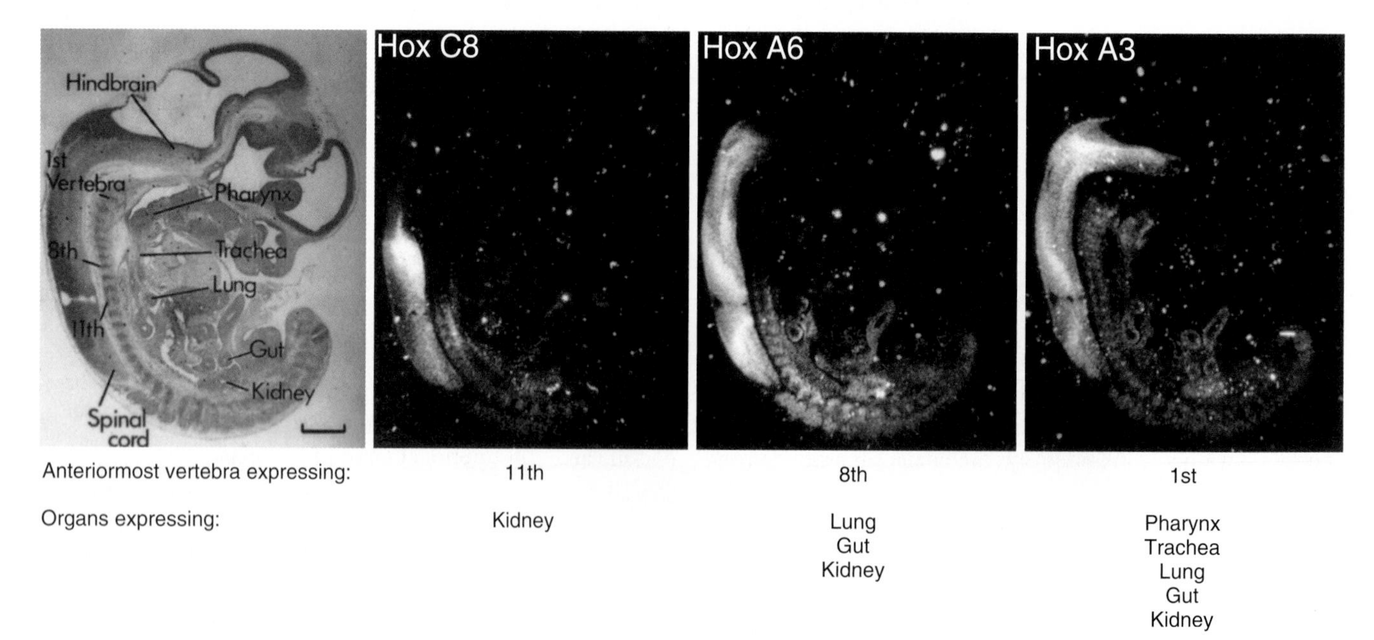

Figure 16-34 Photomicrographs showing the RNA expression patterns of three mouse Hox genes in the vertebral column of a sectioned 12.5-day-old mouse embryo. Note that the anterior limit of each of the expression patterns is different. *(From S. J. Gaunt and P. B. Singh,* Trends in Genetics *6, 1990, 208.)*

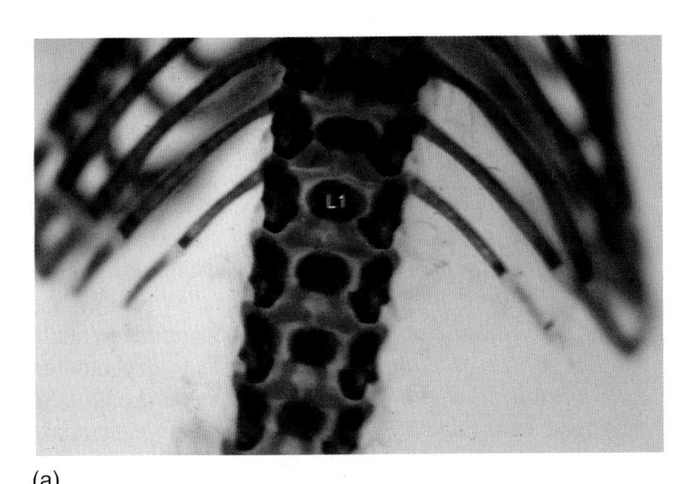

(a)

(b)

Figure 16-35 The phenotype of a homeotic mutant mouse. Mice homozygous for a targeted knockout of the *Hox C8* gene were created by using cultured ES cells. (a) An enlargement of the thoracic and lumbar vertebrae of a homozygous *Hox C8*$^-$ mouse. Note the ribs coming from L1, the first lumbar vertebra. L1 in wild-type mice had no ribs. (b) An unexpected second phenotype of the *Hox C8*$^-$ knockout. Note that the homozygous mutant mouse on the right has clenched fingers, whereas the wild-type mouse on the left has normal fingers. *(From H. Le Mouellic, Y. Lallemand, and P. Brulet,* Cell *69, 1992, 251.)*

tionary conservation of the HOM-C and Hox genes is not a singular occurrence. Indeed, as we are beginning to compare whole genomes, we are finding that such evolutionary and functional conservation seems to be the norm rather than the exception. For example, 60 percent of human genes associated with a heritable disease have related genes in *Drosophila.*

MESSAGE

Developmental strategies in animals are quite ancient and highly conserved. In essence, a mammal, a worm, and a fly are put together with the same basic building blocks and regulatory devices. *Plus ça change, plus c'est la même chose!*

DO THE LESSONS OF ANIMAL DEVELOPMENT APPLY TO PLANTS?

The evidence emerging from comparative studies of pattern formation in a variety of animals indicates that many important developmental pathways are ancient inventions conserved and maintained in many, if not all, animal species. The life history, cell biology, and evolutionary origins of plants would, in contrast, argue against the appearance of the same sets of pathways in the regulation of plant development. Plants have very different organ systems from those of animals, depend on inflexible cell walls for structural rigidity, separate germ line from soma very late in development, and are very dependent on light intensity and duration to trigger various developmental events. Certainly, plants use hormones to regulate gene activity, to signal locally between cells by as yet unknown signals, and to create cell-fate differences by means of transcription factors. The general themes for establishing cell fates in animals are likely to be seen in plants as well, but the participating molecules in these developmental pathways are likely to be considerably different from those encountered in animal development.

An active area of plant developmental genetics research utilizes a small flowering plant called *Arabidopsis thaliana* as a model system. The genome of *Arabidopsis* is 120 megabase pairs of DNA, small for a plant, organized into a haploid complement of five chromosomes. Thus, its genome size and complexity compare to those of the fruit fly, *Drosophila melanogaster.* It is easy to grow in the laboratory in culture tubes or on petri plates, and because it is a self-fertilizing plant, F_2 mutagenesis surveys can be done in a straightforward manner. Thus, many mutations with interesting phenotypes that affect a variety of developmental events have been obtained. Finally, the genome of *Arabidopsis* has been recently sequenced, and many tools such as insertional mutagenesis exist for overlaying the genetic and transcriptional maps of this plant.

Perhaps one of the most intensively studied events in *Arabidopsis* development is flower pattern formation. Just as the homeotic gene cluster controls segmental identity in animal development, a series of transcriptional regulators determines the fate of the four layers (whorls) of the flower. The outermost whorl of the flower normally develops into the sepal; the next whorl, the petals; the next, the stamen; and the innermost develops into the carpel (Figure 16-36).

(a)

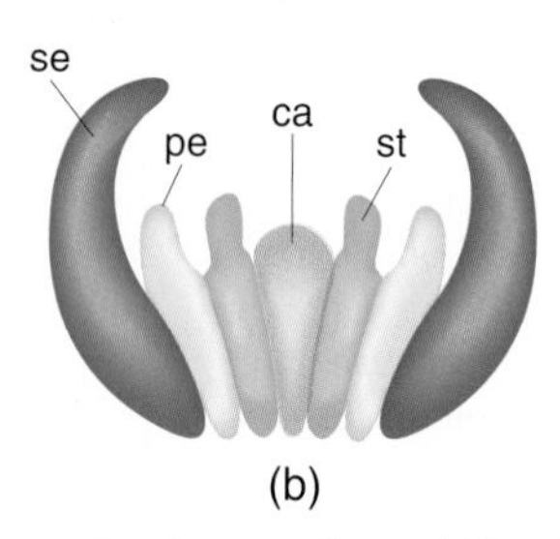

(b)

Figure 16-36 Flower development in *Arabidopsis thaliana.* (a) The mature products of the four whorls of a flower. (b) A cross-sectional diagram of the developing flower, with the normal fates of the four whorls indicated. From outermost to innermost, they are sepal (se), petal (pe), stamen (st) and carpel (ca). *(Photograph courtesy of Vivian F. Irish; from V. F. Irish, "Patterning the Flower,"* Developmental Biology *209, 1999, 211–222.)*

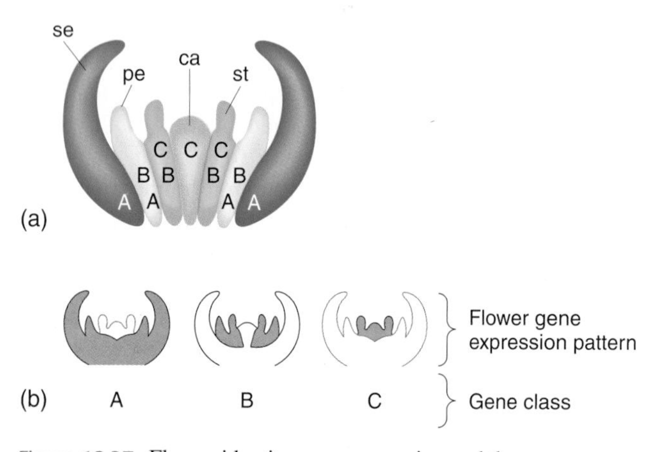

Figure 16-37 Flower-identity gene expression and the establishment of whorl fate. (a) The patterns of gene expression corresponding to the different whorl fates. (b) The shaded regions of the cross-sectional diagrams of the developing flower indicate the gene expression patterns for genes of the A, B, and C classes. Refer to Figure 16-36 for the normal anatomy of the developing flower. *(From V. F. Irish, "Patterning the Flower,"* Developmental Biology *209, 1999, 211–222.)*

Several genes have been identified that, when knocked out or ectopically expressed, transform one or more of these whorls into another. For example, the gene *AP1 (Apetala-1)* causes the homeotic transformation of the outer two whorls into the inner two. Analogously to the homeotic mutants in animals, the number of whorls remains the same (four), but their identities are transformed. The study of the spatial expression patterns and mutant phenotypes of the various flower-identity genes has produced a model in which whorl fate is established through the combinatorial action of multiple transcription factors (Figure 16-37). Thus, sepal (outermost whorl fate) is established through the action of transcription factors expressed by genes of the class A type only. Petal fate is established through the action of transcription factors produced by simultaneous expression of the class A and class B genes. Stamen fate is established through the action of transcription factors produced by simultaneous expression of class B and class C genes. Finally, carpel fate is established through the action of transcription factors expressed by class C genes. Just as the homeotic segment-identity genes in animals encode a series of structurally related (homeodomain-containing) transcription factors, the flower-identity genes encode a series of structurally related (MADS-domain) transcription factors. Thus, although different in detail, the overall strategy of differentially expressed transcription factors is one of the approaches by which plant cell fate is established. With its combination of sophisticated genetics and genomics, studies of *Arabidopsis* pattern formation should reveal much about the ways in which plants develop.

Now that we have examined some developmental pathways in detail, do we have any additional insights into possible candidate genes for Fibrodysplasia Ossificans Progressiva? In fact, we do. It turns out that some proteins called bone morphogenetic proteins (BMPs) were isolated about 15 years ago based on their ability to induce normally patterned bone growth when implanted in experimental animals. These BMPs are members of a family of signaling molecules called the TGF-beta family, which we have seen before in several contexts, including as the molecules that drive many events of pattern formation in flies, worms, and vertebrates. Because of work in these model systems, the receptors and signal transduction molecules for this family of signaling molecules are known, and these are currently being investigated as candidate genes for FOP. Further, with the recent release of the first version of the human genome sequence, most or all sequences related to known elements of the BMP signaling pathways have now been identified. It is possible that the Human Genome Project will help in another way, in that microarray technology can be used to attempt to evaluate the range of alterations in gene regulation that occur in FOP connective tissue. These observations may help localize where the defect resides within the pathway of bone development.

This does not mean the task to identify the right gene is an easy one. The candidate gene approach is second best—a negative result for one gene doesn't get you much closer to the right answer. However, in the absence of sufficient pedigree data to carry out positional cloning, there is not much choice but to take this approach. With good logic and some luck, the right gene is on the list of candidates and the approach will succeed.

FOP is an example of an orphan disease—a disease that has too few sufferers to garner much attention from the medical research and pharmaceutical communities. We should be aware that there are issues of priority and economics that necessarily contribute to decisions about clinical research and drug development, and that a host of value judgments and medical ethical considerations come into play. These are difficult problems with no easy answers, but as genetics and genomics open avenues of research even for the orphan diseases, society will need to wrestle with the appropriate use of its resources to alleviate suffering and cure disease.

SUMMARY

1. A programmed set of instructions in the genome of a higher organism establishes the developmental fates of cells with respect to the major features of the basic body plan. These instructions take totipotent cells and eventually produce a fine-grained mosaic of different cell types deployed in the proper spatial pattern.
2. Developmental pathways are formed by the sequential implementation of various regulatory steps. Within each pathway, there is usually one step that serves as a decision point, with a switch that can be toggled into an "on" or "off" position for binary switches, or for a "one of many" decision for more complex switches.
3. The zygote is totipotent, giving rise to every adult cell type; as development proceeds, successive decisions restrict each cell and its descendants (a lineage) to its particular fate. We see that the first developmental decisions are very coarse, and that the restriction of fates is a gradual process.
4. Gradients of maternally derived regulatory proteins establish polarity along the major body axes; these proteins control the local transcriptional activation of genes encoding master regulatory proteins in the zygote. In all cases that have been well described, the intrinsic polarity of the cytoskeletal system underlies the establishment of the primary positional information within the embryo. It is likely that this reflects exploitation of preexisting architecture (the cytoskeleton) that evolved to control cell shape, motility, and intracellular transport as a means to build higher-order cellular pattern.
5. Many proteins that act as master regulators of early development are transcription factors; others are components of pathways that mediate signaling between cells. The transcription factors ultimately determine cell fate. The signaling pathways provide an avenue for cells to share and coordinate their fate decisions with adjacent cells. Thus, it is not surprising that many fate decisions require communication and collaboration between cells.
6. The same basic set of genes identified in *Drosophila* and the regulatory proteins that they encode are conserved in mammals and appear to govern major developmental events in many—perhaps all—higher animals. It is fair to say that the majority of genes in the metazoan genome are shared among most members of the animal kingdom. The take-home message from this is that the basic pathways that underlie pattern formation are ancient and are exploited in many different ways to produce animals that are superficially very different from one another.
7. The underlying molecules that regulate development in plants are different from those in animals, but many of the themes seen in animal development are used to assign cell fate in plant development. Transcription factors and signaling systems are exploited to create pattern. However, because of the ancient phylogenetic split between plants and animals, and because the life strategies of plants and animals are dramatically different, it is not surprising that different molecules fulfill the parallel roles.

CONCEPT MAP

Draw a concept map interrelating as many of the following terms as possible. Note that the terms are listed in no particular order.

targeted gene knockouts / homeotic genes / totipotency / microtubules / A-P axis / gap genes / paralogous genes / segment-polarity genes / morphogen gradients / lateral inhibition / positional information / feedback loops

SOLVED PROBLEMS

1. In developmental pathways, the crucial events seem to be the activation of master switches that set in motion a programmed cascade of regulatory responses. Identify the master switches and explain how they operate in the two examples of sex determination and differentiation described in this chapter: (a) sex determination in *Drosophila* and (b) sex determination in mammals.

Solution

a. In *Drosophila* sex determination, the master switch is the transcriptional regulation of the *Sxl* gene during early embryogenesis. The properties of the SXL protein and the existence of a constitutive late promoter for *Sxl* ensure that, when active SXL protein is produced, an autoregulatory loop will continue to maintain SXL protein activity in the cell. SXL protein will also initiate a downstream set of regulated RNA splicing events, culminating in the production of *dsx-F* mRNA. In the absence of SXL, the default RNA splicing machinery will produce *dsx-M* mRNAs. The key then is whether the *Sxl* switch is flipped on or stays off. This is controlled by the level of active numerator-encoded transcription factors. The higher level of these transcription factors conferred by an X:A ratio of 1 is necessary to activate early *Sxl* transcription and set the female developmental pathway in motion. In the absence of these higher levels of numerator-encoded transcription factors, development continues along the default pathway leading to male development.

b. In mammalian sex determination, the master switch is the presence or absence of the *SRY* ~ *Sry* gene, which is ordinarily located on the Y chromosome. In the presence of the protein product of this gene, which acts as a DNA-binding protein, certain cells of the gonad (Leydig cells) synthesize androgens, male-inducing steroid hormones. These hormones are secreted into the bloodstream and act on target tissues to induce the transcription-factor activity of the androgen receptors. In the absence of androgen-receptor activation, development proceeds along the default pathway leading to female development. The factors that activate *SRY* ~ *Sry* expression in the testis are not understood. Because the master switch here is the actual presence or absence of the *SRY* ~ *Sry* gene itself, it is likely that the regulatory molecules that activate *SRY* ~ *Sry* are present in the indifferent gonad early in development.

2. In *Drosophila* you have identified a new mutation, *mll (malelike),* that causes flies with an X:A ratio of 1 to develop as phenotypic males. You want to understand how the product of the *mll* gene acts in the developmental pathway for *Drosophila* sex determination and differentiation. You measure the presence of functional SXL, TRA, DSX-F, and DSX-M protein in animals with an X:A ratio of 1. In each of the following cases, propose a role for MLL protein in the sex-determination and differentiation pathway.

a. You observe that functional early SXL, late SXL, TRA, and DSX-M proteins are produced.

b. You observe that functional early SXL and DSX-M are produced. No functional late SXL or TRA is produced.

c. The only functional protein that you observe is DSX-M.

Solution

It is actually through experiments like these that the sequential pathway of sex determination and differentiation in *Drosophila* was elucidated. More generally, it is through observations like these, where the effects of a mutant in one gene on the mRNA or protein expression of another gene are studied, that developmental pathways are pieced together.

a. Given that SXL and TRA proteins are still operating normally in the context of an X:A ratio of 1, whereas the alternative splicing of *dsx* mRNA is leading to the male-specific form of the protein, it is likely that *mll* contributes to *dsx* RNA-splicing regulation.

b. Here, the block seems to be between early and late SXL protein production. This finding may indicate that *mll* plays a role in the autoregulation of late *Sxl* alternative splicing.

c. In this example, the block is before *Sxl* expression in the early embryo. One possible explanation is that MLL protein contributes to the numerator function in interpreting the X:A ratio.

3. In the embryogenesis of mammals, the inner cell mass, or ICM (the prospective fetus), quickly separates from the cells that will serve as enclosing membranes and respiratory, nutritive, and excretory channels between the mother and the fetus.

a. Design experiments using mosaics in mice to determine when the two fates are decided.

b. How would you trace the formation of different fetal membranes?

Solution

a. We must have markers that enable us to distinguish different cell lineages. This can be done with mice by using strains that differ in chromosomal or biochemical markers. (Other ways would be to use differences in sex chromosomes of XX and XY cells or to induce chromosome loss or aberrations by irradiating embryos.)

When you have decided on the marker difference to be used, one way to answer the question is to inject a single cell from one of two strains into embryos of the other at various developmental stages. Another approach is to fuse embryos of defined cell numbers from the two strains. In either case, you would inspect the embryos when the ICM and membranes are distinct and recognizable. When cell insertion or fusion results in membranes and ICM that are exclusively made up of one cell type and never a mosaic of the two, the two developmental fates have been set.

b. Carry out the same injection or fusion experiment on early embryos. Now look for the pattern of mosaicism. Correlate the presence of cells of similar genotype in different membranes. It should be possible to determine the lineage of cells in each set of membranes.

PROBLEMS

Basic Problems

1. In *Drosophila,* individual flies with two X chromosomes and three sets of autosomes (X:A ratio = 0.67) are intersexes. When examined more closely, it turns out that their intersexual phenotype is due to mosaicism. Some cells differentiate with an entirely male phenotype and other cells with an entirely female phenotype. Explain this observation in regard to how the *Sxl* on-off switch is established and maintained.

2. XYY humans are fertile males. XXX humans are fertile females. What do these observations reveal about the mechanisms of sex determination and dosage compensation?

3. Humans who are mosaics of XX and XY tissue occasionally occur. They generally exhibit a uniform sexual phenotype. Some of them are phenotypically female, others male. Explain these observations in regard to the mechanism of sex determination in mammals.

4. How are the gradients of BCD and HB-M established during early embryogenesis in *Drosophila?* What is the role of the cytoskeleton in this process?

5. What are the similarities between DL/CACT in *Drosophila* and Rb/E2F (see Chapter 15)?

6. In *C. elegans* vulva development, one anchor cell in the gonad interacts with six equivalence-group cells (cells with the potential to become parts of the vulva). The six equivalence group cells have three distinct phenotypic fates: primary, secondary, and tertiary. The equivalence group cell closest to the anchor cell develops the primary vulva phenotype. If the anchor cell is ablated, all six equivalence group lineages differentiate into the tertiary state.

a. Set up a model to explain these results.

b. The anchor cell and the six equivalence group cells can be isolated and grown in vitro; design an experiment to test your model.

7. There are two types of muscle cell in *C. elegans:* pharyngeal muscles and body wall muscles. They can be distinguished from each other, even as single cells. In the following diagram, letters designate particular muscle precursor cells, black triangles (▲) are pharyngeal muscles, and white triangles (△) are body wall muscles.

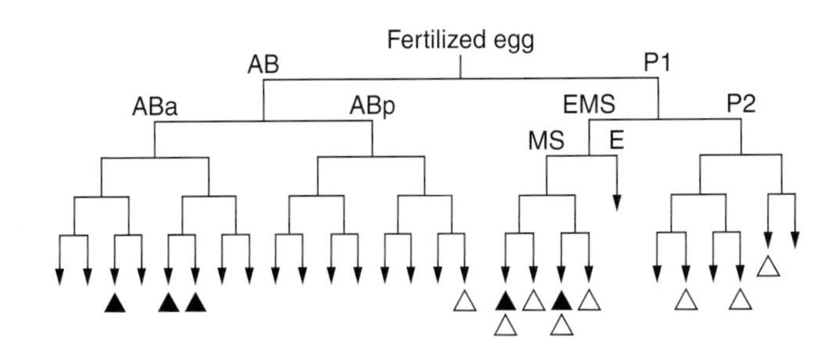

a. It is possible to move these cells around during development. When the positions of ABa and ABp are physically interchanged, the cells develop according to their new position. In other words, the cell that was originally in the ABa position now gives rise to one body wall muscle cell, whereas the cell that was originally in the ABp position now gives rise to three pharyngeal muscle cells. What does that tell us about the developmental processes controlling the fates of ABa and ABp?

b. If EMS is ablated (by heat inactivation with a laser beam aimed through a microscope lens), no AB descendants make muscles. What does that suggest?

c. If P2 is ablated, all AB descendants turn into muscle cells. What does that suggest?

(Diagram from J. Priess and N. Thomson, *Cell* 48, 1987, 241.)

8. In humans, the male steroid hormone testosterone binds to the testosterone receptor. The hormone-receptor complex then activates transcription of male-specific genes. A mutation in the testosterone receptor prevents binding to testosterone. What would be the phenotype of a person who is XY and is homozygous for such a mutation? Explain your conclusion in relation to how steroid receptors act.

9. When an embryo is homozygous mutant for the gap gene *Kr*, the fourth and fifth stripes of the pair-rule gene *ftz* (counting from the anterior end) do not form normally. When the gap gene *kni* is mutant, the fifth and sixth *ftz* stripes do not form normally. Explain these results in regard to how segment number is established in the embryo.

10. Explain why active SXL product from the early promoter is required for active SXL product from the late promoter. How does this create an autoregulatory loop?

11. The *Drosophila* embryo has a polarity that is developed through the action of maternal genes that are expressed in the developing egg follicle in the ovary. Mothers homozygous for a *nanos* produce embryos that lack posterior segments. However, these embryos do not display mirror-image anterior (bicephalic) pattern. In contrast, mothers homozygous for a *bicoid* mutation produce embryos that not only lack anterior segments but also display a mirror-image posterior (bicaudal) pattern. Explain these observations in terms of the roles of *nanos, bicoid,* and *hunchback* in anterior-posterior axis formation.

12. For many of the mammalian Hox genes, it has been possible to determine that some of them are more similar to one of the insect HOM-C genes than to the others. Describe an experimental approach using the tools of molecular biology that would allow such a determination.

Challenging Problems

13. a. When you remove the anterior 20 percent of the cytoplasm of a newly formed *Drosophila* embryo, you can cause a bicaudal phenotype, in which there is a mirror-image duplication of the abdominal segments. In other words, from the anterior tip of the embryo to the posterior tip, the order of segments is A8-A7-A6-A5-A4-A4-A5-A6-A7-A8. Explain this phenotype in regard to the action of the anterior and posterior determinants and how they affect gap-gene expression.

b. Females homozygous for the maternally acting mutation *nanos (nos)* produce embryos in which the abdominal segments are absent and in which the head and thoracic segments are broader. In regard to the action of the anterior and posterior determinants and gap-gene action, explain how *nos* produces this mutant phenotype. In your answer, explain why there is a loss of segments rather than a mirror-image duplication of anterior segments.

14. The three homeodomain proteins ABD-B, ABD-A, and UBX are encoded by genes within the BX-C of *Drosophila.* In wild-type embryos, the *Abd-B* gene is expressed in the posterior abdominal segments, *abd-A* in the middle abdominal segments, and *Ubx* in the anterior abdominal and posterior thoracic segments. When the *Abd-B* gene is deleted, *abd-A* is expressed in both the middle and the posterior abdominal segments. When *abd-A* is deleted, *Ubx* is expressed in the posterior thorax and in the anterior and middle abdominal segments. When *Ubx* is deleted, the patterns of *abd-A* and *Abd-B* expression are unchanged from wild type. When both *abd-A* and *Abd-B* are deleted, *Ubx* is expressed in all segments from the posterior thorax to the posterior end of the embryo. Explain these observations, taking into consideration the fact that the gap genes control the initial expression patterns of the homeotic genes.

i **15.** In considering the formation of the A-P and D-V axes in *Drosophila,* we noted that, for mutations such as *bcd,* homozygous mutant mothers uniformly produce mutant offspring with segmentation defects. This outcome is always true regardless of whether the offspring themselves are bcd^+ / *bcd* or *bcd* / *bcd.* Some other maternal-effect lethal mutations are different, in that the mutant phenotype can be "rescued" by introducing a wild-type allele of the gene from the father. In other words, for such rescuable maternal-effect lethals, mut^+ / *mut* animals are normal, whereas *mut* / *mut* animals have the mutant defect. Rationalize the difference between rescuable and nonrescuable maternal-effect lethal mutations.

16. The anterior determinant in the *Drosophila* egg is *bcd.* A mother heterozygous for a *bcd* deletion has only one copy of the bcd^+ gene. With the use of P elements to insert copies of the cloned bcd^+ gene into the genome by transformation, it is possible to produce mothers with extra copies of the gene. Shortly after the blastoderm has formed, the *Drosophila* embryo develops an indentation called the cephalic furrow that is more or less perpendicular to the longitudinal body axis. In the progeny of bcd^+ monosomics, this furrow is very close to the anterior tip, lying at a position one-sixth of the distance from the anterior to the posterior tip. In the progeny of standard wild-type diploids (disomic for bcd^+), the cephalic furrow arises more posteriorly, at a position one-fifth of the distance from the anterior to the posterior tip of the embryo. In the progeny of bcd^+ trisomics, it is even more posterior. As additional gene doses are added, it moves more and more posteriorly, until, in the progeny of hexasomics, the cephalic furrow is midway along the A-P axis of the embryo.

a. Explain the gene dose effect of bcd^+ on cephalic furrow formation in relation to the contribution that *bcd* makes to A-P pattern formation.

b. Diagram the relative expression patterns of mRNAs from the gap genes *Kr* and *kni* in blastoderm embryos derived from *bcd* monosomic, trisomic, and hexasomic mothers.

17. In *Drosophila,* homozygous mutations in the *tra* gene transform XX flies into phenotypic males with regard to somatic secondary sexual characteristics. The gonad in *Drosophila* forms from somatic mesoderm tissue and germ-line cells. XX ; *tra* homozygotes are sterile and have rudimentary gonads. You suspect that the reason for this sterility is that the somatic tissue is sexually transformed into male by *tra* but the germ-line cells remain female. Design an experiment that tests this prediction.

18. There are dominant mutations of the *Sxl* gene, called Sxl^M alleles, that transform XY animals into females but do not affect XX animals. Reversions of these Sxl^M alleles can be readily induced with mutagen treatment. These reversions, called Sxl^f alleles, yield normal individuals in XY animals, but, in homozygotes, they transform XX animals into phenotypic males. Provide a possible explanation for these observations, keeping in mind that *Sxl* is ordinarily dispensable for male development.

19. In the *Drosophila* embryo, the 3′ untranslated regions (3′ UTR) of the mRNAs [the regions between the translation-termination codons and the poly(A) tails] are responsible for localizing *bcd* and *nos* to the anterior and posterior poles, respectively. Experiments have been done in which the 3′ UTRs of *bcd* and *nos* have been swapped. Suppose that we make P-element transformation constructs with both swaps (*nos* mRNA with *bcd* 3′ UTR and *bcd* mRNA with *nos* 3′ UTR) and transform them into the *Drosophila* genome. We then make a female that is homozygous mutant for *bcd* and *nos* and carries both swap constructs. What phenotype would you expect for her embryos in regard to A-P axis development?

20. a. If you had a mutation affecting anterior-posterior patterning of the *Drosophila* embryo in which every other segment of the developing mutant larva was missing, would you consider it a mutation in a gap gene, a pair-rule gene, a segment-polarity gene, or a segment-identity gene?

b. You have cloned a piece of DNA that contains four genes. How could you use the spatial-expression pattern of their mRNA in a wild-type embryo to identify which represents a candidate gene for the mutation described in part a?

c. Assume you have identified the candidate gene. If you now examine the wild-type spatial-expression pattern of its mRNA in an embryo homozygous mutant for the gap gene *Krüppel,* would you expect to see a normal expression pattern? Explain.

21. You have in your possession wild-type and *bicoid* mutant strains. You also have cloned cDNAs for *nanos* and *bicoid.* These plasmids are as follows:

plasmid 1	full-length *nanos* cDNA
plasmid 2	full-length *bicoid* cDNA
plasmid 3	*bicoid* 5′ UTR + *bicoid* ORF + *nanos* 3′ UTR
plasmid 4	*nanos* 5′ UTR + *bicoid* ORF + *bicoid* 3′ UTR
plasmid 5	*nanos* 5′ UTR + *bicoid* ORF + *nanos* 3′ UTR

where UTR = untranslated region and ORF = open reading frame (protein-coding region)

The plasmids are constructed so that you can generate a synthetic mRNA corresponding to the sequences described in each cDNA. Describe how you could use these mRNAs to determine that *bicoid* mRNA localization is due to its 3′ UTR.

22. In mice, the *tfm* (testicular feminization) gene, located on the X chromosome, encodes the testosterone receptor.

a. XY mice hemizygous for a loss-of-function *Tfm* mutation are phenotypic females. Explain the basis of this mutant phenotype.

b. An otherwise wild-type XXY individual is phenotypically male. In each of the somatic cells, one of the two X chromosomes is inactivated by the normal process of dosage compensation. Recall that X-inactivation occurs at random early in development and that the decision which X is inactivated in a given cell is passed on to all descendants of that cell. If an XXY individual is heterozygous for the same loss-of-function *Tfm* mutation described above, what would be the sexual phenotype of this individual?

POPULATION GENETICS 17

Key Concepts

1. The goal of population genetics is to understand the genetic composition of a population and the forces that determine and change that composition.
2. In any species, a great deal of genetic variation within and between populations arises from the existence of various alleles at different gene loci.
3. In a randomly mating population in which no forces are acting to change allelic frequencies, an equilibrium of genotype frequencies is reached in one generation.
4. The frequency of a given allele in a population can be changed by recurrent mutation, selection or migration, or by random-sampling effects.
5. Natural selection results in changes in allele frequencies and may lead to the replacement of one allele by another in the population or to a stable equilibrium of allele frequencies.
6. Most changes in the frequencies of phenotypes are the consequence of changes in the allele frequencies at several loci.
7. Genetic variation in populations is a consequence of the interaction of the forces of selection, migration, and mutation.
8. Changes in the frequencies of alleles in populations occur even in the absence of natural selection as a result of random differences in the reproduction of different genotypes.

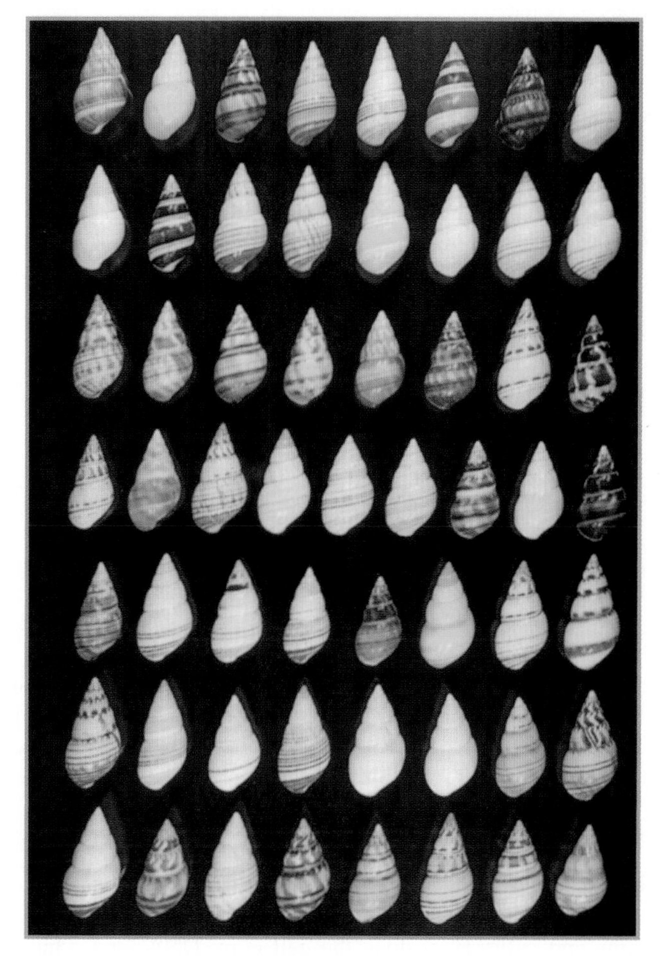

Shell-color polymorphism in *Liguus fascitus*. *(From David Hillis,* Journal of Heredity, *July-August 1991.)*

In western and central Africa, in parts of southern India, and in the Arabian peninsula, between 1 and 2 percent of people suffer from an extreme form of an inherited blood disorder, called sickle-cell anemia, which is the consequence of the crystallization of their hemoglobin and the breakdown of their red blood cells (Figure 17-1). The resulting anemia is severe enough to cause death. People with this disease are homozygous for an aberrant allele of the gene that specifies the amino acid sequence of hemoglobin. About 25 to 30 percent of people in these regions suffer a much milder form of anemia, and these people turn out to be heterozygous for the same sickle-cell allele that is so deleterious to homozygotes. The descendants of West African slaves brought to America also have a high frequency of these disorders, although not as high a frequency as their ancestral populations in Africa. How are we to explain such high frequencies of a serious inherited blood disorder in the African, Indian, and Arabian populations? What are the various forces that have reduced that frequency in North Americans of African ancestry? What is the likely future of this deleterious allele in human populations?

(b)

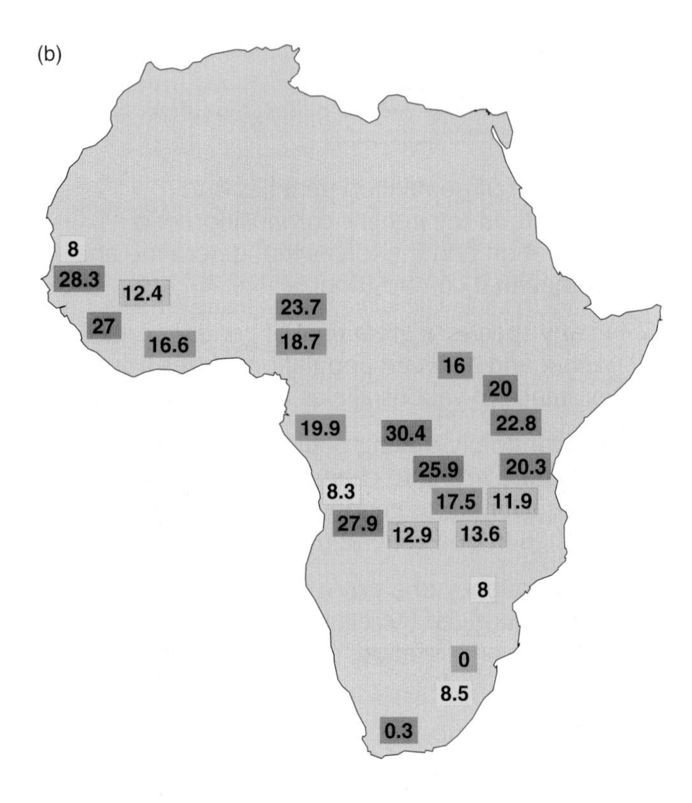

(a)

Figure 17-1 (a) Red blood cells of a person with sickle-cell anemia, showing a few normal disc-shaped red blood cells, surrounded by distorted sickle-shaped cells. (b) Percentages of the population with sickle cells in various localities in Africa, including both heterozygotes and homozygotes for the abnormal allele. *(Dr. Marion I. Barnhart.)*

CHAPTER OVERVIEW

So far in our investigation of genetics, we have been concerned with processes that take place in individual organisms and cells. How does the cell copy DNA, and what causes mutations? How do the mechanisms of segregation and recombination affect the kinds and proportions of gametes produced by an individual organism? How is the development of an organism affected by the interactions among its DNA, the cellular machinery of protein synthesis, cellular metabolic processes, and the external environment? But the members of a species do not live as isolated individuals. They interact with one another in groups, called **populations.** There are questions about the genetic composition of populations that cannot be answered only from a knowledge of the basic individual-level genetic processes. Why are mutant alleles of the genes that encode the proteins Factor VIII and Factor IX, which cause a failure of normal blood clotting—hemophilia—so rare in all human populations, whereas the allele of the hemoglobin β gene that causes sickle-cell anemia is so common in some parts of Africa? What changes in the frequency of sickle-cell anemia are to be expected in the descendants of Africans in North America as a consequence of the change in environment and of interbreeding between Africans and Europeans and Native Americans? What genetic changes occur in a population of insects subjected to insecticides generation after generation? What is the consequence of an increase or decrease in the rate of mating between close relatives? All are questions about what determines the genetic composition of populations and how that composition may be expected to change over time. These questions are the domain of **population genetics.**

MESSAGE

Population genetics relates the processes of individual heredity and development to the genetic composition of populations and to changes in that composition over time and space.

To relate the basic individual-level genetic processes to the genetic composition of populations, we must investigate the following phenomena (Figure 17-2):

1. The rate of introduction of new genetic variation into the population by *mutation,* which generates new alleles.
2. Changes in the population composition due to *migration* of individuals between populations.
3. The effect of *mating patterns* on the population. Individuals may mate at random, they may mate preferentially with close relatives *(inbreeding),* or they may mate preferentially on the basis of genotypic or phenotypic similarity *(assortative mating).*

CHAPTER OVERVIEW figure

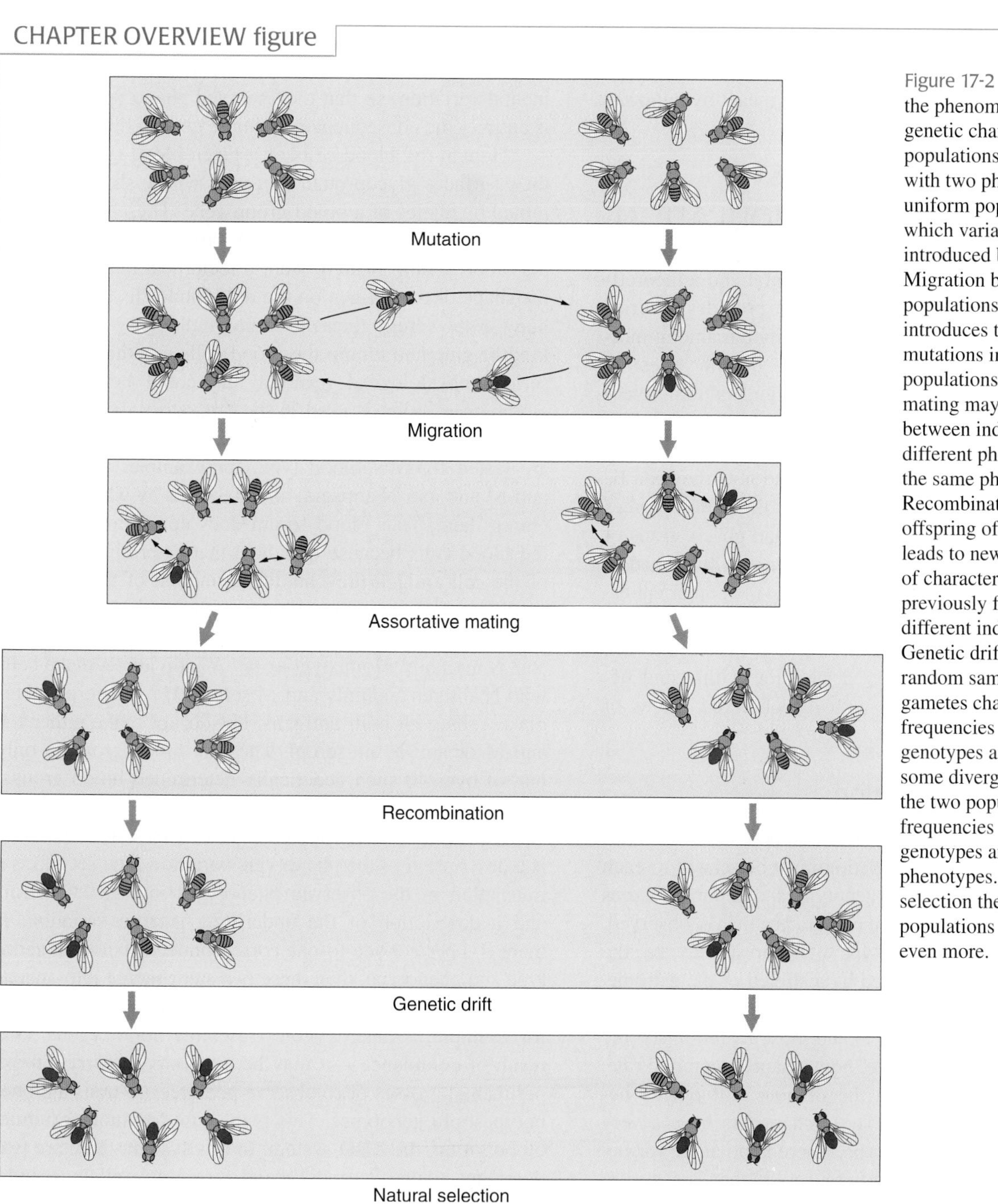

Figure 17-2 Overview of the phenomena that cause genetic change in populations. We begin with two phenotypically uniform populations, into which variation is introduced by mutation. Migration between populations then introduces these separate mutations into both populations. Assortative mating may occur between individuals of different phenotypes or the same phenotype. Recombination in the offspring of these matings leads to new combinations of characters that were previously found in different individuals. Genetic drift due to random sampling of gametes changes the frequencies of the genotypes and causes some divergence between the two populations in the frequencies of their genotypes and phenotypes. Natural selection then causes the populations to diverge even more.

4. The production of new combinations of characters by *recombination,* which re-sorts combinations of alleles at different loci.
5. Random fluctuations in the actual reproductive rates of different genotypes. Because any given individual has only a few offspring, and because the total population size is limited, the theoretical expectations of genetic ratios from meiosis are never realized exactly in real families and real populations. This random fluctuation causes *genetic drift* in allele frequencies from generation to generation.
6. Differential rates of reproduction by different genotypes and the differential chances of survival of genetically different offspring of these matings. These changes in population composition that occur because of these differential rates are what is meant by *natural selection.*

VARIATION AND ITS MODULATION

Population genetics is both an experimental and a theoretical science. On the experimental side, it provides descriptions of the actual patterns of genetic variation among individuals in populations and estimates of the rates of reproduction, mutation, recombination, and natural selection, as well as random variation in reproductive rates. On the theoretical side, it makes predictions of what the genetic composition of populations would be, and how they can be expected to change, as a consequence of the various forces operating on them. Experimental work and observations of natural populations provide the measurements that need to be explained by theory, and they also provide numerical estimates of the quantities that appear in the theory. Reciprocally, the theoretical formulations make it possible to ask which explanations for observed variation within and between species are consistent with the observed values of various population parameters.

Observations of Variation

As in other areas of genetic research, in population genetics phenotypic and genotypic variation must be related to each other. The relation between phenotype and genotype varies in complexity depending on the character that is observed. At one extreme, the phenotype of interest may be the mRNA or polypeptide encoded by a stretch of the genome. At the other extreme lie the bulk of the characters of interest to plant and animal breeders and most evolutionists: the variations in yield, growth rate, body shape, metabolic rate, and behavior that constitute the obvious differences between varieties and species. These characters have a very complex relation to the genotype. There is no allele for being 5′8″ or 5′4″ tall. Such differences, if they are consequences of genetic variation, will be affected by several or many genes and by environmental variation as well. We must use the methods that will be introduced in Chapter 18 if we are to say anything at all about the genotypes underlying the phenotypic variation. But, as we shall see in Chapter 18, it is not possible to make very precise statements about the genotypic variation underlying quantitative characters. For that reason, experimental population genetics has concentrated on characters with simple relations to the genotype, ones in which different phenotypes can be shown to be the result of different allelic forms of a single gene. A favorite object of study for human population geneticists, for example, has been the various human blood groups. The qualitatively distinct phenotypes of a given blood group—say, the MN group—are encoded by alternative alleles at a single locus, and the phenotypes are insensitive to environmental variation, so that the observed phenotypic variation is entirely the consequence of simple genetic differences.

Human red blood cells of a given blood type carry on their surfaces glycoprotein *antigens,* whose shape is determined by alleles of a blood group gene. These same alleles also govern the shape of antibody proteins that circulate in the blood serum. If there were a reciprocal match between the shape of the antigen on the red blood cells and the antibody in the serum, then the serum antibody would act as a kind of glue and clump the blood cells together, causing a serious physiological reaction that could be fatal. This clumping reaction is used in the laboratory to determine a person's blood type. The red blood cells of an individual are tested for MN blood type, for example, against both anti-M and anti-N antisera, and classified by which antisera clump them. Your blood serum does not clump your own red blood cells because the allele that determines the form of the cell antigen prevents the formation of the matching antibody. For example, homozygous *M / M* individuals make red blood cells with M antigen and serum with only anti-N antibody. Homozygous *N / N* individuals make cells with N antigen and only anti-M serum. *M / N* heterozygotes make cells with both antigens, but do not make either the anti-M or anti-N antiserum. The MN blood group is only one of over 40 such genetically determined blood groups, each coded by a different gene.

The study of variation consists of two stages. The first is a description of the phenotypic variation. The second is a translation of the observed phenotypes into genetic terms and a description of the underlying genetic variation. If there is a perfect one-to-one correspondence between genotype and phenotype, then these two steps merge into one, as in the MN blood group. If the relation is more complex—for example, if heterozygotes resemble homozygotes as a result of dominance—it may be necessary to carry out experimental crosses or to observe pedigrees to translate phenotypes into genotypes. This is the case for another human blood group, the ABO system. In this system, there are two dominant alleles, I^A and I^B, and a recessive allele, *i*. Individuals with type A or type B blood may be homozygous

TABLE 17-1 Frequencies of Genotypes for Alleles at MN Blood Group Locus in Various Human Populations

	GENOTYPE			ALLELE FREQUENCIES	
Population	***M / M***	***M / N***	***N / N***	***p(M)***	***q(N)***
Eskimo	0.835	0.156	0.009	0.913	0.087
Australian Aborigine	0.024	0.304	0.672	0.176	0.824
Egyptian	0.278	0.489	0.233	0.523	0.477
German	0.297	0.507	0.196	0.550	0.450
Chinese	0.332	0.486	0.182	0.575	0.425
Nigerian	0.301	0.495	0.204	0.548	0.452

SOURCE: W. C. Boyd, *Genetics and the Races of Man*. D. C. Heath, 1950.

for their respective alleles (I^A / I^A or I^B / I^B), or they may be heterozygous for their type allele and the recessive allele (I^A / i or I^B / i).

The simplest description of single-gene variation is a list of the frequencies of genotypes in a population. Table 17-1 shows this frequency distribution for the three genotypes at the MN locus in several human populations. Note that there is variation between individuals in each population, because different genotypes are present, and that there is also variation in the frequencies of these genotypes from population to population. For example, most people in the Eskimo population are *M* / *M*, while this genotype is quite rare among Australian Aborigines.

More typically, instead of the frequencies of the diploid genotypes, the frequencies of the alternative alleles themselves are given. The frequency of an allele is simply the proportion of that allelic form of the gene among all the copies of the gene in the population. Each individual organism in the population is counted as contributing one diploid set of genes. Thus, there are twice as many copies of the gene in the population as there are individuals, because every individual is diploid. Homozygotes for an allele have two copies of that allele, whereas heterozygotes have only one copy. So the frequency of an allele is the frequency of homozygotes plus half the frequency of heterozygotes. Thus, if the frequency of *A* / *A* individuals were, say, 0.36, and the frequency of *A* / *a* individuals were 0.48, the **allele frequency** of *A* would be 0.36 + 0.48/2 = 0.60. Box 17-1 gives the general form of this calculation. Table 17-1 shows the values of ***p*** and ***q***, the allele frequencies of the two alleles in the different populations.

A *measure* of genetic variation (in contrast to its *description* in terms of allele frequencies) is the amount of **heterozygosity** for a gene in a population, which is given by the total frequency of heterozygotes for the gene. If one allele has a very high frequency and all others have frequencies near zero, then there is very little heterozygosity because most individuals are homozygous for the common allele. We expect heterozygosity to be greatest when there are many alleles of a gene, all at equal frequencies. In Table 17-1, the heterozygosity of each population is simply equal to the frequency of the *M* / *N* genotype in that population.

When more than one locus is considered, there are two possible ways of calculating heterozygosity. The *S* gene

BOX 17-1

Calculation of Allele Frequencies

If $f_{A/A}$, $f_{A/a}$, and $f_{a/a}$ are the frequencies of the three genotypes at a locus with two alleles, then the frequency *p* of the *A* allele and the frequency *q* of the *a* allele can be obtained by counting alleles. Because each homozygote *A* / *A* consists only of *A* alleles, and because half the alleles of each heterozygote *A* / *a* are *A* alleles, the total frequency of *A* alleles in the population, *p*, is calculated as

$$p = f_{A/A} + \tfrac{1}{2} f_{A/a} = \text{frequency of } A$$

Similarly, the frequency *q* of the *a* allele is given by

$$q = f_{a/a} + \tfrac{1}{2} f_{A/a} = \text{frequency of } a$$

Therefore

$$p + q = f_{A/A} + f_{a/a} + f_{A/a} = 1.00$$

$$q = 1 - p$$

If there are more than two different allelic forms, the frequency for each allele is simply the frequency of its homozygote plus half the sum of the frequencies for all the heterozygotes in which it appears.

TABLE 17-2 Frequencies of Gametic Types for MNS System in Various Human Populations

Population	HAPLOTYPE *M S*	*M s*	*N S*	*N s*	HETEROZYGOSITY (*H*) Haplotype Diversity	Allelic Heterozygosity
Ainu	0.024	0.381	0.247	0.348	0.672	0.438
Ugandan	0.134	0.357	0.071	0.438	0.658	0.412
Pakistani	0.177	0.405	0.127	0.291	0.704	0.455
English	0.247	0.283	0.080	0.290	0.700	0.469
Navaho	0.185	0.702	0.062	0.051	0.467	0.286

SOURCE: A. E. Mourant, *The Distribution of the Human Blood Groups*. Blackwell Scientific Pub., 1954.

(which encodes the secretor factor, determining whether the M and N proteins are also contained in the saliva) is closely linked to the MN blood group gene in humans. Table 17-2 shows the frequencies of the four possible combinations of the two alleles for the two genes (*M S*, *M s*, *N S*, and *N s*) in various populations. One way of measuring heterozygosity is to look at **haplotypes,** which are combinations of alleles of different genes on the same chromosomal homolog. (To determine whether two different genes are associated on the same chromosomal homolog, it is necessary either to sequence the DNA from individuals or to have information about their parents or offspring.) If we consider each haplotype as a unit, as in Table 17-2, we can calculate the proportion of all individuals who carry two different haplotypic or gametic forms. This form of heterozygosity is also referred to as *haplotype diversity* or *gametic diversity*. Alternatively, we can calculate the frequency of heterozygotes at each locus separately and average them, which yields the values shown in the last column of Table 17-2 (allelic heterozygosity). Note that the haplotype diversity is always greater than the average heterozygosity of the separate loci, because an individual is a haplotypic heterozygote if *either* of the two genes is heterozygous.

Genetic variation that could serve as the basis for evolutionary change is ubiquitous. Simple variation can be observed within and between populations of any species at various levels, from external morphology down to the amino acid sequences of proteins. Indeed, genotypic variation can be directly characterized by sequencing DNA from the same gene or for intergenic regions in multiple individuals. Every species ever examined has revealed considerable genetic variation, or **polymorphism,** manifested at one or more levels of observation within populations, between populations, or both. A gene or a phenotypic trait is said to be *polymorphic* if there is more than one form of the gene, or more than one phenotype for the character, in a population. In some cases, nearly the entire population is characterized by one form of the gene or character, with rare individuals carrying an unusual variant. The extremely common form is called the **wild type,** in contrast to the rare mutants. In other cases two or more forms are common, and it is not possible to pick out a wild type.

The tasks of population genetics are to describe heritable variation quantitatively and to build a theory of evolutionary change that can use those observations to predict what changes will occur in the genetic composition of populations over time. But the structure of population genetics as an explanatory and predictive study requires that we be able to describe variation in terms of specific alleles at specific loci—alleles whose frequency of occurrence in populations can be observed. As a consequence, population genetics necessarily is concerned with simple variation: variation that can be directly observed in the genome itself or at a phenotypic level that can be directly related to allelic variation in genes. Essentially, this means that the observations of variations with which population genetic formulations can deal must be variations in

1. molecules, such as proteins and RNA, that are coded directly by genes
2. DNA structure or sequence, including chromosomal structure
3. physiological or morphological characters whose variant forms have a simple relationship to allelic variation in a particular gene—for example, flower colors, or many human diseases

There is a great deal of morphological variation in size and shape in every species, but most of it is a result of the interaction between environmental and genetic variation that is difficult to ascribe to distinct gene loci. The analysis of such variation requires the techniques of quantitative genetics (see Chapter 18) and does not allow for the kind of analysis and prediction that is possible for simple genetic variation.

It is impossible to provide an adequate picture in this text of the immense richness of even the simple genetic variation that exists in species. We can consider only a few examples of different kinds of variation to gain a sense of this diversity. Each of these examples can be multiplied many times over in other species and with other characters.

Protein Polymorphisms

Immunologic polymorphism. A number of loci in vertebrates encode antigenic specificities such as the ABO blood types. More than 40 different specificities are known in human red blood cells, and several hundred are known in domesticated cattle. Table 17-3 gives the allele frequencies for the ABO blood group locus in some very different human populations. The polymorphism of the HLA system of cellular antigens, which is implicated in tissue graft compatibility, is vastly greater. There appear to be two main loci in this system, each with five distinguishable alleles. Thus, there are $5^2 = 25$ different possible gametic types, making 25 different homozygous forms and $(25)(24)/2 = 300$ different heterozygotes possible. All of these genotypes are not phenotypically distinguishable, however, so only 121 phenotypic classes can be seen. In one study of a sample of only 100 Europeans, 53 of the 121 possible phenotypes were actually observed!

Amino acid sequence polymorphism. Studies of genetic polymorphism have been carried down to the level of the polypeptides encoded by the coding regions of the genes themselves. If there is a nonsynonymous codon change in a gene (say, GGU to GAU), the result is an amino acid substitution in the polypeptide produced at translation (in this case, aspartic acid is substituted for glycine). Variation in the amino acid sequence of a particular protein can be detected by sequencing the DNA that codes for that protein from a large number of individuals. This is the method that would be used if the one wished to know exactly which amino acids in the protein sequence were varying, but it is extremely time-consuming and expensive to carry out such DNA sequencing projects for many different protein-coding genes.

There is, however, a practical substitute for DNA sequencing that can be used if one is interested only in detecting variant forms of a protein, without knowing the particular amino acid changes involved. This method makes use of the changes in the physical properties of a protein that result when an amino acid is substituted. Five amino acids (glutamic acid, aspartic acid, arginine, lysine, and histidine) have ionizable side chains that give a protein a characteristic net charge, depending on the pH of the surrounding medium. An amino acid substitution may directly replace one of these charged amino acids, or a noncharged substitution near one of them in the polypeptide chain may affect the degree of ionization of the charged amino acid, or a substitution at the junction between two α helices may cause a slight shift in the three-dimensional packing of the folded polypeptide. In all these cases, the net charge on the polypeptide is altered because the net charge on a protein is not simply the sum of all the individual charges on its amino acids, but depends on their exposure to the liquid medium surrounding them.

The change in net charge on a variant protein can be detected by gel electrophoresis. Figure 17-3 shows the outcome of such an electrophoretic separation of variants of an esterase enzyme in *Drosophila pseudoobscura,* where each track is the protein from a different individual fly. Figure 17-4 on the next page shows a similar gel for different human hemoglobin A variants. In this case, most individuals are heterozygous for variant and normal hemoglobin A. Table 17-4 on the next page shows the frequencies of different alleles for three different enzyme-coding genes in *D. pseudoobscura* from several populations: a nearly monomorphic locus (malic dehydrogenase), a moderately polymorphic locus (α-amylase), and a highly polymorphic locus (xanthine dehydrogenase).

TABLE 17-3 Frequencies of the Alleles I^A, I^B, and i at the ABO Blood Group Locus in Various Human Populations

Population	I^A	I^B	i
Eskimo	0.333	0.026	0.641
Sioux	0.035	0.010	0.955
Belgian	0.257	0.058	0.684
Japanese	0.279	0.172	0.549
Pygmy	0.227	0.219	0.554

SOURCE: W. C. Boyd, *Genetics and the Races of Man.* D. C. Heath, 1950.

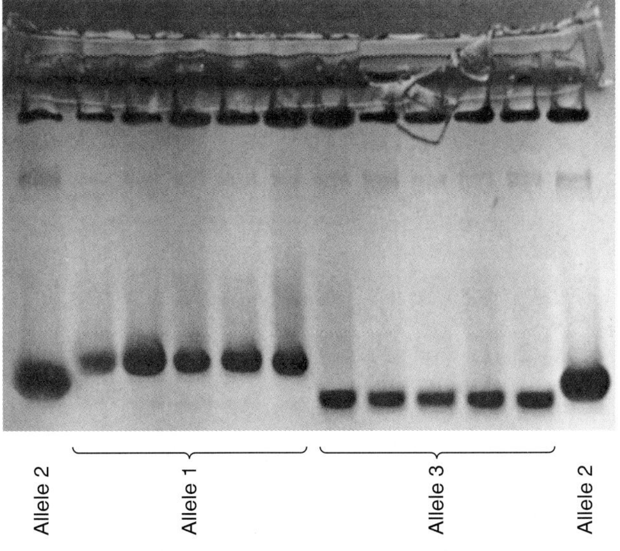

Figure 17-3 Electrophoretic gel of the proteins encoded by homozygotes for three different alleles at the *esterase-5* locus in *Drosophila pseudoobscura.* Each lane is the protein from a different individual fly. Replicate samples of the protein encoded by each allele give identical patterns, but there are repeatable differences between the patterns for different alleles. *(Richard C. Lewontin.)*

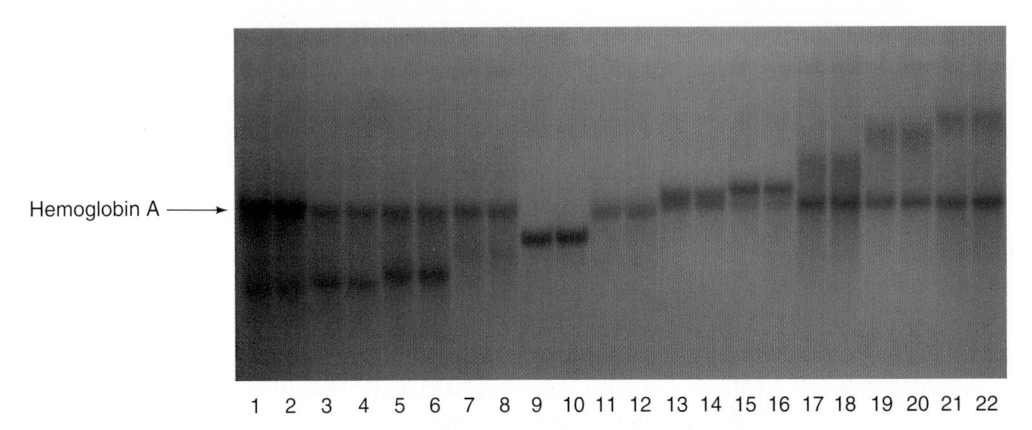

Figure 17-4 Electrophoretic gel showing normal hemoglobin A and a number of variant hemoglobin alleles. Each lane represents a different individual. One of the dark-staining bands is marked as normal hemoglobin A. The other dark-staining band seen in most of the lanes (seen most clearly in lanes 3 and 4) represents a variant hemoglobin derived from the second allele of a heterozygote. Hemoglobin A is missing from lanes 9 and 10 because these individuals are homozygous for the variant allele. *(Richard C. Lewontin.)*

The technique of protein gel electrophoresis (like DNA sequencing) differs fundamentally from other methods of genetic analysis in allowing the study of genes that are not actually varying in a population, whereas classic methods of studying genes require that these be different genotypes whose phenotypes differ. The presence of a polypeptide is prima facie evidence of a DNA sequence encoding a protein. Thus, it has been possible to ask what proportion of all structural genes in the genome of a species are polymorphic and what the average heterozygosity is in a population. Very large numbers of species have been sampled using this method, including bacteria, fungi, higher plants, vertebrates, and invertebrates. The results are remarkably consistent over species. About one-third of genes coding for proteins are detectably polymorphic at the protein level, and the average heterozygosity in a population over all loci sampled is about 10 percent. This means that scanning the genome of virtually any species would show that about 1 in every 10 structural genes in an individual is in heterozygous condition and that about one-third of all structural genes have two or more alleles segregating in any population. Thus the potential for evolution is immense. The disadvantage of the electrophoretic technique is that it can detect variation only in protein-coding regions of genes and misses the important changes in regulatory elements that underlie much of the evolution of form and function.

DNA Structure and Sequence Polymorphism

DNA analysis makes it possible to examine variation in genome structure among individuals and between species. There are three levels at which such studies can be done.

TABLE 17-4 Frequencies of Various Alleles at Three Enzyme-Coding Loci in Four Populations of *Drosophila pseudoobscura*

	POPULATION			
Allele of Locus (Enzyme Encoding)	**Berkeley**	**Mesa Verde**	**Austin**	**Bogotá**
Malic dehydrogenase				
A	0.969	0.948	0.957	1.00
B	0.031	0.052	0.043	0.00
α-Amylase				
A	0.030	0.000	0.000	0.00
B	0.290	0.211	0.125	1.00
C	0.680	0.789	0.875	0.00
Xanthine dehydrogenase				
A	0.053	0.016	0.018	0.00
B	0.074	0.073	0.036	0.00
C	0.263	0.300	0.232	0.00
D	0.600	0.581	0.661	1.00
E	0.010	0.030	0.053	0.00

SOURCE: R. C. Lewontin, *The Genetic Basis of Evolutionary Change*. Columbia University Press, 1974.

TABLE 17-5 Frequencies of Plants with Supernumerary Chromosomes and of Translocation Heterozygotes in a Population of *Clarkia elegans* from California

No Supernumeraries or Translocations	Supernumeraries	Translocations	Both Translocations and Supernumeraries
0.560	0.265	0.133	0.042

SOURCE: H. Lewis, *Evolution* 5, 1951, 142–157.

Examining variation in chromosome number and morphology provides a large-scale view of reorganizations of the genome. Studying variation at the sites recognized by restriction enzymes provides a coarse view of base-pair variation. At the finest level, methods of DNA sequencing allow variation to be observed base pair by base pair.

Chromosomal polymorphism. Although the karyotype is often regarded as a distinctive characteristic of a species, in fact numerous species are polymorphic for chromosome number and morphology. Extra chromosomes (supernumeraries), reciprocal translocations, and inversions are observed in many populations of plants, insects, and even mammals. As an example, Table 17-5 gives the frequencies of supernumerary chromosomes and translocation heterozygotes in a population of the plant *Clarkia elegans* from California. The "typical" species karyotype would be hard to identify in this plant, in which only 56 percent of the individuals lacked either supernumerary chromosomes or translocations.

Restriction-site variation. An inexpensive and rapid way to observe variation in DNA sequences is to use restriction enzymes (see Chapter 8). There are many different restriction enzymes, each of which will recognize a different base sequence and cut the DNA at the site of that sequence. The result will be two DNA fragments, whose length will be determined by the location of the restriction site in the original uncut molecule. A restriction enzyme that recognizes six-base sequences (a "six-cutter") will recognize an appropriate sequence approximately once every $4^6 = 4096$ base pairs along a DNA molecule (determined from the probability that a specific base, of which there are four, will be found at each of the six positions). If there is polymorphism in the population for one of the six bases at the recognition site, then there will be a **restriction fragment length polymorphism (RFLP)** in the population, because in one variant the enzyme will recognize and cut the DNA, whereas in the other variant it will not. A panel of, say, eight different six-cutter enzymes will then sample every $4096/8 \cong 500$ base pairs for such polymorphisms. However, when one is found, we do not know which of the six base pairs at the recognition site is polymorphic.

If we use enzymes that recognize four-base sequences ("four-cutters"), there is a recognition site every $4^4 = 256$ base pairs, so a panel of eight different enzymes can sample about once every $256/8 = 32$ base pairs. In addition to single-base-pair changes that destroy restriction enzyme recognition sites, there are insertions and deletions of stretches of DNA that occur along the DNA strand between the locations of restriction sites, and these will also cause restriction fragment lengths to vary.

A variety of different restriction enzyme studies of different regions of the X chromosome and the two large autosomes of *Drosophila melanogaster* have found between 0.1 and 1.0 percent heterozygosity per nucleotide site, with an average of 0.4 percent.

Variable number tandem repeats. Another form of DNA sequence variation that can be revealed by restriction fragment studies arises from the occurrence of multiply repeated DNA sequences. In the human genome, there are a variety of different short DNA sequences dispersed throughout the genome, each one of which is multiply repeated in a tandem row (see Chapter 9). The number of repeats may vary from a dozen to more than a hundred in different individual genomes. Such sequences are known as **variable number tandem repeats (VNTRs).** If restriction enzymes are used to cut the sequences that flank either side of such a tandem array, a fragment will be produced whose size is proportional to the number of repeated elements. These different-sized fragments will migrate at different rates in an electrophoretic gel. Although the individual copies of the repeated sequence elements are too short to allow distinguishing between, say, 64 and 68 repeats, size classes that include a range of repeat numbers *(bins)* can be established, and a population can be assayed for the frequencies of the different classes. Table 17-6 on the next page shows the results of such an assay for two different VNTRs sampled in two American Indian groups from Brazil. In one case, that of D14S1, the Karitiana are nearly homozygous, whereas the Surui are very variable; in the other case, that of D14S13, both populations are variable but have different frequency patterns.

Complete sequence variation. A ubiquitous form of genetic variation is **single-nucleotide polymorphisms (SNPs).** Studies of variation at the level of single base pairs, carried out by DNA sequencing, can provide information of two kinds. First, translating the sequences of DNA coding regions obtained from different individuals in a population, or from different species, allows the exact amino acid sequence differences in the encoded proteins to be determined. Electrophoretic studies of a protein from different individuals can show that there is variation in amino acid sequences but cannot identify how many or

TABLE 17-6 Size Class Frequencies for Two Different VNTR Sequences in Two South American Indian Groups from Brazil

	D14S1		D14S13	
Size Class	**Karitiana**	**Surui**	**Karitiana**	**Surui**
3–4	105	4	0	0
4–5	0	3	3	14
5–6	0	11	1	4
6–7	0	2	1	2
7–8	0	1	1	2
8–9	3	3	8	16
9–10	0	11	28	9
10–11	0	2	22	0
11–12	0	4	18	8
12–13	0	0	13	18
13–14	0	0	13	3
>14	0	0	0	2
	108	78	108	78

SOURCE: Data from J. Kidd and K. Kidd, *American Journal of Physical Anthropology* 81, 1992, 249.

which amino acids differ between individuals. So, when DNA sequences were obtained for the various electrophoretic variants of esterase-5 in *Drosophila pseudoobscura* (see Figure 17-3), the electrophoretic variants were found to differ from one another by an average of 8 amino acids, and the 20 different kinds of amino acids were involved in polymorphisms at about the frequency at which they were present in the protein. Such studies also show that different regions of the same protein may have different amounts of polymorphism. In the esterase-5 protein, which consists of 545 amino acids, 7 percent of the amino acid positions are polymorphic, but the last few amino acids at the carboxyl terminus of the protein are totally invariant between individuals, probably because of functional constraints on these amino acids.

DNA sequence variation can also be studied for those base pairs that do not determine or change protein sequences. Such base-pair variation can be found in 5′-flanking sequences that may be regulatory. The importance of studying variation in regulatory sequences cannot be overemphasized. If most of the evolution of shape, physiology, and behavior rests on changes in regulatory sequences, then much of the sequence variation in coding regions and in the amino acid sequences they encode is irrelevant to evolution. There is also variation in introns, in nontranscribed DNA 3′ to the gene, and in those nucleotide positions in codons (usually third positions) whose variation does not result in amino acid substitutions. These so-called *silent* or *synonymous* base-pair polymorphisms are much more common than are those that result in changes in amino acid sequences, presumably because many amino acid changes interfere with the normal function of the protein and are eliminated by natural selection. An examination of the codon translation table (see Figure 3-20) shows that if base-pair changes were random, approximately 25 percent of them would be synonymous, giving an alternative codon for the same amino acid, whereas 75 percent of random changes would change the amino acid encoded. For example, a codon that changes from AAT to AAC still encodes asparagine, but a change to ATT, ACT, AAA, AAG, AGT, TAT, CAT, or GAT—all single-base-pair changes from AAT—changes the amino acid encoded. So, if mutations of base pairs were random and if the substitution of an amino acid made no difference to function, we would expect a 3:1 ratio of amino acid replacement to silent polymorphisms. The actual ratios found in *Drosophila,* however, vary from 2:1 to 1:10. Clearly, there is a great excess of synonymous polymorphism, showing that most amino acid changes do make a difference to function and therefore are subject to natural selection.

It should not be assumed, however, that silent sites in coding sequences are entirely free from constraints. Different alternative codons for the same amino acid may differ in the speed and accuracy of their transcription, and the mRNA corresponding to different alternative codons may have different accuracy and speed of translation because of limitations on the pool of tRNAs available. Evidence for the latter effect is that the alternative synonymous codons for an amino acid are not used equally (the codon bias we discussed in Chapter 9) and that the inequality of use is much more pronounced for genes that are transcribed at a very high rate. There are also constraints on 5′ and 3′ noncoding sequences and on intron sequences. Both 5′ and 3′ noncoding sequences contain signals for transcription, and introns may contain enhancers of transcription.

Genetically Simple Morphological Variation

There are a number of morphological variations in populations that turn out to be caused by allelic variation at a single locus. For example, a subject of intense study by population geneticists is the pattern of shell color and banding in the European land snail *Cepaea nemoralis.* This widely distributed species shows regional variation in shell characteristics. Shells may be either pink or yellow, depending on two alleles of a single gene, with pink dominant to yellow. In addition, colored bands may be present or absent on the shell, as a result of the segregation of alleles of a second gene, with unbanded dominant to banded. Table 17-7 shows the variation of these phenotypes in several European colonies of the snails. (The populations are also polymorphic for the height of the shell and the number of bands, but these characteristics have a complex genetic basis.) Because shell color and banding pattern have a simple genetic basis, they can be used to test various theories about the forces governing their variation among populations.

TABLE 17-7 Frequencies of Snails *(Cepaea nemoralis)* with Different Shell Colors and Banding Patterns in Three Populations in France

	YELLOW		PINK	
Population	**Banded**	**Unbanded**	**Banded**	**Unbanded**
Guyancourt	0.440	0.040	0.337	0.183
Lonchez	0.196	0.145	0.564	0.095
Peyresourde	0.175	0.662	0.100	0.062

SOURCE: Maxime Lamotte, *Bulletin Biologique de France et Belgique* 35 (suppl.), 1951.

MESSAGE

Within species, there is great genetic variation. This variation is manifest at the morphological level, at the level of chromosome form and number, and at the level of DNA segments that may have no observable developmental effects.

THE EFFECT OF SEXUAL REPRODUCTION ON VARIATION

Before Mendel, blending inheritance was the standard model of inheritance. This model held that the mechanism of heredity was some physical blending of characteristics from both parents, as, for example, by the mixing of blood. Such a system of inheritance would have powerful consequences for population variation. Suppose that some trait (say, height) had a certain distribution in a population and that individuals mated more or less at random. If individuals of intermediate height mated with each other, they would produce only intermediate offspring if the mechanism of heredity were by some physical blending. The mating of a tall with a short individual also would produce only intermediate offspring. Only the mating of tall with tall individuals and short with short individuals would preserve extreme types. The net result of all matings would be an increase in intermediate types and a decrease in extreme types. The variation in the population would shrink, simply as a result of sexual reproduction, so that the population would be essentially uniformly intermediate in height before very many generations had passed.

The particulate nature of inheritance changes this picture completely. Because of the discrete nature of genes and the segregation of alleles at meiosis, a cross of intermediate with intermediate individuals does *not* result in all intermediate offspring. On the contrary, extreme types (homozygotes) segregate out of the cross between the two intermediate heterozygotes.

To see the consequence of meiotic transmission of particulate genes for genetic variation, consider a population in which males and females mate with each other at random with respect to some gene locus A (that is, individuals do not choose their mates preferentially with respect to the genotype at that locus). Such random mating is equivalent to mixing all the sperm and all the eggs in the population together and then matching randomly drawn sperm with randomly drawn eggs. The outcome of such a random pairing of sperm and eggs is easy to calculate. If the allele frequency of A is 0.60 in sperm and eggs, then the chance that a randomly chosen sperm and a randomly chosen egg are both A is $0.60 \times 0.60 = 0.36$. Thus, in a randomly mating population with this allele frequency, offspring will be 36 percent A/A. By the same calculation, the frequency of a/a offspring will be $0.40 \times 0.40 = 0.16$. Heterozygotes will be produced by the fusion either of an A sperm with an a egg or of an a sperm with an A egg. If gametes pair at random, then the chance of an A sperm and an a egg is 0.60×0.40, and the reverse combination has the same probability, so the frequency of heterozygous offspring is $2 \times 0.6 \times 0.4 = 0.48$. Moreover, the process of random mating does nothing to change *allele* frequencies, as can be easily checked by calculating the frequencies of the alleles A and a among the offspring in this example (using the method described in Box 17-1). Therefore, the proportions of homozygotes and heterozygotes in each successive generation will remain the same.

MESSAGE

Meiotic segregation in randomly mating populations results in an equilibrium distribution of genotypes after only one generation, so that genetic variation is maintained.

Box 17-2 on the next page gives a general form of this equilibrium result, which is called the **Hardy-Weinberg equilibrium** after two of the people who independently discovered it. (A third independent discovery was made by the Russian geneticist Sergei Chetverikov.) The equilibrium distribution of alleles in a randomly mating population is

A/A	A/a	a/a
p^2	$2pq$	q^2

BOX 17-2

The Hardy-Weinberg Equilibrium

If the frequency of allele *A* is p in both the sperm and the eggs, and the frequency of allele *a* is $q = 1 - p$, then the consequences of random unions of sperm and eggs are as shown in the adjoining diagram. The probability that both the sperm and the egg in any mating will carry *A* is

$$p \times p = p^2$$

so this will be the frequency of *A* / *A* homozygotes in the next generation. Likewise, the chance of heterozygotes *A* / *a* will be

$$(p \times q) + (q \times p) = 2pq$$

and the chance of homozygotes *a* / *a* will be

$$q \times q = q^2$$

The three genotypes, after a generation of random mating, will be at the frequencies

$$p^2 : 2pq : q^2$$

The frequency of *A* in the F_1 will not change (it will still be p) because, as the diagram shows, the frequency of *A* in the zygotes is the frequency of *A* / *A* plus half the frequency of *A* / *a*, or

$$p^2 + pq = p(p + q) = p$$

Therefore, in the F_2, the frequencies of the three genotypes will again be

$$p^2 : 2pq : q^2$$

and so on, forever. These are the Hardy-Weinberg equilibrium frequencies.

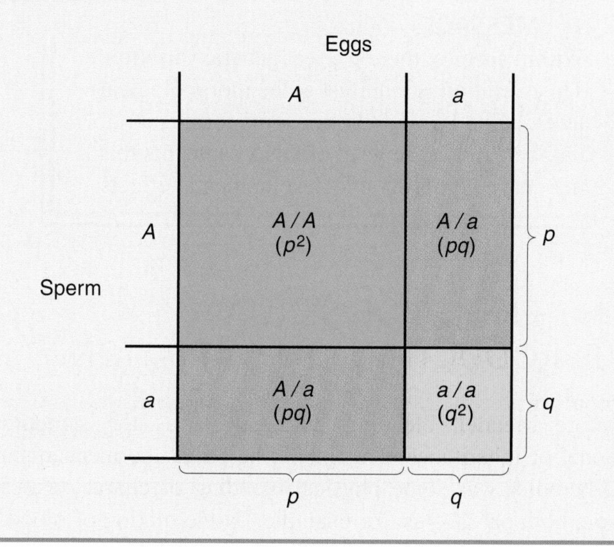

where p is the frequency of the *A* allele, q is the frequency of the *a* allele, and

$$p + q + 1$$

The Hardy-Weinberg equilibrium means that sexual reproduction does not cause a constant reduction in genetic variation in each generation; on the contrary, the amount of variation remains constant generation after generation in the absence of other disturbing forces. The equilibrium is the direct consequence of the segregation of alleles at meiosis in heterozygotes.

Numerically, the equilibrium shows that, irrespective of the particular mixture of genotypes in the parental generation, the genotypic distribution after one round of random mating is completely specified by the allele frequency p. For example, consider three hypothetical populations:

Population	$f_{A/A}$	$f_{A/a}$	$f_{a/a}$
I	0.3	0.0	0.7
II	0.2	0.2	0.6
III	0.1	0.4	0.5

The allele frequency p of *A* in the three populations is:

I $\quad p = f_{A/A} + f_{A/a} = 0.3 + 1/2(0) = 0.3$

II $\quad p = 0.2 + 1/2(0.2) = 0.3$

III $\quad p = 0.1 + 1/2(0.4) = 0.3$

So, despite their very different genotypic compositions, they have the same allele frequency. After one generation of random mating within each population, however, each of them will have the same genotypic frequencies:

$$A/A: p^2 = (0.3)^2 = 0.09$$

$$A/a: 2pq = 2(0.3)(0.7) = 0.42$$

$$a/a: q^2 = (0.7)^2 = 0.49$$

and they will remain so indefinitely.

One consequence of the Hardy-Weinberg proportions is that rare alleles are virtually never in homozygous condition. An allele with a frequency of 0.001 is present in homozygotes at a frequency of only one in a million; most

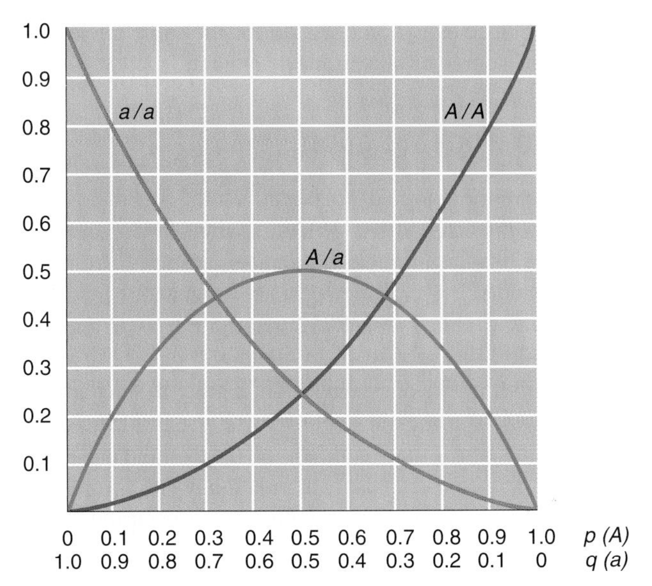

Figure 17-5 Curves showing the proportions of homozygotes *A / A*, homozygotes *a / a*, and heterozygotes *A / a* in populations with different allele frequencies if the populations are at Hardy-Weinberg equilibrium.

copies of such rare alleles are found in heterozygotes. Because two copies of any allele are found in homozygotes, but only one copy of that allele is found in each heterozygote, the relative frequency of the allele in heterozygotes (in contrast with homozygotes) is, from the Hardy-Weinberg equilibrium frequencies,

$$\frac{2pq}{2q^2} = \frac{p}{q}$$

which for $q = 0.001$ is a ratio of 999 : 1. The general relation between homozygote and heterozygote frequencies as a function of allele frequencies is shown in Figure 17-5.

In our derivation of the equilibrium, we assumed that the allele frequency p is the same in sperm and eggs. The Hardy-Weinberg equilibrium theorem does not apply to sex-linked genes if males and females start with unequal gene frequencies.

The Hardy-Weinberg equilibrium was derived on the assumption of "random mating," but we must carefully distinguish two meanings of that term. First, we may mean that individuals do not choose their mates on the basis of some heritable character. Humans mate randomly with respect to blood groups in this first sense, because they generally do not know the blood type of their prospective mates, and even if they did, it is unlikely that they would use it as a criterion for mate choice. In this first sense, random mating will occur with respect to genes that have no effect on appearance, behavior, smell, or other characteristics that directly influence mate choice.

The second sense of random mating is relevant when there is any division of a species into subgroups. If there is genetic differentiation between subgroups so that the frequencies of alleles differ from group to group, and if individuals tend to mate within their own subgroup **(endogamy),** then, with respect to the species as a whole, mating is not at random, and frequencies of genotypes will depart more or less from Hardy-Weinberg frequencies. In this sense, humans do not mate randomly, because ethnic and racial groups and geographically separated populations differ from one another in genotype frequencies, and people show high rates of endogamy not only within major races but also within local ethnic groups. Spaniards and Russians, for example, differ in their ABO blood group frequencies. Spaniards usually marry Spaniards and Russians usually marry Russians, so there is unintentional nonrandom mating with respect to ABO blood groups.

Table 17-8 shows the results of random mating in the first sense and nonrandom mating in the second sense for the MN blood group. Within Eskimo, Egyptian, Chinese, and Australian Aborigine subpopulations, people do not choose their mates by MN type, and, thus, Hardy-Weinberg equilibrium exists *within* the subpopulations. But Egyptians do not often mate with Eskimos or Australian Aborigines, so the nonrandom associations in the human species *as a whole* result in large differences in genotype frequencies from group to group. It follows that if we could take the

TABLE 17-8 Comparison between Observed Frequencies of Genotypes for the MN Blood Group Locus and the Frequencies Expected from Random Mating

	OBSERVED			EXPECTED		
Population	***M / M***	***M / N***	***N / N***	***M / M***	***M / N***	***N / N***
Eskimo	0.835	0.156	0.009	0.834	0.159	0.008
Egyptian	0.278	0.489	0.233	0.274	0.499	0.228
Chinese	0.332	0.486	0.182	0.331	0.488	0.181
Australian Aborigine	0.024	0.304	0.672	0.031	0.290	0.679

Note: The expected frequencies are computed according to the Hardy-Weinberg equilibrium, using the values of p and q computed from the observed genotypic frequencies.

human species as a whole and calculate the average allele frequency for the entire species, we would observe a departure from Hardy-Weinberg equilibrium. To perform such a calculation, however, we would need to know the population size and allele frequencies of every local population. To illustrate the effect, however, suppose we brought together an equal number of Eskimos and Australian Aborigines. From the observed genotype frequencies given in Table 17-8, we can use the method described in Box 17-1 to calculate the allele frequencies in the two subgroups and then average them to get the values for the merged group:

	$p(M)$	$q(N)$
Eskimos	0.915	0.085
Australian Aborigines	0.178	0.822
Merged average	0.546	0.454

If the merged group were really a single randomly mating population, we would expect the Hardy-Weinberg proportions given by the average allele frequencies:

$p^2\ (M/M)$	$2pq\ (M/N)$	$q^2\ (N/N)$
0.298	0.496	0.206

What we find in the merged group, however, is the averaged proportion of homozygotes and heterozygotes from the two original parental populations:

(M/M)	(M/N)	(N/N)
0.430	0.230	0.340

SOURCES OF VARIATION

For any given population, there are three sources of genetic variation: mutation, recombination, and immigration of genes. However, recombination between genes by itself does not produce variation unless there is allelic variation segregating at the different loci; otherwise there is nothing to recombine. Similarly, migration cannot provide variation if the entire species is homozygous for the same allele. Ultimately, the source of all variation is mutation.

Variation from Mutation

Mutations are the *source* of variation, but the *process* of mutation does not itself drive genetic change in populations. The rate of change in allele frequencies resulting from the mutation process is very low because spontaneous mutation rates are low (Table 17-9). The mutation rate is defined as the probability that a copy of an allele will change to some other allelic form in one generation. Thus, the increase in the frequency of a mutant allele will be the product of the mutation rate times the frequency of the nonmutant allele. Suppose, for example, that a population were completely homozygous for *A* and that mutations to *a* occurred at the rate of 1 per 100,000 gene copies. In the next generation, the frequency of *a* alleles would be only $1.0 \times 1/100{,}000 = 0.00001$, and the frequency of *A* alleles would be 0.99999. After yet another generation of mutation, the frequency of *a* would be increased by $0.99999 \times 1/100{,}000 = 0.000009$ to a new frequency of 0.000019, whereas the original *A* allele would be reduced in frequency to 0.999981. It is obvious that the rate of increase of the new allele is extremely slow, and that *it gets slower every generation,* because there are fewer copies of the old allele still left to mutate. A general formula for the change in allele frequencies by the process of mutation is given in Box 17-3.

MESSAGE

Mutation rates are so slow that mutation alone cannot account for the rapid genetic changes seen in populations and species.

If we look at the mutation process from the standpoint of the increase of a particular new allele rather than the decrease of the old form, the process is even slower. Most mutation rates that have been determined are the sum of all mutations of *A* to any mutant form with a detectable effect. Any *specific* base substitution is likely to be at least two or-

TABLE 17-9 Point Mutation Rates in Different Organisms

Organism	Gene	Mutation Rate per Generation
Bacteriophage	Host range	2.5×10^{-9}
Escherichia coli	Phage resistance	2×10^{-8}
Zea mays (corn)	*R* (color factor)	2.9×10^{-4}
	Y (yellow seeds)	2×10^{-6}
Drosophila melanogaster	Average lethal	2.6×10^{-5}

SOURCE: T. Dobzhansky, *Genetics and the Origin of Species,* 3d ed., rev. Columbia University Press, 1951.

BOX 17-3

The Effect of Mutation on Allele Frequencies

Let μ be the **mutation rate** from allele *A* to some other allele *a* (the probability that a copy of gene *A* will become *a* during the DNA replication preceding meiosis). If p_t is the frequency of the *A* allele in generation *t*, if $q_t = 1 - p_t$ is the frequency of the *a* allele in generation *t*, and if there are no other causes of allele frequency change (no natural selection, for example), then the change in allele frequency in one generation is

$$\Delta p = p_t - p_{t-1} = (p_{t-1} - \mu p_{t-1}) - p_{t-1} = -\mu p_{t-1}$$

where p_{t-1} is the frequency in the preceding generation. This tells us that the frequency of *A* decreases (and the frequency of *a* increases) by an amount that is proportional to the mutation rate μ and to the proportion *p* of all the genes that are still available to mutate. Thus Δp gets smaller as the frequency of *p* itself decreases, because there are fewer and fewer *A* alleles to mutate into *a* alleles. We can make an approximation that, after *n* generations of mutation,

$$p_n = p_0 e^{-n\mu}$$

where *e* is the base of natural logarithms.

This relation of allele frequency to number of generations is shown in the adjoining figure for $\mu = 10^{-5}$. After 10,000 generations of continued mutation of *A* to *a*,

$$p = p_0 e^{-(10^4) \times (10^5)} = p_0 e^{-0.1} = 0.904 p_0$$

If the population started with only A alleles ($p_0 = 1.0$), it would still have only 10 percent *a* alleles after 10,000 generations at this rather high mutation rate and would require 60,000 additional generations to reduce *p* to 0.5.

Even if mutation rates were doubled (say, by environmental mutagens), the rate of change would be very slow. For example, radiation levels of sufficient intensity to double the mutation rate over the reproductive lifetime of an individual human are at the limit of occupational safety regulations, and a dose of radiation sufficient to increase mutation rates by an order of magnitude would be lethal, so rapid genetic change in our species would not be one of the effects of increased radiation. Although we have many things to fear from environmental radiation pollution, turning into a species of monsters is not one of them.

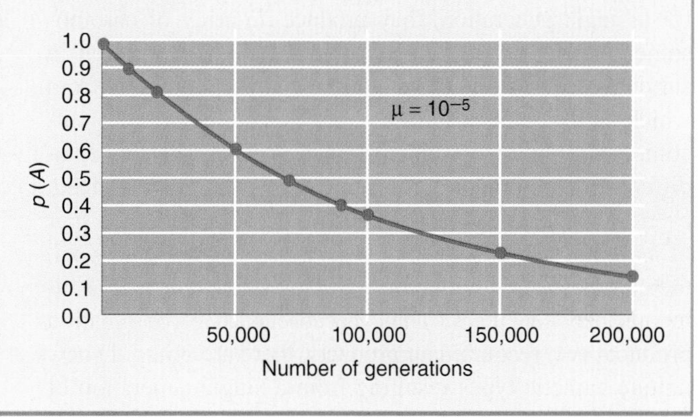

ders of magnitude lower in frequency than the sum of all such changes.

It is not possible to measure locus-specific mutation rates for continuously varying characters (see Chapter 18), but the rate of accumulation of genetic variation can be determined. For example, beginning with a completely homozygous line of *Drosophila* derived from a natural population, 1/1000 to 1/500 of the genetic variation in abdominal bristle number in the original population is restored each generation by spontaneous mutation.

Variation from Recombination

When a new mutation of a gene arises in a population, it occurs as a single event on a particular copy of a chromosome carried by some individual. But that chromosome copy also has a particular allelic composition for all the other polymorphic genes on the chromosome. So, if a mutant allele *a* arose at the *A* locus on a chromosome copy that already had the allele *b* at the *B* locus, then without recombination, all gametes carrying the *a* allele would also carry the *b* allele in future generations. The population would then contain only the original *AB* haplotype and the new *ab* haplotype that arose from the mutation. Recombination between the *A* gene and the *B* gene in the double heterozygote *AB* / *ab*, however, would produce two new haplotypes, *Ab* and *aB*.

The consequence of repeated recombination between genes is to randomize combinations of alleles of different genes. If the frequency of allele *a* at locus *A* is, say 0.2, and the allele frequency of allele *b* at locus *B* is, say, 0.4, then the frequency of *a b* would be (0.2)(0.4) = 0.08 if the combinations were randomized. This randomized condition is called **linkage equilibrium.**

But recombination between genes will not produce linkage equilibrium in a single generation if the alleles at the different genes began in nonrandom association with each other. This original association, called **linkage disequilibrium,** decays only slowly from generation to generation at a rate that is proportional to the amount of recombination between the genes. This fact can be used to find the locations of unknown genes on chromosomes and to provide evidence that some phenotypic variant is, in fact, influenced by an unknown gene. Suppose that people who suffered from some disease—say, diabetes—also turned out to carry an allele of some marker gene, which has nothing to do with insulin formation, more often than would be expected if the association between diabetes and the marker allele were random. This finding would be evidence that diabetes is influenced by some gene on the same chromosome as the marker gene and, if the linkage disequilibrium were quite strong, that the diabetes-related gene is fairly close to the marker. The existence of such a linkage disequilibrium would presumably be the accidental result of the mutational origin of the marker allele on the same chromosome copy as the diabetes-associated allele.

The creation of genetic variation by recombination can be a much faster process than its creation by mutation. When just two homologous chromosomes, carrying a variety of alleles that together confer "normal" survival, from a natural population of *Drosophila* are allowed to recombine for a single generation, they produce an array of chromosomes with 25 to 75 percent as much genetic variation in survival as was present in the entire natural population from which the parent chromosomes were sampled. This outcome is simply a consequence of the very large number of different recombinant chromosomes that can be produced, even if we take into account only single crossovers. If a pair of homologous chromosomes is heterozygous at n loci, then a crossover can take place in any one of the $n - 1$ intervals between them. Thus, because each recombination produces two recombinant products, there are $2(n - 1)$ new unique gametic types resulting from a single generation of crossing-over, even considering only single crossovers. If the heterozygous loci are well spread out along the chromosome, these new gametic types will be frequent, and considerable variation will be generated. Asexual organisms or organisms, such as bacteria, that very seldom undergo sexual recombination do not have this source of variation, so new mutations are the only way in which a change in gene combinations can be achieved. As a result, populations of asexual organisms may change more slowly than sexual organisms.

Variation from Migration

A further source of variation in populations is migration into the population of individuals from other populations with different genotype frequencies. The resulting mixed population will have an allele frequency that is somewhere intermediate between its original value and the frequency in the donor population.

Suppose, for example, that a population receives a group of migrants whose number is equal to, say, 10 percent of its population size. Then the newly formed mixed population will have an allele frequency that is a 0.90:0.10 mixture between its original allele frequency and the allele frequency of the donor population. If its original frequency of allele *A* were, say, 0.70, whereas the donor population had an *A* allele frequency of only, say, 0.40, the new mixed population would have a frequency of $0.70 \times 0.90 + 0.40 \times 0.10 = 0.67$. As shown in Box 17-4, the change in allele frequency is proportional to the difference in frequency between the recipient population and the donor population. Unlike the mutation rate, the migration rate (m) can be large, so if the difference in allele frequency between the donor and recipient population is large, the resulting change in frequency may be substantial.

We must understand *migration* as meaning any form of the introduction of genes from one population into another. So, for example, genes from Europeans have "migrated" into the population of African origin in North America steadily since the Africans were introduced there as slaves. We can determine the amount of this "migration" by looking at the frequency of an allele that is found only in Europeans but not in Africans and comparing its frequency among blacks in North America.

We can use the formula for the change in gene frequency from migration in Box 17-4 if we modify it slightly to account for the fact that several generations of admixture have taken place. If the rate of admixture has not been too great, then (to a close order of approximation) the sum of the single-generation migration rates over several generations (let's call this M) will be related to the total change in the recipient population after these several generations by

BOX 17-4

Calculating the Effect of Migration on Allele Frequency

If p_t is the frequency of an allele in a recipient population in generation t, P is the frequency of that allele in a donor population (or the average over several donor populations), and m is the proportion of the recipient population that is made up of new migrants from the donor population, then the allele frequency in the recipient population in the next generation, p_{t+1}, is the result of mixing $1 - m$ genes from the recipient population with m genes from the donor population. Thus

$$p_{t+1} = (1 - m)p_t + mP = p_t + m(P - p_t)$$

and

$$\Delta p = p_{t+1} - p_t = m(P - p_t)$$

the same expression as the one used for a single generation in Box 17-4. If P is the allele frequency in the donor population and p_0 is the original frequency among the recipient population, then

$$\Delta p_{\text{total}} = M(P - p_0)$$

so

$$M = \frac{\Delta p_{\text{total}}}{(P - p_0)}$$

For example, the Duffy blood group allele Fy^a is absent in Africa but has a frequency of 0.42 in whites from the state of Georgia. Among blacks from Georgia, the Fy^a frequency is 0.046. Therefore, the total migration of genes from whites into the black population since the introduction of slaves in the eighteenth century is

$$M = \frac{\Delta p_{\text{total}}}{(P - p)} = \frac{0.046 - 0}{0.42 - 0} = 0.1095$$

That is, on average, about 11 percent of the alleles of Americans of African ancestry in Georgia have been derived from a European ancestor. This is only an average, however, and different individuals have different proportions of European and African ancestry. When the same analysis is carried out on American blacks from Oakland (California) and Detroit, M is 0.22 and 0.26, respectively, showing either greater admixture rates in these cities than in Georgia or differential movement into these cities by American blacks who have more European ancestry. In any case, the genetic variation at the Fy locus has been increased by this admixture.

Nonrandom Mating

Random mating with respect to a genetic locus is common within populations, but it is not universal. Two kinds of deviation from random mating must be distinguished. First, individuals may mate with each other nonrandomly because of their degree of common ancestry—that is, their degree of genetic relatedness. If relatives mate with each other more often than would occur by pure chance, then the population is said to be undergoing **inbreeding.** If mating between relatives is less common than would occur by chance, then the population is said to be undergoing **enforced outbreeding,** or **negative inbreeding.**

Second, individuals may choose each other as mates not because of their degree of genetic relatedness but because of their degree of resemblance to each other in some phenotypic trait. Bias toward mating of like with like is called **positive assortative mating.** Bias toward mating with unlike partners is called **negative assortative mating.** Assortative mating is never complete, so in any population some matings will be random and some the result of assortative mating. Assortative mating for some traits is common. In humans, for example, there is a positive assortative mating bias for skin color and height.

Inbreeding and assortative mating are not the same. Close relatives resemble each other, on average, more than unrelated individuals do, but not necessarily for any particular phenotypic trait. So inbreeding can result in the mating of quite dissimilar individuals. On the other hand, individuals who resemble each other for some trait may do so because they are relatives, but unrelated individuals also may have such resemblances. Brothers and sisters, for example, do not all have the same eye color, and blue-eyed people are not all related to one another.

An important difference between assortative mating and inbreeding is that the former is specific to a particular phenotype, whereas the latter applies to the entire genome. Unrelated individuals may mate assortatively with respect to height but at random with respect to blood group. Cousins, on the other hand, resemble each other genetically to the same degree, on average, at all loci.

Under both positive assortative mating and inbreeding, the consequence for population structure is the same: there is an increase in homozygosity above the level predicted by the Hardy-Weinberg equilibrium. If two individuals are related to each other, they have at least one common ancestor. Thus, there is some chance that an allele carried by one of them and an allele carried by the other are both descended from the identical DNA molecule. The result is that there is an extra chance of this **homozygosity by descent** that must be added to the chance of homozygosity $(p^2 + q^2)$ that arises from the random mating of unrelated individuals. The probability of this extra homozygosity by descent is called the **inbreeding coefficient** (F). Figure 17-6 and Box 17-5, both on the next page, illustrate the calculation of the probability of homozygosity by descent.

Individuals I and II in Figure 17-6 are full sibs because they share both parents. We shall label each allele in the parents uniquely to keep track of them. Individuals I and II mate to produce individual III. Suppose individual I is A_1 / A_3 and the gamete that it contributes to III contains the allele A_1. We can calculate the probability that the gamete produced by II is also A_1. The chance is 1/2 that II has received A_1 from its father, and, if it has, the chance is 1/2 that II will pass A_1 on to the gamete in question. Thus, the probability that III will receive A_1 from II is $1/2 \times 1/2 = 1/4$, and this is the chance that III—the product of a full-sib mating—will be homozygous A_1 / A_1 by descent from the original ancestor.

Such close inbreeding can have deleterious consequences. Let's consider a rare deleterious allele a that, when homozygous, causes a metabolic disorder. If the frequency of the a allele in the population is p, then the probability that a random couple will produce a homozygous offspring is only p^2 (from the Hardy-Weinberg equilibrium). Thus, if p is, say, 1/1000, the frequency of homozygotes

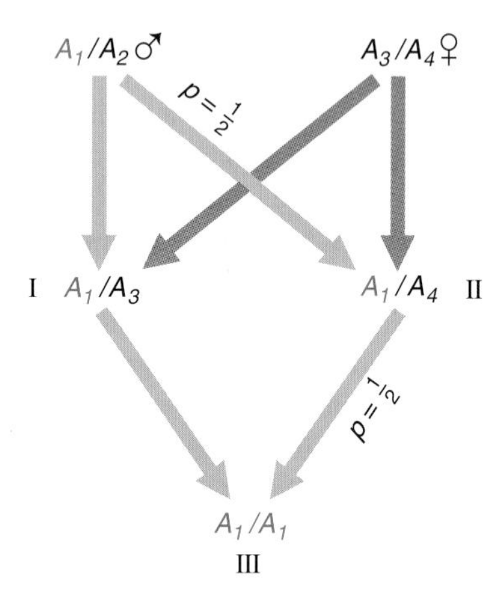

Figure 17-6 Calculation of homozygosity by descent for an offspring (III) of a brother-sister (I-II) mating. Assume that individual III has received one copy of A_1 from its grandfather through individual I. The probability that II has received A_1 from its father is 1/2; if it has, the probability that II will pass A_1 on to III is 1/2. Thus, the probability that III will receive an A_1 from II is $1/2 \times 1/2 = 1/4$.

will be 1 in 1,000,000. Now suppose instead that the couple are brother and sister. If one of their common parents is a heterozygote for *a*, they may both receive it and may both pass it on to their offspring. As the calculation in Box 17-5 shows, the rarer the gene, the worse their *relative* risk—compared with that of a random couple—of producing a defective offspring. For matings between more distant relatives, the chance of homozygosity by descent is less, but still substantial. For first cousins, for example, the relative risk is $1/16p$ compared with random mating.

Systematic inbreeding between close relatives eventually leads to complete homozygosity of the population, at a rate depending on the degree of relationship. If two alleles are present in an inbreeding population, one will eventually be lost, and the other will have a frequency of 1.0—in other words, it will become **fixed.** Which allele will become fixed is a matter of chance. Suppose, for example, that several groups of individuals are taken from a population and subjected to inbreeding. If, in the original population from which the inbred lines are taken, allele *A* has frequency *p* and allele *a* has frequency $q = 1 - p$, then a proportion *p* of the homozygous lines established by inbreeding will be homozygous *A / A* and a proportion *q* of the lines will be *a / a*. Inbreeding takes the genetic variation present *within* the original population and converts it into variation *between* homozygous inbred lines sampled from the population (Figure 17-7).

The same consequences may occur in a population that is established from only a few individuals. Suppose that a population is founded by some small number of individuals, who then mate at random to produce the next generation, and no further immigration into the population occurs. (These conditions would apply to the European rabbits now living in Australia, which are probably descended from a single introduction of a few animals in the nineteenth century.) In later generations, everyone will be related to everyone else, because their family trees will have common ancestors here and there in their pedigrees. Such a population is then inbred, in the sense that there is some probability of a gene's being homozygous by descent. Because the population is, of necessity, finite in size, some of the originally introduced family lines will become extinct in every generation, just as family names disappear in a human population because, by chance, no male offspring are left. As original family lines disappear, the population comes to be made up of descendants of fewer and fewer of the original founder

BOX 17-5

The Effect of Mating between Close Relatives on Homozygosity

Let's assume that allele *a* is rare enough in a population that it is carried in heterozygous condition by only one of the parents of a brother and sister. The probability of a homozygous *a / a* offspring from a mating between that brother and sister is

probability that one or the other parent is *A / a*
× probability that *a* is passed to male sib
× probability that *a* is passed to female sib
× probability of a homozygous *a / a* offspring from *A / a* × *A / a*

$$= (2pq + 2pq) \times 1/2 \times 1/2 \times 1/4$$
$$= pq/4$$

We assume that the chance that both parents of the sibling pair are *A / a* is negligible. If *p* is very small, then *q* is nearly 1.0, and the chance of a homozygous offspring is close to *p*/4. For example, if $p = 1/1000$, there is 1 chance in 4000 of a homozygous child—a small chance, but much greater than the one-in-a-million (p^2) chance from a random mating. In general, for full sibs, the ratio of probabilities of homozygosity will be

$$\frac{p/4}{p^2} = \frac{1}{4p}$$

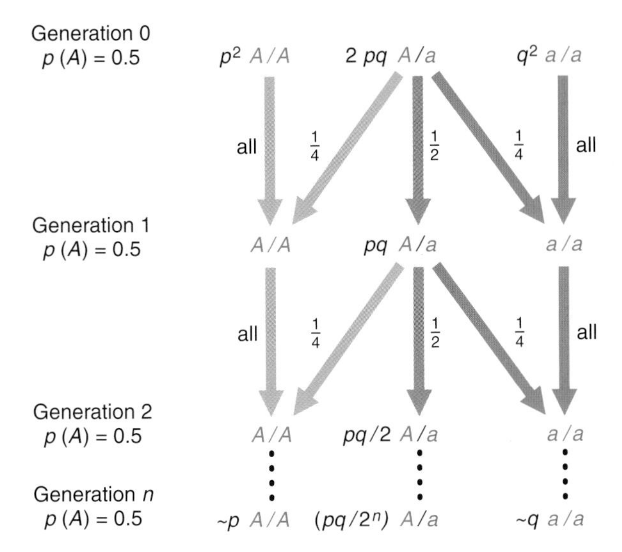

Figure 17-7 Repeated generations of inbreeding (or self-fertilization) will eventually split a heterozygous population into a series of completely homozygous lines. The frequency of A / A lines among the homozygous lines will be equal to the frequency of allele A in the original heterozygous population (p), while the frequency of a / a lines will be equal to the original frequency of a (q).

individuals, and all the members of the population become more and more likely to carry the same alleles by descent. In other words, the inbreeding coefficient F increases, and the heterozygosity decreases, over time until finally F reaches 1.00 and heterozygosity reaches 0.

The rate of loss of heterozygosity per generation in such a closed, finite, randomly mating population is inversely proportional to the total number ($2N$) of haploid genomes, where N is the number of diploid individuals in the population. In each generation, $1/2N$ of the remaining heterozygosity is lost, so

$$H_t = H_0 (1 - 1/2N)^t \cong H_0 e^{-t/2N}$$

where H_t and H_0 are the proportions of heterozygotes in the tth and original generations, respectively. As the number t of generations becomes very large, H_t approaches zero.

SELECTION

So far in this chapter, we have considered changes in a population arising from the forces of mutation, migration, recombination, and nonrandom mating. But these changes are random with respect to the way in which organisms make a living in the environments in which they live. Changes in a species in response to a changing environment occur because the different genotypes produced by mutation and recombination have different abilities to survive and reproduce. These differential rates of survival and reproduction are what is meant by **selection,** and the process of selection results in changes in the frequencies of the various genotypes in the population. Darwin called the process of differential survival and reproduction of different types **natural selection,** by analogy with the **artificial selection** carried out by animal and plant breeders when they deliberately select some individuals of a preferred type.

The relative probability of survival and rate of reproduction of a phenotype or genotype is now called its **Darwinian fitness.** Although geneticists sometimes speak of the fitness of an individual, this Darwinian concept of fitness really applies to the average survival probability and reproductive rate of individuals in some phenotypic or genotypic class. Because of chance events in the life histories of individuals, even two organisms with identical genotypes and identical environments will differ in their survival and reproductive rates. It is the fitness of a genotype on average over all its possessors that matters.

Fitness is a consequence of the relation between the phenotype of an organism and the environment in which the organism lives, so the same genotype will have different fitnesses in different environments. In part, this is because exposure to different environments during development may result in different phenotypes for the same genotype. But even if the phenotype is the same in every environment, the success of the organism depends on the environment. Having webbed feet is fine for paddling in water but a positive disadvantage for walking on land, as a few moments spent observing a duck will reveal. No genotype is unconditionally superior in fitness to all others in all environments.

Furthermore, the environment is not a fixed set of circumstances that is experienced passively by an organism. The environment of an organism is defined by the activities of the organism itself. For example, dry grass is part of the environment of a junco, so juncos that are the most efficient at gathering it may waste less energy in nest building and thus have a higher reproductive fitness. But dry grass is part of a junco's environment *because juncos gather it to make nests.* The rocks among which the grass grows are not part of the junco's environment, although the rocks are physically present there. But the rocks are part of the environment of thrushes, because these birds use the rocks to break open snails. Moreover, the environment that is defined by the life activities of an organism evolves as a result of those activities. The structure of soil, which in part determines the kinds of plants that will grow on it, is altered by the growth of those very plants. On a larger scale, as primitive plants evolved photosynthesis, they changed the Earth's atmosphere from one that had essentially no free oxygen and a high concentration of carbon dioxide to the atmosphere that we know today, which contains 21 percent oxygen and only 0.03 percent carbon dioxide. Plants that evolve today must do so in an environment created by the evolution of their own ancestors. Environment is both the cause and the result of the evolution of organisms.

Darwinian fitness is not to be confused with "physical fitness" in the everyday sense of the term, although they may be related. No matter how strong, healthy, and mentally alert the possessor of a genotype may be, that genotype has a fitness of zero if for some reason its possessors leave no offspring. Thus such statements as "the unfit are outreproducing the fit" are meaningless in population genetics. The fitness of a genotype is a consequence of all the phenotypic effects of the genes involved. Thus, an allele that doubles the fecundity of its carriers but at the same time reduces the average life span of its possessors by 10 percent will have greater fitness than its alternatives, despite its life-shortening property. The most common example is parental care. An adult bird that expends a great deal of its energy gathering food for its young will have a lower probability of survival than one that keeps all the food for itself. But a totally selfish bird will leave no offspring, because its young cannot fend for themselves. As a consequence, parental care is favored by natural selection.

Two Forms of Selection

Because the differences in reproduction and survival between genotypes depend on the environments in which the genotypes live and develop, and because organisms may alter their own environments, there are two fundamentally different forms of selection. In the simplest case, the fitness of an individual does not depend on the composition of the population; rather, it is a fixed property of the individual's phenotype and the external physical environment. For example, the relative ability of two plants that live at the edge of the desert to get sufficient water will depend on how deep their roots grow and on how much water they lose through their leaf surfaces. These characteristics are a consequence of their developmental patterns and are not sensitive to the composition of the population in which they live. The fitness of a genotype in such a case does not depend on how rare or how common it is in the population. Under these circumstances, fitness is **frequency independent.**

In contrast, consider organisms that are competing to catch prey or to avoid being captured by a predator. Under these circumstances, the relative abundances of two different genotypes will affect their relative fitnesses. An example is Müllerian mimicry in butterflies. Some species of brightly colored butterflies (such as monarchs and viceroys) are distasteful to birds, which learn, after a few trials, to avoid attacking butterflies with that color pattern. Therefore, if two butterfly species differ in pattern, there will be selection to make them more similar because both will be protected from predation and they will share the burden of the birds' initial learning period. The less frequent pattern will be at a disadvantage with respect to the more frequent one, because birds will learn less often to avoid a pattern they encounter more rarely. Within a species, rarer patterns will be selected against for the same reason. The rarer the pattern, the greater is its selective disadvantage, because birds will be unlikely to have had a prior experience with a low-frequency pattern and therefore will not avoid it. This selection to blend in with the crowd is an example of **frequency-dependent fitness,** in which the fitness of a type changes as it becomes more or less common in the population.

For reasons of mathematical convenience, most models of natural selection are based on frequency-independent fitness. In fact, however, a very large number of selective processes (perhaps most) are frequency dependent. The kinetics of the change in allele frequency depends on the exact form of frequency dependence, and for that reason alone, it is difficult to make any generalizations about these processes. The result of *positive* frequency dependence (in which the fitness of a genotype increases with increasing frequency) is quite different from that of *negative* frequency dependence (in which fitness declines with increasing frequency). For the sake of simplicity, and as an illustration of the main qualitative features of selection, we deal only with models of frequency-independent selection in this chapter, but convenience should not be confused with reality.

Measuring Fitness Differences

For the most part, the differential fitness of different genotypes can be most easily measured when the genotypes differ at many loci. In very few cases—such as laboratory mutants, horticultural varieties, and organisms with major metabolic disorders—does an allelic substitution at a single locus make enough difference to the phenotype to be reflected in measurable fitness differences.

Figure 17-8 shows the probability of survival from egg to adult—that is, the **viability**—of a number of different lines made homozygous for the second chromosome in *D. pseudoobscura* at three different temperatures. As is generally the case, the fitness (or, in this case, viability, which is a component of the total fitness) of each genotype is different in different environments. The homozygous state is lethal, or nearly so, in a few cases at all three temperatures, whereas a few genotypes have consistently high viability. Most genotypes, however, are not consistent in viability between temperatures, and no genotype is unconditionally the most fit at all temperatures. The fitnesses of these chromosomal homozygotes were not measured in competition with each other; all were measured in competition with a common standard genotype, so we do not know whether they are frequency dependent.

Examples of clear-cut fitness differences associated with single-gene substitutions are the many "inborn errors of metabolism" in which a recessive allele interferes with a metabolic pathway and is lethal to homozygotes. As we saw in Chapter 3, one example in humans is phenylketonuria, in which tissue degeneration is the result of the accumulation of a toxic intermediate in the pathway of phen-

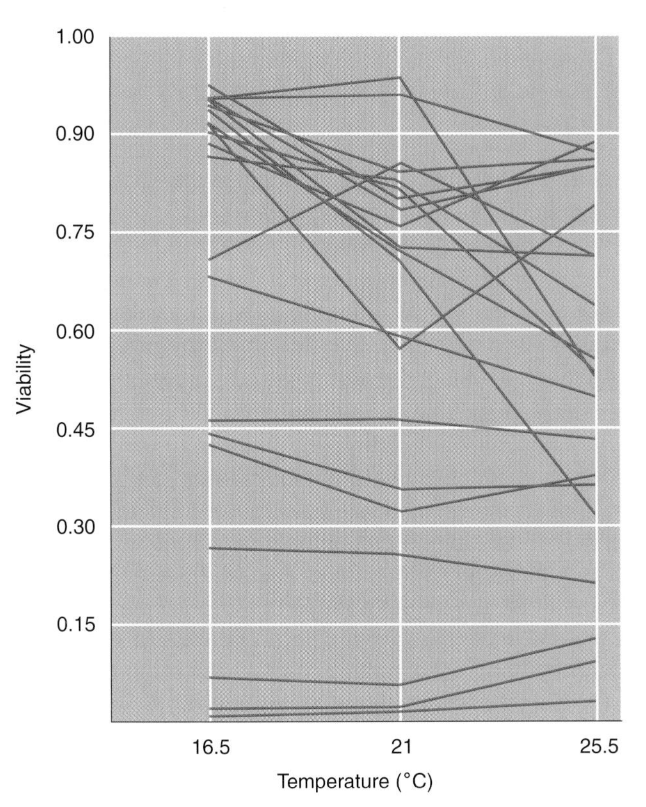

Figure 17-8 Viabilities of various chromosomal homozygotes of *Drosophila pseudoobscura* at three different temperatures.

ylalanine metabolism. In most cases, however, it has not been possible to measure fitness differences for single-locus polymorphisms, even in humans, even though so much is known about our physiology, metabolism, and reproductive rates. The evidence for differential net fitness for different ABO or MN blood types, for example, is shaky at best. The extensive enzyme polymorphism present in all sexually reproducing species has for the most part not been connected with measurable fitness differences, although in *Drosophila,* clear-cut differences in the fitness of different genotypes have been demonstrated in the laboratory for a few loci such as those encoding α-amylase and alcohol dehydrogenase. If fitness differences among genotypes do exist for most polymorphisms, these differences are too small to measure by observations of natural populations or by experiments that are practical to perform.

How Selection Works

The simplest way to see the effect of selection is to consider an allele, *a*, that is completely lethal before reproductive age in homozygous condition (such as the allele that leads to Tay-Sachs disease). Suppose that, in some generation, the allele frequency of this gene is 0.10. In a randomly mating population, the proportions of the three genotypes after fertilization forming the zygotes that begin the next generation will be

A/A	A/a	a/a
0.81	0.18	0.01

At reproductive age, however, the homozygotes a/a will have already died, leaving the proportions at this stage as

A/A	A/a	a/a
0.81	0.18	0.00

But these proportions add up to only 0.99, because only 99 percent of the population is still alive. For the actual surviving reproducing population, the proportions must be recalculated by dividing by 0.99 so that the total proportions add up to 1.00. After this readjustment, we have

A/A	A/a	a/a
0.818	0.182	0.00

The frequency of the lethal *a* allele among the gametes produced by these survivors is then

$$0.00 + 0.182/2 = 0.091$$

and the change in allele frequency in one generation, expressed as the new value minus the old one, has been

$$0.091 - 0.100 = -0.019$$

We can repeat this calculation in each successive generation to obtain the predicted frequencies of the lethal and normal alleles in a succession of future generations. The same kind of calculation can be carried out if each genotype is not simply lethal or normal, but rather has some relative probability of survival to reproductive age. This general calculation is shown in Box 17-6 on the next page. After one generation of selection, the new value of the frequency of A (p') is equal to the old value (p) multiplied by the ratio of the mean fitness of A alleles, $\overline{W}_A$, to the mean fitness of the whole population, $\overline{W}$. If the fitness of A alleles is greater than the average fitness of all alleles, then $\overline{W}_A / \overline{W}$ will be greater than unity and p' will be larger than p. Thus, the allele A increases in the population. Conversely, if $\overline{W}_A / \overline{W}$ is less than unity, A decreases. But the mean fitness of the population ($\overline{W}$) is the average fitness of the A alleles and of the a alleles. So if $\overline{W}_A$ is greater than the mean fitness of the population, it must be greater than $\overline{W}_a$, the mean fitness of a alleles. Thus the allele with the higher mean fitness increases in frequency.

It should be noted that the fitnesses $W_{A/A}$, $W_{A/a}$, and $W_{a/a}$ may be expressed as absolute probabilities of survival and absolute reproductive rates, or they may all be rescaled relative to one of the fitnesses, which is given the standard value of 1.0. This rescaling has absolutely no effect on the formula for p' because it cancels out in the numerator and denominator.

BOX 17-6

The Effect of Selection on Allele Frequencies

Suppose that a population is mating at random with respect to a given locus that has two alleles, and that the population is so large that (for the moment) we can ignore inbreeding. Just after the eggs have been fertilized, the genotypes of the zygotes will be in Hardy-Weinberg equilibrium:

Genotype	A/A	A/a	a/a
Frequency	p^2	$2pq$	q^2

and

$$p^2 + 2pq + q^2 = (p + q)^2 = 1.0$$

where p is the frequency of A.

Further, suppose that the three genotypes have the relative probabilities of survival to adulthood (viabilities) of $W_{A/A}$, $W_{A/a}$, and $W_{a/a}$. (For the sake of simplicity, we shall assume that all fitness differences are differences in survivorship between the fertilized egg and the adult stage. Differences in fertility give rise to much more complex mathematical formulations.) Among the progeny, once they have reached adulthood, the frequencies will be

Genotype	A/A	A/a	a/a
Frequency	$p^2W_{A/A}$	$2pqW_{A/a}$	$q^2W_{a/a}$

These adjusted frequencies do not add up to unity because the Ws are all fractions smaller than 1. However, we can readjust them so that they do, without changing their relation to each other, by dividing each frequency by the sum of the frequencies after selection ($\overline{W}$):

$$\overline{W} = p^2W_{A/A} + 2pqW_{A/a} + q^2W_{a/a}$$

So defined, $\overline{W}$ is called the **mean fitness** of the population because it is, indeed, the mean of the fitnesses of all individuals in the population. After this adjustment, we have

Genotype	A/A	A/a	a/a
Frequency	$p^2\dfrac{W_{A/A}}{\overline{W}}$	$2pq\dfrac{W_{A/a}}{\overline{W}}$	$q^2\dfrac{W_{a/a}}{\overline{W}}$

We can now determine the frequency p' of the allele A in the next generation by summing up genes:

$$p' = A/A + (1/2)\,A/a = p^2\frac{W_{A/A}}{\overline{W}} + \frac{pqW_{A/a}}{\overline{W}}$$

$$= p\,\frac{pW_{A/A} + qW_{A/a}}{\overline{W}}$$

Finally, we note that the expression $pW_{A/A} + qW_{A/a}$ is the mean fitness of A alleles because, from the Hardy-Weinberg frequencies, a proportion p of all A alleles are present in homozygotes with another A, in which case they have a fitness of $W_{A/A}$, whereas a proportion q of all the A alleles are present in heterozygotes with a and have a fitness of $W_{A/a}$. Using $\overline{W}_A$ to denote $pW_{A/A} + qW_{A/a}$, the mean fitness of the allele A, yields the final new allele frequency:

$$p' = p\frac{\overline{W}_A}{\overline{W}}$$

An alternative way to look at the process of selection is to solve for the *change* in allele frequencies in one generation:

$$\Delta p = p' - p = \frac{p\overline{W}_A}{\overline{W}} - p = \frac{p(\overline{W}_A - \overline{W})}{\overline{W}}$$

But $\overline{W}$, the mean fitness of the population, is the average of the allele fitnesses $\overline{W}_A$ and $\overline{W}_a$, so

$$\overline{W} = p\overline{W}_A + q\overline{W}_a$$

where W_a is the mean fitness of a alleles. Substituting this expression for $\overline{W}$ in the formula for Δp and remembering that $q = 1 - p$, we obtain (after some algebraic manipulation)

$$\Delta p = \frac{pq(\overline{W}_A - \overline{W}_a)}{\overline{W}}$$

MESSAGE

As a result of selection, the allele with the highest mean fitness relative to the mean fitnesses of other alleles increases in frequency in the population.

An increase in the frequency of the allele with the higher fitness means that the average fitness of the population as a whole increases, so selection can also be described as a process that *increases mean fitness.* This rule is strictly true only for frequency-independent genotypic fitnesses, but it is close enough to a general rule to be used as a fruitful generalization. This maximization of fitness does not necessarily lead to any optimal property for the species as a whole, however, because fitnesses are defined only relative to one another within a population. It is relative (not absolute) fitness that is increased by selection. The population does not necessarily become larger or grow faster, nor is it less likely to become extinct. For example, suppose that an allele causes its carriers to lay more eggs than do other genotypes in the population. This higher-fecundity allele will increase in the population. But the population size at the adult stage may depend on the total food supply available to the immature stages, so the higher fecundity of the population may not result in an increase in the total population size, but only an increase in the number of immature individuals that starve to death before adulthood.

Rate of Change in Allele Frequencies

The general expression for the change in allele frequencies derived in Box 17-6 is particularly illuminating. It says that Δp will be positive (*A* will increase) if the mean fitness of *A* alleles is greater than the mean fitness of *a* alleles, as we saw before. But it also shows that the speed of the change depends not only on the difference in fitness between the two alleles, but also on the factor pq, which is proportional to the frequency of heterozygotes ($2pq$). For a given difference in fitness of alleles, their frequency will change most rapidly when the alleles *A* and *a* are at intermediate frequencies, so that pq is large. If p is near 0 or 1 (that is, if *A* or *a* is nearly fixed at a frequency of 0 or 1), then pq is nearly 0, and selection will proceed very slowly.

Figure 17-9 shows the S-shaped curve that represents the course of selection of a new favorable allele *A* that has recently entered a population of homozygotes *a* / *a*. At first, the change in frequency is very small because p is still close to 0. Then it accelerates as *A* becomes more frequent, but it slows down again as *A* takes over and *a* becomes rare (q approaches 0). This is precisely what is expected from a selection process. When most of the population is of one type, there is nothing to select. For change by natural selection to occur, there must be genetic variation; the more variation, the faster the process.

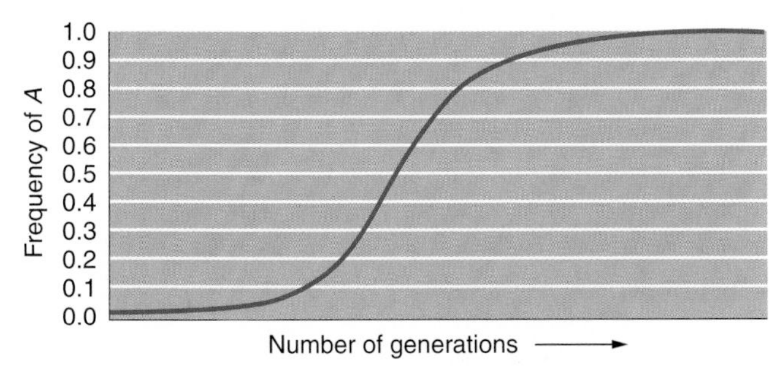

Figure 17-9 The pattern of increasing frequency of a new favorable allele *A* that has entered a population of *a* / *a* homozygotes.

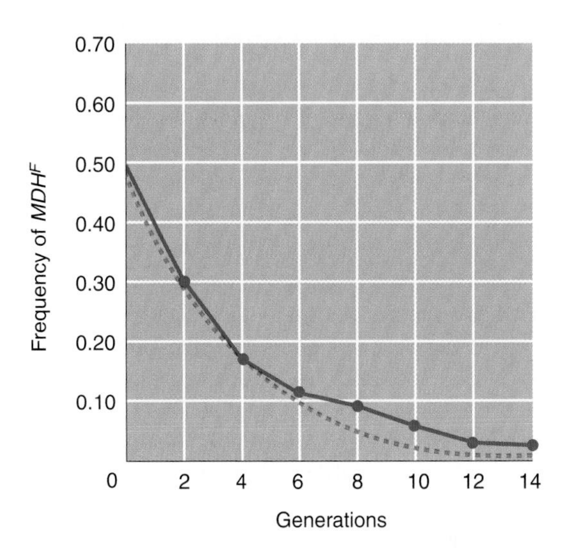

Figure 17-10 The loss of allele MDH^F at the malic dehydrogenase locus due to selection in a laboratory population of *Drosophila melanogaster*. The red dashed line shows the theoretical curve of change computed for fitnesses $W_{A/A} = 1.0$, $W_{A/a} = 0.75$, $W_{a/a} = 0.4$; the blue line shows the observed changes. *(From R. C. Lewontin,* The Genetic Basis of Evolutionary Change. *© 1974 by Columbia University Press. Data courtesy of E. Berger.)*

One consequence of the dynamics shown in Figure 17-9 is that it is extremely difficult to reduce significantly the frequency of an allele that is already rare in a population. Thus, eugenic programs designed to eliminate deleterious recessive alleles from human populations by preventing the reproduction of affected persons do not work. Of course, if all heterozygotes could be prevented from reproducing as well, the alleles could be eliminated (except for new mutations) in a single generation. Because every human is heterozygous for a number of different deleterious alleles, however, no one would be allowed to reproduce.

When alternative alleles are not rare, selection can cause rapid changes in allele frequencies. Figure 17-10 shows the course of elimination of a malic dehydrogenase allele that had an initial frequency of 0.5 in a laboratory population of *D. melanogaster*. The fitnesses in this case are

$$W_{A/A} = 1.0 \qquad W_{A/a} = 0.75 \qquad W_{a/a} = 0.40$$

The frequency of allele *a* declines rapidly but is not reduced to 0, and any further reduction in its frequency would require longer and longer times.

MESSAGE

Selection is dependent on genetic variation. Unless alternative alleles are present in intermediate frequencies, selection (especially against recessives) is quite slow.

BALANCED POLYMORPHISM

Let's reexamine the general formula for allele frequency change (see Box 17-6):

$$\Delta p = pq \frac{\overline{W}_A - \overline{W}_a}{\overline{W}}$$

Under what conditions will the process stop? That is, when is $\Delta p = 0$? Two answers are: when $p = 0$ or when $q = 0$ (that is, when either allele *A* or allele *a* has been eliminated from the population). One of these events will eventually occur if $\overline{W}_A - \overline{W}_a$ is consistently positive or negative, so that Δp is always positive or negative irrespective of the value of *p*.

The condition for such unidirectional selection is that the fitness of the heterozygote be somewhere between the fitnesses of the two homozygotes. If *A* / *A* homozygotes are the most fit, then *A* alleles will be more fit than *a* alleles in both the heterozygous and the homozygous conditions. Then the mean allelic fitness of *A*, $\overline{W}_A$, will be greater than the mean allelic fitness of *a*, $\overline{W}_a$, no matter what the frequencies of the genotypes may be. In this case, $\overline{W}_A - \overline{W}_a$ is positive, and *A* always increases until it reaches $p = 1$. If, on the other hand, *a* / *a* homozygotes are the most fit, then $\overline{W}_A - \overline{W}_a$ is negative, irrespective of allele frequencies, and *a* always increases until it reaches $q = 1$. But there is another possibility for $\Delta p = 0$, even when *p* or *q* is not 0:

$$\overline{W}_A = \overline{W}_a$$

which can occur if the heterozygote is not intermediate between the homozygotes but has a fitness that is more extreme than either homozygote. In this case, selection will lead to an intermediate equilibrium allele frequency, symbolized by $\hat{p}$ as calculated in Box 17-7.

There are, in fact, two qualitatively different possibilities for $\hat{p}$. One possibility is that $\hat{p}$ is an *unstable* equilibrium. There will be no change in allele frequencies if the population has exactly this value of *p*, but the frequencies will move *away* from the equilibrium (toward $p = 0$ or $p = 1$) if the slightest perturbation of frequencies occurs by chance. This unstable equilibrium will exist when the heterozygote is *lower* in fitness than either homozygote; such a condition is an example of **underdominance.**

The alternative possibility is a *stable* equilibrium, or **balanced polymorphism,** in which slight perturbations from the value of $\hat{p}$ will result in a return to $\hat{p}$. The condition for this balance is that the heterozygote be *greater* in fitness than either homozygote—a condition termed **overdominance.** (The mathematical proof of these conditions for stability is beyond the scope of this book.)

The best-studied cases of balanced polymorphism in nature and in the laboratory are the chromosomal inversion polymorphisms found in several species of *Drosophila.* Figure 17-11 shows the course of frequency change for the inversion *ST* (Standard) in competition with the alternative chromosomal type *CH* (Chiricahua) in a laboratory population of *D. pseudoobscura.* The inversions *ST* and *CH* are part of a chromosomal polymorphism found in natural populations of this species (see Figure 11-32 for some exam-

BOX 17-7

Natural Selection Leading to Equilibrium of Allele Frequencies

In Box 17-6, the average fitness of an allele is defined as the average of all the fitnesses of the genotypes that carry that allele:

$$\overline{W}_A = (pW_{A/A} + qW_{A/a})$$

$$\overline{W}_a = (pW_{A/a} + qW_{a/a})$$

Suppose that the fitness of the heterozygote, $W_{A/a}$, is greater than the fitnesses of both homozygotes, $W_{A/A}$ and $W_{a/a}$. When *p* is small, most of the *A* alleles are in heterozygotes, so the mean fitness of *A* alleles is large, whereas most of the *a* alleles are in homozygotes, and therefore the mean fitness of *a* alleles is small. The consequence is that when *p* is small, $\overline{W}_A > \overline{W}_a$. Conversely, when *p* is large, most of the *A* alleles are in homozygous condition, whereas most of the *a* alleles are in heterozygotes. In this range of *p*, then, $\overline{W}_A < \overline{W}_a$. Just between those ranges is a value of *p* (denoted by $\hat{p}$) for which the mean fitnesses of the two alleles are equal. A little algebraic manipulation of

$$\overline{W}_A - \overline{W}_a = 0 = (\hat{p}W_{A/A} + \hat{q}W_{A/a}) - (\hat{p}W_{A/a} + \hat{q}W_{a/a})$$

gives us the solution for the equilibrium value $\hat{p}$:

$$\hat{p} = \frac{W_{a/a} - W_{A/a}}{(W_{a/a} - W_{A/a}) + (W_{A/A} - W_{A/a})}$$

The equilibrium value is a ratio of the differences in fitness between the homozygotes and the heterozygote. As an example, suppose the fitnesses are

$W_{A/A}$	$W_{A/a}$	$W_{a/a}$
0.9	1.0	0.8

The equilibrium value will be

$$\hat{p} = 2/3.$$

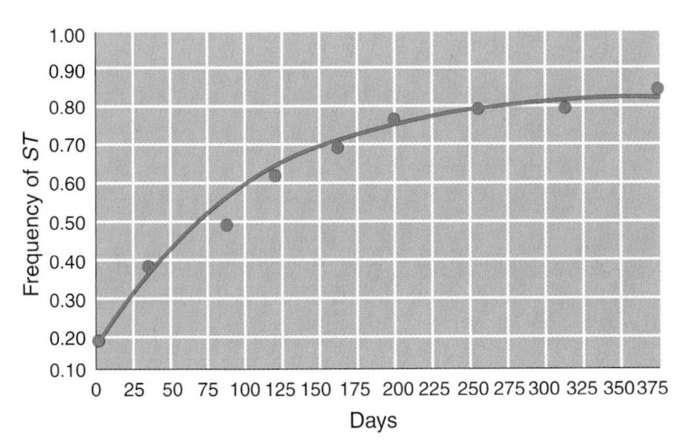

Figure 17-11 Changes in the frequency of the inversion Standard (*ST*) in competition with Chiricahua (*CH*) in a laboratory population of *Drosophila pseudoobscura.* The solid line shows the theoretical course of change for the estimated fitnesses of the three genotypes: $W_{ST/ST} = 0.89$; $W_{ST/CH} = 1.0$; $W_{CH/CH} = 0.41$. The points show the actual frequencies observed in successive generations.

ples of such chromosomes). The fitnesses estimated for the three genotypes in the laboratory are

$$W_{ST/ST} = 0.89 \qquad W_{ST/CH} = 1.0 \qquad W_{CH/CH} = 0.41$$

Applying the formula for the equilibrium value $\hat{p}$, we obtain $\hat{p} = 0.85$, which agrees quite well with the observations in Figure 17-11.

In nature, the chance that genotype frequencies will remain balanced on the knife edge of an unstable equilibrium is negligible. In any generation in a real population made up of a finite number of individuals, there will be chance fluctuations in allele frequencies, so we should not expect to find naturally occurring polymorphisms in which heterozygotes are less fit than homozygotes. On the contrary, the observation of a long-lasting polymorphism in nature might be taken as evidence of a superior heterozygote.

Unfortunately, life confounds theory. The locus that encodes the Rh blood group in humans has a widespread polymorphism, with Rh^+ and Rh^- alleles. In Europeans, for example, the frequency of the Rh^- allele is about 0.4, whereas in Africans, it is about 0.2, and such a polymorphism is found in all human groups. Thus, this human polymorphism must be very old, antedating the origin of modern geographical races. But this polymorphism causes a maternal-fetal incompatibility when an Rh^- mother (homozygous Rh^- / Rh^-) produces an Rh^+ fetus (heterozygous Rh^- / Rh^+). This incompatibility may result in hemolytic anemia (from a destruction of red blood cells) and death for the fetus if the mother has been previously sensitized to the Rh^+ antigen by an earlier pregnancy with an incompatible fetus. Thus, there is selection against heterozygotes. This polymorphism is therefore predicted to be unstable and should have disappeared from the species, yet it exists in most human populations. Many hypotheses have been proposed to explain its apparent stability, but the mystery remains.

In contrast, no fitness difference at all can be demonstrated for many polymorphisms of blood groups (and for the ubiquitous polymorphism of enzymes revealed by electrophoresis). It has been suggested that such polymorphisms are not under selection at all, but that

$$W_{A/A} = W_{A/a} = W_{a/a}$$

This situation of **selective neutrality** would also satisfy the requirement that $\overline{W}_A = \overline{W}_a$. Instead of a stable equilibrium, however, it gives rise to a **passive (neutral) equilibrium** such that any allele frequency p is as good as any other. This leaves unanswered the question of how populations became highly polymorphic in the first place.

A balance of selective forces is not the only situation in which an equilibrium of allele frequencies may arise. Another cause of polymorphic genetic equilibrium in populations is a balance between the introduction of new alleles by repeated mutation and their removal by natural selection. This balance is probably the cause of many low-level polymorphisms for genetic diseases in human populations. New deleterious mutations are constantly arising spontaneously or as the result of the action of mutagens. These mutations may be completely recessive or partly dominant. Selection removes them from the population, but there will be an equilibrium between their appearance and removal.

The general equation for this equilibrium is given in detail in Box 17-8 on the next page. It shows that the frequency of the deleterious allele at equilibrium depends on the ratio of the mutation rate, μ, to the intensity of selection, s, against the deleterious genotype. For a completely recessive deleterious allele whose fitness in the homozygous state is $1 - s$, the equilibrium frequency is

$$q = \sqrt{\frac{\mu}{s}}$$

So, for example, a recessive lethal allele ($s = 1$) mutating at the rate of $\mu = 10^{-6}$ will have an equilibrium frequency of 10^{-3}. Indeed, if we knew that an allele was a recessive lethal and had no heterozygous effects, we could estimate its mutation rate as the square of its allele frequency. But the biological basis for the assumptions behind such calculations must be firm. Sickle-cell anemia was once thought to be a recessive lethal with no heterozygous effects, which led to an estimated mutation rate in Africa of 0.1 for this locus. But we now know that its equilibrium is a result of higher fitness of heterozygotes.

A similar result can be obtained for a deleterious allele with some effect in heterozygotes. If we let the fitnesses be $W_{A/A} = 1.0$, $W_{A/a} = 1 - hs$, and $W_{a/a} = 1 - s$ for a partly dominant allele a, where h is the degree of

BOX 17-8

The Balance between Selection and Mutation

If we let q be the frequency of the deleterious allele a and $p = 1 - q$ be the frequency of the normal allele A, then the change in allele frequency due to the mutation rate μ is

$$\Delta q_{\text{mut}} = \mu p$$

A simple way to express the fitnesses of the genotypes in the case of a recessive deleterious allele a is $W_{A/A} = W_{A/a} = 1.0$ and $W_{a/a} = 1 - s$, where s is the loss of fitness in the recessive homozygotes. We now can substitute these fitnesses in our general expression for allele frequency change (see Box 17-6) and obtain

$$\Delta q_{\text{sel}} = \frac{-pq(sq)}{1 - sq^2} = \frac{-spq^2}{1 - sq^2}$$

Equilibrium means that the increase in the allele frequency due to mutation must exactly balance the decrease in the allele frequency due to selection, so

$$\Delta \hat{q}_{\text{mut}} + \Delta \hat{q}_{\text{sel}} = 0$$

Remembering that $\hat{q}$ at equilibrium will be quite small, so that $1 - s\hat{q}^2 \approx 1$, and substituting the terms for $\Delta \hat{q}_{\text{mut}}$ and $\Delta \hat{q}_{\text{sel}}$ in the preceding formula, we have

$$\mu \hat{p} - \frac{s\hat{p}\hat{q}^2}{1 - s\hat{q}^2} \approx \mu \hat{p} - s\hat{p}\hat{q}^2 = 0$$

or

$$\hat{q}^2 = \frac{\mu}{s} \quad \text{and} \quad \hat{q} = \sqrt{\frac{\mu}{s}}$$

dominance of the deleterious allele, then a similar calculation gives us

$$\hat{q} = \frac{\mu}{hs}$$

Thus, if $\mu = 10^{-6}$ and the lethal allele is not totally recessive, but has a 5 percent deleterious effect in heterozygotes ($s = 1.0$, $h = 0.05$), then

$$\hat{q} = \frac{10^{-6}}{5 \times 10^{-2}} = 2 \times 10^{-5}$$

which is smaller by two orders of magnitude than the equilibrium frequency for the purely recessive case. In general, then, we can expect deleterious, completely recessive alleles to have frequencies much higher than those of partly dominant alleles, because the recessive alleles are protected in heterozygotes.

ARTIFICIAL SELECTION

In contrast with the difficulties of finding simple, well-behaved cases in nature that exemplify the simple formulas of natural selection, there is a vast record of the effectiveness of artificial selection in changing populations phenotypically. Many such changes have been produced by laboratory selection experiments and by selection of animals and plants in agriculture (as, for example, selection for increased milk production in cows and for rust resistance in wheat). No analysis of these experiments in terms of allele frequencies is possible because individual loci have not been identified and followed. Nevertheless, it is clear that genetic changes have occurred in the populations because the populations maintain their characteristics even after the selection has been terminated. Figure 17-12 shows, as an example, the magnitude of the changes in average bristle number that are achieved in a selection experiment with

Figure 17-12 Changes in average bristle number obtained in two laboratory populations of *Drosophila melanogaster* through artificial selection for high bristle number in one population and for low bristle number in the other. The dashed segment in the curve for the upwardly selected line indicates a period of five generations in which no selections were performed. *(From K. Mather and B. J. Harrison, "The Manifold Effects of Selection,"* Heredity *3, 1949, 1.)*

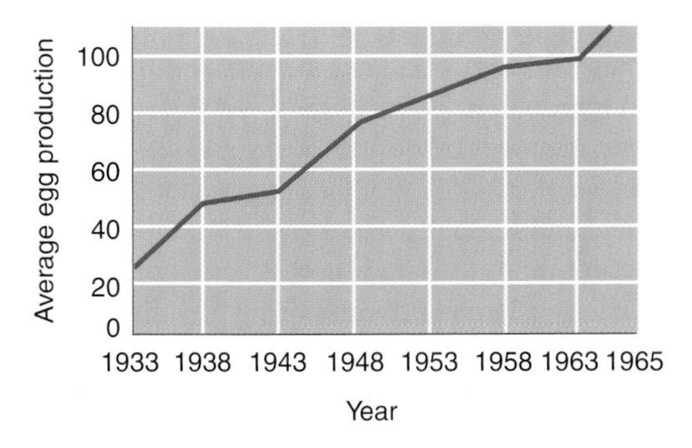

Figure 17-13 Changes in average egg production in a chicken population selected for a higher egg-laying rate over a period of 30 years. *(From I. M. Lerner and W. J. Libby,* Heredity, Evolution, and Society, *2d ed. © 1976 by W. H. Freeman and Company. Data courtesy of D. C. Lowry.)*

D. melanogaster. Figure 17-13 shows the changes in the number of eggs laid per chicken as a consequence of 30 years of selection.

The usual method of selection for a continuously varying trait is **truncation selection.** The individuals in a given generation are pooled (irrespective of their families), a sample is measured, and only those individuals above (or below) a given phenotypic value (the truncation point) are chosen as parents for the next generation. This phenotypic value may remain the same over successive generations; in this case, selection is by **constant truncation.** We might, for example, select as parents of the next generation only those chickens whose eggs have shells thick enough to withstand a standard breaking pressure. More commonly, a fixed percentage of the population representing the highest (or lowest) value of the selected phenotype is chosen; in this case, selection is by **proportional truncation.** As an example, when selecting for egg weight, we might select the 20 percent of the chickens whose eggs weighed the most. With constant truncation, the intensity of selection decreases with time as more and more of the population exceeds the fixed truncation point. With proportional truncation, the intensity of selection is constant, but the truncation point moves upward as the population distribution changes. Figure 17-14 illustrates these two types of truncation.

A common experience in artificial selection programs is that, as the population becomes more and more extreme for the selected phenotype, its viability and fertility decrease. As a result, eventually no further progress under selection is possible, despite the presence of genetic variation for the character, because the selected individuals do not reproduce. The loss of fitness may be a direct phenotypic effect of the genes for the selected character, in which case nothing much can be done to improve the population further. Often, however, the loss of fitness is tied to alleles at loci linked to the genes that are under selection and that are carried along with the selected genes. In such cases, a number of generations without selection may allow recombinants to be formed between the two sets of genes, and selection can then be continued, as in the upwardly selected line in Figure 17-12.

We must be very careful in our interpretation of long-term agricultural selection programs. In the real world of agriculture, changes in cultivation methods, machinery, fertilizers, insecticides, herbicides, and so forth are occurring along with the production of genetically improved varieties. Increases in average yields are consequences of all of these changes. The average yield of corn in the United States, for example, increased from 40 bushels to 80 bushels per acre between 1940 and 1970. But experiments comparing old and new varieties of corn in common environments show that only about half this increase is a direct result of new

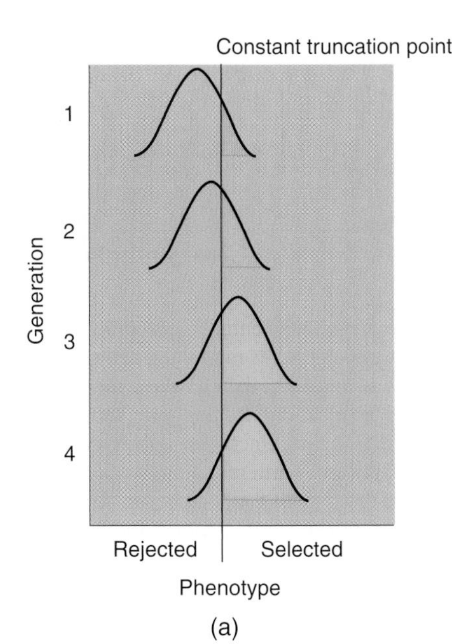

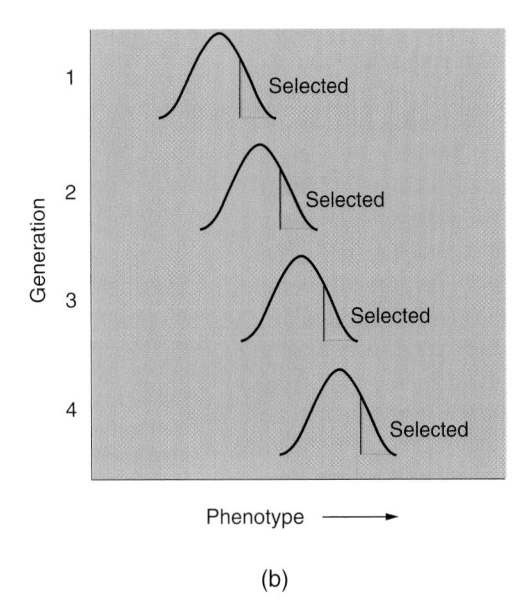

Figure 17-14 Two schemes of truncation selection for a continuously varying trait. (a) Constant truncation, in which individuals with a phenotype greater than some fixed value are selected. (b) Proportional truncation, in which some percentage of the population with the greatest value for a phenotype is selected each generation.

corn varieties (the other half being a result of improved farming techniques). Furthermore, the new varieties are superior to the old ones only at the high densities of modern planting for which they were selected.

RANDOM EVENTS

If a population consists of a finite number of individuals (as all real populations do), and if a given pair of parents has only a small number of offspring, then, even in the absence of all selective forces, the frequency of an allele will not be exactly reproduced in the next generation, simply because of sampling error. If, for example, in a population of 1000 individuals, the frequency of *a* is 0.5 in one generation, it may be 0.493 or 0.505 in the next generation, just because of the chance production of slightly more or slightly fewer progeny of each genotype. In the second generation, there will be another sampling error based on the new allele frequencies, so the frequency of *a* may go from 0.505 to 0.511 or back to 0.498. This process of random fluctuation continues generation after generation. There is no force pushing the frequency back to its initial state because the population has no "genetic memory" of its state many generations ago; each generation is an independent event. The final result of this random change in allele frequencies, known as **genetic drift,** is that the population eventually drifts to $p = 1$ or $p = 0$. At this point, no further change is possible; the allele is fixed and the population has become homozygous. A different population of the same species, isolated from the first, also undergoes this random genetic drift, but it may become homozygous for allele *A*, whereas the first population has become homozygous for allele *a*. As time goes on, isolated populations diverge from each other, each losing heterozygosity. The variation originally present *within* populations now appears as variation *among* populations.

One form of genetic drift occurs when a small group breaks off from a larger population to found a new colony. This acute drift, called the **founder effect,** results from a single generation of sampling of a small number of colonizers from the original large population, followed by several generations during which the new population remains small in number. Even if the population grew large after some time, it would continue to drift, but at a slower rate. The founder effect is probably responsible for the virtually complete lack of blood group B in Native Americans, whose ancestors migrated in very small numbers across the Bering Strait at the end of the last Ice Age, about 20,000 years ago, but whose ancestral population in northeastern Asia had an intermediate frequency of group B.

The process of genetic drift should sound familiar. It is, in fact, another way of looking at the inbreeding effect in small populations discussed earlier. Populations that are the descendants of a very small number of ancestral individuals have a high probability that all the copies of a particular allele will be identical by descent from a single common ancestor (see Figure 17-7). Whether regarded as inbreeding or as random sampling of genes, the effect is the same. Populations do not reproduce their genetic constitutions exactly; there is always a random component of allele frequency change.

One result of random sampling is that most new mutations, even if they are not selected against, never succeed in becoming part of the long-term genetic composition of the population. Suppose that a single individual is heterozygous for a new mutation. There is some chance that the individual in question will have no offspring at all. Even if it has one offspring, there is a chance of 1/2 that the new mutation will not be transmitted to that offspring. If the individual has two offspring, the probability that neither offspring will carry the new mutation is 1/4, and so forth. Even if the new mutation is successfully transmitted to an offspring, the lottery is repeated in the next generation, and again the allele may be lost. In fact, if a population has a size of *N*, the chance that a new mutation will eventually be lost by chance is $(2N - 1)/2N$. (For a derivation of this result, which is beyond the scope of this book, see Chapters 2 and 3 of Hartl and Clark, *Principles of Population Genetics,* 3d ed., Sinauer Associates, 1997.) But, if the new mutation is not lost, then the only thing that can happen to it in a finite population is that eventually it will sweep through the population and become fixed. This event has the probability

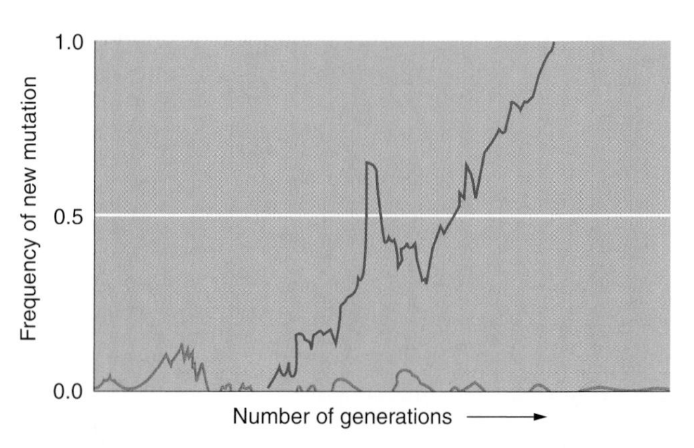

Figure 17-15 The appearance, loss, and eventual incorporation of new mutations in the life of a population. If random genetic drift does not cause the loss of a new mutation, then it must eventually become fixed (in the absence of selection). In the figure, ten mutations arise, of which nine (red lines) increase slightly in frequency and then die out. Only the fourth mutation to occur (blue line) eventually spreads throughout the population. *(After J. Crow and M. Kimura,* An Introduction to Population Genetics Theory. *© 1970 by Harper & Row.)*

of $1/2N$. In the absence of selection, then, the history of a population looks something like Figure 17-15. For some period of time, the population is homozygous; then a new mutation appears. In most cases, the new mutant allele will be lost immediately or very soon after it appears. Occasionally, however, a new mutant allele drifts through the population, and the population becomes homozygous for the new allele. The process then begins again.

A striking example of the effect of genetic drift in human populations is the variation in frequencies of the VNTR length variants among populations of South American Indians that we saw illustrated in Table 17-6. For one VNTR, D14S1, the Surui are very variable, but the Karitiana, living several hundred miles away in the Brazilian rain forest, are nearly homozygous for one allele, presumably because of genetic drift in these very small isolated populations. For the other VNTR, D14S13, neither population has become homozygous, but the pattern of allele frequencies is very different between the two.

Even a new mutation that is slightly favorable selectively will usually be lost in the first few generations after it appears in the population, a victim of genetic drift. If a new mutation has a selective advantage of s in the heterozygote in which it appears, then the chance is only $2s$ that the mutation will ever succeed in becoming fixed. So a mutant allele that is 1 percent better in fitness than the standard allele in the population will be lost 98 percent of the time by genetic drift. It is even possible for a very slightly deleterious mutation to rise in frequency and become fixed in a population by genetic drift.

MESSAGE

New mutations can become established in a population even though they are not favored by natural selection, simply by a process of random genetic drift. Even new favorable mutations are often lost, and occasionally a slightly deleterious mutation can become fixed in a population by genetic drift.

Another consequence of the interaction of random and selective forces is that the effectiveness of selection in driving population composition depends on population size. The magnitude of the random effect is proportional to the reciprocal of population size, $1/N$. Selection is effective only if the selection coefficient, s, is large enough so that $s \geq 1/N$, or, equivalently, $Ns \geq 1$.

When Ns is small because selection is weak or population size is small, then mutations are *effectively neutral,* even though there is some selection on them. Small populations will be less affected by selection than large populations even under otherwise identical conditions. Human populations, for example, were very small for nearly all the history of our species, having grown large only in the past few hundred generations. Thus, we may expect to find that many mutations that are now under selection were effectively neutral for a long time and may have reached high frequency by chance.

The same dependency on population size holds for the effects of migration or mutation. Thus, we can say, roughly, that migration and mutation will drive the genetic composition of a population only if the migration rate, m, or the mutation rate, μ, is high enough relative to the reciprocal of population size. The requirement is that $m \geq 1/N$ or $\mu \geq 1/N$, or, in other words, $Nm \geq 1$ or $N\mu \geq 1$.

We now return to the strange case with which we began this chapter, the very high frequency in West Africa of an inherited severe blood disorder, sickle-cell anemia. The primary molecular cause of sickle-cell anemia is an allele, Hb^S, with a nucleotide substitution in the coding region of the hemoglobin β chain gene, which results in the substitution of valine for the normal glutamic acid at polypeptide chain position 6. The abnormal hemoglobin crystallizes at low oxygen pressure, and the red cells deform and hemolyze. Homozygotes Hb^S / Hb^S have severe anemia, and their survivorship is low. Heterozygotes (Hb^A / Hb^S) have a mild anemia and under ordinary circumstances exhibit the same or only slightly lower fitness than normal homozygotes Hb^A / Hb^A. However, in regions of Africa with a high incidence of malaria, which is caused by the parasite *Plasmodium falciparum,* heterozygotes have a *higher* fitness than normal homozygotes because the presence of some sickling hemoglobin apparently protects them from the parasite. Where malaria is absent, as in North America, the fitness advantage of heterozygosity is lost. Thus, in the descendants of the slaves brought to North America from Africa, there should be a decrease in the frequency of the sickling allele, for two reasons. First, mating between Africans and non-Africans in North America should reduce the frequency of the Hb^S allele in the African-American population, since neither European nor Native American populations originally had the allele. Second, natural selection operating against the allele in North America through the premature death of homozygotes, with no countervailing advantage to heterozygotes, would cause a rapid decrease in the frequency of the Hb^S allele. From the estimates of the proportion of non-African ancestry in African-Americans given above, it appears that roughly half the decrease in the frequency of the sickle-cell anemia gene in North America has been the result of this natural selection.

SUMMARY

1. Population genetics relates changes in the genetic composition of populations to the underlying individual processes of inheritance and development. Population genetics is the study of inherited variation and its modification in time and space.
2. Identifiable inherited variation within a population can be studied by observing morphological differences between individuals, by examining differences in the amino acid sequences of specific proteins, and, most recently, by examining the differences in nucleotide sequences within the DNA. These kinds of observations have led to the conclusion that there is considerable polymorphism at many loci within populations. A measure of this variation is the amount of heterozygosity in a population.
3. One result of Mendelian segregation is that random mating results in an equilibrium distribution of genotypes after one generation. However, inbreeding can convert genetic variation within a population into differences between populations by making each separate population homozygous for a randomly chosen allele. On the other hand, for most populations, a balance is reached between the forces of inbreeding, mutation, and migration.
4. The ultimate source of all variation is mutation. Within a population, however, the frequencies of specific genotypes can be changed by recombination, migration of genes, continued mutational events, and chance.
5. "Directed" changes of allele frequencies within a population occur through natural selection. In many cases, such changes lead to homozygosity at a particular locus. On the other hand, the heterozygote may be more suited to a given environment than either of the homozygotes, leading to a balanced polymorphism.
6. Selection of specific genotypes is rarely so simple, however. More often than not, phenotypes are determined by several interacting genes, and alleles at these different loci will be selected for at different rates. Furthermore, closely linked loci, unrelated to the phenotype in question, may have specific alleles carried along during the selection process that affect its outcome.
7. In general, genetic variation is the result of the interaction of several forces. For instance, a recessive, deleterious mutation may never be totally eliminated from a population because mutation continues to resupply it or because migration reintroduces it into the population.
8. Unless alternative alleles are at intermediate frequencies, selection (especially against recessives) is very slow, requiring many generations. In many populations, especially those of small size, new mutations can become established even though they are not favored by natural selection, simply by a process of random genetic drift.

CONCEPT MAP

Draw a concept map interrelating as many of the following terms as possible. Note that the terms are listed in no particular order.

allele frequency / heterozygosity / polymorphism / mutation / selection / Hardy-Weinberg equilibrium / migration / inbreeding / genetic drift / Mendelian ratios

SOLVED PROBLEM

1. The polymorphisms for shell color (yellow or pink) and for the presence or absence of shell banding in the snail *Cepaea nemoralis* are each the result of a pair of segregating alleles at a separate locus. Design an experimental program that would reveal the forces that determine the frequency and geographical distribution of these polymorphisms.

Solution

a. Describe the frequencies of the different morphs for samples of snails from a large number of populations covering the geographical and ecological range of the species. Each snail must be scored for *both* polymorphisms. At the same time, record a description of the habitat of each population. In addition, estimate the number of snails in each population.

b. Measure migration distances by marking a sample of snails with a spot of paint on the shell, replacing them in the population, and then resampling at a later date.

c. Raise broods from eggs laid by individual snails so that the genotype of male parents can be inferred and nonrandom mating patterns can be observed. The segregation frequencies *within* each family will reveal differences between genotypes in probability of survivorship of early developmental stages.

d. Seek further evidence of selection from (1) geographical patterns in the frequencies of the alleles; (2) correlation between allele frequencies and ecological variables, including population density; (3) correlation between the frequencies of the two different polymorphisms (are populations with, say, high frequencies of pink shells also characterized by, say, high frequencies of banded shells?); and (4) nonrandom associations *within* populations of the alleles at the two loci, indicating that certain combinations may have a higher fitness.

e. Seek evidence of the importance of random genetic drift by comparing the variation in allele frequencies among small populations with the variation among large populations. If small populations vary more from each other than do large ones, random drift is implicated.

SOLVED PROBLEM

2. About 70 percent of all white North Americans can taste the chemical phenylthiocarbamide, and the remainder cannot. The ability to taste this chemical is determined by the dominant allele *T*, and the inability to taste is determined by the recessive allele *t*. If the population is assumed to be in Hardy-Weinberg equilibrium, what are the genotype and allele frequencies in this population?

Solution

Because 70 percent are tasters (*T* / *T* and *T* / *t*), 30 percent must be nontasters (*t* / *t*). This homozygous recessive frequency is equal to q^2; so, to obtain q, we simply take the square root of 0.30:

$$q = \sqrt{0.30} = 0.55$$

Because $p + q = 1$, we can write

$$p = 1 - q = 1 - 0.55 = 0.45$$

Now we can calculate

$$p^2 = 0.45^2 = 0.20, \text{ the frequency of } T/T$$

$$2pq = 2 \times 0.45 \times 0.55 = 0.50, \text{ the frequency of } T/t$$

$$q^2 = 0.3, \text{ the frequency of } t/t$$

SOLVED PROBLEM

3. In a large natural population of *Mimulus guttatus*, one leaf was sampled from each of a large number of plants. The leaves were crushed and subjected to gel electrophoresis. The gel was then stained for a specific enzyme, X. Six different banding patterns were observed, as shown in the accompanying diagram.

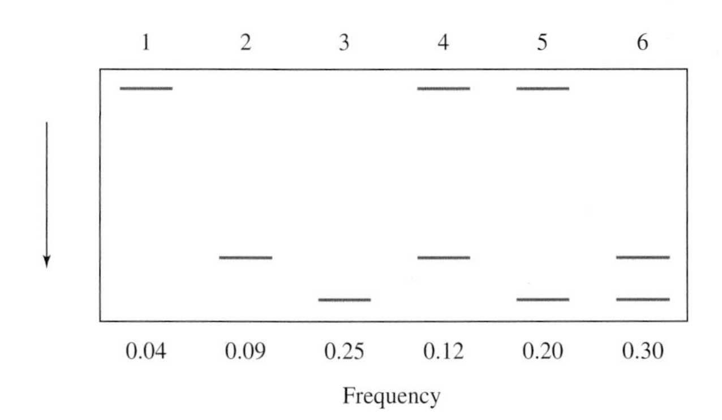

a. Assuming that these patterns are produced by a single locus, propose a genetic explanation for the six types.

b. How can you test your hypothesis?

c. What are the allele frequencies in this population?

d. Is the population in Hardy-Weinberg equilibrium?

Solution

a. Inspection of the gel reveals that there are only three band positions: we shall call them slow, intermediate, and fast, according to how far each has migrated in the gel. Furthermore, any individual can show either one band or two. The simplest explanation is that there are three alleles of one locus (let's call them *S*, *I*, and *F*) and that the individuals with two bands are heterozygotes. Hence, lane 1 = *S* / *S*, 2 = *I* / *I*, 3 = *F* / *F*, 4 = *S* / *I*, 5 = *S* / *F*, and 6 = *I* / *F*.

b. The hypothesis can be tested by making controlled crosses. For example, from a self of type 5, we can predict 1/4 *S* / *S*, 1/2 *S* / *F*, and 1/4 *F* / *F*.

c. The frequencies can be calculated by a simple generalization from the two-allele formulas. Hence:

$$f_S = 0.04 + 1/2(0.12) + 1/2(0.20) = 0.20 = p$$

$$f_I = 0.09 + 1/2(0.12) + 1/2(0.30) = 0.30 = q$$

$$f_F = 0.25 + 1/2(0.20) + 1/2(0.30) = 0.50 = r$$

d. The Hardy-Weinberg genotypic frequencies are

$$(p + q + r)^2 = p^2 + q^2 + r^2 + 2pq + 2pr + 2qr$$
$$= 0.04 + 0.09 + 0.25 + 0.12 + 0.20 + 0.30$$

which are precisely the observed frequencies. So it appears that the population is in equilibrium.

SOLVED PROBLEM

4. In a large experimental *Drosophila* population, the fitness of a recessive phenotype is calculated to be 0.90, and the mutation rate to the recessive allele is 5×10^{-5}. If the population is allowed to come to equilibrium, what allele frequencies can be predicted?

Solution

Here mutation and selection are working in opposite directions, so an equilibrium is predicted. Such an equilibrium is described by the formula

$$\hat{q} = \sqrt{\frac{\mu}{s}}$$

In the present question, $\mu = 5 \times 10^{-5}$ and $s = 1 - W = 1 - 0.9 = 0.1$. Hence

$$\hat{q} = \sqrt{\frac{5 \times 10^{-5}}{10^{-1}}} = 2.2 \times 10^{-2} = 0.022$$

$$\hat{p} = 1 - 0.022 = 0.978$$

PROBLEMS

Basic Problems

1. What are the forces that can change the frequency of an allele in a population?

2. In a population of mice, there are two alleles of the *A* locus (A_1 and A_2). Tests showed that in this population there are 384 mice of genotype A_1 / A_1, 210 of A_1 / A_2, and 260 of A_2 / A_2. What are the frequencies of the two alleles in the population?

3. In a randomly mating laboratory population of *Drosophila,* 4 percent of the flies have black bodies (encoded by the autosomal recessive *b*) and 96 percent have brown bodies (the wild type, encoded by *B*). If this population is assumed to be in Hardy-Weinberg equilibrium, what are the allele frequencies of *B* and *b* and the genotypic frequencies of *B* / *B* and *B* / *b*?

4. In a wild population of beetles of species X, you notice that there is a 3:1 ratio of shiny to dull wing covers. Does this ratio prove that the *shiny* allele is dominant? (Assume that the two states are caused by two alleles of one gene.) If not, what does it prove? How would you elucidate the situation?

5. The fitnesses of three genotypes are $W_{A/A} = 0.9$, $W_{A/a} = 1.0$, and $W_{a/a} = 0.7$.

a. If the population starts at the allele frequency $p = 0.5$, what is the value of p in the next generation?

b. What is the predicted equilibrium allele frequency?

i **6.** *A* / *A* and *A* / *a* individuals are equally fertile. If 0.1 percent of the population is *a* / *a*, what selection pressure exists against *a* / *a* if the $A \rightarrow a$ mutation rate is 10^{-5}?

7. In a survey of Native American tribes in Arizona and New Mexico, albinos were completely absent or very rare in most tribes (there is 1 albino per 20,000 North American Caucasians). However, in three Native American populations, albino frequencies are exceptionally high: 1 per 277 Native Americans in Arizona; 1 per 140 Jemez in New Mexico; and 1 per 247 Zuni in New Mexico. All three of these populations are culturally but not linguistically related. What possible factors might explain the high incidence of albinos in these three tribes?

Challenging Problems

8. In a population, the $D \rightarrow d$ mutation rate is 4×10^{-6}. If $p = 0.8$ today, what will p be after 50,000 generations?

9. You are studying protein polymorphism in a natural population of a certain species of a sexually reproducing haploid organism. You isolate many strains from various parts of the test area and run extracts from each strain on electrophoretic gels. You stain the gels with a reagent specific for enzyme X and find that in the population there is a total of, say, five electrophoretic variants of enzyme X. You speculate that these variants represent various alleles of the structural gene for enzyme X.

a. How could you demonstrate that your speculation is correct, both genetically and biochemically? (You can make crosses, make diploids, run gels, test enzyme activities, test amino acid sequences, and so forth.) Outline the steps and conclusions precisely.

b. Name at least one other possible way of generating the different electrophoretic variants, and explain how you would distinguish this possibility from your speculation above.

10. A study made in 1958 in the mining town of Ashibetsu in Hokkaido, Japan, revealed the frequencies of MN blood type genotypes (for individuals and for married couples) shown in the following table:

Genotype	Number of individuals or couples
Individuals	
L^M / L^M	406
L^M / L^N	744
L^N / L^N	332
Total	1482
Couples	
$L^M / L^M \times L^M / L^M$	58
$L^M / L^M \times L^M / L^N$	202
$L^M / L^N \times L^M / L^N$	190
$L^M / L^M \times L^N / L^N$	88
$L^M / L^N \times L^N / L^N$	162
$L^N / L^N \times L^N / L^N$	41
Total	741

a. Show whether the population is in Hardy-Weinberg equilibrium with respect to MN blood types.

b. Show whether mating is random with respect to MN blood types.

(Problem 10 is from J. Kuspira and G. W. Walker, *Genetics: Questions and Problems.* © 1973 by McGraw-Hill.)

11. Consider the populations that have the genotypes shown in the following table:

Population	A / A	A / a	a / a
1	1.0	0.0	0.0
2	0.0	1.0	0.0
3	0.0	0.0	1.0
4	0.50	0.25	0.25
5	0.25	0.25	0.50
6	0.25	0.50	0.25
7	0.33	0.33	0.33
8	0.04	0.32	0.64
9	0.64	0.32	0.04
10	0.986049	0.013902	0.000049

a. Which of the populations are in Hardy-Weinberg equilibrium?

b. What are p and q in each population?

c. In population 10, it is discovered that the $A \rightarrow a$ mutation rate is 5×10^{-6} and that reverse mutation is negligible. What must be the fitness of the a / a phenotype?

d. In population 6, the a allele is deleterious; furthermore, the A allele is incompletely dominant, so that A / A is perfectly fit, A / a has a fitness of 0.8, and a / a has a fitness of 0.6. If there is no mutation, what will p and q be in the next generation?

12. Colorblindness results from a sex-linked recessive allele. One in every 10 males is colorblind.

a. What proportion of all women are colorblind?

b. By what factor is colorblindness more common in men (or, how many colorblind men are there for each colorblind woman)?

c. In what proportion of marriages would colorblindness affect half the children of each sex?

d. In what proportion of marriages would all children be normal?

e. In a population that is not in equilibrium, the frequency of the allele for colorblindness is 0.2 in women and 0.6 in men. After one generation of random mating, what proportion of the female progeny will be colorblind? What proportion of the male progeny?

f. What will the allele frequencies be in the male and in the female progeny in part e?

(Problem 12 courtesy of Clayton Person.)

13. It seems clear that most new mutations are deleterious. Why?

14. Most mutations are recessive to the wild type. Of those rare mutations that are dominant in *Drosophila,* for example, the majority turn out either to be chromosomal mutations or to be inseparable from chromosomal mutations. Explain why the wild type is usually dominant.

15. Ten percent of the males of a large and randomly mating population are colorblind. A representative group of 1000 people from this population migrates to a South Pacific island, where there are already 1000 inhabitants and where 30 percent of the males are colorblind. Assuming that Hardy-Weinberg equilibrium applies throughout (in the two original populations before the migration and in the mixed population immediately after the migration), what fraction of males and females can be expected to be colorblind in the generation immediately after the arrival of the migrants?

16. Using pedigree diagrams, find the probability of homozygosity by descent of the offspring of (a) parent-offspring matings; (b) first-cousin matings; (c) aunt-nephew or uncle-niece matings.

17. In an animal population, 20 percent of the individuals are *A* / *A*, 60 percent are *A* / *a*, and 20 percent are *a* / *a*.

a. What are the allele frequencies in this population?

b. In this population, mating is always with *like phenotype* but is random within phenotype. What genotype and allele frequencies will prevail in the next generation?

c. Another type of assortative mating is that which takes place only between *unlike* phenotypes. Answer the preceding question with this restriction imposed.

d. What will the end result be after many generations of mating of each type?

18. A *Drosophila* stock isolated from nature has an average of 36 abdominal bristles. By the selective breeding of only those flies with the most bristles, the mean is raised to 56 bristles in 20 generations.

a. What is the source of this genetic flexibility?

b. The 56-bristle stock is infertile, so selection is relaxed for several generations and the bristle number drops to about 45. Why doesn't it drop back to 36?

c. When selection is reapplied, 56 bristles are soon attained, but this time the stock is *not* infertile. How can this situation arise?

19. Allele *B* is a deleterious autosomal dominant. The frequency of affected individuals is 4.0×10^{-6}. The reproductive capacity of these individuals is about 30 percent that of normal individuals. Estimate μ, the rate at which *b* mutates to its deleterious allele *B*.

20. Of 31 children born of father-daughter matings, 6 died in infancy, 12 were very abnormal and died in childhood, and 13 were normal. From this information, calculate roughly how many recessive lethal genes we have, on average, in our human genomes. (HINT: If the answer were 1, then a daughter would stand a 50 percent chance of carrying the lethal allele, and the probability of the union's producing a lethal combination would be $1/2 \times 1/4 = 1/8$. So 1 is not the answer.) Consider also the possibility of undetected fatalities in utero in such matings. How would they affect your result?

21. If we define the **total selection cost** to a population of deleterious recessive genes as the loss of fitness per individual affected (s) multiplied by the frequency of affected individuals (q^2), then

$$\text{selection cost} = sq^2$$

a. Suppose that a population is at equilibrium between mutation and selection for a deleterious recessive allele, where $s = 0.5$ and $\mu = 10^{-5}$. What is the equilibrium frequency of the allele? What is the selection cost?

b. Suppose that we start irradiating individual members of the population, so that the mutation rate doubles. What is the new equilibrium frequency of the allele? What is the new selection cost?

c. If we do not change the mutation rate, but we lower the selection intensity to 0.3 instead, what happens to the equilibrium frequency and the selection cost?

QUANTITATIVE GENETICS 18

Key Concepts

1. As an experimental laboratory science, genetics has concentrated on clear-cut, qualitative phenotypic differences between mutant and wild-type genotypes. In populations of organisms in nature, however, variation in most characters takes the form of a continuous phenotypic range, rather than discrete phenotypic classes. In other words, the variation is quantitative, not qualitative.
2. Statistical techniques are usually employed to study quantitative traits.
3. A major task of quantitative genetics is to determine the ways in which genes interact with the environment to contribute to the formation of a given quantitative character distribution.
4. The genetic variation underlying a continuous character distribution can be the result of segregation at a single genetic locus or at numerous interacting loci that produce cumulative effects on the phenotype.
5. Estimates of genetic and environmental variance are specific to the single population and the particular set of environments in which the estimates are made.
6. The estimated ratio of genetic to environmental variance is a measure of the relative amounts of variation in a population that arise from genetic and environmental sources, respectively. It is *not* a measure of the relative contribution of genes and environment to an individual's phenotype.

Quantitative variation in flower color, flower diameter, and number of flower parts in the composite flowers of *Gaillardia pulchella.* (*J. Heywood,* Journal of Heredity, *May/June 1986.*)

Skin color is a physical feature of human beings of which people in many societies are acutely conscious and which has powerful social consequences. It is obvious to us that genes have an important role in the development of skin color. For example, the children of African slaves, born in North America, were black, like their African-born parents. It is also obvious that skin color does not fall into a few clear-cut classes, but varies continuously from the very pale northern Europeans to the very dark central Africans, and that children born to two parents whose skin colors are very different from each other are intermediate between them (Figure 18-1). Finally, we also know that environment affects skin color, because light-skinned people who spend a month sunning themselves on tropical beaches come home a great deal darker than when they left. In these ways, skin color, like many other characters such as height, weight, and metabolic rate, is rather different from the simple phenotypic contrasts that we have considered up to this point, and a different approach is required to study such characters.

Figure 18-1 A sample of the variety of human skin colors that have arisen in the evolution of our species and in the subsequent matings between individuals from different geographical populations. *(Cleo/PhotoEdit.)*

CHAPTER OVERVIEW

Ultimately, the goal of genetics is the analysis of how the phenotypes of organisms are influenced by their genotypes. But generally the genotype can be identified—and therefore studied—only through its phenotypic effect. We recognize two genotypes as different from each other because the phenotypes of their carriers are different. Basic genetic experiments, then, depend on the existence of a simple relation between genotype and phenotype. That is why studies of DNA sequences are so important: for the first time, we can look at the genotype of the individual directly.

For the most part, the study of genetics presented in the preceding chapters has been the study of allelic substitutions that cause *qualitative* differences in phenotype. In general, we hope to find a uniquely distinguishable phenotype for each genotype and only a simple genotype for each phenotype. At worst, when one allele is completely dominant, it is necessary to perform a simple genetic cross to distinguish the heterozygote from the homozygote. Where possible, geneticists avoid studying genes that have only partial penetrance and incomplete expressivity because of the difficulty of making genetic inferences from such phenotypes.

However, most actual variation among organisms is *quantitative,* not qualitative. Even the simple phenotypic classes that are commonly observed in laboratory mutants have some variation within them, but this continuous variation can be ignored for the purposes of the experiment if there is no overlap between the phenotypic classes corresponding to the different genotypes. The *Bar* mutation in *Drosophila,* for example, reduces the size of the eye by different amounts in different individuals, but the reduction is always so large that there is no confusion between Bar and wild-type individuals. On the other hand, wheat plants in a cultivated field or wild asters at the side of the road are not neatly sorted into categories of "tall" and "short," any more than humans are neatly sorted into categories of "black" and "white." Height, weight, shape, color, metabolic activity, reproductive rate, and behavior are characteristics that vary more or less continuously over a wide range. Even when the character is intrinsically countable (such as eye facet or bristle number in wild-type *Drosophila*), the number of distinguishable classes may be so large that the variation is nearly *continuous.* If we consider extreme individuals—say, a corn plant 8 feet tall and another one 3 feet tall—a cross between them will not produce a uniform result in the F_1 generation. Rather, such a cross will produce plants about 5.5 feet tall, with some clear variation among the siblings. Furthermore, the F_2 from selfing the F_1 will not fall into two or three discrete height classes in ratios of 3:1 or 1:2:1. Instead, the F_2 will be continuously distributed in height from one parental extreme to the other.

The analysis of a continuously varying character can be carried out by an array of investigations, shown schematically in Figure 18-2:

- Norm of reaction studies, in which different genotypes are allowed to develop in an array of different environments in order to determine the interaction of genotype and environment in the development of the character.
- Selection studies, in which successive generations are produced from the extreme individuals in the previous generation (as, for example, by crossing the two shortest corn plants and the two tallest corn plants in our example above). If repeated generations of selection result in divergence between the populations selected for the opposite extremes of phenotype, then the divergent populations must differ genetically at one or more loci influencing the character.
- Heritability studies, in which the variation in the progeny of crosses is analyzed statistically to estimate the proportion of the variation in the original population that is a consequence of genetic and environmental differences affecting the character.
- Quantitative trait locus (QTL) studies, which associate phenotypic differences with alleles of a marker gene of known chromosomal location. Such an association with the marker gene reveals the approximate location of a gene affecting the quantitative character.

CHAPTER OVERVIEW figure

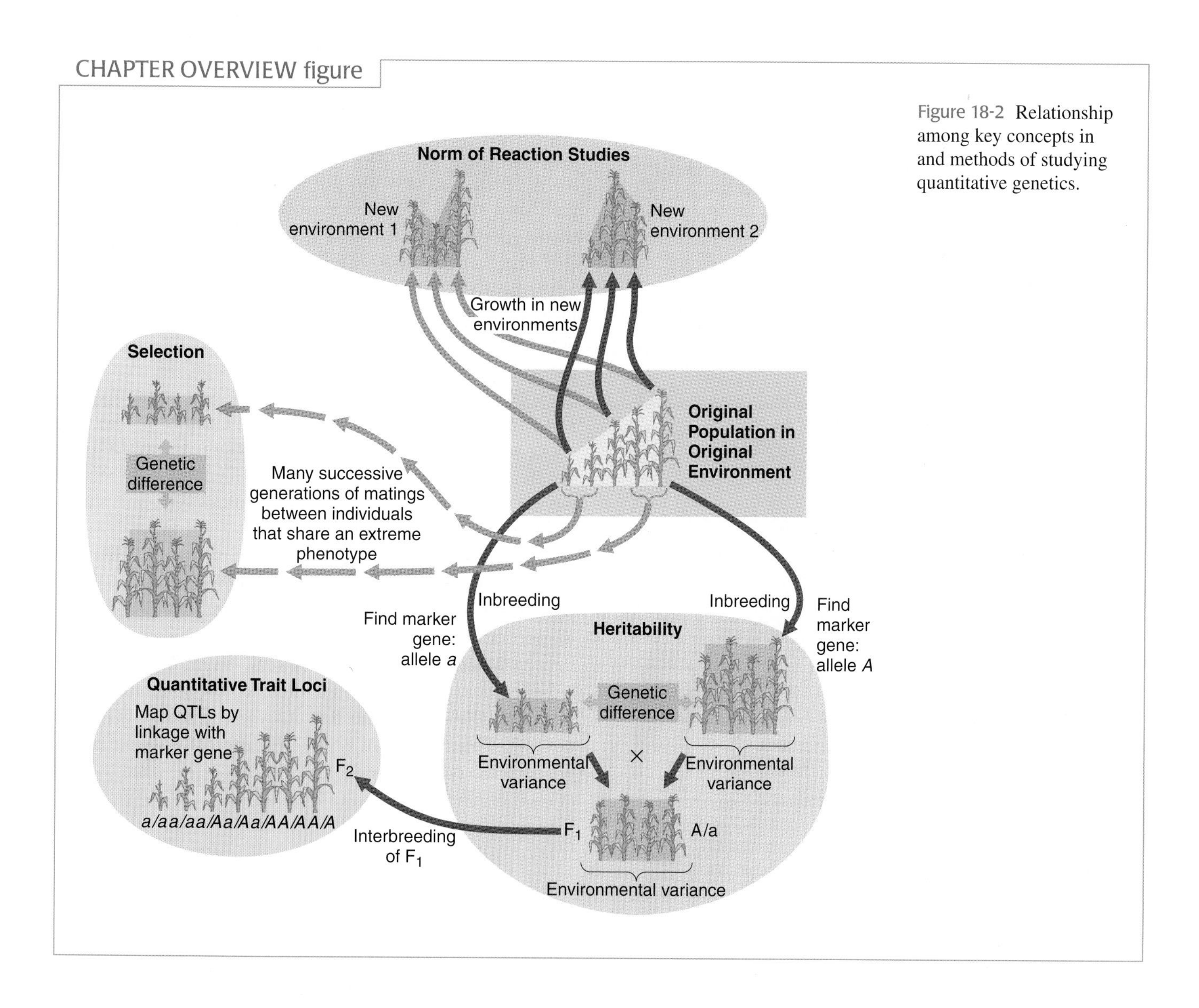

Figure 18-2 Relationship among key concepts in and methods of studying quantitative genetics.

GENES AND QUANTITATIVE TRAITS

A classic example of the outcome of crosses between strains that differ in a quantitative character is the experiment shown in Figure 18-3. The length of the corolla (flower tube) was measured in a number of individual plants from two true-breeding lines of *Nicotiana longiflora,* a relative of tobacco. The distribution of corolla lengths in the two parental lines is shown in the top panel of Figure 18-3. The difference between these two lines is genetic, but the variation among individuals within each line is a result of uncontrolled environmental variation and developmental noise (see Chapter 1). The F_1 plants, whose mean corolla length is very close to halfway between those of the two parental lines, also vary because of environmental and developmental variation. In the F_2, the mean corolla length remains essentially unchanged from the F_1, but there is a large increase in the variation, because there is now segregation of the genetic differences that were introduced from the two original parental lines. Evidence that at least part of this variation is the result of genetic differences among the F_2 individuals can be seen in the F_3. Different pairs of parents were chosen from four different parts of the F_2 distribution and crossed to produce the next, F_3, generation. In each case, the F_3 mean is close to the value of that part of the F_2 distribution from which its parents were sampled.

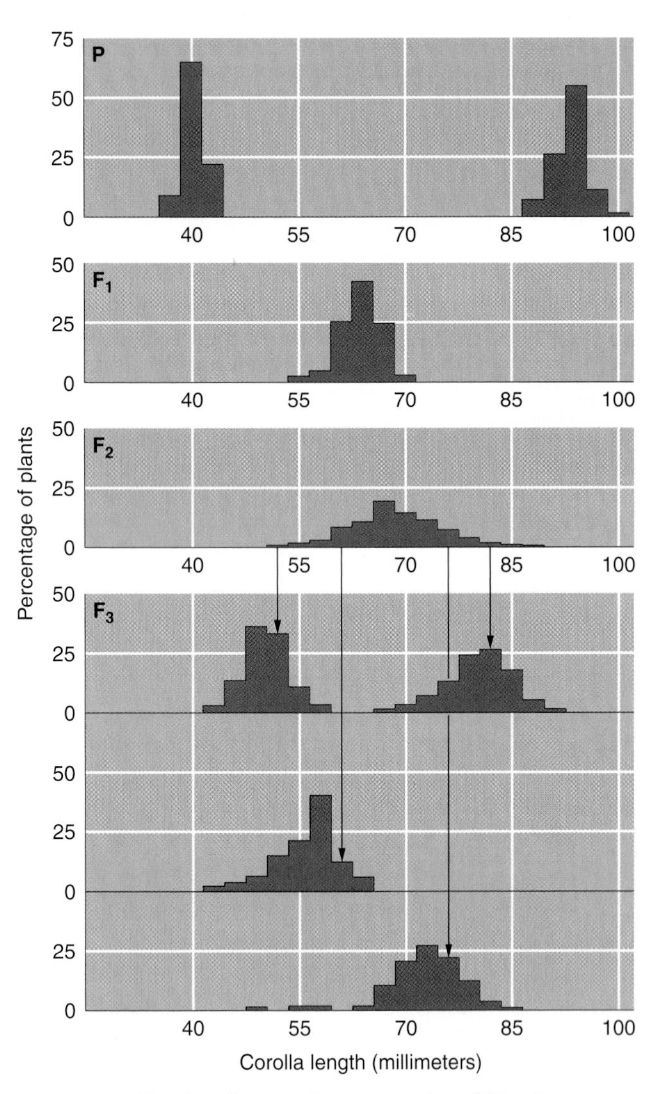

Figure 18-3 Results of crosses between strains of *Nicotiana longiflora* that differ in corolla length. The graphs show (from top to bottom) the frequency distribution of corolla lengths in the two parental strains (P); the distribution of corolla lengths in the F_1; the frequency distribution in the F_2; and the distribution of corolla lengths in four F_3 crosses made by taking parents from the four indicated parts of the F_2 distribution. *(Adapted from K. Mather,* Biometrical Genetics, *Methuen, 1959; data from E. M. East, 1916,* Genetics, *1:164–176.)*

This behavior of crosses is not an exception; it is the rule for most characters in most species. Mendel obtained his simple results because he worked with horticultural varieties of the garden pea that differed from one another by single allelic differences that had drastic phenotypic effects. If Mendel had conducted his experiments on the natural variation of the weeds in his garden, instead of on abnormal pea varieties, he would never have discovered any of his hereditary laws. The phenotypic differences of most organisms are a consequence of complex interactions among many genes and between genes and environment. In general, size, shape, color, physiological activity, and behavior do not assort in a simple way in crosses.

The fact that most characters vary continuously does not mean that their variation is the result of genetic mechanisms different from those that apply to the Mendelian characters that we have studied in earlier chapters. The continuity of a phenotype is a result of two phenomena. First, each genotype does not have a single phenotypic expression, but rather a norm of reaction (see Chapter 1) that covers a wide phenotypic range. As a result, the phenotypic differences between genotypic classes become blurred, and we are not able to assign a particular phenotype unambiguously to a particular genotype.

Second, many segregating loci may have alleles that make a difference to the phenotype being observed. Suppose, for example, that five equally important loci affect the number of flowers that will develop in an annual plant and that each locus has two alleles (call them $+$ and $-$). For simplicity, suppose also that there is no dominance, and that a $+$ allele adds one flower, whereas a $-$ allele adds nothing. Thus, there are $3^5 = 243$ different possible genotypes [three possible genotypes ($+/+$, $+/-$, and $-/-$) at each of five loci], ranging from

$$\frac{+++++}{+++++} \quad \text{through} \quad \frac{+++++}{-----} \quad \text{to} \quad \frac{-----}{-----}$$

but there are only 11 phenotypic classes (10 flowers, 9 flowers, . . . 0 flowers) because many of the genotypes have the same numbers of + and − alleles. For example, although there is only one genotype with 10 + alleles and therefore an average phenotypic value of 10, there are 51 different genotypes with 5 + alleles and 5 − alleles, such as

$$\frac{++++-}{+----} \quad \text{and} \quad \frac{++-+-}{++---}$$

Thus, many different genotypes may have the same average phenotype. At the same time, because of environmental variation, two individuals of the same genotype may *not* have the same phenotype. This lack of a one-to-one correspondence between genotype and phenotype obscures the underlying Mendelian mechanism.

If we cannot study the behavior of the Mendelian factors influencing such phenotypes directly, then what can we learn about their genetics? Using current experimental techniques, geneticists can answer the following questions about the genetics of a continuously varying character in a population (say, height in a human population). These questions constitute the study of **quantitative genetics**—the genetics of continuously varying characters.

1. Is the observed variation in the character influenced *at all* by genetic variation? Are there alleles segregating in the population that produce some differential effect on the character, or is all the variation simply the result of environmental variation and developmental noise?
2. If there is genetic variation, what are the norms of reaction of the various genotypes?
3. How important is genetic variation as a source of total phenotypic variation? Are the norms of reaction and the environments such that nearly all the variation is a consequence of environmental variation and developmental noise, or does genetic variation predominate?
4. Do many loci (or only a few) contribute to the variation in the character? How are they distributed over the genome?
5. How do the different loci interact with one another to influence the character? Is there dominance, and is there any epistasis (interaction between genes at different loci)?
6. Is there any nonnuclear inheritance (for example, any maternal effect)?

In the end, the purpose of asking these questions is to be able to predict what kinds of offspring will be produced by crosses of different phenotypes by mapping the traits genetically and by understanding the mechanistic basis of the genetic contributions to phenotype.

The precision with which these questions can be framed and answered varies greatly. In experimental organisms, on the one hand, it is relatively simple to determine whether there is any genetic influence on a character at all, but extremely laborious experiments are required to localize the genes involved (even approximately). In humans, on the other hand, it is extremely difficult to answer even the question of the presence of genetic influence for most traits, because it is almost impossible to separate environmental from genetic effects in an organism that cannot be manipulated experimentally. As a consequence, we know a lot about the genetics of bristle number in *Drosophila* but virtually nothing about the genetics of complex human characteristics; a few (such as skin color) clearly are influenced by genes, whereas others (such as the specific language spoken) clearly are not. It is the purpose of this chapter to develop the basic statistical and genetic concepts needed to answer these questions and to provide some examples of their application to particular characters in particular species.

SOME BASIC STATISTICAL NOTIONS

To consider the answers to these questions about the most common kinds of genetic variation, we must first examine a number of statistical tools that are essential in the study of quantitative variation.

Statistical Distributions

In the case of simple variation that depends only on the allelic differences at a single locus, the outcome of a cross can be described in terms of the proportions of the offspring that fall into several distinct phenotypic classes, or often simply in terms of the presence or absence of a class. For example, a cross between a red-flowered plant and a white-flowered plant might be expected to yield all red-flowered plants or, if it were a backcross of an F_1 individual to the white-flowered parent, 1/2 red-flowered plants and 1/2 white-flowered plants. However, we require a different mode of description for quantitative characters. If the heights of a large number of male undergraduates are measured to the nearest 5 centimeters (cm), they will vary (say, between 145 and 195 cm), but many more individuals will fall into the middle measurement classes (say, 170, 175, and 180 cm) than into the classes at the two extremes. Such a description of a set of quantitative measurements is known as a **statistical distribution.**

We can graph such measurements by representing each measurement class as a bar, with its height proportional to the number of individuals in that class, as shown in Figure 18-4a on the next page. Such a graph of numbers of individuals observed against measurement class is called a **frequency histogram.** Now suppose that we measure five times as many individuals, each to the nearest centimeter, so that we divide them into even smaller measurement classes, producing a histogram like the one shown in Figure

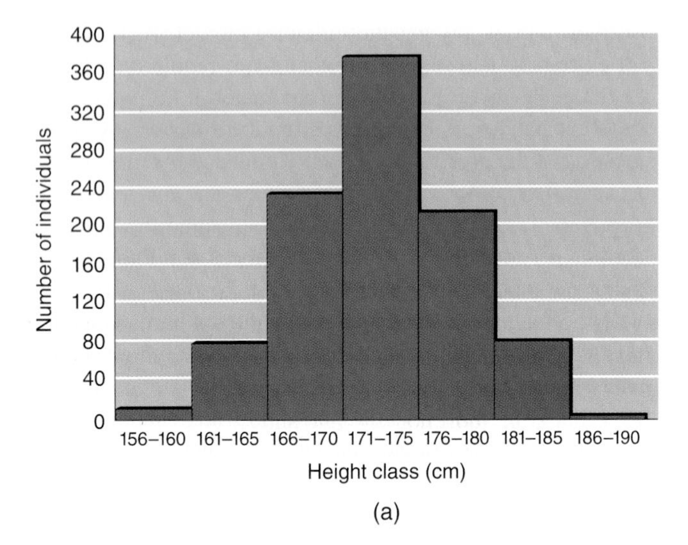

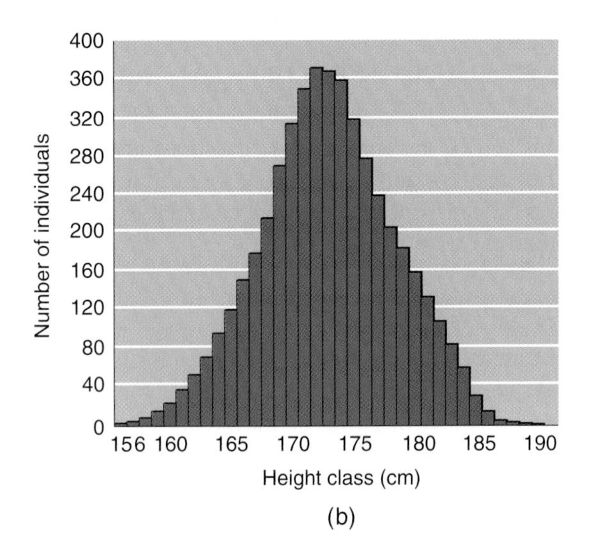

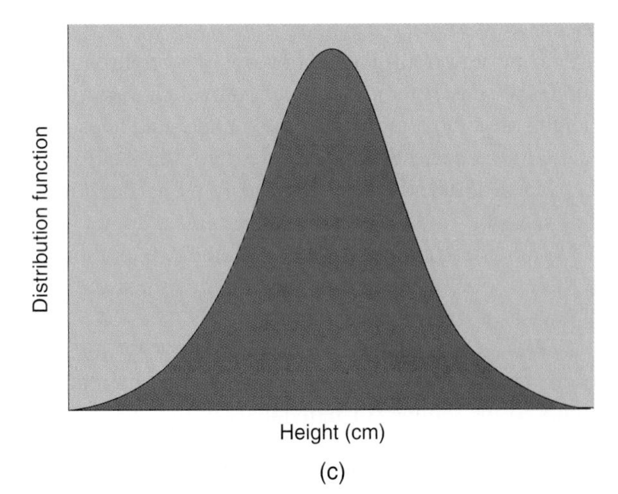

Figure 18-4 Statistical distributions for the height of male undergraduates. (a) A frequency histogram with 5-cm class intervals. (b) A frequency histogram with 1-cm class intervals. (c) The limiting continuous distribution function.

18-4b. If we continue this process, making each measurement finer but proportionately increasing the number of individuals measured, the histogram eventually takes on the continuous appearance of Figure 18-4c. Such a continuous curve is called the **distribution function** of the measure in the population.

The distribution function is an idealization of the actual frequency distribution of a measurement in any real population, because no measurement can be taken with infinite accuracy or on an unlimited number of individuals. Moreover, the measured character itself may be intrinsically discontinuous because it is the count of some number of discrete objects, such as eye facets or bristles. It is sometimes convenient, however, to develop concepts by using this slightly idealized curve as shorthand for the more cumbersome observed frequency histogram.

Statistical Measures

Although a statistical distribution contains all of the information we need about a set of measurements, it is often useful to distill this information into a few characteristic numbers that convey the necessary information about the distribution without giving it in detail. There are several questions about the height distribution for male undergraduates, for example, that we might like to answer:

1. Where is the distribution located along the range of possible values? Are our observed values of height, for example, closer to 100 or to 200 cm? This question can be answered with a measure of **central tendency.**
2. How much variation is there among the individual measurements? Are they all concentrated around the central measurement, or do they vary widely across a large range? This question can be answered with a measure of **dispersion.**
3. If we are considering more than one measured quantity, how are the values of the different quantities related? Do taller parents, for example, have taller sons? If they do, we would regard this as evidence that genes influence height. Thus, we need measures of **relation** between measurements.

Among the most commonly used measures of central tendency are the **mode,** which is the most frequent observation, and the **mean,** which is the arithmetic average of the observations. The dispersion of a distribution is almost always measured by the **variance,** which is the average squared deviation of the observations from their mean. The relation between different variables is measured by their **correlation,** which is the average product of the deviation of one variable from its own mean times the deviation of the other variable from its own mean. These common measures are discussed in detail in the Specialized Topics section on statistical analysis at the end of this chapter. The detailed discussion of these statistical concepts is placed in a separate section in order not to interrupt the flow of logic as we consider quantitative genetics. It should not be assumed, however, that an understanding of these statistical concepts is somehow secondary. A proper understanding of quantitative genetics requires a grasp of the basics of statistical analysis.

GENOTYPES AND PHENOTYPIC DISTRIBUTIONS

Using the concepts of distribution, mean, and variance, we can understand the difference between quantitative phenotypic variation and discrete segregation ratios such as those discussed in Chapters 5 and 6. Suppose that a population of plants contains three genotypes, *a* / *a*, *A* / *a*, and *A* / *A*, each of which has some differential effect on growth rate. Furthermore, assume that there is some environmental variation from plant to plant because the soil in which the population is growing is not homogeneous and that there is some developmental noise. For each genotype, there will be a separate distribution of phenotypes, with a mean and a variance that depend on the genotype and the set of environments. Assume that these distributions look like the three height distributions in Figure 18-5a. Finally, assume that the population consists of a mixture of the three genotypes, but in the unequal proportions 1:2:3 (*a* / *a* : *A* / *a* : *A* / *A*).

Under these circumstances, the phenotypic distribution of individuals in the population as a whole will look like the black line in Figure 18-5b, which is the result of summing the three underlying separate genotypic distributions, weighted by their frequencies in the population. This weighting by frequency is indicated in Figure 18-5b by the different heights of the component distributions. The mean of the total distribution is the average of the three genotypic means, again weighted by the frequencies of the genotypes in the population. The variance of the total distribution is produced partly by the environmental variation within each genotype and partly by the slightly different means of the three genotypes.

Two features of the total distribution are noteworthy. First, there is only a single mode. Despite the existence of three separate genotypic distributions underlying it, the population distribution as a whole does not reveal the separate modes. Second, any individual whose height lies between the two arrows in Figure 18-5b could have any one of the three genotypes, because the phenotypes of the three overlap so much. For these reasons, we cannot carry out a simple breeding analysis like those carried out by Mendel to determine the genotype of an individual plant. For example, suppose that the three genotypes are the two homozygotes and the heterozygote for a pair of alleles at a locus. Let *a* / *a* be the short homozygote and *A* / *A* be the tall one, with the heterozygote being of intermediate height. Because there is so much overlap of the phenotypic distributions, we cannot know to which genotype a given individual belongs. Conversely, if we cross a homozygote *a* / *a* and a heterozygote *A* / *a*, the offspring will not fall into two discrete phenotypic classes, *A* / *a* and *a* / *a*, in a 1:1 ratio, but will cover almost the entire range of phenotypes smoothly. Thus, we cannot know from looking at the offspring that the cross was, in fact, *a* / *a* × *A* / *a* rather than *a* / *a* × *A* / *A* or *A* / *a* × *A* / *a*.

Suppose we grew the hypothetical plants in Figure 18-5 in an environment that exaggerates the differences between genotypes—for example, by doubling the growth rate of all genotypes. At the same time, we were very careful to provide all the plants with exactly the same environment. Under these conditions, the phenotypic variance of each separate genotype would be reduced, because all the plants were grown under identical conditions; at the same time, the phenotypic differences between the genotypes would be exaggerated by the more rapid growth (Figure 18-6a on the next page). The result (Figure 18-6b) would be a separation of the population into three nonoverlapping phenotypic distributions, each characteristic of one genotype. We could now carry out a perfectly conventional Mendelian analysis of plant height. A "quantitative" character has been converted into a "qualitative" one. This conversion was accomplished by finding a way to make the

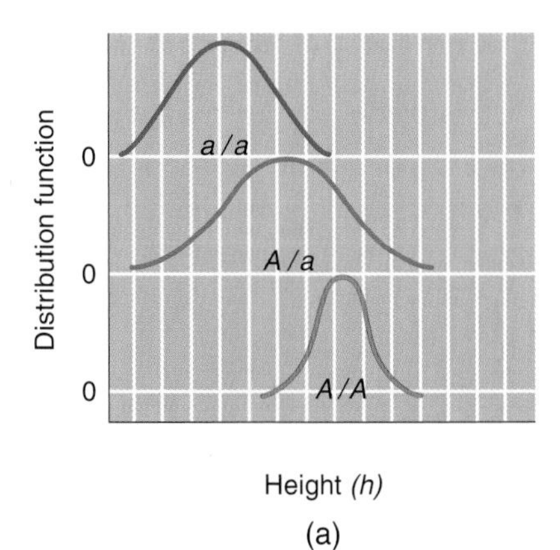

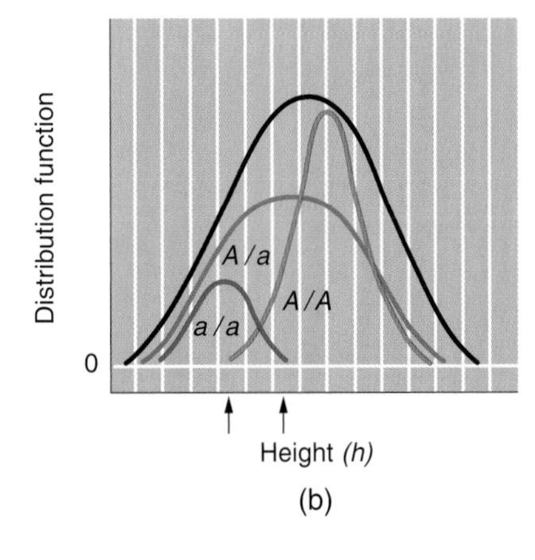

Figure 18-5 (a) Phenotypic distributions of three plant genotypes. (b) A phenotypic distribution for the total population (black line) can be obtained by summing the three genotypic distributions in the proportion 1:2:3 (*a* / *a* : *A* / *a* : *A* / *A*) in which they occur in the population (shown by the heights of the individual distribution curves).

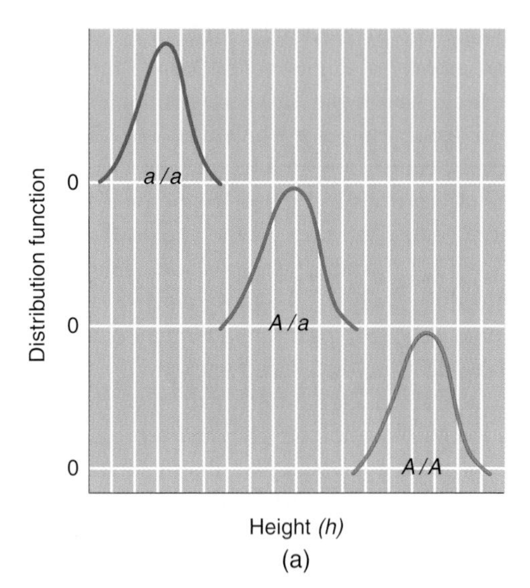

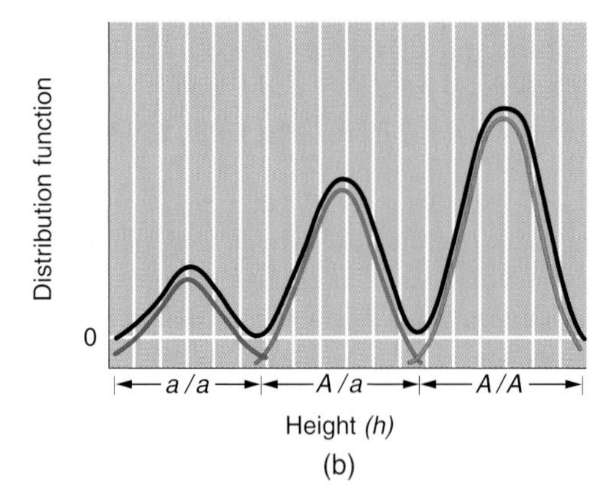

Figure 18-6 (a) Phenotypic distributions of the same three plant genotypes shown in Figure 18-5 when grown in carefully controlled environments. (b) The result is smaller phenotypic variation within each genotype and a greater difference between genotypes. The heights of the individual distributions are proportional to the frequencies of the genotypes in the population.

difference between the means of the genotypes large compared with the variation within each genotype.

The difference between the outcomes shown in Figures 18-5 and 18-6 can also be described in terms of differences in penetrance and expressivity (see Chapter 14). The expressivity of the genotypes—the intensity with which the different genotypes are manifested phenotypically—has changed. The *A* / *A* plants in Figure 18-6 are taller than those in Figure 18-5, and the height variation within the *A* / *a* plants has become much less than in Figure 18-5. This change in expressivity has resulted in changes in penetrance. The penetrance of a genetic difference—the proportion of individuals whose genetic difference can be observed as a distinct phenotypic difference—is 100 percent in Figure 18-6, but is only partial in Figure 18-5. We see that penetrance and expressivity are two facets of the same dependence of phenotype on other factors beside allelic differences at a particular locus.

MESSAGE

A *quantitative* character is one for which the average phenotypic differences between genotypes are small compared with the variation between individuals within genotypes.

It is sometimes assumed that continuous variation in a character is necessarily caused by a large number of segregating genes, so continuous variation is taken as evidence for control of the character by many genes. But, as we have just shown, this is not necessarily true. If the difference between genotypic means is small compared with the environmental variance, then even a simple one-gene–two-allele case can result in continuous phenotypic variation.

If the range of a character is limited, and if many segregating loci influence it, then we expect the character to show continuous variation, because each allelic substitution must account for only a small difference in phenotype. This **multiple-factor hypothesis** (that large numbers of genes, each with a small effect, are segregating to produce quantitative variation) has long been the basic model of quantitative genetics. It is important to remember, however, that the *number* of segregating loci that influence a trait is not what separates quantitative and qualitative characters. Even in the absence of large environmental variation, it takes only a few genetically varying loci (or even a single locus, as we saw in Figure 18-5) to produce variation that is indistinguishable from that caused by many loci of small effect.

As an example, let us consider one of the earliest experiments in quantitative genetics, that of Wilhelm Johannsen on pure lines. Pure lines—that is, lines that are true-breeding from generation to generation—have a number of uses in quantitative genetics. They are produced by inbreeding, the repeated mating of close relatives generation after generation. In corn, for example, in which self-fertilization is possible, a single individual is chosen and self-pollinated. Then, in the next generation, a single one of its offspring is chosen and self-pollinated. In the third generation, a single offspring is chosen and self-pollinated, and so forth. Suppose that the original individual chosen for selfing is already a homozygote at some locus. Then all of its offspring will also be homozygous and identical at that locus. Future generations of self-pollination will simply preserve this homozygosity. If, on the other hand, the original individual is a heterozygote, then the selfing of *A* / *a* × *A* / *a* will produce offspring that are 1/4 *A* / *A* homozygotes and 1/4 *a* / *a* homozygotes. If a single offspring is chosen to propagate the line, then there is a 50 percent chance that it will be a homozygote. If, by bad luck, the chosen individual should be a heterozygote, there is another 50 percent chance that the selected individual in the third generation will be homozygous, and so forth. Of the ensemble of all heterozygous loci, then, after one generation of selfing, only 1/2 will still be heterozygous; after two generations, 1/4; after three, 1/8. In the nth generation,

$$\text{Het}_n = \frac{1}{2^n}\text{Het}_0$$

where Het_n is the proportion of heterozygous loci in the nth generation and Het_0 is the proportion in the 0th generation. When selfing is not possible, brother-sister mating will accomplish the same end, although more slowly (see Figure 17-7).

Through inbreeding, Johannsen produced 19 homozygous lines of bean plants from an originally genetically heterogeneous population. Each line had a characteristic average seed weight. These weights ranged widely, from 0.64 g per seed for the heaviest line to 0.35 g per seed for the lightest line. Suppose that these lines are all genetically different. In that case, Johannsen's results would be incompatible with a simple one-locus–two-allele model of gene action. If the original population were segregating for the two alleles A and a, all inbred lines derived from that population would have to fall into one of two classes: A / A or a / a. If, in contrast, there were, say, 100 loci, each of small effect, segregating in the original population, then a vast number of different inbred lines could be produced, each with a different combination of homozygotes at different loci.

However, we do not need such a large number of loci to obtain the results observed by Johannsen. One possibility is that there was only a single variable gene, but that there were a large number of alleles of that gene segregating in the population. On the other hand, if the observed variation was the consequence of variation at several loci, each with only a few alleles, we would not need many such loci to produce the results observed by Johannsen. If there were only five loci, each with three alleles, then $3^5 = 243$ different kinds of homozygotes could be produced by the inbreeding process. If we make 19 inbred lines at random, there is a good chance (about 50 percent) that each of the 19 lines will belong to a different one of the 243 classes. So Johannsen's experimental results can be easily explained by the segregation of alleles at a relatively small number of genes (or even at only one gene with multiple alleles).

Thus, there is no real dividing line between "multigenic" traits and other traits. It is safe to say that no phenotypic character above the level of the amino acid sequence of a polypeptide is influenced by only one gene. The phenotypic manifestation of almost any single-gene mutation in *Drosophila,* for example, can be made more or less extreme by selective breeding in a population that is homozygous for the mutation but varies at a large number of other loci. Moreover, traits influenced by many genes are not equally influenced by all of them. Some genes will have major effects, others, minor effects.

MESSAGE

The critical difference between qualitative and quantitative character differences is not the number of segregating loci, but the size of phenotypic differences between genotypes compared with the individual variation within genotypic classes.

NORMS OF REACTION

Norms of Reaction and Phenotypic Distribution

The phenotype of an organism depends not only on its genotype but also on the environment that it has experienced at various critical stages in its development. For a given genotype, different phenotypes will develop in different environments. The relationship between environment and phenotype for a given genotype is called the genotype's **norm of reaction.** The norm of reaction of a genotype with respect to some environmental variable—say, temperature—can be visualized by a graph showing phenotype as a function of that variable, as exemplified in Figure 18-7.

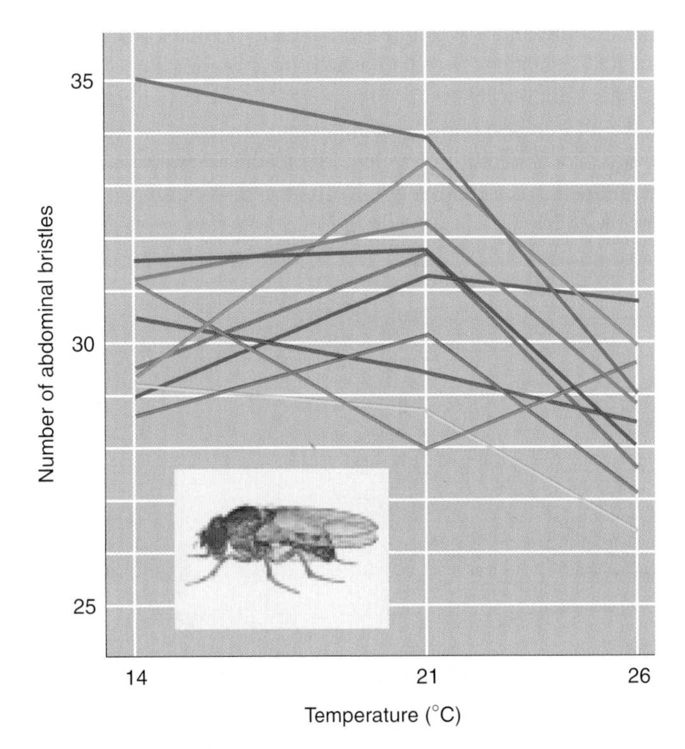

Figure 18-7 Norms of reaction for the number of abdominal bristles in different second-chromosome homozygotes of *Drosophila pseudoobscura* developing at three different temperatures. Each colored line represents a different genotype. *(Data courtesy of A. P. Gupta. Photo, Plate IV, University of Texas Publication 4313,* Studies in the Genetics of Drosophila III, The Drosophilidae of the Southwest, *by J. T. Patterson. Courtesy of the Life Sciences Library, University of Texas, Austin.)*

The phenotypic distribution of a character, as we have seen, is a function of the average phenotypic differences between genotypes and of the phenotypic variation among genotypically identical individuals. But, as the norms of reaction in Figure 18-7 show, both are functions of the environments in which the organisms develop and live. For a given genotype, each environment will result in a given phenotype (for the moment, ignoring developmental noise). Thus, for any given genotype, a *distribution of environments* will be reflected biologically as a *distribution of phenotypes.*

The way in which the environmental distribution is transformed into the phenotypic distribution depends on the norm of reaction, as shown in Figure 18-8, in which the horizontal axis represents environment (say, temperature) and the vertical axis represents phenotype (say, plant height). The norm of reaction curve for a genotype shows how each particular temperature results in a particular plant height. This norm of reaction converts a distribution of environments into a distribution of phenotypes. Thus, for example, the dashed line from the 18°C point on the horizontal environment axis is reflected off the norm of reaction curve to a corresponding plant height on the vertical phenotype axis, and so forth for each temperature. If a large number of individuals develop at, say, 20°C, then a large number of individuals will have the phenotype that corresponds to 20°C, as shown by the dashed line from the 20°C point; if only small numbers develop at 18°C, few individuals will have the corresponding plant height. In other words, the frequency distribution of developmental environments will be reflected as a frequency distribution of phenotypes as determined by the shape of the norm of reaction curve. It is as if an observer, standing at the vertical phenotype axis, were seeing the environmental distribution, not directly, but reflected in the curved mirror of the norm of reaction. The shape of its curvature will determine how the environmental distribution is distorted on the phenotype axis. Thus, the norm of reaction in Figure 18-8 falls very rapidly at lower temperatures (the phenotype changes dramatically with small changes in temperature) but flattens out at higher temperatures, showing that plant height is much less sensitive to temperature differences at the higher temperatures. The result is that the symmetrical environmental distribution is converted into an asymmetrical phenotypic distribution with a long tail at the larger plant heights, corresponding to the lower temperatures.

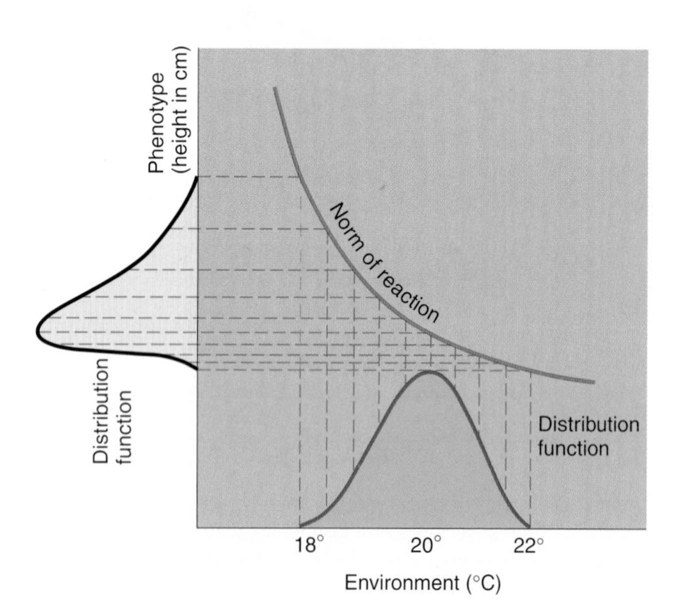

Figure 18-8 The distribution of environments on the horizontal axis is converted into the distribution of phenotypes on the vertical axis by the norm of reaction of a genotype.

Determining Norms of Reaction

Remarkably little is known about the norms of reaction for any quantitative trait in any species. This is partly because determining a norm of reaction requires the testing of many individuals of identical (or near identical) genotype. It is for this reason, for example, that we do not have a norm of reaction for any genotype for any human quantitative character, although we do have basic knowledge about how such characters are influenced by the environment. Thus, we know that malnutrition can cause reductions in height and muscle mass, and that sunlight darkens light skins, and we understand the basic cellular mechanisms that are responsible for these effects. What we lack is knowledge of the differential reactions of different genotypes to these environmental factors.

Three methods are available for replicating a genotype for norm of reaction studies:

1. For a few experimental organisms, special methods make it possible to replicate genotypes and thus to determine norms of reaction. Such studies are particularly easy in plants that can be propagated vegetatively (that is, by cuttings). The pieces cut from a single plant all have the same genotype; they are, in fact, clones. All offspring produced in this way have identical genotypes. Such a study was performed using the yarrow plant, *Achillea millefolium* (Figure 18-9a). Many plants were collected, and three cuttings were taken from each plant. One cutting from each plant was planted at low elevation (30 meters above sea level), one at medium elevation (1400 meters), and one at high elevation (3050 meters). Figure 18-9b shows the mature individuals that developed from the cuttings of seven plants; each set of three plants of identical genotype is aligned vertically in the figure for comparison. As the figure shows, the ordering of plants from tallest to shortest at the low elevation is completely changed at the two higher elevations. It is not possible to say which genotype results in the tallest

(a)

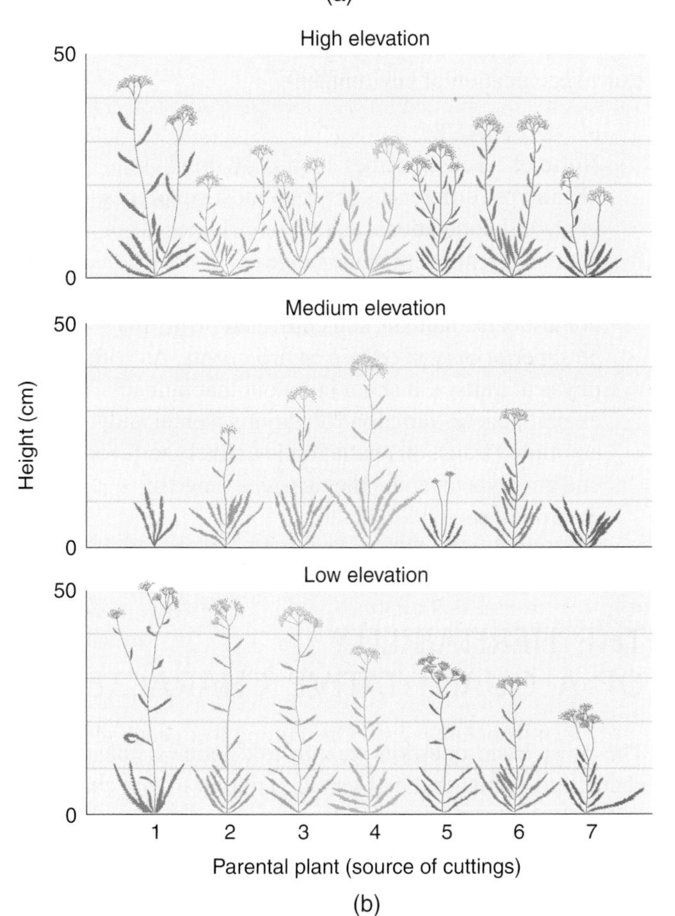

(b)

Figure 18-9 (a) *Achillea millefolium.* (b) Norms of reaction of plant height to elevation for seven different individual *Achillea* plants (seven different genotypes). A cutting from each plant was grown at low, medium, and high elevations. *(Part a, Harper Horticultural Slide Library; part b, Carnegie Institution of Washington.)*

or shortest plant unless the environment is also specified. Cloning of animals is now possible, so this method, previously restricted to plants, is now available for some animal studies.

2. With domesticated species, inbred lines can be created by mating close relatives. By selfing (as with corn plants) or by mating brother and sister repeatedly generation after generation, a **segregating line** (one that contains both homozygotes and heterozygotes at a locus) can be made homozygous, as we saw above.
3. In species such as *Drosophila,* in which the necessary dominant markers and crossover suppressors are available, it is possible in a few generations of crosses to produce lines that are homozygous for one or two entire chromosomes.

The purpose of creating homozygous lines is to produce groups of organisms within which all individuals are genetically identical. These homozygous lines can then be allowed to develop in different environments to determine a norm of reaction for the genotype. Alternatively, two different homozygous lines can be crossed and the heterozygous F_1 offspring, all genetically identical with one another, can be tested in different environments.

Very few norm of reaction studies have been carried out for quantitative characters found in natural populations, but many have been carried out for domesticated species such as corn, which can be self-pollinated, or strawberries, which can be clonally propagated. The outcomes of such studies resemble those given in Figure 18-7. No genotype consistently produces a phenotypic value above or below that of the others under all environmental conditions. Instead, there are small differences between genotypes, and the direction of these differences varies over a wide range of environments.

These features of norms of reaction have important consequences. One consequence is that selection for "superior" genotypes in domesticated animals and cultivated plants will result in varieties adapted to very specific conditions, which may not show their superior properties in other environments. To some extent, this problem can be overcome by deliberately testing genotypes in a range of environments (for example, over several years and in several locations). It would be even better, however, if plant breeders could test their selections in a variety of controlled environments in which different environmental factors could be separately manipulated.

The consequences of actual plant-breeding practices can be seen in Figure 18-10 on the next page, in which the yields of two varieties of corn are shown as a function of different farm environments. Variety 1 is an older variety of hybrid corn; variety 2 is a later "improved" hybrid. Their performances are compared at a low planting density, which was usual when variety 1 was developed, and at a high planting density, characteristic of farming practice

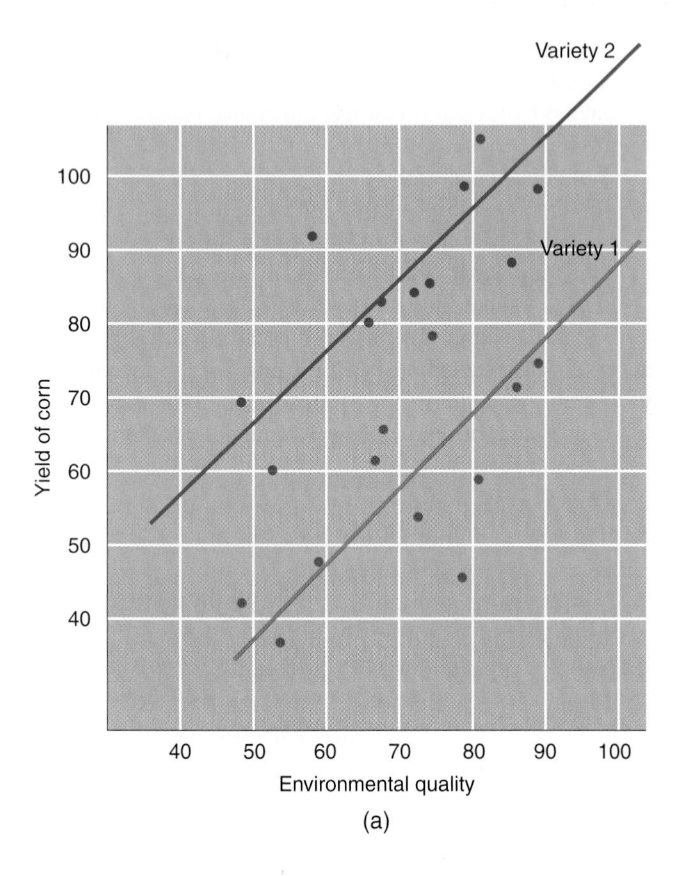

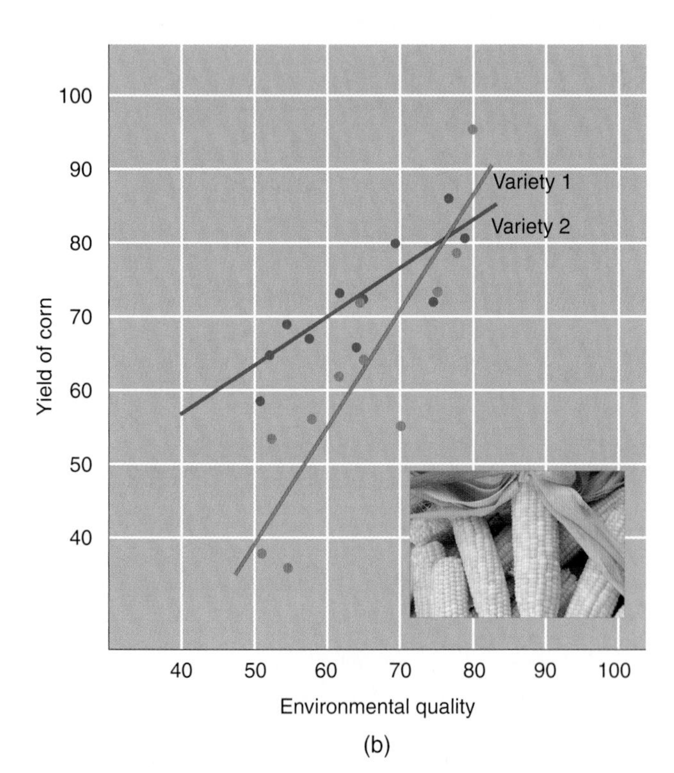

Figure 18-10 Yields of grain of two varieties of corn in different environments (a) at a high planting density and (b) at a low planting density. *(Data courtesy of W. A. Russell,* Proceedings of the 29th Annual Corn and Sorghum Research Conference, *1974. Photo © Bonnie Sue/Photo Researchers.)*

when variety 2 was created. At the high planting density, variety 2 is clearly superior to variety 1 in all environments (Figure 18-10a). At the low planting density (Figure 18-10b), however, the situation is quite different. First, note that the new variety is less sensitive to environmental variation than the older hybrid, as evidenced by its flatter norm of reaction. Second, the new "improved" variety actually performs more poorly than the older variety under the best farm conditions. Third, the yield improvement of the new variety does not occur under the low planting densities characteristic of earlier agricultural practice.

The nature of norms of reaction also has implications for human social relations and policy. Even if it should turn out that there is genetic variation for various mental and emotional traits in the human species—which is by no means clear—this variation is unlikely to favor one genotype over another across a range of environments. We must beware of hypothetical norms of reaction for human cognitive traits that show one genotype being unconditionally superior to another. Even putting aside all questions of moral and political judgment, there is simply no basis for describing different human genotypes as "better" or "worse" on any scale, unless the investigator is able to make a very exact specification of environment.

MESSAGE

Norm of reaction studies show only small phenotypic differences between most genotypes in natural populations, and these differences are not consistent over a wide range of environments. Thus, "superior" genotypes in domesticated animals and cultivated plants may be superior only in certain environments. As with physical traits, if it should turn out that humans exhibit genetic variation for various mental and emotional traits, this variation is unlikely to favor one genotype over another across a range of environments.

THE HERITABILITY OF A QUANTITATIVE CHARACTER

The most basic question we can ask about a quantitative character is whether the observed variation in that character is influenced by genes at all. It is important to note that this is not the same as asking whether genes play any role in the character's development. Gene-mediated developmental processes lie at the base of every character, but *variation* in a character from individual to individual is not necessarily the result of *genetic variation.* For example, the possibility of an individual speaking any language at all depends critically on the structures of the central nervous system as well as on the vocal cords, tongue, mouth, and ears, which depend in turn on many genes in the human genome. There is no environment in which cows will speak. But, although the particular language that is spoken by humans varies from nation to nation, that variation is not genetic. A character is said to show **heritability** only if there is genetic variation in that character.

MESSAGE

The question of whether a trait is heritable is a question about the role that genetic differences play in the phenotypic differences between individuals or groups.

Familiality and Heritability

In principle, it is easy to determine whether any genetic variation influences the phenotypic variation in a particular trait. If genes are involved, then (on average) biological relatives should resemble one another more than unrelated individuals do. This resemblance would be seen as a positive correlation in the values of a trait between parents and offspring or between siblings (offspring of the same parents). Parents who are larger than the average, for example, would have offspring who are larger than the average; the more seeds a plant produces, the more seeds its siblings would produce. Such correlations between relatives, however, are evidence for genetic variation *only if the relatives do not share common environments more than nonrelatives do.* It is absolutely fundamental to distinguish *familiality* from *heritability.* Character states are **familial** if members of the same family share them, for whatever reason. They are **heritable** only if the similarity arises from shared genotypes.

There are two general methods for establishing the heritability of a trait as distinct from its familiality. The first depends on *phenotypic similarity* between relatives. For most of the history of genetics, this method has been the only one available, so nearly all the evidence about heritability for most traits in experimental organisms and in humans has been established using this approach. The second method, using *marker-gene segregation,* depends on showing that genotypes carrying different alleles of certain marker genes also differ in their average phenotype for a quantitative character. If the marker genes (which have nothing to do with the character under study) are seen to vary in relation to the character, then presumably they are linked to genes that *do* influence the character and its variation. Thus, heritability is demonstrated even if the actual genes causing the variation in the character are not known. This method requires that the organism being studied have large numbers of detectable, genetically variable marker loci spread throughout its genome. Such marker loci can be observed through variants in DNA sequence, electrophoretic studies of protein variation or, in vertebrates, immunological studies of blood group proteins. Within flocks, for example, chickens with different blood groups show some difference in egg weight, but as far as is known, the blood group antigens and antibodies do not themselves cause the difference in egg size. Presumably, genes that do influence egg weight are linked to the loci determining blood group.

Since the introduction of molecular methods for studying DNA sequences, a great deal of variation has been discovered in a great variety of organisms. This variation consists either of substitutions at single nucleotide positions or of variable numbers of insertions or repeats of short sections of DNA. These variations are usually detected by the gain or loss of recognition sites for restriction enzymes or by length variation in DNA sequences between two fixed restriction sites, both of which are forms of restriction fragment length polymorphism (RFLP; see Chapter 17). In tomatoes, for example, strains carrying different RFLP variants differ in fruit characteristics. It is assumed that the DNA sequences involved in these RFLPs do not themselves influence fruit characteristics, but rather are landmarks located near genes that do, and therefore show high levels of cosegregation for these characteristics.

Because so much of what is known or claimed about heritability still depends on phenotypic similarity between relatives, however, especially in human genetics, we shall begin our examination of the problem of heritability by analyzing phenotypic similarity.

Phenotypic Similarity between Relatives

In experimental organisms, there is no problem in separating environmental from genetic similarities. The offspring of a cow producing milk at a high rate and the offspring of a cow producing milk at a low rate can be raised together in the same environment to see whether, despite the environmental similarity, each resembles its own parent. In natural populations, however, and especially in humans, this kind of study is difficult to perform. Because of the nature of human societies, members of the same family share not only genes, but also similar environments. Thus, the observation of simple familial similarity of phenotype is genetically uninterpretable. In general, people who speak Hungarian have Hungarian-speaking parents and people who speak Japanese have Japanese-speaking parents. Yet the massive experience of immigration to North America has demonstrated that these linguistic differences, although familial, are nongenetic. The highest correlations between parents and offspring for any social trait in the United States are those for political party and religious affiliation, but these traits are not heritable. The distinction between familiality and heredity is not always so obvious, however. The U.S. Public Health Commission, when it studied the vitamin-deficiency disease pellagra in the southern United States in 1910, came to the conclusion that it was genetic because it ran in families. However, pellagra is now well understood to have been prevalent in southern U.S. populations because of their poor diet.

To determine whether a human trait is heritable, we must use studies of certain adopted individuals to avoid the usual environmental similarity between biological relatives. The ideal experimental subjects are monozygotic (identical) twins reared apart because they are genetically identical but experience different environments. Such adoption studies must be so contrived that there is no correlation between the social environments of the adopting families and those of the biological families, or else the similarities between the twins' environments will not have been eliminated by the adoption. These requirements are exceedingly difficult to meet; therefore, in practice, we know very little about whether human quantitative characters that are familial are also heritable.

Skin color is clearly heritable, as is adult height—but even for characters such as these we must be very careful. We know that skin color is affected by genes, both from studies of cross-racial adoptions and from observations that the offspring of black African slaves were black even when they were born and reared in North America. But are the differences in height between Japanese and Europeans affected by genes? The children of Japanese immigrants who are born and reared in North America are taller than their parents but shorter than the North American average, so we might conclude that there is some influence of genetic difference. However, second-generation Japanese Americans are even taller than their American-born parents. It appears that some environmental-cultural influence, possibly nutritional, or perhaps an effect of maternal inheritance, is still felt in the first generation of births in North America. We cannot yet say anything definitive about genetic differences that might contribute to the height differences between North Americans of, say, Japanese and Swedish ancestry.

Personality traits, temperament, cognitive performance (including IQ scores), and a whole variety of behaviors such as alcoholism and mental disorders such as schizophrenia have been the subject of heritability studies in human populations. Many of these traits show familiality (that is, familial similarity). There is a positive correlation, for example, between the IQ scores of parents and the scores of their children (the correlation is about 0.5 in white American families), but this correlation does not distinguish familiality from heritability. To make that distinction requires that the environmental correlation between parents and children be broken, so studies on adopted children are common. Because it is difficult to randomize environments, even in cases of adoption, evidence of heritability for human personality and behavior traits remains equivocal despite the very large number of studies that exist.

Figure 18-11 summarizes the usual method of testing for heritability in experimental organisms. Individuals from both extremes of the phenotypic distribution are mated with other individuals of their own extreme group, and the offspring are raised in a common controlled environment. If there is an average difference between the two offspring groups, the phenotypic difference is heritable. Most morphological characters in *Drosophila,* for example, turn out to be heritable—but not all of them. If flies with right wings that are slightly longer than their left wings are mated with each other, their offspring have no greater tendency to be "right-winged" than do the offspring of "left-winged" flies. As we shall see below, this method can also be used to obtain quantitative information about heritability.

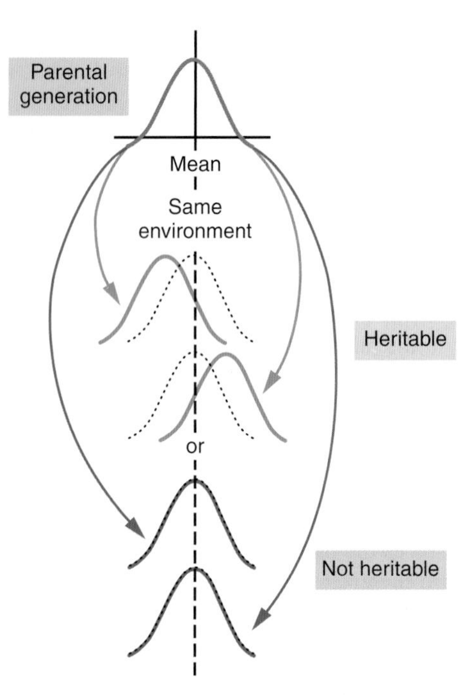

Figure 18-11 Standard method of testing for heritability in experimental organisms. Crosses are performed within two populations of individuals selected from the extremes of the phenotypic distribution in the parental generation. If the phenotypic distributions of the two groups of offspring are significantly different from each other (red curves), then the character difference is heritable. If both offspring distributions resemble the distribution for the parental generation (blue curves), then the phenotypic difference is not heritable.

MESSAGE

In experimental organisms, environmental similarity can often be readily distinguished from genetic similarity (or heritability). In humans, however, it is very difficult to determine whether a particular trait is heritable.

QUANTIFYING HERITABILITY

If a character is shown to have some heritability in a population, then it is possible to quantify the degree of heritability. In Figure 18-3, we saw that the variation between phenotypes in a population arises from two sources. First, there are average differences between the genotypes; second, each genotype exhibits phenotypic variation because of environmental variation. The total phenotypic variance of the population (s_p^2) can thus be broken into two parts: the variance between genotypic means (s_g^2) and the remaining variance (s_e^2). The former is called the **genetic variance,** and the latter is called the **environmental variance;** however, as we shall see, these names are quite misleading. Moreover, the breakdown of the phenotypic variance into environmental and genetic variances leaves out the possibility of some covariance between genotype and environment. For example, suppose it were true (we do not know this) that there were genes that influence musical ability in humans. Parents with such genes might themselves be musicians, who would create a more musical environment for their children, who would then have both the genes and the environment promoting musical performance. The result would be an increase in the phenotypic variance of musical ability in the human population and an erroneous estimate of genetic and environmental variances. If the phenotype is the sum of a genetic and an environmental effect, $p = g + e$, then the variance of the phenotype is the sum of the genetic variance, the environmental variance, and twice the covariance between the genotypic and environmental effects:

$$s_p^2 = s_g^2 + s_e^2 + 2 \text{ cov } ge$$

If genotypes are not distributed randomly across environments, there will be some covariance between genotypic and environmental values, and that covariance will be hidden in the genetic and environmental variances.

The heritability of a character can be defined as that part of the total phenotypic variance that is due to genetic variance:

$$H^2 = \frac{s_g^2}{s_p^2} = \frac{s_g^2}{s_g^2 + s_e^2}$$

H^2, so defined, is called the **broad heritability** of the character.

It must be stressed that this measure of "genetic influence" tells us what part of the population's *variation* in phenotype can be attributed to *variation* in genotype. It does not tell us what parts of an *individual's* phenotype can be ascribed to its genotype and to its environment. This latter distinction is not a reasonable one. An individual's phenotype is a consequence of the interaction between its genes and the sequence of environments it experiences as it develops. It would be silly to say that 60 inches of your height were produced by your genes and 10 inches were then added by your environment. All measures of the "importance" of genes are framed in terms of the proportion of phenotypic variance ascribable to their variation. This approach is a special application of the more general technique of **analysis of variance,** used for apportioning relative weight to contributing causes. The technique was, in fact, invented originally to deal with experiments in which different environmental and genetic factors were influencing the growth of plants. (For a sophisticated but accessible treatment of the analysis of variance written for biologists, see R. Sokal and J. Rohlf, *Biometry,* 3d ed., W. H. Freeman and Company, 1995.)

Methods of Estimating H^2

Genetic variance and heritability in a population can be estimated in several ways. Most directly, we can obtain an estimate of the environmental variance in the population, s_e^2, by making a number of homozygous lines, crossing them in pairs to reconstitute individual heterozygotes typical of the population, and measuring the phenotypic variance *within* each heterozygous genotype. Because there is no genetic variance within a genotypic class, these variances will (when averaged) provide an estimate of s_e^2. This value can then be subtracted from the value of s_p^2 in the original population to give s_g^2. With the use of this method, any covariance between genotype and environment in the original population will be hidden in the estimate of genetic variance and will inflate it. So, for example, if individuals with genotypes that would make them taller on average over random environments were also given better nutrition than individuals with genotypes that would make them shorter, then the observed difference in heights between the two genotypic groups would be exaggerated.

Other estimates of genetic variance can be obtained by considering the genetic similarities between relatives. Using simple Mendelian principles, we can see that half the genes of two full siblings will (on average) be identical. For identification purposes, we can label the alleles at a locus carried by the parents uniquely—say, as A_1 / A_2 and A_3 / A_4. The older sibling has a probability of 1/2 of getting A_1 from its father, as does the younger sibling, so the two siblings have a chance of $1/2 \times 1/2 = 1/4$ of both carrying A_1. On the other hand, they might both receive A_2 from their father; so, again, they have a probability of 1/4 of sharing that allele. Thus, the chance is $1/4 + 1/4 = 1/2$ that both siblings will inherit the same allele from their father. The other half of the time, one sibling will inherit an A_1 and the other will inherit an A_2. So, as far as paternally inherited genes are concerned, full siblings have a 50 percent chance of carrying the same allele. But the same reasoning applies to their maternally inherited allele. Averaging over their paternally and maternally inherited genes, half the genes of full siblings will be identical between them. Their **genetic correlation,** which is equal to the chance that they carry the same allele, will be 1/2.

If we apply this reasoning to half-siblings, say, with a common father but with different mothers, we get a different result. Again, the two siblings have a 50 percent chance of inheriting an identical gene from their father, but this time they have no way of inheriting the same gene from their mothers because they have two different mothers. Averaging the maternally inherited and paternally inherited genes thus gives a probability of (1/2 + 0)/2 = 1/4 that these half-siblings will carry the same gene.

We might be tempted to use this theoretical correlation between relatives to estimate H^2. If the observed phenotypic correlation between siblings were, for example, 0.4, and we expected, on purely genetic grounds, a correlation of 0.5, then our estimate of heritability would be 0.4/0.5 = 0.8. But such an estimate fails to take into account the fact that the environments of siblings may also be correlated. Unless we are careful to raise the siblings in independent environments, our estimate of H^2 will be too large, and could even exceed 1 if the observed phenotypic correlation were greater than 0.5.

All these estimates, as well as others based on correlations between relatives, depend *critically* on the assumption that environmental correlations between individuals are the same for all degrees of relationship—which is unlikely to be the case. Full sibs, for example, are usually raised by the same pair of parents, while half-sibs are likely to be raised in circumstances with only one parent in common. If closer relatives have more similar environments, as they do in humans, these estimates of heritability will be biased. It is reasonable to assume that most environmental correlations between relatives are positive, in which case the heritabilities would be overestimated. But negative environmental correlations also can exist. For example, if the members of a litter must compete for food that is in short supply, there could be negative correlations in growth rates among siblings.

The difference in phenotypic correlation between monozygotic and dizygotic twins is commonly used in human genetics to estimate H^2 for cognitive or personality traits. Here the problem of degree of environmental similarity is very severe. Monozygotic (identical) twins are generally treated more similarly to each other than are dizygotic (fraternal) twins. Parents often give their identical twins names that are similar, dress them alike, treat them the same way, and, in general, accentuate their similarities. As a result, heritability is overestimated.

The Meaning of H^2

Attention to the problems of estimating broad heritability distracts from the deeper questions about the meaning of the ratio even when it can be estimated. Despite its widespread use as a measure of how "important" genes are in influencing a character, H^2 actually has a special and limited meaning.

There are two alternative conclusions that can be drawn from the results of a properly designed heritability study. First, if there is a nonzero heritability, we can conclude that, in the population measured and in the environments in which the organisms have developed, genetic differences have influenced the phenotypic variation among individuals, so genetic differences do matter to the trait. This is not a trivial finding, and it is a first step in a more detailed investigation of the role of genes.

It is important to notice that the reverse is not true. If zero heritability for the trait is found, that is not a demonstration that genes are irrelevant to the trait, but only that, in the particular population studied, either there is no genetic variation at the relevant loci, or the environments in which the population developed were such that different genotypes resulted in the same phenotype. In other populations or other environments, the character might be heritable.

MESSAGE

The heritability of a character difference is different in each population and in each set of environments; it cannot be extrapolated from one population and set of environments to another.

Moreover, we must distinguish between *genes* being relevant to a trait and *genetic differences* being relevant to *differences* in a trait. The natural experiment of immigration to North America has proved that the ability to pronounce the sounds of North American English, rather than French, Swedish, or Russian, is not a consequence of genetic differences between our immigrant ancestors. But, without the appropriate genes, we could not speak any language at all.

Second, the value of H^2 provides a limited prediction of the effect of environmental modification under particular circumstances. If all the relevant environmental variation is eliminated *and the new constant environment is the same as the mean environment in the original population,* then H^2 estimates how much phenotypic variation will still be present. So, if the heritability of performance on an IQ test were found to be, say, 0.4, then we could predict that, if all children had the same developmental and social environment as the "average child," about 60 percent of the variation in IQ test performance would disappear and 40 percent would remain.

The requirement that the new constant environment be at the mean of the old environmental distribution is absolutely essential to this prediction. If the environment is shifted toward one end or the other of the environmental distribution present in the population used to determine H^2, or if a new environment is introduced, nothing at all can be predicted. In the example of IQ test performance, the heritability gives us no information at all about how variable performance would be if children's developmental and social environments were generally enriched. To understand why this is so, we must return to the concept of the norm of reaction.

The separation of phenotypic variance into genetic and environmental components, s_g^2 and s_e^2, does not really separate the genetic and environmental causes of variation. Consider the results we saw in Figure 18-10b. When the environment is poor (an environmental quality of 50), corn variety 2 has a much higher yield than variety 1, so a population made up of a mixture of the two varieties would have a lot of genetic variance for yield in that environment. But in a richer environment (scoring 75), there is no difference in yield between varieties 1 and 2, so a mixed population would have no genetic variance at all for yield in that environment. Thus, *genetic* variance has been changed by changing the *environment.* On the other hand, variety 2 is less sensitive to environment than variety 1, as shown by the slopes of the two lines. So a population made up mostly of variety 2 would have a lower environmental variance than one made up mostly of variety 1. So, *environmental* variance in the population is changed by changing the proportion of *genotypes.*

As a consequence of the argument just given, knowledge of the heritability of a character difference does not permit us to predict how the distribution of variation in the character will change if either genotypic frequencies or environmental factors change markedly.

MESSAGE

A high heritability does not mean that a character is unaffected by the environment. Because genotype and environment interact to produce phenotype, no partition of variation into its genetic and environmental components can actually separate causes of variation.

This argument has immense importance for public policy, especially since studies have claimed that variation in some human characteristics, such as IQ and propensity toward violence, may have a high heritability. It is often argued that a high heritability means that environmental changes can have only a limited effect in changing the phenotype, so that major efforts in education or public health cannot have any important effect on such characteristics. According to this view, demands for social equality are necessarily unrealistic because the inequalities are based on heritable, and therefore unchangeable, biological properties. Therefore, to alter the incidence of violence or low IQ, it would be necessary to influence the reproductive rates and prevent the immigration of "undesirables." The error of this claim is in its misunderstanding of the meaning of heritability. All that high heritability means is that, for the particular population developing in the particular distribution of environments in which the heritability was measured, average phenotypic differences between genotypes are large compared with environmental variation within genotypes. If the environment is changed, large changes in phenotype may occur. Thus, the analysis of variance, which is the technique that is used to estimate the genetic and environmental variance, does not allow us to predict what changes in phenotype will occur if the distribution of environments is changed. Only a study of reaction norms can do that.

Perhaps the best-known example of the erroneous use of heritability arguments to make claims about the changeability of a trait is the case of human IQ test performance and social success. In 1969, an educational psychologist, A. R. Jensen, published a long paper in the *Harvard Educational Review,* asking the question (in its title) "How much can we boost IQ and scholastic achievement?" Jensen's conclusion was "not much." As evidence of this unchangeability, he offered a claim of high heritability for IQ test performance. A great deal of criticism has been directed at the evidence offered by Jensen for the high heritability of IQ scores. But, irrespective of the value of H^2 for IQ test performance, the real error of Jensen's argument lies in his assumption that high heritability of IQ means that IQ is unchangeable by environmental modification. In fact, the heritability of IQ is *irrelevant* to the question raised in the title of his article.

To see why this is so, let us consider the usual results of IQ studies on children that have been separated from their biological parents in infancy and reared by adoptive parents. Although these results vary quantitatively from study to study, they have three characteristics in common. First, because adoptive parents usually come from a better-educated population than the biological parents who offer their children for adoption, they generally have higher IQ scores than the biological parents. Second, the adopted children have higher IQ scores than their biological parents. Third, the adopted children show a higher correlation of IQ scores with their biological parents than with their adoptive parents. The following table is a hypothetical data set that shows all these characteristics, in idealized form, to illustrate the concepts involved. The scores given for parents are meant to be the average of mother and father.

	Children	Biological parents	Adoptive parents
	110	90	118
	112	92	114
	114	94	110
	116	96	120
	118	98	112
	120	100	116
Mean	115	95	115

First, we can see that the scores of the children have a high correlation with those of their biological parents but a low correlation with those of their adoptive parents. In fact, in our hypothetical example, the correlation of children with biological parents is $r = 1.00$, but with adoptive parents it is $r = 0$. (Remember that a correlation between two sets of numbers does not mean that the two sets are identical, but that, for each unit of increase in one set, there is a constant proportional increase in the other set—see the Special Topics section on statistical analysis at the end of this chapter.) This perfect correlation with biological parents and zero correlation with adoptive parents means that $H^2 = 1$, given the arguments developed above. All the variation in IQ score among the children is explained by the variation in IQ score among the biological parents.

Second, however, we notice that the IQ score of each child is 20 points higher than that of its biological parents, and that the mean IQ of the children is equal to the mean IQ of the adoptive parents. Thus, adoption has raised the average IQ of the children 20 points higher than the average IQ of their biological parents, so that, as a *group*, the children resemble their adoptive parents. So we have perfect heritability, yet high plasticity in response to environmental modification.

An investigator who is seriously interested in knowing how genes might constrain or influence the course of development of any character in any organism must study directly the norms of reaction of the various genotypes in the population over the range of projected environments. No less detailed information will do. Summary measures such as H^2 are not valuable in themselves.

MESSAGE

Heritability is not the opposite of phenotypic plasticity. A character may have perfect heritability in a population and still be subject to great changes resulting from environmental variation.

LOCATING GENES

It is not possible to identify all the genes that influence the development of a given character using purely genetic techniques. In a given population, only a subset of the genes that contribute to the development of any given character will be genetically variable. Hence, only some of the possible variation will be observed. This is true even for genes that determine simple qualitative traits—for example, the genes that determine the total antigenic configuration of the membrane of the human red blood cell. About 40 loci determining human blood groups are known at present; each has been discovered by finding at least one person with an immunological specificity that differs from the specificities of other people. Many other loci that determine red blood cell membrane structure may remain undiscovered because all the individuals studied are genetically identical at these loci. *Genetic* analysis detects genes only when there is some allelic variation. In contrast, *molecular* analysis deals directly with DNA and its translated information and so can identify genes as stretches of DNA coding for certain products, even when they do not vary—provided the gene products can be identified.

Even though a character may show continuous phenotypic variation, the genetic basis for the differences may be allelic variation at a single locus. Most of the classic mutations in *Drosophila* are phenotypically variable in their expression, and in many cases the mutant class differs little from wild type, so that many individuals that carry the mutation are indistinguishable from normal individuals. Even the genes of the bithorax gene complex, which have dramatic homeotic mutations that turn halteres into wings (see Figure 16-26), also have weak alleles that increase the size of the halteres only slightly on average, so that individuals of the mutant genotype may appear to be wild type.

It is sometimes possible to use prior knowledge of the biochemistry and development of an organism to guess that variation at a known locus is responsible for at least some of the variation in a certain character. Such a locus is a **candidate gene** for the investigation of continuous phenotypic variation. The variation in activity of the enzyme acid phosphatase in human red blood cells was investigated in this way. Because we are dealing with variation in enzyme activity, a good hypothesis would be that there is allelic variation at the locus that codes for this enzyme. When H. Harris and D. Hopkinson sampled an English population, they found that there were, indeed, three allelic forms, *A*, *B*, and *C*, that resulted in enzymes with different activity levels. Table 18-1 shows the mean activity, the variance in activity, and the population frequency of the six genotypes. Figure 18-12 shows the distribution of activity for the entire population and how it is composed of the distributions of the different genotypes. As Table 18-1 shows, of the vari-

TABLE 18-1 Activity of Different Genotypes of Acid Phosphatase in Red Blood Cells in the English Population

Genotype	Mean Activity	Variance of Activity	Frequency in Population
A / A	122.4	282.4	.13
A / B	153.9	229.3	.43
B / B	188.3	380.3	.36
A / C	183.8	392.0	.03
B / C	212.3	533.6	.05
C / C	~240	—	.002
Grand average	166.0	310.7	
Total distribution	166.0	607.8	

Note: Averages are weighted by frequency in population.
SOURCE: H. Harris, *The Principles of Human Biochemical Genetics*, 3d ed. North-Holland, 1980.

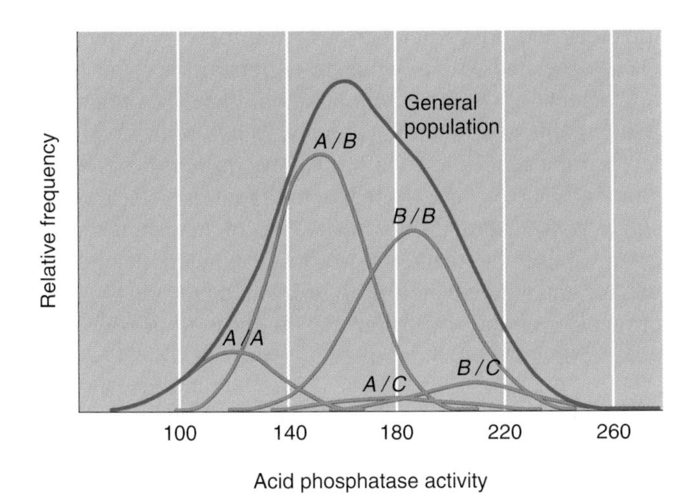

Figure 18-12 Acid phosphatase activity in red blood cells for different genotypes (red curves) and the distribution of activity in an English population made up of a mixture of these genotypes (green curve). *(H. Harris,* The Principles of Human Biochemical Genetics, *3d ed. © 1970 by North Holland.)*

ance in activity in the total distribution (607.8), about half is explained by the average variance within genotypes (310.7), so half (607.8 − 310.7 = 297.1) must be accounted for by the variance between the means of the six genotypes. Although so much of the variation in activity is explained by the mean differences between the genotypes, there remains variation within each genotype that may be the result of environmental influences or of the segregation of other, as yet unidentified, genes.

This explanation of a part of the variation in a population by the presence of different alleles at a single identified locus is typical of what is found by the candidate gene method, but the proportion of variance associated with the single locus is usually less than what was found for acid phosphatase activity. For example, the three common alleles for the gene *apoE,* which encodes the protein apolipoprotein E, account for only about 16 percent of the variance in blood levels of the low-density lipoproteins that carry cholesterol and are implicated in excess cholesterol levels. The remaining variance is a consequence of some unknown combination of genetic variation at other loci and environmental variation.

Marker-Gene Segregation

The specific loci whose allelic differences are responsible for the genetic variation in a quantitative trait—so-called **quantitative trait loci,** or **QTLs**—cannot be individually identified in most cases. It is possible, however, to locate those regions of the genome in which the relevant loci lie and to estimate how much of the total variation is accounted for by QTL variation in each region. This analysis can be done in experimental organisms by crossing two lines that differ markedly in the quantitative trait and which also differ in alleles at well-known loci, called **marker genes.** The marker genes used for such analyses are ones for which the different genotypes can be distinguished by some visible phenotypic effect that cannot be confused with the quantitative trait (for example, eye color in *Drosophila*), or by the electrophoretic mobility of the proteins that they encode, or by the DNA sequence of the genes themselves. The F_1 resulting from the cross between the two lines may then be crossed with itself to make a segregating F_2, or it may be backcrossed to one of the parental lines. If there are QTLs closely linked to a marker gene, then the different marker genotypes in the segregating generation will also carry the QTL alleles that were linked to them in the original parental lines. Thus different marker genotypes in the F_2 or backcross will have different average phenotypes for the quantitative character.

Linkage Analysis

The localization of QTLs to small regions within chromosomes requires the presence of closely spaced marker loci along the chromosome. Moreover, it must be possible to create parental lines that differ from each other in the alleles carried at these marker loci. With the advent of molecular techniques that can detect genetic polymorphism at the DNA level, a very high density of variant loci has been discovered along the chromosomes of all species. Especially useful are restriction fragment length polymorphisms (RFLPs), tandem repeats, and single-nucleotide polymorphisms (SNPs) in DNA. Such polymorphisms are so common that any two lines selected for a difference in quantitative traits are also sure to differ from each other at some known molecular marker loci spaced a few crossover units from each other along each chromosome.

An experimental protocol for localizing genes, shown in Figure 18-13 on the next page, uses groups of individuals that differ markedly in the quantitative character of interest as well as at marker loci. These groups may be created by several generations of divergent selection to create extreme lines, or advantage may be taken of existing varieties or family groups that differ markedly in the trait. These lines must then be surveyed for marker loci that differ between them. A cross is made between the two lines, and the F_1 is then crossed with itself to produce a segregating F_2 or is crossed back to one of the parental lines to produce a segregating backcross. A large number of offspring from the segregating generation are then measured for the quantitative phenotype and characterized for their genotype at the marker loci. A marker locus that is unlinked, or very loosely linked, to any QTLs affecting the quantitative trait of interest will have the same average value of the quantitative trait for all its genotypes, whereas one that is closely linked to some QTLs will differ in its mean quantitative phenotype from one marker genotype to another.

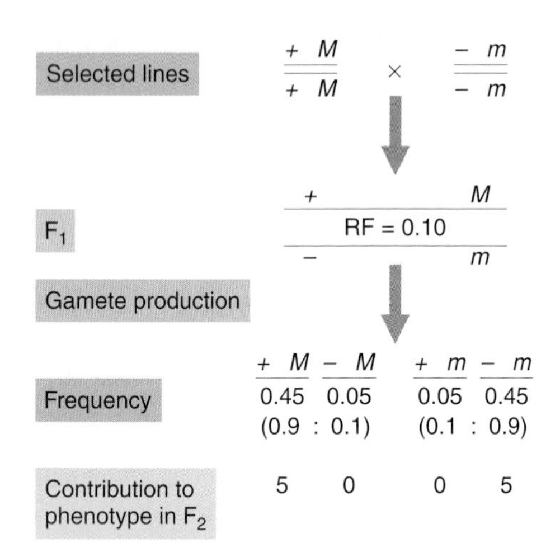

Average phenotypic effect of *M* class = 5 (0.9) + 0 (0.1) = 4.5
Average phenotypic effect of *m* class = 5 (0.1) + 0 (0.9) = 0.5
Difference between *M*-carrying gametes and *m*-carrying gametes = 4.5−0.5 = 4
Difference between average F_2 *M*/*M* homozygotes and average F_2 *m*/*m* homozygotes = 8

Figure 18-13 Results of a cross between two selected lines that differ at a QTL (+ or −) and at a molecular marker locus 10 crossover units (recombination fraction = 0.10) away from the QTL (*M* or *m*). The QTL + allele adds 5 units of difference to the phenotype.

How much difference there is in the mean quantitative phenotype between the different marker genotypes depends both on the strength of the effect of the QTL and on the tightness of linkage between the QTL and the marker locus. Suppose, for example, that there are two selected lines that differ by a total of 100 units in some quantitative character. The line with the high value is homozygous + / + at a particular QTL, whereas the line with the low value is homozygous − / −, and each + allele at this QTL accounts for 5 units of the total difference between the lines. Further, suppose that the high line is *M* / *M* and the low line is *m* / *m* at a marker locus 10 crossover units away from the QTL. Then, as shown in Figure 18-13, there are 4 units of difference between the average gamete carrying an *M* allele and the average gamete carrying an *m* allele in the segregating F_2. We can therefore calculate that 8 units of the difference between an *M* / *M* homozygote and an *m* / *m* homozygote are attributable to that QTL. Thus, we have accounted for 8 percent of the average difference between the original selected lines. The QTL actually accounts for 10 percent of the difference; the discrepancy comes from the recombination between the marker gene and the QTL. We could then repeat this process using marker loci at other locations along the chromosome and on different chromosomes to account for yet further fractions of the quantitative difference between the original selected lines.

This technique has been used to locate chromosomal segments associated with such traits as fruit weight in tomatoes and bristle number in *Drosophila.* Typically, any QTL associated with any marker region will account for a few percent of the effect, but taking all the QTLs together, it is possible to account for as much as half of the total difference between lines. Unfortunately, in human genetics, although marker gene segregation can be used to localize single-gene disorders, the small size of human pedigree groups makes the marker segregation technique inapplicable for quantitative trait loci because there are too few progeny from any particular marker cross to provide any accuracy.

MORE ON ANALYZING VARIANCE

Knowledge of the broad heritability (H^2) of a character in a population is not very useful in itself, as we have seen, but a finer subdivision of phenotypic variance can provide important information for plant and animal breeders. The genetic variance and the environmental variance can themselves each be further subdivided to provide information about gene action and the possibility of shaping the genetic composition of a population.

Additive and Dominance Variance

When a heterozygote *A* / *a* is not exactly intermediate in phenotype between the two homozygotes *A* / *A* and *a* / *a*, then not all the genetic variance in the population can be accounted for by differences in the *average* effect of the *A* and *a* alleles. Some of the genetic variance is the result of the deviation of the heterozygote from the phenotype that a simple averaging of the two allelic effects would be expected to produce. In this case, the total genetic variance in the population can be subdivided into two components: **additive genetic variance,** s_a^2, the variance that arises because there is an average difference between the carriers of *a* alleles and the carriers of *A* alleles, and **dominance variance,** s_d^2, the variance that results from the fact that heterozygotes are not exactly intermediate between the homozygotes. Thus

$$s_g^2 = s_a^2 + s_d^2$$

The total phenotypic variance can now be written as

$$s_p^2 = s_g^2 + s_e^2 = s_a^2 + s_d^2 + s_e^2$$

We can define a new kind of heritability, the **heritability in the narrow sense** (h^2), as

$$h^2 = \frac{s_a^2}{s_p^2} = \frac{s_a^2}{s_a^2 + s_d^2 + s_e^2}$$

It is this heritability, not to be confused with H^2, that is useful in determining whether a program of selective breeding will succeed in changing a population. Heritability in the broad sense, H^2, is the proportion of the phenotypic variance that is accounted for by all of the genetic variance, while heritability in the narrow sense, h^2, is the proportion of the phenotypic variance that is accounted for by the additive genetic variance alone. The greater h^2 is, the greater the fraction of the difference between selected parents and the population as a whole that will be preserved in the offspring of those parents.

To see why it is h^2 that matters for response to selection, consider an extreme case in which the two homozygotes are equal in phenotype—say, with a value of 10—but the heterozygote is overdominant, with a value of 12. Further, suppose the population mates randomly and that the proportions of the three genotypes in the population are

$$0.25\ A\,/\,A \qquad 0.5\ A\,/\,a \qquad 0.25\ a\,/\,a$$

If we now select the individuals with the highest values to breed and produce the next generation, they will all be heterozygotes, and their offspring will exactly reproduce the 1:2:1 distribution of genotypes in the parental generation. Selection will have been completely ineffective in making any change in the F_1, despite the existence of genetic variance in the population, because all the variance was due to dominance and there was no additive variance.

MESSAGE

The effect of selection depends on the amount of *additive* genetic variance in a population and not on the genetic variance in general. Therefore, the narrow heritability, h^2, not the broad heritability, H^2, is relevant for a prediction of response to selection.

What has been described here as the "dominance variance" actually arises from more than the deviation of heterozygotes from exact intermediacy between the homozygotes. It is all the genetic variation that cannot be explained by the average effect of substituting A for a. If there is more than one locus affecting the character, then any epistatic interactions between loci will appear as variance not associated with the average effect of substituting alleles at the A locus. In principle, we could separate this **interaction variance** (s_i^2) from the dominance variance (s_d^2). In practice, however, this cannot be done with any semblance of accuracy, so all the nonadditive variance is attributed to "dominance variance."

Estimating the Components of Genetic Variance

The different components of genetic variance can be estimated from covariance between relatives, but the derivation of these estimates is beyond the scope of this text. There is, however, another way to estimate h^2 that reveals its real meaning. If, in two generations of a population, we plot the phenotype—say, height—of offspring against the average phenotype of their two parents (the **midparent value**), we may observe a relation like the one illustrated by the red line in Figure 18-14. The regression line will pass through the mean height of all the parents and the mean height of all the offspring, which will be equal to each other because no change has occurred in the population between generations. Moreover, taller parents have taller children and shorter parents have shorter children, so that the slope of the line is positive. But the slope is not unity; very short parents on average have children who are somewhat taller, and very tall parents on average have children who are somewhat shorter, than they themselves are. This slope of less than unity arises because heritability is less than perfect. If the phenotype were additively inherited with complete fidelity, then the heights of the offspring would be identical to the midparent values and the slope of the regression line would be 1. On the other hand, if the offspring had no heritable similarity to their parents, all parents would have offspring of the same average height and the slope of the line would be 0. This reasoning suggests that the slope of the regression line of the offspring value on the midparent value provides an estimate of additive heritability. In fact, the slope of the line can be shown mathematically to be a correct estimate of h^2.

The fact that the slope of the regression line estimates additive heritability allows us to use h^2 to predict the effects of artificial selection. Suppose that we select parents for the next generation who are, on average, 2 units above the general mean of the population from which they were chosen. If $h^2 = 0.5$, then the offspring of those selected parents will

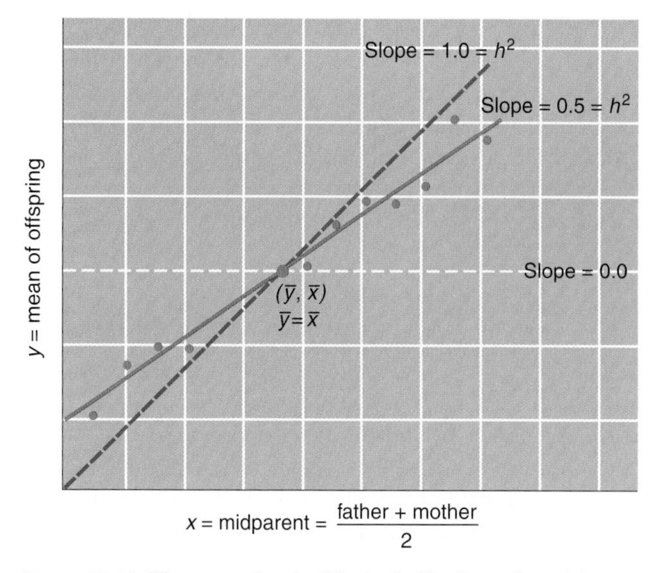

Figure 18-14 The regression (red line) of offspring values (y) on midparent values (x) for a character with a heritability (h^2) of 0.5. The blue line would be the regression slope if the trait were perfectly heritable.

lie 0.5(2.0) = 1.0 unit above the mean of the parental population, because the slope of the regression line predicts how much increase in y will result from a unit increase in x. We can define the **selection differential** as the difference between the selected parents and the mean of the entire population in their generation, and the **selection response** as the difference between the offspring of the selected parents and the mean of the parental generation. Thus

$$\text{Selection response} = h^2 \times \text{selection differential}$$

or

$$h^2 = \frac{\text{selection response}}{\text{selection differential}}$$

The second expression provides us with yet another way to estimate h^2: by carrying out selective breeding for one generation and comparing the response with the selection differential. Usually this process is carried out for several generations using the same selection differential, and the average response is used as an estimate of h^2.

Remember that any estimate of h^2, just as for H^2, depends on the assumption of no correlation between the similarity of the individuals' environments and the similarity of their genotypes. Moreover, h^2 in one population in one set of environments will not be the same as h^2 in a different population in a different set of environments. To illustrate this principle, Figure 18-15 shows the range of heritabilities reported in various studies for a number of characters in chickens. For most traits for which a substantial heritability has been reported in some population, there are big differences from study to study, presumably because different populations have different amounts of genetic variation and because the different studies were carried out in different environments. Thus, breeders who want to know whether selection will be effective in changing some character in their chickens cannot count on the heritabilities found in previous studies but must estimate the heritability in the particular population and particular environment in which the selection program is to be carried out.

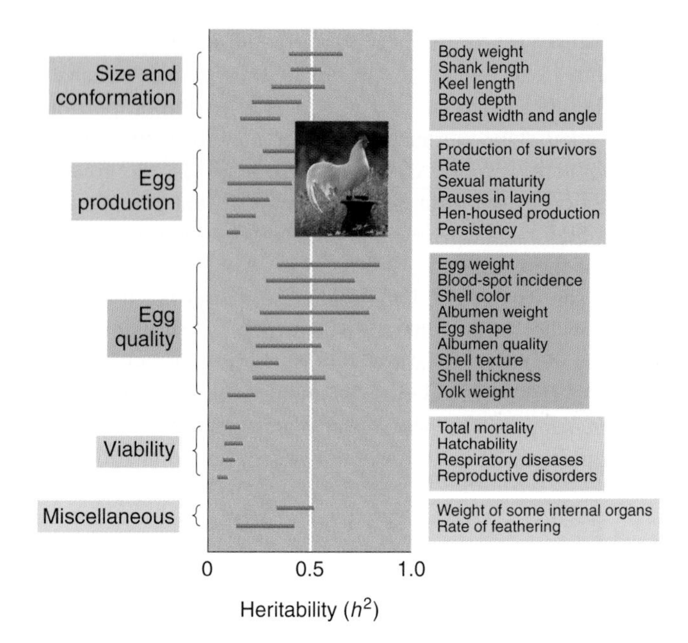

Figure 18-15 Ranges of heritabilities reported for a variety of characters in chickens. *(From I. M. Lerner and W. J. Libby,* Heredity, Evolution and Society. *© 1976 by W. H. Freeman and Company. Photo © Kenneth Thomas/Photo Researchers.)*

The Use of h^2 in Breeding

Even though h^2 is a number that applies only to a particular population and a given set of environments, it is still of great practical importance to breeders. A poultry geneticist interested in increasing, say, the growth rate of chickens is not concerned with the genetic variance over all possible flocks and all environmental distributions. Given a particular flock (or a choice between a few particular flocks) under the environmental conditions approximating present husbandry practice, the question becomes: Can a selection scheme be devised to increase growth rate and, if so, how rapidly can it be increased? If one flock has a lot of genetic variance for growth rate and another only a little, the breeder will choose the former flock to carry out selection. If the heritability in the chosen flock is very high, then the mean of the population will respond quickly to the selection imposed, because most of the superiority of the selected parents will appear in the offspring. The higher h^2 is, the higher the parent-offspring correlation is. If, on the other hand, h^2 is low, then only a small fraction of the superiority of the selected parents will appear in the next generation.

If h^2 is very low, some alternative scheme of selection or husbandry may be needed. In this case, H^2 together with h^2 can be of use to the breeder. Suppose that h^2 and H^2 are both low, which means that there is a large proportion of environmental variance compared with genetic variance. Some scheme of reducing s_e^2 must be used. One method is to change the husbandry conditions so that environmental variance is lowered. Another is to use **family selection.** Rather than selecting the best individuals, the breeder allows pairs to produce several trial progeny, and parental pairs are selected to produce the next generation on the basis of the average performance of those progeny. Averaging over progeny allows uncontrolled environmental variation and developmental noise to be canceled out, and a better estimate of the genotypic difference between pairs can be made so that the best pairs can be chosen as parents of the next generation.

If, on the other hand, h^2 is low but H^2 is high, then there is not much environmental variance. The low h^2 is the result of a small proportion of additive genetic variance compared with dominance and interaction variance. Such a situation calls for special breeding schemes that make use of nonadditive variance. One such scheme is the **hybrid-inbred method,** which is used almost universally for corn. A large number of inbred lines are created by selfing. These inbred lines are then crossed in many different combinations (all possible combinations, if this is economically feasible), and the cross that gives the best hybrid is chosen. Then new inbred lines are developed from this best hybrid, and again crosses are made to find the best hybrid cross. This process is continued cycle after cycle. This scheme selects not only for additive effects but also for dominance effects, because it selects the best heterozygotes; it has been the basis of major genetic advances in hybrid maize yield in North America since 1930. Yield in corn does not appear to have large amounts of nonadditive genetic variance, however, so it is debatable whether this technique *ultimately* produces higher-yielding varieties than those that would have resulted from years of simple selection techniques based on additive variance.

The hybrid-inbred method has been introduced into the breeding of all kinds of plants and animals. Tomatoes and chickens, for example, are now almost exclusively hybrids. Attempts also have been made to breed hybrid wheat, but thus far the wheat hybrids obtained do not yield consistently better than do the nonhybrid varieties now used.

MESSAGE
The subdivision of genetic variation and environmental variation provides important information about gene action that can be used in plant and animal breeding.

The case of human skin color with which we began this chapter provides an example of the difficulty of analyzing the genetic basis of quantitative characters in a species for which neither deliberately controlled crosses nor experimental manipulation of development is possible. Everyday observation makes it clear that differences in skin color must be strongly influenced by genotypic differences. Moreover, we know that the environment also affects skin color. Yet, beyond these two general observations, virtually nothing is known about the relation of phenotypic to genotypic variation for skin pigmentation. Although many matings have occurred between people of different skin colors, and people of different skin colors have been exposed to different environments, we do not even have quantitative estimates of the genetic and environmental components of variance for skin color because it is not possible to control carefully the types of matings that occur or to arrange that the offspring of different matings be subjected to the same environments. Nor do we know which genes influence skin color differences between Africans and Europeans, or between Asians and Africans, or between individuals within these major geographical groups. No one so far has carried out the observations on the segregation of marker loci in the offspring of the matings of various phenotypes that would be needed to identify such genes by the techniques of quantitative trait locus (QTL) analysis. Nor is it clear what would be learned by doing so, beyond the already obvious fact that skin color differences depend on more than a single gene and are also susceptible to some environmental modification. It must be borne in mind that genetics, like all other sciences, requires resources of time and money, so that decisions must always be made about the priorities to be assigned to different investigations.

SPECIALIZED TOPICS
STATISTICAL ANALYSIS

Complete information about the distribution of a phenotype in a population can be given only by specifying the frequency of each measured class, but a great deal of information can be summarized in just two statistics. First, we need some measure of the location of the distribution along the axis of measurement (for example, do the individual measurements of height for male undergraduates tend to cluster around 100 cm or around 200 cm)? Second, we need some measure of the amount of variation within the distribution (for example, are the heights of the male undergraduates all concentrated around the central measurement, or do they vary widely across a large range?).

Measures of Central Tendency

The mode. One possibility for describing the central tendency of the distribution is to give the measurement of the most common class, the **mode.** In Figure 18-4b, the mode is 172 cm (for females, the mode would be about 6 cm less). Most distributions of phenotypes look roughly like those in Figure 18-4: a single mode is located near the middle of the distribution, with frequencies decreasing on either side. There are exceptions to this pattern, however. Figure 18-16a on the next page shows the very asymmetrical distribution of seed weights in the plant *Crinum longifolium.* Figure 18-16b shows the **bimodal** (two-mode) distribution of larval survival probabilities for several different second-chromosome homozygotes in *Drosophila willistoni.* A bimodal distribution may indicate that the population being studied could be better considered a mixture of two populations, each with its own mode. In Figure 18-16b, the left-hand mode probably represents a subpopulation of severe single-locus mutations that are extremely deleterious when homozygous but whose effects are not felt in the heterozygous state in which they usually exist in natural populations. The right-hand mode is probably part of the distribution of "normal" viability modifiers of small effect.

The mean. A more common measure of central tendency is the arithmetic average, or the **mean.** The mean of the

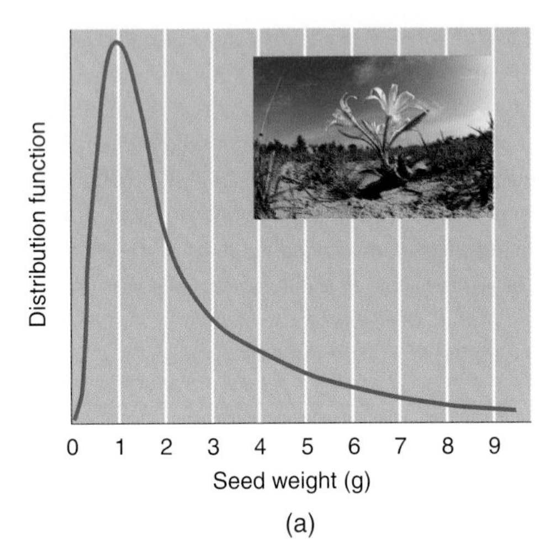

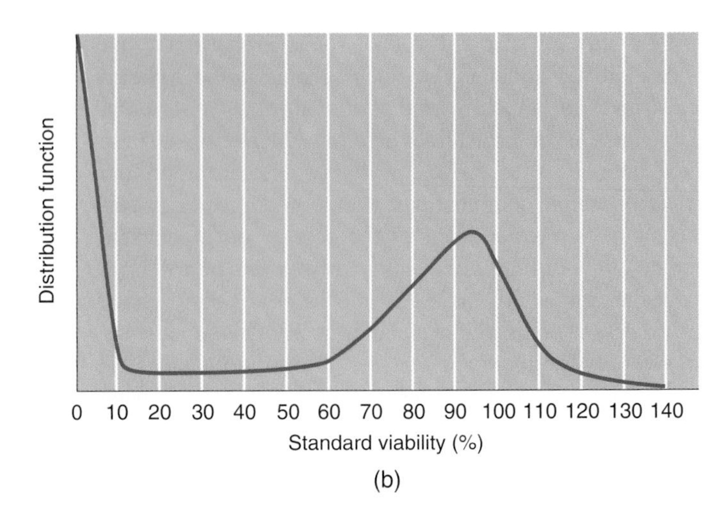

Figure 18-16 Asymmetrical distribution functions. (a) Asymmetrical distribution of seed weight in *Crinum longifolium*. (b) Bimodal distribution of survival of *Drosophila willistoni* expressed as a percentage of standard survival. *(Adapted from S. Wright,* Evolution and the Genetics of Populations, *Vol. 1. © 1968 by University of Chicago Press, 1968. Photo: Earth Scenes/© Thompson GOSF.)*

measurement x, denoted $\bar{x}$, is simply the sum of all the individual measurements divided by the number of measurements in the sample (N):

$$\text{Mean} = \bar{x} = \frac{x_1 + x_2 + x_3 + \cdots + x_N}{N} = \frac{1}{N}\Sigma x_i$$

where Σ represents the operation of summing over all values of i from 1 to N, and x_i is the ith measurement.

In a typical large sample, the same measured value will appear more than once, because several individuals will have the same value within the accuracy of the measuring instrument. In such a case, $\bar{x}$ can be rewritten as the sum of all measurement values, each weighted by how frequently it occurs in the population. From a total of N individuals measured, suppose that n_1 fall into the class with value x_1, that n_2 fall in the class with value x_2, and so forth, so that $\Sigma n_i = N$. If we let f_i be the **relative frequency** of the ith measurement class, so that

$$f_i = \frac{n_i}{N}$$

then we can rewrite the mean as

$$\bar{x} = f_1 x_1 + f_2 x_2 + \cdots + f_k x_k = \Sigma f_i x_i$$

where x_i equals the value of the ith measurement class.

Let us apply these calculation methods to the data in Table 18-2, which gives the numbers of toothlike bristles in the sex combs on the right (x) and left (y) front legs and on both legs ($T = x + y$) of 20 *Drosophila*. Looking for the moment only at the sum of the two legs, T, we find the mean number of sex comb teeth, $\bar{T}$, to be

$$\bar{T} = \frac{11 + 12 + 12 + 12 + 13 + \cdots + 15 + 16 + 16}{20}$$
$$= \frac{274}{20}$$
$$= 13.7$$

Alternatively, using the relative frequencies (f_i) of the different measurement values, we find that

$$\bar{T} = 0.05(11) + 0.15(12) + 0.20(13) + 0.35(14) + 0.15(15) + 0.10(16)$$
$$= 13.7$$

Measures of Dispersion: The Variance

A second characteristic of a distribution is the width of its spread around the central class. Two distributions with the same mean might differ a great deal in how closely the measurements are concentrated around the mean. The most common measure of variation around the center of a distribution is the **variance,** which is defined as the average squared deviation of the observations from the mean, or

$$\text{Variance} = s^2$$
$$= \frac{(x_1 - \bar{x})^2 + (x_2 - \bar{x})^2 + \cdots + (x_N - \bar{x})^2}{N}$$
$$= \frac{1}{N}\Sigma(x_i - \bar{x})^2$$

TABLE 18-2 Number of Teeth in the Sex Combs on the Right (x) and Left (y) Legs and the Sum of the Two (T) for 20 *Drosophila* Males

x	y	T	n_i	$f_i = n_i/N$
6	5	11	1	$\frac{1}{20} = 0.05$
6	6	12		
5	7	12	3	$\frac{3}{20} = 0.15$
6	6	12		
7	6	13		
5	8	13	4	$\frac{4}{20} = 0.20$
6	7	13		
7	6	13		
8	6	14		
6	8	14		
7	7	14		
7	7	14	7	$\frac{7}{20} = 0.35$
7	7	14		
6	8	14		
8	6	14		
8	7	15		
7	8	15	3	$\frac{3}{20} = 0.15$
6	9	15		
8	8	16	2	$\frac{2}{20} = 0.10$
7	9	16		
$N = 20$				
$\bar{x} = 6.65$		$s_x^2 = 0.8275$		$s_x = 0.9096$
$\bar{y} = 7.05$		$s_y^2 = 1.1475$		$s_y = 1.0722$
$\bar{T} = 13.70$		$s_T^2 = 1.71$		$s_T = 1.308$
		$\text{cov } xy = -0.1325$		
		$r_{xy} = -0.1360$		

When more than one individual has the same measured value, the variance can be written

$$s^2 = f_1(x_1 - \bar{x})^2 + f_2(x_2 - \bar{x})^2 + \cdots + f_k(x_k - \bar{x})^2$$
$$= \Sigma f_i(x_i - \bar{x})^2$$

To avoid subtracting every value of x separately from the mean, we can use an alternative computing formula that is algebraically identical with the preceding equation:

$$s^2 = \frac{1}{N}\Sigma x_i^2 - \bar{x}^2$$

Because the variance is in squared units (square centimeters, for example), it is common to take the square root of the variance, which then has the same units as the measurement itself. This square-root measure of variation is called the **standard deviation** of the distribution:

$$\text{Standard deviation} = s = \sqrt{\text{variance}} = \sqrt{[s^2]}$$

The data for sex comb teeth in Table 18-2 can be used to exemplify these calculations:

$$s_T^2 = \frac{(11 - 13.7)^2 + (12 - 13.7)^2 + (12 - 13.7)^2}{20}$$
$$\frac{+ \cdots + (15 - 13.7)^2 + (16 - 13.7)^2}{20}$$
$$= \frac{34.20}{20} = 1.71$$

We can also use the computing formula that avoids calculating all the individual deviations:

$$s_T^2 = \frac{1}{N}\Sigma T_i^2 - \bar{T}^2 = \frac{3788}{20} - 187.69 = 1.71$$

and

$$s_T = \sqrt{1.71} = 1.308$$

Figure 18-17 shows two distributions that have the same mean but different standard deviations (curves A and B) and two distributions that have the same standard deviation but different means (curves A and C).

The mean and the variance of a distribution do not describe it completely. They do not distinguish a symmetrical distribution from an asymmetrical one, for example. There are even symmetrical distributions that have the same mean and variance but still have somewhat different shapes. Nevertheless, for the purposes of dealing with most quantitative genetics problems, the mean and variance suffice to characterize a distribution.

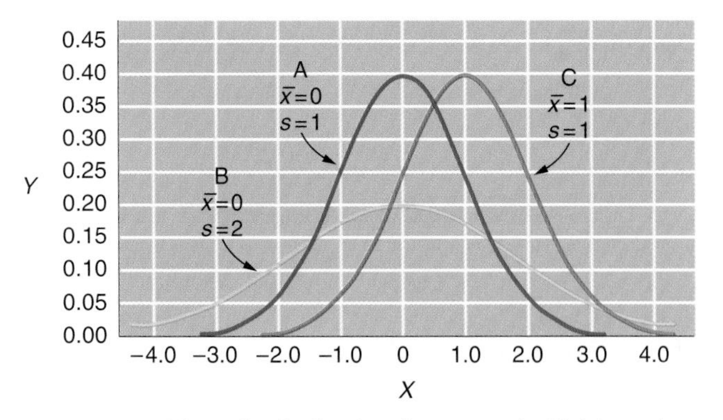

Figure 18-17 Three distribution functions, two of which have the same mean (A and B) and two of which have the same standard deviation (B and C).

Measures of Relationship

Covariance and correlation. Another statistical notion that is useful in the study of quantitative genetics is the association, or **correlation,** between variables. As a result of complex paths of causation, many variables in nature vary together, but in an imperfect or approximate way. Figure 18-18a provides an example, showing the lengths of two particular teeth in several individual specimens of a fossil mammal, *Phenacodus primaevis.* There is a rough trend such that individuals with longer first molars tend to have longer second molars, but there is considerable scatter of the data points around this trend. In contrast, Figure 18-18b shows that body length and tail length in individual snakes *(Lampropeltis polyzona)* are quite closely related to each other, with all the points falling close to a straight line that could be drawn through them from the lower left of the graph to the upper right.

The usual measure of the precision of a relation between two variables x and y is the **correlation coefficient** (r_{xy}). It is calculated in part from the product of the deviation of each observation of x from the mean of the x values and the deviation of each observation of y from the mean of the y values—a quantity called the **covariance** of x and y (cov xy):

$$\text{cov } xy = \frac{(x_1 - \bar{x})(y_1 - \bar{y}) + (x_2 - \bar{x})(y_2 - \bar{y}) + \cdots}{N} \frac{+ (x_N - \bar{x})(y_N - \bar{y})}{N}$$

$$= \frac{1}{N} \Sigma (x_i - \bar{x})(y_i - \bar{y})$$

A formula that is exactly algebraically equivalent but that makes computation easier is

$$\text{cov } xy = \frac{1}{N} \Sigma x_i y_i - \overline{xy}$$

Using this formula, we can calculate the covariance between the right (x) and the left (y) leg counts in Table 18-2:

$$\text{cov } xy = \frac{1}{N} \Sigma x_i y_i - \overline{xy}$$

$$= \frac{(6)(5) + (6)(6) + \cdots + (8)(8) + (7)(9)}{20}$$

$$-(6.65)(7.05)$$

$$= -0.1325$$

The correlation, r_{xy}, is defined as

$$\text{Correlation} = r_{xy} = \frac{\text{cov } xy}{s_x s_y}$$

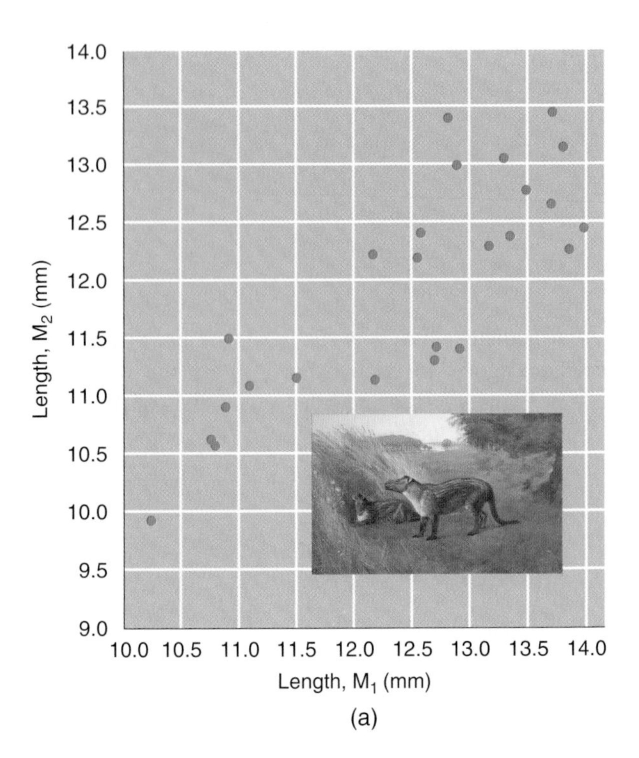

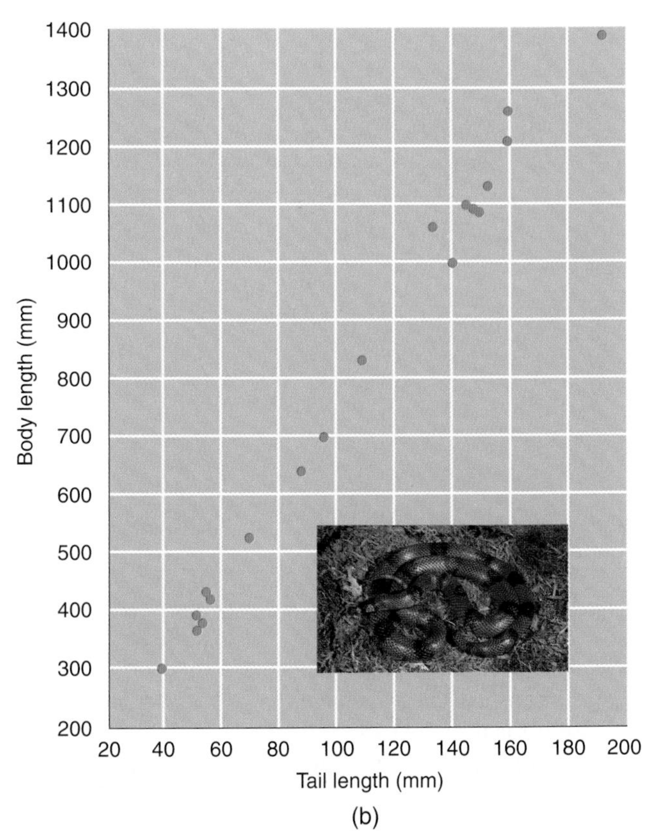

Figure 18-18 Scatter diagrams of relations between pairs of variables. (a) Relation between the lengths of the first and second lower molars (M_1 and M_2) in the extinct mammal *Phenacodus primaevis.* Each point gives the M_1 and M_2 measurements for one individual. (b) Tail length and body length of 18 individuals of the snake *Lampropeltis polyzona.* *(Photos: part a, negative 2430,* Phenacodus, *painting by Charles Knight; courtesy Dept. of Library Services, American Museum of Natural History; part b, Zig Leszczynski/Animals Animals.)*

In the formula for correlation, the products of the deviations are divided by the product of the standard deviations of x and y (s_x and s_y). This normalization by the standard deviations has the effect of making r_{xy} a dimensionless number that is independent of the units in which x and y are measured. So defined, r_{xy} will vary from -1, which signifies a perfectly linear negative relation between x and y, to $+1$, which indicates a perfectly linear positive relation between x and y. If $r_{xy} = 0$, there is no linear relation between the variables. Intermediate values between 0 and $+1$ or -1 indicate intermediate degrees of relationship between the variables. The data in Figure 18-18a and b have r_{xy} values of 0.82 and 0.99, respectively. In the example of the sex comb teeth in Table 18-2, the correlation between left and right legs is

$$r_{xy} = \frac{\text{cov } xy}{\sqrt{s_x^2 s_y^2}} = \frac{-0.1325}{\sqrt{(0.8275)(1.1475)}} = -0.1360$$

a very small value.

It is important to notice that sometimes when there is no *linear* relation between two variables, but there is a regular *nonlinear* relation between them, one variable may be perfectly predicted from the other. Consider, for example, the parabola shown in Figure 18-19. The values of y are perfectly predictable from the values of x; yet $r_{xy} = 0$, because, on average over the whole range of x values, larger x values are not associated with either larger or smaller y values.

Correlation and equality. It is important to notice that correlation between two sets of numbers is not the same as numerical identity. For example, two sets of values can be perfectly *correlated,* even though the values in one set are very much larger than the values in the other set. Consider the following pairs of values:

x	y
1	22
2	24
3	26

The variables x and y in the pairs are perfectly correlated ($r = +1.0$), although each value of y is about 20 units greater than the corresponding value of x. Two variables are perfectly correlated if, for a fixed value increase in one, there is a constant linear increase in the other (or a constant decrease if r is negative).

The importance of the difference between correlation and identity arises when we consider the effect of environment on heritable characters. Parents and offspring might be perfectly correlated in some character such as height, yet, because of an environmental difference between generations, every child might be taller than its parents. This phenomenon appears in adoption studies, in which children may be correlated with their biological parents but, on average, may be quite different from those parents as a result of the change in their social situation.

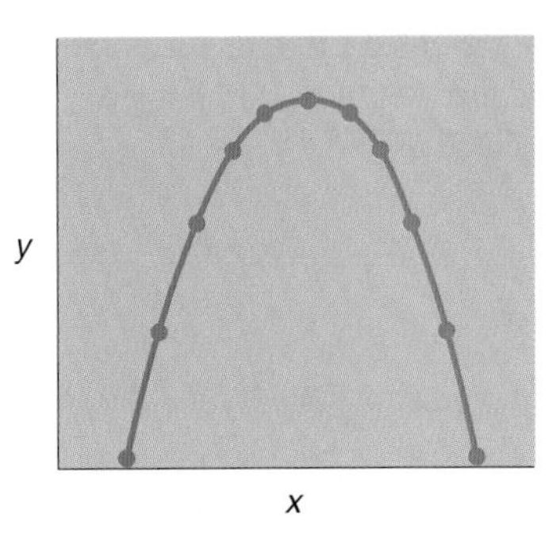

Figure 18-19 A parabola. Each value of y is perfectly predictable from the value of x, but there is no linear correlation.

Regression. The correlation coefficient provides us with only an estimate of the *precision* of the relation between two variables. A related problem is predicting the value of one variable given the value of the other: If x increases by two units, for example, by how much will y increase? If the two variables are linearly related, then that relation can be expressed as

$$y = bx + a$$

where b is the slope of the line relating y to x and a is the y intercept of that line.

Figure 18-20 shows a scatter diagram of points for two variables, y and x, together with a straight line expressing the general linear trend of y with increasing x. This line, called the **regression line of y on x,** has been positioned so that the deviations of the points from the line are as small as possible. Specifically, if Δy is the distance of any point from the line in the y direction, then the line has been chosen so that the sum of $(\Delta y)^2$ equals a minimum. Any other straight line passed through the points on the scatter diagram will have a larger total squared deviation of the points from it.

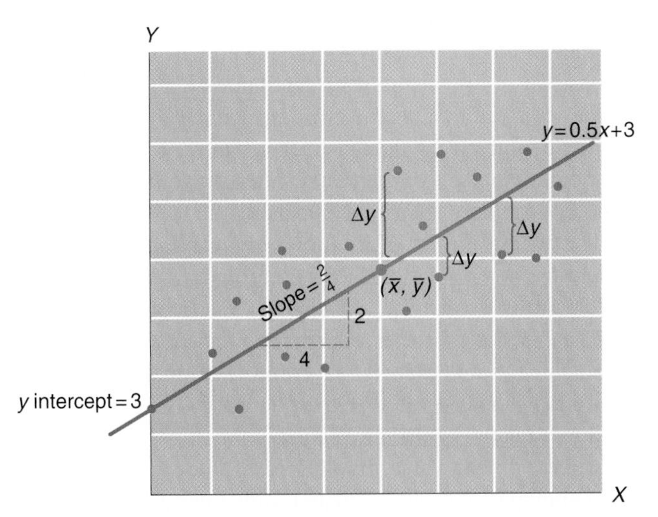

Figure 18-20 A scatter diagram showing the relation between two variables, x and y, with the regression line of y on x. This line, with a slope of 2/4, minimizes the squares of the deviations (Δy).

Obviously, we cannot find this **least-squares regression line** by trial and error. It turns out, however, that, if slope b of the line is calculated as

$$b = \frac{\text{cov } xy}{s_x^2}$$

and if a is then calculated as

$$a = \bar{y} - b\bar{x}$$

so that the line passes through the point $\bar{x}, \bar{y}$, then these values of b and a will yield the linear equation of the least-squares regression line.

Note that the prediction equation given above cannot predict y exactly for a given x, because there is scatter around the least-squares regression line. The equation predicts the *average* y for a given x, if large enough samples are taken.

Samples and Populations

The preceding sections have described the distributions of, and some statistics for, particular assemblages of individuals that have been collected in experiments or sets of observations. For some purposes, however, we are not really interested in the particular 100 male undergraduates or 18 snakes that have been measured. Instead, we are interested in the wider world of phenomena of which those particular individuals are representative. For example, we might want to know the average height, *in general,* of male undergraduates in the United States. That is, we are interested in the characteristics of a **universe,** of which our small collection of observations is only a **sample.** The characteristics of any particular sample are not identical with those of the universe but vary from sample to sample.

We can use the sample mean to estimate the true mean of the universe, but the sample variance and covariance will be, on average, a little smaller than the true value in the universe. That is because the deviations from the sample mean are not all independent of one another. In fact, by definition, the sum of all the deviations of the observations from the mean is 0! (Try to prove this as an exercise.) Therefore, if we are told $N - 1$ of the deviations from the mean in a sample of N observations, we can calculate the missing deviation, since all the deviations must add up to zero.

It is simple to correct for this bias in our estimate of the variance. Whenever we are interested in the variance of a set of measurements—not as a characteristic of the particular sample, but as an estimate of a universe that the sample represents—then the appropriate quantity to use, rather than s^2 itself, is

$$\frac{N}{N-1} s^2$$

Note that this new quantity is equivalent to dividing the sum of the squared deviations by $N - 1$ instead of N in the first place, so

$$\left(\frac{N}{N-1}\right) s^2 = \left(\frac{N}{N-1}\right) \frac{1}{N} \Sigma (x_i - \bar{x})^2$$

$$= \frac{1}{N-1} \Sigma (x_i - \bar{x})^2$$

All these considerations about bias also apply to the sample covariance. In the formula for the correlation coefficient above, however, the factor $N/(N - 1)$ would appear in both the numerator and the denominator and therefore cancel out, so we can ignore it for the purposes of computation.

SUMMARY

1. Many—perhaps most—of the phenotypic characters we observe in organisms vary continuously. In many cases, the variation in a character is determined by more than a single segregating locus. Each of these loci may contribute equally to a particular phenotype, but it is more likely that they contribute unequally.

2. The measurement of continuously varying phenotypes and the determination of the contributions of specific alleles to their distribution must be made on a statistical basis. Some of these variations in phenotype (such as height in some plants) may show a normal distribution around a mean value. Others (such as seed weight in some plants) will have an asymmetrical distribution around a mean value. In other cases, the variation in one phenotype may be correlated with the variation in another. A correlation coefficient may be calculated for two such variables.

3. The distribution of environments is reflected biologically as a distribution of phenotypes for individuals of the same genotype. The transformation of environmental distribution into phenotypic distribution is determined by the norm of reaction. Norms of reaction can be characterized in organisms in which large numbers of genetically identical individuals can be produced.

4. A quantitative character is one for which the average phenotypic differences between genotypes are small compared with the variation between individuals within the genotypes. This may be true even for characters that are influenced by alleles at only one locus. With the use of marker genes, it is possible to determine the relative contributions of different chromosomes to variation in a quantitative trait, to observe dominance and epistasis from whole chromosomes, and, in some cases, to map genes that are segregating for a trait.

5. Characters are familial if they are common to members of the same family, for whatever reason. They are heritable, however, only if the similarity arises from common genotypes. In experimental organisms, similarities resulting from common environments may be readily distinguished from heritable ones. In humans, however, it is very difficult to determine whether a particular trait is heritable. Norm of reaction studies show only small differences between most genotypes in most organisms, and these differences are not consistent over a wide range of environments. Thus, "superior" genotypes in domesticated animals and cultivated plants may be superior only in certain environments. If it should turn out that humans exhibit genetic variation for various mental and emotional traits, this variation is unlikely to favor one genotype over another across a range of environments.
6. The attempt to quantify the influence of genes on a particular character has led to the determination of heritability in the broad sense (H^2). In general, the heritability of a character is different in each population and each set of environments and cannot be extrapolated from one population and set of environments to another. Because H^2 characterizes present populations in present environments only, it is fundamentally flawed as a predictive device. Heritability in the narrow sense, h^2, measures the proportion of phenotypic variance that results from substituting one allele for another. This quantity, if large, predicts that selection for a character will succeed rapidly. If h^2 is small, special forms of selection are required.

CONCEPT MAP

Draw a concept map interrelating as many of the following terms as possible. Note that the terms are listed in no particular order.

quantitative variation / Mendel's laws / norm of reaction / environment / heritability / variance / selection response / additive variance / QTL

SOLVED PROBLEM

1. In some species of songbirds, populations living in different geographical regions sing different "local dialects" of their species' song. These differences are characterized by quantitative differences in the frequency spectrum of emitted sounds. Some persons believe that these differences in dialect are the result of genetic differences between populations, whereas others believe that these differences arose from purely individual idiosyncrasies in the founders of the populations and have been passed on from generation to generation by learning. Outline an experimental program that would determine the importance of genetic and nongenetic factors and their interaction in this dialect variation. If there is evidence of genetic differences, what experiments could be done to provide a detailed description of the genetic system, including the number of segregating genes, their linkage relations, and their additive and nonadditive phenotypic effects?

Solution

This example has been chosen because it illustrates the very considerable experimental difficulties that arise when we try to examine claims that observed differences in quantitative characters have a genetic basis. To be able to say anything at all about the roles of genes and environment during development requires, at minimum, that the organisms be raised from fertilized eggs in a controlled laboratory environment. To be able to make more detailed statements about the genotypes underlying variation in the character requires, further, that the results of crosses between parents of known phenotype and known ancestry be observable and that the offspring of some of those crosses be, in turn, crossed with other individuals of known phenotype and ancestry. Very few animal species can satisfy this requirement, although it is much easier to carry out such controlled crosses in plants. We shall assume that the songbird species in question can indeed be raised and crossed in captivity, but that is a big assumption.

a. To determine whether there is any genetic difference underlying the observed phenotypic difference in dialect between songbird populations, we need to raise birds of each population, from the egg, in the absence of auditory input from their own ancestors and in various combinations of auditory environments of other populations. This is done by raising birds under the following conditions:

1. In isolation
2. Surrounded by other hatchlings consisting only of birds from their own population
3. Surrounded by other hatchlings consisting of birds from other populations
4. In the presence of singing adults from other populations
5. In the presence of singing adults from their own population (as a control on the rearing conditions)

If there are no genotypic differences in dialect between populations and all dialect differences are learned, then birds from group 5 will sing their own population's dialect and those from group 4 will sing the foreign dialect. Groups 1, 2, and 3 may sing a generalized song not corresponding to any of the dialects; or they may all sing the same dialect—this dialect would then represent the "intrinsic" developmental program unmodified by learning. It is also possible that in the absence of adult models, they will not sing at all.

If dialect differences are totally determined by genetic differences, birds from groups 4 and 5 will sing the same dialect—that of their own population. Birds from groups 1, 2, and 3, if they sing at all, will each sing the dialect of their own population, irrespective of the other birds in their group. There are then the possibilities of less clear-cut results, indicating that both genetic and learned differences influence the trait. For example, birds in group 4 might sing a song with elements from both populations. Note that if the birds in group 5, the control group, do not sing their own population's dialect, the rest of the results are uninterpretable, because the conditions of artificial rearing are interfering with the normal developmental program.

b. If the results of the first experiments show some heritability in the broad sense, then a further analysis is possible. This analysis requires a genetically segregating population, made from crosses between two dialect populations—say, A and B. A cross between males from population A and females from population B, and the reciprocal cross, will give an estimate of the average degree of dominance of genes influencing the character and show whether there is any sex linkage. (Remember that in birds the female is the heterogametic sex.) The offspring of this cross and all subsequent crosses *must* be raised in conditions that do not confuse the learned and the genetic components of the dialect differences, as revealed in the experiments in part a. If learned effects cannot be separated out, this further genetic analysis is impossible.

c. To localize the genes influencing dialect differences would require a large number of segregating genetic markers. These markers could be morphological mutants or molecular variants such as restriction fragment length polymorphisms. Families segregating for the quantitative trait differences in song dialect would be examined to see if there were cosegregation of any of the marker loci with the quantitative trait. These cosegregated loci could then be tested for linkage to the quantitative trait loci. Further crosses between individuals with and without the markers and measurement of the quantitative trait values in F_2 individuals would establish whether there was actual linkage between the marker and QTLs. In practice, it is very unlikely that such experiments could be carried out on a songbird species, because of the immense time and effort required to establish lines carrying a large number of different marker genes and molecular polymorphisms.

SOLVED PROBLEM

2. Two inbred lines of beans are intercrossed. In the F_1, the variance in bean weight is measured at 1.5. The F_1 is selfed; in the F_2, the variance in bean weight is 6.1. Estimate the broad heritability of bean weight in the F_2 population of this experiment.

Solution

The key here is to recognize that all the variance in the F_1 population must be environmental, because all the F_1 individuals must have the same genotype, since they are all heterozygotes from a cross between two inbred lines. Furthermore, the F_2 variance must be a combination of environmental and genetic components, because all the genes that are heterozygous in the F_1 will segregate in the F_2 to give an array of different genotypes that relate to bean weight. Hence, we can estimate

$$s_e^2 = 1.5$$

$$s_e^2 + s_g^2 = 6.1$$

Therefore

$$s_g^2 = 6.1 - 1.5 = 4.6$$

and broad heritability is

$$H^2 = \frac{4.6}{6.1} = 0.75\ (75\%)$$

SOLVED PROBLEM

3. In an experimental population of *Tribolium* (flour beetles), body length shows a continuous distribution with a mean of 6 mm. A group of males and females with body lengths of 9 mm are removed and interbred. The body lengths of their offspring average 7.2 mm. From these data, calculate the heritability in the narrow sense for body length in this population.

Solution

The selection differential is $9 - 6 = 3$ mm, and the selection response is $7.2 - 6 = 1.2$ mm. Therefore, the heritability in the narrow sense is

$$h^2 = \frac{1.2}{3} = 0.4\ (40\%)$$

PROBLEMS

Basic Problems

1. Distinguish between continuous and discontinuous variation in a population, and give some examples of each.

2. The following table shows a distribution of bristle number in *Drosophila:*

Bristle number	Number of individuals
1	1
2	4
3	7
4	31
5	56
6	17
7	4

Calculate the mean, variance, and standard deviation of this distribution.

3. A book on the problem of heritability of IQ makes the following three statements. Discuss the validity of each statement and its implications about the authors' understanding of h^2 and H^2.

a. "The interesting question, then, is . . . . 'How heritable?' The answer [0.01] has a very different theoretical and practical application from the answer [0.99]." [The authors are talking about H^2.]

b. "As a rule of thumb, when education is at issue, H^2 is usually the more relevant coefficient, and when eugenics and dysgenics (reproduction of selected individuals) are being discussed, h^2 is ordinarily what is called for."

c. "But whether the different ability patterns derive from differences in genes . . . is not relevant to assessing discrimination in hiring. Where it could be relevant is in deciding what, in the long run, might be done to change the situation."

(From J. C. Loehlin, G. Lindzey, and J. N. Spuhler, *Race Differences in Intelligence.* © 1975 by W. H. Freeman and Company.)

4. Using the concepts of norms of reaction, environmental distribution, genotypic distribution, and phenotypic distribution, try to restate the following statement in more exact terms: "80 percent of the difference in IQ performance between the two groups is genetic." What would it mean to talk about the heritability of a difference between two groups?

Challenging Problems

i **5.** In a large herd of cattle, three different characters showing continuous distribution are measured, and the variances in the following table are calculated:

	CHARACTERS		
Variance	Shank length	Neck length	Fat content
Phenotypic	310.2	730.4	106.0
Environmental	248.1	292.2	53.0
Additive genetic	46.5	73.0	42.4
Dominance genetic	15.6	365.2	10.6

a. Calculate the broad-sense *and* narrow-sense heritabilities for each character.

b. In the population of animals studied, which character would respond best to selection? Why?

c. A project is undertaken to decrease mean fat content in the herd. The mean fat content is currently 10.5 percent. Animals with 6.5 percent fat content are interbred as parents of the next generation. What mean fat content can be expected in the descendants of these animals?

6. Suppose that two triple heterozygotes A / a ; B / b ; C / c are crossed. Assume that the three loci are on different chromosomes.

a. What proportions of the offspring are homozygous at one, two, and three loci, respectively?

b. What proportions of the offspring carry 0, 1, 2, 3, 4, 5, and 6 alleles represented by capital letters, respectively?

7. Among the offspring of the cross in Problem 6, suppose that the average effects of the three genotypes at the A locus on some phenotypic trait are $A / A = 4$, $A / a = 3$, $a / a = 1$, and that similar effects exist for the B and C loci. Moreover, suppose that the effects of the loci add to each other. Calculate and graph the distribution of phenotypes in this population (assuming no environmental variance).

8. In Problem 7, suppose that the character in question is bristle number in *Drosophila.* Suppose that there is a phenotypic effect threshold such that when the phenotypic value is above 9, the individual fly has three bristles; when it is between 5 and 9, the individual has two bristles; and when the value is 4 or less, the individual has one bristle. Describe the outcome of crosses within and between bristle classes. Given the result, could you infer the underlying genetic situation?

9. Suppose that the general form of a distribution of a character for a given genotype is

$$f = 1 - \frac{(x - \bar{x})^2}{s_e^2}$$

over the range of x where f is positive.

a. On the same scale, plot the distributions for three genotypes with the following means and environmental variances:

Genotype	$\bar{x}$	s_e^2	Approximate range of phenotype
1	0.20	0.3	$x = 0.03$ to $x = 0.37$
2	0.22	0.1	$x = 0.12$ to $x = 0.24$
3	0.24	0.2	$x = 0.10$ to $x = 0.38$

b. Plot the phenotypic distribution that would result if the three genotypes were equally frequent in a population. Can you see distinct modes? If so, what are they?

10. The following sets of hypothetical data represent paired observations on two variables (x, y). Plot each set of data pairs as a scatter diagram. Look at the plot of the points, and make an intuitive guess about the correlation between x and y. Then calculate the correlation coefficient for each set of data pairs, and compare this value with your estimate.

a. (1, 1); (2, 2); (3, 3); (4, 4); (5, 5); (6, 6)

b. (1, 2); (2, 1); (3, 4); (4, 3); (5, 6); (6, 5)

c. (1, 3); (2, 1); (3, 2); (4, 6); (5, 4); (6, 5)

d. (1, 5); (2, 3); (3, 1); (4, 6); (5, 4); (6, 2)

11. Describe an experimental protocol for studies of relatives that could estimate the broad heritability of alcoholism. Remember that you must make an adequate observational definition of the trait itself.

12. A line selected for high bristle number in *Drosophila* has a mean of 25 sternopleural bristles, whereas a line selected for low bristle number has a mean of only 2. Stocks with marker genes on the two large autosomes II and III are used to create stocks with various mixtures of chromosomes from the high (h) and low (l) lines. The mean number of bristles for each chromosomal combination is as follows:

$$\frac{h\ \ h}{h\ \ h}\ 25.1 \qquad \frac{h\ \ h}{l\ \ h}\ 22.2 \qquad \frac{l\ \ h}{l\ \ h}\ 19.0$$

$$\frac{h\ \ h}{h\ \ l}\ 23.0 \qquad \frac{h\ \ h}{l\ \ l}\ 19.9 \qquad \frac{l\ \ h}{l\ \ l}\ 14.7$$

$$\frac{h\ \ l}{h\ \ l}\ 11.8 \qquad \frac{h\ \ l}{l\ \ l}\ 9.1 \qquad \frac{l\ \ l}{l\ \ l}\ 2.3$$

What conclusions can you reach about the distribution of genetic factors and their actions from these data?

13. Suppose that number of eye facets is measured in a population of *Drosophila* raised at various temperatures. Further, suppose that it is possible to estimate the total genetic variance (s_g^2) as well as the phenotypic distribution. Finally, suppose that there are only two genotypes in the population. Draw pairs of norms of reaction that would lead to the following results:

a. An increase in mean temperature decreases the phenotypic variance.

b. An increase in mean temperature increases H^2.

c. An increase in mean temperature increases s_g^2 but decreases H^2.

d. An increase in temperature *variance* changes a unimodal into a bimodal phenotypic distribution (one norm of reaction is sufficient here).

14. Francis Galton compared the heights of male undergraduates with the heights of their fathers, with the results shown in the accompanying graph.

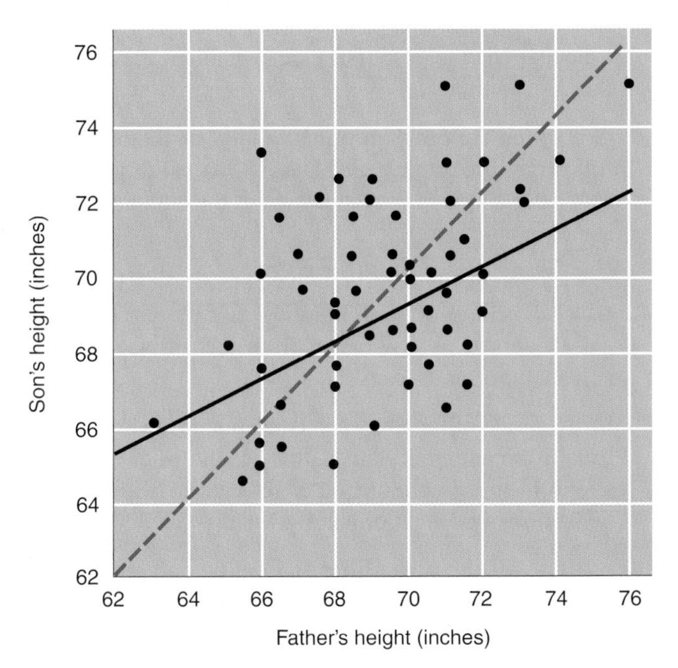

The average height of all fathers is the same as the average height of all sons, but the individual height classes are not equal across generations. The very tallest fathers had somewhat shorter sons, whereas the very shortest fathers had somewhat taller sons. As a result, the best line that can be drawn through the points on the scatter diagram has a slope of about 0.67 *(solid line),* rather than 1.00 *(dashed line).* Galton used the term *regression* to describe this tendency for the phenotype of the sons to be closer than the phenotype of their fathers to the population mean.

a. Propose an explanation for this regression.

b. How are regression and heritability related here?

(Graph after W. F. Bodmer and L. L. Cavalli-Sforza, *Genetics, Evolution, and Man.* © 1976 by W. H. Freeman and Company.)

EVOLUTIONARY GENETICS 19

Key Concepts

1. Evolution consists of the continuous heritable change of organisms within a single line of descent (phyletic evolution) and the differentiation between different lines of descent to form different species (diversification). The Darwinian mechanism of evolution rests on three principles: (1) organisms within a species vary from one another, (2) the variation is heritable, and (3) different types leave different numbers of offspring in future generations.
2. Natural selection is the differential reproduction of different genotypes that is a consequence of their different physiological, morphological, and behavioral traits.
3. Both phyletic change and diversification are the results of the interaction between the directional force of natural selection and random events. Random effects include the sampling of gametes each generation in finite populations and the random occurrence of mutations.
4. A consequence of the random factors in evolution is that the same forces of natural selection do not lead to the same evolutionary result in independent lines of descent.
5. Because all organisms are related to each other through common ancestors at different times in the past, their genes have similar DNA sequences, the similarities being smaller the farther in the past their common ancestor.
6. Some genes have diverged very rapidly in DNA sequence, but others have maintained important parts of their sequence through hundreds of millions of years of evolution, depending on the selective importance of these parts of the sequence.
7. Species are reproductively isolated populations of organisms that can exchange genes within the group but not with other species, because the groups that belong to different species are physiologically, behaviorally, or developmentally incompatible.
8. Evolutionary novelties are possible because new DNA is acquired either by the duplication and subsequent differentiation of DNA already present in the species or by the introduction of novel DNA from other species.
9. The genes underlying the embryonic development and basic body plan of animals have retained similarity of DNA sequence through hundreds of millions of years of evolution.

Charles Darwin. *(Corbis/Bettmann.)*

Humans, chimpanzees, and gorillas are all descended from a common ancestral form that separated into the three distinct evolutionary branches between about 4 million and 5 million years ago. In the time since then, a number of different species of each of these branches have arisen. In the case of chimpanzees and gorillas, there are several species of each now living. There is evidence from the fossil record that there have been several species in the human lineage (the exact number is a matter of dispute), but only one human species, *Homo sapiens,* has survived. Chimpanzees and gorillas differ from each other and from humans in their size, shape, hairiness, behavior, and the range of environments in which they live. The most striking difference for us, however, is the immense difference in cognitive ability between us and our close simian relatives. Only we humans have speech and culture. Although people have been able to teach chimpanzees and gorillas a limited ability to engage in symbolic exchange (Figure 19-1), no chimpanzee or gorilla will ever be able to write or read a book on genetics. What are the evolutionary forces that have led to this divergence among the various species of apes and humans and to the unique ability of *Homo sapiens* to communicate and to conceptualize and manipulate the world?

Figure 19-1 Dr. Sue Savage-Rumbaugh communicates with a bonobo, a close relative of the chimpanzee, by using a specially invented keyboard called a Lexigram. *(Courtesy of Georgia State University Language Research Center.)*

CHAPTER OVERVIEW

The modern theory of evolution is so completely identified with the name of Charles Darwin (1809–1882) that many people think the concept of organic evolution was first proposed by Darwin, which is certainly not the case. Most scholars had abandoned the notion of fixed species, unchanged since their origin in a grand creation of life, long before publication of Darwin's *The Origin of Species* in 1859. By that time, most biologists agreed that new species arise through some process of evolution from older species; the problem was to explain *how* this evolution could occur.

Darwin's theory of the mechanism of evolution begins with the variation that exists among organisms within a species. Individual organisms of one generation are qualitatively different from one another. Evolution of the species as a whole results from the differential rates of survival and reproduction of the various types; so the relative frequencies of the types change through time. Evolution, in this view, is a sorting process, in which types that are better

MESSAGE

Darwin proposed a new explanation to account for the accepted phenomenon of evolution. He argued that the population of a given species at a given time includes individual members of varying characteristics. The population of the next generation will contain a higher frequency of those types that most successfully survive and reproduce under the existing environmental conditions. Thus, the frequencies of various types within the species will change through time.

suited by their anatomy, physiology, and behavior to the environment in which the organisms live leave more offspring than do those that are more poorly suited. For Darwin, evolution of the group resulted from the differential survival and reproduction of individual variants *already existing* in the group—variants arising in a way unrelated to the environment whose chance of survival and reproduction do depend on the environment (Figure 19-2).

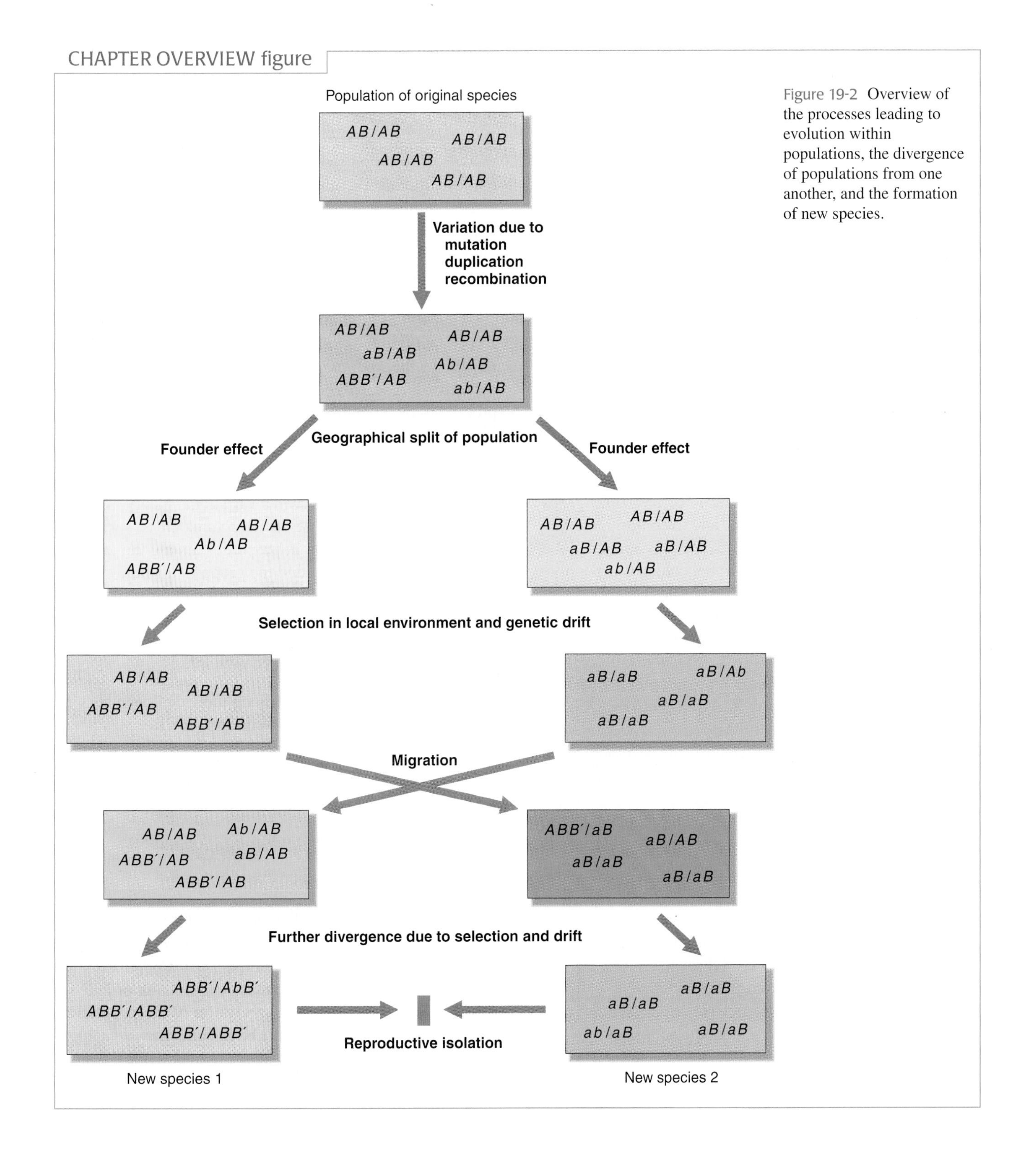

Figure 19-2 Overview of the processes leading to evolution within populations, the divergence of populations from one another, and the formation of new species.

CONSERVATION AND CHANGE IN EVOLUTION

Although the study of the processes of evolution emphasizes the mechanisms of change of species and the origin of new species, the evidence that all living species derive from common ancestors in the past comes from their similarities, rather than their differences. So, classical studies of the morphology and development of vertebrates showed that structures as different as the wings of birds, the front flippers of seals, and the front legs of frogs are all developmental variations of the same basic structure common to all vertebrates. Such features that are variants on the same developmental scheme inherited from a common ancestor are said to be **homologous.** On the other hand, the wings of birds and the wings of flies are derived from quite different embryonic structures and are not developmental variations of a common ancestral wing. Such unrelated structures are said to be only **analogous.** Before the development of modern molecular genetics and the biophysical analysis of molecular structure, the judgment that two features were homologous or analogous had to be based either on their basic anatomical structure or on similarities in their development or, more rarely, on fossil forms that were intermediate between them. The present ability to locate and sequence genes and to visualize the three-dimensional structure of proteins now makes it possible to make exact statements about the similarities and divergences of particular genes and proteins from different species and to confirm that they are indeed homologous. When this analysis is applied to the genes and proteins involved in morphogenesis, it is possible to determine if common elements underlie what appear to be very different anatomical structures in different species. One consequence has been that some structures judged to be only analogous on the basis of their anatomical development have turned out to be based on genes that are derived from common ancestral genes in the distant past. So, although the wings of birds and the wings of insects are not derived from a common ancestral wing, the development of both of them depends on genes controlling body segmentation that are very similar in vertebrates and insects. Thus, the neat division that was made between homologous and analogous structures when only anatomical evidence was available has become less clear-cut now that evidence from molecular genetics is considered.

MESSAGE

All living organisms are related through remote common ancestors at various times in the past. Similarities and differences in DNA and protein sequences allow inferences about the common evolutionary origin of molecules and structures that differ between species.

THE BASIC PRINCIPLES OF DARWINIAN EVOLUTION

There is an obvious similarity between the process of evolution as Darwin described it and the process by which the plant or animal breeder improves a domestic stock. The plant breeder selects the highest-yielding plants from the current population and (as far as possible) uses them as the parents of the next generation. If the characteristics causing the higher yield are heritable, then the next generation should produce a higher yield. It was no accident that Darwin chose the term **natural selection** to describe his model of evolution through differential rates of reproduction of different variants in the population. As a model for this evolutionary process, he had in mind the selection that breeders exercise on successive generations of domestic plants and animals. This **variational** model of an evolutionary process is fundamentally different from the other processes of change through time that are also referred to as "evolution" but are **transformational** in nature. In a transformational evolutionary process, the collection of all the objects in the system changes through time because each object undergoes a change. The best example of such a process is the evolution of the cosmos, in which each star has been passing through a set of changes in temperature and density that began with the Big Bang and will end as each star rapidly expands and then finally collapses into a very dense cold mass. Darwin's variational theory of evolution is quite different in its structure. In a variational scheme, there is variation in properties among the different objects in the population, and the group as a whole changes through time because objects with different properties leave different numbers of descendants.

We can summarize Darwin's theory of evolution through natural selection in three principles:

1. **Principle of variation.** Among individual members within any population, there is variation in morphology, physiology, and behavior.
2. **Principle of heredity.** Offspring resemble their parents more than they resemble individuals to which they are unrelated.
3. **Prinicple of selection.** Some forms are more successful at surviving and reproducing than other forms in a given environment.

Clearly, a selective process can produce change in the population composition only if there are some variations among which to select. If all members of a population are identical, no amount of differential reproduction of individual members can affect the composition of the population. Furthermore, the variation must be in some part heritable if differential reproduction is to alter the population's genetic composition. If large animals within a population have more offspring than do small ones but their offspring are no larger on average than those of small animals, then there can be no change in population composition from one generation to another. Moreover, the variation must not be destroyed by the process of reproduction itself, which accounts for the importance of the particulate nature of heredity. If inheritance were by a mixture of blood from the two parents, then the offspring would be intermediate between the parents and, through time, all the variation in the population would be lost, just as the mixing of paints of different colors leads to a uniform intermediate color. Finally, if all variant types leave, on average, the same number of offspring, then we can expect the population to remain unchanged.

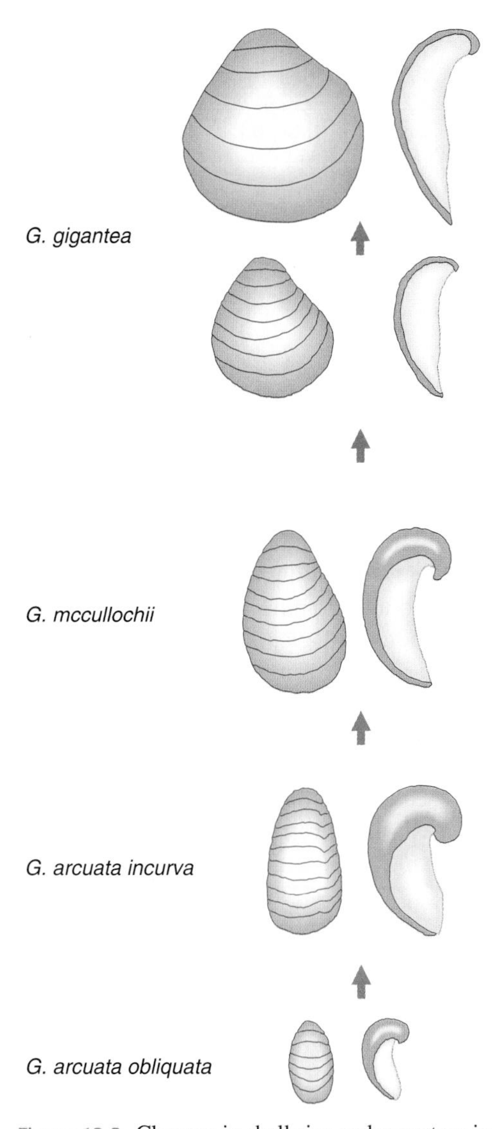

Figure 19-3 Changes in shell size and curvature in the bivalve mollusc *Gryphaea* in the course of its phyletic evolution in the early Jurassic. Only the left shell is shown. In each case, the shell back and a longitudinal section through it are illustrated. *(After A. Hallam, "Morphology, Palaeoecology and Evolution of the Genus* Gryphaea *in the British Lias,"* Philosophical Transactions of the Royal Society of London Series B *254, 1968, 124.)*

MESSAGE

Darwin's principles of variation, heredity, and selection must hold true if there is to be evolution by a variational mechanism.

The Darwinian explanation of evolution must apply to two different aspects of the history of life. One aspect is the successive change of form and function that takes place in a single continuous line of descent time, **phyletic evolution.** Figure 19-3 shows such a continuous change in the size and curvature of the left shell of the oyster, *Gryphaea,* over a period of 40 million years. The other is **diversification** among species: in the history of life on Earth, there are many different contemporaneous species having quite different forms and living in different ways. Figure 19-4 shows some of the variety of bivalve mollusc forms that existed at various times in the past 130 million years. Every species eventually becomes extinct and more than 99.9 percent of all the species that have ever existed are already extinct, yet the number of species and the diversity of their forms and functions have increased in the past billion years. Thus species not only must be changing, but must give rise to new and different species in the course of evolution. Both of these processes are the consequences of heritable variation within populations. Heritable variation provides the raw material for successive changes within a species and for the multiplication of new species. The basic mechanisms of those changes (as discussed in Chapter 17) are the origin of new variation by various kinds of mutational mechanisms, the change in frequency of alleles by selective and random processes, the possibility of divergence of isolated local populations because the selective forces are different or because of random drift, and the reduction of variation between populations by migration. From those basic mechanisms, population genetics, as discussed in Chapter 17,

Figure 19-4 A variety of bivalve mollusc shell forms that have appeared in the past 300 million years of evolution. *(After C. L. Fenton and M. A. Fenton,* The Fossil Book, *Doubleday, 1958.)*

derives a set of principles governing changes in the genetic composition of populations. The application of these principles of population genetics provides an articulated theory of evolution.

MESSAGE

Evolution, under the Darwinian scheme, is the conversion of heritable variation between individual members within populations into heritable differences between populations in time and in space, by population genetic mechanisms.

A SYNTHESIS OF FORCES: VARIATION AND DIVERGENCE OF POPULATIONS

In evolution, the various forces of breeding structure, mutation, migration, and selection are all acting simultaneously in populations. We need to consider how these forces, operating together, mold the genetic composition of populations to produce both variation within local populations and differences between them.

The genetic variation within and between populations is a result of the interplay of the various evolutionary forces (Figure 19-5). Generally, as Table 19-1 shows, forces that increase or maintain variation within populations prevent the differentiation of populations from each other, whereas the divergence of populations is a result of forces that make each population homozygous. Thus, random drift (or inbreeding) produces homozygosity while causing different populations to diverge. This divergence and homozygosity are counteracted by the constant flux of mutation and the migration between localities, which introduce variation into the populations again and tend to make them more like one another.

When Darwin arrived in the Galápagos Islands in 1835, he found a remarkable group of finchlike birds that provided a very suggestive case for the development of his

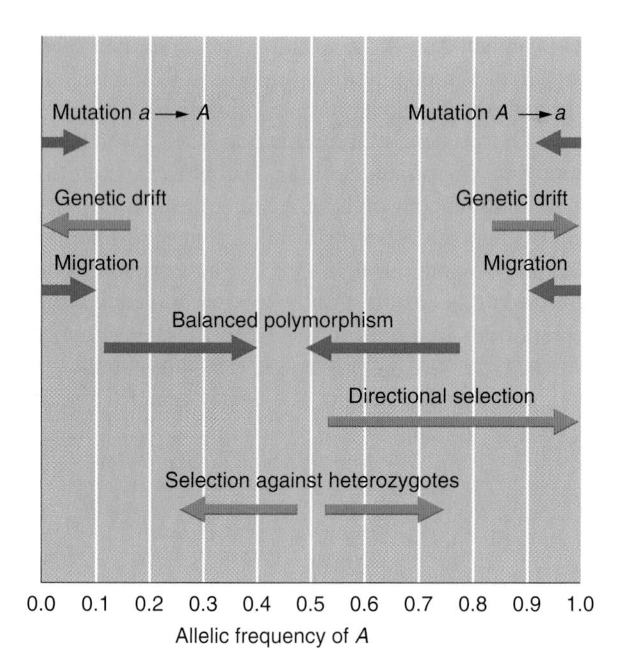

Figure 19-5 The effects on gene frequency of various forces of evolution. The blue arrows show a tendency toward increased variation within the population; the red arrows, decreased variation.

theory of evolution. The Galápagos archipelago is a cluster of 29 islands and islets of different sizes lying on the equator about 600 miles off the coast of Ecuador. Figure 19-6 shows the 13 Galápagos finch species. Finches are generally ground-feeding seed eaters with stout bills for cracking the tough outer coats of the seeds. The Galápagos species, though clearly finches, have an immense variation in how they make a living and in their bill shapes and their behaviors, which underly these ecological differences. For example, a vegetarian tree finch eats fruits, leaves, and buds; an insectivorous finch has a bill with a biting tip for eating large insects; and, most remarkable of all, the woodpecker finch, with its long bill for probing crevices in the bark, can grasp a twig in its beak and use it to obtain insect prey by probing holes in trees. This diversity of species arose from

TABLE 19-1 How the Forces of Evolution Increase (+) or Decrease (−) Variation within and between Populations

Force	Variation within Populations	Variation between Populations
Inbreeding or genetic drift	−	+
Mutation	+	−
Migration	+	−
Selection		
Directional	−	+/−
Balancing	+	−
Incompatibility	−	+

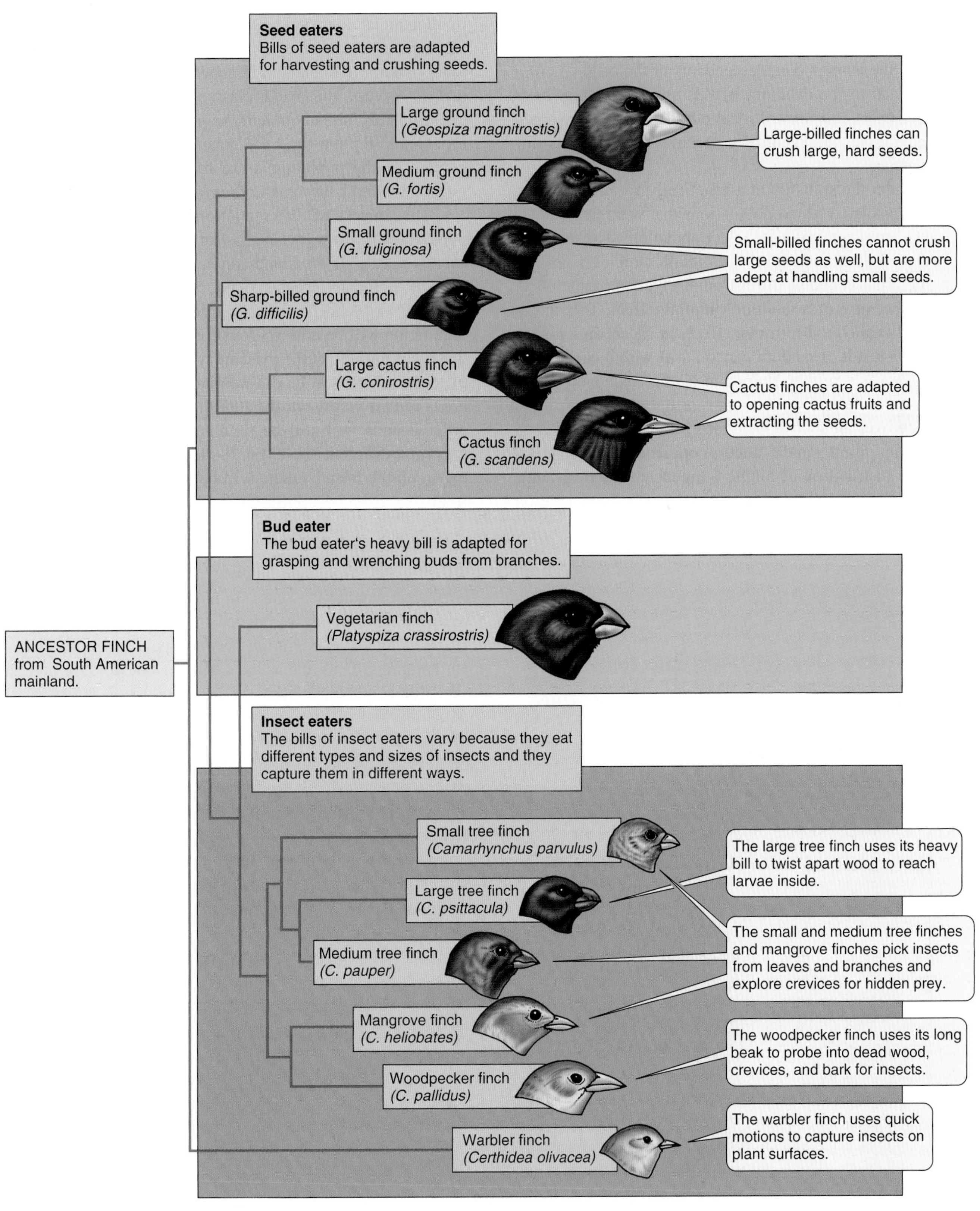

Figure 19-6 The 13 species of finches found in the Galápagos Islands. *(After W. K. Purves, G. H. Orians, and H. C. Heller,* Life: The Science of Biology, *4th ed. New York: Sinauer Associates/W. H. Freeman and Company, 1995, Figure 20.3, p. 450.)*

an original population of a seed-eating finch that arrived in the Galápagos from the mainland of South America and populated the islands. The descendants of the original colonizers spread to the different islands and to different parts of large islands and formed local populations that diverged from each other and eventually became established as different species.

Consider the situation at a genetically variable locus in a group of isolated island populations that were founded by migrants from an initial single population. The original founders of each population are small samples from the donor population and so differ from each other in allele frequencies because of a random-sampling effect. This initial variation is called the **founder effect.** In succeeding generations, as a result of random genetic drift within each population, there is a further change in allelic frequency, p_i, of each of the i alleles toward either 1 or 0, but average allelic frequency over all such populations remains the same as it was in the initial single donor population. Figure 19-7 shows the distribution of allelic frequencies among islands in successive generations, where $p(A_1) = 0.5$. In generation 0, all populations are identical. As time goes on, the gene frequencies among the populations diverge and some become fixed. After about $2N$ generations, every allelic frequency except the fixed classes ($p = 0$ and $p = 1$) is equally likely, and about half the populations are totally homozygous. By the time $4N$ generations have gone by, 80 percent of the populations are fixed, half being homozygous A / A and half being homozygous a / a.

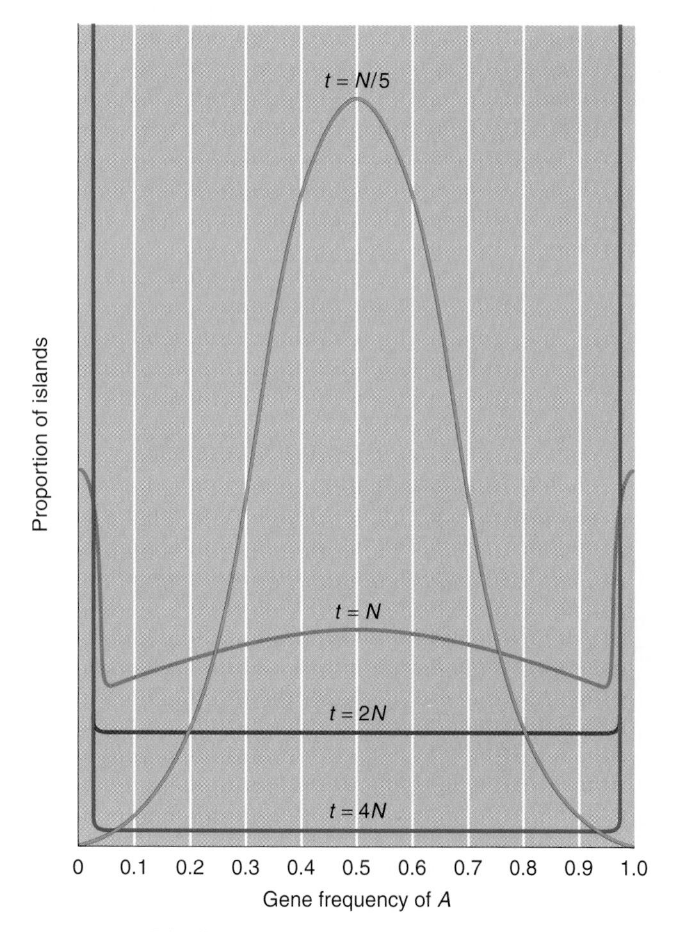

Figure 19-7 Distribution of gene frequencies among island populations after various numbers of generations of isolation, where the number of generations that have passed (t) is given in multiples of the population size (N).

The process of differentiation by inbreeding in island populations is slow, but not on an evolutionary or geological time scale. If an island can support, say, 10,000 members of a rodent species, then, after 20,000 generations (about 7000 years, assuming 3 generations per year), the population will be homozygous for about half of all the loci that were initially at the maximum of heterozygosity. Moreover, the island will be differentiated from other similar islands in two ways. For the loci that are fixed, many of the other islands will still be segregating, and others will be fixed at a different allele. For the loci that are still segregating in all the islands, there will be a large variation in gene frequency from island to island, as shown in Figure 19-7.

Any population of any species is finite in size; so all populations should eventually become homozygous and differentiated from one another as a result of inbreeding. Evolution would then cease. In nature, however, new variation is always being introduced into populations by mutation and by some migration between populations at different localities. Thus, the actual variation available for natural selection is a balance between the introduction of new variation and its loss through local inbreeding. The rate of loss of heterozygosity in a closed population is $1/(2N)$ per generation; so any effective differentiation between populations will be negated if new variation is introduced at this rate or a higher rate. If m is the migration rate into a given population and μ is the rate of mutation to new alleles per generations, then roughly (to an order of magnitude) a population will retain most of its heterozygosity and will not differentiate much from other populations by local inbreeding if

$$m \geq 1/N \quad \text{or} \quad \mu \geq 1/N$$

or if

$$Nm \geq 1 \quad \text{or} \quad N\mu \geq 1$$

For populations of intermediate and even fairly large size, it is unlikely that $N\mu \geq 1$. For example, if the population size is 100,000, then the mutation rate must exceed 10^{-5}, which is somewhat on the high side for known mutation rates, although it is not an unknown rate. On the other hand, a migration rate of 10^{-5} per generation is not unreasonably large. In fact,

$$m = \frac{\text{number of migrants}}{\text{total population size}} = \frac{\text{number of migrants}}{N}$$

TABLE 19-2 Examples of Extreme Differentiation and Close Similarity in Blood Group Allelic Frequencies in Three Geographical Races

Gene	Allele	POPULATION		
		Caucasoid	Negroid	Mongoloid
Duffy	*Fy*	0.0300	0.9393	0.0985
	Fy^a	0.4208	0.0000	0.9015
	Fy^b	0.5492	0.0607	0.0000
Rhesus	R_0	0.0186	0.7395	0.0409
	R_1	0.4036	0.0256	0.7591
	R_2	0.1670	0.0427	0.1951
	r	0.3820	0.1184	0.0049
	r′	0.0049	0.0707	0.0000
	Others	0.0239	0.0021	0.0000
P	P_1	0.5161	0.8911	0.1677
	P_2	0.4839	0.1089	0.8323
Auberger	Au^a	0.6213	0.6419	No data
	Au	0.3787	0.3581	No data
Xg	Xg^a	0.67	0.55	0.54
	Xg	0.33	0.45	0.46
Secretor	*Se*	0.5233	0.5727	No data
	se	0.4767	0.4273	No data

SOURCE: A summary is provided in L. L. Cavalli-Sforza and W. F. Bodmer, *The Genetics of Human Populations* (W. H. Freeman and Company, 1971), pp. 724–731. See L. L. Cavalli-Sforza, P. Menozzi, and A. Piazza, *The History and Geography of Human Genes* (Princeton University Press, 1994), for detailed data.

Thus, the requirement that $Nm \geq 1$ is equivalent to the requirement that

$$Nm = N \times \text{number of migrants}/N \geq 1$$

or that

$$\text{number of migrant individuals} \geq 1$$

regardless of population size. For many populations, more than a single migrant individual per generation is quite likely. Human populations (even isolated tribal populations) have a higher migration rate than this minimal value, and, as a result, no locus is known in humans for which one allele is fixed in some populations and an alternative allele is fixed in others, although a particular allele may be common in one population and very rare or even absent in another (Table 19-2).

The effects of selection are more variable. **Directional selection** pushes a population toward homozygosity, rejecting most new mutations as they are introduced but occasionally (if the mutation is advantageous) spreading a new allele through the population to create a new homozygous state. Whether such directional selection promotes differentiation of populations depends on the environment and on chance events. Two populations living in very similar environments may be kept genetically similar by directional selection, but, if there are environmental differences, selection may direct the populations toward different compositions.

Selection favoring heterozygotes (balancing selection) will, for the most part, maintain more or less similar polymorphisms in different populations. However, again, if the environments are different enough between them, then the populations will show some divergence. The opposite of balancing selection is selection against heterozygotes, which produces unstable equilibria. Such selection will cause homozygosity and divergence between populations.

MULTIPLE ADAPTIVE PEAKS

We must avoid taking an overly simplified view of the consequences of selection. At the level of the gene—or even at the level of a particular aspect of the phenotype—the outcome of selection for a trait in a given environment is not unique. Selection to alter a trait (say, to increase size) may be successful in a number of ways. In 1952, Forbes Robertson and Eric Reeve successfully selected to change wing size in *Drosophila* in two different populations. However, in one case, the number of cells in the wing changed, whereas, in the other case, the *size* of the wing cells changed. Two different genotypes had been selected, both causing a change in wing size. The initial state of the population at the outset of selection determined which of these selections occurred.

The way in which the same selection can lead to different outcomes can be most easily illustrated by a simple hypothetical case. Suppose that the variation of two loci (there will usually be many more) influences a character and that (in a particular environment) intermediate phenotypes have the highest fitness. (For example, newborn babies have a higher chance of surviving birth if they are neither too big nor too small.) If the alleles act in a simple way in influencing the phenotype, then the three genetic constitutions $A\ B\ /\ a\ b$, $A\ b\ /\ A\ b$, and $a\ B\ /\ a\ B$ will produce a high fitness because they will all be intermediate in phenotype. On the other hand, very low fitness will characterize the double homozygotes $A\ B\ /\ A\ B$ and $a\ b\ /\ a\ b$. What will the result of selection be? We can predict the result by using the mean fitness, $\overline{W}$, of a population. As discussed in Chapter 17, selection acts in most simple cases to increase the value of $\overline{W}$ Therefore, if we calculate $\overline{W}$ for every possible combination of gene frequencies at the two loci, we can determine which combinations yield high values of $\overline{W}$. Then we should be able to predict the course of selection by following a curve of increasing $\overline{W}$.

The plot of mean fitness for all possible combinations of allelic frequency is called an **adaptive surface** or an **adaptive landscape** (Figure 19-8). The plot is like a topographic map. The frequency of allele A at one locus is plotted on one axis, and the frequency of allele B at the other locus is plotted on the other axis. The height above the plane (represented by topographic lines) is the value of $\overline{W}$ that the population would have for a particular combination of frequencies of A and B. According to the rule of increasing fitness, selection should carry the population from a low-fitness "valley" to a high-fitness "peak." However, Figure 19-8 shows that there are two adaptive peaks, corresponding to a fixed population of $A\ b\ /\ A\ b$ and a fixed population of $a\ B\ /\ a\ B$, with an adaptive valley between them. Which peak the population will ascend—and therefore what its final genetic composition will be—depends on whether the initial genetic composition of the population is on one side or the other of the dashed "fall line," inasmuch as the surface rises on both sides of this line.

MESSAGE

Under identical conditions of natural selection, two populations may arrive at two different genetic compositions as a direct result of natural selection.

It is important to note that nothing in the theory of selection requires that the different adaptive peaks be of the same height. The kinetics of selection is such that the value of $\overline{W}$ increases, not that it necessarily reaches the highest possible peak in the field of gene frequencies. Suppose, for example, that a population is near the peak $a\ B\ /\ a\ B$ in Figure 19-8 and that this peak is lower than the $A\ b\ /\ A\ b$ peak. Selection alone cannot carry the population to $A\ b\ /\ A\ b$, because that would require a temporary decrease in the value of $\overline{W}$ as the population descended the $a\ B\ /\ a\ B$ slope, crossed the saddle, and ascended the other slope. Thus, the force of selection is myopic. It drives the population to a *local* maximum value of $\overline{W}$ in the field of gene frequencies—not to a *global* one.

The existence of multiple adaptive peaks for a selective process means that some differences between species are the result of history and not of environmental differences. For example, African rhinoceroses have two horns, and Indian rhinoceroses have one (Figure 19-9). We need not invent a special story to explain why it is better to have two horns on the African plains and one in India. It is much

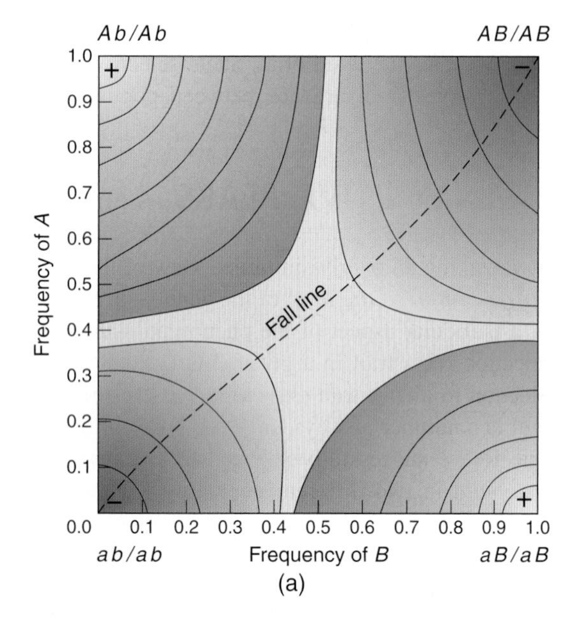

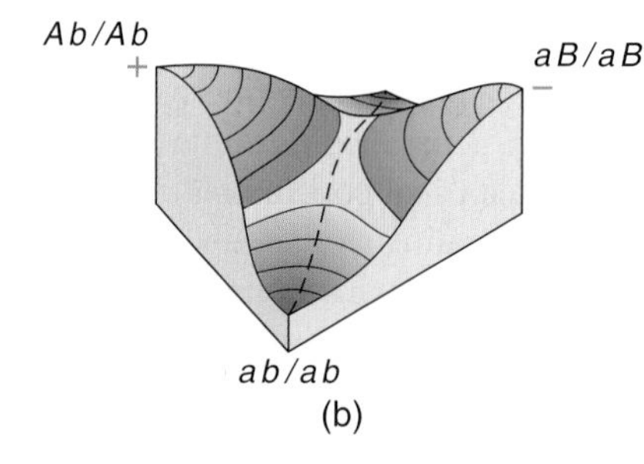

Figure 19-8 An adaptive landscape with two adaptive peaks (red), two adaptive valleys (blue), and a topographic saddle in the center of the landscape. The topographic lines are lines of equal mean fitness. If the genetic composition of a population always changes in such a way as to move the population "uphill" in the landscape (to increasing fitness), then the final composition will depend on where the population began with respect to the fall (dashed) line. (a) Topographic map of the adaptive landscape. (b) A perspective sketch of the surface shown in the map.

(a) (b)

Figure 19-9 Differences in horn morphology in two geographically separated species of rhinoceros: (a) the African rhinoceros; (b) the Indian rhinoceros. *(Part a from Anthony Bannister/NHPA; part b from K. Ghani/NHPA.)*

more plausible that the trait of having horns was selected but that two long, slender horns and one short, stout horn are simply alternative adaptive features, and historical accident differentiated the species. Explanations of adaptations by natural selection do not require that every difference between species be differentially adaptive.

Exploration of Adaptive Peaks

Random and selective forces should not be thought of as simple antagonists. Random drift may counteract the force of selection, but it can enhance it as well. The outcome of the evolutionary process is a result of the simultaneous operation of these two forces. Figure 19-10 illustrates these possibilities. Note that there are multiple adaptive peaks in this landscape. Because of random drift, a population under selection does not ascend an adaptive peak smoothly. Instead, it takes an erratic course in the field of gene frequencies, like an oxygen-starved mountain climber. Pathway I shows a population history where adaptation has failed. The random fluctuations of gene frequency were sufficiently great that the population by chance became fixed at an unfit genotype. In any population, some proportion of loci are fixed at a selectively unfavorable allele because the intensity of selection is insufficient to overcome the random drift to fixation. The existence of multiple adaptive peaks and the random fixation of less-fit alleles are integral features of the evolutionary process. Natural selection cannot be relied on to produce the best of all possible worlds.

Pathway II in Figure 19-10, on the other hand, shows how random drift may improve adaptation. The population was originally in the sphere of influence of the lower adaptive peak ($a\ B\ /\ a\ B$); however, by random fluctuation in gene frequency, its composition passed over the adaptive saddle, and the population was captured by the higher, steeper adaptive peak. This passage from a lower to a higher adaptive stable state could never have occurred by selection in an infinite population, because, by selection

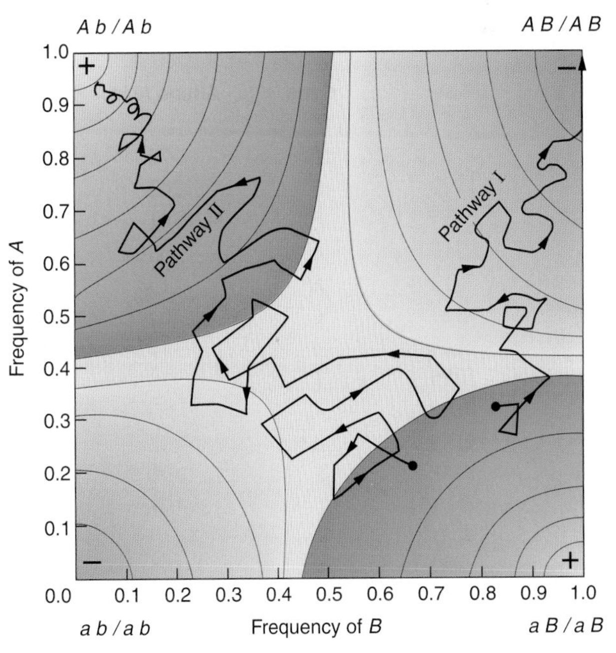

Figure 19-10 Selection and random drift can interact to produce different changes in gene frequency in an adaptive landscape. Without random drift, both populations would have moved toward $a\ B\ /\ a\ B$ as a result of selection alone.

alone, $\overline{W}$ could never decrease temporarily to cross from one slope to another.

An important source of indeterminacy in the outcome of a long selective process is the randomness of the mutational process. After the initial genetic variation has been exhausted by the selective and random fixation of alleles, new variation arises from mutation that can be the source of yet further evolutionary change. The particular direction of this further evolution depends on the particular mutations that occur and the temporal order in which they take place. A very clear illustration of this historical contingency of the evolutionary process is the selection experiment of Holly Wichman and her colleagues to allow the bacteriophage ΦX174 to reproduce at high temperatures and to change its host from *Escherichia coli* to *Salmonella typhimurium.* Two independent selection lines were established and each evolved changes in its temperature tolerance and its host. In one of the two lines, the ability to reproduce on *E. coli* still existed, but, in the other line, that ability was lost. The bacteriophage has only 11 genes, and the successive changes in the DNA for all these genes and in the proteins encoded by them were recorded during the selection process. The result for the two strains is shown in Table 19-3. There were 15 DNA changes in strain TX, located in 6 different genes; in strain ID, there were 14 changes in 4 of the genes. The strains had identical changes in 7 cases, including a large deletion, but the temporal order of these identical changes differed between the lines. So, for example, the change at DNA site 1533 (dark blue type in Table 19-3), causing a substitution of isoleucine for threonine, was the third change in the ID strain, but the fourteenth change in the TX strain. What these results show is that, when there is selection for the same change in physiology in two independent evolutionary lines, even when those two lines were initially genetically identical, the sequence of genetic events leading to adaptation in the two lines will be very different. Evolution never repeats itself.

HERITABILITY OF VARIATION

The first rule of any reconstruction or prediction of evolution is that the phenotypic variation must be heritable. It is easy to construct stories of the possible selective advantage of one form of a trait over another, but it is a matter of considerable experimental difficulty to show that the variation in the trait corresponds to genotypic differences (see Chapter 18).

It should not be supposed that all variable traits are heritable. Certain metabolic traits (such as resistance to high salt concentrations in *Drosophila*) show individual variation but no heritability. In general, behavioral traits have lower heritabilities than do morphological traits, especially in organisms with more complex nervous systems that exhibit immense individual flexibility in the connectivity of their central nervous systems. Before any judgment

TABLE 19-3 Molecular Changes in Two Replicated Selection Lines, TX and ID, to Change the Temperature Tolerance and Host Range of Bacteriophage ΦX174

	TX				ID			
Order	**Site**	**Gene**	**Amino Acid**	**Change**	**Site**	**Gene**	**Amino Acid**	**Change**
1	782	*E*	72	T → I	2167	*F*	388	H → Q
2	1727	*F*	242	L → F	1613	*F*	204	T → S
3	2085	*F*	361	A → V	**1533**	*F*	**177**	T → I
4	319	*C*	63	V → F	1460	*F*	153	Q → E
5	2973	*H*	15	G → S	1300	*F*	99	Silent
6	323	*C*	64	D → G	1305	*F*	101	G → D
7	4110	*A*	44	H → Y	1308	*F*	102	Y → C
8	1025	*F*	8	E → K	4110	*A*	44	H → Y
9	3166	*H*	79	A → V	4637	*A*	219	Silent
10	5185	*A**	402	T → M	**965–991**		**deletion**	
11	1305	*F*	101	G → D	5365	*A**	462	M → T
12	**965–991**		**deletion**		**4168**	*A*	**63**	**Q → R**
13	5365	*A**	462	M → T	3166	*H*	79	A → V
14	**1533**	*F*	**177**	T → I	1809	*F*	269	K → R
15	**4168**	*A*	**63**	Q → R				

Note: The temporal order of mutations is in the left-hand column. The changes are given as the nucleotide site number, the gene name (A–H, A*), the amino acid residue number, and the nature of the amino acid substitution (amino acid abbreviations are as given in Table 3-1). Changes identical in the two strains are in colored type.

SOURCE: Data from H. A. Wichman, M. R. Badgett, L. A. Scott, C. M. Boulianne, and J. J. Bull, *Science* 285, 1999, 422–424.

can be made about the evolution of a particular quantitative trait, it is essential to determine if there is genetic variance for it in the population whose evolution is to be predicted. Thus, suggestions that such traits in the human species as performance on IQ tests, temperament, and social organization are in the process of evolving or have evolved at particular epochs in human history depend critically on evidence about whether there is genetic variation for these traits. Reciprocally, traits that appear to be completely phenotypically invariant in a species may nevertheless evolve. Such evolution may occur when there is genetic variation that is relevant to a trait, but the variation only appears in some environments—as, for example, in the presence of an environmental stress.

One of the most important findings in evolutionary genetics has been the discovery of substantial genetic variation underlying characters that show no morphological variation. These characters are called **canalized characters,** because the final outcome of their development is held within narrow bounds despite disturbing forces. Development is such that all the different genotypes for canalized characters have the same constant phenotype in the range of environments that is usual for the species. The genetic differences are revealed only if the organisms are put in an unusual environment, such as a stressful environment, or if a severe mutation stresses the developmental system. For example, all wild-type *Drosophila* have exactly four scutellar bristles (Figure 19-11). If the recessive mutant *scute* is present, the number of bristles is reduced, but, in addition, there is variation from fly to fly. This variation is heritable, and lines with no bristles or one bristle and lines with three or four bristles can be obtained by selection in the presence of the *scute* mutation. When the mutation is removed, these lines now have two and six bristles, respectively. Similar experiments have been performed by using extremely stressful environments instead of mutations. A consequence of such hidden genetic variation is that a character that is phenotypically uniform in a species may nevertheless undergo rapid evolution if a stressful environment uncovers the genetic variation.

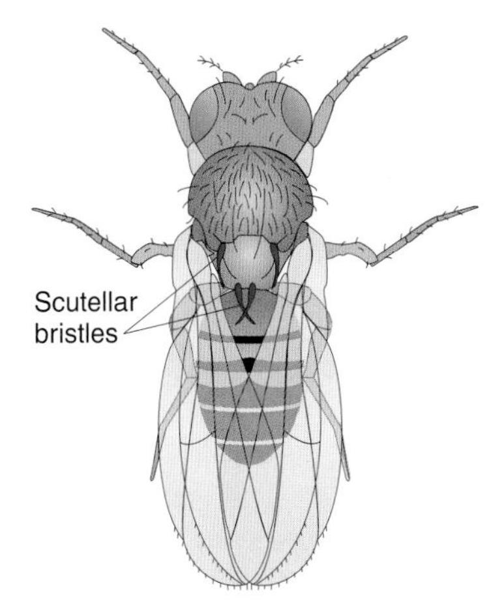

Figure 19-11 The scutellar bristles of the adult *Drosophila,* shown in blue. This is an example of a canalized character; all wild-type *Drosophila* have four scutellar bristles in a very wide range of environments.

OBSERVED VARIATION WITHIN AND BETWEEN POPULATIONS

In Chapter 17, the existence of genetic variation within populations at the levels of morphology, karyotype, proteins, and DNA was documented. The general conclusion is that about one-third of all protein-encoding loci are polymorphic and that all classes of DNA, including exons, introns, regulatory sequences, and flanking sequences, show nucleotide diversity among individual members within populations. Several of these examples also documented some differences in the genotype frequencies between populations (see Tables 17-1, 17-2, 17-4, 17-5, and 17-6). The relative amounts of variation within and between populations vary from species to species, depending on history and environment. In humans, some gene frequencies (for example, those for skin color or hair form) are well differentiated between populations and major geographical groups (so-called geographical races). If, however, we look at single structural genes identified immunologically or by electrophoresis rather than by these outward phenotypic characters, the situation is rather different. Table 19-2 lists the three loci for which Caucasians, Negroids, and Mongoloids are known to be most different from one another (Duffy and Rhesus blood groups and the P antigen) compared with the three polymorphic loci for which the races are most similar (Auberger blood group and Xg and secretor factors). Even for the most divergent loci, no race is homozygous for one allele that is absent in the other two races.

In general, different human populations show rather similar frequencies for polymorphic genes. Figure 19-12 on the next page is a **triallelic diagram** for the three main allelic classes, I^A, I^B, and i, of the ABO blood group. Each point represents the allelic composition of a population, where the three allelic frequencies can be read by taking the lengths of the perpendiculars from each side to the point. The diagram shows that all human populations are bunched together in the region of high i, intermediate I^A, and low I^B frequencies. Moreover, neighboring points (enclosed by dashed lines) do not correspond to geographical races; so such races cannot be distinguished from one another by characteristic allelic frequencies for this gene. The results of studies of polymorphic blood groups and enzyme loci in a variety of human populations have shown that about 85 percent of total human genetic diversity is found within local populations, about 7 percent is found among local populations within major geographical races, and the remaining

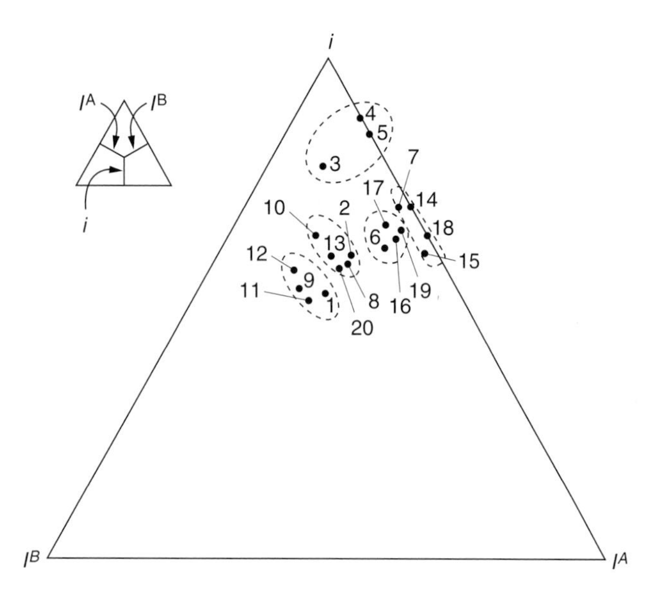

Figure 19-12 Triallelic diagram of the ABO blood group allelic frequencies for human populations. Each point represents a population; the perpendicular distances from the point to the sides represent the allelic frequencies, as indicated in the small triangle. Populations 1 through 3 are African, 4 through 7 are Native American, 8 through 13 are Asian, 14 and 15 are Australian Aborigine, and 16 through 20 are European. Dashed lines enclose arbitrary classes with similar gene frequencies; these groupings do not correspond to "racial" classes. *(After A. Jacquard,* Structures génétiques des populations. *©1970 by Masson et Cie.)*

8 percent is found among major geographical races. Clearly, the genes influencing skin color, hair form, and facial form that are well differentiated among races are not a random sample of structural gene loci.

PROCESS OF SPECIATION

When we examine the living world, we see that individual organisms are usually clustered into collections that resemble each other more or less closely and are clearly distinct from other clusters. A close examination of a sibship of *Drosophila* will show differences in bristle number, eye size, and details of color pattern from fly to fly, but an entomologist has no difficulty whatsoever in distinguishing *Drosophila melanogaster* from, say, *Drosophila pseudoobscura.* One never sees a fly that is halfway between these two kinds. Clearly, in nature at least, there is no effective interbreeding between these two forms. A group of organisms that exchanges genes within the group but cannot do so with other groups is what is meant by a **species.** Within a species there may also exist local populations that are easily distinguished from one another by some phenotypic characters, but it is also the case that genes can be easily exchanged between them. For example, no one has any difficulty distinguishing a "typical" Senegalese from a "typical" Swede, but, as a consequence of the migration and mating history of humans in North America in the past 300 years, an immense number of people exist of every degree of intermediacy between these local geographical types. They are not separate species. A geographically defined population that is genetically distinguishable from other local populations but is capable of exchanging genes with those other local populations is sometimes called a **geographical race.** In general, there is some difference in the frequency of alleles of various genes in different geographical populations of a species; so the marking out of a particular population as a distinct race is arbitrary and, as a consequence, the concept of race is no longer much used in biology. In human affairs, the term *race* is still commonly used to denote groups with different skin colors, but this usage reflects a social reality rather than a significant biological differentiation between groups.

MESSAGE

A species is a group of organisms that can exchange genes among themselves but are genetically unable to exchange genes in nature with other such groups. A geographical race is a phenotypically distinguishable local population within a species that is capable of exchanging genes with other races within that species. Because nearly all geographical populations are different from others in the frequencies of some genes, race is a concept that makes no clear biological distinction.

All the species now existing are related to each other by common ancestors at various times in the evolutionary past. That relatedness means that each of these species has separated out from a previously existing species and has become genetically distinct and genetically isolated from its ancestral line. In extraordinary circumstances, the founding of such a genetically isolated group might occur by a single mutation, but the carrier of that mutation would need to be capable of self-fertilization or vegetative reproduction. Moreover, that mutation would have to cause complete mating incompatibility between its carrier and the original species and to allow the new line to compete successfully with the previously established group. Although not impossible, such events must be rare.

The usual pathway to the formation of new species is through geographical isolation. As stated earlier in this chapter, populations that are geographically separated will diverge from one another genetically as a consequence of a combination of unique mutations, selection, and genetic drift. Migration between populations will prevent them from diverging too far, however. As shown on pages 626–627, even a single migrant per generation will be sufficient to prevent populations from fixing at alternative alleles by genetic drift alone, and even selection toward different adaptive peaks will not succeed in causing complete divergence unless it is extremely strong. As a consequence, populations that diverge enough to become new, reproductively isolated species must first be virtually totally isolated from each other by some mechanical barrier. This isolation almost always requires some spatial separation, and the separation must be great enough or the natural barriers to the passage of migrants must be strong enough to prevent any effective migration. Such populations are referred to as **allopatric.** The isolating barrier might be, for example, the extending tongue of a continental glacier during glacial epochs that forces apart a previously continuously distributed population, the drifting apart of continents that become separated by water, or the infrequent colonization of islands that are far from shore. The critical point is whether the mechanisms of dispersal of the original species will make further migration between the separated populations a very rare event. If so, then the populations are now genetically independent and will continue to diverge by mutation, selection, and genetic drift. Eventually, the genetic differentiation between the populations becomes so great that the formation of hybrids between them would be physiologically, developmentally, or behaviorally impossible **(hybrid breakdown)** even if the geographical separation were abolished. These *biologically* isolated populations are now new species, formed by the process of **allopatric speciation.**

MESSAGE

Allopatric speciation occurs through an initial geographical and mechanical isolation of populations that prevents any gene flow between them, followed by genetic divergence of the isolated populations sufficient to make it biologically impossible for them to exchange genes in the future.

Biological isolating mechanisms between species may arise to prevent even the formation of hybrid zygotes **(prezygotic isolation)** or they may come into play after zygote formation to prevent the passage of the hybrid genome on to future generations **(postzygotic isolation).** Prezygotic mechanisms include separation in times or places of sexual activity or in behavioral and physical incompatabilites between males and females of the different species. Examples of prezygotic isolating mechanisms are well known in plants and animals. The two species of pine growing on the Monterey peninsula, *Pinus radiata* and *P. muricata,* shed their pollen in February and April, respectively, and so do not exchange genes. The light signals that are emitted by male fireflies and attract females differ in intensity and timing between species. In the tsetse fly, *Glossina,* mechanical incompatibilities in the copulatory apparatus of the two sexes cause severe injury and even death if males of one species mate with females of another. The pollen of different species of *Nicotiana,* the genus to which tobacco belongs, either fail to germinate or cannot grow down the style of other species.

Postzygotic isolation includes both the failure of hybrid embryos to reach adulthood and the sterility of one or both sexes of hybrids if they do develop to the usual age of reproduction. Postzygotic isolation is more common in animals than in plants, apparently because the development of many plants is much more tolerant of genetic incompatibilities and chromosomal variations. When the eggs of the leopard frog, *Rana pipiens,* are fertilized by sperm of the wood frog, *R. sylvatica,* the embryos do not succeed in developing. Horses and asses can easily be crossed to produce mules, but, as is well known, these hybrids are sterile.

Genetics of Species Isolation

Usually, it is not possible to carry out any genetic analysis of the isolating mechanisms between two species for the simple reason that, by definition, they cannot be crossed with each other. It is possible, however, to make use of very closely related species in which the isolating mechanism is an incomplete hybrid sterility and hybrid breakdown. Then, segregating progeny of hybrid F_2 or backcross generations can be analyzed by using genetic markers. An example is shown in Figure 19-13 on the next page. *Drosophila pseudoobscura* and *D. persimilis* are closely related species that never exchange genes in nature but can be crossed in the laboratory. The F_1 males are completely sterile, but the F_1 females have normal fertility and can be backcrossed to males of the parental species. A manifestation of hybrid male sterility is that, in the cross between *D. persimilis* females and *D. pseudoobscura* males, the testes of F_1 males are about one-fifth normal size, presumably indicative of a severe reduction in the amount of sperm-forming tissue. Genetically marking the chromosomes with visible mutants and backcrossing F_1 females to males of either species permits every combination of X chromosomes and autosomes to be identified and their effects on testis size to be determined. As Figure 19-13 shows, when an X chromosome from *D. persimilis* is present together with a complete diploid set of autosomes from the other species, the testes are at a minimum size. As individual autosomes of the species to which the X chromosome belongs are substituted, the testis size increases, up to a complete haploid set of compatible autosomes but not beyond that. There is also evidence (not shown) of an interaction between the source of the cytoplasm in the egg and the source of the X chromosome. These variations in testis size of different hybrid genotypes correspond roughly to observed differences in the fertility of the males.

When such marker experiments have been performed on other species, mostly in the genus *Drosophila,* the general conclusions are that gene differences responsible for hybrid inviability are on all the chromosomes more or less

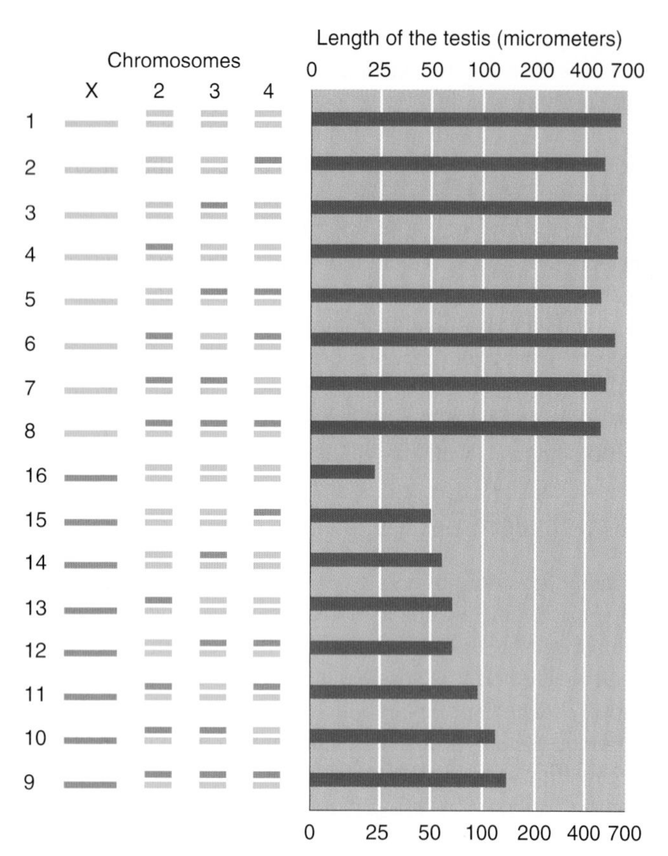

Figure 19-13 Testis size in backcross hybrids between *Drosophila pseudoobscura* and *D. persimilis*. Chromosomes of *D. pseudoobscura* are represented in orange, and those of *D. persimilis* in red.

equally and that, for hybrid sterility, there is some added effect of the X chromosome. For behavioral sexual isolation, the results are variable. In *Drosophila,* genes on all the chromosomes are involved, but, in Lepidoptera, the genes are much more localized, apparently because specific pheromones, chemicals that serve as sexual attraction signals, have a role. The sex chromosome has a very strong effect in butterflies; in the European corn borer, only three loci, one of which is on the sex chromosome, account for the entire isolation between pheromonal types that do not mate with each other.

ORIGIN OF NEW GENES

It is clear that evolution consists of more than the substitution of one allele for another at loci of defined function. New functions have arisen that have resulted in major new radiations of ways of making a living. Many of these new functions—for example, the development of the mammalian inner ear from a transformation of the reptilian jaw bones—result from continuous transformations of shape for which we do not have to invoke totally new genes and proteins. But qualitative novelties arise at the level of genes and proteins, such as the origination of photosynthesis in plants, of cell walls, of contractile proteins, of a variety of cell and tissue types, of oxygenation molecules such as hemoglobin, of the immune system, of chemical detoxification cycles, and of digestive enzymes. Older metabolic functions must have necessarily been maintained while new ones were being developed, which in turn means that old genes had to be preserved while new genes with new functions evolved. Where does the DNA for new functions come from?

Polyploidy

As we saw in Chapter 11, one process for the provision of new DNA is the duplication of the entire genome by polyploidization, much more common in plants than in animals. Evidence that polyploids have played a major role in the evolution of plant species was presented in that chapter. As shown in Figure 11-3, even haploid numbers are much more common than odd numbers, a consequence of frequent polyploidy followed by differentiation of the duplicate chromosome complements, giving rise to new species.

Duplications

Chapter 11 also discussed a second process for an increase in DNA—that is, the duplication of small sections of the genome as a consequence of unequal exchange at meiosis (see Figure 11-19b). At first, after a duplicated segment has arisen, there is the possibility of an increase in the production of the polypeptide encoded by the duplicated segment, but then functional differentiation between the duplicate sequences may occur in one of two directions. In one direction, the general function of the original sequence is maintained in the new DNA, but there is some differentiation of the sequences by accumulated mutations so that variations on the same protein theme are produced, allowing a somewhat more complex molecular structure. In the other direction, one of the duplicate sequences may evolve to take on a wholly new function.

A classic example of differentiation in the first direction, where there is variation on the same functional scheme, is the set of gene duplications and divergences that underlie the production of human hemoglobin. Adult hemoglobin is a tetramer consisting of two α polypeptide chains and two β chains, each with its bound heme molecule. The gene encoding the α chain is on chromosome 16 and the gene for the β chain is on chromosome 11, but the two chains are about 49 percent identical in their amino acid sequences, an identity that clearly points to their common origin. However, in fetuses, until birth, about 80 percent of β chains are substituted by a related γ chain. These two polypeptide chains are 75 percent identical. Furthermore, the gene for the γ chain is close to the β-chain gene on chromosome 11, and both have an identical intron–exon structure, again pointing to a common ancestral origin.

This developmental change in globin synthesis is part of a larger set of developmental changes that are shown in Figure 19-14. The early embryo begins with α, γ, ε, and ζ chains and, after about 10 weeks, the ε and ζ are lost, there is a large increase in α and γ, and the production of β begins. Near birth, β replaces γ, and a small amount of yet a sixth globin, δ, is produced. Table 19-4 shows the percentage of amino acid identity among these chains, and Figure 19-15 shows the chromosomal locations and intron-exon structures of the genes encoding them. The story is remarkably consistent. The β, δ, γ, and ε chains all belong to a "β-like" group; they have very similar amino acid sequences and are encoded by genes of identical intron-exon structure that are all contained in a 50-kb stretch of DNA on chromosome 11, indicating their origin by duplication. The α and ζ chains belong to an "α-like" group and are encoded by genes contained in a 30-kb region on chromosome 16. Two slightly different forms of the α chain are encoded by neighboring genes with identical intron-exon structure, as are two forms of the ζ chain. In addition, Figure 19-15 shows that on both chromosome 11 and chromosome 16 are pseudogenes, labeled $\Psi_{\alpha 1}$ and $\Psi_{\beta 1}$. These pseudogenes are duplicate copies of the genes that did not acquire new functions but accumulated random mutations that render them nonfunctional. At every moment in development, hemoglobin molecules consist of two chains from the α-like group and two from the β-like group, but the specific members of the groups change in embryonic, fetal, and newborn life. What is even more remarkable is that the order of genes on each chromosome is the same as the temporal order of appearance of the globin chains in the course of development. This complexity of chain replacement evolved in mammals and is not present in fish, reptiles, birds, and monotremes, which have only the basic α and β system. Figure 19-16 on the next page shows the order of origin of the different components of the β-like system.

TABLE 19-4 Percentage of Similarity in Amino Acid Sequences among Globin Chains in Humans

	α	ζ	β	γ	ε
α		58	42	39	37
ζ			34	38	37
β				73	75
γ					80

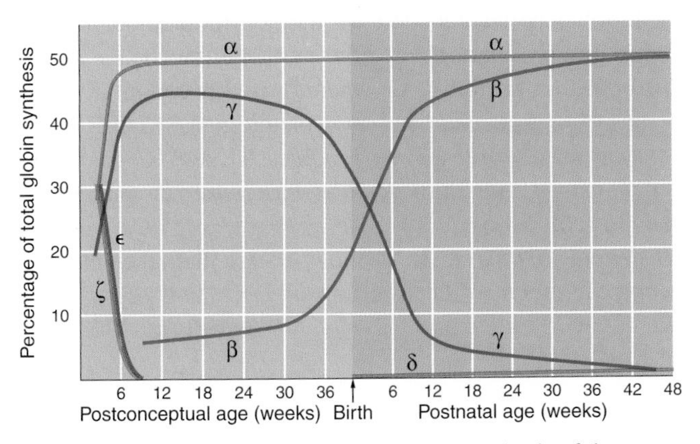

Figure 19-14 Developmental changes in the synthesis of the α-like and β-like globins that make up human hemoglobin.

In the evolution of hemoglobin, the duplicated DNA encodes a function closely related to that served by the original gene from which it arose. The other possibility for evolution of duplicated DNA is a complete qualitative divergence resulting in the appearance of a new function. An example of such a divergence is shown in Figure 19-17 on the next page. Birds and mammals, like other eukaryotic organisms, have a gene encoding lysozyme, a protective enzyme that breaks down the cell wall of bacteria. This gene has been duplicated in mammals to produce a second sequence that encodes a completely different, nonenzymatic protein, α-lactalbumin. Figure 19-17 shows that the duplicated gene has the same intron-exon structure as that of the lysozyme gene, whose array of four exons and three introns itself suggests an earlier multiple duplication event in the origin of lysozyme.

Imported DNA

New DNA that is the basis of new functions does not arise only from the duplication of DNA at an adjacent chromosomal location. Repeatedly in evolution, "extra" DNA has

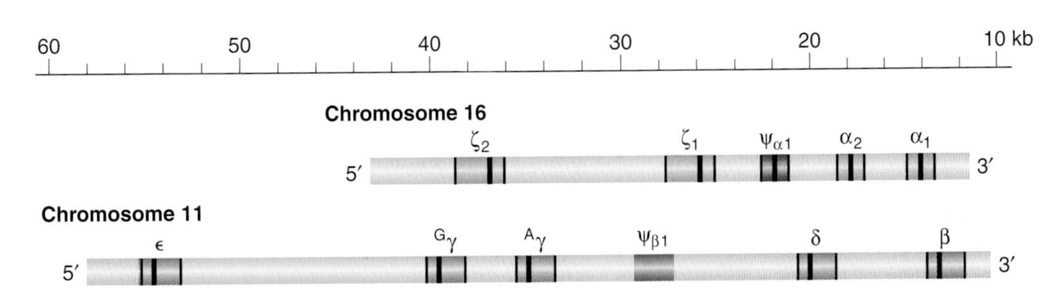

Figure 19-15 Chromosomal distribution of the genes for the α family of globins on chromosome 16 and the β family of globins on chromosome 11 in humans. Gene structure is shown by black bars (exons) and colored bars (introns).

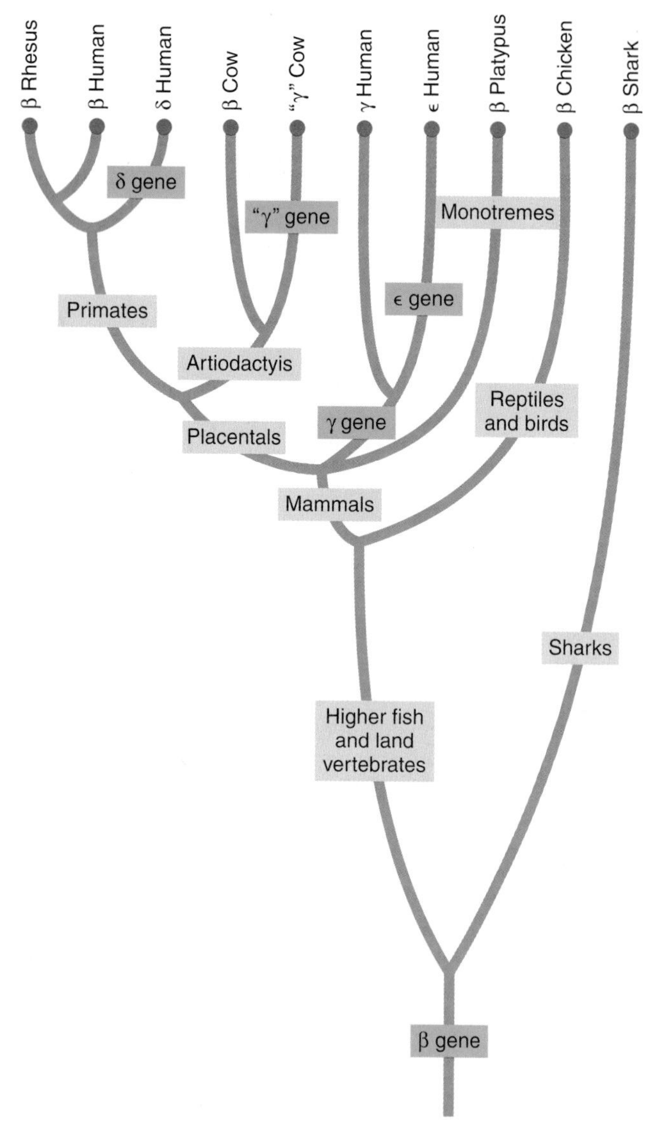

Figure 19-16 Reconstruction of the diversification of the β-globin-gene family in the evolution of the vertebrates. The cow "γ" gene is a γ-like gene that originated independently in the line leading to the artiodactyls.

been imported into a place in the genome from outside sources by mechanisms other than normal sexual reproduction. DNA can be inserted into a chromosome from other chromosomal locations and from other species, and genes from totally unrelated organisms can become incorporated into cells to become part of their heredity and function.

Cellular organelles. Eukaryotic cells contain cellular organelles such as mitochondria or the chloroplasts of photosynthetic organisms. Both these organelles are the descendants of prokaryotes that entered eukaryotic cells either by infection or by having been ingested. These prokaryotes became symbionts, transferring much of their genomes to the nuclei of their eukaryotic hosts but retaining genes that are essential to cellular functions. Mitochondria have retained about three dozen genes concerned with cellular respiration as well as some tRNA genes, whereas chloroplast genomes have about 130 genes encoding enzymes of the photosynthetic cycle as well as ribosomal proteins and tRNAs.

Important evidence for the extracellular origin of mitochondria is in their genetic code. There is no "universal" code, although most eukaryotic organisms have the same genetic code in their nuclear genomes. The usual DNA-RNA code of nuclear genes differs in a few codons, however, from that in mitochondria. Moreover, mitochondria in different organisms differ from one another in these coding elements, providing evidence that the invasion of eukaryotic cells by prokaryotes may have taken place at least five times, each time by a prokaryote with a different coding system. In the nuclear genome, for example, isoleucine is the only threefold redundantly encoded amino acid, with methionine being encoded by the fourth member of the codon group, separated by a purine-to-purine change in the third position from A to G; whereas, in this codon group in in mitochondria, there are two methionines whose third-position bases are the pyrimidines U and C and two isoleucines whose third-position bases are the purines A and G (Table 19-5).

Horizontal transfer. It is now clear that the nuclear genome is open to the insertion of DNA both from other

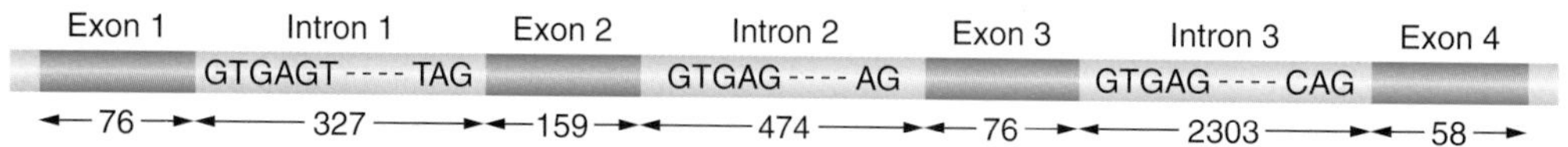

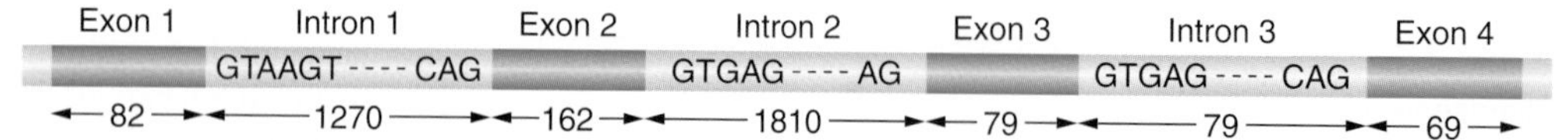

Figure 19-17 Structural homology of the gene for hen lysozyme and mammalian α-lactalbumin. Exons and introns are indicated by dark green bars and light green bars, respectively. Nucleotide sequences at the beginning and end of each intron are indicated, and the numbers refer to the nucleotide lengths of each segment. *(After I. Kumagai, S. Takeda, and K.-I. Miura, "Functional Conversion of the Homologous Proteins α-Lactalbumin and Lysozyme by Exon Exchange,"* Proceedings of the National Academy of Sciences USA *89, 1992, 5887–5891.)*

TABLE 19-5 Comparison of the Universal Nuclear DNA Code with Several Mitochondrial Codes for Five Triplets in Which They Differ

	TRIPLET CODE				
	TGA	**ATA**	**AGA**	**AGG**	**AAA**
Nuclear	Stop	Ile	Arg	Arg	Lys
Mitochondrial					
Mammalia	Trp	Met	Stop	Stop	Lys
Aves	Trp	Met	Stop	Stop	Lys
Amphibia	Trp	Met	Stop	Stop	Lys
Echinoderms	Trp	Ile	Ser	Ser	Asn
Insecta	Trp	Met	Ser	Stop	Lys
Nematodes	Trp	Met	Ser	Ser	Lys
Platyhelminthes	Trp	Met	Ser	Ser	Asn
Cnidaria	Trp	Ile	Arg	Arg	Lys

parts of the same genome and from outside (see Chapter 9). The chromosomes of an individual *Drosophila,* for example, contain a large variety of families of transposable elements with multiple copies of each distributed throughout the genome. As much as 25 percent of the DNA of *Drosophila* may be of transposable origin. It is not clear at present what role this mobile DNA plays in functional evolution. The transposition that takes place when transposable elements are introduced into zygotes at mating, as in the P elements of *Drosophila,* results in an explosive proliferation of the elements in the recipient genome. When a mobile element becomes inserted into a gene, the resulting mutation usually has a drastic deleterious effect on the organism, but this effect may be an artifact of the laboratory methods used to detect the presence of such elements, because such methods usually cannot detect small changes in physiology and survival. On the other hand, the results of laboratory selection experiments on quantitative characters have shown that transposition can act as an added source of selectable variation. Finally, there is the possibility that genes are transferred from the nuclear genome of one species to the nuclear genome of another by the phenomenon of retrotransposition mediated by retroviruses (see Chapters 9 and 10). This possibility could be a powerful source of the acquisition of new functions by a species because such retroviruses could be carried between very distantly related species by common disease vectors such as insects or by bacterial infections. The sequencing of the human genome has provided suggestive evidence of a transfer of DNA into humans from the bacterium *E. coli.*

Relation of Genetic Change to Functional Change

There is no simple relation between the amount of change in DNA and how much functional change takes place in the encoded protein. At one extreme, almost the entire amino acid sequence of a protein can be replaced while maintaining the original function. Eukaryotes, from yeast to humans, produce an enzyme, lysozyme, that breaks down bacterial cell walls. In the evolutionary divergence that has occurred in the yeast and vertebrate lines since their descent from an ancient common ancestor, virtually every amino acid in this protein has been replaced; so an alignment of their two protein or DNA sequences would not reveal any similarity. The evidence that they are descended from an original common ancestral gene comes from comparisons of evolutionarily intermediate forms that show more and more divergence of sequence as species are more divergent. The maintenance of the function despite the replacement of the amino acids has been the result of the maintenance of the three-dimensional structure of the enzyme by the selective substitutions of just the right amino acids to maintain shape.

In contrast, it is possible to change the function of an enzyme by a single amino acid substitution. The sheep blow fly, *Lucilia cuprina,* has evolved resistance to organophosphate insecticides used widely to control it. This resistance is the consequence of a single substitution of aspartic acid for a glycine residue in the active site of an enzyme that ordinarily is a carboxylesterase. The mutation causes complete loss of the carboxylesterase activity and its replacement by phosphatase activity. A three-dimensional modeling of the molecule indicates that the amino acid substitution causes a small change in the distance between two amino acid residues in the protein. This change results in

MESSAGE

There is no simple regular relation between how much DNA change takes place in evolution and how much functional change results.

the acquisition of the ability of the substituted protein to bind a water molecule close to the site of attachment of the organophosphate, which is then hydrolyzed by the water.

When more than one mutation is required for the origin of a new function, the order in which these mutations occur in the evolution of the molecule may be critical. Barry Hall experimentally changed a gene's function into a new one in *E. coli* by a succession of mutations and selection. In addition to the *lacZ* genes specifying the usual lactose-fermenting activity in *E. coli,* another structural gene locus, *ebg,* specifies another β-galactosidase that does not ferment lactose, although it is induced by lactose. The natural function of this second gene is unknown. Hall was able to mutate and select this extra gene to enable *E. coli* to live, without any lactose, on a wholly new substrate, lactobionate. To do so, he first had to mutate the regulatory sequence of *ebg* so that the *ebg* gene became constitutive and no longer required lactose to induce its translation. Next, he tried to select mutants that would ferment lactobionate, but he failed. First, it was necessary to select a form that would ferment a related substrate, lactulose, and then the lactulose fermenters could be mutated and selected to operate on lactobionate. Moreover, only some of the independent mutants from lactose fermentation to lactulose utilization could be further mutated to forms that could be selected to ferment lactobionate. The others were dead ends. Thus, the sequence of evolution had to be (1) from an inducible to a constitutive enzyme, followed by (2) just the right mutation from lactose to lactulose fermentation, followed by (3) a mutation to lactobionate fermentation.

MESSAGE

In the evolution of new functions by mutation and selection, particular pathways through the array of mutations must be followed. Other pathways come to dead ends that do not allow further evolution.

RATE OF MOLECULAR EVOLUTION

Although it is possible that only one or a few mutations lead to a change in the specificity of a protein, the more usual situation is that DNA accumulates substitutions over long periods of evolution without making a qualitative change in the functional properties of the proteins that are encoded. There may, however, be smaller effects influencing the kinetic properties, timing of production, or quantities of the encoded proteins that, in turn, will affect the fitness of the organism that carries them. Mutations of DNA can have three effects on fitness. First, they may be deleterious, reducing the probability of survival and reproduction of their carriers. All of the laboratory mutants used by the experimental geneticist have some deleterious effect on fitness. Second, they may actually increase fitness by providing increased efficiency or by expanding the range of environmental conditions in which the species can make a living or by enabling the individual organism to adapt to changes in the environment. Third, they may have no effect on fitness, leaving the probability of survival and reproduction unchanged; they are the so-called **neutral mutations.** For the purposes of understanding the rate of molecular evolution, however, we need to make a slightly different distinction—that between *effectively neutral* and *effectively selected* mutations. In Chapter 17, we learned that, in a finite population of N individuals, the process of random genetic drift will not be materially altered if the intensity of selection, s, on an allele is of lower order than $1/N$. That means that the class of evolutionarily neutral mutations includes both those that have absolutely no effect on fitness and those whose fitness effects are less than the reciprocal of population size, so small as to be effectively neutral.

We would like to know how much of molecular evolution is a consequence of new, favorable adaptive mutations sweeping through a species, the picture presented by a simplistic Darwinian view of evolution, and how much is simply the accumulation of the random fixation of effectively neutral mutations. Mutations that are effectively deleterious need not be considered, because they will be kept at low frequencies in populations and will not contribute to evolutionary change. If a newly arisen mutation is effectively neutral then, as pointed out in Chapter 17, there is a probability of $1/2N$ that it will replace the previous allele at that locus because of random genetic drift. If the rate of appearance of new effectively neutral mutations at a locus per gene copy per generation is μ, then the absolute number of new mutational copies that will appear in a population of N diploid individuals is $2N\mu$. Each one of these new copies has a probability of $1/2N$ of eventually taking over the population. Thus, the absolute rate of replacement of old alleles by new ones at a locus per generation is their rate of appearance multiplied by the probability that any one of them will eventually take over by drift:

$$\text{Rate of neutral replacement} = 2N\mu \times 1/2N = \mu$$

That is, we expect that in every generation there will be μ substitutions of a new allele for an old one at each locus in the population, purely from genetic drift of effectively neutral mutations.

MESSAGE

The rate of replacement in evolution resulting from the random genetic drift of effectively neutral mutations is equal to the mutation rate to such alleles, μ.

The constant rate of neutral substitution predicts that, if the number of nucleotide differences between two species

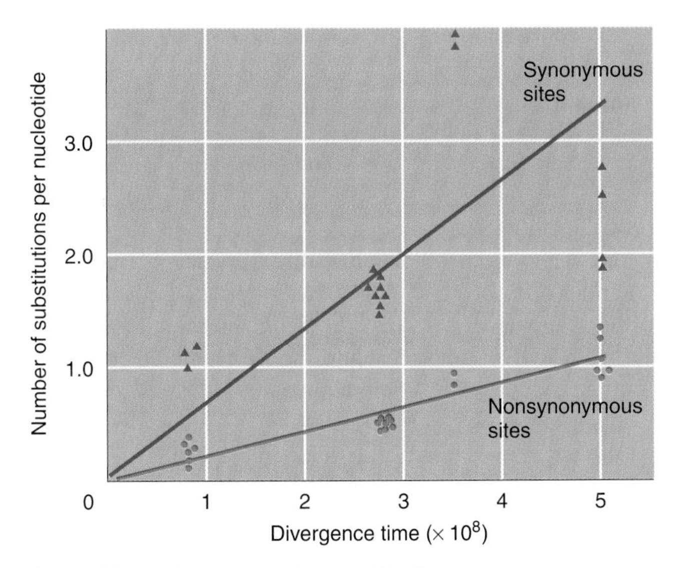

Figure 19-18 The amount of nucleotide divergence at synonymous and nonsynonymous sites of the β-globin gene as a function of time since divergence.

is plotted against the time since their divergence from a common ancestor, the result should be a straight line with a slope equal to μ. That is, evolution should proceed according to a **molecular clock** that is ticking at the rate μ. Figure 19-18 shows such a plot for the β-globin gene. The results are quite consistent with the claim that nucleotide substitutions have been effectively neutral over the past 500 million years. Two sorts of nucleotide substitutions are plotted: (1) **synonymous substitutions,** which are from one alternative codon to another, making no change in the amino acid, and (2) **nonsynonymous substitutions,** which result in an amino acid change. Figure 19-18 shows a much lower slope for nonsynonymous substitutions than that for synonymous changes, which means that the mutation rate to neutral nonsynonymous substitutions is much lower than that to synonymous ones. This is precisely what we expect. The mutation rate to neutral alleles, μ, is the product of the intrinsic nucleotide mutation rate, *M*, and the proportion of all mutations that are neutral, *f*. That is,

$$\mu = M \times f$$

It is reasonable that mutations that cause an amino acid substitution should more often have a deleterious effect, *s*, above the threshold for neutral evolution than do synonymous substitutions and therefore the proportion of neutral changes, *f*, will be smaller for nonsynonymous than for synonymous changes. It is important to note that these observations do not show that synonymous substitutions have *no* selective constraints on them; rather they show that these constraints are, on the average, not as strong as those for mutations that change amino acids. In fact, synonymous changes do have effects on probabilities of correct splicing, on the stability and lifetime of mRNA, on use by the translation apparatus of the available pool of tRNA molecules (and thus on the rate of translation), and on the folding of the translated polypeptide.

Another prediction of neutral evolution is that different proteins will have different clock rates, because the metabolic function of some proteins will be much more sensitive to changes in their amino acid sequence. Proteins in which every amino acid makes a difference will have smaller values of the effectively neutral mutation rate, *Mf*, than will proteins that are more tolerant of substitution. Figure 19-19 compares the clocks for fibrinopeptides, hemoglobin, and cytochrome *c*. That fibrinopeptides have a much higher proportion of neutral mutations is reasonable because these peptides are merely a nonmetabolic safety catch, cut out of fibrinogen to activate the blood-clotting reaction. From *a priori* considerations, why hemoglobins are less sensitive to amino acid changes than is cytochrome *c* is less obvious.

MESSAGE

The rate of neutral evolution for the amino acid sequence of a protein depends on the sensitivity of a protein's function to amino acid changes.

The demonstration of the molecular clock argues that most nucleotide substitutions are neutral, but it does not tell us how much of nonneutral molecular evolution is adaptive. One way of detecting the adaptive evolution of a protein is by comparing the synonymous and nonsynonymous polymorphisms within species with the synonymous and nonsynonymous changes between species. Under the operation

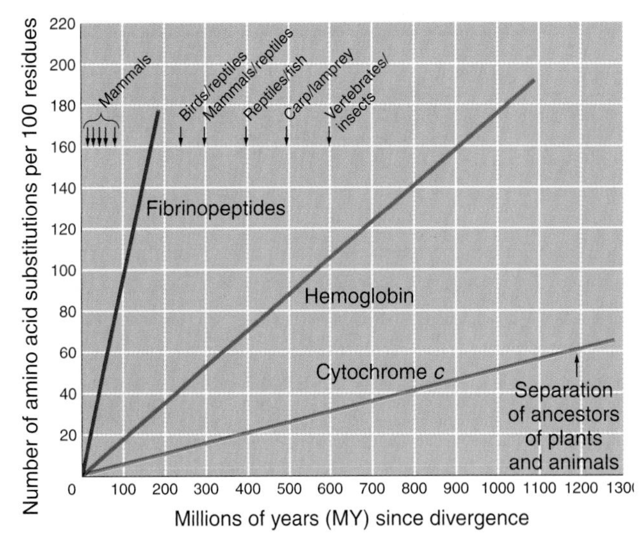

Figure 19-19 Number of amino acid substitutions in the evolution of the vertebrates as a function of time since divergence. The three proteins—fibrinopeptides, hemoglobin, and cytochrome *c*—differ in rate because different proportions of their amino acid substitutions are selectively neutral.

TABLE 19-6 Synonymous and Nonsynonymous Polymorphisms and Species Differences for Alcohol Dehydrogenase in Three Species of *Drosophila*

	Species Differences*	Polymorphisms
Nonsynonymous	7	2
Synonymous	17	42
Ratio	0.29:0.71	0.05:0.95

*For all the species differences, one of the species differed from the other two, which were the same as each other.
SOURCE: J. McDonald and M. Kreitman, *Nature* 351, 1991, 652–654.

of evolution by random genetic drift, polymorphism within a species is simply a stage in the eventual fixation of a new allele; so the ratio of nonsynonymous to synonymous polymorphisms within a species should be the same as the ratio of nonsynonymous to synonymous differences between species. On the other hand, if some amino acid changes between species have been driven by a positive adaptive selection, there ought to be an excess of nonsynonymous changes between species compared with those within species. This excess arises because amino acid polymorphisms are a mixture of neutral and positively selected changes, and most neutral polymorphisms will never be fixed between species. Table 19-6 shows an application of this principle by John MacDonald and Martin Kreitman to the alcohol dehydrogenase gene in three closely related species of *Drosophila.* Clearly, there is an excess of amino acid replacements between species over what is expected from the polymorphisms.

GENETIC EVIDENCE OF COMMON ANCESTRY IN EVOLUTION

When we think of evolution, we think of change. The species living at any particular time have changed in their various characteristics from those possessed by their ancestors, changes that have taken place by the mechanisms reviewed so far in our consideration of the genetics of the evolutionary process. But there is a second feature of the diverse forms of life that is a consequence of their evolutionary history, a feature that was taken by Darwin as an important argument for the reality of evolution. Not only have present organisms descended from earlier, different, organisms; but, if we go back in time, organisms that are at present very different are descended from a single ancestral form. Indeed, if we go back far enough in time to the origin of life, all the organisms on Earth are descended from an original ancestral form that is common to them all.

A consequence of this common ancestry of apparently different species is that we expect to find that they possess some attributes of their common ancestor that have been conserved over evolutionary time despite all the changes that have taken place. Before the tools of modern biochemistry and genetics were available, the chief evidence of underlying similarity of apparently different structures in different species was taken from anatomical observations of adult and embryonic forms. So, the detailed bone structure of the wings of bats and the forelimbs of running mammals makes it evident that these structures were derived evolutionarily from a common mammalian ancestor. Moreover, the anatomy of the wings of birds points to the common ancestry of mammals and birds (Figure 19-20). It was even

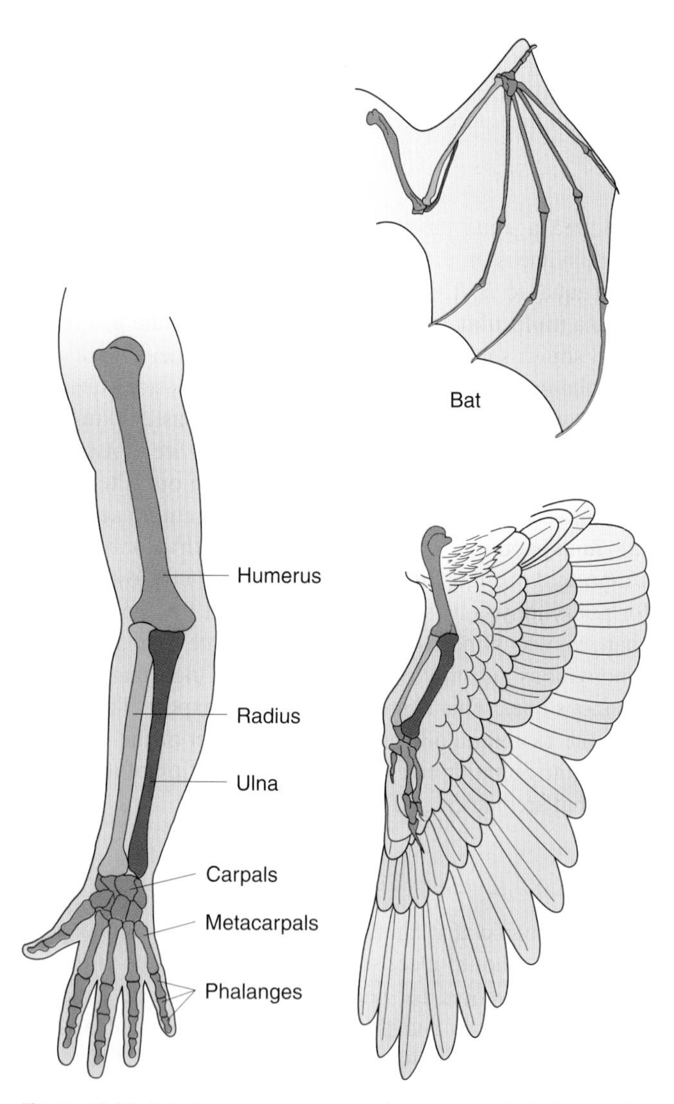

Figure 19-20 The bone structures of a bat wing, a bird wing, and a human arm and hand, showing the underlying anatomical similarity between them and the way in which different bones have become relatively enlarged or diminished to produce these different structures. *(After W. T Keeton and J. L. Gould.* Biological Science. *W. W. Norton & Company, 1986.)*

argued that the basic segmentation of the bodies of insects and of vertebrates were evolutionary variants on a common ancestral pattern derived from the common ancestor of invertebrates and vertebrates. Although this argument may seem to have pushed the claim of evolutionary conservation too far, it turns out, as discussed in Chapter 16, that the insect Hox and the vertebrate HOM-C genes, which regulate the formation of the body plan, are closely related in structure and function. Thus, genetic analysis of the patterns of development of insects and vertebrates provides a powerful demonstration of the common ancestry of animals as different as insects and mammals.

We saw in Chapter 16 that such disparate organisms as the fly, mouse, and human have similar sequences in the genes controlling the development of body form. (The same is true for the worm *C. elegans.*) The simplest interpretation to explain this similarity is that the Hox and HOM-C genes are the insect and vertebrate descendants of a homeobox gene cluster present in a common ancestor some 600 million years ago. The evolutionary conservation of the Hox and HOM-C genes is not a singular occurrence. Many examples of strong evolutionary and functional conservation of genes and entire pathways have been uncovered. For example, the pathways for activating the *Drosophila* DL and mammalian $NF_{\kappa}B$ transcription factors are essentially completely conserved (Figure 19-21). The *Drosophila* protein at any step in the DL activation pathway is similar in amino acid sequence to its counterpart in the mammalian $NF_{\kappa}B$ activation pathway. (Don't worry about what the particular proteins do; just appreciate the incredible conservation of cellular and developmental pathways as demonstrated by the similarity between the corresponding components of the two pathways indicated by similarly shaped objects in the diagrams. We do indeed know that DL and $NF_{\kappa}B$ participate in some equivalent developmental decisions.) Indeed, a survey of the known examples suggests that such evolutionary and functional conservation seems to be the norm rather than the exception. What has made developmental genetics into an extraordinarily exciting field of biological inquiry is the demonstration, by means of genetic analysis, of the conservation of basic developmental pathways and their genetic basis through hundreds of millions of years of evolution.

At a second, deeper, level, we can observe the common evolutionary origin of organisms in the structure of their proteins and of their genomes. The advantage of direct observation of the protein and DNA sequences is that we do not have to depend on similarity of the function of the proteins or anatomical structures that result from the possession of particular genes. We have already seen that a single amino acid change can alter the function of a protein from an esterase to an acid phosphatase. Yet, despite this change in function, we have no difficulty in determining from DNA sequences that the two enzymes are produced by reading genes that are virtually identical, one of which was derived by a single mutational step from the other in the course of the evolution of insecticide resistance by natural selection.

Through evolutionary time, genes that have descended from a common ancestor will diverge in DNA sequence and in their physical position in the genome, as a result of mutations and chromosomal rearrangements. If enough time elapsed and there were no counteracting force of natural selection, this divergence would finally result in the loss of any observable similarity in genes or proteins between different species, even if they were descended from a common ancestor. In fact, however, even the time that has elapsed since the existence of the common ancestor of present-day vertebrates and invertebrates has not erased the similarity of DNA and amino acid sequences between *Drosophila* and mice. This is partly because mutation rates are not high

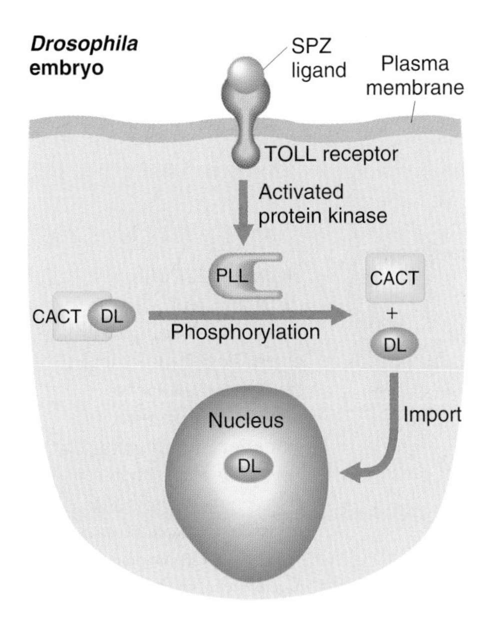

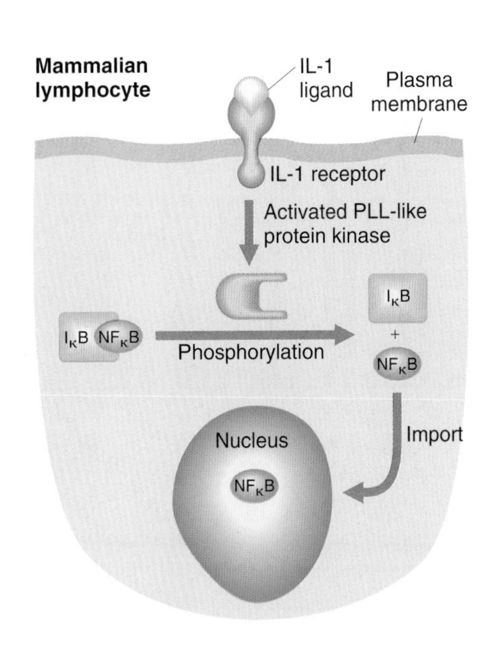

Figure 19-21 The signaling pathway for activation of the *Drosophila* DL morphogen parallels a mammalian signaling pathway for activation of $NF_{\kappa}B$, the transcription factor that activates the transcription of genes encoding antibody subunits. There are structural protein similarities between SPZ and IL-1, TOLL and IL-1R, CACT and $I_{\kappa}B$, and DL and $NF_{\kappa}B$. *(After H. Lodish, D. Baltimore, A. Berk, S. L. Zipursky, P. Matsudaira, and J. Darnell,* Molecular Cell Biology, *3d ed. © Scientific American Books, 1995.)*

enough to cause complete loss of similarity even over hundreds of millions of years and partly because most new mutations result either in a deleterious loss or change of function of a protein or in the control of the time and place of protein production. Therefore, the amount of divergence that has been preserved in evolution has been limited by natural selection.

MESSAGE

In the evolutionary divergence of different species from a common ancestor in the remote past, some DNA sequences have diverged so much that no residual similarity can be detected between them, but others have been conserved, sometimes very strongly, even when the proteins that they specify have changed function.

COMPARATIVE GENOMICS AND PROTEOMICS

As we saw in Chapter 9, a major effort of molecular genetics is the determination of the complete DNA sequence of a variety of different species. At the time this paragraph was written, the genomes had been sequenced from 39 species of bacteria; a yeast; the nematode *C. elegans;* one insect, *Drosophila melanogaster;* and a plant, *Arabidopsis.* By the time you read these lines, the genomes of many more species will have been sequenced. The availability of such data make it possible to reconstruct the evolution of the genomes of widely diverse species from their common ancestors. Moreover, comparison of the gene sequences in various species with gene sequences that code for the amino acid sequences of proteins with known function makes it possible to infer the similarities and differences in the proteomes of these species.

Comparing the Proteomes of Distant Species

With our current state of knowledge, we can suggest functions for about half the proteome of each eukaryote whose genome has been sequenced, by using the similarity of its sequence with proteins of known function. Figure 19-22 depicts the distribution of this half of each proteome into general functional categories. Strikingly, the group of proteins participating in defense and immunity has expanded greatly in humans compared with the other species. For other functional categories, though there are greater numbers of proteins in the human lineage, there is no case in which the difference between humans and other eukaryotes is as pronounced. As discussed in Chapter 13, a great deal of the

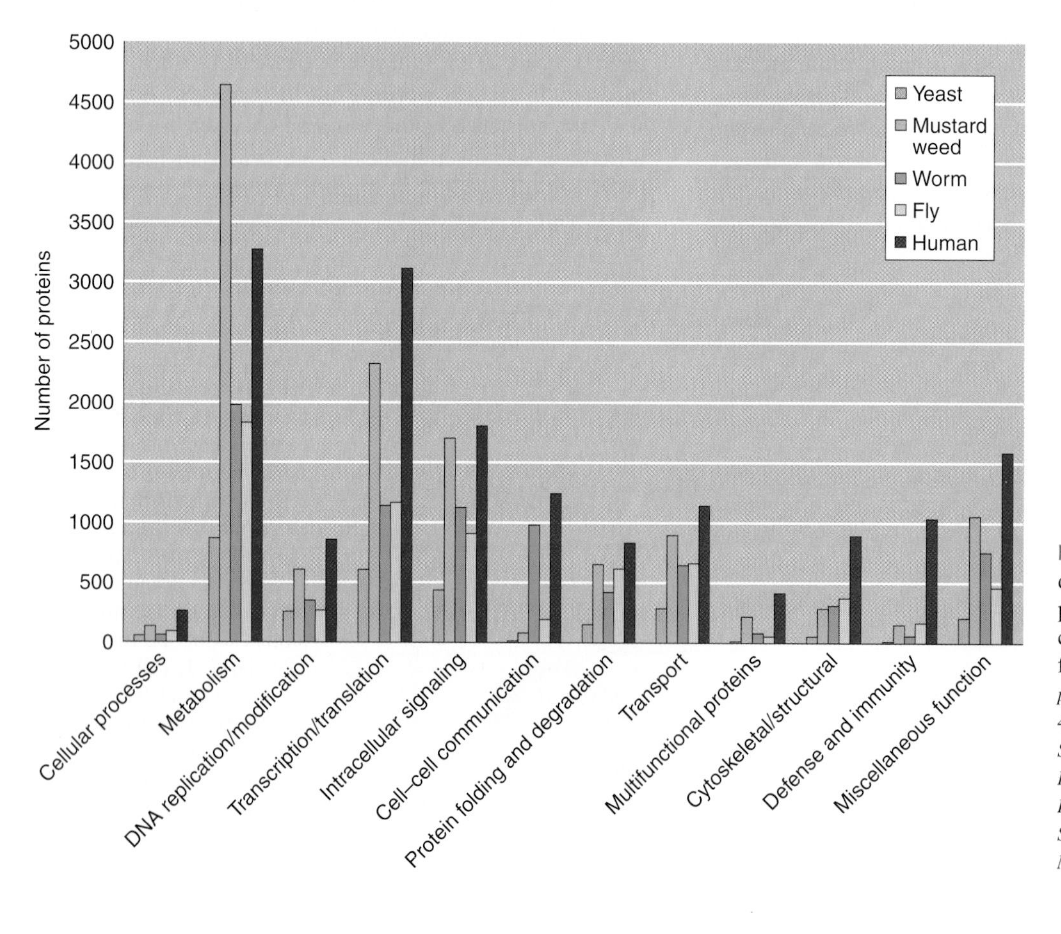

Figure 19-22 The distribution of eukaryotic proteins according to broad categories of biological function. *(Reprinted by permission from* Nature, *vol. 409:902, 15 Feb. 2001, "Initial Sequencing and Analysis of the Human Genome," The International Human Genome Sequencing Consortium, © 2001 Macmillan Magazines Ltd.)*

control of gene expression is through the regulation of transcription by proteins called transcription factors. Perhaps because of the many cell types that differentiate in humans, the size and distribution of the families of specific transcription factors in humans far exceed the numbers for the other sequenced eukaryotes, with the exception of mustard weed, *Arabidopsis thaliana* (see Figure 19-22).

The distribution of proteins described in the preceding paragraph refers to only half of each proteome. What about the other half? It can be broken down into two components. One component, comprising about 30 percent of each proteome, consists of proteins that have relatives among the different genomes, but all of the relatives are functionally anonymous—that is, none of the related proteins have had a function ascribed to them. The other component, comprising the remaining 20 percent or so of each proteome, consists of proteins that are unrelated by amino acid sequence to any protein known in any other branch of the eukaryotic evolutionary tree. We can imagine two possible explanations for these novel polypeptides in the various sequenced eukaryotes. One possibility is that some of these polypeptides first evolved after the descendants of the common ancestor to any of the species that have been sequenced diverged from one another. Because none of these species are evolutionarily closer than a few hundred million years, it is perhaps not surprising to find this frequency of newly evolved proteins. The other possibility is that some of these proteins are very rapidly evolving, and so their ancestry has essentially been erased by the overlay of new mutations that have accumulated. It is almost certain that both possibilities are correct for a subset of these novel polypeptides.

Finally, we can ask where the protein-coding genes in the human genome come from. Figure 19-23 depicts the distribution of human genes in other species. About a fifth of the known human genes have been found only in vertebrates. Another fifth seem to be ubiquitous in eukaryotes and prokaryotes. About a third are found throughout eukaryotes but not in bacteria. Curiously, a few hundred genes (less than 1 percent) appear to be found only in humans and in prokaryotes. Either these genes were present in a common ancestor of prokaryotes and eukaryotes and have disappeared from most other eukaryotes in the course of their evolution or else these genes that we and prokaryotes uniquely have in common have been passed on to us from prokaryotes through horizontal gene transfer.

Comparing the Genomes of Near Neighbors: Human-Mouse Comparative Genomics

It is thought that genomes evolved in part by a process of chromosome rearrangement—that is, the breaking and rejoining of the backbones of double-stranded DNA molecules, thereby producing new gene orders and new chromosomes. (See Chapter 11 for a discussion of chromosome rearrangements.) The extent to which chromosome rearrangements have accumulated during evolution can be assessed by looking for common gene orders between diverged species. As an example of this approach, we will compare the genomes of humans and mice, two species that diverged from a common ancestor about 50 million years ago. There is now sufficient genome sequencing in the mouse such that their placement and relative gene order on chromosomes can be determined. It turns out that large blocks of conserved gene order are easily recognized. Through systematic comparisons of this type, one can make **synteny maps,** in which the chromosomal origin of one species is essentially painted onto the karyotype of the other. Figure 19-24 on the next page depicts a color representation of the syntenic mouse-human genome. In this illustration, 21 different colors represent the mouse X and Y sex chromosomes and the 19 mouse autosomes. For example, most of mouse chromosome 14 can be found in three blocks on human chromosome 13, but, in addition, small segments can also be found on human chromosomes 3, 8, 10, and 14. Similar block-by-block distributions in the human genome are observed for each of the other mouse chromosomes. Thus, we can conclude that many chromosomal rearrangements have occurred between man and mouse, but not enough to have completely scrambled the two genomes relative to each other.

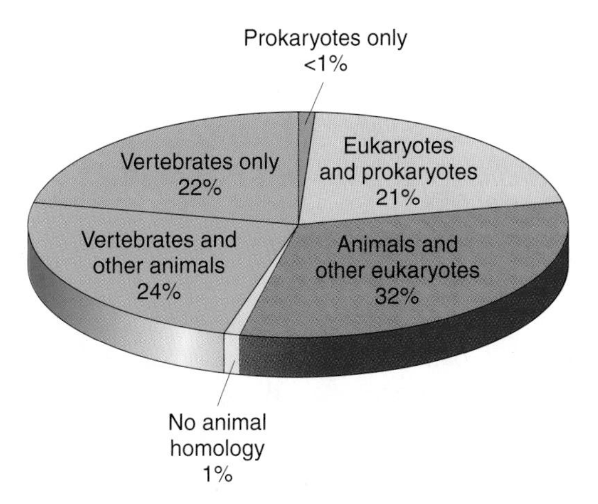

Figure 19-23 The distribution of human proteins according to the identification of significantly related proteins in other species. Note that about one-fifth of human proteins have been identified only within the vertebrate lineage; whereas, at the other extreme, one-fifth have been identified in all of the major branches of the evolutionary tree. *(Reprinted by permission from* Nature, *vol. 409:902, 15 Feb. 2001, "Initial Sequencing and Analysis of the Human Genome," The International Human Genome Sequencing Consortium, © 2001 Macmillan Magazines Ltd.)*

MESSAGE

Comparative genomics can be a source of insight into gene-level and chromosome-level changes that take place in the process of evolution.

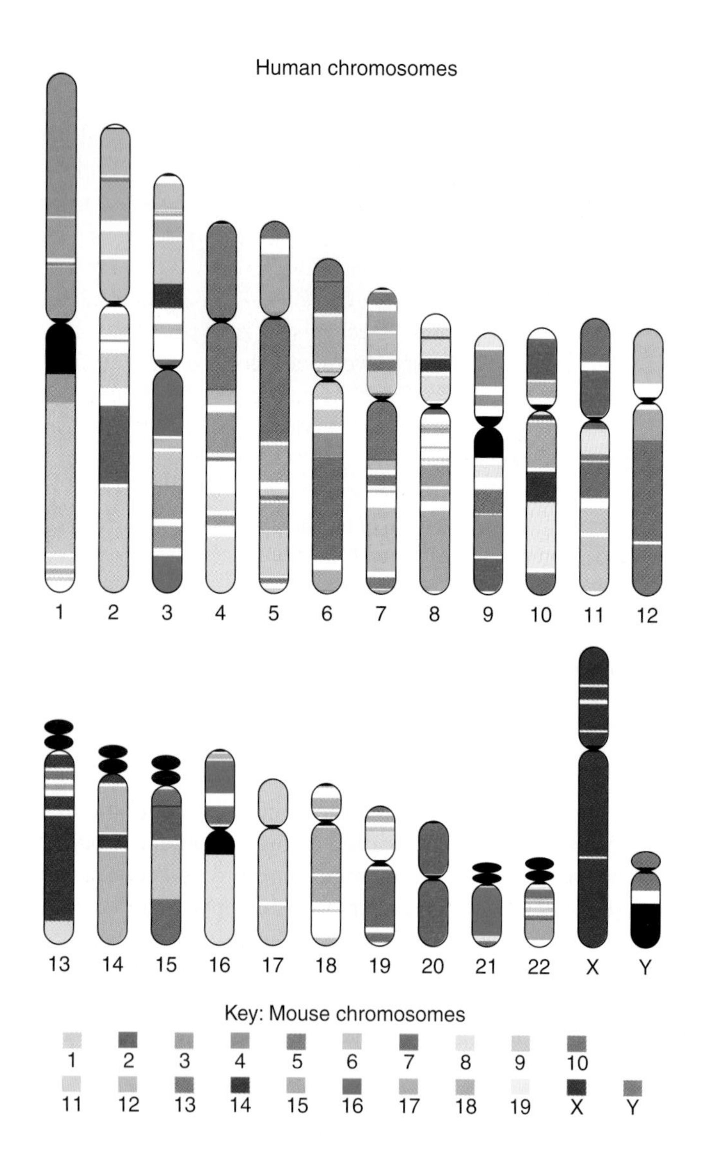

Figure 19-24 A synteny map of the human genome on which color coding is used to depict regional matches of each bloc of the human genome to the corresponding sections of the mouse genome. Each color represents a different mouse chromosome, as indicated in the key. *(Reprinted by permission from* Nature, *vol. 409:910, 15 Feb. 2001, "Initial Sequencing and Analysis of the Human Genome," The International Human Genome Sequencing Consortium, © 2001 Macmillan Magazines Ltd.)*

At the beginning of this chapter, we noted that humans, chimpanzees, and gorillas are all descended from a common ancestral form. We asked what forces are responsible for the evolutionary divergence among these various species of apes and humans and how we might account for the acquisition of human cognition and especially of a true novelty, human grammatically structured language, for which no equivalent exists in any other species. In this chapter, we have examined the various processes implicated in the acquisition of evolutionary novelties and in the splitting of species. Although we do not know, and probably will never know, the specific details of the forces of selection and random events that led to humans, we can be confident that the basic repertoire of forces described in this chapter must have taken part.

SUMMARY

1. The Darwinian theory of evolution is a variational scheme that explains the changes that take place in populations of organisms as being the result of changes in the relative frequencies of different variants in the population. If there is no variation within a species for some trait, there can be no evolution. Moreover, that variation must be influenced by genetic differences. If differences are not heritable, they cannot evolve, because the differential reproduction of the different variants will not carry across generational lines. Thus, all hypothetical evolutionary reconstructions depend critically on whether the traits in question are, in fact, heritable.
2. The processes that give rise to the genetic variation within a population are causally independent of the processes that are responsible for the differential reproduction of the various types. It is this independence that is meant when it is said that mutations are "random." The process of mutation supplies undirected variation.
3. The process of natural selection culls the variation produced by genetic mechanisms, increasing the frequency of those variants that by chance are better able to survive and reproduce. Many are called, but few are chosen.
4. The evolutionary divergence of populations in space and time is not only a consequence of natural selection. Natural selection is not a globally optimizing process that finds the "best" organisms for a particular environment. Instead, it finds one of a set of alternative "good" solutions to adaptive problems, and the particular outcome of selective evolution in a particular case is subject to chance historical events.
5. Random factors such as genetic drift and the chance occurrence or loss of new mutations may result in radically different outcomes of an evolutionary process even when the force of natural selection is the same. The metaphor usually employed is that there is an "adaptive landscape" of genetic combinations and that

natural selection leads the population to a "peak" in that landscape, but only to one of several alternative local peaks.

6. Not all of evolution is impelled by natural selective forces. If the selective difference between two genetic variants is small enough, less than the reciprocal of population size, there may be a replacement of one allele by another purely by genetic drift. A great deal of molecular evolution seems to be the replacement of one protein sequence by another one of equivalent function. The evidence for this neutral evolution is that the number of amino acid differences between two different species in some molecule—for example, hemoglobin—is directly proportional to the number of generations since their divergence from a common ancestor in the evolutionary past. Such a "molecular clock" with a constant rate of change would not be expected if the selection of differences were dependent on particular changes in the environment.
7. We expect the molecular clock to run faster for proteins such as fibrinopeptides, in which the amino acid composition is not critical for the function, and this difference in clock rate is, in fact, observed. Thus we cannot assume without evidence that evolutionary changes are the result of adaptive natural selection. Overall, genetic evolution is a historical process that is subject to historical contingency and chance, but it is constrained by the necessity of organisms to survive and reproduce in a constantly changing world.
8. The vast diversity of different living forms that have existed is a consequence of independent evolutionary histories in separate populations. For different populations to diverge from each other, they must not exchange genes; so the independent evolution of large numbers of different species requires that these species be reproductively isolated from each other. Indeed, we define a species as a population of organisms that exchange genes among themselves and reproductively isolated from other populations. The mechanisms of reproductive isolation may be prezygotic or postzygotic. Prezygotic isolating mechanisms are those that prevent the union of gametes of two species. These mechanisms may be behavioral incompatibility of the males and females of the different species, differences in timing or place of their sexual activity, anatomical differences that make mating mechanically impossible, or physiological incompatibility of the gametes themselves. Postzygotic isolating mechanisms include the inability of hybrid embryos to develop to adulthood, the sterility of hybrid adults, and the breakdown of later generations of recombinant genotypes. For the most part, the genetic differences responsible for the isolation between closely related species are spread throughout all the chromosomes, although in species with chromosomal sex determination there may be a concentration of incompatibility genes on the sex chromosome.
9. If new functions are to arise in evolution without the sacrifice of previously existing functions, new DNA must be made available for the evolution of added genes. This new DNA may arise by duplication of the entire genome (polyploidy) followed by a slow evolutionary divergence of the extra chromosomal set. Such duplication has been a frequent occurrence in plants. An alternative is the duplication of single genes followed by selection for differentiation. Yet another source of DNA, recently discovered, is the entry into the genome of DNA from totally unrelated organisms by infection followed by integration of the foreign DNA into the nuclear genome or by the establishment of extranuclear cell organelles with their own genomes. Mitochondria and chloroplasts in higher organisms have arisen by this route.
10. The fact that the diverse forms of living organisms trace back to a common ancestor means that structures that appear to be quite different between species can be shown to be based on the same genetic pathways, inherited from their common ancestor in remote time.

CONCEPT MAP

Draw a concept map interrelating as many of the following terms as possible. Note that the terms are listed in no particular order.

mutation / founder effect / species / alternate adaptive peak / migration / reproductive isolation / genetic drift / geographical race / natural selection / hybrid sterility / neutral evolution / duplication / horizontal transfer / evolutionary novelty

SOLVED PROBLEM

1. An entomologist who studies insects that feed on rotting vegetation has discovered an interesting case of diversification of fungus gnats on several islands in an archipelago. Each island has a gnat population that is extremely similar in morphology, although not identical, to those on the other islands, but each lives on a different kind of rotting vegetation that is not present on the other islands. The entomologist postulates that these gnats are closely

related species that have diverged by adapting to feeding on slightly different rot conditions.

To support this hypothesis, he carries out an electrophoretic study of the alcohol dehydrogenase enzyme in the different populations. He discovers that each population is characterized by a different electrophoretic form of the alcohol dehydrogenase, and he then reasons that each of these alcohol dehydrogenase forms is specifically adapted to the particular alcohols that are produced in the fermentation of the vegetation characteristic of a particular island. There is, in addition, some polymorphism of alcohol dehydrogenase within each island, but the frequency of variant alleles is low on each island and can be easily explained as the result of an occasional mutation or rare migrant from another island. These fungus gnats then become a textbook example of how species diversity can come about by natural selection adapting each newly forming species to a different environment.

A skeptical population geneticist reads about the case in a textbook, and she immediately has some doubts. It seems to her that, given the evidence, an equally plausible explanation is that these populations are not species at all but just local geographic races that have become slightly differentiated morphologically by random genetic drift. Moreover, the different electrophoretic forms of the alcohol dehydrogenase protein may be physiologically equivalent variants of a gene undergoing neutral molecular evolution in isolated populations.

Outline a program of investigation that could distinguish between these alternative explanations. How could you test whether the different populations are indeed different species? How could you test the hypothesis that the different forms of the alcohol dehydrogenase have diverged selectively?

Solution

To test whether the different populations of gnats are distinct species, it is necessary to be able to manipulate and breed them in captivity. If they cannot be bred in the laboratory or greenhouse, then their species distinctness cannot be established. The mating-behavior compatibility of the different forms can be tested by placing a mixture of males of two different populations with females of one of the forms to see whether there are any female mating preferences. The same experiment can then be repeated with mixed females and one sort of male and with mixtures of males and females of both forms. From such experiments, patterns of mating preference can be observed. A small amount of mating of different forms may occur only because of the unnatural conditions in which the test is being carried out. On the other hand, no mating of any kind may occur, even between the same forms, because the necessary cues for mating are missing, in which case nothing can be concluded.

If matings between different forms do occur, the survivorship of the interpopulation hybrids can be compared with that of the intrapopulation matings. If hybrids survive, their fertility can be tested by attempting to backcross them to the two different parental strains. As in the mating tests, under the unnatural conditions of the laboratory or greenhouse, some survivorship or fertility of species hybrids is possible even though the isolation in nature is complete. Any clear reduction in observed survivorship or fertility of the hybrids is strong presumptive evidence that they belong to different species.

To test whether the different amino acid sequences underlying the electrophoretic mobility differences are the result of selective divergence, a program of DNA sequencing of the alcohol dehydrogenase (*Adh*) locus is necessary. Replicated samples of *Adh* sequences from each of the island populations must be obtained. The number of such sequences needed from each population depends on the degree of nucleotide polymorphism that is present in the populations, but results from many loci in many species suggest that, as a rule of thumb, at least 10 sequences should be obtained from each population. The polymorphic sites within populations are classified into nonsynonymous (a) and synonymous (b) sites. The fixed nucleotide differences between populations are also classified into nonsynonymous (c) and synonymous (d) differences. If the divergence between the populations is purely the result of random genetic drift, then we expect a/b to be equal to c/d. If, on the other hand, there has been selective divergence, there should be an excess of fixed nonsynonymous differences, and so a/b should be less than c/d. The equality of these ratios can be tested by a 2 × 2 contingency χ^2 test of the form

	POLYMORPHISMS Nonsynonymous	Synonymous
POPULATION	a	b
DIFFERENCES	c	d

$$\chi^2 = \frac{(a+b+c+d)(ad-bc)^2}{(a+c)(b+d)(a+b)(c+d)}$$

If the χ^2 test shows that the difference between a/b and c/d is significant and if the ratio of nonsynonymous to synonymous changes between the species is greater than that ratio within species, then there is strong evidence for natural selection causing the change between the species, as the entomologist postulated.

SOLVED PROBLEM

2. How could the molecular evolution of a set of different proteins be used to provide evidence of the relative importance of exact amino acid sequence to the function of each protein?

Solution

Obtain DNA sequences from the genes for each protein from a wide variety of very divergent species whose approximate time to a common ancestor is known from the fossil record. Translate the DNA sequences into amino acid sequences. For each protein, plot the observed amino acid difference for each pair of species against the estimated time of divergence for those species. The line plotted for each protein will have a slope that is proportional to the amount of functional constraint on amino acid substitution in that protein. Highly constrained proteins will have very low rates of substitution, whereas more tolerant proteins will have higher slopes.

PROBLEMS

Basic Problems

1. What is the difference between a transformational and a variational scheme of evolution? Give an example of each (not including the Darwinian theory of organic evolution).

2. What are the three principles of Darwin's theory of variational evolution?

3. Why is the Mendelian explanation of inheritance essential to Darwin's variational mechanism for evolution? What would the consequences for evolution be if inheritance were by the mixing of blood? What would the consequence for evolution be if heterozygotes did not segregate exactly 50 percent of each of the two alleles at a locus but were consistently biased toward one or the other allele?

4. What is a geographical race? What is the difference between a geographical race and a separate species? Under what conditions will geographical races of a species become new species?

Challenging Problems

5. If the mutation rate to a new allele is 10^{-5}, how large must isolated populations be to prevent differentiation in the frequency of this allele from developing among them by chance?

6. Suppose that a number of local populations of a species are each about 10,000 members in size and that there is no migration between them. Suppose, further, that they were originally established from a large population with the frequency of an allele A at some locus equal to 0.4. Show by approximate sketches what the distribution of allele frequencies among the local populations would be after 100, 1000, 5000, 10,000, and 100,000 generations of isolation.

7. Show the results for the populations described in Problem 6 if there were an exchange of migrants among the populations at the rate of (a) one migrant individual per population every 10 generations; (b) one migrant individual per population every generation.

8. Suppose that a population is segregating for two alleles at each of two loci and that the relative probabilities of survival to sexual maturity of zygotes of the nine genotypes are as follows:

	A/A	A/a	a/a
B/B	.95	.90	.80
B/b	.90	.85	.70
b/b	.90	.80	.65

Calculate the mean fitness, $\overline{W}$, of the population if the allele frequencies are $p(A) = .8$ and $p(B) = .9$. What direction of change do you expect in allele frequencies in the next generation? Make the same calculation and prediction for the allele frequencies $p(A) = .2$ and $p(B) = .2$. From inspection of the genotypic fitnesses, how many adaptive peaks are there? What are the allele frequencies at the peak(s)?

9. Suppose the genotypic fitnesses in Problem 8 were:

	A/A	A/a	a/a
B/B	.9	.8	.9
B/b	.7	.9	.7
b/b	.9	.8	.9

Calculate the mean fitness, $\overline{W}$, for allelic frequencies $p(A) = .5$ and $p(B) = .5$. What direction of change do you expect for the allele frequencies in the next generation? Repeat the calculation and prediction for $p(A) = .1$ and $p(B) = .1$. From inspection of the genotype fitnesses, how many adaptive peaks are there and where are they located?

10. What is the evidence that polyploid formation has been important in plant evolution?

11. What is the evidence that gene duplication has been the source of the α and β gene families in human hemoglobin?

12. The human blood group allele I^B has a frequency of about 0.10 in European and Asian populations but is almost entirely absent in Native American populations. What explanations can account for this difference?

13. *Drosophila pseudoobscura* and *D. persimilis* are now considered separate species, but originally they were classified as Race A and Race B of a single species. They are morphologically indistinguishable from each other, except for a small difference in the genitalia of the males. When they are crossed in the laboratory, abundant adult F_1 progeny of both sexes are produced. Outline what program of observations and experiments you would undertake to test the claim that the two forms are different species.

14. Using the data on amino acid similarity of the α-, β-, γ-, ζ-, and ε-globin chains given in Table 19-4, draw a branching tree of the evolution of these chains from an original ancestral sequence in which the order of branching in time is as consistent as possible with the observed amino acid similarity on the assumption of a molecular clock.

15. DNA-sequencing studies for a gene in two closely related species produce the following numbers of sites that vary:

Synonymous polymorphisms	50
Nonsynonymous species differences	2
Synonymous species differences	18
Nonsynonymous polymorphisms	20

Does this result support a neutral evolution of the gene? Does it support an adaptive replacement of amino acids? What explanation would you offer for the observations?

16. Outline a program of investigation to test whether the tails of frog tadpoles and the tails of mice are evolutionarily derived from a structure in a common vertebrate ancestor.

Exploring Genomes: A Web-Based Bioinformatics Tutorial

Measuring Phylogenetic Distance

Sequence data allow us to estimate the evolutionary distance among organisms based on the extent of sequence divergence. In the Genomics tutorial at www.whfreeman.com/mga, we will use sequence comparisons to generate or support our conclusions regarding the structure of the evolutionary tree.

Appendix A

Genetic Nomenclature

There is no universally accepted set of rules for naming genes, alleles, protein products, and associated phenotypes. At first, individual geneticists developed their own symbols for recording their work. Later, groups of people working on any given organism met and decided on a set of conventions that all would use. Because *Drosophila* was one of the first organisms to be used extensively by geneticists, most of the systems now being used are variants of the *Drosophila* system. However, there has been considerable divergence. Some scientists now advocate a standardization of this symbolism, but standardization has not been achieved. Indeed, the situation has been made more complex by the advent of DNA technology. Whereas most genes had previously been named for the phenotypes produced by mutations within them, the new technology has shown the precise nature of the products of many of these genes. Hence it seems more appropriate to refer to them by their cellular function. However, the old names are still in the literature, so many genes have two parallel sets of nomenclature.

The following examples by no means cover all the organisms used in genetics, but most of the nomenclature systems follow one of these types:

Drosophila melanogaster (insect)

ry	A gene that when mutated causes rosy eyes
ry^{502}	A specific recessive mutant allele producing rosy eyes in homozygotes
ry^+	The wild-type allele of *rosy*
ry	The rosy mutant phenotype
ry^+	The wild-type phenotype (red eyes)
RY	The protein product of the *rosy* gene
XDH	Xanthine dehydrogenase, an alternative description of the protein product of the *rosy* gene; named for the enzyme that it encodes
D	*Dichaete;* a gene that when mutated causes a loss of certain bristles and wings to be held out laterally in heterozygotes and causes lethality in homozygotes
D^3	A specific mutant allele of the *Dichaete* gene
D^+	The wild-type allele of *Dichaete*
D	The Dichaete mutant phenotype
D^+	The wild-type phenotype
D	(Depending on context) the protein product of the *Dichaete* gene (a DNA-binding protein)

Neurospora crassa (fungus)

arg	A gene that when mutated causes arginine requirement
arg-1	One specific *arg* gene
arg-1	An unspecified mutant allele of the *arg* gene
arg-1 (1)	A specific mutant allele of the *arg-1* gene
$arg\text{-}1^+$	The wild-type allele
arg-1	The protein product of the $arg\text{-}1^+$ gene
Arg^+	A strain not requiring arginine
Arg^-	A strain requiring arginine

Saccharomyces cerevisiae (fungus)

ARG	A gene that when mutated causes arginine requirement
ARG1	One specific *ARG* gene
arg1	An unspecified mutant allele of the *ARG1* gene
arg1-1	A specific mutant allele of the *ARG1* gene
$ARG1^+$	The wild-type allele
ARG1p	The protein product of the $ARG1^+$ gene
Arg^+	A strain not requiring arginine
Arg^-	A strain requiring arginine

Homo sapiens (mammal)

ACH	A gene that when mutated causes achondroplasia
ACH^1	A mutant allele (dominance not specified)
ACH	Protein product of *ACH* gene; nature unknown
FGFR3	Recent name for gene for achondroplasia
$FGFR3^1$ or *FGFR3**1 or *FGFR3*<1>	Mutant allele of *FGR3* (unspecified dominance)
FGFR3 protein	Fibroblast growth factor receptor 3

Mus musculus (mammal)

Tyrc	A gene for tyrosinase
$+^{Tyrc}$	The wild-type allele of this gene
$Tyrc^{ch}$ or *Tyrc-ch*	A mutant allele causing chinchilla color
Tyrc	The protein product of this gene
+TYRC	The wild-type phenotype
TYRCch	The chinchilla phenotype

Escherichia coli (bacterium)

lacZ	A gene for utilizing lactose
lacZ$^+$	The wild-type allele
lacZ1	A mutant allele
LacZ	The protein product of that gene
Lac$^+$	A strain able to use lactose (phenotype)
Lac$^-$	A strain unable to use lactose (phenotype)

Arabidopsis thaliana (plant)

YGR	A gene that when mutant produces yellow-green leaves
YGR1	A specific *YGR* gene
YGR1	The wild-type allele
ygr1-1	A specific recessive mutant allele of *YGR1*
ygr1-2D	A specific dominant (D) mutant allele of *YGR1*
YGR1	The protein product of *YGR1*
Ygr$^-$	Yellow-green phenotype
Ygr$^+$	Wild-type phenotype

Appendix B

Bioinformatic Resources for Genetics and Genomics

"You certainly usually find something, if you look, but it is not always quite the something you were after." — *The Hobbit,* J. R. R. Tolkien

The field of bioinformatics encompasses the use of computational tools to distill complex data sets. Genetic and genomic data are so diverse that it has become a considerable challenge to identify the authoritative site(s) for a specific type of information. Furthermore, the landscape of Web-accessible software for analyzing this information is constantly changing as new and more powerful tools are developed. This appendix is intended to provide *some* valuable starting points for exploring the rapidly expanding universe of online resources for genetics and genomics.

1. Finding Genetic and Genomic Web Sites

Here are listed several central resources that contain large lists of relevant Web sites.

- The scientific journal called *Nucleic Acids Research (NAR)* publishes a special issue every January listing a wide variety of online database resources at: http://nar.oupjournals.org/
- The Virtual Library has Model Organisms and Genetics subdivisions with rich arrays of Internet resources at: http://ceolas.org/VL/mo/ and http://www.ornl.gov/TechResources/Human_Genome/organisms.html
- The National Human Genome Research Institute (NHGRI) maintains a list of genome Web sites at: http://www.nhgri.nih.gov/Data/
- The Department of Energy (DOE) maintains a Human Genome Project site at the Oak Ridge National Laboratory at: http://public.ornl.gov/hgmis/
- SwissProt maintains Amos' WWW links page at: http://www.expasy.ch/alinks.html

2. General Databases

Nucleic Acid and Protein Sequence Databases By international agreement, three groups collaborate to house the primary DNA and mRNA sequences of all species: The National Center for Biotechnology Information (NCBI) houses GenBank; The European Bioinformatics Institute (EBI) houses the European Molecular Biology Laboratory (EMBL) Data Library; and the National Institute of Genetics in Japan houses the DNA DataBase of Japan (DDBJ).

Primary DNA sequence records, called accessions, are submitted by individual research groups. In addition to providing access to these DNA sequence records, these sites provide many other data sets. For example, NCBI also houses RefSeq, a summary synthesis of information on the DNA sequences of fully sequenced genomes and the gene products that are encoded by these sequences.

Many other important features can be found at the NCBI, EBI, and DDBJ sites. Home pages and some other key Web sites are:

- **NCBI** http://www.ncbi.nlm.nih.gov/
- **NCBI-Genomes** http://www.ncbi.nlm.nih.gov/Genomes/index.html
- **NCBI-RefSeq** http://www.ncbi.nlm.nih.gov/LocusLink/refseq.html
- **EBI** http://www.ebi.ac.uk/
- **DDBJ** http://www.nig.ac.jp/

The harsh reality is that, with so much biological information, the goal of making these online resources "transparent" to the user is not fully achieved. Thus, exploration of these sites will entail familiarizing yourself with the contents of each site and exploring some of the ways the site helps you to focus your queries so you get the right answer(s) to your queries. For one example of the power of these sites, consider a search for a nucleotide sequence at NCBI. Databases typically store information in separate bins called "fields." By using queries that limit the search to the appropriate field, a more directed question can be asked. Using the "Limits" option, a query phrase can be used to identify or locate a specific species, type of sequence (genomic or mRNA), gene symbol, or any of several other data fields. Query engines usually support the ability to join multiple query statements together. For example: retrieve all DNA sequence records that are from the species *Caenorhabditis elegans* **AND** that were published after January 1, 2000. Using the "History" option, the results of multiple queries can be joined together, so that only those hits common to multiple queries will be retrieved. By proper use of the available query options on a site, a great many false positives can be computationally eliminated while not discarding any of the relevant hits.

Because protein sequence predictions are a natural part of the analysis of DNA and mRNA sequences, these same sites provide access to a variety of protein databases. One important protein database is SwissProt/TrEMBL. TrEMBL sequences are automatically predicted from DNA and/or mRNA sequences. SwissProt sequences are curated, meaning that an expert scientist reviews the output of computational analysis and makes expert decisions about which results to accept or reject. In addition to the primary protein sequence records, SwissProt also offers databases of protein domains and protein signatures (amino acid sequence strings that are characteristic of proteins of a particular type). The SwissProt home page is http://www.ebi.ac.uk/swissprot/.

Protein Domain Databases The functional units within proteins are thought to be local folding regions called domains. Prediction of domains within newly discovered proteins is one way to guess at their function. Numerous protein domain databases have emerged that predict domains in somewhat different ways. Some of the individual domain databases are Pfam, PROSITE, PRINTS, SMART, ProDom, TIGRFAMs, BLOCKS, and CDD. InterPro allows querying multiple protein domain databases simultaneously and presents the combined results. Web sites for some domain databases are:

- **InterPro** http://www.ebi.ac.uk/interpro/
- **Pfam** http://www.sanger.ac.uk/Software/Pfam/index.shtml
- **PROSITE** http://www.expasy.ch/prosite/
- **PRINTS** http://www.bioinf.man.ac.uk/dbbrowser/PRINTS/
- **SMART** http://smart.embl-heidelberg.de/
- **ProDom** http://prodes.toulouse.inra.fr/prodom/doc/prodom.html
- **TIGRFAMs** http://www.tigr.org/TIGRFAMs/
- **BLOCKS** http://www.blocks.fhcrc.org/
- **CDD** http://www.ncbi.nlm.nih.gov/Structure/cdd/cdd.shtml

Protein Structure Databases The representation of three-dimensional protein structures has become an important aspect of global molecular analysis. Three-dimensional structure databases are available from the major DNA/protein sequence database sites and from independent protein structure databases, notably The Protein DataBase (PDB). NCBI has an application called Cn3D that helps in viewing PDB data.

- **PDB** http://www.rcsb.org/pdb/
- **Cn3D** http://www.ncbi.nlm.nih.gov/Structure/CN3D/cn3d.shtml

3. Specialized Databases

Organism-Specific Genetic Databases In order to mass some classes of genetic and genomic information, especially phenotypic information, expert knowledge of a particular species is required. Thus, MODs (model organism databases) have emerged to fulfill this role for the major genetic systems. These include databases for *Saccharomyces cerevisiae* (SGD), *Caenorhabiditis elegans* (WormBase), *Drosophila melanogaster* (FlyBase), the zebrafish *Danio rerio* (ZFIN), the mouse *Mus musculus* (MGI), the rat *Rattus norvegicus* (RGD), *Zea mays* (MaizeDB), and *Arabidopsis thaliana* (TAIR). Home pages for these MODs can be found at:

- **SGD** http://genome-www.stanford.edu/Saccharomyces/
- **WormBase** http://www.wormbase.org/
- **FlyBase** http://flybase.bio.indiana.edu/
- **ZFIN** http://zfin.org/
- **MGI** http://www.informatics.jax.org/
- **RGD** http://rgd.mcw.edu/
- **MaizeDB** http://www.agron.missouri.edu/top.html
- **TAIR** http://www.arabidopsis.org/

Human Genetics and Genomics Databases Because of the importance of human genetics in clinical as well as basic research, a diverse set of human genetic databases have emerged. Among them are a human genetic disease database called Online Mendelian Inheritance in Man (OMIM), a database of brief descriptions of human genes called GeneCards, a compilation of all known mutations in human genes called Human Gene Mutation Database (HGMD), a database of the current sequence map of the human genome called the Golden Path, and some links to human genetic disease databases:

- **OMIM** http://www3.ncbi.nlm.nih.gov/Omim/
- **GeneCards** http://mach1.nci.nih.gov/cards/index.html
- **HGMD** http://www.hgmd.org
- **Golden Path** http://genome.ucsc.edu/goldenPath/hgTracks.html
- **Online Genetic Support Groups** http://www.mostgene.org/support/index.html
- **Genetic Disease Information** http://www.geneticalliance.org/diseaseinfo/search.html

Genome Project Databases The individual genome projects also have Web sites, where they display their results, often including information that doesn't appear on any other Web site in the world. The largest of the publicly funded genome centers include:

- **Whitehead Institute/MIT Center for Genome Research** http://www-genome.wi.mit.edu/
- **Washington University School of Medicine Genome Sequencing Center** http://genome.wustl.edu/gsc/
- **Baylor College of Medicine Human Genome Sequencing Center** http://www.hgsc.bcm.tmc.edu/
- **Sanger Institute** http://www.sanger.ac.uk/
- **DOE Joint Genomics Institute** http://www.jgi.doe.gov/

4. Relationships of Genes within and between Databases

Gene products may be related by virtue of sharing a common evolutionary origin, sharing a common function, or participating in the same pathway.

BLAST: Identification of Sequence Similarities Evidence for a common evolutionary origin comes from the identification of sequence similarities between two or more sequences. One of the most important tools for identifying such similarities is BLAST (Basic Local Alignment Search Tool), which was developed by NCBI. BLAST is really a suite of related programs and databases in which local matches between long stretches of sequence can be identified and ranked. A query for similar DNA or protein sequences through BLAST is one of the first things that a researcher does with a newly sequenced gene. Different sequence databases can be accessed and organized by type of sequence (reference genome, recent updates, nonredundant, ESTs, etc.), and a particular species or taxonomic group can be specified. One BLAST routine matches a query nucleotide sequence translated in all six frames to a protein sequence database. Another matches a protein query sequence to the six-frame translation of a nucleotide sequence database. Other BLAST routines are customized to identify short sequence pattern matches or pairwise alignments, to screen genome-sized DNA segments, and so forth, and can be accessed through the same top-level page:

- **NCBI-BLAST** http://www.ncbi.nlm.nih.gov/BLAST/

Function Ontology Databases Another approach to developing relationships among gene products is by assigning these products to functional roles based on experimental evidence or prediction. Having a common way of describing these roles, regardless of the experimental system, is then of great importance. A group of scientists from different databases are working together to develop a common set of hierarchically arranged terms—an ontology—for *function* (biochemical event), *process* (the cellular event to which a protein contributes), and *subcellular location* (where a product is located in a cell) as a way of describing the activities of a gene product. This particular ontology is called the Gene Ontology (GO), and many different databases of gene products now incorporate GO terms. A full description can be found at:

- http://www.geneontology.org/

Pathway Databases Still another way to relate products to each other is by assigning them to steps in biochemical or cellular pathways. Pathway diagrams can be used as organized ways of presenting relationships of these products to one another. Some of the more advanced attempts at producing such pathway databases include Kyoto Encyclopedia of Genes and Genomes (KEGG), Signal Transduction Database (TRANSPATH), and What Is There—Interactive Metabolic Database (WIT):

- **KEGG** http://www.genome.ad.jp/kegg/
- **TRANSPATH** http://transpath.gbf.de/
- **WIT** http://wit.mcs.anl.gov/WIT2/

Glossary

A Adenine or adenosine.

abortive transduction The failure of a transducing DNA segment to be incorporated into the recipient chromosome.

AC element *See* **activator element.**

acentric chromosome A chromosome having no centromere.

achondroplasia A type of dwarfism in humans, inherited as an autosomal dominant phenotype.

acridine orange A mutagen that tends to produce frameshift mutations.

acrocentric chromosome A chromosome having the centromere located slightly nearer one end than the other.

activator A protein that, when bound to a cis-regulatory DNA element such as an operator or an enhancer, activates transcription from an adjacent promoter.

activator (AC) element A transposable element in corn with **short inverted terminal repeat sequences.**

active site The part of a protein that must be maintained in a specific shape if the protein is to be functional—for example, in an enzyme, the part to which the substrate binds.

adaptation In the evolutionary sense, some heritable feature of an individual's phenotype that improves its chances of survival and reproduction in the existing environment.

adaptive landscape A surface plotted in a three-dimensional graph with all possible combinations of allele frequencies for different loci plotted in the plane and with mean fitness for each combination plotted in the third dimension.

adaptive peak A high point (perhaps one of several) on an adaptive landscape. Selection tends to drive the genotype composition of the population toward a genotype corresponding to an adaptive peak.

adaptive surface *See* **adaptive landscape.**

adaptor protein A protein that binds to certain specific phosphorylated amino acid sequences on a second protein, often a transmembrane receptor, and associates with still other proteins, thereby allowing a complex of proteins to "dock" to the receptor. This docking brings proteins of the complex into proximity with one another and, by doing so, permits propagation of an intracellular signal in a signal transduction pathway.

addition *See* **indel mutation.**

additive genetic variance Genetic variance associated with the average effects of substituting one allele for another.

adenine A purine base that pairs with thymine in the DNA double helix.

adenine methylase An enzyme that methylates adenines before replication, thereby distinguishing old strands from the newly synthesized complementary strands.

adenosine The nucleoside containing adenine as its base.

adenosine triphosphate *See* **ATP.**

adjacent segregation In a reciprocal translocation, the passage of a translocated and a normal chromosome to each of the poles.

ADP Adenosine diphosphate.

AFB_1 *See* **aflatoxin B_1**.

aflatoxin B_1 (AFB_1) A mutagen that cleaves the bond between guanine and deoxyribose, producing an apurinic site.

Ala Alanine (an amino acid).

albino A pigmentless "white" phenotype, determined by a mutation in a gene coding for a pigment-synthesizing enzyme.

alkylating agent A chemical agent that can add alkyl groups (for example, ethyl or methyl groups) to another molecule; many mutagens act through alkylation.

allele One of the different forms of a gene that can exist at a single locus.

allele frequency A measure of the commonness of an allele in a population; the proportion of all alleles of that gene in the population that are of this specific type.

allopatric speciation The formation of new species from populations that are geographically isolated from each other.

allopolyploid *See* **amphidiploid.**

allosteric effector The small molecule that binds to an allosteric site.

allosteric site The site on a protein to which a small molecule binds, causing a change in the conformation of the protein that modifies the activity of its active site.

allosteric transition A change from one conformation of a protein to another.

alternate segregation In a reciprocal translocation, the passage of both normal chromosomes to one pole and both translocated chromosomes to the other pole.

alternation of generations The alternation of gametophyte and sporophyte stages in the life cycle of a plant.

alternative splicing The process by which different mRNAs are produced from the same primary transcript, through variations in the splicing pattern of the transcript. Multiple mRNA "isoforms" can be produced in a single cell, or the different isoforms can display different tissue-specific patterns of expression. If the alternative exons fall within the open reading frames of the mRNA isoforms, different proteins will be produced by the alternative mRNAs.

amber codon The codon UAG, a nonsense codon.

amber suppressor A mutant allele encoding a tRNA whose anticodon is altered in such a way that the tRNA inserts an amino acid at an amber codon in translation.

Ames test A widely used test to detect possible chemical carcinogens; based on mutagenicity in the bacterium *Salmonella.*

amino acid A peptide; the basic building block of proteins (or polypeptides).

amniocentesis A technique for testing the genotype of an embryo or fetus in utero with minimal risk to the mother or the child.

AMP Adenosine monophosphate.

amphidiploid An allopolyploid; a polyploid formed from the union of two separate chromosome sets and their subsequent doubling.

analogous structures Structures of different species that are functionally similar but that are developed from different embryonic structures as, for example, the wings of birds and the wings of insects.

anaphase An intermediate stage of nuclear division during which chromosomes are pulled to the poles of the cell.

androgen A family of related steroid hormones that promote male development, such as testosterone and 5-hydroxy-testosterone.

aneuploid cell A cell having a chromosome number that differs from the normal chromosome number for the species by a small number of chromosomes.

animal breeding The practical application of genetic analysis for development of lines of domestic animals suited to human purposes.

annealing Spontaneous alignment of two single DNA strands to form a double helix.

antibody A protein (immunoglobulin) molecule, produced by the immune system, that recognizes a particular substance (antigen) and binds to it.

anticodon A nucleotide triplet in a tRNA molecule that aligns with a particular codon in mRNA under the influence of the ribosome, so that the amino acid carried by the tRNA is inserted into a growing protein chain.

antigen A molecule that is recognized by antibody (immunoglobulin) molecules. Generally, multiple antibody molecules can recognize a given antigen.

antiparallel A term used to describe the opposite orientations of the two strands of a DNA double helix; the 5′ end of one strand aligns with the 3′ end of the other strand.

AP endonuclease An enzyme that removes apurinic sites in DNA so that repair synthesis can replace them with appropriate complementary strands.

AP site Apurinic or apyrimidinic site resulting from the loss of a purine or pyrimidine residue from the DNA.

apoptosis The cellular pathways responsible for cell death and subsequent removal of the remains of the dead cell. Cell death may be induced by intracellular damage or by signals from neighboring cells.

apurinic site A DNA site that has lost a purine residue.

apyrimidinic site A DNA site that has lost a pyrimidine residue.

Arg Arginine (an amino acid).

ascospore A sexual spore from certain fungus species in which spores are found in a sac called an ascus.

ascus In fungi, a sac that encloses a tetrad or an octad of ascospores.

asexual spore *See* **spore.**

Asn Asparagine (an amino acid).

Asp Aspartate (an amino acid).

ATP (adenosine triphosphate) The "energy molecule" of cells, synthesized mainly in mitochondria and chloroplasts; energy from the breakdown of ATP drives many important cell reactions.

attached X A pair of *Drosophila* X chromosomes joined at one end and inherited as a single unit.

autofeedback loop *See* **autoregulatory loop.**

autonomous controlling element A controlling element that seems to have both regulator and receptor functions combined in a single unit, which enters a gene and causes an unstable mutation.

autonomous phenotype A genetic trait in multicellular organisms in which only genotypically mutant cells exhibit the mutant phenotype. Conversely, a *nonautonomous* trait is one in which genotypically mutant cells cause other cells (regardless of their genotype) to exhibit a mutant phenotype.

autonomous replication sequence (ARS) A segment of a DNA molecule needed for the initiation of its replication; generally a site recognized and bound by the proteins of the replication system.

autophosphorylation The process by which a protein kinase phosphorylates specific amino acid residues on itself.

autopolyploid A polyploid formed from the doubling of a single genome.

autoradiogram A pattern of dark spots in a developed photographic film or emulsion, in the technique of autoradiography.

autoradiography A process in which radioactive materials are incorporated into cell structures, which are then placed next to a film or photographic emulsion, thus forming a pattern on the film (an autoradiogram) corresponding to the location of the radioactive compounds within the cell.

autoregulatory loop The process by which the expression of a gene is controlled by its own product.

autosome Any chromosome that is not a sex chromosome.

auxotroph A strain of microorganisms that will proliferate only when the medium is supplemented with some specific substance not required by wild-type organisms (*compare* **prototroph**).

BAC Bacterial artificial chromosome; an F plasmid engineered to act as a cloning vector that can carry large inserts.

back mutation *See* **reversion.**

bacteriophage (phage) A virus that infects bacteria.

balanced polymorphism Stable genetic polymorphism maintained by natural selection.

balancer A chromosome with multiple inversions, used to retain favorable allele combinations in the uninverted homolog.

Balbiani ring A large chromosome puff.

Barr body A densely staining mass that represents an inactivated X chromosome.

base analog A chemical whose molecular structure mimics that of a DNA base; because of the mimicry, the analog may act as a mutagen.

base correction The production (possibly by excision and repair) of a properly paired nucleotide pair from a sequence of hybrid DNA that contains an illegitimate pair.

base-excision repair One of several excision repair pathways. In this pathway, subtle base-pair distortions are repaired by creation of apurinic sites followed by repair synthesis.

base-pair substitution *See* **nucleotide-pair substitution.**

basic helix-loop-helix proteins A family of transcription regulatory proteins in which a part of the polypeptide forms two alpha helices separated by a loop (the HLH domain); this domain acts as a sequence-specific DNA-binding domain. Many of these proteins have a positively charged (basic) domain as well.

bead theory The disproved hypothesis that genes are arranged on the chromosome like beads on a necklace, indivisible into smaller units of mutation and recombination.

behavior mutation A mutation affecting some aspect of the behavior of the mutant individual.

bHLH proteins *See* **basic helix-loop-helix proteins.**

bimodal distribution A statistical distribution having two modes.

binary fission The process in which a parent cell splits into two daughter cells of approximately equal size.

bioinformatics Computational information systems and analytical methods applied to biological problems such as genomic analysis.

blastoderm In an insect embryo, the stage in which there is a single layer of nuclei *(syncitial blastoderm)* or cells *(cellular blastoderm)* surrounding the central yolk.

blastula An early developmental stage of lower vertebrate embryos, in which the embryo consists of a single layer of cells surrounding the central yolk.

blending inheritance A discredited model of inheritance suggesting that the characteristics of an individual result from the smooth blending of fluidlike influences from its parents.

brachydactyly A human phenotype of unusually short digits, generally inherited as an autosomal dominant.

branch migration The process by which a single "invading" DNA strand extends its partial pairing with its complementary strand as it displaces the resident strand.

broad heritability (H^2) The proportion of total phenotypic variance at the population level that is contributed by genetic variance.

bud A daughter cell formed by mitosis in yeast; one daughter cell retains the cell wall of the parent, and the other (the bud) forms a new cell wall.

buoyant density A measure of the tendency of a substance to float in some other substance; large molecules are distinguished by their different buoyant densities in some standard fluid. Measured by density-gradient ultracentrifugation.

Burkitt's lymphoma A cancer of the lymphatic system manifested by tumors in the jaw, associated with a chromosomal translocation bringing a specific oncogene next to regulatory elements of one of the immunoglobulin genes.

C Cytosine or cytidine.

callus An undifferentiated clone of plant cells.

cAMP *See* **cyclic adenosine monophosphate.**

cancer The class of disease characterized by rapid and uncontrolled proliferation of cells within a tissue of a multitissued eukaryote. Cancers are generally thought to be genetic diseases of somatic cells, arising through sequential mutations that create oncogenes and inactivate tumor-suppressor genes.

candidate gene approach An approach to the molecular identification of a gene (and its encoded gene products) defined by one or more mutant alleles of that gene, based upon plausible molecular functions, gene expression patterns, or other properties that might account for the mutant phenotypes or disease syndrome.

canalized characters Characters that are constant in development despite environmental and genetic disturbances.

CAP *See* **catabolite activator protein.**

carbon source A nutrient (such as sugar) that provides carbon "skeletons" needed in an organism's synthesis of organic molecules.

carcinogen A substance that causes cancer.

cardinal gene Those pattern-formation genes in *Drosophila* that are the zygotically acting genes directly responding to the gradients of the anterior-posterior and dorsal-ventral positional information created by the maternally expressed pattern-formation genes. *See* **gap gene.**

carrier An individual who possesses a mutant allele but does not express it in the phenotype because of a dominant allelic partner; thus, an individual of genotype *A* / *a* is a carrier of *a* if there is complete dominance of *A* over *a*.

caspase A member of a family of related proteases that take part in initiating and carrying out the cell death response.

catabolite activator protein A protein that unites with cAMP at low glucose concentrations and binds to the *lac* promoter to facilitate RNA polymerase action.

catabolite repression The inactivation of an operon caused by the presence of large amounts of the metabolic end product of the operon.

cation A positively charged ion (such as K^+).

CDK *See* **cyclin-dependent protein kinase.**

cDNA *See* **complementary DNA.**

cDNA library A library composed of cDNAs, not necessarily representing all mRNAs.

cell cycle The set of events that take place in the divisions of mitotic cells. The cell cycle oscillates between mitosis (M phase) and interphase. Interphase can be subdivided in order into G_1, S phase, and G_2. DNA synthesis occurs during S phase. The length of the cell cycle is regulated through a special option in G_1, in which G_1 cells can enter a resting phase called G_0.

cell division The process by which two cells are formed from one.

cell fate The ultimate differentiated state to which a cell has become committed.

cell lineage A set of cells derived from a common ancestral cell by mitotic divisions.

centimorgan (cM) *See* **map unit.**

central dogma The hypothesis that information flows only from DNA to RNA to protein; although some exceptions are now known, the rule is generally valid.

centromere A specialized region of DNA on each eukaryotic chromosome that acts as a site for binding of the kinetochore proteins.

character Some attribute of individuals within a species for which various heritable differences can be defined.

character difference Alternative forms of the same attribute within a species.

chase *See* **pulse-chase experiment.**

checkpoints Stages of the cell cycle at which the completion of certain events of the cell cycle, such as chromosome replication, must have been successfully completed in order for the cell cycle to progress to the next stage.

chemical genetics The use of libraries of small chemical ligands to interfere specifically with individual protein functions, thereby phenocopying the effects of mutations in the genes encoding these functions.

chi-square (χ^2) test A statistical test used to determine the probability of obtaining observed proportions by chance, under a specific hypothesis.

chiasma (plural, **chiasmata**) A cross-shaped structure commonly observed between nonsister chromatids in meiosis; the site of crossing-over.

chimera *See* **mosaic.**

chloroplast A chlorophyll-containing organelle in plants that is the site of photosynthesis.

chorionic villus sampling (CVS) A placental sampling procedure for obtaining fetal tissue for chromosome and DNA analysis to assist in prenatal diagnosis of genetic disorders.

chromatid One of the two side-by-side replicas produced by chromosome division.

chromatid conversion A type of gene conversion that is inferred from the existence of identical sister-spore pairs in a fungal octad that shows a non-Mendelian allele ratio.

chromatid inference A situation in which the occurrence of a crossover between any two nonsister chromatids can be shown to affect the probability of those chromatids being involved in other crossovers in the same meiosis.

chromatin The substance of chromosomes; now known to include DNA, chromosomal proteins, and chromosomal RNA.

chromocenter The point at which the polytene chromosomes appears to be attached together.

chromomere A small beadlike structure visible on a chromosome during prophase of meiosis and mitosis.

chromosome A linear end-to-end arrangement of genes and other DNA, sometimes with associated protein and RNA.

chromosome aberration Any type of change in the chromosome structure or number.

chromosome loss Failure of a chromosome to become incorporated into a daughter nucleus at cell division.

chromosome map *See* **linkage map.**

chromosome mutation Any type of change in the chromosome structure or number.

chromosome puff A swelling at a site along the length of a polytene chromosome; the site of active transcription.

chromosome rearrangement A chromosome mutation in which chromosome parts are in new juxtaposition.

chromosome set The group of different chromosomes that carries the basic set of genetic information of a particular species.

chromosome theory of inheritance The unifying theory stating that inheritance patterns may be generally explained by assuming that genes are located in specific sites on chromosomes.

chromosome walking A method for the dissection of large segments of DNA, in which a cloned segment of DNA, usually eukaryotic, is used to screen recombinant DNA clones from the same genome bank for other clones containing neighboring sequences.

cis-acting element A site on a DNA (or RNA) molecule that functions as a binding site for a sequence-specific DNA- (or RNA-) binding protein. The term *cis-acting* indicates that protein binding to this site affects only nearby DNA (or RNA) sequences on the same molecule.

cis conformation In a heterozygote involving two mutant sites within a gene or within a gene cluster, the arrangement $++/a_1a_2$.

cis dominance The ability of a gene to affect genes next to it on the same chromosome.

cis-regulatory element *See* **cis-acting element.**

clone (1) A group of genetically identical cells or individuals derived by asexual division from a common ancestor. (2) *(colloquial)* An individual formed by some asexual process so that it is genetically identical to its "parent." (3) *See* **DNA clone.**

clone-based sequencing An approach to genome sequencing on a clone-by-clone basis.

cloning (1) In recombinant DNA research, the process of creating and amplifying specific DNA segments. (2) The production of genetically identical organisms from the somatic cells of an individual.

cloning vector In cloning, the plasmid or phage chromosome used to carry the cloned DNA segment.

cM (centimorgan) *See* **map unit.**

code dictionary A listing of the 64 possible codons and their translational meanings (the corresponding amino acids).

codominance The situation in which a heterozygote shows the phenotypic effects of both alelles equally.

codon A section of DNA (three nucleotide pairs in length) that encodes a single amino acid.

codon bias The skewed synonymous codon usage characteristic of a given species.

coefficient of coincidence The ratio of the observed number of double recombinants to the expected number.

cohesive ends Single-stranded ends of DNA that are cut in a staggered pattern and can then hydrogen-bond with complementary base sequences from other similarly formed ends.

cointegrate The product of the fusion of two circular transposable elements to form a single, larger circle during replicative transposition.

colinearity The correspondence between the location of a mutant site within a gene and the location of an amino acid substitution within the polypeptide translated from that gene.

colony A visible clone of cells.

comparative genomics The analysis of the relationships of the genome sequences of two or more species.

competent Able to take up exogenous DNA and thereby be transformed.

complementary DNA (cDNA) Synthetic DNA transcribed from a specific RNA through the action of the enzyme reverse transcriptase.

complementation The production of a wild-type phenotype when two different mutations are combined in a diploid or a heterokaryon.

complementation test *See* **cis-trans test.**

complex trait Discontinuous variant whose inheritance can be explained by the interaction of several genes plus the environment.

conditional mutation A mutation that has the wild-type phenotype under certain (permissive) environmental conditions and a mutant phenotype under other (restrictive) conditions.

conjugation The union of two bacterial cells, during which chromosomal material is transferred from the donor to the recipient cell.

conjugation tube *See* **pilus.**

consensus sequence The inferred most likely real sequence of a segment of DNA based on multiple sequence reads of the same segment.

conservative replication A disproved model of DNA synthesis suggesting that one-half of the daughter DNA molecules should have both strands composed of newly polymerized nucleotides.

conservative substitution Nucleotide-pair substitution within a protein-coding region that leads to replacement of an amino acid with one of similar chemical properties.

conservative transposition A mechanism of transposition that moves the mobile element to a new location in the genome as it removes it from its previous location.

constitutive A biological function that is always in one state, regardless of environment conditions. (1) With reference to gene expression, it refers to a gene that is always expressed. (2) With reference to chromosome structure, it refers to portions of the chromosome that are always heterochromatic.

constitutive expression Constant expression of a gene product, regardless of environmental conditions.

constitutive heterochromatin Specific regions of heterochromatin always present and in both homologs of a chromosome.

contig A set of ordered overlapping clones that constitute a chromosomal region or a genome.

continuous variation Variation showing an unbroken range of phenotypic values.

***copia*-like element** Retroviral-like transposable element, flanked by long terminal repeats and typically encoding a reverse transcriptase.

core promoter In eukaryotes, the RNA polymerase II–binding region of a gene.

correlation coefficient A statistical measure of the extent to which variations in one variable are related to variations in another.

cosegregation Parallel inheritance of two genes due to their close linkage on a chromosome.

cosmid A cloning vector that can replicate autonomously like a plasmid and be packaged into a phage.

cotransduction The simultaneous transduction of two bacterial marker genes.

cotransformation The simultaneous transformation of two bacterial marker genes.

coupling conformation Linked heterozygous gene pairs in the arrangement *AB* / *ab*.

covariance A statistical measure used in computing the correlation coefficient between two variables; the covariance is the mean of $(x - \bar{x})(y - \bar{y})$ over all pairs of values for the variables x and y, where $\bar{x}$ is the mean of the x values and $\bar{y}$ is the mean of the y values.

cpDNA Chloroplast DNA.

cri du chat syndrome A lethal human condition in infants caused by deletion of part of one homolog of chromosome 5.

crisscross inheritance Transmission of a gene from male parent to female child to male grandchild—for example, X-linked inheritance.

cross The deliberate mating of two parental types of organisms in genetic analysis.

crossing-over The exchange of corresponding chromosome parts between homologs by breakage and reunion.

crossover suppressor An inversion (usually complex) that makes pairing and crossing-over impossible.

cruciform configuration A region of DNA with palindromic sequences in both strands, so that each strand pairs with itself to form a helix extending sideways from the main helix.

culture Tissue or cells multiplying by asexual division, grown for experimentation.

CVS *See* **chorionic villus sampling.**

cyclic adenosine monophosphate (cAMP) A molecule containing a diester bond between the 3' and 5' carbons of the ribose portion of the nucleotide. This modified nucleotide cannot be incorporated into DNA or RNA. It plays a key role as an intracellular signal in the regulation of various processes.

cyclin A family of labile proteins that are synthesized and degraded at specific times within each cell cycle and regulate the progression of the cell cycle through their interactions with specific cyclin-dependent protein kinases.

cyclin-dependent protein kinase A family of protein kinases that, on activation by cyclins and an elaborate set of positive and negative regulatory proteins, phosphorylate certain transcription factors whose activity is necessary for a particular stage of the cell cycle.

Cys Cysteine (an amino acid).

cystic fibrosis A potentially lethal human disease of secretory glands; the most prominent symptom is excess secretion of lung mucus; inherited as an autosomal recessive.

cytidine The nucleoside containing cytosine as its base.

cytochromes A class of proteins, found in mitochondrial membranes, whose main function is oxidative phosphorylation of ADP to form ATP.

cytogenetics The cytological approach to genetics, mainly consisting of microscopic studies of chromosomes.

cytoplasm The material between the nuclear and cell membranes; includes fluid (cytosol), organelles, and various membranes.

cytoplasmic inheritance Inheritance through genes found in cytoplasmic organelles.

cytosine A pyrimidine base that pairs with guanine.

cytoskeleton The protein cable systems and associated proteins that together form the architecture of a eukaryotic cell.

cytosol The fluid part of the cytoplasm, outside the organelles.

Darwinian fitness The relative probability of survival and reproduction for a genotype.

daughter cells Two identical cells formed by asexual division of a cell.

daughter chromatids Two identical chromatids formed by the replication of one chromosome.

deamination The removal of amino groups from compounds. Deamination of cytosine produces uracil. If this is not repaired, C to T transitions can occur.

default state The developmental state of a cell (or group of cells) in the absence of activation of a developmental regulatory switch.

degenerate code A genetic code in which some amino acids may be encoded by more than one codon each.

deletion Removal of a chromosomal segment from a chromosome set.

deletion mutation *See* **indel mutation.**

denaturation The separation of the two strands of a DNA double helix or the severe disruption of the structure of any complex molecule without breaking the major bonds of its chains.

denaturation map A map of a stretch of DNA showing the locations of local denaturation loops, which correspond to regions of high AT content.

deoxyribonuclease *See* **DNase.**

deoxyribonucleic acid *See* **DNA.**

determinant A spatially localized molecule that causes cells to adopt a particular fate or set of related fates.

determination The process of commitment of cells to particular fates.

development The process whereby a single cell becomes a differentiated organism.

developmental field A group of developmentally equivalent cells that together organize and form a particular part of the body plan.

developmental pathway The chain of molecular events that take a set of equivalent cells and produce the assignment of different fates among those cells.

dicentric chromosome A chromosome with two centromeres.

dideoxy sequencing The most popular method of DNA sequencing. It uses dideoxynucleotide triphosphates mixed in with standard nucleotide triphosphates to produce a ladder of DNA strands whose synthesis is blocked at different lengths. This method has been incorporated into automated DNA-synthesis machines. Also called **Sanger sequencing** after its inventor, Sir Fred Sanger.

differentiation The changes in cell shape and physiology associated with the production of the final cell types of a particular organ or tissue.

dihybrid cross A cross between two individuals identically heterozygous at two loci—for example, $AB / ab \times AB / ab$.

dimorphism A "polymorphism" with only two forms.

dioecious plant A plant species in which male and female organs are on separate plants.

diploid A cell having two chromosome sets or an individual organism having two chromosome sets in each of its cells.

directional selection Selection that changes the frequency of an allele in a constant direction, either toward or away from fixation for that allele.

discontinuous variation Variation having distinct classes of phenotypes for a particular character.

dispersive replication Disproved model of DNA synthesis suggesting more or less random interspersion of parental and new segments in daughter DNA molecules.

distribution *See* **statistical distribution.**

distribution function A graph of some precise quantitative measure of a character against its frequency of occurrence.

diversification Increase in evolution in the diversity of the forms of a character among related species.

DNA (deoxyribonucleic acid) A double chain of linked nucleotides (having deoxyribose as their sugars); the fundamental substance of which genes are composed.

DNA-binding domain The site on a DNA-binding protein that directly interacts with specific DNA sequences.

DNA breakpoint The sequence location of the breakpoint of a chromosome aberration.

DNA clone A section of DNA that has been inserted into a vector molecule, such as a plasmid or a phage chromosome, and then replicated to form many copies.

DNA fingerprint The autoradiographic banding pattern produced when DNA is digested with a restriction enzyme that cuts outside a family of VNTRs and a Southern blot of the electrophoretic gel is probed with a VNTR-specific probe. Unlike true fingerprints, these patterns are not unique to each individual.

DNA glycosylase Enzymes that remove altered bases, leaving apurinic sites.

DNA polymorphism A naturally occurring variation in DNA sequence at a given location in the genome.

DNase (deoxyribonuclease) An enzyme that degrades DNA to nucleotides.

dominance variance Genetic variance at a single locus attributable to dominance of one allele over another.

dominant allele An allele that expresses its phenotypic effect even when heterozygous with a recessive allele; thus, if A is a dominant over a, then A / A and A / a have the same phenotype.

dominant phenotype The phenotype of a genotype containing the dominant allele; the parental phenotype that is expressed in a heterozygote.

donor DNA Any DNA to be used in cloning or in DNA-mediated transformation.

donor organism An organism that provides DNA for use in recombinant DNA technology or DNA-mediated transformation.

dosage compensation The process in organisms using a chromosomal sex determination mechanism (such as XX versus XY) that allows standard structural genes on the sex chromosome to be expressed at the same levels in females and males, regardless of the number of sex chromosomes. In mammals, dosage compensation operates by maintaining only a single active X chromosome in each cell; in *Drosophila,* it operates by hyperactivating the male X chromosome.

dose *See* **gene dose.**

double crossover Two crossovers in a chromosomal region under study.

double helix The structure of DNA first proposed by Watson and Crick, with two interlocking helices joined by hydrogen bonds between paired bases.

double infection Infection of a bacterium with two genetically different phages.

double-stranded RNA interference (dsRNAi) The use of double-stranded RNA to interfere with the stability and translation of homologous mRNAs, thereby phenocopying the effects of mutations in the genes that encode the mRNAs.

Down syndrome An abnormal human phenotype, including mental retardation, due to a trisomy of chromosome 21; more common in babies born to older mothers.

drift *See* **random genetic drift.**

drug resistance A mutant phenotype in which mutant individuals are able to survive on doses of a specific antibiotic or other drug that are toxic to wild-type individuals.

dsRNAi *See* **double-stranded RNA interference.**

Duchenne muscular dystrophy A lethal muscle disease in humans caused by mutation in a huge gene coding for the muscle protein dystrophin; inherited as an X-linked recessive phenotype.

duplicate genes Two identical allele pairs in one diploid individual.

duplication More than one copy of a particular chromosomal segment in a chromosome set.

dyad A pair of sister chromatids joined at the centromere, as in the first division of meiosis.

ecdysone A molting hormone in insects.

ectopic expression The occurrence of gene expression in a tissue in which it is normally not expressed. It can be caused by the juxtaposition of novel enhancer elements with a gene.

ectopic integration In a transgenic organism, the insertion of an introduced gene at a site other than its usual locus.

electrophoresis A technique for separating the components of a mixture of molecules (proteins, DNAs, or RNAs) in an electric field within a gel.

embryonic (or tissue) polarity The production of axes of asymmetry in a developing embryo or tissue primordium.

embryonic stem cells (ES cells) Cultured cell lines that are established from very early embryos and that are essentially totipotent. That is, these cells can be implanted into a host embryo and populate many or all tissues of the developing animal, most importantly including the germ line. Manipulations of these embryonic stem cells (ES cells) are used extensively in mouse genetics to produce targeted gene knockouts.

endocrine signal The hormone (ligand) secreted into the circulatory system from a gland (that is, endocrine organ). This hormone will bind to receptors within or on the surface of target cells.

endocrine system The organs in the body that secrete hormones into the circulatory system.

endogenote *See* **merozygote.**

endonuclease An enzyme that cleaves the phosphodiester bonds within a nucleotide chain.

endopolyploidy An increase in the number of chromosome sets caused by replication without cell division.

endosperm Triploid tissue in a seed, formed from the fusion of two haploid female nuclei and one haploid male nucleus.

enforced outbreeding Deliberate avoidance of mating between relatives.

enhanceosome The macromolecular assembly responsible for interaction between enhancer elements and the promoter regions of genes.

enhancer element A cis-regulatory sequence that can elevate levels of transcription from an adjacent promoter. Many tissue-specific enhancers can determine spatial patterns of gene expression in higher eukaryotes. Enhancers can act on promoters over many tens of kilobases of DNA and can be 5′ or 3′ to the promoter that they regulate.

enhancer mutation A modifier mutation that makes the phenotype of a mutation of another gene more severe.

enhancer trap A transgenic construction inserted into a chromosome and used to identify tissue-specific enhancers in the genome. In such a construct, a promoter sensitive to enhancer regulation is fused to a reporter gene, such that expression patterns of the reporter gene identify the spatial regulation conferred by nearby enhancers.

enol form *See* **tautomeric shift.**

enucleate cell A cell having no nucleus.

environment The combination of all the conditions external to an organism that could affect the expression of its genome.

environmental variance The variance due to environmental variation.

enzyme A protein that functions as a catalyst.

epigenetic inheritance Processes by which heritable modifications in gene function occur but are not due to changes in the base sequence of the DNA of the organism. Examples of epigenetic inheritance are paramutation, X chromosome inactivation, and parental imprinting.

epistasis A situation in which the differential phenotypic expression of a genotype at one locus depends on the genotype at another locus.

equational division A nuclear division that maintains the same ploidy level of the cell.

equivalence group A set of immature cells that all have the same developmental potential. In many cases, cells of an equivalence group end up adopting different fates.

ES cells *See* **embryonic stem cells.**

EST *See* **expressed sequence tag.**

ethidium A molecule that can intercalate into DNA double helices when the helix is under torsional stress.

euchromatin A chromosomal region that stains normally; thought to contain the normally functioning genes.

eugenics State-controlled human breeding in an attempt to improve future generations.

eukaryote An organism having eukaryotic cells.

eukaryotic cell A cell containing a nucleus.

euploid A cell having any number of complete chromosome sets or an individual composed of such cells.

excision repair The repair of a DNA lesion by removal of the faulty DNA segment and its replacement with a wild-type segment.

exconjugant A female bacterial cell that has just been in conjugation with a male and that contains a fragment of male DNA.

exogenote *See* **merozygote.**

exon Any nonintron section of the coding sequence of a gene; together, the exons correspond to the mRNA that is translated into protein.

exonuclease An enzyme that cleaves nucleotides one at a time from an end of a polynucleotide chain.

expressed sequence tag (EST) A sequence-tagged site derived from a cDNA clone; used to position and identify genes in genomic analysis.

expression library A library in which the vector carries transcriptional signals to allow any cloned insert to produce mRNA and ultimately a protein product.

expression vector A vector with the appropriate bacterial regulatory regions located 5' to the insertion site, allowing transcription and translation of a foreign protein in bacteria.

expressivity The degree to which a particular genotype is expressed in the phenotype.

F^- cell In *E. coli,* a cell having no fertility factor; a female cell.

F^+ cell In *E. coli,* a cell having a free fertility factor; a male cell.

F factor *See* **fertility factor.**

F′ factor A fertility factor into which a part of the bacterial chromosome has been incorporated.

F_1 generation The first filial generation, produced by crossing two parental lines.

F_2 generation The second filial generation, produced by selfing or intercrossing the F_1.

FACS *See* **fluorescence-activated chromosome sorting.**

facultative heterochromatin Heterochromatin located in positions that are composed of euchromatin in other individuals of the same species or even in the other homolog of a chromosome pair.

familial trait A trait shared by members of a family.

family selection A breeding technique of selecting a pair on the basis of the average performance of their progeny.

fate map A map of an embryo showing areas that are destined to develop into specific adult tissues and organs.

fate refinement The process by which decisions are fine-tuned so that all necessary cell types are fated in the proper spatial locations and cell numbers.

feedback loop *See* **autoregulatory loop.**

fertility factor (F factor) A bacterial episome whose presence confers donor ability (maleness).

filial generations Successive generations of progeny in a controlled series of crosses, starting with two specific parents (the P generation) and selfing or intercrossing the progeny of each new (F_1, F_2, . . .) generation.

filter enrichment A technique for recovering auxotrophic mutants in filamentous fungi.

fingerprint The characteristic spot pattern produced by electrophoresis of the polypeptide fragments obtained by denaturing a particular protein with a proteolytic enzyme.

first-division segregation pattern A linear pattern of spore phenotypes within the ascus for a particular allele pair, produced when the alleles go into separate nuclei at the first meiotic division, showing that no crossover has occurred between the allele pair and the centromere.

FISH *See* **fluorescence in situ hybridization.**

fitness *See* **Darwinian fitness.**

fixed allele An allele for which all members of the population under study are homozygous, so that no other alleles for this locus exist in the population.

fixed breakage point According to the heteroduplex DNA recombination model, the point from which unwinding of the DNA double helices begins, as a prelude to formation of heteroduplex DNA.

flow sorting *See* **fluorescence-activated chromosome sorting.**

fluctuation test A test used in microbes to establish the random nature of mutation or to measure mutation rates.

fluorescence-activated chromosome sorting (FACS) Using specific fluorescence signals of stained chromosomes in droplets to activate deflector plates that sort chromosomes into individual tubes of uniform types.

fluorescence in situ hybridization (FISH) In situ hybridization using a probe coupled with a fluorescent molecule.

fMet *See* **formylmethionine.**

foreign DNA DNA from another organism.

formylmethionine (fMet) A specialized amino acid that is the very first one incorporated into the polypeptide chain in the synthesis of proteins.

forward genetics The classical approach to genetic analysis, in which genes are first identified by mutant alleles and mutant phenotypes and later cloned and subjected to molecular analysis.

forward mutation A mutation that converts a wild-type allele into a mutant allele.

founder effect A random difference from the parental population in the frequency of a genotype in a new colony that results from a small number of founders.

frameshift mutation The insertion or deletion of a nucleotide pair or pairs, causing a disruption of the translational reading frame.

frequency-dependent fitness Fitness differences whose intensity changes with the relative frequency of genotypes in the population.

frequency-dependent selection Selection that depends on frequency-dependent fitness.

frequency histogram A "step curve" in which the frequencies of various arbitrarily bounded classes are graphed.

frequency-independent fitness Fitness that does not depend on interactions with other individuals of the same species.

frequency-independent selection Selection in which the fitnesses of genotypes are independent of their relative frequency in the population.

fruiting body In fungi, the organ in which meiosis takes place and sexual spores are produced.

functional complementation The use of a cloned fragment of wild-type DNA to transform a mutant into wild type; used in identifying a clone containing one specific gene.

functional genomics Studying the patterns of transcript and protein expression and molecular interactions at a genome-wide level.

G Guanine or guanosine.

G-protein A member of a family of proteins that contribute to signal transduction through protein-protein interactions that take place when the G-protein binds GTP but not when the G-protein binds GDP.

G_0 phase The resting phase that can occur in G_1 of interphase of the cell cycle.

G_1 phase The part of interphase of the cell cycle that precedes S phase.

G_2 phase The part of interphase of the cell cycle that follows S phase.

gain-of-function mutation A mutation that results in a new functional ability for a protein, detectable at the phenotypic level. These are often dominant, because only one copy of the novel gene is needed to produce the new function.

gamete A specialized haploid cell that fuses with a gamete from the opposite sex or mating type to form a diploid zygote; in mammals, an egg or a sperm.

gametophyte The haploid gamete-producing stage in the life cycle of plants; prominent and independent in some species but reduced or parasitic in others.

gap gene In *Drosophila,* a class of cardinal genes that are activated in the zygote in response to the anterior-posterior gradients of positional information. Through regulation of the pair-rule and homeotic genes, the patterns of expression of the various gap gene products lead to the specification of the correct number and types of body segments. Gap mutations cause the loss of several adjacent body segments.

gastrulation The first process of movements and infoldings of the single-layered cell sheet in early animal embryos.

gel electrophoresis A method of molecular separation in which DNA, RNA, or proteins are separated in a gel matrix according to molecular size, using an electric field to draw the molecules through the gel in a predetermined direction.

gene The fundamental physical and functional unit of heredity, which carries information from one generation to the next; a segment of DNA composed of a transcribed region and a regulatory sequence that makes transcription possible.

gene amplification The process by which the number of copies of a chromosomal segment is increased in a somatic cell.

gene conversion A meiotic process of directed change in which one allele directs the conversion of a partner allele into its own form.

gene disruption Inactivation of a gene by the integration of a specially engineered introduced DNA fragment.

gene dose The number of copies of a particular gene present in the genome.

gene family A set of genes in one genome, all descended from the same ancestral gene.

gene frequency *See* **allele frequency.**

gene fusion A novel gene that is produced by juxtaposition of DNA sequences of two separate genes. Gene fusions can be the result of chromosomal rearrangements, transposable element insertion, or genetic engineering.

gene interaction The collaboration of several different genes in the production of one phenotypic character (or related group of characters).

gene knockout Inactivation of a gene by the integration of a specially engineered introduced DNA fragment. In some systems, this occurs randomly using transgenic constructs that insert at many different locations in the genome. In other systems, it can be carried out in a directed fashion. *See* **targeted gene knockout.**

gene locus The specific place on a chromosome where a gene is located.

gene map A linear designation of mutant sites within a gene, based on the various frequencies of interallelic (intragenic) recombination.

gene mutation A change in the DNA that is entirely contained within a single gene, such as a **point mutation** or an **indel mutation** of several nucleotide pairs.

gene rearrangement The process of programmed changes in the DNA structure of somatic cells, leading to changes in gene number or in the structural and functional properties of the rearranged gene.

gene replacement The insertion of a genetically engineered transgene in place of a resident gene; often achieved by a double crossover.

gene tagging The use of a piece of foreign DNA or a transposon to tag a gene so that a clone of that gene can be identified readily in a library.

gene therapy The correction of a genetic deficiency in a cell by the addition of new DNA and its insertion into the genome. Different techniques have the potential to carry out gene therapy only in somatic tissues, or alternatively by correcting the genetic deficiency in the zygote, thereby correcting the germ line as well.

generalized transduction The ability of certain phages to transduce any gene in the bacterial chromosome.

genetic code The set of correspondences between nucleotide-pair triplets in DNA and amino acids in protein.

genetic dissection The use of recombination and mutation to piece together the various components of a given biological function.

genetic engineering The process of producing modified DNA in a test tube and reintroducing that DNA into host organisms.

genetic markers Alleles used as experimental probes to keep track of an individual, a tissue, a cell, a nucleus, a chromosome, or a gene.

genetic screen *See* **screen.**

genetic selection *See* **selective system.**

genetic variance The phenotypic variance associated with the average difference in phenotype among different genotypes.

genetically modified organism (GMO) A popular term for a transgenic organism, especially applied to transgenic agricultural organisms.

genetics (1) The study of genes. (2) The study of inheritance.

genome The entire complement of genetic material in a chromosome set.

genome projects Large-scale, often multilaboratory efforts required to sequence complex genomes.

genomic library A library encompassing an entire genome.

genomics The cloning and molecular characterization of entire genomes.

genotype The specific allelic composition of a cell—either of the entire cell or, more commonly, of a certain gene or a set of genes.

geographical race A geographically distinct population that differs from other populations of the species in the frequency of a genotype.

germ layer One of the primary embryonic cell sheets that are formed as a result of gastrulation. The primary germ layers in higher animals are the endoderm, mesoderm, and ectoderm.

germ line The cell lineage in a multitissued eukaryote from which the gametes derive.

germ-line gene therapy *See* **gene therapy.**

germinal mutation Mutations occurring in the cells that are destined to develop into gametes.

Gln Glutamine (an amino acid).

Glu Glutamate (an amino acid).

Gly Glycine (an amino acid).

GMO *See* **genetically modified organism.**

gradient A gradual change in some quantitative property over a specific distance.

growth factor A class of paracrine signaling molecules, usually secreted polypeptides, that are mitogenic, that is, that promote cell division in cells receiving these signals.

guanine A purine base that pairs with cytosine.

guanosine The nucleoside having guanine as its base.

gynandromorph An individual that is a mosaic of male and female structures. The underlying cause is frequently sex chromosome mosaicism, in which some cells are chromosomal females and others are chromosomal males.

half-chromatid conversion A type of gene conversion that is inferred from the existence of nonidentical sister spores in a fungal octad showing a non-Mendelian allele ratio.

haplo-insufficient Description of a gene that, in a diploid cell, cannot promote wild-type function in only one copy (dose).

haplo-sufficient Description of a gene that, in a diploid cell, can promote wild-type function in only one copy (dose).

haploid A cell having one chromosome set or an organism composed of such cells.

haploidization Production of a haploid from a diploid by progressive chromosome loss.

Hardy-Weinberg equilibrium The stable frequency distribution of genotypes A/A, A/a, and a/a, in the proportions of p^2, $2pq$, and q^2, respectively (where p and q are the frequencies of the alleles A and a), that is a consequence of random mating in the absence of mutation, migration, natural selection, or random drift.

harlequin chromosomes Sister chromatids that stain differently, so one appears dark and the other light.

helicase An enzyme that breaks hydrogen bonds in DNA and unwinds it during movement of the replication fork.

helix-loop-helix protein *See* **HLH proteins.**

hemizygous gene A gene present in only one copy in a diploid organism—for example, X-linked genes in a male mammal.

hemoglobin (Hb) The oxygen-transporting blood protein in most animals.

hemophilia A human disease in which the blood fails to clot, caused by a mutation in a gene encoding a clotting protein; inherited as an X-linked recessive phenotype.

hereditary nonpolyposis colorectal cancer (HNPCC) One of the most common hereditarily predisposed cancers.

heredity The biological similarity of offspring and parents.

heritability in the broad sense *See* **broad heritability.**

heritability in the narrow sense (h^2) The proportion of phenotypic variance that can be attributed to additive genetic variance.

hermaphrodite (1) A plant species in which male and female organs are in the same flower of a single individual (*compare* **monoecious plant**). (2) An animal with both male and female sex organs.

heterochromatin Densely staining condensed chromosomal regions, believed to be for the most part genetically inert.

heterodimer A protein consisting of two nonidentical polypeptide subunits.

heteroduplex A DNA double helix formed by annealing single strands from different sources; if there is a structural difference between the strands, the heteroduplex may show such abnormalities as loops or buckles.

heteroduplex DNA model A model that explains both crossing-over and gene conversion by assuming the production of a short stretch of heteroduplex DNA (formed from both parental DNAs) in the vicinity of a chiasma.

heterogametic sex The sex that has heteromorphic sex chromosomes (for example, XY) and hence produces two different kinds of gametes with respect to the sex chromosomes.

heterogeneous nuclear RNA (HnRNA) A diverse assortment of RNA types found in the nucleus, including mRNA precursors and other types of RNA.

heterokaryon A culture of cells composed of two different nuclear types in a common cytoplasm.

heterokaryon test A test for cytoplasmic mutations, based on new associations of phenotypes in cells derived from specially marked heterokaryons.

heteromorphic chromosomes A chromosome pair with some homology but differing in size, shape, or staining properties.

heteromultimer A protein consisting of at least two polypeptide subunits, with at least two of the subunits being nonidentical.

heteroplasmon A cell containing a mixture of genetically different cytoplasms and generally different mitochondria or different chloroplasts.

heterozygosity A measure of the genetic variation in a population; with respect to one locus, stated as the frequency of heterozygotes for that locus.

heterozygote An individual having a heterozygous gene pair.

heterozygous gene pair A gene pair having different alleles in the two chromosome sets of the diploid individual—for example, A/a or A^1/A^2.

hexaploid A cell having six chromosome sets or an organism composed of such cells.

high-frequency recombination (Hfr) cell In *E. coli,* a cell having its fertility factor integrated into the bacterial chromosome; a donor (male) cell.

high-throughput Highly-automated methods used to generate large-scale data collections.

Himalayan A mammalian temperature-dependent coat phenotype, generally albino with pigment only at the cooler tips of the ears, feet, and tail.

His Histidine (an amino acid).

histocompatibility antigens Antigens that determine the acceptance or rejection of a tissue graft.

histocompatibility genes The genes that encode the histocompatibility antigens.

histone A type of basic protein that forms the unit around which DNA is coiled in the nucleosomes of eukaryotic chromosomes.

HLH proteins A family of proteins in which a part of the polypeptide forms two helices separated by a loop (the HLH domain); this structure acts as a sequence-specific DNA-binding domain. HLH proteins are thought to act as transcription factors.

homeodomain A highly conserved family of sequences, 60 amino acids in length and found within a large number of transcription factors, that can form helix-turn-helix structures and bind DNA in a sequence-specific manner.

homeosis The replacement of one body part by another. Homeosis can be caused by environmental factors leading to developmental anomalies or by mutation.

homeotic gene A gene that controls the fate of segments along the anterior-posterior axis of higher animals.

homeotic gene complex One of the tightly linked clusters of genes controlling A/P segment determination in segmented animals. These genes all encode homeodomain transcription factors.

homeotic mutation A mutation that leads to the replacement of one part of the body by another.

homodimer A protein consisting of two identical polypeptide subunits.

homogametic sex The sex with homologous sex chromosomes (for example, XX).

homolog A member of a pair of homologous chromosomes.

homologous chromosomes Chromosomes that pair with each other at meiosis or chromosomes in different species that have retained most of the same genes during their evolution from a common ancestor.

homologous structures Structures of different species that develop from the same embryonic structures whether or not they have the same function, as, for example, the wings of birds, the wings of bats, and the forelimbs of humans.

homomultimer A protein consisting of two or more identical polypeptide subunits.

homozygote An individual having a homozygous gene pair.

homozygous gene pair A gene pair having identical alleles in both copies—for example, A / A or A^1 / A^1.

hormone A molecule that is secreted by an endocrine organ into the circulatory system and that acts as a long-range signaling molecule by activating receptors on or within target cells.

hormone response element (HRE) For hormones that act by binding to receptors that can act as transcription factors, an HRE is a cis-regulatory DNA sequence that is a binding site for a hormone-receptor complex.

host range The spectrum of strains of a given bacterial species that a given strain of phage can infect.

hot spot A part of a gene that shows a very high tendency to become a mutant site, either spontaneously or under the action of a particular mutagen.

HRE *See* **hormone response element.**

Huntington disease A lethal human disease of nerve degeneration, with late-age onset. Inherited as an autosomal dominant phenotype; new mutations are rare.

hybrid (1) A heterozygote. (2) A progeny individual from any cross between parents of different genotypes.

hybrid breakdown The loss of viability or fertility in the offpsring of crosses between different species.

hybrid dysgenesis A syndrome of effects including sterility, mutation, chromosome breakage, and male recombination in the hybrid progeny of crosses between certain laboratory and natural isolates of *Drosophila.*

hybridization in situ Finding the location of a gene by adding specific radioactive probes for the gene and detecting the location of the radioactivity on the chromosome after hybridization.

hybridize (1) To form a hybrid by performing a cross. (2) To anneal nucleic acid strands from different sources.

hydrogen bond A weak bond in which an atom shares an electron with a hydrogen atom; hydrogen bonds are important in the specificity of base pairing in nucleic acids and in the determination of protein shape.

hydroxyapatite A form of calcium phosphate that binds double-stranded DNA.

hypermorphic mutation A mutation that confers a mutant phenotype through the production of more gene product by mutant individuals than is produced by wild-type individuals.

hyperploid Aneuploid containing a small number of extra chromosomes.

hypha (plural, **hyphae**) A threadlike structure (composed of cells attached end to end) that forms the main tissue in many fungus species.

hypomorphic mutation A mutation that confers a mutant phenotype but still retains a low but detectable level of wild-type function.

hypoploid Aneuploid with a small number of chromosomes missing.

ICR compound One of a series of quinacrine-mustard compounds synthesized at the Institute for Cancer Research (Fox Chase, Pennsylvania) that act as **intercalating agents.**

Ig (immunoglobulin) *See* **antibody.**

Ile Isoleucine (an amino acid).

imago An adult insect.

immune system The animal cells and tissue that recognize and attack foreign substances within the body.

immunoglobulin *See* **antibody.**

immunohistochemistry The use of antibodies or antisera as histological tools for identifying patterns of protein distribution within a tissue or an organism. An antibody (or mixture of antibodies) that binds to a specific protein is tagged with an enzyme that can convert a substrate into a visible dye. The tagged antibody is incubated with the tissue, unbound antibody is washed off, and the enzymatic substrate is then added, revealing the pattern of protein (actually, antigen) localization.

in situ "In place"; *see* **hybridization in situ.**

in vitro In an experimental situation outside the organism (literally "in glass").

in vitro mutagenesis The production of either random or specific mutations in a piece of cloned DNA. Typically, the DNA will then be repackaged and introduced into a cell or an organism to assess the results of the mutagenesis.

in vivo In a living cell or organism.

inborn errors of metabolism Hereditary metabolic disorders, particularly in reference to human disease.

inbreeding Mating between relatives.

inbreeding coefficient The probability of homozygosity that results because the zygote obtains copies of the *same* ancestral gene.

incomplete dominance The situation in which a heterozygote shows a phenotype quantitatively (but not exactly) intermediate between the corresponding homozygote phenotypes. (Exact intermediacy is no dominance).

indel mutation A mutation involving the **addition** or **deletion** of one or more nucleotide pairs.

independent assortment *See* **Mendel's second law.**

induced mutation A mutation that arises after mutagen treatment.

inducer An environmental agent that triggers transcription from an operon.

induction (1) The relief of repression for a gene or set of genes under negative control. (2) An interaction between two or more cells or tissues that is required for one of those cells or tissues to change its developmental fate.

inductive interaction The interaction between two groups of cells in which a signal passed from one group of cells causes the other group of cells to change their developmental state (or fate).

infectious transfer The rapid transmission of free episomes (plus any chromosomal genes that they may carry) from donor to recipient cells in a bacterial population.

inosine A rare base that is important at the wobble position of some tRNA anticodons.

insertion sequence (IS) element A mobile piece of bacterial DNA (several hundred nucleotide pairs in length) that is capable of inactivating a gene into which it inserts.

insertional mutation A mutation arising by the interuption of a gene by foreign DNA, such as from a transgenic construct or a transposable element.

insertional translocation The insertion of a segment from one chromosome into another nonhomologous one.

interactome The entire set of physical interactions among the gene products and nucleotide sequences of the genome.

intercalating agent A mutagen that can insert itself between the stacked bases at the center of the DNA double helix, causing an elevated rate of **indel mutations.**

interchromosomal recombination Recombination resulting from independent assortment.

interference A measure of the independence of crossovers from each other, calculated by subtracting the coefficient of coincidence from 1.

intermediate filaments A heterogeneous class of cytoskeletal elements characterized by an intermediate cable diameter, larger than that of microfilaments but smaller than that of microtubules.

internal resolution site (IRS) A region of replicative transposable elements necessary for cointegrate recombination.

interphase The cell cycle stage between nuclear divisions, when chromosomes are extended and functionally active.

interrupted mating A technique used to map bacterial genes by determining the sequence in which donor genes enter recipient cells.

interstitial region The chromosomal region between the centromere and the site of a rearrangement.

intervening sequence An intron; a segment of largely unknown function within a gene. This segment is initially transcribed, but the transcript is not found in the functional mRNA.

intrachromosomal recombination Recombination resulting from crossing-over between two gene pairs.

intron *See* **intervening sequence.**

inversion A chromosomal mutation consisting of the removal of a chromosome segment, its rotation through 180 degrees, and its reinsertion in the same location.

inverted repeat (IR) sequence A sequence found in identical (but inverted) form—for example, at the opposite ends of a transposon.

IR sequence *See* **inverted repeat sequence.**

IRS *See* **internal resolution site.**

IS element *See* **insertion sequence element.**

isoaccepting tRNAs The various types of tRNA molecules that carry a specific amino acid.

isotope One of several forms of an atom having the same atomic number but differing atomic masses.

karyotype The entire chromosome complement of an individual or cell, as seen during mitotic metaphase.

keto form *See* **tautomeric shift.**

kilobase 1000 nucleotide pairs.

kinase An enzyme that adds one or more phosphate groups to its substrates. If the substrates are specific proteins, the enzyme is called a **protein kinase.**

kinetochore A complex of proteins to which a nuclear spindle fiber attaches.

Klinefelter syndrome An abnormal human male phenotype due to an extra X chromosome (XXY).

knockout Inactivation of one specific gene. Same as gene disruption.

lagging strand In DNA replication, the strand that is synthesized apparently in the 3′-to-5′ direction, by the ligation of short fragments synthesized individually in the 5′-to-3′ direction.

λ (lambda) phage One kind ("species") of temperate bacteriophage.

λdgal A λ phage carrying a *gal* bacterial gene and defective *(d)*for some phage function.

lampbrush chromosomes Large chromosomes found in amphibian eggs, with lateral DNA loops that produce a brushlike appearance under the microscope.

lateral inhibition The signal produced by one cell that prevents adjacent cells from adopting the same developmental fate as that of the first cell.

lawn A continuous layer of bacteria on the surface of an agar medium.

leader An untranslated segment at the 5′ end of mRNA between the transcriptional and translational start sites.

leader sequence The sequence at the 5′ end of an mRNA that is not translated into protein.

leading strand In DNA replication, the strand that is made in the 5′-to-3′ direction by continuous polymerization at the 3′ growing tip.

leaky mutation A mutation that confers a mutant phenotype but still retains a low but detectable level of wild-type function.

lethal mutation A gene whose expression results in the death of the individual expressing it.

lesion A damaged area in a gene (a mutant site), a chromosome, or a protein.

lethal gene A gene whose expression results in the death of the individual expressing it.

Leu Leucine (an amino acid).

library A collection of DNA clones obtained from one DNA donor.

ligand *See* **ligand-receptor interaction.**

ligand-receptor interaction The interaction between a molecule (usually of extracellular origin) and a protein on or within a target cell. One type of ligand-receptor interaction can be between steroid hormones and their cytoplasmic or nuclear receptors. Another can be between secreted polypeptide ligands and transmembrane receptors.

ligase An enzyme that can rejoin a broken phosphodiester bond in a nucleic acid.

LINE *See* **long interspersed element.**

linear tetrad A tetrad that results from the occurrence of the meiotic and postmeiotic nuclear divisions in such a way that sister products remain adjacent to one another (with no passing of nuclei).

linkage The association of genes on the same chromosome.

linkage group A group of genes known to be linked; a chromosome.

linkage map A chromosome map; an abstract map of chromosomal loci, based on recombinant frequencies.

locus (plural, **loci**) *See* **gene locus.**

long interspersed element (LINE) A type of large repetitive DNA segment found throughout the genome.

long terminal repeat Sequence identical regions at the 5' and 3' ends of retroviruses and copialike elements.

loss-of-function mutation A mutation that partly or fully eliminates normal gene activity. Hypomorphic and null mutations are loss-of-function.

LTR *See* **long terminal repeat.**

Lys Lysine (an amino acid).

lysis The rupture and death of a bacterial cell on the release of phage progeny.

lysogen *See* **lysogenic bacterium.**

lysogenic bacterium A bacterial cell capable of spontaneous lysis due, for example, to the uncoupling of a prophage from the bacterial chromosome.

M phase The mitotic phase of the cell cycle.

macromolecule A large polymer such as DNA, a protein, or a polysaccharide.

Manx Tailless phenotype in cats, caused by an autosomal dominant mutation that is lethal when homozygous.

map unit (m.u.) The "distance" between two linked gene pairs where 1 percent of the products of meiosis are recombinant; a unit of distance in a linkage map.

mapping function A formula expressing the relation between distance in a linkage map and recombinant frequency.

Marfan syndrome A human disorder of the connective tissue expressed as a range of symptoms, including heart defects and very long limbs and digits; inherited as an autosomal dominant phenotype.

marker *See* **genetic markers.**

marker retention A technique used to test the degree of linkage between two mitochondrial mutations in yeast.

maternal effect The environmental influence of the mother's tissues on the phenotype of the offspring.

maternal effect lethal mutation A mutation that is viable in zygotes, but mothers having the mutation produce inviable offspring.

maternal inheritance A type of uniparental inheritance in which all progeny have the genotype and phenotype of the parent acting as the female.

maternally expressed gene A gene that contributes to the phenotype of an offspring on the basis of its expression in the mother.

mating types The equivalent in lower organisms of the sexes in higher oganisms; the mating types typically differ only physiologically, not in the physical form.

matroclinous inheritance Inheritance in which all offspring have the nucleus-determined phenotype of the mother.

mean The arithmetic average.

medium Any material on (or in) which experimental cultures are grown.

megabase One million nucleotide pairs.

meiocyte Cell in which meiosis takes place.

meiosis Two successive nuclear divisions (with corresponding cell divisions) that produce gametes (in animals) or sexual spores (in plants and fungi) that have one-half of the genetic material of the original cell.

meiospore Cell that is one of the products of meiosis in plants.

melting Denaturation of DNA.

Mendelian ratio A ratio of progeny phenotypes corresponding to the operation of Mendel's laws.

Mendel's first law The two members of a gene pair segregate from each other in meiosis; each gamete has an equal probability of obtaining either member of the gene pair.

Mendel's second law The law of independent assortment; unlinked or distantly linked segregating gene pairs assort independently at meiosis.

merozygote A partly diploid *E. coli* cell formed from a complete chromosome (the endogenote) plus a fragment (the exogenote).

messenger RNA *See* **mRNA.**

Met Methionine (an amino acid).

metabolism The chemical reactions that take place in a living cell.

metacentric chromosome A chromosome having its centromere in the middle.

metameres Segmental repeat units in higher animals.

metaphase An intermediate stage of nuclear division when chromosomes align along the equatorial plane of the cell.

methylation Modification of a molecule by the addition of a methyl group.

microfilaments The smallest-diameter cable system of the cytoskeleton. Microfilament cables are composed of actin polymers.

microsatellite DNA A type of repetitive DNA based on very short repeats such as dinucleotides.

microsatellite marker A DNA difference at the equivalent location in two genomes due to different repeat lengths of a microsatellite.

microtubule The largest-diameter cable system of the cytoskeleton. Microtubules are composed of polymerized tubulin subunits forming a hollow tube.

microtubule organizing center (MTOC) The part of the microtubule cytoskeleton in which all of the minus ends of the microtubules are clustered. Ordinarily, this is near the center of the cell.

midparent value The mean of the values of a quantitative phenotype for two specific parents.

minimal medium A medium containing only inorganic salts, a carbon source, and water.

minimum tiling path The minimal set of clones (with the least overlap) in a physical map that recapitulate the entirety of a genome.

minisatellite DNA A type of repetitive DNA sequence based on short repeat sequences with a unique common core; used for DNA fingerprinting.

misexpression *See* **ectopic expression.**

mismatch-repair A repair system for repairing damage to DNA that has already been replicated.

missense mutation Nucleotide-pair substitution within a protein-coding region that leads to replacement of one amino acid by another amino acid.

mitochondrial cytopathies Human disorder caused by point mutations or deletions in mitochondrial DNA; inherited maternally.

mitochondrion A eukaryotic organelle that is the site of ATP synthesis and of the citric acid cycle.

mitogen *See* **growth factor.**

mitosis A type of nuclear division (occurring at cell division) that produces two daughter nuclei identical with the parent nucleus.

mitotic crossover A crossover resulting from the pairing of homologs in a mitotic diploid.

mobile genetic element *See* **transposable element.**

mode The single class in a statistical distribution having the greatest frequency.

modifier gene A gene that affects the phenotypic expression of another gene.

modifier mutation A gene that affects the phenotypic expression of another gene.

molecular clock The constant rate of substitution of amino acids in proteins or nucleotides in nucleic acids over long evolutionary time.

molecular genetics The study of the molecular processes underlying gene structure and function.

monocistronic mRNA An mRNA that encodes one protein.

monoecious plant A plant species in which male and female organs are found on the same plant but in different flowers (for example, corn).

monohybrid cross A cross between two individuals identically heterozygous at one gene pair—for example, $A / a \times A / a$.

monoploid A cell having only one chromosome set (usually as an aberration) or an organism composed of such cells.

monosomic A cell or individual that is basically diploid but that has only one copy of one particular chromosome type and thus has chromosome number $2n + 1$.

morphogen A molecule that can induce the acquisition of different cell fates according to the level of morphogen to which a cell is exposed.

morphological mutation A mutation affecting some aspect of the appearance of an individual.

mosaic A chimera; a tissue containing two or more genetically distinct cell types or an individual composed of such tissues.

motif A short DNA sequence that is associated with a particular functional role.

motor protein Proteins that are able to move unidirectionally along a specific type of cytoskeletal cable. Kinesins and dyneins are microtubule-based and myosins are microfilament-based motor proteins. By attaching to other subcellular components, motor proteins are capable of directed movement of these components within the cell.

mRNA (messenger RNA) An RNA molecule transcribed from the DNA of a gene and from which a protein is translated by the action of ribosomes.

mtDNA Mitochondrial DNA.

MTOC *See* **microtubule organizing center.**

m.u. *See* **map unit.**

***mu* phage** A kind ("species") of phage with properties similar to those of insertion sequences, being able to insert, transpose, inactivate, and cause rearrangements.

multimer A protein consisting of two or more subunits.

multimeric structure A structure composed of several identical or different subunits held together by weak bonds.

multiple allelism The existence of several known alleles of a gene.

multiple-cloning site *See* **polylinker.**

multiple-factor hypothesis A hypothesis that explains quantitative variation by assuming the interaction of a large number of genes (polygenes), each with a small additive effect on the character.

multiple-hit hypothesis The proposal that a single cell must receive a series of mutational events to become malignant or cancerous.

multiplicity of infection The average number of phage particles that infect a single bacterial cell in a specific experiment.

mutagen An agent that is capable of increasing the mutation rate.

mutagenesis An experiment in which experimental organisms are treated with a mutagen and their progeny are examined for specific mutant phenotypes.

mutant An organism or cell carrying a mutation.

mutant allele An allele differing from the allele found in the standard, or wild type.

mutant hunt The process of collecting different mutants showing abnormalities in a certain structure or in a certain function, as a preparation for mutational dissection of that function.

mutant rescue *See* **functional complementation.**

mutant sector *See* **sector.**

mutant site The damaged or altered area within a mutated gene.

mutation (1) The process that produces a gene or a chromosome set differing from the wild type. (2) The gene or chromosome set that results from such a process.

mutation breeding Use of mutagens to develop variants that can increase agricultural yield.

mutation event The actual occurrence of a mutation in time and space.

mutation frequency The frequency of mutants in a population.

mutation rate The number of mutation events per gene copy in a population per unit of time (for example, per cell generation).

mutational dissection The study of the components of a biological function through a study of mutations affecting that function.

mutational specificity The constellation of mutational damage that characterizes a particular mutagen.

muton The smallest part of a gene that can be involved in a mutation event; now known to be a nucleotide pair.

myeloma A cancer of the bone marrow.

nanometer 10^{-9} meters.

narrow heritability *See* **heritability in the narrow sense.**

natural selection The differential rate of reproduction of different types in a population as the result of different physiological, anatomical, or behavioral characteristics of the types.

negative assortative mating Preferential mating between phenotypically unlike partners.

negative control Regulation mediated by factors that block or turn off transcription.

neomorphic mutation A mutation with phenotypic effects due to the production of a novel gene product or novel pattern of gene expression.

neoplasia *See* **cancer.**

neurofibromatosis A human disease with tumors of nerve cells and *café au lait* spots, both in the skin. The allele generally arises from germinal mutation, but it is inherited as an autosomal dominant.

Neurospora A pink mold, commonly found growing on old food.

neutral DNA sequence variation Variation in DNA sequence that is not under natural selection.

neutral mutations Mutations that have no effect on the survivorship and reproductive rate of the organisms that carry them.

neutral petite A petite that produces all wild-type progeny when crossed with wild type.

neutrality *See* **selective neutrality.**

nicking Nuclease action to sever the sugar-phosphate backbone in one DNA strand at one specific site.

nitrocellulose filter A type of filter used to hold DNA for hybridization.

nitrogen bases Types of molecules that form important parts of nucleic acids, composed of nitrogen-containing ring structures; hydrogen bonds between bases link the two strands of a DNA double helix.

nonautonomous *See* **autonomous phenotype.**

nonconservative substitution Nucleotide-pair substitution within a protein-coding region that leads to replacement of an amino acid with one of different chemical properties.

nondisjunction The failure of homologs (at meiosis) or sister chromatids (at mitosis) to separate properly to opposite poles.

non-Mendelian ratio An unusual ratio of progeny phenotypes that does not conform to the simple operation of Mendel's laws; for example, mutant : wild ratios of 3 : 5, 5 : 3, 6 : 2, or 2 : 6 in tetrads indicate that gene conversion has occurred.

nonsense codon A codon for which no formal tRNA molecule exists; the presence of a nonsense codon causes termination of translation (ending of the polypeptide chain). The three nonsense codons are called amber, ocher, and opal.

nonsense mutation Nucleotide-pair substitution within a protein-coding region that changes a codon for an amino acid into a termination codon.

nonsense suppressor A mutation that produces an altered tRNA that will insert an amino acid in translation in response to a nonsense codon.

nonsynonymous substitution Mutational replacement of an amino acid with one of different chemical properties

norm of reaction The pattern of phenotypes produced by a given genotype under different environmental conditions.

Northern blotting Transfer of electrophoretically separated RNA molecules from a gel onto an absorbent sheet, which is then immersed in a labeled probe that will bind to the RNA of interest.

nu body *See* **nucleosome.**

nuclease An enzyme that can degrade DNA by breaking its phosphodiester bonds.

nucleoid A DNA mass within a chloroplast or mitochondrion.

nucleolar organizer A region (or regions) of the chromosome set physically associated with the nucleolus and containing rRNA genes.

nucleolus An organelle found in the nucleus, containing rRNA and amplified multiple copies of the genes encoding rRNA.

nucleoside A nitrogen base bound to a sugar molecule.

nucleosome A nu body; the basic unit of eukaryotic chromosome structure; a ball of eight histone molecules wrapped about by two coils of DNA.

nucleotide A molecule composed of a nitrogen base, a sugar, and a phosphate group; the basic building block of nucleic acids.

nucleotide pair A pair of nucleotides (one in each strand of DNA) that are joined by hydrogen bonds.

nucleotide excision-repair system An excision-repair pathway that breaks the phosphodiester bonds on either side of the damaged base, removing that base and several on either side. This is followed by repair replication.

nucleotide-pair substitution The replacement of a specific nucelotide pair by a different pair; often mutagenic.

null allele An allele whose effect is either an absence of normal gene product at the molecular level or an absence of normal function at the phenotypic level.

null mutation A mutation that results in complete absence of function for the gene.

nullisomic A cell or individual with one chromosomal type missing, with a chromosome number such as $n - 1$ or $2n - 2$.

nurse cells The sister cells of the oocyte in insects. The nurse cells produce the bulk of the cytoplasmic contents of the mature oocyte.

ocher codon The codon UAA, a nonsense codon.

octad An ascus containing eight ascospores, produced in species in which the tetrad normally undergoes a postmeiotic mitotic division.

Okazaki fragment A small segment of single-stranded DNA synthesized as part of the lagging strand in DNA replication.

oligonucleotide A short segment of synthetic DNA.

oncogene A gain-of-function mutation that contributes to the production of a cancer.

opal codon The codon UGA, a nonsense codon.

open reading frame (ORF) A section of a sequenced piece of DNA that begins with a start codon and ends with a stop codon; it is presumed to be the coding sequence of a gene.

operator A DNA region at one end of an operon that acts as the binding site for repressor protein.

operon A set of adjacent structural genes whose mRNA is synthesized in one piece, plus the adjacent regulatory signals that affect transcription of the structural genes.

ordered clone sequencing The sequencing of a set of clones such as a **minimum tiling path** that corresponds to a genomic segment, a chromosome, or a genome.

ORF *See* **open reading frame.**

organelle A subcellular structure having a specialized function—for example, the mitochondrion, the chloroplast, or the spindle apparatus.

organogenesis The production of organ systems in animal embryogenesis.

origin of replication The point of a specific sequence at which DNA replication is initiated.

overdominance A phenotypic relation in which the phenotypic expression of the heterozygote is greater than that of either homozygote.

oxidatively damaged base One origin of spontaneous DNA damage.

P element A transposable *Drosophila* **short-inverted-repeat element** that has been used as a tool for insertional mutagenesis and for germ-line transformation.

P granules The cytoplasmic granules associated with germ-line formation in *C. elegans.*

PAC (P1-based artificial chromosome) A derivative of phage P1 engineered as a cloning vector for carrying large inserts.

pair-rule gene In *Drosophila,* a member of a class of zygotically expressed genes that act at an intermediary stage in the process of establishing the correct numbers of body segments. Pair-rule mutations have half of the normal number of segments, owing to the loss of every other segment.

paired-end sequences The sequences corresponding to the two ends of a cloned insert.

palindrome A DNA sequence that has 180-degree rotational symmetry.

paracentric inversion An inversion not involving the centromere.

paracrine signal The process by which a secreted molecule binds to a receptor on or within nearby cells, thereby inducing a signal transduction pathway within the receiving cell.

paralogous genes Two genes in the same species that have evolved by gene duplication.

paramutation An epigenetic phenomenon in plants, in which the genetic activity of a normal allele is heritably reduced by virtue of that allele having been heterozygous with a special "paramutagenic" allele.

parental imprinting An epigenetic phenomenon in which the activity of a gene depends on whether it was inherited from the father or the mother. Some genes are maternally imprinted, others paternally.

parthenogenesis The production of offspring by a female with no genetic contribution from a male.

partial diploid. *See* **merozygote.**

particulate inheritance The model proposing that genetic information is transmitted from one generation to the next in discrete units ("particles"), so that the character of the offspring is not a smooth blend of essences from the parents (*compare* **blending inheritance**).

pathogen An organism that causes disease in another organism.

patroclinous inheritance Inheritance in which all offspring have the nucleus-based phenotype of the father.

pattern formation The developmental processes resulting in the complex shape and structure of higher organisms.

PCR *See* **polymerase chain reaction.**

pedigree A "family tree," drawn with standard genetic symbols, showing inheritance patterns for specific phenotypic characters.

penetrance The proportion of individuals with a specific genotype who manifest that genotype at the phenotype level.

peptide *See* **amino acid.**

peptide bond A bond joining two amino acids.

pericentric inversion An inversion that involves the centromere.

permissive conditions Those environmental conditions under which a conditional mutant shows the wild-type phenotype.

petite A yeast mutation producing small colonies and altered mitochondrial functions. In cytoplasmic petites (neutral and suppressive petites), the mutation is a deletion in mitochondrial DNA; in segregational petites, the mutation occurs in nuclear DNA.

phage *See* **bacteriophage.**

Phe Phenylalanine (an amino acid).

phenocopy An environmentally induced phenotype that resembles the phenotype produced by a mutation.

phenome The description of the phenotypes elicited by mutations in all of the genes in the genome, singly and in combination.

phenotype (1) The form taken by some character (or group of characters) in a specific individual. (2) The detectable outward manifestations of a specific genotype.

phenotypic sex determination Sex determination by nongenetic means.

phenylketonuria (PKU) A human metabolic disease caused by a mutation in a gene encoding a phenylalanine-processing enzyme, which leads to mental retardation if not treated; inherited as an autosomal recessive phenotype.

Philadelphia chromosome A translocation between the long arms of chromosomes 9 and 22, often found in the white blood cells of patients with chronic myeloid leukemia.

phosphodiester bond A bond between a sugar group and a phosphate group; such bonds form the sugar-phosphate backbone of DNA.

photolyase An enzyme that cleaves thymine dimers.

phyletic evolution The formation of new related species as a result of the splitting of previously existing species.

physical map The ordered and oriented map of cloned DNA fragments on the genome.

piebald A mammalian phenotype in which patches of skin are unpigmented because of lack of melanocytes; generally inherited as an autosomal dominant.

pilus (plural, **pili**) A conjugation tube; a hollow hairlike appendage of a donor *E. coli* cell that acts as a bridge for transmission of donor DNA to the recipient cell during conjugation.

plant breeding The application of genetic analysis to the development of plant lines better suited to human purposes.

plaque A clear area on a bacterial lawn, left by lysis of the bacteria through progressive infections by a phage and its descendants.

plasmid Autonomously replicating extrachromosomal DNA molecule.

plate (1) A flat dish used to culture microbes. (2) To spread cells over the surface of solid medium in a plate.

pleiotropic mutation A mutation that has effects on several different characters.

ploidy The number of chromosome sets.

point mutation A mutation that can be mapped to one specific locus.

poky A slow-growing mitochondrial mutant in *Neurospora.*

polar granules Cytoplasmic granules localized at the posterior end of a *Drosophila* oocyte and early embryo. These granules are associated with the germ-line and posterior determinants.

polar mutation A mutation that affects the transcription or translation of the part of the gene or operon on only one side of the mutant site—for example, nonsense mutations, frameshift mutations, and IS-induced mutations.

poly(A) tail A string of adenine nucleotides added to mRNA after transcription.

polyacrylamide A material used to make electrophoretic gels for separation of mixtures of macromolecules.

polycistronic mRNA An mRNA that codes for more than one protein.

polydactyly More than five fingers or toes or both. Inherited as an autosomal dominant phenotype.

polygenes *See* **multiple-factor hypothesis.**

polylinker A vector DNA sequence containing multiple unique restriction-enzyme-cut sites, convenient for inserting foreign DNA. Sometimes also referred to as an MCS (multiple cloning site).

polymerase chain reaction (PCR) A method for amplifying specific DNA segments that exploits certain features of DNA replication.

polymorphism The occurrence in a population (or among populations) of several phenotypic forms associated with alleles of one gene or homologs of one chromosome.

polypeptide A chain of linked amino acids; a protein.

polyploid A cell having three or more chromosome sets or an organism composed of such cells.

polysaccharide A biological polymer composed of sugar subunits—for example, starch or cellulose.

polytene chromosome A giant chromosome produced by an endomitotic process in which the multiple DNA sets remain bound in a haploid number of chromosomes.

position effect Used to describe a situation in which the phenotypic influence of a gene is altered by changes in the position of the gene within the genome.

positional cloning The identification of the DNA sequences encoding a gene of interest based on knowledge of its genetic or cytogenetic map location.

positional information The process by which chemical cues that establish cell fate along a geographic axis are established in a developing embryo or tissue primordium.

position-effect variegation Variegation caused by the inactivation of a gene in some cells through its abnormal juxtaposition with heterochromatin.

positive assortative mating A situation in which like phenotypes mate more commonly than expected by chance.

positive control Regulation mediated by a protein that is required for the activation of a transcription unit.

postzygotic isolation The failure of two species to exchange genes because the zygotes formed by the mating between them are either inviable or infertile.

prezygotic isolation The failure of two species to exchange genes because of barriers either to mating between them or to fertilization of the gametes of one species by the gametes of the other species.

primary structure of a protein The sequence of amino acids in the polypeptide chain.

primase An enzyme that makes RNA primers during DNA replication.

primer A short single-stranded RNA or DNA that can act as a start site for 3′ chain growth when bound to a single-stranded template.

primer walking The use of a primer based on a sequenced area of a genome to sequence into a flanking unsequenced area.

prion Proteinaceous infectious particle that causes degenerative disorders of the central nervous system, such as "scrapie" in sheep and Creutzfeldt-Jacob disease in humans.

Pro Proline (an amino acid).

probe Defined nucleic acid segment that can be used to identify specific DNA molecules bearing the complementary sequence, usually through autoradiography.

product of meiosis One of the (usually four) cells formed by the two meiotic divisions.

product rule The probability of two independent events occurring simultaneously is the product of the individual probabilities.

proflavin A mutagen that tends to produce frameshift mutations.

programmed cell death *See* **apoptosis.**

prokaryote An organism composed of a prokaryotic cell, such as a bacterium or a blue-green alga.

prokaryotic cell A cell having no nuclear membrane and hence no separate nucleus.

promoter A regulator region a short distance from the 5' end of a gene that acts as the binding site for RNA polymerase.

promoter-proximal element The series of transcription-factor binding sites located near the core promoter.

prophage A phage "chromsome" inserted as part of the linear structure of the DNA chromosome of a bacterium.

prophase The early stage of nuclear division during which chromosomes condense and become visible.

propositus In a human pedigree, the person who first came to the attention of the geneticist.

protein kinase An enzyme that phosphorylates specific amino acid residues on specific target proteins. One major class of protein kinases phosphorylates tyrosines, and the other phosphorylates serines and threonines on target proteins.

proteome The complete set of protein-coding genes in a genome.

proto-oncogene The normal cellular counterpart of genes that can be mutated to become dominant oncogenes.

protoplast A plant cell whose wall has been removed.

prototroph A strain of organisms that will proliferate on minimal medium (*compare* **auxotroph**).

provirus The chromosomally inserted DNA genome of a retrovirus.

pseudodominance The sudden appearance of a recessive phenotype in a pedigree, due to deletion of a masking dominant gene.

pseudogene A mutationally inactive gene for which no functional counterpart exists in wild-type populations.

puff *See* **chromosome puff.**

pulse-chase experiment An experiment in which cells are grown in radioactive medium for a brief period (the pulse) and then transferred to nonradioactive medium for a longer period (the chase).

pulsed field gel electrophoresis An electrophoretic technique in which the gel is subjected to electrical fields alternating between different angles, allowing very large DNA fragments to "snake" through the gel and hence permitting efficient separation of mixtures of such large fragments.

Punnett square A grid used as a graphic representation of the progeny zygotes resulting from different gamete fusions in a specific cross.

pure-breeding line or strain A group of identical individuals that always produce offspring of the same phenotype when intercrossed.

purine A type of nitrogen base; the purine bases in DNA are adenine and guanine.

pyrimidine A type of nitrogen base; the pyrimidine bases in DNA are cytosine and thymine.

quantitative trait locus (QTL) A gene affecting the phenotypic variation in continuously varying traits such as height, weight, and so forth.

quantitative variation The existence of a range of phenotypes for a specific character, differing by degree rather than by distinct qualitative differences.

quaternary structure of a protein The multimeric constitution of the protein.

R plasmid A plasmid containing one or several transposons that bear resistance genes.

radiation hybrid A type of human-mouse hybrid cell in which human chromosomes have been fragmented by radiation to determine which markers are inherited together and therefore linked.

radiation hybrid map A map of cloned DNA segments based on an analysis of hybridization of those DNA segments to a panel of radiation hybrid clones.

random genetic drift Changes in allele frequency that result because the genes appearing in offspring are not a perfectly representative sampling of the parental genes.

random mating Mating between individuals in which the choice of a partner is not influenced by the genotypes (with respect to specific genes under study).

randomly amplified polymorphic DNA (RAPD) A set of several genomic fragments amplified by a single PCR primer; somewhat variable from individual to individual; + / − heterozygotes for individual fragments can act as markers in genome mapping.

RAPD *See* **randomly amplified polymorphic DNA.**

reading frame The codon sequence that is determined by reading nucleotides in groups of three from some specific start codon.

realized heritability The ratio of the single-generation progress of selection to the selection differential of the parents.

reannealing Spontaneous realignment of two single DNA strands to re-form a DNA double helix that had been denatured.

receptor *See* **ligand-receptor interaction.**

receptor tyrosine kinase (RTK) A transmembrane receptor whose cytoplasmic domain includes a tyrosine kinase enzymatic activity. In normal situations, the kinase is activated only on binding of the appropriate ligand to the receptor.

recessive allele An allele whose phenotypic effect is not expressed in a heterozygote.

recessive phenotype The phenotype of a homozygote for the recessive allele; the parental phenotype that is not expressed in a heterozygote.

reciprocal crosses A pair of crosses of the type genotype A ♀ × genotype B ♂ and B ♀ × A ♂.

reciprocal translocation A translocation in which part of one chromosome is exchanged with a part of a separate nonhomologous chromosome.

recombinant An individual or cell with a genotype produced by recombination.

recombinant DNA A novel DNA sequence formed by the combination of two nonhomologous DNA molecules.

recombinant frequency (RF) The proportion (or percentage) of recombinant cells or individuals.

recombination (1) In general, any process in a diploid or partly diploid cell that generates new gene or chromosomal combinations not found in that cell or in its progenitors. (2) At meiosis, the process that generates a haploid product of meiosis whose genotype is different from either of the two haploid genotypes that constituted the meiotic diploid.

recombinational repair The repair of a DNA lesion through a process, similar to recombination, that uses recombination enzymes.

reduction division A nuclear division that produces daughter nuclei each having one-half as many centromeres as the parental nucleus.

redundant DNA *See* **repetitive DNA.**

regression A term coined by Galton for the tendency of the quantitative traits of offspring to be closer to the population mean than are their parents' traits. It arises from dominance, gene interaction, and nongenetic influences on traits.

regression coefficient The slope of the straight line that most closely relates two correlated variables.

regulatory genes Genes that are involved in turning on or off the transcription of structural genes.

regulatory region Upstream (5′) end of a gene to which bind various proteins that cause transcription of the gene at the correct time and place.

repetitive DNA Redundant DNA; DNA sequences that are present in many copies per chromosome set.

replication DNA synthesis.

replication fork The point at which the two strands of DNA are separated to allow replication of each strand.

replicative transposition A mechanism of transposition that generates a new insertion element integrated elsewhere in the genome while leaving the original element at its original site of insertion.

replicon A chromosomal region under the influence of one adjacent replication-initiation locus.

reporter gene A gene whose phenotypic expression is easy to monitor; used to study tissue-specific promoter and enhancer activities in transgenes.

repressor A protein that binds to a cis-acting element such as an operator or a silencer, thereby preventing transcription from an adjacent promoter.

repressor protein A molecule that binds to the operator and prevents transcription of an operon.

repulsion conformation Two linked heterozygous gene pairs in the arrangement of *Ab* / *aB*.

resolving power The ability of an experimental technique to distinguish between two genetic conditions (typically discussed when one condition is rare and of particular interest).

restriction enzyme An endonuclease that will recognize specific target nucleotide sequences in DNA and break the DNA chain at those points; a variety of these enzymes are known, and they are extensively used in genetic engineering.

restriction fragment length polymorphism (RFLP) A DNA sequence difference between individuals or haplotypes recognized as different restriction fragment lengths. For example, a nucelotide-pair substitution can cause a restriction-enzyme-recognition site to be present in one allele of a gene and absent in another. Consequently, a probe for this DNA region will hybridize to different-sized fragments within restriction digests of DNAs from these two alleles.

restriction map A map of a chromosomal region showing the positions of target sites of one or more restriction enzymes.

restrictive conditions Environmental conditions under which a conditional mutant shows the mutant phenotype.

retinoblastoma A childhood cancer of the retina.

retrotransposon A transposable element that utilizes reverse transcriptase to transpose through an RNA intermediate.

retrovirus An RNA virus that replicates by first being converted into double-stranded DNA.

reverse genetics An experimental procedure that begins with a cloned segment of DNA or a protein sequence and uses it (through directed mutagenesis) to introduce programmed mutations back into the genome to investigate function.

reverse mutation *See* **reversion.**

reverse transcriptase An enzyme that catalyzes the synthesis of a DNA strand from an RNA template.

reversion The production of a wild-type gene from a mutant gene.

RFLP *See* **restriction fragment length polymorphism.**

RFLP mapping A technique in which DNA restriction fragment length polymorphisms are used as reference loci for mapping in relation to known genes or other RFLP loci.

rho A protein factor required to recognize certain transcription termination signals in *E. coli.*

ribonucleic acid *See* **RNA.**

ribosomal RNA *See* **rRNA.**

ribosome A complex organelle that catalyzes the translation of messenger RNA into an amino acid sequence. Composed of proteins plus rRNA.

ribozymes RNAs with enzymatic activities—for instance, the self-splicing RNA molecules in *Tetrahymena.*

RNA (ribonucleic acid) A single-stranded nucleic acid similar to DNA but having ribose sugar rather than deoxyribose sugar and uracil rather than thymine as one of the bases.

RNA in situ hybridization A technique that is used to identify the spatial pattern of expression of a particular transcript (usually an mRNA). In this technique, the DNA probe is labeled, either radioactivity or by chemically attaching an enzyme that can convert a substrate into a visible dye. A tissue or organism is soaked in a solution of single-stranded labeled DNA under conditions that allow the DNA to hybridize to complementary RNA sequences in the cells; unhybridized DNA is then removed. Radioactive probe is detected by autoradiography. Enzyme-labeled probe is detected by soaking the tissue in the substrate; the dye develops in sites where the transcript of interest was expressed.

RNA localization A mechanism of exploiting asymmetry within a developing oocyte or within a somatic cell to create asymmetrically localized proteins. Typically, mRNA asymmetry is achieved through interactions with polarized cytoskeletal elements.

RNA polymerase An enzyme that catalyzes the synthesis of an RNA strand from a DNA template. In eukaryotes, there are several classes of RNA polymerase; structural genes for proteins are transcribed by RNA polymerase II.

robotics The application of automation technology to large-scale biological data collection efforts such as **genome projects.**

rRNA (ribosomal RNA) A class of RNA molecules, coded in the nucleolar organizer, that have an integral (but poorly understood) role in ribosome structure and function.

RTK *See* **receptor tyrosine kinase.**

S (Svedberg unit) A unit of sedimentation velocity, commonly used to describe molecular units of various sizes (because sedimentation velocity is related to size).

S phase The part of interphase of the cell cycle in which DNA synthesis occurs.

SARs Scaffold attachment regions; the positions along DNA where it is anchored to the central scaffold of the chromosome.

Sanger sequencing *See* **dideoxy sequencing.**

satellite A terminal section of a chromosome, separated from the main body of the chromosome by a narrow constriction.

satellite chromosome A chromosome that seems to be an addition to the normal genome.

satellite DNA Any type of highly repetitive DNA; formerly defined as DNA forming a satellite band after cesium chloride density-gradient centrifugation.

saturation mutagenesis Induction and recovery of large numbers of mutations in one area of the genome or in one function, in the hope of identifying all the genes in that area or affecting that function.

scaffold (1) The central framework of a chromosome to which the DNA solenoid is attached as loops; composed largely of topoisomerase. (2) In genome projects, an ordered set of contigs in which there may be unsequenced gaps connected by paired-end sequence reads.

SCE *See* **sister-chromatid exchange.**

screen A mutagenesis procedure in which essentially all mutagenized progeny are recovered and are individually evaluated for mutant phenotype.

secondary mutational screen A mutational detection screen in which individuals with phenotypes more severe or more mild than a starting mutant phenotype are identified.

second-division segregation pattern A pattern of ascospore genotypes for a gene pair showing that the two alleles separate into different nuclei only at the second meiotic division, as a result of a crossover between that gene pair and its centromere; can be detected only in a linear ascus.

second-site mutation The second mutation of a double mutation within a gene; in many cases, the second-site mutation suppresses the first mutation, so the double mutant has the wild-type phenotype.

secondary sexual characteristics The sex-associated phenotypes of somatic tissues in sexually dimorphic animals.

secondary structure of a protein A spiral or zigzag arrangement of the polypeptide chain.

sector An area of tissue whose phenotype is detectably different from the surrounding tissue phenotype.

sedimentation The sinking of a molecule under the opposing forces of gravitation and buoyancy.

segment One of the repeating units along the anterior-posterior axis of the body of higher animals such as annelids, arthropods, and chordates.

segment identity The process by which the anterior-posterior fates of the segments are established.

segment-polarity gene In *Drosophila,* a member of the class of genes that contribute to the final aspects of establishing the correct number of segments. Segment-polarity mutations cause a loss of a comparable part of each of the body segments.

segmentation The process by which the correct number and types of segments are established in a developing higher animal.

segregation (1) Cytologically, the separation of homologous structures. (2) Genetically, the production of two separate phenotypes, corresponding to two alleles of a gene, either in different individuals (meiotic segregation) or in different tissues (mitotic segregation).

selection coefficient (*s*) The proportional excess or deficiency of fitness of one genotype in relation to another genotype.

selection differential The difference between the mean of a population and the mean of the individuals selected to be parents of the next generation.

selection progress The difference between the mean of a population and the mean of the offspring in the next generation born to selected parents.

selective neutrality A situation in which different alleles of a certain gene confer equal fitness.

selective system A mutational selection technique that enriches the frequency of specific (usually rare) genotypes by establishing environmental conditions that prevent growth or survival of other genotypes.

self To fertilize eggs with sperms from the same individual.

self-assembly The ability of certain multimeric biological structures to assemble from their component parts through random movements of the molecules and formation of weak chemical bonds between surfaces with complementary shapes.

semiconservative replication The established model of DNA replication in which each double-stranded molecule is composed of one parental strand and one newly polymerized strand.

semisterility (half sterility) The phenotype of individuals heterozygotic for certain types of chromosome aberration; expressed as a reduced number of viable gametes and hence reduced fertility.

sequence assembly The compilation of thousands or millions of independent DNA sequence reads into a set of contigs and scaffolds.

sequence similarity The level of relationship of two nucleotide or amino acid sequences to one another.

sequence-tagged site (STS) A type of small repetitive DNA sequence found throughout a eukaryotic genome.

Ser Serine (an amino acid).

sex chromosome A chromosome whose presence or absence is correlated with the sex of the bearer; a chromosome that plays a role in sex determination.

sex determination The genetic or environmental process by which the sex of an individual is established.

sex linkage The location of a gene on a sex chromosome.

sex reversal A syndrome known in humans and mice in which chromosomally XX individuals develop as males. In some cases, sex reversal is now known to be due to the translocation of the testis-determining region of the Y chromosome to the tip of the X chromosome in such individuals.

sexduction Sexual transmission of donor *E. coli* chromosomal genes on the fertility factor.

sexual spore *See* **spore.**

short interspersed element (SINE) A type of small repetitive DNA sequence found throughout a eukaryotic genome.

short-inverted-repeat (SIR) element An element such as the P element or Ac element that has identical short sequences present in inverted orientation at the two ends of the element.

short-sequence-length polymorphism (SSLP) The presence of different numbers of short repetitive elements (mini- and microsatellite DNA) at one particular locus in different homologous chromosomes; heterozygotes represent useful markers for genome mapping.

shotgun technique Cloning a large number of different DNA fragments as a prelude to selecting one particular clone type for intensive study.

shuttle vector A vector (for example, a plasmid) constructed in such a way that it can replicate in at least two different host species, allowing a DNA segment to be tested or manipulated in several cellular settings.

sickle-cell anemia Potentially lethal human disease caused by a mutation in a gene encoding the oxygen-transporting molecule hemoglobin. The altered molecule causes red blood cells to be sickle shaped. Inherited as an autosomal recessive.

signal sequence The N-terminal sequence of a secreted protein, which is required for transport through the cell membrane.

signal transduction cascade A series of sequential events, such as protein phosphorylations, that pass a signal received by a transmembrane receptor through a series of intermediate molecules until final regulatory molecules, such as transcription factors, are modified in response to the signal.

silencer element A cis-regulatory sequence that can reduce levels of transcription from an adjacent promoter.

silent mutation A mutation that has no effect on the function of a gene product.

SINE *See* **short interspersed element.**

single-copy DNA DNA sequences present in only one copy per haploid genome.

single-nucleotide polymorphism (SNP) A nucleotide-pair difference at a given location in the genomes of two or more naturally occurring individuals.

SIR element *See* **short-inverted-repeat element.**

sister-chromatid exchange (SCE) An event similar to crossing-over that can take place between sister chromatids at mitosis or at meiosis; detected in harlequin chromosomes.

site-directed mutagenesis The alteration of some specific part of a cloned DNA segment followed by reintroduction of the modified DNA back into an organism for assay of the mutant phenotype or for production of the mutant protein.

site-specific recombination Recombination between two specific sequences that need not be homologous; mediated by a specific recombination system.

SNP *See* **single-nucleotide polymorphism.**

solenoid structure The supercoiled arrangement of DNA in eukaryotic nuclear chromosomes produced by coiling the continuous string of nucleosomes.

somatic cell A cell that is not destined to become a gamete; a "body cell," whose genes will not be passed on to future generations.

somatic cell genetics Asexual genetics, including the study of somatic mutation, assortment, crossing-over, and cell fusion.

somatic gene therapy *See* **gene therapy.**

somatic mutation A mutation that arises in a somatic cell and consequently is not passed through the germ line to the next generation.

somatostatin A human growth hormone.

SOS repair system The error-prone process whereby gross structural DNA damage is circumvented by allowing replication to proceed past the damage through imprecise polymerization.

Southern blot Transfer of electrophoretically separated fragments of DNA from a gel to an absorbent sheet such as paper. This sheet is then immersed in a solution containing a labeled probe that will bind to a fragment of interest.

spacer DNA DNA found between genes; its function is unknown.

specialized (restricted) transduction The situation in which a particular phage will transduce only specific regions of the bacterial chromosome.

species A group of organisms that can exchange genes among themselves but are reproductively isolated from other such groups.

specific-locus test A system for detecting recessive mutations in diploids. Normal individuals treated with mutagen are mated to testers that are homozygous for the recessive alleles at a number of specific loci; the progeny are then screened for recessive phenotypes.

spindle The set of microtubular fibers that appear to move eukaryotic chromosomes during division.

splicing The reaction that removes introns and joins together exons in RNA.

spontaneous mutation A mutation occuring in the absence of exposure to mutagens.

spore (1) In plants and fungi, sexual spores are the haploid cells produced by meiosis. (2) In fungi, asexual spores are somatic cells that are cast off to act either as gametes or as the initial cells for new haploid individuals.

sporophyte The diploid sexual-spore-producing generation in the life cycle of plants—that is, the stage in which meiosis occurs.

SSLP *See* **short-sequence-length polymorphism.**

stacking The packing of the flattish nitrogen bases at the center of the DNA double helix.

staggered cuts The cleavage of two opposite strands of duplex DNA at points near one another, giving rise to short terminal single-stranded overhangs.

standard deviation The square root of the variance.

statistic A computed quantity characteristic of a population, such as the mean.

statistical distribution The array of frequencies of different quantitative or qualitative classes in a population.

stem cell A cell that divides, generally asymmetrically, to give rise to two different progeny cells. One is a blast cell just like the parental cell and the other is a cell that enters a differentiation pathway. In this manner, a continuously propagating cell population can maintain itself and spin off differentiating cells.

steroid hormone A class of hormones synthesized by glands of the endocrine system that, by virtue of their nonpolar nature, are able to pass directly through the plasma membrane of cells. Steroid hormones act by binding to and activating transcription factors called steroid hormone receptors.

steroid hormone receptor A family of related proteins that act as transcription factors when bound to their cognate hormones. Not all members of this family actually bind to steroids; the name derives from the first family member to have been discovered, which was indeed a steroid hormone receptor.

strain A pure-breeding lineage, usually of haploid organisms, bacteria, or viruses.

structural gene A gene encoding the amino acid sequence of a protein.

structural genomics The large-scale analysis of three-dimensional protein structures.

STS *See* **sequence-tagged site.**

STS content map A physical map of clones based on overlaps determined by the presence of common STSs.

subvital gene A gene that causes the death of some proportion (but not all) of the individuals that express it.

sum rule The probability that one or the other of two mutually exclusive events will occur is the sum of their individual probabilities.

supercoil A closed, double-stranded DNA molecule that is twisted on itself.

superinfection Phage infection of a cell that already harbors a prophage.

suppression The production of a phenotype closer to wild type by the addition of another mutation to a phenotypically abnormal genotype.

suppressor A secondary mutation that can cancel the effect of a primary mutation, resulting in wild-type phenotype.

suppressor mutation A mutation that counteracts the effects of another mutation. A suppressor maps at a different site from the mutation that it counteracts, either within the same gene or at a more distant locus. Different suppressors act in different ways.

survivor factor A ligand that, when bound to a receptor, blocks the activation of the apoptosis pathway.

synapsis Close pairing of homologs at meiosis.

synaptonemal complex A complex structure that unites homologs during prophase of meiosis.

syncytial blastoderm In insects, the stage of blastoderm preceding the formation of cell membranes around the individual nuclei of the early embryo.

syncytium A single cell with many nuclei.

synonymous substitution A nucleotide-pair substitution mutation within a protein-coding region that converts one codon for an amino acid into another codon for that same amino acid.

syntenic Description of DNA segments in which the gene order is identical in different related species.

T (1) Thymine or thymidine. (2) *See* **tetratype.**

T-DNA A part of the Ti plasmid that is inserted into the genome of the host plant cell.

tRNA (transfer RNA) A class of small RNA molecules that bear specific amino acids to the ribosome during translation; the amino acid is inserted into the growing polypeptide chain when the anticodon of the tRNA pairs with a codon on the mRNA being translated.

tagging *See* **gene tagging.**

tandem duplication Adjacent identical chromosome segments.

targeted gene knockout The introduction of a null mutation in a gene by a designed alteration in a cloned DNA sequence that is then introduced into the genome through homologous recombination and replacement of the normal allele.

tautomeric shift The spontaneous isomerization of a nitrogen base from its normal keto form to an alternative hydrogen-bonding enol form.

telocentric chromosome A chromosome having the centromere at one end.

telomerase An enzyme that adds repetitive units to the ends of linear chromosomes to prevent shortening after replication, by using a special small RNA as a template.

telomere The tip (or end) of a chromosome.

telophase The late stage of nuclear division when daughter nuclei re-form.

temperate phage A phage that can become a prophage.

temperature-sensitive mutation A conditional mutation that produces the mutant phenotype in one temperature range and the wild-type phenotype in another temperature range.

template A molecular "mold" that shapes the structure or sequence of another molecule; for example, the nucleotide sequence of DNA acts as a template to control the nucleotide sequence of RNA during transcription.

teratogen An agent that interferes with normal development.

terminal redundancy In phage, a linear DNA molecule with single-stranded ends that are longer than is necessary to close the DNA circle.

tertiary structure of a protein The folding or coiling of the secondary structure to form a globular molecule.

testcross A cross of an individual of unknown genotype or a heterozygote (or a multiple heterozygote) with a tester individual.

tester An individual homozygous for one or more recessive alleles; used in a testcross.

testicular feminization syndrome A human condition, caused by loss-of-function *Tfm* mutations in the gene encoding androgen receptors. Chromosomally XY individuals develop as phenotypic females.

tetrad (1) Four homologous chromatids in a bundle in the first meiotic prophase and metaphase. (2) The four haploid product cells from a single meiosis.

tetrad analysis The use of tetrads (definition 2) to study the behavior of chromosomes and genes in meiosis.

tetramer A protein consisting of four polypeptide subunits.

tetraploid A cell having four chromosome sets; an organism composed of such cells.

Thr Threonine (an amino acid).

three-point testcross A testcross in which one parent has three heterozygous gene pairs.

thymidine The nucleoside having thymine as its base.

thymine dimer A pair of chemically bonded adjacent thymine bases in DNA; the cellular processes that repair this lesion often make errors that create mutations.

Ti plasmid A circular plasmid of *Agrobacterium tumifaciens* that enables the bacterium to infect plant cells and produce a tumor (crown gall tumor).

tissue-specific gene expression The expression of a gene in a higher eukaryote in a specific and reproducible subset of tissues and cells during development.

topoisomerase An enzyme that can cut and re-form polynucleotide backbones in DNA to allow it to assume a more relaxed configuration.

totipotent The state of a cell lineage that is able to give rise to all possible cell fates found within a given organism.

trailer An untranslated segment at the 3′ end of mRNA.

trans-acting factor A diffusible regulatory molecule (almost always a protein) that binds to a specific cis-acting element.

trans conformation In a heterozygote with two mutant sites within a gene or gene cluster, the arrangement $a_1+ / +a_2$.

transcription The synthesis of RNA from a DNA template.

transcription factor A protein that binds to a cis-regulatory element (for example, an enhancer) and thereby, directly or indirectly, affects the initiation of transcription.

transcriptome The entire constellation of transcripts encoded by a genome.

transduction The movement of genes from a bacterial donor to a bacterial recipient with a phage as the vector.

transfection The process by which exogenous DNA in solution is introduced into cultured cells.

transfer RNA *See* **tRNA.**

transformation (1) The directed modification of a genome by the external application of DNA from a cell of different genotype. (2) Conversion of normal higher eukaryotic cells in tissue culture into a cancerlike state of uncontrolled division.

transformational evolution An evolutionary process that results from successive changes in the properties of individuals during their own lifetimes.

transgene A gene that has been modified by externally applied recombinant DNA techniques and reintroduced into the genome by germ-line transformation.

transgenic organism One whose genome has been modified by externally applied new DNA.

transient diploid The stage of the life cycle of predominantly haploid fungi (and algae) during which meiosis occurs.

transition A type of nucleotide-pair substitution involving the replacement of a purine with another purine or of a pyrimidine with another pyrimidine—for example, G·C to A·T.

translation The ribosome-mediated production of a polypeptide whose amino acid sequence is derived from the codon sequence of an mRNA molecule.

translocation The relocation of a chromosomal segment in a different position in the genome.

transmembrane receptor A protein that spans the plasma membrane of a cell, with the extracellular part of the protein having the ability to bind to a ligand and the intracellular part having an activity (such as protein kinase) that can be induced on ligand binding.

transmission genetics The study of the mechanisms involved in the passage of a gene from one generation to the next.

transposable element A general term for any genetic unit that can insert into a chromosome, exit, and relocate; includes insertion sequences, transposons, some phages, and controlling elements.

transposase The enzyme encoded by transposable elements that undergo conservative transposition.

transposition The process by which mobile genetic elements move from one location in the genome to another.

transposon A mobile piece of DNA that is flanked by terminal repeat sequences and typically bears genes coding for transposition functions.

transversion A type of nucleotide-pair substitution in which a pyrimidine replaces a purine or vice versa—for example, G·C to T·A.

triallelic diagram An equilateral triangle in which the frequencies of three different alleles at a locus are given by the distances of a point from the three sides of the triangle.

trinucleotide repeat disease The expansion of a 3-bp repeat from a relatively low number of copies to a high number of copies that is responsible for a number of genetic diseases, such as Huntington's disease.

triplet The three nucleotide pairs that compose a codon.

triplet expansion The expansion of a 3-bp repeat from a relatively low number of copies to a high number of copies that is responsible for a number of genetic diseases, such as fragile X syndrome.

triploid A cell having three chromosome sets or an organism composed of such cells.

trisomic Basically a diploid with an extra chromosome of one type, producing a chromosome number of the form $2n + 1$.

tritium A radioactive isotope of hydrogen.

Trp Tryptophan (an amino acid).

true-breeding line or strain *See* **pure-breeding line or strain.**

truncation selection A breeding technique in which individuals in whom quantitative expression of a phenotype is above or below a certain value (the truncation point) are selected as parents for the next generation.

tumor-suppressor gene A gene encoding a protein that suppresses tumor formation. The wild-type alleles of tumor-suppressor genes are thought to function as negative regulators of cell proliferation.

tumor virus A virus that is capable of inducing a cancer.

Turner syndrome An abnormal human female phenotype produced by the presence of only one X chromosome (XO).

twin spot A pair of mutant sectors within wild-type tissue, produced by a mitotic crossover in an individual of appropriate heterozygous genotype.

two-hybrid system A pair of *Saccharomyces cerevisiae* (yeast) vectors used for detecting protein-protein interaction. Each vector carries the gene for a different foreign protein under test; if these vectors unite physically, a reporter gene is transcribed.

2-μm (2-micrometer) plasmid A naturally occurring extragenomic circular DNA molecule found in some yeast cells, with a circumference of 2 μm. Engineered to form the basis for several types of gene vectors in yeast.

Tyr Tyrosine (an amino acid).

U Uracil, or uridine.

underdominance A phenotypic relation in which the phenotypic expression of the heterozygote is less than that of either homozygote.

unequal crossover A crossover between homologs that are not perfectly aligned.

uniparental inheritance The transmission of certain phenotypes from one parental type to all the progeny; such inheritance is generally produced by organelle genes.

unstable mutation A mutation that has high frequency of reversion; a mutation caused by the insertion of a controlling element whose subsequent exit produces a reversion.

uracil A pyrimidine base that appears in RNA in place of the thymine found in DNA.

URF (unassigned reading frame) An open reading frame (ORF) whose function has not yet been determined.

uridine The nucleoside having uracil as its base.

Val Valine (an amino acid).

variable A property that may have different values in various cases.

variable number tandem repeat (VNTR) A chromosomal locus at which a particular repetitive sequence is present in different numbers in different individuals or in the two different homologs in one diploid individual.

variable region A region in an immunoglobulin molecule that shows many sequence differences between antibodies of different specificities. The variable regions of the light and heavy chains of an immunoglobulin bind antigen.

variance A measure of the variation around the central class of a distribution; the average squared deviation of the observations from their mean value.

variant An individual organism that is recognizably different from an arbitrary standard type in that species.

variate A specific numerical value of a variable.

variation The differences among parents and their offspring or among individuals in a population.

variational evolution An evolutionary process that results from changes over time in the frequencies of different types that are present in the population as a result of different rates of survival or reproduction of the different types.

variegation The occurrence within a tissue of sectors with differing phenotypes.

vector *See* **cloning vector.**

viability The probability that a fertilized egg will survive and develop into an adult organism.

viral transforming gene A gene within a viral genome that can induce abnormal proliferation of cells in culture and, similarly, can induce tumors in infected whole animals.

virulent phage A phage that cannot become a prophage; infection by such a phage always leads to lysis of the host cell.

VNTR *See* **variable number tandem repeat.**

Western blot Membrane carrying an imprint of proteins separated by electrophoresis. Can be probed with a labeled antibody to detect a specific protein.

whole genome shotgun (WGS) sequencing The sequencing of ends of clones without regard to any information about the location of the clones.

wild type The genotype or phenotype that is found in nature or in the standard laboratory stock for a given organism.

wobble The ability of certain bases at the third position of an anticodon in tRNA to form hydrogen bonds in various ways, causing alignment with several different possible codons.

X : A ratio The ratio between the X chromosome and the number of sets of autosomes.

X chromosome inactivation The process by which the genes of an X chromosome in a mammal can be completely repressed as part of the dosage compensation mechanism (*see* **dosage compensation** and **Barr body**).

X hyperactivation In *Drosophila,* the process by which the structural genes of the male X chromosome are transcribed at the same rate as the two X chromosomes of the female combined.

X linkage The inheritance pattern of genes found on the X chromosome but not on the Y.

X-and-Y linkage The inheritance pattern of genes found on both the X and Y chromosomes (rare).

xeroderma pigmentosum (XP) A human genetic disease syndrome due to mutation in one of several genes encoding products contributing to excision repair.

XP *See* **xeroderma pigmentosum.**

X-ray crystallography A technique for deducing molecular structure by aiming a beam of X rays at a crystal of the test compound and measuring the scatter of rays.

Y linkage The inheritance pattern of genes found on the Y chromosome but not on the X (rare).

YAC *See* **yeast artificial chromosome.**

yeast artificial chromosome (YAC) A cloning vector system in *Saccharomyces cerevisiae* employing yeast centromere and replication sequences.

yeast two-hybrid system *See* **two-hybrid system.**

zygote The cell formed by the fusion of an egg and a sperm; the unique diploid cell that will divide mitotically to create a differentiated diploid organism.

zygotic induction The sudden release of a lysogenic phage from an Hfr chromosome when the prophage enters the F^2 cell, and the subsequent lysis of the recipient cell.

zygotically acting gene A gene whose product is expressed only in the zygote and not included in the maternal contribution to the oocyte.

zymogen The inactive precursor form of an enzyme. Zymogens are typically activated by proteolytic cleavage.

Further Readings

STUDENTS INTERESTED in pursuing genetics further should start reading original research articles in scientific journals. Some important journals are *Cell, Chromosoma, Current Genetics, Evolution, Gene, Genetic Research, Genetics, Genomics, Heredity, Human Genetics, Journal of Medical Genetics, Journal of Molecular Biology, Molecular and General Genetics, Mutation Research, Nature, Plasmid, Proceedings of the National Academy of Sciences of the United States of America,* and *Science.* Useful review articles may be found in the review sections of the major journals listed above, as well as in special review-only journals such as *Trends in Genetics, Nature Reviews: Genetics, Current Opinion in Genetics and Development, Bioessays,* and *Current Biology.* In addition, there are many useful sites dealing with genetics on the Internet. Some particularly useful references, mostly general reviews, are listed below under the chapters to which they relate.

Chapter 1

Carlson, E. A. 1996. *The Gene: A Critical History.* Philadelphia: Saunders. A readable history of genetics.

Clausen, J., D. D. Keck, and W. W. Hiesey. 1940. *Experimental Studies on the Nature of Species.* Vol. 1: *The Effect of Varied Environments on Western North American Plants.* Carnegie Institute of Washington, Publ. No. 520, 1–452. This publication and the following one by the same authors are the classic studies of norms of reaction of plants from natural populations.

Clausen, J., D. D. Keck, and W. W. Hiesey. 1958. *Experimental Studies on the Nature of Species.* Vol. 3: *Environmental Responses of Climatic Races of* Achiella. Carnegie Institute of Washington, Publ. No. 581, 1–129.

Horgan, J. 1993. "Eugenics Revisited." *Scientific American* (June). A discussion of the selective regulation of human reproduction.

Milunsky, A., and G. J. Annas, eds. 1975. *Genetics and the Law.* New York: Plenum Press. Interesting accounts of the ramifications of genetics in the lives of individuals.

Moore, J. A. 1985. *Science as a Way of Knowing—Genetics. American Zoologist* 25:1–165. One short book from an excellent series that focuses on the modus operandi of science.

Schmalhausen, I. I. 1949. *Factors of Evolution: The Theory of Stabilizing Selection.* Philadelphia: Blakiston. The most general discussion of the relation of genotype and environment in the formation of phenotypic variation.

Silver, L. 1998. *Remaking Eden: Cloning and Beyond in a Brave New World.* New York: Avon Books. A popular book on the effects of genetic technology, by one of the leading experts in mammalian genetics.

Sturtevant, A. H. 1965. *A History of Genetics.* New York: Harper & Row. Another useful historical test.

Chapter 2

Brown, S. W. 1966. "Heterochromatin." *Science* 151: 417–425. A nice review of the classic cytological observations.

Chambon, P. 1981. "Split Genes." *Scientific American* (May). The discovery of intervening sequences is described.

Dickerson, R. E. 1983. "The DNA Helix and How It Is Read." *Scientific American* (December). An article with some beautiful color models of DNA structures.

Dupraw, E. J. 1970. *DNA and Chromosomes.* New York: Holt, Rinehart & Winston. A useful book on chromosome substructure, containing excellent photographs by the author.

Felsenfeld, G. 1985. "DNA." *Scientific American* (October).

Greider, C. W. 1990. "Telomeres, Telomerase, and Senescence." *BioEssays* 12: 363–369. A good review written for nonexperts.

Lewin, B. 1994. *Genes V.* Cambridge, MA: Cell Press. A large comprehensive text on gene structure and function and on genome structure.

Lodish, H., et al. 2000. *Molecular Cell Biology,* 4th ed. New York: Scientific American Books. A useful book that does a good job of translating many of the concepts of genetics into cellular processes, including a particularly strong section on chromosome structure and types of repetitive DNA.

Petrov, D. A. 2001. "Evolution of Genome Size: New Approaches to an Old Problem." *Trends in Genetics* 17: 23–28. A discussion of the causes of variation in genome size in the light of new sequence data.

Rick, C. M. 1978. "The Tomato." *Scientific American* (August). Includes an account of tomato genes and chromosomes and their role in breeding.

Venter, J. C., et al. 2001. "The Sequence of the Human Genome." *Science* 291: 1304–1351. The first view of the human sequence, showing maps of genes and their structures and functions. Long and technical but quite readable and worth a look.

Watson, J. D. 1981. *The Double Helix.* New York: Atheneum. An enjoyable personal account of Watson and Crick's discovery, including the human dramas.

Chapter 3

Chambon, P. 1981. "Split Genes." *Scientific American* (May). A popular article on introns and exons.

Crick, F. H. C. 1962. "The Genetic Code." *Scientific American* (October). This article and the following one are popular accounts of code-cracking experiments.

Crick, F. H. C. 1966. "The Genetic Code: III." *Scientific American* (October).

Darnell, J. E., Jr. 1985. "RNA." *Scientific American* (October).

Doolittle, R. F. 1985. "Proteins." *Scientific American* (October).

Dressler, D., and H. Potter. 1991. *Discovering Enzymes.* New York: Scientific American Library. A delightful book that describes the discovery of enzymes and shows many of their features in well-planned easy-to-read illustrations.

Lake, J. A. 1981. "The Ribosome." *Scientific American* (August). Three-dimensional model of the ribosome.

Lane, C. 1976. "Rabbit Hemoglobin from Frog Eggs." *Scientific American* (August). This article describes experiments that illustrate the universality of the genetic system.

Lawn, R. M., and G. A. Vehar. 1986. "The Molecular Genetics of Hemophilia." *Scientific American* (March).

Miller, O. L. 1973. "The Visualization of Genes in Action." *Scientific American* (March). A discussion of electron microscopy of transcription and translation.

Moore, P. B. 1976. "Neutron-Scattering Studies of the Ribosome." *Scientific American* (October). This article gives the details of ribosome substructure.

Nirenberg, M. W. 1963. "The Genetic Code: II." *Scientific American* (March). Another account of early code-cracking experiments.

Radman, M., and R. Wagner. 1988. "The High Fidelity of DNA Duplication." *Scientific American* (August).

Rich, A., and S. H. Kim. 1978. "The Three-Dimensional Structure of Transfer RNA." *Scientific American* (January). A presentation of the experimental evidence for the structure described in this chapter.

Steitz, J. A. 1988. "Snurps." *Scientific American* (June). The mechanism of intron splicing.

Watson, J. D., et al. 1987. *The Molecular Biology of the Gene,* 4th ed. Menlo Park, CA: Benjamin/Cummings. A superb development of the subject, written in a highly readable style and well illustrated.

Weinberg, R. A. 1985. "The Molecules of Life." *Scientific American* (October).

Yanofsky, C. 1967. "Gene Structure and Protein Structure." *Scientific American* (May). This article gives the details of colinearity at the molecular level.

Chapter 4

Greider, C. W., and E. H. Blackburn. 1996. "Telomeres, Telomerase and Cancer." *Scientific American* (February).

Kornberg, A., and T. Baker. 1991. *DNA Replication,* 2d ed. New York: W. H. Freeman and Company.

McIntosh, J. R., and K. L. McDonald. 1988. "The Mitotic Spindle." *Scientific American* (October).

McLeish, J., and B. Snoad. 1958. *Looking at Chromosomes.* New York: Macmillan. A short classic book consisting of many superb photographs of mitosis and meiosis.

Radman, M., and R. Wagner. 1988. "The High Fidelity of DNA Duplication." *Scientific American* (August).

Wang, J. C. 1982. "DNA Topoisomerases." *Scientific American* (July). Diagrams different topological forms of DNA.

Chapter 5

Crow, J. F. 1983. *Genetics Notes,* 8th ed. New York: Macmillan. A short and concise review of genetics from Mendel to populations.

Finchman, J. R. S., P. R. Day, and A. Radford. 1979. *Fungal Genetics,* 3d ed. London: Blackwell. A large, standard technical work. Good for tetrad analysis.

Grivell, L. A. 1983. "Mitochondrial DNA." *Scientific American* (March). An excellent and up-to-date review including a discussion of the unique mitochondrial translation system and of splicing of mitochondrial introns.

Hutt, F. B. 1964. *Animal Genetics.* New York: Ronald Press. A standard text on the subject with many interesting examples.

Jegalian, K., and B. T. Lahn. 2001. "Why the Y Chromosome Is So Weird." *Scientific American* (February). A discussion of the genes of the Y chromosome, and where they originated, in the light of the full DNA sequence.

Jennings, P. R. 1976. "The Amplification of Agricultural Production." *Scientific American* (September). A discussion of genetics and the green revolution.

Jobling, M. A., and Tyler-Smith, C. 2000. "New Uses for New Haplotypes: The Human Y Chromosome, Disease, and Selection." *Trends In Genetics* 16: 356–361. An analysis of variation in the sequence of the Y chromosome.

Mange, A. P., and E. J. Mange. 1990. *Genetics: Human Aspects,* 2d ed. Sunderland, MA: Sinauer. A useful and up-to-date text for a course in human genetics.

McKusick, V. A., et al. 1990. *Mendelian Inheritance in Man,* 9th ed. Baltimore and London: Johns Hopkins Press. The "bible" of medical genetics; a 2000-page compendium of all the known human inherited disorders, both autosomal and X-linked. Also accessible on the Internet.

McLaren, A. 1988. "Sex Determination in Mammals." *Trends in Genetics* 4:153–157.

Olby, R. C. 1966. *Origins of Mendelism.* London: Constable. An enjoyable account of Mendel's work and the intellectual climate of his time.

Robertson, S. C., J. A. Tynan, and D. J. Donoghue. 2000. "Receptor Tyrosine Kinases and Human Syndromes." *Trends In Genetics* 16: 2565–2571. Important cellular signaling enzymes, and the disorders resulting from mutations in their genes. Included are conditions described in this book, for example, piebald spotting and dwarfism.

Rousseau, F., et al. 1994. "Mutations in the Gene Encoding Fibroblast Growth Factor Receptor in Achondroplasia." *Nature* 371: 253–254. A short research paper describing the molecular identification of the abnormal allele that causes achondroplasia.

Sager, R. 1965. "Genes Outside Chromosomes." *Scientific American* (January). A popular description of early *Chlamydomonas* experiments.

Sager, R. 1972. *Cytoplasmic Genes and Organelles.* New York: Academic Press.

Stern, C., and E. R. Sherwood. 1966. *The Origin of Genetics: A Mendel Source Book.* New York: W. H. Freeman and Company. A short collection of important early papers, including Mendel's papers and correspondence.

Wallace, D. C. 1992. "Mitochondrial Genetics: A Paradigm for Aging and Degenerative Diseases?" *Science* 256: 628–632. A review article on the various types of mtDNA mutations in humans and their disease outcomes.

Chapter 6

Alberts, B., D. Bray, J. Lewis, M. Raff, K. Roberts, and J. D. Watson. 1994. *Molecular Biology of the Cell,* 3d ed. New York: Garland. Includes an excellent description of recombination at the molecular level, with nice figures.

O'Brien, S. J., ed. 1984. *Genetic Maps.* Cold Spring Harbor, NY: Cold Spring Harbor Laboratory Press. A compendium of the detailed maps of 80 well-analyzed organisms.

Peters, J. A., ed. 1959. *Classic Papers in Genetics.* Englewood Cliffs, NJ: Prentice-Hall. A collection of important papers in the history of genetics.

Stahl, F. W. 1979. *Genetic Recombination: Thinking about It in Phage and Fungi.* New York: W. H. Freeman and Company. A rather technical short book on recombination models.

Stahl, F. W. 1994. "The Holliday Junction on Its Thirtieth Anniversary." *Genetics* 138:241–246. An easy-to-read review of the Holliday model and subsequent variations that are used today to explain recombination.

White, R., et al. 1985. "Construction of Linkage Maps with DNA Markers for Human Chromosomes." *Nature* 313: 101–104. An extension of the techniques of this chapter to DNA markers.

Whitehouse, H. L. K. 1973. *Towards an Understanding of the Mechanism of Heredity,* 3d ed. London: Edward Arnold. An excellent general introduction to genetics stressing the historical approach and the pivotal experiments. It includes a good section on recombination models.

Chapter 7

Adelberg, E. A. 1966. *Papers on Bacterial Genetics.* Boston: Little, Brown.

Birge, E. A. 1981. *Bacterial and Bacteriophage Genetics.* New York: Springer-Verlag.

Brock, T. D. 1990. *The Emergence of Bacterial Genetics.* Cold Spring Harbor, NY: Cold Spring Harbor Laboratory Press. The definitive treatise on the beginnings and development of the field of bacterial genetics.

Donachie, W. D. 1993. "The Cell Cycle of *Escherichia coli.*" *Annual Review of Microbiology* 47: 199–230.

Hayes, W. 1968. *The Genetics of Bacteria and Their Viruses,* 2d ed. New York: Wiley. The standard and classic text, written by a pioneer in the subject.

Holloway, B. W. 1993. "Genetics for All Bacteria." *Annual Review of Microbiology* 47: 659–684. A modern review.

Lewin, B. 1977. *Gene Expression.* Vol. 1: *Bacterial Genomes.* New York: Wiley. An excellent set of volumes, all of which are relevant to various sections of this text.

Lewin, B. 1977. *Gene Expression.* Vol. 3: *Plasmids and Phages.* New York: Wiley.

Miller, J. H. 1992. *A Short Course in Bacterial Genetics.* Cold Spring Harbor, NY: Cold Spring Harbor Laboratory Press. A laboratory manual with extensive historical introductions to each area of bacterial genetics.

Nicolau, K. C., and C. N. Boddy. 2001. "Behind Enemy Lines." *Scientific American* (May). A discussion of bacterial drug resistance.

Stent, G. S., and R. Calendar. 1978. *Molecular Genetics,* 2d ed. New York: W. H. Freeman and Company. A lucidly written account of the development of our present understanding of the subject, based mainly on experiments in bacteria and phage.

Chapter 8

Capecchi, M. R. 1994. "Targeted Gene Replacement." *Scientific American* (March). A popular article describing details of the production of knockout mice by gene replacement.

Estruch, J. J. 1997. "Transgenic Plants: An Emerging Approach to Pest Control." *Nature Biotechnology* 15: 137–141. A recent account of the use of Bt toxin in transgenic plants.

Fernández-Cañón, J. M., et al. 1996. "The Molecular Basis of Alkaptonuria." *Nature Genetics* 14: 19–24. A description of the cloning of the *aku* gene.

Mullis, K. B. 1990. "The Unusual Origin of the Polymerase Chain Reaction." *Scientific American* (April). An account by the discoverer of the PCR.

Rommens, J. M., et al. 1989. "Identification of Cystic Fibrosis Gene by Chromosome Walking and Jumping." *Science* 245: 1059–1065.

Scriver, C. R. 1996. "Alkaptonuria: Such a Long Journey." *Nature Genetics* 14: 5–6. An interesting commentary on the genetics of this inborn error of metabolism.

Sinsheimer, R. L. 1977. "Recombinant DNA." *Annual Review of Biochemistry* 46: 415–438. A provocative article by a leading molecular biologist who expressed concern about potential hazards of DNA manipulation.

Somia, N., and I. M. Verma. 2000. "Gene Therapy: Trials and Tribulations." *Nature Reviews Genetics* 1: 91–99.

Watson, J. D., M. Gilman, J. Witkowski, and M. Zoller. 1992. *Recombinant DNA,* 2d ed. New York: W. H. Freeman and Company. A well-written account of recombinant DNA and its applications, with many wonderful illustrations.

White, R., and J. M. Lalouel. 1988. "Chromosome Mapping with DNA Markers." *Scientific American* (February).

Wolfenbarger, L. L., and P. R. Phifer. 2000. "The Ecological Risks and Benefits of Genetically Engineered Plants." *Science* 290: 2088–2093.

Chapter 9

Brenner, S. E. 2001. "A Tour of Structural Genomics." *Nature Reviews: Genetics* 2: 801–809.

Cronk, Q. C. B. 2001. "Plant Evolution and Development in a Post-Genomic Context." *Nature Reviews: Genetics* 2: 607–619.

Fleischmann, M. D., et al. 1995. "Whole-Genome Random Sequencing and Assembly of *Haemophilus influenzae* Rd." *Science* 269: 496–512. The article reports the first full genomic sequence for a free-living organism.

Goffeau, A., et al. 1996. "Life with 6000 Genes." *Science* 274: 546–567. An analysis of the complete genomic sequence of yeast.

Green, E. D. 2001. "Strategies for the Systematic Sequencing of Complex Genomes." *Nature Reviews: Genetics* 2: 573–583.

Green, E. D., and M. V. Olson. 1990. "Chromosomal Region of the Cystic Fibrosis Gene in Yeast Artificial Chromosomes: A Model for Human Genome Mapping." *Science* 250: 94–98. An easy-to-understand article on the principles of genome mapping with the use of YACs.

Hudson, T. J., et al. 1995. "An STS-Based Map of the Human Genome." *Science* 270: 1945–1954. A technical description of the use of STSs.

Jeffreys, A. J., V. Wison, and S. L. Thein. 1985. "Hypervariable 'Minisatellite' Regions in Human DNA." *Nature* 314: 66–73. One of the first papers on the subject of DNA fingerprinting.

Kim, S. K. 2001. "http://C.elegans: Mining the Functional Genomics Landscape." *Nature Reviews: Genetics* 2: 681–690.

Kumar, A., and M. Snyder. 2001. "Emerging Technologies in Yeast Genomics." *Nature Reviews: Genetics* 2: 302–312.

Nature, Vol. 409, No. 6822, 2001. "The Human Genome." A landmark issue with many papers devoted to an analysis of the compiled draft sequence of the human genome.

Prak, E. T., and, H. H. Kazazian Jr. 2000. "Mobile Elements and the Human Genome." *Nature Reviews: Genetics* 1: 134–144.

Quackenbush, J. 2001. "Computational Analysis of Microarray Data." *Nature Reviews: Genetics* 2: 418–427.

Ramsay, G. 1998. "DNA Chips: State of the Art." *Nature Biotechnology* 16: 40–44.

Schuler, G. D., et al. 1996. "A Gene Map of the Human Genome." *Science* 274: 540–546. An account of the use of ESTs.

Science, Vol. 287, No. 5461, 2000. "The *Drosophila* Genome." A landmark issue with many papers devoted to an analysis of the first eukaryotic genome sequenced by whole-genome shotgun sequence assembly.

Stein, L. 2001. "Genome Annotation: From Sequence to Biology." *Nature Reviews: Genetics* 2: 493–503.

Trends in Genetics, Vol. 11, No. 12, 1995. The entire issue deals with the analysis of complex traits and quantitative traits by using molecular markers. Quite readable.

Venter, J. C., et al. 1998. "Shotgun Sequencing of the Human Genome." *Science* 280: 1540–1542. Discussion of the whole-genome shotgun strategy for large-scale sequencing.

Walter, M., et al. 1994. "A Method for Constructing Radiation Hybrid Maps of Whole Genomes." *Nature Genetics* 7: 22–28. An easy-to-understand research paper describing the technique of radiation hybrid mapping.

White, K. P. 2001. "Functional Genomics and the Study of Development, Variation, and Evolution." *Nature Reviews: Genetics* 2: 528–537.

Chapter 10

Auerbach, C. 1976. *Mutation Research.* London: Chapman & Hall. Standard text by a pioneer researcher.

Caldecott, K. W. 2001. "Mammalian DNA Single-Strand Break Repair: an X-Ra(y)ted Affair." *Bioessays* 23: 447–455.

Dianov, G. L., P. O'Neill, and D. T. Goodhead. 2001. "Securing Genome Stability by Orchestrating DNA Repair: Removal of Radiation-Induced Clustered Lesions in DNA." *Bioessays* 23: 745–749.

Drake, J. W. 1970. *The Molecule Basis of Mutation.* San Francisco: Holden-Day. One of the few standard texts on the subject.

Eisenstadt, E. 1987. "Analysis of Mutagenesis in *Escherichia coli* and *Salmonella typhimurium.*" *Cellular and Molecular Biology,* F. C. Neidhardt, ed. Washington, DC: American Society for Microbiology. A comprehensive review of studies of mutagenesis in bacteria.

Fedoroff, N. V. 1984. "Transposable Genetic Elements in Maize." *Scientific American* (June). An account of the early experiments in maize, with later results on the molecular basis of transposition.

Friedberg, E. C., G. C. Walker, and W. Siede. 1995. *DNA Repair and Mutagenesis.* Washington, DC: American Society for Microbiology. A complete review text covering mutagenesis and repair in both bacteria and higher cells. An excellent source on human genetic diseases and cancer in relation to repair genes.

Jiricny, J. 2000. "Mismatch Repair: The Praying Hands of Fidelity." *Current Biology* 10: 788–790.

Kanaar, R., J. H. Hoeijmakers, and D. C. van Gent. 1998. "Molecular Mechanisms of DNA Double-Strand Break Repair." *Trends in Cell Biology* 8: 483–489.

Michelson, R. J., and T. Weinert. 2000. "Closing the Gaps among a Web of DNA Repair Disorders." *Bioessays* 22: 966–969.

Modesti, M, and R. Kanaar. 2001. "DNA Repair: Spot(light)s on Chromatin." *Current Biology* 11: 229–232.

Ninio, J., 2000. "Illusory Defects and Mismatches: Why Must DNA Repair Always Be (Slightly) Error Prone?" *Bioessays* 22: 396–401.

Prak, E. T., and H. H. Kazazian Jr. 2000. "Mobile Elements and the Human Genome." *Nature Reviews: Genetics* 1: 134–144.

Scharer, O. D., and J. Jiricny. 2001. "Recent Progress in the Biology, Chemistry and Structural Biology of DNA Glycosylases." *Bioessays* 23: 270–281.

Schull, W. J., et al. 1981. "Genetic Effect of the Atomic Bombs: A Reappraisal." *Science* 213: 1220–1227. A summary of all the indicators of potential genetic effects of the Hiroshima and Nagasaki explosions, concluding that, "In no instance is there a statistically significant effect of parental exposure; but for all indicators the observed effect is in the direction suggested by the hypothesis that genetic damage resulted from the exposure."

Sutton, M. D., B. T. Smith, V. G. Godoy, and G. C. Walker. 2000. "The SOS Response: Recent Insights into umuDC-Dependent Mutagenesis and DNA Damage Tolerance." *Annual Review of Genetics* 34: 479–497.

van Gent, D. C., J. H. Hoeijmakers, and R. Kanaar. 2001. "Chromosomal Stability and the DNA Double–Stranded Break Connection." *Nature Reviews: Genetics* 2: 196–206.

Chapter 11

deGrouchy, J., and C. Turleau. 1984. *Clinical Atlas of Human Chromosomes.* New York: Wiley. A systemic examination of all the human chromosomes and the aberrations associated with them.

Dellarco, V. L., P. E. Voytek, and A. Hollaender. 1985. *Aneuploidy: Etiology and Mechanisms.* New York: Plenum. A collection of research summaries on aneuploidy in humans and experimental organisms.

Epstein, C. J., et al. 1983. "Recent Developments in Prenatal Diagnosis of Genetic Diseases and Birth Defects." *Annual Review of Genetics* 17: 49–83. Includes amniocentesis.

Feldman, M. G., and E. R. Sears. 1981. "The Wild Gene Resources of Wheat." *Scientific American* (January). A general discussion of the genomes of wheat and its relatives and how new genes can be introduced.

Friedmann, T. 1971. "Prenatal Diagnosis of Genetic Disease." *Scientific American* (November). An early article on amniocentesis and its uses.

Fuchs, F. 1980. "Genetic Amniocentesis." *Scientific American* (August).

German, J., ed. 1974. *Chromosomes and Cancer.* New York: Wiley. A large technical work, but readable, describing the relation of chromosome changes and cancer.

Hassold, T., and Hunt, P., 2001 "To Err (Meiotically) Is Human: The Genesis of Human Aneuploidy". *National Review of Genetics* 2(4) (April): 280–291.

Hassold, T. J., and P. A. Jacobs. 1984. "Trisomy in Man." *Annual Review of Genetics* 18: 69–98. A comprehensive summary of trisomy, including a discussion of the maternal age effect.

Hulse, J. H., and D. Spurgeon. 1984. "Triticale." *Scientific American* (August). An account of the development and possible benefits of this wheat-rye amphidiploid.

Lawrence, W. J. C. 1968. *Plant Breeding.* London: Edward Arnold (*Studies in Biology,* No. 12). A short introduction to the subject.

Mangelsdof, P. C. 1986. "The Origins of Corn." *Scientific American* (August).

Maniatis, T. E., et al. 1980. "The Molecular Genetics of Humam Hemoglobins." *Annual Review of Genetics* 14: 145–178. A useful summary, which could be profitably read at this point in the course or after reading the material on molecular genetics.

Mills, A. A., and A. Bradley. 2001. "From Mouse to Man: Generating Megabase Chromosome Rearrangements." *Trends In Genetics* 17: 331–338. A rather technical but readable account of synthesizing large chromosome mutations.

Patterson, D. 1987. "The Causes of Down Syndrome." *Scientific American* (August).

Reeves, R. H., L. L. Baxter, and J. T. Richtsmeier. 2001. "Too Much of a Good Thing: Mechanisms of Gene Action in Down Syndrome." *Trends In Genetics* 17: 83–88. Possible ways in which extra gene copies result in Down syndrome.

Shepherd, J. F. 1982. "The Regeneration of Potato Plants from Protoplasts." *Scientific American* (May). A review by one of the leaders in this field.

Swanson, C. P., T. Mertz, and W. J. Young. 1967. *Cytogenetics.* Englewood Cliffs, N.J.: Prentice-Hall.

Chapter 12

Barstead, R. 2001. "Genome-Wide RNAi." *Current Opinions in Chemical Biology* 5: 63–66.

Brown, S. D., and R. Balling. 2001. "Systematic Approaches to Mouse Mutagenesis." *Current Opinion in Genetics and Development* 11: 268–273.

Forsburg, S. L. 2001. "The Art and Design of Genetic Screens: Yeast." *Nature Reviews: Genetics* 2: 659–668.

Hammond, S. M, A. A. Caudy, and G. J. Hannon. 2001. "Post-Transcriptional Gene Silencing by Double-Stranded RNA." *Nature Reviews: Genetics* 2: 110–119

Hartwell, L. H, P. Szankasi, C. J. Roberts, A. W. Murray, and S. H. Friend. 1997. "Integrating Genetic Approaches into the Discovery of Anticancer Drugs." *Science* 278: 1064–1068. A forward-looking discussion of the role of genetic analysis in drug screening. The lead author is one of the recipients of the 2001 Nobel Prize in physiology and medicine for his work on the cell cycle (see Chapter 15).

Justice, M. J. 2000. "Capitalizing on Large-Scale Mouse Mutagenesis Screens." *Nature Reviews: Genetics* 1: 109–115.

Kumar, A., and M. Snyder. 2001. "Emerging Technologies in Yeast Genomics." *Nature Reviews: Genetics* 2: 302–312.

Lekven, A. C, K. A. Helde, C. J. Thorpe, R. Rooke, and R. T. Moon. 2000. "Reverse Genetics in Zebrafish." *Physiological Genomics* 2: 37–48.

Lewandoski, M. 2001. "Conditional Control of Gene Expression in the Mouse." *Nature Reviews: Genetics* 2: 743–755.

Lindsay, E. A. 2001. "Chromosome Microdeletions: Dissecting del22q11 Syndrome." *Nature Reviews: Genetics* 2: 858–868.

Malicki, J. 2000. "Harnessing the Power of Forward Genetics—Analysis of Neuronal Diversity and Patterning in the Zebrafish Retina." *Trends in Neurosciences* 11: 531–541.

Nadeau, J. H. 2001. "Modifier Genes in Mice and Humans." *Nature Reviews: Genetics* 2: 165–174.

Parinov, S., and V. Sundaresan. 2000. "Functional Genomics in *Arabidopsis:* Large-Scale Insertional Mutagenesis Complements the Genome-Sequencing Project." *Current Opinion in Biotechnology* 11: 157–161.

Sharp, P. A. 2001. "RNA Interference." *Genes and Development* 15: 485–490.

Sijen, T., and J. M. Kooter. 2000. "Post-Transcriptional Gene-Silencing: RNAs on the Attack or on the Defense?" *Bioessays* 22: 520–531.

Sokolowski, M. B. 2001. "*Drosophila:* Genetics Meets Behaviour." *Nature Reviews: Genetics* 2: 878–890.

Stanford, W. L., J. B. Cohn, and S. P. Cordes. 2001. "Gene-Trap Mutagenesis: Past, Present, and Beyond." *Nature Reviews: Genetics* 2: 756–768.

Stockwell, B. R. 2000. "Chemical Genetics: Ligand-Based Discovery of Gene Function". *Nature Reviews: Genetics* 1: 116–125.

Chapter 13

Buratowski, S. 2000. "Snapshots of RNA Polymerase II Transcription Initiation." *Current Opinions in Cell Biology* 12: 320–325.

Dillon, N., and P. Sabbattini. 2000. "Functional Gene Expression Domains: Defining the Functional Unit of Eukaryotic Gene Regulation." *Bioessays* 22: 657–665.

Dvir, A., J. W. Conaway, and R. C. Conaway. 2001. "Mechanism of Transcription Initiation and Promoter Escape by RNA Polymerase II." *Current Opinion in Genetics and Development* 11: 209–214.

Fiedler, U., and H. T. Marc Timmers. 2000. "Peeling by Binding or Twisting by Cranking: Models for Promoter Opening and Transcription Initiation by RNA Polymerase II." *Bioessays* 22: 316–326.

Jacob, F., and J. Monod. 1961. "Genetic Regulatory Mechanisms in the Synthesis of Proteins." *Journal of Molecular Biology* 3: 318–356. A classic paper setting forth the elements of an operon and the experimental evidence.

Kelley, R. L, and M. I. Kuroda. 2000. "The Role of Chromosomal RNAs in Marking the X for Dosage Compensation." *Current Opinion in Genetics and Development* 10: 555–561.

Malik, S., and R. G. Roeder. 2000. "Transcriptional Regulation through Mediator-Like Coactivators in Yeast and Metazoan Cells." *Trends in Biochemical Sciences* 25: 277–283.

Maniatis, T., and M. Ptashne. 1976. "A DNA Operator-Repressor System." *Scientific American* (January). This article discusses the molecular structures of the components of the *lac* operon.

Mann, M. R, and M. S. Bartolomei. 1999. "Towards a Molecular Understanding of Prader-Willi and Angelman Syndromes." *Human Molecular Genetics* 8: 1867–1873.

Nicholls, R. D., S. Saitoh, and B. Horsthemke. 1998. "Imprinting in Prader-Willi and Angelman Syndromes." *Trends in Genetics* 14: 194–200.

Pennacchio, L. A., and E. M. Rubin. 2001. "Genomic Strategies to Identify Mammalian Regulatory Sequences." *Nature Reviews: Genetics* 2: 100–109.

Ptashne, M. 1987. *A Genetic Switch: Gene Control and Phage λ.* Cambridge, MA: Cell Press. A detailed account of regulation in this *E. coli* phage.

Ptashne, M. 1989. "How Gene Activators Work." *Scientific American* (January).

Ptashne, M., and W. Gilbert. 1970. "Genetic Repressors." *Scientific American* (June). The exciting story of how repressors were identified and purified, thereby confirming the predictions of Jacob and Monod.

Rachez, C., and L. P. Freedman. 2001. "Mediator Complexes and Transcription." *Current Opinions in Cell Biology* 13: 274–280.

Rideout, W. M. 3rd, K. Eggan, and R. Jaenisch. 2001. "Nuclear Cloning and Epigenetic Reprogramming of the Genome." *Science* 293: 1093–1098.

Tijan, R. 1995. "Molecular Machines That Control Genes." *Scientific American* (February). An up-to-date review of eukaryotic gene transcription regulation.

Chapter 14

Bodmer, W. F., and L. L. Cavalli-Sforza. 1976. *Genetics, Evolution, and Man.* New York: W. H. Freeman and Company. A very readable, well-illustrated book, including a clear account of HLA genetics.

Griffiths, A. J. F., and F. R. Ganders. 1984. *Wildflower Genetics.* Vancouver: Flight Press. A field guide to plant variation in natural populations and its genetic basis, including examples relevant to this chapter.

Griffiths, A. J. F., and J. McPherson. 1989. *One Hundred Principles of Genetics.* New York: W. H. Freeman and Company. A short book that distills genetics to its basic key concepts; useful for rapid review of the whole subject.

Hutt, W. B. 1979. *Genetics for Dog Breeders.* New York: W. H. Freeman and Company. A short book that will make genetics immediately relevant to all dog owners and breeders.

Scriver, C. R., and P. J. Waters. 1999. "Monogenic Traits Are Not Simple: Lessons from PKU." *Trends In Genetics* 15: 267–272. A complete discussion of the molecular action of the genes involved in the pathways leading to the disease PKU.

Searle, A. G. 1968. *Comparative Genetics of Coat Color in Mammals.* New York: Academic Press. A classic treatment of the subject with many examples relevant to this and other chapters.

Silvers, W. K. 1979. *The Coat Colors of Mice.* New York: Springer-Verlag. A standard handbook on the subject, including many examples of gene interaction.

Todd, N. B. 1977. "Cats and Commerce." *Scientific American* (November). Includes some genetics of domestic cat coat colors and the use of this information to study cat migration throughout history.

Wright, M., and S. Walters, eds. 1981. *The Book of the Cat.* New York: Summitt Books. A fascinating book that has an excellent chapter on gene interaction in determining the coat colors of domestic cats.

Chapter 15

Clarke, D. J., and J. F. Gimenez-Abian. 2000. "Checkpoints Controlling Mitosis." *Bioessays* 22: 351–363.

Daujat, S., H. Neel, and J. Piette. 2001. "MDM2: Life without p53." *Trends in Genetics* 17: 459–464.

Ekholm, S. V., and S. I. Reed. 2000. "Regulation of G_1 Cyclin-Dependent Kinases in the Mammalian Cell Cycle." *Current Opinion in Cell Biology* 12: 676–684.

Evan, G. I., and K. H. Vousden. 2001. "Proliferation, Cell Cycle, and Apoptosis in Cancer." *Nature* 411: 342–348.

Fishel, R., and R. D. Kolodner. 1995. "Identification of Mismatch Repair Genes and Their Role in the Development of Cancer." *Current Opinions in Genetics and Development* 5: 382–396.

Futcher, B. 2000. "Microarrays and Cell Cycle Transcription in Yeast." *Current Opinion in Cell Biology* 12: 710–715.

Hartwell, L. H., and M. B. Kastan. 1994. "Cell Cycle Control and Cancer." *Science* 266: 1821–1828. A thoughtful review of this field. The lead author is one of the recipients of the 2001 Nobel Prize in physiology and medicine for his work on the cell cycle.

Jiricny, J., and M. Nystrom-Lahti. 2000. "Mismatch Repair Defects in Cancer." *Current Opinion in Genetics and Development* 10: 157–161.

Kaelin, W. G. Jr. 1999. "Functions of the Retinoblastoma Protein." *Bioessays* 21: 950–958.

Karin, M., and T. Hunter. 1995. "Transcriptional Control by Protein Phosphorylation: Signal Transmission from the Cell Surface to the Nucleus." *Current Biology* 5: 747–757.

Kaufmann, S. H., and G. J. Gores. 2000. "Apoptosis in Cancer: Cause and Cure." *Bioessays* 22: 1007–1017.

Kinzler, K. W., and B. Vogelstein. 1996. "Lessons from Hereditary Colorectal Cancer." *Cell* 87: 159–170.

Knudson, A. G. 2000. "Chasing the Cancer Demon." *Annual Review of Genetics* 34:1–19.

Lengauer, C., K. W. Kinzler, and B. Vogelstein. 1998. "Genetic Instabilities in Human Cancers." *Nature* 396: 643–649.

Liotta, L., and E. Petricoin. 2000. "Molecular Profiling of Human Cancer." *Nature Reviews: Genetics* 1: 48–56.

Nurse, P. 2000. "A Long Twentieth Century of the Cell Cycle and Beyond." *Cell* 100: 71–78. A prospective and retrospective by one of the recipients of the 2001 Nobel Prize in physiology and medicine.

Paulovich, A. G., D. P. Toczyski, and L. H. Hartwell. 1997. "When Checkpoints Fail." *Cell* 88: 315–321.

Ryan, K. M., A. C. Phillips, and K. H. Vousden. 2001. "Regulation and Function of the p53 Tumor Suppressor Protein." *Current Opinion in Cell Biology* 13: 332–337.

Sherr, C. J., and J. D. Weber. 2000. "The ARF/p53 Pathway." *Current Opinion in Genetics and Development* 10: 94–99.

van Gent, D. C., J. H. Hoeijmakers, and R. Kanaar. 2001. "Chromosomal Stability and the DNA Double-Stranded Break Connection." *Nature Reviews: Genetics* 2: 196–206.

Walworth, N. C. 2000. "Cell-Cycle Checkpoint Kinases: Checking in on the Cell Cycle." *Current Opinion in Cell Biology* 12: 697–704.

Welcsh, P. L., and M. C. King. 2001. "BRCA1 and BRCA2 and the Genetics of Breast and Ovarian Cancer." *Human Molecular Genetics* 10: 705–713.

What You Need to Know about Cancer. 1997. A special issue of *Scientific American.* New York: W. H. Freeman and Company. An excellent, easy-to-read source on all aspects of cancer, including genetics.

Chapter 16

Affolter, M., and B. Z. Shilo. 2000. "Genetic Control of Branching Morphogenesis during *Drosophila* Tracheal Development." *Current Opinion in Cell Biology* 12: 731–735. An example of the role of cell-cell communication during organogenesis.

Avner, P., and E. Heard. 2001. "X-Chromosome Inactivation: Counting, Choice, and Initiation." *Nature Reviews: Genetics* 2: 59–67.

Cronk, Q. C. B. 2001. "Plant Evolution and Development in a Post-Genomic Context." *Nature Reviews: Genetics* 2: 607–619.

Curtis, D., R. Lehmann, and P. D. Zamore. 1995. "Translational Regulation in Development." *Cell* 81: 171–178.

Davidson, D., and R. Baldock. 2001. "Bioinformatics beyond Sequence: Mapping Gene Function in the Embryo." *Nature Reviews: Genetics* 2: 409–417.

DiNardo, S., J. Heemskerk, S. Dougan, and P. H. O'Farrell. 1994. "The Making of a Maggot: Patterning the *Drosophila* Embryonic Epidermis." *Current Opinion in Genetics and Development* 4: 529–534.

Ferrier, D. E., and P. W. Holland. 2001. "Ancient Origin of the Hox Gene Cluster." *Nature Reviews: Genetics* 2: 33–38.

Hülskamp, M., and D. Tautz. 1991. "Gap Genes and Gradients: The Logic behind the Gaps." *Bioessays* 13: 261–268.

Kelley, R. L., and M. I. Kuroda. 2000. "The role of Chromosomal RNAs in Marking the X for Dosage Compensation." *Current Opinion in Genetics and Development* 10: 555–561.

Kenyon, C. 1995. "A Perfect Vulva Every Time: Gradients and Signaling Cascades in *C. elegans.*" *Cell* 82: 171–174.

Kimbrell, D. A., and B. Beutler. 2001. "The Evolution and Genetics of Innate Immunity." *Nature Reviews: Genetics 2: 256–267.*

Kuwabara, P. E., and M. D. Perry. 2001. "It Ain't Over Till It's Ova: Germline Sex Determination in *C. elegans.*" *Bioessays* 23: 596–604.

Lovell-Badge, R. 1992. "Testis Determination: Soft Talk and Kinky Sex." *Current Opinion in Genetics and Development* 2: 596–601.

Lu, C. C., J. Brennan, and E. J. Robertson. 2001. "From Fertilization to Gastrulation: Axis Formation in the Mouse Embryo." *Current Opinion in Genetics and Development* 11: 384–392.

McGinnis, W., and M. Kuziora. 1994. "The Molecular Architecture of Body Design." *Scientific American* (February).

McKeown, M. 1992. "Sex Differentiation: The Role of Alternative Splicing." *Current Opinion in Genetics and Development* 2: 299–304.

Ng, M., and M. F. Yanofsky. 2000. "Three Ways to Learn the ABCs." *Current Opinion in Plant Biology* 3: 47–52. A clear review of the genetic control of flower development.

Ng, M., and M. F. Yanofsky. 2001. "Function and Evolution of the Plant MADS-Box Gene Family." *Nature Reviews: Genetics* 2: 186–195. A discussion of the plant gene family that controls flower development and other events.

Rougvie, A. E. 2001. "Control of Developmental Timing in Animals." *Nature Reviews: Genetics* 2: 690–701.

Saga, Y., and H. Takeda. 2001. "The Making of the Somite: Molecular Events in Vertebrate Segmentation." *Nature Reviews: Genetics* 2: 835–845.

Schier, A. F. 2001. "Axis Formation and Patterning in Zebrafish." *Current Opinion in Genetics and Development* 11: 393–404.

Sommer, R. J. 2000. "Evolution of Nematode Development." *Current Opinion in Genetics and Development* 10: 443–448.

St. Johnston, R. D. 1994. "RNA Localization: Getting to the Top." *Current Opinion in Genetics and Development* 4: 54–56.

Sullivan, W., and W. E. Theurkauf. 1995. "The Cytoskeleton and Morphogenesis of the Early *Drosophila* Embryo." *Current Opinions in Cell Biology* 7: 18–22.

Tabata, T. 2001. "Genetics of Morphogen Gradients." *Nature Reviews: Genetics* 2: 620–630.

Tautz, D. 2000. "Evolution of Transcriptional Regulation." *Current Opinion in Genetics and Development* 10: 575–579.

Tautz, D., and R. J. Sommer. 1995. "Evolution of Segmentation Genes in Insects." *Trends in Genetics* 11: 23–27.

Vaiman, D., and E. Pailhoux. 2000. "Mammalian Sex Reversal and Intersexuality: Deciphering the Sex-Determination Cascade." *Trends in Genetics* 11: 488–494.

Wawersik, S., and R. L. Maas. 2000. "Vertebrate Eye Development as Modeled in *Drosophila.*" *Human Molecular Genetics* 9: 917–925.

White, K. P. 2001. "Functional Genomics and the Study of Development, Variation and Evolution." *Nature Reviews: Genetics* 2: 528–537.

Chapter 17

Bodmer, W. F., and L. L. Cavalli-Sforza. 1976. *Genetics, Evolution, and Man.* New York: W. H. Freeman and Company.

Cavalli-Sforza, L. L., P. Menozzi, and A. Piazza. 1994. *The History and Geography of Human Genes.* Princeton, NJ: Princeton University Press. A complete discussion of human genetic variation, its history, and the forces molding it, with recent and complete tables of human gene frequencies.

Clarke, B. 1975. "The Causes of Biological Diversity." *Scientific American* (August). A popular account emphasizing genetic polymorphism.

Crow, J. F. 1979. "Genes That Violate Mendel's Rules." *Scientific American* (February). An interesting article on a topic called segregation distortion (not treated in this text) and its effects in populations.

Dobzhansky, T. 1951. *Genetics and the Origin of Species.* New York: Columbia University Press. The classic synthesis of population genetics and the processes of evolution. The most influential book on evolution since Darwin's *Origin of Species.*

Ford, E. B. 1971. *Ecological Genetics,* 3d ed. London: Chapman & Hall. A nonmathematical treatment of the role of genetic variation in nature, stressing morphological variation.

Futuyma, D. J. 1986. *Evolutionary Biology,* 2d ed. Sunderland, MA: Sinauer. The best modern discussion of population genetics, ecology, and evolution from both a theoretical and experimental point of view.

Hartl, D. L., and A. G. Clark. 1989. *Principles of Population Genetics,* 2d ed. Sunderland, MA: Sinauer.

Lerner, I. M., and W. J. Libby. 1976. *Heredity, Evolution, and Society,* 2d ed. New York: W. H. Freeman and Company. A text meant for nonscience students.

Lewontin, R. C. 1974. *The Genetic Basis of Evolutionary Change.* New York: Columbia University Press. A discussion of the prevalence and role of genetic variation in natural populations. Both morphological and protein variations are considered.

Lewontin, R. C. 1982. *Human Diversity.* New York: Scientific American Books. Applies concepts of population genetics to human diversity and evolution.

Li, W. H., and D. Graur. 1991. *Fundamentals of Molecular Evolution.* Sunderland, MA: Sinauer. An excellent review of what is known about the forces and rates of evolution of proteins and DNA sequences.

Scientific American (September) 1978. "Evolution." This issue contains several articles relevant to population genetics.

Wilson, A. C. 1985. "The Molecular Basis of Evolution." *Scientific American* (October).

Chapter 18

Bodmer, W. F., and L. L. Cavalli-Sforza. 1970. "Intelligence and Race." *Scientific American* (October). A popular treatment, including a discussion of heritability.

Falconer, D. S. 1981. *Introduction to Quantitative Genetics,* 2d ed. New York: Ronald Press. A widely read text with a strong mathematical emphasis.

Feldman, M. W., and R. C. Lewontin. 1975. "The Heritability Hangup." *Science* 190: 1163–1168. A discussion of the meaning of heritability and its limitations, especially in relation to human intelligence.

Lewontin, R. C. 1974. "The Analysis of Variance and the Analysis of Causes." *American Journal of Human Genetics.* 26: 400–411. A discussion of the meaning of the analysis of variance in genetics as a method for determining the roles of heredity and environment in determining phenotype.

Lewontin, R. C. 1982. *Human Diversity.* New York: Scientific American Books. Includes a discussion of quantitative variation in human populations.

Chapter 19

Dobzhansky, T. 1951. *Genetics and the Origin of Species.* New York: Columbia University Press. The classic synthesis of population genetics and the processes of evolution. The most influential book on evolution since Darwin's *Origin of Species.*

Futuyma, D. J. 1998. *Evolutionary Biology,* 3d ed. Sunderland, MA: Sinauer. The best modern discussion of population genetics, ecology, and evolution from both a theoretical and an experimental point of view.

Li, W. H., and D. Grauer. 1991. *Fundamentals of Molecular Evolution.* Sunderland, MA: Sinauer. An excellent review of what is known about the forces and rates of evolution of proteins and DNA sequences.

Scientific American (September) 1978. "Evolution." This issue contains several articles relevant to population genetics.

Wilson, A. C. 1985. "The Molecular Basis of Evolution." *Scientific American* (October).

Answers to Selected Problems

Chapter 1

2. DNA determines all the specific attributes of a species (shape, size, form, behavioral characteristics, biochemical processes, etc.) and sets the limits for possible variation that is environmentally induced.

4. There are 4^{10}, or 1,048,576, possible DNA molecules!

6. Human somatic cells are diploid ($2n$) and the haploid number is 23 ($n = 23$). Each chromosome is a double-stranded DNA molecule, so there are 46 molecules of DNA. It is generally stated that the haploid number represents the number of different types of DNA molecules, but that does not take into account the difference between the X and Y chromosome in males. So females have 23 different types (called 1–22 and X) and males have 24 (1–22, X and Y).

8. Yes.

10. For the following, normal typeface represents previously polymerized nucleotides and *italic* typeface represents newly polymerized nucleotides.

3′—TTGGCACGTCGTAAT—5′

5′—*AACCGTGCAGCATTA*—3′

3′—*TTGGCACGTCGTAAT*—5′

5′—AACCGTGCAGCATTA—3′

12. Sulfur.

Chapter 2

2. In each cell there are two copies of the nuclear genome (plants are diploid) and many copies of the mitochondrial and chloroplast genomes.

4. A **chromosome** is a single DNA molecule containing many genes, often with associated protein and RNA.

A **chromomere** is a small, beadlike structure visible on a chromosome during prophase of mitosis and meiosis.

A **chromocenter** is the point at which polytene chromosomes appear to be attached together.

Chromatin is the substance of chromosomes; it includes the DNA and associated proteins and RNA.

6. The DNA double helix is held together by two types of bonds: covalent and hydrogen. Covalent bonds occur within each linear strand and strongly bond the bases, sugars, and phosphate groups (both within each component and between components). Hydrogen bonds occur between the two strands and link a base from one strand with a base from the second in complementary pairing. These hydrogen bonds are individually weak but collectively quite strong.

8. If the DNA is double-stranded, the G content should equal the C content. Since the total content cannot equal more than 100%, it is not possible to have 55% G. (However, it is possible if the DNA is single-stranded.)

10.
- **a.** Since diploid organisms usually have two copies of each of their chromosomes, an odd number of chromosomes likely indicates the organism is haploid.
- **b.** Although an odd number of chromosomes likely indicates that the organism is haploid, an even number does not necessarily indicate the organism is diploid. Either this organism is haploid with 6 different chromosomes or it is diploid with 3 pairs of chromosomes.

12. No. The average size of genes varies from one organism to another depending on the number and length of introns; the amount of repetitive DNA varies greatly as well.

14. If the DNA is double stranded, $G = C = 24\%$ and $A = T = 26\%$.

16. Human somatic cells are diploid ($2n$) and the haploid number is 23 ($n = 23$).

18. 3′—TAACCGAGA—5′

20.
- **a.** Sulfur is not found in DNA.
- **b.** Neither nitrogen nor sulfur is found in the backbone of DNA.

22.
- **a.** 1830 kb/1703 genes = 1074 base pairs
- **b.** $1074 - 1000 = 74$ base pairs
- **c.** These DNA sequences will predominantly be regulatory regions.
- **d.** 0%: bacteria do not have introns.
- **e.** The same as part b, 74 base pairs.

24. *Neurospora*—7 bands
pea—7 bands
housefly—6 bands

26. $(1.7\%)\ 3 \times 10^9 = 51 \times 10^6$ nucleotide differences.

28. The polarity of the sugar-phosphate backbone in DNA is defined by the orientation of the deoxyribose sugar. By convention, the carbons are numbered as primes to differentiate the atoms of the base from those of the sugar. It is useful to use 5′ and 3′ to define the orientation of DNA (or RNA) strands. In this way, the two strands of the double helix are said to be antiparallel (run in opposite orientations). Also, as you learn more about the enzymology of replication, transcription, and translation, the importance and concept of strand orientation can easily be conveyed.

30. There are roughly 10 base pairs per turn.

32. Herpes virus > *E. coli* > *Arabidopsis thaliana* > *Homo sapiens*

34. Taking the average number of genes from chromosomes 21 and 22 and assuming this to be representative of all human chromosomes, the total number of genes in the human genome would be: (225 + 545)/2 = 385 genes per 1% of the genome or 38,500 for the entire genome. (This is in line with the predicted 30,000 to 40,000 genes suggested by the preliminary completion of the sequencing of the human genome.) An obvious source of error in this calculation is the simple assumption that genes are spread proportionally over the entire genome. Given that chromosome 22 contains more than twice as many genes as chromosome 21, it is likely that this assumption is too simple.

36. The average amount of DNA per gene is 33.46 mb/545 genes = 61.4 kb. Assuming the gene has 5 introns, it would have 6 exons of average length of 266 bp or approximately 1.6 kb of coding DNA (6 × 266 = 1596 bp) and each intron would be, on average, 3.52 kb long (19.2 kb − 1.6 kb = 17.6 kb of introns divided by 5 = 3.52 kb per intron). The average spacing between genes would be 61.4 kb (the average amount of DNA per gene) − 19.2 kb (the average gene size) = 42.4 kb.

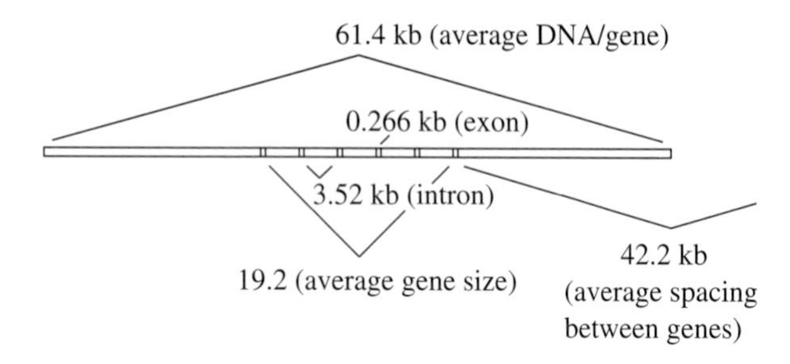

Chapter 3

2.

3′ CGT ACC ACT GCA 5′	DNA double helix (transcribed strand)
5′ GCA TGG TGA CGT 3′	DNA double helix
5′ GCA UGG UGA CGU 3′	mRNA transcribed
3′ CGU ACC ACU GCA 5′	Appropriate tRNA anticodon
NH_3 − Ala − Trp − (stop) − COOH	Amino acids incorporated

4. **a.** *his*⁻, (since it cannot make histidine)
b. *y*, (lowercase because it is recessive)
c. *Pr*, (uppercase because it is dominant)
d. *H*, (uppercase because it is dominant). Although most null mutations are recessive, some are dominant because of haplo-insufficiency.
e. y^+ / y; Pr / Pr^+; H / H^+

6. There are many: polymerases, nucleases, transcription factors, ligases, restriction enzymes, ribosomal proteins, histones, topoisomerases, etc.

8. **a.** The main use is in detecting carrier parents and in diagnosing the disorder in the fetus.
b. Because the values for normal individuals and carriers overlap for galactosemia, there is ambiguity if a person has 25 to 30 units. That person could be either a carrier or normal.
c. These wild-type genes are phenotypically dominant but are incompletely dominant at the molecular level. Apparently a minimal level of enzyme activity is enough to ensure normal function and phenotype.

10. **a.** White.
b. Blue.

12. **a.** Recessive. The normal allele provides enough enzyme to be sufficient for normal function (the definition of haplo-sufficient).
b. There are many ways to mutate a gene to destroy enzyme function. One possible mutation might be a frameshift mutation within an exon of the gene. Assuming that a single base pair was deleted, the mutation would completely alter the translational product 3′ to the mutation.
c. Hormone replacement could be given to the patient.
d. If the hormone is required before birth, it can be supplied by the mother.

14. DNA has often been called the "blueprint of life," but how does it actually compare to a blueprint used for house construction? Both are abstract representations of instructions for building three-dimensional forms, and both require interpretation for their information to be useful. But real blueprints are two-dimensional renderings of the various views of the final structure drawn to scale. There is one-to-one correlation between the lines on the drawing and the real form. The information in the DNA is encoded in a linear array—a one-dimensional set of instructions which only becomes three-dimensional as the encoded linear array of amino acids fold into their many forms. Included in the informational content of the DNA are also all the directions required for its "house" to maintain and repair itself, respond to change, and replicate. Let's see a real blueprint do that!

16. There are three codons for isoleucine: 5′ AUU 3′, 5′ AUC 3′, and 5′ AUA 3′. Possible anticodons are 3′ UAA 5′ (complementary), 3′ UAG 5′ (complementary), and 3′ UAI 5′ (wobble). 5′ UAU 3′, although complementary, would also base-pair with 5′ AUG 3′ (methionine) due to wobble and therefore would not be an acceptable alternative.

18. **a., b.**

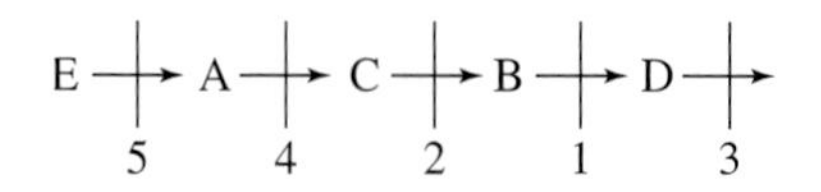

Vertical lines indicate step where each mutant is blocked.

20. Protein function can be destroyed by a mutation that causes the substitution of a single amino acid even though the protein has the same immunological properties. For example, enzymes require very specific amino acids in exact positions within their active site. A substitution of one of these key amino acids might have no effect on overall size and shape of the protein while completely destroying the enzymatic activity.

22. a. The allele *sn* will show dominance over *sf* because there will be only 40 units of square factor in the heterozygote.

b. No. Here the functional allele is recessive.

c. The allele *sf* may become dominant over time by several mechanisms: it could mutate in such a way that it produces 50 units; other genes might mutate to lower the cell's need to have 50 units; or other genes might mutate to increase the production or activity of *sf*.

24. a., b.

Base deleted

Old: $\mathrm{AA}^{\mathrm{A}}_{\mathrm{G}}$ (A)GU CCA UCA CUU AAU GCN GCN $\mathrm{AA}^{\mathrm{A}}_{\mathrm{G}}$

New: $\mathrm{AA}^{\mathrm{A}}_{\mathrm{G}}$ GUC CAU CAC UUA AU(G) GCN GCN $\mathrm{AA}^{\mathrm{A}}_{\mathrm{G}}$

Base added

Chapter 4

2. The choices are many, but the following is one example. Because of the antiparallel nature of DNA, the synthesis of new DNA is continuous for the leading strand and discontinuous for the lagging strand.

4. Yes, both strands serve as templates along their entire length. For each, however, telomerase adds additional sequences to the 3′ end using its own RNA template. These added sequences serve as templates for the newly synthesized lagging strands.

6. Six.

8.

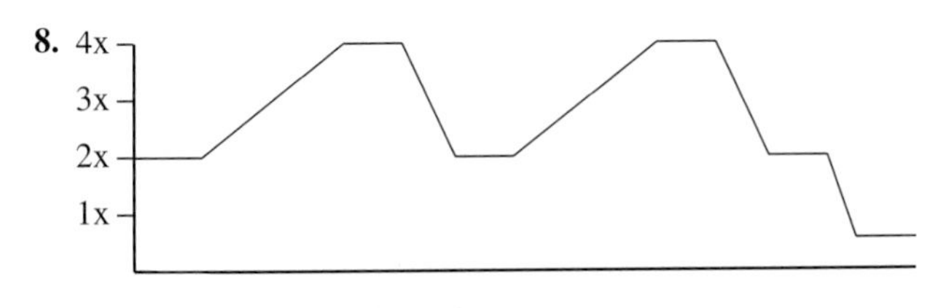

10. As cells divide mitotically, each chromosome consists of identical sister chromatids that are separated to form genetically identical daughter cells. Although the second division of meiosis appears to be a similar process, the "sister" chromatids are likely to be different. Recombination during earlier meiotic stages has swapped regions of DNA between sister and nonsister chromosomes such that the two daughter cells of this division are typically not genetically identical.

12.

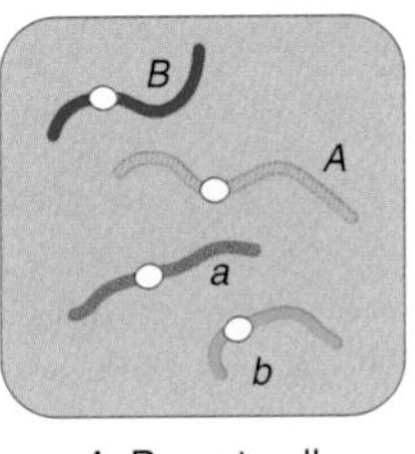

1 Parent cell

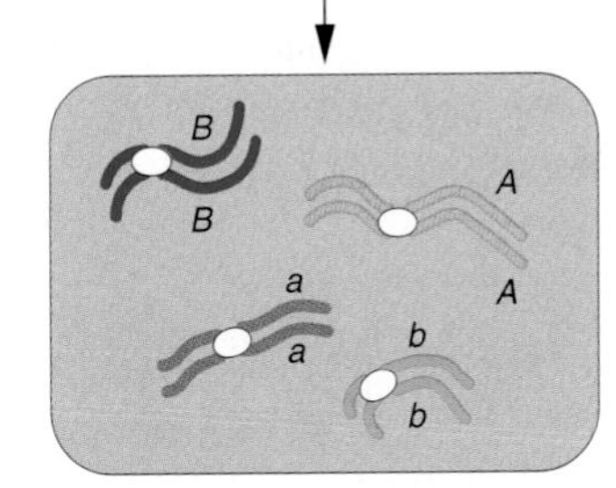

2 Chromosome duplication

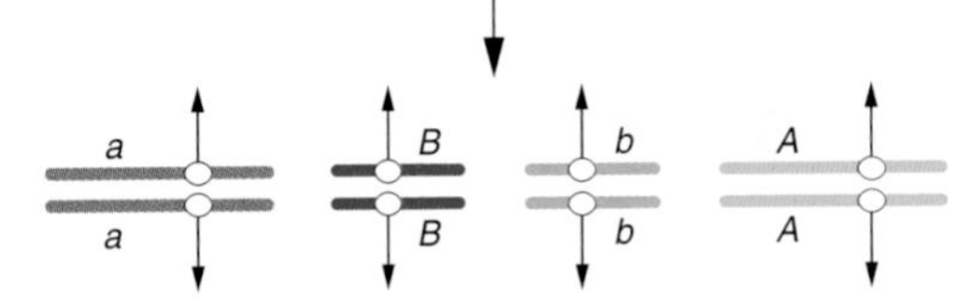

3 Segregation

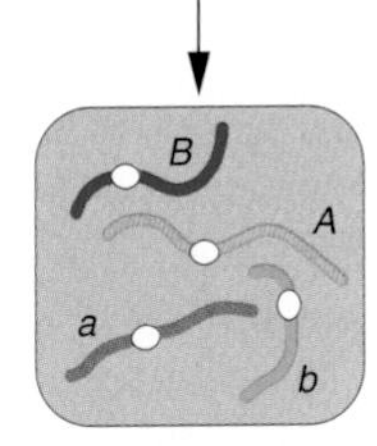

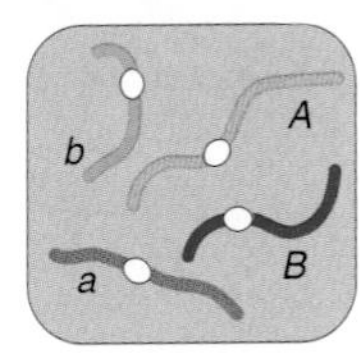

4 Daughter cells

14. The key function of mitosis is to generate two daughter cells genetically identical to the original parent cell. It could be argued that this "sameness" is conservative. The key functions of meiosis are to halve the DNA content and to reshuffle the genetic content of the organism to generate genetic diversity among the progeny. It could be argued that this "purposeful diversity" is liberal.

16. The nucleus contains the genome and separates it from the cytoplasm. However, during cell division, the nuclear envelope dissociates (breaks down) and it is the job of the microtubule-based spindle to actually separate the chromosomes (divide the genetic material) around which nuclei re-form during telophase. In this sense, it can be viewed as a passive structure that is divided by the cell's cytoskeleton.

18. Chromosome pairing is unique to prophase I of meiosis.

20. 3′ . . . .GGAATTCTGATTGATGAATGACCCTAG. . . .5′

22. Without functional telomerase, the telomeres would shorten at each replication cycle, leading to eventual loss of essential coding information and death. In fact, there are some current observations that decline or loss of telomerase activity plays a role in the mechanism of aging in humans.

24. $\frac{1}{8}$.

26. It is likely that the failure of the spindle to attach to the kinetochore of just one chromatid would result in mitotic nondisjunction or chromosome loss.

Chapter 5

2. a. Orange phenotype is dominant to the red phenotype. Let r = allele for red and R = allele for orange.

The cross is:	Parents	R / R × r / r
	Progeny	all R / r

b.

The cross is:	Parents	R / r × R / r	
	Progeny	25% R / R	(orange)
		50% R / r	(orange)
		25% r / r	(red)

4. First cousins share grandparents. If either grandparent was heterozygous for the recessive galactosemia allele (noted by shading in half of the symbol), the cousins could each inherit the allele from that grandparent.

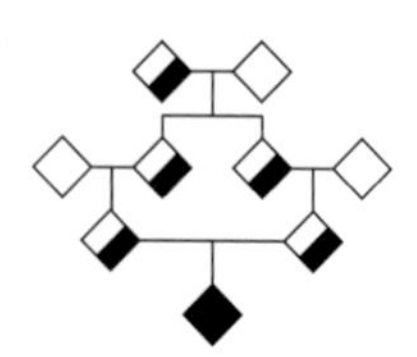

6. Black (B) is dominant to white (b).

Parents: B / b × B / b

Progeny: 3 black : 1 white (1 B / B : 2 B / b : 1 b / b)

8. a. Let A = blotched, a = unblotched.

P A / a (blotched) × A / a (blotched)

F_1 1 A / A : 2 A / a : 1 a / a

3 A / – (blotched) : 1 a / a (unblotched)

b. All unblotched plants should be pure-breeding in a testcross with an unblotched plant (a / a), and one-third of the blotched plants should be pure-breeding.

10. a. dominant

b. I: d / d, D / d
II: D / d, d / d, D / d, d / d
III: d / d, D / d, d / d, D / d, d / d, d / d, D / d, d / d
IV: D / d, d / d, D / d, d / d, d / d, d / d, d / d, D / d, d / d

c. $\frac{1}{16}$.

12. a. leu^+ / leu.

b. 1 leu^+ : 1 leu.

c. Certain amino acids are essential to protein structure and function, and a change of even one of these could totally destroy an enzyme's activity. There are many ways to change a DNA sequence that encodes an enzyme that will result in an altered amino acid sequence. For the following, a small DNA sequence has arbitrarily been chosen to show three classes of mutation that could cause loss of enzymatic activity. The first, a missense mutation, is the result of a single base-pair change that alters a single amino acid. The second, a deletion of three base pairs, causes the loss of one amino acid. The third, a frameshift mutation caused by the addition of one base pair, alters the amino acid sequence from the site of the addition until the end of translation is reached. Other mutations, not shown, would also lead to null alleles: mutations in the promoter for this gene preventing transcription, mutations that alter or prevent proper splicing, insertions of DNA into the coding region of the gene, etc.

Wild-type allele

. . . . 5′-ATTCGTACGATCGAC-3′ . . . .
. . . . 3′-TAAGCATGCTAGCTG-5′ . . . . DNA

. . . . 5′-AUU CGU ACG AUC GAC-3′ . . . . mRNA

. . . . NH_2-Ile Arg Thr Ile Asp-COOH . . . . Protein

Missense allele

. . . . 5′-ATTCATACGATCGAC-3′ . . . .
. . . . 3′-TAAGTATGCTAGCTG-5′ . . . . DNA

. . . . 5′-AUU CAU ACG AUC GAC-3′ . . . . mRNA

. . . . NH_2-Ile His Thr Ile Asp-COOH . . . . Protein

Small deletion allele (3 base pairs)

CGT
GCA

. . . . 5′-ATTACGATCGAC-3′ . . . .
. . . . 3′-TAATGCTAGCTG-5′ . . . . DNA

. . . . 5′-AUU ACG AUC GAC-3′ . . . . mRNA

. . . . NH_2-Ile Thr Ile Asp-COOH . . . . Protein

Frameshift allele (+ 1)

. . . . 5′-ATTCGTAACGATCGAC-3′ . . . .
. . . . 3′-TAAGCATTGCTAGCTG-5′ . . . . DNA

. . . 5′-AUU CGU AAC GAU CGA C-3′ . . . . mRNA

. . . . NH_2-Ile Arg Asn Asp Arg-COOH . . . . Protein

14. Horizontal lines (H) is dominant to vertical lines (h).

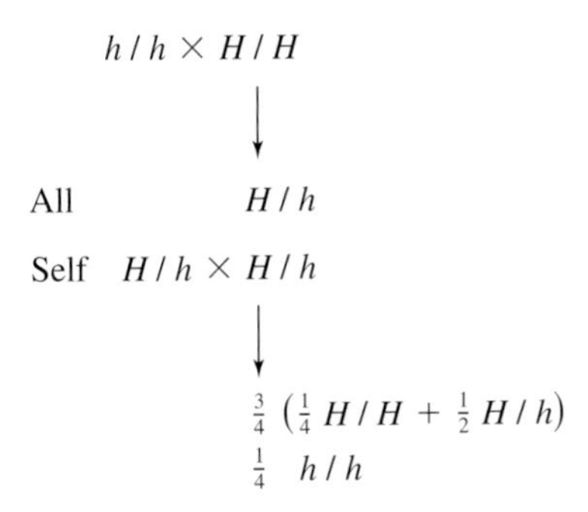

16. Vertical line (h^+) is dominant to horizontal line (h).

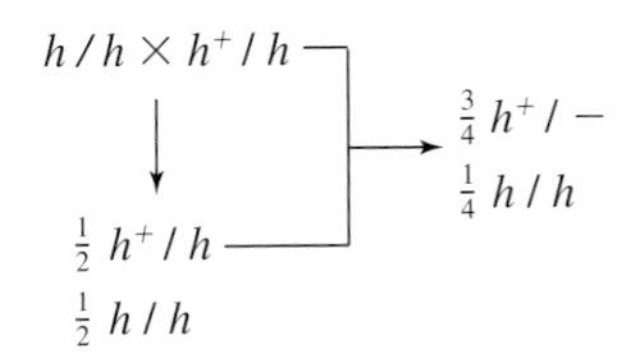

18. Star (s) is recessive to no star (s^+) and is X-linked.

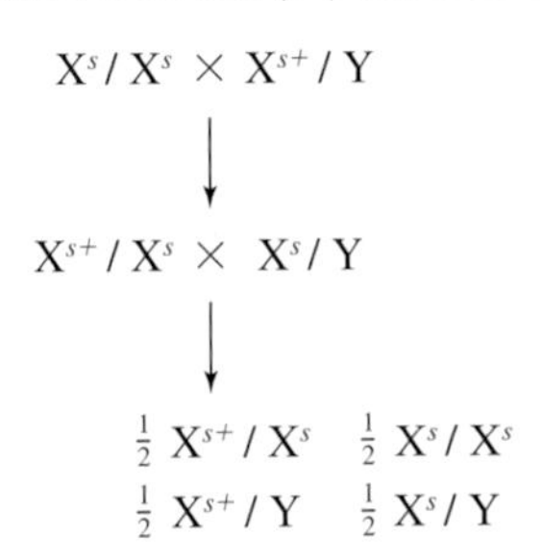

20. **a.** X-linked recessive allele.

b. Generation I: $X^+ / Y, X^+ / X^m$

Generation II: $X^+ / X^+, X^m / Y, X^+ / Y, X^+ / -, X^+ / X^m, X^+ / Y$

Generation III: $X^+ / X^+, X^+ / Y, X^+ / X^m, X^+ / X^m, X^+ / Y, X^+ / X^+, X^m / Y, X^+ / Y, X^+ / -$

c. The first couple: 0%. The second couple: 50% chance of having affected sons and no chance of having affected daughters. The third couple: 0%.

22. **a., b.** The "stopper" or "continuous" phenotype of the maternal parent is inherited by all its offspring. This is typical of maternal inheritance (or organelle-based inheritance) and would be expected if this trait mapped to a mitochondrial gene. Nuclear genes that differ between the two parental strains should segregate in the normal Mendelian manner and produce 1:1 ratios in the progeny. (Remember, *Neurospora* is haploid and the progeny of these crosses are actually the haploid products of meiosis.) This is observed for the orange/yellow trait, suggesting that this trait maps to a nuclear gene.

24. **a.** Favism pedigree.

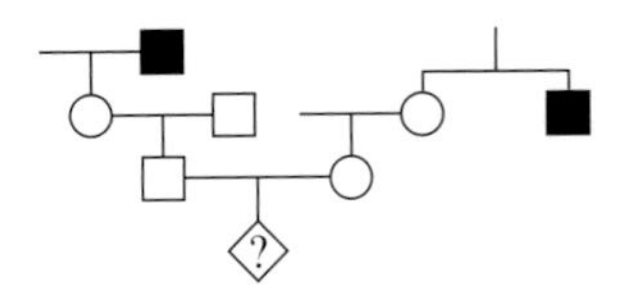

b. $\frac{1}{16}$.

c. $\frac{1}{4}$.

26. H = hairy allele; h = smooth allele.

plant 1: H / h

plant 2: H / H

plant 3: H / h

plant 4: h / h

28. **a.** Pedigree 1: X-linked recessive. Pedigree 2: autosomal recessive.

b. $\frac{1}{16}$; 0%.

Chapter 6

2. **a., b.** Parents: dilute; Black × Intense; brown

d / d ; B / b D / d ; b / b

d B × D b

d b d b

Progeny:	dilute ; Black	d / d ; B / b
	dilute ; brown	d / d ; b / b
	Intense ; Black	D / d ; B / b
	Intense ; brown	D / d ; b / b

4. 45%.

6. Since only parental types are recovered, the two genes must be tightly linked and recombination must be very rare.

8. **a.** 25%.
b. 0%.
c. 6%.
d. 12%.

10. **a.** C / c ; S / s × C / c ; S / s
b. C / C ; S / s × C / C ; s / s
c. C / c ; S / S × c / c ; S / S
d. c / c ; S / s × c / c ; S / s
e. C / c ; s / s × C / c ; s / s
f. C / C ; S / s × C / C ; S / s
g. C / c ; S / s × C / c ; s / s

12. Cross 1: nonsignificant; hypothesis cannot be rejected.
Cross 2: nonsignificant; hypothesis cannot be rejected.
Cross 3: significant; hypothesis must be rejected.
Cross 4: significant; hypothesis must be rejected.

14. **a.** The three genes are linked.

b. The map units between v and p = 100%(2200)/10,000 = 22 m.u., and the map units between p and b = 100%(1500)/10,000 = 15 m.u. The map is

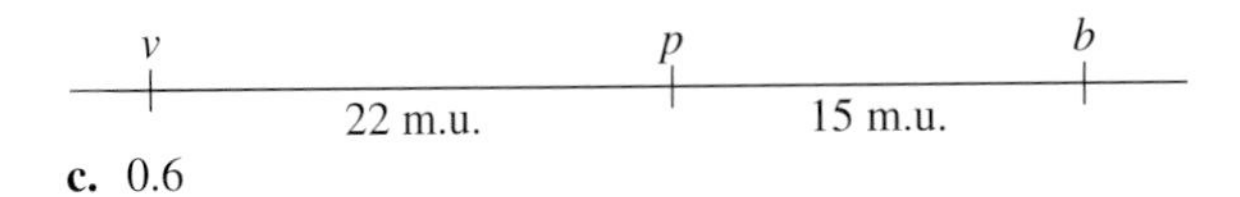

c. 0.6

16. Dear Brother Mendel:

I have recently read your most engrossing manuscript detailing the results of your most wise experiments with garden peas. I salute both your curiosity and your ingenuity in conducting said experiments, thereby opening up for scientific exploration an entire new area of our Maker's universe. Dear Sir, your findings are extraordinary!

While I do not pretend to compare myself to you in any fashion, I beg to bring to your attention certain findings I have made with the aid of that most fascinating and revealing instrument, the microscope. I have been turning my attention to the smallest of worlds with an instrument that I myself have built, and I have noticed some structures that may parallel in behavior the factors that you have postulated in the pea.

I have worked with grasshoppers, however, not your garden peas. Although you are a man of the cloth, you are also a man of science, and I pray that you will not be offended when I state that I have specifically studied the reproductive organs of male grasshoppers. Indeed, I did not limit myself to studying the organs themselves; instead, I also studied the smaller units that make up the male organs and have beheld structures most amazing within them.

These structures are contained within numerous small bags within the male organs. Each bag has a number of these structures, which are long and threadlike at some times and short and compact at other times. They come together in the middle of a bag, and then they appear to divide equally. Shortly thereafter, the bag itself divides, and what looks like half of the threadlike structures goes into each new bag. Could it be, Sir, that these threadlike structures are the very same as your factors? I know, of course, that garden peas do not have male organs in the same way that grasshoppers do, but it seems to me that you found it necessary to emasculate the garden peas in order to do some crosses, so I do not think it too far-fetched to postulate a similarity between grasshoppers and garden peas in this respect.

Pray, Sir, do not laugh at me and dismiss my thoughts on this subject even though I have neither your excellent training nor your astounding wisdom in the Sciences. I remain your humble servant to eternity!

18. Two unlinked genes; "no dots" (d) is recessive to "dots" (d^+) and "no lines" (l) is recessive to "lines" (l^+).

no dots, Lines × Dots, no lines

$d\,/\,d\ ;\ l^+\,/\,l^+ \qquad d^+\,/\,d^+\ ;\ l\,/\,l$

↓

All: Dots, Lines

$d^+\,/\,d\ ;\ l^+\,/\,l$

↓

Self: $\frac{9}{16}$ Dots, Lines

$d^+\,/\,-\ ;\ l^+\,/\,-$

$\frac{3}{16}$ Dots; no lines

$d^+\,/\,-\ ;\ l\,/\,l$

$\frac{3}{16}$ no dots, Lines

$d\,/\,d\ ;\ l^+\,/\,-$

$\frac{1}{16}$ no dots, no lines

$d\,/\,d\ ;\ l\,/\,l$

20. Two X-linked genes, 10 m.u. apart; "white" (b) is recessive to "black" (b^+) and "wavy tail" (s) is recessive to "straight tail" (s^+).

P $\quad b^+\,s^+\,/\,b^+\,s^+ \times b\,s\,/\,\mathrm{Y}$

↓

$F_1 \quad b^+\,s^+\,/\,b\,s$ and $b^+\,s^+\,/\,\mathrm{Y}$

↓

F_2 Females: Although all phenotypically $b^+\,s^+$, genotypically they are

45% $b^+\,s^+\,/\,b^+\,s^+$

45% $b\,s\,/\,b^+\,s^+$

5% $b\,s^+\,/\,b^+\,s^+$

5% $b^+\,s\,/\,b^+\,s^+$

Males: 45% $b^+\,s^+\,/\,\mathrm{Y}$

45% $b\,s\,/\,\mathrm{Y}$

5% $b\,s^+\,/\,\mathrm{Y}$

5% $b^+\,s\,/\,\mathrm{Y}$

22. **a.** For (1) = $\frac{9}{128}$. For (2) = $\frac{9}{128}$. For (3) = $\frac{9}{64}$. For (4) = $\frac{55}{64}$.

b. For (1) = $\frac{1}{32}$. For (2) = $\frac{1}{32}$. For (3) = $\frac{1}{16}$. For (4) = $\frac{15}{16}$.

24. **a.** 3.36%.

b. 18.2%.

26. (1) Impossible
(2) Meiosis II
(3) Meiosis II
(4) Meiosis II
(5) Mitosis
(6) Impossible
(7) Impossible
(8) Impossible
(9) Impossible
(10) Meiosis I
(11) Impossible
(12) Impossible

28. **a.** Long (s^+) is dominant to short (s). Large (v^+) is dominant to vestigial (v).

b. Cross 1: $v^+\,/\,v\ ;\ s^+\,/\,s^+ \times v\,/\,v\ ;\ s\,/\,s$
Cross 2: $v^+\,/\,v\ ;\ s^+\,/\,s \times v\,/\,v\ ;\ s^+\,/\,s$
Cross 3: $v^+\,/\,v\ ;\ s^+\,/\,s \times v^+\,/\,v\ ;\ s\,/\,s$
Cross 4: $v^+\,/\,v\ ;\ s\,/\,s \times v\,/\,v\ ;\ s^+\,/\,s$

30. **a.** The two genes are 100% (21 + 19)/total = 4 map units apart.

Tall, Red, Wide × short, white, narrow

$$\frac{T \quad r}{t \quad R} \cdot \frac{W}{W} \times \frac{t \quad r}{t \quad r} \cdot$$

b. 0.04%.

Chapter 7

2. All cultures of F^+ strains have a small proportion of cells in which the F factor is integrated into the bacterial chromosome and are, by definition, Hfr cells. These Hfr cells transfer markers from the host chromosome to a recipient during conjugation.

4. Generalized transduction occurs with lytic phages that enter a bacterial cell, fragment the bacterial chromosome, and then, while new viral particles are being assembled, improperly incorporate some bacterial DNA within the viral protein coat. Because the amount of DNA, not the information content of the DNA, is what governs viral particle formation, any bacterial gene can be included within the newly formed virus. In contrast, specialized transduction occurs with improper excision of viral DNA from the host chromosome in lysogenic phages. Because the integration site is fixed, only those bacterial genes very close to the integration site will be included in a newly formed virus.

6. —*M*—*Z*—*X*—*W*—*C*—*N*—*A*—*L*—*B*—*R*—*U*—

8. Prototrophic strains of *E. coli* will grow on minimal media, while auxotrophic strains will only grow on media supplemented with the required molecule(s). Thus, strain 3 is prototrophic (wild-type), strain 4 is met^-, strain 1 is arg^-, and strain 2 is arg^- met^-.

10. a. The two genotypes being cultured are pro^+ thi^- and pro^+ thi^+.

b. Two recombination events must occur, one on either side of *pro*.

c. 11.1%.

12. The best explanation is that the integrated F factor of the Hfr looped out of the bacterial chromosome abnormally and is now an F′ that contains the pro^+ gene. This F′ is rapidly transferred to F^- cells, converting them to pro^+ (and F^+).

14. First carry out a cross between the Hfr and F^-, and then select for colonies that are ala^+ str^r. If the Hfr donates the *ala* region late, then redo the cross but now interrupt the mating early and select for ala^+. This selects for an F′, since this Hfr would not have transferred the *ala* gene early. If the Hfr instead donates this region early, then use a Rec^- strain that cannot incorporate a fragment of the donor chromosome by recombination. Any ala^+ colonies from the cross should then be used in a second mating to another ala^- strain to see whether they can donate the *ala* gene easily, which would indicate that there is F′ *ala*. (This would also require another marker to differentiate the donor and recipient strains. For example, the ala^- strain could be tetracycline^r and selection would be for ala^+ tet^r.)

16. a. The two genes are located close enough together to be cotransformed at a rate of 0.17%.

b. Here, when the two genes must be contained on separate pieces of DNA, the rate of cotransformation is much lower, confirming the conclusion in part a.

18. a. 26%.

b. 46%.

c. *pdx* is closer, as determined by cotransduction rates.

d. The gene order is *pur pdx nad*.

20. The most straightforward way would be to pick two Hfr strains that are near the genes in question but are oriented in opposite directions. Then, measure the time of transfer between two specific genes, in one case when they are transferred early and in the other when they are transferred late. For example,

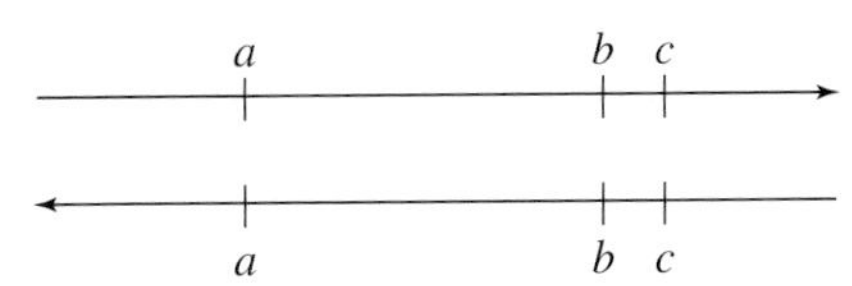

22. In a small percent of the cases, gal^+ transductants can arise by recombination between the gal^+ DNA of the λdgal transducing phage and the gal^- gene on the chromosome. This will generate gal^+ transductants without phage integration.

24. a., b.

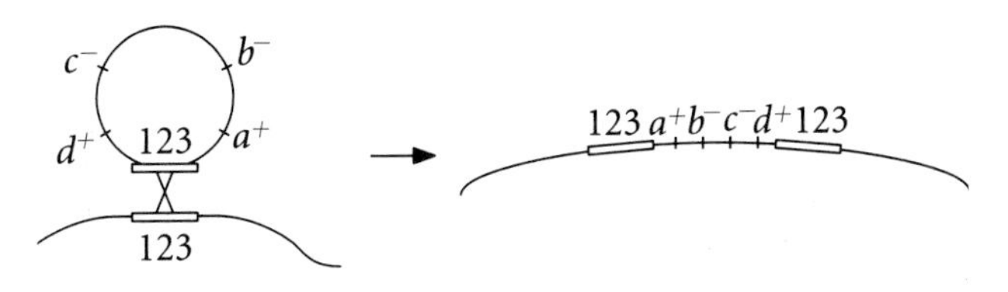

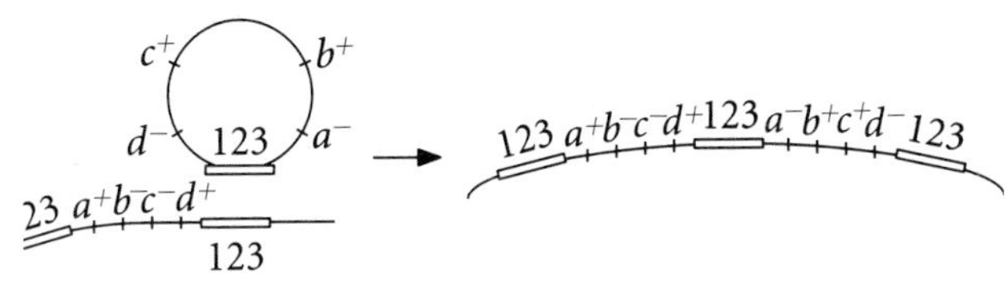

c.

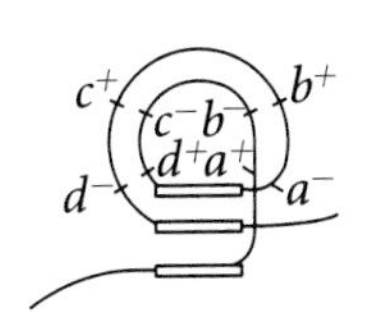

d.

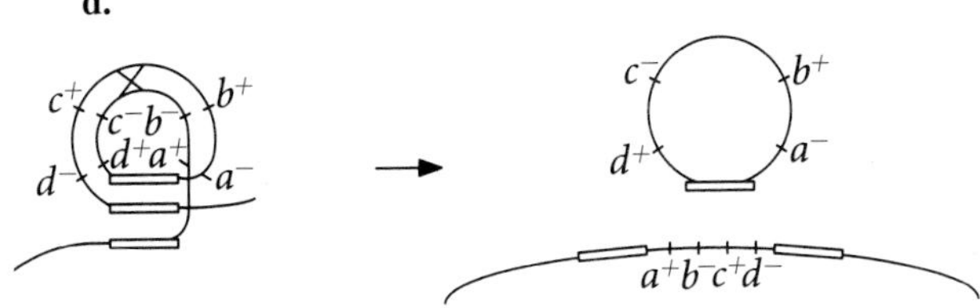

26. a. Owing to the medium used, all colonies are cys^+ but either + or − for the other two genes.

b. (1) cys^+ leu^+ thr^+ and cys^+ leu^+ thr^+ (supplemented with threonine)

(2) cys^+ leu^+ thr^+ and cys^+ leu^- thr^+ (supplemented with leucine)

(3) cys^+ leu^+ thr^+ (no supplements)

c. 39% of the colonies.

d.

Chapter 8

2. a.

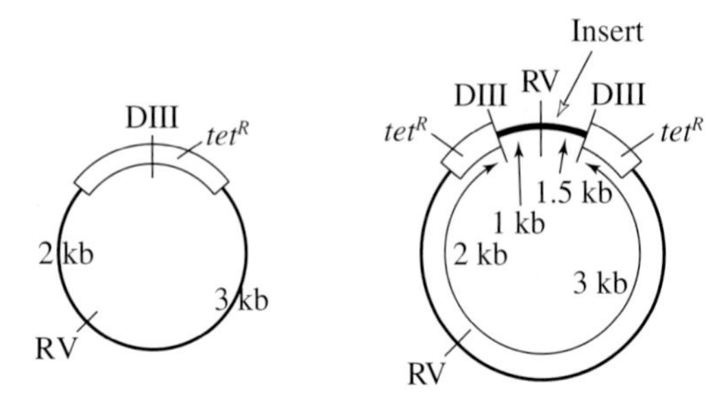

b. For the clone 15 digests, the *Hin*dIII 5-kb band will be radioactive and both *Eco*RV bands will be radioactive. For the *Hin*dIII + *Eco*RV double digest, the 3-kb and 2-kb bands will be radioactive.

c. No bands will be radioactive in the control lanes, and the clone 15 lanes will all have at least one radioactive band. For *Hin*dIII, the 2.5-kb band (the insert) will be radioactive. For *Eco*RV, the 4.5-kb and 3.0-kb bands will be radioactive. For the *Hin*dIII + *Eco*RV double digest, the 1.5-kb and 1-kb bands will be radioactive.

4. a. Since the actin protein sequence is known, a probe could be synthesized by "guessing" the DNA sequence based on the amino acid sequence. Alternatively, the gene for actin cloned in another species can be used as a probe to find the homologous gene in *Drosophila.* If an expression vector was used, it might also be possible to detect a clone coding for actin by screening with actin antibodies.

b. Hybridization using the specific tRNA as a probe could identify a clone coding for itself.

6. a.

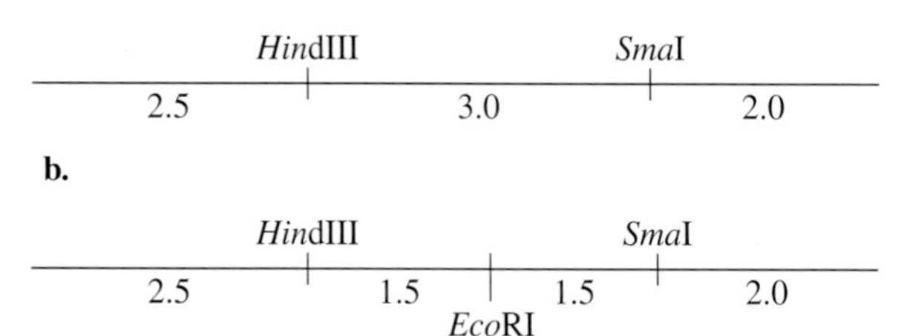

8. a. The transformed phenotype would map to the same locus. If gene replacement occurred by a double crossing-over event, the transformed cells would not contain vector DNA. If a single crossing-over took place, the entire vector would now be part of the linear *Neurospora* chromosome.

b. The transformed phenotype would map to a different locus than that of the auxotroph if the transforming gene was inserted ectopically (i.e., at another location).

10. This DNA mostly represents the introns that must be correctly spliced out of the primary transcript during RNA processing for correct translation. (There are also comparatively very small amounts of both 5′ and 3′ untranslated regions of the final mRNA that are necessary for correct translation encoded by this 60 kb of DNA.)

12. a. There is one *Bgl*II site, and the plasmid is 14 kb.

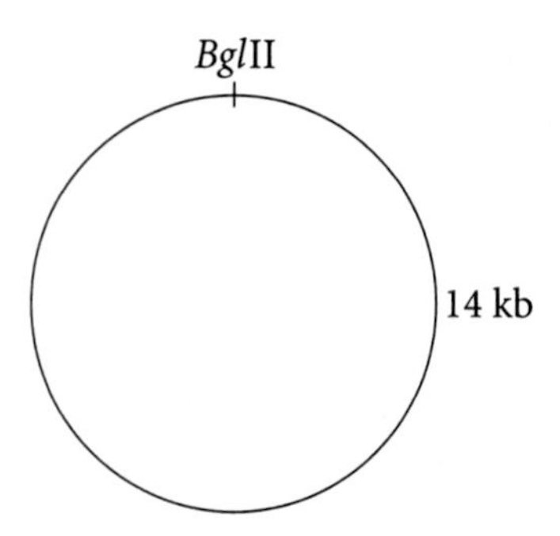

b. There are two *Eco*RV sites.

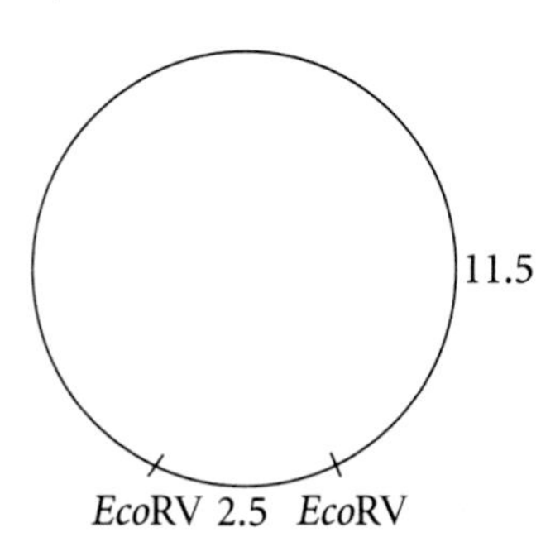

c. The 11.5-kb RV fragment is cut by *Bgl*II. The arrangement of the sites must be as indicated below.

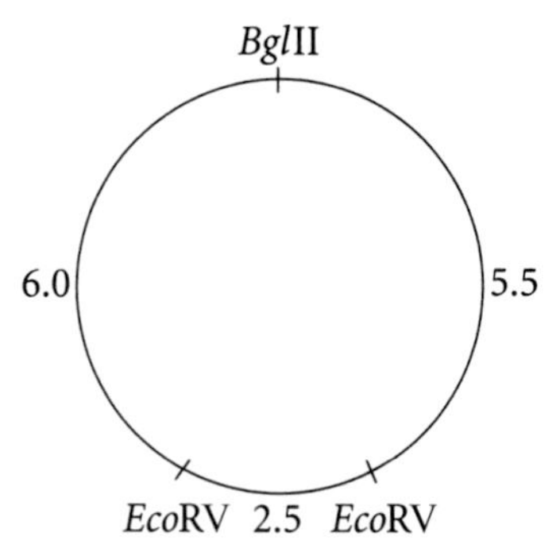

d. The *Bgl*II site must be within the *tet* gene.

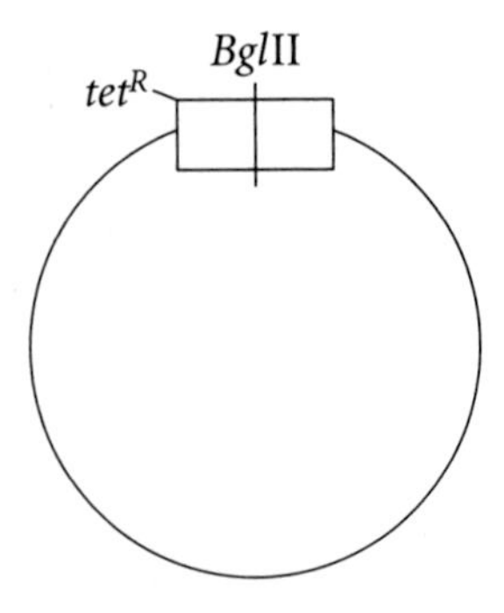

e. There was an insert of 4 kb.

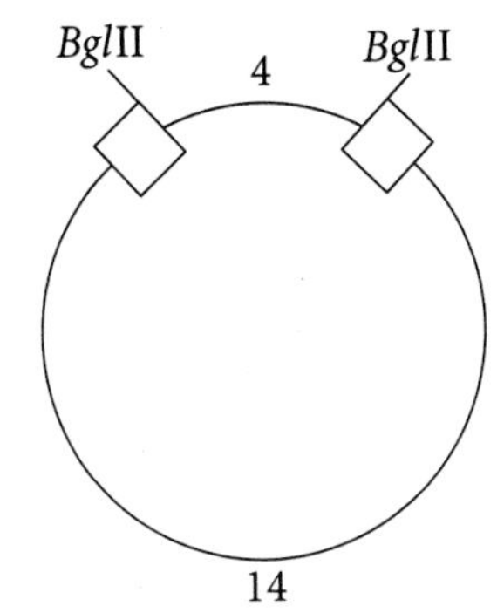

f. There was an *Eco*RV site within the insert.

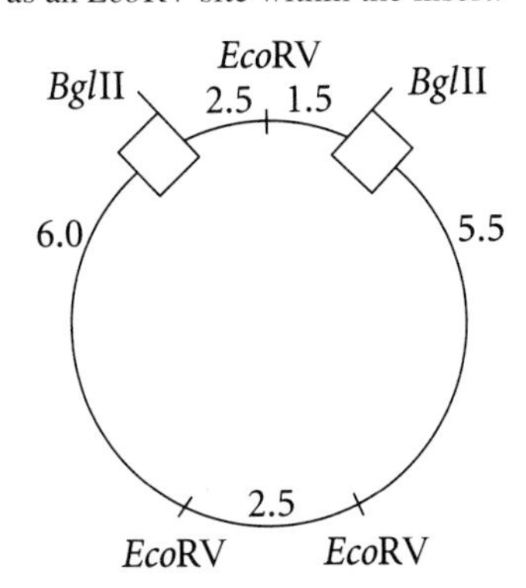

14. a. To ensure that a colony is not, in fact, a prototrophic contaminant, the prototrophic line should be sensitive to a drug to which the recipient is resistant. A simple additional marker would also achieve the same end.

b. Use a nonrevertible auxotroph as the recipient (such as one containing a deletion).

16. a. 5′ TTCGAAAGGTGACCCCTGGACCTTTAGA 3′

b. 3′ AAGCTTTCCACTGGGGACCTGGAAATCT 5′

c. 5′ TTCGAAAGGTGACCCCTGGACCTTTAGA 3′
3′ AAGCTTTCCACTGGGGACCTGGAAATCT 5′

d. There are a total of four open reading frames of the six possible.

18. Plant 1 shows the typical inheritance for a dominant gene that is heterozygous. Assuming kanamycin resistance is dominant to kanamycin sensitivity, the cross can be outlined as follows:

$$kan^R \,/\, kan^S \quad \times \quad kan^S \,/\, kan^S$$
$$\downarrow$$
$$\tfrac{1}{2}\ kan^R \,/\, kan^S$$
$$\tfrac{1}{2}\ kan^S \,/\, kan^S$$

Plant 2 shows a 3 : 1 ratio in the progeny of the backcross. This suggests that there have been two unlinked insertions of the kan^R gene and presumably the gene of interest as well.

20. The promoter and control regions of the plant gene of interest must be cloned and joined in the correct orientation with the glucuronidase gene. This places the reporter gene under the same transcriptional control as the gene of interest. Figure 8-33 in the text discusses the methodology used to create transgenic plants. Transform plant cells with the reporter gene construct, and, as shown in the figure, grow into transgenic plants. The glucuronidase gene will now be expressed in the same developmental pattern as the gene of interest and its expression can easily be monitored by bathing the plant in an X-Gluc solution and assaying for the blue reaction product.

22. a., b. During Ti plasmid transformation, the kanamycin gene will insert randomly into the plant chromosomes. Colony A, when selfed, has $\frac{3}{4}$ kanamycin-resistant progeny, and colony B, when selfed, has $\frac{15}{16}$ kanamycin-resistant progeny. This suggests that there was a single insertion into one chromosome in colony A and two independent insertions on separate chromosomes in colony B.

24. a. If the plasmid never integrates, the linear plasmid will be cut once by *Xba*I and two fragments will be generated that will both hybridize to the *Bgl*II probe. The autoradiogram will show two bands whose combined length will equal the full length of the plasmid.

b. If the plasmid integrates occasionally, most cells will still have free plasmids and these will be indicated by the two bands mentioned above. However, when the plasmid is integrated, two bands will still be generated, but their sizes will vary based on where other genomic *Xba*I sites are relative to the insertion point. If integration is random, many other bands will be observed, but if it is at a specific site, only two other bands will be detected.

Chapter 9

2. a.

YAC A:				1	4	3
YAC B:		5	1			
YAC C:				4	3	7
YAC D:	(6 2)		5			
YAC E:					3	7

b.

(6 2) — 5 — 1 — 4 — 3 — 7

A: 1 to 3; B: 5 to 1; C: 4 to 7; D: (6 2) to 5; E: 3 to 7

4. a. The following stylized schematic of a reciprocal translocation between chromosomes 3 and 21 is arbitrarily chosen to show the salient details. Band 3.1 of the q arm of chromosome 3 is split by the translocations that are correlated to the *N* disease allele. Probe c hybridizes to the region of 3q3.1 that remains with chromosome 3; and probes a, b, and d hybridize to the region of 3q3.1 that is translocated in this case to chromosome 21.

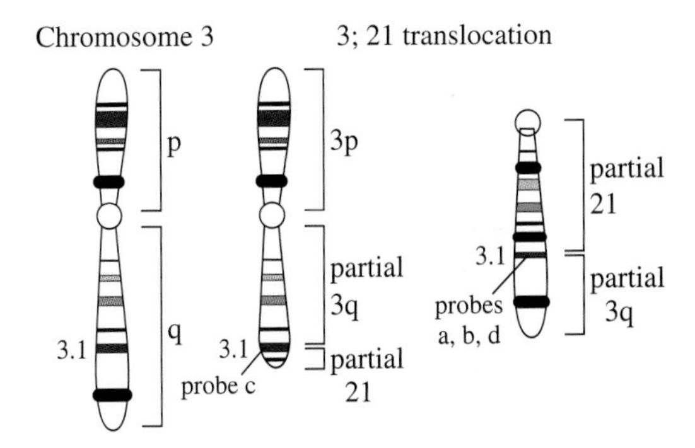

b. Since translocations of chromosome 3 that break band 3q3.1 are correlated to the disease, it is reasonable to assume that these rearrangements split the normal gene (*n*) in two, separating vital coding or regulatory regions. Therefore analysis and cloning of this specific region should be attempted.

c. Once *n* is cloned, it can be used to clone the various alleles from individuals who have the disease but not a translocation. The various alleles could then be compared with *n* by sequence, regulation, etc.

6. e, b, h, g

8.

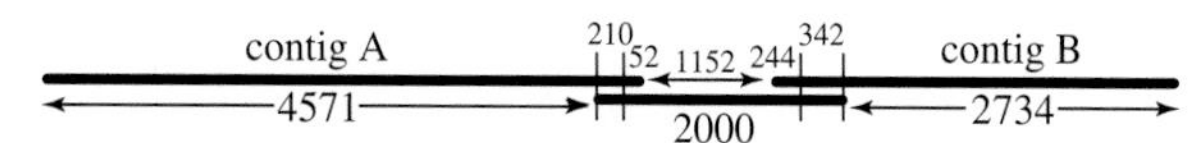

10. a. Since the triplet code is redundant, changes in the DNA nucleotide sequence (especially at those nucleotides coding for the third position of a codon) can occur without change to its encoded protein.

b. It can be expected that protein sequences will evolve and diverge more slowly than the genes that encode them.

12. a. The sequence is 2 1 3 (or 3 1 2).

b. A diagram of these results is shown below. In the diagram, there is no way to know the exact location of the ends of each YAC.

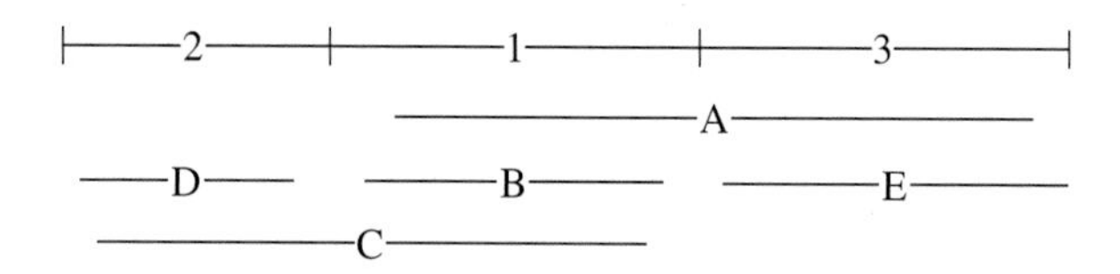

14. a. The *cys-1* locus is in this region of chromosome 5. If it were not in this region, linkage to either of the RFLP loci would not be observed.

b. *cys-1* to *RFLP-1* = 5 map units
cys-1 to *RFLP-2* = 12 map units
RFLP-1 to *RFLP-2* = 17 map units

c. A number of strategies could be tried. Since this is an auxotrophic mutant, functional complementation can be attempted. Positional cloning or chromosome walking from the RFLPs is also a very common strategy.

16. a. Of the regions of overlap for cosmids C, D, and E, region 5 is the only region in common. Thus, gene *x* is localized to region 5.

b. The common region of cosmids E and F, or the location of gene *y*, is region 8.

c. Both probes are able to hybridize with cosmid E because the cosmid is long enough to contain part of genes *x* and *y*.

18. a. Clone fingerprinting: clones are digested with restriction enzyme(s) to generate a set of bands whose number and positions are a unique "fingerprint" of each clone; the different bands generated from separate clones can be aligned to determine if there is any overlap between the inserted DNAs; the overlap is used to generate contigs. STS content mapping: amass a large set of random clones with small genomic inserts; sequence short regions of each clone and design pairs of PCR primers based on these sequences to amplify the short DNA sequence flanked by the primers (these short sequences are called STSs, or sequence-tagged sites); characterize clones of large genomic inserts to identify the contained STSs; clones that are shown to have specific STSs in common must be overlapping and can therefore be aligned into contigs. Radiation hybrid mapping: irradiate cells to fragment chromosomes; fuse irradiated cells with rodent cells to form a panel of different hybrids; each hybrid will have a random assortment of fragments integrated into the rodent chromosomes; analyze different radiation hybrids for co-occurrence of markers that may indicate linkage.

b. If two different clones have repetitive transposable element sequences in common, they will also share restriction fragments (bands) from those elements. If the randomly chosen STS is from a repeated element, its presence in various clones will not necessarily indicate that the clones overlap. Co-occurrence of markers that by chance are repetitive will not necessarily indicate linkage in radiation hybrid mapping.

c. Attempt to use restriction enzymes that do not cut within repeated elements. Make certain that the STSs are unique (present just once in the genome). Make certain that markers being assessed for co-occurrence are not repeated.

20. Identification of consensus donor and acceptor splice site sequences and the use of comparative genomics. That is, the conservation of the predicted amino acid encoded by the micro-exon in the same or other genomes.

Chapter 10

2. Frameshift mutations arise from addition or deletion of one or more bases in other than multiples of three. When translated, this will alter the reading frame and therefore the amino acid sequence from the site of the mutation to the end of the protein product. Also, frameshift mutations often result in premature stop codons in the new reading frame, leading to shortened protein products. A missense mutation changes only a single amino acid in the protein product.

4. Depurination, deamination, oxidatively damaged bases, such as 8-OxodG (8-oxo-7-hydrodeoxyguanosine), and errors during DNA replication (see Figure 10-12 in the text) can all lead to spontaneous mutations.

6. An AP site is an apurinic or apyrimidinic site. AP endonucleases introduce chain breaks by cleaving the phosphodiester bonds at the AP sites. Some exonuclease activity follows, so that a number of bases are removed. The resulting gap is filled by DNA pol I and then sealed by DNA ligase. General excision repair is used to remove damaged DNA, including photodimers. It cleaves the phosphodiester backbone on either side of the damage, removing 12 or 13 nucleotides in bacteria and 27 to 29 nucleotides in eukaryotes. The resulting gap is filled by DNA pol I and then sealed by DNA ligase.

8. Yes. It will cause CG-to-TA transitions.

10. In replicative transposition, transposable elements move to a new location by replicating into the target DNA, leaving behind a copy of the transposable element at the original site. If, on the other hand, the transposable element excises from its original position and inserts into a new position, this is called conservative transposition.

12. **a.** Because 5′-UAA-3′ does not contain G or C, a transition to a GC pair in the DNA cannot result in 5′-UAA-3′. 5′-UGA-3′ and 5′-UAG-3′ have the DNA antisense-strand sequence of 3′-ACT-5′ and 3′-ATC-5′, respectively. A transition to either of these stop codons occurs from the nonmutant 3′-ATT-5′, respectively. However, a DNA sequence of 3′-ATT-5′ results in an RNA sequence of 5′-UAA-3′, itself a stop codon.
b. Yes. An example is 5′-UGG-3′, which codes for Trp, to 5′-UAG-3′.
c. No. In the three stop codons the only base that can be acted upon is G (in UAG, for instance). Replacing the G with an A would result in 5′-UAA-3′, a stop codon.

14. a., b. Mutant 1: Most likely a deletion. It could be caused by radiation. Mutant 2: Most likely a frameshift mutation by an intercalating agent. Mutant 3: Most likely a GC-to-AT transition. It could have been caused by base analogs. Mutant 4: Most likely a transversion. X-irradiation or oxidizing agents could have caused the original mutation. Mutant 5: Most likely an AT-to-GC transition, which could be caused by base analogs.
c. The suggestion is a second-site reversion linked to the original mutant by 20 map units and therefore most likely in a second gene. Note that auxotrophs equal half the recombinants.

16. O^{-6}-Methyl G leads predominantly to high levels of GC → AT transitions. 8-Oxo dG gives rise predominantly to high levels of G → T transversions. Finally, C-C photodimers will most often cause C → T transitions, but some transversions are also possible.

18. Yes. Because DNA is a double-stranded molecule, replication of the DNA strand with a T to T* (altered T that base pairs with G) change produces an A-to-G transition in the newly replicated complementary DNA strand. If the mRNA is transcribed from the strand with the A-to-G change (the template strand), a U-to-C change is produced in the corresponding mRNA.

20. The movement of nonautonomous P elements can be controlled by doing crosses in which a source of the transposase is introduced into a genome containing only deleted elements. Their controlled transposition create mutations by insertion, and these interrupted and mutated genes can then be cloned with the use of P element segments as a probe—a method called "tagging".

Chapter 11

2.

a. paracentric inversion

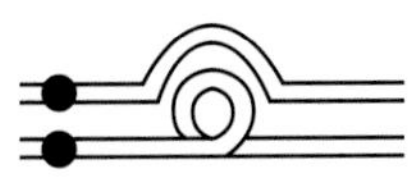

b. deletion

c. pericentric inversion

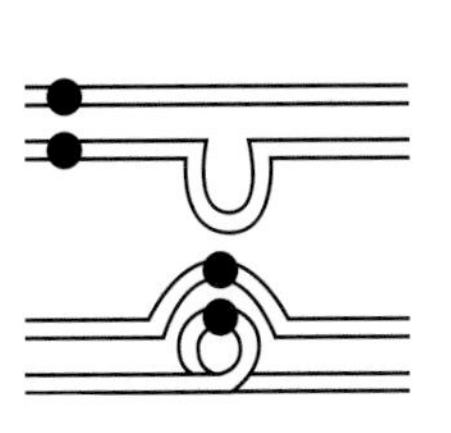

d. duplication

4. A deletion has occurred; half the progeny will be w^+ / w and half will be $w^+ / w^{deletion}$. If $w^+ / w^{deletion}$ are intercrossed, 25% of the progeny will not develop (the homozygous deletion would likely be lethal), and no waltzers will be observed.

6. The data suggest that one or both breakpoints of the inversion are located within an essential gene, causing a recessive lethal mutation.

8. **a.** The progeny are not in the 1:1:1:1 ratio expected for independent assortment; instead, the data indicate close linkage. And half the progeny did not develop, indicating semisterility.
b. These observations are best explained by a translocation that brought the two loci close together.
c. Parents: $T\,R / t\,r \times t / t\ ;\ r / r$

Progeny:

98	$T\,R / t\ ;\ r$
104	$t\,r / t\ ;\ r$
3	$T\,r / t\ ;\ r$
5	$t\,R / t\ ;\ r$

d. Assume a translocation heterozygote in coupling. If pairing is as diagrammed below, then you would observe the following:

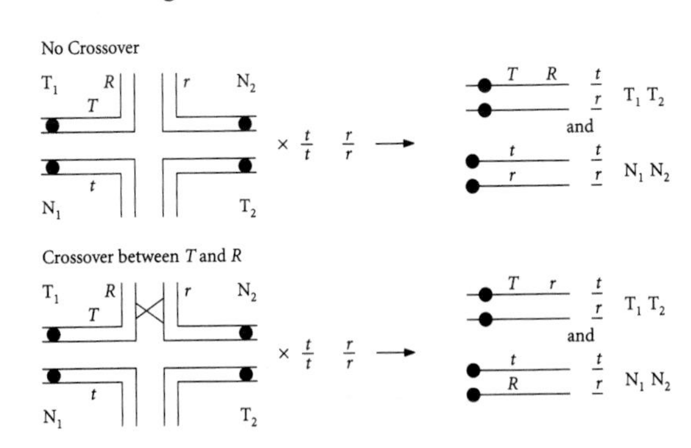

e. The two recombinant classes result from a recombination event followed by proper segregation of chromosomes, as diagrammed above.

10. 18.17 m.u.

12. Create a hybrid by crossing the two plants and then double the chromosomes with a treatment that disrupts mitosis, such as colchicine treatment. Alternatively, diploid somatic cells from the two plants could be fused and then grown into plants through various culturing techniques.

14. **a., b., c.** One of the parents of the woman with Turner syndrome (XO) must have been a carrier for colorblindness, an X-linked recessive disorder. Because her father has normal vision, she could not have obtained her only X from him. Therefore, nondisjunction occurred in her father. A sperm lacking a sex chromosome fertilized an egg with the X chromosome carrying the colorblindness allele. The nondisjunctive event could have occurred during either meiotic division.

d. If the colorblind patient had Klinefelter syndrome (XXY), then both X's must carry the allele for colorblindness. Therefore, nondisjunction had to occur in the mother. Remember that, during meiosis I, given no crossover between the gene and the centromere, allelic alternatives separate from each other. During meiosis II, identical alleles on sister chromatids separate. Therefore, the nondisjunctive event had to occur during meiosis II because both alleles are identical.

16. Type a: the extra chromosome must be from the mother. Because the chromosomes are identical, nondisjunction had to have occurred at meiosis II. Type b: the extra chromosome must be from the mother. Because the chromosomes are not identical, nondisjunction had to have occurred at meiosis I. Type c: the mother correctly contributed one chromosome, but the father did not contribute any chromosome 4. Therefore, nondisjunction occurred in the male during either meiotic division.

18. *y* is on chromosome 1, *cot* is on chromosome 7, and *h* is on chromosome 10. Genes *d* and *c* do not map to any of these chromosomes.

20. Cross 1: Independent assortment of the 2 genes (expected for genes on separate chromosomes). Cross 2: The 2 genes now appear to be linked (the observed RF is 1%); also, half of the progeny are inviable. These data suggest that a reciprocal translocation occurred and both genes are very close to the breakpoints. Cross 3: The viable spores are of two types: half contain the normal, nontranslocated chromosomes, and half contain the translocated chromosomes.

22. **a.** Among the progeny of this cross, the phenotypic ratio will be 5 wild type (a^+) : 1 *a*.

b. Among the progeny of this cross, the phenotypic ratio will be 1 wild type (a^+) : 1 *a*.

c. Among the progeny of this cross, the phenotypic ratio will be 3 wild type (a^+) : 1 *a*.

d. Among the progeny of this cross, the phenotypic ratio will be 11 wild type (a^+) : 1 *a*.

24. **a.** Loss of one X in the developing fetus after the two-cell stage.

b. Nondisjunction leading to Klinefelter syndrome (XXY), followed by a nondisjunctive event in one cell for the Y chromosome after the two-cell stage, resulting in XX and XXYY.

c. Nondisjunction of the X at the one-cell stage.

d. Fused XX and XY zygotes (from the separate fertilizations either of two eggs or of an egg and a polar body by one X-bearing and one Y-bearing sperm).

e. Nondisjunction of the X at the two-cell stage or later.

26. **a.** For the first ascus, the most reasonable explanation is that nondisjunction occurred at the first meiotic division. Second-division nondisjunction and chromosome loss are two explanations of the second ascus. Crossing-over best explains the third ascus.

b.

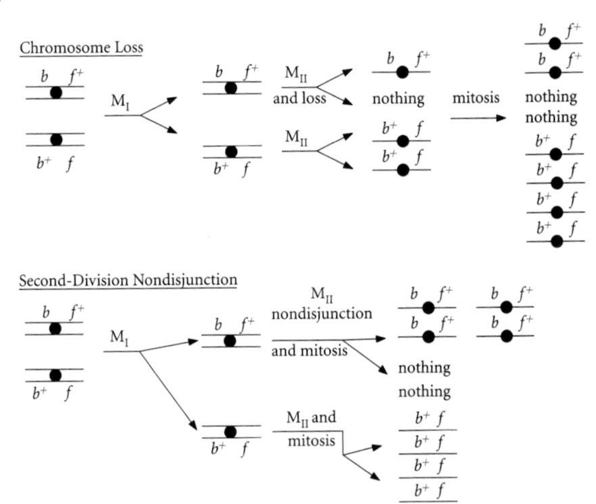

Chapter 12

2. Starting with a yeast strain that is *pro-1*, plate the cells on a medium lacking proline. Only those cells that are able to synthesize proline will form colonies. Most of these will be revertants; however, some will have second-site suppressors. (Treating the cells with a mutagen before plating them would significantly increase the yield.)

4. The commission was looking for induced recessive X-linked lethal mutations, which would show up as a shift in the sex ratio. A shift in the sex ratio is the first indication that a population has sustained lethal genetic damage.

6. **a.** To select for a nerve mutation that blocks flying, place *Drosophila* at the bottom of a cage and place a poisoned food source at the top of the cage.

b. Make antibodies against flagellar protein and expose mutagenized cultures to the antibodies.

c. Do filtration through membranes with various-size pores.

d. Screen visually.

e. Go to a large shopping mall and set up a rotating polarized disk. Ask the passersby to look through the disk for a free evaluation of their vision and their need for sunglasses. People with normal vision will see light with a constant intensity through the disk. Those with polarized vision will see alternating dark and light.

f. Set up a Y tube (a tube with a fork giving the choice of two pathways) and observe whether the flies or unicellular algae move to the light or the dark pathway.

g. Set up replica cultures and expose one of the two plates to low doses of UV.

8. Phenocopying is the mimicking of a mutant phenotype by inactivating the gene product rather than the gene itself. Three methods of producing phenocopies are: using RNA complementary to the gene-specific mRNA; double-stranded RNA interference; and inhibition of a specific protein through high-affinity binding of compound(s) identified through chemical genetics.

10. A forward mutation is any change away from the wild-type allele; any change back to the wild-type allele is called a reverse mutation.

12. a., b. The pattern of growth for Prototroph 1 suggests that it is a reversion of the original mutation. When crossed with wild type, a reversion would be expected to produce all met^+ progeny, and when backcrossed, it would be expected to produce a 1 : 1 ratio. The pattern of growth for Prototroph 2 suggests that a suppressor at another, unlinked locus is responsible for its prototrophic growth.

14. Neomorphic mutations result in novel gene activity and are dominant. Reversion of the dominant phenotype will commonly be the result of introducing another mutation in the already mutant gene that now eliminates its function completely. Most gene knockout mutations are recessive, so it is likely that most "revertants" will actually be recessive loss-of-function mutations.

16. a. A reverse genetic approach would be used.
b. The two general approaches would be: directed mutations in the gene of interest, or the generation of phenocopies of the mutant phenotype by inactivating the gene product rather than the gene itself.

18. a. There are three genes represented.
b. As stated above, 1 and 4 map to the same gene. Of the other combinations, only 2 and 3 show linkage. In this case, the testcross of the 2 × 3 F_1 produces 10% wild-type progeny, not the expected 25% if the genes were unlinked. You can also use these data to determine the map distance between genes 2 and 3. The percentage of wild-type progeny from the testcross will be equal to half that of the recombinants (the other half will be mutant for both genes). Thus genes 2 and 3 are 20 map units apart.
c. 1 × 2: $m_1 / m_1 \; ; \; + / + \times + / + \; ; \; m_2 / m_2$

↓

$m_1 / + \; ; \; m_2 / + \times m_1 / m_1 \; ; \; m_2 / m_2$ (testcross)

↓

25% $m_1 / m_1 \; ; \; m_2 / +$
25% $m_1 / + \; ; \; m_2 / m_2$
25% $m_1 / m_1 \; ; \; m_2 / m_2$
25% $m_1 / + \; ; \; m_2 / +$ (wild-type)

1 × 3: $m_1 / m_1 \; ; \; + / + \times + / + \; ; \; m_3 / m_3$

↓

$m_1 / + \; ; \; m_3 / + \times m_1 / m_1 \; ; \; m_3 / m_3$ (testcross)

↓

25% $m_1 / m_1 \; ; \; m_3 / +$
25% $m_1 / + \; ; \; m_3 / m_3$
25% $m_1 / m_1 \; ; \; m_3 / m_3$
25% $m_1 / + \; ; \; m_3 / +$ (wild-type)

1 × 4: $m_1 / m_1 \times m_4 / m_4$

↓

$m_1 / m_4 \times m_1 / m_1$ or m_4 / m_4 (testcross)

↓

50% m_1 / m_4
50% m_1 / m_1

2 × 3: $m_2 + / m_2 + \times + m_3 / + m_3$

↓

$m_2 + / + m_3 \times m_2 m_3 / m_2 m_3$ (testcross)

↓

40% $m_2 + / m_2 m_3$
40% $+ m_3 / m_2 m_3$
10% $m_2 m_3 / m_2 m_3$
10% $+ + / m_2 m_3$ (wild-type)

2 × 4: as in 1 × 2

3 × 4: as in 1 × 2

20. a., b. It is likely that the observed abnormalities are the result of mitotic recombination.

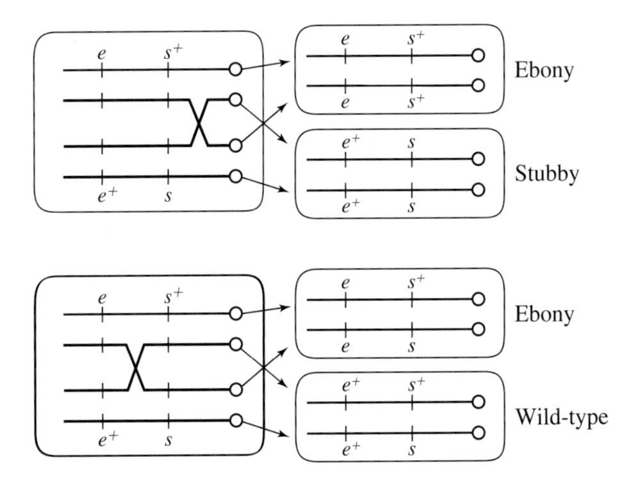

Chapter 13

2. An operator controls only the genes on the same DNA strand, those genes that it is adjacent or "cis" to.

4.

	β-Galactosidase		Permease	
Part	**No lactose**	**Lactose**	**No lactose**	**Lactose**
a	+	+	−	+
b	+	+	−	−
c	−	−	−	−
d	−	−	−	−
e	+	+	+	+
f	+	+	−	−
g	−	+	−	+

6. The *lacY* gene produces a permease that transports lactose into the cell. A cell containing a *lacY*⁻ mutation cannot transport lactose into the cell, so β-galactosidase will not be induced.

8. Bacterial operons contain a promoter region that extends approximately 35 bases upstream of the site where transcription is initiated. Within this region is the promoter. Activators and repressors, both of which are trans-acting proteins that bind to the promoter region, regulate transcription of associated genes in cis only. The eukaryotic gene has the same basic organization. However, the promoter region is somewhat larger. Also, enhancers up to several thousand nucleotides upstream or downstream can influence the rate of transcription. A major difference is that eukaryotes have not been demonstrated to have polycistronic messages.

10. The *araC* product has two conformations, which are determined by the presence and absence of arabinose. When it has bound arabinose, the *araC* product can bind to the initiator site *(araI)* and activate transcription. When it is not bound to arabinose, the *araC* product binds to both the initiator *(araI)* and the operator *(araO)* sites, forming a loop of the intermediary DNA. When both sites are bound to the *araC* product, transcription is inhibited. The *araC* product is trans-acting. Many eukaryotic trans-acting protein factors also bind to promoters, enhancers, or both that are upstream from the protein-encoding gene. These factors are required for the initiation of transcription. Additionally, some bind to other proteins, such as RNA polymerase II, in order to initiate transcription. Like their counterparts in the *ara* operon, the eukaryotic trans-acting protein factors can bind DNA at two sites, with the intermediary DNA forming a loop between the binding sites.

12. Construct a set of reporter genes with the promoter region, the introns, and the region 3′ to the transcription unit of the gene in question containing different alterations that do not disrupt transcription or processing. Use these reporter genes to make transgenic animals by germ-line transformation. Assay for expression of the reporter gene in various tissues and the kidney of both sexes.

14. Because very small amounts of the repressor are made, the system as a whole is quite responsive to changes in lactose concentration. In the heterodiploids, repressor tetramers may form by association of polypeptides encoded by I^- and I^+. The operator binding site binds two subunits at a time. Therefore, the repressors produced may reduce operator binding, which in turn would result in some expression of the *lac* genes in the absence of lactose.

16. The *S* mutation is an alteration in *lacI* such that the repressor protein binds to the operator regardless of whether inducer is present or not. The constitutive reverse mutations that map to *lacI* are mutational events that inactivate the ability of this repressor to bind to the operator. The constitutive reverse mutations that map to the operator alter the operator DNA sequence such that it will not permit binding to any repressor molecules (wild-type or mutant repressor).

18.

	Glucose	Lactose	Lactose + Glucose
wild type	0	100	1
lacI⁻	1	100	1
*lacI*ˢ	0	0	0
lacO⁻	1	100	1
crp⁻	0	1	1

20. a. Between the 3′ end of exon 1 [E1(f)] and the 5′ end of exon 3 [E3(f)].

b. In the 3′ region of E3(f) or the 3′ flanking sequences of this gene.

c. To the 3′ side of E3(g) or the 3′ flanking regions of the globin gene.

Chapter 14

2. Two facts are important: (1) the parents consist of only one phenotype, yet the offspring have three phenotypes, and (2) the progeny appear in an approximate ratio of 1 : 2 : 1. These facts should tell you immediately that you are dealing with a heterozygous × heterozygous cross involving one gene and that the erminette phenotype must be the heterozygous phenotype. To test the hypothesis that the erminette phenotype is a heterozygous phenotype, you could cross an erminette with either, or both, of the homozygotes. You should observe a 1 : 1 ratio in the progeny of both crosses.

4. a. The data indicate that there is a single gene with multiple alleles. The order of dominance is:

Black > sepia > cream > albino.

	Cross Parents	Progeny	Conclusion
Cross 1	*b / a* × *b / a*	3 *b* / − : 1 *a* / *a*	Black is dominant to albino.
Cross 2	*b / s* × *a / a*	1 *b* / *a* : 1 *s* / *a*	Black is dominant to sepia; sepia is dominant to albino.
Cross 3	*c / a* × *c / a*	3 *c* / − : 1 *a* / *a*	Cream is dominant to albino.
Cross 4	*s / a* × *c / a*	1 *c* / *a* : 2 *s* / − : 1 *a* / *a*	Sepia is dominant to cream.
Cross 5	*b / c* × *a / a*	1 *b* / *a* : 1 *c* / *a*	Black is dominant to cream.
Cross 6	*b / s* × *c* / −	1 *b* / − : 1 *s* / −	"−" can be *c* or *a*.
Cross 7	*b / s* × *s* / −	1 *b* / *s* : 1 *s* / −	"−" can be *s*, *c*, or *a*.
Cross 8	*b / c* × *s / c*	1 *s* / *c* : 2 *b* / − : 1 *c* / *c*	"−" can be *s* or *c*.
Cross 9	*s / c* × *s / c*	3 *s* / − : 1 *c* / *c*	"−" can be *s* or *c*.
Cross 10	*c / a* × *a / a*	1 *c* / *a* : 1 *a* / *a*	Cream is dominant to albino.

b. The progeny of the cross *b / s* × *b / c* will be $\frac{3}{4}$ black ($\frac{1}{4}$ *b / b*, $\frac{1}{4}$ *b / c*, $\frac{1}{4}$ *b / s*) : $\frac{1}{4}$ sepia (*s / c*).

6. a. The sex ratio is expected to be 1 : 1.

b. The female parent was heterozygous for an X-linked recessive lethal allele. This would result in 50% fewer males than females.

c. Half of the female progeny should be heterozygous for the lethal allele and half should be homozygous for the nonlethal allele. Individually mate the F_1 females and determine the sex ratio of their progeny.

8. a. The most obvious control is to cross the two pure-breeding lines. The cross would be $A/A\ ;\ b/b \times a/a\ ;\ B/B$. The progeny will be $A/a\ ;\ B/b$, and all should be reddish purple.

b. The most likely explanation is that the red pigment is produced by the action of at least two different gene products.

c. The genotypes of the two lines should be $A/A\ ;\ b/b$ and $a/a\ ;\ B/B$.

d. The F_1 would be all be pigmented, $A/a\ ;\ B/b$. This is an example of complementation. The mutants are defective for different genes. The F_2 would be

9	$A/-\ ;\ B/-$	Pigmented
3	$a/a\ ;\ B/-$	White
3	$A/-\ ;\ b/b$	White
1	$a/a\ ;\ b/b$	White

10. a. Complementation refers to gene products within a cell, which is not what is happening here. Most likely, what is known as cross-feeding is occurring, whereby a product made by one strain diffuses to another strain and allows growth of the second strain. This is equivalent to supplementing the medium. Because cross-feeding seems to be taking place, the suggestion is that the strains are blocked at different points in the metabolic pathway.

b. For cross-feeding to occur, the growing strain must have a block that occurs earlier in the metabolic pathway than does the block in the strain from which it is obtaining the product for growth.

c. The metabolic pathway is

trpE ⟶ *trpD* ⟶ *trpB*

d. Without some tryptophan, no growth at all would occur, and the cells would not have lived long enough to produce a product that could diffuse.

12. a. The best explanation is that Marfan syndrome is inherited as a dominant autosomal trait.

b. The pedigree shows both pleiotropy (multiple affected traits) and variable expressivity (variable degree of expressed phenotype). Penetrance is the percentage of individuals with a specific genotype who express the associated phenotype. There is no evidence of decreased penetrance in this pedigree.

c. Pleiotropy indicates that the gene product is required in a number of different tissues, organs, or processes. When the gene is mutant, all tissues needing the gene product will be affected. Variable expressivity of a phenotype for a given genotype indicates modification by one or more other genes, random noise, or environmental effects.

14. a. Let A = wild type, a = white, B = wild type, and b = pink.

Cross 1:	P	blue × white	$A/A\ ;\ B/B \times a/a\ ;\ B/B$
	F_1	All blue	All $A/a\ ;\ B/B$
	F_2	3 blue: 1 white	$3\ A/-\ ;\ B/B{:}1\ a/a\ ;\ B/B$
Cross 2:	P	blue × pink	$A/A\ ;\ B/B \times A/A\ ;\ b/b$
	F_1	All blue	All $A/A\ ;\ B/b$
	F_2	3 blue: 1 pink	$3\ A/A\ ;\ B/-{:}1\ A/A\ ;\ b/b$
Cross 3:	P	pink × white	$A/A\ ;\ b/b \times a/a\ ;\ B/B$
	F_1	All blue	All $A/a\ ;\ B/b$
	F_2	9 blue	$9\ A/-\ ;\ B/-$
		4 white	$3\ a/a\ ;\ B/-{:}1\ a/a\ ;\ b/b$
		3 pink	$3\ A/-\ ;\ b/b$

b. The white parent had to have been heterozygous, and the F_2 cross was $A/a\ ;\ B/b \times a/a\ ;\ B/b$.

16. $A/a\ ;\ R/r \times A/a\ ;\ r/r$

18. a. This type of gene interaction is called *epistasis*. The phenotype of e/e is epistatic to the phenotypes of $B/-$ or b/b.

b. The following are the inferred genotypes:

I	1 $(B/b\ \ E/e)$	2 $(B/b\ \ E/e)$	
II	1 $(b/b\ \ E/e)$	2 $(B/b\ \ E/e)$	3 $(-/-\ \ e/e)$
	4 $(b/b\ \ E/-)$	5 $(B/b\ \ E/e)$	6 $(b/b\ \ E/e)$
III	1 $(B/b\ \ E/-)$	2 $(-/b\ \ e/e)$	3 $(b/b\ \ E/-)$
	4 $(B/b\ \ E/-)$	5 $(b/b\ \ E/-)$	6 $(B/b\ \ E/-)$
	7 $(-/b\ \ e/e)$		

20. a. The following genotypes can be inferred:

I-1 and I-2	$A/a\ ;\ B/B$
I-3 and I-4	$A/A\ ;\ B/b$
II-(1, 3, 4, 5, 6)	$A/-\ ;\ B/B$
II-(9, 10, 12, 13, 14, 15)	$A/A\ ;\ B/-$
II-2 and II-7	$a/a\ ;\ B/B$
II-8 and II-11	$A/A\ ;\ b/b$

b. Generation III shows complementation. All are $A/a\ ;\ B/b$.

22. Cross 1:	P	$A/A\ ;\ B/B \times a/a\ ;\ B/B$	Wild type × Platinum
	F_1	$A/a\ ;\ B/B$	All wild type
	F_2	$3\ A/-\ ;\ B/B$: $1\ a/a\ ;\ B/B$	3 wild type : 1 Platinum
Cross 2:	P	$A/A\ ;\ B/B \times A/A\ ;\ b/b$	Wild type × Aleutian
	F_1	$A/A\ ;\ B/b$	All wild type
	F_2	$3\ A/A\ ;\ B/-$: $1\ A/A\ ;\ b/b$	3 wild type : 1 Aleutian

Cross 3:	P	$a / a\ ;\ B / B\ \times$ $A / A\ ;\ b / b$	Platinum × Aleutian
	F_1	$A / a\ ;\ B / b$	All wild type
	F_2	$9\ A / -\ ;\ B / -$	Wild type
		$3\ A / -\ ;\ b / b$	Aleutian
		$3\ a / a\ ;\ B / -$	Platinum
		$1\ a / a\ ;\ b / b$	Sapphire

b.

	Sapphire × Platinum		Sapphire × Aleutian	
P	$a / a\ ;\ b / b\ \times\ a / a\ ;\ B / B$		$a / a\ ;\ b / b\ \times\ A / A\ ;\ b / b$	
F_1	$a / a\ ;\ B / b$	Platinum	$A / a\ ;\ b / b$	Aleutian
F_2	$3\ a / a\ ;\ B / -$	Platinum	$3\ A / -\ ;\ b / b$	Aleutian
	$1\ a / a\ ;\ b / b$	Sapphire	$1\ a / a\ ;\ b / b$	Sapphire

24. a. There are three genes indicated by this complementation test.

b.

$$\text{AIR} \xrightarrow{4} \text{CAIR} \xrightarrow{1} \text{SAICAR} \xrightarrow{3} \text{Adenine}$$

Where a mutant is blocked, it is indicated by the vertical line through the arrow.

c. 1.1%

26. a. The data from the crosses indicate that the mutations are in different, unlinked genes and both are recessive.

b. Assume p^+ = normal P protein; r^+ = normal regulatory protein; both are required for normal function.

Lane 1:	$p^+ / p^+\ ;\ r^+ / -$
Lane 2:	$p^+ / p^+\ ;\ r / r$
Lane 3:	$p^+ / p\ ;\ r^+ / -$
Lane 4:	$p^+ / p\ ;\ r / r$
Lane 5:	$p / p\ ;\ r^+ / -$
Lane 6:	$p / p\ ;\ r / r$

c. Type 4 is $p^+ / p\ ;\ r / r$. $\frac{1}{2}$ of the progeny will be p^+ / p, and $\frac{1}{4}$ of the progeny will be independently r / r, so $\frac{1}{2} \times \frac{1}{4} = \frac{1}{8}$.

d. Lane 1 ($p^+ / p^+\ ;\ r^+ / -$) and lane 3 ($p^+ / p\ ;\ r^+ / -$) will be phenotypically wild type.

e. Parents, $p / p\ ;\ r^+ / r^+\ \times\ p^+ / p^+\ ;\ r / r$ will look like lane 5 and lane 2, respectively. F_1, $p^+ / p\ ;\ r^+ / r$, will look like lane 3.

28. These data suggest that three alleles of one gene are being studied and that the order of dominance is black (B) > red (r) > blue (bl).

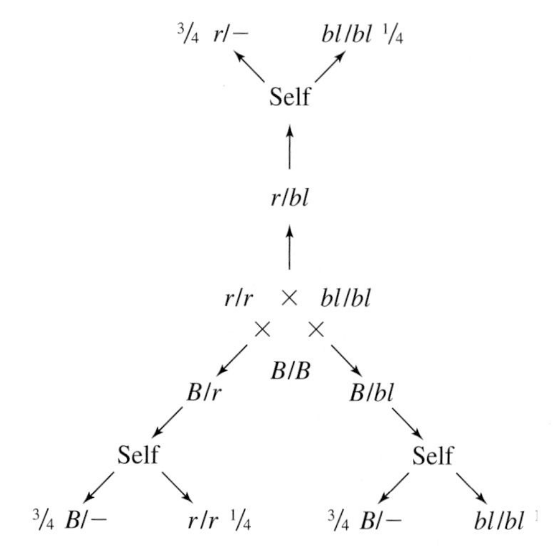

30. There is a 13:3 ratio of F_2 progeny, which suggests two genes segregating independently and epistasis.

Pure-breeding × Pure-breeding

$$A / A\ ;\ b / b\ \times\ a / a\ ;\ B / B$$

$$\downarrow$$

$$\text{All } A / a\ ;\ B / b$$

$$\downarrow$$

Self

$$\downarrow$$

$$131\quad 9\ A / -\ ;\ B / - : 3\ a / a\ ;\ B / - : 1\ a / a\ ;\ b / b$$

$$29\quad 3\ A / -\ ;\ b / b$$

32. These data suggest that three alleles of one gene are being studied. Both codominance (=) and classical dominance (>) are present in this multiple allelic series: l^v (vertical lines) = l^h (horizontal lines), and both l^v and $l^h > l^0$ (no lines).

$$l^v / l^v \times l^0 / l^0 \longrightarrow \text{all } l^v / l^0 \longrightarrow F_2\ 3\ l^v / - : 1\ l^0 / l^0$$

$$l^h / l^h \times l^0 / l^0 \longrightarrow \text{all } l^h / l^0 \longrightarrow F_2\ 3\ l^h / - : 1\ l^0 / l^0$$

$$l^v / l^v \times l^h / l^h \longrightarrow \text{all } l^v / l^h \longrightarrow F_2\ 1\ l^v / l^v : 2\ l^v / l^h : 1\ l^h / l^h$$

Chapter 15

2. Phosphorylation/dephosphorylation; protein-protein interactions; and proteolysis.

4. Point mutations; gene fusions; and deletions of key regulatory domains.

6. Misexpression or overexpression of nonmutated protein and gene fusions.

8. Translocation and fusion of an immunoglobin enhancer to the *bcl2* gene causes large amounts of Bcl2 to be expressed in B lymphocytes. Since Bcl2 is a negative regulator of apoptosism, this overexpression essentially blocks apoptosis in these mutant B lymphocytes and allows them to accumulate cell proliferation-promoting mutations over their unusually long lifetime.

10. **a.** The v-*erbB* oncogene encodes a mutated form of the epidermal growth factor receptor (EGFR). Unlike wild type, the mutant EGFR oncoprotein lacks key regulatory regions as well as the extracellular ligand-binding domain. These deletions allow the mutant protein to constitutively dimerize, leading to continuous autophosphorylation, which results in continuous transduction of a signal from the receptor.
b. The *ras* oncogene is caused by a missense mutation that allows the mutated Ras protein to always bind GTP, even in the absence of the signals normally required for such binding by wild-type Ras protein. As a result, the Ras oncoprotein continually promotes cell proliferation.
c. The Philadelphia chromosome is a translocation between chromosomes 9 and 22 that results in the fusion of two genes, *bcr1* and *abl*. The Bcr1-Abl fusion protein has an activated kinase activity that is responsible for causing chronic myelogenous leukemia.

12. **a.** Overproduction of one of the cyclins could disrupt the orderly process of cell division, but it would be limited by the amount of CDK present as well as the state of the p21 "brake."
b. This would likely be recessive and would slow cell proliferation, not accelerate it.
c. This would be dominant and it would lead to excess cell death, not proliferation.
d. This would likely be recessive and would slow cell proliferation, not accelerate it.
e. This chromosomal rearrangement would be dominant.

14. Gain-of-function mutations effect mitogenic pathways to increase cell proliferation or cell survival pathways to decrease apoptosis. Loss-of-function mutations promote tumors by loss of growth-inhibitor pathways, p53 pathways, or apoptosis-activator pathways.

16. Normal Ras is a G-protein that activates a protein kinase, which in turn phosphorylates a transcription factor. If it were simply deleted, no cancer could develop, because cell division would not occur. If it were simply duplicated, an excess of the G-protein could not cause cancer, because it must be activated before it can activate the protein kinase, and presumably the enzyme that activates normal Ras is closely regulated and would not activate too many copies. However, if it were to have a point mutation, it might now bind GTP, even in the absence of normal control signals, and be in a state of permanent activation. As a positive regulator of cell growth, this mutant Ras would continually promote cell proliferation. In contrast, normal *c-myc* is a transcription factor. If the gene were to be duplicated, too much transcription factor could lead to malignancy.

18. **a.** Mutations in a tumor-suppressor gene are recessive and caused by loss of function. That function can be restored by the introduction of a wild-type allele.
b. Mutations in an oncogene are dominant and caused by gain of function (overexpression or misexpression). The normal function will not inhibit these mutants, and the introduced gene would be ineffective in restoring the normal phenotype.

20. p53 detects and is activated by DNA damage. When activated, p53 activates p21, an inhibitor of the cyclin-CDK complex necessary for the progression of the cell cycle. If the DNA damage is repairable, this system will eventually deactivate p53 and allow cell division. However, if the damage is irreparable, p53 would stay active and would activate the apoptosis pathway, ultimately leading to cell death. It is for this reason that the "loss" of p53 is often associated with cancer.

22. **a.** In the absence of functional Rb protein, E2F will be in the nucleus.
b. No. The absence of functional E2F would be epistatic to the absence of functional Rb protein.
c. A mutation that causes permanent sequestering of E2F in the cytoplasm will likely inhibit the cell cycle.

Chapter 16

2. In humans, a single copy of the Y chromosome is sufficient to shift development toward normal male phenotype. The extra copy of the X chromosome is simply inactivated. Both mechanisms seem to be all-or-none rather than based on concentration levels.

4. The *bcd* mRNA is tethered to the minus ends of microtubules that are located at the anterior pole of the egg. Upon translation, BCD protein diffuses in the common cytoplasm creating the observed gradient. The *hb-m* mRNA is uniformly distributed throughout the oocyte. However, translation of *hb-m* mRNA is blocked by the NOS protein product. The *nos* mRNA is localized to the posterior pole through its association with the plus ends of microtubules and, upon translation, a posterior-to-anterior gradient of NOS is generated.

6. **a.** There must be a diffusible substance produced by the anchor cell that affects development of the six cells.
b. Remove the anchor cell and the six equivalent cells. Arrange the six cells in a circle around the anchor cell. All six cells will develop the same phenotype, which will depend on the distance from the anchor cell.

8. Because the receptor is defective, testosterone cannot signal the cell and initiate the cascade of developmental changes that will switch the embryo from the "default" female development to male development. Therefore, the phenotype will be female.

10. The early promoter of the *Sxl* gene is activated by the NUM-NUM transcription factors and is active only early in embryogenesis. Later in embryogenesis and for the remainder of the life cycle, the *Sxl* gene is transcribed in both sexes from the late promoter. Subject to alternative splicing, the processed transcript from the late promoter encodes active SXL protein only if spliced in the presence of preexisting SXL protein. Therefore, the presence of active SXL protein ensures the further production of more active SXL protein.

12. A number of experiments could be devised. A comparison of amino acid sequence between mammalian gene products and insect gene products would indicate which genes are most similar to each other. Using cloned cDNA sequences from mammalian genes for hybridization to insect DNA would also indicate which genes are most similar to each other.

14. Deletion of an anterior gene does not allow extension of the next-most-posterior segment in an anterior direction. The gap genes activate *Ubx* in both thoracic and abdominal segments, whereas the *abd-A* and *Abd-B* genes are activated only in the middle and posterior abdominal segments. The functioning of the *abd-A* and *Abd-B* genes in those segments somehow prevents *Ubx* expression. However, if the *abd-A* and *Abd-B* genes are deleted, *Ubx* can be expressed in these regions.

16. a. The determination of anterior-posterior portions of the embryo is governed by a concentration gradient of *bcd*. The furrow develops at a critical concentration of *bcd*. As bcd^+ gene dosage (and, therefore, BCD concentration) decreases, the furrow shifts anteriorly; as the gene dosage increases, the furrow shifts posteriorly.

b.

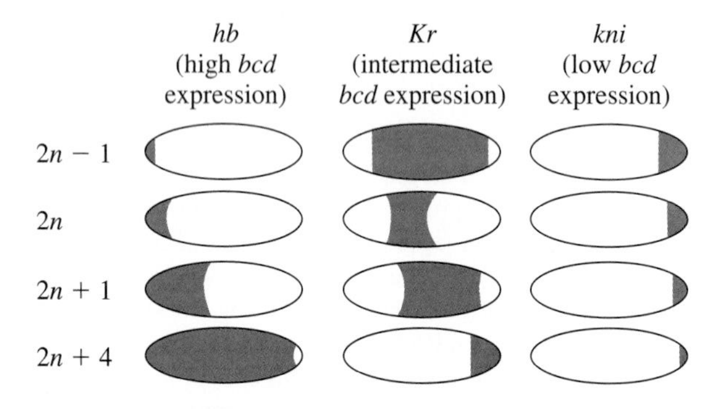

18. The concentration of *Sxl* is crucial for female development and dispensable for male development. The dominant Sxl^M male-lethal mutations may not actually kill all males but simply produce an excessive amount of gene product so that only females (fertile XX and XY) result. The reversions may eliminate all gene product, resulting in XX (sterile) and XY males. The reversions would be recessive because, presumably, a single normal copy of the gene may produce enough gene product to "toggle the switch" in development to female.

20. a. A pair-rule gene.

b. Look for expression of the mRNA from the candidate gene in repeating pattern of seven stripes along the A–P axis of developing embryo.

c. No. An embryo mutant for the gap gene *Krüppel* would be missing many anterior segments, and this effect would be epistatic to the expression of a pair-rule gene.

22. a. Developmentally, the default sexual pathway in mice is female unless testosterone is able to trigger male development. In the absence of the testosterone receptor, the presence of testosterone cannot be detected by the organism.

b. Mosaic. Cells that derived from progenitors that had inactivated the wild-type X chromosome would develop along the female pathway, while those derived from progenitors that had inactivated the *Tfm* mutation-carrying X would follow the male pathway.

Chapter 17

2. The frequency of *A1* is 978/1708 = 0.57, and the frequency of *A2* is 730/1708 = 0.43.

4. The frequency of a phenotype in a population is a function of the frequency of alleles that lead to that phenotype in the population. To determine dominance and recessiveness, do standard Mendelian crosses.

6. $s = 0.01$

8. $p_{50,000} = 0.65$

10. a. The genotypes should be distributed as follows if the population is in equilibrium:

$L^M / L^M = p^2\,(1482) = 401$ Actual: 406

$L^M / L^N = 2pq\,(1482) = 740$ Actual: 744

$L^N / L^N = q^2\,(1482) = 341$ Actual: 332

This compares well with the actual data, so the population is in equilibrium.

b. If mating is random with respect to blood type, then the following frequency of matings should occur:

$L^M / L^M \times L^M / L^M = (p^2)(p^2)(741) = 54$
Actual: 58

$L^M / L^M \times L^M / L^N$ or $L^M / L^N \times L^M / L^M = (2)(p^2)(2pq)(741) = 200$
Actual: 202

$L^M / L^N \times L^M / L^N = (2pq)(2pq)(741) = 185$
Actual: 190

$L^M / L^M \times L^N / L^N$ or $L^N / L^N \times L^M / L^M = (2)(p^2)(q^2)(741) = 92$
Actual: 88

$L^M / L^N \times L^N / L^N$ or $L^N / L^N \times L^M / L^N = 2(2pq)(q^2)(741) = 170$
Actual: 162

$L^N / L^N \times L^N / L^N = (q^2)(q^2)(741) = 39$
Actual: 41

Again, this compares nicely with the actual data, so the mating is random with respect to blood type.

12. a. 0.01

b. There would be 10 colorblind men for every colorblind woman (q / q^2).

c. 0.018

d. 0.81

e. The frequency of colorblind females will be 0.12 and colorblind males 0.2.

f. 0.2 in males; 0.4 in females.

14. Wild-type alleles are usually dominant because most mutations result in lowered or eliminated function. To be dominant, the heterozygote has approximately the same phenotype as the dominant homozygote. This will typically be true when the wild-type allele produces a product and the mutant allele does not. Chromosomal rearrangements are often dominant mutations because they can cause gross changes in gene regulation or even cause fusions of several gene products. Novel activities, overproduction of gene products, etc., are typical of dominant mutations.

16. a. For a parent-sib mating, the pedigree can be represented as follows

In this example, it is only the chance of the incestuous parent's being heterozygous that matters. Thus, the chance of the descendant being homozygous is

$$2pq(\tfrac{1}{2})(\tfrac{1}{4}) = \tfrac{pg}{4} \text{ or, for rare alleles, approximately } \tfrac{q}{4}.$$

b. The probability of inheriting the recessive allele if *either* grandparent is heterozygous can be represented as follows:

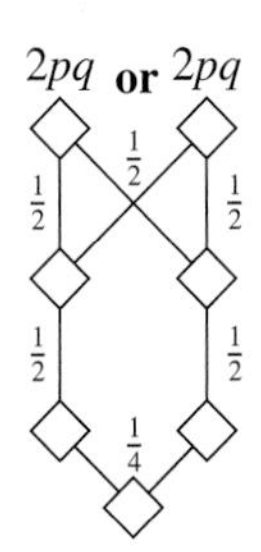

Thus the chance of this child's being affected is

$$2pq(\tfrac{1}{2})(\tfrac{1}{2})(\tfrac{1}{2})(\tfrac{1}{2})(\tfrac{1}{4}) + 2pq(\tfrac{1}{2})(\tfrac{1}{2})(\tfrac{1}{2})(\tfrac{1}{2})(\tfrac{1}{4}) = (\tfrac{pq}{16})$$

or, for rare alleles, approximately $\frac{q}{16}$.

c. An aunt-nephew (or uncle-niece) mating can be represented as:

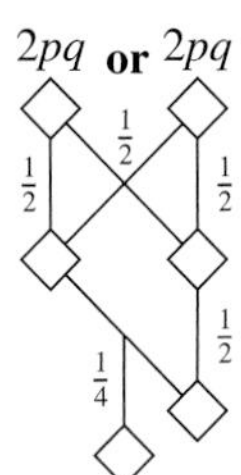

The chance of this child's being homozygous is

$$2pg(\tfrac{1}{2})(\tfrac{1}{2})(\tfrac{1}{2})(\tfrac{1}{4}) + 2pg(\tfrac{1}{2})(\tfrac{1}{2})(\tfrac{1}{2})(\tfrac{1}{4}) = \tfrac{pq}{8}$$

or, for rare alleles approximately $\frac{q}{8}$.

18. Many genes affect bristle number in *Drosophila*. The artificial selection resulted in lines with mostly high-bristle-number alleles. Some mutations may have occurred during the 20 generations of selective breeding, but most of the response was caused by alleles present in the original population. Assortment and recombination generated lines with more high-bristle-number alleles. Fixation of some alleles causing high bristle number would prevent complete reversal. Some high-bristle-number alleles would have no negative effects on fitness, so there would be no force pushing bristle number back down because of those loci. The low fertility in the high-bristle-number line could have been a result of pleiotropy or linkage. Some alleles that caused high bristle number may also have caused low fertility (pleiotropy). Chromosomes with high-bristle-number alleles may also carry alleles at different loci that cause low fertility (linkage). After artificial selection was relaxed, the low-fertility alleles would have been selected against through natural selection. A few generations of relaxed selection would have allowed low-fertility-linked alleles to recombine away, producing high-bristle-number chromosomes that did not contain low-fertility alleles. When selection was reapplied, the low-fertility alleles had been reduced in frequency or separated from the high-bristle loci, so this time there was much less of a fertility problem.

20. The probability of not getting a recessive lethal genotype for one gene is $1 - \frac{1}{8} = \frac{7}{8}$. If there are n lethal genes, the probability of not being homozygous for any of them is $(\frac{7}{8})^n = \frac{13}{31}$. Solving for n, an average of 6.5 recessive lethals are predicted. If the actual percentage of "normal" children is less owing to missed in utero fatalities, the average number of recessive lethals would be higher.

Chapter 18

2. Mean $= x = 4.7$ average number of bristle/individual

$$\text{Variance} = s^2 = 0.26$$

$$\text{Standard deviation} = s = 0.51$$

4. The following are unknown: (1) norms of reaction for the genotypes affecting IQ; (2) the environmental distribution in which the individuals developed; and (3) the genotypic distributions in the populations. Even if the above were known, because heritability is specific to a specific population and its environment, the difference between two different populations cannot be given a value of heritability.

6. a. $p(\text{homozygous at 1 locus}) = 3(\frac{1}{2})^3 = \frac{3}{8}$

$p(\text{homozygous at 2 loci}) = 3(\frac{1}{2})^3 = \frac{3}{8}$

$p(\text{homozygous at 3 loci}) = 2(\frac{1}{2})^3 = \frac{2}{8}$

b. $p(\text{0 capital letters}) = \frac{1}{64}$

$p(\text{1 capital letter}) = \frac{3}{32}$

$p(\text{2 capital letters}) = \frac{15}{64}$

$p(\text{3 capital letters}) = \frac{10}{32}$

$p(\text{4 capital letters}) = \frac{15}{64}$

$p(\text{5 capital letters}) = \frac{3}{32}$

$p(\text{6 capital letters}) = \frac{1}{64}$

8. The population described would be distributed as follows:

3 bristles	$\frac{19}{64}$
2 bristles	$\frac{44}{64}$
1 bristle	$\frac{1}{64}$

The 3-bristle class would contain 7 different genotypes, the 2-bristle class would contain 19 different genotypes, and the 1-bristle class would contain only 1 genotype. It would be very difficult to determine the underlying genetic situation by doing controlled crosses and determining progeny frequencies.

10. a.

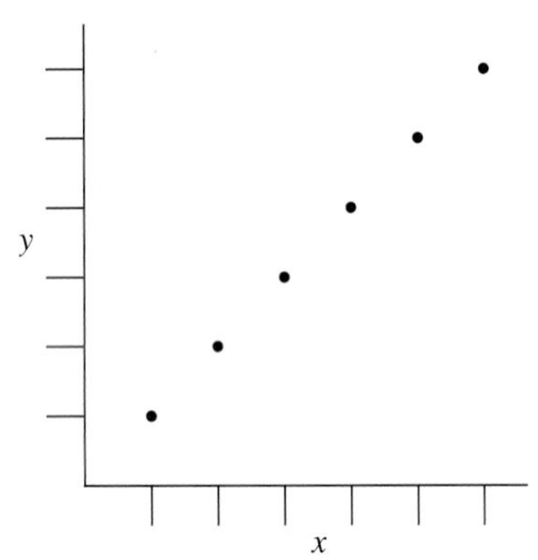

b.

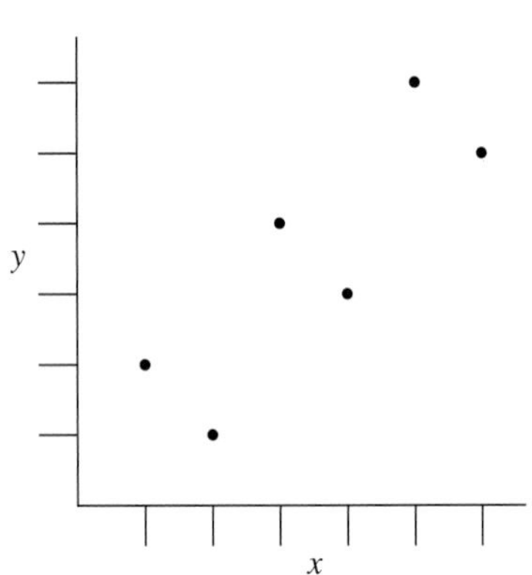

c.

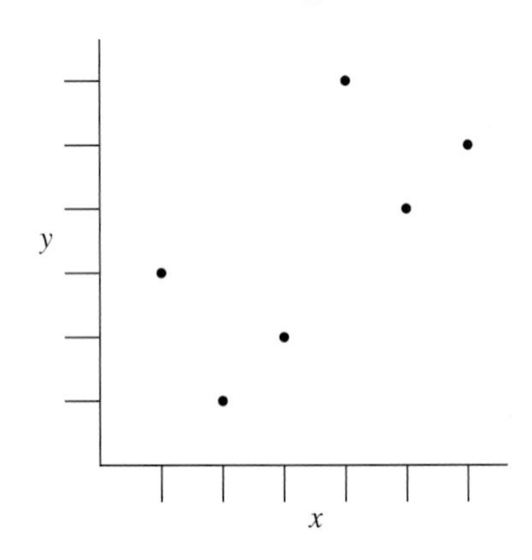

d.

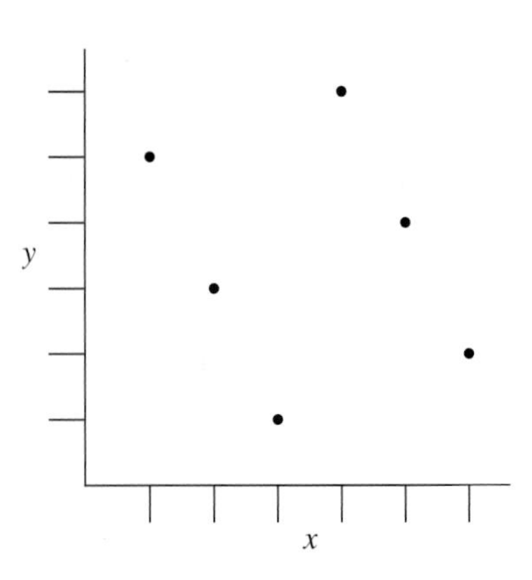

a. cov xy = 2.92
r_{xy} = 2.92/(1.71)(1.71) = 1.0.
b. 0.83
c. 0.66
d. − 0.20

12. Here is a summary of these results:

	Chromosome II	**Chromosome III**
h / h to *h / l*	2.9	2.9
h / l to *l / l*	5.1	11.5

Each set of alleles for both chromosomes is expressed in the phenotype, but that expression varies with the chromosome. Chromosome III appears to have a stronger effect on the phenotype than does chromosome II. (Compare *h / h* *h / h* with both *l / l* *h / h* and *h / h* *l / l*. The difference in the first case is 6.1, and in the second case, 13.3.) Finally, there is partial dominance of *h* over *l* for both chromosomes. The change from *h / h* to *h / l* is less than the change from *h / l* to *l / l*.

14. a. The regression line shows the relationship between the two variables. It attempts to predict one (the son's height) from the other (the father's height.) If the relationship is perfectly correlated, the slope of the regression line should approximate 1. If you assume that individuals at the extreme of any spectrum are homozygous for the genes responsible for these phenotypes, then their offspring are more likely to be heterozygous than are the original individuals. That is, they will be less extreme. Also, there is no attempt to include the maternal contribution to this phenotype.

b. For Galton's data, regression is an estimate of heritability (h^2), *assuming* that there were few environmental differences between all fathers and all sons both individually and as a group. However, there is no evidence given to determine whether the traits are familial but not heritable. This data would indicate genetic variation only if the relatives do not share common environments more than nonrelatives do.

Chapter 19

2. The three principles are: (1) organisms within a species vary from one another, (2) the variation is heritable, and (3) different types leave different numbers of offspring in future generations.

4. A geographical race is a population that is genetically distinguishable from other local populations but is capable of exchanging genes with those other local populations. A species is a group of organisms that exchanges genes within the group but cannot do so with other groups. Populations that are geographically separated will diverge from each other as a consequence of a combination of unique mutations, selection, and genetic drift. For populations to diverge enough to become reproductively isolated, spatial separation sufficient to prevent any effective migration is usually necessary.

6. The rate of loss of heterozygosity in a closed population is $1/(2N)$ per generation.

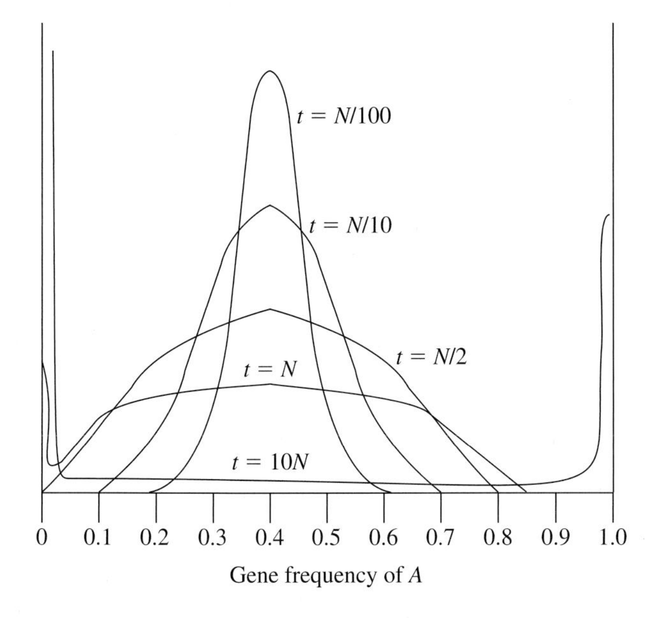

8. The mean fitness of Population 1 [$p(A) = 0.8, p(B) = 0.9$] is 0.92. Since selection acts to increase the mean fitness, the frequency of both A and B should increase in the next generation (the $A / A\ .\ B / B$ class has a fitness of 0.95). The mean fitness of Population 2 [$p(A) = 0.2, p(B) = 0.2$] is 0.73. Again, the frequency of both A and B should increase. There is a single adaptive peak at $A / A\ .\ B / B$. By inspection, the fitness is lowest at $a / a\ b / b$ and highest at $A / A\ .\ B / B$. The allelic frequencies at the peak is 1.0 for both A and B.

10. Above a chromosome number of 12, even numbers are much more common than odd numbers. This is evidence of frequent polyploidization during plant evolution. For example, if a species of plant with an odd haploid chromosome number undergoes a "doubling" event, the chromosome number becomes even.

12. All human populations have high i, intermediate I^A, and low I^B frequencies. The variations that do exist among the different geographical populations are most likely a result of genetic drift. There is no evidence that selection plays any role regarding these alleles.

14.

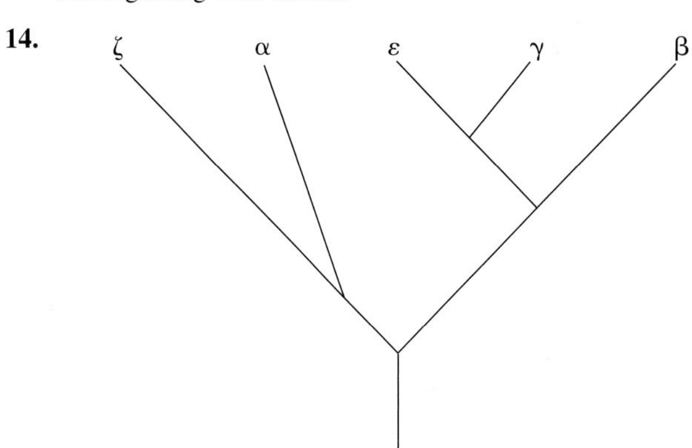

16. Morphological, anatomical, and developmental studies would help indicate whether the tails derived from the same basic structures. Also, comparison of the genes and proteins involved in the morphogenesis of the tails will determine whether common elements underlie their development.

Index

Note: Page numbers followed by f indicate figures; those followed by t indicate tables.

Animations

Forty-five animations were developed by Tony Griffiths and are fully integrated with the content and figures in the text chapters.

Chapter 1
Regulation of the Lactose System in *E. coli* (Figure 1-11)

Chapter 2
Three-Dimensional Structure of Nuclear Chromosomes (Figures 2-22 and 2-26)

Chapter 3
Transcription (Figure 3-5)
Translation: Peptide-Bond Formation (Figure 3-17)
Translation: The Three Steps of Translation (Figure 3-24)

Chapter 4
DNA Replication: The Nucleotide Polymerization Process (Figure 4-4)
DNA Replication: Coordination of Leading- and Lagging-Strand Synthesis (Figure 4-6)
DNA Replication: Replication of a Chromosome (Figure 4-8)
Mitosis (Figure 4-21)
Meiosis (Figure 4-22)

Chapter 6
Meiotic Recombination between Unlinked Genes by Independent Assortment (Figure 6-4)
Meiotic Recombination between Linked Genes by Crossing-Over (Figure 6-10)
A Mechanism of Crossing-Over: Meselson-Radding Heteroduplex Model (Figure 6-18)
A Mechanism of Crossing-Over: Genetic Consequences of the Meselson-Radding Model (Figure 6-20)

Chapter 7
Bacterial Conjugation and Mapping by Recombination (Figures 7-8 and 7-12)

Chapter 8
Polymerase Chain Reaction (Figure 8-8)
Finding Specific Cloned Genes by Functional Complementation: Functional Complementation of the Gal^- Yeast Strain and Recovery of the Wild-Type GAL Gene
Finding Specific Cloned Genes by Functional Complementation: Making a Library of Wild-Type Yeast DNA (Figure 8-9)
Finding Specific Cloned Genes by Functional Complementation: Using the Cloned GAL Gene as a Probe for GAL mRNA (Figure 8-18)

Chapter 9
RFLP Analysis and Gene Mapping (Figures 9-3 and 9-5)
DNA Microarrays: Using an Oligonucleotide Array to Analyze Patterns of Gene Expression (Figure 9-38)
DNA Microarrays: Synthesizing an Oligonucleotide Array (Figure 9-39)

Chapter 10
Molecular Mechanism of Mutation: The Streis Model (Figure 10-12)
Replicative Transposition (Figures 10-30 and 10-31)
UV-Induced Photodimers and Excision Repair (Figure 10-37)

Chapter 11
Autotetraploid Meiosis (Figure 11-7)
Meiotic Nondisjunction at Meiosis I (Figure 11-13)
Meiotic Nondisjunction at Meiosis II (Figure 11-13)
Chromosome Rearrangements: Paracentric Inversion, Formation of Paracentric Inversions (Figure 11-21)
Chromosome Rearrangements: Paracentric Inversion, Meiotic Behavior of Paracentric Inversions (Figure 11-22)
Chromosome Rearrangements: Reciprocal Translocation, Formation of Reciprocal Translocations (Figure 11-24)
Chromosome Rearrangements: Reciprocal Translocation, Meiotic Behavior of Reciprocal Translocations (Figure 11-24)
Chromosome Rearrangements: Reciprocal Translocation, Pseudolinkage of Genes by Reciprocal Translocations (Figure 11-26)

Chapter 12
Screening and Selecting for Mutations (Figure 12-26)

Chapter 13
Regulation of the Lactose System in *E. coli*: Assaying Lactose Presence/Absence through the Lac Repressor (Figure 13-7)
Regulation of the Lactose System in *E. coli*: O^c *lac* Operator Mutations (Figure 13-10)
Regulation of the Lactose System in *E. coli*: I^- Lac Repressor Mutations (Figure 13-11)
Regulation of the Lactose System in *E. coli*: I^s Lac Super-Repressor Mutations (Figure 13-12)

Chapter 14
Interactions between Alleles at the Molecular Level, *RR*: Wild-Type Interactions between Alleles at the Molecular Level, *rr*: Homozygous Recessive, Null Mutation
Interactions between Alleles at the Molecular Level, $r'r'$: Homozygous Recessive, Leaky Mutation
Interactions between Alleles at the Molecular Level, *Rr*: Heterozygous, Complete Dominance (Figure 14-5)
Nonsense Suppression at the Molecular Level: The rod^{ns} Nonsense Mutation
Nonsense Suppression at the Molecular Level: The tRNA Nonsense Suppressor
Nonsense Suppression at the Molecular Level: Nonsense Suppression of the rod^{ns} Allele

Chapter 16
Gene Interactions in *Drosophila* Embryogenesis: Cell Signaling in Dorsal-Ventral Axis Formation (Figure 16-20)